MW00844887

PATTY'S TOXICOLOGY

Fifth Edition
Volume 6

EULA BINGHAM
BARBARA COHRSSEN
CHARLES H. POWELL
Editors

CONTRIBUTORS

Robert A. Barter
C. Bevan
Michael S. Bisesi
John H. Butala
Raymond M. David
Martin Kayser
Richard H. McKee
David A. Morgott
John L. O'Donoghue
Lynn H. Pottenger
Jon B. Reid
Douglas C. Topping
Gauke E. Veenstra
John M. Waechter, Jr.

A Wiley-Interscience Publication
JOHN WILEY & SONS, INC.
New York / Chichester / Weinheim / Brisbane / Singapore / Toronto

DISCLAIMER: Extreme care has been taken in preparation of this work. However, neither the publisher nor the authors shall be held responsible or liable for any damages resulting in connection with or arising from the use of any of the information in this book.

This book is printed on acid-free paper. ♾

For ordering and customer service, call 1-800-CALL-WILEY.

Library of Congress Cataloging in Publication Data:

Patty's toxicology / [edited by] Eula Bingham, Barbara Cohrssen, Charles H. Powell.— 5th ed.
p. ; cm.
"A Wiley-Interscience publication."
Includes bibliographical references and index.
ISBN 0-471-31939-2 (cloth: v. 6 : alk.paper); 0-471-31943-0 (set)
1. Industrial toxicology—Encyclopedias. I. Bingham, Eula. II. Cohrssen, Barbara. III. Powell, Charles. H. IV. Patty's industrial hygiene and toxicology
[DNLM: 1. Occupational Medicine. 2. Occupational Diseases. 3. Poisons. 4. Toxicology. WA 400 P3222 2000]
RA1229 .P38 2000
613.6′2—dc21 99-053898

Printed in the United States of America.

10 9 8 7 6 5 4 3 2 1

Contributors

Robert A. Barter, Ph.D., DABT, Eastman Kodak Company, Rochester, New York

C. Bevan, Ph.D., DABT, AMOCO Corporation, Warrenville, Illinois

Michael S. Bisesi, Ph.D., CIH, Medical College of Ohio, Toledo, Ohio

John H. Butala, MS, DABT, Gibsonia, Pennsylvania

Raymond M. David, Ph.D., DABT, Eastman Kodak Company, Rochester, New York

Martin Kayser, MD, Eastman Kodak Company, Rochester, New York

Richard H. McKee, Ph.D., DABT, Eastman Kodak Company, Rochester, New York

David A. Morgott, Ph.D., DABT, CIH, Eastman Kodak Company, Rochester, New York

John L. O'Donoghue, VMD, Ph.D., DABT, Eastman Kodak Company, Rochester, New York

Lynn H. Pottenger, Ph.D., Health and Environmental Sciences, The Dow Chemical Company, Midland, Michigan

Jon B. Reid, Ph.D., DABT, University of Cincinnati, Cincinnati, Ohio

Douglas C. Topping, Ph.D., DABT, Eastman Kodak Company, Rochester, New York

Gauke E. Veenstra, Ph.D., Shell Chemicals Ltd. CHSE-GV, Shell Centre London SE1 7NA

John M. Waechter, Jr., Ph.D., DABT, Health and Environmental Sciences, The Dow Chemical Company, Midland, Michigan

Preface

In this Preface to the Fifth Edition, we acknowledge and note that it has been built on the work of previous editors. We especially need to note that Frank Patty's words in the preface of the second edition are cogent:

> This book was planned as a ready, practical reference for persons interested in or responsible for safeguarding the health of others working with the chemical elements and compounds used in industry today. Although guidelines for selecting those chemical compounds of sufficient industrial importance for inclusion are not clearly drawn, those chemicals found in carload price lists seem to warrant first consideration.
>
> Where available information is bountiful, an attempt has been made to limit the material presented to that of a practical nature, useful in recognizing, evaluating, and controlling possible harmful exposures. Where the information is scanty, every fragment of significance, whether negative or positive, is offered the reader. The manufacturing chemist, who assumes responsibility for the safe use of his product in industry and who employs a competent staff to this end, as well as the large industry having competent industrial hygiene and medical staffs, are in strategic positions to recognize early and possibly harmful exposures in time to avoid any harmful effects by appropriate and timely action. Plant studies of individuals and their exposures regardless of whether or not the conditions caused recognized ill effects offer valuable experience. Information gleaned in this manner, though it may be fragmentary, is highly important when interpreted in terms of the practical health problem.

While we have not insisted that chemical selection be based on carload quantities we have been most concerned about agents (chemical and physical) in the workplace that are of toxicological concern for workers. We have attempted to follow the guide as expressed by Frank Patty in 1962 regarding practical information.

The expansion of this edition to include biological agents, e.g., wood dust, Histoplasma, not previously covered, reflects our concern with their toxicology and potential for adverse health effects in workers. In the workplace of the new century, physical agents and human factors appear to be of more concern. Traditionally, these agents or factors, ergonomics, biorhythms, vibration, and heat and cold stress were centered on how one measures them. Today, understanding the toxicology of these agents (factors) is of great importance because it can assist in the anticipation, recognition, evaluation, and control of the physical agent. Their mechanisms of actions and the assessment of adverse health effects are as much a part of toxicology as dusts and the heavy metals.

Chapters on certain topics such as reproduction and development, and neurotoxicology reflect the importance of having at hand for practical use such information to help those persons who are responsible for helping to safeguard health to better understand toxicological information and tests reported for the various chemicals. As noted in Chapter One, the trend in toxicology is increasingly focused on molecular biology and, for this "decade of the genome," molecular genetics. Therefore, it seemed crucial to have a chapter that would help to explain the dogma of our teachers in industrial toxicology that, frequently, there are two workers side by side, and one develops an occupational disease and the other does not. Hence the chapter on genetics was authored by an expert in environmental genetics.

The thinking and planning of this edition was a team effort by us: Charlie, Barbara, and Eula. Over many months we worked on the new framework and selected the contributors. When Charlie died in September, 1998, we (Barbara and Eula) knew that we had a road map and, with the help of our expert contributors, many of whom the three of us have known for 10, 20, or even 30 years, would complete this edition. The team effort was fostered among the current editors by many of the first contributors to Patty's such as Robert A. Kehoe, Francis F. Heyroth, William B. Deichmann, and Joseph Treon, all of whom were at Kettering Laboratory at the University of Cincinnati sometime during their professional lives. The three of us have a long professional association with the Kettering Laboratory: Charles H. Powell received a ScD., Barbara Cohrssen received a MS, and Eula Bingham has been a lifetime faculty member. Many of the authors were introduced to us through this relationship and association.

The authors have performed a difficult task in a short period of time for a publication that is as comprehensive as this one is. We want to express our deep appreciation and thanks to all of them.

Kettering Laboratory, Cincinnati, Ohio EULA BINGHAM, Ph.D.

San Francisco, California BARBARA COHRSSEN, MS

CHARLES H. POWELL, ScD.

Contents

USEFUL EQUIVALENTS AND CONVERSION FACTORS

1 kilometer = 0.6214 mile
1 meter = 3.281 feet
1 centimeter = 0.3937 inch
1 micrometer = 1/25,4000 inch = 40 microinches = 10,000 Angstrom units
1 foot = 30.48 centimeters
1 inch = 25.40 millimeters
1 square kilometer = 0.3861 square mile (U.S.)
1 square foot = 0.0929 square meter
1 square inch = 6.452 square centimeters
1 square mile (U.S.) = 2,589,998 square meters = 640 acres
1 acre = 43,560 square feet = 4047 square meters
1 cubic meter = 35.315 cubic feet
1 cubic centimeter = 0.0610 cubic inch
1 cubic foot = 28.32 liters = 0.0283 cubic meter = 7.481 gallons (U.S.)
1 cubic inch = 16.39 cubic centimeters
1 U.S. gallon = 3,7853 liters = 231 cubic inches = 0.13368 cubic foot
1 liter = 0.9081 quart (dry), 1.057 quarts (U.S., liquid)
1 cubic foot of water = 62.43 pounds (4°C)
1 U.S. gallon of water = 8.345 pounds (4°C)
1 kilogram = 2.205 pounds
1 gram = 15.43 grains
1 pound = 453.59 grams
1 ounce (avoir.) = 28.35 grams
1 gram mole of a perfect gas ≈ 24.45 liters (at 25°C and 760 mm Hg barometric pressure)
1 atmosphere = 14.7 pounds per square inch
1 foot of water pressure = 0.4335 pound per square inch
1 inch of mercury pressure = 0.4912 pound per square inch
1 dyne per square centimeter = 0.0021 pound per square foot
1 gram-calorie = 0.00397 Btu
1 Btu = 778 foot-pounds
1 Btu per minute = 12.96 foot-pounds per second
1 hp = 0.707 Btu per second = 550 foot-pounds per second
1 centimeter per second = 1.97 feet per minute = 0.0224 mile per hour
1 footcandle = 1 lumen incident per square foot = 10.764 lumens incident per square meter
1 grain per cubic foot = 2.29 grams per cubic meter
1 milligram per cubic meter = 0.000437 grain per cubic foot

To convert degrees Celsius to degrees Fahrenheit: °C (9/5) + 32 = °F
To convert degrees Fahrenheit to degrees Celsius: (5/9) (°F − 32) = °C
For solutes in water: 1 mg/liter ≈ 1 ppm (by weight)
Atmospheric contamination: 1 mg/liter ≈ 1 oz/1000 cu ft (approx)
For gases or vapors in air at 25°C and 760 mm Hg pressure:
To convert mg/liter to ppm (by volume): mg/liter (24,450/mol. wt.) = ppm
To convert ppm to mg/liter: ppm (mol. wt./24,450) = mg/liter

CONVERSION TABLE FOR GASES AND VAPORS[a]

(Milligrams per liter to parts per million, and vice versa; 25°C and 760 mm Hg barometric pressure)

Molecular Weight	1 mg/liter ppm	1 ppm mg/liter	Molecular Weight	1 mg/liter ppm	1 ppm mg/liter	Molecular Weight	1 mg/liter ppm	1 ppm mg/liter
1	24,450	0.0000409	39	627	0.001595	77	318	0.00315
2	12,230	0.0000818	40	611	0.001636	78	313	0.00319
3	8,150	0.0001227	41	596	0.001677	79	309	0.00323
4	6,113	0.0001636	42	582	0.001718	80	306	0.00327
5	4,890	0.0002045	43	569	0.001759	81	302	0.00331
6	4,075	0.0002454	44	556	0.001800	82	298	0.00335
7	3,493	0.0002863	45	543	0.001840	83	295	0.00339
8	3,056	0.000327	46	532	0.001881	84	291	0.00344
9	2,717	0.000368	47	520	0.001922	85	288	0.00348
10	2,445	0.000409	48	509	0.001963	86	284	0.00352
11	2,223	0.000450	49	499	0.002004	87	281	0.00356
12	2,038	0.000491	50	489	0.002045	88	278	0.00360
13	1,881	0.000532	51	479	0.002086	89	275	0.00364
14	1,746	0.000573	52	470	0.002127	90	272	0.00368
15	1,630	0.000614	53	461	0.002168	91	269	0.00372
16	1,528	0.000654	54	453	0.002209	92	266	0.00376
17	1,438	0.000695	55	445	0.002250	93	263	0.00380
18	1,358	0.000736	56	437	0.002290	94	260	0.00384
19	1,287	0.000777	57	429	0.002331	95	257	0.00389
20	1,223	0.000818	58	422	0.002372	96	255	0.00393
21	1,164	0.000859	59	414	0.002413	97	252	0.00397
22	1,111	0.000900	60	408	0.002554	98	249.5	0.00401
23	1,063	0.000941	61	401	0.002495	99	247.0	0.00405
24	1,019	0.000982	62	394	0.00254	100	244.5	0.00409
25	978	0.001022	63	388	0.00258	101	242.1	0.00413
26	940	0.001063	64	382	0.00262	102	239.7	0.00417
27	906	0.001104	65	376	0.00266	103	237.4	0.00421
28	873	0.001145	66	370	0.00270	104	235.1	0.00425
29	843	0.001186	67	365	0.00274	105	232.9	0.00429
30	815	0.001227	68	360	0.00278	106	230.7	0.00434
31	789	0.001268	69	354	0.00282	107	228.5	0.00438
32	764	0.001309	70	349	0.00286	108	226.4	0.00442
33	741	0.001350	71	344	0.00290	109	224.3	0.00446
34	719	0.001391	72	340	0.00294	110	222.3	0.00450
35	699	0.001432	73	335	0.00299	111	220.3	0.00454
36	679	0.001472	74	330	0.00303	112	218.3	0.00458
37	661	0.001513	75	326	0.00307	113	216.4	0.00462
38	643	0.001554	76	322	0.00311	114	214.5	0.00466

CONVERSION TABLE FOR GASES AND VAPORS (*Continued*)
(*Milligrams per liter to parts per million, and vice versa;*
25°C and 760 mm Hg barometric pressure)

Molecular Weight	1 mg/liter ppm	1 ppm mg/liter	Molecular Weight	1 mg/liter ppm	1 ppm mg/liter	Molecular Weight	1 mg/liter ppm	1 ppm mg/liter
115	212.6	0.00470	153	159.8	0.00626	191	128.0	0.00781
116	210.8	0.00474	154	158.8	0.00630	192	127.3	0.00785
117	209.0	0.00479	155	157.7	0.00634	193	126.7	0.00789
118	207.2	0.00483	156	156.7	0.00638	194	126.0	0.00793
119	205.5	0.00487	157	155.7	0.00642	195	125.4	0.00798
120	203.8	0.00491	158	154.7	0.00646	196	124.7	0.00802
121	202.1	0.00495	159	153.7	0.00650	197	124.1	0.00806
122	200.4	0.00499	160	152.8	0.00654	198	123.5	0.00810
123	198.8	0.00503	161	151.9	0.00658	199	122.9	0.00814
124	197.2	0.00507	162	150.9	0.00663	120	122.3	0.00818
125	195.6	0.00511	163	150.0	0.00667	201	121.6	0.00822
126	194.0	0.00515	164	149.1	0.00671	202	121.0	0.00826
127	192.5	0.00519	165	148.2	0.00675	203	120.4	0.00830
128	191.0	0.00524	166	147.3	0.00679	204	119.9	0.00834
129	189.5	0.00528	167	146.4	0.00683	205	119.3	0.00838
130	188.1	0.00532	168	145.5	0.00687	206	118.7	0.00843
131	186.6	0.00536	169	144.7	0.00691	207	118.1	0.00847
132	185.2	0.00540	170	143.8	0.00695	208	117.5	0.00851
133	183.8	0.00544	171	143.0	0.00699	209	117.0	0.00855
134	182.5	0.00548	172	142.2	0.00703	210	116.4	0.00859
135	181.1	0.00552	173	141.3	0.00708	211	115.9	0.00863
136	179.8	0.00556	174	140.5	0.00712	212	115.3	0.00867
137	178.5	0.00560	175	139.7	0.00716	213	114.8	0.00871
138	177.2	0.00564	176	138.9	0.00720	214	114.3	0.00875
139	175.9	0.00569	177	138.1	0.00724	215	113.7	0.00879
140	174.6	0.00573	178	137.4	0.00728	216	113.2	0.00883
141	173.4	0.00577	179	136.6	0.00732	217	112.7	0.00888
142	172.2	0.00581	180	135.8	0.00736	218	112.2	0.00892
143	171.0	0.00585	181	135.1	0.00740	219	111.6	0.00896
144	169.8	0.00589	182	134.3	0.00744	220	111.1	0.00900
145	168.6	0.00593	183	133.6	0.00748	221	110.6	0.00904
146	167.5	0.00597	184	132.9	0.00753	222	110.1	0.00908
147	166.3	0.00601	185	132.2	0.00757	223	109.6	0.00912
148	165.2	0.00605	186	131.5	0.00761	224	109.2	0.00916
149	164.1	0.00609	187	130.7	0.00765	225	108.7	0.00920
150	163.0	0.00613	188	130.1	0.00769	226	108.2	0.00924
151	161.9	0.00618	189	129.4	0.00773	227	107.7	0.00928
152	160.9	0.00622	190	128.7	0.00777	228	107.2	0.00933

CONVERSION TABLE FOR GASES AND VAPORS (*Continued*)
(*Milligrams per liter to parts per million, and vice versa;*
25°C and 760 mm Hg barometric pressure)

Molecular Weight	1 mg/liter ppm	1 ppm mg/liter	Molecular Weight	1 mg/liter ppm	1 ppm mg/liter	Molecular Weight	1 mg/liter ppm	1 ppm mg/liter
229	106.8	0.00937	253	96.6	0.01035	227	88.3	0.01133
230	106.3	0.00941	254	96.3	0.01039	278	87.9	0.01137
231	105.8	0.00945	255	95.9	0.01043	279	87.6	0.01141
232	105.4	0.00949	256	95.5	0.01047	280	87.3	0.01145
233	104.9	0.00953	257	95.1	0.01051	281	87.0	0.01149
234	104.5	0.00957	258	94.8	0.01055	282	86.7	0.01153
235	104.0	0.00961	259	94.4	0.01059	283	86.4	0.01157
236	103.6	0.00965	260	94.0	0.01063	284	86.1	0.01162
237	103.2	0.00969	261	93.7	0.01067	285	85.8	0.01166
238	102.7	0.00973	262	93.3	0.01072	286	85.5	0.01170
239	102.3	0.00978	263	93.0	0.01076	287	85.2	0.01174
240	101.9	0.00982	264	92.6	0.01080	288	84.9	0.01178
241	101.5	0.00986	265	92.3	0.01084	289	84.6	0.01182
242	101.0	0.00990	266	91.9	0.01088	290	84.3	0.01186
243	100.6	0.00994	267	91.6	0.01092	291	84.0	0.01190
244	100.2	0.00998	268	91.2	0.01096	292	83.7	0.01194
245	99.8	0.01002	269	90.9	0.01100	293	83.4	0.01198
246	99.4	0.01006	270	90.6	0.01104	294	83.2	0.01202
247	99.0	0.01010	271	90.2	0.01108	295	82.9	0.01207
248	98.6	0.01014	272	89.9	0.01112	296	82.6	0.01211
249	98.2	0.01018	273	89.6	0.01117	297	82.3	0.01215
250	97.8	0.01022	274	89.2	0.01121	298	82.0	0.01219
251	97.4	0.01027	275	88.9	0.01125	299	81.8	0.01223
252	97.0	0.01031	276	88.6	0.01129	300	81.5	0.01227

[a]A. C. Fieldner, S. H. Katz, and S. P. Kinney, "Gas Masks for Gases Met in Fighting Fires," *U.S. Bureau of Mines, Technical Paper No. 248*, 1921.

PATTY'S TOXICOLOGY

Fifth Edition

Volume 6

Ketones
Alcohols
Esters
Epoxy Compounds
Organic Peroxides

CHAPTER SEVENTY-FOUR

Acetone

David A. Morgott, Ph.D., DABT, CIH

Acetone is a clear, colorless liquid that is highly flammable and infinitely soluble in water. Years of clinical study, laboratory testing, and practical experience have shown that acetone can be used safely and without harm in many industrial and commercial applications. The long-standing interest in the biochemical, pharmacological, and toxicological properties of acetone can be traced to three important characteristics of the chemical:

1. Acetone is a normal by-product of mammalian metabolism; levels within the body can, however, be altered by changes in nutrition or energy balance.
2. Acetone is a highly volatile organic solvent that is miscible with water; thus large amounts of the vapor can be absorbed through the lungs and quickly distributed throughout the body.
3. Acetone is manufactured and used in large amounts in a variety of commercial and industrial applications; thus the potential for exogenous human exposure is widespread.

1.0 Acetone

1.0.1 CAS Number: [67-64-1]

1.0.2 Synonyms: 2-Propanone; β-ketopropane; dimethyl ketone; dimethyl formaldehyde; methyl ketone; propanone; pyroacetic acid; pyroacetic ether; allylic alcohol; dimethylketal; ketone propane; and acetone oil

Patty's Toxicology, Fifth Edition, Volume 6, Edited by Eula Bingham, Barbara Cohrssen, and Charles H. Powell.
ISBN 0-471-31939-2

1.0.3 Trade Names: NA

1.0.4 Molecular Weight: 58.08

1.0.5 Molecular Formula: $(CH_3)_2CO$

1.0.6 Molecular Structure:

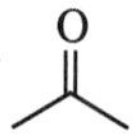

1.1 Chemical and Physical Properties

The commercial grade of acetone is generally 99.5% pure and contains less than 0.4% water and 0.1% organic matter (1). Other important chemical and physical properties of the material are listed in Table 74.1 (2–8).

1.1.1 General

The greatest potential hazard of acetone arises from its flammability. The high vapor pressure and low flash point of acetone can combine to make it a potentially lethal

Table 74.1. Important Chemical and Physical Properties of Acetone

Property	Value	Reference
Empirical formula	C_3H_6O	
Freezing point	−94.7°C	2
Boiling point	56.2°C at 760 mmHg	2
Density	0.790 g/cm^3 at 20°C	1
	0.784 g/cm^3 at 25°C	
	0.780 g/cm^3 at 30°C	
Vapor pressure	70 mmHg at 0°C	3
	185 mmHg at 20°C	
	410 mmHg at 40°C	
Partition coefficient	−0.24 (log $K_{octanol/water}$)	4
	−0.50 (log $K_{oil/water}$)	5
	−2.82 (log $K_{air/water}$)	6
	−2.34 (log $K_{air/saline}$)	5
	−2.44 (log $K_{air/blood}$)	7
Henry's law constant	2.05 atm	6
Water solubility	Infinite	1
Vapor density	2.0 (air = 1.0)	8
Flash point	Cleveland open cup: −9°C	1
	Tag closed cup: −17°C	
Autoignition temperature	465°C	8
Flammability	Lower limit: 2.5% (v/v) at 25°C	8
	Upper limit: 13.0% (v/v) at 25°C	
Hazard identification code	Health: 1 (slight)	8
	Flammability: 3 (high)	
	Reactivity: 0 (stable)	

incendiary under some conditions (9). Acetone easily ignites at room temperature, and its vapors can accumulate in closed containers at levels higher than the lower explosive limit. The vapors are heavier than air and may travel long distances along the ground to an ignition source. Acetone may react violently with strong oxidizing agents such as hydrogen peroxide or nitrosyl chloride. Mixtures of acetone and chloroform form highly reactive intermediates in the presence of alkali (2).

1.1.2 Odor and Warning Properties

Acetone vapors have an intense and characteristically sweet and fruity odor at low concentrations (10). Patte et al. reviewed the odor detection threshold data from six early studies of acetone and reported a normalized value of 24 ppm (11). Devos et al. reviewed an even larger set of work and calculated a standardized odor detection threshold of about 15 ppm (12). The American Industrial Hygiene Association proposed an odor detection threshold of 62 ppm and a recognition threshold of 130 ppm based on their review of the literature (10). A recent review of all available published odor threshold information for acetone revealed considerable variability among the different studies, and the threshold values range from 0.5 to 11,600 ppm (13). When unreliable and poorly documented studies were removed from consideration, the range of threshold values decreased to levels between 20 and 400 ppm.

People adapt quickly to the odor of acetone both during an exposure and between exposure sessions. In a recent study, Wysocki et al. determined the odor detection thresholds for occupationally exposed and nonexposed volunteers and reported mean values of 414 and 50 ppm, respectively (14). Likewise, Dalton et al. reported that the odor detection thresholds for a sample of nonadapted volunteers were not normally distributed and that a 20-min exposure to 800 ppm of acetone could shift the median threshold from a preexposure value of 84 ppm to a postexposure level of 278 ppm (15). Odeigah reported that the recognition thresholds for acetone was bimodally distributed in a group of 970 Nigerian students and that differences in sensitivity could be observed as a function of sex and age (16). Odor threshold data for acetone and other homologous ketones have been used to develop a quantitative structure–activity relationship for a large series of ketonic solvents (17).

Numerous studies have examined the intensity of the human olfactory stimulus response to acetone using different models of intensity perception (18–21). In each case, a relatively steep intensity response function was observed, indicating that the human olfactory nerve is quite sensitive to the odor of acetone and to changes in the odorant concentration.

1.2 Production and Use

Worldwide production capacity of acetone was nominally 3.8 million metric tons (tonnes) in 1995, but the actual volume produced was somewhat less at 3.7 million tonnes (22). Production capacity in the United States constituted about 33% (1.3 million tonnes) of the global capacity, and the capacities in Western Europe and Asia (including Japan) were about 31% (1.2 million tonnes) and 19% (0.7 million tonnes), respectively. The average annual production of acetone is expected to rise at a global rate of 3.3% until the year 2000.

Major end uses of acetone can be divided into three categories. These include use as (*1*) a chemical feedstock, (*2*) a formulating solvent for commercial products, and (*3*) an industrial process solvent. The majority of worldwide production is used as a feedstock to prepare methyl methacrylate/methacrylic acid and bisphenol A (22). Several aldol chemicals, such as methyl isobutyl ketone, methyl isobutyl carbinol, isophorone, and diacetone alcohol are also prepared directly from nascent acetone. Acetone has many favorable properties that make it a preferred formulating solvent for a variety of paints, inks, resins, varnishes, lacquers, surface coatings, paint removers, and automotive care products. As an industrial process solvent, acetone is used to manufacture cellulose acetate yarn, polyurethane foam, vitamin C, and smokeless gun powder. At least 75% of the acetone consumed in 1995 was used in captive processes for preparing downstream chemicals, and only about 12% was used as a formulating solvent for commercial products.

Large-scale commercial production of acetone is generally accomplished by one of two processes. The first, and by far the most common, is through the acid-catalyzed hydrolytic cleavage of cumene hydroperoxide (22). Acetone and phenol are formed as coproducts in this reaction at a ratio of 0.61 to 1.00. The second process, catalytic dehydrogenation of isopropyl alcohol, accounted for about 6% of the U.S. production in 1995. Other methods, such as biofermentation, propylene oxidation, and diisopropylbenzene oxidation, are either experimental or account for a very small percentage of worldwide production.

1.3 Exposure Assessment

Exposure to acetone is inescapable. In addition to its release from many exogenous emission sources, acetone is produced normally within the body from the breakdown and utilization of stored fats and lipids (23). Acetone has routinely been detected in the expired air of humans and in air samples from many different occupied environments (Table 74.2) (24–29). The levels in these samples can vary greatly and range from a few (μg/m^3 to nearly 25 mg/m^3. Cigarette smoking, emissions from furnishings and construction materials, and excretion by the lung are perhaps the greatest contributors to indoor acetone levels. The acetone levels in indoor air are generally higher than those found outdoors (30).

Release by chemical manufacturers' and end users accounts for a very small percentage (1%) of the estimated 40 million tonnes of acetone that is emitted annually to the

Table 74.2. Comparison of Breath Acetone Concentration with the Levels in Different Living Environments

Sample Type	Airborne Concentration (μg/m^3)	Reference
Inside office building	7.1–28.5	24
Inside home	9.5–81	25
Urban street	2.4–306	26
Nonsmoker home	4.7–415	27
Inside aircraft cabin	7.1–560	28
Smoking workplace	9.5–21	27
Human breath	230–11,285	29

Table 74.3. Estimated Average Annual Emissions of Acetone from Different Sources[a]

	Global Annual Emissions (tonnes × 10^{-6})	
Acetone Source	Average	Range
Primary Anthropogenic		
Stationary sources	0.5	0.4–0.7
Mobile sources	0.3	0.2–0.3
Primary Biogenic		
Vegetation	9	4–18
Secondary Anthropogenic & Biogenic		
Propane oxidation	17	15–20
Isobutane & isopropane oxidation	2	1–3
Isobutene & isopropene oxidation	1	1–2
Myrcene oxidation	0.2	0.2–0.3
Biomass Burning	10	8–12
Total	40	30–46

[a]Taken from Singh et al. (31).

environment (Table 74.3) (31). Vegetative releases, forest fires, and other natural events account for nearly half (47%) of the estimated annual emissions of acetone, and another 50% results from the tropospheric photooxidation of propane and other alkanes and alkenes (31). Since 1993, U.S. industries have not been required to report their emissions of acetone as stated under SARA Title III, Section 313. The 1993 Toxic Release Inventory for acetone reveals that 2510 facilities reported a total environmental release of 60,688 tonnes, 58,550 tonnes emitted to the air, 448 tonnes to water, and 215 tonnes to land (32). Investigative, clinical, and environmental methods have been developed for analyzing acetone (33).

1.3.1 Air

Background levels of acetone in outdoor air have been assessed from both ground level and airborne monitoring stations located throughout the world. The average acetone concentrations at rural ground level sites are generally lower than the values reported for urban areas. The concentration of acetone in urban areas can show large unpredictable variations that are likely to be related to vehicle traffic and the emission of precursor alkanes and alkenes (34, 35). Airborne measurements of acetone in the upper troposphere and lower stratosphere show an average concentration range of 190–285 ng/m^3 (36).

Measurable amounts of acetone can be found in the emissions from both mobile and stationary sources (Table 74.4) (37–42). Although the concentrations of acetone emitted from municipal landfills and cigarette smoke can be relatively high, they are minor

Table 74.4. Concentration of Acetone in Various Mobile and Stationary Emissions

Emission source	Airborne Acetone Concentration (mg/m^3)	Reference
Fuel or crude oil fire	0.02–0.16	37
Automobile exhaust	0.09–4.50	38
Factory fence line	1.9–9.7	39
Tree foliage	7.8–12.6	40
Municipal landfill	15.7–77.1	41
Cigarette smoke	498–869	42

contributors to the total global mass. The direct release of acetone from vegetation is an important source that can often be overlooked in an emissions profile. In a qualitative evaluation, acetone was emitted from each of the 22 forest plant species examined in a laboratory study (43).

1.3.2 Background Levels

Acetone can be found as an ingredient in many consumer products ranging from cosmetics and remedies to processed and unprocessed foods. Acetone has been rated as a GRAS (Generally Recognized as Safe) substance when present in beverages, baked goods, desserts, and preserves at concentrations ranging from 5–8 mg/L (44). It can also be detected in measurable amounts in onions, grapes, cauliflower, tomatoes, milk, cheese, beans, peas, and other natural foods. Milk from dairy cattle may contain very high levels of acetone, that range as high as 225 mg/L for cows that have hyperketonemia (45). Acetone has also been identified, but not quantified, in air samples from numerous microorganisms. Acetone is eliminated in the expired air of all mammals, and is excreted as a metabolic end product by some bacteria (*Clostridium butylicium*), molds, fungi (*Paecilaomyces variotii*), and algae (*Cryptomonas ovata palustris*) (46–48).

Acetone is often detected as an end product of thermal combustion and biological decomposition. Except for tree foliage, the release of acetone from living vegetation has been poorly quantified (40). Emissions from poultry manure (530 g/kg), backyard waste incinerators (4.0 g/kg), pine wood combustion (2.8 g/kg), neoprene combustion (990 mg/kg), and wood-burning stoves (145 mg/kg) have all been measured and reported (49–52).

Although dissolution into rainwater is not an important process for removing acetone vapors from the air, small amounts are routinely detected in water samples from various sources. Low levels of acetone have reportedly been found in Los Angeles rainwater ($<$ 50 mg/L), Philadelphia municipal drinking water (1.0 mg/L), and surface seawater off the coast of Florida (22 mg/L) (53–55). Volatilization and biodegradation are the most important processes for removing acetone from water, whereas, photolysis and hydroxyl radical mediated photooxidation are the predominate processes for removing atmospheric acetone (56, 57).

Table 74.5. Concentration of Acetone in Various Polluted and Unpolluted Waters

Sample type	Aqueous Acetone Concentration (μg/L)	Reference
Residential well water	2–7	58
Seawater	5–53	59
Groundwater	12–25	60
Lake water	1–50	61
Storm water runoff	0–100	62
Cloud water	0–17,300	63
Industrial wastewater	138–37,709	64
Landfill leachate	50–62,000	65

The amount of acetone in surface and ground water samples is highly dependent on the source of the sample and the extent of contamination (Table 74.5) (58–65). Ambient background levels of acetone are the result of both natural and commercial releases and generally reflect the physical processes that affect absorption from air, movement through soil, and microbial biodegradation. A search of nearly 2000 entries in the USEPAs STORET database revealed that acetone levels in natural water and industrial monitoring wells rarely exceeded 1 mg/L. A USEPA-sponsored survey of the acetone levels in discharges from 4000 industrial and publicly owned wastewater treatment plants revealed an average water concentration ranging from 0.006 to 2.5 mg/L (64). Potable water in plastic pipe may be contaminated by acetone from the surrounding soil. Water levels of 15 μg/L were observed when the concentration in the surrounding soil was 4.45 g/kg (66).

1.3.3 Workplace Methods

High airborne concentrations of acetone have been found in a variety of occupational environments (Table 74.6) (67–79). These levels reflect the high volatility and low intrinsic toxicity that have combined to make acetone an attractive industrial process solvent. The predominant route of both occupational and consumer exposure to acetone is through inhalation. Oral and dermal uptake can occur, but the body burden from these exposure routes is relatively small compared to respiratory absorption. NIOSH Method 1300, for ketones, is recommended for determining workplace exposures to acetone (79a).

1.3.4 Community Methods

Acetone is found in a wide variety of consumer and commercial products, but only a few contain appreciably high percentages by volume (80). These include paints and paint-related products, such as paint thinners, finger nail polish removers, automotive waxes, and tar removers (Table 74.7). Consumer exposures are most likely to occur by inhalation and will be greatest for products containing a high percentage of acetone such as adhesives, automotive products, and paints.

Table 74.6. Concentration of Acetone in Different Work Environments

Location	Time-Weighted Average Conc (mg/m^3)	Reference
Glue spraying	1–40	67
Automotive repair shop	12–77	68
Hospital EEG lab	1–160	69
Print shop	6–235	70
Shoe factory	25–393	71
Automotive assembly	0.01–460	72
Electronics plant	2–648	73
Coin and medal mint	415–888	74
Decontamination unit	440–1090	75
Fiberglass fabrication	40–1580	76
Varnish production	5–1448	77
Boatyards	30–1700	78
Cellulose acetate plant	12–2876	79

1.3.5 Biomonitoring/Biomarkers

Virtually every tissue and organ in the human body contains some acetone, which is one of three biochemicals collectively referred to as the "ketone bodies" (the others are β-hydroxybutyrate and acetoacetate). Acetone is produced within the body from the breakdown and utilization of stored fats and lipids as a source of energy. Consequently, conditions such as strenuous physical exercise and prolonged dieting, which lead to a breakdown of fat within the body, may result in higher than average amounts of acetone in the bloodstream. Measurable amounts of acetone are continuously excreted in the breath and urine of humans because of its high volatility and solubility in water (Table 74.8) (29, 81–97).

Table 74.7. Average Acetone Concentration in Various Consumer Product Categories[a]

Product Category	No. Products Surveyed	Acetone Prevalence (%)	Avg. Acetone Conc (%)
Oils, greases and lubricants	71	5.3	0.2
Cleaners for electronic equipment	111	16.1	0.3
Household cleaners and polishers	463	10.8	0.3
Miscellaneous products	76	17.2	7.4
Fabric and leather treatments	91	14.6	12.9
Adhesive-related products	69	24.3	18.8
Automotive products	111	22.7	28.1
Paint-related products	167	51.5	29.3

[a]Taken from Sack et al. (80).

Table 74.8. Normal Values for the Acetone Concentration in Human Blood and Urine Specimens

Type of Specimen	No. of Subjects	Average Conc mg/L	Average Conc μg/L	Std. Dev. or Range (mg/L or μg/L)	Reference
Plasma	20	4.35	—	1.31	81
Plasma	20	1.74	—	11.6	82
Plasma	31	0.41	—	0.17	83
Serum	11	2.9	—	0.3	84
Whole blood	6	0.93	—	0.06	85
Whole blood	216	1.25	—	0.0–17.4	86
Whole blood	88	0.84	—	0.56	87
Whole blood	16	1.56	—	?–5.21	88
Whole blood	1062	1.8	—	0.64– >6.0	89
Whole blood	288	1.59	—	0.15–15.4	90
Spot urine	20	3.02	—	1.25	81
Spot urine	49	0.84	—	0.13–9.35	91
Spot urine	15	0.76	—	0.63	92
Spot urine	10	0.8	—	0.2	84
Expired air	9	—	1.52	0.36	93
Expired air	88	—	0.71	0.02–3.32	91
Expired air	187	—	1.45	0.29–8.25	29
Expired air	13	—	1.19	0.52–2.07	94
Expired air	23	—	1.04	0.29	95
Expired air	67	—	1.10	0.88	96
Expired air	40	—	1.1	0.5	97
Expired air	14	—	0.97	0.07	84

Detectable amounts of acetone have been found in a variety of biological specimens including whole blood, cerebrospinal fluid, urine, exhaled air, vitreous humor, and breast milk (98–102). Because endogenous acetone formation is closely linked with the utilization of stored fats as a source of energy, levels can fluctuate, depending on an individual's health, nutrition, and level of physical activity. The following variables may influence the amount of acetone found in a normal blood or plasma specimen: (1) the duration and degree of fasting (103), (2) the point in the circadian ketogenic cycle (104); (3) the type of test specimen collected and the length of time before analysis (105), and (4) the procedures used for separating and analyzing the specimen.

Analytical techniques have been developed to monitor exposure from both endogenous and exogenous acetone. The methods have been valuable for monitoring the weight loss programs of diabetics, assessing occupational exposures, evaluating community exposures to contaminated air or water, and tracking the effectiveness of drugs (106–110). Other areas of use for endogenous acetone measurements have been in treating diabetes, determining alcohol consumption, and evaluating congestive heart failure (111–113). Urinary levels of acetone have also been used to monitor occupational exposures to isopropanol (114).

Perhaps the greatest potential application of acetone as a biomonitoring tool is in the use of blood, urine, and expired air levels as indicators of occupational exposure. In a study of 104 workers from three different factories (paint, plastics, and artificial fibers), Pezzagno et al. found a close linear relationship ($r = 0.91$) between 4-h urine acetone levels and the 4-h time-weighted average (TWA) exposure concentration (92, 115). The data indicated that urinary acetone levels could be used to monitor acetone exposures biologically in the workplace and that a urine acetone concentration of 55 mg/L was equivalent to a 4-h TWA exposure concentration of 1000 ppm. Kawai et al. examined the relationship between urine and ambient air levels of acetone in a group of 28 workers who had 8-h TWA exposures ranging from about 1–45 ppm. They found a good correlation ($r = 0.90$) between the exposure concentration and the amount of acetone in the urine (116). The authors, however, noted that repetitive exposures at concentrations greater than 15 ppm caused daily accumulation of acetone in the urine (117). Fujino et al. measured the urine, expired air, and blood concentrations of acetone in 110 workers exposed to 8-hr time-weighted exposures ranging from about 5–1200 ppm (118). The best statistical correlations were observed when the urinary acetone concentrations and ambient exposure levels were compared ($r = 0.71$). The acetone concentrations corresponding to an 8-h TWA exposure of 750 ppm were 118 mg/L in blood, 177 ppm in expired air, and 76.6 mg/L in urine. In subsequent work, Satoh et al. examined whether the day-to-day carryover of acetone in the urine could distort the ambient relationship between exposure concentration and urinary excretion by measuring urine acetone levels at the beginning of a second day of work (119). Although the average urinary concentration of acetone did not completely return to normal during the 16 h away from work, the change did not affect the strength of the statistical relationship between exposure concentration and urine level.

The strong statistical relationship that exists between inhaled acetone levels and the concentrations in expired air and blood have led to several alternate approaches for monitoring occupational exposure to acetone (120, 121). Brugnone et al. measured the alveolar air and venous blood acetone levels of shoe factory employees exposed to airborne concentrations ranging up to 21.1 ppm (122, 123). The acetone concentration in venous blood was linearly related to the alveolar air concentration but was not related to the levels in ambient air. The acetone blood levels (0.5 to 3.0 mg/L) of the employees were quite low, however, and within physiologically normal limits. Baumann and Angerer reported that the acetone concentration in the blood of four workers exposed to 30 ppm for 2 h only increased slightly and rose from 1.0 mg/L to 3.3 mg/L by the end of the shift (124).

Wang measured the concentration of acetone in the blood, urine, and expired air of shoe, printing, and plastic workers and found good correlations for the relationship between inspired air concentrations and the levels in blood ($r = 0.77$) and expired air ($r = 0.80$) (91). The biological monitoring of workers exposed to acetone was not affected by the endogenous production of acetone or by exposure to isopropanol. These results agreed with those of Mizunuma et al. who found a strong relationship between the ambient air concentration of acetone and levels in the blood ($r = 0.86$) and serum ($r = 0.90$) of 41 workers who were coexposed to styrene (125). Acetone exposures, however, alter the biological monitoring of occupational styrene exposures by impeding the excretion of key metabolites into the urine (126, 127). In an unusual study, Tomita and Nishimura proposed using saliva acetone measurements as an exposure monitoring tool based on the observed relationship between salivary concentration and expired air levels in a volunteer exposed to 600 or 2500 ppm of acetone for 15 min (128). Considerable attention has also been given to the potential impact of endogenous acetone exhalation on the accuracy of breath alcohol measurements by law enforcement groups (129–131).

Acetone may be used as a biochemical marker of past exposures to any chemical that can cause oxidative stress through lipid peroxidation. Acetone has repeatedly been detected at high concentrations in the urine and exhaled air of laboratory animals previously exposed to a chemical that can be metabolized to a free radical intermediate (Table 74.9) (132–149). Once formed, the radical intermediates react with molecular oxygen to produce peroxy radicals which then oxidize polyunsaturated fatty acids located in cell membranes. Acetone and several low molecular weight aldehydes are products of this peroxidation reaction and are released into the blood for elimination. Pretreating rats with antioxidants such as vitamin E and ellagic acid prevented lipid peroxidation and acetone release caused by administration of endrin (150). Surprisingly, treating mice with a 10% concentration of ethanol in drinking water for six months did not affect their blood levels of acetone compared to untreated control mice (151).

The structural requirements for halocarbon-induced acetonemia were investigated in a series of inhalation experiments performed by Filser et al. (149, 152). Male rats were exposed to 14 different solvent vapors for 50 h, during which the acetone production rate

Table 74.9. Chemicals that Affect the Formation of Acetone in Laboratory Animals by Causing Oxidative Stress

Test Chemical	Species and Treatment	Type of Specimen	Acetone Conc. Change Observed	Reference
Endrin	Rat—oral	Urine	2.7-fold increase	132
Ethanol	Rat—oral	Urine	2.8-fold increase	133
Paraquat	Rat—oral	Urine	10.7-fold increase	134
Carbon tetrachloride	Rat—oral	Urine	4-fold increase	135
Alachlor	Rat—oral	Urine	4.5-fold increase	136
Malondialdehyde	Rat—oral	Urine	50% decrease	137
Menadione	Rat—oral	Urine	3.2-fold increase	138
2,3,7,8-TCDD	Rat—oral	Serum	8.7-fold increase	139
2,3,7,8-TCDD	Mouse—oral	Amniotic fluid	58% increase	140
Lindane	Mouse—oral	Serum	2.4-fold increase	141
Sodium dichromate	Rat—oral	Urine	3.3-fold increase	142
Cadmium chloride	Rat—oral	Urine	1.5-fold increase	143
Smokeless tobacco	Rat—oral	Urine	1.6-fold increase	144
Chloral hydrate	Mouse—*in vitro*	Microsomes	5.1-fold increase	145
1,2-Trichloroethylene	Rat—inhalation	Exhaled air	39-fold increase	146
Diquat	Rat—i.p.	Urine	5-fold increase	147
1,3-Butadiene	Rat—inhalation	Exhaled air	50-fold increase	148
1,1,2-Trichloroethane	Rat–inhalation	Exhaled air	60-fold increase	149

was measured. Halogenated ethylenes generally resulted in greater acetone production than halogenated ethanes or methanes. The only solvents that did not result in any appreciable exhalation of acetone were 1,1,1-trichloroethane and *n*-hexane. Simon et al. initially postulated that acetone formation involved a complex compensatory increase in the rate of fatty acid utilization which resulted in above normal production of the ketone bodies (153). Attempts by Buchter et al. to use acetone production as a biomonitor for tetrachloroethylene exposure in humans were not successful (154). Although acetone exhalation in the exposed workers tended to be higher than the unexposed controls, high individual variation prevented any meaningful application of the results.

Disulfiram, a therapeutic agent used to treat chronic alcoholics and a potent inhibitor of hepatic cytochrome P450IIE1 and alcohol dehydrogenase activity, also reportedly caused acetonemia in both humans and rats (155, 156). The mechanism, however, does not involve lipid peroxidation but is related instead to the suicidal inactivation of a key enzyme required for metabolizing acetone. Compounds of similar structure such as diallyl sulfide and diethyldithiocarbamate can similarly cause irreversible inactivation of cytochrome P450IIE1 that is partially prevented when acetone is incorporated into the reaction medium as a competitive inhibitor (157). Filser and Bolt observed that diethyldithiocarbamate, a major metabolite of disulfiram, caused acetonemia in rats when used experimentally to inhibit hepatic metabolism (158). DeMaster and Nagasawa showed that nonfasting male rats given daily oral doses of disulfiram responded with acetone blood levels that were elevated up to 25-fold (159). A single subcutaneous dose of disulfiram or intraperitoneal dose of cyanamide could also increase blood acetone levels up to 16-fold in rats (160, 161). Disulfiram-induced acetonemia was attributed to the enzymatic inhibition of acetone metabolism. The acetonemia observed following cyanamide administration occurred by a different mechanism because parallel increases in blood acetoacetate and β-hydroxybutyrate levels also occurred.

Humans who received disulfiram therapy for at least a month showed an average increase of 15-fold in their expired air levels of acetone (160). These results were mirrored by those of Stowell et al. who found that nonalcoholic volunteers given disulfiram for two days responded with a rapid and marked rise in blood acetone that lasted for 3 to 5 days (162). Disulfiram-induced acetonemia was not observed, however, when another group of test subjects was treated subcutaneously with the drug (163).

1.3.5.1 Blood. The normal limits for blood acetone are presented in Table 74.10 along with the levels observed under extreme conditions (164). The acetone levels in the body at any time reflect acetoacetate production which is in turn affected by the use of free fatty acids by the liver. Consequently, altered physiological states, such as those described in Table 74.11, can appreciably increase ketogenesis and the body burden of acetone (165–171). For instance, infants, pregnant women, and exercising humans can have ketone body levels that are elevated two- to 20-fold above normal due to the ketogenesis from their higher energy requirements (172, 173). Likewise, children can have appreciably higher blood acetone blood levels than adults due to their higher energy expenditure and the possible ketogenic influence of growth hormone (174). The average blood acetone levels in infants 2 to 5 days old were reportedly threefold higher than in adolescents 10 to 15 years of age (175). Blood acetone levels as high 100 mg/L have been observed in obese

Table 74.10. Reference Values for Human Plasma Acetone under Different Conditions[a]

	Acetone Plasma Conc Limits		
Condition	(mg/L)	(mg%)	(mM)
Normal	< 10	< 1.0	< 0.17
Occupational exposure	< 100	< 10.0	< 1.72
Diabetic ketacidosis	100–700	10.0–70.0	1.72–12.04
Overt toxicity	> 200	> 20.0	> 3.44

[a]Taken from Teitz (164).

individuals who participated in a comprehensive weight loss program (176). The average blood levels of acetone are not different for smokers and nonsmokers (91).

Trotter et al. reported that the acetone concentration in plasma is 8–11% greater than the level in whole blood, whereas, Gavino et al. found no difference in these two types of specimens (82, 177). The average blood acetone concentration in nonfasting rats was nearly equivalent to the level found in nonfasting human subjects (0.99 mg/L vs. 0.93 mg/L, respectively). Others have reported that the average blood acetone level in nonfasting Wistar rats were 1.20 mg/L (178). Recent data have also shown that acetone levels can

Table 74.11. Human Physiological and Clinical Conditions that Lead to an Increase in Acetone Production[a]

Physiological Conditions

- Pregnancy
- Postnatal growth
- High fat consumption
- Dieting
- Lactation
- Vigorous physical exercise
- Perinatal development
- Physical exertion

Disease States

- Starvation
- Alcoholism
- Diabetes mellitus
- Hypoglycemia
- Eating disorders
- Prolonged vomiting
- Prolonged fasting
- Acute trauma
- Inborn errors of metabolism

[a]Refs. 165–171.

Table 74.12. Blood and Urine Acetone Levels in Alcoholics, Diabetics, and Fasting Humans

Subject Description	Sample Type	No. of Subjects	Average Conc mg/L	Average Conc μg/L	Reference
Controlled diabetics	Plasma	50	5.8	—	81
Ketotic diabetics	Plasma	10	23.2	—	180
Ketoacedotic diabetics	Plasma	27	424	—	100
Controlled diabetics	Plasma	55	5.4	—	181
Uncontrolled diabetics	Serum	8	45.5	—	84
Nonobese adults fasted 3 days	Plasma	6	46.5	—	180
Obese adults fasted 21 days	Plasma	3	81.3	—	180
Alcohol intoxication	Blood	5	5.3	—	182
Autopsy (hyperglycemics)	Blood	15	395	—	183
Autopsy (chronic alcoholics)	Blood	9	183	—	184
Controlled diabetics	Spot urine	9	6.8	—	84
Uncontrolled diabetics	Spot urine	8	14.0	—	84
Diabetics	Spot urine	6	3.1	—	185
Autopsy (diabetics)	Spot urine	14	599	—	98
Healthy adults 36-h fast	Expired air	6	—	63.1	93
Controlled diabetics	Expired air	10	—	2.3	84
Uncontrolled diabetics	Expired air	7	—	7.9	84
Controlled diabetics fasting overnight	Expired air	129	—	30.1	97
Juvenile diabetics fasting overnight	Expired air	33	—	104.7	97
Juvenile diabetics	Expired air	49	—	4.4	96
Diabetics reduced carbohydrate diet	Expired air	19	—	2.78	95

increase after sample collection due to the decarboxylation of blood acetoacetate. This reaction is catalyzed by plasma proteins, particularly albumin, and is active at temperatures as low as $-20°C$ (179).

Aside from cases of accidental or intentional exposure, the most dramatic increases in blood ketone bodies occur in uncontrolled diabetes mellitus. As shown in Table 74.12 (180–185), patients who have severe diabetic ketoacidosis can have plasma acetone levels as high as 750 mg/L which is more than 300 times the normal limit (180, 186). Clinical findings in cases of acute acetone intoxication suggest that acetone blood levels in excess of 1000 mg/L are necessary to cause unconsciousness in humans (187, 188). Brega et al. reported that occupationally exposed workers can have plasma acetone levels that range from 6.0 to 74.9 mg/L (81). Similarly, Kobayashi et al. indicated that acetone intoxicated patients can display urine levels that range from 31.0 to 650.9 mg/L (185).

Other clinical conditions, such as diabetic ketosis and prolonged fasting, lead to more moderate increases in the acetone level. In each of these conditions, the elevations in blood acetone are typically accompanied by even larger increases in the two remaining ketone bodies; however, unlike acetone, acetoacetate and β-hydroxybutyrate are ionizable organic acids that can disrupt normal acid–base balance when formed in sufficient amounts. In contrast, acetone is nonionic and is produced along with carbonic acid during acetoacetate breakdown. Some authors have postulated that the formation of acetone and carbonic acid in the body represents a control mechanism for diabetics, whereby the acid–base imbalance from the strongly acidic acetoacetate is ameliorated by the metabolic formation of a weakly acidic carbonic acid metabolite (186).

1.3.5.2 Urine. Values for the normal background concentration of acetone in spot urine specimens from humans are presented in Table 74.6. Although studies have shown that the acetone concentration in repetitive urine collections from resting humans does not increase appreciably following light physical exercise, the studies did reveal a consistent diurnal trend; urine concentrations were higher in the late evening and early morning than during the daytime (92). Kobayashi et al. measured the urine acetone levels in several human volunteers and found that the levels were generally 100- to 1000-fold lower than those from autointoxicated individuals (185).

1.3.5.3 Other. A large number of investigators have published normal values for the concentration of acetone in human expired air. The values listed in Table 74.6 show that the average acetone concentration ranges between 1.0 and 1.5 mg/L, regardless of the degree of fasting. Sex-related differences in acetone exhalation existed in a large group of nonfasting healthy subjects (189). Significantly higher exhaled air concentrations were found in normal females than in males, but the sex difference largely disappeared when diabetic outpatients were examined. Exhaled acetone levels in normal subjects also (1) increased or decreased in response to weight loss or gain; (2) increased rapidly and dramatically when carbohydrates were removed from the diet, and (3) were relatively constant for the different age groups examined. The rate of acetone exhalation can vary over a broad range in healthy human volunteers, values typically fall between 29 and 230 mg/h (190).

Numerous studies have documented slight to moderate increases in breath acetone for diabetics and fasting adults. As shown in Table 74.12, elevations in breath acetone are generally less severe than those observed for plasma acetone; there are, however, several notable exceptions. Stewart and Boettner found that breath acetone levels in cases of juvenile diabetes were, on average, nearly 100 times greater than normal. Likewise, diabetics who tested strongly positive for glucosuria had breath levels that averaged about 340 times the normal level (97). Rooth and Østensen reported that an individual who consumed a paint thinning solvent had a breath acetone level of 2200 mg/L two days after the event (96). Several reports have shown that a 36-hour fast can result in a 40-fold elevation in breath acetone level and that the increase could be immediately reduced by consuming small amounts of ethanol (93, 191, 192).

Kernbach et al. reported that acetone was detected in postmortem cerebrospinal fluid and that the levels did not exceed 5 mg/L, except where diabetic coma was the cause of death (193). In cases of ketoacidotic coma, the cerebrospinal fluid levels averaged 143 mg/L.

1.4 Toxic Effects

An examination of all available information on the biological activity of acetone indicates that the vapors are mildly toxic after both direct contact or systemic absorption. The primary effect of acute high-level exposure is central nervous system depression. The data indicate that acetone does not pose a neurotoxic, carcinogenic, or reproductive health hazard at the concentrations found anywhere in the environment. Information obtained from animal feeding studies indicate that the kidney may be the most sensitive target tissue. The mild effects of acetone have allowed using it as a carrier solvent for dissolving and testing less soluble chemicals in many *in vitro* toxicity assays.

1.4.1 Experimental Studies

1.4.1.1 Acute Toxicity. Acute toxicity studies of acetone in laboratory animals have focused primarily on three endpoints, lethality, narcosis, and sensory irritation. The acute effects of acetone have been examined in mice, rats, rabbits, guinea pigs, dogs, cats, and monkeys. Early interest in the anesthetic properties of acetone also inspired some investigators to examine the cardiovascular and pulmonary effects of the material. The lethality and irritancy potential of acetone have been repeatedly examined using both *in vivo* and *in vitro* techniques. Although, the oral route was typically used to assess the acute lethality of acetone, a considerable amount of data have been obtained following vapor inhalation, dermal application, and intraperitoneal injection (Table 74.13) (194–204). The oral LD_{50} values for male and female rats from different age groups reveal that acetone is more acutely toxic to newborn (LD_{50} is 1.8 g/kg) rats than to adults (LD_{50} is 6.8 g/kg). The LD_{50} values for rats aged 14 days were midway between these two values (198).

Vapor inhalation has been the preferred route of exposure for most neurotoxicological and sensory irritation studies of acetone. The ocular and dermal routes have typically been

Table 74.13. Acute Lethality of Acetone to Laboratory Animals by Different Routes of Exposure

Strain & Species	Sex	Exposure route	LD_{50} or LC_{50} Value (ppm or g/kg)	Reference
Carworth–Nelson rat	F	Inhalation	21,150	194
Carworth–Wistar rat	F	Oral	8.5	195
Carworth–Wistar rat	F	Oral	10.0	194
Rat	Unk	Oral	9.8	196
Sprague–Dawley rat	M	Oral	9.75	197
Sprague–Dawley rat	M & F	Oral	7.3	198
Rabbit	Unk	Oral	5.3	199
ddY mouse	M	Oral	5.2	200
New Zealand rabbit	M	Dermal	>15.8	195
Rabbit	Unk	Dermal	20	201
Harley guinea pig	M	Dermal	>7.4	202
CF-1 mouse	M	Intraperitoneal	3.1	203
CR rat	Unk	Intraperitoneal	0.62	204
Rat	Unk	Intraperitoneal	1.3	196

used to examine the local irritancy of acetone and its activity as a contact sensitizer. The intravenous and subcutaneous routes have generally been used to collect data on pulmonary and cardiovascular effects. The hallmark of many acetone inhalation studies has been the extremely high vapor concentrations or long exposures needed to cause a serious adverse effect. Smyth et al. reported that five of six rats survived a 4-h exposure to 16,000 ppm of acetone but that the remaining rats died when the concentration was doubled to 32,000 ppm (195). Likewise, Mashbitz et al. reported that mice exposed to acetone at concentrations ranging from 42,200 to 84,400 ppm became unconscious after about 35 min of exposure (205). The following gross abnormalities were observed in female guinea pigs that succumbed to acetone exposure ranging from 10,000 to 50,000 ppm: pulmonary congestion and edema, splenic congestion and hemorrhage, renal congestion, and glomerular distension (206). Safronov et al. determined the LC_{50} of acetone in rats and mice for exposures lasting 15, 30, 60, 120, and 240 min (207). The LC_{50} values ranged from 604,000 mg/m^3 (15 min) to 44,000 mg/m^3 (4 h) for male mice and 724,000 mg/m^3 (15 min) to 71,000 mg/m^3 (4 h) for male rats. Kennedy and Graepel reviewed much of the acute lethality data and concluded that acetone possesses a low order of toxicity to rats (208).

The acute data summarized in Table 74.14 (209–212) indicate that acetone concentrations higher than 8,000 ppm are required to elicit any sign of intoxication in laboratory animals, regardless of the exposure duration. The data also indicate that the degree of CNS involvement intensifies when either the exposure concentration or the exposure duration is increased. The narcotic effects of acetone proceed through several distinct phases that include drowsiness, incoordination, loss of autonomic reflexes, unconsciousness, respiratory failure, and death. Using rats, Haggard et al. demonstrated that the onset of narcosis was closely related to the acetone levels in the blood and that many of the stages of toxicity could be induced by manipulating either the exposure concentration or the exposure duration (210). Blood levels of 1000 to 2000 mg/L were found necessary for slight incoordination, 2910 to 3150 mg/L for loss of the righting reflex, and 9100 to 9300 mg/L for respiratory failure. Albertoni reported that a minimum oral acetone dose of 4 g/kg caused narcosis in dogs and that 8 g/kg caused death (213). Similarly, Walton et al. reported that the minimum narcotic and lethal doses of acetone for rabbits were 7 and 10 mL/kg, respectively (214).

Anderson reported that the intracranial administration of acetone at a dose of about 550 mg/kg caused convulsive seizures and spasms in rabbits (215). Frantík et al. examined the acute neurotoxic effects of acetone and many other solvents by determining the inhalation concentration needed to cause a 30% depression in an electrically induced seizure response in rats (4-h exposure) and mice (2-h exposure) (216). The isoeffective concentration (EC_{30} value) of acetone was 3500 ppm in rats and 5000 ppm in mice. A subsequent structure–activity analysis revealed that hydrophobicity was one of several factors needed to adequately describe the relationship between chemical structure and neurotropic potency (217). Kanada et al. found that an oral acetone dose of 2.44 g/kg caused an increase in the concentration of some brain neurotransmitters in rats (197). Larsen et al. examined the ability of acetone and other solvents to block the action potential propagation in isolated frog sciatic nerve and found that a concentration of 618 mM (35.9 g/L) was needed to produce a 50% decrease in the intensity of the response

Table 74.14. Acute Toxicity of Acetone to Laboratory Animals Following Inhalation Exposure

Species	Exposure conc (ppm)	Duration (h)	Observed effects	Reference
Mice	8,440	4.0–7.75	Loss righting reflex	209
	8,440	7.75	Narcosis in some animals	
	20,256	1.5	Deep narcosis	
	20,256	1.0–1.2	Loss righting reflex	
	46,420	0.6–1.0	Deep narcosis	
	46,420	1.0	Lethal	
Rats	2,100	8.0	None	210
	4,220	8.0	None	
	10,550	1.7–4.2	Incoordination	
	21,100	2.2–2.7	Loss righting reflex	
	42,200	1.75–1.9	Loss corneal reflex	
	42,200	4.5–5.5	Respiratory failure	
	84,400	2.5–3.0	Respiratory failure	
	84,400	0.35–0.83	Loss corneal and righting reflex	
	126,600	1.75–2.25	Respiratory failure	
	126,600	0.17–0.42	Loss corneal and righting reflex	
Cats	16,880	3.75–4.0	Loss righting reflex	211
	48,108	1.5	Loss righting reflex	
	75,116	1.0–1.25	Deep narcosis	
Cats	3,375–4,220	5.0	Eye and nose irritation	212
	8,440–21,100	3.0–4.0	Drowsiness and stupor	
	33,760–42,200	4.0	Narcosis with convulsions	
	52,750	1.5	Narcosis with convulsions	
Guinea pigs	10,000	47–48	Some lethality	206
	20,000	9.0	Narcosis	
	50,000	3.0–4.0	Lethal	

(218). Using synaptosomes isolated from mouse brain, Huang et al. found that acetone concentrations of 1080 μM and 656 μM caused 50% inhibition in beta-adrenergic receptor binding and sodium potassium ATPase activity, respectively (219). The anesthetic potency of ketonic solvents was determined in mice by Tanii and reported as the intraperitoneal dose needed to eliminate the righting reflex (220). The AD_{50} value for acetone was 59.6 mmol/kg (3.46 g/kg). Ghoniem et al. developed an animal model for human irritable bowel syndrome by intravesical installation of 20 mL of a 50% acetone solution into the bladders of African green monkeys (221). When measured one week after treatment, micturation frequency, urine volume, peak urine flow, urea absorption, and bladder capacity were affected by the acetone administration. Acetone caused interstitial cystitis in the bladder by destroying the mucin layer. Similar findings were also observed in rabbits treated with a 50% acetone solution (222). The functional changes observed in the rabbit bladder were reversible within two months of treatment.

Panson and Winek examined the effects of acetone aspiration in anesthetized members of each sex by examining the degree of lung hemorrhage and changes in the lung weight/body weight ratio (223). Acetone was hazardous when aspirated, so the standard approach (vomiting and gastric lavage) to oral acetone poisoning was judged dangerous and inappropriate. Enhorning et al. studied the effects of acetone on the surfactant film that lines the bronchi and alveoli of rabbit lung and reported that a 25% vapor concentration did not alter the surface tension of the film (224).

Bruckner and Petersen examined the anesthetic properties of acetone in male rats exposed for 3 h to inhalation of 12,600, 19,000, 25,300, and 50,600 ppm (225). A dose-related decline in unconditioned performance and involuntary reflex was observed at all but the highest acetone exposure level at which death occurred within 2 h of initiating the exposure. Performance scores returned to baseline levels after a 24-h recovery period. Goldberg et al. examined the avoidance and escape behavior of female rats exposed to 3000, 6000, 12,000, or 16,000 ppm of acetone vapors for 10 days for 4 h/day (226). The 3000 ppm exposures had no effect on all exposure days, the 6000 ppm exposure initially inhibited the conditioned avoidance response but not the unconditioned escape response, and the two highest exposures inhibited both responses. Normal responses were obtained after three days of exposure to 6000 and 12,000 ppm, indicating that adaptive changes develop upon repeated exposure.

Geller et al. reported that continuous (24 h/day) exposure to 500 ppm acetone for seven days acetone caused minimal effects in male baboons that were taught to perform a complex operant discrimination task (227). In other studies, operant behavior was monitored in male rats exposed to 150 ppm of acetone for times ranging from 30 min to 4 hours (228). The animals were trained to respond on a fixed-ratio (FR) or a fixed-interval (FI) schedule of reinforcement (food reward) following an auditory stimulus. The results were highly variable, the 30-min exposure caused no effect, the 1-h exposure caused an increase in FR and FI values, the 2-h exposure resulted in a decrease in both values, and the 4-h exposure caused inconsistent changes among the animals.

Garcia et al. studied the lever-pressing activity of rats operantly trained on a FR schedule of reinforcement and found that acetone exposures of 25 to 100 ppm for 3 h produced extremely variable results (229). Preexposure baseline values measured two days before the exposure was initiated varied over a seven-fold range. Most of animals were reportedly unaffected by the acetone exposure; however, several showed either a relative increase or decrease in their response rate for up to five days postexposure. Glowa and Dews examined the effects of acetone on schedule-controlled operant behavior in mice serially exposed to six nominal concentrations of acetone ranging from 100 to 56,000 ppm (230). The mice received a food reward on a FI schedule after correctly responding to a specific presentation of two feeding stimuli (blinking lights and white noise). The correct response rate was not affected by acetone concentrations less than 1000 ppm, whereas, 56,000 ppm completely eliminated the response. The response rate returned to the normal baseline levels 30 min after terminating the 56,000 ppm exposure. The EC_{50} value for acetone-induced changes in schedule-controlled operant behavior was 10,964 ppm.

More recent studies of this type by Christoph and Stadler yielded better data but also showed some variability (231). No consistent or permanent changes were observed in the schedule-controlled operant performance of rats exposed 6 h/day to 1000, 2000, or

4000 ppm of acetone for 13 weeks. Animals were trained on a multiple FR-FI schedule of food presentation and were tested daily at the beginning of the exposure session. During the last few weeks of the exposure, the authors observed an increase in the variability of the FR post-reinforcement pause duration relative to control values that occasionally caused a significant difference. All other FR and FI performance measurements were unchanged during or after the 13-week exposure period.

De Ceaurriz et al. subjected mice to a behavioral despair swimming test following a 4-h inhalation exposure to acetone at levels ranging from about 2000 to 3000 ppm and found that concentrations ranging from 2600 to 3000 ppm caused the swimming lag time to decrease up to 59% (232). The ID_{50} value for acetone in this test was 2800 ppm. Kane et al. examined the irritancy of acetone vapors in mice by using a short-term test that measured the reflexive decline in respiration rate after a 10-min inhalation exposure (233). By this technique, the RD_{50} value of acetone found was 77,516 ppm. Acetone was also one of only two chemicals examined that caused the mice to develop a rapid and complete tolerance to any reflexive decline in respiration resulting from the exposure. Using male mice of a different strain and a 5-min exposure period, De Ceaurriz et al. reported that the acetone irritancy RD_{50} value was 23,480 ppm (234). Comparison of the RD_{50} values for nearly forty chemical vapors and gases showed that acetone was the weakest irritant in the entire group (235). Numerous structure–activity relationships have been developed using the irritancy data collected for acetone and other ketonic solvents (236–240). These relationships are based on the correlations observed between the irritancy RD_{50} values in mice and various physiochemical properties of the chemical tested.

The parenteral routes of treatment most frequently used in acute studies of acetone toxicity have been intramuscular, intravenous, intraperitoneal, and subcutaneous injection. The lethality of acetone by secondary parenteral routes is not substantially different from the potency observed by oral and inhalation exposure. The minimum lethal subcutaneous dose of acetone for both dogs and guinea pigs was reportedly 5 g/kg (209, 213). A 4.0-mL/kg intravenous dose of acetone was immediately lethal to rabbits, whereas 2.0 mL/kg caused delayed mortality in several of the animals (214). The maximum tolerated and minimum lethal intravenous doses of acetone for rats were 5 mL/kg and about 7 mL/kg, respectively. DiVincenzo and Krasavage reported that two of four guinea pigs given a 3.0-g/kg intraperitoneal dose of acetone died and that all animals survived a 1.5-g/kg dose (241). Sanderson estimated that the average lethal intraperitoneal dose of acetone was 0.5 g/kg for female rats (242). This value is somewhat less than the intraperitoneal LD_{50} values measured in rats by other investigators (Table 74.13).

Parenteral routes of administration have been commonly used to examine the effects of acetone on respiration, circulation, and the CNS. Early interest in the physiological effects of acetone was sparked by its potential use as an anesthetic and by the acetonemia observed in diabetics. The studies summarized in Table 74.15 (243–245) show that the clinical signs that accompany the intravenous and intramuscular administration of acetone are similar to those observed following inhalation exposure. Saliant and Kleitman studied the circulatory and respiratory effects of acetone in some detail following intravenous administration to anesthetized cats and dogs at doses ranging from 125 to 500 mg/kg (246). In both species, the treatment caused an immediate 15 to 40% drop in blood pressure that was followed by either respiratory stimulation or respiratory depression,

Table 74.15. Acute Toxicity of Acetone to Laboratory Animals Following Parenteral Exposure

Species	Exposure Route	Dose (g/kg)	Treatment Regimen	Observations	Reference
Dog	i.v.	0.47	Single bolus dose	Listlessness, weakness, incoordination, recovery	243
	i.v.	2.1	Single 15-min infusion	Excitation, listlessness, weakness, atoxia, recovery	
	i.v.	0.29–0.36	Four injections over 7-h	Vomiting, slight weakness, listlessness, recovery	
	i.v.	1.62	Single bolus dose	Prostration, incoordination, labored breathing, recovery	
	i.v.	2.55	Two injections an hour apart	Prostration, coma, convulsions, recovery	
	i.v.	0.38–0.76	Six injections over 4-h	Uneasiness, atoxia, weakness, convulsions, dyspnea, cauce, recovery	
Rabbit	i.v.	0.80	Single 60-min infusion	CNS and respiratory depression, recovery	214
	im	4.0	Single 40-min infusion	Paralysis at injection site, CNS and respiratory depression, recovery	
	i.v.	3.2	Single 50-min infusion	Incoordination, drowsiness, slightly increased respiration, recovery	
	im	2.4	Single 70-min infusion	No effect	
	i.v.	0.5	Single 40-min infusion	No effect	244
Dog	i.v.	1.14, 0.38	Two injections 10-min apart	Unsteadiness, drowsiness, dyspnea, recovery	245
	i.v.	0.62–3.11	Seven injections at 1-h intervals	Excitement, dyspnea, prostration, coma, death	
	s.c.	0.76	Five injections over 1-h period	Excitement, dyspnea, prostration, coma, death	

depending on the size of the dose, the concentration of the dose solution, the rate of administration, and the number of treatments. Schlomovitz and Seybold administered acetone by intravenous infusion to four species of laboratory animals and calculated the blood concentration that would cause death (247). Continuous intravenous administration of acetone to dogs at rates ranging from 0.045 to 1.19 mL/min/kg was lethal in less than one hour. Respiratory depression, irregular heart beat, and a drop in blood pressure often preceded death. The average lethal blood level of acetone for rabbits, guinea pigs, cats, and dogs, was 72,300, 42,000, 46,000, and 3000 mg/L, respectively. When the rate of increase in blood acetone exceeded 30,000 mg/L/min, the primary cause of death in the animals was respiratory paralysis. Using isolated and perfused heart-lung preparations from dogs, Bagoury reported that a minimum acetone concentration of 250–400 mg/L caused heart dilatation and that 8000 to 9000 mg/L caused cardiac failure (248).

Studies of rabbits showed that acetone is a severe eye irritant when applied undiluted and left in contact with the cornea for an extended period of time (Table 74.16) (249–259). Ocular damage and irritancy is much less severe when the exposures involve brief contact or diluted aqueous solutions of acetone. To determine whether undiluted acetone could be used as a solvent to remove cyanoacrylate adhesive from the cornea of rabbits, Turss et al. performed gross and histopathological examinations of the ocular tissue from rabbits treated with acetone following the adhesive repair of an artificial corneal injury (260). Several minutes of acetone contact harmed the corneal epithelium; however, the stroma was not permanently affected. All injury to the corneal epithelium was reversible within 4 to 6 days, and adhesive removal using acetone was reportedly acceptable and a preferable technique for adhesive removal. Kennah et al. compared the Draize irritation method for scoring eye injury to an ocular swelling technique that entailed corneal thickness measurements using acetone solutions that ranged in strength from 1 to 100% (251). The diluted acetone solutions (1, 3, 10, and 30%) caused minimal injury; undiluted acetone,

Table 74.16. Ocular and Skin Irritation Studies with Acetone

Assay	Concentration (v/v)	Eye or Skin Irritation Score	Reference
Draize eye	100%	5 (severe)	249
Draize eye	100%	<4 (moderate)	250
Draize eye	100%	66 (severe)	251
	10%	3 (minimal)	
Draize eye	100%	>2 (moderate)	252
Draize eye	10%	3.3 (slight)	253
Corneal swelling	100%	215 (severe)	251
	10%	110 (none)	
Corneal thickness ratio	100%	1.37 (severe)	254
Cornea thickness ratio	100%	1.1 (minimal)	255
Bovine lens opacity	2.0%	5 (mild)	256
Corneal opacity & swelling	100%	slight	257
Draize skin	10%	0 (no irritation)	253
Guinea pig skin	100%	0 (no irritation)	258
Rabbit skin irritation	100%	0 (no irritation)	259

however, was severely irritating by both techniques. Eye irritation measurements for acetone and forty-one other chemicals were used to establish a quantitative structure–activity relationship by using principal component analysis (261). Guthrie and Seitz found that contact lenses could protect against the ocular damage seen in rabbits treated with a 50% solution of acetone (262).

Bolková and Cejková used enzymatic activity measurements to detect damage in rabbit corneal tissue briefly treated with undiluted acetone (263). Acid phosphatase activity in the corneal epithelium was minimally affected by the treatment; alkaline phosphatase activity in the corneal epithelial tissue increased, however, beginning 7 days after treatment. The acetone-induced damage was limited to the ocular epithelium, and no permanent corneal damage was detected. A large number of *in vitro* quantitative assays have been developed to measure the ocular and dermal irritancy potential of acetone. These assays used a wide variety of cell types and damage indices to measure and report potency as a function of the dose that causes a 50% change in response (Table 74.17) (264–268).

The ability of acetone to dehydrate and delipidate unprotected skin is well known from industrial and laboratory experience. Laboratory animal studies confirmed this observation and showed a low potential for acute lethality following dermal exposure (Table 74.13). Animal studies also showed that acetone is neither a skin irritant nor a contact allergen. Smyth et al. found that the application of 10 mL of undiluted acetone to the depilated skin of rabbits did not result in any sign of irritation within 24 hours (195). Likewise, Anderson et al. found that acetone was not irritating when applied to the skin of guinea pigs (258). Skin irritation was measured on three scales: visual erythema and edema, inflammatory cell count in histology specimens, and epidermal thickness in biopsy specimens.

Grubauer et al. examined the effects of acetone treatment on the barrier function of hairless mouse skin by measuring the rate of transcutaneous water loss and the amount of lipid removal that followed the application of undiluted acetone (269, 270). Perturbations in the barrier function of skin were related primarily to the amount of sphingolipids and

Table 74.17. Quantitative Determinations of the Irritancy Potential of Acetone

Test System	Determination	50% Response Conc (mg/mL)	Reference
Chinese hamster lung cells	Neutral red uptake	0.4	253
Human skin fibroblasts	MTT dye reduction	2	253
Rat erthrocytes	Cell hemolysis	1	253
Balb/3T3 mouse fibroblasts	Neutral red uptake	39	264
Human keratinocytes	Neutral red uptake	22	265
Balb/3T3 mouse fibroblasts	Uridine uptake	31.7	266
Balb/3T3 mouse fibroblasts	Growth inhibition	25.5	251
Rabbit corneal epithelial cells	Leucine incorporation	34.6	267
Rabbit corneal epithelial cells	MTT dye reduction	41.8	267
Rabbit corneal epithelial cells	Neutral red uptake	69.8	267
Chinese hamster lung cells	Leucine incorporation	15.2	267
Chinese hamster lung cells	MTT dye reduction	14.4	267
Anesthetized rabbits	Ocular edema	227	268

free sterols removed from the stratum corneum. If the mouse skin was allowed contact with ambient air following treatment, the barrier function of the skin returned to normal after 2 days. In contrast, if the treatment site was occluded with a water impermeable membrane, recovery was prevented. The rate of water loss was linearly related to the amount of lipid removed by the acetone treatment. The effects of acetone on cutaneous water permeability in mouse skin has been associated with the production of proinflammatory cytokines and increases in Langerhans cell responsiveness (271, 272).

The repair of the acetone-induced dermal barrier disruption was accelerated by treating of mice with a mixture of lipids normally found in the stratum corneum (273). Commercial creams and ointments, however, varied considerably in their ability to repair the barrier disruption caused by acetone (274). When applied to acetone-treated mouse skin, partial mixtures of topical lipids could under some circumstances inhibit normal barrier repair (275). Studies showed that cholesterol, sphingolipid, and ceremide synthesis is stimulated in the epidermal layers of mice treated with acetone and that the resynthesis of these substances was essential to restoring the skin's barrier function (276–278). Iversen et al. examined the cell-cycle kinetics and epidermal cell histology in mice treated dermally with 200 μL of acetone (279). The number of suprabasal cells increased slightly 6–12 h after application but quickly returned to normal. Labeling of the DNA with tritiated thymidine revealed a minor reduction in the DNA labeling index and in the fraction of basal cells in the S phase 12–24 h after treatment.

The ability of acetone to affect the toxicity of other solvents was examined in female rats by both the oral and inhalation routes of exposure (194). When administered in combination with nine other different solvents, only the acetone–acetonitrile mixture displayed LC_{50} and LD_{50} values that were significantly less than the predicted additive value (Table 74.18). Smyth et al. expanded upon these findings by determining the rat oral LD_{50} values for 26 different solvent pairs that included acetone as one of the solvents (195). A synergistic response was again seen with the acetone–acetonitrile pair, which produced an LD_{50} value that was nearly fourfold lower than the predicted result. Similar

Table 74.18. Toxic Interactions Between Acetone and Other Solvents Coadministered to Rats by Either the Inhalation or Oral Route[a]

	Inhalation LC_{50} (mg/L)		Oral LD_{50} (mL/kg)	
Solvent Pair	Predicted	Observed	Predicted	Observed
Acetone–hexane	100.7	123.4	—	—
Acetone–dioxane	61.2	61.6	8.28	7.13
Acetone–ethyl acetate	53.0	68.1	11.7	12.9
Acetone–carbon tetrachloride	50.8	61.2	4.78	3.73
Acetone–acetonitrile	39.7	14.6	9.99	2.75
Acetone–toluene	30.4	37.3	10.59	7.96
Acetone–propylene oxide	14.0	20.5	1.04	2.38
Acetone–epichlorohydrin	2.46	1.84	0.32	0.26

[a]Taken from Pozzani et al. (194).

results were obtained when drinking water containing 1% acetone was given to rats 24-hours before a 1 mmol/kg intraperitoneal dose of acetonitrile (280).

The mechanistic basis for the acute toxic synergism between acetone and acetonitrile was investigated by Freeman and Hayes in female rats (281). The appearance of severe signs (tremors and convulsions) of acetonitrile toxicity was delayed for more than a day when 2 g/kg acetone was orally coadministered. Compared to acetonitrile alone, the oral administration of a 25% solution of an acetone–acetonitrile mixture delayed the time when the peak concentration of cyanide appeared in the blood. In addition, the extent of acetonitrile metabolism to cyanide was increased by the acetone cotreatment. Subsequent *in vitro* studies found that acetone competitively inhibited then induced a specific liver microsomal cytochrome P450 isozyme responsible for the metabolism of acetonitrile to cyanide (282). Rats treated with 1% acetone in their drinking water for 7 days also metabolized acetonitrile to an active epoxide intermediate five times faster than untreated rats (283). The metabolism was mediated by hepatic microsomal cytochrome P450 IIE1.

Acetone affects the toxicity of other chemicals, especially when administered several days before treatment with a test chemical. Acetone pretreatment of rats or mice increased barbiturate-induced narcosis (284, 285), thiobenzamide hepatotoxicity (286), cisplatin nephrotoxicity (287), hydrazine cytotoxicity (288), cephaloridine nephrotoxicity (289), and *N,N*-dimethylformamide hepatotoxicity (290). The mechanism responsible for the increase in toxic potency often centers on the ability of acetone to alter metabolic elimination by inducing key enzymes in the activation pathway of the test chemical. In fact, acetone is routinely used as a model inducing agent to study the influence of cytochrome P450IIE1 on the pharmacokinetics and metabolism of various chemicals (Table 74.19) (291–307). Cytochrome P450IIE1 has a wide range of enzymatic activities toward a diverse group of substrates. Hepatic microsomes from acetone-treated rodents display a two- to ten-fold increase in aniline hydroxylase (308, 309), *p*-nitrophenol hydroxylase (310–312, 314), diethylnitrosamine *N*-deethylase (313, 314), dimethylnitrosamine *N*-demethylase (315–317), ethoxycoumarin *O*-deethylase (318, 319), benzene hydroxylase (320, 321), and pentoxyresorufin *O*-dealkylase activity (322, 323).

Elovaara et al. found that acetone pretreatment potentiated the lung damage observed in rats given styrene by inhalation (324, 325). The consumption of drinking water containing 1% acetone for one week before a 488-ppm inhalation exposure to styrene resulted in larger than expected changes in lung glutathion content, urinary metabolite excretion, and lung P450-dependent mixed-function oxidase activity. The acetone pretreatment also enhanced the metabolism of styrene to mandelic and phenylglyoxylic acids by inducing liver cytochrome P450IIE1 activity. Liver microsomes from rats administered 1% acetone in their drinking water for one week formed significantly larger amounts of styrene oxide but not mandelic acid (326). Metabolic interactions were also observed in liver homogenates from rats exposed by inhalation to 1000 ppm of acetone and 300 ppm of styrene for 5 days (327).

Lorr et al. reported that *N*-nitrosodimethylamine hepatotoxicity was potentiated in rats by administering a 2.5-mL/kg oral dose of acetone (328). Male rats given a single 2.5-mL/kg (2.0 g/kg) dose of acetone before NDMA were observed histologically. The acetone pretreatment caused a two-fold elevation in plasma alanine aminotransferase activity and

Table 74.19. The Effects of Acetone Pretreatment on the *In vivo* and *In vitro* Metabolism of Chemicals

Test Chemical	Test Species	Acetone Pretreatment	Evaluation Matrix	Changes Observed	Reference
Lauric acid	Rabbits	1% drinking water 7 days	Liver microsomes	↑in rate metabolite formation	291
Propranolol	Rats	10% drinking water 7 days	Liver microsomes	↑in rate metabolite formation	292
Benzo[*a*]pyrene	Rabbits	10–50 μL in the reaction medium	Lung microsomes	↓in rate metabolite formation	293
Aflatoxin B_1	Rats	0.3–1.2 mM in the reaction medium	Liver microsomes	↓in rate metabolite formation	294
Benzene	Mice	1% drinking water 8 days	Urine specimens	↑in mass metabolite excreted	295
Ethyl chloride	Rats & mice	Oral dose 730 mg/kg	Liver microsomes	↑in rate compound oxidation	296
1,1-Dichloroethylene	Mice	1% drinking water 8 days	Liver microsomes	↑in rate metabolite formation	297
Dimethyl sulfoxide	Rats	1% drinking water 10–12 days	Liver microsomes	↑in rate metabolite formation	298
Glycerol	Rats	1% drinking water 10 days	Liver microsomes	↑in rate compound oxidation	299
Ethyl carbamate	Mice	2 daily intraperitoneal doses 2 g/kg	Blood specimens	↑in blood clearance rate	300
Azoxymethane	Rat	Oral dose 987 mg/kg	Liver microsomes	↑in rate compound oxidation	301
Diethyl ether	Rat	Oral dose 493 mg/kg	Liver microsomes	↑in rate metabolite formation	302
Chloroform	Rat	Oral dose 871 mg/kg	Liver microsomes	↑in rate compound oxidation	303
Methyl *t*-butyl ether	Rat	Oral dose 395 mg/kg	Liver microsomes	↑in rate metabolite formation	304
Ethanol	Rat	2 daily oral doses 1.3 g/kg	Liver microsomes	↑in rate compound oxidation	305
1,1-Dichloroethylene	Mice	Oral dose 2.0 g/kg	Liver microsomes	No effect	297
Carbon disulfide	Rats	1% drinking water 6 days	Urine specimens	No effect	306
Dichlobenil	Mice	1% drinking water 9 days	Olfactory microsomes	No effect	307

an increase in centrilobular necrosis severity. Haag and Sipes examined the mutagenicity of *N*-nitrosodimethylamine in the Ames assay using the liver microsomal fraction from mice treated with a 3-mL/kg intraperitoneal dose of acetone and found a dramatic increase in the number of revertant colonies compared to the effects observed with microsomes from untreated control mice (329). The cytotoxicity of *N*-nitrosodimethylamine to isolated hepatocytes was, however, significantly reduced when the incubations were performed in the presence of a 2 mM acetone solution (330).

Subchronic pretreatment of rats with acetone affects the toxicity of 2,5-hexanedione. Fertility and testicular changes were observed in male rats treated with a combination of 0.5% acetone and 0.13–0.5% 2,5-hexanedione in their drinking water for six weeks (331). The number of pregnancies was significantly lower, and the morphological changes in the testis significantly higher in the animals that received the combined treatment, compared to those that received only 2,5-hexanedione. The treatment of rats with a combination of 0.5% acetone and 0.5% 2,5-hexanedione in their drinking water for six weeks significantly reduced nerve conduction velocities and balance time in the rotarod test compared to the animals given only the dione (332). A similar treatment regimen also caused a decline in open field behavior and a change in the size distribution of sciatic and tibial nerve cross-sectional areas (333). On closer examination, the combined treatment with acetone and 2,5-heaxanedione did not, however, potentiate the dione-induced changes in the neocortical volume or neocortical neuron number observed in the brains of treated rats (334). Minimal effects of combined treatment were also observed in rats for maze performance, brain-swelling, and synaptosomal viability (335). The subcutaneous administration of a single combined dose of acetone and 2,5-hexanedione retarded the elimination of 2,5-hexanedione from the serum and sciatic nerves of rats (336, 337). A similar decrease in body clearance was also observed in rabbits administered both chemicals intravenously (338).

Acetone also antagonizes the adverse effects of some xenobiotics. Lo et al. reported that acetone reduced the nephrotoxicity of *N*-(3,5-dichlorophenyl)-succinimide in male rats given a 10 mmol/kg dose of acetone 16 h before the test chemical (339). Acetone doses less than 10 mmol/kg were, however, ineffective and did not antagonize the renal damage. Jenney and Pfeiffer reported that acetone was effective in preventing the seizures and convulsions that accompanied the treatment of mice with semicarbizide (340). Kohli et al. observed that acetone prevented the seizures caused by either electroshock treatment or the administration of isoniazid (341). The acetone ED_{50} value for preventing electroshock-induced convulsions was 220 mg/kg. Price and Jollow found that rats pretreated with 2.5 mL/kg of acetone exhibited appreciably less liver necrosis following the administration of acetaminophen (342). Metabolic and pharmacokinetic data suggested that the mechanism of action resided in a decrease in the rate of formation of a reactive mercapturate metabolite. The following studies, however, were in conflict and indicated that acetone could potentiate the toxicity of acetaminophen. The administration of 1% acetone to nursing dams in their drinking water also increased the toxicity of a 500-mg/kg intraperitoneal dose of acetaminophen to 14-day old nursing offspring (343). Moldéus and Gergely also reported that rat hepatocyte cultures pretreated with phenobarbital rapidly lost viability when 100 mM of acetone and 5 mM of acetaminophen were added to the incubation mixture (344).

Acetone pretreatment, it has also repeatedly been shown, potentiates the hepatotoxicity and nephrotoxicity of some halogenated hydrocarbons. Acetone shares its activity as an inducing agent with other polar solvents, such as ethanol, isopropanol, and 1,3-butanediol (345–347). The ability of these solvents to potentiate halocarbon-induced organ damage is associated with the induction of a microsomal enzyme that metabolizes the chemicals to toxic intermediates (348–351). The specific isoform responsible for potentiating halocarbon toxicity in rats is cytochrome P450IIE1 (303, 352–353).

Traiger and Plaa were the first to demonstrate that the oral pretreatment of male rats with a 1.0-mL/kg dose of acetone increased the hepatocellular damage from a single intraperitoneal dose of carbon tetrachloride (CCl_4) (345, 354). Large increases in serum alanine aminotransferase activity, liver glucose-6-phosphatase activity, and liver triglyceride content were observed 24 h after treatment with the haloalkane. No liver damage was observed without acetone pretreatment. The oral administration of a 2.5-mL/kg dose of acetone to male mice 18 h before an intraperitoneal dose of chloroform ($CHCl_3$) or 1,1,2-trichloroethane also exacerbated the hepatocellular toxicity of these solvents (355). Little or no potentiation was observed, however, when 1,1,1-trichloroethane or perchloroethylene was administered.

Hewitt et al. examined the structure activity and dose–response relationships for the potentiation caused by a homologous series of ketone solvents and found a positive correlation between the number of carbon atoms in the solvent and the increase in plasma enzymatic activities (356, 357). The five ketonic solvents (acetone, 2-butanone (MEK), 2-pentanone (MPK), 2-hexanone (MnBK), and 2-heptanone (MnAK) were administered to rats 18 h before an intraperitoneal dose of chloroform ($CHCl_3$). An acetone pretreatment dose of 1.0 mmol/kg (58 mg/kg) did not potentiate $CHCl_3$ nephrotoxicity or hepatotoxicity. Several other ketonic solvents (methyl *n*-butyl ketone, or 2,5-hexanedione) also potentiated $CHCl_3$ toxicity in rats (358). Raymond determined the dose of acetone needed to affect the hepatotoxicity of CCl_4 and the nephrotoxicity of $CHCl_3$ in 50% of the treated rats (359). The acetone ED_{50} values of 2.4 mmol/kg (139 mg/kg) for CCl_4 and 3.4 mmol/kg (197 mg/kg) for $CHCl_3$ were appreciably higher than the corresponding ED_{10} values of 1.1 and 0.8 mmol/kg, respectively. Hepatic and renal microsomes isolated from rats treated with either 6.8 or 13.6 mmol/kg of acetone did not show any increase in aminopyrine *N*-demethylase activity, benzphetamine *N*-demethylase activity, or cytochrome P450 content (360). Aniline hydroxylation, however, increased in the microsomes from both the liver and kidney.

The relationship between the pretreatment time interval and the severity of $CHCl_3$-induced hepatobiliary dysfunction was examined in rats given a 15-mmol/kg oral dose of acetone 10 to 90 h before an oral dose of $CHCl_3$ (361, 362). The administration of acetone 18–24 h before $CHCl_3$ treatment caused the greatest increase in plasma bilirubin and the greatest decrease in bile flow rate. There was no potentiation of the hepatobiliary damage when the time interval between pretreatment and $CHCl_3$ administration was longer than 24 h. Liver microsomal fractions from the acetone pretreated rats contained slightly higher cytochrome C reductase and ethoxycoumarin *O*-deethylase activity. The liver cytochrome P-450 level, aminopyrine *N*-demethylase activity, and aniline hydroxylase activity were unaffected, however.

Plaa et al. reported that a threshold or minimally effective dosage (MED) could be determined for the potentiation of CCl_4 liver toxicity by acetone pretreatment (363). Dose-related increases in serum enzymatic activity were observed when male rats were orally administered graded doses of acetone ranging from 0.025 to 2.5 mL/kg before CCl_4. Using these data, the MED of acetone determined was 0.25 mL/kg (198 mg/kg), and the noneffective dose (NED) was 0.1 mL/kg (79 mg/kg). Oral pretreatment with acetone twice per day for three days at the NED did not change the threshold value. When a similar multiday treatment schedule was used with a MED of 0.25 mL/kg, the degree of potentiation was greater than that observed after a single treatment. The effects of acetone pretreatment on CCl_4 hepatotoxicity were limited to the severity of the response and did not alter the time to the appearance of the liver injury or the pattern of repair (364). A chronic pretreatment/treatment regimen with acetone and CCl_4 for 4 to 12 weeks potentiated CCl_4-induced liver cirrhosis and renal damage (365).

The effect of the dosing schedule on the degree of potentiation was further examined by administering a total acetone dose of 1.5 mL/kg at the following four rates before CCl_4 treatment: (1) 1.5 mL/kg orally in a single bolus dose, (2) 0.25 mL/kg every 12 h, (3) 0.125 mL/kg every 6 h, and (4) continuous intravenous infusion at 0.0208 mL/kg/h (363). A single oral dose of acetone elicited the greatest effect on CCl_4 hepatotoxicity, whereas graded dosages produced far less potentiation, and the intravenous infusion produced none. Pharmacokinetic studies indicated that the potentiating effects of acetone correlated with the peak concentration in the blood and not with the area under the blood-acetone concentration versus time curve.

In a detailed study, Charbonneau et al. examined the influence of the exposure regimen, the treatment vehicle, and the route of administration on the dose required for CCl_4 potentiation (366). Preliminary experiments established that the 4-h inhalation NED and MED for acetone potentiation of CCl_4 hepatotoxicity were 1000 and 2500 ppm, respectively. Using plasma enzyme and bilirubin measurements, it was found that the inhalation NED (1000 ppm) and MED (2500 ppm) were independent of the number of exposure sessions. When compared to water as a treatment vehicle, corn oil resulted in a slower acetone absorption rate and lower peak blood levels. Acetone blood levels below approximately 100 mg/L did not show any potentiation of CCl_4 hepatotoxicity, regardless of the vehicle or the exposure route. The degree of potentiation was also greater when acetone was administered orally rather than by inhalation, which was attributed to the first-pass liver metabolism that occurred following oral, but not inhalation exposure. A pharmacokinetic analysis suggested that the degree of potentiation was related to the area under the blood concentration versus time curve which was above the threshold blood concentration.

Hewitt and Plaa found that pretreating male rats with a 15-mmol/kg (871 mg/kg) oral dose of acetone resulted in a significant potentiation of the liver but not the kidney toxicity from bromodichloromethane and dibromochloromethane (367). MacDonald et al. reported that orally administered acetone had a biphasic effect on the hepatocellular damage caused by the intraperitoneal administration of 1,1,2-trichloroethane to rats (368). Unlike the dose-related increase observed when CCl_4 was administered, acetone dosages greater than 0.5 mL/kg did not cause a further increase in 1,1,2-trichloroethane hepatotoxicity but protected against further liver damage. The effects were not, however, associated with any changes in the binding of radiolabeled 1,1,2-trichloroethane to hepatic proteins. A similar biphasic type of response was observed by Hewitt and Plaa while investigating the potentiation of 1,1-dichloroethylene toxicity by acetone (367). Brondeau et al. observed a biphasic response in rats and mice exposed to acetone and 1,2-dichlorobenzene by inhalation (369). Depending upon the exposure concentration employed, a 4-h inhalation pretreatment of male rats or mice with acetone either potentiated or antagonized 1,2-dichlorobenzene hepatotoxicity. Liver cytochrome P-450 level and glutathione transferase activity were uniformly elevated by about 50 to 100% in all of the rats preexposed to acetone.

Charbonneau et al. examined the effects of an acetone pretreatment on the hepatotoxicity of trichloroethylene (TCE), CCl_4, or a TCE–CCl_4 mixture in male rats given an oral dose of acetone at levels of 0.25, 0.75, and 1.5 mL/kg (370). Regardless of the acetone dose, the pretreatment had no effect on either the serum enzyme or bilirubin levels of rats administered an intraperitoneal dose of TCE. Acetone pretreatment caused a marked dose-related increase in the liver damage caused by the TCE–CCl_4 mixture. The potentiating effects of acetone were most severe when the 0.75-mL/kg (592 mg/kg) dose was administered. In follow-up studies, Charbonneau et al. determined the MED for the liver damage caused by TCE, CCl_4, or a TCE–CCl_4 mixture by orally pretreating male rats at each of five dose levels ranging from 0.05 to 0.25 mL/kg (371). Regardless of the dose level, the acetone pretreatment did not cause any TCE-induced hepatotoxicity. Using plasma ALT as the most sensitive end point for the liver damage, the acetone MED for CCl_4 hepatotoxicity was 0.25 mL/kg (197 mg/kg), and the MED for the TCE–CCl_4 mixture was at least 0.05 mL/kg (40 mg/kg).

Charbonneau et al. continued their investigation into the effects of acetone pretreatment on the hepatotoxicity of halocarbon mixtures and recently reported on the results of using 28 different binary mixtures (372). Male rats were examined after receiving a 0.75-mL/kg dose of acetone in corn oil before an oral treatment with all possible binary combinations of CCl_4, $CHCl_3$, TCE, 1,1-dichloroethylene, tetrachloroethylene, 1,1,1-trichloroethane, 1,1,2-trichloroethane, and 1,1,2,2-tetrachloroethane. When there was no acetone pretreatment of the rats, 26 mixtures showed an additive effect and two were supra-additive. When the acetone pretreatment was used, 10 mixtures showed infra-additive effects, 17 showed additivity, and one showed supra-additivity. The supra-additive effect was observed only with the TCE–CCl_4 combination. The mechanism responsible for the protective effects of the acetone pretreatment on the hepatotoxicity of the solvent mixtures was not investigated.

1.4.1.2 Chronic and Subchronic Toxicity. Information on the health effects resulting from long-term, high-level exposures to acetone is available from several types of studies. The subchronic toxicity of acetone has been examined in rodents following oral, dermal, and inhalation exposure. These studies involved treating animals at a rate of 5 days/week for periods of time from 2 to 13 weeks. In addition, there have been several mixture studies, in which acetone was combined with other chemicals and administered to mice or rats via inhalation or their drinking water for periods lasting up to 4 months (373, 374). Solllmann conducted a chronic feeding study with acetone that involved the administering

2.5% acetone in the drinking water of three female albino rats for 18 weeks (375). The average dose of acetone was 1.8 mL/kg/day. No deaths resulted, but, there was a sharp decrease in water consumption, a 23% decrease in food consumption, and a 34% drop in the expected weight gain over the course of the study. Unfortunately, the study did not go beyond these general measures of toxicity, and nothing further was reported.

Skutches et al. examined the effects of acetone on insulin-stimulated glucose use in male rats given 0.5, 1.0, 2.0, or 3.0% in their drinking water for 7 days (376). The rate of glucose oxidation by isolated epididymal adipose tissue was significantly lower in the animals treated at the 3.0% level and disappeared when the rats were allowed four days of recovery. Sinclair et al. found that male mice administered 1% acetone together with 5-aminolevulinic acid in their drinking water had high uroporphyrin levels in their livers and urine after 35 days of treatment (377). The changes, however, were observed only in iron-loaded mice that were treated with both chemicals in combination. Bruckner and Petersen exposed rats for 3 h per day to 19,000 ppm of acetone for up to eight weeks and found little evidence of toxicity (378). Blood and tissue (lung, liver, brain, kidney, and heart) specimens were examined biochemically and histologically at intervals for evidence of damage. Other than a slight, statistically significant decrease in the absolute weight of the brain and kidney, there were no observable effects of the treatment relative to control animals. No organ weight changes were observed in a 2-week postexposure recovery group.

Acetone has been administered subchronically to laboratory animals both by gavage and through drinking water. The organs most sensitive to acetone are generally the liver and kidney, which show increases in relative or absolute organ weight that are not accompanied by any observable morphological alterations. Some differences in toxicity were reported in two studies, however, particularly with regard to the hematopoietic system.

In the gavage study, acetone was administered to 30 male and female Sprague–Dawley rats at three dose levels (100, 500, and 2500 mg/kg) for 90 consecutive days (379, 380). No toxicologically significant differences were noted in food consumption or body weight gain at any dosage. Animals of both sexes treated at 2500 mg/kg demonstrated an increase in several hematological parameters (hemoglobin concentration, hematocrit, and erythrocyte indices) and an increase in the activity of three serum enzymes (alanine aminotransferase, aspartate aminotransferase, and alkaline phosphatase). Male rats also showed an increase in serum cholesterol at the two highest dosages and a decrease in serum glucose at the 2500-mg/kg dosage.

Female rats showed increases in the absolute liver and kidney weight at the two highest dosages; however, parallel increases in the relative organ weight ratios were observed only at the highest dose level. Male rats administered 2500 mg/kg displayed an increase in the relative organ-to-body weight ratios for the liver and kidney, but the absolute weights of these organs were not affected. Histopathological examination of tissue sections from the liver did not reveal any structural abnormalities, and those observed in renal tubular cells were also seen in control animals. The acetone treatment accentuated the renal nephropathy that is typically observed in aging rats (381).

Dietz et al. administered acetone in the drinking water of male and female Fisher 344 rats and B6C3F$_1$ mice for either 14 days or 13 weeks (382, 383). The drinking water

Table 74.20. Time-Weighted Average Daily Dose of Acetone to Fisher 344 Rats and B6C3/F1 Mice Used in a Subchronic Drinking Water Study[a]

	14-Day Average Dose (mg/kg/day)				13-Week Average Dose (mg/kg/day)			
	Rats		Mice		Rats		Mice	
Water Concentration (%)	Male	Female	Male	Female	Male	Female	Male	Female
0.125	—	—	—	—	—	—	380	—
0.25	—	—	—	—	200	200	611	892
0.5	714	751	965	1569	400	600	1353	2007
1.0	1616	1485	1579	3023	900	1200	2258	4156
2.0	2559	2328	3896	5481	1700	1600	4858	5945
5.0	4312	4350	6348	8804	3400	3100	—	11,298
10.0	6942	8560	10,314	12,725	—	—	—	—

[a]Taken from Dietz (383).

concentrations and the calculated TWA daily doses to the mice and rats are shown in Table 74.20. No mortality was seen in the rats or mice used in either the 14-day or 13-week study. Overt clinical signs of toxicity were observed only in rats treated at the 10% dose level in the 14-day study and were described as an emaciated appearance from decreased weight gain. Acetone-induced increases in relative kidney weights were observed at the 2% dose level in female rats and at the 5% dose level in male rats. The kidney weight changes were reportedly associated with a nephropathy that occurred spontaneously in untreated control rats.

Male and female rats also exhibited an increase in relative liver weight at the 2% level and an increase in the relative testicular weight at the 5% level. The liver weight changes were not associated with treatment-related morphological alterations. No ophthalmic lesions occurred at any treatment level in either the rats or the mice. A decline was noted in some hematologic indexes for male rats treated at the 0.5% level and higher that was consistent with macrocytic anemia. The anemia was not apparent in female rats, although some mild hematologic changes were noted in female rats treated with 5% acetone. Minimal to mild splenic pigmentation (hemosiderosis) was observed exclusively in male rats treated at the 2% and 5% levels. The most notable finding in the mice was an increase in the absolute liver and a decrease in the absolute spleen weight of female mice given 5% acetone. Centrilobular hepatocellular hypertrophy was noted histologically in female mice treated at the 5% level. In contrast to the rats, mice showed no change in hematology. The maximally tolerated dose of acetone in the drinking water was 2% for male rats and male mice and 5% for female mice. No toxic effects were identified for female rats.

Many of the organ weight changes observed in the subchronic studies with acetone may be related to the autoinductive effects of the compound which can increase the turnover of microsomal proteins. Hètu and Joly examined liver microsomal mixed-function oxidase activities in the liver of female rats given 1% acetone in their drinking water for two weeks and found elevated levels of cytochrome P-450, as well as increases in *p*-nitrophenol hydroxylase, aniline hydroxylase, and 7-ethoxycoumarin *O*-deethylase activity (319). The treatment did not, however, cause any change in body weight, liver weight, or liver microsomal protein content.

Studies by Rengstorff et al. suggested that acetone was cataractogenic in guinea pigs treated either cutaneously or subcutaneously with a 5% or 50% aqueous solution 3 days per week for three weeks (384). Male and female and guinea pigs treated dermally (0.5 mL) or subcutaneously (0.05 mL) with acetone developed cataracts in both eyes by the third month posttreatment (15 mL total volume). When a 1% acetone solution was applied dermally to guinea pigs or rabbits twice daily, 5 days per week for either four or eight weeks, several of the guinea pigs, but neither of the rabbits, developed cataracts during the six-month examination period. In additional studies, Rengstorff et al. reported that male and female rabbits treated with 1 mL of acetone for 3 days per week for three weeks showed no lens abnormalities during the six-month observation period (385, 386). Male and female guinea pigs treated with 0.5 mL of acetone for 5 days per week for six weeks, however, showed evidence of cataract development by 3 months posttreatment. Ascorbate levels in aqueous humor specimens from the test animals that had cataracts were appreciably lower than the levels found in control animals. The authors concluded that development of cataracts was species-specific and intimately related to ascorbate synthesis.

In an attempt to develop a animal model, Taylor et al. conducted additional studies and found that acetone was not cataractogenic in guinea pigs even under extreme conditions (387). Hairless guinea pigs were treated dermally with 0.5 mL of acetone for 5 days/week for 6 months (65 mL total volume) and then monoitored during a 2-year period. Except for some mild erythematous skin irritation, the acetone treatment caused no other adverse effects. Treating guinea pigs on a low ascorbate diet with acetone also failed to cause any cataracts to develop. The authors did not speculate on possible causes for the inconsistent results when comparing their findings with earlier studies.

1.4.1.3 Pharmacokinetics, Metabolism, and Mechanisms. The rates and routes of acetone metabolism have been extensively examined in both humans and laboratory animals. Although research in this area has been somewhat clouded, a clear and consistent picture of acetone anabolism and catabolism has emerged. Both the formation of endogenously created acetone and the breakdown of exogenously administered acetone have received qualitative and quantitative attention. Acetone is one of three ketone bodies that arises from the production of acetyl coenzyme A within the liver. The enzyme systems that catalyze the formation of acetone reside primarily within the mitochondria of hepatocytes and, to a much smaller extent, renal cells. Because acetone is nonionic and miscible with water, the chemical can passively diffuse across cell membranes and distribute throughout the body fluids. The remaining ketone bodies, acetoacetate and β-hydroxybutyrate, are organic acids that can cause metabolic acidosis in large amounts. Endogenous acetone is derived from both the spontaneous and catalytic breakdown of acetoacetate. Endogenous and exogenous acetone are both eliminated from the body by excretion into the urine and exhaled air or by enzymatic metabolism. Under normal circumstances, metabolism is the predominate route of elimination; however, with high blood levels, metabolic saturation occurs that makes the pulmonary and urinary excretion pathways much more important.

The first step in the enzymatic metabolism of acetone is cytochrome P450-dependent oxidation of acetone to acetol by acetone monooxygenase. Then, the acetol is biotransformed by either of two pathways, an extrahepatic propanediol pathway or an intrahepatic methylglyoxal pathway. Recent studies have also shown that a third metabolic pathway that leads to acetate formation can be recruited when the primary metabolic pathways become overloaded. The intrahepatic oxidation of acetol to methyl glyoxal is also cytochrome P450-dependent and is catalyzed by acetol monooxygenase. Studies have shown that acetone monooxygenase and acetol monooxygenase activities are associated with a specific isoform of cytochrome P450 in rat and rabbit liver (388, 389). P450IIE1 enzymatic activity can induced to function at a much higher rate when large amounts of acetone are circulating through the body. This autoinductive effect allows acetone to regulate its own metabolism by increasing the rate of enzymatic elimination when the need arises.

Metabolism studies with acetone have shown that it is an important biochemical constituent of the human body that can be broken down to two- and three-carbon intermediates that are basic building blocks for much larger molecules. Acetone belongs to a class of chemicals that are gluconeogenic, or capable of causing glucose biosynthesis. Amino acids, glycogen, cholesterol, and other biochemicals also contain the carbon atoms from metabolized acetone. The basic building block for the biosynthesis of glucose and other macromolecules from acetone is pyruvic acid which is the product of acetone metabolism by both the intrahepatic and extrahepatic pathways.

The primary source of endogenous acetone is acetyl coenzyme A (CoA), and smaller amounts arise directly from the breakdown of certain amino acids (390). As shown in Figure 74.1, the production of acetyl CoA is controlled by several biochemical processes, including lipolysis, glycolysis, and oxidation of ketogenic amino acids. Of the three processes, lipolysis is considered the most active and exerts the most influence on the available pool of acetyl CoA within the liver. Consequently, factors that affect the formation and release of long-chain fatty acids from adipose tissue during lipolysis can profoundly affect the circulating levels of acetone within the body.

The five enzymatic reactions that lead to the biosynthesis of acetone from acetyl CoA are depicted in Figure 74.2. The sequence involves the condensation of three molecules of acetyl CoA by the enzyme acetoacetyl CoA thiolase to yield β-hydroxymethyl-β-methylglutaryl CoA, which is next cleaved by a lyase to produce acetoacetate and a molecule of the starting material (391). There is still some confusion about the last step in the reaction scheme. Many researchers have shown that endogenous acetone is formed spontaneously from acetoacetate (390, 392); however, others have demonstrated that some proteins can catalyze the decarboxylation of acetoacetate to acetone (393). A third group of researchers argues that a specific enzyme, acetoacetate decarboxylase, catalyzes the formation of acetone in mammals and that enzyme is associated with the albumin fraction of human blood, as well as with proteins from rat liver and kidney cells and rat placental tissue (394, 395). By metabolizing acetoacetate to acetone, acetoacetate decarboxylase activity may regulate metabolic pH when excess amounts of acetoacetate are produced (396, 397).

The rate of acetone production can be quite high., especially during times of high energy demand. Hetenyi and Ferrarotto examined the rate and extent of acetone turnover in male rats fasting for 3 days and administered a small tracer amount (0.17 to 0.20 mg/kg)

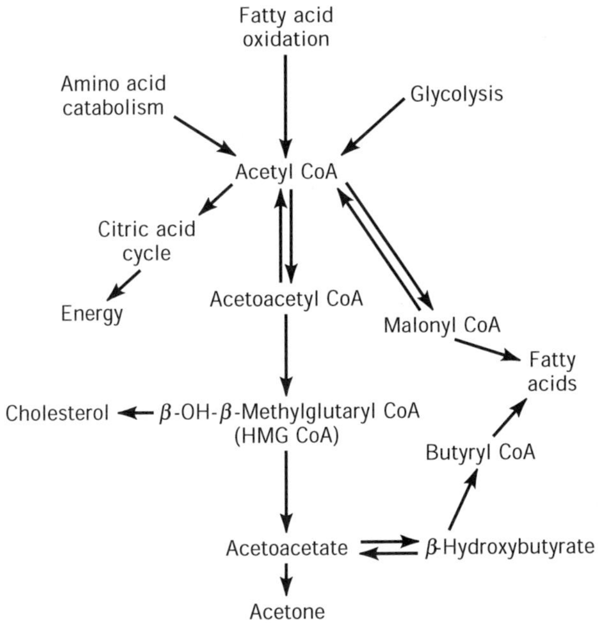

Figure 74.1. Endogenous biosynthesis of acetone (390) copyright © 1982, reprinted with permission.

of radiolabeled acetone ([U-^{14}C]-acetone) by arterial injection (398). The calculated metabolic clearance and turnover rates of acetone were 2.25 mL/min/kg (1.78 g/min/kg) and 0.74 mmol/kg/min (2.58 g/kg/h), respectively. By comparison, Filser and Bolt reported that the endogenous rate of acetone production was 5 mmol/kg/h (290 mg/kg/hr) for nonfasting rats (399). Haggard et al. calculated the rate of acetone metabolism by subtracting the rate of excretion from the total rate of elimination and found that the rate of metabolism began to plateau and exhibit evidence of saturation when the blood concentration exceeded about 400–500 mg/L (210). Once acetone blood levels reached about 800–900 mg/L, the rate of metabolism was at a maximum of about 12 mg/kg/h. The rate of acetone metabolism in fasting rats was also about 20 to 30% greater than in fed animals.

Sakami and Lafaye initially proposed that acetone was metabolized by separate carbon cleavage and carbon oxidation to yield pyruvate, a major metabolite (400, 401). These conclusions were based on a determination of the isotopic distribution and labeling patterns in the liver glycogen isolated from rats administered 2[^{14}C]-acetone. Rudney then isolated 1,2-propanediol phosphate as a metabolite from the livers of rats given a subcutaneous injection of 2[^{14}C]-acetone (402). These studies also indicated that 1,2-propanediol can metabolized to one- and two-carbon intermediates by a rat liver preparation. Mourkides et al. amplified further on the two pathways for acetone metabolism using male rats given a single intraperitoneal injection of 2[^{14}C]-acetone (403). At dose levels less than 25 mg/kg, approximately 25–45% of the dose could be recovered as [^{14}C]CO_2 within 3 h of treatment; however, at dose levels greater than about

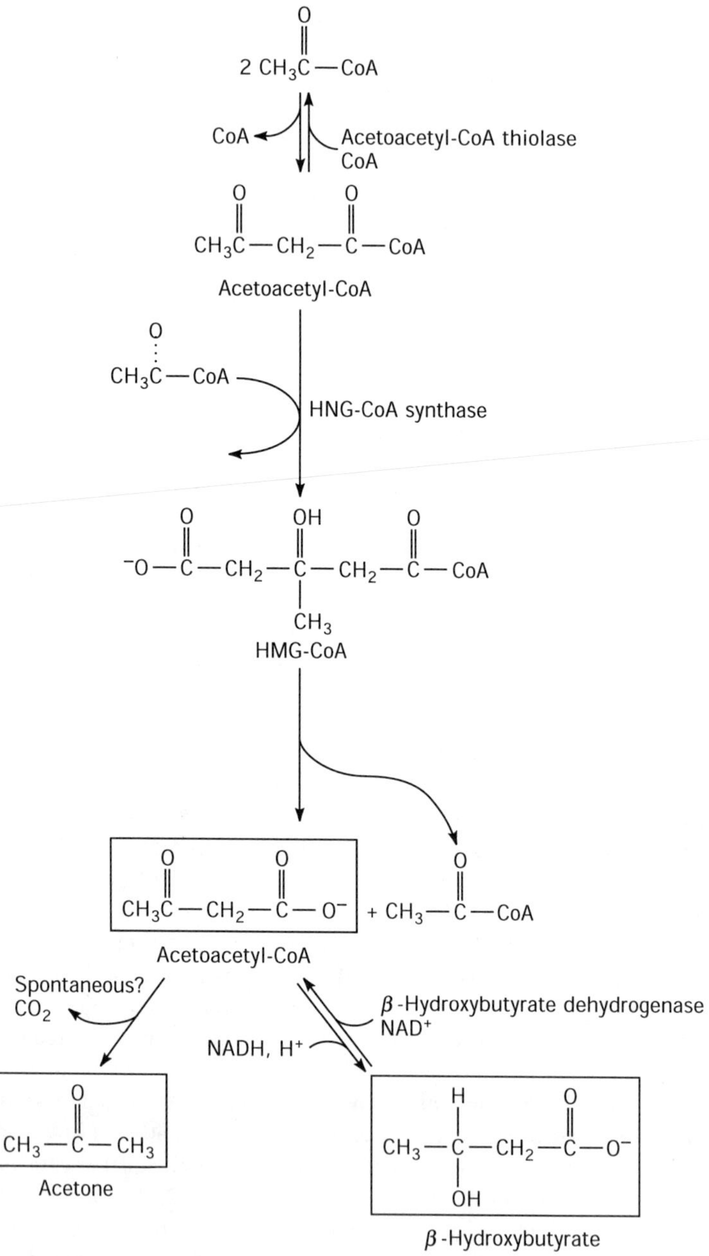

Figure 74.2. Biosynthesis of ketone bodies (391) copyright © 1982, reprinted with permission.

260 mg/kg, the recovery of [^{14}C]CO_2 declined to about 2–6% of the dose. Isolation and purification of alanine, glutamic acid, and aspartic acid from the liver and carcass of the treated animals revealed a carbon labeling pattern that was consistent with the metabolism of acetone to pyruvate and acetate by two separate pathways. Acetone was less efficiently metabolized at high-dose levels.

Casazza et al. proposed two alternate pathways for the biotransformation of acetone in fasting male rats given 1% acetone in their drinking water for 3 days before the administration of 5 mmol/kg of 2[^{14}C]-acetone by the intraperitoneal route (404). Two metabolic pathways proceeding through same common intermediate, acetol, operate with acetone. The acetol intermediate was either oxidized to methylglyoxal by the enzyme acetol monooxygenase or, alternatively, reduced to the levorotatory (L) isomer of 1,2-propanediol by an unknown extrahepatic mechanism. Thus, the methylglyoxal pathway led to the formation of pyruvate through D-lactate, and the 1,2-propanediol pathway led to the formation of pyruvate through L-lactate. The use of metabolic inhibitors suggested that the methylglyoxal pathway was slightly more active than the 1,2-propanediol pathway in rats.

Perinado et al. provided supportive evidence for the biotransformation of acetone by two separate pathways in pregnant and virgin rats administered a 100-mg/kg intravenous dose of acetone (405). Variable amounts of acetol, 1,2-propanediol, and methylglyoxal were detected in plasma and liver specimens from the pregnant dams, term fetuses, and virgin rats. The metabolite levels were generally higher in virgin than in pregnant animals, and the levels of acetol and methylglyoxal were many fold higher in fed virgin versus fasting virgin rats. Fasting pregnant rats eliminated acetone at a much faster rate than either the fed pregnant rats or virgin rats.

Kosugi et al. obtained strong evidence for the biotransformation of acetone by a third pathway involving acetate as an intermediate (406). Following either a diabetes-inducing streptozocin injection or a pretreatment with 1% acetone in the drinking water for 3–4 days, rats were intravenously infused over a 3-h period with a dose of 2[^{14}C]-acetone between 323–1113 mg/kg. The resulting very high blood acetone concentrations ranged from 4.1 to 9.7 mM. The labeling pattern in the isolated glucose was different from the pattern observed with the administration of 2[^{14}C]-pyruvate but was consistent with the formation of acetate as a metabolic intermediate. Gavino et al. provided unequivocal evidence for the formation of acetate during acetone metabolism (393). The liver perfusates of male rats given 4.5 mM (260 mg/L) of 2[^{14}C]-acetone contained a large percentage of 1[^{14}C]-acetate and small amounts of labeled lactate and 1,2-propanediol.

Kosugi et al. established that a third metabolic pathway appeared when large doses of acetone were administered and that the pathway involved the formation of both acetate and formate metabolites (407). Fasting, fed, and diabetic (streptozocin-induced) rats were given a 3-h intravenous infusion of 2[^{14}C]-acetone at low trace levels (0.17–0.23 mg/kg) or at high saturating levels (313–383 mg/kg). The labeling pattern in the isolated glucose was consistent with the metabolism of acetone to methylglyoxal and lactate at the low dose; however, at the high dose, the carbon atom labeling indicated the formation of acetate as a metabolite.

Black et al. examined the incorporation of radiolabeled carbon into milk casein and lactose from normal and spontaneously ketotic cows given a single bolus intravenous dose of 2[C^{14}]- acetone (408). The labeling intensity in the amino acids from casein increased in the following order: alanine > glutamic acid > aspartic acid > serine > glycine. The results indicated that acetone was gluconeogenic for cows and that 40–70% of the endogenous acetone was metabolized in the citric acid cycle through a common precursor. Coleman found that the synthesis of gluconeogenic precursors from acetone was induced by fasting or prior acetone administration (409). Normal mice and two variant strains (obese and diabetic) were given acetone in their drinking water for up to three days and then administered 0.55 mmol/kg of 2[^{14}C]-acetone. Liver homogenates from the normal and mutant mouse strains that were administered 1.0–2.5% acetone in drinking water for up to 3 days showed a five-fold increase in the rate of acetone metabolism. The degree of induction was dose- and time-related in all types of mice, but significantly greater induction was observed in the genetic variants than in the normal mice.

Lindsay and Brown assessed the incorporation of acetone into plasma glucose by intravenously infusing normal male and ketotic female (2-day fast during pregnancy) sheep with small amounts of 1,3[^{14}C]-acetone (410). Endogenous acetone accounted for less than 2.2% of the plasma glucose in ketotic sheep and about 0.5% in normal animals. The rate of acetone formation from acetoacetate decarboxylation was estimated at less than 1 mg/min in normal sheep and 2.4–6.5 mg/min in ketotic sheep. Studies of ketotic and normal cows given a small tracer amount of 2[^{14}C]-acetone by bolus intravenous injection produced dissimilar results (411). Five labeled constituents were isolated from the cow milk: lactose, casein, fat, albumin, and citrate. The level of radioactivity incorporated into the carbon atoms from these milk constituents was 12–29% for normal cows and 2.4–3.5% for ketotic animals.

Argilés has compiled much of the available metabolic information on acetone and constructed a detailed description of mammalian acetone metabolism (412). Figure 74.3 shows that the pivotal step in the reaction sequence occurs with the initial oxidation of acetone to acetol by acetone monooxygenase. This cytochrome P450IIE1-dependent reaction is considered the rate-limiting process that controls the overall accumulation and elimination of acetone from the body.

Acetone administration induces a specific cytochrome P450 protein that has enzymatic activity toward a wide variety of chemical substrates. Cytochrome P450IIE1, known also as P450j, $P450_{ac}$, or P450 3a, is a constitutive isozyme in the P450 subfamily that can be induced severalfold following high-dose treatments with acetone. Acetone induces the P450IIE1 activity in rat and rabbit liver by posttranscriptional stabilization of the enzyme protein against autophagosomal/lysosomal degradation (413–417). In addition to acetone, a wide variety of chemical exposures and physiological conditions induce P450IIE1 activity in rat and rabbit liver. These include ethanol (418), MEK (419), isoniazid (420), 95% oxygen (421), trichloroethylene (422), high fat/low carbohydrate diets (423), diabetes (424), and fasting (425). The P450IIE1 induction observed with ethanol, fasting, and diabetes is unrelated to the secondary production of endogenous acetone (426–428). Acetone-induced P450IIE1 activity has been detected in the lung, liver, and kidney of rabbits and hamsters (429–431). Likewise, chicken and mouse liver, rabbit bone marrow and olfactory mucosa, and rat Kupffer cells possess acetone-inducible P450IIE1 activity (432–436). The acetone-induced P450IIE1 from rabbit and rat liver is immunochemically and catalytically similar to the constitutive P450IIE1 from human liver (437, 438).

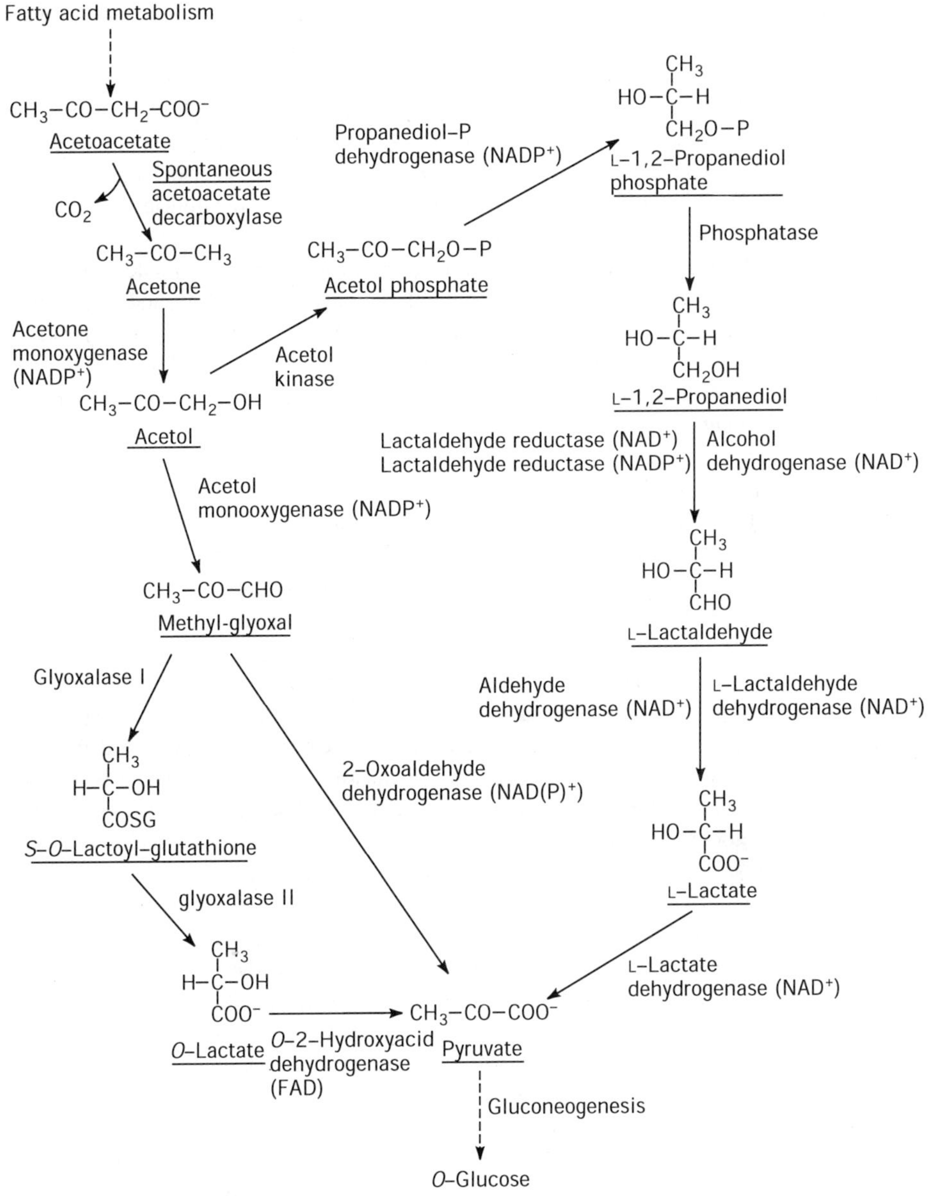

Figure 74.3. Mammalian metabolism of acetone (412) copyright © 1982, reprinted with permission.

The induction of hepatic P450IIE1 typically requires repeated administration of acetone either by gavage or via drinking water. The induction of P450IIE1 for experimental purposes is generally accomplished by giving animals a 1–10% solution of acetone in their drinking water for at least 7 days. Enzyme induction has also been achieved by administering up to four intragastric dosages of acetone at 1–3 g/kg. Under these conditions, a two- to sixfold elevation in hepatic P450IIE1 may be observed 18–24 h after the final treatment (415, 422, 430, 439). The minimum drinking water regimen needed to induce hepatic P450IIE1 activity in hamsters was 0.5% for 7 days (439), whereas, the minimum treatment needed to induce hepatic P450IIE1 activity in Sprague–Dawley rats was a single oral dose of 0.20 g/kg (440). High acetone doses also sharply induce cytochrome P450IIB1 (phenobarbital-induced form), glutathione transferase, and glucuronosyltransferase in rat liver (413, 414, 441, 442).

In contrast to the enzyme induction observed upon *in vivo* treatment with acetone, the *in vitro* addition to hepatic microsomal preparations from rats, mice, rabbits, and dogs causes catalytic enhancement. The mechanism for the catalytic enhancement has not been clearly elucidated; however, the increase in activity was not associated with P450IIE1 induction by acetone. Acetone enhancement has been associated with specific configurational or conformational changes in cytochrome P450 that arise when the chemical binds at a specific site on the cytochrome protein (443–446). Acetone enhanced aniline hydroxylation two to fourfold in rat liver microsomes (447–449). Enhancement of acetanilide and *N*-butylaniline hydroxylation was also observed (450), along with the *O*-dealkylation of *p*-anisidine and *p*-phenetidine (451) and the omega hydroxylation of prostaglandin (452). The addition of acetone inhibited the microsomal hydroxylation of coumarin and benzo[*a*]pyrene (453, 454) and the esteratic hydrolysis of procaine (455).

1.4.1.3.1 Absorption. The absorption, distribution, and excretion of acetone have been evaluated pharmacokinetically under a variety of test conditions. Studies of mice, rats, rabbits, and dogs provided additional evidence that acetone elimination is dose-dependent. Egle examined the uptake of acetone vapors by anesthetized dogs exposed to acetone vapor concentrations ranging from 360–800 mg/m^3 (456). The uptake of acetone vapors by the lungs was about 52% at ventilation rates between 4 and 18 breaths/min but tended to drop to about 42% when the ventilation rate was greater than 20 breaths/min. The pulmonary uptake increased from about 52% to 59% when the concentration of acetone increased from 460 to 1750 mg/m^3. Aharonson et al. investigated further and found that the uptake of acetone vapors by the nose and throat of anesthetized dogs more than doubled when the flow rate of inspired air was increased from 2 to 5 L/min (457). The increase in uptake, it was thought, is associated with an increase in the perfusion of blood to the lung tissues, an increase in the transfer of vapors into the tissues, or an increase in the effective area for vapor uptake by the lung.

Morris et al. conducted studies on the regional deposition of acetone vapors in the upper respiratory tract (URT) of many different species (458). Anesthetized male Fisher 344 rats and Hartley guinea pigs that had tracheostomies were exposed for less than 20 min to acetone vapor concentrations ranging between 810 and 1800 mg/m^3. The deposited fraction did not depend on the inspired acetone concentration, which indicated that URT metabolism was not an important factor in the overall uptake. Furthermore, the deposited fraction was inversely related to the inspiratory flowrate which indicated that the acetone was rapidly absorbed into the bloodstream. The URT deposition in the rat was about twofold higher than in the guinea pig. Additional studies of anesthetized Sprague–Dawley rats exposed at vapor concentrations from 1000 to 1300 mg/m^3 produced absorption rates that were 2–3 times greater than those for either the Fisher rat or the Hartley guinea pig (459). The absolute deposition of acetone vapors at inspiratory flow rates of 70, 150, 300, and 500 mL/min were 0.66, 0.96, 1.18, and 1.76 mmol/min, respectively. The uptake of acetone vapors has also been examined in male Syrian golden hamsters (460) and B6C3F1 mice (461).

When the uptake efficiencies were compared at ventilation rates that were 2.5- to threefold higher than the normal, the results from the deposition studies varied as follows for each species: Sprague–Dawley rat (21%), B6C3F1 mouse (14%), Fisher 344 rat (12%), Hartley guinea pig (7%), and Syrian hamster (5%). The inefficient uptake of acetone vapors by the hamster was attributed to the slow perfusion of blood through the nasal tissue, whereas anatomical or physiological factors were thought to be responsible for differences between rat and guinea pig.

1.4.1.3.2 Distribution. Early studies by Widmark showed a logarithmic (i.e., first-order) decrease in acetone blood levels following the absorption of a 700-mg/kg subcutaneous dose by rabbits (462). Peak blood levels of acetone ranged from about 675 to 728 mg/g and typically occurred 50 to 90 min following treatment. The average volume of distribution (V_d) was 1.8 L and the average blood elimination half-life ($t_{1/2}$) was 7.5 hr. A rabbit was given a 350-mg/kg dose of acetone by intravenous injection that had an average elimination half-life of 7.4 h. When rabbits were given the same dose of acetone by either oral, intraperitoneal, or rectal administration, the postexposure blood levels were described by an exponential equation.

DiVincenzo et al. found that the rate of acetone elimination in the breath and blood of dogs decreased in a first-order manner following a 2-h exposure to vapor concentrations of 100, 500, or 1000 ppm (463). The calculated half-life for the blood elimination of acetone was about 3 h for each of the exposure concentrations examined, and the amount of acetone excreted into the expired air was approximately proportional to the exposure concentration.

Haggard et al. examined the pharmacokinetics of acetone in rats following intraperitoneal, oral, and pulmonary administration (210). Rats exposed by inhalation at concentrations of 5000, 25,000, or 50,000 mg/m^3 for 8 h displayed peak acetone blood levels of 250, 2000, and 4400 mg/L, respectively. Appreciable day-to-day accumulation of acetone occurred in rats exposed for nine consecutive days to 5350 mg/m^3 of acetone. When continuous 24-h exposures were performed at the same concentration, the blood levels plateaued at about 1000 mg/L three days later. Rats given an intraperitoneal dose of acetone at levels from 25 to 2000 mg/kg had an average blood-to-tissue partition coefficient of 0.82. The oral administration of a 3-g/kg dose of acetone produced peak blood concentrations that tended to remain elevated at about 2300 mg/L for nearly 48 hours. The elimination of acetone at blood concentrations between 500 and 2300 mg/L was appreciably slower than at lower blood concentrations.

Mast et al. exposed pregnant rats to 440, 2200, or 11,000 ppm of acetone for 6 h per day on days 6 through 19 of gestation and collected blood specimens for acetone analysis on three of the exposure days (464). The acetone blood levels at 30-min postexposure showed a disproportionate increase relative to the rise in exposure concentration, indicating that the metabolic elimination of acetone was saturated when the animals were

Table 74.21. Acetone Plasma Levels Following Daily Inhalation Exposure of Pregnant Rats[a]

	Acetone Level 30 min Postexposure (mg/L)			Acetone Level 17 h Postexposure (mg/L)		
Exposure Conc (ppm)	Day 7	Day 14	Day 19	Day 7	Day 14	Day 19
0	< 0.06	< 0.6	1.7	< 0.6	< 0.6	< 0.6
440	38.3	43.0	52.9	< 4.1	< 0.6	< 1.7
2,200	296.2	273.0	296.2	< 0.6	0.6	23.2
11,000	2149.0	2090.9	1974.7	174.2	174.2	348.5

[a]Taken from Mast et al. (464).

exposed at the highest concentration (Table 74.21). Blood level measurements at 17 h postexposure also revealed that an exposure concentration of 11,000 ppm caused day-to-day accumulation of acetone in the body. Blood levels of the two remaining ketone bodies, acetoacetate and β-hydroxybutyrate, were not, however, affected by inhalation exposure to acetone at any level.

Hallier et al. examined the pharmacokinetics of acetone vapor in rats using a closed, recirculating, exposure chamber (465). Male rats were exposed in a closed chamber at concentrations as high as 62,000 ppm, and the rate of acetone loss from the exposure chamber was monitored for up to about 30 h. Acetone elimination exhibited apparent first-order kinetics when the chamber concentration was 100 ppm or less. After an absorption and distribution phase lasting about 8 h, acetone disappeared from the chamber at a rate equal to its rate of metabolism within the animal. The calculated Michaelis constant (K_m) and maximum velocity (V_{max}) for the processes were 160 ppm and 230 mmol/kg/h, respectively. The average blood-to-air and urine-to-air partition coefficients for acetone were 210 and 330, respectively.

Wigaeus et al. examined the distribution and respiratory excretion of radiolabeled acetone in mice that received either single or multiple exposures at 1200 mg/m^3 (500 ppm) (466). The concentration of unmetabolized acetone reached a steady-state level in most tissues and organs after 6 h of exposure. Exposures longer than 6 h caused further accumulation in the liver and brown adipose tissue. The highest concentration of radioactivity was observed in the liver after 24 h of continuous exposure, but only about 10% of the radioactivity in the liver was composed of unmetabolized acetone. The rate of organ removal of radioactivity was highest for the blood, kidneys, lungs, muscle, and brain, and half-life values ranged from 2 to 3 h. The slowest removal occurred in subcutaneous adipose tissue where the $t_{1/2}$ was slightly greater than 5 h. The acetone concentration in all tissues returned to normal levels within 24 h after exposure stopped. The results indicated that acetone was evenly distributed in the body water and that it did not accumulate in the body after repeated exposures to 500 ppm.

Plaa et al. examined the elimination of acetone in male rats administered a single oral dose at 0.25, 1.00, or 2.50 mL/kg (363). As shown in Table 74.22, the maximum plasma concentration (Cp_{max}), the volume of distribution (V_d), and the blood half-life ($t_{1/2}$)

Table 74.22. Pharamacokinetics of Acetone in Male Rats Following the Oral Administration of Either Acetone or Isopropanol[a]

Dose (mL/kg)	Maximum Plasma Conc (mg/L)	Area Under the Curve (mg/h/L)	Total Body Clearance (mL/h)	Volume of Distribution (mL)	Elimination Half-life (h)
			Acetone		
0.25	311	3,049	64	227	2.4
1.00	871	9,143	86	608	4.9
2.50	1292	26,195	75	779	7.2
			Isopropanol		
0.25	207	1,850	62	361	4.0
1.00	634	6,978	84	372	3.1
2.50	917	21,120	75	779	7.2

[a]Taken from Plaa et al. (363).

showed disproportionate changes relative to the dose level, which indicated metabolic saturation and Michaelis–Menten type kinetics. In contrast to these parameters, values for the area under the blood concentration versus time curve (AUC) increased in proportion to the dose, and the total body clearance estimate (CL_{TB}) remained relatively constant. A separate determination of systemic bioavailability (F) by comparing intravenous infusion and oral gavage kinetics revealed that acetone was completely absorbed from the GI tract ($F = 1.02$). The data indicated that acetone elimination was saturated when the plasma levels rose above 300–400 mg/L. Rats repeatedly exposed by inhalation to 1000, 2000, or 4000 ppm of acetone for either 5 days or 13 weeks showed evidence of metabolic saturation at the highest exposure concentration when the blood AUC values at the two exposure durations were compared (467).

Once absorbed and metabolized, the carbon atoms of acetone are incorporated into cellular macromolecules and distributed throughout the body. Price and Rittenberg detected the radioactivity from orally administered 1,3[^{14}C]-acetone in the following tissue macromolecules: cholesterol, heme, glycogen, fatty acids, urea, tyrosine, leucine, arginine, aspartic acid, and glutamic acid (468). Acetate and another two-carbon fragment from acetone, it was thought, were used in the citric acid cycle or in the synthesis of amino acids. Similar result were obtained by Zabin and Bloch after feeding rats 0.5 mM/100 g/day of 1,3[^{14}C]-acetone in their diet (469). Radiolabeled fatty acids and cholesterol were isolated from the livers of the treated animals. Labeled glycogen, serine, choline, and methionine were isolated from the liver and gastrointestinal tracts following the sacrifice of male rats given subcutaneous injections of 1,3[^{14}C]-acetone during a 10-h period (470). Degradation of the labeled products revealed that the radioactivity in choline and methionine was confined to the carbon atom in the terminal methyl group.

1.4.1.3.3 Excretion. The urinary and pulmonary excretion of acetone is normally a secondary route of elimination. When metabolic saturation occurs, however, a much larger percentage of the body burden is disposed of in the urine and exhaled air. Haggard et al. observed that about 75% of an absorbed acetone dose was excreted in the urine or expired air when the blood levels were higher than 1000 mg/L (210). When the blood levels declined to 150 mg/L or less, excretion in the urine and expired air declined to about 40%. Price and Rittenberg orally administered 1.2 mg/kg of labeled acetone to a female rat and found 47% of the radioactivity in the expired air as radiolabeled carbon dioxide within about 14 h (468). Further studies of a rat given 7.1-mg/kg oral doses of acetone for 7 days revealed that only about 7% of the administered dose was in the expired air as acetone and that an additional 76% appeared as carbon dioxide. Sakami and Lafaye found that male albino rats orally administered 2.9 mmol/kg of labeled acetone excreted 27% of the dose in expired air as carbon dioxide (400). Coleman examined the excretion of acetone in normal, obese, and diabetic mice that fasted were then administered 0.55 mmol/kg of labeled acetone (409). The obese and diabetic mice converted 20–40% of the dose into carbon dioxide within 3 h, whereas normal mice converted 15–25%. Approximately 10–20% of the acetone was excreted unmetabolized in the expired air, and an average of 1.5% was excreted the urine. Bergman et al. administered a single radiolabeled dose of acetone to guinea pigs by intracardiac injection at dose levels from about 1 to 210 mg/kg (471). The percentage of acetone excreted as carbon dioxide after 6 h decreased from about 53% at the low dose to about 4% at the high dose.

1.4.1.4 Reproductive and Developmental. Acetone treatment has generally produced mild reproductive and developmental changes in laboratory animals or cultured embryos. Because of its favorable properties as a treatment vehicle, acetone has been tested in a variety of *in vitro* screening assays designed to detect changes in growth or structural development of developing embryos. The assays have revealed either no adverse effects or some adverse effects at extremely high treatment concentrations. Acetone has also been examined for its effects in developmental screening assays that have used chicken and rodent embryos, frog eggs and embryos, mallard embryos, and the hydra (472–477).

High doses of acetone were also needed to cause any treatment-related effects in laboratory animals. Dietz et al. detected some mild adverse spermatogenic effects in male rats, but not mice, that consumed 5% acetone (3.1 g/kg/day) in their drinking water for 13 weeks (382, 383). Rats showed a relative decrease in testicular weight, along with depressed sperm motility, increased epididymal weight, and an increased incidence of abnormal sperm at this dose. No histopathological changes were observed in the testis of the test animals. Larsen et al. examined the reproductive effects of acetone on male Wistar rats given 0.5% acetone in their drinking water for six weeks (331). On the fifth week of treatment, the rats were allowed to mate with untreated females, and the number of matings was recorded together with the number of pregnancies and the number of fetuses per pregnancy. The absolute weight of the testis was measured, along with the diameter of the seminiferous tubules and histopathological changes in the testis. The acetone treatment did not affect any of the measures of reproductive and testicular toxicity.

Using Chernoff and Kavlock's screening assay for reproductive toxicants, acetone caused developmental effects in mice (478). Acetone was administered orally at a dosage of 3500 mg/kg/day to a group of 50 female mice on days 6 through 15 of gestation. The weight and survival of the pups in each litter were determined postnatally on days 1 and 3.

Fetal weight and survival were evaluated, along with maternal mortality, using a ranking system that scaled the severity of each adverse effect. The changes caused by acetone yielded a total score of 19 out of a possible 22 points, which indicated a need for additional testing in both rats and mice.

The additional testing was performed on virgin and pregnant Sprague–Dawley rats and Swiss CD-1 mice (464). Groups of positively mated rats were exposed by inhalation for 6 h/day to 440, 2200, or 11,000 ppm of acetone on days 6 through 19 of gestation (7 days/week). The mice were exposed similarly at concentrations of 0, 440, 2200, or 6600 ppm of acetone on days 6 through 17 of gestation. Maternal uterine and fetal body weights were recorded upon sacrifice, and the live fetuses were examined for gross, visceral, skeletal, and craniofacial defects. No deaths were observed in the treated rats, and the mean pregnancy rate was greater than or equal to 93% in all treatment groups.

Clinical signs of toxicity were confined to the 11,000-ppm treatment group and included a statistically significant reduction in maternal body weight gain that started on day 14 of gestation and a decrease in the uterine and extragestational weight gain of the dams. The fetal weights were significantly lower for the 11,000-ppm group relative to the control group. No effect was seen in the mean liver or kidney weights of pregnant dams, the organ to body weight ratios, the number of implantations, the mean percentage of live pups per litter, the mean percentage of resorptions per litter, or the fetal sex ratio. The incidence of rat fetal malformations was not significantly increased by gestational exposure to acetone vapors, although the percentage of litters with at least one pup that exhibited a malformation was reportedly greater for the 11,000-ppm group (11.5%) than for the control group (3.8%).

No treatment-related effects on maternal or virgin body weight or on maternal uterine weight occurred in the mice. The mean pregnancy rate in treated mice was greater than or equal to 85%. Treatment at 6600 ppm of acetone produced a statistically significant reduction in fetal weight and a slight, but statistically significant increase in the percentage incidence of late resorptions. The incidence of fetal malformations was not altered at any exposure level; however, a treatment-related increase was observed in the liver to body weight ratio of the pregnant dams. The authors concluded that 2200 ppm of acetone was the no-observable-effect level (NOEL) for developmental toxicity in both rats and mice. Minimal maternal toxicity was seen at levels of 11,000 ppm of acetone in rats and 6600 ppm in mice.

Kitchin and Ebron evaluated the use of acetone as a treatment vehicle for a teratogenic screening assay that relied on a rat whole-embryo culture system (479). Acetone was added to the culture medium at concentrations of 0.1, 0.5, 2.5, and 10.0%. After a 48-h incubation, acetone concentrations of 0.5% and greater caused structural abnormalities and a high rate of embryo lethality. Picard and Van Maele-Fabry calculated an ELC_{50} embryo lethality value of 0.20 M (11.6 mg/mL) using the results from this screening assay (480). Schmid evaluated the effects of acetone and 38 other chemicals in a similar postimplantation rat embryo culture (481). Acetone did not cause any structural malformations when incubated with embryo cultures from rats at concentrations of 0.1, 0.3, 0.6, or 1.0%; however, evidence of embryo toxicity was observed at acetone concentrations of 0.6% or greater. An acetone level of 3% inhibited the growth and differentiation of all of the embryos tested. Guntakatta et al. showed that acetone did not cause any effects in a mouse embryo limb-bud cell culture system designed to estimate the teratogenic potential of chemicals by measuring DNA and protein synthesis (482). The mouse embryos were incubated with acetone at concentrations from 10 to 100 mg/mL for periods up to 96 h.

Acetone did not decrease the number of fertile eggs, the number of viable embryos, or the percentage of live embryos when 0.8 mL/kg was given to developing chick embryos (483). Additional studies with chick embryos showed that the use of acetone as a treatment vehicle did not alter the liver weights or the RNA and DNA contents of viable embryos (484). McLaughlin et al. injected acetone into the yolk sacs of fresh fertile chicken eggs and found minimal toxicity and no teratogenicity following a 20-day incubation at dose levels of 39 and 78 mg/kg (485, 486). Walker used a similar yolk sac injection procedure and found that the treatment of chicken eggs with 0.05 to 0.1 mL of acetone resulted in a survival rate of 20 to 37%, depending on the time of treatment (487). Embryo mortality was associated with the formation of a proteinaceous coagulum immediately following the injection. Acetone volumes less than 0.05 mL did not cause any increase in mortality. Korhonen et al. found that injecting 5 mL of acetone into the eggs of developing chick embryos did not significantly increase the number of early or late deaths or the percentage of malformations in dead embryos (488, 489). Swartz treated chicken embryos with 0.05 mL of acetone for either 5 or 12 consecutive days and did not observe any increase in mortality relative to untreated controls (490). DiPaolo et al. added acetone at concentrations of 0.02% or less to cell cultures derived from hamster embryos and morphologically evaluated the cell clones (491). Cloning efficiency and induced transformants were unaffected by the treatment. Intraperitoneal treatment of pregnant hamsters with a 3.0-mL/kg dose of acetone 3 days before isolation and preparation of the embryo cell cultures did not result in any morphological evidence of cell transformation (492). Acetone concentrations of 0.1–1000 mg/L failed to cause treatment-related effects on progesterone release from human luteinizing granulosa cells (493).

1.4.1.5 Carcinogenesis. The carcinogenicity of acetone was not formally examined in a two-year rodent bioassay. However, tests for mutagenicity, chromosomal damage, and DNA interaction have shown little interaction with genetic material. Because of its utility as a treatment vehicle for dissolving and applying water-insoluble chemicals, acetone was examined in both dermal carcinogenicity and initiation-promoting studies. The tumorigenicty of acetone was also examined in an alternative cancer bioassay using Tg.AC transgenic mice that were genetically altered to increase their susceptibility to tumor-promoting agents (494). In the transgenic assay, male and female mice treated dermally for 7 days per week for 26 weeks with 200 μL of acetone did not show a statistically significant increase in the incidence of dermal papillomas relative to untreated control mice.

In other studies, Van Duuren et al. found that 0.1 mL acetone did not act as a tumor-promoting agent when applied to the shaved dorsal skin of mice three times a week for 441 days (495). The initiator, 7,12-dimethylbenz[*a*]anthracene, was applied for 2 weeks before beginning the acetone treatment. Moreover, Weiss et al. reported that acetone inhibited tumor formation when used in a mouse two-stage initiation-promoting experiment (496, 497). A 56% decrease in the total number of tumors per mouse was observed when 7,12-dimethylbenz[*a*]anthracene-initiated CD-1 mice were treated with 0.1–0.4 mL of acetone

3 to 5 min before treatment with a promoter (phorbol-12-myristate-13-acetate) twice a week for 10 weeks.

In a dermal oncogenicity study reported by Van Duuren et al., mice were treated with 0.1 mL of acetone or 0.1 mL of an acetone–water mixture (9:1) three times a week for up to 424 days (498). Histopathological analysis of all major organs revealed a total of 14 lung tumors, one liver tumor, one forestomach tumor, but no skin tumors in the acetone-treated and acetone/water-treated groups. Except for one undifferentiated malignant liver tumor, which was not cited as a remarkable finding, the incidence of systemic tumors in acetone-treated and acetone–water-treated mice was not different from the background incidence in untreated mice. Park and Koprowska saw no increase in the incidence of dyspasia or cervical squamous cell carcinoma in mice treated dermally with acetone for up to 5 months (499). In two separate studies, Barr-Nea and Wollman treated male mice dermally with acetone twice a week for a periods of 5, 7, or 12 months and examined excised tissues and organs for signs of damage (500, 501). No tumors were observed in the treated animals, however, a significant increase in amyloid deposition was observed in the tissues and organs of animals treated with acetone for 12 months. The application of acetone to the buccal pouch of male and female hamsters three times per week for 6 weeks caused some inflammation and hyperkeratosis, but no dysplasia (502).

Iversen reported extremely low malignant skin tumor incidence rates in hairless mice treated with 0.1 mL of acetone once or twice a week for 20 weeks (503). Zakova et al. applied 0.2 mL of acetone to the shaved dorsal skin of male and female CF1 mice once a week for two weeks and found that acetone had no effect on the survival of the 300 animals tested (504). The acetone treatment caused dermal inflammatory reactions (dermal fibrosis and focal acanthosis) in 6% of the animals. DePass et al. reported that the mean survival time was 504 days in a group of 40 male C3H/HeJ mice treated with 25 mL of acetone three times a week for life (505). No carcinomas and two sarcomas (fibrosarcoma and lymphosarcoma) were found on or below the skin in the acetone-treated control group. The historical incidence of sarcomas in the acetone-treated mice was reportedly 5/724.

Peristianis et al. conducted a two-year dermal carcinogenicity in which a group of 99 male and 100 female CF1 mice were treated with 0.2 mL of acetone twice a week for up to 103 weeks (506). The percentages of male and female mice that survived until the end of the study were 26% and 22%, respectively. The most common causes of death or illness were systemic neoplasia, renal failure, urethral obstructions, and suppurative lesions. One skin tumor was found in a male mouse from the acetone-treated control group, and about 15% of the male and female mice had irritant lesions on their skin. The incidence of each tumor type in the acetone-treated control group was not statistically different from historical control values.

Ward et al. provided a historical review of the organ pathology in the mice used in two previous initiation-promoting experiments (507). A combined treatment group of 60 female Sencar mice were treated with 0.2 mL of acetone once or twice a week for at least 88 weeks. Fifty percent of the animals survived past 96 weeks of age, 15 of the mice died due to neoplastic lesions, and 27 died due to nonneoplastic lesions. The two major contributing causes of death were glomerulonephritis and histocytic sarcoma. Based on a consideration of the background incidence in untreated mice, the authors determined that there was no evidence that the lesions were associated with the acetone treatment.

1.4.1.6 Genetic and Related Cellular Effects Studies. Acetone has been repeatedly tested in a variety of prokaryotic and eukaryotic systems without causing genotoxic effects [Tables 74.23 (508–526) and 74.24 (527–550)]. Studies using the *Salmonella* assay showed that acetone is nonmutagenic and is an acceptable vehicle for dissolving and delivering water-insoluble chemicals to the test strains (551). EPA-sponsored studies showed that acetone is negative in *Salmonella* strains TA97, TA98, TA100, and TA1535 at levels up to 1 mg/plate (552). Subsequent studies found that acetone was negative in strains TA92, TA94, TA98, TA100, TA1535, and TA1537 at a concentration of 10 mg/plate (511). Acetone was not genotoxic to *Schizosaccharomyces pombe* with or without metabolic activation (523). Acetone induced aneuploidy, but not mitotic recombination or point mutations, in *Saccharomyces cerevisiae* at concentrations greater than 40 mg/mL using a cold-interruption procedure (519). These effects were not observed, however, when *S. cerevisiae* was tested by the standard overnight incubation procedure (521).

Acetone did not produce genotoxic effects in an embryo cell transformation assay performed on rats and mice and was also negative in a micronucleus assay using hamsters (550, 553). Acetone did not cause chromosomal aberrations or sister chromatid exchanges in Chinese hamster ovary cells treated at concentrations up to 5 mg/mL (537). Acetone concentrations from 10.5 to 20.9 mM (0.6 to 1.2 mg/mL) also did not cause chromosomal aberrations or sister chromatid exchanges in cultured human lymphocytes (538). Acetone did not cause point mutations at the thymidine kinase locus in L5178Y mouse lymphoma cells treated at a level of 10 mg/mL (539).

1.4.1.7 Other: Neurological, Pulmonary, Skin Sensitization. Goldberg et al. recorded mild neurobehavioral changes in female Carworth rats exposed for two weeks (5 days/wk, 4 h/day) to 3000, 6000, 12,000, or 16,000 ppm of acetone (226). Repeated exposures to 6000 ppm of acetone inhibited avoidance behavior but did not produce any signs of motor imbalance. Acetone concentrations of 12,000 and 16,000 ppm produced ataxia in several animals on the first exposure day; however, rapid tolerance developed, and ataxia was not seen on subsequent days. Kurnayeva et al. studied the combined effect of vapors and noise on the immune, cardiovascular, endocrine, and nervous systems of Wistar rats exposed for 1.5–2.0 months (554). The animals were exposed (5 days/wk, 4 h/day) to 2000 mg/m^3 (843 ppm) of acetone and 85 Db of noise. The combined exposure was reportedly without effect on the various measures of physiological function.

Spencer et al. compared the neurotoxicity of various ketonic solvents and reported that repeated acetone treatments did not cause neuropathological lesions in Sprague–Dawley rats (555). Acetone was initially administered in drinking water for 8 weeks at 0.5%; the concentration was then increased to 1.0% for an additional 4 weeks. Tissue specimens were collected from the spinal cord and spinocerebellar area, as well as from the plantar, tibial, and sciatic nerves. Histological examination of nerve tissue from acetone-treated rats showed no evidence of peripheral distal axonopathy.

A synergistic relationship was demonstrated in rats cotreated with acetone and 2,5-hexanedione (HD) for 6 weeks. Ladefoged et al. maintained groups of male Wistar rats on drinking water that contained either HD, acetone, or a combination of the two chemicals at concentrations of 0.5% (332). Caudal motor nerve conduction velocities were measured weekly during the last 4 weeks of treatment, and the balance time on a rotarod were

Table 74.23. Genotoxicity Studies with Acetone Using Prokaryotic and Eukaryotic Organisms

Assay	Indicator System	Highest Conc Tested	Metabolic Activation	Results (with/without S9)	Reference
	Prokaryotic organisms—In Vitro Tests				
Reverse mutation (Ames assay)	*S. typhimurium* TA98, TA100, TA1535, & TA1537	10 mg/plate	Rat liver S9	–/–	508
Reverse mutation (Ames assay)	*S. typhimurium* TA98, TA100, TA1535, & TA1537	10 mg/plate	Rat & hamster liver S9	–/–	509
Reverse mutation (Ames assay)	*S. typhimurium* TA98, TA100, TA1535, TA1537, & TA 1538	73 mg/plate	None	/–	510
Reverse mutation (Ames assay)	*S. typhimurium* TA92, TA94, TA98, TA100, TA1535, & TA1537	10 mg/plate	Rat liver S9	–/	511
Lambda prophage WP2s(λ) induction (microscreen assay)	*E. coli* TH-008	10% (v/v)	Rat liver S9	–/–	512
Lambda prophage WP2s(λ) induction (microscreen assay)	*E. coli* SR714	10% (v/v)	Rat liver S9	–/–	513
β-Galactosidase activation (SOS chromotest)	*E. coli* PQ37	100 mM	Rat liver S9	–/–	514
Colitis phage DNA transfection assay	*E. coli* CR63	0.1 mL	Rat liver S9	–/	515
DNA binding assay	*E. coli* Q13	0.05% (v/v)	Rat liver S9	–/–	516
Recombination assay	*B. subtilis* H-17 & M-45	10 mg/well	Rat liver S9	–/–	517
β-Galactosidase activation (SOS chromotest)	*S. typhimurium* TA1535/pSK1002	33 mg/mL	Rat liver S9	–/–	518
	Eukaryotic organisms—In Vitro Tests				
Chromosomal malsegregation	*S. cerevisiae* D61.M	7.8% (v/v)	None	+	519
Point mutations & mitotic recombination	*S. cerevisiae* D61.M	7.8% (v/v)	None	–	519
Chromosomal malsegregation	*S. cerevisiae* D61.M	50 mg/mL	None	±	520
Chromosomal malsegregation	*S. cerevisiae* D61.M	8% (v/v)	None	±	521
Reverse mutation	*S. cerevisiae* D7	10% (v/v)	None	±	522
Forward mutation	*S. pombe* P_1	3.7% (v/v)	Mouse liver S10	–/	523
Forward mutation	*S. cerevisiae* D4	5% (v/v)	Rat liver S9	–/	524
Plant mitotic index	*A. cepa*	1%	None	–	525
Plant seed gene mutation	A. thaliana	500 mM	–	–	526

Table 74.24. Genotoxicity Studies with Acetone in Mammalian Cell Systems

Assay	Indicator System	Highest Conc Tested	Metabolic Activation	Results (with/ without S9)	Reference
	Eukaryotic Organisms—In Vitro Tests				
Cell transformation assay	Syrian hamster embryo cells	135 μg/m^3	None	–	527
Cell transformation assay	Syrian hamster embryo cells	8% (v/v)	None	–	528
Cell transformation assay	Rat embryo cells	100 μg/mL	None	–	529
Cell transformation assay	Rat embryo cells	0.1% (v/v)	Rat liver S9	–/–	530
Transformation assay	Asynchronous mouse embryo fibroblsts	0.5% (v/v)	None	–	531
Cell transformation assay	Mouse embryo fibroblasts	0.5% (v/v)	None	–	532
Cell transformation assay	Mouse prostate fibroblasts	0.5% (v/v)	None	–	533
Sister chromatid exchange (SCE)	Chinese hamster lung fibroblasts	100 mM	Rat liver S9	–/–	534
Chromosomal aberration	Chinese hamster fibroblasts	5% (v/v)	None	+	511
Sister chromatid exchange (SCE)	Chinese hamster lung fibroblasts	8.6 mM	None	/–	535
Chromosomal aberration & SCE	Chinese hamster ovary cells	1 mg/mL	Rat liver S9	–/–	536
Chromosomal aberration & SCE	Chinese hamster ovary cells	5 mg/mL	Rat liver S9	–/–	537
Chromosomal aberration & SCE	Human lymphocytes	20.9 mM	None	–	538
Mouse lymphoma mutation assay	L5178Y mouse lymphoma cells	470 mM	None	–	539
Mouse lymphoma mutation assay	L5178Y mouse lymphoma cells	1% (v/v)	Rat liver S9	–/	540
Mouse lymphoma mutation assay	S49 mouse lymphoma cells	140 mM	Rat liver S9	–/	541
Reverse mutation Ouabain resistance	Chinese hamster lung fibroblasts	0.2% (v/v)	None	/–	542
Forward mutation thioguanine resistance	Chinese hamster lung fibroblasts	0.5% (v/v)	Rat liver S9	/–	543
Micronucleus test	Human lymphocytes	5 mM	Rat liver S9	–	544
Unscheduled DNA synthesis	Bovine lymphocytes	0.4 mg/mL	None	–	545
Unscheduled DNA synthesis	Human skin cells	10% (v/v)	None	–	546
Metabolic cooperation assay	Chinese hamster lung fibroblasts	5% (v/v)	None	+	547
Alkaline elution assay	Rat hepatocytes	1% (v/v)	None	–	548
Two-stage cell tranformation assay	Mouse 3T3 cells	0.5% (v/v)	None	–	549
	Eukaryotic Cells—In Vivo Tests				
Micronucleus test	Chinese hamster bone marrow cells	865 mg/kg	—	–	550
Host-mediated assay	Hamster fetal cells	2.3 g/kg	—	–	492

recorded weekly for all 6 weeks of treatment. The body weight gain and water consumption significantly decreased in the animals cotreated with acetone and HD. HD and the HD–acetone mixture also caused a significant reduction in the rotarod balance and the nerve conduction velocities. The difference in balance time between HD-treated rats and those given HD and acetone was statistically significant during the last three treatment weeks. Likewise, the difference in conduction velocities for these two groups was significant during the fourth and sixth weeks of treatment. Acetone exposures alone had no effect on the balance times observed throughout the study. The acetone cotreatment potentiated the neurophysiological and neurobehavioral effects of HD; however, the authors were not certain whether acetone caused a purely neurotoxic interaction or interfered with the metabolism and elimination of HD. Previous studies by Ladefoged and Perbellini had shown that the whole body clearance of HD was retarded in New Zealand white rabbits that were administered an i.v. dose of HD and acetone (338).

Misumi and Nagano performed a neurophysiological study of mice treated subcutaneously with acetone (556). Male mice treated for 15 weeks (5 days/week) with a 400-mg/kg/day dose of acetone showed no evidence of neurological dysfunction relative to control animals. The acetone treatment did not cause any difficulty in walking or dullness in movement, and there were no significant changes in motor or sensory nerve conduction velocities. The authors concluded that acetone was not neurotoxic to the peripheral nervous system.

Studies showed that acetone does not cause skin hypersensitization and is not a contact allergen in laboratory animals (Table 74.25) (557–562). Several testing approaches have been taken to identify the allergic potential of acetone; these include ear swelling measurements in mice, the appearance of erythema and edema (maximization test) in guinea pigs, and variations on the local lymph node assay in both species. These assays have generally involved subcutaneous administration of an initial sensitizing dose of acetone followed by a dermal challenge dose several days later. Some tests used the well known sensitizer, 2,4-dinitrochlorobenzene, for initial sensitization, and then the effects of an acetone challenge dose were evaluated. In a variation of this approach, Czerniecki et al. reported that acetone pretreatment did not affect the development of contact hypersensitivity in mice sensitized and challenged by the well-known sensitizer, 2,4-dinitrochlorobenzene (563). In another newer approach, Brouland et al. reported that acetone was

Table 74.25. Contact Hypersensitivity Tests with Acetone on Laboratory Animals

Test species	Test Method	Acetone Concentration (%)	Result	Reference
Mouse	Ear swelling test	100	Negative	557
Guinea pig	Maximization test (erythema & edema)	100	Negative	558
Mouse	Local lymph node assay	100	Negligible	559
Guinea pig	Lymph node size and histology	100	Negative	560
Mouse	Lymph node cell proliferation	100	Negative	561
Mouse	Ear swelling test	100	Negative	562
Guinea pig	Maximization test (erythema & edema)	100	Negative	562

negative in popliteal lymp node assay, which includes an examination of treated mice for a graft-versus-host immune response and a morphological evaluation of excised lymph nodes for changes in cellular composition (564). Singh et al. reported that dermal application of up to 300 μL of acetone twice a week for 8 weeks affected humoral immunity development in mice (565). The acetone treatment did not affect antibody titers, lymphocyte proliferation, or lymphocytic phenotyping in single-cell suspensions from the spleens of the treated animals. Significant acetone-related reductions in humoral response were observed, however, when the animals were given a mitogen or alloantigen to induce cell proliferation.

1.4.2 Human Experience

Acetone toxicity is characterized by nonspecific local and systemic effects following acute exposure to relatively large amounts of vapor or liquid. The local effects of acetone are primarily irritative and are observed in the mucous membranes of the eyes, nose, and throat. Except for prolonged liquid contact or high vapor exposures, acetone irritation is not particularly severe or debilitating compared with the effects of other primary sensory irritants. Respiratory tract irritation is the most sensitive indicator of acetone overexposure in humans and is the basis for limiting occupational exposure. The systemic effects of acetone are its anesthetic properties and its ability to disrupt normal intermediary metabolism at high concentrations. The most noticeable effect of systemic overexposure is a generalized central nervous system (CNS) depression, which can range from light-headedness to narcosis, depending on the magnitude and length of the exposure. Accidental exposures to extremely large amounts of acetone vapor are necessary for the development of any overt signs of acetone-induced toxicity. The signs and symptoms of acetone overexposure are perhaps most notable for their lack of specificity; the clinical picture in human acetone intoxication can be easily confused with other medical conditions, such as diabetes and nonspecific mixed-solvent-induced narcosis.

1.4.2.1 General Information. The following section describes instances of accidental or intentional exposures to large amounts of acetone. The 1997 yearly summary of the American Association of Poison Control Centers listed 1505 incidents of human acetone poisoning throughout the United States (566). Approximately 94% of these incidents involved accidental ingestion of pure acetone, and about 34% involved children less than six-years old. A total of 462 cases required treatment in a health-care facility, and only one case was described as a major medical event; no fatalities were reported in any of these incidents. Minor organ damage was occasionally reported in cases of acute acetone-induced narcosis, and was likely to be due to the oxygen deprivation and tissue hypoxia that accompany very high vapor exposures. Hyperglycemia is characteristically observed in severe acetone intoxication and is the result of acetone's gluconeogenic effect, whereby a portion of the absorbed dose is metabolized to glucose. Available information indicates that acetone causes coma and possibly seizures following acute intoxication; however, no deaths have ever been documented. Acetone has a low potential for causing chronic or delayed health effects following an acute exposure.

1.4.2.2 Clinical Cases. Nine iatrogenic cases of acetone poisoning were reported in the literature. Patient descriptions and clinical observations are presented in Table 74.26

Table 74.26. Reports of Acute Acetone Intoxication of Hospital Patients Fitted with a Synthetic Cast Material

Case Reports Traits and Signs	Cossman (567)	Strong (568)	Chatterton and Elliott (569)	Fitzpatrick and Claire (570)	Pomerantz (571)	Anonymous (572)	Harris and Jackson (573)	Renshaw and Mitchell (574)	Hift and Patel (575)
Circumstances									
Age (yrs)	12	42	10	21	32	3	10	1.5	7
Sex	Male	Male	Female	Female	Female	Male	Male	Female	Male
Onset (hrs)	< 24	10	10	1?	12	1	9	12	6?
Recovery (days)	—	5	3	1?	1.5	2?	4	1	3.5
Type of cast	Hip cast	Body cast	Leg cast	Hip cast	Body cast	Leg cast	Hip cast	Hip cast	Leg cast
Exposure route	Dermal?	Inhalation	Inhalation	Inhalation	Inhalation	Inhalation	Inhalation	Inhalation	Dermal
Signs and Symptoms									
Unconscious	Yes	Yes	No	Yes	No	Yes	No	Yes	No
Drowsinwss	Yes	Yes	Yes	Yes	Yes	Yes	Yes	Yes	Yes
Vomiting	Yes	Yes	Yes	Yes	No	—	Yes	Yes	Yes
Hematemesis	—	Yes	Yes	Yes	No	—	Yes	Yes	Yes
Acetonuria	Yes	Yes	Yes	Yes	Yes	Yes	Yes	Yes	Yes
Glycosuria	—	Yes	Yes	—	Yes	—	Yes	Yes	Yes
Tachycardia	Yes	Yes	Yes	Yes	Yes	Yes	Yes	Yes	Yes
Breath acetone	Yes	Yes	Yes	Yes	Yes	—	Yes	Yes	Yes
Labored breathing	Yes	Yes	Yes	No	—	Yes	Yes	No	Yes
Throat irritation	—	Yes	Yes	—	—	—	No	—	Yes
Dermal irritation	—	Yes	No	—	—	—	No	No	—

(567–575) for each case report. The incidents generally involved hospital patients who required joint immobilization for a broken hip or leg. An outdated synthetic plaster substitute, containing large amounts of acetone as a setting fluid, was used in each instance. Males and female from about 2 to 42 years of age were affected. The exposures typically occurred via vapor inhalation; however, in some cases, skin absorption was considered the portal of entry. The onset of symptoms was typified by initial lethargy and drowsiness within 1 to 12 h of the exposure and nausea and vomiting later. Many patients lapsed into unconsciousness, and glycosuria, acetonuria, hematemesis, labored breathing, tachycardia, and throat irritation were often observed by the attending physician. A complete summary and analysis of the signs and symptoms of acetone intoxication in a hospital setting is presented in the case described by Hift and Patel (575).

In addition to the iatrogenic cases of acetone intoxication, there have been several occupationally related mishaps (Table 74.27) (576–584). Ross described an instance of overt acetone intoxication in an industrial environment that involved five individuals who were removing standing water from an unventilated pit (576). Water had been diverted to the pit following rupture of a water main. Two members of a crew entered the pit to shovel the water into buckets that were hauled to the surface by the others. Before taking a lunch break, both members of the pit-crew experienced some sensory irritation, and one of the members reportedly felt inebriated. Upon returning to the pit, the later worker became unconscious, whereupon another member of the crew entered the pit. Nearby employees, who came to provide assistance, were also affected by the vapors. Five of the six workers exposed to acetone vapor during the rescue operation reported feeling symptoms such as dizziness, eye and throat irritation, and leg weakness. The acetone vapor concentration in the pit was greater than 12,000 ppm. Foley reported an unusual case of acetone poisoning in an employee who attempted suicide by inhaling acetylene gas from a cylinder where acetone was used a carrier (577). The employee attached a hose from a cylinder of acetylene gas to a paper cup which he fastened to his face with a rubber band. The marked acidosis observed in the patient was attributed to the hypotension and hypoxia that accompanied unconsciousness.

Sack documented an acetone overexposure in an employee who was cleaning a large kettle that previously contained a synthetic fiber dissolved in acetone (578). The employee wore a respirator while working in the kettle; however, a poor fit resulted in severe overexposure. Blood acetone levels 9 and 11 hours after the mishap were reportedly 430 and 302 mg/L, respectively. Acetone was detected in the urine together with urobilin, red and white blood cells, and some albumin. These findings, together with a rise in serum glucose and bilirubin levels, suggested that slight liver and kidney damage had occurred. Smith and Mayers reported two cases of acute poisoning in a raincoat manufacturing operation where a waterproofing resin was applied that contained acetone and MEK (579). Samples of the workroom air showed that the acetone vapor concentration ranged from 330 to 495 ppm, and the MEK concentration varied between 398 and 561 ppm.

The most severe cases of acetone intoxication have occurred following attempted suicides. Sakata et al. described a male patient who attempted suicide by ingesting approximately 100 mL of an adhesive cement that contained 39% cyclohexanone, 28% MEK, 18% acetone, and 15% polyvinyl chloride (580). The individual reportedly drank about 720 mL of sake (10% ethanol) 30 min before drinking the adhesive solution.

Table 74.27. Case Reports of Human Acetone Intoxication from the Literature

Circumstances	No. People	Acetone Exposure	Symptoms	Reference
Workplace accident	5	12,000 ppm in air	Coma, dizziness, eye and throat irritation, leg weakness, confusion, and drowsiness	576
Attempted suicide	1	None	Coma, rapid breathing and heart beat, cyanosis, acidosis, elevated serum lactic acid, hyperglycemia, hematuria, glycosuria, and ketonuria	577
Workplace accident	1	430 mg/L in blood	Coma, vomiting, salivation, hyperactivity, hematuria, proteinuria, hyperglycemia, and increased serum bilirubin	578
Workplace overexposure	2	495 ppm in workroom air	Coma, stomach distress, watery eyes, acetosis, confusion, and headache	579
Attempted suicide	1	100 mL adhesive cement ingestion (18% acetone)	Coma, hyperglycemia, and increased serum transaminases	580
Attempted suicide	1	200 mL pure acetone ingestion	Coma, shallow respiration, proteinuria, acetonuria, hyperglycemia, and gait disturbances	581
Ethanol substitute	1	2500 mg/L in blood	Lethargy, ketonuria, and acetonemia	187
Accidental ingestion by an infant	1	4450 mg/L in blood	Coma, seizure, acetonuria, acetonemia, metabolic acidosis, respiratory depression, hypothermia, and hyperglycemia	188
Intentional administration to an infant	1	4.48 mL/kg oral dose	Abdominal distension and tenderness, portal venous gas, diarrhea, decreased hematocrit, bloody stool, gastric hemorrhage, and eukocytosis	582
Attempted suicide	1	2000 mg/L in blood; 800 mL pure acetone ingestion	Coma, ketosis, acetonuria, acetonemia, hyperglycemia, leukocytosis, and elevated lactate dehydrogenase	583
Theraputic application to abraded skin	1	465 mg/L in blood; 40 mL dermal application	Coma, weakness, dizziness, fatigue, acidosis, ketonuria, and elevated serum β-hydroxybutyrate	584

Gitelson et al. described a case involving an adult male who consumed 200 mL of pure acetone in an attempted suicide (581). The elevated glucose levels observed in this patient returned to normal only after two months of dietary restriction. The authors noted that the duration of the hyperglycemia was more prolonged than most cases but did not speculate on the etiology. Ramu et al. described a case of acetone intoxication that involved the ingestion of nail polish remover (187). The patient denied drinking from a half-empty bottle of nail polish remover found in her home but had a history of alcohol abuse and was known to have suffered from liver cirrhosis, peripheral neuropathy, cerebral atrophy, and throat inflammation. Extremely high acetone blood levels (2500 mg/L) were observed upon admission, but no unconsciousness, hyperglycemia, or glycosuria was reported.

In perhaps one of the most severe case reports of acute acetone intoxication ever reported, Gamis and Wasserman documented the accidental ingestion of acetone by a young child (188). A 2.5-year-old child reportedly consumed most of the contents of a 6-ounce bottle of fingernail polish remover that contained 65% acetone and 10% isopropanol. The child was unconscious when found in his home and began having a seizure while being taken to a hospital. The acetone blood levels observed at 1, 18, 48, and 72 h after the onset of symptoms were 4450, 2650, 420, and 40 mg/L, respectively. The initial acetone blood level of nearly 4.5 g/L is the highest ever reported for a human. The decline in blood acetone closely followed the course of recovery, and the patient was discharged on the fourth day after admission when a neurological examination showed no abnormalities. A six-month follow-up examination of the patient also showed no signs of neurodevelopmental complications. Acetone potentiation was responsible for the death of an individual who simultaneously ingested acetone and acetonitrile with suicidal intent (585). The patient's physical and clinical appearance was unremarkable for 18 h and then rapidly deteriorated with convulsions, acidosis, and cardiovascular depression. The patient's death 30 h after ingestion of the mixture was attributed to the delayed development of acetonitrile toxicity.

1.4.2.2.1 Acute Toxicity. Albertoni reported that the consumption of 15 to 20 g of acetone per day for several days caused no adverse effects other than slight drowsiness (213). Likewise, Frerichs found no apparent effects in diabetics who consumed 20 g of acetone per day for 5 days (586). Koehler et al. showed that the intravenous administration of 200 mL of a 0.5% solution of acetone in normal saline over 2 h to healthy and diabetic test subjects resulted in a small decrease in blood pressure and a slight transient drowsiness (587). Widmark measured acetone blood levels following the administration of 8 to 16 g of acetone in aqueous solution by the oral or peranal route. No apparent adverse effects followed the treatment (462). Haggard et al. reported no ill effects for several test subjects who drank 4.7 to 5.4 g of acetone (210).

In self-exposure trials, Kagan reported that 22 mg/L (9300 ppm) of acetone vapor could not be tolerated for longer than 5 min due to throat irritation (211). DiVincenzo et al. reported that human volunteers exposed to either 200 or 500 ppm of acetone for 2 h did not experience any subjective symptoms of irritation other than odor awareness at 500 ppm (463). Additional hematological and biochemical measurements on blood specimens collected before and after treatment did not reveal any exposure-related alterations. An inhalation study performed by Haggard et al. revealed that a single resting individual exposed to either 211 or 2110 ppm for 8 h did not experience any loss of judgment or coordination (210). These results were not altered when moderate exertion was required of the test subject during the 2110-ppm exposure. When the exposure duration was extended to three days at 2110 ppm (8 h/day), a similar negative finding was reported.

Using several different treatment regimens, Stewart et al. studied the local and systemic effects of repetitive exposures in two series of experiments with male and female volunteers (121). In the first, two small groups of male subjects were exposed to each of four vapor concentrations (0, 200, 1000, and 1250 ppm) for either 3.0 or 7.5 h per day for 4 days/week. The groups were exposed to progressively higher vapor levels of acetone in each succeeding week of treatment. Following the fourth week of exposure at 1250 ppm of acetone, the two groups were given a fifth week of exposure at 0 ppm and then a final week where the vapor concentration was allowed to fluctuate between 750 and 1250 ppm (average of 1000 ppm) on each of four exposure days. The second series of studies was performed in groups of female subjects who were exposed to 1000 ppm of acetone for either 1.0, 3.0, or 7.5 h per day for 4 days/week.

All of the subjects in the investigation were given complete medical and physical exams at the beginning and end of the study. The exams included weekly collection of blood for a complete blood count and a 23-element clinical chemistry analysis. Blood pressure, temperature, subjective responses, clinical signs and symptoms, and urinalysis were recorded daily. Electrocardiograms were recorded continuously during each exposure session. Cardiopulmonary testing (heart rate, minute ventilation, expiratory flow rate, alveolar-capillary gas exchange, and vital capacity) was performed on male subjects shortly before the end of each weekly exposure session. Using a separate group of male subjects exposed for 7.5 h, metabolic, hematologic, and cardiopulmonary measurements were conducted before and after physical exercise. Urinalysis, complete blood counts, and clinical chemistry values were within normal ranges for both the male and female subjects. No statistically significant changes were noted in the electrocardiograms, and no trends or consistent changes were noted in the pulmonary function tests.

As mentioned earlier, subjective complaints of slight to mild eye and throat irritation were often reported for workers exposed to acetone vapor. Three different experimental methods were used to study the sensory irritation potential of acetone in the eyes, nose, and throat: physiological methods, psychophysical methods, and symptom questionnaires. The studies listed in Table 74.28 (588–597) were conducted in the workplace using acetone-exposed employees and in the laboratory using naive volunteers exposed to acetone under controlled conditions. The studies that used objective physiological and psychophysical techniques tended to show that acetone is an extremely weak sensory irritant. Subjective symptom questionnaires, in contrast, indicated that acetone was a sensory irritant at much lower vapor concentrations. Recent research indicates that the irritancy responses obtained from subjective symptom questionnaires are likely to be caused by the odor of acetone. Dalton et al. showed that both acetone and phenyl ethyl alcohol, a known nonirritant that has a strong odor, could produce subjective irritancy responses in humans following a 20-min inhalation exposure at 800 ppm (598). Ninety male and female subjects were divided into three groups and given positive, negative, or neutral information about the consequences of the exposure. Using psychophysical methods to obtain ratings of odor

Table 74.28. Human Sensory Irritation Studies of Acetone

Test Method	Type of Subjects	No Effect Level (mg/m^3)	Reference
		Subjective	
Questionnaire	Naive	475	588
Questionnaire	Workers	<595	589
Questionnaire	Naive	595	590
Questionnaire	Naive	1185	463
Questionnaire	Workers	1900	591
Questionnaire	Both	2375	592
Questionnaire	Naive	2850	121
Questionnaire	Workers	3560	593
		Objective	
Acoustic rhinometry	Naive	7120	594
Spirometry	Naive	18,985	595
Psychophysics	Naive	>23,730	596
Psychophysics	Naive	>23,730	597
Laterialization	Workers	>35,600	14
Laterialization	Naive	>83,070	14

and irritation intensity, the study revealed that the subjects given positive information experienced significantly less odor and irritation than those given neutral or negative information. In addition, the irritation intensity rating strongly correlated with odor intensity in all of the groups but particularly in the group given neutral hazard information at the beginning of the exposure. The irritancy rating for acetone also strongly correlated with the irritancy rating for phenyl ethyl alcohol, which was used as nonirritaing control odorant.

MacGregor et al. used several sensitive and specific tests to detect the nasal irritancy from short-term acetone exposures in six to eight male and female volunteers (599). The subjects were either given consecutive 20-minute whole body exposures at 750 and 1500 ppm of acetone or consecutive 3–5 min face mask exposures at 500, 3000, and 6000 ppm. Three types of tests were conducted on each volunteer: (1) physiological measurements using respiratory spirometry, laser Doppler flowmetry, and acoustic rhinometry, (2) biochemical measurements of total protein, albumin, lysozyme, and mucin in nasal lavage secretions; and (3) subjective assessments of illness using a 17-item symptom questionnaire that allowed four types of response (none, mild, moderate, and severe) in five categories (nose, eyes, throat, head, and chest). Neither protocol resulted in any statistically significant physiological changes. The whole body exposures caused a statistically significant increase in total protein and lysozyme content of the lavage fluid, whereas the face mask exposures caused a significant increase in the mucin content. The biochemical changes were not observed following the high-level exposures and did not increase in a concentration-related fashion. The biochemical changes observed at 1500 ppm were not associated with any complaints of congestion, rhinorrhea, or changes in mucosal blood flow. The whole body exposures also produced mild complaints of respiratory or ocular irritation once the exposures began, but they did not worsen when the concentration was increased to 1500 ppm but caused a mild runny nose in one subject and a sore throat in another.

Cometto-Muñiz and Cain determined the odor and sensory irritation thresholds for a homologous series of ketones, alcohols, and acetates (596). Solvent vapors were inhaled briefly by a group of four normosmic and four anosmic subjects, who were exposed to headspace vapors from squeeze bottles that contained acetone in solution. The testing used a rigorous protocol that involved progressively higher concentrations of acetone vapor, along with procedures to minimize false responses. Acetone vapors had irritancy thresholds of approximately 100,000 ppm in anosmic subjects and an odor threshold of 10,000 ppm in normosmics. Below the irritancy threshold concentration, anosmic subjects could not detect the presence of acetone vapors. In a subsequent study, the same authors examined the ocular irritancy threshold of acetone vapors using a similar headspace technique (597). Ten male and female volunteers delivered a brief puff of vapor to their eyes using custom-fitted squeeze bottles. Using the two-alternative forced-choice method of obtaining psychophysical response data, the ocular irritation for acetone was in excess of 100,000 ppm.

Matsushita et al. presented data indicating that acetone exposures of 500 ppm or greater caused elevations in white blood cell and eosinophil counts (590, 600). The studies were performed on groups of male university students who received either a single exposure or six consecutive daily exposures to acetone vapor. The single day exposures were performed on groups of five subjects at concentrations of 0, 100, 250, 500, or 1000 ppm, whereas the multiple day exposures were conducted on groups of six students at concentrations of 250 and 500 ppm. In both studies, the exposures lasted 6 h, and a 45-min lunch break separated the morning and afternoon segments. The authors reported that exposures at 500 or 1000 ppm resulted in a temporary decrease in the phagocytic activity of neutrophils and an increase in the eosinophil and leukocyte counts in peripheral blood specimens collected at 3, 7, 24, and 32 h postexposure. All of the these measurements returned to normal after 48 h. The authors attributed the changes in the white blood cell count to an inflammatory reaction caused by the irritating effects of acetone vapor.

Douglas and Coe used constant volume whole body plethysmography to measure the bronchoconstrictive activity of acetone and eight other gases or vapors (595, 601). The experiments with acetone were terminated after testing just two subjects because of the weak irritancy response that was observed. The first individual inhaled 6000 ppm of acetone and did not experience any throat irritation; however, the subject experienced nausea, suffocation, and slight dizziness from the exposure. The second volunteer exposed to 8000 ppm of acetone felt nausea, a mild anesthetic feeling, and peripheral vasodilatation. The second subject also failed to feel any throat irritation, and the plethysmographic results showed little change in either airway resistance or thoracic gas volume. Acetone was not a bronchoconstrictor at any concentration or a sensory irritant, except at uncommonly high exposure concentrations.

1.4.2.2.2 Chronic and Subchronic Toxicity. Although the acetone levels characteristic of diabetic ketoacidosis may cause persistent drowsiness and mild proteinuria, the prevailing opinion is that acetone is not responsible for diabetic coma or any of the symptoms that accompany diabetic shock (100, 602).

1.4.2.2.3 Pharmacokinetics, Metabolism, and Mechanisms. Although most exposures to acetone occur by inhalation, the overall absorption and elimination of acetone has been examined in humans following many routes of absorption, including oral, inhalation, and dermal contact. Evaluation of the pulmonary uptake and elimination of acetone have demonstrated that the vapors may be absorbed into the aqueous film that lines the respiratory tract during inspiration and are then partially desorbed upon exhalation. Although the high blood-to-air partition coefficient for acetone suggest that a large percentage of inhaled acetone is absorbed into the body, the occurrence of this peculiar wash-in/wash-out effect effectively reduces uptake into the blood. This unusual characteristic and the difficulties associated with its unambiguous determination provide a likely explanation for the wide range of pulmonary retention values reported for acetone (Table 74.29) (603–609).

Landahl and Hermann investigated the nasal retention of acetone vapors in two volunteers exposed at concentrations of 300 and 3000 mg/m^3 (126 and 1264 ppm) and found that values varied between 18 and 40% (603). In addition, the nasal retention was independent of the acetone vapor concentration. Dalhamn et al. found that the average retention of acetone was 56% for the mouth and 86% for the lung; however, these studies did not take into consideration the reversible absorption of acetone into the fluid layer that lines the inside surface of the nose and lung (605, 610). Nomiyama and Nomiyama examined the respiratory retention, alveolar uptake, and alveolar excretion of acetone in male and female subjects (five of each sex) exposed to about 130 ppm for 4 h and found a statistically significant sex difference for body retention and alveolar uptake; men displayed higher percentages than women (605). In a follow-up report, the same authors reported that the rate of respiratory excretion of acetone was very slow and that less was expired by the women than the men; however, they also reported that up to 76% more

Table 74.29. Respiratory Retention of Acetone in Humans Exposed under Controlled and Uncontrolled Conditions

Type of Setting	No. of Subjects (sex)	Exposure Level (ppm)	Respiratory Retention (%)	Reference
Laboratory	2	337	53	603
Laboratory	16 (M & F)	430	86	604
Laboratory	10 (M & F)	130	14.4	605
Laboratory	15 (M & F)	24–211	55	92
Occupational	4	30	80	124
Occupational	20 (M & F)	0.5–21.1	81.4	122
Laboratory	9 (M)	100	77.5	463
Laboratory	15	20–238	54	606
Laboratory	5 (M)	85	43	607
Laboratory	33 (M)	180–690	65	608
Laboratory	4 (M)	50–200	52.6	609

acetone was exhaled than was retained, which can be explained only by the respiratory removal of metabolically produced acetone (611).

Cander and Forster first suggested that the acetone absorbed in the upper respiratory tract tissues during inspiration could be reentrained into the alveolar air upon expiration (612). Two subjects inspired 20,000 ppm of acetone in a single breath and forcibly exhaled after holding their breath for a period of time ranging from 1.5 to 15 s. The acetone concentration in the air at the beginning and end of the breath-holding period provided an actual measurement of the acetone removed by the capillary blood flowing to the lung. When the theoretical and actual values for residual alveolar acetone were compared, it was found that the rate of acetone removal by the lung was about eight times slower than predicted. The authors concluded that a substantial portion of the acetone adsorbed into the pulmonary tissue upon inspiration was removed during expiration, thus contaminating the expired alveolar air sample used for analysis.

Schrikker et al. provided additional data to support the concept that highly soluble vapors, such as acetone, are partially desorbed from epithelial tissues in the respiratory tract upon exhalation (613, 614). Data were collected from four to seven male subjects who inspired 0.01% (100 ppm) or 0.1% (1000 ppm) of acetone at rest or while exercising under workloads of 25–150 watts. The excretion of acetone from venous blood accounted for only about 50 to 60% of the acetone in the expired air samples, and this amount increased only slightly as the workload was raised. There was an appreciable difference in the amount of acetone expired at the beginning and end of any single breath during the wash-in cycle. In addition, the total amount of acetone found during each wash-out breath decreased linearly. This was ascribed to the initial absorption of acetone into the nonperfused tissues that line the upper airways (i.e., the nose, pharynx, and bronchi) during the wash-in period. This zone of high acetone was gradually displaced into the deeper and more distal regions of the respiratory tract. Then, as breathing continued after the exposure, the high concentration zone moved back up toward the mouth and was reentrained in the relatively acetone-free expired air of the wash-out cycle.

Pezzagno et al. examined the pharmacokinetics of acetone in humans and derived linear mathematical equations that related inhaled acetone to the concentration in blood and urine (92, 115). The amount of acetone taken up by resting and exercising subjects exposed to intentionally varying vapor concentrations of acetone ranging from 56–500 mg/m^3 (21–211 ppm) was strongly dependent on the rate of pulmonary ventilation. Jakubowski and Wieczorek investigated the effect of ventilation rate on the pulmonary uptake of acetone by five male subjects exposed on four different occasions to about 200 mg/m^3 (84 ppm) of acetone vapor for 2 h (607). Inspired and expired air samples were used to calculate the percentage of acetone retained in the body, the intake rate, and the uptake rate. The results indicated that the percentage of acetone retained by the body remained relatively constant as pulmonary ventilation increased higher workloads. It was concluded that the rate of acetone uptake by the body is directly proportional to the ventilation rate.

Wigaeus et al. examined the relative uptake and excretion of acetone in resting and exercising humans exposed for 2 h at vapor concentrations of 300, 311, or 522 ppm (7). As shown in Table 74.30, the uptake of acetone remained relatively constant at about 44% of the exposure concentration, regardless of the exposure regimen. The blood acetone

Table 74.30. Uptake and Elimination of Acetone Vapor in Groups of Human Subjects Who Received a Two-Hour Inhalation Exposure[a]

Activity Level (watts)	Exposure Conc (mg/m^3)	Pulmonary Uptake (%)	Respiratory Excretion (%)	Urinary Excretion (%)	Total Excretion (%)
Rest	1309	44	15	1	16
50	712	44	19	1	20
150	737	44	26	1	27

[a]Taken from Wigaeus et al. (7).

concentration increased continuously during exposure and did not reach an apparent steady state. The calculated half-life ($t_{1/2}$) for acetone excretion by the lung was 4.3 h, and the $t_{1/2}$ for acetone elimination from venous and arterial blood was 3.9 and 6.1 h, respectively. The amount of acetone taken up in the body was greater in the exercising subjects because of their higher ventilation rates. Likewise, the lower than expected values for respiratory uptake were, it was thought, due to the nonspecific absorption of acetone vapors into the mucosa that lined the nose and throat.

Haggard et al. studied the pharmacokinetics of acetone in resting and exercising male volunteers exposed for 8 h to 1, 3, and 5 mg/L (422, 1266, and 2110 ppm) of acetone (210). The end-exposure blood concentrations of acetone were 30, 99, and 165 mg/L at the three exposure concentrations, and the increase in blood concentrations were approximately proportional to the rise in the exposure concentration. A twofold increase in the acetone blood level occurred when a subject was exposed to 422 ppm under moderate exercise conditions. The accumulation of acetone in the body was examined in a single subject repeatedly exposed to 2110 ppm for 8 h/day on each of 3 days. The end-exposure blood levels of acetone were 162 mg/L, 180 mg/L, and 182 mg/L for the first, second, and third day, respectively. The blood level obtained 24 h after the start of the last exposure was 91 mg/L. Exposures less than 1266 ppm for a resting individual or 422 ppm for a moderately active person were estimates of the exposure limits that would not cause any accumulation of acetone after repeated 8-h exposures.

A large number of studies have focused on the direct measurement (615) or mathematical calculation (616) of acetone partition coefficients in various biological fluids and tissues (Table 74.31) (617–622). The earliest of these studies was performed by Widmark, who used both *in vivo* and *in vitro* techniques to measure the blood-to-air partitioning of acetone (617). The *in vitro* experiments were repeated several times using defibrinated calf serum and blood, and the results were compared to the water-to-air partition coefficient for acetone. Using a different approach, Haggard et al. determined the *in vivo* blood-to-air and urine-to-blood partition coefficients in four subjects who consumed a 50-mg/kg dose of acetone (210). The average blood-to-air and urine-to-blood partition values were reportedly 330 and 334, respectively. Sato and Nakajima used a more modern and precise vial equilibration procedure to measure the water-to-air, oil-to-air, and blood-to-air distributions of acetone and 16 other solvents (619). The average *in vitro* water-, oil-, and blood-to-air values for acetone were 395, 86, and 245, respectively. Acetone had the highest partition coefficients of the 17 solvents tested. Fiserova-Bergerova

Table 74.31. Partition Coefficients for Acetone in Human Tissue

Partition Coefficient	Method of Determination	Ratio	Reference
Blood/air	*In vivo*	334	617
Blood/air	*In vivo*	301	618
Blood/air	*In vitro*	245	619
Blood/air	*In vitro*	268	620
Blood/air	*In vitro*	196	621
Blood/air	*In vivo*	370	7
Blood/air	*In vivo*	330	210
RBC/air	*In vitro*	170	621
RBC/air	*In vitro*	210	620
Urine/blood	*In vivo*	1.34	210
Urine/blood	*In vivo*	1.24	617
Plasma/air	*In vitro*	315	620
Plasma/air	*In vitro*	217	621
Serum/air	*In vivo*	355	622
Muscle/air	*In vitro*	151	621
Kidney/air	*In vitro*	146	621
Lung/air	*In vitro*	160	621
Fat/air	*In vitro*	86	621
Brain/air	*In vitro*	130	621
Liver/air	Calculated	113	621

and Diaz used a slightly modified method to measure the distribution of acetone between air and organ tissues obtained from five autopsied humans (621). The tissue-to-air partition coefficients were determined for the following specimens: muscle, kidney, lung, cerebral white matter, cerebral gray matter, fat, whole blood, packed erythrocytes, and plasma. A liver-to-air partition coefficient was not determined because of the metabolic consumption that occurred during incubation. The highest tissue-to-air partitioning was observed when using whole blood and the smallest using fat.

Stewart et al. exposed male and female volunteers to acetone for either 3.0 h or 7.5 h per day for 4 days/week (the first day of each week was a control exposure at 0 ppm) (121). The groups were exposed to progressively higher vapor levels of acetone in each succeeding week of treatment. Following the fourth week of exposure at 1250 ppm of acetone, the two groups were given a fifth week of exposure at 0 ppm and then a final week where the vapor concentration was allowed to fluctuate between 750 and 1250 ppm (average of 1000 ppm). As shown in Table 74.32, the acetone average blood level 30-min after ending the last 7.5-h exposure to 1000 ppm was 69.3 mg/L for male subjects and 97.7 mg/L for females. A maximum blood acetone level of 117.6 mg/L was obtained for an individual exposed for 7.5 h to vapor concentrations that fluctuated between 750 and 1250 ppm. Baumann and Angerer reported that the acetone concentration in the urine of four workers exposed to 30 ppm for 2 h increased from about 0.75 mg/L at the beginning of the work shift to about 2.0 mg/L by the end of the shift (124). Similarly, the acetone in

Table 74.32. Average Acetone Blood Concentration in Humans Following Inhalation Exposures[a]

Exposure Conc (ppm)	Exposure Duration (h)	No. of Subjects	Avg. Blood Conc (mg/L)	
			Male	Female
0	7.5	4	12.9	28.6
200	7.5	4	19.6	
1000	7.5	3–4	69.3	97.7
1000	7.5	3	90.2	
1250	7.5	4	69.5	
0	3.0	2–4	12.3	35.8
200	3.0	4	21.8	
1000	3.0	2–3	28.7	72.7
1000	3.0	2	49.1	
1250	3.0	3	37.2	

[a]Taken from Stewart et al. (121).

venous blood increased from 1.0 mg/L at the start of the shift to 3.3 mg/L by the end. The urine and blood acetone levels returned to normal within 24 h.

The blood elimination of acetone in humans displays Michaelis–Menten saturation kinetics. At blood concentrations below the K_m, the elimination kinetics are first-order and are governed strictly by the rate of liver metabolism. At higher blood levels, the elimination of acetone becomes zero-order, and respiratory and urinary excretion pathways predominate. Under normal first-order conditions where metabolic saturation does not occur, respiratory and urinary excretion do not represent important pathways of elimination. In addition, a direct linear relationship was observed between the rate of endogenous acetone production and the observed levels in the blood (Table 74.33) (623–625). Reichard et al. measured the rates of acetone production and elimination in obese and nonobese humans during starvation-induced ketonemia (623). Following a prolonged fast, three groups of human volunteers were given a small i.v. dose of 2[^{14}C]-acetone (1.01 to 2.72 mmol) and were monitored for pulmonary and urinary excretion, turnover rate, rate of metabolism, and metabolic incorporation into plasma constituents (Table 74.34). Radioactivity from the administered acetone was detected in plasma glucose, lipids, and proteins, but not in plasma free fatty acids, acetoacetate, or β-hydroxybutyrate. A linear

Table 74.33. Acetone Blood Levels and Production Rates in Normal and Ketotic Humans

Subject Type	Blood Level (mg/L)	Production Rate (mg/kg/day)	Production Mass (g/day)	Reference
Normal adult	11	41	2.9	623
Fasting adult	44	105	7.4	623
Moderate diabetic	90	81	5.7	624
Severe diabetic	189	637	44.6	625

Table 74.34. Elimination of Acetone in Fasting Obese and Nonobese Human Volunteers Given a Small Tracer Dose of 2[^{14}C]-Acetone and Monitored for Six Hours[a]

Body Weight (kg)	Length of Fast (days)	Avg. Blood Conc (mg/L)	Avg. Urine Conc (mg/L)	Pulmonary Excretion (%)	Urinary Excretion (%)	Metabolic Breakdown (%)	Glucose Formation (%)	Turnover Rate (mmol/m^2/min)
108.7	3	16.8	2.3	5.3	0.6	94.1	3.1	31
72.3	3	44.1	69.7	14.7	1.4	83.9	4.2	51
124.1	21	79.6	151.0	25.2	1.3	73.5	11.0	61

[a]Taken from Reichard et al. (623).

first-order decline in the plasma radioactivity was observed in all three test groups, but the percentage of endogenous acetone excreted in the expired air and eliminated by metabolism increased in concert with the rise in blood levels.

Owen et al. examined the pharmacokinetics of endogenous acetone in patients who had moderate to severe diabetic ketoacidosis and who were administered a small tracer dose (0.75 to 1.56 mmol) of 2[^{14}C]-acetone by bolus i.v. injection (624). The initial mean plasma acetone concentration in the patients was 288 mg/L, and the values ranged from 90 to 517 mg/L. Acetone turnover rates ranged from 68 to 581 mmol/min/1.73 m^2 (values normalized to the body surface area of a standard human), and the rates were unrelated to the initial plasma acetone concentration. Distinct nonlinearities in the rates of acetone production and elimination were observed in the study. When the plasma acetone concentration was less than 5 mM (290 mg/L), there was a direct linear relationship between the rate of formation and the amount in the plasma; however, at higher plasma levels there was a marked decrease in acetone production. The excretion of acetone in expired air accounted for about 20% of the production rate at plasma levels less than 5 mM, but then increased to about 80% of the production rate when the plasma concentration was higher than this level. Corresponding differences were described for the amount of acetone metabolized. The data indicated that acetone metabolism was capacity-limited and that deviations from normal first-order kinetics could be expected when acetone levels in the blood were approximately 300 mg/L.

In a follow-up study, Reichard et al. examined the elimination of acetone by ketoacidotic diabetics who were administered 5.7 to 6.7 mmol of 2[^{14}C]-acetone at a constant i.v. infusion rate rather than bolus injection (625). The radiolabeled acetone was infused during a 4-h period to attain steady-state levels of radioactivity in the plasma. The initial average concentration of acetone in the plasma was measured at 3.26 mM (189 mg/L) and ranged from 0.50 to 6.02 mM (29 to 349 mg/L) for individual subjects. A minimum of 0.5–4.1% of the plasma glucose from treated patients was derived from endogenously produced acetone. The authors reported that the acetone turnover rate was linearly related to the plasma concentration up to a level of 7.61 mM (442 mg/L).

Haggard et al. compared acetone pharmacokinetics in both resting and exercising humans given oral dosages ranging from 40 to 70 mg/kg (210). The percentage of absorbed acetone eliminated by excretion decreased as blood levels declined from a high of 36% to a low of 7%. The calculated percentage of acetone metabolized increased from 64 to 94% blood levels fell from 73 to 2 mg/L. Periodic estimates of the metabolic rate showed that the value decreased from about 2.1 to 1.1 mg/kg/h during a 24-h period and that the magnitude was related to the amount of acetone in the blood (i.e., the rate was apparently first-order). When the metabolic rates were estimated before and during exercise for an individual given a 70-mg/kg dose of acetone, they increased from about 2.7 to 6.0 mg/kg/h.

In series of pharmacokinetics experiments, Widmark measured the absorption and excretion of acetone following oral administration of a 137-mg/kg dose (462). Following a fasting diet, the blood levels of acetone rose rapidly, and a peak blood level of 310 mg/g was observed at 10 min posttreatment. The absorption was much slower after a meal, and a peak blood level of 190 mg/g was observed 42 min after ingestion. Koehler et al. examined the disappearance of acetone from the blood of diabetic and healthy individuals administered 10 g (ca. 140 mg/kg) by i.v. infusion during a 2-h period (587). The average peak blood level of acetone at the end of the infusion was 230 mg/L in the healthy subjects and 195 mg/L in the diabetic patients. The elimination of acetone from the blood was extremely slow for both groups, and the rate of decline was somewhat faster in the diabetic patients. The initially slow disappearance of acetone from the blood was, it was thought, due to a derangement of acetone metabolism that occurred when the blood concentrations were excessively high.

Parmeggiani and Sassi examined the dermal absorption of acetone in human volunteers who were exposed below the neck in a static chamber (626). Acetone-saturated pads were placed on the skin of the subjects for 30 min to saturate the air with acetone vapors. After 90 min of exposure, acetone concentrations in the blood and urine were reportedly substantially elevated above preexposure levels. Fukabori et al. used human volunteers and applied acetone to 12.5 cm^2 of skin on four consecutive days (627). The acetone concentration ranges in the blood, urine, and expired air were 5–12 mg/L, 8–14 mg/L, and 5–12 ppm, respectively. These values increased by three- to fivefold when the exposure duration was increased from 2 to 4 h/day.

Six different physiological-pharmacokinetic models were constructed to predict tissue levels of acetone in humans following inhalation exposure (628–633). Each of the models uses slightly different physiological biochemical constants to describe the uptake and disposition of acetone in the body. Likewise, each of the models has different constraints on its use in risk assessment. Kumagai et al. used their model to predict how variations in workload and workplace exposure concentrations would affect the concentration of acetone in blood, urine, and expired air (634).

1.4.2.2.4 Reproductive and Developmental. Much of the information on the reproductive and developmental effects of acetone in human populations has been collected in groups coexposed to other chemicals. Jelnes examined the semen quality in 25 workers who produced reinforced plastic and were exposed to acetone and styrene at concentrations of 52–1334 mg/m^3 (22–560 ppm) and 162–961 mg/m^3, respectively (635). When compared to group of controls, statistically significant differences were noted in the percentage of live sperm, immotile sperm, and normally shaped sperm. The workplace population showed a higher percentage of live sperm and a decreased percentage of immotile sperm compared to controls. The workers, however, had an increased incidence of amorphous and pyriform sperm heads. Agnesi et al. evaluated the risk of spontaneous abortion in 108 women exposed to 12 different solvents during shoe manufacturing (636). Workplace concentrations of acetone averaged 31 mg/m^3 (13 ppm). Using a standard questionnaire and case matching techniques to control confounding risk factors, the solvent-exposed women had a cumulative risk of spontaneous abortion that was significantly higher than nonexposed controls.

In contrast, Axelsson et al. reported that the miscarriage rate in women exposed to acetone during the first trimester of pregnancy was 12.5% (637). When 1160 pregnancies were divided into those where laboratory solvent exposure occurred and those who had no exposure, no statistical difference was observed in perinatal death rates, prevalence of malformations, or the rate of miscarriage. Taskinen et al. evaluated the risk of spontaneous abortion in a group of 68 laboratory employees, who were divided into groups exposed to acetone 1 or 2 days a week versus 3 to 5 days a week (638). When compared to a group of matched controls and adjusted for confounding, there was no statistically significant difference in the odds ratio for spontaneous abortion. Stewart et al. exposed a group female volunteers to 1000 ppm of acetone for 1.0, 3.0, or 7.5 hours per day for 4 days/week and found that three of the four women began their menstrual cycle earlier than normal (121). Acetone concentrations of 0.1 to 1000 μg/mL did not affect the *in vitro* production of progesterone by human luteinized granulosa cells held in culture (493).

1.4.2.2.5 Carcinogenesis. Ott et al. examined the cancer and mortality rates for 948 employees of a cellulose fiber production plant who were exposed to median TWA acetone exposures of 380, 770, or 1070 ppm during a span of 23 years (639–641). These workers were compared to the U.S. general population and used as a reference group for a larger methylene chloride epidemiology study. As shown in Table 74.35, no statistical differences were found for either men or women in the acetone-exposed groups and the control group. The observed deaths for all causes, cardiovascular disease, and total malignant neoplasms were below expectation by 55%, 61%, and 43%, respectively. Likewise, there was no indication that occupational acetone exposures up to 1070 ppm had an adverse affect on selected hematologic and clinical chemistry determinations.

1.4.2.2.6 Genetic and Related Cellular Effects Studies. Lake et al. reported that acetone concentrations up to 10% inhibited DNA synthesis but did not cause unscheduled DNA synthesis in human epithelial cell cultures (546). Kabarity reported that human leukocytes treated with high concentrations of acetone could cause a nonspecific effect on cell division (642). Incubation of the cells with 25 or 50 mM (1.4 or 2.9 g/L) of acetone caused the formation of abnormal multipolar anaphase and star metaphase cells. Zarani et al. found that an acetone concentration of 5 mM did not induce the formation of micronuclei in whole blood lymphocyte cultures from two donors (544). Similar results were obtained with and without metabolic activation using a rat liver S9 fraction. Acetone was also inactive in a human skin culture assay used to determine cytokine mRNA expression as a result of dermal contact with a primary irritant (643). Dermal application of neat acetone to excised human skin maintained in a culture medium for up to 24 h did not result in the mRNA expression for tumor necrosis factor-α or interleukin-1. Müller-Decker et al. measured the release of the imflammatory mediators arachidonic acid and

Table 74.35. Observed and Expected Mortality Rates for Men and Women Occupationally Exposed to Acetone[a]

	Male Mortality Ratio		Female Mortality Ratio	
Cause of death	Observed	Expected	Observed	Expected
All causes	24	53.8	3	6.7
Malignant neoplasm	5	10.0	2	2.3
Cardiovascular disease	15	40.4	2	2.8

[a]Taken from Ott et al. (639–641).

interleukin-1α in human keratinocytes grown in culture and reported half-maximal stimulatory (SC_{50}) and cell viability concentrations (IC_{50}) for a variety of chemicals (644). The 12-h SC_{50} and IC_{50} values for acetone were very high at $\geq$ 1.4 M and 1.0 M, respectively. Mrowietz et al. evaluated chemotaxis and superoxide anion generation in isolated human monocytes treated with 0.1% acetone and found no effect (645).

1.4.2.2.7 Other: Neurological, Pulmonary, Skin Sensitization. There have been several reports of direct ocular contact with liquid acetone, and although the effects were more serious than with acetone vapor, the data show that prompt treatment prevents permanent corneal damage. In fact, acetone has been used with no apparent difficulty to remove polymerized cyanoacrylate adhesives from the eyelid following accidental fusion (646). Acetone has also been used to prevent recurrent vaginal hemorrhage associated with gynecologic cancer (647), to dissolve Styrofoam impactions in the ear (648), to sterilize ophthalmic instruments before use (649), to degrease the skin before treatment of actinic keratosis with a chemical peel (650), and to remove Superglue from the external ear (651).

The ocular damage from an accidental splash of acetone has often been limited to the corneal epithelium, which healed completely after several days. Grant reported that small amounts of acetone in the eye caused an immediate stinging sensation, but when promptly irrigated, the damage was confined to the epithelium (652). Victims of ocular contact with acetone often reported a "foreign-body feeling" in the eye. McLaughlin reported three cases of human corneal burns from liquid acetone that healed within 48 h of irrigation and removal of the damaged corneal epithelium (653). Mayou once speculated that the cataract and lens opacity observed in a male patient may have resulted from his occupational exposures to acetone (654). The assertion was based on the similarity between the patient's cataract and those previously observed in diabetics. Weill reported a case of severe eye injury in a worker exposed to the vapors from a methanolic solution that contained 5% acetone (655). Permanent corneal opacity has been reported in a single incident involving an employee who was splashed by a solution of cellulose acetate in acetone; a second similarly affected patient recovered completely (656). A tenacious film of cellulose acetate was deposited on the surface of the affected employee's cornea following evaporation of the acetone. After reviewing many reports, Grant and Schuman concluded that there was no substantive evidence of cataract formation or optic nerve damage in humans systemically exposed to acetone (657).

An ultrastructural examination of acetone-treated human skin has shown that topical exposure can affect the architectural integrity of the epidermis. Lupulescu et al. obtained biopsy specimens of the epidermis from the forearms of six male volunteers who were treated with acetone for periods of 30 or 90 min (658). The biopsies were collected immediately after treatment and at 72 h posttreatment for analysis by light and electron microscopy. Acetone contact caused cellular damage in both the stratum corneum and stratum spinosum which was more severe after the 90-min exposure. Tissue specimens collected at 72 h posttreatment showed a high degree of repair and restoration, but some evidence of damage was still present. In a follow-up study using seven subjects, Lupulescu and Birmingham found that pretreatment of the skin with a protective gel substantially reduced the severity of the cellular and structural damage caused by 90 min of acetone exposure (659).

Tosti et al. reported a single rare case of acute contact dermatitis from acetone in a female laboratory technician who was sensitized on the job (660). Drexler et al. reported that skin prick tests with 1 and 5% acetone in 136 subjects failed to show any evidence of a dermal hypersensitivity reaction or immunoglobulin E induction (661). Gad et al. reported that acetone did not cause dermal sensitization in 50 volunteers repeatedly patch tested with nine 0.2 mL treatments during a 3-week period and a subsequent challenge 10–14 days after the final treatment (562). Abrams et al. examined transepidermal water loss (TEWL) as a measure of skin barrier disruption in human cadaver skin treated with acetone for up to 12 min (662). The acetone treatment did not affect TEWL relative to water; however, a considerable quantity of surface lipids (triglycerides, squalene, and cholesterol esters) was detected in the extracts. In contrast, Zhai et al. showed that the application of acetone to the upper backs of six volunteers caused a rapid and severe increase in the rate of transcutaneous water loss (663). Recovery of barrier function was complete after a 96-h repair period. Ghadially et al. examined barrier repair in the skins of fifteen young and six old volunteers treated with acetone and found much faster repair in the young versus the aged skin (664). The young skin showed 70–80% barrier recovery in 48 h, whereas the aged skin required 5–6 days for the same level of recovery. Proksch and Brasch reported that epidermal permeability barrier disruption by acetone in six volunteers who had known sensitization to specific chemicals resulted in an significant increase in Langerhans cells and epidermal proliferation after 24 h (665). Based on clinical experience, Kechijian concluded that the overzealous use of acetone-containing nail polish removers could cause onychoschizia, a condition characterized by split and peeling nails (666).

The scientific literature contains eight different studies that have measured either the neurobehavioral performance or neurophysiological response of humans exposed to acetone (Table 74.36) (667–672). Many of the early neurotoxicity studies with acetone were not amenable to reliable statistical analysis because of the variability in the data and the inability to reproduce the results. A close inspection of these early investigations also reveals many problems in design, conduct, or interpretation that hinder their use. Among

Table 74.36. Minimal Effect Vapor Concentration Reported in Various Human Nuerobehavioral Studies with Acetone

Type of Setting	No. of Subjects	Minimal Effect Level (ppm)	End Point Affected	Reference
Laboratory	6 (M)	250	Visual reaction time	600
Laboratory	8 (M)	250	Skin response and vasoactivity	667
Laboratory	4 (M & F)	> 170	Time estimation	668
Occupational	5 (?)	200	Visual reaction time	669
Laboratory	22 (M & F)	250	Auditory reaction time	670
Occupational	8 (M)	> 1000	Memory and reaction time	671
Laboratory	16 (M)	> 1000	Memory and reaction time	672
Laboratory	4 (M)	1250	Visual evoked response	121
Occupational	110	> 1000	Reaction time and digit span	79

more recent studies of acetone, NOAELs ranging from vapor concentrations of 600 mg/m^3 to greater than 2375 mg/m^3 have been reported. The wide range in effect levels are likely to be due to statistical errors caused by large numbers of independent variables, analytical problems, and the failure to use multiple concentrations to evaluate dose–response characteristics.

Matsushita et al. performed a neurobehavioral examination of subjects repeatedly exposed to 250 or 500 ppm for six consecutive days at 6 h/day (600). An additional group of subjects was exposed to 250 ppm of acetone while doubling their metabolic rate through physical exercise. The average relative reaction times to a visual stimulus were longer on each of the six exposure days for the 500 ppm group and on two of the six exposure days for the both of the 250-ppm exposure groups (resting and exercising). The absolute response times obtained during each exposure session varied considerably among individuals and were not statistically different from the untreated control group. The data did not reveal any consistent dose- or time-related trends in the magnitude of the response. Seeber et al. examined reaction time, memory, and spontaneous motor activity in a group of 16 male volunteers who received either a 1000-ppm acetone exposure or a combination exposure to 500 ppm acetone and 200 ppm ethyl acetate for 4 hours (672). Simple reaction time and choice reaction time were determined at 2-hour intervals before, during, and after the exposure. Spontaneous motor activity was measured continuously using an actinometer that recorded each subject's arm movement in three dimensions. An examination of the results did not show any significant changes in the values obtained from four tests used in the study.

Dick et al. performed a series of neurobehavioral studies on groups of about 20 male and female volunteers who were exposed to either 250 ppm of acetone for 4 h or to a combination of 125 ppm of acetone and 200 ppm of methyl ethyl ketone (MEK) for 4 h (670, 673). Four psychomotor tests, one sensorimotor test, and one psychological test were performed on the subjects before, during, and after the exposure session. Acetone was caused an effect on the responses obtained in two of these tests, a dual auditory tone discrimination compensatory tracking test and a profile of mood states (POMS) test. Relative to preexposure control values, the 250-ppm acetone exposure caused an increase in both the response time and the percentage of incorrect responses in the auditory tone portion of the dual discrimination task when the stimuli were presented in series. The response measurements were not affected by the exposure when both portions of the dual task were presented simultaneously. Male subjects who took the POMS test showed an increase in the anger-hostility score. Except for a small change in the percentage of incorrect responses in the dual auditory discrimination test, none of these effects were noted when the subjects were exposed to the acetone-MEK mixture. The authors noted that the results of their study needed careful interpretation and that additional research was needed to detect more distinct declines in human performance.

Suzuki performed several neurophysiological tests on groups of eight or nine male student volunteers exposed to acetone at concentrations of 250–270 ppm or 500–750 ppm for two 3-h sessions with a one-hour break (667). Spontaneous and evoked changes in galvanic skin response (GSR), vasoconstriction, heart beat interval, respiration interval, and cerebral activity were measured. The following were noted for the exposed groups: (*1*) a decrease in the spontaneous GSR and an increase in the evoked GSR, (*2*) a decrease in evoked vasoconstrictive activity, (*3*) a decrease in mean time for ten heart beats at the high exposure concentration, and (*4*) an increase in cerebral activity. An increase in air temperature within the exposure chamber was positively correlated with several of the observed response trends. Nakaaki performed a neurobehavioral time estimation test on two male and two female student volunteers exposed to acetone vapors for 4 h on a single day (668). Because of fluctuating chamber concentrations of acetone, the exposure sessions were described as either 170 and 450 ppm or between 450 and 690 ppm. The results from a time estimation test indicated a prolonged sense of time passage at both concentration ranges relative to control values.

Seeber et al. examined the neuropsychological relationship between an individual's inherent susceptibility for a particular chronic symptom and the acute subjective response to an organic solvent exposure (674). Groups of 16 male volunteers received either a 1000-ppm acetone exposure or a combination exposure to 500 ppm acetone and 200 ppm ethyl acetate. A chronic symptom questionnaire was completed by each subject 1 week before the 4-h exposure. The responses on the questionnaire were grouped into one of six chronic symptom categories: psycho-/neurovegetative lability, neurological symptom, lack of activity, excitability, lack of concentration, and special symptoms. Information from the chronic symptom questionnaire was subsequently compared to the results obtained from a second set of ranked-response questions that dealt with four acute symptoms: tension, tiredness, complaints, and annoyance. A correlation between any of the six chronic symptom categories and any of the four acute symptoms suggested a possible temporal relationship between the chronic symptom and the acute response. No statistically significant cross-correlations were observed in the subjects exposed to 1000 ppm of acetone; however, the group that received a combined exposure to ethyl acetate and acetone displayed a cross-correlation for psycho-/neurovegetative lability and their scores for both complaints and annoyance.

Stewart et al. examined the neurotoxic effects of repetitive exposures to acetone vapors in male and female volunteers using a variety of treatment regimens (121). Two series of experiments were performed. In the first series, two small groups of male subjects were exposed to each of four vapor concentrations (0, 200, 1000, or 1250 ppm) for either 3.0 or 7.5 hours per day for 4 days/week (the first day of each week was a control exposure at 0 ppm). The groups were exposed to progressively higher vapor levels of acetone in each succeeding week of treatment. Following the fourth week of exposure at 1250 ppm of acetone, the two groups were given a fifth week of exposure at 0 ppm and then a final week where the vapor concentration was allowed to fluctuate between 750 and 1250 ppm (1000 ppm, average) on each of four exposure days. The second series of studies was performed on groups of female subjects who were exposed to 1000 ppm of acetone for either 1.0, 3.0, or 7.5 hours per day for 4 days/week. A battery of neurophysiological and neurobehavioral tests was performed at various times throughout the exposures. The neurophysiological tests included spontaneous electroencephalograms, visually evoked response using a strobe light, and a Romberg heel-to-toe equilibrium examination. Cognitive neurobehavioral testing included an arithmetic test, a coordination test, and a visual inspection test. Male subjects exposed to 1250 ppm of acetone for 7.5 hours/day showed a statistically significant increase in the amplitude of the visually evoked response compared to background values.

Table 74.37. Occupational Health Surveys in Acetone-Exposed Worker Populations

Factory Location	No. of Subjects	Employment (years)	Exposure Conc (mg/m^3)	Clinical Measurements	Reference
United States	948	< 23	900–2540	Hematology, urinalysis, and mortality	640
Italy	60	> 5	1305–2490	Hematology, urinalysis, and clinical chemistry	106
United States	800	Unknown	1425–5100	Hematology and urinalysis	593
United States	1271	> 0.25	0–3800	Mortality and neoplasms	675
Japan	110	15	48–2415	Hematology, immunology, and clinical chemistry	79

1.4.2.3 Epidemiology Studies. The following section describes the results of many of the medical surveillance and epidemiology investigations available for acetone. Most, but not all, of the data have been collected in the workplace under mixed-solvent conditions where the exposures involved any of a variety of oxygenated, aromatic, and hydrocarbon solvent vapors. Because of the technical difficulties encountered in this type of investigation, it is often difficult to make a definite statement about the potential health effects of a chemical on the basis of the results of any single epidemiological study. When multiple studies are available, however, a clearer picture begins to emerge, and the impact of confounding variables begins to diminish. Five health monitoring or epidemiology studies stand out from those involving acetone contact in the workplace (Table 74.37) (675). They are distinguished by the number of people involved, the magnitude of the acetone exposures, and the diverse number of health-related indicators examined. Taken together, these studies indicate that the daily inhalation of acetone vapors does not cause systematic toxicity or chronic health effects at workplace concentrations.

1.4.2.3.1 Acute Toxicity. Raleigh and McGee performed two occupational health surveys on filter press operators and support personnel exposed to acetone vapor in a cellulose fiber facility (591). In the first study, nine employees were monitored for seven 8-h workdays and were asked to rate (slight, mild, or strong) any symptoms of sensory irritation following each sample collection. Breathing zone samples produced a mean daily TWA exposure of 1006 ppm (950 to 1060 ppm range) for the 7-day survey period. Reports of irritation were recorded as follows for the nine employees: eye irritation in seven, throat irritation in four, headache and light-headedness in three, and nasal irritation in two employees. The symptoms were transient and generally occurred when the vapor concentrations exceeded 1000 ppm. Of the 31 individual reports of eye irritation, 21 occurred when the acetone concentration was higher than 1500 ppm. Medical examinations were essentially normal in all respects, except for a slight redness in the nasal mucosa of one individual and slight congestion in the nose and throat of another. Four filter press operators were involved in the second study. Half of them were monitored for three 8-h work shifts and the other half for two 8-h shifts. Air samples collected from the breathing zones of each employee revealed a TWA exposure concentration of

2070 ppm during the 3-h monitoring period (155 to 6596 ppm range). In this study, two of the four employees complained of eye irritation, three of nasal irritation, and one of throat irritation. Physical examinations were negative except for one individual whose throat was slightly congested. Reports of eye irritancy were highly variable in each survey, and the subjective responses showed only a general concentration-related increase in severity. Dalton et al. evaluated the perceived irritancy response to an 800-ppm acetone exposure in 27 male and female volunteers who worked in a cellulose fiber production plant and compared the results with 32 subjects who had no prior contact with acetone vapor (598). The naive control subjects gave greater estimates of irritative intensity than the workers when psychophysical techniques were used to grade the response. In contrast, the workers, responded with a higher frequency of general complaints (irritative and otherwise) than the control subjects when symptom questionnaires were used to assess irritancy. The perceived intensity of irritation correlated strongly with the odor of acetone and the response bias of the test group.

Kiesswetter et al. examined acute subjective complaints and states of well-being in a group of employees from a cellulose acetate plant located in Germany (671). The concentration of acetone in the breathing zone and in the urine was measured at 4-h intervals in a group of eight employees on three different work shifts. The study was performed on three consecutive days during a midsummer heat wave when temperatures within the facility were higher than 40°C (104°F) throughout most of the day. Analysis of 122 samples revealed a mean 4-h TWA exposure concentration of about 940 ppm and a standard deviation of 340 ppm. Two questionnaires were administered to the subjects before, during, and after each work shift. The first requested information on the intensity of 17 different acute subjective symptoms (severity ranked from 0 to 6) that included headache, tiredness, dizziness, and irritation of the eyes, nose, and throat, whereas the second questionnaire inquired about the following four states of well-being: tension, tiredness, complaints, and annoyance. The subjective symptom survey revealed that the total score for all complaints increased during each work shift; the results, however, were poorly correlated with the exposure concentration. When the results from the 17-symptom questionnaire were arbitrarily divided into four complaint categories, two categories (discomfort and irritation) dominated the severity scores and correlated with the total amount of acetone excreted in the urine during the 8-h shift. The acetone-exposed employees also responded with a high frequency of subjective complaints at the beginning of each work shift before any vapor exposure occurred compared to a control group that was not exposed to acetone and did not experience the high ambient temperatures on the work floor. These data suggest that factors other than the acetone exposure (e.g., high ambient temperature, shift start time, etc.) contributed to the employees' tendency to respond adversely. Similar results were obtained from the questionnaire on well-being. The scores for each of the four states increased during the work shift. The scores for annoyance were the only values that correlated with the ambient airborne concentration of acetone.

Sietmann and Seeber performed a detailed statistical analysis of the responses from a 17-item acute symptom questionnaire that was previously administered by Kiesswetter et al. to occupationally exposed individuals (676). Instead of grouping all of the results from the symptom questionnaire into one of four categories (discomfort, irritation, tiredness, and difficulties in breathing), the most relevant responses from the questionnaire were placed into one of the following three categories: (1) odor, which included responses of an unpleasant odor and bad taste; (2) irritation, which included responses to questions about irritation of the eyes, nose, throat, and tearing of the eyes; and (3) distress, which included responses to questions of headache, dizziness, nausea, and weakness. The pooled results from the three symptom categories were compared to the results from the four ratings of well-being (tension, tiredness, complaints, and annoyance) recorded by each subject during the course of the workplace exposure. Using a statistical technique known as bipartialling, which allowed filtering out and statistically controlling interfering and masking effects, the annoyance response from the previous field study correlated with odor and not with irritation or distress.

Nasterlack et al. examined serum alanine aminotransferase (ALT), aspartate aminotransferase (AST), and γ-glutamyl transpeptidase (GGT) activity in a group of 33 printers and found no exposure-related changes in either of two studies conducted 4 years apart (70). The printers were exposed primarily to the following six solvents: acetone, MEK, ethyl acetate, white spirits, toluene, xylenes, and trichloroethylene. Acetone exposures ranged from 2.5–157 ppm during the first study and to 2.5–99 ppm during the second. Franco et al. measured serum AST, ALT, GGT, direct bilirubin, and bile acids in a group of 30 workers exposed to toluene, xylene, acetone, *n*-butyl acetate, *n*-butanol, ethyl acetate, and several other solvents in a factory that produced fillers and varnishes (77). Acetone concentrations in the workplace ranged from 5–1448 mg/m^3 during a 6-year period. Except for the bile acid determinations, which were significantly higher in the exposed workers than in the controls, all clinical measurements were similar for the workers and the reference group. Tan and Wong found evidence of renal and hepatic damage in a group of 50 workers exposed to methyl ethyl ketone, toluene, and acetone in a chemical drum filling operation (677). A comparison of workers exposed to low and high solvent concentrations with a control group composed of production and office workers revealed an apparent exposure-related increase in mean values for total bilirubin and alkaline phosphatase. In addition, individual employees showed sporadic increases in urine or blood chemistry values that were outside normal limits. Tomei et al. examined liver function in a group of 33 shoe repairers exposed to acetone and five other solvents for about 9 years (678). The 8-h TWA acetone concentration reportedly ranged from 20–25 mg/m^3. When compared to controls, a significant increase was observed in blood ALT, AST, alkaline phosphatase, and conjugated bilirubin. Likewise, a significantly higher number of exposed workers had individual clinical test values above the normal limit than were observed in the control group.

1.4.2.3.2 Chronic and Subchronic Toxicity. Oglesby et al. summarized eighteen years of medical surveillance of employees in a cellulose acetate production facility where more than 21 million man-hours of acetone exposure had accumulated (593). Using medical department visits, lost-time records, and comprehensive medical examinations, the authors did not detect any increased incidence of illness relative to an appropriate control population. Mild transient symptoms of irritation were recorded when the average exposure concentrations exceeded 2500 ppm. The analysis indicated that acetone concentrations up to 1500 ppm would be entirely without injurious or objectionable effects for continuous exposure up to 8 h.

Parmeggiani and Sassi examined workers at three separate manufacturing sites where acetone was used in combination with other solvents (626). The first work site involved a group of six employees in a plant manufacturing acetate rayon who were exposed to acetone at concentrations ranging from 309–918 ppm for up to 3 h during a 7- to 15-year period. The employees complained of drowsiness, eye and throat irritation, dizziness, inebriation, and headache. Muscular weakness, vertigo, and chronic inflammation of the stomach, duodenum, and air passages were also reported. A physical examination showed signs of pharyngeal, conjunctival, and lung irritation in five of the six employees. The second work site employed four workers in a bottle lacquering operation where acetone concentrations ranged from 84–147 ppm. Workers at this site experienced nausea, abdominal pain, headache, vertigo, loss of appetite, vomiting, and other debilitating symptoms. The final site was a polyvinyl chloride fiber production plant with eleven employees who were exposed to acetone vapors at concentrations ranging from 13–86 ppm. The exposures reportedly caused irritation to eyes, nose, throat, and bronchi, as well as some severe CNS disturbances. High concentrations of carbon disulfide, a potent CNS toxicant, were also found in this plant.

Grampella et al. considered the possibility that long-term occupational exposures to acetone could cause systemic organ damage by examining a group of 60 volunteers employed for at least five years in an acetate fiber manufacturing facility (106). The high-exposure group received personal TWA exposures ranging from 2251–2488 mg/m^3 and had an average urine acetone level of 93 mg/L, whereas, the low-exposure group had TWA acetone exposures that ranged from 1303–1550 mg/m^3 and a midshift urine acetone value that averaged 62 mg/L. Clinical measurements included determination of glucose, ALT, AST, GGT, protein electrophoresis, blood urea nitrogen, creatinine, platelet count, and red and white blood cell counts. After taking risk factors, past medical histories, and age into consideration, no statistically significant difference was noted between the exposed and unexposed groups in any of the test results.

Soden summarized the health files of employees who participated in a health monitoring program at a triacetate fiber plant to determine whether occupational exposures to methylene chloride, acetone, and methanol adversely affected hematology or blood chemistry test results (679). The test values for 150 acetone-exposed employees who had an average 8-h TWA exposure of 900 ppm were compared with the results from a group of 260 nonexposed controls. A comparison of the frequency distributions for the exposed and nonexposed populations failed to show any significant difference in ALT, AST, total bilirubin, or hematocrit in the two groups. Likewise, no differences were found in the response rate for symptoms such as loss of memory, headache, and dizziness.

In another study, Rösgen and Mamier focused specifically on the question of acetone-induced blood dyscrasias (680). The authors examined 45 men and 39 women in a factory where acetone was extensively used in the manufacture of plasticizers. Blood specimens from the employees were analyzed for hemoglobin concentration, coagulation time, sedimentation rate, red and white blood cell counts, and white blood cell differential count. The clinical measurements revealed a below normal hemoglobin concentration in several of the women, which was attributed to their poor nutritional status. The authors concluded that the acetone exposures did not cause any hematological abnormalities.

Mitran et al. conducted a study that examined the effects of acetone on a group of 71 workers who used acetone to clean enamel from trophy medals (74). The 8-h time weighted average exposures to acetone ranged from 988–2114 mg/m^3. The acetone exposures caused an increase in the percentage of reported neurotoxic symptoms (mood disorders, memory difficulties, sleep disturbances, and headache) and rheumatic symptoms (bone pain, joint pain, and muscular pain). Symptoms of ocular and respiratory tract irritation were also more common, but tests of statistical significance were not performed in any instance. Measurement of motor nerve conduction revealed a statistically significant increase in the latency period for proximal and distal regions of the median, ulnar, or peronal nerve and a significant decrease in pulse duration for the same regions of each nerve. Nerve conduction velocity was also significantly decreased in these nerves relative to controls. The authors noted, however, that nerve conduction velocity experiments were relatively insensitive to any underlying CNS abnormalities.

1.4.2.3.3 Pharmacokinetics, Metabolism, and Mechanisms. Dolara et al. looked for evidence of enzymatic induction in a group of 19 workers exposed to styrene and acetone in a plant that manufactured polystyrene plastics (681). The 8-h TWA exposure concentration of acetone in the workroom air ranged from 55–571 mg/m^3. Two markers of monooxygenase induction were significantly increased in the exposed employees. Urinary glucaric acid concentrations and the ratio of urinary 6-β-OH-cortisol to 17-OH-corticosteriods were both increased relative to nonexposed controls.

1.4.2.3.4 Reproductive and Developmental Toxicity. Swan et al. performed a retrospective cohort study of spontaneous abortion in fabrication workers from the semiconductor industry who were exposed to twelve different chemical agents (682). The acetone concentrations in the work place were not measured; however, low- and high-exposure groups were identified from calculated exposure scores during the first trimester of pregnancy. The unadjusted relative risk of spontaneous abortion in the acetone-exposed groups increased in a exposure-related fashion and was statistically significant compared to unexposed controls. When the relative risk ratios were adjusted for coexposure to other solvents and the type of work being performed, no increase in risk was observed for the acetone-exposed employees. Khattak et al. conducted a prospective case-controlled study of pregnancy outcome and fetal malformations in a group of 125 women occupationally exposed to a mixture of at least seven different solvents (683). Although, no exposure data were collected during the study, a statistically significant increase in major fetal malformations was observed in the solvent-exposed group relative to the controls. In addition, a significantly greater number of women reported previous miscarriages while working with solvents than was recorded for the controls.

1.4.2.3.5 Carcinogenesis. Spirtas et al. examined the causes of death in a retrospective cohort study of 14,457 male and female workers at an aircraft maintenance facility (684). Standardized mortality ratios (SMR) were calculated for workers exposed to a large number of individual organic solvents during their employment. Using the population of Utah as a referent, the acetone exposures did not increase the SMR for deaths due to non-Hodgkin's lymphoma in either male or female study groups. An increase was noted, however, in the SMR for multiple myeloma in the female, but not the male cohort. Information was not provided on the level of workplace exposure to acetone. In a follow-up study, Blair et al. reported on the cancer incidence in the same cohort of solvent-exposed workers from the Air Force maintenance facility (685). Although mortality rate ratios were determined for the workers exposed to 13 different solvents, most exposures were actually to a mixture of solvent types. The acetone-exposed group did not show any statistically significant excesses in the rate ratios for mortality from non-Hodgkin's lymphoma (male and female), multiple myeloma (female), or breast cancer (female).

Lanes et al. examined the mortality of 1271 workers in a cellulose fiber production plant who were exposed to methylene chloride, acetone, and methanol (675). The workers were employed for at least 3 months and the 8-h TWA exposure to acetone was 1600 ppm in 1977 when the cohort was assembled. Comparing the results to county residents, no difference was found in the standardized mortality ratio for all causes. The standard mortality ratio increased for neoplasms of buccal cavity and pharynx, biliary passages and liver, and for melanoma. The standard mortality ratio less than unity for cancer of the respiratory system, breast, and pancreas. Walker et al. performed a retrospective cohort mortality study on 7814 male and female workers employed for at least one month in shoe manufacturing at two different plants (686). The workers were exposed to wide variety of solvents, but monitoring data was only available for toluene, methyl ethyl ketone, acetone, and hexane. The 8-h TWA acetone concentrations in the plants ranged from 25–270 mg/m^3. Calculation of SMR values using the U.S. population as a reference revealed a statistically significant increase in cancer of the trachea, bronchus, and lung in the total cohort and in the men. The SMRs for malignant neoplasms in tissues and organs such as kidney, bladder, buccal cavity, lymphatics, and digestive tract were not appreciably different from unity.

1.4.2.3.6 Genetic and Related Cellular Effects Studies. Pitarque et al. used an alkaline single-cell gel electrophoresis assay to evaluate the degree of DNA damage in 34 female shoe industry employees exposed to gasoline, toluene, acetone, ethyl acetate, and diisocyanate in two plants (687). Acetone concentrations at the two work sites ranged from 382.0–926.87 mg/m^3. Compared to the nonexposed control group, the level of DNA damage in peripheral blood mononuclear leukocytes was not affected by solvent exposure, regardless of the glutathione *S*-transferase M1 genotype of the individual. Hagmar et al. performed a micronucleus assay on cultured blood leukocytes from 20 employees of a reinforced plastics plant where styrene, acetone, and methylene chloride were used (688). The acetone exposures ranged from 15–137 mg/m^3 during the last year monitoring took place. No statistically differences were observed in micronucleus size ratios, micronucleus frequencies, or chromosomal breaks between the solvent-exposed employees and a control population. Chromosomal gaps were detected at statistically higher frequency in the control group than in the solvent-exposed workers.

1.4.2.3.7 Neurological, Pulmonary, Skin Sensitization. Kiesswetter et al. examined reaction time and vigilance in a group of eight employees of a cellulose acetate plant (671). The concentration of acetone was measured in the breathing zone and in the urine of the employees at 4-hour intervals during three different work shifts. The study was performed during a midsummer heat wave when temperatures within the facility were higher than 40°C (104°F) throughout much of the day. Analysis of 122 samples revealed a mean 4-hour exposure concentration of 940 ppm. The mean 4-hour exposure values for the individual employees ranged from 475 to 1500 ppm, and some individual 4-hour results ranged as high as 5050 ppm. Simple reaction time and color word vigilance tests were performed three times per shift using a standardized test method. No significant differences were noted in any of the neurobehavioral measurements as a result of the acetone exposures.

Satoh et al. conducted an occupational health survey during a two-day period on 110 male workers exposed to acetone concentrations ranging from 5 to 1212 ppm (average value 350 ppm) for a mean of 14.9 years (79, 589). The employees worked in three factories that produced cellulose acetate fibers. They were divided into three categories groups according to their level of exposure: (*1*) low, less than 250 ppm; (*2*) medium, 250 to 500 ppm; and (*3*) high, higher than 500 ppm. A large subjective symptom questionnaire administered to each employee solicited information on symptoms experienced before and after work and during the previous 6 months of employment. Five neurobehavioral tests were also performed along with tests of autonomic nervous system function. Except for an occasional decrease in simple reaction time and digit span scores in some age groups, none of the neurobehavioral or neurophysiological tests showed any clear exposure–response relationship relative to an unexposed control population.

Israeli et al. conducted neurobehavioral testing of a group of five employees who worked on an office appliance production line where acetone was used in a solvent-based glue and in a cleaning solution (669). The instantaneous workplace concentration of acetone was reportedly about 200 ppm. Test results were obtained both before and after an 8-h work shift and were compared to the results obtained for the same group of employees after a 2-day confinement in an acetone-free work area. A statistically significant increase was observed in the reaction time to a light stimulus when the mean values for each individual were averaged for the five subjects. The response measurements, however, were highly variable

Kishi et al. investigated the relationship between organic solvent exposure and alterations in central nervous system function in a group of 81 industrial painters (689). A neurobehavioral test battery consisting of eight separate tests was administered to the painters who were exposed to 12 different solvents that included acetone to a small extent. The solvent-exposed group showed poorer performance on vocabulary and digit symbol tests than the control population. When the duration of exposure was taken into consideration, significantly poorer performance was observed on vocabulary, block design, and digit span tests than in the referent controls. Baird administered a color vision discrimination test and several cognitive tests to 82 print shop workers exposed to acetone and five other solvents (690). The 8-h TWA exposure to acetone ranged from 1.1–11.6 mg/m^3 in different areas of the factory. Color vision test performance using the Lanthony D-15 color panel was similar in the solvent-exposed group and in the controls. The number of subjects who made ≥ 1 error on the test was also higher in the low-exposure group than in the high-exposure group. Results from the two tests of cognitive function (Symbol Digit Modality test and Trials A and B test) also failed to show a difference between the exposed employees and the control group. Calabrese et al.

conducted an otoneurological study of 20 workers exposed to styrene and acetone in the fiberglass industry (691). The 8-h TWA concentration of acetone ranged from 70–277 mg/m^3 in the workroom air. Tests of audiometric function and auditory brainstem response showed no statistically significant differences between the exposed workers and a control group. The examination of vestibular function, however, revealed some disturbance in performance in the caloric test, and only 3 of the 20 workers responded normally.

Socie et al. surveyed workers from four plastic manufacturing plants to assess risk factors for the development of occupationally related skin disorders (692). A self-administered questionnaire was completed by 122 workers, 26 of whom had skin problems. The workers were grouped according to their exposure to any of 17 different chemicals. When the data were examined relative to the acetone exposure, a crude odds ratio of 1.69 (95% CI, 0.49–4.48) was obtained, which was not statistically significant compared to nonexposed controls. A prospective 2.5-year evaluation of the pulmonary function in a group of 40 employees exposed to organic solvents did not reveal any statistically significant changes in forced vital capacity or the 1-s forced expiratory volume relative to nonexposed controls (693). Solvents in use included acetone, *n*-butyl acetate, isopropyl alcohol, methanol, toluene, xylene, and MEK. The acetone exposure concentration in the work area averaged 1.2 ppm.

1.5 Standards, Regulations, or Guidelines of Exposure

Occupational exposure standards have been established for acetone by many countries worldwide. The 8-h limits are shown in Table 74.38.

Table 74.38. Worldwide Occupational Exposure Limits for Acetone

Country and Organization	Year Established	8-hour Exposure Limit (ppm)	Short-term or Ceiling Limit (ppm)
United States—NIOSH	1978	250	—
United States—OSHA	1989	1000	—
United States—ACGIH	1997	500	750 (15 min)
Mexico	1994	1000	1260 (15 min)
Canada (Ontario)	1990	750	1000 (15 min)
Canada (Quebec)	1992	750	1000 (15 min)
European Union—SCOEL	1997	500	1000 (15 min)
Germany—MAK	1996	500	1000 (5 min)
United Kingdom—HSE	1992	750	1500 (10 min)
Netherlands	1986	750	1500 (15 min)
Sweden	1985	250	500 (15 min)
Denmark	1985	—	250 (30 min)
Finland	1981	500	625 (15 min)
China	1979	—	168 (ceiling)
Japan	1972	200	—
Australia	1990	500	1000 (15 min)
Russia	1960	84	—

Table 74.39. Comparison of Environmental Removal Processes for Acetone

Acetone Removal Process	Approximate Half-life (days)	Reference
Soil biodegradation	7	694
Total tropospheric removal	23	695
Aqueous photolysis	40	696
Aqueous biodegradation	0.6	56
Hydroxyl radical reaction	31	57
Atmospheric photolysis	80	57
Volatilization river	6	56
Volatilization lake	100	64

1.6 Studies on Environmental Impact

Two processes govern the photochemical removal of acetone from the troposphere: reaction with hydroxyl radicals and photolysis. The two processes occur at about equal rates in clear unpolluted skies and yield a total tropospheric lifetime of about 32 days (57). The reaction with hydroxyl radicals predominates over photolysis in urban areas where hydroxyl radical concentrations are greater and during cloudy wintertime conditions where photodecomposition is minimal. Rain out and other forms of wet deposition are considered minor tropospheric removal processes (35). Calculated and measured rate constants have been used to estimate the elimination half-life ($t_{1/2} = 0.693/k_{\text{calc}}$) of acetone through various environmental processes (Table 74.39) (694–696). These data show that acetone is rapidly biodegraded in water and that this is the dominant removal process in the environment. The slow removal of acetone from the troposphere indicates that it is relatively un-reactive and a minor contributor to urban ozone and peroxyacyl nitrate (PAN) concentrations (697).

BIBLIOGRAPHY

1. J. A. Riddick, W. B. Bunger, and T. K. Sakano, eds., Physical properties and methods of purification. In *Organic Solvents*, 4th ed., Vol. II, Wiley, New York, 1986, pp. 336–337.
2. S. Sifniades, Acetone. In W. Gerhartz et al., eds., *Ullmann's Encyclopedia of Industrial Chemistry*, 5th ed., Vol. A1, VCH, Deerfield Beach, FL, 1985, pp. 79–96.
3. W. L. Howard, Acetone. In J. I. Kroschwitz and M. Howe-Grant, eds., *Kirk-Othmer Encyclopedia of Chemical Technology*, 4th ed., Vol. 1, Wiley, New York, 1991, p. 76.
4. G. G. Cash and R. G. Clements, Comparison of structure-activity relationships derived from two methods for estimating octanol-water partition coefficients. *QSAR Environ. Res.* **5**, 113–124 (1996).
5. P. Poulin and K. Krishnan, Molecular structure-based prediction of the partition coefficients of organic chemicals for physiological pharmacokinetic models. *Toxicol. Methods* **6**, 117–137 (1996).
6. H.-J. Benkelberg, S. Hamm, and P. Warneck, Henry's law coefficient for aqueous solutions of acetone, acetaldehyde and acetonitrile, and equilibrium constants for the addition compounds of acetone and acetaldehyde with bisulfite. *J. Atmos. Chem.* **20**, 17–34 (1995).

7. E. Wigaeus, S. Holm, and I. Astrand, Exposure to acetone. Uptake and elimination in man. *Scand. J. Work Environ. Health* **7**, 84–94 (1981).
8. National Fire Protection Association (NFPA), Fire hazard properties of flammable liquids, gases, and volatile solids. In *Fire Protection Guide on Hazardous Materials*, 12th ed., NFPA, Quincy, MA, 1997, pp. 325–310.
9. D.A. Boiston, Acetone oxidation, *Br. Chem. Eng.* **13**, 85–88 (1968).
10. American Industrial Hygiene Association (AIHA), *Odor Thresholds for Chemicals with Established Occupational Health Standards*, AIHA, Fairfax, VA, 1997, pp. 42–43.
11. F. Patte, M. Etcheto, and P. Laffort, Selected and standardized values of supra-threshold odor intensities for 110 substances. *Chem. Senses Flavour* **1**, 283–305 (1975).
12. M. Devos et al., in *Standardized Human Olfactory Thresholds*, Oxford University Press, Oxford, UK, 1990, p. 145.
13. J. H. E. Arts et al., *An Analysis of Human Response to the Irritancy of Acetone Vapours*, TNO Rep. V98.357s, Nutrition and Food Research Institute, 1998.
14. C. J. Wysocki et al., Odor and irritation thresholds for acetone in acetone-exposed factory workers and control (occupationally-nonexposed) subjects. *Am. Ind. Hyg. Assoc. J.* **58**, 704–712 (1997).
15. P. Dalton et al., Perceived odor, irritation and health symptoms following short-term exposure to acetone. *Am. J. Ind. Med.* **31**, 558–569 (1996).
16. P. G. C. Odeigah, Smell acuity for acetone and its relationship to taste ability to phenylthiocarbamide in a Nigerian population. *East Afr. Med. J.* **71**, 462–466 (1994).
17. K. M. Hau and D. W. Connell, Quantitative structure-activity relationships (QSARs) for odor thresholds of volatile organic compounds (VOCs). *Indoor Air* **8**, 23–33 (1998).
18. R. W. Moncrief, Olfactory adaptation and odor intensity. *Am. J. Psychol.* **70**, 1–20 (1957).
19. W. S. Cain, Odor intensity: Differences in the exponent of the psychophysical function. *Percept. Psychophys.* **6**, 349–354 (1969).
20. B. Berglund et al., Individual psychophysical functions for 28 odorants. *Percept. Psychophys.* **9**, 379–384 (1971).
21. M. Chastrette, T. Thomas-Danguin, and E. Rallet, Modeling the human olfactory stimulus-response function. *Chem. Senses* **23**, 181–196 (1998).
22. S. N. Bizzari, *CEH Marketing Research Report Acetone*, Chemical Economics Handbook, SRI International, Menlo Park, CA, 1996.
23. O. Wieland, Ketogenesis and its regulation. *Adv. Metab. Disord.* **3**, 1–47 (1968).
24. J. M. Daisey et al., Volatile organic compounds in twelve california office buildings: Classes, concentrations and sources. *Atmos. Environ.* **28**, 3557–3562 (1994).
25. C. W. Lewis and R. B. Zweidinger, Apportionment of residential indoor aerosol VOC and aldehyde species to indoor and outdoor sources, and their source strengths. *Atmos. Environ.* **26A**, 2179–2184 (1992).
26. A. Jonsson, K. A. Persson, and V. Grigoriadis, Measurements of some low molecular-weight oxygenated, aromatic, and chlorinated hydrocarbons in ambient air and in vehicle emissions. *Environ. Int.* **11**, 383–392 (1985).
27. D. L. Heavner, W. T. Morgan, and M. W. Ogden, Determination of volatile organic compounds and respirable suspended particulate matter in New Jersey and Pennsylvania homes and workplaces. *Environ. Int.* **22**, 159–183 (1996).

28. M. Dechow, H. Sohn, and J. Steinhanses, Concentrations of selected contaminants in cabin air of airbus aircrafts. *Chemosphere* **35**, 21–31 (1997).

29. O. B. Crofford et al., Acetone in breath and blood. *Trans. Am. Clin. Climatol. Assoc.* **88**, 128–139 (1977).

30. F. H. Jarke, A. Dravnieks, and S. M. Gordon, Organic contaminants in indoor air and their relation to outdoor contaminants. *ASHRAE Trans.* **87**, 153–166 (1981).

31. H. B. Singh et al., Acetone in the atmosphere: Distribution, sources, and sinks. *J. Geophys. Res.* **99**, 1805–1819 (1994).

32. National Library of Medicine (NLM), *Toxicology Data Network—TOXNET*, NLM, Washington, DC, 1999. Toxic Releases (TRI) Search: http://toxnet. nlm. nih. gov/servlets/simple-search?

33. D. A. Morgott, Acetone. In G. D. Clayton and F. E. Clayton, eds., *Patty's Industrial Hygiene and Toxicology*, 4th ed., Vol. II, Pt. A, Wiley, New York, 1993, pp. 149–281.

34. R. B. Zweidinger et al., Detailed hydrocarbon and aldehyde mobile source emissions from roadway studies. *Environ. Sci. Technol.* **22**, 956–962 (1988).

35. R. B. Chatfield, E. P. Gardner, and J. G. Calvert, Sources and sinks of acetone in the troposphere: Behavior of reactive hydrocarbons and a stable product. *J. Geophys. Res.* **92** 4208–4216 (1987).

36. H. B. Singh et al., High concentrations and photochemical fate of oxygenated hydrocarbons in the global troposphere. *Nature (London)* **378**, 50–54 (1995).

37. L. E. Booher and B. Janke, Air emissions from petroleum hydrocarbon fires during controlled burning. *Am. Ind. Hyg. Assoc. J.* **58** 359–365 (1997).

38. F. Grimaldi et al., Study of air pollution by carbonyl compounds in automobile exhaust. *Pollut. Atmos.* **149** 68–76 (1996).

39. Y. Hoshika, Y. Nihei, and G. Muto, Pattern display for characterization of trace amounts of odorants discharged from nine odour sources. *Analyst (London)* **106**, 1187–1202 (1981).

40. M. A. K. Khalil and R. A. Rasmussen, Forest hydrocarbon emissions: Relationships between fluxes and ambient concentrations. *J. Air Waste Manage. Assoc.* **42**, 810–813 (1992).

41. J. Brosseau and M. Heitz, Trace gas compound emissions from municipal landfill sanitary sites. *Atmos. Environ.* **28**, 285–293 (1994).

42. D. E. Euler, S. J. Davé, and H. Guo, Effect of cigarette smoking on pentane excretion in alveolar breath. *Clin. Chem. (Winston-Salem, N.C.)* **42**, 303–308 (1996).

43. V. A. Isidorov, I. G. Zenkevich, and B. V. Ioffe, Volatile organic compounds in the atmosphere of forests. *Atmos. Environ.* **19**, 1–8 (1985).

44. B. L. Oser and R. A. Ford, Recent progress in the consideration of flavoring ingredients under the food additives amendment 6. GRAS substances. *Food Technol.* **27**, 64–67 (1973).

45. L. Andersson and K. Lundström, Milk and blood ketone bodies, blood isopropanol and plasma glucose in dairy cows; methodological studies and diurnal variations. *Zentralbl. Veterinaer. Med. A.* **31**, 340–349 (1984).

46. H. A. George et al., Acetone, isopropanol, and butanol production by *Clostridium Beijerinckii* (Syn. *Clostridium Butylicium*) and *Clostridium Aurantibutyricum*. *Appl. Environ. Microbiol.* **45**, 1160–1163 (1983).

47. A.-L. Sunesson et al., Volatile metabolites produced by two fungal species cultivated on building materials. *Ann. Occup. Hyg.* **40**, 397–410 (1966).

48. R. P. Collins and K. Kalnins, Carbonyl compounds produced by cryptomonas ovata Var. *Palustris. J. Protozool.* **13**, 435–437 (1966).
49. M. S. Smith, A. J. Francis, and J. M. Duxbury, Collection and analysis of organic gases from natural ecosystems: Application to poultry manure. *Environ. Sci. Technol.* **11**, 51–55 (1977).
50. J. E. Yocom, G. M. Hein, and H. W. Nelson, A study of the effluents from backyard incinerators. *J. Air Pollut. Control Assoc.* **6**, 84–89 (1956).
51. A. M. Hartstein and D. R. Forshey, *Coal Mine Combustion Products. Neoprenes, Polyvinyl Chloride Compositions, Urethane Foam, and Wood*, Rep. No. 7977, U.S. Department of the Interior, Washington, DC, 1974.
52. F. Lipari, J. M. Dasch, and W. F. Scruggs, Aldehyde emissions from woodburning fireplaces. *Environ. Sci. Technol.* **18**, 326–330 (1984).
53. D. Grosjean and B. Wright, Carbonyls in urban fog, ice fog, cloudwater and rainwater. *Atmos. Environ.* **17**, 2093–2096 (1983).
54. L. H. Keith et al., Identification of organic compounds in drinking water from thirteen U. S. cities. In *Identification and Analysis of Organic Pollutants in Water*, 1979, pp. 329–373.
55. K. Mopper and W. L. Stahovec, Sources and sinks of low molecular weight organic carbonyl compounds in seawater. *Mar. Chem.* **19**, 305–321 (1986).
56. R. E. Rathbun et al., Fate of acetone in water. *Chemosphere* **11**, 1097–1114 (1982).
57. H. Meyrahn et al., Quantum yields for the photodissociation of acetone in air and an estimate for the life time of acetone in the lower troposphere. *J. Atmos. Chem.* **4**, 277–291 (1986).
58. F. B. Dewalle and E. S. K. Chian, Detection of trace organic's in well water near a solid waste landfill. *J. Am. Water Works Assoc.* **73**, 206–211 (1981).
59. J. F. Corwin, Volatile oxygen-containing organic compounds in sea water: Determination. *Bull. Mar. Sci.* **19**, 504–509 (1969).
60. G. V. Sabel and T. P. Clark, Volatile organic compounds as indicators of municipal solid waste leachate contamination. *Waste Manage. Res.* **2**, 119–130 (1984).
61. G. A. Jungclaus, V. Lopez-Avila, and R. A. Hites, Organic compounds in an industrial wastewater: A case study of their environmental impact. *Environ. Sci. Technol.* **12**, 88–96 (1978).
62. D. E. Line et al., Water quality of first flush runoff from 20 industrial sites. *Water Environ. Res.* **69**, 305–310 (1997).
63. V. P. Aneja, Organic compounds in cloud water and their deposition at a remote continental site. *J. Air Waste Manage. Assoc.* **43**, 1239–1244 (1993).
64. P. H. Howard et al., Acetone. In *Handbook of Environmental Fate and Exposure Data for Organic Chemicals*, 1990, pp. 9–18.
65. K. W. Brown and K. C. Donnelly, An estimation of the risk associated with the organic constituents of hazardous and municipal waste landfill leachates. *Hazard. Waste Hazard. Mater.* **5**, 1–30 (1988).
66. T. M. Holsen et al., Contamination of potable water by permeation of plastic pipe. *J. Am Water Works Assoc.* **83**, 53–56 (1991).
67. L. W. Whitehead et al., Solvent vapor exposures in booth spray painting and spray glueing, and associated operations. *Am. Ind. Hyg. Assoc.* **45**, 767–772 (1984).
68. C. Winder and P. J. Turner, Solvent exposure and related work practices amongst apprentice spray painters in automotive body repair workshops. *Ann. Occup. Hyg.* **36**, 385–394 (1993).

69. B. Young et al., Vapors from colloidon and acetone in an EEG laboratory. *J. Clin. Neurophysiol.* **10**, 108–110 (1993).

70. M. Nasterlack, G. Triebig, and O. Stelzer, Hepatotoxic effects of solvent exposure around permissible limits and alcohol consumption in printers over a 4-year period. *Int. Arch. Occup. Environ. Health* **66**, 161–165 (1994).

71. I. Ahonen and R. W. Schimberg, 2,5-Hexanedione excretion after occupational exposure to *n*-Hexane. *Br. J. Ind. Med.* **45**, 133–136 (1988).

72. N. A. Nelson et al., Historical characterization of exposure to mixed solvents for an epidemiologic study of automotive assembly plant workers. *Appl. Occup. Environ. Hyg.* **8**, 693–702 (1993).

73. M. F. Hallock et al., Assessment of task and peak exposures to solvents in the microelectronics fabrication industry. *Appl. Occup. Environ. Hyg.* **8**, 945–954 (1993).

74. E. Mitran et al., Neurotoxicity associated with occupational exposure to acetone, methyl ethyl ketone, and cyclohexanone. *Environ. Res.* **73**, 181–188 (1997).

75. M. Schultz, J. Cisek, and R. Wabeke, Simulated exposure of hospital emergency personnel to solvent vapors and respirable dust during decontamination of chemically exposed patients. *Ann. Emerg. Med.* **26**, 324–329 (1995).

76. E. De Rosa et al., The importance of sampling time and co-exposure to acetone in the biological monitoring of styrene-exposed workers. *Appl. Occup. Environ. Hyg.* **11**, 471–475 (1996).

77. G. Franco et al., Serum bile acid concentrations as a liver function test in workers occupationally exposed to organic solvents. *Int. Arch. Occup. Environ. Health* **58**, 157–164 (1986).

78. A. A. Jensen et al., Occupational exposures to styrene in Denmark 1955–88. *Am. J. Ind. Med.* **17**, 593–606 (1990).

79. T. Satoh et al., Relationship between acetone exposure concentration and health effects in acetate fiber plant workers. *Int. Arch. Occup. Environ. Health* **68**, 147–153 (1996).

79a. NIOSH. *Manual of Analytical Methods*, 4th ed., U.S. Govt Printing Office, Supt. of Docs, Washington, DC, 1994.

80. T. M. Sack et al., A survey of household products for volatile organic compounds. *Atmos. Environ.* **26A**, 1063–1070 (1992).

81. A. Brega et al., High-performance liquid chromatographic determination of acetone in blood and urine in the clinical diagnostic laboratory. *J. Chromatogr.* **553**, 249–254 (1991).

82. M. D. Trotter, M. J. Sulway, and E. Trotter, The rapid determination of acetone in breath and plasma. *Clin. Chim. Acta* **35**, 137–143 (1971).

83. M. Kimura et al., Head-space gas-chromatographic determination of 3-hydroxybutyrate in plasma after enzymic reactions, and the relationship among the three ketone bodies. *Clin. Chem.* (*Winstan Salem N.C.*) **31**, 596–598 (1985).

84. S. Levey et al., Studies of metabolic products in expired air. II. Acetone. *J. Lab. Clin. Med.* **63**, 574–584 (1964).

85. V. C. Gavino et al., Determination of the concentration and specific activity of acetone in biological fluids. *Anal. Biochem.* **152**, 256–261 (1986).

86. S. Felby and E. Nielsen, Congener production in blood samples during preparation and storage. *Blutalkohol* **32**, 50–58 (1995).

87. F. Brugnone et al., Blood and urine concentrations of chemical pollutants in the general population. *Med. Lav.* **85**, 370–389 (1994).

88. T. J. Buckley et al., Environmental and biomarker measurements in nine homes in the Lower Rio Grande Valley: Multimedia results for pesticides, metals, PAHs, and VOCs. *Environ. Int.* **23**, 705–732 (1997).

89. D. L. Ashley et al., Blood concentrations of volatile organic compounds in a nonoccupationally exposed US population and in groups with suspected exposure. *Clin. Chem. (Winston Salem, N.C.)* **40**, 1401–1404 (1994).

90. A. W. Jones et al., Concentrations of acetone in venous blood samples from drunk drivers, type-I diabetic outpatients, and healthy blood donors. *J. Appl. Toxicol.* **17**, 182–185 (1993).

91. G. Wang et al., Blood acetone concentration in "normal people" and in exposed workers 16 h after the end of the workshift. *Int. Arch. Occup. Environ. Health* **65**, 285–289 (1994).

92. G. Pezzagno et al., Urinary concentration, environmental concentration, and respiratory uptake of some solvents: Effect of the work load. *Am. Ind. Hyg. Assoc. J* **49**, 546–552 (1988).

93. A. W. Jones, Breath acetone concentrations in fasting male volunteers: Further studies and effect of alcohol administration. *J. Anal. Toxicol.* **12**, 75–79 (1988).

94. B. O. Jansson and B. T. Larsson, Analysis of organic compounds in human breath by gas chromatography-mass spectrometry. *J. Lab. Clin. Med.* **74**, 961–966 (1969).

95. C. N. Tassopoulos, D. Barnett, and T. R. Fraser, Breath-acetone and blood-sugar measurements in diabetes. *Lancet* **2**, 1282–1286 (1969).

96. G. Rooth and S. Östenson, Acetone in alveolar air, and the control of diabetes. *Lancet* **2**, 1102–1105 (1966).

97. R. D. Stewart and E. A. Boettner, Expired-air acetone in diabetes mellitus. *N. Engl. J. Med.* **270**, 1035–1038 (1964).

98. J. E. Smialek and B. Levine, Diabetes and decomposition. A case of diabetic ketoacidosis with advanced postmortem change. *Am. J. Forensic Med. Pathol.* **19**, 98–101 (1998).

99. B. J. Dowty, J. L. Laseter, and J. Storer, The transplacental migration and accumulation in blood of volatile organic compounds. *Pediatr./Res.* **10**, 696–701 (1976).

100. M. J. Sulway et al., Acetone in uncontrolled diabetes. *Postgrad. Med. J.* **47**, 383–387 (1971).

101. A. Zlatkis et al., Profile of volatile metabolites in urine by gas chromatography mass chromatography. *Anal. Chem.* **45**, 763–767 (1973).

102. E. D. Pellizzari et al., Purgeable organic compounds in mother's milk. *Bull. Environ. Contam. Toxicol.* **28**, 322–328 (1982).

103. H. Göschke and T. Lauffenburger, Aceton in der Atemluft und Ketone im Venenblut bei Vollständigem Fasten Normal und Übergewichtiger Personen. *Res. Exp. Med.* **165**, 233–244 (1975).

104. K. E. Wildenhoff, Diurnal variations in the concentrations of blood acetoacetate, 3-hydroxybutyrate and glucose in normal persons. *Acta Med. Scand.* **191**, 303–306 (1972).

105. E. M. P. Widmark, The kinetics of the ketonic decomposition of acetoacetic acid. *Acta Med. Scand.* **53**, 393–421 (1920).

106. C. Grampella et al., Health surveillance in workers exposed to acetone. *Proc. 7th Int. Symp. Occup. Health Prod. of Arti. Org. Fibres, 1987*, pp. 137–141.

107. D. Dyne, J. Cocker, and H. K. Wilson, A novel device for capturing breath samples for solvent analysis. *Sci. Total Environ.* **199**, 83–89 (1997).

108. S. K. Kundu and A. M. Judilla, Novel solid-phase assay of ketone bodies in urine. *Clin. Chem. (Winston-Salem, N.C.)* **37**, 1565–1569 (1991).

109. J. Schuberth, A full evaporation headspace technique with capillary GC and ITD: A means for quantitating volatile organic compounds in biological samples. *J. Chromatogr. Sci.* **34**, 314–319 (1996).

110. S. C. Connor et al., Antidiabetic efficacy of BRL 49653, a potent orally active insulin sensitizing agent, assessed in the C57BL/KsJ *db/db* diabetic mouse by non-invasive ^{1}H NMR Studies of Urine. *J. Pharm. Pharmacol.* **49**, 336–344 (1997).

111. O. E. Owen et al., Acetone metabolism during diabetic ketoacidosis. *Diabetes* **31**, 242–248 (1982).

112. H. Crato, G. Walther, and A. Hermann, Das Vorkommen Von Aceton in Zur Alkohol-Bestimmung Eingesandten Blutproben. *Beitr. Gerichtl. Med.* **36**, 275–279 (1978).

113. M. Kupari et al., Breath acetone in congestive heart failure. *Am. J. Cardiol.* **76**, 1076–1078 (1995).

114. F. Brugnone et al., Isopropanol exposure: Environmental and biological monitoring in a printing works. *Br. J. Ind. Med.* **40**, 160–168 (1983).

115. G. Pezzagno et al., Urinary elimination of acetone in experimental and occupational exposure. *Scand. J. Work Environ. Health*, **12**, 603–608 (1986).

116. T. Kawai et al., Urinary excretion of unmetabolized acetone as an indicator of occupational exposure to acetone. *Int. Arch. Occup. Environ. Health* **62**, 165–169 (1990).

117. T. Kawai et al., Curvilinear relationship between acetone in breathing zone air and acetone in urine among workers exposed to acetone vapor. *Toxicol. Lett.* **62**, 85–91 (1992).

118. A. Fujino et al., Biological monitoring of workers exposed to acetone in acetate fibre plants. *Br. J. Ind. Med.* **49**, 654–657 (1992).

119. T. Satoh et al., Acetone excretion into urine of workers exposed to acetone in acetate fiber plants. *Int. Arch. Occup. Environ. Health* **67**, 131–134 (1995).

120. R. R. Vangala et al., Biomonitoring of human experimental exposures to acetone and ethyl acetate. In N. V. C. Swamy and M. Singh, eds., *Physiological Fluid Dynamics III*, Nora Publishing House, New Delhi, India, 1992, pp. 354–359.

121. R. D. Stewart et al., Acetone: Development of a biologic standard for the industrial worker by breath analysis. U.S. C.F.S.T.I., *PB Rep.* PB82-172917 (1975).

122. F. Brugnone et al., Solvent exposure in a shoe upper factory. I. *n*-hexane and acetone concentration in alveolar and environmental air and in blood. *Int. Arch. Occup. Environ. Health* **42**, 51–62 (1978).

123. F. Brugnone et al., Biomonitoring of industrial solvent exposures in workers' alveolar air. *Int. Arch. Occup. Environ. Health* **47**, 245–261 (1980).

124. K. Baumann and J. Angerer, Untersuchungen zur Frage der beruflichen Lösungsmittel-belastung mit Aceton. *Krebsgefaehrd. Arbeitsplatz Arbeitsmed.* **19**, 403–408 (1979).

125. K. Mizunuma et al., Exposure-excretion relationship of styrene and acetone in factory workers: A comparison of a lipophilic solvent and a hydrophilic solvent. *Arch. Environ. Contam. Toxicol.* **25**, 129–133 (1993).

126. P. Apostoli et al., Metabolic interferences in subjects occupationally exposed to binary styrene-acetone mixtures. *Int. Arch. Occup. Environ. Health* **71**, 445–452 (1998).

127. D. Marhuenda et al., Biological monitoring of styrene exposure and possible interference of acetone co-exposure. *Int. Arch. Occup. Environ. Health* **69**, 455–460 (1997).

128. M. Tomita and M. Nishimura, Using saliva to estimate human exposure to organic solvents. *Bull. Tokyo Dent. Coll.* **23**, 175–188 (1982).

129. D. Mebs, J. Gerchow, and K. Schmidt, Interference of acetone with breath-alcohol testing. *Blutalkohol* **21**, 193–198 (1984).

130. A. W. Jones, L. Andersson, and K. Berglund, Interfering substances identified in the breath of drinking drivers with intoxilyzer 5000S. *J. Anal. Chem.* **20**, 522–527 (1996).

131. J. F. Frank and A. L. Flores, The likelihood of acetone interference in breath alcohol measurement. *Alcohol Drugs Driving* **3**, 1–8 (1987).

132. D. Bagchi et al., Endrin-induced urinary excretion of formaldehyde, acetaldehyde, malondialdehyde and acetone in rats. *Toxicology* **75**, 81–89 (1992).

133. J. Moser et al., Excretion of malondialdehyde, formaldehyde, acetaldehyde, and acetone in the urine of rats following acute and chronic administration of ethanol. *Alcohol Alcohol.* **28**, 287–295 (1993).

134. M. A. Shara et al., Excretion of formaldehyde, malondialdehyde, acetaldehyde and acetone in the urine of rats in response to 2,3,7,8-tetrachlorodibenzo-*p*-dioxin, paraquat, endrin and carbon tetrachloride. *J. Chromatogr. Biomed. Appl.* **576**, 221–233 (1992).

135. L. L. de Zwart et al., Evaluation of urinary biomarkers for radical–induced liver damage in rats treated with carbon tetrachloride. *Toxicol. Appl. Pharmacol.* **148**, 71–82 (1998).

136. P. I. Akubue and S. J. Stohs, Effect of alachlor on the urinary excretion of malondialdehyde, formaldehyde, acetaldehyde, and acetone by rats. *Bull. Environ. Contam. Toxicol.* **50**, 565–571 (1993).

137. P. I. Akubue et al., Excretion of malondi–aldehyde, formaldehyde, acetaldehyde, acetone, and methyl ethyl ketone in the urine of rats given an acute dose of malondialdehyde. *Arch. Toxicol.* **68**, 338–341 (1994).

138. D. Bagchi et al., Effects of carbon tetrachloride, menadione, and paraquat on the urinary excretion of malondialdehyde, formaldehyde, acetaldehyde, and acetone in rats. *J. Biochem. Toxicol.* **8**, 101–106 (1993).

139. D. Bagchi et al., Time-dependent effects of 2,3,7,8-tetrachlorodibenzo-*p*-dioxin on serum and urine levels of malondialdehyde, formaldehyde, acetaldehyde, and acetone in rats. *Toxicol. Appl. Pharmacol.* **123**, 83–88 (1993).

140. E. A. Hassoun, D. Bagchi, and S. J. Stohs, Evidence of 2,3,7,8-tetrachloro-dibenzo-*p*-dioxin (TCDD)-induced tissue damage in fetal and placental tissues and changes in amniotic fluid lipid metabolites of pregnant CF_1 mice. *Toxicol. Lett.* **76**, 245–250 (1995).

141. E. A. Hassoun, D. Bagchi, and S. J. Stohs, TCDD, endrin, and lindane induced increases in lipid metabolites in maternal sera and amniotic fluids of pregnant C57Bl/6J and DBA/2J mice. *Res. Commun. Mol. Pathol. Pharmacol.* **94**, 157–169 (1996).

142. D. Bagchi et al., Oxidative stress induced by chronic administration of sodium dichromate [Cr(VI)] to rats. *Comp. Biochem. Physiol.* **110C**, 281–287 (1995).

143. D. Bagchi et al., Induction of oxidative stress by chronic administration of sodium dichromate [chromate VI] and cadmium chloride [cadmium II] to rats. *Free Radical Biol. Med.* **22**, 471–478 (1997).

144. D. Bagchi et al., Subchronic effects of smokeless tobacco extract (STE) on hepatic lipid peroxidation, DNA damage and excretion of urinary metabolites in rats. *Toxicology* **127**, 29–38 (1998).

145. Y.-C. Ni et al., Mouse liver microsomal metabolism of chloral hydrate, trichloroacetic acid, and trichloroethanol leading to induction of lipid peroxidation via a free radical mechanism. *Drug Metab. Dispos.* **24**, 81–90 (1996).

146. J. G. Filser et al., Exhalation of acetone by rats on exposure to trans-1,2-Dichloroethylene and Related Compounds. *Toxicol. Lett.* **2**, 247–252 (1978).

147. L. L. de Zwart et al., Urinary excretion of biomarkers for radical-induced damage in rats treated with NDMA or diquat and the effects of calcium carbimide co-administration. *Chem. -Biol. Interact.* **117**, 151–172 (1999).

148. H. M. Bolt et al., Biological activation of 1,3-butadiene to vinyl oxirane by rat liver microsomes and expiration of the reactive metabolite by exposed rats. *J. Cancer Res. Clin. Oncol.* **106**, 112–116 (1983).

149. J. G. Filser, P. Jung, and H. M. Bolt, Increased acetone exhalation induced by metabolites of halogenated C_1 and C_2 compounds. *Arch. Toxicol.* **49**, 107–116 (1982).

150. D. Bagchi et al., Protective effects of antioxidants against endrin-induced hepatic lipid peroxidation, DNA damage, and excretion of urinary lipid metabolites. *Free Radical Biol. Med.* **15**, 217–222 (1993).

151. P. Ninfali et al., Acetaldehyde, ethanol and acetone concentrations in blood of alcohol-treated mice receiving aldehyde dehydrogenase-loaded erythrocytes. *Alcohol Alcohol.* **27**, 19–23 (1992).

152. H. M. Bolt, Covalent binding of haloethylenes. *Adv. Exp. Med. Biol.* **136**, 667–683 (1981).

153. P. Simon, H. M. Bolt, and J. G. Filser, Covalent interaction of reactive metabolites with cytosolic coenzyme A as mechanism of haloethylene-induced acetonemia. *Biochem. Pharmacol.* **34**, 1981–1986 (1985).

154. A. V. Buchter et al., Studie zur Perchlorethylen-(Tetrachlorethen-) Exposition unter Berücksichtigung von Expositions-Spitzen und Aceton-Bildung. *Zentralbl. Arbeitsmed.* **34**, 130–136 (1984).

155. B. T. Langeland and J. S. McKinley-McKee, The effects of disulfiram and related compounds on equine hepatic alcohol dehydrogenase. *Comp. Biochem. Physiol.* **117C**, 55–61 (1997).

156. J. F. Brady et al., Effects of disulfuram on hepatic P450IIE1, other microsomal enzymes, and hepatotoxicity in rats. *Toxicol. Appl. Pharmacol.* **108**, 366–373 (1991).

157. V. V. Lauriault, S. Khan, and P. J. O'Brien, hepatocyte cytotoxicity induced by various hepatotoxins mediated by cytochrome P-450_{IIE1}: Protection with diethyldithiocarbamate administration. *Chem. -Biol. Interact.* **81**, 271–289 (1992).

158. J. G. Filser and H. M. Bolt, Characteristics of haloethylene-induced acetonemia in rats. *Arch. Toxicol.* **45**, 109–116 (1980).

159. E. G. DeMaster and H. T. Nagasawa, Disulfuram-induced acetonemia in the rat and man. *Res. Commun. Chem. Pathol. Pharmacol.* **18**, 361–364 (1977).

160. E. G. DeMaster and J. M. Stevens, Acute effects of the aldehyde dehydrogenase inhibitors, disulfiram, paragyline and cyanamide, on circulating ketone body levels in the rat. *Biochem. Pharmacol.* **37**, 229–234 (1988).

161. P. Pronko, S. Zimatkin, and A. Kuzmich, Effect of aldehyde dehydrogenase and alcohol dehydrogenase inhibitors and ethanol on blood and liver ketone bodies in the rat. *Pol. J. Pharmacol.* **46**, 445–449 (1994).

162. A. Stowell et al., Disulfiram-induced acetonemia. *Lancet*, **1**, 882–883 (1983).

163. J. Johnsen et al., A double-blind placebo controlled study of healthy volunteers given a subcutaneous disulfiram implantation. *Pharmacol. Toxicol.* **66**, 227–230 (1990).

164. N. W. Teitz, *Clinical Guide to Laboratory Tests*, Saunders, Philadelphia, PA, 1983, pp. 2–3.

165. P. Paterson et al., Maternal and foetal ketone concentration in plasma and urine. *Lancet* **2**, 862–865 (1967).

166. J. H. Koeslag, T. D. Noakes, and A. W. Sloan, Post-exercise ketosis. *J. Physiol. (London)* **301**, 79–90 (1980).

167. R. E. Kimura and J. B. Warshaw, Metabolic adaptations of the fetus and newborn. *J. Pediatr. Gastroenterol. Nutr.* **2**, S12–S15 (1983).

168. S. B. Lewis et al., Effect of diet composition on metabolic adaptations to hypocaloric nutrition: Comparison of high carbohydrate and high fat isocaloric diets. *Am. J. Clin. Nutr.* **30**, 160–170 (1977).

169. L. J. Levy et al., Ketoacidosis associated with alcoholism in nondiabetic subjects. *Ann. Intern. Med.* **78**, 213–219 (1973).

170. R. Smith et al., Initial effect of injury on ketone bodies and other blood metabolites. *Lancet* **1**, 1–3 (1975).

171. G. Rooth and S. Carlstrom, Therapeutic fasting. *Acta Med. Scand.* **187**, 455–463 (1970).

172. E. H. Williamson and E. Whitelaw, Physiological aspects of the regulation of ketogenesis. *Biochem. Soc. Symp.* **43**, 137–161 (1978).

173. A. Walther and G. Neumann, Über den Verlauf Azetonausscheidung in der Ausatmungsluft bei Radsportlern vor, während und nach einer spezifischen Laborbelastung. *Acta Biol. Med. Ger.* **22**, 117–121 (1969).

174. P. Felig, E. B. Marliss, and G. F. Cahill, Metabolic response to human growth hormone during prolonged starvation. *J. Clin. Invest.* **50**, 411–421 (1971).

175. V. H. Peden, Determination of individual serum 'ketone bodies,' with normal values in infants and children. *J. Lab. Clin. Med.* **63**, 332–343 (1964).

176. P. G. Lindner and G. L. Blackburn, Multidisciplinary approach to obesity utilizing fasting modified by protein-sparing therapy. *Obes. Bariatric Med.* **5**, 198–216 (1976).

177. V. C. Gavino et al., Determination of the concentration and specific activity of acetone in biological fluids. *Anal. Biochem.* **152**, 256–261 (1986).

178. F. J. López-Soriano and J. M. Argilés, Simultaneous determination of ketone bodies in biological samples by gas chromatographic headspace analysis. *J. Chromatogr. Sci.* **23**, 120–123 (1985).

179. M. Kimura et al., Acetoacetate decarboxylase activity of plasma components. *Jap. J. Clin. Chem.* **15**, 37–43 (1986).

180. A. C. Haff and G. A. Reichard, A method for estimation of acetone radioactivity and concentration in blood and urine. *Biochem. Med.* **18**, 308–314 (1977).

181. M. F. Mason and D. Hutson, The range of concentrations of free acetone in the plasma and breath of diabetics and some observations on its plasma/breath ratio. In *Proc. 6th Int. Conf. Alcohol, Drugs, Traffic Saf.* Toronto, Canada, 1975.

182. M. Kitazawa et al., Concentrations of ethanol metabolites and symptoms of acute alcohol-intoxicated patients. *Jpn. J. Alcohol Drug Depend.* **29**, 31–39 (1994).

183. C. Péclet, P. Picotte, and F. Jobin, The use of vitreous humor levels of glucose, lactic acid and blood levels of acetone to establish antemortem hyperglycemia in diabetes. *Forensic Sci. Int.* **65**, 1–6 (1994).

184. B. Brinkmann et al., Ketoacidosis and lactic acidosis — frequent causes of death in chronic alcoholics?. *Int. J. Leg. Med.* **111**, 115–119 (1998).

185. K. Kobayashi et al., A gas chromatographic method for the determination of acetone and acetoacetic acid in urine. *Clin. Chim. Acta* **133**, 223–226 (1983).

186. M. J. Sulway and J. M. Mullins, Acetone in diabetic ketoacidosis, *Lancet* **1**, 736–740 (1970).

187. A. Ramu, J. Rosenbaum, and T. F. Blaschke, Disposition of acetone following acute acetone intoxication. *West. J. Med.* **129**, 429–432 (1978).

188. A. S. Gamis and Wasserman, Acute acetone intoxication in a pediatric patient. *Pediatr. Emerg. Care* **4**, 24–26 (1988).

189. M. J. Henderson, B. A. Karger, and G. A. Wrenshall, Acetone in breath. A study of acetone exhalation in diabetic and nondiabetic human subjects. *Diabetes* **1**, 188–200 (1952).

190. J. P. Conkle, B. J. Camp, and B. E. Welch, Trace composition of human respiratory gas. *Arch. Environ. Health* **30**, 290–295 (1975).

191. A. W. Jones, Breath-acetone concentrations in fasting healthy men: Response of infrared breath-alcohol analyzers. *J. Anal. Toxicol.* **11**, 67–69 (1987).

192. J. Neiman et al., Combined effect of a small dose of ethanol and 36 hr fasting on blood-glucose response, breath-acetone profiles and platelet function in healthy men. *Alcohol Alcoholism* **22**, 265–270 (1987).

193. G. Kernbach, E. Koops, and B. Brinkmann, Biochemische Parameter bei 31 tödlichen diabetischen Stoffwechselentgleisungen. *Beitr. Gerichtl. Med.* **42**, 301–306 (1984).

194. U. C. Pozzani, C. S. Weil, and C. P. Carpenter, The toxicological basis of threshold limit values: 5. The experimental inhalation of vapor mixtures by rats, with notes upon the relationship between single dose inhalation and single dose oral data. *Am. Ind. Hyg. Assoc. J.* **20**, 364–369 (1959).

195. H. F. Smyth et al., An exploration of joint toxic action: Twenty-seven industrial chemicals intubated in rats in all possible pairs. *Toxicol. Appl. Pharmacol.* **14**, 340–347 (1969).

196. R. H. Clothier et al., Comparison of the *in vitro* cytotoxicities and acute *in vivo* toxicities of 59 chemicals. *Mol. Toxicol.* **1**, 571–577 (1987).

197. M. Kanada et al., Neurochemical profile of effects of 28 neurotoxic chemicals on the central nervous system in rats. (1) Effects of oral administration on brain contents of biogenic amines and metabolites. *Ind. Health* **32**, 145–164 (1994).

198. E. T. Kimura, D. M. Ebert, and P. W. Dodge, Acute toxicity and limits of solvent residue for sixteen organic solvents. *Toxicol. Appl. Pharmacol.* **19**, 699–704 (1971).

199. W. J. Krasavage, J. L. O'Donoghue, and G. D. DiVincenzo, Ketones. In G. D. Clayton and F. E. Clayton, eds., *Patty's Industrial Hygiene and Toxicology*, 3rd ed., Vol. IIC, Wiley, New York, 1982, pp. 4720–4727.

200. H. Tanii, H. Tsuji, and K. Hashimoto, Structure-toxicity relationship of monoketones. *Toxicol. Lett.* **30**, 13–17 (1986).

201. H. Nishimura et al., Analysis of acute toxicity (LD_{50}-value) of organic chemicals to mammals by solubility parameter (δ). *Jpn. J. Ind. Health* **36**, 428–434 (1994).

202. R. L. Roudabush et al., Comparative acute effects of some chemicals on the skin of rabbits and guinea pigs. *Toxicol. Appl. Pharmacol.* **7**, 559–565 (1965).

203. S. Zakhari, Acute oral, intraperitoneal, and inhalational toxicity of methyl isobutyl ketone in the mouse. In L. Goldberg, ed., *Isopropanol and Ketones in the Environment*, CRC Press, Cleveland, OH, 1977, pp. 101–104.

204. P. Mikolajczak et al., The toxic and pharmacokinetic interactions between ethanol and acetone in rats. *Vet. Hum. Toxicol.* **35**, 363 (abstr.) (1993).

205. L. M. Mashbitz, R. M. Sklianskaya, and F. I. Urieva, The relative toxicity of acetone, methylalcohol and their mixtures: II. Their action on white mice. *J. Ind. Hyg. Toxicol.* **18**, 117–122 (1936).

206. H. Specht, J. W. Miller, and P. J. Valaer, Acute response of guinea pigs to the inhalation of dimethyl ketone (acetone) vapor in air. *Public Health Rep.* **52**, 944–954 (1939).

207. G. A. Safronov et al., Comparative acute inhalation toxicity of aliphatic aldehydes and ketones according to exposure time. *Curr. Toxicol.* **1**, 47–51 (1993).

208. G. L. Kennedy and G. J. Graepel, Acute toxicity in the rat following either oral or inhalation exposure. *Toxicol. Lett.* **56**, 317–326 (1991).

209. E. Browning, Ketones. In *Toxicity of Industrial Organic Solvents*, Chem. Pub. Co., New York, 1953, pp. 320–339.

210. H. W. Haggard, L. A. Greenberg, and J. M. Turner, The physiological principles governing the action of acetone together with the determination of toxicity. *J. Ind. Hyg. Toxicol.* **26**, 133–151 (1944).

211. E. Kagan, Experimentelle Studien Über den Einfluss technisch und hygienisch wichtiger Gase und Dämfpe auf den Organismus. XXXVI. Aceton. *Arch. Hyg.* **94**, 41–53 (1924).

212. F. Flury and W. Wirth, Zur Toxikologie der Lösungsmittel (Verschiedene Ester, Aceton, Methylalkohol). *Arch. Gewerbepathol. Gewerhyg.* **5**, 1 (1934).

213. P. Albertoni, Die Wirkung und die Verwandlungen Einiger Stoffe im Organismus in Beziehung zur Pathogenese der Acetonämie und des Diabetes. *Arch. Exp. Pathol. Pharmakol.* **18**, 219–241 (1884).

214. D. C. Walton, E. F. Kehr, and A. S. Lovenhart, A comparison of the pharmacological action of diacetone alcohol and acetone. *J. Pharmacol. Exp. Ther.* **33**, 175–183 (1928).

215. C. L. Anderson, Experimental production of convulsive seizures. *J. Nerv. Ment. Dis.* **109**, 210–219 (1949).

216. E. Frantík, M. Hornychová, and M. Horváth, (1994). Relative acute neurotoxicity of solvents: Isoeffective air concentrations of 48 compounds evaluated in rats and mice. *Environ. Res.* **66**, 173–185 (1994).

217. M. T. D. Cronin, Quantitative structure-activity relationship (QSAR) analysis of the acute sublethal neurotoxicity of solvents. *Toxicol. In Vitro* **10**, 103–110 (1996).

218. J. Larsen, K. Gasser, and R. Hahin, An analysis of dimethylsulfoxide-induced action potential block: A comparative study of DMSO and other aliphatic water soluble solutes. *Toxicol. Appl. Pharmacol.* **140**, 296–314 (1996).

219. J. Huang et al., Structure-toxicity relationship of monoketones: *In vitro* effects on beta-adrenergic receptor binding and Na^+-K^+-ATPase activity in mouse synaptosomes. *Neurotoxicol. Teratol.* **15**, 345–352 (1993).

220. H. Tanii, Anesthetic activity of monoketones in mice: Relationship to hydrophobicity and *in vivo* effects on Na^+/K^+-ATPase activity and membrane fluidity. *Toxicol. Lett.* **85**, 41–47 (1996).

221. G. M. Ghoniem, A. M. Shaaban, and M. R. Clarke, Irritable bladder syndrome in an animal model: A continuous monitoring study. *Neurourol. Urodyn.* **14**, 657–665 (1995).

222. K. Kato et al., Time-course of alterations of bladder function following acetone-induced cystitis. *J. Urol.* **144**, 1272–1276 (1990).

223. R. D. Panson and C. L. Winek, Aspiration toxicity of ketones. *Clin. Toxicol.* **17**, 271–317 (1980).

224. G. Enhorning et al., Pulmonary surfactant films affected by solvent vapors. *Anesth. Analg.* **65**, 1275–1280, (1986).

225. J. V. Bruckner and R. G. Petersen, Evaluation of toluene and acetone inhalant abuse. i. pharmacology and pharmacodynamics. *Toxicol. Appl. Pharmacol.* **61**, 27–38 (1981).

226. M. E. Goldberg et al., Effect of repeated inhalation of vapors of industrial solvents on animal behavior. I. Evaluation of nine solvent vapors on pole-climb performance in rats. *Am. Ind. Hyg. Assoc. J.* **25**, 369–375 (1964).

227. I. Geller et al., Effects of acetone, methyl ethyl ketone and methyl isobutyl ketone on a match-to-sample task in the baboon. *Pharmacol. Biochem. Behav.* **11**, 401–406 (1979).

228. I. Geller et al., Effects of acetone and toluene vapors on multiple schedule performance of rats. *Pharmacol., Biochem. Behav.* **11**, 395–399 (1979).

229. C. R. Garcia, I. Geller, and H. L. Kaplan, Effects of ketones on lever-pressing behavior of rats. *Proc. West. Pharmacol. Soc.* **21**, 433–438 (1978).

230. J. R. Glowa and P. B. Dews, Behavioral toxicology of volatile organic compounds. IV. Comparison of the rate-decreasing effects of acetone, ethyl acetate, methyl ethyl ketone, toluene, and carbon disulfide on schedule-controlled behavior of mice. *J. Am. Coll. Toxicol.* **6**, 461–469 (1987).

231. G. R. Christoph and J. C. Stadler, Subchronic inhalation exposure to acetone vapor in rats and schedule-controlled operant performance. *Neurotoxicology* (1999)(submitted for publication).

232. J. De Ceaurriz et al., Quantitative evaluation of sensory irritating and neurobehavioral properties of aliphatic ketones in mice. *Food Chem. Toxicol.* **22**, 545–549 (1984).

233. L. E. Kane, B. S. Dombroske, and Y. Alarie, Evaluation of sensory irritation from some common industrial solvents. *Am. Ind. Hyg. Assoc. J.* **41**, 451–455 (1980).

234. J. C. De Ceaurriz et al., Sensory irritation caused by various industrial airborne chemicals. *Toxicol. Lett.* **9**, 137–143 (1981).

235. Y. Alarie and J. I. Luo, Sensory irritation by airborne chemicals: A basis to establish acceptable levels of exposure. In C. S. Barrow, ed., *Toxicology of the Nasal Passages*, Hemisphere Publishing, Washington, DC, 1986, pp. 91–100.

236. J. Muller and G. Greff, Recherche de relations entre toxicité de molécules d'intérêt industriel et propriétés physio-chimiques: Test d'irritation des voies aérïennes supérieures appliqué à quatre familles chimiques. *Food Chem. Toxicol.* **22**, 661–664 (1984).

237. D. W. Roberts, QSAR for upper-respiratory tract irritants. *Chem. -Biol. Interact.* **57**, 325–345 (1986).

238. M. H. Abraham et al., Hydrogen bonding 12. A new QSAR for upper respiratory tract irritation by airborne chemicals in mice. *Quant. Struct. -Act. Relat.* **9**, 6–10 (1990).

239. Y. Alarie et al., Structure-activity relationships of volatile organic chemicals as sensory irritants. *Arch. Toxicol.* **72**, 125–140 (1998).

240. Y. Alarie et al., Physicochemical properties of nonreactive volatile organic chemicals to estimate RD_{50}: Alternatives to animal studies. *Toxicol. Appl. Pharmacol.* **134**, 92–99 (1995).

241. G. D. DiVincenzo and W. J. Krasavage, Serum ornithine carbamyl transferase as a liver response test for exposure to organic solvents. *Am. Ind. Hyg. Assoc. J.* **35**, 21–29 (1975).

242. D. M. Sanderson, A note on glycerol formal as a solvent in toxicity testing. *J. Pharm. Pharmacol.* **11**, 150–156 (1959).

243. A. R. J. Dungan, Experimental observations on the acetone bodies. *J. Metab. Res.* **4**, 229–295 (1924).

244. R. Schneider and H. Droller, The relative importance of ketosis and acidosis in the production of diabetic coma. *Q. J. Exp. Physiol.* **28**, 323–333 (1938).

245. F. M. Allen and M. B. Wishart, Experimental studies in diabetes. Series V. Acidosis. 9. Administration of acetone bodies and related acids. *J. Metab. Res.* **4**, 613–648 (1923).

246. W. Saliant and N. Kleitman, Pharmacological studies on acetone. *J. Pharmacol. Exp. Ther.* **19**, 293–306 (1922).

247. B. H. Schlomovitz and E. G. Seybold, The toxicity of the 'acetone bodies.' *Am. J. Physiol.* **70**, 130–139 (1924).

248. M. M. Bagoury, The action of acetone and of the ketone bodies present in diabetic blood upon the heart. *Br. J. Exp. Pathol.* **16**, 25–33 (1935).

249. C. P. Carpenter and H. F. Smyth, Jr., Chemical burns of the rabbit cornea. *Am. J. Ophthalmol.* **29**, 1363–1372 (1946).

250. T. Märtins, J. Pauluhn, and L. Machemer, Analysis of alternate methods for determining ocular irritation. *Food Chem. Toxicol.* **30**, 1061–1067 (1992).

251. H. E. Kennah et al., An objective procedure for quantitating eye irritation based upon changes of corneal thickness. *Fundam. Appl. Toxicol.* **12**, 258–268 (1989).

252. G. A. Jacobs, OECD eye irritation tests on two ketones. *J. Am Coll. Toxicol.* **1**, 190–191 (1992).

253. H. Kojima et al., Evaluation of seven alternative assays on the main ingredients in cosmetics as predictors of draize eye irritation scores. *Toxicol. In Vitro* **9**, 333–340 (1995).

254. R. L. Morgan, S. S. Sorenson, and T. R. Castles, Prediction of ocular irritation by corneal pachymetry. *Food Chem. Toxicol.* **8**, 609–613 (1987).

255. M. Berry and D. L. Easty, Isolated human and rabbit eye: Models of corneal toxicity. *Toxicol. In Vitro* **7**, 461–464 (1993).

256. J. G. Sivak and K. L. Herbert, Optical damage and recovery of the *in vitro* bovine ocular lens for alcohols, surfactants, acetates, ketones, aromatics, and some consumer products: A review. *J. Toxicol. Cutaneons. Ocu. Toxicol.* **16**, 173–187 (1997).

257. M. York, R. S. Lawerence, and G. B Gibson, An *in vitro* test for the assessment of eye irritancy in consumer products — preliminary findings. *Int. J. Cosmet. Sci.* **4**, 223–234 (1982).

258. C. Anderson, K. Sundberg, and O. Groth, Animal model for assessment of skin irritancy, *Contact Dermatitis* **15**, 143–151 (1986).

259. H. F. Smyth et al., Range finding toxicity data: List VI. *Am. Ind. Hyg. Assoc. J.* **23**, 95–107 (1962).

260. U. Turss, R. Turss, and M. F. Refojo, Removal of isobutyl cyanoacrylate adhesive from the cornea with acetone. *Am. J. Ophthalmol.* **70**, 725–728 (1970).

261. M. D. Barratt, QSARS for the eye irritation potential of neutral organic chemicals. *Toxicol. In vitro* **11**, 1–8 (1997).

262. J. W. Guthrie and G. F. Seitz, An investigation of the chemical contact lens problem. *J. Occup. Med.* **17**, 163–166 (1975).

263. A. Bolková and J. Cejková, Changes in alkaline and acid phosphatases of the rabbit cornea following experimental exposure to ethanol and acetone: a biochemical and histochemical study. *Graefe's Arch. Clin. Exp. Ophthalmol.* **220**, 96–99 (1983).

264. F. Bartnik and K. Künstler, *Validierung von Ersatzmethoden für Tierversuche zur Prüfung auf lokale Veträglichkeit.* I. *In vitro Untersuchungen mit Zellkulturen.* BMFT Proj. No. 03 8640 9, Bundesministerium für Forschung und Technologie, Bonn, Germany, 1987.

265. A. Hoh, K. Meier, and R. M. Dreher, Multilayered keratinocyte culture used for *in vitro* toxicology. *Mol. Toxicol.* **1**, 537–546 (1987).

266. C. Shopsis and S. Sathe, Uridine uptake inhibition as a cytotoxicity test: Correlations with the draize test. *Toxicology* **29**, 195–206 (1984).

267. J. F. Sina et al., Assessment of cytotoxicity assays as predictors of ccular irritation of pharmaceuticals. *Fundam. Appl. Toxicol.* **18**, 515–521 (1992).

268. P. S. Larson, J. K. Finnegan, and H. B. Haag, Observations on the effect of chemical configuration on the edema-producing potency of acids, aldehydes, ketones and alcohols. *J. Pharmacol. Exp. Ther.* **116**, 119–122 (1956).

269. G. Grubauer et al., Lipid content and lipid type as determinants of the epidermal permeability barrier. *J. Lipid Res.* **30**, 89–96 (1989).

270. G. Grubauer, P. M. Elias, and K. R. Feingold, Transepidermal water loss: The signal for recovery of barrier structure and function. *J. Lipid Res.* **30**, 323–333 (1989).

271. N. Katoh et al., Acute cutaneous barrier perturbation induces maturation of langerhan's cells in hairless mice. *Acta Derm. Venereol.* **77**, 365–369 (1997).

272. L. C. Wood et al., Cutaneous barrier perturabation stimulates cytokine production in the epidermis of mice. *J. Clin. Invest.* **90**, 482–487 (1992).

273. L. Yang et al., Topical stratum corneum lipids accelerate barrier repair after tape stripping, solvent treatment and some but not all types of detergent treatment. *Br. J. Dermatol.* **133**, 679–685 (1995).

274. C. G. Mortz, K. E. Andersen, and L. Halkier-Sørensen, The efficacy of different moisturizers on barrier recovery in hairless mice evaluated by non-invasive bioengineering methods. *Contact Dermatitis* **36**, 297–301 (1997).

275. M. Mao-Qiang, K. R. Feingold, and P. M. Elias, Exogenous lipids influence permeability barrier recovery in acetone-treated murine skin. *Arch. Dermatol.* **129**, 728–738 (1993).

276. K. R. Feingold et al., Cholesterol synthesis is required for cutaneous barrier function in mice. *J. Clin. Invest.* **86**, 1738–1745 (1990).

277. W. M. Holleran et al., Permeability barrier requirements regulate epidermal β-glucocerebrosidase. *J. Lipid Res.* **35**, 905–912 (1994).

278. W. M. Holleran et al., Localization of epidermal sphingolipid synthesis and serine palmitoyl transferase activity: Alterations imposed by permeability barrier requirements. *Arch. Dermatol. Res.* **287**, 254–258 (1995).

279. O. H. Iversen, S. Ljunggren, and W. M. Olsen, The early effects of a single application of acetone and various doses of 7,12-dimethylbenz(α)anthracene on CD-1 and hairless mouse epidermis. *Acta Pathol. Microbiol. Immunol. Scand., Suppl.* **27**, 7–80 (1988).

280. R. K. Felton, D. B. DeNicola, and G. P. Carlson, Minimal effects of acrylonitrile on pulmonary and hepatic cell injury enzymes in rats with induced cytochrome P450. *Drug Chem. Toxicol.* **21**, 181–194 (1998).

281. J. J. Freeman and E. P. Hayes, Acetone potentiation of acute acetonitrile toxicity in rats. *J. Toxicol. Environ. Health* **15**, 609–621 (1985).

282. J. J. Freeman and E. P. Hayes, Acetone potentiation of acute acetonitrile toxicity in rats. *J. Toxicol. Environ. Health* **15**, 609–621 (1985).

283. G. L. Kedderis, R. Batra, and D. R. Koop, Epoxidation of acrylonitrile by rat and human cytochromes P450. *Chem. Res. Toxicol.* **6**, 866–871 (1993).

284. V. Postolache et al., Die Einwirkung des Acetons auf die Barbiturat-Narkose. *Arch. Toxikol.* **25**, 333–337 (1969).

285. B. D. Astill and G. D. DiVincenzo, Some factors involved in establishing and using biological threshold limit values (BTLV's). *Proc. 3rd Annu. Conf. Environ. Toxicol.* Paper No. 7, Aerospace Medical Research Laboratory, Wright-Patterson Air Force Base, OH, 1972, pp. 115–135.

286. E. Chieli et al., Possible role of the acetone-inducible cytochrome P-450IIE1 in the metabolism and hepatotoxicity of thiobenzamide. *Arch. Toxicol.* **64**, 122–127 (1990).

287. L. A. Scott, E. Maden, and M. A. Valentovic, Influence of streptozotocin (STZ)-induced diabetes, dextrose diuresis and acetone on cisplatin nephrotoxicity in Fisher 344 (F344) rats. *Toxicology* **60**, 109–125 (1990).

288. J. Delaney and J. A. Timbrell, Role of cytochrome P450 in hydrazine toxicity in isolated hepatocytes *in vitro. Xenobiotica* **25**, 1399–1410 (1995).

289. M. Valentovic, J. G. Ball, and D. Anestis, Contribution of acetone and osmotic diuresis by streptozotocin-induced diabetes in attenuation of cephaloridine nephrotoxicity. *Toxicology* **71**, 245–255 (1992)

290. E. Chieli t al., Hepatotoxicity and P-4502E1-dependent metabolic oxidation of *N,N*-dimethylformamide in rats and mice. *Arch. Toxicol.* **69**, 165–17 (1995).

291. X. Guan Cytochrome P450-dependent desaturation of lauric acid: Isoform selectivity and mechanism of formation of 11-dodecanoic acid. *Chem. -Biol. Interact.* **110**, 103–121 (1998).

292. S. Narimatsu et al., Cytochrome P450 enzymes involved in enhancement of propranolol *N*-deisopropylation after repeated administration of propranolol in rats. *Chem. -Biol. Interact.* **101**, 207–224 (1996).

293. D. M. Kontir et al., Effects of organic solvent vehicles on benzo[*a*]pyrene metabolism in rabbit lung microsomes. *Biochem. Pharmacol.* **35**, 2569–2575 (1986).

294. H. L. Gurtoo and R. P. Dahms, Effects of inducers and inhibitors on the metabolism of aflatoxin B_1 by rat and mouse. *Biochem. Pharmacol.* **28**, 3441–3449 (1979).

295. E. M. Kenyon et al., Influence of gender and acetone pretreatment on benzene metabolism in mice exposed by nose-only inhalation. *J. Toxicol. Environ. Health* **55**, 421–443 (1998).

296. N. Fedtke et al., Species differences in the biotransformation of ethyl chloride. I. Cytochrome P450-dependent metabolism. *Arch. Toxicol.* **68**, 158–166 (1994).

297. T. F. Dowsley et al., Reaction of glutathione with the electrophilic metabolites of 1,1-dichloroethylene. *Chem. -Biol. Interact.* **95**, 227–244 (1995).

298. S. Puntarulo and A. I. Cederbaum, Increased microsomal interaction with iron and oxygen radical generation after chronic acetone treatment. *Biochim. Biophys. Acta* **964**, 46–52 (1988).

299. D. K. Winters and A. I. Cederbaum, Oxidation of glycerol to formaldehyde by rat liver microsomes. Effects of cytochrome P-450 inducing agents. *Biochem. Pharmacol.* **39**, 697–705 (1990).

300. R. A. Kurata et al., Studies on inhibition and induction of metabolism of ethyl carbamate by acetone and related compounds. *Drug Metab. Dispos.* **19**, 388–393 (1991).

301. O. S. Sohn et al., Metabolism of azoxymethane, methylazoxymethanol and *N*-nitrosodimethylamine by cytochrome P450IIE1. *Carcinogenesis (London)* **12**, 127–131 (1991).

302. J. F. Brady et al., Diethyl ether as a substitute for acetone/ethanol-inducible cytochrome P-450 and as an inducer for cytochrome(s) P-450. *Mol. Pharmacol.* **33**, 148–154 (1988).

303. J. F. Brady et al., Induction of cytochromes P450IIE1 and P450IIB1 by secondary ketones and the role of P450IIE1 in chloroform metabolism. *Toxicol. Appl. Pharmacol.* **100**, 342–349 (1989).

304. J. F. Brady et al., Metabolism of methyl *tertiary*-butyl ether by rat hepatic microsomes. *Arch. Toxicol.* **64**, 157–160 (1990).

305. E. Albano et al., Role of ethanol-inducible cytochrome P450 (P450IIE1) in catalyzing the free radical activation of aliphatic alcohols. *Biochem. Pharmacol.* **41**, 1895–1902 (1991).

306. H. Kivistö et al., Effect of cytochrome P450 isozyme induction and glutathione depletion on the metabolism of CS2 to TTCA in rats. *Arch. Toxicol.* **69**, 185–190 (1995).

307. S. P. Eriksen and E. B. Brittebo, Metabolic activation of the olfactory toxicant dichlobenil in rat olfactory microsomes: Comparative studies with *p*-nitrophenol. *Chem. -Biol. Interact.* **94**, 183–196 (1995).

308. H. Clark and G. Powis, Effect of acetone administration *in vivo* upon hepatic microsomal drug metabolizing activity in the rat. *Biochem. Pharmacol.* **23**, 1015–1019 (1974).

309. C. J. Patten et al., Acetone-inducible cytochrome P-450: Purification. Catalytic activity, and Interaction with cytochrome b5. *Arch. Biochem. Biophys.* **251**, 629–638 (1986).

310. L. E. Rikans, Effects of ethanol on microsomal drug metabolism in aging female rats. I. induction. *Mech. Ageing Dev.* **48**, 267–280 (1989).

311. G. Bànhegyi et al., Accumulation of phenols and catechols in isolated hepatocytes in starvation or after pretreatment with acetone. *Biochem. Pharmacol.* **21**, 4157–4162 (1988).

312. S. Kinsler, P. E. Levi, and E. Hodgson, Relative contributions of the cytochrome P450 and flavin-containing monooxygenases to the microsomal oxidation of phorate following treatment of mice with phenobarbital, hydrocortisone, acetone, and piperonyl butoxide. *Pestic. Biochem. Physiol.* **37**, 174–181 (1990).

313. R. Puccini et al., High affinity diethylnitrosamine-deethylase in liver microsomes from acetone-induced rats. *Carcinogenesis (London)* **10**, 1629–1634 (1989).

314. R. Puccini et al., Effects of acetone administration on drug-metabolizing enzymes in mice: Presence of a high-affinity diethylnitrosamine de-ethylase. *Toxicol. Lett.* **54**, 143–150 (1990).

315. I. G. Sipes, M. L. Slocumb, and G. Holtzman, Stimulation of microsomal dimethylnitrosamine-*N*-demethylase by pretreatment of mice with acetone. *Chem. -Biol. Interact.* **21**, 155–166 (1978).

316. M. F. Argus et al., Induction of dimethyl-nitrosamine-demethylase by polar solvents. *Proc. Soc. Exp. Biol. Med.* **163**, 52–55 (1980).

317. A. D. Ross et al., Effect of propyl-thiouracil treatment on NADPH-cytochrome P450 reductase levels, oxygen consumption and hydroxyl radical formation in liver microsomes from rats fed ethanol or acetone chronically. *Biochem. Pharmacol.* **49**, 979–989 (1995).

318. Y. Y. Tu et al., Induction of a high affinity nitrosamine demethylase in rat liver microsomes by acetone and isopropanol. *Chem. -Biol. Interact.* **44**, 247–260 (1983).

319. C. Hètu and J. G. Joly, Effects of chronic acetone administration of ethanol-inducible monooxygenase activities in the rat. *Biochem. Pharmacol.* **37**, 421–426 (1988).

320. D. R. Koop, C. L. Laethem, and G. G. Schnier, Identification of ethanol-inducible P450 isozyme 3a (P450IIE1) as benzene and phenol hydroxylase. *Toxicol. Appl. Pharmacol.* **98**, 278–288 (1989).

321. I. Johansson and M. Ingelman-Sundberg, Benzene metabolism by ethanol-, acetone-, and benzene-inducible cytochrome P-450 (IIE1) in rat and rabbit liver microsomes. *Cancer Res.* **48**, 5387–5390 (1988).

322. M. C. Canivenc-Lavier et al., Comparative effects of flavanoids and model inducers in drug-metabolizing enzymes in rat liver. *Toxicology* **114**, 19–27 (1996).

323. D. Lucas et al, Ethanol-inducible cytochrome P-450: Assessment of substrates' specific chemical probes in rat liver microsomes. *Alcohol.: Clin. Exp. Res.* **14**, 590–594 (1990).

324. H. Elovaara, H. Vainio, and A. Aitio, Pulmonary toxicity of inhaled styrene in acetone-, phenobarbital- and 3-methylcholanthrene-treated rats. *Arch. Toxicol.* **64**, 365–369 (1990).

325. H. Elovaara et al., Metabolism of inhaled styrene in acetone-, phenobarbital- and 3-methylcholanthrene-pretreated rats: Stimulation and stereochemical effects by induction of cytochrome P450IIE1, P450IIB, and P450IA. *Xenobiotica* **21**, 651–661 (1991).

326. B. Mortensen, and O. G. Nilsen, Optimization and application of the head space liver S9 equilibration technique for metabolic studies of organic solvents. *Pharmacol. Toxicol.* **82**, 142–146 (1998).

327. H. Vainio and A. Zitting, Interaction of styrene and acetone with drug biotransformation enzymes in rat liver. *Scand. J. Work Environ. Health* **4**(Suppl. 2), 47–52 (1978).

328. N. A. Lorr et al., Potentiation of the hepatotoxicity of *N*-nitrosodimethylamine by fasting, diabetes, acetone, and isopropanol. *Toxicol. Appl. Pharmacol.* **73**, 423–431 (1984).

329. S. M. Haag and I. G. Sipes, Differential effects of acetone or aroclor 1254 pretreatment in the microsomal activation of dimethylnitrosamine to a mutagen. *Mutat. Res.* **74**, 431–438 (1980).

330. Z. Quan, S. Khan, and P. J. O'Brien, Role of cytochrome P-450IIE1 in *N*-nitroso-*N*-methylamine induced hepatocyte cytotoxicity. *Chem. -Biol. Interact.* **83**, 221–233 (1992).

331. J. J Larsen, M. Lykkegaard, and O. Ladefoged, Infertility in rats induced by 2,5-hexanedione in combination with acetone. *Pharmacol. Toxicol.* **69**, 43–46 (1991).

332. O. Ladefoged, H. Hass, and L. Simonsen, Neurophysiological and behavioral effects of combined exposure to 2,5-hexanedione and acetone or ethanol in rats. *Pharmacol. Toxicol.* **65**, 372–375 (1989).

333. O. Ladefoged, K. Roswall, and J. J. Larsen, Acetone potentiation and influence on the reversibility of 2,5-hexanedione-induced neurotoxicity studied with behavioral and morphometric methods in rats. *Pharmacol. Toxicol.* **74**, 294–299 (1994).

334. P. Strange et al., Total number and mean cell volume of neocortical neurons in rats exposed to 2,5-hexanedione with and without acetone. *Neurotoxicol. Teratol.* **13**, 401–406 (1991).

335. H. R. Lam et al., Effects of 2,5-hexanedione alone and in combination with acetone on radial arm maze behavior, the brain-swelling reaction and synaptosomal functions. *Neurotoxicol. Teratol.* **13**, 407–412 (1991).

336. W. Zhao et al., Effects of methyl ethyl ketone, acetone, or toluene coadministration on 2,5-hexanedione concentration in the sciatic nerve, serum, and urine or rats. *Int. Arch. Occup. Environ. Health* **71**, 236–244 (1998).

337. K. Aoki et al., Changes in 2,5-hexanedione concentration in the sciatic nerve, serum and urine of rats induced by combined administration of 2,5-hexanedione with acetone or methyl ethyl ketone. *J. Occup. Health* **38**, 30–35 (1996).

338. O. Ladefoged and L. Perbellini, Acetone-induced changes in the toxicokinetics of 2,5-hexanedione in rabbits. *Scand. J. Work Environ. Health* **12**, 627–629 (1986).

339. H. H. Lo et al., Acetone effects on *N*-(3,5-dichlorophenyl) succinimide-induced nephrotoxicity. *Toxicol. Lett.* **38**, 161–168 (1987).
340. E. H. Jenney and C. C. Pfeiffer, The convulsant effect of hydrazides and the antidotal effect of anticonvulsants and metabolites. *J. Pharmacol. Exp. Ther.* **122**, 110–123 (1958).
341. R. P. Kohli et al., Anticonvulsant activity of some carbonyl containing compounds. *Ind. J. Med. Res.* **55**, 1221–1225 (1967).
342. V. F. Price and D. J. Jollow, Mechanism of ketone-induced protection from acetaminophen hepatotoxicity in the rat. *Drug Metab. Dispos.* **11**, 451–457 (1983).
343. J. Llamas et al., Increase in the renal damage induced by paracetamol in rats exposed to ethanol translactationally. *Biol. Neonate.* **74**, 385–392 (1998).
344. P. Moldéus and V. Gergely, Effect of acetone on the activation of acetaminophen. *Toxicol. Appl. Pharmacol.* **53**, 8–13 (1980).
345. G. J. Traiger and G. L. Plaa, Differences in the potentiation of carbon tetrachloride in rats by ethanol and isopropanol pretreatment. *Toxicol. Appl. Pharmacol.* **20**, 105–112 (1971).
346. G. J. Traiger and G. L. Plaa, Chlorinated hydrocarbon toxicity. Potentiation by isopropyl alcohol and acetone. *Arch. Environ. Health* **28**, 276–278 (1974).
347. W. R. Hewitt et al., Dose-response relationships in 1,3-butanediol-induced potentiation of carbon tetrachloride toxicity. *Toxicol. Appl. Pharmacol.* **64**, 529–540 (1982).
348. T. H. Ueng et al., Isopropanol enhancement of cytochrome p-450-dependent monooxygenase activities and its effects on carbon tetrachloride intoxication. *Toxicol. Appl. Pharmacol.* **71**, 204–214 (1983).
349. I. G. Sipes et al., Enhanced hepatic microsomal activity by pretreatment of rats with acetone or isopropanol. *Proc. Soc. Exp. Biol. Med.* **142**, 237–240 (1973).
350. H. M. Mailing et al., Enhanced hepatotoxicity of carbon tetrachloride, thioacetamide, and dimethylnitrosamine by pretreatment of rats with ethanol and some comparisons with potentiation by isopropanol. *Toxicol. Appl. Pharmacol.* **33**, 291–308 (1975).
351. A. B. Kobush, G. L. Plaa, and P. du Souich, Effects of acetone and methyl *n*-butyl ketone on hepatic mixed-function oxidase. *Biochem. Pharmacol.* **38**, 3461–3467 (1989).
352. A. Sato and T. Nakajima, Enhanced metabolism of volatile hydrocarbons in rat liver following food deprivation, restricted carbohydrate intake, and administration of ethanol, phenobarbital, polychlorinated biphenyl and 3-methylcholanthrene: a comparative study. *Xenobiotica* **15**, 67–75 (1985).
353. I. Johansson and M. Ingelman-Sundberg, Carbon tetrachloride-induced lipid peroxidation dependent on an ethanol-inducible form of rabbit liver microsomal cytochrome P450. *FEBS Lett.* **183**, 265–269 (1985).
354. G. J. Traiger and G. L. Plaa, Effect of aminotriazole on isopropanol- and acetone-induced potentiation of CCl_4 hepatotoxicity. *Can. J. Physiol. Pharmacol.* **51**, 291–296 (1973).
355. W. R. Hewitt et al., Dose-response relationships in 1,3-butanediol-induced potentiation of carbon tetrachloride toxicity. *Toxicol. Appl. Pharmacol.* **64**, 529–540 (1982).
356. W. R. Hewitt, E. M. Brown, and G. L. Plaa, Relationship between the carbon skeleton length of ketonic solvents and potentiation of chloroform-induced hepatotoxicity in rats. *Toxicol. Lett.* **16**, 297–304 (1983).
357. E. M. Brown and W. R. Hewitt, Dose-response relationships in ketone-induced potentiation of chloroform hepato- and nephrotoxicity. *Toxicol. Appl. Pharmacol.* **76**, 437–453 (1984).

358. W. R. Hewitt et al., Acute alteration of chloroform-induced hepato- and nephrotoxicity by *n*-hexane, methyl *n*-butyl ketone, and 2,5-hexanedione. *Toxicol. Appl. Pharmacol.* **53**, 230–248 (1980).

359. P. Raymond, and G. L. Plaa, Ketone potentiation of haloalkane-induced hepato- and nephrotoxicity: I. Dose-response relationships. *J. Toxicol. Environ. Health* **45**, 465–480 (1995).

360. P. Raymond, and G. L. Plaa, Ketone potentiation of haloalkane-induced hepato- and nephrotoxicity: II. Implications of monooxygenases. *J. Toxicol. Environ. Health* **46**, 317–328 (1995).

361. L. A. Hewitt, P. Ayotte, and G. L. Plaa, Modifications in rat hepatobiliary function following treatment with acetone, 2-butanone, 2-hexanone, mirex, or chlordecone and subsequently exposed to chloroform. *Toxicol. Appl. Pharmacol.* **83**, 465–473 (1986).

362. L. A. Hewitt, C. Valiquette, and G. L. Plaa, The role of biotransformation-detoxification in acetone-, 2-butanone-, and 2-hexanone-potentiated chloroform-induced hepatotoxicity. *Can. J. Physiol. Pharmacol.* **65**, 2313–2318 (1987).

363. G. L. Plaa et al., Isopropanol and acetone potentiation of carbon tetrachloride-induced hepatotoxicity: Single versus repetitive pretreatment in rats. *J. Toxicol. Environ. Health* **9**, 235–250 (1982).

364. M. Charbonneau et al., Temporal analysis of rat liver injury following potentiation of carbon tetrachloride hepatotoxicity with ketonic or ketogenic compounds. *Toxicology* **35**, 95–112 (1985).

365. M. Charbonneau, B. Tuchweber, and G. L. Plaa, Acetone potentiation of chronic liver injury induced by repetitive administration of carbon tetrachloride. *Hepatology.* **6**, 694–700 (1986).

366. M. Charbonneau et al., Correlation between acetone-potentiated CCl_4-induced liver injury and blood concentrations after inhalation or oral administration. *Toxicol. Appl. Pharmacol.* **84**, 286–294 (1986).

367. W. R. Hewitt and G. L. Plaa, Dose-dependent modification of 1,1-dichloroethylene toxicity by acetone. *Toxicol. Lett.* **16**, 145–152 (1983).

368. J. R. MacDonald, A. J. Gandolfi, and I. G. Sipes, Acetone potentiation of 1,1,2-trichloroethane hepatotoxicity. *Toxicol. Lett.* **13**, 57–69 (1982).

369. M. T. Brondeau et al., Acetone compared to other ketones in modifying the hepatotoxicity of inhaled 1,2-dichlorobenzene in rats and mice. *Toxicol. Lett.* **49**, 69–78 (1989).

370. M. Charbonneau et al., Acetone potentiation of rat liver injury induced by trichloroethylene-carbon tetrachloride mixtures. *Fundam. Appl. Toxicol.* **6**, 654–661 (1986).

371. M. Charbonneau et al., Assessment of the minimal effective dose of acetone for potentiation of the hepatotoxicity induced by trichloroethylene-carbon tetrachloride mixtures. *Fundam. Appl. Toxicol.* **10**, 431–438 (1988).

372. M. Charbonneau et al., Influence of acetone on the severity of the liver injury induced by haloalkane mixtures. *Can. J. Physiol. Pharmacol.* **69**, 1901–1907 (1991).

373. M. Naruse, Effects on mice of long-term exposure to organic solvents in adhesives. *Nagoya Med. J.* **28**, 183–120 (1984).

374. J. E. Simmons et al., Toxicology studies of a chemical mixture of 25 groundwater contaminants: Hepatic and renal assessment, response to carbon tetrachloride challenge, and influence of treatment-induced water restriction. *J. Toxicol. Environ. Health* **43**, 305–325 (1994).

375. T. Sollmann, Studies of chronic intoxications on albino rats. II. Alcohols (ethyl, methyl and wood) and acetone. *J. Pharmacol. Exp. Ther.* **16**, 291–309 (1921).

376. L. Skutches, O. E. Owen, and G. A. Reichard, Acetone and acetol inhibition of insulin-stimulated glucose oxidation in adipose tissue and isolated adipocytes. *Diabetes* **39**, 450–455 (1990).

377. P. R. Sinclair et al., Uroporphyrin caused by acetone and 5-aminolevulinic acid in iron-loaded mice. *Biochem. Pharmacol.* **23**, 4341–4344 (1989).

378. J. V. Bruckner and R. G. Petersen, Evaluation of toluene and acetone inhalant abuse. II. Model development and toxicology. *Toxicol. Appl. Pharmacol.* **61**, 302–312 (1981).

379. B. Sonawane et al., Estimation of reference dose (RfD) for oral exposure to acetone. *J. Am. Coll. Toxicol.* **5**, 605(abstr.) (1986).

380. D. A. Mayhew and L. D. Morrow, *Ninety Day Gavage Study in Albino Rats Using Acetone*, EPA Contract No. 68-01-7075, U. S. Environmental Protection Agency, Washington, DC, 1988.

381. G. N. Rao, J. Edmondson, and M. R. Elwell, Influence of dietary protein concentration on severity of nephropathy in fischer-344 (F-344/N) rats. *Toxicol. Pathol.* **21**, 353–361 (1993).

382. D. D. Dietz et al., Toxicity studies of acetone administered in the drinking water of rodents. *Fundam. Appl. Toxicol.* **17**, 347–360 (1991).

383. D. Dietz, *Toxicity Studies of Acetone in F344/N Rats and B6C3F$_1$ Mice (Drinking Water Studies)*, NTP TOX 3, NIH 91-3122, U.S.Department of Health and Human Services, National Toxicology Program, Research Triangle Park, NC, (1991), pp. 1–38.

384. R. H. Rengstorff, J. P. Petrali, and V. M. Sim, Cataracts induced in guinea pigs by acetone, cyclohexanone, and dimethyl sulfoxide. *Am. J. Optom.* **49**, 308–319 (1972).

385. R. H. Rengstorff, J. P. Petrali, and V. M. Sim, Attempt to induce cataracts in rabbits by cutaneous application of acetone. *Am. J. Opt. Physiol. Opt.* **53**, 41–42 (1975).

386. R. H. Rengstorff and H. I. Khafagy, Cutaneous acetone depresses aqueous humor ascorbate in guinea pigs. *Arch. Toxicol.* **58**, 64–66 (1985).

387. A Taylor et al., Relationship between acetone, cataracts, and ascorbate in hairless guinea pigs. *Ophthalmol. Res.* **25**, 30–35 (1993).

388. I. Johansson et al., Hydroxylation of acetone by ethanol- and acetone-inducible cytochrome P-450 in liver microsomes and reconstituted membranes. *FEBS Lett.* **196**, 59–64 (1986).

389. D. R. Koop and J. P. Casazza, Identification of ethanol-inducible P-450 isozyme 3a as the acetone and acetol monooxygenase of rabbit microsomes. *J. Biol. Chem.* **260**, 13607–13612 (1985).

390. F. N. LeBaron, Lipid metabolism. I. Utilization and storage of energy in lipid form. In T. M. Devlin, ed., *Textbook of Biochemistry with Clinical Correlations*, Wiley, New York, 1982, pp. 467–479.

391. D. E. Vance, Metabolism of fatty acids and triacylglycerols. In G. Zubay, ed., *Biochemistry*, Addison-Wesley, Reading, MA, 1984, pp. 471–503.

392. O. Wieland, Ketogenesis and its regulation. *Adv. Metab. Disord.* **3**, 1–47 (1967).

393. V. C. Gavino et al., Production of acetone and conversion of acetone to acetate in the perfused rat liver. *J. Biol. Chem.* **262**, 6735–6740 (1987).

394. G. J. Van Stekelenburg and G. Koorevaar, Evidence for the existence of mammalian acetoacetate decarboxylase: With special reference to human blood serum. *Clin. Chim. Acta.* **39**, 191–199 (1972).

395. G. Koorevaar and G. J. Van Stekelenburg, Mammalian acetoacetate decarboxylase activity. Its distribution in subfractions of human albumin and occurrence in various tissues of the rat. *Clin. Chim. Acta.* **71**, 173–183 (1976).

396. F. J. Lopez-Soriano and J. M. Argilés, Rat acetoacetate decarboxylase: Its role in the disposal of 4C-ketone bodies by the fetus. *Horm. Metab. Res.* **18**, 446–449 (1986).

397. F. J. Lopez-Soriano, M. Alemany, and J. M. Argilés, Rat acetoacetic acid decarboxylase inhibition by acetone. *Int. J. Biochem.* **17**, 1271–1273 (1985).

398. G. Hetenyi and C. Ferrarotto, Gluconeogenesis from acetone in starved rat. *Biochem. J.* **231**, 151–155 (1985).

399. J. G. Filser and H. M. Bolt, Inhalation pharmacokinetics based on gas uptake studies. IV. The endogenous production of volatile compounds. *Arch. Toxicol.* **52**, 123–133 (1983).

400. W. Sakami and J. M. Lafaye, The metabolism of acetone in the intact rat. *J. Biol. Chem.* **193**, 199–203 (1951).

401. W. Sakami and H. Rudney, The metabolism of acetone and acetoacetate in the mammalian organism. In *The Major Metabolic Fuels*, Brookhaven Symp. Biol, Brookhaven National Laboratory, Upton, NY, 1952, pp. 176–190.

402. H. Rudney, Propanediol phosphate as a possible intermediate in the metabolism of acetone. *J. Biol. Chem.* **210**, 361–371 (1954).

403. G. A. Mourkides, D. C. Hobbs, and R. E. Koeppe, The metabolism of acetone-2-C_{14} by intact rats. *J. Biol. Chem.* **234**, 27–30 (1959).

404. J. P. Casazza, M. E. Felver, and R. L. Veech, The metabolism of acetone in rat. *J. Biol. Chem.* **259**, 231–236 (1984).

405. J. Perinado, F. J. Lopez-Soriano, and J. M. Argilés, The metabolism of acetone in the pregnant rat. *Biosci. Rep.* **6**, 983–989 (1986).

406. K. Kosugi et al., Pathways of acetone's metabolism in the rat. *J. Biol. Chem.* **261**, 3952–3957 (1986).

407. K. Kosugi et al., Determinants in the pathways followed by the carbons of acetone in their conversion to glucose. *J. Biol. Chem.* **261**, 13179–13181 (1986).

408. A. L. Black et al., Glucogenic pathway for acetone metabolism in the lactating cow. *Am. J. Physiol.* **222**, 1575–1580 (1972).

409. D. L. Coleman, Acetone metabolism in mice: Increased activity in mice heterozygous for obesity genes. *Proc. Natl. Acad. Sci. U.S.A.* **77**, 290–293 (1980).

410. D. B. Lindsay and R. E. Brown, Acetone Metabolism in Sheep. *Biochem. J.* **100**, 589–592 (1966).

411. J. R. Luick et al., Acetone metabolism in normal and ketotic cows. *J. Dairy Sci.* **50**, 544–549 (1967).

412. J. M. Argilés, Has acetone a role in the conversion of fat to carbohydrate in mammals? *Trends Biochem. Sci.* **11**, 61–63 (1986).

413. I. Johansson et al., Ethanol-, fasting-, and acetone-inducible cytochromes P-450 in rat liver: Regulation and characteristics of enzymes belonging to the IIB and IIE gene subfamilies. *Biochemistry* **27**, 1925–1934 (1988).

414. M. J. J. Ronis et al., Acetone-regulated synthesis and degradation of cytochrome P4502E2 and cytochrome P4502B1 in rat liver. *Eur. J. Biochem.* **198**, 383–389 (1991).

415. M. J. Ronis and M. Ingelman-Sundberg, Acetone-dependent regulation of cytochrome P-450J IIE1 and P-450B IIB1 in rat liver. *Xenobiotica* **19**, 1161–1166 (1989).

416. B.-J. Song et al., Induction of rat hepatic *N*-nitrosodimethylamine demethylase by acetone is due to protein stabilization. *J. Biol. Chem.* **264**, 3568–3572 (1989).

417. C. M. Hunt et al., Regulation of rat hepatic cytochrome P450IIE1 in primary monolayer hepatocyte culture. *Xenobiotica* **21**, 1621–1631 (1991).

418. E. T. Morgan, D. R. Koop, and J. M. Coon, Catalytic activity of cytochrome P-450 isozyme 3a isolated from liver microsomes of ethanol-treated rabbits. Oxidation of alcohols. *J. Biol. Chem.* **257**, 13951–13957 (1982).

419. S. Imaoka and Y. Funae, Induction of cytochrome P450 isozymes in rat liver by methyl *n*-alkyl ketones and *n*-alkylbenzenes. *Biochem. Pharmacol.* **42**, S143–S150 (1991).

420. D. E. Ryan et al., Characterization of a major form of rat hepatic microsomal cytochrome P-450 induced by isoniazid. *J. Biol. Chem.* **260**, 6385–6393 (1985).

421. N. Tindberg and M. Ingelman-Sundberg, Cytochrome P-450 and oxygen toxicity. Oxygen-dependent induction of ethanol-inducible cytochrome P-450 (IIE1) in rat liver and lung. *Biochemistry* **28**, 4499–4504 (1989).

422. D. R. Koop et al., Immunochemical evidence for induction of the alcohol-oxidizing cytochrome P-450 of rabbit liver microsomes by diverse agents: Ethanol, imidazole, trichloroethylene, acetone, pyrazole, and isoniazid. *Proc. Natl. Acad. Sci. U.S.A.* **82**, 4065–4069 (1985).

423. J.-S. H. Yoo et al., Regulation of hepatic microsomal cytochrome P450IIE1 level by dietary lipids and carbohydrates in rats. *J. Nutr.* **121**, 959–965 (1991).

424. M. R. Past and D. E. Cook, Effect of diabetes on rat liver cytochrome P-450. Evidence for a unique diabetes-dependent rat liver cytochrome P-450. *Biochem. Pharmacol.* **31**, 3329–3334 (1982).

425. J.-Y. Hong et al., The induction of a specific form of cytochrome P-450 (P450j) by fasting. *Biochem. Biophys. Res. Commun.* **142**, 1077–1083 (1987).

426. C. S. Lieber et al., Role of acetone, dietary fat and total energy intake in induction of hepatic microsomal ethanol oxidizing system. *J. Pharmacol. Exp. Ther.* **247**, 791–795 (1988).

427. J.-Y. Hong et al., Regulation of *N*-nitrosodimethylamine demethylase in rat liver and kidney. *Cancer Res.* **47**, 5948–5953 (1987).

428. Z. Dong et al., Mechanism of induction of cytochrome P-450ac (P-450j) in chemically-induced and spontaneous diabetic rats. *Arch. Biochem. Biophys.* **263**, 29–35 (1988).

429. T. D. Porter, S. C. Khani, and M. J. Coon, Induction and tissue-specific expression of rabbit cytochrome P450IIE1 and IIE2 genes. *Mol. Pharmacol.* **36**, 61–65 (1989).

430. S. Menicagli et al., Effect of acetone administration on renal, pulmonary and hepatic monooxygenase activities in hamster. *Toxicology* **64**, 141–153 (1990).

431. T.-H. Ueng et al., Effects of acetone administration on cytochrome P-450-dependent monooxygenases in hamster liver, kidney, and lung. *Arch. Toxicol.* **65**, 45–51 (1991).

432. F. Sinclair et al., Isolation of four forms of acetone-induced cytochrome P-450 in chicken liver by HPLC and their enzymic characterization. *Biochem. J.* **269**, 85–91 (1990).

433. G. Forkert et al., Distribution of cytochrome CYP2E1 in murine liver after ethanol and acetone administration. *Carcinogenesis (London)* **12**, 2259–2268 (1991).

434. X. Ding and M. J. Coon, Induction of cytochrome P-450 isozyme 3a (P-450IIE1) in rabbit olfactory mucosa by ethanol and acetone. *Drug Metab. Dispos.* **18**, 742–745 (1990).

435. G. Schnier, C. L. Laethem, and D. R. Koop, Identification and induction of cytochromes P450, P450IIE1 and P450IA1 in rabbit bone marrow. *J. Pharmacol. Exp. Ther.* **251**, 790–796 (1989).

436. R. Koop, A. Chernosky, and E. P. Brass, Identification and induction of cytochrome P450 2E1 in rat kupffer cells. *J. Pharmacol. Exp. Ther.* **258**, 1072–1076 (1991).

437. R. Robinson et al., Human liver cytochrome P-450 related to a rat acetone-inducible, nitrosamine-metabolizing cytochrome P-450: Identification and isolation. *Pharmacology* **39**, 137–144 (1989).

438. J. L. Raucy et al., Identification of a human liver cytochrome P-450 exhibiting catalytic and immunochemical similarities to cytochrome P-450 3a of rabbit liver. *Biochem. Pharmacol.* **36**, 2921–2926 (1987).

439. J. -Y. Hong et al., Roles of pituitary hormones in the regulation of hepatic cytochrome P450IIE1 in rats and mice. *Arch. Biochem. Biophys.* **281**, 132–138 (1990).

440. K. W. Miller and C. S. Yang, Studies on the mechanisms of induction of *N*-nitrosodimethylamine demethylase by fasting, acetone, and ethanol. *Arch. Biochem. Biophys.* **229**, 483–491 (1984).

441. H. Sippel, K. E. Penttilä, and K. O. Lindros, Regioselective induction of liver glutathione transferase by ethanol and acetone. *Pharmacol. Toxicol.* **68**, 391–393 (1991).

442. L. Braun et al., Molecular basis of bilirubin UDP-glucuronosyltransferase induction in spontaneously diabetic rats, acetone-treated rats and starved rats. *Biochem. J.* **336**, 587–592 (1998).

443. M. W. Anders and J. E. Gander, Acetone enhancement of cumene hydroperoxide supported microsomal aniline hydroxylation. *Life Sci.* **25**, 1085–1090 (1979).

444. M. Kitada et al., Effect of acetone on aniline hydroxylation by a reconstituted system. *Biochem. Pharmacol.* **21**, 3151–3155 (1983).

445. R. Bidlack and G. L. Lowery, Multiple drug metabolism: *p*-nitroanisole reversal of acetone enhanced aniline hydroxylation. *Biochem. Pharmacol.* **31**, 311–317 (1982).

446. G. Powis and A. R. Boobis, The effect of pretreating rats with 3-methylcholanthrene upon the enhancement of microsomal aniline hydroxylation by acetone and other agents. *Biochem. Pharmacol.* **24**, 424–426 (1975).

447. M. W. Anders, Acetone enhancement of microsomal aniline parahydroxylase activity. *Arch. Biochem. Biophys.* **126**, 269–275 (1968).

448. H. Vainio and O. Hänninen, Enhancement of aniline *p*-hydroxylation by acetone in rat liver microsomes. *Xenobiotica* **2**, 259–267 (1972).

449. W. Anders, Effect of phenobarbital and 3-methylcholanthrene administration on the *in vitro* enhancement of microsomal aromatic hydroxylation. *Arch. Biochem. Biophys.* **153**, 502–507 (1972).

450. C. A. Lee et al., Activation of acetaminophen-reactive metabolite formation by methylxanthines and known cytochrome P-450 activators. *Drug Metab. Dispos.* **19**, 966–971 (1991).

451. M. Kitada, T. Kamataki, and H. Kitagawa, NADH-synergism of NADPH-dependent *O*-dealkylation of type II compounds, *p*-anisidine and *p*-phenetidine, in rat liver. *Arch. Biochem. Biophys.* **178**, 151–157 (1977).

452. Y. Kikuta et al., Purification and characterization of hepatic microsomal prostaglandin ω-hydroxylase cytochrome P-450 from pregnant rabbits. *J. Biochem. (Tokyo)* **106**, 468–473 (1989).

453. A. J. Draper, A. Madan, and A. Parkinson, (1997). Inhibition of coumarin 7-hydroxylase activity in human liver microsomes. *Arch. Biochem. Biophys.* **341**, 47–61.

454. D. M. Kontir et al., Effects of organic solvent vehicles on benzo[*a*]pyrene metabolism in rabbit lung microsomes. *Biochem. Pharmacol.* **35**, 2569–2575 (1986).

455. W. E. Luttrell and M. C. Castle, Enhancement of hepatic microsomal esterase activity following soman pretreatment in guinea pigs. *Biochem. Pharmacol.* **46**, 2083–2092 (1993).

456. J. L. Egle, Jr., Retention of inhaled acetone and ammonia in the dog. *Am. Ind. Hyg. Assoc. J.* **34**, 533–539 (1973).

457. E. F. Aharonson et al., Effect of respiratory airflow rate on removal of soluble vapors by the nose. *J. Appl. Physiol.* **37**, 654–657 (1974).

458. J. B. Morris, R. J. Clay, and D. G. Cavanagh, Species differences in upper respiratory tract deposition of acetone and ethanol vapors. *Fundam. Appl. Toxicol.* **7**, 671–680 (1986).

459. J. B. Morris and D. G. Cavanagh, Deposition of ethanol and acetone vapors in the upper respiratory tract of the rat. *Fundam. Appl. Toxicol.* **6**, 78–88 (1986).

460. J. B. Morris and D. G. Cavanagh, Metabolism and deposition of propanol and acetone vapors in the upper respiratory tract of the hamster. *Fundam. Appl. Toxicol.* **9**, 34–40 (1987).

461. J. B. Morris, Deposition of acetone in the upper respiratory tract of the B6C3F1 mouse. *Toxicol. Lett.* **56**, 187–196 (1991).

462. E. M. P. Widmark, Studies in the concentration of indifferent narcotics in blood and tissues. *Acta Med. Scand.* **52**, 87–164 (1919).

463. G. E. DiVincenzo, F. J. Yanno, and B. D. Astill, Exposure of man and dog to low concentrations of acetone vapor. *Am. Ind. Hyg. Assoc. J.* **34**, 329–336 (1973).

464. T. J. Mast et al., *Inhalation Developmental Toxicology Studies: Teratology Study of Acetone in Mice and Rats*, Contract DE-AC06–76RLO 1830, National Institute of Environmental Health Sciences, National Toxicology Program, Pacific Northwest Laboratory, Battelle Memorial Institute, Columbus, OH, 1988.

465. E. Hallier, J. G. Filser, and H. M. Bolt, Inhalation pharmacokinetics based on gas uptake studies. II. Pharmacokinetics of acetone in rats. *Arch. Toxicol.* **47**, 293–304 (1981).

466. E. Wigaeus, A. Löf, and M. B. Nordqvist, Distribution and elimination of 2-[$_{14}$C]- acetone in mice after inhalation exposure. *Scand. J. Work Environ. Health* **8**, 121–128 (1982).

467. G. R. Christoph, D. A. Keller, and J. C. Stadler, Subchronic inhalation of acetone vapor: Schedule-controlled operant behavior and time–course of blood acetone concentration in rats. *Toxicologist* **17**, 63(abstr.) (1997).

468. T. D. Price and D. Rittenberg, The metabolism of acetone. I. Gross aspects of catabolism and excretion. *J. Biol. Chem.* **185**, 449–459 (1950).

469. I. Zabin and K. Bloch, The utilization of isovaleric acid for the synthesis of cholesterol. *J. Biol. Chem.* **185**, 131–138 (1950).

470. W. Sakami, Formation of formate and labile methyl groups from acetone in the intact rat. *J. Biol. Chem.* **187**, 369–377 (1950).

471. E. N. Bergman, A. F. Sellers, and F. A. Spurrell, Metabolism of C_{14}-labeled acetone, acetate and palmitate in fasted pregnant and nonpregnant guinea pigs. *Am. J. Physiol.* **198**, 1087–1093 (1960).

472. S. Ameenuddin and M. L. Sunde, Sensitivity of chick embryo to various solvents used in egg injection studies. *Proc. Soc. Exp. Biol. Med.* **175**, 176–178 (1984).

473. K. T. Kitchin and M. T. Ebron, Combined use of a water-insoluble chemical delivery system and a metabolic activation system in whole embryo culture. *J. Toxicol. Environ. Health* **13**, 499–509 (1984).

474. J. R. Rayburn et al., Altered developmental toxicity caused by three carrier solvents. *J. Appl. Toxicol.* **11**, 253–260 (1991).

475. D. Marchal-Segault and F. Ramade, The effects of lindane, an insecticide, on hatching and postembryonic development of *Xenopus laevis* (Daudin) anurian amphibian. *Environ. Res.* **24**, 250–258 (1981).

476. D. J. Hoffman and W. C. Eastin, Effects of industrial effluents, heavy metals, and organic solvents on mallard embryo development. *Toxicol. Lett.* **9**, 35–40 (1981).

477. E. M. Johnson et al., An analysis of the hydra assay's applicability and reliability as a developmental toxicity prescreen. *J. Am. Coll. Toxicol.* **7**, 111–126 (1988).

478. M. Pereira, P. Barnwell, and W. Bailes, *Screening of Priority Chemicals for Reproductive Hazards: Benzethonium Chloride, 3-Ethoxy-1-propanol, and Acetone*, Proj. No. ETOX-85-1002, Environmental Health Research and Testing, 1987.

479. K. T. Kitchin and M. T. Ebron, Further development of rodent whole embryo culture: Solvent toxicity and water insoluble compound delivery system. *Toxicology* **30**, 45–57 (1984).

480. J. J. Picard and G. Van Maele-Fabry, A new method to express embryotoxic data obtained *in vitro* on whole murine embryos. *Teratology* **35**, 429–437 (1987).

481. B. P. Schmid, Xenobiotic influences on embryonic differentiation, growth and morphology in vitro. *Xenobiotica* **15**, 719–726 (1985).

482. E. J. Guntakatta, E. J. Matthews, and J. O. Rundell, Development of a mouse embryo limb bud cell culture system for the estimation of chemical teratogenic potential. *Teratog. Carcinog. Mutagen.* **4**, 349–364 (1984).

483. J. R. Strange, P. M. Allred, and W. E. Kerr, Teratogenic and toxicological effects of 2,4,5-trichlorophenoxyacetic acid in developing chick embryos. *Bull. Environ. Contam. Toxicol.* **15**, 682–688 (1976).

484. P. M. Allred and J. R. Strange, The effects of 2,4,5-trichlorophenoxyacetic acid and 2,3,7,8-tetrachlorodibenzo-*p*-dioxin on developing chicken embryos. *Arch. Environ. Contam. Toxicol.* **6**, 483–489 (1977).

485. J. McLaughlin et al., The injection of chemicals into the yolk sac of fertile eggs prior to incubation as a toxicity test. *Toxicol. Appl. Pharmacol.* **5**, 760–771 (1963).

486. J. McLaughlin et al., Toxicity of fourteen volatile chemicals as measured by the chick embryo method. *Am. Ind. Hyg. J.* **25**, 282–284 (1964).

487. N. E. Walker, Distribution of chemicals injected into fertile eggs and its effect upon apparent toxicity. *Toxicol. Appl. Pharmacol.* **10**, 290–299 (1967).

488. A. Korhonen, K. Hemminki, and H. Vainio, Embryotoxicity of industrial chemicals on the chicken embryo: Thiourea derivatives. *Acta Pharmacol. Toxicol.* **51**, 38–44 (1982).

489. A. Korhonen, K. Hemminki, and H. Vainio, Embryotoxic effects of acrolein, methacrylates, guanidines and resorcinol on three day chicken embryos. *Acta Pharmacol. Toxicol.* **52**, 95–99 (1983).

490. W. J. Swartz, Long- and short-term effects of carbaryl exposure in chick embryos. *Environ. Res.* **26**, 463–471 (1981).

491. J. A. DiPaolo, P. Donovan, and R. Nelson, Quantitative studies of *in vitro* transformation by chemical carcinogens. *J. Natl. Cancer Inst.* **42**, 867–876 (1969).

492. J. M. Quarles et al., Transformation of hamster fetal cells by nitrosated pesticides in a transplantation assay. *Cancer Res.* **39**, 4525–4533 (1979).

493. D. Badeaux et al., Detection of toxins to human female reproduction using human luteinized-granulosa cells in culture. *Primary Care Update Obstet. Gynaecol.* **4**, 267–271 (1997).

494. H. E. Holden et al., Hemizygous Tg.AC transgenic mouse as a potential alternative to the two-year mouse carcinogenicity bioassay: Evaluation of husbandry and housing factors. *J. Appl. Toxicol.* **18**, 19–24 (1998).

495. B. L. Van Duuren et al., Cigarette smoke carcinogenesis: Importance of tumor promoters. *J. Natl. Cancer Inst.* **47**, 235–240 (1971).

496. H. S. Weiss et al., Inhibitory effect of toluene on tumor promotion in mouse skin. *Proc. Soc. Exp. Biol. Med.* **181**, 199–204 (1986).

497. H. S. Weiss, K. S. Pacer, and C. A. Rhodes, Tumor inhibiting effect of pre-promotion with acetone. *Fed. Am. Soc. Exp. Biol. J.* **2**, A1153(abstr.) (1988).

498. B. L. Van Duuren et al., Mouse skin carcinogenicity tests of the flame retardants tris(2,3-dibromopropyl)phosphate, tetrakis(hydroxymethyl)phosphonium chloride, and polyvinyl bromide. *Cancer Res.* **38**, 3236–3240 (1978).

499. H.-Y. Park and I. Koprowska, A comparative *in vitro* and *in vivo* study of induced cervical lesions of mice. *Cancer Res.* **28**, 1478–1489 (1968).

500. L. Barr-Nea and M. Wolman, Tumors and amyloidosis in mice painted with crude oil found on bathing beaches. *Bull. Environ. Contam. Toxicol.* **18**, 385–391 (1977).

501. L. Barr-Nea and M. Wolman, The effects of cutaneous applications of beach-polluting oil on mice. *Isr J. Med. Sci.* **8**, 1745–1749 (1972).

502. O. Odukoya and G. Shklar, Initiation and promotion in experimental oral carcinogenesis. *Oral Surg.* **58**, 315–320 (1984).

503. O. H. Iversen, The skin tumorigenic and carcinogenic effects of different doses, numbers of dose fractions and concentrations of 7,12-dimethylbenz[a]anthracene in acetone applied on hairless mouse epidermis. Possible implications for human carcinogenesis. *Carcinogenesis (London)* **12**, 493–502 (1991).

504. N. Zakova et al., Evaluation of skin carcinogenicity of technical 2,2-bis(*p*-glycidyloxyphenyl)propane in CF1 mice. *Food Chem. Toxicol.* **23**, 1081–1089 (1985).

505. L. R. DePass et al., Dermal oncogenicity studies on two methoxysilanes and two ethoxysilanes in male C3H mice. *Fundam. Appl. Pharmacol.* **12**, 579–583 (1989).

506. G. C. Peristianis et al., Two-year carcinogenicity study on three aromatic epoxy resins applied cutaneously to CF1 mice. *Food Chem. Toxicol.* **26**, 611–624 (1988).

507. J. M. Ward et al., Pathology of aging female SENCAR mice used as controls in skin two-stage carcinogenesis studies. *Environ. Health Perspect.* **68**, 81–89 (1986).

508. J. E. McCann et al., Detection of carcinogens as mutagens in the *Salmonella*/microsome test: Assay of 300 chemicals. *Proc. Natl. Acad. Sci. U.S.A.* **72**, 5135–5139 (1975).

509. E. Zeiger et al., Salmonella mutagenicity tests: V. Results from the testing of 311 chemicals. *Environ. Mol. Mutagen.* **19**(Suppl. 21), 2–141 (1992).

510. S. De Flora et al., Genotoxic activity and potency of 135 compounds in the ames reversion test and in a bacterial DNA-repair test. *Mutat. Res.* **133**, 161–198 (1984).

511. M. Ishidate et al., Primary mutagenicity screening of food additives currently used in Japan. *Food Chem. Toxicol.* **22**, 623–636 (1984).

512. D. M. DeMarini et al., Compatibility of organic solvents with the microscreen prophage-induction assay: Solvent-mutagen interactions. *Mutat. Res.* **263**, 107–113 (1991).

513. T. G. Rossman et al., Performance of 133 compounds in the lambda prophage induction endpoint of the microscreen assay and a comparison with *S. typhimurium* mutagenicity and rodent carcinogenicity assays. *Mutat. Res.* **260**, 349–367 (1991).

514. W. Von der Hude et al., Evaluation of the SOS chromotest. *Mutat. Res.* **203**, 81–94 (1988).

515. H. A. Vasavada and J. D. Padayatty, Rapid transfection assay for screening mutagens and carcinogens. *Mutat. Res.* **91**, 9–14 (1981).

516. H. Kubinski, G. E. Gutzke, and Z. O. Kubinski, DNA-cell-binding (DCB) assay for suspected carcinogens and mutagens. *Mutat. Res.* **89**, 95–136 (1981).

517. N. E. McCarroll, B. H. Keech, and C. E. Piper, A microsuspension adaptation of the *Bacillus subtilis* "rec" assay. *Environ. Mutagen.* **3**, 607–616 (1981).

518. S. Nakamura et al., SOS-inducing activity of chemical carcinogens and mutagens in *Salmonella typhimurium* TA1535/pSK1002: Examination with 151 chemicals. *Mutat. Res.* **192**, 239–246 (1987).

519. F. K. Zimmermann et al., Acetone, methyl ethyl ketone, ethyl acetate, acetonitrile and other polar aprotic solvents are strong inducers of aneuploidy in *Saccharomyces cerevisiae*. *Mutat. Res.* **149**, 339–351 (1985).

520. W. G. Whittaker et al., Detection of induced mitotic chromosome loss in *Saccharomyces cerevisiae*—An interlaboratory study. *Mutat. Res.* **224**, 31–78 (1989).

521. S. Albertini, Reevaluation of the 9 compounds reported conclusive positive in yeast *Saccharomyces cerevisiae* aneuploidy test systems by the gene-tox program using strain D61. M of *Saccharomyces cerevisiae*. *Mutat. Res.* **260**, 165–180 (1991).

522. A. S. Yadav, R. K. Vashishat, and S. N. Kakar, Testing endosulfan and fenitrothion for genotoxicity in *Saccharomyces cerevisiae*. *Mutat. Res.* **105**, 403–407 (1982).

523. A. Abbondandolo et al., The use of organic solvents in mutagenicity testing. *Mutat. Res.* **79**, 141–150 (1980).

524. R. Barale et al., The induction of forward gene mutation and gene conversion in yeasts by treatment with cyclophosphamide *in vitro* and *in vivo*. *Mutat. Res.* **111**, 295–312 (1983).

525. G. Fiskesjö, Benzo(*a*)pyrene and *N*-methyl-*N*-nitroguanidine in the *Allium* Test. *Hereditas* **95**, 155–162 (1981).

526. G. P. Rédei, Mutagen assay with *Arabidopsis*. A report of the U. S. Environmental Protection Agency Gene-Tox Program. *Mutat. Res.* **99**, 243–255 (1982).

527. G. G. Hatch et al., Chemical enhancement of viral transformation in syrian hamster embryo cells by gaseous and volatile chlorinated methanes and ethanes. *Cancer Res.* **43**, 1945–1950 (1983).

528. R. J. Pienta, Transformation of Syrian hamster embryo cells by diverse chemicals and correlation with their reported carcinogenic and mutagenic activities. In F. J. de Serres and A. Hollaender, eds., *Chemical Mutagen: Principles and Methods for their Detection*, Plenum, New York, NY, 1980, pp. 175–202.

529. A. E. Freeman et al., Transformation of cell cultures as an indication of the carcinogenic potential of chemicals. *J. Natl. Cancer Inst.* **51**, 799–808 (1973).

530. N. K. Mishra et al., Simultaneous determination of cellular mutagenesis and transformation by chemical carcinogens in fischer rat embryo cells. *J. Toxicol. Environ. Health* **4**, 79–91 (1978).

531. A. R. Peterson et al., Oncogenic transformation and mutation of C3H/10T$\frac{1}{2}$ clone 8 mouse embryo fibroblasts by alkylating agents. *Cancer Res.* **41**, 3095–3099 (1981).

532. J. R. Lillehaug and R. Djurhuus, Effect of diethylstilbestrol on the transformable mouse embryo fibroblast C3H/10T1/2Cl8 cells. Tumor promotion, cell growth, DNA synthesis, and ornithine decarboxylase. *Carcinogenesis (London)* **3**, 797–799 (1982).

533. E. B. Gehly and C. Heidelberger, The induction of Oua^{r}-mutations in nontransformable CVP3SC6 mouse fibroblasts. *Carcinogenesis (London)* **3**, 963–967 (1982).

534. W. von der Hude et al., Genotoxicity of three-carbon compounds evaluated in the SCE test *in vitro*. *Environ. Mutagen.* **9**, 401–410 (1987).

535. S. A. Latt et al., Sister-chromatid exchanges: A report of the gene-tox program. *Mutat. Res.* **87**, 17–62 (1981).

536. A. D. Tates and E. Kriek, Induction of chromosomal aberrations and sister chromatid exchanges in chinese hamster cells *in vitro* by some proximate and ultimate carcinogenic arylamide derivatives. *Mutat. Res.* **88**, 397–410 (1981).

537. K. S. Loveday et al., Chromosome aberration and sister chromatid exchange tests in chinese hamster ovary cells *in vitro*. V: Results with 46 chemicals. *Environ. Mol. Mutagen.* **16**, 272–303 (1990).

538. H. Norppa, The *in vitro* induction of sister chromatid exchanges and chromosome aberrations in human lymphocytes by styrene derivatives. *Carcinogenesis (London)* **2**, 237–242 (1981).

539. D. E. Amacher et al., Point mutations at the thymidine kinase locus in L5178Y mouse lymphoma cells. II. Test validation and interpretation. *Mutat. Res.* **72**, 447–474 (1980).

540. D. B. McGregor et al., Studies of an S9-based metabolic activation system used in the mouse lymphoma L5178Y cell mutation assay. *Mutagenesis* **3**, 485–490 (1988).

541. U. Friedrich and G. Nass, Evaluation of a mutation test using S49 mouse lymphoma cells and monitoring simultaneously the induction of dexamethasone resistance, 6-thioguanine resistance and ouabain resistance. *Mutat. Res.* **110**, 147–162 (1983).

542. G. R. Lankas, Effect of cell growth rate and dose fractionation on chemically-induced ouabain-resistant mutations in chinese hamster V79 cells. *Mutat. Res.* **60**, 189–196 (1979).

543. S. J. Cheng et al., Promoting effect of Roussin's red identified in pickled vegetables from linxian china. *Carcinogenesis (London)* **2**, 313–319 (1981).

544. F. Zarani, P. Papazafiri, and A. Kappas, Induction of micronuclei in human lymphocytes by organic solvents *in vitro*. *J. Environ. Pathol. Toxicol. Oncol.* **18**, 21–28 (1999).

545. S. P. Targowski and W. Klucinski, Reduction in mitogenic response of bovine lymphocytes by ketone bodies. *Am. J. Vet. Res.* **44**, 828–830 (1983).

546. R. S. Lake et al., Chemical induction of unscheduled DNA synthesis in human skin epithelial cell cultures. *Cancer Res.* **38**, 2091–2098 (1978).

547. T.-H. Chen et al., Inhibition of metabolic cooperation in chinese hamster cells by various organic solvents and simple compounds. *Cell Biol. Toxicol.* **1**, 155–171 (1984).

548. J. F. Sina et al., Evaluation of the alkaline elution/rat hepatocyte assay as a predictor of carcinogenic/mutagenic potential. *Mutat. Res.* **113**, 357–391 (1983).

549. A. Sakai and M. Sato, Improvement of carcinogen identification in BALB/3T3 cell transformation by application of a 2-stage method. *Mutat. Res.* **214**, 285–296 (1989).

550. A. Basler, Aneuploidy-inducing chemicals in yeast evaluated by the micronucleus test. *Mutat. Res.* **174**, 11–13 (1986).

551. D. Anderson and D. B. MacGregor, The effect of solvents upon the yield of revertants in the *Salmonella*/activation mutagenicity assay. *Carcinogenesis (N.Y.)* **1**, 363–366 (1980).

552. National Toxicology Program (NTP), *Cellular and Genetic Toxicology*, Fiscal Year 1987 Annual Plan, NTP-87-001, U.S.Department of Health and Human Services, Research Triangle Park, NC, 1987, pp. 43–96.

553. J. S. Rhim et al., Evaluation of an *in vitro* assay system for carcinogens based on prior infection of rodent cells with nontransforming RNA tumor virus. *J. Natl. Cancer Inst.* **52**, 1167–1173 (1974).

554. V. P. Kurnayeva et al., Experimental study of combined effect of solvents and noise. *J. Hyg. Epidemiol. Microbiol. Immunol.* **30**, 49–56 (1986).

555. P. S. Spencer, M. C. Bischoff, and H. H. Schaumburg, On the specific molecular configuration of neurotoxic aliphatic hexacarbon compounds causing central-peripheral distil axonopathy. *Toxicol. Appl. Pharmacol.* **44**, 17–28 (1978).

556. J. Misumi and M. Nagano, Neurophysiological studies on the relation between the structural properties and neurotoxicity of aliphatic hydrocarbon compounds in rats. *Br. J. Ind. Med.* **41**, 526–532 (1984).

557. J. Descotes, Identification of contact allergens: The mouse ear sensitization assay. *J. Toxicol. Cutaneous Ocul. Toxicol.* **7**, 263–272 (1988).

558. A. Nakamura et al., A new protocol and criteria for quantitative determination of sensitization potencies of chemicals by guinea pig maximization test. *Contact Dermatitis* **31**, 72–85 (1994).

559. J. Montelius et al., The murine lymph node assay: Search for an alternative, more adequate, vehicle than acetone/olive oil (4:1). *Contact Dermatitis* **34**, 428–430 (1996).

560. M. H. Friedlaender, F. V. Chisari, and H. Baer, The role of the inflammatory response of skin and lymph nodes in the induction of sensitization to simple chemicals. *J. Immunol.* **111**, 164–170 (1973).

561. J. R. Heylings et al., Sensitization to 2,4-dinitrochlorobenzene: Influence of vehicle on absorption and lymph node activation. *Toxicology* **109**, 57–65 (1996).

562. S. C. Gad et al., Development and validation of an alternative dermal sensitization test: The mouse ear swelling test (MEST). *Toxicol. Appl. Pharmacol.* **84**, 93–114 (1986).

563. B. Czerniecki et al., The development of contact hypersensitivity in mouse skin is supressed by tumor promotors. *J. Appl. Toxicol.* **8**, 1–8 (1988).

564. J.-L. Brouland et al., Morphology of popliteal lymph node responses in brown-Norway rats. *J. Toxicol. Environ. Health* **41**, 95–108 (1994).

565. K. P. Singh et al., Modulation of the development of humoral immunity by topically applied acetone, ethanol, and 12-*O*-tetradecanoyl-phorbol-13-acetate. *Fundam. Appl. Toxicol.* **33**, 129–139 (1996).

566. T. L. Litovitz et al., 1997 annual report of the American Association of Poison Control Centers toxic exposure surveillance system. *Am. J. Emerg. Med.* **16**, 443–497 (1998).

567. T. Cossmann, Acetonvergiftung nach Anlegung eines Zelluloid-Mullverbandes. *Muench. Med. Wochenschr.* **50**, 1556–1557 (1903).

568. G. F. Strong, Acute acetone poisoning. *Can. Med. Assoc. J.* **51**, 359–362 (1944).

569. C. C. Chatterton and R. B. Elliott, Acute acetone poisoning from leg casts of a synthetic plaster substitute. *J. Am. Med. Assoc.* **130**, 1222–1223 (1946).

570. L. J. Fitzpatrick and D'D. C. Claire, Acute acetone poisoning (resulting from synthetic plaster substitute in spica cast). *Curr. Res. Anesth. Analg.* **26**, 86–87 (1947).

571. R. B. Pomerantz, Acute acetone poisoning from castex. *Am. J. Surg.* **80**, 117–118 (1950).

572. Anonymous, Acetone poisoning and immobilizing casts. *Br. Med. J.* **2**, 1058 (1952).

573. L. C. Harris and R. H. Jackson, Acute acetone poisoning caused by setting fluid for immobilizing casts. *Br. Med. J.* **2**, 1024–1026 (1952).

574. P. K. Renshaw and R. M. Mitchell, Acetone poisoning following the application of a lightweight cast. *Br. Med. J.* **1**, 615 (1956).

575. W. Hift and P. L. Patel, Acute acetone poisoning due to a synthetic plaster cast. *S. Afr. Med. J.* **35**, 246–250 (1961).

576. D. S. Ross, Acute acetone intoxication involving eight male workers. *Ann. Occup. Hyg.* **16**, 73–75 (1973).

577. R. J. Foley, Inhaled industrial acetylene. A diabetic ketoacidosis mimic. *J. Am. Med. Assoc.* **254**, 1066–1067 (1985).

578. G. Sack, Ein Fall von gewerblicher Azetonvergiftung. *Arch. Gewerbepathol. Gewerbehyg.* **10**, 80–86 (1940).

579. A. R. Smith and M. R. Mayers, Study of poisoning and fire hazards of butanone and acetone. *Ind. Bull. NY State Dep. Labor* **23**, 174–176 (1944).

580. M. Sakata, J. Kikuchi, and M. Haga, Disposition of acetone, methyl ethyl ketone and cyclohexanone in acute poisoning. *Clin. Toxicol.* **27**, 67–77 (1989).

581. S. Gitelson, A. Werczberger, and J. B. Herman, Coma and hyperglycemia following drinking of acetone. *Diabetes* **15**, 810–811 (1966).

582. J. F. Knapp et al., Case records of the LeBonheur Children's Medical Center: A 17-month-old girl with abdominal distension and portal vein gas. *Pediatr. Emerg. Care.* **13**, 237–242 (1997).

583. G. Zettinig et al., Überlebte Vergiftung nach Einnahme der zehnfachen Letaldosis von Aceton. *Dtsch. Med. Wochenschr.* **122**, 1489–1492 (1997).

584. M. Wijngaarden et al., Coma and metabolic acidosis related to the use of muscle liniment. *Crit. Care Med.* **23**, 1143–1145 (1995).

585. M. D. Boggild, R. W. Peck, and C. V. R. Tomson, Acetonitrile ingestion: Delayed onset of cyanide poisoning due to concurrent ingestion of acetone. *Postgrad. Med. J.* **66**, 40–41 (1990).

586. F. T. Frerichs, Über den plötzlichen Tod und über das Coma bei Diabetes (diabetische Intoxication). *Zr. Klin. Med.* **6**, 3–53 (1883).

587. A. E. Koehler, E. Windsor, and E. Hill, Acetone and acetoacetic acid studies in man. *J. Biol. Chem.* **140**, 811–825 (1941).

588. K. W. Nelson et al., Sensory response to certain industrial solvent vapors. *J. Ind. Hyg. Toxicol.* **25**, 282–285 (1943).

589. T. Satoh et al., Cross-sectional study of effects of acetone exposure on workers' health. *Proc. 9th Int. Symp. Epidemiol. Occup. Health*, Cincinnati, OH, 1994, pp. 407–412.

590. T. Matsushita et al., Experimental studies for determining the MAC value of acetone. 1. Biologic reactions in the one-day exposure to acetone. *Jpn. J. Ind. Health* **11**, 477–485 (1969).

591. R. L. Raleigh and W. A. McGee, Effects of short, high-concentration exposures to acetone as determined by observation in the work area. *J. Occup. Med.* **14**, 607–610 (1972).

592. A. Seeber, E. Kiesswetter, and M. Blaszkewicz, Correlations between subjective disturbances due to acute exposure to organic solvents and internal dose. *Neurotoxicology* **13**, 265–270 (1992).

593. F. L. Oglesby, J. L. Williams, and D. W. Fassett, Eighteen-year experience with acetone. *Annu. Meet. Am. Ind. Hyg. Assoc.* Detroit, MI, 1949.

594. D. N. Roberts et al., Monitoring of the nasal response to industrial and environmental stimuli. *Soc. Occup. Med. Annu. Sci. Meet.*, Birmingham, UK, 1996.

595. R. B. Douglas, Inhalation of irritant gases and aerosols. In J. G. Widdicombe, ed., *International Encyclopedia of Pharmacology and Therapeutics*, Pergamon, Oxford, U. K., 1981, pp. 297–333.

596. J. E. Cometto-Muñiz and W. S. Cain, Efficacy of volatile organic compounds in evoking nasal pungency and odor. *Arch. Environ. Health* **48**, 309–314 (1993).

597. J. E. Cometto-Muñiz and W. S. Cain, Relative sensitivity of the ocular trigeminal, nasal trigeminal and olfactory systems to airborne chemicals. *Chem. Senses* **20**, 191–198 (1995).

598. P. Dalton et al., The influence of cognitive bias on the perceived odor, irritation and health symptoms from chemical exposure. *Int. Arch. Occup. Environ. Health* **69**, 407–417 (1996).

599. F. B. MacGregor et al., *The Effects of Short-Term Exposures to Acetone and Ammonia on the Human Nose*, Royal Postgraduate Medical School, Department of Medicine, London, 1996 (unpublished).

600. T. Matsushita et al., Experimental studies for determining the MAC value of acetone. 2. Biologic reactions in the six-day exposure to acetone. *Jpn. J. Ind. Health* **11**, 507–515 (1969).

601. R. B. Douglas and J. E. Coe, The relative sensitivity of the human eye and lung to irritant gases. *Ann. Occup. Hyg.* **31**, 265–267 (1987).

602. P. Fisher, The role of the ketone bodies in the etiology of diabetic coma. *Am. J. Med. Sci.* **221**, 384–387 (1951).

603. H. D. Landahl and R. G. Hermann, Retention of vapors and gases in the human nose and lung. *Arch. Ind. Hyg.* **1**, 36–45 (1950).

604. T. Dalhamn, M.-L. Edfors, and R. Rylander, Retention of cigarette smoke components in human lungs. *Arch. Environ. Health* **17**, 746–748 (1968).

605. K. Nomiyama and H. Nomiyama, Respiratory retention, uptake and excretion of organic solvents in man. Benzene, toluene, *n*-hexane, trichloroethylene, acetone, ethyl acetate and ethyl alcohol. *Int. Arch. Arbeitsmed.* **32**, 75–83 (1974).

606. M. Imbriani et al., Esposizione as acetone. Studio sperimentale dell'assorbimento e della eliminazione polmonare in soggetti normali. *Med. Lav.* **7**, 223–229 (1985).

607. M. Jakubowski and H. Wieczorek, The effects of physical effort on pulmonary uptake of selected organic compound vapours. *Pol. J. Occup. Med.* **1**, 62–71 (1988).

608. K. Teramoto et al., Initial uptake of organic solvents in the human body by short-term exposure. *J. Sci. Labour.* **63**, 13–19 (1987).

609. S. Kumagai et al., Uptake of 10 polar solvents during short-term respiration. *Toxicol. Sci.* **48**, 255–263 (1999).

610. T. Dalhamn, M.-L. Edfors, and R. Rylander, Mouth absorption of various compounds in cigarette smoke. *Arch. Environ. Health* **16**, 831–835 (1968).

611. K. Nomiyama and H. Nomiyama, Respiratory elimination of organic solvents in man. Benzene, toluene, *n*-hexane, trichloroethylene, acetone, ethyl acetate and ethyl alcohol. *Int. Arch. Arbeitsmed.* **32**, 85–91 (1974).

612. L. Cander and R. E. Forster, Determination of pulmonary parenchymal volume and pulmonary capillary blood flow in man. *J. Appl. Physiol.* **59**, 541–551 (1959).

613. A. C. M. Schrikker et al., Uptake of highly soluble gases in the epithelium of the conducting airways. *Pflügers Arch.* **405**, 389–394 (1985).

614. A. C. M. Schrikker et al., The excretion of highly soluble gases by the lung in man. *Pflügers Arch.* **415**, 214–219 (1989).

615. G. Pezzagno et al., La misura dei coefficienti di solbilitá degli aeriformi nel sangue. *G. Ital. Med. Lav.* **5**, 49–63 (1983).
616. S. Paterson and D. Mackay, Correlation of tissue, blood and air partition coefficients of volatile organic chemicals. *Br. J. Ind. Med.* **46**, 321–328 (1989).
617. E. M. P. Widmark, XXXII. Studies in the acetone concentration in blood, urine, and alveolar air. III. The elimination of acetone through the lungs. *Biochem. J.* **14**, 379–394 (1920).
618. R. L. Dills et al., Interindividual variability in blood/air partitioning of volatile organic compounds and correlation with blood chemistry. *J. Expos. Anal. Environ. Epidemiol.* **4**, 229–245 (1994).
619. A. Sato and T. Nakajima, Partition coefficients of some aromatic hydrocarbons and ketones in water, blood and oil. *Br. J. Ind. Med.* **36**, 231–234 (1979).
620. I. H. Young and P. D. Wagner, Effect of intrapulmonary hematocrit maldistribution on O_2, CO_2, and inert gas exchange. *J. Appl. Physiol.* **46**, 240–248 (1979).
621. V. Fiserova-Bergerova and M. L. Diaz, Determination and prediction of tissue-gas partition coefficients. *Int. Arch. Occup. Environ. Health* **58**, 75–87 (1986).
622. A. P. Briggs and P. A. Shaffer, The excretion of acetone from the lungs. *J. Biol. Chem.* **48**, 413–428 (1921).
623. G. A. Reichard et al., Plasma acetone metabolism in the fasting human. *J. Clin. Invest.* **63**, 619–626 (1979).
624. O. E. Owen et al., Acetone metabolism during diabetic ketoacidosis. *Diabetes* **31**, 242–248 (1982).
625. G. A. Reichard et al., Acetone metabolism in humans during diabetic ketoacidosis. *Diabetes* **35**, 668–674 (1986).
626. L. Parmeggiani and C. Sassi, Occupational poisoning with acetone-clinical disturbances, investigations in workrooms and physiopathological research. *Med. Lav.* **45**, 431–468 (1954).
627. S. Fukabori, K. Nakaaki, and O. Tada, On the cutaneous absorption of acetone. *J. Sci. Labour* **55**, 525–532 (1979).
628. V. Fiserova-Bergerova, Toxicokinetics of organic solvents. *Scand. J. Work Environ. Health* **11**, 7–21 (1985).
629. W. D. Brown et al., Body burden profiles of single and mixed solvent exposures. *J. Occup. Med.* **29**, 877–883 (1987).
630. J. Brodeur et al., Le problème de l'ajustement des Valeurs Limites d'Exposition (VLE) pour des horaires de travail non-conventionnels: Utilit, de la modélisation pharmacocinétique à base physiologique. *Trav. Santé* **6**, S11–S16 (1991).
631. G. Johanson, Modelling of respiratory exchange of polar solvents. *Ann. Occup. Hyg.* **35**, 323–339 (1991).
632. S. Kumagai and I. Matsunaga, Physiologically based pharmacokinetic model for acetone. *Occup. Environ. Med.* **52**, 344–352 (1995).
633. R. Gentry et al., Development of a physiologically-based pharmacokinetic (PB-PK) model for isopropanol and acetone. *Toxicologist* **48**, 144(abstr.) (1999).
634. S. Kumagai, I. Matsunaga, and T. Tabuchi, Effects of variation in exposure to airborne acetone and difference in work load on acetone concentrations in blood, urine, and exhaled air. *Am. Ind. Hyg. Assoc. J.* **59**, 242–251 (1998).
635. J. E. Jelnes, Semen quality in workers producing reinforced plastic. *Reprod. Toxicol.* **2**, 209–212 (1988).

636. R. Agnesi, F. Valentini, and G. Mastrangelo, Risk of spontaneous abortion and maternal exposure to organic solvents in the shoe industry. *Int. Arch. Occup. Environ. Health* **69**, 311–316 (1997).

637. G. Axelsson, C. Lütz, and R. Rylander, Exposure to solvents and outcome of pregnancy in university laboratory employees. *Br. J. Ind. Med.* **41**, 305–312 (1984).

638. H. Taskinen et al. Laboratory work and pregnancy outcome. *J. Occup. Med.* **36**, 311–319 (1994).

639. M. G. Ott et al., Health evaluation of employees occupationally exposed to methylene chloride. General study design and environmental considerations. *Scand. J. Work Environ. Health* **9**(Suppl. 1), 1–7 (1983).

640. M. G. Ott et al., Health evaluation of employees occupationally exposed to methylene chloride. Mortality. *Scand. J. Work Environ. Health* **9**(Suppl. 1), 8–16 (1983).

641. M. G. Ott et al., Health evaluation of employees occupationally exposed to methylene chloride. Clinical laboratory evaluation. *Scand. J. Work Environ. Health* **9**(Suppl. 1), 17–25 (1983).

642. A. Kabarity, Wirkung von Aceton auf die Chromosomen-Verteilung bei menschlichen Leukocyten. *Virchows Arch. B* **2**, 163–170 (1969).

643. T. Matsunaga et al., Epidermal cytokine mRNA expression induced by hapten differs from that induced by primary irritant in human skin organ culture system. *J. Dermatol.* **25**, 421–428 (1998).

644. K. Müller-Decker, G. Fürstenberger, and F. Marks, Keratinocyte-derived proinflammatory key mediators and cell viability as *in vitro* parameters of irritancy: A possible alternative to the draize skin irritation test. *Toxicol. Appl. Pharmacol.* **127**, 99–108 (1994).

645. U. Mrowietz et al., Inhibition of human monocyte functions by anthralin. *Br. J. Dermatol.* **127**, 382–386 (1992).

646. A. M. Mindlin, Acetone used as a solvent in accidental tarsorrhaphy. *Am. J. Ophthalmol.* **83**, 136–137 (1977).

647. B. Pastner, Topical acetone for control of life-threatening vaginal hemorrhage from recurrent gynecologic cancer. *Eur. J. Gynaecol. Oncol.* **14**, 33–35 (1993).

648. S. J. White and S. Broner, The use of acetone to dissolve a styrofoam impaction of the ear. *Ann. Emerg. Med.* **23**, 580–582 (1994).

649. V. Agrawal and S. Sharma, The efficacy of acetone in the sterilization of ophthalmic instruments. *Indian J. Ophthalmol.* **41**, 20–22 (1993).

650. J. M. Peikert et al., The efficacy of various degreasing agents used trichloroacetic acid peels. *J. Dermatol. Surg. Oncol.* **20**, 724–728 (1994).

651. W. F. Abadir, V. Nakhla, and P. Chong, Removal of superglue from the external ear using acetone: Case report and literature review. *J. Laryngol. Otol.* **109**, 1219–1221 (1995).

652. W. M. Grant, Acetone. In *Toxicology of the Eye*, 1st ed., Thomas, Springfield, IL, 1962, pp. 9–10.

653. R. S. McLaughlin, Chemical burns of the human cornea. *Am. J. Ophthalmol.* **29**, 1355–1362 (1946).

654. M. S. Mayou, Cataract in an acetone Worker. *Proc. R. Soc. Med.* **25**, 475 (1932).

655. M. G. Weill, Severe ocular disturbance from inhalation of impure acetone. *Bull. Soc. Fr. Ophthalmol.* **47**, 412–414 (1934).

656. J. J. Gomer, Corneal opacities from pure acetone. *Albrecht von Graefe's Arch. Ophthalmol.* **150**, 622–624 (1950).

657. W. M. Grant and J. S. Schuman, Acetone. In *Toxicology of the Eye*, 4th ed., Thomas, Springfield, IL, 1993, pp. 55–56

658. A. P. Lupulescu, D. J. Birmingham, and H. Pinkus, Electron microscopic study of human epidermis after acetone and kerosene administration. *J. Invest. Dermatol.* **60**, 33–45 (1973).

659. A. P. Lupulescu and D. J. Birmingham, Effect of protective agent against lipid-solvent-induced damages. *Arch. Environ. Health* **31**, 29–32 (1976).

660. A. Tosti, F. Bardazzi, and P. Ghetti, Unusual complication of sensitizing therapy for alopecia areata. *Contact Dermatitas.* **18**, 322 (1988).

661. H. Drexler et al., Skin prick tests with solutions of acid anhydrides in acetone. *Int. Arch. Allergy Immunol.* **100**, 251–255 (1993).

662. K. Abrams et al., Effect of organic solvents on *in vitro* human skin water barrier function. *J. Invest. Dermatol.* **101**, 609–613 (1993).

663. H. Zhai, Y.-H. Leow, and H. I. Maibach, Human barrier recovery after acute acetone perturbation: An irritant dermatitis model. *Clin. Exp. Dermatol.* **23**, 11–13 (1998).

664. R. Ghadially et al., The aged epidermal permeability barrier. Structural, functional, and lipid biochemical abnormalities in humans and a senescent murine model. *J. Clin. Invest.* **95**, 2281–2290, (1995).

665. E. Proksch and J. Brasch, Influence of epidermal permeability barrier disruption and Langerhan's cell density on allergic contact dermatitis. *Acta Derm.-Venereol.* **77**, 102–104.

666. P. Kechijian, Nail polish removers: Are they harmful? *Semin. Dermatol.* **10**, 26–28 (1991).

667. H. Suzuki, An experimental study on physiological functions of the autonomic nervous system of man exposed to acetone gas. *Jpn. J. Ind. Health* **15**, 147–164 (1973).

668. K. Nakaaki, An experimental study on the effect of exposure to organic solvent vapor in human subjects. *J. Sci. Labour* **50**, 89–96 (1974).

669. R. Israeli et al., Reaktionszeit als Mittel zur Aceton-TLV-(MAK)-Wertbestimmung. *Zentralbl Arbeitsmed.* **27**, 197–199 (1977).

670. R. B. Dick et al., Neurobehavioural effects of short duration exposures to acetone and methyl ethyl ketone. *Br. J. Ind. Med.* **46**, 111–121 (1989).

671. E. Kiesswetter et al., Acute exposure to acetone in a factory and ratings of well-being. *Neurotoxicology* **15**, 597–602 (1994).

672. A. Seeber et al., Neurobehavioral effects of experimental exposure to acetone and ethyl acetate. *Proc. 23rd Int. Cong. Occup. Health*, Montréal, Can., *1990*.

673. R. B. Dick et al., Effects of short duration exposures to acetone and methyl ethyl ketone. *Toxicol. Lett.* **43**, 31–49 (1988).

674. A. Seeber et al., Combined exposure to organic solvents: An experimental approach using acetone and ethyl acetate. *Appl. Psychol. Int. Rev.* **41**, 281–292 (1992).

675. S. F. Lanes et al., Mortality of cellulose fiber production workers. *Scand. J. Work Environ. Health* **16**, 247–251 (1990).

676. B. Sietmann and A. Seeber, Irritationen und Lästigkeitserleben bei Expositionen Gegen Aceton. Eine Statistische Re-Analyse Zeier Laborexperimente und Einer Feldstudie als Ergebnis eines Forschungs-Projektes. *Irritationen und Lästigkeitserleben bei Expositionen gegeniiber Aceton*, Rhone-Poulenc Rhodia, AG, Freiburg. Institute of Occupational Health., University of Dortmund, Dortmund, Germany, 1996.

677. T. C. Tan and L. L. T. Wong, Health effects of exposure to methyl ethyl ketone, toluene and acetone in a chemical drum filling plant in Hong Kong. *J. Occup. Med. Singapore* **2**, 9–13 (1990).

678. F. Tomei et al., Liver damage among shoe repairers. *Am. J. Ind. Med.* **36**, 541–547 (1999).

679. K. J. Soden, An evaluation of chronic methylene chloride exposure. *J. Occup. Med.* **35**, 282–286 (1993).

680. H. Rösgen and G. Mamier, Sind Azetongase blutschädigend? *Öeff. Gesundheitsdienst* **10**, A83–A86 (1944).

681. P. Dolara et al., Enzyme induction in humans exposed to styrene. *Ann. Occup. Hyg.* **27**, 183–188 (1983).

682. S. H. Swan et al., Historical cohort study of spontaneous abortion among fabrication workers in the semiconductor health study: Agent-level analysis. *Am. J. Ind. Med.* **28**, 751–769 (1995).

683. S. Khattak et al., Pregnancy outcome following gestational exposure to organic solvents. A prospective controlled study. *J. Am. Med. Assoc.* **281**, 1106–1109 (1999).

684. R. Spirtas et al., Retrospective cohort mortaility study of workers at an aircraft maintenance facility. I. Epidemiological results. *Br. J. Ind. Med.* **48**, 515–530 (1991).

685. A. Blair et al., Mortality and cancer incidence of aircraft maintenace workers exposed to trichloroethylene and other organic solvents and chemicals: Extended follow up. *Ocuup. Environ. Med.* **55**, 161–171 (1998).

686. J. T. Walker et al., Mortality of workers employed in shoe manufacturing. *Scand. J. Environ. Med.* **19**, 89–95 (1993).

687. M. Pitarque et al., Evaluation of DNA damage by the comet assay in shoe workers exposed to toluene and other organic solvents. *Mutat. Res.* **441**, 115–127 (1999).

688. L. Hagmar et al., Cytogenetic and hematological effects in plastic workers exposed to styrene. *Scand. J. Work Environ. Health* **15**, 136–141 (1989).

689. R. Kishi et al., Neurobehavioral effects of chronic occupational exposure to organic solvents among japanese industrial painters. *Environ. Res.* **62**, 303–313 (1993).

690. B. Baird et al., Solvents and color discrimination ability. *J. Occup. Med.* **36**, 747–751 (1994).

691. G. Calabrese et al., Otoneurological study in workers exposed to styrene in the fiberglass industry. *Int. Arch. Occup. Environ. Health* **68**, 219–223 (1996).

692. E. M. Socie et al., Work-related skin disease in the plastics industry. *Am. J. Ind. Med.* **31**, 545–550 (1997).

693. F. Akbar-Khanzadeh, and R. D. Rivas, Exposure to isocyanates and organic solvents, and pulmonary-function changes in workers in a polyurethane molding process. *J. Occup. Environ. Med.* **38**, 1205–1212 (1996).

694. P. F. Sanders, Calculation of soil cleanup criteria for volatile organic compounds as controlled by the soil-to-groundwater pathway: Comparison of four unsaturated soil zone leaching models. *J. Soil Contam.* **4**, 1–24 (1995).

695. T. Gierczak et al., Photochemistry of acetone under tropospheric conditions. *Chem. Phys.* **231**, 229–244 (1998).

696. E. A. Betterton, The partitioning of ketones between the gas and aqueous phases. *Atmos. Environ.* **25A**, 1473–1477 (1991).

697. R. G. Derwent, M. E. Jenkin, and S. M. Saunders, Photochemical ozone creation potentials for a large number of reactive hydrocarbons under European conditions. *Atmos. Environ.* **30**, 181–199 (1996).

CHAPTER SEVENTY-FIVE

Ketones of Four or Five Carbons

David A. Morgott, Ph.D., DABT, CIH,
Douglas C. Topping, Ph.D., DABT, and
John L. O'Donoghue, VMD, Ph.D., DABT

INTRODUCTION

A ketone is an organic compound containing a carbonyl group (C=O) attached to two carbon atoms and can be represented by the general formula

$$\mathrm{R-\overset{\overset{\Large O}{\|}}{C}-R'}$$

Several billion pounds of ketones are produced annually for industrial use in the United States. Those with the highest production volumes include acetone, methyl ethyl ketone, methyl isobutyl ketone, cyclohexanone, 4-hydroxy-4-methyl-2-pentanone, isophorone, mesityl oxide, and acetophenone. Common methods used to manufacture ketones include aliphatic hydrocarbon oxidation, alcohol dehydration with subsequent oxidation, dehydrogenation of phenol, alkyl aromatic hydrocarbon oxidation, and condensation reactions.

Ketones are used because of their ease of production, low manufacturing cost, excellent solvent properties, and desirable physical properties such as low viscosity, moderate vapor pressure, low to moderate boiling points, high evaporation rates, and a wide range of miscibility with other liquids. The low-molecular-weight aliphatic ketones are miscible with water and organic solvents, whereas the high-molecular-weight aliphatic and aromatic ketones are generally immiscible with water. Most ketones are chemically stable.

Patty's Toxicology, Fifth Edition, Volume 6, Edited by Eula Bingham, Barbara Cohrssen, and Charles H. Powell.
ISBN 0-471-31939-2 © 2001 John Wiley & Sons, Inc.

The exceptions are mesityl oxide, which can form peroxides, and methyl isopropenyl ketone, which polymerizes. Most ketones are generally of low flammability.

Ketones are commonly used in industry as solvents, extractants, chemical intermediates, and to a lesser extent, flavor and fragrance ingredients. Ketones have also been reported in the ambient air, in wastewater treatment plants (1), and in oil field brine discharges (2).

Occupational Exposures

In an occupational setting, the primary routes of exposure to ketones are inhalation and skin contact. Ingestion is rare. Since most ketones have a significant vapor pressure at room temperature, exposure by inhalation in the workplace is likely to occur. The principal hazard associated with exposure to ketone vapors is irritation of the eyes, nose, and throat. Many ketones have excellent warning properties and can be easily detected by their odor. Accidental overexposure should be relatively rare provided warning properties are not ignored and olfactory fatigue does not occur. The classic symptoms produced by an overexposure to ketones include, progressively, irritation of the eyes, nose, and throat, headache, nausea, vertigo, uncoordination, central nervous system depression, narcosis, and cardiorespiratory failure. Recovery is usually rapid and without residual toxic effects. In the case of accidental spills, personnel should wear protective clothing including respiratory protection. Contaminated clothing should be removed promptly, and the exposed areas of the body should be thoroughly flushed with water. Many ketones are absorbed through the skin; therefore, caution should be exercised to avoid repeated or prolonged skin contact. The vapors produced by accidental spills may present a fire or explosion hazard.

Toxic Effects

Although the relative toxicity of most ketones is low and the effects of acute exposures are well recognized, the effects of chronic exposure have received less study. In some cases, metabolic studies have helped to elucidate the toxic effects of several ketones. Generally, when ketones are absorbed into the bloodstream, they may be eliminated unchanged in the expired air, or metabolized by a variety of metabolic pathways to secondary alcohols, hydroxyketones, diketones, and carbon dioxide. Recent studies indicate that carbonyl reduction, α and $\omega - 1$ oxidation, decarboxylation, and transamination play important roles in the metabolism of aliphatic ketones. Aromatic ketones and ketones such as cyclohexanone and isophorone may undergo oxidative metabolism by dehydrogenation, ring hydroxylation, or substituent group oxidation. In addition, aromatic and aliphatic ketones may be conjugated with glucuronic acid, sulfuric acid, or glutathione prior to excretion in the urine. Glucuronic and sulfuric acid conjugation usually occur after a ketone is reduced to a secondary alcohol or oxidized to a carboxylic acid. Of the various conjugation mechanisms that occur, glucuronic acid conjugation appears to be the predominant pathway.

Ketone exposure may alter the toxicity of other chemicals, including other ketones that are metabolized by cytochrome P450 enzymes. Under certain circumstances non-neurotoxic ketones may potentiate the neurotoxicity of other ketones or organophosphates or the hepatotoxicity and renal toxicity of haloalkanes. Data on these interactions are referred to in sections on specific ketyones in Chapters 74, 75, and 76.

1.0 Methyl Ethyl Ketone

1.0.1 CAS Number: *[78-93-3]*

1.0.2 Synonyms: 2-butanone, 2-oxobutane, ethyl methyl ketone, butane-2-one, methyl acetone, and MEK

1.0.3 Trade Names: Methyl ethyl ketone is sold under ten different trade names, including: Acryloid A-21LV, AroCy F-40S, Black-Out Black, CMD 2493, D.E.R. 684-EK40, and Electrodag 442 (3).

1.0.4 Molecular Weight: 72.11

1.0.5 Molecular Formula: C_4H_8O

1.0.6 Molecular Structure:

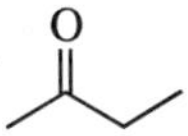

1.1 Chemical and Physical Properties

Methyl ethyl ketone is miscible with water and a variety of other solvents including: ethanol, diethyl ether, acetone, and benzene. It has a boiling point of 79.6°C, a melting point of −86.6°C, a density of 0.807 g/mL at 20°C, and a vapor pressure of 77.5 torr at 25°C (4). The flash point is −9°C by the closed-cup method and 10°C with an open-cup (5). The autoignition temperature is 404°C and the flammability limits at 93°C are 1.4 and 11.4%. Additional physical–chemical properties are summarized in Table 75.1.

1.1.2 Odor and Warning Properties

The odor detection thresholds for MEK reportedly range from about 2 to 10 ppm, with 6 ppm being the best estimate for an odor recognition threshold (6–8). Devos et al. summarized and standardized the results from eight different threshold studies and reported an odor detection threshold of 7.8 ppm (9). A substantially lower odor detection threshold of 0.25 ppm was reported by Ruth (10). After critically analyzing all available data, the American Industrial Hygiene Association reported that the range of acceptable odor thresholds concentrations for MEK seemed to vary from 2 to 85 ppm and that the odor quality for MEK could be described as sharp and sweet (11). The frequency distribution of odor threshold values for 187 male and female students exposed to MEK showed a distinct trimodal distribution, with no significant differences noted for the two sexes (12).

1.2 Production and Use

Methyl ethyl ketone is commercially manufactured from *n*-butene in a metal-catalyzed hydrogenation reaction that proceeds through the intermediate formation of 2-butanol

Table 75.1. Physical–Chemical Properties of Ketones

Compound	Molecular Formula	Mol. Wt.	Boiling Point (°C)	Melting Point (°C)	Specific Gravity[a]	Refractive Index (20°C)	Vapor Pressure (mmHg)	Vapour Density (Air = 1)
Acetone	C_3H_6O	58.08	56.2	– 95.4	0.791	1.3588	180	2.0
Methyl ethyl ketone	C_4H_8O	72.11	79.6	– 86.6	0.807 (20/4)	1.3788	77.5 (20)	2.4
3-Butyn-2-one	C_4H_4O	68.08	85.0	—	0.879	1.4024	40.0	
Methyl *n*-propyl ketone	$C_5H_{10}O$	86.17	102.2	– 76.9	0.809 (20/4)	1.3895	16.0	3.0
Methyl isopropyl ketone	$C_5H_{10}O$	86.14	93.0	– 92.0	0.803 (20/0)	1.3879		
3-Pentyn-2-one	C_5H_6O	82.10	133.0	– 28.7	0.910	1.141		
Methyl isopropenyl ketone	C_5H_8O	84.12	97.7	– 53.7	0.855	1.4220	42.0	
2,4-Pentanedione	$C_5H_8O_2$	100.12	138.3	– 23.2	0.976	1.4494	7.0	3.5

[a]Specific gravity is at 20/20°C unless otherwise noted.
[b]Vapor pressure is at 25°C unless otherwise noted.
[c]Closed cup unless otherwise noted, [O.C.] open cup. Figures in parentheses are °C.
[d]S = readily soluble, Sl = slightly soluble, I = insoluble.
[e]At 25°C, 760 mm Hg.

(13). A second method of synthesis involves the liquid-phase oxidation of *n*-butane with the formation of acetic acid as a co-product. MEK is widely used as a coating and formulating solvent and as a feedstock for the synthesis of methyl isopropenyl ketone, MEK peroxide, 2,3-butanedione, and MEK oxime. MEK use as a formulating solvent includes products such as rubber cements, printing inks, urethane lacquers, cleaning fluids, and lubricant dewaxing agents. The U.S. consumption of MEK was estimated to be 442 million pounds for the year 1996, with 61% of the use going into the preparation of paints and other surface coatings (13).

1.3 Exposure Assessment

1.3.1 Air

MEK has been found in a wide range of air samples from both polluted and unpolluted sources such as grasses, compost piles, automobile exhaust, and office furnishings. The average emission of MEK from the clover grown in a pasture lot was found to be up to 50% of the total emissions (14). MEK was also emitted at rates as high as 50–100 μg/kg/h for some types of tree foliage and was routinely detected in the air surrounding forest trees at concentrations of 2–3 ppb (15). Appreciable amounts of MEK were shown to be emitted from winter cover crops, including legumes such as Berseem clover, hairy vetch, and crimson clover (16). MEK levels of 11.2 mg/kg were found in ground tobacco leaves (17). The concentration in cigarette smoke (235–250 μg/L) was shown to be similar for filtered and nonfiltered cigarettes (18). MEK was found in the headspace from biodegradable household waste (19). The composting of municipal solid wastes such as newspapers, food and yard wastes, and cardboard can result in the emission of MEK at concentrations as high as 36 mg/m^3 (20).

Max. Vapor Concen. (ppm @20°C)	Flash Point[c] °F (°C)	Flammability (%)		Solubility[d]			Conversion Factors[e]	
		Lower Limit	Upper Limit	Water	Alcohol	Ether	1 ppm⇔mg/m^3	1 mg/L⇔ppm
23.700	0 (18)	2.15	13.0	S	S	S	2.37	422
102,000	50 (9)	1.4	11.4	S	S	S	2.94	340
	90 (32)	1.3	8.8				2.78	360
21,000	45 (7)	1.6	8.2	S	S	S	3.52	284
52,600	43 (6)			S			3.52	284
							3.35	298
54,300	49 (9)			S	S	S	3.43	290
4,900	(34)	1.2	8.0	S	S	S	4.10	244

MEK is present in the air from private homes at levels as high as 38 μg/m^3 (21). These levels can be traced in part to the use of products that contain MEK, such as caulks, paints, and particleboard (22). Products containing the highest concentrations of MEK appear to be adhesives and automobile-related cleaners and lubricants (23), although it was found in 6 out of 20 paint thinners at levels ranging from 5 to 45% (24). The weekly average emission of MEK from the gypsum wallboard, chipboard, and ceramic tile used in home construction ranged from about 100 to 250 ng/L (25). Microbial growth on ceramic tile was found to nearly double the emission of MEK into the surrounding air, raising the level from 118 to 205 ng/L (26). The level of MEK in dust from mold-infested offices and schools was generally less than 70 μg/m^3 (19).The concentration of MEK found in a variety of additional environmental samples is listed in Table 75.4 (27–40).

Li et al. showed that substantial amounts (5.1×10^4 mole/d) of MEK could be produced in the troposphere as a breakdown product from the photooxidation of other higher-molecular-weight compounds (41). Grosjean et al. found MEK at an average level of 1.2 ppb in ambient air samples from four locations in and around Los Angeles (42). MEK has also been shown to constitute about 0.04% of the total volatile organic carbon emitted in the exhaust of light-duty vehicles (43). The exhaust emission factors for MEK from gasoline and diesel-powered vehicles were shown to average 0.08 and 1.24 mg/km, respectively (44). MEK also constituted 0.06% of the volatile nonmethane hydrocarbons on a busy roadway, 0.18% of the hydrocarbons in a bus garage, and 0.20% of the emissions from a lead smelter (45). Verhoeff et al. reported indoor air concentrations ranging from about 0.1–1.5 mg/m^3 for 27 homes in the vicinity of a car-body repair shop that used MEK (46).

Occupational exposures to MEK vary considerably depending upon the factory operation being examined and the work habits of the employees being monitored. A survey

Table 75.2. Comparison of the MEK Concentrations Found in Different Environmental Samples

Sample Type	Airborne Concentration ($\mu g/m^3$)	(ppb)	Reference
Composting effluent	7800		27
Poultry manure dryers		8–260	28
Inside home	0–19		29
Indoor air		4.1–14.3	30
Summer indoor air		1.4–6.9	31
Summer outdoor air		0.8–2.7	31
Outside ambient air	0–3		29
Urban air		1.9–8.5	32
City air		0.2–58	33
Inside office area		2.4	33
Ventilation return air	5.7–40.9		34
Inside machine shop		1.2	33
Outside furniture factory		0.7–1.2	33
Outside paint incinerator		0.9	33
Automobile exhaust		90	35
Oil fire	10–170		36
Chemical waste site	1.5–33.0		37
Factory exhaust gas	20–680		38
Outside laboratory oven	1800		39
Municipal land fill gas		3092–5200	40

of the chemical products used in 19 different industry groups revealed that the largest volumes of MEK were used in the manufacture of fabricated metal products, in the manufacture of transport equipment, and in the printing and publishing industry (47). The 8-h time-weighted average exposure for a laborer in a woodworking shop was reported to be 0.12 ppm (48). Whitehead et al. reported a mean concentration of 0.3 ppm for 89 employees involved in spray painting and spray gluing operations at three different plants (49). Similarly, Winder and Ng reported a mean airborne concentration of 12.2 mg/m^3 (4.1 ppm) for 20 of the 70 spray painters surveyed from several automobile repair shops (24). In a study of paint stripping and paint spraying operations in the aeronautical industry, Vincent et al. found daily mean MEK levels of 14.7–31.2 mg/m^3 for the 5 to 9 workers examined (50). The MEK levels in different areas of an aircraft degreasing operation varied from nondetectable to 6.33 mg/m^3 (51). An analysis of the levels found in 31 shoe factories over a 10-y period showed average levels ranging from 41 to 67 mg/m^3 (52). In a survey of the MEK levels associated with different types of manufacturing operations, De Rosa et al. reported ranges of 2–152 ppm for shoe factories, 1–376 ppm for painting operations, and 3–277 ppm for printing plants (53). The range of values in an organic solvent recycling plant were considerably lower, averaging 1–38 ppm (54). Likewise, the 8-h time-weighted-average exposure levels for waste receivers at a chemical waste incinerator ranged from 0.02 to 64 ppm (55). Nelson et al. reported MEK exposures ranging from 0 to 630 ppm for 215 workers in automobile and truck assembly plants (56). An analysis of the information in a database of exposure measurements for pulp, paper, and paper products industries revealed that only 6 of 137 values exceeded the 8-h occupational threshold limit value of 200 ppm (57).

1.3.2 Background Levels

MEK has been detected, and in many cases quantified, in a wide variety of naturally occurring materials such as foodstuffs and agricultural crops, bacteria, and fungi. It has been found in many dairy products, including milk, cream, and cheese at concentrations up to 0.3 ppm (58). It was also detected, but not quantified, in bread, honey, oranges, tea, chicken, tomatoes, potato chips, mushrooms, coffee, eggs, yogurt, and smoked fish (59–61). MEK is produced by several different types of fungi, including *Penicillium commune* and *Phialophora fastigata* (62, 63). When cultivated on gypsum board, the fungus *Streptomyces albidoflavus* produced MEK at a rate of 300 ng/h (64). MEK has been shown to be present in some types of soils purged with air (65) and in the gaseous headspace above many types of marine algae (66).

MEK has been found in water samples from a variety of sources. Sea water samples from the Straits of Florida contained up to 220 μg/L of MEK (67), whereas sea water from Antarctica contained less than 4 μg/L (68). MEK was detected in 1 of 9 samples of household water at a concentration of 1.2 μg/L (69). It was also found in the potable water delivered by polyvinyl chloride (PVC) pipe at concentrations up to 4.5 mg/L when the water had a residence time of at least 64 h in the pipe (70). In another study with drinking water, MEK did not permeate through the PVC pipe when the concentration in the surrounding soil was 1700 μg/kg (71). MEK has been found in rainwater at a concentration of 4.6 ppb, in the wastewater from a chemical plant at 8–20 ppm, and in the leachate from a municipal landfill at levels up to 5.2 ppm (72–74). Sabel and Clark reported that MEK was present at a concentration range of 110–27,000 μg/L in all six of the municipal landfill leachates sampled (75). It was also detected in ground water samples from the monitoring wells around a landfill at a range of 6.8–6200 μg/L. The levels in ground water from uncontaminated sites ranged from 5.1 to 1100 μg/L. Brown and Donnelly summarized data from the literature and reported MEK concentrations up to 53 mg/L for industrial landfill leachates and 27 mg/L for municipal landfill leachates (76).

1.3.3 Workplace Methods

The determination of workplace concentrations of MEK is generally accomplished by adsorption onto a high-affinity medium, such as activated carbon, followed by desorption in carbon disulfide and quantitation by gas chromatography (NIOSH Method 2500) (77). Porapak N, coconut charcoal, and Anasorb have been shown through testing to be useful adsorbents for MEK (78–80). A systematic evaluation of the adsorption capacity and thermal desorption efficiency of four different solid adsorbents for long-term personal monitoring showed that Carboseive gave the best overall performance with MEK (81). Baltussen et al. showed that the best overall adsorbent for MEK vapors in the air was perhaps polydimethylsiloxane (82). More recently, passive organic vapor monitors have replaced sampling tubes as the method of choice for sample collection (83). The airborne limit of detection using passive samplers is about 50 mg/m^3. Care is required in their use, however, especially in multisolvent environments where the MEK sampling rate can decrease, due to saturation and in high-humidity environments where the MEK sampling rate can be affected (84, 85). Remote-sensing Fourier transform infrared spectroscopy has also been shown to be a useful workplace method that correlates well with gas chromatographic techniques (86). Rowell developed a thermal desorption–gas chromatography method to determine the MEK concentration in charcoal-containing adhesive plasters used to monitor glove permeation and dermal exposure to solvents (87).

1.3.4 Community Methods

Many different approaches have been developed to search for MEK in the community environment. A very sensitive method has been developed to identify the formation of MEK as a product of atmospheric photooxidation (96). The method involves the formation of an oxime derivative using *O*-(2,3,4,5,6-pentafluorobenzyl)-hydroxylamine (PFBHA) and subsequent separation by ion trap mass spectrometry. Ciccioli et al. used three-stage carbon adsorption traps and high-resolution gas chromatography–mass spectrometry to determine the concentration of MEK in polluted and unpolluted air (97). The MEK concentration in diesel exhaust has been measured following derivatization with 2,4-dinitrophenylhydrazine (DNPH) and separation by high-pressure liquid chromatography (98). Vacuum distillation and gas chromatography–mass spectrometry has been used to determine the MEK in fresh and spoiled seafood samples (99). An automated headspace gas chromatography has been used to assess the production and excretion of MEK by bacteria (100). A gas chromatography–mass spectrometry technique was specifically developed to analyze for MEK and other volatile materials in air at concentrations of 10 ppm or greater (101). Residual levels of MEK and other solvents in commercially manufactured drug products were determined using a thermal desorption cold trap–gas chromatography procedure (102). Headspace gas chromatography with both flame ionization and electron capture detection was a useful approach for assessing the level of MEK and other volatile substances in biological fluids when inhalant abuse was suspected (103). Dual-column headspace gas chromatography has been used to analyze for the MEK in blood specimens when other interfering solvents are also present (104). The concentration of MEK in hazardous waste samples was monitored by gas chromatography–mass spectrometry using a purge and trap apparatus (105). A differential scanning calorimetry method is under development for the trace gas analysis of very low MEK levels in samples such as soil or solid waste (106).

Analysis of effluent samples for MEK has been accomplished with a limit of detection down to about 5 ppb by direct aqueous injection into a gas chromatograph (107). Lower levels (1.6 ppb) of MEK have been determined in drinking water by adsorption onto a Zeolite resin followed by derivatization of the extract with 2,4-dinitrophenylhydrazine and separation by high-pressure liquid chromatography (108). MEK can be detected in aqueous samples at the low part-per-trillion level by cryotrap membrane introduction mass spectrometry (109).

1.3.5 Biomonitoring/Biomarkers

MEK has been detected in the serum, urine, and expired air of healthy adults(110, 111). When compared to normal individuals, serum levels of MEK were reportedly unchanged in patients with long-standing diabetes mellitus (112). Urinary levels of MEK can, however, be elevated in children with congenital errors in isoleucine catabolism (113, 114).

The analysis of breath samples for MEK requires the use of a method with greater sensitivity and selectivity. This is typically accomplished by high-resolution gas chromatography using fused capillary columns for separation of the volatile chemicals in alveolar air (88). Air samples can be collected by any of a number of techniques, including evacuated Teflon bags, glass sampling tubes, or adsorbent resin sampling devices (89). Dyne et al. reported on the use of a method that collected expired air containing MEK onto Tenax resin, which was analyzed with an automated thermal desorption device attached directly to a gas chromatograph (90).

1.3.5.1 Blood. An evaluation of the MEK blood levels in 1101 non occupationally exposed participants of the Third National Health and Nutrition Examination Survey (NHANES III) revealed an average concentration of 7.1 μg/L, a median concentration of 5.4 μg/L, and a 95% upper confidence limit concentration of 16.9 μg/L (115). The blood levels of MEK in a group of 16 individuals concerned about chemical exposures from nearby manufacturing plants were not appreciably different from the values reported in the NHANES III study (116). The blood concentrations in the citizen survey was 5.9 μg/L and the concentration at the upper 95% confidence limit was 18 μg/L. In contrast, a survey of 100 residents living near an industrial complex with high levels of soil contamination revealed an MEK blood concentration of 6.11 μg/L, which was higher than the value of 5.05 μg/L for 106 control subjects (117).

Imbriani et al. evaluated the relationship between blood levels and ambient exposure to MEK in a group of fifteen volunteers who were exposed to concentrations ranging from 11.9–621.8 mg/m^3 (118). Three groups of five individuals were either exposed for 2 h while at rest, for 4 h while at rest, or for 2 h while exercising. Peak blood concentrations of MEK ranged from 206–2306 μg/L for the three treatment groups and the blood values showed a good correlation with the mass of MEK excreted in the urine ($r = 0.88$). Yoshikawa et al. examined the relationship between occupational exposures to MEK and the amount present in the urine and blood of 72 workers in a printing plant (119). Workroom MEK concentrations of 1.3–223.7 ppm resulted in end-of-shift blood levels of 0.01–6.68 mg/L and end-of-shift urine levels of 0.20–8.08 mg/L. The correlation coefficient for the relationship between air and blood was 0.82, whereas the coefficient for air and urine was 0.89. The BEI level for a 200-ppm MEK exposure was calculated to be 5.1 mg/L.

1.3.5.2 Urine. Miyasaka et al. monitored the urinary excretion of MEK in 62 male employees exposed to average ambient air levels ranging up to about 85 ppm (120). Ten of the employees with an 8-h time-weighted-average MEK exposure of about 20 ppm tended to rapidly excrete MEK in the urine at the beginning of the workshift, but the rate declined and became slower as the exposure continued. Only about 0.1% of the absorbed MEK was excreted in the urine of the exposed employees. A positive correlation was observed between the MEK concentrations in the urine, which ranged up to 3000 μg/L, and the concentration in the workroom air ($r = 0.774$).

Perbellini et al. examined the elimination of MEK in two groups of occupationally exposed workers concentrations ranging from 8 to 272 μg/L (2.7–92.5 ppm) and found end-of-shift urine MEK levels of 120–1120 μg/L and 3-hydroxy-2-butanone levels of 0–1780 μg/L (121). Taken together, the two urinary metabolites were calculated to account for 0.1% of the pulmonary uptake of MEK. The range of end-exposure blood and expired air MEK levels were 842–9573 and 4–54 μg/L, respectively. An *in vitro* blood-to-air partition coefficient of 183 was used along with tissue-to-blood values that were close to unity to construct a physiological model of MEK kinetics in humans. Positive linear correlations were reported for the relationship between the ambient air concentrations of MEK and the concentration of MEK in the urine ($r = 0.688$) and the concentration of 3-hydroxy-2-butanone in the urine ($r = 0.818$) (122). The regression analysis led to the proposal for a Biological Equivalent Exposure (BEI) Limit of 2 mg/L for the urinary excretion of MEK after a 4-h occupational exposure (123).

Urinary excretion was again studied in a group of 78 workers exposed to MEK concentrations ranging from 6 to 790 mg/m^3 during the manufacture of leather suitcases (118). A good correlation was observed between the workroom concentration and the excretion of MEK in the urine ($r = 0.93$). Similar results were obtained in a group of fifteen volunteers who were exposed to MEK concentrations of 11.9–621.8 mg/m^3 for 2–4 h either at rest or during light physical exercise. The correlation coefficient for the urinary excretion in the experimentally exposed volunteers ranged from 0.94 to 0.99. Jang etal. examined the urinary excretion of MEK in 14 workers in a surface coating operation and found that workroom concentrations of 34 to 356 ppm resulted in urinary MEK levels of 0.2–2.5 mg/L (124). A linear correlation was observed between the exposure concentration and the urinary excretion of MEK ($r = 0.65$). The data indicated a BEI level of 1.4 mg/L for a 4-h exposure at 200 ppm. This value was noted to be less than the 2.0 mg/L value recognized by the American Conference of Governmental Industrial Hygienists.

Urine specimens have been analyzed for MEK using a similar gas chromatographic approach that utilized solid-phase extraction media for initial isolation and preconcentration of the sample (91). The analysis of blood specimens for MEK has been accomplished using purge and trap gas chromatography–mass spectrometry, gas chromatography coupled with Fourier transform infrared spectroscopy, and gas chromatography accompanied by ion trap detection (92–94). Petrarulo et al. reported that the MEK in the urine of occupationally exposed workers could be analyzed by high-pressure liquid chromatography after derivatization with 2,4-dinitrophenylhydrazine (95). Using this method, the MEK concentration in the urine of 9 nonexposed control subjects was less than 0.1 mg/L, whereas the concentration range in 60 rotogravure workers was 0.3–3.6 mg/L at the end of the work week.

1.3.5.3 Other. Ong et al. evaluated potential biomonitors for MEK exposure in a group 59 men with 8-h time-weighted average exposures ranging from 10 to 300 ppm (125). About 10% of the absorbed MEK was estimated to be excreted in the breath, and the exposures all resulted in blood concentrations of MEK that were less than 10 μg/mL. The evaluation of end-of-shift blood, breath, and urine specimens revealed better exposure correlations for the concentration in blood than for the levels in expired air due to the marked variability in the breath excretion data. The best correlations were, however, observed between the workroom air concentration and the level in urine ($r = 0.89$). The data indicated that an MEK exposure of 200 ppm would result in 9.7 ppm of MEK in the expired air and 3.62 mg/L in the urine.

Caughlin reported that the post mortem blood and vitreous humor specimens from homicide victims were found contain MEK as a result of using a new fingerprinting procedure that employed a fluorescent dye dissolved in 20% MEK (126). Jones et al. found that the MEK in expired air could interfere with the breath analyzers used to identify drunk drivers (127). Jones et al. also found that MEK was present in 77 of 21,153 blood specimens collected from individuals suspected of driving while intoxicated in Sweden. The average blood concentration of MEK was 49.2 mg/L and the range was 5–144 mg/L. A good correlation existed between the blood levels of MEK and the levels of acetone, which indicated the possible consumption of a widely available denatured alcohol solution that contained 5% MEK and 2% acetone (128).

1.4 Toxic Effects

Methyl ethyl ketone is slightly toxic and may be irritating to the skin, eyes, and nasal passages. It may produce transient corneal injury. Irritation is common at atmospheric concentrations above 200 ppm. Exposure to high concentration can produce central nervous system depression.

1.4.1 Experimental Studies

1.4.1.1 Acute Toxicity. MEK is weakly toxic following acute administration by either the enteral or parenteral routes. Pozzani et al. determined the oral LD_{50} value to be 6.86 mL/kg [95% confidence interval (CI), 5.59–8.45 mL/kg] in female Nelson rats (129). Tanii et al. reported that the LD_{50} of MEK in male ddY mice was 56.16 mmol/kg (4.05 g/kg) with a 95% confidence interval of 44.14–70.98 mmol/kg (3.18–5.12 g/kg) (130). The LD_{50} value for MEK was 73.35 mmol/kg when the mice were previously treated with a single intraperitoneal (IP) dose of carbon tetrachloride. Zakhari and Snyder reported that the 24-h oral LD_{50} of MEK in male CF-1 mice was 43.5 mmol/kg (3.14 g/kg) (131). Kimura et al. reported that the 7-day oral LD_{50} values for immature,young, and old male Sprague–Dawley rats was 3.1 mL/kg (95% CI, 2.5–3.9 mL/kg), 3.6 mL/kg (95% CI, 2.9–4.4 mL/kg), and 3.4 mL/kg (95% CI, 2.6–4.4 mL/kg), respectively (132). Kennedy and Graepel ranked MEK as "slightly toxic" based upon a summary of the acute lethality studies conducted by the oral and pulmonary routes (133). Kohli et al. reported that a single 80.5 mg/kg oral dose of MEK to albino rats of either sex was able to delay the onset of isoniazid-induced convulsions and appreciably reduce the percentage of rats that convulsed and died from the treatment. The same dose of MEK also provided 100% protection against the seizures produced by electroshock treatments and 40% protection against the convulsions caused by the administration of metrazol (134).

DiVincenzo and Krasavage administered a single IP dose of 750, 1500, or 2000 mg/kg of MEK to male guinea pigs and observed that the highest two dose levels caused an elevation in serum ornithine carbamyl transferase activity 24 h after treatment (135). Liver histology revealed lipid accumulation, but no necrosis, at the two highest dose levels. Lundberg et al. reported an IP LD_{50}value of 607 mg/kg (95% CI,486–758 mg/kg) in female Sprague–Dawley rats (136), whereas Zakhari and Snyder found the 24-h IP LD_{50} of MEK to be 23.0 mmol/kg (1.66 g/kg) in male CF-1 mice (131). Terhaar reported acute IP LD_{50} values of 1.28 mL/kg (95% CI, 0.81–1.98 mL/kg) and 0.73 mL/kg (forced regression) in male Carworth rats and male guinea pigs (Duncan–Hartley), respectively (137). When a mixture of MEK:MIBK (9:1 ratio) was administered IP, the LD_{50} values increased to 1.64 mL/kg (95% CI, 1.00–2.80 mL/kg) and 1.06 mL/kg (forced regression) for rats and guinea pigs, respectively. Likewise, the LD_{50} values increased to 1.41 mL/kg (95% CI, 0.91–2.20 mL/kg) and 0.94 mL/kg (forced regression) when a mixture of MEK:MnBK (9:1 ratio) was administered to rats and guinea pigs, respectively. Tham et al. studied the effects of MEK on the vestibulo-ocular reflex of female Sprague–Dawley rats following continuous intravenous (IV) infusion of 0.1–10% for 60 min (138). The vestibulo-ocular reflex was depressed (decreased eye movement in response to rapid rotation) at an MEK infusion rate of 70 μM/min/kg and an MEK blood level of 1.4 mM/L. The depressive effects were noted to occur at blood levels that did not cause CNS depression. Tanii examined the anesthetic potency of MEK by determining the concentration required to inhibit the righting reflex in male ddY mice by 50% (139). The AD_{50} value following IP administration was 16.0 mmol/kg. As a measure of potential neurotoxicity, Huang et al. evaluated the *in vitro* effects of MEK on the binding affinity of ^{3}H-dihydroalprenolol for beta-adrenergic receptors and on the enzymatic activity of Na^{+}-K^{+}-adenosone triphosphatase from male ddY mice synaptosomes (140). The concentration causing a 50% inhibition in response (IC_{50}) was 478 μM in the receptor assay and 211 μM in the enzyme assay. When evaluated with other ketones, the results were shown to correlate well with lipophilicity.

Opdyke reported the dermal LD_{50} value for rabbits to be greater than 5 g/kg (141), and Smyth et al. reported the value to be greater than 10 g/kg(142). Panson and Winek quoted a rabbit dermal LD_{50} value of 13 g/kg for MEK (143). Wahlberg quantified the erythema (laser Doppler flowmetry) and edema (skin-fold thickness) that developed at the site where 0.1 mL of MEK was applied daily for ten consecutive days (144, 145). Erythema and edema appeared 24–72 h after the first treatment and were more frequent and more severe in New Zealand rabbits than in Hartley guinea pigs. Anderson et al. assessed the erythema and edema (subjective macroscopic observations), epidermal thickness (micrometer measurements), and inflammatory cell infiltration (histologic examinations) in female Duncan–Hartley guinea pigs that were treated with 10 μL of MEK three times daily for three days (146). A slight redness was observed on the second day of treatment, and the epidermal thickness was slightly higher on the third day; however, no change was noted in the inflammatory cell response.

Smyth et al. rated the skin damage of MEK for albino rabbits that were treated with varying volumes or concentrations of the solvent (142). On a scale of 1 to 10, MEK was given a rating of 2 for skin irritation. Weil and Scala summarized the findings from an interlaboratory comparison of the skin irritancy of a large group of solvents in male albino rabbits (147). The skin test was conducted with 0.5 mL of MEK placed under an occlusive wrap. The primary skin irritation score (erythema, edema, and necrosis) for MEK ranged from 0.0 to 3.2 (maximum value of 30) for the 22 laboratories that conducted the skin test. Twelve percent of the laboratories rated MEK as a skin irritant.

Smyth et al. rated the eye damage of MEK for albino rabbits that were treated with varying volumes or concentrations of the solvent (142). On a scale of 1 to 10, MEK was given a rating of 5 for corneal injury. Larsen et al. measured the ocular edema in anesthetized male albino rabbits 1 h after instillation of MEK under the eyelid (148). An MEK concentration of 1.8 M (129.8 g/L) was needed to produce a 50% increase in the moist-to-dry-weight ratio of the treated eye. Weil and Scala summarized the findings from an interlaboratory comparison of the eye irritancy of a large group of solvents in male albino rabbits (147). The eye test was performed with 0.1 mL of MEK and scored according to the Draize procedure. The median 24 h and 7 d scores from the 24 laboratories that examined the eye irritancy of MEK were 19.2 and 0.8 (maximum score of 110), respectively. Seventy-one percent of the laboratories rated MEK as an eye irritant. Sivak and Herbert cultured the ocular lens from cows and examined the effects of MEK on optical performance (focal length and transmittance) using a scanning laser beam (149). Treatment of bovine lenses with a 2.0% solution of MEK was found to cause a moderate change in the treatment time necessary to cause a 100% change in focal-length variability. Reboulet et al. used cultured rabbit corneal fibroblasts to assess the ocular irritancy potential of MEK relative to results from the Draize assay (150). Using cytolysis as the endpoint, MEK volumes up to 25 μL were judged to be nonirritating. Kennah et al. compared the Draize irritation method for scoring ocular injury to an ocular swelling technique that entailed corneal thickness measurements (151). Ocular irritation (Draize) and swelling (corneal thickness) were determined in rabbits following treatment with MEK solutions ranging from 1 to 100% in strength. A maximum Draize score of 50 (moderately irritating) was obtained at 24 h post-treatment with undiluted MEK. Corneal swelling, however, was severe, increasing over two-fold relative to controls at the highest level. The diluted MEK solutions (3, 10, and 30%) caused minimal injury in both procedures.

Safronov et al. determined the LC_{50} of MEK to be 59.8 mg/m^3 in male albino rats and 124 mg/m^3 in male albino mice following a 1 h exposure (152). When the exposure duration was increased to 4 h, the LC_{50} values for rats and mice decreased to 12.7 and 32.0 mg/m^3, respectively. La Belle and Brieger found that adult white mice could survive for 43 min when exposed to 103,000 ppm of MEK (153). The 14-d LC_{50} value for a 4-h exposure was 11,700 ppm and the standard error of the mean 2400 ppm.The LC_{50} for MEK was not appreciably different from the value obtained when a mixture containing about 56% MEK, and an unknown percentage of seven other solvents was used to conduct the inhalation exposures. Smyth et al. reported that rats exposed to 2000 ppm of MEK vapor for 2 h showed no apparent toxicity. An 8-h exposure to 8000 ppm of MEK,however, killed half the animals, and doubling the exposure to 16,000 ppm caused the death of all animals in the exposure group (142). Carpenter et al. reported that a 4-h exposure to 2000 ppm of MEK killed 2–4 (exact number not stated) of the six Sherman rats that were exposed (154). The authors used these data to rank the MEK inhalation hazard as "moderate."

Zakhari and Snyder reported that the 45-min inhalation LC_{50} value for male CF-1 mice was 2.84 mmol/L (69,400 ppm) (131). Pozzani et al. reported an inhalation LC_{50} concentration (observation period not stated) of 23.5 mg/L (7970 ppm) and a 95% confidence interval of 12.8–43.2 mg/L (4340–14,640 ppm) in female Nelson rats receiving a single 8-h exposure to MEK (129). When MEK was co-administered with equal amounts of MIBK and butyl alcohol, the inhalation LC_{50} was greater than additive, which suggested an antagonistic interaction with the solvent mixture. When the mixture was administered by the oral route, however, the LD_{50} was approximately equal to the value predicted from a mathematical equation that assumed an additive relationship.

Patty et al. exposed guinea pigs to MEK vapor concentrations of 0.33, 1.0, 3.3, and 10% (3300, 10,000,33,000, and 100,000 ppm) for periods up to 13.5 h (155). The acute effects of MEK advanced through several distinct phases as the exposure proceeded with nose irritation, eye irritation, lacrimation, incoordination,narcosis, labored breathing, and death occurring in succession. None of these signs or symptoms was observed in the animals exposed to 3300 ppm MEK. Vapor concentrations of 33,000 ppm or greater caused death in the animals after 45–260 min of exposure. Narcosis was observed after 10–280 min of exposure at the three highest concentrations. Gross pathology revealed slight to marked congestion in the brain, lungs, liver, and kidneys of those animals that succumbed during the exposure. Animals surviving an exposure to 100,000 ppm of MEK for at least 30 min showed severe corneal opacity that was barely present 8 d after the exposure. Specht exposed two female outbred guinea pigs to the vapors of MEK and documented the clinical signs of toxicity and their temporal development (156). Animals were exposed to MEK concentrations of 1.0, 2.5, or 5.0% (10,000, 25,000, or 50,000 ppm) for periods of 4–14 h. Respiration rate, heart rate, and rectal temperature were found to decrease as the exposure duration or exposure concentration was increased. An exposure to 25,000 ppm of MEK for about 5 h caused the following overt signs of intoxication: salivation, lacrimation, incoordination, loss of auditory and corneal reflex, and death. Frantík et al. determined the isoeffective concentrations necessary to produce a percentage decrease in the maximum intensity of an electrically evoked seizure in both male Wistar rats and female H strain mice (157). The MEK concentration needed for a 30% decrease in response for rats was 1250 ppm (90% CI, 1010–1490 ppm), whereas the concentration required for a 37% decrease in mice was 860 ppm (90% CI, 590–1130 ppm).

Panson and Winek found that four out of six Sprague–Dawley rats died instantly when they were administered 1 mg/kg of undiluted MEK directly into their lungs (143). Gross findings included some lung hemorrhage that was not severe; death was attributed to asphyxiation, cardiac arrest, or respiratory arrest. DeCeaurriz et al. studied the CNS effects of MEK in a "behavioral despair" swimming test that determined the vapor concentration necessary to cause a 50% decrease in the length of time that male Swiss mice remained immobile in a water tank (158). The concentration of MEK that produced this effect was 2065 ppm (95% CI, 1841–2553 ppm), which was among the highest values recorded for the group of solvents tested. Garcia et al. reported neurobehavioral changes in schedule-controlled operant behavior(SCOB) (variable interval schedule) of Sprague–Dawley rats exposed to MEK vapor concentrations ranging from 25 to 800 ppm for 2–6 h (159). The increase in lever-pressing behavior was shown to persist for upto 11 d in some of the animals exposed to 25 ppm for 6 h; the effect, however, was extremely variable with some animals showing no change in the response rate.

DeCeaurriz et al. measured the irritancy of MEK in a mouse bioassay that quantified the sudden reflexive drop in respiration caused by a short duration vapor exposure (160). Male Swiss OF_1 mice were exposed to at least four vapor concentrations of MEK for 5 min. The concentration-related decline in the respiration rate allowed for the calculation of an RD_{50} value, which corresponded to the irritant concentration needed to produce a 50% decrease in the initial respiration rate. The RD_{50} value for MEK was very high, 10,745 ppm, and was superseded by only one compound, acetone. The RD_{50} value was presented as a reliable means of estimating the irritation threshold of chemical vapor, and the results for MEK have been used to develop several quantitative structure–activity relationships for establishing occupational exposure limits (161–165). Muller and Black examined the sensory irritation caused by a Spanished wall covering that emitted a relatively large amount (514 $\mu g/m^3$) of MEK at room temperature and found less than a 5% depression in respiratory rate in Swiss–Webster mice (166). Hansen et al. examined the receptor activation mechanism responsible for the irritancy effects of MEK in male CF-1 mice and found that the thresholds for the decrease in respiratory rate and tidal volume were affected by the length of exposure (167). The concentration threshold for respiratory rate depression decreased from 3589 ppm at the beginning of an exposure to 2518 ppm after 21–30 min of inhalation. The authors concluded that at low, but not high, concentrations MEK could desensitize the trigeminal receptor responsible for the irritancy effects of the chemical.

An examination of the toxicity and blood kinetics of both 2-butanol and MEK led Traiger and Bruckner to report that MEK was likely responsible for the potentiation of carbon tetrachloride (CCl_4) hepatotoxicity observed when male Sprague–Dawley rats were treated with either of these compounds (168). Dietz and Traiger further suggested that the oral treatment of male Sprague–Dawley rats with either MEK (2.1 mL/kg) or the MEK metabolite 2,3-butanediol (2.12 mL/kg) potentiated the hepatotoxicity of CCl_4 as measured by the increase in serum ALT activity and liver triglyceride content (169). Traiger et al. found that a 1.87-mL/kg oral dose of MEK or a 2.2-mL/kg dose of 2-butanol produced the greatest potentiation of CCl_4 hepatotoxicity when they were administered 16 h prior to the CCl_4 treatment (170). In addition, the liver of the rats treated with MEK displayed an increase in acetanilide hydroxylase, but no meaningful change in liver cytochrome c reductase activity, cytochrome P450 content, or cytochrome b_5 content. Examination of the hepatocytes from the treated rats by electron microscopy revealed an increase in the smooth endoplasmic reticulum. The authors suggested that the potentiation of CCl_4 hepatotoxicity by MEK could be related to the stimulation of the hepatic mixed function oxidase system.

Hewitt et al. examined the relationship between carbon chain length and the capacity of various ketones to potentiate chloroform ($CHCl_3$) hepatotoxicity (171, 172). Five ketonic solvents [acetone, MEK, 2-pentanone (MPK), MnBK, and 2-heptanone (MnAK)] were administered to male Sprague–Dawley rats 18 h before an IP dose of $CHCl_3$. A positive correlation was observed between the number of carbon atoms in the solvent and the increase in plasma ALT and ornithine carbamyl transferase activities. None of the ketones caused any hepatotoxicity when given alone. Brown and Hewitt investigated the dose–response and structure–activity relationships for ketone solvent-induced potentiation of $CHCl_3$ hepatotoxicity and nephrotoxicity (173). A challenge dose of $CHCl_3$ was administered 18 h after oral treatment of male Fisher 344 rats with 1.0–15.0 mmol/kg of the following five ketonic solvents: acetone, MEK, MPK, MnBK, and MnAK. An MED dose of 12.5 mmol/kg caused a slight increase in mortality, and a dose of 5.0 mmol/kg or greater caused a statistically significant increases in plasma creatinine and ALT activity when challenged with $CHCl_3$; however, the degree of damage showed no apparent dose–response relationship. MEK pretreatment to $CHCl_3$ challenge also caused a decrease in *p*-aminohippurate (PAH) uptake by kidney slices and an increase in the severity of renal tubular necrosis when the dosages exceeded 1.0 mmol/kg. The results indicated that pretreatment with each of the five ketones was equally effective at potentiating the nephrotoxicity and hepatotoxicity of $CHCl_3$ when the dose was between 5 and 10 mmol/kg. The authors concluded that the correlation between the degree of $CHCl_3$ potentiation and length of the carbon chain in the ketonic solvent was only apparent at a dose of 1.0 mmol/kg.

Hewitt et al. also studied the relationship between the time interval for the $CHCl_3$ and MEK treatment and the severity of $CHCl_3$-induced hepatobiliary dysfunction (174). Male Sprague–Dawley rats were given a 15-mmol/kg oral dose of MEK 10–90 h before an oral dose of $CHCl_3$. The administration of MEK 18 h prior to $CHCl_3$ treatment was found to cause the greatest increase in plasma bilirubin and the greatest decrease in bile flow rate. There was no potentiation of the hepatobiliary damage when the time interval between MEK treatment and $CHCl_3$ administration was greater than 24 h. No hepatobiliary dysfunction was noted when MEK was administered alone. Hewitt et al. reported that MEK did not potentiate $CHCl_3$ hepatotoxicity by altering hepatic lysosomal fragility or the ability of the hepatic mitochondria and microsomes to sequester calcium (175). Hewitt et al. also studied the activation mechanism that was responsible for the potentiation of $CHCl_3$ hepatotoxicity by MEK (176). The studies were designed to determine if increased mixed function oxidase-mediated metabolism or decreased glutathione-mediated detoxification was responsible for the potentiating effects of MEK. Male Sprague–Dawley rats were administered an oral dose of $CHCl_3$ at various times following a 15-mmol/kg oral dose of MEK. Liver microsomal fractions from MEK-treated animals contained slightly higher cytochrome reductase and 7-ethoxycoumarin *O*-deethylase activities; however, cytochrome P450 levels, aminopyrine *N*-demethylase activity, and aniline hydroxylase activity were unaffected. MEK pretreatment significantly affected hepatic glutathione levels, but only at a single time point (24 h). The authors suggested that the potentiating effects of MEK were mostly attributable to increased rates of $CHCl_3$ metabolism by the monooxygenase system; however, a specific mechanism was not proposed.

Raymond and Plaa examined the dose–response relationships for the potentiation of CCl_4 hepatotoxicity and $CHCl_3$ nephrotoxicity in male Sprague–Dawley rats orally pretreated with a daily dose of MEK for 3 d (177). The minimum effective dose of MEK necessary to increase plasma ALT activity in the CCl_4-treated animals was 5.4 mmol/kg. When a 6.8-mmol/kg pretreatment regimen was used, the oral dose of CCl_4 needed to affect ALT activity in 50% of the animals (ED_{50}) was 4.4 mmol/kg, which was greater than the value for the other two ketones tested (acetone and MIBK). An MEK daily dose of 13.6 mmol/kg was necessary to potentiate the nephrotoxic effects of $CHCl_3$.

The ED_{50} value for $CHCl_3$ was found to be 2.0 mmol/kg when PAH uptake was used as the indicator of toxicity. In supplemental studies, *in vitro* enzyme activities were evaluated using the hepatic and renal microsomes isolated from male Sprague–Dawley rats treated with either 6.8 or 13.6 mmol/kg of MEK (178). The low dose of MEK significantly increased liver aniline hydroxylase activity, whereas the high dose significantly increased kidney and liver aniline hydroxylase activity but did not affect aminopyrine *N*-demethylation or benzphetamine *N*-demethylation activities. In addition, treatment at the two MEK dose levels did not significantly increase the rate of irreversible binding of radiolabeled CCl_4 or $CHCl_3$ to liver and renal microsomal protein. Raymond and Plaa also examined whether the potentiation of CCl_4 hepatotoxicity was the result of changes in cellular membranes (179). Treatment of male Sprague–Dawley rats with a daily oral dose of MEK (6.8 mmol/kg) for 3 d did not affect membrane fluidity in isolated basal canalicular membranes, or the activities of three plasma membrane enzymes: 5′-nucleotidase, leucine aminopeptidase, or alkaline phosphatase.

Cunningham et al. examined the effects that an IP dose of MEK ranging from 5to15 mmol/kg had on the pharmacologic and metabolic activity of ethanol in male CD-1 mice (180). MEK was found to prolong the loss of the righting reflex due to ethanol in a dose-related manner and to reduce the blood elimination rate of ethanol when an MEK dose of 15 mmol/kg was administered. The mechanism appeared to involve the competitive inhibition of liver alcohol dehydrogenase activity, which was reduced when MEK was added *in vitro*.

1.4.1.2 Chronic and Subchronic Toxicity. Egan et al. reported that male Sprague–Dawley rats continuously exposed 22 h/d to 500 ppm of MEK vapors for up to 6 mo failed to show any significant clinical or histopathological evidence of neurological dysfunction (181). The lumbar cord, dorsal and ventral spinal roots, spinal ganglia, sciatic notch, and tibial nerve were included in the histological examination. Duckett et al. exposed rats (sex and strain not stated) to either 200 ppm of MnBK or 2200 ppm of an MEK:MnBK mixture (10:1 ratio) for 6 wk at 8 h/d (5 d/wk) and found clinical (muscular weakness) and histopathological (axonal hypertrophy and degeneration in the sciatic nerve) evidence of neurological damage in both treatment groups (182).

Saida et al. reported that MEK vapors could potentiate the neurotoxic effects of methyl *n*-butyl ketone (MnBK) in rats when the animals were simultaneously exposed to the two chemicals (183). Quantitative histopathology was performed on sciatic nerves and intramuscular nerves of intrinsic foot muscles taken from Sprague–Dawley rats exposed for up to 5 mo (24 h/d) to either MEK (1125 ppm), MnBK(400 ppm), or a mixture of MEK:MnBK (5:1 ratio). Noperipheral neuropathy occurred in the animals treated with MEK alone. Co-exposure to MEK:MnBK caused hind-limb paralysis after 25 d of exposure, compared to 42 d with MnBK alone. Electron microscopy revealed a marked increase in the number neurofilaments, neurotubules, denuded fibers, swollen axons, and myelin sheath inpouchings in the nerves from animals simultaneously treated with both solvents. The data indicated that an MEK co-exposure could shorten the latency period for the onset of MnBK-induced neuropathy.

Altenkirch et al. reported that MEK could potentiate the neurotoxicity of *n*-hexane in male Wistar rats exposed by inhalation for 15 wk (184, 185). The animals were exposed to

6000 ppm of MEK, 10,000 ppm of *n*-hexane, or 10,000 ppm of an MEK:*n*-hexane mixture (1.1:8.9 ratio) for 8 h/d and 7 d/wk. The animals exposed to MEK all died of bronchopneumonia and showed no signs of peripheral neuropathy. The rats co-exposed to MEK and *n*-hexane developed severe hind-limb paralysis after about 5 wk of exposure, compared to about 9 wk for those exposed to *n*-hexane alone. Paranodal axonal swelling, neurofilament formation, and myelin retraction were observed in tibial nerves and in nerve tracts from the spinal cord and medulla of rats co-exposed to MEK and *n*-hexane for 4 wk. In subsequent studies, Altenkirch et al. explored the concentration–response relationship for the potentiation of *n*-hexane neuropathy by MEK (186). Male Wistar rats were exposed by inhalation to either a 500 or 700 ppm mixture of MEK:*n*-hexane blended at several different ratios. Two continuous (7 d/wk) exposure regimens were employed: 22 h/d for up to 9 wk and 8 h/d for 40 wk. All of the animals exposed for 9 wk survived the treatment; however, narcosis and an orangish-red discoloration of the skin and fur were commonly observed. Severe hind-limb paralysis occurred about 1 wk earlier in animals exposed to combinations of MEK and *n*-hexane (1:4 and 2:3 ratios) than in animals receiving a 500-ppm *n*-hexane exposure. The rats exposed to a 700-ppm mixture, in contrast, showed no difference in the time to onset relative to *n*-hexane alone. The rats exposed to the 700-ppm combinations for 8 h/d failed to show any overt clinical signs of neurotoxicity during the 40-wk study.

Spencer and Schaumburg treated cats with combinations of MEK, MnBK, and methyl isobutyl ketone (MIBK) and looked for synergistic interactions leading to solvent-induced neuropathy (187). Plantar nerves, interosseous muscle, posterior tibial nerves, and pacinian corpuscles were excised from cats treated twice daily with 150 mg/kg SC injections (5 d/wk) of MnBK, MEK, MIBK, MEK:MIBK (9:1 ratio), or MEK:MnBK (9:1 ratio) for up to 8.5 mo. Narcosis, excessive salivation, and a generalized weakness were observed in the animals treated with MEK or MEK:MnBK. Overt neurological effects were confined to those cats treated with MnBK alone. Tissue damage was observed in the animals treated with either MnBK or the MEK:MnBK mixture and was more severe for the cats receiving MnBK alone. Tibial nerve branches from the animals treated with the MEK:MnBK mixture had an abnormal number of nerve fibers with segmental remyelination. In a similar series of studies, O'Donoghue and Krasavage administered 150 and 300 mg/kg doses (5 d/wk) of MnBK, MEK, MIBK, MEK:MIBK (9:1 ratio), or MEK:MnBK (9:1 ratio) by SC injection to male beagle dogs for approximately 11 mo (188,189). The dose was decreased from 300 to 150 mg/kg after 1 wk due to severe irritation at the injection site. Clinical signs of neurotoxicity were confined to those animals treated with MnBK alone. Animals treated with MEK-containing mixtures showed statistically significant increases in the white blood cell count and in the distribution of white blood cells in the differential count; these changes, however, were attributed to the inflammatory reaction that developed at the injection site. Monthly analysis of plasma and cerebrospinal fluid specimens for changes in clinical chemistry did not show any statistically significant differences that could be specifically related to the treatment. Except for the animals treated with MnBK, electromyographic examinations of fore-limb and hind-limb muscles were normal for each treatment group. Histologic examination of tissues and organs other than the skin failed to reveal any treatment-related lesions for animals given MEK or an MEK-containing mixture.

Krasavage and O'Donoghue also studied the neurotoxicity of MEK and MEK-containing mixtures in male Sprague–Dawley rats treated by SC injection for 35 wk (190). The rats were treated daily (5 d/wk) with MEK, MIBK, MnBK, an MEK:MIBK mixture (9:1 ratio), and an MEK:MnBK mixture (9:1 ratio). Dose levels of 10, 30, or 100 mg/kg were doubled to 20, 60, or 200 mg/kg after the first week of treatment. Body weight gain was decreased at various times in each of the treatment groups; however, there were no clinical signs of neuropathy in any of the treated animals. Histopathology on the sciatic, tibial, peroneal, and sural nerves failed to show any treatment-related alterations in morphology. O'Donoghue et al. also found no clinical or histological evidence of a neuropathy in male Hartley guinea pigs treated dermally with 2 mL of MEK, MIBK, MnBK, MEK:MIBK (9:1 ratio) or MEK:MnBK (9:1 ratio) twice a day for 31 wk (191).

Schmidt et al. (192) and Schnoy et al. (193) examined MEK:*n*-hexane solvent interactions in the lung tissue of male Wistar rats exposed by inhalation. Three 8 h/d exposure regimens were employed: (1) 500 ppm of *n*-hexane or an MEK:*n*-hexane mixture (1:4 ratio) for up to 89 d, (2) 700 ppm of *n*-hexane or an MEK:*n*-hexane mixture (2:5 ratio) for up to 86 d, (3) 10,000 ppm of *n*-hexane or an MEK:*n*-hexane mixture (1:9 ratio) for up to 14 d. Intrapulmonary nerves from the hilus, central, and peripheral sections of the lung were examined by light and electron microscopy for damage to the axon cylinder, medullary sheath, and the ganglion cells. In addition, alveolar epithelium and type I and type II pneumocytes were examined for evidence of fatty degeneration and lamellar inclusions. The 10,000 ppm *n*-hexane exposures failed to alter the ultrastructure of the pulmonary nerves, but did cause the appearance of numerous lamellar inclusions in type II cells. The latter effects were also noted in the cells from animals exposed to MEK:*n*-hexane mixtures. After 2 wk of exposure to the MEK:*n*-hexane mixtures, a focal fatty degeneration of type II cells appeared along with an increase in alveolar brush cells that was not apparent in the animals treated with *n*-hexane alone. These latter alterations were only observed in the *n*-hexane-treated animals after 7 wk of exposure. Axonal swelling and neurofilament proliferation of medullary nerve fibers were observed together with focal fatty degeneration of Schwann cells. These morphological alterations of the intrapulmonary nerves occurred after about 7 wk of exposure to *n*-hexane versus 6 wk of exposure to the MEK:*n*-hexane mixtures. The authors attributed the histopathological changes to an interference in lipid metabolism with subsequent effects on nerve morphology. Co-exposures to MEK:*n*-hexane were judged to be more potent than *n*-hexane alone at causing these effects.

Takeuchi et al. performed monthly neurophysiological evaluations on rats exposed by inhalation to MEK and *n*-hexane (194). Male Wistar rats were exposed to MEK (200 ppm), *n*-hexane(100 ppm), or 300 ppm of MEK:*n*-hexane (2:1 ratio) 12 h/d (7 d/wk) for 6 mo. Following 1 mo of exposure to MEK, the dorsal tail nerve of the rats showed a significant increase in motor nerve conduction velocity and mixed nerve conduction velocity and a significant decrease in distal latency that was not observed during subsequent evaluation sessions. Statistically significant decreases in the motor nerve conduction velocity and mixed nerve conduction velocity were observed after 5 and 6 mo of exposure to the MEK:*n*-hexane mixture. On termination of the study, there were no histopathologic abnormalities in the proximal or distal nerves of the tail from any of the solvent-exposed animals. Using the results from previously reported experiments, the authors reported that a concentration-related decrease in the motor nerve conduction velocity occurred with subchronic *n*-hexane exposures of 100, 200, and 500 ppm and that the effects for the MEK:*n*-hexane–treated rats were more severe than in those rats exposed to 200 ppm of *n*-hexane.

In a similar set of experiments, Iwata et al. performed monthly electrophysiological examinations with male Wistar rats exposed to MEK and *n*-hexane vapors for 33 wk (195). Animals received an 8-h/d inhalation exposure to 500 ppm of *n*-hexane, 1000 ppm of an MEK:*n*-hexane mixture (1:1 ratio), or 1000 ppm of a toluene:*n*-hexane mixture (1:1 ratio). None of the three exposure regimens caused a statistically significant change in motor nerve conduction velocity or distal latency in the dorsal tail nerve of the rat. Likewise, no histological changes were observed in tail nerve morphology in any of the exposure groups. Regular measurement of the urinary metabolites of *n*-hexane revealed a large initial decrease in the amount of 2,5-hexanedione excreted in the urine of rats exposed to MEK:*n*-hexane compared to the rats exposed to *n*-hexane alone. The changes in metabolite excretion were more severe and persistent following exposure to toluene:*n*-hexane, which led the authors to conclude that the mechanism responsible for the potentiation of *n*-hexane neurotoxicity by MEK may not involve alterations in metabolism. Ichihara et al. exposed male Wistar rats to 2200 ppm of MEK:*n*-hexane (1:10 ratio) or 4000 ppm MEK:*n*-hexane (1:1 ratio) for 12 h/d 6 d/wk for 20 wk and measured nerve conduction velocity, distal latency, and the urinary excretion of 2,5-hexanedione at 4-wk intervals (196). When compared to treatments with *n*-hexane alone, the 4000-ppm combined treatment caused a significant increase in distal latency and a significant decrease in the motor nerve conduction velocity at weeks 12, 16, and 20. The 2200-ppm combined treatment also caused some decrement in performance at weeks 16 and 20, but the changes were not as severe. Compared to *n*-hexane alone, the amount of 2,5-hexanedione excreted in the urine was significantly lower on the first day of exposure to 4000 ppm of MEK:*n*-hexane, but was significantly higher at all subsequent time points. The 2200-ppm combined exposure did not, however, affect the urinary elimination of 2,5-hexanedione.

O'Donoghue et al., in contrast to the findings reported, found evidence of metabolic induction and chemical potentiation inmale Sprague–Dawley rats administered MEK, together with ethyl *n*-butyl ketone (EnBK) by either the oral or pulmonary route (197). Rats were given daily (5 d/wk) oral treatments of EnBK (0.25–4.0 g/kg) either alone or together with 0.75 or 1.5 g/kg of MEK for 14 wk. In separate experiments, rats were exposed to 700 ppm of EnBK or to 770, 1400, or 2100 ppm of an MEK:EnBK mixture (1:10, 1:1, and 2:1 ratios, respectively) for 4 d at 16–20 h/d. In some animals, narcosis and death occurred several days after beginning treatment with 2 or 4 g/kg of EnBK and 1.5 g/kg of MEK. Clinical signs of neurotoxicity (hindquarter weakness) developed during the third week of treatment with 2 g/kg of EnBK after the fourteenth week of treatment with 1 g/kg EnBK and 1.5 g/kg MEK. Neuropathologic lesions were limited to these same two treatment groups and consisted of enlarged or "giant" axons in the sciatic and tibial nerves, the dorsal spinal roots, and the intramuscular nerves of calf and quadriceps. The central and peripheral nervous systems were both affected, but the former to a lesser degree. Neurogenic muscle lesions (angular or atrophied myofibers) were observed in the animals treated with 1 g/kg EnBK and 1.5 g/kg MEK. The urinary levels of the EnBK metabolites 2,5-heptanedione and 2,5-hexanedione were increased in the animals given 1 g/kg EnBK together with 1.5 g/kg MEK relative those animals given 1 g/kg of EnBK alone. A similar increase in EnBK metabolites was observed in the plasma of rats following exposure to 1400 or 2100 ppm of the EnBK:MEK mixture relative to the animals exposed to 700 ppm of EnBK. The authors concluded that MEK was capable of potentiating EnBK-induced neurotoxicity by inducing its metabolism to toxicologically active metabolites.

Cavender et al. conducted a 90-d inhalation study in male and female Fischer-344 and found no histopathologic evidence of neurotoxicity and little evidence of organ toxicity due to MEK (198). Animals were exposed to MEK concentrations of 1250, 2500, or 5000 ppm 5 d/wk for 6 h/d. No signs of nasal irritation were observed during the study and ophthalmologic examinations were negative at all exposure concentrations. Body weight gain was not significantly affected, but the male rats showed statistically significant increases in liver (absolute and relative) and kidney (relative to body weight) weight following the 5000-ppm exposure. Female rats showed statistically significant increases in absolute liver weight at all three exposure concentrations. Relative or absolute kidney- and brain-weight changes were also noted in the female rats exposed to 5000 ppm of MEK. Serum chemistry, urine chemistry, and blood hematology values were not affected, except in the case of female rats receiving a 5000-ppm exposure. These animals showed significant elevations in serum potassium, glucose, and alkaline phosphatase activity, and a significant decrease in alanine aminotransferase (ALT) activity. Histological examination of the organs failed to show any treatment-related lesions. Neurohistological examination of the sciatic and tibial nerves and of the medulla revealed several nonspecific changes in isolated animals.

Bernard et al. examined the nephrotoxic potential of MEK and 16 other solvents in female Sprague–Dawley rats treated intraperitoneally (199). The rats were administered daily 340 mg/kg injections of MEK (5 d/wk) for 2 wk. MEK did not cause renal damage, as determined from measurements of urine albumin.

Li et al. exposed female Wistar rats to 300 ppm of MEK for 7 d (8 h/d) and found that leukocyte and serum alkaline phosphatase activities were unaffected by the exposures (200). Geller et al. continuously (24 h/d) exposed male baboons to 100 ppm of MEK, 50 ppm MIBK, or 150 ppm of an MEK:MIBK mixture (2:1 ratio) for 7 d and measured any changes in operant performance (201). The animals were trained to perform an operant discrimination task that reportedly gave a collective measurement of associative learning, short-term memory, stimuli differentiation, psychomotor function, and reaction time. The performance measurements (percentage of correct responses) were not affected by either of the exposure regimens; however, the number of extra responses recorded during the initial delay period were increased for some of the daily test sessions. In addition, the mean response time for animals exposed to 100 ppm of MEK was appreciably slower than the control values for two of the four animals tested. Despite the variable and inconsistent results, the authors proposed pharmacokinetic and metabolic mechanisms to account for the differences between MEK-exposed animals and those exposed to the MEK:MIBK mixture. In preliminary operant behavior studies with two baboons exposed continuously to 20 or 40 ppm of MEK for 7 d, neither animal showed any impairment in a perceptual acuity/discrimination task (202). One of the two animals exposed to 20 ppm did, however, showed a decrease in the extra responses observed during the pre-exposure and post-exposure control sessions.

1.4.1.3 Pharmacokinetics, Metabolism, and Mechanisms. Ralston et al. studied the mechanism responsible for the MEK-induced potentiation of γ-diketone neuropathy by examining the toxicity, total neural tissue accumulation, and pharmacokinetics of 2,5-hexanedione following coadministration with MEK to male Fischer-344 rats (203). When a 2.2–4.4 mmol/kg/d dose of MEK:2,5-hexanedione (1:1ratio) was given by gavage for up to 85 d (5 d/wk),there was a reduction in time necessary to observe neurobehavioral performance decrements in sensorimotor tasks. This decrement was found to be correlated with higher blood values for 2,5-hexanedione when coadministered with MEK. The reduced blood clearance of 2,5-hexanedione led the authors to conclude that MEK acted by competitively inhibiting 2,5-hexanedione metabolism.

Couri et al. reported that hepatic microsomal enzyme activities were significantly enhanced in male Wistar rats exposed to 750 ppm MEK or to 975 ppm mixture of MEK:MnBK (3:1 ratio) intermittently (7 h/d) or continuously (24 h/d) for 7 d (204). Hexobarbital sleeping times were reduced following exposure to both the mixture and MEK alone. Aniline hydroxylase, aminopyrine demethylase, *p*-nitrobenzoate reductase, and neoprontosil reductase activities were found to be increased 2–3-fold in liver microsomal preparations from rats exposed under the two treatment regiment. Toftgård et al. reported that the exposure of male Sprague–Dawley rats to 800 ppm of MEK for 4 wk (5 d/wk) for 6 h/d resulted in an increase in absolute and relative (to body weight) liver weight together with an increase in liver microsomal biphenyl, benzo(a)pyrene, and steroid hydroxylase activities; however, there was no concomitant increase in liver microsomal P450 content (205, 206). Nedelcheva exposed male Wistar rats to an airborne MEK concentration of 4 mg/L for 4 h or to a 6 mg/ L combination of MEK:styrene (2:1 ratio) and measured cytochrome P450-associated enzyme activities in isolated lung, liver, and kidney microsomes (207). The MEK exposures caused a marked increase in hepatic ethoxyresorufin deethylation, benzyloxyresorufin deethylation, penthoxyresorufin deethylation, and chlorzoxazone hydroxylation. Lung and liver activities of ethoxyresorufin deethylation and chlorzoxazone hydroxylation were also appreciably affected by the MEK exposures. The combined exposure to styrene and MEK noticeably enhanced the activities of benzyloxyresorufin deethylation and penthoxyresorufin deethylation beyond what was observed with styrene or MEK alone. These data indicated that MEK had its greatest effect on the enzyme activities associated with cytochrome CYP2E1 from lung, liver, and kidney.

Raunio et al. reported that the oral administration of a 1.4-mL/kg d dose of MEK alone or in combination with a 1.0-mL/kg/d dose of *m*-xylene for 3 d increased the total microsomal cytochrome P450 content in the livers of male Wistar rats (208). The increase in cytochrome P450 was associated with proteins in the P450IIB and P450IIE subfamilies. The enzymatic activities of liver microsomal 7-ethoxyresorufin *O*-deethylase, 7-ethoxycoumarin *O*-deethylase, 7-pentoxyresorufin *O*-depentylase, *N*-nitrosodimethylamine *N*-demethylase, and aniline hydroxylase were all increased at least several fold by the treatments. The effects of MEK and *m*-xylene on the enzyme activities appeared to be additive in most cases. Mortensen et al. pretreated male Sprague–Dawley rats with a daily oral dose of 1.87 mL/kg MEK for 3 d and found that the *in vitro* metabolism of *n*-hexane to 2,5-hexanedione was significantly increased in the isolated hepatic S9 fraction (209). The levels of the three other metabolites, 2-hexanol, 2,5-hexanediol, and MnBK, were not affected. The *in vitro* addition of MEK to nontreated rat liver S9 inhibited the metabolism of *n*-hexane to 2,5-hexanedione and MnBK. These data indicated that MEK specifically induced the activity of the enzyme pathway responsible for the conversion of *n*-hexane to 2,5-hexanedione.

Aarstad et al. reported that vapor exposures to 500 or 2000 ppm of 2-butanol for 3–5 d resulted in high serum concentrations of MEK that were likely responsible for the increase in liver or kidney cytochrome P450 content and the increase in the *in vitro* hydroxylation of *n*-hexane (210). Carlson found that male Sprague–Dawley rats given a single IP injection of MEK at a dose of 0.75 or 1.5 mL/kg caused an increase in butanol oxidase activity in the liver and the lung microsomes, but there was no clear time-related or dose-related pattern to the observed changes (211). Harada et al. showed that an SC injection of 4.0 mmol/kg MEK or an MEK:MnBK mixture (1:1 ratio) for 3 d caused an increase in liver microsomal P450 content, along with an increase in aniline hydroxylase and aminopyrine *N*-demethylase activities in male Donryu rats (212). Similar treatment by SC injection, with 0.8 mmol/kg of MEK, had no effect on the liver microsomal P450 content or enzyme activities.

Zhao et al. administered a single subcutaneous dose of MEK:2,5-hexanedione in either a 1:1 or 5:1 ratio and examined how the combined treatment affected the levels of 2,5-hexanedione in the blood, urine, and sciatic nerve of male Wistar rats (213, 214). When compared to treatments with 2,5-hexanedione alone, the combined treatment with 13.0 mmol/kg MEK and 2.6 mmol/kg of 2,5-hexanedione resulted in a marked increase in the serum concentration, sciatic nerve concentration, and urinary amount of 2,5-hexanedione. Similar, although less dramatic, effects were also observed when the combined dose was reduced to 2.6 mmol/kg MEK and 2.6 mmol/kg of 2,5-hexanedione. The elevated levels of 2,5-hexanedione were related to a decrease in the elimination rate for the serum and sciatic nerve.

Liira et al. studied the behavior of MEK in male Wistar rats exposed to 600 ppm for 6–10 h/d either once or for eight consecutive days and found that the animals exposed repeatedly showed a slight reduction in liver glutathione content and a slight increase in the microsomal monooxygenase activity for one (7-ethoxyresorufin *O*-deethylase) of the four substrates examined (215). There was, however, no change in the liver microsomal cytochrome P450 content compared to untreated controls. The authors reported that a coexposure to 900 ppm of MEK:*m*-xylene mixture (2:1 ratio) resulted in higher blood concentrations of *m*-xylene and reduced urine concentrations of xylene metabolites when compared to the results for *m*-xylene alone. An 8-d coexposure to the solvents also resulted in a synergistic effect on all of the liver microsomal enzyme activities examined. The inhibition of *m*-xylene metabolism by MEK was found to diminish rapidly during the 18-h monitoring period that followed the coexposure. Ukai et al. exposed female Wistar rats for 8 h to 50 or 100 ppm of toluene either alone or in combination with 50, 100, 200, and 400 ppm of MEK and examined the excretion of two toluene metabolites in the urine (216). At both toluene exposure levels,the MEK coexposures resulted in a decrease in the urinary elimination of hippuric acid at exposure concentrations of 200 ppm or higher.

There was no effect, however, on the urinary elimination of *o*-cresol under any of the treatment conditions.

Shibata et al. performed similar studies in male Wistar rats and found that a single 8-h coexposure to MEK:*n*-hexane at concentrations of 2200 ppm (1:10 ratio), 2630 ppm (1:3.2 ratio), and 4000 ppm (1:1 ratio) reduced the urinary excretion of *n*-hexane metabolites when compared to an exposure to *n*-hexane alone (217, 218). In addition, the rate of elimination of *n*-hexane metabolites in the serum was examined at the highest coexposure concentration and found that the rates of production and clearance of 2,5-hexanedione were reduced. The authors concluded that MEK was capable of inhibiting the metabolism of *n*-hexane in a concentration-related manner and that the potentiation of *n*-hexane neurotoxicity by MEK could not be explained by the amount of 2,5-hexanedione being formed.

Robertson et al. administered a 1.87 mL/kg/d dose of MEK by oral gavage to male F344 rats for 1–7 d and found an increase in total cytochrome P450 content and 7-ethoxycoumarin *O*-deethylase activity after only a single treatment (219). When the MEK was administered orally for 4 d prior to a single 6-h inhalation exposure to 1000 ppm *n*-hexane, there was an increase in the concentration of the metabolites 2,5-hexanedione and 2,5-dimethylfuran in the blood, sciatic nerve, and testis relative to the animals exposed to *n*-hexane alone. The authors concluded that MEK could potentiate the neurotoxicity of *n*-hexane by inducing the enzymes responsible for the metabolic activation of *n*-hexane.

1.4.1.3.1 Absorption. Kinoshita et al. found that the ethanol metabolite, acetaldehyde, could inhibit the intestinal absorption of ethanol and MEK when both compounds were administered in combination to anesthetized male Wistar rats by insitu intestinal perfusion (220). Animals were perfused for 30 min with a solution containing 4% ethanol and 2% MEK. The rate constant for the intestinal absorption of MEK in the absence of any inhibitor of ethanol metabolism was 2.4 h^{-1}. This value decreased to 1.7 h^{-1} when the animals were pretreated with cyanamid to inhibit liver alcohol dehydrogenase activity and increase the level of blood acetaldehyde. Dietz et al. reported a peak MEK blood concentration of 0.95 mg/mL 4 h after treatment and a blood elimination profile that appeared to be zero order when Sprague–Dawley rats were given a 1.69 g/kg oral dose of MEK (221). The blood time-course data for MEK and its metabolites were used to construct a physiologically based pharmacokinetic model for 2-butanol. Using the gas uptake kinetics from a closed recirculating chamber, Kessler et al. determined that the pulmonary uptake of MEK was limited by the rate ventilation at vapor concentrations below 180 ppm (222). The pulmonary uptake of MEK was determined to be 40% in male Wistar rats. Only about 14% of the inhaled MEK was excreted unchanged by metabolism.

1.4.1.3.2 Distribution. Liira et al. examined the kinetics of MEK in male Wistar rats exposed to 600 ppm for 6–10 h/d either once or for eight consecutive days (215). Peak blood levels of MEK averaged 1041 and 1138 μM in the rats receiving the single and repeated exposures, respectively. Relative to a single exposure, multiple exposures to MEK did not affect the end-exposure perirenal fat concentration of MEK or the excretion of thioether metabolites in the urine. Using the data collected from gas uptake studies, Kessler et al. reported an MEK body-to-air partition coefficient of 103 for Wistar rats

(222). Ostwald's partition coefficients of water/air and olive oil/air were very similar, which suggested that MEK would equally distribute in the aqueous and fatty compartments of the body.

1.4.1.3.3 Excretion. MEK has been shown to be a pivotal metabolite in the metabolic elimination of 2-butanol in the rat and rabbit following either oral or inhalation exposure (221, 223,224). Akubue et al. (225) showed that MEK is excreted in the urine of female Sprague–Dawley rats treated with malondialdehyde as a means of causing oxidative stress and lipid peroxidation. The urinary excretion of MEK increased quickly from nondetectable levels at the start of the experiment to about 490 nmol/kg in the 12-h specimen. Mathews et al. found that MEK was excreted in the expired air of male Fisher 344 rats administered *trans*-1,2-dichloroethylene as a metabolic inhibitor of cytochrome P450 2E1 activity (226). Levels of MEK in the expired air increased from 20 pmol/100 g rat/min before treatment to 253 pmol/100 g rat/min 4–6 h after treatment. DiVincenzo et al. examined the metabolism and pharmacokinetics of MEK in guinea pigs given a single dose of 450 mg/kg by IP injection (227). The serum half-life for MEK was calculated to be 270 min, and the following three MEK metabolites were identified in the serum of the treated animals: 2-butanol, 3-hydroxy-2-butanone, and 2,3-butanedione. Dietz et al. detected the same three metabolites in male Sprague–Dawley rats given an oral dose of 1.69 g/kg MEK (221). Walter et al. reported that MEK displayed saturation kinetics in male Wistar rats at concentrations above 150 ppm with the maximum rate of metabolism being 600 μmol/h/kg (228). Rats exposed to 11,000 ppm of a MEK:*n*-hexane mixture (1:10 ratio) for 5 d (8 h/d) showed an increase in the maximum rate for the *in vivo* metabolism of *n*-hexane.

1.4.1.4 Reproductive and Developmental. Schwetz et al. exposed groups of pregnant Sprague–Dawley rats to MEK vapors for 7 h/d on days 6–15 of gestation (229). Exposure concentrations of 1000 or 3000 ppm did not produce any appreciable change in maternal body weight gain, liver weight, or plasma alanine aminotransferase activity. There were also no statistically significant effects on the number of implantations or fetal resorptions, the average litter size, or the conception rate. The total number of litters with fetuses showing skeletal anomalies was significantly increased following the 1000-, but not the 3000-ppm exposure. The 3000-ppm exposure appeared to cause a statistically significant increase in the total number of gross (acaudia, brachygnathia, and imperforated anus) and soft tissue (subcutaneous edema and dilated ureters) abnormalities. The authors concluded that MEK was embryotoxic, fetotoxic, and potentially teratogenic to rats.

In a followup study, Deacon et al. re-examined the developmental effects observed at an MEK vapor concentration of 3000 ppm (230, 231). Pregnant Sprague–Dawley rats were exposed to 400, 1000, or 3000 ppm of MEK for 7 h/d on days 6–15 of gestation. There were no effects on maternal liver weight; however, a significant decrease in maternal body weight gain was observed in the animals exposed to 3000 ppm of MEK. In addition, none of the exposures affected the pregnancy rate, the number of implantations or fetal resorptions, the litter size, or the fetal body weight. Unlike the initial study, there were no gross or soft-tissue malformations at any exposure level; however, the 3000-ppm exposure caused significant increase in the incidence of delayed skeletal ossifications (skull and neck) and an increase in the incidence of extra lumbar ribs. The authors concluded that MEK was not embryotoxic or teratogenic and only slightly fetotoxic in the rat at exposure concentrations of 3000 ppm.

Schwetz et al. expanded further on these studies and more recently reported on the developmental effects of MEK in Swiss mice (232). Groups of pregnant and nonpregnant mice were exposed to MEK vapor concentrations of 400, 1000, or 3000 ppm for 7 h/d for 10 consecutive days (for pregnant mice, days 6–15 of gestation). The nonpregnant and pregnant mice showed no differences in sensitivity to the MEK exposures. The maternal body weight gain, extragestational weight gain, uterine weight, pregnancy rate, and implantation ratio were not affected by exposure to MEK vapors relative to control. The 3000-ppm exposure did, however, result in a relative increase in maternal liver and kidney weight. The percentage of live fetuses, resorptions, and dead fetuses per litter was also unaffected by the treatment. A concentration-related decrease in fetal weight was observed for both the male and females. Although there were no statistically significant increases in the number of malformed fetuses per litter, there were several atypical malformations (cleft-palate, fused ribs, missing vertebrae, and syndactyly) not often found in control litters. A statistically significant increase in the incidence of misaligned sternebrae was observed in the fetuses exposed to 3000 ppm of MEK. No developmental or maternal effects were observed at vapor concentrations of 1000 ppm or less. After considering the results from the present study together with results obtained in rats, the authors concluded that MEK vapors caused mild developmental toxicity at concentrations that caused maternal toxicity.

Stoltenburg-Didinger et al. studied the effects of inhaled MEK and *n*-hexane vapors in Wistar rats exposed continuously (23 h/d) throughout gestation and postnatal development (233). Beginning on day 1 of pregnancy, groups of rats were exposed to (1) 800 ppm of MEK or *n*-hexane for up to 42 d, (2) 1000–1500 ppm of MEK or *n*-hexane for up to 51 d, or (3) 1500 ppm of an MEK:*n*-hexane mixture (1:4 ratio) for 51 d. Half of the dams in each exposure group were terminated on day 21 to assess the prenatal effects of the solvent exposures. Pregnancy and resorption rates decreased in a concentration-related manner following exposure to MEK vapors. The treatment groups exposed to 1000–1500 ppm of MEK, *n*-hexane, or MEK:*n*-hexane all showed resorption rates of 50%. Relative to control animals, the newborn pups in the 1000–1500 ppm exposure groups displayed reductions in body weight and increases in the brain-to-body-weight ratio that became more severe as the exposures continued during postnatal development. The decrease in body weight was greatest for the animals exposed to the MEK:*n*-hexane mixture than for those exposed to *n*-hexane. Pups exposed for 3 wk during gestation and 3 wk postnatally to 1000–1500 ppm of MEK:*n*-hexane failed to show any clinical or histopathological evidence of neurological damage. The dams, however, showed distinct evidence of peripheral neuropathy when exposed to the mixture for the same length of time (i.e., 6 wk). The authors concluded that all the solvent exposures caused embryotoxicity and fetotoxicity in the rat and that the adult animals appeared more sensitive to the neurotoxic effects of the MEK:*n*-hexane mixture than the immature animals.

1.4.1.5 Carcinogenesis. Although MEK has not been specifically examined in a rodent 2-y bioassay, there is little to suggest that the material is carcinogenic. When used as a delivery vehicle in a dermal carcinogenicity bioassay for organic sulfur compounds, Horton et al.found that the application of benzyl disulfide or phenylbenzylthiophene in a 25–29% solution of MEK in dodecylbenzene together failed to increase the incidence of benign papillomas in male C3H/HeJ mice (234). The mice used in the experiments were treated twice a week for 52 wk with the MEK-containing test solution.

1.4.1.6 Genetic and Related Cellular Effects Studies. Zeiger et al. reported that MEK levels ranging from 100 to 10,000 μg/plate were not mutagenic to *Salmonella typhimurium* strains TA97, TA98, TA100, TA1535, and TA1537 when tested with and without metabolic activation in a preincubation-type assay. Metabolic activation was accomplished with up to 30% of a liver S9 fraction from rats and hamsters treated with Arochlor 1254 (235). Likewise, Shimizu et al. noted that MEK was not mutagenic to *S. typhimurium* strains TA98, TA100, TA1535, TA1537, and TA 1538 or *Escherichia coli* strain WP2uvrA when tested at levels of 5–5000 μg/plate with and without metabolic activation by liver S9 from polychlorinated biphenyl-induced rats (236). Nestmann et al. also obtained negative results with levels up to 10,000 μg/plate using the same five *S. typhimurium* tester strains and the same source of liver S9 for metabolic activation (237). Marnett et al. showed that MEK at levels up to 3 μmol/plate (216 μg/plate), was not mutagenic to TA104, a special *S.typhimurium* strain sensitive to mutagens containing a carbonyl group (238). Muller and Black reported that MEK was not mutagenic to *S. typhimurium* strain TA102 when tested in three different laboratories using water, ethanol, or dimethylsulfoxide as the delivery solvent (166).

Zimmermann et al. examined the ability of MEK to induce either chromosomal malsegregation, mitotic recombination, or point mutations in *Saccharomyces cerevisiae* strain D61.M (239). The authors reported that an MEK concentration of 3.54% induced chromosomal malsegregation, characterized as aneuploidy, but did not induce mitotic recombinations or point mutations. An overnight incubation of the cells in ice-cold water was needed consistently to observe the aneuploidy. The authors proposed that MEK acted directly on tubuli during the growth phase of the cells, causing the formation of labile microtubules that functioned poorly during mitosis. The studies of Mayer and Goin revealed that MEK concentrations ranging from 0.5 to 1.96% were capable of causing chromosome loss (i.e., aneuploidy) in *S. cerevisiae* strain D61.M following a 16-h incubation at 4°C and that these levels were capable of dramatically potentiating the aneuploidy caused either by nocodazole or ethyl acetate (240, 241). Further studies with strain D61.M and the cold-shock regimen revealed that chromosome loss only occurred at MEK concentrations above 20 mg/mL; however, an ethyl acetate–propionitrile mixture was capable of significantly increasing the potency of MEK (242). In mixture studies using MEK with *n*-hexane, 2-hexanone, or 2,5-hexanedione, it was shown that MEK was able to potentiate the chromosomal loss observed in strain D61.M of *S. cerevisiae* at high concentrations (243). The most pronounced effects were observed when 2,5-hexanedione concentrations of 105 mM or greater were combined with MEK concentrations greater than 138 mM.

Yeast cell aneuploidy was observed with MEK at incubation temperatures ranging from 0 to 14°C;however, peak activity was observed when the cold-shock temperature was 10–14°C (244). Studies using microtubular protein from porcine brain revealed that those solvents capable of inducing aneuploidy in yeast cells were also capable of interfering with the *in vitro* assembly of brain tubuli (245). MEK amounts ranging from 10–20 μL inhibited tubule assembly in a concentration-related fashion.

Basler examined the clastogenicity of MEK in the micronucleus assay by administering 10 mL/kg of MEK intraperitoneally to male and female Chinese hamsters and determining the number of micronuclei in the polychromatic erythrocytes obtained from bone marrow (246). Specimens collected for up to 72 h failed to show any mutagenic effect due to the MEK treatment. Chen et al. reported that 5–20 μL of MEK caused a marginal inhibition of gap junction-mediated metabolic cooperation between V79 Chinese hamster lung fibroblasts (247). The metabolic cooperation assay examined the ability of MEK to alter the *in vitro* transfer of 6-thioguanine across gap junctions in normal and hypoxanthine-guanine phosphoribosyltransferase-deficient cells. Perocco et al. found that MEK concentrations ranging from 0.1 to 10 mM did not affect the viability or uptake of tritiated thymidine by cultured human lymphocytes incubated in the presence or absence of a rat liver S9 fraction (248). Using a structure–activity analysis based on the biological activity associated with the structural fragments from 212 chemicals (CASE methodology), Yang et al. predicted that MEK would not be genotoxic in the bone marrow micronucleus assay (249).

1.4.1.7 Other: Neurological, Pulmonary, Skin Sensitization. Holmberg et al. observed that MEK, unlike many of the solvents studied, was anti-hemolytic only at high concentrations (56.25 mM), when incubated *in vitro* with the red blood cells from outbred albino rats (250). Holmberg and Malmfors assessed the cytotoxicity of MEK and 32 other organic solvents using Ehrlich-Landschütz diploid ascites tumor cells (251). A 5-h incubation with MEK concentrations of 50 and 100 μg/mL caused a 10–14% decrease in cell survival, which the authors rated as being moderately toxic. Ebert et al. found that an MEK concentration of 86 mM was able to stimulate hemoglobin synthesis 3–13-fold in murine erythroleukemia cells and tetraploid erythroleukemia cells when incubated with the cultured cells for 5 d (252). MEK was found to be among the best inducers of this activity, and the presence of a carbonyl function in the molecule was judged to be an important mechanistic factor for promoting the effect.

Graham et al. studied the *in vitro* covalent interaction of radiolabelled MEK (2-[^{14}C]butanone) with spinal cord neurofilaments from Sprague–Dawley rats (253). Unlike the γ-diketone examined in parallel, MEK (20 mM) did not react with the neurofilaments and cause protein crosslinking, which was hypothesized to be the mechanism for solvent-induced peripheral neuropathies. Veronesi et al. treated dorsal root ganglia from fetal mouse spinal cord with mixtures of MEK and *n*-hexane (254). The incubation of the nerve cultures with MEK at concentrations ranging from 10 to 100 μg/mL for 49 d did not cause any change in morphology; higher concentrations of 200–600 μg/mL, however, caused the appearance of distinct changes in axonal appearance. Using the time to appearance of axonal swellings as an endpoint, MEK was found to potentiate the neurotoxic effects of *n*-hexane when mixtures of the compounds were added to the cultures at concentrations that were nontoxic when tested separately. The authors postulated that the mechanism responsible for the potentiation of *n*-hexane-induced neurotoxicity could be related to effects on *n*-hexane metabolism or binding. In a subsequent paper, Veronesi described the ultrastructural appearance of the nerve tissue lesions that developed when 300 μg/mL MEK was incubated with the dorsal root ganglia from the fetal mouse spinal cord for up to 7 wk (255). After 5 wk of incubation, Nissl bodies and a swollen granular endoplasmic reticulum were observed in the cytoplasm of sensory and motor neurons. Other observations included an excessive agranular reticulum in the peripheral and central nerves that was associated with foci of axoplasmic debris and the appearance of membrane bound vacuoles. The author concluded that MEK could affect neuronal metabolism in culture after a prolonged incubation. In a related study, Stoltenburg-Didinger et al. examined the neurotoxic effects of MEK on cultures of dorsal root ganglia isolated from chick embryos (256). MEK concentrations from 5 to 200 mg/mL caused a distinctive swelling of the cell body and somatic spines that was unlike the changes observed with the neurotoxicants *n*-hexane and 2,5-hexanedione. Differences in the appearance, time course, and location of the lesions led the authors to postulate that histopathological changes observed *in vitro* may be unrelated to the *in vivo* development of a peripheral neuropathy.

Gad et al. reported that MEK did not cause dermal hypersensitization in the CF-1 mouse ear swelling test (MEST) when the animals were treated dermally with 100 μL for 4 d then challenged with 20 μL 7 d after the final treatment (257). The MEK treatment caused no sensitization or swelling in the test area around the ear 24 or 48 h after treatment with the challenge dose. Similar findings were reported in Hartley guinea pigs intradermally treated twice in one day with MEK in Freund's complete adjuvant followed by a 48-h dermal treatment 7 d later. The challenge dose was administered topically 14 d after the induction regimen was complete. Descotes found that MEK was not a contact allergen in a bioassay that measured ear swelling of male and female Swiss mice that were sensitized with 0.05 mL of the solvent 3–14 days before topical and subcutaneous challenge (258).

1.4.2 Human Experience

1.4.2.1 General Information. At moderate concentrations, MEK is only slightly irritating to the mucous membranes. Because of its odor threshold, it can be safely detected well below the recommended exposure levels. High atmospheric concentrations of MEK are irritating to the eyes, nose, and throat, and prolonged exposure may produce central nervous system (CNS) depression and narcosis. Other symptoms of exposure include headache, dizziness, vomiting, or numbness of the extremities.

1.4.2.2 Clinical Cases. Smith and Mayers reported two cases of acute poisoning in a raincoat manufacturing operation where the seams were waterproofed with a resin dissolved in either acetone or MEK (259). Samples of the workroom air indicated that the MEK concentration ranged from 398 to 561 ppm, whereas the acetone concentration ranged from 330 to 495 ppm. Two female employees were reportedly affected by the solvent vapors on different occasions. The first complained of stomach distress and watery eyes one morning, and later was found unconscious in a rest room. The employee had a strong acetone odor on her breath, but no acetonuria,upon admission to the local hospital. She was treated with a CNS stimulant and awoke immediately, but complained of a severe headache.The patient was discharged from the hospital on the day following admission.

The second case followed the first by one day and was apparently less severe. The individual fainted and convulsed at the work site but regained consciousness immediately afterwards. At the hospital she was confused and had a headache, but after one hour of observation she was allowed to return home.

Sakata et al. presented a case study in which a male patient attempted suicide by ingesting approximately 100 mL of an adhesive cement that contained 39% cyclohexanone, 28% MEK, 18% acetone, and 15% polyvinyl chloride (260). The individual reportedly drank about 720 mL of sake (10% ethanol) 30 min prior to drinking the adhesive solution. The patient was unconscious when taken to a hospital about 2 h after ingestion of the solvent mixture. Gastric lavage, plasma exchange, and charcoal hemoperfusion were all performed, and the patient regained consciousness about 7 h after ingestion; however, signs of systemic toxicity persisted. High blood concentrations of the cyclohexanone metabolite, cyclohexanol, were thought to be responsible for the coma, whereas the hyperglycemia was attributed to acetone. The authors were not certain what caused the increased serum transaminase levels.

Berg reported a case of retrobulbar neuritis in a sailor who was using large quantities of MEK as a paint remover (261). The affected individual complained of blurred vision, light-headedness, and nausea after being exposed to MEK vapor for 1.5 h. Upon admission to a military hospital, the patient's near- and far-field visual acuity was reduced, and blind spots were markedly enlarged when a white target was viewed. An ophthalmoscopic examination, however, did not reveal any abnormalities. Clinical testing on the day of admission revealed elevations in blood methanol and formaldehyde. Visual acuity returned to normal within two days of admission; however, the blood methanol levels did not return to normal until the sixth day of hospitalization. The author tentatively concluded that MEK was the causative agent for the optic nerve damage that was observed. The MEK was assumed to have been systemically absorbed both dermally and through the lungs and then partially metabolized to methanol, which subsequently affected the optic nerve and cerebral cortex. The report contained no mention of possible methanol consumption by the patient. Neither methanol nor formaldehyde has been identified as a major metabolite of MEK. An independent evaluation of the preceding case study found the results to be inconclusive (262). Kopelman and Kalfayan reported a case of severe metabolic acidosis in an individual that unintentionally consumed MEK placed in a rum bottle (263). The patient was unconscious and hyperventilating when first examined but regained consciousness 12 h after admission to a hospital. A plasma MEK level of 950 mg/L was reported, and the urine MEK level was noted to be elevated. Liver and renal function tests remained normal during the 1 wk hospitalization; however, a severe lactic acidosis was apparent on admission. The authors were not certain whether MEK or circulatory insufficiency caused metabolic acidosis observed in the patient.

Welch et al. described a case toxic encephalopathy with dementia and cerebellar ataxia in a 38-year-old male who spray painted a truck in an unventilated garage with a paint that contained toluene, MEK, and propylene glycol (264). As time passed following the exposure, the individual began to experience memory problems along with headaches, shortness of breath, appetite and sleep disturbances, ataxia, and suicidal depression. Neuropsychological evaluations over a 2.5-yr period revealed distinct cognitive, motor, and behavioral changes.The chronic nature of the toxicity led the authors to postulate a possible preexisting neurological disease. Price et al. reported an incident involving a 42-year-old male who ingested approximately 240 mL of a cleaning solution that contained 47% MEK,45% methanol, and 3% isopropanol (265). On arrival at the hospital the patient showed depressed mental activity (was comatose), tachycardia, and a minimal respiratory acidosis with a large osmolar gap. Whole blood analysis revealed an MEK concentration of 124 mg/dL and a 2-butanol concentration of 24 mg/dL on admission. The 2-butanol metabolite of MEK was believed to have possibly affected the time course and early clinical course of the toxicity. The patient recovered completely without any permanent effects.

1.4.2.2.1 Acute Toxicity. Nakaki performed time-estimation tests on human volunteers exposed to MEK for 4 h at a vapor concentration that ranged between 90 and 270 ppm (266). When the subjects estimated the passage of time for intervals ranging from 5 to 30 sec, the MEK-exposed male subjects tended to overestimate the timed interval and the female subjects tended to underestimate the interval, relative to control values. The time-estimation values from this study were highly variable and no statistical differences were noted between or among the exposure groups. In a series of large-scale studies using hundreds of male and female volunteers, Dick et al. found that a controlled 4-h exposure to either MEK at 200 ppm, MEK:toluene at 150 ppm (2:1 ratio), MEK:acetone at 225 ppm (1:1.25 ratio), or MEK:MIBK at 150 ppm (2:1 ratio) failed to show any signs of a performance decrement in a large array of behavioral and psychomotor tests conducted before, during, and after the exposures (267–270). The types of tests used in the various studies included: visual vigilance, choice reaction time, simple reaction time, profile of moods, postural sway, eye blink, dual-task performance, and short-term memory. The primary effects observed with the MEK exposures were limited to subjective complaints of sensory irritation.

Ukai et al. performed a wide variety of hematology, liver function, renal function, and subjective symptom tests on a group of 303 workers exposed to toluene, xylene, ethylbenzene, isopropyl alcohol, ethyl acetate, and MEK and found no exposure-related effects except for the prevalence of self-reported symptoms from a questionnaire (271). The geometric mean for the MEK exposures in the workplace was 15.9 ppm. Exposure to MEK was not related to appearance of those symptoms (e.g., eye and nose irritation, sore throat, unusual taste and smell, dizziness, and rough skin), which showed a statistically significant increase in prevalence in the four solvent-exposed subgroups that were created. In addition, MEK did not affect the excretion of hippuric acid in the urine of the toluene-exposed individuals. Nasterlack et al. examined serum ALT, AST, and γ-glutamyl transpeptidase activities in a group of 33 printers and found no exposure-related changes in either of two studies conducted 4 years apart (272). The printers were primarily exposed to the following six solvents: acetone, MEK, ethyl acetate, white spirits, toluene, xylenes, and trichloroethylene. MEK exposures ranged from 3.3 to 907 ppm during the first study and 23.1–759 ppm during the second. The average MEK blood levels were 1.9 and 6.2 mg/L for the two studies. Shamy et al., in contrast, reported that a group of 32 workers exposed to phenol, benzene, toluene, and MEK showed elevations in many of the biochemical and hematological test values examined (273). The majority of these elevations were also observed in a test group exposed only to phenol. Elevations in prothrombin time, platelet count, and eosinophil count were confined, however, to the mixed solvent group. The time-weighted average MEK exposure concentration for the solvent-exposed group was 90 ppm and the average MEK concentration in the urine was 2.75 mg/g creatinine. Franco summarized the results from two mixed solvent studies involving MEK exposures and reported that increased concentrations of serum bile acids, cholic acid and chenodeoxycholic acid, were observed in one study involving xylene, toluene, acetone, and MEK, but not in another where trichloroethane, heptane, and MEK exposures occurred (274).

Nelson et al. determined the irritation threshold in subjects exposed for 3–5 min to MEK concentrations ranging from 100 to 350 ppm (275). The degree of sensory irritation from the MEK exposure was subjectively rated as not irritating, slightly irritating, or very irritating. A vapor concentration of 350 ppm irritated the eyes, nose, and throat of most subjects, and a majority of the subjects estimated that a vapor concentration of 200 ppm would be a satisfactory 8-h exposure limit. Douglas and Coe reportedly determined the eye and lung irritancy threshold of MEK and other ketones in human volunteers; however, the results for MEK were not presented (276). Wayne and Orcutt reported that fewer than half of the 9–11 volunteers exposed to 5 or 20 ppm of warm MEK vapors experienced any eye irritation during the 90 sec of exposure (277). Irradiation of the MEK vapor by mercury vapor lamps resulted in large increase in the response by the subjects. Doty reported that male and female volunteers with multiple chemical sensitivity (MCS) were not more sensitive to the odor of MEK than control subjects (278). The odor threshold values for males and females with MCS were 5.7 and 7.6 ppm, whereas the corresponding values in male and female controls were 8.2 and 8.1 ppm.

1.4.2.2.2 Chronic and Subchronic Toxicity. Dyro described three cases of polyneuropathy following repeated occupational exposure to MEK in combination with other solvents (279). All three cases involved the use of MEK-containing cements or glues by workers in a shoe factory. In the first case, a woman who worked for 6.5 years with a cement that contained 51% MEK and 27% toluene displayed parasthesias in the feet and complained weakness, fatigue, heaviness of the chest, and vertigo. Neurological abnormalities included decreased motor nerve conduction velocities, distal motor latencies, glove and stocking hyperesthesia, and a decreased number of motor units with fasciculations and occasional fibrillations. Conduction velocities and distal motor latencies returned to normal after approximately 8 mo of observation. The second and third cases were less severe and generally involved numbness in the hands and feet, headaches or other CNS complaints, and some reduction in peroneal nerve conduction velocities. The latter two individuals were exposed to a mixture of MEK and toluene or MEK and acetone. The etiologic agent in these case reports was not determined.

Altenkirch et al. reported that juveniles who had intentionally inhaled a commercially available glue thinner developed "glue sniffers' neuropathy," which was characterized by rapidly progressing damage in the ascending nerves along with motor atrophy and vegetative alterations (280). The polyneuropathies occurred after the composition of the glue thinner was altered and began containing about 11% MEK, along with 29% toluene, 18% ethyl acetate, and 16% *n*-hexane. Vallat et al. described a case of polyneuropathy in a 39-year-old woman who worked with a glue that contained 20% MEK and 8% *n*-hexane

(281). A biopsy specimen from the leg revealed giant axons, numerous neurofilaments, and a marked accumulation of glycogen granules in unmyelinated axons. Binaschi et al. documented a decrement in performance in several neuropsychological and neurophysiological tests administered to eight employees in a paint factory where MEK was used together with 5 other solvents (282). Although the air concentration of each individual solvent was below the TLV, the mixture TLV was exceeded by nearly 75%.

Vaider et al. suggested that MEK may have produced peripheral neuropathy in a 55-year-old employee working with MEK, tetrahydrofuran, and polyester (283). Callender described a case involving a 31-year-old employee in a quality-control laboratory who worked with MEK and was exposed to fumes generated when fiberglass was pyrolyzed (284). The patient complained of central and peripheral neurologic disturbances, respiratory tract irritation, cardiac arrhythmias, and a skin rash. Test results for neuropsychology, nerve conduction velocity, electronystagmography, and cognitive evoked potential indicated significant central and peripheral nervous system impairment that was consistent with a diagnosis of chronic progressive encephalopathy. Brain scan imaging indicated small ischemic areas in both the right and left cerebral hemisphere. In another case, a 27-year-old male exposed exclusively to MEK for 2 yr complained of dizziness, asthenia, anorexia, and weight loss prior to hospital admission (285). The patient was ataxic with multifocal myoclonus and postural tremors when first observed, but the condition improved dramatically after a single oral dose of ethanol. The results from blood and urine analysis, serum biochemistry, thyroid function, bacterial blood cultures, brain scans, nerve conduction velocity, and toxicity scans were all normal.

1.4.2.2.3 Pharmacokinetics, Metabolism, and Mechanisms. Imbriani et al. determined the relative pulmonary uptake of MEK in a group of fifteen volunteers exposed to concentrations ranging from 11.9 to 621.8 mg/m^3 (118). Three groups of five individuals were either exposed for 2 h while at rest, for 4 h while at rest, or for 2 h while exercising. Peak concentrations of MEK for the three treatment groups ranged from 206 to 2306 μg/L for blood and 113–1757 μg/L for urine. The average pulmonary uptake for the three treatment groups was calculated to be 53%.Brown et al. described the pharmacokinetics of MEK in 70 male and female volunteers following a 4-h inhalation exposure either alone (200 ppm) or in combination with acetone (225 ppm,1:1.25 ratio) (286). Blood and expired air elimination data for theMEK-exposed individuals revealed on average that (1) MEK steady-statebreath levels of about 10–12 ppm were attained, (2)non-steady-state peak blood levels of 3–4 μg/mL occurred, (3) no significant difference in uptake or elimination were observed for males and females, and (4) no appreciable interaction affecting elimination took place with the combined exposures. Liira et al. reported that the average relative uptake of MEK was 53% in eight male volunteers exposed for 4 h to 200 ppm. The uptake of MEK was not appreciably altered for a group of volunteers who exercised during the exposure (287). A two-compartment model of the blood elimination data revealed a terminal half-life of about 81 min for MEK. Pulmonary excretion was calculated to account for about 2–3% of the absorbed dose and another 2% was excreted in the urine as the metabolite, 2,3-butanediol. The blood levels of MEK were routinely found to be higher in exercising individuals. Subsequent kinetic studies with 2 male volunteers exposed for 4 h to 25,200, or 400 ppm of MEK were performed by Liira et al. to obtain the blood elimination data necessary to develop a physiologic-based pharmacokinetic model of MEK disposition in the body. Peak blood levels of about 5, 95, and 275 μM were observed for the three exposure concentrations, and the authors concluded from the kinetic model that MEK metabolism was dose dependent, with saturation kinetics occurring at exposure concentrations of 50–100 ppm, depending on the level of physical exercise (288). Tada et al. presented some excretion data for four test subjects exposed to MEK vapor concentrations ranging from 60 to 300 ppm for 2–4 h on three consecutive days (289). Using a nonspecific method for ketonic compounds, the authors found no change in the rate or amount of ketones excreted in the urine over the 3-d exposure period. MEK could routinely be detected in the expired air of the exposed individuals and was regarded as a good biological indicator of MEK exposure.

In a study of the kinetic interactions between solvents, Liira et al. determined that a coexposure to 300 ppm of MEK:*m*-xylene (2:1 ratio) for 4 h significantly inhibited the metabolism of *m*-xylene, but not MEK, when compared to the disposition of each solvent separately (215). The metabolic effects were determined by examining the blood elimination profiles along with metabolite levels in the urine. The 4-h inhalation exposure to 200 ppm of MEK resulted in an average pulmonary retention value of 41.1% of the absorbed dose, a terminal blood elimination half-life of 95 min, and a large inter-individual difference in urinary excretion of the metabolite 2,3-butanediol. In contrast, Liira et al. found that the ingestion of a 0.8-g/kg dose of ethanol before or after a 200 ppm exposure to MEK for 4 h inhibited the oxidative metabolism of MEK to 2,3-butanediol causing an increase in the blood concentrations of MEK, a decrease in the blood clearance, and an increase in the elimination of MEK in the urine and expired air (290). The blood levels of the primary metabolite, 2-butanol, were also increased upto 10-fold when ethanol was ingested.

Van Engelen et al. examined the effects of an MEK coexposure on the elimination of *n*-hexane and its metabolite 2,5-hexanedione in 16 male and female volunteers (291). Three different types of exposure regimens were employed which involved 15.5-min coexposures to *n*-hexane(60 ppm) and MEK (200 or 300 ppm) that either preceded or followed a pure hexane exposure by 4 h. The time course was followed for the elimination of *n*-hexane in the expired air and 2,5-hexanedione in the serum of each treatment group. Compared to an exposure to *n*-hexane alone, the combined exposures did not affect the excretion of *n*-hexane. The elimination of 2,5-hexanedione from the serum was appreciably affected, however, with the combined exposures causing a decrease in the peak concentration and a delay in the time until the peak concentration was observed. Compared to animals exposed to *n*-hexane alone, the differences in 2,5-hexanedione serum elimination were statistically different when the MEK concentration was increased from 200 to 300 ppm. The effects were also observed when the 300-ppm MEK:*n*-hexane coexposure either preceded or followed the pure *n*-hexane exposure by 4 h.Ashley and Prah exposed 5 human volunteers for 4 h to either 3.2 or 6.4-ppm of a 21-component mixture of solvents (292). The MEK concentration in the 6.4-ppm mixture was 0.0264 ppm and the mean pre-exposure blood levels were 3.0 ppb. The mean end-exposure blood levels of MEK for the 3.2- and 6.4-ppm exposures were 2.7 and 2.9 ppb, respectively.

Wurster and Munies added MEK to an absorption cell affixed to the palmar surface (91.5 cm^2) of the forearm of 12 volunteers (293). Shortly after contact with normal skin,

MEK was detected in the expired air with steady-state concentrations ranging from 3–13 μg/L occurring within 2–3 h. Hydration of the stratum corneum with water prior to the application of the MEK was observed to cause a large initial rise in the expired air concentrations of MEK than a slow decline to the level observed with normal skin. Further more, the use of a strong desiccant, magnesium perchlorate, to dehydrate the stratum corneum was shown to retard absorption and delay the attainment of steady-state expired air levels for 1–2 h compared to normal skin (294). The first-order elimination of MEK in expired air following percutaneous absorption was calculated to be 0.011–0.018 min^{-1}. Comparable rate constants were found for individuals who ingested a capsule containing 375 mg of MEK (i.e., 0.016–0.018 min^{-1}). DiVincenzo et al. measured the skin absorption rate for a 9:1 (v/v) mixture of radiolabeled MEK:MnBK that was applied to the forearm of two volunteers for 1 h (295). The cumulative excretion of radioactivity derived from MnBK in the breath and urine indicated that the skin absorption for the two test subjects had been 4.2 and 5.6 μg/min/cm^2. Brooke et al. examined the difference MEK uptake for 4 volunteers exposed to MEK vapor either by whole body or dermal only contact (296). Subjects were exposed to 200 ppm of MEK for 4 h with the dermal-only group wearing T-shirts and shorts, and breathing through a supplied-air respirator. MEK adsorption levels immediately following the whole body exposure ranged from 54 to 81 μmol/L for blood, 517–1242 nmol/L for breath, and 7.6–34.1 μmol/L for urine. Dermal exposure was shown on average to account for 3.1–3.8% of the MEK found in blood, breath, and urine. The half-life for elimination of MEK in the urine appeared to be greater for the dermal only group ($t_{1/2} = 2.7$ h) than for the whole-body group ($t_{1/2} = 1.5$ h).

Fiserova-Bergerova and Diaz determined the *in vitro* tissue–gas partition coefficients for human tissues obtained by autopsy from 5 subjects (297) . The average tissue/air partition coefficients for muscle, kidney, lung,and brain ranged from 96 to 111, whereas the fat/air value was 162. The average partition coefficient values for blood/air, plasma/air, and red blood cells/air were 126,133, and 106, respectively. Fisher et al. developed a physiologically based pharmacokinetic model of lactational transfer and calculated the amount of MEK that would be ingested by a nursing infant (298). Using human specimens, blood/air and milk/air partition coefficients of 195 and 221 were determined experimentally and used to calculate a milk/blood value of 1.13. The partition coefficient data were used in the model, along with a scenario that assumed 7 h of occupational exposure to MEK at a concentration of 200 ppm and a 2–3 h nursing schedule. The amount of MEK ingested by a nursing infant under these conditions was determined to be 2.09 mg/d. Poulin and Krishnan used a biologically based algorithm to calculate human tissue/blood partition coefficients for MEK that utilized information on chemical solubility along with the lipid content and water content of tissues (299). The predicted tissue/blood partition coefficients for liver, muscle, lung, kidney, and brain ranged from 0.95 to 1.18. The predicted fat/blood partition coefficient was 1.98, which was considerably higher than experimentally determined values ranging from 0.75 to 1.30.

1.4.2.2.4 Reproductive and Developmental Studies. Jankovic and Drake developed a screening method for assessing the occupational reproductive health risks from solvent exposures by constructing a toxicological database and comparing the observed health effect levels with the current occupational exposure limits (300). Using this approach, an occupational reproductive guideline (ORG) level of 590 mg/m^3 was established for MEK.

1.4.2.2.7 Other: Neurological, Pulmonary, Skin Sensitization. Varigos and Nurse described a single case of contact urticaria in a painter who began using an epoxy–polyamide paint 18 mo earlier (301). Dermatitis initially appeared on the individual's hands and then spread to his forearms, face, and neck before he discontinued use of the paint. Patch tests revealed a positive reaction to epoxy resins. The painter also used MEK as a solvent to some degree, and he noted that his hands became severely irritated when they came into contact with the solvent. A small amount of MEK reportedly caused erythema but no edema within 15 min of application to the patient's forearm. The test was repeated after 2 d with the same results. No reaction was observed in five medical staff members at the hospital. The authors concluded that the patient was sensitized to MEK and that it could produce contact urticaria.

Malten et al. reported that the application of MEK to the forearm skin of two volunteers for 1 h/d on six successive days damaged only the stratum corneum and that the dermal toxicity was similar to ethanol (302). Opdyke indicated that the application of 20% MEK in petrolatum to human forearm skin for 48 h did not cause any irritation and that no sensitization reactions occurred when the same concentration was tested on 24 volunteers (141).

1.4.2.3 Epidemiology Studies. Alderson and Rattan reported that there was no increase in mortality or clear evidence of a cancer hazard in 262 MEK-exposed workers who were examined in a prospective epidemiology study (303). The workers, who were employed in a lubricant dewaxing plant, showed a lower than expected death rate from all causes and from all types of neoplasms. For the seven specific cancer types examined, there was an excess in death due to cancer of the buccal cavity and pharynx but a deficiency in deaths from lung cancer.

1.4.2.3.1 Acute Toxicity. Chia et al. examined a group of 19 employees in a video tape manufacturing plant who were exposed to cyclohexanone, toluene, tetrahydrofuran, and MEK (304). The mean time-weighted average exposures to MEK ranged from 5 to 426 ppm in three different areas of the plant. The authors noted a statistically significant increase in the prevalence of headache, eye and nose irritation, coughing, and irritability from a symptom survey that was administered. Kishi et al. obtained symptom complaints from a group of 81 industrial painters who were exposed to 12 different solvents, including MEK (305). MEK exposure concentrations were not provided, and only 5 of the 81 workers reported any contact with the solvent. Reports of subjective symptoms such as dizziness, difficulties in hearing, dry and scaly skin, being easily irritated, and loss of appetite were significantly more common in the solvent exposed group than in the controls.

1.4.2.3.2 Chronic and Subchronic Toxicity. Mitran et al. conducted a study that examined the effects of MEK in a group of 41 workers in a cable factory who were exposed while using a lacquer coating (306). The 8-h time-weighted average exposures to

MEK ranged from 149 to 342 mg/m^3. The MEK exposures were found to cause an increase in the percentage of reported neurotoxic symptoms (mood disorders, memory difficulties, sleep disturbances, and headache) and rheumatic symptoms (bone pain, joint pain, and muscular pain). Symptoms of ocular and respiratory tract irritation also appeared to be more common, but tests of statistical significance were not performed. Measurement of motor nerve conduction velocity revealed a statistically significant increase in the latency period for proximal and distal regions of the median, ulnar, or peroneal nerves. Nerve conduction velocity was also significantly decreased in these nerves relative to controls. The authors noted, however, that the nerve conduction velocity experiments were relatively insensitive to any underlying CNS abnormalities.

1.4.2.3.4 Reproductive and Developmental Studies. Agnesi et al. conducted a case-control investigation of 108 spontaneous abortion reports for individuals in a health district where about 8000 people were employed in shoe manufacturing (52). A 10-y historical exposure profile showed that shoe industry employees were exposed to MEK concentrations ranging from an average of 41 to 67 mg/m^3. These same individuals were also exposed to 11 other oxygenated and aliphatic solvents. Removing confounding variables such as previous abortions and coffee intake, the relative risk of spontaneous abortion in groups with high versus no solvent exposure was 3.85, and the 95% confidence interval was 1.24–11.9. The relative risk values for the high-exposure, but not the low-exposure, group were significantly elevated compared to nonexposed women. Solvent exposure was assessed from personal interviews concerning work habits and the use of solvent-containing glues.

1.4.2.3.5 Carcinogenesis. Several retrospective cohort studies have been conducted on employees exposed to MEK along with other solvents. Wen et al. performed such a study on a group of 1008 male oil refinery workers exposed to MEK, toluene, benzene, hexane, xylene, and MIBK (307). The cohort was composed of two subcohorts in a lubricating department, a dewaxing subgroup that was exposed to the majority of MEK, and a subgroup involved in other lubricating department operations. MEK exposures ranged from a mean of 0.20 ppm for area samples to a mean of 4.38 ppm for personal samples (full range, 0.10–24.99 ppm). The standardized mortality ratio (SMR) for all cancers was found to be 0.86 for the cohort when compared to the U.S. population. SMR ratios were statistically higher for cancer of the bone and nonstatistically higher for pancreatic, lymphatic, and hematopoietic cancer. When the two subcohorts were examined separately, the excess of bone cancers was apparent in the other lubricating subcohort.

Spirtas et al. examined the causes of death in retrospective cohort study of 14,457 male and female workers at an aircraft maintenance facility (308). Standardized mortality ratios were calculated for workers exposed to a large number of individual organic solvents during their employment. An increase was noted in the SMR values for deaths due to multiple myeloma and non-Hodgkin's lymphoma for female, but not male, workers previously exposed to MEK. Information was not provided on the level of workplace exposure to MEK. In a followup study, Blair et al. reported on the cancer incidence in the same cohort of solvent-exposed workers from the Air Force maintenance facility (309). Although mortality rate ratios were determined for the workers exposed to 13 different solvents, most exposures were actually to a mixture of solvent types. The MEK-exposed group showed nonstatistically significant excesses in the rate ratios for mortality from non-Hodgkin's lymphoma (male and female), multiple myeloma (female), and breast cancer (female). Rate ratios were calculated using the mortality experience of Utah residents.

1.4.2.3.6 Genetic and Related Cellular Effects Studies. Lemasters et al. studied a group of 50 solvent- and jet-fuel-exposed individuals from a military installation to determine the frequency that micronuclei and sister-chromatid exchanges appeared in cultured blood lymphocyte specimens relative to a group of eight control individuals (310). The total average exposure to all solvents, except jet fuel and benzene, was less than 1.6 ppm, and the average expired air level was 8.6 ppb. The solvent exposures included 1,1,1-trichloroethane, MEK, xylenes, toluene, and methylene chloride. In addition, workers were also exposed to JP-4 jet fuel and benzene at low ppm or ppb levels. When examined after 30 wk of exposure, the lymphocytes from the solvent-exposed group showed a statistically significant increase in the frequency of sister chromatid exchanges, but no change in the frequency of micronuclei occurrence. Individuals employed in the paint and sheet metal shops showed the greatest change in sister chromatid exchange values.

1.4.2.3.7 Other: Neurological, Pulmonary, Skin Sensitization, etc. A relatively large number of epidemiology studies have examined the relationship between solvent exposure and the potential neurotoxic effects of MEK. Most, however, have involved mixed solvent exposures where MEK was one out a large number of solvents being examined. Lee et al. administered a Neurobehavioral Core Test Battery (6 different tests) to a group of 40 female workers from shoe manufacturing factory where toluene, MEK, *n*-hexane, cyclohexane, dichloroethylene, trichloroethylene, benzene, and xylene were used (311). The average MEK exposures in the frame making and adhesive application areas of the plant were 17.5 and 67.0 ppm, respectively. Comparing three different measures of exposure, the authors found that the solvent-exposed individuals performed significantly poorer on the Santa Ana dexterity test when cumulative exposure estimates (equal to a summation of the products of exposure concentration and exposure duration) were used to evaluate the data. Performance was also noticeably, but not significantly, worse on the Benton visual retention test compared to the control group.

White et al. performed a 2-y prospective study on 30 male and female workers in a screen printing plant (312). The employees were exposed to toluene, MEK, mineral spirits, diacetone alcohol, methylene chloride, and acetic acid. MEK exposures appeared to range from about 1.3 to 24 ppm, but only several instantaneous room values were provided. The study population was divided into two subgroups having either high acute or high chronic exposures based on the duration of their contact with solvents. Neurobehavioral testing (13 separate tests) revealed that the high acute exposure subgroup performed significantly worse on the Santa Ana dexterity test and the visual reproduction test when the effects of age, gender, and education were factored into the analysis. The high-chronic-exposure group also showed significantly poorer performance on the visual reproduction test when confounding variables were taken into consideration. No neurological impairment was noted in either subgroup relative to controls and performance was found to improve when the tests were administered a year later. Chia et al. examined a group of 19 employees in a video tape manufacturing plant who were exposed to cyclohexanone, toluene, tetrahydrofuran, and MEK (304). The mean time-weighted average exposures to MEK ranged from 5 to 426 ppm in three different areas of the plant. A neurobehavioral test battery consisting of four tests revealed significantly poorer performance on the Santa Ana dexterity, digit span, and visual reproduction tests when the results were adjusted for age, drinking, and smoking history. There were no dose-related trends in any of the test scores, however, when the results were evaluated against individual measures of total exposure. Triebig et al. performed a cross-sectional evaluation of motor and sensory nerve conduction velocities on a group of 66 workers with exposures to ethyl acetate, MEK, toluene, xylene, and trichloroethylene (313). MEK concentrations in the workroom air ranged from less than 10 to 400 mg/m^3. The results revealed a statistically significant reduction in the sensory nerve conduction velocity of the ulnar, but not the medial, nerve relative to a control group. Motor nerve conduction velocity of the ulnar nerve was likewise unaffected. The effects observed in the ulnar nerve appeared to be greater in those subgroups with medium-and long-term histories of solvent exposure.

Kishi et al. investigated the relationship between organic solvent exposure and alterations in central nervous system function in a group of 81 industrial painters (305). A neurobehavioral test battery consisting of eight separate tests was administered to the painters who were exposed to 12 different solvents that included MEK to a small extent. The solvent-exposed group showed poorer performance on vocabulary and digit symbol tests than the control population. When the duration of exposure was taken into consideration, significantly poorer performance was observed on vocabulary, block design, and digit span tests than in the referent controls. Williamson and Winder conducted a 3-yr prospective cohort evaluation of the neurobehavioral effects of solvent exposure on a group of 25 male vehicle spray painters (314). The group was exposed to twelve different solvents, with the MEK exposures ranging from 2 to 14 mg/m^3. The study did not reveal a decrement in performance, relative to controls, for any of the six neurobehavioral tests administered to the spray painters each year.

A prospective 2.5-y evaluation of the pulmonary function in a group of 40 employees exposed to organic solvents did not reveal any statistically significant changes in forced vital capacity or the forced expiratory volume in 1 sec relative to nonexposed controls (315). Solvents in use included acetone, *n*-butyl acetate, isopropyl alcohol, methanol, toluene, xylene, and MEK. The MEK exposure concentrations in the work area averaged 5.3 ppm. Oleru and Onyek were found that a subgroup of 43 employees from an area of a shoe factory where leather, dyes, and MEK were present displayed lower than expected forced expiratory volume (FEV_1) and a normal forced vital capacity (316). The ratio of these respiratory parameters was shown to correlate positively with the average age and duration of employment of the subgroup. The prevalence of obstructive and restrictive lung disease was 4.7 and 49.7 cases per 1000 person-years, respectively. A medical questionnaire revealed higher than expected reports of five symptoms: headache, dizziness, sleep disorders, dizziness, and chest pain. The MEK levels in the workroom air were not determined.

Socie et al. conducted a survey of workers from four plastic manufacturing plants to assess risk factors for the development of occupationally related skin disorders (317). A self-administered questionnaire was completed by 122 workers, 26 of whom had a skin problem. When the data were examined relative to MEK exposure, a crude odds ratio of

Table 75.3. Toxicologic Properties of Ketones

Compound	Approximate Oral Rat LD_{50} (mL/kg)	Lowest Reported Lethal Air Conc. Rat (ppm/h)	Skin Irritation[a]	Ocular Injury[a]
Acetone	8–11	16,000/4	Sl	M
Methyl ethyl ketone	3–7	2,000/4	Sl	Sl
3-Butyn-2-one	0.01	10/4	SV	SV
Methyl *n*-propyl ketone	3.7	30,000/1	Sl	M
Methyl isopropyl ketone	4–7	5,700/4	Sl	Sl
3-Pentyn-2-one	0.1	Sat'd./0.1	SV	SV
Methyl isopropenyl ketone	0.2	125/4	M	M
2,4-Pentanedione	1	1,000/4	Sl	M

[a] Sl, slight; M, moderate; SV, severe; skin irritation and ocular injury ratings are for direct application of liquids.

1.49 (95% CI, 0.49–4.48) was obtained, which was not statistically significant compared to nonexposed controls. Morata et al. examined the effects of organic solvent exposure on hearing using a group of 39 individuals from a paint manufacturing plant (318). Solvents in use at the paint plant included toluene, xylene, benzene, MIBK, ethanol, and MEK. MEK was present in the air at concentrations ranging from 0 to 32 ppm. Pure tone and intermittence audiometry tests revealed that the prevalence of high-frequency hearing loss in the solvent-exposed group was more than two times higher than in an unexposed control group. Compared to controls, the relative risk of some hearing loss was 5.0 (95% CI, 1.5–17.5) in the solvent-exposed group.

1.5 Standards, Regulations, or Guidelines of Exposure

Hygienic standards for MEK are listed in Table 75.3.

1.6 Studies on Environmental Impact

MEK does not possess the chemical and physical properties that are characteristic of a persistent environmental pollutant. The environmental fate and transport of MEK has been evaluated following its release to air and water (319). Tropospheric removal will occur by both photolysis and reaction with hydroxyl radicals. Reactions with hydroxyl radicals will predominate in the air and result in a half-life of 14 d. The high vapor pressure of MEK will result in rapid volatilization from soil; however, migration through soil may also occur to an appreciable degree. MEK is expected to volatilize rapidly from water with an estimated half-life of 19 h. Biodegradation will occur to a large degree with over 80% removed over a 24 h period.

2.0 3-Butyn-2-one

2.0.1 CAS Number: *[1423-60-5]*

2.0.2 Synonyms: Methyl ethynyl ketone, acetylacetylene, 1-butyn-3-one, butynone, and acetylethyne

2.0.3 Trade Names: NA

2.0.4 Molecular Weight: 68.075

2.0.5 Molecular Formula: C_4H_4O

2.0.6 Molecular Structure: 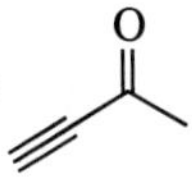

2.1 Chemical and Physical Properties

2.1.1 General

3-Butyn-2-one is a clear, colorless liquid that is chemically reactive. It has a boiling point of 85.0°C. The autoignition temperature for 3-butyn-2-one is 274°C. Additional physical–chemical properties are summarized in Table 75.4.

2.1.2 Odor and Warning Properties

3-Butyn-2-one has a penetrating odor and is a lacrimator.

2.2 Production and Use

Production data for 3-butyn-2-one are not available.

2.3 Exposure Assessment

No data on exposure to 3-butyn-2-one are available. Due to its strong odor and irritating nature, exposure to low levels of 3-butyn-2-one is expected to be readily apparent.

Table 75.4. Hygienic Standards for Ketones

	ACGIH TLV[a]		OSHA PEL[d]		NIOSH REL[e]		DFG MAK[f]
Compound	TWA[b]	STEL[c]	TWA	STEL	TWA	STEL	TWA
Methyl ethyl ketone	200	300	200	—	200	300	200
3-Butyn-2-one	—	—	—	—	—	—	—
Methyl *n*-propyl ketone	200	250	200	—	150	—	200
Methyl isopropyl ketone	200	—	—	—	200	—	—
3-Pentyn-2-one	—	—	—	—	—	—	—
Methyl isopropenyl ketone	—	—	—	—	—	—	—
2,4-Pentanedione	—	—	—	—	—	—	—

[a]American Conference of Governmental Industrial Hygienists threshold limit values.
[b]Time-wighted average (ppm).
[c]Short-term exposure limit (ppm).
[d]Occupational Safety and Health Administration permissible exposure limits.
[e]U.S. National Institute for Occupational Safety and Health recommended exposure limit.
[f]Deutsche Forschungsgemeinschaft (Federal Republic of Germany) maximum concentration values in the workplace.

2.4 Toxic Effects

2.4.1 Experimental Studies

3-Butyn-2-one is an extremely toxic material. It is a strong lacrimator, and both the liquid and vapor states are very irritating to the eyes, skin, and mucous membranes. It is readily absorbed percutaneously and is highly toxic both dermally and orally.

2.4.1.1 Acute Toxicity. The oral LD_{50} of 3-butyn-2-one was between 6.3 and 12.6 mg/kg for rats. Clinical signs of toxicity included tremors, diarrhea, and depression (320).

Application of the undiluted material to rabbit eyes produced a severe irritant response. Lethal amounts may be absorbed through the conjunctival membranes. When the material was diluted to a 1% solution with propylene glycol, severe irritation was still evident (320).

Undiluted and 10% solutions of 3-butyn-2-one were severe skin irritants and were lethal when applied for a few hours. One percent solutions were also strong dermal irritants, particularly to abraded skin. When applied as a 10% solution in propylene glycol under an impervious cuff, the LD_{50} was 40–50 mg/kg, with deaths occurring within hours of application (320).

Rats exposed to 10 ppm of 3-butyn-2-one vapors died between 4 and 7 h of exposure. At 25 ppm, deaths occurred between 0.5 and 1 h of exposure. At 50 ppm, deaths occurred between 12 min and 2 h of exposure. At 100 ppm, all deaths occurred between 12 min and 1 h of exposure. At 200 ppm, all deaths occurred between 3 and 30 min of exposure. At atmospheric saturation, all rats died within 6 min (320).

The vapors produced immediate, severe eye and respiratory irritation at all concentrations studied. Lung congestion was observed in all test animals, and liver and kidney damage was apparent in animals receiving longer exposures (320).

2.4.1.2 Chronic and Subchronic Toxicity. No repeated dose or long-term studies are available for this material.

2.4.1.3 Pharmacokinetics, Metabolism, and Mechanisms. No pharmacokinetic or metabolism studies are available for this material.

2.4.1.4 Reproductive and Developmental Studies. No reproductive or developmental toxicity studies are available for this material.

2.4.1.5 Carcinogenesis. No carcinogenicity studies are available for this material.

2.4.1.6 Genetic and Related Cellular Effects Studies. No genotoxicity studies are available for this material.

2.4.1.7 Other: Neurological, Pulmonary, Skin Sensitization. No studies on specific organ systems are available for this material.

2.4.2 *Human Experience*

2.4.2.1 General Information. Industrial exposures to 3-butyn-2-one may occur by inhalation, skin, or eye exposures. High oral toxicity, the ease of skin and eye absorption, and the highly irritating nature of vapors and solutions demand extreme caution. Limited human experience suggests that man is at least as sensitive to 3-butyn-2-one as experimental animals. 3-Butyn-2-one has been characterized as a strong lacrimator.

2.4.2.2 Clinical Cases. No clinical case studies reporting adverse health effects are available for this material.

2.4.2.3 Epidemiology Studies. No epidemiology studies reporting adverse health effects are available for this material.

2.5 Standards, Regulations, or Guidelines of Exposure

There are no recommended hygienic standards for 3-butyn-2-one.

2.6 Studies on Environmental Impact

Specific studies on environmental impacts for this material are not available. Due to the irritating nature of 3-butyl-2-one, spills to bodies of water may result in acute toxicity to fish, however, long-term effects would not be expected, as this material should biodegrade.

3.0 Methyl *n*-Propyl Ketone

3.0.1 CAS Number: *[107-87-9]*

3.0.2 Synonyms: Methyl propyl ketone, 2-pentanone, and MPK.

3.0.3 Trade Names: NA

3.0.4 Molecular Weight: 86.17

3.0.5 Molecular Formula: $C_5H_{10}O$

3.0.6 Molecular Structure:

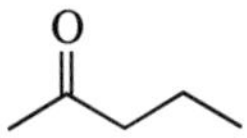

3.1 Chemical and Physical Properties

MPK is miscible with alcohol and ether and its water solubility is 4.3×10^{-4} mg/L at 25°C. It has a boiling point of 102.2°C and a melting point of −76.9°C. The autoignition point for this compound is 505°C. Its moderate flammability indicates a possible fire hazard. Additional physical–chemical properties are summarized in Table 75.1.

3.1.2 *Odor and Warning Properties*

Methyl *n*-propyl ketone is a clear, colorless liquid with a powerful, ethereal, fruity odor. The odor threshold concentration for MPK has been reported to be 8 ppm (321). The odor threshold for MPK is significantly below its threshold for nasal irritation, as measured by nasal pungency techniques (322).

3.2 Production and Use

MPK can be manufactured by oxidation of 2-pentanol, from ethylene and methyl acetoacetate, or by distillation of a mixture of calcium acetate and calcium butyrate. Commercial purity can be 90% MPK, however, some commercial materials are a mixture of MPK and diethylketone with small amounts of *sec*-amyl acetate. MPK is used as a solvent in lacquers and lacquer removers, as a chemical intermediate, and as a synthetic flavor ingredient.

3.3 Exposure Assessment

Under industrial conditions, MPK exposure may be expected to occur via inhalation or dermal contact. Since MPK is approved for food use by the USFDA, small amounts may appear in the diet.

3.3.1 Air

MPK has been detected in the air emissions resulting from composting operations (323). Indoor air levels of MPK may be derived from fungal growth on common pine or other surfaces (324).

3.3.2 Background Levels

Background concentrations in the environment can be expected as ketones such as MPK produced by naturally occurring metabolic pathways in the microorganisms. Normal F344 rats have been reported to exhale MPK. The amount of MPK exhaled can be increased by inhibition of cytochrome P450 2E1 by *trans*-1,2-dichloroethylene (226).

3.3.3 Workplace Methods

NIOSH Method 1300 is recommended for determining workplace exposures to MPK (77).

3.3.5 Biomonitoring/Biomarkers

3.3.5.3 Other. MPK has been detected in samples of human milk (325).

3.4 Toxic Effects

3.4.1 Experimental Studies

MPK is moderately toxic orally and can produce slight to moderate irritation of the skin and eyes. Repeated doses show no evidence of cumulative toxicity.

3.4.1.1 Acute Toxicity. The oral LD_{50} for rats has been calculated to be 3730 mg/kg by Smyth et al. (142). Unpublished data indicated the oral LD_{50} for both rats and mice was between 1600 and 3200 mg/kg (326). The intraperitoneal LD_{50} for methyl *n*-propyl ketone was 800 mg/kg for rats and 1600 mg/kg for mice (326).

Giroux et al. (327) injected mice intraperitoneally and found that doses of 250–500 mg/kg produced agitation and hyperesthesia. Doses of 2000 mg/kg produced agitation, loss of balance, staggering, paresis, loss of sensitivity, and deaths in 20% of the group. A dose of 2500 mg/kg caused 100% mortality. Haggard et al. (328) reported respiratory failure and death in rats after an intraperitoneal dose of 2530 mg/kg. The plasma concentration of MPK in these rats was 156 mg/dL. Tanii (139) reported the dose to produce a 50% loss of righting reflex in mice was 8.78 mmol/kg.

When MPK was placed in the conjunctival sac of rabbit eyes, slight to moderate irritation was observed (326).

Single doses of MPK applied to the backs of guinea pigs under an impervious cuff produced only slight irritation (326). Doses up to 20 mL/kg on the skin of guinea pigs did not produce clinical abnormalities, although percutaneous absorption appeared to occur. Repeated application for 10 d to the backs of guinea pigs resulted in slight to moderate erythema and slight edema (326). The dermal LD_{50} of MPK for rats has been reported to be 8 mL/kg (142).

Smyth et al. (142) studied the inhalation toxicity of MPK and reported that none of six rats died within a 2-wk period following a 30-min exposure to a saturated vapor (142). The LC_{50} for rats receiving a 4-h exposure was greater than 2000 ppm.

Yant et al. (329) reported that exposures to 30,000–50,000 ppm of MPK for 30–60 min produced lethality in guinea pigs. Guinea pigs exposed to 5000 ppm for 1 h, 2000 ppm for 8 h, or 1500 ppm for several hours showed slight or no abnormal clinical signs (329). Specht et al. (330) reported that MPK produced narcosis, depression of cardiac and respiratory rates, and loss of cranial nerve reflexes. Generalized vascular congestion was noted at levels of 2500–40,000 ppm.

Respiratory depression to 72% or 37% of the normal frequency has been reported in rats at concentrations of 3820 or 8355 ppm, respectively. Respiratory depression to 50% of the normal frequency was obtained at a concentration of 5915 ppm MPK (331). Inhalation of 976–1965 ppm of MPK for 4 h resulted in significantly decreased duration of the immobility phase (from 28 to 82%, respectively) in a "behavioral despair" swimming test. The differences were statistically significant only at exposure concentrations of 1180 ppm or greater. (331).

Hansen and Nielsen (332) reported that MPK reduced the respiratory rate in mice through activation of trigeminal receptors in the upper airways. Pulmonary irritation had little impact on respiratory rate.

3.4.1.2 Chronic and Subchronic Toxicity. MPK was administered in the drinking water to rats at concentrations of 0.25, 0.5, and 1.0%. The mean daily doses at the 1.0% level were calculated to be 144 mg/kg for 10 mo and 250 and 454 mg/kg for 13 mo. Adverse effects due to 1.0% MPK were limited to a slight reduction (9%) in weight gain compared to controls after 67 d of exposure. Clinical signs, organ weights, and histology were normal. No neuropathologic changes were noted in the central or peripheral nervous systems (322).

When a dose of 400 mg/kg of MPK was injected subcutaneously, 5 d/wk for a period of 9 wk, abnormal clinical signs were limited to salivation during the latter stages of the dosing regimen and a slight reduction in activity. No significant effects were noted in motor or sensory nerve conduction velocity or motor distal latency in treated animals (333). When MPK (345 mg/kg) was administered with equimolar doses of methyl *n*-butyl ketone (MnBK) by subcutaneous injection 5 d/wk for 20 wk, the neurotoxicity of MnBK, as measured by motor fiber conduction velocity and motor distal latency, was enhanced. The combined effect was, in general, less than the effect seen with methyl *n*-hexyl ketone together with MnBK but greater than or equal to that seen with combinations of MnBK and either methyl *n*-amyl ketone or MEK (334).

Five rats were exposed to MPK at a calculated level of 305 ppm to determine its neurotoxic potential. The exposures were given during two 16-h and two 20-h periods on four consecutive days per week over approximately a 17.5-wk period for a total of 1240 h of exposure. Histopathologic screening of a broad sample of tissue taken from these animals was normal except for the possibility of slight hepatocyte enlargement. Central and peripheral nerves embedded in epoxy resins did not reveal any neurologic damage (326).

3.4.1.3 Pharmacokinetics, Metabolism, and Mechanisms. Absorption can occur through the skin. MPK eliminated (38–54%) unchanged in expired air in 25–35 h. The rate of elimination is less than 2-pentanol, and a portion of the absorbed dose is excreted as a glucuronide (335). MPK is the major metabolite of 2-pentanol (335). Partition coefficients for blood, water, and air have been determined for MPK (336, 337). The rate of excretion of MPK through the lungs can be increased in normal F344 rats by inhibition of cytochrome P450 2E1 (226).

3.4.1.4 Reproductive and Developmental Studies. A reproductive and developmental screening test was recently conducted on MPK. In this study, groups of male and female rats were exposed to 0, 1, 2.5, or 5 mg/L MPK. Exposures were conducted 6 h/d, 7 d/wk. The study consisted of a 14-d premating phase, a mating phase lasting from 1 to 14 d, a gestation phase, and a lactation phase (days 0–4 postpartum). Females received 35–48 exposures, and males received 51 exposures. In this study no effect of exposure was seen in body weights, reproductive organ weights, sperm motility, sperm counts, gross pathology or histopathology of the reproductive organs, or reproductive indices. The only test-substance-related effect was reduced activity during exposure at the 5.0 mg/L exposure level. The NOEL for this study was determined to be 2.5 mg/L, while 5.0 mg/L was a NOEL for reproductive effects (338).

3.4.1.5 Carcinogenesis. No carcinogenicity studies are available for this material.

3.4.1.6 Genetic and Related Cellular Effects Studies. Few data are available on the genetic toxicity of MPK. In an unconventional assay using *Saccharomyces cerevisiae* as a test organism, MPK was a weak inducer of aneuploidy. This effect was noted in organisms subjected to a 17-h period in ice between two periods of incubation at 28°C in the presence of media containing 1.36% MPK (240). When the test organisms were incubated with MPK and propionitrile, MPK strongly potentiated the action of concentrations of propionitrile that by themselves did not induce aneuploidy (233). The basis of this test is the disruption of microtubular assembly required for the segregation of chromosomes during mitosis. Since no direct action on genetic material occurs, however, the validity of this test as a predictor of genetic damage is questionable.

3.4.1.7 Other: Neurological, Pulmonary, Skin Sensitization. Chloroform-induced hepato- and nephrotoxicity were evaluated in rats pretreated with oral doses of from 86 to 1292 mg/kg MPK. Animals were challenged by injection of 0.5 or 0.75 chloroform/kg body weight 18 h after oral doses of MPK. Potentiation of chloroform-induced liver and kidney toxicity, as measured by mortality; plasma creatinine, glutamic–pyruvic transaminase, or ornithine carbamoyl transferase activity; or uptake of *p*-aminohippurate by renal cortical slices were noted, though the magnitude of the potentiated response was nonlinear (171–173).

3.4.2 Human Experience

3.4.2.1 General Information. MPK has a low odor threshold and pungent odor. The primary effects expected following exposures are irritation to the eyes, nose, and throat. The low degree of systemic toxicity of MPK would indicate its relative safety, provided skin and eye protection are available. Inhalation should not be hazardous if adequate ventilation is provided. The greatest hazard associated with its use is its moderate flammability.

3.4.2.2 Clinical Cases. Yant et al. (329) reported that exposure to 1500 ppm of MPK in the air was associated with a strong odor and caused irritation to the nasal passages and eyes. No reports of ill effects resulting from use in industry have been reported (151).

MPK was one of a series of chemicals used to determine the relative sensitivity of the human eye and lung to irritant gases. Eye irritation was based upon subjective descriptions following 15 sec of MPK exposure inside close-fitting goggles. Respiratory irritation was evaluated by measuring reflex bronchoconstriction after inspiration of 10 breaths of 1 L each. In this study, the threshold for bronchoconstriction and for subjective responses based on ocular irritation and were within the 1000–1500 ppm range, with the threshold for eye irritation the lower of the two (276).

3.4.2.2.1 Acute Toxicity. There are no case reports of systemic toxicity following acute exposures to this material.

3.4.2.3 Epidemiology Studies. There are no epidemiological studies reporting adverse effects after exposure to this material.

3.5 Standards, Regulations, or Guidelines of Exposure

Hygienic standards for MPK are listed in Table 75.4.

3.6 Studies on Environmental Impact

There are no published environmental impact studies for MPK. MPK is not expected to present any unusual hazard to the environment. Since MPK would be expected to biodegrade, long-term environmental effects are not anticipated.

4.0 Methyl Isopropyl Ketone

4.0.1 CAS Number: [563-80-4]

4.0.2 Synonyms: 3-Methyl-2-butanone, methyl butanone, 2-acetyl propane, 2-methyl-3-butanone, and MIPK

4.0.3 Trade Names: NA

4.0.4 Molecular Weight: 86.14

4.0.5 Molecular Formula: $C_5H_{10}O$

4.0.6 Molecular Structure:
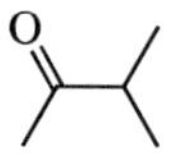

4.1 Chemical and Physical Properties

4.1.1 General

Methyl isopropyl ketone (MIPK) is a colorless, slightly water-soluble, flammable liquid with an acetonelike odor. It has a boiling point of 93.0°C and a melting point of −92.0°C. Additional physical–chemical properties are summarized in Table 75.1.

4.1.2 Odor and Warning Properties

MIPK has an odor similar to acetone.

4.2 Production and Use

Production volume and use information is not available for this material.

4.3 Exposure Assessment

Likely exposure pathways for this material are inhalation and dermal contact. Specific exposure information is not available for MIPK. Background levels in the environment are possible due to microbial metabolism.

4.3.3 Workplace Methods: NA

4.4 Toxic Effects

4.4.1 Experimental Studies

Methyl isopropyl ketone is slightly toxic by the oral and inhalation routes. Contact with the eyes produces slight irritation, and skin contact may produce slight to moderate irritation.

4.4.1.1 Acute Toxicity. Carpenter et al. (339) reported the oral LD_{50} for methyl isopropyl ketone to be 5.66 mL/kg in rats. In other studies, the oral LD_{50} values for male rats, male mice, and female rats were 3078–3200 mg/kg (322). Clinical signs observed included weakness and prostration. Other workers reported oral LD_{50} values in rats of 4100 (340) or 2572 mg/kg (130). The intraperitoneal LD_{50} was reported to be 800 mg/kg for male rats and between 200 and 400 mg/kg for male mice (322).

Slight ocular injury resulted when methyl isopropyl ketone was tested in the rabbit eye (322, 341). Washed and unwashed eyes appeared clinically normal by 72 h and 7 d, respectively, after dosing (322).

Slight irritation was observed when methyl isopropyl ketone was applied to the uncovered skin of rabbits. Moderate irritation was observed when it was applied undiluted to intact and abraded rabbit skin for 24 h, under an occlusive dressing (340). The acute dermal LD_{50} for rabbits was reported to be 6.35 mL/kg (339) or greater than 5 g/kg (340). Nine of ten rabbits survived exposure to a dermal dose of 20 mL/kg (326).

Methyl isopropyl ketone, applied to the abdomens of guinea pigs at doses of 5 to 20 mL/kg under an occlusive dressing for 24 h, produced moderate skin irritation as evidenced by edema and necrosis at 24 h, and residual eschars and scarring after 2 wk (326). When methyl isopropyl ketone was applied topically 9 times over a period of 11 d to the clipped skin of the backs of guinea pigs, no irritant response developed (326).

Exposure to an airborne concentration of 5700 ppm MIPK for 4 h resulted in the death of one of six rats (339). In groups of rats exposed for 6 h to 4026, 5708, or 8270 ppm methyl isopropyl ketone, clinical signs included dose-dependent CNS depression, lachrymation, hypoventilation, and death. The combined LC_{50} value for males and females was 6377 ppm. Clinical signs of toxicity in survivors resolved by 72 h (326). Rats were also exposed to target concentrations of up to 3000 ppm methyl isopropyl ketone 6 h/d, 5 d/wk for a total of 22 exposures. Animals exhibited lethargy, narcosis, and tearing during exposures, but all abnormal clinical signs resolved prior to the following day's exposure. Exposure-related findings included increased adrenal gland weights in 1500 and 3000 ppm males. Hyalin droplet formation was seen in the renal epithelium of all male exposure groups, with increased severity in the 1500 and 3000 ppm groups. Although a no-effect level for CNS depression and irritation was not obtained, there was no clear evidence of organ toxicity at doses as high as 3000 ppm (326).

4.4.1.2 Chronic and Subchronic Toxicity. Subchronic and chronic toxicity reports are not available for MIPK.

4.4.1.3 Pharmacokinetics, Metabolism, and Mechanisms. When a dose of 1000 mg/kg methyl isopropyl ketone was administered to rats, 38–54% was eliminated unchanged in 25–35 h, predominantly in the expired air (326). Williams (341) reported that methyl isopropyl ketone increases glucuronic acid output in urine.

Methyl isopropyl ketone appears to be biologically active primarily through its irritant properties and by the induction of narcosis.

4.4.1.4 Reproductive and Developmental Studies. Reproductive and developmental toxicity studies are not available for this material.

4.4.1.5 Carcinogenesis. Carcinogenicity studies are not available for this material.

4.4.1.6 Genetic and Related Cellular Effects Studies. Few data are available on the genetic toxicity of methyl isopropyl ketone. In an assay using *Saccharomyces cerevisiae* as a test organism, cells were subjected to a 17-h period in ice between two periods of incubation at 28°C in the presence of media containing 1.23% methyl isopropyl ketone. Under the test conditions, the type and number of colonies that grew were considered to be indicative of changes related to point mutation, mitotic recombination, or deletion of chromosomal fragments (240). The significance of this result is difficult to assess, however, given the unconventional nature of the test method.

4.4.1.7 Other: Neurological, Pulmonary, Skin Sensitization. *Sensitization.* Methyl isopropyl ketone did not cause a dermal sensitization response in guinea pigs (326). Studies on specific organ systems are not available for this material.

4.4.2 Human Experience

4.4.2.1 General Information. The main hazards associated with methyl isopropyl ketone are slight eye irritation, slight to moderate skin irritation, inhalation of its vapors, and its flammability. Eye and skin protection plus adequate ventilation would be expected to prevent serious injury.

4.4.2.2 Clinical Cases. Methyl isopropyl ketone was one of a series of chemicals used to determine the relative sensitivity of the human eye and lung to irritant gases. Eye irritation was based upon subjective descriptions following 15 sec of ketone exposure inside close-fitting goggles. Respiratory irritation was evaluated by measuring reflex bronchoconstriction after inspiration of 10 breaths of 1 L each. In this study, the threshold for bronchoconstriction and for subjective responses based on ocular irritation and were within the 300–500 ppm range, with the threshold for eye irritation the lower of the two (276).

4.4.2.2.1 Acute Toxicity. Skin sensitization patch tests on 25 volunteers, using a 10% concentration of methyl isopropyl ketone, produced neither irritation nor sensitization (342).

There are no case reports of systemic toxicity following exposure to MIPK.

4.4.2.3 Epidemiology Studies. There are no epidemiological studies of adverse health effects following exposure to MIPK.

4.5 Standards, Regulations, or Guidelines of Exposure

Hygienic standards for methyl isopropyl ketone are listed in Table 75.4

4.6 Studies on Environmental Impact

There are no published environmental impact studies for MIPK. MIPK is not expected to present any unusual hazard to the environment. Since MIPK would be expected to biodegrade, long-term environmental effects are not anticipated.

5.0 3-Pentyn-2-one

5.0.1 CAS Number: *[7299-55-0]*

5.0.2 Synonyms: No commonly used synonyms were identified.

5.0.3 Trade Names: NA

5.0.4 Molecular Weight: 82.102

5.0.5 Molecular Formula: C_5H_6O

5.0.6 Molecular Structure:

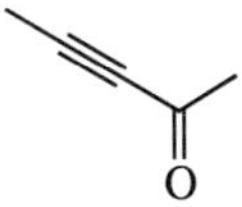

5.1 Chemical and Physical Properties

5.1.1 General

5.1.2 Odor and Warning Properties

3-Pentyn-2-one is a water-clear, chemically reactive liquid with a pronounced unpleasant odor. It has a boiling point of 133°C and a melting point of −28.7°C. Additional physical–chemical properties are summarized in Table 75.1.

5.2 Production and Use

No production and use information was identified for this material.

5.3 Exposure Assessment

Industrial exposure is expected to be via inhalation or dermal contact. Due to the irritating nature of this chemical, exposure is expected to be readily apparent.

3-Pentyn-2-one is moderately toxic orally, extremely irritating to the skin and eyes, appears to be readily absorbed, and is highly toxic percutaneously. Since inhalation of the vapors and skin contact are the primary routes of exposure, this compound presents a definite irritant hazard if handled without proper protective equipment.

5.3.3 Workplace Methods: NA

5.4 Toxic Effects

5.4.1 Experimental Studies

5.4.1.1 Acute Toxicity. Range-finding studies indicated an oral LD_{50} for 3-pentyn-2-one of between 63 and 128 mg/kg for rats (320). Clinical signs of toxicity included tremors and convulsions, with death occurring within 2 d. Necropsy of survivors revealed severe liver and kidney damage and marked irritation of the stomach mucosa.

When applied undiluted or as a 10% solution, 3-pentyn-2-one produced severe eye irritation (320).

Both the undiluted compound and a 10% solution were severely irritating to the skin. One to two mL of the neat compound, applied to the skin of rabbits, caused mortality within a few hours (320).

Three rats exposed for 6 min to a saturated atmosphere of the vapor died over a 5-d period. Immediate irritation of the eyes and nose was evident (320).

5.4.1.2 Chronic and Subchronic Toxicity. No repeated dose or long-term studies are available for this material.

5.4.1.3 Pharmacokinetics, Metabolism, and Mechanisms. No pharmacokinetic or metabolism studies are available for this material.

5.4.1.4 Reproductive and Developmental Studies. No reproductive or developmental toxicity studies are available for this material.

5.4.1.5 Carcinogenesis. No carcinogenesis studies are available for this material.

5.4.1.6 Genetic and Related Cellular Effects Studies. No geneticity studies are available for this material.

5.4.1.7 Other: Neurological, Pulmonary, Skin Sensitization. No studies on specific organ systems are available for this material.

5.4.2 Human Experience

5.4.2.1 General Information. The limited range-finding studies in animals reported indicate that this compound is moderately toxic by both the oral and dermal routes. It is very irritating to the skin, eyes, and mucous membranes. Precautions should be taken to avoid all possible contact. Protective clothing and respiratory equipment should be worn when handling 3-pentyn-2-one.

5.4.2.2 Clinical Cases. No clinical case studies reporting adverse health effects following exposure to this material have been published.

5.4.2.3 Epidemiology Studies. No epidemiological studies reporting adverse health effects following exposure to this material have been published.

5.5 Standards, Regulations, or Guidelines of Exposure

There are no recommended hygienic standards for 3-pentyn-2-one.

5.6 Studies on Environmental Impact

Specific studies on environmental impacts for this material are not available. Due to the irritating nature of 3-pentyn-2-one, spills to bodies of water may result in acute toxicity to fish; however, long-term effects would not be expected, as this material should biodegrade.

6.0 Methyl Isopropenyl Ketone

6.0.1 CAS Number: [814-78-8]

6.0.2 Synonyms: 2-Methyl-1-butene-3-one, 3-methyl-3-butene-2-one, isopropenyl methyl ketone, methyl butenone, and propen-2-yl methyl ketone.

6.0.3 Trade Names: NA

6.0.4 Molecular Weight: 84.118

6.0.5 Molecular Formula: C_5H_8O

6.0.6 Molecular Structure:

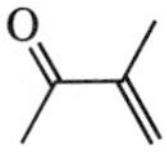

6.1 Chemical and Physical Properties

Methyl isopropenyl ketone is a chemically reactive, clear, colorless liquid with moderate flammability. The boiling point is 98°C, the melting point is -54°C, and it is soluble in alcohol. It is stored at subzero temperatures to prevent polymerization. Additional physical–chemical properties are summarized in Table 75.1.

6.1.2 Odor and Warning Properties

Methyl isopropenyl ketone has a pungent odor. A threshold odor concentration of 0.291 ppm has been reported (320).

6.2 Production and Use

This ketone is manufactured by catalytic hydration of isopropyl acetylene, by condensation of acetylene with acetone to form dimethyl ethylene carbinol, which requires further processing, or by condensation of methyl ethyl ketone with formaldehyde. Commercial samples may be stabilized with hydroquinone. The ketone is used as a solvent or as a monomer in copolymers.

6.3 Exposure Assessment

Methyl isopropenyl ketone is highly toxic by oral, dermal, and inhalation routes. Topical exposure to the skin may be moderately irritating while topical or vapor exposure to the eye may be severely irritating. Enclosed processing vessels, ventilation, or personal protective equipment are recommended to limit exposure.

6.4 Toxic Effects

6.4.1 Experimental Studies

6.4.1.1 Acute Toxicity. The oral LD_{50} for methyl isopropenyl ketone was reported to be 180 mg/kg for rats (343). The oral LD_{50} for guinea pigs was reported to be between 60 and 250 mg/kg, indicating a moderate to high degree of toxicity (320).

When applied to rabbit eyes, methyl isopropenyl ketone was found to be severely irritating (343). Vapors of methyl isopropenyl ketone may also be irritating to the eye.

Methyl isopropenyl ketone was moderately irritating to rabbit skin, particularly when applied under an occlusive wrap. The percutaneous LD_{50} has been calculated to be 230 mg/kg for the rabbit (343).

Smyth et al. (343) reported that a 4-h exposure to 125 ppm killed five of six rats, while exposure to a saturated atmosphere for 2 min killed all test animals. Rats survived a 90-min exposure to nominal vapor concentrations of 524 ppm, but exhibited marked irritation of ocular and nasal tissues (320). At a nominal vapor concentration of 1455 ppm for 30 min, methyl isopropenyl ketone was very irritating, and deaths occurred within minutes. Animals dying during these exposures were cyanotic and died in convulsions. Delayed deaths were related to irritation of the respiratory system.

6.4.1.2 Chronic and Subchronic Toxicity. Exposure of rats, guinea pigs, and rabbits, 7 h per day for 20 exposures over a period of 28 d at a concentration of 30 ppm, resulted in marked ocular and nasal irritation in all species, death and weight loss in rats, and decreased growth in guinea pigs (320). In rats, the lungs were severely affected and unspecified changes were found in the kidneys and spleen. One hundred 7-h exposures to 15 ppm of methyl isopropenyl ketone over a 140-d period produced ocular and nasal irritation in rats, guinea pigs, and rabbits. Mortality, leukocytosis, and slight renal damage (characterized by increased kidney weight and slight tubular damage) occurred in the rats (320).

6.4.1.3 Pharmacokinetics, Metabolism, and Mechanisms. Pharmacokinetics and metabolism studies with this material are not available.

6.4.1.4 Reproductive and Developmental Studies. No reproductive and developmental toxicity studies are available for this material.

6.4.1.5 Carcinogenesis. No carcinogenicity studies are available for this material.

6.4.1.6 Genetic and Related Cellular Effects Studies. Standard Ames/Salmonella plate assays using strains TA98, TA100, TA1535, TA1537, and TA1538 with and without metabolic activation (rat and hamster liver) were negative for mutagenicity (344).

6.4.1.7 Other: Neurological, Pulmonary, Skin Sensitization. No studies on specific organ systems are available for this material.

6.4.2 Human Experience

6.4.2.1 General Information. Methyl isopropenyl ketone is a highly toxic, highly irritating liquid or vapor. Industrial exposures are likely to be by inhalation, skin, and eye contact. Its strong irritating and lacrimatory properties may not prevent toxicity because these effects are often not immediately sensed.

6.4.2.2 Clinical Cases. Clinical case reports of adverse health effects following exposure to this material have not been published.

6.4.2.3 Epidemiology Studies. Short-term exposures to concentrations of methyl isopropenyl ketone ranging from 0.291 to 14.5 ppm have been reported (320). There was a questionable odor but no irritation at 0.291 ppm. At 0.77 ppm, there was an immediate definite odor, and effects of vapors were noted in eyes after a few minutes. At 1.45 and 2.91 ppm, eye irritation was apparent. At 14.5 ppm, there was an immediate strong odor with eye irritation apparent after 2 min. Ocular irritation rapidly increased in severity with time.

Epidemiological studies reporting adverse health effects following exposure to this material have not been published.

6.5 Standards, Regulations, or Guidelines of Exposure

There are no recommended hygienic standards for methyl isopropenyl ketone.

6.6 Studies on Environmental Impact

Specific studies on environmental impacts for this material are not available. Due to the irritating nature of methyl isopropenyl ketone, spills to bodies of water may result in acute toxicity to fish; however, long-term effects would not be expected, as this material should biodegrade.

7.0 2,4-Pentanedione

7.0.1 CAS Number: [123-54-6]

7.0.2 Synonyms: Acetylacetone, acetoacetone, diacetyl methane, acetyl 2-propanone, 2,4-pentanedione, and pentanedione

7.0.3 Trade Names: NA

7.0.4 Molecular Weight: 100.12

7.0.5 Molecular Formula: $C_5H_8O_2$

7.0.6 Molecular Structure

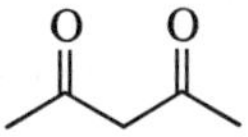

7.1 Chemical and Physical Properties

7.1.1 General

2,4-Pentanedione is a clear liquid with a boiling point of 138.3°C and a melting point of – 23.2°C. It presents a moderate fire hazard. Additional physical–chemical properties are summarized in Table 75.3.

7.1.2 Odor and Warning Properties

2,4-Pentanedione has an unpleasant rancid odor. Its odor threshold level is reported to be 0.010 ppm (345).

7.2 Production and Use

Published production information is not available for this material. 2,4-Pentanedione has been used as a fungicide, a bactericide, and a wood preservative.

7.3 Exposure Assessment

Exposure to 2,4-pentanedione may produce toxicity by the oral, dermal, or inhalation routes. Although ingestion would not appear to be a common route of exposure, accidental ingestion could be lethal or could produce irreversible neurologic impairment. Single high-level exposures dermally or by inhalation should be avoided. Repeated exposures to low concentrations may be cumulative.

7.4 Toxic Effects

7.4.1 Experimental Studies

High doses of 2,4-pentanedione produce dyspnea, severe central nervous system depression, and death in experimental animals. Similar clinical effects are produced by lower repeated doses, except that some animals survive and develop a central nervous system disorder characterized by an irreversible cerebellar syndrome. Thymic necrosis and atrophy may accompany central nervous system damage in experimental animals. Slight decreases in male fertility have been reported in experimental animals with exposure to 2,4-pentanedione.

7.4.1.1 Acute Toxicity. Smyth et al. (346) reported an oral LD_{50} of 1000 mg/kg for rats. Others reported an approximate oral LD_{50} of 800 mg/kg for rats and 951 mg/kg for mice (326). A study by Ballantyne et al. (347) determined the LD_{50} to be 0.78 mL/kg in male rats and 0.59 mL/kg in female rats. Signs of acute toxicity were a sluggish and unsteady gait, prostration, lacrimation, and convulsions. No changes in body weight occurred, and no gross pathology was observed at necropsy. The intraperitoneal LD_{50} was 750 mg/kg for mice (326).

Smyth and Carpenter (348) reported that 2,4-pentanedione was as irritating to the eye as acetone. A more recent study indicated that 2,4-pentanedione was a mild ocular irritant producing slight to moderate chemosis and mild iritis, which healed by 24 h (347). No corneal injury was noted.

2,4-Pentanedione applied to the depilated abdomens of guinea pigs under an occlusive wrap for 24 h produced moderate skin irritation (333). In a standard primary dermal irritation test in which rabbits received 2,4-pentanedione under an occluded patch for 4 h, slight erythema and edema developed by 24 h (347) and persisted in half the animals for up to 3 d. No effects other than desquamation were observed at 7 d. The acute dermal toxicity in rabbits was 1.41 mL/kg for males and 0.81 mL/kg for females. Signs of toxicity included dilated pupils and salivation. No signs of toxicity were observed at 0.5 mL/kg. Dermal irritation after the 24-h exposure persisted for up to 7 d. Four guinea pigs receiving 20 mL/kg under a wrap died within 24–72 h of application (326), but only one of five guinea pigs showed a weak sensitization reaction (326).

Smyth and Carpenter (348) reported that 30 min was the maximum time that rats could be exposed to a saturated atmosphere of 2,4-pentanedione vapor before death occurred. The same level was lethal to all animals exposed for 1 h. Ballantyne et al. determined the 4-h LC_{50} to be 1224 ppm in rats (347). Signs of toxicity included ataxia, perinasal and perioral wetness, tremors, and breathing difficulties. No signs of toxicity were observed at 628 ppm.

Dodd et al. conducted repeated exposure studies for 9 d and 14 wk in rats in which animals were exposed to concentrations of 197, 418, or 805 ppm for 9 d, and 101, 307, or 650 ppm for 14 wk (349). Exposure for 9 d resulted in significantly lower body weights for male and female rats at 805 ppm, and slightly lower weight gain for male rats at 418 ppm. No mortality occurred. At 805 ppm, signs of sensory irritation were evident, and a mild leukocytosis was noted. Lower concentrations produced no effect. Inflammation of the nasal tissues with lymphocyte infiltration was noted in animals exposed to 805 ppm, and to some extent 418 ppm. These results are consistent with irritation of the mucosa. In addition, thymus-to-body weight ratios in animals exposed to 805 ppm were significantly lower than in controls, but no histopathologic findings were observed. Exposure to 605 ppm for 14 wk resulted in 100% mortality of female rats and 33% (10 of 30) mortality of male rats. Animals that died in the test had degenerative changes in the brain and thymus. Survivors had gliosis and malacia in the brain, squamous metaplasia of the nasal mucosa, and lymphocytosis. The body weights of survivors were significantly reduced after 14 wk of exposure, but rebounded during the 4-wk recovery period. Body weights of animals exposed to 307 ppm were only slightly reduced, and there was no effect at 101 ppm.

7.4.1.2 Chronic and Subchronic Toxicity. Groups of five rats were administered 0, 100, 500, or 1000 mg/kg 2,4-pentanedione by gavage for 15 d (326). The 1000-mg/kg dose produced a rapid onset of dyspnea and respiratory depression followed by death within 1 h of dosing. Animals receiving 500 mg/kg also showed signs of respiratory depression in addition to tremors and ataxia. Death occurred within a few days. Animals receiving 100 mg/kg showed signs of slight respiratory depression but were otherwise normal. Subsequent studies evaluated the effects of various dose levels on survival and overt clinical signs of toxicity. No clinical signs of toxicity were observed at 100 mg/kg, and no adverse effects on weight gain, feed consumption, hematology, clinical pathology, organ weight, or histopathology were observed. A dose of 250 mg/kg twice a day by gavage was lethal within 4 d. A dose of 150 mg/kg given twice daily produced head tilt and increased muscle tone in one animal. Another animal developed extreme weakness and tremors. The dose was lowered to 100 mg/kg and treatment continued for 6 wk without any adverse effects. After 6 wk, the dose was increased to 125 mg/kg twice daily. One of four animals developed quadraparesis. Histologic examination of the brain from all treated animals indicated that acute changes were characterized by perivascular edema, hemorrhage into the Virchow–Robbin spaces, and endothelial cell swelling, all primarily localized in the brainstem and cerebellum. Chronic central nervous system lesions were also present as bilateral, symmetrical areas of malacia and gliosis centered on the cerebellar peduncles, olivary nuclei, and lower brainstem.

To exclude the possibility that these changes were peculiar to the rat, groups of two male New Zealand rabbits were administered 250, 500, or 1000-mg/kg of 2,4-

pentanedione once a day for 5 d/wk by gavage (326). At 1000 mg/kg, both rabbits died within 24 h after a single dose. At autopsy, rabbits at the 1000 mg/kg level showed congestion of the brain, lungs, and thymus. Histologically, both rabbits at the 1000-mg/kg level showed congestion and hemorrhage in the thymus and in one of the rabbits the thymus was very small. At 500 mg/kg, one rabbit died on the ninth day of the study, and the second rabbit died on the twelfth day following severe central nervous system depression. Gross changes in one rabbit at 500 mg/kg included hemorrhage in the brain, an atrophic thymus, pulmonary congestion, and gastric mucosal hemorrhages. Both rabbits in the 500 mg/kg group showed congestion of the brain. Hemorrhage into the spinal cord was observed in one rabbit, and the one rabbit in which the thymus and stomach were examined showed hemorrhage into the mediastinal fat, marked thymic atrophy with heavy macrophage infiltration, and gastric mucosal hemorrhage. One rabbit receiving 250 mg/kg survived 14 d. Gross changes in rabbits at 250 mg/kg were absent. Animals in the 250 mg/kg group showed no compound-related histologic lesions.

7.4.1.3 Pharmacokinetics, Metabolism, and Mechanisms. Little information about the *in vivo* mechanisms of action of 2,4-pentanedione exists, but *in vitro* studies indicate that 2,4-pentanedione may inactivate enzyme activity in several ways. The first involves its metal chelating properties. 2,4-Pentanedione binds iron in peroxidase and thus prevents the oxidation of human serum proteins (350). Iron chelation may also be involved in the degradation of heme in cytochrome P450. 2,4-Pentanedione suppresses cytochrome P450 without lowering other hepatic microsomal enzymes (351). Enzymatic inactivation can also occur through reactions involving arginine and lysine. Reaction with the arginine residues in proteins leads to the formation of stable pyrimidines, whereas reaction with lysine residules leads to enamines (352). Of the two amine reactions, that of arginine is the slower and more irreversible one (352). In correlating the *in vitro* activity of 2,4-pentanedione with its toxicity following repeated administration to rats and rabbits, it appears that the neurologic and lymphoid damage may be due to the inactivation of B vitamins or their coenzymes. Neuropathologic lesions due to 2,4-pentanedione closely resemble those produced by thiamine deficiency in experimental animals (326). Antilymphoid effects of folate antagonists are well known. In vitro, 2,4-pentanedione inactivates dihydrofolate reductase activity, which is a target of antifolate drugs for cancer chemotherapy (353). It also inactivates the pyridoxal phosphate and pyridoxamine phosphate forms of porcine heart aspartate aminotransferase. The similarity between the morphologic damage produced by 2,4-pentanedione and acute vitamin B deficiency diseases and the known ability of 2,4-pentanedione to inactivate the lysyl residues of vitamin B coenzymes suggest that the toxicity of 2,4-pentanedione is due to its ability to produce deficiencies of thiamine, folic acid, pyridoxine, or their combinations.

7.4.1.4 Reproductive and Developmental Studies. The effect of 2,4-pentanedione exposure on reproductive capability and development were evaluated. The reproductive capacity of male rats exposed to 99, 412, or 694 ppm for 5 consecutive days were tested by Tyl et al. (354). Concentrations of 694 and 412 produced slight decreases in viable fetuses 2–4 wk after exposure, suggesting a slight decrease in male fertility. However, there were no histopathologic changes in the testes of exposed males. Exposure of pregnant female rats to concentrations of 53, 202, or 398 ppm did not produce any malformations in offspring (355). A concentration of 398 ppm reduced maternal body weight, and a concentration of 202 ppm reduced fetal body weight slightly. No effects were observed at 53 ppm.

7.4.1.5 Carcinogenesis. No carcinogenicity studies are available for this material.

7.4.1.6 Genetic and Related Cellular Effects Studies. Kato et al. (356) determined that 2,4-pentanedione was mutagenic in *Salmonella* strain TA104 without enzymic activation of the ketone and in *E. coli* with enzymic activation, but not in *Salmonella* strains TA98 and TA100. Guzzie et al. (357) also evaluated the genotoxicity of 2,4-pentanedione *in vitro* using the Ames/*Salmonella* bacterial assay and Chinese hamster ovary HGPRT mutation assay. 2,4-Pentanedione was not mutagenic in either assay whether in the presence or absence of S9 enzymic activation. Clastogenic changes in the form of sister chromatid exchange did occur in the Chinese hamster ovary assay, however. Micronuclei were observed in Swiss–Webster mice that were injected IP with 200, 400, or 640 mg/kg of 2,4-pentanedione. The numbers of micronuclei were significantly increased in both treated males and females. When 2,4-pentanedione was incubated with yeast cells, no chromosome loss occurred even at a concentration of 1.96% (v/v) and cold shock (0EC) (240), but the addition of propionitrile to the incubation mixture stimulated chromosomal loss (243).

7.4.1.7 Other: Neurological, Pulmonary, Skin Sensitization. See Section 7.4.1.2 for information about neurotoxicity. Nagano et al. (358) and Misumi and Nagano (333) reported slowing of motor and sensory nerve conduction velocity in rats dosed subcutaneously with 2,4-pentadione at 200 or 415 mg/kg/d, 5 d/wk for up to 40 wk.

7.4.2 Human Experience

7.4.2.2 Clinical Cases

7.4.2.2.1 Acute Toxicity. 2,4-Pentanedione (2 to 14 ppm) has been reported to produce nausea and headaches in several persons, and may have been associated with the occurrence of hives in one worker (326). Sterry and Schmoll (359) reported a case involving an individual who had developed contact dermatitis to Cu(II)–acetyl acetonate and who also showed a cross reaction to 2,4-pentanedione.

7.4.2.3 Epidemiology Studies. Epidemiological studies of adverse health effects following exposure to this material have not been published.

7.5 Standards, Regulations, or Guidelines of Exposure

There are no recommended hygienic standards for 2,4-pentanedione.

7.6 Studies on Environmental Impact

There are no published environmental impact studies for this material. This material is not expected to present any unusual hazard to the environment. Since the material would be expected to biodegrade, long-term environmental effects are not anticipated.

BIBLIOGRAPHY

1. V. S. Dunovant et al., Volatile organics in the wastewater and airspace of three wastewater treatment plants. *J. Water Pollut. Control Fed.* **58**, 866–895 (1986).
2. T. C. Sauer, Jr., Volatile liquid hydrocarbon characterization of underwater hydrocarbon vents and formation waters from offshore production operations. *Environ. Sci. Technol.* **15**, 917–923 (1981).
3. M. Ash and I. Ash, Chemical to tradename reference. In *Industrial Chemical Thesaurus*, 2nd ed., Vol. 1, VCH Publishers, Inc., New York, 1992, p. 363.
4. J. A. Riddick, W. B. Bunger, and T. K. Sakano, *Organic Solvents Physical Properties and Methods of Purification*, 4th ed., John Wiley & Sons, Inc., New York, 1986, pp. 338–339.
5. National Fire Protection Association, Fire hazard properties of flammable liquids, gases, and volatile solids. In *Fire Protection Guide on Hazardous Materials*, 12th ed., NFPA, Quincy, MA, 1997, pp. 325–369.
6. J. E. Amoore and E. Hautala, Odor as an aid to chemical safety: Odor thresholds compared with threshold limit values and volatilities for 214 industrial chemicals in air and water dilution. *J. Appl. Toxicol.* **3**, 272–290 (1983).
7. G. Leonardos, D. Kendall, and N. Barnard, Odor threshold determinations of 53 odorant chemicals. *J. Air Pollut. Control Assoc.* **19**, 91–95 (1969).
8. T. M. Hellman and F. H. Small, Characterization of the odor properties of 101 petrochemicals using sensory methods. *J. Air Pollut. Control Assoc.* **24**, 979–982 (1974).
9. M. Devos et al., *Standardized Human Olfactory Thresholds*, Oxford University Press, Oxford, UK, 1990, p. 145.
10. J. H. Ruth, Odor thresholds and irritant levels of several chemical substances: A review. *Am. Ind. Hyg. Assoc. J.* **47**, A142–A151 (1986).
11. American Industrial Hygiene Association (AIHA), *Odor Thresholds for Chemicals with Established Occupational Health Standards*, AIHA, Fairfax, VA, 1997, pp. 42–43.
12. G. Forrai et al., Studies on the sense of smell to ketone compounds in a Hungarian population. *Hum. Genet.* **8**, 348–353 (1970).
13. S. N. Bizzari, A. E. Leder, and K. Miyashita, CEH data summary–Methyl ethyl ketone (MEK). *Stanford Res. Inst., Rep.* 675.5000 A, 1–15 (1977).
14. W. Kirstine et al., Emissions of volatile organic compounds (primarily oxygenated species) from pasture. *J. Geophys. Res.* **103**(10), 605–610, 619 (1998).
15. M. A. K. Khalil and R. A. Rasmussen, Forest hydrocarbon emissions: Relationships between fluxes and ambient concentrations. *J. Air Waste Manage. Assoc.* **42**, 810–813 (1992).
16. J. M. Bradow and W. J. Connick, Volatile seed germination inhibitors from plant residues. *J. Chem. Ecol.* **16**, 645–666 (1990).
17. J. A. Weybrew and R. L. Stephens, Survey of the carbonyl contents of tobacco. *Tob. Sci.* **6**, 53–57 (1962).
18. J. R. Newsome, V. Norman, and C. H. Keith, Vapor phase analysis of tobacco smoke. *Tob. Sci.* **9**, 102–110 (1965).
19. K. Wilkins, Volatile organic compounds from household waste. *Chemosphere* **29**, 47–53 (1994).
20. B. D. Eitzer, Emissions of volatile organic chemicals from municipal solid waste composting facilities. *Environ. Sci. Technol.* **29**, 896–902 (1995).

21. S. K. Brown et al., Concentrations of volatile organic compounds in indoor air–A review. *Indoor Air* **4**, 123–134 (1994).
22. B. A. Tichenor and M. A. Mason, Organic emissions from consumer products and building materials to the indoor environment. *J. Air Pollut. Control Assoc.* **38**, 264–268 (1988).
23. T. M. Sack et al., A survey of household products for volatile organic compounds. *Atmos. Environ.* **26A**, 1063–1070 (1992).
24. C. Winder and S. K. Ng, The problem of variable ingredients and concentrations in solvent thinners. *Am. Ind. Hyg. Assoc. J.* **56**, 1225–1228 (1995).
25. A. Korpi, A.-L. Pasanen, and P. Pasanen, Volatile compounds originating from mixed microbial cultures on building materials under various humidity conditions. *Appl. Environ. Microbiol.* **64**, 2914–2919 (1998).
26. A.-L. Pasanen et al., Critical aspects on the significance of microbial volatile metabolites as indoor air pollutants. *Environ. Int.* **24**, 703–712 (1998).
27. E. Smet, H. Van Langenhove, and I. De Bo, The emission of volatile compounds during the aerobic and the combined anaerobic/aerobic composting of biowaste. *Atmos. Environ.* **33**, 1295–1303 (1999).
28. Y. Hoshika, Y. Nihei, and G. Muto, Pattern display for characterisation of trace amounts of odorants discharged from nine odour sources. *Analyst (London)* **106**, 1187–1202 (1981).
29. C. C. Chan, L.Vainer, and J. W. Martin, Determination of organic contaminants in residential indoor air using an absoption-thermal desorption technique. *J. Air Waste Manage. Assoc.* **40**, 62–67 (1990).
30. J. J. Shah and H. B. Singh, Distribution of volatile organic chemicals in outdoor and indoor air. *Environ. Sci. Technol.* **22**, 1381–1388 (1988).
31. R. Reiss et al., Measurement of organic acids, aldehydes, and ketones in residential environments and their relation to ozone. *J. Air Waste Manage. Assoc.* **45**, 811–822 (1995).
32. A. Jonsson, K. A. Persson, and V. Grigoriadis, Measurements of some low molecular-weight oxygenated, aromatic, and chlorinated hydrocarbons in ambient air and in vehicle emissions. *Environ. Int.* **11**, 383–392 (1985).
33. T. J. Kelly and P. J. Callahan, Method development and field measurements for polar volatile organic compounds in ambient air. *Environ. Sci. Technol.* **27**, 1146–1153 (1993).
34. A. T. Hodgson and J. M. Daisey, Sources and source strengths of volatile organic compounds in a new office building. *J. Air Waste Manage. Assoc.* **41**, 1461–1468 (1991).
35. T. A. Bellar and J. E. Sigsby, Direct gas chromatographic analysis of low molecular weight substituted organic compounds in emissions. *Environ. Sci. Technol.* **4**, 150–156 (1970).
36. L. E. Booher and B. Janke, Air emissions from petroleum hydrocarbon fires during controlled burning. *Am. Ind. Hyg. Assoc. J.* **58**, 359–365 (1997).
37. E. D. Pellizzari, Analysis for organic vapor emissions near industrial and chemical waste disposal sites. *Environ. Sci. Technol.* **16**, 781–789 (1982).
38. K. J. James, M. Cherry, and M. A. Stack, Assessment of chemical plant emissions on an urban environment: A new approach to air quality measurements. *Chemosphere* **31**, 3741–3751 (1995).
39. D. Rotival et al., The emissions of carbonyl compounds during thermal processing of an acrylic polymer. *Indoor Environ.* **3**, 79–82 (1994).
40. J. Brosseau and M. Heitz, Trace gas compound emissions from municipal landfill sanitary sites. *Atmos. Environ.* **28**, 285–293 (1994).

41. S.-M. Li et al., Emission ratios and photochemical production efficiencies of nitrogen oxides, ketones, and aldehydes in the lower Fraser Valley during the summer Pacific 1993 oxidant study. *Atmos. Environ.* **31**, 2037–2048 (1997).

42. E. Grosjean et al., Air quality model evaluation data for organics. 2. C_1-C_{14} carbonyls in Los Angeles air. *Environ. Sci. Technol.* **30**, 2687–2703 (1996).

43. T. W. Kirchstetter, B. C. Singer, and R. A. Harley, Impact of oxygenated gasoline use on California light-duty vehicle emissions. *Environ. Sci. Technol.* **30**, 661–670 (1996).

44. K. Staehelin et al., Emissions factors from road traffic from a tunnel study (Gubrist Tunnel, Switzerland). Part III: Results of organic compounds, SO_2 and speciation of organic exhaust emissions. *Atmos. Environ.* **32**, 999–1009 (1998).

45. P. V. Doskey et al., Source profiles for nonmethane organic compounds in the atmosphere of Cairo, Egypt. *J. Air Waste Manage. Assoc.* **49**, 814–822 (1999).

46. A.P. Verhoeff et al., Organic solvents in the indoor air of two small factories and surrounding houses. *Int. Arch. Occup. Environ. Health* **59**, 153–163 (1987).

47. N. P. Brandorff et al., National survey on the use of chemicals in the working environment: Estimated exposure events. *Occup. Environ. Med.* **52**, 454–463 (1995).

48. B. C. Doney, Solvent, wood dust, and noise exposure in a decoy shop. *Appl. Occup. Environ. Hyg.* **10**, 446–448 (1995).

49. L. W. Whitehead et al., Solvent vapor exposures in booth spray painting and spray glueing, and associated operations. *Am. Ind. Hyg. Assoc. J.* **45**, 767–772 (1984).

50. R. Vincent et al., Occupational exposure to organic solvents during paint stripping and painting operations in the aeronautical industry. *Int. Arch. Occup. Environ. Health* **65**, 377–380 (1994).

51. M. Kiefer et al., Investigation of d-Limonene use during aircraft maintenance degreasing operations. *Appl. Occup. Environ. Hyg.* **9**, 303–311 (1994).

52. R. Agnesi, F. Valentini, and G. Mastrangelo, Risk of spontaneous abortion and maternal exposure to organic solvents in the shoe industry. *Int. Arch. Occup. Environ. Health* **69**, 311–316 (1997).

53. De Rosa et al., The industrial use of solvents and risk of neurotoxicity. *Ann. Occup. Hyg.* **29**, 391–397 (1985).

54. L. L. Kupferschimid and J. L. Perkins, Organic solvent recycling plant exposure levels. *Appl. Ind. Hyg.* **1**, 122–124 (1986).

55. D. W. Decker et al., Worker exposure to organic vapors at a liquid chemical waste incinerator. *Am. Ind. Hyg. Assoc. J.* **44**, 296–300 (1983).

56. N. A. Nelson et al., Historical characterization of exposure to mixed solvents for an epidemiologic study of automotive assembly plant workers. *Appl. Occup. Environ. Hyg.* **8**, 693–702 (1993).

57. T. Kauppinen et al., International data base of exposure measurements in the pulp, paper and paper product industries. *Int. Arch. Occup. Environ. Health* **70**, 119–127 (1997).

58. S. S. Lande et al., Investigation of selected potential environmental contaminants: Ketonic solvents. *Environ. Prot. Agency* EPA-560/2-76-003 (1976).

59. S. Zakhari et al., Acute oral, intraperitoneal and inhalational toxicity in the mouse. In L. Goldberg, ed., *Isopropanol and Ketones in the Environment*, CRC Press, Boca Raton, FL, 1977, pp. 67–69.

60. I. Laye, D. Karleskind, and C. V. Morr, Chemical, microbiological and sensory properties of plain nonfat yogurt. *J. Food Serv.* **58**, 991–995 (1993).

61. H. Sakakibara et al., Volatile flavor compounds of some kinds of dried and smoked fish. *Agric. Biol. Chem.* **54**, 9–16 (1990).
62. T. O. Larsen and J. C. Frisvad, Characterization of volatile metabolites from 47 penicillium taxa. *Mycol. Res.* **99**, 1153–1100 (1995).
63. A.-L. Sunesson et al., Identification of volatile metabolites from five fungal species cultivated on two media. *Appl. Environ. Microbiol.* **61**, 2911–2918 (1995).
64. A.-L. Sunesson et al., Production of volatile metabolites from *Streptomyces albidoflavus* cultivated on gypsum board and tryptone glucose extract agar—Influence of temperature, oxygen and carbon dioxide levels. *Ann. Occup. Hyg.* **41**, 393–413 (1997).
65. D. A. Pavlica et al., Volatiles from soil influencing activities of soil fungi. *Ecol. Epidemiol.* **68**, 758–765 (1978).
66. J. K. Whelan, M. E. Tarafa, and J. M. Hunt, Volatile C_1-C_8 organic compounds in macroalgae. *Nature (London)*, **299**, 50–52 (1982).
67. J. F. Corwin, Volatile oxygen-containing organic compounds in sea water: Determination. *Bull. Mar. Sci.* **19**, 504–509 (1969).
68. A. B. Crockett, Water and wastewater quality monitoring, McMurdo Station, Antarctica. *Environ. Monit. Assess.* **47**, 39–57 (1997).
69. M. R. Berry, L. S. Johnson, and K. P. Brenner, dietary characterizations in a study of human exposure in the Lower Rio Grande Valley: II. Household waters. *Environ. Int.* **23**, 693–703 (1997).
70. T. C. Wang and J. L. Bricker, 2-Butanone and tetrahydrofuran contamination in the water supply. *Bull. Environ. Contam. Toxicol.* **23**, 620–623 (1979).
71. T. M. Holsen et al., Contamination of potable water by permeation of plastic pipe. *J. Am. Water Works Assoc.* **83**, 53–56 (1991).
72. R. Snider and G. A. Dawson, Tropospheric light alcohols, carbonyls, and acetonitrile: Concentrations in the southwestern United States and Henry's Law data. *J. Geophys. Res.* **90**, 3797–3805 (1985).
73. G. A. Jungclaus, V. Lopez-Avila, and R.A. Hites, Organic compounds in an industrial waste-water: A case study of their environmental impact. *Environ. Sci. Technol.* **12**, 88–96 (1978).
74. B. L. Sawhney and J. A. Raabe, Groundwater contamination: Movement of organic pollutants in the Granby Landfill. *Bull. Conn. Agric. Exp. Stn.* **833**, 1–9 (1986).
75. G. B. Sabel and T. P. Clark, Volatile organic compounds as indicators of municipal solid waste leachate contamination. *Waste Manage. Res.* **2**, 119–130 (1984).
76. K. W. Brown and K. C. Donnelly, An estimation of the risk associated with the organic constituents of hazardous and municipal waste landfill leachates. *Hazard. Waste Hazard. Mater.* **5**, 1–30 (1988).
77. NIOSH, *Manual of Analytical Methods*, 4th ed., National Institute for Occupational Safety and Health, Washington, DC, 1994.
78. V. B. Stein and R. S. Narang, Determination of aldehydes, ketones, esters, and ethers, in air using Porapak N and charcoal. *Arch. Environ. Contam. Toxicol.* **30**, 476–480 (1996).
79. P. W. Langvardt and R. G. Melcher, Simultaneous determination of polar and non-polar solvents in air using a two-phase desorption from charcoal. *Am. Ind. Hyg. Assoc. J.* **40**, 1006–1012 (1979).
80. M. Harper et al., An evaluation of sorbents for sampling ketones in workplace air. *Appl. Occup. Environ. Health* **8**, 293–304 (1993).

81. N. Vahdat et al., Adsorption capacity and thermal desorption efficiency of selected adsorbents. *Am. Ind. Hyg. Assoc. J.* **56**, 32–38 (1995).

82. E. Baltussen et al., Sorption tubes packed with polydimethylsiloxane: A new and promising technique for the preconcentration of volatiles and semi-volatiles from air and gaseous samples. *J. High Resolut. Chromatogr.* **21**, 332–340 (1998).

83. S. T. Rodriguez, D. W. Gosselink, and H. E. Mullins, Determination of desorption efficiencies in the 3M 3500 organic vapor monitor. *Am. Ind. Hyg. Assoc. J.* **43**, 569–574 (1982).

84. P. P. Ballesta, E. G. Ferradas, and A. M. Aznar, Simultaneous passive sampling of volatile organic compounds. *Chemosphere* **25**, 1797–1809 (1992).

85. S.-H. Byeone et al., Evaluation of an activated carbon felt passive sampler in monitoring organic vapors. *Ind. Health* **35**, 404–414 (1997).

86. H. Xiao, Analysis of organic vapors in the workplace by remote sensing fourier transform infrared spectroscopy. *Am. Ind. Hyg. Assoc. J.* **54**, 545–556 (1993).

87. F. J. Rowell, A. Fletcher, and C. Packham, Recovery of some common solvents from an adhesive commercial skin adsorption pad by thermal desorption-gas chromatography. *Analyst (London)* **122**, 793–796 (1997).

88. P. Clair, M. Tua, and H. Simian, Capillary columns in series for GC analysis of volatile organic pollutants in atmospheric and alveolar air. *J. High Resolut. Chromatogr.* **14**, 383–387 (1991).

89. H. K. Wilson, Breath analysis. *Scand. J. Work Environ. Health* **12**, 174–192 (1986).

90. D. Dyne, J. Cocker, and H. K. Wilson, A novel device for capturing breath samples for solvent analysis. *Sci. Total Environ.* **199**, 83–89 (1997).

91. S. Kezic and A. C. Monster, Determination of methyl ethyl ketone and its metabolites in urine using capillary gas chromatography. *J. Chromatogr.* **428**, 275–280 (1988).

92. L. Dunemann and H. Hajimiragha, Development of a screening method for the determination of volatile organic compounds in body fluids and environmental samples using purge and trap gas chromatography-mass spectrometry. *Anal. Chim. Acta* **283**, 199–206 (1993).

93. I. Ojanpera, R. Hyppola, and K. Vuori, Identification of volatile organic compounds in blood by purge and trap PLOT-capillary gas chromatography coupled with Fourier transform infrared spectroscopy. *Forensic Sci. Int.* **80**, 201–209 (1996).

94. J. Schuberth, A full evaporation headspace technique with capillary GC and ITD: A means for quantitating volatile organic compounds in biological samples. *J. Chromatogr. Sci.* **34**, 314–319 (1996).

95. M. Petrarulo, S. Pellegrino, and E. Testa, High-performance liquid chromatographic microassay for methyl ethyl ketone in urine as the 2,4-dinitrophenyl-hydrazone derivative. *J. Chromatog.* **579**, 324–328 (1992).

96. J. Yu, H. E. Jeffries, and R. M. Le Lacheur, Identifying airborne carbonyl compounds in isoprene atmospheric photooxidation products by their PFBHA oximes using gas chromatography/ion trap mass spectrometry. *Environ. Sci. Toxicol.* **29**, 1923–1932 (1995).

97. P. Ciccioli et al., Evaluation of organic pollutants in the open air and atmospheres in industrial sites using graphatized carbon black traps and gas chromatographic-mass spectrophotometric analysis with specific detectors. *J. Chromatog.* **126**, 757–770 (1976).

98. J. Lange and S. Eckhoff, Determination of carbonyl compounds in exhaust gas by using a modified DNPH-method. *Fresenius' J. Anal. Chem.* **356**, 385–389 (1996).

99. A. Yasuhara, Comparison of volatile components between fresh and rotten mussels by gas chromatography-mass spectrometry. *J. Chromatogr.* **409**, 251–258 (1987).

100. J. M. Zechman, S. Aldinger, and J. N. Labows, Characterization of pathogenic bacteria by automated headspace concentration-gas chromatography. *J. Chromatogr.* **377**, 49–57 (1986).

101. J. D. Pleil and M. L. Stroupe, Measurement of vapor-phase organic compounds at high concentrations. *J. Chromatogr.* **676**, 399–408 (1994).

102. M. Sugimoto et al., Determination of residual solvents in drug substances by gas chromatography with thermal desorption cold trap inject. *Chem. Pharm. Bull.* **43**, 2010–2013 (1995).

103. P. J. Streete et al., Detection and identification of volatile substances by headspace capillary gas chromatography to aid the diagnosis of acute poisoning. *Analyst (London)* **117**, 1111–1127 (1992).

104. B. K. Logan, G. A. Case, and E. L. Kiesel, Differentiation of diethyl ether/acetone and ethanol/acetonitrile solvents pairs, and other common volatiles by dual column headspace gas chromatography. *J. Forensic Sci.* **39**, 1544–1551 (1994).

105. D. F. Gurka et al., Interim method for determination of volatile organic compounds in hazardous wastes. *J. Assoc. Off. Anal. Chem.* **47**, 776–782 (1984).

106. H. Hartkamp, J. Rottmann, and M. Schmitz, Trace gas analysis using thermoanalytical methods. *Fresenius' J. Anal. Chem.* **337**, 729–739 (1990).

107. B. S. Middleditch et al., Trace analysis of volatile polar organics by direct aqueous injection gas chromatography. *Chromatographia* **23**, 273–277 (1987).

108. I. Ogawa and J. S. Fritz, Determination of low-molecular-weight aldehydes and ketones in aqueous samples. *J. Chromatogr.* **329**, 81–89 (1985).

109. M. A. Mendes et al., A cryotrap membrane introduction mass spectrometry system for analysis of volatile organic compounds in water at the low parts-per-trillion level. *Anal. Chem.* **68**, 3502–3506 (1996).

110. J. P. Conkle, B. J. Camp, and B. E. Welch, Trace composition of human respiratory gas. *Arch. Environ. Health* **30**, 290–295 (1975).

111. M. U. Tsao and E. L. Pfeiffer, Isolation and identification of a new ketone body in normal human urine. *Proc. Soc. Exp. Biol. Med.* **94**, 628–629 (1957)

112. A. Zlatkis et al., Volatile metabolites in sera of normal and diabetic patients. *J. Chromatog.* **182**, 137–145 (1980).

113. H. Przyrembel et al., Propionyl-CoA carboxylase deficiency with overflow of metabolites of isolucine catabolism at all levels. *Eur. J. Pediat.* **130**, 1–14 (1979).

114. D. Gompertz et al., A defect in L-isoleucine metabolism associated with α-methyl-β-hydroxybutyric acid and α-methylacetoacetic aciduria: Quantitative *in vivo* and *in vitro* studies. *Clin. Chim. Acta* **57**, 269–281 (1974).

115. D. L. Ashley et al., Blood concentrations of volatile organic compounds in a nonoccupationally exposed US population and in groups with suspected exposure. *Clin. Chem.* **40**, 1401–1404 (1994).

116. T. J. Buckley et al., Environmental and biomarker measurements in nine homes in the lower Rio Grande Valley: Multimedia results for pesticides, metals, PAHs, and VOCs. *Environ. Int.* **23**, 705–732 (1997).

117. G. B. Hamar et al., Volatile organic compounds testing of a population living near a hazardous waste site. *J. Exposure Anal. Environ. Epidemiol.* **6**, 247–255 (1996).

118. M. Imbriani et al., Methyl Ethyl Ketone (MEK) in urine as biological index of exposure. *G. Ital. Med. Lav.* **11**, 255–261 (1989).

119. M. Yoshikawa et al., Biological monitoring of occupational exposure to methyl ethyl ketone in Japanese workers. *Arch. Environ. Contam. Toxicol.* **29**, 135–139 (1995).

120. M. Miyasaka et al., Biological monitoring of occupational exposure to methyl ethyl ketone by means of urinalysis for methyl ethyl ketone itself. *Int. Arch. Occup. Environ. Health* **50**, 131–137 (1992).

121. L. Perbellini et al., Methyl ethyl ketone exposure in industrial workers. Uptake and kinetics. *Int. Arch. Occup. Environ. Health* **54**, 73–81 (1984).

122. F. Brugnone et al., Environmental and biological monitoring of occupational methyl ethyl ketone exposure. In A.W. Hayes, R. C. Schnell, and T.S. Miya, eds., *Developments in the Science and Practice of Toxicology*, Vol. 11, Elsevier, New York, 1983, pp. 571–574.

123. S. Ghittori et al., The urinary concentration of solvents as a biological indicator of exposure: Proposal for the biological equivalent exposure limit for nine solvents. *Am. Ind. Hyg. Assoc. J.* **48**, 786–790 (1987).

124. J.-Y. Jang, S.-K. Kang, and H. K. Chung, Biological exposure indices of organic solvents for Korean workers. *Int. Arch. Occup. Environ. Health* **65**, S219–S222 (1993).

125. C. N. Ong et al., Biological monitoring of occupational exposure to methyl ethyl ketone. *Int. Arch. Occup. Environ. Health* **63**, 319–324 (1991).

126. J. D. Caughlin, An unusual source for postmortem findings of methyl ethyl ketone and methanol in two homicide victims. *Forensic Sci. Int.* **67**, 27–31 (1997).

127. A. W. Jones, L. Andersson, and K. Berglund, Interfering substances identified in the breath of drinking drivers with intoxilyzer 5000S. *J. Anal. Toxicol.* **20**, 522–527 (1996).

128. A. W. Jones, M. Lund, and E. Andersson, Drinking drivers in Sweden who consume denatured alcohol preparation: An analytical-toxicological study. *J. Anal. Toxicol.* **13**, 199–203 (1989).

129. U. C. Pozzani, C. S. Weil, and C. P. Carpenter, The toxicological basis of threshold limit values: 5. The experimental inhalation of vapor mixtures by rats, with notes upon the relationship between single dose inhalation and single dose oral data. *Am. Ind. Hyg. Assoc. J.* **20**, 364–369 (1959).

130. H. Tanii, H.Tsuji, and K. Hashimoto, Structure-toxicity relationship of monoketones. *Toxicol. Lett.* **30**, 13–17 (1986).

131. S. Zakhari, and D. L. Snyder, Hemodynamic effects of methyl ethyl ketone inhalation in the dog. In L. Goldberg, ed., *Isopropanol and Ketones in the Environment*, CRC Press, Cleveland, OH, 1977, pp. 79–86.

132. E. T. Kimura, D. M. Ebert, and P. W. Dodge, Acute toxicity and limits of solvent residue for sixteen organic solvents. *Toxicol. Appl. Pharmacol.* **19**, 699–704 (1971).

133. G. L. Kennedy and G. J. Graepel, Acute toxicity in the rat following either oral or inhalation exposure. *Toxicol. Lett.* **56**, 317–326 (1991).

134. R. P. Kohli et al., Anticonvulsant activity of some carbonyl containing compounds. *Indian J. Med. Res.* **55**, 1221–1225 (1967).

135. G. D. DiVincenzo and W.J. Krasavage, Serum ornithine carbamyl transferase as a liver response test for exposure to organic solvents. *Am. Ind. Hyg. Assoc. J.* **35**, 21–29 (1974).

136. I. Lundberg et al., Relative hepatotoxicity of some industrial solvents after intraperitoneal injection or inhalation exposure in rats. *Environ. Res.* **40**, 411–420 (1986).

137. C. J. Terhaar, *Acute Intraperitoneal Toxicity of Methyl Ethyl Ketone, Methyl n-Butyl Ketone and Methyl Iso-butyl Ketone to Rats and Guinea Pigs*, Unpublished Report TL-76-17, Eastman Kodak Company, Rochester, NY, 1976.

138. R. Tham et al., Vestibulo-ocular disturbances in rats exposed to organic solvents. *Acta Pharmacol. Toxicol.* **54**, 58–63 (1984)

139. H. Tanii, Anesthetic activity of monoketones in mice: Relationship to hydrophobicity and *in vivo* effects on Na^+/K^+-ATPase activity and membrane fluidity. *Toxicol. Lett.* **85**, 41–47 (1996).

140. J. Huang et al., Structure-toxicity relationship of monoketones: *In vitro* effects on beta-adrenergic receptor binding and Na^+-K^+-ATPase activity in mouse synaptosomes. *Neurotoxicol. Teratol.* **15**, 345–352 (1993).

141. D. L. J. Opdyke, Monographs on fragrance raw materials. *Food Cosmet. Toxicol.* **15**, 611–638 (1977).

142. H. F. Smyth, Jr. et al., Range finding toxicity data: List VI. *Am. Ind. Hyg. Assoc. J.* **23**, 95–107 (1962).

143. R. D. Panson and C. L. Winek, Aspiration toxicity of ketones. *Clin. Toxicol.* **17**, 271–317 (1980).

144. J. E. Walberg, Edema-inducing effects of solvents following topical administration. *Dermatosen* **32**, 91–94 (1984).

145. J. E. Wahlberg, Assessment of skin irritancy: Measurement of skin fold thickness. *Contact Dermatitis* **9**, 21–26 (1983).

146. C. Anderson, K. Sundberg, and O. Groth, Animal model for assessment of skin irritancy. *Contact Dermatitis* **15**, 143–151 (1986).

147. C. S. Weil and R. A. Scala, Study of intra- and interlaboratory variability in the results of rabbit eye and skin irritation tests. *Toxicol. Appl. Pharmacol.* **19**, 276–360 (1971).

148. P. S. Larson, J. K. Finnegan, and H. D. Haag, Observations on the effect of the edema-producing potency of acids, aldehydes, ketones and alcohols. *J. Pharmacol. Exp. Ther.* **116**, 119–122 (1956).

149. J. G. Sivak and K. L. Herbert, Optical damage and recovery of the *in vitro* bovine ocular lens for alcohols, surfactants, acetates, ketones, aromatics, and some consumer products: A review. *J. Toxicol.—Cutaneous Ocul. Toxicol.* **16**, 173–187 (1997).

150. J. T. Reboulet et al., The agar diffusion cytolysis method: An alternative *in vitro* screen for the prediction of a severe ocular response. *Toxicol. Methods* **4**, 234–242 (1994).

151. H. E. Kennah, II et al., An objective procedure for quantitating eye irritation based upon changes of corneal thickness. *Fundam. Appl. Toxicol.* **12**, 258–268 (1989).

152. G. A. Safronov et al., Comparative acute inhalation toxicity of aliphatic aldehydes and ketones according to exposure time. *Curr. Toxicol.* **1**, 47–51 (1990).

153. C. W. La Belle and H. Brieger, The vapor toxicity of a composite solvent and its principal components. *Arch. Ind. Health* **12**, 623–627 (1955).

154. C. P. Carpenter, H. F. Smyth, Jr., and U. C. Pozzani, The assay of acute vapor toxicity an the grading and interpretation of results on 96 chemical compounds. *J. Ind. Hyg. Toxicol.* **31**, 343–346 (1949).

155. F. A. Patty, H. H. Schrenk, and W. P. Yant, Butanone. *Public Health Rep.* **50**, 1217–1228 (1935).

156. H. Specht, Acute response of guinea pigs to inhalation of methyl isobutyl ketone. *Public Health Rep.* **53**, 292–300 (1938).

157. E. Frantik, M. Hornychova, and M. Horvath, Relative acute neurotoxicity of solvents: Isoeffective air concentrations of 48 compounds evaluated in rats and mice. *Environ. Res.* **66**, 173–185 (1994).

158. J. DeCeaurriz et al., Concentration-dependent behavioral changes in mice following short-term inhalation exposure to various industrial solvents. *Toxicol. Appl. Pharmacol.* **67**, 383–389 (1983).

159. C. R. Garcia, I. Geller, and H. L. Kaplan, Effects of ketones on lever pressing behavior of rats. *Proc. West. Pharmacol. Soc.* **21**, 433–438 (1978).

160. J. C. DeCeaurriz et al., Sensory irritation caused by various industrial airborne chemicals. *Toxicol. Lett.* **9**, 137–143 (1981).

161. M. H. Abraham et al., Hydrogen bonding 12. A new QSAR for upper respiratory tract irritation by airborne chemicals in mice. *Quant. Struct.-Act. Relat.* **9**, 6–10 (1990).

162. J. Muller and G. Greff, Recherche de relations entre toxicité de molecules d'interet industriel et propriétés physio-chimiques: test d'irritation des voies aeriennes superiéures appliqué à quatre familles chimiques. *Food Chem. Toxicol.* **22**, 661–664 (1984).

163. D. W. Roberts, QSAR for upper-respiratory tract irritants. *Chem.-Biol. Interact.* **57**, 325–345 (1986).

164. Y. Alarie et al., Physicochemical properties of nonreactive volatile organic chemicals to estimate RD_{50}: Alternatives to animal studies. *Toxicol. Appl. Pharmacol.* **134**, 92–99 (1995).

165. Y. Alarie et al., Structure-activity relationships of volatile organic chemicals as sensory irritants. *Arch. Toxicol.* **72**, 125–140 (1998).

166. W. J. Muller and M. S. Black, Sensory irritation in mice exposed to emissions from indoor products. *Am. Ind. Hyg. Assoc. J.* **56**, 794–803 (1995).

167. L. F. Hansen, A. Knudsen, and G. D. Nielsen, Sensory irritation effects of methyl ethyl ketone and its receptor activation mechanism. *Pharmacol Toxicol* **71**, 201–208 (1992).

168. G. J. Traiger and J. V. Bruckner, The participation of 2-butanone in 2-butanol-induced potentiation of carbon tetrachloride hepatotoxicity. *J. Pharmacol Exp. Ther.* **196**, 493–500 (1976).

169. F. K. Dietz and G. J. Traiger, Potentiation of CCl_4 Hepatotoxicity in rats by a metabolite of 2-butanone: 2,3-butanediol. *Toxicology* **14**, 209–215 (1979).

170. G. J. Traiger et al., Effect of 2-butanol and 2-butanone on rat hepatic ultrastructure and drug metabolizing enzyme activity. *J. Toxicol. Environ. Health* **28**, 235–248 (1989).

171. W. R. Hewett, E. M. Brown, and G. L. Plaa, Relationship between the carbon skeleton length of ketonic solvents and potentiation of chloroform-induced hepatotoxicity in rats. *Toxicol. Lett.* **16**, 297–304 (1983).

172. W. R. Hewitt and E. M. Brown, Nephrotoxic interactions between ketonic solvents and halogenated aliphatic chemicals. *Fundam. Appl. Toxicol.* **4**, 902–908, (1984).

173. E. M. Brown and W. R. Hewitt, Dose-response relationships in ketone-induced potentiation of chloroform hepato- and nephrotoxicity. *Toxicol. Appl. Pharmacol.* **76**, 437–453 (1984).

174. L. A. Hewitt, P. Ayotte, and G. L. Plaa, Modifications in rat hepatobiliary function following treatment with acetone, 2-butanone, 2-hexanone, mirex, or chlordecone and subsequently exposed to chloroform. *Toxicol. Appl. Pharmacol.* **83**, 465–473 (1986).

175. L. A. Hewitt et al., Evidence for the involvement of organelles in the mechanism of ketone-potentiated chloroform-induced hepatotoxicity. *Liver* **10**, 35–48 (1990).

176. L. A. Hewitt, C. Valiquette, and G. L. Plaa, The role of biotransformation detoxification in acetone-, 2-butanone-, and 2-hexanone-potentiated chloroform-induced hepatotoxicity. *Can. J. Physiol. Pharmacol.* **65**, 2313–2318 (1987).

177. P. Raymond and G. L. Plaa, Ketone potentiation of haloalkane-induced hepato- and nephrotoxicity. I. Dose-response relationships. *J. Toxicol. Environ. Health* **45**, 465–480 (1995).

178. P. Raymond and G. L. Plaa, Ketone potentiation of haloalkane-induced hepato-and nephrotoxicity. ii. implication of monooxygenases. *J. Toxicol. Environ. Health* **46**, 317–328 (1995).

179. P. Raymond and G. L. Plaa, Ketone potentiation of haloalkane-induced hepatotoxicity: CCl_4 and ketone treatment on hepatic membrane integrity. *J. Toxicol. Environ. Health* **49**, 285–300 (1996).

180. J. Cunningham, M. Sharkawi, and G. L. Plaa, Pharmacological and metabolic interactions between ethanol and methyl *n*-butyl ketone, methyl isobutyl ketone, methyl ethyl ketone, or acetone in mice. *Fundam. Appl. Toxicol.* **13**, 102–109 (1989).

181. G. Egan et al., *n*-Hexane-"free" hexane mixture fails to produce nervous system damage. *Neurotoxicology* **1**, 515–524 (1980).

182. S. Duckett, N. Williams, and S. Frances, Peripheral neuropathy associated with inhalation of methyl *n*-butyl ketone. *Experientia* **30**, 1283–1284 (1974).

183. K. Saida, J. R. Mendell, and H. F. Weiss, Peripheral nerve changes induced by methyl *n*-butyl ketone and potentiation by methyl ethyl ketone. *J. Neuropathol. Exp. Neurol.* **35**, 207–225 (1976).

184. H. Altenkirch, G. Stoltenburg, and H. M. Wagner, Experimental studies on hydrocarbon neuropathies induced by methyl-ethyl-ketone (MEK). *J. Neurol.* **219**, 159–170 (1978).

185. H. Altenkirch, G. Stoltenburg-Didinger, and H. M. Wagner, Experimental data on the neurotoxicity of methyl ethyl ketone (MEK). *Experientia* **35**, 503–504 (1979).

186. H. Altenkirch et al., Nervous system responses of rats to subchronic inhalation of *n*-hexane and *n*-hexane + methyl-ethyl-ketone mixtures. *J. Neurol. Sci.* **57**, 209–219 (1982).

187. P. S. Spencer and H. H. Schaumburg, Feline nervous system response to chronic intoxication with commercial grades of methyl *n*-butyl ketone, methyl iso-butyl ketone, and methyl ethyl ketone. *Toxicol. Appl. Pharmacol.* **37**, 301–311 (1976).

188. J. L. O'Donoghue and W. J. Krasavage, *Neurotoxicity Studies on Methyl n-Butyl Ketone, Methyl Isobutyl Ketone, Methyl Ethyl Ketone and their Combinations*, Unpublished Report TX-81-62, Eastman Kodak Company, Rochester, NY, 1981.

189. J. L. O'Donoghue, *Electromyographic Examination of Dogs Treated with Ketone Solvents*, Unpublished Report TX-76-11, Eastman Kodak Company, Rochester, NY, 1977.

190. W. J. Krasavage and J. L. O'Donoghue, *Chronic Intraperitoneal Administration of MnBK, MEK, MIBK and Combinations of these Ketones to Rats*, Unpublished Report TL-77-48, Eastman Kodak Company, Rochester, NY, 1977.

191. J. L. O'Donoghue et al., *Chronic Skin Application of MnBK, MEK, MIBK, MEK/MIBK and MEK/MnBK*, Unpublished Report, Eastman Kodak Company, Rochester, NY, 1978.

192. R. Schmidt et al., Ultra-structural alteration of intrapulmonary nerves after exposure to organic solvents. A contribution to 'sniffers disease'. *Respiration* **46**, 362–369 (1984).

193. N. Schnoy et al., Ultrastructural alteration of the alveolar epithelium after exposure to organic solvents. *Respiration* **43**, 221–231 (1982).

194. Y. Takeuchi et al., An experimental study of the combined effects of *n*-hexane and methyl ethyl ketone. *Br. J. Ind. Med.* **40**, 199–203 (1983).

195. M. Iwata et al., Changes of *n*-hexane metabolites in urine of rats exposed to various concentrations of *n*-hexane and to its mixture with toluene or MEK. *Int. Arch. Occup. Environ. Health* **53**, 1–8 (1983).

196. G. Ichihara et al., Urinary 2,5-hexanedione increases with potentiation of neurotoxicity in chronic coexposure to *n*-hexane and methyl ethyl ketone. *Int. Arch. Occup. Environ. Health* **71**, 100–104 (1998).

197. J. L. O'Donoghue et al., Further studies on ketone neurotoxicity and interactions. *Toxicol. Appl. Pharmacol.* **72**, 201–209 (1984).

198. F. L. Cavender et al., A 90-day vapor inhalation toxicity study of methyl ethyl ketone. *Fundam. Appl. Toxicol.* **3**, 264–270 (1983).

199. A. M. Bernard et al., Evaluation of the subacute nephrotoxicity of cyclohexane and other industrial solvents. *Toxicol. Lett.* **45**, 271–280 (1989).

200. G. L. Li et al., Benzene-specific increase in leukocyte alkaline phosphatase activity in rats to vapors of various organic solvents. *J. Toxicol. Enviorn. Health* **19**, 581–589 (1986)

201. I. Geller et al., Effects of acetone, methyl ethyl ketone, and methyl isobutyl ketone on a match to sample task in the baboon. *Pharmacol., Biochem. Behav.* **11**, 401–406 (1979).

202. I. Geller, J. R. Rowlands, and H. L. Kaplan, *DHEW Publ.* (ADM.) *(U.S.)* ADM79–779, 363–376 (1979).

203. W. H. Ralston et al., Potentiation of 2,5-hexanedione neurotoxicity by methyl ethyl ketone. *Toxicol. Appl. Pharmacol.* **81**, 319–327 (1985).

204. D. Couri et al., The influence of inhaled ketone solvent vapors on hepatic microsomal biotransformation activities. *Toxicol. Appl. Pharmacol.* **41**, 285–289 (1977).

205. R. Toftgård, O. G. Nilsen, and J.Å. Gustafsson, Changes in rat liver microsomal cytochrome P-450 and enzymatic activities after the inhalation of *n*-hexane, xylene, methyl ethyl ketone and methylchloroform for four weeks. *Scand. J. Work Environ. Health* **7**, 31–37 (1981).

206. O. G. Nilsen and R. Toftgård, The influence or organic solvents on cytochrome P-450-mediated metabolism of biphenyl and benzo(*a*)pyrene. In M. J. Coon et al., eds., *Microsomes, Drug Oxidations, and Chemical Carcinogenesis*, Vol. 2, Academic Press, New York, 1980, pp. 1235–1238.

207. V. Nedelcheva, Interaction of styrene and ethylmethylketone in the induction of ctyochrome P450 enzymes in rat lung, kidney and liver after separate and combined inhalation exposures. *Cent. Eur. J. Public. Health* **4**, 115–118 (1995).

208. H. Raunio et al., Cytochrome P450 isozyme induction by methyl ethyl ketone and *m*-xylene in rat liver. *Toxicol. Appl. Pharmacol.* **103**, 175–179 (1990).

209. B. Mortensen, K. Zahlsen, and O. G. Nilsen, Metabolic interaction of *n*-hexane and methyl ethyl ketone *in vitro* in a head space rat liver S9 vial equilibration system. *Pharmacol. Toxicol.* **82**, 67–73 (1998).

210. K. Aarstad, K. Zahlsen, and O. G. Nilsen, Inhalation of butanols: Changes in the cytochrome P-450 enzyme system. *Arch. Toxicol. Suppl.* **8**, 418–421 (1985).

211. G. P. Carlson, Effect of ethanol, carbon tetrachloride, and methyl ethyl ketone on butanol oxidase activity in rat lung and liver. *J. Toxicol. Environ. Health* **27**, 255–261 (1989).

212. K. Harada et al., Combined effects of methyl *n*-butyl ketone and methyl ethyl ketone on activities of microsomal drug metabolizing enzymes in rats. *Jpn. J. Ind. Health* **31**, 156–157 (1989).

213. W. Zhao et al., Effects of methyl ethyl ketone, acetone, or toluene coadministration on 2,5-hexanedione concentration in the sciatic nerve, serum, and urine of rats. *Inte. Arch. Occup. Environ. Health* **71**, 236–244 (1998).

214. W. Zhao et al., Relationships between 2,5-hexanedione concentrations in nerve, serum, and urine alone or under co-treatment with different doses of methyl ethyl ketone, acetone, and toluene. *Neurochemi. Res.* **23**, 837–843 (1998).

215. J. Liira et al., Metabolic interaction and disposition of methyl ethyl ketone and *m*-xylene in rats at single and repeated inhalation exposures. *Xenobiotica* **21**, 53–63 (1991).

216. H. Ukai et al., Dose-dependent suppression of toluene metabolism by isopropyl alcohol and methyl ethyl ketone after experimental exposure of rats. *Toxicol. Lett.* **81**, 229–234 (1995).

217. E. Shibata et al., Effects of MEK on kinetics of *n*-hexane metabolites in serum. *Arch. Toxicol.* **64**, 247–250 (1990).

218. E. Shibata et al., Changes in urinary *n*-hexane metabolites by co-exposure to various concentrations of methyl ethyl ketone and fixed *n*-hexane levels. *Arch. Toxicol.* **64**, 165–168 (1990).

219. P. Robertson, E. L. White, and J. S. Bus, Effects of methyl ethyl ketone on hepatic mixed-function oxidase activity and on *in vivo* metabolism of *n*-hexane. *Xenobiotica* **19**, 721–729 (1989)

220. H. Kinoshita et al., Inhibitory mechanism of intestinal absorption induced by high acetaldehyde concentrations: Effect of intestinal blood flow and substance specificity. *Alcohol.: Clin. Exp. Res.* **20**, 510–513 (1996).

221. F. K. Dietz et al., Pharmacokinetics of 2-butanol and its metabolites in the rat. *J. Pharmacokinet. Biopharm.* **9**, 553–576 (1981).

222. W. Kessler, B. Denk, and J. G. Filser, Species-specific inhalation of 2-nitropropane, methyl ethyl ketone, and *n*-hexane. In C.C. Travis, ed., *Biologically Based Methods for Cancer Risk Assessment*, Plenum, New York, 1989, pp. 123–139.

223. K. Aarstad, K. Zahlsen, and O. G. Nilsen, Effekter Etter Inhalasjon av Forskjellige Butanolisomerer. *Färg Och Lack Scand.* **4**, 69–74 (1986).

224. I. A. Kamil, J. N. Smith, and R.T. Williams, Studies on detoxication. 46. The metabolism of aliphatic alcohols. The glucuronic acid conjugation of acyclic aliphatic alcohols. *Biochem. J.* **53**, 129–136 (1953).

225. P. I. Akubue et al., Excretion of malondialdehyde, formaldehyde, acetaldehyde, acetone and methyl ethyl ketone in the urine of rats given an acute dose of malondialdehyde. *Arch. Toxicol.* **68**, 338–341 (1994).

226. J. M. Mathews et al., Do endogenous volatile organic chemicals measured in breath reflect and maintain CYP2E1 levels *in vivo? Toxicol. Appl. Pharmacol.* **146**, 255–260 (1997).

227. G. D. DiVincenzo, C. J. Kaplan, and J. Dedinas, Characterization of the metabolites of methyl *n*-butyl ketone, methyl *iso*-butyl ketone and methyl ethyl ketone in guinea pig serum and their clearance. *Toxicol. Appl. Pharmacol.* **36**, 511–522 (1976).

228. G. Walter et al., Toxicokinetics of the inhaled solvents *n*-hexane, 2-butanone, and toluene in the rat: Alone and in combination. *Arch. Pharmacol.* **334**(Suppl.), R22 (abstr.)(1986).

229. B. A. Schwetz, B. K. Leong, and P. J. Gehring, Embryo and feto toxicity of inhaled carbon tetrachloride, 1,1-dichloroethane and methyl ethyl ketone in rats. *Toxicol. Appl. Pharmacol.* **28**, 452–464 (1974).

230. J. A. John et al., Teratologic evaluation of methyl ethyl ketone in the rat. *Teratology* **21**, 47A (abstr.)(1980).

231. M. M. Deacon et al., Embryo- and fetotoxicity of inhaled methyl ethyl ketone in rats. *Toxicol. Appl. Pharmacol.* **59**, 620–622 (1981).

232. B. A. Schwetz et al., Developmental toxicity of inhaled methyl ethyl ketone in Swiss mice. *Toxicol. Appl. Pharmacol.* **16**, 742–748 (1991).

233. G. Stoltenburg-Didinger, H. Altenkirch, and M. Wagner, Neurotoxicology of organic solvent mixtures: Embryotoxicity and fetotoxicity. *Neurotoxicol. Teratol.* **12**, 585–589 (1990).

234. A. W. Horton et al., Carcinogenesis of the skin. III. The contribution of elemental sulfur and of organic sulfur compounds. *Cancer Res.* **25**, 1759–1763 (1965).

235. E. Zeiger et al., *Salmonella* mutagenicity tests: V. Results from the testing of 311 chemicals. *Environ. Mol. Mutagen.* **19**(Suppl. 21), 2–141 (1992).

236. H. Shimizu et al., The results of microbial mutation test for forty-three industrial chemicals. *Jpn. J. Ind. Health* **27**, 400–419 (1985).

237. E. R. Nestmann et al., Mutagenicity of constitutents identified in pulp and paper mill effluents using the *Salmonella*/mammalian-microsome assay. *Mutat. Res.* **79**, 203–212 (1980).

238. L. J. Marnett et al., Naturally occurring carbonyl compounds are mutagens in *Salmonella* tester strain TA104. *Mutat. Res.* **148**, 25–34 (1985).

239. F. K. Zimmermann et al., Acetone, methyl ethyl ketone, acetonitrile, and other polar aprotic solvents are strong inducers of aneuploidy in *Saccharomyces cerevisiae. Mutat. Res.* **149**, 339–351 (1985).

240. V. W. Mayer and C. J. Goin, Effects of chemical combinations on the induction of aneuploidy in *Saccharomyces cerevisiae. Mutat. Res.* **187**, 21–30 (1987).

241. V. W. Mayer and C. J. Goin, Investigations of aneuploidy-inducing chemical combinations in *Saccharomyces cerevisiae. Mutat. Res.* **201**, 413–421 (1988).

242. F. K. Zimmermann, I. Scheel, and M.A. Resnick, Induction of chromosome loss by mixtures of organic solvents including neurotoxins. *Mutat. Res.* **224**, 287–303 (1989).

243. V. W. Mayer and C. J. Goin, Induction of chromosome loss in yeast by combined treatment with neurotoxic hexacarbons and monoketones. *Mutat. Res.* **341**, 83–91 (1994).

244. F. K. Zimmermann et al., Aprotic polar solvents that affect porcine brain tubulin aggregation *in vitro* induce aneuploidy in yeast cells growing at low temperatures. *Mutat. Res.* **201**, 431–442 (1988).

245. U. Gröschel-Stewart et al., Aprotic solvents inducing chromosomal malsegregation in yeast interfere with the assembly of porcine brain tubulin *in vitro. Mutat. Res.* **149**, 333–338 (1985).

246. A. Basler, Aneuploidy-inducing chemicals in yeast evaluated by the micronucleus test. *Mutat. Res.* **174**, 11–13 (1986).

247. T.-H. Chen et al., Inhibition of metabolic cooperation in Chinese hamster cells by various organic solvents and simple compounds. *Cell Biol. Toxicol.* **1**, 155–171 (1984).

248. P. Perocco, S. Bolognesi, and W. Alberghini, Toxic activity of seventeen industrial solvents and halogenated compounds on human lymphocytes cultured *in vitro. Toxicol. Lett.* **16**, 69–75 (1983).

249. W.-L. Yang, G. Klopman, and H. S. Rosenkranz, Structural basis of the *in vivo* induction of micronuclei. *Mutat. Res.* **272**, 111–124 (1992).

250. B. Holmberg, I. Jakobson, and T. Malmfors, The effect of organic solvents on erythrocytes during hypotonic hemolysis. *Environ. Res.* **7**, 193–205 (1974).

251. B. Holmberg, and T. Malmfors, The cytotoxicity of some organic solvents. *Environ. Res.* **7**, 183–192 (1974).

252. P. S. Ebert, H. L. Bonkowsky, and I. Wars, Stimulation of hemoglobin synthesis in murine erthyroleukemia cells by low molecular weight ketones, aldehydes, acids, alcohols and ethers. *Chem.-Biol. Interact.* **36**, 61–69 (1981).

253. D. G. Graham et al., *in vitro* evidence that covalent crosslinking of neurofilaments occurs in γ-diketone neuropathy. *Proc. Natl. Acad. Sci. U. S. A.* **81**, 4979–4982, (1984).

254. B. Veronesi, A. Lington, and P. S. Spencer, A tissue culture model of methyl ethyl ketone's potentiation of *n*-hexane neuropathy. *Neurotoxicology* **5**, 43–52 (1984).

255. B. Veronesi, An ultrastructural study of methyl ethyl ketone's effect on cultured nerve cells. *Neurotoxicology* **5**, 31–43 (1984).

256. G. Stoltenburg-Didinger et al., Specific neurotoxic effects of different organic solvents on dissociated cultures of the nervous system. *Neurotoxicology* **13**, 161–164 (1992).

257. S. C. Gad et al., Development and validation of an alternative dermal sensitization. *Toxicol. Appl. Pharmacol.* **84**, 93–114 (1986).

258. J. Descotes, Identification of contact allergens: The mouse ear sensitization assay. *J. Toxicol. — Cutaneous. Ocu. Toxicol.* **7**, 263–272 (1988).

259. A. R. Smith and M. R. Mayers, Study of poisoning and fire hazards of butanone and acetone. *Ind. Bull., NY State Dep Labor* **23**, 174–176 (1944).

260. M. Sakata, J. Kikuchi, and M. Haga, Disposition of acetone, methyl ethyl ketone and cyclohexanone in acute poisoning. *Clin. Toxicol.* **27**, 67–77 (1989).

261. E. Berg, Retrobulbar neuritis: A case report of presumed solvent toxicity. *Ann. Ophthalmol*, **3**, 1351–1353 (1971).

262. W. M. Grant, Methyl ethyl ketone. In *Toxicology of the Eye*, 3rd ed., Thomas, Springfield, IL, 1986, pp. 617–618.

263. P. G. Kopelman and P. Y. Kalfayan, Severe metabolic acidosis after ingestion of butanone. *Br. Med. J.* **286**, 21–22 (1983).

264. L. Welch et al. Chronic neuropsychological and neurological impairment following acute exposure to a solvent mixture of toluene and methyl ethyl ketone (MEK). *Clin. Toxicol.* **29**, 435–445 (1991).

265. E. A. Price et al; Osmolar gap with minimal acidosis in combined methanol and methyl ethyl ketone ingestion. *Clin. Toxicol.*, **32**, 79–84 (1994).

266. K. Nakaaki, An experimental study on the effect of exposure to organic solvent vapor in human subjects. *J. Sci. Labour* **50**, 89–96 (1974).

267. R. B. Dick et al; Neurobehavioral effects from acute exposures to methyl isobutyl ketone and methyl ethyl ketone. *Fundam. Appl. Toxicol.* **19**, 453–473 (1992).

268. R. B. Dick et al., Neurobehavioural effects of short duration exposures to acetone and methyl ethyl ketone. *Br. J. Ind. Med.* **46**, 111–121 (1989).

269. R. B. Dick et al., Effects of short duration exposures to acetone and methyl ethyl ketone. *Toxicol. Lett.* **43**, 31–49 (1988).

270. R. B. Dick et al., Effects of acute exposure of toluene and methyl ethyl ketone on psychomotor performance. *Int. Arch. Occup. Environ. Health* **54**, 91–109 (1984).

271. H. Ukai et al., Occupational exposure to solvent mixtures: Effects on health and metabolism. *Occup. Environ. Med.* **51**, 523–529 (1994).

272. M. Nasterlack, G. Triebig, and O. Stelzer, Hepatotoxic effects of solvent exposure around permissible limits and alcohol consumption in printers over a 4-year period. *Int. Arch. Occup. Environ. Health* **66**, 161–165 (1994).

273. M. Y. Shamy et al., Study of some biochemical changes among workers occupationallly exposed to phenol, alone or in combination with other organic solvents. *Ind. Health* **32**, 207–217 (1994).

274. G. Franco, New perspectives in biomonitoring liver function by means of serum bile acids: Experimental and hypothetical biochemical basis. *Br. J. Ind. Med.* **48**, 557–561 (1991).

275. K. W. Nelson et al., Sensory response to industrial solvent vapors. *J. Ind. Hyg. Toxicol.* **25**, 282–285 (1943).

276. R. B. Douglas and J. E. Coe, The relative sensitivity of the human eye and lung to irritant gases. *Ann. Occup. Hyg.* **31**, 265–267 (1987).

277. L. G. Wayne and H. H. Orcutt, The relative potentials of common organic solvents as percusors of eye-irritants in urban atmospheres. *J. Occup. Med.* **2**, 383–388 (1960).

278. R. L. Doty, Toxicology and industrial health *Olfaction Mult. Chem Sensitivity* **10**, 359–368 (1994).

279. F. M. Dyro, Methyl ethyl ketone polyneuropathy in shoe factory workers. *Clin. Toxicol.* **13**, 1371–1376 (1978).

280. H. Altenkirch et al., Toxic polyneuropathies after sniffing a glue thinner. *Acta neuropathol.* **214**, 137–152 (1977).

281. J. M. Vallat et al., *N*-hexane- and methylethylketone-induced polyneuropathy abnormal accumulation of glycogen in unmyelinated axons. Report of a case. *Acta Neuropathol.* **55**, 275–279 (1981).

282. S. Binaschi, G. Gazzaniga, and E. Crovato, Behavioral toxicology in the evaluation of the effects of solvent mixtures. In M. Horváth, ed., *Adverse Effects of Environmental Chemicals and Psychotropic Drugs*, Vol. 2, Elsevier, New York, 1976, pp. 91–98.

283. F. Vaider, B. Lechevalier, and T. Morin, Polyneurite toxique chez un travailleur du plastique. Rle possible du méthyl-éthyl-cétone. *Nouv Presse Med.* **4**, 1813–1814 (1975).

284. T. J. Callender, Neurotoxic impairment in a case of methylethyl-ketone exposure. *Arch. Environ. Health* **50**, 392(abstr.) (1995).

285. M. Orti-Pareja et al., Reversible myoclonus, tremor, and ataxia in a patient exposed to methyl ethyl ketone. *Neurology* **46**, 272 (1996).

286. W. D. Brown et al., Body burden profiles of single and mixed solvent exposures. *J. Occup. Med.* **29**, 877–883 (1987).

287. J. Liira, V. Riihimäki, and P. Pfäffli, Kinetics of methyl ethyl ketone in man: Absorption, distribution and elimination in inhalation exposure. *Int. Arch. Occup. Environ. Health* **60**, 195–200 (1988).

288. J. Liira, G. Johanson, and V. Riihimäki, Dose-dependent kinetics of inhaled methylethylketone in man. *Toxicol. Lett.* **50**, 195–201 (1990).

289. O. Tada, K. Nakaaki, and S. Fukabori, An experimental study on acetone and methyl ethyl ketone concentrations in urine and expired air after exposure to those vapors. *J. Sci. Labour* **48**, 305–336 (1972).

290. J. Liira, V. Riihimäki, and K. Engström, Effects of ethanol on the kinetics of methyl ethyl ketone in man. *Br. J. Ind. Med.* **47**, 325–330 (1990).

291. J. G. M. Van Engelen et al., Effects of coexposure to methyl ethyl ketone (MEK) on *n*-hexane toxicokinetics in human volunteers. *Toxicol. Appl. Pharmacol.* **144**, 385–395 (1997).

292. D. L. Ashley and J. D. Prah, Time dependence of blood concentrations during and after exposure to a mixture of volatile organic compounds. *Arch. Environ. Health* **52**, 26–33 (1997).

293. D. E. Wurster and R. Munies, Factors influencing percutaneous absorption II. Absorption of methyl ethyl ketone. *J. Pharm. Sci.* **54**, 554–556 (1965).

294. R. Munies and D. E Wurster, Investigation of some factors influencing percutaneous absorption III. Absorption of methyl ethyl ketone. *J. Pharm. Sci.* **54**, 1281–1284 (1965).

295. G. D. DiVincenzo et al., Studies on the respiratory uptake and excretion and the skin absorption of methyl *n*-butyl ketone in humans and dogs. *Toxicol. Appl. Pharmacol.* **44**, 593–604 (1978).

296. I. Brooke et al., Dermal uptake of solvents from the vapour phase: An experimental study in humans. *Ann. Occup. Hyg.* **42**, 531–540 (1998).

297. V. Fiserova-Bergerova and M. L. Diaz, Determination and prediction of tissue-gas partition coefficients. *Int. Arch. Occup. Environ. Health* **58**, 75–87 (1986).

298. J. Fisher et al., Lactational transfer of volatile chemicals in breast milk. *Am. Ind. Hyg. Assoc. J.* **58**, 425–431 (1997).

299. P. Poulin and K. Krishnan, A biologically-based algorithm for predicting human tissue: Blood: partition coefficients of organic chemicals. *Hum. Exp. Toxicol.* **14**, 273–280 (1995).

300. J. Jankovic and F. Drake, A screening method for occupational reproductive health risk. *Am. Ind. Hyg. Assoc. J.* **57**, 641–649 (1996).

301. G. A. Varigos and D. S. Nurse, Contact uticaria from methyl ethyl ketone.*Contact Dermatitis.* **15**, 259–260 (1986).

302. K. E. Malten et al., Horny layer injury by solvents. *Berufs-Dermatosen* **16**, 135–147 (1968).

303. M. R. Alderson and N. S. Rattan, Mortality of workers in an isopropyl alcohol plant and two MEK dewaxing plants. *Br. J. Ind. Med.* **37**, 85–89 (1980).

304. S. E. Chia et al., Neurobehavioral effects on workers in a video tape manufacturing factory in Singapore. *Neurotoxicology* **14**, 51–56 (1993).

305. R. Kishi et al., Neurobehavioral effects of chronic occupational exposure to organic solvents among Japanese industrial painters. *Environ. Res.* **62**, 303–313 (1993).

306. E. Mitran et al., Neurotoxicity associated with occupational exposure to acetone, methyl ethyl ketone, and cyclohexanone. *Environ. Res.* **73**, 181–188 (1997).

307. C. P. Wen et al., Long-term mortality study of oil refinery workers. IV. Exposure to the lubricating-dewaxing process. *J. Natl. Cancer Inst.* **74**, 11–18 (1985).

308. R. Spirtas et al., Retrospective cohort mortality study of workers at an aircraft maintenance facility. I. Epidemiological Results. *Br. J. Ind. Med.* **48**, 515–530 (1991).

309. A. Blair et al., Mortality and cancer incidence of aircraft maintenance workers exposed to trichloroethylene and other organic solvents and chemicals: Extended follow up. *Occup. Environ. Med.* **55**, 161–171 (1998).

310. G. K. Lemasters et al., Genotoxic changes after low-level solvent and fuel exposure on aircraft maintenance personnel. *Mutagenesis* **12**, 237–243 (1997).

311. D. H. Lee et al., Neurobehavioral changes in shoe manufacturing workers. *Neurotoxicol. Teratol.* **20**, 259–263 (1998).

312. R. F. White et al., Neurobehavioral effects of acute and chronic mixed-solvent exposure in the screen printing industry. *Am. J. Ind. Med.* **28**, 221–231 (1995).

313. G. Triebig et al., Untersuchungen zur Neurotoxizität von Arbeitsstoffen. *Int. Arch. Occup. Environ. Health* **52**, 139–150 (1983).

314. A. M. Williamson and C. Winder, A prospective cohort study of the chronic effects of solvent exposure. *Environ. Res.* **62**, 256–271 (1993).

315. F. Akbar-Khanzadeh and R. D. Rivas, Exposure to isocyanates and organic solvents, and pulmonary-function changes in workers in a polyurethane molding process. *J. Environ. Med.* **38**, 1205–1212 (1996).

316. U. G. Oleru and C. Onyekwere, Exposures to polyvinyl chloride, methyl ketone and other chemicals. The pulmonary and non-pulmonary effect. *Int. Arch. Occup. Environ. Health* **63**, 503–507 (1992).

317. E. M. Socie et al., Work-related skin disease in the plastics industry. *Am. J. Ind. Med.* **31**, 545–550 (1997).

318. T. C. Morata et al., Effects of occupational exposure to organic solvents and noise on hearing. *Scand. J. Work Environ. Health* **19**, 245–254 (1993).

319. Hazardous Substances Data Bank (HSDB), *Methyl Ethyl Ketone*, National Library of Medicine, Bethesda, MD (CD-ROM) MICROMEDIX, Englewood, Co. 1999, pp. 1–50.

320. Biochemical Research Laboratory, Dow Chemical Co., Midland, MI, unpublished data.

321. K. Verscheuren, *Handbook of Environmental Data on Organic Chemicals*, 2nd ed., Van Nostrand-Reinhold, New York, 1981.

322. J. E. Cometto-Muiz and W.S. Cain, Efficacy of volatile organic compounds in evoking nasal pungency and odor. *Arch. Environ. Health* **48**, 309–314 (1993).

323. M. Day et al., An investigation of the chemical and physical changes occurring during commercial composting. *Compost. Sci. Util.* **6**, 44–66 (1998).

324. A.-L. Sunesson et al., Volatile metabolites produced by two fungal species cultivated on building materials. *Ann. Occup. Hyg.* **40**, 397–410 (1996).

325. E. D. Pellizzari et al., Purgeable organic compounds in mother's milk. *Bull. Environ. Contam. Toxicol.* **28**, 322–328 (1982).

326. Health and Environment Laboratories, Eastman Kodak Co., Rochester, NY, unpublished data.

327. J. Giroux, R. Granger, and P. Monnier, Comparative study of 2-pentanone and 2-ethyl butanone in the mouse. *Trav. Soc. Pharm. Montpellier* **14**, 342–346 (1954).

328. H. W. Haggard, D. P. Miller, and L. A. Greenberg, The amyl alcohols and their ketones: Their metabolic fates and comparative toxicities. *J. Ind. Hyg. Toxicol.* **27**, 1–14 (1945).

329. W. P. Yant, F. A. Patty, and H. H. Schrenk, Pentanone. *Public Health Rep.* **51**, 392 (1936).

330. H. Specht et al., *Acute Response of Guinea Pigs to the Inhalation of Ketone Vapors*, NIH Bull. No. 176, U.S. Public Health Service, Div. Ind. Hyg. Washington, DC, 1940, pp. 1–66.

331. J. DeCeaurriz et al., Quantitative evaluation of sensory irritating and neurobehavioral properties of aliphatic ketones in mice. *Food Chem. Toxicol.* **22**, 545–549 (1984).

332. L. F. Hansen and G. D. Nielsen, Sensory irritation and pulmonary irritation of *n*-methyl ketones: receptor activation mechanisms and relationships with threshold limit values. *Arch. Toxicol.* **68**, 192–202 (1994).

333. J. Misumi and M. Nagano, Neurophysiological studies on the relation between the structural properties and neurotoxicity of aliphatic hydrogen compounds in rats. *Br. J. Ind. Med.* **41**, 526–532 (1984).

334. J. Misumi and M. Nagano, Experimental study on the enhancement of the neurotoxicity of methyl *n*-butyl ketone by non-neurotoxic aliphatic monoketones. *Br. J. Ind. Med.* **42**, 155–161 (1985).

335. E. Browning, Methyl propyl ketone. In *Toxicity and Metabolism of Industrial Solvents*, Elsevier, New York, 1965, pp. 425–427.

336. A. Sato and T. Nakajima, Partition coefficients of some aromatic hydrocarbons and ketones in water, blood and oil. *Br. J. Ind. Med.* **36**, 231–234 (1979).
337. P. Poulin and K. Krishnan, An algorithm for predicting tissue: blood partition coefficients of organic chemicals from *n*-Octanol: Water partition coefficient data. *J. Toxicol. Environ. Health* **46**, 117–129 (1995).
338. L. G. Bernard, *Reproduction/Developmental Toxicity Screening Test in the Rat*, Unpublished Report TX-99-075, Eastman Kodak Company, Rochester, NY, 1999.
339. C. P. Carpenter, C. S. Weil, and H. F. Smyth, Jr., Range finding toxicity data: List VIII. *Toxicol. Appl. Pharmacol.* **28**, 313–319 (1974).
340. O. M. Moreno, Report to Research Institute for Fragrance Materials, August 1976, cited in D.L.J. Opdyke, Fragrance raw materials monographs, methyl isopropyl ketone. *Food Cosmet. Toxicol.* **16**, 819 (1978).
341. R. Williams, Detoxication mechanisms. In *The Metabolism and Detoxication of Drugs, Toxic Substances and Other Organic Compouncts*, 2nd ed., Chapman & Hall, London, 1959, p. 96.
342. A. M. Kligman, Report to Research Institute for Fragrance Materials, July 1976, in D. L. J. Opdyke, Methyl isopropyl ketone, fragrance raw materials monographs. *Food Cosmet. Toxicol.* **16**, 819–820 (1978).
343. H. F. Smyth, Jr., C. P. Carpenter, and C. S. Weil, Range finding toxicity data: List IV. *Arch. Ind. Hyg. Occup. Med.* **4**, 119–122 (1951).
344. National Cancer Institute (NCI), *Chemical Carcinogenesis Research Information System*, NCI, Washington, DC, 1994.
345. K. Verschueren, *Handbook of Environmental Data on Organic Chemicals*, 3rd ed., Van Nostrand-Reinhold, New York, 1983.
346. H. F. Smyth, Jr., C. P. Carpenter, and C. S. Weil, Range finding toxicity data: List III. *J. Ind. Hyg. Toxicol.* **31**, 60–62 (1949).
347. B. Ballantyne et al., The acute toxicity and primary irritancy of 2,4-pentanedione. *Drug Chem. Toxicol.* **9**, 133–146 (1986).
348. H. F. Smyth, Jr. and C. P. Carpenter, The place of the range finding test in the industrial toxicology laboratory. *J. Ind. Hyg. Toxicol.* **26**, 269 (1944).
349. D. E. Dodd et al., 2,4-pentanedione: 9-day and 14-week vapor inhalation studies in Fischer-344 rats. *Fundam. Appl. Toxicol.* **7**, 329–339 (1986).
350. A. Germent, Inhibition of oxidation by peroxidase of human serum proteins. *Mol. Biol. Rep.* **3**, 283–287 (1977).
351. K. M. Ivanetich et al., Organic compounds: Their interaction with and degradation of hepatic microsomal drug-metabolizing enzymes *in vitro*. *Drug Metab. Dispos.* **6**, 218–225 (1978).
352. H. F. Gilbert, III and M. H. O'Leary, Modification of arginine and lysine in proteins with 2,4-pentanedione. *Biochemistry* **14**, 5194-5198 (1975).
353. H. B. Otwell, K.L. Cipollo, and R. B. Dunlap, Modification of lysyl residues of dihydrofolate reductase with 2,4-pentanedione *Biochim. Biophys. Acta* **568**, 297–306 (1979).
354. R. W. Tyl et al., Dominant lethal assay of 2,4-pentanedione vapor in Fischer 344 rats. *Toxicol. Ind. Health* **5**, 463–477 (1989).
355. R. W. Tyl et al., An evaluation of the developmental toxicity of 2,4-pentanedione in Fischer 344 rat by vapour exposure. *Toxicol. Ind. Health* **6**, 461–474 (1990).
356. F. Kato et al., Mutagenicity of aldehydes and diketones. *Mutat. Res.* **216**, 366(abstr.) (1989).

357. P. J. Guzzie, R. S. Slesinski, and B. Ballantyne, An *in vitro* and *in vivo* evaluation of the genotoxic potential of 2,4-pentanedione. *Environ. Mutagen.* **9**(Suppl. 8), 44(abstr.) (1987).

358. M. Nagano, J. Misumi, and S. Nomura, An electrophysiological study on peripheral neurotoxicity of 2,3-butanedione, 2,4-pentanedione and 2,5-hexanedione in rats. *Sangyo Igaku* **25**, 471–482 (1983).

359. W. Sterry and M. Schmoll. Contact urticaria and dermatitis from self-adhesive pads. *Contact Dermatitis* **13**, 284–285 (1985).

CHAPTER SEVENTY-SIX

Ketones of Six to Thirteen Carbons

Douglas C. Topping, Ph.D., DABT, David A. Morgott, Ph.D., DABT, CIH, and John L. O'Donoghue, VMD, Ph.D., DABT

INTRODUCTION

Ketones of carbon number 6–13 are important commercial and industrial materials. Their primary use is as solvents that find use in numerous products and industrial applications. Due to their volatility, environmental regulations have been directed at restricting emissions, particularly to the atmosphere. A number of the ketones discussed in this chapter can undergo photochemical transformations that contribute to their abiotic degradation but may also contribute to the formation of smog. Regulations limiting or prohibiting release of materials that may contribute to smog formation are leading to reductions in the use of some of these materials.

Occupational Exposures

As for the short-chain ketones discussed in Chapter 75, the ketones covered in this chapter are mainly of concern due to inhalation and dermal exposure routes. Acute exposure to high vapor concentrations of these materials may result in narcosis; however, such exposures are rare except in cases of accidents.

Low levels of exposure to many of these ketones can be expected in the environment and through endogenous exposure because ketones are common substrates for many of the enzymes associated with intermediary metabolism in organisms from bacteria to man.

Patty's Toxicology, Fifth Edition, Volume 6, Edited by Eula Bingham, Barbara Cohrssen, and Charles H. Powell.
ISBN 0-471-31939-2

Toxic Effects

Acute exposures to ketone vapors may result in irritation to the eyes and throat. Repeated dermal exposures may result in defatting of the skin, resulting in dryness, cracking, peeling, and inflammation of the epidermis.

The more serious effect of exposure to some of the ketones covered in this chapter is peripheral neuropathy, which has been reported to occur in occupational environments. Other effects including hematologic effects and altered activity levels of various enzyme systems have been reported in experimental animal systems but not in human clinical cases.

Structure–Activity Relationships

Alkanes, primary and secondary alcohols, carboxylic acids, glycols, diketones, epoxides, hydroxy acids, and ketones are metabolically related in many biologic systems. Thus a knowledge of the structure–activity relationships of these various compounds adds to our understanding of their individual and/or combined toxicities. Much of our present knowledge about the structure–activity relationships of ketones has been developed in response to an occupationally related outbreak of neurotoxicity. Since this incident, the emphasis on ketone toxicity has been directed primarily toward neurotoxicity. It must be realized, however, that these same ketones produce effects other than neurotoxicity.

Table 76.1 lists ketones and related compounds that have been examined for neurotoxicity (1–18). Those indicated as positive are substances that showed a specific anatomic and morphologic types of nerve degeneration characterized by large multifocal axonal swellings, often referred to as "giant axonal" neuropathy. These swellings are filled with masses of disorganized neurofilaments and other organelles. Myelin damage

Table 76.1. Neurotoxicity of Ketones and Related Substances

Chemical	Structure	Neurotoxicity[a]	Ref.
	Six-Carbon Structures		
n-Hexane	$CH_3(CH_2)_4CH_3$	+	1
Practical-grade hexanes	Mixed hexanes	+	1
1-Hexanol	$HOCH_2(CH_2)_4CH_3$	−	1, 2
2-Hexanol	$CH_3CHOH(CH_2)_3CH_3$	+	1, 2
6-Amino-1-hexanol	$HOCH_2(CH_2)_5NH_2$	−	3
Methyl *n*-butyl ketone	$CH_3CO(CH_2)_3CH_3$	+	1, 3–7
Methyl isobutyl ketone	$CH_3COCH_2CH(CH_3)_2$	−	3, 6, 7
2,5-Hexanediol	$CH_3CHOH(CH_2)_2CHOHCH_3$	+	1, 8
1,6-Hexanediol	$HOCH_2(CH_2)_4CH_2OH$	−	8
5-Hydroxy-2-hexanone	$CH_3CO(CH_2)_2CHOHCH_3$	+	1
2,3-Hexanedione	$CH_3COCO(CH_2)_2CH_3$	−	3, 8, 9
2,4-Hexanedione	$CH_3COCH_2COCH_2CH_3$	−	3, 8, 9
2,5-Hexanedione	$CH_3CO(CH_2)_2COCH_3$	+	1, 3, 8–10

Table 76.1. (*Continued*)

Chemical	Structure	Neurotoxicity[a]	Ref.
	Seven-Carbon Structures		
n-Heptane	$CH_3(CH_2)_5CH_3$	–	3
Methyl *n*-amyl ketone	$CH_3CO(CH_2)_4CH_3$	–	11
Methyl isoamyl ketone	$CH_3CO(CH_2)_2CH(CH_3)_2$	–	3
Ethyl *n*-butyl ketone[b]	$CH_3CH_2CO(CH_2)_3CH_3$	+	3
Di-*n*-propyl ketone	$CH_3(CH_2)_2CO(CH_2)_2CH_3$	–	3
2,5-Heptanedione	$CH_3CO(CH_2)_2COCH_2CH_3$	+	3, 9
2,6-Heptanedione	$CH_3CO(CH_2)_3COCH_3$	–	3, 9
3,5-Heptanedione	$CH_3CH_2COCH_2COCH_2CH_3$	–	8
3-Methyl-2,5-hexanedione	$CH_3COCHCH_3CH_2COCH_3$	+	12
	Eight-Carbon Structures		
3,6-Octanedione	$CH_3CH_2CO(CH_2)_2COCH_2CH_3$	+	3, 9
5-Methyl-3-heptanone	$CH_3CH_2COCH_2CHCH_3CH_2CH_3$	+	13
3-Acetyl-2,5-hexanedione	$CH_3COC(CH_3CO)CH_2COCH_3$	–	14
3,4-Dimethyl-2,5-hexanedione	$CH_3COCHCH_3CHCH_3COCH_3$	+	14–17
3,3-Dimethyl-2,5-hexanedione	$CH_3COC(CH_3)_2CH_2COCH_3$	–	15
	Nine-Carbon Structures		
5-Nonanone	$CH_3(CH_2)_3CO(CH_2)_3CH_3$	+	3
5-Methyl-2-octanone	$CH_3CO(CH_2)_2CHCH_3(CH_2)_2CH_3$	[c]	3
Diisobutyl ketone	$(CH_3)_2CHCH_2COCH_2CH(CH_3)_2$	–	3
	Ten-Carbon Structure		
3,4-Diethyl-2,5-hexanedione	$CH_3COCH(CH_3CH_2)CH(CH_3CH_2)$-COCH	–	16
	Eleven-Carbon Structures		
Diisoamyl ketone	$(CH_3)_2CH(CH_2)_2CO(CH_2)_2CH$-$(CH_3)_2$	–	3
	Twelve-Carbon Structure		
3,4-Diisopropyl-2,5-hexanedione	$CH_3COCH(CH_3CH_3CH_2)CH$-$(CH_3CH_3CH_2)COCH_3$	–	18

[a] – Indicates that the material was tested experimentally and found not to be neurotoxic; + indicates the material may produce giant axonal neuropathy.
[b] Ethyl *n*-butyl ketone is metabolized to 2,5-heptanedione, which is neurotoxic.
[c] Commercial samples of 5-methyl-2-octanone may contain 5-nonanone, which is neurotoxic. 5-Methyl-2-octanone enhances 5-nonanone neurotoxicity.

also occurs but is generally considered to be a secondary effect. Clinical symptomatology in man includes bilaterally symmetrical paresthesia, best described as a "pins and needles" feeling, and muscle weakness, primarily in the legs and arms.

The metabolic interrelationships of some of these neurotoxins are shown in Figure 76.1. Initially, studies of *n*-hexane and methyl *n*-butyl ketone neurotoxicity revealed that the γ-diketone 2,5-hexanedione was a neurotoxin. Subsequently, a series of diketones were examined for their ability to produce "giant axonal" neuropathy in rats. Table 76.2 lists these compounds and further emphasizes the necessity of the γ-diketone spacing for the production of neuropathy. These findings have led to the theory that neurotoxicity is related to a common metabolic pathway leading to the formation of a γ-diketone, which is the toxic metabolite that produces the neuropathy. Except for 2,5-heptanedione and 3,6-octanedione, all metabolic interconversions are oxidation of the ω-1 carbon(s), first to an alcohol or diol, then to a γ-diketone. In the case of *n*-heptane, where ω-1 oxidation may occur, the ketone formed would be a δ-diketone such as 2,6-heptanedione, which is not neurotoxic. When the ω carbon is oxidized in preference to the ω-1 carbon, as when *n*-hexane is converted to 1-hexanol, no γ-diketone is formed.

These data also suggest that, as chain length increases, the neurotoxicity of the diketone decreases, possibly owing to steric hindrance. However, chain length may not be as important for some materials such as 5-nonanone. The neurotoxicity of 5-nonanone appears to involve two metabolic pathways, one to 2,5-nonanedione, a mechanism similar to that of the other compounds shown in Figure 76.1, and the other to methyl *n*-butyl ketone via a series of oxidative and decarboxylative pathways (Fig. 76.2).

A third modifying factor affecting the neurotoxic potential of these substances is the number and size of substituent groups located between the γ-spaced carbonyls. Single methyl groups on the carbons located between the carbonyl groups increase the potential neurotoxicity of the γ-diketone (i.e., 3-methyl-2,5-hexanedione or 3,4-dimethyl-2,5-hexanedione, Table 76.1). Two methyl groups positioned on one of the methyl groups between the carbonyls (i.e., 3,3-dimethyl-2,5-hexanedione) eliminate neurotoxicity. Metabolism to a substituted γ-diketone may play a role in the neurotoxicity of 5-methyl-3-heptanone (Fig. 76.1).

Summary

A summary of the acute toxicologic properties of the ketones is presented in Table 76.3. During the past 30 years, a significant amount of data has been accumulated on the biologic and toxicologic effects of ketones in experimental animals and man. With the exception of certain studies that have shown that ketones with a particular structure produce a toxic polyneuropathy, these findings support the existing concepts of the relatively innocuous biologic effects of most ketones. The most widely and extensively used ketones appear to be the least toxic.

1.0 Methyl *n*-Butyl Ketone

1.0.1 CAS Number: [591-78-6]

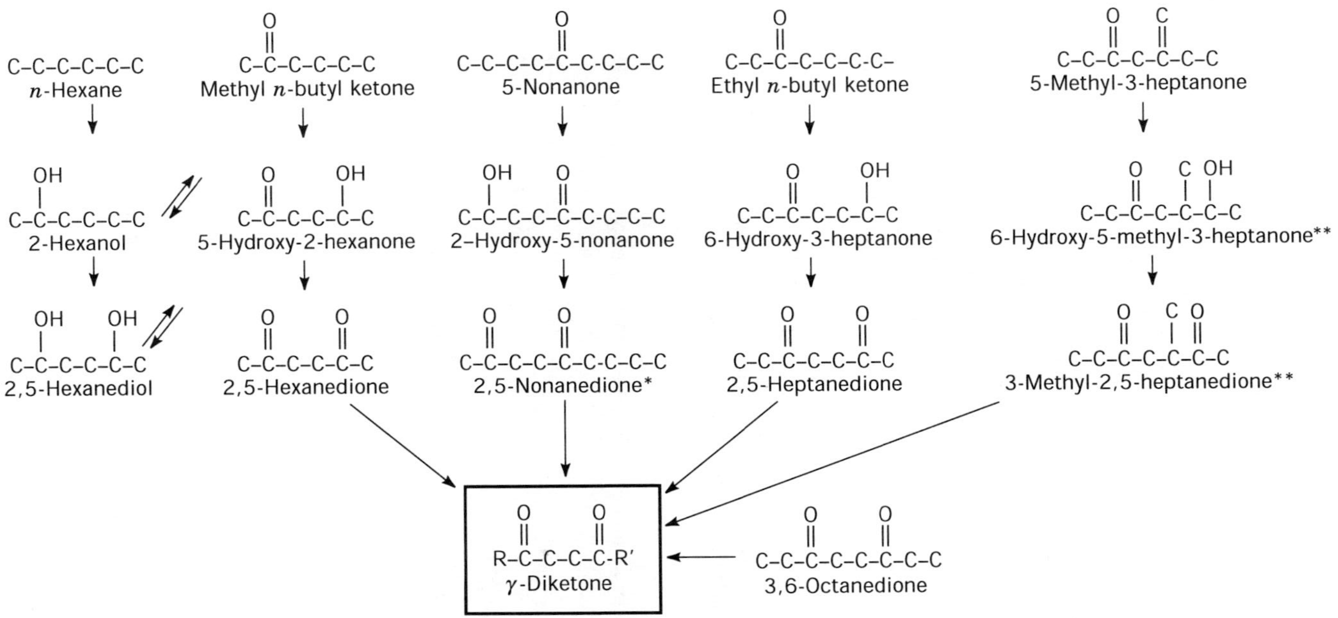

Figure 76.1. Relationships of alkanes, alcohols, and ketones that produce "giant axonal" neuropathy. Hydrogen atoms are included only when present as hydroxyl ions. * Further oxidative and decarboxylative pathways lead to the formation of methyl *n*-butyl ketone and 2,5-hexanedione (see Fig. 76.2). **The 5-methyl-3-heptanone metabolic scheme is an hypothesized pathway that is consistent with the observed neurotoxicity of this material.

Table 76.2. Structure–Activity Relationships of Diketones

Ketone Structure	Spacing	Ketone Neuropathy	Giant Axonal
2,4-Pentanedione	$CH_3COCH_2COCH_3$	β	–[a]
2,3-Hexanedione	$CH_3COCO(CH_2)_2CH_3$	α	–
2,4-Hexanedione	$CH_3COCH_2COCH_2CH_3$	β	–
2,5-Hexanedione	$CH_3CO(CH_2)_2COCH_3$	γ	+
3-Methyl-2,5-hexanedione	$CH_3COCHCH_3CH_2COCH_3$	γ	+
3,4-Dimethyl-2,5-hexanedione	$CH_3COCHCH_3CHCH_3COCH_3$	γ	+
3,3-Dimethyl-2,5-hexanedione	$CH_3COC(CH_3)_2CH_2COCH_3$	γ	–
2,3-Heptanedione	$CH_3CH_2COCH_2COCH_2CH_3$	β	–
2,5-Heptanedione	$CH_3CO(CH_2)_2COCH_2CH_3$	γ	+
2,6-Heptanedione	$CH_3CO(CH_2)_3COCH_3$	δ	–
3,6-Octanedione	$CH_3CH_2CO(CH_2)_2COCH_2CH_3$	γ	+
3-Acetyl-2,5-hexanedione	$CH_3COC(CH_3CO)CH_2COCH_3$	γ	–
3,4-Diethyl-2,5-hexanedione	$CH_3COCH(CH_3CH_2)CH$-$(CH_3CH_2)COCH_3$	γ	–
3,4-diisopropyl-2,5-hexanedione	$CH_3COCH(CH_3CH_3CH_2)CH$-$(CH_3CH_3CH_2)COCH_3$	γ	–

[a]2,4-Pentanedione produces central nervous system damage, which is clinically, anatomically, and morphologically distinguishable from "giant" axonal neuropathy.

1.0.2 Synonyms: 2-Hexanone, hexan-2-one, methyl butyl ketone, MBK, MnBK; and propylacetone.

1.0.3 Trade Names: NA

1.0.4 Molecular Weight: 100.16

1.0.5 Molecular Formula: $CH_3COCH_2CH_2CH_2CH_3$

1.0.6 Molecular Structure:

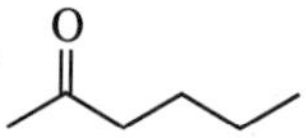

1.1 Chemical and Physical Properties

1.1.1 General

MnBK is a water-clear liquid. It has a relatively low cool flame ignition temperature of 188°C, and an autoignition temperature of 424°C. Additional physical–chemical data are summarized in Table 76.4.

1.1.2 Odor and Warning Properties

MnBK has an odor that resembles acetone but is more pungent. The odor threshold for MnBK in air is 0.076 ppm and in water is 0.25 mg/L (19).

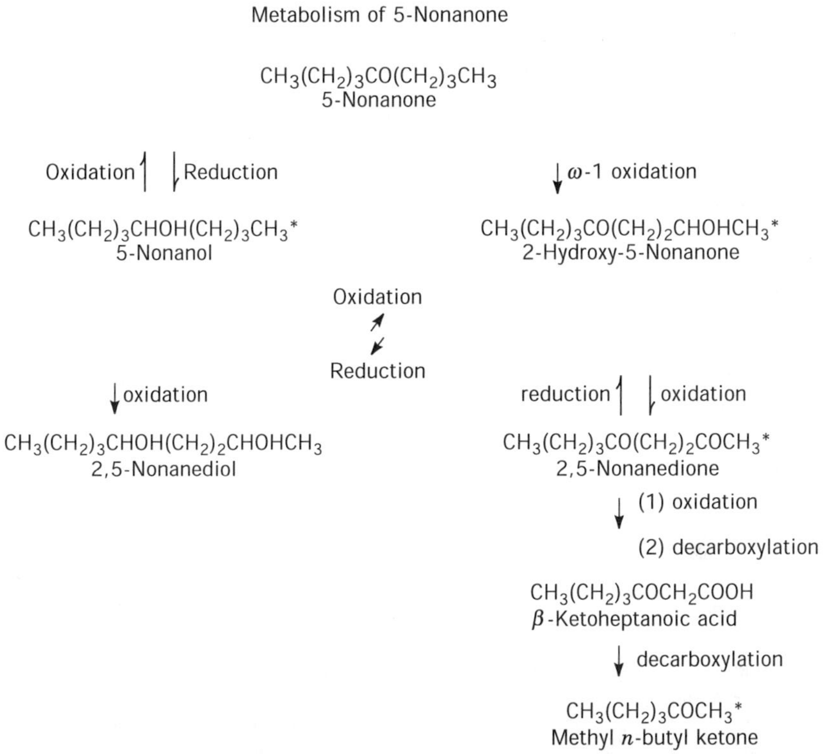

Figure 76.2. Metabolism of 5-Nonanone. *Designates actual metabolites found in blood or urine.

1.2 Production and Use

Production of MnBK in the United States ceased in 1979. Commercial production and importation in the United States was discontinued following the publication of a significant new use rule by the U.S. EPA. MnBK can be produced by a reaction between acetic acid and ethylene under pressure with a catalyst. MnBK has been used as a solvent for laquers, varnishes, adhesives, inks, oils, fats, waxes, and resins. Commercial grades of MnBK vary in purity from 70 to 96%; the more dilute grades include predominantly methyl isobutyl ketone (MiBK).

1.3 Exposure Assessment

1.3.1 Air

Occupational exposure to MnBK has been primarily associated with inhalation of vapor, although dermal exposures have also been reported. Nonoccupational exposures are most likely to occur through the diet as low levels of MnBK have been identified in food. Since MnBK is a metabolite of *n*-hexane and 2-hexanol, exposure to these materials provide an internalized dose of MnBK.

Table 76.3. Toxicologic Properties of Ketones

Compound	Approximate Oral Rat LD_{50} (mL/kg)	Lowest Reported Lethal Air Conc. Rat (ppm/h)	Skin Irritation[a]	Ocular Injury[a]
Methyl *n*-butyl ketone	3	8000/4	Sl	Sl
Methyl isobutyl ketone	5–6	4000/4	Sl	Sl
Mesityl oxide	1	500/8	Sl	M
4-Hydroxy-4-methyl-2-pentanone	4	> Sat'd./8	Sl	M
Methyl *n*-amyl ketone	2	4000/4	M	Sl
Methyl isoamyl ketone	4	3813/6	Sl	Sl
Ethyl *n*-butyl ketone	3	4000/4	M	Sl
Di-*n*-propyl ketone	4	2670/6	Sl	Sl
Diisopropyl ketone	4	> 2765/6	Sl	Sl
2-Octanone	3	> 1673/6	M	Sl
3-Octanone	> 5	—	M	—
5-Methyl-3-heptanone	4	3484/4	Sl	M
5-Nonanone	> 2	—	—	—
Diisobutyl ketone	6	Sat'd./8	Sl	Sl
Trimethyl nonanone	9	> Sat'd./4	Sl	Sl
2,5-Hexanedione	3	> Sat'd./1	Sl	M
Cyclohexanone	2	2000/4	M	M
Methyl cyclohexanones	2	2800/4	M	M
Acetophenone	3	> Sat'd./8	M	SV
Propiophenone	> 4	> Sat'd./8	Sl	Sl
Isophorone	2– > 3	1840/4	Sl	SV
Benzophenone	2–3	—	Sl	Sl

[a]Sl, slight; M, moderate; SV, severe. Skin irritation and ocular injury ratings are for direct application of liquids.

1.3.2 Background Levels

Background levels of MnBK occur in several foods as MnBK is a natural volatile component of blue and Beaufort cheeses, nectarines, roasted filberts, chicken muscle, milk, cream, and bread (20). MnBK levels in milk and cream have been reported in the range of 7–18 ppb (20). As materials such as *n*-hexane and 2-hexanol can be metabolized by organisms in the environment, low levels of MnBK may be encountered at sites in which *n*-hexane and 2-hexanol are found. MnBK has been identified in ground water, surface water, and soil at 15 National Priorities List Waste Sites (20). Concentrations of 7–202 ppb MnBK have been detected in the effluent from an oil shale processing plant (20). MnBK has also been detected in the effluents from a coal gasification plant and a chemical plant, and leachate (0.148 ppm MnBK) from a municipal landfill (20).

1.3.3 Workplace Methods

Gas chromatography using mass spectroscopy or a flame ionization detection is the method of choice for precise analysis of MnBK. MnBK vapor is trapped and concentrated from a large volume of air on a sorbent such as activated carbon and then thermally

Table 76.4. Physical–Chemical Properties of Ketones

Compound	CAS Number	Molecular Formula	Mol. Wt.	Boiling Point (°C)	Melting Point (°C)	Specific Gravity[a]	Refractive Index (20°C)	Vapor Pressure (mmHg)[b]	Vapor Density (Air = 1)	Max. Vapor Concen. (ppm @ 20°C)	Flash Point[c] °F (°C)	Flammability (%) Lower Limit	Flammability (%) Upper Limit	Solubility[d] Water	Solubility[d] Alcohol	Solubility[d] Ether	Conversion Factors[e] 1 ppm O (mg/m³)	Conversion Factors[e] 1 mg/L O (ppm)
Methyl *n*-butyl ketone	[*591-78-6*]	$C_6H_{12}O$	100.16	127.5	−59.6	0.821	1.4007	3.8	3.5	4,900	73 (23)	1.2	8.0	S	S	S	4.10	244
Methyl isobutyl ketone	[*108-10-1*]	$C_6H_{12}O$	100.16	115.8	−84.7	0.802	1.3959	18.8	3.5	9,700	64 (18)	1.2	8.0	S	S	S	4.10	244
Mesityl oxide	[*141-79-7*]	$C_6H_{10}O$	98.14	129.6	−53	0.857	1.4440	9.5	3.4	12,300	90 (32)	1.3	8.8	Sl	S	S	4.02	249
4-Hydroxy-4-methyl-2-pentanone	[*123-42-2*]	$C_6H_{12}O_2$	116.16	168.0	−43	0.941	1.4242	1.2	4.0	1,550	136 (58)	1.8	6.9	S	S	S	4.74	211
2,5-Hexanedione	[*110-13-4*]	$C_6H_{10}O_2$	114.14	194.0	−5.5	0.973	1.4230	1.6	3.9	2,100	(78)			S	S	S	4.66	214
Cyclohexanone	[*108-94-1*]	$C_6H_{10}O$	98.14	155.7	−32.1	0.948 (20/4)	1.4507	4.8	3.4	6,700	111 (44)	1.1	9.4	Sl	S	S	4.02	249
Methyl *n*-amyl ketone	[*110-43-0*]	$C_7H_{14}O$	114.19	151.1	−35	0.817	1.4073	2.1	3.9	2,700	102 (39)	1.1	7.9	Sl	S	S	4.66	214
Methyl isoamyl ketone	[*110-12-3*]	$C_7H_{14}O$	114.19	144.0	−73.9	0.813	1.4062	4.5 (20)	3.9	5,900	110 (43)	1.1	8.2	Sl	S	S	4.66	214
Ethyl *n*-butyl ketone	[*106-35-4*]	$C_7H_{14}O$	114.19	147.4	−39.0	0.818	1.3994	5.6	3.9	1,800	98 (37)	1.4	8.8	Sl	S	S	4.66	214
Di-*n*-propyl ketone	[*123-19-3*]	$C_7H_{14}O$	114.19	143.7	−32.6	0.817 (20/4)	1.4069	5.5 (20)		7,200	120 (49)			I	S	S	4.66	214
Diisopropyl ketone	[*565-80-0*]	$C_7H_{14}O$	114.19	124.0	−69	0.803 (20/4)		10 (37)	3.95		68 (15)			I			4.66	214
2-Methylcyclohexanone	[*583-60-8*]	$C_7H_{12}O$	112.17	165	−13.9	0.925	1.4440 (25)		3.9		118 (46)			I	S	S	4.58	218
3-Methylcyclohexanone	[*591-24-2*]	$C_7H_{12}O$	112.17	169	−73.5	0.914 (20/4)	1.4449		3.9		(51)			I	S	S	4.58	218
4-Methylcyclohexanone	[*589-92-4*]	$C_7H_{12}O$	112.17	170	−40.6	0.914 (20/4)	1.4451	10 (54)	3.9	11,800	(40)	1.1	7.4	I	S	S	4.58	218
Acetophenone	[*98-86-2*]	C_8H_8O	120.13	202.0	19.6	1.0281 (20/4)	1.5363	0.37 (25)	4.1	430	180 (77)	1.1	6.7	Sl	S	S	4.95	204

Table 76.4. *(Continued)*

Compound	CAS Number	Molecular Formula	Mol. Wt.	Boiling Point (°C)	Melting Point (°C)	Specific Gravity[a]	Refractive Index (20°C)	Vapor Pressure (mmHg)[b]	Vapor Density (Air = 1)	Max. Vapor Concen. (ppm @ 20°C)	Flash Point[c] °F (°C)	Flammability (%) Lower Limit	Flammability (%) Upper Limit	Solubility[d] Water	Solubility[d] Alcohol	Solubility[d] Ether	Conversion Factors[e] 1 ppm O (mg/m³)	Conversion Factors[e] 1 mg/L O (ppm)
2-Octanone	*[111-13-7]*	$C_8H_{16}O$	128.21	172.9	− 16.0	0.819 (20/4)	1.4151	1.2	4.4	1,550	(62)			Sl	S	S	5.24	191
3-Octanone	*[106-68-3]*	$C_8H_{16}O$	128.21	167.0		0.822	1.4150	2.0		2,600	(43)			I	S	S	5.24	191
5-Methyl-3-heptanone	*[541-85-5]*	$C_8H_{16}O$	128.21	160.5	− 56.7	0.820 (0/4)	1.4160	2.0		2,600	110 (43)			I	S	S	5.24	191
Propiophenone	*[93-55-0]*	$C_9H_{10}O$	134.19	218.0	18.6	1.010	1.5269	1.5 (25)		180	(88)			I	S	S	5.48	183
Isophorone	*[78-59-1]*	$C_9H_{14}O$	138.21	215.2	− 8.0	0.923	1.478	0.26 (20)	4.77	340	184 (84)	0.8	3.5	Sl	S	S	5.71	175
5-Nonanone	*[502-56-7]*	$C_9H_{18}O$	142.24	188.4	− 50	0.822	1.4195				(65)			SI	S	S	5.81	172
Diisobutyl ketone	*[108-83-8]*	$C_9H_{18}O$	142.24	168.1	− 46.0	0.807	1.4210	1.7	4.9	3,100	(49)	0.8	7.1	Sl	S	S	5.81	172
Trimethyl nonanone	*[123-18-2]*	$C_{12}H_{24}O$	184.32	207–228		0.817	1.4273				195 (91) [O.C.]			I	S	S	7.54	133
Benzophenone	*[119-61-9]*	$C_{13}H_{10}O$	182.22	305.4	48.5	1.1108 (18/4)	1.5975 (45.2)	1 (108)			155			I	Sl	Sl	7.44	134

[a]Specific gravity is at 20/20°C unless otherwise noted.
[b]Vapor pressure is at 25°C unless otherwise noted.
[c]Closed cup unless otherwise noted, [OC] open cup. Figures in parentheses are °C.
[d]S, readily soluble; Sl, slightly soluble; I, insoluble.
[e]At 25°C, 760 mm Hg.

released or eluted with a solvent. MnBK in water may be collected by purging with an inert gas or cryogenic trapping.

NIOSH Method 1300 is recommended for determining workplace exposures to methyl *n*-butyl ketone (19a).

1.3.4 Community Methods

Environmental samples can be analyzed by methodologies that are similar to those used for occupational exposure assessments. However, these methods may not be adequate for detection of background levels of MnBK. Several methods that have been used for environmental sampling have detection limits in the range of $< 10-50$ µg/L (20).

1.3.5 Biomonitoring/Biomarkers

1.3.5.1 Blood. MnBK and its metabolites can be measured in the blood; however, these are not specific biomarkers for exposure to MnBK, since the metabolites of *n*-hexane and 2-hexanol are shared with MnBK and *n*-hexane and 2-hexanol can be metabolized to MnBK itself.

1.3.5.2 Urine. As for blood metabolites, MnBK metabolites can be detected in the urine, but the metabolites are not specific markers for exposure to MnBK. Protein-bound pyrroles in the hair have been suggested as a possible biomarker to 2,5-hexandione, a metabolite of MnBK (21).

1.4 Toxic Effects

The acute oral toxicity of MnBK is low. Topical damage to the skin and eyes is also low, but transient eye irritation can occur. Percutaneous absorption may occur and with concomitant inhalation exposure may contribute to MnBK toxicity. High vapor concentrations of MnBK may produce ocular and respiratory irritation followed by CNS depression and narcosis.

Subchronic to chronic inhalation produces effects that are significantly different from those produced by acute exposure. These include degenerative axonal changes primarily in the peripheral nerves and long spinal cord neural tracts, atrophy of testicular germinal cell epithelium, and depression of circulating white blood cells. Neurologic and hematologic changes occur in experimental animals at low dose levels, while high dose levels are required to produce germinal cell damage. The principal hazard in industry has been peripheral neuropathy, which is associated with repeated inhalation and dermal exposures.

The mechanism of acute MnBK toxicity is by progressive depression of the CNS resulting in coma and cardiorespiratory failure. The mechanism(s) associated with subchronic and chronic toxicity are associated with significant binding of MnBK metabolites to axonal proteins. The precise nature of the binding and the details of the mechanism(s) of action are yet to be completely understood.

1.4.1 Experimental Studies

1.4.1.1 Acute Toxicity. Smyth et al. (22) reported an oral LD_{50} for MnBK of 2590 mg/kg for rats. The oral LD_{50} for mice was 1000 mg/kg. A second study indicated that the oral

LD_{50} in mice was 2426 mg/kg (23). The intraperitoneal LD_{50} for rats and guinea pigs was 1142 and 1140 mL/kg, respectively (3).

Smyth et al. (22) reported that MnBK caused mild eye irritation and minor transient corneal injury.

Tests on rabbit skin indicated that MnBK produced no significant irritation. The dermal LD_{50} was 5.99 mL/kg (22). Repeated skin contact may be irritating due to the ability of MnBK to defat the skin, resulting in dermatitis.

Smyth et al. (22) reported that more than 30 min of exposure to a saturated atmosphere of MnBK was required to kill rats. Exposure to 8000 ppm for 4 h resulted in the death of all exposed rats, whereas 4000 ppm for 4 h did not kill any of the animals. Schrenk et al. (24) exposed guinea pigs to a maximum of 1000 ppm for several hours and produced slight or no abnormal clinical signs. A maximum of 1500 ppm for several hours had no serious effect; and a maximum of 3000 ppm for 60 min did not cause serious disturbances. Exposures to 4000–6000 ppm were potentially lethal after several hours, 10,000–20,000 ppm was potentially lethal in 30–60 min, and >20,000 ppm killed guinea pigs within a few minutes. The acute effects of MnBK exposure were eye and nasal irritation followed by narcosis and death. Pathologic examination revealed vascular congestion in major organs. Specht et al. (25) reported that acute inhalation exposure of guinea pigs to MnBK produced depression of body temperature and respiratory and heart rates; loss of corneal, auditory, and equilibrium reflexes; narcosis; coma; and death. Mortality in mice has been reported following inhalation exposures to 7000 ppm MnBK vapor (26).

1.4.1.2 Chronic and Subchronic Toxicity. Since 1974, a series of reports have been published describing neurotoxicity after repeated exposure to MnBK. MnBK has been used as a model compound for exploring mechanisms of neurotoxicity and used as a positive control in neurotoxicity test batteries. Therefore, the scientific literature includes a number of studies that provide redundant information about repeated exposure to MnBK. Since 2,5-hexanedione is considered the proximate active metabolite of MnBK and because MnBK is a metabolite of *n*-hexane, the scientific literature for 2,5-hexanedione, *n*-hexane, and MnBK are closely linked.

Spencer and Schaumburg (7) injected cats subcutaneously with 150 mg/kg MnBK or a 9:1 mixture of MEK and MnBK twice a day, 5 d/wk for 8.5 mo. MnBK alone produced clinical signs of peripheral neuropathy after 8–10 wk of exposure. Weakness was first evident in the hind limbs and later in the fore limbs. Pathologic changes included axonal damage with secondary myelin changes in the peripheral and central nervous systems. The long spinal tracts in the central nervous systems were particularly affected. Neuropathologic changes but not clinical neurotoxicity were produced by the 9:1 mixture of MEK and MnBK. Beagle dogs receiving treatments similar to the cats had a similar clinical course and developed similar neuropathologic lesions, except that the 9:1 mixture of MEK and MnBK did not produce evidence of neurotoxicity (3).

Groups of 10 rats were administered MnBK in their drinking water at concentrations of 1.0% (560 mg/kg), 0.5% (266 mg/kg), or 0.25% (143 mg/kg) daily for 10–13 mo (27). MnBK produced a dose-dependent reduction in body weight gain that was present by the second week in the 1.0 and 0.5% groups and by the third week in the 0.25% group. Clinical neurotoxicity consisting of hind limb weakness was evident after 6 wk in the 1.0% group and after 10 wk in the 0.5% group. The 0.25% group was clinically normal. Axonal degeneration was found in the cerebellum, brainstem, spinal cord, peripheral nerves, and skeletal muscle of all groups.

Misumi et al. (28) reported that rats given 415 mg/kg/d MnBK 5 d/wk for several weeks had decreased motor sensory nerve conduction times, and decreased motor and sensory nerve action potential amplitudes that correlated with pathologic findings. Eben et al. (29) reported no overt signs of neurotoxicity when doses of 400 mg/kg/d MnBK were administered by stomach tube over a 40-wk period. Reduced body weight gain was present after the second week of treatment, and transient hind limb weakness was observed between the seventeenth and twenty-eighth weeks. Rats given intraperitoneal injections of 100–200 mg/kg of MnBK, or 9:1 mixture of MEK and MnBK for 35 wk showed no clinical or neuropathologic changes (3). Guinea pigs given 0.25 or 0.1% MnBK in their drinking water had an increased body weight gain, depressed locomotor activity, and impaired pupillary response (30), whereas studies with other species have consistently shown decreased body weight gain. Johnson et al. (31) found that 68 mg/kg MnBK given by gavage produced a reduction in bar pressing response rate, which was attributed to mild narcosis.

The relative neurotoxicity of MnBK, *n*-hexane, practical-grade hexane, and their metabolites was studied by Krasavage et al. (1). They found that MnBK (660 mg/kg for 90 d) was less neurotoxic than its metabolites 2,5-hexanediol, 5-hydroxy-2-hexanone, and 2,5-hexanedione, but more neurotoxic than 2-hexanol, *n*-hexane, and practical-grade hexane. In addition to neurotoxicity, testicular germinal cell atrophy was evident. MnBK produced body weight depression and neurotoxicity after repeated doses as low as 143 mg/kg. Testicular atrophy was seen in animals given doses of between 560 and 660 mg/kg for 90 d.

Spencer et al. (6) reported that rats exposed to 1300 ppm MnBK 6 h/d, 5 d/wk showed slight narcosis by 4 h of exposure, incoordination after 5.5 h, decreased weight gain, slow progressive weight loss beginning after the seventy-third exposure, and symmetrical hind limb weakness between the third and fourth months of exposure. Pathologic changes included "giant" axonal swellings, paranodal myelin retraction or loss, and axonal degeneration in peripheral nerves, spinal cord, medulla oblongata, and cerebellum. Electron microscopy showed masses of neurofilaments, numerous mitochondria, and sparse neurotubules in the swellings of myelinated and unmyelinated axons. Mendell et al. (5) described an identical neuropathy in chickens, rats, and cats exposed to 100–200 ppm, 24 h/d, 7 d/wk. Clinical signs of weakness were observed in chickens at 4–5 wk and in cats at 5–8 wk. Electrodiagnostic findings and pathologic changes comparable to those described in humans exposed to methyl *n*-butyl ketone were found in experimental animals.

Duckett et al. (32) described a neuropathy in rats exposed to 200 ppm MnBK or 200 ppm MnBK/2000 ppm MEK, 8 h/d, 5 d/wk for 6 wk. These findings are difficult to integrate with the other studies of MnBK neurotoxicity because (1) the method used to determine the concentration of MnBK was not specified, (2) the rats were weak after each exposure, which is an unlikely effect from an exposure to only 200 ppm, and (3) the pathologic changes were not characteristic of MnBK. Similar difficulties were also encountered when trying to evaluate a report of subclinical neuropathologic changes in rats exposed to 50 ppm MnBK (33).

Johnson et al. (31) reported that exposure of rats and monkeys to 1000 ppm MnBK for 6 h/d, 5 d/wk for 25 wk produced decreased body weight, clinical signs of neurotoxicity, and abnormal electrodiagnostic parameters in the peripheral nerves and visual system. Rats exposed for 29 wk and monkeys exposed for 41 wk to 100 ppm of MnBK had nerve conduction velocities and evoked muscle action potentials that differed significantly from the controls. Recovery of nerve conduction velocities occurred in monkeys 6 mo after cessation of exposure to 1000 ppm and 2 mo after exposure to 100 ppm. Recovery from clinical neuropathy has also been observed in dogs (3). Chronic inhalation of MnBK at 100 and 330 ppm for 6 h/d, 5 d/wk for 72 wk (rats) or 2 yr (cats) produced subclinical neuropathy in cats at 330 ppm but not at 100 ppm, and a dose-dependent reduction of body weight gain in rats after 4 mo (330 ppm) and 9 mo (100 ppm) (3). Other effects due to MnBK exposure include testicular germinal cell atrophy (1, 34) and a reduction in total circulating white blood cells (35).

Enhancement of MnBK neurotoxicity by MEK was reported by Saida et al. (36). These investigators compared continuous exposures of 225 ppm MnBK, 1125 ppm MEK, and 225 ppm MnBK plus 1125 ppm MEK in rats, and demonstrated that the addition of high levels of MEK to a neurotoxic dose of MnBK shortened the onset of paralysis from 66 to 25 d. It is possible that exposure of animals to a subneurotoxic dose of MnBK plus a high dose of MEK might also produce neurotoxicity, as has been reported for *n*-hexane and MEK (37).

Application of 2 mL MnBK or a 9:1 mixture of MEK:MnBK, with or without dimethyl sulfoxide, to the backs of guinea pigs twice a day, 5 d/wk for 31 wk did not produce clinical or histologic evidence of toxic neuropathy. The only adverse effect noted was desquamation due to chronic defatting of the skin (3).

1.4.1.3 Pharmacokinetics, Metabolism, and Mechanisms

1.4.1.3.1 Absorption. MnBK is readily absorbed by oral, inhalation, and dermal routes. Oral gavage administration of radiolabeled MnBK to rats indicated that 98% of the dose was absorbed (4).

Beagle dogs that inhaled 50 or 100 ppm MnBK for 6 h absorbed 65–68% of the vapor (38). Absorption of radiolabeled MnBK through dog skin was observed to be slow at first but increased dramatically after 20 min of exposure (38). Based on these findings, it was calculated that approximately twice as much MnBK would be absorbed during a 1-h dermal exposure as would be absorbed by inhalation of 25 ppm of MnBK vapor for 1 h. Therefore, repeated dermal exposure to MnBK may contribute to its toxicity. Concomitant exposure to methyl ethyl ketone (MEK) may increase absorption of MnBK.

1.4.1.3.2 Distribution. Distribution of radiolabeled MnBK was widespread following an oral gavage dose of 200 mg/kg, with the highest counts in liver and blood (4). Subcellular localization of counts in the liver, brain, and kidney indicated that the highest counts were associated with the crude lipid fraction and protein. Partition coefficients for air, water, blood, and oil mixtures indicate that MnBK partitions readily from air into oil $>$ blood $\geq$ water (39).

1.4.1.3.3 Excretion. Rats given a single oral dose of MnBK excreted 45% of the dose via inhalation (40% as CO_2 and 5% as MnBK), 35% in urine, 1.5% in the feces, and about 15% remained in the carcass (4). Beagle dogs exposed to MnBK via inhalation at 50 or 100 ppm for 6 h excreted 32 or 35% of the vapor in the expired breath (38). MnBK was not detected in the breath 3–5 h postexposure. MnBK is eliminated from the body primarily as carbon dioxide in expired air, and as 2-hexanol, 5-hydroxy-2-hexanone, 2,5-hexanedione, 2,5-dimethyl furan, γ-valerolactone, norleucine, and urea in the urine (4, 30, 38, 40).

Metabolism of hydrocarbons and methyl ketones is a common capability of many forms of life and has many similarities from microbes to man (4, 38, 41–45). Since 1974, much work has been done on the metabolic interrelationships of *n*-hexane, MnBK, and their neurotoxic metabolites, particularly 2,5-hexanedione (1, 2, 4, 29, 38, 40, 41, 46, 47). This work clearly indicates that the toxicity of *n*-hexane and MnBK are linked through their oxidation to secondary alcohols (2-hexanol, 2,5-hexanediol, and 5-hydroxy-2-hexanone) and to the -diketone, 2,5-hexanedione. Uptake of the parent compounds is rapid through pulmonary, gastrointestinal, and percutaneous tissues, as is their metabolic conversion to neurotoxic metabolites. The principal neurotoxic metabolite, 2,5-hexanedione, is slowly excreted from the body, giving rise to its possible accumulation in target organs. The concentration of 2,5-hexanedione is significantly higher in the lungs and plasma after inhalation as compared to oral gavage with a comparable dose level, suggesting lung metabolism of MnBK (48).

The proximate neurotoxicant associated with exposure to MnBK is generally considered to be 2,5-hexanedione. See Section 5 of this chapter for information on 2,5-hexanedione.

1.4.1.4 Reproductive and Developmental. Atrophy of the testicular germinal epithelium has been reported in rats exposed to 700 ppm MnBK vapor for 11 wk (35) or 660 mg/kg/d (5 d/wk) MnBK in a 90-d oral gavage study (1).

Pretreatment of F344 rats with three doses of 10 mM/kg MnBK by gavage 12 h apart followed by treatment with 120 mg/kg 1,2-dibromo-3-chloropropane 6 h later significantly reduced testicular 17-α-hydroxylase activity and thus reduced steroidogenesis (49). Exposure to MnBK alone did not produce a similar effect.

Testicular atrophy has been studied in much greater depth following exposure of animals to 2,5-hexanedione, a metabolite of MnBK (See section on 2,5-hexanedione).

Peters et al. (50) evaluated the effects of MnBK on gestation and postnatal development in rats. Pregnant females exposed to 1000 or 2000 ppm for 6 h/d throughout gestation had decreased body weight, and fewer and smaller offspring. Exposure to 500 ppm resulted in normal offspring. Weight gains of offspring from mothers exposed to 1000 or 2000 ppm were normal in the first few weeks, but lower thereafter. Evaluation of postnatal development indicated enhanced grip strength, but a general decrease in motor activity and avoidance learning in young adult or adult offspring.

1.4.1.5 Carcinogenesis. No carcinogenicity studies on MnBK could be found in the literature.

1.4.1.6 Genetic and Related Cellular Effects Studies. Few data are available on the genetic toxicity of MnBK except for an unconventional assay of aneuploidy described by

Zimmermann et al. (51). The basis for this test is the disruption of microtubule assembly required for the segregation of chromosomes during mitosis. Polar solvents such as propionitrile disrupt tubulin interactions resulting in incomplete segregation of chromosomes, but they can also act synergistically with other solvents to produce the same effect. Thus concentrations of 3–7 mg/mL MnBK incubated with D61.M yeast (*Saccharomyces cerevisiae*) did not induce chromosomal loss except when incubated in the presence of propionitrile. Since no direct action of genetic material occurs, however, the validity of this test as a predictor of genetic damage is questionable.

Mayer and Goin (52) also examined the ability of MnBK, *n*-hexane, and 2,5-hexanedione to induce chromosome loss in strain D61.M of *S. cerevisiae*. MnBK and *n*-hexane induce only marginally positive results whereas 2,5-hexanedione was strongly positive.

1.4.1.7 Other: Neurological, Pulmonary, Skin Sensitization. Simultaneous exposure to other chemicals can alter the rate at which neurotoxicity is observed following exposure to MnBK. MEK, in particular, can accelerate the onset of neurotoxicity (36). Enhancement may be due to the induction of the hepatic mixed-function oxidase system. Inhalation of MEK/MnBK shortened hexobarbital sleeping time and increased *in vitro* production of *n*-aminophenol, formaldehyde, *n*-aminobenzoate, and sulfanilamide by rat hepatocyte microsomes (53). On the other hand, Abdel-Rahman et al. (46) reported that phenobarbital, also a microsomal inducer, protects against MnBK neurotoxicity. Mathews et al. (54) reported that *trans*-1,2-dichloroethylene, an inhibitor of P450 2E1, increased the exhaled MnBK concentration in rats not exposed to MnBK, presumably by decreasing the metabolism of endogeneous MnBK.

Just as exposure to other materials may impact the toxic properties of MnBK, exposure to MnBK has been reported to alter the effects of other toxicants in experimental animals.

While MnBK itself does produce liver or kidney toxicity, exposure to MnBK is reported to enhance the activity of materials that are hepatotoxic or nephrotoxic. Prior exposure to an oral dose of MnBK followed by chloroform has been reported to enhance the toxicity, resulting in increased liver enzyme and urea nitrogen levels in the plasma, altered bile flow, and degenerative changes in hepatic and renal tissue (55, 56). These effects are postulated to be due to alteration in P450 levels by MnBK (55). MnBK exposure may also alter the activity of other enzyme systems. For example, the inhibition of alcohol dehydrogenase by MnBK results in slower elimination of ethanol from the blood and prolongation of ethanol-induced loss of the righting reflex (57, 58).

1.4.2 Human Experience

1.4.2.1 General Information. The concern for human health effects following MnBK exposure has been focused on peripheral neuropathy and weight loss. Acute effects and lethality have not been reported for humans. Other effects such as those observed in experimental animals (i.e., testicular atrophy, hematologic effects, enhancement of hepatotoxicity and nephrotoxicity) have not been observed in humans exposed to MnBK.

1.4.2.2 Clinical Cases

1.4.2.2.1 Acute Toxicity. There are no case reports of acute toxicity or lethality following exposures to MnBK.

1.4.2.2.2 Chronic and Subchronic Toxicity. Gilchrist et al. (59), Billmaier et al. (60), and Allen et al. (61) reported an outbreak of toxic neuropathy that resulted when MnBK was substituted for methyl *iso*butyl ketone (MiBK) in a mixture with MEK. The clinical course in affected workers included distal symmetrical sensory changes and weakness, most severely affecting the legs, and an unexplained weight loss. The amount of time from the introduction of the ketone change to the onset of neuropathy was approximately 7 mo. Although ambient solvent levels were measured, it is questionable whether they reflect the actual exposure levels. Similar case reports involving painters (62, 63), cabinet finishers (64), and a screen cleaner (65) are reported in the literature. Other cases of peripheral neuropathy have been reported for shipyard workers (64) and leather workers (66) although exposure was not restricted to MnBK. Changes in lymphocyte populations associated with reduced immune function have been reported in floor-layer workers (67), but functional changes in immunocompetence have not been demonstrated. In each case, exposure was probably by both the inhalation and the dermal routes.

1.4.2.2.3 Pharmacokinetics, Metabolism, and Mechanisms

ABSORPTION. MnBK is readily absorbed following oral ingestion or inhalation by humans. Ingestion of 0.1 mg/kg of radiolabeled MnBK resulted in absorption of at least 66% of the dose based on observed excretion amount (4). Inhalation of 10, 50, or 100 ppm MnBK vapor resulted in absorption of 75–92% of the inhaled dose (38). Application of radiolabeled MnBK to the shaved forearms of two volunteers for 60 min under an occlusive wrap resulted in absorption of 4.8 or 8.0 μg MnBK/min/cm^2 (38).

DISTRIBUTION. Specific distribution studies have not been conducted with humans exposed to MnBK. However, the distribution of MnBK would be expected to be widespread based on physiochemical properties and the data available from studies with experimental animals.

EXCRETION. Following ingestion of 0.1 mg/kg of radiolabeled MnBK, humans excreted 40% of the radiolabel in the breath and 26% in the urine (38). Following inhalation of 10 or 50 ppm MnBK for 7.5 or 100 ppm for 4 h, humans excreted 75–92% of the radiolabel in the breath (38). Application of radiolabeled MnBK to the skin under an occlusive wrap for 60 min resulted in excretion of the label in breath and urine (38).

METABOLISM. DiVincenzo et al. (38) observed metabolism of MnBK to carbon dioxide and 2,5-hexanedione suggesting that metabolism of MnBK was similar in man and experimental animals.

1.4.2.2.4 Reproductive and Developmental Toxicity. There are no case reports of reproductive or developmental toxicity in humans following exposure to MnBK.

1.4.2.2.5 Carcinogenesis. There are no case studies of carcinogenicity in humans following exposure to MnBK.

1.4.2.2.6 Genetic and Related Cellular Effects Studies. There are no case studies of genetic toxicity in humans following exposure to MnBK.

1.4.2.2.7 Other: Neurological, Pulmonary, Skin Sensitization, etc. There are no case reports of pulmonary toxicity or skin sensitization following exposure to MnBK. For neurological effects see Section 1.4.2.2.2.

1.4.2.3 Epidemiology Studies. Epidemiologic studies and controlled exposure studies are limited to acute controlled exposures to MnBK. These studies examine the acute effects of MnBK on eye and nasal irritation and absorption, excretion, and metabolism of MnBK following acute exposure. Epidemiologic studies of subchronic and chronic toxicity, cacinogenicity, reproductive and developmental toxicity, genetic effects, or other forms of acute or repeated dose toxicity have not been reported.

1.4.2.3.1 Acute Toxicity. Schrenk et al. (24) reported that volunteers exposed for several minutes to 1000 ppm MnBK experienced moderate eye and nasal irritation.

1.4.2.3.2 Chronic and Subchronic Toxicity. There are no epidemiology studies reporting adverse health effects following chronic or subchronic exposure to MnBK.

1.4.2.3.3 Pharmacokinetics, Metabolism, and Mechanisms. Available data indicate that MnBK is readily absorbed by oral and inhalation routes. Human volunteers who inhaled 10 or 50 ppm MnBK for 7.5 h or 100 ppm for 4 h absorbed 75–92% of the inhaled vapor (38). Humans who ingested capsules containing 0.1 mg/kg of radiolabeled MnBK excreted 40% of the dose in expired air and 26% of the dose in their urine over the following 8 d (38). Skin absorption rates of 4.8 and 8.0 μg/min/cm^2 MnBK were calculated following application of radiolabeled MnBK to the shaved forearms of two human volunteers (38). The dermal exposure site was occluded, and the exposure time was 60 min. The metabolism of MnBK is similar in experimental animals and humans. 2,5-Hexanedione has been detected in humans following administration of MnBK to humans (38).

1.4.2.3.4 Reproductive and Developmental. There are no epidemiology studies reporting reproductive or developmental effects following exposure to MnBK.

1.4.2.3.5 Carcinogenesis. There are no epidemiology studies reporting carcinogenic effects following exposure to MnBK.

1.4.2.3.6 Genetic and Related Cellular Effects Studies. There are no epidemiology studies reporting genetic effects following exposure to MnBK.

1.4.2.3.7 Other: Neurological, Pulmonary, Skin Sensitization, etc. There are no epidemiology studies reporting other systemic effects following exposure to MnBK.

1.5 Standards, Regulations, or Guidelines of Exposure

Industrial exposure to MnBK by the oral route is unlikely in the normal work environment. The greatest hazards of MnBK are by topical (eye and skin) and inhalation routes. Contact with the eyes should be avoided by wearing appropriate eye protection. Skin contact should be avoided to prevent dermatitis due to the defatting action of MnBK and to prevent percutaneous absorption. Inhalation exposure, particularly repeated exposures, should be kept to a minimum to prevent peripheral neuropathy.

Hygienic standards for MnBK are listed in Table 76.5.

1.6 Studies on Environmental Impact

Significant environmental exposure to MnBK is not expected in the United States, since MnBK is not manufactured, imported, or used. Availability of MnBK in other countries is not reported. The most likely anthropogenic sources of MnBK in the environment are as a waste product from wood pulping, coal gasification, or oil shale processing, or as a leachate from hazardous waste sites. MnBK can be expected to occur naturally in the environment from microbiological metabolism of other hexacarbons. Due to the potential sources for MnBK and its solubility in water, it is expected that it may be found in water. As MnBK is volatile, partitioning from water into air is expected. Degeneration in the environment may be biotic through microbial metabolism or abiotic through photodegradation. Bioconcentration and biomagnification are not expected to occur to a significant extent (68).

2.0 Methyl Isobutyl Ketone

2.0.1 CAS Number: *[108-10-1]*

2.0.2 Synonyms: 4-Methyl-2-pentanone, 4-methylpentane-2-one, isopropyl acetone, hexone, MiBK, and MBK [reference to MiBK as MBK has caused some confusion in scientific literature because methyl *n*-butyl ketone is also (and more commonly) referred to as MBK]

2.0.3 Trade Names: Araldite GZ, 571 KX-75, CMD 9015, Epi-Rez 2036, Filmex A-1, and Propsolv III-1 (69)

2.0.4 Molecular Weight: 100.16

2.0.5 Molecular Formula: $C_6H_{12}O$

2.0.6 Molecular Structure: $CH_3\overset{\overset{O}{\|}}{C}CH_2CH(CH_3)_2$

2.1 Chemical and Physical Properties

Methyl isobutyl ketone is highly soluble in water and miscible with many organic solvents, including ethanol, diethyl ether, benzene, and chloroform. MiBK has a boiling point of 117.8°C, a melting point of −84.7°C, a density of 0.8017 g/cm^3 at 20°C, and a vapor pressure of 18.8 torr at 25°C (70). The flash point of MiBK is 18°C by the closed-cup method and 23°C (73°F) with an open-cup method (71). The autoignition temperature is 448°C, and the flammability limits at 93°C are 1.2 and 8.0%.

Table 76.5. Hygienic Standards for Ketones

Compound	CAS No.	ACGIH TLV[a]		OSHA PEL[d]		NIOSH REL[e]		DFG MAK[f]
		TWA[b]	STEL[c]	TWA	STEL	TWA	STEL	TWA
Methyl *n*-butyl ketone	[*591-78-6*]	5(S)[g]	10	100	—	1	—	5
Methyl isobutyl ketone	[*108-10-1*]	50	75	100	—	50	75	20
Mesityl oxide	[*147-79-7*]	15	25	25	—	10	—	25
4-Hydroxy-4-methyl-2-pentanone	[*123-42-2*]	50	—	50	—	50	—	50
2,5-Hexanedione	[*110-13-4*]	—	—	—	—	—	—	—
Cyclohexanone	[*108-94-1*]	25(S)[g]A4[j]	—	50	—	25(S)	—	—
Methyl *n*-amyl ketone	[*110-43-0*]	50	—	100	—	100	—	—
Methyl isoamyl ketone	[*110-12-3*]	50	—	100	—	50	—	—
Ethyl *n*-butyl ketone	[*106-35-4*]	50	75	50	—	50	—	—
Di-*n*-propyl ketone	[*123-19-3*]	50	—	—	—	50	—	—
Diisopropyl ketone	[*565-80-0*]	—	—	—	—	—	—	—
Methyl cyclohexanone	[*583-60-8*]	50(S)[g]	75	100(S)[g]	—	50(S)[g]	75	50
Acetophenone	[*98-86-2*]	10	—	—	—	—	—	—
2-Octanone	[*111-13-7*]	—	—	—	—	—	—	—
3-Octanone	[*106-68-3*]	—	—	—	—	—	—	—
5-Methyl-3-heptanone	[*541-85-5*]	25	—	25	—	25	—	—

Propiophenone	*[93-55-0]*	—	—	—	—	—	—	—
Isophorone	*[78-59-1]*	—	5(CL)[h]A3[i]	25	—	4	—	2
5-Nonanone	*[502-56-7]*	—	—	—	—	—	—	—
Diisobutyl ketone	*[108-83-8]*	25	—	50	—	25	—	50
Trimethyl nonanone	*[123-18-2]*	—	—	—	—	—	—	—
Benzophenone	*[119-61-9]*	—	—	—	—	—	—	—

[a]American Conference of Governmental Industrial Hygienists Threshold Limit Values.
[b]Time-weighted average (ppm).
[c]Short-term exposure limit (ppm).
[d]Occupational Safety and Health Administration Permissible Exposure Limits.
[e]U.S. National Institute for Occupational Safety and Health Recommended Exposure Limit.
[f]Deutsche Forschungsgemeinschaft (Federal Republic of Germany) maximum concentration values in the workplace.
[g]Indicates a skin notation applicable to TLV.
[h]Indicates ceiling limit applies.
[i]A3 Confirmed animal carcinogen with unknown relevance to humans.
[j]A4 Not classifiable as a human carcinogen.

2.1.1 General

Additional physical–chemical properties are summarized in Table 76.4.

2.1.2 Odor and Warning Properties

Methyl isobutyl ketone is a clear liquid with a sweet, sharp odor. AIHA reviewed and critiqued all available odor threshold studies and rejected 3 of the 11 published values (72). Odor detection threshold determinations ranging from 0.4 to 32 ppm were deemed acceptable, as were odor recognition thresholds ranging from 1.1 to 64 ppm. Amoore and Hautala calculated the geometric average for all reported air-derived and water-derived odor threshold determinations for MiBK and reported values of 0.68 and 1.3 ppm, respectively (19). Devos et al. summarized and standardized the results from 6 different studies and reported an MiBK odor threshold of 0.54 ppm (73). Others have reported odor threshold concentrations ranging between 0.10 and 0.68 ppm (74–76). Gagnon et al. have shown that volunteers exposed to MiBK for 7 h at concentrations of 20 or 40 ppm exhibited a ninefold average increase in their relative odor threshold at the end of the exposure period (77). Recovery to pre-exposure threshold levels was dependent on the exposure concentration and was not complete 95 min after exposure termination.

2.2 Production and Use

Methyl isobutyl ketone can be manufactured by either of two processes (78). The first is a mixed ketone process where MiBK, diisobutyl ketone, and acetone are co-produced in a single reaction using isopropanol as a starting material. The second method is used to produce the majority of MiBK and involves a three-step reaction sequence in which diacetone alcohol and mesityl oxide are formed as intermediates. Most (79%) of the MiBK produced in the United States is used as an extracting solvent in the mining and pharmaceutical industries and in the preparation of high-solids coatings such vinyl and acrylate paints and nitrocellulose-based lacquers. Approximately 17% of the MiBK is used to synthesize an antiozanant that is found in products made of rubber. A small amount of MiBK is also used to synthesize acetylenic surfactant glycols, which are found in paints, inks, and pesticide formulations. The U.S. consumption of MiBK was 155 million pounds in 1995.

2.3 Exposure Assessment

2.3.1 Air

An examination of VOC emissions from grass and clover fields showed that MiBK accounted for 0.5–1.1% of the total organic carbon emitted from the grass field, but less than 0.1% of the organic carbon from the clover (79). Furia and Bellanca cited data indicating that oranges and grapes contain MiBK; the levels, however, were not provided (80). A survey of flavoring ingredients showed that MiBK is used at a maximum level of 6.3 ppm in beverages, ice cream, candy, and baked goods (81). A survey of 1039 household products showed that MiBK was present in 6.2% of the products examined (82). MiBK was detected in low amounts ($<6.9\%$) in several product categories, including paint-related materials, fabric and leather treatments, adhesives, and automotive care products.

MiBK was found in the plume from a fuel oil fire at a concentration 14 μg/m^3 (83). It was also detected, but not quantified, in the building materials and dust from mold-infested rooms (84). It was not, however, identified as a volatile metabolite for three of the mold strains cultured from the dust samples. An MiBK concentration of 600 ng/L was measured in a school classroom where the average concentration in the expired air from 12 of the students attending the school was 150 ng/L (85).

Two of the four air samples collected around a chemical waste disposal site contained MiBK at concentrations of 2.1 and 6.0 μg/m^3 (86). The MiBK concentration in an air sample collected on the dump site was 440 μg/m^3. Dunovant et al. found MiBK in air samples from 2 of 3 municipal wastewater treatment plants at levels ranging from 2.1 to 13 ppm (87). Similarly, the average MiBK concentration in the air being vented from a municipal solid waste landfill was 5020 μg/m^3 (88). The Hazardous Substance Release and Health Effects Database contains an extensive listing of the MiBK contaminant levels found at landfills, surface impoundments, recycling plants, and hazardous waste sites throughout the United States (89). Although much of the monitoring data is for the ground water, surface water, and soil samples collected at these locations, some air emissions levels are available for unspecified air samples in and around the sites. Outdoor air levels ranging from less than 1 ppb to over 30 ppm were recorded, with most values ranging from 0.1 to 12 ppb. Soil gas samples contained MiBK at appreciably higher concentrations, whereas indoor air samples were at the low end of the stated concentration range. Using emissions modeling data from 60,803 census tracts throughout the United States, the highest individual airborne concentration of MiBK was estimated to be 9.9 μg/m^3 (90). Frostling et al. showed that MiBK was emitted when polypropylene was thermally oxidized at temperatures (220°C) used during industrial molding operations (91).

An evaluation of solvent use in Japan indicated that MiBK was one of the ten most commonly found solvents in the workplace (92). A survey of MiBK use in 95 Japanese factories showed a median airborne concentration of 0.5 ppm and a maximum concentration of 2.7 ppm (93). An examination of ten different manufacturing operations indicated that the highest prevalence of MiBK use occurred at facilities where painting operations were performed. Similar results were reported in a survey of MiBK use in Belgian industries (94). An analysis of the information in an IARC database of exposure levels for the pulp, paper, and paper products industries revealed that 30 of the 38 recorded MiBK measurements occurred in the manufacture of paper products (95). Eitzer determined the MiBK concentrations in eight different solid waste composting facilities and found average concentrations ranging from 0 to 1500 μg/m^3 for various operations within the plant (96). The highest concentrations of MiBK were generally located at the shredding operation, where the municipal waste was cut into smaller pieces before being composted. A survey of apprentice spray painters at automotive repair shops revealed an average MiBK exposure concentration of 6.8 mg/m^3 in 4 of the 70 workers examined (97). Whitehead et al. reported a mean concentration of 0.3 ppm for 89 employees involved in spray painting and spray gluing operations at three different plants (98). The highest exposure concentrations were associated with the use of low-aromatic spray paints where the 8-h time-weighted average exposures were 0.6 ppm. MiBK was present at a concentration of 29.9 mg/m^3 in the air at 1 out of 12 auto paint shops surveyed for organic solvent exposures (99). The MiBK concentration in the breathing zone of 38 workers applying polyurethane coatings reportedly ranged from 0.50 to 58.0 ppm (100). Nelson et al. summarized personal monitoring data for employees working in automotive assembly plants and found MiBK exposure concentrations ranging from 0.5 to 13 ppm for 26 employees (101).

2.3.2 Background Levels

MiBK has repeatedly been shown to be a contaminant of ground water, surface water, soil, and sediment. The ATSDR Hazardous Substance Release and Health Effects Database contains monitoring data showing MiBK concentrations that range up to 47,000 mg/L for ground water, 35 mg/L for surface water, and 97 ppm for soil (89). Atlas et al. detected MiBK in raw and filtered sediment from a dump site for municipal and industrial waste in the Gulf of Mexico (102). Little if any MiBK could be found in four species of seaweed grown in the laboratory (103). MiBK was present at a concentration of 190 μg/L in the formation water that was generated with the gas and oil being pumped at an offshore oil well (104). An examination of groundwater monitoring data from 479 waste disposal site investigations showed that MiBK was present in measurable amounts at 10 different sites (105). MiBK was measured in the leachate from 2 of 5 municipal landfill sites at concentrations of 35 and 200 μg/L (106). Brown and Donnelly examined and summarized published information on levels of groundwater contamination and reported MiBK concentrations ranging from 2 to 10 mg/L for industrial landfill leachates and 0.18–0.74 mg/L for municipal landfill leachates (107). Canter and Sabatini examined the monitoring data from three Superfund sites located in or around the Biscayne aquifer and found a maximum MiBK contaminant concentration of 90 μg/L (108). Francis et al. reported the presence of MiBK in leachates from two low-level radioactive disposal sites in the United States (109).

Surface water from the Adige river in Italy contained MiBK levels ranging from 0.01 to 332 μg/L (110). Hall et al. detected MiBK in water samples from the Potomac River at concentrations less than 40 μg/L (111). MiBK was not detected, however, in a survey of organic compounds in water from the Delaware river (112). Three of the eleven domestic wells near a municipal landfill in Connecticut contained MiBK at concentrations ranging from 25 to 150 ppb (113). Sabel and Clark reported that MiBK was present the leachate from 6 municipal landfills at a concentration range of 10–640 μg/L (114). It was also detected in groundwater samples from 4 of 13 monitoring wells around a landfill at a concentration range of 5.6–3000 μg/L. Groundwater from 1 of 7 uncontaminated sites contained MiBK at a concentration 9.0 μg/L. The estimated discharge of MiBK into the wastewater leaving a chemical/pharmaceutical plant ranged from 10 to 340 kg over a 2-wk monitoring period (115).

2.3.3 Workplace Methods

Either of two NIOSH methods can be used to determine MiBK concentrations in workroom air (116). The first method (NMAM 2549) is a screening method for volatile organic compounds that uses thermal desorption tubes containing graphitized carbon and molecular sieve adsorbents for sampling and gas chromatography–mass spectrometry for analysis. The second method (NMAM 1300) is more specific for ketones and uses a solid sorbent tube with coconut shell charcoal and gas chromatography with flame ionization detection. Levin and Carleborg evaluated the performance of 5 different adsorbents for the collection of MiBK vapors and found that charcoal and Ambersorb XE-348 performed better than XAD resins (117). Tomczyk and Rogaczewska published and validated a similar method that could also be used to monitor methyl isobutyl carbinol, toluene, and acetone levels in the air (118). Gazzaniga et al. described an automated method for monitoring MiBK concentrations in the workplace at 15-min intervals using gas chromatography with flame ionization detection (119). Smith and Wood described a colorimetric method for the determination of MiBK in air that involved the collection of an aqueous sample using impingers and the formation of the 2,4-dinitrophenylhydrazine derivative, which formed a red color under basic conditions (120). Eitzer used thermal desorption gas chromatography–mass spectrometry and an unknown type of adsorbent to determine MiBK concentrations in the air at a composting facility (106).

2.3.4 Community Methods

Dunemann and Hajimiragha outlined a method for determining MiBK in body fluids and environmental samples that used purge and trap gas chromatography–mass spectrometry (121). A combination of four absorbents was used in the sampling trap: silica gel, Tenax-TA, Carbotrap, and SP-2100 on Chromosorb. Dreisch and Munson showed that a similar purge and trap method could be used to detect MiBK in environmental samples and that the method was generally preferable to an EPA recommended method (122). Ojanperä et al. also used purge and trap gas chromatography with a Tenax adsorbent to analyze for MiBK in blood, but combined it with Fourier transform infrared spectroscopy (123). Brown and Purnell described the sampling and desorption properties of MiBK on a Tenax adsorbent used for airborne vapor collection (124). XAD-2 resin was shown to be a good adsorbent for collecting MiBK from a potable water sample when it was used with a gas chromatographic or spectrophotometric analysis technique (125). Pleil and Stroupe developed a gas chromatography–mass spectrometry technique to analyze for MiBK and other volatile materials at airborne concentrations of 10 ppm or greater (126). Marley and Gaffney described a reactive hydrocarbon analyzer with a chemiluminescent detector that could be used to analyze for MiBK in the atmosphere (127). The method was compared with a gas chromatographic technique using flame ionization detection and shown to be comparable in performance.

2.3.5 Biomonitoring/Biomarkers

2.3.5.1 Blood. Hjelm et al. measured breath, blood, and urine levels of MiBK in eight volunteers exposed at different times to 10, 100, or 200 mg/m^3 MiBK and to a combination of 200 mg/m^3 MiBK and 150 mg/m^3 toluene for 2 h under conditions of light exercise (128). Upon termination of the exposure, the concentration of MiBK in the urine was somewhat higher than the levels in arterial blood. Blood levels at the end of the

100 mg/m^3 exposure were about 6 μmol/L; slightly lower blood levels were observed with the toluene co-exposure.

MiBK was detected in the blood of two people who died as a result of spraying inside a storage tank with an epoxy-based paint that contained MiBK, MEK, xylenes, and toluene (129). Blood concentrations of 0.04 and 0.14 mg/dL were observed in autopsy specimens. Higher levels were noted in brain, liver, lung, kidney, and vitreous humor specimens. Blood samples from normal human volunteers contained an MiBK concentration of 0.06 μg/L (121).

2.3.5.2 Urine. Hjelm et al. showed a good correlation between the urinary levels of MiBK and the pulmonary uptake of vapors by eight volunteers exposed to 10, 100, or 200 mg/m^3 (128). The levels of two MiBK metabolites in urine were below detection limits (5 nmol/L) after each of these exposures. Gobba et al. cited a report that showed a good correlation ($r = 0.81$) between the airborne exposure of 28 workers to MiBK and the concentration in the urine (130). Ogata et al. examined the excretion of MiBK in the urine of 38 workers co-exposed to toluene and xylenes (131). The time-weighted average exposure to 14.6 ppm resulted in the excretion of 0.54 mg/L of MiBK in the urine. A regression analysis showed a high degree of correlation ($r = 0.90$) between the airborne concentration and the level in urine. A comparison of these results with those from the previous studies indicated that the combined exposure did not affect the rate of MiBK excretion into the urine. A study of 20 workers exposed to MiBK, toluene, ethyl benzene, and xylene indicated that measurable amounts of the MiBK metabolite, 4-methyl-2-pentanol, were excreted into the urine (132). The 8-h time-weighted average exposures to MiBK were 21.9 ppm, and the correlation coefficient for the urinary excretion of the MiBK metabolite was 0.63. Dunemann and Hajimiragha measured 0.97 μg/L of MiBK in the urine of an unstated number of human volunteers (121). Zlatkis and Liebich identified MiBK as well as other substituted 2-pentanones in the 24-h urine samples of unexposed men and women (133).

2.3.5.3 Other. MiBK could not be detected in the breath of 8 volunteers being monitored for personal exposure to volatile organic compounds (134). An MiBK level of 0.52 μg/L was found in human breast milk (135). Trace levels of MiBK have been detected in the expired breath (3.9–24.0 μg/h) and urine of normal adult humans and in both maternal and umbilical cord blood (136, 137).

2.4 Toxic Effects

2.4.1 Experimental Studies

2.4.1.1 Acute Toxicity. Smyth et al. reported a 14-d oral LD_{50} of 2.08 g/kg (95% CI, 1.91–2.27 g/kg) for rats when MiBK was administered as a 20% emulsion in Tergitol 7 surfactant (138). Smyth also reported an oral LD_{50} of 5.7 mL/kg for rats when the compound was given undiluted (139). Oral LD_{50} values of 1600–3200 mg/kg have been reported for both rats and guinea pigs (3). Zakhari reported a 24-h oral LD_{50} for MiBK of 1.90 g/kg (95% CI, 1.22–2.58 g/kg) for male CF-1 mice (140). Batyrova conducted acute oral studies in mice and rats and found the lethal dose to be 2.85 and 4.6 g/kg, respectively (141). In another study, groups of rats were given single oral doses of 900, 1350, 2025, 3038, or 5000 mg/kg. The low-dosed animals exhibited reduced activity levels that cleared within 1 d, while higher-dosed animals exhibited more severe signs of CNS depression that cleared within 2 d; the LD_{50} was 3200 mg/kg (142). Tanii et al. reported the oral LD_{50} for MiBK in male ddY mice, treated with olive oil by intraperitoneal injection 24 h previously, was 2671 mg/kg (95% CI, 2111–3372 mg/kg) (23). Pretreatment with carbon tetrachloride lowered the oral LD_{50} slightly to 1977 mg/kg. Nishimura et al. determined the MiBK LD_{50} value to be 2671 and 2080 mg/kg for mice and rats, respectively (143).

Exposure to 19,500 ppm of MiBK produced anesthesia in 7 of 10 mice within 30 min (144). When the animals were removed to fresh air, four recovered immediately, and three were awake within 5 min. Concentrations above 20,000 ppm produced anesthesia within 30 min with subsequent death of most of the animals. Gross examination at necropsy revealed congestion of the lungs. Daily exposures to 20,000 ppm for 20 min for 15 d killed 6 of 10 mice. Smyth et al. reported that 15 min in a saturated atmosphere of MiBK was the minimum time necessary to kill the first animal (138). Six rats survived a 4-h exposure to 2000 ppm MiBK, but 4000 ppm killed all six animals in 4 h. Specht et al. exposed guinea pigs to concentrations of 1000, 16,800, and 28,000 ppm MiBK (25, 145). The 1000-ppm level caused little or no irritation of the eyes and nose of the animals; however, the operator did experience eye and nose irritation. Guinea pigs showed a decreased respiratory rate during the first 6 h of exposure, which the authors attributed to a low-grade narcosis. The 16,800-ppm level caused immediate signs of eye and nose irritation followed by salivation, lacrimation, ataxia, and death. Nine of 10 guinea pigs died within 6 h of exposure. The highest concentration used (28,000 ppm) killed 50% of the animals within 45 min. Only a few guinea pigs survived 60 min of exposure. Fatty livers and congestion of the brain, lungs, and spleen were noted. The heart and kidneys were not affected. Exposures of 21,000 ppm MiBK vapor killed all rats within 53 min, whereas exposures of 4000 ppm for 6 h caused loss of coordination and prostration followed by recovery with no residual adverse signs (3). Zakhari exposed male CF-1 mice to a series of nominal concentrations of MiBK for 45 min (146). An exposure to 10,000 produced no mortality, while an exposure to a saturated atmosphere (30,000 ppm) resulted in 80% mortality. The 45-min LC_{50} was calculated to be 18,105 ppm. Aspiration of MiBK into the lungs may produce significant irritation and lethality (147). Using a model system, aspiration of 1 mL/kg of liquid MiBK into the lungs of Sprague–Dawley rats resulted in the rapid death of all six treated animals.

Zakhari reported an intraperitoneal LD_{50} value of 0.59 g/kg (95% CI, 0.36–0.82 g/kg) in male CF-1 mice given undiluted MiBK (140). In CD-1 mice, the estimated intraperitoneal LD_{20} of MiBK has been reported to be 0.73 mL/kg (148). Terhaar et al. determined the intraperitoneal LD_{50} for rats and guinea pigs to be 1.14 and 0.92 mL/kg, respectively (3). Cunningham et al. reported that 752 mg/kg was lethal to male CD-1 mice when given in corn oil by IP injection, but 500 mg/kg caused no deaths (58). Single doses of 500 or 250 mg/kg in corn oil administered by IP injection did not result in ataxia or loss of righting reflex in male CD-1 mice. At the 500-mg/kg dose level, MiBK did prolong the loss of righting reflex induced by ethanol administration. When MiBK was administered in combination with MEK (9:1 MEK:MiBK), the intraperitoneal LD_{50} for these species were essentially unchanged, 1.64 mL/kg for the rat and 1.07 mL/kg for the guinea pig (3).

Undiluted MiBK (0.1 mL) produced some irritation within 10 min when instilled in the rabbit eye. Inflammation and swelling occurred in 8 h, and inflammation, swelling, and exudate were present at 24 h (144). MiBK has been included in several interlaboratory eye irritation validation studies using both *in vivo* (Draize) and *in vitro* (BCOP) procedures (143, 149). Using the *in vivo* Draize approach with 0.1 mL of undiluted material, Weil and Scala (149) reported that the majority of the laboratories did not find MiBK to be an eye irritant. In 6 of the 24 participating laboratories, irritation responses were reported. The ocular irritancy of pure MiBK was also evaluated in an *in vitro* bovine corneal opacity and permeability (BCOP) assay and found to be a mild irritant (150). These results were compared to the findings from an *in vivo* Draize test, where damage to rabbit cornea, iris, and conjunctiva were scored on days 1 and 14. The Draize study also indicated that MiBK was mildly irritating and that recovery from the damage occurred in 4 d.

Gilleron et al. evaluated the irritancy potential of MiBK in the hen's egg test–chorioallantoic membrane (HET–CAM) assay and showed it to be an ocular irritant (151). Sivak and Herbert cultured the ocular lens from cows and examined the effects of MiBK on optical performance (focal length and transmittance) using a scanning laser beam (152). Treatment of bovine lenses with a 2.0% solution of MEK was found to cause a mild–moderate change in the treatment time necessary to cause a 100% change in focal length variability. Kennah et al. compared the Draize irritation method for scoring ocular injury to an ocular swelling technique that entailed corneal thickness measurements (153). Ocular irritation (Draize) and swelling (corneal thickness) were determined in rabbits following treatment with MiBK at concentrations of 30 or 100%. A maximum Draize score of 5 (mildly irritating) was obtained at 24 h post-treatment with undiluted MiBK. Corneal swelling was also mild, increasing less than 20% relative to control.

A single application of MiBK to the skin of rabbits produced only transient erythema, but daily applications of 10 mL for 7 d caused drying and flaking of the skin (154). Undiluted MiBK (5 or 10 mL) held in contact with the depilated skin of guinea pigs under an occlusive wrap for 24 h produced slight irritation with no clinical evidence of absorption (3). In a collaborative study involving 23 laboratories, 0.5 mL of MiBK was held against the skin of rabbits for 24 h and scored for irritation at 30 min and 48 h after removal of an occlusive dressing. The overall results indicated that MiBK was relatively nonirritating to the skin (149). Other studies showed that 500 mg of MiBK produced moderate irritation of rabbit skin after 24 h (154). MiBK or MEK:MiBK (9:1), with or without dimethyl sulfoxide, dropped on the backs of guinea pigs in amounts up to 2 mL twice daily for 31 wk produced only desquamation with no clinical or histologic evidence of toxic neuropathy (3). Huang et al. evaluated the *in vitro* effects of MiBK on the binding affinity of ^{3}H-dihydroalprenolol for beta-adrenergic receptors and on the enzymatic activity of Na^{+}–K^{+}–adenosine triphosphatase from male ddY mice synaptosomes (155). The concentration causing a 50% inhibition in response (IC_{50}) was 46 μM in the receptor assay and 43 μM in the enzyme assay. When evaluated with other ketones, the results were shown to correlate well with lipophilicity.

Tham et al. infused MiBK intravenously into female Sprague–Dawley rats at a rate of 30 μM/min/kg until the animal's vestibulo-ocular reflex was depressed (156). Reflex depression occurred at a blood concentration of 0.2 mM/L, which was below the level at which general CNS depression occurred. DeCeaurriz et al. measured the irritancy of MiBK in a mouse bioassay that quantified the sudden reflexive drop in respiration caused by a short-duration vapor exposure (157). The concentration-related decline in the respiration rate allowed for the calculation of an RD_{50} value, which corresponded to the irritant concentration needed to produce a 50% decrease in the initial respiration rate. The RD_{50} value for MiBK male Swiss OF_1 mice exposed to at least four vapor concentrations of MiBK for 5 min was 3195 ppm. The RD_{50} value was presented as a reliable means of estimating the irritation threshold of chemical vapor and the results for MiBK have been used to develop several quantitative structure–activity relationships for establishing occupational exposure limits (158–162). Muller and Black examined the sensory irritation caused by a Spanished wall covering that emitted a relatively large amount (701 $\mu g/m^3$) of MiBK at room temperature and found less than a 5% depression in respiratory rate in Swiss–Webster mice (163). In another study, DeCeaurriz et al. calculated that an exposure of mice to 803 ppm MiBK for 4 h decreased the duration of immobility in a 3-min forced swimming test (164). A 4-h exposure to 662 ppm MiBK decreased the duration of immobility by 25%. Interpretation of data from this unconventional test designed to detect the efficacy of antidepressant drugs is uncertain as the endpoint measured is enhanced performance.

In anesthetized, surgically prepared, open-chest dogs and cats, Aviado reported that MiBK inhalation induced pulmonary vascular hypertension (165). The author attributed the change in pressure to effects on the caliber of large pulmonary vessels. The changes noted did not significantly alter pulmonary arterial blood flow. Sharkawi et al. showed that a single IP treatment with MiBK or either of its two primary metabolites could potentiate the loss of a righting reflex in ethanol treated CD-1 mice (166). The prolonged loss in righting reflex from ethanol was observed at 5 mmol/kg of MiBK and 2.5 mmol/kg of either 4-hydroxymethyl-2-pentanone or 4-methyl-2-pentanol.

Brondeau et al. exposed male Sprague–Dawley rats to MiBK vapor at levels of 0, 595, 1280, or 3020 ppm and OF_1 mice to levels of 0, 664, 1477, or 3260 ppm for 4 h (167) followed 18 h later by a 4-h exposure to 1,2-dichlorobenzene (DCB) at hepatotoxic vapor levels (167). Exposure to the ketone alone increased hepatic glutathione-*S*-transferase activity (40–65%) and cytochrome P450 levels but did not alter serum glutamate dehydrogenase activity in rats or liver glucose-6-phosphatase activity in mice. Following pretreatment with MiBK, DCB-induced increases in glutamate dehydrogenase and glucose-6-phosphatase activities were observed. The potentiation of DCB hepatotoxicity appears to be related to a ketone-induced enhancement of hepatic cytochrome P450 levels which increases the formation of toxic DCB metabolites.

2.4.1.2 Chronic and Subchronic Toxicity. Phillips et al. conducted 2- and 14-wk inhalation studies with Fischer-344 rats and $B6C3F_1$ mice of both sexes (168). Exposure concentrations were 0, 100, 500, or 2000 ppm MiBK vapor for two weeks or 0, 50, 250, or 1000 ppm for 14 wk administered 6 h/d, 5 d/wk. At 2000 ppm, low incidences of lethargy and lacrimation were observed as well as increased liver weight in all groups except the male mice. Increased kidney weights were observed in male rats and female mice. Hyalin droplet formation was present in the renal epithelium of male rats exposed to

2000 or 500 ppm MiBK. No effects were observed at 100 ppm. At 1000 ppm (14-wk exposure), liver weights were increased in both male rats and mice. Hematologic changes were limited to an increased platelet count in male rats (1000 ppm) and decreased eosinophils in female rats (1000 ppm). Serum cholesterol was increased in male rats (1000 and 250 ppm). Urinary glucose was increased in male (1000 and 250 ppm) and female rats (1000 ppm) while urinary protein was increased only in male rats (1000 ppm). Hyalin droplet formation was present in only in the kidneys of male rats (1000 and 250 ppm), as has been observed with certain other hydrocarbons.

Oral administration of MiBK and its metabolites (4-hydroxymethyl-2-pentanone and 4-methyl-2-pentanol) in corn oil can potentiate chemically induced cholestasis in rats. The oral pretreatment of male Sprague–Dawley rats with 7.5 mmol/kg (3 d) or 3.75 mmol/kg (7 d) of MiBK caused a significant increase in the cholestatic effects from an IV injection of taurolithocholate (169). There was no decrease in bile flow from MiBK treatment in the absence of taurolithocholate administration. MiBK was also capable of potentiating the decrease in bile flow that was observed when Sprague–Dawley rats were treated with an IV dose of either manganese or manganese–bilirubin in combination (170). Dose–response analysis revealed that a 3-d oral pretreatment regimen at a minimum dose of 3.75 was necessary to cause an effect. When the pretreatment regimen was reduced to a single dose 18 h before the cholestatic agent, the minimum dose increased to 15 mmol/kg. Single or repetitive oral treatments with the MiBK metabolites, 4-hydroxymethyl-2-pentanone and 4-methyl-2-pentanol, were capable of potentiating the cholestasis induced in anesthetized Sprague–Dawley rats by IV co-administration of manganese and bilirubin (171). Pretreatment at metabolite doses ranging from 1.88 to 15 mmol/kg showed that 4-methyl-2-pentanol was a more potent inducer than 4-hydroxymethyl-2-pentanone when the manganese–bilirubin combination was used to induce cholestasis. The opposite was true, however, when manganese was used alone to induced a cholestatic response. The exposure of male Sprague–Dawley rats to 200–600 ppm of MiBK vapors for 3 d (4 h/d) was also capable of intensifying the cholestatic effects from both taurolithocholate and a manganese–bilirubin mixture (172). A minimum exposure concentration of 400 ppm was necessary to observe any additional decrease in bile flow from the cholestatic agents when they were administered by IV injection 18 h after the last inhalation exposure. Mechanistic studies further revealed that the effects of a 3-d oral pretreatment with 7.5 mmol/kg MiBK on taurolithocholate-induced cholestasis were not caused by any change in taurolithocholate kinetics (173). The blood clearance, distribution volume, protein binding, and elimination rates were not affected by the MiBK pretreatment. Joseph et al. subsequently examined the effects of MiBK pretreatment on bile salt formation and secretion rates following an IV dose of taurolithocholate (174). An oral MiBK dose of 7.5 μmol/kg once a day for 3 d was shown significantly to decrease the secretion rates of bile salts, cholesterol, and phospholipids even in the absence of taurolithocholate. When taurolithocholate was administered by IV infusion following the MiBK pretreatment, hepatocellular transport rate was apparently affected.

The ability of MiBK to potentiate the cholestatic reaction from either taurolithocholate or the manganese–bilirubin mixture was not affected by the route of administration in male Sprague–Dawley rats (175). The minimally effective dose for an oral pretreatment schedule was estimated to be 3 mmol/kg for 3 d, whereas the corresponding minimum for a 4-h inhalation exposure was 400 ppm for 3 d. The severity of the decrease in bile flow from the pretreatment was found to be dependent on the plasma concentration of MiBK, irrespective of the route of exposure. More recent studies indicated that a 3-d pretreatment to 600 ppm for 4 h/d affected the cholesterol content in the bile canalicular membranes following the IV administration of a manganese–bilirubin mixture (176). Cholesterol analysis on the hepatic microsomes, mitochondria, cytosol, and plasma membranes from treated Sprague–Dawley rats revealed an increase in cholesterol in canalicular but not sinusoidal plasma membranes was apparently associated with a corresponding increase in de novo synthesis.

MiBK has also been examined for its ability to potentiate the hepatotoxicity of chloroform and carbon tetrachloride. Pilon et al. observed that single oral dose of MiBK at concentrations ranging from 0.3 to 20 mmol/kg could affect the hepatotoxicity of an oral dose of CCl_4 given 18 h later (177). The minimally effective dose of MiBK needed significantly to increase the level of serum alanine aminotransferase (ALT) was dependent on the dose of CCl_4 administered. An MiBK dose of 12 mmol/kg could potentiate the effects from a 0.01-ml/kg dose of CCl_4 and a 1.5-mmol/kg dose was effective with 0.1 mL/kg of CCl_4. Vézina et al. evaluated the ability of MiBK, 4-hydroxymethyl-2-pentanone, and 4-methyl-2-pentanol to potentiate the hepatotoxic effects of an IP dose of chloroform in male Sprague–Dawley rats (178). Treatment for 3 d at concentrations ranging from 3.75 to 7.5 mmol/kg revealed that a minimum dose of 5.6 mmol/kg of all three chemicals was necessary significantly to increase to serum enzyme levels of alanine aminotransferase and ornithine carbonyl transferase following the $CHCl_3$ treatment. Multiple treatments with 5.0 or 7.0 mmol/kg MiBK alone were also found to increase the liver microsomal enzyme activities and cytochrome P450 content. Aniline hydroxylase, 7-ethoxycoumarin *O*-deethylase, and aminopyrine *N*-demethylase activities were all significantly increased by the MiBK treatment.

Raymond and Plaa examined the dose–response relationships for the potentiation of CCl_4 hepatotoxicity and $CHCl_3$ nephrotoxicity in male Sprague–Dawley rats orally pretreated with a daily dose of MiBK for 3 d (179). The minimum effective dose of MiBK necessary to increase plasma ALT activity in the CCl_4-treated animals was 1.4 mmol/kg. When a 6.8-mmol/kg pretreatment regimen was used, the oral dose of CCl_4 needed to affect ALT activity in 50% of the animals (ED_{50}) was 1.8 mmol/kg, which was lower than the value for the other two ketones tested. An MiBK daily dose of 13.6 mmol/kg was necessary to potentiate the nephrotoxic effects of $CHCl_3$. The ED_{50} value for $CHCl_3$ was found to be 2.2 mmol/kg when PAH uptake was used as the indicator of toxicity. In supplemental studies, *in vitro* enzyme activities were evaluated using the hepatic and renal microsomes isolated from male Sprague–Dawley rats treated with either 6.8 or 13.6 mmol/kg of MiBK (180). The low dose of MiBK significantly increased liver aniline hydroxylase activity, whereas the high dose significantly increased kidney and liver aniline hydroxylase, aminopyrine *N*-demethylase, and benzphetamine *N*-demethylase activities. In addition, treatment at the two MiBK dose levels significantly increased the rate of irreversible binding of radiolabeled CCl_4 or $CHCl_3$ to liver and renal microsomal proteins. Raymond and Plaa also examined whether the potentiation of CCl_4 hepatotoxicity was the result of changes in cellular membranes (181). Treatment of male Sprague–Dawley rats with a daily oral dose of MiBK (6.8 mmol/kg) for 3 d did not significantly affect membrane fluidity in isolated basal canalicular membranes, or the activities of three plasma membrane enzymes: 5′-nucleotidase, leucine aminopeptidase, or alkaline phosphatase. The authors concluded that MiBK did not potentiate the effects of CCl_4 by increasing the susceptibility of cellular membranes to damage. Potentiation of haloalkane-induced hepatotoxicity may in large part be associated with enhancement of cytochrome P450 bioactivation pathways (182). For potentiation to occur, there is a critical time interval between the exposure to MiBK and the exposure to the haloalkane. The enhancement of hepatotoxicity also appears to be greater for brominated alkanes than for chlorinated alkanes. Male rats given 3 daily oral gavage doses of 750 mg/kg of MiBK in corn oil had liver weights that were heavier than with control animals (174). Vernot et al. (183) and MacEwen et al. (184) reported increased kidney weights and kidney/body weight ratios in rats exposed to 100 ppm MiBK continuously for 2 wk. Kidney and liver weights and the organ/body weight ratios were also increased after exposure to 200 ppm for 2 wk and to 100 ppm for 90 d. Similar effects were not seen in dogs, monkeys, or mice after the 2-wk exposures. MacKenzie described kidney damage at 2 wk after 100 ppm MiBK as hyaline droplet tubular nephrosis (185). This damage was reversible, even after 90 d of exposure, when rats were removed from the exposure environment.

2.4.1.3 Pharmacokinetics, Metabolism, and Mechanisms. DiVincenzo et al. examined the metabolism of MiBK in male guineas pigs given a single 450-mg/kg IP injection (41). MiBK was metabolized in guinea pigs by ω-1 oxidation to the corresponding hydroxy ketone, 4-hydroxymethyl-2-pentanone, and by carbonyl reduction to the secondary alcohol, 4-methyl-2-pentanol. The latter compound was identified by mass spectroscopy, but the blood levels were too low for quantitation. The half-life for MiBK in the blood was 66 min, and the clearance time was 6 h.

Plasma levels of MiBK and its two metabolites were monitored after the oral or inhalation exposure of male Sprague–Dawley rats (172). Oral exposure at dose levels of 1.5, 3, or 6 mmol/kg resulted in MiBK blood concentrations of 5.27, 8.45, and 16.06 μg/mL, respectively. The corresponding peak blood levels after a 4-h inhalation exposure at 200, 400, or 600 ppm were 5.02, 8.11, and 14.30 μg/mL. Blood levels of 4-hydroxymethyl-2-pentanone ranged from 1.10 to 13.27 μg/mL after the oral exposure, and from 5.03 to 7.10 μg/mL after inhalation. Detectable amounts of 4-methyl-2-pentanol were only observed after inhalation exposure at the two highest concentrations.

2.4.1.3.1 Absorption. The dermal absorption of MiBK was evaluated in 8 anesthetized female guinea pigs exposed to the neat liquid for 2.5 h (186). The average maximum blood concentration was 26.7 μmol/L after 23.1 min of exposure. The dermal uptake of MiBK decreased from an average value of 0.861 μmol/cm^2/min after 15–75 min of dermal contact to 0.564 μmol/cm^2/min after 75–135 min exposure. The apparent volume of distribution averaged 1.5 L/kg.

2.4.1.3.2 Distribution. The tissue distribution of MiBK and its two primary metabolites was determined in male Sprague–Dawley rats treated for 3 d either by inhalation or oral gavage (48). Oral exposure at dose levels of 1.5, 3, or 6 mmol/kg resulted in MiBK liver concentrations of 1.0–15.6 μg/mL and lung concentrations ranging from

1.0–5.1 μg/mL. Inhalation exposures at 200, 400, or 600 ppm for 4 h resulted in MiBK liver concentrations of 0.7–9.1 μg/mL and lung levels of 4.1–5.3 μg/mL. For both routes of exposure, the tissue levels of 4-hydroxymethyl-2-pentanone were generally lower than for the parent compound, but only slightly. Detectable tissue quantities of 4-methyl-2-pentanol were only observed following the inhalation exposures. All tissue and plasma levels of MiBK and its metabolites were proportional to the exposure level, and there was no evidence of metabolic saturation.

Poulin and Krishnan published *in vitro* tissue/air partition coefficients for MiBK and used the data to develop an algorithm to predict the partition coefficients for untested chemicals (187, 188). The muscle/blood, liver/blood, and fat/blood partition coefficients for MiBK in rat tissues were 0.72, 1.97, and 6.72.

MiBK distribution was also examined in guinea pigs treated by continuous IV infusion at rates up to 3.5 μmol/min/kg (186). The highest infusion rate produced a maximum blood level of 30 μmol/L. Infusion rates of 0.680–0.928 μmol/min/kg resulted in a steady-state blood concentration of 3.94 μmol/L and an average clearance rate of 130 mL/min. The average half-life for MiBK in the blood was 3.29 min.

2.4.1.4 Reproductive and Developmental. Pregnant Fischer-344 rats and CD-1 mice were exposed to MiBK vapor (0, 300, 1000, or 3000 ppm) by inhalation on gestational days 6 through 15 (189). On gestation day 21, the fetuses were collected and examined for external, visceral, and skeletal abnormalities. The 3000-ppm exposure was associated with maternal toxicity and fetotoxicity in both species, but embryotoxicity and fetal malformations were not seen. Effects at 3000 ppm were generally more severe in mice. The 1000- and 300-ppm exposures were not associated with maternal toxicity, embryotoxicity, or fetotoxicity. Teratogenicity was not observed at any exposure level.

Call et al. assessed the effects of MiBK on the growth and development of fathead minnows after continuous exposure of embryos, larvae, and juveniles to concentrations of 57, 105, 169, 246, 418 mg/L for 31–33 d (190). An evaluation of egg hatchability, incidence of developmental abnormalities, survival, and growth indicated that the maximum acceptable toxicant concentration that produced no adverse effects was 77.4 mg/L. An MiBK concentration of 105 mg/L was found to cause a significant reduction in the wet weight of the juvenile fish and a reduction in their standard length.

2.4.1.5 Carcinogenesis. No published studies of carcinogenicity following MiBK exposure were identified.

2.4.1.6 Genetic and Related Cellular Effects Studies. In general, MiBK exposure has not been associated with genotoxicity either *in vitro* or *in vivo*. MiBK was negative in standard Ames/*Salmonella* tests using a preincubation procedure and a plate incorporation method with and without metabolic activation (191, 192). Zeiger et al. reported that MiBK levels ranging from 100 to 6667 μg/plate were not mutagenic to *S. typhimurium* strains TA97, TA98, TA100, and TA1535 when tested with and without metabolic activation in a preincubation-type assay (191). Zimmermann et al. tested MiBK alone and in combination with propionitrile and cold shock for induction of chromosome loss in *Saccharomyces cerevisiae* strain D61.M (51). While chromosome loss was reported with the mixture,

MiBK alone did increase chromosome loss over a range of concentrations that inhibited cell growth or caused lethality (4.8–7.3 mg/mL). Brooks et al. tested MiBK in several mutagenicity assays with and without metabolic activation using methods appropriate for volatile chemicals (34). The following assays gave negative results: *Saccharomyces* mitotic gene conversion assay, rat liver *in vitro* chromosome assay, Chinese hamster ovary *in vitro* chromosome assay, and bacterial cell mutagenicity assays using *S. typhimurium* and *Escherichia coli* WP_1 and WP_2 uvr A.

O'Donoghue et al. reported the results of a battery of tests including: Ames/*Salmonella* assay, L5178Y/$TK^{+/-}$ mouse lymphoma assay, BALB/3T3 cell transformation assay, unscheduled DNA synthesis assay, and an *in vivo* micronucleus assay (148). These assays were conducted with and without metabolic activation as appropriate for the test, and all were conducted under conditions appropriate for volatile chemicals. The Ames/*Salmonella* assay, the unscheduled DNA synthesis assay, and the micronucleus assays were negative. The mouse lymphoma assay gave equivocal results when tested at high concentrations without metabolic activation. The BALB/3T3 assay without activation resulted in an increased number of Type III foci and a reduced cloning efficiency. These results were not duplicated in a repeat assay. Brooks et al. evaluated the effects of MiBK on cell survival in Chinese hamster ovary fibroblast and micronuclei formation in rat lung epithelial cells (193). MiBK levels ranging from 0.01 to 6.25 mM caused a concentration-related decrease in the cloning efficiency of the CHO cells due to cell death. The same levels did not affect micronuclei formation in the lung epithelial cells. The authors also showed that MiBK did not increase the genotoxic effects of ionizing radiation in the lung epithelial cells.

2.4.1.7 Other: Neurological, Pulmonary, Skin Sensitization. The neurotoxic potential of MiBK has been studied in rats, cats, and dogs following repeated exposures (3). Rats were given intraperitoneal injections of MiBK or a mixture of MEK and MiBK (9:1 by volume) five times weekly for 35 wk. The 10-, 30-, and 100-mg/kg dose levels were increased to 20, 60, and 200 mg/kg after 2 wk of treatment. Except for body weight suppression after 3–4 wk of treatment, the only other effect noted was transient anesthesia during the first month of treatment in the 200-mg/kg animals. No toxic neuropathy was noted. Electromyographic examination of dogs administered subcutaneous doses of MiBK (300 mg/kg) daily for 11 mo revealed no evidence of neurotoxicity (3). Spencer and Schaumburg also treated cats subcutaneously with 150 mg/kg MiBK or MEK:MiBK (9:1) twice daily, five times a week for up to 8.5 mo (7). They found no detectable nervous system damage attributable to MiBK. Beagle dogs receiving similar treatments as the cats also did not show neurotoxic changes due to MiBK or the combination of MEK and MiBK (9:1 by volume) (3).

Spencer et al. reported on groups of male rats exposed to 1300 ppm MnBK for 4 mo or 1500 ppm MiBK for 5 mo (6). MnBK produced a toxic distal axonopathy, whereas MiBK exposure was associated with minimal distal axonal changes. However, this may have been due to the 3% MnBK present as a contaminant in the MiBK or, more likely, was associated with compression neuropathy due to the type of cages used in housing the animals (7).

In a series of experiments, domestic hens were exposed to 250 or 1000 ppm of MiBK vapor continuously for 29 (1000 ppm), 30 (250 ppm), or 90 (1000 ppm) d (194–196).

Exposure to 1000 ppm vapor for 29 d resulted in the hens stopping production of eggs (194). Continuous exposure to 1000 ppm for 90 d resulted in a slight reduction in weight gain and transient weakness after 44 d of exposure, but no histopathological evidence of neurotoxicity (196). Exposure to 250 ppm vapor for 30 d was not associated with changes in body weight, or alterations of brain acetylcholinesterase and neuropathy target esterase levels (195); hepatic cytochrome P450 activity was increased. Enhanced P450 levels may also explain the potentiation of MnBK, *n*-hexane, and *O*-ethyl-*O*-4-nitrophenyl phenylphosphonothioate-induced neurotoxicity in association with exposure to MiBK (194–196).

Batyrova reported that 20–30 ppm of MiBK 4 h/d for 4.5 mo caused disturbances in the conditioned reflexes of rats and interfered with the detoxifying function of the liver; the eosinophil count was also elevated (141). Geller et al. reported that rats exposed to 25 ppm MiBK showed a minimal statistical increase in pressor lever response, but the discriminatory behavior of baboons was not impaired by exposures of 20–40 ppm (197). Geller et al. reported delayed behavioral response times in baboons exposed to 50 ppm of MiBK alone, but no alteration of response was seen when MiBK was combined with MEK (100 ppm) (198). Garcia et al. also reported subtle behavior alterations in rats exposed to low levels of MiBK (199). David et al. examined the effects of MiBK on the schedule-controlled operant behavior of male Sprague–Dawley rats exposed 6 h/d at 250, 750, or 1500 ppm for 13 wk (200). Using a multiple fixed ratio/fixed interval schedule of reinforcement, the food-restricted animals showed no significant differences in fixed-ratio run rate, fixed-ratio pause duration, fixed-interval response rate, or index of curvature values at any exposure concentration.

2.4.2 Human Experience

2.4.2.1 General Information. Smyth has stated that the most important manifestation of MiBK toxicity was its narcotic effect, which may be associated with its solubility in lipids of the CNS (201). MiBK itself is normally metabolized by cytochrome P450 enzyme pathways, which may be enhanced when exposures to MiBK exceed the body's ability to clear MiBK from the blood. Enhancement of cytochrome P450 activity is a likely explanation for the potentiation observed when MiBK exposures have been combined with exposures to chemicals that themselves are hepatotoxic or neurotoxic.

The most likely exposure routes for MiBK in the workplace are by inhalation of the vapors and by skin and eye contact. Levels 20 times the recommended TLV (i.e., 1000 ppm) produce central nervous system depression and narcosis. Exposure to 100 ppm appears to present a problem only because of the objectionable odor, whereas exposure to 200–400 ppm is irritating to the eyes, nose, and throat. However, reported industrial exposures show that levels from 100 to 500 ppm may also be associated with symptoms of gastrointestinal effects such as nausea, vomiting, loss of appetite, and diarrhea. MiBK should be handled where adequate ventilation is available. Skin contact should be avoided because the defatting property of this ketone may produce dermatitis. Eye contact may produce painful irritation, but immediate flushing of the eyes with water should be palliative. If exposure to high concentrations of MiBK is likely to occur, proper respiratory protection and protective clothing should be used.

2.4.2.2 Clinical Cases

2.4.2.2.1 Acute Toxicity. Elkins reported that one group of workers exposed to 100 ppm MiBK developed headache and nausea, whereas another group complained only of respiratory tract irritation (202). Tolerance seemed to develop during the work week but was lost over the weekend. Most of these effects were not noted at 20 ppm. Silverman et al. reported that 12 persons exposed to MiBK for 15-min periods indicated that 100 ppm was the highest concentration they considered satisfactory for an 8-h exposure period, and that 200 ppm had an objectionable odor and definitely caused eye irritation (203). Others have reported that 5-min exposures of unconditioned volunteers revealed that the odor threshold for all subjects was $<$ 100 ppm; 200–400 ppm produced eye irritation in 50% of the subjects, and 400 ppm caused nasal irritation in 50% (144). Armeli et al. reported on 19 workers exposed to concentrations of 80–500 ppm MiBK and other chemicals for 30 min/d (204). Common complaints were weakness, loss of appetite, headache, eye and throat irritation, nausea, vomiting, and diarrhea. When examined 5 yr later after improved ventilation reduced exposure levels to 50–105 ppm, these signs were greatly reduced.

May conducted studies that showed that the first perceptible concentration of MiBK detected by a test subject was 8 ppm, whereas 15 ppm was reported as being definitely detectable (205). Wayne and Orcutt directed MiBK concentrations of 5 or 20 ppm at 8–11 subjects wearing eye masks for up to 90 sec and found that MiBK was somewhat more irritating to the eyes than several other solvents tested (206). The time until a subjective irritancy response was obtained from the volunteers was about 23 sec at 5 ppm and 28 sec at 20 ppm. Iregren et al. evaluated heart rate, neurophysiological performance and self-reported symptoms in 6 male and 6 female volunteers exposed to 10 or 200 mg/m^3 for 2 h (207). The subjects were asked to perform light physical exercise (50 W) during the first 90 min of the exposure. There were no statistically significant effects on heart rate, reaction time, or arithmetic ability. An increase was observed in the reporting frequency for symptoms related to sensory irritation and CNS function at both exposure concentrations. Gagnon et al. (77) noted that 4 subjects exposed to 20 or 40 ppm of MiBK for 7 h reportedly experienced headaches and eye, nose, or throat irritation at various times during the exposure. Dalton et al. (208) examined the odor and irritation sensitivity to MiBK in 40 male and female human volunteers exposed for 1–2 sec using squeeze bottles containing a known concentration of MEK. Odor threshold determinations using alternative forced choice techniques produced a mean value of 10 ppm. The sensory irritation threshold for MiBK was found to be 8874 ppm using a lateralization procedure. When a subjective scaling procedure was used to determine the perceived irritation threshold of MiBK, the response was more intense than the corresponding reports obtained with acetone. This difference was subsequently shown to involve a perceptual bias that was associated with the unfamiliar odor of MiBK. These data are consistent with sensory irritation studies in six human volunteers exposed for 7 min at 4–9 concentrations of MiBK (209). The sensory irritation threshold for MiBK was determined to be 340 ppm in volunteers asked to indicate the presence or absence of eye, nose, and throat irritation during the vapor exposure.

Hjelm et al. exposed eight volunteers for 2 h to 10, 100, or 200 mg/m^3 MiBK or to a combination of 100 mg/m^3 MiBK and 150 mg/m^3 toluene at four different times (128).

All participants in this study exercised lightly (50 W) during the ketone exposures. No separate control group or control exposure was included in this study. No more than a third of the participants reported symptoms of irritation to the eyes, nose, or throat or headache, nausea, and vertigo. Assessment of a simple reaction time task, a mental arithmetic test, and a mood scaling test did not reveal any alterations due to the ketone exposures. Dick et al. measured the performance of volunteers exposed for 4 h to 100 ppm MiBK or a combined exposure to 50 ppm MiBK and 100 ppm MEK (210). Five psychomotor tests including choice reaction time, simple reaction time, visual vigilance, auditory tone discrimination and tracking, memory scanning, postural sway, and a profile of mood states did not detect effects directly attributable to either exposure situation. Responses to questionnaires attempting to detect self-reported sensory or irritant effects indicated that approximately 20–30% of the volunteers reported symptoms of odor, headache, nausea, throat irritation, or tearing. Among these symptoms only strong odor was statistically increased, as chemically unexposed volunteers also reported similar symptoms.

2.4.2.2.2 Chronic and Subchronic Toxicity. Two case reports of peripheral neuropathy potentially associated with MiBK exposure have been recorded in the literature. In one case, a young man was exposed to a mixture of solvents during spray paint application (211). The second case was associated with solvent abuse (212). In neither case was there a strong indication that the cases were etiologically related to MiBK exposure.

2.4.2.2.3 Pharmacokinetics, Metabolism, and Mechanisms. In humans breathing 100 ppm MiBK, steady-state breath concentrations were reached after 2 h of exposure (133, 210). Unlike MEK, MiBK blood levels did not increase significantly between the 2- and 4-h exposure periods. When exposure was to 50 ppm MiBK and 100 ppm MEK, MiBK breath levels increased < 1 ppm between the 2- and 4-h exposure points. MiBK was largely cleared from the body by 90 min after a 4-h vapor exposure.

Hjelm et al. studied the kinetics of breath, blood, and urine levels of MiBK in eight volunteers exposed at different times to 10, 100, or 200 mg/m^3 MiBK and to a combination of 200 mg/m^3 MiBK and 150 mg/m^3 toluene for 2 h under conditions of light exercise (128). Levels of metabolites (4-methyl-2-pentanol and 4-hydroxy-4-methyl-2-pentanone) in urine were below detection limits (5 nmol/L) after these exposures. Relative pulmonary uptake of MiBK was 60%. Total uptake of MiBK after 100 ppm ketone exposures was slightly less after the combined exposure with toluene than after exposure to the ketone alone. Ketone blood levels increased rapidly after the onset of exposure and decreased slightly with continued exposure but did not reach a plateau within 2 h. The mean blood clearance of MiBK was approximately 1.6 L/h/kg body weight under all exposure conditions. Total urinary excretion of MiBK was 10.04% of total dose within 3 h post exposure.

Sato and Nakajima reported *in vitro* blood/air and oil/air partition coefficients of 90 and 926, respectively, indicating a high degree of blood and lipid solubility (39). In the blood, MiBK partitions approximately equally between the red blood cells (RBC) and plasma (213). In both RBCs and plasma, MiBK is primarily associated with protein. In RBCs, only 6.4% of the MiBK in the RBCs is distributed to the membrane fraction. MiBK is probably taken up primarily by the hydrophobic sites of blood proteins (213).

2.4.2.2.4 Reproductive and Developmental. No clinical cases reporting reproductive or developmental toxicity following MiBK exposure have been reported.

2.4.2.2.5 Carcinogenesis. No clinical cases of carcinogenicity following MiBK exposure have been reported.

2.4.2.2.6 Genetic and Related Cellular Effects Studies. No cases of genetic toxicity following exposure to MiBK have been reported.

2.4.2.2.7 Other: Neurological, Pulmonary, Skin Sensitization, etc. See Section 2.4.2.2.2 for information about neurotoxicity. No clinical cases of pulmonary toxicity or skin sensitization have been reported.

2.4.2.3 Epidemiology Studies

2.4.2.3.1 Acute Toxicity. The incidence of chronic respiratory symptoms was shown to be higher for children from 42 elementary schools located in a highly industrialized area of West Virginia, where volatile organic compounds were detected in the air (209). Health symptom questionnaires were administered to 4334 children who were exposed to a mixture of 15 different VOCs in the classroom. The maximum airborne concentration of MiBK was 0.90 mg/m^3. Responses from the questionnaire indicated that the children exposed to organic solvents had a higher rate of doctor-diagnosed asthma and a higher incidence of five chronic lower-respiratory symptoms (cough, phlegm, bronchitis, wheezing, and shortness of breath) relative to matched controls. Reports of acute irritancy symptoms (eye, nose, and throat) did not show any relationship to organic vapor exposure.

2.4.2.3.7 Other: Neurological, Pulmonary, Skin Sensitization, etc. Morata et al. examined the effects of organic solvent exposure on hearing using a group of 39 individuals from a paint manufacturing plant (214). Solvents in use at the paint plant included toluene, xylene, benzene, MEK, ethanol, and MiBK. MiBK was present in the air at concentrations ranging from 0 to 20 ppm. Pure tone and intermittence audiometry tests revealed that the prevalence of high-frequency hearing loss in the solvent-exposed group was more than two times higher than in an unexposed control group. Compared to controls, the relative risk of some hearing loss was 5.0 (95% CI, 1.5–17.5) in the solvent-exposed group.

Triebig et al. performed a cross-sectional evaluation of the neurological status in 105 spray painters from ten different plants who were exposed to a variety of aromatic, aliphatic, and oxygenated solvents (215, 216). MiBK was detected in the air from one of the ten worksites at a concentration of 6.4 mg/m^3. The results from neurotoxic symptom questionnaires (e.g., loss of memory and headache), neurologic examinations (e.g., pyramidal and cerebellar signs), psychological examinations (e.g., reaction time and self-rating scales), and cranial computerized axial tomography (CAT scan) did not show any statistically significant effects in the spray painters. Psychiatric examinations did, however, reveal a statistically significant difference in some measures of depression and the loss of mental concentration for the spray painters relative to nonexposed controls. In another study, Triebig et al. reported on the results from a retrospective study of 105 house painters exposed to the solvents in paints and lacquers (217). The workers were exposed to ethyl acetate, toluene, butyl acetate, xylene, ethyl benzene, and MiBK. Personal samplers indicated a maximum 6–8 h MiBK exposure concentration of 11 ppm. Neurophysiological, neuroradiological, and neurobehavioral testing did not reveal any signs of polyneuropathy or encephalopathy in the solvent-exposed painters relative to controls.

Wang et al. looked at the prevalence of acute and chronic neurologic symptoms in 196 workers from 27 companies in China that manufactured or used paints (218). The employees were exposed to eight different aromatic or oxygenated solvents. The 8-h time-weighted average exposures to MiBK ranged from 0 to 68 ppm. No overt signs of neurotoxicity were observed; however, there was a statistically significant increase in the number of acute and chronic symptoms reported on a questionnaire. These included reports of problems such as headache, dizziness, and tiredness. Tsai et al. evaluated 325 workers at 6 paint factories in northern Taiwan for any neurobehavioral effects from solvent exposures (219). The workers were exposed to toluene, xylene, *n*-hexane, *n*-butyl acetate, and acetone. MiBK exposures in the plants ranged from 0 to 36.6 ppm. Eleven neurobehavioral tests were performed on the workers, who were subdivided into white collar and blue collar categories to control for differences in demographic characteristics and socioeconomic status. A statistically significant change was observed in tests that examined either visual attention or perception. For blue collar workers, tests of continuous performance, pattern comparison, and pattern memory were affected, whereas only the continuous performance test was affected in the white collar workers.

Elofsson et al. performed a cross-sectional epidemiology study on 80 spray painters exposed to as many as 17 different solvents at very low levels. MiBK exposure concentrations ranged from 2.1 to 8.4 ppm (220). Psychiatric symptom evaluations were conducted on all subjects along with neurobehavioral performance tests, neurophysiological tests, and neurological tests. Psychometric tests revealed a statistically significant difference between the solvent-exposed painters and the controls for reaction time, manual dexterity, perceptual speed, and short-term memory. Neurophysiological tests also indicated a loss of function in the peripheral and central nervous systems using multiple measures of activity.

Hänninen et al. studied the behavioral effects of long-term exposures to a mixture of organic solvents in a group of 100 automobile painters (221). The workers were exposed to toluene, xylene, butyl acetate, white spirit, isopropanol, ethyl acetate, acetone, ethanol, and MiBK. The mean airborne concentration of MiBK in the workroom air was 1.7 ppm. Fifteen different tests for memory, intelligence, personality, and psychomotor performance were administered to the test subjects and a group of controls. The test results indicated impairments in performance, personality, visual intelligence, verbal memory, and emotional reactivity for the solvent-exposed group relative to controls.

2.5 Standards, Regulations, or Guidelines of Exposure

Hygienic standards for MiBK are listed in Table 76.5.

2.6 Studies on Environmental Impact

MiBK does not possess the chemical and physical properties characteristic of a persistent environmental pollutant. The environmental fate and transport of MiBK has been evaluated following its release to air and water (222). Tropospheric removal will occur by both photolysis and reaction with hydroxyl radicals. Reactions with hydroxyl radicals will predominate in the air and result in a half-life of 27 h. The high vapor pressure of MiBK will result in rapid volatilization from soil; however, migration through soil may also occur to an appreciable degree. MiBK is expected to rapidly volatilize from water with an estimated half-life of 9 h in a model river. MiBK is not expected to adsorb to suspended solids or sediment. Biodegradation will occur to a large degree under both aerobic and anaerobic conditions.

3.0 Mesityl Oxide

3.0.1 CAS Number: *[141-79-7]*

3.0.2 Synonyms: Methyl isobutenyl ketone, 4-methyl-3-penten-2-one, 2-methyl-2-penten-4-one, and isopropylidene acetone, isobutenyl methyl ketone; 1-isobutenyl methyl ketone; 4-methyl-3-penten-2-one; Mesityl oxide, mixt. of alpha- and beta isomers (ca. 93% alpha)

3.0.3 Trade Names: NA

3.0.4 Molecular Weight: 98.14

3.0.5 Molecular Formula: $(CH_3)_2CCHCOCH_3$

3.0.6 Molecular Structure:

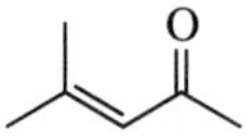

3.1 Chemical and Physical Properties

Mesityl oxide (MO) is a colorless to light yellow, oily liquid that may darken upon standing. It is flammable and may present a moderate fire hazard. The boiling point is 129.6°C, and the melting point is -53°C. Chemical and physical properties of mesityl oxide are listed in Table 76.4.

3.1.1 General

MO can produce marked irritation and transient corneal injury to the eye. Occasional skin contact may produce some irritation; prolonged contact may produce dermatitis and, if the dose is high enough, systemic injury. MO, like many other ketones, can produce narcosis via inhalation exposure. Sublethal concentrations of the vapors may result in vascular congestion, which has been reported to occur primarily in the kidneys. The liver and lung are affected to a lesser degree. Death is generally attributed to its narcotic action.

3.1.2 Odor and Warning Properties

Mesityl oxide has a pungent odor that has been described as peppermint, spearmint, or honeylike. The absolute perception limit has been reported to be 0.017 ppm (223). The penetrating odor and eye and nose irritation that occurs following exposure at low levels should prevent overexposure.

3.2 Production and Use

Mesityl oxide has been used as a solvent for cellulose resins, gums, rubbers, vinyl resins, inks, cleaners, lacquers, varnishes, and enamels. Current use is as a site-limited chemical intermediate in the manufacture of solvents, particularly methyl isobutyl ketone (MIBK), although a small amount of MO has been imported into the United States in recent years for use as an inert in a herbicide. All other known uses are industrial. Outside the United States, MO uses include the manufacture of drying oils, drugs, and plasticisers.

3.3 Exposure Assessment

3.3.1 Air

Mesityl oxide may be released to the air during manufacture of MIBK, or in the formulation of various drugs, insecticides, resins, inks, or cleaners. It has been shown to occur at a concentration of up to 1.5 ppm in gasoline exhaust (224). Atmospheric MO will be degraded rapidly, with an estimated half-life of 4.8 h, by reaction with photochemically produced hydroxyl radicals (225).

3.3.2 Background Levels

Background levels of MO are expected to be low. MO is estimated to exist entirely in the vapor phase in ambient atmospheres. Vapor-phase MO is rapidly degraded by reaction with photochemically formed hydroxyl radicals. Volatilization from water is expected to occur over a period of days. MO can also be expected to evaporate from dry soil surfaces. Bioconcentration and adsorption to sediment are not expected to be significant. The estimated K_{oc} of 15.3 suggests high mobility and leaching in soil (226). Low levels of MO have been measured in baked potatoes (227) and nectarines (228).

3.3.3 Workplace Methods

MO is an intermediate in the production of solvents, particularly MIBK. In this application, MO is not isolated, but is converted directly to the final product in closed systems. No specific workplace monitoring methods have been defined, largely due to the low probability of exposure. In the event of a spill, its penetrating odor and eye and nose irritation, which occur following exposure to even low levels, should prevent overexposure. Workplace monitoring can be accomplished through the use of NIOSH Method 1301. This method uses gas chromatography with flame ionization detection following elution from solid sorbent tubes with isopropyl alcohol (116).

3.4 Toxic Effects

3.4.1 Experimental Studies

3.4.1.1 Acute Toxicity. The acute oral LD_{50} has been reported to be 1120 (154) or 655 mg/kg (229) for the rat, 923 mg/kg for the mouse (229), 1000 mg/kg for the rabbit

(144), and 800 mg/kg for the guinea pig (154). The single-dose intraperitoneal LD_{50} was 354 mg/kg for mice (144).

MO has been reported to cause severe irritation when 0.5 mL was placed in a rabbit eye (230). A similar reaction has been reported following instillation of 20 mg into rabbit eyes (231, 232).

When applied to the uncovered skin of rabbits, 430 mg of MO produced only mild irritation. The acute dermal LD_{50} was reported to be 5.99 mL/kg for the rabbit (154). When doses of 0.5 mL were dropped on the skin of mice, marked irritation occurred within a few minutes and the animals became ataxic and narcotized within 15 min. This dose killed all ten animals in 3–9 h. Applications of 0.1 mL to the backs of mice produced local irritation and excitement within 5 min; one of 10 animals died in 12 h, but the others recovered (233). Repeated contact with MO may cause dermatitis because of its defatting properties (144). The 14-d dermal LD_{50} for guinea pigs is reported to be 1.9 g/kg. However, the conditions of exposure may have caused an underestimation of the LD_{50}. Gross pathology lesions included mottled liver, pale spleen and kidneys, and congestion of the gastrointestinal tract (139).

Vapor concentrations of 2300, 5000, or 10,000 ppm for up to 8 h produced a dose-dependent narcotic response in guinea pigs. Clinical signs of toxicity included respiratory irritation, hypothermia, reduction in respiratory and heart rates, loss of reflexes, coma, and death at all exposure levels (25). Smyth et al. (234) found that 12,000 ppm MO killed rats and guinea pigs after 1 h of exposure. Eight-hour exposures of rats to 2500, 1000, and 500 ppm killed 100, 68, and 30% of the test animals, respectively. Hart et al. (233) exposed mice to concentrations of 6000–24,000 ppm MO in air. Clinical signs of toxicity were ocular and nasal irritation, labored breathing, convulsions, narcosis, vasodilation, cyanosis, and death. The time to death was concentration dependent, and ranged from 23 to 135 min. Carpenter et al. (235) subjected groups of six male or female rats to acute 4-h exposures of increasing concentrations of MO until two to four of the animals died within a 14-d observation period. They found that a concentration of 1000 ppm was necessary to produce this mortality. Other investigators determined the 4-h LC_{50} in rats and mice to be 1130 and 2000–4000 ppm, respectively (229). Respiratory depression to 50% of the normal frequency in rats was obtained at a concentration of 61 ppm (164). Inhalation of from 347 to 540 ppm of MO for 4 h resulted in significantly decreased duration of the immobility phase (to 31–67%, respectively) in a "behavioral despair" swimming test. The differences between the exposed and control rats were statistically significant at all exposure concentrations (164). Rabbits exposed to 13,000 ppm MO for 30 or 90 min exhibited only eye and nose irritation (233).

Hart et al. (233) also exposed mice and rabbits repeatedly to 13,000 ppm MO. Five daily exposures of 15 min each produced no deaths in mice, whereas 11 exposures killed 3/10 animals. Increasing the exposure time to 30 min/d killed all 10 mice within 6 d. Rabbits exposed to 13,000 ppm MO for 30 min/d showed only eye and nose irritation after 15 d, whereas exposures of 60 min duration produced paralysis and death within 21 d. Smyth et al. (234) exposed 10 male rats and 10 guinea pigs of both sexes to MO at concentrations of 50, 100, 250, and 500 ppm for 8 h/d, 5 d/wk for 6 wk. No deaths occurred in the 50-, 100-, or 250-ppm exposed groups. The 500-ppm group was terminated after 10 d due to high mortality (65%). Pathologic changes, observed at all concentrations except 50 ppm, included poor growth, congestion of the liver, dilated Bowman's capsules and swollen convoluted tubular epithelium in the kidney, and often congestion of the lungs. The frequency and intensity of these changes were concentration dependent.

3.4.1.2 Chronic and Subchronic Toxicity. When rats were offered drinking water containing 5000 or 10,000 ppm MO for 14 d, water consumption was markedly reduced and animals lost weight (154).

Rats exposed to 300 ppm MO for 2 h/d for 30 d showed leucocytosis and hypertrophy of the liver, kidney, and spleen (233). When male and female rats were exposed for up to 49 exposures, 6 h/d, 7 d/wk, to 31, 103, or 302 ppm MO in a "Combined Repeat Dose and Reproductive/Developmental Toxicity Screening Test" conducted under OECD guideline 422, no mortality was observed. Exposure to MO caused an exposure-dependent reduction in feed consumption and body weight in both male and female rats. The irritating odor of MO was indicated by postexposure sialorrhea and porphyrin nasal discharge. No significant exposure-related organ weight changes or clinical chemistry alterations were associated with exposure. Histopathologic examination revealed exposure-related increases in sero-cellular exudate, chronic inflammation, and focal metaplasia of the respiratory and olfactory epithelium of the nasal passages (236). Leucopenia without any change in differential or red blood cell counts was observed following exposure for 4 h to either 86 or 130 ppm MO (237), but the effect on leukocyte count was not seen in adrenalectomized, MO-exposed rats.

3.4.1.3 Pharmacokinetics, Metabolism, and Mechanisms. Information on the metabolism of MO is fragmentary. It has been reported that metabolism occurs through formation of a glucuronide (238).

3.4.1.4 Reproductive and Developmental. The effect of MO on reproductive toxicity in rats by the inhalation route has recently been conducted using the Combined Repeat Dose and Reproductive/Developmental Toxicity Screening Test detailed in OECD Guideline 422 (236). Animals were exposed to 31, 103, or 302 ppm of the test substance during premating, mating, gestation, and lactation phases for a total of up to 49 exposures. Fourteen days of exposure to 302 ppm MO reduced the number of litters produced by the mating pairs. Reproductive performance of male rats exposed to 302 ppm of the test material for a total of 42 d and bred to nonexposed females was comparable to control values. The number of litters produced by mating pairs exposed to 103 or 31 ppm MO was unaffected by treatment. The NOEL for reproductive toxicity was 103 ppm under the test conditions employed (236).

3.4.1.5 Carcinogenesis. No carcinogenicity studies are available in the published literature on MO.

3.4.1.6 Genetic and Related Cellular Effects Studies. MO was negative in the Ames/*Salmonella* assay in the presence and absence of S-9 fraction, and was also negative in a mouse micronucleus assay (236). In a series of studies designed to test the effect of chlorinated water on the mutagenicity of MO, the addition of submillimolar quantities of

MO to aqueous solutions of hypochlorite at pH 8.5 produced a mixture that was positive in the Ames test. The specific mutagenic species was not identified in these test (239).

3.4.2 Human Experience

3.4.2.2 Clinical Cases

3.4.2.2.1 Acute Toxicity. Silverman et al. (203) reported that 25 ppm MO produced eye irritation in humans, and 50 ppm also produced nasal irritation and an objectionable odor and taste. The subjects exposed to MO reported that the bad taste lasted 3–6 h. Other studies using unconditioned subjects revealed that, when exposed for 5 min, half of the people could detect the odor of MO at 12 ppm and all detected it at 25 ppm. Exposure to 50 to 100 ppm caused eye irritation and half of the test subjects experienced nasal irritation and pulmonary discomfort at 25 ppm (238). Ito (240) reported the possibility of anemia and leucopenia in workers due to MO exposure.

3.4.2.2.2 Chronic and Subchronic Toxicity. No case reports of subchronic or chronic toxicity to humans resulting from the industrial handling of MO were found. Brief skin exposure may produce irritation, and prolonged or repeated contact may result in dermatitis. Systemic injury may occur due to percutaneous absorption. However, MO has a distinct odor; its irritant properties at concentrations below the NIOSH recommended exposure limit should prevent overexposure.

3.4.2.2.3 Pharmacokinetics, Metabolism, and Mechanisms. There are no case reports on pharmacokinetics, metabolism, or mechanisms in humans following exposure to MO.

3.4.2.2.4 Reproductive and Developmental. There are no case reports of reproductive or developmental toxicity in humans following exposure to MO.

3.4.2.2.5 Carcinogenesis. There are no case studies of carcinogenicity in humans following exposure to MO.

3.4.2.2.6 Genetic and Related Cellular Effects Studies. There are no case studies of genetic toxicity in humans following exposure to MO.

3.4.2.2.7 Other: Neurological, Pulmonary, Skin Sensitization, etc. There are no case reports of neurotoxicity, pulmonary toxicity, or skin sensitization following exposure to MO.

3.4.2.3 Epidemiology Studies. No epidemiology studies following exposure to MO have been published.

3.5 Standards, Regulations, or Guidelines of Exposure

Hygienic standards for MO are listed in Table 76.5.

3.6 Studies on Environmental Impact

There are no published environmental impact studies for MO. MO is not expected to present any unusual hazards to the environment. Spills of MO in the water environment may result in acute toxicity; however, no such effects have been observed. MO is not expected to be persistent in the environment; therefore, no long-term environmental impacts are expected.

4.0 4-Hydroxy-4-methyl-2-pentanone

4.0.1 CAS Number: *[123-42-2]*

4.0.2 Synonyms: Diacetone alcohol, 2-methyl-2-pentanol-4-one, and 2-hydroxy-2-methyl-4-pentanone, 4-hydroxy-2-keto-4-methylpentane; acetonyldimethyl carbinol; diketone alcohol; 4-methyl-4-hydroxy-2-pentanone; Tyranton; diacetone; 4-hydroxy-4-methylpentan-2-one

4.0.3 Trade Names: NA

4.0.4 Molecular Weight: 116.16.

4.0.5 Molecular Formula: $CH_3COCH_2C(CH_3)_2OH$

4.0.6 Molecular Structure:

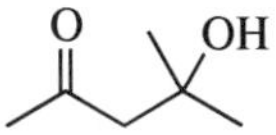

4.1 Chemical and Physical Properties

4-Hydroxy-4-methyl-2-pentanone is a colorless, flammable liquid that becomes yellow with age. The boiling point is 168°C, and the melting point is −43°C. Additional physical–chemical properties are summarized in Table 76.4.

4.1.1 General

4-Hydroxy-4-methyl-2-pentanone has a low degree of oral toxicity. It is only slightly irritating to the skin, but prolonged contact may defat the skin and cause dermatitis. Eye contact may produce moderate to severe irritation. 4-Hydroxy-4-methyl-2-pentanone is slightly toxic by dermal exposure. High concentrations of the vapor may produce narcosis and systemic injury. Eye, nose, and throat irritation occur at levels well below those required to produce systemic injury.

4.1.2 Odor and Warning Properties

4-Hydroxy-4-methyl-2-pentanone has an absolute odor threshold of about 0.28 ppm. Fifty percent odor recognition has been reported at 1.1 ppm, and 100% recognition at 1.7 ppm. It has a characteristically pleasant odor that has been described as sweet to mintlike (223).

4.2 Production and Use

4-Hydroxy-4-methyl-2-pentanone is manufactured through the action of barium hydroxide, potassium hydroxide, or calcium hydroxide on acetone. Commercial materials may contain up to 15% acetone. Major uses include the manufacture of mesityl oxide and hexylene glycol. It has also been reported as a solvent for cellulosic products, fats, oils, and waxes, and as a component of antifreeze and hydraulic fluids. Additional uses include coatings, ink formulations, and adhesives.

4.3 Exposure Assessment

4.3.1 Air

Exposure to vapors of 4-hydroxy-4-methyl-2-pentanone is the most likely route of human contact. The time-weighted average concentration of the compound measured in the breathing zone of workers in a screen printing plant ranged up to 14 ppm in printing press operations. The concentrations were somewhat lower in the general air of the workplace except in the manual dryer operations. The overall average concentration found in the general air of the plant was 3.2 ppm (241).

4.3.2 Background Levels

Information on background levels of 4-hydroxy-4-methyl-2-pentanone has not been published.

4.3.3 Workplace Methods

Workplace monitoring can be accomplished through the use of NIOSH Method 1402. In this method, airborne vapors are collected on solid sorbent tubes; materials eluted are measured through the use of gas chromatography with flame ion detection. Alternatively, a field test for ketones exists in which vapor can be collected in water and allowed to react with 2,4-dinitrophenylhydrazine, which is subsequently reacted with methanolic potassium hydroxide, which results in the formation of quantifiable red coloration (120). Exposure to this substance outside the workplace is unlikely; exposure could occur via ingestion of contaminated drinking water or from ingestion of foods contaminated due to contact with surfaces freshly painted with acetone-containing paints (242).

4.3.4 Community Methods

Specific methods for analysis of environmental samples have not been published.

4.3.5 Biomonitoring/Biomarkers

Biomonitoring and biomarker studies have not been published for this ketone.

4.4 Toxic Effects

4.4.1 Experimental Studies

4.4.1.1 Acute Toxicity. The oral LD_{50} of 4-hydroxy-4-methyl-2-pentanone has been reported to be 4.0 g/kg for rats (243). Other investigators (144, 244) reported that 2 mL/kg given orally produced transient liver damage in rats, and that 4 mL/kg was lethal. A dose of 2 mL/kg was reported to cause destruction of circulating erythrocytes (244). Single doses of 2.4–4.0 mL/kg produced narcosis in rabbits, while 5 mL/kg caused death (245). The intraperitoneal LD_{50} in mice was reported to be 933 mg/kg (144). Intravenous injections of 1.0–1.5 mL/kg produced narcosis in rabbits, and 3.25 mg/kg was the lowest dose to produce lethality. The lowest intramuscular dose that produced lethality in rabbits was between 3 and 4 mL/kg (245).

Undiluted 4-hydroxy-4-methyl-2-pentanone has been reported to produce significant eye irritation and transient corneal damage in the rabbit (230).

At most, only mild skin irritation was produced by application of up to 500 mg of undiluted 4-hydroxy-4-methyl-2-pentanone to the skin of rabbits (144, 154, 243). The acute dermal LD_{50} was 14.5 mL/kg for rabbits (243).

Rats survived 8-h exposures to atmospheres containing 1500 ppm 4-hydroxy-4-methyl-2-pentanone (243). Mice, rats, rabbits, and cats exposed to 2100 ppm for 1–3 h exhibited restlessness, irritation of the mucous membranes, excitement, and later, somnolence (246). Effects noted following exposure for 6 h/d, 5 d/wk for 6 wk to 973 ppm 4-hydroxy-4-methyl-2-pentanone included slight increases in kidney and liver weights, and histopathologic changes in the kidneys. Increases in liver weight were also seen in the 224-ppm group, but no effect was seen at 50 ppm (247, 248).

4.4.1.2 Chronic and Subchronic Toxicity. Renal effects were seen in rats ingesting 40 mg/kg/d of 4-hydroxy-4-methyl-2-pentanone in the drinking water for 30 d. No effects were seen in rats ingesting 10 mg/kg/d (243). Oral administration of 2 mL daily for 12 d produced narcosis, kidney damage, and death in three of four rabbits (245). Repeated subcutaneous injections of 0.08 mL of 4-hydroxy-4-methyl-2-pentanone produced narcosis followed by recovery in rats (245).

4.4.1.3 Pharmacokinetics, Metabolism, and Mechanisms. In a study designed to assess the effect of metabolites of methyl *n*-butyl ketone and methyl isobutyl ketone on ethanol-induced loss of righting reflex in mice, 4-hydroxy-4-methyl-2-pentanone was administered IP to animals 30 min prior to administration of ethanol. Ethanol-induced loss of righting reflex was significantly prolonged by 2.5 mmol/g, but the rate of ethanol elimination was unaffected (166).

4.4.1.4 Reproductive and Developmental. No reproductive or development studies were found in the literature for this substance.

4.4.1.5 Carcinogenesis. No carcinogenicity studies were found in the literature for this substance.

4.4.1.6 Genetic and Related Cellular Effects Studies. 4-Hydroxy-4-methyl-2-pentanone was negative in both a bacterial mutation assay (Ames test) and a yeast mitotic gene conversion assay (34). However, in the rat liver chromosome assay, a small increase in chromatid damage was observed in the concentration range just below that causing growth inhibition of the cells. The effect, though marginal and not quantitatively dose related, was observed in two separate experiments (34).

4.4.2 Human Experience

The primary systemic effect of 4-hydroxy-4-methyl-2-pentanone is narcosis. High concentrations produce irritation and pulmonary discomfort. Death may result from respiratory failure due to the depression of the respiratory center. Additional effects include renal and hepatic injury and decreased blood pressure.

Occupational exposure to 4-hydroxy-4-methyl-2-pentanone is most likely to be by inhalation and skin contact. It presents a low degree of hazard if good work practices are observed. Appropriate protective clothing and eye protection should be made available as prolonged exposure may defat the skin and cause dermatitis. The occurrence of eye, nose, and throat irritation and a recognizable odor at low concentrations should protect against overexposure to 4-hydroxy-4-methyl-2-pentanone.

4.4.2.2 Clinical Cases

4.4.2.2.1 Acute Toxicity. There is a single report of the development of a nephrotic syndrome in a 59-year-old man 40 d after a 3-d exposure to diacetone alcohol and ethanol paint solvents. In this case, renal biopsy gave evidence of a subacute proliferative glomerulonephritis (249).

4.4.2.2.2 Chronic and Subchronic Toxicity. There are no case reports of chronic or subchronic toxicity in humans following exposure to 4-hydroxy-4-methyl-2-pentanone.

4.4.2.2.3 Pharmacokinetics, Metabolism, and Mechanisms. There are no case reports on pharmacokinetics, metabolism, or mechanisms in humans following exposure to 4-hydroxy-4-methyl-2-pentanone.

4.4.2.2.4 Reproductive and Developmental. There are no case reports of reproductive or developmental toxicity in humans following exposure to 4-hydroxy-4-methyl-2-pentanone.

4.4.2.2.5 Carcinogenesis. There are no case studies of carcinogenicity in humans following exposure to 4-hydroxy-4-methyl-2-pentanone.

4.4.2.2.6 Genetic and Related Cellular Effects Studies. There are no case studies of genetic toxicity in humans following exposure to 4-hydroxy-4-methyl-2-pentanone.

4.4.2.2.7 Other: Neurological, Pulmonary, Skin Sensitization, etc. There are no case reports of neurotoxicity, pulmonary toxicity, or skin sensitization following exposure to 4-hydroxy-4-methyl-2-pentanone.

4.4.2.3 Epidemiology Studies. Controlled studies in humans reported by Silverman et al. revealed that 100 ppm 4-hydroxy-4-methyl-2-pentanone produced eye, nose, and throat irritation (203). This level of exposure also produced a bad taste and an objectionable odor. Nasal irritation occurred only at levels greater than 100 ppm. Other investigators (144) reported that humans exposed to 400 ppm for 15 min experienced pulmonary discomfort and eye, nose, and throat irritation. A level of 100 ppm produced the same results seen by Silverman et al. (203).

No epidemiological studies have been published following exposure to 4-hydroxy-4-methyl-2-pentanone.

4.5 Standards, Regulations, or Guidelines of Exposure

Hygienic standards for 4-hydroxy-4-methyl-2-pentanone are listed in Table 76.5.

4.6 Studies on Environmental Impact

There are no published environmental impact studies for 4-hydroxy-4-methyl-2-pentanone. 4-Hydroxy-4-methyl-2-pentanone is not expected to present any unusual hazards to the environment. 4-Hydroxy-4-methyl-2-pentanone is not expected to be persistent in the environment; therefore, no long-term environmental impacts are expected.

5.0 2,5-Hexanedione

5.0.1 CAS Number: *[110-13-4]*

5.0.2 Synonyms: Acetonyl acetone, α,β-diacetyl ethane, 2,5-diketohexane, 1,2-diacetylethane, and diacetonyl

5.0.3 Trade Names: NA

5.0.4 Molecular Weight: 114.14

5.0.5 Molecular Formula: $CH_3COCH_2CH_2COCH_3$

5.0.6 Molecular Structure:

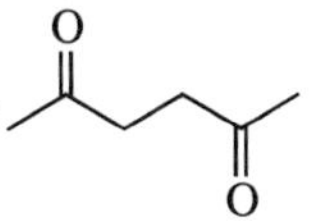

5.1 Chemical and Physical Properties

5.1.1 General

2,5-Hexanedione (2,5-HD) is a clear liquid. It is a relatively reactive chemical and has an autoignition temperature of 493°C. Its boiling point is 194.0°C and its melting point is – 5.5°C. Additional physical–chemical properties are summarized in Table 76.4.

5.1.2 Odor and Warning Properties

2,5-HD has a sweet aromatic odor.

5.2 Production and Use

No information on the production and use of 2,5-hexanedione was identified.

2,5-Hexanedione has been studied extensively as a model neurotoxicant, and it is the most active neurotoxic metabolite of *n*-hexane and methyl *n*-butyl ketone. Use of this material, other than as a research material, appears to be low.

5.3 Exposure Assessment

5.3.1 Air

No specific information on levels of 2,5-hexanedione in ambient air was identified.

5.3.2 Background Levels

Information on background levels of 2,5-hexanedione in the environment were not identified.

2,5-Hexanedione was detected in the urine of 12 people who had not been exposed to it or *n*-hexane (250). The amount of 2,5-hexanedione in the urine depended on the pH used for acid hydrolysis of the urine samples. The 2,5-hexanedione concentration in the urine ranged from 0.12 to 0.78 mg/L (0.45 ± 0.20 mg/L; mean ± SD). A reference value for excretion of 2,5-hexanedione for the Italian general population was reported to be 0.795 mg/L for men and 0.627 mg/L for women (251). Background levels of 2,5-hexanedione in blood of unexposed individuals was 6–30 μg/L (252). Low levels of 2,5-hexanedione in the blood and urine are likely produced by endogenous metabolism of hexacarbons or substances metabolized to hexacarbons.

5.3.3 Workplace Methods

Specific methods for analysis of 2,5-hexanedione in the workplace were not identified.

5.3.4 Community Methods

Specific methods for analysis of 2,5-hexanedione in the environment were not identified.

5.3.5 Biomonitoring/Biomarkers

Occupational exposure to 2,5-hexanedione appears to be minimal; hence biomonitoring methods and biomarkers associated with exposure to 2,5-hexanedione itself have not been developed. Based on the work of DiVincenzo et al. (41), which showed that 2,5-hexanedione was a metabolite of *n*-hexane and methyl *n*-butyl ketone, and the suggestion by Perbellini et al. (253) that it might be a useful biomarker for exposure to *n*-hexane, 2,5-hexanedione itself has become an important biomarker for *n*-hexane exposure.

5.3.5.1 Blood. Sensitive techniques for the analysis of 2,5-hexanedione in blood are available. Perbellini et al. (252) reported that the serum concentration of total and free 2,5-hexanedione in workers exposed to *n*-hexane ranged from 33 to 418 and 14 to 283 μg/L, respectively. For subjects not exposed to *n*-hexane, 2,5-hexanedione blood concentrations were 6–30 μg/L. However, relatively little work has been done using blood levels of 2,5-hexanedione as a biomarker. The principal reasons for this situation appear to be the good correlation between urine 2,5-hexanedione levels and levels of *n*-hexane exposure and the fact that urine measurement is convenient and noninvasive.

5.3.5.2 Urine. Urine level of 2,5-hexanedione has become a well-accepted standard as a biomarker for exposure to *n*-hexane (254–257). The American Conference of Governmental Industrial Hygienists recommends a biological exposure index of 5 mg/L 2,5-hexanedione in urine.

In a review of several cases of *n*-hexane toxicity, Takeuchi (258) suggested that workplace exposure to 40 ppm *n*-hexane was equivalent to 2.2 mg/L of 2,5-hexanedione in the urine.

In addition to being a biomarker of exposure to *n*-hexane, 2,5-hexanedione levels have shown significant correlations with mild electrophysiologic changes in peripheral nerves and skeletal muscle, suggesting that urinary 2,5-hexanedione levels can be used as a biomarker of effect (259, 260).

Levels of 2,5-hexanedione in the urine from *n*-hexane can be affected by coexposure to other solvents. For example, toluene coexposure tends to lower 2,5-hexanedione levels, while MEK has been reported to both increase and decrease urinary 2,5-hexanedione levels (261, 262).

5.3.5.3 Other. Pyrrole adducts in hair have been proposed as a biomarker for 2,5-hexanedione and *n*-hexane (21). Pyrrole adducts may serve as an indicator of chronic exposure to *n*-hexane, since the adducts are stable and hair is a less labile sample source compared to urine.

Polymorphonuclear (PMN) leukocyte chemotaxis has been studied as a biomarker to effect following *n*-hexane exposure. Inhibition of chemotaxis was observed in the PMNs of some *n*-hexane–exposed subjects, suggesting that biologic effects may be seen at 2,5-hexanedione levels below the ACGIH biological exposure index of 5 mg/L (263).

5.4 Toxic Effects

5.4.1 Experimental Studies

Acute exposures to 2,5-hexanedione pose a low hazard, but chronic exposures may produce severe neurologic damage and testicular germinal cell damage. There is also the possibility that prolonged systemic exposure can effect immunocompetence. 2,5-Hexanedione may produce moderate to marked eye irritation when applied directly to the eye. Repeated skin contact may produce immunologically mediated skin sensitization and dermatitis. Acute toxicity by inhalation of 2,5-hexanedione is unlikely because its irritant properties should preclude significant exposures. Repeated skin and inhalation exposure may lead to neurologic damage.

5.4.1.1 Acute Toxicity. Smyth and Carpenter (264) reported an approximate oral LD_{50} of 2.7 g/kg for rats. Other studies (3) report the approximate oral LD_{50} to be 1600 mg/kg for male rats and to be between 1600 and 3200 mg/kg for male mice. More recent studies in female rats and male mice indicate an oral LD_{50} of 2.14 mL/kg for rats (265) and 4.90 mL/kg in mice (266). Treatment of rats with approximately 1 mL/kg produced a decrease in thymus weight and size, and an increase in spleen weight. Hematological parameters were unchanged. In mice, thymic weights were increased, but spleen weights were decreased following a single dose of approximately 1 mL/kg 2,5-hexanedione. In addition, delayed-type hypersensitivity and phagocytic activity were suppressed. When 2,5-hexanedione was administered intraperitoneally, the LD_{50} was approximately 800 mg/kg for rats and 1600 mg/kg for mice (3).

When placed on the cornea of rabbits, 2,5-hexanedione produced moderate to marked eye irritation with transient corneal injury (230).

Skin irritation was slight when 5–20 mL/kg was applied to the abdominal skin of guinea pigs under an occlusive wrap for 24 h. Systemic toxicity due to percutaneous absorption was not evident, but the compound had a tanning effect and stained the skin (3). Repeated application to the uncovered, shaved backs of guinea pigs was not irritating (3). In a standardized skin sensitization test, four of five guinea pigs developed a weak positive skin sensitization reaction (3). Smyth and Carpenter (264) reported a dermal LD_{50} of 6.6 mL/kg for guinea pigs.

Smyth and Carpenter (264) exposed rats to a saturated (1700 ppm calculated) atmosphere for a maximum of 1 h and did not produce deaths. Specht et al. (25) stated that guinea pigs exposed to a saturated atmosphere at 25°C (400 ppm) showed a slight decrease in respiratory rate and slight nasal and ocular irritation. Five male rats exposed to 650 ppm 2,5-hexanedione for four 16-h periods and three 20-h periods over 10 d developed orange-brown staining of the hair and lost body weight. Limited histopathologic examination showed testicular germinal cell degeneration but no neuropathologic changes (3).

5.4.1.2 Chronic and Subchronic Toxicity. Spencer and Schaumburg (267) administered 2,5-hexanedione by subcutaneous injection to six rats, 5 d/wk for 19–23 wk at a dose of 97–388 mg/kg/d. The mean dose level was 340 mg/kg/d. Axonal degeneration similar to the "giant" axonal degeneration associated with MnBK was found in the peripheral nerves and medulla oblongata. O'Donoghue et al. (10) exposed groups of 10 male rats to 0.25, 0.5, and 1.0% (v/v) of 2,5-hexanedione in the drinking water for up to 165 d. Mean calculated doses based on water consumption and body weight were 240, 450, and 750 mg/kg/d, respectively. Seven of 10 rats ingesting the 1.0% solution died between 14 and 43 d. All survivors showed clinical neurologic defects in the hind limbs by the forty-sixth day of exposure. Hind limb weakness was evident in the 0.5% group at 52 d, and in the 0.25% group at 101 d. Neuropathologic changes identical to those produced by MnBK were seen in the cerebellum, medulla oblongata, spinal cord, and peripheral nerves. Additional toxic responses included a dose-dependent body weight loss or reduced body weight gain, testicular germinal cell degeneration in all test groups, and hypocellularity of the bone marrow in the 1.0 and 0.5% groups. Intraperitoneal injections of rats 5 d/wk up to 7.5 mo produced axonal damage in the hind limb nerves and the medulla oblongata in 1.5 mo. The morphologic damage in the peripheral nerve progressed with time but not to an extent that produced clinical signs of neurotoxicity. Testicular atrophy was not observed.

Treatment of rats with 1% 2,5-hexanedione in the drinking water for 5 wk reduced the number of rods in the retina of treated animals (268). Recovery for 13 wk did not improve the conditions and all rods were lost while only 50% of the cones were lost. Four cats were exposed to 0.5% 2,5-hexanedione in the drinking water for up to 136 d to examine the effects on the visual system (269). Hind limb weakness was evident in 60–75 d; followed by quadriparesis. Electron microscopy revealed widespread axonal degeneration in the optic tracts, mammillary bodies, lateral geniculate nucleus, and superior colliculus, indicating that chronic exposure to relatively high levels of 2,5-hexanedione can result in degeneration in the visual pathways. Treatment of cats with 0.5% 2,5-hexanedione in the drinking water for 38 consecutive days selectively damaged the large retinal ganglion cell with loss of flicker resolution but retention of visual acuity (270). The reduction in large ganglion cells persisted through 8 mo of recovery. Monkeys orally dosed with 73 mg/kg/d for 15–17 wk lost contrast sensitivity, but not necessarily visual acuity (271). Histopathologic examination indicated swollen axons typical of 2,5-hexanedione neurotoxicity. This condition was reversible after 5 mo of recovery. Auditory brain stem responses were reduced in rats given 350 mg/kg 2,5-hexanedione, 5 d/wk for 3 wk (272).

Dogs exposed to approximately 110 mg/kg 2,5-hexanedione developed a neurotoxic syndrome similar to that seen in cats and rats (273). Powell et al. (274) emphasized that Schwann cell damage may occur *pari passu* with axonal damage. Misumi et al. (28) demonstrated that changes in nerve conduction velocity coincide with neuropathologic abnormalities.

Krasavage et al. (1), in a comparative study with 2,5-hexanedione and other metabolites of *n*-hexane and MnBK, showed that the ability of *n*-hexane and MnBK to produce neurotoxicity was related to their quantitative conversion to 2,5-hexanedione. 2,5-Hexanedione was 3.3 times as neurotoxic as methyl *n*-butyl ketone and 38 times as neurotoxic as *n*-hexane.

Compared to other diketones, such as 2,5-heptanedione and 3,6-octanedione, 2,5-hexanedione produced neurotoxicity at lower doses (9).

5.4.1.3 Pharmacokinetics, Metabolism, and Mechanisms

5.4.1.3.1 Absorption. Based on observed effects, 2,5-hexanedione is well absorbed via the oral route. 2,5-Hexanedione is also absorbed through the skin (275). The half-life of 2,5-hexanedione applied to the skin was 6 h (275).

5.4.1.3.2 Distribution. A limited amount of data is available on the distribution of 2,5-hexanedione. Based on observed effects, it is clearly widely distributed in the body. Following application of a dermal dose of 50 mg/kg radiolabeled 2,5-hexanedione, the liver and kidneys contained the highest tissue levels, while the brain, spinal cord, and peripheral nerves had lower concentrations (275).

5.4.1.3.3 Excretion. A limited amount of work has been done studying excretion pathways with 2,5-hexanedione. However, it is clear from studies of *n*-hexane and MnBK that 2,5-hexanedione is excreted in the urine following its formation. Application of 50 mg/kg of radiolabeled 2,5-hexanedione to the skin of the back of hens resulted in excretion of 35% of the radiolabel via exhalation, as volatile material, largely 2,5-hexanedione, 11.9% as CO_2, and 15% as combined urine and feces (275). The highest concentration of radiolabel was in the bile.

5.4.1.3.4 Metabolism. The pharmacokinetics of 2,5-hexanedione have been studied primarily in relation to the metabolism of *n*-hexane, MnBK, and their metabolites. In each of these cases the effects produced by the parent compound depend on the amount of 2,5-hexanedione produced when they are metabolized (1). DiVincenzo et al. (41) showed that guinea pigs given an intraperitoneal injection of 450 mg/kg 2,5-hexanedione converted a fraction of the dose to 5-hydroxy-2-hexanone and that both compounds were cleared from the serum in 16 h. The serum half-life of 2,5-hexanedione was 100 min. Rats given 0.5% 2,5-hexanedione in the drinking water for 49 d had a mean 2,5-hexanedione serum level of 17.4 μg/mL (0–74.05 μg/mL). A level of 4.49 μg/mL 5-hydroxy-2-hexanone was detected in the serum of one animal. A concentration of 0.25% 2,5-hexanedione in the drinking water did not produce detectable serum levels ($\geq$ 0.1 μg/mL) of 2,5-hexanedione, 5-hydroxy-2-hexanone, or 2,5-hexanediol in the rats (10). The chronic toxicity of 2,5-hexanedione is at least partially due to its prolonged retention in the body; small doses accumulate and produce adverse effects. Suwita et al. (275) studied the pharmacokinetics of a 50-mg/kg dermal dose of 2,5-hexanedione in hens and found that the half-life of absorption was 6 h based on the disappearance of 2,5-hexanedione from the application site. Excretion was primarily through expired air, which contained 35% volatile organics, most of which was parent compound, and 12% carbon dioxide. Only 15% of the administered dose was excreted in the urine and feces, mostly as 5-hydroxy-2-hexanone. Nervous tissue contained small amounts of 2,5-hexanedione relative to the kidneys and liver, but tissue half-lives ranged from 12 h for fat to 71 h for muscle.

The ability of methyl ethyl ketone, acetone, and ethanol to potentiate the effects of 2,5-hexanedione has been studied in rats. Repeated oral doses of MEK produced a more rapid onset of 2,5-hexanedione-induced neuromotor dysfunction as determined by hind-limb grasp and rotorod performance. Repeated doses of MEK also prolonged the blood clearance of 2,5-hexanedione (276), perhaps by inhibiting the excretion of a toxic metabolite. Acetone, but not ethanol, in the drinking water enhanced the neuromuscular dysfunction caused by 2,5-hexanedione alone (277), as indicated by substantially lower balance times on a rotorod. A concentration of 0.5% acetone in the drinking water for up to 6 wk reduced the balance time by 94% when given in combination with 0.5% 2,5-hexanedione, which had minimal effect alone. A concentration of 5.0% ethanol for the same period of time had no effect either alone or in combination with 2,5-hexanedione.

Diketones without a γ spacing between carbonyl groups do not produce the same type of neurotoxicity as 2,5-hexanedione (9). Available data suggest that among aliphatic dicarbonyls a γ-diketone is necessary for neurotoxicity and that among γ-diketones, a six-carbon chain length is more neurotoxic than a seven- or eight-carbon chain length (9).

5.4.1.3.5 Mechanisms. Local effects of 2,5-hexanedione are probably related to its ability to interact with surface proteins as it does in commercial tanning processes. Reaction with epidermal proteins by cross-linking or by pyrrole formation may be the reason skin sensitization was observed in guinea pigs. The ability of 2,5-hexanedione to form pyrrole rings in the presence of protein has led to the study of the interaction of this ketone with proteinaceous filaments in the axon. Graham et al. (278) demonstrated that 2,5-hexanedione crosslinks neurofilaments to produce the distal axonal swelling characteristic of intoxication. The action of 2,5-hexanedione on neurofilaments has been confirmed in cell culture of neurons (279) with the additional finding that 2,5-hexanedione is cytotoxic to glial cells in cell culture (280). The ability of 2,5-hexanedione to crosslink protein filaments has also been investigated as a mechanism of testicular toxicity. Assembly of microtubules from the brain and testes of rats treated with 1% of 2,5-hexanedione in the drinking water was altered after 4 wk of treatment (281) or 5 wk of treatment followed by 17 wk of recovery (282).

The ability of γ-diketones to induce neurotoxicity is highly dependent upon the ease with which the dicarbonyl can form a pyrrole ring and eventually form protein crosslinks. Structural modifications that make it easier for the dicarbonyl moiety to form pyrroles and cross links lead to enhanced neurotoxicity (14–17, 283).

5.4.1.4 Reproductive and Developmental. The testicular effects of 2,5-hexanedione have been studied by several investigators. Oral treatment of male rats with 1% 2,5-hexanedione in the drinking water produced azospermic testes after 6 wk of treatment without alteration of gonadotropic hormones (284). Aspermia was not reversed after 10 wk of recovery (285) or even after 75 wk of recovery (286). The rate of intoxication rather than the total exposure dose appears to influence the toxicity (18). No testicular effects were observed at an intoxication rate of 217 mg/kg/d, but intermediate effects were observed at a rate of 433 mg/kg/d suggesting that a protective mechanism was overwhelmed by this rate of exposure.

5.4.1.5 Carcinogenesis. No carcinogenicity studies with 2,5-hexanedione have been published.

5.4.1.6 Genetic and Related Cellular Effects Studies. Kato et al. (287) determined that 2,5-hexanedione was not mutagenic in *Salmonella typhimurium* strains TA98, TA100, or TA104 or in *E. coli*. Aeschbacher et al. (288) reported negative results in *S. typhimurium* strains TA98, TA100, and TA102 with and without rat S9 preincubation. Zimmermann et al. (51) found 2,5-hexanedione to be a weak inducer of chromosomal loss in yeast cells, and the addition of propionitrile to the incubation mixture enhanced the induction of aneuploidy. Mayer and Goin (52) reported that 2,5-hexanedione alone or in combination with acetone or methyl ethyl ketone was a strong inducer of chromosomal loss in strain D61.M of *Saccharomycs cerevisiae*.

5.4.1.7 Other: Neurological, Pulmonary, Skin Sensitization. Repeated oral doses of approximately 1 mL/kg 2,5-hexanedione for 1 wk suppressed delayed-type hypersensitivity and phagocytic activity in mice, and increased their susceptibility to endotoxin (266). Treated mice did not demonstrate a sensitization reaction to sheep red blood cells as did control animals, and treated mice had higher mortality following endotoxin challenge compared with control mice. Molinari et al. (289) reported that *in vitro* exposure of a human carcinoma cell to 2,5-hexanedione modified the expression of HLA histocompatibility antigens on the surface of the cell. Governa et al. (290) demonstrated that 2,5-hexanedione inhibits neutrophil chemotaxis *in vitro*. The authors suggest that rearrangement of the cytoskeleton of the cell membrane caused by 2,5-hexanedione alters the cell-to-cell communication required for chemotaxis. Altered cell-to-cell communication may also influence development of other immune responses such as delayed-type hypersensitization. Upreti and Shanker (266) gave Swiss albino mice single or seven consecutive doses of 0.98 mL/kg (0.2 × LD_{50}) 2,5-hexanedione and then examined the mice for morphological and function changes in the immune system. A single dose caused atrophy of the thymus while multiple doses did not. Both dosing regimens caused reduced cellularity in the spleen, thymus, and mesenteric lymph nodes. In addition, the delayed-type hypersensitivity reaction, plaque-forming cell assay, phagocytosis by adherent peritoneal exudate cells, and resistance to endotoxic shock were impaired.

5.4.2 *Human Experience*

5.4.2.1 General Information. No reports of adverse effects in humans due to the handling of 2,5-hexanedione were found. The discoloration of the skin produced by 2,5-hexanedione may serve as an indicator that skin contact has occurred.

Any toxicity due to 2,5-hexanedione exposure is expected to be due to eye or skin contact, or from inhalation of vapors. Topically, 2,5-hexanedione may be a slight skin irritant and a moderate eye irritant. Chronic exposure to 2,5-hexanedione by either the skin or inhalation routes could result in severe impairment of the central and peripheral nervous systems, the testes, and the circulating elements of the blood. Skin and eye contact should be prevented and inhalation exposures minimized.

5.4.2.2 Clinical Cases. No clinical case reports of adverse health effects following exposure to 2,5-hexanedione were identified.

5.4.2.3 Epidemiology Studies. No epidemiology studies are available for 2,5-hexanedione.

5.5 Standards, Regulations, or Guidelines of Exposure

No threshold limit value has been established for 2,5-hexanedione.

5.6 Studies on Environmental Impact

No information about the environmental impact of 2,5-hexanedione was identified. 2,5-Hexanedione is a reactive material and can be metabolized through common pathways for intermediary metabolism. Therefore, it is not expected to persist in the environment or cause long-term environmental effects.

6.0 Cyclohexanone

6.0.1 CAS Number: *[108-94-1]*

6.0.2 Synonyms: Cyclohexyl ketone, ketohexamethylene, pimelic ketone, pimelin ketone, sextone, and anone

6.0.3 Trade Names: Nadone (69)

6.0.4 Molecular Weight: 98.14

6.0.5 *Molecular Formula:* $C_6H_{10}O$

6.0.6 *Molecular Structure:*

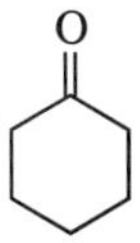

6.1 Chemical and Physical Properties

Cyclohexanone is slightly soluble in water and miscible in most organic solvents, including ethanol, benzene, *n*-hexane, and diethylamine. It has a melting point of −32.1°C, a boiling point of 155.7°C, a density of 0.945 g/cm^3 at 20°C, and a vapor pressure of 24.8 torr at 20°C (70). The flash point (closed-cup) is 44°C, and the autoignition temperature is 420°C (71). The lower and upper flammability limits at 100°C are 1.1 and 9.4.

6.1.1 *General*

Additional physical–chemical properties are summarized in Table 76.4.

6.1.2 *Odor and Warning Properties*

Hellman and Small reported that 50% of the people on an odor panel could detect the odor of cyclohexanone at a concentration of 0.12 ppm and the scent was described as pleasantly sharp and sweet (75). Summarizing the results from five different studies, the standardized odor detection threshold for cyclohexanone was reported to be 0.71 ppm (73). The American Industrial Hygiene Association lists the range of acceptable detection threshold values to be 0.12–100 ppm (72).

6.2 Production and Use

Cyclohexanone is primarily manufactured by the catalytic oxidation of cyclohexane with smaller amounts produced by the hydrogenation of phenol (291). Production in the United States was estimated to be 1100 million pounds in 1994. Cyclohexanone is used as a feedstock for the synthesis of caprolactone, which is used to manufacture thermoplastic and polyurethane elastomers and others polymers, and as a process solvent to manufacture magnetic disks. Smaller amounts are used to synthesize cyclohexylamine derivatives, formulate herbicides, and manufacture surface coatings such as paints, lacquers, and inks (291).

6.3 Exposure Assessment

6.3.1 *Air*

Cyclohexanone was detected in the cabin air from 1 of 28 space shuttle missions at a concentration between 0.01 and 0.1 mg/m^3 (292). It was also reportedly found in an unspecified air sample from a Superfund site at a concentration of 33.1 μg/m^3 (89). The maximum concentration of cyclohexanone in the workroom air of eight different screen printing shops ranged from 0.18 to 8.03 mg/m^3 (293). Air from the vulcanization area of a factory manufacturing shoe soles contained cyclohexanone at a concentration ranging from 0 to 10 μg/m^3 (294). The indoor air from 2 of 6 homes in northern Italy was found to contain cyclohexanone at concentrations of 1 and 6 μg/m^3 (295). An analysis of ambient air samples from four locations in and around Los Angeles revealed an average cyclohexanone concentration of 0.20 ppb (296). Cyclohexanone was emitted from new polyvinyl chloride flooring at a rate of 23 μg/m^2/h after 4 wk and a rate of 20 μg/m^2/h after 26 wk of being applied (297). Air samples from forests in the Sierra Nevada Mountains of California and the Black Forest of southern Germany were also found to contain cyclohexanone, but the amounts were not determined (298, 299).

6.3.2 Background Levels

Cyclohexanone was detected in 10 of 29 samples of printing ink and showed a concentration ranging from 2.7 to 17.0% by weight (300). It was also detected in 4 of 20 paint thinners at a level of 5–20% (301). Cyclohexanone was not detected in any of the 26 types of model or hobby glue examined (302).

Cyclohexanone was detected in the drinking water from 1 of 10 U.S. cities at a concentration of 0.1 μg/L (303). It was also detected, but not measured, in a drinking water sample that was disinfected with chlorine dioxide and stabilized with ferric chloride (304). Cyclohexanone was found at concentrations ranging from 2.9 to 15.9 mg/L in saline, glucose, and Ringer intravenous solutions stored in polyvinyl chloride bags for 2–3.5 yr (305). It was similarly detected, but not quantified, in a single secondary effluent sample from 1 of 10 wastewater treatment plants (306) and in a groundwater leachate sample from a municipal landfill (307). Cyclohexanone was detected at a concentration of 35 mg/L in the wastewater stream from a shale oil recovery operation (308). The highest concentration of cyclohexanone found in contaminated groundwater samples from The Netherlands was 30 μg/L (309). Paxeus reported that cyclohexanone was detected in the effluent from one of three wastewater treatment plants at a concentration less than 0.1 μg/L (310). Clark et al. found measurable quantities of cyclohexanone at all three of the publicly owned wastewater treatment plants surveyed (311). Concentrations at the sites, which treated both municipal and industrial wastewater, ranged from 0.3 to 10 ppb. The wastewater influent from a textile finishing plant was shown to contain cyclohexanone, but the levels were not quantified (312).

6.3.3 Workplace Methods

OSHA has published a recommended method for assessing workplace exposures to cyclohexanone (313). The method uses Chromosorb 106 instead of activated charcoal for sample collection and gas chromatography with flame ionization detection for quantitation. Previous studies have shown that cyclohexanone can only be partially recovered when a charcoal adsorbent tube is used for personal monitoring. The recovery efficiency declined with storage time and ranged from a high of 80% after 2 d to a low of about 50% after 23 d (314). Levin and Carlborg evaluated different types of solid adsorbents for sampling workroom air and found that cyclohexanone was only retained on charcoal and Ambersorb XE-348. The recovery efficiencies from these two media were up to 80–90% if the samples were analyzed immediately (118). Samimi reported on the calibration and use of an inferred gas analyzer to measure airborne vapor concentrations of cyclohexanone

(120). Using charcoal sampling tubes and GC-FID analysis, the time-weighted average breathing zone concentration of cyclohexanone for the employees in different locations of a screen printing plant were found to range from 6 to 28 ppm. The workroom concentrations in the same areas ranged from about 3 to 23 ppm. NIOSH Method 1300 may also be used to determine workplace exposures to cyclohexanone (116).

6.3.4 *Community Methods*

The cyclohexanone in atmospheric samples may analyzed by reverse-phase liquid chromatography after separation and quantitation of the 2,4-dinitrophenyl-hydrazone derivative (315). Applying this method, the analytical detection limit for ambient air samples of cyclohexanone was found to be 3.7 ng/m^3 (316). The purity of cyclohexanone may be determined by gas liquid chromatography using a DC-710 or Carbowax 20 M column (317). When other carbonyl compounds are absent, cyclohexanone may be derivitized with hydroxylamine and assayed as the oxime.

6.3.5 *Biomonitoring/Biomarkers*

Cyclohexanone can be detected in the serum of normal healthy adults and in the urine of certain patients with diabetes mellitus (318, 319). The presence of cyclohexanone in the urine of diabetics appears to dependent on the form, stage, and severity of the disease.

6.3.5.1 Blood. Development of specific biomonitoring methods and biomarkers for cyclohexane in blood has not been reported.

6.3.5.2 Urine. Urinary cyclohexanol has been shown to correlate well with occupational cyclohexanone exposures (320). Urine samples from 27 workers with cyclohexanone time-weighted exposures (8-h) ranging from 2 to 30 ppm were found to contain cyclohexanol concentrations ranging up to 14 mg/g creatinine. Regression analysis revealed a correlation coefficient of 0.88. Fiserova-Bergerova reported that ethanol consumption could competitively increase the urinary levels of cyclohexanol from a cyclohexanone exposure (321). Cyclohexanone was detected in the urine of workers exposed to cyclohexane, but the levels were too low to be used as an exposure monitor (322).

6.3.5.3 Other. Cyclohexanone breath levels did not correlate well ($r = 0.51$) with the occupational exposure levels of 59 workers with time-weighted average exposures ranging from 1 to 40 ppm (323). Breath levels ranged from 0 to 4.5 ppm and showed a large interindividual variability that was deemed to be too excessive for exposure monitoring.

6.4 Toxic Effects

Cyclohexanone is slightly toxic and may be irritating to the skin, eyes, and nasal passages. It may produce transient corneal injury. Irritation is common at atmospheric concentrations above 50 ppm. Exposure to high concentration can produce central nervous system depression.

6.4.1 *Experimental Studies*

6.4.1.1 Acute Toxicity. Smyth et al. reported that the oral LD_{50} for cyclohexanone in Wistar rats was 1.62 mL/kg with a 95% confidence interval (CI) of 1.20–2.20 mL/kg (324, 325). Deichmann and LeBlanc reported a rat oral LD_{50} value of 1.84 (326). Treon et al. examined the acute toxicity of cyclohexanone and five related alicyclic compounds following oral administration to outbred albino rabbits (327). The minimum lethal dose of cyclohexanone ranged from 1.6 to 1.9 g/kg, whereas animals administered 0.9–1.6 g/kg of cyclohexanone survived the treatment. Survival time was found to be inversely related to the administered dose of cyclohexanone. Narcosis without convulsions was noted in the animals that succumbed to the treatment. At high lethal dosages, the alicyclic compounds used in the study were found to cause widespread vascular damage and coagulative necrosis in many of the tissues. Cyclohexanone was notable for the degree of lung damage observed on gross pathology. Animals that survived treatment with cyclohexanone showed some evidence of myocardial fibrosis, chronic interstitial pneumonitis, bronchitis, hepatic necrosis, and diffuse glomerulonephritis.

Smyth et al. reported that a 4-h inhalation exposure to 2000 ppm (of cyclohexanol) was lethal to one of the six rats exposed, whereas doubling the exposure concentration to 4000 ppm was lethal to all six of the treated rats (324, 328). Guinea pigs exposed to 4000 ppm for 6 h showed signs of lacrimation, salivation, decreased body temperature, decreased heart rate, opacity of the cornea, and central nervous system depression (25). Specht et al. exposed guinea pigs of mixed stock to 0.4% (4000 ppm) of cyclohexanone for nearly 6 h and recorded the signs and symptoms that developed during the exposure (25). Lacrimation, salivation, ataxia, and narcosis were noted along with a time-dependent decrease in body temperature, respiratory rate, and heart rate. Three of the ten treated animals died within 4 d of the exposure and the remainder recovered within several weeks.

Klimisch et al. summarized the results from an interlaboratory study where cyclohexanone and seven other solvents were evaluated in the Inhalation Hazard Test (OECD Method No. 403) (329). The Inhalation Hazard Test was designed to determine the maximum exposure duration (LT_0) to an atmosphere saturated with solvent vapors that could be tolerated without any lethality during a 14-d observation period. The laboratories exposed male and female rats of a Sprague–Dawley or Wistar strain to cyclohexanone vapors for periods of time ranging from 3 min to 7 h. The average nominal cyclohexanone exposure concentration was 17 mg/L (4250 ppm) for the six laboratories participating in the study. The LT_0 value for cyclohexanone was found to be 1 h in five of the six laboratories. The remaining laboratory reported a cyclohexanone LT_0 value of 3 h. A similar 14-d study by Smyth et al. revealed that Wistar rats could tolerate a saturated atmosphere of cyclohexanone vapors (concentration not stated) for up to 30 min with no observed mortality (324, 325). Gupta et al. determined the 7-d LT_{50} for ICR mice exposed to approximately 19.0 mg/L (4750 ppm) of cyclohexanone vapors and reported a value of 99.9 min (95% CI, 91.5–109.2 min) (330). Histological analysis of the dead and moribund animals revealed congestion, edema, and focal to diffuse hemorrhage in the lung tissue. Splenic hyperplasia of the white pulp was noted in the animals that survived the cyclohexanone exposure.

Gupta et al. determined the IP LD_{50} values for male ICR mice, Sprague–Dawley rats, New Zealand rabbits, and an unstated strain of guinea pig (330). The following LD_{50} values were calculated: mouse, 1.23 g/kg (95% CI, 1.12–1.35 g/kg); rat, 1.13 g/kg (95% CI, 0.91–1.39 g/kg); rabbit, 1.54 g/kg (95% CI, 1.00–2.37 g/kg); and guinea pig, 0.93 g/kg (95% CI, 0.61–1.42 g/kg). The animals exhibited acute "hypnotic" effects and labored respiration prior to death, which typically occurred within 24 h of treatment. Tremors and peritoneal irritation were exclusively noted in those animals treated by the IP route.

Gupta et al. evaluated the acute local and systemic effects cyclohexanone in an extensive study that involved six alternate modes of treatment routes with four different animal species (330). The 7-d oral LD_{50} for cyclohexanone in male and female ICR mice was calculated to be 2.07 g/kg (95% CI, 1.83–2.38 g/kg) and 2.11 g/kg (95% CI, 2.02–2.34 g/kg), respectively. The LD_{50} value in both sexes of Sprague–Dawley rat was 1.80 g/kg; however, the confidence intervals were slightly greater for the female rats (95% CI, 1.47–2.20 g/kg) than for the male rats (95% CI, 1.58–2.05 g/kg).

Smyth et al. found that cyclohexanone produced a grade 5 corneal injury (scale of 1 to 10) when placed in the eyes of New Zealand rabbits (324, 325). A grade 5 injury indicated that 0.005 mL of the undiluted solvent caused a damage score of at least 5 (range of 1 to 20) and that 0.02 mL of the solvent caused a damage score greater than 5 (323). Gupta et al. reported that a cyclohexanone concentration of 5% or less caused little or no irritation to the eyes of New Zealand rabbits, whereas concentrations greater than 40% caused a marked irritation (330). A graded irritative response was observed at cyclohexanone concentrations ranging from 5 to 40%. The ocular irritancy of pure cyclohexanone was evaluated in an *in vitro* bovine corneal opacity and permeability (BCOP) assay and found to be a severe irritant (151). These results were compared to the findings from an *in vivo* Draize test, where damage to rabbit cornea, iris, and conjunctiva were scored on days 1 and 14. The Draize study indicated that cyclohexanone was moderately irritating and that recovery from the damage occurred in 14 d. Cyclohexanone was also examined in the hen's egg test–chorioallantoic membrane (HET–CAM) assay and shown to be an ocular irritant (152).

Cyclohexanone was moderately irritating when held in contact with guinea pig skin for 24 h and mildly irritating when applied to the skin of rabbits. Smyth et al. reported that the acute dermal LD_{50} for New Zealand rabbits was 1.00 mL/kg (95% CI, 0.63–1.63 mL/kg) and that 0.01 mL of a cyclohexanone solution was not a skin irritant when applied to the shaved abdomen of rabbits (unoccluded) and examined 24 h later (324, 325). Deichmann and LeBlanc reported that the subcutaneous LD_{50} values of cyclohexanone in rats was 2.17 g/kg (326). Treon et al. determined the concentration of cyclohexanone that was lethal to outbred albino rabbits by applying the solvent to the shaved abdominal skin of the animals and then removing the unabsorbed liquid after 1 h (327). Using this procedure, a topically applied cyclohexanone dose of 10.2–23.0 g/kg was lethal to all rabbits, with marked hypothermia, convulsions, and narcosis observed in the affected animals.

Gupta et al. found that cyclohexanone concentrations of 12.4 and 24.8% in cottonseed oil caused little or no irritation when applied to shaved backs of New Zealand rabbits (occluded) for 24 h and observed daily thereafter for 7 d (330). Increasing the cyclohexanone concentration to 49.5% caused moderate irritation on the first day of observation and little or no irritation on the following six days. A cyclohexanone concentration of 99% caused a marked irritation on days 1 and 2 of the observation period, but the damage gradually subsided and disappeared by the seventh day. When administered intradermally to the rabbits, cyclohexanone concentrations of 31.1% or greater in cottonseed oil caused marked irritation, whereas concentrations between 4 and 31.1% produced a concentration-related increase in the degree of irritation. The irritation threshold for intradermally administered cyclohexanone was calculated to be 12.8%.

DeCeaurriz et al. determined the irritancy of cyclohexanone and 21 other solvent vapors by measuring the vapor concentration needed to cause a 50% decline in respiration rate (RD_{50} value) relative to the control value (164). Male Swiss OF_1 were exposed for 5 min to cyclohexanone vapor concentrations ranging from about 250 to 1000 ppm to obtain an RD_{50} value of 750 ppm (confidence interval not stated). According to empirically derived guidelines, the authors concluded that the RD_{50} value in mice was consistent with the occupational exposure standards established for humans. The results from this study have been used to develop several quantitative structure–activity relationships for predicting the irritancy potential of ketones and establishing occupational exposure limits of chemicals in this class (159–163).

In a separate study, DeCeaurriz et al. assessed the neurological effects of cyclohexanone and a variety of other solvents in a "behavioral despair" swimming test that followed a 4-h vapor exposure (331). Male Swiss OF_1 mice were exposed to cyclohexanone at five vapor concentrations ranging from 184 to 577 ppm. The animals were placed in a water tank immediately following the exposure and the length of time they remained immobile during a 3-min test period was recorded. These data were used to calculate an ID_{50} value that corresponded to the vapor concentration that caused a 50% decrease in the length of time the animals remained motionless in the water relative to the control values. The ID_{50} for cyclohexanone was 308 ppm (95% CI, 279–341 ppm), which ranked second in potency for the 13 solvents tested.

6.4.1.2 Chronic and Subchronic Toxicity. Treon et al. exposed groups of young albino rabbits and a single Rhesus monkey to cyclohexanone vapors for periods ranging from 3 to 10 wk (332). The monkey exposed 6 h/d for 10 wk (5 d/wk) showed extensive damage to the heart, lungs, liver, and kidneys; however, these results were compromised by a chronic bronchopulmonary infection that developed. Rabbits were exposed 6 h/d for 10 wk to either 190, 309, 773, or 1414 ppm of cyclohexanone and 6 h/d for 3 wk to 3082 ppm of cyclohexanone. The 3082-ppm exposure caused narcosis, incoordination, salivation, and conjunctival irritation and congestion. Two of the four animals in the high-exposure group failed to survive, but no convulsions or tremors were observed. Vapor exposures of 309–1414 ppm generally resulted in eye irritation and salivation. An atmospheric concentration of 3082 ppm produced light narcosis and central nervous system effects. Weekly measurements of blood hemoglobin, the red blood cell count, and the leukocyte count failed to show any change for any of the treatment groups. Gross and microscopic tissue examination apparently revealed some degree liver and kidney damage at all exposure levels; however, specific details were not provided.

Greener et al. assessed the toxicity of cyclohexanone in two strains of rats following IV treatment for 28 consecutive days (333). Male Wistar and Gunn (congenital hyperbilirubinemia) rats were administered cyclohexanone doses of 50 or 100 mg/kg/d via their tail veins. The latter strain of rat was selected for use because of their deficiency in UDP-glucuronosyl transferase, an enzyme needed to conjugate and excrete many phenolic compounds, including bilirubin. Necropsies and histopathological analysis of selected tissues (spleen, skeletal muscle, kidney, lung, liver, stomach, duodenum, pancreas, bladder, brain, eyes) did not reveal any gross or microscopic changes as a result of cyclohexanone treatment in either rat strain. Likewise, no changes in body-weight gain or organ-to-body weight ratios were recorded relative to the control groups, and no changes in lens opacity were noted following ophthalmoscopic examination. Except for a small dose-related decrease in serum calcium values, no statistically significant changes were observed in blood hematology or clinical chemistry values.

Koeferl et al. investigated the toxicity of cyclohexanone to rats, dogs, and monkeys following intravenous administration (334). Dogs and monkeys (strain or type not stated) were given a daily dose of cyclohexanone for 21 d at either of two dose levels (142 or 284 mg/kg), whereas the rats were treated at three dose levels (142, 284, or 568 mg/kg). Nearly all the rats treated at the high dose died immediately after the first treatment apparently because of respiratory distress. Several dogs and one monkey were similarly affected at lower dose levels. Signs of toxicity included lacrimation, salivation, apnea, tremors, hypokinesis, ataxia, and narcosis. The severity of the response was dose-related and decreased in intensity from the first to the last treatment. Increased numbers of nucleated red blood cells and increased hematocrit values were observed in the circulating blood, and these were accompanied by changes in the myeloid-to-erythroid ratios in the bone marrow.

In a subsequent study, Koeferl et al. examined how the toxicity profile of cyclohexanone was affected by alterations in the dose solution concentration and in the IV administration rate (335). Beagle dogs were treated for 18–21 d with a 284-mg/kg/d dose of cyclohexanone at concentrations of 0.75% or 6%. Each concentration was administered at rates of 5 or 75 mL/min into either the cephalic or saphenous veins of the dogs. The dogs treated with a 6% solution of cyclohexanone at a rate of 75 mL/min were euthanatized after 18 d of treatment due to their waning health; all other treatment groups received 21 continuous days of treatment before being examined. A wide variety of behavioral, autonomic, and CNS-mediated clinical signs were noted in the treated animals. The severity of these effects was related strongly to the treatment regimen employed and was least severe in the animals given a 0.75% solution at 5 mL/min. Blood gas analysis in the most severely affected dogs, 6% cyclohexanone at 75 mL/min, showed a rapid and progressive decline in blood pH with corresponding changes in carbonate and bicarbonate levels that were indicative of a severe metabolic acidosis. Statistically significant increases in absolute and relative organ weight (adrenal glands and liver) were only noted in the most severely affected animals. Hematologic measurements revealed a statistically significant decrease in packed cell volume, red blood cell counts, and total hemoglobin and a statistically significant increase in the white blood cell count in several of the treatment groups. Gross and microscopically observed tissue damage was generally limited to the animals treated with 6% cyclohexanone and included inflammation, edema, and venous thrombi at the injection sites, hemosiderin deposits and extramedullary hematopoiesis in the spleen and liver, and erythroid or myeloid hyperplasia in the bone marrow. The authors demonstrated that the toxicity of cyclohexanone was affected by both the administration rate and the concentration of cyclohexanone in the dose solution; however, the changes in the latter appeared to have the greatest effect.

Rengstorff et al. reported that the repeated administration of cyclohexanone to guinea pigs was capable of causing cataracts (336). Male and female albino guinea pigs were treated with cyclohexanone either topically (0.5 mL/day, neat) or by subcutaneous injection (0.05 mL/day, diluted) 3 d/wk for the 3 wk and then examined monthly for 6 mo. The treated animals were examined using an ophthalmoscope and slit-lamp for evidence of lens vacuolization, but no vehicle-treated control groups were included for comparison. Bilateral cataracts were observed in 25% of the guinea pigs treated dermally with cyclohexanone and 12% of the animals treated by subcutaneous injection.

Using a nearly identical treatment regimen, Greener and Youkilis repeated the studies of Rengstorff et al. and found that lens vacuolization occurred in a high percentage of both the cyclohexanone-treated and vehicle-treated animals (337). Male and female Hartley guinea pigs and New Zealand rabbits were treated by intravenous injection at two dose levels (0.5 or 5.0 mg/kg) and topically (0.5 mL, neat) 3 d/wk for 3 wk. Positive and negative control groups were treated in parallel with known cataractogenic substances and a saline vehicle, respectively. Histopathological examinations were performed on the eyes at monthly intervals for 6 mo following treatment. No lenticular changes were observed in any of the treated rabbits used in the study. In guinea pigs, there was a 50–83% incidence of lenticular damage (subcapsular droplets, opacification, and vacuolization) in both the test and control animals, and no statistically significant differences were found among the various treatment groups. The authors concluded that lenticular changes occurred spontaneously in the guinea pig and that this species was an inappropriate model for assessing the cataractogenic potential of a chemical.

Mayhew et al. also attempted to verify the findings of Rengstorff et al. by dermally treating male and female Hartley guinea pigs and Sprague–Dawley rats with cyclohexanone (338, 339). The guinea pigs were treated 3 d/wk for 3 wk with 0.05 mL of either a 2% solution of cyclohexanone in saline or undiluted cyclohexanone, whereas the rats were treated 3 d/wk for 3 wk with neat cyclohexanone and 3 d/wk for 13 wk with diluted (2%, v/v) and undiluted cyclohexanone. Positive (dimethyl sulfoxide or galactose) and negative (saline) controls were administered to both species, and all animals were observed for 6 mo. In guinea pigs, the incidence of lenticular vacuolization was similar for the saline control group (40%), the group given 2% cyclohexanone (53%), and the group given undiluted cyclohexanone (47%). Rats treated for either 3 or 13 wk showed no cyclohexanone treatment-related changes in body weight gain, hematological values, or ophthalmoscopic appearance. The authors concluded that lenticular vacuolization developed spontaneously in the guinea pig.

Lijinsky and Kovatch assessed the subchronic and chronic effects of cyclohexanone supplied via the drinking water (340). In the subchronic study, groups of male and female Fisher 344 rats received tap water containing 190, 400, 800, 1600, 3300, 4700, or 6500 ppm of cyclohexanone for 25 wk. The only effect noted in rats was slight reduction in weight gain (10%) relative to control animals at the 6500-ppm level. Male and female $B6C3F_1$ mice were treated repeatedly with cyclohexanone concentrations of 400, 2300, 6500, 13,000, 25,000 34,000, or 47,000 ppm for 13 wk. Increased mortality, focal coagulative liver necrosis, and thymic hyperplasia were observed with 47,000 ppm of cyclohexanone. The only recorded effect of cyclohexanone at 25,000 and 34,000 ppm was a reduction in weight gain relative to controls.

6.4.1.3 Pharmacokinetics, Metabolism, and Mechanisms. James and Waring found small amounts of a mercapturate conjugate in the urine of rats and rabbits treated with cyclohexanone (341). Urine specimens from female New Zealand rabbits and Wistar rats treated by gavage with 2.5 mmol/kg were found to contain small amounts of hydroxycyclohexylmercapturic acid, *cis*-2-hydroxycyclo-hexylmercapturic acid, and a third unidentified sulfur-containing metabolite. Boyland and Chasseaud reported that a 0.4-mL/kg IP dose of cyclohexanone caused a slight reduction in liver glutathione levels in rats (342). Female Chester–Beatty rats showed a 12% decline in liver glutathione 30 min after treatment and a 28% decline after 2 h.

Parmer and Burka found that male Fisher rats treated orally with up to 30 mg/kg of cyclohexanone oxime rapidly hydrolyzed the chemical to cyclohexanone as the first step in its overall metabolism (343).

Greener et al. reported that male Sprague–Dawley and Gunn rats excreted a large percentage of cyclohexanol glucuronide in their urine when treated with 50 or 100 mg/kg/d of cyclohexanone for 28 consecutive days (333). The weekly collection of 24-h urine specimens revealed small amounts (less than 1% of the dose) of cyclohexanone and cyclohexanol. The cyclohexanol glucuronide conjugates, in contrast, accounted for 15–25% of the dose at 50 mg/kg and 19–34% of the dose at 100 mg/kg, regardless of the rat strain examined. Sulfate conjugates of cyclohexanol were not detected in urine specimens collected on days 1 and 7 of the treatment regimen. Elliott et al. administered about 250 mg/kg of cyclohexanone to rabbits (sex and strain not stated) and isolated an average of 66% of the dose as conjugated glucuronides that were composed primarily of cyclohexylglucuronide (344). Cronholm reported that ethanol was capable of increasing the rate of biliary excretion of the cyclohexanone metabolite, cyclohexylglucuronide, in male Sprague–Dawley rats when 5 μmol, but not 500 μmol, of cyclohexanone was administered by IP injection (345). Sakata et al. showed that the oral co-administration of a 4.8-mmol/kg dose of ethanol and cyclohexanone could affect the elimination of cyclohexanone in domestic white rabbits of both sexes (346). When compared to cyclohexanone alone, the co-administration with significantly increased the peak blood concentration of cyclohexanol at 1 h post-treatment but did not affect the rates of elimination of cyclohexanone or cyclohexanol from the blood.

6.4.1.3.1 Absorption. In a study which examined the metabolism of cyclohexanone in rats exposed by inhalation, male Sprague–Dawley rats received a single 6-h inhalation exposure to either 400 or 1600 ppm of cyclohexanone (347). The high concentration reportedly caused pronounced sedation of the rats throughout the exposure. Terminal blood specimens were collected from the animals together with three 24-h postexposure urine specimens. The average levels of cyclohexanone and cyclohexanol in the terminal blood specimens were approximately equal at both the 400-ppm (26 and 20 μg/mL) and 1600-ppm (122 and 140 μg/mL) exposures.

6.4.1.3.2 Distribution. Martis et al. examined the pharmacokinetics of cyclohexanone in dogs following intravenous administration (348). Male beagle dogs were treated for 18 or 21 consecutive days with a 284-mg/kg/d dose of cyclohexanone at solution concentrations of 0.75% or 6%. Infusion rates of 35.5 (group IV) and 532.5 mL/min (group III) were used with the 0.75% solution, whereas infusion rates of 280 (group II) and 4200 mg/min (group I) were used with the 6% solution. The cyclohexanone and cyclohexanol detected in post-treatment blood samples were shown to decline in a logarithmic fashion and were described mathematically by a two-compartment and one-compartment model, respectively. The plasma concentration of cyclohexanol showed a peak value ranging from 140 to 220 μg/mL within 5–30 min of cyclohexanone administration to the group I dogs. The conversion of cyclohexanone to cyclohexanol was calculated to range between 74 and 100% of the dose. Renal clearance measurements for cyclohexanone and cyclohexanol showed that urinary excretion did not contribute greatly to the overall elimination of these compounds. The distribution half-life for cyclohexanone in group I–IV dogs ranged from 6.6 to 10.7 min and did not change appreciably for the first and last day of treatment. In contrast, the mean elimination half-life for cyclohexanone in group I–IV dogs showed a statistically significant increase between the first (102 min) and last day (123 min) of treatment. Under the treatment conditions employed, the study showed no evidence of cyclohexanone accumulation or enzyme induction upon repeated administration.

6.4.1.3.3 Excretion. The total cumulative excretion of cyclohexanone and cyclohexanol in the urine specimens of male Sprague–Dawley rats receiving a single 6-h inhalation exposure was 16 and 15 μg, respectively, for the rats exposed at 400 ppm, and 143 and 264 μg for those exposed the animals exposed at 1600 ppm (347). A large amount of cyclohexanol conjugates was excreted in the urine (72-h total) from the rats exposed to 400 ppm (13 mg recovered) and 1600 ppm (72 mg recovered of cyclohexanone). Treon et al. found that albino rabbits administered 890 mg/kg of cyclohexanone by gavage excreted 45–50% of the dose into the urine as a glucuronide conjugate (327).

The analysis of 24-h urine specimens in male beagle dogs treated for 18 or 21 consecutive days with a 284-mg/kg/d dose of cyclohexanone revealed that approximately 60% of the administered cyclohexanone was excreted as the glucuronide conjugate of cyclohexanol with less than 1% of the dose appearing as free cyclohexanone or free cyclohexanol (348).

6.4.1.4 Reproductive and Developmental. Gray et al. evaluated the effects of forty chemicals, including cyclohexanone, on the postnatal behavioral development of rats by examining figure-eight maze activity (349). Pregnant CD-1 mice were administered 800 mg/kg/d of cyclohexanone by gavage on days 8–12 of gestation and allowed to deliver their pups. On days 22, 58, and 200 of post-parturition, the offspring were placed in a figure-eight maze for 1 h and monitored for the level of motor activity. The cyclohexanone treatment did not affect motor activity at any stage of development relative to controls. Greener et al. looked for evidence of systemic toxicity in neonatal rats administered cyclohexanone for 18 d (333). Three-day-old neonatal rats (strain not specified) were given daily IV injections of cyclohexanone at doses of 1.0, 10.0, or 25.0 mg/kg/d. No statistically significant changes were found in hematology and clinical chemistry values, body weight gain, or organ weights relative to controls. Likewise, no gross or microscopic pathological changes were noted in the animals treated at each dose level.

Hall et al. examined cyclohexanone and over 50 other substances in a mouse fertility screening assay (350). Female CF-1 mice were administered a 50 mg/kg/d IP dose of cyclohexanone for 28 d with the matings occurring on the tenth day of treatment. Cyclohexanone was judged to have no effect on fertility since the number of pregnancies, viable fetuses, and resorptions in the treated groups was close to the values from vehicle-treated mice. Weller and Griggs exposed chick embryos to cyclohexanone vapors and evaluated the effects on chick development (351). The eggs were exposed to an unstated concentration of cyclohexanone vapor for 3, 6, or 12 h on day 0 or day 4 of incubation. Although the embryos showed no gross pathology, the newly hatched chicks began developing a severe motor deficit that manifested itself as an inability to walk, or as a spastic extension of their legs. The chicks died within several days due to dehydration. The location of the neurological damage responsible for these clinical effects was not determined.

Samini et al. determined the teratogenicity of cyclohexanone following inhalation exposure (352). Pregnant Sprague–Dawley rats were exposed to cyclohexanone vapor at concentrations of 100, 250, or 500 ppm for 7 h/d on days 5 through 20 of gestation. Examination of the dams and near-term fetuses on day 21 of gestation failed to reveal any cyclohexanone-related changes in fetal weight, implantations, resorptions, fetal death, sex ratio, or number of corpora lutea. Few external, visceral, or skeletal malformations were observed in the cyclohexanone-exposed animals. Cyclohexanone exposures up to 500 ppm were considered to be free of developmental effects in rats. Gondry et al. reported that 1% cyclohexanone in the diet of mice for several generations resulted in an increase in fetal mortality for several generations and a decrease in the growth weight of animals in the first generation (353).

Seidenberg et al. evaluated the potential developmental effects of cyclohexanone and 54 other chemicals in the Chernoff/Kavlock developmental toxicity screening assay (354, 355). A large daily dose (2200 mg/kg) of cyclohexanone, calculated to produce overt maternal toxicity, was administered orally to pregnant ICR/SIM mice during a critical period of organogenesis (days 8–12 of gestation). The treated mice were allowed to deliver their pups, and all dams and neonates were examined for evidence of treated-related toxicity relative to the vehicle-treated control group. Six of the 28 mice failed to survive the treatment, and the remaining mice showed a statistically significant decrease in body weight gain. The live-born litter size, total litter size, number of dead neonates, and 2-d survival of the neonates was unaffected by the treatment; however, a statistically significant decrease in birth weight was reported. Consequently, the authors concluded that cyclohexanone produced a positive result in the screening assay and that more extensive testing was necessary.

An unpublished teratology study subsequently revealed that inhaled cyclohexanone was only embryotoxic or fetotoxic to rats and mice at concentrations that adversely affected the dams (356, 357). Pregnant Sprague–Dawley rats were exposed to cyclohexanone vapor concentrations of 300, 650, or 1400 ppm for 6 h/d on days 6–19 of gestation, whereas the CD-1 mice were exposed to 1400 ppm for 6 h/d on days 6–17 of gestation. Rats were terminated on day 20 and the mice on day 18 of gestation. The cyclohexanone exposure did not affect maternal body weight in dams treated at 300 or 650 ppm. At the highest exposure concentration, however, there was a statistically significant decrease in maternal body weight gain and an increased prevalence of lacrimation, lethargy, and nasal discharge during the exposure. Relative to controls, there were no statistically significant changes in the mean number of implantations, viable fetuses, or resorptions at any exposure concentration. Rat fetuses from the high-exposure group showed a reduction in mean body weight and incomplete ossification of cranial bones, sternebrae, and fore-limb bones. No other treatment-related alterations were observed in the incidence of external, visceral, or skeletal malformations at any exposure concentration. Similar maternal and fetal effect were observed in the mice treated at 1400 ppm. There were, however, statistically significant increases in the mean number of resorption sites and in the incidence of visceral malformations, and a significant decrease in the mean number of viable fetuses per pregnant female. A cyclohexanone exposure concentration of 1400 ppm was concluded to be maternally toxic and embryotoxic to both rats and mice, and fetotoxic to just mice. No teratogenicity was, however, observed in either species at any exposure concentration.

A second unpublished report documented the results from a two-generation reproduction study designed to ascertain the effects of inhaled cyclohexanone on the reproductive performance, growth, and development of rats (358, 359). Male Sprague–Dawley rats in the first (F_0) generation were exposed to 250, 500, or 1000 ppm of cyclohexanone for 6 h/d for at least 10-wk. The exposures to the female rats in the first generation were continued beyond the 10-wk period and included the mating period and days 0 through 20 of gestation. For the second (F_1) generation, the exposure concentrations were 250, 500, or 1400 ppm. Exposures of the male and female F_1 animals began immediately after weaning and were continued for 12 wk when the animals were again mated to produce an F_{2a} generation, and after a 2 wk rest for the females, an F_{2b} generation. Exposure of the F_1 male and female rats continued until the F_{2b} generation was weaned. The exposures of the F_0 animals failed to cause any mortality or changes in body weight; however, numerous clinical signs, some mortality, and significant body weight reductions were observed in the F_1 male or female animals that received a 1400-ppm exposure. The F_1 animals exposed to 1400 ppm also showed a decreased fertility in males, decreased survival of the progeny, and decreased body weights in the progeny. The infertility of the male animals at the highest exposure ceased after the exposures were terminated and the animals allowed to recover. All other evaluations for neurodevelopmental effects, reproductive nervous system histopathology, reproductive performance, and survivability failed to show any consistent treatment-related effects for remaining treatment groups or their progeny.

6.4.1.5 Carcinogenesis. On the basis of findings from a 13-wk drinking water study, a 2-yr (104 wk) carcinogenicity bioassay was performed at cyclohexanone concentrations of 3300 or 6500 ppm for male and female rats, 6500 or 13,000 for male mice, and 6500, 13,000, or 25,000 ppm for female mice (340). Male and female rats used in the chronic study showed a distinct reduction in weight gain at both treatment levels and a decreased survival at the high exposure level. On terminating the study, histopathology revealed a decreased incidence of mammary gland neoplasms and uterine stromal polyps in female rats and an increased incidence of adenomas in the cortex of the adrenal glands of male rats treated at the low exposure level. Male and female mice treated at the high exposure level displayed a 15–20% reduction in body weight that was evident throughout most of the study. The female mice treated at 13,000 or 25,000 ppm showed poor survival, with less than half of the animals surviving more than 90 wk of treatment. Male mice showed an increased incidence of hepatocellular neoplasms (adenomas and carcinomas) at the low exposure level and a decreased incidence of alveolar–bronchiolar neoplasms at both exposure levels relative to the control animals. Female mice showed an increased incidence of malignant lymphoma at the low exposure level. The failure to observe any concentration-related increases in tumor incidence led the authors to conclude that their was only marginal evidence for cyclohexanone carcinogenicity in rats and mice and that the carcinogenic activity, if present, was weak.

6.4.1.6 Genetic and Related Cellular Effects Studies. Haworth et al. reported that cyclohexanone was not mutagenic to *Salmonella typhimurium* strains TA98, TA100, TA1535, or TA1537 in an Ames assay (328). Preincubation of cyclohexanone with the tester strains at concentrations ranging from 100 to 10000 μg/plate failed to increase the number of revertant colonies above background in the presence or absence of a rat liver S9 from Arochlor 1254–treated Sprague–Dawley rats. Likewise, Florin et al. found that 3 μmol/plate of cyclohexanone was not mutagenic to *S. typhimurium* strains TA98 or TA100 in a spot test conducted with and without metabolic activation by a liver S9 from Arochlor-treated rats (360). In an abstract containing few details, Massoud et al. claimed that cyclohexanone severely affected *Bacillus subtilis* survival and caused forward mutations in several types of amino acid requiring bacterial strains (361). In addition, the authors reported that cyclohexanone produced a large number of revertants in *S. typhimurium* strain TA98; however, the concentration range and test conditions were not stated. Rosenkranz and Leifer summarized a personnel communication in which cyclohexanone purportedly yielded a positive result with DNA-polymerase-deficient *Escherichia coli* (362). Growth inhibition was apparently obtained when an unknown concentration of cyclohexanone was incubated with the tester strain in the absence of metabolic activation from an S9 fraction.

McGregor et al. found that cyclohexanone concentrations up to 5000 μg/mL did not affect cell survival or mutation fraction (number mutants at thymidine kinase locus/million clonable cells) in the L5178Y tk^{+}/tk^{-} mouse lymphoma forward point mutation assay (363). The negative findings were found both in the presence and absence of an S9 from Arochlor 1254–treated Fischer-344 rats. Aaron et al. examined three genotoxicity endpoints following incubation of 7.5 μg/mL of cyclohexanone with Chinese hamster ovary cells entering the DNA synthesis phase (S phase) of multiplication (364). In the absence of an S9 activation system, cyclohexanone increased the sister chromatid exchange frequencies and HGPRT gene mutations, but did not increase the number of chromosomal aberrations. The genetic effects were not observed when an S9 fraction was incorporated into the study design. The authors concluded that cyclohexanone was a weak genotoxicant. The genetic effects of cyclohexanone on the germ cells of *Drosophila melanogaster* have been described in an unpublished report (365). The exposure of

Canton-S wild-type males to approximately 1900 ppm of cyclohexanone for 4 h did not cause statistically significant number of sex-linked recessive lethal mutations in the postmeiotic germ cells. Malcolm and Mills showed that a cyclohexanone concentration of 3 mg/L could inhibit metabolic cooperation in Chinese hamster V79 lung fibroblasts and prevent the formation of gap junctions that would normally facilitate the cytotoxic effects of 6-thioguianine (366).

Holmberg et al. determined the antihemolytic effects of cyclohexanone and 59 other organic solvents in an *in vitro* test that was purportedly indicative of anesthetic potency (367). Rat erythrocytes were incubated hypotonic saline and cyclohexanone to determine the solvent concentration that 50% (ED_{50}) and 100% (ED_{max}) protection against hemolysis. An ED_{50} value could not be determined for cyclohexanone, and the ED_{max} value was 25.2 mM. The ED_{50} values obtained for many of the solvents were shown to be linearly related to the *iso*-octane/water partition coefficients of the compounds; however, cyclohexanone was not included in the activity analysis.

Gupta et al. examined the effects of cyclohexanone on the growth of mouse fibroblasts (strain L-929) in culture (330). A cyclohexanone concentration of 19.5 mM was found to produce a 50% decrease in cell growth (ID_{50} value) during a 72-h incubation. The same group of authors also examined the effects of cyclohexanone on muscular contractility in the isolated perfused rabbit heart. A 30-min perfusion with cyclohexanone at concentrations ranging from about 2 to 20 mM caused a concentration-related decrease in the contractile force in the cardiac muscle (negative inotropic effect).

6.4.1.7 Other: Neurological, Pulmonary, Skin Sensitization. Cyclohexanone did not cause dermal hypersensitization in the mouse ear swelling test (MEST) or in the guinea pig maximization test (368). The MEST was performed in CF-1 mice treated dermally with 100 μL of pure cyclohexanone for 4 d and then challenged with 20 μL 7 d after the final treatment. The cyclohexanone treatment caused no sensitization and little swelling in the test area around the ear. These results were consistent with the findings from Hartley guinea pigs intradermally treated twice in one day with MEK in Freund's complete adjuvant followed by a continuous 48-h dermal treatment 7 d later. The challenge dose was administered topically 14 d after the induction regimen was finished. Bruze et al. also found that cyclohexanone did not cause dermal sensitization potential in the guinea pig maximization test when induction occurred either by intradermal injection at a concentration of 8% or by epidermal application at a concentration of 25% (369).

The acute neurological effects of cyclohexanone were evaluated in a bioassay designed to assess the acute neurotropic effects following a 4-h inhalation of rats and 2-h exposure of mice (370). The isoeffective concentration necessary for a 30% depression in an electrically evoked seizure discharge was determined to be 440 ppm (95% CI, 342–538 ppm) in rats and 490 ppm (95% CI, 340–640 ppm) in mice. Similarly, Kanada et al. evaluated the effects of an oral 738-mg/kg dose of cyclohexanone on neurotransmitter levels in different regions of male Sprague–Dawley rat brain (371). Levels of acetylcholine, 3,4-dihydroxyphenylalanine (DOPA), norepinephrine, 3-methoxy-4-hydroxyphenylglycol, and serotonin were not significantly affected. Levels of dopamine in the midbrain and hypothalamus and 5-hydroxyindoleacetic acid (5HIAA) in the hypothalamus were significantly decreased by the treatment.

Holland et al. screened a group of unsubstituted and monoalkyl-substituted cyclohexanones and cyclopentanones for their anticonvulsant activity following pentylenetetrazol or maximum electroshock treatment (372). The *in vivo* convulsant activity and neurotoxicity of the compounds were also determined along with their ability to displace [^{35}S]-butylbicyclophosphorothionate, a potent convulsant, from the picrotoxin receptor isolated from rat brain. Convulsant activity was not detected in the female CF-1 mice administered a single IP dose of cyclohexanone at levels in excess of 500 mg/kg (dose range not stated). The value calculated to cause half of the cyclohexanone-treated animals to fall from the rotorod twice during a 10-min test period was 473 mg/kg (95% CI, 431–536 mg/kg). The cyclohexanone dose level that prevented seizures in the pentylenetetrazol-treated animals was calculated to be 338 mg/kg (95% CI, 203–275 mg/kg), whereas the corresponding anticonvulsion value in the animals given an electroshock treatment was 397 mg/kg (95% CI, 327–532 mg/kg). The value for the displacement of TBPS from the picrotoxin receptor isolated from the cerebral cortex of Sprague–Dawley rats was 29.8 mM, which was appreciably higher than the value obtained with the monoalkyl-substituted cyclohexanones used in the study.

Panhuber performed odor-aversion learning experiments with cyclohexanone in rats to determine whether the odor acted as a learning stimulus for avoiding a toxic effect (373). Vapors of cyclohexanone and six other solvents were passed by the noses of male Sprague–Dawley rats while they were allowed access to drinking water. After establishing baseline consumption with untreated tap water, the animals were exposed to four cyclohexanone vapor concentrations (nominal) ranging from about 3 to 1250 ppm and simultaneously presented with drinking water for 10 min. The drinking water session was immediately followed with an IP injection of lithium chloride, which was used to induce malaise. After several days recovery to reestablish baseline drinking behavior, the experiment was repeated to determine whether the vapors caused a conditioned aversion as measured by the volume of water consumed. Cyclohexanone was one of only two compounds that did not cause an aversion over the range of exposure concentrations examined. Panhuber et al. also found that the continuous exposure of male Sprague–Dawley rats to 8 ppm of cyclohexanone for 10 wk caused a 21% reduction in the size but not the number of cells in their olfactory bulbs (374). A smaller change was observed in young rats exposed to cyclohexanone and in rats exposed to deodorized air for a similar length of time. The changes were attributed to olfactory deprivation rather than to the direct effects of cyclohexanone.

Perbellini et al. reported that the twice daily IP administration of 200 mg/kg of cyclohexanone to Sprague–Dawley rats for up to 13 wk did not produce damage to the peripheral nervous system (375). Electrophysiological (motor conduction velocities) measurements and neuropathological (posterior tibial nerve) examinations were similar for the treated and control animals.

6.4.2 Human Experience

6.4.2.1 General Information. Occupational exposure to cyclohexanone is likely to occur by inhalation, or dermal or eye contact. Because of its odor, cyclohexanone can be detected at concentrations well below the recommended exposure limits, and it is therefore treated as a chemical with adequate warning properties. Accidental overexposure can cause dizziness and narcosis, and irritation of the eyes, nose, and throat. Dermal exposure can result in dryness, irritation, and inflammation of the skin.

6.4.2.2 Clinical Cases

6.4.2.2.1 Acute Toxicity. Sakata et al. described a male patient who attempted suicide by ingesting approximately 100 mL of an adhesive cement that contained 39% cyclohexanone, 28% MEK, 18% acetone, and 15% polyvinyl chloride (376). The individual reportedly drank about 720 mL of sake (10% ethanol) 30 min prior to drinking the adhesive solution. The patient was unconscious when taken to a hospital about 2 h after ingestion, then regained consciousness about 5 h later. Increases in serum glucose and transaminase values were observed at various times during recovery. Cyclohexanone and cyclohexanol were detected in plasma and urine specimens for about 25 h following the incident. The cyclohexanone concentrations in plasma were barely above detection limits; however, the plasma cyclohexanol concentration was initially high, decreasing from 220 μg/mL at 5 h post-ingestion to about 10 μg/mL at 20 h. Cyclohexanol glucuronide concentrations as high as 440 μg/mL were detected in the urine specimens and the metabolite could be found in the urine for up to about 2 d following ingestion. The initial coma and drowsiness observed in the patient were attributed to the cyclohexanone that was consumed, whereas the hyperglycemia was attributed to acetone. The authors could not ascertain what caused the increased serum transaminase levels.

Zuckerman reported an altered mental status, shock, metabolic acidosis, hepatotoxicity, and renal insufficiency in a 15-year-old boy who attempted to commit suicide by ingesting an unknown amount of a paint thinner that contained pure cyclohexanone (377). Hematuria and proteinuria was noted on day 1 of admission. Liver enzyme (AST and ALT) levels continued to rise and peaked on day 6 before returning to normal. On day 6, the patient complained of muscle cramps, weakness, and difficulty walking. Blood creatine phosphokinase levels were also elevated and myoglobin was detected in the urine. These delayed signs were consistent with a diagnosis of rhadomyolysis, from which the patient recovered after about a month of hospitalization.

Passeron et al. described a poisoning case that involved the ingestion of 200 mL of an organophosphate pesticide solution that contained 50% cyclohexanone (378). Hepatorenal damage was observed in the affected individual with elevations in creatinine, AST, ALT, and bilirubin observed on admission and for several days thereafter. A liver biopsy showed areas of ballooned hepatocytes and inflammatory infiltration. The concentration of cyclohexanone in the urine was 5 mg/L. The patient recovered slowly and was discharged 25 d after admission.

Nelson et al. evaluated the irritancy potential of sixteen volatile solvents in humans by recording the subjective responses following a short-term inhalation exposure (379). Approximately ten volunteers of both sexes were exposed to cyclohexanone for 3–5 min then asked to qualitatively describe the effects on the eyes, nose, and throat. Three responses were possible: no reaction, slightly irritating, and very irritating. The majority of subjects considered 25 ppm to be the highest acceptable vapor concentration for a continuous 8-h exposure. A vapor concentration of 50 ppm caused throat irritation in some of the individuals (number not stated), and 75 ppm caused eye, nose, and throat irritation in the many of subjects. Somewhat different were the results obtained in more recent studies with six human volunteers exposed for 7 min at 4–9 concentrations of cyclohexanone (380). The sensory irritation threshold for cyclohexanone was determined to be 160 ppm in volunteers asked to indicate the presence or absence of eye, nose, and throat irritation during the vapor exposure.

6.4.2.2.2 Chronic and Subchronic Toxicity. Jacobsen et al. described a case where an individual exposed to cyclohexanone, white spirits, and isopropanol was affected by weekly to monthly epileptic seizures for the 28 yr he was employed as a silk screen printer (381). The individual often experienced headache, nausea, vertigo, and a feeling of drunkenness while on the job. The seizures were preceded by periods of irritability, visual hallucinations, and depressed consciousness several days before onset. The seizures disappeared when the printer left for other employment, but briefly reappeared when he returned to a solvent environment. The individual did not show any of the signs or symptoms associated with chronic toxic encephalopathy, and his cerebral tomography test results were normal.

6.4.2.2.3 Pharmacokinetics, Metabolism, and Mechanisms. Mráz et al. examined the uptake and urinary excretion of cyclohexanone and its metabolites in 8 male and female human volunteers exposed by inhalation for 8 h at concentrations of 101, 207, or 406 mg/m^3 (382). The respiratory retention of cyclohexanone was about 58% at all three exposure levels. Analysis of hydrolyzed 72-h urine specimens revealed the presence of three metabolites: cyclohexanol, 1,2-cyclohexanediol, and 1,4-cyclohexane. Approximately 40% of the dose was recovered as the 1,2-diol, 19% as the 1,4-diol, and 1% as the alcohol. The parent cyclohexanone was not detected in any of the urine specimens. Ong et al. reported finding cyclohexanol in the urine of workers occupationally exposed to cyclohexanone in a video tape manufacturing plant (383). Cyclohexanol was isolated from the glucuronide conjugate present in spot urine specimens collected at the end of the shift for 27 employees with a time-weighted average exposure to cyclohexanone of 2–30 ppm. The urine cyclohexanol values ranged up to about 14 mg/g of creatinine and showed a good linear relationship with the cyclohexanone exposure concentration ($r = 0.88$).

Mills and Walker detected a variety of cyclohexanediols in the urine of infants in a hospital neonatal unit and associated their appearance to the metabolism of cyclohexanone that was present in the dextrose solutions used for intravenous feeding (384). Using gas chromatography–mass spectrometry, unconjugated *trans*-1,2-cyclohexanediol, *cis*-1,3-cyclohexanediol, *trans*-1,4-cyclohexanediol, and *cis*-1,2-cyclohexanediol were detected in 101 of 584 urine specimens collected from a total of 278 infants. Cyclohexanol and glucuronide conjugates of cyclohexanediols were not detected in any of the specimens. All the diol-containing urine specimens were obtained from infants that received intravenous feeding. Subsequent analysis revealed that 150 mL of dextrose infused over 24-h period would contain an average of about 0.9 mg of cyclohexanone. Cyclohexanone had previously been shown to leach into solution from the polyvinyl chloride plastic used to construct the infusion bags (305, 385). The failure to find glucuronide conjugates of cyclohexanol was ascribed to the inability of neonates to carry out this metabolic reaction for several weeks following birth. Mutti et al. found that small amounts of cyclohexanone ($\leq$1%) were excreted in the urine of human volunteers exposed to cyclohexane for 4 h at concentrations ranging from 53 to 266 mg/m^3 (386).

6.4.2.2.4 Reproductive and Developmental. Jankovic and Drake developed a screening method for assessing the occupational reproductive health risks from solvent exposures by constructing a toxicological database and comparing the observed health effect levels with the current occupational exposure limits (387). Using this approach, an occupational reproductive guideline (ORG) level of 100 mg/m^3 was established for cyclohexanone.

6.4.2.2.5 Carcinogenesis. No case studies reporting carcinogenesis following exposure to cyclohexanone were identified.

6.4.2.2.6 Genetic and Related Cellular Effects Studies. Perocco et al. examined the cytotoxicity and genotoxicity of cyclohexanone using cultured human lymphocytes (388). A cyclohexanone concentration of 10 mM caused a marked decrease in tritiated thymidine incorporation without affecting cell viability. The response was not observed at lower concentrations of cyclohexanone (0.1 or 1 mM) or when a rat liver S9 was incorporated into the assay. Lederer et al. observed some chromosomal abnormalities in the karyotypes from cultured human lymphocytes incubated with cyclohexanone concentrations of 0.1, 1.0, or 10 mM (389).

6.4.2.2.7 Other: Neurological, Pulmonary, Skin Sensitization, etc. A case of allergic contact dermatitis was reported in a woman who worked with a solution of polyvinyl chloride in cyclohexanone (390). Patch testing with a 10% aqueous solution of cyclohexanone showed some redness in the treatment area after 2 days that further developed into an itchy papular rash 4 days after treatment. Subsequent testing of 20 naive subjects using a 50% aqueous solution of cyclohexanone produced essentially negative results.

Likewise, Mitran et al. examined the neurotoxic effects in a group of 75 workers in a furniture factory who were exposed while coating wood with cyclohexanone (391). The 8-h time-weighted average exposures ranged from 162 to 368 mg/m^3. The cyclohexanone exposures were observed to cause an increase in the percentage of reported neurotoxic symptoms (mood disorders, irritability, memory difficulties, sleep disturbances, and headache) and rheumatic symptoms (bone pain, joint pain, and muscular pain). Symptoms of ocular and respiratory tract irritation also appeared to be more common, but tests of statistical significance were not performed. In addition, measurement of the motor nerve conduction revealed a statistically significant increase in the latency period at distal regions of the median, ulnar, and peronal nerves. Other measures of nerve conduction were also occasionally affected relative to controls. The authors noted, however, that the nerve conduction velocity experiments were relatively insensitive to any underlying CNS abnormalities.

6.4.2.3 Epidemiology Studies. Very few epidemiology studies have been published that involve exposure to cyclohexanone.

6.4.2.3.1 Acute Toxicity. Chia et al. reported an increase in the prevalence of acute symptoms in a group of 19 employees exposed to cyclohexanone and 3 other solvents (toluene, tetrahydrofuran, and MEK) in a video tape manufacturing plant (392). The mean time-weighted average exposure to cyclohexanone ranged from about 1 to 17 ppm in different regions of the plant. Statistically significant increases in headache, eye and nose irritation, coughing, and irritability were reported the solvent-exposed group relative to controls.

6.4.2.3.2 Chronic and Subchronic Toxicity. No case reports of subchronic or chronic toxicity to humans following exposure to cyclohexanone are available.

6.4.2.3.3 Pharmacokinetics, Metabolism, and Mechanisms. There are no case reports on pharmacokinetics, metabolism, or mechanisms in humans following exposure to cyclohexanone.

6.4.2.3.4 Reproductive and Developmental. There are no case reports of reproductive or developmental toxicity in humans following exposure to cyclohexanone.

6.4.2.3.5 Carcinogenesis. There are no case studies of carcinogenicity in humans following exposure to cyclohexanone.

6.4.2.3.6 Genetic and Related Cellular Effects Studies. There are no case studies of genetic and related cellular effects following exposure to cyclohexanone.

6.4.2.3.7 Other: Neurological, Pulmonary, Skin Sensitization, etc. Chia et al. reported a decrease in neurobehavioral test scores for a group of 19 employees exposed to cyclohexanone, toluene, tetrahydrofuran, and MEK in a video tape manufacturing plant (392). The mean time-weighted average exposure to cyclohexanone ranged from about 1 to 17 ppm in different regions of the plant. A statistically significant change was found in the neurobehavioral test results for dexterity, digit span memory, and visual reproduction when age, drinking, and smoking history were taken into consideration. After examining the results as function of total exposure, the effects failed to show any dose relationship.

6.5 Standards, Regulations, or Guidelines of Exposure

Hygienic standards for cyclohexanone are listed in Table 76.5.

6.6 Studies on Environmental Impact

The environmental fate and transport of cyclohexanone following its release to air, water, and soil have been reported (223). Cyclohexanone is expected to have high mobility in soil and volatilize from the surface layers. If emitted to the air, cyclohexanone will degrade relatively rapidly (half-life $\leq$ 4.3 d) due its oxidation by photolysis and reaction with hydroxyl radicals. If released into water, cyclohexanone will not volatilize rapidly, except in rapidly moving streams and rivers, where the half-life will be 4.1 d. The aerobic biodegradation of cyclohexanone is relatively high.

7.0 Methyl *n*-Amyl Ketone

7.0.1 CAS Number: [110-43-0]

7.0.2 Synonyms: 2-Heptanone, methyl pentyl ketone, MnAK, MAK, amyl methyl ketone, methyl amyl ketone, *n*-amyl methyl ketone, butyl acetone; heptan-2-one, and ketone C-7

7.0.3 Trade Names: NA

7.0.4 Molecular Weight: 114.19

7.0.5 Molecular Formula: $CH_3CO(CH_2)_4CH_3$

7.0.6 Molecular Structure:

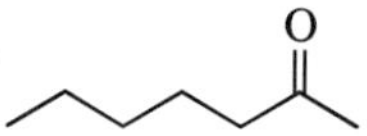

7.1 Chemical and Physical Properties

Methyl *n*-amyl ketone is a clear, colorless liquid of low volatility. Its boiling point is 151.1°C, and its melting point is −35°C. Additional physical–chemical properties are summarized in Table 76.4.

7.1.2 Odor and Warning Properties

MnAK has a penetrating but fruity odor. Its odor threshold is about 0.15 (223) to 0.35 ppm (393).

7.2 Production and Use

MnAK is used as a solvent for lacquers, paints, and synthetic resin coatings with a high solids content and as a fragrance and flavoring additive. Specific information about production and use was not identified. MnAK is produced by the condensation of acetone with butyraldehyde.

7.3 Exposure Assessment

7.3.1 Air

Information about ambient levels of MnAK in air was not identified.

7.3.2 Background Levels

MnAK is a natural product of intermediary metabolism of organisms from bacteria to mammals. MnAK has been identified in human mother's milk, indoor air in spaces contaminated with molds, in decomposing meat and compost, and in exhaled breath for unexposed humans and animals (54, 394–398). It occurs naturally in oil of cloves and cinnamon bark oil.

7.3.3 Workplace Methods

Either of two NIOSH methods can be used to determine MnAK concentrations in workroom air (116). The first method (NMAM 2549) is a screening method for volatile organic compounds that uses thermal desorption tubes containing graphitized carbon and molecular sieve adsorbents for sampling and gas chromatography–mass spectrometry for analysis. The second method (NMAM 1301) is a more specific method for ketones that uses a solid sorbent tube with coconut shell charcoal and gas chromatography with flame ionization detection.

7.3.4 Community Methods

Specific community methods were not identified.

7.3.5 Biomonitoring/Biomarkers

No biomonitoring methods or biomarkers for MnAK were identified.

7.4 Toxic Effects

Single acute and subchronic toxicity studies and repeated inhalation studies indicate that methyl *n*-amyl ketone has a low degree of toxicity. Evidence is available that shows that it is not neurotoxic. Although high concentrations are capable of producing serious narcotic effects, the penetrating odor and eye and nose irritation noted at low concentrations should provide good warning signs.

7.4.1 Experimental Studies

7.4.1.1 Acute Toxicity. The acute oral LD_{50} in rats has been reported to be 1670 (324), 2407 (23), and approximately 1600 mg/kg (3). The acute oral LD_{50} in mice has been reported as 730 mg/kg (399), while others have reported survival following doses of as high as 1600 mg/kg (3). The intraperitoneal LD_{50} in rats is 800 mg/kg, while the intraperitoneal LD_{50} in mice was between 400 and 800 mg/kg (3).

Respiratory depression to 80% or 37% of the normal frequency in rats was obtained at a concentration of 535 or 1225 ppm, respectively. Depression to 50% of normal breathing frequency was attained at a concentration of 895 ppm (164). Inhalation of 345–545 ppm of methyl *n*-amyl ketone for 4 h resulted in significantly decreased duration of the immobility phase (from 27 to 65%, respectively) in a "behavioral despair" swimming test. The differences were statistically significant at all exposure concentrations (164). The concentration of MnAK that depressed respiration by 50% (RD_{50}) in mice was 3790 ppm (26).

Smyth et al. (325) reported mild eye irritation in rabbits. Other investigators reported mild (Draize score 11–18) to minimal (Draize score 1–3) irritation at concentrations of 10–100% and 1–3%, respectively (153).

Smyth et al. (324) reported that methyl *n*-amyl ketone applied uncovered to the abdomen of rabbits produced moderate skin irritation. The acute dermal LD_{50} was 12.6 mL/kg. MnAK applied undiluted to the intact or abraded skin of the back of rabbits under an occlusive wrap for 24 h was not irritating (400). The undiluted compound in quantities from 5 to 20 mL/kg when held in contact with the depilated skin of guinea pigs under an occlusive wrap for 24 h produced slight to moderate irritation. No deaths occurred, and there was no evidence of percutaneous absorption (3).

Single exposures of guinea pigs to concentrations of 1500 ppm produced irritation of the mucous membranes; 2000 ppm was strongly narcotic; and 4800 ppm produced narcosis and death in 4–8 h of exposure (25). Rats survived 4-h exposures to 2000 ppm, whereas 4000 ppm killed all six animals (324). The maximum length of inhalation exposure to a saturated atmosphere of methyl *n*-amyl ketone with no deaths was 30 min (324). A concentration of 5100 ppm killed all three rats exposed within 4 h, and 4100 ppm killed all three rats within 6 h of exposure. All rats survived 6-h exposures to concentrations of 830–2000 ppm. Clinical signs of toxicity at all concentrations were piloerection, vasodilation, hyperpnea, ataxia, prostration, and dyspnea (3). When rats were exposed for 6–8 h to 1500 ppm methyl *n*-amyl ketone or lower, behavioral changes were not observed, while like exposures to 1575 ppm or higher resulted in effects similar to those seen after intraperitoneal administration of 175 mg/kg (401). Rats and monkeys were exposed by inhalation to 131 and 1025 ppm methyl *n*-amyl ketone 6 h/d, 5 d/wk for 9 mo. No alterations of pulmonary function, electrocardiographic, or clinical chemistry parameters were noted in either species. Inhalation of methyl *n*-amyl ketone did not induce microsomal enzymes, and no liver pathology was observed (402). In the same series of studies, no neurologic deficits, as measured by motor nerve conduction velocity or evoked muscle action potential, were noted at the highest level tested (11). Exposure of rats for a total of 2300 h (regimen not described) to 278 ppm methyl *n*-amyl ketone resulted in no signs of neurological impairment, and no histologic changes were noted in the peripheral nerves (403). Tison et al. (404) reported that exposure to 650 ppm of methyl *n*-amyl ketone did not produce peripheral neuropathy in rats or chickens.

7.4.1.2 Chronic and Subchronic Toxicity. Gaunt et al. (405) gave doses of 20, 100, and 500 mg/kg methyl *n*-amyl ketone in oil by gavage to male and female rats for 13 wk. No untoward effects were seen in the animals given 20 mg/kg. Liver weights were increased in both sexes, and kidney weights were increased in the males only in the 100 or 500 mg/kg dose groups. No statistically significant effects were found on body weight gain, feed consumption, hematology, or histopathology. Urinalysis was unremarkable except for increased numbers of cells observed in urine from animals administered 100 or 500 g/kg/d. Rats given 0.5% methyl *n*-amyl ketone in their drinking water for 12 wk failed to develop clinical or histologic signs of peripheral neuropathy (8). Anger et al. (401) exposed rats to methyl *n*-amyl ketone through a sequence of 24 daily intraperitoneal injections ranging from 18 to 175 mg/kg. The animals were tested for behavioral changes daily for 1 h on a multiple fixed ratio–fixed interval schedule of reinforcements. Tests were conducted 15 min after the injections. Little or no change was seen in the fixed interval response after 18 mg/kg was given intraperitoneally; moderate decreases were seen after 37 and 74 mg/kg; and near-cessation of response was found after 175 mg/kg. Subcutaneous injections of 400 mg/kg methyl *n*-amyl ketone 5 d/wk for 15 wk did not have an effect on growth, did not produce clinical signs of toxicity, and did not affect conduction velocity of motor or sensory nerves or motor distal latency (406). When methyl *n*-amyl ketone (456 mg/kg) was administered with equimolar doses of MnBK by subcutaneous injection 5 d/wk for 20 wk, the neurotoxicity of MnBK, as measured by motor fiber conduction velocity and motor distal latency, was enhanced ($p < 0.01$). The combined effect was, in general, less than the effect seen with methyl *n*-hexyl ketone and MnBK but less than or equal to that seen with MnBK and either MPK or methyl ethyl ketone (407).

7.4.1.3 Pharmacokinetics, Metabolism, and Mechanisms. MnAK undergoes carbonyl reduction to a secondary alcohol and o-1 oxidation to a hydroxyketone, which is further oxidized to 2,6-heptanedione (3). 2,6-Heptanedione has been tested and shown not to be neurotoxic in rats (3). Kamil et al. had earlier reported that 41% of an orally administered dose of 0.95 g/kg in rabbits was excreted as 2-heptyl glucuronide (408). MnAK is excreted in the breath of unexposed rats (54). Inhibition of P450 2EL increases the amount of MnAK exhaled (54).

7.4.1.4 Reproductive and Developmental. A reproductive and developmental screening test was conducted with MnAK. In this study, groups of male and female rats were exposed to 0, 80, 400, or 1000 ppm MnAK. Exposures were conducted 6 h/d, 7 d/wk. The study consisted of a 14-d premating phase, a mating phase lasting from 1 to 14 d, a gestation phase, and a lactation phase (days 0–4 postpartum). Females received 34–47 exposures and males received 50 exposures. In this study, there were transient reductions in feed consumption and transient body weight change, but no changes in reproductive organ weights, gross pathology, or histopathology of the reproductive organs. The only statistically significant change noted in any of the reproductive indices was a slightly increased growth rate for the pups from the low-concentration group between days 0 and 1 postpartum. This was not dose-related and was not considered to be an exposure-related effect. The no-observed-effect level (NOEL) for this study for reproductive and developmental toxicity was determined to be 1000 ppm (409).

7.4.1.6 Genetic and Related Cellular Effects Studies. MnAK has been shown to be negative in the Ames *Salmonella* bacterial mutation assay (Shell Chemical Company, unpublished results, 1994) and in an *in vitro* chromosomal aberration assay (Eastman chemical company, unpublished results, 1998). Albro et al. (410) reported that MnAK was able to bind to rat liver DNA *in vitro* to the extent of 400 pmol/mg DNA, and that MnAK given by gavage was also able to bind to rat liver *in vivo*. In a more recent study using direct, sensitive techniques (411), radiolabeled MnAK was added *in vitro* to rat liver DNA and incubated for 4 h. No binding to DNA was detectable. In a subsequent study, female F344 rats were exposed to 0, 80, 400, or 1000 ppm MnAK, 6 h/d for 10 d. Liver DNA was purified from the control and high-dose animals and ^{32}P-Postlabeling was used to assay for adducts. No adducts were detected using this technique.

In brain synaptosomes prepared from MnAK anesthetized mice, membrane 1,6-diphenyl-1,3,5-hexatriene fluidity was decreased (412).

7.4.1.7 Other: Neurological, Pulmonary, Skin Sensitization. Chloroform-induced hepato- and nephrotoxicity were evaluated in rats pretreated with oral doses of from

114 to 1710 mg/kg MnAK. Animals were challenged by injection of 0.5 or 0.75 mL chloroform/kg body weight 18 h after oral doses of MnAK. Potentiation of chloroform-induced liver and kidney toxicity, as measured by mortality; plasma creatinine, glutamic-pyruvic transaminase, or ornithine carbomoyl transferase activity; or uptake of *p*-aminohippurate by renal cortical slices were noted, though the magnitude of the potentiated response was nonlinear (56, 413, 414).

7.4.2 Human Experience

7.4.2.1 General Information. The principal route of exposure in the occupational setting is by inhalation. Skin and eye contact may also occur. MnAKs low odor threshold, its penetrating odor, and its irritant properties at levels below those that produce serious effects should preclude injury due to this ketone.

7.4.2.2 Clinical Cases. No clinical case reports of adverse health effects following exposure to MnAK have been published.

7.4.2.3 Epidemiology Studies. Skin sensitization studies were conducted with 26 human volunteers. At a concentration of 4% MnAK in petrolatum, no positive reactions were observed (415).

No epidemiology studies reporting adverse health effects following exposure to MnAK have been published. A cross-sectional study of paint makers designed to assess cognitive performance and mental health of workers exposed to solvents including MnAK below current exposure levels found no adverse effects (417, 418).

7.5 Standards, Regulations, or Guidelines of Exposure

Hygienic standards for MnAK are listed in Table 76.5.

7.6 Studies on Environmental Impact

Studies on the environmental impact of MnAK are not available; however, MnAK is not expected to present any unusual hazard to the environment. Since MnAK would be expected to biodegrade, long-term environmental effects are not anticipated.

8.0 Methyl Isoamyl Ketone

8.0.1 CAS Number: *[110-12-3]*

8.0.2 Synonyms: 5-Methyl-2-hexanone, 2-methyl-5-hexanone, methyl isopentyl ketone, MIAK, and methyl hexanone

8.0.3 Trade Names: NA

8.0.4 Molecular Weight: 114.19

8.0.5 Molecular Formula: $CH_3COCH_2CH_2CH(CH_3)_2$

8.0.6 *Molecular Structure:* 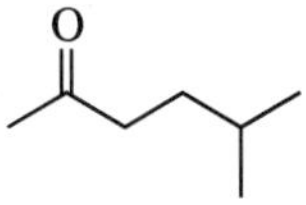

8.1 Chemical and Physical Properties

MIAK is miscible with most organic solvents. It is soluble in water to 5.4 g/L at 20°C. Its boiling point is 144°C, and its melting point is −73.9°C; log K_{ow} is 1.88. Additional physical–chemical properties are summarized in Table 4.

8.1.2 *Odor and Warning Properties*

MIAK is a clear, colorless liquid with a sharp but pleasant, sweet odor. The odor threshold level for MIAK is 0.012 ppm (223).

8.2 Production and Use

MIAK is primarily used as a solvent for nitrocellulose, cellulose acetate, butyrate, acrylics, and vinyl copolymers. Commercial material is typically 97–98% pure and is manufactured by oxidation of 5-methyl-2-hexanol.

8.3 Exposure Assessment

8.3.1 *Air*

The principal routes of exposure to MIAK are expected to be inhalation and dermal contact. Ingestion is not an expected route of exposure for industrial use. Due to low volatility, significant airborne concentrations are not expected unless the solvent is heated.

8.3.2 *Background Levels*

Ketones are naturally present in the environment and can be released to the air during composting operations, or any activity that disturbs soils. MIAK has been detected in fumes from cooking meat (416).

8.3.3 *Workplace Methods*

Either of two NIOSH methods can be used to determine MIAK concentrations in workroom air (116). The first method (NMAM 2549) is a screening method for volatile organic compounds that uses thermal desorption tubes containing graphitized carbon and molecular sieve adsorbents for sampling and gas chromatography–mass spectrometry for analysis. The second method (NMAM 1300) is a more specific method for ketones that uses a solid sorbent tube with coconut shell charcoal and gas chromatography with flame ionization detection.

8.3.4 *Community Methods*

Specific community methods for measurement of MIAK were not identified.

8.3.5 Biomonitoring/Biomarkers

Specific biomonitoring methods and biomarkers for MIAK were not identified.

8.4 Toxic Effects

MIAK is a material of low volatility with a sharp but pleasant odor. Exposure is expected to be via inhalation or possibly dermal contact. Ingestion is expected to be rare and has not been reported to occur. The principal hazards associated with exposure are eye, nose, and throat irritation. Overall, MIAK has a low degree of systemic toxicity. However, repeated exposure to large amounts of this ketone may lead to hepatic toxicity.

8.4.1 Experimental Studies

8.4.1.1 Acute Toxicity. The oral LD_{50} for MIAK in rats has been reported to be 3870 (324), 3200 (3), and 2542 mg/kg (23). The LD_{50} in mice is between 3200 and 6400 mg/kg (3). The intraperitoneal LD_{50} is between 400 and 800 mg/kg for rats and approximately 800 mg/kg for mice (23).

Respiratory depression to 73% or 39% of the normal frequency has been reported in rats exposed at concentrations of 416 or 1515 ppm, respectively. Respiratory depression to 50% of normal frequency was estimated to occur at a concentration of 1222 ppm (164). Inhalation of 270–637 ppm of MIAK for 4 h resulted in significantly decreased duration of the immobility phase (from 26 to 68%, respectively) of a "behavioral despair" swimming test. This enhanced performance was statistically significant at all exposure concentrations (164).

MIAK produced slight eye irritation when dropped into the conjunctival sacs of six rabbits' eyes, three of which were washed with water promptly after instillation of the material (3).

When undiluted MIAK (5–20 mL/kg) was held in contact with the abdomen of guinea pigs under an occlusive wrap for 24 h, slight irritation developed. Percutaneous toxicity may have occurred at 20 mL/kg since weight gain was reduced (3). Smyth et al. (324) have reported a dermal LD_{50} of 10 mL/kg for rabbits. Repeated daily applications to the backs of guinea pigs by an open rub-on technique resulted in exacerbation of the irritative response. After 10 applications, all animals developed eschars (3).

Single, 6-h exposures of rats to 802 ppm MIAK produced no effects. At a concentration of 1603 ppm, a decreased response to noise was observed. A concentration of 3207 ppm produced eye irritation and a decreased respiratory rate, narcosis, and death in one of four rats. Four rats exposed to 5678 ppm died within 2 h. The 6-h LC_{50} was calculated to be 3813 ppm for rats (3). Smyth et al. (324) reported that the lowest dose to produce an effect was 2000 ppm for 4 h of exposure.

8.4.1.2 Chronic and Subchronic Toxicity. Groups of three rats were administered 1000, 2000, or 4000 mg/kg of MIAK by gavage, 5 d/wk for 3 wk. At 4000 mg/kg, all rats died within 1.5 h of dosing due to central nervous system depression and subsequent cardiorespiratory failure; pathologic lesions were limited to vascular congestion. At 2000 mg/kg, all rats survived, and effects noted included reduced weight gain, reduced feed consumption, chronic irritation of the gastric mucosa, hepatocyte hypertrophy, and renal hyaline droplet formation. At 1000 mg/kg, the only effect noted was evidence of minor chronic gastric irritation (3). Subsequent doses of 2000 mg/kg MIAK were given by gavage 5 d/wk for 13 wk to a group of eight rats to determine its neurotoxic potential. Feed consumption was reduced only the first week of exposure, but this produced an early body weight gain depression that was not fully recovered by the end of the study. Hematologic determinations were comparable to controls, but serum clinical chemistries showed slight elevations of hepatic enzymes. Absolute and relative liver and adrenal gland weights and relative kidney weights were increased. Histopathologic changes included chronic gastric irritation, individual hepatocyte degeneration, diffuse hepatocyte hypertrophy, and microfoci of hepatocyte hyperplasia (3).

Five rats exposed to 400 ppm MIAK, 5 d/wk for 12 exposures showed no effects on body weight, organ weights, hematology, clinical chemistries, or gross and histopathology (3). Rats were also exposed for 6 h/d, 5 d/wk to target exposures of 0, 1000, or 2000 ppm for a total of 12 exposures over 16 d or to target exposures of 0, 200, 1000, or 2000 ppm for a total of 69 exposures over 96 d (419). Body weights, hematology, and serum clinical chemistries were comparable to controls in both studies. Clinical signs of toxicity under both exposure regimes included lethargy, decreased aural response, and nasal and eye irritation. Excretion of gel-like casts was seen in seminal fluid in males exposed to 2000 or 1000 ppm MIAK in both studies. Additional effects noted in both studies in both male and female animals exposed to 2000 or 1000 ppm MIAK included increased absolute and relative liver and kidney weights. Regeneration of tubular epithelium and hyalin droplet formation were noted in males in the 2-wk and 90-d studies. Minor tubular regeneration was noted in females exposed to 2000 ppm for 90 d. No compound-related changes were noted in males or females following exposure to 200 ppm (419).

8.4.1.3 Pharmacokinetics, Metabolism, and Mechanisms. No pharmacokinetics and metabolism studies are available for MIAK.

8.4.1.4 Reproductive and Developmental. No reproductive and developmental toxicity studies are available for MIAK.

8.4.1.5 Carcinogenesis. No carcinogenicity studies are available for MIAK.

8.4.1.6 Genetic and Related Cellular Effects Studies. *In vitro* studies with MIAK indicate that it can penetrate cell membranes due to its lipophilicity. This type of *in vitro* exposure may result in conformational changes in cell membranes (155).

8.4.1.7 Other: Neurological, Pulmonary, Skin Sensitization. Slight skin sensitization was observed in one of five guinea pigs immunized with MIAK and Freund's complete adjuvant (3).

8.4.2 Human Experience

8.4.2.1 General Information. Exposure to MIAK in the workplace is most likely to occur by skin contact, eye contact, or inhalation, particularly at elevated temperatures. Repeated skin exposure may result in contact dermatitis of both the allergic and nonallergic types. Good industrial hygiene practices should include skin and eye protection and adequate ventilation. The warning properties associated with MIAK vapor should prevent overexposure.

8.4.2.2 Clinical Cases. No clinical case studies reporting adverse health effects associated with exposure to MIAK were identified.

8.4.2.3 Epidemiology Studies. No epidemiology studies reporting adverse health effects associated with exposure to MIAK were identified.

8.5 Standards, Regulations, or Guidelines of Exposure

Hygienic standards for MIAK are listed in Table 76.5.

8.6 Studies on Environmental Impact

No studies on the environmental impact of MIAK are available; however, MIAK is not expected to present any unusual hazard to the environment. Since MIAK would be expected to biodegrade, long-term environmental effects are not anticipated.

9.0 Ethyl *n*-Butyl Ketone

9.0.1 CAS Number: [106-35-4]

9.0.2 Synonyms: 3-Heptanone, heptan-3-one, and EnBK

9.0.3 Trade Names: NA

9.0.4 Molecular Weight: 114.19

9.0.5 Molecular Formula: $CH_3CH_2CO(CH_2)_3CH_3$

9.0.6 Molecular Structure:

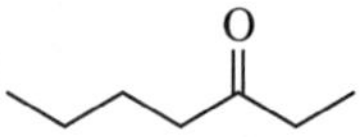

9.1 Chemical and Physical Properties

9.1.1 General

EnBK has a boiling point of 147.4°C and a melting point of −39°C. It is miscible in alcohol and ether and slightly soluble to 0.43% by weight at 20°C in water. Additional physical–chemical properties are summarized in Table 76.4.

9.1.2 Odor and Warning Properties

EnBK is a clear, colorless liquid with a strong fruity odor.

9.2 Production and Use

EnBK is produced by catalytic dehydrogenation of 3-heptanol or by hydrogenation of the mixed alcohol condensation product of propionaldehyde and methyl ethyl ketone. Commercial samples can be 95% pure. EnBK is used as a flavoring agent in food and beverages and as a solvent for lacquers, polyvinyl resins, and nitrocellulose resins.

9.3 Exposure Assessment

9.3.1 Air

Information on airborne levels if EnBK were not identified.

9.3.2 Background Levels

EnBK has been detected in breast milk (395) and the expired air of nonsmokers (394). EnBK has been identified in urine of unexposed humans (133).

9.3.3 Workplace Methods

Either of two NIOSH methods can be used to determine EnBK concentrations in workroom air (116). The first method (NMAM 2549) is a screening method for volatile organic compounds that uses thermal desorption tubes containing graphitized carbon and molecular sieve adsorbents for sampling and gas chromatography–mass spectrometry for analysis. The second method (NMAM 1301) is a more specific method for ketones that uses a solid sorbent tube with coconut shell charcoal and gas chromatography with flame ionization detection.

9.3.4 Community Methods

Specific community methods for monitoring EnBK were not identified.

9.3.5 Biomonitoring/Biomarkers

Specific biomonitoring methods and biomarkers for EnBK were not identified.

9.4 Toxic Effects

9.4.1 Experimental Studies

9.4.1.1 Acute Toxicity. The acute oral LD_{50} of EnBK is reported to be 2.76 g/kg for rats (418). Smyth et al. (420) and Marhold (231) reported that EnBK produced mild skin irritation in rabbits. Studies by Moreno (421) indicated that EnBK applied undiluted to the intact or abraded skin of rabbits under an occlusive wrap for 24 h was moderately irritating. The dermal LD_{50} was greater than 20 mL/kg for rabbits (420). EnBK was tested for its potential to produce contact dermatitis in the guinea pig and was found to be negative (422).

Smyth et al. (420) reported that rats survived 4-h exposures to 2000 ppm of EnBK, but all rats exposed for 4 h to 4000 ppm died.

A single 6-h exposure to 5800 ppm EnBK killed all three rats exposed. Six-hour exposures to 420 or 3000 ppm did not produce death. Ataxia, prostration, and narcosis were seen at the higher level (3). Rats and guinea pigs survived 6-h daily exposures to 400–500 ppm for 10 d. Clinical signs included piloerection, vasodilation, and lacrimation. No pathologic lesions were found in the lungs, livers, or kidneys (3).

Katz et al. (35) studied the neurotoxic potential of EnBK in rats. Exposures to 700-ppm EnBK vapors for 24 wk produced no clinical signs of systemic toxicity or neurotoxicity. No significant effects were seen on body weight gain, hematology, clinical chemistries, and histopathology except for a significant depression in total white blood cells. It was postulated that the lack of neurotoxicity may have been due to the low concentration of the known neurotoxic metabolite, 2,5-heptanedione, found in the serum of these animals.

9.4.1.2 Chronic and Subchronic Toxicity. Homan and Maronpot (423) reported that EnBK did not produce discernible neurotoxic effects when administered in the drinking water of rats for 120 d. Other investigators administered EnBK by gavage, 5 d/wk for 14 wk at doses of 0.25, 0.5, 1, 2, or 4 g/kg. Animals given 4 g/kg died after 1 or 2 doses, while animals given doses of 2 g/kg developed peripheral neuropathy and microscopically distinguishable neuropathologic lesions, including "giant" axonal neuropathy (424). Doses of 1 g/kg EnBK did not produce neurotoxicity. EnBK was also administered to rats using the same dosing regime described previously but followed by doses of 1.5 g/kg EnBK, 0.75 g/kg EnBK, 1.5 g/kg 5-methyl-2-octanone. When either 4 or 2 g/kg EnBK was given in conjunction with methyl ethyl ketone or 5-methyl-2-octanone, all animals died. Neurotoxicological signs and neuropathological lesions were evident when EnBK was dosed along with 1.5 g/kg methyl ethyl ketone. In the group exposed to 1 g/kg EnBK followed by 1.5 g/kg 5-methyl-2-octanone, slight weakness of the hind limbs was noted during weeks 4 to 6, but these animals later appeared normal. An unequivocal neuropathy was not observed in animals from this group. Other test groups did not develop either clinically apparent neurotoxicity or neuropathological lesions (424). Rats exposed to *n*-heptane excrete EnBK in their urine (425).

9.4.1.3 Pharmacokinetics, Metabolism, and Mechanisms. Analysis of the sera of rats exposed to 700 ppm EnBK revealed the presence of 6-hydroxy-3-heptanone and 2,5-heptanedione (35). These compounds are the seven-carbon atom counterparts of the neurotoxic metabolites of MnBK, that is, 5-hydroxy-2-hexanone and 2,5-hexanedione. These data suggest that ethyl *n*-butyl ketone follows a metabolic pathway similar to that of MnBK (see Fig. 76.1). Urine from animals gavaged with EnBK or EnBK together with methyl ethyl ketone or 5-methyl-2-octanone contained the neurotoxic metabolites 2,5-heptanedione and 2,6-hexanedione, although inhalation exposures to high concentrations of EnBK failed to produce levels of neurotoxic metabolites high enough to produce neuropathy in rats. Methyl ethyl ketone appeared to increase excretion of both 2,5-heptanedione and 2,6-hexanedione, while 5-methyl-2-octanone did not affect the excretion of either metabolite (424). Rats exposed to *n*-heptane excrete EnBK in their urine (425).

9.4.2 Human Experience

9.4.2.1 General Information. EnBK has been generally recognized as safe and is approved by the FDA for use as an artificial flavoring substance in foods. Its toxicity is low by the oral and percutaneous routes. It produces only mild irritation when in contact with the skin and eyes, but repeated or prolonged skin exposure may result in dermatitis. EnBK may be metabolized by rats to 2,5-heptanedione, which, like its six-carbon counterpart, 2,5-hexanedione, has been shown to be neurotoxic. However, repeated inhalation exposures to high concentrations of EnBK failed to produce serum levels of 2,5-heptanedione high enough to produce neuropathy in rats. The low volatility of this ketone along with good industrial hygiene practices should prevent the occurrence of serious toxic effects.

Exposure in the industrial setting would be principally by inhalation and by skin and eye contact. Appropriate protective equipment and good industrial hygiene practices should preclude injury to the skin and eyes, and the low volatility of EnBK should prevent injury due to inhalation of the vapors.

9.4.2.2 Clinical Cases. No clinical cases reports of adverse health effects associated with exposure to EnBK are available.

9.4.2.3 Epidemiology Studies. In controlled studies, ethyl *n*-butyl ketone (4% in petrolatum) produced no irritation to human skin after 48 h under an occlusive patch. This same concentration used in maximization tests on 25 volunteers failed to produce sensitization (415).

No epidemiology studies reporting adverse health effects associated with exposure to EnBK are available.

9.5 Standards, Regulations, or Guidelines of Exposure

The hygienic standards for ethyl *n*-butyl ketone are listed in Table 76.5.

9.6 Studies on Environmental Impact

Studies on the environmental impact of EnBK are not available; however, EnBK is not expected to present any unusual hazard to the environment. Since EnBK would be expected to biodegrade, long-term environmental effects are not anticipated.

10.0 Di-*n*-propyl Ketone

10.0.1 CAS Number: *[123-19-3]*

10.0.2 Synonyms: Dipropyl ketone, 4-heptanone, propyl ketone, butyrone, heptane-4-one, and DnPK

10.0.3 Trade Names: NA

10.0.4 Molecular Weight: 114.19

10.0.5 Molecular Formula: $(CH_3CH_2CH_2)_2CO$

10.0.6 Molecular Structure:

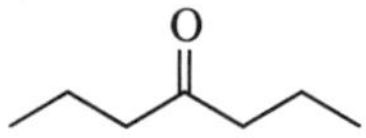

10.1 Chemical and Physical Properties

DnPK has a boiling point of 143.7°C and a melting point of −32.1°C; its vapor pressure is 5.5 mm Hg. Additional physical–chemical properties are summarized in Table 76.4.

10.1.1 General

DnPK is a stable, colorless, liquid that is essentially insoluble in water. It is categorized as a DOT flammable liquid.

10.1.2 Odor and Warning Properties

DnPK has a pleasant but penetrating odor and a burning taste.

10.2 Production and Use

DnPK is used as a solvent for cellulosics, oils, resins, and polymers; it is also used in the manufacture of optical gratings, and is approved for use as a flavorant.

10.3 Exposure Assessment

Under industrial conditions, DnPK exposure may be expected to occur via inhalation or skin contact. Since it is used as a flavorant, small amounts may appear in the diet. Information on airborne and background levels of DnPK are not available.

10.3.3 Workplace Methods

Either of two NIOSH methods can be used to determine DnBK concentrations in workroom air (116). The first method (NMAM 2549) is a screening method for volatile organic compounds that uses thermal desorption tubes containing graphitized carbon and molecular sieve adsorbents for sampling and gas chromatography–mass spectrometry for analysis. The second method (NMAM 1300) is a more specific method for ketones that uses a solid sorbent tube with coconut shell charcoal and gas chromatography with flame ionization detection.

10.3.4 Community Methods

Specific community methods for measurement of DnPK were not available.

10.3.5 Biomonitoring/Biomarkers

Specific biomonitoring methods and biomarkers for DnPK were not identified.

10.4 Toxic Effects

10.4.1 Experimental Studies

In general, DnPK has a relatively low degree of toxicity. Acute studies have shown it is only slightly irritating. Short exposure to relatively high levels may cause transient narcosis. Repeated exposures to high concentrations may lead to hepatic enlargement and may lower blood glucose levels.

10.4.1.1 Acute Toxicity. The LD_{50} of DnPK was found to be 3.73 mL/kg for rats (426) and > 3200 mg/kg for mice (3). The intravenous LD_{50} was 271 mg/kg, while an intravenous dose of 151 mg/kg caused anesthetic effects (loss of righting reflex) in mice (427).

When DnPK was applied to the eyes of rabbits, only slight irritation was observed (3, 231).

Single applications of DnPK to the skin of guinea pigs (3) or rabbits (231) produced slight irritation. Repeated applications of 0.5 mL for a 10-d period produced slight erythema (3). Although absorption through the skin may occur, doses of 5, 10, and 20 mL/kg were not lethal when applied to the skin of guinea pigs under an occlusive wrap for 24 h (3). The dermal LD_{50} has been reported to be 5.66 mL/kg for rabbits (426).

The LC_{50} of a single 6-h exposure of rats to DnPK was 2690 ppm. A concentration of 400 ppm DnPK decreased respiration and 825 ppm produced depression. Narcosis occurred at 1600 ppm, and 3200 ppm killed three of four rats (3). Respiratory depression to 50% of the normal frequency has been reported in rats exposed to a concentration of 1098 ppm (159). Four rats exposed to a saturated atmosphere (4970 ppm) died (3). Exposure to 2000 ppm for 4 h did not kill rats, whereas 4000 ppm for 4 h killed six of six animals (425). Six-hour exposures to 1200 ppm of DnPK 5 d/wk for 2 wk resulted in a slightly decreased response to stimulation during exposure (3). In this study, marginal liver enlargement, but no change in hematology, clinical chemistries, or pathology, was seen at the end of the 2-wk exposure period.

10.4.1.2 Chronic and Subchronic Toxicity. DnPK administered by gavage to eight rats at a dose of 2000 mg/kg 5 d/wk produced severe central nervous system depression and reduced weight gain. One rat died after 1 wk of treatment due to cardiorespiratory failure. Lowering the dose to 1000 mg/kg resulted in improved weight gain and the absence of deleterious clinical effects over a 12-wk period. Hematologic determinations and serum clinical chemistries, except for a reduction in glucose, were unaffected. Relative liver and kidney weights were increased, and histologically there was hepatocyte hypertrophy. Repeated contact with the stomach resulted in hyperkeratosis and evidence of chronic irritation of the nonglandular gastric epithelium (3).

10.4.1.3 Pharmacokinetics, Metabolism, and Mechanisms. No pharmacokinetic or metabolism studies are available for DnPK.

10.4.1.4 Reproductive and Developmental. No reproductive and developmental toxicity studies are available for DnPK.

10.4.1.5 Carcinogenesis. No carcinogenicity studies are available for DnPK.

10.4.1.6 Genetic and Related Cellular Effects Studies. No genotoxicity studies are available for DnPK.

10.4.1.7 Other: Neurological, Pulmonary, Skin Sensitization. No other systemic toxicity studies are available for DnPK.

10.4.2 Human Experience

10.4.2.1 General Information. Industrial exposures to DnPK are expected to be primarily by inhalation, and by skin and eye contact. Single topical exposures may be slightly irritating, but repeated exposures may exacerbate the irritation. Oral exposures are not expected to be hazardous unless large quantities of DnPK are consumed. Repeated exposure to high levels may lead to liver enlargement and may lower blood glucose levels.

10.4.2.2 Clinical Cases. No published reports of toxicity following human exposure to DnPK were found in the literature.

10.4.2.3 Epidemiology Studies. No epidemiology studies reporting adverse health effects associated with exposure to DnPK are available.

10.5 Standards, Regulations, or Guidelines of Exposure

Hygienic standards for DnPK are listed in Table 76.5.

10.6 Studies on Environmental Impact

Studies on the environmental impact of DnPK are not available; however, DnPK is not expected to present any unusual hazard to the environment. Since DnPK would be expected to biodegrade, long-term environmental effects are not anticipated.

11.0 Diisopropyl Ketone

11.0.1 CAS Number: *[565-80-0]*

11.0.2 Synonyms: 2,4-Dimethyl-3-pentanone, isopropyl ketone, isobutyrone, and DIPK

11.0.3 Trade Names: NA

11.0.4 Molecular Weight: 114.19

11.0.5 Molecular Formula: $(CH_3)_2CHCOCH(CH_3)_2$

11.0.6 Molecular Structure:

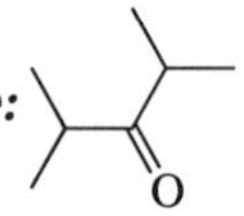

11.1 Chemical and Physical Properties

11.1.1 General

DIPK is a colorless, flammable liquid with negligible solubility in water. It has a boiling point of 124°C. Additional physical–chemical properties are summarized in Table 76.4.

11.1.2 Odor and Warning Properties

DIPK is a colorless liquid with a characteristic ketone odor.

11.2 Production and Use

DIPK has been used in automotive assembly plants as a ketonic solvent for fuel injectors and as a photoinitiator for IV-curable acrylic coatings in optical fiber gratings.

11.3 Exposure Assessment

Exposure in the industrial setting would be expected to be by inhalation and through skin and eye contact. Appropriate care should be exercised to prevent eye contact. Vapor concentrations should be kept below 50 ppm to preclude eye, nose, and throat irritation.

11.3.3 Workplace Methods

Either of two NIOSH methods can be used to determine DIPK concentrations in workroom air (116). The first method (NMAM 2549) is a screening method for volatile organic compounds that uses thermal desorption tubes containing graphitized carbon and molecular sieve adsorbents for sampling and gas chromatography–mass spectrometry for analysis. The second method (NMAM 1300) is a more specific method for ketones that uses a solid sorbent tube with coconut shell charcoal and gas chromatography with flame ionization detection.

11.3.4 Community Methods

Specific community methods for measurement of DIPK were not identified.

11.3.5 Biomonitoring/Biomarkers

Specific biomonitoring methods and biomarkers for DIPK were not identified.

11.4 Toxic Effects

11.4.1 Experimental Studies

11.4.1.1 Acute Toxicity. The oral LD_{50} in rats was reported to be 3535 mg/kg. Signs of toxicity included salivation, CNS depression, and weakness (3). The oral LD_{50} in mice was approximately 1600–3200 mg/kg. The intraperitoneal LD_{50} values in mice and rats were between 100 and 400 mg/kg (3).

DIPK was a slight skin irritant, causing slight erythema following exposure of guinea pigs to 0.5 mL for 24 h under an occlusive wrap. Repeated topical application of the neat material to the clipped backs of guinea pigs over an 11-d period caused slight exacerbation of the dermal response. The dermal LD_{50} in rats was greater than 20 mL/kg (3).

DIPK was a slight eye irritant. It caused erythema of the adnexal structures and slight discharges, but all eyes appeared normal by 48 h after dosing (3).

The LC_{50} for DIPK for a 6-h exposure was greater than 2765 ppm. Clinical signs included concentration-dependent depression of activity and ocular and respiratory irritation. All exposure-related abnormalities resolved within 48 h (3).

11.4.1.2 Chronic and Subchronic Toxicity. When rats were exposed to DIPK at concentrations of up to 1221 ppm for 6 h/d, 5 d/wk for a 4-wk period, clinical signs were mild and reversible. Intracellular hyalin droplets were noted in the proximal convoluted tubules of the kidneys of all male animals exposed to diisopropyl ketone. Absolute and relative liver weights were significantly greater for the 1221- and 386-ppm groups, but hepatocyte hypertrophy was observed only in the 1221-ppm group. The no-effect level was close to, but below, 128 ppm (3).

11.4.1.3 Pharmacokinetics, Metabolism, and Mechanisms. No pharmacokinetic or metabolism studies are available for DIPK.

11.4.1.4 Reproductive and Developmental. No reproductive or developmental toxicity studies are available for DIPK.

11.4.1.5 Carcinogenesis. No carcinogenicity studies are available for DIPK.

11.4.1.6 Genetic and Related Cellular Effects Studies. No genotoxicity studies are available for DIPK.

11.4.1.7 Other: Neurological, Pulmonary, Skin Sensitization. DIPK did not cause a dermal sensitization response when tested in guinea pigs using a standardized test method (3).

11.4.2 Human Experience

11.4.2.1 General Information. No studies on humans are reported in the literature.

11.4.2.2 Clinical Cases. No case studies reporting adverse health effects associated with DIPK exposure were identified.

11.4.2.3 Epidemiology Studies. No epidemiology studies reporting adverse health effects associated with exposure to DIPK were identified.

11.5 Standards, Regulations, or Guidelines of Exposure

There are no recommended hygienic standards for diisopropyl ketone.

11.6 Studies on Environmental Impact

No studies on the environmental impacts of DIPK are available; however, DIPK is not expected to present any unusual hazard to the environment. Since DIPK would be expected to biodegrade, long-term environmental effects are not anticipated.

12.0 2-Methylcyclohexanone

12.0.1 CAS Number: 2-Methylcyclohexanone *[583-60-8]*, 3-methylcyclohexanone *[591-24-2]*, 4-methylcyclohexanone *[589-92-4]*, and methylcyclohexanone mixtures *[1331-22-2]*

12.0.2 Synonyms: 2-Methylcyclohexanone is also known as *o*-methylcyclohexanone, cyclohexanone, 2-methyl or α-methylcyclohexanone. 3-Methylcyclohexanone is also known as *m*-methylcyclohexanone or cyclohexanone, 3-methyl. 4-Methylcyclohexanone is also known as *p*-methylcyclohexanone and cyclohexanone, 4-methyl

12.0.3 Trade Names: NA

12.0.4 Molecular Weight: 112.17

12.0.5 Molecular Formula: $C_7H_{12}O$

12.0.6 Molecular Structure:

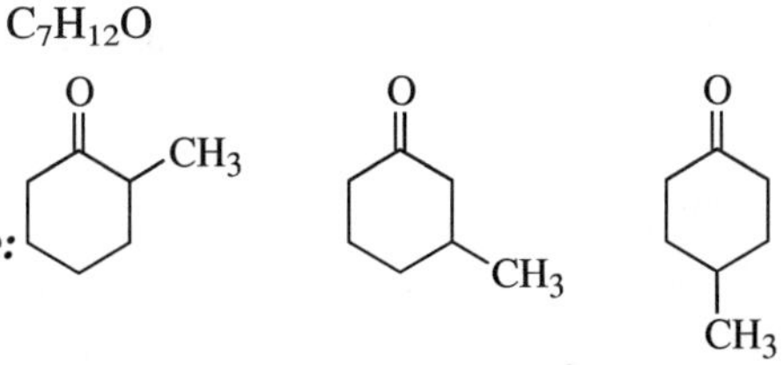

ortho *meta* *para*

12.1 Chemical and Physical Properties

Methylcyclohexanones are generally insoluble in water, but readily soluble in alcohols and ethers. 2-Methylcyclohexanone (±) has a boiling point of 165°C, a melting point of −13.9°C, and a density of 0.9250 g/cm^3 at 20°C. 3-Methylcyclohexanone (±) has boiling point of 169°C, a melting point of −73.5°C, and a density of 0.9136 g/cm^3 at 20°C. 4-Methylcyclohexanone has boiling point of 170°C, a melting point of −40.6°C, and a density of 0.9138 g/cm^3 at 20°C (428). Additional physical–chemical properties are summarized in Table 76.4. The National Fire Protection Association lists the flash point of methylcyclohexanone at 48°C (71). The ChemFinder database on the World Wide Web lists flash points of 46°C, 51°C, and 40°C for the *ortho*, *meta*, and *para* isomers of methylcyclohexanone (429).

12.1.1 General

Methylcyclohexanones are chemically stable, but may darken when exposed to light. Although methylcyclohexanones may be obtained as pure compounds, they are usually found as isomeric mixtures.

12.1.2 Odor and Warning Properties

The isomers of methylcyclohexanone are colorless liquids of low volatility with a weak peppermintlike odor (430).

12.2 Production and Use

2-Methylcyclohexanone is manufactured by the catalytic hydrogenation of *o*-cresol or by the dehydrogenation of *o*-methylcyclohexanol (431). Methylcyclohexanones are used in small amounts as general-purpose solvents for cellulose-based finishes and coatings. It can be found in commercial-grade nitrocellulose-containing lacquers at levels of about 5%.

12.3 Exposure Assessment

An examination of the organic solvent use in small- to medium-sized industries in Japan revealed that methylcyclohexanone represented less than 1% of the solvents surveyed (92).

12.3.1 Air

In a comprehensive search for carbonyls, Grosjean et al. was unable to identify 2-methylcyclohexanone in ambient air samples from four locations in and around the city of Los Angeles (296). Ciccioli et al. found 4-methylcyclohexanone in an air sample from the city of Rome (Italy) at a concentration of 0.89 ng/L (432).

12.3.2 Background Levels

A 3-yr study of the extractable organic compounds in the surface water from a network of canals fed by the Rhône river revealed the presence of 2-methylcyclohexanone at the ng/L level (433). Praxéus detected 4-methylcyclohexanone at a concentration of 1 μg/L in the effluent from a municipal sewage treatment plant in Sweden that processed wastewater from industries producing foods and beverages, fabricated metal, equipment and machinery, paper and chemicals, and plastic matting (310). Methylcyclohexanone was not detected the other two sewage treatment plants examined. 2-Methyl and 4-methylcyclohexanone were detected in the influent, but not the effluent, of an industrial wastewater treatment plant in Japan that primarily served the needs of a petrochemical plant. Several other factories producing machinery and metal products, electrical power, and foodstuffs also relied on the treatment plant (434).

12.3.3 Workplace Methods

Methylcyclohexanes can be measured in the workplace using NIOSH Method 2521, which utilizes a solid sorbent tube containing Porapak Q for sampling and gas chromatography with flame ionization detection for quantification (116). OSHA notes that detector tubes and infrared gas analyzers may also be used to monitor for methylcyclohexanone in the workplace. Andrew et al. described a simple colorimetric test for determining the airborne concentration of methylcyclohexanone that relied on the red azo dye produced when 4-amino-5-hydroxynaphthalene-2,7-disulfonic acid was added to a sample collected in water (435). Wright published a method for monitoring workplace exposures to 2-methylcyclohexanone that used thermal desorption gas chromatography together with a dual-capillary column system to enhance resolution (436).

12.3.4 Community Methods

Paxéus analyzed wastewater for methylcyclohexanones using tandem C18 bonded phase cartridges for the solid-phase adsorption of the organics, and gas chromatography–mass spectrometry in the electron-impact mode for separation and quantification (310). Deroux et al. used liquid–liquid extraction for the isolation of organics in surface water samples followed by gas chromatography–mass spectrometry in both the chemical ionization and electron impact modes for identification and quantification (437). Grosjean et al. used DPNH-coated C18 cartridges to sample and derivatize carbonyl-containing urban air samples (296). The eluted samples were separated by liquid chromatography–mass spectrometry in the chemical-ionization mode and quantitation was made possible through the application of a response factor for 2-methylcyclohexanone. Ciccioli et al. used three-stage carbon adsorbent traps along with a cryofocusing desorption unit and high-resolution gas chromatograph with flame ionization detector to measure 4-methylcyclohexanone in an urban air sample (432).

12.3.5 Biomonitoring/Biomarkers

Specific biomonitoring methods and biomarkers have not been identified for these ketones.

12.4 Toxic Effects

12.4.1 Experimental Studies

The isomers of methylcyclohexanone do not present a high degree of health hazard in the workplace. They are moderately toxic to laboratory animals and may cause moderate skin irritation. Low concentrations are characterized by a strong persistent odor and by irritation to the eyes, nose, and throat. High concentrations produce central nervous system depression and narcosis.

12.4.1.1 Acute Toxicity. Smyth et al. reported an oral LD_{50} value of 2.14 mL/kg (95% CI, 1.48–3.10 mL/kg) for 2-methylcyclohexanone in female Carworth–Wister for rats (325). Treon et al. reported that 1.0–1.2 g/kg was the range of lethal doses for white rabbits (327). Death was preceded by twitching, convulsions, and narcosis. 4-Methylcyclohexanone had an oral LD_{50} of 800–1600 mg/kg for rats and 1600–3200 mg/kg for mice (3). 2-Methylcyclohexanone was moderately irritating when held in contact with the skin of guinea pigs for 24 h. The lethal intravenous dose of 2-methyl, 3-methyl, and 4-methylcyclohexanone for dogs was 270, 310, and 370 mg/kg, respectively (438).

Smyth et al. reported that 2-methylcyclohexanone had a dermal LD_{50} of 1.77 mL/kg in white rabbits (325). It was also mildly irritating to rabbit skin, producing a grade 3 response (range 1–10) when left in contact with the skin for 24 h. 2-Methylcyclohexanone also caused a severe grade 5 (scale of 1 to 10) corneal burn when instilled into the eyes of white rabbits. Exposures to 2800-ppm 2-methylcyclohexanone vapor for 4 h produced lethality in three of six female rats. Treon et al. reported that the minimum lethal dose by skin application ranged from 4.9 to 7.2 g/kg for white rabbits (327). Anesthesia, lacrimation, dermal irritation, and hypothermia were observed in the animals that survived.

Lehmann and Flury reported that 450-ppm methylcyclohexanone was irritating to the eyes and respiratory tract of mice (439). Mice, rats, and guinea pigs exposed to 3500-ppm methylcyclohexanone vapor for 30 min resulted in prostration. Rabbits and cats exposed to 2500 ppm became tired within an hour and showed labored breathing and an unsteady gait. An intradermal injection of 0.5 mL/kg to rabbits caused an increase in the respiration rate, but not apparent damage to the circulation.

Frantík et al. determined the isoeffective concentrations necessary to produce a percentage decrease in the maximum intensity of an electrically evoked seizure in both male Wistar rats and female H strain mice (370). Methylcyclohexanone was examined along with 47 other chemicals. The concentration needed for a 30% decrease in response was 660 ppm (90% CI, 460–860 ppm) in rats, whereas the concentration required for a 37% decrease in mice was 390 ppm (90% CI, 190–590 ppm).

12.4.1.2 Chronic and Subchronic Toxicity. Treon et al. exposed white rabbits to methylcyclohexanone concentrations of 182, 514, 1139, or 1822 ppm for 6 h/d (5 d/wk) for either 10 wk (two lowest concentrations) or 3 wk (two highest concentrations) (327). None of the animals died during the study, and the animals all gained weight. Exposure at the two highest concentrations for a 3-wk period caused a few minor effects, including lethargy, salivation, lacrimation, and eye irritation. Slight conjunctival congestion was noted in the animals exposed to 514 ppm for 10 wk. Liver and kidney histology, hemoglobin concentration, erythrocyte count, and leukocyte count were unaffected by treatment at any concentration. Extending the study duration from 3 to 10 wk at the two highest exposure concentrations appeared to cause slight evidence of hepatic and renal damage. The no-observed-effect level for the inhalation of methylcyclohexanone by rabbits was reported to be 182 ppm.

12.4.1.3 Pharmacokinetics, Metabolism, and Mechanisms. Methylcyclohexanone may be produced as metabolic products for several different chemicals. Spillane et al. found small amounts of 4-methylcyclohexanone in the urine of rats treated with non-nutritive sweetener, 4-methylcyclohexylsulfamate (440). Madyastha and Raj detected 3-methylcyclohexanone in the urine of rats treated with menthofuran, which was thought to be the metabolite responsible for the hepatotoxic effects of pulegone (441). Pulegone is a major constituent of pennyroyal oil, which is a widely used flavoring agent extracted from the plant *Mentha pulegium.*

Treon et al. reported that the exposure of rabbits to cyclohexanone vapor at concentrations ranging from 182 to 1822 ppm for up to 10 wk at 6 h/d resulted in a large dose-related increase in the urinary excretion of glucuronic acid (327). This increase was also observed when rabbits were exposed to methylcyclohexanol. Tao and Elliot found that rabbits metabolize racemic (dextro and levo-rotatory) mixtures of 2-methylcyclohexanone to *trans*-2-methylcyclohexanol and that all 2-, 3-, and 4-methylcyclohexanone were all reduced *in vitro* to their corresponding methylcyclohexanol by NADH and horse-liver alcohol dehydrogenase (442). Elliott et al. studied the stereospecific metabolism of 2-, 3-, and 4-methylcyclohexanones in greater detail by isolating the optically active *cis* and *trans* forms of methylcyclohexanol from the urine of treated rabbits (443). Onishi et al. also observed that the fungus *Aspergillus repens* was also capable of preferential stereoselective reduction of 2-methylcyclohexanone to specific isomeric forms of 2-methylcyclohexanol (444). Nakajina et al. showed that 2-methylcyclohexanol was a good substrate for the *in vitro* carbonyl reductase activity exhibited by pig testicular 20β-hydroxysteroid dehydrogenase (445).

12.4.1.3.3 Excretion. All available evidence indicates that methylcyclohexanones are metabolized to methylcyclohexanols and excreted in the urine as sulfate and glucuronate conjugates (327).

12.4.1.4 Reproductive and Developmental. No reproductive or developmental studies were identified for these ketones.

12.4.1.5 Carcinogenesis. No carcinogenesicity studies were identified for these ketones.

12.4.1.6 Genetic and Related Cellular Effects Studies. Dierickx determined the EC_{50} concentration of 2-methylcyclohexanone that caused a 50% reduction in the total protein content in cultured FHM (fathead minnow) cells (446). The EC_{50} value for 2-methylcyclohexanone was found to be 28.5 mg/mL after 24 h of incubation.

12.4.1.7 Other: Neurological, Pulmonary, Skin Sensitization. No other toxicity studies were identified for these ketones.

12.4.2 Human Experience

12.4.2.1 General Information. 3-Methylcyclohexanone is a minor constituent of pennyroyal oil, which is a plant-derived oil used as a flavoring agent, a fragrance in cosmetics, and a herbal remedy. Pennyroyal oil is distilled from the plant *Mentha pulgegium*, which has been shown contain 3-methylcyclohexanone at concentrations ranging from 70 to 140 ppm. The primary ingredient of pennyroyal oil is pulegone, which is a derivative of 3-methylcyclohexanone, which contains an isopropylidene substituent on the ring. Pennyroyal is hepatotoxic to humans and has been the cause of death due to massive hepatocellular necrosis (447). Studies in mice have indicated that pulegone is responsible for the hepatotoxic effects observed in humans. Removal of the isopropylidene from puglegone and administration of 3-methylcyclohexanone reportedly eliminated the hepatotoxic effects of this compound (448). Pennyroyal oil is a GRAS (generally recognized as safe) compound and has been approved for human consumption in small amounts as a food flavoring agent (81).

Occupational exposure is likely to occur by inhalation and eye and skin contact. Irritation may occur following exposure of mucous membranes to methylcyclohexanone. Methylcyclohexanones have good warning properties and are considered to be relatively nonhazardous in the workplace provided good hygienic standards are followed.

12.4.2.2 Clinical Cases

12.4.2.2.1 Acute Toxicity. Lehmann and Flury reported, with no supporting data, that methylcyclohexanone caused no effect following skin application and that it was a relatively strong irritant for mucous membranes (439). No other clinical cases were identified in the literature.

12.4.2.3 Epidemiology Studies. No epidemiological studies on methylcyclohexanone were identified.

12.5 Standards, Regulations, or Guidelines of Exposure

Hygienic standards for 2-methylcyclohexanone are listed in Table 76.5. There are no recommended hygienic standards for 3- and 4-methylcyclohexanone.

12.6 Studies on Environmental Impact

The environmental fate and transport of methylcyclohexanones have not been well studied.

13.0 Acetophenone

13.0.1 CAS Number: [98-86-2]

13.0.2 Synonyms: Acetylbenzene, phenyl methyl ketone, 1-phenylethanone, benzoyl methide, and hypnone

13.0.3 Trade Names: NA

13.0.4 Molecular Weight: 120.13

13.0.5 Molecular Formula: C_8H_8O

13.0.6 Molecular Structure:

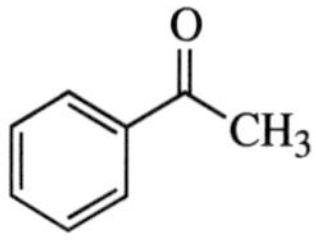

13.1 Chemical and Physical Properties

Acetophenone is a colorless liquid at room temperature that is slightly soluble in water and highly soluble solvents such as ethanol, diethyl ether, chloroform, and glycerol (449). It has a boiling point of 202.0°C, a melting point of 19.6°C, a density of 1.0281 g/cm^3 at 20°C, and a vapor pressure of 0.37 mm Hg at 25°C (70). The flash point of acetophenone is 77°C by the closed-cup method, and the autoignition temperature is 570°C (71).

13.1.1 General

Additional physical–chemical properties are summarized in Table 76.4.

13.1.2 Odor and Warning Properties

Hellman and Small reported that acetophenone vapors have a sweet almondlike smell and that the odor recognition threshold was 0.60 ppm for a least half of the individuals on an odor panel (75). Devos et al. summarized and standardized the results from two odor studies with acetophenone and reported an odor detection threshold of 0.36 ppm (73).

13.2 Production and Use

Acetophenone can be prepared by the Friedel–Crafts reaction between benzene and acetyl chloride in the presence of aluminum chloride or by the catalytic oxidation of ethyl benzene (450, 451). Acetophenone finds use as a chemical intermediate for dyes and pharmaceuticals, as a solvent for plastics and resins, as a fragrance in soaps, detergents, and lotions, and as a flavorant in nonalcoholic beverages, chewing gum, ice cream, candy, and baked goods.

13.3 Exposure Assessment

13.3.1 Air

Acetophenone has been often detected, but rarely quantified, in air samples from polluted and unpolluted sources. Acetophenone has been found in solvent extracts from aerosol samples collected in a polluted urban environment (452). Yokouchi et al. identified acetophenone in the solvent extract of atmospheric aerosol samples collected in a area of Japan that contained numerous red pine forests (453). Similarly, acetophenone was found as a volatile component in air samples from forests in the Sierra Nevada Mountains of California and the Black Forest of southern Germany (298, 299). Samples of air from a remote location in Hawaii (Mauna Loa Observatory) also contained acetophenone as a trace volatile component (454).

In their comparison of indoor and outdoor air pollution for 36 homes in the Chicago area, Jarke et al. found that acetophenone was present 17 of the indoor samples and only 9 of the outdoor samples (455). Sunesson et al. reported that acetophenone could be detected in the background emissions from a mixture of gypsum board and mineral wool bought at a local retail store (456). A survey of volatile organic exposures for eight New Jersey residents showed that measurable amounts of acetophenone were present on the personal monitors worn by four of the volunteers (134). Using emissions modeling data from 60,803 census tracts throughout the United States, Caldwell et al. estimated that the highest individual airborne concentration of acetophenone would be 0.00045 $\mu g/m^3$ (90). An airborne acetophenone concentration of 6 $\mu g/m^3$ was found in outdoor air samples from a hazardous waste landfill site in the U.S. Virgin Islands (89).

Leary et al. identified acetophenone in chloroform extracts of the exhaust smoke from a residential oil burner using No. 2 fuel oil (457). Acetophenone has repeatedly been shown to be present in the exhaust of automobiles using either diesel fuel or gasoline (458). Acetophenone was emitted at a rate 5100 μg/km in the gas phase, but not the particle phase of the exhaust from a medium-duty diesel truck (459). Cocheo et al. measured acetophenone at levels ranging from 0 to 30 $\mu g/m^3$ in air samples from the extrusion area of a factory that manufactured electrical cables (294). It was not, however, detected in air samples collected from a shoe-sole factory or from the vulcanization and extrusion areas of a tire retreading plant.

13.3.2 Background Levels

Acetophenone is found in the essential oils castoreum and labdanum (80). It is also found in dried beans, split peas, and lentils at the ppb level (460). Acetophenone has been added as a flavoring agent to the following products in limited amounts; nonalcoholic beverages at a maximum concentration of 0.98 ppm, ice cream at 2.8 ppm, candy at 3.6 ppm, baked goods at 5.6 ppm, gelatins and puddings at 7.0 ppm, and chewing gum at levels ranging from 0.60 to 20 ppm (81).

Acetophenone was present in the drinking water for 1 of 10 U.S. cities at a concentration of 1.0 μg/L (303). Suffet et al. identified acetophenone in the extracts from 6 of 11 drinking water samples collected in the Philadelphia, PA, area over a 2-yr period (461). Wallace et al. found measurable amounts of acetophenone in the drinking water used by one of the New Jersey residents participating in a survey of VOC exposures (134). Acetophenone was found at a concentration of 5.4 ppb in extracts of drinking water samples collected in Japan (462).

Zoeteman et al. reported that acetophenone was found at a maximum concentration of 10 μg/L in groundwater samples collected at 232 pumping stations in The Netherlands (309). An analysis of monitoring data from disposal-site groundwater investigations showed that acetophenone was present in one of the 479 sites examined (463). Acetophenone concentrations ranging from 0.038 to 0.053 μg/L were found in the secondary effluent from a wastewater treatment plant at a military base (464). Ellis et al. detected acetophenone in the secondary effluent from 2 of 10 municipal and industrial wastewater treatment plants examined in Illinois (306). Surface water collected downstream of a tire fire was shown to contain acetophenone (465). Acetophenone was detected at a concentration of 10 mg/L in the wastewater stream from a shale oil recovery operation (308). Groundwater samples from a oil refining plant were found to contain acetophenone at a level of 170 μg/L (89). Extracts from water samples collected on the river Waal in The Netherlands were shown to contain acetophenone (466). Ehrhardt identified acetophenone in seawater collected in Hamilton Harbor, Bermuda (467). Acetophenone was detected in sediment collected from the Calcasieu River, which lies in the vicinity of a heavily industrialized area of Louisiana (12). Likewise, it was also found in sediment samples from Tobin Lake on the North Saskatchewan River (468).

13.3.3 Workplace Methods

OSHA has published a partially validated method for assessing workplace exposures to acetophenone method uses 5/95 mixture of isopropanol/carbon disulfide to elute the chemical from an adsorbent of Tenax GC (469). Analysis was accomplished by gas chromatography with flame ionization detection. Cocheo et al. used charcoal containing adsorbent tubes and gas chromatography/mass spectrometry to measure the airborne levels of acetophenone in an electrical cable insulation plant (294).

13.3.4 Community Methods

Acetophenone analysis is most often accomplished by gas chromatography/mass spectrometry used in combination with several different sampling techniques. This approach was used with either of 5 different adsorbent media to detect the presence of acetophenone in ambient air at a remote location (298). Suffet et al. used XAD-2 resin and a continuous liquid–liquid extractor to trap the acetophenone in drinking water prior to GC/MS analysis (461). Shinohara et al. measured the concentration of acetophenone in XAD-2 extracts of drinking water by GC/MS and mass fragmentography (462). Ellis et al. identified acetophenone in the secondary effluent from a waste water treatment plant by initial adsorption onto Tenax resins before final analysis by GC/MS in a total ionization mode (306). Steinheimer et al. used a Soxlet extractor and Kuderna–Danish concentrator to isolate acetophenone from samples of river sediment prior to a GC/MS determination (12). A similar approach was used to isolate and analyze for the acetophenone in airborne particulate matter (462). Schauer et al. used a slightly different approach to analyze diesel exhaust vapors (459). A DNPH (2,4-dinitrophenylhydrazone) impregnated C18 cartridge was used to trap the acetophenone vapors before final analysis by liquid chromatography with an ultraviolet detector. An interlaboratory evaluation of an EPA-validated GC/MS procedure (Method 8270) for the analysis of semivolatile material in groundwater revealed that 56% of the samples were falsely reported to contain no acetophenone (470).

13.3.5 Biomonitoring/Biomarkers

There are very few reports of biological or clinical monitoring for acetophenone in human body fluids or excreta. The only information available was reported by Wallace et al. in a survey of volatile organic exposures for a group of New Jersey residents (134). The study showed that measurable amounts of acetophenone could be detected in the expired air from 5 of the 12 volunteers examined. In their review of breast feeding, Giroux et al. reported that acetophenone had been detected in human breast milk in the absence of any known exposure, but few details were provided (471).

13.3.5.1 Blood. No information on biomarkers in the blood was available.

13.3.5.2 Urine. Rhodes et al. detected acetophenone in the urine of untreated male Sprague–Dawley rats and showed that the levels increased when the animals were treated with alloxan and streptozotocin to induce diabetes (472). In an evaluation of a new breath collection system for the study of endogenous and exogenous volatile compounds, Raymer et al. did not find acetophenone in the expired air of either untreated or CCl_4-treated rats Fisher-344 rats (473).

13.4 Toxic Effects

13.4.1 Experimental Studies

Acetophenone has a low to moderate degree of acute toxicity. The principal hazard associated with acetophenone exposure is irritation of the skin and eyes. Eye contact may produce transient corneal injury. The low volatility of acetophenone limits any exposure by the pulmonary route. Ingestion of large amounts may produce anesthesia.

13.4.1.1 Acute Toxicity. Smyth and Carpenter reported an approximate oral LD_{50} value 3.0 g/kg for male Wistar rats in a 14-d range finding test (264), and an LD_{50} value of 0.90 g/kg (95% CI, 0.81–1.00 g/kg) in a more definitive test using Sherman rats (243). Jenner et al. used male and female Osborne–Mendel rats and determined an oral LD_{50} value of 3.20 g/kg (95% CI, 2.46–4.16 g/kg) with narcosis reported a toxic sign (474). Grübner et al. reported an oral LD_{50} of 900 mg/kg for Wistar rats and a decrease in hexobarbital sleeping time and ascorbic acid excretion in the urine of rats administered 10% of the LD_{50} (475). Smyth et al. determined an intraperitoneal LD_{50} value of 1.07 g/kg for male mice and a maximum tolerated dose of 0.7 g/kg (234).

Smyth and Carpenter (264) reported that the dermal LD_{50} was greater than 20 mL/kg for guinea pigs. Rats exposed to a saturated atmosphere of acetophenone (ca. 430 ppm) for 8 h survived the treatment (264). Carpenter and Smyth reported that acetophenone produced grade 8 (severe on a scale of 1 to 10) ocular damage in albino rabbits based on the total point score 24 h after treatment (230). Points were accumulated according to the degree of corneal opacity, necrosis, and iritis observed in the treated eye, with the maximum total score being 20 points. A grade 8 injury corresponded to a point score of 5 or less when 5 μL of a 5% solution of acetophenone was applied, and a point score greater than 5 with a 15% solution. Muir reported that acetophenone incubated with isolated ileum at a concentration of 0.01% (v/v) caused a 50% decrease in the spontaneous contractile activity of the isolated terminal ileum *in vitro*. The results of the assay were related to ocular irritancy of acetophenone (476).

Muller and Greff measured the irritancy potential of acetophenone in a mouse bioassay that quantified the sudden reflexive drop in respiration caused by a short-duration vapor exposure (159). The concentration-related decline in the respiration rate allowed for the calculation of an RD_{50} value, which corresponded to the irritant concentration needed to produce a 50% decrease in the initial respiration rate. The RD_{50} value for acetophenone in male Swiss OF_1 mice exposed for 5 min was 500 mg/m^3 (477). A correlation has been observed between the irritancy data from the bioassay and physico-chemical properties such as polarizability, boiling point, and gas–liquid partition coefficient. Several different quantitative structure–activity relationships have been developed for use in establishing occupational exposure limits (158, 161, 162).

13.4.1.2 Chronic and Subchronic Toxicity. Smyth and Carpenter found that acetophenone levels ranging from 1 to 102 mg/kg/d failed to cause any reduction in body weight or any histopathologic abnormalities in the liver, kidney, spleen, or testis when incorporated in the diet of Sherman rats for 30 d (243). Hagan et al. fed acetophenone in the diet of male and female Osborne–Mendel rats at levels 1000, 2500, or 10,000 ppm (0.1, 0.25, or 1.0%) for 17 wk and found no toxic effects on body weight, hematological indices (red and white blood cell counts, hemoglobin, and hematocrit), nor histopathological abnormalities of the liver, kidney, spleen, heart, testis, muscle, or bone marrow (478).

Zissu exposed male Swiss CF-1 mice to 100 or 300 ppm acetophenone for 4, 9, or 14 d (6 h/d) and histopathologically examined the respiratory tissue for evidence of damage (479). In all cases, the acetophenone treatment did not cause any observable damage to the trachea, lungs, or nasal passages. Dalland and Doving continuously exposed male Wistar rats to acetophenone vapor concentrations of 2.0×10^{-9} or 1.2×10^{-7} M (ca. 0.05 or 3.0 ppm) for up to 230 d and found selective degeneration of mitral cells in the olfactory bulb (480). Despite these changes, the animals were still capable of distinguishing the odor of acetophenone in a passive-avoidance behavioral test that used an olfactory conditioning stimulus. Laing and Panhuber exposed rats to 0.0003 ppm of acetophenone and showed a decrease in the sensitivity to cyclohexanone odors (481).

13.4.1.3 Pharmacokinetics, Metabolism, and Mechanisms. Much of what is known of acetophenone metabolism comes from studies conducted with ethylbenzene, which has been shown to be metabolized through an acetophenone intermediate (482, 483). Leibman showed that the cytosolic fraction from a rabbit liver homogenate was capable of reducing about 20% of acetophenone to 1-phenylethanol after *in vitro* incubation; however, the microsomal fraction showed little corresponding activity (484). The authors also found that phenobarbital pretreatment did not affect the rates of acetophenone reduction and that the cytosolic fractions from kidney, heart, and lung were also capable of reducing acetophenone to its carbinol.

Kiese and Lenk showed that the rate of hydroxylation of acetophenone to ω-hydroxyacetophenone by rabbit liver microsomes was increased over seven-fold following pretreatment of the rabbits with phenobarbital (485). Weiner et al. determined that acetophenone was a moderately good reversible inhibitor of glyceraldehyde reduction *in vitro* by horse liver aldehyde dehydrogenase (486). Mahy et al. found that acetophenone was produced *in vitro* when microsomal prostaglandin H synthase from sheep seminal vesicles was incubated with the phenylhydrazone derivative of acetophenone (487).

13.4.1.3.3 Excretion. Early studies identified 1-phenylethanol, benzoic acid, and mandelic acid as urinary metabolites of acetophenone in rabbits and dogs. Smith et al. found that rabbits administered acetophenone by gavage excreted 47% of the dose as glucuronide conjugates of 1-phenylethanol and about 20% as hippuric acid (488). Kiese and Lenk, in contrast, reported that 1-phenylethanol and its glucuronide conjugate constituted only about 4% of the dose for rabbits treated by the intraperitoneal route (489). These authors also detected *m*-hydroxyacetophenone, *p*-hydroxyacetophenone, and ω-hydroxyacetophenone as minor urinary metabolites of acetophenone (less than 1% of the dose) in rabbits. Sullivan et al. found that 10% of a 100 mg/kg intraperitoneal dose of radiolabeled acetophenone was excreted as CO_2 after 4 h and that the amount increased to 30% after 13 h (447). The authors also showed that mendelic acid was present in the urine of rats treated intraperitoneally with acetophenone and that this metabolite most likely arose from ω-hydroxyacetophenone.

13.4.1.4 Reproductive and Developmental. Application of 480 mg/kg of acetophenone to the skin of pregnant rats on days 10 through 15 of pregnancy did not cause any change in the gestation period, size of litter, weight of the offspring, time for appearance of teeth or hair, opening of the eyes, or appearance of reflexes (490).

13.4.1.5 Carcinogenesis. No carcinogenicity studies were identified for acetophenone.

13.4.1.6 Genetic and Related Cellular Effects Studies. Elliger et al. determined that acetophenone was not mutagenic in the Ames assay at levels up to 3000 nmoles/plate. The negative results were obtained using *Salmonella typhimurium* strains TA98, TA100, and TA1537 either with or without metabolic activation with a liver S9 fraction from Arochlor 1254–induced rats (491). Fluck et al. incubated 50 μL of acetophenone with *E. coli* strains Pol A^+ (DNA polymerase positive) and Pol A^- (DNA polymerase negative) and found little or no growth inhibition (492). Acetophenone did not exhibit antimutagenic activity toward 2-amino-3-methy-imidazo[4,5-*f*]-quinoline in *Salmonella* strain TA98 (493). Önfelt et al. presented data on the ability of acetophenone and 29 other chemicals to induce narcosis in tadpoles and mitotic spindle disturbances in *Allium cepa* root tips and used the information in an attempt to a establish a quantitative structure–activity relationship to predict chromosomal abnormalities in humans (494).

Acetophenone may be photoactivated and used as a tool to study the mechanisms responsible for UV-induced DNA damage. Its photoreactivity occurs when molecules absorb photons of light in the UVA and UVB regions of the ultraviolet spectrum and move into a highly reactive and relatively stable triplet energy state. Single-strand chain breaks and thymine dimers increased in a concentration-related manner when acetophenone and tritiated DNA from *E. coli* were irradiated at a wavelength 313 nm (495). The formation of thymine dimers appeared to be the result of an energy transfer from the triplet state of benzophenone to the pyrimidine nucleotides in DNA. The single-strand breaks appeared to be the result of the intermediate formation of singlet oxygen, which preferentially attacked particular regions of the DNA strand. The irradiation (λ 295 nm) of 0.01 M acetophenone with DNA prepared from trout sperm resulted in a 16-fold increase in the thymine dimerization and a fourfold increase in single-strand breaks (496). Chouini-Lalanne et al. used acetophenone as a positive control in their study of dimers and breaks by four anti-inflammatory drugs. A 40-sec irradiation (λ 313 nm) of supercoiled DNA and 25 μM acetophenone resulted in a quantum yield of 6.0×10^{-3} for thymine dimers and 1.4×10^{-3} for single-strand breaks (497).

13.4.1.7 Other: Neurological, Pulmonary, Skin Sensitization. Acetophenone was not a skin sensitizer in Hartley guinea pigs when tested using a modified Draize procedure that involved 4 intradermal injections at levels 2.5 times the minimally irritating concentration (422). The test involved the use of both an intradermal and a dermal challenge dose 14 d after intradermal sensitization. Preliminary studies indicated that the minimally irritating intradermal concentration of acetophenone was 0.25% and that the dermal challenge concentration causing no local erythema was 20%.

13.4.2 Human Experience

13.4.2.1 General Information. Occupational exposure to acetophenone may occur by inhalation and eye and skin contact. Its low volatility and good warning properties should reduce the likelihood of overexposure.

13.4.2.2 Clinical Cases. No clinical case studies reporting adverse health effects associated with exposure to acetophenone were identified.

13.4.2.3 Epidemiology Studies. No skin sensitization was noted when 2% acetophenone in petrolatum was tested on humans (451).

No other epidemiology studies on acetophenone were identified.

13.5 Standards, Regulations, or Guidelines of Exposure

There are no recommended hygienic standards for acetophenone.

13.6 Studies on Environmental Impact

The environmental fate and transport of acetophenone has been evaluated following its release to air, water, and soil (498). Acetophenone is expected to be highly mobile in soil and may evaporate. The major removal processes from water will be volatilization and biodegradation. The half-life for biodegradation in ground, river, and lake water is estimated to range from 4.5 to 32 d, whereas the half-life for volatilization from a river is calculated to be about 3.8 d. Acetophenone will have a relatively short lifetime in the troposphere, with a half-life of 2.2 d from its reaction with hydroxyl radicals. Aqueous hydrolysis and sediment adsorption are not expected to be important removal processes for acetophenone.

14.0 2-Octanone

14.0.1 CAS Number: [111-13-7]

14.0.2 Synonyms: Methyl *n*-hexyl ketone, hexyl methyl ketone, and 2-oxooctane

14.0.3 Trade Names: NA

14.0.4 Molecular Weight: 128.21

14.0.5 Molecular Formula: $CH_3CO(CH_2)_5CH_3$

14.0.6 Molecular Structure:

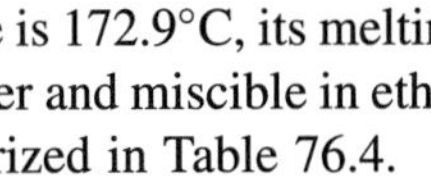

14.1 Chemical and Physical Properties

14.1.1 General

The boiling point for 2-octanone is 172.9°C, its melting point is − 16°C, and its log K_{ow} is 2.37. It is slightly soluble in water and miscible in ether and alcohol. Additional physical–chemical properties are summarized in Table 76.4.

14.1.2 Odor and Warning Properties

2-Octanone is a colorless liquid with a pleasant fruitlike odor resembling the essence of apple. 2-Octanone is low in volatility and has an odor threshold of 248 ppm (223). There is little fire or explosion hazard associated with the use of 2-octanone in an industrial setting.

14.2 Production and Use

2-Octanone can be produced by oxidation of methyl hexyl carbinol, 2-octanol, or 1-octene or by reductive condensation of acetone with pentanol. Commercial samples can have a purity of 98%. 2-Octanol is used as a flavoring substance in foods and beverages, an odorant in fragrances, an antibushing agent in nitrocellulose lacquers, and a chemical intermediate in the production of fragrance substances. Production volume is low; in the United States production is probably on the order of 2–3 ton/yr.

14.3 Exposure Assessment

Occupational exposure is expected to be by inhalation and skin contact. 2-Octanone is relatively innocuous and does not pose a hazard in the workplace when good industrial hygiene practices are followed. Specific methods for measurement of 2-octanone, background levels, biomonitoring methods, and biomarkers were not identified.

14.4 Toxic Effects

14.4.1 Experimental Studies

14.4.1.1 Acute Toxicity. 2-Octanone is slightly toxic by the oral route The acute oral LD_{50} in rats has been reported to be 3.2 g/kg (3), greater than 5 g/kg (35), or 9.2 g/kg (499). The oral LD_{50} in mice has been reported to be 3.2 g/kg (3), 3.1 g/kg (499), 1.6 g/kg (500), or 3.87 g/kg (23). Intraperitoneal LD_{50} values were between 800 and 1600 mg/kg for both rats and mice (3).

2-Octanone produced slight irritation when applied undiluted to rabbit eyes (3).

2-Octanone applied to the skin of guinea pigs produced slight to moderate skin irritation. Guinea pigs lost weight during a 2-wk period following occluded application to the skin, suggesting that 2-octanone may have been absorbed percutaneously (3). When 500 mg of 2-octanone was applied to the skin of rabbits or rodents for 24 h, slight irritation was noted (501). The acute dermal LD_{50} was greater than 5 g/kg for rabbits (501).

Rats exposed to calculated vapor concentrations of 8.9 mg/L (1673 ppm) for 6 h exhibited signs of mild eye irritation (3). Guinea pigs exposed to saturated atmospheres of 2-octanone developed immediate signs of eye and nasal irritation. Exposure for 10 h produced evidence of central nervous system depression, and after 12 h of exposure, the animals were comatose (25). Respiratory depression to 50% of the normal frequency has been reported in mice exposed to a concentration of 472 ppm (159).

14.4.1.2 Chronic and Subchronic Toxicity. No repeated dose or long-term studies on 2-octanone were identified.

14.4.1.3 Pharmacokinetics, Metabolism, and Mechanisms. Studies conducted in rats and mice indicate that 2-octanone significantly lowers serum cholesterol in both species, and serum triglyceride and glycerol levels in rats. Rats given 20 mg/kg of 2-octanone per day for 16 d showed significantly elevated serum lipase activity (500, 502). The conduction velocity of motor and sensory fibers of the tail in rats were not affected following repeated subcutaneous injections of 2-octanone alone (406), but injections of 2-octanone together with MnBK significantly enhanced the neurotoxic effect seen with MnBK alone (407).

14.4.1.4 Reproductive and Developmental. No reproductive or developmental toxicity studies for 2-octanone were identified.

14.4.1.5 Carcinogenesis. No carcinogenicity studies with 2-octanone were identified.

14.4.1.6 Genetic and Related Cellular Effects Studies. 2-Octanone was negative in the Ames/*Salmonella* bacterial mutagenicity assay when tested using a series of 10 strains of *S. typhimurium* (503).

14.4.1.7 Other: Neurological, Pulmonary, Skin Sensitization. 2-Octanone was negative when tested for potential to induce allergic contact dermatitis using a modified Draize procedure (422).

14.4.2 Human Experience

14.4.2.1 General Information. 2-Octanone has a relatively low toxicity. Direct skin contact may cause defatting and irritation of the skin. Inhalation may produce mild symptoms of eye, nose, and throat irritation at low concentrations, and may cause narcosis at high concentrations.

14.4.2.2 Clinical Cases. No published reports of toxicity following human exposure to 2-octanone were found in the literature.

14.4.2.3 Epidemiology Studies. 2-Octanone was not irritating in a 48-h closed patch test with human subjects, and did not produce skin sensitization in a panel of 25 human volunteers (504).

No epidemiological studies on 2-octanone were identified.

14.5 Standards, Regulations, or Guidelines of Exposure

There are no recommended hygienic standards for 2-octanone.

14.6 Studies on Environmental Impact

No studies on the environmental impacts of 2-octanone are available; however 2-octanone is not expected to present any unusual hazard to the environment. Since 2-octanone would be expected to biodegrade, long-term environmental effects are not anticipated.

15.0 3-Octanone

15.0.1 CAS Number: [106-68-3]

15.0.2 Synonyms: Ethyl *n*-amyl ketone, ethyl pentyl ketone, ethyl amyl ketone, and EAK

15.0.3 Trade Names: NA

15.0.4 Molecular Weight: 128.21

15.0.5 Molecular Formula: $CH_3CH_2CO(CH_2)_4CH_3$

15.0.6 Molecular Structure:

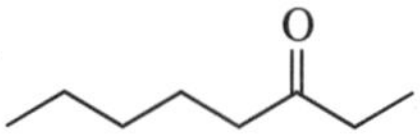

15.1 Chemical and Physical Properties

15.1.1 General

The boiling point of 3-octanone is 167.0°C. It is of low volatility and should not present a fire hazard unless used at high temperatures. Additional physical–chemical properties are summarized in Table 76.4.

15.1.2 Odor and Warning Properties

3-Octanone is a clear liquid with a penetrating, but pleasant, fruity odor.

15.2 Production and Use

3-Octanone can be produced by oxidation of 3-octanol or by heating propionic acid and caproic acid over thorium oxide. 3-Octanone is used as an ingredient in soaps, perfumes, lotions, and creams. It is also used as a flavoring agent in foods.

15.3 Exposure Assessment

15.3.1 Air

No specific information about airborne levels of 3-octanone were identified.

15.3.2 Background Levels

No specific information about background levels for 3-octanone were identified.

15.3.3 Workplace Methods

Either of two NIOSH methods can be used to determine 3-octanone concentrations in workroom air (116). The first method (NMAM 2549) is a screening method for volatile organic compounds that uses thermal desorption tubes containing graphitized carbon and molecular sieve adsorbents for sampling and gas chromatography–mass spectrometry for analysis. The second method (NMAM 1300) is a more specific method for ketones that uses a solid sorbent tube with coconut shell charcoal and gas chromatography with flame ionization detection.

15.3.4 Community Methods

Specific community methods for 3-octanone were not identified.

15.3.5 Biomonitoring/Biomarkers

Specific biomonitoring methods and biomarkers for 3-octanone were not identified.

15.4 Toxic Effects

3-Octanone should present only a low degree of hazard in the occupational setting. It has a low acute oral toxicity. Skin contact may cause moderate irritation, and repeated skin contact may produce dermatitis due to its defatting action.

15.4.1 Experimental Studies

15.4.1.1 Acute Toxicity. An acute oral LD_{50} of greater than 5 g/kg has been reported for rats (505). The intraperitoneal LD_{50} was 406 mg/kg for mice (144).

No reports of eye irritation studies in experimental animals were found. 3-Octanone applied full strength to the intact or abraded skin of rabbits under an occlusive wrap for 24 h produced moderate irritation. The acute dermal LD_{50} was reported to be greater than 5 g/kg (505).

15.4.1.2 Chronic and Subchronic Toxicity. No data are available on the subchronic or chronic effects of 3-octanone or on the effects of inhaling its vapors.

15.4.1.3 Pharmacokinetics, Metabolism, and Mechanisms. No pharmacokinetics, metabolism, or mechanisms studies were identified for 3-octanone.

15.4.1.4 Reproductive and Developmental. No reproductive or developmental studies were identified for 3-octanone.

15.4.1.5 Carcinogenesis. No carcinogenesis studies were identified for 3-octanone.

15.4.1.6 Genetic and Related Cellular Effects Studies. No genetic studies were identified for 3-octanone.

15.4.1.7 Other: Neurological, Pulmonary, Skin Sensitization. Using a modified Draize procedure, Sharp (422) found that 3-octanone was not a skin sensitizer for guinea pigs.

15.4.2 Human Experience

15.4.2.1 General Information. The principal route of exposure in the workplace is by inhalation of the vapors. Occasional skin contact should present no hazard. However, repeated contact may result in dermatitis. The low volatility of 3-octanone and its penetrating odor should minimize the occurrence of overexposures.

15.4.2.2 Clinical Cases. No published reports of toxicity following human exposure to 3-octanone were found in the literature.

15.4.2.3 Epidemiology Studies. Kligman (506) tested 3-octanone at 2.0% in petrolatum with 25 volunteers. It produced no skin irritation after 48 h under an occlusive patch and a maximization test revealed no sensitization reactions.

No epidemiological studies on 3-octanone were identified.

15.5 Standards, Regulations, or Guidelines of Exposure

There are no recommended hygienic standards for 3-octanone.

15.6 Studies on Environmental Impact

No studies on the environmental impacts of 3-octanone are available; however, 3-octanone is not expected to present any unusual hazard to the environment. Since 3-octanone would be expected to biodegrade, long-term environmental effects are not anticipated.

16.0 5-Methyl-3-heptanone

16.0.1 CAS Number: *[541-85-5]*

16.0.2 Synonyms: Ethyl *sec*-amyl ketone, ethyl isoamyl ketone, and methyl heptanone

16.0.3 Trade Names: NA

16.0.4 Molecular Weight: 128.21

16.0.5 Molecular Formula: $CH_3CH_2COCH_2CHCH_3CH_2CH_3$

16.0.6 Molecular Structure:

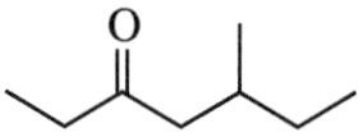

16.1 Chemical and Physical Properties

16.1.1 General

The boiling point for 5-methyl-3-heptanone is 160.5°C, and the melting point is −56.7°C. Additional physical–chemical properties are summarized in Table 76.4.

16.1.2 Odor and Warning Properties

5-Methyl-3-heptanone is a colorless liquid of low volatility. It has an agreeable penetrating odor that resembles the essence of apricots and peaches. Threshold odor concentrations of 6 and < 5 ppm have been reported (144, 223).

16.2 Production and Use

No information on production and use was available.

16.3 Exposure Assessment

No information on exposure assessment was available.

16.3.1 Air

No information on airborne levels of 5-methyl-2-heptanone were identified.

16.3.2 Background Levels

No information on background levels of 5-methyl-2-heptanone were identified.

16.3.3 Workplace Methods

Either of two NIOSH methods can be used to determine 5-methyl-3-heptanone concentrations in workroom air (116). The first method (NMAM 2549) is a screening method for volatile organic compounds that uses thermal desorption tubes containing graphitized carbon and molecular sieve adsorbents for sampling and gas chromatography–mass spectrometry for analysis. The second method (NMAM 1301) is a more specific method for ketones that uses a solid sorbent tube with coconut shell charcoal and gas chromatography with flame ionization detection.

16.3.4 Community Methods

Specific community methods for 5-methyl-2-heptanone were not identified.

16.3.5 Biomonitoring/Biomarkers

Specific biomonitoring methods and biomarkers for 5-methyl-2-heptanone were not identified.

16.4 Toxic Effects

16.4.1 Experimental Studies

16.4.1.1 Acute Toxicity. Oral LD_{50} values of 3.5, 3.8, and 2.5 g/kg for rats, mice, and guinea pigs, respectively, have been reported (3).

5-Methyl-3-heptanone has produced eye irritation with some transient corneal damage in laboratory animals (144). 5-Methyl-3-heptanone is mildly irritating to the skin, and prolonged repeated exposures may cause defatting and dermatitis (144).

Exposure to 3000 ppm 5-methyl-3-heptanone for 4 h was lethal to three of six mice. Similar exposures conducted in rats were not lethal, but obvious signs of irritation of the respiratory tract and eyes were noted. Exposure of rats and mice to 3484 ppm for 4 and 8 h, respectively, was lethal to some of the animals. Exposure to 5888 ppm for 4 h killed four of six rats (144). Respiratory depression to 50% of normal frequency has been reported in mice exposed to a concentration of 760 ppm (159).

16.4.1.2 Chronic and Subchronic Toxicity. When male rats were administered 82, 410, or 820 mg/kg 5-methyl-3-heptanone 5 d/wk for 13 wk, depressed activity and reduced weight gain were seen in the two highest-dose groups. Hind-limb grip strength was reduced at 820 mg/kg/d. Histological evidence of γ-diketone neuropathy was observed at both 820 and 410 mg/kg/d, though effects at 410 mg/kg were minimal. 82 mg/kg/d did not cause adverse effects (13). A proposed metabolic scheme for 5-methyl-3-heptanone is shown is Figure 76.2.

16.4.1.3 Pharmacokinetics, Metabolism, and Mechanisms. No pharmacokinetics, metabolism, or mechanisms studies were identified for 5-methyl-3-heptanone.

16.4.1.4 Reproductive and Developmental. No reproductive or developmental studies were identified for 5-methyl-3-heptanone.

16.4.1.5 Carcinogenesis. No carcinogenesis studies were identified for 5-methyl-3-heptanone.

16.4.1.6 Genetic and Related Cellular Effects Studies. No genetic studies were identified for 5-methyl-3-heptanone.

16.4.1.7 Other: Neurological, Pulmonary, Skin Sensitization. No other systemic toxicity studies were identified for 5-methyl-3-heptanone.

16.4.2 Human Experience

16.4.2.2 Clinical Cases. No clinical case reports of adverse health effects associated with exposure to 5-methyl-3-heptanone were identified.

16.4.2.3 Epidemiology Studies. The odor of 5-methyl-3-heptanone can be detected at 5 ppm by most individuals. Exposure to 25 ppm produced a strong odor and mild irritation of the nasal passages. Exposures to 50 and 100 ppm caused moderate irritation of the eyes, nose, and throat, as well as headache and nausea (144).

No epidemiology studies are available for 5-methyl-3-heptanone.

16.5 Standards, Regulations, or Guidelines of Exposure

Hygienic standards for 5-methyl-3-heptanone are listed in Table 76.5.

16.6 Studies on Environmental Impact

No studies on the environmental impacts of 5-methyl-3-heptanone are available; however, 5-methyl-3-heptanone is not expected to present any unusual hazard to the environment. Since 5-methyl-3-heptanone would be expected to biodegrade, long-term environmental effects are not anticipated.

17.0 Propiophenone

17.0.1 CAS Number: *[93-55-0]*

17.0.2 Synonyms: Phenyl ethyl ketone and 1-phenyl-1-propanone

17.0.3 Trade Names: NA

17.0.4 Molecular Weight: 134.19

17.0.5 *Molecular Formula:* $C_9H_{10}O$

17.0.6 *Molecular Structure:*

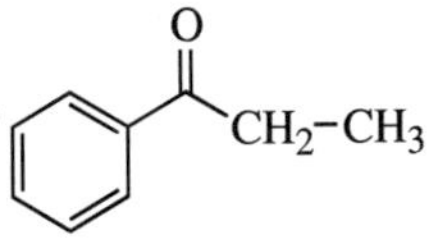

17.1 Chemical and Physical Properties

17.1.1 *General*

Propiophenone is a liquid at room temperature that is soluble in ethanol, benzene, and toluene, but insoluble in water, glycerol, ethylene glycol, and propylene glycol (507). It has a boiling point of 218.0°C, a melting point of 18.6, and a density of 1.010 g/cm^3 at 20°C. The flash point of propiophenone is 88°C by the closed-cup method and 90°C by the open-cup method (508). Propiophenone has a vapor pressure of 1.5 mm Hg at 25°C (509). Additional physical–chemical properties are listed in Table 76.4.

17.1.2 *Odor and Warning Properties*

Propiophenone has a strong agreeable and flowery odor. There are no available data on the odor detection or odor recognition thresholds for propiophenone.

17.2 Production and Use

Propiophenone is produced by the Friedel–Crafts acylation of benzene with propionic acid chloride in the presence of aluminum chloride. It may also be produced by the reaction of benzoic acid and propionic acid using a calcium acetate–aluminum oxide catalyst (510). The chemical is used in the synthesis of ephedrine and other pharmaceutical drugs, as a fragrance enhancer, and as a polymerization sensitizer (508).

17.3 Exposure Assessment

No information on exposure assessment was available.

17.3.1 *Air*

Jüttner detected propiophenone in air samples collected in the Black Forest of southern Germany (299). Propiophenone was detected in test samples collected from three different waste incinerators at concentrations of 200, 220, and 120 μg/mL (511).

17.3.2 *Background Levels*

Background levels other than those discussed in Section 17.3.1 were not identified.

17.3.3 *Workplace Methods*

Recommended workplace exposure monitoring methods for propiophenone have not been published by OSHA or NIOSH.

17.3.4 Community Methods

Propiophenone was measured in breath samples by gas chromatography with high-resolution mass spectrometry after initially preconcentrating the sample on an adsorbent of Tenax GC (512). A similar technique using a Tenax TA adsorbent and a gas chromatography–mass spectrometer was used to identify propiophenone in ambient air samples based on the fragmentation pattern observed (299).

17.3.5 Biomonitoring/Biomarkers

Propiophenone was detected in the expired air of 26.4% of the 54 normal nonsmoking individuals surveyed (512). The median concentration in the breath of these individuals was 0.261 ng/L. Kundu et al. reported that the normal physiological concentration of propiophenone in the expired air of humans was 0.004 nmol/L (513).

17.4 Toxic Effects

17.4.1 Experimental Studies

Propiophenone has a low potential for producing toxicity or irritation. Acute exposure by inhalation or dermal routes would not be expected to cause severe injury.

17.4.1.1 Acute Toxicity. Carpenter et al. (426) reported the oral LD_{50} in rats as 4.49 mL/kg (95% CI, 2.33–8.64 mL/kg), while others indicated that the LD_{50} was between 1.6 and 3.2 g/kg (23). In mice, the oral LD_{50} was found to be >3.2 g/kg, whereas the subcutaneous LD_{50} was 2.25 g/kg (514). Transient signs of weakness, prostration, and vasodilatation were observed in animals treated with high doses of propiophenone.

Carpenter et al. reported that propiophenone was slightly irritating to the eyes of rabbits, and the dermal LD_{50} in rabbits was found to be 4.49 mL/kg (95% CI, 2.11–9.54 mL/kg) (426). Other investigators have reported the dermal LD_{50} of propiophenone to be >10 mL/kg in guinea pigs (3). Carpenter et al. also reported the dermal irritation in rabbits to be barely visible after a 24-h exposure and that all rats survived an 8-h exposure to a saturating concentration (ca. 430 ppm) of propiophenone vapors (426). Studies in guinea pigs have indicated the presence of slight edema and slight to moderate erythema after a 24-h exposure (3). Desquamation was observed in these guinea pigs for 14 d following treatment.

17.4.1.2 Chronic and Subchronic Toxicity. Repeated dose studies with propiophenone were not identified.

17.4.1.3 Pharmacokinetics, Metabolism, and Mechanisms. Studies in rabbits indicated that propiophenone is metabolized to 1-phenyl-1-propanol, which is subsequently conjugated with glucuronic acid and excreted in the urine (515, 516). Coutts et al. confirmed that 60–75% of propiophenone was converted to this metabolite *in vitro* using rabbit liver homogenates (517). The major metabolite of propiophenone was also 1-phenyl-1-propanol when rat liver homogenate was used. However, rat liver metabolized only 37–38% of the propiophenone compared with the 79–95% metabolized by rabbit liver. Rat liver also converted nearly 20% of the alcohol back to the ketone. Little 1-phenyl-1-propanol was further metabolized to 1-phenyl-1,2-propanediol by either species. About 13% of propiophenone was converted to 2-hydroxy-1-phenyl-2-propanone by both rat and rabbit liver homogenates.

17.4.1.4 Reproductive and Developmental. No reproductive or developmental studies were identified for propiophenone.

17.4.1.5 Carcinogenesis. No cancinogenesis studies were identified for propiophenone.

17.4.1.6 Genetic and Related Cellular Effects Studies. The mutagenic activity of propiophenone was investigated by McMahon et al. in a *Salmonella* bacterial assay (503). Frame-shift mutations were investigated using strains D3052, TA1538, TA98, C3076, and TA1537, while point mutations were investigated using strains G46, TA1535, TA100, WP2, and WP2 *uvrA*. Propiophenone was negative in all assays and has no mutagenic activity even in the presence of enzymic activation.

17.4.1.7 Other: Neurological, Pulmonary, Skin Sensitization. No other systemic toxicity studies were identified for propiophenone.

17.4.2 Human Experience

17.4.2.1 General Information. Propiophenone is not considered to be hazardous due to its low toxicity. Exposure by inhalation and dermal routes are likely especially owing to its pleasant odor.

17.4.2.2 Clinical Cases. No clinical case reports of adverse health effects following exposure to propiophenone were identified.

17.4.2.2.1 Acute Toxicity. Two male and one female maintenance worker exposed by inhalation to propiophenone after an industrial spill reported bladder pain and dysfunction after cleaning up the liquid. Urgency, frequency, and void-volume changes were reported by the affected individuals (518).

17.4.2.3 Epidemiology Studies. No epidemiology studies are available for propiophenone.

17.4.2.2.4 Reproductive and Developmental. The intravenous infusion of 8 mg propiophenone to one woman during labor did not affect the number of contractions or the rise in blood concentration of nonesterified fatty acids (519). Nonesterified fatty acids normally increased two- to threefold during labor.

17.4.2.2.5 Carcinogenesis. There are no case studies of carcinogenicity in humans following exposure to propiophenone.

17.4.2.2.6 Genetic and Related Cellular Effects Studies. There are no case studies of genetic toxicity in humans following exposure to propiophenone.

17.4.2.2.7 Other: Neurological, Pulmonary, Skin Sensitization, etc. There are no case reports of pulmonary toxicity or skin sensitization following exposure to propiophenone.

17.5 Standards, Regulations, or Guidelines of Exposure

There are no recommended hygienic standards for propiophenone.

17.6 Studies on Environmental Impact

Limited information on environmental fate and transport of propiophenone has been reported (509). Propiophenone may volatilize from moist and dry soil and photooxidize in the air by reaction with hydroxyl radicals. The half-life in air is about 5 d, whereas the volatilization from still surface water is expected to be about 11 d. The importance of biodegradation as a removal process is unknown. Propiophenone will not bioconcentrate in aquatic organisms and does not photooxidize in air or water.

18.0 Isophorone

18.0.1 CAS Number: *[78-59-1]*

18.0.2 Synonyms: 3,5,5-Trimethyl-2-cyclohexen-1-one, 1,1,3-trimethyl-3-cyclohexen-5-one, trimethylcyclohexenone, isoacetophorone, 3,5,5-trimethyl-2-cyclohexenone, 3,5,5-trimethylcyclohexenone, 3,5,5-trimethylcyclohex-2-enone, alpha-isophorone, isoforon, and 5,5-trimethyl-1-cyclohexen-3-one

18.0.3 Trade Names: NA

18.0.4 Molecular Weight: 138.2

18.0.5 Molecular Formula: $C_9H_{14}O$

18.0.6 Molecular Structure:

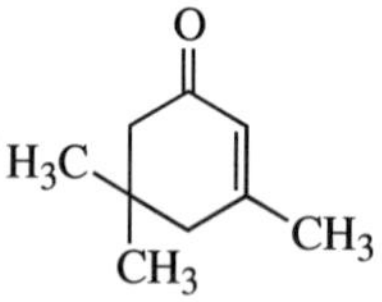

18.1 Chemical and Physical Properties

18.1.1 General

Isophorone is a colorless liquid with a specific gravity of 0.923 and a vapor pressure of 0.26 mmHg. It has a boiling point of 215.2°C and a melting point of −8.0°C. Typical commercial samples may contain 1–3% of the isomer β-isophorone. Additional physical–chemical properties are summarized in Table 76.4.

18.1.2 Odor and Warning Properties

Isophorone has a sharp peppermint- or camphorlike odor. Its odor threshold has been reported to be 0.20 ppm (520).

18.2 Production and Use

Isophorone is used as a solvent for laquers and plastics, oils, fats, nitrocellulose, and in vinyl–resin copolymers. It is also an intermediate in the preparation of trimethylcyclohexanol and xylenols. It is prepared by feeding a mixture of acetone, water, and alkali to a pressure column operated at 200°C and 35 atm of pressure. A vapor-phase process in which acetone vapor is passed over a catalyst bed of mixed oxides has also been described.

18.3 Exposure Assessment

18.3.1 Air

No specific information on levels of isophorone in ambient air was identified.

18.3.2 Background Levels

Information on background levels of isophorone in the environment were not identified.

18.3.3 Workplace Methods

Either of two NIOSH methods can be used to determine isophorone concentrations in workroom air (116). The first method (NMAM 2549) is a screening method for volatile organic compounds that uses thermal desorption tubes containing graphitized carbon and molecular sieve adsorbents for sampling and gas chromatography–mass spectrometry for analysis. The second method NMAM 2508 uses a solid sorbent tube and gas chromatography with flame ionization detection.

18.3.4 Community Methods

Specific community methods for measurement of isophorone were not identified.

18.3.5 Biomonitoring/Biomarkers

Specific biomonitoring methods and biomarkers for isophorone were not identified.

18.4 Toxic Effects

18.4.1 Experimental Studies

18.4.1.1 Acute Toxicity. The oral LD_{50} of isophorone has been reported as 2000 and 2370 mg/kg for mice and rats, respectively (521). Other investigators have reported oral LD_{50} values of >3200 mg/kg for rats and mice (3) and 2150 mg/kg for rats (522). The intraperitoneal LD_{50} was 400 mg/kg for mice and 400–800 mg/kg for rats (3).

Isophorone applied to the eyes of rabbits produced opacity of the cornea, inflammation of the eyelids and conjunctiva, and a purulent discharge (523). Moderate corneal injury and burns (a grade of 4 on a scale of 1 to 10) have been reported by Carpenter and Smyth (230). Other investigators have categorized the irritant response as moderate (231, 521).

Isophorone had a weak irritant action on rabbit (523, 524) and guinea pig skin (521). There was no difference in the irritation observed following 1- and 4-h exposures to 0.5 mL isophorone, when applied under either an occluded or semioccluded wrap (524). The dermal LD_{50} was reported to be 1500 mg/kg for rabbits (154). When isophorone (5–20 mL/kg) was held in contact with guinea pig skin for 24 h under an occlusive covering, moderate skin irritation was noted (3).

The minimum lethal concentration for rats exposed for 4 h was reported as 1840 ppm (525), although at this concentration some of the material must have been in aerosol form. Guinea pigs survived 8-h exposures to a similar atmosphere.

18.4.1.2 Chronic and Subchronic Toxicity. Groups of male and female rats were fed diets containing 0.075, 0.15, or 0.3% isophorone for 90 d. No effects were seen at the 0.075 or 0.15% dose levels. The only untoward finding was a reduction in body weight gain in rats fed the 0.3% diet (526). In another series of studies, male and female rats and mice were dosed with isophorone by gavage for 16 d at doses up to 2000 mg/kg/d, and for 13 wk at doses up to 1000 mg/kg/d. In the 16-d studies, all 2000-mg/kg mice and three of ten 2000-mg/kg rats died. Reduced weight gain was noted at 1000 mg/kg. In the 13-wk studies, compound-related deaths were seen in one of ten female rats and three of ten female mice at the 1000-mg/kg dose. These treatments did not produce gross or histopathological changes (526). No toxicologically significant effects were seen in beagle dogs given 90 oral doses of up to 150 mg/kg/d (527).

Rats and guinea pigs exposed to 100–500 ppm isophorone 8 h/d, 5 d/wk for 6 wk had decreased weight gain. Exposure to 500 ppm produced chronic conjunctivitis and pulmonary inflammation. Kidney damage occurred at the higher concentrations. There was a concentration-related increase in mortality. However, no effects were noted after exposure to 25 ppm (234). When male and female rats were exposed to 500 ppm isophorone 8 h/d, 5 d/wk for 6 and 4 mo, respectively, a total of three of twenty animals died, but the only other reported adverse effects consisted of irritation of the eyes and nose (528). Eye and nose irritation were also seen in a limited 18-mo study in which rats and rabbits were exposed to 250 ppm isophorone for 6 h/d, 5 d/wk. Microvacuolation of the liver was attributed to exposure to isophorone (528). Leucopenia without any change in differential or red blood cell counts was observed following exposure for 4 h to either 67 or 90 ppm isophorone (237). Respiratory depression to 50% of the normal frequency has been reported in rats exposed to a concentration of approximately 30 ppm (157). Inhalation of 89–137 ppm of isophorone for 4 h resulted in significantly decreased duration of the immobility phase (from 35 to 70%, respectively) in a "behavioral despair" swimming test. The differences between the exposed and control rats were statistically significant at all exposure concentrations (164).

18.4.1.3 Pharmacokinetics, Metabolism, and Mechanisms. Rabbits dosed by gavage with 1 g/kg isophorone excreted 5,5-dimethylcyclohexen-3-one, 5,5-dimethylcyclohexene-1-carboxylic acid, and glucuronic acid conjugates in the urine (529). These studies demonstrate that isophorone undergoes methyl group oxidation and subsequent conjugation with glucuronic acid. Rabbits partially eliminated isophorone unchanged in the expired air and urine (530, 531). The remainder was oxidized to a carboxylic acid and reduced to isophorol, which was eliminated in the urine as a glucuronide conjugate. Male rats treated with 0.5 or 1.0 g isophorone exhibited a significant increase in hyalin droplets in the kidney. Isophorone was identified in the α2-μglobulin fraction of the droplets. Biological alterations resembled those occurring following administration of trimethyl pentane and 1,4-dichlorobenzene, suggesting a similar mechanism of nephrotoxicity among these compounds (532). Administration of isophorone reduced glutathione both in the liver and male reproductive tract of the rat (533). Perturbation of reproductive tract glutathione by isophorone also enhanced the binding of ethyl methanesulfonate to spermatocytes. The temporal pattern of the ethylations was consistent with the stage of sperm development known to be susceptible to ethylation by EMS (533).

18.4.1.4 Reproductive and Developmental. The developmental toxicity of isophorone in rats and mice was studied using atmospheres containing isophorone at concentrations up to 115 ppm. At a concentration of 115 ppm, there was slight maternal toxicity in both rats and mice. No effects were seen at 50 ppm. No developmental effects were noted in either species (534). Dutertre-Catella (528) also reported that exposure of rats 6 h/d, 5 d/wk to 500 ppm isophorone for 3 mo did not cause adverse reproductive or developmental effects.

18.4.1.5 Carcinogenesis. In a 2-yr bioassay in rats and mice (conducted by gavage on a 5-d/wk treatment schedule) at dose levels of 0, 250, and 500 mg/kg isophorone, there was decreased survival of male rats and slight nephrotoxicity in female rats. There was no evidence of carcinogenicity in female rats or mice. In male rats, there was an increase in renal tumors in animals given either 250 or 500 mg/kg/d, and a low incidence of preputial gland tumors at 500 mg/kg. Other proliferative lesions in male rats included hyperplasia of the renal pelvis and tubular cell hyperplasia. Mineralization of the medullary collecting ducts was also noted in the isophorone-dosed male rats. In general, the incidence of nephropathy was increased in female rats, but the incidence was similar in dosed and control male rats. Responses in male mice included slight increases in liver and mesenchymal tumors at 500 mg/kg, and malignant lymphomas at 250 mg/kg. The tumors in male rats were judged to have provided some evidence of carcinogenicity, while the results were judged equivocal in male mice (526, 535).

18.4.1.6 Genetic and Related Cellular Effects Studies. Isophorone was negative in a battery of genetic toxicity tests, including the BALB/3T3 cell transformation assay, unscheduled DNA synthesis, and the mouse micronucleus assay (148). Others have shown isophorone to be positive in the UDS assay, whether the method of evaluation was by flow cytometry or by the more conventional measurement of radioactivity (536). Isophorone was also negative in the Ames/*Salmonella* assay in the presence and absence of S9 fraction, but induced sister chromatid exchanges in the absence, but not the presence, of S9 in Chinese hamster ovary cells. It did not induce chromosomal aberrations in Chinese hamster ovary cells in the presence or absence of S9 (537). Isophorone was negative in the mouse lymphoma assay in one study (148), but was weakly mutagenic in the absence of S9 in another (363). DNA-binding studies showed no covalent binding of isophorone or its metabolites to DNA from liver or kidney from Fischer 344 or $B6C3F_1$ mice, supporting the view that isophorone is may exert its affects through a nongenotoxic mechanism (538). In a series of studies designed to test the effect of chlorinated water on the mutagenicity of isophorone, the addition of millimolar quantities of isophorone to aqueous solutions containing hypochlorite at pH 8.5 produced a mixture that was positive in the Ames/*Salmonella* test. The specific mutagenic species were not identified in these tests (239).

18.4.1.7 Other: Neurological, Pulmonary, Skin Sensitization. Isophorone was negative in a Magnusson–Kligman guinea pig maximization test to detect the potential for dermal sensitization (539).

18.4.2 Human Experience

18.4.2.1 General Information. The most common route of exposure to isophorone is by inhalation. The sharp odor of isophorone may induce olfactory fatigue. Its vapors are also irritating to the eyes and mucous membranes. Prolonged skin contact may cause irritation.

Isophorone vapor (25 ppm) produced irritation to the eyes, nose, and throats of unacclimatized volunteers following a 15-min exposure (203). Workers exposed to 5–8 ppm isophorone for 1 mo complained of fatigue and malaise. No complaints were reported after exposure to 1–4 ppm. Workers exposed to 40, 85, 200, and 400 ppm experienced eye, nose, and throat irritation (525). Some complained of nausea, headache, dizziness, faintness, inebriation, and a feeling of suffocation at the higher concentrations. Symptoms of irritation and narcotic action decreased with time at 40 and 85 ppm. Useful warning properties existed only at 200 and 400 ppm. In another study in which six adult volunteers were exposed to graded concentrations of isophorone for 7-min periods, irritation was noted at 77 ppm (540).

Isophorone is not considered to be hazardous in the workplace owing to its low volatility. However, in the case of an accidental spill, caution should be exercised. Workers exposed to relatively low vapor concentrations of isophorone may experience irritation of the eyes, nose, and throat. Central nervous system effects may also occur at higher vapor concentrations.

18.4.2.3 Epidemiology Studies. No epidemiological studies on isophorone were identified.

18.5 Standards, Regulations, or Guidelines of Exposure

Hygienic standards for isophorone are listed in Table 76.5.

18.6 Studies on Environmental Impact

Isophorone does not posses the chemical and physical properties of a persistent environmental pollutant. Based on its vapor pressure, isophorone will exist primarily in the vapor state. Modeling predicts that the half-life for reaction with ambient ozone and photochemically generated hydroxyl radicals will be only 32 min (541, 542). Following release to water, isophorone may either volatilize or biodegrade. Its high solubility in water and its partition coefficient of 2.22 suggest that absorption to sediment will be unlikely to influence its fate in aquatic systems. (543). Its estimated soil adsorption coefficient (K_{oc}) of 25 suggests that isophorone would be mobile in the soil and that adsorption to suspended solids and sediment in water would be insignificant (544).

19.0 5-Nonanone

19.0.1 CAS Number: [502-56-7]

19.0.2 Synonyms: Di-*n*-butyl ketone, dibutyl ketone, 5-oxononane, and nonan-5-one

19.0.3 Trade Names: NA

19.0.4 Molecular Weight: 142.24

19.0.5 Molecular Formula: $CH_3(CH_2)_3CO(CH_2)_3CH_3$

19.0.6 Molecular Structure:

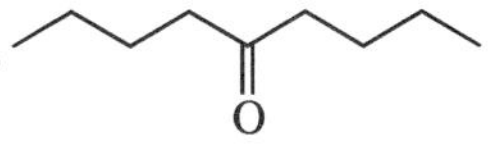

19.1 Chemical and Physical Properties

19.1.1 General

5-Nonanone has a boiling point of 188.4°C and a melting point of − 50°C. It is soluble in ethanol and very soluble in ether and chloroform. Its water solubility is 363 mg/L at 30°C. 5-Nonanone is also a stable, colorless to light yellow liquid. Additional physical–chemical properties are summarized in Table 76.4.

19.1.2 Odor and Warning Properties

No odor threshold was identified for 5-nonanone.

19.2 Production and Use

No information on production and use of 5-nonanone was identified.

19.3 Exposure Assessment

No information on exposure assessment was available for 5-nonanone.

19.4 Toxic Effects

19.4.1 Experimental Studies

5-Nonanone is moderately toxic orally. Repeated exposure has produced neurotoxicity in experimental animals, which may be enhanced by concomitant exposure to other ketones.

19.4.1.1 Acute Toxicity. The oral LD_{50} of 5-nonanone was >2000 mg/kg for rats (3). The intravenous LD_{50} has been reported to be 138 mg/kg for mice (545), while the anesthetic dose by intravenous administration, as measured by loss of righting reflex, was 118 mg/kg (427).

19.4.1.2 Chronic and Subchronic Toxicity. Repeated administration of 5-nonanone by gavage at doses of 1000 and 2000 mg/kg, 5 d/wk for 13 and 4 wk, respectively, produced a neuropathy in rats that was indistinguishable from that produced by MnBK. An oral dose of 233 mg/kg 5-nonanone administered to rats 5 d/wk for 13 wk produced a minor subclinical neuropathy. When the same amount of 5-nonanone (233 mg/kg) was present as an 11% contaminant in a commercial lot of 5-methyl-2-octanone (sometimes referred to as methyl heptyl ketone), neuropathy developed in less than 90 d. Clinical neurotoxicity similar in degree to that seen in the commercial mixture of 5-methyl-2-octanone could be achieved only with a 4.3-fold increase in 5-nonanone. The neuropathy following 5-nonanone administration results in axonal damage in the peripheral nerves, spinal cord, brainstem, and cerebellum (546). Other investigators were able to produce a clinical neuropathy that progressed to clinical paralysis within 1 wk in rats given daily doses of, sequentially, 150 mg/kg for 3 wk, 750 mg/kg for 8 wk, and 1500 mg/kg for 3 wk (547). Respiratory depression to 50% of the normal frequency has been reported in mice at a vapor concentration of 270 ppm (159).

19.4.1.3 Pharmacokinetics, Metabolism, and Mechanisms. 5-Nonanone is absorbed from the gastrointestinal tract of the rat. Metabolic transformations include ω-1 oxidation to 2,5-nonanedione with subsequent oxidative and decarboxylative steps to produce MnBK and 2,5-hexanedione (Fig. 2). In addition, 5-nonanone is oxidized to carbon dioxide (38% of the dose) (548). Radioactivity excreted in the urine accounted for approximately 50% of the dose after 72 h. No unchanged 5-nonanone was detected in the urine, but both MnBK and 2,5-hexanedione were present (548). The half-life of 5-nonanone in the blood of rats dosed with 500 mg/kg was 1.6 h (3).

The toxicity of 5-nonanone appears to be related to its metabolic conversion to 2,5-hexanedione; thus its mechanism of action is expected to be similar to γ-diketones. By combining 5-nonanone and purified 5-methyl-2-octanone in the same proportion found in a commercial preparation of 5-methyl-2-octanone, the degree of neurotoxicity of the commercial product was reproduced, suggesting that 5-methyl-2-octanone potentiates the neurotoxicity of 5-nonanone approximately sixfold. This occurred although 5-methyl-2-octanone did not cause neurotoxicity when administered alone (546). Shifman et al. (547) also found that administration of twice weekly administration of methyl ethyl ketone shortened the time period required for the onset of signs of neurotoxicity in rats administered 5-nonanone. Animals administered methyl ethyl ketone and 5-nonanone developed hind-limb paralysis after 11 wk, while animals administered 5-nonanone alone developed paralysis after 14 wk. Morphological observations were identical to those described for hexane, MnBK, and 2,5-hexanedione (547).

19.4.2 Human Experience

19.4.2.1 General Information. No adverse effects involving humans associated with exposure to 5-nonanone have been reported in the literature.

Any exposure in the workplace is expected to be by inhalation or by skin or eye contact. Repeated skin exposures could result in dermatitis and the accumulation of neurotoxic metabolites. Although exposures to 5-nonanone are not expected to occur frequently, 5-nonanone may be found as a contaminant in other solvents. Other ketones may potentiate the neurotoxic potential of 5-nonanone.

19.4.2.2 Clinical Cases. No clinical cases associated with exposure to 5-nonanone were identified.

19.4.2.3 Epidemiology Studies. No epidemiological studies on 5-nonanone were identified in the published literature.

19.5 Standards, Regulations, or Guidelines of Exposure

There are no recommended hygienic standards for 5-nonanone.

19.6 Studies on Environmental Impact

No studies on the environmental impacts of 5-nonanone are available; however 5-nonanone is not expected to present any unusual hazard to the environment. Since 5-nonanone would be expected to biodegrade, long-term environmental effects are not anticipated.

20.0 Diisobutyl Ketone

20.0.1 CAS Number: [108-83-8]

20.0.2 Synonyms: 2,6-Dimethyl-4-heptanone, isobutyl ketone, DiBK, isovalerone, valerone, *sym*-diisopropyl acetone, *s*-diisopropylacetone, 2,6-dimethylheptan-4-one, diisopropylacetone, and 4,6-dimethyl-2-heptanone

20.0.3 Trade Names: Isovalerone

20.0.4 Molecular Weight: 142.24

20.0.5 Molecular Formula: $(CH_3)_2CHCH_2COCH_2CH(CH_3)_2$

20.0.6 Molecular Structure:

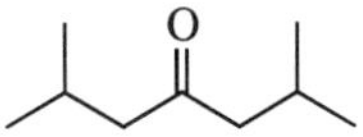

20.1 Chemical and Physical Properties

20.1.1 General

Diisobutyl ketone has a boiling point of 168.1°C and a melting point of −46°C. Its vapor pressure is 1.7 mm Hg, its specific gravity at 20°C is 0.807, and its flash point is 49°C. Diisobutyl ketone is a colorless stable liquid of low volatility that is poorly soluble in water but is soluble in alcohol and ether. NFPA has graded it Grade 2 for flammability. Additional physical–chemical properties are summarized in Table 76.4.

20.1.2 Odor and Warning Properties

Diisobutyl ketone has a mild ketone odor that has been described as sweet or peppermintlike. The threshold odor concentration has been reported to be below 0.11 ppm (520).

20.2 Production and Use

Diisobutyl ketone is produced by the hydrogenation of phorone or by metal-catalyzed decomposition of isovaleric acid. It is also a byproduct in the manufacture of methyl isobutyl ketone. It has use as a solvent, usually as a coating solvent for cellulosics, rubber, vinylite, and resins. It is also used as a solvent for paint, dyes, and adhesives, and as an intermediate in the production of pharmaceuticals and insecticides.

20.3 Exposure Assessment

20.3.1 Air

Under industrial conditions, exposure to diisobutyl ketone may be expected to occur by dermal or inhalation contact.

20.3.2 Background Levels

No information about background levels of diisobutyl ketone was identified.

20.3.3 Workplace Methods

Exposure in the workplace can be monitored by the use of NIOSH analytical method 1300 (116). Vapor can be collected via solid sorbent tube, is then desorbed with carbon disulfide, and is measured by GC/FID.

20.3.4 Community Methods

Specific community methods for measurement of diisobutyl ketone were not identified.

20.3.5 Biomonitoring/Biomarkers

Specific biomonitoring methods and biomarkers for diisobutyl ketone were not identified.

20.4 Toxic Effects

20.4.1 Experimental Studies

20.4.1.1 Acute Toxicity. Smyth et al. (420) reported an oral LD_{50} of 5800 mg/kg for rats. Single oral doses as high as 3200 mg/kg failed to produce death in rats (3). An oral LD_{50} of 1416 mg/kg has been reported for mice (144). Intraperitoneal doses as high as 1600 mg/kg caused no deaths in rats (3).

Diisobutyl ketone instilled undiluted into the rabbit eye produced essentially no irritation (3, 230, 420). Application of the undiluted liquid to covered and uncovered guinea pig skin caused moderate irritation (3). Ten milligrams per kilogram held in contact with the skin of rabbits for 24 h under an occlusive wrap, and 500 mg/kg applied to the covered skin of rabbits, produced only mild irritation (154, 420). There was no difference in the irritation observed following 1- and 4-h exposures to 0.5 mL diisobutyl ketone, when applied under either an occluded or semioccluded wrap (524). Smyth et al. (420) have reported a dermal LD_{50} of greater than 20 mL/kg for the rabbit.

McOmie and Anderson (549) reported no deaths of rats or guinea pigs exposed for 7.5–16 h to saturated vapors of diisobutyl ketone. A single 6-h exposure of rats to 1500 ppm caused ataxia and drowsiness but no deaths (3). Smyth et al. (420) reported a higher degree of toxicity for rats; an 8-h exposure to an essentially saturated atmosphere (2000 ppm) caused the deaths of five of six animals. In another report, seven of twelve female rats succumbed during an 8-h exposure to 2000 ppm (550). Four hours was reported to be the shortest exposure time that produced a toxic effect in rats exposed to 2000 ppm (235). Respiratory depression to 50% of the normal frequency has been reported in rats exposed to vapor concentrations of 320 (159) or 290 ppm (164). Inhalation of from 243 to 415 ppm of diisobutyl ketone for 4 h resulted in significantly decreased duration of the immobility phase (from 33 to 71%, respectively) of a "behavioral despair" swimming test. The differences between the exposed and control rats were statistically significant at all exposure concentrations (331). McOmie and Anderson (549) reported that mice survived twelve 3-h exposures to a saturated atmosphere. Carpenter et al. (550) exposed guinea pigs and rats to concentrations of 125–1650 ppm for thirty 7-h exposures. The 125-ppm concentration was a no-effect level for both species. The 250-ppm concentration increased liver and kidney weights of female rats, but decreased the liver weight of male guinea pigs. Rats exposed to intermediate concentrations of 530 or 920 ppm had increased liver and kidney weights. Increased mortality and liver and kidney injury were seen at 1650 ppm. Dodd et al. (551) exposed rats to concentrations of up to 905 ppm for nine 6-h exposures. Evidence of toxicity was noted only in male rats and included increased absolute and relative kidney weights, increased urine volumes, hyaline droplets in the proximal renal tubules, increased serum proteins, and increased liver weights. Similar but lesser effects were seen at 300 ppm. The effects either disappeared or lessened following a 2-wk recovery period (551). In another study in which male rats were exposed to 400 ppm for twelve 6-h exposures, no treatment-related changes were seen in weight gain, clinical signs, clinical chemistries, liver or kidney weights, or gross or microscopic tissue evaluation (3).

Daily doses of 4000 mg/kg given by gavage killed all rats after two or three treatments (3). Deaths were attributed to severe depression of the central nervous system, hepatotoxicity, and dehydration. Pulmonary congestion and edema and renal toxicity were contributing factors in the deaths. Doses of 2000 and 1000 mg/kg administered daily for 3 wk did not cause mortality. The 2000-mg/kg dose produced central nervous system depression initially, but a tolerance developed after several doses. Body weight gain and feed consumption were not affected by the ketone exposure. Hematologic and serum clinical chemistry determinations were comparable to the control values. Absolute and relative liver and kidney weights were slightly increased. Histologically, only minor changes were seen in the livers and kidneys (3).

20.4.1.2 Chronic and Subchronic Toxicity. When diisobutyl ketone was tested in a 90-d gavage study at doses as high as 2000 mg/kg (3), abnormalities noted included reduced glucose levels, increased absolute and relative liver and adrenal gland weights, increased relative kidney weights, and decreased absolute brain and heart weights. Compound-related histopathology included minor changes in the stomach, liver, and kidneys. No neurotoxicity was evident.

20.4.1.3 Pharmacokinetics, Metabolism, and Mechanisms. No pharmacokinetics, metabolism, or mechanism studies were identified for diisobutyl ketone.

20.4.1.4 Reproductive and Developmental. No reproductive or developmental studies were identified for diisobutyl ketone.

20.4.1.5 Carcinogenesis. No carcinogenesis studies were identified for diisobutyl ketone.

20.4.1.6 Genetic and Related Cellular Effects Studies. Diisobutyl ketone was negative in the Ames/*Salmonella* test in the presence and absence of rat or hamster liver S9 fractions (192). Other investigators confirmed the lack of activity of diisobutyl ketone in bacterial mutation assays (34). Diisobutyl ketone was also negative in a yeast assay for mitotic gene conversion and an *in vitro* assay for chromosome damage using cultured rat liver cells (34).

20.4.1.7 Other: Neurological, Pulmonary, Skin Sensitization. No other systemic toxicity studies were identified for diisobutyl ketone.

20.4.2 Human Experience

20.4.2.1 General Information. Limited controlled human studies indicate that diisobutyl ketone presents a low degree of hazard in the usual industrial setting. Oral toxicity is low, and eye and skin contact may produce slight to moderate irritation. Concentrations of the vapor well below those that cause irritation and narcosis have an odor and may cause discomfort, and thus would serve as warnings to preclude overexposure. However, the principal hazard to health would be inhalation of the vapors; so precautions should be taken to provide proper ventilation.

Silverman et al. (203) studied the sensory response of humans to diisobutyl ketone vapors and reported that concentrations higher than 25 ppm had an objectionable odor and caused eye irritation. No nose and throat irritation was observed at 50 ppm. Carpenter et al. (550) found that a 3-h exposure to 50 and 100 ppm caused slight eye, nose, and throat irritation in three volunteers. The 100-ppm level was judged to be "unsatisfactory," whereas 50 ppm was "satisfactory."

20.4.2.2 Clinical Cases. A single occupational case history associated with exposure to diisobutyl ketone was identified. This case concerned a laboratory technician whose job consisted of hydraulic stripping experiments on iron ore. During these experiments, diisobutyl ketone was heated to 700°C and placed under 2300 pounds of pressure. After 1 mo on the job, the subject complained of headaches and loss of vision. MRIs revealed multiple small foci in the subcortical region, and some deficits were noted during psychological testing (552).

20.4.2.3 Epidemiology Studies. No epidemiology studies are available for diisobutyl ketone.

20.5 Standards, Regulations, or Guidelines of Exposure

Hygienic standards for diisobutyl ketone are listed in Table 76.5.

20.6 Studies on Environmental Impact

No studies on the environmental impacts of diisobutyl ketone are available; however, diisobutyl ketone is not expected to present any unusual hazard to the environment. Since diisobutyl ketone would be expected to biodegrade, long-term environmental effects are not anticipated.

21.0 Trimethyl Nonanone

21.0.1 CAS Number: *[123-18-2]*

21.0.2 Synonyms: 2,6,8-Trimethyl-4-nonanone and isobutyl heptyl ketone.

21.0.3 Trade Names: NA

21.0.4 Molecular Weight: 184.32

21.0.5 Molecular Formula: $CH_3CHCH_3CH_2COCH_2CHCH_3CH_2CH(CH_3)_2$

21.0.6 Molecular Structure:

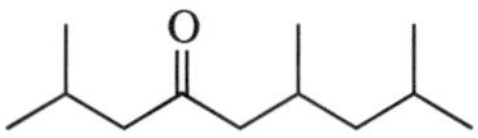

21.1 Chemical and Physical Properties

21.1.1 General

Trimethyl nonanone has a boiling point of 207–228°C. Additional physical–chemical properties are summarized in Table 76.4.

21.1.2 Odor and Warning Properties

Trimethyl nonanone is a colorless liquid with a pleasant fruity odor.

21.2 Production and Use

No information on production or use of trimethyl nonanone was identified.

21.4 Toxic Effects

21.4.1 Experimental Studies

Trimethyl nonanone is slightly toxic orally and is a mild dermal and ocular irritant.

21.4.1.1 Acute Toxicity. An oral LD_{50} of 8.47 g/kg has been reported for trimethyl nonanone in rats (138). Trimethyl nonanone was a mild irritant when applied undiluted to rabbit eyes (22). Ten milligrams applied to the skin of rabbits for 24 h under a cover and 500 mg/kg applied uncovered produced mild skin irritation (138). Repeated application may produce irritation by defatting the skin. The dermal LD_{50} was 11 mL/kg for rabbits (138). Saturated vapor exposures of 4 h caused no deaths in rats (138).

21.4.1.2 Chronic and Subchronic Toxicity. No long-term studies on trimethyl nonanone were identified.

21.4.1.3 Pharmacokinetics, Metabolism, and Mechanisms. No pharmacokinetics, metabolism, or mechanisms studies were identified for trimethyl nonanone.

21.4.1.4 Reproductive and Developmental. No reproductive or developmental studies were identified for trimethyl nonanone.

21.4.1.5 Carcinogenesis. No carcinogenesis studies were identified for trimethyl nonanone.

21.4.1.6 Genetic and Related Cellular Effects Studies. No genetic studies were identified for trimethyl nonanone.

20.4.1.7 Other: Neurological, Pulmonary, Skin Sensitization. No other systemic toxicity studies are available for trimethyl nonanone.

21.4.2 Human Experience

21.4.2.1 General Information. No published reports of toxicity to humans were found in the literature. Industrial exposures are expected to be primarily by inhalation or skin and eye contact. Single topical exposures may be slightly irritating, but repeated exposures may exacerbate the irritation. Oral exposures should not be hazardous unless large volumes of trimethyl nonanone are swallowed. Trimethyl nonanone has a relatively high flash point and should not present a fire hazard.

21.5 Standards, Regulations, or Guidelines of Exposure

There are no recommended hygiene standards for trimethyl nonanone.

21.6 Studies on Environmental Impact

No studies on the environmental impact of trimethyl nonanone are available; however, trimethyl nonanone is not expected to present any unusual hazard to the environment.

Since trimethyl nonanone would be expected to biodegrade, long-term environmental effects are not anticipated.

22.0 Benzophenone

22.0.1 *CAS Number: [119-61-9]*

22.0.2 Synonyms: Diphenyl ketone, α-oxodiphenylmethane, α-oxoditane, benzoylbenzene, diphenyl ketone, diphenylmethanone, and phenyl ketone

22.0.3 Trade Names: NA

22.0.4 Molecular Weight: 182.22

22.0.5 Molecular Formula: $C_{13}H_{10}O$

22.0.6 Molecular Structure:

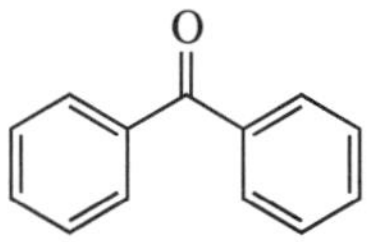

22.1 Chemical and Physical Properties

22.1.1 *General*

Benzophenone is a white crystalline solid with a low vapor pressure. It is insoluble in water, but soluble to various degrees in acetone, diethyl ether, chloroform, and benzene (553). The melting and boiling points of benzophenone are 48.5 and 305.4°C, respectively. Benzophenone has a vapor pressure of 1 mm Hg at 108.2°C, a density of 1.1108 g/cm^3 at 20°C, and a flash point of approximately 155°C (428, 510). Additional physical–chemical properties are summarized in Table 76.4.

22.1.2 *Odor and Warning Properties*

Benzophenone is used as a fragrance in perfumes and has a geranium or roselike odor. No odor detection threshold has been published for benzophenone.

22.2 Production and Use

Benzophenone is commercially synthesized by the atmospheric oxidation of diphenylmethane using a catalyst of copper naphthenate. Alternatively, it can be produced by a Friedel–Crafts acylation of benzene using either benzoyl chloride or phosgene in the presence of aluminum chloride (510). Benzophenone is used as a synthetic intermediate for manufacture of pharmaceuticals and agricultural chemicals. It is also used as a photoinitiator in UV-curable printing inks, as a fragrance in perfumes, as a flavor enhancer in foods. (508, 555). Benzophenone can be added as a UV-absorbing agent to plastics, lacquers, and coatings at concentrations of 2–8% (556). In 1973, Opdyke reported that about 100,000 lb/yr of benzophenone was being used in the United States as a flavoring agent (557). Lande et al. reported that 741,000 lb of benzophenone was produced in the United States in 1973 (68).

22.3 Exposure Assessment

22.3.1 Air

Benzophenone was detected in air samples collected inside a spruce tree forest in northern Germany (557). Benzophenone was also detected, but not quantified, in the exhaust from a residential oil burner using No. 2 fuel oil (457). Likewise, it was found in flue gas emissions from a municipal waste incinerator at a level of 1.16 $\mu g/m^3$ (558). Cocheo et al. measured benzophenone levels ranging from 0 to 1 $\mu g/m^3$ in air samples from the extrusion area of a factory that manufactured electrical cables (294). It was not, however, detected in air samples collected from a shoe-sole factory or from the vulcanization and extrusion areas of a tire retreading factory. Nylander-French et al. used a sampling tube that collected both aerosols and vapors to measure the benzophenone levels in a factory that manufactured a UV-cured acrylate coating (560). Airborne concentrations of benzophenone ranging from < 1 to 30 $\mu g/m^3$ were measured in the breathing zones of 10 workers. The highest levels were observed in the employees who performed a "hand-filling" operation.

22.3.2 Background Levels

Benzophenone has been identified in food, salt water sediment, groundwater, and wastewater either as an intentional additive or as a contaminant. Benzophenone may be added to soap, detergents, lotions, and perfumes as a fragrance at maximum concentrations of 0.15, 0.015, 0.015, and 0.3%, respectively (557). Because of its status as a GRAS substance (generally recognized as safe), benzophenone may be added as a flavoring agent to the following products in limited amounts: nonalcoholic beverages at a maximum concentration of 0.50 ppm, ice cream at 0.61 ppm, candy at 1.70 ppm, or baked goods at 2.40 ppm (81). Gilbert et al. reported that benzophenone was present in the cardboard packaging used for frozen and shelf-stable foods at a level of 1.1 mg/dm^2 (561). Use of the packaging as a splatter shield during microwave preparation of frozen lasagna resulted in the appearance of benzophenone in the food at 150 μg/kg. The sunlight irradiation of benzyl benzoate, a commonly used fragrance in skin care products, resulted in the formation of benzophenone as a breakdown product (562). Benzophenone was found at a concentration of 8.8 ppb in a drinking water sample collected in Japan (308). Fernández et al. identified benzophenone in extracts from ocean coastal sediment collected near a major urban area in Spain (563). Benzophenone was also found in soil, sediment, and sludge samples taken from inside a chemical plant at concentrations of 15,000–30,000 ppm (89). Ehrhardt et al. identified benzophenone in seawater samples taken at different depths, but the results were not quantified (467). Ellis et al. detected benzophenone in the secondary effluent from 3 of 10 municipal and industrial wastewater treatment plants examined in Illinois (306). Clark et al. surveyed three publicly owned wastewater treatment plants and found benzophenone in the effluent from two of the sites at concentrations of 0.5 and 2 ppb (311). Similarly, Praxéus showed that benzophenone was present in the effluent from 2 of 3 large wastewater treatment plants in Sweden that treated water from a variety of industries (310). The benzophenone concentration in the water effluent at the two sites was 11 and 48 μg/L. The wastewater influent from a textile finishing plant was shown to contain benzophenone, but the levels were not quantified (312).

Zoeteman et al. reported that benzophenone was found at a maximum concentration of 0.03 μg/L in groundwater samples collected at 232 pumping stations in the Netherlands (309). Ahel found benzophenone in groundwater samples from 1 of 3 sites in Yugoslavia at concentrations of 10–100 ng/L (564). The site was located near a canal that received untreated wastewater from a pharmaceutical plant and a factory producing baker's yeast. The concentration of benzophenone in the canal wastewater was 100–1000 ng/L.

Benzophenone was identified in groundwater samples collected at 3 out of 4 military camps using a tertiary wastewater treatment system for handling secondary effluents from a wastewater treatment plant (565). The benzophenone concentrations in the groundwater at the 3 sites ranged from 0.05 to 2.13 μg/L. The secondary effluent concentration of benzophenone was determined to be 0.038 μg/L at one site (Fort Polk, LA) and 5.17 μg/L at another (Fort Devens, MA) of the sites where groundwater contamination was present (462, 566). Further studies indicated that raw sewage from the Fort Devens site contained an average benzophenone concentration of 0.73 μg/L and that the level was reduced to an average of 0.40 μg/L in the secondary effluent that was sent to large soil-lined basins for final tertiary treatment by rapid infiltration (567). The soil infiltrate from the catch basins also showed an average benzophenone concentration of 0.40 μg/L, which indicated that benzophenone was not removed by the soil column and could migrate into the groundwater. Average groundwater concentrations of benzophenone from monitoring wells near the infiltration basins were 0.20 μg/L.

22.3.3 Workplace Methods

OSHA has published a recommended method for determining airborne concentrations of benzophenone in the workplace that relies on the use of an OVS (OSHA versatile sampler) containing XAD-7 for sample collection, and gas chromatography with a flame ionization detector for quantitation (568). Nylander-French et al. have developed a unique method for sampling and analyzing benzophenone aerosols and vapors in the workplace that also used an OVS that that contained a glass fiber filter along with an adsorbent of Tenax (560). Samples were quantified by capillary gas chromatography with flame ionization detection.

22.3.4 Community Methods

Geary assessed the recovery of benzophenone from cured acrylic coatings by derivitizing an extract with *p*-nitrophenylhydrazone and measuring the concentration spectrophotometrically at a wavelength of 253 nm (569). Kringstad et al. used reverse-phase high-performance liquid chromatography to determine the octanol–water partition coefficient of benzophenone (570). Bulterman et al. was able to determine benzophenone at the ppb level in surface water by combining gas chromatography–mass spectrometry with a liquid chromatographic trace-enrichment procedure (571). Benzophenone was identified in sediment using gel permeation chromatography for isolation, normal-phase liquid chromatography for separation, and gas chromatography–mass spectrometry in the electron-impact mode for final identification (563).

Several approaches have been taken to analyze for benzophenone in environmental samples. Franz et al. developed an analytical method that looked at the migration of benzophenone into and out of bottles made from recycled polyethylene terephthalate (572). The migration of benzophenone out of the bottle was determined using acid or ethanol as a solvent and either gas chromatography with flame ionization detection or gas chromatography–mass spectrometry in the selective ion mode for quantitation. A similar gas chromatography–mass spectrometry approach was used by Monteiro et al. to examine the migration of UV stabilizers in polyethylene terephthalate bottles. The authors, however, used benzophenone as an internal standard for determining the concentration of chromatographically similar test chemicals (573).

22.3.5 Biomonitoring/Biomarkers

22.3.5.1 Blood. No information on biomarkers in blood was identified.

22.3.5.2 Urine. Goenechea published a method that specifically focused on the analysis of benzophenone in the urine (574). The method involves hexane extraction of the urine followed by spectrophotometric analysis of the dried and purified extract at a wavelength of 247 nm.

22.4 Toxic Effects

22.4.1 Experimental Studies

22.4.1.1 Acute Toxicity. Caprino et al. reported the 7-d oral LD_{50} of benzophenone in male Swiss mice to be 2895 mg/kg (95% CI, 2441–3433 mg/kg) (575). No tissue or organ lesions were noted on necropsy; however, animals that died showed sedation, depressed motor activity, unstable gait, tremors, and respiratory depression. Another unpublished study reported an LD_{50} of 1600 mg/kg in mice (555). In rats, one report indicated that the oral LD_{50} was $>10{,}000$ mg/kg (363), whereas another study reported the LD_{50} to be approximately 1900 mg/kg with clinical signs of slight to severe weakness, unkempt haircoats, and ataxia (555). The intraperitoneal LD_{50} in male Swiss mice was reported to be 727 mg/kg (95% CI, 634–833 mg/kg) (575).

Benzophenone produced slight irritation of the conjunctiva of rabbit eyes (555). Three animals had crystals of benzophenone instilled into one eye, and the eyes were monitored for irritation for 14 d. Slight to moderate erythema of the conjunctiva and nictitating membrane occurred at 1 h and persisted for 24 h. By 48 h, slight erythema was observed in only one animal, and by 14 d all eyes were normal. Irrigation of the eye immediately after instillation was beneficial in that only slight irritation was seen at 1 h and no irritation was observed at 24 h.

Benzophenone was slightly irritating to the skin of guinea pigs when applied for 24 h under an occlusive patch (555). Slight erythema, desquamation, and slight-to-moderate edema were noted. The dermal LD_{50} was reported to be >1 g/kg in an unstated strain of guinea pig (555) and 3.5 g/kg in an unstated strain of rabbit (557).

22.4.1.2 Chronic and Subchronic Toxicity. Several studies have assessed the toxicity of repeated exposure to benzophenone. Male rats given 1.0% benzophenone in the diet for 11 d (average dose of 661 mg/kg/d) had slightly decreased weight gain and feed consumption compared with controls. The liver enzyme alanine aminotransferase was slightly increased in the blood, and liver weights were greatly increased. Histologic changes in these animals included mild degeneration of the liver. Diffuse bone marrow vacuolization and erythroid hyperplasia were also noted, indicating that the liver and the bone marrow were potential sites of action. Animals treated with 0.1% benzophenone in the diet (81 mg/kg/d) gained weight normally. Liver weights in these animals were moderately increased, and kidney weights were slightly increased (555).

Similar effects were noted in male guinea pigs given a 5-mg/kg intraperitoneal dose of benzophenone once a day for 15 d (576). Liver enlargement and nodular cirrhotic lesions were observed in the treated animals 24 h after the last treatment. Histologic examination of the liver revealed hepatic degeneration and mild fatty infiltration that was associated with areas of perilobular and centrilobular necrosis. Obstructive fibrinous thrombi and leukocytic infiltrates appeared within the lumen of hepatic blood vessels. Regenerative changes, including the growth of connective tissue and the proliferation of bile duct endothelial epithelial cell, were also noted. Histochemical analysis revealed a depletion of liver glycogen and an increase in mitotic activity in biliary epithelial cells (Feulgen reaction).

In longer-term studies, male and female Sprague–Dawley rats were treated with oral doses of benzophenone at levels of 0, 20, 100, or 500 mg/kg/d in the diet for 28 (all dose groups) or 90 d (0- and 20-mg/kg dose groups). Male and female animals at the two highest dose levels showed a decrease in body weight gain relative to controls. At the mid- and high doses, female rats showed signs of hemolysis, and both male and female rats had increased absolute and relative liver weights (577). Blood albumin was increased at all dose levels for both males and females, whereas total protein levels were increased in the high-dose males and mid- and high-dose females. Microscopic hepatocellular changes occurred only in the mid- and high-dose groups (both male and female). Kidney weights were also increased at 100 and 500 mg/kg, but there was no associated histopathology. Low-dose female rats had increased liver weight after 28 d of treatment, but not after 90 d. No histopathology was noted after 90 d of treatment.

In a more recent study benzophenone was administered in the diet of male and female Fischer 344 rats and B6C3F$_1$ mice at levels of 0.125, 0.25, 0.5, 1.0, or 2.0% for 13 wk (578). A decrease in food consumption and body weight gain was observed in rats treated at dose levels greater than 0.25% and mice treated at levels greater than 0.5% in the diet. All high-dose male and female rats were sacrificed after 6 wk due to excessive weight loss. By the end of the study, there was a statistically significant decrease in body weight for all rat treatment groups, except the male rats treated with 0.125%. A similar decrease in body weight was observed in female mice treated at dose levels of 0.5% or higher. Body weight in male mice was normal except at the 1.0% dose level. Both rats and mice showed an increase in serum bile salts, relative and absolute liver weight, and hepatocellular hypertrophy incidence. Rats also showed cellular atrophy in the bone marrow and papillary necrosis in the kidney. The data indicated that the palatability of benzophenone in the diet played a role in the toxicity.

Repeated topical applications of benzophenone to guinea pig skin over 10 d initially produced a slight to moderate erythema that was not exacerbated by additional treatment (555).

22.4.1.3 Pharmacokinetics, Metabolism, and Mechanisms. The metabolism of benzophenone has been studied in rabbits and rats. The results of these studies indicate that the metabolic pathways differ slightly in these two species. Rabbits given an oral dose of 1 g (2 mmol/kg) benzophenone excreted 50% of the dose as a glucuronide conjugate of benzhydrol within 24 h (579). No sulfate conjugate, free benzophenone, or *p*-hydroxybenzophenone was detected in the urine. The percentage of the dose excreted in the urine of rabbits as the glucuronide conjugate of benzhydrol can range from 48 to 77% (580). In a subsequent study with Sprague–Dawley rats given an oral dose of either 5.0 or 6.2 mmol (approx. 15.5 or 19.1 mmol/kg), 1% of the benzophenone dose was converted to the glucuronide or sulfate conjugate of *p*-hydroxybenzophenone (581).

Leibman showed that benzophenone was rapidly reduced by the S9 fraction (microsomes and cytosol) from rabbit liver (582). The incubation of the S9 supernatant from male New Zealand rabbits 8 mM of benzophenone resulted in the formation of benzhydrol at a rate of 7.6 μmol/h. The reaction required the addition of an NADP-generation system to the supernatant mixture and did not occur after heat inactivation. Benzophenone may be produced as an *in vitro* metabolic product of the cyclooxygenase activity of prostaglandin H synthase on the phenylhydrazone derivative of benzophenone (487). Benzophenone was capable of activating or inhibiting the bioluminescent activity of several bacterial luciferase reactions *in vitro* as a result of its oxidation reduction potential (583).

22.4.1.3.1 Absorption. The dermal absorption of benzophenone was determined in four female rhesus monkeys under occluded patch and at unoccluded sites (584). A dose of 4 $\mu g/cm^2$ of radiolabeled benzophenone was applied for 24 h. Evaporative loss from the unoccluded site was substantial since only 43.8% of the applied dose was absorbed compared with 68.7% absorption from the occluded site. Of the dose absorbed, 92.6% was excreted in the urine over 4 d. No determination of metabolites was performed.

22.4.1.4 Reproductive and Developmental. Benzophenone did not affect limb regeneration in Japanese newts that were treated following amputation of the forelimb (585). The direct application of 5 μg of benzophenone to the regeneration blastema 7 d after amputation caused abnormal limb development in 1 of the 10 animals treated. Call et al. assessed the effects of benzophenone on the growth and development of fathead minnows after continuous exposure of embryos, larvae, and juveniles to concentrations of 0.54, 0.99, 1.76, 3.31, 6.38, or 8.66 mg/L for 31–33 days (190). An evaluation of egg hatchability, incidence of developmental abnormalities, survival, and growth indicated that the maximum acceptable toxicant concentration that produced no adverse effects was 0.73 mg/L. A benzophenone concentration of 0.99 mg/L was found to cause a significant reduction in the wet weight of the juvenile fish.

The estrogenic activity of benzophenone was examined in an *in vitro* assay that measured the inhibition of 17β-estradiol to an estrogenic receptor from rainbow trout (586). Benzophenone was weakly active in the assay, with a concentration of 1×10^{-3} M causing an 80% reduction in binding. A benzophenone concentration of 1 μM also failed to inhibit human progesterone receptor binding in transformed DY150 yeast cells grown in culture (587).

22.4.1.5 Carcinogenesis. Lifetime dermal carcinogenicity studies in mice and rabbits did not show any tumor excess in the treated animals (588, 589). Female Swiss mice and New Zealand White rabbits of both sexes were treated dermally with 0, 5, 25, or 50% of benzophenone (0.02 mL) twice a week for 120 or 180 wk, respectively. Weekly examination of the rabbits did not reveal any reduction in survival or appearance of tumors. Mice treated with benzophenone did not show any excess in the number of tumor-bearing animals or in total number of tumors relative to untreated control animals. Although three skin tumors were observed in the benzophenone-treated mice (1 squamous cell carcinoma and 2 squamous cell papillomas), there were also three tumors (1 carcinoma and 2 papillomas) observed in the control animals.

22.4.1.6 Genetic and Related Cellular Effects Studies. The mutagenic activity of benzophenone was evaluated in the Ames/*Salmonella typhimurium* bacterial assay at levels up to 1000 μg/plate using strains TA98, TA100, TA1535, and TA1537. No mutagenic activity was found in the presence or absence of enzymatic activation using Arochlor-induced rat liver S9 (590). In a nearly identical study, benzophenone was not mutagenic to TA98, TA100, TA1537, or TA 1538 with and without activation at levels up to 2000 μg/plate (555). Benzophenone did not exhibit antimutagenic activity towards 2-amino-3-methy-imidazo[4,5-*f*]-quinoline in *Salmonella* strain TA98 (493). Fluck et al. incubated 500 μg of benzophenone with *Escherichia coli* strains Pol A^+ (DNA polymerase positive) and Pol A^- (DNA polymerase negative) and found little or no growth inhibition (492). Benzophenone was not genotoxic to mouse lymphoma cells (L51178Y (TK^+/TK^-) at concentrations up to 90 μg/mL without enzyme activation and concentrations up to 145 μg/mL with liver S9 activation (555). A concentration of 5×10^{-5} M was also negative under both black light (330–380 nm) and sunlamp (293–325 nm) conditions in an *in vitro* phototoxicity assay using cultured Ehrlich ascites cells (591).

Benzophenone is a known photosensitizer that can be photoactivated and used as a tool to study the mechanisms responsible for UV-induced DNA damage. Benzophenone is a photoreactive substance because of its ability to absorb photons of light in the UVA and UVB regions of the ultraviolet spectrum and move into a highly reactive and relatively stable triplet energy state (592). Flash photolysis experiments have revealed that benzophenone molecules in the triplet energy state can be identified in solution and that they have a high affinity for interacting with the purine and pyrimidine bases in DNA (593). Single-strand chain breaks and thymine dimers have been observed when benzophenone and tritiated DNA from *E. coli* were irradiated at a wavelength 313 nm (495). The formation of thymine dimers appeared to be the result of an energy transfer from the triplet state of benzophenone to pyrimidine nucleotides in DNA. The single-strand breaks, in contrast, appeared to be the result of the intermediate formation of hydroxyl radicals, which preferentially attacked particular regions of the DNA strand (594). The free radicals are the result of natural decay and dissociation of the triplet-state molecule and have been shown to be responsible for the peroxidation of linoleic acid, a fatty acid found in biological membranes (595). Benzophenone photoactivation has been used mechanistically to study and evaluate DNA–protein crosslinks in cell-free systems, inhibitors of photooxidative damage in calf thymus DNA, and the mode of action of therapeutic drugs that operate through free radical intermediates (596–598).

22.4.1.7 Other: Neurological, Pulmonary, Skin Sensitization. Using four different methods to evaluate contact hypersensitivity in Himalayan spotted guinea pigs, Klecak et al. reported that benzophenone was not a sensitizer in any of the tests employed (599). The four tests included an open percutaneous test, a Draize test, a maximization test, and a Freund's complete adjuvant test. The open percutaneous test relied on an initial determination of a minimally irritating concentration, which yielded a 1-d acute value of 30% and a 21-d subchronic value of 3%. A challenge dose of 30 or 60% benzophenone was dermally applied on day 21 and day 35 of the test protocol. In the Draize test, the animals were treated every other day with an intradermal injection of 0.1% solution of benzophenone, then intradermally challenged with the same dose on days 35 and 49. Animals in the maximization test were treated intradermally treated with a 5% solution, both with and without adjuvant, for 7 d, and then dermally with a 25% solution for 2 d. The challenge dose was applied on day 21 under an occlusive patch. The Freund's complete adjuvant (FCA) test utilized five alternate-day intradermal injections of 250 mg of benzophenone in FCA. The challenge dose was administered dermally on days 21 and 35 of the test.

Benzophenone was not a skin sensitizer in Hartley guinea pigs when tested using a modified Draize procedure that involved 4 intradermal injections at levels 2.5 times the minimally irritating concentration (422). The test involved the use of both an intradermal and a dermal challenge dose 14 d after intradermal sensitization. Preliminary studies indicated that the minimally irritating intradermal concentration of benzophenone was 0.25% and that the dermal challenge concentration causing no local erythema was 20%.

In another separate study, the irritancy and dermal sensitization potential was assessed using the Draize test in rabbits and the maximization test in guinea pigs (600). The treatment of abraded or unabraded skin of New Zealand white rabbits with a 20% solution of benzophenone caused slight to medium erythema with focal necrosis with and without dyskeratosis. The primary cutaneous irritation index value of 2.0 indicated a medium irritancy potential. The maximization test was performed in female Hartley guinea pigs. The induction phase occurred in two steps with the initial intradermal injections (2×0.05 mL) of 1% benzophenone then the topical application of 10% benzophenone at the same site after 1 wk. A challenge concentration of either 1 or 5% benzophenone was applied after another 2-wk wait. Under these conditions, benzophenone did not cause an observable sensitization reaction in any of animals, and biopsy specimens did not reveal any histopathological reactions at the treatment site.

22.4.2 Human Experience

22.4.2.1 General Information. Benzophenone was given a GRAS ranking by the Flavor and Extract Manufacturers Association due to its low acute toxicity. Although approved by the FDA for use in foods, the most common route of benzophenone exposure is expected to be dermal because of the widespread use of benzophenone in the cosmetic industry. Benzophenone derivatives have sunscreen properties, which has led to their use in cosmetics and suntanning products. Inhalation exposure to benzophenone is unlikely because of its low volatility.

22.4.2.2 Clinical Cases

22.4.2.2.1 Acute Toxicity. There are no case reports of acute toxicity or lethality following exposures to benzophenone.

22.4.2.2.2 Chronic and Subchronic Toxicity. No case reports of subchronic or chronic toxicity to humans resulting from exposure to benzophenone were identified.

22.4.2.2.3 Pharmacokinetics, Metabolism, and Mechanisms. There are no case reports on pharmacokinetics, metabolism, or mechanisms in humans following exposure to benzophenone.

22.4.2.2.4 Reproductive and Developmental. There are no case reports of reproductive or developmental toxicity in humans following exposure to benzophenone.

22.4.2.2.5 Carcinogenesis. There are no case studies of carcinogenicity in humans following exposure to benzophenone.

22.4.2.2.6 Genetic and Related Cellular Effects Studies. Jobling et al. reported that benzophenone was not mitogenic and did not enhance the growth of breast cancer cells in culture when tested at a concentration of 1×10^{-5} M (586).

22.4.2.2.7 Other: Neurological, Pulmonary, Skin Sensitization. Patch testing and phototesting of 12 patients with a history of photocontact dermatitis from oral ketoprofen or topical tiaprofenic acid therapy revealed a high degree of cross-sensitivity to benzophenone (601). Only 1 of the 12 patients tested with 1% benzophenone displayed a positive patch test; however, 11 out of 12 showed a definite reaction in the photopatch test where the treated skin was irradiated with either UVA or full-spectrum ultraviolet light 24 h after removal of the test patch. Phototesting with 1% benzophenone in 15 unsensitized control subjects did not cause any positive reactions. In another study with three photosensitized subjects who worked in a plant that manufactured UV-cured inks that contained benzophenone, photopatch testing did not produce a positive reaction (591). The phototesting was performed at a benzophenone concentration of 5% using sunlight to irradiate the treated skin area.

Opdyke reported that benzophenone failed to produce a reaction in human volunteers participating in a maximization test (557). A total of 25 people were involved in a study utilizing a challenge concentration of 6%. Mitchell et al. summarized the results of previous standardized patch testing with benzophenone in humans (602). Positive reactions were noted in 1–2% of the 1191 patients tested with a benzophenone concentration of 1%. When a test concentration of 4–5% was employed, positive reactions were observed in 1.7% of the 147 patients examined.

22.4.2.3 Epidemiology Studies. A health survey was performed of upper airway symptoms in 123 workers from a wood products manufacturing plant that used UV-curable coatings containing benzophenone as a photoinitiator (603). Workers were exposed to wood dust, multifunctional acrylates, various organic solvents, and benzophenone. The 8-h time-weighted average exposure concentration for benzophenone was 26 $\mu g/m^3$. A symptom questionnaire revealed a higher level of nasal complaints in all four of the work areas examined, but the increase was not statistically significant. A cell differential count on nasal lavage fluid indicated a statistically significant increase in the number of macrophages and cylinder cells when all of the exposed employees were compared to controls. There were, however, no differences relative to control when the cell counts were examined for the individuals in each of the four work areas. The concentration of eosinophil cationic protein in the lavage fluid was also significantly elevated for all exposed employees and for UV line workers and UV finishers. Albumin levels in the lavage fluid were also significantly elevated when the values for all employees were combined and compared to controls. The increase in ECP and albumin levels was not correlated with any other clinical sign or reported symptom. The authors concluded that an inflammatory response was present in the nasal mucosa of the workers exposed to UV-curing acrylate coatings.

22.5 Standards, Regulations, or Guidelines of Exposure

There are no recommended hygienic standards for benzophenone.

22.6 Studies on Environmental Impact

The environmental fate and transport of benzophenone has been evaluated following its release to air, water, and soil (604). Benzophenone emitted to the water will degrade relatively slowly by photooxidation ($t_{1/2} > 91$ d) or hydrolysis. Volatilization from water will also be unimportant because of the low vapor pressure it exhibits. Under some aerobic (water) and anaerobic (soil) conditions biodegradation may occur; however, the rates are not high. Benzophenone has a relatively short half-life in the troposphere ($t_{1/2} > 5.4$ d) due to its reaction with hydroxyl radicals. Benzophenone has low to medium soil mobility, and leaching will be an important environmental process.

BIBLIOGRAPHY

1. W. J. Krasavage et al., The relative neurotoxicity of methyl *n*-butyl ketone and *n*-hexane and their metabolites. *Toxicol. Appl. Pharmacol.* **52**, 433–441 (1980).
2. L. Perbellini et al., An experimental study on the neurotoxicity of *n*-hexane metabolites: Hexanol-1 and hexanol-2. *Toxicol. Appl. Pharmacol.* **46**, 421–427 (1978).
3. Health and Environment Laboratories, Eastman Kodak Co., Rochester, NY (unpublished data).
4. G. D. DiVincenzo, C. J. Kaplan, and J. Dedinas, Metabolic fate and disposition of ^{14}C-labeled methyl *n*-butyl ketone in the rat. *Toxicol. Appl. Pharmacol.* **41**, 547–560 (1977).
5. J. R. Mendell et al., Toxic polyneuropathy produced by methyl *n*-butyl ketone. *Science* **185**, 787–789 (1974).
6. P. S. Spencer et al., Nervous system degeneration produced by the industrial solvent methyl *n*-butyl ketone. *Arch. Neurol. (Chicago)* **32**, 219–222 (1975).

7. P. S. Spencer and H. H. Schaumburg, Feline nervous system response to chronic intoxication with commercial grades of methyl *n*-butyl ketone, methyl *Iso*-butyl ketone, and methyl ethyl ketone. *Toxicol. Appl. Pharmacol.* **37**, 301–311 (1976).
8. P. S. Spencer, M. C. Bischoff, and H. H. Schaumburg, On the specific molecular configuration of neurotoxic aliphatic hexacarbon compounds causing central-peripheral distal axonopathy. *Toxicol. Appl. Pharmacol.* **44**, 17–28 (1978).
9. J. L. O'Donoghue and W. J. Krasavage, Hexacarbon neuropathy: A γ-diketone neuropathy? *J. Neuropathol. Exp. Neurol.* **38**, 333 (1979).
10. J. L. O'Donoghue, W. J. Krasavage, and C. J. Terhaar, Toxic effects of 2,5-hexanedione. *Toxicol. Appl. Pharmacol.* **45**, 269 (1978).
11. B. L. Johnson et al., An electrodiagnostic study of the neurotoxicity of methyl *n*-amyl ketone. *Am. Ind. Hyg. Assoc. J.* **39**, 866–872 (1978).
12. T. R. Steinheimer, W. E. Pereira, and S. M. Johnson, Application of capillary gas chromatography mass spectrometry/computer techniques to synoptic survey of organic material in bed sediment. *Anal. Chim. Acta* **129**, 57–67 (1981).
13. C. B. Salocks et al., Subchronic neurotoxicity of 5-methyl-3-heptanone in rats: Correlation of behavioral symptoms and neuropathology. *Toxicologist* **10**, 121 (abstr.) (1990).
14. M. B. G. St. Clair et al., Pyrrole oxidation and protein cross-linking as necessary steps in the development of gamma-diketone neuropathy. *Chem. Res. Toxicol.* **1**, 179–185 (1988).
15. L. M. Sayre et al., Structural basis of gamma-diketone neurotoxicity: Non-neurotoxicity of 3,3-dimethyl-2,5-hexanedione, a gamma-diketone incapable of pyrrole formation. *Toxicol. Appl. Pharmacol.* **84**, 36–44 (1986).
16. M. G. Genter et al., Evidence that pyrrole formation is a pathogenetic step in gamma-diketone neuropathology. *Toxicol. Appl. Pharmacol.* **87**, 351–362 (1987).
17. M. B. St. Clair et al., Neurofilament protein cross linking in gamma-didketone neuropathy: *in vitro* and *in vivo* studies using the seaworm *Myxicola infundibulum. Neurotoxicology* **10**, 743–756 (1990).
18. K. Boekelheide and J. Eveleth, The rate of 2,5-hexanedione intoxication, not total dose, determines the extent of testicular injury and altered microtubule assembly in the rat. *Toxicol. Appl. Pharmacol.* **94**, 76–83 (1988).
19. J. E. Amoore and E. Hautala, Odor as an aid to chemical safety: odor thresholds compared with threshold limit values and volatilities for 214 industrial chemicals in air and water dilution. *J. Appl. Toxicol.* **3**, 272–290 (1983).

19a. NIOSH. *Manual of Analytical Methods*, 4th ed., United States Government Printing Office, Supt. of Docupments, Washington, DC, 1994.

20. Agency for Toxic Substances and Disease Registry (ATSDR), *Toxicological Profile for 2-Hexanone*, TP-91/18, U. S. Public Health Service, Washington, DC, 1992.
21. D. J. Johnson et al., Protein-bound pyrroles in rat hair following subchronic intraperitoneal injections of 2,5-hexanedione. *J. Toxicol. Environ. Health* **45**, 313–324 (1995).
22. H. F. Smyth, Jr. et al., Range finding toxicity data: List V. *Arch. Ind. Hyg. Occup. Med.* **10**, 61–68 (1954).
23. H. Tanii, H. Tsuji, and K. Hashimoto, Structure-toxicity relationship of monoketones. *Toxicol. Lett.* **30**, 13–17 (1986).
24. H. H. Schrenk, W. P. Yant, and F. A. Patty, Hexanone. *Public Health Rep.* **51**, 624–631 (1936).
25. H. Specht et al., *Acute Response of Guinea Pigs to the Inhalation of Ketone Vapors*, NIH Bull. No. 176, U.S. Public Health Service, Div. Ind. Hyg., Washington, DC, 1940, pp. 1–66.

26. L. F. Hansen and G. D. Nielsen, Sensory irritation and pulmonary irritation of *n*-methyl ketones: Receptor activation mechanisms and relationships with threshold Limit values. *Arch. Toxicol.* **68**, 193–202 (1994).
27. W. J. Krasavage, J. L. O'Donoghue, and C. J. Terhaar, Oral chronic toxicity of methyl *n*-propyl ketone, methyl *n*-butyl ketone and hexane in rats. *Toxicol. Appl. Pharmacol.* **48**(Part 2), A205 (1979).
28. J. Misumi et al., Effects of *n*-hexane, methyl *n*-butyl ketone and 2,5-hexanedione on the conduction velocity of motor and sensory nerve fibers in rats' tail. *Sangyo Igaku* **21**, 180–181 (1979).
29. A. Eben et al., Toxicological and metabolic studies of methyl *n*-butyl ketone, 2,5-hexanedione, and 2,5-hexanediol in male rats. *Ecotoxicol. Environ. Saf.* **3**, 204–217 (1979).
30. M. S. Abdel-Rahman et al., The effect of 2-hexanone and 2-hexanone metabolites on pupillmotor activity and growth. *Am. Ind. Hyg. Assoc. J.* **39**, 94–99 (1978).
31. B. L. Johnson et al., Neurobehavioral effects of methyl *n*-butyl ketone and methyl *n*-amyl ketone in rats and monkeys: A summary of NIOSH investigators. *J. Environ. Pathol. Toxicol.* **2**, 123–133 (1979).
32. S. Duckett, N. Williams, and S. Frances, Peripheral neuropathy associated with inhalation of methyl *n*-butyl ketone. *Experientia* **30**, 1283–1284 (1974).
33. S. Duckett et al., 50 ppm MnBK subclinical neuropathy in rats. *Experientia* **35**, 1365–1366 (1979).
34. T. M. Brooks, A. L. Meyer, and D. H. Hutson, The genetic toxicology of some hydrocarbon and oxygenated solvents. *Mutagenesis* **3**, 227–232 (1988).
35. G. V. Katz et al., Comparative neurotoxicity and metabolism of ethyl *n*-butyl ketone and methyl *n*-butyl ketone in rats. *Toxicol. Appl. Pharmacol.* **52**, 153–158 (1980).
36. K. Saida, J. R. Mendell, and H. F. Weiss, Peripheral nerve changes induced by methyl *n*-butyl ketone and potentiation by methyl ethyl ketone. *J. Neuropathol. Exp. Neurol.* **35**, 207–225 (1976).
37. H. Altenkirch, G. Stoltenburg-Didinger, and H. M Wagner, Experimental data on the neurotoxicity of methyl ethyl ketone (MEK). *Experientia* **35**, 503–504 (1979).
38. G. D. DiVincenzo et al., Studies on the respiratory uptake and excretion and the skin absorption of methyl *n*-butyl ketone in humans and dogs. *Toxicol. Appl. Pharmacol.* **44**, 593–604 (1978).
39. A. Sato and T. Nakajima, Partition coefficients of some aromatic hydrocarbons and ketones in water, blood and oil. *Br. J. Ind. Med.* **36**, 231–234 (1979).
40. D. Couri, M. S. Abdel-Rahman, and L. B. Hetland, Biotransformation of *n*-hexane and methyl *n*-butyl ketone in guinea pigs and mice. *Am. Ind. Hyg. Assoc. J.* **39**, 295–300 (1978).
41. G. D. DiVincenzo, C. J. Kaplan, and J. Dedinas, Characterization of the metabolites of methyl *n*-butyl ketone, methyl *iso*-butyl ketone and methyl ethyl ketone in guinea pig serum and their clearance. *Toxicol. Appl. Pharmacol.* **36**, 511–522 (1976).
42. C. T. Hou et al., Microbial oxidation of gaseous hydrocarbons: Production of methyl ketones from their corresponding secondary alcohols by methane and methanol-grown microbes. *Appl. Environ. Microbiol.* **38**, 135–142 (1979).
43. R. D. King and G. H. Clegg, The metabolism of fatty acids, methyl ketones and secondary alcohols by *Penicillium roqueforti* in blue cheese slurries. *J. Sci. Food Agric.* **30**, 197–202 (1979).

44. H. B. Lukins and J. W. Foster, Methyl ketone metabolism in hydrocarbon utilizing microbacteria. *J. Bacteriol.* **85**, 1074–1086 (1963).

45. A. J. Markovetz, Intermediates from the microbial oxidation of aliphatic hydrocarbons. *J. Am. Oil Chem. Soc.* **55**, 430–434 (1978).

46. M. S. Abdel-Rahman, L. B. Hetland, and D. Couri, Toxicity and metabolism of methyl *n*-butyl ketone. *Am. Ind. Hyg. Assoc. J.* **37**, 95–102 (1976).

47. D. Couri and J. P. Nachtman, Biochemical and biophysical studies of 2,5-hexanedione neuropathy. *Neurotoxicology* **1**, 269–283 (1979).

48. A. B. Duguay and G. L. Plaa, Tissue concentrations of methyl *iso*butyl ketone, methyl *n*-butyl ketone and their metabolites after oral or inhalation exposure. *Toxicol. Lett.* **75**, 51–58 (1995).

49. W. R. Kelce, F. M. Raisbeck, and V. K. Ganjam, Gonadotoxic effects of 2-hexanone and 1,2-dibromo-3-chloropropane on the enzymatic activity of rat testicular 17-alpha-hydroxylase/C17,20-lyase. *Toxicol. Lett.* **52**, 331–338 (1990).

50. M. A. Peters, P. M. Hudson, and R. L. Dixon, The effect totigestational exposure to methyl *n*-butyl ketone has on postnatal development and behavior. *Ecotoxicol. Environ. Saf.* **5**, 291–306 (1981).

51. F. K. Zimmermann, I. Scheel, and M. A. Resnick, Induction of chromosome loss by mixtures of organic solvents including neurotoxins. *Mutat. Res.* **224**, 287–303 (1989).

52. V. W. Mayer and C. J. Goin, Induction of chromosome loss in yeast by combined treatment with neurotoxic hexacarbons and monoketones. *Mutat. Res.* **341**, 83–91 (1994).

53. D. Couri et al., The influence of inhaled ketone solvent vapors on hepatic microsomal biotransformation activities. *Toxicol. Appl. Pharmacol.* **41**, 285–289 (1977).

54. J. M. Mathews et al., Do endogenous volatile organic chemicals measured in breath reflect and maintain CYP2E1 levels *in vivo*? *Toxicol Appl. Pharmacol.* **146**, 255–260 (1997).

55 R. V. Branchflower et al., Comparison of the effects of methyl *n*-butyl ketone and phenobarbital on rat liver cytochromes P-450 and the metabolism of chloroform to phosgene. *Toxicol. Appl. Pharmacol.* **71**, 414–421 (1983).

56. W. R. Hewitt and E. M. Brown, Nephrotoxic interactions between ketonic solvents and halogenated aliphatic chemicals. *Fundam. Appl. Toxicol.* **4**, 902–908 (1984).

57. M. Sharkawi and B. Elfassy, Inhibition of mouse liver alcohol dehydrogenase by methyl *n*-butyl ketone. *Toxicol. Lett.* **25**, 185–190 (1985).

58. J. Cunningham, M. Sharkawi, and G. L. Plaa, Pharmacological and metabolic interactions between ethanol and methyl *n*-butyl ketone, methyl *Iso*butyl ketone, methyl ethyl ketone, or acetone in mice. *Fundam. Appl. Toxicol.* **13**, 102–109 (1989).

59. M. A. Gilchrist et al., Toxic peripheral neuropathy. *Morbid. Mortal. Wkly. Rep.* **23**, 9–10 (1974).

60. D. J. Billmaier et al., Peripheral neuropathy in a coated fabrics plant. *J. Occup. Med.* **16**, 665–671 (1974).

61. N. Allen et al., Toxic polyneuropathy due to methyl *n*-butyl ketone. *Arch. Neurol. (Chicago)* **32**, 209–218 (1975).

62. J. S. Mallov, Methyl *n*-butyl ketone neuropathy among spray painters. *J. Am. Med. Assoc.* **235**, 1445–1457 (1976).

63. J. G. Davenport, D. F. Farrell, and S. M. Sumi, 'Giant axonal neuropathy' caused by industrial chemicals: Neurofilamentous axonal masses in man. *Neurology* **26**, 919–923 (1976).

64. P. Halonen et al., Vibratory perception threshold in shipyard workers exposed to solvents. *Acta Neurol. Scand.* **73**, 561–565 (1986).
65. C. W. Wickersham, III and E. J. Fredricks, Toxic polyneuropathy secondary to methyl *n*-butyl ketone. *Conn. Med.* **40**, 311–312 (1976).
66. C. H. Hawkes, J. B. Cavanagh, and A. J. Fox, Motorneuron disease: A disorder secondary to solvent exposure? *Lancet* January 14, 73–75 (1989).
67. W. Denkhaus et al., Lymphocyte subpopulations in solvent-exposed workers. *Int. Arch. Occup. Environ. Health* **57**, 109–115 (1986).
68. S. S. Lande et al., *Investigation of Selected Potential Environmental Contaminants: Ketonic Solvents*, EPA-560/2-76-003, NTIS No. TR 76-5001-330 U.S. Environmental Protection Agency, Office of Toxic Substances, Washington, DC, 1976.
69. M. Ash and I. Ash, Chemical to tradename reference. In *Industrial Chemical Thesaurus*, 2nd ed., Vol. 1, VCH Publishers, Inc., New York, 1992, pp. 1–831.
70. J. A. Riddick, W. B. Bunger, and T. K. Sakano, *Organic Solvents Physical Properties and Methods of Purification*, 4th ed., John Wiley & Sons, Inc., New York, 1986, pp. 1–1325.
71. National Fire Protection Association, Fire Hazard properties of flammable liquids, gases, and volatile solids. In *Fire Protection Guide on Hazardous Materials*, 12th ed., NFPA, Quincy, MA, 1997, pp. 1–94.
72. AIHA, *Odor Thresholds for Chemicals with Established Occupational Health Standards*, AIHA, Washington, DC, 1933, pp. 1–80.
73. M. Devos et al., *Standardized Human Olfactory Thresholds*, Oxford University Press, Oxford, 1990, pp. 1–165.
74. G. Leonardos, D. Kendall, and N. Barnard, Odor threshold determinations of 53 odorant chemicals. *J. Air Pollut. Control Assoc.* **19**, 91–95 (1969).
75. T. M. Hellman and F. H. Small, Characterization of the odor properties of 101 petrochemicals using sensory methods. *J. Air Polluy. Control Assoc.* **24**, 979–982 (1974).
76. J. H. Ruth, Odor thresholds and irritant levels of several chemical substances: A review. *Am. Ind. Hyg. Assoc. J.* **47**, A142–A151 (1986).
77. P. Gagnon, D. Mergler, and S. LaPare, Olfactory adaptation, threshold shift and recovery at low levels of exposure to methyl *Iso*butyl ketone (MIBK). *Neurotoxicology* **15**, 637–642 (1994).
78. S. N. Bizzari, Methyl isobutyl ketone and methyl isobutyl carbinol — United States, CEH Research Report, Methyl Ketones, *Chemical Economics Handbook — SRI International*, Menlo Park, CA, 1996, pp. 1–9.
79. W. Kirstine et al., Emissions of volatile organic compounds (primarily oxygenated species) from pasture. *J. Geophys. Res.* **103**, 10, 605–10,619 (1998).
80. G. A. Burdock, *Fenaroli's Handbook of Flavor Ingredients*, 3rd ed., Vol. II, CRC Press, Boca Raton, FL, 1995 pp. 1–990.
81. R. L. Hall and B. L. Oser, Recent progress in the consideration of flavoring ingredients under the food additives amendment, III. GRAS substances. *Food Technol.* **253**, 151–197 (1965).
82. T. M. Sack et al., A survey of household products for volatile organic compounds. *Atmos. Environ.* **26A**, 1063–1070 (1992).
83. L. E. Booher and B. Janke, Air emissions from petroleum hydrocarbon fires during controlled burning. *Am. Ind. Hyg. Assoc. J.* **58**, 359–365 (1997).
84. K. Wilkins, E. M. Nielsen, and P. Wolkoff, Patterns in volatile organic compounds in dust from moldy buildings. *Indoor Air* **7**, 128–134 (1997).

85. A. Cailleux et al., Volatile organic compounds in indoor air and in expired air as markers of activites. *Chromatographia* **37**, 57–59 (1993).
86. E. D. Pellizzari, Analysis of organic vapor emissions near industrial and chemical waste disposal sites. *Environ. Sci. Technol.* **16**, 781–785 (1982).
87. V. S. Dunovant et al., Volatile organics in the wastewater and airspace of three wastewater treatment plants. *J. Water Pollut. Control Fed.* **58**, 866–895 (1986).
88. D. Manca, B. Birmingham, and D. Raba, Toxicological screening of chemical emissions from municipal solid waste landfills. Application of predictive framework to a state-of the-art facility. *Hum. Ecol. Risk Assess.* **3**, 257–286 (1997).
89. Agency for Toxic Substances and Disease Registry, (ATSDR), HazDat Database, ATSDR, Washington, DC. Available: *http://www.atsdr.cbc.gov/hazdat.html*
90. J. C. Caldwell et al., Application of health information to hazardous air pollutants modeled in EPA's cumulative exposure projects. *Toxicol. Ind. Health* **14**, 429–454 (1998).
91. H. Frostling et al., Analytical, occupational and toxicologic aspects of the degradation products of polypropylene plastics. *Scand. J. Work Environ. Health* **10**, 163–169 (1984).
92. H. Ukai et al., Types of organic solvents used in small-to-medium-scale industries in Japan; A nationwide field survey. *Int. Arch. Occup. Environ. Health* **70**, 382–392 (1997).
93. T. Yasugi et al., Types of organic solvents used in workplaces and work environment conditions with special references to reproducibility of work environment classification. *Ind. Health* **36**, 223–233 (1998).
94. H. Veulemans et al., Survey of ethylene glycol ether exposures in Belgian industries and workshops. *Am. Ind. Hyg. Assoc. J.* **48**, 671–676 (1987).
95. T. Kauppinen et al., International data base of exposure measurements in the pulp, paper and paper product industries. *Int. Arch. Occup. Environ. Health* **70**, 119–127 (1997).
96. B. D. Eitzer, Emissions of volatile organic chemicals from municipal solid waste composting facilites. *Environ. Sci. Technol.* **29**, 896–902 (1995).
97. C. Winder and P. J. Turner, Solvent exposure and related work practices amongst apprentice spray painters in automotive body repair workshops. *Ann. Occup. Hyg.* **36**, 385–394 (1992).
98. L. W. Whitehead et al., Solvent vapor exposures in booth spray painting and spray glueing, and associated operations. *Am. Ind. Hyg. Assoc. J.* **45**, 767–772 (1984).
99. J. De Medinilla and M. Espigares, Contamination by organic solvents in auto paint shops. *Ann. Occup. Hyg.* **32**, 509–513 (1988).
100. H. E. Myer, S. T. O'Block, and V. Dharmarajan, A survey of airborne HDI, HDI based polysiocyanate and solvent concentrations in the manufacture and application of polyurethane coatings. *Am. Ind. Hyg. Assoc. J.* **54**, 663–670 (1993).
101. N. A. Nelson et al., Historical characterization of exposure to mixed solvents for an epidemiologic study of automotive assembly plant workers. *Appl. Occup. Environ. Hyg.* **8**, 693–702 (1993).
102. E. Atlas et al., Environmental aspects of ocean dumping in the western Gulf of Mexico. *J. Water Pollut. Control Fed.* **52**, 329–350 (1980).
103. J. K. Whelan, M. E. Tarafa and J. M. Hunt, Volatile C_1-C_8 organic compounds in macroalgae. *Nature (London)* **299**, 50–52 (1982).
104. T. C. Sauer, Volatile liquid hydrocarbon characterization of underwater hydrocarbon vents and formation waters from offshore production operations. *Environ. Sci. Technol.* **15**, 917–923 (1981).

105. R. H. Plumb, The occurrence of appendix IX organic constituents in disposal site ground water. *Ground Water Rep.* **11**, 157–164 (1990).

106. B. L. Sawhney and R. P. Kozloski, Organic pollutants in leachates from landfill sites. *J. Environ. Qual.* **13**, 349–352 (1984).

107. K. W. Brown and K. C. Donnelly, An estimation of the risk associated with the organic constituents of hazardous and municipal waste landfill leachates. *Hazard. Waste Hazard. Mater.* **5**, 1–30 (1988).

108. L. W. Canter and D. A. Sabatini, Contamination of public groundwater supplies by superfund sites. *Int. J. Environ. Stud.* **46**, 35–57 (1994).

109. A. J. Francis et al., Characterization of organics in leachates from low-level radioactive waste disposal sites. *Nucl. Technol.* **50**, 158–163 (1980).

110. E. Benfenati et al., Characterization of organic and inorganic pollutants in the Adige River (Italy). *Chemosphere* **25**, 1665–1674 (1992).

111. L. W. Hall et al., *In situ* striped bass (*Morone saxatilis*) contaminant and water quality studies in the potomac river. *Aquat. Toxicol.* **10**, 73–99 (1987).

112. L. S. Sheldon and R. A. Hites, Organic compounds in the Delaware River. *Environ. Sci. Technol.* **12**, 1188–1194 (1978).

113. B. L. Sawhney and J. A. Raabe, Groundwater contamination: Movement of organic pollutants in the granby landfill. *Bull—Conn. Agric. Exp. Stn. New Haven* **833**, 1–9 (1986).

114. G. V. Sabel and T. P. Clark, Volatile organic compounds as indicators of municipal solid waste leachate contamination. *Waste Manage. Res.* **2**, 119–130 (1984).

115. T. Brorson et al., Comparison of two strategies for assessing exotoxicological aspects of complex wastewater from a chemical-pharmaceutical plant. *Environ. Toxicol. Chem.* **13**, 543–552 (1994).

116. In NIOSH *Manual of Analytical Methods* 4th ed., P. M. Eller ed. National Institute for Occupational Safety and Health, Washington, DC, 1994.

117. J.-O. Levin and L. Carleborg, Evaluation of solid sorbents for sampling ketones in work-room air. *Ann. Occup. Hyg.* **31**, 31–38 (1987).

118. H. Tomczyk and T. Rogaczewska, Oznaczanie W Powietrzu Par Ketonu Methyloizobutylowego, Metyloizobutylokarbinolu, Actonu, Toluenu I o-Ksylenu Metoda Chromatografii Gazowej. *Med. Pr.* **30**, 417–424 (1979).

119. G. Gazzaniga et al., Analisi Gascromatografica Automatica Di Solventi Neel'Ambiente Di Lavoro. *Med. Lav.* **69**, 232–247 (1978).

120. A. F. Smith and R. Wood, A field test for the determination of some ketone vapours in air. *Analyst (London)* **97**, 363–371 (1972).

121. L. Dunemann and H. Hajimiragha, Development of a screeing method for the determiination of volatile organic compounds in body fluids and environmental samples using purge and trap gas chromatography-mass spectrometry. *Anal. Chim. Acta* **283**, 199–206 (1993).

122. F. A. Dreisch and T. O. Munson, Purge-and-trap analysis using fused silica capillary column GC/MS. *J. Chromatogr. Sci.* **21**, 111–118 (1983).

123. I. Ojanperä, R. Hyppola, and E. Vuori, Identification of volatile organic compounds in blood by purge and trap PLOT-capillary gas chromatography coupled with fourier transform infrared spectroscopy. *Forensic Sci. Int.* **80**, 201–209 (1996).

124. R. H. Brown and C. J. Purnell, Collection and analysis of trace organic vapour pollutants in ambient atmospheres. *J. Chromatog.* **178**, 79–80 (1979).

125. A. K. Burnham et al., Identification and estimation of neutral organic contaminants in potable water. *Anal. Chem.* **44**, 139–142 (1972).

126. J. D. Pleil and M. L. Stroupe, Measurement of vapor-phase organic compounds at high concentration. *J. Chromatog.* **676**, 399–408 (1994).

127. N. A. Marley and J. S. Gaffney, A comparison of flame ionization and ozone chemiluminescence for the determination of atmospheric hydrocarbons. *Atmos. Environ.* **32**, 1435–1444 (1998).

128. E. W. Hjelm et al., Exposure to methyl isobutyl ketone: Toxicokinetics and occurrence of irritative and CNS symptoms in man. *Int. Arch. Occup. Environ. Health* **62**, 19–26 (1990).

129. J. A. Bellanca et al., Detection and quantitation of multiple volatile compounds in tissues by GC and GC/MS. *J. Anal. Toxicol.* **6**, 238–240 (1982).

130. F. Gobba et al., The urinary excretion of solvents and gases for the biological monitoring of occupational exposure: A review. *Sci. Total Environ.* **199**, 3–12 (1997).

131. M. Ogata, T. Taguchi, and T. Horike, Evaluation of exposure to solvents from their urinary excretions in workers coexposed to toluene, xylene, and methyl *Iso*butyl ketone. *Appl. Occup. Environ. Hyg.* **10**, 913–920 (1995).

132. N. Hirota, Methyl *Iso*butyl ketone. *Okayama Tgakkai Zaiski* **103**, 327–336 (1991).

133. A. Zlatkis and H. M. Liebich, Profile of volatile metabolites in human urine. *Clin. Chem. (Winstor-Salem, N.C.)* **17**, 592–594 (1971).

134. L. A. Wallace et al., Personal exposure to volatile organic compounds. I. Direct measurements in breathing-zone air, drinking water, food, and exhaled breath. *Environ. Res.* **35**, 293–319 (1984).

135. H. Nishimura et al., Analysis of acute toxicity (LD_{50} Value) of organic chemicals to mammals by solubility parameter (8) (1) Acute oral toxicity to rats. *Jpn. J. Ind. Health* **36**, 314–323 (1994).

136. J. P. Conkle, B. J. Camp, and B. E. Welch, Trace composition of human respiratory gas. *Arch. Environ. Health* **30**, 290–295 (1975).

137. B. J. Dowty, J. L. Laseter, and J. Storer, The transplacental migration and accumulation in blood of volatile organic compounds. *Pediatr. Res.* **10**, 696–701 (1976).

138. H. F. Smyth, Jr., C. P. Carpenter, and C. S. Weil, Range finding toxicity data: List IV. *Arch. Ind. Hyg. Occup. Med.* **4**, 119–122 (1951).

139. Chemical Hygiene Fellowship, Mellon Institute, Pittsburgh, PA (unpublished data).

140. S. Zakhari, Acute oral, intraperitoneal and inhalational toxicity of methyl isobutyl ketone in the mouse. In L. Goldberg, ed., *Isopropanol and Ketones in the Environment*, CRC Press, Cleveland, OH, 1977, pp. 101–104.

141. T. F. Batyrova, Materials for substantiation of maximal permissible concentration of methyl *Iso*butyl ketone in the atmospheres of a work zone. *Gig. Tr. Prof. Zabol.* **17**, 52–53 (1973).

142. Exxon Chemical Company, *Acute Oral Toxicity—Male Albino Rats*, USEPA/OTS Doc. 878210455, TSCA 8(d) Submission, Exxon Chemial Company, 1982.

143. H. Nishimura et al., Analysis of acute toxicity (LD_{50} value) of organic chemicals to mammals by solubility parameter (8) (2) Acute oral toxicity to mice. *Jpn. J. Ind. Health* **36**, 421–427 (1994).

144. Shell Chemical Company, Houston, TX (unpublished data).

145. H. Specht, Acute response of guinea pigs to inhalation of methyl isobutyl ketone. *Public Health Rep.* **53**, 292 (1938).

146. S. Zakhari, Acute oral, intraperitoneal and inhalational toxicity in the mouse. In L. Goldberg, ed., *Isopropanol and Ketones in the Environment*, CRC Press, Cleveland, OH, 1977, pp. 67–69.
147. R. D. Panson and C. L. Winek, Aspiration toxicity of ketones. *Clin. Toxicol.* **17**, 271–317 (1980).
148. J. L. O'Donoghue et al., Mutagenicity studies on ketone solvents: Methyl ethyl ketone, methyl *Iso*butyl ketone, and isophorone. *Mutat. Res.* **206**, 149–161 (1988).
149. C. S. Weil and R. A. Scala, Study of intra- and interlaboratory variability in the results of rabbit eye and skin irritation tests. *Toxicol. Appl. Pharmacol.* **19**, 276–360 (1971).
150. P. Gautheron et al., Interlaboratory assessment of the bovine corneal opacity and permeability (BCOP) assay. *Toxicol. In Vitro* **8**, 381–392 (1994).
151. L. Gilleron et al., Evaluation of a modified HET-CAM assay as a screening test for eye irritancy. *Toxicol. In Vitro* **10**, 431–446 (1996).
152. J. G. Sivak and K. L. Herbert, Optical damage and recovery of the *in vitro* bovine ocular lens for alcohols, surfactants, acetates, ketones, aromatics, and some consumer producs: A review. *J. Toxicol. — Cutaneous Ocul. Toxicol.* **16**, 173–187 (1997).
153. H. E. Kennah, II et al., An objective procedure for quantitating eye irritation based upon changes of corneal thickness. *Fundam. Appl. Toxicol.* **12**, 258–268 (1989).
154. Union Carbide Corp., *Med. and Toxicol. Dept.*, New York (unpublished data).
155. J. Huang et al., Structure-toxicity relationship of monoketones: *in vitro* effects on beta-adrenergic receptor binding and Na^+-K^+-ATPase activity in mouse synaptosomes. *Neurotoxicol. Teratol.* **15**, 345–352 (1993).
156. R. Tham et al., Vestibulo-ocular disturbances in rats exposed to organic solvents. *Acta Pharmacol. Toxicol.* **54**, 58–63 (1984)
157. J. C. DeCeaurriz et al., Sensory irritation caused by various industrial airborne chemicals. *Toxicol. Lett.* **9**, 137–143 (1981).
158. M. H. Abraham et al., Hydrogen bonding 12. A new QSAR for upper respiratory tract irritation by airborne chemicals in mice. *Quant. Struct.-Act. Relat.* **9**, 6–10 (1990).
159. J. Muller and G. Greff, Recherche de relations entre toxicité de molécules d'interêt industriel et propriétés physio-chimiques: Test d'irritation des voies aérïennes supérieures appliqué à quatre familles chimiques. *Food Chem. Toxicol.* **22**, 661–664 (1984).
160. D. W. Roberts, QSAR for upper-respiratory tract irritants. *Chem.-Biol. Interact.* **57**, 325–345 (1986).
161. Y. Alarie et al., Physicochemical properties of nonreactive volatile organic chemicals to estimate RD50: alternatives to animal studies. *Toxicol. Appl. Pharmacol.* **134**, 92–99 (1995).
162. Y. Alarie et al., Structure-activity relationships of volatile organic chemicals as sensory irritants. *Arch. Toxicol.* **72**, 125–140 (1998).
163. W. J. Muller and M. S. Black, Sensory irritation in mice exposed to emissions from indoor products. *Am. Ind. Hyg. Assoc. J.* **56**, 794–803 (1995).
164. J. DeCeaurriz et al., Quantitative evaluation of sensory irritating and neurobehavioral properties of aliphatic ketones in mice. *Food Chem. Toxicol.* **22**, 545–549 (1984).
165. D. M. Aviado, Summary of the effects of ketones, chlorinated solvents, fluorocarbons, and isopropanol on the pulmonary circulation. In L. Goldberg, ed., *Isopropanol and Ketones in the Environment*, CRC Press, Cleveland, OH, 1977, pp. 129–131.

166. M. Sharkawi et al., Pharmacodynamic and metabolic interactions between ethanol and two industrial solvents (methyl *n*-butyl ketone and methyl isobutyl ketone) and their principal metabolites in mice. *Toxicology* **94**, 187–195 (1994).

167. M. T. Brondeau et al., Acetone compared to other ketones in modifying the hepatotoxicity of inhaled 1,2-dichlorobenzene in rats and mice. *Toxicol. Lett.* **49**, 69–78 (1989).

168. R. D. Phillips et al., A 14-week vapor inhalation study of methyl isobutyl ketone. *Fundam. Appl. Toxicol.* **9**, 380–388 (1987).

169. G. L. Plaa and P. Ayotte, Taurolithocholate-induced intrahepatic cholestasis: Potentiation by methyl isobutyl ketone and methyl *n*-butyl ketone in rats. *Toxicol. Appl. Pharmacol.* **80**, 228–234 (1985).

170. M. Vézina and G. L. Plaa, Potentiation by methyl isobutyl ketone of the cholestasis induced in rats by a manganese-bilirubin combination or manganese alone. *Toxicol. Appl. Pharmacol.* **91**, 477–483 (1987).

171. M. Vezina and G. L. Plaa, Methyl isobutyl ketone metabolites and potentiation of the cholestasis induced in rats by a manganese-bilirubin combination or manganese alone. *Toxicol. Appl. Pharmacol.* **92**, 419–427 (1988).

172. A. B. Duguay and G. L. Plaa, Ketone potentiation of intrahepatic cholestasis: Effect of two aliphatic isomers. *J. Toxicol. Environ. Health* **50**, 41–52 (1997).

173. L. Dahlström-King et al., The influence of severity of bile flow reduction, cycloheximide, and methyl isobutyl ketone pretreatment on the kinetics of taurolithocholic acid disposition in the rat. *Toxicol. Appl. Pharmacol.* **104**, 312–321 (1990).

174. L.-D. Joseph et al., Potentiation of lithocholic acid-induced cholestasis by methyl isobutyl ketone. *Toxicol. Lett.* **61**, 39–47 (1992).

175. A. B. Duguay and G. L. Plaa, Plasma concentrations in methyl isobutyl ketone potentiated experimental cholestasis after inhalation or oral administration. *Fundam. Appl. Toxicol.* **21**, 222–227 (1993).

176. A. Duguay and G. L. Plaa, Altered cholesterol synthesis as a mechanism involved in methyl isobutyl ketone-potentiated experimental cholestasis. *Toxicol. Appl. Pharmacol.* **147**, 281–288 (1997).

177. D. Pilon, J. Brodeur, and G. L.Plaa, Potentiation of CCl_4-induced liver injury by ketonic and ketogenic compounds: Role of the CCl_4 dose. *Toxicol. Appl. Pharmacol.* **94**, 183–190 (1988).

178. M. Vézina et al., Potentiation of chloroform induced hepatotoxicity by methyl isobutyl ketone and two metabolites. *Can. J. Physiol. Pharmacol.* **68**, 1055–1061 (1990).

179. P. Raymond and G. L. Plaa, Ketone potentiation of haloalkane-induced hepato- and nephrotoxicity. I. Dose-response relationships. *J. Toxicol. Environ. Health* **45**, 465–480 (1995).

180. P. Raymond and G. L. Plaa, Ketone potentiation of haloalkane-induced hepato- and nephrotoxicity. II. Implication of monooxygenases. *J. Toxicol. Environ. Health* **46**, 317–328 (1995).

181. P. Raymond and G. L. Plaa, Ketone potentiation of haloalkane-induced hepatotoxicity: CCl_4 and ketone treatment on hepatic membrane integrity. *J. Toxicol. Environ. Health* **49**, 285–300 (1996).

182. G. L. Plaa, Experimental evaluation of haloalkanes and liver injury. *Fundam. Appl. Toxicol.* **10**, 563–570 (1988).

183. E. H. Vernot, J. D. MacEwen, and E. S. Harris, *Continuous Exposure of Animals to Methyl Isobutyl Ketone Vapors.* AD Rep., ISS No. 751443. U.S. Natl. Tech. Inf. Serv., Springfield, VA, 1971, p. 11.

184. J. D. MacEwen, E. H. Vernot, and C. C. Haun, *Effect of 90-day Continuous Exposure to Methylisobutyl Ketone on Dogs, Monkeys, and Rats*, AD Rep., ISS No. 730291, U.S. Natl. Tech. Inf. Serv., Springfield, VA, 1971.
185. W. F. MacKenzie, *Pathological Lesions Caused by Methyl Isobutyl Ketone*. AD Rep., ISS No. 751444, U.S. Natl. Tech. Inf. Serv., Springfield, VA, 1971, pp. 311–322.
186. E. W. Hjelm et al., Percutaneous uptake and kinetics of methyl isobutyl ketone (MIBK) in the guinea pig. *Toxicol. Lett.* **56**, 79–86 (1991).
187. P. Poulin and K. Krishnan, A tissue composition-based algorithm for predicting tissue: air partition coefficients of organic chemicals. *Toxicol. Appl. Pharmacol.* **136**, 126–130 (1996).
188. P. Poulin and K. Krishnan, An algorithm for predicting tissue: blood partition coefficients of organic chemicals from *n*-octanol: Water partition coefficient data. *J. Toxicol. Environ. Health* **46**, 117–129 (1995).
189. R. W. Tyl et al., Developmental toxicity evaluation of inhaled methyl isobutyl ketone in Fischer 344 rats and CD-1 mice. *Fundam. Appl. Toxicol.* **8**, 310–327 (1987).
190. D. J. Call et al., Fish subchronic toxicity chemicals that produce narcosis. *Environ. Toxicol. Chem.* **4**, 335–341 (1985).
191. E. Zeiger et al., *Salmonella* mutagenicity tests: V. Results from the testing of 311 chemicals. *Environ. Mol. Mutagen.* **19**(Suppl. 21), 2–141 (1992).
192. Goodyear Tire and Rubber Company, *Mutagenicity Evaluation of Methyl Isobutyl Ketone*, USEPA/OTS Doc. 878210382, TSCA 8(d) Submission, Goodyear Tire and Rubber Company, Akron, OH, 1982.
193. A. L. Brooks et al., The combined genotoxic effects of radiation and occupational pollutants. *Appl. Occup. Environ. Hyg.* **11**, 410–416 (1996).
194. D. M. Lapadula et al., Induction of cytochrome P450 *iso*zymes by simultaneous inhalation exposure of hens to *n*-hexane and methyl *iso*-butyl ketone (MiBK). *Biochem. Pharmacol.* **41**, 877–883 (1991).
195. M. B. Abou-Donia et al., Mechanisms of joint neurotoxicity of *n*-hexane, methyl isobutyl ketone and *O*-ethyl *O*-4-nitrophenyl phenylphosphonothioate in hens. *J. Pharmacol. Exp. Ther.* **257**, 282–289 (1991).
196. M. B. Abou-Donia et al., The synergism of *n*-hexane-induced neurotoxicity by methyl isobutyl ketone following subchronic (90 days) inhalation in hens: Induction of hepatic microsomal cytochrome P450. *Toxicol. Appl. Pharmacol.* **81**, 1–16 (1985).
197. I. Geller, J. R., Rowlands, and H. L. Kaplan, *Effects of Ketones on Operant Behavior of Laboratory Animals*, DHEW Publ. Adm- 79A-A779, United States Government Printing Office, Washington, DC, 1978, pp. 363–376.
198. I. Geller et al., Effects of acetone, methyl ethyl ketone, and methyl isobutyl ketone on a match to sample task in the baboon. *Pharmacol., Biochem. Behav.* **11**, 401–406 (1979).
199. C. R. Garcia, I. Geller, and H. L. Kaplan, Effects of ketones on lever pressing behavior of rats. *Proc. West. Pharmacol. Soc.* **21**, 433–438 (1978).
200. R. M. David et al., The effect of repeated methyl *Iso*-butyl ketone vapor exposure on schedule-controlled operant behavior in rats. *Neurotoxicology* **20**, 583–594 (1999).
201. H. F. Smyth, Jr., Hygienic standards for daily inhalation. *Am. Ind. Hyg. Assoc. Q.* **17**, 129–185 (1956).
202. H. B. Elkins, *The Chemistry of Industrial Toxicology*, 2nd ed., Wiley, New York, 1959.

203. L. Silverman, H. F. Schulte, and M. W. First, Further studies on sensory response to certain industrial solvent vapors. *J. Ind. Hyg. Toxicol.* **28**, 262–266 (1946).

204. G. Armeli, F. Linari, and G. Martorano, Rilievi clinici ed ematochimici in operai eposti all'azione di un chetone superiore (MIBK) ripetuti a distanza Di 5 anni. *Lav. Um.* **20**, 418–424 (1968).

205. J. May, Odor thresholds of solvents for evaluating solvent odors in air. *Staub.* **26**, 385 (1966).

206. L. G. Wayne and H. H. Orcutt, The relative potentials of common organic solvents as percusors of eye-irritants in urban atmospheres. *J. Occup. Med.* **2**, 383 (1960).

207. A. Iregren, M. Tesarz, and E. Wigaeus-Hjelm, Human experimental MIBK exposure: Effects on heart rate, performance, and symptoms. *Environ. Res.* **63**, 101–108 (1993).

208. P. H. Dalton, D. D. Dilks, and M. L. Banton, Evaluation of odor and sensory irritation thresholds for methyl *iso*-butyl ketone (MIBK) in humans. *Am. Ind. Hyg. J.* **61**, 340–350 (2000).

209. J. H. Ware et al., Respiratory and irritant health effects of ambient volatile organic compounds. *Am. J. Epidemiol.* **137**, 1287–1301 (1993).

210. R. B. Dick et al., Neurobehavioral effects from acute exposures to methyl isobutyl ketone and methyl ethyl ketone. *Fundam. Appl. Toxicol.* **19**, 453–473 (1992).

211. J. AuBuchon, H. I. Robins, and C. Viseskul, Letter to the editor: Peripheral neuropathy after exposure to methyl-isobutyl ketone in spray paint. *Lancet* **8138**, 363–364 (1979).

212. S. J. Oh and J. M. Kim, Giant axonal swelling in "Huffer's" neuropathy. *Arch. Neurol. (Chicago)* **33**, 583–586 (1976).

213. C.-W. Lam et al., Mechanism of transport and distribution of organic solvents in blood. *Toxicol. Appl. Pharmacol.* **104**, 117–129 (1990).

214. T. C. Morata et al., Effects of occupational exposure to organic solvents and noise on hearing. *Scand. J. Work Environ. Health* **19**, 245–254 (1993).

215. G. Triebig, K. H. Schaller, and D. Weltle, Neurotoxicity of solvent mixtures in spray painters. I. Study design, workplace exposure, and questionnaire. *Int. Arch. Occup. Environ. Health* **64**, 353–359 (1992).

216. G. Triebig et al., Neurotoxicity of solvent mixtures in spray painters. II. Neurologic, psychiatric, psychological, and neuroradiologic findings. *Int. Arch. Occup. Environ. Health* **64**, 361–372 (1992).

217. G. Triebig et al., Cross-sectional epidemiological study on neurotoxicity. *Int. Arch. Occup. Environ. Health* **60**, 233–241 (1998).

218. J.-D. Wang and J.-D. Chen, Acute and chronic neurological symptoms among paint workers exposed to mixtures of organic solvents. *Environ. Res.* **61**, 107–116 (1993).

219. S.-Y. Tsai et al., Neurobehavioral effects of occupational exposure to low-level organic solvents among Taiwanese workers in paint factories. *Environ. Res.* **73**, 146–155 (1997).

220. S.-A. Elofsson et al., A cross-sectional epidemiologic investigation on occupationally exposed car and industrial spray painters with special reference to the nervous system. *Scand. J. Work Environ. Health* **6**, 239–273 (1980).

221. H. Hänninen et al., Behavioral effects of long-term exposure to a mixture of organic solvents. *Scand. J. Work Environ. Health* **4**, 240–255 (1976).

222. Hazardous Substances Data Bank (HSDB), *Methyl Isobutyl Ketone*, National Library of Medicine, Bethesda, MD, (CD-ROM) MICROMEDIX, Englewood, Co., 1997, pp. 1–30.

223. K. Verschueren, *Handbook of Environmental Data on Organic Chemicals*, Van Nostrand-Reinhold, New York, 1983.

224. E. D. Setzinger and B. Deimitriades, Oxygenates in exhaust from simple hydrocarbon fuels. *J. Air Pollut. Control Assoc.* **22**, 42–51 (1972).

225. R. Atkinson, Estimation of the gas phase hydroxyl radical rate constants for organic chemists. *Environ. Toxicol. Chem.* **7**, 435–442 (1988).

226. W. J. Lyman, W. F. Reehl, and E. F. Rosenblatt, *Handbook of Chemical Estimation Methods*, McGraw-Hill, New York, 1990.

227. E. C. Coleman, C.-T. Ho, and S. S. Chang, Isolation and identification of volatile compounds from baked potatoes. *J. Agric. Food Chem.* **29**, 42–48 (1981).

228. G. R. Takeoka et al., Nectarine volatiles: Vacuum steam distillation versus headspace sampling. *J. Agric. Food Chem.* **36**, 553–560 (1988).

229. C. H. Hine et al., *Acute Toxicity of Mesityl Oxide and Isomesityl Oxide*, Univ. Calif. Rep. No. 290, Shell Development Company, Emeryville, CA, 1960.

230. C. P. Carpenter and H. F. Smyth, Jr., Chemical burns of the rabbit cornea. *Am. J. Ophthalmol.* **29**, 1363–1372 (1946).

231. J. Marhold, *Prehled Prumyslove Toxikologie; Organicke Latky*, Prague, Czechoslovakia, 1986, cited in Ref. 232.

232. National Institute for Occupational Safety and Health (NIOSH), *Registry of Toxic Effects of Chemical Substances*, NIOSH, Washington, DC, 1990.

233. E. R. Hart, J. N. Shick, and C. D. Leake, The toxicity of mesityl oxide. *Univ. Calif., Berkeley, Publ. Pharmacol.* **1**, 161–173 (1939).

234. H. F. Smyth, Jr., J. Seaton, and L. Fischer, Response of guinea pigs and rats to repeated inhalation of vapors of mesityl oxide and isophorone. *J. Ind. Hyg. Toxicol.* **24**, 46–50 (1942).

235. C. P. Carpenter, H. F. Smyth, Jr., and U. C. Pozzani, The assay of acute vapor toxicity and the grading and interpretation of results on 96 chemical compounds. *J. Ind. Hyg. Toxicol.* **31**, 343–346 (1949).

236. *Testing Completed in Compliance with a USEPA TSCA Testing Consent Order* (56 FR 43878–43881; Chemical Manufacturers Association, Arlington, VA, 1991 (unpublished data).

237. M. T. Brondeau et al., Adrenal-dependent leucopenia after short-term exposure to various airborne irritants in rats. *J. Appl. Toxicol.* **10**, 83–86 (1990).

238. R. T. Williams, *Detoxication Mechanisms*, 2nd ed., Chapman & Hall, London, 1959, pp. 95–96.

239. A. M. Cheh, Mutagen production by chlorination of methylated α,β-unsaturated ketones. *Mutat. Res.* **169**, 1–9 (1986).

240. S. Ito, Industrial toxicological studies of mesityl oxide. *Yokohama Igaku* **20**, 253–365 (1969).

241. B. Samimi, Exposure to isophorone and other organic solvents in a screen printing plant. *Am. Ind. Hyg. J.* **43**, 43–48 (1982).

242. S. S. Lande, *Investigation of Potential Environmental Contaminants: Haloalkyl Phosphates*, USEPA-560/2-76-007, NTIS PN257910, USEPA, Washington, DC, 1976, pp. 101, 126.

243. H. F. Smyth, Jr. and C. P. Carpenter, Further experience with the range finding Test in the industrial toxicology laboratory. *J. Ind. Hyg. Toxicol.* **30**, 63–68 (1948).

244. H. M. Keith, Effect of diacetone alcohol on the liver of the rat. *Arch. Pathol. Lab. Med.* **13**, 707–712 (1932).

245. D. C. Walton, E. F. Kehr, and A. S. Loevenhart, A comparison of the pharmacological action of diacetone alcohol and acetone. *J. Phamacol. Exp. Ther.* **33**, 175–183 (1928).

246. W. F. von Oettingen, Aliphatic alcohols. *Public Health Bull.* **281**, 138 (1943).

247. Shell Research, Ltd., London, 1979, (unpublished data), in Ref. 248.

248. V. K. Rowe and S. B. McCollister, Alcohols. In G. D. Clayton and F. E. Clayton, eds., *Patty's Industrial Hygiene and Toxicology*, 3rd ed., Vol. 2C, Wiley, New York, 1982, pp. 4527–4708.

249. C. von Scheele et al., Nephrotic syndrome Due to subacute glomerulonephritis — Association with hydrocarbon exposure? *Acta Med. Scand.* **200**, 427–429 (1976).

250. N. Fedtke and H. M. Bolt, Detection of 2,5-hexanedione in the urine of persons not exposed to *n*-hexane, *Int. Arch. Occup. Environ. Health* **57**, 143–148 (1986).

251. P. Bavazzano et al., Determination of urinary 2,5-hexanedione in the general italian population. *Int. Arch. Occup. Environ. Health* **71**, 284–288 (1998).

252. L. Perbellini et al., Biochemical and physiological aspects of 2,5-hexanedione: Endogenous or exogenous product? *Int. Arch. Occup. Environ. Health* **65**, 49–52 (1993).

253. L. Perbellini, F. Brugnone, and G. Faggionato, Urinary excreton of the metabolites of *n*-hexane and its isomers during occupational exposure. *Br. J. Ind. Med.* **38**, 20–26 (1981).

254. T. Kawai et al., Comparative evaluation of blood and urine analysis as a tool for biological monitoring of *n*-hexane and toluene. *Int. Arch. Occup. Environ. Health* **65**, 123–126 (1993).

255. L. Perbellini, F. Brugnone, and E. Gaffuri, Urinary metabolite excretion in the exposure to technical hexane compounds. In M. H. Ho and H. K. Dillon, eds., *Organic Compounds*, Wiley, New York, 1987, pp. 197–205.

256. J. G. van Engelen et al., Determination of 2,5-hexanedione, a metabolite of *n*-hexane, in urine: Evaluaiton and application of three analytical methods. *J. Chromatogr. B.* **667**, 233–240 (1995).

257. J. F. Periago et al., Biological monitoring of occupational exposure to *n*-hexane by exhaled air analysis and urinalysis. *Int. Arch. Occup. Environ. Health* **65**, 275–278 (1993).

258. Y. Takeuchi, *n*-hexane polyneuroathy in Japan: A review of *n*-hexane poisoning and its preventive measures. *Environ. Res.* *62*, 76–80 (1993).

259. M. Governa et al., Urinary excretion of 2,5-hexanedione and peripheral polyneuropathies workers exposed to hexane. *J. Toxicol. Environ. Health* **20**, 219–228 (1987).

260. C. Pastore et al., Early diagnosis of *n*-hexane-caused neuropathy. *Muscle Nerve* **17**, 981–986 (1994).

261. A. Cardona et al., Biological monitoring of occupational exposure to *n*-hexane by measurement of urinary 2,5-hexanedione. *Int. Arch. Occup. Environ. Health* **65**, 71–74 (1993).

262. J. G. M. van Engelen et al., Effect of coexposure to methyl ethyl kentone (MEK) on *n*-hexane toxico-kinetics in human volunteers. *Toxicol. Appl. Pharmacol.* **144**, 385–395 (1997).

263. M. Governa et al., Human polymorphonuclear leukocyte chemotaxis as a tool in detecting biological early effects in workers occupationally exposed to low levels of *n*-hexanes. *Hum. Exp. Toxicol.* **13**, 663–670 (1994).

264. H. F. Smyth, Jr. and C. P. Carpenter, The place of the range finding test in the industrial toxicology laboratory. *J. Ind. Hyg. Toxicol.* **26**, 269 (1944).

265. R. K. Upreti et al., Effect of 2,5-hexanedione on lymphoid organs of rats: A preliminary report. *Environ. Res.* **39**, 188–198 (1986).

266. R. K. Upreti and R. Shanker, 2,5-hexanedione-induced immunomodulatory effect in mice. *Environ. Res.* **43**, 48–59 (1987).

267. P. S. Spencer and H. H. Schaumburg, experimental neuropathy produced by 2,5-hexanedione — A major metabolite of the neurotoxic solvent methyl *n*-butyl ketone. *J. Neurol., Neurosurg. Psychol.* **38**, 771–775 (1975).

268. B. Bäckström and V. P. Collins, The effects of 2,5-hexanedione on rods and cones of the retina of albino rats. *Neurotoxicology* **13**, 199–202 (1992).

269. H. H. Schaumburg and P. S. Spencer, Environmental hydrocarbons produced degeneration in cat hypothalamus and optic tract. *Science* **199**, 199–200 (1978).

270. T. Pasternak et al., Selective damage to large cells in the cat retinogeniculate pathway by 2,5-hexanedione. *J. Neurosci.* **5**, 1641–1652 (1985).

271. J. L. Lynch, III, W. H. Merigan, and T. A. Eskin, Subchronic dosing with 2,5-hexanedione causes long-lasting motor dysfunction but reversible visual loss. *Toxicol. Appl. Pharmacol.* **98**, 166–180 (1989).

272. M. Hirata, Reduced conduction function in central nervous system by 2,5 hexanedione. *Neurotoxicol. Teratol.* **12**, 623–626 (1990).

273. G. Krinke et al., Clinoquinol and 2,5-hexanedione induced different types of distal axonopathy in the dog. *Acta Neuropathol.* **47**, 213–221 (1979).

274. H. C. Powell et al., Schwann cell abnormalities in 2,5-hexanedione neuropathy. *J. Neurocytol.* **7**, 517–528 (1978).

275. E. Suwita, A. A. Nomeir, and M. B. Abou-Donia, Disposition, pharmacokinetics, and metabolism of a dermal dose of [^{14}C]2,5-hexanedione in hens. *Drug Metab. Dispos.* **15**, 779–785 (1987).

276. W. H. Ralston et al., Potentiation of 2,5-hexanedione neurotoxicity by methyl ethyl ketone. *Toxicol. Appl. Pharmacol.* **81**, 319–327 (1985).

277. O. Lagefoged, U. Hass, and L. Simonsen, Neurophysiological and behavioural effects of combined exposure to 2,5-hexanedione and acetone or ethanol in rats. *Pharmacol. Toxicol.* **65**, 372–375 (1989).

278. D. G. Graham et al., Covalent crosslinking of neurofilaments in the pathogenesis of *n*-hexane neuropathy. *Neurotoxicology* **6**, 55–64 (1985).

279. P. R. Sager and D. W. Matheson, Mechanisms of neurotoxicity related to selective disruption of microtubules and intermediate filaments. *Toxicology* **49**, 479–492 (1988).

280. F. Boegner et al., 2,5-Hexanedione is a potent gliatoxin in *in vitro* cell cultures of the nervous system. *Neurotoxicology* **13**, 151–154 (1992).

281. K. Boekelheide, Rat testis during 2,5-hexanedione intoxication and recovery. *Toxicol. Appl. Pharmacol.* **92**, 28–33 (1988).

282. K. Boekelheide, 2,5-hexanedione alters microtubule assembly. I. Testicular atrophy, not nervous system toxicity, correlates with enhanced tubulin polymerization. *Toxicol. Appl. Pharmacol.* **88**, 370–382 (1987).

283. D. C. Anthony et al., The effect of 3,4-dimethyl substitution on the neurotoxicity of 2,5-hexanedione. II. Dimethyl substitution accelerates pyrrole formation and protein crosslinking. *Toxicol. Appl. Pharmacol.* **71**, 372–382 (1983).

284. R. E. Chapin et al., The effects of 2,5-hexanedione on reproductive hormones and testicular enzyme activities in the F-344 rat. *Toxicol. Appl. Pharmacol.* **62**, 262–272 (1982).

285. J.-J. Larsen, M. Lykkegaard, and O. Ladefoged, Infertility in rats induced by 2,5-hexanedione in combination with acetone. *Pharmacol. Toxicol.* **69**, 43–46 (1991).

286. K. Boekelheide and S. J. Hall, 2,5-hexanedione exposure in the rat results in long-term testicular atrophy despite the presence of residual spermatogonia. *J. Androl.* **12**, 18–29 (1991).

287. F. Kato et al., Mutagenicity of aldehydes and diketones. *Mutat. Res.* **216**, 336 (1989).

288. H. U. Aeschbacher et al., Contribution of coffee aroma constituents to the mutagenicity of coffee. *Food Chem. Toxicol.* **27**, 227–232 (1989).

289. A. Molinari, G. Formisano, and W. Malorni, Modification of the cell surface expression of histocompatibility antigens induced by the neurotoxin 2,5-hexanedions. *Cell Biol. Toxicol.* **3**, 417–430 (1987).

290. M. Governa et al., Impairment of human polymorphonuclear leukocyte chemotaxis by 2,5-hexanedione. *Cell Biol. Toxicol.* **2**, 33–39 (1986).

291. F. Stahl, Cyclohexanol and cyclohexanone-United States. *Chemical Economics Handbook—SRI International*, Stanford Res. Inst., Stanford, CA, 1998, 638.7000A–638.7000M.

292. J. T. James et al., Volatile organic contaminants found in the habitable environment of the space shuttle: STS-26 to STS-55. *Aviat. Space, Environ. Med.* **65**, 851–857 (1994).

293. A. P. Verhoeff et al., Organic solvents in the indoor air of two small factories and surrounding houses. *Int. Arch. Occup. Environ. Health* **59**, 153–163 (1987).

294. V. Cocheo, M. L. Bellomo, and G. G. Bombi, Rubber manufacture: Sampling and identification of volatile pollutants. *Am. Ind. Hyg. Assoc. J.* **44**, 521–527 (1983).

295. M. De Bortoli et al., Concentrations of selected organic pollutants in indoor and outdoor air in northern Italy. *Environ. Int.* **12**, 343–350 (1986).

296. E. Grosjean et al., Air quality model evaluation data for organic. 2. C1-C14 carbonyls in Los Angeles air. *Environ. Sci. Technol.* **30**, 2687–2703 (1996).

297. Lundgren, B. Jonsson, and B. Ek-Olausson, Materials emission of chemical-PVC flooring materials. *Indoor Air* **9**, 202–208 (1999).

298. D. Helmig and J. Arey, Organic chemicals in the air at whitaker's Forest/Sierra Nevada Mountains, California. *Sci. Total Environ.* **112**, 233–250 (1992).

299. F. Juttner, Analysis of organic compounds (VOC) in the forest air of the southern Black Forest. *Chemosphere* **15**, 985–992 (1986).

300. S. C. Rastogi, Levels of organic solvents in printer's inks. *Arch. Environ. Contam. Toxicol.* **20**, 543–547 (1991)

301. C. Winder and K. Steven, The problem of variable ingredients and concentrations in solvent thinners. *Am. Ind. Hyg. Assoc. J.* **56**, 1225–1228 (1995).

302. S. C. Rastogi, Organic solvent levels in model and hobby glues. *Bull. Environ. Contam. Toxicol.* **51**, 501–507 (1993).

303. L. H. Keith et al., Identification of organic compounds in drinking water from thirteen U.S. cities. *Identification and Analysis of Organic Pollutants in Water*, Ann Arbor Press, Ann Arbor, MI, 1976, pp. 329–373.

304. S. D. Richardson et al., Multispectral identification of chlorine dioxide disinfection byproducts in drinking water. *Environ. Sci. Technol.* **28**, 592–599 (1994).

305. G. A. Ulsaker and R. M. Korsnes, Determination of cyclohexanone in intravenous solutions stored in PVC bags by gas chromatography. *Analyst (London)* **102**, 882–883 (1977).

306. D. D. Ellis et al., Organic constituents of mutagenic secondary effluents from wastewater treatment plants. *Arch. Environ. Contam. Toxicol.* **11**, 373–382 (1982).

307. M. Reinhard and N. L. Goodman, Occurrence and distribution of organic chemicals in two landfill leachate plumes. *Environ. Sci. Technol.* **18**, 953–961 (1984).

308. K. R. Dobson et al., Identification and treatability of organics in oil shale retort water. *Water Res.* **19**, 849–856 (1985).

309. B. C. J. Zoeteman, E. De Greek, and F. J. J. Brinkmann, Persistency of organic contaminants in groundwater, lessons from soil pollution incidents in the Netherlands. *Sci. Total Environ.* **21**, 187–202 (1981).

310. N. Paxeus, Organic pollutants in the effluents of large wastewater treatment plants in Sweden. *Water Res.* **30**, 1115–1122 (1996).

311. L. B. Clark et al., Determination of nonregulated pollutants in three New Jersey publicly owned treatment works (POTWs). *J. Water Pollut. Control Fed.* **63**, 104–113 (1991).

312. A. W. Gordon and M. Gordon, Analysis of volatile organic compounds in a textile finishing plant effluent. *Trans. Ky. Acad. Sci.* **42**, 149–157 (1981).

313. *Occupational Safety and Health Administration*, (OSHA), *Cyclohexanone* OSHA, Washington, DC, Available: *http://www.osha-slc.gov/SLTC/analy. . .l–methods/organic/org_1/org_1.html*

314. J. Folke, I. Johansen, and K.-H. Cohr, The recovery of ketones from gas-sampling charcoal tubes. *Am. Ind. Hyg. Assoc. J.* **45**, 231–235 (1984).

315. A. Vairavamurthy, J. M. Roberts, and L. Newman, Methods for determination of low molecular weight carbonyl compounds in the atmosphere: A review. *Atmos. Environ.* **26a**, 1965–1993 (1992).

316. K. Fung and D. Grosjean, Determination of nanogram amounts of carbonyls as 2,4-dinitrophenylhydrazones by high-performance liquid chromatography. *Anal. Chem.* **53**, 168–171 (1981).

317. W. B. Fisher and J. F. VanPeppen, Cyclohexanol and cyclohexanone. In *Encyclopedia of Chemical Technology*, 4th ed., Vol. 7, Wiley, New York, 1993, pp. 851–859.

318. A. Zlatkis et al., Analysis of trace volatile metabolites in serum and plasma. *J. Chromatog.* **91**, 379–383 (1974).

319. H. M. Liebich and O. Al-Babbili, Gas chromatographic-mass spectrometric study of volatile organic metabolites in urines of patients with *Diabetes mellitus. J. Chromatogr.* **112**, 539–550 (1975).

320. C. N. Ong, G. L. Sia, and S. E. Chia, Determination of cyclohexanol in urine and its use in envionmental monitoring of cyclohexanone exposure. *J. Anal. Toxicol.* **15**, 13–16 (1991).

321. V. Fiserova-Bergerova, Biological monitoring VIII: Interference of alcoholic beverage consumption with biological monitoring of occupational exposure to industrial chemicals. *Appl. Occup. Environ. Hyg.* **8**, 757–760 (1993).

322. T. Yasugi et al., Exposure monitoring and health effect studies of workers occupationally exposed to cyclohexane vapor. *Arch. Occup. Environ. Health* **65**, 343–350 (1994).

323. C. N. Ong et al., Monitoring of exposure to cyclohexanone through the analysis of breath and urine. *Scand. J. Work Environ. Health* **17**, 430–435 (1991).

324. H. F. Smyth, Jr. et al., Range finding toxicity data: List VI. *Am. Ind. Hyg. Assoc. J.* **23**, 95–107 (1962).

325. H. F. Smyth, Jr. et al., Range finding toxicity data: List VII. *Am. Ind. Hyg. Assoc. J.* **30**, 470–476 (1969).

326. W. B. Deichmann and T. V. Leblanc, Determination of the approximate lethal dose with about six animals. *J. Ind. Hyg. Toxicol.* **25**, 415–417 (1943).

327. J. F. Treon, W. E. Crutchfield, and K. V. Kitzmiller, The physiological response of animals to cyclohexane, methylcyclohexane, and certain derivatives of these compounds. Oral administration and cutaneous application. *J. Ind. Hyg. Toxicol.* **25**, 199–214 (1943).

328. S. Haworth et al., *Salmonella* mutagenicity test results for 250 chemicals. *Environ. Mutagen.* **5**(Suppl.1), 3–142 (1983).

329. H. J. Klimisch et al., Inhalation hazard test. Interlaboratory trial with OECD method 403. *Arch. Toxicol.* **61**, 318–320 (1988).

330. P. K. Gupta et al., Toxicological aspects of cyclohexanone. *Toxicol. Appl. Pharmacol.* **49**, 525–533 (1979).

331. J. DeCeaurriz et al., Concentration-dependent behavioral changes in mice following short-term inhalation exposure to various industrial solvents. *Toxicol. Appl. Pharmacol.* **67**, 383–389 (1983).

332. J. F. Treon, W. E. Crutchfield, and K. U. Kitzmiller, The physiological response of animals to cyclohexane, methylcyclohexane and certain derivatives of these compounds. II. Inhalation. *J. Ind. Hyg. Toxicol.* **25**, 323–347 (1943).

333. Y. Greener et al., Assessment of the safety of chemicals administered intravenously in the neonatal rat. *Teratology* **35**, 187–194 (1987).

334. M. T. Koeferl et al., Subacute toxicity of cyclohexanone in rats, dogs, and monkeys. *Toxicol. Appl. Pharmacol.* **37**, 115 (Abstr.) (1976).

335. M. T. Koeferl et al., Influence of concentration and rate of intravenous administration on the toxicity of cyclohexanone in beagle dogs. *Toxicol. Appl. Pharmacol.* **59**, 215–219 (1981).

336. R. H. Rengstorff, J. P. Petrali, and V. M. Sim, Cataracts induced in guinea pigs by acetone, cyclohexanone, and dimethyl sulfoxide. *Am. J. Optom.* **49**, 308–319 (1972).

337. Y. Greener and E. Youkilis, Assessment of the cataractogenic potential of cyclohexanone in guinea pigs and rabbits. *Fundam. Appl. Toxicol.* **4**, 1055–1066 (1984).

338. D. A. Mayhew et al., Cataractogenesis studies in guinea pigs with cyclohexanone. *Toxicol. Lett.* **31**, 51 (abstr.)(1986).

339. Wil Research Laboratories, *Cataractogenic Potential of Cyclohexanone Dosed Dermally to Guinea Pigs and Rats*, Proj. Nos. WIL-81154, WIL-81155, Will Research Laboratories, Cincinnati, OH, 1983 (unpublished report).

340. W. Lijinsky and R. M. Kovatch, Chronic toxicity study of cyclohexanone in rats and mice. *J. Natl. Cancer Inst.* **77**, 941–949 (1986).

341. S. P. James and R. H. Waring, The metabolism of alicyclic ketones in the rabbit and rat. *Xenobiotica* **1**, 573–580 (1971).

342. E. Boyland and L. F. Chasseaud, The effect of some carbonyl compounds on rat liver glutathione levels. *Biochem. Pharmacol.* **19**, 1526–1528 (1970).

343. D. Parmer and L. T. Burka, Metabolism and disposition of cyclohexanone oxime in male F-344 rats. *Drug Metab. Dispos.* **19**, 1101–1107 (1991).

344. T. H. Elliott, D. V. Parke, and R. T. Williams, Studies on detoxication, T 79. The metabolism of cyclo[^{14}C] hexane and its derivatives. *Biochem. J.* **72**, 193–200 (1959).

345. T. Cronholm, Isotope effects and hydrogen transfer during simultaneous metabolism of ethanol and cyclohexanone in rats. *Eur. J. Biochem.* **43**, 189–196 (1975).

346. M. Sakata et al., Metabolic interaction of ethanol and cyclohexanone in rabbits. *J. Toxicol. Environ. Health* **38**, 33–32 (1993).

347. Tegeris Laboratories, *Acute Inhalation Metabolism Study of Cyclohexanone in Rats*, Proj. No. 86009, Tegeris Laboratories, Temple Hills, MD, 1987 (unpublished report).

348. L. Martis et al., Disposition kinetics of cyclohexanone in beagle dogs. *Toxicol. Appl. Pharmacol.* **55**, 545–553 (1980).

349. L. E. Gray et al., An evaluation of figure-eight maze activity and general behavioral development following prenatal exposure to forty chemicals: Effects of cytosine arabinoside, dinocap, nitrofen, and vitamin A. *Neurotoxicology* **7**, 449–462 (1986)

350. I. H. Hall et al., Cycloalkanones. 4. Antifertility activity. *J. Med. Chem.* **17**, 1253–1257 (1974).

351. E. M. Weller and J. H. Griggs, Hazards of embryonic exposures to chemical vapors. *Anat. Rec.* **184**, 561 (1976).

352. B. S. Samini, S. B. Harris, and A. De Peyster, Fetal effects of inhalation exposure to cyclohexanone vapors in pregnant rats. *Toxicol. Ind. Health* **5**, 1035–1043 (1989).

353. E. Gondry, Recherches sur la toxicité de la cyclohexylamine, de la cyclohexanone et du cyclohexanol, métabolites du cyclamate. *J. Eur. Toxicol.* **5**, 227–238 (1972).

354. J. M. Seidenberg, D. G. Anderson, and R. A. Becker, Validation of an *in vivo* developmental toxicity screen in the mouse. *Teratog., Carcinog., Mutagen.* **6**, 361–374 (1986).

355. J. M. Seidenberg and R. A. Becker, A summary of the results of 55 chemicals screened for developmental toxicity in mice. *Teratog., Carcinog., Mutagen.* **7**, 17–28 (1987).

356. Biodynamics Inc., *An Inhalation Teratolgy Study in Rats with Cyclohexanone*, Proj. No. 83-2719, Biodynamics, Inc., East Millstone, NJ, 1984 (unpublished report).

357. Biodynamics Inc., *An Inhalation Teratogenicity Study in the Mouse with Cyclohexanone*, Proj. No. 83-2766, Biodynamics Inc., East Millstone, NJ, 1984 (unpublished report).

358. American Biogenics Corp., *Two-generation Reproduction Study via Inhalation (with a Neurotoxicology/Pathology Component) in Albino Rats using Cyclohexanone*, Study No. 450-1587, American Biogenics Corp., Decatur, IL, 1986 (unpublished report).

359. American Biogenics Corp., *Assessment of Male Reproductive Performance During a Post-Exposure Recovery Period of Second Generation Males from a Two Generation Reproduction Study with Cyclohexanone via Inhalation*, Study No. 450-2326, American Biogenics Corp., Decatur, IL, 1986 (unpublished report).

360. I. Florin et al., Screening of tobacco smoke constituents for mutagenicity using the Ames test. *Toxicology* **15**, 219–232 (1980).

361. A. Massoud, A. Aly, and H. Shafik, Mutagenicity and carcinogenicity of cyclohexanone. *Mutat. Res.* **74**, 174 (abstr.) (1980).

362. H. S. Rosenkranz and Z. Leifer, Determining the DNA-modifying activity of chemicals using DNA-polymerase-deficient *Escherichia coli.* In F. J. de Serres and A. Hollaender, eds., *Chemical Mutagens: Principles and Methods for Their Detection*, Vol. 6, Plenum, New York, 1980, pp. 109–147.

363. D. B. McGregor et al., Response of the L5178Y tk^{+}/tk^{-} mouse lymphoma cell forward mutation assay: III. 72 coded chemicals. *Environ. Mol. Mutagen.* **12**, 85–154 (1988).

364. C. S. Aaron et al., Comparative mutagenesis in mammalian cells (CHO) in culture: Multiple genetic endpoint analysis of cyclohexanone *in vitro*. *Environ. Mutagen.* **7**(Suppl. 3), 60 (abstr) (1985).

365. University of Wisconsin, Drosophilia melanogaster *Sex-linked Recessive Lethal Test of Cyclohexanone*, Proj. No. 116, University of Wisconsin, Madison, 1986 (unpublished report).

366. A. R. Malcolm and L. J. Mills, The potential role of bioactivation in tumor promotion: Indirect evidence from effects of phenol, sodium cyclamate and their metabolites on metabolic cooperation *in vitro*. *Biochem. Mech. Regul. Intercell. Commun.* **14**, 237–249 (1987).

367. B. Holmberg, I. Jakobson, and T. Malmfors, The effect of organic solvents on erythrocytes during hypotonic hemolysis. *Environ. Res.* **7**, 193–205 (1974).

368. S. C. Gad et al., Development and validation of an alternative dermal sensitization test: The mouse ear swelling test (MEST). *Toxicol. Appl. Pharmacol.* **84**, 93–114 (1986).

369. M. Bruze et al., Contact allergy to a cyclohexanone resin in humans and guinea pigs. *Contact Dermatitis* **18**, 46–49 (1988).

370. E. Frantik, M. Hornychova, and M. Horvath, Relative acute neurotoxicity of solvents: Isoeffective air concentration of 48 compounds evaluated in rats and mice. *Environ. Res.* **66**, 173–185 (1994).

371. M. Kanada et al., Neurochemical profile of effects of 28 neurotoxic chemicals on the central nervous system in rats. 1. Effects of oral administration on brain contents of biogenic amines and metabolites. *Ind. Health* **32**, 145–164 (1994).

372. K. D. Holland et al., Convulsant and anticonvulsant cyclopentanones and cyclohexanones. *Mol. Pharmacol.* **37**, 98–103 (1989).

373. H. Panhuber, Effect of odor quality and intensity on conditioned odor aversion learning in the rat. *Physiol. Behav.* **28**, 149–154 (1982).

374. H. Panhuber, A. Mackay-Sim, and D. G. Laing, Prolonged odor exposure causes severe cell shrinkage in the adult rat olfactory bulb. *Dev. Brain Res.* **31**, 307–311 (1987).

375. L. Perbellini et al., Studio sperimentale sulla neurotossicità del cicloesanolo e del cicloesanone. *Med. Lav.* **2**, 102–107 (1981).

376. M. Sakata, J. Kikuchi, and M. Haga, Disposition of acetone, methyl ethyl ketone and cyclohexanone in acute poisoning. *Clin. Toxicol.* **27**, 67–77 (1989).

377. G. B. Zuckerman, S. C. Lam, and S. M. Santo, Rhabdomyolysis following oral ingestion of the hydrocarbon cyclohexanone in an adolescent. *J. Environ. Pathol. Toxicol. Oncol.* **17**, 11–15 (1998).

378. D. Passeron et al., Insuffisance hepato-renale aigue apres ingestion de cyclohexanone. *J. Toxicol. Clin. Exp.* **12**, 207–211 (1992).

379. K. W. Nelson et al., Sensory response to industrial solvent vapors. *J. Ind. Hyg. Toxicol.* **25**, 282–285 (1943).

380. Esso Research and Engineering Co., *Acute Inhalation (LC50) and Human Sensory Irritation Studies on Cyclopentanone, Isophorone, Dihydroisophorone, Cyclohexanone, and Methyl Isobutyl Ketone*, OTS Doc., EPA/OTS: 86960000031, ESSO Research and Engineering Co., Baton Rauge, LA, 1965, pp. 1–7.

381. M. Jacobsen, J. Baelum, and J. P. Bonde, Temporal epileptic seizures and occupational exposure to solvents. *Occup. Environ. Med.* **51**, 429–430 (1994).

382. J. Mráz et al., Uptake, metabolism and elimination of cyclohexanone in humans. *Arch. Occup. Environ. Health* **66**, 203–208 (1994).

383. C. N. Ong et al., Determination of cyclohexanol in urine and its use in environmental monitoring of cyclohexanone exposure. *J. Anal. Toxicol.* **15**, 13–16 (1991).

384. G. A. Mills and V. Walker, Urinary excretion of cyclohexanediol, a metabolite of the solvent cyclohexanone, by infants in a special care unit. *Clin Chem.* **36**, 870–874 (1990).

385. R. P. Snell, Capillary GC analysis of compounds leached into parenteral solutions packaged in plastic bags. *J. Chromatogr. Sci.* **27**, 524–528 (1989).

386. A. Mutti et al., Absorption and alveolar exretion of cyclohexane in workers in a shoe factory. *J. Appl. Toxicol.* **1**, 220–223 (1981).

387. J. Jankovic and F. Drake, A screening method for occupational reproductive health risk. *Am. Ind. Hyg. Assoc. J.* **57**, 641–649 (1996).

388. P. Perocco, S. Bolognesi, and W. Alberghini, Toxic activity of seventeen industrial solvents and halogenated compounds on human lymphocytes cultured *in vitro*. *Toxicol. Lett.* **16**, 69–75 (1983).

389. J. Lederer et al., L'action cytogénétique et tératogéne du cyclamate et de ses métabolites. *Thérapeutique* **47**, 357–363 (1971).

390. O. Sanmartin and J. De La Cuadra, Occupational contact dermatitis from cyclohexanone as a PVC adhesive. *Contact Dermatitis* **27**, 189–190 (1992).

391. E. Mitran et al., Neurotoxicity associated with occupational exposure to acetone, methyl ethyl ketone, and cyclohexanone. *Environ. Res.* **73**, 181–188 (1997).

392. S. E. Chia et al., Neurobehavioral effects on workers in a video tape manufacturing factory in singapore. *Neurotoxicology* **14**, 51–56 (1993).

393. U.S. Department of Health and Human Services, Occupational safety and health guideline for methyl (*n*-amyl) ketone. In *Occupational Health Guidelines for Chemical Hazards*, Public Health Service, Washington, DC, 1988, pp. 1–5.

394. B. K. Krotoszynski and H. J. O'Neill, Involuntary bioaccumulation of environmental pollutants in nonsmoking heterogeneous human population. *J. Environ. Sci. Health* **A17**, 855–883 (1982).

395. E. D. Pellizzari et al., Purgeable organic compounds in mother's milk. *Bull. Environ. Contam. Toxicol.* **28**, 322–328 (1982).

396. S. Dewey et al., Microbial volatile organic compounds: A new approach in assessing health risks by indoor mould? *Zentralbl. Hyg. Umweltmed.* **197**, 504–515 (1995).

397. M. Day et al., An investigation of the chemical and physical changes occurring during commercial composting. *Compost Sci. Util.* **6**, 44–66 (1998).

398. K.-O. Intarapichet and M. E. Bailey, Volatile compounds produced by meat pseudomonas grown on beef at refrigeration temperatures. *Asean Food J.* **8**, 14–21 (1993).

399. H. Srepel and B. Akacic, Testing the antihelmintic effectiveness of volatile oils from plants of the genus Ruta. *Acta Pharm. Jugosl.* **12**, 79–87 (1962).

400. I. Levinstein, Report to research institute for fragrance materials, in D. L. J. Opdyke, Methyl *n*-amyl ketone, fragrance raw materials monographs. *Food Cosmet. Toxicol.* **13**, 847–848 (1975).

401. W. K. Anger, M. K. Jordan, and D. W. Lynch, Effect of inhalation exposure and intraperitonal injections of methyl *n*-amyl ketone on multiple fixed ratio, fixed-interval response rates in rats. *Toxicol. Appl. Pharmacol.* **49**, 407–416 (1979).

402. D. W. Lynch et al., Inhalation toxicity of methyl *n*-amyl ketone (2-heptanone) in rats and monkeys. *Toxicol. Appl. Pharmacol.* **58**, 341–352 (1981).

403. L. Prockop and D. Couri, Nervous system damage from mixed organic solvents. In C. W. Sharp and M. L. Brehm, eds., *Review of Inhalants: Euphoria to Dysfunction*, National Institute on Drug Abuse, Rockville, MD, 1977, pp. 185–198.

404. J. H. Tison, L. D. Prockup, and E. D. Means, In S. S. Lande et al., Investigation of selected potential environmental contaminants; Ketonic solvents, *U.S. NTIS PB Rep.* PB 252970 (1976).

405. I. F. Gaunt et al., Short term toxicity of methyl amyl ketone in rats. *Food Cosmet. Toxicol.* **10**, 625–636 (1972).

406. J. Misumi and M. Nagano, Neurophysiological studies on the relation between the structural properties and neurotoxicity of aliphatic hydrogen compounds in rats. *Brit. J. Ind. Med.* **41**, 526–532 (1984).

407. J. Misumi and M. Nagano, Experimental study on the enhancement of the neurotoxicity of methyl *n*-butyl ketone by non-neurotoxic aliphatic monoketones. *Br. J. Ind. Med.* **42**, 155–161 (1985).

408. I. A. Kamil, J. N. Smith, and R. T. Williams, Studies on detoxication. 46. The metabolism of aliphatic alcohols. The glucuronic and conjugation of acrylic aliphatic alcohols. *Biochem. J.* **53**, 129–136 (1953).

409. L. G. Bernard, *Methyl n-Amyl Ketone: Reproduction/Developmental Toxicity Screening Test in the Rat*, Rep. TX-96-11, Eastman Kodak Company, Rochester, NY, 1996.

410. P. W. Albro, J. T. Corbett, and J. L. Schroeder, Metabolism of methyl *n*-amyl ketone (MAK) and its binding to DNA of rat liver *in vivo* and *in vitro*. *Chem.-Biol. Interact.* **51**, 295–308 (1984).

411. E. D. Barber et al., The lack of binding of methyl *n*-amyl ketone (MAK) to rat liver DNA as demonstrated by direct binding measurements and ^{32}P-postlabeling techniques. *Mutat. Res.* **442**, 133–147 (1999).

412. H. Tanii, Anesthetic activity of monoketones in mice: Relationship to hydrophobicity and *in vivo* effects on Na^+/K^+-ATPase activity and membrane fluidity. *Toxicol. Lett.* **85**, 41–47 (1996).

413. W. R. Hewitt, E. M. Brown, and G. L. Plaa, Relationship between the carbon skeleton length of ketonic solvents and potentiation of chloroform-induced hepatotoxicity in rats. *Toxicol. Lett.* **16**, 297–304 (1983).

414. E. M. Brown and W. R. Hewitt, Dose-response relationships in ketone-induced potentiation of chloroform hepato- and nephrotoxicity. *Toxicol. Appl. Pharmacol.* **76**, 437–453 (1984).

415. W. L. Epstein, Report to research institute of fragrance materials. April, 1974, in G. L. J. Opdyke, *Food Cosmet. Toxicol.* **16**, 731 (1978).

416. H. P. Thiebaud et al., Mutagenicity and chemical analysis of fumes from cooking meat. *J. Agric. Food Chem.* **42**, 1502–1510 (1994).

417. A. Spurgeon et al., Investigation of dose related neurobehavioral effects in paint makers exposed to low levels of solvents. *Occup. Environ. Med.* **51**, 626–630 (1994).

418. D. C. Glass et al., Retrospective assessment of solvent exposure in paint manufacturing. *Occup. Environ. Med.* **51**, 617–625 (1994).

419. G. V. Katz, E. R. Renner, Jr., and C. J. Terhaar, Subchronic inhalation toxicity of methyl *iso*amyl ketone in rats. *Fundam. Appl. Toxicol.* **6**, 498–505 (1986).

420. H. F. Smyth, Jr., C. P. Carpenter, and C. S. Weil, Range finding toxicity data. List III. *J. Ind. Hyg. Toxicol.* **31**, 60–62 (1949).

421. O. M. Moreno, Report to research institute for fragrance materials, August 1976, cited in D. L. J. Opdyke, Fragrance Raw Materials Monographs, Ethyl Butyl Ketone. *Food Cosmet. Toxicol.* **16**, 731 (1978).

422. D. W. Sharp, The sensitization potential of some perfume ingredients tested using a modified draize procedure. *Toxicology* **9**, 261–271 (1978).

423. E. R. Homan and R. R. Maronpot, Neurotoxic evaluation of some aliphatic ketones. *Toxicol. Appl. Pharmacol.* **45**, 312 (1978).

424. J. L. O'Donoghue et al., Further studies on ketone neurotoxicity and interactions. *Toxicol. Appl. Pharmacol.* **72**, 201–209 (1984).

425. L. Perbellini et al., Identification of *n*-heptane metabolites in rat and human urine. *Arch. Toxicol.* **58**, 229–234 (1986).

426. C. P. Carpenter, C. S. Weil, and H. F. Smyth, Jr., Range finding toxicity data: List VIII. *Toxicol. Appl. Pharmacol.* **28**, 313–319 (1974).

427. R. Jeppsson, Parabolic relationship between lipophilicity and biological activity of aliphatic hydrocarbons, ethers and ketones after intravenous injections of emulsion fortifications into mice. *Acta Pharmacol. Toxicol.* **37**, 56–64 (1975).

428. M. Weast, in D. R. Lide, ed., *CRC Handbook of Chemistry and Physics*, 80th ed., CRC Press, Boca Raton, FL, 1999.

429. Chemfinder.com, 2-Methylcyclohexanone. Available: *http://www.chemfinder.com/database and Internet searching.*

430. E. Browning, *Toxicity of Industrial Organic Solvents*, Chemical Publishing Co., New York, 1953, pp. 334–339.

431. R. J. Lewis, Ed., 6-methyl-3-cyclohexene carboxal. *Hawley's Condensed Chemical Dictionary* 13th ed., Van Nostrand-Reinhold, New York, 1997, p. 733.

432. P. Ciccioli et al., Use of carbon adsorption traps combined with high resolution gas chromatography-mass spectrometry for the analysis of polar and non-polar C_4-C_{14} hydrocarbons involved in photochemical smog formation. *J. High Resolut. Chromatogr.* **15**, 75–84 (1992).

433. J. M. Deroux et al., Long-term extractable compounds screening in surface water to prevent accidental organic pollution. *Sci. Total Environ.* **203**, 261–274 (1997).

434. T. Kozawa et al., Management of toxics for the Fukashiba industrial wastewater treatment plant of the kashima petrochemical complex. *Water Sci. Technol.* **25**, 247–254 (1992).

435. P. Andrew, A. F. Smith, and R. Wood, A simple field test for the determination of cyclohexanone and methylcyclehexanone vapours in air. *Analyst (London)* **96**, 528–534 (1971).

436. M. D. Wright, A dual-capillary column system for automated analysis of workplace contaminants by thermal desorption-gas chromatography. *Anal. Proc.* **24**, 309–311 (1987).

437. J. M. Deroux et al., Analysis of extractable organic compounds in water by gas chromatography/mass spectrometry: Applications to surface water. *Talanta* **43**, 365–380 (1996).

438. F. Caujolle et al., Toxicité de la cyclohexanone et de quelques cétones homologues, *C. R. Hebd. Seances Acad. Sci.* **236**, 633 (1953).

439. K. B. Lehmann and F. Flury, eds., Methyl cyclohexanone. *Toxicology and Hygiene of Industrial Solvents*, Williams & Wilkins, Baltimore, MD, 1943, p. 247.

440. W. J. Spillane, G. A. Benson, and G. McGlinchey, Metabolic studies with nonnutritive sweeteners cyclooctylsulfamate and 4-methylcyclohexysulfamate. *J. Pharm. Sci.* **68**, 372–374 (1979).

441. K. M. Madyastha and C. P. Raj, Metabolic fate of menthofuran in rats. *Drug Metab. Dispos.* **20**, 295–301 (1992).

442. C. C. Tao and T. H. Elliott, The stereospecific metabolism of methylcyclohexane derivatives. *Biochemistry* **84**, 38–39 (1962).

443. T. H. Elliott, E. Jacob, and C. C. Tao, The *in vitro* and *in vivo* metabolism of optically active methylcyclohexanols and methycyclohexanones. *J. Pharm. Pharmacol.* **21**, 561–572 (1969).

444. H. Onishi et al., Optical analysis of reduction products of 2-methylcyclohexanone by *Aspergillus repens* MA0197. *Biosci. Biotech. Biochem.* **60**, 486–487 (1996).

445. S. Nakajina et al., Carbonyl reductase activity exhibited by pig testicular 20β-hydroxysteriod dehydrogenase. *Biol. Pharm. Bull.* **20**, 1215–1218 (1997).

446. P. J. Dierickx, Comparison between fish lethality data and the *in vitro* cytotoxicity of lipophilic solvents to cultured fish cells in a two-compartment model. *Chemosphere* **27**, 1511–1518 (1993).

447. H. R. Sullivan, W. M. Miller, and R. E. McMahon, Reaction pathways of *in vivo* stereoselective conversion of ethylbenzene to (−)mandelic acid. *Xenobiotica* **6**, 49–54 (1976).

448. W. P. Gordon et al., Hepatotoxicity and pulmonary toxicity of pennyroyal oil and its constituent terpenes in the mouse. *Toxicol. Appl. Pharmacol.* **65**, 413–424 (1982).

449. S. Budavari et al., eds., Acetorphenone. *The Merck Index* 12th ed., Merck & Co., Rahway, NJ, 1996, p. 72.

450. A. J. Papa and P. D. Sherman, Ketones. in H. F. Mark et al., eds., *Kirk-Othmer Encyclopedia of Chemical Technology*, 3rd ed., Vol. 13, Wiley, New York, 1980, pp. 894–941.

451. D. L. J. Opdyke, Monographs on fragrance materials. *Food Cosmet. Toxicol.* **11**, 95–115 (1973).

452. W. Cautreels and K. Van Cauwenberghe, Experiments on the distribution of organic pollutants between airborne particulate matter and the corresponding gas phase. *Atmos. Environ.* **12**, 1133–1141 (1978).

453. Y. Yokouchi, T. Ito, and T. Ambe, Identification of C_6-C_{15} γ-lactones in atmospheric aerosols. *Chemosphere* **16**, 1143–1147 (1987).

454. D. Helmig et al., Gas chromatography mass spectrometry analysis of volatile organic trace gases at Mauna Loa Observatory, Hawaii. *J. Geophy. Res.* **101**, 14,697–14,710 (1996).

455. F. H. Jarke, A. Dravnieks, and S. M. Gordon, Organic contaminants in indoor air and their relation to outdoor contaminants. *ASHRAE Trans.* **87**, 153–166 (1981).

456. A.-L. Sunesson et al., Volatile metabolites produced by two fungal species cultivated on building materials. *Ann. Occup. Hyg.* **40**, 397–410 (1996).

457. J. A. Leary et al., Chemical and toxicological characterization of residential oil burner emissions: I. Yields and chemical characterization of extractables from combustion of No. 2 fuel oil at different Bacharach smoke number and firing cycles. *Environ. Health Perspect* **73**, 223–234 (1987).

458. C. V. Hampton et al., Hydrocarbon gases emitted from vehicles on the road. 1. A literative review. *Environ. Sci. Technol.* **16**, 287–298 (1982).

459. J. J. Schauer et al., Measurement of emissions from air pollution sources. 2. C_1 through C_{30} organic compounds from medium duty diesel trucks. *Environ. Sci. Technol.* **33**, 1578–1587 (1999).

460. N. V. Lovegren et al., Volatile constituents of dried legumes. *J. Agric. Food Chem.* **27**, 851–853 (1979).

461. I. H. Suffet, L. Brenner, and P. R. Cairo, GC/MS identification of trace organics in Philadelphia drinking waters during a 2-year period. *Water Res.* **14**, 853–867 (1980).

462. R. Shinohara et al., Identification and determination of trace organic substances in tap water by computerized gas chromatography-mass spectrometry and mass fragmentography. *Water Res.* **15**, 535–542 (1981).

463. R. H. Plumb, Jr., The occurrence of appendix IX organic constituents in disposal site ground water. *Ground water Monit. Rev.* **13**, 157–164 (1991).

464. S. R. Hutchins, M. B. Tomson, and C. H. Ward, Trace organic contamination of ground water from a rapid infiltration site: a laboratory-field coordinated study. *Environ. Toxicol. Chem.* **2**, 195–216 (1983).

465. J. C. Peterson, D. F. Clark, and P. S. Sleevl, Monitoring a new environmental pollutant. *Anal. Chem.* **58**, 71A-74A.

466. A. P. Meijers and R. C. Van Der Leer, The occurrence of organic micropollutants in the River Rhine and the River Maas in 1974. *Water Res.* **10**, 597–604 (1976).

467. M. Ehrhardt, Photo-oxidation products of fossil fuel components in the water of Hamilton Harbour, Bermuda. *Mar. Chem.* **22**, 85–94 (1987).

468. M. R. Samolloff et al., Combined bioassay-chemical fractionation scheme for the determination and ranking of toxic chemicals in sediments. *Environ. Sci. Technol.* **17**, 329–334 (1983).

469. Occupational Safety and Health Administration (OSHA), *Acetophenone*, cas **98-86-2**, OSHA, Washington, DC. Available: *http://www.osha-slc.gov/sltc/analy. . .ial/acetophenone/acetophenone.html*

470. G. H. Stanko and P. E. Fortini, Inter- and intralaboratory assessment of SW-846 methods manual for analysis of appendix VIII compounds in groundwater. *Hazard. Wastes Hazard. Mater.* **2**, 67–97 (1985).

471. D. Giroux, G. Lapointe, and M. Baril, Toxicological index and the presence in the workplace of chemical hazards for workers who breast-feed infants. *Am. Ind. Hyg. Assoc. J.* **53**, 471–474 (1992).

472. G. Rhodes et al., Structural relationships between the endogenous volatile urinary metabolites of experimentally diabetic rats and certain neurotoxins. *Experientia* **38**, 7577 (1982).

473. J. H. Raymer et al., A nonrebreathing breath collection system for the study of exogenous and endogenous compounds in the Fisher-344 rat. *Toxicol. Methods* **4**, 243–258 (1994).

474. P. M. Jenner et al., Food flavorings and compounds of related structure. I. Acute oral toxicity. *Food Cosmet. Toxicol.* **2**, 327 (1964).

475. I. Grübner, W. Klinger, and H. Ankermann, Untersuchung Verschiedener Stoffe und Stoffklassen auf Induktoreigenschaften. II. Mitteilung. *Arch. Int. Pharmacodyn. Ther.* **196**, 288–297 (1972).

476. C. K. Muir, The toxic effect of some industrial chemicals on rabbit ileum *in vitro* compared with eye irritancy *in vivo. Toxicol. Lett.* **19**, 309–312 (1983).

477. M. Schaper, Development of a database for sensory irritants and its use in establishing occupational exposure limits. *Am. Ind. Hyg. Assoc. J.* **54**, 488–544 (1993).

478. E. C. Hagan et al., Food flavourings and compounds of related structure. II. Subacute and chronic toxicity. *Food Cosmet. Toxicol.* **5**, 141–157 (1967).

479. D. Zissu, Histopathological changes in the respiratory tract of mice exposed to ten families of airborne chemicals. *J. Appl. Toxicol.* **15**, 207–213 (1995).

480. T. Dalland and K. B. Doving, Reaction to olfactory stimuli in odor-exposed rats. *Behav. Neurol. Biol.* **32**, 79–88 (1981).

481. D. G. Laing and H. Panhuber, Neuronal and behavioral changes in rats following continuous exposure to odour. *J. Comp. Physiol.* **124**, 259–265 (1978).

482. K. M. Engström, Metabolism of inhaled ethylbenzene in rats. *Scand. J. Work Environ. Health* **10**, 83–87 (1984).

483. K. M. Engström, Urinalysis of minor metabolites of ethylbenzene and *m*-xylene. *Scand. J. Work Environ. Health* **10**, 75–81 (1984).

484. K. C. Leibman, Reduction of ketones in liver cytosol. *Xenobiotica* **1**, 97–104 (1971).

485. M. Kiese and W. Lenk, ω-and (ω-1)-hydroxylation of 4-chloropropionanilide in liver microsomes of rabbits treated with phenobarbital or 3-methylcholanthrene. *Biochem. Pharmacol.* **22**, 2575–2580 (1973).

486. H. Weiner et al., Reversible inhibitors of aldehyde dehydrogenase. *Prog. Clin. Biol. Res.* **114**, 91–102 (1983).

487. J. P. Mahy, S. Gaspard, and D. Mansuy, Phenylhydrazones as new good substrates for the dioxygenase and peroxidase reactions of prostaglandin synthase: Formation of Iron(III)-*o*-phenyl complexes. *Biochemistry* **32**, 4014–4021 (1993).

488. J. N. Smith, R. H. Smithies, and R. T. Williams, Studies in detoxication. 56. The metabolism of alkylbenzenes. Stereochemical aspects of the biological hydroxylation of ethylbenzene to methylphenylcarbinol. *Biochem. J.* **56**, 320–324 (1954).

489. M. Kiese and W. Lenk, Hydroxyacetophenones: Urinary metabolites of ethylbenzene and acetophenone in the rabbit. *Xenobiotica* **4**, 337–343 (1974).

490. T. S. Lagno and G. Z. Bakhtizina, The skin resorptive action of acetophenone. *Tr. — Ufim. Nauchno-Issled. Gig. Prof. Zabol.* **5**, 90–94 (1969).

491. C. A. Elliger, P. R. Henika, and J. T. MacGregor, Mutagenicity of flavones, chromones and acetophenones in *Salmonella typhimurium*. *Mutat. Res.* **135**, 77–86 (1984).

492. E. R. Fluck, F. A. Poirier, and H. W. Ruelius, Evaluation of a DNA polymerase-deficient mutant of *E.coli* for the rapid detection of carcinogens. *Chem.-Biol. Interac.* **15**, 219–231 (1976).

493. R. Edenharder, I. von Petersdorff, and R. Rauscher, Antimutagenic effects of flavonoids, chalcones and structurally related compounds on the activity of 2-amino-3-methylimidazo[4,5-*f*]quinoline (IQ) and other heterocyclic amine mutagens from cooked food. *Mutat. Res.* **287**, 261–274 (1993).

494. A. Önfelt, S. Hellberg, and S. Wold, Relationships between induction of anesthesia and mitotic spindle disturbances studied by means of principal component analysis. *Mutat. Res.* **174**, 109–113 (1986).

495. R. O. Rahn, L. C. Landry, and W. L. Carrier, Formation of chain breaks and thymine dimers in DNA upon photosensitization at 313 nm with acetophenone, acetone, or benzophenone. *Photochem. Photobiol.* **19**, 75–78 (1974).

496. B. E. Zierenberg et al., Effects of sensitized and unsensitized longwave U.V.-irradiation on the solution properties of DNA. *Photochem. Photobiol.* **14**, 515–520 (1971).

497. N. Chouini-Lalanne, M. Defais, and N. Pailous, Nonsteroidal antiinflammatory drug-photosensitized formation of pyrimidine dimer in DNA. *Biochem. Pharmacol.* **55**, 441–446 (1998).

498. Hazardous Substances Data Bank (HSDB), *Acetophenone*, National Library of Medicine, Bethesda, MD, (CD-ROM), MICROMEDIX, Englewood, CO, 1991, pp. 1–17.

499. B. M. Abbasov, K. G. Zeinaloba, and I. A. Safarova, Toxicology of some ketones. *Tr. Azerb. Nauchno-Issled. Inst. Gig. Pr. Prof. Zabol.* **11**, 145–148 (1977).

500. I. H. Hall and G. L. Carlson, Cycloalkanones. 9. Comparison of analogs which inhibit cholesterol and fatty acid synthesis. *J. Med. Chem.* **19**, 1257–1261 (1976).

501. D. L. J. Opdyke, Monographs on fragrance raw materials. Methyl hexyl ketone. *Food Cosmet. Toxicol.* **13**, 861 (1975).

502. G. L. Carlson, I. H. Hall, and C. Piantadosi, Cycloalkanones. Part 7: Hypocholesterolemic activity of aliphatic compounds related to 2,8-dibenzylcyclooctanone. *J. Med. Chem.* **18**, 1024–1026 (1975).

503. R. E. McMahon, J. C. Cline, and C. Z. Thompson, Assay of 855 test chemicals in 10 tester strains using a new modification of the Ames test for bacterial mutagens. *Cancer Res.* **39**, 682–693 (1979).

504. A. M. Kligman, Report to the Research Institute for Fragrance Materials, October 1973, in D. L. J. Opdyke, Methyl hexyl ketone, fragrance raw materials monographs. *Food Cosmet. Toxicol.* **13**, 861 (1975).

505. M. V. Shelanski, Report to Research Institute for Fragrance Materials, January 1973, in D. L. J. Opdyke, Fragrance raw materials, ethyl amyl ketone. *Food Cosmet. Toxicol.* **12**, 715 (1974).

506. A. M. Kligman, Report to Research Institute for Fragrance Materials, November 1972, in D. L. J. Opdyke, Fragrance raw materials, ethyl amyl ketone. *Food Cosmet. Toxicol.* **12**, 715 (1974).

507. S. Budavari et al., eds., Propiophenone. *The Merck Index*, 12th ed., Merck & Co., Rahway, NJ, 1996, pp. 8015–8016.

508. J. Braithwaite, Ketones. In *Encyclopedia of Chemical Toxicology*, 4th ed., Vol. 14, Wiley, New York, 1995, pp. 978–1021.

509. Hazardous Substances Data Bank (HSDB), *Phenyl Ethyl Ketone*, National Library of Medicine, Bethesda, MD, (CD-ROM), MICROMEDIX, Englewood, CO, 1996, pp. 1–9.

510. H. Siegel and M. Eggersdorfer, Ketones. In B. Elvers, S. Hawkins, and G. Schulz, Eds., *Ullmann's Encyclopedia of Industrial Chemistry*, 5th ed., Vol. A15, VCH Publishers, New York, 1990, pp. 77–95.

511. R. H. James et al., Evaluation of analytical methods for the determination of POHC in combustion products. *Proc. 77th Annu. Meet., Air Pollut. Control Assoc.* San Francisco, CA, 1984, Paper 84-18.5 (1984), pp. 1–25.

512. B. K. Krotoszynski, G. M. Bruneau, and H. J. O'Neill, Measurement of chemical inhalation exposure in urban population in the presence of endogenous effluents. *J. Anal. Toxicol.* **3**, 225–234 (1979).

513. S. K. Kundu et al., Breath acetone analyzer: diagnostic tool to monitor dietary fat loss. *Clin. Chem. (Winston-Salem, N.C.)* **39**, 87–92 (1993).

514. E. Hannig, Chemische Konstitution und Pharmakologische Wirkung des 4-Propoxy-β-Piperidinopropiophenon-Hydrochlorid, Seiner Bausteine und Derivate. *Arznei.-Forsch.* **10**, 559–566 (1955).

515. J. N. Smith, R. H. Smithies, and R. T. Williams, Studies in detoxication. 59. The metabolism of alkylbenzene. The biological reduction of ketones derived from alkylbenzenes. *Biochem. J.* **57**, 74–76 (1954).

516. D. B. Prelusky, R. T. Coutts, and F. M. Pasutto, Stereospecific metabolic reduction of ketones. *J. Pharm. Sci.* **71**, 1390–1393 (1982).

517. R. T. Coutts, D. B. Prelusky, and G. R. Jones, The effects of cofactor and species differences on the *in vitro* metabolism of propiophenone and phenylacetone. *Can. J. Physiol. Pharmacol.* **59**, 195–201 (1981).

518. U.S. Environmental Protection Agency (USEPA), *Letter from Union Carbide Corporation to USEPA Regarding Information on Employees Who Have Been Exposed to Propiophenone*, EPA/OTS 86-8000065, USEPA, Washington, DC, 1986.

519. F. Jaisle, K. Krautstrunk and L. Dinter, Die Wehenauslosung Durch Lipide. *Geburtshilfe Frauenheilkd.* **29**, 541–553 (1969).

520. K. Verscheuren, *Handbook of Environmental Data on Organic Chemicals*, 2nd ed., Van Nostrand-Reinhold, New York, 1981.

521. A. A. Bukhalovskii and V. V. Shugaev, Toxicity and hygienic standardization of isophorone, dihydroisophorone and dimethylphenylcarbinol. *Prom-st. Sint. Kauch.* **2**, 4–5 (1976).

522. H. F. Smyth, Jr. et al., An exploration of joint toxic action: Twenty-seven industrial chemicals intubated in rats in all possible pairs. *Toxicol. Appl. Pharmacol.* **14**, 340–347 (1969).

523. R. Truhaut, H. Dutertre-Catella, and P. Nguyen, Study of the toxicity of an industrial solvent, isophorone. Irritating capacity with regard to the skin and mucous membranes. *Eur. J. Toxicol.* **5**, 31–37 (1972).

524. M. Potokar et al., Studies on the design of animal tests for the corrosiveness of industrial chemicals. *Food Cosmet. Toxicol.* **23**, 615–617 (1985).

525. H. F. Smyth, Jr. and J. Seaton, Acute response of guinea pigs and rats to inhalation of the vapors of isophorone. *J. Ind. Hyg. Toxicol.* **22**, 477–483 (1940).

526. National Toxicology Program (NTP), *Toxicology and Carcinogenicity Studies of Isophorone in F344/N Rats and B6C3F1 Mice*, NIH Publ. 86–2547, Washington, DC, 1986.

527. Rohm & Haas Company, *90-Day Subchronic Toxicity of Isophorone in the Dog*, USEPA/OTS Doc. 878212178, TSCA 8(d) Submission, Rohm & Haas Company, Philadelphia, PA, 1972.

528. H. Dutertre-Catella, Contribution to the analytical toxicological and bio-chemical study of isophorone. Thesis for Doctorate in Pharmacology, Université Rene Descartes, Paris, in *Joint Assessment of Commodity Chemicals* No. 10, Isophorone, ECETOC, Brussels, 1989.

529. R. Truhaut, H. Dutertre-Catella, and P. Nguyen, Metabolic study of an industrial solvent, isophorone, in the rabbit. *C. R. Hebd. Seances Acad. Sci., Ser. D* **271**, 1333–1336 (1970).

530. H. Dutertre-Catella et al., Metabolic transformations of the trimethyl-3,5,5-cyclohexene-2-one (isophorone). *Toxicol. Eur. Res.* 109–216 (1978).

531. R. Truhaut et al., Metabolic transformations of 3,5,5-trimethylcyclohexanone (dihydroisophorone). New metabolic pathway dismutation. *C. R. Hebd. Seances Acad. Sci., Ser. D* **276**, 2223–2228 (1973).

532. J. Strasson, Jr. et al., Renal protein droplet formation in male Fischer 344 rats after isophorone (IPH) treatment. *Toxicologist* **8**, 136 (abstr.) (1988).

533. J. Gandy et al., Effects of selected chemicals on the glutathione status in the male reproductive system of rats. *J. Toxicol. Environ. Health* **29**, 45–57 (1990).

534. Chemical Manufacturers Association, *Inhalation Teratology Study in Rats and Mice—Isophorone*, USEPA/OTS Doc. 408555049, TSCA Negotiated Testing Agreement Submission, Chemical Manufacturers Association, Arlington, VA, Ketone Program Panel, 1985.

535. J. R. Bucher, J. Huff and W. M. Kluwe, Toxicology and carcinogenesis studies of isophorone in F344 rats and B6C3F$_1$ mice. *Toxicology* **39**, 207–219 (1986).

536. J. R. Selden et al., Validation of a flow cytometric *in vitro* DNA repair (UDS) assay in rat hepatocytes. *Mutat. Res.* **35**, 147–167 (1994).

537. D. K. Gulati et al., Chromosome aberration and sister chromatid exchange test in Chinese hamster ovary cells *in vitro* III: Results with 27 chemicals. *Environ. Mol. Mutagen.* **13**, 133–193 (1989).

538. R. Thier et al., DNA binding study of isophorone in rats and mice. *Arch. Toxicol.* **64**, 684–685 (1990).

539. Hüls Corporation, *Test of the Skin Sensitizing Effect of Isophorone in the Guinea Pig*, Internal Rep. No. 1278, Hüls Corporation Marl, Germany (unpublished data), cited in *Joint Assessment of Commodity Chemicals*, No. 10, Isophorone, ECETOC, Brussels, 1989.

540. Exxon Chemical Americas, *Acute Inhalation (LC50) and Human Sensory Irritation Studies on Cyclopentanone, Isophorone, Diydroisophorone, Cyclohexanone, and Methyl Isobutyl Ketone*. USEPA/OTS Doc 86960000031, TSCA 8(d) Submission, Exxon Biomedical Science, East Millstone, NJ, 1995.

541. U.S. Environmental Protection Agency (USEPA), *Graphical Exposure Modeling System (GEMS). Fate of Atmospheric Pollutants*, USEPA, Washington, DC, (1986).

542. U.S. Environmental Protection Agency (USEPA), *Health Effects Assessment for Isophorone*, 600/8-88/044, NTIS PB88-179916, USEPA, Washington, DC, 1987.

543. M. A. Callahan, M. W. Slimak, and N. W. Gabel, *Water-related Environmental Fate of 129 Priority Pollutants*. 440/4-79-029B, USEPA, Washington, DC, 1979.

544. R. L. Swann et al., A rapid method for the estimation of the environmental parameters octanol/water partition coefficient, soil sorption constant, water to air ratio, and water solubility. *Res. Rev.* **85**, 17–28 (1983).

545. T. DiPaolo, Molecular connectivity in quantitative structure-activity relationship study of anesthetic and toxic activity of aliphatic hydrocarbons, ethers, and ketones. *J. Pharm. Sci.* **67**, 566–568 (1978).

546. J. L. O'Donoghue et al., Commercial-grade methyl heptyl ketone (5-methyl-2-octanone) neurotoxicity: Contribution of 5-nonanone. *Toxicol. Appl. Pharmacol.* **62**, 307–316 (1982).

547. M. A. Shifman et al., The neurotoxicity of 5-nonanone: Preliminary report. *Toxicol. Lett.* **8**, 283–288 (1981)

548. G. D. DiVincenzo et al., Possible role of metabolism in 5-nonanone neurotoxicity. *Neurotoxicology* **3**, 55–63 (1982).

549. W. A. McOmie and H. H. Anderson, Comparative toxicologic effects of some isobutyl carbinols and ketones. *Univ. Calif., Berkeley, Publ. Pharmacol.* **2**, 217 (1949).

550. C. P. Carpenter, U. C. Pozzani, and C. S. Weil, Toxicity and hazard of diisobutyl ketone vapors. *Arch. Ind. Hyg. Occup. Med.* **8**, 377 (1953).

551. D. E. Dodd et al., Hyalin droplet nephrosis in male Fischer-344 rats following inhalation of diisobutyl ketone. *Toxicol. Ind. Health* **3**, 443–457 (1987).

552. R. B. White et al., Magnetic resonance imaging (MRI), neurobehavioral testing, and toxic encephalopathy: Two cases. *Environ. Res.* **61**, 117–123. (1993).

553. S. Budavari et al., eds., Benzophenone. *The Merck Index*, 12th ed., Merck & Co., Rahway, NJ, 1996, p. 184.

554. H. Siegel and M. Eggersdorfer, Ketones. In *Ullmann's Encyclopedia of Industrial Chemistry*, Vol. A15, VCH Publishers, New York, 1990, pp. 77–95.

555. A. D. Little, Benzophenone, National Toxicology Program, CAS Number 119-61-9, http://ntp-server.niehs.nih.gov/ht...kground/ExecSumm/Benzophenone.html (1991).

556. M.-L. Henriks-Eckerman and L. Lasse Kanerva, Product analysis of acrylic resins compared to information given in material safety data sheets. *Contact Dermatitis* **36**, 164–165 (1997).

557. D. L. J. Opdyke, Monographs on fragrance raw materials. *Food Cosmet. Toxicol.* **11**, 855–876 (1973).

558. D. Helmig, J. Muller, and W. Klein, Volatile organic substances in a forest atmosphere. *Chemosphere* **19**, 1399–1412 (1989).

559. K. Jay and L. Stieglitz, Identification and quantification of volatile organic components in emissions of waste incineration plants. *Chemosphere* **30**, 1249–1260 (1995).

560. L. A. Nylander-French et al., A method for monitoring worker exposure to airborne multifunctional acrylates. *Appl. Occup. Environ.* **9**, 977–983 (1994).

561. J. Gilbert et al., Current research on food contact materials undertaken by the UK Ministry of Agriculture, Fisheries and Food. *Food Addit. Contam.* **11**, 231–240 (1994).

562. T. Shibamoto and K. Umano, Photochemical products of benzyl benzoate: Possible formation of skin allergens. *J. Toxicol.—Cutaneous Ocul. Toxicol.* **4**, 97–103 (1985).

563. P. Fernández et al., Bioassay-directed chemical analysis of genotoxic components in coastal sediments. *Environ. Sci. Technol.* **26**, 817–829 (1992).

564. M. Ahel, Infiltration of organic pollutants into groudwater: Field studies in the alluvial aquifer of the Sava River. *Bull. Environ. Contam. Toxicol.* **47**, 586–593 (1991).

565. S. R. Hutchins and C. H. Ward, A predictive laboratory study of trace organic contamination of groundwater: Preliminary results. *J. Hydrol.* **67**, 223–233 (1984).

566. P. B. Bedient et al., Ground-water transport from wastewater infiltration. *J. Environ. Eng.* **109**, 485–501 (1983).

567. S. R. Hutchins et al., Fate of trace organics during rapid infiltration of primary wastewater at Fort Devens, Massachusetts. *Water Res.* **18**, 1025–1036 (1984).

568. Occupational Safety and Health Administration (OSHA), *Benzophenone*, OSHA Chem. Sampling Inf., CAS No. 119-61-9, OSHA, Washington, DC. Available: *http://www.osha-slc.gov/ChemSamp_data/CH_220325.html*

569. J. T. Geary, Qualitative and quantitative analysis of UV initiators in cured acrylic coatings. *J. Coat. Technol* **49**, 25–28 (1977).

570. K. P. Kringstad, F. De Sousa, and M. Stomberg, Evaluation of lipophilic mutagens present in the spent chlorination liquor from pulp bleaching. *Environ. Sci. Technol.* **18**, 200–203 (1984).

571. A.-J. Bulterman et al., Selective and sensitive detection of organic contaminants in water samples by on-line trace enrichment-gas chromatography-mass spectrometry. *J. High Resolut. Chromatogr.* **16**, 397–403 (1993).

572. R. Franz et al., Study of functional barrier properties of multilayer recycled poly (ethylene terephthalate) bottles for soft drinks. *J. Agric. Food Chem.* **44**, 892–897 (1996).

573. M. Monteiro et al., A GC/MS method for determining UV stablizers in polyethyleneterephthalate bottles. *J. High Resolut. Chromatogr.* **21**, 317–320 (1998).

574. S. Goenechea, Spektrophotometrische Bestimmung von Benzophenon im Urin. *Fresenius' Z. Anal. Chem.* **300**, 48 (1980).

575. L. Caprino, G. Togna, and M. Mazzei, Toxicological studies of photosensitizer agents and photodegradable polyolefins. *Eur. J. Toxicol.* **9**, 99–103 (1976).

576. K. Dutta, M. Das, and T. Rahman, Toxicological impacts of benzophenone on the liver of guinea pig *(Cavia Porcellus). Bull. Environ. Contam. Toxicol.* **50**, 282–285 (1993).

577. G. A. Burdock, D. H. Pence, and R. A. Ford, Safety evaluation of benzophenone. *Food Chem. Toxicol.* **29**, 741–750 (1991).

578. G. B. Freeman et al., 13-week dosed feed toxicity study of benzophenone in F344 rats and B6C3F mice. *Toxicologist* **14**, 221 (1994).

579. D. Robinson, Studies in detoxication. 74. The metabolism of benzhydrol, benzophenone and p-hydrobenzophenone. *Biochem. J.* **68**, 584–586 (1958).

580. D. Robinson and R. T. Williams, The metabolism of benzophenone. *Biochem. J.* **66**, 46–47 (1957).

581. A. W. Stocklinski, O. B. Ware, and T. J. Oberstat, Benzophenone metabolism. I. Isolation of *p*-hydroxybenzophenone from rat urine. *Life Sci.* **26**, 365–369 (1980).

582. K. C. Leibman, Reduction of ketones in liver cytosol. *Xenobiotica* **1**, 97–104 (1971).

583. N. S. Kudryasheva, V. A. Kratasyuk, and P. I. Belobrov, Bioluminescent analysis. The action of toxicants: Physical-chemical regularities of the toxicants effects. *Anal. Lett.* **27**, 2931–2947 (1994).

584. R. L. Bronaugh et al., *In vivo* percutaneous absorption of fragrance ingredients in rhesus monkeys and humans. *Food Chem. Toxicol.* **28**, 369–373 (1990).

585. P. A. Tsonis and G. Eguchi, Abnormal limb regeneration without tumor production in adult newts directed by carcinogens, 20-methylcholanthrene and benzo (*a*) pyrene. *Dev. Growth Differ.* **24**, 183–190 (1982).

586. S. Jobling et al., A variety of environmentally persistent chemicals, including some phthalate plasticizers, are weakly estrogenic. *Environ. Health Perspect.* **103**, 582–587 (1995).

587. D. Q. Tran et al., Inhibition of progesterone receptor activity in yeast by synthetic chemicals. *Biochem. Biophys. Res. Commun.* **229**, 518–523 (1996).

588. F. Stenbäck, Local and systemic effects of commonly used cutaneous agents: Lifetime studies of 16 compounds in mice and rabbits. *Acta Pharmacol. Toxicol.* **41**, 417–431 (1977).

589. F. Stenbäck and P. Shubik, Lack of toxicity and carcinogenicity of some commonly used cutaneous agents. *Toxicol. Appl. Pharmacol.* **30**, 7–13 (1974).

590. K. Mortelmens et al., *Salmonella* mutagenicity tests: II. Results from the testing of 270 chemicals. *Environ. Mutagen.* **8**, 1–119 (1986).

591. E. A. Emmett, B. R. Taphorn, and J. R. Kominshky, Phototoxicity occurring during the manufacture of ultraviolet-cured ink. *Arch. Dermatol.* **113**, 770–775 (1977).

592. F. Bosca and M. A. Miranda, Photosensitizing drugs containing the benzophenone chromophore. *J. Photochem. Photobiol. B* **43**, 1–26 (1998).

593. M. Charlier and C. Helene, Photochemical reactions of aromatic ketones with nucleic acids and their components 1. Purine and pyrimidine bases and nucleosides. *Photochem. Photobiol.* **15**, 71–87 (1972).

594. M. Charleir and C. Helene, Photochemical reactions of aromatic ketones with nucleic acids and their components. III. Chain breakage and thymine dimerization in benzophenone photosensitiezed DNA. *Photochem. Photobiol.* **15**, 527–536 (1972).

595. D. Z. Markovic and L. K. Patterson, Benzophenone-sensitized lipid peroxidation in linoleate micelles. *Photochem. Photobiol.* **58**, 329–334 (1993).

596. B. Morin and J. Cadet, Type I benzophenone-mediated nucleophilic reaction of 5′-amino-2′,5′-dideoxyguanosine. A model system for the investigation of photosensitized formation of DNA-protein cross-links. *Chem. Res. Toxicol.* **8**, 792–799 (1995).

597. W. Adam et al., Inhibitory effect of ethyl oleate hydroperoxide and alcohol in photosensitized oxidative DNA damage. *J. Photochem. Photobiol. B* **34**, 51–58 (1996).

598. R. H. Bisby, S. A. Johnson, and A. W. Parker, Quenching of reactive oxidative species by probucol and comparison with other antioxidants. *Free Radical Biol. Med.* **20**, 411–420 (1996).

599. G. Klecak, H. Geleick, and J. R. Frey, Screening of fragrance materials for allergenicity in the guinea pig I. Comparison of four testing methods. *J. Soc. Cosmet. Chem.* **28**, 53–64 (1977).

600. E. Calas et al., Allergic contact dermatitis to a photopolymerizable resin used in printing. *Contact Dermatitis* **3**, 186–194 (1977).

601. C. J. Le Coz et al., Photocontact dermatitis from ketoprofen and tiaprofenic acid: Cross-reactivity study in 12 consecutive patients. *Contact Dermatitis* **38**, 245–252 (1998).

602. J. C. Mitchell et al., Results of standard patch tests with substances abandoned. *Contact Dermatitis* **8**, 336–337 (1982).

603. P. Granstrand, L. Nylander-French, and M. Holmstrom, Biomarkers of nasal inflammation in wood-surface coating industry workers. *Am. J. Ind. Med.* **33**, 392–399 (1998).

604. Hazardous Substances Data Bank (HSDB), *Benzophenone*, National Library of Medicine, Bethesda, MD, (CD-ROM), MICROMEDIX, Englewood, CO, 1993, pp. 1–9.

CHAPTER SEVENTY-SEVEN

Monohydric Alcohols—C_1 to C_6

C. Bevan, Ph.D., DABT

This chapter reviews both linear and branched monohydric aliphatic C_1 to C_6 alcohols. The C_7 to C_{20} monohydric alcohols are covered in Chapter 78.

Chemical and Physical Properties

The physical and chemical properties for the C_1 to C_6 monohydric alcohols are listed in Table 77.1. At ambient temperature, the vapor pressure decreases with increasing carbon number as shown in Table 77.1. The water solubility also decreases with an increasing carbon number. The National Fire Protection Association (NFPA) has prepared a rating system to assess the physical and chemical hazards of chemicals with respect to flammability, health, and reactivity (1, 2). The C_1 to C_6 monohydric alcohols are flammable, but not reactive.

Production and Use

The C_1 to C_6 alcohols represent an important class of industrial chemicals with a wide number of uses. Based on production volume, the monohydric alcohols represent the most important group of the alcohol family (3). Methanol was in the top 50 chemicals produced in the United States in 1995 with a production volume of 11.3 billion pounds (4). In general, the alcohols of commercial significance are produced synthetically, although some alcohols are made from natural products or by fermentation. The most important

Patty's Toxicology, Fifth Edition, Volume 6, Edited by Eula Bingham, Barbara Cohrssen, and Charles H. Powell.
ISBN 0-471-31939-2

Table 77.1. Chemical and Physical Properties of C_1 to C_6 Monohydric Alcohols

Compound	CAS #	Mol. formula	Mol. wt.	Boiling point (°C)	Melting point (°C)	Sp. gr.	Refractive index (20°C)	Vapor pressure (mmHg) (°C)	Maximum vapor concn. % (°C)	Flammability (%) Lower Limit	Flammability (%) Uper Limit	Solubility in water (%) (°C)	1 ppm =mg/m^3	1 mg/L =ppm
Methanol	*[67-56-1]*	CH_4O	32.0	65	– 97.8	0.792	1.3285	160 (30)	21.05 (30)	6.0	36.5	Misc.[a]	1.31	764
Ethanol	*[64-17-5]*	C_2H_6O	46.1	79	– 114.1	0.789	1.3614	50 (25)	6.58 (25)	3.3	19.0	Misc.[a]	1.88	532
1-Propanol	*[71-23-8]*	C_3H_8O	60.1	97	– 126.2	0.804	1.3850	21 (25)	2.7 (25)	2.0	13.0	Misc.[a]	2.45	408
Isopropanol	*[67-63-0]*	C_3H_8O	60.1	83	– 88.5	0.785	1.3777	44 (25)	5.8 (25)	2.5	11.8	Misc.[a]	2.45	408
1-Butanol	*[71-36-3]*	$C_4H_{10}O$	74.1	118	– 90.0	0.810	1.3991	6.5 (25)	0.86 (25)	1.4	11.2	7.3 (25)	3.03	330
Isobutanol	*[78-83-1]*	$C_4H_{10}O$	74.1	108	– 108.0	0.803	1.3959	12.2 (25)	1.61 (25)	1.7	10.9	9.8 (20)	3.03	330
2-Butanol	*[78-92-2]*	$C_4H_{10}O$	74.1	100	– 111.7	0.807	1.3972	23.9 (30)	31.4 (30)	1.7	9.8	12.5 (20)	3.03	330
tert-Butyl alcohol	*[75-65-0]*	$C_4H_{10}O$	74.1	82	25.6	0.787	1.3841	42.0 (25)	5.53 (25)	2.4	8.0	Misc.[a]	3.03	330
1-Pentanol	*[71-41-0]*	$C_5H_{12}O$	88.2	138	– 79.0	0.815	1.4100	10 (44.9)	0.77 (37)	1.2	9.0	2.7 (22)	3.60	278
2-Pentanol	*[6032-29-7]*	$C_5H_{12}O$	88.2	119	—	0.809	1.4053	—	—	1.5	9.7	16.6 (20)	3.60	278
3-Pentanol	*[584-02-1]*	$C_5H_{12}O$	88.2	116	– 75.0	0.822	1.4098	2 (20)	—	1.2	9.0	55 (30)	3.60	278
2-Methyl-1-butanol	*[137-32-6]*	$C_5H_{12}O$	88.2	128	< – 70.0	0.816	1.4098	3.4 (25)	—	1.4	9.0	2.2	3.60	278
3-Methyl-1-butanol	*[123-51-3]*	$C_5H_{12}O$	88.2	131	– 117.2	0.812	1.4078	2.8 (20)	1.02 (35)	1.2	9.0	2.4	3.60	278
tert-Amyl alcohol	*[75-85-4]*	$C_5H_{12}O$	88.2	102	– 11.9	0.809	1.4052	10 (17.2)	—	1.2	9.0	8.0 (25)	3.60	278
3-Methyl-2-butanol	*[598-75-4]*	$C_5H_{12}O$	88.2	113	—	0.819	1.4095	—	—	1.2	9.0	7.8 (30)	3.60	278
2,2-Dimethyl-1-propanol	*[75-84-3]*	$C_5H_{12}O$	88.2	114	52.0	0.812	—	—	—	1.5	9.1	—	3.60	278
1-Hexanol	*[111-27-3]*	$C_6H_{14}O$	102.2	157	– 51.6	0.814	1.4178	0.98 (20)	0.32 (37)	1.2	7.7	0.58	4.25	239
2-Hexanol	*[626-93-7]*	$C_6H_{14}O$	102.2	—	—	—	—	—	—	—	—	—	4.25	239
2-Methyl-1-pentanol	*[105-30-6]*	$C_6H_{14}O$	102.2	148	—	0.825	1.4190	1.5 (20)	—	—	—	0.31	4.25	239

3-Methyl-1-pentanol	[*589-35-5*]	$C_6H_{14}O$	102.2	153	—	0.822	1.4182	—	—	—	—	—	4.25	239
4-Methyl-1-pentanol	[*626-89-1*]	$C_6H_{14}O$	102.2	152	—	0.821	1.4134	—	—	1.4	9.0	1.7	4.25	239
Methyl Iso-butyl-carbinol	[*108-11-2*]	$C_6H_{14}O$	102.2	132	−90.0	0.807	1.4113	3.52 (20)	0.46 (20)	—	—	—	4.25	239
2-Ethyl-1-butanol	[*97-95-0*]	$C_6H_{14}O$	102.2	149	< −50.0	0.835	1.4224	1.8 (20)	0.22 (20)	1.9	8.0	0.43	4.25	239
2,2-Dimethyl-1-butanol	[*1185-33-7*]	$C_6H_{14}O$	102.2	—	—	—	—	—	—	—	—	—	4.25	239
3,3-Dimethyl-2-butanol	[*464-07-3*]	$C_6H_{14}O$	102.2	—	—	—	—	—	—	—	—	—	4.25	239

[a]Miscible at 25°C, 760 mmHg.

industrial processes are the methanol process and the oxo process. The oxo process can be used to produce alcohols in the C_3 to C_{20} range by using alkenes as starting materials (3).

The uses of alcohols are numerous and can vary depending on their chemical and physical properties. In general, alcohols are used as solvents, cosolvents, and chemical intermediates. Among the higher alcohols, those containing six carbons or more, the C_6 to C_{11} alcohols are used in the manufacture of plasticizers. Some of the alcohols exist as pure chemicals, but higher monohydric alcohols (C_6 to C_{18}) can also exist as complex isomeric mixtures (3, 5, 6). The major routes of industrial or occupational exposure to alcohols is by dermal contact and/or inhalation. The extent of the exposure pathways depends on the use of the chemical and physical properties of the alcohol.

Metabolism and Disposition

A comparative uptake study has been conducted using human abdominal skin *in vitro* for the C_1 to C_{10} linear alcohols (7). The rate of dermal uptake for the neat material decreases with increasing carbon number.

Williams wrote an early general review on the metabolism and disposition of alcohols (8). Primary alcohols are readily oxidized to the corresponding aldehydes, which are further converted to the corresponding acids. Secondary alcohols are converted to ketones. Alcohols can be conjugated either directly or as a metabolite with glucuronic acid, sulfuric acid, or glycine, and excreted. Tertiary alcohols are more resistant to metabolism and are generally conjugated more readily than secondary or primary alcohols. Biomonitoring methods exist for several of the alcohols, including methanol, ethanol, isopropanol, and *n*-butanol (9). It is important to note that there are endogenous sources for some low molecular weight alcohols that must be considered in evaluating biomonitoring data for those alcohols.

Health Effects in Animals

The acute toxicity data for the monohydric alcohols indicate a low order of acute toxicity by oral, dermal, or inhalation routes of exposure. The C_1 to C_6 monohydric alcohols produce central nervous system depression. Based on the inhalation data, the secondary and tertiary alcohols are more biologically active than the primary alcohols. The C_1 to C_6 linear alcohols have been studied for their aspiration hazard (10). Except for methanol, the linear alcohols are aspiration hazards. Although not studied, the branched isomeric alcohols may be aspiration hazards. Sensory irritation increases with increasing carbon number (11).

Nelson et al. (12) investigated the developmental toxicity of aliphatic alcohols (C_1 to C_{10}) administered by inhalation to rats. Several of the alcohols (methanol, 1-propanol, isopropanol, 1-butanol) produced developmental toxicity but at concentrations that are at least an order of magnitude higher than existing occupational exposure standards and generally in the presence of maternal toxicity. Developmental toxicity did not, as predicted, increase as the carbon chain length was increased from six to eight carbons, after which toxicity would expected to decrease, although sufficiently high vapor concentrations of the longer chain alcohols could not be generated to produce maternal or fetal toxicity. Furthermore, behavioral teratogenic effects were not observed at concentrations lower than those that produced embryo toxicity (e.g., malformations) as revealed by traditional assessment. Although sporadic deviations in behavioral and neurochemical end points were observed with the alcohols, no pattern of effect was seen with any of the alcohols examined. The C_1 to C_6 monohydric alcohols are typically inactive in genotoxicity assays, but on occasion, some weak activity has been noted for methanol and ethanol.

Health Effects in Humans

This review discusses primarily dermal and inhalation routes of exposure, which are the major routes of occupational exposure to alcohols. Many of the high-production alcohols such as methanol, ethanol, propanols, and butanols cause adverse effects when ingested; ingestion, however, is not a major route of occupational exposure. There are some alcohols that have produced adverse effects in humans, including death, in an occupational environment. Nevertheless, alcohols have been used extensively in the workplace generally with few or minor problems. Occasionally, methanol, ethanol, and the propanols produce a skin sensitizing response in humans. In some, but not all, cases the sensitization response was considered to be due to contaminants and not to the alcohol itself.

A common property of some of the alcohols is to produce local irritation to the skin, eyes, and respiratory tract, and the effect or potency varies for the type of alcohol. Many alcohols produce minimal or no adverse effects in humans, possibly because of low exposure combined with the low toxicity potential of the alcohol.

Few alcohols produce neuropathic effects in humans. Abuse of products containing methanol and ethanol has produced some indications of neurotoxicity in humans, but nothing has been reported in an occupational environment. 2-Hexanol produces neurotoxicity by the oral and intraperitoneal routes in animals, but there is no evidence of such an effect having occurred in the workplace.

There is no clear evidence that occupational exposures to alcohols represent a carcinogenic risk to humans. Based on epidemiological data, there is an association between the manufacture of ethanol and isopropanol by the strong acid process (a process no longer used in the United States) and an excess of upper respiratory tract cancer in humans (13). The effect has been attributed to by-products such as dialkyl sulfates (14) and sulfuric acid (15), not the alcohols themselves.

Some of these alcohols such as ethanol and isopropanol can enhance the toxic effects of various chemicals, particularly hepatotoxins. It is thought that the effects may be due largely to an inductive effect of the alcohol on microsomal enzymes, particularly the cytochrome P450 system, which may allow a greater metabolic conversion of the hepatotoxin to its toxic metabolite.

A C_1 TO C_3 ALCOHOLS

The low molecular weight alcohols, including methanol, ethanol, 1-propanol, and isopropanol are used extensively in industry (16). These alcohols exist as volatile liquids at ambient temperatures, and exposure can occur in both industrial and nonindustrial environments.

The best studied monohydric alcohol is ethanol, although an extensive database exists for methanol and isopropanol. These alcohols have a low order of acute toxicity in animals, and the principal effects from inhalation exposure are local irritation and central nervous system depression. Inhalation studies in animals indicate a high no-observed-adverse-effect-level for developmental effects (12).

1.0 Methanol

1.0.1 CAS Number: *[67-56-1]*

1.0.2 Synonyms: Methyl alcohol, carbinol, wood spirits, wood alcohol, methylol, wood, columbian spirits, colonial spirit, columbian spirit, methyl hydroxide, monohydroxymethane, pyroxylic spirit, wood naphtha

1.0.3 Trade Names: NA

1.0.4 Molecular Weight: 32.0

1.0.5 Molecular Formula: CH_4O

1.0.6 Molecular Structure:

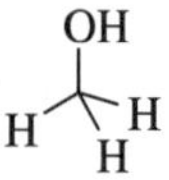

1.1 Chemical and Physical Properties

1.1.1 General

Methanol is a clear, water-white liquid with a mild odor at ambient temperatures. The physical and chemical properties are listed in Table 77.1.

1.1.2 Odor and Warning Properties

The air odor threshold for methanol has been reported as 100 ppm (17). Others have reported that 2000 ppm or 5900 ppm methanol is barely detectable (18).

1.2 Production and Use

Modern industrial-scale methanol production is based exclusively on synthesis from pressurized mixtures of hydrogen, carbon monoxide, and carbon dioxide gases in the presence of catalysts. Based on production volume, methanol has become one of the largest commodity chemicals produced in the world (16).

Methanol is largely used to produce formaldehyde, acetic acid, and methyl *tert*-butyl ether or MTBE (16). MTBE, obtained by the reaction of methanol and isobutylene, has quickly become a high-volume chemical because of its use as an oxygenated fuel additive (16). Methanol itself is also being used as an oxygenated fuel additive, as well as an alternate transportation fuel. Methanol is also used as a chemical intermediate, as an animal feed additive, as a solvent, and in various miscellaneous applications (16).

1.3 Exposure Assessment

1.3.3 Workplace Methods

NIOSH Method No. 2000 has been recommended for the analysis of methanol in air (19). This method involves drawing a known volume of air through silica gel to trap the organic vapors present (the recommended sample is 1 to 5 liters at a rate of 0.02 to 0.2 L/min). The analyte is desorbed with water. The sample is separated by injection into a gas chromatograph equipped with a flame ionization detector and quantified. The working range is 25 to 900 mg/m^3 (19 to 690 ppm) for a 5-liter sample. Other methods of collection and analysis by gas chromatography have been published (20, 21).

1.3.5 Biomonitoring/Biomarkers

A number of gas chromatographic methods for determining methanol in urine have been described and validated in field studies: head space analysis (22) and direct injection (23). Methanol in blood can be measured by gas chromatography with a detection limit of 0.6 mg/L (24) and 2 mg/L (25). A gas chromatographic method has also been developed to measure the metabolite, formic acid, in the urine. Formic acid is methylated and converted to *N,N*-dimethylformamide and analyzed with a capillary gas chromatograph using a flame ionization detector (26).

Biological exposure indexes have been proposed for methanol exposure because monitoring levels in the air may underestimate the absorbed dose. Attention has focused on various approaches, including methanol in the expired air (22, 27, 28), methanol in the blood (24, 28–30), methanol in urine (22–24, 26, 28, 29), formic acid in blood (27), and formic acid in urine (24, 26, 27, 29, 31). It is important to remember that methanol and formic acid can be found in the human body when there has been no occupational exposure.

1.4 Toxic Effects

1.4.1 Experimental Studies

Methanol exhibits a low order of acute toxicity by the oral, dermal, and inhalation routes of exposure, as measured by lethality. Sublethal doses, however, produce central nervous system effects and ocular injury that may result in blindness. This effect has been seen in primates, but not in rodents, and has been attributed to the differences in blood levels of the metabolite, formic acid. Methanol is somewhat irritating to the eyes. The vapor does not produce sensory irritation unless exposures are extremely high ($>$25,000 ppm). Prolonged exposure by inhalation to methanol has not resulted in any systemic toxicity in either rodents or monkeys. There is no evidence that methanol is carcinogenic in animals. It is generally inactive in a variety of genotoxicity assays; a weak response, however, has been occasionally noted. Conflicting results have been obtained concerning the effect of methanol on testicular hormones in rats. Methanol causes developmental effects at very high exposure levels in both rats ($\geq$ 10,000 ppm) and mice ($\geq$ 5000 ppm), but it is not a developmental neurobehavioral toxicant. High concentrations of methanol can potentiate the toxicity of other chemicals, particularly hepatotoxins. This effect has been attributed to the induction of certain cytochrome P450 isozymes.

1.4.1.1 Acute Toxicity. Data on the acute oral toxicity of methanol to animals are given in Table 77.2 (32–36). The acute dermal LD_{50} for rabbits was reported to be 20 mL/kg (Carnegie Mellon Institute of Research, unpublished data). The acute inhalation toxicity of methanol is listed in Table 77.3 (37–39a).

Exposure of animals to methanol vapor may produce increased respiration rates, nervous system depression followed by excitation, irritation of the mucous membranes, ataxia, prostration, deep narcosis, convulsions, and weight loss. Lethal concentrations resulted in death from respiratory failure (18). The lethal dose for nonprimates was two to three times higher than for monkeys. Primates, but not other species, developed signs of early intoxication, then a day later, acidosis and eye toxicity, which preceded death. The other species developed an initial narcosis from which they either survived or died, but acidosis was not observed (36, 40). Aspiration of 0.2 mL of methanol by rats resulted in the death of only 1 of 10 animals (10).

Methanol is irritating to the eye and causes conjunctivitis, chemosis, some iritis, and corneal opacity (41). Methanol was reported to be a mild eye irritant (42). However, in another study, undiluted methanol caused moderate corneal opacity in three of six rabbits and conjunctival redness in all six rabbits (18). A 50% aqueous methanol solution caused minimal to no effects, and a 25% aqueous solution caused no effects (18). In a sensory irritation (Alarie) test, methanol produced sensory irritation in mice at very high vapor concentrations with a RD_{50} (concentration that decreased respiration rate by 50%) of 25,300 ppm (43) and 41,514 ppm (44).

1.4.1.2 Chronic and Subchronic Toxicity. Cynomolgus monkeys or Sprague–Dawley rats were exposed by inhalation to 0, 500, 2000, or 5000 ppm of methanol 6 h/day, 5 days/week for 4 weeks. No treatment-related effects in the nasal turbinates, trachea, lungs,

Table 77.2. Single-Dose Oral Toxicity Values for Methanol in Animals[a]

Species	LD_{50} values (g/kg)	Ref.
Rat	6.2	32
	9.1	32a
	12.9	33
	13.0	34
Rabbit	14.4	35
Monkey	2–3[b]	36
	7.0[b]	

[a]Taken from Rowe and McCollister (18).
[b]Minimal lethal dose.

Table 77.3. Results of Single Inhalation Exposures of Animals to Vapors of Methanol[a]

Animal	Concentration ppm	mg/L	Duration of exposure (h)	Signs of intoxication	Outcome	Ref.
Mouse	72,600	95.0	54	Narcosis	Died	37
	72,600	95.0	28	Narcosis	Died	37
	54,000	70.7	54	Narcosis	Died	37
	48,000	62.8	24	Narcosis	Survived	37
	10,000	13.1	230	Ataxia	Survived	37
	152,800	200.0	94 min	Narcosis	Overall	38
	101,600	133.0	91 min	Narcosis	mortality	38
	91,700	120.0	95 min	Narcosis	45%	38
	76,400	100.0	89 min	Narcosis		38
	61,100	80.0	134 min	Narcosis		38
	45,800	60.0	153 min	Narcosis		38
	30,600	40.0	190 min	Narcosis		38
Rat	60,000	78.5	2.5	Narcosis, convulsions		39
	31,600	41.4	18–20		Died	39
	22,500	29.5	8	Narcosis		39
	13,000	17.0	24	Prostration		39
	8,800	11.5	8	Lethargy		39
	4,800	6.3	8	None		39
	3,000	4.0	8	None		39
	50,000	65.4	1	Drowsiness	Survived	39a
Dog	37,000	48.4	8	Prostration, incoordination		39
	13,700	17.9	24	None		39
	2,000	2.6	24	None		39

[a]Taken from Rowe and McCollister (18).

esophagus, liver, eye, or optic nerve were observed. Increased discharges around the eyes and nose were observed and were attributed to upper respiratory tract irritation (45). Groups of four male rats were exposed by inhalation to 0, 200, 2000, or 10,000 ppm of methanol 6 h/day, 5 days/week for 1, 2, 4, or 6 weeks. Evaluation of a number of parameters including lung weights, surfactant levels, and enzyme activities did not reveal any adverse effects on the lung. No histopathological examinations were performed (46). Newborn stumptail macaques received aspartame in the formula daily for 9 months, starting between 17 and 42 days of age. Aspartame hydrolyzes in the gut to aspartate, phenylalanine, and methanol. The methanol accounts for 10% of the molecular weight of aspartame. The exposed animals ate 1, 2, or 2.5 to 2.7 g/kg/day aspartame, which is equivalent to 100, 200, and 250 to 270 mg methanol/kg/day. No adverse effects were noted; however, no histopathological examinations were performed (47).

In an oral study, Sprague–Dawley rats were dosed by oral gavage with 0, 100, 500, or 2500 mg/kg of methanol for 90 days. There were no differences between dosed animals and controls in body weight gain and food consumption. Brain weights were decreased in both sexes in the 2500-mg/kg dose group. Elevated serum glutamic pyruvate transaminase and alkaline phosphatase were noted in the 2500-mg/kg dose group, but there were no adverse treatment-related effects in the gross pathology and histopathological evaluation (48).

Rats given 1% of methanol in their drinking water daily for 6 months did not show values significantly different from those of untreated controls (49). In a Russian study, rats that received oral doses of 10, 100, or 500 mg/kg/day of methanol for 1 month showed liver changes characterized by focal proteinic degeneration of the hepatocytic cytoplasm, changes in the activity of some microsomal enzymes, and enlarged hepatic cells (18). In another Russian study, rabbits were exposed by inhalation to 61 mg/m^3 (46.6 ppm) of methanol for 6 months (duration of exposure not given). Electron microscopic examination of the retina showed ultrastructural changes in the photoreceptor cells and Mueller fibers (18). Rats exposed to 16.8 ppm of methanol vapor 4 h/day for 6 months and simultaneously administered orally 0.7 mg/kg of methanol daily showed changes in blood morphology, oxidation–reduction processes, and liver function. The effects of methanol given simultaneously by the two routes were additive (49a).

1.4.1.3 Pharmacokinetics, Metabolism, and Mechanisms. The metabolism and pharmacokinetics of methanol have been previously reviewed (50–52). The metabolism of methanol is important because the acute toxicity of methanol is linked to its metabolites, not to the alcohol itself.

Methanol is absorbed following inhalation or ingestion, and inhalation is the major route of absorption in the occupational environment. There is no agreement on the potential risk of dermal exposure to methanol. Methanol toxicity due to dermal exposure is controversial. Methanol is uniformly distributed according to the relative water content of the tissue (53). Pulmonary retention in humans after 8 h of inhalation exposure to methanol was 58% of the inhaled amount, regardless of the exposure level, duration of exposure, or pulmonary ventilation (22). The methanol blood concentrations in dogs exposed repeatedly to vapor concentrations of 450 or 500 ppm, 8 h/day, for 379 continuous days varied between 10 and 15 mg/100 mL at the end of an 8-h period, but occasional values as high as 52 mg/dL were obtained (54). Following oral administration of 4 g/kg of methanol, 100 to 700 mg/100 mL methanol was found in the blood of rats (54a).

The total uptake of methanol that penetrated guinea pig skin *in vitro* over a period of 19 h was on the order of 1% of the total dose (55). An absorption rate of 0.015 $mL/cm^2/h$ was obtained from dermal exposure of rabbits to methanol for 12 to 30 min. With an exposure of 60 min, the rate decreased to 0.01 $mL/cm^2/h$. The rate of absorption in the dog was similar to that in the rabbit (56).

The first step in the metabolism of methanol is oxidation to formaldehyde. In rats, a catalase–peroxidase system is primarily responsible for the initial step, whereas in humans and monkeys, alcohol dehydrogenase plays a major role. Despite the difference, the initial metabolic step proceeds at similar rates. The third possible pathway in methanol oxidation is hepatic microsomal oxidation involving the cytochrome P450 system.

Formaldehyde is converted to formic acid, which is converted to formate and a hydrogen ion. Conversion to formic acid is a two-step process, the second step is irreversible. In the first reaction, formaldehyde combines with reduced glutathione (GSH) to form *S*-formylglutathione. This is mediated by an NAD-dependent formaldehyde dehydrogenase. In the second reaction, thiolase catalyzes the hydrolysis of *S*-formylglutathione to form formic acid and GSH. A folate-dependent pathway in the liver is responsible for formate metabolism in both rats and primates. Formate first forms a complex with tetrahydrofolate (THF) that is sequentially converted to 10-formyl-THF (by formyl-THF synthetase) and then to CO_2 (by formyl-THF dehydrogenase). THF is derived from folic acid in the diet and is also regenerated in the folate pathway. Although the folate pathway metabolizes formate in both rats and monkeys, rats use the pathway more efficiently.

Following a 2-mg/kg oral dose of methanol to primates and rats, about 90% is metabolized (50). Even after a high dose of 1 g/kg of radiolabeled methanol to monkeys, 78% of the activity was recovered within 24 h as exhaled CO_2 (57). Following a 2-mL/kg oral dose of methanol to rabbits, 5.3% was excreted in the expired air and 7.8% in the urine (58). Urinary excretion of methanol following an oral dose of 2 g/kg to rats was 1.81% of the dose administered (59). Rabbits given a single oral dose of methanol (2 to 10 g/kg) excreted 0.1 to 1.1% of it as formate and 13 to 20% of it as methanol in the urine within 47 to 143 h (60). Dogs given 1 to 2 g/kg of methanol orally excreted 5 to 15% in the urine as formate and 5 to 8% as unchanged methanol (60). Dogs excrete more formate than rabbits after ingestion of methanol. The rabbit excreted 10% of the dose (2.38 g/kg) unchanged in the urine within 95 h but the urinary formate was only questionably higher than the normal amount. From two dogs given 1.7 and 1.97 g/kg, respectively, 6 and 8.7% were recovered in the urine as methanol, and 22.4 and 23.7% were found in the urine as formate within 100 h (60a, 60b).

The rate of eliminating of methanol from the blood is much lower than that of ethanol (52). The rate of methanol disappearance from the blood also depends on dose. Elimination follows first-order kinetics (52). The monkey and the rat have similar capabilities for eliminating methanol at low inhalation exposures (200 ppm) but not at higher exposure levels. In rats exposed by inhalation to 200, 1200, and 2000 ppm methanol for 6 h, the peak methanol concentrations in the blood were 3.3, 32.2, and 90.8 mg/mL, respectively, indicating a nonlinear response. The blood half-lives were 0.8, 2.3, and 2.4 h at 200, 1200, and 2000 ppm, respectively (61). In rhesus monkeys exposed to 200, 1200, and 2000 ppm of methanol for 6 h, the peak methanol concentration in the blood concentration was 7.7, 39.3, and 66.1 mg/mL, respectively, a directly proportional response. The primate half-lives after 200, 1200, and 2000 ppm were 2.4, 3.3, and 3.5 h, respectively (62). Exposures did not cause an elevation in blood formate levels that could be attributed to methanol metabolism.

Although formate metabolism follows the folate pathway in both species, rats use the pathway more efficiently than monkeys. Both formate clearance from the blood and its metabolism to CO_2 proceed about 2 to 2.5 times faster in rats than in primates. Following an intravenous infusion at doses less than 100 mg/kg, formate is eliminated from rats with a half-life of 12 min and from primates with a half-life of 31 min (63). Rats metabolize formic acid at about twice the rate of that seen in monkeys (64). This difference between rats and primates in eliminating formic acid is thought to be due to the differences in the hepatic THF levels in primates (and humans), which are only half those in rats (65). The activity of 10-formyl-THF dehydrogenase, the enzyme that catalyzes the final step of formic acid oxidation, markedly lower in the livers of monkeys and humans, is 20 to 25% of that in rat liver (66). It has been suggested that this may explain to some extent why rats are resistant to the ocular effects of methanol intoxication via the toxic metabolite formic acid, whereas primates (and humans) are sensitive.

Female Long–Evans rats were exposed by inhalation to 4500 ppm of methanol 6 h/day from gestational day 6 to birth, and dams and their pups were exposed until postnatal day 21. The average methanol concentrations were 0.35, 0.499, and 1.3 mg/mL for pregnant dams, nursing dams, and neonates, respectively (67). In pregnant female Long–Evans rats exposed to 2% methanol in drinking water on gestational day 17 and euthanized the next day (approximately 2 g/kg consumed), there was a rapid equilibration of methanol across placental and blood–brain barriers (68).

The pharmacokinetics of short-term, low level exposure to methanol was studied in cynomolgus monkeys. Exposure was to 0, 10, 45, 200, and 900 ppm of [^{14}C]methanol for 2 hours. Monkeys were then placed on a folate-deficient diet until folate concentrations consistent with moderate deficiency (29–107 ng/mL) developed in red blood cells and then reexposed to 900 ppm of methanol for 2 hours. The average end-of-exposure blood [^{14}C] methanol concentrations were 0.65, 3.0, 21, 106, and 211 μM for the 10, 45, 200, 900, and the reexposed folate-deficient 900 ppm exposure groups, respectively. The respective [^{14}C]folate blood levels were 0.07, 0.25, 2.3, 2.8, and 9.5 μM. The blood concentrations of [^{14}C]methanol-derived formate from all exposures were 10 to 1000 times lower than the endogenous blood formate concentration (0.1 to 0.2 mM) reported for monkeys (69).

A physiologically-based pharmacokinetic (PBPK) model of inhaled methanol has been developed (70). A PBPK model has also been developed to describe the disposition of methanol in pregnant rats and mice (71).

1.4.1.4 Reproductive and Developmental

Developmental Toxicity

Female Sprague–Dawley rats were exposed by inhalation to 0, 5000, and 10,000 ppm for 7 h/day during gestational days 1 to 19 and to 20,000 ppm 7 h/day during gestational days 7 to 15. At 20,000 ppm, methanol produced slight maternal toxicity (unsteady gait only). Fetal body weights were decreased in the 10,000- and 20,000-ppm groups with a high incidence of malformations. The malformations were predominantly extra or rudimentary cervical ribs and urinary or cardiovascular defects. Similar malformations were observed at 10,000 ppm, but the incidence was not statistically significant from controls. No maternal or developmental effects were noted at 5000 ppm. The blood methanol concentrations in female rats at the end of the first 7 h of exposure were 1000 mg/mL at 5000 ppm, 2240 mg/mL at 10,000 ppm, and 8650 mg/mL at 20,000 ppm (72).

Female CD-1 mice were exposed by inhalation to 0, 2000, 5000, and 15,000 ppm of methanol 7 h/day during gestational days 6 to 15. A food-deprived nonexposed group was added. Methanol exposure did not produce maternal toxicity, but all exposure groups gained less weight than fed or food-deprived controls. Most of the litters of dams exposed to 15,000 ppm were completely resorbed and 38% of fetuses that survived to day 17 had exencephaly. Exposure to 5000 ppm resulted in exencephaly in about one-third of the litters and in 5 to 10% of all fetuses. At 2000 ppm, 1 of 220 fetuses had exencephaly, and there was none in the controls. The maternal plasma methanol level in exposed pregnant female CD-1 mice was 2000 and 8000 mg/mL after the first 7 h of exposure at 5000 and 15,000 ppm, respectively (73).

Methanol inhaled 6 h/day by pregnant female CD-1 mice at exposures of 10,000 and 15,000 ppm produced neural tube defects and ocular lesions when exposure was between gestational days 7–9, limb anomalies only during gestational days 9–11, and cleft palate and hydronephrosis after exposure during either period. No effects occurred in either period at an exposure level of 5000 ppm (74).

In a single oral exposure, pregnant female Long–Evans rats were dosed by gavage with 0, 1.3, 2.6, and 5.2 mL/kg of methanol on gestational day 10. At 5.2 mL/kg, maternal weight loss was $>10\%$. A significant decrease in fetal body weight (11 to 21%) was noted. Both internal and external examination of the fetuses showed a dose-related increase in total anomalies of 0.6, 4.8, 6.7, and 17.3% at doses of 0, 1.3, 2.6, and 5.2 mL/kg of methanol, respectively. The dose-related anomalies involved undescended testes and the eye (75).

In whole-rat embryo culture, day 9 embryos were explanted and cultured in 0, 4000, 6000, 8000, 12,000, or 16,000 mg of methanol/mL serum for 24 h and then transferred to rat serum alone for 24 h. Methanol concentrations of 8000 mg/mL and higher resulted in a dose-related decrease in somite number and overall development. The 12,000-mg/mL concentration resulted in embryo lethality, as well as dysmorphogenesis, and 16,000 mg/mL was embryolethal. Head length and crown-rump length were not affected by methanol exposure below embryolethal concentrations (76).

Behavioral Developmental Toxicity

Pregnant Long–Evans rats were exposed to 2% methanol in drinking water on gestational days 15 to 17 or 17 to 19. The animals were allowed to deliver, and the postnatal behavior of the pups was assessed. Methanol consumption averaged 2.5 g/kg/day. Methanol had no effect on maternal weight gain, gestational duration, water consumption, litter size, birth weight, neonatal mortality, postnatal growth, or the time until eye opening. However, compared to nonexposed rats, prenatal exposure impaired suckling behavior, as well as homing behavior, as determined by increased time to return to the nest area (77). No significant behavioral effects were noted in the offspring of female rats exposed to 15,000 ppm of methanol for 7 h/day during gestational days 7 to 19. Maternal blood methanol levels, measured from samples taken within 15 minutes after removal from the exposure chamber, declined from about 3800 μg/mL to 3100 μg/mL on the 12th day of exposure (78).

In another study using a somewhat different study design, pregnant Long–Evans rats were exposed to 4500 ppm of methanol vapor for 6 h/day beginning on gestational day 6. Exposure continued for both dams and pups through postnatal day 21. There was no alteration of suckling or olfactory conditioned behavior, but the level of motor activity in the older neonates decreased. Methanol effects were also detected in the two operant behaviors studied in the adults, but the results were subtle (79).

Reproductive Toxicity

There have been conflicting reports on the effect of methanol on testicular hormones in rats. Six weeks exposure (6 h/day) to 200 ppm of methanol resulted in decreased serum testosterone levels (80). The elimination rate of testosterone from the blood was not changed, and there were no significant changes in luteinizing hormone (LH) or follicle-stimulating hormone (FSH) (80). Exposure to 200 ppm of methanol for 1 week did not affect testosterone synthesis (81). A 6-h exposure to methanol at 200 ppm decreased serum testosterone by 59% (81); however, neither serum nor interstitial fluid testosterone concentrations in rats exposed to 200, 5000, or 10,000 ppm of methanol for 6 h differed from those observed in control rats (82). Exposure to 200 ppm of methanol 8 h/day, 5 days/week for 6 weeks did not reduce serum testosterone, nor was testosterone synthesis affected (83). In rats exposed to 800 ppm of methanol 20 h/day, 7 days/week for 90 days, the testes-to-body weight ratio did not differ from control rats, nor were there any adverse effects on testicular morphology in either normal rats or folate-reduced methanol-sensitive rats (when they were 10 months old). However, a greater incidence of testicular degeneration was noticed in the 18-month-old, folate-reduced rats exposed for 90 days to 800 ppm of methanol, suggesting that methanol may have a potential to accelerate the age-related degeneration of the testes (83). In a follow-up study, folate reduced-rats were exposed to 200 or 800 ppm of methanol, 20 h/day for 18 months. There were no signs of altered spermatogenesis. Increased incidence of multiple small hyperplastic nodules or interstitial cell masses were noted in the testes of these folated-reduced mice compared to air controls and folate-sufficient rats. There was no difference in pituitary weights or histological appearances or in serum testosterone concentrations. LH concentration decreased in the 800-ppm rats compared to controls. Prolactin level increased in the 200-ppm animals but decreased in the 800-ppm animals (84).

In Long–Evans rats dosed by oral gavage with 0, 0.8, or 1.6 g/kg of methanol for 21 days, serum testosterone levels were unaffected, but serum LH was increased. Testis weights were decreased as were the number of morphologically normal caput and cauda epididymal sperm (85).

1.4.1.5 Carcinogenesis. Chronic inhalation was studied by the Japan New Energy Development Organization (86). Only summary reports of these studies with limited information are available. Groups of eight monkeys were exposed to 0, 10, 100, or 1000 ppm of methanol 22 h/day for up to 2.5 years. Slight changes in the liver and kidney at 1000 ppm and some pathological changes in the nervous system of all animals at 1000 ppm were observed but were considered transient in nature and probably reversible. Rats and mice were exposed to 0, 10, 100, or 1000 ppm of methanol 20 h/day for 18 months (mice) or 24 months (rats). There was no evidence of carcinogenicity at 1000 ppm. Lung papillary adenomas and adrenal pheochromocytomas were observed and reported as "biologically insignificant" (86).

In another study, dogs were exposed to 450 to 500 ppm of methanol 8 h/day for 379 consecutive days. No unusual behavior or loss of weight was observed. Ophthalmologic, hematologic, and clinical chemistry values were unaffected. There were no noteworthy gross or histopathological findings (54).

1.4.1.6 Genetic and Related Cellular Effects Studies. Methanol was not mutagenic to *Salmonella* strains TA97, TA98, TA100, TA1535, TA1537, and TA1538 in Ames tests with or without metabolic activation (87–90). Equivocal results were obtained with *Salmonella* strain TA102 in the presence of metabolic activation (88). Methanol was not mutagenic in a DNA-repair test using various strains of *Escherichia coli* WP2 (87) and in a forward mutation assay using *Schizosaccharomyces pombe* (91).

Methanol did not induce micronuclei in Chinese hamster lung V79 cells *in vitro* (92). Methanol was mutagenic in the mouse lymphoma assay, in a Basc test, or in *Drosophila*, sex-linked, recessive lethal mutation assay (90). Treatment of primary cultures of Syrian golden hamster embryo cells with methanol did not lead to cell transformation (93).

In mice exposed by inhalation to 800 or 4000 ppm of methanol for 5 days, there were no increased frequencies of micronuclei in blood cells; sister chromatid exchanges, chromosomal aberrations, or micronuclei in lung cells; or synaptosomal complex damage in spermatocytes (94). The frequency of micronucleated erythrocytes in normal or folate-deficient mice treated with four daily intraperitoneal injections of up to 2500 mg/kg of methanol (95) was not increased. No increase in micronuclei was observed in the bone marrow of mice exposed to methanol (90).

1.4.1.7 Other. In rats, inhalation exposure to 200, 2000, and 10,000 ppm of methanol 6 h/day for 5 days resulted in a dose-related increase in the cytochrome P450 system in the livers, lungs, and kidneys (96).

Acute exposure to methanol potentiates the hepatotoxicity of carbon tetrachloride. In one study, methanol was administered in a single oral 320-mg/kg dose to Sprague–Dawley rats 18 h before a single oral dose of 0.1, 0.5, or 1.0 mL/kg of carbon tetrachloride. Massive liver injury was observed, as determined by increased plasma alanine aminotransferase and aspartate aminotransferase levels and histopathological damage was observed (97). In another study, F344 rats were exposed by inhalation to 1000, 2500, 5000, or 10,000 ppm of methanol for 6 h and gavaged with 0.075 mL carbon tetrachloride/kg body weight on days 1, 2, 3, and 6 after methanol exposure. When carbon tetrachloride was administered the first day, methanol markedly increased carbon tetrachloride hepatotoxicity in a dose-related manner. Little or no potentiation was observed when carbon tetrachloride was administered after day 1 (98). This potentiation of carbon tetrachloride toxicity by methanol has been associated with cytochrome P450 2E1 induction (99).

1.4.2 Human Experience

1.4.2.1 General Information. Nearly all of the available information on methanol toxicity in humans is from acute exposure. Cases of intoxication following oral consumption are the most frequently reported and, for this reason, this route of intake has been examined more thoroughly than the other routes of exposure. The case reports of methanol toxicity have been reviewed by others (100, 101).

1.4.2.2 Clinical Cases

1.4.2.2.1 Acute Toxicity. The minimal lethal dose of methanol in humans has not been determined. It has been suggested that ingestion of about 1 g/kg can cause death if the patient is untreated and has not consumed ethanol (102). In 10 adults acutely intoxicated with methanol, a latency period before treatment exceeding 10 h and a blood formate level above 0.5 g/L was reported to be predictive of severe methanol poisoning (103).

Acute methanol toxicity in humans is characterized in a well-defined pattern. Depending on the amount consumed, a narcotic effect varying from mild to severe results from central nervous system depression. This is followed by an asymptomatic interval lasting 10 to 48 h, although 12 to 24 h is most common. The latent period is then followed by headaches, nausea, dizziness, and vomiting, followed by severe abdominal pain and difficult, periodic breathing (Kussmaul breathing). The individual also experiences blurred vision, photophobia, and pains in the eyes. Depending on the amount of methanol consumed, the individual susceptibility and the time at which treatment began, these visual disturbances may either recede or develop within a few days into visual impairment or total blindness. The appearance of these symptoms coincides with increasing acidosis caused by an accumulation of formic acid. In severe cases, this leads to coma and eventually death from respiratory failure.

From ophthalmoscopic and histological evidence, both the retina and the optic nerve have been postulated as the primary sites of the toxic lesion produced by formate in methanol poisoning. Permanent visual damage in methanol-poisoned humans is associated with exposures usually greater than 20 h to blood formate concentrations in excess of 7 mM or 322 mg/dL (104). The initiating effect in the ocular toxicity of formic acid is believed to involve the inhibition of cytochrome oxidase, which is located in the mitochondria and is involved in oxidative phosphorylation.

Reports of cases of poisoning due to methanol inhalation are relatively rare, although in one report there had been about 100 known cases of failing sight or death due to inhalation of "wood alcohol" (105). Usually exposure was in more confined spaces with poor ventilation. There was one case reported of a woman who died after inhaling 4000 to 13,000 ppm methanol over 12 h (106). In 23 workers who, partly due to unfavorable working conditions (blackout precautions during wartime), had been exposed to methanol concentrations up to about 8300 ppm, no symptoms were seen apart from one case of "temporary" blindness. Another case involved a painter who spent three days painting the engine room of a submarine with a methanol-containing varnish. No details were given of the varnish composition or the methanol concentration in the air. On the first day, the painter experienced dizziness; on the second day, he showed exuberance; and on the third day, increased irritability. He also complained of abdominal pain, insomnia, double vision, and ptosis (drooping eyelids). The patient's sight became severely restricted in the left eye, the pupils dilated, the optic disks pallid, and the retinal veins engorged. His gait became uncoordinated, and he suffered acidosis. Under treatment, the patient's condition improved gradually over three weeks, and his sight eventually returned to normal (100).

A study of the wooden heel industry in Massachusetts showed average methanol vapor concentrations ranging from 160 to 780 ppm, with no definite evidence of injury to the exposed workers (106). There was no mention of symptoms or complaints in spirit duplicating processes in which exposures ranged from 400 to 1000 ppm (106). In small, poorly ventilated rooms, headaches occurred among employees using methanol in duplicating machines. The air concentrations were calculated for two unventilated sizes of compartment and two types of Ditto fluid (one containing 5% and the other 45% methanol, though some fluids contained up to 100% methanol). Air concentrations from evaporation of 1 pint of fluid were as follows: 510 and 1000 ppm with 5% solution for 1000- and 512-ft^3 rooms, respectively; 4500 and 8870 ppm with 45% methanol solution, respectively. The human tolerance values for exposure for methanol were then determined. For single exposures, the estimated tolerance value was 1000 ppm for 1 h, 500 ppm for 8 h, 200 ppm for 24 h, and 200 ppm for 40 h, based on five 8-h working days (101).

NIOSH evaluated the health hazard to determine if vapors from duplicating fluid (99% methanol) used in direct-process spirit duplicating machines were causing adverse health effects among teacher's aides. A self-administered symptom questionnaire was distributed to current teacher's aides (exposed group) and a comparison group of teachers. The aides reported significantly more blurred vision, headaches, dizziness, and nausea than the comparison group. The concentration of airborne methanol ranged from 365 to 3080 ppm; 15 of 21 measurements exceeded 800 ppm (107).

There have been reports of diminished vision or blindness from skin application of methanol or preparations that contain methanol for the relief of pain, for removal of varnish from skin, as well as following accidental exposure (101). In some of these cases, it is hard to isolate the skin exposure from possible inhalation effects.

Direct contact with methanol can be mildly irritating to the skin. There have been occasional reports of allergic contact dermatitis from dermal exposure to methanol (108–110).

1.4.2.2.2 Pharmacokinetics, Metabolism, and Mechanisms. For a general review of the metabolism of methanol, see section 1.4.1.3. When methanol was administered orally to three human volunteers at doses of 2.4, 4.0, or 5.6 g, elimination of methanol from the blood followed first-order kinetics with a half-time of about 3 h (111). Urinary methanol levels decreased exponentially with a half-life of about 2.5 to 3 h in four volunteers exposed by inhalation to 102, 205, or 300 mg/m^3 (78, 157, or 229 ppm) for 8 h (22).

Low-level inhalation exposures to methanol cause small increases in blood and urine formate levels. A study was conducted of 20 workers in a printing office who were exposed to an estimated methanol concentration between 111 and 174 mg/m^3 (85 and 133 ppm) throughout the workday. During the day, the blood level of formate increased an average of 4.7 mg/L (3.2 mg/L before the work shift to 7.9 mg/L when work ended), and urinary formate increased an average of 7.1 mg/L. A control group maintained relatively stable levels throughout the day of 5.3 mg/L of blood and 11.8 mg/L of urine (27). In another study, 20 workers were exposed throughout the day to 120 mg/m^3 (91.5 ppm) of methanol. At the end of the day, blood and urine levels of methanol were 8.9 and 21.8 mg/L, respectively; a control group had a mean blood and urine level of < 0.6 and 1.1 mg/L, respectively. Urinary formic acid was significantly higher in the workers (29.9 mg/L) than in the controls (12.7 mg/L) (24).

In an inhalation study, six human volunteers were exposed to 200 ppm methanol for 6 h. At the end of the exposure, the blood methanol level was increased from a mean of 1.8 mg/mL to 7.0 mg/mL. With light exercise, the total amount of methanol inhaled during the exposure period was 1.8 times higher that inhaled at rest; however, no statistically significant increase in blood methanol was observed. Formate did not accumulate in the blood above the background level in subjects at rest or during exercise (112).

The dermal uptake rate of liquid methanol applied to the forearm of human volunteers was 11.5 $mg/cm^2/h$ (28). An absorption rate of 0.145 $mg/cm^2/min$ was observed after application to the forearm for 15 min, increasing to 0.22 and then decreasing to 0.185 $mg/cm^2/min$ after 35 and 60 min, respectively (113). When 15 mL of methanol was applied to the forearm skin of human volunteers, there was some evidence of uptake based on increased blood and urine methanol levels (114). The dermal flux for methanol in human skin (epidermis) *in vitro* is 8.29 $mg/cm^2/h$ (7).

Background levels of methanol and formic acid in the body are derived mainly from dietary and metabolic processes. A mean methanol blood level of 0.73 mg/L in 31 unexposed subjects (range 0.32 to 2.61 mg/L) was reported (22); in another study, 26 unexposed workers averaged 1.1 mg/L, with a range of $<$ 0.6 to 2.9 mg/L (24). A mean of 0.25 mg/L methanol in expired breath of nine "normal" people (range 0.06 to 0.45 mg/L) was reported (115). Formic acid is present in the blood at background levels that range from 3 to 19 mg/L (27, 50). Urine concentrations of methanol of up to 5 mg/L have been measured after ingestion of alcoholic beverages by subjects not otherwise exposed to methanol (29).

1.5 Standards, Regulations, or Guidelines of Exposure

The ACGIH TLV and the OSHA PEL are 200 ppm (260 mg/m^3) as an 8-h TWA and a short-term exposure limit (STEL) at 250 ppm (116,117). Both ACGIH and OSHA warn against skin contact. Many other countries have set their occupational standards at 200 ppm (118).

2.0 Ethanol

2.0.1 CAS Number: *[64-17-5]*

2.0.2 Synonyms: Ethyl alcohol, grain alcohol, methyl carbinol, ethyl hydrate, anhydrol, alcohol, denatured alcohol, ethyl hydroxide, ethanol, algrain, cologne spirit, fermentation alcohol, molasses alcohol, potato alcohol, spirit, spirits of wine, Tecsol, alcohol dehydrated, ethanol 200 proof, ethanol absolute

2.0.3 Trade Names: NA

2.0.4 Molecular Weight: 46.1

2.0.5 Molecular Formula: C_2H_6O

2.0.6 Molecular Structure: 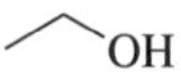

2.1 Chemical and Physical Properties

2.1.1 General

Ethanol is a colorless, flammable, volatile liquid. Its physical and chemical properties are listed in Table 77.1.

2.1.2 Odor and Warning Properties

An odor threshold of 84 ppm for ethanol has been reported (17).

2.2 Production and Use

Ethanol is produced synthetically from ethylene, as a by-product of certain industrial operations, or by the fermentation of sugar, cellulose, or starch (119).

There are two main processes for synthesizing ethanol from ethylene. The earliest method developed was the indirect hydration (strong-acid) process. The other synthetic process is called the direct hydration process and was designed to eliminate the use of sulfuric acid. Since the early 1970s, this has completely supplanted the old sulfuric acid process (119). The current production of ethanol from ethylene far exceeds its production from fermentation. Anhydrous ethanol is manufactured on a large scale by azeotropic distillation (18).

Industrial ethanol is one of the largest volume organic chemicals used in industrial and consumer products. The main uses are as an intermediate in the production of other chemicals and as a solvent. As a solvent, ethanol is second only to water. Ethanol is a key raw material in the manufacture of drugs, plastics, lacquers, polishes, plasticizers, perfumes, and cosmetics (119). In recent years, ethanol's role as a solvent has increased sharply. Its use as an intermediate has lost ground as other materials have replaced ethanol for making acetaldehyde, butyraldehyde, and ethylhexanol (119). Ethanol was used in the United States to a considerable extent as an antifreeze, but it has largely been replaced for such use by ethylene glycol (119). Ethanol is also used for preparations such as aerosol sprays and mouthwashes, cleaning fluids, liquid detergents, and as an automotive fuel additive (119). A number of denaturants can be added to ethanol to make it unpalatable as a beverage (119).

Because of the diverse of uses of ethanol, there is considerable opportunity for exposure, particularly by inhalation of the vapors. Workers are also likely to be exposed from dermal contact. Because there are strict regulations governing the accounting of its use, oral ingestion is unlikely, except from alcoholic beverages.

2.3 Exposure Assessment

2.3.3 Workplace Methods

NIOSH Method No. 1400 has been recommended for analyzing ethanol in air (19). It involves drawing a known volume of air through coconut shell charcoal to trap the organic vapors present (recommended sample is 0.1 to 1 liter at a rate of 0.05 l/min). The analyte is desorbed with carbon disulfide containing 1% of 2-butanol. The sample is separated and quantified by a gas chromatograph with a flame ionization detector. NIOSH Method No. 8002 is also available to measure ethanol in blood by gas chromatography with flame ionization detection (19). Other gas chromatographic methods have been recommended to measure ethanol in air (20, 21).

2.3.5 *Biomonitoring/Biomarkers*

A gas chromatographic method (headspace analysis) with flame ionization detection is available to measure ethanol in blood. The detection limit is 0.05 mg/mL (120).

2.4 Toxic Effects

2.4.1 *Experimental Studies*

Ethanol exhibits a low order of acute toxicity by oral, dermal, and inhalation exposure. At very high vapor concentrations, it is irritating to the eyes, nose, and respiratory tract. Hence the effects of inhalation are not likely to be serious under good hygienic conditions of handling and use. Ethanol is irritating to the eyes of rabbits, but it is not appreciably irritating to the skin. Repeated exposure to high doses of ethanol causes liver injury. There is no clear evidence that ethanol is carcinogenic to laboratory animals; it is, however, a tumor promoter. Ethanol is typically inactive in genotoxicity assays, but on some occasions, a weak response has been noted. Studies in rats and mice have shown the effects of ethanol on the testis and on other reproductive tissues at high oral doses, but generally there have been no effects on reproductive performance. Oral exposure to ethanol produces malformations and developmental toxicity in rats and mice at maternally toxic doses. No developmental effects were observed in rats from inhalation at doses up to 20,000 ppm. Exposure to ethanol by inhalation does not cause developmental neurobehavioral effects in rats. Ethanol can also affect the metabolism of a variety of toxicants, particularly hepatotoxins. This effect has been attributed to the induction of certain cytochrome P450 isozymes.

2.4.1.1 Acute Toxicity. The acute oral LD_{50} values are listed in Table 77.4 (32, 34, 35, 120a, 120b) and the effects of inhalation of ethanol at various concentrations are given in Table 77.5 (39, 122, 123). Aspiration of 0.2 mL of 100% ethanol caused deaths in 5 out of 10 rats tested, whereas 0.2 mL of 70% ethanol in water caused death in only 1 out of 10 rats (10).

Ethanol is not appreciably irritating to the skin of rabbits (41, 124, 125). Ethanol (95%) produced irritation to the eyes of rabbits (124). In a sensory irritation (Alarie) test, ethanol produced sensory irritation in mice, with RD_{50} values of 13,675 ppm (43) and 27,314 ppm (44).

2.4.1.2 Chronic and Subchronic Toxicity. The effects on renal function were studied in dogs given a single or 7-day repeated 3-g/kg oral dose of ethanol. Glomerular filtration rate was elevated, and tubular reabsorption of sodium and potassium excretion increased (126).

Rats given 3.26 M ethanol in drinking water for 12 weeks (equivalent to a dose of 10.2 g/kg/day by the end of the study) exhibited a decrease in weight gain and fatty livers (127). When rats were dosed by oral gavage with 8 to 15 g/kg/day over 4 months and fed a diet containing 25% of total calories as fat, focal necrosis, inflammation, and fibrosis were observed in the liver. Serum levels of transaminases were also elevated. Alcoholic hepatitis and Mallory bodies were not detected (128). When ethanol was given to rats continuously by oral gavage at doses that resulted in blood concentrations of 2160 mg/L, together with a

Table 77.4. Single-dose Oral Toxicity Values for Ethanol in Animals[a]

Species	LD_{50} values (g/kg)	Ref.
Rat	13.7	34
	17.8[b]	32
	6.2[c]	32
	11.5[d]	32
Mouse	9.5	120a
	8.3	120b
Guinea pig	5.6	34
Rabbit	9.9	35
Rabbit	9.9[e]	121
	7.0[e]	121a
Dog	5.5–6.6[f]	120a

[a]Taken from Rowe and McCollister (18).
[b]Young adults.
[c]14 days old.
[d]Older adults.
[e]Minimal lethal dose.
[f]Lethal dose.

low-fat diet (4.9% of total calories), progressive fatty infiltration of the liver was detected. After 30 days, one-third of the animals also showed focal necrosis with infiltration of macrophages in the liver (129). Nine baboons fed ethanol at 50% of total calories developed fatty liver, and four animals developed hepatitis within 9 to 12 months (130).

Rabbits exposed to saturated vapors of ethanol for periods ranging from 25 to 365 days developed cirrhosis of the liver (130a). Guinea pigs exposed to 3000 ppm of ethanol 4 h/day, 6 days/week for 10.5 weeks showed no adverse effects (131). Rats, guinea pigs, rabbits, squirrel monkeys, and beagle dogs were exposed continuously to 86 mg/m^3 (46 ppm) of ethanol for 90 days with no effects on behavior, mortality, hematologic parameters, or gross pathology (132).

2.4.1.3 Pharmacokinetics, Metabolism, and Mechanisms. The absorption, distribution, excretion, and metabolism of ethanol have been reviewed extensively (133, 134).

Ethanol is absorbed by the oral, dermal, and inhalation routes of administration. Most of the absorption data are on the oral route, and little information is available on the dermal and inhalation routes. Ethanol is absorbed from the gastrointestinal tract by simple diffusion. The rate of absorption is decreased by delayed gastric emptying and by the intestinal contents. It has been demonstrated in several animal species that food delays absorption, producing a slower rise and lower peak value of blood ethanol in fed than in fasting animals.

The total amount of ethanol that penetrated guinea pig skin *in vitro* during a period of 19 h was of the order of 1% of the total dose. There was no apparent increase in penetration with increasing dose volume, but penetration was significantly enhanced by occlusion (55). The concentration of alcohol in the blood of dogs that breathed an

Table 77.5. Results of Single Acute Inhalation Exposures of Animals to Vapors of Ethanol[a]

Animal	Concentration ppm	Concentration mg/L	Duration of exposure (h)	Signs of intoxication	Outcome	Ref.
Mouse	31,900	70.0	0.33	Ataxia		122
	29,300	55.0	7.0	Narcosis	Died	122
	23,900	45.0	1.25	Narcosis		122
	13,300	25.0	1.33	Ataxia		122
Guinea pig	45,000	84.6	3.75	Incoordination		39
	44,000	82.7	7.5	Deep narcosis		39
	50,170	94.3	10.2	Deep narcosis	Died	39
	19,260	36.2	3.75	None		39
	20,000	37.6	6.5	Incoordination		39
	21,900	41.2	9.8	Deep narcosis	Died	39
	9,080	17.1	5.25	None		39
	12,850	24.2	8.75	Incoordination		39
	13,300	25.0	24.0	Light narcosis		39
	6,400	12.0	8.0	None	Survived	39
Rat	32,000	60.1	8.0	—	Some died	123
	16,000	30.1	8.0	—	Some died	123
	45,000	84.6	3.75	Deep narcosis		39
	44,000	82.7	6.5	Deep narcosis	Died	39
	19,260	36.2	2.0	Light narcosis		39
	21,960	41.2	9.8	Deep narcosis	Died	39
	18,200	34.2	1.0	Excitation		39
	18,200	34.2	1.75	Incoordination		39
	22,800	42.9	8.0	Deep narcosis		39
	22,100	41.5	15.0	Deep narcosis	Died	39
	10,750	20.2	0.5	None		39
	10,750	20.2	2.0	Incoordination		39
	12,400	23.3	8.5	Deep narcosis		39
	12,700	23.8	21.75	Deep narcosis	Died	39
	5,660	10.6	1.75	Incoordination		39
	6,400	12.3	12.0	Light narcosis	Survived	39
	3,260	6.1	6.0	None		39
	3,260	6.1	8.0	Drowsiness		39
	4,580	8.6	21.13	Ataxia	Survived	39

[a]Taken from Rowe and McCollister (18).

undetermined concentration of ethanol rose from 0.8 g/kg of blood after 2 h to 4.0 g/kg of blood after 6 h (135).

Following absorption, ethanol distributes into total body water and easily penetrates the blood–brain barrier because it is both water- and lipid-soluble. Ethanol was reported present in the amniotic fluid of near-term pregnant rats following a single oral dose (136). When pregnant guinea pigs were given four 1-g/kg doses of ethanol by oral gavage in the third trimester, maternal and fetal blood ethanol concentration–time curves were superimposable, indicating that ethanol crosses the placenta bidirectionally (137).

Ethanol is oxidized in the liver to acetaldehyde by alcohol dehydrogenase within the cytosol. Acetaldehyde is then converted by aldehyde dehydrogenase to acetic acid mainly in the mitochondria. Acetic acid is then released into the blood and is oxidized by peripheral tissues to CO_2 and water. A cytochrome P450-dependent pathway oxidizes ethanol to acetaldehyde in microsomes (138). Chronic ethanol treatment increases *in vivo* ethanol metabolism. This metabolic tolerance to ethanol has been attributed either to enhanced mitochondrial reoxidation of NADH produced in the alcohol dehydrogenase pathway or to metabolism by the microsomal ethanol oxidizing system, perhaps by induction of cytochrome P450 enzymes (139).

Ethanol is eliminated from the body mainly by metabolism in the liver and only minimally by urinary excretion and pulmonary exhalation. Other tissues such as the kidney (140), stomach, and intestine oxidize ethanol to a small extent (141, 142). After a 2-ml/kg oral dose to rabbits, 0.8% of the dose was excreted in breath and 3.28% of the dose was excreted in urine (58). After a 2-g/kg oral dose to rats, 2.02% of the dose was excreted in urine (59). The total amount eliminated by dogs in the expired air following ingestion of 4 g/kg amounted to 4% of the total dose during an 8-h period (18).

After oral dosage, ethanol disappeared from the blood of dogs linearly in the postabsorption phase, irrespective of the concentration of ethanol in the body; zero-order kinetics was also found in dogs, cats, rabbits, pigeons, and chickens after intravenous administration (143). The rat metabolizes ethanol more slowly than the mouse and more rapidly than the dog (144). In rats following a 2-g/kg oral dose, blood levels peaked at 2 h and dropped rapidly. Urinary excretion peaked after 1.5 h (18). After an oral dose of 20% ethanol (amount not specified) to rabbits, the concentration in blood peaked at 1 h and then decreased. The concentration in the urine reached a maximum in 2 h, and the concentration in breath reached a maximum in 1 h (18). In female ferrets, a 6-mL/kg oral dose resulted in a peak blood alcohol level of 202 mg% in 1.5 h and 50 mg% after 12 h. Metabolism of ethanol in this range was not a first-order process (145).

Inhalation pharmacokinetics of ethanol was characterized in male and female $B6C3F_1$ mice and F344 rats. During exposure to 600 ppm ethanol for 6 h, steady-state blood ethanol concentrations (BEC) were reached within 30 min in rats and within 5 min in mice. The maximum BEC ranged from 71 μM in rats to 105 μM in mice. Exposure to 200 ppm of ethanol for 30 min resulted in peak BEC of approximately 15 μM in rats and approximately 25 μM in mice. Peak BEC of about 10 μM were measured following exposure to 50 ppm in female rats and male mice, and blood ethanol was undetectable in male rats. No sex-dependent differences in peak BEC at any exposure level were observed (146).

A physiologically-based pharmacokinetic (PBPK) model was developed for ethanol ingestion in the mouse (147). Subsequently, Pastino et al. (146), using inhalation pharmacokinetic data from $B6C3F_1$ mice and F344 rats, extended the PBPK model to account for the inhalation of ethanol in mice, rats, and humans. This PBPK model accurately similated BEC in rats and mice at all exposure levels used in the pharmacokinetic studies, as well as BEC reported in human males in all previously reported studies. Simulated peak BEC in human males following exposure to 50 to 600 ppm ranged from 7 to 23 μM and 86 and 293 μM, respectively.

2.4.1.4 Reproductive and Developmental Toxicity

Developmental Toxicity

Pregnant female Sprague–Dawley rats were exposed by inhalation to 0, 10,000, 16,000, and 20,000 ppm 7 h/day during gestational days 1 to 19. Dams in the 20,000-ppm group were narcotized by the end of the exposure, and maternal weight gain and food consumption were decreased during the first week of exposure. At 16,000 ppm, dams had a slight but statistically significant decreased weight gain during the first week of exposure. There were no treatment-related effects on uterine implantation parameters. No developmental adverse effects were noted at any dose level (72).

In a drinking water study, pregnant CF-1 mice, Sprague–Dawley rats, and New Zealand white rabbits were given 15% ethanol in their drinking water on gestational days 6 to 15 (rats and mice) and 6 to 18 (rabbits). Maternal toxicity in the form of decreased liquid intake, maternal body weight, and food consumption occurred in all species. Fetal body weights significantly decreased in rats and mice, but not in rabbits. There was a significant increase in the incidence of minor skeletal variants among the litters of rats and mice, but not in rabbits. No teratogenic effects were observed (148).

CBA and C3H female mice were maintained on liquid diets containing 15 to 35% ethanol-derived calories for at least 30 days before and during gestation until day 18. A dose-related increase in the number of resorptions and a decrease in fetal body weight were noted. Deficient occiput ossification and neural anomalies were observed in the low ethanol diets, and cardiac and eye-lid dysmorphology were noted in the higher ethanol diets (149). Male Long–Evans rats were given 20% ethanol in drinking water for 60 days and mated with three untreated females once per week for three consecutive weeks. Implantation sites, live birth rates, total litter weights, and pup weights decreased, whereas fetal deaths increased. Malformations were noted in 55% of the offspring of treated rats compared to 12% in the controls. Absolute and relative testes weights decreased significantly in the treated males compared to controls (150).

Ethanol, and not acetaldehyde, has been implicated as the causative agent of the teratogenic effects in laboratory animals. Oral coadministration of 100 mg/kg of 4-methylpyrazole, an inhibitor of alcohol dehydrogenase, with 6 g/kg of ethanol intraperitoneally on gestation day 10 dramatically increased the embryotoxicity of ethanol in CD-1 mice. Five 200-mg/kg intraperitoneal treatments of acetaldehyde at 2-h intervals on gestational day 10 did not significantly increase the percentage of resorptions and malformed fetuses or decrease fetal body weight, whereas these effects as well as malformations were increased with ethanol. This suggests that ethanol is the proximate teratogen (151). Both ethanol and acetaldehyde were accessible to the embryo on gestational day 10 (152).

Ethanol was added to the culture medium bathing rat embryos at 200 to 800 mg% for 48 h (0 to 24 somite stage), 600 to 800 mg% for 24 h, or 6 h during the organogenesis. Exposure to 600 and 800 mg% ethanol throughout the entire 48-h culture period or the first 24-h period (0 to 12 somites) produced marked growth retardation, particularly of the head region in a dose-dependent manner, but did not prevent neural tube closure. Exposure to high levels of ethanol during specific 6-h periods of early organogenesis (three to nine somites) prevented closure of the neural tube in 30% of the cultured embryos (153). A 24-h exposure to 300, 450, 600, and 800 mg/dL of ethanol produced a dose-dependent increase in neural tube defects and concomitant growth retardation in mouse whole embryo culture. However, when the exposure period *in vitro* was reduced to mimic the half-life *in vivo*, no adverse effects on embryonic growth or development were produced (154). Rat embryos exposed to 300 to 900 mg/dL of ethanol on gestational day 9.5 or 10 showed a dose-related growth retardation associated with a high frequency of malformations. When day 10 embryos were exposed to 300 mg/dL, no effects were observed, but 900 mg/dL produced moderate growth retardation and morphological defects (155).

Behavioral Developmental Toxicity
Groups of 18 male and 15 female Sprague–Dawley rats were exposed 7 h/day for 6 weeks or throughout gestation, respectively, to 0, 10,000, or 16,000 ppm by inhalation and then mated with untreated rats. Litters were culled to four males and four females and were fostered within 16 h after birth to untreated dams that had delivered their litters within 48 h previously. No adverse effects were observed in either the exposed males or females. Offspring from paternally or maternally exposed animals performed as well as controls on days 10 to 90 in tests of neuromotor coordination, activity levels, and learning ability. Norepinephrine levels in the brain were altered in paternally exposed offspring, and 5-hydroxytryptamine (cerebellum) and Met-enkephalin levels were altered in both maternally and paternally exposed rats (156, 157).

Reproductive Toxicity
A few studies in mice and rats have shown effects on the testis and on other reproductive tissues but generally have not shown an effect in reproductive performance (133). No significant effects were observed in reproductive function or pup development in female C57Bl/Crgl mice given 10% ethanol (v/v) in drinking water before mating and during gestation and lactation (128). When female Wistar rats were given 20 to 25% of the calories consumed as 12% ethanol in a sucrose solution as drinking fluid before mating, during gestation, and during lactation, there was no effect on reproductive performance or on development of the offspring (158). Male C57Bl/6J male mice were given 5 or 6% ethanol in a liquid diet for 70 or 35 days, respectively. There was a significant decrease in testicular weight and in seminal vesicle/prostate weight, an increase in the frequency of germ-cell desquamation, inactive seminiferous tubules, and a significant decrease in the total number of motile sperm (159). Male Sprague–Dawley rats maintained on a liquid diet containing 6% ethanol for 1 week followed by 4 weeks on a 10% ethanol liquid diet showed adverse effects on the testes, seminal vesicles, and ductules, as well as a significant decrease in serum testosterone levels (160). In a continuous breeding protocol, CD-1 mice were exposed to 0, 5, 10, or 15% ethanol in water. Other than a slight decrease in the number of live pups at the high dose, there were no fertility effects. The second generation mating resulted only in decreased adjusted pup weight (161).

2.4.1.5 Carcinogenesis. In 1987, the International Agency for Research on Cancer (IARC) evaluated the cancer data on ethanol and alcoholic beverages in humans and animals (133). The IARC concluded that there was inadequate evidence for the carcinogenicity of ethanol and of alcoholic beverages in experimental animals, but there was sufficient evidence for the carcinogenicity of alcoholic beverages in humans. The IARC classified alcoholic beverages as a Group 1 carcinogen based on the occurrence of malignant tumors of the oral cavity, pharynx, larynx, esophagus, and liver that have been causally related to the consumption of alcoholic beverages.

Ethanol has been studied in rats, mice, and hamsters for carcinogenicity using various protocols; the IARC extensively reviewed these studies and found them inadequate (133). A few of these studies will be summarized. A group of 108 male and female CF1 mice was given 43% ethanol in drinking water 5 days/week for up to 1020 days. A group of 44 male CF1 mice was given 14% ethanol similarly for up to 735 days. Two mice given 43% ethanol developed papillomas of the forestomach. A few other tumors, four malignant lymphomas, and three lung adenomas were also found in the high-dose group. A further group of 100 male ddN mice was given 19.5% ethanol in drinking water intermittently for 664 days; one mouse developed a papilloma of the stomach (162). A group of 15 female C3H/St mice (20 to 35 days old) received 12% ethanol (v/v) in drinking water for 80 weeks. Mammary tumors developed between 6 and 11 months of age in 8/11 (73%) mice in the ethanol-treated group and between 12 and 16 months of age in 22/27 (82%) control mice. The tumor incidence in ethanol-treated mice was not statistically different from that in the controls, but the shorter median time to tumor appearance in the ethanol-treated mice (8 months versus 14.2 months of age) was statistically significant (163). Groups of 40 Sprague–Dawley rats were given 0.5 mL of 30 or 50% ethanol (v/v) daily by oral gavage. The average life span of rats treated with 30% ethanol was 500 days, and that of the group given 50% ethanol was 396 days. No esophageal, stomach, or hepatic tumors were found in the treated groups (164).

Male and female Sprague–Dawley rats received 0, 1 or 3% ethanol or an equicaloric amount of glucose in a semisynthetic diet. There was no increased incidence in tumors in the male rats. For females, it was concluded that there was also no increased incidence in tumors, although different frequency comparisons resulted in conflicting results in the incident rate for mammary tumors in the 1% dose group. Liver and bile duct injury was observed in the male rats. Inflammatory reactions were seen in the pancreas of males and in the clitoral gland of females. There was also hyperplasia in the thyroid gland in both sexes and in the adrenal glands in the females. Peripheral nerve degeneration was reported for the low-dose males and in the high-dose females (165).

A number of studies have been conducted where ethanol was used as part of a study of modifying effects (133). In one study, 25 male and female BDVI rats were given 40% ethanol (amount unspecified) by oral gavage twice weekly for 78 weeks. The average life span for treated animals was 98 weeks (males) and 105 weeks (females) and for control animals 107 and 113 weeks, respectively. No tumors that could be related to ethanol treatment were observed in these animals (166). In another study, groups of 48 Sprague–Dawley rats were given 25% ethanol as the drinking fluid five times a week until their natural deaths. Man survival times were 780 and 730 days in the treated and control groups, respectively. No statistically significant increase in tumor incidence was observed (167).

Experimental data in animals indicate that ethanol consumption can enhance the carcinogenic activity of a broad spectrum of organ-specific carcinogens in several animal species. This cocarcinogenic effect may be influenced, however, by the dose and timing of the ethanol exposure (168).

Rats were given a single intraperitoneal dose of diethylnitrosamine followed by treatment with ethanol in the drinking water for 12 to 18 months. Ethanol was an effective promoter of liver tumors (169, 170).

2.4.1.6 Genetic and Related Cellular Effects Studies. The genotoxicity of ethanol has been extensively reviewed (133, 171). Ethanol is not mutagenic in *Salmonella typhimurium* strains TA97, TA98, TA100, TA1535, TA1537, or TA1538 (87–89, 172–174) in the presence or absence of metabolic activation. In the presence of a metabolic activation system, ethanol is slightly mutagenic to *Salmonella* strain TA102, a strain considered to respond to the presence of oxygen radicals (87, 88). In DNA repair tests with different strains of *E. coli* WP2, ethanol has been reported to give both very weak positive results (87) as well as negative results (175). In *Aspergillus nidulans*, 5% ethanol produced nondisjunction, mitotic crossing-over, and chromosomal male segregation (176, 177), but was inactive in *Neurospora crassa* (178).

Ethanol did not induce mutations in mouse lymphoma L5178Y TK+/– cells (179) and did not induce micronuclei in Chinese hamster V79 cells in the absence of metabolic activation (92). No chromosomal aberrations or sister chromatid exchanges were observed in Chinese hamster ovary cells treated with ethanol (180). However, in one study a slight increase in sister chromatid frequency was reported that became slightly higher in the presence of a metabolic activation system (181). Treatment of HeLa cells with ethanol did not lead to an elevation of exchange-type aberrations or micronuclei in the absence of a metabolic activation system (180). Treatment of primary cultures of Syrian golden hamster embryo cells with ethanol did not lead to cell transformation (93).

Oral or subcutaneous administration of ethanol to male ddY mice or given to male Swiss mice in drinking water up to 40% for 26 days did not increase the frequency of micronuclei in the bone marrow (182, 183). However, an elevation in sister chromatid exchanges was noted in male CBA mice given 10 and 20% ethanol as their only liquid supply for up to 16 weeks (184), but oral administration of 10% ethanol to male Swiss Webster mice once daily for 4 days did not (185). Feeding CD rats for 6 weeks with a liquid diet containing 36% of the calories as ethanol led to an elevation in micronuclei in the bone marrow (186), but male Wistar rats that received 10 and 20% ethanol as the only liquid supply for 3 or 6 weeks did not show an increase in the frequency of micronuclei in bone marrow or hepatocytes (187). Furthermore, the treatment did not lead to chromosomal aberrations in bone marrow cells or in lymphocytes; although the frequencies of sister chromatid exchanges were elevated in the peripheral lymphocytes but not in bone marrow cells (187).

Administration of up to 30% ethanol as the only liquid supply to male Wistar rats for 35 days did not lead to dominant lethal mutations (188). Feeding male Sprague–Dawley rats a liquid diet containing 6% ethanol for 1 week, followed by 10% ethanol for 4 weeks (189), or administering 20% ethanol in water as the drinking water fluid to male Long–Evans rats for 60 days (150) led to the induction of dominant lethal mutations. Dominant lethal mutations have been reported in male CBA mice after intubation on three consecutive days with 0.1 mL of 40 to 60% ethanol(190).

2.4.1.7 Other. Ethanol had no effect in the mouse ear sensitization assay, which is used to determine the sensitizing potential of compounds in Balb/c mice (191).

Pretreatment of animals with ethanol enhances the metabolism of a variety of agents, including alcohols, ketones, halogenated alkanes, alkenes, ethers, aromatic compounds, aromatic amines, and nitrosoamines (133, 192, 193). Alcohol enhances the activity of many hepatotoxic agents (194). Long-term ethanol consumption results in proliferation of smooth hepatic endoplasmic reticulum and an increase in the level of cytochrome P450. The induction of cytochrome P450 by ethanol is associated with an increase in metabolism and, in some instances, in the toxicity of several compounds, including *N*-nitrosodimethylamine, benzene, and carbon tetrachloride.

Chronic ingestion of high doses of ethanol leads to nutritional imbalances in the body such as zinc deficiency and depressed vitamin A levels (195). Ethanol administration in animals depressed hepatic levels of vitamin A, even when administered with diets containing large amounts of the vitamin. The hepatic depletion of vitamin A was associated with decreased detoxification of chemical carcinogens, suggesting induced microsomal metabolism of ethanol (195). Zinc deficiency potentiated the teratogenic effects of ethanol in CBA mice (196, 197). Folate deficiency is frequently associated with alcoholism in humans (198, 199).

The effect of ethanol exposure on the immune and hematopoietic system of rats was studied (200). Exposure to 25 mg/L (approximately 13,300 ppm) of ethanol vapor for 14 days produced decreased cellularity in the spleen, thymus, and bone marrow. There was a significant decline in the number of erythroid progenitor cells but no effect on any other hematopoietic cell population. Splenic lymphocytes, although fewer in number, showed no significant difference in the ability to proliferate when stimulated by nonspecific mitogens. Mice exposed to 10 to 25 mg/L (5320 to 13,300 ppm) for 20 to 43 days developed thrombocytopenia but not anemia or macrocytosis (201).

2.4.2 Human Experience

2.4.2.1 General Information. Few adverse effects have been reported in humans from industrial exposure to ethanol by either inhalation or dermal contact. The effects of chronic alcoholism from the excessive use of alcoholic beverages are well known and are not extensively discussed in this chapter. It is important to note that ethanol ingestion from alcoholic beverages can affect the metabolism of other chemicals (133, 193, 202). The enhanced hepatotoxic effects of carbon tetrachloride in alcoholics have been known for more than 50 years (193).

2.4.2.2 Clinical Cases

2.4.2.2.1 Acute Toxicity. In human subjects exposed to 10 to 20 mg/L (5000 to 10,000 ppm) of ethanol, there was some coughing and smarting of the eyes and nose, and the symptoms disappeared within a few minutes. Although the atmosphere was not exactly comfortable, it was felt that there would be no difficulty in tolerating these levels. At 30 mg/L (about 15,000 ppm), there was continuous lacrimation and coughing. At levels higher than 40 mg/L (about 20,000 ppm), it was impossible to tolerate the atmosphere even for a short time. The subjective symptoms increased when the breathing rate was increased by two or three times the resting level (203).

In a nontolerant human subject, inhalation exposure to 1380 ppm ethanol for 39 min resulted in no effects at 28 min, but headaches and slight numbness after 33 min. At

3340 ppm for 100 min, sensations of warmth and coldness, nasal irritation, headaches, and numbness were reported. When exposed to 8840 ppm for 64 min, the subjects complained of momentary intolerable odor and difficulty in breathing, conjunctival and nasal irritation, a feeling of warmth, headache, drowsiness, and fatigue (39). In tolerant individuals, the symptoms are less severe, and the time required to produce them is greater than in nontolerant individuals. For instance, a human subject tolerant to alcohol reported slight headaches after 20 min exposure to 5030 ppm for 120 min (39). Intoxication has been seen among humans subjected to inhalation of vapors from hot alcohol (122).

The clinical features of ethanol intoxication in a nontolerant individual are related to blood alcohol levels: at 50 to 150 mg/dL (0.05 to 0.15%), there is mild intoxication: slight impairment of visual acuity, muscular incoordination, and reaction time; and mood, personality, and behavioral changes; at 150 to 300 mg/dL (0.15 to 0.30%), moderate intoxication occurs, resulting in visual impairment, sensory loss, muscular incoordination, slowed reaction time, and slurred speech; at 300 to 500 mg/dL (0.30 to 0.50%), there is severe intoxication characterized by marked muscular incoordination, blurred or double vision, sometimes stupor and hypothermia, vomiting and nausea, and occasional hypoglycemia and convulsions; and at $>$ 400 mg/dL ($>$ 0.40%), there are coma, depressed reflexes, respiratory depression, hypotension and hypothermia, and death from respiratory or circulatory failure or as the result of aspiration of stomach contents in the absence of a gag reflex (204, 205).

Human subjects reported no apparent skin irritation when ethanol was applied to the forearm of human subjects in a modified Draize test. No irritation was noted when ethanol was applied to the forearm openly for 21 days, whereas a 21-day occlusive test caused erythema and induration toward the end of the exposure period (125). There have been infrequent reports of skin sensitization reactions attributed to ethanol (110). Ethanol is a weak sensitizer in a patch test (110, 206).

2.4.2.2.3 Pharmacokinetics, Metabolism, and Mechanisms. In human subjects exposed to ethanol vapors from 11 to 20 mg/L (5000 to 10,000 ppm), the average absorption was reported to be 62% and was independent of concentration (203). Human subjects were exposed to 13 to 16 mg/L of ethanol vapors for 3 to 6 h (207). The highest blood level obtained in any of the subjects was 47 mg of alcohol/100 mL of blood after a 6-h exposure to 16 mg/L (8500 ppm) of alcohol at a high ventilation rate of 22 L/min. When the ventilation rate was reduced to 15 L/min with an alcohol concentration in the air of 15 mg/L (8000 ppm), there was definitely a plateau reached in the blood level of about 7 to 8 mg/100 mL of blood. The rate of alcohol uptake was equal to the rate of metabolism. It was pointed out by the authors that a 70-kg human exposed to 1000 ppm of alcohol would have to breathe at a rate of 117 L/min to obtain any continuous rise in alcohol concentration (207).

Few data exist on the dermal uptake of ethanol. The dermal flux for ethanol in human skin (epidermis) *in vitro* is 0.57 mg/cm^2/h (7).

In humans, 80% of an ingested dose of ethanol is absorbed in the small intestine, and the remainder is absorbed in the stomach. About 80 to 90% of the ethanol is absorbed within 30 to 60 min, but food may delay complete absorption for 4 to 6 h (204).

The rate of ethanol metabolism varies among individuals. Studies of twins indicate that interindividual variability in the rate of ethanol metabolism may be genetically controlled (208). The main pathway for ethanol oxidation in humans is to acetaldehyde via the alcohol dehydrogenase pathway. Acetaldehyde is oxidized further to acetic acid by aldehyde dehydrogenase. Disulfiram (Antabuse) is an inhibitor of aldehyde dehydrogenase and produces higher blood levels of acetaldehyde during alcohol consumption. Genetic differences have been identified in the metabolism of ethanol. Some Orientals are known to be more sensitive to the health effects of ethanol; the sensitivity has been attributed to different forms of the enzyme acetaldehyde dehydrogenase (209). Acetaldehyde concentrations during ethanol oxidation in the blood following ingestion of 0.8 g/kg were low in healthy male Caucasians at 0.1 to 1.0 mg/L in hepatic venous blood and $<$ 0.1 mg/L in peripheral venous blood (210). In contrast, alcohol ingestion by Orientals resulted in marked elevations of blood acetaldehyde levels ranging from 0.4 to 3 mg/L (211), and individuals developed facial flushing and tachycardia as a direct consequence of elevated blood acetaldehyde levels (211–213).

As reviewed by Ellenhorn and Barceloux (204), the elimination of ethanol from the blood follows zero-order kinetics except at very low or very high concentrations. First-order elimination occurs only for blood alcohol levels of less than 20 mg/dL or 0.02% (205). The kidney and lungs excrete only 5 to 10% of an absorbed dose unchanged. The maximum rate of metabolism is 100 to 125 mg/kg/h, although, tolerant individuals can increase their metabolic rates by enzymatic induction, to 175 mg/kg/h. The average adult metabolizes 7 to 10 g/h and reduces the ethanol level 15 to 20 mg/dL/h. Chronic alcoholics have metabolic rates as high as 30 to 40 mg/dL/h. Children have higher metabolic rates than adults, ranging up to 28 mg/dL/h (204).

2.4.2.2.4 Reproductive and Developmental. Prenatal exposure to ethanol (as alcoholic beverages) is associated with a distinct pattern of congenital malformations that have been collectively termed the fetal alcohol syndrome (214, 215). Among the characteristics of this syndrome are intrauterine and postnatal growth deficiency, a distinctive pattern of physical malformations, including microcephaly; shortened palpebral fissures; joint, limb, and cardiac anomalies; and behavioral/cognitive impairment such as fine motor dysfunction and mental retardation. Not all affected children have all of the features of the syndrome. This syndrome has been associated with alcoholic women who drank heavily and chronically during pregnancy. There have been no reports of fetal alcohol syndrome as a result of industrial exposure by the oral, dermal, or inhalation routes.

2.5 Standards, Regulations, or Guidelines of Exposure

The ACGIH TLV and the OSHA PEL are 1000 ppm as an 8-h TWA (116,117). The TLV recommendation was based on the lack of eye and upper respiratory tract irritation at levels below 5000 ppm and on widespread and long industrial hygiene experience with human exposures to ethanol. In many other countries, the occupational exposure standard is 1000 ppm (118).

3.0 1-Propanol

3.0.1 CAS Number: *[71-23-8]*

3.0.2 Synonyms: *n*-Propanol, *n*-propyl alcohol, propyl alcohol, 1-propanol, ethyl carbinol, propan-1-ol

3.0.3 Trade Names: NA

3.0.4 Molecular Weight: 60.1

3.0.5 Molecular Formula: C_3H_8O

3.0.6 Molecular Structure: 

3.1 Chemical and Physical Properties

3.1.1 General

1-Propanol is a clear, colorless liquid with a typical alcohol odor. Its physical and chemical properties are listed in Table 77.1.

3.1.2 Odor and Warning Properties

The odor threshold for 1-propanol is 2.6 ppm (17).

3.2 Production and Use

1-Propanol is produced commercially by the oxo process by reacting ethylene with carbon monoxide and hydrogen in the presence of a catalyst to give propionaldehyde, which is then hydrogenated (216). 1-Propanol is used mainly as a solvent and as a chemical intermediate (216). Its chief use historically has been as a specialty solvent in flexographic printing inks, particularly for printing on polyolefin and polyamide film. It is used in producing *n*-propyl acetate, in water-based inks, and as a ruminant feed supplement, and it reportedly improves the water tolerance of motor fuels (216).

3.3 Exposure Assessment

3.3.3 Workplace Methods

NIOSH Method 1401 has been recommended for analyzing 1-propanol in air (19). It involves drawing a known volume of air (recommended sample is 1 to 10 liters at a rate of 0.01 to 0.2 L/min) through coconut shell charcoal to trap the vapors present. The analyte is desorbed with carbon disulfide containing 1% 2-propanol. The sample is separated and quantified by gas chromatography with flame ionization detection. The working range is 50 to 900 mg/m^3 for a 10-liter sample.

3.3.5 Biomonitoring/Biomarkers

Other analytical methods for determining 1-propanol in air, water, and biological media have been summarized (217). These methods mostly use gas chromatography with flame ionization detection and are quite sensitive with limits of detection of 0.05 mg/m^3 in air, 0.1 mg/L in water, and 2 mg/L in serum or urine. Endogenous levels of 1-propanol can be measured in serum and urine using gas chromatography–mass fragmentography (218).

3.4 Toxic Effects

3.4.1 Experimental Studies

1-Propanol has a low order of acute toxicity. It is not very irritating to the skin, but is moderately irritating to the eyes of rabbits. 1-Propanol produces sensory irritation but only at very high concentrations. There are no adequate subchronic or carcinogenicity studies on 1-propanol, but it is not genotoxic. In rats, 1-propanol produces developmental toxicity but only at maternally toxic doses. No developmental behavioral effects were noted in the offspring of female rats exposed to 1-propanol. It may affect the male reproductive organ at very high vapor concentrations.

3.4.1.1 Acute Toxicity. The acute oral LD_{50} has been reported to be 1.87 g/kg (219), 5.4 g/kg (18), and 6.5 g/kg (220) in rats; 4.5 g/kg in mice (37); and 2.82 g/kg (35) in rabbits. The acute dermal LD_{50} is 4 g/kg (219) and 6.7 g/kg (18) in rabbits. No deaths occurred in rats exposed to saturated vapors for 2 h; however, two out of six rats died after a 4-h exposure at 4000 ppm (219). No mortality resulted from exposure of rats to 20,000 ppm for 1 h (18).

Exposures of mice to 13,120 ppm (32.2 mg/L) for 160 min or 19,680 ppm (48.2 mg/L) for 120 min were fatal (37). Groups of two mice were exposed to 3250, 4100, 8150, 12,250, 16,300, and 24,500 ppm for 480, 240, 135, 120, 90, and 60 min, respectively. The length of time required for the appearance of ataxia, prostration, and deep narcosis was inversely proportional to the concentration to which the mice were exposed. Ataxia appeared in 10 to 14 min at 24,500 ppm and in 90 to 120 min at 3250 ppm. Prostration was noted in 19 to 23 min at 24,500 ppm and in 165 to 180 min at 3250 ppm. Deep narcosis was manifest in 60 min at 24,500 ppm and in 240 min at 4100 ppm. Only 1 of 12 mice that showed these clinical signs died. Mice exposed for 480 min to 2050 ppm showed no adverse effects (18). Aspiration of 0.2 mL 1-propanol caused deaths in 9 out of 9 rats (10).

1-Propanol is very slightly irritating to the skin of rabbits (219), and it is moderately irritating to the eyes of rabbits (219). In a sensory irritation (Alarie) test, 1-propanol produced sensory irritation in mice, and the RD_{50} values were 4800 ppm (43), 12,704 ppm (44), and 22,080 ppm (221).

3.4.1.2 Chronic and Subchronic Toxicity. A group of six male rats was given 60.1 g/L of 1-propanol in drinking water for 4 months. There were no differences in body weight gain, food consumption, and liver pathology compared to the control animals (222). In another drinking water study, Wistar rats were exposed to 320 g/L (approximately 16 g/kg) of 1-propanol for 13 weeks. The exposed rats gradually became weak, lost their appetites, and showed decreased body weight gain. Electron microscopic studies of the liver showed irregularly shaped megamitochondria with few cristae and normally sized but irregularly shaped mitochondria with a decreased number of cristae (223).

3.4.1.3 Pharmacokinetics, Metabolism, and Mechanisms. 1-Propanol is readily absorbed orally. Oral administration of 3 g/kg of 1-propanol to Wistar rats resulted in a maximum blood concentration of 1860 mg/L of 1-propanol in 1.5 h (224).

1-Propanol is oxidized to propionaldehyde primarily by alcohol dehydrogenase, which is then oxidized to propionic acid (225). 1-Propanol is a better substrate for alcohol dehydrogenase than ethanol (226). It has also been shown *in vitro* that rat and rabbit liver cytochrome P450 enzymes can metabolize 1-propanol to propionaldehyde (227, 228). The relative affinity of the cytochrome P450 system for 1-propanol is about three times higher than for ethanol (227, 228). The isozymes responsible for metabolizing 1-propanol are inducible by ethanol (227).

1-Propanol, which is water-soluble, is rapidly distributed throughout the body (217). Rabbits dosed orally with 2 mL/kg excreted 1.65% in the expired air and 0.7% in urine (58). In rats dosed orally with 2 g/kg, 0.13% of the dose was excreted in urine (59). Following an 800-mg/kg oral dose to rabbits, only 0.9% was found in urine as propylglucuronide and none as a sulfate conjugate (229).

The elimination of 1-propanol was reported to be independent of dose above a single oral dose of 1 g/kg in rats and above a single intraperitoneal dose of 1.2 g/kg in rabbits (224–226). In rats that received a single 3 g/kg oral dose, the disappearance of 1-propanol from the blood followed zero-order kinetics at a rate of 510 mg/kg/h (224). At lower doses, the elimination rate followed first-order kinetics. The half-life of 1-propanol in the blood was 45 min in rats given a single 1-g/kg oral dose (226). Blood levels reached a maximum at 90 min and disappeared by 8 h after a 2-g/kg oral dose to rats (59). In pregnant female rats exposed to 3500, 7000, or 10,000 ppm of 1-propanol or isopropanol 7 h/day during gestational days 1 to 19, isopropanol blood levels were consistently much higher than those of 1-propanol (230).

3.4.1.4 Reproductive and Developmental

Developmental Toxicity

In an inhalation study, pregnant female Sprague–Dawley rats were exposed to 0, 3500, 7000, or 10,000 ppm of 1-propanol 7 h/day on days 1 to 19 of gestation. Food consumption was significantly reduced throughout gestation in the mid- and high-dose groups, but maternal body weight was affected only at the end of gestation in the high-dose group. Resorptions were increased at the high dose, and fetal body weights were significantly reduced in the mid- and high-dose groups. Significantly more litters had malformations following exposure to the mid and high doses; no effects were observed at 3500 ppm (230).

3.4.1.4.2 Behavioral Developmental Toxicity. Pregnant female Sprague–Dawley rats were exposed by inhalation to 0, 3500, or 7000 ppm of 1-propanol 7 h/day throughout gestation. Male rats were also exposed to the same concentrations for 6 weeks. Pregnant animals exposed to 7000 ppm showed reduced weight gain, and female offspring also had reduced weight gain through 3 weeks of age; the authors reported a low incidence of crooked tails, a minimal teratological effect, at 7000 ppm. Exposed males were mated with unexposed females, and fertility was reduced in males exposed to 7000 ppm of 1-

propanol (2 viable litters from 17 matings). There were no consistent effects seen in behavioral or neurochemical tests (231, 232). At 7000 ppm, the male rats exhibited reduced fertility (only 2 of 17 produced litters) compared to controls.

Reproductive Toxicity
In the developmental behavioral study mentioned previously (232), male Sprague–Dawley rats were exposed to 0, 3500, or 7000 ppm of 1-propanol for 6 weeks. Beginning on the third day after the last exposure, males were mated for a maximum of 5 days with unexposed females (232).

3.4.1.5 Carcinogenesis. Eighteen Wistar rats were dosed by oral gavage with 0.3 mL/kg twice weekly. The average survival time was 570 days. In addition to severe liver injury and hyperplasia of the hematopoietic parenchyma, five malignant tumors (two myeloid leukemias, two liver sarcomas, one liver cell carcinoma), and 10 benign tumors were observed. Three benign but no malignant tumors were found in the controls given saline (233). In the same study, 31 Wistar rats were given 0.06 mL/kg subcutaneous injections of 1-propanol twice weekly. The average survival time was 666 days. Fifteen malignant tumors were noted (five liver sarcomas, four myeloid leukemias, one local sarcoma, two spleen sarcomas, one kidney pelvis carcinoma, one bladder carcinoma, and one uterine carcinoma), and seven benign tumors were identified. Severe liver injury and marked hematopoietic effects were noted in the treated rats. Only two benign tumors were found in the controls given saline (233). Although there was an apparent increase in the incidence of liver sarcomas, the study is inadequate to assess carcinogenicity.

3.4.1.6 Genetic and Related Cellular Effects Studies. 1-Propanol was not mutagenic to *S. typhimurium* in an Ames test with or without metabolic activation (234, 235). 1-Propanol inactivated *E. coli* CA 274 in a concentration-dependent manner and increased the reversion rate (236). 1-Propanol did not increase the number of sister chromatid exchanges in Chinese ovary hamster cells (180) or in V79 Chinese hamster lung fibroblasts with or without metabolic activation (237). It did not induce micronuclei in V79 Chinese hamster lung fibroblasts (92).

3.4.1.7 Other. No sensitization was noted in an ear-swelling test in CF1 mice (238).

3.4.2 Human Experience

3.4.2.1 General Information. There have been no reports of adverse health effects in humans from 1-propanol exposure. There was one case, however, in which a positive patch test reaction was elicited in a person sensitive to isopropanol, 1-butanol, and 2-butanol (but not to methanol and ethanol), suggesting that cross-sensitization may have occurred (239).

3.4.2.2.3 Pharmacokinetics, Metabolism, and Mechanisms. When 10 human volunteers drank 1-propanol in ethanolic orange juice in 3.75-mg doses of 1-propanol and 1200 mg ethanol/kg body weight over a period of 2 h, 1-propanol was detected in the blood and in the urine, partly as glucuronide. The total urinary excretion of 1-propanol was 2.1% of the dose. The urinary levels of 1-propanol were lower when the amount of simultaneously ingested ethanol was less (240, 241).

The dermal flux for 1-propanol in human skin (epidermis) *in vitro* is 0.096 mg/cm^2/h (7).

3.5 Standards, Regulations, or Guidelines for Exposure

Both the ACGIH TLV and the OSHA PEL are 200 ppm. The ACGIH STEL is 250 ppm (116). Other countries have set identical exposure standards (118).

4.0 Isopropanol

4.0.1 CAS Number: [67-63-0]

4.0.2 Synonyms: Isopropyl alcohol, 2-propanol, sec-propyl alcohol, dimethyl carbinol, IPA, rubbing alcohol (70% isopropanol), propan-2-ol, i-propanol, 2-hydroxypropane, *n*-propan-2-ol

4.0.3 Trade Names: NA

4.0.4 Molecular Weight: 60.1

4.0.5 Molecular formula: C_3H_8O

4.0.6 Molecular Structure:

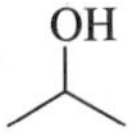

4.1 Chemical and Physical Properties

4.1.1 General

Isopropanol is a colorless, volatile liquid. Its physical and chemical properties are listed in Table 77.1.

4.1.2 Odor and Warning Properties

The air odor threshold for isopropanol has been reported as 22 ppm (17) and 40 ppm (18).

4.2 Production and Use

Three processes can be used to produce isopropanol: a strong-acid process, a weak-acid process, and a nonacid process (242). Historically, isopropanol was manufactured by the strong-acid process (or indirect hydration) process; however, the strong-acid process has been replaced by the weak-acid and the nonacid processes (Chemical Manufacturers Association Isopropanol Panel, unpublished data). Isopropanol is used as a solvent and is a component of numerous industrial and consumer products (242). Isopropanol is also used to produce acetone (242). The major potential route of occupational exposure is by inhalation.

4.3 Exposure Assessment

4.3.3 *Workplace Methods*

NIOSH Method No. 1400 has been recommended for analyzing isopropanol in air (19). It involves drawing a known volume of air through coconut shell charcoal to trap the vapors present (recommended sample is 0.2 to 3.0 liters at a rate of 0.01 to 0.2 L/min). The analyte is desorbed with carbon disulfide containing 1% of 2-butanol. The sample is separated and quantified by a gas chromatograph with a flame ionization detector. Other gas chromatographic methods have been recommended (20, 21).

4.3.5 *Biomonitoring/Biomarkers*

Analytic methods for isopropanol have been reviewed (243, 244). These methods are available for detecting isopropanol in various media, including air, water, blood, serum, and urine with detection limits of 0.02 mg/m^3, 0.04 mg/L, and 1 mg/L for air, water, and blood, respectively.

4.4 Toxic Effects

4.4.1 *Experimental Studies*

Isopropanol has a low order of acute and chronic toxicity. At high exposure levels, isopropanol is irritating to the eyes, nose, and throat and may cause transient central nervous system depression. It does not produce adverse effects on reproduction, and it is not a teratogen, a selective developmental toxicant, or a developmental neurotoxicant. Isopropanol is not genotoxic in a variety of tests, and it is not an animal carcinogen. It has produced effects to several rodent toxicity end points at high dose levels, such as motor activity, male mating index, and exacerbated renal disease.

4.4.1.1 Acute Toxicity. The acute oral LD_{50} has been reported as 4.7 g/kg (32), 5.3 g/kg (32), 5.5 g/kg (243), and 5.84 g/kg in rats (245); 4.5 g/kg in mice (243); and 5.03 g/kg (246), 7.8 g/kg (121), and 7.9 g/kg (35) in rabbits. The acute dermal LD_{50} in rabbits has been reported to be 12.9 g/kg (245). The acute inhalation 8-h LC_{50} in rats was 19,000 ppm in females and 22,500 ppm in males (157). Exposure of rats to 16,000 ppm for 8 h resulted in four deaths out of six animals (245). Deaths occurred in 6 of 10 rats following aspiration of 0.2 mL of 100% isopropanol and in 1 of 10 rats following aspiration of 70% isopropanol (10).

Isopropanol applied to the intact or abraded skin of rabbits and guinea pigs produced negligible irritation (247). Liquid isopropanol is moderately irritating to the eyes of rabbits (18, 243, 248). In a sensory irritation (Alarie) test, isopropanol produced sensory irritation in mice, and RD_{50} values were 5000 ppm (249) and 17,693 ppm (44).

In an acute study, F344 rats were exposed to 0, 500, 1500, 5000, or 10,000 ppm of isopropanol for 6 h. Neurobehavioral function was evaluated. A spectrum of behavioral effects indicative of narcosis, defined as a generalized loss of neuromotor and reflex function, was observed in animals of the 10,000-ppm group and to a lesser extent in the 5000-ppm animals. Recovery from these effects was observed by 24 h for the 10,000-ppm animals and by 6 h for the 5000-ppm animals. A concentration-dependent decrease in motor activity was observed for males exposed to ≥1500 ppm and for females exposed to ≥ 5000 ppm. The results showed that exposure of rats to isopropanol vapor produces transient, concentration-related narcosis and/or central nervous system sedation (250).

4.4.1.2 Chronic and Subchronic Toxicity. In an inhalation study, F-344 rats and CD-1 mice were exposed to 0, 100, 500, 1500, or 5000 ppm isopropanol 6 h/day, 5 days/week for 13 weeks. There were no deaths during the study. During and immediately following exposure to 5000 ppm, ataxia, narcosis, hypoactivity, and a lack of startle reflex were observed in some rats and mice exposed to 5000 ppm. Narcosis was not observed in rats during exposure following week 2, suggesting some adaptation to isopropanol. During exposures to 1500 ppm, narcosis, ataxia, and hypoactivity were observed in some mice, whereas only hypoactivity was observed in rats. Immediately following exposures, ataxia and/or hypoactivity were observed in a few rats or mice exposed to 5000 ppm. Overall, the 1500- and 5000-ppm rats and the 5000-ppm female mice showed increased body weights and/or body weight gain during the study. The liver was the only organ weight affected by isopropanol exposure. An increase in liver weight relative to body weight was observed in rats of both sexes and female mice exposed to 5000 ppm. No corresponding microscopic changes were noted in the liver. Histopathological evaluation showed a slight increase in the size and frequency of hyaline droplets in the kidneys of the isopropanol-exposed rats (251).

In a subchronic neurotoxicity study, F344 rats were exposed by inhalation to 0, 100, 500, 1500, or 5000 ppm for 13 weeks. Neurobehavioral evaluations included a functional observation battery (FOB), motor activity, and neuropathology. Effects of narcosis were observed in the 5000-ppm groups only. There were no changes in FOB, but increased motor activity was noted in female rats of the 5000-ppm group at weeks 9 and 13. Neuropathological examination revealed no exposure-related lesions in the nervous system (251).

An additional subchronic neurotoxicity study was conducted to clarify the increased motor activity findings. Female F344 rats were exposed to 0 or 5000 ppm isopropanol vapor, 6 h/day, 5 days/week. Half of the animals in each group were exposed for 9 consecutive weeks and the other half for 13 consecutive weeks. After 9 weeks of exposure, the motor activity effect was reversible within 2 days after the last exposure. Subtle differences in the shape of the motor activity versus test session time curve were noted in both the 9-week- and the 13-week-exposed animals, although it was unclear whether these changes were treatment-related. Complete reversibility of these changes did not occur until 1 and 6 weeks after the last exposure in the 9- and 13-week exposure groups, respectively (252).

In a drinking water study, rats ingested 0.5 to 10% of isopropanol for 27 weeks and showed decreased body weight gain but no gross or microscopic tissue abnormalities (246). Increased formation of hyaline droplets in the proximal tubules was reported in male rats given 1–4% isopropanol in drinking water for 12 weeks (253). In another study, daily application of a 50% solution to the heads of rats for 187 days produced no apparent injury to the skin (254). Isopropanol was administered in drinking water to three dogs for 7 months. The alcohol concentration was 4% from the end of the first month until the conclusion of the experiment. Tolerance to the alcohol developed as manifested by an increased degree of neuromuscular coordination at similar blood levels in habituated versus control animals and by increased elimination of the alcohol. The only significant histopathological changes were noted in the kidneys of one dog that died (255).

4.4.1.3 Pharmacokinetics, Metabolism, and Mechanisms. Isopropanol is readily absorbed by the oral and inhalation exposure. Following single oral doses of up to 2.9 g/kg isopropanol to dogs, maximum blood levels of up to 2.95 mg/L isopropanol were found within 0.5 to 2 h. The blood concentration was directly related to the dose level (255). Following an oral dose of 2 g/kg or 6 g/kg isopropanol to rats, the peak blood concentrations were 1.08 mg/L after 1 h and 4.8 to 6.0 mg/L after 8 h, respectively (243, 256). When rats inhaled 8000 ppm of isopropanol for 4 or 8 h, the blood concentrations of isopropanol attained at the end of the exposure period were 0.503 and 1.742 mg/mL (157).

Groups of three rabbits received isopropanol orally at 2 or 4 mg/kg or were exposed to isopropanol on towels, one applied to the chest and others on the floor of the inhalation chamber, with or without a plastic layer to prevent skin contact. The highest blood levels of isopropanol were produced by the oral exposures, followed by the combined dermal and inhalation exposure. Inhalation alone was of little significance (257).

It is well established that isopropanol is metabolized to acetone in the rat, dog, and rabbit (157, 256, 258). Pretreatment of rats with an inhibitor of alcohol dehydrogenase decreased isopropanol clearance and delayed the rate of acetone production (256). Acetone is further metabolized to CO_2. After inhalation exposure of rats to isopropanol for 4 h, blood levels of isopropanol and its metabolite acetone were directly related to airborne concentrations in the range of 500 to 8000 ppm. Following inhalation, the acetone/isopropanol ratio in blood decreased with increasing isopropanol concentrations, indicating saturation of the oxidative metabolic pathway above 4000 ppm (157).

As a minor pathway, isopropanol is conjugated by glucuronic acid and excreted in the urine. In rabbits dosed orally with 3 g/kg of isopropanol, 10.2% of the dose was excreted in the urine as the glucuronide conjugate within 48 h. Acetone was also identified in the expired air (229). A cytochrome P450-dependent pathway for the oxidation of isopropanol has been identified in rat liver microsomes (259).

The major route of excretion is via the lungs. Urinary excretion of both isopropanol and its metabolite acetone is minor and in each case does not exceed 4% of the dose in the rat, rabbit, and dog (258). The disappearance of isopropanol from the blood of experimental animals was found to be a first-order process at oral doses less than 1500 mg/kg (258). In rats, the half-life of isopropanol was 1.5 h at a 500-mg/kg intraperitoneal dose and increased to 2.5 h at a 1500-mg/kg dose (258). Simultaneous administration of ethanol to rats increased the half-life of isopropanol in the blood approximately fivefold following acute exposure (226). A half-life of 4 h was determined in dogs administered 1 g/kg isopropanol intravenously (258). After prolonged administration of isopropanol to dogs (255) and rats (260), elimination of isopropanol was increased.

In some recent studies, male and female rats and mice were exposed to 300 mg/kg of isopropanol intravenously and to 500 and 5000 ppm isopropanol by inhalation for 6 h. Additionally, isopropanol was given to rats by gavage (300 and 3000 mg/kg) in single and multiple doses. Exhalation was the major route of excretion, and acetone was the major metabolite along with unmetabolized isopropanol and CO_2. Urinary excretion accounted for 5 to 8% of the administered dose, which included isopropanol, acetone, and the glucuronide conjugate of isopropanol. A small amount was excreted in the feces. Excretion and distribution patterns were similar by all routes of administration. High doses exceeded the metabolic threshold as evidenced by a greater proportion of isopropanol excretion. There was no bioaccumulation, and distribution was similar for single and repeated administration. No major differences in absorption, distribution, metabolism, or excretion between species were observed (261).

Isopropanol, as a 70% aqueous solution, was applied under occluded conditions to the shaved backs of male and female Fischer 344 rats for a period of 4 h. Maximum analyzed blood concentrations of isopropanol were attained at 4 h and decreased steadily following removal from the skin. Blood concentrations were below the limit of detection at 8 h. Acetone blood levels rose steadily during the 4-h exposures and continued to rise after removal of isopropanol from the skin, reaching peak analyzed levels at 4.5–5 h. Basic pharmacokinetic parameters were similar for male and female rats, and mean, first-order elimination half-lives for isopropanol and acetone were 0.8–0.9 h and 2.1–2.2 h, respectively. [^{14}C]Isopropanol studies (intravenous and dermal) were conducted to determine dermal absorption rates. Using two independent methods, the values obtained were 0.78 ± 0.03 and 0.85 ± 0.04 mg/cm^2/h for males and 0.77 ± 0.13 and 0.78 ± 0.16 mg/cm^2/h for females. The calculated permeability coefficients are $1.37–1.50 \times 10^{-3}$ cm/h for males and $1.35–1.37 \times 10^{-3}$ cm/h for females (262, 263).

4.4.1.4 Reproductive and Developmental

Developmental Toxicity

Pregnant female Sprague–Dawley rats were exposed to 0, 3500, 7000, or 10,000 ppm of isopropanol 7 h/day during gestational days 1 to 19. The animals showed unsteady gait and narcotization during initial exposures in the mid- and high-dose groups; reduced food consumption and reduced weight gain were also noted in both the mid- and high-dose groups. Fetal body weights per litter were reduced in all dose groups. Exposure to 10,000 ppm also resulted in failure of implantation, fully resorbed litters, increased resorptions per litter, and increased incidence of cervical ribs (230). In another study, rats were exposed to isopropanol in drinking water for three generations (average intakes were 1470, 1380, and 1290 mg/kg/day for the first, second, and third generations, respectively). The growth of the first generation was retarded initially but returned to normal by the 13th week. There were no other effects on growth and no effects on reproduction (255). In a developmental toxicity study, reported in abstract form only, rats were dosed with 2% of isopropanol in drinking water. Reduced fetal body weight and delayed ossification were observed, but there was little evidence of "frank" malformations (264).

In a rat developmental study, female Sprague–Dawley rats were dosed by oral gavage with either 0, 400, 800, or 1200 mg/kg of isopropanol during gestational days 6 to 15. Two dams (8%) died at 1200 mg/kg and one dam (4%) died at 800 mg/kg. At 1200 mg/kg, maternal body weights were reduced throughout gestation (GS 0-20; 89.9% of control value), associated with reduced gravid uterine weight. There were no other treatment-related effects on the dams. Fetal body weights per litter were also significantly reduced at the 800 and 1200 mg/kg dose levels, but there were no teratogenic effects. There were no adverse maternal or developmental effects at 400 mg/kg (265). In a rabbit developmental study, female New Zealand white rabbits were dosed by oral gavage with either 0, 120, 240, or 480 mg/kg of isopropanol during gestational days 6 to 18. At 480 mg/kg, isopropanol was unexpectedly toxic to pregnant female rabbits, resulting in the deaths of four does (26%). Maternal body weights were significantly reduced during treatment (gestational days 6–18) and were associated with reduced maternal food consumption during this period. Profound clinical signs were noted at 480 mg/kg and included flushed and/or warm ears, cyanosis, lethargy, and labored respiration. No adverse maternal effects were noted at 120 or 240 mg/kg. There were no developmental or teratogenic effects at any dose tested (265).

Behavioral Developmental Toxicity
Isopropanol was given by oral gavage to Sprague–Dawley rats from gestational days 6 to 21 in doses of 0, 200, 700, or 1200 mg/kg. The dams were allowed to deliver, litters were culled on postnatal day (PND) 4, pups were weaned on PND 22, and their dams were killed. Weaned pups were assessed for day of testes descent or vaginal opening, motor activity, auditory startle, and active avoidance. The pups were killed on PND 68. Some of the pups were taken from each dose group and were perfused *in situ* for pathological examination of the central nervous system. There were no biologically significant findings in the behavioral tests, no changes in organ weights, and no pathological findings of note. Thus, there was no evidence of developmental neurotoxicity from isopropanol exposure (266).

Reproductive Toxicity
In a two-generation reproductive toxicity study, Sprague–Dawley rats were dosed by oral gavage with 0, 100, 500, or 1000 mg/kg isopropanol. There were seven parental deaths that were considered treatment-related: two high-dose F_0 females, two F_1 high-dose females, one mid-dose F_0 female, and two low-dose F_1 males. Lactation body weight gain was increased in the 500- and 1000-mg/kg females in both generations, and liver and kidney weights were increased in the 500- and 1000-mg/kg groups in both sexes. Centrilobular hepatocyte hypertrophy was noted in some 1000 mg/kg F_1 males. There were some kidney effects in the 500- and 1000-mg/kg F_0 males and in all treated F_1 male rats. The kidney effects were characterized by an increased number of hyaline droplets in the convoluted proximal tubular cells, epithelial degeneration and hyperplasia, and proteinaceous casts. Increased mortality occurred in the high-dose F_1 offspring during the early postnatal period; no other clinical signs of toxicity were observed in the offspring from either generation. Offspring body weight, however, in the 1000-mg/kg group was reduced during the early postnatal period. There was significant mortality in the F_1 weanlings (18/70) before the selection of the F_1 adults. A statistically significant reduction was observed in the F_1 male mating index of the 1000-mg/kg group (73 versus 97% in the controls). There were no other treatment-related effects on reproduction, including fertility and gestational indices, or histopathology of the reproductive organs (267).

In a single-generational reproductive toxicity study, rats were given 0, 0.5, 1.0, or 5.0% isopropanol in their drinking water. Parental animals dosed with 2% isopropanol had decreased body weights and increased liver and kidney weights. Offspring of these animals had reduced pup weight gain and decreased survival compared with the controls. In the offspring, there was also a dose-related increase in relative liver weights. There were no histopathological changes from isopropanol treatment (268).

The effects of isopropanol (2.5% in drinking water) on the reproduction and growth of rats was assessed in a multigenerational study. No reproductive toxicity was observed (255).

Isopropanol was administerd as a 3% solution in drinking water to Wistar rats. Reduced parental body weight gain, food, and water consumption was observed in the treated animals compared with the controls. Fertility, litter size, and pup weights at postnatal days 4 and 21 were reduced in treated animals compared with the controls. In the second generation, the isopropanol concentration was reduced to 2%, and there were essentially no effects (264).

4.4.1.5 Carcinogenesis. CD-1 mice were exposed by inhalation to 0, 500, 2500, or 5000 ppm of isopropanol vapor for 6 h/day, 5 days/week for 18 months. An additional group of mice (all exposure levels) were assigned to a recovery group which were exposed to isopropanol for 12 months and then retained until study termination at 18 months. There was no increased frequency of neoplastic lesions in any of the isopropanol-exposed animals. Nonneoplastic lesions were limited to the testes (males) and the kidney. In the testes, enlargement of the seminal vesicles occurred in the absence of associated inflammatory or degenerative changes. The kidney effects included tubular proteinosis and/or tubular dilatation. The incidence of testicular and kidney effects was not increased in the isopropanol-exposed recovery animals (269).

Fischer 344 rats were exposed to 0, 500, 2500, or 5000 ppm of isopropanol vapor for 6 h/day, 5 days/week for 24 months. The mortality rates for all male rats were 82, 83, 91, and 100% for the 0-, 500-, 2500-, and 5000-ppm groups, respectively. The corresponding values for the female rats were 54, 48, 55, and 69%. The main cause of death for the 5000-ppm rats (both sexes), as well as for much of the mortality of the 2500-ppm male rats, was chronic progressive nephropathy. Isopropanol exposure resulted in impaired kidney function, as indicated by various urine chemistry changes in male (2500- and 5000-ppm) and female (5000-ppm) rats. Animals in these groups also exhibited histopathological effects in the kidneys which appeared to be an exacerbated form of chronic progressive nephropathy. The only neoplastic lesion noted was increased interstitial (Leydig) cell adenomas in male rats. The frequency of these tumors, although elevated above the control animals, was within the historical control range of the testing facility and within the range reported for control animals from the National Toxicology Program carcinogenicity studies (269).

The carcinogenic potential of isopropanol was evaluated via inhalation using three strains of mice. Male mice were exposed to 7.5 ppm of isopropanol for 3 to 7 h/day, 5 days/week for 5 to 8 months. Animals were killed at either 8 or 12 months. There was no significant increase in the number of lung tumors observed. Similarly, no increases in lung tumors were observed in the same strains of mice that received subcutaneous injections of isopropanol once weekly for 20 to 40 weeks (13). This study is considered inadequate for cancer risk assessment (270).

4.4.1.6 Genetic and Related Cellular Effects Studies. All genotoxicity assays conducted so far with isopropanol have been negative. Isopropanol was not mutagenic in the *Salmonella* microsomal assay using the spot test, in strains TA98, TA100, TA1535, and TA1537 with and without S9 from the livers of Aroclor-induced rats (91). Isopropanol was not mutagenic in *Salmonella* strains TA97, TA98, TA100, TA102, TA104, TA1535, TA1537, and TA1538 with and without metabolic activation, when tested using a plate-incorporation modification of this assay (174).

Isopropanol was inactive in mutagenicity tests in *Neurospora crassa* (178), and isopropanol did not enhance adenovirus (SA7) transformation using Syrian hamster embryo cells (93). Isopropanol was inactive in a sister chromatid exchange assay with and without S9 metabolic activation (237). Isopropanol was inactive in an *in vitro* CHO/HGPRT gene mutation assay and in an *in vivo* bone marrow micronucleus assay in mice (271).

4.4.1.7 Other. Acute or chronic treatment of rats with isopropanol caused a significant increase in hepatic and renal cytochrome P450 content (272, 273). Acute oral pretreatment of rats with isopropanol resulted in a dose-related potentiation of carbon tetrachloride hepatotoxicity. The minimum effective dose was 0.25 mL/kg (274).

4.4.2 Human Experience

4.4.2.1 General Information. Reports of adverse effects on humans are relatively few despite the widespread use of isopropanol. Although there have been a few cases of dermal irritation and/or sensitization, no recorded cases of systemic poisonings due to industrial exposure have been found.

4.4.2.2 Clinical Cases. Exposure to 400 ppm of isopropanol vapors for 3 to 5 min caused mild irritation of the eyes, nose, and throat of human volunteers. At 800 ppm, the effects were not severe, but the majority felt that the atmosphere was unsuitable (275). Although isopropanol produced little irritation when tested on the skin of six human subjects, there have been reports of isolated cases of dermal irritation and/or skin sensitization (111, 239, 276–279). Except for three case reports (239, 277), the positive reactions were observed on patch testing patients with contact dermatitis due to ethanol. These patients also had a positive reaction to ethanol.

The use of isopropanol as a sponge treatment to control fever has resulted in cases of intoxication, probably the result of both dermal absorption and inhalation (18). Blood levels as high as 130 mg/100 mL were noted 95 min after admission for treatment (18). Unfortunately, there are no data on the amount used or the duration of exposure. In all cases where high blood levels were noted, a comatose condition existed. Recovery occurred within 34 h in all of these cases.

There have been a number of cases of poisoning due to the ingestion of isopropanol, particularly among alcoholics or suicide victims (18). The most common observation is a comatose condition. Pulmonary difficulty, nausea, vomiting, and headache accompanied by various degrees of central nervous system depression are typical. In the absence of shock, recovery usually occurred.

Acute carbon tetrachloride poisoning has been reported when workers were simultaneously exposed to isopropyl alcohol. In one incident, fourteen workers in an isopropyl alcohol packaging plant became ill after accidental exposure to carbon tetrachloride (280). In another outbreak, three workers from a color printing factory were admitted to a hospital with signs of acute hepatitis. One of the three had superimposed acute renal failure and pulmonary edema. Seventeen of 25 workers from the plant had abnormal liver function tests 10 days after the incident. It was determined that the incident occurred following inadvertent use of carbon tetrachloride to clean a pump in a printing machine that used isopropyl alcohol (281).

4.4.2.2.3 Pharmacokinetics, Metabolism, and Mechanisms. Lung uptake was studied in 12 printing workers exposed to workplace levels of isopropanol in the range of 8 to 647 mg/m^3 (3.3 to 264 ppm). The alveolar isopropanol concentration was highly correlated with the exposure level at any time of exposure. Acetone, but not isopropanol, was detected in the blood or urine. The acetone concentration ranged between 0.76 and 15.6 mg/L in the blood and between 3 and 93 mg/m^3 in the alveolar air. The acetone levels in alveolar air and blood increased with the increasing exposure period and were linearly related to the alveolar isopropanol levels. Elimination of acetone was mainly via the lungs, varying from 10.7 to 39.8% of the uptake and was inversely related to the exposure level (282).

Human data support the importance of the alcohol dehydrogenase pathway for the oxidation of isopropanol, as observed in experimental animals. When human subjects drank orange juice containing 3.75 mg/kg of isopropanol and 1.2 g/kg of ethanol over a period of 2 h, the average peak blood concentration of isopropanol at the end of this period was 0.83 mg/L (240). Isopropanol was detected in the blood only when ethanol was ingested simultaneously, indicating retarded elimination. Among 74 persons intoxicated by isopropanol, significantly lower acetone values were found in the blood of those who had also been exposed to ethanol (283).

The elimination of isopropanol in human beings also follows first-order kinetics. In two alcoholics who had ingested rubbing alcohol (70% isopropanol), isopropanol was eliminated from the blood with half-lives of 155 and 187 min, respectively. The levels of acetone declined slowly over the next 30 h. The initial blood concentrations of isopropanol were 1.7 and 1.0 g/L, and no ethanol was detected (284). A 46-year-old woman who was not an alcoholic had ingested rubbing alcohol. The half-life of isopropanol was 6.4 h, and the acetone level in the blood reached a maximum 30 h after admission to the hospital. The half-life of acetone was 22.4 h. Elimination of both isopropanol and acetone followed first-order kinetics (285). In six isopropanol ingestion episodes among five patients, serum isopropanol levels on admission and peak acetone concentrations ranged from 16.5 to 220 mg/dL and 141 to 585 mg/dL, respectively. The isopropanol and acetone apparent half-lives ranged from 2.9 to 16.2 h and 7.6 and 26.2 h, respectively (286). In another case report of an acute isopropanol overdose, the calculated half-life of isopropanol was 7.3 h (287).

The relationship of isopropanol vapor exposure to urinary excretion of acetone and unmetabolized isopropanol was studied in 99 printers of both sexes, who were exposed to levels up to 66 ppm of isopropanol (time-weighted average). Acetone and isopropanol concentrations were also studied in 34 nonexposed subjects. Acetone was detected in the urine of most of the nonexposed, and urinary acetone concentration increased in proportion to isopropanol exposure. Isopropanol itself was not found in the urine of nonexposed subjects and was detectable in the urine of only those who were exposed to isopropanol levels higher than 5 ppm (244).

Acute isopropanol intoxication has been observed in humans, and serum concentrations ranged from 128 to 480 mg/dL, whereas lethal serum concentrations have ranged as low as 100 to 240 mg/dL. The highest isopropanol serum concentrations among surviving ingestors were reported as single values of 467 and 560 mg/dL obtained from two patients (286).

Endogenous formation of isopropanol has been demonstrated at autopsies of individuals not previously exposed to this alcohol. This observation and the results of additional studies on rats show that isopropanol can result from the reduction of acetone by liver alcohol dehydrogenase, especially when high levels of acetone and high NADH/NAD ratios occur. Such conditions are found in diabetes mellitus, starvation, high fat feeding, chronic alcoholism, and dehydration (288, 289).

4.4.2.3 Epidemiology Studies

4.4.2.3.5 Carcinogenesis. Epidemiological evidence indicates that the manufacture of isopropanol by the strong-acid process is associated with an excess of upper respiratory tract cancer in workers (13). The International Agency for the Research on Cancer (IARC) concluded that there is sufficient evidence for carcinogenicity to humans in the manufacture of isopropanol by this process (270). Although the use of this strong-acid process has raised concerns of carcinogenicity, they are related to sulfuric acid (15) and to by-products such as dialkyl sulfates that are formed during this manufacture process, not to isopropanol itself (14).

4.5 Standards, Regulations, or Guidelines of Exposure

The ACGIH TLV and the OSHA PEL are set at 400 ppm as an 8-hr TWA, with a STEL of 500 ppm (116,117). The TLV is expected to minimize the potential for inducing narcotic effects or significant irritation of the eyes or upper respiratory tract in workers. This standard is the same for Australia and the United Kingdom (118). The occupational exposure standards in two other countries are 400 ppm in Germany and 150 ppm in Sweden (118).

B BUTANOLS

Butanols consist of four structural isomers of the molecular formula $C_4H_{10}O$: two primary, one secondary, and one tertiary alcohol. There are two stereoisomers of 2-butanol due to an asymmetric carbon atom in the secondary alcohol. All the butanols exist as liquids at room temperature, except *tert*-butanol (melting point 25°C). A relatively good toxicity data base exists for these alcohols. Occupational exposure standards exist for all four butanols.

5.0 1-Butanol

5.0.1 CAS Number: [71-36-3]

5.0.2 Synonyms: *n*-Butyl alcohol, propyl carbinol, butyl hydroxide, 1-hydroxybutane, alcool butylique, *n*-butanol, butan-1-ol, *n*-butan-1-ol, butanolen, butanolo, 1-butyl alcohol, butylowy alkohol, normal primary butyl alcohol, Methylopropane, propylmethanol

5.0.3 Trade Names: NA

5.0.4 Molecular Weight: 74.1

5.0.5 Molecular Formula: $C_4H_{10}O$

5.0.6 Molecular Structure: 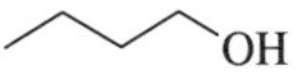

5.1 Chemical and Physical Properties

5.1.1 General

1-Butanol is a colorless, volatile liquid with a rancid sweet odor. Its physical and chemical properties are listed in Table 77.1.

5.1.2 Odor and Warning Properties

The air odor threshold of 1-butanol was reported to be 0.83 ppm (17); others have identified the minimum concentration with identifiable odor as 11 and 15 ppm (18).

5.2 Production and Use

The principal commercial source of 1-butanol is *n*-butyraldehyde obtained from the oxo reaction of propylene, followed by hydrogenation in the presence of a catalyst (290). 1-Butanol has also been produced from ethanol via successive dehydrogenation to acetaldehyde, followed by an aldol process (290). The earliest commercial route to 1-butanol, which is still used extensively in many Third World countries, employs fermentation of molasses or corn products with *Clostridium acetobutylicum* (290).

The largest volume commercial derivatives of 1-butanol are *n*-butyl acrylate and methacrylate (290). Other important derivatives of 1-butanol are 2-butoxyethanol and *n*-butyl acetate (290). 1-Butanol is used as a solvent for paints, coatings, natural resins, waxes, gums, synthetic resins, dyes, alkaloids, and camphor (18, 291). It is used as an extractant in the manufacture of antibiotics, hormones, vitamins, hops, and vegetable oils. It is also used as an intermediate in the manufacture of dibutyl phthalate, dibutyl sebacate, and herbicides (18, 291). Other miscellaneous applications of 1-butanol are as a swelling agent in textiles, as a component of brake fluids, cleaning formulations, degreasers, repellents, as flotation agents, and as a protective coating for glass objects (18, 291).

The production or, in some cases, use of the following substances may result in exposure to 1-butanol: artificial leather, butyl esters, rubber cement, dyes, fruit essences, lacquers, motion picture and photographic films, raincoats, perfumes, pyroxylin plastics, rayon, safety glass, shellac varnish, and waterproof cloths (18, 291).

5.3 Exposure Assessment

5.3.3 Workplace Methods

NIOSH Method No. 1401 has been recommended for the analysis of 1-butanol in air. It involves drawing a known volume of air through coconut shell charcoal. The recommended sample is 1 to 10 liters at a rate of 0.01 to 0.2 L/min. The analyte is desorbed with carbon disulfide containing 1% of 2-propanol. The sample is separated and quantified by a gas chromatograph with a flame ionization detector. The working range is about 50 to 900 mg/m^3 (16 to 297 ppm) for a 10-liter sample (19).

5.3.5 Biomonitoring/Biomarkers

Endogenous levels of 1-butanol can also be measured by gas chromatography–mass fragmentography in serum and urine (218).

5.4 Toxic Effects

5.4.1 Experimental Studies

1-Butanol has a low order of acute toxicity. It is severely irritating to the eyes but less so to the skin. Exposure by inhalation to 1-butanol results in central nervous system depression and irritation to the mucous membranes. No systemic toxicity has been observed from repeated exposure, and 1-butanol is not mutagenic. 1-Butanol is a developmental toxicant to rats at very high doses but only at maternally toxic doses. 1-Butanol is not a behavioral developmental toxicant.

5.4.1.1 Acute Toxicity. The acute oral LD_{50} has been reported to be 4.36 g/kg (292) and 2.5 g/kg (220) for the rat; and 3.4 g/kg (121) and 3.5 g/kg (35) in rabbits. The acute dermal LD_{50} in rabbits is 5.3 and 4.2 g/kg (Carnegie Mellon Institute of Research, unpublished data; 293). Rats survived exposure to 8000 ppm for 4 h (123). No deaths were observed in rats exposed to saturated vapors for 8 h (292). No evidence of intoxication was observed in mice exposed to 3300 ppm (10 mg/L) or 1650 ppm (5 mg/L) for 420 min (294). Exposures of mice to 6600 ppm produced giddiness after 1 h, prostration after 1.5 to 2 h, deep narcosis with loss of reflexes after 3 h, and the deaths of some animals (294a). Aspiration of 0.2 mL of 1-butanol caused deaths in 9 out of 10 rats (10).

1-Butanol is slightly to moderately irritating to the skin of rabbits (18). 1-Butanol was moderately to severely irritating to the eyes of rabbits (292). Severe corneal irritation resulted from instillation of 0.005 mL of undiluted 1-butanol and from an excess of a 40% solution in propylene glycol. A 15% aqueous solution caused minor corneal injury (42). 1-Butanol was also caused conjunctivitis, chemosis, and iritis, and corneal opacity (41). In a sensory irritation (Alarie) test, 1-butanol produced sensory irritation in mice, and reported RD_{50} values were 1268 (249), 4784 (44), and 11,696 ppm (156).

5.4.1.2 Chronic and Subchronic Toxicity. In a 13-week oral gavage study, rats were dosed with 0, 30, 125, or 500 mg/kg of 1-butanol. Ataxia and hypoactivity were observed in the high-dose animals during the final six weeks of the study. There were no significant differences between treated and control animals in body and organ weights, food consumption, hematologic and serum chemistry values, or histopathology (295). In a drinking water study, rats were given 6.9% of 1-butanol in drinking water (which also contained 25% sucrose) for 13 weeks. Control rats were given tap water. Electron microscopic studies showed that within five weeks irregularly shaped megamitochondria were formed in the liver (223). When three guinea pigs were exposed to 100 ppm of 1-butanol daily for 2 weeks (exposure unspecified) and then to 100 ppm 1-butanol 4 h/day, 6 days/week for about 2.5 months, the number of red blood cells and relative and absolute lymphocytes decreased. There was some evidence of lung hemorrhage, albuminuria, and early degenerative changes of the liver and kidney (131).

There are no neurotoxicity studies conducted with 1-butanol. However, there is a 13-week inhalation rat study of *n*-butyl acetate, which is hydrolyzed to 1-butanol and acetic acid. Animals were exposed to 0, 500, 1500, or 3000 ppm, 6 h/day, 5 days/week for 65 exposures. Functional observation battery (FOB) and motor activity were conducted during weeks 1, 4, 8, and 13 with neuropathology at termination. Scheduled-controlled operant behavior (SCOB), using a 4FR20:2FI120 schedule, was analyzed for week 1, 4, 8, and 13. There was no evidence of neurotoxicity during the FOB examinations. Mean total motor activity for the 3000-ppm male group was significantly higher than for the control group only during week 4, but motor activity was at control levels during weeks 8 and 13. No significant differences were seen in the SCOB at any concentration, and no treatment-related neuropathological effects were observed in the control and the 3000-ppm groups (296).

5.4.1.3 Pharmacokinetics, Metabolism, and Mechanisms. 1-Butanol is readily absorbed through the lungs, skin, and intestinal tract. Dogs exposed by inhalation to 1-butanol vapor at 50 ppm for more than 6 h absorbed 55% of the inhaled vapor (297). 1-Butanol was absorbed through the skin of dogs at a rate of 8.8 $mg/cm^2/min$ (297). When 12 human volunteers were exposed to 600 mg/m^3 of 1-butanol for 2 h, the difference in the inspired and expired air indicated an uptake of 47% of 1-butanol (298).

Once absorbed, 1-butanol disappears rapidly from the blood and is distributed to various tissues without evidence of bioaccumulation. Oral administration of 2.0 g/kg to rats resulted in the rapid appearance of 1-butanol in the blood after about 15 min; the maximum blood concentration was 51 mg% after 2 h and was reduced to 15 mg% at 8 h (59). Following an oral 2-mL/kg dose to rabbits, blood levels reached a peak after 30 to 60 min and became a trace after 10 h (58). In rats given 450-mg/kg oral doses of [^{14}C]1-butanol, the blood alcohol level reached a peak after 1 h (71 mg/mL) and disappeared rapidly; at 4 h it was below the detection limit (297). After a 500-mg/kg oral dose, 1-butanol reached a peak blood concentration of 240 ppm in 45 to 50 min and disappeared from the blood in 2 to 3 h (299).

Radioactivity was found in the liver, kidneys, small intestine, and lungs 1 h after oral dosing of [^{14}C]1-butanol to rats. Three hours later, the amounts of radioactivity in these organs had decreased. Over the next 3 days, 95% of the ^{14}C was excreted from the body; 2.8% of the radioactivity was eliminated in the urine and feces combined (300).

1-Butanol is a substrate for alcohol dehydrogenase, but not for catalase (228, 301). 1-Butanol is successively oxidized to *n*-butyraldehyde, *n*-butyric acid, and then to CO_2 and water. A 500-mg/kg oral dose of 1-butanol to rats resulted in detectable blood levels of *n*-butyl aldehyde (299). Inhibition of alcohol dehydrogenase with pyrazole produced a significant increase in peak blood concentration and persistence of 1-butanol (299). It is oxidized more rapidly than ethanol, presumably due to the high substrate affinity of 1-butanol for alcohol dehydrogenase (300). 1-Butanol can also be oxidized by the microsomal fraction of rat liver homogenate by the cytochrome P450 system (228, 301). Although 1-butanol is readily oxidized, it was also reported to be conjugated and excreted in the urine as glucuronide and sulfate conjugates. Rabbits administered 16 mmol of 1-butanol by oral gavage excreted 1.8% of the dose as a glucuronide conjugate over 24 h (229). In rats dosed with 450 mg/kg of 1-butanol by oral gavage, 83.3% of the dose was excreted as CO_2 in 24 h. Less than 1% was eliminated in the feces, 4.4% was excreted in the urine, and 12.3% remained in the carcass. Similar excretion patterns were observed in rats dosed with 45 and 4.5 mg/kg of 1-butanol. About 75% of 1-butanol was excreted conjugated to sulfate (44%) or to glucuronic acid (30%) (297).

5.4.1.4 Reproductive and Developmental

Developmental Toxicity

In an inhalation study, Sprague–Dawley rats were exposed to 0, 3500, 6000, and 8000 ppm of 1-butanol 7 h/day on days 1 to 19 of gestation. At 8000 ppm, narcosis was observed in approximately one-half of the maternal animals, and 2 out of 18 died. Maternal weight gain and food consumption decreased, and fetal body weights decreased by about 25% compared with controls. At 8000 ppm, the incidence of skeletal malformations (primarily rudimentary cervical ribs) was also greater than in controls, and a smaller percentage of the fetuses was judged normal (85 vs. 100% in controls). At 6000 ppm, maternal food consumption decreased, but maternal body weights were not affected. Fetal body weights were reduced by 12% from controls, but there were no significant increases in malformations compared to controls. No effects were observed at 3500 ppm. Thus 1-butanol is not a selective developmental toxicant (302).

Behavioral Developmental Toxicity

Sprague–Dawley rats were exposed by inhalation to 0, 3000, and 6000 ppm of 1-butanol, 7 h/day throughout gestation; male rats were similarly exposed for 7 h/day for 6 weeks and mated to unexposed females. Litters were culled and fostered to untreated controls. A number of behavioral and neurochemical tests were conducted in the offspring from days 10 to 90. There were few behavioral or neurochemical alterations detected in the offspring following parental exposure to 3000 or 6000 ppm. Thus inhalation exposure to 1-butanol does not result in any behavioral teratological effects in rats (303).

Reproductive Toxicity

1-Butanol was given by oral gavage to young male Sprague–Dawley rats at doses equimolar to 2.0 g/kg dibutyl phthalate for 4 days. Dibutyl phthalate produced testicular injury and reduced testicular weight, whereas 1-butanol, a hydrolysis product of dibutyl phthalate, had no effect (304).

5.4.1.6 Genetic and Related Cellular Effects Studies. 1-Butanol is not mutagenic to *S. typhimurium* in an Ames test with and without metabolic activation (173). 1-Butanol did not induce sister chromatid exchanges or chromosomal breakage in the "chick embryo cytogenetic test" (305). There were no sister chromatid exchanges in Chinese hamster ovary cells treated with 1-butanol (180), and no micronuclei were formed in V79 Chinese hamster cells treated with 1-butanol (92).

5.4.2 *Human Experience*

5.4.2.2.3. Pharmacokinetics, Metabolism, and Mechanisms. When 12 human volunteers were exposed to 600 mg/m^3 of 1-butanol for 2 h, the difference in the inspired air and expired air indicated an uptake of 47% of 1-butanol (298). The *in vitro* rate of dermal uptake of neat 1-butanol in human skin (epidermis) is 0.048 mg/cm^2/h (7).

5.4.2.3 Epidemiology Studies. A study was conducted of workers who used 1-butanol alone or in combination with other solvents at six plants where 1-butanol concentrations in the air ranged from 5 to 115 ppm. The major complaints were eye irritation when the air concentration was greater than 50 ppm, disagreeable odor, slight headache and vertigo, slight irritation of the nose and throat, and dermatitis of the fingers and hands (306). Mild irritation of the nose and throat was reported in subjects exposed for 3 to 5 min to 25 ppm. Exposure to 50 ppm was objectionable because it produced pronounced irritation of the eyes, nose, and throat in all subjects, and some experienced mild headaches (275). A 10-year study was conducted of men exposed to 1-butanol in an industrial setting. At the beginning of the study, the concentration of 1-butanol was 200 ppm or more, and corneal inflammation was occasionally observed. Ocular symptoms included a burning sensation, blurring of vision, lacrimation, and photophobia. Symptoms became more severe toward the end of the work week. Later in the study after the average exposure concentration of 1-butanol was reduced to 100 ppm, no systemic effects were observed and complaints of eye irritation were rare (307).

Audiological impairment was reported in workers exposed to 1-butanol. The 1-butanol concentration was measured at 80 ppm in combination with unprotected noise exposure. Of 11 workers exposed for 3 to 11 years without personal protective equipment from noise, nine experienced greater hearing loss (hypoacousia) in direct relation to the exposure time compared with a control group of 47 workers exposed to measured amounts of industrial noise at 90 to 100 dB but without butyl alcohol exposure. The ages of the affected workers ranged from 20 to 39 years (308, 309).

5.5 Standards, Regulations, or Guidelines of Exposure

The OSHA PEL for 1-butanol is 100 ppm (300 mg/m^3) as an 8-hr TWA (117). The ACGIH TLV, and the National Institute for Occupational Safety and Health (NIOSH) recommended exposure level (REL) are 50 ppm or 152 mg/m^3 as a ceiling value. There is a skin notation as well (116, 117). The TLV and PEL were recommended to protect against vestibular and auditory nerve injury, as well as headaches and irritation. Many other countries have set identical occupational exposure limits (118).

6.0 Isobutanol

6.0.1 CAS Number: [78-83-1]

6.0.2 Synonyms: Isobutyl alcohol, 2-methyl-1-propanol, isopropyl carbinol, 1-hydroxy-2-methylpropane, alcohol isobutylique, fermentation butyl alcohol, isobutyl-alkohol, 2-methyl propanol, 2-methylpropan-1-ol, 2-methylpropyl alcohol, IBA, *i*-butyl alcohol, butanol-iso

6.0.3 Trade Names: NA

6.0.4 Molecular Weight: 74.1

6.0.5 Molecular Formula: $C_4H_{10}O$

6.0.6 Molecular Structure:

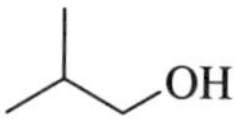

6.1 Chemical and Physical Properties

6.1.1 General

Isobutyl alcohol is a colorless, volatile liquid. Its physical and chemical properties are listed in Table 77.1.

6.1.2 Odor and Warning Properties

The limit of odor detection for isobutanol was reported as 1.6 (17) and 40 ppm (309a).

6.2 Production and Use

Isobutanol is commercially produced almost exclusively by the hydrogenation of isobutyraldehyde obtained from propylene using the oxo process (290). The primary industrial application for isobutanol is as a direct solvent replacement for 1-butanol (290). It is also used as an intermediate in the flavor and fragrance, pharmaceutical, and pesticide industries (290). The major routes of industrial exposure are dermal contact and inhalation.

6.3 Exposure Assessment

6.3.3 Workplace Methods

NIOSH Method No. 1401 has been recommended for the analysis of isobutyl alcohol in air. It involves drawing a known volume of air through coconut shell charcoal to trap the organic vapors present (recommended sample is 1 to 10 liters at a rate of 0.01 to 0.2 L/min). The analyte is desorbed with carbon disulfide containing 1% of 2-propanol, and the analyte is separated and quantified by gas chromatography with flame ionization detection. The working range is about 50 to 900 mg/m^3 (16 to 297 ppm) for a 10-liter sample (19).

6.3.2 *Biomonitoring/Biomarkers*

Endogenous levels of isobutanol can be measured by using gas chromatography–mass fragmentography in serum and urine (218).

6.4 Toxic Effects

6.4.1 *Experimental Studies*

Isobutanol has a low order of acute toxicity. It is not very irritating to skin, but it is severely irritating to the eye. Isobutanol is a sensory irritant at high concentrations. No systemic toxicity has been observed from repeated exposure to isobutanol in drinking water. It is not a developmental toxicant, and the data are inadequate to evaluate its carcinogenic potential.

6.4.1.1 Acute Toxicity. The acute oral LD_{50} has been reported as 2.46 g/kg (219) and 3.1 g/kg (291) in rats, 3.5 g/kg in mice (291), and 3.4 g/kg in rabbits (35). The acute dermal LD_{50} in rabbits is 3.4 g/kg (219). Exposure of mice to 10,600 ppm (32.2 mg/L) for 300 min or 15,950 ppm (48.3 mg/L) for 250 min resulted in deaths (37). No deaths occurred in rats exposed to saturated vapors for 2 h; however, two out of six deaths occurred when rats were exposed to 8000 ppm for 4 h (219). Aspiration of 0.2 mL of isobutanol caused deaths in 10 out of 10 rats. A solution of 85% isobutanol in water caused deaths in 7 out of 10 rats (10).

Isobutanol was very slightly irritating to the skin of rabbits, but it was moderately to severely irritating to rabbits' eyes (219). In a sensory irritation (Alarie) test, isobutanol produced sensory irritation in mice, and the RD_{50} was 1818 ppm (249).

In an acute neurotoxicity study, Sprague–Dawley rats were exposed by inhalation to 0, 1500, 3000, or 6000 ppm of isobutanol for 6 h. Functional observational battery and motor activity tests were conducted. There were marked signs of narcosis in the 3000- and 6000-ppm groups that disappeared after the end of the exposure period. Hypoactivity was seen in the 1500-ppm rats during the exposure period (310).

6.4.1.2 Chronic and Subchronic Toxicity. Male and female Wistar rats were administered isobutanol in drinking water at doses of 0, 1000, 4000, or 16,000 ppm for 3 months. These doses are approximately equivalent to 0, 80, 340, and 1450 mg/kg of isobutanol. There were no adverse effects on clinical observations, body weights, hematologic and serum chemistry parameters, organ weights, or histopathology (311).

In a subchronic neurotoxicity study, Sprague–Dawley rats were exposed by inhalation to 0, 250, 1000, or 2500 ppm isobutanol, 6 h/day, 5 days/week for 13 weeks. Neurobehavioral evaluations included a functional observation battery (FOB), motor activity, and neuropathology. The FOB and motor activity tests were conducted before study initiation and during the 4th, 8th, and 13th weeks of exposure. No histopathological or behavioral effects were observed that indicated a persistent or progressive effect of isobutanol on the nervous system at any exposure level. There was a slight decrease in response to external stimuli during the exposures in all groups, consistent with the acute transient central nervous system depression of isobutanol (310).

The effect of subchronic exposure on the schedule-controlled operant behavior (SCOB) was evaluated using Sprague–Dawley rats. The rats were trained to perform a multiple fixed ratio 20 (FR 20), fixed interval 120 second (FI120s) schedule of food reinforcement. This schedule is sensitive to chemically induced increases and decreases in rates of reinforcement and changes in index of curvature. After performance on this schedule was stabilized, the animals were exposed to 0, 250, 1000, or 2500 ppm, 6 h/day, 5 days/week for 13 weeks. On each exposure day, the rats were required to perform the multiple schedule of reinforcement for 47 minutes at approximately the same time each day just before exposure. Subchronic exposure up to 2500 ppm had no effects on the F1 index of curvature, FR rate, FI rate, or FR postreinforcement pause. All exposure levels caused a slight reduction in responsiveness to external stimuli during exposure, which is likely to be due to the transient effects of acute central nervous system depression. It was concluded that subchronic exposure to high vapor concentrations of isobutanol does not affect the performance of a complex behavior dependent on learning and memory (312).

6.4.1.3 Pharmacokinetics, Metabolism, and Mechanisms. Isobutanol can be readily absorbed via the lungs and gastrointestinal tract. Peak blood levels of 0.5 mg/mL of isobutanol were observed 1 h after administering a 2-mL/kg oral dose to rabbits. After 6 h, isobutanol was no longer detectable in the blood (58).

Isobutanol is oxidized to isobutyraldehyde by alcohol dehydrogenase, which is then oxidized to isobutyric acid. Isobutyric acid is thought to react with coenzyme A to form succinate, which enters the tricarboxylic acid cycle and subsequently, CO_2 is liberated (58).

The urinary metabolites from rabbits administered a 2-mL/kg oral dose followed by drinking water saturated with isobutanol (concentration unspecified) were acetaldehyde, acetic acid, isobutyraldehyde, and isovaleric acid. A small amount of unchanged isobutanol was also found in the urine. The rate of isobutanol oxidation was reported to be between those of 1-propanol and ethanol (313). When approximately 618 mg/kg of isobutanol was administered orally to rabbits, 4.4% of the dose was excreted as a glucuronide conjugate within 24 h in the urine. No aldehydes or ketones were detected in the expired air of a rabbit dosed orally with 6 mL (approximately 1600 mg/kg) of isobutanol (229). Following a 2-g/kg oral dose to rats, 0.27% was excreted in the urine within 8 h (59).

Isobutanol is rapidly cleared from the blood following systemic absorption. A 2-g/kg oral dose of a 20% aqueous solution of isobutanol given to rats resulted in isobutanol in the blood 15 min after dosing that achieved a peak concentration of 25 mg/dL blood at 90 min. This peak time coincided with the presence of isobutanol in the urine (52 mg/100 mL urine), after which levels dropped to 27 mg/100 mL urine after 8 h (207). In another study, 35 ppm of isobutanol was detected in the blood 45 to 50 min after a 500-mg/kg oral dose to rats; no alcohol was detected after 2 to 3 h. The metabolites isobutyraldehyde, isobutyric acid, and acetone were detected in the blood (299).

The rate of isobutanol oxidation was studied using both rat liver homogenate and a liver perfusion system. Addition of the alcohol dehydrogenase inhibitor pyrazole inhibited the *in vitro* oxidation by 65% (313).

6.4.1.4 Reproductive and Developmental

Developmental Toxicity
Pregnant female rabbits were exposed by inhalation to 0, 0.5, 2.5, or 10 mg/L (0, 165, 825, 3300 ppm) of isobutanol 6 h/day during gestational days 7 to 19. Maternal toxicity was observed at the high dose as evidenced by a slight body weight suppression, especially during the first phase of the exposure period. No adverse fetal or developmental effects were noted at any dose level (311). Pregnant female rats were exposed by inhalation to 0, 0.5, 2.5, or 10 mg/L of isobutanol 6 h/day during gestational days 6 to 15. No maternal or fetal toxicity or developmental effects were noted at any dose level (311).

6.4.1.5 Carcinogenesis. Nineteen Wistar rats were dosed with 0.2 mL of isobutanol twice weekly by oral intubation. The average survival time was 495 days. It was reported that malignant tumors developed in three animals; one had a forestomach carcinoma and a liver cell carcinoma, another had a forestomach carcinoma and myelogenous leukemia, and the third, a myelogenous leukemia (233). In the same study, 24 rats were injected subcutaneously with 0.05 mL/kg twice weekly. The average survival time was 544 days. A total of eight malignant tumors developed, two forestomach carcinomas, two liver sarcomas, one spleen sarcoma, one mesothelioma, and two retroperitoneal sarcomas (233). Increased incidences of total tumors were observed by both routes of administration, but there was no significant increased incidence of any tumor type at any site. This study is considered inappropriate for cancer risk assessment.

6.4.1.6 Genetic and Related Cellular Effects Studies. Isobutanol was not mutagenic in an Ames test when tested in the absence or presence of a S9 metabolic activation system (314). Isobutanol increased the reverse mutation rate in *E. coli* CA274 without metabolic activation (236).

6.4.2 Human Experience

6.4.2.1 General Information

6.4.2.2 Clinical Cases. Slight erythema without the formation of wheals was observed following the application of isobutanol to human skin (315). Irritation of the eyes and throat, formation of vacuoles in the superficial layers of the cornea, and loss of appetite and weight were reported in workers subjected to undetermined, but apparently high concentrations of isobutanol and butyl acetate from the lacquering of cables under conditions of crowding, ineffectual ventilation, and oppressive heat (89 to 107°F). When the conditions were improved, the symptoms disappeared (315a). In another report, no evidence of eye irritation was noted with repeated 8-h exposure to about 100 ppm of isobutanol (207).

Seven histories of cases that occurred between 1965 and 1971 described workers who had been exposed to 1-butanol and isobutanol in an unventilated photographic laboratory. They handled the alcohols under intense, hot light without any precautions. Exposure levels were not quantified, but they appear to have been excessive; exposure time ranged from 1.5 months to 2 years. Two workers had transient vertigo with nausea, vomiting, and/or headaches (291).

6.5 Standards, Regulations, or Guidelines of Exposure

The ACGIH TLV for isobutanol is 50 ppm as an 8-h TWA (116, 117) and the OSHA PEL is 100 ppm (300 mg/m^3). The TLV is based on the slight acute toxic potential of isobutanol versus 1-butanol. OSHA concluded that the PEL would reduce the significant risk of skin irritation. The occupational exposure standards in other countries are 50 ppm, Australia; 100 ppm, Germany; and 50 ppm, United Kingdom (118).

7.0 2-Butanol

7.0.1 CAS Number: [78-92-2]

7.0.2 Synonyms: sec-Butyl alcohol, *s*-butanol; methyl ethyl carbinol; 2-Hydroxybutane; butylene hydrate, *sec*-butanol; ethyl methyl carbinol; 1-methylpropyl alcohol; butan-2-ol; methylethylcarbinol; 1-methyl-1-propanol; methyl-1-propanol

7.0.3 Trade Names: NA

7.0.4 Molecular Weight: 74.1

7.0.5 Molecular Formula: $C_4H_{10}O$

7.0.6 Molecular Structure:

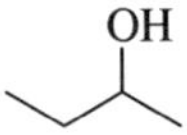

7.1 Chemical and Physical Properties

7.1.1 General

2-Butanol is a colorless liquid, whose physical and chemical properties are listed in Table 77.1.

7.1.2 Odor and Warning Properties

The air odor detection limit for 2-butanol is 2.6 ppm (17).

7.2 Production and Use

2-Butanol is produced commercially by the indirect hydration of *n*-butenes (290). The major use of 2-butanol is as a chemical intermediate for conversion to methyl ethyl ketone (316). It is also used to some extent as a solvent for lacquers, enamels, vegetable oils, gums, and natural resins (316). It is also used in hydraulic brake fluids, industrial cleaning compounds, polishes, paint removers, penetrating oils, and in the preparation of ore-flotation agents and perfumes (316).

7.3 Exposure Assessment

7.3.3 Workplace Methods

NIOSH Method No. 1401 has been recommended for the analysis of *sec*-butyl alcohol in air. It involves drawing a known volume of air through coconut shell charcoal to trap the organic vapors present (recommended sample is 1 to 10 liters at a rate of 0.01 to 0.2 L/min). The analyte is desorbed with carbon disulfide containing 1% of 2-propanol, and the sample is separated and quantified by gas chromatography with flame ionization detection. The working range is about 50 to 900 mg/m^3 for a 10-liter sample (19).

7.3.5 Biomonitoring/Biomarkers

A method involving solid-phase extraction and capillary gas chromatography has been reported for measuring 2-butanol in urine. The detection limit is in the range of 0.1 to 0.15 mg/L (317).

7.4 Toxic Effects

7.4.1 Experimental Studies

2-Butanol has a low order of acute toxicity. It is very slightly irritating to the skin and moderately irritating to the eyes. 2-Butanol is a central nervous system depressant, that is more potent than ethanol. It is not a reproductive toxicant. It can cause developmental effects such as delayed development but only at maternally toxic doses. 2-Butanol is not a teratogen, and it is not mutagenic.

7.4.1.1 Acute Toxicity. The acute oral LD_{50} is 6.5 g/kg in the rat (219) and 4.9 g/kg in the rabbit (35,121). Five of six rats died following inhalation exposure for 4 h to 16,000 ppm (48.5 mg/L) of 2-butanol (219). Inhalation exposures of 10,670 ppm for 225 min and 16,000 ppm for 160 min were fatal for mice (37). Mice subjected repeatedly to 5330 ppm for a total of 117 h were narcotized but survived (37).

2-Butanol was very slightly irritating to the skin of rabbits, but it was moderately irritating to the eyes of rabbits (219).

7.4.1.3 Pharmacokinetics, Metabolism, and Mechanisms. 2-Butanol is absorbed through the lungs, gastrointestinal tract, and skin. It is oxidized via alcohol dehydrogenase to the corresponding ketone, 2-butanone or methyl ethyl ketone, which is excreted either in the breath or the urine. 2-Butanol can also be conjugated directly with glucuronic acid. An inhibitor of alcohol dehydrogenase caused an increase in the peak concentrations and persistence of 2-butanol in the blood of rats given an oral dose of 2-butanol (299). In rabbits, 2-butanol was oxidized to methyl ethyl ketone, which could be detected in the expired air; it was also conjugated to form 2-butyl glucuronide and excreted in the urine (8). In rats dosed orally with 500 mg/kg of 2-butanol, a large portion was exhaled as 2-butanone and a smaller portion was excreted in the urine as glucuronides (299). 2-Butanone, 2,3-butanediol, and 3-hydroxy-2-butanone were found in the blood of rats administered a 1776-mg/kg oral dose of 1-butanol (318).

In rabbits given a 2-mL/kg oral dose of 2-butanol, the blood concentration of 2-butanol peaked within an hour at about 1 g/L and disappeared to a trace after 10 h. Unchanged 2-butanol was excreted to the extent of 3.3% of the dose in the breath and 2.6% in the urine. Methyl ethyl ketone (or 2-butanone) was detected in the blood and reached a maximum level after 6 h; it was excreted in amounts equivalent to 22.3% of the dose in the breath and 4% in the urine (58).

A pharmacokinetic model was developed to describe the biotransformation of 2-butanol and its metabolites 2-butanone, 3-hydroxy-2-butanone, and 2,3-butanediol using *in vivo* experimental blood concentrations (318). Male Sprague–Dawley rats were dosed by oral gavage with 2.2 mL/kg or 1776 mg/kg of 2-butanol after an overnight fast. Blood concentrations of 2-butanol reached a maximum of 0.59 g/L within 2 h and declined to less than 0.05 g/L after 16 h. As the blood concentration of 2-butanol fell, the concentration of 2-butanone, 3-hydroxy-2-butanone, and 2,3-butanediol rose to maximum levels of 0.78, 0.04, and 0.21 g/L at 8, 12, and 18 h, respectively. Approximately 97% of the 2-butanol dose was converted to 2-butanone by alcohol dehydrogenase; the calculated clearance constant for 2-butanol was 0.40 mL/min.

In rats given 1776-mg/kg oral doses of 2-butanol, the apparent blood elimination half-life was 2.5 h in rats. The blood levels of 2-butanol peaked after 1 h at 800 mg/L and declined; 2-butanone levels were 430 mg/L at 1 h and reached a maximum of 1050 mg/L by 4 h (319).

7.4.1.4 Reproductive and Developmental

Developmental Toxicity
Sprague–Dawley rats were exposed to 0, 3500, 5000, or 7000 ppm of 2-butanol 7 h/day on gestational days 1 to 19. At 7000 ppm, narcosis was observed in all animals; at 5000 ppm, animals were partially narcotized, and locomotor activity was impaired. Maternal weight gain and food consumption were significantly reduced in all dose groups. The number of live fetuses was significantly reduced, and resorptions increased only in the high-dose group. Fetal body weights were significantly reduced in the mid- and high-dose groups. There was no evidence of teratogenic effects (302).

Reproductive Toxicity
2-Butanol was administered to rats at 0.3, 1.0, and 2.0% in drinking water to the P1 and P2 generations. The 2% level produced several changes that represented mild toxicity reminiscent of stress lesions. No effect on reproductive parameters was noted. At the 2% dose level, 2-butanol produced a significant reduction in the growth of weanlings with evidence of retarded skeletal maturation (264).

7.4.1.6 Genetic and Related Cellular Effects Studies. 2-Butanol was not mutagenic to *Salmonella* strains TA98, TA100, TA1535, TA1537, and TA1538 in an Ames plate incorporation assay with or without metabolic activation (320). 2-Butanol was not mutagenic to yeast *S. pombe* with or without metabolic activation (91) or to *Saccharomyces cerevisiae* in a mitotic gene conversion assay with or without metabolic activation (320). No chromosomal effects were observed in Chinese hamster ovary cells treated with 2-butanol with or without metabolic activation (320).

7.4.1.7 Other. Cytochrome P450 enzymes in the liver and kidney were induced in Sprague–Dawley rats exposed by inhalation for 3 days (2000 ppm) and 5 days (500 ppm) (321).

2-Butanol at high concentrations potentiated carbon tetrachloride liver toxicity in rodents, probably due to the enhancement of the biotransformation of carbon tetrachloride to cytotoxic metabolites by activation of selective cytochrome P450 isozymes (322, 323).

7.4.2 Human Experience

7.4.2.1 General Information. Irritation of the eyes, nose, and throat, headache, nausea, fatigue, and dizziness have been experienced from excessive exposure to 2-butanol (18). A few cases of skin sensitization to 2-butanol have been reported (110, 324).

7.5 Standards, Regulations, or Guidelines of Exposure

The ACGIH TLV for 2-butanol is 100 ppm (305 mg/m^3) as an 8-h TWA and the OSHA PEL is 150 ppm (450 mg/m^3) (116, 117). The TLV was set to protect against the narcotic and irritative effects of 2-butanol. Many other countries have set identical occupational exposure standards (118).

8.0 *tert*-Butyl Alcohol

8.0.1 CAS Number: *[75-65-0]*

8.0.2 Synonymns: 2-Methyl-2-propanol, *tert*-butyl alcohol, *tert*-butanol, *tert*-butanol, trimethyl carbinol, *t*-butanol, 1,1-dimethylethanol; 2-methylpropan-2-ol, TBA, *t*-butyl hydroxide, trimethyl methanol, dimethylethanol, methyl-2-propanol

8.0.3 Trade Names: NA

8.0.4 Molecular Weight: 74.1

8.0.5 Molecular Formula: $C_4H_{10}O$

8.0.6 Molecular Structure: 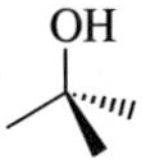

8.1 Chemical and Physical Properties

8.1.1 General

t-Butyl alcohol is a liquid that has a camphor-like odor. The physical and chemical properties are listed in Table 77.1.

8.1.2 Odor and Warning Properties

An air odor threshold of 47 ppm has been reported (17).

8.2 Production and Use

t-Butyl alcohol is produced as a by-product from the isobutane oxidation process for manufacturing propylene oxide (290). *t*-Butyl alcohol is also produced by the acid-catalyzed hydration of isobutylene, a process no longer used in the United States (290). *t*-Butyl alcohol is used in the manufacture of drugs, perfumes, paint removers, plastics, lacquers, and oil-soluble polyester resins. *t*-Butyl alcohol is a component of industrial cleaning compounds and insecticidal formulations and is an important additive in unleaded gasoline (291). It has been approved by the FDA as a defoaming agent for use in components and coatings with food contact (325).

Industrial exposures are likely to be by dermal contact and inhalation. Indirect exposure to *t*-butyl alcohol may occur by metabolism of methyl *t*-butyl ether, an oxygenated additive in gasoline (326).

8.3 Exposure Assessment

8.3.1 *Workplace Methods*

NIOSH Method 1400 has been recommended for the analysis of *t*-butyl alcohol in air. It involves drawing a known volume of air through coconut shell charcoal (recommended sample is 0.5 to 10 liters at a rate of 0.01 to 0.2 L/min). The analyte is desorbed with 1% 2-butanol in carbon disulfide, and the sample is analyzed by gas chromatography with flame ionization detection (19). Other gas chromatography methods have been published (20, 21).

8.3.5 *Biomonitoring/Biomarkers*

A method has been reported for measuring *t*-butyl alcohol in the blood (327).

8.4 Toxic Effects

8.4.1 *Experimental Studies*

t-Butyl alcohol shows a low order of acute oral toxicity. The primary acute effects are signs of alcohol intoxication. *t*-Butyl alcohol is a more potent narcotic agent than ethanol and other butanol isomers. No published data exist for skin and eye irritation. A subchronic drinking water study in rodents indicates that the urinary tract is a target organ for *t*-butyl alcohol toxicity. Chronic exposure to *t*-butyl alcohol in a two-year drinking water study resulted in increased kidney tumors in male rats and thyroid tumors in female mice. *t*-Butyl alcohol is not teratogenic to rodents, but fetotoxicity has been observed at maternally toxic doses. There is no evidence that *t*-butyl alcohol adversely effects the reproductive organs. *t*-Butyl alcohol was not active in a variety of genotoxicity assays.

8.4.1.1 Acute Toxicity. The oral LD_{50} is 3.5 g/kg for rats (328) and 3.6 g/kg for rabbits (35). It has been reported that *t*-butyl alcohol has a stronger narcotic effect on mice and rats than other butanol isomers (12, 37). The oral narcotic dose (ND_{50}) for rabbits is 19 mM or 1.4 g/kg (35). The potency for intoxication from *t*-butyl alcohol is also greater than that of ethanol (291, 329). In a performance test on an incline plane, the effect of an oral dose of different alcohols on the angle at which the rat slides down was determined. A 0.0163-mol/kg dose of *t*-butyl alcohol exhibited a strong intoxicating effect which was 4.8 times that of ethanol and was exceeded only by 1-butanol. Intoxication was for a greater period of time than the other alcohols tested; recovery was slow, and there was no improvement in performance 7 h after administration (329a).

8.4.1.2 Chronic and Subchronic Toxicity. F344/N rats and B6C3F_1 mice were exposed by inhalation to 0, 450, 900, 1750, 3500, or 7000 ppm of *t*-butyl alcohol, 6 h/day, 5 days/week for 12 exposure days. All rats and mice exposed to 7000 ppm were moribund after a single 6-h exposure and were killed. One 3500-ppm male mouse died on day 3. Final mean body weights of the 3500-ppm males and female rats were significantly lower than those of the controls. Final mean body weights and body weight gains of all other exposed groups were similar to those of the controls. The thymus weights of the male and female rats and female mice were less than those of the controls. The liver weights of the male and female mice exposed to 3500 ppm were greater than those of the controls. No treatment-related gross or microscopic lesions were noted in either the rats or the mice (330).

F344 rats and B3C3F_1 mice were exposed by inhalation to 0, 135, 270, 540, 1080, or 2100 ppm, 6 h/day, 5 days/week for 13 weeks. One 2100-ppm and five 1080-ppm exposed male mice died before the end of the studies. Adverse clinical signs in the male mice that died in the 1080-ppm group were rough coats and emaciated appearance, hypoactivity, and prostration. Mean body weights of the 2100-ppm female mice and mean body weigh gains of the 1080- and 2100-ppm female mice were significantly lower than those of the controls. Hematocrit values, hemoglobin concentrations, and erythrocyte counts decreased minimally in the 1080- and 2100-ppm male rats. Hemoglobin and/or hematocrit values also minimally decreased in the male rats in the lower exposure groups. Urine pH decreased minimally in the 1080-ppm and 2100-ppm male and female rats. Neutrophilia occurred in the 2100-ppm male mice. Kidney weights of the 1080- and 2100-ppm rats (both sexes) increased and liver weights decreased in the 1080- and 2100-ppm female rats. There was an exposure concentration-dependent increase in the severity of chronic nephropathy in the male rats. Splenic lymphoid depletion occurred in the male mice that died during the study, presumably secondary to stress (330).

In a 13-week drinking water study, F-344 rats and B6C3F_1 mice were given 0, 0.25, 0.5, 1.0, 2.0, and 4.0% (w/v) of *t*-butyl alcohol. Lethality was observed in the 4% dose group of both species. Body weights were significantly decreased in male rats in all dose groups, in female rats at 4%, in male mice at 1% or higher, and in the 2 and 4% female mouse groups. Water consumption increased at lower doses in male rats and decreased in the higher dose groups of both sexes of rats and female mice. The urinary tract was the target organ for *t*-butyl alcohol toxicity. In rats, urinary volumes were reduced in association with crystallinuria. Urinary tract lesions were observed in both sexes of rats and mice and included hyperplasia of the transitional epithelium and inflammation of the urinary bladder. No urinary tract effects were observed at 1% in male rats and mice and at 2% in the female rats and mice. In male rats, microscopic kidney changes suggestive of α2u-globulin nephropathy was noted. Evidence of hyaline droplets, and hyaline crystals in tubular epithelium was observed in all treated groups of male rats except the 4% group (331).

8.4.1.3 Pharmacokinetics, Metabolism, and Mechanisms. *t*-Butyl alcohol is readily absorbed by animals through the lungs and gastrointestinal tract; no information is available on dermal absorption.

t-Butyl alcohol is not a substrate for alcohol dehydrogenase (58, 59, 332, 333), but it can be conjugated with glucuronic acid (8, 229, 299). Following an oral dose of *t*-butyl alcohol, 24% of the dose was detected as a glucuronic acid conjugate in the urine of rabbits (229). Glucuronic acid conjugates of *t*-butyl alcohol have not been detected in dogs (334). *t*-Butyl alcohol is metabolized to formaldehyde (335) and acetone (336) by an oxidative demethylation reaction. Oxidative demethylation is probably a minor pathway, and *t*-butyl alcohol is mainly excreted unmetabolized in the air or urine (229, 299) or conjugated to glucuronic acid. *t*-Butyl alcohol is slowly eliminated from the blood of rats and mice. The elimination of *t*-butyl alcohol from the blood of rats given an intraperitoneal 1-g/kg dose of *t*-butyl alcohol showed first-order kinetics and a half-life of 9.1 h (336). Mice given an intraperitoneal 770-mg/kg dose of *t*-butyl alcohol (337) and pregnant mice dosed by oral gavage with 777 mg/kg of *t*-butyl alcohol (338) showed zero-order kinetics. *t*-Butyl alcohol was found in the blood of rabbits 70 h after oral administration of 2 mL/kg body weight (58).

Several studies have looked at *t*-butyl alcohol in the blood or various tissues. Rats given an oral dose of *t*-butyl alcohol (500 mg/kg) showed peak blood levels of 450 ppm of *t*-butyl alcohol at 45 min postexposure (299). *t*-Butyl alcohol was still detectable in the blood 24 h later. Rats exposed to inhaled levels of *t*-butyl alcohol (500 and 2000 ppm) 6 h/day for 3 to 5 days had serum concentrations of *t*-butyl alcohol of 74 and 666 ppm for the low- and high-dose group, respectively (321). In rats, *t*-butyl alcohol was found in greater amounts in brain tissue compared to fat tissue (326). In pregnant mice dosed by oral gavage with 777 mg/kg *t*-butyl alcohol, absorption and distribution were completed 1.5 h after dosing; elimination was complete by 12 h. There was little difference in the elimination rate between pregnant and nonpregnant mice (338).

Increased cytochrome P450 levels have been demonstrated in the livers and kidneys of rats exposed by inhalation to 500 or 2000 ppm of *t*-butyl alcohol for 3 to 5 days (321). *t*-Butyl alcohol elimination in mice can be substantially increased by pretreatment with *t*-butyl alcohol (329); however, *t*-butyl alcohol pretreatment does not seem to induce *t*-butyl alcohol metabolism in rats (336).

8.4.1.4 Reproductive and Developmental

Developmental Toxicity

Pregnant female Sprague–Dawley rats were exposed to 0, 2000, 3500, or 5000 ppm of *t*-butyl alcohol 7 h/day during days 1 to 19 of gestation. *t*-Butyl alcohol produced narcosis and unsteady gait at all exposure levels and a significant decrease in body weight and food consumption at 5000 ppm. Although fetal body weights significantly decreased at all exposure levels, there was no evidence of teratogenic effects (302). In another study, Swiss–Webster mice were given 0.5, 0.75, or 1.0% of *t*-butyl alcohol-derived calories in a liquid diet on gestational days 5 to 19 and all groups were pair-fed to the 1.0% *t*-butyl alcohol group. Maternal intake of 1.0% *t*-butyl alcohol reduced food consumption and maternal weight gain, and the number of stillborn pups increased in a dose-related manner. Interpretation of this study is difficult because of the small numbers of litters and the use of the number of pups rather than the number of litters as the basis for the statistical analyses (339). Two strains of mice (CBA/J and C57BL/6J) were dosed by oral gavage with 770 mg/kg of *t*-butyl alcohol on gestational days 6 through 18. No teratogenic effects were observed (338).

Behavioral Neurotoxicity Studies
Pregnant Sprague–Dawley rats were exposed to 6000 and 12,000 mg/m^3 (1975 and 3950 ppm) *t*-butyl alcohol for 7 h/day on gestational days 1 to 19; groups of male rats were similarly exposed for 7 h/day for 6 weeks and mated to unexposed females. Litters were culled at birth and fostered to untreated controls. From days 10 to 90, offspring were tested for neuromotor coordination, activity, and learning. In addition neuropathology was performed on selected offspring at 10 days of age. The results showed that the high dose was maternally toxic and reduced maternal body weight and food consumption. There were few differences from controls in the behavioral measures; however, the high dose produced elevations in a number of the neurotransmitters in brain tissue (340).

Reproductive Toxicity
In a 13-week inhalation study, F344 rats and B3C3F$_1$ mice were exposed to 0, 135, 270, 540, 1080, or 2100 ppm of *t*-butyl alcohol for 6 h/day, 5 days/week. No adverse effects on reproductive parameters were noted in either the rats or the mice (330).

8.4.1.5 Carcinogenesis. Male rats were given doses of 0, 1.25, 2.5, or 5 mg/mL (approximate average daily dose of 85, 195, or 420 mg/kg) and female rats were given doses of 0, 2.5, 5, or 10 mg/mL (approximate average daily dose of 175, 330, or 650 mg/kg) for two years. Male and female mice received doses of 0, 5, 10, or 20 mg/mL resulting in an average daily dose of approximately 535, 1035, or 2065 mg/kg for males and 510, 1015, or 2105 mg/kg for females for two years. Survival was significantly reduced in rats (both sexes) and male rats that received the high dose. There were increased incidences of renal tubule adenoma and carcinoma in male rats, transitional epithelia hyperplasia of the kidney in male and female rats, follicular cell adenoma of the thyroid in female mice, and follicular cell hyperplasia of the thyroid and inflammation and hyperplasia of the urinary bladder in male and female mice. In addition, a slight increase in follicular cell adenoma or carcinoma of the thyroid (combined) in male mice may have been related to exposure to *t*-butyl alcohol (341).

t-Butyl alcohol was inactive on mouse skin as a complete carcinogen or as a tumor promoter (342).

8.4.1.6 Genetic and Related Cellular Effects Studies. *t*-Butyl alcohol was not mutagenic in the *Salmonella* Ames test with and without metabolic activation (343). *t*-Butyl alcohol was inactive in the mouse lymphoma assay both with and without metabolic activation (344), and in the *Neurospora crassa* reversion test (178). *t*-Butyl alcohol did not produce chromosomal aberrations or sister chromatid exchanges in Chinese hamster ovary cells (345). NTP reported induction of micronucleated erythrocytes in the bone marrow of cells of rats administered *t*-butyl alcohol by intraperitoneal injection (330). However, there were no increased micronuclei in erythrocytes from the blood of mice administered *t*-butyl alcohol in drinking water for 13 weeks (341).

8.4.1.7 Other. *t*-Butyl alcohol at a high oral dose can enhance the hepatotoxic effect of carbon tetrachloride (346). Although this effect is not completely understood, *t*-butyl alcohol induces cytochrome P450 enzymes that may enhance the metabolism of carbon tetrachloride to its toxic metabolites.

8.4.2 Human Experience

8.4.2.1 General Information. There is no evidence that *t*-butyl alcohol poses a serious health concern in an occupational environment. Prolonged exposure to high concentrations may produce narcotic effects such as headaches, dizziness, drowsiness, and stupor. Overexposure to *t*-butyl alcohol vapors can also produce irritation of the eyes, nose, and throat. *t*-Butyl alcohol is slightly irritating to the skin. No reaction other than slight erythema and hyperemia was noted after application of *t*-butyl alcohol to human skin (315).

8.5 Standards, Regulations, or Guidelines of Exposure

The ACGIH TLV and the OSHA PEL for *t*-butyl alcohol are 100 ppm (300 mg/m^3) as an 8-h TWA, and the STEL is 150 ppm or 450 mg/m^3 (116, 117)). The TLV was set to protect against narcosis, the major effect of *t*-butyl alcohol by inhalation. A number of other countries have adopted an occupational exposure standard of 100 ppm for *t*-butyl alcohol (118).

C AMYL ALCOHOLS

This group consists of saturated aliphatic alcohols that contain five carbons. It includes three pentanols, four substituted butanols, and a disubstituted propanol, eight structural isomers that have the empirical formula $C_5H_{12}O$. There are four primary, three secondary, and one tertiary alcohol (347). In addition, 2-pentanol, 2-methyl-1-butanol, and 3-methyl-2-butanol are optically active.

The odd-numbered carbon structure and the extent of branching provide amyl alcohols with unique physical and solubility properties and often offer ideal characteristics for solvent, surfactant, extraction, gasoline additive, and fragrance applications. The original source of amyl alcohols was fusel oil, which is a by-product of the ethanol fermentation industry. Refined amyl alcohol from this source contained 85% 3-methyl-1-butanol and about 15% 2-methyl-1-butanol (347). Amyl alcohols were first produced synthetically from pentane by chlorination and hydrolysis (347). Today the most important industrial process for producing amyl alcohol is the oxo process based on butenes (347). Mixtures of isomeric amyl alcohols such as 1-pentanol and 2-methyl-1-butanol are often preferred because the different degree of branching imparts a more desirable combination of properties (347).

Inhalation of amyl alcohol vapors by humans have been reviewed by Rowe and McCollister (18). The vapors cause marked irritation of the eyes and respiratory tract, headache, vertigo, dyspnea and cough, nausea, vomiting, and diarrhea. Double vision, deafness, delirium, and occasionally fetal poisoning, preceded by severe nervous symptoms, have been attributed to the absorption of amyl alcohol. Coma, glycosuria, and sometimes methemoglobinemia have been reported as characteristic of amyl alcohol intoxication. A few cases of industrial poisoning were caused by amyl alcohol, although the presence of other solvents confounded the issue (18). Workers engaged in producing smokeless powder reported cough, eye irritation, colic, diarrhea, vomiting, heart palpitation, nervous symptoms, headache, vertigo, vision disturbances, forgetfulness, insomnia, somnolence, weakness, and one fatality. Although other alcohols and ether were employed, the signs increased as the use of amyl alcohol (probably fusel oil) increased (348). Nonindustrial cases of poisoning from drinking fusel oil were characterized by coma, glycosuria, and methemoglobinuria (349). An occupational exposure standard exists only for 3-methyl-1-butanol.

9.0 1-Pentanol

9.0.1 CAS Number: *[71-41-0]*

9.0.2 Synonyms: *n*-Amyl alcohol, *n*-butyl carbinol, *n*-pentyl alcohol, *n*-pentanol, pentyl alcohol, 1-pentanol, 1-pentol, pentan-1-ol, pentanol-1, pentasol, primary amyl alcohol, normal amyl alcohol, *n*-pentan-1-ol

9.0.3 Trade Name: NA

9.0.4 Molecular Weight: 88.2

9.0.5 Molecular Formula: $C_5H_{12}O$

9.0.6 Molecular Structure: 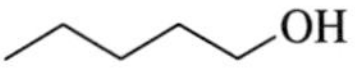

9.1 Chemical and Physical Properties

9.1.1 General

1-Pentanol is a clear, colorless liquid at ambient temperatures. Its physical and chemical properties are listed in Table 77.1.

9.1.2 Odor and Warning Properties

The air odor detection level (lowest perceptible level) was reported as 10 ppm for 1-pentanol (18).

9.2 Production and Use

1-Pentanol is made primarily by the oxo process, which involves the reaction of butenes with carbon monoxide and hydrogen in the presence of a catalyst, followed by hydrogenation (347, 350). 1-Pentanol is used as a solvent, as a chemical intermediate for esters, and as a food additive and flavoring substance (347, 350, 353). The primary routes of industrial exposure are by dermal contact and inhalation.

9.3 Exposure Assessment

9.3.1 Workplace Methods

A NIOSH method exists for isoamyl alcohol (specifically 3-methyl-1-pentanol) that may also be suitable for 1-pentanol (19).

9.4 Toxic Effects

9.4.1 Experimental Studies

1-Pentanol has a low order of acute toxicity. It is severely irritating to eyes and skin. Repeated oral exposure to 1-pentanol resulted in no systemic toxicity. It did not produce developmental toxicity at saturated vapor concentrations, and it was not mutagenic to bacteria.

The studies that have been conducted on a mixture consisting of 74% 1-pentanol, 25% 2-methyl-1-butanol, and 1% 3-methylbutanol, and not on pure 1-pentanol, have been identified in the following with an asterisk.

9.4.1.1 Acute Toxicity. The acute oral LD_{50} is 3.03 g/kg (18) and 2.69* g/kg in rats (351). Central nervous system depression and gastrointestinal irritation were observed. The acute dermal LD_{50} in rabbits is 2.0 g/kg and 4.5 g/kg (18) and greater than 3.2* g/kg (351). In an acute inhalation study, groups of 10 mice, rats, and guinea pigs were exposed to the aerosolized mixture* calculated to be 14 mg/L for 6 h (351). Two rats and seven mice died during the exposure; all other animals survived. Histological examination showed the lung and kidney as the principal target organs. Appreciable lung edema was observed in the mice. Aspiration of 0.2 mL *n*-amyl alcohol caused deaths in 10 out of 10 rats. The deaths were instantaneous and were attributed to cardiac and respiratory arrest (10).

The mixture* was severely irritating to the skin of rabbits when applied for 24 h under occlusive conditions (351). The mixture* also caused severe irritation to rabbit eyes and caused conjunctival erythema, chemosis, discharge, iritis, and varying degrees of corneal injury (351). In a sensory irritation (Alarie) test, 1-pentanol produced sensory irritation, with an RD_{50} of 610 ppm (43), 3000 ppm (352), and 4039 ppm (44).

9.4.1.2 Chonic and Subchonic Toxicity. 1-Pentanol, dissolved in corn oil, was administered to ASH/CSE rats by oral gavage at dose levels of 0, 50, 150, or 1000 mg/kg for 13 weeks. There were no treatment-related effects on body weight gain, food and water consumption, hematologic and serum chemistry values, urinalysis, renal function, organ weights, or histopathology (353).

9.4.1.3 Pharmacokinetics, Metabolism, and Mechanisms. Rats given 1 g/kg of 1-pentanol (0.25 g at 15-min intervals) by intraperitoneal injection showed a peak blood concentration of about 21 mg% 1 h after dosing started and disappearance from the blood after 3.5 h. Only 0.88% and 0.29% were excreted in the expired air and urine, respectively (135). Rats given a 2-g/kg oral dose of 1-pentanol had a peak blood concentration of 20 to 25 mg% in 1 h and essentially none after 4 h. No alcohol was detected in the urine (59).

After an oral dose of 0.5, 1, or 2 g/kg of 1-pentanol, the maximal blood levels in mice occurred at 10 min and were nonlinear. The blood levels were 7, 11, and 16 mg/dL, respectively. No 1-pentanol could be found in the blood after 4 h (354). When rats were exposed by inhalation to 100, 300, or 600 ppm 6 h/day, 5 days/week for 7 weeks, 1-pentanol was found in the brain at 0.2, 1.4, and 3.2 ppm, respectively. Similar but smaller values were found in rats exposed to the same concentrations for 14 weeks. A dose-dependent blood concentration of 1-pentanol was also observed at 7 and 14 weeks (355). 1-Pentanol is oxidized by alcohol dehydrogenase to valeric acid in rats (135), and it can undergo further oxidation or be excreted in the urine.

9.4.1.4 Reproductive and Developmental Toxicity

Developmental Toxicity
Pregnant female Sprague–Dawley rats were exposed to saturated vapors (3900 ppm) of 1-pentanol 7 h/day on days 1 to 19 of gestation. Food consumption significantly decreased, and body weight gain was reduced but was not statistically significant in the treated dams. There was no evidence of developmental toxicity (356).

9.4.1.6 Genetic and Related Cellular Effects Studies. 1-Pentanol was not mutagenic to *S. typhimurium* in an Ames test with and without metabolic activation (357).

9.4.1.7 Other. 1-Pentanol (0.1 mmol) did not produce hepatic peroxisome induction in rat hepatocytes (358).

9.4.2 Human Experience

9.4.2.1 General Information. Both industrial and nonindustrial cases of 1-pentanol poisoning have been reported. These cases, however, involved mixed exposures; see the introductory section for amyl alcohols.

9.4.2.2.3 Pharmacokinetics, Metabolism, and Mechanisms. The *in vitro* dermal flux of 1-pentanol in human skin (epidermis) is 0.041 mg/cm^2/hr (7).

9.5 Standard, Regulations, or Guidelines of Exposure

An ACGIH TLV or OSHA PEL has not been established for 1-pentanol. Russia has set an exposure standard of 10 mg/m^3 (2.8 ppm); Bulgaria and Poland, 100 mg/m^3 (28 ppm); and Hungary, 200 mg/m^3 (56 ppm) (118). There was no firm documentation given for these levels.

10.0 Isoamyl Alcohols

10.0a 2-Methyl-1-butanol

10.0b 3-Methyl-1-butanol

10.0c 2,2-Dimethyl-1-propanol

***10.0.1a Cas Number:** [137-32-6]*

***10.0.1b Cas Number:** [123-51-3]*

***10.0.1c Cas Number:** [75-84-3]*

10.0.2a Synonyms: DL-2-Methyl-1-butanol, 2-methylbutan-1-ol

10.0.2b Synonyms: 3-Methyl-l-butanol, fermentation amyl alcohol, isobutylcarbinol, 3-methylbutan-1-ol, 3-methylbutanol, 1-hydroxy-3-methylbutane

10.0.2c Synonyms: *t*-Butyl carbinol, 2,2-dimethylpropanol, 2,2-dimethylpropan-1-ol

10.0.3 Trade Name: NA

10.0.4 Molecular Weight: 88.2

10.0.5 Molecular Formula: $C_5H_{12}O$

10.0.6a Molecular Structure:

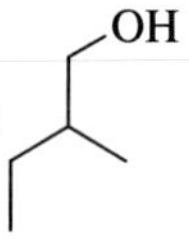

10.0.6b Molecular Structure:

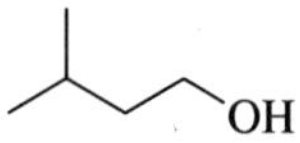

10.0.6c Molecular Structure:

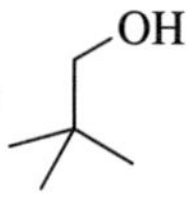

10.1 Chemical and Physical Properties

10.1.1 General

Except for 2,2-dimethyl-1-propanol, which is a crystalline solid, purified isoamyl alcohols are colorless liquids with a mild odor. Their physical and chemical properties are listed in Table 77.1.

10.1.2 Odor and Warning Properties

The air odor threshold for 3-methyl-1-butanol was reported as 0.042 ppm (17), which provides some acute warning for exposure to this chemical. No air odor threshold data were located for 2-methyl-1-butanol and 2,2-dimethyl-1-propanol.

10.2 Production and Use

Isoamyl alcohols are made principally by the oxo process, in which butenes are reacted with carbon monoxide and hydrogen in the presence of a catalyst, followed by hydrogenation (347, 350). Small amounts of 3-methyl-1-butanol and 2-methyl-1-butanol are refined from ethanol production as fusel oil (347, 350). Isoamyl alcohols are used as solvents for oils, fats, resins, and waxes, in the plastics industry in spinning polyacrylonitrile, and in manufacturing lacquers, chemicals, and pharmaceuticals (18). They are also used as flavoring agents and in fragrances (359). Industrial exposure is principally by the dermal contact and inhalation.

10.3 Exposure Assessment

10.3.1 Workplace Methods

NIOSH specifically recommends Method 1402 for measuring 3-methyl-1-butanol in air. This method involves drawing a known volume of air through coconut shell charcoal (recommended volume of 1 to 10 liters at a flow rate of 0.01 to 0.2 L/min). The analyte is desorbed with carbon disulfide containing 5% of 2-propanol. The sample is separated and quantified by gas chromatography with flame ionization detection. The working range is 15 to 150 mg/m^3 (4.2 to 42 ppm) for a 10-liter sample (19). Other gas chromatographic methods have been published (20, 21) that may be applicable to isoamyl alcohol.

10.3.5 Biomonitoring/Biomarkers

Endogenous levels of isoamyl alcohol can be measured by gas chromatography–mass fragmentography in serum and urine (218).

10.4 Toxic Effects

10.4.1 Experimental Studies

3-Methyl-1-butanol is the most studied of the isoamyl alcohols. Only acute toxicity data could be found for 2-methyl-1-butanol, and no data could be found for 2,2-dimethyl-1-propanol.

Both 2-methyl-1-butanol and 3-methyl-1-butanol have a low order of acute toxicity. They are slightly irritating to the skin but severely irritating to the eye. No systemic toxicity has been observed from subchronic oral exposure. 3-Methyl-1-butanol is not a developmental toxicant. The carcinogenicity toxicity data are considered inadequate.

10.4.1.1 Acute Toxicity. The acute oral LD_{50} for 2-methyl-1-butanol is 4.01 g/kg in rats (18). The acute oral LD_{50} for 3-methyl-1-butanol has been reported to be 5.77 g/kg (360) and 1.3 to 4.0 g/kg (359) in rats and 3.44 g/kg in rabbits (35). The acute dermal LD_{50} in rabbits was reported to be 2.89 g/kg for 2-methyl-1-butanol (18) and 3.24 g/kg for 3-methyl-1-butanol (360). No deaths occurred in rats exposed to concentrated vapors of either alcohol for up to 8 h (18, 360).

Both 2-methyl-1-butanol and 3-methyl-1-butanol produced slight irritation to rabbit skin and were severely irritating to the eyes of rabbits (18, 360). In a sensory irritation test, isopentanol produced sensory irritation and had an RD_{50} of 4452 ppm (44) and 731 ppm (43).

10.4.1.2 Chronic and Subchronic Toxicity. In an oral gavage study, ASH/CSE rats were dosed with 0, 150, 500, or 1000 mg/kg isoamyl alcohol for 17 weeks. Body weight gain was reduced slightly, but not statistically significantly, at 1000 mg/kg, but no other adverse effects were noted (361).

10.4.1.3 Pharmacokinetics, Metabolism, and Mechanisms. 2-Methyl-1-butanol and 3-methyl-1-butanol are oxidized to their corresponding aldehydes, which are further oxidized to the corresponding acids (135). Valeric acid was formed during perfusion of rabbit liver with 3-methyl-1-butanol (362). The oxidation of the alcohols to aldehydes occurs mainly in the liver, for this reaction was largely inhibited in partially hepatectomized rats (135). As a minor pathway, these alcohols can be conjugated directly with glucuronic acid. In rabbits dosed with isoamyl alcohol, 9% of the dose was excreted in the urine within 24 h as the glucuronide conjugate (229).

Small amounts of 2-methyl-1-butanol and 3-methyl-1-butanol are excreted in the air (0.97 to 5.6% of total dose) or urine (0.27 to 2.0% of total dose) following 1 g/kg intraperitoneal injections to rats. The maximum blood concentrations ranged from 14 to 55 mg%, and they disappeared from the blood in 4 to 9 h (135). Similar findings were reported following an 2-g/kg oral dose of 3-methyl-1-butanol to rats (59). The alcohol was not detectable in the urine; a maximum blood level of 17 mg/100 mL was found after 1 h, and only trace amounts were detected after 4 h. Because 3-methyl-1-butanol has a higher reaction rate with alcohol dehydrogenase than ethanol (18), these findings indicate that rapid metabolism probably accounts for the low blood levels reported.

10.4.1.4 Reproductive and Developmental Toxicity. Pregnant female Himalayan rabbits were exposed by inhalation to 0, 0.5, 2.5, or 10 mg/mL (0, 165, 825, 3300 ppm) of 3-methyl-1-butanol 6 h/day during gestational days 7 to 19. Maternal toxicity was observed in the high dose as evidenced by a slight retardation of body weight during the first days of the exposure period. No adverse fetal or developmental effects were noted at any dose level (311). Pregnant female rats were exposed by inhalation to 0, 0.5, 2.5, or 10 mg/mL 3-methyl-1-butanol 6 h/day during gestational days 6 to 15. Maternal toxicity was observed in the high dose as evidenced by a slight retardation of body weight during the first days of the exposure period. No adverse fetal or developmental effects were noted at any dose level (311).

In the chick embryo test, isoamyl alcohol did not produce any deformities in hatched chicks at doses up to 16 mg/egg (363).

10.4.1.5 Carcinogenesis. 3-Methyl-1-butanol was administered orally (0.1 mg/kg twice weekly; mean total dose was 27 ml) to 10 Wistar rats or subcutaneously (0.04 mg/kg once weekly; mean total dose was 3.8 ml) to 24 Wistar rats. It was not stated whether the alcohol was administered undiluted or in a preparation. All animals were observed until they died naturally; the average survival time was 527 days. In the oral study, four malignant tumors were reported: a myeloid leukemia, two liver cell carcinomas, and one antestomach carcinoma. In the subcutaneous study, a total of 10 malignant tumors were reported occurring in the liver, spleen, antestomach, and retroperitoneal region (233). This study is considered inadequate for cancer risk assessment.

10.4.2 Human Experience

10.4.2.2 Clinical Cases. The majority of test subjects found that isoamyl alcohol is irritating to the nose and throat at 100 ppm and to the eyes at 150 ppm within 3 to 5 min

(275). Isoamyl alcohol (8% in petrolatum) did not produce skin irritation and was not a skin sensitizer in human volunteers (359). In a single case, a positive reaction in a 48-h covered patch test with 10% 3-methyl-1-butanol was reported (109). Signs of isoamyl alcohol poisoning in humans who had ingested 50 to 100 mL included central nervous system depression, weakness, pain, a burning sensation in the chest and stomach, nausea, headache, sleep within 10 to 15 min, and terminal coma and death within 1 h to 6 days (364).

10.5 Standards, Regulations, or Guidelines of Exposure

The ACGIH TLV and the OSHA PEL for 3-methyl-1-butanol are 100 ppm (360 mg/m^3) as an 8-h TWA, and the STEL is 125 ppm (116,117). A number of other countries have recommended 100 ppm as an occupational exposure standard (118). No adequate documentation was given. Neither an ACGIH TLV nor an OSHA PEL has been established for 2-methyl-1-butanol or 2,2-dimethyl-1-propanol.

11.0 Secondary Amyl Alcohols

11.0a 2-Pentanol

11.0b 3-Pentanol

11.0c 3-Methyl-2-Butanol

11.0.1a CAS Number: *[6032-29-7]*

11.0.1b CAS Number: *[584-02-1]*

11.0.1c CAS Number: *[598-75-4]*

11.0.2a Synonyms: 2-pentanol, DL-2-pentanol, DL-sec-amyl alcohol, methyl propyl carbinol; pentan-2-ol

11.0.2b Synonyms: Diethyl carbinol, pentanol-3, pentan-3-ol

11.0.2c Synonyms: DL-3-Methyl-2-butanol; 3-methylbutan-2-ol

11.0.3 Trade Name: NA

11.0.4 Molecular Weight: 88.2

11.0.5 Molecular Formula: $C_5H_{12}O$

11.0.6a Molecular Structure: 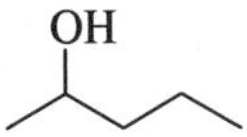

11.0.6b Molecular Structure: 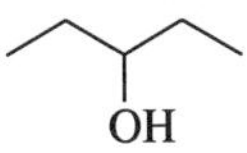

11.0.6c Molecular Structure: 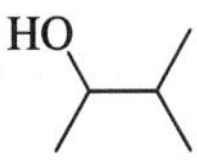

11.1 Chemical and Physical Properties

11.1.1 General

These alcohols exist as liquids at room temperature. Their physical and chemical properties are listed in Table 77.1.

11.1.2 Odor and Warning Properties

No information was located.

11.2 Production and Use

The secondary amyl alcohols are produced mainly by hydrating pentenes (350). They can also be produced by hydrogenating the corresponding ketones: 2-pentanone, 3-pentanone, and 3-methyl-2-butanone for 2-pentanol, 3-pentanol, and 3-methyl-2-butanol, respectively (350). The secondary amyl alcohols are used to produce catalyst systems for dimerization and polymerization and for synthesizing methanol from carbon monoxide and hydrogen (350). They may also be used as solvents, stabilizing agents, chemical intermediates, lubricant additives, cleaning agents, defrosting agents, herbicides, and fragrances (350). Industrial exposure is likely to be by dermal contact and inhalation.

11.3 Exposure Assessment

11.3.3 Workplace Methods

A NIOSH method exists for isoamyl alcohol (specifically 3-methyl-1-pentanol) that may be suitable for the secondary amyl alcohols (19). For other methods that may be applicable, see the section on 1-pentanol.

11.4 Toxic Effects

11.4.1 Experimental Methods

11.4.1.1 Acute Toxicity. The acute oral LD_{50} in rats for 3-pentanol is 1.87 g/kg (219), and in rabbits for 2-pentanol is 2.8 g/kg (121). For unspecified isomers, the acute oral LD_{50} is 1.47 g/kg in rats (18) and 2.82 g/kg in rabbits (35). The acute dermal LD_{50} for 3-pentanol is 2.0 g/kg in rabbits (219).

The secondary amyl alcohols are reported to have the highest narcotic activity of the amyl alcohols, followed by tertiary and the primary alcohols (121). The narcotic dose of *sec*-amyl alcohol (isomeric composition not given) in rabbits was found to be 5 mmol or 0.44 g/kg (35). This is defined as the dose that produces stupor and loss of voluntary movements in 50% of the animals.

3-Pentanol was slightly irritating to the skin of rabbits and moderately irritating to rabbit eyes (219).

11.4.1.3 Pharmacokinetics, Metabolism, and Mechanisms. As shown in Table 77.6, appreciable amounts of a given dose of a secondary amyl alcohol were excreted as ketones in the expired air following an intraperitoneal dose to rats (135).

Table 77.6. Pattern of Excretion of Secondary Amyl Alcohols by Rats after Intraperitoneal Injection (135)[a]

	Percent of administered dose				
	Expired air		Urine		
Isomer	Alcohol	Ketone	Alcohol	Ketone	Total Excreted
2-Pentanol	5.4	36.8	1.2	2.0	45.4
3-Pentanol	0.3	50.0	0.1	4.6	55.0
3-Methyl-2-butanol	8.1	48.0	2.9	2.4	61.4

[a]Taken from Rowe and McCollister (18).

Secondary amyl alcohols are oxidized to ketones, which were found in the blood in measurable quantities after administration of the alcohols to animals (135). The ketones also had longer half-lives in the blood than their respective alcohols. The secondary amyl alcohols can form glucuronic acid conjugates. Small amounts of conjugated glucuronic acid were found in the urine following oral administration of 2-pentanol to rabbits (365, 366). Rabbits excreted more glucuronic acid than dogs (365, 366).

Amyl alcohols were excreted from the body in the order primary > secondary > and tertiary (135). Following an 1-g/kg intraperitoneal injection to rats, the concentration of secondary amyl alcohols in the blood was 51 to 65 mg%.These alcohols disappeared from the blood in 13 to 16 h (135).

The minimum concentrations of 2-pentanol, 3-pentanol, and 3-methyl-2-butanol in the jugular vein blood of rats necessary to cause death from respiratory failure were 86, 87, and 90 mg/100 mL of blood, respectively (135).

11.4.2 Human Experience

11.4.2.1 General Information. There have been no reports of adverse effects from industrial exposure to these alcohols. Dermal exposure to any of the secondary amyl alcohols did not result in erythema, hyperemia, or any local wheal formation (315).

11.5 Standards, Regulations, or guidelines of Exposure

No occupational exposure standards have been established for these alcohols.

12.0 *tert*-Amyl Alcohol

12.0.1 CAS Number: *[75-85-4]*

12.0.2 Synonyms: 2-Methyl-2-butanol, *tert*-pentyl alcohol, dimethyl ethyl carbinol, *tert*-Pentanol, amylene hydrate, 2-methylbutan-2-ol

12.0.3 Trade Names: NA

12.0.4 Molecular Weight: 88.2

12.0.5 Molecular Formula: $C_5H_{12}O$

12.0.6 Molecular Structure:

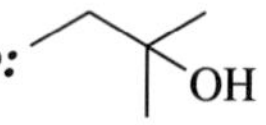

12.1 Chemical and Physical Properties

tert-Amyl alcohol is a volatile liquid. Its physical and chemical properties are listed in Table 77.1.

12.1.1 General

12.1.2 Odor and Warning Properties

2-Methyl-2-butanol has a sour odor, a threshold value of 8.2 mg/m^3 (2.3 ppm), an absolute perception limit of 0.04 ppm, and a 100% recognition level of 0.23 ppm (367).

12.2 Production and Use

tert-Amyl alcohol is prepared by hydrating 2-methyl-2-butenes (347). It can also be prepared by reducing pivalic acid (350). *tert*-Amyl alcohol is used as a solvent for coating materials based on epoxy resins and polyurethanes, in the oxidation of olefins to the corresponding carbonyl compounds, and as a stabilizing agent (350). No exposure data were located on *tert*-amyl alcohol.

12.3 Exposure Assessment

12.3.5 Biomonitoring/Biomarkers

A method has been developed for determining *tert*-amyl alcohol in plasma with a detection limit of 0.1 mg/mL. The method involves direct injection of heparinized plasma, separation by gas–liquid–solid chromatography, and quantification by selective ion monitoring mass spectrophotometry. About 50 mL of blood is required (368).

12.4 Toxic Effects

12.4.1 Experimental Effects

tert-Amyl alcohol has a low order of acute toxicity. It is not very irritating to the skin but markedly irritating to the eyes. *tert*-Amyl alcohol is not a skin sensitizer. No systemic toxicity has been observed from subchronic inhalation exposure, and it is not genotoxic.

12.4.1.1 Acute Toxicity. The acute oral LD_{50} is 1 g/kg (328) and 1.0 to 2.0 g/kg (Dow Chemical Company, unpublished data) for rats and 2.0 g/kg for rabbits (35). The acute dermal LD_{50} 1.72 g/kg for rabbits (Dow Chemical Company, unpublished data). Rats exposed to 5700 ppm of *tert*-amyl alcohol vapor for 6 h were unconscious by the end of exposure and died within 24 h; rats exposed to 3000 ppm were unconscious by the end of exposure but survived (Dow Chemical Company, unpublished data).

The narcotic action of a series of amyl alcohols in rabbits decreased in the order secondary > tertiary > primary (121). The narcotic dose of *tert*-amyl alcohol in rabbits is 8 mmol or 0.7 g/kg (369). This is defined as the dose that produces stupor and loss of voluntary movements in 50% of the animals.

No irritation was observed to the uncovered skin of rabbits when *tert*-amyl alcohol was applied repeatedly (Dow Chemical Company, unpublished data). However, if the skin was covered with a cloth, redness and exfoliation resulted. Confinement of the material under an impervious covering resulted in severe necrosis (Dow Chemical Company, unpublished data). *tert*-Amyl alcohol was markedly irritating to the rabbit eye (Dow Chemical Company, unpublished data).

12.4.1.2 Chronic and Subchronic Toxicity. Rats were exposed to 0, 150, 500, and 1500 ppm of *tert*-amyl alcohol 6 h/day for 7 days. The high-dose animals showed motor incoordination after the first two exposures but not by study termination, suggesting that some tolerance may have occurred. They also appeared lethargic for 3 to 4 h after exposure. There were increased absolute and relative liver and kidney weights and decreased blood glucose level; gross pathology showed no treatment-related changes (Dow Chemical Company, unpublished data).

Rats, mice, and dogs were exposed to 0, 50, 225, 1000 ppm of *tert*-amyl alcohol 6 h/day for a total of 59 to 60 exposures in 3 months. Tearing was noted in rats at 225 and 1000 ppm and in one dog at 1000 ppm. Impaired motor coordination was very apparent in the dogs during and for a short time after exposure to 1000 ppm. The female rats also exhibited motor incoordination but only after the first exposure. Tolerance developed in both species. The only serum chemistry and hematologic change noted was increased serum alkaline phosphatase activity in the dogs at 1000 ppm. There were no treatment-related histological changes in the rats and mice, and the livers of all of the dogs appeared normal except that in each exposure group one out of four dogs had hepatocellular cytoplasmic inclusions (Dow Chemical Company, unpublished data).

12.4.1.3 Pharmacokinetics, Metabolism, and Mechanisms. *tert*-Amyl alcohol is not a substrate for alcohol dehydrogenase (18). It is poorly metabolized, an appreciable amount of absorbed *tert*-amyl alcohol is excreted unchanged in the expired air, and the amount varies between species. A small amount of conjugated glucuronic acid was found in the urine following oral administration of *tert*-amyl alcohol to rabbits (365, 366). Dogs did not excrete glucuronic acid following administration of *tert*-amyl alcohol (365, 366). It is unknown whether *tert*-amyl alcohol undergoes a demethylation reaction similar to *tert*-butyl alcohol.

In general, rats and rabbits exhale less *tert*-amyl alcohol in expired air than do cats and dogs. A dog injected subcutaneously with 0.1 mL of *tert*-amyl alcohol exhaled 65% of the dose within 5.75 h (207). Another dog given a slightly larger dose of *tert*-amyl alcohol intravenously exhaled 52% within 6 h (207). When rabbits were injected intravenously with a dose approximately the same as that given to dogs, 21% was found unchanged in the air within 4 h (18). A rat given a 1-g/kg intraperitoneal dose eliminated 26.4% unchanged in the expired air and 8.9% in the urine within 50 h (135).

tert-Amyl alcohol did not disappear from the blood of rats until 50 h after a 1-g/kg intraperitoneal dose (135). Immediately after a single 6h inhalation exposure to 150, 500, or 1500 ppm, the concentrations of *tert*-amyl alcohol in rat plasma were 10, 75, and 350 mg/mL, respectively. The clearance of *tert*-amyl alcohol from the plasma of rats exposed to 150 ppm was first-order ($t_{1/2} = 100$ min) but exhibited saturation kinetics following acute exposures of 500 and 1500 ppm of *tert*-amyl alcohol. Upon repeated exposures to 50 ppm, *tert*-amyl alcohol was cleared by rats ($t_{1/2} = 47$ min) and dogs ($t_{1/2} = 69$ min) in first-order manner. End-exposure plasma concentrations after repeated exposure to 1000 ppm of *tert*-amyl alcohol were approximately 110 mg/mL in rats and mice and 480 mg/mL in dogs. Following repeated exposure to 1000 ppm, the plasma clearance of *tert*-amyl alcohol was first-order ($t_{1/2} = 29$ min) in mice but exhibited Michaelis–Menten kinetics in rats and dogs (370).

12.4.1.6 Genetic and Related Cellular Effects Studies. *tert*-Amyl alcohol was not mutagenic to *Salmonella* strains TA1535, TA1537, and TA1538 in an Ames assay with or without metabolic activation (Dow Chemical Company, unpublished data). It was also not mutagenic to *S. cerevisiae* with or without metabolic activation (Dow Chemical Company, unpublished data).

12.4.1.7 Other. *tert*-Amyl alcohol was not a skin sensitizer to guinea pigs (Dow Chemical Company, unpublished data).

12.4.2 Human Experience

12.4.2.1 Clinical Cases. Two case reports from nonindustrial exposures to *tert*-amyl alcohol were previously summarized (18). A patient died following an enema of 35 g of *tert*-amyl alcohol. A woman who intentionally drank about 27 g of *tert*-amyl alcohol experienced coma, dyspnea, irregular pulse, and dilation, then contraction, of the pupils. Application of *tert*-amyl alcohol to the skin of men had no effect on the nerves, no local wheal formation, erythema, or hyperemia (315). No adverse effects have been reported in humans from occupational exposure.

12.5 Standards, Regulations, or Guidelines of Exposure

No occupational exposure standards have been established for *tert*-amyl alcohol.

C HEXANOLS

The C_6 alcohols consist of a number of primary, secondary, and tertiary structural isomers. Enantiomers can exist for 2-hexanol. The most important commercial members are 1-hexanol, 2-ethyl-1-butanol, and isohexyl alcohols (a mixture of branched C_6 alcohols). The most important member of the secondary alcohols, is methyl isobutyl carbinol. The major use of these alcohols is in the production of esters, such as plasticizers. No occupational exposure standards exist for the hexanols except for methyl isobutyl carbinol.

There is toxicological interest in 2-hexanol because it is a metabolite of *n*-hexane and methyl-*n*-butyl ketone, which are known neuropathic agents in humans and animals.

13.0 1-Hexanol

13.0.1 CAS Nomber: [111-27-3]

13.0.2 Synonyms: *n*-Hexyl alcohol, amyl carbinol, pentyl carbinol, 1-hydroxyhexane

13.0.3 Trade Name: NA

13.0.4 Molecular Weight: 102.2

13.0.5 Molecular Formula: $C_6H_{14}O$

13.0.6 Molecular Structure:

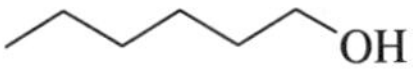

13.1 Chemical and Physical Properties

13.1.1 General

1-Hexanol is a liquid at room temperature. Its physical and chemical properties are listed in Table 77.1.

13.1.2 Odor and Warning Properties

The absolute perceived concentration has been reported as 0.01 ppm, and the recognition level is 0.09 ppm (367).

13.2 Production and Use

1-Hexanol is commercially prepared from the addition of ethylene to triethylaluminum followed by oxidation (3). It is also produced from natural products derived from coconut or palm oils (3). 1-Hexanol is used in the manufacture of antiseptics, fragrances, and perfumes, as a solvent, and in the production of plasticizers with higher n-alcohols (3, 371).

13.3 Exposure Assessment

13.3.1 Workplace Methods

No NIOSH specific method exists to measure 1-hexanol in the air; other gas chromatographic methods have been suggested that may be applicable (20, 21).

13.3.2 Biomonitoring/Biomarkers

A capillary gas chromatographic method with flame ionization detection is available for measuring 1-hexanol in urine (372). A method involving a capillary gas chromatographic method with normal phase high-performance liquid chromatography has been developed for detecting 1-hexanol in plasma (373).

13.4 Toxic Effects

13.4.1 Experimental Studies

1-Hexanol has a low order of acute toxicity. It is moderately irritating to the skin and severely irritating to the eye. 1-Hexanol can produce sensory irritation. No systemic toxicity has been observed from repeated oral exposure to 1-hexanol. 1-Hexanol is not a skin tumor promoter in mice, and it is not a developmental toxicant in rats. The major health effect of 1-hexanol is local irritation and possibly narcosis.

13.4.1.1 Acute Toxicity. The acute oral LD_{50} in rats is 4.87 g/kg (374), 4.59 g/kg (292) in rats, and 4 g/kg in mice (375). The acute dermal LD_{50} in rabbits is 2.53 g/kg (292) and greater than 5 g/kg (371). No deaths occurred in rats exposed to saturated vapors for 8 h (292). Aspiration of 0.2 mL of 1-hexanol caused deaths in 10 out of 10 rats. Death was instantaneous due to respiratory arrest (10).

1-Hexanol was moderately irritating to rabbit skin when applied for 24 h under occlusive conditions (292, 371). 1-Hexanol was severely irritating to the eyes of rabbits (292). In a sensory irritation (Alarie) test, 1-hexanol produced sensory irritation in mice with an RD_{50} of 240 ppm (43).

13.4.1.2 Chronic and Subchronic Toxicity. 1-Hexanol did not affect serum lipids or induce peroxisome proliferation in male F344 rats fed 2% 1-hexanol (about 1.5 g/kg) in the diet for three weeks (376, 377). In a Russian inhalation study, ultrastructural eye effects were reported in rabbits exposed to 28 ppm of hexyl alcohol (1-hexanol was not specifically mentioned) for 6 months (18).

13.4.1.3 Pharmacokinetics, Metabolism, and Mechanisms. Metabolic studies in rabbits indicate that oxidation to hexanoic acid is the major pathway, mediated by alcohol dehydrogenase and aldehyde dehydrogenase (8). Direct conjugation with glucuronic acid is a minor pathway (8).

13.4.1.4 Reproductive and Developmental

Developmental Toxicity

Sprague–Dawley rats were exposed by inhalation to the maximally attainable vapor concentration of 1-hexanol (3500 mg/m^3 or 824 ppm) for 7 h/day on days 1 to 19 of gestation. Resorptions increased slightly, but no malformations were observed (12,356). In an oral gavage teratology screening study, CD rats were dosed with 0, 200, and 1000 mg/kg of 1-hexanol (in corn oil) on days 6 to 15 of gestation. Maternal toxicity occurred at the 1000 mg/kg level as indicated by clinical signs and decreased body weight gain. No embryotoxic or teratogenic effects were observed at either dose level (378).

13.4.1.5 Carcinogenesis. 1-Hexanol was not a skin tumor promoter when applied three times a week for 60 weeks to mice skin that had been initiated with dimethylbenz[a]anthracene (379).

13.4.1.6 Other. 1-Hexanol showed equivocal signs of peripheral neuropathy in rats treated intraperitoneally for 8 months (380). In rat hepatocytes, 0.1 mM of 1-hexanol did not induce peroxisome proliferation (381).

13.4.2 Human Experience

13.4.2.1 Clinical Cases. In a human patch test, 1-hexanol (1% in petrolatum) was neither a skin irritant nor a skin sensitizer (371). Overexposure to 1-hexanol is likely to lead to irritation of the eyes and respiratory tract and possible narcosis. In another human patch test, 1-hexanol was applied to the skin for up to 4 h. The irritation response was significantly lower than that of the positive control (20% sodium dodecyl sulfate). The authors concluded that 1-hexanol would not be classified as a skin irritant using the EU criteria (382).

13.4.2.2.3. Pharmacokinetics, Metabolism, and Mechanisms. The *in vitro* dermal flux in human skin (epidermis) was reported to be 0.044 mg/cm^2/h, indicating a low rate of dermal uptake (7).

13.5 Standards, Regulations, or Guidelines of Exposure

Neither ACGIH nor OSHA have set permissible exposure levels for 1-hexanol. Russia recommends an occupational standard of 10 mg/m^3 (2.4 ppm), but the basis for this standard has not been documented (118).

14.0 2-Ethyl-1-Butanol

14.0.1 CAS Number: *[97-95-0]*

14.0.2 Synonyms: 2-Ethylbutyl alcohol, 2-ethylbutanol

14.0.3 Trade Name: NA

14.0.4 Molecular Weight: 102.2

14.0.5 Molecualr Formula: $C_6H_{14}O$

14.0.6 Molecular Structure:

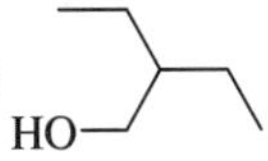

14.1 Chemical and Physical Properties

2-Ethylbutanol exists as a colorless, pleasant-smelling liquid. Its physical and chemical properties are listed in Table 77.1.

14.2 Production and Use

2-Ethylbutanol can be prepared commercially by the aldol condensation of acetaldehyde and 1-butanal and subsequent hydrogenation (3). The alcohol is used as a solvent and flow improver for paints and varnishes, as a component in the manufacture of penetrating oils and corrosion inhibitors, as a cleaning agent for printed circuits, and in the manufacture of plasticizers (3). The primary routes of occupational exposure are dermal contact and inhalation.

14.4 Toxic Effects

14.4.1 Experimental Studies

14.4.1.1 Acute Toxicity. The acute oral LD_{50} in rats is 1.85 g/kg, and the dermal LD_{50} in rabbits is 1.05 g/kg (219) and 4.16 g/kg for guinea pigs (219). No deaths were observed among rats exposed to saturated vapors for 8 h (219).

2-Ethylbutanol is slightly irritating to rabbit skin, but it is severely irritating to the eyes of rabbits (219).

14.4.1.3 Pharmacokinetics, Metabolism, and Mechanisms. 2-Ethylbutanol is oxidized to the corresponding acid, 2-ethylbutyric acid, followed by conjugation with glucuronic acid (229). Rabbits excrete 40% of an ingested dose of 2-ethylbutanol as the urinary glucuronide, diethyl acetyl glucuronide (229).

14.5 Standards, Regulations, or Guidelines of Exposure

No occupational exposure standards exist for 2-ethylbutanol. No adverse effects have been reported in humans from occupational exposure.

15.0 Methyl Isobutyl Carbinol

15.0.1 CAS Number: [108-11-2]

15.0.2 Synonyms: 4-Methyl-2-pentanol, methylamyl alcohol, *sec*-hexyl alcohol, MIBC methylpentanol, 4-methylpentanol-2, 4-methylpentan-2-ol, *sec*-hexyl alcohol, 2-methyl-4-pentanol, isobutylmethylcarbinol

15.0.3 Trade Name: NA

15.0.4 Molecular Weight: 102.2

15.0.5 Molecualr Formula: $C_6H_{14}O$

15.0.6 Molecular Structure:

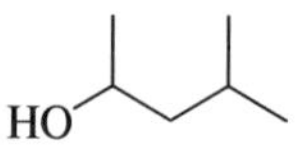

15.1 Chemical and Physical Properties

15.1.1 General

Methyl isobutyl carbinol exists as a colorless liquid which is flammable but not reactive. Other physical and chemical properties are listed in Table 77.1.

15.1.2 Odor and Warning Properties

The air odor threshold for methyl isobutyl carbinol is reported to be 0.07 ppm (17), which may provide some warning of acute exposure.

15.2 Production and Use

Methyl isobutyl carbinol is prepared commercially as a by-product of the synthesis of methyl isobutyl ketone (3). It is used as a solvent in the paint industry, as a brake fluid, as a flotation aid, and as a fungicide (3). No exposure information on methyl isobutyl carbinol was found; however, the main routes of industrial exposure are likely to be skin contact and inhalation. Indirect exposure to methyl isobutyl carbinol may occur from exposure to methyl isobutyl ketone, which can be reduced *in vivo* to methyl isobutyl carbinol.

15.3 Exposure Assessment

15.3.3 Workplace Methods

NIOSH Method No. 1402 is recommended for measuring methyl isobutyl carbinol in air. The method involves drawing vapors from a known volume of air through coconut shell charcoal (recommended sample is 1 to 10 liters at a flow rate of 0.01 to 0.2 L/min). The analyte is desorbed with carbon disulfide containing 5% of 2-propanol, and separated and quantified by gas chromatography with flame ionization detection. The working range is 15 to 150 mg/m^3 (3.5 to 35 ppm) for a 10-liter sample (19).

15.4 Toxic Effects

15.4.1 Experimental Studies

Methyl isobutyl carbinol has a low order of acute toxicity. It is slightly irritating to the skin but is severely irritating to the eyes and respiratory tract. It is not genotoxic.

15.4.1.1 Acute Toxicity. The acute oral LD_{50} is 2.59 g/kg in rats (292) and 1.2 g/kg in mice (383). The acute dermal LD_{50} in rabbits is 2.88 g/kg (292). Five out of six rats died when exposed to 2000 ppm for 8 h (292). There were no deaths among rats exposed to saturated vapors (about 4600 ppm or 20 mg/L) for 2 h (292).

Groups of 10 mice were exposed to 20 mg/L (4600 ppm) of methyl isobutyl carbinol for 1, 4, 8.5, 10, or 15 h. No deaths occurred among the animals exposed up to 8.5 h; however, six and eight mice died after 10 and 15 h of exposure, respectively. Exposure of 1 h caused somnolence and after 4 h, anesthesia. Microscopic examination showed respiratory tract irritation and congestion of the lungs (18, 383).

Methyl isobutyl carbinol is slightly irritating to rabbit skin, but it is moderately irritating to the eyes of rabbits (292). In a sensory irritation (Alarie) test, methyl isobutyl carbinol produced sensory irritation in mice, and had a RD_{50} (the concentration that decreased respiration rate by 50%) of 425 ppm (43).

15.4.1.2 Chronic and Subchronic Toxicity. No deaths were observed among mice exposed 4 h/day for 12 inhalation exposures to 20 mg/L (4600 ppm), although irritation of the respiratory tract was noted (18). Five repeated dermal applications of about 2.5 g/kg to rabbits (five applications during a 15 to 21 day period) produced "severe drying of the skin with some sloughing and cracking," but failed to produce any overt systemic effects or microscopic changes in organs (18, 383).

15.4.1.3 Pharmacokinetics, Metabolism, and Mechanisms. Methyl isobutyl carbinol is oxidized to the corresponding ketone, methyl isobutyl ketone (384). Methyl isobutyl carbinol is also a metabolite of methyl isobutyl ketone (384). No data exist on the absorption or excretion of methyl isobutyl carbinol.

15.4.1.6 Genetic and Related Cellular Effects Studies. Methyl isobutyl carbinol was not mutagenic to either *S. typhimurium* in an Ames test or to *E. coli* in the presence or absence of a metabolic activation system (385).

15.4.1.7 Other. High doses of methyl isobutyl carbinol can enhance the toxic effects of other chemicals. The cholestatic effect of a manganese–bilirubin combination on the livers of rats was enhanced by methyl isobutyl carbinol (386), and chloroform-induced hepatotoxicity was potentiated by methyl isobutyl carbinol in rats (387). This effect has been attributed to methyl isobutyl ketone, the ketone metabolite of methyl isobutyl carbinol, not the alcohol.

15.4.2 Human Experience

15.4.2.1 General Information

15.4.2.2 Clinical Cases. Exposure of human subjects to 50 ppm of methyl isobutyl carbinol vapor for 15 min resulted in nose and throat irritation; 25 ppm was estimated to be the highest concentration acceptable for 8 h (388).

15.5 Standards, Regulations, or Guidelines of Exposure

The ACGIH TLV and the OSHA PEL are 25 ppm (8-h TWA) for methyl isobutyl carbinol (116, 117). The ACGIH has also recommended a skin notation to call attention to a potential hazard from absorption through the skin. A number of other countries have also set occupational exposure standards at 25 ppm (118).

16.0 Isohexyl Alcohol

16.0.2 C_6 Oxo Alcohols

16.1 Chemical and Physical Properties

Isohexyl alcohol is a liquid at room temperature.

16.2 Production and Use

Isohexyl alcohol is produced in the oxo process by the reaction of pentenes with carbon monoxide and hydrogen in the presence of a catalyst, followed by hydrogenation (3). The commercial product typically consists of three isomers with the following composition: 44% 1-hexanol, 53% methylpentanols, and 3% 2-ethylbutanol (3). It is used as a solvent and as a chemical intermediate in the production of esters such as acetates and phthalates

(3). Industrial exposure may occur in handling and use, and indirect exposure may occur by hydrolysis of esters containing these alcohols.

16.4 Toxic Effects

16.4.1 Experimental Studies

16.4.1.1 Acute Toxicity. The acute oral LD_{50} in rats has been reported to be 3.67 g/kg, and the dermal LD_{50} in rabbits is greater than 2.6 g/kg (351). No deaths occurred among mice, rats, or guinea pigs exposed to saturated vapors of 1060 ppm for 6 h (351). However, the animals showed moderate irritation of the upper respiratory tract, slight congestion of the lungs, and a questionable effect on the central nervous system.

Isohexyl alcohol is moderately irritating to the skin of rabbits when applied for 24 h under occlusive conditions (351). The material is severely irritating to the eyes of rabbits and causes persistent iritis, corneal opacity, and, in two animals, corneal vascularization (351).

16.4.1.2 Chronic and Subchronic Toxicity. A 2-week dermal study in rabbits at doses of 0.5 and 2.5 mL/kg of a C_6 oxo alcohol showed evidence of skin irritation but no systemic effects (Exxon Chemical Company, unpublished data).

16.4.2 Human Experience

16.4.2.1 General Information. No data were found.

16.5 Standards, Regulations, or Guidelines of Exposure

No occupational exposure standards have been established for the C_6 oxo alcohols.

17.0 Methyl-1-Pentanols

17.0a 2-Methyl-1-Pentanol

17.0b 3-Methyl-1-Pentanol

17.0c 4-Methyl-1-Pentanol

17.0.1a CAS Number: *[105-30-6]*

17.0.1b CAS Number: *[589-35-5]*

17.0.1c CAS Number: *[626-89-1]*

17.0.2 Synonyms: 3-Methylpentanol, 2-Ethyl-4-butanol

17.0.3 Trade Name: NA

17.0.4 Molecular Weight: 102.2

17.0.5 Molecular Formula: $C_6H_{14}O$

17.0.6a Molecular Structure:

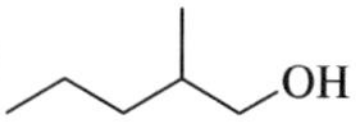

17.0.6b Molecular Structure:

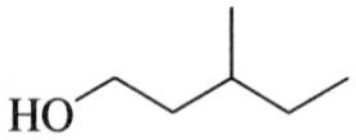

17.0.6c Molecular Structure:

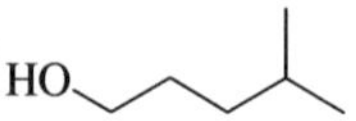

17.1 Chemical and Physical Properties

Methyl-1-pentanols exist as liquids at room temperature. Their physical and chemical properties are listed in Table 77.1.

17.2 Production and Use

2-Methyl-1-pentanol is prepared by the aldol condensation of propionaldehyde and subsequent hydrogenation of the intermediate 2-methyl-2-pentanol (3). Commercially, the methyl-1-pentanols are rarely used alone but as a mixture. For instance, isohexyl alcohol contains a mixture of methyl-1-pentanols, as well as 1-hexanol (3). The uses of these alcohols are in the production of esters such as phthalates and acetates. 2-Methyl-1-pentanol is also used as a solvent (3). Industrial exposure to these alcohols is by dermal contact and by inhalation.

17.4 Toxic Effects

17.4.1 Experimental Studies

17.4.1.1 Acute Toxicity. For 2-methyl-1-pentanol, the acute oral LD_{50} in the rat is 1.41 g/kg, and the acute dermal LD_{50} in rabbits is 3.56 mL/kg (219). For 4-methyl-1-pentanol, the acute oral LD_{50} in the rat is 6.5 mL/kg, and the acute dermal LD_{50} in rabbits is 3.97 mL/kg (389). No deaths occurred among rats exposed to saturated vapors of either 2-methyl-1-pentanol or 4-methyl-1-pentanol for 8 h (219,389). No toxicity data were found for 3-methyl-1-pentanol.

2-Methyl-1-pentanol is slightly irritating to the skin of rabbits and is severely irritating to rabbit eyes (219). 4-Methyl-1-pentanol is slightly irritating to the skin of rabbits and is moderately irritating to rabbit eyes (389).

17.4.2 Human Experience

No data were found.

17.5 Standards, Regulations, or guidelines of Exposure

No exposure standards exist for these alcohols.

18.0 Secondary Alcohols

18.0a 2-Hexanol

18.0b 3,3-Dimethyl-2-butanol

18.0c 2,2-Dimethyl-1-butanol

18.0.1a CAS Number: *[626-93-7]*

18.0.1b CAS Number: *[464-07-3]*

18.0.1c CAS Number: *[1185-33-7]*

18.0.2a Synonyms:

18.0.2b Synonyms: 3,3-Dimethylbutan-2-ol, 3,3-dimethyl-2-butanol, *tert*-butyl methyl carbinol, pinacolyl alcohol

18.0.2c Synonyms: 2,2-Dimethylbutanol

18.0.3 Trade Name: NA

18.0.4 Molecular Weight: 102.2

18.0.5 Molecular Formula: $C_6H_{14}O$

18.0.6a Molecular Structure:

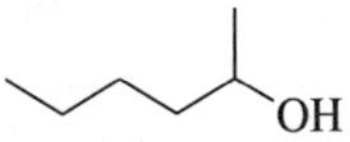

18.0.6b Molecular Structure:

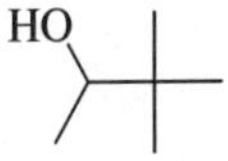

18.0.6c Molecular Structure:

18.1 Chemical and Physical Properties

The physical and chemical properties are listed in Table 77.1.

18.2 Production and Use

The secondary hexanols are of low commercial importance. However, there is some toxicological interest because 2-hexanol can produce peripheral neuropathy in animals, and 3,3-dimethyl-2-butanol is nephrotoxic in male rats. No exposure data exists for 2-hexanol; this lack may be due to its limited commercial use. Indirect exposure to 2-hexanol may occur by exposure to *n*-hexane which is known to be metabolized to 2-hexanol (390, 391). 3,3-Dimethyl-2-butanol is also known as pinacolyl alcohol (392). No commercial uses were found for this material.

18.3 Exposure Assessment

18.3.5 Biomonitoring/Biomarkers

A method involving gas chromatography and normal-phase high-performance liquid chromatography is available to measure 2-hexanol in plasma (373). There is also an analytical method for measuring 2-hexanol in urine involving gas chromatography with flame ionization detection. The detection limit was in the range of 0.2 to 0.7 mg/L and between 0.05 to 0.1 mg/L when a packed or capillary column was used (393). Another gas chromatography method is also available to measure 2-hexanol in urine (372).

18.4 Toxic Effects

18.4.1 Experimental Methods

18.4.1.1 Acute Toxicity. No data were found for 2-hexanol. For 3,3-dimethyl-2-butanol, no deaths occurred among rats exposed to 5 mg/L (1200 ppm) of 3,3-dimethyl-2-butanol vapor for 6 h (394). For 2,2-dimethyl-1-butanol, the acute oral LD_{50} in rats is 2.33 mL/kg, and the acute dermal LD_{50} in rabbits is 1.77 mL/kg (389). 2,2-Dimethyl-1-butanol is reported to be a slight skin irritant and a severe eye irritant in rabbits (389).

18.4.1.2 Chronic and Subchronic Toxicity. 2-Hexanol has been tested in several subchronic studies. In an oral gavage study, male rats were dosed with 2-hexanol for 90 days. Clinical and histological evidence of peripheral neuropathy was observed. Additionally, atrophy of the testicular germinal epithelium was observed (395). When rats were given 101 mg/kg/day of 2-hexanol by intraperitoneal injection 6 days/weeks for 20 weeks and subsequently in doses of 152 mg/kg/day in two injections for another 14 weeks, peripheral neuropathy was observed, which was characterized by reduced nerve conduction velocity and huge axonal swelling due to accumulation of packed filaments (380). It is believed that the neurotoxicity of 2-hexanol results from metabolism to 2,5-hexanedione because the peripheral neuropathy produced by both chemicals is similar.

3,3-Dimethyl-2-butanol has been tested in an inhalation subchronic study. Sprague–Dawley rats were exposed to 0, 48, 240, and 1200 ppm, 6 h/day, 5 days/week for 13 weeks. Clinical signs were limited to the high-dose groups and included ataxia and lacrimation. Kidney and ovary weights were significantly increased in the high-dose males and females, respectively. Histological effects were observed in the kidneys of the mid- and high-dose male rats. These lesions consisted of tubular regeneration in the cortex and outer medulla with occasional epithelial necrosis and, in high-dose, proteinaceous casts in the outer medulla (392).

18.4.1.3 Pharmacokinetics, Metabolism, and Mechanisms. 2-Hexanol is oxidized to methyl-*n*-butyl ketone, which can be further metabolized to 2,5-hexanedione, the putative neuropathic form of *n*-hexane, and methyl-*n*-butyl ketone (390, 391). No metabolism data were found for 3,3-dimethyl-2-butanol.

18.5 Standards, Regulations, or Guidelines of Exposure

No occupational standards exist for these chemicals, and there have been no adverse effects reported in humans from occupational exposure.

BIBLIOGRAPHY

1. National Fire Protection Association (NFPA), *Fire Protection Guide on Hazardous Materials*, 9th ed., NFPA, 1986.
2. National Fire Protection Association (NFPA), *Fire Protection Guide on Hazardous Materials*, 10th ed., NFPA, 1991.
3. J. Falbe et al., in *Ullman's Encyclopedia of Industrial Chemistry*, 5th rev. ed., Vol. A1, VCH Publishers, New York, 1984, pp. 279–303.
4. *Chem. Eng. News*, June 24 (1996).
5. M. F. Gautreaux, W. T. Davis, and E. D. Travis, in M. Grayson and D. Eckroth, eds., *Kirk-Othmer Encyclopedia of Chemical Technology*, 3rd ed., Vol. 1, Wiley, New York, 1982, pp. 740–752.
6. R. A. Peters, in M. Grayson and D. Eckroth, eds., *Kirk-Othmer Encyclopedia of Chemical Technology*, 3rd ed., Vol. 1, Wiley, New York, 1978, pp. 716–739.
7. R. J. Schueplein and I. H. Blank, *Physiol. Rev.* **51**, 702–747 (1971).
8. R. T. Williams, in *The Metabolism and Detoxication of Drugs, Toxic Substances and Other Organic Compounds*, 2nd ed., Chapman & Hall, London, 1959.
9. R. C. Baselt, *Biological Monitoring Methods for Industrial Chemicals*, PSG Publishing, Littleton, ME, 1988.
10. H. W. Gerarde, and D. B. Ahlstrom, *Arch. Environ. Health* **13**, 457–461 (1966).
11. P. M. J. Bos et al., *Crit. Rev. Toxicol.* **21**(6), 423–450 (1992).
12. B. K. Nelson, W. S. Brightwell, and E. F. Krieg, Jr., *Toxicol. Ind. Health* **6**, 373–387 (1990).
13. C. S. Weil, H. F. Smyth, Jr., and T. W. Nale, *Arch. Ind. Hyg. Occup. Med.* **5**, 535–547 (1952).
14. J. Lynch et al., *J. Occup. Med.* **21**, 333–341 (1979).
15. C. L. Soskolne et al., *Am. J. Epidemiol.* **120**, 358–369 (1984).
16. A. English, J. Rovner, and S. Davies, in J. Kroschwitz ed., *Kirk-Othmer Encyclopedia of Chemical Technology*, 4th ed., Vol. 16, Wiley, New York, 1995, pp. 537–554.
17. J. E. Amoore and E. Hautala, *J. Appl. Toxicol.* **3**, 272–290 (1983).
18. V. K. Rowe and S. B. McCollister, in G. D. Clayton and F. E. Clayton, eds., *Patty's Industrial Hygiene and Toxicology*, 3rd rev. ed., Wiley, New York, 1982, pp. 4527–4708.
19. National Institute for Occupational Safety and Health (NIOSH), *Manual of Analytical Methods*, 3rd ed., Vols. 1 and 2, U.S. Department of Health and Human Services, Cincinnati, OH, 1984.
20. P. W. Langvardt and R. G. Melcher, *Am. Ind. Hyg. Assoc. J.* **40**, 1006–1012 (1979).
21. F. X. Mueller and J. A. Miller, *Am. Ind. Hyg. Assoc. J.* **40**, 380–386 (1979).
22. V. Sedivec, M. Mraz, and J. Flek, *Int. Arch. Occup. Environ. Health* **48**, 257–271 (1981).
23. T. Kawai, J. Hirashima, and S. Horiguchi, *J. Sci. Labour* **63**, 21–24 (1987).
24. R. Heinrich and J. Angerer, *Int. Arch. Occup. Environ. Health* **50**, 341–349 (1982).
25. G. M. Pollack and J. L. Kawagoe, *J. Chromatogr.* **570**, 406–411 (1991).
26. J. Liesivuori and H. Savolainen, *Am. Ind. Hyg. Assoc. J.* **48**, 32–34 (1987).
27. K. Baumann and J. Angerer, *Int. Arch. Occup. Environ. Health* **42**, 241–249 (1979).
28. B. Dutkiewicz, H. Korczalik, and W. Karwacki, *Int. Arch. Occup. Environ. Health* **47**, 81–88 (1980).
29. D. G. Ferry, W. A. Temple, and E. G. McQueen, *Int. Arch. Occup. Environ. Health* **47**, 155–163 (1980).

30. T. Kawai et al., *Bull. Environ. Contain. Toxicol.* **47**, 797–803 (1991).
31. A. Franzblau et al., *App. Occup. Environ. Hyg.* **7**, 467–471 (1992).
32. E. T. Kimura, D. M. Ebert, and P. W. Dodge, *Toxicol. Appl. Pharmacol.* **19**, 699–703 (1971).
32a. H. Welch and G. G. Slocum, *J. Lab. Chem. Med.* **28**, 1440 (1943).
33. W. B. Deichman and E. G. Mergard, *J. Ind. Hyg. Toxicol.* **30**, 373–378 (1948).
34. H. F. Smyth, Jr., J. Seaton, and L. Fisher, *J. Ind. Hyg. Toxicol.* **23**, 259–268 (1941).
35. J. C. Munch, *Ind. Med. Surg.* **41**, 31–33 (1972).
36. A. P. Gilger and A. M. Potts, *Am. J. Ophthalmol.* **39**, 63–85 (1955).
36a. J. R. Cooper and P. Felig, *Toxicol. Appl. Pharmacol.* **3**, 202 (1961).
37. H. Weese, *Arch. Exp. Pathol. Pharmakol.* **135**, 118–130 (1928).
38. L. M. Mashbitz, R. M. Sklianskaya, and F. M. Urieva, *J. Ind. Hyg. Toxicol.* **18**, 117 (1936).
39. A. Loewy and R. von der Heide, *Biochem. Z.* **86**, 125 (1918).
39a. R. Muller, *Z. Angew. Chem.* **23**, 351 (1910).
40. A. P. Gilger, A. M. Potts, and L. V. Johnson, *Am. J. Ophthalmol.* **35**(Pt. II), 113–126 (1952).
41. G. A. Jacobs, *J. Am. Coll. Toxicol., Part B* **1**(1), 56 (1990).
42. C. P. Carpenter and H. F. Smyth, Jr., *Am. J. Ophthalmol.* **29**, 1363–1372 (1946).
43. J. Muller and G. Greff, *Food Chem. Toxicol.* **22**, 661–664 (1984).
44. L. E. Kane, R. Dombroske, and Y. Alarie, *Am. Ind. Hyg. Assoc. J.* **41**, 451–455 (1980).
45. L. S. Andrews et al. *J. Toxicol. Environ. Health* **20**, 117–124 (1987).
46. L. R. White et al., *Toxicol. Lett.* **17**, 1–5 (1983).
47. W. A. Reynolds et al., in L. D. Stegink and L. J. Filer, Jr., eds., *Aspartame: Physiology and Biochemistry*, Dekker, New York, 1984, p. 405.
48. U.S. Environmental Protection Agency (USEPA), *Rat Oral Subchronic Study on Methanol*, USEPA, Washington, DC, 1986; cited in IRIS Database for Methanol, Ref. 298.
49. H. P. Kloecking, *Fortschr. Wasserchem. Ihre Grenzgeb.* **14**, 189 (1972).
49a. S. M. Pavlenko, *Gig. Sanit.* **37**(1), 40 (1972).
50. R. Kavet and K. M. Nauss, *CRC Crit. Rev. Toxicol.* **20**, 21–50 (1990).
51. J. Liesivuori and H. Savolainen, *Pharmacol. Toxicol.* **69**, 157–163 (1991).
52. T. R. Tephly, *Life Sci.* **48**, 1031–1041 (1991).
53. W. P. Yant and H. H. Schrenk, *J. Ind. Hyg. Toxicol.* **19**, 337–345 (1937).
54. R. R. Sayers et al., *J. Ind. Hyg. Toxicol.* **26**, 255–259 (1944).
54a. H. W. Haggard and L. A. Greenberg, *J. Pharmacol.* **66**, 479 (1939).
55. C. L. Gummer and H. I. Maibach, *Food Chem. Toxicol.* **24**, 305–309 (1986).
56. H. H. Euler and K.-H. Gedicke, *Arch. Toxikol.* **15**, 409–414 (1955).
57. J. T. Eells et al., *J. Pharmacol. Exp. Ther.* **227**, 349–353 (1983).
58. M. Saito, *Nichidai Igaku Zasshi* **34**, 569–585 (1975).
59. D. Gaillard and R. Derache, *Trav. Soc. Pharm. Montpellier* **25**, 51–62 (1965).
60. J. T. Bastrup, *Acta Pharmacol.* **3**, 303 (1947).
60a. A. Lund, *Acta Pharmacol.* **4**, 99 (1948).
60b. A. Lund, *Acta Pharmacol.* **4**, 108 (1948).
61. V. L. Horton et al., *Toxicologist* **6** (Abstr. No. 1044) (1986).

62. V. L. Horton, M. A. Higuchi, and D. E. Richert, *Toxicologist* **7**, 233 (Abstr. No. 932) (1987).
63. K. L. Clay, R. C. Murphy, and W. D. Watkins, *Toxicol. Appl. Pharmacol.* **34**, 49–61 (1975).
64. K. E. McMartin et al., *J. Pharmacol. Exp. Ther.* **201**, 564–572 (1977).
65. K. A. Black et al., *Proc. Natl. Acad. Sci. U.S.A.* **82**, 3854–3858 (1985).
66. F. C. Johlin et al., *Mol. Pharmacol.* **31**, 557–561 (1987).
67. A. Sharma et al., *Toxicologist* **12**, 102 (Abstr. No. 321) (1992).
68. R. Infurna and G. G. Berg, *Toxicologist* **2**, 73 (Abstr. No. 260) (1982).
69. D. C. Dorman et al., *Toxicol. Appl. Pharmacol.* **128**, 229–238 (1994).
70. V. L. Horton, M. A. Higuchi, and D. E. Rickert, *Toxicol. Appl. Pharmacol.* **117**, 26–36 (1992).
71. K. W. Ward et al., *Toxicol. Appl. Pharmacol.* **145**, 311–322 (1997).
72. B. K. Nelson et al., *Fundam. Appl. Toxicol.* **5**, 727–736 (1985).
73. J. M. Rogers, N. Chernoff, and M. L. Mole, *Toxicologist* **11**, 344 (Abstr. No. 1350) (1991).
74. B. Bolon et al., *Fundam. Appl. Toxicol.* **21**, 508–516 (1993).
75. A. F. Youssef et al., *Teratology* **43**, 467 (Abstr.) (1991).
76. J. E. Andrews, M. Ebron-McCoy, and J. M. Rogers, *Teratology* **43** 461 (Abstr.) (1991).
77. R. Infurna and B. Weiss, *Teratology* **33**, 259–265 (1986).
78. M. E. Stanton et al., *Fundam. Appl. Toxicol.* **28**, 100–110 (1995).
79. S. Stern et al., *Fundam. Appl. Toxicol.* **36**, 163–176 (1997).
80. A. M. Cameron et al., *Arch. Toxicol. Suppl.* **7**, 441–443 (1984).
81. A. M. Cameron et al., *Arch. Toxicol. Suppl.* **18**, 422–424 (1985).
82. R. L. Cooper et al., *Toxicology* **71**, 69–81 (1992).
83. E. Lee et al., *Toxicol. Ind. Health* **7**, 261–275 (1991).
84. E. W. Lee, A. N. Brady, and R. A. Hess, *Toxicologist* **14**, 75 (Abstr. No. 207) (1994).
85. R. L. Cooper et al., *Toxicologist* **10**, 211 (Abstr. No. 843) (1990).
86. New Energy Development Organization (NEDO), *Summary Report of the Workshop on Methanol Vapors and Health Effects: What We Know and What We Need to Know*, JOSI Risk Science Institute, 1989.
87. S. De Flora et al., *Mutat. Res.* **133**, 161–198 (1984).
88. S. De Flora et al., *Mutat. Res.* **134**, 159–165 (1984).
89. I. Florin et al., *Toxicology* **18**, 219–232 (1980).
90. E. Gocke et al., *Mutat. Res.* **90**, 91–109 (1981).
91. A. Abbondandolo et al., *Mutat. Res.* **79**, 141–150 (1980).
92. C. Lasne et al., *Mutat. Res.* **130**, 273–282 (1984).
93. C. Heidelberger et al., *Mutat. Res.* **114**, 283–385 (1983).
94. J. A. Campbell et al., *Mutat. Res.* **260**, 257–264 (1991).
95. K. O'Loughlin et al., *Environ. Mutagen Soc.* **47** (Abstr.) (1992).
96. K. Aarsted, K. Zahlsen, and O. G. Nilsen, *Arch. Toxicol. Suppl.* **7**, 295–297 (1984).
97. S. D. Ray and H. M. Mehendale, *Fundam. Appl. Toxicol.* **15**, 429–440 (1990).
98. ,A. McDonald et al., *Toxicologist* **12**, 239 (Abstr. No. 891) (1992).
99. J. W. Allis et al., *Toxicologist* **12**, 85 (Abstr. No. 254) (1992).

100. W.-R. Delbrück, A. Kluge, and U. Täuber, DGMK-Project 260-07, Germany Society for Petroleum Sciences and Coal Chemistry, 1982.

101. H. S. Posner, *J. Toxicol. Environ. Health* **1**, 153–171 (1975).

102. Ö. Roe, *CRC Crit. Rev. Toxicol.* **10**, 275–286 (1982).

103. P. Mahieu, A. Hassourn, and R. Lauwerys, *Hum. Toxicol.* **8**, 135–137 (1989).

104. D. Jacobsen and K. E. McMartin, *Med. Toxicol.* **1**, 309–334 (1986).

105. H. H. Tyson and M. J. Schoenberg, *J. Am. Med. Assoc.* **63**, 915 (1914).

106. American Conference of Governmental Industrial Hygienists (ACGIH), *Documentation of the Threshold Limit Values and Biological Exposure Indices*, ACGIH, Cincinnati, OH, 1986.

107. L. J. Frederick, P. A. Schulte, and A. Apol, *Am. Ind. Hyg. Assoc. J.* **45**, 51–55 (1984).

108. A. A. Fisher, in *Contact Dermatitis*, 3rd ed., Lea & Febiger, Philadelphia, 1986, pp. 531–545.

109. S. Fregert et al., *J. Allergy* **34**, 404–408 (1963).

110. S. Fregert et al., *Acta Derm. Venereol.* **49**, 493–497 (1969).

111. G. Leaf and L. J. Zatman, *Br. J. Ind. Med.* **9**, 19–31 (1952).

112. E. W. Lee et al., *Am. Ind. Hyg. Assoc. J.* **53**, 99–104 (1992).

113. H. S. Posner, *Toxicologist* **2**, 89 (Abstr. No. 314) (1982).

114. K. Nakaaki, S. Fukabori, and O. Tada, *J. Sci. Labour Part 2* **56**, 1–9 (1980).

115. S. P. Ericksen and A. B. Kulkarni, *Science* **141**, 539–540 (1963).

116. American Conference of Governmental Industrial Hygienists (ACGIH), *Threshold Limit Values for Chemical Substances and Physical Agents and Biological Exposure Indices*, ACGIH, Cincinnati, OH, 1997.

117. U.S. Department of Labor, Occupational Safety and Health Administration (OSHA), *Fed. Regist.* **62**, 42018 (1997).

118. International Labour Office (ILO), *Occupational Exposure Limits for Airborne Toxic Substances; Values of Selected Countries Prepared from the ILO-CIS Data Base of Exposure Limits*, 3rd ed., Occup. Saf. Health Ser. No. 37, ILO, Geneva, 1991.

119. J. E. Logsdan, in J. Kroschwitz, ed., *Kirk-Othmer Encyclopedia of Chemical Technology*, 4th ed., Vol. 9, Wiley, New York, 1991, pp. 812–860.

120. T. J. Kneip and J. V. Crable, eds., *Methods for Biological Monitoring: A Manual for Assessing Human Exposure to Hazardous Substances*, American Public Health Association, Washington, DC, 1988, p. 327.

120a. W. S. Specter, ed., *Handbook of Toxicology*, Vol. 1, Sanders, Philadelphia and London, 1956, p. 128; cited by H. Maling, *Int. Encycl. Pharmacol. Ther.* **20**, 277 (1970).

120b. W. Bartsch et al., *Arzneim.-Forsch.* **26**(8), 1581 (1976).

121. J. C. Munch and E. W. Schwartze, *J. Lab. Clin. Med.* **10**, 985 (1925).

121a. I. Takeda, *Nichidai Igaku Zasshi* **31**(6), 518 (1972).

122. K. B. Lehmann and F. Flury, *Toxikologie und Hygiene der Technischen Losungsmittel*, Springer, Berlin, 1938, p. 152.

123. H. F. Smyth, Jr., *Am. Ind. Hyg. Assoc. Q.* **17**, 129–185 (1956).

124. W. L. Guess, *Toxicol. Appl. Pharmacol.* **16**, 382–390 (1970).

125. L. Phillips, II et al., *Toxicol. Appl. Pharmacol.* **21**, 369–382 (1972).

126. W. Q. Sargeant, J. R. Simpson, and J. D. Beard, *J. Pharmacol. Exp. Ther.* **188**, 461–471 (1974).

127. L. Kager and J. Ericsson, *Acta Pathol. Microbiol. Scand., Sect. A* **82A**, 534 (1974).
128. H. Tsukamoto et al., *Hepatology* **6**, 814–822 (1986).
129. H. Tsukamoto et al., *Hepatology* **5**, 224–232 (1985).
130. C. S. Lieber, L. M. deCarli, and E. Rubin, *Proc. Natl. Acad. Sci. U.S.A.* **72**, 437–441 (1975).
130a. H. Mertens, *Arch. Int. Pharmacodyn.* **2**, 127 (1896).
131. H. F. Smyth and H. F., Smyth, Jr., *J. Ind. Hyg.* **10**, 261–271 (1928).
132. R. A. Coon et al., *Toxicol. Appl. Pharmacol.* **16**, 646–655 (1970).
133. International Agency for Research on Cancer, IARC Monogr. Vol. 44, IARC, Geneva, 1988.
134. C. S. Lieber, *Rev. Biochem Toxicol* **5**, 267–311 (1983).
135. H. W. Haggard, D. P. Miller, and L. A. Greenberg, *J. Hyg. Toxicol.* **27**, 1–14 (1945).
136. M. Hayashi et al., *Bull. Environ. Contam. Toxicol.* **47**, 184–189 (1991).
137. D. W. Clarke et al., *Can. J. Physiol. Pharmacol.* **64**, 1060–1067 (1986).
138. C. S. Lieber and L. M. DeCarli, *J. Pharmacol. Exp. Ther.* **181**, 279–287 (1972).
139. J. Alderman, S. Kato, and C. S. Lieber, *Arch. Biochem. Biophys.* **271**, 33–39 (1989).
140. L. F. Leloir, and J. M. Muñoz, *Biochem. J.* **32**, 299–307 (1938).
141. E. A. Carter and K. J. Isselbacher, *Proc. Soc. Exp. Biol. Med.* **138**, 817–819 (1971).
142. Y. Lamboeuf, G. de Saint Blanquat, and R. Derache, *Biochem. Pharmacol.* **30**, 542–545 (1981).
143. H. W. Newman and A. J. Lehman, *Arch. Int. Pharmacodyn.* **55**, 440–446 (1937).
144. J. S. Aull, Jr., W. J. Roberts, Jr., and F. W. Kinard, *Am. J. Physiol.* **186**, 380–382 (1956).
145. D. E. McLain, D. A. Roe, and J. G. Babish, *Toxicologist* **4**, 143 (Abstr. No. 572) (1984).
146. G. M. Pastino et al., *Toxicol. Appl. Pharmacol.* **145**, 147–157 (1997).
147. G. M. Pastino, L. G. Sultatos, and E. J. Flynn, *Alcohol Alcohol.* **31**, 365–374 (1996).
148. B. A. Schwetz, F. A. Smith, and R. E. Staples, *Teratology* **18**, 385–392 (1978).
149. G. F. Chernoff, *Teratology* **15**, 223–230 (1977).
150. R. F. Mankes et al., *J. Toxicol. Environ. Health* **10**, 871–878 (1982).
151. P. M. Blakley and W. J. Scott, Jr., *Toxicol. Appl. Pharmacol.* **72**, 355-363 (1984).
152. P. M. Blakley and W. J. Scott, Jr., *Toxicol. Appl. Pharmacol.* **72**, 364–371 (1984).
153. J. M. Wynter et al., *Teratog., Carcinog., Mutagen.* **3**, 421–428 (1983).
154. E. S. Hunter et al., *Toxicologist* **12**, 332 (Abstr. No. 1289) (1992).
155. E. Giavini et al., *Teratology* **44**, 30A (Abstr.) (1991).
156. U. Kristiansen, A. M. Vinggaard, and G. D. Nielsen, *Arch. Toxicol.* **61**, 229–236 (1988).
157. S. Laham et al., *Drug Chem. Toxicol.* **3**, 343–360 (1980).
158. J. F. Oisund, A. E. Fjorden, and J. Morland, *Acta Pharmacol. Toxicol.* **43**, 145–155 (1978).
159. R. A. Anderson, Jr., B. R. Willis, and C. Oswald, *Alcohol* **2**, 479–484 (1985).
160. R. W. Klassen and T. V. N. Persaud, *Int. J. Fertil.* **23**, 176–184 (1978).
161. R. E. Morrissey et al., *Fundam. Appl. Toxicol.* **13**, 747–777 (1989).
162. A. Horie, S. Kohchi, and M. Kuratsume, *Gann* **56**, 429–441 (1965).
163. G. N. Schrauzer et al., *J. Stud. Alcohol* **40**, 240–246 (1979).
164. W. Gibel, *Arch. Geschwulstforsch.* **30**, 181–189 (1967).

165. B. Homberg and Y. Ekström, *Toxicology* **96**, 133–145 (1996).
166. L. Griciute et al., *Cancer Lett.* **31**, 267–275 (1986).
167. D. Schmähl, *Cancer Lett.* **1**, 215–218 (1976).
168. A. J. Garro and C. S. Lieber, *Annu. Rev. Pharmacol. Toxicol.* **30**, 219–249 (1990).
169. H. E. Driver and A. E. M. McClean, *Food Chem. Toxicol.* **24**, 241–245 (1986).
170. S. Mitjavila, J. Faccini, and A. E. M. McLean, *Hum. Exp. Toxicol.* **10**, 69 (Abstr.) (1991).
171. G. Obe and D. Anderson, *Mutat. Res.* **186**, 177–200 (1987).
172. R. D. Blevins and D. E. Taylor, *J. Environ. Sci. Health, Part A* **A17**(2), 217–239 (1982).
173. J. McCann et al., *Proc. Natl. Acad. Sci. U.S.A.* **72**, 5135–5139 (1975).
174. E. Zeiger et al., *Environ. Mol. Mutagen.* **19**(Suppl. 21), 2–141 (1992).
175. L. Hellmer and G. Bolcsfoldi, *Mutat. Res.* **272**, 145–160 (1992).
176. R. Crebelli et al., *Mutat. Res.* **215**, 187–195 (1989).
177. Z. Harsanyi, J. A. Granek, and D. W. R. MacKenzie, *Mutat. Res.* **48**, 51–74 (1977).
178. H. E. Brockman et al., *Mutat. Res.* **133**, 87–134 (1984).
179. D. E. Amacher et al., *Mutat. Res.* **72**, 447–474 (1980).
180. G. Obe and H. Ristow, *Mutat. Res.* **56**, 211–213 (1977).
181. W. K. De Raat, P. B. Davis, and G. L. Bakker, *Mutat. Res.* **124**, 85–90 (1983).
182. R. C. Chaubey et al., *Mutat. Res.* **43**, 441–444 (1977).
183. M. Watanabe et al., *Mutat. Res.* **97**, 43–48 (1982).
184. G. Obe et al., *Mutat. Res.* **68**, 291–294 (1979).
185. B. N. Nayak and H. S. Buttar, *Teratog., Carcinog. Mutagen.* **6**, 83–91 (1989).
186. E. Baraona, M. Guerra, and C. S. Lieber, *Life Sci.* **29**, 1797–1802 (1981).
187. A. D. Tates, N. de Vogel, and I. Neuteboom, *Mutat. Res.* **79**, 285–288 (1980).
188. P. S. Chauhan et al., *Mutat. Res.* **79**, 263–275 (1980).
189. R. W. Klassen and T. V. N. Persaud, *Exp. Pathol.* **12**, 38–45 (1976).
190. F. M. Badr and R. S. Badr, *Nature (London)* **253**, 134–136 (1975).
191. J. Descotes, *J. Toxicol. Cutaneous Ocul. Toxicol.* **7**, 263–272 (1988).
192. D. R. Koop and M. J. Coon, *Alcohol.: Clin. Exp. Res.* **10**, 445–495 (1986).
193. H. J. Zimmerman, *Alcohol.: Clin. Exp. Res.* **10**, 3–15 (1986).
194. O. Strubelt, *Fundam. Appl. Toxicol.* **4**, 144–151 (1984).
195. C. S. Lieber, *J. Am. Coll. Nutr.* **10**, 602–632 (1991).
196. L. D. Keppen, T. Pysher, and O. M. Rennert, *Pediatr. Res.* **19**, 944–947 (1985).
197. S. I. Miller et al., *Pharmacol., Biochem. Behav.* **18**(Suppl. 1), 311–315 (1983).
198. E. R. Eichner and R. S. Hillman, *J. Clin. Invest.* **52**, 584–591 (1973).
199. V. Hebert, R. E. Zalusky, and C. S. Davidson, *Ann. Intern. Med.* **58**, 977–988 (1963).
200. C. A. Marietta et al., *Alcohol.: Clin. Exp. Res.* **12**, 211 (1988).
201. F. Malik and S. N. Wickramasinghe, *Br. J. Exp. Pathol.* **67**, 831–838 (1986).
202. B. W. Hills, and H. L. Venable, *Am. J. Ind. Med.* **3**, 321–333 (1982).
203. D. Lester and L. A. Greenberg, *Q. J. Stud. Alcohol* **12**, 167–178 (1951).
204. M. J. Ellenhorn and D. G. Barceloux, in *Medical Toxicology: Diagnosis and Treatment of Human Poisoning*, Elsevier, New York, 1988, pp. 781–814.

205. T. A. Gossel and J. D. Bricker, in *Principles of Clinical Toxicology*, 2nd ed., Raven Press, New York, 1990, pp. 65–75.
206. J. Stotts and W. J. Ely, *J. Invest. Dermatol.* **69**, 219–222 (1977).
207. J. F. Treon, in D. W. Fassett and D. D. Irish, eds., *Patty's Industrial Hygiene and Toxicology*, 2nd rev. ed., Vol. 2, Interscience, New York, 1963, pp. 1409–1496.
208. M. Kopun and P. Propping, *Eur. J. Clin. Pharmacol.* **11**, 337–344 (1977).
209. D. P. Agarwal, S. Harada, and H. W. Goedde, *Alcohol. Clin. Exp. Res.* **5**, 12–16 (1981).
210. H. Nuutinen, K. O. Lindros, and M. Salaspuro, *Alcohol. Clin. Exp. Res.* **7**, 163–168 (1983).
211. I. Ijiri, *Jpn. J. Stud. Alcohol* **9**, 35–59 (1974).
212. K. Inoue, M. Fukunaga, and K. Yamasawa, *Pharmacol., Biochem. Behav.* **13**, 295–297 (1980).
213. Y. Mizoi et al., *Pharmacol., Biochem. Behav.* **10**, 303–311 (1979).
214. C. L. Randall, *Alcohol, Suppl.* **1**, 125–132 (1987).
215. J. L. Schardein and K. A. Keller, *Crit. Rev. Toxicol.* **19**, 251–339 (1989).
216. J. D. Unruh and D. Pearson, in J. Kroschwitz, ed., *Kirk-Othmer Encyclopedia of Chemical Technology*, 4th ed., Vol. 20, Wiley, New York, 1996, pp. 241–248.
217. World Health Organization, *Environmental Health Criteria* 102; WHO, Geneva, 1990.
218. H. M. Liebich, H. J. Buelow, and R. Kallmayer, *J. Chromatogr.* **239**, 343–349 (1982).
219. H. F. Smyth, Jr. et al., *Arch. Ind. Hyg. Occup. Med.* **10**, 61–68 (1954).
220. P. M. Jenner et al., *Food Cosmet. Toxicol.* **2**, 327–343 (1943).
221. U. Kristiansen et al., *Acta Pharmacol. Toxicol.* **59**, 60–72 (1986).
222. M. E. Hillbom, K. Franssila, and O. A. Forsander, *Res. Commun. Chem. Pathol. Pharmacol.* **9**, 177–180 (1974).
223. T. Wakabayashi et al., *Acta Pathol. Jpn.* **34**, 471–480 (1984).
224. F. Beauge et al., *Chem.-Biol. Interact.* **26**, 155–166 (1979).
225. S. I. Oerskov, *Acta Physiol. Scand.* **20**, 258–262 (1950).
226. U. Abshagen and N. Rietbrock, *Naunyn-Schmiedebergs Arch. Pharmakol.* **265**, 411–424 (1970).
227. E. T. Morgan, D. R. Koop, and M. J. Coon, *J. Biol. Chem.* **257**, 13951–13957 (1982).
228. R. Teschke, Y. Hasumura, and C. S. Lieber, *J. Biol. Chem.* **250**, 7397–7404 (1975).
229. I. A. Kamil, J. N. Smith, and R. T. Williams, *Biochem. J.* **53**, 129–136 (1953).
230. B. K. Nelson et al., *Food Chem. Toxicol.* **26**, 247–254 (1988).
231. B. K. Nelson, W. S. Brightwell, and J. R. Berg, *Neurobehav. Toxicol. Teratol.* **7**, 779–783 (1985).
232. B. K. Nelson et al., *Neurotoxicol. Teratol.* **11**, 153–159 (1989).
233. W. Gibel, Kh. Lohs, and G. P. Wildner, *Arch. Geschwulstforsch.* **45**, 19–24 (1975).
234. V. V. Hudolei, I. V. Mizgirev, and G. B. Pliss, *Vopr. Onkol.* **32**, 73–80 (1987).
235. S. J. Stolzenberg and C. H. Hine, *J. Toxicol. Env. Health* **5**, 1149–1158 (1979).
236. H. Hilscher et al., *Acta Biol. Med. Ger.* **23**, 843–852 (1969).
237. W. von der Hude et al., *Environ. Mutagen.* **9**, 401–410 (1987).
238. S. C. Gad et al., *Toxicol. Appl. Pharmacol.* **84**, 93–114 (1986).
239. E. Ludwig and B. M. Hausen, *Contact Dermatitis* **3**, 240–244 (1977).

240. W. Bonte et al., *J. Blutalkohol* **18**, 399–411 (1981).
241. W. Bonte et al., *J. Blutalkohol*, **18**, 412–426 (1981).
242. J. E. Logsdon and R.A. Loke, in J. Kroschwitz, ed., *Kirk-Othmer Encyclopedia of Chemical Technology*, 4th ed., Vol. 20, Wiley, New York, 1996, pp. 216–240.
243. World Health Organization, *Environ. Health Criteria* 103; WHO, Geneva, 1990.
244. T. Kawai et al., *Int. Arch. Occup. Environ. Health* **62**, 409–413 (1990).
245. H. F. Smyth, Jr. and C. P. Carpenter, *J. Ind. Hyg. Toxicol.* **30**, 63–68 (1948).
246. A. J. Lehman and H. F. Chase, *J. Lab. Clin. Med.* **29**, 561–567 (1944).
247. G. A. Nixon, C. A. Tyson, and W. C. Wertz, *Toxicol. Appl. Pharmacol.* **31**, 481–490 (1975).
248. J. F. Griffith et al., *Toxicol. Appl. Pharmacol.* **55**, 501–513 (1980).
249. J. C. De Ceaurriz et al., *Toxicol. Lett.* **9**, 137–143 (1981).
250. M. W. Gill et al., *J. Appl. Toxicol.* **15**, 77–84 (1995).
251. H. D. Burleigh-Flayer et al., *Fundam. Appl. Toxicol.* **23**, 412–428 (1994).
252. H. D. Burleigh-Flayer et al., *J. Appl. Toxicol.* **18**, 373–381 (1998).
253. K. Pilegaard and O. Ladefoged, *In Vivo* **7**, 325–330 (1993).
254. L. I. Boughton, *J. Am. Pharm. Assoc.* **33**, 111–113 (1944).
255. A. J. Lehman, H. Schwerma, and E. Rickards, *J. Pharmacol. Exp. Ther.* **85**, 61–69 (1945).
256. R. Nordmann et al., *Life Sci.* **13**, 919–932 (1973).
257. T. T. Martinez et al., *Vet. Hum. Toxicol.* **28**, 233–236 (1986).
258. U. Abshagen and N. Rietbrock, *Naunyn-Schmeidebergs Arch. Pharmakol* **264**, 110–118 (1969).
259. A. I. Cederbaum, A. Qureshi, and P. Messenger, *Biochem. Pharmacol.* **30**, 825–831 (1981).
260. H. Savolainen, K. Pekari, and H. Helojoki, *Chem.-Biol. Interact.* **28**, 237–248 (1979).
261. R. W. Slaughter et al., *Fundam. Appl. Toxicol* **23**, 407–420 (1994).
262. R. J. Boatman et al., *Toxicologist* **30**, 41 (Abstr. No. 210) (1996).
263. R. J. Boatman et al., *Drug Metab. Dispos.* **26**, 197–202 (1998).
264. M. A. Gallo et al., *Toxicol. Appl. Pharmacol.* **41**, 135 (1977).
265. R. W. Tyl et al., *Fundam. Appl. Toxicol.* **22**, 139–151 (1994).
266. H. K. Bates et al., *Fundam. Appl. Toxicol.* **22**, 152–158 (1994).
267. C. Bevan et al., *J. Appl. Toxicol.* **15**, 117–123 (1995).
268. British Industrial Biological Research Association (BIBRA), Rep. No. 0570/3/86, BIBRA, Carshalton, England, 1998.
269. H. Burleigh-Flayer et al., *Fundam. Appl. Toxicol.* **36**, 95–111 (1997).
270. International Agency for Research on Cancer, IARC Monogr. Suppl. 7, IARC, Geneva, 1987.
271. R. W. Kapp Jr. et al., *Environ. Mol. Mutagen.* **22**, 93–100 (1993).
272. T.-H. Ueng et al., *Toxicol. Appl. Pharmacol.* **71**, 204–214 (1983).
273. K. Zahlsen, K. Aasted, and O. G. Nilsen, *Toxicology* **34**, 57–66 (1985).
274. G. L. Plaa et al., *J. Toxicol. Environ. Health* **9**, 235–250 (1982).
275. K. W. Nelson et al., *J. Ind. Hyg. Toxicol.* **25**, 282–285 (1943).
276. O. Jensen, *Contact Dermatitis* **7**, 148–150 (1981).
277. C. Wasilewski, Jr., *Arch. Dermatol.* **98**, 502–504 (1968).
278. W. G. Van Ketel and K. N. Tan-Lim, *Contact Dermatitis* **1**, 7–10 (1975).

279. Y. Keith et al., *Arch. Toxicol., Suppl.* **12**, 274–277 (1988).
280. D. S. Folland et al., *J. Am. Med. Assoc.* **236**, 1853–1856 (1976).
281. J.-F. Deng et al., *Am. J. Ind. Med.* **12**, 11–19 (1987).
282. F. Brugnone et al., *Br. J. Ind. Med.* **40**, 160–168 (1983).
283. M. Kelner and D. N. Bailey, *J. Toxicol. Clin. Toxicol.* **20**, 497–505 (1983).
284. D. R. Daniel, B. H. McAnalley, and J. C. Garriot, *J. Anal. Toxicol.* **5**, 110–112 (1981).
285. M. Natowicz et al., *Clin. Chem. (Winston-Salem, N.C.)* **31**, 326–328 (1985).
286. A. A. Pappas et al., *Clin. Toxicol.* **29**, 11–21 (1991).
287. M. P. Gaudet and G. L. Fraser, *Am. J. Emerg. Med.* **7**, 297–299 (1989).
288. P. L. Davis, L. A. Dal Cortivo, and J. Maturo, *J. Anal. Toxicol.* **8**, 209–212 (1984).
289. G. D. Lewis et al., *J. Forensic Sci.* **29**, 541–549 (1984).
290. E. Billig, in J. Kroschwitz , ed., *Kirk-Othmer Encyclopedia of Chemical Technology*, 4th ed., Vol. 4, Wiley, New York, 1992, pp. 691–700.
291. World Health Organization, *Environ. Health Criteria 65*, WHO, Geneva, 1987.
292. H. F. Smyth, Jr., C. P. Carpenter, and C. S. Weil, *Arch. Ind. Hyg. Occup. Med.* **4**, 119–122 (1951).
293. Y. L. Egorov, *Toksikol. Gig. Neftekhim. Proizvod.* **98**, 102 (1972).
294. E. Starrek, Dissertation, Wurzburg, 1938.
294a. E. Starrek, in K. B. Lehmann and F. Flury, eds., *Toxikologie and Hygiene der Technischen Losungsmittel*, Springer, Berlin, 1938.
295. U.S. Environmental Protection Agency (USEPA), *Rat Oral Subchronic Toxicity Study on 1-Butanol* USEPA, Washington, DC, 1986, cited in IRIS Database for 1-Butanol, Ref. 137.
296. R. M. David et al., *Toxicologist* **36**, 61 (Abstr. No. 314) (1997).
297. G. D. DiVincenzo and M. L. Hamilton, *Toxicol. Appl. Pharmacol.* **48**, 317–326 (1979).
298. I. Ästrand et al., *Scand. J. Work Environ. Health* **2**, 165–175 (1976).
299. D. Bechtel and H. Cornish, *Toxicol. Appl. Pharmacol.* **33**, 179 (Abstr.) (1975).
300. K. R. Brandt, *J. Am. Coll. Toxicol.* **6**(3), 403–424 (1987).
301. A. I. Cederbaum, E. Dicker, and G. Cohen, *Biochemistry* **17**, 3058–3064 (1978).
302. B. K. Nelson et al., *Fundam. Appl. Toxicol.* **12**, 469–479 (1989).
303. B. K. Nelson et al., *Neurotoxicol. Teratol.* **11**, 313–315 (1989).
304. B. R. Cater et al., *Toxicol. Appl. Pharmacol.* **41**, 609–614 (1977).
305. S. E. Bloom, in T. C. Hsu, ed., *Cytogenetic Assays for Environmental Mutagens*, Allanheld, Osmun, Montclair, NJ, 1982, pp. 137–159.
306. I. R. Tabershaw, J. P. Fahy, and J. B. Skinner, *J. Ind. Hyg. Toxicol.* **26**, 328–331 (1944).
307. J. H. Sterner et al., *Am. Ind. Hyg. Assoc. J.* **10**, 53–59 (1949).
308. J. Velasquez, *Med. del Trabajo* **1**, 43–45 (1964).
309. J. Velasquez, R. Escobar, and A. Almaraz, *Proc. 16th Int. Congr. Occup. Health*, Tokyo, 1969, pp. 231–234; cited in Ref. 6.
309a. J. May, *Staub* **26**(9), 385 (1966).
310. A. A. Li et al., *Toxicologist* **36**, 62 (Abstr. No. 317) (1997).
311. H.-J. Klimisch, and J. Hellwig, *Fundam. Appl. Toxicol.* **3**, 77–89 (1995).
312. T. A. Kaempfe et al., *Toxicologist* **36**, 62 (Abstr. No. 316) (1997).

313. S. G. Hedlund and K. H. Kiessling, *Acta Pharmacol. Toxicol.* **27**, 381–396 (1969).
314. E. Zeiger et al., *Environ. Mol. Mutagen.* **11**(Suppl. 12), 1–158 (1988).
315. H. Oettel, *Arch. Exp. Pathol. Pharmacol.* **183**, 641–691 (1936).
315a. D. Steinkoff, *Zenfralbl. Arbeitsmed. Arbeitsschutz* **2**, 13 (1952).
316. H.-D. Hahn and N. Rupprich, in *Ullman's Encyclopedia of Industrial Chemistry*, 5th rev. ed., Vol. A4, VCH Publishers, New York, 1985, pp. 463–474.
317. S. Kezic and A. C. Monster, *J. Chromatogr.* **428**, 275–280 (1988).
318. F. K. Dietz et al., *J. Pharmacokinet. Biopharmacol.* **9**, 553–576 (1981).
319. G. J. Traiger and J. V. Bruckner, *J. Pharmacol. Exp. Ther.* **196**, 493–500 (1976).
320. T. M. Brooks, A. L. Meyer, and D. H. Hutson, *Mutagenesis*, **3**, 227–232 (1988).
321. K. Aarsted, K. Zahlsen, and O. G. Nilsen, *Arch. Toxicol. Suppl.* **8**, 418–421 (1985).
322. H. H. Cornish and J. Adefuin, *Arch. Environ. Health* **14**, 447–449 (1967).
323. G. J. Traiger et al., *J. Toxicol. Environ. Health* **28**, 235–248 (1989).
324. S. Fregert et al., *Acta Derm.-Venereol.* **51**, 271–272 (1971).
325. K. R. Brandt, *J. Am. Coll. Toxicol.* **8**(4), 627–641 (1989).
326. H. Savolainen, P. Pfåffi, and E. Elovaara, *Eur. Arch. Toxicol.* **57**, 285–288 (1985).
327. J. M. Wood and R. Laverty, *Pharmacol., Biochem. Behav.* **10**, 113–119 (1979).
328. M. W. Schaffarzick and B. J. Brown, *Science*, **116**, 663–666 (1952).
329. J. A. McComb and D. B. Goldstein, *J. Pharmacol. Exp. Ther.* **208**, 113–117 (1979).
329a. H. Wallgren, *Acta Pharmacol. Toxicol.* **16**, 217 (1960).
330. National Toxicology Program (NTP), *Toxicity Studies of t-Butyl Alcohol (CAS No. 75-65-0) Administered by Inhalation to F344/N Rats and B6C3F$_1$* Mice, NTP TOX-53, NTIS No. PB98-108905, NTP, Washington, DC, 1997.
331. C. Lindamood, III et al., *Fundam. Appl. Toxicol.* **19**, 91–100 (1992).
332. M. J. Arslanian, E. Pascoe, and J. G. Reinhold, *Biochem. J.* **125**, 1039–1047 (1971).
333. R. Derache, in J. Tremolieves, ed., *Alcohols and Derivatives*, Vol. II, Sect. 20, Pergamon, London, 1970, pp. 507–522.
334. E. Browning, *Toxicology and Metabolism of Industrial Solvents*, Chapter 7, Elsevier, Amsterdam, 1965, pp. 310–411.
335. A. I. Cederbaum and G. Cohen, *Biochem. Biophys. Res. Commun.*, **97**, 730–736 (1980).
336. R. C. Baker, S. M. Sorensen, and R. A. Deitrich, *Alcohol. Clin. Exp. Res.* **6**, 247–251 (1982).
337. T. P. Faulkner and A. S. Hussain, *Res. Commun. Chem. Pathol. Pharmacol.* **64**, 31–39 (1989).
338. T. P. Faulkner et al., *Life Sci.* **45**, 1989–1995 (1989).
339. M. A. Daniel and M. A. Evans, *J. Pharmacol. Exp. Ther.* **222**, 294–300 (1982).
340. B. K. Nelson et al., *Teratology* **39**, 504 (Abstr.) (1989).
341. J. D. Cirvello et al., *Toxicol. Ind. Health* **11**, 151–165 (1995).
342. H. Hoshino, G. Chihara, and F. Fukuoka, *Gann* **61**, 121–124 (1970).
343. E. Zeiger et al., *Environ. Mol. Mutagen.* **9**(Suppl. 9), 1–110 (1987).
344. D. B. McGregor et al., *Environ. Mol. Mutagen* **11**, 91–118 (1988).
345. National Toxicology Program (NTP), *Review of Current DHHS, DOE and EPA Research Related to Toxicology*, NTP-85-056, NTP, Washington, DC, 1985.

346. U. Rannug, R. Gothe, and C. A. Wachtmeister, *Chem.-Biol. Interact.* **12**, 251–263 (1976).
347. A. J. Papa, in J. Kroschwitz, ed., *Kirk-Othmer Encyclopedia of Chemical Technology*, 4th ed., Vol. 2, Wiley, New York, 1991, pp. 709–728.
348. Eyquem, *Ann. Hyg. Publigue. Med. Leg.* **3**, 71 (1905).
349. T. B. Fuchter, *Am. Med.* **2**, 210 (1901).
350. P. Lappe and T. Hoffmann, in *Ullman's Encyclopedia of Industrial Chemistry*, 5th rev. ed., Vol. A19, VCH Publishers, New York, 1985, pp. 49–60.
351. R. A. Scala and E. G. Burtis, *Am. Ind. Hyg. Assoc. J.* **34**, 493–499 (1973).
352. L. F. Hansen and D. Nielsen, *Toxicology* **88**, 81–99 (1994).
353. K. R. Butterworth et al., *Food Cosmet. Toxicol.* **16**, 203–207 (1978).
354. R. P. Maickel and J. F. Nash, Jr., *Neuropharmacology* **24**, 83–89 (1985).
355. H. Savolainen, P. Pfäffli, and E. Elovaara, *Acta Pharmacol. Toxicol.* **56**, 260–264 (1985).
356. B. K. Nelson et al., *J. Am. Coll. Toxicol.* **8**(2), 405–410 (1989).
357. National Toxicology Program (NTP), *Review of Current DHHS, DOE, and EPA Research Related to Toxicology*, NTP, Washington, DC, 1986.
358. T. J. B. Gray et al., *Toxicology* **28**, 167–179 (1983).
359. D. L. J. Opdyke, *Food Cosmet. Toxicol.* **16**, 785–788 (1978).
360. H. F. Smyth Jr. et al., *Am. Ind. Hyg. Assoc. J.* **30**, 470–476 (1969).
361. F. M. B. Carpanini et al., *Food Cosmet. Toxicol.* **11**, 713–724 (1973).
362. M. Guggenheim and W. Loffler, *Biochem. Z.* **72**, 325 (1916).
363. J. McLaughlin J. et al., *Ind. Hyg. J.* **25**, 282–284 (1964).
364. Y. I. Ardeev, *Nauchn. Tr.—Omsk. Med. Inst. im M. I. Kalinina* **69**, 146 (1966).
365. O. Neubauer, *Arch. Exp. Pathol. Pharmakol.* **46**, 133 (1901).
366. H. Thierfelder and J. V. Mering, Hoppe-Seyler's *Z. Physiol. Chem.* **9**, 511 (1985).
367. K. Verschueren, *Handbook of Environmental Data on Organic Chemicals*, Van Nostrand-Reinhold, New York, 1977.
368. P. W. Langvardt and W. H. Braun, *J. Anal. Toxicol.* **2**, 83–85 (1978).
369. J. M. Robinson, *J. Am. Med. Assoc.* **148** (1918).
370. R. J. Nolan, P. W. Langvardt, and G. E. Blau, *Toxicol. Appl. Pharmacol.* **48**, A164 (Abstr.) (1979).
371. D. L. J. Opdyke, *Food Cosmet. Toxicol.* **13**, 695–696 (1975).
372. N. Fedtke and H. M. Bolt, *Int. Arch. Occup. Environ. Health* **57**, 149–158 (1986).
373. A. A. Nomeir and M. B. Abou-Donia, *Anal. Biochem.* **151**, 381–388 (1985).
374. F. U. Bär and F. Griepentrog, *Med. Ernehr.* **8**, 244 (1967).
375. G. N. Zaeva and V. I. Fedorova, *Toksikol. Nov. Prom. Khim. Veshchestv* **5**, 51 (1963).
376. D. E. Moody and J. K. Reddy, *Toxicol. Appl. Pharmacol.* **45**, 497–504 (1978).
377. D. E. Moody and J. K. Reddy, *Toxicol. Lett.* **10**, 379–383 (1982).
378. D. E. Rodwell et al., *Toxicologist* **8**, 213 (Abstr. No. 848) (1988).
379. Sice, J. *Toxicol. Appl. Pharmacol.* **9**, 70–74 (1966).
380. L. Perbellini et al., *Toxicol. Appl. Pharmacol.* **46**, 421–427 (1978).
381. T. J. B. Gray et al., *Toxicol. Lett.* **10**, 273–279 (1982).
382. H. A. Griffiths et al., *Food Chem. Toxicol.* **35**, 255–260 (1997).

383. British Industrial Biological Research Association (BIBRA), *Toxicity Profile for Methyl Isobutyl Carbinol*, BIBRA, Carshalton, England, 1988.

384. G. D. DiVincenzo, C. J. Kaplan, and J. Dedinas, *Toxicol. Appl. Pharmacol.* **36**, 511–522 (1976).

385. H. Shimuzi et al., *Jpn. J. Ind. Health* **27**, 400–419 (1985).

386. M. Vezina and G. L. Plaa, *Toxicol. Appl. Pharmacol.* **92**, 419–427 (1988).

387. M. Vezina et al., *Can. J. Physiol. Pharmacol.* **68**, 1055–1061 (1990).

388. L. Silverman, H. F. Schulte, and M. W. First, *J. Ind. Hyg. Toxicol.* **28**, 262–268 (1946).

389. H. F. Smyth, Jr. et al., *Am. Ind. Hyg. Assoc. J.* **23**, 95–107 (1962).

390. J. L. O'Donoghue, in *Neurotoxicology of Industrial and Commercial Chemicals*, Vol. II, CRC Press, Boca Raton, FL, 1985, pp. 61–97.

391. P. S. Spencer et al., *Crit. Rev. Toxicol.* **7**, 279–356 (1980).

392. J. T. James et al., *J. Appl. Toxicol.* **7**, 135–142 (1987).

393. D. L. J. Opdyke and C. Letizia, *Food Chem. Toxicol.* **20**, 839–840 (1982).

394. J. T. James, B. P. Infiesto, and M. R. Landauer, *J. Appl. Toxicol.* **7**, 307–312 (1987).

395. W. J. Krasavage et al., *Toxicol. Appl. Pharmacol.* **52**, 433–441 (1980).

CHAPTER SEVENTY-EIGHT

Monohydric Alcohols — C_7 to C_{18}, Aromatic, and Other Alcohols

C. Bevan Ph.D., DABT

This chapter reviews linear and branched C_7 to C_{18} monohydric aliphatic alcohols as well as aromatic, alicyclic, aliphatic unsaturated, and aliphatic halogenated alcohols. The CAS registry number and molecular structures have been provided for all of the alcohols, except for the oxo alcohols. These alcohols are mixtures of isomeric alcohols with the same molecular formula, with the composition and CAS registry number dependent on the olefin feedstock.

CHEMICAL AND PHYSICAL PROPERTIES

The physical and chemical properties for these alcohols are listed in Table 78.1. The National Fire Protection Association (NFPA) has prepared a rating system to assess the physical and chemical hazards of chemicals with respect to flammability, health, and reactivity (1, 2). In general, these alcohols are not reactive chemicals, except for the unsaturated alcohols.

PRODUCTION AND USE

Alcohols represent an important class of industrial chemicals with a wide number of uses. Based on production volume, monohydric alcohols represent the most important group of

Patty's Toxicology, Fifth Edition, Volume 6, Edited by Eula Bingham, Barbara Cohrssen, and Charles H. Powell.
ISBN 0-471-31939-2

Table 78.1. Chemical and Physical Properties of Alcohols

Compound	CAS Number	Mol. Formula	Mol. Wt.	Boiling Point (°C)	Melting Point (°C)	Sp. Gr.	Refractive Index (20°C)	Vapor Pressure (mmHg) (°C)	Maximum Vapor Concn. % (°C)	Flammability Lower Limit	Flammability Upper Limit	Solubility in Water (%) (°C)	Conversion Factors[a] 1 ppm = mg/m³	Conversion Factors[a] 1 mg/L = ppm
1-Heptanol	*117-70-6*	$C_7H_{16}O$	116.2	176	− 35.0	0.824	1.4233	—	0.085 (37)	—	—	—	4.75	210.4
2-Heptanol[b]	*543-47-7*	$C_7H_{16}O$	116.2	159	—	0.817	1.4213	—	—	—	—	0.35	4.75	210.4
3-Heptanol[b]	*3913-02-8*	$C_7H_{16}O$	116.2	156	− 70.0	0.821	1.4222	—	—	—	—	—	4.75	210.4
2,3-Dimethyl-1-pentanol	*143-23-4*	$C_7H_{16}O$	116.2	—	—	—	—	—	—	—	—	—	4.75	210.4
1-Octanol	*111-87-5*	$C_8H_{18}O$	130.2	195	− 16.3	0.827	1.4300	—	0.034 (37)	—	—	0.01–0.05	5.32	187.8
2-Octanol	*123-96-6*	$C_8H_{18}O$	130.2	179	− 38.6	0.821	1.4260	1.0 (32.8)	—	—	—	—	5.32	187.8
3-Octanol	*589-29-1*	$C_8H_{18}O$	130.2	—	—	—	—	—	—	—	—	—	5.32	187.8
2-Ethyl-1-hexanol	*104-76-7*	$C_8H_{18}O$	130.2	185	< − 75.0	0.833	1.4315	0.05 (20)	0.0066 (20)	—	—	0.07	5.32	187.8
2,2,4-Trimethyl-1-pentanol	*123-44-4*	$C_8H_{18}O$	130.2	168	− 70.0	0.839	1.4300	—	—	—	—	—	5.32	187.8
2-Ethyl-4-methyl-1-pentanol	*106-67-2*	$C_8H_{18}O$	130.2	—	—	—	—	—	—	—	—	—	5.32	187.8
1-Nonanol	*143-08-8*	$C_9H_{20}O$	144.3	214	− 5.0	0.827	1.4323	0.3 (20)	—	2.8	6.1	—	5.90	169.6
3,5,5-Trimethyl-1-hexanol[b]	*3452-97-9*	$C_9H_{20}O$	144.3	194	− 70.0	0.824	1.4330	—	—	—	—	—	5.90	169.6
1-Decanol	*112-30-1*	$C_{10}H_{22}O$	158.3	234	6.4	0.832	1.4359	10 (695)	—	—	—	Insol.	6.47	154.5
Isodecanol	*2533-17-7*	$C_{10}H_{22}O$	158.3	226	—	—	—	—	—	—	—	—	6.47	154.5
1-Undecanol[b]	*112-42-5*	$C_{11}H_{24}O$	172.3	245	14.3	0.830	1.4392	—	—	—	—	Insol.	7.05	141.8
1-Dodecanol	*112-53-8*	$C_{12}H_{26}O$	186.3	259	23.8	0.831	1.4428	—	—	—	—	Insol.	7.62	131.3
1-Tetradecanol[b]	*112-72-1*	$C_{14}H_{30}O$	214.4	—	38.0	0.817	1.4358	—	—	—	—	Insol.	8.76	114.2
1-Hexadecanol	*36653-82-4*	$C_{16}H_{34}O$	242.5	—	49.0	0.816	1.4392	1 (127.3)	—	—	—	Insol.	9.91	100.9
1-Octadecanol	*112-92-5*	$C_{18}H_{38}O$	270.5	—	58.0	0.812	—	—	—	—	—	Insol.	11.06	90.4
Eicosanol	*629-96-9*	$C_{20}H_{42}O$	298.6	—	66.0	—	—	—	—	—	—	Insol.	12.21	81.9
Benzyl alcohol	*100-51-6*	C_7H_8O	108.1	205	15.2	—	1.5404	0.15 (25)	0.02	—	—	4.0	4.42	226.1
2-Phenylethanol	*60-12-8*	$C_8H_{10}O$	122.2	220	− 25.8	—	1.5300	1.0 (58)	0.13 (52)	—	—	2.0	5.00	200.2
1-Phenylethanol	*98-85-1*	$C_8H_{10}O$	122.2	202	20.1	—	—	—	—	—	—	—	5.00	200.2
2-Phenyl-2-propanol	*1123-85-9*	$C_9H_{12}O$	134.2	—	—	—	—	—	—	—	—	—	5.49	182.1

p-tolyl alcohol	*589-18-4*	$C_8H_{10}O$	122.2	—	—	—	—	—	—	—	—	3.6	5.00	200.2
Cyclohexanol	*108-93-0*	$C_6H_{12}O$	100.2	161	24.0	—	1.4656	3.5 (34)	—	—	—	—	4.10	244.1
Methylcyclo-hexanol	*25639-42-3*	$C_7H_{14}O$	114.2	174	– 50.0	0.913	1.4610	1.5 (30)	0.2 (30)	—	—	3.4	4.67	214.2
3,5,5-Trimethyl-cyclohexanol	*116-02-9*	$C_9H_{18}O$	142.0	—	—	—	—	—	—	—	—	—	5.81	172.1
Furfuryl alcohol	*98-00-0*	$C_5H_6O_2$	98.1	170	– 14.6	—	1.4840	1.0 (31.5)	0.13 (30.2)	1.8	16.3	Misc[c]	4.01	249.4
Tetrahydrofuran methanol	*97-99-4*	$C_5H_{10}O_2$	102.1	178	< – 80.0	1.050	1.4520	2.3 (40)	—	1.5	9.7	Misc[c]	4.18	239.5
Allyl alcohol	*107-18-6*	C_3H_6O	58.1	94	– 50.0	0.848	1.4135	23.8 (25)	3.13 (25)	2.5	18.6	Misc[c]	2.37	422
Propargyl alcohol	*107-19-7*	C_3H_4O	56.1	114	– 50.0	0.972	1.4306	11.6 (20)	—	—	—	Misc[c]	2.29	437
Hexynol	*105-31-7*	$C_6H_{10}O$	98.1	142	– 80.0	0.882	—	13.0 (25)	—	—	—	38	4.01	249
Butynol[b]	*2028-69-9*	C_4H_8O	70.1	107	—	—	—	—	—	—	—	—	2.87	348
Methylbutynol[b]	*115-9-5*	C_5H_8O	84.1	—	—	—	—	—	—	—	—	—	3.44	291
Methylpentynol[b]	*77-75-8*	$C_6H_{10}O$	98.1	121	– 197.2	—	—	—	—	—	—	—	4.01	249
Ethyloctynol[b]	*5877-42-9*	$C_{10}H_{18}O$	154.2	—	—	0.871	—	—	—	—	—	—	6.31	158
2-Chloroethanol	*107-07-3*	C_2H_5ClO	80.5	129	—	1.205	1.4419	4.9 (20)	—	0.64	4.9	Misc[c]	3.29	303.8
1-Chloro-2-propanol	*127-00-4*	C_3H_7ClO	94.5	—	—	—	—	—	—	—	—	—	3.90	258
2-Chloro-1-propanol	*78-89-7*	C_3H_7ClO	94.5	—	—	—	—	—	—	—	—	—	3.90	258

[a] At 25°C, 760 mm Hg.
[b] Toxicity of these chemicals is formed in Table 78.4.
[c] Miscible.

the alcohol family (3). The most important alcohols in this chapter are the plasticizer alcohols (C_7 to C_{11}) and the detergent range alcohols (C_{12} to C_{18}). In 1997, the production volume of 2-ethylhexanol reached 7.68 billion pounds in the United States (4). In general, alcohols of commercial significance are produced synthetically, although some alcohols are made from natural products of fermentation. The most significant industrial process is the oxo process that can be used to produce alcohols in the C_3 to C_{20} range by using alkenes as starting materials (3).

The uses of alcohols are numerous and can vary depending on their chemical and physical properties. In general, alcohols are used as solvents, cosolvents, and chemical intermediates. Among the higher alcohols, defined as containing six carbons or more, the C_6 to C_{11} alcohols are used to manufacture plasticizers. Some of the alcohols are pure chemicals, but higher monohydric alcohols (C_6 to C_{18}) are also complex isomeric mixtures (3, 5, 6). The major routes of industrial or occupational exposure to alcohols is by dermal contact and/or inhalation. The extent of the exposure pathways depends on the use of the chemical and the physical properties of the alcohol.

METABOLISM AND DISPOSITION

A comparative uptake study has been conducted using human abdominal skin *in vitro* for C_1 to C_{10} linear alcohols (7). The rate of dermal uptake for the neat material decreases with increasing carbon number.

Williams prepared an early general review on the metabolism and disposition of alcohols (8). Primary alcohols are readily oxidized to the corresponding aldehydes, which are further converted to the corresponding acids. Secondary alcohols are converted to ketones. Alcohols can be conjugated either directly or as a metabolite with glucuronic acid, sulfuric acid, or glycine, and excreted. Tertiary alcohols are more resistant to metabolism and are generally conjugated more readily than secondary or primary alcohols.

HEALTH EFFECTS

A common property of some of the alcohols is to produce local irritation to the skin, eyes, and respiratory tract, and the effect or potency varies with the type of alcohol. Many alcohols produce minimal or no adverse effects in humans, possibly because of low exposure combined with the low toxicity potential of the alcohol.

A HEPTANOLS

The most important commercial member of this group is isoheptyl alcohol, which is a mixture of branched C_7 alcohols. This alcohol is used for the manufacture of esters such as phthalate plasticizers. 1-Heptanol has little commercial value. Other C_7 alcohols are 2,3-dimethyl-1-pentanol and the secondary alcohols, 2-heptanol, 3-heptanol, 4-heptanol, and 2,4-dimethyl-3-pentanol. 2-Heptanol and 3-heptanol can exist as enantiomers.

The available toxicity data indicate that heptanols have a low order of acute toxicity and no occupational exposure standards exist for them.

1.0 1-Heptanol

1.0.1 CAS Number: [117-70-6]

1.0.2 Synonyms: *n*-Heptanol; *n*-heptyl alcohol, 1-hydroxyheptane; heptyl alcohol; Alcohol C-7; Enanthyl alcohol; Enanthic alcohol

1.0.3 Trade Names: NA

1.0.4 Molecular Weight: 116.2

1.0.5 Molecular Formula: $C_7H_{16}O$

1.0.6 Molecular Structure: 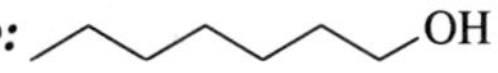

1.1 Chemical and Physical Properties

1-Heptanol is a colorless liquid; other physical and chemical properties are listed in Table 78.1.

1.2 Production and Use

1-Heptanol is made by the oxo process by reacting hexenes with carbon monoxide (3) or by the catalytic reduction of heptaldehyde (9). It has little commercial value except in fragrances and as an artificial flavoring agent (9).

1.4 Toxic Effects

1.4.1 Experimental Studies

1.4.1.1 Acute Toxicity. The acute oral LD_{50} was reported as 6.2 g/kg (male rats) and 5.5 g/kg (female rats) (10) and 3.25 mg/kg (9) in rats and 4.3 g/kg (11) and 6 g/kg (12) in mice. The dermal LD_{50} in rabbits was reported as about 2 g/kg (10) and greater than 5 g/kg (9). The acute inhalation LC_{50} was reported as 6.6 mg/L (1390 ppm) in mice (11). No deaths occurred in rats exposed to saturated vapors for 4 h (10). Aspiration of 0.2 mL 1-heptanol caused deaths in 10 out of 10 rats. Death was instant and due to respiratory arrest (13).

1-Heptanol was moderately irritating to intact or abraded rabbit skin when applied for 24 h under occlusive conditions (9). It was moderately irritating to the eyes of rabbits (10). Exposure of mice to 1-heptanol in a sensory irritation (Alarie) test produced sensory irritation with a RD_{50} of 100 ppm (14) and 700 ppm (15).

1.4.1.2 Chronic and Subchronic Toxicity. In a Russian study, subacute inhalation exposure in rabbits and rats led to conjunctivitis and a decrease in blood cholinesterase activity (11). In another Russian study, inhalation of 0.18 to 0.35 mg/L (38 to 74 ppm) 2 h/day for 4.5 months caused minor hematologic changes and some unspecified histological changes (12).

1.4.1.3 Pharmacokinetics, Metabolism, and Mechanisms. 1-Heptanol is primarily oxidized in the rabbit to heptanoic acid, which either undergoes further oxidation to CO_2 or forms an ester glucuronide (8). There is also a lesser metabolic pathway using direct conjugation with glucuronic acid to form an ether glucuronide conjugate (8).

1.4.2 Human Experience

1.4.2.1 General Information. No adverse effects have been reported in humans from occupational exposure.

1.4.2.2.1 Acute Toxicity. 1-Heptanol (1% in petrolatum) was not irritating after a 48-h closed skin patch test, and it did not produce a skin sensitization reaction in 20 human volunteers (9).

1.4.2.2.3 Pharmacokinetics, Metabolism, and Mechanisms. The dermal flux for 1-heptanol in human skin (epidermis) *in vitro* is 0.021 mg/cm^2/h (7).

1.5 Standards, Regulations, or Guidelines of Exposure

No occupational health standards exist for 1-heptanol. The United States does not have an occupational standard for 1 heptanol. The Russian occupational exposure limit, set in 1993, is 10 mg/m^3 and a skin notation (16).

2.0 Isoheptyl Alcohol

2.0.1 CAS Number:

2.0.2 Synonyms: C_7 Oxo Alcohols

2.0.3 Molecular Weight: 116.20

2.0.4 Molecular Formula: $C_7H_{16}O$

2.1 Chemical and Physical Properties

Isoheptyl alcohol is a colorless liquid.

2.2 Production and Use

Isoheptyl alcohol is produced in the oxo process by reacting hexenes with carbon monoxide and hydrogen, followed by hydrogenation (3). The commercial product typically consists of methyl-1-hexanols. A major use is in producing ester compounds such as phthalate plasticizers and esters of dicarboxylic acids (3). Isoheptyl alcohols are also used as solvents or solubilizers in the paint and printing inks, as components in textile auxiliaries and pesticides, for hormone extraction, and in the surfactant field as foam boosters or antifrothing agents (3).

2.4 Toxic Effects

2.4.1 *Experimental Studies*

2.4.1.1 Acute Toxicity. The acute oral LD_{50} was reported as 3.9 g/kg in rats, and the acute dermal LD_{50} was greater than 3.16 g/kg in rabbits (17). There were no deaths or treatment-related effects in rats, mice, and guinea pigs exposed to 0.72 mg/L isoheptyl alcohol vapor for 6 h (17). Isoheptyl alcohol was severely irritating to both the skin and eyes of rabbits (17).

2.4.2 *Human Experience*

2.4.2.1 General Information. Prolonged exposure to excessive concentrations of isoheptyl alcohol is expected to be very irritating to the eyes, nose, and respiratory tract.

2.5 Standards, Regulations, or Guidelines of Exposure

No occupational exposure standards have been established for isoheptanol.

3.0 2,3-Dimethyl-1-Pentanol

3.0.1 *CAS Number: [10143-23-4]*

3.0.2 *Synonym:* 1-Hydroxy-2,3-dimethylpentane; 2,3-dimethylpentanol

3.0.3 *Trade Names:* NA

3.0.4 *Molecular Weight:* 116.2

3.0.5 *Molecular Formula:* $C_7H_{16}O$

3.0.6 *Molecular Structure:*

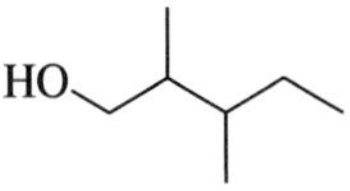

3.1 Chemical and Physical Properties

2,3-Dimethylpentanol is a liquid; other chemical and physical properties are listed in Table 78.1.

3.4 Toxic Effects

3.4.1 *Experimental Studies*

3.4.1.1 Acute Toxicity. The acute oral LD_{50} in rats is 2.38 ml/kg, and the acute dermal LD_{50} in rabbits is 2.5 ml/kg (18). There were no deaths among rats exposed to saturated vapors for 8 h (18). 2,3-Dimethylpentanol was slightly irritating to the skin of rabbits, but it was severely irritating to rabbit eyes (18).

3.4.1.6 Genetic and Related Cellular Effects Studies. 2,3-Dimethylpentanol was not mutagenic in *Salmonella typhimurium* in an Ames test with or without metabolic activation (19).

B OCTANOLS

The most important commercial C_8 alcohols are 2-ethylhexanol and a mixture of branched C_8 alcohols referred to as isooctyl alcohol. Other octanols of lesser commercial interest are 2-octanol, 1-octanol, 3,5-dimethyl-1-hexanol, 2,2,4-trimethyl-1-pentanol, and 2-ethyl-4-methyl-1-pentanol. These alcohols are liquids at ambient temperature and are used primarily in producing esters, such as plasticizers. No occupational exposure standards exist for octanols except for isooctyl alcohol.

4.0 1-Octanol

4.0.1 CAS Number: *[111-87-5]*

4.0.2 Synonyms: *n*-Octyl alcohol, *n*-octanol; caprylic alcohol; octanol; *n*-capryl alcohol; 1-octyl alcohol; capryl alcohol; octan-1-ol; *n*-caprylic alcohol; alcohol C-8; heptyl carbinol

4.0.3 Trade Names: NA

4.0.4 Molecular Weight: 130.2

4.0.5 Molecular Formula: $C_8H_{18}O$

4.0.6 Molecular Structure:

4.1 Chemical and Physical Properties

1-Octanol is a colorless liquid with a fresh orange-rose odor. Its chemical and physical properties are listed in Table 78.1.

4.1.2 Odor and Warning Properties

The odor and taste threshold for 1-octanol was reported at about 24 ppm (20).

4.2 Production and Use

1-Octanol is made commercially by sodium reduction or high-pressure catalytic hydrogenation of the esters of naturally occurring caprylic acid or by oligomerization of ethylene using aluminum alkyl technology (3, 21). 1-Octanol is used in the perfume industry and in food products (3, 21). No exposure data were found on 1-octanol in the workplace.

4.4 Toxic Effects

4.4.1 Experimental Studies

1-Octanol has a low order of acute toxicity. It is a slight skin irritant and a sensory irritant. 1-Octanol is not teratogenic and may be a weak tumor promoter in mice.

4.4.1.1 Acute Toxicity. The acute oral LD_{50} is greater than 5 g/kg in rats, and the acute dermal LD_{50} is greater than 5 g/kg in rabbits (21). No deaths were observed among rats exposed to 6400 mg/m^3 (1203 ppm) 1-octanol for 1 h; no lung lesions were noted (21a). Three out of ten rats died within 2 days following a 4-h exposure to 5600 mg/m^3 (1053 ppm) 1-octanol. Lung lesions were noted (21a) and consisted of necrosis of the bronchial epithelium with alveolar edema and an accumulation of alveolar macrophages. Aspiration of 0.2 mL 1-octanol to rats produced deaths in 10 out of 10 rats, and death occurred after a few breaths (13). 1-Octanol was slightly irritating to the skin of rabbits (22) and is considered an eye irritant using the EU criteria (23). In a sensory irritation (Alarie) test, 1-octanol produced sensory irritation in mice, with an RD_{50} of 50 ppm (14).

4.4.1.2 Chronic and Subchronic Toxicity. A Russian study indicated no cumulative effects in mice after oral administration of 179 mg/kg 1-octanol for 1 month (24).

4.4.1.4 Reproductive and Developmental. Pregnant female Sprague–Dawley rats were exposed by inhalation to a saturated vapor (65 ppm) of 1-octanol 7 h/day during gestational days 6 to 15. No maternal or developmental toxicity was observed (25, 26).

In an oral gavage study, pregnant female Wistar rats were dosed with 0, 130, 650, 975, or 1300 mg/kg 1-octanol during gestational days 6 through 15. Maternal toxicity occurred with a dose-related increase in severity of clinical signs, including lateral and abdominal position, unsteady gait, salivation, piloerection, nasal discharge, and pneumonia. Food consumption and body weight gain decreased slightly in all but the lowest dose group. There was no developmental toxicity (27).

4.4.1.5 Carcinogenesis. There was no evidence of tumors in the cancer screening lung adenoma study in which mice were injected intraperitoneally with 100 and 500 mg/kg 1-octanol three times a week for 8 weeks (28). This assay has not been validated as a reliable screen for cancer.

1-Octanol is a weak skin tumor promoter when applied three times a week for 60 weeks to mice skin that had been initiated with dimethylbenz[a]anthracene (29).

4.4.1.7 Other: Neurological, Pulmonary, Skin Sensitization, etc. 1-Octanol (0.1 mmol) did not induce peroxisome proliferation in rat hepatocyte culture (30).

4.4.2 Human Experience

4.4.2.2.1 Acute Toxicity. In a human patch test, 1-octanol in 2% petrolatum was neither a skin irritant nor a skin sensitizer (21, 22). In another human patch test, 1-octanol was applied to the skin for up to 4 h. The irritation response was significantly lower compared to that of the positive control (20% sodium dodecyl sulfate). The authors concluded that 1-octanol would not be classified as a skin irritant using the EU criteria (31).

4.4.2.2.3 Pharmacokinetics, Metabolism, and Mechanisms. In vitro dermal flux in human skin (epidermis) was reported as 0.008 mg/cm^2/h (7), suggesting a low rate of penetration.

4.5 Standards, Regulations, or Guidelines of Exposure

An ACGIH TLV or an OSHA PEL has not been established for 1-octanol. A workplace environmental exposure level (WEEL) value of 50 ppm as an 8-h TWA has been recommended (32). Occupational standards for 1-octanol exist in other countries, but there is no documentation to support these values (33). These standards are 10 mg/m^3 or 1.9 ppm (Russia), 100 mg/m^3 or 19 ppm (Yugoslavia), 150 mg/m^3 or 28 ppm (Romania).

5.0 2-Ethylhexanol

5.0.1 CAS Number: *[104-76-7]*

5.0.2 Synonyms: 2-Ethylhexanol-1; 2-ethyl-*n*-hexyl Alcohol; 2-ethylhexan-1-ol

5.0.3 Trade Names: NA

5.0.4 Molecular Weight: 130.2

5.0.5 Molecular Formula: $C_8H_{18}O$

5.0.6 Molecular Structure:

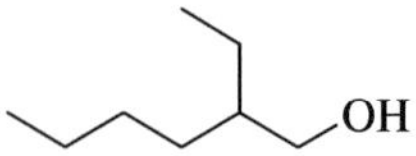

5.1 Chemical and Physical Properties

2-Ethylhexanol is a clear, colorless liquid. Other chemical and physical properties are listed in Table 78.1.

5.2 Production and Use

2-Ethylhexanol is the most important C_8 alcohol and is used mainly in manufacturing plasticizers (3). Other minor uses include the manufacturing of 2-ethylhexyl acrylate, as a dispersing agent and wetting agent, as a solvent for gums and resins, as a cosolvent for nitrocellulose, and in ceramics, paper coatings, rubber latex, textiles, and fragrances (34, 35). The primary routes of industrial exposure are dermal contact and inhalation. Indirect exposure can occur by hydrolysis of esters containing 2-ethylhexanol which can then be absorbed.

5.3 Exposure Assessment

5.3.3 Workplace Methods

A NIOSH method does not exist for specifically measuring 2-ethylhexanol in air. However, the NIOSH method for measuring isooctyl alcohol should be applicable to 2-

ethylhexanol (36). A method has been published for measuring 2-ethylhexanol in air that involves gas chromatography with flame ionization detection (37).

5.3.5 Biomonitoring/Biomarkers

Other methods have been published to measure 2-ethylhexanol in air and in biological media (38–40).

5.4 Toxic Effects

5.4.1 Experimental Studies

The toxicity of 2-ethylhexanol has been reviewed elsewhere (41). 2-Ethylhexanol displayed a low order of acute toxicity in animals. It is moderately irritating to the skin but severely irritating to the eyes, and it is a sensory irritant. Subchronic oral studies revealed that the liver is a target organ in rats and mice. High oral doses ($\geq$ 5 mmol/kg) can produce liver enlargement and hepatic peroxisome induction in rodents. 2-Ethylhexanoic acid, the major metabolite of 2-ethylhexanol, is viewed as the putative toxic form of 2-ethylhexanol for both peroxisome induction and developmental effects. 2-Ethylhexanol is not carcinogenic to rats or male mice; a possible increase in liver tumors was observed in female mice dosed with 750 mg/kg. 2-Ethylhexanol was consistently inactive in a battery of *in vitro* and *in vivo* genotoxicity assays. No developmental effects were observed in the dermal and inhalation studies; but in oral studies, 2-ethylhexanol produced developmental effects including teratogenicity at maternally toxic doses. 2-Ethylhexanol was not a selective developmental toxicant. There is no evidence that 2-ethylhexanol is a reproductive toxicant.

5.4.1.1 Acute Toxicity. The acute oral LD_{50} is 3.73 g/kg (42), 2.05 g/kg (43), and 3.25 g/kg (44) in rats, and 3.2 to 6.4 g/kg in mice (45). Gastrointestinal irritation was reported at gross necropsy following an acute oral dose (42). The acute dermal LD_{50} is 2 g/kg (43) and greater than 2.6 g/kg (42) for rabbits, and greater than 8.3 g/kg for guinea pigs (45). There were no deaths among rats, mice, or guinea pigs exposed to nearly saturated vapors (227 ppm) of 2-ethylhexanol for 6 h (42). No deaths occurred among rats exposed to saturated vapors for 6 h or to rats exposed for 8 h (43). Signs of central nervous system depression were observed, including labored breathing. Irritation involving the mucous membranes of the eyes, nose, throat, and respiratory tract was considered moderate (42).

2-Ethylhexanol was reported to be slightly irritating (43) and moderately irritating (42) to rabbit skin when applied for 24 h as an occlusive patch. 2-Ethylhexanol produced moderate irritation (43) and severe irritation, including burns and corneal damage, to the eyes of rabbits (42). In a sensory irritation test, 2-ethylhexanol produced sensory irritation (Alarie) in mice, and had an RD_{50} of 44 ppm (14).

5.4.1.2 Chronic and Subchronic Toxicity

Dermal
During an 11-day period, F344 rats received nine dermal applications of 2-ethylhexanol at doses of 0.41 or 0.83 g/kg. At 0.83 g/kg, treatment-related effects included lymphopenia,

decreased spleen weight, and minor local skin irritation (46). When applied daily to the skin of 10 rabbits at a dose of 1.66 g/kg for 12 days, 2-ethylhexanol produced only slight skin effects. Decreased body weights were observed after 9 days (47).

Diet

Hens were fed 2% 2-ethylhexanol for an unspecified period. A depression in serum lipids was noted, but there was no significant effect on liver weights (48). Mice were fed 1.0% 2-ethylhexanol (about 1500 mg/kg) for 4 days. Increases in hepatic peroxisomes were observed (49). In male Sprague–Dawley rats fed 2.0% 2-ethylhexanol for 2 weeks, no significant effects on peroxisome enzymes in the liver were observed (50). In contrast, male F344 rats fed 2.0% (about 1000 mg/kg) 2-ethylhexanol for 3 weeks showed decreased serum lipids, increased liver weights, and an increase in liver peroxisomes (51, 52).

Rats (strain unspecified) were fed 0, 100, 500, 2500, or 12,500 ppm 2-ethylhexanol for 13 weeks. The highest dose was about 650 mg/kg/day. There were no effects on mortality, body weight, food consumption, or kidney weight. Histopathological lesions included cortical degeneration in the kidneys of the high-dose male rats and focal liver congestion or swelling in the high-dose female rats. No effects were observed in the 2500-ppm group (52a).

Drinking Water

F344 rats received 2-ethylhexanol in drinking water at concentrations of 0, 308, or 636 ppm for 9 days. The high dose is the maximum limit of water solubility and is equivalent to about 160 mg/kg/day. No adverse effects were observed (46).

Gavage

Mice (strain unspecified) were dosed by oral gavage with 0, 143, 351, 702, 1053, or 1755 mg/kg for 2 weeks. Significant increases in liver weight and hepatic peroxisomes were observed at doses $\geq$702 mg/kg (53). B6C3F1 mice were dosed by gavage (aqueous suspension with an emulsifier) with 0, 25, 125, 250, or 500 mg/kg 2-ethylhexanol 5 days/week for 13 weeks. Treatment-related effects included increased stomach weight ($>$250 mg/kg) and liver weights (125 and 250 mg/kg). The gross pathological examination revealed forestomach lesions in the high-dose animals (54). Male Wistar rats were intubated at a dose of 130 mg/kg/day for 2 weeks. There were no significant effects on the testes, serum lipids, or various liver end points, including peroxisome induction (55). Wistar rats were dosed by gavage with doses of 143, 351, 702, or 1053 mg/kg 5 days/week for 2 weeks. Significant increases in liver weight and hepatic peroxisomes were observed at 702 and 1053 mg/kg (53).

F344 rats dosed with 1350 mg/kg 2-ethylhexanol for 7 days showed increased liver weights and liver peroxisomes (56). Over an 11-day period, F344 rats received nine daily doses of 2-ethylhexanol by oral gavage at 0, 83, 249, 830, or 1245 mg/kg. The major findings were decreased food consumption and body weight at $\geq$ 249 mg/kg; decreased leukocytes, lymphocytes, and spleen weights at $\geq$830 mg/kg; and increased liver and stomach weights at $\geq$830 mg/kg. Histological findings included decreased extramedullary hematopoiesis, thymic atrophy, and evidence of forestomach irritation at $\geq$249 mg/kg (46). In another study, F344 rats were dosed with 100, 320, or 950 mg/kg 5 days/week for

4 weeks. Liver weights and liver peroxisomes significantly increased at 320 and 950 mg/kg. There was no evidence of testicular damage (57). F344 rats were dosed with 0, 25, 125, 250, or 500 mg/kg 2-ethylhexanol 5 days/week for 13 weeks. Body weights decreased in the high-dose group of both sexes. Relative liver, kidney, and stomach weights increased at 250 and 500 mg/kg. Gross pathological examination revealed forestomach lesions in the 500-mg/kg animals (54).

5.4.1.3 Pharmacokinetics, Metabolism, and Mechanisms. The rate and extent of dermal uptake of 2-ethylhexanol are low. The *in vitro* dermal flux for 2-ethylhexanol was calculated as 0.22 $mg/cm^2/hr$ in full thickness rat skin (58).

The metabolic fate of 2-ethylhexanol has been studied in rats and rabbits. It is readily converted to 2-ethylhexanoic acid, which can be oxidized to a hydroxy acid and a diacid. Within 28 h following a single oral dose of [^{14}C]ethylhexanol to rats, about 81% of the radioactivity was recovered in the urine, about 6% was recovered in the air as CO_2, and about 8% was in the feces (44). The chief metabolites in the urine were 2-ethylhexanoic acid, 2-ethylhexanoic acid glucuronide, 2-ethyl-1,6-hexanedioic acid, and 2-ethyl-5-hydroxyhexanoic acid (44). In the rabbit given an oral dose of 2-ethylhexanol, the major urinary metabolite was 2-ethylhexanoyl glucuronide; 2-ethyl-2,3-dihydroxyhexanoic acid and a trace amount of unmetabolized 2-ethylhexanol were also present (59). Excretion studies on 2-ethylhexanol were conducted in female Fischer 344 rats following single oral doses of [^{14}C]2-ethylhexanol (50 or 500 mg/kg), following repeated oral dosing with unlabeled 2-ethylhexanol at 50 mg/kg, following dermal exposure for 6 h with a 1 g/kg applied dose of [^{14}C]2-ethylhexanol and following a 1 mg/kg intravenous dose of [^{14}C]2-ethylhexanol. Similar excretion balance profiles were observed in the low, high, and repeated oral dose studies with 2-ethylhexanol, and there was some evidence of metabolic saturation at 500 mg/kg. All of the oral doses were eliminated rapidly, predominantly in the urine during the first 24 hours following dosing. There was no evidence of metabolic induction following the repeated low oral dosing. Only about 5% of the 1 g/kg dose was absorbed by the dermal route during the 6-h period. Finally, urinary metabolites eliminated by both routes were predominately glucuronides of oxidized metabolites of 2-ethylhexanol, including 2-ethyladipic acid, 2-ethylhexanoic acid, 5-hydroxy-2-ethylhexanoic acid, and 6-hydroxy-2-ethylhexanoic acid (60).

5.4.1.4 Reproductive and Developmental

Developmental Toxicity

No developmental toxicity was observed in dermal and inhalation developmental studies. In oral studies, 2-ethylhexanol produced developmental effects, including teratogenicity at maternally toxic doses, but it was not a selective developmental toxicant.

Pregnant female Wistar rats were dosed by oral gavage with 0, 130, 650, or 1300 mg/kg 2-ethylhexanol on days 6 to 15 of gestation. The 1300 mg/kg dose group was maternally toxic based on deaths, clinical symptoms (nasal discharge, salivation, and signs of central nervous system depression), reduced food consumption and body weight loss during the entire treatment period, and gross necropsy findings in the animals that died during the study (liver and lung effects). Developmental effects occurred in the 1300 mg/kg group which included increased number of early resorptions and a high implantation loss. Mean fetal body weights were markedly reduced and an increased frequency of fetuses that had skeletal malformations was observed. The number of fetuses that had skeletal variations and retardations and dilated renal pelvis increased in the high-dose group. Slight maternal toxicity was noted in the 650-mg/kg group. Fetal body weights were slightly reduced and an increased number of fetuses that had skeletal variations was observed. There were no maternal or developmental effects in the 130-mg/kg dose group (27).

In a range-finding study, Wistar rats received 1 or 2 mL/kg 2-ethylhexanol by oral gavage on day 12 of gestation. The incidence of malformations was none in controls, 2% in the low-dose group and 22% in the high-dose group (61). In another range-finding study, pregnant CD-1 mice were given oral doses of 1525 mg/kg/day 2-ethylhexanol on days 7 to 14 of gestation. Severe maternal toxicity was observed as evidenced by lethality (34%) and decreased body weight gain, and neonatal effects were observed (62). In an inhalation study, Sprague–Dawley rats were exposed to 850 mg/m^3 (about 190 ppm) 2-ethylhexanol 7 h/day during gestational days 1 to 19. There were limited maternal effects but no teratogenicity or developmental toxicity (63). In a dermal study, 2-ethylhexanol was administered by occluded dermal applications to pregnant female F344 rats at doses of 0.3, 1.0, or 3.0 mL/kg for 6 h/day from days 6 to 15 of gestation. Maternal effects (depression in body weight gain) were observed at the mid and high doses. There was no evidence of treatment-related developmental effects (64).

In another developmental study, CD-1 mice were given 2-ethylhexanol by microencapsulation at doses of 0.3, 0.49, and 1.49 mmol/kg on days 0 to 17 of gestation. There were no maternal or developmental effects at any doses up to 1.49 mmol/kg or 194 mg/kg/day (65).

Reproductive Toxicity

2-Ethylhexanol did not produce adverse effects on reproductive organs of rats and mice in any subchronic or chronic studies (54, 66). In addition, 2-ethylhexanol (0.2 mmol) did not produce any adverse effects on Sertoli cells *in vitro* (67–69). 2-Ethylhexanol was also inactive in a dominant lethal assay in mice dosed by oral gavage with 250, 500, or 1000 mg/kg for 5 days (70).

5.4.1.5 Carcinogenesis. Male and female F344 rats were dosed by oral gavage with 0, 50, 150, or 500 mg/kg 2-ethylhexanol (0.005% in aqueous Cremophor EL, a polyoxyl-35 castor oil), 5 days/week for 2 years. There were no differences of biological importance between the vehicle control and a water control group that was included in the study. Reduced body weight gain occurred in the mid- and high-dose groups and increased incidence of lethargy and unkemptness. There were dose-related increases in relative liver, stomach, brain, kidney, and testis weights at study termination. Mortality was significantly increased among the high-dose females, and marked aspiration-induced bronchopneumonia in the high-dose animals. There was no evidence of treatment-related neoplastic lesions in any of the exposed groups (66).

Male and female B6C3F1 mice were dosed by oral gavage with 0, 50, 200, or 750 mg/kg 2-ethylhexanol (0.005% in aqueous Cremophor EL, a polyoxyl-35 castor oil), 5 days/week for 18 months. There were no differences of biological importance between the vehicle control and a water control group that was also included in the study. All treatment-related effects occurred only in the high-dose animals. Mortality was increased and body weight gain was reduced, and there was a slight increase in nonneoplastic focal hyperplasia in the forestomach. Relative liver and stomach weights occurred in the high-dose animals. There was a weak adverse trend in hepatocellular carcinoma incidence in the 750-mg/kg dose group which may have been associated with toxicity. The time-adjusted incidence of hepatocellular carcinomas in male mice (18.8%) was within the historical control range at the testing facility (0–22%), but it lay outside the normal range of 0–2% for the female mice (13.1%) (66).

2-Ethylhexanol was studied for tumor promotion in two assays. It was inactive in mouse JB6 epidermal cells (71). At a dietary dose of 1700 ppm (about 90 mg/kg), it was not a liver tumor-promoting agent in rats previously treated with diethylnitrosamine (72).

5.4.1.6 Genetic and Related Cellular Effects Studies. 2-Ethylhexanol was not mutagenic to *S. typhimurium* in an Ames test with or without metabolic activation (73–76). It was also inactive in the mouse lymphoma assay (74) and did not produce chromosomal aberrations in two *in vitro* cytogenetic assays (77, 78). 2-Ethylhexanol was inactive in an in vivo micronucleus assay in which mice were injected intraperitoneally with 460 mg/kg 2-ethylhexanol (73). 2-Ethylhexanol did not induce chromosomal aberrations in F344 rats dosed by oral gavage with up to 0.21 mg/kg for 5 days (79). There was no significant covalent binding to liver DNA of mice treated with an oral dose of 100 mg/kg (80). 2-Ethylhexanol was also inactive in the Balb 3T3 cell transformation assay (73). In a dominant lethal assay, mice were dosed by oral gavage with 0, 250, 500, or 1000 mg/day for 5 days and no effects were observed (70). There was no evidence that mutagenic substances were excreted in the urine by rats dosed with 2 g/kg 2-ethylhexanol for 15 days when tested in a modified Ames test with or without metabolic activation (81).

5.4.1.7 Other: Neurological, Pulmonary, Skin Sensitization, etc. The liver is a target organ of 2-ethylhexanol, but this effect depends on species, dose, route, and duration of treatment. The liver effects include liver enlargement, which is associated with an increase in microsomal and peroxisomal enzymes. On a mg/kg basis, 2-ethylhexanol is viewed as a weak peroxisome inducer. Based on hepatocyte work, the putative peroxisomal inducing form of 2-ethylhexanol is 2-ethylhexanoic acid (30, 82, 83). *In vitro* (hepatocyte culture) and/or *in vivo* studies indicate that rats and mice are sensitive to the peroxisome effects of 2-ethylhexanol (or 2-ethylhexanoic acid), but guinea pigs and marmosets are relatively insensitive to these effects (83).

5.4.2 Human Experience

5.4.2.2.1 Acute Toxicity. No serious adverse effects in humans have been observed from occupational exposure to 2-ethylhexanol. Exposure to unspecified concentrations of 2-ethylhexanol reportedly cause headaches, dizziness, fatigue, intestinal disorders, and a slight decrease in blood pressure (41). In a patch test, a 48-h closed patch of 4% 2-ethylhexanol in petrolatum was neither irritating nor sensitizing to the skin of human volunteers (41). Overexposure to 2-ethylhexanol may produce irritation of the skin, eyes, and mucous membranes of the respiratory tract.

5.4.2.2.3 Pharmacokinetics, Metabolism, and Mechanisms. The rate and extent of dermal uptake of 2-ethylhexanol is low. The *in vitro* dermal flux for 2-ethylhexanol was calculated as 0.038 $mg/cm^2/h$ in human skin (stratum corneum) (58).

5.5 Standards, Regulations, or Guidelines of Exposure

No occupational exposure standards exist for 2-ethylhexanol.

6.0 Isooctyl Alcohol

6.0.1 CAS Number: NA

6.0.2 Synonym: C_8 oxo alcohols; isooctanol; 6-methyl-1-heptanol

6.0.3 Trade Names: NA

6.0.4 Molecular Weight: 130.2

6.0.5 Molecular Formula: $C_8H_{18}O$

6.1 Chemical and Physical Properties

6.1.1 General

Isooctyl alcohol is a colorless liquid.

6.2 Production and Use

Isooctyl alcohols or isooctanols are mixtures of isomeric C_8 alcohols that are made by the oxo process in which heptenes are reacted with carbon monoxide and hydrogen in the presence of a catalyst, followed by hydrogenation (3). The commercial product typically consists of methyl-1-heptanols and/or dimethyl-1-hexanols; the composition and CAS registry number depend on the olefin feedstock. The principal use of these alcohols is in preparing plasticizers, mainly diisoctyl phthalate, but also esters of adipic, sebacic, azelaic, and trimellitic acids (3). Isooctyl alcohol is also used as a solvent for fats, oils, and waxes, as well as various rubber formulations and resins (3). Industrial exposure may occur from dermal contact in handling; indirect exposure may occur from the hydrolysis of esters containing isooctyl alcohol.

6.4 Toxic Effects

6.4.1 Experimental Studies

6.4.1.1 Acute Toxicity. The acute oral LD_{50} in rats has been reported as 1.48 g/kg (42) and greater than 2 g/kg (17). The dermal LD_{50} in rabbits is greater than 2.6 g/kg (42). No deaths were observed among rats, mice, or guinea pigs exposed to saturated vapors (200 ppm) for 6 h (42).

Isooctyl alcohols are moderately irritating to rabbit skin when applied under an occlusive wrap for 24 h (42) or under 4-h semiocclusive conditions (17). Isooctyl alcohols are severely irritating to the eyes of rabbits (42).

6.4.1.2 Chronic and Subchronic Toxicity. A C_8 oxo alcohol was administered dermally to rabbits for 2 weeks. No systemic effects were noted at doses of 0.5 or 2.5 mL/kg/day (17). In a 2-week oral study, rats dosed with 130 mg/kg/day isooctanol did not show testicular atrophy, liver enlargement, hepatic peroxisome induction, or hypolipidemia (55). An inhalation study in rats exposed to saturated vapors of 180 ppm isooctyl alcohol 6 hr/day for 13 days revealed no toxic signs and no gross pathological effects at autopsy (84).

6.4.1.4 Reproductive and Developmental. A C_8 oxo alcohol was administered by oral gavage to pregnant female CRL:CDBR rats at doses of 0, 100, 500, and 1000 mg/kg during days 6 through 15 of gestation. Adverse clinical signs, reduced body weight gains and food consumption, were noted in the high-dose. Fetal toxicity and malformations were not observed. Increases in fetal skeletal variations were seen in the mid- and high-dose groups compared to controls. However, these incidences were considered as alternative normal patterns of development (85).

6.4.2 Human Experience

6.4.2.1 General Information. No serious industrial effects were observed from exposure to isooctyl alcohols.

6.5 Standards, Regulations, or Guidelines of Exposure

The ACGIH TLV for isooctyl alcohol (methyl-1-heptanols) CAS Number: *[26952-21-6]* is 50 ppm as an 8-h TWA (86).

7.0 2-Octanol

7.0.1 CAS Number: *[123-96-6]*

7.0.2 Synonyms: *sec*-Caprylic alcohol, capryl alcohol, methyl hexyl carbinol; *sec*-octyl alcohol

7.0.3 Trade Names: NA

7.0.4 Molecular Weight: 130.2

7.0.5 Molecular Formula: $C_8H_{18}O$

7.0.6 Molecular Structure:

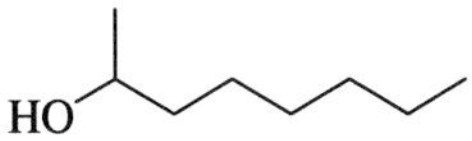

7.1 Chemical and Physical Properties

2-Octanol is a colorless liquid at room temperature. Its chemical and physical properties are listed in Table 78.1.

7.2 Production and Use

2-Octanol is produced commercially by heating a soap of castor oil with sodium hydroxide (3, 34). The resulting commercial product may contain 2 to 14% methyl hexyl ketone as a contaminant (34). It is used in the paint industry, as a wetting agent in the textile industry, and as a component of brake fluids (3).

7.4 Toxic Effects

7.4.1 Experimental Studies

7.4.1.1 Acute Toxicity. The acute oral LD_{50} is reportedly greater than 3.2 g/kg in rats (45) and 4.0 g/kg (12) in mice. The dermal LD_{50} in guinea pigs is greater than 0.5 g/animal (45). Immersion of the tails of mice in liquid 2-octanol for 3 h caused some mortality due to absorption of 2-octanol (12). 2-Octanol is slightly irritating to guinea pig skin (45), and application of 2-octanol to the eyes of rabbits caused pronounced erythema (12).

7.4.1.2 Chronic and Subchronic Toxicity. In a Russian study, rats were exposed to 34 to 56 ppm 2-octanol, 2 h/day, 6 days/week for 4.5 months. Mild reversible central nervous system effects (not described) were reported, as well as hematologic changes (decreased hemoglobin and erythrocyte numbers) and minor changes in the heart, liver, and kidney (12). 2-Octanol (2 mL/day) applied for 6 days to the shaven skin of rabbits caused erythema, inflammation, and cracking of the skin, which healed in 10 to 12 days (12).

7.4.2 Human Experience

7.4.2.1 General Information. No adverse effects were reported in humans.

7.5 Standards, Regulations, or Guidelines of Exposure

No occupational standards exist for 2-octanol.

8.0 3-Octanol

8.0.1 CAS Number: *[589-98-0]*

8.0.2 Synonyms: Amylethylcarbinol, ethyl *n*-amyl carbinol

8.0.3 Trade Names: NA

8.0.4 Molecular Weight: 130.2

8.0.5 Molecular Formula: $C_8H_{18}O$

8.0.6 Molecular Structure:

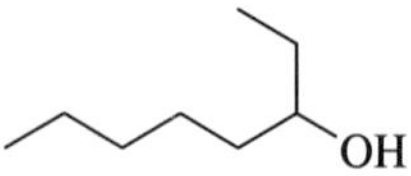

8.1 Chemical and Physical Properties

The chemical and physical properties are listed in Table 78.1

8.2 Production and Use

3-Octanol is a liquid produced by hydrogenation of ethyl *n*-amyl ketone (87). It is used in fragrances and as an artificial flavoring in foods (87).

8.4 Toxic Effects

8.4.1 Experimental Studies

8.4.1.1 Acute Toxicity. The acute oral LD_{50} of 3-octanol was reportedly greater than 5 g/kg in rats, and the acute dermal LD_{50} is greater than 5 g/kg in rabbits (87). 3-Octanol is moderately irritating when applied to intact or abraded rabbit skin under a 24-h occlusive wrap (87).

8.4.2 Human Experience

8.4.2.1 General Information. There were no adverse effects reported in humans from exposure to 3-octanol.

8.4.2.2 Clinical Cases. 3-Octanol (12% in petrolatum) did not produce any skin irritation after a 48-h closed patch test on human subjects, nor did 3-octanol produce skin sensitization in a maximization test (87).

9.0 Other C_8 Alcohols

***9.0.1a 2,2,4-Trimethyl-1-Pentanol (CAS Number:** [123-44-4])*

***9.0.1b 2-Ethyl-4-Methyl-1-Pentanol (CAS Number:** [106-67-2])*

9.0.2b Synonyms: 4-methyl-2-ethyl-1-pentanol

9.0.3 Trade Names: NA

9.0.4 Molecular Weight: 130.2

9.0.5 Molecular Formula: $C_8H_{18}O$

9.0.6a Molecular Structure:

9.0.6b Molecular Structure:

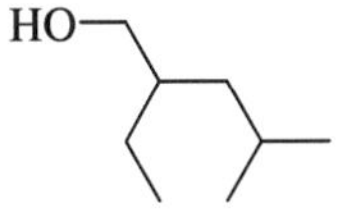

9.1 Chemical and Physical Properties

9.1.1 General

The chemical and physical properties are listed in Table 78.1.

9.2 Production and Use

These alcohols alone have little commercial value compared to other octanols. Some of them can exist as components in mixtures of octanols.

9.4 Toxic Effects

9.4.1 Experimental Studies

9.4.1.1 Acute Toxicity. The acute oral LD_{50} of 2,2,4-trimethylpentanol in rats is 3.73 mL/kg (18) and 2.70 mL/kg (88), and the acute dermal LD_{50} in rabbits is 6.30 mL/kg (18) and 3.88 mL/kg (88). No deaths occurred among rats exposed to saturated or near saturated vapors for 6 h (88) or 8 h (18). 2,2,4-Trimethylpentanol is slightly irritating to the skin of rabbits (18, 88), and it produces moderate to severe irritation to the eyes of rabbits (18). Minor corneal injury, iritis, and minor to moderate conjunctivitis were observed in rabbits at 1 h. All eyes healed by 7 days (88).

The acute oral LD_{50} of 2-ethyl-4-methylpentanol in rats is 4.29 mL/kg (18) and 3.35 g/kg (47); the dermal LD_{50} in rabbits is greater than 5 mL/kg (18). No deaths were reported among rats exposed to saturated vapors for 8 h (18). 2-Ethyl-4-methylpentanol produced a slight to moderate irritation to rabbit skin and was moderately irritating to the eyes of rabbits (18).

9.4.1.3 Pharmacokinetics, Metabolism, and Mechanisms. The limited data suggest that 2,2,4-trimethylpentanol is oxidized to 2,2,4-trimethylpentanoic acid (89).

Rats were given a single dose of 4.4 mmol/kg 2,2,4-trimethylpentanol by oral gavage. Another group of rats was implanted with osmotic minipumps containing [^{3}H]thymidine 1 day before dosing with 2,2,4-trimethylpentanol, and incorporation of [^{3}H]thymidine into kidney DNA was measured 5 days later to quantify cell proliferation. An increase in renal protein droplet accumulation and α-2u-globulin concentrations was observed, but [^{3}H]thymidine was not incorporated into the DNA (90).

2,2,4-Trimethylpentanol also reversibly binds to α-2u-globulin, a low molecular weight protein that is excreted in large quantities in male rat urine (91). This interaction is unique to α-2u-globulin (92), and is involved in the development of a male rat-specific nephrotoxicity manifested acutely as excessive accumulation of protein phagolysosomes of renal proximal tubular cells (93, 94). Although humans excrete trace amounts of proteins similar to α-2u-globulin (the protein content is 1% that of male rats) (95), humans are not considered at risk for this type of nephropathy (96).

9.4.2 Human Experience

9.4.2.1 General Information. No adverse effects in humans have been reported.

C NONANOLS

The most important commercial members of this subgroup of alcohols are the C_9 oxo alcohols, which are a mixture of predominantly C_9 branched alcohols, diisobutyl carbinol, and 2,6-dimethyl-4-heptanol. Two C_9 alcohols of lesser commercial importance are 1-nonanol and 3,5,5-trimethyl-1-hexanol. All of these alcohols are liquids at ambient temperatures.

Acute studies in animals indicate a low order of toxicity. These alcohols are irritating to the skin, eyes, and respiratory tract. They are also aspirations hazard. No serious adverse effects from industrial exposure were reported in humans. Prolonged or excessive exposure to the alcohols can produce local irritation and narcosis. No occupational exposure standards have been established for any of the nonanols.

10.0 1-Nonanol

10.0.1 CAS Number: [143-08-8]

10.0.2 Synonyms: *n*-Nonyl alcohol, 1-hydroxynonane, pelargonic alcohol; nonyl alcohol; alcohol C-9

10.0.3 Trade Names: NA

10.0.4 Molecular Weight: 144.3

10.0.5 Molecular Formula: $C_9H_{20}O$

10.0.6 Molecular Structure:

10.1 Chemical and Physical Properties

1-Nonanol is found in a number of citrus oils (97). It is a liquid; other chemical and physical properties are listed in Table 78.1.

10.2 Production and Use

1-Nonanol is produced by the high-pressure catalytic reduction of esters of pelargonic acid (97). A major use of 1-nonanol is as a chemical intermediate in the production of esters and in fragrances (97). Industrial exposure is associated with contact in handling and is not expected to be significant unless the material is handled hot.

10.4 Toxic Effects

10.4.1 Experimental Studies

10.4.1.1 Acute Toxicity. The acute dermal LD_{50} of 1-nonanol in rabbits is 2.96 g/kg (98). The acute inhalation LC_{50} is 5.5 mg/liter (duration not specified) in mice (98). Based on the eye irritation scores that were reported, 1-nonanol would be considered an eye irritant using the EU criteria (99). Aspiration of 0.2 mL of 1-nonanol produced deaths among 10 out of 10 rats; death was instant and was due to respiratory arrest (13).

The acute oral LD_{50} of nonyl alcohol containing 2% 2-propylheptanol was reported as 3.2 to 6.4 g/kg for rats and 6.4 to 12.8 g/kg for mice (45). The acute dermal LD_{50} is greater than 10 mL/kg in guinea pigs (45). There were no deaths among rats exposed to either 215 or 730 ppm for 6 h (45).

10.4.1.4 Reproductive and Developmental. Exposure of female rats to saturated vapors of 1-nonanol (25 ppm or 150 mg/m^3) 7 h/day on days 1 to 19 of gestation did not result in maternal or developmental toxicity (25, 26). Embryotoxicity and delayed fetal development (including retardation of ossification) was reported in a study in which female rats were dosed orally with a 40% solution of 1-nonanol during days 1 through 15 of pregnancy (100).

10.4.2 Human Experience

10.4.2.1 General Information. No serious adverse effects have been reported in humans.

10.4.2.2 Clinical Cases

10.4.2.2.1 Acute Toxicity. 1-Nonanol (2% in petrolatum) was reportedly neither a skin irritant nor a skin sensitizer to humans (22, 97).

10.4.2.2.3 Pharmacokinetics, Metabolism, and Mechanisms. Skin absorption is low; the dermal flux of 1-nonanol in human skin (epidermis) *in vitro* is 0.003 mg/cm^2/h (7).

11.0 Isononyl Alcohol

11.0.1 CAS Number:

11.0.2 Synonym: C_9 oxo alcohols

11.0.3 Trade Names: NA

11.0.4 Molecular Weight: 144.26

11.0.5 Molecular Formula: $C_9H_{20}O$

11.1 Chemical and Physical Properties

The chemical and physical properties are listed in Table 78.1.

11.2 Production and Use

Isononyl alcohol is made in the oxo process by reacting olefins with carbon monoxide and hydrogen in the presence of a catalyst, followed by hydrogenation (3). The commercial product typically consists of dimethyl-1-heptanols and methyl-1-octanols; the composition and CAS registry number depend on the olefin feedstock. A major use of isononyl alcohol is as a chemical intermediate in producing esters, such as phthalates, acetates, and adipates (3). Isononyl alcohol is frequently used in the paint industry. Industrial exposure may occur in handling and indirect exposure can occur from hydrolysis of esters containing these alcohols.

11.4 Toxic Effects

11.4.1 Experimental Studies

11.4.1.1 Acute Toxicity. The acute oral LD_{50} of an isononyl alcohol (composition typically consisted of 75 to 85% dimethyl-1-heptanols, 5 to 10% methyl-1-octanols, and 10 to 20% other homologous primary alcohols) was reportedly 2.98 g/kg in rats (42), and the acute dermal LD_{50} is greater than 3.2 g/kg in rabbits (42). No deaths were observed among rats, mice, or guinea pigs exposed to an aerosol (21.7 mg/L) for 6 h (42). Central nervous system depression and some irritation of the eyes and nose was noted.

This material is markedly irritating to rabbit skin when applied under an occlusive wrap for 24 h (42), but only moderately irritating when applied for 4 h under semiocclusive conditions (17). This material produced marked irritation in the eyes of rabbits (42).

11.4.1.2 Chronic and Subchronic Toxicity. In a 2-week oral study, rats dosed with 1 mmol/kg isononyl alcohol did not develop testicular atrophy, liver enlargement, hepatic peroxisome induction, or hypolipidemia (55).

11.4.1.4 Reproductive and Developmental. In an oral developmental toxicity study, Wistar rats were dosed with 1.0, 5.0, and 10.0 mmol/kg isononyl alcohol (144, 720, 1440 mg/kg) on days 6 to 15 of gestation. Exposure to the high dose resulted in 20% mortality of the dams and produced a weak teratogenic effect in the fetuses. Developmental toxicity occurred only at maternally toxic doses. No maternal or developmental effects were noted at 1.0 mmol/kg (27).

Pregnant female Wistar rats were dosed by oral gavage with 0, 144, 720, or 1440 mg/kg isononanol CAS Number: [*68515-81-1*]. This alcohol had a set of isomers with a medium degree of branching, including about 16% isodecanol. It consisted of roughly equivalent amounts of 3,4-, 4,6-, 3,6-, 3,5-, 4,5-, and 5,6-dimethylheptanol-1. All high-dose dams died during the study. At 720 mg/kg, there were reduced body weight gain, clinical signs of apathy, and nasal discharge. There was also increased number of skeletal variations and retardations. No maternal or developmental effects were observed in the 144 mg/kg dose group (27). In a supplementary study, maternal toxicity occurred in dams dosed with 1080 mg/kg. The incidence of malformations (mainly related to the heart) and retardations increased significantly. There was also a higher incidence of rudimentary cervical rib(s) (27). In another developmental study, pregnant female Wistar rats were dosed by oral gavage with 0, 144, 720 or 1440 mg/kg of another isononanol CAS Number: [*685515-81-1*]. This alcohol had a set of isomers with a low degree of branching. The main components were 4,5-dimethylheptanol-1 ($\sim$ 23%), 4-methyloctanol-1 (29%), 3-ethylheptanol-1 (3%), 6-methyloctanol-1 (15%), and 3-ethyl-4-methylhexanol-1 (1%). Exposure to the high dose resulted in 30% mortality of the dams. Fetal weights were also reduced in the high-dose and the number of fetuses that had common variations and skeletal retardations also increased. At 720 mg/kg, there were adverse clinical signs in the dams and slightly reduced body weight gain. Resorptions rates increased slightly. In a supplemental study, maternal toxicity occurred in dams dosed with 1080 mg/kg. There was a marginal increase in the number of resorptions, higher postimplantation losses, and an elevated number of fetuses that were malformed (27).

11.4.2 Human Experience

11.4.2.1 General Information. No serious industrial intoxication was observed in using isononyl alcohol.

12.0 Secondary C_9 Alcohols

12.0.1a 2,6-Dimethyl-4-Heptanol (Diisobutyl Carbinol), CAS Number: *[108-82-7]*

12.0.1b 2,6-Dimethyl-2-Heptanol (Dimetol, Freesiol, Lolitol), CAS Number: *[13254-34-7]*

12.0.2a Synonyms: Diisobutylcarbinol

12.0.3 Trade Names: NA

12.0.4 Molecular Weight: 144.3

12.0.5 Molecular Formula: $C_9H_{20}O$

12.0.6a Molecular Structure: 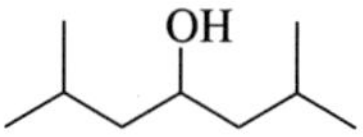

12.0.6b Molecular Structure: 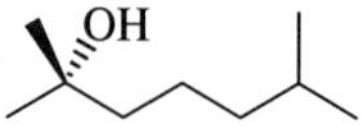

12.1 Chemical and Physical Properties

The physical and chemical properties of diisobutyl carbinol only are listed in Table 78.1.

12.2 Production and Use

Diisobutyl carbinol is prepared by aldol condensation of acetone and subsequent hydrogenation (3). Diisobutyl carbinol is used as a reaction medium for preparing hydrogen peroxide and as an effective defoaming agent (3). 2,6-Dimethyl-2-heptanol is prepared from the reaction of methyl heptenone and methyl magnesium halide, followed by hydrogenation (101).

12.4 Toxic Effects

12.4.1 Experimental Studies

12.4.1.1 Acute Toxicity. The acute oral LD_{50} of diisobutyl carbinol is 3.56 g/kg in rats, and the acute dermal LD_{50} in rabbits is greater than 5.66 ml/kg (102). No deaths were observed among rats exposed to saturated vapors for 8 h (102). Diisobutyl carbinol was slightly irritating to both the skin and eyes of rabbits (101).

The acute oral LD_{50} of 2,6-dimethyl-2-heptanol in rats is greater than 5 g/kg based on 2 out of 10 deaths (101), and 6.8 g/kg for a 31.6 to 50% emulsion in olive oil (101). The acute dermal LD_{50} was greater than 5 g/kg in rabbits (101). 2,6-Dimethyl-2-heptanol was reportedly both a severe skin and eye irritant to rabbits (101). As part of a dermal LD_{50} study, 5 g/kg of undiluted 2,6-dimethyl-2-heptanol produced slight to moderate erythema and edema on rabbit skin after an occluded application for 24 h (101).

12.4.2 *Human Experience*

12.4.2.2 Clinical Cases. In a maximization test, 2,6-dimethyl-2-heptanol (10% in petrolatum) did not produce any irritation or sensitization to the skin of 25 volunteers (101).

13.0 3,5,5-Trimethylhexanol

13.0.1 CAS Number: *[3452-97-9]*

13.0.2 Synonyms: Nonylol; 3,5,5-trimethylhexyl alcohol; 3,5,5-trimethylhexan-1-ol; 3,5,5-trimethyl-1-hexanol

13.0.3 Trade Names: NA

13.0.4 Molecular Weight: 144.3

13.0.5 Molecular Formula: $C_9H_{20}O$

13.0.6 Molecular Structure:

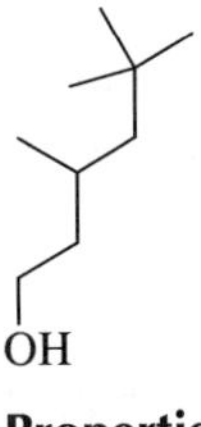

13.1 Chemical and Physical Properties

3,5,5-Trimethylhexanol is a liquid at ambient temperatures. Its chemical and physical properties are listed in Table 78.1.

13.2 Production and Use

3,5,5-Trimethylhexanol is prepared from diisobutene by the oxo process (3). This alcohol is also a minor component of isononyl alcohol (3). The uses of 3,5,5-trimethylhexanol are similar to those of isononyl alcohol, principally in the production of plasticizers (3).

13.4 Toxic Effects

13.4.1 *Experimental Studies*

All toxicity studies, except for peroxisome induction work, were carried out on material defined as a nonyl alcohol rich in trimethylhexanol (45).

13.4.1.1 Acute Toxicity. The acute oral LD_{50} was reportedly 1.4 to 1.75 ml/kg in rats and 1.4 to 2.1 ml/kg in rabbits (45). The acute dermal LD_{50} is less than 3.6 ml/kg in rabbits (45). 3,5,5-Trimethylhexanol reportedly produced erythema with slight necrosis in rabbit skin after a 24-h application (45).

13.4.1.2 Chronic and Subchronic Toxicity. In a 2-week oral study, rats dosed with 1 mmol/kg 3,5,5-trimethylhexanol showed some liver enlargement but no testicular atrophy, hepatic peroxisome induction, or hypolipidemia (55).

An oral dose of 0.148 g/kg given to rabbits on each of 67 days over a period of 83 days resulted normal growth, uniform survival, and caused no signs of intoxication (45). Contact of 5 mL (1.6 to 2.0 g/kg) of nonyl alcohol for 1 h/day with the skin of rabbits on each of 50 days over a period of 75 days resulted in retarded growth and erythema but no mortality (45). In rabbits treated with nonyl alcohol rich in trimethylhexanols (presumably by oral administration; no information given), there were degenerative changes in the neurons in all parts of the brain and brain stem. In addition, there was hepatocellular degeneration in the central regions and kidney damage, which consisted of severe degeneration and frequent necrosis of the epithelium of the proximal and loop tubules, as well as the epithelium of the glomeruli. Slight degenerative changes of the myocardium were also reported (45, 103).

D DECANOLS

The decanols consist of more than 20 structural isomers, including a number of enantiomers. The most important commercial members are the C_{10} oxo alcohols, which exist as a mixture of C_{10} branched alcohols. Many of these alcohols are liquids. Unlike the lower alcohols, the decanols are less volatile and flammable (Table 78.1). Toxicity studies indicate that these alcohols have a low order of acute toxicity but they are irritating to both the skin and eyes. No serious industrial intoxication has been reported for the decanols. No occupational exposure standards exist for the decanols.

14.0 1-Decanol

14.0.1 CAS Number: *[112-30-1]*

14.0.2 Synonyms: *n*-Decyl alcohol, 1-hydroxydecane; decyl alcohol; decanol; decan-1-ol; capric alcohol; alcohol C-10; nonyl carbinol

14.0.3 Trade Names: NA

14.0.4 Molecular Weight: 158.3

14.0.5 Molecular Formula: $C_{10}H_{22}O$

14.0.6 Molecular Structure: 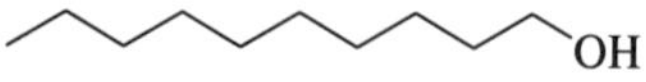

14.1 Chemical and Physical Properties

The chemical and physical properties are listed in Table 78.1.

14.1.2 Odor and Warning Properties

The threshold odor concentration in air for decyl alcohol (isomer not specified) was reportedly 6.3 ppb (20).

14.2 Production and Use

1-Decanol is prepared commercially by sodium reduction or by the high-pressure catalytic reduction of coconut oil, coconut fatty acids, or esters (3, 104). It is also produced by the Ziegler process, which involves oxidation of trialkylaluminum compounds (3, 104). 1-Decanol is used as a chemical intermediate in the production of esters and fragrances and as an artificial flavor in foods, as well as in detergents and as a defoaming agent (3, 104).

14.4 Toxic Effects

14.4.1 Experimental Studies

14.4.1.1 Acute Toxicity. The acute dermal LD_{50} for 1-decanol is 18.8 mL/kg (34). The acute inhalation LC_{50} is 525 ppm (duration not specified) for 1-decanol (11). Aspiration of 0.2 mL 1-decanol produced deaths among nine out of nine rats (13). Decanol reportedly caused corneal injury when instilled into rabbit eyes (11). A mixture of 1-decanol and *sec*-decanol was moderately irritating to the skin of guinea pigs (45).

The acute oral LD_{50} of a mixture of 1-decanol and *sec*-decanol was reported as 12.8 to 25.6 g/kg in the rat and 6.4 to 12.8 g/kg in the mouse (45). The acute dermal LD_{50} is greater than 10 mL/kg in guinea pigs (45). There were no deaths among rats exposed to either 65 or 906 ppm for 6 h (45).

14.4.1.4 Reproductive and Developmental. Exposure of pregnant female rats to a saturated vapor of 1-decanol (15 ppm or 100 mg/m^3) 7 h/day on gestational days 1 to 19 did not result in maternal or developmental toxicity (25, 26). Embryotoxicity was reported in a study in which female rats were dosed orally with a 40% solution of 1-decanol during days 1 through 15 of pregnancy (100).

14.4.1.5 Carcinogenesis. 1-Decanol showed weak to moderate tumor-promoting activity when applied three times a week for 60 weeks to the skin of female Swiss mice that previously received an initiating dose of dimethylbenz[a]anthracene (29).

14.4.2 Human Experience

14.4.2.1 General Information. 1-Decanol has not reportedly had adverse effects on workers.

14.4.2.2 Clinical Cases. It was not a skin irritant or a skin sensitizer to humans at a dose of 3% in petrolatum (22, 104).

14.5 Standards, Regulations, or Guidelines of Exposure

The ACGIH has not established a TLV for 1-decanol. Russia has a MAC value of 10 mg/m^3 (1.5 ppm) for 1-decanol, and Yugoslavia has a MAC value of 200 mg/m^3 (31 ppm); no documentation was given (33).

15.0 Isodecyl Alcohol

15.0.2 Synonym: C_{10} oxo alcohols

15.0.4 Molecular Weight: 158.3

15.1 Chemical and Physical Properties

These alcohols are colorless liquids.

15.2 Production and Use

Isodecyl alcohol, or the C_{10} oxo alcohols, are made in the oxo process by reacting nonenes with carbon monoxide and hydrogen in the presence of a catalyst, followed by hydrogenation (3, 5). The commercial product typically consists of trimethyl-1-heptanols and dimethyl-1-octanols; the composition and CAS registry number depend on the olefin feedstock. A major use of this alcohol is in the production of plasticizers. Industrial exposure may occur in handling the alcohol.

15.4 Toxic Effects

15.4.1 Experimental Studies

Different investigators used different compositions of isodecyl alcohol. Studies by Scala and Burtis (42) used an isodecyl alcohol mixture (oxo) consisting typically of 95% by weight of trimethylheptanols and other homologous primary alcohols; studies by Smyth et al. (105) used mixed isomers of decyl oxo alcohols.

15.4.1.1 Acute Toxicity. The acute oral LD_{50} of isodecyl alcohol is 4.72 g/kg (42) and 9.8 g/kg (105) in rats. The acute dermal LD_{50} is greater than 2.6 g/kg (42) and 3.56 mL/kg (105) in rabbits. No deaths were observed in rats, mice, or guinea pigs exposed to 95 ppm for 6 h (42). There was, however, severe irritation to the mucous membrane of the eyes, nose, throat, and respiratory tract. There were no deaths in rats exposed to saturated vapors for 8 h (105).

Isodecyl alcohol is moderately irritating (42) and severely irritating (105) to the skin of rabbits when applied as an occlusive patch for 24 h. It is a severe eye irritant to rabbits (42).

15.4.1.2 Chronic and Subchronic Toxicity. In a 2-week oral study, male Wistar rats dosed with 1 mmol/kg (158 mg/kg) isodecyl alcohol did not show testicular atrophy, liver enlargement, hepatic peroxisome induction, or hypolipidemia (55).

15.4.1.4 Reproductive and Developmental. Pregnant female Wistar rats were dosed by oral gavage with 0, 158, 790, or 1580 mg/kg isodecanol, CAS Number [*25339-17-7*], during gestational days 6 through 15. In the 790- and 1580-mg/kg groups, maternal toxicity was dose-dependent with clinical symptoms (nasal discharge, salivation and signs of central nervous system depression) and reduced food consumption. Deaths occurred in the 1580 mg/kg dose group. At 1580 mg/kg, mean uterine and fetal weights were reduced, the number of resorptions increased as did the postimplantation loss. Fetal skeletal retardations and malformations were noted. Two fetuses (in two litters) lacked external genitalia, a rare malformation. There was no developmental toxicity in the 790- and 158-mg/kg dose groups (27).

15.4.2 Human Experience

15.4.2.1 General Information. No adverse effects in humans were reported in the use of the C_{10} oxo alcohols.

16.0 3,7-Dimethyl-1-Octanol

16.0.1 CAS Number: *[106-21-8]*

16.0.2 Synonym: Tetrahydrogeraniol; dihydrocitronellol

16.0.3 Trade Names: NA

16.0.4 Molecular Weight: 158.3

16.0.5 Molecular Formula: $C_{10}H_{22}O$

16.0.6 Molecular Structure:

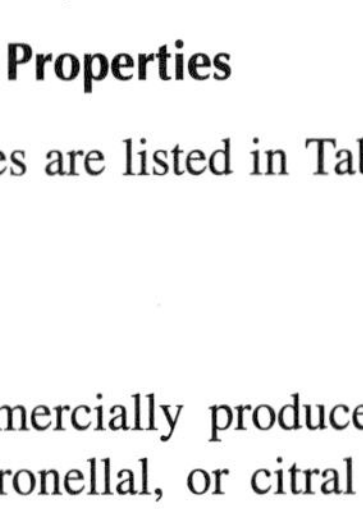

16.1 Chemical and Physical Properties

Chemical and physical properties are listed in Table 78.1.

16.2 Production and Use

3,7-Dimethyl-1-octanol is commercially produced by the reduction of geraniol or by the reduction of citronellol, citronellal, or citral (106). It is used in fragrances and in foods (106).

16.4 Toxic Effects

16.4.1 Experimental Studies

16.4.1.1 Acute Toxicity. The acute oral LD_{50} in rats is greater than 5 g/kg, and the acute dermal LD_{50} in rabbits is 2.4 g/kg (106). 3,7-Dimethyl-1-octanol applied to intact or abraded rabbit skin produced irritation (106).

16.4.2 Human Experience

16.4.2.1 General Information. No adverse effects in humans have been reported.

16.4.2.2 Clinical Cases. 3,7-Dimethyl-1-octanol was not a human skin irritant in a 48-h closed patch test or a sensitizer at a concentration of 8% in petrolatum (106).

E DODECANOLS

These alcohols consist of more than 20 structural isomers, including a number of enantiomers. The two most prominent members of this group of alcohols are 1-dodecanol and isodecyl alcohol, a mixture of predominantly C_{12} branched alcohols. Toxicity studies indicate that dodecanols have a low order of acute toxicity. 1-Dodecanol is the most studied C_{12} alcohol and is a tumor promoter in mice. There have been no reports of adverse effects in humans. No occupational exposure standards have been established for any of the decanols.

17.0 1-Dodecanol

17.0.1 CAS Number: [112-53-8]

17.0.2 Synonyms: *n*-Dodecyl alcohol; lauryl alcohol; 1-hydroxydodecane; dodecanol; dodecan-1-ol; alcohol C12; undecyl carbinol

17.0.3 Trade Names: NA

17.0.4 Molecular Weight: 186.3

17.0.5 Molecular Formula: $C_{12}H_{26}O$

17.0.6 Molecular Structure:

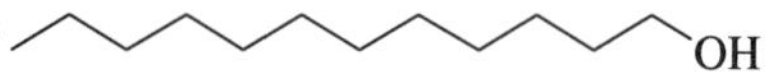

17.1 Chemical and Physical Properties

1-Dodecanol is a crystalline solid which has a melting point of 24°C. Its chemical and physical properties are listed in Table 78.1.

17.1.2 Odor and Warning Properties

The air odor threshold for dodecyl alcohol (isomer not specified) is reported as 7.1 ppb (20).

17.2 Production and Use

1-Dodecanol is produced commercially by the oxo process and from ethylene by the Ziegler process, which involves oxidation of trialkylaluminum compounds (3). It can also be produced by sodium reduction or high-pressure hydrogenation of esters of naturally occurring lauric acid (104). 1-Dodecanol is used mostly in manufacturing detergents and soaps (3). It is used to a lesser extent in wetting, emulsifying, and foaming agents (3). It is also used in fragrances (104) and is approved by the Food and Drug Administration (FDA) for food use.

17.4 Toxic Effects

17.4.1 Experimental Studies

17.4.1.1 Acute Toxicity. The acute oral LD_{50} in rats is reportedly greater than 10.6 g/kg (45) and 12.8 g/kg (107), and greater than 36 mL for rabbits (45). The animals that survived either 19.9 or 29.9 g/kg technical lauryl alcohol demonstrated no significant gross microscopic changes (108). The acute dermal LD_{50} is greater than 8.31 g/kg in guinea pigs (45). The 6-h LC_{50} of 1-dodecanol dispersed as an aerosol is greater than 1050 mg/m^3 (138 ppm) in rats (109).

Aspiration of 0.2 mL of 1-dodecanol produced death among 9 out of 10 rats. The deaths were caused by pulmonary edema rather than cardiac arrest or respiratory failure as with the C_3 to C_{10} alcohols. The lungs were dark red, seven rats died within 7 to 30 min, and two rats died 5 h or longer after dosing (13).

Slight to moderate irritation was noted when 1-dodecanol was applied for 24 h under occlusive conditions to the skin of rabbits and mice (110). No irritation was observed when 1-dodecanol was applied to the skin of guinea pigs (45).

17.4.1.3 Pharmacokinetics, Metabolism, and Mechanisms. After 24-h covered contact with the skin of mice, about 95% of a 100-μL dose of 0.5% 1-dodecanol in triethyl citrate remained on the skin. A small proportion, 0.1%, was recovered in the feces an urine, 0.13% was recovered from the body, and 2.61% was excreted in the air (110). These data indicate a low amount of dermal uptake.

17.4.1.6 Carcinogenesis. 1-Dodecanol showed weak tumor-promoting activity when applied three times a week for 60 weeks to the skin of mice that had previously received an initiating dose of dimethylbenz[a]anthracene. Papillomas developed in 2 of 30 mice after 39 and 49 weeks of treatment (29).

17.4.1.6 Genetic and Related Cellular Effects Studies. 1-Dodecanol was not mutagenic to *S. typhimurium* in the Ames assay with and without metabolic activation, or to *E. coli* without metabolic activation (111).

17.4.2 Human Experience

17.4.2.1 General Experience. No adverse effects in humans were reported.

17.4.2.2 Clinical Cases. A 48-h contact with 4% 1-dodecanol in petrolatum was not irritating to 25 human volunteers (20), but marked skin irritation was noted when 25% 1-dodecanol in mineral oil was given in open contact with scarified skin of 5 to 10 volunteers once a day for 3 days (109). There was no skin sensitization in 25 human volunteers at a concentration of 4% in petrolatum (104).

18.0 Isodecyl Alcohol

18.0.1 CAS Number:

18.0.2 Synonyms: C_{12} oxo alcohols, Isodecanol

18.0.3 Trade Names: NA

18.0.4 Molecular Weight: 186.3

18.0.5 Molecular Formula: $C_{10}H_{22}O$

18.1 Chemical and Physical Properties

Isododecyl alcohols are solid or liquids.

18.2 Production and Use

Isodecyl alcohols are a mixture of alcohols produced by the oxo process in which undecenes are reacted with carbon monoxide and hydrogen in the presence of a catalyst, followed by hydrogenation (5, 6). They are also prepared by other processes, including the Ziegler process which involves oxidation of trialklylaluminum compounds (3). Two of the major isomers found in the commercial product are trimethyl-1-nonanols and dimethyl-1-decanols; the composition and CAS registry number depend on the olefin feedstock. These alcohols are used in producing phthalates and in detergents (6).

18.4 Toxic Effects

18.4.1 Experimental Studies

18.4.1.1 Acute Toxicity. The acute oral LD_{50} in rats is greater than 2 g/kg (17). These alcohols are moderately irritating to both the skin and eyes of rabbits (17).

F TRIDECANOLS

19.0 Isotridecyl Alcohol

19.0.1 CAS Number:

19.0.2 Synonym: C_{13} oxo alcohols

19.0.3 Trade Names: NA

19.0.4 Molecular Weight: 200.3

19.0.5 Molecular Formula: $C_{13}H_{28}O$

19.1 Chemical and Physical Properties

Isotridecyl alcohols are liquids or solids.

19.2 Production and Use

Isotridecyl alcohol is produced by the oxo process in which dodecenes are reacted with carbon monoxide and hydrogen in the presence of a catalyst, followed by hydrogenation (3). A major isomer of a commercial-grade product is tetramethyl-1-nonanol. Because of their low volatility these alcohols are used to produce plasticizers, they are also used as surfactant raw materials, as lubricant intermediates, and as solvents (3).

19.4 Toxic Effects

19.4.1 Experimental Studies

19.4.1.1 Acute Toxicity. The acute oral LD_{50} is 4.75 g/kg in rats (42), and the dermal LD_{50} in rabbits is greater than 2.6 g/kg (42). No deaths were observed in rats, mice, or guinea pigs exposed to saturated vapors (12 ppm) for 6 h (42). Animals exhibited slight irritation to the eyes, nose, and throat.

These alcohols are reportedly moderately irritating to rabbit skin when applied for 24-h under an occlusive wrap, but slightly irritating when applied under 4-h semiocclusive conditions (17). Isotridecyl alcohol was moderately irritating to eyes of rabbits (42).

19.4.2 Human Experience

19.4.2.1 General Information. No adverse effects in humans were reported from exposure to these alcohols.

G HIGHER ALCOHOLS

Most of the toxicity data on the higher alcohols exist on the C_{16} and the C_{18} alcohols. Limited data exist for 1-tetradecanol and eicosnol (see Table 78.4). No occupational exposure standards have been established for the higher alcohols.

20.0 C_{16} Alcohols

20.0.1 CAS Number:

20.0.2 Synonyms: Hexyldecanol; C_{16} oxo alcohols; cetyl alcohol; cetanol; 1-hexadecanol; palmityl alcohol; hexadecan-1-ol; hexadecyl alcohol; hexadecanol; alcohol; C16

20.0.4 Molecular Weight: 242.5

20.0.5 Molecular Formula: $C_{16}H_{34}O$

20.1 Chemical and Physical Properties

These alcohols may be solids or liquids. The chemical and physical properties are listed in Table 78.1.

20.2 Production and Use

This section includes both linear and branched C_{16} alcohols. The linear C_{16} alcohol, also known as 1-hexadecanol CAS Number: [*36653-82-4*] and cetyl alcohol, is prepared commercially by the catalytic reduction of fats containing palmitic acid and by the saponification of spermaceti wax (112). Cetyl alcohol is used extensively in skin lotions and creams, in cosmetics, perfumes, toilet articles, and medicinals, and in manufacturing detergents (112).

Branched C_{16} alcohols, also known as isohexadecyl alcohol or C_{16} oxo alcohols, are prepared by the oxo process and by aldol-like condensation of isooctyl alcohol and subsequent hydrogenation (3,6). These alcohols are used in lubricants, as a component of detergents, and in wetting agents, textile aids, softeners, evaporation preventers, and foam breakers (3).

20.3 Exposure Assessment

Gas chromatographic methods are available to measure these alcohols (6, 112).

20.4 Toxic Effects

20.4.1 Experimental Studies

20.4.1.1 Acute Toxicity. The oral LD_{50} for a C_{16} oxo alcohol is greater than 8.42 g/kg, and the dermal LD_{50} is greater than 2.6 g/kg in rabbits (42). No deaths were observed among rats, mice, and guinea pigs exposed to 26 ppm for 6 h (42). There were no deaths or pathological effects in rats and guinea pigs exposed to an aerosol of 9.6 mg/L for 10 min of every 30 min during a 4-h period (80 min total) (42). The oral LD_{50} of unidentified C_{16} alcohol(s) was reported as 6.4 to 12.8 g/kg for rats and 3.2 to 6.4 g/kg for mice, and the dermal LD_{50} was reported as less than 10 g/kg in guinea pigs (45). All rats died when exposed to a calculated concentration of 2.22 mg/L for 6 h, but all survived at 0.41 mg/L (45).

The skin and eye irritation of cetyl alcohol was extensively reviewed (112). Cetyl alcohol is slightly irritating to both the skin and eyes of rabbits. A C_{16} oxo alcohol was slightly irritating to the skin and eyes of rabbits (42). Cetyl alcohol is not sensitizing to the skin of guinea pigs (112).

20.4.1.3 Pharmacokinetics, Metabolism, and Mechanisms. These alcohols are poorly absorbed by the skin (110). Cetyl alcohol is oxidized in rats to the corresponding fatty acid, palmitic acid (8). When fed to rats at a dose level of 2 g/kg, it is partly absorbed and metabolized, about 20% of the dose is recovered in the feces (113). A small rise in urinary glucuronic acid excretion was observed (113).

20.4.1.5 Carcinogenesis. Cetyl alcohol demonstrated weak tumor-promoting activity when applied three times a week for 60 weeks to the skin of mice that had received an initiating dose of dimethylbenz[a]anthracene (29).

20.4.1.6 Genetic and Related Cellular Effects Studies. Cetyl alcohol was not mutagenic in *S. typhimurium* strains TA98, TA100, TA1535, TA1537, and TA1538 in the spot test (114).

20.4.2 Human Experience

20.4.2.2 Clinical Cases. Clinical studies indicate a low order of skin irritation and sensitization in humans (112).

20.5 Standards, Regulations, or Guidelines of Exposure

No occupational health standards exist for these alcohols.

21.0 C_{18} Alcohols

21.0.1 CAS Number:

21.0.2 Synonyms: Octyldecanol; C_{18} oxo alcohols; stearic alcohol; stearyl alcohol; 1-octadecanol; 1-hydroxyoctadecane; *n*-octadecanol; octadecyl alcohol; octadecan-1-ol

21.0.3 Trade Names: NA

21.0.4 Molecular Weight: 270.5

21.0.5 Molecular Formula: $C_{18}H_{38}O$

21.1 Chemical and Physical Properties

These alcohols may be solids or liquids. Their chemical and physical properties are listed in Table 78.1.

21.2 Production and Use

This section includes both linear and branched C_{18} alcohols. The linear C_{18} alcohol, also known as 1-octadecanol, CAS Number *[112-92-5]* and stearyl alcohol, is prepared commercially via Ziegler aluminum alkyl hydrolysis or the catalytic, high-pressure hydrogenation of stearyl acid, followed by filtration and distillation (115). It may also be derived from natural fats and oils (115). Stearyl alcohol is used in surface-active agents, lubricants, emulsions, resins, and USP ointments (115). Synthetic stearyl alcohol has been approved as a direct and indirect food additive ingredient and as an ingredient in over-the-counter drugs (115).

Branched C_{18} alcohols, also known as C_{18} oxo alcohols, are prepared by the oxo process and by an aldol-like condensation of isononyl alcohol (3, 6). They are used for manufacturing synthetic lubricants and hydraulic fluids. Because of the low vapor pressure, they are often applied to water surfaces to prevent evaporation (3).

21.3 Exposure Assessment

Methods are available to measure these alcohols (6, 115).

21.4 Toxic Effects

21.4.1 Experimental Studies

21.4.1.1 Acute Toxicity. The acute oral LD_{50} in rats is greater than 8 g/kg, and the acute dermal LD_{50} in rabbits is greater than 3 g/kg (115). Stearyl alcohol is slightly irritating to both the skin and eyes of rabbits (115). Stearyl alcohol is not sensitizing to the skin of guinea pigs (115).

21.4.1.3 Pharmacokinetics, Metabolism, and Mechanisms. Stearyl alcohol is found naturally in various mammalian tissues. It is used in the biosynthesis of lipids and other naturally occurring cellular constituents and enters metabolic pathways for energy production. This fatty alcohol is readily converted to stearic acid, another common constituent of mammalian tissue. Results from several studies indicate that stearyl alcohol is poorly absorbed from the gastrointestinal tract (115).

21.4.1.5 Carcinogenesis. Stearyl alcohol was a weak tumor promoter when applied three times a week for 60 weeks to the skin of mice that had received an initiating dose of dimethylbenz[a]anthracene (29).

21.4.1.6 Genetic and Related Cellular Effects Studies. Stearyl alcohol was not mutagenic to *S. typhimurium* strains TA98, TA100, TA1535, and TA1537 in an Ames test with or without metabolic activation (116).

21.4.2 Human Experience

21.4.2.1 General Information. No adverse effects were reported in an occupational environment.

21.4.2.2 Clinical Cases. The results of single-insult clinical patch testing indicate a very low order of skin irritation potential for stearyl alcohol with a sensitization rate of 19 of 3740 (0.51%) individuals (115).

21.5 Standards, Regulations, or Guidelines of Exposure

No occupational health standards exist for these alcohols.

H AROMATIC ALCOHOLS

There are at least five aromatic alcohols of commercial interest; three primary alcohols (benzyl alcohol, 2-phenylethanol, and *p*-tolyl alcohol) and two secondary alcohols (1-phenylethanol and 2-phenyl-2-propanol). All of these alcohols are liquids at ambient temperatures. No valid NIOSH methods exist to measure these alcohols in air.

The toxicities of benzyl alcohol and 2-phenylethanol are the most studied of the aromatic alcohols. As a group, these alcohols have a low to moderate order of acute toxicity and have been both active and inactive in genotoxicity assays. No occupational exposure standards have been set for these alcohols.

22.0 Benzyl Alcohol

22.0.1 CAS Number: [100-51-6]

22.0.2 Synonyms: Benzene methanol; phenyl carbinol; phenyl methanol; α-hydroxytoluene; phenylmethyl alcohol; benzoyl alcohol; hydroxytoluene; benzenecarbinol; alpha-toluenol; (hydroxymethyl)benzene

22.0.4 Molecular Weight: 108.1

22.0.5 Molecular Formula: C_7H_8O

22.0.6 Molecular Structure:

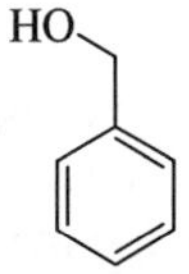

22.1 Chemical and Physical Properties

Benzyl alcohol is a colorless liquid with a faint aromatic odor and a sharp, burning taste. The chemical and physical properties are listed in Table 78.1.

22.2 Production and Use

Benzyl alcohol is manufactured commercially from benzyl chloride by refluxing with sodium carbonate (117). A significant use of benzyl alcohol is in the photographic and textile industries (117). Benzyl alcohol is used in the pharmaceutical industry, and can be found in cough syrups and drops, ophthalmic solutions, and burn, dental, and insect-repellent solutions and ointments. It is also used in cosmetics and in acne treatment preparations, in rug cleaners as a degreasing agent, and in the polymer industry (117). In agriculture, it is used in insect repellents, as a stabilizer in insecticidal formulations, and in treating fruits and vegetables (117).

22.3 Exposure Assessment

22.3.3 Workplace Methods

There are no methods available for determinining benzyl alcohol in air in the workplace.

22.4 Toxic Effects

22.4.1 Experimental Studies

Benzyl alcohol displayed a low to moderate order of acute toxicity. It produced neurotoxic effects in rat and mouse oral subchronic toxicity studies. Benzyl alcohol was not carcinogenic to rats and mice; it has produced both positive and negative effects in *in vitro* genotoxicity studies. There is no indication that benzyl alcohol produces adverse reproductive effects.

22.4.1.1 Acute Toxicity. The acute oral LD_{50} has been reported as 1.23 g/kg (107), 3.1 g/kg (105), and 2.08 g/kg (118) in rats; 1.58 g/kg in mice (119); and 1.94 g/kg (118) in rabbits. The acute dermal LD_{50} was less than 5 ml/kg in guinea pigs (118). No deaths occurred in rats exposed to a saturated vapor of benzyl alcohol for 2 h (105). The inhalation 8-h LC_{50} in rats was 1000 ppm (105).

Undiluted benzyl alcohol was moderately irritating when applied to the depilated skin of guinea pigs for 24 h (45). It was moderately irritating when applied to rabbit skin (105). Benzyl alcohol was severely irritating to the eyes of rabbits (105).

22.4.1.2 Chronic and Subchronic Toxicity. Fischer 344 rats were given oral doses of 50, 100, 200, 400, and 800 mg/kg for 13 weeks. The high dose produced clinical signs indicative of neurotoxicity, including staggering, respiratory difficulty, and lethargy. Reduction in weight gain was noted in males at 800 mg/kg and females at $\geq$200 mg/kg. The high-dose animals also showed hemorrhages around the mouth and nose and histological lesions in the brain, thymus, skeletal muscle, and kidney (120). B6C3F1 mice were given oral doses of 50, 100, 200, 400, and 800 mg/kg for 13 weeks. The high dose produced clinical signs of neurotoxicity. Reduction in weight gain was noted in males at $\geq$400 mg/kg and females at $\geq$200 mg/kg. No treatment-related histopathological effects were noted (120).

22.4.1.3 Pharmacokinetics, Metabolism, and Mechanisms. Benzyl alcohol is readily absorbed from the gastrointestinal tract and rapidly oxidized to benzoic acid, which is conjugated with glycine and excreted as hippuric acid in the urine. Higher doses result in excretion of benzyl alcohol conjugated with glucuronide (8). Benzyl alcohol is oxidized by alcohol dehydrogenase (121, cited in Ref. 122). Pyrazole, an inhibitor of alcohol dehydrogenase, and disulfiram, an inhibitor of aldehyde dehydrogenase resulted in increased levels of benzyl alcohol and benzaldehyde, respectively (123). Within 6 h after the oral administration of 0.40 g benzyl alcohol/kg body weight, rabbits eliminated 65.7% of the dose as hippuric acid (124).

22.4.1.5 Carcinogenesis. In an NTP study, F344 rats were dosed by oral gavage with 0, 200, and 400 mg/kg, 5 days/week for 2 years. Benzyl alcohol had no effect on the survival of male rats; female rats had reduced survival, and many of the early deaths were considered related to the gavage procedure. There were no treatment-related effects on nonneoplastic or neoplastic lesions in either sex treated with benzyl alcohol. It was concluded that under the conditions of the study, there was no evidence of carcinogenic activity (120). In the same NTP study, B6C3F1 mice were dosed by oral gavage with 0, 100, and 200 mg/kg, 5 days/week for 2 years. No effects on survival or body weight gain were observed. There were no treatment-related effects on nonneoplastic or neoplastic lesions in either sex. It was concluded that under the conditions of the study, there was no evidence of carcinogenic activity (120).

22.4.1.6 Genetic and Related Cellular Effects Studies. Benzyl alcohol was not mutagenic to *S. typhimurium* when tested in an Ames test with or without metabolic activation system (116). When tested by the preincubation protocol, benzyl alcohol was not mutagenic to *S. typhimurium* strains TA98, TA100, TA1535, and TA1537 in an Ames test with or without metabolic activation (120, 125). Benzyl alcohol was mutagenic in *Bacillus subtilis* M45 (rec-) and H17 (rec+) (126). In the L5178Y/TK+/ − mouse lymphoma assay, benzyl alcohol was active without, but not with a metabolic activation system; the effect was associated with toxicity (120). In cytogenetic assays with Chinese hamster ovary cells, treatment with benzyl alcohol produced an increase in sister chromatid exchanges, which was judged equivocal both with and without an S9 metabolic activation system. A significant increase in chromosomal aberrations was observed after exposure to benzyl alcohol with, but not without a metabolic activation system (120, 127). No cytogenetic effects were observed in Chinese hamster lung fibroblast cells treated with benzyl alcohol without a metabolic activation system (77).

22.4.1.7 Other: Neurological, Pulmonary, Skin, Sensitization. Benzyl alcohol was not a skin sensitizer in guinea pigs (45).

22.4.2 Human Experience

22.4.2.1 General Information. No serious intoxication has been linked with occupational exposure to benzyl alcohol.

22.4.2.2 Clinical Cases

22.4.2.2.3 Pharmacokinetics, Metabolism, and Mechanisms. The dermal flux for benzyl alcohol across human skin *in vitro* was reported as 0.073 mg/cm^2/hr, indicating a low rate of dermal uptake (128). The percentage of the applied dose that penetrated through human skin *in vitro* in 6 h was 1.42% for adult skin and 0.73% for full-term infant skin (129).

Human subjects eliminated 75 to 85% of the dose in the urine as hippuric acid within 6 h after taking 1.5 g of benzyl alcohol orally (130).

22.2.2.7 Other: Neurological, Pulmonary, Skin sensitization, etc. Local reactions indicating sensitization were seen in humans with various skin complaints in patch studies. These studies have involved 24- or 48-h covered patches with 5 to 10% benzyl alcohol in petrolatum (131–133). In the largest study conducted, about 1% of a total of 2261 volunteers gave positive reactions (132). The exposure regime that initially sensitized these individuals was not reported, but a number of reported case studies have involved repeated contact with perfumes or medications containing benzyl alcohol (134). There were no sensitization reactions in 25 volunteers exposed to 10% benzyl alcohol in five 48-h patch tests during a 10-day period (134). Cross-sensitization to benzyl alcohol has been reported in subjects sensitized to Peru balsam (118). In a covered patch test with 0.05% benzyl alcohol in either ethanol or a cream base, irritation was produced in 18 of 614 subjects (134).

Benzyl alcohol, which is used as a preservative in intravascular flush solutions, caused neurological deterioration and deaths in very low birth weight infants (134). Preterm infants who received large volumes of fluids containing 0.9% benzyl alcohol via catheter developed "gasping baby syndrome." Estimated intakes of 99 to 405 mg benzyl alcohol/kg body weight for 2 to 28 days caused effects, including severe metabolic acidosis, gasping, neurological deterioration, blood abnormalities, skin breakdown, liver and kidney failure, lowered blood pressure, heart failure, and death (135, 136). No effects were seen for intakes of 27 to 99 mg/kg body weight over similar periods (136), although there is one report of 32 to 105 mg/kg body weight for 7 days that caused breathing difficulty (134).

Benzyl alcohol poisoning was reported in a case of a 5-yr-old girl following treatment with a continuous intravenous infusion of diazepam for 36 h. Severe metabolic acidosis occurred in which the patient received 180 mg/kg/day benzyl alcohol, resulting in benzoic acid serum and urine levels of 18 mg/mL and 120 mg/dL, respectively (137).

22.5 Standards, Regulations, or Guidelines of Exposure

No occupational standards have been established for benzyl alcohol.

23.0 2-Phenylethanol

23.0.1 CAS Number: *[60-12-8]*

23.0.2 Synonyms: β-Phenylethyl alcohol, phenethyl alcohol, benzyl carbinol, β-hydroxy ethylbenzene; phenylethyl alcohol; 2-Phenylethanol; benzeneethanol; hydroxy-ethylbenzene; 1-phenyl-2-ethanol

23.0.3 Trade Names: NA

23.0.4 Molecular Weight: 122.2

23.0.5 Molecular Formula: $C_6H_{10}O$

23.0.6 Molecular Structure:

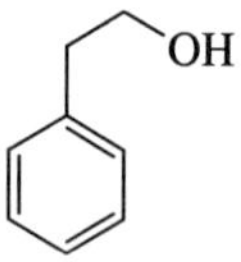

23.1 Chemical and Physical Properties

2-Phenylethanol is a low volatility liquid with a characteristic odor of roses. Its chemical and physical properties are listed in Table 78.1.

23.2 Production and Use

2-Phenylethanol is produced from benzene and ethylene oxide by the Friedel–Crafts reaction (138). It is used primarily as a fragrance, and also as a synthetic food flavoring and antimicrobial agent (138–140). The primary routes of occupational exposure are dermal contact and inhalation. Exposure from ingestion may occur from foods or cosmetic products. No information was found concerning actual exposure levels in the workplace.

23.3 Exposure Assessment

A number of analytical methods are available to measure 2-phenylethanol: infrared spectrophotometry, gas chromatography, gas–liquid chromatography, plasma chromatography, and ion-exchange chromatography. Gas chromatography–mass spectrometry has been used to identify 2-phenylethanol in urine by conversion to the pentafluoropropionyl derivative (140).

23.4 Toxic Effects

23.4.1 Experimental Studies

2-Phenylethanol produces slight to moderate acute toxicity by the oral, dermal, or inhalation routes. It is severely irritating to the eyes and slightly irritating to the skin of rabbits and guinea pigs. It was not a skin sensitizer in guinea pigs. A subchronic dermal rat study showed little systemic toxicity. 2-Phenylethanol was not genotoxic in *in vitro* genotoxicity tests. High dermal and oral doses of 2-phenylethanol in rats produced maternal and teratogenic effects.

23.4.1.1 Acute Toxicity. The acute oral LD_{50} is 1.79 g/kg (119) and 2.46 mL/kg (141) in rats, 0.8 to 1.5 g/kg for mice, and 0.4 to 0.8 g/kg in guinea pigs (45). The acute dermal LD_{50} is from 5 to 10 mL/kg in guinea pigs (45) and 0.79 mL/kg in rabbits (141). No deaths occurred among rats exposed for 8 h to a saturated vapor (141).

2-Phenylethanol was slightly irritating to the skin of guinea pigs and rabbits (45, 141). When instilled into the eyes of rabbits, 0.005 mL of undiluted material or 0.5 mL of 5 to 15% solutions in propylene glycol caused severe corneal irritation and iritis (141). A 1% solution caused irritation of the conjunctiva and transient clouding of the corneal epithelium (142).

23.4.1.2 Chronic and Subchronic Toxicity. 2-Phenylethanol was applied to the skin of Sprague–Dawley rats at doses of 0, 0.25, 0.5, 1.0, and 2.0 ml/kg for 13 weeks. Weight gain was depressed in both sexes in the 1.0- and 2.0-mL/kg groups, but no effect on food intake was observed. Decreased hemoglobin concentration and white blood cell counts were noted in the high-dose males at week 6 and 13. There were no differences in organ weights and no treatment-related histopathological effects. No other adverse treatment-related effects were observed (143).

23.4.1.3 Pharmacokinetics, Metabolism, and Mechanisms. 2-Phenylethanol is oxidized almost entirely to the corresponding acid, which is conjugated with glycine to form phenylaceturic acid (8). Phenylaceturic acid is the major urinary metabolite of 2-phenylethanol in animals. Other urinary metabolites that have been found, but at much lower concentrations, include 2-phenylethyl glucuronide, phenylacetyl glucuronide, and hippuric acid (140).

23.4.1.4 Reproductive and Developmental. Pregnant female Long–Evans rats were dosed by oral gavage with 0, 4.3, 43, or 432 mg/kg 2-phenylethanol on days 6 to 15 of gestation. Embryo lethality was noted in rats treated with 4.3 and 43 mg/kg, and intrauterine growth retardation and skeletal variations were observed in the offspring of rats treated with 4.3 and 432 mg/kg, but not at 43 mg/kg. The incidence of malformations was dose related, and there were increased incidences of malformed eyes, neural-tube defects, hydronephrosis, and limb defects (144).

Pregnant female Sprague–Dawley rats were fed dietary levels of 1000, 3000, and 10,000 ppm 2-phenylethanol on days 6 to 15 of gestation. At the high dose, slight body weight loss and decreased food consumption were observed during the first 2 days of treatment. No treatment-related teratogenic effects were observed. There were no apparent treatment effects on skeletal variants, number of live young, embryolethality, implants, litter weight, mean fetal weights, or sex ratio (145). Pregnant female rats were treated with a dermal application of 2-phenylethanol under an occlusive patch at doses of 0.14, 0.43, or 1.4 ml/kg for 24 h/day on days 6 to 15 of gestation. The high dose resulted in maternal toxicity and developmental effects such as morphological abnormalities in virtually all fetuses. The middle dose produced no maternal toxicity but a slightly increased incidence in cervical rib buds was noted in 30/129 fetuses. No maternal or developmental effects were observed at the low dose (146).

23.4.1.6 Genetic and Related Cellular Effects Studies. 2-Phenylethanol was not mutagenic to *S. typhimurium* in an Ames test with or without metabolic activation (116). 2-Phenylethanol inhibited DNA synthesis in *E. coli* (140) but was not active in a DNA repair assay of *E. coli* (140). There was no increase in the frequency of sister chromatid exchanges in human lymphocytes treated with 2-phenylethanol (147).

23.4.1.7 Other: Neurological, Pulmonary, Skin Sensitization. 2-Phenylethanol was not a skin sensitizer in guinea pigs (139).

23.4.2 Human Experience

23.4.2.1 General Information. No adverse effects were reported from occupational exposure to 2-phenylethanol by inhalation.

23.4.2.2 Clinical Cases

23.4.2.2.1 Acute Toxicity. 2-Phenylethanol is irritating to the eyes. Aqueous solutions of 0.75% and 0.6% produced irritation and eye watering (140, 142). No skin irritation was observed in humans exposed for 24 h to a covered application of neat 2-phenylethanol (142).

23.4.2.2.3 Pharmacokinetics, Metabolism, and Mechanisms. The dermal flux of 2-phenylethanol in human skin *in vitro* was reported as 0.26 mg/cm^2/h (128). This value was about 3.5-fold higher than the dermal flux as reported for benzyl alcohol (128) under identical test conditions. Phenylacetyl glutamine was identified as a urinary metabolite in humans (140).

23.4.2.2.7 Other: Neurological, Pulmonary, Skin Sensitization etc. An 8% solution of 2-phenylethanol in petrolatum failed to produce skin sensitization in 25 subjects (148). However, local reactions indicative of sensitization were seen in patients who had various skin complaints in patch studies (149).

23.5 Standards, Regulations, or Guidelines of Exposure

No occupational exposure standards exist for 2-phenylethanol.

24.0 1-Phenylethanol

24.0.1 CAS Number: [98-85-1]

24.0.2 Synonyms: 1-Phenyl ethyl alcohol; alpha-methyl benzyl alcohol; styrallyl alcohol; phenylmethylcarbinol; alpha-methylbenzenemethanol; beta-hydroxyethylbenzene; beta-phenethyl Alcohol; methylphenylcarbinol; methyl phenyl methanol; alpha-phenylethyl alcohol; 1-phenylethan-1-ol; *sec*-phenethyl alcohol; DL-*sec*-phenethyl alcohol; (*S*)-1-phenylethanol; (*S*)-1-phenylethyl alcohol

24.0.3 Trade Names: NA

24.0.4 Molecular Weight: 122.2

24.0.5 Molecular Formula: $C_8H_{10}O$

24.0.6 Molecular Structure:

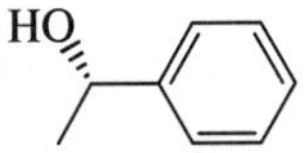

24.1 Chemical and Physical Properties

1-Phenylethanol is a colorless liquid with a floral odor. Its chemical and physical properties are listed in Table 78.1.

24.2 Production and Use

1-Phenylethanol is coproduced with propylene oxide by reaction of α-peroxyethylbenzene (formed by the oxidation of ethylbenzene) with propylene (150). It is used as a fragrance additive in cosmetics such as perfumes, creams, and soaps and is an intermediate in styrene production (151). 1-Phenylethanol is also added to foods as a flavoring agent (151). Industrial exposure may occur from dermal contact and ingestion.

24.4 Toxic Effects

24.4.1 Experimental Studies

1-Phenylethanol displayed a low to moderate order of acute and subchronic oral toxicity in rodents. It was not carcinogenic to mice or female rats, but kidney tumors were observed in male rats. 1-Phenylethanol displayed both negative and positive findings in various genotoxicity assays.

24.4.1.1 Acute Toxicity. The acute oral LD_{50} was reported as 400 mg/kg in rats (152). The acute dermal LD_{50} in rabbits is greater than 15 mL/kg (152) and greater than 2.5 g/kg (153). No deaths occurred among rats exposed to saturated vapors for 8 h (152).

1-Phenylethanol was irritating to rabbit skin (152). Undiluted 1-phenylethanol was moderately irritating to intact or abraded rabbit skin when applied under occluded conditions for 24 h (153). 1-Phenylethanol was irritating to rabbit eyes (154).

24.4.1.2 Chronic and Subchronic Toxicity. Male and female F344 rats were dosed by gavage with 93, 187, 375, 750, and 1500 mg/kg 1-phenylethanol for 13 weeks. The high dose produced some deaths and reduction in body weight gain but no histopathological lesions. Ataxia and lethargy were noted at the two highest doses. Relative liver weights increased in all treated female rats and at the three highest doses in male rats (151). Male and female B6C3F1 mice were dosed by oral gavage with 47, 94, 188, 375, and 750 mg/kg 1-phenylethanol for 13 weeks. No deaths were observed, and there were no treatment-related body weight effects or histopathological lesions. Ataxia and lethargy were noted at the two highest doses (151).

24.4.1.3 Pharmacokinetics, Metabolism, and Mechanisms. A single oral dose of 460 mg/kg 1-phenylethanol was rapidly excreted by rabbits, and 82% of the dose appeared as urinary metabolites. Fifty percent of the material was the glucuronide conjugate of 1-phenylethanol, 30% was hippuric acid, and 1 to 2% was mandelic acid (151).

24.4.1.5 Carcinogenesis. In an NTP study, both sexes of F344 rats were dosed by gavage with 0, 375, and 750 mg/kg 1-phenylethanol 5 days/week for 2 years. Survival was poor in low- and high-dose male and high-dose female rats. Renal toxicity characterized by severe nephropathy and related secondary lesions was observed in the dosed male rats. There was increased incidence of nonneoplastic and neoplastic kidney tumors in the high-dose male rats, but no evidence of carcinogenicity in the female rats (151). In the same NTP study, both sexes of B6C3F1 mice were dosed by oral gavage with 0, 375, and 750 mg/kg 1-phenylethanol 5 days/week for 2 years. Survival rates were similar among groups and a reduction in body weight gain was evident in the high-dose animals. There was no evidence that 1-phenylethanol was carcinogenic to mice in this study (151).

24.4.1.6 Genetic and Related Cellular Effects Studies. 1-Phenylethanol was not mutagenic to *S. typhimurium* strains TA98, TA100, TA1535, and TA1537 in an Ames assay with and without activation (151, 155). It was active in a L5178/TK +/− mouse lymphoma assay without metabolic activation; it was not tested with activation (151). No sister chromatid exchanges were observed in Chinese hamster ovary cells treated with 1-phenylethanol with and without activation (151). In Chinese hamster ovary cells treated with 1-phenylethanol, chromosomal aberrations were observed with but not without metabolic activation (151).

24.4.2 Human Experience

24.4.2.2.7 Other: Neurological, Pulmonary, Skin Sensitization, etc. In a maximization test carried out on 25 human volunteers, 1-phenylethanol (8% in petrolatum) did not produce any skin sensitizing reaction (153). However, a 25% solution of 1-phenylethanol produced a sensitization reaction in 1 out of 179 individuals (149).

24.5 Standards, Regulations, or Guidelines of Exposure

No occupational exposure standards were established for 1-phenylethanol.

25.0 2-Phenyl-1-Propanol

25.0.1 CAS Number: *[1123-85-9]*

25.0.2 Synonyms: Hydratropyl alcohol, α-methyl phenylethyl alcohol, 2-phenyl-propan-1-ol

25.0.3 Trade Names: NA

25.0.4 Molecular Weight: 134.2

25.0.5 Molecular Formula: $C_9H_{12}O$

25.0.6 Molecular Structure:

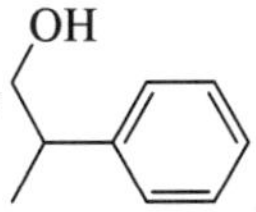

25.1 Chemical and Physical Properties

2-Phenyl-1-propanol is a colorless liquid. Its chemical and physical properties are listed in Table 78.1.

25.2 Production and Use

2-Phenyl-1-propanol is made by the catalytic hydrogenation of 2-phenylpropanal (156). It is used as a component of fragrances and as a food flavoring agent (156).

25.4 Toxic Effects

25.4.1 Experimental Studies

25.4.1.1 Acute Toxicity. The acute oral LD_{50} in rats was reported as 2.3 g/kg, and the acute dermal LD_{50} in rabbits exceeded 5 g/kg (156). This material was not irritating to intact or abraded rabbit skin when applied for 24 h under occluded conditions (156).

25.4.1.2 Chronic and Subchronic Toxicity. A 13-week subchronic oral study was conducted in rats at doses of 10, 40, and 160 mg/kg 2-phenyl-1-propanol. Increased liver weights at the highest dose in both sexes and increased kidney weights at the middle and high doses in males were noted, but there were no pathological lesions (157).

25.4.1.3 Pharmacokinetics, Metabolism, and Mechanisms. 2-Phenyl-1-propanol is oxidized to hydratropic acid, the corresponding acid, which is excreted as a glucuronide conjugate (8). A smaller pathway (10 to 20% of the dose) involves direct glucuronide conjugation of the alcohol.

25.4.1.6 Genetic and Related Cellular Effects Studies. 2-Phenyl-1-propanol was inactive in an Ames test, a Basc test on *Drosophila melanogaster*, and in a mouse bone marrow micronucleus assay (158).

25.4.2 Human Experience

No adverse effects have been reported in humans.

25.4.2.2 Clinical Cases. 2-Phenyl-1-propanol at 6% in petrolatum was not a skin irritant or a sensitizer in humans (156).

25.5 Standards, Regulations, or Guidelines of Exposure

No occupational exposure limits exist for 2-phenyl-1-propanol.

26.0 *p*-Tolyl Alcohol

26.0.1 CAS Number: *[589-18-4]*

26.0.2 Synonyms: 4-methylbenzyl alcohol, *p*-tolyl carbinol; 4-methyl benzenemethanol; 4-(hydroxymethyl)toluene; *p*-toluyl alcohol

26.0.3 Trade Names: NA

26.0.4 Molecular Weight: 122.2

26.0.5 Molecular Formula: $C_8H_{10}O$

26.0.6 Molecular Structure:

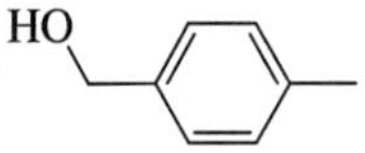

26.1 Chemical and Physical Properties

p-Tolyl alcohol is a white crystalline powder. Its chemical and physical properties are listed in Table 78.1.

26.2 Production and Use

p-Tolyl alcohol is commercially prepared by reducing *p*-tolyl aldehyde (159). It is used mainly in fragrances (159), and the major route of exposure is likely to be dermal contact.

26.4 Toxic Effects

26.4.1 Experimental Studies

26.4.1.1 Acute Toxicity. The acute oral LD_{50} is 3.9 g/kg for rats and 1.4 g/kg for mice (159). The acute dermal LD_{50} in rabbits exceeds 5 g/kg (159). *p*-Tolyl alcohol was moderately to severely irritating to intact or abraded rabbit skin (159) when applied for 24 h under occluded conditions.

26.4.1.6 Genetic and Related Cellular Effects Studies. *p*-Tolyl alcohol was not mutagenic to *S. typhimurium* in an Ames test with and without metabolic activation (19).

26.4.2 Human Experience

26.4.2.2 Clinical Cases. When tested at 4% in petrolatum, *p*-tolyl alcohol was not irritating and did not produce a sensitization reaction in human volunteers (159).

26.5 Standards, Regulations, or Guidelines of Exposure

No occupational standards exist for *p*-tolyl alcohol.

M ALICYCLIC ALCOHOLS

There are at least four alicyclic alcohols of commercial interest, two from the cyclohexyl family (cyclohexanol, methylcyclohexanols) and two from the furan family (furfuryl alcohol and tetrahydrofuran methanol). These alcohols are liquids. Cyclohexanol and furfuryl alcohol are the best studied alcohols of this group. Occupational exposure standards exist for these alcohols, except for tetrahydrofuran methanol.

27.0 Cyclohexanol

27.0.1 CAS Number: *[108-93-0]*

27.0.2 Synonyms: Cyclohexyl alcohol; hexalin; hexahydrophenol; hydrophenol; hydralin; hydroxycyclohexane; 1-cyclohexanol; cyclohexyl alcohol; hydrophenol; adronol anol

27.0.3 Trade Names: NA

27.0.4 Molecular Weight: 100.2

27.0.5 Molecular Formula: $C_6H_{12}O$

27.0.6 Molecular Structure:

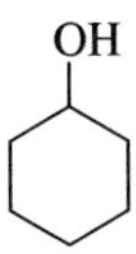

27.1 Chemical and Physical Properties

Cyclohexanol is a colorless, viscous liquid with a camphor-like odor. Its chemical and physical properties are listed in Table 78.1.

27.1.2 Odor and Warning Properties

The air odor detection limit for cyclohexanol is 0.15 ppm (160).

27.2 Production and Use

Cyclohexanol is prepared by the catalytic air oxidation of cyclohexane or by the catalytic hydrogenation of phenol (161). The most important use of cyclohexanol is in producing adipic acid used to the manufacture caprolactam. Cyclohexanol is used in the manufacture of esters for use as plasticizers; it is also used as a chemical intermediate, a stabilizer, a homogenizer for various soap and detergent emulsions, and as a solvent for lacquers and varnishes (161). The primary routes of occupational exposure are dermal and inhalation.

27.3 Exposure Assessment

27.3.3 Workplace Methods

NIOSH Method No. 1402 has been recommended for measuring cyclohexanol in air with a working range of 15 to 150 mg/m^3 (3.7 to 37 ppm). This method involves drawing a known volume of air through charcoal to trap the organic vapors present (the recommended sample is 10 liters at a rate of 0.2 L/min). The analyte is desorbed with carbon disulfide containing 5% 2-propanol. The sample is separated by injection into a gas chromatograph equipped with a flame ionization detector (36).

27.3.5 Biomonitoring/Biomarkers

A gas chromatographic method also exists to measure cyclohexanol in urine (162, 163).

27.4 Toxic Effects

27.4.1 Experimental Studies

Cyclohexanol has a low order of acute oral and inhalation toxicity. It is markedly irritating to the eyes but only slightly irritating to the skin; however, it can be absorbed through the skin in toxic amounts if exposures are severe. Excessive inhalation exposure is irritating to the eyes, nose, and throat and can cause narcosis. Repeated inhalation exposure may result in adverse effects on the heart, liver, kidney, and brain. Cyclohexanol is not mutagenic, but it may affect reproduction.

27.4.1.1 Acute Toxicity. The acute oral LD_{50} for rats is 2.06 g/kg (108) and the minimum lethal dose for rabbits is 2.2 to 2.6 g/kg (164). No deaths occurred in rats exposed to saturated vapor for 8 h (18).

Cyclohexanol is slightly irritating to rabbit skin (165) and can be absorbed through the skin in toxic amounts which, at high concentrations and if applied to extensive skin areas, result in tremors, narcosis, hypothermia, and death (164). It produces moderately severe irritation and reversible corneal injury when instilled into the eyes of the rabbits (34, 165).

27.4.2 Chronic and Subchronic Toxicity

Exposure of rabbits to 997 and 1229 ppm cyclohexanol vapor 6 hr/day, 5 days/week for 5 or 11 weeks, respectively, induced an intoxication characterized by conjunctival congestion and irritation, lacrimation, salivation, lethargy, incoordination, narcosis, mild convulsions, and 50% mortality (166). One monkey and eight rabbits survived concentrations below 700 ppm for 300 h (166). Toxic degenerative changes were found in the brain, heart, liver, and kidneys of the rabbits exposed to 997 and 1229 ppm. Similar but less severe changes were seen in the myocardium, liver, and kidneys of rabbits exposed to 272 ppm cyclohexanol. Rabbits exposed to 145 ppm suffered only slight degenerative changes in the liver and kidneys (166).

Temporary erythema and superficial sloughing of skin were noted when cyclohexanol as a 15% ointment was applied to the skin of rabbits for 1 h/day during a period of 15 days (164). Necrosis, exudative ulceration, and thickening of the skin were observed following application of 10 mL cyclohexanol to the intact skin of a single rabbit for 1 h/day for 10 consecutive days. These 10-ml applications also induced narcosis, tremors, athetoid movements, and hypothermia, and the rabbit died the day after the 10th treatment (164).

27.4.1.3 Pharmacokinetics, Metabolism, and Mechanisms. Cyclohexanol can be absorbed via the skin or lungs or by ingestion. Cyclohexanol is not aromatized *in vivo*; large amounts are excreted as urinary glucuronides (8). No urinary metabolites were detected when cyclohexanol was administered to dogs as a subcutaneous dose of 0.29 g/kg (34), but glucuronic acid was found in the urine of a dog following oral administration of cyclohexanol (165). Following oral administration to rabbits or by inhalation, cyclohexanol was excreted in the urine conjugated with sulfuric and glucuronic acids (164, 166). When 1.2 g/kg was given orally to rabbits, 45 to 50% was excreted with glucuronic acid, and the ratio of urinary inorganic sulfates to total sulfates decreased (164). Similar results were obtained when rabbits were exposed to repeated inhalation exposures of cyclohexanol (166). In general, the increased elimination of conjugated sulfates could be correlated with the concentration of cyclohexanol in air. However, if rabbits were exposed to lower concentrations of cyclohexanol, such as 145 ppm in air, an increase in the conjugation of urinary sulfates was not observed but urinary glucuronic acid levels were five times higher than in unexposed rabbits (166). When rabbits received 50 mg cyclohexanol as a single oral dose, the urine contained 25% of the unchanged alcohol and $< 1\%$ as the glucuronide (167).

No cyclohexanol was recovered from the urine of a rabbit given 33% of the minimum lethal oral dose (164). More than 65% of a dose of 0.25 g/kg cyclohexanol was excreted by rabbits as glucuronides, chiefly cyclohexyl glucuronide, and an additional 6% was excreted as conjugated *trans*-cyclohexane-1,2-diol (8).

27.4.1.4 Reproductive and Developmental. Pregnant and nonpregnant TB and MNRI mice were given 0.1, 0.5, or 1.0% cyclohexanol in the diet through gestation, lactation, and weaning for several generations. The 1.0% concentration produced a significant increase in mortality and growth of the offspring decreased during the 21 days after birth. In strain TB mice, the percentage mortality in the first- and second-generation treated offspring was 14.1 and 53.3%, respectively, compared to 11.9% for controls. The percentage mortality of the first generation offspring of the treated MNRI mice was 43.1% compared to 12.2% for controls (168).

27.4.1.6 Genetic and Related Cellular Effects Studies. Cyclohexanol was not mutagenic to *S. typhimurium* in the Ames test with or without metabolic activation (169). Cyclohexanol produced questionable or equivocal results in a *Salmonella* Ames test; no details were given (170). In a mouse bone marrow micronucleus assay, a single oral dose of 500, 1000, or 1500 mg/kg cyclohexanol did not induce micronuclei (171).

27.4.2 Human Experience

27.4.2.2.1 Acute Toxicity. Data on the toxic effects of cyclohexanol on humans are very limited. The estimated acceptable concentration in air for 8 h was reported by volunteer subjects as less than 100 ppm (172). At 100 ppm, human subjects complained of eyes, nose, and throat irritation (172). In a human patch test (4% in petrolatum), cyclohexanol failed to produce irritation or sensitization (173).

27.4.2.2.3 Pharmacokinetics, Metabolism, and Mechanisms. Four men and four women were exposed to 236 mg/m^3 cyclohexanol in an exposure chamber for 8 h. During the 72 h following exposure, 1.1, 19.1, and 8.4% of the dose was excreted as cyclohexanol, 1,2-cyclohexanediol, and 1,4-cyclohexanediol, respectively. These amounts included both free and conjugated metabolites (174).

27.5 Standards, Regulations, or Guidelines for Exposure

The ACGIH TLV is currently 50 ppm (8-h TWA), and a skin notation is recommended (86). Other countries have generally adopted 50 ppm, except for Japan, where 25 ppm is the standard (33). NIOSH REL and OSHA PEL are 50 ppm.

28.0 Methylcyclohexanol

28.0.1 CAS Number: *[25639-42-3]*

28.0.2 Synonyms: 2-, 3-, 4-Methylcyclohexanol; methyl hexalin; hexahydrocresols; hexahydromethyl phenol; methyl adronal; methyl anol; sextol

28.0.3 Trade Names: NA

28.0.4 Molecular Weight: 114.2

28.0.5 Molecular Formula: $C_7H_{14}O$

28.0.6 Molecular Structure:

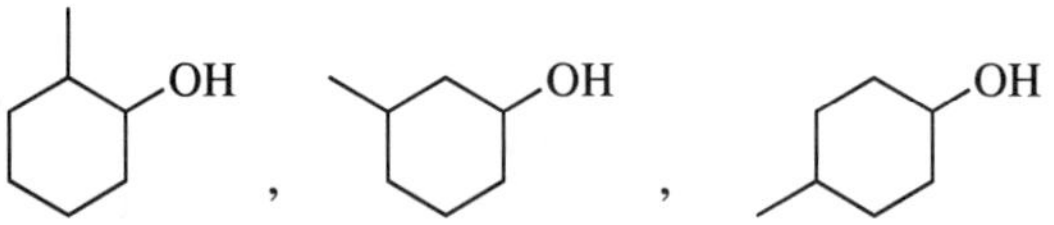

28.1 Chemical and Physical Properties

Methylcyclohexanol exists as a colorless liquid. Its chemical and physical properties are listed in Table 78.1.

28.1.2 Odor and Warning Properties

Methylcyclohexanol vapor in air can be detected and recognized by its odor when present to the extent of 500 ppm, a concentration capable of causing upper respiratory irritation (179a). The odor threshold for *cis*-3-methylcyclohexanol has been reported as 500 ppm (160).

28.2 Production and Use

Methylcyclohexanol is produced by hydrogenating *m*- and *p*-cresols (35). Methylcyclohexanol can exist as three separate isomers, ortho, meta, and para, or as a mixture of isomers. The commercial product consists essentially of the meta and para isomers (35). Methylcyclohexanol is used as a solvent in lacquers, a blending agent in textile soaps, and an antioxidant in lubricants. In addition, methylcyclohexanol dissolves gums, oils, resins, and waxes.

28.3 Exposure Assessment

28.3.3 Workplace Methods

NIOSH Method No. 1404 has been recommended for measuring methylcyclohexanol in air. It involves drawing a known volume of air through charcoal to trap the vapors. The analyte is desorbed with methylene chloride and separated with a gas chromatograph equipped with a flame ionization detector (36).

28.3.5 Biomonitoring/Biomarkers

There are also methods to measure methylcyclohexanol in urine (162).

28.4 Toxic Effects

28.4.1 Experimental Studies

Methylcyclohexanol displayed a low order of acute toxicity by the oral route. It is slightly irritating to the skin. Methylcyclohexanol is not particularly toxic when inhaled. Excessive exposure to vapors may cause headaches and irritation of the eyes, nose, and throat.

28.4.1.1 Acute Toxicity. The acute oral LD_{50} in the rat is 1.66 g/kg (175), and the minimum lethal dose has been reported as 1.75 to 2.0 g/kg for the rabbit (164). There were no signs of intoxication in a dog exposed for 10 min daily on seven consecutive days to air saturated with methylcyclohexanol (165).

Application on the intact skin of rabbits of 5 g of an ointment consisting of 15% methylcyclohexanol in potassium oleate for 1 h/day during a period of 15 days produced only temporary erythema and superficial sloughing of the skin (164).

28.4.1.2 Chronic and Subchronic Toxicity. The application of 10 mL of methylcyclohexanol on the intact skin of rabbits for 1 h for six consecutive days resulted in tremors, narcosis, hypothermia, and death. The minimum lethal dose was from 6.8 to 9.4 g/kg. At various stages of the treatment, weakness, deep anesthesia, and local petechiae, gross hemorrhage, and thickening of the skin were also noted (164).

Rabbits exposed by inhalation to 503 ppm methylcyclohexanol 6 h/day, 5 days/week for 10 weeks showed lethargy, irritation, salivation, and conjunctival congestion but no hematologic effects. There were no adverse effects at 121 and 232 ppm under similar exposure conditions (166).

28.4.1.3 Pharmacokinetics, Metabolism, and Mechanisms. Methylcyclohexanol can be absorbed via the dermal, oral, and inhalation routes of exposure. Methylcyclohexanol is a substrate for alcohol dehydrogenase (121, cited in Ref. 122). In rabbits, conjugation of methylcyclohexanol or a metabolite with both glucuronic acid and sulfuric acids has been demonstrated in the urine after oral administration of methylcyclohexanol (164). A conjugation product with sulfuric acid was also found in the urine of animals exposed to 232 and 503 ppm methylcyclohexanol in air (166). The rate of excretion of glucuronic acid in the urine of rabbits is correlated with the concentration of methylcyclohexanol in air (166). Rabbits exposed to 121 ppm methylcyclohexanol had twice the normal quantity of glucuronic acid in the urine (166).

28.4.2 Human Experience

28.4.2.1 General Information. Prolonged exposure to excessive concentrations of methylcyclohexanol vapors may lead to headaches and irritation of the eyes, nose, and throat.

28.4.2.2 Clinical Cases. Several workers exposed to a cellulose solvent containing methylcyclohexanol showed some hematologic effects, including slight relative lymphosis and a diminished number of leukocytes in the blood (176). In this mixed exposure environment, it remains unclear whether these hematologic effects were entirely due to methylcyclohexanol.

28.5 Standards, Regulation, or Guidelines of Exposure

The ACGIH TLV is 50 ppm. The OSHA PEL is 100 ppm (177). Occupational exposure standards for methylcyclohexanol are either 50 ppm or 100 ppm for a number of countries (16).

29.0 Furfuryl Alcohol

29.0.1 CAS Number: *[98-00-0]*

29.0.2 Synonyms: 2-Furanmethanol; 2-furylcarbinol; 2-furan carbinol; 2-hydroxymethylfuran

29.0.3 Trade Names: NA

29.0.4 Molecular Weight: 98.1

29.0.5 Molecular Formula: $C_5H_6O_2$

29.0.6 Molecular Structure:

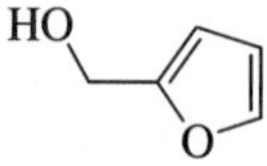

29.1 Chemical and Physical Properties

Furfuryl alcohol is a colorless liquid that darkens on exposure to light. Its chemical and physical properties are listed in Table 78.1.

29.1.2 Odor and Warning Properties

The odor threshold for furfuryl alcohol has been reported as 8 ppm (160).

29.2 Production and Use

Furfuryl alcohol is produced by the high-pressure catalytic hydrogenation of furfural (178). Because of its reactivity with acids, furfuryl alcohol, is used as a monomer in manufacturing furfuryl resins (178). Furfuryl alcohol is used alone or in combination with other solvents for various cleaning and paint-removing operations (178). It is used in manufacturing dark-colored thermosetting resins and phenolic resins and as a solvent in manufacturing abrasive wheels (35, 178). It is employed as a solvent and dispersant for dyes in the textile industry and in the manufacture of wetting agents (179). The major routes of industrial exposure are by dermal contact and inhalation.

29.3 Exposure Assessment

29.3.3 Workplace Methods

NIOSH Method 2505 has been recommended with a working range of 20 to 600 mg/m^3 (4 to 150 ppm). It involves drawing a known volume of air through a glass tube containing Porapak Q to trap furfuryl alcohol vapors (recommended sample size is 6 liters). The analyte is desorbed with acetone and analyzed by gas chromatography with flame ionization detection (36). Other gas chromatographic methods have been reported which may be applicable to furfuryl alcohol (39, 40).

29.4 Toxic Effects

29.4.1 Experimental Methods

Furfuryl alcohol was moderately acutely toxic by the oral, dermal, and inhalation routes of administration. It is an eye irritant and may be a weak skin sensitizer in guinea pigs. Subchronic inhalation studies in rats and mice revealed nasal lesions at very low doses (2 ppm). Furfuryl alcohol has shown both positive and negative findings in genotoxicity tests.

29.4.1.1 Acute Toxicity. The acute oral LD_{50} has been reported as 132 mg/kg (179a) and 275 mg/kg in the rat (179b). The acute dermal LD_{50} is 657 mg/kg in rabbits and 4920 mg/kg in mice (179a). No overt adverse effects were reported following application of 8.5 g/kg furfuryl alcohol to the skin of three guinea pigs (179a). Several animal species have been exposed to various concentrations of vapor for various periods of time. No deaths were observed in mice exposed to 243 or 597 ppm for 6 h, in rabbits exposed to 416 ppm for 6 h, in dogs exposed to 349 ppm for 6 h, or in a monkey exposed to 260 ppm for 6 h (179a). The inhalation 4-h LC_{50} in rats has been reported as 233 ppm (181), although another study reported that all rats died at a vapor concentration of 243 ppm (179a).

Slight redness was observed when about 0.1 g/day was applied 5 days/week for 4 weeks to the rabbit skin (179c). Instillation of about 0.02 or more neat furfuryl alcohol into the eyes of four rabbits resulted in inflammation, thick secretions, corneal clouding, and eyelid swelling, which were all reversible (179a).

29.4.1.2 Chronic and Subchronic Toxicity. Dogs exposed by inhalation for 6 h/day, 5 days/week for 4 weeks to 239 ppm furfuryl alcohol showed no changes in behavior and no gross pathology. Microscopic examination showed minimal chronic inflammation of the bronchi (179a). A monkey exposed to 239 ppm furfuryl alcohol for 6 h/day for 3 consecutive days showed no irritation or toxicity (179a). When B6C3F_1 mice and F344 rats were exposed by inhalation to 0, 16, 31, 63, 125, or 250 ppm furfuryl alcohol for 14 days, all animals in the high-dose group died. Animals that survived developed lesions in the nasal respiratory epithelium and/or olfactory epithelium, and the severities of these lesions generally increased with increasing exposure concentration (182–184).

In a 13-week inhalation study, B6C3F_1 mice and F344 rats were exposed to 0, 2, 4, 8, 16, or 32 ppm furfuryl alcohol 6 h/day, 5 days/week. No treatment-related effects were observed on survival, body weights, or organ weights. Exposure-related histopathological changes were present in the nose of both rats and mice at all exposure concentrations. These lesions consisted of inflammation of the nasal turbinates accompanied by necrosis and squamous metaplasia of the respiratory epithelium and necrosis and degeneration of the olfactory epithelium. In addition, squamous metaplasia and goblet cell hyperplasia of the respiratory epithelium, squamous metaplasia of the transitional epithelium and degeneration, and hyperplasia, and some respiratory metaplasia of the olfactory epithelium were observed in rats. In mice, hyaline droplets were observed in the respiratory epithelium and chronic inflammation and respiratory metaplasia in the olfactory epithelium (182).

Furfuryl alcohol reportedly produced lesions in the thyroid, spleen, kidneys, and liver, but few details were supplied on dose, species, and duration of treatment (180).

29.4.1.3 Pharmacokinetics, Metabolism, and Mechanisms. Studies with laboratory animals have shown that furfuryl alcohol is absorbed by the oral, dermal, and inhalation routes of exposure. Furfuryl alcohol is first metabolized by oxidation to the aldehyde (furfural) and then to the corresponding acid (furoic acid), followed by decarboxylation to CO_2, excreted as the unchanged acid, conjugated with glycine, or condensed with acetic acid (185, 186). Furoylglycine was identified as the major urinary metabolite in rats after an oral dose of furfuryl alcohol (185, 186). Furoic acid and furanacrylic acid were also identified as minor metabolites (185, 186). Following oral administration of 0.275, 2.75, and 27.5 mg/kg [^{14}C]furfuryl alcohol in rats, at least 86 to 89% was absorbed. The major route of excretion was in urine, where 83 to 88% of the dose was excreted, whereas 2 to 4% was excreted in the feces. No furfuryl alcohol was identified in the urine. Furoylglycine was the major urinary metabolite (73 to 80% of dose), and furoic acid (1 to 6%) and furanacrylic acid (3 to 8%) were the minor metabolites (186). At 72 h following treatment, the liver and kidney contained the highest concentrations of radioactivity, and the concentration was proportional to the dose (186).

29.4.1.5 Carcinogenesis. The NTP conducted a 2-year inhalation study on furfuryl alcohol. F344 rats and B6C3F_1 mice were exposed to 0, 2, 8, or 32 ppm furfuryl alcohol for 6 h/day, 5 days/week. All rats exposed to 32 ppm died by week 99; survival of all other animals was similar to control animals. There were increased incidences of nasal tumors in the male rats and increased incidences of kidney tubule tumors in male mice. Increased incidences of nonneoplastic lesions of the nose and increased severities of nephropathy were observed in male and female rats and male mice. Nonneoplastic lesions of nose and corneal degeneration occurred in female mice (187).

29.4.1.6 Genetic and Related Cellular Effects Studies. Furfuryl alcohol was not mutagenic to *S. typhimurium* in an Ames test (116, 125). It was, however, reported to be mutagenic to *Bacillus subtilis* in a Rec assay (188). In an *in vitro* cytogenetics test, furfuryl alcohol produced chromosomal aberrations in Chinese hamster ovary cells and without metabolic activation (189). There was no increase in the frequency of sister chromatid exchanges in human lymphocytes treated with furfuryl alcohol (190), nor was there any evidence of genotoxicity in *Drosophila melanogaster* in assays for sex-linked recessive lethal mutations and sex chromosome loss (191).

Furfuryl alcohol was not mutagenic in *S. typhimurium* strain TA98, TA100, TA1535, or TA1537 with or without metabolic activation (187). It did induce sister chromatid exchanges in cultured Chinese hamster ovary cells, but not with metabolic activation (187). No induction of chromosomal aberrations was noted in cultured Chinese hamster ovary cells treated with furfuryl alcohol without metabolic activation, but equivocal results were obtained with metabolic activation (187). There was no induction of chromosomal aberrations, sister chromatid exchanges, or micronuclei in bone marrow cells from mice treated with furfuryl alcohol (187).

29.4.1.7 Other: Neurological, Pulmonary, Skin Sensitization. In an inadequately described test, furfuryl alcohol produced signs of skin sensitization in guinea pigs (180).

29.4.2 Human Experience

29.4.2.2 Clinical Cases. Three workers exposed for 15 min up to 43 mg/m^3 (11 ppm) did not report discomfort. Acid-resistant cement containing furfuryl alcohol has been associated with eye and respiratory tract irritation and dermatitis, but it is unclear whether this was specifically due to furfuryl alcohol or to other chemicals that were present (180).

29.5 Standards, Regulations, or Guidelines of Exposure

The OSHA PEL is 50 ppm (177). The ACGIH TLV is 10 ppm with a STEL of 15 ppm with a skin notation (86). A few countries have set lower occupational standards for furfuryl alcohol (16).

30.0 Tetrahydro-2-Furanmethanol

30.0.1 CAS Number: *[97-99-4]*

30.0.2 Synonyms: Tetrahydrofurfuryl alcohol, tetrahydro-2-furancarbinol, tetrahydro-2-furylmethanol

30.0.3 Trade Names: NA

30.0.4 Molecular Weight: 102.1

30.0.5 Molecular Formula: $C_5H_{10}O_2$

30.0.6 Molecular Structure:

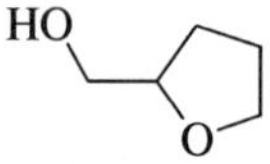

30.1 Chemical and Physical Properties

Tetrahydro-2-furanmethanol is a colorless liquid with a mild, pleasant odor. Its chemical and physical properties are listed in Table 78.1.

30.2 Production and Use

Tetrahydro-2-furanmethanol is made by hydrogenating furfuryl alcohol (178). Tetrahydro-2-furanmethanol is used as a specialty organic solvent. A major use of this alcohol is as an ingredient in proprietary stripping formulations. It is also used in formulations for crop sprays, cleaners, and water-based paints, and in dyeing and finishing textiles and leathers (178).

30.4 Toxic Effects

30.4.1 Experimental Studies

30.4.1.1 Acute Toxicity. The acute oral LD_{50} is reported as 1.6 to 3.2 g/kg (45) and 2.50 g/kg (34) in the rat, 2.3 g/kg in the mouse (34), and 0.8 to 1.6 g/kg (45) and 3.0 g/kg (34) in the guinea pig. The acute dermal LD_{50} in guinea pigs is less than 5 mL/kg (45). No deaths occurred among any of the three rats exposed by inhalation to 655 ppm tetrahydro-2-furanmethanol for 6 h (45); loss of coordination, prostration, and vasodilation of the ears and feet were noted.

Tetrahydro-2-furanmethanol was moderately irritating when tested on guinea pig skin (45).

30.4.1.7 Other: Neurological, Pulmonary, Skin Sensitization. Tetrahydro-2-furanmethanol was not a skin sensitizer in guinea pigs (45).

30.4.2 Human Experience

No data were found.

30.5 Standards, Regulations, or Guidelines of Exposure

No occupational standards have been established for tetrahydro-2-furanmethanol.

N UNSATURATED ALCOHOLS

Unsaturated alcohols can be divided into olefinic (double-bond) and acetylenic (triple-bond) alcohols. Allyl alcohol is the most important olefinic alcohol as well as the most studied in this subset; propargyl alcohol and hexynol alcohol are the most commercially important acetylenic alcohols. There are also a number of other olefinic alcohols (C_5) and acetylenic alcohols (C_4 to C_{10}) with toxicity data (see Table 78.4). In general, these alcohols are liquids and are quite reactive. Occupational exposure standards exist for allyl alcohol and propargyl alcohol.

31.0 Allyl Alcohol

31.0.1 CAS Number: [107-18-6]

31.0.2 Synonyms: 2-Propen-1-ol, vinyl carbinol, 1-propenol-3

31.0.3 Trade Names: NA

31.0.4 Molecular Weight: 58.1

31.0.5 Molecular Formula: C_3H_6O

31.0.6 Molecular Structure:

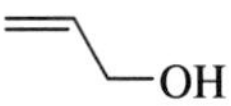

31.1 Chemical and Physical Properties

Allyl alcohol is a colorless liquid with a pungent mustard odor. Its chemical and physical properties are listed in Table 78.1.

31.1.2 Odor and Warning Properties

The air odor threshold level is reported as 1.1 ppm (160). Allyl alcohol is also a potent sensory irritant (194). The warning properties may be adequate to prevent voluntary exposure to acutely dangerous concentrations but inadequate to prevent excessive prolonged and/or repeated exposure.

31.2 Production and Use

Allyl alcohol is prepared by several different processes (195), the original is alkaline hydrolysis of allyl chloride by steam injection at high temperatures. A more recent commercial process used oxidation of propylene to acrolein, which in turn reacts with a secondary alcohol to yield allyl alcohol and a ketone. In this process, allyl alcohol is not isolated but its aqueous stream is converted directly to glycerol. The most recent commercial process is isomerization of propylene oxide over a lithium phosphate catalyst (195). Most of the allyl alcohol produced is used in synthesizing glycerol (195). It is also used as a flavoring agent and in preparing allyl resins and plastics, pharmaceutical, and chemicals (33). Industrial exposure is mainly from dermal contact and inhalation from the transfer and maintenance of equipment.

31.3 Exposure Assessment

31.3.3 Workplace Methods

NIOSH Method No. 1402 has been recommended (36). The method involves drawing a known volume of air through charcoal to trap the organic vapors present (the recommended sample is 10 liters at a rate of 0.02 to 0.2 L/min). The analyte is desorbed with carbon disulfide containing 5% isopropanol. The sample is separated and quantified by a gas chromatograph with a flame ionization detector.

31.3.5 Biomonitoring/Biomarkers

A gas chromatographic method (head-space analysis) exists to measure allyl alcohol in blood (196).

31.4 Toxic Effects

31.4.1 Experimental Methods

Allyl alcohol is quite toxic by the oral, dermal, and inhalation routes of exposure. Even at concentrations of a few parts per million, its vapors cause eye, nose, and respiratory tract irritation. It is a potent sensory irritant. The liquid material can be irritating to the skin and is readily absorbed through the skin. Allyl alcohol causes liver and kidney damage following repeated exposure. No adequate data exist to assess whether allyl alcohol is teratogenic. The data indicate that allyl alcohol is not carcinogenic to rodents. It is active in several *in vitro* genotoxicity assays apparently due to its metabolite, acrolein.

31.4.1.1 Acute Toxicity. The acute oral LD_{50} was reported as 105 mg/kg (197), 70 mg/kg (119), 99 mg/kg (197), and 64 mg/kg (105) for rats; 96 mg/kg for mice (197); and 71 mg/kg for rabbits (197). The acute dermal LD_{50} has been reported as 45 mg/kg (198) and 89 mg/kg (197) in rabbits. The acute inhalation toxicity data for allyl alcohol are summarized in Table 78.2 (199–201).

Allyl alcohol was slightly irritating to intact and abraded skin of rabbits (198). In rabbits, 0.02 mL of the undiluted material produced severe injury, including corneal

Table 78.2. Results of Single or Short-term Exposures of Animals to Vapors of Allyl Alcohol[a]

Concentration (ppm)	Animal	Duration of Exposure (h)	Outcome	Ref.
1000	Monkey	4	Death	199
1000	Rabbit	4	100% lethal	199
200	Rabbit	18 × 7	100% lethal	199
1000	Rat	4	100% lethal	199
1060	Rat	1	LC_{50}	200
1000	Rat	1	LC_{67}	198
500	Rat	1	Survived	201
250	Rat	4	Some deaths	201
165	Rat	4	LC_{50}	200
76	Rat	8	LC_{50}	200
200	Rat	2 × 7	100% lethal	199

[a]Taken from Ref. 34.

necrosis (154). In another study, 0.05 mL was only a slight irritant and produced some conjunctivitis and corneal opacity that disappeared after 7 days (197). Allyl alcohol produced sensory irritation in mice with an RD_{50} of 1.6 ppm (14) and 3.9 ppm (202).

31.4.1.2 Chronic and Subchronic Toxicity. In a drinking water study, Wistar rats were exposed to 0, 50, 100, 200, or 800 ppm allyl alcohol in drinking water for 15 weeks. Based on water consumption data, these concentrations were equivalent to dosages of 0, 4.8, 8.3, 14.0, and 48.2 mg/kg/day for males and 0, 6.2, 6.9, 17.1, and 58.4 mg/kg/day for females, respectively. Food intake and growth were depressed at ≥100-ppm dose groups. Relative liver, kidney, and spleen weights were significantly increased in ≥100-ppm dose groups. Tests of renal function indicated impairment in males at doses of 100 ppm and higher, and in females at 200 and 800 ppm. There were no noteworthy adverse histopathological findings (203).

In an inhalation study, rats were exposed to 0, 1, 2, 5, 20, 40, 60, 100, and 150 ppm allyl alcohol 7 h/day for 90 days. Mortality occurred at 150 ppm, and rats exhibited gasping, depression, nasal discharge, and eye irritation. Pathological changes were noted in the lungs and liver. Similar signs, lesions, and pathological findings were noted at 100, 60, and 40 ppm, but they were less severe. Except for lack of weight gain at 20 ppm, there were no effects at exposures of ≥20 ppm (197). In another inhalation study, mild reversible changes were seen in the liver (dilation of the sinusoids, cloudy swelling, and necrosis) and kidneys (necrosis of the epithelium of the convoluted tubules and proliferation of the interstitial tissue) of male and female rats, male guinea pigs, and female rabbits exposed to 7 ppm for 7 h/day for 5 weeks. No effects were found on growth, mortality, or body or organ weights. No histopathological lesions were noted in animals exposed to 2 ppm for 6 months (204).

31.4.1.3 Pharmacokinetics, Metabolism, and Mechanisms. There are no quantitative data on the uptake of allyl alcohol, but animal studies have shown that it can be absorbed via the skin or lungs or by ingestion. In rats, allyl alcohol is metabolized to acrolein by alcohol dehydrogenase (205). Inhibitors of alcohol dehydrogenase prevent periportal liver necrosis in animals treated with allyl alcohol. Acrolein can react with glutathione to form the corresponding thiol ether, which can be further metabolized to mercapturic acids and excreted in the urine (206). In the presence of NADPH and liver and lung microsomes, allyl alcohol and acrolein were oxidized to the corresponding epoxides, glycidol, and glycidaldehyde, respectively (205).

The toxic effects of allyl alcohol are caused by its metabolism to acrolein. Acrolein reacts with sulfhydryl groups, primarily glutathione. In the absence of glutathione or when glutathione is depleted, acrolein can react with other macromolecules, including protein and DNA.

31.4.1.4 Reproductive and Developmental. Litters sired by male rats treated with a dose of 0.86% allyl alcohol 7 days/week to week 12 and 5 days/week from week 13 to 33 did not develop any malformations. No adverse reproductive effects were observed (207).

31.4.1.5 Carcinogenesis. Male and female F344 rats (20/group) were given allyl alcohol in the drinking water at a concentration of 0 or 300 mg/L for 106 weeks. The incidence of tumors was similar to that in controls (208). Male and female hamsters (20/group) were dosed by oral gavage with 2 mg allyl alcohol/week for 60 weeks. The incidence of tumors did not increase significantly compared to controls (208).

31.4.1.6 Genetic and Related Cellular Effects Studies. Allyl alcohol was not mutagenic to *S. typhimurium* strains TA98, TA100, TA1535, TA1537, and TA1538 in an Ames test with or without metabolic activation (209). In a liquid suspension modification of the Ames test, allyl alcohol showed a strong mutagenic effect which was diminished by the presence of a metabolic activation system (210). Allyl alcohol was also mutagenic to V79 mammalian cells *in vitro* (211). It has been suggested that the mutagenic activity is due to its metabolism to acrolein (206).

31.4.2 Human Experience

31.4.2.2 Clinical Cases. Skin irritation has been frequently reported from exposure with allyl alcohol, and absorption through the skin leads to deep pain (197). Accidental splashes produced corneal burns in six workers; in all but one case there was prompt healing (212). In air moderately contaminated with allyl alcohol (concentration unspecified), men complained of excessive secretion of tears, pain behind the eyes, sensitivity to light, and some blurring of vision (197). Volunteers have complained that 12.5 ppm was moderately irritating to the nose; only slight nasal irritation was reported at 0.8 ppm, the lowest concentration tested (197). Allyl alcohol vapor reportedly blinded one man temporarily by delayed corneal necrosis (201).

31.4.2.3 Epidemiology Studies. There was no evidence of liver damage or kidney dysfunction in a group of employees worked with allyl alcohol for 10 years (197).

31.5 Standards, Regulations, or Guidelines of Exposure

The ACGIH TLV is 0.5 ppm (1.25 mg/m^3) and the OSHA PEL is 2 ppm (5 mg/m^3) as an 8-h TWA, with a recommendation for a skin notation because dermal absorption can be significant, leading to systemic toxicity (86, 177). Many other countries have adopted the same standard (33).

32.0 Propargyl Alcohol

32.0.1 CAS Number: *[107-19-7]*

32.0.2 Synonyms: 2-Propyn-1-ol; ethynol carbinol; propiolic alcohol; acetylenyl carbinol; ethynyl methanol

32.0.3 Trade Names: NA

32.0.4 Molecular Weight: 56.1

32.0.5 Molecular Formula: C_3H_4O

32.0.6 Molecular Structure:

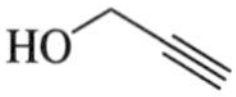

32.1 Chemical and Physical Properties

Propargyl alcohol is a moderately volatile, clear to slightly straw-colored liquid with a mild geranium-like odor. Its chemical and physical properties are listed in Table 78.1.

32.2 Production and Use

Propargyl alcohol is the major commercially available acetylenic primary alcohol (213). Propargyl alcohol is a by-product of butynediol manufacture. In the usual high-pressure butynediol process, about 5% of the product is propargyl alcohol. Some processes give higher proportions of propargyl alcohol (213). It is used as an inhibitor for the corrosion of steel by hydrochloric acid, as a stabilizer in certain chlorinated hydrocarbon formulations, as a soil fumigant, and as a chemical intermediate in organic chemical synthesis (179). The primary routes of occupational exposure are from dermal contact and inhalation.

32.3 Exposure Assessment

32.3.3 Workplace Methods

OSHA has developed an analytical method for propargyl alcohol (method #97). It involves drawing a known volume of air through petroleum-based charcoal which has been coated with hydrobromic acid, to trap the organic vapors present (recommended sample is 6 liters at a rate of 0.05 liters/min). The analyte is desorbed with toluene, and the analyte is separated and quantified by gas chromatography with an electron capture detector. The reliable quantitation limit is 10.08 ng/sample, which corresponds to an air concentration of 0.73 ppb or 1.68 $\mu g/m^3$ (213a).

32.4 Toxic Effects

32.4.1 Experimental Methods

Propargyl alcohol has a high order of acute toxicity by the oral, dermal, and inhalation routes of exposure. It is irritating to the eyes, skin, and respiratory tract. Repeated oral or inhalation exposure has produced liver and kidney damage in rats. Propargyl alcohol is not mutagenic to *Salmonella* in an Ames test, except for a weak effect in strain D3052.

32.4.1.1 Acute Toxicity. The acute oral LD_{50} is 93 mg/kg (214) and 110 mg/kg (215) for male rats and 54 mg/kg (214) and 55 mg/kg (215) for female rats. Other oral LD_{50} values for rat have been reported as 20 to 50 mg/kg (216) and 70 mg/kg (217). The acute oral LD_{50} for mice is 50 mg/kg (218), and for guinea pigs, 60 mg/kg (217). The acute dermal LD_{50} for rabbits is 16 mg/kg (216) and 88 mg/kg (102). A 1-h acute inhalation LC_{50} is 1040 to 1200 ppm for rats (214). A concentration of 874 ppm was lethal to mice exposed for 2 h (218). Two out of three rats died after a 6-min exposure to a saturated vapor of propargyl alcohol, and exposures of 12 min or longer were fatal to all exposed animals (216).

Undiluted propargyl alcohol caused irritation and some superficial necrosis to the skin of rabbits. A 10% aqueous solution produced mild irritation, but a 1% aqueous solution had no effect (216). Instillation of undiluted propargyl alcohol into the eyes of rabbits caused marked pain and irritation, and the resulting injury to the cornea was judged permanent (216). A 10% aqueous solution was slightly irritating, but the eyes cleared within a few days. A 1% solution was without effect (216).

32.4.1.2 Chronic and Subchronic Toxicity. In a dermal study, propargyl alcohol was applied to young adult rabbits at daily doses of 1, 3, or 10 mg/kg/day during a 63-day period, and 20 mg/kg/day during a 28-day period. No systemic effects were noted (218a). In a preliminary inhalation study, rats, mice, guinea pigs, rabbits, and cats exposed to 100 ppm for up to 75 days were observed with irritation of the mucous membranes (219). In an inhalation study, rats were exposed to 80 ppm propargyl alcohol 7 h/day, 5 days/week for 59 days during a period of 89 days. Increased liver weights were seen in male rats, and increased liver and kidney weights in female rats. Histopathological examination revealed degenerative damage in the kidneys and liver, which was more marked in the females (216). In an oral gavage study, rats were dosed with 0, 5, 15, or 50 mg/kg propargyl alcohol for 13 weeks. Hematologic changes and some enzyme changes characteristic of liver damage were seen in the mid- and high-dose groups. Increased liver and kidney weights were noted in the mid- and high-dose groups. Histological examination revealed kidney and liver lesions in the 15- and 50 mg/kg groups. The kidney lesion was karyomegaly of renal tubular epithelial cells, and the liver effects were megalocytosis and cytoplasmic vacuolation of the liver cells. No effects were observed at the low dose of 5 mg/kg (220, 221).

32.4.1.3 Pharmacokinetics, Metabolism, and Mechanisms. Male Sprague–Dawley rats were dosed orally with 40 mg/kg mixture of [1,2,3-^{13}C]propargyl alcohol and [1,2-^{14}C]-propargyl alcohol. Approximately 60% of the dose was excreted in the urine by 96 h. Major metabolites were identified in the urine by 1- and 2-D NMR and confirmed by isolation and purification of the individual metabolites followed by ^{13}C FT-NMR and mass spectometry. The proposed pathway involves oxidation of propargyl alcohol to 2-propynoic acid and glutathione conjugation, the first example of multiple glutathione additions to a triple bond. The following final products were identified: 3-{[2-(acetylamino)-2-carboxyethyl]thio}-2-propenoic acid, *S*-*S*′-(3-hydroxypropylidene)-bis[*N*-acetylcysteine], and 3-[[2-(acetylamino)-2-carboxyethyl]-sulfinyl]-3-[[2-(acetylamino)-2-carboxyethyl]thio]1-propanol (222).

32.4.1.6 Genetic and Related Cellular Effects Studies. Propargyl alcohol was not mutagenic to *S. typhimurium* strains TA 98, TA100, TA 1535, TA 1537, or TA1538 in the Ames assay with and without metabolic activation (219). However, propargyl alcohol was weakly mutagenic to one unusual strain of *S. typhimurium*, D3052 (223).

In another study, propargyl alcohol was not mutagenic to *S. typhimurium* strains TA97, TA98, TA100, and TA102 in the Ames assay with and without metabolic activation (224). Propargyl alcohol induced chromosomal aberrations in Chinese hamster ovary (CHO) cells with and without metabolic activation (224). It was, however, inactive in a mouse bone marrow micronucleus assay (224).

32.4.1.7 Other. Propargyl alcohol has been reported to not be a skin sensitizer to guinea pigs (217).

32.4.2 Human Experience

No adverse effects in humans were reported.

32.5 Standards, Regulations, or Guidelines of Exposure

The ACGIH TLV and the OSHA PEL for propargyl alcohol are 1 ppm (2.3 mg/m^3), with a skin notation (86, 177). The ACGIH TLV was based on the structural and apparent toxicological similarity to allyl alcohol. The NIOSH REL is 1 ppm. Many other countries have an identical occupational standard as the TLV for propargyl alcohol (33).

33.0 Hexynol

33.0.1 CAS Number: *[105-31-7]*

33.0.2 Synonym: 1-Hexyn-3-ol

33.0.3 Trade Names: NA

33.0.4 Molecular Weight: 98.1

33.0.5 Molecular Formula: $C_6H_{10}O$

33.0.6 Molecular Structure:

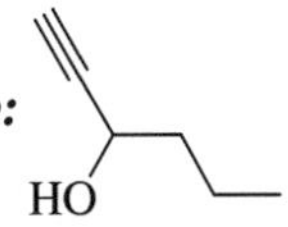

33.1 Chemical and Physical Properties

is a pale yellow liquid. Its chemical and physical properties are listed in Table 78.1.

33.1.2 Odor and Warning Properties

The odor is described as musty and terpene-like, but no air odor threshold data exist on hexynol.

33.2 Production and Use

Hexynol is prepared commercially by the reaction of acetylene with butyraldehyde (213). It is used as a corrosion inhibitor for steel in the presence of mineral acids, as an inhibitor for the acidification of oil wells, and in electroplating (225). Industrial exposure may occur by direct contact with the liquid or with solutions containing it.

33.4 Toxic Effects

33.4.1 Experimental Studies

33.4.1.1 Acute Toxicity. The acute oral LD_{50} in rats was reported as 88 to 170 mg/kg (226) and 130 to 250 mg/kg (216). The acute dermal LD_{50} from 24-h exposure was 30 to

60 mg/kg (216), and less than 200 mg/kg for rabbits (34). A dose of 252 mg/kg applied to 81 cm^2 of rabbit skin for 1 h and then washed off resulted in death (216). An acute 1-h inhalation LC_{50} of greater than 20 mg/L was reported for rats (34). Rats exposed to saturated vapors for 12 min had liver and kidney injury, but no deaths were reported (216). All rats died when exposure was for 30 or 60 min (216).

Hexynol was slightly to moderately irritating to rabbit skin (216). Undiluted hexynol caused moderate irritation to rabbit eyes which was characterized by conjunctival irritation, corneal injury, and iritis; recovery was not complete within a week (216).

33.4.2 Human Experience

33.4.2.2 Clinical Cases. One death was resulted from dermal exposure to hexynol. A workman spilled hexynol on his trousers, did not remove the material, and delayed cleansing himself. Death occurred from kidney failure within 24 h (216).

33.5 Standards, Regulations, or Guidelines of Exposure

No occupational standards exist for hexynol. It would be prudent to minimize skin contact because the material can be rapidly absorbed through intact skin in toxic amounts.

O HALOGENATED ALCOHOLS

The most important commercial members of the halogenated alcohol series are 2-chloroethanol, or ethylene chlorohydrin, and chloropropanols, which are two isomers. These three alcohols exist as volatile liquids. Some of the other halogenated alcohols not covered in this chapter have been reviewed by Rowe and McCollister (34).

34.0 2-Chloroethanol

34.0.1 CAS Number: *[107-07-3]*

34.0.2 Synonyms: Ethylene chlorohydrin, β-chloroethyl alcohol, glycol chlorohydrin, glycol monochlorohydrin, 2-monochloroethanol, glycomonochlorohydrin, 2-chloro-1-ethanol

34.0.3 Trade Names: NA

34.0.4 Molecular Weight: 80.5

34.0.5 Molecular Formula: C_2H_2ClO

34.0.6 Molecular Structure:

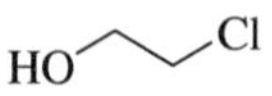

34.1 Chemical and Physical Properties

The chemical and physical properties are listed in Table 78.1.

34.2 Production and Use

2-Chloroethanol is manufactured by the reaction of ethylene gas with dilute hydrochlorous acid (227). The principal use of 2-chloroethanol was formerly in producing of ethylene oxide, in which ethylene was reacted with hypochlorous acid. However, the current production of ethylene oxide does not use this procedure. Facilities for ethylene chlorohydrin have, in many cases, been converted to the production of propylene chlorohydrin, the dehydrochlorination of which yields propylene oxide (227).

34.3 Exposure Assessment

34.3.3 Workplace Method

NIOSH recommends Method No. 2513, which has a working range of 1.5 to 50 mg/m^3 (0.5 to 15 ppm). This method involves drawing a known volume of air through charcoal (the recommended sample is 20 liters at a rate of 0.2 L/min). The analyte is desorbed with 5% 2-propanol in carbon disulfide, and the sample is separated by injection into a gas chromatograph equipped with a flame ionization detector (36). Other methods of collection and analysis by gas chromatography have been developed (39, 40) that have given excellent results and are applicable in when a variety of other materials may be present. These methods are expected to be adaptable for 2-chloroethanol.

34.4 Toxic Effects

34.4.1 Experimental Studies

2-Chloroethanol has a moderate to high order of acute toxicity. It is extremely irritating to the eyes but not to the skin, and it is not a skin sensitizer. Repeated oral exposure resulted in histopathological effects in the liver and lungs. 2-Chloroethanol is not carcinogenic to laboratory animals. Although 2-chloroethanol is a weak base-pair mutagen in bacteria, it has been both active and inactive in a variety of genotoxicity tests. Fetotoxicity and maternal toxicity in mice (but not rabbits) have been observed from 2-chloroethanol exposure. Teratogenicity was observed only in mice given 2-chloroethanol intravenously at a maternally toxic dose.

34.4.1.1 Acute Toxicity. The acute oral LD_{50} in the rat was reported as 72 mg/kg (227), 95 mg/kg (228), and 71 mg/kg (229, 230); in the mouse, 81 mg/kg (229) and 91 mg/kg (230); and in the guinea pig, 110 mg/kg (228). The acute dermal LD_{50} is reported as 68 mg/kg for rabbits (229), 84 mg/kg for rats (230), and 70 ml/kg for guinea pigs (231). The acute inhalation toxicity of 2-chloroethanol is listed in Table 78.3.

Undiluted 2-chloroethanol is not significantly irritating to the skin of rabbits (229, 234), but it is very irritating to the eyes of rabbits (229, 234).

34.4.1.2 Chronic and Subchronic Toxicity. 2-Chloroethanol was administered to dogs in their diet, for 90 days by oral gavage, and to monkeys by syringe. In the rats, no adverse findings were seen with dose levels of 30 and 45 mg/kg/day; but with 67.5 mg/kg/day,

Table 78.3. Results of Single Exposures of Animals to the Vapors of 2-Chloroethanol[a]

Animal	Dose (mg/L)	(ppm)	Duration of Exposure (h)	Outcome	Ref.
Guinea pig	18.0	5468	0.25	Death	235
	5.0	1544	0.9	Survived	227
	3.6	1094	1.0	Death	235
	3.0	911	1.8	Death	227
	3.0	911	0.5	Survived	227
Mouse	7.0	2430	2.0	Death	227
	4.5	1367	0.5	Death	227
	4.0	1215	0.25	LC_{67}	227
	3.0	911	1.0	Death	227
	1.2	365	2	LC_{17}	232
	1.0	304	2	Survived	227
	0.38	115	?	LC_{50}	230
Rat	4.0	1215	0.5	Death	227
	3.4	1033	0.25 × 3, 6, or 11	Death	227
	3.0	911	0.25	Survived	227
	0.29	88	?	LC_{50}	230
	0.11	33	4	LC_{50}	233

[a]Taken from Rowe and McCollister (34).

growth was depressed in both sexes and mortality was high. In dogs, ingestion of 2-chloroethanol was followed by severe emesis, which limited the highest intake that could be retained to about 18 to 20 mg/kg/day regardless of dietary level. The treated dogs did not grow, but all survived. The monkeys all failed to increase in weight at treatment levels up to 62.5 mg/kg, but no other differences were noted. Gross and histopathological examination showed no consistent dose-related effects in any of the the species (236). Rats maintained on diets containing 2-chloroethanol (0.01 to 0.08%) for at least 220 days showed no significant toxic effects; however, in rats diets of 0.12% showed dose-related growth retardation (237). In an inhalation study, rats were exposed to 0.31 or 3.1 ppm 4 h/day for 4 months. Decreased body weights and histopathological effects in the liver and lungs were observed. Both concentrations produced nervous system effects (230).

34.4.1.3 Pharmacokinetics, Metabolism, and Mechanisms. 2-Chloroethanol can be absorbed by the oral, dermal, or inhalation routes of exposure but quantitative data are limited. When rats were given an oral dose of 5 or 50 mg/kg radio-labeled 2-chloroethanol, the radioactivity was rapidly eliminated, mainly in the urine. Twenty-four hours after administration of the low dose, about 77% of the dose was found in the urine, 2% in the feces, and 1% as CO_2 in the expired air. Similar results were obtained with the high dose. The major urinary metabolites identified were thiodiactic acid and thionyldiactic acid. At the low dose, both metabolites were excreted in equal amounts but at the high dose, thiodiacetic acid was the predominant metabolite (238).

34.4.1.4 Reproductive and Developmental. Fetotoxicity and maternal toxicity were produced when 2-chloroethanol was administered by oral gavage to pregnant female Swiss CD-1 mice on days 4 to 12 of gestation (239). 2-Chloroethanol administered by oral gavage to pregnant CD-1 on days 6 through 16 of gestation produced a significant reduction in maternal weight gain and a decrease in fetal body weight and liver weight. A higher dose of 150 mg/kg was maternally lethal, and a lower dose of 50 mg/kg had no consistent effect (240). No effect on the dams or offspring occurred when 2-chloroethanol was administered in drinking water at doses up to 227 mg/kg to Swiss CD-1 mice on days 6 to 16 of gestation (240). No teratogenic effects and no significant maternal or embryo-fetal effects were noted in New Zealand white rabbits administered 2-chloroethanol intravenously at doses up to 36 mg/kg (241). In CD-1 mice administered 120 mg/kg 2-chloroethanol intravenously, malformations (treated on gestational days 8 to 10) and resorptions (treated on gestational days 4 to 6 and 10 to 12) were significantly increased compared to control animals, but only at maternally toxic doses (241).

34.4.1.5 Carcinogenesis. No evidence of carcinogenicity was observed in rats that ingested doses of 4, 8, or 16 mg/kg 2-chloroethanol in drinking water for up to 2 years. The study was not reported in detail, and small numbers of animals were used (242). F344 rats were given subcutaneous injections twice weekly at doses of 0.3, 1, 3, or 10 mg/kg for 1 year, followed by a 6-month observation period. An increased incidence of pituitary gland adenomas was observed in female rats; the incidence in the the dosed female rats (all doses combined) was 7/100 and the control incidence was 1/50 (243). 2-Chloroethanol was not carcinogenic in NMRI mice given subcutaneous injections once weekly for 70 weeks at doses of 0.3, 1.0, or 3 mg for 70 weeks (244). In an NTP study, 2-chloroethanol was given to F-344 rats and CD-1 mice at a dermal dose of 50 or 100 mg/kg for 2 years. There was no evidence of carcinogenicity (239).

34.4.1.6 Genetic and Related Cellular Effects Studies. The genotoxicity of 2-chloroethanol was investigated in a wide variety of short-term studies. 2-Chloroethanol is a direct-acting mutagen to *S. typhimurium* in the Ames test that causes base-pair substitution mutations (170, 239, 245). 2-Chloroethanol was mutagenic in *S. typhimurium* strains TA100 and TA135 (but not TA1537 or TA98) in either the presence or absence of Aroclor 1254-induced rat or hamster liver S9 (239). The addition of metabolic activation enhanced the mutagenicity of 2-chloroethanol, presumably mediated by the metabolite, 2-chloroacetaldehyde, a known mutagen. 2-Chloroethanol was not mutagenic in yeast *Saccharomyces cerevisiae* or *Schizosaccharomyces pombe* (247), and it did not induce sex-linked recessive-lethal mutations in *Drosophila* (239, 248). It was mutagenic in *E. coli* (249) and fungi (245). 2-Chloroethanol produced chromosomal aberrations and sister chromatid exchanges in Chinese hamster ovary cells *in vitro* (250). As cited in the NTP study, 2-chloroethanol increased the frequency of chromosomal aberrations in rat bone marrow after inhalation exposure, but neither chromosomal aberrations nor micronuclei were found in mouse bone marrow cells after exposure to 2-chloroethanol by either the oral or intraperitoneal injection routes. 2-Chloroethanol did not induce dominant-lethal mutations (251) or heritable translocations in the mouse (252).

34.4.1.7 Other: Neurological, Pulmonary, Skin Sensitization. 2-Chloroethanol was not a skin sensitizer to guinea pigs (253).

34.4.2 Human Experience

Human fatalities have resulted from ingestion, inhalation, or dermal contact with 2-chloroethanol.

34.4.2.2 Clinical Cases. The effects of acute exposure to 2-chloroethanol were reviewed (254). The following signs of intoxication were usually reported: nausea, vomiting, incoordination of the legs, vertigo, weakness, weak irregular pulse, and respiratory failure. Vomiting of bile, profuse perspiration, headache, visual disturbance, decreased blood pressure, hematuria, and spastic contracture of the hands were also reported. A 24-year-old man was reported to have accidentally drunk 20 mL (345 mg/kg) 2-chloroethanol and died 3 days later (255). In a fatal case, the individual had been using 2-chloroethanol for 2 h to clean trays on which rubber strips were stored (232). Analysis of the air revealed a concentration of about 1 mg/L. One fatal and several nonfatal cases of 2-chloroethanol poisoning in industrial workers occurrred at exposure concentrations (for unspecified periods) estimated at between 300 and 500 ppm (256). The survivors had nausea, vomiting, an irritation of the eyes, nose, and lungs. The average concentration of 2-chloroethanol in a plant was 18 ppm at the time of nine cases of nonfatal intoxication. A single fatality occurred from apparent cutaneous absorption of 2-chloroethanol (257). Rubber gloves have been reported to offer little protection against 2-chloroethanol. Laboratory results have shown that dangerous amounts of 2-chloroethanol or its aqueous solutions pass rapidly through rubber (231). Death occurred in less than 12 h following ingestion of approximately 2 mL of 2-chloroethanol by a 2-yr-old child (258).

34.4.2.3 Epidemiology Studies. An excess of deaths due to pancreatic cancer and leukemias was reported for 278 men assigned two years or more to the chlorohydrin unit of Union Carbide's South Charleston plant in the Kanawha Valley, West Virginia, from 1925 to 1957 (259). There were six deaths due to pancreatic cancer (0.7 expected) and three deaths due to leukemia (0.4 expected). Statistically significant trends with duration of assignment to the chlorohydrin unit were found for both diseases. The chlorohydrin unit primarily produced ethylene chlorohydrin from ethylene and chlorine and produced ethylene dichloride (1,2-dichloroethane) and bis-chloroethyl ether as by-products. Ethylene oxide was intermittently used in this unit to produce these needed and by-products for special production of ethylene chlorohydrin by an alternative process.

In a 10-year update of these chlorohydrin production workers, there were two additional cases of pancreatic cancer, bringing the total to eight (1.6 expected) and a standard mortality (SMR) ratio of 492. There were no additional deaths due to leukemia, but there was an excess risk for lymphopoietic and hematopoietic cancers (eight observed, 2.7 expected, SMR 294). Most of the cases were first assigned to the unit in the 1930s when chemical manufacturing was in its infancy and exposures were less controlled. Based on the review of the toxicology, medical records which indicated accidental overexposures for three cases of pancreatic cancer, and available industry hygiene data, it was suggested by the authors that likely past high exposure to ethylene dichloride, perhaps in combination with other chlorinated hydrocarbons, was a likely explanation for these findings (260).

In another study, mortality from pancreatic and lymphopoietic cancers were reported for male workers assigned to the Dow Chemical Company's ethylene and propylene chlorohydrin production processes. The cohort consisted of 1361 workers during 1940 and 1992. There was no increased risks of pancreatic cancer and lymphopoietic and hematopoietic cancer (261). The primary difference between Dow Chemical and Union Carbide ethylene chlorohydrin operations was that the Dow plants involved the production of ethylene chlorohydrin and also its subsequent epoxidation to ethylene oxide, in another part of the plant. Union Carbide's unit produced ethylene chlorohydrin which was subsequently pumped to a different unit to produce ethylene oxide.

34.5 Standards, Regulations, or Guidelines of Exposure

The ACGIH TLV is 1 ppm (3 mg/m^3) ceiling; The OSHA PEL is 5 ppm. Both have a skin notation (86, 177). The OSHA PEL was set to reduce the significant risk of central nervous system effects and other systemic effects associated with potential workplace exposure (177). The standard is similar for many countries (33).

35.0 Propylene Chlorohydrin (Mixture of 1-Chloro-2-Propanol and 2-Chloro-1-Propanol)

35.0.1a CAS Number: *[127-00-4]* (1-Chloro-2-propanol)

35.0.1b CAS Number: *[78-89-7]* (2-Chloro-1-propanol)

35.0.2a Synonyms: *sec*-Propylene chlorohydrin; α-propylene chlorohydrin; 1-chloroisopropyl alcohol; 1-chloro-2-hydroxypropane

35.0.2b Synonyms: 2-Chloropropyl alcohol; 1-hydroxy-2-chloropropane; β-chloropropyl alcohol

35.0.3 Trade Names: NA

35.0.4 Molecular Weight: 94.5

35.0.5 Molecular Formula: C_3H_7ClO

35.0.6a Molecular Structure: 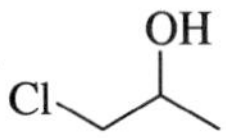

35.0.6b Molecular Structure: 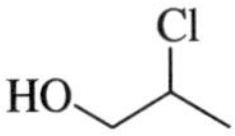

35.1 Chemical and Physical Properties

The chemical and physical properties are listed in Table 78.1.

35.2 Production and Use

Propylene chlorohydrin is used as an intermediate for producing propylene oxide and as a starting material for producing polyurethane polyols and propylene glycol (226, 262). Propylene chlorohydrins are a mixture of 1-chloro-2-propanol and 2-chloro-1-propanol. Both isomers are formed from propylene oxide during sterilization of foodstuffs because propylene oxide reacts with naturally occurring inorganic chloride in foods.

35.4 Toxic Effects

35.4.1 Experimental Studies

Chloropropanols have a moderate to severe order of acute toxicity. They are not very irritating to the skin but severely irritating to the eyes. These alcohols do not show any systemic toxicity by the oral route, and they are not reproductive toxicants. The chloropropanols have been shown to be genotoxic both *in vitro* and *in vivo*.

35.4.1.1 Acute Toxicity. The acute oral LD_{50} in the rat is reported as 100 to 300 mg/kg (179b), 381 mg/kg (263), and 0.22 ml/kg for propylene chlorohydrin (43). The acute dermal LD_{50} in rabbits for both isomers is 480 mg/kg (263), 0.48 ml/kg (43), and about 0.5 g/kg (179b). One out of six rats died when exposed to 500 ppm propylene chlorohydrin for 4 h (201). No deaths were observed when rats were exposed to saturated vapors for 15 min (43).

35.4.1.2 Chronic and Subchronic Toxicity. Rats given 15 exposures by inhalation (6 h each) of 250 ppm 1-chloro-2-propanol displayed lethargy, irregular weight gains, congestion, and perivascular edema in the lungs. At 100 ppm, the lungs were congested with perivascular edema; no effects were observed at 30 ppm (84). In a 21-day oral study, rats were dosed with either 7.6, 25.4, or 76 mg/kg/day propylene chlorohydrin (75 : 25 mixture of 1-chloro-2-propanol and 2-chloro-1-propanol). No systemic toxicity was observed (264).

In a 13-week study, Sprague–Dawley rats were dosed by oral gavage with 1, 7, or 50 mg/kg/day propylene chlorohydrin (75:25 1-chloro-2-propanol and 2-chloro-1-propanol). There were no findings of note except for histopathological changes in the acinar cells of the pancreas (265). In a 13-week study, dogs were dosed by oral gavage with 1, 7, or 50 mg/kg/day propylene chlorohydrin (75:25 1-chloro-2-propanol and 2-chloro-1-propanol). The only effect noted was decreased body weights in the high-dose animals (265).

In NTP studies, F344 rats and $B6C3F_1$ mice were given 0, 33, 100, 1000, or 3300 ppm technical grade propylene chlorohydrin (75 : 25 1-chloro-2-propanol and 2-chloro-1-propanol) in drinking water (266).

35.4.1.3 Pharmacokinetics, Metabolism, and Mechanisms. Propylene chlorohydrin is absorbed, widely distributed, and rapidly metabolized and eliminated. In rats dosed orally with 1-chloro-2-propanol, the major urinary metabolites were identified as *N*-acetyl-*S*-(2-hydroxypropyl)cysteine and β-chlorolactate. Four percent of the administered dose was excreted unchanged (267, 268). When rats were exposed nose-only for 6 h to 8 or 77 ppm of radiolabeled 1-chloro-2-propanol, there were two major routes of elimination of [^{14}C], urinary and exhalation of CO_2, which together accounted for about 80% of the total radioactivity. The half-times for elimination of [^{14}C] in urine and as [^{14}C]CO_2 were between 3 and 7 h. After the end of exposure, kidneys, livers, trachea, and nasal turbinates contained high concentrations of radioactivity. Major metabolites detected in the urine and tissues were *N*-acetyl-*S*-(hydroxypropyl)cysteine and/or *S*-(2-hydroxypropyl)-cysteine (269).

35.4.1.4 Reproductive and Developmental. Technical grade propylene chlorohydrin (75:25 1-chloro-2-propanol and 2-chloro-1-propanol) was administered in drinking water to Sprague–Dawley rats and tested for its effect on fertility and reproduction in a continuous breeding protocol. There was no effect on fertility or any of the reproductive parameters, including mean number of litters, litte size, or days to deliver. However, it was concluded that propylene chlorohydrin may adversely affect the production of morphologically normal sperm at maximally tolerated doses (270).

35.4.1.5 Carcinogenesis. No evidence of carcinogenicity was observed in an NTP study in which F344 rats were administered a technical grade of propylene chlorohydrin (75 : 25 1-chloro-2-propanol and 2-chloro-1-propanol) in drinking water at 0, 150, 325, 650 ppm for two years. These concentrations equated to average daily doses of approximately 15, 30, or 65 mg/kg during the first several months of the study and 8, 17, or 34 mg/kg (266). In the same study, there was no evidence of carcinogenicity in $B6C3F_1$ administered technical grade propylene chlorohydrin in drinking water at 0, 250, 500, or 1000 ppm for two years. These concentrations equated to average daily doses of approximately 44, 74, or 149 mg/kg to males and 58, 104, or 210 mg/kg to females during the first several months of the study and 25, 50, or 100 mg/kg for the remainder of the study (266).

35.4.1.6 Genetic and Related Cellular Effects Studies. 1-Chloro-2-propanol was mutagenic to *S. typhimurium* in an Ames test with and without metabolic activation (155, 271). A 75 : 25 mixture of 1-chloro-2-propanol and 2-chloro-1-propanol was mutagenic in the Ames test with and without metabolic activation, in the mouse lymphoma assay, and in the rat bone marrow cytogenic assay (272).

A technical grade of propylene chlorohydrin (75 : 25 1-chloro-2-propanol and 2-chloro-1-propanol) put through a battery of tests sponsored by the NTP program. It was weakly mutagenic in *S. typhimurium* strain TA100 in the presence of hamster or liver S9 activation enzymes and was positive with and without S9 in TA1535. No mutagenic activity was detected with strain TA97, TA98, or TA1537, with or without S9 (155, 266). Propylene chlorohydrin induced sister chromatid exchanges and chromosomal aberrations *in vitro* in Chinese hamster ovary (CHO) cells both with and without metabolic activation. Induction of sex-linked recessive lethal mutations in germ cells of male *Drosophila melanogaster* occurred by injection, but not by feed. No effects were observed in a subsequent germ cell reciprocal translocation test in *Drosophila melanogaster.* Furthermore, propylene chlorohydrin was inactive in a mouse bone marrow micronucleus assay in which mice were given propylene chlorohydrin in drinking water for 13 weeks (266).

Table 78.4. Toxicity Data on Miscellaneous Alcohols

Alcohol	Acute Oral Rat LD_{50}	Acute Dermal Rabbit LD_{50}	Acute Inhalation Rat LC_{50}	Skin Irritation	Eye Irritation	Skin Sensitization	Ref.
2-Heptanol (CAS #*543-47-7*)	2.58 g/kg	1.78 mL/kg	No deaths, 8 hr satd. vapor	Slight	Severe	—	273
3-Heptanol (CAS #*3913-02-8*)	1.87 g/kg	4.36 mL/kg	No deaths, 4 hr satd. vapor	Slight	Moderate	—	105
1-Undecanol (CAS #*112-42-5*)	3 g/kg	> 5 g/kg	—	Moderate	—	—	274
1-Tetradecanol (CAS #*112-72-1*)	> 5 g/kg	> 5 g/kg	—	No	Slight	—	275
Eicosanol (mixed isomers) (CAS #*629-96-9*)	> 64 mL/kg	> 20 mL/kg	No deaths, 8 hr satd. vapor	Slight	Slight	—	43
3,5,5-Trimethyl cyclohexanol (CAS #*116-02-9*)	3.45 g/kg	2.8 mL/kg	—	Moderate	Very severe	—	102
3-Butyn-2-ol (CAS #*2028-69-9*)	34 mg/kg	32–36 mg/kg	1 hr LC_{50}; std. vapor 6 min exposure, 100% lethal	—	Very severe	—	216
Methyl butynol (CAS #*115-9-5*)	1.9 g/kg	> 5 g/kg	1 hr LC_{50}; > 20 mg/L[a]	None	Severe (10% in H_2O not irritating)	Possibly	216
Methyl pentynol (CAS #*77-75-8*)	0.8 g/kg	> 1 g/kg	1 hr LC_{50}; > 20 mg/L[b]	Slight	Severe	Possibly	216
Ethyl octynol (CAS #*5877-42-9*)	2.1 g/kg	0.2–1.0 g/kg	—	—	—	Possibly	34
3,7-Dimethyl-1-octonol (CAS #*106-21-8*)	> 5 g/kg	2.4 g/kg	—	Yes. unspec.	—	—	106

[a] A single 4-hr exposure of rats to 3000 ppm methyl butynol caused death, but a 7-hr exposure to 2000 ppm did not. Both exposures caused liver and kidney injury. A single 7-hr exposure to 1000 ppm did not cause grossly apparent injury. Rats that received 81 7-hr exposures to 76 ppm for 115 day exhibited no adverse effects.

[b] A single 4-hr or 7-hr exposure of rats to 4600 ppm methyl pentynol caused anesthesia, considerable injury to the lungs, liver, and kidney, and death. A 2-hr exposure did not cause death but did cause kidney injury and weight loss. A few deaths occurred after 4- and 7-hr exposures to 2000 ppm, whereas all survived a 7-hr exposure to 1000 ppm, although kidney injury was noted. Groups of rats tolerated 67 7-hr exposures to 100 ppm for 98 days without adverse effects.

35.4.2 Human Experience

35.4.2.3 Epidemiology Studies. The mortality from pancreatic and lymphopoietic cancers was reported for male workers assigned to the Dow Chemical Company's ethylene and propylene chlorohydrin production processes. The cohort consisted of 1361 workers during 1940 and 1992 (261). There was no increased risk of pancreatic, lymphopoietic or hematopoietic cancer.

35.5 Standards, Regulations, or Guidelines of Exposure

No occupational exposure standards have been established for proylene chlorohydrin.

P MISCELLANEOUS ALCOHOLS

Limited toxicity data exist on a number of alcohols of lesser industrial importance, which are listed in Table 78.4. These alcohols include six monohydric alcohols, one alicyclic alcohol, and four acetylenic alcohols. It should be noted that a number of these alcohols exist as stereoisomers with different CAS registry numbers. The CAS registry numbers listed in Table 78.4 represent the mixed product. It is not known whether biological differences exist among the different isomers.

BIBLIOGRAPHY

1. National Fire Protection Association (NFPA), *Fire Protection Guide on Hazardous Materials*, 9th ed., NFPA, 1986.
2. National Fire Protection Association (NFPA), *Fire Protection Guide on Hazardous Materials*, 10th ed., NFPA, 1991.
3. J. Falbe et al., in *Ullman's Encyclopedia of Industrial Chemistry*, 5th rev. ed., Vol. A1, VCH Publishers, New York, 1984, pp. 279–303.
4. Chem. Eng. News, June 29 (1998).
5. M. F. Gautreaux, W. T. Davis, and E. D. Travis, in *Kirk-Othmer's Encyclopedia of Chemical Technology*, 3rd ed., Vol. 1, Wiley, New York, 1982, pp. 740–752.
6. J. D. Wagner, G. R. Lappin, and J. R. Zietz, in *Kirk-Othmer's Encyclopedia of Chemical Technology*, 4th ed., Vol. 1, Wiley, New York, 1991, pp. 865–913.
7. R. J. Schueplein and I. H. Blank, *Physiol. Rev.* **51**, 702–747 (1971).
8. R. T. Williams, in *The Metabolism and Detoxication of Drugs, Toxic Substances and Other Organic Compounds*, 2nd ed., Chapman & Hall, London, 1959.
9. D. L. J. Opdyke, *Food Cosmet. Toxicol.* **13**, 697–698 (1975).
10. R. Truhaut et al., *Arch. Mal. Prof. Med. Trav. Secur. Soc.* **35**, 501–509 (1974).
11. Y. L. Egoros and L. A. Andrianov, *Uch. Zap.—Mosk. Nauchno.-Issled. Inst. Gig. im F. F. Erismana* **9**, 47 (1961).
12. G. N. Zaeva and V. I. Fedorova, *Toksikol. Nov. Prom. Khim. Veshchestv* **5**, 51 (1963).
13. H. W. Gerarde and D. B. Ahlstrom, *Arch. Environ. Health* **13**, 457–461 (1966).

14. J. Muller and G. Greff, *Food Chem. Toxicol.* **22**, 661–664 (1984).
15. L. F. Hansen and D. Nielsen, *Toxicology* **88**, 81–99, 1994.
16. National Institute for Occupational Safety and Health (NIOSH), *Registry of Toxic Effects of Chemical Substances (RTECS)*, CCOHS CD-ROM, NIOSH, Washington, DC, 1999.
17. Exxon Chemical Company, unpublished data.
18. H. F. Smyth. Jr. et al., *Am. Ind. Hyg. Assoc. J.* **23**, 95–107 (1962).
19. E. Zeiger et al., *Environ. Mol. Mutagen.* **19**(Suppl. 21), 2–141 (1992).
20. K. Verschueren, *Handbook of Environmental Data on Organic Chemicals*, Van Nostrand-Reinhold, New York, 1977.
21. D. L. J. Opdyke, *Food Cosmet. Toxicol.* **11**, 101–102 (1973).
21a. Amoco Corporation, unpublished data.
22. D. L. J. Opdyke, *Food Cosmet. Toxicol.* **11**, 1079 (1973).
23. G. A. Jacobs, *J. Am. Coll. Toxicol.* **11**(6), 726 (1992).
24. V. B. Voskobinikova, *Gig. Sanit.* **31**, 310–315 (1966).
25. B. K. Nelson et al., *J. Am. Coll. Toxicol.* **9**, 93–97 (1990).
26. B. K. Nelson et al., *Toxicol. Ind. Health* **6**, 373–387 (1990).
27. J. Hellwig and R. Jäckh, *Food. Chem. Toxicol.* **35**, 489–500, 1997
28. G. D. Stoner et al., *Cancer Res.* **33**, 3069–3085 (1985).
29. Sice, *J. Toxicol. Appl. Pharmacol.* **9**, 70–74 (1966).
30. T. J. B. Gray et al., *Toxicology* **28**, 167–179 (1983).
31. H. A. Griffiths et al., *Food Chem. Toxicol.* **35**, 255–260, 1997.
32. American Industrial Hygiene Association, *Workplace Environmental Exposure Level Guide (WEEL) for 1-Octanol*, AIHA, Akron, OH, 1986.
33. International Labour Office, *Occupational Exposure Limits for Airborne Toxic Substances; Values of Selected Countries Prepared from the ILO-CIS Data Base of Exposure Limits*, 3rd ed., Occup. Saf. Health Ser. No. 37, ILO, Geneva, 1991.
34. V. K. Rowe and S. B. McCollister, In G. D. Clayton and F. E. Clayton, Eds., *Patty's Industrial Hygiene and Toxicology*, 3rd rev. ed., Wiley, New York, 1982, pp. 4527–4708.
35. I. Mellan, *Industrial Solvents*, Reinhold, New York, 1950.
36. National Institute for Occupational Safety and Health (NIOSH), *Manual of Analytical Methods*, 4th ed., NIOSH Pub. 94–113, U.S. Government Printing Office, Superintendent of Documents, Washington, DC, 1994.
37. H. Russo and S. S. Que Hee, *Anal. Chem.* **55**, 400–403 (1983).
38. T. Gorski et al., *J. Chromatogr.* **509**, 383–389 (1990).
39. P. W. Langvardt and R. G. Melcher, *Am. Ind. Hyg. Assoc. J.* **40**, 1006–1012 (1979).
40. F. X. Mueller and J. A. Miller, *Am. Ind. Hyg. Assoc. J.* **40**, 380–386 (1979).
41. BIBRA, *Toxicity Profile for 2-Ethyl-1-Hexanol*, BIBRA, 1990.
42. R. A. Scala, and E. G. Burtis, *Am. Ind. Hyg. Assoc. J.* **34**, 493–499 (1973).
43. H. F. Smyth, Jr. et al., *Am. Ind. Hyg. Assoc. J.* **30**, 470–476 (1969).
44. P. W. Albro, *Xenobiotica* **5**, 625–636 (1975).
45. D. W. Fassett, personal communication to J. F. Treon, in D. W. Fassett and D. D. Irish, eds., *Patty's Industrial Hygiene and Toxicology*, 2nd rev. ed., Vol. 2, Interscience, New York, 1963, pp. 1409–1496.

46. E. V. Weaver et al., *Toxicologist* **9**, 247 (abstr. 989) (1989).
47. P. Schmidt, R. Gohlke, and R. Rothe, *Z. Gesamte. Hyg. Grenzgeb.* **19**, 485–490 (1973).
48. D. L. Wood and J. Bitman, *Poult. Sci.* **63**, 469–474 (1984).
49. B. Lundgren et al., *Chem.-Biol. Interact.* **68**, 219–240 (1988).
50. A. E. Ganning et al., *Acta Chem. Scand. Ser.* β **B36**, 563–565 (1982).
51. D. E. Moody and J. K. Reddy, *Toxicol. Appl. Pharmacol.* **45**, 497–504 (1978).
52. D. E. Moody and J. K. Reddy, *Toxicol. Lett.* **10**, 379–383 (1982).
52a. Union Carbide, unpublished data.
53. Y. Keith et al., *Arch. Toxicol.* **66**, 321–326 (1992).
54. B. D. Astrill et al., *Fundam. Appl. Toxicol.* **29**, 31–39 (1996).
55. C. Rhodes et al., *Toxicol. Lett.* **21**, 103–109 (1984).
56. B. G. Lake et al., *Toxicol. Appl. Pharmacol.* **32**, 355–367 (1975).
57. J. R. Hodgson, *Toxicol. Ind. Health* **3**(2), 49–61 (1987).
58. E. D. Barber et al., *Fundam. Appl. Toxicol.* **19**, 493–497 (1992).
59. I. A. Kamil, J. N. Smith, and R. T. Williams, *Biochem. J.* **53**, 137 (1953).
60. P. J. Deisinger, R. J. Boatman, and D. Guest, *Xenobiotica* **24**, 429–440 (1994).
61. E. J. Ritter et al., *Teratology* **35**, 41–46 (1987).
62. B. D. Hardin et al., *Teratog., Carcinog., Mutagen.* **7**, 29–48 (1987).
63. B. K. Nelson et al., *J. Am. Coll. Toxicol.* **8**, 405–410 (1989).
64. R. W. Tyl et al., *Fundam. Appl. Toxicol.* **19**, 176–185 (1992).
65. C. J. Price et al., *Teratology* **43**, 457 (abstr.) (1991).
66. B. D. Astrill et al., *Fundam. Appl. Toxicol.* **31**, 29–41 (1996).
67. T. J. B. Gray and J. A. Beamond, *Food Chem. Toxicol.* **22**, 123–131 (1984).
68. E. J. Moss et al., *Toxicol. Lett.* **40**, 77–84 (1988).
69. P. Sjöberg et al., *Acta Pharmacol. Toxicol.* **58**, 225–233 (1986).
70. C. J. Bushbrook, T. A. Jorgenson, and J. R. Hodgson, *Environ. Mutagen.* **4**, 387 (abstr.) (1982).
71. J. M. Ward et al., *Environ. Health Perspect.* **65**, 279–291 (1986).
72. A. B. DeAngelo et al., *Toxicology* **41**, 279–288 (1986).
73. E. D. Barber et al., *Toxicologist* **5**, 211 (abstr. No. 841) (1985).
74. P. E. Kirby et al., *Environ. Mutagen.* **5**, 657–663 (1983).
75. E. Zeiger et al., *Environ. Mutagen.* **7**, 213–232 (1985).
76. E. Zeiger et al., *Environ. Mol. Mutagen.* **11**(Suppl. 12), 1–158 (1988).
77. M. Ischidate, Jr., M. C. Harnois, and T. Sofuni, *Mutat. Res.* **195**, 151–213 (1988).
78. B. J. Phillips, T. E. B. James, and S. D. Gangolli, *Mutat. Res.* **102**, 297–304 (1982).
79. D. L. Putman et al., *Environ. Mutat.* **5**, 227–231 (1983).
80. A. von Däniken et al., *Toxicol. Appl. Pharmacol.* **73**, 373–387 (1984).
81. G. D. DiVincenzo et al., *Toxicology* **34**, 247–259 (1985).
82. C. R. Elcombe and A. M. Mitchell, *Environ. Health Perspect.* **70**, 211–219 (1986).
83. Y. Keith et al., *Arch. Toxicol. Supp.* **12**, 274–277 (1988).
84. J. C. Gage, *Br. J. Ind. Med.* **27**, 1–18 (1970).

85. S. B. Harris, L. H. Keller, and A. I. Nikiforov, *Toxicologist* **30**, 195 (Abstr. No. 995), 1996.
86. American Conference of Governmental Industrial Hygienists (ACGIH), *Threshold Limit Values for Chemical Substances and Physical Agents and Biological Exposure Indices*, 2000, ACGIH, Washington, DC, 2000.
87. D. L. J. Opdyke, *Food Cosmet. Toxicol.* **17**, 881 (1979).
88. R. C. Myers and R. R. Tyler, *J. Am. Coll. Toxicol., Part B* **1**(3), 194–195 (1992).
89. C. T. Olson et al., *Biochem. Biophys. Res. Commun.* **130**, 313–316 (1985).
90. M. Charbonneau et al., *Toxicologist* **7**, 89 (Abstr. No. 356) (1987).
91. E. A. Lock et al., *Toxicol. Appl. Pharmacol.* **91**, 182–192 (1987).
92. L. D. Lehman-McKeeman and D. Caudill, *Toxicol. Appl. Pharmacol.* **116**, 170–176 (1992).
93. S. J. Borghoff, B. G. Short, and J. A. Swenberg, *Annu. Rev. Pharmacol. Toxicol.* **30**, 349–367 (1990).
94. W. G. Flamm and L. D. Lehman-McKeeman, *Regul. Toxicol. Pharmacol.* **13**, 70–86 (1991).
95. M. J. Olson, J. T. Johnson, and C. A. Reidy, *Toxicol. Appl. Pharmacol.* **102**, 524–536 (1990).
96. K. Baetake et al., *Alpha-2u-globulin: Association with Renal Toxicity and Neoplasia in the Male Rat*, EPA/625/3-91/019F, U.S. Environmental Protection Agency, Risk Assessment Forum, Washington, DC, 1991.
97. D. L. J. Opdyke, *Food Cosmet. Toxicol.* **11**, 103–104 (1973).
98. Yu. L. Egorov, *Toksikol. Gig. Prod. Ncftekhim.*, Neftekhim. Proizvol. Vses. Knof. (Dkol.) 2nd, 98 (1972); *Chem. Abstr.* **80**, 91721b (1974).
99. J. A. Guido and M. A. Martens, *J. Am. Coll. Toxicol.* **11**(6), 736 (1992).
100. Barilyak, V. I. Korkach, and L. D. Spitkovskaya, *Sov. J. Dev. Biol.* **22**(1), 36–39 (1990).
101. D. L. J. Opdyke, *Food Chem. Toxicol.* **30**, 23S (1990).
102. H. F. Smyth, Jr., C. P. Carpenter, and C. S. Weil, *J. Ind. Hyg. Toxicol.* **31**, 60–62 (1949).
103. K. V. Kitzmiller, personal communication. Cited in Ref. 45.
104. D. L. J. Opdyke, *Food Cosmet. Toxicol.* **11**, 109 (1973).
105. H. F. Smyth, Jr., C. P. Carpenter, and C. S. Weil, *Arch. Ind. Hyg. Occup. Med.* **4**, 119–122 (1951).
106. D. L. J. Opdyke, *Food Cosmet. Toxicol.* **12**, 535 (1974).
107. F. U. Bär and F. Griepentrog, *Med. Ernehr.* **8**, 244 (1967).
108. F. E. Shaffer, personal communication. Cited in Ref. 45.
109. BIBRA, *Toxicity Profile on 1-Dodecanol*, BIBRA, 1990.
110. Y. Iwata, Y. Moriya, and T. Kobayashi, *Cosmet. Toxicol.* **102**, 53–68 (1987).
111. H. Shimuzi et al., *Jpn. J. Ind. Health* **27**, 400–419 (1985).
112. W. Johnson, Jr., *J. Am. Coll. Toxicol.* **7**(3), 359–413 (1988).
113. W. M. McIaac and R. T. Williams, *W. Afr. J. Biol. Chem.* **2**, 42 (1958).
114. R. D. Blevins and D. E. Taylor, *J. Environ. Sci. Health, Part A* **A17**(2), 217–239 (1982).
115. J. Moore, *J. Am. Coll. Toxicol.* **4**(5), 1–29 (1985).
116. I. Florin et al., *Toxicology* **18**, 219–232 (1980).
117. B. D. Mookherjee, in *Kirk-Othmer Encyclopedia of Chemical Technology*, 4th ed., Vol. 4, Wiley, New York, 1992, pp. 116–125.
118. D. L. J. Opdyke, *Food Cosmet. Toxicol.* **11**, 1011–1013 (1973).

119. P. M. Jenner et al., *Food Cosmet. Toxicol.* **2**, 327–343 (1943).

120. National Toxicology Program (NTP), *Toxicology and Carcinogenesis Studies of Benzyl Alcohol*, NTP Tech. Rep. No. 343, NIH Pub. No. 89-2599, NTP, Washington, DC, 1989.

121. A. D. Winer, *Acta Chem. Scand.* **12**, 1695 (1958).

122. R. Derache, *Int. Encycl. Pharmacol. Ther.* **20**, 507 (1970).

123. S. E. McCloskey et al., *Toxicologist* **6**, (Abstr. No. 1037) (1986).

124. S. L. Diack and H. B. Lewis, *J. Biol. Chem.* **77**, 89 (1928).

125. K. Mortelsmans et al., *Environ. Mutagen.* **8**(Suppl. 7), 1–119 (1986).

126. K. Kuroda, Y. S. Yoo, and T. Ishibashi, *Mutat. Res.* **130**, 369 (abstr.) (1984).

127. B. E. Anderson et al., *Environ. Mol. Mutagen.* **16**(Suppl. 18), 55–137 (1990).

128. P. H. Dugard et al., *Environ. Health Perspect.* **57**, 193–197 (1984).

129. L. B. Fisher, in R. L. Bronaugh and H. I. Maibach, eds., *Percutaneous Absorption. Mechanism—Methodology—Drug Delivery*, New York, 1989, pp. 135–145.

130. I. Snapper, A. Grunbaum, and S. Sturkop, *Biochem. Z.* **155**, 163 (1925).

131. K. E. Malten et al., *Contact Dermatitis* **11**, 1–10 (1984).

132. J. C. Mitchell et al., *Contact Dermatitis* **8**, 336–337 (1982).

133. J. R. Nethercott, *Contact Dermatitis* **8**, 389–395 (1982).

134. BIBRA, *Toxicity Profile on Benzyl Alcohol*, BIBRA, 1989.

135. W. J. Brown et al., *Lancet* **1**, 1250 (1982).

136. J. Gershanik et al., *N. Engl. J. Med.* **307**, 1384–1388 (1982).

137. J. Lopez-Herce et al., *Ann. Pharmacother.* **29**, 632 (1995).

138. W. Ringk and E. T. Theimer, in *Kirk-Othmer's Encyclopedia of Chemical Tehnology*, 3rd ed., Vol. 3, Wiley, New York, 1978, pp. 848–864.

139. K. R. Brandt, *J. Am. Coll. Toxicol.* **9**(2), 165–183 (1990).

140. *Chemical Hazard Information Profile*, Draft Report on 2-Phenylethanol and 2-Phenylethyl Acetate, prepared by Oak Ridge National Laboratory for the US EPA, Washington, DC, 1985.

141. C. P. Carpenter, C. S. Weil, and H. F. Smyth, Jr., *Toxicol. Appl. Pharmacol.* **28**, 313–319 (1974).

142. BIBRA, *Toxicity Profile on Phenylethyl Alcohol*, BIBRA, 1988.

143. E. Owston, R. Lough, and D. L. Opdyke, *Food Cosmet. Toxicol.* **19**, 713–715 (1981).

144. C. A. Mankes et al., *J. Toxicol. Environ. Health* **12**, 235–244 (1983).

145. G. A. Burdock et al., *Toxicologist* **7**, 176 (Abstr. No. 702) (1987).

146. R. A. Ford, A. M. Api, and T. M. Palmer, *Toxicologist* **7**, 175 (Abstr No. 701) (1987).

147. H. Norppa and H. Vainio, *Mutat. Res.* **116**, 379–387 (1983).

148. D. L. J. Opdyke, *Food Cosmet. Toxicol.* **13**, 903–904 (1975).

149. A. C. De Groot et al., *Contact Dermatitis* **12**, 87–92 (1985).

150. M. R. Schoenberg, J. W. Blieszner, and C. G. Papadopoulos, in *Kirk-Othmer's Encyclopedia of Chemical Technology*, 3rd ed., Vol. 19, Wiley, New York, 1982, pp. 228–246.

151. National Toxicology Program (NTP), *Toxicology and Carcinogenesis Studies of Alpha Methylbenzyl Alcohol*, NTP Tech. Rep. TR 369, NIH Publ. 89-2824, NTP, Washington, DC, 1990.

152. H. F. Smyth, Jr. and C. P. Carpenter, *J. Ind. Hyg. Toxicol.* **26**, 269–273 (1944).

153. D. L. J. Opdyke, *Food Cosmet. Toxicol.* **12**, 995–996 (1974).
154. C. P. Carpenter and H. F. Smyth, Jr., *Am. J. Ophthalmol.* **29**, 1363–1372 (1946).
155. E. Zeiger et al., *Environ. Mol. Mutagen* **9**(Suppl. 9), 1–110 (1987).
156. D. L. J. Opdyke, *Food Cosmet. Toxicol.* **13**, 547 (1975).
157. I. F. Gaunt, M. G. Wright, and R. Cottrel, *Food Cosmet. Toxicol.* **20**, 519–525 (1982).
158. D. Wild et al., *Food Chem. Toxicol.* **21**, 707–719 (1983).
159. D. L. J. Opdyke and C. Letizia, *Food Chem. Toxicol.* **20**, 839–840 (1982).
160. J. E. Amoore and E. Hautala, *J. Appl. Toxicol.* **3**, 272–290 (1983).
161. W. B. Fisher and J. F. Van Peppen, in *Kirk-Othmer's Encyclopedia of Chemical Technology*, Vol. 7, 4th ed., Wiley, New York, 1993, pp. 851–859.
162. L. Perebellini, F. Brugnone, and I. Pavan, *Toxicol. Appl. Pharmacol.* **53**, 220–229 (1980).
163. L. Perbellini et al., *Int. Arch. Occup. Environ. Health* **48**, 99–106 (1981).
164. J. F. Treon, W. E. Crutchfield, and K. V. Kitzmiller, *J. Ind. Hyg. Toxicol.* **25**, 199–214 (1943).
165. J. Pohl, *Zentralbl. Gewerbehyg. Unfallverhuet.* **12**, 91 (1925).
166. J. F. Treon, W. E. Crutchfield and K. V. Kitzmiller, *J. Ind. Hyg. Toxicol.* **25**, 323–347 (1943).
167. H. Ichibagase et al., *Chem. Pharm. Bull.* **20**, 175–180 (1972).
168. E. Gondry, *Eur. J. Toxicol.* **5**, 227–238 (1972).
169. J. P. Collin, *Diabete* **19**, 215 (1971).
170. S. Haworth et al., *Environ. Mutagen. Suppl.* **1**, 3–142 (1983).
171. BASF (1991), Submitted to USEPA under TSCA 8(d), EPA/OTS; Doc No. 8695000035 (1995).
172. K. W. Nelson, et al., *J. Ind. Hyg. Toxicol.* **25**, 282–285 (1943).
173. D. L. J. Opdyke, *Food Cosmet. Toxicol.* **13**, 777 (1975).
174. J. Mráz et al., *Clin. Chem. (Winston-Salers, N.C.)*, **40**, 146–148 (1994).
174a. J. F. Treon, In D. W. Fassett and D. D. Irish, Eds., *Patty's Industrial Hygiene and Toxicology*, 2nd rev. ed., Interscience, New York, 1963, pp 1459.
175. BIBRA, *Toxicity Profile for Methylcyclohexanol, BIBRA*, 1988.
176. E. Browning, *Toxicology and Metabolism of Industrial Solvents*,Elsevier, New York, 1965, pp. 310–411.
177. U.S. Department of Labor, Occupational Safety and Health Administration (OSHA), 29CFR, p. 1910., 1000, Jaule Z-I, Limits for Air Contaminants, 1997.
178. W. J. McKillip, and E. Sherman, in *Kirk-Othmer's Encyclopedia of Chemical Technology*, 3rd ed., Vol. 11, Wiley, New York, 1978, pp. 499–527.
179. M. W. Windholz, ed., *The Merck Index*, 9th ed., Merck & Co., Rahway, N J, 1976.
179a. The Quaker Oats Company, unpublished data. Cited in Ref. 34 and 180.
179b. J. Gajewship and W. Alsdrof, *Fed. Proc.* **8**, 294 (1949).
180. BIBRA, *Toxicity Profile on Furfuryl Alcohol*, BIBRA, 1989.
181. K. H. Jacobson et al., *Am. Ind. Hyg. Assoc. J.* **19**, 91–100 (1958).
182. R. D. Irwin, et al., *J. Appl. Toxicol.* **17**, 159–169 (1997).
183. P. W. Mellick et al., *Toxicologist* **11**, 183 (Abstr. No. 670) (1991).
184. R. A. Miller et al., *Toxicologist* **11**, 183 (Abstr. No. 669) (1991).
185. R. D. Irwin, S. B. Enke, and J. D. Prejean, *Toxicologist* **5**, 240 (Abstr. No. 960) (1985).

186. A. A. Nomeir et al., *Drug Metab. Dispos.* **20**, 198–204 (1992).
187. National Toxicology Program (NTP), *Toxicology and Carcinogenesis Studies of Furfuryl Alcohol (CAS No. 98-00-0) in F344/N Rats and B6C3F$_1$* Mice (Inhalation Studies). NTP TR-482, NIH Publ. No. 99-3972, NTP, Washington, DC, 1999.
188. Z.-L. Kong, et al., *Mutat. Res.* **203**, 376 (Abstr.) (1988).
189. H. F. Stich et al., *Cancer Lett.* **13**, 89–95 (1981).
190. S. Gomez-Arroyo and V. S. Souza, *Mutat. Res.* **156**, 233–238 (1985).
191. R. Rodriguez-Arnaiz, et al., *Mutat. Res.* **223**, 309–311 (1989).
192. H. F. Smyth, Jr., C. P. Carpenter, C. S. Weil, and U. C. Pozzani, *Arch. Ind. Hyg. Occup. Med.* **10**, 61–68 (1954).
193. D. L. J. Opdyke, *Food Cosmet. Toxicol.* **16**, 641 (1978).
194. P. M. J. Bos et al., *CRC Crit. Rev. Toxicol.* **21**(6), 423–450 (1992).
195. N. Nagato, in *Kirk-Othmer's Encyclopedia of Chemical Technology*, 4th ed., Vol. 2, Wiley, New York, 1992, pp. 144–160.
196. W. M. Doizaki and M. D. Levitt, *J. Chromatogr.* **176**, 11–18 (1983).
197. M. K. Dunlap et al., *AMA Arch. Ind. Health* **18**, 303–311 (1958).
198. H. F. Smyth, Jr. and C. P. Carpenter, *J. Ind. Hyg. Toxicol.* **30**, 63–68 (1948).
199. C. P. McCord, *JAMA, J. Am. Med. Assoc.* **98**, 2269–2270 (1932).
200. H. E. Driver and A. E. M. McClean, *Food Chem. Toxicol.* **24**, 241–245 (1986).
201. H. F. Smyth, Jr., *Am. Ind. Hyg. Assoc. Q.* **17**, 129–185 (1956).
202. G. D. Nielsen, J. C. Bakbo, and E. Holst, *Acta Pharmacol. Toxicol.* **54**, 292–298 (1984).
203. F. M. B. Carpanini et al., *Toxicology*, **9**, 29–45 (1978).
204. T. R. Torkelson et al., *Am. Ind. Hyg. Assoc. J.* **20**, 224–229 (1959).
205. J. M. Patel et al., *Drug Metab. Dispos.* **11**, 164–166 (1983).
206. R. O. Beauchamp, et al., *CRC Crit. Rev. Toxicol.* **14**, 309–380 (1991).
207. P. C. Jenkinson and D. Anderson, *Mutat. Res.* **229**, 173–184 (1990).
208. W. Lijinsky, and M. D. Reuber, *Toxicol. Ind. Health* **3**, 337–345 (1987).
209. W. Lijinsky and A. W. Andrews, *Teratog. Carcinog. Mutagn.* **1**, 259–267 (1990).
210. D. Lutz et al., *Mutat. Res.* **93**, 305–315 (1982).
211. R. A. Smith, S. M. Cohen, and T. A. Lawson, *Carcinogenesis (London)* **11**, 497–498 (1990).
212. R. S. McLaughlin, *Am. J. Ophthalmol.* **29**, 1355 (1946).
213. E. V. Hort, in *Kirk-Othmer's Encyclopedia of Chemical Technology*, 3rd ed., Vol. 1, Wiley, New York, 1978, pp. 244–276.
213a. OSHA Computerized Information System Database, SLCAC Chemical Sampling Information, Propargyl Alcohol, OSHA SLTC, Salt Lake City, UT.
214. E. H. Vernot et al., *Toxicol. Appl. Pharmacol.* **42**, 417–423 (1977).
215. T. E. Archer, *J. Environ. Sci. Health, Part B* **B20**, 593–596 (1985).
216. The Dow Chemical Company, unpublished data.
217. General Aniline and Film Corporation, Bull. IM-7-61, Ap-51, General Aniline and Film Corp., cited in Ref. 34.
218. K. P. Stasenkova and T. A. Kochekova, *Toksikol, Nov. Prom. Khim. Veschestv* **8**, p. 97, 1966, *Chem. Abstr.* **67**, 89293b (1967).

218a. General Aniline and Film Corporation, unpublished data.

219. BASF, unpublished data, cited in *BG Chemie Toxicological Evaluations*, Vol. 2, Springer-Verlag, Berlin, 1991.

220. R. Rubenstein et al., *Toxicologist* **9**, 248 (Abstr. No. 993) (1989).

221. U.S. Enviornmental Protection Agency (USEPA), *Rat Oral Subchronic Toxicity Study with Propargyl Alcohol*, USEPA, Washington DC, 1989, cited in IRIS Database for Propargyl Alcohol.

222. A. R. Banijamali, et al., *J. Agri. Food Chem.* **47**, 1717–1729 (1999).

223. A. K. Basu and L. J. Marnett, *Cancer Res.* **44**, 2848–2854 (1984).

224. D. H. Blakey, et al., *Mutat. Res.* **320**, 273–283 (1994).

225. Air Products and Chemicals, Inc., *Product Bulletin on Hexynol*, No. 120-042.16. Air Products and Chemicals, 1980.

226. G. H. Riesser, in *Kirk-Othmer's Encyclopedia of Chemical Technology*, 3rd ed., Vol. 5, Wiley, New York, 1978, pp. 848–864.

227. M. W. Goldblatt, *Br. J. Ind. Med.* **1**, 213–223 (1944).

228. H. F. Smyth, Jr., J. Seaton, and L. Fisher, *J. Ind. Hyg. Toxicol.* **23**, 259–268 (1941).

229. W. H. Lawrence, J. E. Turner, and J. Autian, *J. Pharm. Sci.* **60**(4), 568–575 (1971).

230. V. N. Semenova, S. S. Kazanina, and B. Y. Ekshtat, *Hyg. Sanit.* **36**, 376 (1971).

231. H. F. Smyth, Jr. and C. P. Carpenter, *J. Ind. Hyg. Toxicol.* **27**, 93 (1945).

232. H. Dierker, and P. Brown, *J. Ind. Hyg. Toxicol.* **26**, 277–279 (1944).

233. C. P. Carpenter, H. F. Smyth, Jr., and U. C. Pozzani, *J. Ind. Hyg. Toxicol.* **31**, 343–436 (1949).

234. W. L. Guess, *Toxicol. Appl. Pharmacol.* **16**, 382–390 (1970).

235. F. Koelsch, *Zbl. Gewerbehyg. N. F.* 4, 312 (1927). Cited by K. B. Lehmann and F. Flury, *Toxicology and Hygiene of Industrial Solvents*, Williams and Wilkins, Baltimore, 1948, pp 215.

236. B. L. Oser, et al., *Food Cosmet. Toxicol.* **13**, 313–315 (1975).

237. A. M. Ambrose, *Arch. Ind. Hyg. Occup. Med.* **2**, 591–597 (1950).

238. I. Grunow, and H. J. Altmann, *Arch. Toxicol.* **49**, 275–284 (1982).

239. National Toxicology Program (NTP), *Toxicology and Carcinogenesis Studies of 2-Chloroethanol (Ethylene Chlorohydrin) (CAS No. 107-07-3) in F344 Rats and Swiss CD-1 Mice: Dermal Studies*, NTP TR 275, NIH Publ. No. 86-2531, NTP, Washington, DC, 1985.

240. K. D. Courtney, J. E. Andrews, and M. Grady, *J. Environ. Sci. Health, Part B* **B17**(4), 381–391 (1982).

241. J. B. Laborde, et al., *Toxicologist* **2**(1), 71 (Abstr. No. 251) (1982).

242. M. K. Johnson, *Food Cosmet. Toxicol.* **5**, 449 (1967).

243. M. M. Mason, C. C. Cate, and J. Baker, *Clin. Toxicol.* **4**, 185–204 (1971).

244. H. Dunkelberg, *Zentralbl. Bakteriol. Mikrobiol. Hyg. B*, **177**, 269–281 (1983).

245. M. Bignami et al., *Chem. Biol. Interact.* **30**, 9–23 (1980).

246. D. L. J. Opdyke, *Food Cosmet. Toxicol.* **13**, 699 (1975).

247. N. Loprieno et al., *Cancer Res.* **37**, 253–254 (1977).

248. R. Valencia et al., *Environ. Mutagen.* **7**, 325–348 (1985).

249. S. Rosenkrantz, H. S. Carr, and H. S. Rosenkrantz, *Mutat. Res.* **26**, 367–370 (1974).

250. J. L. Ivett et al., *Environ. Mol. Mutagen.* **14**, 165–187 (1989).

251. S. Epstein et al., *Toxicol. Appl. Pharmacol.* **23**, 288–325 (1972).
252. C. W. Sheu et al., *J. Am. Coll. Toxicol.* **3**(2), 221–223 (1983).
253. W. H. Lawrence, et al., *J. Pharm. Sci.* **60**, 1163–1168 (1971).
254. Goldblatt, M. W. and W. E. Chiesman, *Br. J. Ind. Med.* **1**, 207–213 (1944).
255. A. O. Saitanov and A. M. Kononova, *Gig. Tr. Prof. Zabol.* **2**, 49–50 (1976).
256. A. F. Bush, H. K. Abrams, and H. V. Brown, *J. Ind. Hyg. Toxicol.* **31**, 352–358 (1949).
257. E. L. Middleton, *J. Ind. Hyg. Toxicol.* **12**, 265–280 (1930).
258. V. Miller, R. J. Dobbs, and S. I. Jacobs, *Arch. Dis. Child.* **45**, 589–590 (1970).
259. H. L. Greenberg Doctoral Disseration at the New York University Graduate School of Arts and Science, New York, 1988 (University Microfilms International, Ann Arbor, MI).
260. L. O. Bensen and M. J. Teta, *Br. J. Med.* **50**, 710–716 (1993).
261. G. W. Olsen, et al., *Occup. Environ. Med.* **54**, 592–598 (1997).
262. R. S. H. Yang, *Rev. Environ. Contam. Toxicol.* 99, 47–59 (1987).
263. H. Savolainen, K. Pekari, and H. Helojoki, *Chem. Biol. Interact.* **28**, 237–248 (1979).
264. R. D. Phillips, and R. DuBois, *Toxicologist* **1**, 150 (Abstr. No. 544) (1981).
265. I. W. Daly, and R. D. Phillips, *Toxicologist* **1**, 13 (Abstr. No. 48) (1981).
266. National Toxicology Program (NTP), *Toxicology and Carcinogenesis Studies of 1-Chloro-2-propanol (Technical Grade) (CAS NO. 127-004) in F344/N Rats and B6C3F$_1$* Mice (Drinking Water Studies), NTP TR 477, NIH Pub. 98-3967 NIFS PB 99-119240, NTP, Washington, DC, 1998.
267. E. A. Barnsley, *Biochem. J.* **100**, 362–372 (1966).
268. A. R. Jones and J. Gibson, *Xenobiotica* **10**, 835–846 (1980).
269. J. A. Bond, et al., *Toxicol. Appl. Pharmacol.* **95**, 444–455 (1988).
270. National Toxicology Program (NTP) *Final Report on the Reproductive Toxicity of 1-Chloro-2-propanol (CAS No. 127-00-4) in Sprague-Dawley Rats*, Rep. No. NTP-T0197, Order No. PB91-158469, NTP, Washington,DC, 1990.
271. H. S. Carr and H. S. Rosenkranz, *Mutat. Res.* **57**, 381–384 (1978).
272. R. W. Biles and C. E. Piper, *Fundam. Appl. Toxicol.* **3**, 27–33 (1983).

CHAPTER SEVENTY-NINE

Esters of Mono- and Alkenyl Carboxylic Acids and Mono- and Polyalcohols

Michael S. Bisesi, Ph.D., CIH

A INTRODUCTION

Overview

This volume contains three chapters reviewing 12 classes of the organic compounds called *esters*. Chapter 79, this chapter, reviews (*1*) esters of monocarboxylic acids and mono- and polyalcohols and (*2*) esters of alkenyl carboxylic acids and monoalcohols; Chapter 80 reviews (*3*) esters of aromatic monocarboxylic acids and monoalcohols, (*4*) esters of monocarboxylic acids and di-, tri-, and polyalcohol; (*5*) dicarboxylic acid esters; (*6*) alkenyl dicarboxylic esters; (*7*) esters of aromatic diacids; (*8*) tricarboxylic acid esters; and, Chapter 81 covers (*9*) esters of carbonic and orthocarbonic acid; (*10*) esters of organic phosphorous compounds; (*11*) esters of monocarboxylic halogenated acids, alkanols, or haloalcohols; and (*12*) organic silicon esters.

The sequence of the compounds has been organized according to the chemical structure of the major functional metabolites. This involves the ester hydrolyzates, primarily the acid and secondarily the alcohol. The reason for this sequence was the general observation that the degree of toxic effect, in addition to that of the original material, more often was the result of the toxicity of the acid rather than the response of the alcohol.

Esters are important from an industrial hygiene perspective since exposure can occur during the process of manufacturing esters, the process of manufacturing materials

Patty's Toxicology, Fifth Edition, Volume 6, Edited by Eula Bingham, Barbara Cohrssen, and Charles H. Powell.
ISBN 0-471-31939-2

containing or composed of esters, handling and use of products containing or composed of esters, and treatment of wastes containing esters. In turn, exposure to esters is important from a toxicological perspective because of the correlated observations of adverse physiological responses exhibited by laboratory animals and humans.

Overviews of the physical, chemical and toxicologic (*i.e.*, physiologic responses) properties of many subclasses of esters and/or of specific compounds are provided. In addition, summaries of relative manufacturing and use information are also included for many compounds.

General Properties of Some Esters

Chemically, esters are organic compounds commonly formed via the combination of an acid, typically an organic (–COOH) mono- or polyacid, plus a hydroxyl (–OH) group of a mono-or polyalcohol or phenol; water (H–OH) is generated as a by-product of the reaction. For example, the reaction (esterfication) of acetic acid (CH_3COOH) with methyl alcohol (CH_3OH) to form methyl acetate proceeds [$CH_3(C{=}O)OCH_3$] as follows:

$$CH_3COOH + CH_3OH \leftrightarrow CH_3(C{=}O)OCH_3 + H_2O$$

The forward reaction (right; k_1) is known as *esterification*; the reverse (left; k_2) as *hydrolysis*. The occurrence of esterification or hydrolysis reactions depends on the differential reaction rates k_1 *versus* k_2 and the physical properties of the respective final products. The ratio of the concentrations of products $[C]$ of k_1 divided by k_2 will give a reaction constant K_s indicative of the stability of the final product:

$$K_s = \frac{[C_{ester}] \times [C_{H_2O}]}{[C_{alcohol}] \times [C_{acid}]}$$

The esters are widely used in industry and commerce. They can be prepared by the reactions of acids with alcohols, by reacting metal salts of acids with alkyl halides, acid halides with alcohols, or acid anhydrides with alcohols by the interchange of radicals between esters. Most esters exist in liquid form at ambient temperatures, but some possess lower boiling points than their original starting materials. They are relatively water-insoluble, except for the lower molecular weight members. Their flash points are in the flammable range. The monocarboxylic acid esters have high volatility and pleasant odors, whereas the di- and polyacid esters are relatively nonvolatile and exhibit essentially no odor. The monocarboxylic esters occur frequently in natural products, as, for example, in fruits, to which they lend their pleasant odor and taste. Because of the different properties of esters from the original acids and alcohols, esterification can be used for their isolation or to chemically protect specific carboxy or hydroxy functions.

General Toxicity of Some Esters

Absorbed esters and/or metabolites derived from biotransformed esters can initiate toxic effects in some mammalian systems, including humans, and cause adverse physiological responses. Indeed, the underlying causes of physiological responses are due to initial interactions biochemically within a system. Within these chapters, a summary of reviewed literature will reveal that, in general, toxic effects associated with exposure to various esters include primary irritation to ocular, upper and lower respiratory, and dermal systems; depression of the central nervous system (CNS) (*e.g.*, anesthesia, narcosis); dermal hypersensitization; impact to the gastrointestinal (GI), hepatic, and renal systems; abnormal cardiac rhythm; and carcinogenesis. Indeed, these and some additional effects, are based predominantly on rodent studies. A review of the literature reported here, however, indicates that the most commonly reported effects in animals and humans are irritation and, to some extent, CNS depression. Data are reported in this chapter for several classes of esters, including formates (1), acetates (2), acrylates and methacrylates (3), propionates (4), and lactates (5).

Ocular, dermal, respiratory, and even GI irritation is reportedly associated with both the parent ester compounds and the corresponding acid metabolites produced via hydrolytic cleavage reactions (1). Some compounds, such as the aliphatic esters used as lacquer solvents, may cause CNS depression when inhaled in sufficiently high concentrations (6). As expected from Overton theories substantiated by Munch (7) in experiments using rabbits and tadpoles, the more highly water-soluble, lower molecular weight derivatives, such as the methyl and ethyl formates and acetates, are less potent than butyl and amyl acetates. Thus, when Munch used tadpoles to evaluate the potency of aliphatic alcohols and their alkyl esters, he observed a direct relationship between CNS depression and increase in homologous series (7). He concluded that the intact ester was the primary causative agent.

Their anesthetic potency is weaker than that of lower chlorinated hydrocarbons and usually less than that of ethyl ether, but greater than that of ethanol, acetone or pentane. When inhaled, aliphatic esters readily pass through the alveoli, owing to their relatively high solubility in plasma fluid. Those materials with higher water solubility have higher blood–air distribution coefficients and thus presumably reach saturation more slowly. This group of esters appears to be readily hydrolyzed and the resulting alcohols and acids rapidly metabolized.

Most of the aliphatic esters possess some degree of irritation on exposed surfaces. The formates are especially irritating to the eyes and respiratory tract. Ethyl acetates may be irritating at concentrations of 400–800 ppm, whereas ethanol is devoid of effects up to 1000 ppm. The irritant range of butyl acetate, however, coincides with that of its corresponding alcohol. The irritant effect of the higher homolog is a function of the esterified rather than the hydrolyzed material. The local skin effects resemble those of other solvents; namely, defatting and cracking may occur.

Practically all the common aliphatic and aromatic esters, except for some phosphates used as plasticizers, are inert. At the most, minor degrees of irritation may follow inhalation of heated vapors or prolonged skin exposure. Some of the literature also suggests that reported skin sensitization appears more likely in the presence of impurities or side products. Many of the materials are so inert that any LD_{50} value is impractical to determine. Specific pathology is usually absent, even when the materials is fed in massive quantities to the point of nutritional deprivation. Oily or watery excretion products, sometimes observed at high feeding levels, indicate lack of absorption. The apparent nontoxicity may also be a sign of rapid hydrolysis, metabolism, and excretion. The resins are completely inert, unabsorbed in the gastrointestinal tract, and nonirritant at the surface of the skin and pulmonary system.

Industrial Hygiene Evaluation

One part of industrial hygiene evaluation of esters involves collecting and analyzing air samples to determine their airborne concentrations. To date, published industrial hygiene air sampling and analytical methods are available for relatively few esters compared to the number of ester compounds. In addition, although methods for biological monitoring of laboratory animals and exposed humans can be conducted to determine levels of absorbed esters and metabolites that originate from biotransformed esters, there are few established limits for comparing biological monitoring data.

Contemporary air sampling methods most commonly involve the use of solid absorbents, such as charcoal, carbosieve, and XAD (8, 9). Following desorption of the solid adsorbents using a specified solvent, samples are typically analyzed using gas chromatography with flame ionization detection (GC-FID). The concentration of a given sample is determined by dividing the mass of the ester detected and measured using GC-FID by the volume of air sampled. Concentrations of air samples may be subsequently used to calculate time-weighted averages (TWAs) for comparison to applicable occupational exposure limits (OELs).

Presently, there also are relatively few OELs compared to the number of ester compounds. There are threshold limit values (TLVs) and permissible exposure limits (PELs) established by the American Conference of Governmental Industrial Hygienists (ACGIH) and the Occupational Safety and Health Administration (OSHA), respectively, for some of the ester compounds discussed in this chapter (9–11). In addition, other applicable agencies, such as the National Institute for Occupational Safety and Health (NIOSH), provide published recommendations concerning limits for occupational exposure to some esters (9).

Industrial hygiene sampling and analytical methods for some ester compounds are presented in Table 79.1 along with their respective OELs. Since sampling and analytical methods and occupational exposure limits are subject to periodic revision; however, the reader is encouraged to refer to current publications of ACGIH, OSHA, and NIOSH.

B ESTERS OF MONOCARBOXYLIC ACIDS AND MONO- AND POLYALCOHOLS

Grouped into this category are the naturally occurring fatty acid esters of C1–C24 acids and C1–C30 alcohols.

Formates

The esters called *formates* are alkyl or aryl derivatives of formic acid HCOOH. These esters have various uses ranging from flavoring agents to industrial solvents. As is the case for most esters of toxicological significance, ocular, respiratory, and dermal irritation, and, at higher concentrations, CNS depression, are the major effects associated with exposure. The inhalation hazard decreases with increased molecular weight due to an observed progressive decrease in vapor pressure. There also is an observed decrease in water solubility with increasing molecular weight. von Oettingen reports that the decreased water

Table 79.1. Summary of Occupational Exposure Limits (OELs) and Monitoring Methods for Some Esters (8–11)

	OSHA (ppm)			ACGIH (ppm)			NIOSH (ppm)				
Compound	PEL	STEL	C	TLV	STEL	C	REL	STEL	C	OEL Notations	Monitoring
Methyl formate	100	—	—	100	150	—	100	150	—	—	Carbosieve B tube GC-FID
Ethyl formate	100	—	—	100	—	—	100	—	—	—	Charcoal tube GC-FID
Methyl acetate	200	—	—	200	250	—	200	250	—	—	Charcoal tube GC-FID
Ethyl acetate	400	—	—	400	—	—	400	—	—	—	Charcoal tube GC-FID
n-Propyl acetate	200	—	—	200	250	—	200	250	—	—	Charcoal tube GC-FID
Isopropyl acetate	250	—	—	250	310	—	—	—	—	—	Charcoal tube GC-FID
n-Butyl acetate	150	—	—	150	200	—	150	200	—	—	Charcoal tube GC-FID
sec-Butyl acetate	200	—	—	200	—	—	200	—	—	—	Charcoal tube GC-FID
tert-Butyl acetate	200	—	—	200	—	—	200	—	—	—	Charcoal tube GC-FID
Isobutyl acetate	150	—	—	150	—	—	150	—	—	—	Charcoal tube GC-FID
n-Amyl acetate	100	—	—	50[a]	100[a]	—	100	—	—	—	Charcoal tube GC-FID
sec-amyl acetate	125	—	—	125	—	—	125	—	—	—	Charcoal tube GC-FID
Isoamyl acetate	100	—	—	100[a]	—	—	100	—	—	—	Charcoal tube GC-FID
sec-Hexyl acetate	50	—	—	50	—	—	50	—	—	—	Charcoal tube GC-FID
Vinyl acetate	—	—	—	10	15	—	—	—	4	NIOSH carcinogen ACGIH A3	Carbosieve B tube GC-FID
Methyl acrylate	10	—	—	2	—	—	10	—	—	Skin ACGIH A4	Charcoal tube GC-FID
Ethyl acrylate	25	—	—	5	15	—	—	—	—	Skin ACGIH A4	Charcoal tube GC-FID
n-Butyl acrylate	—	—	—	2	—	—	10	—	—	Sensitizer ACGIH A4	Charcoal tube GC-FID
Methyl methacrylate	100	—	—	100	—	—	100	—	—	ACGIH A4	XAD-2 tube GC-FID
Butyl lactate	—	—	—	5	—	—	5	—	—	—	—

[a]= Notice of intended change, changed as of 2000.

solubility is associated with increased resistance to acid hydrolysis of the higher homologues. Accordingly, the lowest homologe, methyl formate, may pose a higher hazard potential because of its high vapor pressure and water solubility, but the higher homologs exhibit higher potency as central nervous system depressants (1). A summary of physical and chemical properties is found in Table 79.2 (12) and summaries of toxicologic data are shown in Tables 79.3 (13–16) and 79.4 (17).

1.0 Methyl Formate

1.0.1 CAS Number: [107-31-3]

1.0.2 Synonyms: Formic acid methyl ester, methyl methanoate, R611

1.0.3 Trade Names: NA

1.0.4 Molecular Weight: 60.05

1.0.5 Molecular Formula: $C_2H_4O_2$

1.0.6 Molecular Structure:

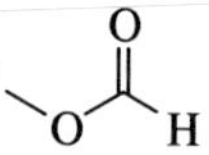

1.1 Chemical and Physical Properties

1.1.1 General

Boiling point: 31.5°C
Freezing point: – 100°C
Specific gravity: 0.987 at 15°C
Vapor pressure: 476 torr at 20°C
Flash point: – 19°C (closed cup)
Solubility: Soluble in water at 30 g/100 g

At ambient temperatures, methyl formate is a colorless, highly volatile water at flammable liquid. Methyl formate can react violently with oxidizing agents (9).

1.1.2 Odor and Warning Properties

Methyl formate has an agreeable odor, but may cause olfactory fatigue (13). One study reported an odor threshold of 600 ppm (18).

1.2 Production and Use

The compound is manufactured by esterification of formic acid and ethanol; heating methanol with sodium formate and hydrochloric acid; or reaction of methanol, carbon dioxide, and steam over a catalyst at high temperature and pressure. Relatively recently, biological synthesis of methyl formate via methylotrophic yeasts that produce methyl formate synthase that catalyses synthesis of the ester from methanol and formaldehyde (19). It is widely used in chemical processes.

Table 79.2. Summary of Physical and Chemical Properties of Some Formates (12)

Compound	CAS Number	Molecular Formula	Molecular Weight	Boiling Point (°C)	Melting Ponint (°C)	Specific Gravity (at 25°C)	Solubility[a] in Water (at 68°F)	Refractive Index (at 20°C)	Vapor Density (Air = 1)	Vapor Pressure [mm Hg (at 20°C)]	Flash Point (°C)	LEL % by Vol. Room Temp.	UEL % by Vol. Room Temp.
Methyl formate	[*107-31-3*]	$C_2H_4O_2$	60.05	31.5	−100	0.987	v	1.3433	2.07	400 (16)	−19	5.0	23
Ethyl formate	[*109-94-4*]	$C_3H_6O_2$	74.08	54.3	−80	0.923	s	1.3598	2.56	200 (21)	−20	2.7	16.0
n-Propyl formate	[*110-74-7*]	$C_4H_8O_2$	88.12	81.3	−92.9	0.91	d	1.3779	3.03	85 (25)	−3	2.3	up
Isopropyl formate	[*625-55-8*]	$C_4H_8O_2$	88.12	68.2	—	0.873	d	1.3678	3.0	—	−5.6	—	—
n-Butyl formate	[*592-84-7*]	$C_5H_{10}O_2$	102.13	106.8	−91.9	0.911	d	1.3912	3.52	30 (25)	17.8	1.7	8.0
Isobutyl formate	[*542-55-2*]	$C_5H_{10}O_2$	102.13	98.4	−95.8	0.89	d	1.3857	—	—	21.0	—	—
n-Amyl formate	[*638-49-3*]	$C_6H_{12}O_2$	116.16	132.1	−73.5	0.893	d	1.3922	4.0	9.6 (25)	26.7	—	—
Isoamyl formate	[*110-45-2*]	$C_6H_{12}O_2$	116.15	124.2	−93.5	0.89	d	1.3976	4.0	—	26.0	—	—
Vinyl formate	[*692-45-5*]	$C_3H_4O_2$	72.08	—	—	—	—	—	2.41	—	21	—	—
Cyclohexyl formate	[*4351-54-6*]	$C_7H_{12}O_2$	128.17	162.5	—	—	i	1.4430	4.4	—	51.1	—	—
Benzyl formate	[*104-57-4*]	$C_8H_8O_2$	136.16	203[b]	—	1.081	i	1.5154	4.7	10 (84)	—	—	—

[a]Solubility in water: v = very soluble; s = soluble; d = slightly soluble; i = insoluble.
[b]At 747 mm Hg.

Table 79.3. Summary of Inhalation Toxicity Data for Some Formates

Compound	Species	Exposure Mode	Approximate Dose or Concentration	Treat-ment Regimen	Observed Effect	Ref.
Methyl-formate	Human	Inhalation	1,500 ppm	1 min	No symptoms	13
	Guinea pig	Inhalation	50,000 ppm	30 min	Lethal	13
	Guinea pig	Inhalation	25,000 ppm	60 min	Lethal	13
	Guinea pig	Inhalation	10,000 ppm	3–4 h	Lethal	13
	Guinea pig	Inhalation	3,500 ppm	8 h	No deaths	13
Ethyl formate	Human	Inhalation	330 ppm	5 min	Eye and nose irritation	13
	Rat	Inhalation	Satd. vap.	5 min	No deaths	14
	Rat	Inhalation	8,000 ppm	4 h	5/6 deaths	1
	Rat	Inhalation	4,000 ppm	4 h	No deaths	15
	Mouse	Inhalation	10,000 ppm	20 min	Eye irritation, dyspnea	16
	Mouse	Inhalation	5,000 ppm	20 min	Eye irritation, dyspnea	16
	Cat	Inhalation	10,000 ppm	80 min	Eye irritation, dyspnea	16
	Cat	Inhalation	5,000 ppm	20 min	Eye irritation, dyspnea	16
	Dog	Inhalation	10,000 ppm	4 h	Pulmonary edema, death	16
n-Butyl-formate	Human	Inhalation	10,000 ppm	< 1 min	Intolerable irritation	16
	Dog	Inhalation	10,000 ppm	60 min	Irritation, narcosis, and recovery	16
	Cat	Inhalation	10,000 ppm	60 min	Irritation, narcosis and death	16

It also has been used as a food additive. In large-scale operations it is a fumigant and larvicide for food and tobacco crops. Other uses include high boiling point refrigerant and solvent.

1.3 Exposure Assessment

Samples of airborne methyl formate can be collected on carbosieve B and desorbed via ethyl acetate. Analysis requires gas chromatography with a flame ionization detector (NIOSH Method S291) (8, 9).

Methyl formate is a metabolic marker of methanol poisoning and is readily identified via gas chromatographic analyses (20).

1.4 Toxic Effects

1.4.1 Experimental Studies

1.4.1.1 Acute Toxicity. Studies observing guinea pigs exposed to 1500–50,000 ppm methyl formate indicated symptoms ranging from ocular and respiratory irritation, narcosis, and death (13). A dose–response relationship was observed in the study and suggested that concentration of 1500–2000 ppm was the maximum tolerated for

Table 79.4. Summary of Oral and Dermal Toxicity Data for Some Formates

Compound	Parameter	Species	Dose (g/kg)	Effect or Time to Death	Ref.
Methyl formate	Oral LD_{LO}	Human	0.500	20–30 min	17
	Oral LD_{50}	Rabbit	1.622	—	7
Ethyl formate	Oral LD_{50}	Rat	4.3	Lowest reported lethal dose	14
	Oral LD_{50}	Rat	1.850	15 min–2h	4
	Oral LD_{50}	Rabbit	2.075	—	7
	Oral LD_{50}	Guinea pig	1.110	—	4
	Dermal LD_{50}	Rabbit	> 20 mL	No absorption	14
Propyl formate	Oral LD_{50}	Rat	3.980	4–18 h	4
	Oral LD_{50}	Mouse	3.4	Few min–6 h	4
n-Butyl formate	Oral LD_{50}	Rabbit	2.656	—	7
Isoamyl formate	Oral LD_{50}	Rat	9.84	Depression immediately following administration	4
		Rabbit	3.0	4 h–4 days	7
Vinyl formate	Oral LD_{50}	Rat	2.820	—	14
	Dermal LD_{50}	Rabbit	3.170	—	14
Allyl formate	Oral LD_{50}	Rat	0.124	—	4

several hours; 5000 ppm, for 1 h; 15,000–25,000 ppm resulted in adverse effects in 30–60 min; and 50,000 was lethal in 20–30 min. (13). The LD_{50} for rabbits was reported at 1600 mg/kg (7).

1.4.1.2 Chronic and Subchronic Toxicity: NA

1.4.1.3 Pharmacokinetics, Metabolism, and Mechanisms: NA

1.4.1.4 Reproductive and Developmental: NA

1.4.1.5 Carcinogenesis: NA

1.4.1.6 Genetic and Related Cellular Effects Studies. Testing of methyl formate via the *Salmonella* assay was reportedly negative for mutagenic activity, and no further testing was conducted (21).

1.4.2 Human Experience

1.4.2.1 General Information. Methyl formate can generate highly irritating vapors at elevated temperatures (1). Physiologically, it is of moderate toxicity when inhaled, but has a narcotic potential (7). It is irritant to the skin and mucous membranes, including the eyes, conjunctiva, and lungs, and it may be directly absorbed. Inhalation of 1500 ppm methyl formate vapor for one minute did not cause irritation (13), but others reported that increasingly higher concentrations cause irritation and may eventually cause convulsions and death (22).

1.4.2.2 Clinical Cases. There is case report of a 19-month-old child who died within 20–30 min after 1 oz of liquid had been applied under a bathcap as a treatment of pediculosis capitis. Pathological findings were nonspecific; however, on autopsy, methyl formate, methanol, and formic acid were detected in the brain and other organs (17). From the industrial use of solvent mixtures containing methyl formate, some CNS effects have been reported, but the corresponding air concentrations were not determined. Inhalation may cause nasal conjunctival irritation, retching, narcosis, and death from pulmonary effects. Visual disturbances have also been recorded, such as amblyopia and nystagmus (13). Exposure of humans in a volunteer study revealed no observed or reported effects at concentrations of 1500 ppm (13).

1.4.2.2.1 Acute Toxicity: NA

1.4.2.2.2 Chronic and Subchronic Toxicity: NA

1.4.2.2.3 Pharmacokinetics, Metabolism, and Mechanisms. The compound can be directly absorbed into the bloodstream via the GI, respiratory, and dermal systems (1). Methyl formate is biotransformed via hydrolysis in the body to yield methanol and formic acid (1).

1.4.2.3 Epidemiology Studies

1.4.2.3.1 Acute Toxicity. Workers exposed to 30% methyl formate combined with ethyl formate and methyl and ethyl acetate exhibited reported and/or exhibited signs of irritation of mucous membranes, dyspnea, some with CNS exhiliration and others with depression, and some with visual disturbances (23).

1.5 Standards, Regulations, or Guidelines of Exposure

OSHA PEL 100 ppm; ACGIH TLV 100 ppm; ACGIH STEL/CEIL (ceiling limit) 150 ppm; NIOSH REL 100 ppm; NIOSH STEL 150 ppm.

1.6 Studies on Environmental Impact: NA

2.0 Ethyl Formate

2.0.1 CAS Number: *[109-94-4]*

2.0.2 Synonyms: Formic acid ethyl ester; ethyl methanoate, formic ether

2.0.3 Trade Names: NA

2.0.4 Molecular Weight: 74.08

2.0.5 Molecular Formula: $C_3H_6O_2$

2.0.6 Molecular Structure:

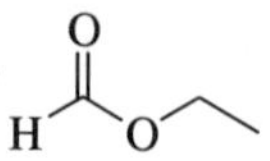

2.1 Chemical and Physical Properties

2.1.1 General

Specific gravity: 0.917 at 20 °C
Melting point: − 80 °C
Boiling point: 54.3 °C
Flash point: − 20 °C closed cup
Vapor pressure: 194 torr at 20 °C
Vapor derisity: 2.56
Solubility: Water solubility 13.6 g/100 mL; very soluble in alcohol, benzene, and other compounds

Ethyl formate is a colorless, combustible liquid. The compound is dangerous when exposed to flame, heat, or oxidizing agents (9).

2.1.2 Odor and Warning Properties

Ethyl formate has a characteristic fruity odor (a pleasant aromatic odor) and bitter taste.

2.2 Production and Use

Ethyl formate is manufactured via esterfication of formic acid and ethanol in presence of sulfuric acid; alternatively, by distillation of ethyl acetate and formic acid in presence of sulfuric acid. It is widely used as a flavoring in foods and beverages. Industrially, it is used in organic syntheses. The larvicidal and fungicidal properties lend applicability to agricultural fumigants and sprays for food crops, cereals, dried fruit, and tobacco. Additional uses of ethyl formate include the manufacture of safety glass, shoes, and artificial silk; a solvent for oils and greases; and as a substitute for acetone.

2.3 Exposure Assessment

2.3.1 Air: NA

2.3.2 Background Levels: NA

2.3.3 Workplace Methods

A sampling procedure (Method 1452) has been recommended by NIOSH (8, 9), that includes sample collection on charcoal, desorption with carbon disulfide, and quantitation by gas chromatography.

2.4 Toxic Effects

2.4.1 Experimental Studies

2.4.1.1 Acute Toxicity. CNS depression is apparent in the lethal dose range for the rat, the mouse, and the cat (13–15). The major response is irritation of mucous membranes, as was shown in experiments using rabbits and guinea pigs (1). Rats survived exposure via inhalation under saturated conditions for 5 min, but exposure was fatal to most rats when exposed for 4 h to 8000 ppm (1). Guinea pigs exposed to airborne ethyl formate vapors exhibited tremors, increased CNS depression, and eventually death (23).

2.4.1.2 Chronic and Subchronic Toxicity: NA

2.4.1.3 Pharmacokinetics, Metabolism, and Mechanisms: NA

2.4.1.4 Reproductive and Developmental: NA

2.4.1.5 Carcinogenesis. A comparison of possible tumorigenic materials when tested by skin application to mice showed no activity with ethyl formate in 10 weeks (24). Intraperitoneal injections of ethyl formate in mice did not cause observable carcinogenic changes in mice (25).

2.4.1.6 Genetic and Related Cellular Effects Studies. There is no evidence of mutagencity based on testing using the yeast *Saccharomyces cerevisiae* (26).

2.4.2 Human Experience

2.4.2.1 General Information. Physiologically, it is an irritant to skin, eyes, and the respiratory system and causes CNS depression, but apparently less CNS effects than does methyl formate (1). Concentrations in excess of 10,500 ppm cause moderate but progressive irritation of the eyes and nose lasting 4 h subsequent to exposure (16).

2.4.2.2 Clinical Cases

2.4.2.2.1 Acute Toxicity: NA

2.4.2.2.2 Chronic and Subchronic Toxicity: NA

2.4.2.2.3 Pharmacokinetics, Metabolism, and Mechanisms. Ethyl formate is highly water-soluble and readily absorbs into the blood via the alveoli of the respiratory system (1). It also is absorbed via the GI system with slight absorption via the dermal system (1). The compound hydrolyzes to irritating formic acid.

2.5 Standards, Regulations, or Guidelines of Exposure

OSHA PEL 100 ppm; ACGIH TLV 100 ppm; NIOSH REL 100 ppm.

2.6 Studies on Environmental Impact: NA

3.0 C4 and Higher Formate Homologs

3.0a Propyl Formate

3.0.1a CAS Number: [110-74-7]

3.0.2a ***Synonyms:*** *n*-Propyl formate, formic acid propyl ester, propyl methonate

3.0.3a ***Trade Names:*** NA

3.0.4a ***Molecular Weight:*** 88.106

3.0.5a ***Molecular Formula:*** $C_4H_8O_2$

3.0.6a ***Molecular Structure:*** 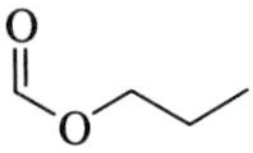

3.0b Isopropyl Formate

3.0.1b ***CAS Number:*** *[625-55-8]*

3.0.2b ***Synonyms:*** Formic acid 1-methylethyl ester, formic acid isopropyl ester

3.0.3b ***Trade Names:*** NA

3.0.4b ***Molecular Weight:*** 88.106

3.0.5b ***Molecular Formula:*** $C_4H_8O_2$

3.0.6b ***Molecular Structure:*** 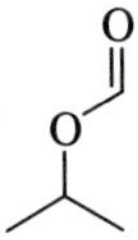

3.0c *n*-Amyl Formate

3.0.1c ***CAS Number:*** *[638-49-3]*

3.0.2c ***Synonyms:*** Amyl formate, pent-1-yl methanoate

3.0.3c ***Trade Names:*** NA

3.0.4c ***Molecular Weight:*** 116.16

3.0.5c ***Molecular Formula:*** $C_6H_{12}O_2$

3.0.6c ***Molecular Structure:*** 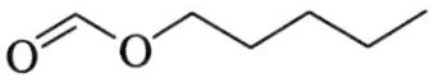

3.0d *n*-Butyl Formate

3.0.1d ***CAS Number:*** *[592-84-7]*

3.0.2d ***Synonyms:*** Butyl formate, formic acid butyl ester

3.0.3d ***Trade Names:*** NA

3.0.4d ***Molecular Weight:*** 102.13

3.0.5d ***Molecular Formula:*** $C_5H_{10}O_2$

3.0.6d ***Molecular Structure:*** 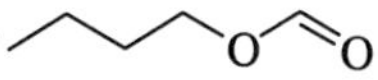

3.0e Isoamyl Formate

3.0.1e CAS Number: *[110-45-2]*

3.0.2e Synonyms: Isopentyl formate, 1-butanol, 3-methyl-, formate

3.0.3e Trade Names: NA

3.0.4e Molecular Weight: 116.16

3.0.5e Molecular Formula: $C_6H_{12}O_2$

3.0.6e Molecular Structure:

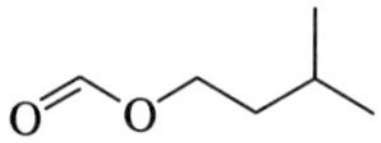

3.0f Vinyl Formate

3.0.1f CAS Number: *[692-45-5]*

3.0.2f Synonyms: Tindal, ethene formate

3.0.3f Trade Names: NA

3.0.4f Molecular Weight: 72.063

3.0.5f Molecular Formula: $C_3H_4O_2$

3.0.6f Molecular Structure:

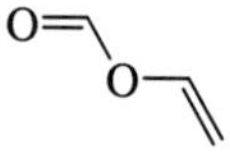

3.0g Allyl Formate

3.0.1g CAS Numbers: *[1838-59-1]*

3.0.2g Synonyms: 2-Propenyl methanoate

3.0.3g Trade Names: NA

3.0.4g Molecular Weight: 86.090

3.0.5g Molecular Formula: $C_4H_6O_2$

3.0.6g Molecular Structure:

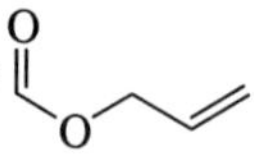

3.0h Isobutyl Formate

3.0.1h CAS Numbers: *[542-55-2]*

3.0.2h Synonyms: Tetryl formate, 2-methylpropyl formate, 2-methyl-1-propyl formate

3.0.3h Trade Names: NA

3.0.4h Molecular Weight: 102.13

3.0.5h Molecular Formula: $C_5H_{10}O_2$

3.0.6h Molecular Structure:

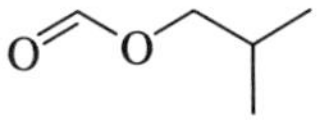

3.1 Chemical and Physical Properties

3.1.1 General

n-Propyl formate is a liquid at ambient temperature and is flammable. Isopropyl is a colorless liquid with a pleasant odor and has physicochemical characteristics similar to those of its normal homolog. Vinyl formate is a clear, flammable liquid. The esters are dangerous when exposed to flame or heat and can react with oxidizing agents. *n*-Butyl and isobutyl formate are clear, flammable liquids. Both chemicals have pleasant odors. The *n*-amyl and isoamyl or pentyl formates are clear liquids. The pentyl formates have pleasant, fruity odors. Additional physicochemical properties are listed for these and other related formate esters in Table 79.2.

3.1.2 Odor and Warning Properties

The C4 and higher formate homologs are characterized as having pleasant odors.

3.2 Production and Use

n-Propyl and isopropyl formates are used as fumigants, and both chemicals are used as raw materials in production. *n*-Butyl and isobutyl formate are used as solvents.

3.3 Exposure Assessment: NA

3.4 Toxic Effects

3.4.1 Experimental Studies

3.4.1.1 Acute Toxicity. Rabbits orally fed *n*-butyl formate via intubation exhibited signs of stupor and loss of voluntary movements. Increasingly higher doses caused disappearance of corneal reflexes, nystagmus, dyspnea, bradycardia, and ultimately death. The dose causing the signs of narcosis in half of the group (ND_{50}) was approximately 1700 mg/kg, and the LD_{50} was reported as approximately 2600 mg/kg (7). From the animal study using rabbits, *n*-propyl formate appeared more irritating and narcotic than its lower homologs (7). An oral LD_{50} for isoamyl formate rat is 9840 mg/kg. The C5 esters appear more highly toxic by ingestion and inhalation than the lower members. High concentrations by acute or chronic exposure may cause narcosis and low CNS effects (7). The allyl formate appears to be much more toxic than its lower homolog and may cause hepatic damage through microvascularization of the liver parenchyma, leading to necrosis of the hepatocytes and progressing to terminal hepatic venule (27). This may be due to the action of acrolein, one of its metabolic products, and the related inhibited incorporation of leucine into hepatic proteins (28). Exposure to the compound may result in elevated levels of amine oxidase (29).

Animal studies using rats and mice were conducted because of concern regarding possible toxicity of chemicals used as food flavorings. Five male and five female rats fasted for approximately 18 h, and mice with full stomachs were treated via intubation with *n*-propyl formate. Treatments caused depression soon after dosing. The LD_{50} for rat was between 3350 and 4750 mg/kg in 14–18 h and for mice 3060–3780 mg/kg in minutes to 6 h (4). Cats exposed via inhalation to 10,418 ppm *n*-butyl formate exhibited increased ocular irritation and salivation, depressed CNS at 20 min, and death due to pulmonary edema at 70 min (16). Cats exposed to a lower concentration of 3941 ppm exhibited increased salivation and irritation followed by staggering in one hour (16). The same researchers reported that a dog exposed to the higher concentration, exhibited ocular irritation, salivation, vomiting, and staggering, but recovered postexposure.

3.4.2 Human Experience

3.4.2.1 General Information. The physiological properties of isopropyl formate are expected to resemble those of *n*-propyl and ethyl formates, except that it may possess increased dermal and eye irritancy (7). Also, as reported in von Oettingen, human experiments by Flury and Neuman showed severe ocular irritation, spasmodic winking, and intolerance after a few inhalations (16).

3.5 Standards, Regulations, or Guidelines of Exposure: NA

3.6 Studies on Environmental Impact: NA

Acetates

The saturated aliphatic acetates, especially the ethyl and butyl acetates, serve as important solvents in the lacquer industry. The aromatic and cyclic acetates are used as flavoring agents for food and scenting of perfumes, soap, and similar articles.

Physiological effects of some are relatively low, since some acetates are, resemble, or convert into natural body metabolites. In both humans and experimental animals, however, administration of excessive quantities produce effects that consist of eye, throat, and nose irritation, followed by gradual onset of narcosis and slow recovery after termination of the exposure (2,6)). Orally administered high concentrations of acetates to rabbits appeared to cause loss of coordination in decreasing order: ethyl = isopropyl > butyl > methyl = isoamyl acetate (30). This may be due to the rapid hydrolysis into acetic acid and the corresponding alcohols, causing a simultaneous decrease of the blood P_{CO_2} and P_{O_2} (30). There is a tendency to acidosis, especially with high concentrations of methyl acetate (30).

No anesthetic symptoms developed in men exposed to 400–600 ppm of ethyl or butyl acetate for 2–3 hrs. Eye irritation has been reported with 200–300 ppm exposure concentrations, but not the characteristic temporary corneal edema, which is caused by the corresponding alcohol (31). An eye injury healed promptly with the C4 but more slowly with the C3 acetate (32). No skin sensitization or dryness has been reported to date. The alkyl acetates possess increased narcotic potential (7), and the C3 and C8 members may have neurotoxic tendencies (33). The cyclic and aromatic acetates produce narcosis and death more readily than do the aliphatic esters. This may be due to the resulting hydrolytic products, which, however, are normally rapidly hydroxylated and excreted (2). For example, phenylacetyglutamine has been found to be excreted daily in 250–500-mg quantities in human urine (34).

A study indicated that thresholds for nasal pungency, odor, and eye irritation decreased logarithmically with the length of carbon chains for acetates, as observed for homologous series of alcohols, and also as seen with narcotic and other toxic responses (35). Data from another study that rated ocular irritation based on corneal thickness suggest that the rating potential for irritation is alcohols > acetates or ketones > aromatics (36).

Physical and chemical properties are summarized in Table 79.5, and toxicologic data are summarized in Tables 79.6 (37–43) and 79.7 (44, 45).

4.0 Methyl Acetate

4.0.1 CAS Number: [79-20-9]

4.0.2 Synonyms: Acetic acid methyl ester; methyl acetic ester, devoton, tereton, methyl ester acetic acid, acetic acid methyl ester

4.0.3 Trade Names: NA

4.0.4 Molecular Weight: 74.08

4.0.5 Molecular Formula: CH_3COOCH_3

4.0.6 Molecular Structure:

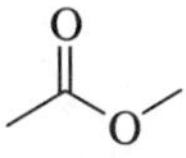

4.1 Chemical and Physical Properties

4.1.1 General

Methyl acetate is a colorless, combustible liquid with an agreeable odor. Methyl acetate is dangerous when exposed to flame, heat or oxidizing agents (9).

4.1.2 Odor and Warning Properties

Methyl acetate has an agreeable odor.

4.2 Production and Use

Methyl acetate occurs naturally in mint, fungus, grapes, bananas, coffee, and nectarines. The compound is manufactured via the esterfication of acetamide with methyl alcohol catalyzed by an acid. Distilled pyroligneous acid derived from wood yields wood spirits, which when refined yield methyl acetate. Uses include solvent for cellulose nitrate, cellulose acetates, resins, and oils in the manufacture of artificial leathers and organic synthesis. Industrially, it is used as a solvent for nitrocellulose and resins, as a catalyst for the biodegradation of organic materials.

Table 79.5. Summary of Physical and Chemical Properties of Some Acetates (12)

Compound	CAS Number	Molecular Formula	Molecular Weight	Boiling Point (°C)	Melting Point (°C)	Specific Gravity at (25 °C)	Solubility[a] in Water at (68 ° F)	Refractive Index at (20 °C)	Vapor density (Air = 1)	Vapor Pressure [(mm Hg) (20°C)]	Flash Point (°C)	LEL % by Vol. Room Temp.	UEL % by Vol. Room Temp.
Methyl acetate	*[79-20-9]*	$C_3H_6O_2$	74.08	56.9	− 98	0.9342	v	1.3593	2.55	100 (9.4)	− 10	3.1	16
Ethyl acetate	*[141-78-6]*	$C_4H_8O_2$	88.10	77	− 83.6	0.902	s	1.3723	3.04	100 (27)	− 4.44	2.2	11.0
n-Propyl acetate	*[109-60-4]*	$C_5H_{10}O_2$	102.13	101.6	− 95	0.887	d	1.3842	3.50	40 (28.8)	14.0	2.0	8
Isopropyl acetate	*[108-21-4]*	$C_5H_{10}O_2$	102.13	89	− 73.4	0.870	s	1.3773	3.52	—	4.44	1.8	8
n-Butyl acetate	*[123-86-4]*	$C_6H_{12}O_2$	116.16	126	− 77	0.882	d	1.3941	4.0	15 (25)	24.0	1.7	7.6
Isobutyl acetate	*[110-19-0]*	$C_6H_{12}O_2$	116.16	116	− 99	0.871	d	1.3902	4.0	—	18.0	2.4	10.5
tert-Butyl acetate	*[540-88-5]*	$C_6H_{12}O_2$	116.16	96	—	—	i	1.3853	4.0	—	—	—	—
n-Amyl acetate	*[628-63-7]*	$C_7H_{14}O_2$	130.18	148.8	− 70	0.879	d	1.4023	4.5	—	25.0	1.1	7.5
Isoamyl acetate	*[624-41-9]*	$C_7H_{14}O_2$	130.19	142	− 78.5	0.870	d	1.4003	4.49	—	25.0	1.0	7.5
2-Amyl acetate	*[626-38-0]*	$C_7H_{14}O_2$	130	134	− 78.5	0.861	i	1.3960	4.5	7 torr at 20°C	26.1	1.0	7.5
n-Hexyl acetate	*[142-92-7]*	$C_8H_{16}O_2$	144.22	338	—	—	i	1.4092	—	—	—	—	—
n-Ocyl acetate	*[112-14-1]*	$C_{10}H_{20}O_2$	172.27	205	− 38.5	—	i	1.4190	—	7.03	82.0	—	—
Vinyl acetate	*[108-05-4]*	$C_4H_6O_2$	86.09	72.3	− 93.2	0.9317	i	1.3959	3.0	115 torr at 25 °C	− 8	2.6	13.4

[a] S = Solubility in water; v = very soluble; s = soluble; d = slightly soluble; i = insoluble.

Table 79.6. Summary of Inhalation Toxicity Data for Some Acetates

Compound	Species	Approximate Dose or Concentration	Treatment Regimen	Observed Effects	Ref.
Methyl acetate	Human	330 ppm	Short	Fruity odor	22
	Human	4,950 ppm	Short	Ocular and respiratory irritation	22
	Human	9,900 ppm	Short	Ocular and respiratory irritation	
	Mouse	55,440 ppm	10–20 min	Immediate irritation, dyspnea, narcosis, lethal from pulmonary edema	37
	Mouse	41,580 ppm	23–35 min	Irritation, dyspnea, convulsion, 1/2 deaths, 3 min postexposure	37
	Mouse	26,400 ppm	31–42 min	Moderate eye irritation, narcosis, 1/2 deaths, 3 h postexposure	37
	Mouse	11,220 ppm	4–5 h	Eye irritation, fatigue, dyspnea, narcosis, lethal 10 h postexposure due to pneumonia	37
	Mouse	7,900 ppm	6 h	1/2 narcosis, irritation, dyspnea, recovery, 1/2 no effects	37
	Mouse	5,000 ppm	20 min	No effect	37
	Cat	53,790 ppm	14–18 min	Irritation, salivation, dyspnea, 1/2 convulsions, narcosis, lethal 1–9 min, later with diffuse pulmonary edema	37
	Cat	34,980 ppm	29–30 min	Irritation, salivation, dyspnea, 1/2 convulsions, narcosis, histology: lateral emphysema or edema	37
	Cat	18,480 ppm	4–4.5 h	Eye irritation, dyspnea, 1/2 vomiting and convulsions, narcosis, slow recovery	37
	Cat	9,900 ppm	10 h	Eye irritation, salivation, somnolence, recovery	37
	Cat	5,000 ppm	20 min	Eye irritation, salivation	37
	Cat	6,600 ppm	6 h / day	Weight loss, weakness, slow recovery	37
	Cat	19,000 ppm	6 h	Narcosis	22
Ethyl acetate	Human	278 ppm	Short	Fruity odor	37
	Human	400 ppm	Short	Irritation of nose and throat	38
	Human	4,170 ppm	Short	Ocular and respiratory irritation	37

Table 79.6. (*Continued*)

Compound	Species	Approximate Dose or Concentration	Treatment Regimen	Observed Effects	Ref.
	Mouse	20,000 ppm	45 min	Toleration of side position	2
	Mouse	12,225 ppm	3 h	10/20 died	39
	Mouse	10,000 ppm	45 min	1/2 lethal, corneal turbidity; 1/4 lethal, immediately. 1/4 in 24 h	2
	Mouse	5,000 ppm	3–4 h	Corneal turbidity	2
	Mouse	2,000 ppm	17 h	Irritation to eyes and nose, dyspnea	2
	Guinea pig	2,000 ppm	65 exposures	No notable effects	6
	Cat	8,000 ppm	20 min	Ocular and respiratory irritation	2
	Cat	9,000 ppm	450 min	Irritation and moderate dyspnea	2
	Cat	43,000 ppm	14–16 min	Deep narcosis, death	2
	Cat	20,000 ppm	45 min	Deep narcosis, recovery	2
	Cat	12,000 ppm	5 h	Lowest narcotic concentration	22
Vinyl acetate	Rat	4,000 ppm	4	Lowest lethal dose	40
n-Propyl acetate	Human	240 ppm	Short	Ocular, nose, pharyngeal irritation	22
	Human	4,595	Short	Ocular and respiratory irritation	22
	Cat	24,000 ppm	0.5 h	Within 5–16 min assumes side position; 13–18 min narcosis; 1/4 death 4 days post-exposure	2
	Cat	7,400 ppm	5.5 h	Staggering within 30–45 min; deep narcosis 4.5–5.5 h; 1/4 death after 5.5 h	2
	Cat	5,300 ppm	6 h/day	Moderate irritation, salivation	2
Isopropyl acetate	Human	200 ppm	—	Eye irritation	41
	Rat	Concentrated vapor	> 30 min	Lethal	14
	Rat	32,000 ppm	4 h	Lethal 5/6 animals	2
n-Butyl acetate	Human	3,300 ppm	Brief	Marked irritation to eyes and nose	42
	Human	200–300 ppm	Brief	Mild irritation to eyes and nose	2
	Mouse	7,400 ppm	3 h	Narcosis, recovery	2
	Guinea pig	14,000 ppm	4 h	Eye irritation, narcosis, lethal	42
	Guinea pig	7,000 ppm	13 h	Eye irriation, deep narcosis, recovery	42

Table 79.6. (*Continued*)

Compound	Species	Approximate Dose or Concentration	Treatment Regimen	Observed Effects	Ref.
	Guinea pig	3,300 ppm	13 h	Eye irritation, no other symptoms	42
	Cat	17,500 ppm	30 min	Narcosis, lethal to some	2
	Cat	12,000 ppm	30 min	Narcosis, recovery	2
	Cat	4,200 ppm	6 h/6 days	Weakness, loss of weight, minor blood changes	2
	Cat	900 ppm	65 experiments, 6 h/day	Weakness	6
Isobutyl acetate	Rat	21,000 ppm	150 min	Narcosis, lethal, 6/6	22
	Rat	3,000 ppm	6 h	No symptoms	22
Isomyl acetate	Human	950 ppm	30 min	Irritation of nose and throat, headache, weakness	2
	Dog	5,000 ppm	1 h	Nasal irritation, drowsiness	2
	Cat	7,200 ppm	24 h	Light narcosis, delayed death due to pneumonia	2
	Cat	4,000 ppm	20 min	Irritation to eyes and nose	2
n-Amyl acetate	Human	200 ppm	30 min	Lowest irrirated dose	38
	Cat	2,182 ppm	215 min	Salivation, no other effects	2
	Cat	10,600 ppm	115 min	Marked salivation, lacrimation, irregular respiration, loss of reflexes after 85 min	2
sec-Amyl acetate	Human	1,000 ppm	1 h	Serious toxic effects	2
	Guinea pig	10,000 ppm	5 h	Eye and nose irritation, narcosis, lethal	2
	Guinea pig	5,000 ppm	13 h	Eye and nose irritation, narcosis recovered	2
	Guinea pig	2,000 ppm	13 h	Eye and nose irritation, no narcosis, recovered	2
n-Hexyl acetate	Rat	4,000 ppm	4 h	Lowest lethal dose	2
Cyclohexyl acetate	Human	516 ppm	Brief	Irritation to eyes and throat	2
	Rabbit	700 ppm	4.8 h	Irritation to nose and eyes, recovery	43
	Rabbit	1,700 ppm	4.8 h	Lethal	43
	Cat	1,700 ppm	10 h	Deep narcosis and death	43
	Cat	860 ppm	9 h	Irritation plus light narcosis	2
	Cat	1,600 ppm	8 h/5 days		2
	Cat	637 ppm	8 h/30 days	No symptoms	2

Table 79.7. Summary of Oral, Dermal, Subcutaneous, and Intraperitoneal Toxicity Data for Some Acetates

Compound	Route of Entry	Parameter Determined	Species Tested	Dose or Concentration	Ref.
Methyl	Oral	LD_{50}	Rabbit	3.7 g/kg	7
		LD_{LO}	Guinea pig	3.0 g/kg	44
	SC	LD_{LO}	Cat	3.0 g/kg	7
		TLm96	Aquatic	1000–100 ppm	17
Ethyl	Oral	LD_{LO}	Rat	11 g/kg	14
		LD_{50}	Rabbit	4.94 g/kg	7
	SC	LD_{50}	Guinea pig	3.0 g/kg	44
		LD_{50}	Cat	3.0 g/kg	44
		TLm96	Aquatic	1000–100 ppm	17
n-Propyl	Oral	LD_{50}	Rabbit	6.64 g/kg	7
	SC	LD_{50}	Guinea pig	3.0 g/kg	44
		LD_{LO}	Cat	3.0 g/kg	44
		TLm96	Aquatic	1000–100 ppm	17
Isopropyl	Oral	LD_{50}	Rabbit	6.95 g/kg	7
		TLm96	Aquatic	>1000 ppm	17
Cyclohexyl	Oral	LD_{50}	Rat	6.73 g/kg	
	Dermal	LD_{50}	Rabbit	10.1 g/kg	
Phenyl	Oral	LD_{50}	Rat	1.63 g/kg	45

4.3 Exposure Assessment

4.3.1 Air: NA

4.3.2 Background Levels: NA

4.3.3 Workplace Methods

Air samples of methyl acetate can be collected on silica gel or activated charcoal. Analysis can be conducted via a gas chromatograph with a flame ionization detector (NIOSH Method 1458) (8, 9).

4.4 Toxic Effects

4.4.1 Experimental Studies

4.4.1.1 Acute Toxicity. Some animal experiments have shown severe effects at high concentrations (Tables 79.6 and 79.7).

4.4.1.2 Chronic and Subchronic Toxicity: NA

4.4.1.3 Pharmacokinetics, Metabolism, and Mechanisms: NA

4.4.1.4 Reproductive and Developmental: NA

4.4.1.5 Carcinogenesis: NA

4.4.1.6 Genetic and Related Cellular Effects Studies. Methyl and ethyl acetate have been shown to induce aneuploidy in yeast cells and porcine brain cells (27) and are negative in *Salmonella* test (Zeiger, E., Anderson, B., Haworth, S., and Mortelman, K. Salmonella Mutagenecity Tests V. Results from the Testing of 311 Chemicals. Environ. Molec. Mutagen. 19 (Suppl 21) (1992): 2–141).

4.4.2 Human Experience

4.4.2.1 General Information. At high concentrations, methyl acetate may cause mild to severe methanol intoxication from ingestion, inhalation, or possibly skin contact (2). The vapor is mildly irritating to the eyes and respiratory system and at high concentrations can cause CNS depression.

4.4.2.2 Clinical Cases

4.4.2.2.1 Acute Toxicity. A teenage girl experienced acute blindness following inhalation of vapors from lacquer thinner. It was determined that methanol and methyl acetate vapors caused optic neuropathy that led to the blindness (28).

4.4.2.2.2 Chronic and Subchronic Toxicity: NA

4.4.2.2.3 Pharmacokinetics, Metabolism, and Mechanisms. Tada et al. (29) have shown with human subjects that the metabolic hydrolysis of methyl acetate to methanol and acetic acid proceeds directly proportional to the exposure level.

4.4.2.3 Epidemiology Studies

4.4.2.3.1 Acute Toxicity: NA

4.4.2.3.2 Chronic and Subchronic Toxicity: NA

4.4.2.3.3 Pharmacokinetics, Metabolism, and Mechanisms. Methyl acetate is readily absorbed through the lungs.

4.5 Standards, Regulations, or Guidelines of Exposure

OSHA PEL 200 ppm; ACGIH TLV-200 ppm; ACGIH STEL 250 ppm; NIOSH REL 200 ppm; NIOSH STEL 250 ppm.

4.6 Studies on Environmental Impact: NA

5.0 Ethyl Acetate

5.0.1 CAS Number: *[141-78-6]*

5.0.2 Synonyms: Acetic acid ethyl ester, ethyl acetic ester, acetoxyethane, acetic ether, vinegar naphtha, acetidin, acetic ester

5.0.3 Trade Names: NA

5.0.4 Molecular Weight: 88.10

5.0.5 Molecular Formula: $C_4H_8O_2$

5.0.6 Molecular Structure:

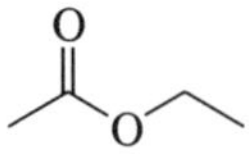

5.1 Chemical and Physical Properties

5.1.1 General

Ethyl acetate is a clear, volatile, and flammable liquid.

5.1.2 Odor and Warning Properties

Ethyl acetate has a fruity odor and an odor threshold of 3.9 ppm (18).

5.2 Production and Use

Ethyl acetate occurs naturally in yeast and sugarcane, where it is photosynthetically produced. A fundamental manufacturing method is a sulfuric acid–catalyzed reaction between ethanol and acetic acid. In the atmosphere it is photolytically degraded. Ethyl acetate has multiple uses including food flavoring agent, fragrance, and solvent. It finds utilization for specialities, such as, contact lenses. The chemical also is released by molds and is present in shoe and leather glues.

5.3 Exposure Assessment

5.3.1 Air: NA

5.3.2 Background Levels: NA

5.3.3 Workplace Methods

An industrial hygiene sampling procedure has been recommended by NIOSH (NIOSH Method 1457) (8). Innumerable analytical determination methods are available that mainly utilize gas chromatographic or spectrometric techniques.

5.4 Toxic Effects

5.4.1 Experimental Studies

Animal experiments showed as acute inhalation LC_{50} of 200.0 mg/L for the rat (46). When ethyl acetate was administered to dogs at blood ethanol concentrations of < 3 mg/mL, the heart rate, systemic arterial pressure, and myocardial contractile force increased slightly, but decreased at higher levels, with a concurrent increase in pulmonary arterial pressure (47). In the guinea pig, the myocardial depressant action was 10 times greater than that of ethanol (48). Ethyl acetate strongly stimulated glucose and sodium transport in the isolated small intestine of the hamster (49). Metabolic studies in the rat have revealed an approximate 2000 ppm no-effect level (50). At higher levels, the rate of hydrolysis of ethyl acetate appeared to exceed ethanol oxidation, leading to its accumulation in the vascular system. Also, when it was injected intraperitoneally at 1.6 g/kg, hydrolysis to acetic acid and ethanol occurred rapidly (50). Intraperitoneal injections of 1 mL/kg to male rats for 8 days increased the blood pyruvic and lactic acid content considerably and also elevated the glycolytic enzymatic activity (51). Ethyl acetate in combination with toluene appeared to produce a mixture of lower toxicity than that of either compound alone. Mixing of ethyl acetate with propylene oxide, propylene glycol, or formalin appears to decrease its LD_{50} value, but its toxicity increases in combination with morpholine, ethylene glycol, or ethyl alcohol (52).

A dose–effect relationship was suggested on the basis of an experimental study measuring various parameters of workers' well-being (*e.g.*, annoyance, tiredness, tension) in relation to exposures to various solvents, including ethyl acetate (53).

5.4.1.1 Acute Toxicity. Ethyl acetate causes ocular and respiratory irritation (2, 54). Acute neurobehavioral effects were observed in mice that exhibited decreased locomotor activity when exposed to 2000 ppm ethyl acetate vapor. In addition, there were marked changes in a functional observational battery profile, including posture, decreased arousal, increased tonic/clonic movements, disturbances in gait, delayed righting reflexes, and increased sensorimotor reactivity (55).

5.4.1.2 Chronic and Subchronic Toxicity: NA

5.4.1.3 Pharmacokinetics, Metabolism, and Mechanisms: NA

5.4.1.4 Reproductive and Developmental: NA

5.4.1.5 Carcinogenesis: NA

5.4.1.6 Genetic and Related Cellular Effects Studies. Cytogenetic toxicological testing was negative for *in vitro* chromosome aberrations but positive for sister chromatid exchange (SCE) (55a) and negative for *Salmonella* (93).

5.4.2 Human Experience

5.4.2.1 General Information. Ocular and respiratory irritation are most common effects (2, 22). At higher concentrations, especially at elevated temperatures, vapors have shown intoxicating properties involving the CNS (2).

5.4.2.2 Clinical Cases

5.4.2.2.1 Acute Toxicity: NA

5.4.2.2.2 Chronic and Subchronic Toxicity: NA

5.4.2.2.3 Pharmacokinetics, Metabolism, and Mechanisms. Ethyl acetate is hydrolyzed to the corresponding alcohol and acid, and ethanol is eliminated via a combination of exhaled air, urination, and biotransformation (2). Human studies using exposure chamber concentrations of 217–770 ppm showed a linear relationship between inhaled ethyl acetate and ethanol in the alveolar air (56). In a comparison of solvents the respiratory retention of ethyl acetate was greater than that of toluene > trichloroethylene > benzene > acetone > ethanol > hexane (57), whereas the respiratory elimination was practically negligible for ethyl acetate when compared with the other above solvents (58).

5.5 Standards, Regulations, or Guidelines of Exposure

OSHA PEL 400 ppm; ACGIH TLV-400 ppm; NIOSH REL 400 ppm.

6.0 Propyl and Isopropyl Acetates

6.0a *n*-Propyl Acetate

6.0.1a CAS Number: *[109-60-4]*

6.0.2a Synonyms: Propyl acetate, 1-acetoxypropane, 1-propyl acetate; acetic acid propyl ester

6.0.3a Trade Names: NA

6.0.4a Molecular Weight: 102.13

6.0.5a Molecular Formate: $C_5H_{10}O_2$

6.0.6a Molecular Structure:

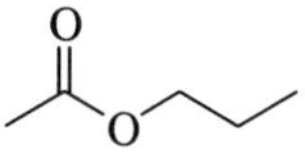

6.0b Isopropyl Acetate

6.0.1b CAS Number: *[108-21-4]*

6.0.2b Synonyms: *sec*-Propyl acetate, acetic acid isopropyl ester, isopropyl ester of acetic acid, 2-acetoxypropane, (2-propyl)acetate

6.0.3b Trade Names: NA

6.0.4b Molecular Weight: 102.13

6.0.5b Molecular Formula: $C_5H_{10}O_2$

6.0.6b Molecular Structure:

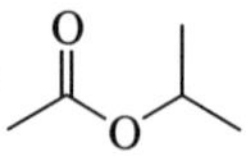

6.1 Chemical and Physical Properties

n-Propyl acetate:

Specific gravity: 0.887 at 20°C
Melting point: −95°C
Boiling point: 101.6°C

Flash point: 14°C (closed cup)
Vapor pressure: 25 torr at 20°C
Solubility: Slightly soluble in water

Isopropyl acetate:

Specific gravity: 0.870
Melting point: −73.4°C
Boiling point: 89°C
Flash point: 11
Vapor density: 3.5
Solubility: 2.90 g/100 mL; soluble in water

6.1.1 General

The isomers are colorless, flammable liquids. *n*-Propyl acetate is dangerous when exposed to heat or flame and reacts with oxidizing agents (9).

6.1.2 Odor and Warning Properties

The isomers have a fruity odor.

6.2 Production and Use

n-Propyl acetate and isopropyl acetate are used as flavoring or fragrant agents and as solvents for resins, cellulose derivatives, and plastics. Other uses for formulating lacquers, insecticides, waxes, and inks. The compound is manufactured via a reaction of acetic acid and a mixture of propene and propane with a zinc catalyst.

6.3 Exposure Assessment

6.3.1 *Air:* NA

6.3.2 *Background Levels:* NA

6.3.3 Workplace Methods

Air sampling is conducted using charcoal for adsorption. Analysis consists of desorption using carbon disulfide followed by gas chromatography using a flame ionization detector (NIOSH Methods 1450 and 1454) (8, 9).

6.4 Toxic Effects

6.4.1 Experimental Studies

6.4.1.1 Acute Toxicity. Animal studies show that at high concentrations the C3 acetates are irritating to the skin, eyes, and mucous membranes (2). Isopropyl acetate may produce corneal burns, but these heal within 3–10 days (32). Takagi and Takyangi observed that exposures caused contractions of the ileum in the guinea pig following release of the neurotransmitter acetylcholine (33). Short-term inhalation exposures to individual solvents, including isopropyl acetate, caused concentration-dependent behavioral changes in mice based on a measurement of reduced immobility in a "behavioral despair" swimming test (59).

6.4.1.2 Chronic and Subchronic Toxicity. Isopropyl acetate was shown to be cytotoxic to local tissues and caused severe, if not fatal, lung damage when infused into a central vein in rats (60).

6.4.2 Human Experience

6.4.2.1 General Information. Reported toxicological effects associated with human exposures to the propyl acetates include irritation of ocular and upper respiratory systems, constriction of lungs and coughing, and CNS depression (22).

6.5 Standards, Regulations, or Guidelines of Exposure

n-Propyl acetate: OSHA PEL 200 ppm, ACGIH TLV 200 ppm; ACGIH STEL 250 ppm; NIOSH REL 200 ppm; NIOSH-STEL 250 ppm. Isopropyl acetate: OSHA PEL 250 ppm; ACGIH TLV-250 ppm; ACGIH STEL 310 ppm. NIOSH does not believe that the limit set by OSHA is protective enough. NIOSH intended changes; TLV-TWA, 100 ppm and STEL/C of 200 ppm.

7.0 Butyl Acetates

7.0a Butyl Acetate

7.0.1a CAS Number: *[123-86-4]*

7.0.2a Synonyms: 1-Butyl Acetate; n-Butyl acetate; acetic acid n-butyl ester

7.0.3a Trade Names: NA

7.0.4a Molecular Weight: 116.16

7.0.5a Molecular Formula: $C_6H_{12}O_2$

7.0.6a Molecular Structure:

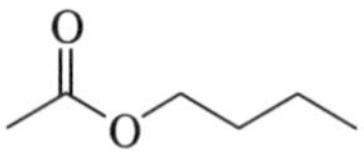

7.0b Isobutyl Acetate

7.0.1b CAS Number: *[110-19-0]*

7.0.2b Synonyms: 2-Methylpropyl acetate; acetic acid isobutyl ester; beta-methylpropyl ethanoate; 2-Methyl-1-propyl acetate; natural isobutyl acetate

7.0.3b Trade Names: NA

7.0.4b Molecular Weight: 116.16

7.0.5b Molecular Formula: $C_6H_{12}O_2$

7.0.6b Molecular Structure: 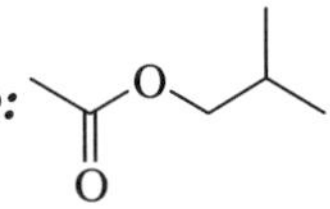

7.0c *sec*-Butyl Acetate

7.0.1c CAS Number: *[105-46-4]*

7.0.2c Synonyms: DL-*sec*-Butyl acetate; *sec*-Butyl Acetate, 98%; acetic acid 1-methylpropyl ester

7.0.3c Trade Names: NA

7.0.4c Molecular Weight: 116.16

7.0.5c Molecular Formula: $C_6H_{12}O_2$

7.0.6c Molecular Structure: 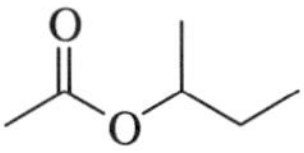

7.0d *tert*-Butyl Acetate

7.0.1d CAS Number: *[540-88-5]*

7.0.2d Synonyms: Acetic acid, 1,1-dimethylethyl ester; *tert*-Butyl acetate, 98%

7.0.3d Trade Names: NA

7.0.4d Molecular Weight: 116.16

7.0.5d Molecular Formula: $C_6H_{12}O_2$

7.0.6d Molecular Structure: 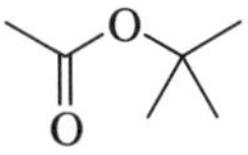

7.1 Chemical and Physical Properties

Butyl acetate:

Specific gravity: 0.8826 at 20°C
Melting point: −77°C
Boiling point: 126°C
Flash point: 24°C (closed cup); 37°C (open cup)
Vapor density: 4.0
Solubility: 0.68 g/100 mL; slightly soluble in water

Isobutyl acetate:

Specific gravity: 0.874
Melting point: −99°C
Boiling point: 116°C
Flash point: 18°C
Vapor density: 4.0
Solubility: 0.67 g/100 mL; slightly soluble in water

sec-Butyl acetate:

Specific gravity: 0.872
Melting point: −99°C
Boiling point: 112°C
Flash point: 16°C
Vapor density: NA
Solubility: 0.80 g/100 mL

tert-Butyl acetate:

Specific gravity: 0.862
Melting point: NA
Boiling point: 96°C
Flash density: 15°C
Vapor density: 4.0
Solubility: Insoluble in water

7.1.1 General

The compounds are colorless, flammable liquids.

7.1.2 Odor and Warning Properties

n-Butyl-, isobutyl, and *tert*-butyl acetate have fruit-like odors. *n*-Butyl acetate has a reported odor threshold between 7 and 20 ppm (21).

7.2 Production and Use

The *n*-butyl, isobutyl, and *tert*-butyl acetates occur naturally among numerous other propionates and butyrates in bananas and other fruit, such as apples. The four isomers are manufactured via the esterification of their respective alcohols with acetic acid or acetic anhydride. *n*-Butyl acetate is also biosynthesized during fermentation processes in yeasts. They are used as flavoring agents, fragrances, synthetic raw materials, and solvents and larvicides. In addition, butyl acetate is used as solvents for nitrocellulose-based lacquers, inks, and adhesives. Other uses include manufacture of artificial leathers, photographic film, safety glass, and plastics. Butyl acetate is also used in fabrication of integrated circuits in the semiconductor industry.

7.3 Exposure Assessment

7.3.1 Air: NA

7.3.2 Background Levels: NA

7.3.3 Workplace Methods

NIOSH has published procedures for the air sampling and determination of several acetates: *n*-butyl, *sec*-butyl, isobutyl, and *tert*-butyl derivatives. Air samples can be collected using charcoal adsorption. Carbon disulfide is used for desorption and gas chromatography with a flame ionization detector for analysis (8, 9).

7.4 Toxic Effects

7.4.1 Experimental Studies

7.4.1.1 Acute Toxicity. Acute neurobehavioral effects were identified in mice following 20-min inhalation exposures to 2000 ppm ethyl acetate and 8000 ppm n-butyl acetate. Effects included changes in posture, decreased arousal, increased tonic/clonic movements, disturbances in gait, delayed righting reflexes, and increased sensorimotor activity. Recovery, however, was rapid and began soon after exposure was discontinued (55). Another study involving exposures of rats and mice to butyl alcohol, *n*-butyl acetate, and a mixture there of caused concentration-dependent effects. The individual solvents and a 5 : 50 mixture caused disturbances in rotarod performances in rats. Median effective concentrations (EC_{50}) were 7559 ppm *n*-butyl alcohol, 8339 ppm *n*-butyl acetate, and 10,672 ppm mixture. The individual solvents and the mixture also caused decreased sensitivity to pain, based on a hot-plate behavior test (61).

7.4.1.2 Chronic and Subchronic Toxicity: NA

7.4.1.3 Pharmacokinetics, Metabolism, and Mechanisms: NA

7.4.1.4 Reproductive and Developmental. An exposure study involving pregnant rabbit and rats exposed to 1500 ppm *n*-butyl acetate vapor during the early days of gestation (1–19 and 1–16, respectively) reportedly caused decreased maternal body weight and suggested some degree of embryotoxicity in rats based on decreased size of rat fetuses (21).

7.4.2 Human Experience

7.4.2.1 General Information. All four compounds exhibit irritating and narcotic effects. It has been reported that n-butyl acetate may be irritating to the ocular and upper respiratory systems at concentrations of 300 ppm and dermatitis also may occur with direct contact (21, 62). Acute human data consist of a report concerning exposure to polyester lacquer, with signs and symptoms manifested as nervous, respiratory, and cardiovascular system disorders. Recovery occured, however, following symptomatic therapy (63). Some experimental data on the guinea pig are presented by Sayers et al. (42). These show the increased lethal effect with rising chamber concentrations and accompanying symptoms occurring at a more rapid rate. Injury from direct skin contact was observed to be of low magnitude and to heal within 1 day (32). However, epidemiologic data are difficult to evaluate, since most reports deal with butyl acetate occuring in solvent mixtures.

7.4.2.2 Clinical Cases

7.4.2.2.1 Acute Toxicity. Rating scales, measures of acute irritation of eyes, and measurement of pulmonary function were conducted following human exposures to airborne concentrations of 350, 700, and 1050 mg/m^3 in 20-min sessions, 70 and 1400 mg/m^3 in 20-min sessions, and 70 and 700 mg/m^3 in 4-h sessions. The results suggested that *n*-butyl acetate caused only very slight irritation (64). Indeed, respiratory and ocular irritation is the major toxicological effect associated with exposures to the four isomers.

7.4.2.2.2 Chronic and Subchronic Toxicity. The USEPA selected *n*-butyl acetate as a chemical of concern relative to their "multisubstance rule for testing of neurotoxicity." Accordingly, a study was conducted involving rats exposed to concentrations of *n*-butyl acetate vapor of 0, 500, 1500, or 3000 ppm for 6 h/day for 65 exposures over 14 weeks. The results indicated transient sedation and hypoactivity at concentrations of 1500 and 3000 ppm. However, on the basis of all tests, including functional observational battery, motor activity, neurohistopathology, and schedule-controlled operant behavior endpoints, there was no indication of cumulative neurotoxicity (65).

7.4.2.3 Epidemiology Studies

7.4.2.3.1 Acute Toxicity. Workers from six paint factories in Taiwan were studied to determine the relationship between exposure to airborne solvent vapors, including toluene, xylene, *n*-hexane, methyl isobutyl ketone, and *n*-butyl acetate, and indications of neurobehavioral effects. Results indicated that the exposures were associated with prolonged response latencies in tests of continuous performance, pattern comparison, and pattern memory (66).

7.4.2.3.2 Chronic and Subchronic Toxicity: NA

7.4.2.3.3 Pharmacokinetics, Metabolism, and Mechanisms. Biotransformation via hydrolysis of the butyl acetates yields acetic acid plus the corresponding alcohol (2). The excreted mammalian metabolite of butyl acetate has been determined as 4-hydroxy-3-methoxy- or vanilemandelic acid, now known to be a normal or a pathological constituent of urine (67).

7.4 Standards, Regulations, or Guidelines of Exposure

n-Butyl acetate: OSHA PEL 150 ppm; ACGIH TLV-150 ppm; ACGIH STEL 200 ppm; NIOSH REL 150 ppm; NIOSH STEL 200 ppm *sec*-Butyl acetate: OSHA PEL 200 ppm; ACGIH TLV 200 ppm; NIOSH REL 200 ppm

tert-Butyl acetate: OSHA PEL 200 ppm; ACGIH TLV-200 ppm; NIOSH REL 200 ppm

Isobutyl acetate: OSHA PEL 150 ppm; ACGIH TLV-150 ppm; NIOSH REL 150 ppm

7.6 Studies on Environmental Impact: NA

8.0 Amyl Acetates

8.0a *n*-Amyl Acetate

8.0.1a CAS Number: *[628-63-7]*

8.0.2a Synonyms: *n*-Pentyl acetate, acetic acid pentyl ester

8.0.3a Trade Names: NA

8.0.4a Molecular Weight: 130.18

8.0.5a Molecular Formula: $C_7H_{14}O_2$

8.0.6a Molecular Structure:

8.0b Isoamyl Acetate

8.0.1b CAS Number: *[624-41-9]*

8.0.2b Synonyms: Isopentyl acetate

8.0.3b Trade Names: NA

8.0.4b Molecular Weight: 130.19

8.0.5b Molecular Formula: $C_7H_{14}O_2$

8.0.6b Molecular Structure:

8.1 Chemical and Physical Properties

8.1.1 General: NA

8.1.2 Odor and Warning Properties

n-Amyl acetate has a low odor threshold ($<$ 1 ppm) and has a pear-like odor (21). Isoamyl acetate has more of a banana-like odor and a reported odor threshold of 7 ppm (21).

8.2 Production and Use

n-Amyl acetate occurs naturally in various fruit, such as the volatile aroma of banana oil, where it is biosynthesized from L-leucine. Manufacturing involves esterification of the alcohol with acetic acid. Isoamyl acetate is used industrially and commercially as an aerosol additive. It is also used in qualitative respirator fit testing in the "banana oil" test. Uses include solvent for paints and lacquers, photographic film, nail polish, a warning odor in glues, printing finishing fabrics; prespotting in dry cleaning, repellent against dogs and cats for ornamental vegetation; and insecticide and miticide to combat wasps and bees.

8.3 Exposure Assessment

8.3.1 Air: NA

8.3.2 Background Levels: NA

8.3.3 Workplace Methods

Charcoal is a common air sampling medium. Gas chromatography is utilized for analysis (NIOSH Method 1450) (8, 9).

8.4 Toxic Effects

8.4.1 Experimental Studies

8.4.1.1 Acute Toxicity. Rare cases of damage to the optic nerve have been reported (62). Amyl acetate was found to be of low toxicity (6). Only guinea pigs, which expired during exposure to very high concentrations of *sec*-amyl acetate, exhibited slight congestion of the brain and marked effect on systemic organs (68). Exposure to 5000 ppm for 13.5 h produced slight congestion of the lungs and 9200 and 18,000 ppm, moderate congestion of brain, lungs, liver, and kidneys immediately after a 30-min exposure (68). No gross pathological effects were noted at 2000 ppm. Additionally, a study using guinea pigs inhaling 320 mg/L showed differential brain waves, which returned to normal within 45 days (69).

8.4.1.2 Chronic and Subchronic Toxicity: NA

8.4.1.3 Pharmacokinetics, Metabolism, and Mechanisms: NA

8.4.1.4 Reproductive and Developmental:. NA

8.4.1.5 Carcinogenesis: NA

8.4.1.6 Genetic and Related Cellular Effects Studies. In a mutagencity assay using the yeast *Saccharomyces cerevisiae*, the methyl esters of propionic and butyric acid and acetic acid ester of *n*- and isopropanol and ethyl propionate were weak inducers of aneuploidy, whereas, *n*-butyl and isoamyl acetate and ethyl formate were not mutagenic (70). Another study involving exposures to mice investigated the acute neurobehavioral effects of ethyl, *n*-butyl, and amyl acetates. The results indicated that ethyl and *n*-butyl acetates caused decreased locomotor activity but amyl acetate did not cause observable effects (55).

8.4.2 Human Experience

8.4.2.1 General Information

8.4.2.2 Clinical Cases

8.4.2.2.1 Acute Toxicity. Inhalation studies indicated that amyl acetate caused eye, nose, and throat irritation in humans (38). Indeed, it appears that irritation is the major toxicological effect associated with exposure to amyl acetates.

8.4.2.2.2 Chronic and Subchronic Toxicity. In a subchronic study, subtle transient hematological effects were observed (71). Chronic worker exposure caused upper respiratory and neurotoxic effects (72).

8.4.2.3 Epidemiology Studies

8.4.2.3.1 Acute Toxicity: NA

8.4.2.3.2 Chronic and Subchronic Toxicity. These compounds also have been found to be a marginal dermal sensitizer (73).

8.5 Standards, Regulations, or Guidelines of Exposure

n-Amyl acetate: OSHA PEL 100 ppm; ACGIH TLV 50 ppm; NIOSH REL 100 ppm
sec-Amyl acetate: OSHA PEL 125 ppm; ACGIH TLV 50 ppm; NIOSH REL 125 ppm
Isoamyl acetate: OSHA PEL 100 ppm; ACGIH TLV 50 ppm; NIOSH REL 100 ppm

For all three compounds ACGIH STEL 100 ppm.

8.6 Studies on Environmental Impact: NA

9.0 Hexyl Acetates

9.0a *n*-Hexyl Acetate

9.0.1a CAS Number: *[142-92-7]*

9.0.2a Synonyms: Hexyl acetate, Acetate C-6, Hexyl Acetate (Nat. C-6 Acetate), 2-methylpentyl acetate

9.0.3a Trade Name: NA

9.0.4a Molecular Weight: 144.21

9.0.5a Molecular Formula: $C_8H_{16}O_2$

9.0.6a Molecular Structure:

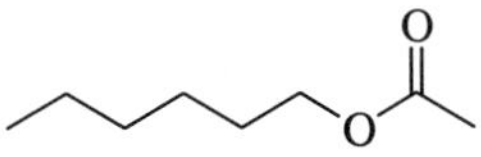

9.0b *sec*-Hexyl Acetate

9.0.1b CAS Number: [108-84-9]

9.0.2b Synonyms: Methyl isoamyl acetate, 2-methylpentyl acetate, 2-methylamyl acetate, methyl isobutyl carbinol acetate, 4-methyl-2-pentanol acetate, 1,3-dimethylbutyl acetate, methyl amyl acetate, 4-methyl-2-pentyl acetate; acetic acid 4-methyl-2-pentyl ester

9.0.3b Trade Name: NA

9.0.4b Molecular Weight: 144.21

9.0.5b Molecular Formula: $C_8H_{16}O_2$

9.0.6b Molecular Structure:

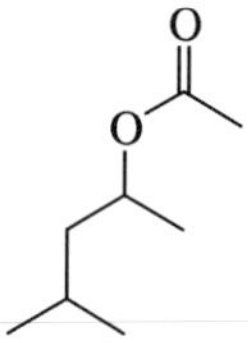

9.1 Chemical and Physical Properties

9.1.1 General

sec-Hexyl acetate, $CH_3COOCH\text{-}(CH_3)(CH_2)_3CH_3$, is a clear liquid with a pleasant odor.

9.2 Production and Use

These compounds also occur naturally in apples and their leaves. They are used as fragrances is the perfume and cosmetics industry, and as components for housefly insecticides, and as chemical components for the large spruce bark beetle attractant (74).

9.3 Exposure Assessment

NIOSH Method 1450 is recommended (8).

9.4 Toxic Effects

9.4.1 Experimental Studies

A range-finding study by Carpenter et al. (95) lists an oral LD_{50} for the rat of 41.5 mL/kg, a dermal LD_{50} of >20 mL/kg for the rabbit, and the maximum no-effect time for concentrated vapor of 8 h for the rat. It was moderately irritant to the rabbit skin and corneal injury was very low (75).

9.4.2.1 General Information

9.5 Standards, Regulations, or Guidelines of Exposure

sec-Hexyl acetate: OSHA PEL 50 ppm; ACGIH TLV 50 ppm; NIOSH REL 50 ppm.

9.6 Studies on Environmental Impact: NA

10.0 Vinyl and Allyl Acetates

10.0a Vinyl Acetate

10.0.1a CAS Number: *[108-05-4]*

10.0.2a Synonyms: Ethenyl acetate, ethenyl ethanoate, acetic acid ethenyl ester, acetic acid ethylene ether, vinyl acetate monomer, acetic acid vinyl ester

10.0.3a Trade Name: NA

10.0.4a Molecular Weight: 86.090

10.0.5a Molecular Formula: $C_4H_6O_2$

10.0.6a Molecular Structure: 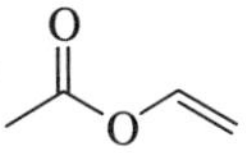

10.0b Allyl Acetate

10.0.1b CAS Number: *[591-87-7]*

10.0.2b Synonyms: 2-Propenyl ethanoate; 3-acetoxypropene, 2-propenyl acetate

10.0.3b Trade Names: NA

10.0.4b Molecular Weight: 100.12

10.0.5b Molecular Formula: $C_5H_8O_2$

10.0.6b Molecular Structure: 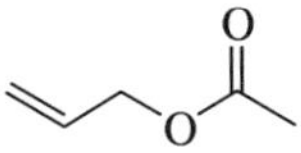

10.1 Chemical and Physical Properties

10.1.1 General

Vinyl acetate is a clear but highly flammable liquid. It may react vigorously with silica gel or alumina and oxidizers such as peroxides (9).

10.2 Production and Use

Vinyl acetate represents the most important member of the unsaturated acetates. The compound is manufactured via a reaction between ethylene and sodium acetate. It polymerizes easily to form inert polyesters, which are widely used as plastic copolymers. Other uses include manufacture of films, lacquers, paints, and safety glass.

10.3 Exposure Assessment

10.3.1 Air: NA

10.3.2 Background Levels: NA

10.3.3 Workplace Methods

Air samples can be collected using a carbosieve adsorbent. Samples can be analyzed via gas chromatography using a flame ionization detector (NIOSH Methods 1453) (8, 9).

10.4 Toxic Effects

10.4.1 Experimental Studies

10.4.1.1 Acute Toxicity. Vinyl acetate exhibits a relatively low toxicity as a monomer and more so as a polymer. An oral LD_{50} of 2.92 g/kg for the rat is reported (76). The inhalation range-finding LD_{50} was about 4000 ppm for a 4-h exposure (76). Vinyl acetate is a low eye irritant, but injuries have been observed to heal within 48 h (32). It defats the skin and is narcotic in high concentrations. Chronic 24-h inhalation of 2.4 mg/m^3 vinyl acetate by rats resulted in subtle hepatic changes of the enzyme rhythms, but were more pronounced with 9.2 and 68 mg/m^3 (77). Allyl acetate, is an irritant and toxic by inhalation, ingestion, eye, and dermal contact (76). The oral LD_{50} for the rat is 130 mg/kg and via percutaneous absorption, in 1100 mg/kg (76). LD_{50} for the rat via inhalation is 1000 ppm in 1 h (76). It is also absorbed through the intact skin. An inhalation LC_{50} of 1000 ppm/h has been recorded (78). Isopropenyl acetate, has been recorded to exhibit an oral LD_{50} in the rat of 3000 mg/kg (78). Exposure of mice to concentration of 2.9 ppm allyl acetate caused sensory irritation marked by 50% decreased respiratory rate (RD_{50}), but did not cause pulmonary irritation. The same study showed other propene derivatives to cause sensory irritation, but acrolein was considered the most potent (79). Allyl alcohol esters, including allyl propionate, allyl formate, and allyl acetate, were shown to be effective as snail fumigants. The relative potencies were allyl propionate $>$ allyl acetate $>$ allyl formate when tested at concentrations of 0–0.03 mM (80).

10.4.1.2 Chronic and Subchronic Toxicity: NA

10.4.1.3 Pharmacokinetics, Metabolism, and Mechanisms: NA

10.4.1.4 Reproductive and Developmental. Rodents were exposed to 0, 200, 1000, or 5000 ppm vinyl acetate in drinking water or exposed to airborne vinyl acetate vapor concentrations of 0, 50, 1200, or 1000 ppm 6 h/day with both routes equivalent to approximately 0, 25, 100, or 500 mg/kg per day. Drinking-water ingestion did not cause observable maternal or developmental toxicity. Inhalation caused maternal toxicity, based on decreased mean weight gain of dams exposed to 1000 ppm. Also, fetal toxicity was suggested on the basis of decreased mean fetal weight, decreased mean crown–rump length, and increased minor skeletal alterations in fetuses exposed to 1000 ppm (81).

10.4.1.5 Carcinogenesis. Another study involving rats consuming drinking water containing 0, 200, 1000, or 5000 ppm vinyl acetate (equivalent to 0, 10, 47, and 202 (mg)(kg)/day for male rats and 16, 76, and 302 (mg) (kg)/day for female rats) for $\leq$ 104 weeks from gestation, however, showed no evidence of target organ toxicity or oncogenic activity (82). The same laboratory conducted another study involving rodents exposed to airborne concentrations of vinyl acetate of 0, 50, 200, and 600 ppm for $\leq$ 104 wk. Oncogenic effects were observed in rats exposed to 200 and 600 ppm and included endo-and exophytic papillomas, squamous-cell carcinoma, carcinoma in olfactory regions, and endophytic papilloma in respiratory regions. One mouse from the 600-ppm exposure group had squamous cell carcinoma in the lung (83). Another drinking water study involving rats indicated that 6 of 20 female rats that consumed water containing 2500 mg/L vinyl acetate for 2 years developed thyroid cancer, and five developed uterine cancer; neither effects were shown in control groups (84).

10.4.1.6 Genetic and Related Cellular Effects Studies. Vinyl acetate caused cytotoxicity in rat nasal tissues following *in vitro* exposure to 50–200 mM, but not 25 mM, for 2 h and DNA–protein crosslinking in rat nasal epithelial cells exposed *in vitro* to 5–75 mM vinyl acetate for 1 to 2 h (85). The toxicity is thought to be related to carboxylesterase hydolysis of vinyl acetate, which generates the active cytotoxic acetic acid and DNA–protein crosslinking acetaldehyde metabolites. Indeed, another study indicated increased sister chromatid exchange and structural chromosome aberrations in cultured human lymphocytes and chinese ovarian cells following treatment with vinyl acetate. The active metabolite of vinyl acetate hydrolysis, acetaldehyde, was thought to be the causative agent (86).

Vinyl esters, including vinyl formate, vinyl chloroformate, vinyl propionate, vinyl crotonate, and vinyl-2-ethylhexanoate, were evaluated for inducing sister chromatid exchange (SCE) in human lymphocytes treated for 48 h. The vinyl formate, vinyl propionate, and vinyl crotonate exhibited dose-dependent SCE at concentrations of 0.125–0.5 mM, vinyl chloroformate at 0.063–1 mM, and vinyl 2-ethylhexanoate at 0.25–4 mM (87).

10.4.2 Human Experience

10.4.2.1 General Information. It has been reported that workers exposed to vinyl acetate exhibited gradual deterioration of cardiac muscle, arrythmias, abnormal ECG, fainting, and chest pain (88).

10.5 Standards, Regulations, or Guidelines of Exposure

Vinyl acetate: ACGIH TLV 10 ppm; ACGIH STEL 15 ppm; ACGIH A3 "Confirmed animal carcinogen with unknown relevance to humans"; NIOSH C 4 ppm.

10.6 Studies on Environmental Impact: NA

11.0 Cyclohexyl Acetate

11.0.1 CAS Number: *[622-45-7]*

11.0.2 Synonyms: Hexalin acetate; acetic acid cyclohexyl ester

11.0.3 Trade Name: NA

11.0.4 Molecular Weight: 142.20

11.0.5 Molecular Formula: $C_8H_{14}O_2$

11.0.6 Molecular Structure:

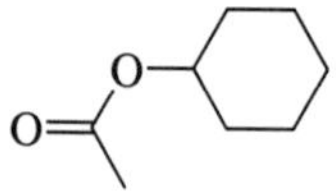

11.1 Chemical and Physical Properties

11.1.1 General

Cyclohexyl acetate is a colorless, oily, flammable liquid.

11.2 Production and Use

The compound is manufactured by direct esterification of cyclohexanol; alternatively, heating alcohol with acetic anhydride in presence of sulfuric acid. It is used as a flavoring agent for foods and an industrial solvent. The C1–C5 allyl homologues occur naturally and some related derivatives are utilized as insect attractants.

11.3 Exposure Assessment: NA

11.4 Toxic Effects

11.4.1 Experimental Studies

11.4.1.1 Acute Toxicity. Lehmann (43) published some human and more detailed animal data in an attempt to relate the data to industrial exposures. He observed 4.8-h exposures at 4 mg/L to be irritant to nose and eyes of the rabbit, from which the animal recovered entirely, although 9.0 mg/L was lethal. To the cat, 10.5 h exposures at 4 mg/L resulted in nose and eye irritation and were also lethal at 9.0 mg/L.

11.4.2 Human Experience

11.4.2.1 General Information. It is a low skin and eye irritant and may have systemic effects when chronically inhaled. A human exposure for 45 min at 3.0 mg/L showed eye and throat irritation a sweet taste following the exposure time (43). It is more highly toxic than the aliphatic C6 and other related derivatives.

11.5 Standards, Regulations, or Guidelines of Exposure: NA

11.6 Studies on Environmental Impact: NA

12.0 Benzyl Acetate

12.0.1 CAS Number: [140-11-4]

12.0.2 Synonyms: Benzyl acetate acetic acid phenyl methyl ester, acetic acid benzyl ester, phenylmethyl acetate, acetic acid phenylmethyl ester, α-acetoxytoluene, acetoxymethyl benzene, acetic acid benzyl ester

12.0.3 Trade Names: NA

12.0.4 Molecular Weight: 150.18

12.0.5 Molecular Formula: $C_9H_{10}O_2$

12.0.6 Molecular Structure:

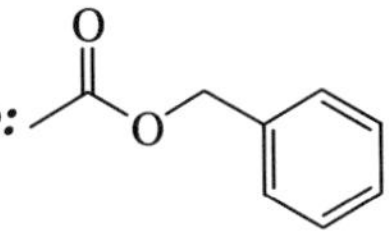

12.1 Chemical and Physical Properties

12.1.1 General

Benzyl acetate is a colorless liquid.

12.1.2 Odor and Warning Properties

Benzyl acetate has an agreeable pear-like odor.

12.2 Production and Use

Benzyl acetate occurs in a number of plants, particularly in jasmine, hyacinth, gardenia, and alfalfa. Accordingly, it is used in the perfume industry and also as a solvent, especially for cellulose acetate and nitrate.

12.3 Exposure Assessment: NA

12.4 Toxic Effects

12.4.1 Experimental Studies

12.4.1.1 Acute Toxicity. Mice exposed to benzyl acetate vapor at airborne concentration of 212 ppm for 7–13 hours developed dyspnea, progressing narcosis, and slowed respiration. One mouse died durign exposure, five postexposure, and only one survived (89). The mice died from hypermia and pulmonary edema. Guinea pigs showed no signs of irritation, but cats showed increased salivation, lacrimation, and tremors when exposed to 10 mg/m^3 benzyl acetate mist (89).

12.4.1.2 Chronic and Subchronic Toxicity: NA

12.4.1.3 Pharmacokinetics, Metabolism, and Mechanisms

12.4.1.3.1 Absorption. Snapper et al. (46) reported that benzyl acetate is rapidly hydrolyzed to acetic acid and benzyl alcohol, and the latter oxidized to benzoic acid and subsequently excreted as hippuric acid (90). Studies with rats showed that significant lung, GI, and dermal absorption of benzyl acetate occurs (2).

12.4.1.5 Carcinogenesis. A 2-year gavage study exposing rats and mice to benzyl acetate showed increased incidence of acinar-cell adenomas of the pancreas in male rats, but the corn oil gavage vehicle may have contributed to the results (91). Increased incidences of hepatocellular adenomas and squamous cell neoplasms of the forestomach were observed in mice (90). A dose–feed study using similar doses as the gavage study did not result in hepatic tumors (92), but both studies revealed necrotic brain lesions in mice. A study comparing gavage against dose feed administration of benzyl acetate to rats and mice was conducted. Gavage administration of benzyl acetate caused much higher plasma concentrations of the metabolite benzoic acid compared to comparable dose fed groups, indicationg that gavage caused saturation of the benzoic acid metabolic pathway whereas dose feed administration did not. The authors concluded that their observations may explain why different toxicity and carcinogenicity responses are observed in 2 y gavage and dose feed studies.

12.4.2 Human Experience

12.4.2.1 General Information. Benzyl acetate, when ingested, can cause general intestinal irritation, and it is also irritating to the eyes, skin, and respiratory system. The compound has been shown to cause decreased blood pressure and depth of respiration, but increased cardiac rate (94–96).

12.5 Standards, Regulations, or Guidelines of Exposure: NA

12.6 Studies on Environmental Impact: NA

13.0 Acetoacetates

13.0a Methyl Acetoacetate

13.0.1a CAS Number: *[105-45-3]*

13.0.2a Synonyms: Butanoic acid, 3-oxo methyl ester, 3-oxobutanoic acid methyl ester; acetic acid methyl ester

13.0.3a Trade Names: NA

13.0.4a Molecular Weight: 116.12

13.0.5a Molecular Formula: $C_5H_8O_3$

13.0.6a Molecular Structure: 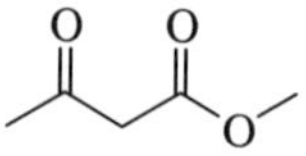

13.0b Ethyl Acetoacetate

13.0.1b CAS Number: *[141-97-9]*

13.0.2b Synonyms: EAA, acetoacetic ester, ethyl 3-oxobutanoate, ethyl acetylacetate, ethyl beta-ketobutyrate, 3-oxobutanoic acid ethyl ester

13.0.3b Trade Names: NA

13.0.4b Molecular Weight: 130.14

13.0.5b Molecular Formula: $C_6H_{10}O_3$

13.0.6b Molecular Structure:

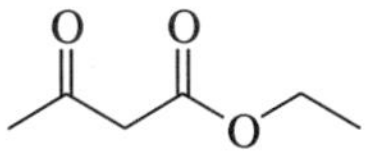

13.1 Chemical and Physical Properties

13.1.1 General

Methyl-and ethyl acetoacetate are colorless liquids.

13.1.2 Odor and Warning Properties

Both acetoacetates have an agreeable odor, and ethyl acetoacetate has a characteristic fruity or rum odor.

13.2 Production and Use

Methyl acetoacetate is manufactured via a reaction of methyl acetate with sodium methoxide. The compound is commonly used as a solvent for cellulose ethers and esters. Ethyl acetoacetate is manufactured through a reaction of high purity ethyl acetate with sodium, followed by neutralization with sulfuric acid. It is used as a flavoring agent for food, and, as a chemical intermediate for yellow pigment in paints, lacquers, inks, and dyes.

13.3 Exposure Assessment: NA

13.4 Toxic Effects

13.4.1 Experimental Studies

13.4.1.1 Acute Toxicity. The toxicity of methyl acetoacetate is low. Like other esters, it is potentially skin-absorbable. An oral LD_{50} in the rat of 3.0 g/kg and a dermal LD_{50} for the rabbit of > 10 ml/mg have been reported (40). An 8-h exposure to saturated vapor was not lethal to rats (40). This level is slightly to moderately irritating to the eyes and skin.

Ethyl acetoacetate is less toxic than its methyl homologue, based on an oral LD_{50} in the rat of 3.98 g/kg and a dermal LD_{50} in the rabbit of > 10 ml/mg (40). Tests conducted on rabbits indicated the potential for ocular irritation, which was especially evident from corneal damage (62). It is also irritating to the respiratory and GI tracts.

13.4.1.2 Chronic and Subchronic Toxicity: NA

13.4.1.3 Pharmacokinetics, Metabolism, and Mechanisms

13.4.1.3.1 Adsorption. Although methyl acetoacetate is similar to other acetates, hydrolysis and metabolic degradation occur rapidly, especially since acetoacetic acid is a common mammalian metabolite.

13.4.2 Human Experience

13.4.2.1 General Information. Methyl-and ethyl acetoacetate may be irritating.

13.5 Standards, Regulations, or Guidelines of Exposure: NA

13.6 Studies on Environmental Impact: NA

13.7 Propionates and Higher Fatty-Acid Esters

In general, these compounds exhibit low toxicity. Physical and chemical properties are summarized in Table 79.8, and toxicologic data are summarized in Table 79.9 (97).

14.0 Propionates

14.0a Methyl Propionate

14.0.1a CAS Number: *[554-12-1]*

14.0.2a Synonyms: Methylester propanoic acid, methyl propylate, propanoic acid methyl ester

14.0.3a Trade Names:

14.0.4a Molecular Weight: 88.106

14.0.5a Molecular Formula: $C_4H_8O_2$

14.0.6a Molecular Structure: 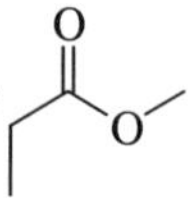

14.0b Ethyl Propionate

14.0.1b CAS Number: *[105-37-3]*

14.0.2b Synonyms: Propanoic acid ethyl ester, ethyl *n*-proponoate, ethyl propionate (Nat C-3 Ethyl Ester)

14.0.3b Trade Names:

14.0.4b Molecular Weight: 102.13

14.0.5b Molecular Formula: $C_5H_{10}O_2$

14.0.6b Molecular Structure: 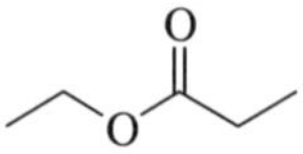

14.0c Ethyl 3-Ethoxypropionate

14.0.1c CAS Number: *[763-69-9]*

Table 79.8. Summary of Physical and Chemical Properties of Some Fatty Acid Esters (12)

Compound	CAS Number	Molecular Formula	Molecular Weight	Boiling Point (°C)	Melting Point (°C)	Specific Gravity at (25°C)	Solubility[a] in Water at (68°F)	Refractive Index at (20°C)	Vapor density (Air = 1)	Vapor Pressure [mm Hg (20°C)]	Flash Point (°C)	LEL % by Vol. Room Temp.	UEL % by Vol. Room Temp.
Methyl propionate	[*554-12-1*]	$C_4H_8O_2$	88.10	79.85	−87.5	0.910	d	1.3775	3.03	40 (11.0)		2.5	13.0
Ethyl propionate	[*105-37-3*]	$C_5H_{10}O_2$	102.13	210	−99.4	0.891	d	1.3839	3.52	40 (27.2)	12	1.9	11.0
Ethyl 3-ethoxy-propionate	[*763-69-9*]		146.19	170.1	−75	0.95	—	—	5.03	—	—	—	—
Methyl butyrate	[*623-42-7*]	$C_5H_{10}O_2$	102.13	102.3	−95	0.8721	d	1.3878	3.53	40 (29.6)	14	—	—
Methyl isobutyrate	[*547-63-7*]	$C_5H_{10}O_2$	102.13	92.3	−84.7	0.8930	d	1.3840	3.5	50 (24)	13	—	—
Ethyl butyrate	[*105-54-4*]	$C_6H_{12}O_2$	116.16	252	−135.4	0.879	d	1.4000	4.0	20 (28)	26	—	—
Ethyl isovalerate	[*108-64-5*]	$C_7H_{14}O_2$	130.19	271	−146.2	0.868	—	1.3962	—	—	—	—	—
Ethyl caproate	[*123-66-0*]	$C_8H_{16}O_2$	144.22	168	−67	0.873	i	1.4073	—	—	—	—	—
Ethyl enanthate	[*106-30-9*]	$C_9H_{18}O_2$	158.24	372	−86.8	0.868	d	1.4100	—	—	—	—	—
Ethyl caprylate	[*106-32-1*]	$C_{10}H_{20}O_2$	172.27	208.5	−47	—	i	1.4178	—	—	—	—	—
Ethyl pelargonate	[*123-29-5*]	$C_{11}H_{22}O_2$	186.30	119	−36.7	0.865	i	1.4220	—	—	—	—	—
Ethyl caprate	[*110-38-3*]	$C_{12}H_{24}O_2$	200.33	245	−20	0.862	i	1.4256	—	—	—	—	—
Ethyl crotonate	[*623-70-1*]	$C_6H_{10}O_2$	114.14	143	+45	0.92	i	1.4243	3.39	—	—	—	—
Vinyl crotonate	[*14861-06-4*]	$C_6H_8O_2$	112.13	134	—	0.94	—	—	4.0	—	25.6	—	—

[a] Solubility in water: v = very soluble; s = soluble; d = slightly soluble; i = insoluble.

Table 79.9 Summary of Toxicity Data for Some Propionates, Butyrates, and Higher Esters

Compound	Species	Route of Entry	Parameter	Result (g/kg)	Time to Death	Ref.
			Propionate			
Methyl	Rabbit	Oral	LD_{50}	2.02		7
Ethyl	Rabbit	Oral	LD_{50}	5.71		7
n-Propyl	Rabbit	Oral	LD_{50}	3.94		7
Isobutyl	Rabbit	Oral	LD_{50}	5.6		7
Isoamyl	Rabbit	Oral	LD_{50}	6.9		7
Ethylethoxy	Rat	Oral	LD_{50}	5.0		97
		Dermal	LD_{50}	>10.0		97
			Butyrate			
Methyl	Rabbit	Oral	LD_{50}	3.38		7
Ethyl	Rat	Oral	LD_{50}	13.0		4
	Rabbit	Oral	LD_{50}	5.23	4–18 h	7
n-Propyl	Rat	Oral	LD_{50}	15.0	1–3 days	4
Amyl	Rat	Oral	LD_{50}	12.2	Few min–2 h	4
	Guinea pig	Oral	LD_{50}	11.95	2 h–6 days	4
Allyl	Rat	Oral	LD_{50}	.25	4 years–5 days	4
Linalyl iso-	Rat	Oral	LD_{LO}	>36.3		4
	Mouse	Oral	LD_{LO}	15.1	4 h–4 days	4
Benzyl	Rat	Oral	LD_{LO}	2.33	4 h–4 days	4
Pentyl pentanoate	Rat	Oral	LD_{50}	>35.4		4
	Guinea pig	Oral	LD_{50}	>17.3	2–6 days	4
Allyl heptanoate	Rat	Oral	LD_{50}	0.50	4–18 h	4
	Guinea pig	Oral	LD_{50}	0.44	4 h–3 days	4
Ethyl caprylate	Rat	Oral	LD_{50}	25.9	4 h–4 days	4
Ethyl nonanoate	Rat	Oral	LD_{50}	>43.0		4
	Guinea pig	Oral	LD_{50}	>24.2		4
Butyl stearate	Rat	Oral	LD_{50}	>32		4
	Rat	IP	LD_{50}	>32		4

14.0.2c Synonyms: Ethyl B-ethoxypropionate

14.0.3c Trade Names: NA

14.0.4c Molecular Weight: 146.19

14.0.5c Molecular Formula: $C_7H_{14}O_3$

14.0.6c Molecular Structure:

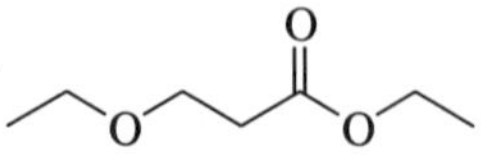

14.1 Chemical and Physical Properties

14.1.1 General

Methyl propionate is a colorless, flammable liquid. Ethyl propionate is a colorless, flammable liquid with a strong rum, pineapple-like odor. Ethyl 3-ethoxypropionate is also a colorless liquid.

14.2 Production and Use

Methyl propionate is prepared by esterification of proprionic acid or proprionic anhydride with methanol. The compound is used as a solvent for cellulose derivatives, coatings, paints, lacquers, varnishes; as a flavoring for food, and also as a fragrance. Methyl propionate is used mainly in organic synthesis. Ethyl propionate is manufactured via esterification of ethanol with proprionic acid or proprionic anhydride. It is used as a solvent for various natural and synthetic resins and for fragrance and food flavoring. In addition, it is used clinically as a solvent to dissolve gallstones (98).

14.3 Exposure Assessment: NA

14.4 Toxic Effects

14.4.1 Experimental Studies

14.4.1.1 Acute Toxicity Propionates are of Low Toxicity. An oral LD_{50} in the rabbit has been determined at 2.02 g/kg (7). Acute toxicity testing using *Daphnia* indicated an 24-h EC_{50} of 516 mg/L and a 21-day no-observed-effect level (NOEL) of 6.3 mg/L (99). Signs and symptoms of ataxia, gasping respiration, and hypothermia occurred at lethal dose levels (7).

Since ethyl propionate is used clinically to dissolve gallstones, there is concern for potential leakage of the solvent to surrounding tissues. A study was conducted using rabbits to determine whether ethyl propionate adversely affected the gallbladder and adjacent tissues. The results indicated that ethyl propionate causes less submucosal inflammation than does methyl *tert*-butyl ether, which is also used, but it did cause more biochemical liver injury when administered intraduodenally (98). In another study involving piglets, ethyl propionate caused moderate injury to the gallbladder and bile ducts, but was not toxic to the liver (100). An oral LD_{50} for ethyl propionate in the rabbit was 5.71 g/kg (8). Symptoms were similar to those observed with methyl propionate, but induced some acidosis, probably due to hydrolytically formed propionic acid. Other reported symptoms following exposure to ethyl propionate include ataxia, gasping respiration, and hypothermia at lethal dose levels. Acute toxicity testing using *Daphnia* indicated an 24-h EC_{50} of 286 mg/L and a 21-day NOEL of 6.3 mg/L (99).

Ethyl 3-ethoxypropionate has a reported oral LD_{50} of 5.0 g/kg in the rat and dermal LD_{50} in the rabbit of about 10 ml/kg, and an 8-h saturated vapor inhalation study in the rat caused no deaths (97). Ethyl-3-ethoxypropionate is metabolized in rat generating 3-ethoxypropionate and monoethyl maleate in urine, in addition to malonic acid and the glycine conjugate of 3-ethoxypropionate (101).

14.5 Standards, Regulations, or Guidelines of Exposure: NA

14.6 Studies on Environmental Impact

Environmental concerns to control volatile organic compounds and toxicity have resulted in development of *n*-butyl-and *n*-pentyl propionate containing compounds for use in the coatings industry because of their high solvency, good odor characteristics, and high electrical resistivity (102).

15.0 Butyrates

15.0a Methyl Butyrate

15.0.1a CAS Number: *[623-42-7]*

15.0.2a Synonyms: Methyl *n*-butanoate, butanoic acid methyl ester

15.0.3a Trade Names: NA

15.0.4a Molecular Weight: 102.13

15.0.5a Molecular Formula: $C_5H_{10}O_2$

15.0.6a Molecular Structure:

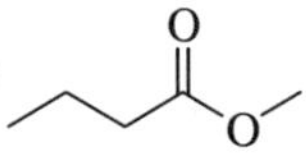

15.0b Methyl Isobutyrate

15.0.1b CAS Number: *[547-63-7]*

15.0.2b Synonyms: Methyl 2-methylpropanoate, methyl isobutrate, isobutyric acid butyl ester

15.0.3b Trade Names: NA

15.0.4b Molecular Weight: 102.13

15.0.5b Molecular Formula: $C_5H_{10}O_2$

15.0.6b Molecular Structure:

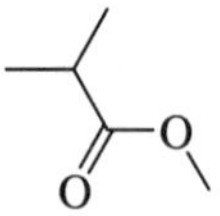

15.0c Ethyl Butyrate

15.0.1c CAS Number: *[105-54-4]*

15.0.2c Synonyms: Ethyl butanoate, butyric ether, butanoic acid ethyl ester, ethyl *n*-butyrate, natural ethyl butyrate

15.0.3c Trade Names: NA

15.0.4c Molecular Weight: 116.16

15.0.5c Molecular Formula: $C_6H_{12}O_2$

15.0.6c Molecular Structure:

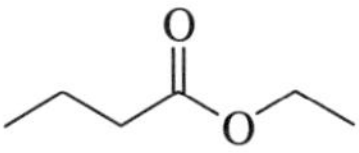

15.1 Chemical and Physical Properties

15.1.1 General

Methyl butyrate is a colorless liquid with a fruity odor. Methyl isobutyrate is a colorless, flammable liquid with a fruity odor. Ethyl butyrate is a colorless, flammable liquid with a pineapple-like odor. Butyl isobutyrate and isobutyl esters are colorless liquids with pleasant odors.

15.2 Production and Use

Methyl butyrate is manufactured from methanol and butyric acid in the presence of sulfuric acid. It is used in the manufacture of rum and essences, as a component for housefly repellent, and as a fumigant. It also is used as a solvent for various celluloses and production of lacquers. Pharmaceutically, it has been reported to enhance the absorption of all types of drugs. Methyl isobutyrate is used as a flavoring agent. Ethyl butyrate is manufactured via esterification of butyric acid with ethanol in the presence of sulfuric acid. It is used as a flavoring agent for foods and beverages, including artifical rum; in fragrance for soaps and perfumes; and as a solvent for lacquers. In a screening program of materials to serve as insect attractants, 2-ethylbutyric acid has been observed to be highly active for some *Diptera* species (103). Ethyl butyrate occurs naturally in brewer's yeast. Amyl butyrates have pleasant odors and occur in banana peel, but presumably have low toxicities. Hexyl-and heptyl butyrates are components of essential oils and serve as insect attractants, especially the heptyl butyrate and also octyl butyrate. Unsaturated esters of butyric acid such as allyl butyrate, $CH_3(CH_2)_2COOCH{:}CH_2$, is used in the cosmetic industry; and, 2,4-hexadienyl isobutyrate, heptyl butyrate, and also propyl 2-heptynoate have been used as synthetic insect attractants, and were not found toxic to mammals. Octyl butyrate is a major component of essential oils of fruits. Some simple fatty-acid butyrate salts exhibit anti-tumor activity. They down-regulate cell growth and promote various differentiated cellular functions. Unfortunately, they are characterized by short-lived biological action since they rapidly diffuse into the blood. Octyl butyrate, however, is hydrolysed by an esterase releasing a biologically active antitumor subunit (104).

15.3 Exposure Assessment: NA

15.4 Toxic Effects

15.4.1 Experimental Studies

15.4.1.1 Acute Toxicity. The toxicity of methyl butyrate is negligible and is similar to that of methyl isobutyrate, which is practically negligible. An oral LD_{50} in rats showed a value of 16 g/kg and intraperitoneally a LD_{50} of 3.2 g/kg (105). At lethal concentrations rats showed labored respiration, vasodilation, slight roughening of the coat, and some muscular twitching, with death occasionally delayed up to 2 days (105). At 25 mg/L about 6400 ppm for 6 h, rats exhibited loss of coordination, and prostration, but recovered with no residual effects (105). The toxicity of ethyl butyrate is low, however, certain metabolic decomposition products may bear an unpleasant odor (106). Ethyl butyrate is reportedly irritating to mucous membranes and is a CNS depressant at elevated concentrations.

The degree of toxicity is about the same for the isomers, butyl isobutyrate and the isobutyl ester. For the isobutyl derivative an oral LD_{100} for both rats and mice was 12.8 g/kg with a LD_1 of 6.4 g/kg (105). An intraperitoneal lethal dose for the rat was 6.3; and for the mouse 1.6 g/kg (107). The dermal LD_{50} value for the rabbit was above 10 cm^3/kg, but such a quantity interfered somewhat with the normal development of the test animal. An inhalation exposure of 5000 ppm in 6 h was lethal to two out of three rats. Symptoms included prostration and complete narcosis. A lower dose of 500 ppm 6-h exposure produced no symptoms or deaths.

Unsaturated esters of butyric acid such as allyl butyrate, 2,4-hexadienyl isobutyrate, heptyl butyrate, and propyl 2-heptynoate were not found toxic to mammals (108).

15.4.2 Human Experience

15.4.2.1 General Information. Other studies involving human subjects indicated no observation of irritation or sensitization after 48 h (111).

15.5 Standards, Regulations, or Guidelines of Exposure: NA

15.6 Studies on Environmental Impact: NA

16.0 Valerates (Pentanoates) and Higher Fatty Acid Esters (Alkanoates)

16.0a Ethyl Caprylate

16.0.1a CAS Number: *[106-32-1]*

16.0.2a Synonyms: Ethyl octanoate, ethyl octylate

16.0.3a Trade Names: NA

16.0.4a Molecular Weight: 172.27

16.0.5a Molecular Formula: $C_{10}H_{20}O_2$

16.0.6a Molecular Structure:

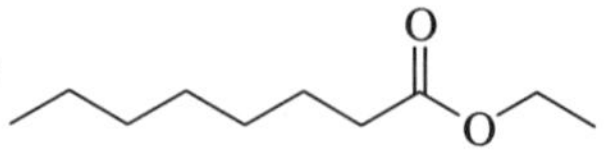

16.0b Ethyl Caproate

16.0.1b CAS Number: *[123-66-0]*

16.0.2b Synonyms: Ethyl hexanoate, hexanoic acid ethyl ester, ethyl n-hexanoate, ethyl caproate (Nat. C-6 ethyl ester), ethyl butyl acetate

16.0.3b Trade Names: NA

16.0.4b Molecular Weight: 144.21

16.0.5b Molecular Formula: $C_8H_{16}O_2$

16.0.6b Molecular Structure:

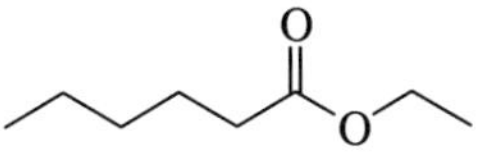

16.0c Ethyl Isovalerate

16.0.1c CAS Number: *[108-64-5]*

16.0.2c Synonyms: Ethyl 3-methylbutanoate, ethyl B-methyl butyrate, ethyl isovalerianate, ethyl isovalerate (Iso C-5 Ethyl Ester), 3-methylbutanoic acid ethyl ester

16.0.3c Trade Names: NA

16.0.4c Molecular Weight: 130.19

16.0.5c Molecular Formula: $C_7H_{14}O_2$

16.0.6c Molecular Structure:

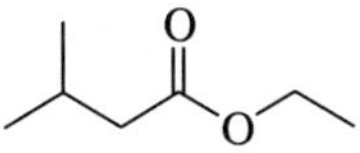

16.0d Ethyl Oleate

16.0.1d CAS Number: *[111-62-6]*

16.0.2d Synonyms: 9-Octadecenoic acid (*Z*)-, ethyl ester, ethyl 9-octadeconate

16.0.3d Trade Names: NA

16.0.4d Molecular Weight: 310.52

16.0.5d Molecular Formula: $C_{20}H_{38}O_2$

16.0.6d Molecular Structure:

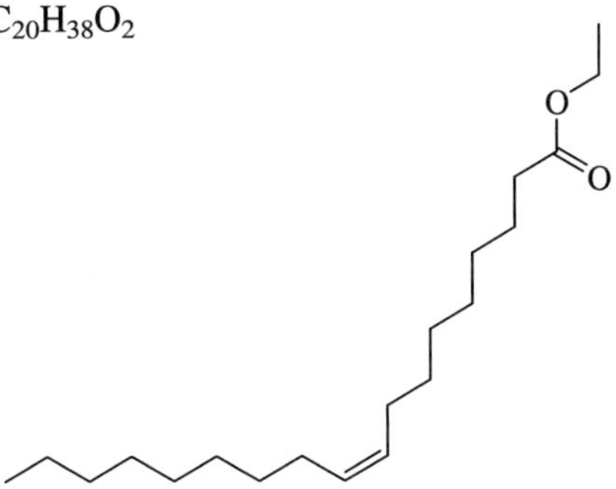

16.1 Chemical and Physical Properties

16.1.1 General

Ethyl isovalerate is a colorless, oily liquid with the odor of apples. The hexanoate, ethyl caproate, is a colorless to yellowish liquid with a pleasant odor. The heptanoate, ethyl enanthate, is a colorless, oily liquid with the odor of grapes. The nonanoate, ethyl pelargonate, is an oily liquid. The dodecanoate, ethyl laurate, is an oily liquid. The decanoate, ethyl caprate, is a colorless liquid. The hexadecanoate, ethyl palmitate, is a white to yellow solid material. The tetradecanoate, ethyl myrisate, is an oily liquid with properties similar to those of the C14 homolog, ethyl laurate. The octadecanoates, methyl stearate and ethyl stearate, are odorless, white crystalline solids. The unsaturated derivative, ethyl oleate, is a yellowish, oily liquid.

16.2 Production and Use

Valerates (pentanoates) and higher fatty-acid esters (alkanoates) are used in the fragrance, food, plastics, pesticide, and insect repellent industries. The higher fatty acids, the butyl laurate, myristate, and stearate, serve principally as plasticizers, and the lower members, the ethyl caproate, heptanoate, caprylate, pelargonate and caprate, mainly as food flavorings. Besides nutritional or fragrant properties, the higher C5–C18 fatty esters have other functions. Such compounds may possess antibacterial action and serve as insect and wildlife attractants or repellents, particularly the C12 and higher members. In a screening program of synthetic insect attractants, the butyric acid, 2-ethyl-, hexyl-, heptyl-, octyl-, and 2,4-hexadienyl esters were highly active (103).

Ethyl isovalerate is used in the fragrance and fod industries. The ethyl ester is a ripening agent for sugarcane. Ethyl caproate is used as a cockroach repellent. Conversely, the 2,4-hexadienyl derivative exhibits insect, such as gnat, attractant properties. The ethyl ester of hexenoic acid, hexenoate, was found to possess bacteriostatic properties (109). It is also utilizable as a pesticidal diffusion vehicle. Commercially, ethyl enanthate is used in the production of artificial flavors such as synthetic cognac oil, manufactured from coconut extracts. The octanoate, ethyl caprylate, is a clear, very mobile liquid, with a pleasant, pineapple-like odor. It is used for the commercial use of fruit flavors and as a derivative and part of the formulation of protective cosmetic creams. Methyl caprylate has antibacterial, antilarval, and miticidal properties. Methyl octanoate occasionally occurs mixed with methyl decanoate. Ethyl pelargonate is used for flavoring cognac-like beverages. Ethyl caprate is used as a wine bouquet, cognac essence, insect attractant, and both in and as stabilizer of pesticide formulations, as is also its methyl homolog. The latter also has bactericidal properties. Ethyl laurate occurs naturally esterified with glycerol and other alcohols in a variety of organisms. Ethyl myrisate is a constituent of the plant *Wisteria floribunda* (110). Ethyl palmitate is a natural and synthetic food antioxidant. The stearate compounds are manufactured via the esterification of the respective alcohols with stearic acid in the presence of a catalyst. Uses include intermediate for stearic acid detergents, emulsifiers, wetting agents, resins, lubricants, textiles, and plasticizers.

16.3 Exposure Assessment: NA

16.4 Toxic Effects

16.4.1 Experimental Studies

16.4.1.1 Acute Toxicity. The low toxicity of ethyl isovalerate may be related to the isovalerate's capability to form acetyl-CoA derivatives and thus expedite its metabolic transformation (111). The C13 unsaturated esters are teratogenic and melanogenic to the housefly (112). The general toxicity toward the mammal is very low except that of special unsaturated or halogenated members. By the oral and intraperitoneal route, the compounds are readily saponified into the corresponding alcohols and carboxylic acids, then nutritionally utilized or further metabolized. Isoacids in the C4–C7 range are readily degraded by *w*-oxidation (78), whereas 2-ethyl butyric acids, for example, tends to be excreted in unchanged form or as an ester glucuronide (90). Stetten and Salcedo (113) studied the effect of the chain lenght of the C4–C18 fatty-acid ethyl esters on the development of fatty liver in choline-deficient rats. At a 35% diet level fed for 2 weeks, the liver fats were normal for the ethyl stearate, and below C12, abnormal for the C16–C14 esters. Gastric hemorrhaging was noted with ethyl butyrate and renal hemorrhaging with ethyl caprylate. No myocarditis was found except for ethyl laurate under simultaneous choline deprivation. Metabolically, fatty-acid esters can be *a*, *B*-, *o*-, or *w*-oxidized, depending on the odd or even number of carbons in the fatty-acid chain. Pentyl pentanoate, (amyl valerate), $CH_3(CH_2)_3COOC_5H_{11}$, is practically nontoxic. A LD_{50} determination revealed a value of >35 g/kg for the rat (5). Ethyl caproate is of low toxicity, since an oral LD_{50} of 218 mg/kg in the rat and 280 mg/kg in the guinea pig (14). The toxicity of ethyl nonanoate is very low, based on an oral LD_{50} in the guinea pig reported as >43 g/kg (4). A diester, decanoic acid trimethylene glycol, has shown as oral LD_{50} in the rate of 7.46 g/kg and a dermal LD_{50} in the rabbit of 11 g/kg (14). For decanoic acid vinyl ester an oral LD_{50} of 6.17 g/kg in the rat and dermal LD_{50} in the rabbit of 14 g/kg has been determined (14). Physiologically, ethyl laurate is multifunctional and is of nutritional and metabolic importance. Conversely, ethyl laurate has produced diffuse myocarditis in cats when fed at high levels: 35–40% in the diets with a lack of choline. If choline was supplied, the myocarditis was absent (114).

Isopropyl myristate has been studied (115) as a possible vehicle for parenteral injection. Its oral toxicity in mice was >100 mL/kg. Repeated intraperitoneal injection of 5 ml/kg in rats daily over 20 days caused some deaths after 5 days. In the other animals, no growth effects or toxic symptoms were noticed. It was not a sensitizer on intracutaneous injection in the guinea pig. No eye irritation was produced on direct contact.

Ethyl palmitate has low toxicity and the same biological properties apply as for laurates. Ethanol consumption is associated with synthesis of ethyl palmitate and ethyl oleate fatty acids (116). An experiment was conducted involving intravenous injection of methyl- and ethyl palmitate into rats. Ethyl palmitate caused selective splenic necrosis and did not appear to affect other organs (117). Another study involving intravenous injection of mice with ethyl palmitate resulted in decreased retention of carbon in both liver and spleen and was cytotoxic to both macrophages and lymphocytes (118). A study conducted *in vitro* resulted in abnormal movement and formation of macrophages treated with ethyl palmitate (119). Ethanol was shown to metabolize to ethyl oleate, ethyl stearate, and ethyl palmitate in rat lung. The researchers concluded that the formation of theses fatty acids may be associated with ethanol-related injuries to the lung (120).

Fitzhugh and Nelson (121), however, determined some unusual toxicity characteristics of ascorbyl palmitate. Rats fed 5% of a 1-ascorbyl-*d*-isoascorbyl palmitate mixture for 9 months developed bladder stones and showed retarded growth. But no effects were noted at the 2% level (121). Ascorbyl palmitate is a natural antioxidant, however, and is considered safer than some synthetic antioxidants added to foods to inhibit or interfere with free-radical formation (122). Ascorbyl palmitate treatment of rats exposed simultaneously to the hepatocarcinogen 3′-methyl-4-dimethylaminoazobenzene resulted in some inhibition of induced hepatocellular carcinomas (74). Ascorbyl palmitate also has been shown to inhibit 12-*O*-tetradecanoylphorbol-13-acetate-induced skin tumors in mice, but was not effective in preventing dysplasia in mouse colon (123). Incubation of ascorbyl esters ascorbyl palmitate and ascorbyl stearate with glioma cells caused decreased cell viability and proliferation, as well as inhibited cytosolic glutathione-*S*-transferase activity (124). The two compounds were also effective in inhibiting glutathione-*S*-transferase activity in human term placenta and fetal liver, which may have clinical implications to improve the effectiveness of anticancer drugs (125).

Methyl oleate was evaluated for carcinogenicity in mice. Oral and subcutaneous administration did not cause carcinogenic changes (126).

Ethyl oleate was shown to damage myocardial mitochondria in rabbits and rats (127). The compound is formed from metabolism of ethanol in the myocardium. Ethanol is partially metabolized to fatty-acid ethyl esters, including ethyl oleate. An *in vitro* study concluded that ethyl oleate and ethyl arachidonate are toxic to human hepatoblastoma cells and, accordingly, may be associated with ethanol-induced liver damage (128). Another study focusing on ethyl oleate concluded that it increases the fragility of rat pancreatic lysosomes which, in turn, may be associated with early pancreatic injury (129). In a study involving rats, ethyl oleate had a potentiating effect on activation of protein kinase C isozymes I, II, and III. It was concluded that interaction of ethyl oleate with a brain-signaling mechanism involving protein kinase C may be associated with the cognitive disturbances related to ethanol (130).

16.4.1.2 Chronic and Subchronic Toxicity. The toxicity of methyl and ethyl stearate is very low, sinc they are common nutritional fat constituents. In a 2-year chronic feeding study, neither 1.25 nor 6.25% of butyl stearate in the diet affected growth of mice or their survival adversely (131). No hematological deviations were observed in the peripheral blood. Differential bone marrow counts after 12 months showed no effects on cellular distribution or myeloid: erythroid ratio. Gross pathological findings at necropsy included some principal lesions, acute and chronic inflammatory changes in the lungs, enlarged kidneys, and fatty changes of the liver. These parameters, however, were not related to the plasticizer administration. Results of a fertility study revealed no adverse effects from feeding 6.25% butyl stearate on litter size or offspring survival. At this concentration the plasticizer retarded growth during the preweaning and postweaning period. However, no gross pathologies were found (131).

16.4.1.3 Pharmacokinetics, Metabolism, and Mechanisms

16.4.1.3.1 Absorption. Weitzel (78) has shown with higher esters that the linear compounds can be almost completely degarded metabolically. Certain branched-chain fatty esters, such as the *n*-propyl stearyl ethyl ester, are similarly catabolized, whereas 2% of *a*-*n*-butyl stearyl derivatives are recovered in the urine in unchanged form (78).

16.4.2 Human Experience

16.4.2.1 General Information. Physiologically, no effects were noted when ethyl heptanoate was applied to human skin for periods from 5 min to 5 h (132). No effects were noted by Oettel (132) when nonanoate was applied to human skin from 5 min to 5 h. Most higher fatty acid esters are cleared to be used in the food industry, such as the laurate polyoxyethylene sorbitan, oleates, palmitates (espicially the ascorbyl derivative), linoleates, or stearates (133). Thus, their general physiological response and toxicity are very low.

16.5 Standards, Regulations, or Guidelines of Exposure: NA

16.6 Studies on Environmental Impact: NA

C ESTERS OF ALKENYLCARBOXYLIC ACIDS AND MONOALCOHOLS

Acrylates, Methacrylates, and Crotonates

Acrylates are esters of propenoic acids and mono- or polyalcohols. Chemically, the acrylic monomers are substituted 2-propene carboxylic acid esters of the type CH_2: CHCOOR. They polymerize readily, with heat or even on standing;the latter reaction is catalyzed by light or oxygen, unless an inhibitor has been added. Uncontrolled polymerization is exothermic and may proceed with explosive force. Methacrylic esters are 2-methyl derivatives of acrylic esters. Chemically, they are of the general structure CH_2: $C(CH_3)COOR$. The higher molecular weight derivatives polymerize to gels or highly viscous liquids. Crotonates are 3-methyl isomers of methacrylates or butanoic acid esters and bear the general structure CH_3CH: CHCOOR. Lower molecular weight acrylic monomers are liquids having relatively higher vapor pressures. The characteristic odors of the monomers can be unpleasant. See Table 79.10 for a summary of physical and chemical properties.

Some acrylates occur naturally in several organisms as intermediates in lipid biosynthesis and degradation. The alkyl monomers exist primarily in liquid form, whereas the formed polymers range from clear, glass-like, brittle masses to highly flexible films, solids, or emulsions. They are widely used in commerce and industry as vinyl, acrylic, or higher molecular weight alkene resins. The acrylic monomers are highly important base components in the manufacture of thermoplastics, acrylic resins, and emulsion polymers. They are also used as solvents, plasticizers, latex coatings, adhesives, fibers, floor finishes, and lubricant additives and serve in medical and dental technology as surgical cement for medical devices and prostheses. The methacrylates sometimes serve as bases for acrylic resins with multifunctional effects. These materials find use in surgical organ repair, in compositing contact lenses, for adhesive dental pretreatment, and for a variety of other applications.

The acrylates can be manufactured by a number of procedures. These include dehydration of the corresponding hydroxyalkanoic acid, saponification of the alkene nitrile, catalytic hydration of acetylene and carbon monoxide, or the reaction of acetone with

Table 79.10. Summary of Physical and Chemical Properties of Some Acrylates, Methacrylates, and Crotonates (12)

Compound	CAS Number	Molecular Formula	Molecular Weight	Boiling Point (°C)	Melting Point (°C)	Specific Gravity at (25°C)	Solubility[a] in Water at (68° F)	Refractive Index at (20°C)	Vapor density (Air = 1)	Vapor Pressure [mm Hg (20°C)]	Flash Point (°C)	LEL % by Vol. Room Temp.	UEL % by Vol. Room Temp.
Methyl acrylate	*[96-33-3]*	$C_4H_6O_2$	86.09	80.5	− 75	0.95	d	1.4040	2.97	100 (28)	− 2.8	2.8	25
Ethyl acrylate	*[140-88-5]*	$C_5H_8O_2$	100.11	99.4	− 71.2	0.92	d	1.4068	3.45	29.3 (20)	15.6	1.8	higher
n-Butyl Acrylate	*[141-32-2]*	$C_7H_{12}O_2$	128.17	146.8	− 64	0.8986	i	1.4185	4.42	10 (35.5)	—	1.5	9.9
2-Ethylbutyl acrylate	*[3953-10-4]*	$C_9H_{16}O_2$	156.22	82	− 70	0.896	—	—	5.4	1.7 (20)	—	—	—
2-Ethyl hexyl acrylate	*[103-11-7]*	$C_{11}H_{20}O_2$	184.28	213.5	− 90	0.887	i	—	6.35	1.0 (50)	—	—	—
Methyl metha-crylate	*[80-62-6]*	$C_5H_8O_2$	100.13	101	− 48	0.945	d	1.4142	3.45	40 (25)	10	2.1	12.5
Ethyl methacry-late	*[97-63-2]*	$C_6H_{10}O_2$	114.14	117	− 75	0.911	d	1.4147	3.94	—	20	1.8	Saturated
n-Butyl metha-crylate	*[97-88-1]*	$C_8H_{14}O_2$	142.20	160	− 75	0.89	i	1.4240	4.8	4.9 (20)	41.1	2	8
Isobutyl metha-crylate	*[97-86-9]*	$C_8H_{14}O_2$	142.20	155	—	0.886	—	—	—	—	—	—	—
2-Ethylisohexyl methacrylate	*[688-84-6]*	$C_{12}H_{22}O_2$	198.30	113	—	—	—	—	6.8	< 1 (20)	—	—	—
Methyl crotonate	*[623-43-8]*	$C_5H_8O_2$	100.12	121	− 42	0.946	i	1.4242	3.5	18 (25)	40.6	—	—

[a] Solubility in water: v = very soluble; s = soluble; d = slightly soluble; i = insoluble.

hydrocyanic acid. The methacrylates can be synthesized by catalytic oxidation of isobutylene and subsequent esterification with the appropriate alcohol, or by reacting acetone with hydrocyanic acid and subsequent esterification in sulfuric acid with the appropriate alcohol.

The low molecular weight monomers are lacrimators and irritants to the eyes, skin, and mucous membranes (3). The main potential for human exposure is by the dermal and respiratory routes;however, the irritating properties of these chemicals may serve as a deterrent to repeated exposures. Nonetheless, chronic inhalation of acrylic acid esters can lead to tissue changes or lesions due to local irritant or inflammatory reactions from acrylic acid or its esters that hydrolyze to form the acid. The acute toxicity of acrylates decreases with increasing molecular weight. For example, results from animal studies indicated that the acute lethal toxicity of methyl acrylate was twice that of ethyl acrylate based on inhalation exposure. Another study, based on 24-h LC_{50} concentrations using rats, indicated that the order of acute toxicity was methyl acrylate > ethyl acrylate > *n*-butyl acrylate > butyl methacrylate > methyl methacrylate. In the same study, rat inhalation exposures to 110 ppm 4 h/day, 5 days/week for 32 days did not cause significant changes in body weight, tissue weight, blood chemistry, gross metabolic performance, or small-intestine motor activities when compared to a control group (134). Structure-toxicity relationships of acrylates, including methacrylates, were analyzed in mice and found to be dependent on the log of the partition coefficient and log of rate order constant (135). In general, the toxicity is theorized to mechanistically involve alkylation of cellular nucelophies via Michael is addition (136).

The introduction of unsaturation into a fatty acid, as, for example, comparing the methyl or ethyl esters of propionic acid with the equivalent propenoic or acrylic acid derivatives, shows a tenfold increase in acute toxicity. Comparing acute toxicity of straight with branched-chain acrylates, ethyl acrylate by the oral route may be only half as toxic as methyl acrylate, but 8 times as toxic as methyl methacrylate, and 13 times as toxic as ethyl methacrylate (137). Acute inhalation of higher concentrations may cause narcosis, salivation, and pronounced nasal, occular, and pulmonary irritation or edema. Prolonged skin or eye contact may result in severe tissue damage. The pathology from single exposure is not particularly characteristic, contrary to repeated exposure effects, which include pulmonary congestion or hemorrhage and cloudy swelling and organ weight changes of the liver and kidney, reported following subchronic exposures to excessive concentrations. Acrylic acid and a series of methacrylates were shown to produce hemangiomas and increase resorptions following intraperitoneal injection of pregnant rats (138). Some compounds, such as the methyl-, ethyl-, *n*-propyl-, or butyl methacrylates, can produce inhibition of barium chloride–induced contraction of the isolated guinea pig ileum (139). Animal studies using rats revealed that some acrylates and methacrylates are embryotoxic. The doses of monomers used in the animal studies, however, were much higher than concentrations likely encountered by workers (3).

Allergic reactions have been reported for some acrylates. Although methyl, ethyl, and butyl methacrylates are potent skin sensitizers, experimental simulation proved rather difficult, owing to the rapid evaporation of the materials tested (140). Despite this, causes of human sensitization may occur. Methyl acrylate, methyl methacrylate, ethyl methacrylate, and butyl methacrylate caused allergic contact dermatitis due to working with artificial nail-bonding chemicals (141). Indeed, a study reviewing 10 y of patch testing data showed that 48 of 275 patients exhibited dermal allergic reactions to at least one acrylate. The most common acrylates that caused allergic reactions were 2-hydroxyenthyl acrylate (12.1%), 2-hydroxypropyl acrylate (12%), and 2-hydroxyethyl methacrylate (11.4%). No allergic response were recorded for 2-ethylhexyl acrylate (141).

Acrylates and methacrylates are detoxified predominantly via conjugation with glutathione via the Michaelis addition reaction or glutathoine-*S*-transferase. They are also likely to be hydrolyzed via carboxylesterases (142). The lower molecular weight esters are rapidly metabolized and eliminated, therefore, will not likely cause cumulative toxicity (3).

The literature revealed much more data regarding methylacrylates than the crotonates. Although the irritant and lacrimatory effects of the crotonates are known, no low level exposure effects in humans have been reported so far. Physiologically, they are somewhat less reactive than the acrylates. They appear to be rapidly metabolized and excreted, possibly because crotonyl and methacrylyl groups are also formed during the normal metabolism of butyryl and isobutyryl derivatives (111). General toxicological data for acrylates, methacrylates, and crotonates are summarized in Tables 79.11 (143, 144) and 79.12 (145–150).

17.0 Methyl Acrylate

17.0.1 CAS Number: [96-33-3]

17.0.2 Synonyms: Methyl acrylate (2-propenoic acid methyl ester), methyl propenoate, methoxycarbonylethylene, acrylic acid methyl ester, 2-propenoic acid methyl ester, methyl ester acrylic acid, curithane 103, methyl 2-propenoate, propenoic acid methyl ester, methyl acrylate, stabilized with 200 ppm MEHQ

17.0.3 Trade Names: NA

17.0.4 Molecular Weight: 86.09

17.0.5 Molecular Formula: $C_4H_6O_2$

17.0.6 Molecular Structure:

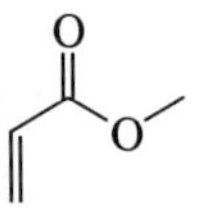

17.1 Chemical and Physical Properties

17.1.1 General

Methyl acrylate is a flammable liquid with an acrid odor. Polymerization may occur spontaneously with an exothermic reaction in uninhibited samples.

17.1.2 Odor and Warning Properties

Human detectability is unreliable, since odor fatigue may occur and vapors may not be irritant until a concentration of $\sim$75 ppm is reached.

Table 79.11. Summary of Inhalation, Oral, Dermal, Subcutaneous, and Intraperitoneal Toxicity Data for Some Acrylates

Acrylate	Oral LD_{50}		Dermal LD_{50} Rabbit (g/kg)	IP LD_{50} Mouse (g/kg)	Inhalation				Saturated Vapor, Rat LC_{00}	Dermal Irritation, Rabbit	Eye Irritation, Rabbit	Ref.
	Species	g/kg			Concen. (ppm)	Time (h)	Parameter	Species				
Methyl	Rat	0.300	1.3	0.256 (143)	1000	4	LC_{50}	Rat		Moderate to severe		40
	Rabbit[b]	0.280			95		TC_{LO}	Rabbit		Nose Irritation	Moderate Irritation	143
Ethyl	Rat	1.02	1.95	0.606 (143)	50,000	0.25	LC_{100}	Rat		Sl. to severe	Moderate	137
	Rabbit	1.0			4000	4	LC_{100}	Rat		Sl. to severe		137
					2000	4	LC_{80}	Rat				137
					1204	7	LC_{LO}	Rabbit				143
		0.4			1204	7	LC_{LO}	Guinea pig				143
					< 1000	4	LC_{50}	Rat				137
					540	19 days	LC_{66}	Rat				105
					70	30	LC_{7}	Rat				105
n-Butyl	Rat (105)	3.7	3.4 (105)	0.853 (143)	1000	4	LC_{80}	Rat		Moderate	Low	105
	Rat (105)	9.05	2.0				LC_{12}	Rat				75
Isobutyl	Rat (75)	7.07	4.49 (75)	0.761 (143)	2000	4	LC_{LO}	Rat		Moderate	Low	75
2-Ethylbutyl	Rat	6.5	5.5						4 h	Moderate	Slight	97
	Rat (105)	5.6	8.5[a] (105)	1.325 (143)					8 h	Moderate to severe	Slight	105
2-Ethylhexyl	Rat	5.66										97
	Rat	6.4–12.8										105
Cyclohexyl	Rat	8.98	2.52						8 h			75
2-Hydroxy	Rat	1.0	1.0		500	4	LC_{80}	Rat	1 h	Slight	Severe	105
	Rat	0.665	1.01 (97)				LC_{LO}	Rat (144)				105
2-Ethoxyethyl	Rat (14)	1.07	1.0 (105)		500	4	LC_{80}	Rat	1 h	Slight	Severe	105
2-Ethoxypropyl	Rat (97)	0.82	1.401 (97)		250	4	LC_{16}	Rat	1 h	Moderate	Moderate	14

[a] Guinea pig.
[b] Rabbit.

Table 79.12. Summary of Toxicity Data for Some Methacrylates and Crotonates

	Oral LD_{50}		Subcutaneous LC_{50}		Intraperitoneal LD_{50}		Inhalation					Irritation (Rabbit)	
Compound	Species	g/kg	Species	g/kg	Species	g/kg	Species	ppm	mg/L	Exposure (h)	Parameter	Skin	Eye
						Methacrylate							
Methyl	Rat (145)	8.4 12.7–18.14	Mouse (146)	6.3	Mouse (39)	1.0						Moderate (105)	
	Rabbit (145)	6.55	Rat (146)	7.5	Mouse (146)	1.1	Human (39)		0.150	Short	TC_{LO}		
	Guinea pig (39)	6.3	Dog (146)	4.5	Rat (147)	1.33	Rat (105)	3750			LC_{50}		
	Dog (39)	5.0	Guinea pig (146)	6.3	Rat (39)	1.8	Rat (147)		15		LC_{50}		
	Rabbit (105)	$> 10.0^{a,b}$			Guinea pig (39)	2.0	Mouse (39)		47.7^{c}		LC_{60}		
							Mouse (39)		61.8^{c}	3	LC_{100}		
							Rabbit (145)		17.5	4.5	LC_{LO}		
							Guinea pig (145)		19.0^{c}	5	LC_{LO}		
Ethyl	Rat (145)	14.8	Rat (145)	25.0^{e}	Rat (147)	1.22	Guinea pig (39)		72.1^{c}	4.5	LC_{100}		
	Rabbit (145)	3.63			Rat (147)	0.37^{f}	Dog (39)		41.2	3	LC_{100}		
	Rabbit (105)	$> 10^{a,d}$			Mouse (146)	1.25	Dog (39)		72.1	1.5	LC_{100}		
					Mouse (148)	1.37	Mouse (39)		41.2	0.5/day×15	LC_{05}		
n-Butyl	Rat (145)	> 20			Rat (147)	2.3			41.2	1.5/day × 15	LC_{10}		
	Rabbit (145)	> 6.3			Mouse (146)	1.49	Guinea pig (39)		39.3	3.0/day × 15	LC_{00}		
									65.5^{c}	3.0/day × 3	LC_{100}		
Isobutyl	Rat (105)	> 6.4			Rat (147)	1.4	Dog (39)		41.2	0.5/day × 15	LC_{00}		
									46.8	0.5/day × 15	LC_{50}		
					Mouse (145)	1.19			46.8	1.5/day × 8	LC_{100}		
						Crotonate							
Methyl	Rat (105)	> 3.2					Rat (105)		15	3	LC_{50}		
	Mouse (105)	2.6–3.2					Rabbit (105)	3300		8	LC_{50}		
	Guinea pig (105)	$10–20^{a,d}$					Rat (150)		6.0				

Ethyl	Rat (149)	3.0	Rat (105)	19,000	6	LC_{00}	Moderate	Slight
	Guinea pig (105)	> 10[a,d]	Rat (149)	Saturated vapor	8	LC_{00}	Moderate	Severe
2-Ethylhexyl	Rat (105)	> 3.2	Rat (105)	2500	6	LD_{00}	Moderate	Slight
	Guinea pig (105)	> 20[a]	Rat (105)	Saturated vapor	6	LD_{LO}		
Vinyl	Rat (105)	6.5	Rat (105)	4000	4	LD_{100}		
	Guinea pig (105)	> 10[a]	Rat (105)	2000	4	LD_{00}		
			Rat (105)	Saturated vapor	1	LD_{LO}	Slight	Slight

[a] Dermal LD_{50}.
[b] mL/kg.
[c] Liver degeneration and focal necrosis.
[d] mL/kg.
[e] TD_{LO}.
[f] Teratogenic TD_{LO}.

17.2 Production and Use

Methyl acrylate is manufactured via a reaction of nickel carbonyl and acetylene with methanol in the presence of an acid; more commonly, however, via oxidation of propylene to acrolein then to acrylic acid. The acid is reacted with methanol to yield the ester. It has wide use in industry and in the dental, medical, and pharmaceutical sciences. It is utilized as a monomer, polymer, or copolymer. Its primary use is for production of acrylic fibers. It may also be used as a microencapsulation mixture component or for the polymerization of radioactive waste into block form for easy transport. It can serve as a resin in the purification and decolorization of industrial effuents or aid in the timed release and disintegration of pesticides.

17.3 Exposure Assessment

17.3.1 Air: NA

17.3.2 Background Levels: NA

17.3.3 Workplace Methods

Air sampling using charcoal adsorbent is a recommended procedure. Gas chromatography using a flame ionization detector is used for analysis (8, 9).

17.4 Toxic Effects

17.4.1 Experimental Studies

17.4.1.1 Acute Toxicity. Animal experiments have shown an oral LD_{50} of 300 mg/kg in the rat (40). The lowest dose to produce a lethal effect in the rabbit was 280 mg/kg (143). The LD_{50} in the mouse by intraperitoneal injection was 254 mg/kg (146). The lowest concentration to have a lethal effect when inhaled by the rat has been observed as 2000 ppm in a 4-h trial (151). Some acrylates produce, and others inhibit, concentrations of the guinea pig ileum. The inhibition of contractions by methyl acrylate may be of myogenic origin (148).

17.4.1.2 Chronic and Subchronic Toxicity. Contact-site degeneration of olfactory epithelium in the nasal turbinates and corneal damage was associated with exposure, but no treatment-relaed chronic systemic toxicity or tumorigenic effects were observed. Indeed, male and female rats exposed in whole-body inhalation chambers to concentrations of methyl or *n*-butyl acrylate vapor at concentration of 0, 15, 45, and 135 ppm for 6 h/day, 5 days/week, for 24 months did not reveal signs of systemic toxicity or carcinogenesis. The rats did, however, show atrophy of neurogenic epithelial cells and hyperplasia of reserve cells in the nasal mucosa and opacity and neovascularization of the cornea (152).

17.4.1.3 Pharmacokinetics, Metabolism, and Mechanisms

17.4.1.3.1 Adsorption. Methyl acrylate, like most acrylates, is more readily absorbed in the respiratory and gastrointestinal tracts than through the skin (139).

17.4.3.2 ***Dsitribution:*** NA

17.4.1.3.3 Excretion. Methyl, ethyl, *n*-butyl, and 2-ethylhexyl acrylate were evaluated for chemical reactivity with glutathione, urinary thioether excretion, and tissue total and nonprotein sulfhydryl (–SH) groups following inhalation of the compounds by rats. The trend for chemical reactivity of acrylates with glutathione was ethyl $>$ *n*-butyl $>$ methyl $>$ 2-ethylhexyl acrylate. All acrylates tested increased the urinary excretion of thioethers, but the protein metabolized to thioethers was only 1.5–8%. Inhalation of ethyl, *n*-butyl, and 2-ethylhexyl acrylates decreased total –SH levels in the liver, ethy acrylate decreased total –SH in blood. Depletion of non-protein-SH was greatest in the liver, less in blood, and moderate in lung and brain tissues. Reaction of the acrylates with –SH groups followed the trend 2-ethylhexyl $>$ ethyl hexyl $>$ *n*-butyl acrylate (153).

17.4.1.4 Reproductive and Developmental

17.4.1.5 Carcinogenesis. Methyl acrylate was not shown to be carcinogenic in male and female rats in a lifetime inhalation study (152).

17.4.1.6 Genetic and Related Cellular Effects Studies. Although micronucleus tests indicated possible mutagenic (clastogenic) activity of methyl acrylate (154), methyl acrylate was negative in a variety of studies for point mutation and *in vivo* clastogenicity (155). Indeed, clastogenic activity was indicated for the series of acrylates methyl acrylate, ethyl acrylate, methyl methacrylate, and ethyl methacrylate using a mouse lymphoma assay. All the compounds induced concentration-dependent increases in mutant frequency (156). Methyl methacrylate was positive for mutagenicity in the mouse lymphoma assay and the AS52-XPRT assay (157).

17.4.2 Human Experience

17.4.2.1 General Information. Methyl acrylate is highly irritating to the skin, eyes, mucous membranes, and GI tract. It may cause serious corneal burns and skin irritation. It is readily absorbed by mucous membranes and through the intact skin. Methyl acrylate may also be an allergen causing dermal sensitization (141).

17.4.2.2 Clinical Cases. Dermal patch testing using methyl acrylate resulted in allergic contact dermatitis after a single exposure (141). This suggested that methyl acrylate is a potent antigen which may induce primary sensitization.

17.5 Standards, Regulations, or Guidelines of Exposure

OSHA PEL 10 ppm (skin); ACGIH TLV-2 ppm (skin); ACGIH A4, "Not classifiable as a human carcinogen"; NIOSH REL 10 ppm (skin)

17.6 Studies on Environmental Impact

Acrylic acid derivatives, including ethyl acrylate, *n*-butyl acrylate, and methyl methacrylate, are components of many common household products that are discharged in municipal wastewater systems. It was determined that the concentrations reaching and discharging from municipal wastewater treatment plants are significantly low and, therefore, are not associated with unreasonable risks to the acquatic environment (158).

18.0 Ethyl Acrylate

18.0.1 CAS Number: [140-88-5]

18.0.2 Synonyms: Ethyl acrylate (acrylic acid ethyl ester; 2-propenoic acid ethyl ester), ethyl propenoate, ethoxycarbonylethylene, ethyl 2-propenoate

18.0.3 Trade Names: NA

18.0.4 Molecular Weight: 100.11

18.0.5 Molecular Formula: $C_5H_8O_2$

18.0.6 Molecular Structure:

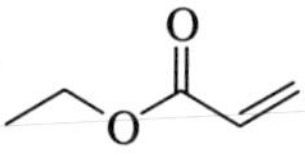

18.1 Chemical and Physical Properties

18.1.1 General

Ethyl acrylate is a clear liquid with an acid, penetrating odor. It is highly flammable and polymerizes exothermically in the presence of light, heat and peroxides above 10°C, unless inhibited. The monomer polymerizes to a transparent, elastic substance that is practically odorless and insoluble in conventional solvents. It is incompatible with peroxides, oxidizing agents, certain acids, and strong alkalies (9).

18.1.2 Odor and Warning Properties

18.2 Production and Use

Ethyl acrylate is manufactured via oxidation of propylene to acrolein then to acrylic acid. The acid is treated with ethanol to yield the ethyl ester. Ethyl acrylate is used as a water emulsion vehicle for paints, textiles, and paper coatings, leather finish, resins, or adhesives, and lends flexibility to hard films. Additional use formerly included additive in food as flavoring and fragrance.

18.3 Exposure Assessment

18.3.1 Air: NA

18.3.2 Background Levels: NA

18.3.3 Workplace Methods

Air sampling is conducted using charcoal adsorbent. Samples are desorbed using carbon disulfide and analyzed using a gas chromatograph (NIOSH Method 1450) (8, 9).

18.4 Toxic Effects

18.4.1 Experimental Studies

18.4.1.1 Acute Toxicity. Irritants, including ethyl acetate and ethyl acrylate, caused leukopenia without altering differential or red blood cell counts in rats exposed to airborne concentrations at the irritant level, suggesting that stress from the irritative effect can confound other hemotological effects (159).

18.4.1.2 Chronic and Subchronic Toxicity. When ethyl acrylate was administered orally to rats at 10% of the LD_{50}, the material appeared to be metabolized promptly and no cumulative effects were observed in the rabbit (143). When it was injected intraperitoneally, a LD_{50} of 606 mg/kg was obtained for the mouse (146).

With a mathematical model and 350 solvent pair mixtures, results demonstrated that ethyl acrylate could be classified slightly more toxic when in combination with other solvents (52). Short-term inhalation studies involving 4-h exposure of rats to 1500 ppm indicated that the ester causes pulmonary tissue damage (134). Lifetime animal inhalation studies, however, indicated contact-site damage only to nasal turbinates of mice and rats at exposures to concentrations of 25 ppm for 6 months; no observed effects at 5 ppm (160). In nasal tissue, ethyl acrylate is rapidly hydrolyzed to acrylic acid by carboxylesterases, but both the ester moiety and the resulting acid contribute to the upper respiratory tract cytotoxicity. Other factors, such as nasal airflow patterns and regional deposition, also are involved (161).

18.4.1.3 Pharmacokinetics, Metabolism, and Mechanisms

18.4.1.3.1 Adsorption: NA

18.4.1.3.2 Distribution: NA

18.4.1.3.3 Excretion. Ethyl acrylate is biotransformed to and excreted as 3-hydroxypropanoic acid, lactic acid, and acetic acid, as well as, two mercapturic acids, *N*-acetyl-*S*-(2-carboxyethyl)cysteine and *N*-acetyl-S-[(2-alkoxycarbonyl)ethyl]cysteine (162).

18.4.1.4 Reproductive and Developmental: NA

18.4.1.5 Carcinogenesis. It has been demonstrated that ethyl acrylate produced contact-site tumors of the forestomach of rats and mice when administered by gavage in corn oil over a major portion of their lifespan (163). Accordingly, NTP and IARC list ethyl acrylate as a "Suspect human carcinogen." However, rats administered 20 mg/kg ethyl acrylate 5 days/week for 6 months developed exhibited squamous-cell proliferation/hyperplasia, which reversed when exposure was discontinued. A time-dependent decrease was observed during a 2–15-month recovery period. Reversal of squamous-cell proliferation/hyperplasia was not observed in groups treated for 12 months (164). Furthermore, this acrylic acid ester did not produce evidence of a carcinogenic response in a 27-month chronic inhalation study of rats and mice (160), in a 2-year water ingestion study conducted in rats and dogs (165), or in a lifetime mouse skin-painting study (166). Ethyl and methyl acrylate were evaluated for tumor induction activity. Although methyl acrylate has been shown to be an inducer in rodent systems, neither methyl or ethyl acrylate induced the human gene HQOR-1 (167).

No evidence of dermal carcinogenic activity was found following treatment of mice with 25-μL application of ethyl acrylate, 1% acrylic acid, or 1% butyl acrylate 3 times weekly for their lifetime. Ethyl acrylate treatments, however, did cause nonneoplastic skin changes, such as dermatitis, dermalfibrosis, epidermal necrosis, and hyperkeratosis, in some mice (169).

18.4.1.6 Genetic and Related Cellular Effects Studies. *In vivo* genotoxicity testing of ethyl acrylate and tripropylene glycol acrylate was conducted using dermal application on transgenic mice. Peripheral blood polychromatic and normochromatic erythrocytes were evaluated and no signs of mutagenicity were indicated, based on the dermal route of exposure (142). Ethyl acrylate has been determined to be clastogeneic when studied *in vitro* (156). Methyl and ethyl acrylate injected intraperitoneally in mice at two doses separated by 24 h of 37.5-300 mg/kg and 225–1800 mg/kg, respectively, induced chromosome damage, forming micronuclei in bone marrow polychromatic erythrocytes (154). Another study of clastogenicity of ethyl acrylate conducted *in vitro* by injecting mice with 0, 125, 250, 500, or 1000 mg/kg indicated no signs of SCE or chromosome aberrations, suggesting that ethyl acrylate is only clastogenic at high concentrations during a specific phase of the cell cycle (168).

18.4.2 Human Experience

18.4.2.1 General Information: NA

18.4.2.2 Clinical Cases

18.4.2.2.1 Acute Toxicity. Irritation and sensitization of workers via dermal contact has been reported (21).

18.4.2.2.2 Chronic and Subchronic Toxicity: NA

18.4.2.2.3 Pharmacokinetics, Metabolism, and Mechanisms: NA

18.4.2.2.4 Reproductive and Developmental: NA

18.4.2.2.5 Carcinogenesis. A retrospective study found an excess of colorectal cancers in one exposed population of workers; however, the data were confounded by other exposures and lack of association of causality and risk in similarly exposed populations from other locations. Therefore, there was inadequate evidence based on the study that ethyl acrylate is a human carcinogen (169). Ethyl acrylate is listed as USEPA group B2, Probable human carcinogen; IARC group B2, Possibly carcinogenic in humans; NIOSH, Carcinogen with no further categorization; NTP group 2, Reasonably anticipated to be a carcinogen, and ACGIH Class A4, Not classifiable as a human carcinogen (10). Dermal studies of acrylic acid, ethyl acrylate, and *n*-butyl acrylate using mice did not result in local carcinogenesis, but several mice in the ethyl acrylate–treated group did exhibit dermatitis, dermal fibrosis, epidermal necrosis, and hyperkeratosis (166).

18.4.2.3 Epidemiology Studies

18.4.2.3.1 Acute Toxicity: NA

18.4.2.3.2 Chronic and Subchronic Toxicity: NA

18.4.2.3.3 Pharmacokinetics, Metabolism, and Mechanisms. A physiologically based pharmacokinetic/pharmacodynamic model has been developed to describe the metabolism of ethyl acrylate in tissues with respect to rate and extent of carboxylesterase catalyzed hydrolysis, conjugation with glutathione, or binding to protein. The model is important for assessing risk associated with dose delivered to target tissues (170, 171).

18.5 Standards, Regulations, or Guidelines of Exposure

OSHA PEL 25 ppm; ACGIH TLV 5 ppm; ACGIH STEL 15 ppm; ACGIH A4, Not classifiable as a human carcinogen.

18.6 Studies on Environmental Impact: NA

19.0 *n*-Butyl Acrylate

19.0.1 CAS Number: [141-32-2]

19.0.2 Synonyms: *n*-Butyl acrylate; butyl 2-propenoate; 2-propenoic acid butyl ester; acrylic acid *n*-butyl ester; propenoic acid *n*-butyl ester; butyl acrylate, stabilized with 20 ppm MEHQ

19.0.3 Trade Names: NA

19.0.4 Molecular Weight: 128.17

19.0.5 Molecualr Formula: $C_7H_{12}O_2$

19.0.6 Molecular Structure:

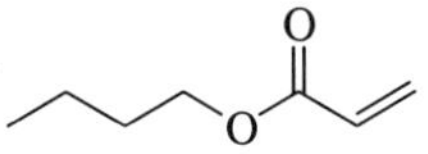

19.1 Chemical and Physical Properties

19.1.1 General

n-Butyl acrylate is a colorless, combustible, highly reactive liquid (20). Isobutyl acrylate is a clear, sharply odorous, and combustible liquid.

19.1.2 Odor and Warning Properties

The compound reportedly has an odor threshold of 0.1 ppb (2).

19.2 Production and Use

n-Butyl acrylate can be manufactured via a reaction of acetylene, butyl alcohol, carbon monoxide, nickel carbonyl, and hydrochloric acid. It is commonly manufactured via an oxidation of propylene to acrolein then to acrylic acid. The acid is reacted with *n*-butanol to yield the butyl ester. The product is used in the manufacture of polymers and resins for leather finishes, paint formulations, and textiles. It is one of the major monomers used to manufacture plastic polymers and resins. Isobutyl acrylate can be manufactured via oxidation of propylene to acrolein, followed by oxidation to acrylic acid, and esterfication of acrylic acid with isobutyl alcohol. It is used as a comonomer in acrylic surface coatings and intermediate for production of polymers; and, monomer in synthetic resin manufacture.

19.3 Exposure Assessment: NA

19.4 Toxic Effects

19.4.1 Experimental Studies

19.4.1.1 Acute Toxicity. *n*-Butyl acrylate is irritating to the skin, eyes and respiratory tract. Delayed hypersensitivity in guinea pigs and possibly humans has been reported (172). Indeed, croos-reactivity of acrylate compounds is also a possibility, suggesting positive sensitivity to more than one type (173). Chinese hamsters and rats exposed to an average concentration of 817 ppm and 820 ppm vapor, respectively, for 4 days resulted in some death and distinct signs of toxicity (174). Isobutyl acrylate is considered only slightly irritating (62). Data from an animal study have suggested the possibility of dermal sensitivity (172).

On the basis of dose–response data, ethylbutyl acrylate, $CH_2{:}CHCOOCH_2CH(C_2H_5)_2$, has relatively lower toxicological activity than the butyl and lower esters. Cyclohexyl acrylate, $CH_2{:}CHCOOC_6H_{11}$, is physiologically in the same activity range as butyl acrylate. However, by inhalation it appears to be of very low toxicity (75). The 2-ethylhexyl acrylate is slighly irritating to the eyes and respiratory tract, and moderately to the skin (146). Intraperitoneally injected 2-ethylhexyl acrylate showed a LD_{50} of 1325 mg/kg in the mouse (146).

Low molecular weight oxygenated acrylates such as 2-hydroxyethyl acrylate ($CH_2{:}CHCOOCH_2CH_2OH$), 2-ethoxyethyl acrylate, ($CH_2{:}CHCOOCH_2CH_2OCH_2CH_2CH_3$), and 3-ethoxypropyl acrylate [$CH_2{:}CHCOO(CH_2)_3$] exhibit physiological properties similar to those of ethyl acrylate, except that the hydroxypropyl monomer appears more acutely toxic following oral administration than its C2 homolog. As a polymer, it is used with urethane as dental filling material. The available data indicate that these compounds are readily hydrolyzed and the products further metabolized by mammals. There is a reported oral LD_{50} for 2-hydroxyethyl acrylate of 1070 mg/kg for rats. These compounds are irritating to eyes, skin, and mucous membranes. Cyclohexyl acrylate, (CH_2:$CHCOOC_6H_{11}$) is physiologically in the same activity range as butyl acrylate. However, by inhalation it appears to be of very low toxicity (75).

19.4.1.2 Chronic and Subchronic Toxicity: NA

19.4.1.3 Pharmacokinetics, Metabolism, and Mechanisms

19.4.1.3.1 Adsorption: NA

19.4.1.3.2 Distribution: NA

19.4.1.3.3 Excretion. As was shown for ethyl acrylate, *n*-butyl acrylate is biotransformed to and excreted as 3-hydroxypropanoic acid, lactic acid, and acetic acid, as well as two mercapturic acids, *N*-acetyl-S-(2-carboxyethyl)cysteine and *N*-acetyl-S-(2-alkoxy-carnonyl)ethyl]cysteine (162). A study revealed that *n*-butyl acrylate introduced intraperitoneally and orally to rats was rapidly metabolized and 70% excreted via exhaled air and 15–22% via urine (175).

19.4.1.4 Reproductive and Developmental. In developmental toxicity studies, inhalation exposure of pregnant mice to levels of butyl acrylate causing maternal toxicity can be embryolethal (176).

19.4.1.5 Carcinogenesis. Male and female rats exposed in whole-body inhalation chambers to concentrations of methyl or n-butyl acrylate vapor at concentrations of 0, 15, 45, and 135 ppm for 6 h/day, 5 days/week, for 24 months did not reveal signs of systemic toxicity or carcinogenesis. The rats did, however, show atrophy of neurogenic epithelial cells and hyperplasia of reserve cells in the nasal mucosa and opacity and neovascularization of the cornea (152).

19.5 Standards, Regulations, or Guidelines of Exposure

n-Butyl acrylate: ACGIH TLV 2 ppm (sensitizer); ACGIH A4 "Not classifiable as a human carcinogen."

19.6 Studies on Environmental Impact: NA

20.0 Methyl Methacrylate

20.0.1 CAS Number: [80-62-6]

20.0.2 Synonyms: Methacrylic acid methyl ester, methyl 2-methyl-2-propenoate, MME, MMA, 2-methylacrylic acid methyl ester, methyl methylacrylate; methyl α-methylacrylate, diakon, methyl 2-methylpropenoate, 2-methylpropenoicate acid methyl ester

20.0.3 Trade Names: NA

20.0.4 Molecular Weight: 100.13

20.0.5 Molecular Formula: $C_5H_8O_2$

20.0.6. Molecula Structure:

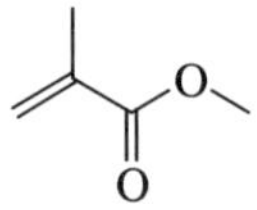

20.1 Chemical and Physical Properties

20.1.1 General

Methyl methacrylate is a clear, flammable liquid with a strong acrid odor. It is incompatible with peroxides, nitrates, oxidizers, alkalies, and even moisture (9).

20.2 Production and Use

Methyl methacrylate can be manufactured via oxidation of *tert-butyl* alcohol to methacrolein, then to methacrylic acid, and subsequent esterification with methanol. It polymerizes easily, especially when heated or in the presence of hydrochloric acid. It forms clear, ceramic-like resins and plastics, some commonly known as *Lucite* or *Plexiglas*. The monomer and polymers have wide applicability in medical technology as bone cement, but have been partially replaced by bucrylate and similar synthetics. The compound also serves as a medicinal spray adhesive or nonirritant bandage solvent, and in dental technology as a ceramic filler or cement, for which various patents have been applied for or granted. The compound can be used as a water-repellent on concrete surfaces.

20.3 Exposure Assessment

20.3.1 Air: NA

20.3.2 Background Levels: NA

20.3.3 Workplace Methods

Air samples can be adsorbed on XAD-2 media. The analyte is desorbed from sampling media using carbon disulfide and analyzed using a gas chromatograph equipped with a flame ionization detector (8, 9).

20.4 Toxic Effects

20.4.1 Experimental Studies

20.4.1.1 Acute Toxicity. The oral lethal dose is 6–9 g/kg in experimental animals and 1.8 mL/kg intraperitoneally (39). Rats receiving 25 mg methylmethacrylate twice weekly showed decreases in serum glycoprotein and albumin levels, but increases in other hepatic enzymatice activity (177). High concentrations that caused CNS effects, such as narcosis, have been shown to be due to the desheathing of the sciatic nerve and hyperpolarization of the resting potential in the amphibian (178).

Methyl methacrylate is classified as a low to moderate toxicant and is irritating to the eyes, skin, and respiratory tract. Methyl methacrylate is a potential sensitizer (140), especially when used in connection with hydroquinone or tertiary amines (179). This observation has been supported by another study where systemic sensitization was observed (180). The compound can also cause contact dermatitis (179).

A rhesus monkey accidentally exposed to methyl methacrylate vapor for 22 h was found comatose and eventually died. Necropsy revealed mottled liver, pulmonary edema, and atelectasis. Blood drawn 1.5 h prior to death exhibited normal hemogram and elevated levels of serum glutamic oxaloacetic transaminase, serum glutamic pyruvic transaminase, lactate dehydrogenase, phosphohexose isomerase, blood urea nitrogen, and serum sodium (181).

20.4.1.2 Chronic and Subchronic Toxicity. Rats and hamsters were exposed to methyl methacrylate vapor concentrations of 0, 25, 100, and 400 ppm 6 h/day 5 days/week, for 24 and 18 months, respectively. The exposures had no observed effects on mortality; hemotology, or blood and urine chemistries. Body weight of female rats and male and female hamsters exposed to 400 ppm were lower than those of controls. Chronic toxicity based on degeneration of nasal tissues was observed in male and female rats exposed to 100 and 400 ppm, but no demonstrable changes were found in hamsters (182).

Rats exposed orally to methyl methacrylate showed maximum accumulation in blood after 10–15 min followed by rapid decline within 1 h. No histopatholgical changes were identified in liver, kidney, heart, spleen, brain, lung, or intestine, nor were there changes in serum enzymes (183).

20.4.1.3 Pharmacokinetics, Metabolism, and Mechanisms

20.4.1.3.1 Adsorption: NA

20.4.1.3.2 Distribution: NA

20.4.1.3.3 Excretion. Oral administration of methyl methacrylate to rats revealed rapid ester hydrolysis with eventual elimination by the liver (184).

20.4.1.4 Reproductive and Development. Following intraperitoneal administration, some teratological effects with relatively high NOELS for resorptions, but lower fetal weight gains, have been observed by Singh et al. (147). A recent teratological study of rats by a relevant exposure route showed no teratogenicity at inhalation exposures of ≤ 2000 ppm (185). Exposure of pregnant rats to concentrations of methyl methacrylate vapor ranging from 0 to 2028 ppm 6 h/day during days 6–15 of gestation did not cause embryo or fetal toxicity despite toxicty to maternal rats (185).

20.4.1.5 Carcinogenesis. In several lifetime animal studies, there was no evidence that methyl methacrylate is carcinogenic. There was no evidence of treatment related tumors in rats or mice exposed for 2 years or hamsters exposed for 18 months by inhalation to concentrations of ≤ 1000 ppm (186, 187). In a 2-year oral study conducted with rats, no treatment related tumors were observed at concentrations of ≤ 2000 ppm in the drinking water (165).

An inhalation study using groups of rats and mice exposed to concentrations of methyl methacrylate vapor up to 5000 ppm for 14-week and 2-year periods did not show evidence of carcinogenicity. Exposure to 5000 ppm was lethal to all rats and most mice, and exposures to 2000 ppm were lethal to some rats and mice. All animals survived exposures to 500–1000 ppm. Inflammation and degeneration of nasal tissues, however, was observed in the surviving animals (187).

20.4.1.6 Genetic and Related Cellular Effects Studies. Bacterial Ames test and mammalian gene mutation gene assays were negative for mutagenicity when methyl methacrylate and 2-hydroxyethyl methacrylate were evaluated (188). In a different study, however, methyl acrylate, ethyl acrylate, methyl methacrylate, and ethyl methacrylate induced concentration-dependent increases in mutant frequency in mouse lymphoma cells suggestive of a clastogenic mechanism (189).

20.4.2 Human Experience

20.4.2.1 General Information

20.4.2.2 Clinical Cases. An individual working with artifical nail bonding chemicals became sensitized to a series of methacrylates, including tripropylene glycol diacrylate and methyl acrylate from photobonded nail gel, ethyl methacrylate, triethylene glycol dimethacrylate, and methyl methacrylate from nail liquid, and butyl methacrylate from nail hardener (190, 191). Following the diagnosis of occupational asthma among six cosmetologists working with artifical fingernail chemicals, NIOSH evaluated the effectiveness of modified table with downdraft ventilation to control methl- and ethyl methacrylate vapors. The results showed a statistically significant reduction of ethyl methacrylate to < 1 ppm and methacrylate to nondetectable levels (192).

20.4.2.2.1 Acute Toxicity: NA

20.4.2.2.2 Chronic and Subchronic Toxicity: NA

20.4.2.2.3 Pharmacokinetics, Metabolism, and Mechanisms: NA

20.4.2.2.4 Reproductive and Developmental: NA

20.4.2.2.5 Carcinogenesis: NA

20.4.2.2.6 Genetic and Related Cellular Effects Studies. No signs of abnormal chromosome aberration rates or SCE frequencies were detected in the peripheral lymphocytes of 38 male workers exposed occupationally to 0.9–71.9 ppm methyl methacrylate vapors during organic glass production. The results based on this human study are in general agreement with results generated fron animal studies and other epidemiological studies (193).

20.5 Standards, Regulations, or Guidelines of Exposure

Methyl methacrylate: OSHA PEL 100 ppm; ACGIH TLV 50 ppm ACGIH A4 "Not classifiable as a human carcinogen," NIOSH-REL 100 ppm.

20.6 Studies on Environmental Impact: NA

21.0 Ethyl Methacrylate

21.0.1 CAS Number: *[97-63-2]*

21.0.2 Synonyms: 2-Methyl-2-propenoic acid ethyl ester; Rhoplex ac-33; methacrylic acid ethyl ester; EMA; ethyl α-methylacrylate; ethyl 2-methyl-2-propenoate

21.0.3 Trade Names: NA

21.0.4 Molecular Weight: 114.14

21.0.5 Molecular Formula: $C_6H_{10}O_2$

21.0.5 Molecular Structure: 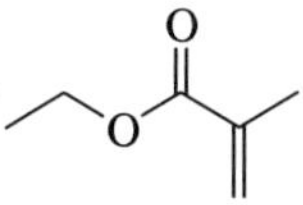

21.1 Chemical and Physical Properties

21.1.1 General

Ethyl methacrylate is a flammable liquid that polymerizes, but at a slower rate than the parent acrylate.

21.2 Production and Use

Ethyl methacrylate can be manufactured via a reaction of methacrylic acid or methyl acrylate with ethanol. It is used primarily for manufacturing polymers and as a component of acrylic polymers for surface coatings and as structural monomer for some artifical fingernail formulations.

21.3 Exposure Assessment: NA

21.4 Toxic Effects

21.4.1 Experimental Studies

21.4.1.1 Acute Toxicity. The Final Report on ethyl methacrylate concluded that the oral LD_{50} for rats ranged from 12.7 to 18.14 g/kg with lesions in the respiratory system and hemoglobinuria (194). Acute inhalation toxicity testing using rats caused ocular and respiratory tract irritation (194).

21.4.1.2 Chronic and Subchronic Toxicity: NA

21.4.1.3 Pharmacokinetics, Metabolism, and Mechanisms: NA

21.4.1.4 Reproductive and Developmental. Pregnant rats injected intraperitoneally exhibited embryotoxic and teratogenic effects (Andersen).

21.4.1.5 Carcinogenesis: NA

21.4.1.6 Genetic and Related Cellular Effects Studies. Both positive and negative mutagenicity testing has been reported (194).

21.4.1.7 Other: Neurological, Pulmonary, Skin Sensitization

21.4.2 Human Experience

21.4.2.1 General Information. Vapors of direct contact may cause mucous membrane irritation (177). The compound causes allergic contact dermatitis in humans (192, 194). Exposure to the monomer also may cause CNS effect (192).

21.5 Standards, Regulations, or Guidelines of Exposure: NA

21.6 Studies on Environmental Impact: NA

22.0 Butyl Methacrylates and Some Higher Homologs

22.0a Butyl Methacrylate

22.0.1a CAS Number: *[97-88-1]*

22.0.2a Synonyms: 2-Methyl-2-propenoic acid butyl ester; butyl 2-methyl-2-propenate; *n*-butyl methacrylate; BMA; 2-methyl butyl acrylate; butyl 2-methyl-2-propenoate; butyl methacrylate, stabilized with 25 ppm methylhydroquinone; methacrylate acid butyl ester

22.0.3a Trade Names: NA

22.0.4a Molecular Weight: 142.20

22.0.5a Molecular Formula: $C_8H_{14}O_2$

22.0.6a Molecular Structure:

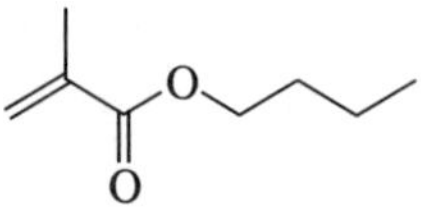

22.0b Isobutyl Methacrylate

22.0.1b CAS Number: *[97-86-9]*

22.0.2b Synonyms: 2-Methylpropyl methacrylate; 2-methyl-2-propenoic acid, 2-methylpropyl ester; 2-methyl-2-propenoic acid isobutyl ester; 2-methylpropyl 2-methylpropenoate; 2-propenoic acid, 2-methyl-, 2-methylpropyl ester; Isobutyl methacrylate, stabilized with 15–20 ppm MEHQ

22.0.3b Trade Names: NA

22.0.4b Molecular Weight: 142.20

22.0.5b Molecular Formula: $C_8H_{14}O_2$

22.0.6b Molecular Structure:

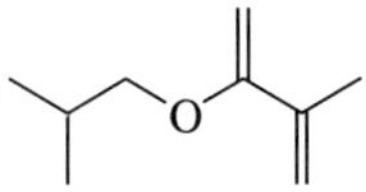

22.1 Chemical and Physical Properties

22.1.1 General

n-Butyl methacrylate is a combustible liquid with a faint characteristic odor of esters. Isobutyl methacrylate is moderately flammable.

22.2 Production and Use

n-Butyl methacrylate is manufactured via a reaction of methacrylic acid or methyl methacrylate with butanol. *n*-Butyl methacrylate monomer and polymers are used in dental technology, as components in oil-dispersible pesticides, and as copolymers, for example, in paraffin embedding media. It is also used for the manufacture of contact lenses and for acrylic surface coatings.

22.3 Exposure Assessment: NA

22.4 Toxic Effects

22.4.1 Experimental Studies

22.4.1.1 Acute Toxicity. *n*-Butyl methacrylate is corrosive to skin and presumably eyes of rabbits (195). Otherwise, it apparently has low systemic toxicity. It has been determined to be practically nontoxic to rats and mice in acute oral and dermal exposures (see Table 79.12). In a 4 week vapor inhalation study, only nasal irritation was shown at concentrations of $\geq$900 ppm (196). Subchronic feeding experiments for 4–6 months with rats showed moderate cumulative properties at 5% of the LD_{50} (197). Physiological, biochemical, and pathomorphological indexes pointed to a NOEL of 3.5 mg/m^3 in air (197). More recent data, however, suggest that possible dermal sensitization may result from exposure (198). Rats exposed for 24 h to a series of acrylates showed LC_{50} values in the following order: methyl acrylate > ethyl acrylate > butyl acrylate > butyl methacrylate > methyl methacrylate > ethyl methacrylate (134). Isobutyl methacrylate appears slightly more toxic, on the basis of some lethality data, than does the *n*-butyl isomer (see Tables 79.11 and 79.12).

2-Ethylisohexyl methacrylate [CH_2:C(CH_3)$COOCH_2CH_2OH$] serves as an important monomer and potential polymer in selected biologic areas. These include applications in optometry, where the products are utilized as basic material for contact lenses and as dental filler. In the pharmaceutical areas it is used as a control led-release drug medium, in medicine as a synthetic wound dressing component, and for surface tension control in bioengineering. Toxicologically, the compound appears almost 3 times as toxic as the methacrylic ethyl ester. A LD_{50} determination by intraperitoneal injection resulted in a value of 497 mg/kg for the mouse (148).

Low molecular weight oxygenated acrylates such as 2-hydroxyethyl acrylate, (CH_2:$CHCOOCH_2CH_2OH$), 2-ethoxyethyl acrylate [(CH_2:$CHCOOCH_2CH_2OCH_2CH_2$ CH_3)], and 3-ethoxypropyl acrylate [CH_2:CHCOO(CH_2)$_3$] exhibit physiological properties similar to those of ethyl acrylate (see Table 79.12), except that the hydroxypropyl monomer appears more acutely toxic following oral administration than its C2 homolog. As a polymer, it is used with urethane as dental filling material. The available data indicate that these compounds are readily hydrolyzed and the products further metabolized by mammals. There is a reported oral LD_{50} for 2-hydroxyethyl acrylate of 1070 mg/kg for rats (97). These compounds are irritating to eyes, skin, and mucous membranes.

2-Ethylhexyl acrylate causes dermatitis, but effective dermal protection is provided by gloves (199). 2-Ethylhexyl acrylate was administered to rats at doses of 3, 20, and 60 mg/kg 5 days/week for 13 weeks. The rats were evaluated for neurotoxicity via a functional observational battery (FOB) and found to exhibit changes in muscular function independent of exposure dose, but no signs of neuropathological changes (200).

22.4.2 Human Experience

22.4.2.1 General Information: NA

22.4.2.2 Clinical Cases

22.4.2.2.1 Acute Toxicity. Facsimile (fax) machine repair technicians repeatedly exposed occupationally to generated emissions containing *n*-butyl methacrylate developed respiratory and blood abnormalities. Respiratory symptoms included sore throat, chest tightness, dry cough, and dyspnea. Chest radiographs were normal, but three of seven technicians had lung crackles and spirometric abnormalities. Blood analyses for these same affected individuals showed increased serum levels of immunoglobulins IgE or IgM (201).

22.5 Standards, Regulations, or Guidelines of Exposure: NA

22.6 Studies on Environmental Impact: NA

23.0 Hydroxyalkyl Esters: Lactates

It is assumed that the low toxicity of the lower lactates stems from their natural occurrence in milk and other food products. Accordingly, most lower hydroxyalkyl esters, owing to their low toxicity, are used as ingredients in the food, biochemical, cosmetic, and pharmaceutical industries. In addition, expanded use now includes applications as solvents for ethyl cellulose gums, oils, dyes, synthetic polymers, and paints (202).

When lactates are ingested, they probably are readily saponified into the basic acid and alcohol, which, in turn, are rapidly metabolized and excreted. In general, on the basis of animal studies and human use, the most significant effects caused by exposure to lactate esters are respiratory, dermal, and ocular irritation (202). Irritation may be associated with the formation of lactic acid, a product of hydrolysis of lactate esters. Lactate esters are not thought to cause dermal sensitization, however, one case is reported of an individual responding to ethyl lactate that was present in an acne medication (202).

A summary of physical and chemical properties is found in Table 79.13, and are summarized in toxicological data, shown in Table 79.14.

23.0a Methyl Lactate

23.0.1a CAS Number: *[547-64-8]*

23.0.2a Synonyms: Methyl (+/−)-lactate; propanoic acid, 2-hydroxy-, methyl ester

23.0.3a Trade Names: NA

23.0.4a Molecular Weight: 104.11

23.0.5a Molecular Formula: $C_4H_8O_3$

23.0.6a Molecular Structure:

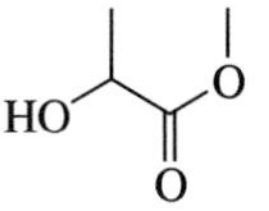

23.0b Ethyl Lactate

23.0.1b CAS Number: *[97-64-3]*

23.0.2b Synonyms: Lactic acid ethyl ester; ethyl 2-hydroxypropionate; propanoic acid, 2-hydroxy-, ethyl ester; ethyl L-lactate

23.0.3b Trade Names: NA

23.0.4b Molecular Weight: 118.13

23.0.5b Molecular Formula: $C_5H_{10}O_3$

23.0.6b Molecular Structure:

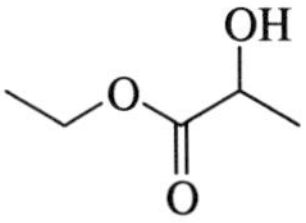

Table 79.13. Summary of Physical and Chemical Properties of Some Lactates and Pyruvates (12, 202)

Compound	CAS Number	Molecular Formula	Molecular Weight	Boiling Point (°C)	Melting Point (°C)	Specific Gravity (at 25 °C)	Solubility[a] in Water at (at 68° F)	Refractive Index (at 20°C)	Vapor density (Air = 1)	Vapor Pressure (mmHg at 20°C)	Flash Point (°C)	LEL % by Vol. Room Temp.	UEL % by Vol. Room Temp.
Methyl lactate	[*547-64-8*]		104.1	35		1.092	s	1.414		2.5	57		
Ethyl lactate	[*97-64-3*]	$C_5H_{10}O_3$	118.13	153		1.033	s	1.413		1.7	61		
Isopropyl lactate	[*617-51-6*]		132.2	157		0.991	s	1.410		1.3	60		
n-Butyl lactate	[*138-22-7*]	$C_7H_{14}O_3$	146.19	188		0.984	s	1.422		0.4	79		
Amyl lactate	[*6382-06-5*]		160.2	207		0.964	s	1.424		0.09	87		
Methyl pyruvate		$C_4H_6O_3$	102.09	134–137			d	1.4046					
Ethyl pyruvate	[*617-35-6*]	$C_5H_8O_3$	116.12	144			d	1.4052					
Allyl acetate	[*591-87-7*]	$C_5H_8O_2$	100.13	103.5		0.928	d	1.4049	3.45		21		
Geranyl acetate	[*105-87-3*]	$C_{12}H_{20}O_2$	196.28	242									
Linalyl acetate	[*115-95-7*]	$C_{12}H_{20}O_2$	196.29	220			i	1.4544					
Cyclohexyl acetate	[*622-45-7*]	$C_8H_{14}O_2$	142.20	173			i	1.4401	4.9	7 (30)	57.8		
Phenyl acetate	[*122-79-2*]	$C_8H_8O_2$	136.16	195.7		1.073	d	1.5033	4.7		80		
Methyl acetoacetate	[*105-45-3*]	$C_5H_8O_3$	116.12	171.7	– 80	1.077	v	1.4184	4.0		76.7	3.1	16
Ethyl acetoacetate	[*141-97-9*]	$C_6H_{10}O_2$	130.14	– 45.4	180.8	1.03	v	1.4194	4.48	0.8 (20)	84.4		

[a] Solubility in water: v = very soluble; s = soluble; d = slightly soluble; i = insoluble.

Table 79.14. Summary of Acute Oral and Inhalation Toxicity of Some Lactates (202)

Compound	Parameter	Species	Dose	Observed Effects
Methyl	Oral LD_{50}	Rat	> 2000 mg/kg	Pilorection ≤ 24 h; absence of gross necroscopy changes
	Inhalation $LC_{50/4\,h}$		> 5030 mg/m^3	During exposure decreased breathing rates and wet nares; postexposure wet fur; gross necropsy showed 7/10 with grayish lungs and two lungs with irregular surfaces
Ethyl	Oral LD_{50}	Rat	> 2000 mg/kg	Pilorection up to 24 h; absence of gross necropsy changes
	Inhalation $LC_{50/4\,h}$		> 5400 mg/m^3	During exposure decreased breathing rates, pilorection, lachrymation, and wet nares; gross necropsy showed pale lungs with spots
Propyl	Oral LD_{50}	Rat	> 2000 mg/kg	Sluggishness ≤4 h; absence of gross necropsy changes
Isoprophyl	Oral LD_{50}	Rat	> 2000 mg/kg	Pilorection ≤24 h; absence of gross necropsy changes
Butyl	Oral LD_{50}	Rat	> 2000 mg/kg	Pilorection ≤24 h and diarrhea; absence of gross necropsy changes
	Inhalation $LC_{50/4\,h}$		> 5140 mg/m^3	During exposure decreased breathing rates and wet head and fur. absence of necropsy changes
Isobutyl	Inhalation $LC_{50/4\,h}$	Rat	> 6160 mg/m^3	During exposure decreased breathing rates, pilorection, and hunched appearance; postexposure apnea; absence of gross necropsy changes

23.0c n-Butyl Lactate

23.0.1c CAS Number: *[138-22-7]*

23.0.2c Synonyms: Lactic acid butyl ester, butyl lactate, 2-propanoic acid, butyl α-hydroxypropionate, butyl 2-hydroxypropanoate, 2-hydroxypropanoic acid butyl ester

23.0.3c Trade Names: NA

23.0.4c Molecular Weight: 146.19

23.0.5c *Molecular Formula:* $C_7H_{14}O_3$

23.0.6c *Molecular Structure:*

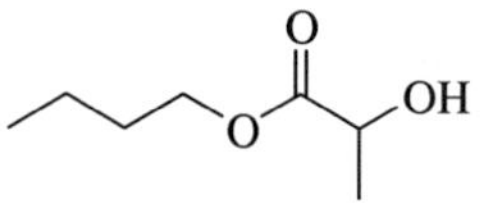

23.1 Chemical and Physical Properties

23.1.1 *General*

Methyl, ethyl, and butyl lactates are a colorless liquids with odors characteristic of most esters.

23.1.2 *Odor and Warning Properties*

The respective odor and odor nuisance thresholds, reported in concentrations expressed as mg/m^3, were 0.89 and 65 for ethyl lactate, 0.095 and 9 for butyl lactate, and 0.45 and 40 for 2-ethylhexyl lacate (202).

23.2 Production and Use

The compounds can be prepared via esterification of lactic acid and the respective alcohol. Methyl lactate is used as a solvent for various forms of cellulose, as a component in lacquers and stains, and as an intermediate for a herbicide. Ethyl lactate can serve to reduce the alkalinity in cosmetic preparations. It is also used as a solvent for basic dyes, in manufacture of lacquers and safety glass; as a solvent for various forms of cellulose, as a component of varnishes, and as a food flavoring agent. Cetyl lactate (hexadecyl 2-hydroxypropionate) [$CH_3CHOOH)COOC_{16}H_{33}$], occurs as a waxy solid. It is used as a nontoxic emollient to improve the feel and texture of cosmetic and pharmaceutical preparations. Cetyl lactate is skin-absorbable and enhances the absorption of other compounds (203).

23.3 Exposure Assessment: NA

23.4 Toxic Effects

23.4.1 *Experimental Studies*

23.4.1.1 Acute Toxicity. It was determined that alkyl lactate esters methyl-, ethyl-, propyl-, and butyl lactate were only slightly more toxic to fish and *Daphnia magna* when compared to nonolar narcotic compounds and that these compounds also rapidly biodegrade (204).

Ingestion studies using rats exposed to 2000 mg/kg methyl-, ethyl-, propyl-, isopropyl, butyl-, isoamyl-, 2-ethylhexyl-, isodecyl, or benzyl lactate did not cause mortality. Nose-only inhalation studies using rats exposed to vapor concentrations as high as 5000 mg/m^3 for 4 h also did not cause mortality (202). LD_{50} values for ethyl lactate in mouse studies have been observed as 25 g/kg orally, 2.5 g/kg subcutaneously, and 0.6 g/kg intravenously (205). The LD_{50} for acute dermal exposure was > 5000 mg/kg ethyl and butyl lactates

(202). It has been reported that the other lactates would be expected to yield a similar response, based on the dermal LD_{50} data for lactic acid and the respective alcohols (202).

23.4.1.2 Chronic and Subchronic Toxicity: NA

23.4.1.3 Pharmacokinetics, Metabolism, and Mechanisms

23.4.1.3.1 Adsorption: NA

23.4.1.3.2 Distribution: NA

23.4.1.3.3 Excretion. When lactates are ingested, they probably are readily saponified into lactic acid and respective alcohols, which, in turn, are rapidly metabolized and excreted.

23.4.1.4 Reproductive and Developmental. Neither dermal nor inhalation studies using rats exposed to ethyl-L-lactate and 2-ethylhexyl-L-lactate, respectively, revealed signs of developmental effects (202).

23.4.1.5 Carcinogenesis: NA

23.4.1.6 Genetic and Related Cellular Effects Studies. Information was found for mutagenicity testing of ethyl-L-lactate using Ames assay. The reported results indicated no indication of mutagenicity in any tester strains evaluated with and without metabolic activation (202).

23.4.2 Human Experience

23.4.2.1 General Information: NA

23.4.2.2 Clinical Cases

23.4.2.2.1 Acute Toxicity. Sensory irritation of the nasopharyngeal region of humans exposed to concentrations of *n*-butyl lactate at levels of > 5 ppm (202).

23.4.2.3 Epidemiology Studies

23.4.2.3.1 Acute Toxicity. Workers exposed to 7 ppm butyl lactate with occasional short peak exposures of 11 ppm reportedly experienced headache and pharyngeal and laryngeal irritation, coughing, and postexposure sleepiness for some (21).

23.5 Standards, Regulations, or Guidelines of Exposure

n-Butyl lactate: ACGIH TLV 5 ppm; NIOSH REL 5 ppm.

23.6 Studies on Environmental Impact

Lactate esters are readily biodegradable in the environment (204).

24.0 Pyruvic Keto Acid Esters

Methyl and ethyl pyruvate [ethyl and methyl 2-oxopropionate; $CH_3COCOOCH_3$ and $CH_3COCOOC_2H_5$] are liquids at ambient temperature and show a very low toxicity rating. According to Asakawa et al. (206), in some cellular systems phenyl pyruvate is enzymatically converted to phenyl acetate for further metabolism. These hydrolytic products are rapidly degraded or reutilized in the mammalian system. Ethyl pyruvate, dimethyl succinate, and aconitic acid are three common flavorings used in candies, beverages, and baked goods. *Salmonella*/mammalian microsome testing at doses of 32, 160, 800, 4000, and 20,000 μg/plate resulted in no observation of mutagenic activity (207).

24.0a Methyl Pyruvate

24.0.1a CAS Number: *[600-22-6]*

24.0.2a Synonyms: Propanoic acid, 2-oxo-, methyl ester; pyruvic acid methylester

24.0.3a Trade Names: NA

24.0.4a Molecular Weight: 102.09

24.0.5a Molecular Formula: $C_4H_6O_3$

24.0.6a Molecular Structure:

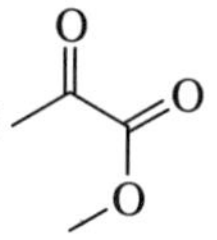

24.0b Ethyl Pyruvate

24.0.1b CAS Number: *[617-35-6]*

24.0.2b Synonyms: Ethyl 2-oxopropionate; pyruvic acid ethyl ester

24.0.3b Trade Names: NA

24.0.4b Molecular Weight: 116.12

24.0.5b Molecular Formula: $C_5H_8O_3$

24.0.6b Molecular Structure:

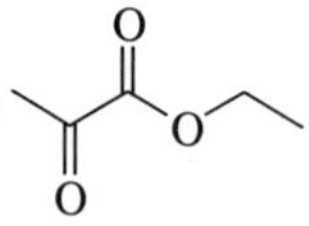

BIBLIOGRAPHY

1. W. F. von Oettingen, The aliphatic acids and their esters: Toxicity and potential dangers — The saturated monobasic aliphatic acids and their esters-toxicity of formic acid and esters. *Arch. Ind. Health* **20**, 517–531 (1959).
2. W. F. von Oettingen, The aliphatic acids and their esters: Toxicity and potential dangers — Toxicity of acetic acid and esters. *Arch. Ind. Health* **21**, 28–65 (1960).

3. J. Autian, Structure-toxicity relationships of acrylic monomers. *Environ. Health Perspect.* **11**, 141–152 (1975).
4. P. M. Janner et al., Food flavourings and compounds of related structure. *Food Cosmet. Toxicol.* **2**, 327 (1964).
5. J. J. Clary, V. J. Feron, and J. A. van Velthuijsen, Safety assessment of lacate esters. *Regul. Toxicol. Pharmacol.* **27**, 88–97 (1998).
6. H. F. Smyth and H. F. Smyth, Jr., Inhalation experiments with certain lacquer solvents. *J. Ind. Hyg.* **10**, 163, 261 (1928).
7. J. C. Munch, Aliphatic alcohols and alkyl esters: Narcotic and lethal potencies to tadpoles and rabbits. *Ind. Med.* **41**(4), 31–33 (1972).
8. National Institute for Occupational Safety and Health (NIOSH), *Manual of Analytical Methods*, 4th ed., U.S. Department of Health and Human Services, Public Health Service, Centers for Disease Control, Cincinnati, OH, 1994.
9. National Institute for Occupational Safety and Health (NIOSH), *Pocket Guide to Chemical Hazards*, U.S. Department of Health and Human Services, Public Health Service, Centers for Disease Control, Cincinnati, OH, 1997.
10. American Conference of Governmental and Industrial Hygienists (ACGIH), *2000 TLVs and BEIs*, ACGIH, Cincinnati, OH, 2000.
11. Occupational Safety and Health Administration *Code of Federal Regulations*, Title 29 1910.1000.
12. D. R. Lide, ed. *Handbook of Chemistry and Physics*, CRC Press, Boca Rotan, FL, 1998.
13. H. H. Schrenk, et al., Acute response of guinea pigs to vapours of some new commercial organic compounds. *Public Health Rep.* **51**, 1329–1337 (1936).
14. H. F. Smyth, Jr. et al., Range-finding toxicity data: List VI. *Am. Ind. Hyg. Assoc. J.* **23**, 95 (1962).
15. H. F. Smyth, Jr., *Am. Ind. Hyg. Assoc.* **17**, 129 (1956).
16. F. Flurry and W. Neuman, (1927), cited in von Oettingen (1, 2).
17. O. Getter, The detection, identification, and quantitative determination of methyl formate in tissues. *Am. J. Clin. Pathol.* **10**, 188 (1940).
18. J. E. Amoore and E. Hautala, Odor as an aid to chemical safety: Odor thresholds compared with threshold limit values and volatilities for 214 industrial chemicals and water dilution. *J. Appl. Toxicol.* **3**(6), 272–290 (1983).
19. A. P. Murdanoto, et al., Ester synthesis by NAD(+)-dependent dehydrogenation of hemiacetal: production of methyl formate by cells of methylotrophic yeasts. *Biosci. Biotechnol. Biochem.* **61**(8), 1391–1393 (1997).
20. A. D. Fraser and W. MacNeil, Gas chromatographic analysis of methyl formate and application in methanol poisoning cases. *J. Anal. Toxicol.* **13**(2), 73–76 (1989).
21. American Conference of Governmental and Industrial Hygienists (ACGIH), *Documentation of the TLVs and BEIs*, ACGIH, Cincinnati, OH 1991.
22. F. Flurry and W. Wirth, (1936), cited in von Oettingen (1, 2).
23. P. Duquenois and P. Revel, (1935), in von Oettingen (1, 2).
24. F. J. Roe and M. H. Salaaman, *Br. J. Cancer* **9**, 177 (1955).
25. G. D. Stoner et al., Test for carcinogenicity of food additives and chemotherapeutic agents by the pulmonary tumor response in strain A mice. *Cancer Res.* **33**, 3069–3085.

26. U. K. Misra, Ethyl acetate. In R. Snyder, ed., *Ethel Browning's Toxicity and Metabolism of Industrial Solvents*, Elsevier, New York, pp. 233–246.
27. U. Groschel-Stewart et al. Aprotic polar solvents inducing chromosomal malsegregation in yeast interfere with the assembly of porcine brain tubulin in vitro. *Mutat. Res.* **149**(3), 333–338 (1985).
28. K. Kohriyama et al., Optic neuropathy induced by thinner sniffing. *J. Univ. Occup. Environ. Health (Jpn.)* **11**(4), 449–454 (1989).
29. O. Tada, K. Nakakaki, and S. Fukabori, *Rodo Kagaku* **50**(4), 239 (1974).
30. S. Tambo, *Nichidai Igaku Zasshi* **32**, 349 (1973).
31. J. H. Sterner et al., *Am. Ind. Hyg. Assoc. Q.* **10**, 53 (1949).
32. R. S. McLaughlin, *Am. J. Ophthalmol.* **29**, 795 (1966).
33. K. Takagi and I. Takyangi, *J. Pharm. Pharmacol.* **18**, 795 (1966).
34. W. H. Stein et al., *J. Am. Chem. Soc.* **76**, 2848 (1954).
35. J. E. Cometto-Muniz and W. S. Cain, *Pharmacol., Biochem. Behav.* **39**(4), 983–989 (1991).
36. H. E. Kennah, II. et al., *Fundam. Appl. Toxicol.* **12**(2), 258–268 (1989).
37. K. J. Reus, Inaugural Dissertation (1993), in von Oettingen (2).
38. K. W. Nelson et al., *J. Ind, Hyg. Toxicol.* **25**, 282 (1943).
39. C. R. Spealman et al., Monomeric methyl methacrylate. *Ind. Med.* **14**, 292 (1945).
40. H. C. Smyth and C. P. Carpenter, *J. Ind. Hyg. Toxicol.* **30**, 63 (1948).
41. L. Silverman, H. F. Schulte, and M. First, *J. Ind. Hyg. Toxicol.* **28**,262 (1946).
42. R. R. Sayers, H. H. Schrenk, and F. A. Patty, *Public Health Rep.* **51**, 1229 (1936).
43. K. B. Lehmann (1913), in von Oettingen (2).
44. I. N. Rymoreva and G. Z. Yakovleva, *Khim. Farm. Zh.* **10**(2), 139–142 (1976)
45. D. Sasse and A. Schenk, *Ger. Acta Anat. (Basel)*, **93**(1), 78–87 (1975).
46. I. Snapper, A. Grunbaum, and S. Sturkop, *Biochem. Z.* **84**, 358 (1945).
47. J. Nakano and J. M. Kessinger, *Eur. J. Pharmacol.* **17**(2), 195 (1972).
48. J. Nakano, S. E. Moore, and C. L. Kesinger, *J. Pharm. Pharmocol.* **25**(12), 1018 (1973).
49. G. Esposito, A. Faellia, and V. Capraro, *Biochim. Biophys. Acta* **426**(3), 489 (1976).
50. E. J. Gallaher and T. A. Loomis, *Toxicol. Appl. Pharmacol.* **34**, 309 (1975).
51. P. K. Seth and S. P. Srivastava, *Bull. Environ. Contam. Toxicol.* **12**(5), 612 (1974).
52. H. F. Smyth et al., An exploration of joint toxic action: Twenty seven industrial chemicals intubated in rats in all possible pairs. *Toxicol. Appl. Pharmacol.* **24**, 340 (1969).
53. M. Seeber et al., Solvent exposure and ratings of well-being: Dose-effect relationships and consistency of data. *Environ. Res.* **73**(1–2), 81–91 (1997).
54. J. G. Sivak, K. L. Herbert, and A. L. Baczmanski, The use of the cultured bovine lens to measure the *in-vitro* ocular irritancy of ketones and acetates. *Alternatives Lab. Anim.* **23**(5), 689–698 (1995).
55. S. E. Bowen and R. L. Balster, A comparison of the acute behavioral effects of inhaled amyl, ethyl, and butyl acetate in mice. *Fundam. Appl. Toxicol.* **35**(2), 189–196 (1997).
55a. K. S. Loveday, B. E. Anderson, M. A. Resnick, and E. Zeiger, Chromosome aberration and sister chromatid exchange tests in Chinese hamster ovary cells *in vitro* V: Results with 46 chemicals. *Environ. Molec. Mutagen.* **16**, 272–303 (1990).

56. J. Fernandez and P. Droz, *Arch. Mal. Prof. Med. Trav. Secur. Soc.* **35**(12), 953 (1965).

57. K. Nomiyana and H. Nomiyana, *Int. Arch. Arbeitsmed.* **32**, 75 (1974).

58. K. Nomiyana and H. Nomiyana, *Int. Arch. Arbeitsmed.* **32**, 85 (1974).

59. J. DeCeaurriz et al., Concentration-dependent behavioral changes in mice following short-term inhalation exposure to various industrial solvents. *Toxicol. Appl. Pharmacol.* **67**(3), 383–389 (1983).

60. R. Akimoto et al., Systemic and local toxicity in the rat of methyl tert-butyl ether A gallstone dissolution agent. *J. Surg. Res.* **53**(6), 572–577 (1992).

61. Z. Korsak and K. Rydzynski, Effects of acute combined inhalation exposure to n-butyl alcohol and n-butyl acetate in experimental animals. *Int. J. Occup. Med. Environ. Health* **7**(3), 273–280 (1994).

62. W. M. Grant, *Toxicology of the Eye*, Thomas, Springfield, IL, 1986.

63. N. N. Titova and R. G. Zakirova, *Kazan. Med. Zh.* **58**(1), 36 (1997).

64. A. Iregren et al., Irritation effects from experimental exposure to *n*-butyl acetate. *Am. J. Ind. Med.* **24**(6), 727–742 (1993).

65. R. M. David et al., Evaluation of subchronic neurotoxicity of *n*-butyl acetate vapor. *Neurotoxicology* **19**(6), 809–822 (1998).

66. S. Y. Tsai et al., Neurobehavirol effects of occupational exposure to low-level organic solvents among Taiwanese workers in paint factories. *Environ. Res.* **73**(1–2), 146–155 (1997).

67. D. Mascia and V. Querci, *Boll. Soc. Ital. Biol. Sper.* **45**(10), 695 (1969).

68. F. A. Patty, W. P. Yant, and H. H. Schrenk, Public *Health Rep.* **51**, 811 (1936).

69. G. Gorgonne et al., *Ann. Ottalmol. Clin. Ocul.* **96**(6), 313 (1970).

70. F. K. Zimmerman et al., Acetone, methyl ethyl ketone, ethyl acetate, acetonitrile and other polar aprotic solvents are strong inducers of aneuploidy in *Saccharomyces cerevisiae. Mutat. Res.* **149**(3), 339–351 (1985).

71. A. Inserra, *Boll. Soc. Ital. Biol. Sper.* **42**(9), 551 (1966).

72. K. Zaikov and G. Bobev, *Khig. Zdraveopaz.* **21**(2), 141 (1978).

73. B. Ballantyne, T. R. Tyler, and C. S. Auletta, *Vet. Hum. Toxicol.* **28**(3), 213–215 (1986).

74. K. Shimpo et al., Inhibition of hepatocellualr carcinomal development and erythrocyte polyamine levels in ODS rats fed on 3′-methyl-4-dimethylaminoazobenzene by hemicalcium ascorbate, 2-O-octadecylasorbic acid and ascorbyl palmitate. *Cancer Detect. prev.* **20**(2), 137–145 (1996).

75. C. P. Carpenter, C. S. Weil, and H. F. Smyth, Jr., *Toxicol. Appl. Pharmocol.* **28**, 313 (1974).

76. H. F. Smyth, C. P. Carpenter, and C. S. Weil, *J. Ind. Hyg. Toxicol.* **31**, 60 (1949).

77. L. V. Tiunova and A. P. Rumiantsev, *B-y-ull. Eksp. Bio. Med.* **79**(4), 101 (1975).

78. G. Weitzel, *Hoppe-seyler's Z. Physiol. Chem.* **287**, 254 (1951).

79. G. D. Nielsen, J. C. Bakbo, and E. Holst, Sensory irritation and pulmonary irritation by airborne allyl acetate, allyl alcohol, and allyl ether compared to acrolein. *Acta Pharmacol. Toxicol.* **54**(4), 292–298 (1984).

80. Y. Ittah and U. Zisman, Evaluation of volatile allyl alcohol derivatives for control of snails on cut roses for export. *Pestic. Sci.* **35**, 183–186 (1992).

81. M. E. Hurtt et al., Developmental toxicity of oral and inhaled vinyl acetate in the rat. *Fundam. Appl. Toxicol.* **24**(2), 198–205 (1995).

82. M. S. Bogdanffy et al., Chronic toxicity and oncogenicity study with vinyl acetate in the rat: In utero exposure to drinking water. *Fundam. Appl. Toxicol.* **23**(2), 206–214 (1994).
83. M. S. Bogdanffy et al., Chronic toxicity and oncogenicity inhalation study with vinyl acetate in that rat and mouse. *Fundam. Appl. Toxicol.* **23**(2), 215–229 (1994).
84. W. Lijinsky amd M. D. Reuber, Chronic toxicity studies of vinyl acetate in Fischer rats. *Toxicol. Appl. Pharmacol.* **68**(1), 43–53 (1983).
85. J. R. Kuykendall, M. L. Taylor, and M. S. Bogdanffy, Cytoxicity and DNA-protein crosslink formation in rat nasal tissues exposed to vinyl acetate are carboxylesterase-mediated. *Toxicol. Appl. Pharmacol.* **123**(2), 283–292 (1993).
86. H. Norppa et al., Chromosome damage induced by vinyl acetate through *in vitro* formation of acetaldehyde in human lymphocytes and Chinese hamster ovary cells. *Cancer Res.* **45**(10), 4816–4821 (1985).
87. P. Sipi, H. Jarventaus, and H. Norppa, Sister-chromatid exchanges induced by vinyl esters and respective carboxylic acids in cultured human lymphocytes. *Muta. Res.* **279**(2), 75–82 (1992).
88. Z. P. Agaronyan and V. A. Amatuni, *Krovoobrashchenie* **13**(4), 31–36 (1980).
89. W. Muller (1932), cited in von Oettingen (2).
90. H. G. Bray, W. V. Thorpe, and K. White, *Biochem. J.* **48**, 88 (1951).
91. National Toxicology Program (NTP), 1993.
92. National Toxicology Program (NTP), 1996.
93. E. Zeiger, B. Anderson, S. Haworth, and K. Mortelman, Salmonella Mutagenicity Tests V. Results from the Testing of 311 Chemicals. *Environ. Molec. Mutagen.* **19**(Suppl 21), 2–141 (1992).
94. C. M. Gruber, *J. Lab. Clin. Med.* **9**, 15 (1923).
95. C. M. Gruber, *J. Lab. Clin. Med.* **9**, 92 (1923).
96. B. E. Graham and M. H. Kuizenga, *J. Pharmacol. Exp. Ther.* **84**, 358 (1945).
97. H. F. Smyth, C. P. Carpenter, and C. S. Weil, *Arch. Ind. Hyg. Occup. Med.* **4**, 119 (1951).
98. C. Clerici et al., Local and systemic effects of intraduodenal exposure to topical gallstone solvents ethyl propionate and methyl tert-butyl ether in the rabbit. *Dig. Dis. Sci.* **42**(3), 497–502 (1997).
99. R. Kuhn et al., *Water Res.* **23**(4), 501–510 (1989).
100. C. Y. Chen et al., Toxic effects of cholelitholytic solvents on gallbladder and liver: A piglet model study. *Dig. Dis. Sci.* **40**(2), 419–426 (1995).
101. P. J. Deisinger, R. J. Boatman, and D. Guest, The metabolism and disposition of ethyl-3-ethoxypropionate in the rat. *Xenobiotica* **20**(10), 989–998 (1990).
102. C. W. Glancy, New solvents for high solids coatings. *In Proceedings of the 14th Water-Borne and Higher-Solids Coatings Symposium*, University of Southern Mississippi, Hattiesburg, pp. 209–229.
103. J. B. Veavers et al., *J. Econ. Entomol.* **65**(6), 1740, (1972).
104. M. Wakselman, I Cerutti, and C. Chany, Anti-tumor protection induced in mice by fatty acid conjugates: Alkyl butyrates and poly(ethylene glycol) dibutyrates. *Int. J. Cancer* **46**(3), 462–467 (1990).
105. F. A. Patty, *Industrial Hygiene and Toxicology*, 2nd rev. ed., Vol. 2, Wiley, New York, 1963.
106. T. Koizumi et al., *Nippon Jozo Kyokai Zasshi* **70**(3), 192 (1975).
107. R. Sugawara, Y. Tominaga, and T. Suzuki, *Insect Biochem.* **7**, 483 (1977).

108. W. M. Rogoff et al., *Ann. Entomol. Soc. Am.* **66**(2), 262 (1973).
109. K. N. Gaind and R. D. Budhiraja, *Indian J. Pharm.* **28**(6), 156 (1966).
110. T. Kurihara and M. Kikuchi, *Yakugaku Zasshi* **95**, 992 (1975).
111. M. J. Coon, *Fed. Proc., Fed. Am. Soc. Exp. Biol.* **14**, 762 (1955).
112. M. S. Quraishi, *Can. Entomol.* **104**(10), 1505 (1972).
113. D. W. Stetten, Jr., and J. Salcedo, Jr., *J. Nutr.* **29**, 167 (1945).
114. H. D. Kesten, J. Salcedo, Jr., and D. W. Stetten, Jr., *J. Nutr.* **29**, 171 (1945)
115. E. L. Platecow and E. Voss, *J. Am. Pharm. Assoc.* **43**, 690 (1954).
116. L. Dan and M. Laposata, Ethyl palmitate and ethyl oleate are the predominant fatty acid ethyl esters in the blood after ethanol ingestion and their synthesis is differentially influenced by the extracellular concentrations of their corresponding fatty acids. *Alcohol.: Clin. Exp. Res.* **21**(2), 286–292 (1997).
117. V. Sebestik and V. Brabec, Experimental elimination of the splenic function by ethyl and methyl palmitate and significance of these substances from an immunological point of view. *Folia Haematol.—Int. Mag. Klin. Morphol. Blutforsch.* **110**(6), 917–923 (1983).
118. A. E. Stuart and I. I. Smith, Histological effects of lipids on the liver and spleen of mice. *J. Pathol.* **115**(2), 63–71 (1975).
119. I. I. Smith and A. E. Stuart, Effects of simple lipids on macrophages *in vitro*. *J. Pathol.* **115**(1), 13–16 (1975).
120. J. E. Manautou and G. P. Carlson, Ethanol-induces fatty acid ethyl ester formation *in-vivo* and *in-vitro* in rat lung. *Toxicology* **70**(3), 303–312 (1991).
121. O. G. Fitzhugh and A. A. Nelson, *Proc. Soc. Exp. Biol. Med.* **61**, 195 (1946).
122. C. Dorko, Antioxidants used in foods. *Food Technol.* **48**, 33 (1994).
123. M. T. Huang et al., Effect of dietary curcumin and ascorbyl palmitate on azoxymethanol-induced colonic epithelial cell proliferation and focal areas of dysplasia. *Cancer Lett.* **64**(2), 117–121 (1992).
124. A. K. Naidu et al., Inhibition of cell proliferation and glutathione S-transferase by ascorbyl esters and interferon in mouse glioma. *J. Neuro-Oncol.* **16**(1), 1–10 (1993).
125. A. Mitra et al., Inhibition of human term placental and fetal liver glutathione-*S*-transferases by fatty acids and fatty acid esters. *Toxicol. Lett.* **60**(3), 281–288 (1992).
126. H. W. Klaer, J. Glavind, and E. Arffmann, Carcinogenicity in mice of some fatty acid methyl esters. 2. Peroral and subcutaneous application. *Acta Pathol. Microbiol. Scand. Sec. A* **83**(5), 550–558 (1975).
127. P. S. Bora et al., Myocardial cell damage by fatty acid ethyl esters. *J. Cardiovasc. Pharmacol.* **27**(1), 1–6 (1996).
128. Z. M. Szczepiorkowski, G. R. Dickersin, and M. Laposata. Fatty acid ethyl esters decrease human hepatoblastoma cell proliferation and protein synthesis. *Gastroenterology* **108**(2), 515–522 (1995).
129. P. S. Haber et al., Fatty acid ethyl esters increase rat pancreatic lysosomal fragility. *J. Lab. Clin. Med.* **121**(6), 759–764 (1993).
130. O. Holian et al., Response of brain protein-kinase-C isozymes to ethyl oleate, an alcohol metabolite. *Brain Res.* **558**(1), 98–100 (1991).
131. C. C. Smith, *Arch. Ind. Hyg. Occup. Med.* **7**, 310 (1953).
132. H. Oettel, *Naunyn-Schniedebergs Arch. Exp. Pathol. Pharmakol.* **183**, 641 (1936).

133. World Health Organization (WHO), *Food Additives*, Ser. No. 5, WHO Geneva, 1974.
134. R. Oberly and M. F. Tansy, LC_{50} values for rats acutely exposed to vapors of acrylic and methacrylic acid esters. *J. Toxicol. Environ. Health* **16**(6), 811–822 (1985).
135. H. Tanii and K. Hashimoto, *Toxicol. Lett.* **11**, 125–129 (1982).
136. T. J. McCarthy and G. Witz, Structure-activity relationships in the hydrolysis of acrylate and methacrylate esters by carboxylesterase in vitro. *Toxicology* **116**(1–3), 153–158 (1997).
137. U. C. Pozzani, C. S. Weil, and C. P. Carpenter, *J. Ind. Hyg. Toxicol.* **31**, 317 (1949).
138. C. L. Doerr, K. H. Brock, and K. L. Dearfield, *Environ. Mol. Mutagen.* **11**(1), 49–63 (1988).
139. E. Seutter and N. V. M. Rigntjes, *Arch. Dermatol. Res.* **270**, 273–284 (1981).
140. C. W. Chung and A. L. Giles, Jr., *J. Invest. Dermatol.* **68**(4), 187 (1977).
141. L. Kanerva, R. Jolanki, and R. Estlander, 10 years of patch testing with the (meth)acrylate series. *Contact Dermatitis* **376**(6), 255–258 (1997).
142. R. R. Tice, L. A. Nylander-French, and J. E. French, Absence of systemic *in vivo* genotoxicity after dermal exposure to ethyl acrylate and tripropylene glycol diacrylate in Tg.AC (v-Ha-ras) mice. *Environ. Mol. Mutagen.* **29**, 240–249 (1997).
143. J. F. Treon et al., *J. Ind. Toxicol.* **31**(6), 311 (1949).
144. Y. Ono and S. Tanaka, *Bunseki Kagaku* **24**(112), 776 (1975).
145. W. Deichmann, *J. Ind. Hyg. Toxicol.* **23**, 343 (1941).
146. W. H. Lawrence, G. E. Bass, and J. Autian, *J. Dent. Res.* **51**, 526 (1972).
147. A. R. Singh, W. H. Lawrence, and J. Autian, *J. Dent. Res.* **51**, 1632 (1972).
148. G. N. Mir, W. H. Lawrence, and J. Autian, *J. Pharm. Sci.* **62**(8), 1258 (1973).
149. K. R. Rees and M. J. Tarlow, The hepatotoxic action of allyl formate. *Biochem. J.* **104**, 757–761 (1967).
150. D. L. J. Opdyke, *Monographs of Fragrance and Raw Materials*, Pergamon, Oxford, 1979.
151. C. P. Carpenter, H. F. Smyth, and U. C. Pozzani, *J. Ind. Hyg. Toxicol.* **31**, 343 (1949).
152. W. Reininghaus, A. Koestner, and H. J. Klimisch, Chronic toxicity and oncogenicity of inhaled methyl acrylate and n-butyl acrylate in Sprague-Dawley rats. *Food Chem. Toxicol.* **29**(5), 329–340 (1991).
153. P. Vodicka, I. Gut, and E. Frantik, Effects of inhaled acrylic acid derivatives in rats. *Toxicology* **65**(1–2), 209–222 (1990).
154. B. Przybojewska, E. Dziubatowska, and Z. Kowalski, Genotoxic effects of ethyl acrylate and methyl acrylate in the mouse evaluated by the micronucleus test. *Mutat. Res.* **135**(3), 189–191 (1984).
155. H. N. Hacmiya, A. Taketani, and Y. Takizawa, *Nippon Koshu Eisei Zasshi* **29**, 236–239 (1982).
156. M. M. Moore et al., Comparison of mutagenicity results for nine compounds evaluated at the HGPRT locus in the standard and suspension CHO assays. *Mutagenesis* **6**(1), 77–86 (1991).
157. T. J. Oberly et al., An evaluation of 6 chromosomal mutagens in the AS52-SPRT mutation assay utilizing suspension culture and soft agar cloning. *Mutat. Res.* **319**(3), 179–187 (1993).
158. J. D. Hamilton, K. H. Reinert, and J. E. McLaughlin, Aquatic risk assessment of acrylates and methacrylates in household consumer products reaching municipal wastewater treatment plants. *Environ. Techno.* **16**(8), 715–727 (1995).

159. M. T. Brondeau et al., Adrenal-dependent leukopenia after short-term exposure to various airborne irritants in rats. *J. Appl. Toxicol.* **10**(2), 83–86 (1990).

160. R. R. Miller et al., Chronic toxicity and oncogenicity bioassay of inhaled ethyl acrylate in Fischer 344 rats and $B6C3F_1$ mice. *Drug Chem. Toxicol.* **8**, 1–42 (1985).

161. C. B. Frederick, J. R. Udinsky, and L. Finch, The regional hydrolysis of ethyl acrylate to acrylic acid in the rat nasal cavity. *Toxicol. Lett. (Shannon)* **70**(1), 49–56 (1994).

162. I. Linhart, M. Vosmanska, and J. Smejkal, Biotransformation of acrylates. excretion of mercapturic acids and changes in urinary carboxylic acid profile in rat dosed with ethyl and 1-butyl acrylate. *Xenobiotica* **24**(10), 1043–1052 (1994).

163. National Toxicology Program (NTP), *Toxicology and Carcinogenesis Studies of Ethyl Acrylate*, Rep. No. 259, NIH Publ. No. 87–2515, NTP, Washington, DC, 1986, p.7.

164. B. I. Ghanayem et al., Demonstration of a temporal relationship between ethyl acrylate-induced forestomach cell proliferation and carcinogenicity. *Toxicol. Pathol.* **22**(5), 497–509 (1994).

165. J. F. Borzelleca et al., *Toxicol. Appl. Pharmacol.* **6**, 29–36 (1964).

166. L. R. DePass et al., Dermal oncogenicity Bioassays of acrylic acid, ethyl acrylate and butyl acrylate. *J. Toxicol. Environ. Health* **14**(2–3), 115–120 1984).

167. E. J. Winner, R. A. Prough, and M. D. Brennan, Human NAD(P)H:quinone oxidoreductase induction in human hepatoma cells after exposure to industrial acrylates, phenolics and metals. *Drug Metabo. Dispos.* **25**(2), 175–181 (1997).

168. A. D. Kligerman et al., Cytogenic studies of ethyl acrylate using C57BL/6 mice. *Mutagenesis* **6**, 137–141 (1991).

169. A. M. Walker et al., Mortality from cancer of the colon or rectum among workers exposed to ethyl acrylate and methyl methacrylate. *Scand. J. Work Environ Health* **17**, 7–19 (1991).

170. C. B. Frederick and I. M. Chang-Mateu, Contact site carcinogenicity: Estimation of an upper limit for risk of dermal dosing site tumors based on oral dosing site carcinogenicity. In T. R. Gerrity and C. J. Henry, eds. *Principles of Route-to-Route Extrapolation for Risk Assessment*, Elsevier, New York, 1990, pp. 237–270.

171. C. B. Frederick et al., *Toxicol. Appl. Pharmacol.* **114**, 246–260 (1992).

172. D. Parker and J. C. Turk, Contact sensitivity to acrylate compounds in guinea pigs. *Contact Dermatitis* **9**, 55–60 (1983).

173. B. B. Levine, Studies on the mechanism of the formation of the penicillin antigen. I. Delayed allergic cross reactions among pencillin G and its degradation products. *J. Exp. Med.* **112**, 1131–1157 (1960).

174. G. Engelhardt and H. J. Klimisch, n-Butyl acrylate: Cytogenetic investigations in the bone marrow of Chinese hamsters and rats after 4-day inhalation. *Fundam. Appl. Toxicol.* **3**(6), 640–641 (1983).

175. A. Sapota, The dynamics of distribution and excretion of butyl-(2,3-14C)-acrylate in male Wistar albino rats. *Pol. J. Occup. Med.* **4**(1), 55–66 (1991).

176. J. Merhle and H. J. Klimisch, n-Butyl acrylate: Prenatal inhalation toxicity in the rat. *Fundam. Appl. Toxicol.* **3**, 443–447 (1983).

177. F. Motoc et al., *Arch. Mal. Prof. Med. Trav. Secur. Soc.* **32**(10–11), 653 (1971).

178. H. G. Boehling, U. Borchard, and H. Drouin, *Arch. Toxicol.* **38**, 307 (1977).

179. A. A. Fisher, *Contact Dermatitis*, 2nd ed., Lea & Febiger, Philadelphia, PA, 1973, p. 358.

180. B. Scolnick and J. Collins, *J. Occup. Med.* **28**(3), 196–198 (1986).

181. M. J. Kessler, J. L. Kupper, and R. J. Brown, Accidental methyl methacrylate toxicity in a rhesus monkey (*Macaca mulatta*). *Lab. Ani. Sci.* **27**(3), 388–390 (1977).

182. L. G. Lomax, N. D. Krivanek, and S. R. Frame, Chronic inhalation toxicity and oncogenicity of methyl methacrylate in rats and hamsters. *Food Chem. Toxicol.* **35**(3–4), 393–407 (1997).

183. Z. Bereznowski, *In vivo* assessment of methyl methacrylate metabolism and toxicity. *Int. J. Biochem. Cell Biol.* **27**(12), 1311–1316 (1995).

184. Z. Bereznowski, Methacrylate uptake by isolated hepatocytes and perfused rat liver. *Int. J. Biochem. Cell Biol.* **29**(4), 675–679 (1997).

185. H. M. Solomon et al., Methyl methacrylate: Inhalation developmental toxicity study in rats. *Teratology* **48**(2), 115–125 (1993).

186. F. E. Reno, *18-Month Vapor Inhalation Safety Evaluation Study in Hamsters*, unpublished, Proj. No. 417–354. Hazelton Laboratories America, Inc., 1979.

187. National Toxicology Program (NTP), *Toxicology and Carcinogenesis Studies of Methyl Methacrylate (Cas No 80-62-6) In F344/N Rats and B6C3F$_1$ Mice (Inhalation Studies)*, Tech. Rep. Ser. 314, NTP, Washington, DC, 1987.

188. H. Schweikl, G. Schmalz, and K. Rackebrandt, The mutagenic activity of unpolymerized resin monomers in *Salmonella typhimurium* and V79 cells. *Muta. Rese.* **415**(1–2), 119–130 (1998).

189. M. M. Moore et al., Genotoxicity of acrylic acid, methyl acrylate, ethyl acrylate, methyl methacrylate, and ethyl methacrylate in L5178Y mouse lymphoma cells. *Environ. Mol. Mutagen.* **11**(1), 49–63 (1988).

190. L. Kanerva et al., A single accidental exposure may result in a chemical burn, primary sensitization and allergic contact dermatitis. *Contact Dermatitis* **31**(4), 229–235 (1994).

191. L. Kanerva et al., Occupational allergic contact dermatitis caused by photobonded sculptured nails and a review of (meth) acrylates in nail cosmetics. *Am. J. Contact Dermatitis* **7**(2), 109–115 (1996).

192. A. B. Spencer et al., Control of ethyl methacrylate exposures during the application of artificial fingernails. *Am. Ind. Hyg. Assoc. J* **58**(3), 214–218 (1997).

193. K. Seiji et al., Absence of mutagenicity in peripheral lymphocytes workers occupationally exposed to methyl methacrylate. *Ind. Health* **32**(2), 97–105 (1994).

194. F. A. Andersen, Final report on ethyl methacrylate. *J. Am. Coll. Toxicol.* **14**(6), 452–467 (1995).

195. J. W. Sarver *Acute Dermal Irritation/Corrosion Study (OECD) with n-Butyl Methacrylate in Rabbits*, Unpublished report, Sponsored by Methacrylate Producers Association, Washington, DC, 1993.

196. J. V. Hagan, *Butyl; Methacrylate: 4-week Vapor Inhalation Study in Rats*, Unpublished report, Sponsored by Methacrylate Producers Association, Washington, DC, 1993.

197. N. R. Shepelskaya, *Gig. Sanit.* **3**, 1107 (1978).

198. P. A. Revell, M. Braden, and M. A. R. Freeman, Review of the biological response to a novel bone cement containing poly(ethyl-methacrylate) and *n*-butyl methacrylate. *Biomaterials* **19**(17), 1579–1856 (1998).

199. M. Tobler and A. U. Freiburghaus, A glove with exceptional protective features minimizes the risks of working with hazardous chemicals. *Contact Dermatitis* **26**(5), 299–303 (1992).

200. V. C. Moser et al., Comparison of subchronic neurotoxicity of 2 hydroxyethylacrylate and acrylamide in rats. *Fundam. Appl. Toxicol.* **18**(3), 343–352 (1992).

201. L. W. Raymond, Pulmonary abnormalities and serum immunoglobulins in facsimile machine repair technicians exposed to butyl methacrylate fume. *Chest* **109**(4), 1010–1018 (1996).

202. J. J. Clary, V. J. Feron, and J. A. van Velthuijsen, Safety assessment of lactate esters. *Regul. Toxicol. Pharmacol.* **27**, 88–97 (1998).

203. F. Kaiho et al., Enhancing effect of cetyl lactate on the percutaneous absorption of indomethacin in rats. *Chem. Pharm. Bull.* **37**(4), 1114–1116 (1989).

204. C. T. Bowner et al., The exotoxicity and the biodegradability of lactic acid, alkyl lactate esters and lactate salts. *Chemosphere* **37**(7), 1317–1333 (1998).

205. A. R. Latven and H. Molitor, *J. Pharmacol. Exp. Ther.* **7**, 159 (1975).

206. T. Asakawa, W. Wada, and T. Yamawo, *Biochim. Biophys. Acta* **170**, 375 (1968).

207. P. H. Andersen and N. J. Jensen, Mutagenic investigation of flavourings: Dimethyl succinate, ethyl pyruvate and aconitic acid are negative in the salmonella/mammalian-microsome test. *Food Addit. Contam.* **1**(3), 283–288 (1984).

CHAPTER EIGHTY

Esters of Aromatic Mono-, Di-, and Tricarboxylic Acids, Aromatic Diacids and Di-, Tri-, or Polyalcohols

Raymond M. David, Ph.D., DABT, Richard H. McKee, Ph.D., DABT, John H. Butala, MS, DABT, Robert A. Barter, Ph.D., DABT, and Martin Kayser, MD

A INTRODUCTION TO ESTERS OF AROMATIC MONOCARBOXYLIC ACIDS AND MONOALCOHOLS

Benzoates

The simple aliphatic esters of benzoic acid are liquids used as solvents, flavors, or perfumes (1). The arylbenzoate benzyl is used as a miticide or plasticizer. For physicochemical data, see Table 80.1. In general, these compounds have a low order of toxicity (2–5) (Table 80.2). The primary effect expected from ingestion of moderate amounts of benzoates is gastrointestinal irritation, gastric pain, nausea, and vomiting. Available data indicate a low order of skin absorbability, and the undiluted materials may be either slight or moderate skin irritants. In rabbits, the degree of skin irritation caused by alkyl benzoates increases with an increase in molecular weight (6).

Salicylates

The salicylates are used as flavorants, perfumes, or analgesics. Physicochemical data are presented in Table 80.1. The most commonly used member of this class of compounds is

Patty's Toxicology, Fifth Edition, Volume 6, Edited by Eula Bingham, Barbara Cohrssen, and Charles H. Powell.
ISBN 0-471-31939-2

Table 80.1. Physical and Chemical Properties of Benzoates, Salicylates, and Parabens

Chemical Name	CAS No.	Molecular Formula	Mol. Wt.	Boiling Point (°C) (mmHg)	Melting Point (°C)	Specific Gravity (Density)	Sol. in Water at 68°F (20°C)	Refractive Index (20°C)	Vapor Pressure (mmHg) (°C)	Flash Point °C (°F)	UEL or LEL % by Vol. Room Temp.
Benzoate											
Methyl	*[93-58-3]*	$C_8H_8O_2$	136.15	199.6	– 12.3	1.0937	< 1 mg/mL[a]	1.5165	1 (39)	83 (181)	—
Ethyl	*[93-89-0]*	$C_9H_{10}O_2$	150.18	212.6	– 34– – 36	1.0937	insol	1.5007	1 (44)	96.1 (205)	—
Isopropyl	*[939-48-0]*	$C_{10}H_{12}O_2$	164.20	218.5	– 26.4	1.010	insol	1.4890	0.12 (44)	98.9 (210)	—
Butyl	*[136-60-7]*	$C_{11}H_{14}O_2$	178.23	250.3	– 26.4	1.000	59 mg/mL	1.4940	0.01 (25)	107.2 (225)	—
Hexyl	*[6789-88-4]*	$C_{13}H_{18}O_2$	206.28	272 (770)	—	(0.979)	insol	—	32 (22)	> 93.3	—
Vinyl	*[769-78-8]*	$C_9H_8O_2$	148.16	—	—	—	—	—	—	—	—
Benzyl	*[120-51-4]*	$C_{14}H_{12}O_2$	212.25	323.4	21	1.114	insol	1.5680	0.0002 (25)	147.8 (298)	—
Salicylate											
Methyl	*[119-36-8]*	$C_8H_8O_3$	152.15	223.3	– 8.6	1.18	0.74%[b]	1.5369	0.034 (25)	96 (205)	—
Ethyl	*[118-61-6*	$C_9H_{10}O_3$	166.18	234	1.02	—	insol	1.5296	—	—	—
Amyl	*]2050-08-0]*	$C_{12}H_{16}O_3$	108.26	266.5	—	—	slight	1.506	—	132.2 (270)	—
Phenyl	*[118-55-8]*	$C_{13}H_{10}O_3$	214.23	173 (12)	41–43	1.2614	insol	—	—	—	—
Paraben											
Methyl	*[99-76-3]*	$C_8H_8O_3$	152.14	270–280	131	—	0.30%[b]	—	—	—	—
Ethyl	*[120-47-8]*	$C_9H_{10}O_3$	166.17	297–280	116–118	—	slight	—	—	—	—
Propyl	*[94-13-3]*	$C_{10}H_{12}O_3$	180.20	—	96.2–98	1.0630[c]	463 mg/L[d]	1.5161	—	—	—
Butyl	*[94-26-8]*	$C_{11}H_{14}O_3$	194.22	—	68–69	—	250 mg/L[d]	—	—	—	—

[a] At 22.5°C.
[b] At 30°C.
[c] At 102°C.
[d] At 25°C.

Table 80.2. Summary of Toxicity of Benzoates

Chemical Name	CAS No.	Species	Exposure Route	Approximate Dose (g/kg)[a]	Treatment Regimen	Observed Effect	Ref.
Benzoate							
Methyl	*[93-58-3]*	Rat	oral	1.35, 2.17, 3.43	Single dose	LD_{50}	3–5
		Mouse	oral	3.33	Single dose	LD_{50} listed. Tremor, excitement	3
		Rabbit	oral	2.17	Single dose	LD_{50} listed. Somnolence, muscle weakness	4
		Guinea pig	oral	4.1	Single dose	LD_{50}	4
		Cat	dermal	10	Unknown	LD_{LO} listed. Tremor, muscle weakness, changes in structure or function of salivary glands.	4
		Rabbit	dermal	0.01 mL	Applied to clipped skin	Irritation score of 3 (10 max)	5
		Rabbit	ocular	0.5 mL	—	Numerical score of 1 (10 max)	5
		Rat	inhal	concentrated vapor	8 h	100% survival	5
Ethyl	*[93-89-0]*	Rat	oral	2.10, 6.48	Single dose	LD_{50} listed. Muscle weakness, dyspnea	4, 5
		Rabbit	oral	2.63	Single dose	LD_{50} listed. Somnolence, muscle weakness	4
		Cat	dermal	10	Unknown	LD_{LO} listed. Tremor, muscle weakness and changes in structure or function of salivary glands.	4
		Rabbit	dermal	0.01 mL	Applied to clipped skin	Irritation score of 4 (10 max)	5
		Rabbit	ocular	0.5 mL	—	Numerical score of 1 (10 max)	5
		Rat	inhal	concentrated vapor	8 h	100% survival	5
Isopropyl	*[939-48-0]*	Rat	oral	3.73	Single dose	LD_{50}	7
		Rabbit	dermal	20 mL/kg	Single dose	LD_{50}	7
Butyl	*[136-60-7]*	Rat	oral	5.14	Single dose	LD_{50}	5
		Mouse	oral	3.45	Single dose	LD_{50} listed. Convulsions, GI changes, hemorrhage	8
		Rat	inhal	concentrated vapor	8 h	100% survival	5
		Rabbit	dermal	>5	—	LD_{50}	9
				0.01 mL	Applied to clipped skin	Irritation score of 5 (10 max)	5
		Rabbit	ocular	0.5 mL	—	Numerical score of 1 (10 max)	5

Table 80.2. *(Continued)*

Chemical Name	CAS No.	Species	Exposure Route	Approximate Dose (g/kg)[a]	Treatment Regimen	Observed Effect	Ref.
Hexyl	*[6789-88-4]*	Rat	oral	12.3	Single dose	LD_{50}	7
		Rabbit	dermal	21 mL/kg	—	LD_{50}	7
Vinyl	*[769-78-8]*	Rat	oral	3.25	Single dose	LD_{50}	5
		Rabbit	dermal	0.01 mL	Applied to clipped skin	Irritation score of 5 (10 max)	5
		Rabbit	ocular	0.5 mL	—	Numerical score of 2 (10 max)	5
		Rat	inhal	concentrated vapor	8 h	100% survival	5
Benzyl	*[120-51-4]*	Mouse	oral	1.4 mL/kg	Single dose, 50 animals	LD_{50} over 6 d	2
		Rat	oral	1.7 mL/kg	Single dose, 40 animals	LD_{50} over 6 d	2
		Cat	oral	2.24	Single dose	LD_{50}	4
		Rabbit	oral	1.68 or 1.80	Single dose, 30 animals	LD_{50} over 6 d	2, 10
		Guinea pig	oral	1.0	Single dose, 50 animals	LD_{50} over 6 d	2
		Dog	oral	> 22.4	Single dose	LD_{50}	4
		Rat	dermal	4.0 mL/kg	Acute	LD_{50} listed. Deaths are delayed. Animals die without exhibiting prior symptoms of systemic effects. Lower doses cause very mild gross skin irritation.	2
		Rat	dermal	2.0 mL/kg	90-d	LD_{50}	2
		Rabbit	dermal	4	—	LD_{50}	10
				0.5 mL	Patch, 3 animals	Primary irritation index score of 0 at 1 h and 1.58 at 2 h (8 max)	11
		Mouse	IP	> 0.5		LD_{50}	12

[a]Unless listed otherwise; LD_{LO} = lowest lethal dose.

methyl salicylate. As noted in Table 80.3, methyl salicylate is more acutely toxic than ethyl- or phenylsalicylate. Ingestion of relatively small quantities of methyl salicylate may cause severe, rapid-onset salicylate poisoning characterized by tinnitus, nausea, vomiting, hyperventilation, hyperthermia, hyperpnea, and hyperpyrexia, central nervous system excitation, confusion, coma, convulsions, respiratory alkalosis, metabolic acidosis, and hyper- or hypoglycemia (105). Pulmonary edema, hemorrhage, acute renal failure, and death may ensue (23). Dimness of vision occurs in 15% of all methyl salicylate poisonings (24). According to Fisher (25), dermal exposure to salicylates is associated with a high incidence of sensitization reactions. Available evidence suggests that salicylates are reproductive and fetal toxicants in animals (26).

4-Hydroxybenzoates (Parabens)

The lower alkyl esters of *p*- or 4-hydroxybenzoic acid (C1 to C4), also named the methyl-, ethyl-, propyl-, and butyl parabens, are high-boiling liquids that decompose on heating (see Table 80.1). They are widely used in the food, cosmetic, and pharmaceutical industries as preservatives, bacteristats, and fungistats (27, 28). Commonly used concentrations range from 0.1 to 0.3% (29). Parabens also have been used therapeutically for the treatment of moniliasis, a *Candida albicans* infection (30).

By the oral route, parabens are rapidly absorbed, metabolized, and excreted. The metabolic reactions and conversions in mammals vary with the chain length of the ester, the animal species, route of administration, and quantity tested. The metabolism of parabens in humans appears to be most closely related to that of dogs (31). The rate of metabolite excretion appears to decrease with increasing molecular weight of the ester (13).

The lower paraben homologues have low potential for acute or chronic systemic toxicity and are therefore approved as human food additives at a dose of 0 to 10 mg/kg by weight for the methyl, ethyl, and propyl derivatives (27). The parabens have a low irritancy potential; however, application of dermatological preparations containing parabens may cause severe and intractable contact dermatitis. Contact allergy to parabens reportedly ranges from 0.3 to 6% (29, 32–34, 142a). The sensitization potential of methyl paraben is approximately twice that of ethyl-, propyl-, and butyl paraben (29). Patients sensitive to one paraben show cross-reactivity to the others (35).

Cinnamates

The cinnamates (phenyl acrylates, phenylpropenoic acid esters) are mainly used as fragrances in the perfume industry (36). Physicochemical data are presented in Table 80.1. An enzyme has been identified that converts *p*-hydroxyphenyl cinnamate to *p*-hydroxyphenyl propionate or *p*-hydroxybenzoate (37). The same type of system may be responsible for benzoic and hippuric acid formation, a common catabolic pathway for all cinnamates. Cinnamates are largely excreted as benzoic and hippuric acid in the urine (38). The cinnamates appear to have low to moderate toxicity in mammals (Table 80.4). In humans, dermal exposure to allyl cinnamate may cause skin irritation (39).

Table 80.3. Summary of Toxicity of Salicylates

Chemical Name	CAS No.	Species	Exposure Route	Approximate Dose (g/kg)[a]	Treatment Regimen	Observed Effect	Ref.
Salicylate							
Methyl	[*119-36-8*]	Rat	oral	0.887	Single dose	LD_{50} listed. Somnolence noted.	3
		Mouse	oral	1.11	Single dose	LD_{50}	14
		Rabbit	oral	1.3, 2.8	Single dose	LD_{50}	3, 15
		Guinea pig	oral	0.7, 1.060	Single dose	LD_{50}	12
		Dog	oral	2.1	Single dose	LD_{50} listed. GI hypermotility, diarrhea and nausea or vomiting noted.	15
		Guinea pig	dermal	0.70 mL/kg	Single dose	LD_{50}	3
		Rabbit	sc	4.25	Single dose	LD_{LO}	15
		Guinea pig	sc	1.5	Single dose	LD_{LO}	16
		Dog	sc	2.25	Single dose	LD_{LO}	15
		Rabbit	dermal	4 mL/kg	90 d	Early death, kidney damage	17
Ethyl	[*118-61-6*]	Rat	oral	1.32	Single dose	LD_{50}	18
		Guinea pig	oral	1.4	Single dose	LD_{LO}	18
		Guinea pig	sc	1.5	Single dose	LD_{LO}	19
		Rabbit	dermal	>5	Single dose	LD_{50}	18
Amyl	[*2050-08-0*]	Dog	IV	0.5–0.8	Single dose	LD_{50}	20
		Rabbit	IV	0.5–0.8	Single dose	LD_{50}	20
Phenyl	[*118-55-8*]	Rat	oral	3	Single dose	LD_{50}	21
		Mouse	IP	>0.5	Single dose	LD_{50}	22

[a]Unless listed otherwise; LD_{LO} = lowest lethal dose.

Table 80.4. Summary of Toxicity of Cinnamates

Chemical Name	CAS No.	Species	Exposure Route	Approximate Dose (g/kg)[a]	Treatment Regimen	Observed Effect	Ref.
Cinnamate							
Methyl	[*103-26-4*]	Rat	oral	2.61	Single dose	LD_{50}	42
		Rabbit	dermal	>5	Single dose	LD_{50}	42
Ethyl	[*103-36-6*]	Rat	oral	4, 7.8	Single dose	LD_{50}	43, 44
		Mouse	oral	4	Single dose	LD_{50}	43
		Guinea pig	oral	4	Single dose	LD_{50}	43
		Rabbit	dermal	>5	Single dose	LD_{50}	44
n-Propyl	[*7778-83-8*]	Mouse	oral	7 mL/kg	Single dose	LD_{50}	2
		Guinea pig	oral	3 mL/kg	Single dose	LD_{50}	2
		Guinea pig	dermal	2 mL/kg	90 d	LD_{50} listed. Mild skin irritation following repeated exposure	2
n-Butyl	[*538-65-8*]	Rat	oral	>5	Single dose	LD_{50}	45
		Mouse	oral	7	Single dose	LD_{50}	46
		Rabbit	dermal	>5	Single dose	LD_{50}	45
Allyl	[*1866-31-5*]	Rat	oral	1.52	Single dose	LD_{50}	39
		Rabbit	dermal	>5	Single dose	LD_{50}	39
Linayl	[*78-37-5*]	Rat	oral	9.96	Single dose	LD_{50}	47
		Rabbit	dermal	>5	Single dose	LD_{50}	47
Benzyl	[*103-41-3*]	Rat	oral	5.53	Single dose	LD_{50}	3

[a]Unless listed otherwise.

p-Aminobenzoates

Some *p*-aminobenzoic acid (PABA) esters occur naturally, since the free compound, PABA, is an intricate part of the vitamin B complex and is utilized for its synthesis. The PABA ethyl ester (benzocaine) and the hydrochloride salt of the 2-diethylamino ethyl ester (procaine) serve as important local anesthetics. A variety of PABA esters are used in sunscreen agents to increase the erythema threshold time (25, 40). Orally or subcutaneously administered PABA esters are rapidly hydrolyzed by the mammalian system to the alcohol and the free acid. Administration of the free acid to experimental animals has been shown to cause methemoglobinemia and leukocytosis (48).

PABA esters exhibit a low order of acute toxicity in experimental animals. In humans, cases of methemoglobinemia after topical benzocaine or procaine use have been reported (49a–53, 203). Sunscreen agents containing PABA esters may occasionally produce allergic photosensitization (25).

o-Aminobenzoates

The ortho-aminobenzoates (anthranilates) are less irritating and less likely to cause sensitization than the para-aminobenzoates, but have less therapeutic usefulness. They are used in some sunscreen lotions. A wide variety of anthranilic esters are utilized as oxidation–corrosion stabilizers for combustion engine lubricants. Readily hydrolyzed by the mammalian system, anthranilic acid presents part of the building block for tryptophan biosynthesis (54, 55). Anthranilates have low toxicity potential.

1.0 Methyl Benzoate

1.0.1 CAS Number: [93-58-3]

1.0.2 Synonyms: Benzoic acid, methyl ester; methyl benzenecarboxylate; Benzoate de methyle [French], Benzoato de metilo [Spanish]; Methylester kyseliny benzoove [Czech]; CCRIS 5851; A13-00525 [NLM];CTFA 05391; EINECS 202-259-7; FEMA No. 2683; HSDB 5283; NSC 9394

1.0.3 Trade Names: Clorius; Essence of Niobe; Niobe Oil; Oil of Niobe; Oniobe Oil; Oxidate Le

1.0.4 Molecular Weight: 136.15

1.0.5 Molecular Formula: $C_8H_8O_2$

1.1 Chemical and Physical Properties

1.1.1 General

The empirical formula of methyl benzoate is $C_8H_8O_2$. The structural formula is $C_6H_5CO_2CH_3$. It is a clear, colorless, oily liquid with a pour point of $-12.3°C$ and a boiling point of 96–98°C (at 24 mmHg). It is soluble in DMSO, 95% ethanol, acetone, ether, and most fixed oils, and miscible with methanol. It reacts with strong oxidizing agents and strong bases and hydrolyzes slowly in water. Its vapor density is 4.68 and its

Henry's Law constant is 3.24×10^{-5} atm cu/m/mole. The flash point is 181°F (83°C). Chemical and physical data are summarized in Table 80.1.

1.1.2 Odor and Warning Properties

Methyl benzoate has a pleasant, distinctive, and persistent odor (56, 57). When heated to decomposition, it emits fumes containing CO and CO_2.

1.2 Production and Use

The compound is manufactured by heating methanol and benzoic acid in the presence of sulfuric acid or by passing dry hydrogen chloride through a solution of benzoic acid in methanol (58). It may also be produced by the alcoholysis of benzonitrile (59). It is a by-product of ozonolysis of water (60).

Methyl benzoate is used as a solvent for cellulose esters and ethers (61), a flavoring agent (58), a carrier for dyeing polyester, an additive for disinfectants or soy sauce, and for manufacture of pesticides, resins, or rubber (61–63). It is also used in perfumes (64). It occurs naturally in plants and is produced by microorganisms (65).

Methyl benzoate has been recognized as a chemical degradation product of cocaine. This fact contributes to a forensic procedure where the characteristic odor serves as a qualitative indicator (57).

1.3 Exposure Assessment

The general population is exposed to methyl benzoate primarily from food (or drinking water in some cities). Occupational exposure occurs via dermal contact and inhalation. No specifics on exposure assessment were located.

1.4 Toxic Effects

1.4.1 Experimental Studies

1.4.1.1 Acute Toxicity. Generally, methyl benzoate is of low to moderate toxicity by ingestion and inhalation. The oral LD_{50} value for the rat ranges from 1.35 to 3.43 g/kg (3–5). The oral LD_{50} values in the mouse, rabbit, and guinea pig are 3.33, 2.17, and 4.1 g/kg, respectively (3, 4). Acute toxicity information is summarized in Table 80.2.

1.4.1.2 Chronic and Subchronic Toxicity. No information found.

1.4.1.3 Pharmacokinetics, Metabolism, and Mechanisms. Methyl benzoate has a low order of skin absorbability (12). It is meteabolized to benzoic acid in rabbits.

1.4.1.4 Reproductive and Developmental. No information found.

1.4.1.5 Carcinogenesis. No information found.

1.4.1.6 Genetic and Related Cellular Effects Studies. At doses of 10–666 μg/plate, methyl benzoate is not a mutagen in Ames *S. typhimurium* strains TA100, TA1535, TA1537, TA97, TA98, or TA100, with or without metabolic activation (66).

1.4.1.7 Other: Neurological, Pulmonary, Skin Sensitization. In nonoccluded skin irritation tests carried out on clipped rabbits, no signs of erythema were observed in any of the animals (14/14), 4 and 24 h after the application of 0.5 mL undiluted methyl benzoate (6). Very slight erythema was observed in the remaining (12/12) rabbits 24 h after a second application. Redness increased with successive treatments. Moderate to severe edema was observed in animals (4/4) receiving 6 applications. In experiments conducted by Smyth and co-workers, methyl benzoate elicited an irritation score of 3 (10 maximum) on clipped rabbit skin (69). Methyl benzoate received a score of 1 (out of 10) in a rabbit corneal necrosis test, indicating that it causes minimal eye injury (5).

1.4.2 Human Experience

Humans using methyl benzoate have experienced irritation to the skin, eyes, mucous membranes, and upper respiratory tract (61). In 25 human volunteers, 4% methyl benzoate in petrolatum did not cause sensitization (67).

1.5 Standards, Regulations, or Guidelines of Exposure

Methyl benzoate is a food additive permitted for direct addition to food for human consumption according to FDA regulations stipulated in 21 CFR 172.515 (68).

1.6 Studies on Environmental Impact

Methyl benzoate may be released into the environment from natural and anthropogenic sources. It is released into wastewater or into the atmosphere, from stack emissions from coal-burning power plants. If released into water, it volatilizes (half-life 1.5 d in a model river) and biodegrades (69). Adsorption to sediment and bioconcentration in fish should not be important fate processes. In the atmosphere, it slowly degrades by reaction with photochemically produced hydroxyl radicals. The estimated half-life in the atmosphere is 18.5 d (70). Washout by rain also contributes to its removal from the atmosphere (71). It is weaekly absorbed by soil, and if released on land, would tend to leach into the soil. Biodegradation rates in soil are not available.

2.0 Ethyl Benzoate

2.0.1 CAS Number: *[93-89-0]*

2.0.2 Synonyms: Benzoic acid, ethyl ester; benzoic ether; benzoyl ethyl ether; ethyl benzenecarboxylate, ethylester kyseliny benzoove [Czech]; AI3-01352 [NLM]; EINECS 202-284-3; FEMA No. 2422 [NLM]; and NSC 8884 [NLM]

2.0.3 Trade Names: Essence of Niobe

2.0.4 Molecular Weight: 150.18

2.0.5 Molecular Formula: $C_9H_{10}O_2$

2.0.6 Molecular Structure:

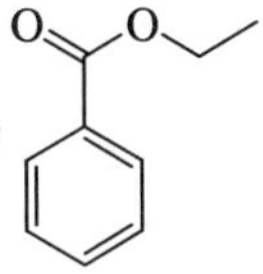

2.1 Chemical and Physical Properties

2.1.1 General

The empirical formula of ethyl benzoate is $C_9H_{10}O_2$. The structural formula is $C_6H_5CO_2C_2H_5$. It is a colorless liquid, with a specific gravity of 1.0937, a melting point of approximately −35°C, and a boiling point of 212.6°C. Chemical and physical data are summarized in Table 80.1.

2.1.2 Odor and Warning Properties

Ethyl benzoate has a pleasant odor.

2.2 Production and Use

According to two patents, ethyl benzoate is used as a sustained-release component for pesticides.

2.3 Exposure Assessment

No information found.

2.4 Toxic Effects

2.4.1 Experimental Studies

2.4.1.1 Acute Toxicity. Available acute toxicity information is summarized in Table 80.2. In two separate studies in the rat, oral LD_{50} values of 2.10 and 6.48 g/kg were obtained (4, 5). For the rabbit, the oral LD_{50} value is 2.63 g/kg (4). In cats, dermal application of 10 g/kg caused lethality (4).

2.4.1.2 Chronic and Subchronic Toxicity. No information found.

2.4.1.3 Pharmacokinetics, Metabolism, and Mechanisms. No information found.

2.4.1.4 Reproductive and Developmental. No information found.

2.4.1.5 Carcinogenesis. No information found.

2.4.1.6 Genetic and Related Cellular Effects. No information found.

2.4.1.7 Other: Neurological, Pulmonary, Skin Sensitization. In nonoccluded skin irritation tests carried out on clipped rabbits, no signs of erythema were observed in any of the animals (10/10), 4-h after the application of 0.5 mL undiluted ethyl benzoate (6). Erythema was observed in the dorsum of 3/10 animals 24-h after the first treatment. Redness gradually increased with successive treatments. Twenty-four hours after five applications, erythema and edema were severe. Histological changes in the dermis were observed at each successive stage of treatment. A mixed inflammatory infiltrate consisting of amphophil granulocytes, lymphocytes, and macrophages was present from the first treatment onwards. Desquamated crusts were present in animals treated two or more times. In animals subjected to four or more treatments, edema, dermal hemorrhage congestion, and epithelial thickening in localized areas were noted. In dermal toxicity studies conducted by Smyth et al. (5), ethyl benzoeate elicited a skin irritation score of 4 (10 maximum).

2.4.2 Human Experience

No information found.

2.5 Standards, Regulations, or Guidelines of Exposure

No information found.

2.6 Studies on Environmental Impact

No information found.

3.0 Isopropyl Benzoate

3.0.1 CAS Number: *[939-48-0]*

3.0.2 Synonyms: Benzoic acid, isopropyl ester; benzoic acid, 1-methylethyl ester; 1-methylethyl benzoate; isopropylester kyseliny bensoove [Czech]; 4-09-00-00289 [Beilstein]; AI3-01132 [NLM]; BRN 2044384 [RTECS]; EINECS 213-361-6; and FEMA No. 2932 [NLM]

3.0.3 Trade Names: NA

3.0.4 Molecular Weight: 164.20

3.0.5 Molecular Formula: $C_{10}H_{12}O_2$

3.0.6 Molecular Structure:

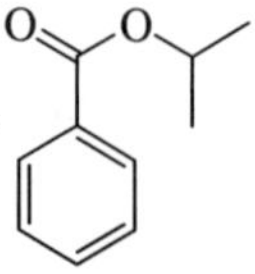

3.1 Chemical and Physical Properties

3.1.1 General

The empirical formula of isopropyl benzoate is $C_{10}H_{12}O_2$. The structural formula is $C_6H_5CO_2C_3H_7$. Its properties are similar to those of ethyl benzoate. Chemical and physical data are summarized in Table 80.1.

3.1.2 Odor and Warning Properties

No information found.

3.2 Production and Use

Isopropyl benzoate is used as an insecticide evaporation control agent.

3.3 Exposure Assessment

No information found.

3.4 Toxic Effects

3.4.1 Experimental Studies

The oral and dermal LD_{50} values in the rat and rabbit are 3.73 g/kg and 20 mL/kg, respectively (7). No other eexperimental toxicity information was located.

3.4.2 Human Experience

No information found.

3.5 Standards, Regulations, or Guidelines of Exposure

No information found.

3.6 Studies on Environmental Impact

No information found.

4.0 Butyl Benzoate

4.0.1 CAS Number: *[136-60-7]*

4.0.2 Synonyms: Benzoic acid, butyl ester; benzoic acid, *n*-butyl ester; *n*-butyl benzoate; butylester kyseliny bensoove [Czech]; 4-09-00-00290 [Beilstein]; AI3-00521 [NLM]; BRN 1867073 [RTECS]; EINECS 205-252-7; HSDB 2089; and NSC 8474 [NLM]

4.0.3 Trade Names: Dai Cari XBN and Anthrapole AZ

4.0.4 Molecular Weight: 178.23

4.0.5 *Molecular Formula:* $C_{11}H_{14}O_2$

4.0.6 *Molecular Structure:*

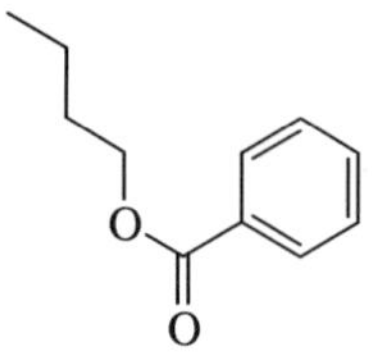

4.1 Chemical and Physical Properties

4.1.1 *General*

The empirical formula of butyl benzoate is $C_{11}H_{14}O_2$. The structural formula is $C_6H_5CO_2C_4H_9$. It is a thick, colorless, oily liquid. It is practically insoluble in water, but is soluble in alcohol, ether, and acetone. It is miscible with oils and hydrocarbons. It is normally stable, but can react with oxidizing materials. Its vapor density is 6.15 and its Henry's Law constant is 3.97×10^{-5} atm cu/m/mol. Its flash point is 225°F (107°C). Chemical and physical data are summarized in Table 80.1.

4.1.2 *Odor and Warning Properties*

When heated to decomposition, butyl benzoate emits acrid smoke and irritating fumes.

4.2 Production and Use

Butyl benzoate is formed by the direct esterification of *n*-butyl alcohol with benzoic acid under azeotropic conditions (72). It is used as a solvent for cellulose ether, a dye carrier for textiles, and a perfume ingredient (73–75). It occurs naturally as a volatile component of various fruits and vegetables (76–79, 88). It has been approved by the FDA as an indirect food additive for use as a component of adhesives (21 CFR 175.105) (80).

4.3 Exposure Assessment

No information found.

4.4 Toxic Effects

4.4.1 *Experimental Studies*

4.4.1.1 Acute Toxicity. The oral LD_{50} values of butyl benzoate for rats and mice are 5.14 and 3.45 g/kg, respectively (5, 8). The acute dermal LD_{50} value for rabbits is >5 g/kg (9). In rats exposed to the concentrated vapor of *n*-butyl benzoate, 8 h is the maximum exposure period that resulted in no deaths within 14 d (5). Acute toxicity data are summarized in Table 80.2.

4.4.1.2 Chronic and Subchronic Toxicity. No information found.

4.4.1.3 Pharmacokinetics, Metabolism, and Mechanisms. No information found.

4.4.1.4 Reproductive and Developmental. No information found.

4.4.1.5 Carcinogenesis. No information found.

4.4.1.6 Genetic and Related Cellular Effects Studies. Butyl benzoate inhibits the growth of *Bacillus subtilis* and some mammalian cell lines (81).

4.4.1.7 Other: Neurological, Pulmonary, Skin Sensitization. In a nonoccluded skin irritation test carried out on clipped rabbits, erythema was observed in all animals (14/14), 4 h after the application of 0.5 mL undiluted butyl benzoate (6). Histological alterations (thickening, loss of elasticity, local increase of temperature, and inflammatory reactions) worsened as the number of treatments increased. After 5 or 6 applications, deep fissures were observed. Studies performed by Smyth indicate that butyl benzoate causes strong erythema, edema, or slight necrosis after application to clipped skin of rabbits (5). It has been referred to as both a "moderate" and a "severe" irritant to rabbit skin by NIOSH (82).

Results of corneal necrosis tests in rabbits indicate that undiluted butyl benzoate causes minimal (score 1/10) eye injury (5). Seizures have been reported to occur in animals receiving lethal doses. All other available toxicity information is presented in Table 80.2.

4.4.2 Human Experience

In 25 human subjects, a preparation of 6% butyl benzoate in petrolatum produced no irritation or sensitization after a 48-h closed-patch test (9).

4.5 Standards, Regulations, or Guidelines of Exposure

No information found.

4.6 Studies on Environmental Impact

No information found.

5.0 Hexyl Benzoate

5.0.1 CAS Number: *[6789-88-4]*

5.0.2 Synonyms: Benzoic acid, hexyl ester; hexyl benzoate; *n*-hexyl benzoate; 1-hexyl benzoate hexylester kyseliny bensoove [Czech]; 4-09-00-00293 [Beilstein]; AI3-02064 [NLM]; BRN 2048117 [RTECS]; EINECS 229-856-5; HSDB 6031; and FEMA No. 3691 [NLM]

5.0.3 Trade Names: NA

5.0.4 Molecular Weight: 206.28

5.0.5 Molecular Formula: $C_{13}H_{18}O_2$

5.0.6 Molecular Structure:

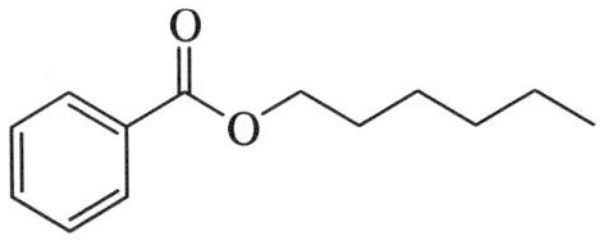

5.1 Chemical and Physical Properties

5.1.1 General

The empirical formula of hexyl benzoate is $C_{13}H_{18}O_2$. The structural formula is $C_6H_5CO_2(CH_2)_5CH_3$. It is a clear, colorless liquid, with a density of 0.979 g/cm^3. It is soluble in DMSO, 95% ethanol, and acetone, but is not very soluble in water. Reactivity data are not available. The flash point is >93.3°C (>200°F). Chemical and physical data are summarized in Table 80.1.

5.1.2 Odor and Warning Properties

When heated to decomposition, hexyl benzoate emits acrid smoke and irritating fumes.

5.2 Production and Use

Hexyl benzoate is used as a fragrance for soaps, perfumes, and creams.

5.3 Exposure Assessment

No information found.

5.4 Toxic Effects

5.4.1 Experimental Studies

The oral and dermal LD_{50} values of hexyl benzoate for the rat and rabbit are 12.3 g/kg and 21 mL/kg, respectively (Table 80.2). No other experimental toxicology information was located.

5.4.2 Human Experience

Exposure may cause mild skin and eye irritation (83).

5.5 Standards, Regulations, or Guidelines of Exposure

No information found.

5.6 Studies on Environmental Impact

No information found.

6.0 Vinyl Benzoate

6.0.1 CAS Number: [769-78-8]

6.0.2 Synonyms: Benzoic acid, vinyl ester; benzoic acid, ethenyl ester; vinylester kyseliny bensoove [Czech]; 4-09-00-00295 [Beilstein]; AI3-24554 [NLM]; BRN 2041125 [RTECS]; EINECS 212-214-3; and NSC 2296 [NLM]

6.0.3 Trade Names: NA

6.0.4 Molecular Weight: 148.16

6.0.5 Molecular Formula: $C_9H_8O_2$

6.0.6 Molecular Structure:

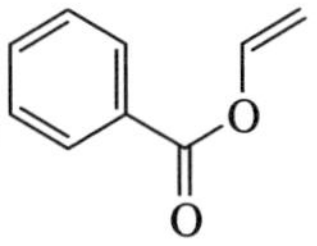

6.1 Chemical and Physical Properties

6.1.1 General

The structural formula is $C_6H_5CO_2CH{:}CH_2$. No other physical or chemical information was located.

6.1.2 Odor and Warning Properties

No information found.

6.1.3 Ionizing Radiation Properties

No information found.

6.2 Production and Use

No information found.

6.3 Exposure Assessment

No information found.

6.4 Toxic Effects

6.4.1 Experimental Studies

Vinyl benzoate is in the same toxicity range as the lower alkyl esters. The oral LD_{50} value for the rat is 3.25 g/kg (5). It is slightly more irritating to rabbit skin than the alkyl benzoates. Acute toxicity data are summarized in Table 80.2.

6.4.2 Human Experience

No information found.

6.5 Standards, Regulations, or Guidelines of Exposure

No information found.

6.6 Studies on Environmental Impact

No information found.

7.0 Benzyl Benzoate

7.0.1 CAS Number: [120-51-4]

7.0.2 Synonyms: Benzoic acid, benzyl ester; benzoic acid, phenylmethyl ester; benzyl alcohol benzoic ester; benzyl benzenecarboxylate; benzyl phenylformate; phenylmethyl benzoate; benzylester kyseliny bensoove [Czech]; 4-09-00-00307 [Beilstein]; AI3-00523 [NLM]; BRN 2049280 [RTECS]; EINECS 204-402-9; Casewell No. 082; CTFA 00287; EPA Pesticide Chemical Code 009501; FEMA No. 2138; HSDB 208; and NSC 8081

7.0.3 Trade Names: Benzylate; Benzylets; Ascabin; Ascabiol; Colebenz; Novoscabin; Scabagen; Scabanca; Scabide; Scabiozon; Scabitox; Scobenol; Vanzoate; and Venzoate

7.0.4 Molecular Weight: 212.25

7.0.5 Molecular Formula: $C_{14}H_{12}O_2$

7.0.6 Molecular Structure:

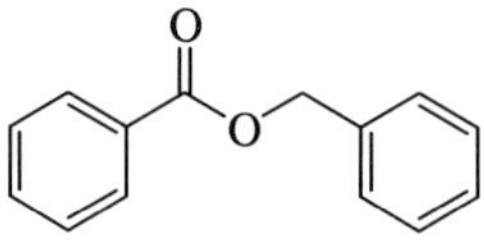

7.1 Chemical and Physical Properties

7.1.1 General

The empirical formula of benzyl benzoate is $C_{14}H_{12}O_2$. The structural formula is $C_6H_5CO_2CH_2C_6H_5$. It is an oily liquid that is insoluble in water or glycerol, soluble in acetone and benzene, and miscible with alcohol, chloroform, ether, and oils. It can react with oxidizing materials and should be stored in light-resistant containers. Its flash point is 148°C (298°F), and its autoignition temperature is 480°C (896°F). It is a slight fire hazard when exposed to heat or flame. Chemical and physical data are summarized in Table 80.1.

7.1.2 Odor and Warning Properties

Benzyl benzoate has a faint, pleasant aromatic odor and a sharp, burning taste.

7.2 Production and Use

Benzyl benzoate is produced from benzyl alcohol and sodium benzoate in the presence of triethylamine, or by the transesterification of methyl benzoate with benzyl alcohol in the presence of an alkali benzyl oxide. It is a by-product in the manufacture of benzoic acid by the oxidation of toluene. In the presence of sodium, benzylaldehyde is condensed to form benzyl benzoate (84). Benzyl benzoate occurs naturally in balsam, concrete, and some flowers (85).

Benzyl benzoate is used in animals and humans as a pediculicide, acaricide, a repellant for chiggers, and a remedy for scabies (86). It is also used as a fixative, plasticizer or dye carrier, a fragrance (synthetic musks), and a flavoring agent for nonalcoholic beverages, ice cream, baked goods, and chewing gum (85).

7.3 Exposure Assessment

The compound can be analyzed using an ultraviolet spectrophotometer.

7.4 Toxic Effects

7.4.1 Experimental Studies

7.4.1.1 Acute Toxicity. Available toxicity data from acute toxicity studies are summarized in Table 80.2. The oral LD_{50} values for the rat, mouse, rabbit, cat, and dog are 1.7 mL/kg, 1.4 mL/kg, 1.68–1.80 g/kg, 2.24 g/kg, and greater than 22.44 g/kg (2, 4, 10). Signs of poisoning in animals receiving toxic oral doses include salivation, piloerection, muscular incoordination, tremors, progressive paralysis of hind limbs, prostration, violent convulsion, dyspnea, and death by respiratory paralysis (87). The dermal LD_{50} values for the rat and rabbit are 4.0 mL/kg and 4.0 g/kg, respectively (2, 4, 11). If applied dermally over too wide an area, or too frequently in small animals (particularly cats), it can cause nausea, vomiting, diarrhea, respiratory and cardiac depression (89).

7.4.1.2 Chronic and Subchronic Toxicity. In a 90-d study in rats, the dermal LD_{50} value of benzyl benzoate was 2.0 mL/kg (2).

7.4.1.3 Pharmacokinetics, Metabolism, and Mechanisms. *In vivo*, benzyl benzoate is hydrolyzed to benzoic acid. Benzoic acid is conjugated with glycine to form benzoylglycine (hippuric acid) and with glucuronic acid to form benzoylglucuronic acid. These conjugates are eliminated in the urine in varying doses, depending on species and dose (87).

7.4.1.4 Reproductive and Developmental. Fetuses from pregnant rats fed benzyl benzoate (0.04 or 1.0%) in the diet from the beginning of gestation to day 21 post-parturition had no external, skeletal, or visceral abnormalities (90).

7.4.1.5 Carcinogenesis. No information found.

7.4.1.6 Genetic and Related Cellular Effects Studies. No information found.

7.4.1.7 Other: Neurological, Pulmonary, Skin Sensitization. In two separate skin irritation studies in 3 rabbits, benzyl benzoate elicited primary irritation index scores of 0 and 1.58 (out of a maximum score of 8) (11).

7.4.2 Human Experience

7.4.2.1 General Information. Benzyl benzoate is relatively nontoxic, but may be irritating to the skin and eyes (24). Increased pruritus and irritation (manifested by burning and stinging) are common with use on the genitalia and scalp (particularly in hot, humid climates) (91). According to Gruber (92, 93), benzyl benzoate has blood pressure lowering capabilities.

7.4.2.2 Clinical Cases

7.4.2.2.1 Acute Toxicity. Convulsions occurred in a 2-mo-old boy 2.5 h after treatment with a scabicide containing benzyl benzoate (43%), soap (0%), ethyl alcohol (20%), and distilled water (17%) (87). An unusual adverse reaction (headache, watering of eyes, rhinorrhea, sweating, malaise, and burning facial erythema) occurred in a healthy 51-year-old man after he had applied benzyl benzoate over his whole body twice a day for 2 d (94). A "disulfeiram-like" reaction was suspected because the man had consumed a "moderate amount of red wine" before symptoms occurred.

7.4.2.2.2 Chronic and Subchronic Toxicity. No information found.

7.4.2.2.3 Pharmacokinetics, Metabolism and Mechanisms. After cutaneous application, 54% of the applied dose penetrates human skin, compared with 69% in the monkey (95).

7.4.2.2.4 Reproductive and Developmental. No information found.

7.4.2.2.5 Carcinogenesis. No information found.

7.4.2.2.6 Genetic and Related Cellular Effects Studies. No information found.

7.4.2.2.7 Other: Neurological, Pulmonary, Skin Sensitization, etc. No information found.

7.4.2.3 Epidemiology Studies. No information found.

7.5 Standards, Regulations, or Guidelines of Exposure

Benzyl benzoate has been approved by the FDA as a direct food additive, when used as a synthetic flavoring substance and adjuvant (96). It has also been approved as an indirect food additive for use as a component of adhesives (80).

7.6 Studies on Environmental Impact

No information found.

8.0 Methyl Salicylate

8.0.1 CAS Number: [119-36-8]

8.0.2 Synonyms: *o*-Anisic acid; benzoic acid, 2-hydroxy-, methyl ester; betula; betula lenta; betula oil; 2-carbomethoxyphenol; gaultheria oil, gaultheriaoel; *o*-hydroxybenzoic acid, methyl ester; 2-hydroxybenzoic acid, methyl ester; methyl 2-hydroxybenzoate;

methyl *o*-hydroxybenzoate; 2-(methoxycarbonyl)phenol; natural wintergreen oil; oil of wintergreen; salicylic acid, methyl ester; spicewood oil; sweet birch oil; synthetic wintergreen oil; teaberry oil; wintergreen oil; wintergruenoel; methylester kyseliny salicylove [Czech]; metylester kyseliny salicylove [Czech]; 4-10-00-00143 [Beilstein]; AI3-00090 [NLM]; BRN 0971516 [RTECS]; EINECS 204-317-7; Casewell No. 577; CCRIS 6259; CTFA 01630; EPA Pesticide Chemical Code 076601; FEMA No. 2745; HSDB 1935; and NSC 8204

8.0.3 Trade Names: Analgit; Exagien; Flucarmint; Theragesic, Panalgesic; and Heet

8.0.4 Molecular Weight: 152.15

8.0.5 Molecular Formula: $C_8H_8O_3$

8.1 Chemical and Physical Properties

8.1.1 General

The empirical formula of methyl salicylate is $C_8H_8O_3$. The structural formula is C_6H_4 (OH) ($COOCH_3$). It is a colorless, yellowish or reddish, oily liquid. It is soluble in chloroform and ether and miscible with alcohol and glacial acetic acid. Its flash point is 96°C (205°F), and its autoignition temperature is 850°F. Chemical and physical data are summarized in Table 80.1.

8.1.2 Odor and Warning Properties

Odor of wintergreen.

8.2 Production and Use

Methyl salicylate occurs naturally in the flowers of *Burnum dilitatum* (97). It is prepared commercially by the esterification of salicylic acid with methanol. It is used in liniments and ointments for pain relief, as a flavorant for foods, chewing gum, candy, beverages, and pharmaceuticals. It is a perfumant for detergents and fragrances, and a mild antiseptic in oral hygiene products. It is also used as a dye carrier, a solvent for insecticides, polishes, and inks, a UV absorber in sunscreen lotions, and a UV light stabilizer in acrylic resins.

8.3 Exposure Assessment

Methyl salicylate has a characteristic odor that can be detected on the breath, and in urine, and vomitus. Analytically, the material can be quantified by thin-layer and gas chromatography (98).

8.4 Toxic Effects

8.4.1 Experimental Studies

8.4.1.1 Acute Toxicity. The oral LD_{50} values for the rat, mouse, and dog are 0.887, 1.11, and 2.1 g/kg, respectively (3, 14, 15). Toxicity data from acute animal studies are summarized in Table 80.3.

8.4.1.2 Chronic and Subchronic Toxicity. In a 17-wk study, ten male and female rats given 1% methyl salicylate in the diet exhibited retarded growth (104). In rats (25/sex) ingesting 0.1, 0.5, 1.0, or 2.0% methyl salicylate in the diet for up to 2 yr, the NOEL was 0.1% (17). Compared to controls, animals ingesting 0.5% had an increased incidence of gross pituitary lesions. Growth retardation and rough coats were noted in animals ingesting 1.0 or 2.0%. An increase in cancellous bone was noted in all bones examined from rats on the 2.0% methyl salicylate diet. All rats ingesting the diet containing 2.0% methyl salicylate died by the fiftieth week.

In a 90-d study, rabbits (3/group) were treated dermally with 0.5, 1.0, 2.0, or 4.0 ml methyl salicylate/kg/d (17). Rabbits in the 4.0 mL/kg/d group exhibited anorexia, weight loss, and depression and died after days 6, 8, or 28 of treatment. Upon autopsy, one of the high-dose rabbits exhibited dilation, desquamation, and formation of atypical epithelium in the distal renal tubules, superficial necrosis and sloughing of the skin, slight hepatitis, vacuolization of pancreatic acinar cells, and histological changes in the bone marrow and voluntary muscles. These changes were not observed in the other high-dose rabbit that was examined. Two rabbits treated with 2.0 mL/kg/d exhibited a slight sloughing of epidermal scales. All treated animals that survived until study termination had very slight to slight dermatitis, and an increased incidence of spontaneous nephritis and mild hepatitis.

In a 59-d oral toxicity study, dogs (1 of each sex/group) were treated with 0, 50, 100, 250, 800, or 1200 mg methyl salicylate/kg/d (17). No adverse effects were noted in animals receiving up to 250 mg/kg/d. Dogs receiving $\geq$500 mg/kg/d more lost weight and were euthanized or died within a month. Livers from animals given 800 or 1200 mg/kg/d had moderate to marked degrees of fatty metamorphosis. In a 2-yr study, dogs (2/sex/group) were given oral doses of 0, 50, 150, or 350 mg methyl salicylate/kg/d (17). Dogs in the 150- and 350-mg/kg/d groups exhibited weight loss or retarded growth, enlarged livers, and hepatocyte hypertrophy. The NOEL was 50 mg/kg/d.

8.4.1.3 Pharmacokinetics, Metabolism and Mechanisms. Salicylates are absorbed rapidly after ingestion or dermal application. Peak plasma levels are reached about 2 h after ingestion. After absorption, salicylate is distributed throughout most body tissues and transcellular fluids. It is metabolized to salicyluric acid via conjugation with glycine, and to salicyl acyl glucuronide and salicylic phenolic glucuronide via conjugation with glucuronic acid. A small fraction is oxidized to gentisic acid, gentisuric acid, 2,3-dihydroxybenzoic acid, and 2,3,5-trihydroxybenzoic acid. Free salicylic acid, salicyluric acid, salicylic phenolic glucuronide, salicylic acyl glucuronides, and a small amount of gentisic acid are excreted in the urine. Excretion of the free salicylate is extremely variable and depends on the dose and urinary pH (105).

8.4.1.4 Reproductive and Developmental. In mice treated with 500 mg/kg methyl salicylate via gavage prior to mating, reproduction was adversely affected. In the treated group, the number of litters per pair and the number of pups per litter were lower than the control group (100). In this study, a single dose of 100 mg/kg or less did not produce any adverse effects on reproduction in mice. However, in mice given 100 mg sodium salicylate/kg/d throughout mating and gestation, the number of live pups per litter, the percentage of live pups, and pup weight decreased (101).

Methyl salicylate has been shown to be teratogenic and/or embryotoxic in mice, rats, rabbits, and hamsters (26). Treatment (IP) of pregnant rats between days 10 and 14 of gestation increases the number of resorptions and fetal mortality (102a). Offspring have decreased weight and an increased incidence of kidney abnormalities (102a, 102b). Seventy-two percent of embryos recovered from hamsters treated orally with 1.75 g/kg methyl salicylate on day 7 of gestation had neural tube defects (103).

In hamsters, topical treatment with 5.25 g/kg methyl salicylate on day 7 of gestation caused a 53% incidence of neural tube defects in embryos (103). Dermal application of undiluted methyl salicylate (1 g/kg/d) to rats during days 6–15 of gestation leads to a 100% incidence of total resortions (104). In contrast, dermal application of up to 6 g/kg/d of a petrolatum preparation containing 3% methyl salicylate to gestating rats caused no signs of maternal toxicity, no alterations in reproductive parameters and no fetal abnormalities (104).

8.4.1.5 Carcinogenesis. Available data suggest that methyl salicylate is not carcinogenic. Similar kinds and numbers of tumors occurred in rats ingesting control diet or a diet containing up to 1.0% methyl salicylate for 2 y (17).

8.4.1.6 Genetic and Related Cellular Effects Studies. At doses of 1,000, 3,300, 10,000, 33,300, 100,000, and 330,300 μg/plate, methyl salicylate was not mutagenic to *S. typhimurium* strains TA1535, TA1537, and TA97 in the presence or absence of rat and hamster liver S-9 and strains TA98 and TA100 in the presence or absence of rat liver S-9 (108). Both positive and negative results have been noted in strains TA98 and TA100 in the presence of hamster liver S-9 (108, 109).

8.4.1.6 Other: Neurological, Pulmonary, Skin Sensitization. No information found.

8.4.2 Human Experience

8.4.2.1 General Information. Refer to introduction.

8.4.2.2 Clinical Cases

8.4.2.2.1 Acute Toxicity. The average lethal dose of methyl salicylate is 10 mL for children and 30 mL (or 0.5 g/kg) for adults (36). The lowest lethal oral dose in adult humans ranges from 0.1 to 1.48 g/kg (3, 110, 111). Effects seen in adults include nausea or vomiting, convulsions, coma, hemorrhage, and respiratory stimulation. In children, the lowest lethal oral dose ranges from 0.17 to 0.7 g/kg (36, 106, 107). In children and infants, dyspnea, nausea or vomiting, respiratory stimulation, and flaccid paralysis have been noted. A 62-year-old man developed salicylate toxicity with metabolic acidosis after applying an ointment containing methyl salicylate to the thigh twice daily for 3 wk (112). Hemorrhage has occurred in patients using both warfarin and topical methyl salicylate (113, 114).

8.4.2.2.2 Chronic and Subchronic Toxicity. No information found.

8.4.2.2.3 Pharmacokinetics, Metabolism, and Mechanisms. In 12 healthy volunteers (ages 21–44) who applied 5 g of an ointment containing 12.5% methyl salicylate to the thigh twice daily, concentrations between 0.31 and 0.91 mg/mL were detected within 1 h of the first application. A C_{max} between 2 and 6 mg/mL was observed on day 4, after 7 applications (115).

8.4.2.2.4 Reproductive and Developmental. No information found.

8.4.2.2.5 Carcinogenesis. No information found.

8.4.2.2.6 Genetic and Related Cellular Effects Studies. No information found.

8.4.2.2.7 Other: Neurological, Pulmonary, Skin Sensitization, etc. Methyl salicylate is reportedly a strong dermal and mucous membrane irritant (14, 116). After spray application to forearms of volunteers, a rubifacient containing methyl and ethyl salicylate caused erythema. Maximum erythema occurred when maximum blood levels were achieved (approximately 30 min after application) (117).

8.4.2.3 Epidemiology Studies. No information found.

8.5 Standards, Regulations, or Guidelines of Exposure

No information found.

8.6 Studies on Environmental Impact

Methyl salicylate is likely released into effluent and the atmosphere. In soil, it is likely to biodegrade. It is fairly soluble in water and may be washed out of soil by rain. If released into water, it should slowly volatilize, biodegrade, hydrolyze (in alkaline water), and be lost as a result of direct photolysis and photooxidation. Its half-life in a model river is 49 d. It should not bioaccumulate in aquatic organisms. In the atmosphere, it reacts with photochemically produced hydroxyl radicals, resulting in an estimated half-life of 1.4 d (118).

9.0 Ethyl Salicylate

9.0.1 CAS Number: [118-61-6]

9.0.2 Synonyms: Benzoic acid, 2-hydroxy-, ethyl ester; *o*-(ethyoxycarbonyl)phenol; ethyl 2-hydroxybenzoate; ethyl *o*-hydroxybenzoate; sal ether; sal ethyl; salicylic acid, ethyl ester; salicylic ether; salicylic ethyl ester; 4-10-00-00149 [Beilstein]; AI3-00513 [NLM]; BRN 0907659 [RTECS]; EINECS 204-265-5; FEMA No. 2458; HSDB 1935; and NSC 8209

9.0.3 Trade Names: Mesotol and Salotan

9.0.4 Molecular Weight: 166.18

9.0.5 Molecular Formula: $C_9H_{10}O_3$

9.0.6 Molecular Structure:

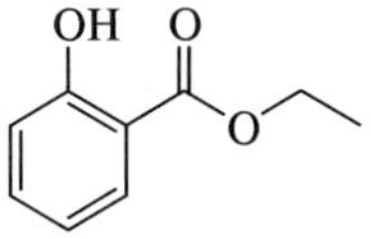

9.1 Chemical and Physical Properties

9.1.1 General

The empirical formula of ethyl salicylate is $C_9H_{10}O_3$. The structural formula is C_6H_4 (OH) ($COOC_2H_5$). It is a liquid at room temperature, with melting and boiling points of 1.02 and 234°C, respectively. Other chemical and physical data are shown in Table 80.1.

9.1.2 Odor and Warning Properties

No information found.

9.2 Production and Use

Ethyl salicylate is found naturally in currants and strawberries (119). It is manufactured commercially by the esterification of salicylic acid with ethyl alcohol. It is used in fragrances.

9.3 Exposure Assessment

No information found.

9.4 Toxic Effects

9.4.1 Experimental Studies

9.4.1.1 Acute Toxicity. Available toxicity information is summarized in Table 80.3. The acute oral LD_{50} value for rats is 1.32 g/kg, and the acute dermal LD_{50} value for rabbits is greater than 5 g/kg (18). In guinea pigs, the minimum lethal oral and subcutaneous doses are 1.4 and 1.5 g/kg, respectively (19).

At dietary levels of 0.35–1.00%, ethyl salicylate induced thymic atrophy in immature rats when administered for 2–3 d (120).

9.4.1.2 Chronic and Subchronic Toxicity. When administered to Leghorn chicks at a concentration of 250 g/ton of food, ethyl salicylate increases food consumption and weight gain, but decreases egg laying and egg weight (121).

9.4.1.3 Pharmacokinetics, Metabolism, and Mechanisms. In contrast to other salicylates, ethyl salicylate penetrates the skin of rats slowly (122). On intact skin, absorption rates of 1.97 and 1.53 μg/mm^2/h were recorded at pH 2 or 3, respectively (123). No absorption was detected at pH 6 or 8. See Section 8.4.1.3 for more data on metabolism.

9.4.1.4 Reproductive and Developmental. No information found.

9.4.1.5 Carcinogenesis. No information found.

9.4.1.6 Genetic and Related Cellular Effects Studies. No information found.

9.4.1.7 Other: Neurological, Pulmonary, Skin Sensitization. When applied full strength to intact or abraded rabbit skin for 24 h under occlusion, ethyl salicylate was moderately irritating (18).

9.4.2 Human Experience

Results of a 48-h closed patch test in 25 human volunteers indicate that application of 12% methyl benzoate in petrolatum does not cause skin sensitization or irritation (142a, 142b). Application of a rubifacient containing both methyl and ethyl salicylate has been shown to cause erythema (117).

9.5 Standards, Regulations, or Guidelines of Exposure

Ethyl salicylate was given GRAS status by FEMA in 1965. It is approved by the FDA for food use (96).

9.6 Studies on Environmental Impact

No information found.

10.0 Amyl Salicylate

10.0.1 CAS Number: *[2050-08-0]*

10.0.2 Synonyms: Benzoic acid, 2-hydroxy-, pentyl ester; 2-hydroxybenzoic acid, pentyl ester; pentyl salicylate; pentyl 2-hydroxybenzoate; salicylic acid, pentyl ester; amylester kyseliny salicylove [Czech]; 4-10-00-00153 [Beilstein]; AI3-00334 [NLM]; BRN 2577253 [RTECS]; Casewell No. 049AA; CTFA 00196; EINECS 218-080-2; NSC 403668; NSC 44877; and NSC 46125

10.0.3 Trade Names: NA

10.0.4 Molecular Weight: 108.26

10.0.5 Molecular Formula: $C_{12}H_{16}O_3$

10.0.6 Molecular Structure:

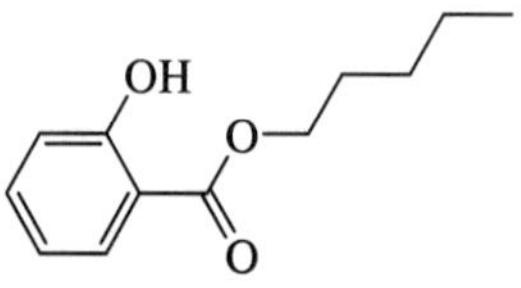

10.1 Chemical and Physical Properties

10.1.1 General

The empirical formula of amyl salicylate is $C_{12}H_{16}O_3$. The structural formula is C_6H_4 (OH) ($COOC_5H_{11}$). Chemical and physical data are summarized in Table 80.1.

10.1.2 Odor and Warning Properties

No information found.

10.2 Production and Use

No information found.

10.3 Exposure Assessment

No information found.

10.4 Toxic Effects

The intravenous LD_{50} values for the dog and rabbit range from 0.5 to 0.8 g/kg (20). Consult Section 8.4.1.3 for information about metabolism.

10.5 Standards, Regulations, or Guidelines of Exposure

No information found.

10.6 Studies on Environmental Impact

No information found.

11.0 Phenyl Salicylate

11.0.1 CAS Number: *[118-55-8]*

11.0.2 Synonyms: Benzoic acid, 2-hydroxy-, phenyl ester; 2-hydroxybenzoic acid, phenyl ester; phenol salicylate; 2-phenoxycarbonylphenol; phenyl 2-hydroxybenzoate; salicylic acid, phenyl ester; salphenyl; fenylester kyseliny salicylove [Czech]; 4-10-00-00154 [Beilstein]; AI3-00195 [NLM]; BRN 0393969 [RTECS]; CCRIS 4859; CTFA 05448; EINECS 204-259-2; and NSC 33406

11.0.3 Trade Names: Musol and Salol

11.0.4 Molecular Weight: 214.23

11.0.5 Molecular Formula: $C_{13}H_{10}O_3$

11.0.6 Molecular Structure:

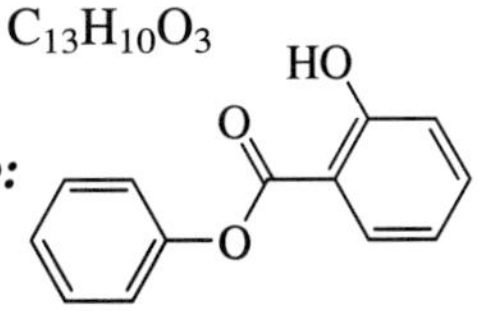

11.1 Chemical and Physical Properties

11.1.1 General

The empirical formula of phenyl salicylate is $C_{13}H_{10}O_3$. The structural formula is $C_6H_5CO_2$ $(CH_2)_5CH_3$. Its physical form is white crystals. Phenyl salicylate melts at 41–43°C and boils at 173°C (12 mmHg). It is soluble in 95% ethanol, acetone, ether, and benzene, but is insoluble in water. Phenyl salicylate is incompatible with bromine water, ferric salts, camphor, phenol, chloral hydrate, monobrominated camphor, thymol, or urethane in trituration. Chemical and physical data are summarized in Table 80.1.

11.1.2 Odor and Warning Properties

No information found.

11.2 Production and Use

Phenyl salicylate is used as an analgesic and antipyretic. It is also used in the manufacture of polymer plastics, lacquers, waxes, polishes, adhesives, and sunscreen products.

11.3 Exposure Assessment

No information found.

11.4 Toxic Effects

11.4.1 Experimental Studies

11.4.1.1 Acute Toxicity. The oral LD_{50} value for rats is 3 g/kg (21). An intraperitoneal dose of >0.5 g/kg is required for lethality in mice (22).

11.4.1.2 Chronic and Subchronic Toxicity. No information found.

11.4.1.3 Pharmacokinetics, Metabolism, and Mechanisms. Consult Section 8.4.1.3 for information about metabolism.

11.4.1.4 Reproductive and Developmental. Reduced fertility (postimplantation mortality and reduced litter size) and increased fetal mortality were observed in rats given 0.6 g/kg from days 7–12 of gestation (125). Developmental abnormalities in the CNS and musculoskeletal system are noted in offspring of rats given 1.2 g/kg from days 7–9 of gestation.

11.4.1.5 Carcinogenesis. No information found.

11.4.1.6 Genetic and Related Cellular Effects Studies. At doses of 1.0–333 μg/plate, phenyl salicylate was not mutagenic in *S. typhimurium* strains TA1535, TA1537, TA98, or TA100 in the presence or absence of rat and hamster liver S-9 (126).

11.4.1.7 Other: Neurological, Pulmonary, Skin Sensitization. No information found.

11.4.2 Human Experience

The oral LD_{LO} of phenyl salicylate in humans is 50 mg/kg (127). Phenyl salicylate is a mild skin and eye irritant. Individual cases of allergy and sensitization have been reported (128). However, in 173 patients referred to the Finnish Institute of Occupational Health for suspected occupational dermatoses, none had allergic or irritant patch test reactions to 1.0% phenyl salicylate (129).

11.5 Standards, Regulations, or Guidelines of Exposure

A NIOSH-approved half-face respirator equipped with an organic vapor/acid gas cartridge and a dust/mist filter should be worn when weighing or diluting neat phenyl salicylate.

11.6 Studies on Environmental Impact

No information found.

12.0 Methyl Paraben

12.0.1 CAS Number: *[99-76-3]*

12.0.2 Synonyms: Benzoic acid, *p*-hydroxy-, methyl ester; benzoic acid, 4-hydroxy-, methyl ester; *p*-carbomethoxyphenol; 4-hydroxybenzoic acid, methyl ester; *p*-hydroxybenzoic acid methyl ester; 4-hydroxybenzoic acid methyl ester; *p*-hydroxybenzoic methyl ester; *p*-(methoxycarbonyl)phenol; 4-(methoxycarbonyl)phenol; methyl 4-hydroxybenzoate; methyl *p*-hydroxybenzoate; methyl *p*-oxybenzoate; methylester kyseliny *p*-hydroxybenzoove [Czech]; *p*-oxybenzoesauremethylester [German]; 4-10-00-00360 [Beilstein]; AI3-01336 [NLM]; BRN 0509801 [RTECS]; EINECS 202-785-7; Casewell No. 573PP; CCRIS 3946; CTFA 01624; EPA Pesticide Chemical Code 061201; FEMA No. 2710; HSDB 1184; and NSC 3827

12.0.3 Trade Names: Abiol, Aseptoform, Maseptol, Metaben, Methaben, Methyl butex, Methyl chemosept, Methyl parasept, Methylben, Metoxyde, Moldex, Nipagin, Nipagin M, Paridol, Preserval, Septos, Solbrol, Solbrol M, and Tegosept M

12.0.4 Molecular Weight: 152.14

12.0.5 Molecular Formula: $C_8H_8O_3$

12.0.6 Molecular Structure:

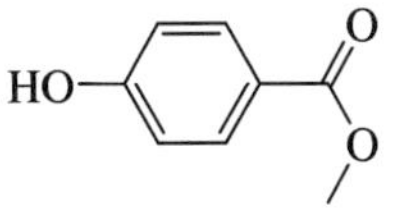

12.1 Chemical and Physical Properties

12.1.1 General

The empirical formula of methyl paraben is $C_8H_8O_3$. The structural formula is $C_6H_4(OH)$-$(COOCH_3)$. At room temperature, methyl paraben is colorless crystals, white needles, or a white crystalline powder. Its melting and boiling points are 131 and 270–280°C, respectively. It is soluble in alcohol, acetone, ether, and benzene. Approximately 1 g dissolves in 40 mL of warm oil or 70 mL warm glycerol. Chemical and physical data are summarized in Table 80.1.

12.1.2 Odor and Warning Properties

When heated to decomposition, methyl paraben emits acrid smoke and fumes.

12.2 Production and Use

Methyl paraben is used as a preservative in parenteral solutions, topical antibiotic or corticosteroid preparations, galencials, cosmetics, foods, and beverages.

12.3 Exposure Assessment

No information found.

12.4 Toxic Effects

12.4.1 Experimental Studies

12.4.1.1 Acute Toxicity. The oral LD_{50} values for the mouse, rabbit, and guinea pig are 8.0, 6.0, and 3.0 mg/kg, respectively (130, 131). In the dog, the LD_{50} and TD_{LO} value are 6.0 and 2.0 mL/kg, respectively (132, 133). Acute toxicity data are summarized in Table 80.5.

12.4.1.2 Chronic and Subchronic Toxicity. The NOELs for methyl paraben in an 80-d study in rats, a 120-d study in guinea pigs, and a 313-d study in mongrel puppies range from 0.5 to 5.0 mg/kg, 11 to 100 mg/kg, and above 1000 mg/kg, respectively (135, 136). Male and female rats that received 1000 or 4000 mg/kg methyl paraben in the diet for 96 wk showed no significant organ changes when examined macroscopically and microscopically (136).

12.4.1.3 Pharmacokinetics, Metabolism, and Mechanisms. After methyl paraben is intravenously infused into the dog, nonhydrolyzed methyl paraben is found in brain, spleen, and pancreas. In liver, kidney, and muscle, it is immediately hydrolyzed to *p*-hydroxybenzoic acid (31). In mice, rats, rabbits, or dogs methyl paraben is excreted in the urine as unchanged benzoate, *p*-hydroxybenzoic acid, *p*-hydroxyhippuric acid (*p*-hydroxybenzoylglycine), ester glucuronides, ether glucuronides, or ether sulfates (31, 137, 138). Six hours after oral administration of 1.0 g/kg to dogs, the peak plasma concentration of free and total methyl paraben (630 and 867 g/cm^3) is reached (31). After 48 h, the vast majority was eliminated.

12.4.1.4 Reproductive and Developmental. No information found.

12.4.1.5 Carcinogenesis. No information found.

Table 80.5. Summary of Toxicity of Parabens

Chemical Name	CAS No.	Species	Exposure Route	Approximate Dose (g/kg)[a]	Treatment Regimen	Observed Effect	Ref.
Paraben							
Methyl	*[99-76-3]*	Mouse	oral	8.0	Single dose	LD_{50} listed. Flaccid paralysis.	130
		Rabbit	oral	6.0	Single dose	LD_{50}	131
		Dog	oral	6.0	Single dose	LD_{50}	132
		Dog	oral	2.0	Single dose	TD_{LO}	133
		Guinea pig	oral	3.0	Single dose	LD_{50}	131
		Rat	sc	>0.5	Single dose	LD_{50}	134
		Mouse	IP	0.96	Single dose	LD_{50} listed. Flaccid paralysis.	130
		Mouse	sc	1.2	Single dose	LD_{50}	13
		Rat	oral	0.5–5 mg	80 d	LD_{00}	135
Ethyl	*[120-47-8]*	Rat	oral	>0.2	Single dose	LD	148
		Mouse	oral	3.0,8.0	Single dose	LD_{50}	149, 150
		Rabbit	oral	5.0	Single dose	LD_{50}	132
		Rabbit	oral	5.0, 4.0	Single dose	LD_{LO}, TD_{LO}	133
		Guinea pig	oral	2.0	Single dose	LD_{50}	131
		Dog	oral	5.0	Single dose	LD_{50}	132
		Dog	oral	5.0, 4.0	Single dose	LD_{LO}, TD_{LO}	133
		Mouse	IP	0.52	Single dose	LD_{50}	150
Propyl	*[94-13-3]*	Mouse	oral	6.0–8.0	Single dose	LD_{50}	150, 151
		Rabbit	oral	6.0	Single dose	LD_{LO}, LD_{50}	133
		Rabbit	oral	3–4	Single dose	TD_{LO}	133
		Dog	oral	6.0	Single dose	LD_{LO}	12, 133
		Dog	oral	3–4	Single dose	TD_{LO}	133
		Mouse	IP	0.4	Single dose	LD_{50}	150
		Mouse	IV	1.65	Single dose	LD_{50}	13
		Mouse	sc	1.65	Single dose	LD_{50}	152
Methyl and propyl mixed		Guinea pig	oral	7.5–75 mg/kg	30 d	LD_{00}	135
Butyl	*[94-26-8]*	Mouse	oral	5.0	Single dose	LD_{50}	150
		Mouse	IP	0.23	Single dose	LD_{50}	150
Amyl	—	Dog	IV	0.5–0.8	Single dose	LD_{50}	12

[a]Unless listed otherwise; LD_{LO} = lowest lethal dose; TD_{LO} = lowest toxic dose; LD_{00} = dose for no lethality.

12.4.1.6 Genetic and Related Cellular Effects Studies. At doses of 0.033–10 mg/plate, methyl paraben is not mutagenic in *S. typhimurium* strains TA98, TA100, TA1535, TA1537, or TA1538 and, *E. coli* strain WP2 in the presence and absence of rat liver S-9 (139). At concentrations of 1.18 and 2.36 mM, methyl paraben reduced beat frequency of cilia in rat tracheal explants (144).

Methyl paraben has been shown to affect cyclic nucleotide metabolism in rats. In rats given 0.4% methyl paraben in feed for 3 wk, cortical cyclic adenosine 3′,5′- monophosphate is decreased and phosphodiesterase IV activity is increased (140).

In vitro, methyl paraben is a potent spermicidal agent against human spermatozoa (141). After treatment with 6 mg/mL methyl paraben, all spermatozoa were immobilized for 30 min. Methyl paraben also inhibits superoxide release from human neutrophils activated with fMet-Leu-Phe and phorbol myristate acetate (143).

12.4.1.7 Other: Neurological, Pulmonary, Skin Sensitization. Subconjunctival injection of methyl paraben at a concentration present in commercially available gentamicin does not cause acute toxic myopathy (147).

12.4.2 Human Experience

12.4.2.1 General Information. Methyl paraben causes allergic contact dermatitis in humans (29, 56, 145). It is moderately irritating to eyes (153).

12.4.2.2 Clinical Cases

12.4.2.2.1 Acute Toxicity. No information found.

12.4.2.2.2 Chronic and Subchronic Toxicity. No information found.

12.4.2.2.3 Pharmacokinetics, Metabolism, and Mechanisms. The excretion and metabolism of methyl paraben have been determined in 6 preterm infants receiving multiple doses of a gentamicin formulation containing methyl paraben. The average recovery of the paraben in urine was 82.6% (150).

12.4.2.2.4 Reproductive and Developmental. No information found.

12.4.2.2.5 Carcinogenesis. No information found.

12.4.2.2.6 Genetic and Related Cellular Effects Studies. No information found.

12.4.2.2.7 Other: Neurological, Pulmonary, and Skin Sensitization. No information found.

12.4.2.3 Epidemiology Studies. In a double-blind study, 200 surgical patients received an IV injection of Citanest (prilocaine containing methyl paraben) or prilocaine alone to produce regional anesthesia in the arm. Seventeen patients in the Citanest group and 4 in the methyl paraben–free group developed erythrematous skin reactions around the injection site (155).

12.5 Standards, Regulations, or Guidelines of Exposure

Methyl paraben is generally recognized as safe when used in accordance with good manufacturing or feeding practice (146).

12.6 Studies on Environmental Impact

No information found.

13.0 Ethyl Paraben

13.0.1 CAS Number: *[120-47-8]*

13.0.2 Synonyms: Benzoic acid, *p*-hydroxy-, ethyl ester; benzoic acid, 4-hydroxy-, ethyl ester; carbethoxyphenol; *p*-carbethoxyphenol; ethyl 4-hydroxybenzoate; ethyl *p*-hydroxybenzoate; ethyl *p*-oxybenzoate; 4-hydroxybenzoic acid, ethyl ester; *p*-hydroxybenzoic acid ethyl ester; 4-hydroxybenzoic acid ethyl ester; *p*-hydroxybenzoate ethyl ester; ethylester kyseliny *p*-hydroxybenzoove [Czech]; *p*-oxybenzoesaureethylester [German]; 4-10-00-00367 [Beilstein]; AI3-30960 [NLM]; BRN 1101972 [RTECS]; EINECS 204-399-4; Casewell No. 447; CTFA 01011; EPA Pesticide Chemical Code 061202; HSDB 938; and NSC 23514

13.0.3 Trade Names: Aseptoform E; Bobomold OE; Easeptol, Ethyl butex, Ethyl parasept, Mycoten, Napagin A, Nipagina A, Nipazin A, Sobrol A, Solbrol A, and Tegosept E

13.0.4 Molecular Weight: 166.17

13.0.5 Molecular Formula: $C_9H_{10}O_3$

13.0.6 Molecular Structure:

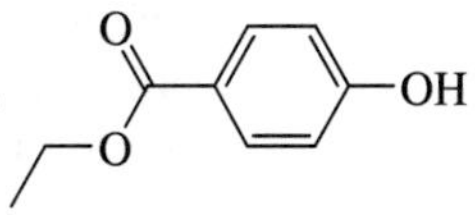

13.1 Chemical and Physical Properties

13.1.1 General

The structural formula is $C_6H_4(OH)(COOC_2H_5)$. At room temperature, it exists as small, colorless crystals or a white powder. Its melting and boiling points are 116–188 and 297–280°C, respectively. It is soluble in alcohol and ether. Chemical and physical properties are summarized in Table 80.1.

13.1.2 Odor and Warning Properties

Ethyl paraben is odorless.

13.2 Production and Use

Ethyl paraben is manufactured via esterification of *p*-hydroxybenzoic acid with ethanol in the presence of an acid. It inhibits the growth of fungi and bacteria and is used as a preservative for pharmaceuticals, adhesives, and various cosmetic preparations.

13.3 Exposure Assessment

No information found.

13.4 Toxic Effects

13.4.1 Experimental Studies

13.4.1.1 Acute Toxicity. The LD_{50} values of ethyl paraben for the mouse, rabbit, guinea pig, and dog are slightly less than those of methyl paraben (Table 80.4). In mice, toxic doses of ethyl paraben cause ataxia, paralysis, and deep depression resembling anesthesia (156).

13.4.1.2 Chronic and Subchronic Toxicity. The NOELs for ethyl paraben in an 80-d study in rats, a 120-d study in guinea pigs, and a 313-d study in mongrel puppies range from 0.5 to 5.0, 11 to 100, and above 1000 mg/kg, respectively (135, 136). After 18 mo of being fed 140 mg/kg of ethyl paraben daily, growth stimulation was noted in rats (150). Growth retardation occurred in rats fed 1600 mg/kg daily (150).

13.4.1.3 Pharmacokinetics, Metabolism, and Mechanisms. After ethyl paraben is intravenously infused into the dog, unhydrolyzed ethyl paraben is found only in the brain. In liver, kidney, and muscle, it is immediately hydrolyzed to *p*-hydroxybenzoic acid. Six hours after oral administration of 1.0 g/kg to dogs, the peak plasma concentration of free and total ethyl paraben (427 and 648 g/cm^3, respectively) is reached (31). After 48 h, all ethyl paraben is completely eliminated.

In mice, rats, rabbits, pigs, or dogs, ethyl paraben is excreted in the urine as unchanged benzoate, *p*-hydroxybenzoic acid, *p*-hydroxyhippuric acid (*p*-hydroxybenzoylglycine), ester glucuronides, ether glucuronides, or ether sulfates (31, 137, 138, 161). Ethyl paraben is also metabolized to *p*-hydoxybenzoic acid by aspergillus (162).

13.4.1.4 Reproductive and Developmental. No teratogenic effects were observed in fetuses of wistar rats which had received ethyl paraben orally (10%, 1%, or 0.01%). However, body weights were lower in some fetuses of rats which had received 10% ethyl paraben (45,600 mg/kg) (148).

13.4.1.5 Carcinogenesis. No information found.

13.4.1.6 Genetic and Related Cellular Effects Studies. At a concentration of 10 mmol/L, ethyl paraben is mutagenic in *E. coli* (158). *In vitro*, ethyl paraben is a potent spermicidal agent against human spermatozoa (141). After treeatment with 8 mg/mL ethyl paraben, all spermatozoa were immobilized for 30 min.

13.4.2 Human Experience

Ethyl paraben is slightly less toxic than the methyl and propyl homologues. It may occasionally cause ocular sensitivity and hypersensitivity (usually manifested as dermatitis) (24, 157). Ointments containing ethyl paraben can cause redness and swelling of eyelids from allergic contact dermatitis (24). Ingestion of a 0.03% aqueous ethyl paraben solution has caused irritation to the intestinal mucosa and a "feltlike" sensation in the mouth (159).

13.5 Standards, Regulations, or Guidelines of Exposure

No information found.

13.6 Studies on Environmental Impact

No information found.

14.0 Propyl Paraben

14.0.1 CAS Number: [94-13-3]

14.0.2 Synonyms: Benzoic acid, *p*-hydroxy-, propyl ester; benzoic acid, 4-hydroxy-, propyl ester; 4-hydroxybenzoic acid, propyl ester; *p*-hydroxybenzoic acid propyl ester; 4-hydroxybenzoic acid propyl ester; *p*-hydroxybenzoic propyl ester; *p*-hydroxy propyl benzoate; propyl 4-hydroxybenzoate; propyl *p*-hydroxybenzoate; propyl *p*-oxybenzoate; propylester kyseliny *p*-hydroxybenzoove [Czech]; *p*-oxybenzoesaurepropylester [German]; 4-10-00-00374 [Beilstein]; AI3-01341 [NLM]; BRN 1103245 [RTECS]; EINECS 202-307-7; Casewell No. 714; CTFA 02642; EPA Pesticide Chemical Code 061203; FEMA No. 2951; HSDB 203; and NSC 23515

14.0.3 Trade Names: Aseptoform P; Betacide P; Betacine P; Bonomold OP; Chemacide pk, Chemocide pk, Nipagin P, Nipasol, Nipasol M, Nipasol P, Nipazol, Paraben, Parasept, Paseptol, Preserval P, Propagin, Propyl aseptoform, Propyl butex, Propyl chemosept, Propyl chemsept, Propyl parasept, Protaben P, Pulvis conservans, Solbrol P, and Tegosept P

14.0.4 Molecular Weight: 180.20

14.0.5 Molecular Formula: $C_{10}H_{12}O_3$

14.0.6 Molecular Structure:

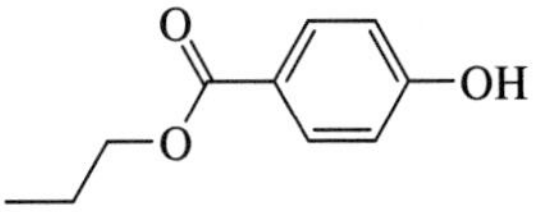

14.1 Chemical and Physical Properties

14.1.1 General

The structural formula is $C_6H_4(OH)(COOC_3H_7)$. Its physical form is white crystals, which melt at 96.2–98°C. It is soluble in alcohol acetone and ether, and slightly soluble in water. Physical properties are summarized in Table 80.1.

14.1.2 Odor and Warning Properties

Propyl paraben is odorless or has a faint odor.

14.2 Production and Use

Propyl paraben is an antifungal agent. It is used as preservative in pharmaceuticals, ophthalmic agents, and foods. It is also used to treat moniliasis (160).

14.3 Exposure Assessment

No information found.

14.4 Toxic Effects

14.4.1 Experimental Studies

14.4.1.1 Acute Toxicity. In experimental animals, the acute toxicity of propyl paraben is similar to that of methyl or ethyl paraben (Table 80.5). In mice, toxic doses of propyl paraben cause ataxia, paralysis, and deep depression resembling anesthesia (156).

14.4.1.2 Chronic and Subchronic Toxicity. Mild growth retardation was noted in rats ingesting 8% propyl paraben in the diet for 96 wk or 16 g/kg/d for 18 mo (150, 156). Weight gains of rats administered 2% propyl paraben in the diet for 96 wk were similar to those of controls (156). Weight gain was noted in rats ingesting 140 mg/kg/d for 18 mo (150). No effect on growth or health was noted in dogs fed 700 mg/kg/d for 90 d (163) or 1 g/kg/d for a year (156).

14.4.1.3 Pharmacokinetics, Metabolism, and Mechanisms. After propyl paraben is intravenously infused into the dog, unhydrolyzed propyl paraben is found only in the brain. In liver, kidney, and muscle, it is immediately hydrolyzed to *p*-hydroxybenzoic acid. Six hours after oral administration of 1.0 g/kg to dogs, the peak plasma concentration of free and total propyl paraben (205 and 370 g/cm^3) is reached (31). After 48 h, all propyl paraben is eliminated. In mice, rats, rabbits, or dogs, propyl paraben is excreted in the urine as unchanged benzoate, *p*-hydroxybenzoic acid, *p*-hydroxyhippuric acid (*p*-hydroxybenzoylglycine), ester glucuronides, ether glucuronides, or ether sulfates (31, 137, 138).

14.4.1.4 Reproductive and Developmental. In female dogs, reproduction was not affected by ingestion of 0.5 g propyl paraben/kg/d for a year (156).

14.4.1.5 Carcinogenesis. In F344 rats initiated with 0.05% *N*-butyl-*N*-(4-hydroxybutyl) nitrosamine, administration of a diet containing 3% propyl paraben for 36 wk did not increase the incidence of preneoplastic lesions, papillary or nodular hyperplasia, or papillomas of the urinary bladder (164). In rats given up to 3% propyl paraben in their diet for 8 wk, food or water consumption and cell proliferation in the forestomach and glandular stomach mucosa were not affected (165). By contrast, Nera and co-workers noted an increase in cell proliferation in epithelial cells of the forestomach after administration of propyl paraben to rats (166).

14.4.1.6 Genetic and Related Cellular Effects Studies. Propyl paraben is not a mutagen in the Ames assay (167). At concentrations of 0.28 and 0.38 mM, propyl paraben is mildly ciliotoxic to epithelial cells in explanted rat tracheal rings (144). *In vitro*, propyl paraben is a potent spermicidal agent against human spermatozoa (141). After treeatment with 3 mg/mL propyl paraben, all spermatozoa were immobilized for 30 min.

14.4.1.7 Other: Neurological, Pulmonary, Skin Sensitization. Propyl paraben blocks nerve conduction in isolated frog peripheral nerve and isolated spinal cord (171).

14.4.2 Human Experience

14.4.2.1 General Information. Saturated aqueous solutions of propyl paraben are moderately irritating to the eye (153). Ingestion of a 0.03% propyl paraben solution causes irritation to the intestinal mucosa and CNS depression (172).

14.4.2.2 Clinical Cases

14.4.2.2.1 Acute Toxicity. No information found.

14.4.2.2.2 Chronic and Subchronic Toxicity. No information found.

14.4.2.2.3 Pharmacokinetics, Metabolism, and Mechanisms. No information found.

14.4.2.2.4 Reproductive and Developmental. No information found.

14.4.2.2.5 Carcinogenesis. No information found.

14.4.2.2.6 Genetic and Related Cellular Effects Studies. No information found.

14.4.2.2.7 Other: Neurological, Pulmonary, Skin Sensitization. A hydrocortisone preparation containing propyl paraben provoked bronchospasm and pruritis when given IV to a 10-year-old asthmatic. Skin tests for immediate hypersensitivity to paraben were positive (168).

14.4.2.3 Epidemiology Studies. No information found.

14.5 Standards, Regulations, or Guidelines of Exposure

Propyl paraben is generally recognized as safe when used in accordance with good manufacturing or feeding practice (146).

14.6 Studies on Environmental Impact

No information found.

15.0 Butyl Paraben

15.0.1 CAS Number: *[94-26-8]*

15.0.2 Synonyms: Aseptoform butyl; benzoic acid, 4-hydroxy-, butyl ester; 4-(butoxycarbonyl) phenol; butyl *p*-hydroxybenzoate; butyl 4-hydroxybenzoate; *n*-butyl *p*-hydroxybenzoate; *n*-butyl parahydroxybenzoate; 4-hydroxybenzoic acid butyl ester; *p*-hydroxybenzoic acid butyl ester; 4-10-00-00375 [Beilstein]; BRN 1103741; and FEMA No. 2203

15.0.3 Trade Names: Butoben, Butyl butex, Butyl chemosept, Butyl parasept, Butyl tegosept, Nipabutyl, Preserval B, Solbrol B, SPF, Tegosept B, and Tegosept butyl

15.0.4 Molecular Weight: 194.22

15.0.5 Molecular Formula: $C_{11}H_{14}O_3$

15.0.6 Molecular Structure:

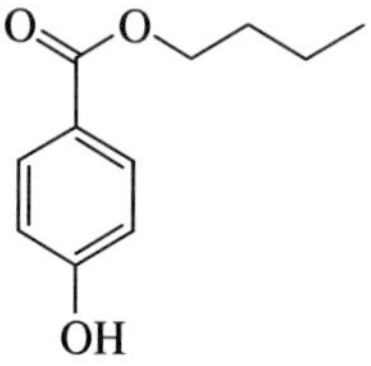

15.1 Chemical and Physical Properties

15.1.1 General

The empirical formula of butyl paraben is $C_{11}H_{14}O_3$. The structural formula is C_6H_4 (OH) ($COOC_4H_9$). At room temperature, its physical form is small, colorless crystals or a powder. It melts at 68–69°C. It is soluble in alcohol, acetone and ether, and chloroform. It is very hygroscopic, but only slightly soluble in water. Physical properties are summarized in Table 80.1.

15.1.2 Odor and Warning Properties

Butyl paraben is odorless.

15.2 Production and Use

Butyl paraben is an antifungal agent. It is used as a preservative in topical antibiotic or corticosteroid preparations and foods.

15.3 Exposure Assessment

No information found.

15.4 Toxic Effects

15.4.1 Experimental Studies

15.4.1.1 Acute Toxicity. The oral and intraperitoneal LD_{50} values for the mouse are 5.0 and 0.23 g/kg, respectively (150).

15.4.1.2 Chronic and Subchronic Toxicity. No information found.

15.4.1.3 Pharmacokinetics, Metabolism, and Mechanisms. After butyl paraben is intravenously infused into the dog, nonhydrolyzed butyl paraben is found in brain, spleen, and pancreas. In liver, kidney, and muscle, it is immediately hydrolyzed to *p*-hydroxybenzoic acid. Six hours after oral administration of 1.0 g/kg to dogs, the peak plasma concentration of free and total butyl paraben (15 and 141 g/cm^3) is reached (31). After 48 h, butyl paraben is eliminated. In mice, rats, rabbits, or dogs, butyl paraben is excreted in the urine as unchanged benzoate, *p*-hydroxybenzoic acid, *p*-hydroxyhippuric acid (*p*-hydroxybenzoylglycine), ester glucuronides, ether glucuronides, or ether sulfates (31, 137, 138).

15.4.1.4 Reproductive and Developmental. No information found.

15.4.1.5 Carcinogenesis. No information found.

15.4.1.6 Genetic and Related Cellular Effects Studies. Butyl paraben is a more potent spermatocide than methyl-, ethyl-, or propyl paraben (141). After treatment with 1 mg/mL propyl paraben, all spermatozoa were immobilized for 30 min.

15.4.1.7 Other: Neurological, Pulmonary, Skin Sensitization. In guinea pigs, mild skin irritation was noted 48 h after dermal application of a preparation containing 5% butyl paraben (169).

15.4.2 Human Experience

Parabens may cause contact dermatitis in humans (29).

16.0 Amyl Paraben

16.0.1 CAS Number: None located

16.0.2 Synonyms: NA

16.0.3 Trade Names: NA

16.0.4 Molecular Weight: NA

16.1 Chemical and Physical Properties

No information found.

16.2 Production and Use

No information found.

16.3 Exposure Assessment

No information found.

16.4 Toxic Effects

The intravenous LD_{50} in the dog ranges from 0.5 to 0.8 mg/kg (12). No other toxicity information was located.

17.0 Methyl Cinnamate

17.0.1 CAS Number: [103-26-4]

17.0.2 Synonyms: Cinnamic acid, methyl ester; 2-propenoic acid, 3-phenyl-, methyl ester; methyl cinnamylate; methyl 3-phenylacrylate; methyl 3-phenylpropenoate; Methyl 3-phenyl-2-propenoate; AI3-00579; EINECS 203-093-8; FEMA No. 2698; and NSC 9411

17.0.3 Trade Names: NA

17.0.4 Molecular Weight: 162.19

17.0.5 Molecular Formula: $C_{10}H_{10}O_2$

17.0.6 Molecular Structure:

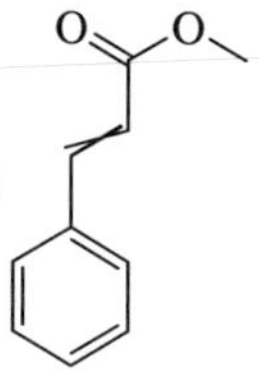

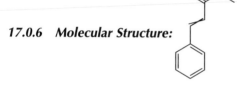

17.1 Chemical and Physical Properties

17.1.1 General

The empirical formula of methyl cinnamate is $C_{10}H_{10}O_2$. The structural formula is $C_6H_5CH{:}CHCOOCH_3$. At room temperature it exists in the crystalline form. Its melting and boiling points are 36.5 and 201.9°C, respectively. It is very soluble in alcohol and ether, and insoluble in water. Chemical and physical properties are summarized in Table 80.6.

17.1.2 Odor and Warning Properties

Methyl cinnamate has a fruity, balsamic odor, similar to strawberries.

17.2 Production and Use

No information found.

17.3 Exposure Assessment

No information found.

17.4 Toxic Effects

The oral and dermal LD_{50} values for the rat and rabbit are 2.61 and >5 g/kg, respectively (42). No other toxicity information was located.

Table 80.6. Physical and Chemical Properties of Cinnamates and Aminobenzoates

Chemical Name	CAS No.	Molecular Formula	Mol. Wt.	Boiling Point (°C) (mmHg)[a]	Melting Point (°C)	Density	Sol. in Water at 20°C (68°F)	Refractive Index (20°C)	Vapor Pressure (mmHg) (°C)	UEL or LEL % by Vol. Room Temp.
Cinnamate										
Methyl	[*103-26-4*]	$C_{10}H_{10}O_2$	162.19	201.9	36.5	1.0911	insol	1.5766	—	—
Ethyl	[*103-36-6*]	$C_{11}H_{12}O_2$	176.22	271.5	12.0	1.0491	insol	1.5598	—	—
n-Propyl	[*7778-83-8*]	$C_{12}H_{14}O_2$	190.24	285	—	1.0435	insol	—	—	—
n-Butyl	[*538-65-8*]	$C_{13}H_{16}O_2$	204.29	145	—	1.012	—	—	—	—
Allyl	[*1866-31-5*]	$C_{12}H_{12}O_2$	188.23	286 (dec.)	—	1.048	insol	1.530	—	—
Linayl	[*78-37-5*]	$C_{19}H_{24}O_2$	284.43	—	—	—	—	—	—	—
Benzyl	[*103-41-3*]	$C_{16}H_{14}O_2$	238.29	350 (dec.)	39	1.109	insol	—	1 (173.8)	—
p-Aminobenzoate										
Methyl	[*619-45-4*]	$C_8H_9NO_2$	151.16	—	114	—	—	—	—	—
Ethyl	[*94-09-7*]	$C_9H_{11}NO_2$	165.19	310	92	—	slight	—	—	—
Propyl	[*94-12-2*]	$C_{10}H_{13}NO_2$	179.22	—	75	—	slight	—	—	—
Butyl	[*94-25-7*]	$C_{11}H_{15}NO_2$	193.25	173.4 (8)	58	—	insol	—	—	—
2-(Diethylamino) ethyl	[*59-46-1*]	$C_{13}H_{20}N_2O_2$	236.32	—	61	—	insol	—	—	—
Procaine	[*51-05-8*]	$C_{13}H_{21}ClN_2O_2$	272.78	—	156	0.707	very	—	—	—
o-Aminobenzoate										
Methyl	[*134-20-3*]	$C_8H_9NO_2$	151.18	256	24–25	1.1682	slight	1.5810	—	—
Methyl *N*-methyl	[*85-91-6*]	$C_9H_{11}NO_2$	165.21	256	18.5–19.5	1.12	insol	—	—	—

[a]dec. = decomposes

18.0 Ethyl Cinnamate

18.0.1 CAS Number: [103-36-6]

18.0.2 Synonyms: Cinnamic acid, ethyl ester; ethylbenzylideneacetate; ethyl beta-phenylacrylate; 2-ethyl cinnamoate; ethyl 3-phenylacrylate; ethyl 3-phenylpropenoate; ethyl 3-phenyl-2-propenoate; ethyl *trans*-cinnamate; 3-phenyl-2-propenoic acid, ethyl ester; propenoic acid, 3-phenyl-, ethyl ester; AI3-00667; BRN 12388804; CTFA 06232; EINECS 203-104-6; FEMA No. 2430; and NSC 6773

18.0.3 Trade Names: NA

18.0.4 Molecular Weight: 176.22

18.0.5 Molecular Formula: $C_{11}H_{12}O_2$

18.0.6 Molecular Structure:

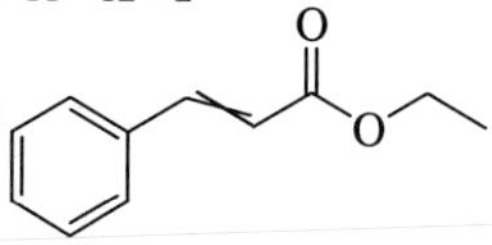

18.1 Chemical and Physical Properties

18.1.1 General

The empirical formula of ethyl cinnamate is $C_{11}H_{12}O_2$. The structural formula is $C_6H_5CH{:}CHCOOC_2H_5$. It is a liquid at room temperature. Its melting and boiling points are 12.0 and 271.5°C, respectively. It is soluble in alcohol and ether, and insoluble in water. Chemical and physical properties are summarized in Table 80.6.

18.1.2 Odor and Warning Properties

Ethyl cinnamate has a fruity, balsamic odor, similar to cinnamon.

18.2 Production and Use

Ethyl cinnamate is found in storax oil, *Kaempferia galanga*, and several other oils. It is produced by the direct esterification of ethanol with cinnamic acid under azeotropic conditions or by Claisen-type condensation of ethyl acetate and benzaldeyde in the presence of sodium metal (170). It is used in fragrances.

18.3 Exposure Assessment

No information found.

18.4 Toxic Effects

18.4.1 Experimental Studies

Acute toxicity information for ethyl cinnamate summarized in Table 80.4. An oral LD_{50} value of 4.0 g/kg was determined for the rat, mouse and guinea pig (43). The dermal LD_{50} value for the rabbit was >5 g/kg (44). When applied full strength to intact or abraded rabbit skin for 24 h under occlusion, it did not cause irritation (44).

18.4.2 Human Experience

When tested at a concentration of 4% in petrolatum, ethyl cinnamate produced no irritation or sensitization reactions in 25 human subjects. When tested at full strength in two separate studies, it caused irritation in 1/22 subjects of one study and none in another (44, 142a, 173).

19.0 *n*-Propyl Cinnamate

19.0.1 CAS Number: *[7778-83-8]*

19.0.2 Synonyms: Cinnamic acid, propyl ester; 2-propenoic acid, 3-phenyl-, propyl ester; propyl 3-phenylpropenoate; propyl 3-phenyl-2-propenoate; propyl beta-phenylacrylate; propylester kyseliny skoricove [Czech]; AI3-02024; EINECS 231-916-0; FEMA No. 2938; and NSC 406146

19.0.3 Trade Names: NA

19.0.4 Molecular Weight: 190.24

19.0.5 Molecular Formula: $C_{12}H_{14}O_2$

19.1 Chemical and Physical Properties

19.1.1 General

The structural formula is $C_6H_5CH{:}CHCOOC_3H_7$. Chemical and physical properties that were located are summarized in Table 80.6.

19.1.2 Odor and Warning Properties

No information found.

19.2 Production and Use

Propyl cinnamate is used as an insect repellant.

19.3 Exposure Assessment

No information found.

19.4 Toxic Effects

All available acute toxicity information for *n*-propyl cinnamate is listed in Table 80.6. The oral LD_{50} values for the mouse and guinea pig are 7 and 3 mL/kg, respectively (2). The 90-d dermal LD_{50} value for the guinea pig is 2 mL/kg (2). The oral LD_{50} value for the isopropyl analog in guinea pigs is 2.7 mL/kg. In guinea pigs, severe skin damage occurs after repeated daily application of 2 mL/kg isopropyl cinnamate (2).

19.5 Standards, Regulations, or Guidelines of Exposure

No information found.

19.6 Studies on Environmental Impact

No information found.

20.0 *n*-Butyl Cinnamate

20.0.1 CAS Number: [538-65-8]

20.0.2 Synonyms: Butyl 3-phenylpropenoate; butyl 3-phenyl-2-propenoate; butyl phenylacrylate; *n*-butyl phenylacrylate; butyl beta-phenylacrylate; cinnamic acid, butyl ester; 2-propenoic acid, 3-phenyl-, butyl ester; cinnamate de *n*-butyl [French]; AI3-02025; BRN 2328067; EINECS 208-699-6; FEMA No. 2192; and NSC 71966

20.0.3 Trade Names: NA

20.0.4 Molecular Weight: 204.29

20.0.5 Molecular Formula: $C_{13}H_{16}O_2$

20.0.6 Molecular Structure:

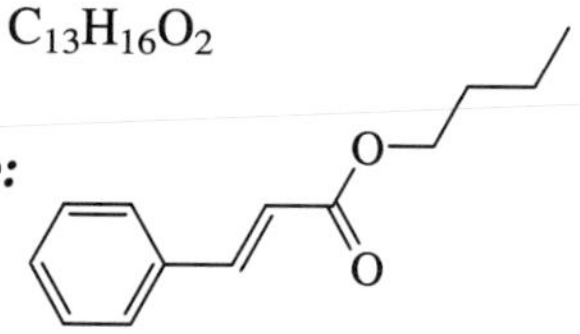

20.1 Chemical and Physical Properties

20.1.1 General

The empirical formula of *n*-butyl cinnamate (butyl cinnamate) is $C_{13}H_{16}O_2$. The structural formula is $C_6H_5CH{:}CHCOOC_3H_7$. It is a colorless, oily liquid. It is incompatible with alkalis, but stable to light, air, and ambient storage temperatures. Chemical and physical properties are summarized in Table 80.6.

20.1.2 Odor and Warning Properties

Butyl cinnamate has an agreeable, ethereal odor.

20.2 Production and Use

Butyl cinnamate is produced by the direct esterification of *n*-butanol with cinnamic acid under azeotropic conditions. It is used in fragrances.

20.3 Exposure Assessment

No information found.

20.4 Toxic Effects

20.4.1 Experimental Studies

All available acute toxicity information for butyl cinnamate is listed in Table 80.4. The oral LD_{50} values for rats and mice and the dermal LD_{50} value for rabbits are all >5 g/kg

(45, 46). When applied full strength to intact or abraded rabbit skin for 24 h under occlusion, it was moderately irritating (174).

20.4.2 Human Experience

At a concentration of 4% in petrolatum, it caused no irritation or sensitization to human skin (45, 142a, 142b).

20.5 Standards, Regulations, or Guidelines of Exposure

Butyl cinnamate was given GRAS status by FEMA in 1965. It is approved by the FDA for food use (68).

20.6 Studies on Environmental Impact

No information found.

21.0 Allyl Cinnamate

21.0.1 CAS Number: *[1866-31-5]*

21.0.2 Synonyms: Allyl beta-phenylacrylate; allyl 3-phenyl-2-propenoate; allyl 3-phenylacrylate; cinnamic acid, allyl ester; 2-propenoic acid, 3-phenyl-, 2-propenyl ester; propenyl cinnamate; 2-propen-1-yl 3-phenyl-2-propenoate; 2-propenyl 3-phenyl-2-propenoate; vinyl carbinyl cinnamate; allylester kyseliny skoricove [Czech]; AI3-02313; EINECS 217-477-8; FEMA No. 2022; and NSC 20972

21.0.3 Trade Names: NA

21.0.4 Molecular Weight: 188.23

21.0.5 Molecular Formula: $C_{12}H_{12}O_2$

21.0.6 Molecular Structure:

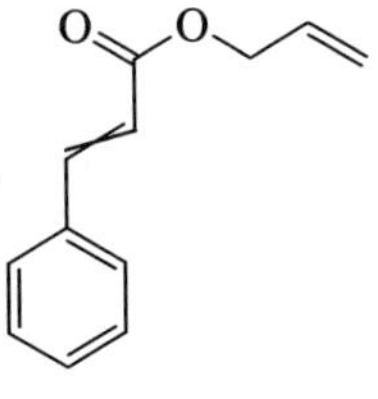

21.1 Chemical and Physical Properties

21.1.1 General

The structural formula is $C_6H_5CH{:}CHCOOCH_2CH{:}CH_2$. It is a colorless, or pale straw-colored liquid, with a density of 1.048. It is very soluble in alcohol and ether, and insoluble in water. Chemical and physical properties are summarized in Table 80.6.

21.1.2 Odor and Warning Properties

No information found.

21.2 Production and Use

Allyl cinnamate is produced by the direct esterification of allyl alcohol with cinnamic acid. It is used in soaps, detergents, creams, lotions, and perfumes.

21.3 Exposure Assessment

No information found.

21.4 Toxic Effects

21.4.1 Experimental Studies

21.4.1.1 Acute Toxicity. Toxicity data for allyl cinnamate are summarized in Table 80.4. The oral LD_{50} value for rats is 1.52 g/kg, and the dermal LD_{50} value for rabbits is >5 g/kg (39). Oral administration of 250–600 mg/kg causes hepatotoxicity in rats, as evidenced by an increase in plasma alanine-α-ketoglutarate transaminase (AKT) activity (174).

21.4.1.2 Chronic and Subchronic Toxicity. No information found.

21.4.1.3 Pharmacokinetics, Metabolism, and Mechanisms. Like other esters, allyl cinnamate is rapidly hydrolyzed *in vivo*. In the rat, allyl compounds are reported to form mercapturic acid, which is excreted in the urine (175).

21.4.1.4 Reproductive and Developmental. No information found.

21.4.1.5 Carcinogenesis. No information found.

21.4.1.6 Genetic and Related Cellular Effects Studies. At concentrations up to 3.6 mg/plate, allyl cinnamate is not mutagenic in *S. typhimurium* strains TA98, TA100, TA1535, TA1537, or TA1538 in the absence or presence of rat liver S-9 (41). Ingestion of 1 mM did not increase the incidence of sex-linked recessive lethal mutations in *Drosophila* (Basc test). Oral treatment of mice with concentrations up to 282 mg/kg did not cause an increase in the number of micronucleated polychromatic erythrocytes from bone marrow (41). *In vitro* tests with *Rhabditis macrocerca* indicate that allyl cinnamate has moderately strong antihelminthic activity (39).

21.4.1.7 Other: Neurological, Pulmonary, Skin Sensitization. When applied full strength to intact or abraded rabbit skin for 24 h under occlusion, allyl cinnamate does not cause irritation (39).

21.4.2 Human Experience

Application of 4.0% allyl cinnamate in petrolatum caused irritation to humans undergoing a 48-h patch test, and as little as 0.1% in petrolatum caused irritation in a multiple irritation test. Dermal application of 4% in petrolatum did not cause sensitization in humans (39).

21.5 Standards, Regulations, or Guidelines of Exposure

Allyl cinnamate was given GRAS status by FEMA in 1965. It is approved by the FDA for food use (176).

21.6 Studies on Environmental Impact

No information found.

22.0 Linayl Cinnamate

22.0.1 CAS Number: [78-37-5]

22.0.2 Synonyms: Cinnamic acid, linayl ester; cinnamic acid, 1,5-dimethyl-1-vinyl-4-hexenyl ester; cinnamic acid, 1,5-dimethyl-1-vinyl-4-hexen-1-yl ester; 3,7-dimethyl-1,6-octadien-3-yl cinnamate; 3,7-dimethyl-1,6-octadien-3-yl beta-phenylacrylate; 3,7-dimethyl-1,6-octadien-3-yl 3-phenylpropenoate; 1,5-dimethyl-1-vinyl-4-hexenyl cinnamate; 1,5-dimethyl-1-vinyl-4-hexen-1-yl cinnamate; 1-ethenyl-1,5-dimethyl-4-hexenyl-3-phenyl-2-propenaote; 4-hexen-1-ol, 1,5 dimethyl-1-vinyl-, cinnamate; linalyl 3-phenylpropenoate; 1,6-octadien-3-ol, 3,7-dimethyl-, cinnamate; 2-propenoic acid, 3-phenyl-1-ethenyl, 1,5-dimethyl-4-hexenyl ester; 2-propenoic acid, 3-phenyl-, 1,5-dimethyl-1-vinyl-4-hexen-1-yl ester; 3-phenyl-2-propenoic acid, 1,5-dimethyl-1-vinyl-4-hexen-1-yl ester; 4-09-00-02010 [Beilstein]; AI3-02451; BRN 3140688; EINECS 201-110-3; FEMA No. 2641; and NSC 46163

22.0.3 Trade Names: NA

22.0.4 Molecular Weight: 284.43

22.0.5 Molecular Formula: $C_{19}H_{24}O_2$

22.0.6 Molecular Structure:

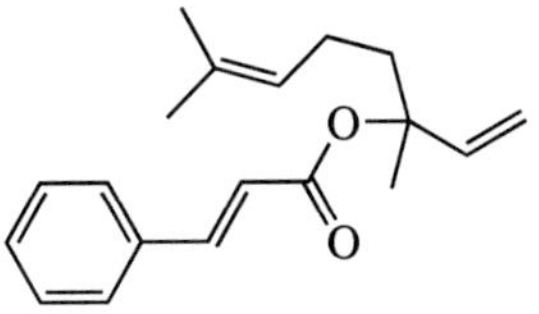

22.1 Chemical and Physical Properties

22.1.1 General

The empirical formula of linayl cinnamate is $C_{19}H_{24}O_2$. The structural formula is $CH_3C(CH_3){:}CH(CH_2)_2(CH_3)C(CH{:}CH_2)COOCH{:}CHC_6H_5$. It is an almost colorless, oily or slightly viscous liquid. No other chemical or physical properties were located.

22.1.2 Odor and Warning Properties

No information found.

22.2 Production and Use

Linayl cinnamate is formed from the reaction of linayl formate with methyl cinnamate and linalool sodium, or from the reaction of dehydrolinalool with cinnamic acid. It is used as a fragrance in soaps, detergents, creams, lotions, and perfumes.

22.3 Exposure Assessment

No information found.

22.4 Toxic Effects

22.4.1 Experimental Studies

22.4.1.1 Acute Toxicity. Acute toxicity information for linayl cinnamate is summarized in Table 80.6. The oral LD_{50} value for rats is 9.96 g/kg, and the dermal LD_{50} value for rabbits is >5 g/kg (3, 47).

22.4.1.2 Chronic and Subchronic Toxicity. Dietary administration of 1000, 2500, or 10,000 ppm linayl cinnamate for 17 wk produced no adverse effects in rats (177).

22.4.1.3 Pharmacokinetics, Metabolism, and Mechanisms. No information found.

22.4.1.4 Reproductive and Developmental. No information found.

22.4.1.5 Carcinogenesis. No information found.

22.4.1.6 Genetic and Related Cellular Effects Studies. No information found.

22.4.1.7 Other: Neurological, Pulmonary Skin Sensitization. When applied full strength to intact or abraded rabbit skin for 24 h under occlusion, linayl cinnamate was mildly irritating (47).

22.4.2 Human Experience

Application of 8.0% linayl cinnamate in petrolatum to human skin does not cause irritation or sensitization (47, 142a, 142b).

22.5 Standards, Regulations, or Guidelines of Exposure

Linayl cinnamate was given GRAS status by FEMA in 1965. It is approved by the FDA for food use (176).

22.6 Studies on Environmental Impact

No information found.

23.0 Benzyl Cinnamate

23.0.1 CAS Number: *[103-41-3]*

23.0.2 Synonyms: Benzyl alcohol, cinnamate; benzyl alcohol, cinnamic ester; benzylcinnamate; benzylcinnamoate; benzyl gamma-phenylacrylate; benzyl 3-phenylpropenoate; cinnamic acid, benzyl ester; phenylmethyl 3-phenyl-2-propenoate; 3-phenyl-2-propenoic acid phenylmethyl ester; 2-propenoic acid, 3-phenyl-, phenylmethyl ester; *trans*-cinnamic acid benzyl ester; benzylester kyseliny skoricove [Czech]; AI3-01268; CTFA 00288; EINECS 203-109-3; FEMA No. 2142; HSDB 359; and NSC 11780

23.0.3 Trade Names: Cinnamein

23.0.4 Molecular Weight: 238.29

23.0.5 Molecular Formula: $C_{16}H_{14}O_2$

23.0.6 Molecular Structure:

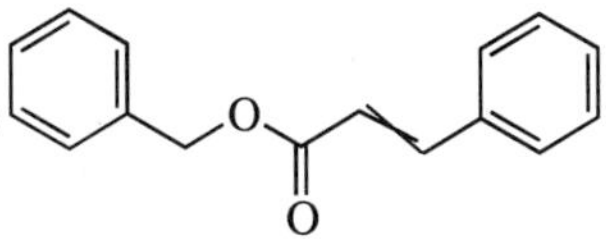

23.1 Chemical and Physical Properties

23.1.1 General

The empirical formula of benzyl cinnamate is $C_{16}H_{14}O_2$. The structural formula is $C_6H_5CH{:}CHCOOCH_2C_6H_5$. Its melting and boiling points are 39 and 350°C, respectively. Like the alkyl cinnamates, benzyl cinnamate is insoluble in water. Chemical and physical properties are summarized in Table 80.6.

23.1.2 Odor and Warning Properties

No information found.

23.2 Production and Use

Benzyl cinnamate is produced by the direct esterification of benzyl alcohol with cinnamic acid. It is used as a flavoring agent, fragrance, and fixative (223).

23.3 Exposure Assessment

No information found.

23.4 Toxic Effects

23.4.1 Experimental Studies

The acute oral LD_{50} value for rats is 5.53 g/kg (3). Dietary administration of 1000 or 10,000 ppm benzyl cinnamate for 19 wk produced no adverse effects in rats (177).

23.4.2 Human Experience

Benzyl cinnamate is a mild dermal irritant (178). Patients allergic to benzyl alcohol may show cross-reactivity to benzyl cinnamate (179).

23.5 Standards, Regulations, or Guidelines of Exposure

No information found.

23.6 Studies on Environmental Impact

No information found.

24.0 Methyl *p*-Aminobenzoate

24.0.1 CAS Number: [619-45-4]

24.0.2 Synonyms: *p*-Aminobenzoic acid methyl ester; 4-aminobenzoic acid methyl ester; benzoic acid, *p*-amino-, methyl ester; benzoic acid, 4-amino-, methyl ester; 4-carbomethoxyaniline; 4-(carbomethoxyl)aniline; *p*-(methoxycarbonyl)aniline; 4-(methoxycarbonyl)aniline; methyl 4-aminobenzoate; methyl aniline-4-carboxylate; methyl PABA; 4-14-00-01128 [Beilstein]; AI3-02437; BRN 0775913; EINECS 210-598-7; and NSC 3783

24.0.3 Trade Names: NA

24.0.4 Molecular Weight: 151.16

24.0.5 Molecular Formula: $C_8H_9NO_2$

24.0.6 Molecular Structure:

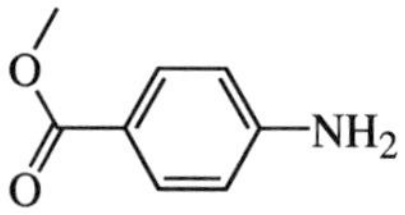

24.1 Chemical and Physical Properties

24.1.1 General

The empirical formula of methyl *p*-aminobenzoate is $C_8H_9NO_2$. The structural formula is $NH_2C_6H_4COOCH_3$, and the boiling point is 114°C. No other chemical and physical data were located.

24.1.2 Odor and Warning Properties

No information found.

24.2 Production and Use

No information found.

24.3 Exposure Assessment

Analytic methods for detection of some PABA esters are available (180).

24.4 Toxic Effects

All available acute toxicity information for methyl *p*-aminobenzoate is listed in Table 80.7. The oral LD_{50} values for rats, mice, and guinea pigs are 2.91, 3.7, and 2.78 g/kg, respectively (3).

24.5 Standards, Regulations, or Guidelines of Exposure

No information found.

24.6 Studies on Environmental Impact

No information found.

25.0 Ethyl *p*-Aminobenzoate

25.0.1 CAS Number: *[94-09-7]*

25.0.2 Synonyms: Amben ethyl ester; 4-aminobenzoic acid ethyl ester; *p*-aminobenzoic acid ethyl ester; benzocaine; benzoic acid, *p*-amino-, ethyl ester; benzoic acid, 4-amino-, ethyl ester; 4-carbethoxyaniline; *p*-carbethoxyaniline; 4-(ethoxycarbonyl)aniline; *p*-(ethoxycarbonyl)aniline; *p*-ethoxycarboxylic aniline; ethyl 4-aminobenzoate; ethyl *p*-aminobenzoate; ethyl *p*-aminophenylcarboxylate; ethyl PABA; aethylium paraminobenzoicum; benzocaina; benzocainum; ethylis aminobenzoas; ethylester kyseliny *p*-aminobenzoove [Czech];4-14-00-01129 [Beilstein]; AI3-02081; BRN 0638434; Casewell No. 430A; CTFA 06234; EINECS 202-303-5; EPA Pesticide Chemical Code 097001; and NSC 41531

25.0.3 Trade Names: Americaine; Anaesthan-syngala; Anaesthesin; Anaesthesinum; Anaesthin; Anestezin; Anestesin; Anesthine; Anesthone; Anbesol; Dermoplast; Ethoforme; Hurricaine; Identhesin; Keloform; Norcain; Norcainum; Ora-jel; Orthesin; Parathesin; Parathesine; Solarcaine; Solu H; and Topcaine

25.0.4 Molecular Weight: 165.19

25.0.5 Molecular Formula: $C_9H_{11}NO_2$

25.0.6 Molecular Structure:

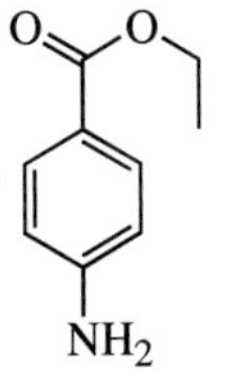

25.1 Chemical and Physical Properties

25.1.1 General

The empirical formula of ethyl *p*-aminobenzoate (PABA) is $C_9H_{11}NO_2$. The structural formula is $NH_2C_6H_4COOC_2H_5$. Its physical form is rhombohedral crystals. Its melting and

Table 80.7. Summary of Toxicity of Aminobenzoates

Chemical Name	CAS No.	Species	Exposure Route	Approximate Dose (g/kg)[a]	Treatment Regimen	Observed Effect	Ref.
p-Aminobenzoate							
Methyl	*[619-45-4]*	Rat	oral	2.91	Single dose	LD_{50}	3
		Mouse	oral	3.7	Single dose	LD_{50}	3
		Guinea pig	oral	2.78	Single dose	LD_{50}	3
		Mouse	IP	0.237	Single dose	LD_{50}	181
Ethyl	*[94-09-7]*	Rat	oral	3.042	Single dose	LD_{50} listed. Ataxia, dyspnea, and cyanosis noted	182
		Mouse	oral	2.5	Single dose	LD_{50} listed. Ataxia, dyspnea, and cyanosis noted	182
		Rabbit	oral	1.15	Single dose	LD_{LO}	183
		Starling	oral	> 0.316	Single dose	LD_{50}	184
		Blackbird	oral	0.056	Single dose	LD_{50}	184
		Mouse	IP	0.216	Single dose	LD_{50}	181
		Guinea pig	dermal	2%	48 h	mild irritation	185
Propyl	*[94-12-2]*	Mouse	IP	0.126	Single dose	LD_{50}	181
Butyl	*[94-25-7]*	Mouse	IP	0.067	Single dose	LD_{50}	181
2-(Diethylamino) ethyl	[59-46-1]	Mouse	oral	0.35	Single dose	LD_{50} listed. Caused tremor, excitement, rigidity	186
		Rat	sc	0.6	Single dose	LD_{50}	187
		Mouse	sc	0.18	Single dose	LD_{50}	188
		Rabbit	sc	0.23	Single dose	LD_{LO}	189
		Guinea pig	sc	0.4	Single dose	LD_{LO}	190
		Cat	sc	0.225	Single dose	LD_{LO}	189
		Dog	sc	0.025	Single dose	LD_{LO}	189
		Frog	sc	0.600	Single dose	LD_{LO}	190
		Rat	IP	0.26	Single dose	LD_{50}	191
		Mouse	IP	0.124	Single dose	LD_{50}	192
		Dog	IP	0.143	Single dose	LD_{LO}	189

		Guinea pig	IP	0.600	Single dose	LD_{LO}	193
		Rat	IV	0.042	Single dose	LD_{50}	191
		Mouse	IV	0.045	Single dose	LD_{50}	194
		Rabbit	IV	0.41	Single dose	LD_{50}	195
		Guinea pig	IV	0.05	Single dose	LD_{LO}	193, 196
		Cat	IV	0.0215	Single dose	LD_{LO}	189
2-(Diethylamino) ethyl	[*59-46-1*]	Rabbit	intraspinal	0.018	Single dose	LD_{LO}	197
		Frog	intraspinal	0.22	Single dose	LD_{LO}	198
o-Aminobenzoate							
Methyl	[*134-20-3*]	Rat	oral	2.91	Single dose	LD_{50} listed. Somnolence and coma noted	199
		Mouse	oral	3.9	Single dose	LD_{50} listed. Somnolence noted	199
		Guinea pig	oral	2.78	Single dose	LD_{50} listed. Somnolence, dyspnea, respiratory stimulation noted	199
		Rabbit	dermal	> 5	Single dose	LD_{50}	200
Methyl *N*-methyl	[*85-91-6*]	Rat	oral	2.25–3.38	Single dose	LD_{50}	201
		Mouse	IV	0.18	Single dose	LD_{LO}	202

[a]Unless listed otherwise; LD_{LO} = lowest lethal dose.

boiling points are 92 and 310°C, respectively. Ethyl *p*-aminobenzoate is soluble in alcohol, ether and lipid media, and slightly soluble in water. Chemical and physical properties are summarized in Table 80.6.

25.1.2 Odor and Warning Properties

No information found.

25.2 Production and Use

Ethyl *p*-aminobenzoate is used as a human topical and veterinary surface anesthetic. It is also used in sunscreens.

25.3 Exposure Assessment

Analytic methods for detection of some PABA esters are available (180).

25.4 Toxic Effects

25.4.1 Experimental Studies

25.4.1.1 Acute Toxicity. All available acute toxicity information for ethyl *p*-aminobenzoate is listed in Table 80.7. The oral LD_{50} values for rats, mice, and wild birds are 3.042, 2.5, and 0.056 g/kg, respectively (182, 183).

25.4.1.2 Chronic and Subchronic Toxicity. No information found.

25.4.1.3 Pharmacokinetics, Metabolism, and Mechanisms. Ethyl *p*-aminobenzoate is rapidly absorbed and metabolized in mammals. The free acid PABA is a normal metabolite.

25.4.1.4 Reproductive and Developmental. No information found.

25.4.1.5 Carcinogenesis. No information found.

25.4.1.6 Genetic and Related Cellular Effects Studies. When incubated with erythrocytes, ethyl *p*-aminobenzoate increases methemoglobin production from 1.40% to 2.66% in 5 h (208).

25.4.1.7 Other: Neurological, Pulmonary, Skin Sensitization. Dermal application of 2% ethyl *p*-aminobenzoate to guinea pigs caused a moderate allergic reaction in 50% of animals tested, which was characterized by slight erythema, slight epidermal thickening, serous parakeratosis, and lymphocyte exocytosis (185).

25.4.2 Human Experience

25.4.2.2 Clinical Cases

25.4.2.2.1 Acute Toxicity. Methemoglobinemia has occurred in patients treated topically with one to three times the recommended dose of benzocaine (49a–52, 203).

Methemoglobinemia and carboxyhemoglobinemia occurred in a 21-d-old infant 2.5 h after administration of half a suppository containing 60 mg benzocaine (53).

25.4.2.2.2 Chronic and Subchronic Toxicity. No information found.

25.4.2.2.3 Pharmacokinetics, Metabolism, and Mechanisms. No information found.

25.4.2.2.4 Reproductive and Developmental. Heinonen et al. have found no association with birth defects and maternal benzocaine use (204).

25.4.2.2.5 Carcinogenesis. No information found.

25.4.2.2.6 Genetic and Related Cellular Effects Studies. No information found.

25.4.2.2.7 Other: Neurological, Pulmonary, Skin Sensitization. Ethyl *p*-aminobenzoate is a mild sensitizer (142a) which may cause contact or photocontact allergy (205a, 205b).

25.4.2.3 Epidemiology Studies. No information found.

25.5 Standards, Regulations, or Guidelines of Exposure

No information found.

25.6 Studies on Environmental Impact

No information found.

26.0 Propyl *p*-Aminobenzoate

26.0.1 CAS Number: *[94-12-2]*

26.0.2 Synonyms: *p*-Aminobenzoic acid propyl ester; benzoic acid, *p*-amino-, propyl ester; benzoic acid, 4-amino-, propyl ester; propyl 4-aminobenzoate; propyl aminobenzoate; propyl *p*-aminobenzoate; 4-(propoxycarbonyl)aniline; propyl PABA; 4-14-00-01130 [Beilstein]; BRN 1211203; EINECS 202-306-1; HSDB 2198; and NSC 23516

26.0.3 Trade Names: Keloform P; Propesin; Propazyl; Propaesin; Propesine; Propylcain; Raythesin; Risocaina; Risocaine; and Risocanium

26.0.4 Molecular Weight: 179.22

26.0.5 Molecular Formula: $C_{10}H_{13}NO_2$

26.0.6 Molecular Structure:

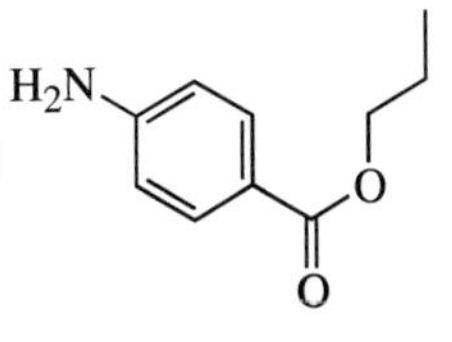

26.1 Chemical and Physical Properties

26.1.1 General

The empirical formula of propyl *p*-aminobenzoate is $C_{10}H_{13}NO_2$. The structural formula is *p*-$NH_2C_6H_4COOC_3H_7$. It exists as crystals or prisms at room temperature. Its melting point is 75°C. Chemical and physical properties are summarized in Table 80.6.

26.1.2 Odor and Warning Properties

No information found.

26.2 Production and Use

Propyl *p*-aminobenzoate is used as local anesthetic and antipruritic.

26.3 Exposure Assessment

Analytic methods for detection of some PABA esters are available (180).

26.4 Toxic Effects

The IP LD_{50} value for the mouse is 126 mg/kg (181). No other toxicity information was located.

26.5 Standards, Regulations, or Guidelines of Exposure

No information found.

26.6 Studies on Environmental Impact

No information found.

27.0 Butyl *p*-Aminobenzoate

27.0.1 CAS Number: *[94-25-7]*

27.0.2 Synonyms: *p*-Aminobenzoic acid butyl ester; 4-aminobenzoic acid butyl ester; benzoic acid, *p*-amino-, butyl ester; benzoic acid, 4-amino-, butyl ester; 4-(butoxycarbonyl) aniline; butyl 4-aminobenzoate; *N*-butyl *p*-aminobenzoate; butyl PABA; butylester kyseliny *p*-aminobenzoove [Czech]; 4-14-00-01130 [Beilstein]; AI3-02284; BRN 1211465; CCRIS 5891; CTFA 00351; EINECS 202-317-1; HSDB 4245; and NSC 128464

27.0.3 Trade Names: Butamben, Butesin, Butesine, Butoform, Butyl keloform, Butylcaine, Planoform, Scuroform, and Scuroforme

27.0.4 Molecular Weight: 193.25

27.0.5 Molecular Formula: $C_{11}H_{15}NO_2$

27.0.6 Molecular Structure:

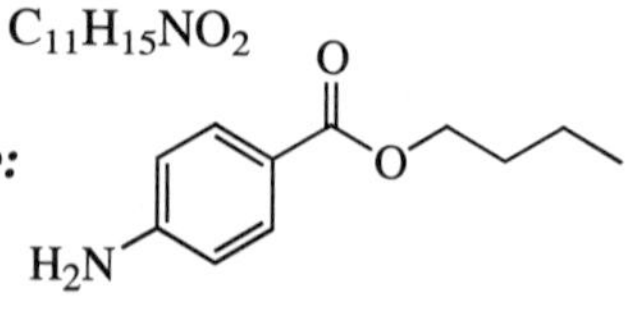

27.1 Chemical and Physical Properties

27.1.1 General

The empirical formula of butyl *p*-aminobenzoate (PABA) is $C_{11}H_{15}NO_2$. The structural formula is *p*-$NH_2C_6H_4COOC_4H_9$. It is a solid at room temperature, with a melting point of 58°C and a boiling point of 173.4°C (at 8 mmHg). Butyl PABA is insoluble in water. Chemical and physical properties are summarized in Table 80.6.

27.1.2 Odor and Warning Properties

No information found.

27.2 Production and Use

Butyl *p*-aminobenzoate is manufactured via esterification of *p*-nitrobenzoic acid with *n*-butyl alcohol, followed by reduction of the nitro group to an amino group. It is used as a local human or veterinary anesthetic. Isobutyl *p*-aminobenzoate, Isocarne, is used as a topical anesthetic (36) and a sunscreen agent (25). No other information about isobutyl *p*-aminobenzoate was located.

27.3 Exposure Assessment

Analytic methods for detection of some PABA esters are available (180).

27.4 Toxic Effects

27.4.1 Experimental Studies

27.4.1.1 Acute Toxicity. The IP LD_{50} value of butyl *p*-aminobenzoate for the mouse is 67 mg/kg (181). A dose of 100 mg/kg fed to starlings or redwing blackbirds caused no adverse effects (184). No other acute toxicity information was located.

27.4.1.2 Chronic and Subchronic Toxicity. No information found.

27.4.1.3 Pharmacokinetics, Metabolism, and Mechanisms. Consult introduction.

27.4.1.4 Reproductive and Developmental. No information found.

27.4.1.5 Carcinogenesis. No information found.

27.4.1.6 Genetic and Related Cellular Effects Studies. At concentrations of 10–1000 μg/plate, butyl *p*-aminobenzoate is not mutagenic to *S. typhimurium* strains TA100, TA1535, TA1537, or TA98, in the presence or absence of rat or hamster liver S-9 (206a).

27.4.1.7 Other: Neurological, Pulmonary, Skin Sensitization. Dogs administered cpidural injections of 10% butyl *p*-aminobenzoate in suspension undergo long-lasting sensory blockade without long-lasting motor effects (206b). Whereas results of a study by

Korsten et al. indicate that epidural injection of 10% butyl *p*-aminobenzoate causes pathomorphologic changes in the spinal cord and spinal nerve roots, results of a study by Shulman et al. indicated that this treatment is not toxic to the spinal cord, meninges, and spinal nerves (206b, 207).

27.4.2 Human Experience

27.4.2.2 Clinical Cases. Upon autopsy, two cancer patients who received multiple injections of a 10% butamben suspension for treatment of cancer pain had no significant pathology in the spinal cord, meninges, and spinal nerves (207).

27.4.2.3 Epidemiology Studies. No information found.

27.5 Standards, Regulations, or Guidelines of Exposure

No information found.

27.6 Studies on Environmental Impact

No information found.

28.0 2-(Diethylamino) Ethyl *p*-Aminobenzoate

28.0.1 CAS Number: *[59-46-1]*

28.0.2 Synonyms: *p*-Aminobenzoic acid 2-diethylaminoethyl ester; 4-aminobenzoic acid diethylaminoethyl ester; *p*-aminobenzoyldiethylaminoethanol; benzoic acid, *p*-amino-, 2-(diethylamino)ethyl ester; Benzoic acid, 4-amino-, 2-(diethylamino)ethyl ester; Diethylaminoethyl *p*-aminobenzoate; 2-diethylaminoethyl *p*-aminobenzoate; 2-diethylaminoethyl 4-aminobenzoate; *b*-diethylaminoethyl 4-aminobenzoate; procaine; 2-diethylaminoethylester kyseliny *p*-aminobenzoove [Czech]; 4-14-00-01138 [Beilstein]; BRN 0913480; and NSC 16947

28.0.3 Trade Names: Allocaine, Duracaine, Gerovital, Jenacain, Jenacaine, Nissocaine, Norocaine, Novocain, Novocaine, Procain, Procaina, Procanium, Scurocaine, and Spinocaine

28.0.4 Molecular Weight: 236.32

28.0.5 Molecular Formula: $C_{13}H_{20}N_2O_2$

28.0.6 Molecular Structure:

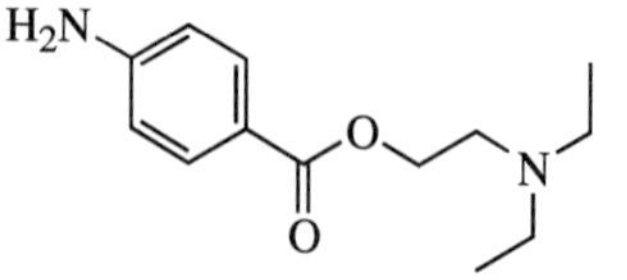

28.1 Chemical and Physical Properties

28.1.1 General

The empirical formula of 2-(diethylamino) *p*-aminobenzoate (PABA) is $C_{13}H_{20}N_2O_2$. The structural formula is *p*-$NH_2C_6H_4COOCH_2CH_2N(C_2H_5)_2$. The hydrochloride salt of

2-(diethylamino) *p*-aminobenzoate ($C_{13}H_{21}ClN_2O_2$ or $N_2C_6H_4COOCH_2CH_2NH(C_2H_5)_2$-HCl) is generally referred to as procaine. The melting points of 2-(diethylamino) PABA and procaine are 61 and 156°C, respectively. Whereas the PABA ester is insoluble in water, the hydrochloride salt is very soluble in water. Chemical and physical properties for both compounds are summarized in Table 80.6.

28.1.2 Odor and Warning Properties

No information found.

28.2 Production and Use

Procaine (Novocaine) is mainly used in dental or medical procedures requiring infiltration anesthesia, peripheral block, or spinal block.

28.3 Exposure Assessment

Analytic methods for detection of some PABA esters are available (180).

28.4 Toxic Effects

28.4.1 Experimental Studies

28.4.1.1 Acute Toxicity. The oral LD_{50} value of 2-(diethylamino) *p*-aminobenzoate for the mouse is 0.35 g/kg (186). The subcutaneous LD_{50} for the rat and mouse are 600 and 180 mg/kg respectively (187, 188). The lowest subcutaneous lethal doses for the frog, guinea pig, rabbit, cat, and dog are 600, 400, 230, 225, and 25 mg/kg, respectively (188–190). Intreaperitoneal, intravenous, and intraspinal toxicity data also are summarized in Table 80.7.

Intravenous injection of 0.5% 2-(diethylamino) *p*-aminobenzoate hydrochloride (Procaine) to exsanguinated hindlimbs of 15 rabbits caused moderate damage to venous epithelium in four animals and severe damage in two (209a).

28.4.1.2 Chronic and Subchronic Toxicity. No information found.

28.4.1.3 Pharmacokinetics, Metabolism, and Mechanisms. No information found.

28.4.1.4 Reproductive and Developmental. Administration of procaine to pregnant rats is associated with an increase in the fetal incidence of cataracts (209b). Exposure of mouse oocytes to 0.1 μg/mL or more procaine adversely affects *in vitro* fertilization and embryo development (209c).

28.4.1.5 Carcinogenesis. No information found.

28.4.1.6 Genetic and Related Cellular Effects Studies. Procaine markedly inhibits the uptake of L-[3*h*] leucine and L-[35*S*]methionine in *E. coli* K1060 cells incubated between 37 and 48°C (209d).

28.4.1.7 Other: Neurological, Pulmonary, Skin Sensitization. Administration of 15 mg/kg procaine (IV) to dogs changed the EEG from an awake pattern to seizure activity (209e). Topical administration of 73.3 mM procaine to urethane-anesthetized cats causes apneustic breathing that is indistinguishable from that produced by 37 mM cocaine (209f). In guinea pigs given 6 IP injections of procaine over a period of 12 d, no systemic or cutaneous anaphylactic reaction occurred after an IV challenge of procaine on day 33 (209g).

28.4.2 Human Experience

28.4.2.1 General Information. The lowest lethal oral dose in humans is 147 mg/kg (210) and the lowest toxic dose given by intramuscular injection is 1.6 mg/kg (211). Cases of methemoglobinemia after procaine use have been reported (52).

28.4.2.2 Clinical Cases. A patient with panhypogammaglobinemia had an anaphylactoid reaction and positive skin test to procaine, despite absent detectable IgE in the serum and a negative RAST test to procaine (212).

28.4.2.3 Epidemiology Studies. No information found.

28.5 Standards, Regulations, or Guidelines of Exposure

No information found.

28.6 Studies on Environmental Impact

No information found.

29.0 Methyl *o*-Aminobenzoate

29.0.1 CAS Number: *[134-20-3]*

29.0.2 Synonyms: 2-Aminobenzoic acid methyl ester; *o*-aminobenzoic acid methyl ester; *o*-aminomethylbenzoate; anthranilic acid, methyl ester; benzoic acid, 2-amino-, methyl ester; carbomethoxyaniline; 2-carbomethoxyaniline; *o*- 2-(methoxycarbonyl)aniline; methyl 2-aminobenzoate; methyl *o*-aminobenzoate; methyl anthranilate; methylester kyseliny anthranilove [Czech]; 4-14-00-01008 [Beilstein]; AI3-01022 [NLM]; BRN 0606965; CCRIS 1349; CTFA06247; EINECS 205-132-4; FEMA No. 2682; HSDB1008; and NSC 3109

29.0.3 Trade Names: Neroli oil, artificial, and Nevoli oil

29.0.4 Molecular Weight: 151.18

29.0.5 Molecular Formula: $C_8H_9NO_2$

29.0.6 Molecular Structure:

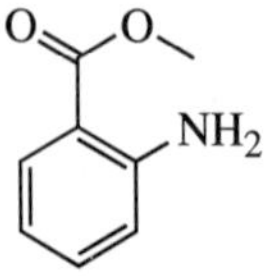

29.1 Chemical and Physical Properties

29.1.1 General

The empirical formula of methyl *o*-aminobenzoate is $C_8H_9NO_2$. The structural formula is *o*-$NH_2C_6H_4COOCH_3$. Methyl *o*-aminobenzoate melts at room temperature (24–25°C) and boils at 256°C. Chemical and physical properties are summarized in Table 80.6.

29.1.2 Odor and Warning Properties

No information found.

29.2 Production and Use

Methyl *o*-aminobenzoate occurs naturally in grapes. It is used in the flavoring, perfume and pharmaceutical industries. It is also used as an oxidation–corrosion stabilizer in lubricants. Several lab and field studies have shown that methyl *o*-aminobenzoate is an effective, non–toxic, and nonlethal bird repellant, with an application for protecting crops, seeds, turf, and fish stocks from bird damage (213).

29.3 Exposure Assessment

Analytic detection methods are available for methyl *o*-aminobenzoate (180).

29.4 Toxic Effects

29.4.1 Experimental Studies

29.4.1.1 Acute Toxicity. The acute oral toxicity of methyl *o*-aminobenzoate in animals is low. The oral LD_{50} values for the rat, mouse, and guinea pig are 2.91, 3.9, and 2.78 mg/kg, respectively (Table 80.6) (199). The dermal LD_{50} for the rabbit is >5 g/kg (17). A 0.3% level in the diet was tolerated by rats without an effect, but a 1% level resulted in slight, histologically identified renal changes and increased liver and kidney weights (12). A 7-h exposure of rats to the saturated vapor at 100°C caused no deaths and only transient weight loss (12). Results of these studies are summarized in Table 80.7.

The acute toxicity of methyl *o*-aminobenzoate also has been tested in a static system against the fry of four species of fish. The LC_{50} values at 24–h for Atlantic salmon, rainbow trout, channel catfish, and bluegill sunfish are 32.3, 23.5, 20.1, and 19.8 mg/L, respectively (213).

29.4.1.2 Chronic and Subchronic Toxicity. No information found.

29.4.1.3 Pharmacokinetics, Metabolism, and Mechanisms. Methyl *o*-aminobenzoate is readily metabolized by hydrolysis into methanol and anthranilic acid, which, in turn, is utilized for anabolic reactions (214, 215).

29.4.1.4 Reproductive and Developmental. In female mice mated with males who were orally treated with 34.8 g/kg methyl *o*-aminobenzoate 8 or 21 d prior to mating, fertility indices (the number of females pregnant/number of sperm-positive females and the number of females pregnant per number of females mated) were reduced (216).

In mice, intraperitoneal injection of 150–300 mg/kg methyl *o*-aminobenzoate on gestational days 10.5 or 11.5 caused a 16% increase in the incidence of cleft lip and/or palate in offspring (217).

29.4.1.5 Carcinogenesis. No information found.

29.4.1.6 Genetic and Related Cellular Effects Studies. At concentrations of 0–5000 μg/plate, methyl *o*-aminobenzoate is nonmutagenic in *S. typhimurium* strains TA98 and TA100, with and without metabolic activation (218). Results of thymidine incorporation studies in UDS rodent hepatocytes incubated with methyl *o*-aminobenzoate are also negative (221).

29.4.1.7 Other: Neurological, Pulmonary, Skin Sensitization. Methyl *o*-aminobenzoate is a slight skin irritant in the rabbit and guinea pig, and may cause eye irritation at high concentrations (200).

29.4.2 Human Experience

No information found.

29.5 Standards, Regulations, or Guidelines of Exposure

Methyl *o*-aminobenzoate is listed as GRAS by the FDA.

29.6 Studies on Environmental Impact

No information found.

30.0 Methyl *N*-Methyl *o*-Aminobenzoate

30.0.1 CAS Number: [85-91-6]

30.0.2 Synonyms: Anthranilic acid, *N*-methyl-, methyl ester; benzoic acid, 2-(methylamino)-, methyl ester; methyl 2-methylaminobenzoate; methyl *N*-methylanthranilate; methyl methanthranilate; MMA; and *N*-methylanthranilic acid, methyl ester

30.0.3 Trade Names: NA

30.0.4 Molecular Weight: 165.21

30.0.5 Molecular Formula: $C_9H_{11}NO_2$

30.0.6 Molecular Structure:

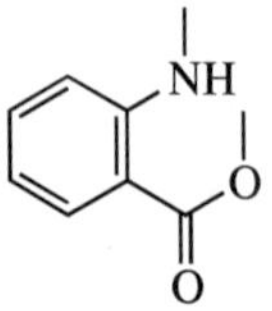

30.1 Chemical and Physical Properties

30.1.1 General

The chemical formula of methyl *N*-methyl *o*-aminobenzoate (methyl *N*-methanthranilate) is $C_9H_{11}NO_2$. The structural formula is $CH_3NHC_6H_4COOCH_3$. At room temperature, it is a pale yellow liquid with a slight bluish fluorescence. Its melting and boiling points are 18.5–19.5 and 256°C, respectively. It is soluble in most organic solvents and is insoluble in water. It has a density of 1.12 (at 15°C). Chemical and physical properties are summarized in Table 80.6.

30.1.2 Odor and Warning Properties

Methyl *N*-methanthranilate has an orange or "mandarin-peel-like" odor, and a musty, grape-like flavor (85).

30.2 Production and Use

Methyl *N*-methanthranilate is a component of mandarin and mandarin-leaves essential oil. It is also found in oil from bulbs of *Kaempferia ethelae L*, hyacinth flowers, and orange petitgrain (85). It is synthesized by methylation of methyl anthranilate or esterification of *N*-methylanthranilic acid. It is used as a flavorant in a wide range of foods (particularly flour and sugar confections).

30.3 Exposure Assessment

In 1965, it was estimated that the daily intake of methyl *N*-methanthranilate in man was 0.2 mg/kg/d (222).

30.4 Toxic Effects

30.4.1 Experimental Studies

30.4.1.1 Acute Toxicity. The oral LD_{50} value for rats is between 2.25 and 3.38 g/kg (201, 202).

30.4.1.2 Chronic and Subchronic Toxicity. In groups of 15 male and female rats were fed diets containing 0, 300, 1200, or 3600 ppm methyl *N*-methanthranilate for 90 d, the NOEL was 300 ppm (20 mg/kg/d) (201). At 6 wk, hematological changes were noted (decreased hemoglobin and red blood cell count in both sexes and decreased white blood cell count in males) These changes were not observed at autopsy. In rats ingesting 1200 or 3600 ppm, there was an increase in absolute (males only) and relative kidney weight (both sexes); however, renal function and histopathology were not altered.

30.4.1.3 Pharmacokinetics, Metabolism, and Mechanisms. Methyl *N*-methanthranilate is rapidly de-esterified in rats and man and excreted as the acid, with little breakdown of the *N*-methyl bond (224).

30.4.1.4 Reproductive and Developmental. In mice, IP injection of 25–75 mg/kg methyl *N*-methanthranilate on gestational days 10.5–12.5 or 11.5–13.5 caused a 16% increase in the incidence of cleft lip and/or palate in offspring (217).

30.4.1.5 Carcinogenesis. No information found.

30.4.1.6 Genetic and Related Cellular Effects Studies. At an unlisted concentration, methyl *N*-methanthranilate increased the rate of protein carboxymethylation in cultured palate cells (217).

30.4.1.7 Other: Neurological, Pulmonary, Skin Sensitization. No information found.

30.4.2 Human Experience

No information found.

30.5 Standards, Regulations, or Guidelines of Exposure

No information found.

30.6 Studies on Environmental Impact

No information found.

B INTRODUCTION TO ESTERS OF MONOCARBOXYLIC ACIDS AND DI-, TRI-, OR POLYALCOHOLS

Glycerol Acetates (Acetins), Propionates and Butyrates

Long-chain fatty acids of glycerides may be replaced by one or more acetyl groups to produce mono-, di-, or triacetin. Glyceryl tripropionate, tributyrate, and triisobutyrate are formed via esterification of glycerol with the corresponding acid. Acetins are resistant to changes in consistency, heat, damage, or oxidative rancidity (257). Acetins, propionates, and butyrates serve as food additives, solvents or plasticizers, and surface-active agents (12). The simple fatty acids (mono-, di-, or triacetin) are readily hydrolyzed by lipases and esterases in serum or other tissues, but most other glycerides are saponified only very slowly, or not at all (214). The rate of hydrolysis decreases with increasing order of acetylation (225). Available evidence indicates that these agents exhibit a low order of toxicity (Table 80.8). Normally, no irritant effects occur upon inhalation or direct dermal contact.

In the mammalian stomach and small intestine glycerides are readily hydrolyzed by glycerolspecific lipases. Lipases are also present at other sites of entry, such as the lung and skin, where glycerides are readily saponified and further metabolized. The simple fatty acids (mono-, di-, or triacetin) are readily hydrolyzed by lipases and esterases in serum or

Table 80.8. Summary of Toxicity of Esters of Acetins

Chemical Name	CAS No.	Species	Exposure Route	Approximate Dose (g/kg)	Treatment Regimen	Observed Effect	Ref.
Monoacetin	[*626446-35-5*]	Rat	sc	5.5 mL/kg	Single dose	LD_{50}	230
		Mouse	sc	3.5 mL/kg	Single dose	LD_{50}	230
		Rabbit	IV	2.0 mL/kg	Single dose	TD_{LO}	12, 231
		Rabbit	IV	4.0 mL/kg	Single dose	LD_{50}	12
		Dog	IV	5.0 mL/kg	Single dose	LD_{50}	12, 231
		Rat	sc	0.1 g/kg/d	70 d	weight gain	232
		Dog	IM	0.1 g/kg/d	69 d	fatty deposits	225
Diacetin	[*25395-31-7*]	Rat	sc	4 mL/kg	Single dose	LD_{50}	230
		Mouse	oral	8.5	Single dose	LD_{50}	233
		Mouse	sc	2.5 mL/kg	Single dose	LD_{50}	230
		Mouse	IP	2.3	Single dose	LD_{50}	233
		Mouse	IV	2.5	Single dose	LD_{50}	234
		Dog	IV	3	Single dose	LD	235
		Rabbit	IV	1.5 mL/kg	Single dose	LD	231
Triacetin	[*102-76-1*]	Rat	oral	3	Single dose	LD_{50}	236
		Mouse	oral	1.1, 8.0 mL/kg	Single dose	LD_{50} listed. Spastic paralysis w/wo sensory change	237, 238
		Frog	oral	0.15	Single dose	LD_{LO}	239
		Rat	sc	2.8 mL/kg	Single dose	LD_{50} listed. Dyspnea and somnolence noted	230
		Mouse	sc	2.3 mL/kg	Single dose	LD_{50} listed. Dyspnea and somnolence noted	230
		Rat	IP	2.1	Single dose	LD_{50}	239
		Mouse	IP	1.4, also 3.65 mL/kg	Single dose		237, 238
		Mouse	IV	1.6	Single dose	LD_{50} listed. Convulsions, respiratory depression.	240
		Dog	IV	1.5	Single dose	LD_{50}	239
		Rabbit	IV	0.75	Single dose	LD_{50}	239
		Guinea pig	IM	1.74	Single dose	LD_{LO} listed.	241

Table 80.8. (*Continued*)

Chemical Name	CAS No.	Species	Exposure Route	Approximate Dose (g/kg)	Treatment Regimen	Observed Effect	Ref.
Tripropionin	[*139-45-7*]	Rat	oral	6.4	Single dose	LD_{50}	242
		Mouse	oral	6.4	Single dose	LD_{50}	242
		Mouse	IV	0.84	Single dose	LD_{50} listed. Respiratory stimulation and somnolence noted.	243
Tributyrin	[*60-01-5*]	Rat	oral	3.2	Single dose	LD_{50}	242
		Mouse	oral	12.8	Single dose	LD_{50}	242
		Mouse	IV	0.32	Single dose	LD_{50} listed. Convulsions, respiratory depression noted.	243
Triisovalerin	—	Mouse	IV	0.082	Single dose	LD_{50}	243
Tricapronin	[*621-70-5*]	Mouse	IV	0.122	Single dose	LD_{50}	243
Triheptylin	[*620-67-7*]	Mouse	IV	0.32	Single dose	LD_{50}	243
Tricaprylin	[*538-23-8*]	Rat	oral	33.3	Single dose	LD_{50}	244
		Mouse	oral	29.6	Single dose	LD_{50}	244
		Rat	sc	> 27.8	Single dose	LD	244
		Mouse	sc	> 27.8	Single dose	LD	244
		Rat	IP	> 27.8	Single dose	LD	244
		Mouse	IP	> 27.8	Single dose	LD	244
		Rat	IV	4	Single dose	LD_{LO}	244
		Mouse	IV	3.7	Single dose	LD_{50}	243

[a]Unless listed otherwise; LD_{LO} = lowest lethal dose; TD_{LO} = lowest toxic dose.

other tissues, but most other glycerides are saponified only very slowly, or not at all (225). The rate of hydrolysis decreases with increasing order of acetylation (226).

Higher Glycerides

The higher glycerides of fatty acids with odd-numbered carbon chains (C_5 to C_{11}) are found naturally in very small quantities in diverse organisms, and the even-numbered (C_{12} to C_{24}) esters are common nutritional constituents. They are used as emulsifiers for foods, industrial raw materials, or nonacid detergent components. The glyceryl oleates are used as emulsifiers in soaps and cosmetics, and the glyceryl stearates as emulsifiers in cosmetics, food products, and medicinals (25). Some glycerol esters (C_8 to C_{12}) are active antimicrobial agents (226). The glycerides are saponified and metabolized in the same manner as other food fats (214). Some toxicity data are available for the C_5 and C_8 compounds. The even-numbered C_{12} to C_{18} glycerides are nontoxic (227). Prolonged feeding of diets containing as much as 25% glyceryl laurate produced no toxic symptoms.

Resorcinol Esters

Of the aromatic dihydroxy compounds, the resorcinol esters resorcinol monoacetate and monobenzoate occur most commonly. Resorcinol monobenzoate is used as a ultraviolet stabilizer in plastic films. Little toxicological information is known about these compounds. Available physical and physiological data are summarized in Tables 80.9 and 80.10.

Gallates

Gallates are chemically trihydroxybenzoic acid esters. They serve generally as antioxidants, and the propyl, octyl, and dodecyl gallates have been approved as food additives.

The gallates exhibit low acute and chronic toxicity in experimental animals. The bulk of evidence suggests that they are not carcinogenic or teratogenic. All gallates are moderate to strong contact sensitizers in guinea pigs. Sensitizing potential increases with increasing side-chain length (228). Gallates have been shown to cause dermatitis in bakers and other workers handling these materials (227).

31.0 Ethylidene Diacetate

31.0.1 CAS Number: *[111-55-7]*

31.0.2 Synonyms: 1,1-Diacetoxyethane and 1,1-ethandiol diacetate

31.0.3 Trade Names: NA

31.0.4 Molecular Weight: 146.14

31.0.5 Molecular Formula: $C_6H_{10}O_4$

31.0.6 Molecular Structure:

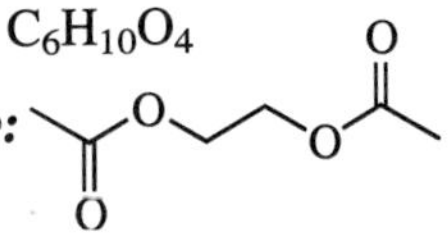

Table 80.9. Physical and Chemical Properties of Esters of Monocarboxylic Acids and Di- or Trialcohols

Chemical Name	CAS No.	Molecular Formula	Mol. Wt.	Boiling Point (°C) (mmHg)	Melting Point (°C)	Density (Sp. Gravity)	Sol. in Water at 20°C (68°F)	Refractive Index (20°C)	Vapor Pressure (mmHg) (°C)	UEL or LEL % by Vol. Room Temp.
Ethylidene diacetate	*[111-55-7]*	$C_6H_{10}O_4$	146.14	190	– 31	1.1063	very	1.4159	1 (38.3)	—
Monoacetin	*[26446-35-5]*	$C_5H_{10}O_4$	134.13	158 (17)	– 78	(1.2060)	soluble	—	3 (130)	—
Diacetin	*[25395-31-7]*	$C_7H_{12}O_5$	176.17	280	– 30	1.1779	very	1.4301	40 (185)	—
Triacetin	*[102-76-1]*	$C_9H_{14}O_6$	218.20	258–260	– 78	(1.1562)	58 g/L[a]	1.4307	0.00248 (25)	—
Tripropionin	*[139-45-7]*	$C_{12}H_{20}O_6$	260.29	175–176	—	1.100	insol	1.4318	—	—
Tributyrin	*[60-01-5]*	$C_{15}H_{26}O_6$	302.36	305–310	– 75	(1.0350)	0.133 g/L[b]	1.4359	—	—
Triisovalerin	—	$C_{18}H_{32}O_6$	344.45	330–335	—	0.9984	insol	1.4354	—	—
Tricapronin	*[621-70-5]*	$C_{21}H_{38}O_6$	386.54	> 200	– 60	0.9867	insol	1.4427	—	
Tricaprylin	*[538-23-8]*	$C_{27}H_{50}O_6$	470.70	233.1	9.8–10.1	0.9540	insol	1.4482	—	—
Resorcinol										
Monoacetate	*[102-29-4]*	$C_8H_8O_3$	152.14	283	—	—	—	—	—	—
Diacetate	*[108-58-7]*	$C_{10}H_{10}O_4$	194.18	178	—	—	insol	—	—	—
Monobenzoate	*[136-36-7]*	$C_{13}H_{10}O_3$	214.21	—	—	—	—	—	—	—
Dibenzoate	*[94-01-9]*	$C_{20}H_{14}O_4$	318.31	—	117	—	—	—	—	—
Propyl gallate	*[121-79-9]*	$C_{10}H_{12}O_5$	212.21	—	130	—	3.5 g/L[a]	—	—	—
Octyl gallate	*[1034-01-1]*	$C_{15}H_{22}O_5$	282.17	—	—	—	—	—	—	—
Dodecyl gallate	*[1166-52-5]*	$C_{19}H_{30}O_5$	338.42	—	—	—	—	—	—	—

[a] At 25°C.
[b] At 37°C.

Table 80.10. Summary of Toxicity of Esters of Resorcinols and Gallates

Chemical Name	CAS No.	Species	Exposure Route	Approximate Dose (g/kg)	Treatment Regimen	Observed Effect	Ref.
Resorcinol							
Monobenzoate	*[136-36-7]*	Rat	oral	0.8–1.6	Single dose	LD_{50}	12, 242
		Mouse	oral	0.8	Single dose	LD_{50}	245
		Rat	IP	0.4–0.8	Single dose	LD_{50}	12
		Mouse	IP	0.71	Single dose	LD_{50} listed. Somnolence, tremor, ataxia noted	246
		Guinea pig	dermal	> 20 mg/kg	Single dose	LD_{50}	12
Dibenzoate	*[94-01-9]*	Mouse	IP	8	Single dose	LD_{50}	246
Propyl gallate	*[121-79-9]*	Rat	oral	2.1–4.0	Single dose	LD_{50}	12, 247–249
		Mouse	oral	1.7–3.5	Single dose	LD_{50}	12, 249, 250
		Rabbit	oral	2.75	Single dose	LD_{50}	251
		Hamster	oral	2.48	Single dose	LD_{50}	251
		Cat	oral	0.4–0.8	Single dose	LD_{50}	12
		Rat	IP	0.38	Single dose	LD_{50} listed. Convulsions or effect on seizure threshold and dyspnea noted	248
Octyl gallate	*[1034-01-1]*	Rat	oral	1.96–2.33, 2.71, 4.7	Single dose	LD_{50}	252–254
		Rat	IP	0.06–0.08	Single dose	LD_{50}	252–254
Dodecyl gallate	*[1166-52-5]*	Rat	oral	6.5	Single dose	LD_{50}	252–254
		Mouse	oral	1.6–3.2	Single dose	LD_{50}	253
		Rat	IP	0.1–0.12	Single dose	LD_{50}	252–255

[a]unless listed otherwise.

31.1 Chemical and Physical Properties

31.1.1 General

The structural formula of ethylidene diacetate is $CH_3CH(OCOCH_3)_2$. It is a liquid at room temperature, with a density of 1.1063, a melting point of $-31°C$ and a boiling point of 190°C. It is very soluble in water, ethanol, and ether. Its flash point is 96.1°C (205°F). Chemical and physical properties are summarized in Table 80.9.

31.1.2 Odor and Warning Properties

Sharp, fruity odor.

31.2 Production and Use

No information found.

31.3 Exposure Assessment

No information found.

31.4 Toxic Effects

Similar to that of diethylmalonate (12).

31.4.1 Experimental Studies

31.4.1.1 Acute Toxicity. The oral LD_{50} values for rats are 3200 mg/kg and 2540 mg/kg for males and females, respectively (242). The dermal LD_{50} value for guinea pigs is >20 mL/kg. Ethylidene diacetate is slightly irritating to the skin and eye.

31.4.1.2 Chronic and Subchronic Toxicity. Groups of male rats were treated with 0, 100, or 1000 mg/kg ethylidene diacetate in distilled water for 2 weeks. There were no effects on body weight, food consumption, hematology, liver weight, kidney weight, or histopathology (242).

32.0 1,3-Propanediol Diacetate

32.0.1 CAS Number: *[628-66-01]*

32.0.2 Synonyms: 1,3-*Bis*(acetyloxy)propane, 1,3-diacetoxypropane, 1,3-propylene-diacetate, 1,3-propylene glycol diacetate, trimethylene acetate, and AI3-07820 [NLM]

32.0.3 Trade Names: NA

32.0.4 Molecular Weight: Not found

32.0.5 Molecular Formula: $C_7H_{12}O_4$

32.0.6 Molecular Structure:

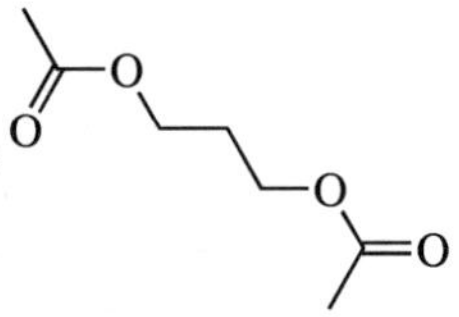

32.1 Chemical and Physical Properties

32.1.1 General

The empirical formula of 1,3-propanediol diacetate is $C_7H_{12}O_4$. Its structural formula is $CH(CH_2OCOCH_3)_2$. No other chemical or physical property data were located.

32.1.2 Odor and Warning Properties

No information found.

32.2 Production and Use

A structurally similar compound, ethynodiol diacetate, has been used as a therapeutic progestin (229).

32.3 Exposure Assessment

No information found.

32.4 Toxic Effects

No information found.

32.5 Standards, Regulations, or Guidelines of Exposure

No information found.

32.6 Studies on Environmental Impact

No information found.

33.0 Glyceryl Monoacetate

33.0.1 CAS Number: *[26446-35-5]*

33.0.2 Synonyms: Acetin, acetin-mono; 1,2,3-propanetriol, monoacetate; acetoglyceride; acetyl monoglyceride; glycerin monoacetate; glycerol acetate; glycerol monoacetate; glyceryl acetate; monoacetin; AI3-24158 [NLM]; CCRIS 5881; EINECS 247-704-6; and HSDB 4285

33.0.3 Trade Names: NA

33.0.4 Molecular Weight: 134.13

33.0.5 Molecular Formula: $C_5H_{10}O_4$

33.0.6 Molecular Structure:

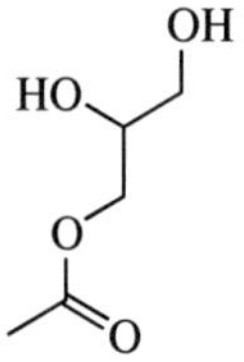

33.1 Chemical and Physical Properties

33.1.1 General

The empirical formula of glyceryl monoacetate (monoacetin) is $C_5H_{10}O_4$. The structural formula is $HOCH_2CH(OH)CH_2OCOCH_3$. It is a clear, highly hygroscopic liquid, with a melting point of −78°C and a boiling point of 158°C. The commercial product is yellow. It is soluble in water and alcohol, slightly soluble in ether, and insoluble in benzene. Physicochemical properties are summarized in Table 80.9.

33.1.2 Odor and Warning Properties

Monoacetin has a "characteristic" odor.

33.2 Production and Use

Monoacetin is manufactured by heating glycerol with acetic acid. It is used in the manufacture of smokeless powder and dynamite, as a solvent for basic dyes, as a food additive, and in the tanning of leather.

33.3 Exposure Assessment

No information found.

33.4 Toxic Effects

33.4.1 Experimental Studies

33.4.1.1 Acute Toxicity. Monoacetin is the least toxic derivative of the acetins. The subcutaneous LD_{50} values for the mouse and rat are 3.5 and 5.5 mL/kg, respectively (230). The intravenous LD_{50} values for the rabbit and dog are 4 and 5 mL/kg, respectively (12, 231). The intravenous lowest toxic dose was 2.0 mL/kg for the rabbit (231). These data are summarized in Table 80.8.

33.4.1.2 Chronic and Subchronic Toxicity. Subchronic subcutaneous injections of 0.1 g/kg/d for 70 d to the rat caused weight gain (225, 232). Intramuscular administration to the dog of 0.1 g/kg/d for 69 d resulted in no cumulative toxicity and no histopathological changes except some fatty deposits (225).

33.4.1.3 Pharmacokinetics, Metabolism, and Mechanisms. Monoacetin is largely, but not completely, hydrolyzed in the bowel to glycerin and acetic acid. In rat intestinal tissue, monacetin is rapid hydrolyzed (225). The intracellularly released active acid appears readily in the serosal and more slowly in the mucosal fluid of the epithelial cell (225).

33.4.1.4 Reproductive and Developmental. No information found.

33.4.1.5 Carcinogenesis. No information found.

33.4.1.6 Genetic and Related Cellular Effects Studies. Sex-linked recessive lethal mutations in *Drosophila* were induced by treatment with 5000 ppm monoacetin (293). In *S. typhimurium* strain 1535, monoacetin is mutagenic at a concentration of 333.3–1000 μg/plate in the absence or presence of rat or hamster liver S-9 (206a). It is nonmutagenic in *S. typhimurium* strains TA98, TA100, or TA1537, in the presence or absence of rat or hamster liver S-9 (206a). It induces sister chromatid exchange in hamster ovary cells at 1.5 g/L (293).

33.4.1.7 Other: Neurological, Pulmonary, Skin Sensitization. No information found.

33.4.2 Human Experience

No information found.

33.5 Standards, Regulations, or Guidelines of Exposure

No information found.

33.6 Studies on Environmental Impact

When released into the air, monoacetin will exist solely as a vapor in the ambient atmosphere. Vapor-phase monoacetin will be degraded in the atmosphere by reacting with photochemically produced hydroxyl radicals. The half-life of this reaction is estimated to be 24 h. If released into the soil, monoacetin is expected to have very high mobility. Volatilization from moist soil is not expected to occur. If released into water, monoacetin is not expected to absorb to suspended solids, sediment in the water column, or volatize. In aquatic systems, approximately 91–84% is expected to degrade over a 4-wk period. The potential for bioconcentration in aquatic organisms is low (124).

34.0 Glyceryl Diacetate

34.0.1 CAS Number: *[25395-31-7]*

34.0.2 Synonyms: Acetin-di; 1,2,3-propanetriol, diacetate; glycerin (e) diacetate; glycerol diacetate; glycerol 1,3-di(acetate); diacetin; diacetylgylcerol; AI3-00676 [NLM]; CTFA 00752; and EINECS 246-941-2

34.0.3 Trade Names: NA

34.0.4 Molecular Weight: 176.17

34.0.5 *Molecular Formula:* $C_{14}H_{24}O_{10}$

34.0.6 *Molecular Structure:*

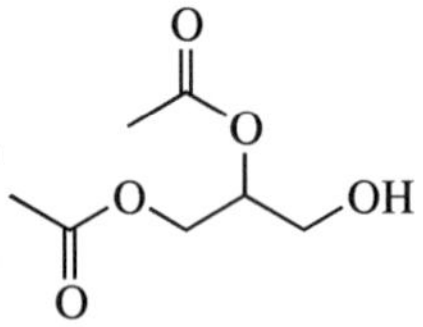

34.1 Chemical and Physical Properties

34.1.1 *General*

The empirical formula of glyceryl diacetate (diacetin) is $C_7H_{12}O_5$. It is a mixture of two isomers, 1,2,3,-propanetriol 1,2-diacetate and 2,3-diacetate [chemically $CH_2(OH)$-$CH(OCOCH_3)CH_2OCOCH_3$ and $(CH_3OCOCH_2)_2CH(OH)$]. It is a colorless, hygroscopic liquid, with a melting point of $-30°C$ and boiling point of 280°C. It is very soluble in water and alcohol, and slightly soluble in ether. Physico-chemical properties are summarized in Table 80.9.

34.1.2 *Odor and Warning Properties*

No information found.

34.2 Production and Use

Diacetin is used as a solvent, plasticizer, and softening agent.

34.3 Exposure Assessment

No information found.

34.4 Toxic Effects

34.4.1 *Experimental Studies*

Acute toxicity information for diacetin is summarized in Table 80.8. Studies by Li et al. (230) have shown a subcutaneous LD_{50} value of 4 mL/kg for the rat. In the mouse, the oral, subcutaneous, intraperitoneal, and intravenous LD_{50} values are 8.5 g/kg, 2.3 mL/kg, 2.3 g/kg, and 2.5 g/kg, respectively (230, 233, 234).

Intravenous lethal doses in the dog and rabbit are 3 g/kg and 1.5 mL/kg, respectively (231, 235). The tolerated dose in the rabbit was 1 mL/kg intravenously (231). Animals near death exhibited severe dyspnea, muscular tremors, and occasional convulsions, as well as various degrees of hemorrhaging in the lungs. However, no pathological effects were observed in the spleen, hepatic, cardiac, or renal tissues (231).

34.4.2 *Human Experience*

If diacetin comes into contact with eyes, it may cause burning pain and redness of the conjunctiva (219).

34.5 Standards, Regulations, or Guidelines of Exposure

No information found.

34.6 Studies on Environmental Impact

No information found.

35.0 Glyceryl Triacetate

35.0.1 CAS Number: [102-76-1]

35.0.2 Synonyms: Acetin-tri; 1,2,3-propanetriol, triacetate; acetic, 1,2,3-propanetriyl ester; glycerin triacetate; glycerol triacetate; triacetin; triacetylglycerol; triacetyl glycerine; 4-02-00-00253 [Beilstein]; triacetina [Spanish]; triacetine [French], triacetinum [Latin]; AI3-00661 [NLM]; BRN 1792353 [RTECS]; CTFA 03234; EINECS 203-051-9; FEMA No. 2007; and HSDB 585

35.0.3 Trade Names: Enzactin; Estol 1581; Fungacetin; Glyped; Kessocoflex TRA; Kadoflex triacetin; Only-Clear Nail; Priacetin 1581; Ujostabil; and Vanay

35.0.4 Molecular Weight: 218.20

35.0.5 Molecular Formula: $C_9H_{14}O_6$

35.0.6 Molecular Structure:

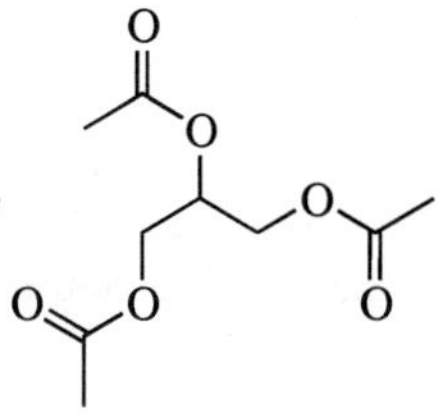

35.1 Chemical and Physical Properties

35.1.1 General

The empirical formula of glyceryl triacetate (triacetin) is $C_9H_{14}O_6$. The structural formula is $CH(OCOCH_3)(CH_2OCOCH_3)_2$. It is a colorless, somewhat oily liquid with a bitter taste. Triacetin melts at −78°C and boils at 258–260°C. It is highly soluble in acetone, alcohol, benzene, and chloroform and slightly soluble in water, carbon disulfide, and carbon tetrachloride. Its flash point and autoignition temperatures are 138°C (280°F) and 433°C (812°F), respectively. Physico-chemical properties are summarized in Table 80.9.

35.1.2 Odor and Warning Properties

Triacetin has a slight fruity or fatty odor.

35.2 Production and Use

Triacetin is formed by the direct reaction of glycerol with acetic acid in the presence of Twitchell reagent; in a benzene solution of glycerol and boiling acetic acid in the presence of cationic resin pretreated with dilute sulfuric acid; or by the reaction of oxygen with a liquid phase mist of allyl acetate and acetic acid in the presence of a bromine catalyst (64, 85). Triacetin is used as a topical antifungal agent, a fixative in perfumes, a solvent for dyes (particularly indulines and tannins), a plasticizer for cellulose acetate or cellulose nitrate, a component of binders for solid rocket fuels, and for mounting microscope slides for analysis of asbestos. It is a camphor substitute in the pyroxalin industries, and can serve to remove carbon dioxide from natural gas. Because parenteral administration of triacetin improves nitrogen balance in rats, its use in parenteral nutrition has been proposed (220, 256).

35.3 Exposure Assessment

No information found.

35.4 Toxic Effects

35.1.1 Experimental Studies

34.4.1.1 Acute Toxicity. The oral LD_{50} value for the rat is 3.0 g/kg (236, 237). For the mouse, oral LD_{50} values of 1.1 g/kg and 8.0 mL/kg have been reported (237, 238). Subcutaneous LD_{50} values are 2.8 mL/kg for the rat and 2.3 mL/kg for the mouse (230). Acute toxicity data are summarized in Table 80.8.

The symptoms of toxicity with triacetin are primarily weakness and ataxia. Animals near death exhibit severe dyspnea, muscular tremors, and occasional convulsions, usually 2–22 min after the injection. Also various degrees of hemorrhage in the lung have been observed. It does not appear to affect the liver, spleen, heart, or kidneys (231).

35.4.1.2 Chronic and Subchronic Toxicity. In the rat, inhalation of 250 ppm triacetin (6 h/d, 5 d/week for 13 wk) or saturated triacetin vapor (6 h/d for 5 d) produced no clinical signs of toxicity or histopathological effects (12).

35.4.1.3 Pharmacokinetics, Metabolism, and Mechanisms. No information found.

35.4.1.4 Reproductive and Developmental. No information found.

35.4.1.5 Carcinogenesis. No information found.

35.4.1.6 Genetic and Related Cellular Effects Studies. No information found.

35.4.1.7 Other: Neurological, Pulmonary, Skin Sensitization. When placed (0.1 mL) undiluted into rabbit eyes, triacetin was slightly to moderately irritating. It was not irritating when immediately irrigated for 6 min (258). Nonaqueous drug vehicles containing triacetin caused local tissue irritation (259). However, it was neither irritating nor sensitizing when absorbed through guinea pig skin (227).

35.4.2 Human Experience

Triacetin appears to be of low toxicity when swallowed, inhaled or placed on skin. In sensitive individuals, it may cause slight irritation (260).

35.5 Standards, Regulations, or Guidelines of Exposure

Triacetin has achieved GRAS status as a food additive (261).

35.6 Studies on Environmental Impact

If released into the atmosphere, triacetin will exist in the vapor phase. It is degraded by reacting with photochemically produced hydroxyl radicals. The half-life of this reaction is estimated to be 2 d (124, 262). If released into alkaline water or moist soil, triacetin can degrade through aqueous hydrolysis. Its half-life at pH 7, 8, or 9 is estimated to be 130, 13, and 1.3 d, respectively. Aquatic volatilization, adsorption sediment, and bioconcentration are not expected to be important fate processes (124, 263).

36.0 Glyceryl Tripropionate

36.0.1 CAS Number: *[139-45-7]*

36.0.2 Synonyms:

Glycerine tripropionate; glycerol tripropionate; glyceryl tripropanoate; 1,2,3-propanetriol, tripropanoate; propionic acid triglyceride; propionin, tri-; tripropionin; tripropionine; tripropionylglycerol; 4-02-00-00717 [Beilstein]; AI3-07939 [NLM]; BRN 1713645; EINECS 205-365-1; FEMA No.3286; and NSC 36744 tripropanoin; tripropanoylglycerol; 1,2,3-propanetriyl tripropionate

36.0.3 Trade Names:

Eastman Tripropionin

36.0.4 Molecular Weight:

260.29

36.0.5 Molecular Formula:

$C_{12}H_{20}O_6$

36.0.6 Molecular Structure:

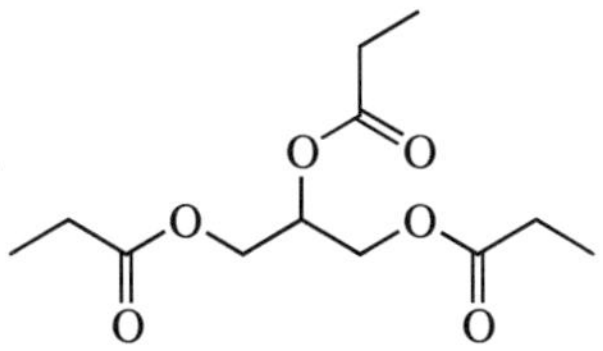

36.1 Chemical and Physical Properties

36.1.1 General

The empirical formula of glyceryl tripropionate (tripropionin) is $C_{12}H_{20}O_6$. The structural formula is $CH_3CH_2COOCH(CH_2OCOCH_2CH_3)_2$. Its boiling point is 175–176°C. Tripropionin is insoluble in water and soluble in ethyl alcohol, ether, and chloroform. Physicochemical properties are summarized in Table 80.9.

36.1.2 Odor and Warning Properties

No information found.

36.2 Production and Use

Tripropionin is used as a plasticizer.

36.3 Exposure Assessment

No information found.

36.4 Toxic Effects

36.4.1 Experimental Studies

36.4.1.1 Acute Toxicity. The oral LD_{50} value for both the rat and mouse is 6.4 g/kg (242). An intravenous LD_{50} value of 840 mg/kg was obtained for the mouse (243).

36.4.1.2 Chronic and Subchronic Toxicity. Subchronic exposure of rats to 750 ppm tripropionin (6 h/d for 90 d) caused no physiological symptoms or abnormal histopathological findings (227).

36.4.1.3 Pharmacokinetics, Metabolism, and Mechanisms. No information found.

36.4.1.4 Reproductive and Developmental. No information found.

36.4.1.5 Carcinogenesis. No information found.

36.4.1.6 Genetic and Related Cellular Effects Studies. No information found.

36.4.1.7 Other: Neurological, Pulmonary, Skin Sensitization. Tripropionin is not a sensitizer in guinea pigs (227).

36.4.2 Human Experience

No information found.

36.5 Standards, Regulations, or Guidelines of Exposure

No information found.

36.6 Studies on Environmental Impact

No information found.

37.0 Glyceryl Tributyrate

37.0.1 CAS Number: *[60-01-5]*

37.0.2 Synonyms: Butanoic acid, 1,2,3-propanetriyl ester; butyric acid triester with glycerin; butyrin; butyrin-tri; butyryl triglyceride; glycerin tributyrate; glycerol tributanoate; glycerol tributyrate; glyceroltributyrin; 1,2,3-propanetriyl butanoate; 1,2,3-propanetriyl tributanoate; tributin; tributyrin; tri-*n*-butyrin; tributyrinine; tributyrl glyceride; tributyroin; tributyrl glyceride; 4-02-00-00799 [Beilstein]; AI3-01776 [NLM]; BRN 1714746 [RTECS]; EINECS 200-451-5; FEMA No. 2223; and HSDB 878

37.0.3 Trade Names: Eastman tributyrin

37.0.4 Molecular Weight: 302.36

37.0.5 Molecular Formula: $C_{15}H_{26}O_6$

37.0.6 Molecular Structure:

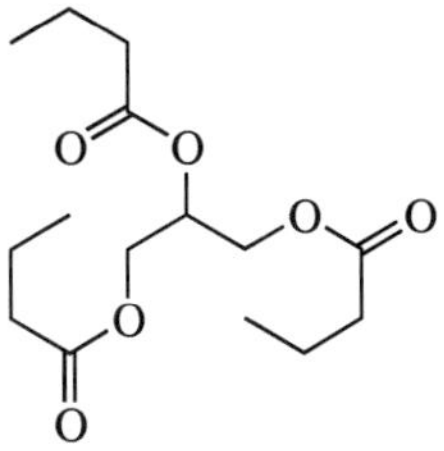

37.1 Chemical and Physical Properties

37.1.1 General

The empirical formula of glyceryl tributyrate (tributyrin) is $C_{15}H_{26}O_6$. The structural formula is $CH_3(CH_2)_2COOCH[CH_2OCO(CH_2)_2CH_3]_2$. Tributyrin is a colorless, oily liquid with a bitter taste. It melts at −75°C and boils at 305–310°C. It is soluble in alcohol and ether and fairly insoluble in water. Its flash point is 356°F. Physico-chemical properties are summarized in Table 80.9.

37.1.2 Odor and Warning Properties

When heated to decomposition, tributyrin emits acrid smoke and fumes.

37.2 Production and Use

Tributyrin is manufactured via esterification of glycerol with butyric acid. It is used as a flavoring agent in foods, a plasticizer; an adjuvant, and a solvent. Irradiation of tributyrin produces various compounds, among them butanetriol triesters, erythritol tetraesters, and polyglycol polyesters (272).

37.3 Exposure Assessment

No information found.

37.4 Toxic Effects

37.4.1 Experimental Studies

37.4.1.1 Acute Toxicity. The oral LD_{50} values for the rat and mouse are 3.2 g/kg and 12.8 mg/kg, respectively (242). Inhalation of 78 ppm for 6 h caused temporary hyperpnea in the rat, but no fatalities or other symptoms were observed (12). An intravenous LD_{50} value for the mouse was 320 mg/kg (243a).

37.4.1.2 Chronic and Subchronic Toxicity. In rats, administration of a diet containing 5% tributyrin for 20 d caused hypertrophy of the forestomach wall, which is characterized by a three- to fourfold increase in the thickness of the mucosa, and an increase in protein, phospholipid, and cholesterol content (243b).

37.4.1.3 Pharmacokinetics, Metabolism, and Mechanisms. Tributyrin is absorbed by the intestinal mucosa of the rat (264) and is not absorbed cutaneously in guinea pigs (227). Crum and co-workers (265) have isolated and characterized tributyrinase, an enzyme specific for the hydrolysis of tributyrins. This enzyme is inhibited by some organic compounds, particularly selected fluorophosphates. Additionally, a fluoride-sensitive tributyrinase has been isolated from rat adipose tissue, which liberates butyric acid from 1-mono-, 1,2-di-, or tributyrin (265).

In mice and rats given 10.3 g/kg tributyrin P.O., plasma butyrate concentrations peaked at 1.75 and 3.07 mM, respectively. They were greater than or equal to 1 mM from 10–60 min after dosing in mice and 30–90 min in rats (266).

37.4.1.4 Reproductive and Developmental. No information found.

37.4.1.5 Carcinogenesis. Administration of sodium butyrate in drinking water potentiates 1,2-dimethylhydrazine-induced colon cancer in rats (267). In mice, dietary administration of 5% tributyrin for 48 wk did not lead to an increase in colonic tumor incidence or focal areas of dysplasia as compared to controls (267). Therefore, it has been suggested that the agent responsible for enhanced tumorigenesis in the rat study was sodium, rather than butyrate.

37.4.1.6 Genetic and Related Cellular Effects Studies. Tributyrin induces differentiation and/or growth inhibition of a number of cell lines *in vitro* (268, 269) and is a potent inhibitor of smooth muscle cell DNA synthesis and cell proliferation (270).

37.4.1.7 Other: Neurological, Pulmonary, Skin Sensitization. No information found.

37.4.2 Human Experience

37.4.2.2 Clinical Cases. Limited human trials have been performed to determine if butyrate has utility in treating cancer. A 10-d infusion of sodium butyrate induced partial remission of acute myelogenous leukemia in a child (271). Similar trials in adults with acute myelogenous leukemia have failed to demonstrate any beneficial effects; however, no severe toxicity was associated with the treatment (273). It has been suggested that butyrate was ineffective in the latter study because the concentration of butyrate achieved in the plasma was less than 10% of that shown to induce differentiation *in vitro* (268).

37.4.2.3 *Epidemiology Studies*

37.4.2.3.1 Acute Toxicity. Once daily PO administration of 50–400 mg/kg tributyrin for 3 wk to 13 cancer patients caused nausea, vomiting, myalgia, diarrhea, headache, abdominal cramping, anemia, constipation, azotemia, lightheadedness, fatigue, rash, alopecia, odor (400 mg/kg only), dysphoria, or clumsiness in some patients (268).

37.4.2.3.2 Chronic and Subchronic Toxicity. No information found.

37.4.2.3.3 Pharmacokinetics, Metabolism, and Mechanisms. In humans administered once daily PO injections of 50–400 mg/kg tributyrin for 3 wk, peak plasma concentrations occurred 0.25–3 h after dosing. The peak plasma concentration of those administered 200 mg/kg ranged from 0.1 to 0.45 mM. Higher plasma concentrations were not observed in those given higher doses (268).

37.5 Standards, Regulations, or Guidelines of Exposure

Triacetin has achieved GRAS status as a food additive (261).

37.6 Studies on Environmental Impact

No information found.

38.0 Glyceryl Triisobutyrate

38.0.1 CAS Number: [14295-64-8]

38.0.2 Synonyms: Glycerol triisobutryate; isobutyric acid, 1,2,3-propanetriyl ester; propanoic acid, 2-methyl-,1,2,3-propanetriyl ester; triisobutyrin; and EINECS 238-224-8

38.0.3 Trade Names: NA

38.0.4 Molecular Weight: 302.27

38.1 Chemical and Physical Properties

38.1.1 General

The empirical formula of glyceryl triisobutyrate (triisobutyrin) is $C_{15}H_{26}O_6$. The structural formula is $(CH_3)_2CHCOOCH[CH_2OCOCH\ (CH_3)_2]_2$. No other chemical or physical data were located.

38.1.2 Odor and Warning Properties

No information found.

38.2 Production and Use

No information found.

38.3 Exposure Assessment

No information found.

38.4 Toxic Effects

38.4.1 Experimental Studies

The oral LD_{50} value of triisobutyrate for the rat is 1.6–3.2 g/kg, with the same symptoms as for tributyrin (12). In rats, inhalation of 145 ppm for 6 h caused a transient increase in depth of respiration (12).

38.4.2 Human Experience

No information found.

38.5 Standards, Regulations, or Guidelines of Exposure

No information found.

38.6 Studies on Environmental Impact

No information found.

39.0 Glyceryl Triisopentanoate

39.0.1 CAS Number: Not found

39.0.2 Synonyms: Triisovalerin

39.0.3 Trade Names: NA

39.0.4 Molecular Weight: 344.45

39.1 Chemical and Physical Properties

39.1.1 General

The empirical formula of glyceryl triisopentanoate (triisovalerin) is $C_{18}H_{32}O_6$. The structural formula is $(CH_3)_2CH(CH_2)COOCH[CH_2OCOCH_2CH\ (CH_3)_2]_2$. It is soluble in alcohol and ether, and insoluble in water. Physical and chemical data that were located are summarized in Table 80.9.

39.1.2 Odor and Warning Properties

No information found.

39.2 Production and Use

No information found.

39.3 Exposure Assessment

No information found.

39.4 Toxic Effects

39.4.1 Experimental Studies

In an intravenously administered trial, a LD_{50} value of 82 mg/kg was determined for the mouse (243a).

39.4.2 Human Experience

No information found.

39.5 Standards, Regulations, or Guidelines of Exposure

No information found.

39.6 Studies on Environmental Impact

No information found.

40.0 Glyceryl Trihexanoate

40.0.1 CAS Number: [621-70-5]

40.0.2 Synonyms: Caproic triglyceride; glycerin tricaproate; glycerin trihexanoate; glycerol tricaproate; glycerol trihexanoate; glyceryl tricaproate; hexanoic acid, 1,2,3-propanetriyl ester; hexanoin, tri-; palmitic acid triglyceride; 1,2,3-propanetriyl hexanoate; tricaproin; tricapronin; tricaproylglycerol; trihexanoin; trihexanoyl glycerol; 4-02-00-00926 [Beilstein]; AI3-07948 [NLM]; BRN 1806732; EINECS 210-701-5; and NSC 406885

40.0.3 Trade Names: NA

40.0.4 Molecular Weight: 386.54

40.0.5 Molecular Formula: $C_{21}H_{38}O_6$

40.0.6 Molecular Structure:

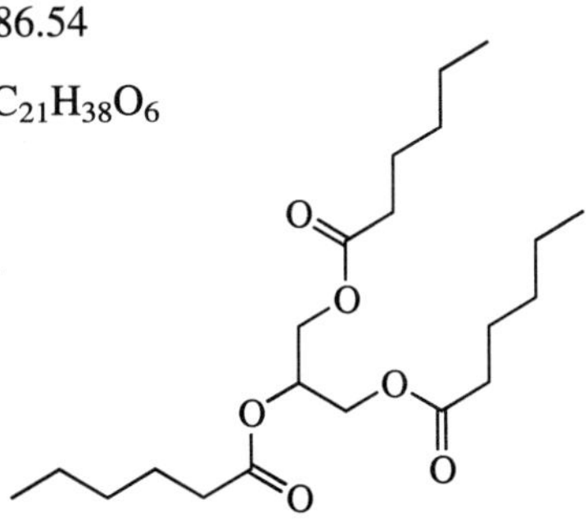

40.1 Chemical and Physical Properties

40.1.1 General

The empirical formula of glyceryl trihexanoate (tricapronin) is $C_{21}H_{38}O_6$. Its structural formula is $CH_3(CH_2)_4COOCH[CH_2OCO\ (CH_2)_4CH_3]_2$. It is very soluble in alcohol and ether and insoluble in water. Its melting and boiling points are $-60°C$ and $>200°C$, respectively. Physical and chemical information is summarized in Table 80.9.

40.1.2 Odor and Warning Properties

No information found.

40.2 Production and Use

No information found.

40.3 Exposure Assessment

No information found.

40.4 Toxic Effects

40.4.1 Experimental Studies

An intravenous LD_{50} value of 122 mg/kg has been determined for the mouse (243a).

40.4.2 Human Experience

No information found.

40.5 Standards, Regulations, or Guidelines of Exposure

No information found.

40.6 Studies on Environmental Impact

No information found.

41.0 Glyceryl Triheptanoate

41.0.1 CAS Number: *[620-67-7]*

41.0.2 Synonyms: Gylcerol triheptanoate; heptanoic acid, 1,2,3-propanetriyl ester; heptanoin, tri-; propane-1,2,3 -triyl trisheptanoate; trienanthoin; triheptanoic glyceride;

triheptanoin; triheptylin; trioenanthoin; 3-02-00-00769 [Beilstein]; BRN 1807724; and EINECS 210-647-2

41.0.3 Trade Names: NA

41.0.4 Molecular Weight: Not found.

41.1 Chemical and Physical Properties

41.1.1 General

The empirical formula of glyceryl triheptanoate (triheptylin) is $C_{24}H_{44}O_6$. Its structural formula is $CH_3(CH_2)_5COOCH[CH_2OCO(CH_2)_5CH_3]_2$. No other chemico-physical information was located.

41.1.2 Odor and Warning Properties

No information found.

41.2 Production and Use

No information found.

41.3 Exposure Assessment

No information found.

41.4 Toxic Effects

41.4.1 Experimental Studies

An intravenous LD_{50} value of 320 mg/kg was determined for the mouse (243a). Changes ienduced included muscle contraction or spasticity and an increase in urine volume.

41.4.2 Human Experience

No information found.

41.5 Standards, Regulations, or Guidelines of Exposure

No information found.

41.6 Studies on Environmental Impact

No information found.

42.0 Glyceryl Trioctanoate

42.0.1 CAS Number: [538-23-8]

42.0.2 Synonyms: Capryilic acid triglyceride; caprylin; caprylic acid, 1,2,3-propanetriyl ester; glycerol tricaprylate; glyceryl tricaprylate; glycerol trioctanoate; octanoic acid, 1,2,3-propanetriyl ester; octanoic acid triglyceride; octanoin, tri-; propane-1,2,3-triyl trioctanoate; 1,2,3-propanetriol trioctanoate; 1,2,3-propanetriol octanoate; tricaprilin; tricaprylin; trioctanoin oil; trioctanoylglycerol; 4-02-00-009991[Beilstein]; BRN 1717202; CTFA 05020; EINECS 208-686-5; and NSC 4059 [NLM]

42.0.3 Trade Names: Captex 8000; Crodamol GTCC; Maceight; NI 01; Panacet 800; RATO; and Sefsol 800

42.0.4 Molecular Weight: 470.70

42.0.5 Molecular Formula: $C_{27}H_{50}O_6$

42.0.6 Molecular Structure:

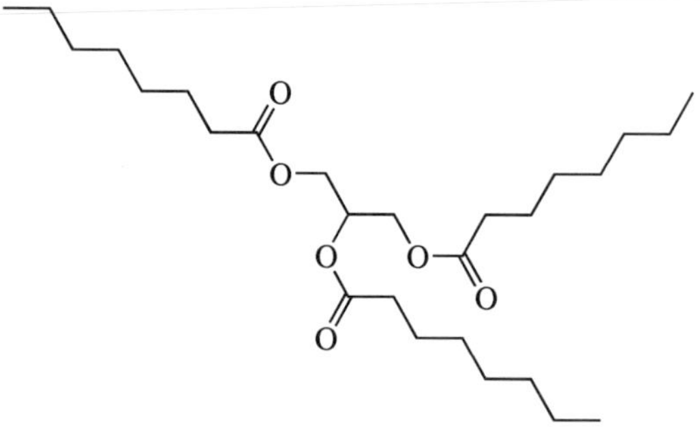

42.1 Chemical and Physical Properties

42.1.1 General

The empirical formula of glyceryl trioctanoate (tricaprylin) is $C_{27}H_{50}O_6$. Its structural formula is $CH_3(CH_2)_6COOCH[CH_2OCO(CH_2)_6CH_3]_2$. It melts around 10°C and boils at 233.1°C. It is very soluble in ether and alcohol and insoluble in water. Chemical and physical data are summarized in Table 80.9.

42.1.2 Odor and Warning Properties

No information found.

42.2 Production and Use

Tricaprylin is used as a suspending agent or vehicle.

42.3 Exposure Assessment

No information found.

42.4 Toxic Effects

42.4.1 *Experimental Studies*

42.4.1.1 Acute Toxicity. The oral LD_{50} values of tricaprylin for the rat and mouse are 33.3 and 29.6 g/kg, respectively (244). Subacute or intraperitoneal injection of greater than 27.8 g/kg was required for lethality in mice (244). These and other data are summarized in Table 80.8.

42.4.1.2 Chronic and Subchronic Toxicity. The effect of daily administration of 2.5, 5.0, and 10 mL/kg tricaprylin by gavage for 2 yr has been studied in male F344 rats. In the high-dose group, survival was less than untreated controls due to moribund kills and deaths that appeared to be related to toxicity (274). Animals tereated with tricaprylin exhibited dose-related increases in pancreatic exocrine hyperplasia and adenomas, and proliferative lesions of the forestomach. In animals treated with the highest dose, the incidence of nephropathy and mononuclear cell leukemia was less than that of controls.

42.4.1.3 Pharmacokinetics, Metabolism, and Mechanisms. No information found.

42.4.1.4 Reproductive and Developmental. A study performed by Ohta and colleagues indicates that at levels up to 10 mL/kg, tricapyrilin is not teratogenic to mice and rabbits (244). Intrperitoneal injection of over 10 gestation days in rabbits or administration by gavage to rats from before mating through lactation also do not cause any teratogenic effects (275, 276). However, results of other studies indicate that oral administration of 10 mL/kg tricaprylin to gestating rats causes a slight increase in the number of resorptions and soft tissue malformations and a decrease in fetal weight (277, 278). Richoloroacetonitrile induces more fetal soft-tissue abnormalities when administered in a tricaprylin vehicle than when administered in a corn oil vehicle (278, 279).

42.4.1.5 Carcinogenesis. In F344 rats given 10 mL/kg/d tricaprylin by gavage for 2 yr, there is significant increase in the incidence of squamous cell papillomas of the forestomach, compared to controls (274).

42.4.1.6 Genetic and Related Cellular Effects Studies. Tricaprylin is mutagenic in *S. typhimurium* strain TA1535 in the presence but not the absence of S-9 (250). It does not induce mutations in strains TA97, TA98, or TA100 with or without S-9.

42.4.1.7 Other: Neurological, Pulmonary, Skin Sensitization. No information found.

42.4.2 *Human Experience*

No information found.

42.5 Standards, Regulations, or Guidelines of Exposure

No information found.

42.6 Studies on Environmental Impact

No information found.

43.0 Glyceryl Trinonanoate

43.0.1 CAS Number: [126-53-4]

43.0.2 Synonyms: Glycerol trinonanoate; glycerol tripelargonate; glyceryl pelargonate; glyceryl tripelargonate; trinonanoate; trinonanoin; trinonylin; tripelargoin; tripelargonin; 4-02-00-01021 [Beilstein]; AI3-22998 [NLM]; BRN 1809503 [RTECS]; EINECS 204-791-5; and NSC 5647 [NLM]

43.0.3 Trade Names: NA

43.0.4 Molecular Weight: Not found

43.1 Chemical and Physical Properties

43.1.1 General

The chemical formula of glyceryl trinonanoate (trinonylin) is $C_{30}H_{56}O_6$. The structural formula is $CH_3(CH_2)_7COOCH[CH_2OCO(CH_2)_7CH_3]_2$. No other chemical or physical data were located.

43.1.2 Odor and Warning Properties

No information found.

43.2 Production and Use

No information found.

43.3 Exposure Assessment

No information found.

43.4 Toxic Effects

43.4.1 Experimental Studies

The intravenous LD_{50} value for the mouse is >10 g/kg (243a).

43.4.2 Human Experience

No information found.

43.5 Standards, Regulations, or Guidelines of Exposure

No information found.

43.6 Studies on Environmental Impact

No information found.

44.0 Glyceryl Tridecanoate

44.0.1 CAS Number: [621-71-6]

44.0.2 Synonyms: Capric acid triglyceride; caprin; decanoic acid, 1,2,3-propanetriyl ester; decanoin, tri-; decanoic acid 1,2,3-propanetriyl ester; glycerin tricaprate; glycerin tridecanoate; glycerol tricaprate; glycerol tricaprin; glycerol tridecanoate; glyceryl tricaprate; 1,2,3-propanol tridecanoate; 1,2,3-propanetriyl decanoate; tricapric glyceride; tricaprin; tricaprinin; tridecanoin; tri (decanoyl)glycerol; 4-02-00-01047 [Beilstein]; AI3-36968 [NLM]; BRN 1717683; CTFA 03833; EINECS 210-702-0; and NSC 147475 [NLM]

44.0.3 Trade Names: Actor M 2; Dynasan 110; Panacet 1000; Triglyceride DDD

44.0.4 Molecular Weight: 554.85

44.0.5 Molecular Formula: $C_{33}H_{62}O_6$

44.0.6 Molecular Structure:

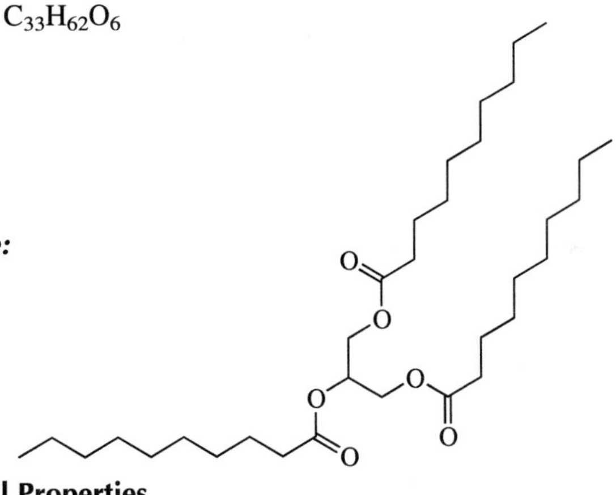

44.1 Chemical and Physical Properties

44.1.1 General

The empirical formula of glyceryl trinonanoate (tricaprinin) is $C_{33}H_{62}O_6$. The structural formula is $CH_3(CH_2)_8COOCH[CH_2OCO(CH_2)_8CH_3]_2$. No other chemical or physical data were located.

44.1.2 Odor and Warning Properties

No information found.

44.2 Production and Use

No information found.

44.3 Exposure Assessment

No information found.

44.4 Toxic Effects

44.4.1 Experimental Studies

The intravenous LD_{50} value for the mouse is > 10 g/kg (243a).

44.4.2 Human Experience

No information found.

44.5 Standards, Regulations, or Guidelines of Exposure

No information found.

44.6 Studies on Environmental Impact

No information found.

45.0 Glyceryl Triundecanoate

45.0.1 CAS Number: *[13552-80-2]*

45.0.2 Synonyms: Glycerol triundecanoate; propane, 1,2,3-triyl triundecanoate; 1,2,3-propanetriyl undecanoate; (*n*)-trihendecanoin; triundecanoin; triundecylin; undecanoic acid 1,2,3-propanetriyl ester; undecanoin, tri-; 3-02-00-00859 [Beilstein]; BRN 1810343; CTFA 01108; and EINECS 236-935-8

45.0.3 Trade Names: NA

45.0.4 Molecular Weight: Not found

45.1 Chemical and Physical Properties

45.1.1 General

The empirical formula of glyceryl triundecanoate (triundecylin) is $C_{36}H_{68}O_6$. The structural formula is $CH_3(CH_2)_9COOCH[CH_2OCO(CH_2)_9CH_3]_2$. No other chemical or physical data were located.

45.1.2 Odor and Warning Properties

No information found.

45.2 Production and Use

No information found.

45.3 Exposure Assessment

No information found.

45.4 Toxic Effects

45.4.1 Experimental Studies

The intravenous LD_{50} value for the mouse is >10 g/kg (243a). Application of triundecylin to rabbit skin does not cause irritation (11).

45.4.2 Human Experience

No information found.

45.5 Standards, Regulations, or Guidelines of Exposure

No information found.

45.6 Studies on Environmental Impact

No information found.

46.0 Resorcinol Monoacetate

46.0.1 CAS Number: [102-29-4]

46.0.2 Synonyms: 3-Acetoxyphenol; acetylresorcinol; 1,3-benzenediol, monoacetate; *m*-hydroxyphenyl acetate; 3-hydroxyphenyl acetate; resorcinol acetate; resorcin acetate; resorcin monoacetate; resorcitate; 4-06-00-05672 [Beilstein]; AI3-02359; BRN 1865490; CTFA 02700; EINECS 203-022-0; and NSC 40511

46.0.3 Trade Names: Euresol and Remonol

46.0.4 Molecular Weight: 152.14

46.0.5 Molecular Formula: $C_8H_8O_3$

46.0.6 Molecular Structure:

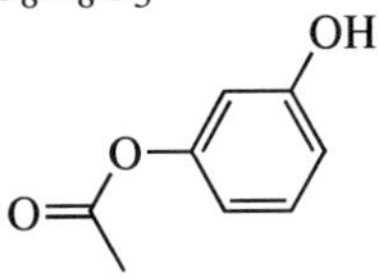

46.1 Chemical and Physical Properties

46.1.1 General

The empirical formula of resorcinol monoacetate is $C_8H_8O_3$. The structural formula is $HOC_6H_4OCOCH_3$. It is a yellow, syrupy, oily liquid with a boiling point of 283°C.

46.1.2 Odor and Warning Properties

No information found.

46.2 Production and Use

Resorcinol monoacetate has been used therapeutically as an antiseborrheic keratolytic agent.

46.3 Exposure Assessment

No information found.

46.4 Toxic Effects

No information found.

46.5 Standards, Regulations, or Guidelines of Exposure

No information found.

46.6 Studies on Environmental Impact

No information found.

47.0 Resorcinol Diacetate

47.0.1 CAS Number: *[108-58-7]*

47.0.2 Synonyms: 1,3-Benzenediol, diacetate; 1,3-diacetoxybenzene; (1,3) dihydroxybenzene diacetate; *m*-phenylene di (acetate); 4-06-00-05673 [Beilstein]; AI3-06314; BRN 1875007; EINECS 203-596-2; and NSC 4885

47.0.3 Trade Names: NA

47.0.4 Molecular Weight: 194.18

47.0.5 Molecular Formula: $C_{10}H_{10}O_4$

47.0.6 Molecular Structure:

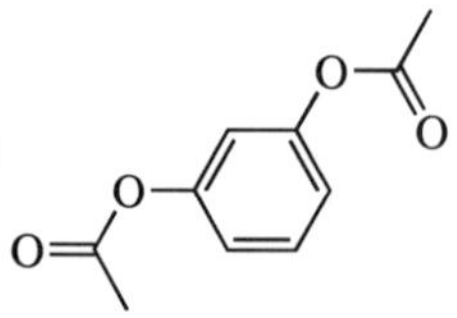

47.1 Chemical and Physical Properties

47.1.1 General

The empirical formula of resorcinol diacetate is $C_{10}H_{10}O_4$. It is very soluble in alcohol and ether and insoluble in water. The boiling point is 178°C.

47.1.2 Odor and Warning Properties

No information found.

47.2 Production and Use

No information found.

47.3 Exposure Assessment

No information found.

47.4 Toxic Effects

No information found.

47.5 Standards, Regulations, or Guidelines of Exposure

No information found.

47.6 Studies on Environmental Impact

No information found.

48.0 Resorcinol Monobenzoate

48.0.1 CAS Number: *[136-36-7]*

48.0.2 Synonyms: 1,3-Benzenediol, monobenzoate; 3-hydroxyphenyl benzoate; benzoic acid, *m*-hydroxyphenyl ester; 4-09-00-00372 [Beilstein]; BRN 1873897; EINECS 205-241-7; and NSC 4807

48.0.3 Trade Names: Eastman Inhibitor RMB

48.0.4 Molecular Weight: 214.21

48.0.5 Molecular Formula: $C_{13}H_{10}O_3$

48.0.6 Molecular Structure:

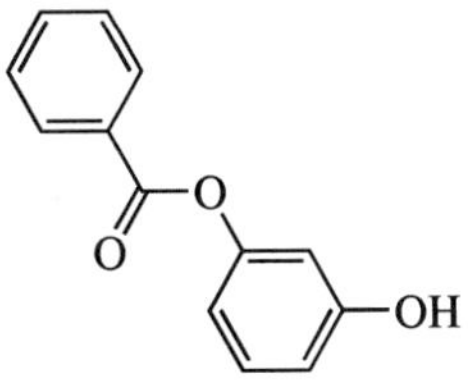

48.1 Chemical and Physical Properties

48.1.1 General

The empirical formula of resorcinol monobenzoate is $C_{13}H_{10}O_3$. The structural formula is $HOC_6H_4OCOC_6H_5$. It is very soluble in alcohol and ether and insoluble in water. No other physical or chemical data were located.

48.1.2 Odor and Warning Properties

No information found.

48.2 Production and Use

Resorcinol monobenzoate is used as an ultraviolet stabilizer in certain types of plastic films.

48.3 Exposure Assessment

No information found.

48.4 Toxic Effects

48.4.1 Experimental Studies

The oral LD_{50} value of resorcinol monobenzoate is 0.8 g/kg for the mouse and ranges from 0.8 to 1.6 g/kg for the rat (12, 242, 245). It is not absorbed through the skin of the guinea pig (227). Experimeental toxicology data are summarized in Table 80.10.

48.4.2 Human Experience

Resorcinol is a very slight primary irritant in humans (12). In 173 patients referred to the Finnish Institute of Occupational Health for suspected occupational dermatoses, 2 (1.2%) had an allergic patch test reaction to 1.0% resorcinol monobenzoate (280).

48.5 Standards, Regulations, or Guidelines of Exposure

No information found.

48.6 Studies on Environmental Impact

No information found.

49.0 Resorcinol Dibenzoate

49.0.1 CAS Number: *[94-01-9]*

49.0.2 Synonyms: 1,3-Benzenediol, dibenzoate; *m*-phenylene dibenzoate; 1,3-*bis* (benzoyloxy)benzene; 1,3-dibenzoyloxybenzene; 4-09-00-00372 [Beilstein]; AI3-00846; BRN 2059467; EINECS 202-294-8; and NSC 33405

49.0.3 Trade Names: NA

49.0.4 Molecular Weight: Not found

49.0.5 Molecular Formula: $C_{20}H_{14}O_4$

49.0.6 Molecular Structure:

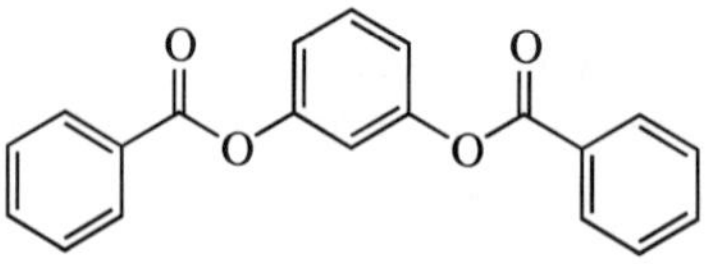

49.1 Chemical and Physical Properties

49.1.1 General

The empirical formula of resorcinol dibenzoate is $C_{20}H_{14}O_4$. At room temperature it is a solid. Its melting point is 117°C. No other chemical or physical information was located.

49.1.2 Odor and Warning Properties

No information found.

49.2 Production and Use

No information found.

49.3 Exposure Assessment

No information found.

49.4 Toxic Effects

The intraperitoneal LD_{50} value of resorcinol dibenzoate for the mouse is 8 g/kg. This dose causes somnolence, tremor, and ataxia (246).

49.5 Standards, Regulations, or Guidelines of Exposure

No information found.

49.6 Studies on Environmental Impact

No information found.

50.0 Propyl Gallate

50.0.1 CAS Number: [121-79-9]

50.0.2 Synonyms: Benzoic acid, 3,4,5,-trihydroxy-, propyl ester; gallic acid, propyl ester; *n*-propyl ester of 3,4,5-trihydroxybenzoic acid; *N*-propyl gallate; *n*-propyl 3,4,5-trihydroxybenzoate; propyl 3,4,5-trihydroxybenzoate; 3,4,5-trihydroxybenzene-1-propyl-carboxylate; 3,4,5-trihydroxybenzoic acid, (*n*) propyl ester; propylester kyseliny gallove [Czech]; AI3-17136; CCRIS 541; CTFA 12641; EINECS 204-498-2; FEMA No. 2947; HSDB 591; NCI-C505888; and NSC 2626

50.0.3 Trade Names: NIPA 49; Nipagallin P Progallin P; and Tenox PG

50.0.4 *Molecular Weight:* 212.21

50.0.5 *Molecular Formula:* $C_{10}H_{12}O_5$

50.0.6 *Molecular Structure:*

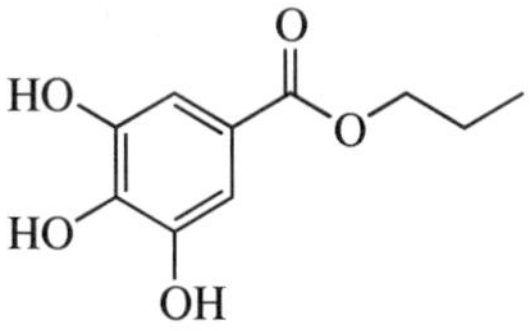

50.1 Chemical and Physical Properties

50.1.1 *General*

The empirical formula of propyl gallate is $C_{10}H_{12}O_5$. The structural formula is $(HO)_3C_6H_2COOC_3H_7$. It is a colorless crystalline material, or fine ivory powder, which melts at 130°C. Its solubilities in alcohol, ether, cottonseed oil (30°C), lard (45°C), and water are 103, 83, 1.23, 1.14, and 0.35 g/100 g, respectively (64). It can react with oxidizers and is a slight fire hazard when exposed to heat or flame. Chemical and physical properties are summarized in Table 80.9.

50.1.2 *Odor and Warning Properties*

When heated to decomposition, it emits toxic smoke and fumes.

50.2 Production and Use

Propyl gallate is manufactured via a reaction of *n*-propanol with 3,4,5-trihydroxybenzoic acid. It is used as an antioxidant for foods and cosmetics; especially fats, oils, emulsions, and waxes. It is also used in transformer oils and as a stabilizer for synthetic vitamin A.

50.3 Exposure Assessment

No information found.

50.4 Toxic Effects

50.4.1 *Experimental Studies*

50.4.1.1 Acute Toxicity. Animal experiments suggest propyl gallate is of low toxicity. The oral LD_{50} values for experimental animals range from 1.7 to 4.0 g/kg (Table 80.9). Symptoms observed at the highest concentrations were gasping respiration and terminal convulsions.

50.4.1.2 Chronic and Subchronic Toxicity. The subchronic toxicity of propyl gallate has been investigated in mice, rats, guinea pigs, and dogs. In these studies, the tested dose levels ranged from 117 to 5000 mg/kg feed. No adverse effects were noted in any of the studies (282).

In a 4-wk oral toxicity study, rats (6 animals/group/sex) ingested 0%, 0.1%, 0.5%, or 2.5% in feed. In rats ingesting the highest dose, a decrease in weight gain of more than 10%, dirty tails, thickening of the stomach wall, necrosis, and ulceration of the stomach mucosa, a moderate to severe granulomatous inflammatory response in the submucosa and muscular wall of the stomach, anemia, hyperplastic tubuli in the outer medulla of the kidneys, and increased activity of several microsomal and cytoplasmic drug-metabolizing enzymes in the liver were observed (281). Increased activity of hepatic drug metabolizing enzymes was also noted in rats treated with 0.5% propyl gallate. No effects were noted in those ingesting 0.1%.

In rats (10/goup/sex) fed a diet containing 0, 0.05, 0.19, or 0.75% propyl gallate for 13 wk, changes in hematological parameters, spleen morphology, and activities of ethoxy-resorufin-*o*-deethylase, glucuronyl transferase, and glutathione-*s*-tranferase (283a) were noted in the high-dose animals. In a 90-d study conducted by Dacre, no toxic effects were noted in mice ingesting diets containing 0.5 or 1.0% propyl gallate (283b). Twenty guinea pigs fed 0.02% propyl gallate in the diet for 14 mo and 7 dogs fed 0.01% for a year showed no signs of toxicity (283b, 284).

In rats (10/sex/group) ingesting diets containing 1.17% or 2.34% propyl gallate for 2 yr, reduced food intake, stunted growth, reduced blood hemoglobin, and renal damage were noted (285). In mice (50/group/sex), ingestion of a diet containing 0.25% or 1.0% propyl gallate for 21 mo did not affect water intake, food consumption, growth, hematologic measurements, or organ pathology. Compared to controls, males treated with 1.0% propyl gallate had decreased relative spleen weights, and males and females ingesting either dose had and an increased survival rate (283a). In both studies, no evidence of toxicity was noted in animals ingesting 0.25% or less propyl gallate in the diet.

In a long-term carcinogenicity study (see the following) rats and mice (50/sex/group) were maintained on a diet containing 0.0%, 0.6%, or 1.2% propyl gallate for 103 wk. Growth retardation and reduced feed utilization efficiency were observed in treated animals. An increased incidence of hepatic cytoplasmic vacuolization and suppurative inflammation of the prostate gland also occurred in treated male rats (327).

50.4.1.3 Pharmacokinetics, Metabolism, and Mechanisms. In rats, nearly 70% of an oral dose of propyl gallate is absorbed in the GI tract (281). *In vivo*, the gallate esters are hydrolyzed to gallic acid and free alcohol. Free alcohol is metabolized through the Krebs cycle, and most of the gallic acid is converted into 4-*o*-methyl gallic acid. Free gallic acid or a conjugated derivative of 4-*o*-methyl gallic acid is excreted in the urine (286a–286c). Significant amounts of unchanged esters are excreted in the feces of rats. In pigs, the metabolism is similar to rats (286b, 286c).

50.4.1.4 Reproductive and Developmental. Available evidence suggests that propyl gallate is not teratogenic in the rat or rabbit (286d, 287). Results of a three-generation study in rats indicate that at concentrations up to 0.5% in the diet, propyl gallate did not affect reproductive performance, indices of reproduction, or fetal morphology (286c). Administration of 2.5% propyl gallate in the diet caused maternal toxicity and slight retardation of fetal development, but no teratogenic effects (288). At dose levels up to 250 mg/kg/d, propyl gallate had no effect on organogenesis in rabbits (289).

50.4.1.5 Carcinogenesis. Results of carcinogenesis tests with propyl gallate in rats and mice suggest that propyl gallate is not carcinogenic (284, 327). In male rats maintained for 103 wk on a diet containing 0.6% propyl gallate, there is an increased incidence of preputial gland tumors, islet cell tumors of the pancreas, and pheochromocytoma of the adrenal glands. These changes were not considered to be relevant because similar changes were not observed in females or the high-dose male rats. In male mice maintained on the high-dose diet (1.2%), there was an increased incidence of malignant lymphoma. This was not considered to be related to propyl gallate because it was not greater than that of historical controls (327).

Inclusion of 0.3–1.0% propyl gallate in the diet of female rats greatly reduced the incidence of mammary gland, ear duct and forestomach tumors induced by 7,12-dimethylbenz[*a*]anthracene (290, 291).

50.4.1.6 Genetic and Related Cellular Effects Studies. At a concentration of 0–0.1 mg/plate, propyl gallate was mutagenic in *S. typhimurium* strain TA102, in the presence, but not the absence, of S-9 (301). It is not mutagenic to *Salmonella* strains G-46, TA97, TA98, TA100, TA1530, TA1537, or TA1538, *E. coli* WP2, or *Saccharomyces* D-3 in the presence or absence of S-9 (139, 283a, 292). At concentrations up to 100 mg/kg, propyl gallate did not alter the cytogenetics of metaphase cells from bone marrow of rats (283a). An intraperitoneal injection of 900 mg/kg propyl gallate caused a positive result in a mouse micronucleus test (295). At concentrations up to 50 μg/mL, propyl gallate does not induce mutations in WI-38 human embryonic lung cells (283a).

At concentrations of 50–500 μM, propyl gallate inhibited liver mixed function oxidase and demethylase activity in rat liver microsomes (283a). However, when fed to rats at a concentration of 0.3% for 1 mo, propyl gallate had no effect on relative liver weight, liver microsomal protein content, cytochrome P450 concentration aniline hydroxylase aminopyrene *N*-demethylase, or cytochrome c reductase activity (283a).

Addition of 50 μmol of propyl gallate to rat testis microsomes in the presence of arachidonate stimulated the production of prostaglandin PGF_2 (283a). Propyl gallate stimulated formation of PGF_2 in mammary gland from rats fed polyunsaturated diets, but inhibits PGF_2 synthesis in rats fed a saturated fat diet (283a).

Propyl geallate (0.5–2.0 mM) caused cytotoxicity, DNA fragmentation, and mitochondrial dysfunction in isolated rat hepatocytes (296, 297). Conversely, it protected hepatocytes and isolated perfused livers against oxygen radical toxicity (298, 299).

50.4.1.7 Other: Neurological, Pulmonary, Skin Sensitization. When injected intradermally, propyl gallate (20 mM) is a moderate skin sensitizer in guinea pigs (300).

50.4.2 Human Experience

Contact dermatitis may occur after dermal exposure to propyl gallate. A 59-year old man presented with an acute lesion of his penis with swelling, erythema, and edema after using Locapred® cream; subsequent patch tests were positive for nickel, neomycin, Locapred®, and propyl gallate (301). Another man who developed acute swelling and erythema of his groins and genitals after applying Timodine® cream for chronic intertrigo patch tested positively for 3 components: propyl gallate, dibutyl phthalate, and hydrocortisone (302). A 49-year-old woman who developed acute eczema after applying traumatocycline (contains 8% propyl gallate) to her arm for a burn patch-tested positively for propyl gallate (303). In 10 patients applying 20% propyl gallate in 70% ethyl alcohol to their forearms once daily for 24 d, 5 patients complained of pruritis and erythema (304). Cases of allergic contact chelitis (an inflammatory reaction of the lips) have been reported in people using lipsticks or lip balms containing propyl gallate (305, 306).

50.5 Standards, Regulations, or Guidelines of Exposure

The recommended ADI for propyl gallate in man is 2.5 mg/kg (307). Propyl gallate is exempted from the requirement of a tolerance when used as an antioxidant in accordance with good agricultural practices as inert (or occasionally active) ingredient in pesticide formulations applied to animals or growing crops, or to raw agricultural commodities after harvest (308).

50.6 Studies on Environmental Impact

No information found.

51.0 Octyl Gallate

51.0.1 CAS Number: [1034-01-1]

51.0.2 Synonyms: Benzoic acid, 3,4,5,-trihydroxy-, octyl ester; gallic acid, octyl ester; *n*-octyl ester of 3,4,5-trihydroxybenzoic acid; *n*-octyl gallate; octyl 3,4,5-trihydroxybenzoate; propyl ester; octylester kyseliny gallove [Czech]; 4-10-00-02005 [Beilstein], BRN 2132305; EINECS 213-853-0; and NSC 97419

51.0.3 Trade Names: Stabilizer GA-8 and Progallin O

51.0.4 Molecular Weight: 282.17

51.1 Chemical and Physical Properties

51.1.1 General

The empirical formula of octyl gallate is $C_{15}H_{22}O_5$. The structural formula is $(HO)_3C_6H_2COOC_8H_{17}$. No other chemical or physical information was located.

51.1.2 Odor and Warning Properties

No information found.

51.2 Production and Use

Octyl gallate is an antioxidant that is used as a preservation agent in a wide variety of foods, oils, and cosmetics.

51.3 Exposure Assessment

No information found.

51.4 Toxic Effects

51.4.1 Experimental Studies

51.4.1.1 Acute Toxicity. Depending on the study consulted, the oral LD_{50} value of octyl gallate for the rat ranges anywhere from 1.96 to 4.7 g/kg (Table 80.9). More recent studies indicate onset of toxic effects in rats at levels >3 g/kg when administered orally (281).

51.4.1.2 Chronic and Subchronic Toxicity. In rats fed diets containing 0.1–0.5% octyl gallate for 13 wk, the only effect observed was a slight elevation of serum glutamic–oxaloacetic transaminase (309). A similar result was observed in dogs ingesting 0.5% octyl gallate for 13 wk (310). In pigs, ingestion of 0.2% in the diet for 13 wk had no effect on growth, hematology, organ weights, or pathology (286c). In a 2-yr study conducted in 1955, no adverse effects were noted in rats given 0.035–0.5% in the diet (286c).

51.4.1.3 Pharmacokinetics, Metabolism, and Mechanisms. Octyl gallate is absorbed and hydrolyzed to a lesser degree than propyl gallate. Metabolism *in vivo* yields *n*-octyl alcohol and gallic acid (311). Unchanged octyl gallate is a major urinary component in rats (286c). See Section 51.4.1.3 for more information about metabolism.

51.4.1.4 Reproductive and Developmental. Results of two-generation reproductive studies indicate that offspring of rats administered 0.25–0.6% octyl gallate in the diet exhibit decreased body weight (312) and survival during weaning (312, 313). A dose-dependent reductions in implantation sites and corpora lutea was noted in P_2 parent animals (312). Ingestion of a diet containing 0.1% octyl gallate had no effect on dams or offspring.

51.4.1.5 Carcinogenesis. No information found.

51.4.1.6 Genetic and Related Cellular Effects Studies. Octyl gallate is more toxic to isolated rat hepatocytes than propyl, butyl, dodecyl, methyl, or ethyl gallate. Addition of 1 mM caused mitochondrial dysfunction, rapid intracellular ATP loss, and cell death (296, 297). Octyl gallate inhibits the activity of 15-lipoxygenase in rabbit reticulocytes (IC_{50} 0.25 μM) and prostaglandin H synthase from sheep vesicular glands (IC_{50} 25 μM) (314). In human polymorphonuclear leukocytes, 10 μM octyl gallate causes complete inhibition of the incorporation of arachidonic acid in triacylgycerols and phospholipids (315).

Application of octyl gallate to mouse skin induced a two fold increase in the activity of glutathione-*S*-transferase (316).

51.4.1.7 Other: Neurological, Pulmonary, Skin Sensitization. When applied to guinea pig skin, octyl gallate (20 mM) is a moderate sensitizer (300).

51.4.2 Human Experience

51.4.2.1 General Information. Allergy may occur after exposure to octyl gallate.

51.4.2.2 Clinical Cases. A 49-year-old female with a 10-y history of "burning mouth" and erythema of the tongue proved to be allergic to octyl gallate (317). A 19-year-old female patch tested positively to octyl gallate after developing itchy swollen hands, face and legs, eyelid edema, and nausea on two occasions after adding a powder containing 97.9% octyl gallate to heated chicken fat (318).

51.4.2.3 Epidemiology Studies. The allergic potential of several different preservatives in 11,485 patients with suspected allergic contact dermatitis was examined by Schnuch and co-workers in 1998 (319). Approximeately 1.2% reacted positively to octyl gallate.

51.5 Standards, Regulations, or Guidelines of Exposure

After being reevaluated in 1987, no ADI was allocated to octyl gallate by the WHO (307).

51.6 Studies on Environmental Impact

No information found.

52.0 Dodecyl Gallate

52.0.1 CAS Number: *[1166-52-5]*

52.0.2 Synonyms: Benzoic acid, 3,4,5,-trihydroxy-, dodecyl ester; dodecyl 3,4,5-trihydroxybenzoate; gallic acid, dodecyl ester; gallic acid lauryl ester; lauryl gallate; lauryl 3,4,5-trihydroxybenzoate; 3,4,5-trihydroxybenzoic acid dodecyl ester; dodecylester kyseliny gallove [Czech]; 4-10-00-02006 [Beilstein], BRN 2701981; CCRIS 5568; CTFA 00948; EINECS 214-620-6; and NSC 133463

52.0.3 Trade Names: Nipagallin LA and Progallin LA

52.0.4 Molecular Weight: 338.42

52.1 Chemical and Physical Properties

52.1.1 General

The empirical formula of dodecyl gallate is $C_{19}H_{30}O_5$. The structural formula is $(OH)_3C_6H_2COOC_{12}H_{25}$. No other chemical or physical information was located.

52.1.2 Odor and Warning Properties

No information found.

52.2 Production and Use

Like the other gallates, dodecyl gallate is used as an antioxidant in food, oils, and cosmetics.

52.3 Exposure Assessment

No information found.

52.4 Toxic Effects

52.4.1 Experimental Studies

52.4.1.1 Acute Toxicity. Dodecyl gallate has a low acute oral toxicity. The oral LD_{50} values for rats and mice are 6.5 and 1.6–3.2 g/kg (252–254) Acute toxicity data are summarized in Table 80.10.

52.4.1.2 Chronic and Subchronic Toxicity. No effects on body weight were noted in rats or pigs given 0.2% dodecyl gallate in the diet for 70 or 91 d, respectively (286c, 320). In a 2-y feeding study, no adverse effects were noted in rats ingesting up to 0.5% in the diet (286c).

52.4.1.3 Pharmacokinetics, Metabolism, and Mechanisms. Dodecyl gallate is absorbed and hydrolyzed to a lesser degree than propyl gallate (286c). See Sectieon 51.4.1.3 for more information about metabolism.

52.4.1.4 Reproductive and Developmental. Growth retardation and loss of litters due to underfeeding and slight hypochromic anemia was noted in rats given 5 or 2 mg/kg over three generations, respectively (286c).

52.4.1.5 Carcinogenesis. In a 2-yr feeding study administration of up to 5 mg/kg feed to rats, no increase in tumor incidence was noted (124). In female mice, a dermal injection of dodecyl gallate (100 nmol in 0.2 mL acetone) inhibits the formation of skin carcinomas and papillomas induced by the tumor promoter phorbol myristate and the initiator 7,12-dimethylbenz[*a*]anthracene (321).

52.4.1.6 Genetic and Related Cellular Effects Studies. *In vitro*, dodecyl gallate protects mammalian and bacterial cells from cytotoxicity induced by hydrogen peroxide (322). Dodecyl gallate is more toxic to isolated rat hepatocytes than propyl, ethyl, or methyl gallate. Addition of 1 mM caused mitochondrial dysfunction, rapid intracellular ATP loss, and cell death (296, 297).

Like octyl gallate, dodecyl gallate inhibits the activity of 15-lipoxygenase in rabbit reticulocytes (IC_{50} 0.25 μM) and prostaglandin H synthase from sheep vesicular glands (IC_{50} 25 μM) (314). In human polymorphonuclear leukocytes, 10 μM dodecyl gallate causes complete inhibition of the incorporation of arachidonic acid in triacylgycerols and phospholipids (315). Application of dodecyl gallate to mouse skin induces a three fold increase in the activity of glutathione-*S*-transferase (316).

52.4.1.7 Other: Neurological, Pulmonary, Skin Sensitization. In guinea pigs, dodecyl gallate (20 mM) is a strong sensitizer (300).

52.4.2 Human Experience

Allergy may occur after exposure to dodecyl gallate. A 23-year-old supermarket cheese counter assistant exhibited a strong positive patch test reaction to 0.1% dodecyl gallate after developing a painful, itchy dermatitis of the hands (359).

52.4.2.3 Epidemiology Studies. No information found.

52.5 Standards, Regulations, or Guidelines of Exposure

After being reevaluated in 1987, no ADI was allocated to dodecyl gallate by the WHO (307).

52.6 Studies on Environmental Impact

No information found.

C INTRODUCTION TO ESTERS OF DICARBOXYLIC ACIDS

Oxalates, malonates, glutarates, and succinates are high-flash, high-boiling fluids. Oxalates and malonates are mainly used as solvents for resins or as chemical intermediates. Some of their physicochemical properties are summarized in Table 80.11. In general, these compounds undergo rapid absorption, metabolism (hydrolysis), and excretion. The general industrial use of these materials has not been associated with any particular toxicity problem. Their low vapor pressure does not warrant any special inhalation hazard precautions. Diethyl oxalate, which can exert typical local solvent and systemic effects (328), may present an exception. In experimental animals, acute ingestion of moderate amounts of diethyl oxalate is associated with deposition of calcium oxalate in the ureter and bladder, which may cause renal obstruction (12). In humans, diethyl oxalate may cause irritation to skin and mucous membranes (62). Acute toxicity data for esters of alkyl dicarboxylic acids are summarized in Table 80.12.

Chemical and physical property data for alkyl and alkoxy adipates, azelates, and sebacates are summarized in Table 80.13. These compounds are important chemical intermediates and are used extensively as plasticizers (324). Some of these agents are used in food packaging materials. They possess low acute toxicities (Table 80.14), and their irritant effects on the skin and eye are very slight (12). Available evidence suggests that the lower alkyl adipates (dimethyl, diethyl, dibutyl) are reproductive and/or fetal toxicants (325). Exposure to products containing 20–30% diethyl sebacate may cause contact dermatitis (326, 329).

53.0 Dimethyl Oxalate

53.0.1 CAS Number: *[553-90-2]*

53.0.2 Synonyms: Dimethyl ethanedioate; ethanedioic acid, dimethyl ester; oxalic acid, dimethyl ester; AI3-21214 [NLM]; EINECS 209-053-6; and NSC 9374

Table 80.11. Physical and Chemical Properties of Esters of Oxalates, Malonates, Succinates, and Glutarates

Chemical Name	CAS No.	Molecular Formula	Mol. Wt.	Boiling Point (°C) (mmHg)	Melting Point (°C)	Density (Specific Gravity	Sol. in Water at 20°C (68°F)	Refractive Index (20°C) (68°F)	Vapor Pressure (mmHg) (°C)	UEL or LEL % by Vol. Room Temp.
Oxalate										—
Dimethyl	*[553-90-2]*	$C_4H_6O_4$	118.09	163.4	52	1.148	slightly	1.379	—	—
Diethyl	*[95-92-1]*	$C_6H_{10}O_4$	146.14	185.7	−41	1.076	slightly	1.4101	10 (84)	
Diisopropyl	*[615-81-6]*	$C_8H_{14}O_4$	174.10	191	—	1.0010	—	1.4100	—	—
Dibutyl	*[2050-60-4]*	$C_{10}H_{18}O_4$	202.25	240	−29	0.986	insol	1.4234	—	—
Diisobutyl	*[2050-61-5]*	$C_{10}H_{18}O_4$	202.25	229 (758)	—	0.9737	insol	1.4180	—	—
Malonate										—
Dimethyl	*[108-59-8]*	$C_5H_8O_4$	132.12	181	−62	1.156	slight	1.41348	—	—
Diethyl	*[105-53-3]*	$C_7H_{12}O_4$	160.17	199	−50	1.055	slight	1.4143	1 (40)	
Succinate										—
Diethyl	*[123-25-1]*	$C_8H_{14}O_4$	174.20	217	−20	1.047	insol	1.4198	1 (54.6)	—
Dipropyl	*[925-15-5]*	$C_{10}H_{18}O_4$	202.25	250.8	−5.9	1.0020	insol	1.4250	—	—
Dibutyl	*[141-03-7]*	$C_{12}H_{22}O_4$	230.30	274.5	−29.0	(0.9768)	insol	1.4299	—	—
Di (2-ethylhexyl)	*[2915-57-3]*	$C_{20}H_{38}O_4$	342.5	257 (50)	−60	0.9346	—	—	—	—
Di (2-hexoxyethyl)	*[10058-20-5]*	$C_{20}H_{38}O_6$	374.58	—	—	—	—	—	—	—
Dioctyl sodium sulfo-	*[577-11-7]*	$C_{20}H_{37}NaO_7S$	445.56	—	—	—	71 g/L [a]	—	—	—
Diethyl glutarate	*[818-38-2]*	$C_9H_{16}O_4$	188.22	236.6	−24.1	1.0220	slight	1.4241	—	—

[a] At 25°C.

Table 80.12. Summary of Toxicity of Esters of Oxalates, Malonates, Succinates and Glutarates

Chemical Name	CAS No.	Species	Exposure Route	Approximate Dose (g/kg)	Treatment Regimen	Observed Effect	Ref.
Oxalate							
Dimethyl	[*553-90-2*]	Rat	oral	> 0.5	Single dose	LD	330
Diethyl	[*95-92-1*]	Rat	oral	0.4–1.6	Single dose	LD_{50}	12
		Mouse	oral	2	Single dose	LD_{50}	333
		Guinea pig	dermal	> 10 mL/kg	Single dose	Slight skin irritation	12
Malonate							
Dimethyl	[*108-59-8*]	Rat	oral	4.62 mL/kg	Single dose	LD_{50}	334
		Rabbit	dermal	> 5 g/kg	Single dose	LD_{50}	334
Diethyl	[*105-53-3*]	Rat	oral	14.9 mL/kg	Single dose	LD_{50}	335
		Rabbit	dermal	> 16 mL/kg	Single dose	LD_{50}	335
Succinate							
Diethyl	[*123-25-1*]	Rat	oral	8.53	Single dose	LD_{50}	7
Dipropyl	[*925-15-5*]	Rat	oral	6.49	Single dose	LD_{50}	5
		Rat	IP	0.29	Single dose	LD_{50}	336
Dibutyl	[*141-03-7*]	Rat	oral	8.0	Single dose	LD_{50}	337
Di (2-hexyloxyethyl)	[*10058-20-5*]	Rat	oral	4.28	Single dose	LD_{50}	5
		Rabbit	dermal	12.3	Single dose	Slight irritation	5
Dioctyl sodium sulfo-	[*577-11-7*]	Rat	oral	1.9	Single dose	LD_{50}	338
		Mouse	oral	2.64	Single dose	LD_{50} listed. Somnolence noted.	339
		Rat	IP	0.59	Single dose	LD_{50} listed. Structural changes in blood vessels noted	340
		Mouse	IV	0.06	Single dose	LD_{50}	341
Diethyl glutarate	[*818-38-2*]	Rat	oral	> 1.6	Single dose	LD_{50} listed. Vasodilation observed.	12
		Guinea pig	dermal	10 mL/kg	Single dose	Moderate skin irritation	12

Table 80.13. Physical and Chemical Properties of Esters of Adipates, Azelates, and Sebacates

Chemical Name	CAS No.	Molecular Formula	Mol. Wt.	Boiling Point (°C) (mmHg)	Melting Point (°C)	Density (Specific Gravity	Sol. in Water at 20°C (68°F)	Refractive Index (20°C)	Vapor Pressure (mmHg) (°C)	UEL or LEL % by Vol. Room Temp.
Adipate										
Dimethyl	[*627-93-0*]	$C_8H_{14}O_4$	174.20	115 (13)	8	(1.0600)	insol	1.4283	—	—
Diethyl	[*141-28-6*]	$C_{10}H_{18}O_4$	202.25	251	– 18	(1.009)	insol	1.4272	—	—
Dibutyl	[*105-99-7*]	$C_{14}H_{26}O_4$	258.36	165 (10)	– 32.4	0.9615	insol	1.4369	—	—
Di(2-ethylbutyl)	[*10022-60-3*]	$C_{18}H_{34}O_4$	314.46	200 (10)	– 15	0.934	insol	1.4434	—	—
Di(2-ethylhexyl)	[*103-23-1*]	$C_{22}H_{42}O_4$	370.57	417	– 67.8	(0.928)	insol	1.4474	4.7×10^{-7} (20)	—
Dioctyl	[*123-79-5*]	$C_{22}H_{42}O_4$	370.57	—	—	—	0.78 mg/L[a]	—	8.5×10^{-7} (20)	—
Diisooctyl	[*1330-86-5*]	$C_{22}H_{42}O_4$	370.57	205–220 (4)	– 70	0.928	—	—	< 0.12 (150)	—
Diisononyl	[*33703-08-1*]	$C_{24}H_{46}O_4$	398.63	232–233 (5)	– 60	(0.92)	—	—	0.9 (200)	—
Didecyl	[*105-97-5*]	$C_{26}H_{50}O_4$	426.68	240 (4)	—	0.919	—	—	—	—
Dibutoxyethyl	[*141-18-4*]	$C_{18}H_{34}O_6$	346.46	208 (4)	—	0.997	—	—	< 0.17 (150)	—
Di(2-hexyloxyethyl)	[*110-32-7*]	$C_{22}H_{42}O_6$	402.57	—	– 6.0	0.9642	insol	—	—	—
Di-2(2-ethyl butoxy) ethyl	[*7790-07-0*]	$C_{22}H_{42}O_6$	402.57	214 (5)	– 60	0.9268	—	—	—	—
Azelate										
Dibutyl	[*2917-73-9*]	$C_{17}H_{32}O_4$	300.44	336	—	—	insol	—	—	—
Di(2-ethylhexyl)	[*103-24-2*]	$C_{25}H_{48}O_4$	412.65	237 (5)	– 78	(0.915)	insol	1.446	5 (237)	—
Sebacate										
Diethyl	[*110-40-7*]	$C_{14}H_{26}O_4$	258.36	312	2	0.963	slight	1.4366	—	—
Dibutyl	[*109-43-3*]	$C_{18}H_{34}O_4$	314.46	180 (3)	– 12	0.9405	40 mg/L	1.4433	4.7×10^{-6} (25)	—
Dioctyl	[*2432-87-3*]	$C_{26}H_{50}O_4$	426.68	248 (4)	– 55	0.916	—	—	—	—
Diisooctyl	[*27214-90-0*]	$C_{26}H_{50}O_4$	426.69	248-255 (4)	—	0.917	—	—	—	—
Di(2-ethylhexyl)	[*122-62-3*]	$C_{26}H_{50}O_4$	426.68	248	– 67	(0.914)	insol	—	—	—

[a] At 22°C.

Table 80.14. Summary of Toxicity of Esters of Esters of Adipates, Azelates and Sebacates

Chemical Name	CAS No.	Species	Exposure Route	Approximate Dose (g/kg)	Treatment Regimen	Observed Effect	Ref.
Adipate							
Dimethyl	[*627-93-0*	Rat	IP	1.809 mL/kg	Single dose	LD_{50}	224
Diethyl	[*141-28-6*]	Mouse	oral	8.1	Single dose	LD_{50}	365
		Rat	IP	2.5119 mL/kg	Single dose	LD_{50}	325
		Mouse	IP	2.19 mL/kg	Single dose	LD_{50}	365
Dibutyl	[*105-99-7*]	Rat	oral	12.9	Single dose	LD_{50}	7
		Mouse	oral	16.89	Single dose	LD_{50}	369
		Rabbit	dermal	20 mL/kg	Single dose	LD_{50}	7
		Rat, mouse	inhal	> 17 mg/m^3	4 h	LC	369
		Rat	IP	5.244 mL/kg	Single dose	LD_{50}	325
Di(2-ethylbutyl)	[*10022-60-3*]	Rat	oral	5.62	Single dose	LD_{50}	5
		Rabbit	dermal	16.8 mL/kg	Single dose	LD_{50}	5
Di(2-ethylhexyl)	[*103-23-1*]	Rat	oral	9.1	Single dose	LD_{50}	5
		Rabbit	dermal	16.3 mL/kg	Single dose	LD_{50}	5
Dioctyl	[*123-79-5*]	Guinea pig	oral	20 mL/kg	Single dose	LC	370, 372
Diisononyl	[*33703-08-1*]	Rat	oral	> 10	Single dose	LD_{50}	371
		Rabbit	dermal	> 3.16	Single dose	LD_{50}	372
Didecyl	[*105-97-5*]	Rat	oral	> 25.6	Single dose	LD_{50}	12
		Mouse	oral	12.8–25.6	Single dose	LD_{50}	12
		Rat, mouse	IP	> 25.0	Single dose	LD_{50}	12
Dibutoxyethyl	[*141-18-4*]	Rat	IP	0.6	Single dose	LD_{50}	12
Di(2-hexyloxyethyl)	[*110-32-7*]	Rat	oral	4.29	Single dose	LD_{50}	5
		Rabbit	dermal	12.3 mL/kg	Single dose	LD_{50}	5
Di-2(2-ethyl butoxy) ethyl	[*7790-07-0*]	Rat	oral	3.25	Single dose	LD_{50}	5
		Rabbit	dermal	4.24 mL/kg	Single dose	LD_{50}	5
Azelate							
Dibutyl	[*2917-73-9*]	Rat, mouse	oral	> 12.8	Single dose	LD_{50}	12
		Rabbit	dermal	> 10 mL/kg	Single dose	LD_{50}	12

Table 80.14. (*Continued*)

Chemical Name	CAS No.	Species	Exposure Route	Approximate Dose (g/kg)	Treatment Regimen	Observed Effect	Ref.
Di(2-ethylhexyl)	[*103-24-2*]	Rat	oral	8.72 mL/kg	Single dose	LD_{50}	373
		Rabbit	dermal	20 mL/kg	Single dose	LD_{50}	373
		Rat	IV	1.06	Single dose	LD_{50}	374
		Rabbit	IV	0.64	Single dose	LD_{50}	374
Sebacate							
Diethyl	[*110-40-7*]	Rat	oral	14.5	Single dose	LD_{50}	3
		Guinea pig	oral	7.28	Single dose	LD_{50}	3
Dibutyl	[*109-43-3*]	Rat	oral	16	Single dose	LD_{50}	375
		Mouse	oral	19.5	Single dose	LD_{50}	369
		Mouse	IP	14.7 mL/kg	Single dose	LD_{50}	238
Di(2-ethylhexyl)	[*122-62-3*]	Mouse	oral	9.5	Single dose	LD_{50}	376
		Rat	IV	0.9	Single dose	LD_{50}	375
		Rabbit	IV	0.54	Single dose	LD_{50}	375

53.0.3 Trade Names: NA

53.0.4 Molecular Weight: 118.09

53.0.5 Molecular Formula: $C_4H_6O_4$

53.0.6 Molecular Structure: 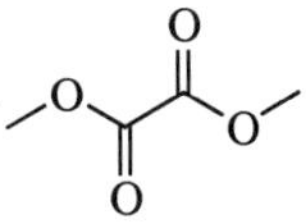

53.1 Chemical and Physical Properties

53.1.1 General

The empirical formula of dimethyl oxalate is $C_4H_6O_4$. It exists as a liquid between its pour point (52°C) and boiling point (163.4°C). It is soluble in ethanol and ether and slightly soluble in water. Chemical and physical data are summarized in Table 80.11.

53.1.2 Odor and Warning Properties

No information found.

53.2 Production and Use

No information found.

53.3 Exposure Assessment

No information found.

53.4 Toxic Effects

The oral lethal dose in the rat is >0.5 g/kg (330). No other toxicological information (in animals or humans) was found.

53.5 Standards, Regulations, or Guidelines of Exposure

No information found.

53.6 Studies on Environmental Impact

No information found.

54.0 Diethyl Oxalate

54.0.1 CAS Number: *[95-92-1]*

54.0.2 Synonyms: Diethyl ethanedioate; ethanedioic acid, diethyl ester; ethyl oxalate; oxalic acid, diethyl ester; oxalic ether; diethylester kyseliny stavelove [Czech]; oxalate d'ethyle [French], oxalato de etilo [Spanish]; 4-02-00-01848 [Beilstein]; BRN 0606350 [RTECS]; EINECS 202-464-1; HSDB 2131; and NSC 8851

54.0.3 Trade Names: NA

54.0.4 Molecular Weight: 146.14

54.0.5 Molecular Formula: $C_6H_{10}O_4$

54.0.6 Molecular Structure:

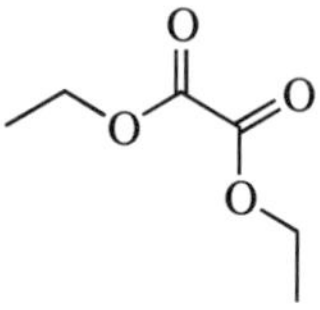

54.1 Chemical and Physical Properties

54.1.1 General

The empirical formula of diethyl oxalate is $C_6H_{10}O_4$. The structural formula is $(COOC_2H_5)_2$. It is an oily, odorous liquid, which decomposes in water. It exists as a liquid between its pour point (−41°C) and boiling point (185.7°C). It is soluble in ethanol and ether. Chemical and physical properties are summarized in Table 80.11.

54.1.2 Odor and Warning Properties

Reaction with moist air or water will release toxic, corrosive, or flammable gases, and may release intense heat that will increase the concentration of fumes in the air.

54.2 Production and Use

Diethyl oxalate is produced via esterification of ethanol and oxalic acid. It is a preferred solvent for cellulose acetate and nitrate. It is used for manufacturing various chemicals, plastics, and dyes, as well as lacquers and perfumes and resins.

54.3 Exposure Assessment

An increase in urinary oxalates may not exclusively indicate oxalate poisoning, because diabetes, cirrhosis, Klinefelter's syndrome, congestive heart failure, renal tubular acidosis, sarcoidosis, and a variety of parasitic diseases similarly affect the renal system (331). Under some pathological conditions, xanthine oxidase may promote the direct oxidation of glyoxylic acid to oxalic acid (332).

54.4 Toxic Effects

54.4.1 Experimental Studies

54.4.1.1 Acute Toxicity. By ingestion this compound appears to be of low to moderate toxicity. Oral LD_{50} values for the rat and mouse are 0.4–1.6 and 2 g/kg, respectively

(Table 80.12). In rats treated with toxic doses of diethyl oxalate, signs of disturbed respiration and muscle twitching were noted (12). These signs are similar to CNS stimulation by oxalic acid, which is indicated by irritability, twitchings, tremors, and convulsions (342). Oxalate crystals in the ureter and bladder may cause pain and hematuria. Death may be due to renal obstruction or to hypocalcemia, leading to cardiac failure. Massive renal oxalate deposits were observed following treatment with 400 mg/kg or higher in the rat (12).

54.4.1.2 Chronic and Subchronic Toxicity. No information found.

54.4.1.3 Pharmacokinetics, Metabolism, and Mechanisms. Diethyl oxalate does not penetrate the intact skin. Ingested or inhaled diethyl oxalate, owing to its water solubility and catabolic properties, presumably undergoes rapid hydrolysis to the stable oxalic acid or sodium oxalate and ethanol. The ethanol is rapidly metabolized and the oxalate conjugated and excreted in the urine. Diethyl oxalate may occur in free or complex form, along with normal metabolites stemming from the dietary intake. Small quantities of diethyl oxalate cause no adverse effects, but larger amounts may increase urinary oxalates and produce calcium oxalate deposits leading to pathogenesis of the renal system (220).

54.4.1.4 Reproductive and Developmental. No information found.

54.4.1.5 Carcinogenesis. No information found.

54.4.1.6 Genetic and Related Cellular Effects Studies. No information found.

54.4.1.7 Other: Neurological, Pulmonary, Skin Sensitization. Dermal application of 10 mL/kg diethyl oxalate to guinea pig skin is slightly irritating (12). The low vapor pressure of diethyl oxalate renders it unlikely to cause toxic effects by inhalation.

54.4.2 Human Experience

In humans, diethyl oxalate is a strong irritant to the skin and mucous membranes (62). Workers exposed to 0.76 mg/L for several months (or a total equivalent of about 128 mg/m^3) have complained of weakness, headache, and nausea (328, 343). In these workers, there was evidence of some slight anemia and leukopenia, neutropenia, and eosinophilia.

54.5 Standards, Regulations, or Guidelines of Exposure

No information found.

54.6 Studies on Environmental Impact

No information found.

55.0 Diisopropyl Oxalate

55.0.1 CAS Number: [615-81-6]

55.0.2 Synonyms: Ethanedioic acid, *bis* (1-methylethyl) ester; AI3-11244 [NLM]; and EINECS 210-448-0

55.0.3 Trade Names: NA

55.0.4 Molecular Weight: 174.10

55.1 Chemical and Physical Properties

The empirical formula of diisopropyl oxalate is $C_8H_{14}O_4$. It is a liquid with a density of 1.001 and a boiling point of 191°C. Chemical and physical properties that were located are summarized in Table 80.11.

55.1.2 Odor and Warning Properties

No information found.

55.2 Production and Use

No information found.

55.3 Exposure Assessment

No information found.

55.4 Toxic Effects

No information found.

55.5 Standards, Regulations, or Guidelines of Exposure

No information found.

55.6 Studies on Environmental Impact

No information found.

56.0 *n*-Dibutyl Oxalate

56.0.1 CAS Number: *[2050-60-4]*

56.0.2 Synonyms: Butyl ethanedioate; dibutyl ethanedioate; di-*n*-butyl oxalate; ethanedioic acid, dibutyl ester; oxalic acid, dibutyl ester; AI3-06011 [NLM]; EINECS 218-092-8; and NSC 8468

56.0.3 Trade Names: NA

56.0.4 Molecular Weight: 202.25

56.0.5 Molecular Formula: $C_{10}H_{18}O_4$

56.0.6 *Molecular Structure:*

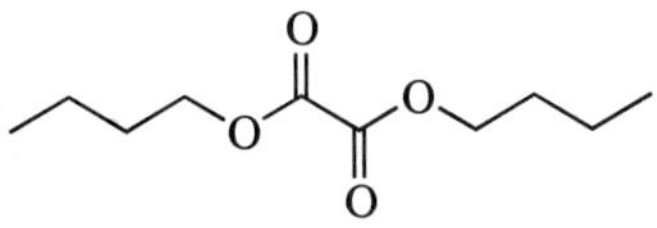

56.1 Chemical and Physical Properties

The empirical formula of dibutyl oxalate is $C_{10}H_{18}O_4$. It exists as a liquid between its pour point (– 29°C) and boiling point (240°C at 3 mmHg). It is insoluble in water. Chemical and physical properties are summarized in Table 80.11.

56.1.2 *Odor and Warning Properties*

No information found.

56.2 Production and Use

No information found.

56.3 Exposure Assessment

No information found.

56.4 Toxic Effects

No information found.

56.5 Standards, Regulations, or Guidelines of Exposure

No information found.

56.6 Studies on Environmental Impact

No information found.

57.0 Diisobutyl Oxalate

57.0.1 *CAS Number: [2050-61-5]*

57.0.2 *Synonyms:* NA

57.0.3 *Trade Names:* NA

57.0.4 *Molecular Weight:* 202.25

57.1 Chemical and Physical Properties

The empirical formula of diisobutyl oxalate is $C_{10}H_{18}O_4$. It is a liquid with a density of 0.9737 and a boiling point of 229°C (at 758 mmHg). Chemical and physical properties are summarized in Table 80.11.

57.1.2 Odor and Warning Properties

No information found.

57.2 Production and Use

No information found.

57.3 Exposure Assessment

No information found.

57.4 Toxic Effects

No information found.

57.5 Standards, Regulations, or Guidelines of Exposure

No information found.

57.6 Studies on Environmental Impact

No information found.

58.0 Dimethyl Malonate

58.0.1 CAS Number: *[108-59-8]*

58.0.2 Synonyms: Dimethyl propanedioate; dimethyl 1,3 -propanedioate; malonic acid, dimethyl ester; propanedioic acid, dimethyl ester; methyl malonate; and EINECS 203-597-8

58.0.3 Trade Names: NA

58.0.4 Molecular Weight: 132.13

58.0.5 Molecular Formula: $C_5H_8O_4$

58.0.6 Molecular Structure:

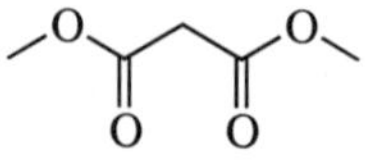

58.1 Chemical and Physical Properties

58.1.1 General

The empirical formula of dimethyl malonate is $C_5H_8O_4$. The structural formula is $CH_2(COOCH_3)_2$. It exists as a liquid between its pour point (−62°C) and boiling point (181°C). Dimethyl malonate is soluble in alcohol and ether, and very slightly soluble in water. It may slowly decompose. Physico-chemical properties are summarized in Table 80.11.

58.1.2 Odor and Warning Properties

No information found.

58.2 Production and Use

Dimethyl malonate is produced via direct esterification of methanol with malonic acid under azeotropic conditions. It is used in fragrances and some artificial flavorings (344).

58.3 Exposure Assessment

No information found.

58.4 Toxic Effects

58.4.1 Experimental Studies

58.4.1.1 Acute Toxicity. The oral LD_{50} value for rats is 4.62 mL/kg and the dermal LD_{50} value for rabbits is >5 g/kg (334).

58.4.1.2 Chronic and Subchronic Toxicity. No information found.

58.4.1.3 Pharmacokinetics, Metabolism, and Mechanisms. No information found.

58.4.1.4 Reproductive and Developmental. No information found.

58.4.1.5 Carcinogenesis. No information found.

58.4.1.6 Genetic and Related Cellular Effects Studies. No information found.

58.4.1.7 Other: Neurological, Pulmonary, Skin Sensitization. When applied full strength to intact or abraded rabbit skin, dimethyl malonate was irritating (334). Treatment of cultured striatal nerve cells with 500 μM dimethyl malonate reduced the ATP/ADP ratio by 30% (345).

58.4.2 Human Experience

Tested at a concentration of 8% in petrolatum, dimethyl malonate was neither irritating or sensitizing to humans (142, 334, 346).

58.5 Standards, Regulations, or Guidelines of Exposure

No information found.

58.6 Studies on Environmental Impact

No information found.

59.0 Diethyl Malonate

59.0.1 CAS Number: [105-53-3]

59.0.2 Synonyms: Carbethoxyacetic ester; dicarbethoxymethane; diethyl propanedioate; ethyl malonate; ethyl methanedicarboxylate; ethyl propanedioate; malonic acid, diethyl ester; malonic ester; methanedicarboxylic acid, diethyl ester; propanedioic acid, diethyl ester; AI3-00656 [NLM]; EINECS 203-305-9; FEMA No. 2375; and NSC 8864

59.0.3 Trade Names: NA

59.0.4 Molecular Weight: 160.17

59.0.5 Molecular Formula: $C_7H_{12}O_4$

59.0.6 Molecular Structure:

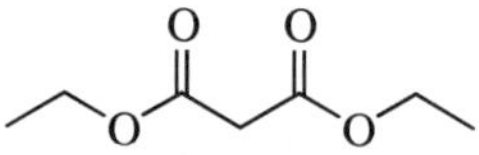

59.1 Chemical and Physical Properties

59.1.1 General

The empirical formula of diethyl malonate is $C_7H_{12}O_4$. The structural formula is $C_2H_5OOCCH_2COOC_2H_5$. It is a clear liquid that is soluble in alcohol and ether and slightly soluble in water. It exists as a liquid between its pour point (−50°C) and boiling point (199°C). Physico-chemical properties are summarized in Table 80.11.

59.1.2 Odor and Warning Properties

Pleasant odor.

59.2 Production and Use

Diethyl malonate is used in the manufacture of some pharmaceutical agents and as an analytic agent.

59.3 Exposure Assessment

Inhalation exposure would be expected to be minimal owing to its low vapor pressure.

59.4 Toxic Effects

59.4.1 *Experimental Studies*

59.4.1.1 Acute Toxicity. Diethyl malonate has a low acute toxicity in the rat and rabbit, with oral and dermal LD_{50} values of 14.9 mL/kg and >16 mL/kg, respectively (335).

59.4.1.2 Chronic and Subchronic Toxicity. No information found.

59.4.1.3 Pharmacokinetics, Metabolism, and Mechanisms. Diethyl malonate is hydrolyzed to ethanol and malonic acid. The ethanol is rapidly catabolized and the malonic acid converted to its sodium or calcium salt, both of which are more highly soluble than calcium oxalate and thus readily metabolized through the C1 carbon unit or fatty acid pathways or directly excreted (227). Should malonic acid occur in excessively high quantities, it is capable of altering the activity of some enzyme systems such as xanthine oxidase in the liver (347), with the consequence of increased urinary diacid excretion.

59.4.1.4 Reproductive and Developmental. No information found.

59.4.1.5 Carcinogenesis. No information found.

59.4.1.6 Genetic and Related Cellular Effects Studies. No information found.

59.4.1.7 Other: Neurological, Pulmonary, Skin Sensitization. Diethyl malonate (10 mL/kg) produces no effects or local irritation to guinea pig skin (12).

59.4.2 *Human Experience*

No information found.

59.5 Standards, Regulations, or Guidelines of Exposure

No information found.

59.6 Studies on Environmental Impact

No information found.

60.0 Diethyl Succinate

60.0.1 *CAS Number: [123-25-1]*

60.0.2 Synonyms: Butanedioic acid, diethyl ester; diethyl butanedioate; diethyl ethanedicarboxylate; ethyl succinate; succinic acid, diethyl ester; diethylester kyseliny jantarove [Czech]; 4-02-00-01914 [Beilstein]; AI3-00682; BRN 0907645; CTFA 05393; EINECS 204-612-0; FEMA No. 2377 [NLM]; and NSC 8875

60.0.3 Trade Names: Clorius

60.0.4 Molecular Weight: 174.20

60.0.5 Molecular Formula: $C_8H_{14}O_4$

60.0.6 *Molecular Structure:*

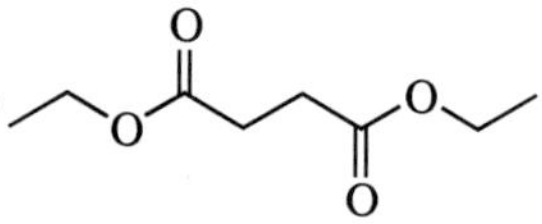

60.1 Chemical and Physical Properties

60.1.1 *General*

The empirical formula of diethyl succinate is $C_8H_{14}O_4$. The structural formula is $C_2H_5OOCCH_2CH_2COOC_2H_5$. It is a colorless, mobile liquid that is soluble in alcohol and ether but insoluble in water. It exists as a liquid between its pour point (−20°C) and boiling point (217°C). Physico-chemical properties are summarized in Table 80.11.

60.1.2 *Odor and Warning Properties*

Diethyl succinate has faint, pleasant odor.

60.2 Production and Use

Alkyl succinates have been used as ashless dispersants and detergent additives for lubricating oils.

60.3 Exposure Assessment

No information found.

60.4 Toxic Effects

60.4.1 *Experimental Studies*

60.4.1.1 Acute Toxicity. The oral LD_{50} value for rats is 8.53 mg/kg (7). Concentrations of greater than 10 g/mL are required to cause irritation to guinea pig skin (12).

60.4.2 *Human Experience*

No information found.

60.5 Standards, Regulations, or Guidelines of Exposure

No information found.

60.6 Studies on Environmental Impact

No information found.

61.0 Dipropyl Succinate

61.0.1 CAS Number: [925-15-5]

61.0.2 Synonyms: Butanedioic acid, dipropyl ester; di-*n*-propyl succinate; succinic acid, dipropyl ester; dipropylester kyseliny jantarove [Czech]; 4-02-00-01916 [Beilstein]; AI3-05512; BRN 1779461; and EINECS 213-114-2

61.0.3 Trade Names: NA

61.0.4 Molecular Weight: 202.25

61.0.5 Molecular Formula: $C_{10}H_{18}O_4$

61.0.6 Molecular Structure:

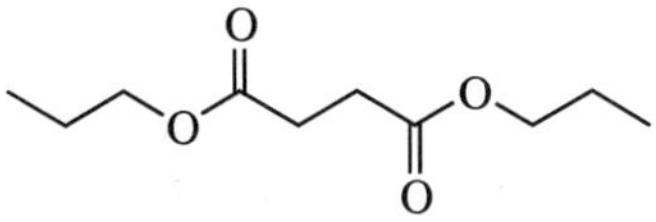

61.1 Chemical and Physical Properties

61.1.1 General

The empirical formula of dipropyl succinate is $C_{10}H_{18}O_4$. The structural formula is $C_3H_7OOCCH_2CH_2COOC_3H_7$. It is a clear, mobile liquid that is soluble in ether but insoluble in water. It exists as a liquid between its pour point (−5.9°C) and boiling point (250.8°C). Physico-chemical properties are summarized in Table 80.11.

61.1.2 Odor and Warning Properties

No information found.

61.2 Production and Use

Alkyl succinates have been used as ashless dispersants and detergent additives for lubricating oils.

61.3 Exposure Assessment

No information found.

61.4 Toxic Effects

61.4.1 Experimental Studies

61.4.1.1 Acute Toxicity. The oral and intraperitoneal LD_{50} values for rats are 6.49 and 0.29 g/kg, respectively (5, 336). Inhalation of concentrated vapor for greater than 8 h is required for lethality in rats (5). Concentrations of greater than 10 g/mL are required to cause irritation to guinea pig skin (12).

61.4.2 Human Experience

No information found.

61.5 Standards, Regulations, or Guidelines of Exposure

No information found.

61.6 Studies on Environmental Impact

No information found.

62.0 *n*-Dibutyl Succinate

62.0.1 CAS Number: [141-03-7]

62.0.2 Synonyms: Butanedioic acid, dibutyl ester; butyl butanedioate; di-*n*-butyl succinate; succinic acid, dibutyl ester; succinic acid di-*n*-butyl ester; di-*n*-butylester kyseliny jantarove [Czech]; 4-02-00-01916 [Beilstein]; AI3-00666; BRN 1786221; Casewell No. 293 [NLM]; EINECS 205-449-8; EPA Pesticide Chemical Code 077802; HSDB 1563; and NSC 1502 [NLM]

62.0.3 Trade Names: B-9; DNBS; and ENT 666

62.0.4 Molecular Weight: 230.31

62.0.5 Molecular Formula: $C_{12}H_{22}O_4$

62.0.6 Molecular Structure:

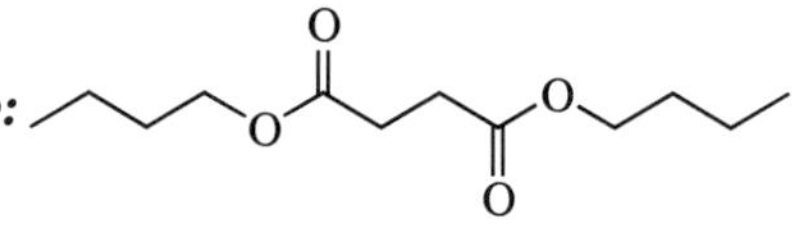

62.1 Chemical and Physical Properties

62.1.1 General

The empirical formula of dibutyl succinate is $C_{12}H_{22}O_4$. The structural formula is $C_4H_9OOCCH_2CH_2COOC_4H_9$. It is a mobile liquid that is soluble in alcohol, benzene, and ether, but insoluble in water. Its melting and boiling points are −29 and 274.5°C, respectively. The flash point is 275°F. Physico-chemical data are summarized in Table 80.11.

62.1.2 Odor and Warning Properties

When heated to decomposition, dibutyl succinate produces acrid smoke and fumes.

62.2 Production and Use

Dibutyl succinate is manufactured via esterification of butanol and succinic acid. It is used as an insect repellant against biting flies, ants, and roaches. It is also a mosquito attractant.

62.3 Exposure Assessment

No information found.

62.4 Toxic Effects

62.4.1 Experimental Studies

62.4.1.1 Acute Toxicity. The oral LD_{50} value for rats is 8 g/kg (337). No other acute toxicity data were located.

62.4.1.2 Chronic and Subchronic Toxicity. Rats fed a diet containing 5% dibutyl succinate suffered no ill effects (348).

62.4.2 Human Experience

Dibutyl succinate is not irritating to the skin or mucous membranes (171).

62.5 Standards, Regulations, or Guidelines of Exposure

No information found.

62.6 Studies on Environmental Impact

No information found.

63.0 Di-(2-ethylhexyl) Succinate

63.0.1 CAS Number: *[2915-57-3]*

63.0.2 Synonyms: *bis*(2-Ethylhexyl) butanedioate; *bis*(2-ethylhexyl) succinate; butanedioic acid, *bis*(2-ethylhexyl)ester; di-(2-ethylhexyl)butanedioate; di-(2-ethylhexyl) succinate; dioctyl succinate; AI3-00998; CTFA 00874; and EINECS 220-836-1

63.0.3 Trade Names: NA

63.0.4 Molecular Weight: 342.5

63.1 Chemical and Physical Properties

63.1.1 General

The empirical formula of di(2-ethylhexyl) succinate is $C_{20}H_{38}O_4$. It exists as a liquid between its pour point (−60°C) and boiling point (257°C at 50 mmHg). Physicochemical properties are summarized in Table 80.11.

63.1.2 Odor and Warning Properties

No information found.

63.2 Production and Use

No information found.

63.3 Exposure Assessment

No information found.

63.4 Toxic Effects

No information found.

63.5 Standards, Regulations, or Guidelines of Exposure

No information found.

63.6 Studies on Environmental Impact

No information found.

64.0 Di(2-hexyloxyethyl) Succinate

64.0.1 CAS Number: [10058-20-5]

64.0.2 Synonyms: Succinic acid, di-2-hexyloxyethyl ester; and succinic acid, *bis* (2-(hexyloxy)ethyl) ester

64.0.3 Trade Names: NA

64.0.4 Molecular Weight: 374.58

64.1 Chemical and Physical Properties

64.1.1 General

The empirical formula of di(2-hexyloxyethyl) succinate is $C_{20}H_{38}O_6$. The structural formula is $(CH_2COOCH_2CH_2OC_6H_{13})_2$. No other chemical or physical property data were located.

64.1.2 Odor and Warning Properties

No information found.

64.2 Production and Use

No information found.

64.3 Exposure Assessment

No information found.

64.4 Toxic Effects

64.4.1 Experimental Studies

The oral and dermal LD_{50} values for the rat and rabbit are 4.28 g/kg and 12.31 mL/kg, respectively (5). Out of a possible score of 10, it induces an irritation score of 3 in rabbits.

64.4.2 Human Experience

No information found.

64.5 Standards, Regulations, or Guidelines of Exposure

No information found.

64.6 Studies on Environmental Impact

No information found.

65.0 Dioctyl Sodium Sulfosuccinate

65.0.1 CAS Number: *[577-11-7]*

65.0.2 Synonyms: *bis* (Ethylhexyl) ester of sodium sulfosuccinic acid; (1,4)-*bis*(2-ethylhexyl) (*S*) sodium sulfosuccinate; *bis*(2-ethylhexyl) sulfosuccinate sodium salt; butanedioic acid sulfo-, 1,4-*bis*(2-ethylhexyl) ester, sodium; di-(2-ethylhexyl) sodium sulfosuccinate; di(2-ethylhexyl)sulfosuccinic acid, sodium salt; dioctyl ester of sodium sulfosuccinate; dioctyl ester of sodium sulfosuccinic acid; dioctyl sodium sulfosuccinate; dioctyl sulfosuccinate sodium; docusate sodium; 2-ethylhexylsufosuccinate sodium; sodium 1,4-*bis*(2-ethylhexyl) sulfosuccinate; sodium *bis*(2-ethylhexyl) sulfosuccinate; sodium di-(2-ethylhexyl)sulfosuccinate; sodium dioctyl sulfosuccinate; sodium dioctyl sulfosuccinate; sodium 2-ethylhexylsulfosuccinate; sodium-sulfodi-(2-ethylhexyl)-sulfo-succinate; succinic acid sulfo-1,4-*bis*(2-ethylhexyl)ester, sodium salt; sulfosuccinic acid *bis*(2-ethylhexyl)ester sodium salt; *bis*-2-ethylhexylester sulfojantaranu sodneho [Czech]; docusate sodique [French]; docusato sodico [Spanish]; docusatum natricum [Latin]; AI3-00239; CTFA 00873; EINECS 209-406-4; and HSDB 3065 aerosol OT; sulfo-butanedioic acid; colace; *bis*(2ethylhexyl) sodium sulfosuccinate; dioctyl sodium sulfonsuccinate

65.0.3 Trade Names: DSS; Aerosol GPG; Aerosol OT; Aerosol OT-B; Aerosol OT 75; Alcopol O Alphasol OT; Berol 478; Celanol DOS 75; Clestol; Colace; Complemix; Constonate; Coprol; Correctol Extra Gentle Tablets; Defilin; Dialose; Dioctylin; Dioctylal; Diomedicone; Diosuccin; Diotilan; Diovac; Diox; Disonate; Doxinate; Doxol; Dulsivac; Duosol; Humifen WT 27G; Konlax; Kosate; Laxinate (100); Manoxol OT; Manoxal OT; Mervamine; Mondane Soft; Molatoc; Molcer; Molofac; Monawet MD 70E, Mo-70 MO-84 R2W or MO-70 RP; Monoxol OT; Nekal WT-27; Nevax; Nikkol OTP 70; Norval; Obston; Rapisol; Regutol; Requtol; Revac; Sanmorin OT 70; SBO; Sobitol; Softil; Soliwax; Solusol (75% or 100%); Sulfimel DOS; SV 102; Tex-Wet 1001; Triton GR-5 or 7; Vastol OT; Velmol; Waxsol; and Wetaid SR

65.0.4 Molecular Weight: 445.56

65.0.5 Molecular Formula: $C_{20}H_{37}NaO_7S$

65.0.6 Molecular Structure:

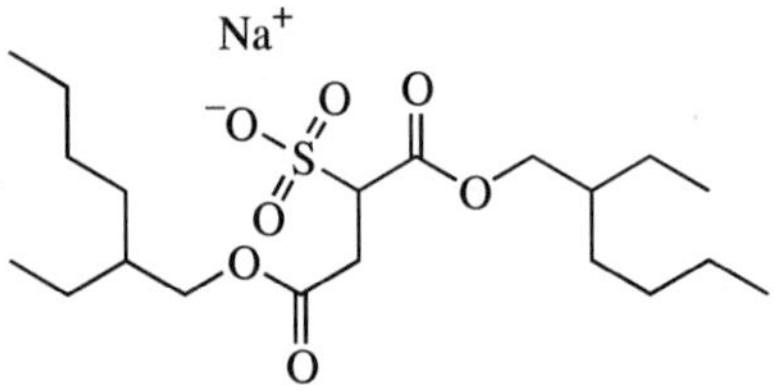

65.1 Chemical and Physical Properties

65.1.1 General

The empirical formula of dioctyl sodium sulfosuccinate is $C_{20}H_{37}NaO_7S$. The structural formula is $C_8H_{17}OOCCH_2CH(SO_3Na)COOC_8H_{17}$. It is a white waxlike solid or clear, viscous liquid. Dioctyl sodium sulfosuccinate is soluble in water (71 g/L), glycerin, carbon tetrachloride petroleum ether, naphtha, xylene, dibutyl phthalate, liquid petrolatum, acetone alcohol, and vegetable oils. It is stable in acid and neutral solutions and hydrolyzes in alkaline ones. No other physiochemical data were located.

65.1.2 Odor and Warning Properties

Dioctyl sodium sulfosuccinate has a "characteristic odor." When heated to decomposition, it emits toxic fumes.

65.2 Production and Use

Dioctyl sodium sulfosuccinate is present in a number of consumer products. It is used as a wetting agent, an adjuvant in tablets, and a dispersing or emulsifying agent in foods and dermatological preparations. It is an ingredient of some adhesives, polymeric coatings, paperboard, detergents, stool softeners, and vitamin preparations.

65.3 Exposure Assessment

No information found.

65.4 Toxic Effects

65.4.1 Experimental Studies

65.4.1.1 Acute Toxicity. Dioctyl sodium sulfosuccinate is of low toxicity orally, dermally, or by inhalation. The oral LD_{50} values for the rat and mouse are 1.9 and 2.64 g/kg, respectively (338, 339). Intraperitoneal and intravenous LD_{50} values are summarized in Table 80.12. Clinical signs of toxicity include anorexia, vomiting, and diarrhea.

65.4.1.2 Chronic and Subchronic Toxicity. In rabbits given 84 mg/kg/d dioctyl sodium sulfosuccinate orally for a period of 2 yr, gastrointestinal hypermotility and diarrhea were observed (349).

65.4.1.3 Pharmacokinetics, Metabolism, and Mechanisms. Dioctyl sodium sulfosuccinate is absorbed from the gastrointestinal tract and a significant amount is excreted unchanged in bile (350). Dioctyl sodium sulfosuccinate acts as a stool softener by lowering the surface tension of the stool. This, in turn, facilitates entry of water and lipids into stool. Dioctyl sodium sulfosuccinate may enhance the absorption of danthron, mineral oil, and phenolphthalein (352).

65.4.1.4 Reproductive and Developmental. Dietary administration of 0.5% and 1.0% dioctyl sodium sulfosuccinate to three successive generations of rats caused a reduction in body weights in males, females, and pups, but did not affect reproductive performance or fetal survival or morphology (353). Dioctyl sodium sulfosuccinate is not teratogenic in mice or rabbits (354). It is teratogenic in dogs at doses that produced maternal toxicity (354). However, use in humans has not demonstrated adverse effects during pregnancy (see Section 66.4.2).

65.4.1.5 Carcinogenesis. No information found.

65.4.1.6 Genetic and Related Cellular Effects Studies. No information found.

65.4.1.7 Other: Neurological, Pulmonary, Skin Sensitization. Dioctyl sodium sulfosuccinate caused mild and reversible damage to rabbit eyes at a concentration of 0.5–2.0%. Concentrations of 10% caused severe damage (355).

65.4.2 Human Experience

In humans, small quantities of dioctyl sodium sulfosuccinate may be hydrolyzed and degraded, partly through the citric acid cycle and thus through general metabolism (55). Use during pregnancy does not appear to be associated with an increased incidence of birth defects (356–358). In ophthalmological formulations, concentrations greater than 0.1% may cause conjunctival irritation. Repeated use may delay healing of corneal lesions (359).

65.5 Standards, Regulations, or Guidelines of Exposure

No information found.

65.6 Studies on Environmental Impact

No information found.

66.0 Diethyl Glutarate

66.0.1 CAS Number: *[818-38-2]*

66.0.2 Synonyms: Diethyl pentanedioate; glutaric acid diethyl ester; pentanedioic acid diethyl ester; AI3-06007 [NLM]; EINECS 212-451-2; and NSC 8890 [NLM]

66.0.3 Trade Names: NA

66.0.4 Molecular Weight: 188.22

66.0.5 *Molecular Formula:* $C_9H_{16}O_4$

66.0.6 *Molecular Structure:*

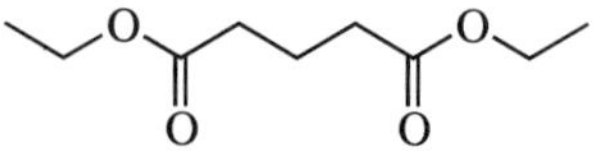

66.1 Chemical and Physical Properties

66.1.1 *General*

The empirical formula of diethyl glutarate is $C_9H_{16}O_4$. The structural formula is CH_2-$(CH_2COOC_2H_5)_2$. It exists as a liquid between its pour point (– 24.1°C) and boiling point (236.6°C). It is soluble in ether and slightly soluble in water and alcohol. Physicochemical properties are summarized in Table 80.11.

66.1.2 *Odor and Warning Properties*

No information found.

66.2 Production and Use

No information found.

66.3 Exposure Assessment

No information found.

66.4 Toxic Effects

66.4.1 *Experimental Studies*

66.4.1.1 Acute Toxicity. The oral LD_{50} value for the rat is > 1.6 g/kg (12). No systemic toxicity was observed in guinea pigs treated dermally with 10 mL/kg (12).

66.4.1.2 Chronic and Subchronic Toxicity. No information found.

66.4.1.3 Pharmacokinetics, Metabolism, and Mechanisms. When ingested, the glutarate is likely to hydrolyze readily into ethanol and glutaric acid. Subsequently, both hydrolyzates will readily enter the metabolism for normal oxidation or catabolism (55). This would then be analogous to ethyl 2-ketoglutarate, $C_2H_5OOCCH_2CH_2COCOOC_2H_5$, which passes through the plasma membrane and is also decarboxylated by yeast (360).

66.4.1.4 Reproductive and Developmental. No information found.

66.4.1.5 Carcinogenesis. No information found.

66.4.1.6 Genetic and Related Cellular Effects Studies. No information found.

66.4.1.7 Other: Neurological, Pulmonary, Skin Sensitization. Diethyl glutarate is a moderate skin irritant in the guinea pig (12).

66.4.2 *Human Experience*

No information found.

66.5 Standards, Regulations, or Guidelines of Exposure

No information found.

66.6 Studies on Environmental Impact

No information found.

67.0 Dimethyl Adipate

67.0.1 *CAS Number:* *[627-93-0]*

67.0.2 *Synonyms:* Hexanedioic acid dimethyl ester

67.0.3 *Trade Names:* NA

67.0.4 *Molecular Weight:* 174.20

67.0.5 *Molecular Formula:* $C_8H_{14}O_4$

67.0.6 *Molecular Structure:*

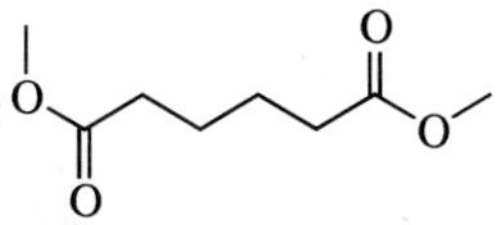

67.1 Chemical and Physical Properties

67.1.1 *General*

The empirical formula of dimethyl adipate is $C_8H_{14}O_4$. Its structural formula is $(CH_2CH_2COOCH_3)_2$. At room temperature, it is a liquid with a density of 1.06. It solidifies at 8°C and boils at 115°C. It is insoluble in water. Physico-chemical properties are summarized in Table 80.13.

67.1.2 *Odor and Warning Properties*

No information found.

67.2 Production and Use

Dimethyl adipate is manufactured via esterification of adipic acid and methanol in the presence of an acid catalyst. It is used as a plasticizer for cellulose-type resins.

67.3 Exposure Assessment

No information found.

67.4 Toxic Effects

67.4.1 Experimental Studies

67.4.1.1 Acute Toxicity. The intraperitoneal LD_{50} value for the rat is 1.809 mL/kg (325).

67.4.1.2 Chronic and Subchronic Toxicity. No information found.

67.4.1.3 Pharmacokinetics, Metabolism, and Mechanisms. No information found.

67.4.1.4 Reproductive and Developmental. Dimethyl adipate (0.0603–0.6028 mL/kg) was administered intraperitoneally to female rats on days 5, 10, and 15 of gestation. Dimethyl adipate had no effect on the number of corporate lutea, fetal viability, or fetal weight at either dose. An increase in resorptions was noted in rats treated with 0.1809 mL/kg dimethyl adipate, but not at other dose levels. Hemangiomas were found in 1.8%, 3.6%, and 8.0% of fetuses from rats treated with 0.1809, 0.3617, or 0.6028 mL/kg dimethyl adipate, respectively. A dose-dependent increase in skeletal abnormalities (mainly elongated and fused ribs) was also noted. Visceral abnormalities (missing left kidney and angulated anal opening) were found in two (8.3%) fetuses from rats treated with the highest dose level. The no-observed-effect level was 0.06028 mL/kg (325).

67.4.1.5 Carcinogenesis. No information found.

67.4.1.6 Genetic and Related Cellular Effects Studies. No information found.

67.4.1.7 Other: Neurological, Pulmonary, Skin Sensitization. Dimethyl adipate has been shown to cause mild degeneration of the olfactory, but not respiratory epithelium in rats (361). The irritant effect of adipic esters on the skin and eyes is very slight (12).

67.4.2 Human Experience

No information found.

67.5 Standards, Regulations, or Guidelines of Exposure

No information found.

67.6 Studies on Environmental Impact

Information on the extent to with dimethyl adipate hydrolyzes in water or soil is not available. It is expected to hydrolyze slowly (half-life of years) (362). After leaching into groundwater, it is not likely to evaporate or absorb strongly to sediments (363). Because of its low vapor pressure, entry into the atmosphere is unlikely (363). That which does enter the atmosphere should degrade with a half-life of 2.7 d (364).

68.0 Diethyl Adipate

68.0.1 CAS Number: [141-28-6]

68.0.2 Synonyms: Adipic acid, diethyl ester; (1,6)-diethyl hexanedioate; ethyl adipate; ethyl delta-carboethoxyvalerate; hexanedioic acid, diethyl ester; diethylester kyseliny adipove [Czech]; 4-02-00-01960 [Beilstein]; AI3-00342 [NLM]; BRN 1780035 [RTECS]; EINECS 205-477-0; HSDB 5413; and NSC 19160 [NLM]

68.0.3 Trade Names: NA

68.0.4 Molecular Weight: 202.25

68.0.5 Molecular Formula: $C_{10}H_{18}O_4$

68.0.6 Molecular Structure:

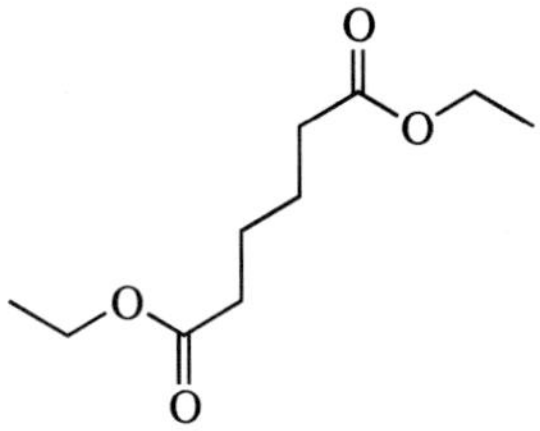

68.1 Chemical and Physical Properties

68.1.1 General

The empirical formula of diethyl adipate is $C_{10}H_{18}O_4$. Its structural formula is $(CH_2CH_2COOC_2H_5)_2$. It exists as a liquid between its pour point (– 18°C) and boiling point (251°C). It is soluble in alcohol in ether and insoluble in water. Physico-chemical properties are summarized in Table 80.13.

68.1.2 Odor and Warning Properties

No information found.

68.2 Production and Use

Diethyl adipate is manufactured by esterifying adipic acid with ethanol. It is used as a plasticizer and intermediate for production of cyclopentanone and putrescine.

68.3 Exposure Assessment

No information found.

68.4 Toxic Effects

68.4.1 Experimental Studies

68.4.1.1 Acute Toxicity. The oral LD_{50} value for the mouse is 8.1 g/kg (365). The IP LD_{50} values for the rat and mouse are 2.5 and 2.19 mL/kg, respectively (325, 366).

68.4.1.2 Chronic and Subchronic Toxicity. No information found.

68.4.1.3 Pharmacokinetics, Metabolism, and Mechanisms. No information found.

68.4.1.4 Reproductive and Developmental. The incidence of gross and skeletal abnormalities in fetuses from rats treated intraperitoneally with 0.8373 mL/kg diethyl adipate on days 5, 10 and 15 of gestation was slightly greater than that of controls (325). The no-effect level was 0.5024 mL/kg. Male mice injected intraperitoneally with 1.1 or 1.46 mL/kg or diethyl adipate prior to mating had decreased fertility and an increased incidence of dominant lethal mutations, as evidenced by an increased number of early fetal deaths (381).

68.4.1.5 Carcinogenesis. No information found.

68.4.1.6 Genetic and Related Cellular Effects Studies. In male rodents, diethyl adipate (1.1 or 1.46 mL/kg) was positive in a dominant lethal test (366).

68.4.1.7 Other: Neurological, Pulmonary, Skin Sensitization. The irritant effect of adipic esters on the skin and eye is very slight (12).

68.4.2 Human Experience

No information found.

68.5 Standards, Regulations, or Guidelines of Exposure

No information found.

68.6 Studies on Environmental Impact

No information found.

69.0 *n*-Dibutyl Adipate

69.0.1 CAS Number: [105-99-7]

69.0.2 Synonyms: Adipic acid, dibutyl ester; butyl adipate; dibutyladipinate; dibutyl hexanedioate; hexanedioic acid, dibutyl ester; dibutylester kyseliny adipove [Czech]; 4-02-00-01961 [Beilstein]; AI3-00671 [NLM]; BRN 1790739; CTFA 00769; EINECS 203-350-4; and NSC 8086 [NLM]

69.0.3 Trade Names: Cetiol B; 3PS; Experimental tick repellant 3; Experimental tick repellant 3PS; and Unitolate B

69.0.4 Molecular Weight: 258.36

69.0.5 Molecular Formula: $C_{14}H_{26}O_4$

69.0.6 Molecular Structure:

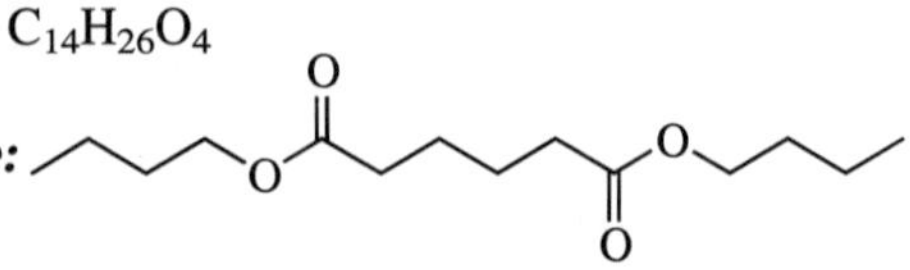

69.1 Chemical and Physical Properties

69.1.1 *General*

The empirical formula of dibutyl adipate is $C_{14}H_{26}O_4$. Its structural formula is $(CH_2CH_2COOC_4H_9)_2$. It exists as a liquid between its pour point (– 32.4°C) and boiling point (165°C at 10 mmHg). Its flash point is > 230°F. Dibutyl adipate is soluble in alcohol and ether and insoluble in water. Physico-chemical properties are summarized in Table 80.13.

69.1.2 *Odor and Warning Properties*

No information found.

69.2 Production and Use

Dibutyl adipate is used in cosmetic formulations as a plasticizer and skin-conditioning agent (367, 368).

69.3 Exposure Assessment

No information found.

69.4 Toxic Effects

69.4.1 *Experimental Studies*

69.4.1.1 Acute Toxicity. The oral LD_{50} values for rats and mice are 12.9 and 16.89 g/kg, respectively (7, 369). The dermal LD_{50} value for rabbits is 20 mL/kg (7). Inhalation of a concentration greater than 17 mg/m^3 was required to cause lethality in rats and mice (369). Acute toxicity data are summarized in Table 80.14.

69.4.1.2 Chronic and Subchronic Toxicity. The subchronic dermal toxicity of dibutyl adipate has been examined in a study employing 10 rabbits/group. Rabbits given topical applications of 1.0 mL/kg/d for 6 wk gained less weight than controls. One rabbit from the 1.0 mL/kg/d group had slight, cloudy hepatic swelling and slight, cloudy swelling of the renal convoluted and loop tubules, and one rabbit from the 0.5 mL/kg/d group had similar renal lesions. When an emulsion containing 6.25% dibutyl adipate was applied to the entire body twice a week for 3 mo, no adverse effects were noted in dogs (377).

69.4.1.3 Pharmacokinetics, Metabolism, and Mechanisms. No information found.

69.4.1.4 Reproductive and Developmental. In a developmental study in which pregnant rats were given intraperitoneal injections of 0.1748, 0.5244, 1.0488, and 1.7480 mL/kg dibutyl adipate, a significant increase in gross fetal abnormalities was observed in offspring from dams treated with the highest dose level only (325). The no-effect level for teratogenicity was 1.0488 mL/kg.

69.4.1.5 Carcinogenesis. No information found.

69.4.1.6 Genetic and Related Cellular Effects Studies. Dibutyl adipate is not toxic to cultured HeLa cells (370).

69.4.1.7 Other: Neurological, Pulmonary, Skin Sensitization. A single application of 0.01 mL undiluted dibutyl adipate to the clipped underside of rabbits caused irritation of grade 2 (out of a maximum of 8) (7). Multiple application of 0.01 mL (8 times over a 4-h period), or 0.025 mL (3 times a day over a 3-d period) caused moderate erythema (370). Application of cloth impregnated with dibutyl adipate was not irritating to rabbit skin (370). Undiluted dibutyl adipate received a primary irritation score of 1 out of 110 in an ocular toxicity study in rabbits (7).

69.4.2 Human Experience

No information found.

69.5 Standards, Regulations, or Guidelines of Exposure

No information found.

69.6 Studies on Environmental Impact

No information found.

70.0 Di(2-ethylbutyl) Adipate

70.0.1 CAS Number: *[10022-60-3]*

70.0.2 Synonyms: Hexanedioic acid, di-2-ethylbutyl ester; adipic acid, di(2-ethylbutyl) ester; adipic acid, *bis*(2-ethylbutyl) ester; and *bis*-(2-ethylbutyl)ester kyseliny adipove [Czech]

70.0.3 Trade Names: NA

70.0.4 Molecular Weight: 314.46

70.0.5 Molecular Formula: $C_{18}H_{34}O_4$

70.0.6 Molecular Structure:

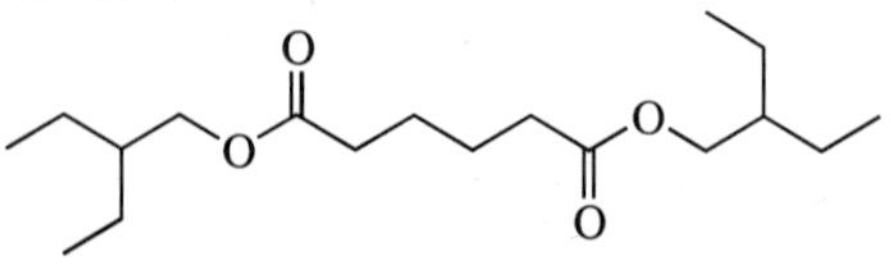

70.1 Chemical and Physical Properties

70.1.1 General

The empirical formula of di(2-ethylbutyl) adipate is $C_{18}H_{34}O_4$. Its structural formula is $[CH_2CH_2COOCH_2CH(C_2H_5)]_2$. It exists as a liquid between its pour point (– 15°C) and boiling point (200°C at 10 mmHg). It is soluble in alcohol and insoluble in water. Physicochemical properties are summarized in Table 80.13.

70.1.2 Odor and Warning Properties

No information found.

70.2 Production and Use

Adipic esters are used as plasticizers and are present in some cosmetics.

70.3 Exposure Assessment

No information found.

70.4 Toxic Effects

70.4.1 Experimental Studies

The oral and dermal LD_{50} values for the rat and rabbit are 5.62 g/kg and 16.8 mL/kg, respectively (5). It is nonirritating to rabbit skin and eyes.

70.4.2 Human Experience

No information found.

70.5 Standards, Regulations, or Guidelines of Exposure

No information found.

70.6 Studies on Environmental Impact

No information found.

71.0 Di(2-ethylhexyl) Adipate

71.0.1 CAS Number: *[103-23-1]*

71.0.2 Synonyms: Hexanedioic acid, *bis*(2-ethylhexyl) ester; dioctyl adipate; and DEHA

71.0.3 Trade Names: Eastman DOA plasticizer; Morflex 310; PX-238; and Plastomoll DOA

71.0.4 Molecular Weight: 370.57

71.0.5 Molecular Formula: $C_{22}H_{42}O_4$

71.0.6 Molecular Structure:

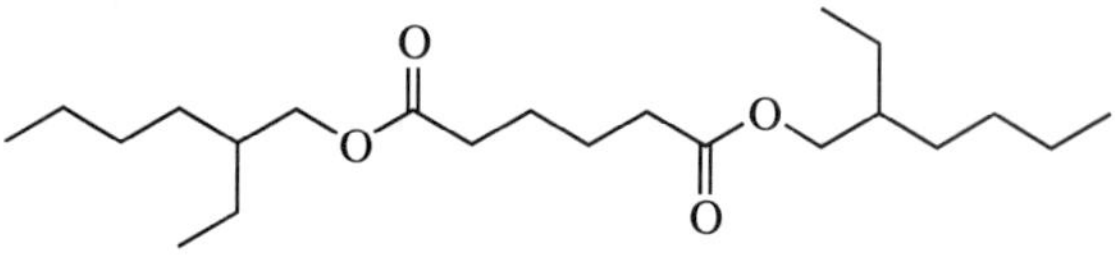

71.1 Chemical and Physical Properties

71.1.1 General

The empirical formula of di(2-ethylhexyl) adipate (DEHA) is $C_{22}H_{42}O_4$. Its structural formula is $[CH_2CH_2COOCH_2CH(C_2H_5)(CH_2)_3CH_3]_2$. It is a colorless to light colored liquid (379) over a wide range of temperatures ranging from $-67.8°C$ (melting point) to 417°C (boiling point). It is virtually insoluble in water (<0.78 mg/L), has a low vapor pressure (4.7×10^{-7} mmHg at 20°C), and has a specific gravity of 0.928. Physicochemical properties are summarized in Table 80.13.

71.1.2 Odor and Warning Properties

No information found.

71.2 Production and Use

DEHA is manufactured by esterification of adipic acid and 2-ethylhexanol. The compound is used as a plasticizer for flexible vinyl foodwraps and as a solvent for aircraft lubricants.

71.3 Exposure Assessment

Values for occupational exposure are not available. Consumer exposure to DEHA via migration into food from cling-film wrap has been reported for human populations in the United States (380), Canada (381), the United Kingdom (382, 383), and elsewhere (384–386). In general, higher levels of DEHA are found in food with high fat content. Estimates of consumer exposure to DEHA are generally low (below the tolerable daily intake levels).

The urinary level of 2-ethylhexanoic acid has been used as a biomarker for exposure (387). It is likely that 2-ethylhexanol and 2-ethylhexanoic acid would also be present in the blood as markers of exposure.

71.4 Toxic Effects

71.4.1 Experimental Studies

71.4.1.1 Acute Toxicity. Oral treatment of rats with DEHA has not resulted in mortality or clinical signs of toxicity except at dose levels greater than the current 2-g/kg limit dose level (388, 389). The oral LD_{50} value for rats is 9.1 g/kg (5). Dermal exposure to doses in excess of the current limit dose level of 2 g/kg does not result in lethality or clinical signs of toxicity (7, 388). The dermal LD_{50} value for rabbits is 16.3 mL/kg (15.029 g/kg) (5).

71.4.1.2 Chronic and Subchronic Toxicity. DEHA has been evaluated for systemic toxicity in 14-, 21-, and 90-day oral feeding studies in rats and mice. The data indicate that repeated exposure of animals to DEHA (up to 90 d) resulted in reduced body weight gain for rats at dose levels of 6300 ppm and higher, and for mice at 3100 ppm and higher. Hepatic hypertrophy and increased peroxisomal enzyme activity occurred in rats and mice within 1 wk of treatment with 12,000 ppm; however, there were no adverse effects on the liver (392–394). No effects were observed at 2500 ppm, which is equivalent to 189 mg/kg for rats and 451 mg/kg for mice (394).

71.4.1.3 Pharmacokinetics, Metabolism, and Mechanisms. DEHA is hydrolyzed almost completely in the gastrointestinal tract to 2-ethylhexanol (2-EH) and adipic acid (395–396). In rodents, peak blood levels of 2-EH or ethylhexanoic acid (EHA) have been observed 3 h after oral exposure (395) and 2 h after oral exposure for primates and humans (397, 398).

Disposition studies in mice indicated that 95–102% of the administered radioactivity from the low- and mid-dose groups was eliminated in the urine, feces, and expired air within 24 h postdosing. Approximately 90% was excreted in the urine and 7–8% in the feces. At the high-dose level, ~12% was recovered in the gastointestinal tract, 75% excreted in the urine, and 4% excreted in the feces. Rats and monkeys also eliminated the majority of radioactivity in the urine, but had a higher fecal elimination than did mice (397). Urinary metabolites consisted of 2-ethylhexanoic acid (EHA), its glucuronide conjugate, 5-hydroxy-EHA, and 2-ethyl-1,6-hexanedioic acid. Rats excreted less glucuronide conjugated metabolite than mice or monkeys. Monkeys excreted more MEHA, EHA, and glucuronide-conjugated metabolites than did rodents.

71.4.1.4 Reproductive and Developmental. The reproductive toxicity of DEHA was evaluated in a one-generation reproduction study using rats (399). In pregnant dams exposed to 12,000 ppm DEHA, the only effect noted was a reduction in body weight gains of pregnant dams and F_1 pups. There were no adverse effects on the reproductive organs of either generation. Singh et al. reported decreased fertility in male mice treated with a single dose of DEHA by IP injection during a dominant lethal study (366). In that study, doses of 5 mL/kg (4610 mg/kg) and 10 mL/kg (9220 mg/kg) resulted in a decrease in fertility, fewer implants, and higher fetal mortality compared with the control group. A dose level of 922 mg/kg had no effect.

The developmental toxicity of DEHA has been evaluated following IP injection and oral exposure in the diet. Singh et al. treated pregnant rats with DEHA (1, 5, and 10 mL/kg) by IP injection on gestation days 5, 10, and 15. There was a slight decrease in the mean fetal weight at the two highest dose levels, and a slight increase in the incidence of gross fetal abnormalities at the highest dose (325). Another study in which pregnant rats were treated with DEHA in the diet throughout gestation also demonstrated reduced body weight gain at the high dose of 1080 mg/kg/d (400). There was evidence of preimplantation fetal loss at 1080 mg/kg/d, but no gross, skeletal, or visceral abnormalities. A dose of 170 mg/kg/d was considered to be slightly fetotoxic because of reduced ossification (which was not statistically significant). A dose of 28 mg/kg/d was determined to be the NOEL.

The estrogenicity of DEHA has been tested *in vitro* (401). At concentrations near the level of solubility, DEHA demonstrated no more than 40% binding to the estrogen receptor. Furthermore, DEHA did not activate the estrogen receptor in breast cancer cells. Based on these results, DEHA would not be expected to act as an estrogen.

The potential for DEHA to cause endocrine modulation was assessed in ten ovariectomized rats treated orally with 1 g/kg/d DEHA for 3 d. Uterine weights and progesterone receptor numbers were not increased, demonstrating that DEHA does not have estrogenic activity (402).

71.4.1.5 Carcinogenesis. DEHA has been evaluated for its carcinogenic potential in rats and mice. No evidence of carcinogenicity was found in rats following oral treatment for 104 wk with 12,000 or 25,000 ppm DEHA. In female mice treated with 12,000 or 25,000 ppm (estimated dose of 3222 or 8623 mg/kg, respectively) there was a significantly higher incidence of hepatocellular neoplasms. In male mice treated with 12,000 or 25,000 ppm DEHA (estimated dose of 2659 or 6447 mg/kg, respectively) evidence was suggestive of carcinogenicity (403). There was no carcinogenic activity when DEHA was applied to the skin of mice (404).

71.4.1.6 Genetic and Related Cellular Effects Studies. Genotoxicity studies *in vitro* have been negative for mutations, unscheduled DNA synthesis, and DNA interaction. No mutations were found when DEHA was incubated with TA98, TA100, TA1535, TA1537, or TA1538 strains of *S. typhimurium* with and without metabolic activation (405, 406). In addition, there was no evidence of mammalian cell mutation in mouse lymphoma assays conducted with and without metabolic activation (407, 408). No chromatid exchange was observed in an SCE assay using Chinese hamster ovary cells with or without activation (409), and there was no induction of unscheduled DNA synthesis in primary rat hepatocytes incubated with DEHA (410).

In vivo studies for genotoxicity also have been negative. Results of two independent mouse micronucleus assays using doses as high as 5 g/kg suggest that DEHA does not interact with DNA (411, 412). In a dominant-lethal study using mice, DEHA did not cause decreases in litter size that might suggest adverse effects on spermatogenesis (366).

Animal data indicate that the liver is a target for DEHA toxicity, and that repeated exposure of rodents to DEHA results in the induction of enzymes associated with the organelles called peroxisomes (392–394, 413, 414). The increased size and number of these organelles (peroxisome proliferation) has been associated with carcinogenesis in rodents (415). This mechanism may be linked to induction of enzymes that produce hydrogen peroxide (oxidative stress), mitogenic responses (cell proliferation), or interference with programmed cell death (apoptosis). In any case, the applicability of rodent liver cancer at high dose levels to humans via this mechanism has been questioned (416) because primates and human cells are refractory to the effects of DEHA and other peroxisome proliferators (396, 417–420). Recent studies have shown that a nuclear receptor (PPARα) plays a central role in the carcinogenic process in rodents (421). The expression of this receptor in human liver is lower than in rodent liver, suggesting that receptor density is a key element for species specificity.

71.4.1.7 Other: Neurological, Pulmonary, Skin Sensitization. Data for acute inhalation exposure are limited, but suggest that there are no adverse effects from prolonged exposure to DEHA vapors (7). Dermal irritation following prolonged exposure (24 h) was slight (424), but a shorter exposure period did not result in any signs of irritation (7). DEHA was not irritating to the eye and was not a dermal sensitizer (424).

71.4.2 Human Experience

Unconjugated EHA was detected in the plasma of 6 human volunteers given oral ^{2}H-DEHA, but there was no evidence of absorption of the parent DEHA. EHA could not be detected after 31 h. In the urine, a conjugated form of EHA was detected as the primary metabolite accounting for 99% of the total deuterium measured, and about 12% of the administered dose. The feces contained only a small amount of the administered dose (398).

A limited population of 112 individuals was given food wrapped in cling-film containing ^{2}H-DEHA, and the amount of urinary EHA measured as an indication of the dose received. Based on the results, the estimated daily intake of 8.2 mg/person/d appears to be correct (387).

71.5 Standards, Regulations, or Guidelines of Exposure

No information found.

71.6 Studies on Environmental Impact

No information found.

72.0 *n*-Dioctyl Adipate

72.0.1 CAS Number: *[123-79-5]*

72.0.2 Synonyms: Adipic acid, dioctyl ester; dicaprylyl adipate; di-*n*-octyl adipate; dioctyl ester hexanedioic acid; dioctyl hexanedioate; hexanedioic acid, dioctyl ester; octyl adipate; AI3-17824 [NLM]; EINECS 204-652-9; HSDB 366; and NSC 16201 [NLM]

72.0.3 Trade Names: Adimoll do

72.0.4 Molecular Weight: 370.57

72.0.5 Molecular Formula: $C_{22}H_{42}O_4$

72.0.6 Molecular Structure:

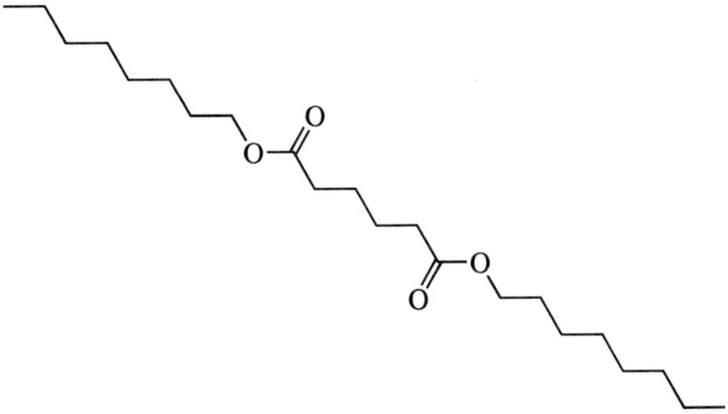

72.1 Chemical and Physical Properties

72.1.1 General

The empirical formula of dioctyl adipate is $C_{22}H_{42}O_4$. Its structural formula is $[CH_2CH_2COO(CH_2)_7CH_3]_2$. It is a clear liquid that is normally stable and does not react with water. Physico-chemical properties that were located are summarized in Table 80.13.

72.1.2 Odor and Warning Properties

No information found.

72.2 Production and Use

Dioctyl adipate is manufactured via esterification of adipic acid with octanol. It is used as a plasticizer for synthetic rubbers, nitrocellulose, and ethyl cellulose.

72.3 Exposure Assessment

No information found.

72.4 Toxic Effects

72.4.1 Experimental Studies

72.4.1.1 Acute Toxicity. In rats, the no-effect level for dioctyl adipate is 6 g/kg (425). Toxic doses resulted in CNS stimulation, then depression, with the signs lasting 5–7 d. In guinea pigs, an oral dose of 20 mL/kg caused death to all animals in ensuing weeks. Death was preceded by progressive infeeblement and muscular atrophy (370).

72.4.1.2 Chronic and Subchronic Toxicity. In rats orally administered 0.4, 1.0, or 2.0 g/kg dioctyl adipate for 6 mo, sulfhydryl compounds in blood were increased. Hepatic detoxification appeared depressed at the beginning, but accelerated after 6 mo (425).

72.4.1.3 Pharmacokinetics, Metabolism, and Mechanisms. No information found.

72.4.1.4 Reproductive and Developmental. No information found.

72.4.1.5 Carcinogenesis. No information found.

72.4.1.6 Genetic and Related Cellular Effects Studies. No information found.

72.4.1.7 Other: Neurological, Pulmonary, Skin Sensitization. No information found.

72.4.2 Human Experience

No information found.

72.5 Standards, Regulations, or Guidelines of Exposure

No information found.

72.6 Studies on Environmental Impact

No information found.

73.0 Diisooctyl Adipate

73.0.1 CAS Number: *[1330-86-5]*

73.0.2 Synonyms: Adipic acid, diisooctyl ester; diisooctyl hexanedioate; dimethyl heptyladipate; di-iso-octyl adipate; hexanedioic acid, diisooctyl ester; isooctyl adipate; EINECS 215-553-5; and HSDB 5813

73.0.3 Trade Names: Adipol 10A and Dioa PX 208

73.0.4 Molecular Weight: 370.57

73.0.5 Molecular Formula: $C_{22}H_{42}O_4$

73.0.6 Molecular Structure:

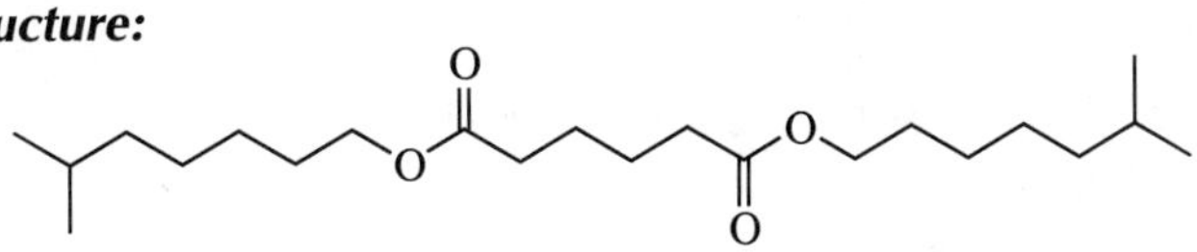

73.1 Chemical and Physical Properties

73.1.1 General

The empirical formula of diisooctyl adipate is $C_{22}H_{42}O_4$. Its structural formula is $[CH_2CH_2COOCH_2CH(CH_3)(C_5H_{11})]_2$. It is a clear liquid between its pour point (− 70°C) and boiling point (205–220°C at 4 mmHg). Physico-chemical properties are summarized in Table 80.13.

73.1.2 Odor and Warning Properties

No information found.

73.2 Production and Use

Diisoctyl adipate is manufactured by esterifying adipic acid with isooctanol and is used as a plasticizer for cellulose nitrate, ethyl cellulose, polymethyl methacrylate, polystyrene, and polyvinyl chloride.

73.3 Exposure Assessment

No information found.

73.4 Toxic Effects

No information found.

73.5 Standards, Regulations, or Guidelines of Exposure

No information found.

73.6 Studies on Environmental Impact

No information found.

74.0 Diisononyl Adipate

74.0.1 CAS Number: *[33703-08-1]* (US), *[90411-51-1]* (Europe). Other CAS numbers that may describe this substance include *[72379-06-7]*, *[68515-75-3]*, *[90411-53-3]*, and *[151-32-6]*

74.0.2 Synonyms: Hexanedioic acid, isononyl ester

74.0.3 Trade Names: JAYFLEX DINA; PX-239; and Plastomoll DNA

74.0.4 Molecular Weight: 398.63

74.0.5 Molecular Formula: $C_{24}H_{46}O_4$

74.0.5 Molecular Structure:

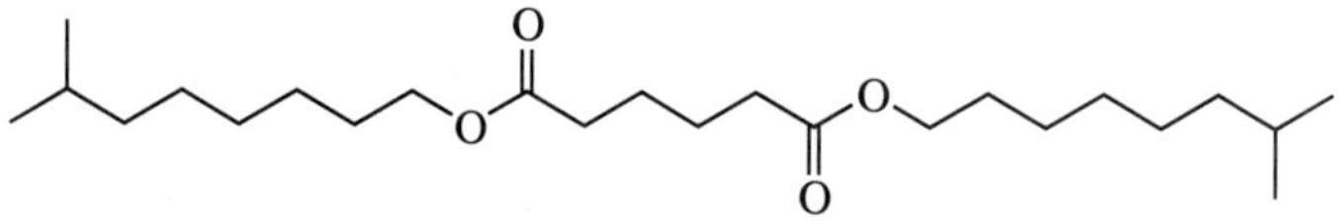

74.1 Chemical and Physical Properties

74.1.1 General

The empirical formula of diisononyl adipate (DINA) is $C_{24}H_{46}O_4$. DINA is a clear liquid. It exists in the liquid state between − 60°C (pour point) and 232–233°C (boiling point at 5 mmHg). The vapor pressure is 0.9 mmHg (200°C). DINA is essentially insoluble in water, but it is soluble in organic solvents. It has a specific gravity of 0.92 (at 20°C). Physico-chemical properties are summarized in Table 80.13.

74.1.2 Odor and Warning Properties

No information found.

74.2 Production and Use

DINA is produced by reacting di-isononyl alcohol with adipic acid in the presence of an acid catalyst. It is used predominantly as a plasticizer in such applications as surfactants, synthetic lubricants, inks, and lube oil additives.

74.3 Exposure Assessment

No information found.

74.4 Toxic Effects

74.4.1 Experimental Studies

74.4.1.1 Acute Toxicity. The rat oral LD_{50} value for DINA is > 10 g/kg and the rabbit dermal LD_{50} value is > 3.16 g/kg (371, 372).

74.4.1.2 Chronic and Subchronic Toxicity. In a 13-wk toxicity study, rats were given DINA by dietary administration at levels equivalent to 0, 50, 150, or 500 mg/kg/d. At the 500-mg/kg level the kidney-to-body-weight ratio was elevated, but the absolute kidney weights were not changed. As there were no histological changes in the kidneys and no other changes of any kind, this effect was not considered to be adverse. The NOAEL was 500 mg/kg/d (426).

DINA was given to beagle dogs for 13 wk at dietary levels of 0, 0.3, 1.0, and 3.0% (increased to 6% at week 9). Effects seen at the high dose level included decreased body weight and food consumption, increased liver weights, elevated enzyme levels, and histologic changes in the liver and kidneys. The NOAEL was 1%, or approximately 274 mg/kg/d (427).

74.4.1.3 Pharmacokinetics, Metabolism, and Mechanisms. No information found.

74.4.1.4 Reproductive and Developmental. No information found.

74.4.1.5 Carcinogenesis. No information found.

74.4.1.6 Genetic and Related Cellular Effects Studies. The genotoxic potential of DINA was evaluated in a battery of *in vitro* assays including the Ames *S. typhimurium* (strains TA98, TA100, TA1535, TA1537, and TA1538)/mammalian microsome mutagenicity assay, the mouse lymphoma test, and two *in vitro* tests for cell transformation, the BALB/c 3T3 and the Syrian hamster embryo test. All these tests were negative (428).

74.4.1.7 Other: Neurological, Pulmonary, Skin Sensitization. DINA produced conjunctival irritation in the eyes of rabbits that cleared completely within 72 h. The maximum Draize score observed was 8 out of a possible 110. Group mean scores at 24, 48, and 72 h for the various indices were: 1.83, 1.0, and 0.17 for conjunctival redness. No chemosis or iridial or corneal effects were noted (429).

74.4.2 Human Experience

No information found.

74.5 Standards, Regulations, or Guidelines of Exposure

No workplace exposure limits have been established for DINA.

74.6 Studies on Environmental Impact

No information found.

75.0 *n*-Didecyl Adipate

75.0.1 CAS Number: *[105-97-5]*

75.0.2 Synonyms: Adipic acid, didecyl ester; di-*n*-decyl adipate; didecyl hexanedioate; hexanedioic acid, didecyl ester; CTFA 000777; EINECS 203-349-9; and NSC 4445 [NLM]

75.0.3 Trade Names: Polycizer 632

75.0.4 Molecular Weight: 426.68

75.0.5 Molecular Formula: $C_{26}H_{50}O_4$

75.0.6 Molecular Structure:

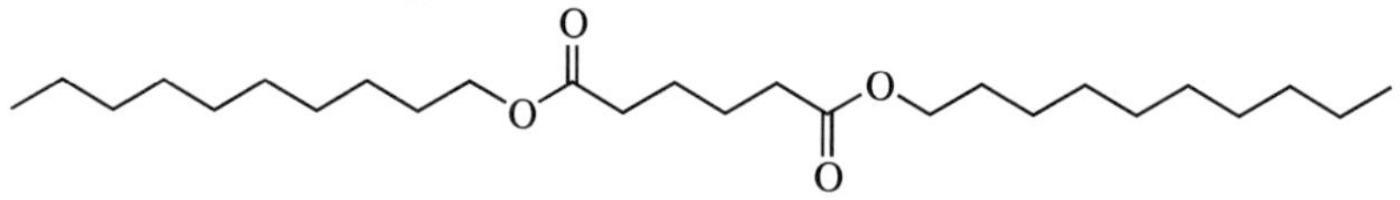

75.1 Chemical and Physical Properties

75.1.1 General

The empirical formula of didecyl adipate is $C_{26}H_{50}O_4$. Its structural formula is $(CH_2CH_2COOC_{10}H_{21})_2$. It is a liquid with a density of 0.919 and a boiling point of 240°C (at 4 mmHg). Physico-chemical properties are summarized in Table 80.13.

75.1.2 Odor and Warning Properties

No information found.

75.2 Production and Use

Plasticizer.

75.3 Exposure Assessment

No information found.

75.4 Toxic Effects

Didecyl adipate is of relatively low oral toxicity. The oral LD_{50} values for the rat and mouse are > 25.6 and 12.8–25.6 g/kg, respectively (12). Dermal application of > 10 mL/kg caused slight irritation in guinea pigs (12). All toxicological information found is summarized in Table 80.14.

75.5 Standards, Regulations, or Guidelines of Exposure

No information found.

75.6 Studies on Environmental Impact

No information found.

76.0 Dibutoxyethyl Adipate

76.0.1 CAS Number: [141-18-4]

76.0.2 Synonyms: Adipic acid, *bis*(2-butoxyethyl) ester; adipic acid, *bis*(ethylene glycol monobutyl ether) ester; adipic acid, dibutoxyethyl ester; *bis*(2-butoxyethyl) adipate; *bis*(2-butoxyethyl) hexanediote; *bis*(ethylene glycol monobutyl ether) adipate; butyl "cellosolve" adipate; di(2-butoxyethyl) adipate; dibutyl cellosolve adipate; hexanedioic acid, *bis*(2-butoxyethyl) ester; 4-02-00-01968 [Beilstein]; AI3-14614 [NLM]; BRN 1803966; EINECS 205-466-0; and NSC 4813 [NLM]

76.0.3 Trade Names: Adipol BCA and Staflex DBEA

76.0.4 Molecular Weight: 346.46

76.0.5 Molecular Formula: $C_{18}H_{34}O_6$

76.0.6 Molecular Structure:

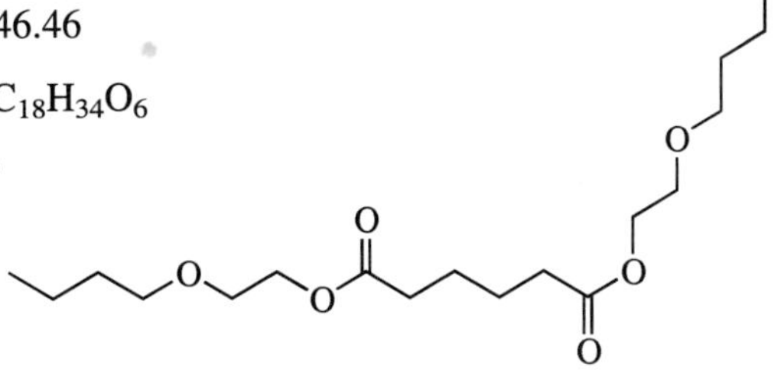

76.1 Chemical and Physical Properties

76.1.1 General

The empirical formula of dibutoxyethyl adipate is $C_{18}H_{34}O_6$. Its structural formula is $(CH_2CH_2COOCH_2CH_2OC_4H_9)_2$. It is a liquid, with a density of 0.997 and a boiling point of 208°C (at 4 mmHg). Physico-chemical properties that were located are summarized in Table 80.13.

76.1.2 Odor and Warning Properties

No information found.

76.2 Production and Use

Dibutoxyethyl adipate is used as a plasticizer for polyvinyl chloride and polyvinyl chloride–acetate copolymers.

76.3 Exposure Assessment

No information found.

76.4 Toxic Effects

The IP LD_{50} value for the rat is 0.6 g/kg (12). No other toxicological information was found.

76.5 Standards, Regulations, or Guidelines of Exposure

No information found.

76.6 Studies on Environmental Impact

No information found.

77.0 Di(2-hexyloxyethyl) Adipate

77.0.1 CAS Number: *[110-32-7]*

77.0.2 Synonyms: Adipic acid, *bis*(2-(hexyloxy)ethyl) ester; adipic acid, di(2-hexyloxyethyl) ester; *bis*(2-(hexyloxy)ethyl)adipate; dihexyloxyethyl adipate; hexanedioic acid di(2-hexyloxy)ethyl) ester; and hexanedioic acid, *bis*(2-hexyloxy)ethyl ester

77.0.3 Trade Names: NA

77.0.4 Molecular Weight: 402.57.

77.0.5 Molecular Formula: $C_{22}H_{42}O_6$

77.0.6 Molecular Structure:

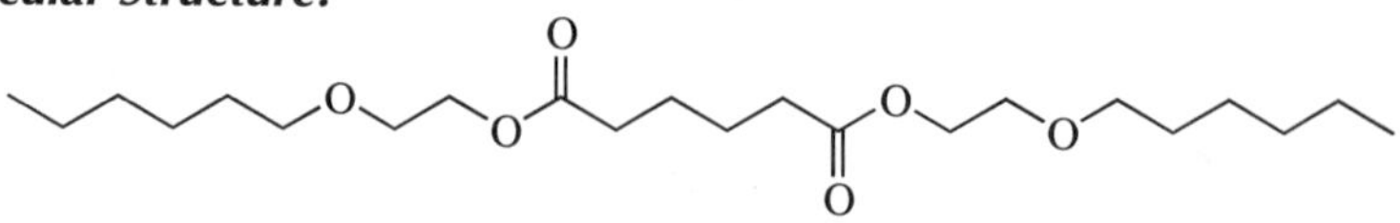

77.1 Chemical and Physical Properties

77.1.1 General

The empirical formula of di (2-hexyloxyethyl) adipate is $C_{22}H_{42}O_6$. Its structural formula is $(CH_2CH_2COOCH_2CH_2OC_6H_{13})_2$. It is a liquid with a pour point of −6.0°C and a density of 0.9642. Physico-chemical properties that were located are summarized in Table 80.13.

77.1.2 Odor and Warning Properties

No information found.

77.2 Production and Use

No information found.

77.3 Exposure Assessment

No information found.

77.4 Toxic Effects

The oral LD_{50} value for the rat is 4.29 g/kg, and the dermal LD_{50} value for the rabbit is 12.31 mL/kg. It is slightly irritating to rabbit skin and eyes.

77.5 Standards, Regulations, or Guidelines of Exposure

No information found.

77.6 Studies on Environmental Impact

No information found.

78.0 Di-2-(2-ethylbutoxy)ethyl Adipate

78.0.1 CAS Number: *[7790-07-0]*

78.0.2 Synonyms: Adipic acid, *bis* (2-(ethylbutoxy)ethyl) ester; adipic acid, di (2-(2-ethylbutoxy)ethyl) ester; di-2-(2-ethylbutoxy)ethyl adipate; hexanedioic acid di-2-(2-ethylbutoxy)ethyl ester; 3-02-00-01717 [Beilstein]; and BRN 1806612 [RTECS]

78.0.3 Trade Names: NA

78.0.4 Molecular Weight: 402.57.

78.0.5 Molecular Formula: $C_{22}H_{42}O_6$

78.0.6 Molecular Structure:

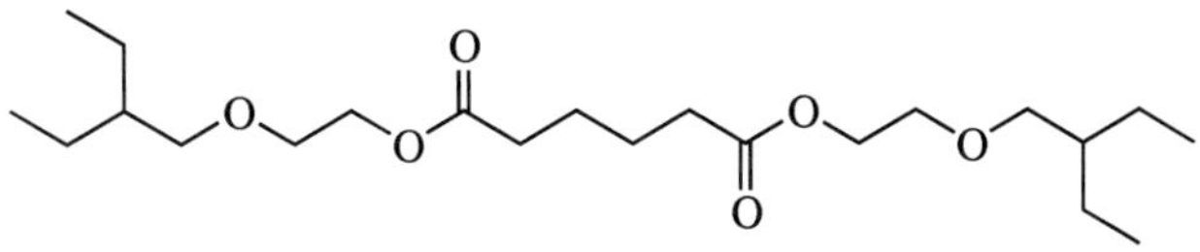

78.1 Chemical and Physical Properties

78.1.1 General

The empirical formula of di-2-(2-ethylbutoxy)ethyl adipate is $C_{22}H_{42}O_6$. Its structural formula is $[CH_2CH_2COOCH_2CH_2OCH_2CH(C_2H_5)_2]_5$. It is a liquid between the pour point (−60°C) and the boiling point (214°C at 5 mmHg). Physico-chemical properties are summarized in Table 80.13.

78.1.2 Odor and Warning Properties

No information found.

78.2 Production and Use

No information found.

78.3 Exposure Assessment

No information found.

78.4 Toxic Effects

The oral LD_{50} value for the rat is 3.25 g/kg and the dermal LD_{50} value for the rabbit is 4.24 mL/kg (5). No other toxicological information was located.

78.5 Standards, Regulations, or Guidelines of Exposure

No information found.

78.6 Studies on Environmental Impact

No information found.

79.0 Dibutyl Azelate

79.0.1 CAS Number: [2917-73-9]

79.0.2 Synonyms: Azelaic acid dibutyl ester; dibutyl nonanedioate; 1,9-di-*n*-butyl nononanedioate; nonanedioic acid dibutyl ester; AI3-01982 [NLM]; EINECS 220-850-8; HSDB 5904; and NSC 93294 [NLM]

79.0.3 Trade Names: Argoplast AZDB

79.0.4 Molecular Weight: 300.44.

79.0.5 Molecular Formula: $C_{17}H_{32}O_4$

79.0.6 Molecular Structure:

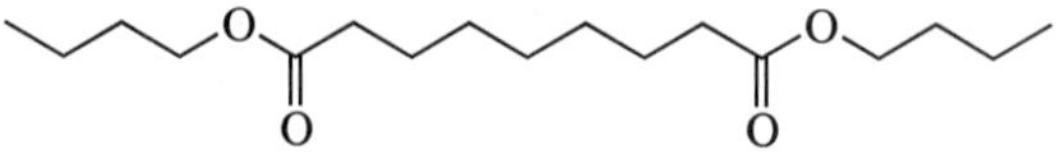

79.1 Chemical and Physical Properties

79.1.1 General

The empirical formula of dibutyl azelate is $C_{17}H_{32}O_4$. Its structural formula is $C_4H_9OOC(CH_2)_7COOC_4H_9$. It is a high-boiling (336°C) liquid that is insoluble in water (Table 80.13).

79.1.2 Odor and Warning Properties

No information found.

79.2 Production and Use

Dibutyl azelate is manufactured by reacting butanol with azelaic acid. It is used as a component of contact lenses and as a plasticizer.

79.3 Exposure Assessment

Air samples can be collected on a filter, extracted, and analyzed using gas chromatography (430).

79.4 Toxic Effects

The oral LD_{50} values for the rat and mouse are > 12.8 g/kg (12). The dermal LD_{50} value for the rabbit is > 10 mL/kg (12). No other toxicological information was located.

79.5 Standards, Regulations, or Guidelines of Exposure

No information found.

79.6 Studies on Environmental Impact

No information found.

80.0 Di(2-ethylhexyl) Azelate

80.0.1 CAS Number: *[103-24-2]*

80.0.2 Synonyms: Azelaic acid, *bis*(2-ethylhexyl) ester; azelaic acid, di(2-ethylhexyl) ester; *bis*(2-ethylhexyl) azelate; *bis*(2-ethylhexyl) nonanedioate; dioctyl azelate; nonanedioic acid, *bis*(2-ethylhexyl) ester; octyl azelate; 3-01-00-01787; AI3-07965 [NLM]; BRN 1806182 [RTECS]; EINECS 203-091-7; and HSDB 2859

80.0.3 Trade Names: DOZ; Plastolein 9058; Plastolein 9058doz; Plastolein 9058DOZ; Staflex DOX; and Truflex DOX

80.0.4 Molecular Weight: 412.65

80.0.5 Molecular Formula: $C_{25}H_{48}O_4$

80.0.6 Molecular Structure:

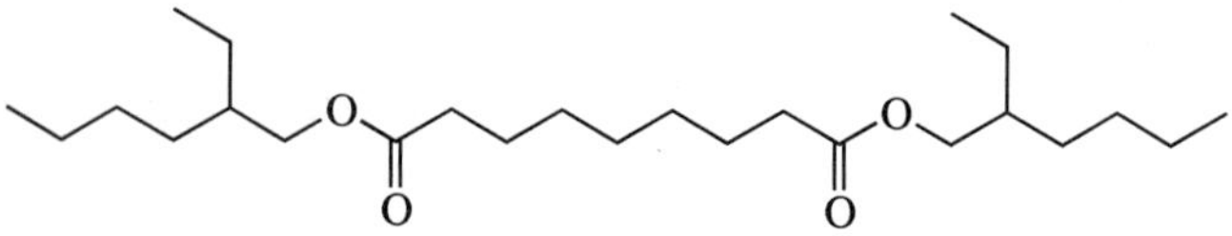

80.1 Chemical and Physical Properties

80.1.1 General

The empirical formula of dibutyl azelate is $C_{25}H_{48}O_4$. Its structural formula is $(C_4H_9)(C_2H_5)CHCH_2COO(CH_2)_7COOCH_2CH(C_2H_5)(C_4H_9)$. It is a high-boiling liquid (237°C at 5 mmHg) that is soluble in alcohol, acetone, and benzene and insoluble in water. Physico-chemical properties are summarized in Table 80.13.

80.1.2 Odor and Warning Properties

No information found.

80.2 Production and Use

Di(2-ethylhexyl) azelate is manufactured by reacting 2-ethylhexanol with azelaic acid in the presence of an acid catalyst. It is used as a plasticizer.

80.3 Exposure Assessment

Air samples can be collected on a filter, extracted, and analyzed using gas chromatography (430).

80.4 Toxic Effects

The oral LD_{50} value for the rat is 8.72 mL/kg and the dermal LD_{50} value for the rabbit is 20 mL/kg (373). The intravenous LD_{50} values for the rat and rabbit are 1.06 and 0.64 g/kg, respectively (374) (Table 80.13).

80.5 Standards, Regulations, or Guidelines of Exposure

No information found.

80.6 Studies on Environmental Impact

No information found.

81.0 Diethyl Sebacate

81.0.1 CAS Number: *[110-40-7]*

81.0.2 Synonyms: Decanedioic acid, diethyl ester; diethyl decanedioate; diethyl 1,10-decanedioate; diethyl 1,8-octanedicarboxylate; ethyl decanedioate; ethyl sebacate; sebacic acid, diethyl ester; 4-02-00-02080 [Beilstein]; AI3-01970 [NLM]; BRN 1790779 [RTECS]; CTFA 00804; EINECS 203-764-5; FEMA No. 2376; and NSC 8911

81.0.3 Trade Names: Bisoflex DES

81.0.4 Molecular Weight: 258.36

81.0.5 Molecular Formula: $C_{14}H_{26}O_4$

81.0.6 Molecular Structure:

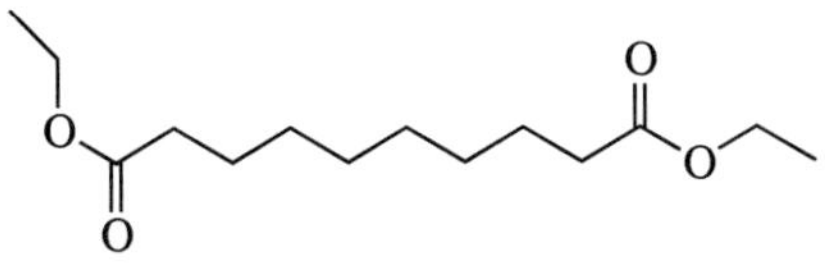

81.1 Chemical and Physical Properties

81.1.1 General

The empirical formula of diethyl sebacate is $C_{14}H_{26}O_4$. Its structural formula is $[(CH_2)_4COOC_2H_5]_2$. It is a liquid between the temperatures of $-2°C$ (melting point) and 312°C (boiling point at 773 mmHg). It is soluble in alcohol and slightly soluble in water. Physico-chemical properties are summarized in Table 80.13.

81.1.2 Odor and Warning Properties

No information found.

81.2 Production and Use

Diethyl sebacate has wide use in the plastics industry and is extensively utilized in the petroleum lubricant manufacture as an antiwear additive.

81.3 Exposure Assessment

No information found.

81.4 Toxic Effects

81.4.1 Experimental Studies

The oral LD_{50} values for the rat and guinea pig are 14.47 and 7.28 g/kg, respectively (3). Oral administration of 10,000 ppm over a 17–18-wk period or 1000 ppm over a 27–29-wk period was not toxic to male or female rats (177).

81.4.2 Human Experience

A few cases of contact dermatitis due to exposure to products containing 20–30% diethyl sebacate have been reported (431, 432).

81.5 Standards, Regulations, or Guidelines of Exposure

No information found.

81.6 Studies on Environmental Impact

No information found.

82.0 Dibutyl Sebacate

82.0.1 CAS Number: [109-43-3]

82.0.2 Synonyms: *bis*(*n*-Butyl) sebacate; butyl sebacate; decanedioic acid dibutyl ester; dibutyl decanedioate; dibutyl 1,8-octanedicarboxylate; di-*n*-butyl sebacate; dibutyl sebacinate; di-*n*-butylsebacate; sebacic acid, dibutyl ester; dibutylester kyseliny sebakove [Czech]; 4-02-00-02081 [Beilstein]; BRN 1798308; CTFA 00773; EINECS 203-672-5; FEMA No. 2373; HSDB 309; and NSC 3893 [NLM] dibutyl sebacate

82.0.3 Trade Names: Monoplex DBS; Polycizer DBS; PX 404; Staflex DBS

82.0.4 Molecular Weight: 314.46.

82.0.5 Molecular Formula: $C_{18}H_{34}O_4$

82.0.6 Molecular Structure:

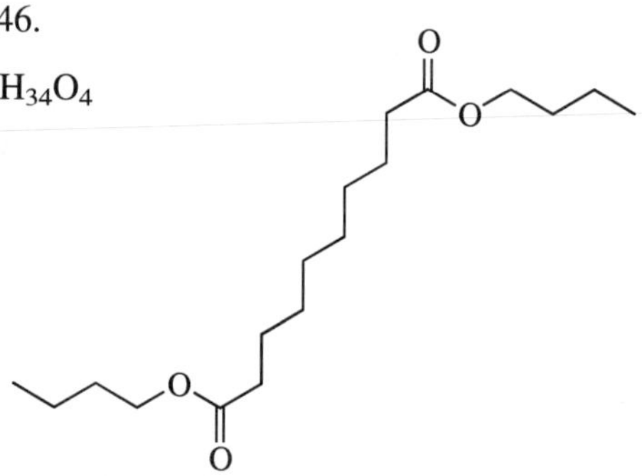

82.1 Chemical and Physical Properties

82.1.1 General

The empirical formula of dibutyl sebacate is $C_{18}H_{34}O_4$. Its structural formula is $[(CH_2)_4COOC_4H_9]_2$. It is a colorless to yellow liquid with a pour point of $-12°C$ and boiling point of 180°C. It is soluble in ether, and fairly insoluble in water (40 mg/L). Physico-chemical properties are summarized in Table 80.13.

82.1.2 Odor and Warning Properties

No information found.

82.2 Production and Use

Dibutyl sebacate is manufactured by reacting butanol with sebacic acid in the presence of a catalyst. It is used as a flavoring agent and as a plasticizer for various plastics and as a lubricating agent in shaving creams.

82.3 Exposure Assessment

No information found.

82.4 Toxic Effects

82.4.1 Experimental Studies

82.4.1.1 Acute Toxicity. Toxicologic data indicate that dibutyl sebacate is practically nontoxic orally or by dermal contact (Table 80.13). In rats and mice, the oral LD_{50} values are 16 and 19.5 g/kg, respectively (369, 375). Inhalation of concentrations greater than 5.4 mg/m^3 is required to cause lethality in mice or rats (369).

82.4.1.2 Chronic and Subchronic Toxicity. In a 2-yr study, dietary levels of 0.25, 1.25, and 6.25% dibutyl sebacate were fed to rats (433). No effects on growth, mortality, gross pathology, hematology or bone marrow were noted at any dose. Isolated inflammatory changes in the lungs, nephrosis, and fatty liver were noted in some animals treated with the highest dose level.

82.4.1.3 Pharmacokinetics, Metabolism, and Mechanisms. Dibutyl sebacate is rapidly hydrolyzed (434). This reaction occurs as rapidly as the saponification of triolein (12).

82.4.1.4 Reproductive and Developmental. In the 2-yr study mentioned in section 82.4.1.2 there was no effect on reproductive ability (433).

82.4.1.5 Carcinogenesis. No information found.

82.4.1.6 Genetic and Related Cellular Effects Studies. At concentrations up to 3.6 mg/plate, dibutyl sebacate was not mutagenic in *S. typhimurium* strains TA98, TA100, TA1535, TA1537, or TA1538 in the absence or presence of rat liver S-9 (41). Ingestion of 19 mM did not increase the incidence of sex-linked recessive lethal mutations in *Drosophila* (Basc test). Oral treatment of mice with concentrations up to 2.829 g/kg did not cause an increase in the number of micronucleated polychromatic erythrocytes from bone marrow (41).

82.4.1.7 Other: Neurological, Pulmonary, Skin Sensitization. No information found.

82.4.2 Human Experience

The human safety factor of dibutyl sebacate, when used in food wrapping, was calculated as greater than 1400 (433).

82.5 Standards, Regulations, or Guidelines of Exposure

No information found.

82.6 Studies on Environmental Impact

No information found.

83.0 *n*-Dioctyl Sebacate

83.0.1 CAS Number: [2432-87-3]

83.0.2 Synonyms: Decanedioic acid, dioctyl ester; di-*n*-octyl sebacate; dioctyl decanedioate; sebacic acid, dioctyl ester; and EINECS 219-411-3

83.0.3 Trade Names: Witamol 500

83.0.4 Molecular Weight: 426.68

83.0.5 Molecular Formula: $C_{26}H_{50}O_4$

83.0.6 Molecular Structure:

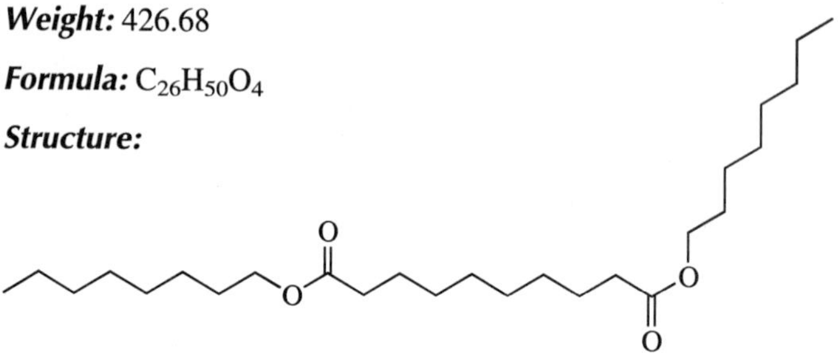

83.1 Chemical and Physical Properties

83.1.1 General

The empirical formula of dioctyl sebacate is $C_{26}H_{50}O_4$. Its structural formula is $[(CH_2)_4COOC_8H_{17}]_2$. It is a liquid between −55°C (pour point) and 248°C (boiling point at 4 mmHg). Physico-chemical properties are summarized in Table 80.13.

83.1.2 Odor and Warning Properties

No information found.

83.2 Production and Use

Dioctyl sebacate is used as a plasticizer.

83.3 Exposure Assessment

Analytic procedures for detection are available (435).

83.4 Toxic Effects

In rats given a diet containing 100 mg/kg dioctyl sebacate for 19 mo, no pathological signs were observed. No abnormalities in reproduction or offspring were noted (436). No other toxicological data were located.

83.5 Standards, Regulations, or Guidelines of Exposure

No information found.

83.6 Studies on Environmental Impact

No information found.

84.0 Diisooctyl Sebacate

84.0.1 CAS Number: [27214-90-0]

84.0.2 Synonyms: Decanedioic acid, diisooctyl ester; diisooctyl decanedioate; and EINECS 248-333-2

84.0.3 Trade Names: NA

84.0.4 Molecular Weight: 426.69

84.1 Chemical and Physical Properties

84.1.1 General

The empirical formula of diisooctyl sebacate is $C_{26}H_{50}O_4$. Its structural formula is $[(CH_2)_4COOCH_2CH(CH_3)(CH_2)_4CH_3]_2$. It is a liquid with a density of 0.917 and a boiling point of 248–255°C (at 4 mmHg). No other chemical or physical data were located.

84.1.2 Odor and Warning Properties

No information found.

84.2 Production and Use

No information found.

84.3 Exposure Assessment

No information found.

84.4 Toxic Effects

No information found.

84.5 Standards, Regulations, or Guidelines of Exposure

No information found.

84.6 Studies on Environmental Impact

No information found.

85.0 Di(2-ethylhexyl) Sebacate

85.0.1 CAS Number: [122-62-3]

85.0.2 Synonyms: *bis*(2-ethylhexyl) decanedioate; *bis*(ethylhexyl) sebacate; *bis*(2-ethylhexyl) sebacate; decanedioic acid, *bis*(2-ethylhexyl) ester; decanedioic acid, di(2-ethylhexyl) ester; di(2-ethylhexyl) sebacate; dioctyl sebacate; 2-ethylhexyl sebacate; 1-

hexanol,2-ethyl-, sebacate; octyl sebacate; sebacic acid, *bis*(2-ethylhexyl)ester; sebacic acid, di(2-ethylhexyl)ester; *bis*-(2-ethylhexyl)ester kyseliny sebakove [Czech]; 4-02-00-02083 [Beilstein]; AI3-09124 [NLM]; BRN 1806504; CCRIS 6191; CTFA 00872; EINECS 204-558-8; HSDB 2898; and NSC 68878 [NLM]

85.0.3 Trade Names: Bisoflex; Bisoflex DOS; DOS; Edenol 888; Edenor DEHS; Ergoplast SDO; Monoplex DOS; Octoil S; Plasthall DOS; Plexol; Plexol 201J; PX 438; Reolube DOS; Reomol DDS; Sansocizer DOS; Staflex DOS; and Uniflex DOS

85.0.4 Molecular Weight: 426.68

85.0.5 Molecular Formula: $C_{26}H_{50}O_4$

85.0.6 Molecular Structure:

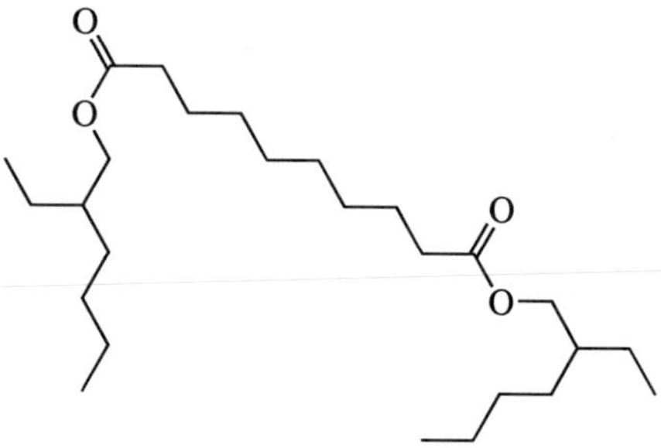

85.1 Chemical and Physical Properties

85.1.1 General

The empirical formula of di(2-ethylhexyl) sebacate is $C_{26}H_{50}O_4$. Its structural formula is $[(CH_2)_4COOCH_2CH(C_2H_5)(C_4H_9)]_2$. It is a pale straw-colored, or colorless, oily liquid which boils at 248°C. Physico-chemical properties that were located are summarized in Table 80.13.

85.1.2 Odor and Warning Properties

Di(2-ethylhexyl) sebacate has a mild odor.

85.2 Production and Use

Di(2-ethylhexyl) sebacate is manufactured by esterifying 2-ethylhexanol with sebacic acid in the presence of an acid catalyst. It is widely used as an antiwear additive in mineral and instrument oils.

85.3 Exposure Assessment

No information found.

85.4 Toxic Effects

85.4.1 Experimental Studies

85.4.1.1 Acute Toxicity. Physiologically, di(2-ethylhexyl) sebacate is of low activity (12). The oral LD_{50} value for mice is 9.5 g/kg (376). The IV LD_{50} values for rats and rabbits are 0.9 and 0.54 g/kg, respectively (374). Rats exposed to sebacate vapor for 6 h, by bubbling air through a liquid column of 100°C, showed no mortality nor any signs of toxicity (12). Vapors heated to 700°F were more toxic than at room temperature to the rat and the rabbit, but not to the pig (437).

85.4.1.2 Chronic and Subchronic Toxicity. No information found.

85.4.1.3 Pharmacokinetics, Metabolism, and Mechanisms. No information found.

85.4.1.4 Reproductive and Developmental. No information found.

85.4.1.5 Carcinogenesis. The ability of di(2-ethylhexyl) sebacate to act as a tumor promoter was examined in Sprague–Dawley rats. Rats were treated three times a week for 12 wk (via gavage) with 200 or 500 mg/kg di(2-ethylhexyl) sebacate or di(2-ethylhexyl) phthalate (DEHP) after initiation with 8 mg/kg diethyl nitrosamine. In contrast to DEHP, di(2-ethylhexyl) sebacate treatment did not lead to an increase in the incidence of liver foci (438).

85.4.1.6 Genetic and Related Cellular Effects Studies. At concentrations of 100–10,000 μg/plate, di(2-ethylhexyl) sebacate was not mutagenic in *S. typhimurium* strains TA98, TA100, TA1535, or TA1537, in the presence or absence of rat or hamster liver S-9 (406).

85.4.1.7 Other: Neurological, Pulmonary, Skin Sensitization. Intratracheal instillation of 37.5 mg/kg di(2-ethylhexyl sebacate) did not cause pulmonary toxicity in Syrian golden hamsters (439).

85.4.2 Human Experience

Di(2-ethylhexyl) sebacate does not appear to cause skin sensitization or irritation in humans (440).

85.5 Standards, Regulations, or Guidelines of Exposure

No information found.

85.6 Studies on Environmental Impact

No information found.

D INTRODUCTION TO ESTERS OF ALKENYL DICARBOXYLIC ACIDS

Maleic acid esters (*cis*-2-butenoates), fumarates (*trans*-2-butenoates), and itaconates have been utilized as plasticizers, raw materials for chemical syntheses, or preservatives for fats and oils (441). Because diethyl maleate causes glutathione depletion in animals (442, 443), it is commonly used to determine the role glutathione plays in toxicity and/or metabolism of chemicals. Physico-chemical properties of esters of alkenyl dicarboxylic acids are summarized in Table 80.15. Maleates are readily hydrolyzed, and small quantities of sodium maleate can easily be metabolized through the citric acid cycle (55). Analytic methods are available for the determination of maleic esters (489). Information about metabolism of fumarates and itaconates was not located.

The esters of alkenyl dicarboxylic acids are of low acute toxicity (Table 80.16). They have a tendency to cause skin or eye irritation in rabbits. Allergic dermatitis has occurred in humans exposed to dibutyl maleate (445). Subacute and chronic toxicity data for these compounds are limited.

86.0 Dimethyl Maleate

86.0.1 CAS Number: [624-48-6]

86.0.2 Synonyms: 2-Butenedioic acid, dimethyl ester; dimethyl *cis*-butenedioate; dimethyl *cis*-ethylenedicarboxylate; maleic acid, dimethyl ester; methyl maleate; dimethylester kyseliny maleinove [Czech]; 4-02-00-02204 [Beilstein]; AI3-07869 [NLM]; BRN 0471705; CTFA 06332; EINECS 210-848-5; and NSC 5161 [NLM]

86.0.3 Trade Names: Sipomer DMM

86.0.4 Molecular Weight: 144.13

86.0.5 Molecular Formula: $C_6H_8O_4$

86.0.6 Molecular Structure:

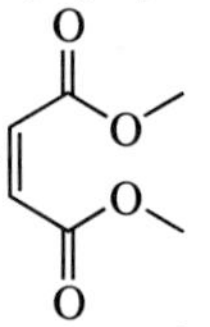

86.1 Chemical and Physical Properties

86.1.1 General

The empirical formula of dimethyl maleate is $C_6H_8O_4$. Its structural formula is *cis*-$CH_2OOCCH{:}CHCOOCH_3$. It is a clear liquid between − 19°C (pour point) and 205°C (boiling point at 762 mmHg). It is soluble in ether and insoluble in water. Physico-chemical properties are summarized in Table 80.15.

86.1.2 Odor and Warning Properties

No information found.

86.2 Production and Use

Dimethyl maleate is widely used in the plastics and chemical industries and as a vaporization retardant of insecticides.

86.3 Exposure Assessment

No information found.

Table 80.15. Physical and Chemical Properties of Alkenyl Dicarboxylic Esters

Chemical Name	CAS No.	Molecular Formula	Mol. Wt.	Boiling Point (°C) (mmHg)	Melting Point (°C)	Density (Specific Gravity	Sol. in Water at 20°C (68°F)	Refractive Index (20°C)	Vapor Pressure (mmHg) (°C)	UEL or LEL % by Vol. Room Temp.
Maleate										—
Dimethyl	[*624-48-6*]	$C_6H_8O_4$	144.13	205	– 19	(1.152)	insol	1.4416	1 (45.7)	—
Diethyl	[*141-05-9*]	$C_8H_{12}O_4$	172.18	225	– 10	(1.064)	insol	1.4416	1 (57.3)	—
Diisopropyl	[*10099-70-4*]	$C_{10}H_{16}O_4$	200.23	444	—	1.00	insol	—	—	—
Dibutyl	[*105-76-0*]	$C_{12}H_{20}O_4$	228.29	129 (4)	– 85	(0.093)	—	—	10 (139)	—
Dipentyl		$C_{14}H_{24}O_4$	256.34	263–300	—	0.981	—	—	—	—
Methyl Amyl	—	$C_{10}H_{16}O_4$	284.40	394 (50)	—	0.90	insol	—	—	—
Dihexyl		$C_{16}H_{28}O_4$	284.40	179 (10)	—	0.9602	insol	—	0.1 (20)	—
Di(2-ethylhexyl)	[*142-16-5*]	$C_{20}H_{36}O_4$	340.50	164 (10)	– 60	(0.9436)	—	—	—	—
Fumarate										
Diethyl	[*623-91-6*]	$C_8H_{12}O_4$	172.18	219	2	(1.052)	—	—	1 (53.2)	—
Diisopropyl	[*7283-70-7*]	$C_{10}H_{16}O_4$	200.23	280	2	—	—	—	—	—
Dibutyl	[*105-75-9*]	$C_{12}H_{20}O_4$	228.29	141 (8)	—	0.987	insol	—	8 (140)	—
Diisobutyl	[*7283-69-4*]	$C_{12}H_{20}O_4$	228.29	—	—	—	—	—	—	—
Di(2-ethylhexyl)	[*141-02-6*]	$C_{20}H_{36}O_4$	340.50	211–220	—	0.942	—	—	—	—
Itaconate										
Dimethyl	[*617-52-7*]	$C_7H_{10}O_4$	158.15	208	37–40	1.124	—	1.4457	—	—
Diethyl	[*2409-52-1*]	$C_9H_{14}O_4$	186.21	228	58–9	1.0467	—	1.4377	—	—

Table 80.16. Summary of Toxicity of Esters of Alkenyl Dicarboxylic Acids

Chemical Name	CAS No.	Species	Exposure Route	LD_{50} (g/kg)	Ref.
Maleate					
Dimethyl	*[624-48-6]*	Rat	oral	1.41 mL/kg	373
		Rabbit	dermal	0.53 mL/kg	373
Diethyl	*[141-05-9]*	Rat	oral	3.2	446
		Rat	dermal	5	447
		Rabbit	dermal	5 mL/kg	447
		Rat	IP	3.07	446
Diisopropyl	*[10099-70-4]*	Rat	oral	2.14	5
Dibutyl	*[105-76-0]*	Rat	oral	3.73	5
		Rabbit	dermal	10.1 mL/kg	5
		Mouse	IP	0.15	448
Dipentyl	*[10099-71-5]*	Rat	oral	4.92 mL/kg	373
		Rabbit	dermal	> 10 mL/kg	373
Dihexyl	*[16064-83-8]*	Rat	oral	7.34 mL/kg	373
Di(2-ethylhexyl)	*[142-16-5]*	Rat	oral	14	446
		Mouse	oral	> 20 mL/kg	449
		Rabbit	dermal	15 mL/kg	446
		Rat	IP	> 5	12
		Mouse	IP	> 1	448
Fumarate					
Diethyl	*[623-91-6]*	Rat	oral	1.78	373
Diisopropyl	*[7283-70-7]*	Rat	oral	3.25	5
		Rabbit	dermal	10 mL/kg	5
Dibutyl	*[105-75-9]*	Rat	oral	8.53	7
		Rabbit	dermal	15.9 mL/kg	7
		Mouse	IP	0.25	448
Diisobutyl	*[7283-69-4]*	Rat	oral	8.12	373
		Rabbit	dermal	7.49	373
Di(2-ethylhexyl)	*[141-02-6]*	Rat	oral	29.2	5
		Rabbit	dermal	> 20 mL/kg	5
		Mouse	IP	0.25	448

86.4 Toxic Effects

86.4.1 Experimental Studies

86.4.1.1 Acute Toxicity. Range-finding studies have listed an oral LD_{50} value of 1.41 mL/kg for the rat and a dermal LD_{50} value of 0.53 mL/kg for the rabbit (373). In rats exposed to saturated vapor, the time to death for the rat was 4 h (373). A single application of 0.5 or 2 g/kg to the back of shaved rats did not cause death or induce any systemic effects (449). Guinea pigs injected intradermally with 1% dimethyl maleate did not show any signs of systemic toxicity (450).

86.4.1.2 ***Chronic and Subchronic Toxicity.*** The effect of subacute dermal exposure to dimethyl maleate has been studied in rats and rabbits. In rats, dose-dependent decreases in food consumption and body weight gain were noted after dermal exposure to 170 and 500 mg/kg, 5 d/wk for 4 wk. Leucocytosis and reduced serum levels of reduced and oxidized glutathione were noted in the high dose group. The dermal no-effect level for systemic effects was 60 mg/kg (450).

86.4.1.3 ***Pharmacokinetics, Metabolism, and Mechanisms.*** No information found.

86.4.1.4 ***Reproductive and Developmental.*** No information found.

86.4.1.5 ***Carcinogenesis.*** No information found.

86.4.1.6 ***Genetic and Related Cellular Effects Studies.*** Concentrations up to 5000 μg/plate did not cause mutations to *S. typhimurium* strains TA98, TA100, TA1535, TA1537, and TA1538, in the presence or absence of rat liver S-9 (450). In mice, oral administration of 1 g/kg did not induce micronuclei (450). A shift from polychromatic to monochromatic erythrocytes was observed, which indicated an effect on bone marrow cell proliferation.

86.4.1.7 ***Other: Neurological, Pulmonary, Skin Sensitization.*** Single- and multiple-dose dermal toxicity studies in rats, rabbits, and guinea pigs have been performed by Ullman and co-workers (449). A single application of 500 or 2000 mg/kg to the back of shaved rats caused local erythema and necrosis. In rabbits, a single dermal application of 0.5 mL caused only slight erythema and edema. All guinea pigs sensitized with 1% dimethyl maleate developed erythema and edema. In rats treated dermally with 60, 170, or 500 mg/kg for 6 h/d, 5 d/wk, for 4 wk, dose-related local effects ranging from slight erythema and scaling (in the lowest dose group) to severe erythema, slight-to-moderate edema and scaling, and slight to severe necrosis (in the highest-dose group) were observed.

86.4.2 Human Experience

No information found.

86.5 Standards, Regulations, or Guidelines of Exposure

No information found.

86.6 Studies on Environmental Impact

No information found.

87.0 Diethyl Maleate

87.0.1 CAS Number: [141-05-9]

87.0.2 Synonyms: 2-Butenedioic acid, diethyl ester; maleic acid, diethyl ester; ethyl maleate; diethylester kyseliny maleinove [Czech]; 4-02-00-02207 [Beilstein]; AI3-000678 [NLM]; BRN 1100825; CCRIS 941; EINECS 205-451-9; and NSC 8394 [NLM]

87.0.3 *Trade Names:* NA

87.0.4 *Molecular Weight:* 172.18

87.0.5 *Molecular Formula:* $C_8H_{12}O_4$

87.0.6 *Molecular Structure:*

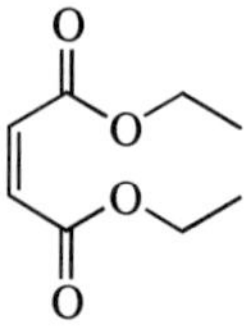

87.1 Chemical and Physical Properties

87.1.1 *General*

The empirical formula of diethyl maleate is $C_8H_{12}O_4$. Its structural formula is *cis*-$C_2H_5OOCCH{:}CHCOOC_2H_5$. It is a clear liquid between the temperatures of − 10°C (pour point) and 225°C (boiling point). Diethyl maleate is soluble in alcohol and ether is insoluble in water. Physico-chemical properties are summarized in Table 80.15.

87.1.2 *Odor and Warning Properties*

No information found.

87.2 Production and Use

Diethyl maleate is formed by the direct esterification of maleic acid with ethyl alcohol. It is used as an experimental tool *in vivo* or *in vitro* to deplete cells of glutathione.

87.3 Exposure Assessment

No information found.

87.4 Toxic Effects

87.4.1 *Experimental Studies*

87.4.1.1 Acute Toxicity. The acute oral and intraperitoneal LD_{50} values for the rat are 3.2 and 3.07 g/kg, respectively (446, 451). Rats treated intraperitoneally with high doses exhibit lacrimation, diarrhea, and hypermotility of the gastrointestinal tract. The acute dermal LD_{50} values for the rat and rabbit are 5 g/kg and 5 mL/kg, respectively (447). In mice treated intragastrically with a dose of diethyl maleate, which causes severe depletion of liver glutathione (12 mmol/kg), lipid peroxidation and necrosis of liver cells occur if the diet is not supplemented with vitamin E (452).

Pretreatment with diethyl maleate enhances the acute toxicity of allyl alcohol, oxygen, potassium bromate, parathion, methyl parathion, methyl paraoxon, fenitrothion, and selenium in experimental animals (453–459).

87.4.1.2 Chronic and Subchronic Toxicity. No information found.

87.4.1.3 Pharmacokinetics, Metabolism, and Mechanisms. Diethyl maleate conjugates directly with reduced glutathione in a reaction catalyzed by glutathione-*S*-transferase (442).

87.4.1.4 Reproductive and Developmental. In gestating mice treated with phenytoin (55 mg/kg, IP), the incidence of cleft palate was increased by pretreatment with diethyl maleate (150 or 300 mg/kg, IP) (460). In an *in vitro* frog embryo teratogenesis assay, the embryo-lethal potential of 4-bromobenzene was increased by addition of diethyl maleate (461). Conversely, fetuses from mice treated IP with diethyl maleate prior to thiabendazole exhibited fewer external or skeletal malformations compared to those from mice treated with thiabendazole alone (462).

87.4.1.5 Carcinogenesis. Compared to rats ingesting a diet containing butylated hydroxyanisole (2%) for 2 yr, rats ingesting a diet containing both diethyl maleate (0.2%) and butylated hydroxyanisole (2%) had a decreased multiplicity and incidence of forestomach epithelial cell hyperplasia and papilloma (463). Diethyl maleate also prevents 3,4-benzopyrene-induced carcinogenesis (464) and increases the latency period for the development of 3-methylcholanthrene-induced tumorigenesis (465).

87.4.1.6 Genetic and Related Cellular Effects Studies. Treatment of cultured cells, embryos or animals with diethyl maleate causes glutathione depletion (442, 443, 454, 456, 466–468). Glutathione plays an important role in metabolism of chemicals and protects cells against oxidative injury. Consequently, diethyl maleate is often used as an experimental tool to help determine whether the mechanism of toxicity of other agents involves active oxygen species or glutathione-containing metabolites.

In addition to causing glutathione depletion, diethyl maleate inhibits protein synthesis (469), affects microsomal drug metabolism (470), and stimulates bile flow (471). By inducing liver metallothionein, it inhibits the hepatotoxicity of cadmium (472).

87.4.1.7 Other: Neurological, Pulmonary, Skin Sensitization. Diethyl maleate is moderately irritating when applied full strength to intact or abraded skin for 24 h under occlusion and is slightly irritating to rabbit eyes (447). Rats treated with 1 mg/kg (IP) diethyl maleate before inspiratory resistive breathing demonstrate a marked impairment of diaphragmatic contractile performance (471).

87.4.2 Human Experience

In a maximation test carried out in 25 volunteers, a concentration of 4% in petrolatum produced no irritation or sensitization reactions (346, 447). In four men working with unsaturated polyester resins, patch testing revealed that diethyl maleate elicited a reaction (474).

87.5 Standards, Regulations, or Guidelines of Exposure

No information found.

87.6 Studies on Environmental Impact

No information found.

88.0 Diisopropyl Maleate

88.0.1 CAS Number: [10099-70-4]

88.0.2 Synonyms: 2-Butenedioic acid, diisopropyl ester; 2-butenedioic acid, *bis*(1-methylethyl) ester; maleic acid, diisopropyl ester; diisopropylester kyseliny maleinove [Czech]; and EINECS 233-242-2

88.0.3 Trade Names: NA

88.0.4 Molecular Weight: 200.23

88.0.5 Molecular Formula: $C_{10}H_{16}O_4$

88.0.6 Molecular Structure:

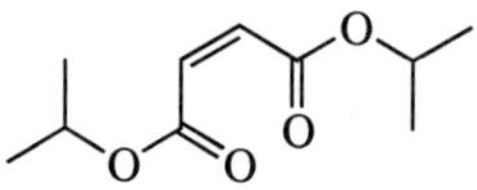

88.1 Chemical and Physical Properties

88.1.1 General

The empirical formula of diisopropyl maleate is $C_{10}H_{16}O_4$. Its structural formula is $(CH_3)_2CHOOCCH{:}CHCOOCH(CH_3)_2$. At ambient temperature, it is a liquid with a density of 1.00. Its boiling point is 444°C. Physico-chemical properties are summarized in Table 80.15.

88.1.2 Odor and Warning Properties

No information found.

88.2 Production and Use

No information found.

88.3 Exposure Assessment

No information found.

88.4 Toxic Effects

88.4.1 Experimental Studies

88.4.1.1 Acute Toxicity. The oral LD_{50} value for the rat is 2.14 g/kg (5).

88.4.1.2 Chronic and Subchronic Toxicity. No information found.

88.4.1.3 Pharmacokinetics, Metabolism, and Mechanisms. No information found.

88.4.1.4 Reproductive and Developmental. No information found.

88.4.1.5 Carcinogenesis. No information found.

88.4.1.6 Genetic and Related Cellular Effects Studies. No information found.

88.4.1.7 Other: Neurological, Pulmonary, Skin Sensitization. In rabbits, dermal application of diisopropyl maleate caused slight (grade 2 out of 10) skin irritation (5). It is slightly irritating (grade 1 out of 10) to rabbit eyes (5).

88.4.2 Human Experience

No information found.

88.5 Standards, Regulations, or Guidelines of Exposure

No information found.

88.6 Studies on Environmental Impact

No information found.

89.0 Dibutyl Maleate

89.0.1 CAS Number: *[105-76-0]*

89.0.2 Synonyms: 2-Butenedioic acid, dibutyl ester; butyl maleate; dibutyl maleinate; maleic acid, dibutyl ester; dibutylester kyseliny maleinove [Czech]; 4-02-00-02209 [Beilstein]; AI3-00644 [NLM]; BRN 1726634; CCRIS 4136; EINECS 203-328-4; and NSC 6711 [NLM]

89.0.3 Trade Names: DBM; PX-504; RC Comonomer DBM; and Staflex DBM

89.0.4 Molecular Weight: 228.29

89.0.5 Molecular Formula: $C_{12}H_{20}O_4$

89.0.6 Molecular Structure:

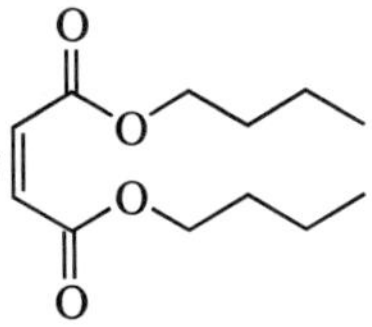

89.1 Chemical and Physical Properties

89.1.1 General

The empirical formula of dibutyl maleate is $C_{12}H_{20}O_4$. Its structural formula is CH_3-$(CH_2)_3OOCCH{:}CHCOO(CH_2)_3CH_3$. It is an oily liquid with a pour point of −85°C and a boiling point of 129°C at 4 mmHg. Physico-chemical properties are summarized in Table 80.15.

89.1.2 Odor and Warning Properties

No information found.

89.2 Production and Use

Dibutyl maleate is used as a copolymer and plasticizer.

89.3 Exposure Assessment

No information found.

89.4 Toxic Effects

89.4.1 Experimental Studies

89.4.1.1 Acute Toxicity. The oral LD_{50} value for the rat is 3.73 g/kg, and the dermal LD_{50} value for the rabbit is 10.1 mL/kg. (5).

89.4.1.2 Chronic and Subchronic Toxicity. Rats inhaling dibutyl maleate (6.9 mg/m^3, 4 h/d) for 4 mo exhibited alterations in hematology and in the physiology of the lungs, liver, heart, adrenal glands, kidneys, and nervous system. Chronic inhalation of 0.54 mg/m^3 caused no pathological or irreversible functional changes in organs (475).

89.4.1.3 Pharmacokinetics, Metabolism, and Mechanisms. No information found.

89.4.1.4 Reproductive and Developmental. No information found.

89.4.1.5 Carcinogenesis. No information found.

89.4.1.6 Genetic and Related Cellular Effects Studies. At a concentration of 100–1000 μg/plate, dibutyl maleate was not mutagenic to *S. typhimurium* strains TA97A, TA98, TA100, or TA102 in the presence or absence of rat liver S-9 (476). Dibutyl maleate also was not mutagenic in a mouse micronucleus test (476).

89.4.1.7 Other: Neurological, Pulmonary, Skin Sensitization. Dermal application of dibutyl maleate caused slight (grade 2/10) skin irritation in rabbits (5). It was mildly irritating (grade 1/10) to rabbit eyes (5).

89.4.2 Human Experience

Ten out of twenty envelope manufacturers exposed to polyvinylacetate glue containing urea and dibutyl maleate developed dermatitis, which upon further investigation was due to allergy to dibutyl maleate (445).

89.5 Standards, Regulations, or Guidelines of Exposure

No information found.

89.6 Studies on Environmental Impact

No information found.

90.0 Dipentyl Maleate

90.0.1 CAS Number:

90.0.2 Synonyms: 2-Butanedioic acid, dipentyl ester; diamyl maleate; maleic acid, dipentyl ester; dibutylester kyseliny maleinove [Czech]; 3-02-00-01925[Beilstein]; AI3-00418 [NLM]; BRN 1727367; and EINECS 233-243-8

90.0.3 Trade Names: NA

90.0.4 Molecular Weight: 256.34

90.0.5 Molecular Formula: $C_{14}H_{24}O_4$

90.0.6 Molecular Structure:

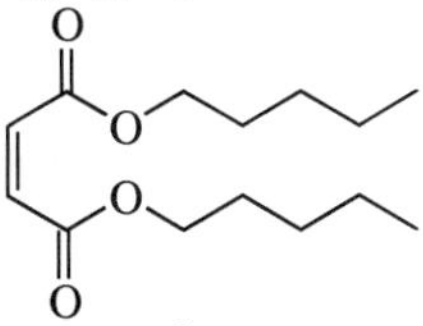

90.1 Chemical and Physical Properties

90.1.1 General

The empirical formula of dipentyl maleate is $C_{14}H_{24}O_4$. Its structural formula is $C_5H_{11}OOCCH{:}CHCOOC_5H_{11}$. It is a liquid with a density of 0.981 and a boiling point of 263–300°C. Physico-chemical properties are summarized in Table 80.15.

90.1.2 Odor and Warning Properties

No information found.

90.2 Production and Use

No information found.

90.3 Exposure Assessment

No information found.

90.4 Toxic Effects

90.4.1 Experimental Studies

90.4.1.1 Acute Toxicity. The oral and dermal LD_{50} values for the rat and rabbit are 4.92 and > 10 mL/kg, respectively (373). In rats, inhalation of saturated vapor proved lethal in an 8-h trial (373).

90.4.1.2 Chronic and Subchronic Toxicity. No information found.

90.4.1.3 Pharmacokinetics, Metabolism, and Mechanisms. No information found.

90.4.1.4 Reproductive and Developmental. No information found.

90.4.1.5 Carcinogenesis. No information found.

90.4.1.6 Genetic and Related Cellular Effects Studies. No information found.

90.4.1.7 Other: Neurological, Pulmonary, Skin Sensitization. Dipentyl maleate causes very little skin or eye irritation in rabbits (373).

90.4.2 Human Experience

No information found.

90.5 Standards, Regulations, or Guidelines of Exposure

No information found.

90.6 Studies on Environmental Impact

No information found.

91.0 Methyl Amyl Maleate

91.0.1 CAS Number: NA

91.0.2 Synonyms: 2-Butenedioic acid, methyl amyl ester

91.0.3 Trade Names: NA

91.0.4 Molecular Weight: 284.40

91.1 Chemical and Physical Properties

91.1.1 General

The empirical formula of methyl amyl maleate is $C_{10}H_{16}O_4$. Its structural formula is $CH_3OOCCH{:}CHCOOC_5H_{11}$. It is a liquid with a density of 0.90 and a boiling point of 394°C (at 50 mmHg). Physico-chemical properties are summarized in Table 80.15.

91.1.2 Odor and Warning Properties

No information found.

91.2 Production and Use

No information found.

91.3 Exposure Assessment

No information found.

91.4 Toxic Effects

No information found.

91.5 Standards, Regulations, or Guidelines of Exposure

No information found.

91.6 Studies on Environmental Impact

No information found.

92.0 *n*-Dihexyl Maleate

92.0.1 CAS Number: *[16064-83-8]*

92.0.2 Synonyms: 2-Butenedioic acid, dihexyl ester; AI3-00997; and EINECS 240-208-0

92.0.3 Trade Names: NA

92.0.4 Molecular Weight: 284.40.

92.1 Chemical and Physical Properties

92.1.1 General

The empirical formula of dihexyl maleate is $C_{16}H_{28}O_4$. Its structural formula is $CH_3(CH_2)_5OOCCH{:}CHCOO(CH_2)_5CH_3$. It is a liquid with a density of 0.9602 and a boiling point of 179°C (at 10 mmHg). Physico-chemical properties are summarized in Table 80.15.

92.1.2 Odor and Warning Properties

No information found.

92.2 Production and Use

No information found.

92.3 Exposure Assessment

No information found.

92.4 Toxic Effects

According to a range-finding study conducted by Smyth and co-workers, dihexyl maleate is less toxic by ingestion than its lower homologues (373). In rabbits, its potential to irritate skin is similar to that of dimethyl maleate (373).

92.5 Standards, Regulations, or Guidelines of Exposure

No information found.

92.6 Studies on Environmental Impact

No information found.

93.0 Di(2-ethylhexyl) Maleate

93.0.1 CAS Number: [142-16-5]

93.0.2 Synonyms: *bis*(2-Ethylhexyl)maleate; 2-butenedioic acid, *bis*(2-ethylhexyl) ester; dioctyl maleate; maleic acid, *bis*(2-ethylhexyl) ester; *bis*-(2-ethylhexyl) ester kyseliny maleinove [Czech]; 4-02-00-02211[Beilstein]; AI3-07870 [NLM]; BRN 1729133 [RTECS]; EINECS 205-524-5; and HSDB 5481

93.0.3 Trade Names: DOM and RC Comonomer DOM

93.0.4 Molecular Weight: 340.50

93.0.5 Molecular Formula: $C_{20}H_{36}O_4$

93.0.6 Molecular Structure:

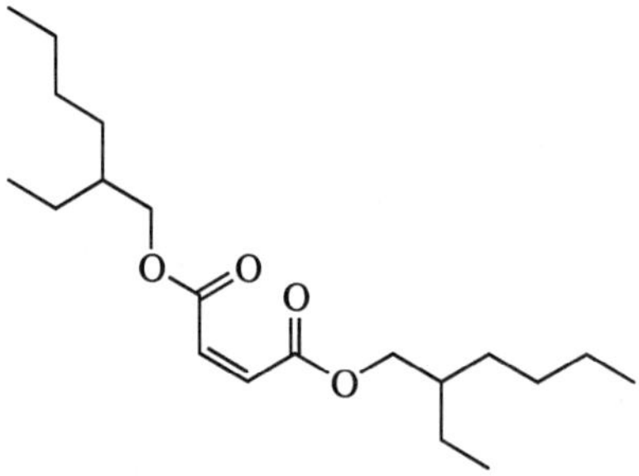

93.1 Chemical and Physical Properties

93.1.1 General

The empirical formula of di(2-ethylhexyl) maleate is $C_{20}H_{36}O_4$. Its structural formula is $[CH_3(CH_2)_3CH(C_2H_5)CH_2OOCCH]_2$. It is a liquid between −60 and 164°C (boiling point at 10 mmHg). Physico-chemical properties are summarized in Table 80.15.

93.1.2 Odor and Warning Properties

No information found.

93.2 Production and Use

No information found.

93.3 Exposure Assessment

No information found.

93.4 Toxic Effects

93.4.1 Experimental Studies

The oral LD_{50} value for the rat is 14 g/kg (446). The oral lethal dose in the mouse is >20 mL/kg (477). The dermal LD_{50} value for the rabbit is 15 mL/kg (446). It is mildly to moderately irritating to rabbit skin (446). Toxicological data that were located are summarized in Table 80.16.

93.4.2 Human Experience

No information found.

93.5 Standards, Regulations, or Guidelines of Exposure

No information found.

93.6 Studies on Environmental Impact

No information found.

94.0 Diethyl Fumarate

94.0.1 CAS Number: [623-91-6]

94.0.2 Synonyms: *trans*-2-Butenedioic acid diethyl ester; 2-butenedioic acid (E)-, diethyl ester; ethyl fumarate; fumaric acid, diethyl ester; diethylester kyseliny fumarove [Czech]; 4-02-00-02207[Beilstein]; AI3-05613 [NLM]; BRN 0775347 [RTECS]; EINECS 210-819-7; HSDB 5722; and NSC 20954 [NLM]

94.0.3 Trade Names: Anti-Psoriaticum

94.0.4 Molecular Weight: 172.18

94.0.5 Molecular Formula: $C_8H_{12}O_4$

94.0.6 Molecular Structure:

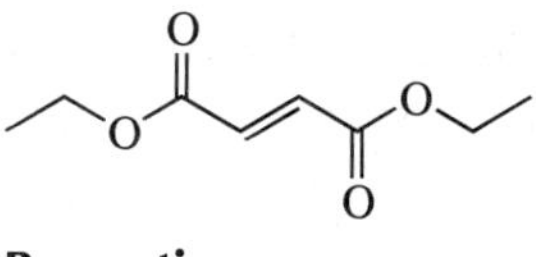

94.1 Chemical and Physical Properties

94.1.1 General

The empirical formula of diethyl fumarate is $C_8H_{12}O_4$. Its structural formula is *trans*-$C_2H_5OOCCH{:}CHCOOC_2H_5$. It occurs as white crystals or in liquid form. Its melting and boiling points are 2 and 219°C, respectively. Physico-chemical properties are summarized in Table 80.15.

94.1.2 Odor and Warning Properties

No information found.

94.2 Production and Use

Diethyl fumarate is manufactured via esterification of ethanol with fumaric acid. It is used as a pesticide or comonomer for production of polystyrene.

94.3 Exposure Assessment

No information found.

94.4 Toxic Effects

94.4.1 Experimental Studies

The oral LD_{50} value for the rat is 1.78 g/kg (373). Lethality occurs in rats exposed to concentrated vapor for 8 h. It is mildly to irritating to rabbit eyes (373). No other information was located.

94.4.2 Human Experience

No information found.

94.5 Standards, Regulations, or Guidelines of Exposure

No information found.

94.6 Studies on Environmental Impact

No information found.

95.0 Diisopropyl Fumarate

95.0.1 CAS Number: *[7283-70-7]*

95.0.2 Synonyms: 2-Butenedioic acid, diisopropyl ester; 2-butenedioic acid (E)-*bis*(1-methylethyl) ester; fumaric acid, diisopropyl ester; diisopropylester kyseliny fumarove [Czech]; 4-02-00-02209 [Beilstein]; AI3-15489 [NLM]; BRN 1726370 [RTECS]; EINECS 230-707-1; and NSC 70161 [NLM]

95.0.3 Trade Names: NA

95.0.4 Molecular Weight: 200.23

95.0.5 Molecular Formula: $C_{10}H_{16}O_4$

95.0.6 Molecular Structure:

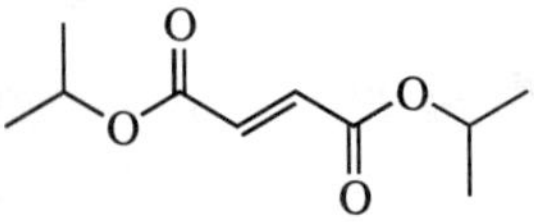

95.1 Chemical and Physical Properties

95.1.1 General

The empirical formula of diisopropyl fumarate is $C_{10}H_{16}O_4$. Its structural formula is *trans*-$(CH_3)_2CHOOCCH:CHCOOCH(CH_3)_2$. Its boiling point is 280°C, and melting point is 2°C. No other chemical or physical property data were located.

95.1.2 Odor and Warning Properties

No information found.

95.2 Production and Use

No information found.

95.3 Exposure Assessment

No information found.

95.4 Toxic Effects

95.4.1 Experimental Studies

The oral and dermal LD_{50} values for the rat and rabbit are 3.25 g/kg and 10 mL/kg, respectively (5). Lethality occurs in rats inhaling concentrated vapor for 8 h. It is moderately irritating (grade 4/10) to rabbit skin and is slightly irritating (grade 2/10) to rabbit eyes (5). No other toxicological information was located.

95.4.2 Human Experience

No information found.

95.5 Standards, Regulations, or Guidelines of Exposure

No information found.

95.6 Studies on Environmental Impact

No information found.

96.0 *n*-Dibutyl Fumarate

96.0.1 CAS Number: *[105-75-9]*

96.0.2 Synonyms: *trans*-2-Butenedioic acid; *n*-butyl diester; 2-butenedioic acid (E), dibutyl ester; butyl fumarate; fumaric acid, di-*n*-butyl ester; dibutylester kyseliny fumarove [Czech]; 4-02-00-02210 [Beilstein]; AI3-09505 [NLM]; BRN 1726635 [RTECS]; EINECS 203-327-9; and NSC 140 [NLM]

96.0.3 Trade Names: RC Comonomer DBF and Staflex DBF

96.0.4 Molecular Weight: 228.29

96.0.5 Molecular Formula: $C_{12}H_{20}O_4$

96.0.6 Molecular Structure:

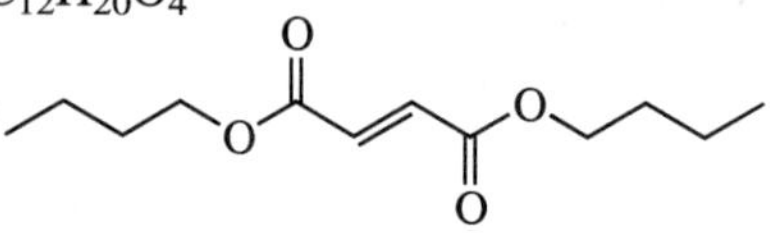

96.1 Chemical and Physical Properties

96.1.1 General

The empirical formula of dibutyl fumarate is $C_{12}H_{20}O_4$. Its structural formula is *trans*-$[CH_3(CH_2)_3OOCCH]_2$. Its density and boiling point are 0.987 and 141°C at 8 mmHg, respectively. It is soluble in alcohol and ether, and insoluble in water. Physico-chemical properties that were located are summarized in Table 80.15.

96.1.2 Odor and Warning Properties

No information found.

96.2 Production and Use

Dibutyl fumarate is used as a comonomer for production of polystyrene.

96.3 Exposure Assessment

No information found.

96.4 Toxic Effects

96.4.1 Experimental Studies

Acute toxicity data are summarized in Table 80.16. The oral and dermal LD_{50} values for the rat and rabbit are 8.53 g/kg, and 15.9 mL/kg, respectively (7). Dibutyl fumarate is moderately irritating to rabbit skin and slightly irritating to rabbit eyes (7). No other toxicological information was located.

96.4.2 Human Experience

No information found.

96.5 Standards, Regulations, or Guidelines of Exposure

No information found.

96.6 Studies on Environmental Impact

No information found.

97.0 Diisobutyl Fumarate

97.0.1 CAS Number: [7283-69-4]

97.0.2 Synonyms: *trans*-2-Butenedioic acid, isobutyl ester; fumaric acid, diisobutyl ester; and diisobutylester kyseliny fumarove [Czech]

97.0.3 Trade Names: NA

97.0.4 Molecular Weight: 228.29

97.0.5 Molecular Formula: $C_{12}H_{20}O_4$

97.0.6 Molecular Structure:

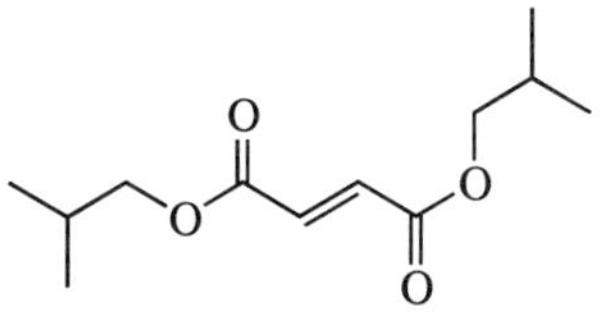

97.1 Chemical and Physical Properties

97.1.1 General

The empirical formula of diisobutyl fumarate is $C_{12}H_{20}O_4$. Its structural formula is *trans*-$[CH_3CH_2CH(CH_3)OOCCH]_2$. No other chemical or physical property data were located.

97.1.2 Odor and Warning Properties

No information found.

97.2 Production and Use

No information found.

97.3 Exposure Assessment

No information found.

97.4 Toxic Effects

The oral and dermal LD_{50} values for the rat and rabbit are 8.12 and 7.49 mL/kg, respectively (373). Diisobutyl fumarate causes a similar degree of skin irritation in rabbits (grade 3/10) as dimethyl or dihexyl maleate (373).

97.5 Standards, Regulations, or Guidelines of Exposure

No information found.

97.6 Studies on Environmental Impact

No information found.

98.0 Diisooctyl Fumarate

98.0.1 CAS Number: *[1330-75-2]*

98.0.2 Synonyms: *trans*-2-Butenedioic acid, diisooctyl ester; EINECS 215-546-7; and NSC 140 [NLM]

98.0.3 Trade Names: NA

98.0.4 Molecular Weight: 340.50

98.0.5 Molecular Formula: $C_{20}H_{36}O_4$

98.0.6 Molecular Structure:

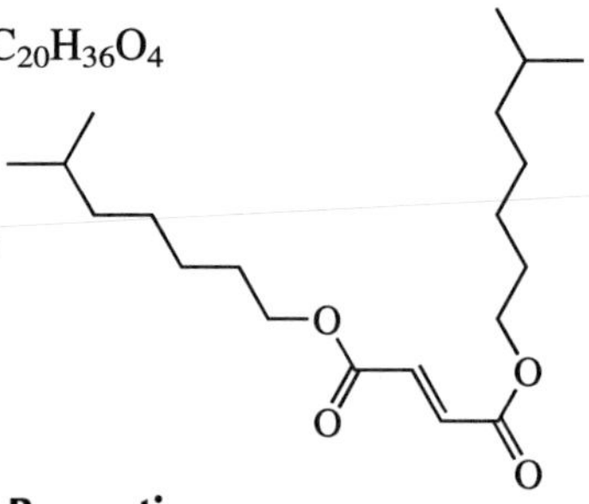

98.1 Chemical and Physical Properties

98.1.1 General

The empirical formula of diisooctyl fumarate is $C_{20}H_{36}O_4$. Its structural formula is *trans*-$[CH_3(CH_2)_4CH(CH_3)CH_2OOCCH]_2$. It is a clear, mobile liquid. No other chemical or physical data were located.

98.1.2 Odor and Warning Properties

Mild odor.

98.2 Production and Use

No information found.

98.3 Exposure Assessment

No information found.

98.4 Toxic Effects

No information found.

98.5 Standards, Regulations, or Guidelines of Exposure

No information found.

98.6 Studies on Environmental Impact

No information found.

99.0 Di(2-ethylhexyl) Fumarate

99.0.1 CAS Number: [141-02-6]

99.0.2 Synonyms: *bis* (2-Ethylhexyl) fumarate; 2-butenedioic acid (E), *bis*(2-ethylhexyl) ester; *trans*-2-butenedioic acid, 2-ethylhexyl ester; dioctyl fumarate; 2-ethylhexyl fumarate; fumaric acid *bis*(2-ethylhexyl) ester; *bis*-(2-ethylhexyl)ester kyseliny fumarove [Czech]; 4-02-00-02211 [Beilstein]; AI3-17415 [NLM]; BRN 1729134 [RTECS]; and EINECS 205-448-2

99.0.3 Trade Names: DOF and RC Comonomer DOF

99.0.4 Molecular Weight: 340.50

99.0.5 Molecular Formula: $C_{20}H_{36}O_4$

99.0.6 Molecular Structure:

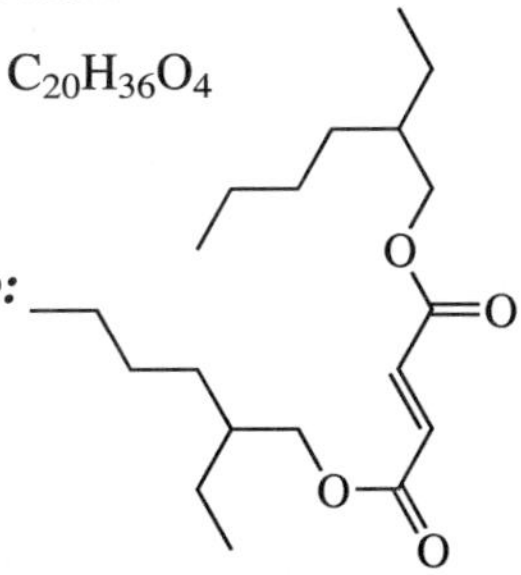

99.1 Chemical and Physical Properties

99.1.1 General

The empirical formula of di(2-ethylhexyl) fumarate is $C_{20}H_{36}O_4$. Its structural formula is $[CH_3(CH_2)_3CH(C_2H_5)CH_2OOCCH]_2$. Its density and boiling point are 0.942 and 211–220°C, respectively. No other chemical or physical data were located.

99.1.2 Odor and Warning Properties

No information found.

99.2 Production and Use

No information found.

99.3 Exposure Assessment

No information found.

99.4 Toxic Effects

Di(2-ethylhexyl) fumarate is of very low toxicity. The oral and dermal LD_{50} values for the rat and rabbit are 29.2 g/kg, and > 20 mL/kg, respectively (5). In rats, lethality occurred after inhalation of saturated vapor for 8 h. Di(2-ethylhexyl) fumarate was moderately irritating to rabbit skin (grade 5/10) and caused minimal eye injury in the Draize test (5).

99.5 Standards, Regulations, or Guidelines of Exposure

No information found.

99.6 Studies on Environmental Impact

No information found.

100.0 Dimethyl Itaconate

100.0.1 CAS Number: *[617-52-7]*

100.0.2 Synonyms: Butanedoic acid, methylene-, dimethyl ester; AI3-16883 [NLM]; CCRIS 7232; and EINECS 210-519-6

100.0.3 Trade Names: NA

100.0.4 Molecular Weight: 158.15

100.0.5 Molecular Formula: $C_7H_{10}O_4$

100.0.6 Molecular Structure:

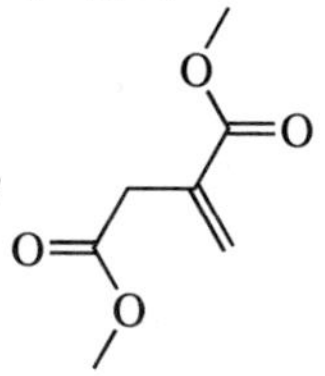

100.1 Chemical and Physical Properties

100.1.1 General

The empirical formula of dimethyl itaconate is $C_7H_{10}O_4$. Its structural formula is $CH_3OOCCH_2C(COOCH_3){:}CH_2$. Its melting and boiling points are 37–40 and 208°C, respectively. It is soluble in alcohol and ether. Physico-chemical properties are summarized in Table 80.15.

100.1.2 Odor and Warning Properties

No information found.

100.2 Production and Use

No information found.

100.3 Exposure Assessment

No information found.

100.4 Toxic Effects

100.4.1 Experimental Studies

100.4.1.1 Acute Toxicity. No information found.

100.4.1.2 Chronic and Subchronic Toxicity. Administration of 0.5% dimethyl itaconate in the drinking water of rats for 37 wk did not cause death or overt toxicity. Upon histological examination, 40% of dosed animals exhibited hyperplastic changes in gastric epithelium (478).

100.4.1.3 Pharmacokinetics, Metabolism, and Mechanisms. No information found.

100.4.1.4 Reproductive and Developmental. No information found.

100.4.1.5 Carcinogenesis. In rats treated with *N*-methylnitrosourea (100 ppm in drinking water for 15 wk) followed by dimethylitaconate (0.5% in drinking water for 37 wk), there was a significant increase in gastric cell proliferation and the incidence of adenocarcinoma, as compared to rats treated with *N*-methylnitrosourea alone (478).

100.4.1.6 Genetic and Related Cellular Effects Studies. Dimethylitaconate is an inducer of the phase II enzymes quinone reductase and glutathione *S*-transferase in mouse liver and stomach (479).

100.4.1.7 Other: Neurological, Pulmonary, Skin Sensitization. No information found.

100.4.2 Human Experience

No information found.

100.5 Standards, Regulations, or Guidelines of Exposure

No information found.

100.6 Studies on Environmental Impact

No information found.

101.0 Diethyl Itaconate

101.0.1 CAS Number: *[2409-52-1]*

101.0.2 Synonyms: Butanedioic acid, methylene-, diethyl ester; diethyl methylenebutanedioate; ethyl itaconate; itaconic acid, diethyl ester; succinic acid methylene–diethyl ester; AI3-11186 [NLM]; NSC1794 [NLM]; and NSC 67389 [NLM]

101.0.3 Trade Names: NA

101.0.4 Molecular Weight: 186.21

101.0.5 Molecular Formula: $C_9H_{14}O_4$

101.0.6 Molecular Structure:

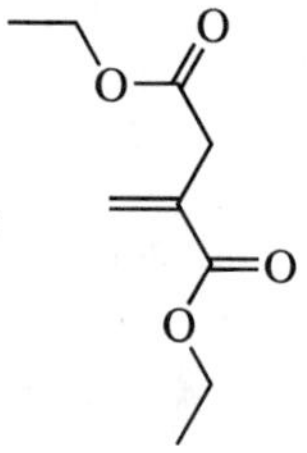

101.1 Chemical and Physical Properties

101.1.1 General

The empirical formula of diethyl itaconate is $C_9H_{14}O_4$. Its structural formula is $C_2H_5OOCCH_2C(COOC_2H_5){:}CH_2$. Its melting and boiling points are 58–59 and 228°C, respectively. It is soluble in alcohol and ether. Physico-chemical properties are summarized in Table 80.15.

101.1.2 Odor and Warning Properties

No information found.

101.2 Production and Use

No information found.

101.3 Exposure Assessment

No information found.

101.4 Toxic Effects

No information found.

101.5 Standards, Regulations, or Guidelines of Exposure

No information found.

101.6 Studies on Environmental Impact

No information found.

E INTRODUCTION TO ESTERS OF AROMATIC DIACIDS

The aromatic *ortho*-dicarboxylic acid (phthalate) esters are among the most important industrial chemicals. They are used as plasticizers for a variety of plastics; those of C8 and above are used to add flexibility to PVC. They also are used with vinyl and cellulose resins to lend toughness and flexibility. They are commonly used in wire and cable coverings, moldings, vinyl consumer products, and medical devices. Some low-molecular-weight phthalate esters (e.g., methyl, ethyl, and butyl) are used as industrial solvents rather than as plasticizers. Occasionally, these low-molecular-weight phthalates have applications for consumer products such as ink and lacquer. Physically, phthalates occur mainly in liquid form (see Table 80.17) with high boiling ranges and very low vapor pressures, with both contributing to the high stability of these materials.

The biological responses to phthalate esters vary based on the alcohol side chain and the animal species tested. In general, phthalate esters have low potential for acute toxicity following oral, dermal, or inhalation exposure (see Table 80.18). They are nonirritating or slightly irritating to the skin and eyes, and they are not sensitizers (480). Repeated oral exposure of rodents to high doses of phthalate esters results in hepatomegaly, changes in enzyme activity, and a burst of cell division in the liver (481–491). The induction of enzymes is associated with organelles called peroxisomes (392–394, 413, 414). The increased size and number of these organelles (peroxisome proliferation) have been associated with carcinogenesis in rodents (415). This mechanism may be linked to induction of enzymes that produce hydrogen peroxide (oxidative stress), mitogenic responses (cell proliferation), or interference with programmed cell death (apoptosis). In any case, the applicability of rodent liver cancer at high dose levels to humans via this mechanism has been questioned (416) because hamsters, guinea pigs, and primates are refractory to the effects on the liver (416–418, 482, 486–488, 493, 494). Recent studies have shown that a nuclear receptor (PPARα) plays a central role in the carcinogenic process in rodents (421). The expression of this receptor in human liver is lower than in rodent liver, suggesting that receptor density is a key element for species specificity (422, 423). Some phthalate esters [generally dibutyl through dihexyl and di(2-ethylhexyl)-phthalate] impair the reproduction of rodents apparently by affecting the testes and decreasing sperm production (481, 495–500). Effects on the testes of primates have not been observed (417, 419).

Some phthalate esters [generally dibutyl through dihexyl and di(2-ethylhexyl)phthalate], have also demonstrated some developmental toxicity (497, 501–507). The effects include craniofacial abnormalities, spina bifida, and urogenital abnormalities. The effects are dose and time dependent. Skeletal anomalies such as extra ribs or fused ribs also have been described. It is not known if these effects may occur in species other than rodents.

Biological effects in rodents occur following oral, but generally not following dermal, IV, or inhalation exposure (508, 509). This difference is believed to be due to the fact that in the gastrointestinal tract, the diester is rapidly metabolized to the half-ester or monoester (only one alcohol), which is an active metabolite. Metabolism to the monoester also occurs outside the gastrointestinal tract via nonspecific lipase activity but is not as rapid or complete.

102.0 Dimethyl Phthalate

102.0.1 CAS Number: [131-11-3]

102.0.2 Synonyms: 1, 2-Benzenedicarboxylic acid, dimethyl ester

Table 80.17. Physical and Chemical Properties of Esters of Aromatic Diacids

Chemical Name	CAS No.	Molecular Formula	Mol. Wt.	Boiling Point (°C) (mm Hg)	Melting Point (°C)	Density (Specific Gravity	Sol. in Water at 20°C (68°F)	Refractive Index (20°C)	Vapor Pressure (mmHg) (°C)	LEL % by Vol. Room Temp.
Phthalate										—
Dimethyl	*[131-11-3]*	$C_{10}H_{10}O_4$	194.19	283.7 (760 ton)	5.5	(1.194)	4.3 g/L	1.5138	< 0.01 (20)	—
Diethyl	*[84-66-2]*	$C_{12}H_{14}O_4$	222.23	298	− 40.5	(1.232)	0.896 g/L	1.5049	4×10^{-4} (25)	—
n-Dibutyl	*[84-74-2]*	$C_{16}H_{22}O_4$	278.34	340	− 35	(1.0459)	0.013 g/L	1.4911	1.4×10^{-5} (20)	—
n-Dihexyl	*[85-75-3]*	$C_{20}H_{30}O_4$	334.46	340–350	− 58	(0.995)	insol	—	—	—
Diisohexyl	*[68515-50-4]*	$C_{20}H_{30}O_4$	334.46	340–350	− 33	(1.01)	insol	—	0.01 (100)	—
n-Diheptyl	*[3648-21-3]*	$C_{22}H_{34}O_4$	362.51	360	—	—	0.1 g/L	—	2.1×10^{-6} (25)	—
Diisoheptyl	*[41451-28-9]*	$C_{22}H_{34}O_4$	362.51	> 300	− 45	(0.994)	insol	—	0.006 (100)	—
n-Dioctyl	*[117-84-0]*	$C_{24}H_{38}O_4$	390.56	220	− 25	(0.978)	insol	0.97	1×10^{-7} (25)	—
Diisooctyl	*[27554-26-3]*	$C_{24}H_{38}O_4$	390.56	> 300	− 46	(0.99)	insol	—	1 (200)	—
Di(2-ethylhexyl)	*[117-81-7]*	$C_{24}H_{38}O_4$	390.54	385 (760 ton)	− 46	(0.9861)	0.34 mg/L[a]	1.485	6.2×10^{-6} (25)	—
Dinonyl	*[84-76-4]*	$C_{26}H_{42}O_4$	418.62	205 (1)	—	(0.97)	insol	—	1 (205)	—
Diisononyl	*[68515-48-0]*; *[28553-12-0]*	$C_{26}H_{42}O_4$	418.62	245–252 (5)	− 48	(0.97)	insol	—	0.5 (200)	—
Diisodecyl	*[68515-49-1]*	$C_{28}H_{46}O_4$	446	256 (5)	− 48	(0.97)	insol	—	0.34 (200)	—
Diundecyl	*[3648-20-2]*	$C_{30}H_{50}O_4$	474.72	270 (5)	− 45	(0.96)	insol	—	< 0.15 (392)	—
Diisoundecyl	*[85507-79-5]*	$C_{30}H_{50}O_4$	474	270 (5)	− 45	(0.96)	insol	—	5.3×10^{-7} (298)	—
Diisotridecyl	*[27253-26-5]*	$C_{34}H_{58}O_4$	530.83	> 250	− 30	(0.957)	insol	—	0.9 (200)	—
Di-C7-11		—	418.6	235–278	− 57	(0.97)	< 0.1 mg/L[a]	1.482 (25)	0.3 (180)	—
Diallyl	*[131-17-9]*	$C_{14}H_{14}O_4$	246.26	290	—	(1.121)	< 0.1 g/L	—	2.4 (150)	—
Butyl Benzyl	*[85-68-7]*	$C_{19}H_{20}O_4$	312.36	370	− 35	(1.1)	2.69 mg/L	—	8×10^{-7} (25)	—

Dimethoxyethyl	[*117-82-8*]	$C_{14}H_{18}O_6$	282.29	340	−45	1.1708	8.5 mg/L	1.502	0.01 (20)	—
Diethoxyethyl	—	$C_{16}H_{22}O_6$	310.35	345	− 34	1.1229	insol	—	—	—
Dibutoxyethyl	[*117-83-9*]	$C_{20}H_{30}O_6$	366.45	270	−55	1.063	300 mg/L	1.486	0.5 (150)	—
Methylphthalyl ethylglycolate	[*85-71-2*]	$C_{13}H_{14}O_6$	266.25	189–195	<−35	1.220	insol	—	—	—
Ethylphthalyl ethyl glycolate	[*84-72-0*]	$C_{14}H_{16}O_6$	280.28	320	—	1.180	insol	—	—	—
Butylphthalyl butylglycolate	[*85-70-1*]	$C_{18}H_{24}O_6$	336.38	220 (10)	<−35	(1.1)	120 ppm [a]	—	—	—
Dimethyl terephthalate	[*120-61-6*]	$C_{10}H_{10}O_4$	194.19	288	141	1.04	<1 g/L at 13°C	—	.0106 (25)	—
Di(2-ethylhexyl) terephthalate	[*6422-86-2*]	$C_{24}H_{38}O_4$	390.56	400	30–34	(0.984)	350 µg/L	—	1 (217)	—

[a] At 25°C.

Table 80.18. Summary of Acute Toxicity of Esters of Aromatic Diacids

Chemical Name	CAS No.	Species	Exposure Route	LD_{50} (g/kg)	Ref.
Phthalate					
Dimethyl	[*131-11-3*]	Rat	oral	6.9 mL/kg	2
		Mouse	oral	7.2 mL/kg	2
		Guinea pig	oral	2.4 mL/kg	2
		Rabbit	oral	4.4 mL/kg	2
		Rabbit	dermal	> 10 mL/kg	48
		Mouse	IP	1.58, 3.6	12, 561
		Cat	inhal	9.3 mg/L	12
Diethyl	[*84-66-2*]	Rat	oral	> 5	536, 537
		Mouse	oral	> 5	536, 537
		Guinea pig	oral	> 5	536, 537
		Rabbit	oral	1.0	536, 537
		Rabbit	dermal	> 20 mL/kg	537
		Mouse	IP	2.8	12, 551
		Guinea pig	sc	3.0	12
n-Dibutyl	[*84-74-2*]	Rat	oral	8.0	549
		Mouse	oral	> 5.0	550
		Rat	IM	8.0	549
		Mouse	IP	4.0	551
		Rabbit	dermal	> 20 mL/kg	550
n-Dihexyl	[*85-75-3*]	Rat	oral	29.6	5
		Rabbit	dermal	> 20 mL/kg	5
Diisoheptyl	[*71888-89-6*]	Rat	oral	> 10	552
		Rabbit	dermal	> 3.16	552
n-Dioctyl	[*117-81-7*]	Rat	oral	13.0	553
		Mouse	oral	53.7	254
		Guinea pig	dermal	75 mL/kg	12
Diisooctyl	[*27554-26-3*]	Rat	oral	> 20	555
Di(2-ethylhexyl)	[*117-81-7*]	Rat	oral	30.6 > 34.5	556, 557
		Mouse	oral	> 25.6	558
		Rabbit	oral	33.9	556
		Rabbit	dermal	> 20 mL/kg	48
		Guinea pig	dermal	> 10 mL/kg	6
		Rat	IP	30.7	12
		Rat	inhalation	4–8.7 mg/L	556
Dinonyl	[*84-76-4*]	Rat	oral	> 2	12
Diisononyl	[*68515-48-0, 28553-12-0*]	Rat	oral	> 10	555
		Rabbit	dermal	> 3.16	485, 605
Diisodecyl	[*68515-49-1, 26761-40-0*]	Rat	oral	> 6	373, 555, 363, 559
		Rabbit	dermal	> 3.16	560
Diundecyl	[*3648-20-2*]	Rat	oral	> 15	561
		Rabbit	dermal	> 7.9	561
		Mouse	IP	> 100	559

Table 80.18. *(Continued)*

Chemical Name	CAS No.	Species	Exposure Route	LD_{50} (g/kg)	Ref.
Diisotridecyl	[*68515-47-9*]	Rat	oral	> 10	555
		Rabbit	dermal	> 3.16	562
Di-C7-11	[*68515-42-4*]	Rat	oral	> 15.8	555, 564
		Mouse	oral	> 20	555
		Rabbit	dermal	> 7.94	564
Diallyl	[*131-17-9*]	Rat	oral	0.896	565
		Mouse	oral	1.0–1.47	566
		Rabbit	oral	1.7	12
		Rabbit	dermal	3.8–3.9	565
		Mouse	IP	0.7	12
Butyl benzyl	[*85-68-7*]	Mouse	oral	6.2 M, 4.2 F	567
		Rat	dermal	6.7	568
		Mouse	dermal	6.7	568
		Rabbit	dermal	> 10	569
Dimethoxyethyl	[*117-82-8*]	Rat	oral	> 4.4	12
		Mouse	oral	3.2–6.4	12
		Guinea pig	oral	1.6–3.2	12
		Guinea pig	dermal	> 10	12
		Rat	IP	3.74 mL/kg	500
		Mouse	IP	2.51	551
Dibutoxyethyl	[*117-83-9*]	Rat	oral	8.4	12
Methylphthalyl ethylglycolate	[*85-71-2*]	Rat	oral	9.04	570
		Guinea pig	dermal	> 10 mL/kg	12
Ethylphthalyl ethyl glycolate	[*84-72-0*]	Mouse	oral	5.66 mL/kg	238
		Mouse	IP	3.65 mL/kg, 4.38 g/kg	238, 559
Butylphthalyl butylgylcolate	[*85-70-1*]	Rat	oral	7.0	571
		Mouse	oral	12.567	572
		Rat	IP	6.889 mL/kg	501
Di (2-ethylhexyl) isophthalate	—	Rat	oral	> 3.2	373
		Rabbit	dermal	7.94 mL/kg	373
		Guinea pig	dermal	> 20 mL/kg	373
Dimethyl terephthalate	[*120-61-6*]	Rat	oral	> 6.59	573
		Rat	IP	3.9	573
Di (2-ethylhexyl) terephthalate	[*6422-86-2*]	Rat	oral	> 5	574
		Mouse	oral	3.2	574
		Guinea pig	dermal	> 20 mL/kg	574
Tetrahydrophthalate					
Dimethyl	—	Rat	oral	0.7	446
		Rabbit	dermal	> 10 mL/kg	446
Di(2-ethylhexyl)	—	Rat	oral	114	12
Didecyl	—	Rat	oral	64	
		Rabbit	dermal	> 20 mL/kg	
Di(1,4-hexadienyl)	—	Rat	oral	45.2	
	—	Rabbit	dermal	> 16 mL/kg	

102.0.3 Trade Names: Palatinol M and Eastman DMP Plasticizer

102.0.4 Molecular Weight: 194.19

102.0.5 Molecular Formula: $C_{10}H_{10}O_4$

102.0.6 Molecular Structure:

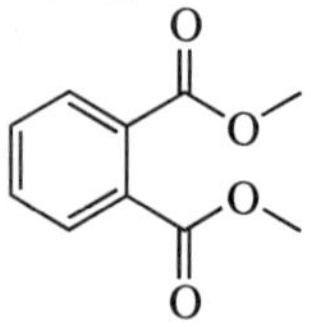

102.1 Chemical and Physical Properties

102.1.1 General

The empirical formula of dimethyl phthalate (DMP) is $C_{10}H_{10}O_4$. Its structural formula is $C_6H_4(COOCH_3)_2$. It is a colorless liquid that has a specific gravity of 1.19. It is very soluble in alcohol and ether and is slightly soluble in water (4.3 g/L). It has a very low vapor pressure, and is a liquid between 5.5°C (melting point) and 283.7°C (boiling point). Physico-chemical properties are summarized in Table 80.17.

102.1.2 Odor and Warning Properties

There is no odor threshold established for DMP, but the substance is described as having a slightly aromatic or sweet odor.

102.2 Production and Use

The compound is manufactured via the reaction of methanol with phthalic anhydride in the presence of sulfuric acid. It serves as a solvent and plasticizer for cellulose acetate, which is used for face shields and tool handles. During World War II, DMP was used as a mosquito and insect repellent (12, 570), but its continued use in these products cannot be confirmed. DMP may also be used for manufacturing lacquers, varnishes, coatings, molding powders, and perfumes. In addition, it may be used as a solvent in hair spray.

102.3 Exposure Assessment

Air samples are collected using charcoal and analyzed via gas chromatography (430).

102.4 Toxic Effects

102.4.1 Experimental Studies

102.4.1.1 Acute Toxicity. Dimethyl phthalate (DMP) has low acute toxicity for all routes of exposure. The oral LD_{50} values for guinea pigs, rabbits, rats, and mice are 2.4, 6.9, and 7.2 mL/kg, respectively (Table 80.17). The dermal LD_{50} value for rabbits is > 10 mL/kg. If ingested, it may irritate the gastrointestinal tract (12).

102.4.1.2 Chronic and Subchronic Toxicity. DMP was applied to the skin of rabbits daily for 90 d at doses of 0.5, 1, 2, or 4 mL/kg. There were “no changes in hemoglobin values and morphology of the blood." However, "edema of the lungs,. . .mild to severe chronic nephritis, and slight to moderate liver damage" were seen at 4 mL/kg/d. It cannot be determined if the animals were allowed to ingest DMP while grooming. The LD_{50} value was >4 mL/kg/d, and the no-observed-effect level (NOEL) was 2 mL/kg/d (511).

In mice inhaling 0.7–1.8 mg/m^3 DMP (4 h/d) for 4 mo, changes in "the frequency of respiration, function of the nervous system, liver, and kidneys, and blood morphology" were observed (512). It is difficult to assess the quality of the data from a transcript of the abstract, since no specifics of the study were given. No other inhalation studies were located.

102.4.1.3 Pharmacokinetics, Metabolism, and Mechanisms. DMP is readily absorbed from the skin, intestinal tract, the peritoneal cavity, and lung. Following absorption, it is rapidly metabolized to the monoester, monomethyl phthalate, and methanol. Monomethyl phthalate can be absorbed through the intestinal wall (513–514). This monoester may undergo further oxidation of the alcohol side chain, but further hydrolysis is minimal (515).

The pharmacokinetics of dermally applied DMP have been examined in rats. Dermal absorption of DMP through rat skin was constant over a 7-d period, with 6–7.5% of the dose excreted in the urine each day. Urine was the primary source of excretion. After 7 d, 19% of the applied dose remained at the application site, but a substantial portion of the dose was unaccounted for (516). Maximum levels of DMP were found in the blood 20 min after oral administration, and the majority of the dose was excreted within 12 h (517).

102.4.1.4 Reproductive and Developmental. In a study of the developmental toxicity of DMP following IP injection, Singh et al. (501) found evidence of resorptions, fetotoxicity, and skeletal anomalies after injecting >1 g/kg into the peritoneal cavity. In a standard rat developmental toxicity (teratology) study, pregnant rats were fed diets containing 0, 0.25, 1.0, or 5.0% DMP on gestational days 6–15 (518). No effects were seen in the fetuses even at dose levels producing significant maternal effects. The no-observed-adverse-effect level (NOAEL) for developmental toxicity was 5% in the diet (3.6 g/kg/d). These data confirm earlier work in which DMP was found to produce no effects in the Chernoff–Kavlock developmental screening test in mice (496, 519).

At a concentration up to 10^{-3} M, DMP was not active in a recombinant yeast screen for estrogenic activity (526). Oral exposure of male rats to 20,000 ppm DMP in the diet for 10 d resulted in decreased levels of testosterone, but no decrease in testicular weight (521). The significance of this finding is not clear. In addition, DMP did not cause testicular lesions in rats or decrease the level of zinc, an early indicator of testicular damage (498). Fredricsson and co-workers found that DMP had little direct effect on human sperm motility (522).

102.4.1.5 Carcinogenesis. In a 2-yr feeding study, 10 female rats per dose level were fed diets containing 2%, 4%, or 8% DMP. There was a slight but significant effect on weight gain at the two higher dose levels. Mortality rates did not differ between treated and control animals. "Chronic nephritis" was seen only at the 8% level. There was no mention of carcinogenicity in this 2-yr feeding study. The NOEL was 2% in the diet (511).

In an initiation–promotion study using mouse skin, DMP was tested for its ability to initiate the development of tumors or promote initiated tumor cells (523). There was no evidence that DMP acted to initiate or promote the development of tumors in mice.

102.4.1.6 Genetic and Related Cellular Effects Studies. The mutagenic potential of DMP has been evaluated in a number of *in vitro* and *in vivo* assays. Studies in *S. typhimurium* strains TA98, TA100, TA1535, and TA1537 with and without S-9 metabolic activation systems indicated that at concentrations up to 5 mg/plate, DMP was nonmutagenic (406, 523). However, Agarwal suggested that DMP was "mildly positive" using TA100 and TA1535 in the absence of S-9 (524). DMP failed to induce dominant lethal mutations in male mice given a single intraperitoneal injection of 250 mg/kg or repeated dermal applications of 1250 mg/kg (525). DMP also had no effect on chromosomal aberrations in Chinese hamster ovary or human leukocyte cell cultures when compared to controls (523, 526, 527). When tested in the Balb/3T3 mouse cell line at concentrations up to 931.6 nL/mL, DMP produced no significantly greater transformation frequencies than controls (528).

DMP induced specific locus mutations in mouse lymphoma cells (mouse lymphoma mutation assay) and sister chromatid exchange in cultured Chinese hamster ovary cells in the presence of S-9 but not in the absence of S-9 (523, 529, 530). Therefore, DMP is mutagenic only in certain *in vitro* studies after metabolism. This is probably due to the formation of a reactive species such as formaldehyde. Since DMP is not mutagenic *in vivo*, any reactive metabolites appear to be quickly detoxified.

102.4.1.7 Other: Neurological, Pulmonary, Skin Sensitization. DMP is slightly irritating to eyes and mucous membranes, but not to the skin (136).

102.4.2 Human Experience

The human skin absorption rate has been reported as 33 $\mu g/cm^2/h$ and as 39.51 $\mu g/cm^2/h$ (531, 532). Both values are consistent with a rating of "slow." Using 39.51 $\mu g/cm^2/h$ for the absorption rate, 720 cm^2 for the skin surface area of both hands, and 60 kg for body weight, contact with the skin of both hands for 1 h can be estimated to produce an absorbed dose of approximately 0.5 mg/kg.

In 144 subjects with suspected occupational dermatoses to plastic or glue allergens, three subjects (2.1%) experienced irritation after patch testing with 5.0% dimethyl phthalate. None of the patients had allergic reactions (129).

102.5 Standards, Regulations, or Guidelines of Exposure

The Cosmetic Ingredient Review Expert Panel has concluded that dimethyl phthalate is safe for use in cosmetics following current practices and at currently used concentrations (5%) (480). The current TLV for DMP set by the ACGIH is 5 mg/m^3 (549). The OSHA PEL and NIOSH REL are also 5 mg/m^3.

102.6 Studies on Environmental Impact

No information found.

103.0 Diethyl Phthalate

103.0.1 CAS Number: [84-66-2]

103.0.2 Synonyms: 1, 2-Benzenedicarboxylic acid, diethyl ester

103.0.3 Trade Names: Palatinol A and Eastman DEP Plasticizer

103.0.4 Molecular Weight: 222.23

103.0.5 Molecular Formula: $C_{12}H_{14}O_4$

103.0.6 Molecular Structure:

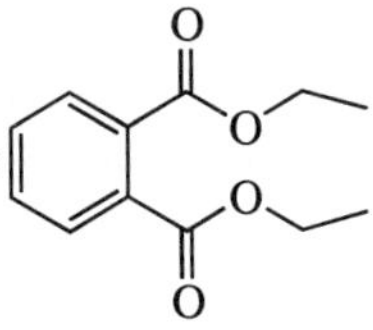

103.1 Chemical and Physical Properties

103.1.1 General

The empirical formula of diethyl phthalate (DEP) is $C_{12}H_{14}O_4$. Its structural formula is $C_6H_4\ (COOC_2H_5)_2$. DEP is a colorless, oily liquid with a bitter, disagreeable taste. It is a liquid from $-40.5°C$ (melting point) to 298°C (boiling point) with a very low vapor pressure (4×10^{-4} mmHg at 25°C). DEP is slightly soluble in water (1.10 g/L) and is soluble in organic solvents. It has a specific gravity of slightly greater than water (1.12). Physico-chemical properties are summarized in Table 80.17.

103.1.2 Odor and Warning Properties

DEP has a slightly aromatic odor.

103.2 Production and Use

DEP is manufactured via reaction of phthalic anhydride with ethanol in the presence of sulfuric acid. It is used as a plasticizer for cellulose acetate. It may be used in cosmetics such as perfume, hair spray, soaps, and nail polish (480, 534).

103.3 Exposure Assessment

Since DEP is used in consumer products, consumer exposure is expected. A survey of human adipose tissue demonstrated that DEP was detectable in fat of various age groups (535). The tissue level was not provided; however, for individuals in the southwestern and northeastern parts of the United States, the relative rank was high compared with other "semivolatile" organics such as trichlorobenzene, organic phosphates, and naphthalene.

103.4 Toxic Effects

103.4.1 Experimental Studies

103.4.1.1 Acute Toxicity. DEP is of low acute toxicity in experimental animals. Oral LD_{50} values for rats, mice, and guinea pigs are >5 g/kg. The oral LD_{50} value for rabbits is 1 g/kg (536, 537). Inhalation exposure of rats to a heated vapor (4.64 mg/L) for 6 h did not produce mortality (537). The dermal LD_{50} value for rabbits is >20 mL/kg (537). Acute toxicity data are summarized in Table 80.18.

A test of the utility of DEP as a solvent for *in vitro* phototoxicity studies indicated that 30% DEP in corn oil was cytotoxic to cells, and that 10% DEP caused 30% cell lethality. These data suggest that DEP is not an appropriate solvent for *in vitro* toxicity studies (538).

103.4.1.2 Chronic and Subchronic Toxicity. The subchronic oral toxicity of DEP has been studied in rats given 0, 0.2, 1.0, or 5.0% DEP in the diet for 16 wk (539). Body weights were consistently lower for the 5.0% group, and occasionally lower for the 1.0% group. There were no adverse effects on hematology parameters. In male rats treated with 5.0%, relative liver and testes weights were higher after 2 and 6 wk of treatment. At necropsy, relative brain, heart, liver, kidney, stomach, adrenal gland, testes, pituitary, and thyroid weights were elevated in males treated with 5.0%; and brain, liver, spleen, kidney, and stomach weights were elevated in females treated with 5.0%. The only changes observed in animals treated with 0.2 or 1.0% DEP was an increase in relative liver weight (females only).

The NTP conducted subchronic and chronic dermal studies with DEP in mice and rats (523). Doses of 0, 37.5, 75, 150, or 300 μL of neat DEP were applied to the shaved skin of F344 rats, and 0, 12.5, 25, 50, or 100 μL were applied to the skin of $B6C3F_1$ mice for 4 wk. Male rats given 300 μL and female rats given 150 or 300 μL had significantly higher relative liver weights. Relative kidney weights for the 150 μL female group were also higher, but kidney weights showed no dose–response trend. There were no clinical signs of toxicity, body weight changes, or dermatotoxicity. Female mice given 25 or 100 μL DEP had higher absolute and relative liver weights. No effect was seen in the 50-μL dose group. There were no clinical signs of toxicity, body weight changes, or dermatotoxicity in any treatment group.

Chronic studies in mice and rats also were conducted by the NTP (523). Doses of 0, 100, or 300 μL neat DEP were applied to the shaved skin of F344 rats for 2 yr. There was a slight decrease in weight gain in male rats dosed with 300 μL between weeks 16 and 90, but there were no clinical signs of toxicity or signs of dermatotoxicity. Differences in organ weight were seen at 4 wk of treatment, but there were no differences in organ weight after 15 mo of treatment. No treatment-related changes in hematology or clinical chemistry were noted. There was no evidence of carcinogenesis, and local effects on the skin at the site of application were considered to be adaptive.

Doses of 0, 7.5, 15, or 30 μL neat DEP were applied to the shaved skin of $B6C3F_1$ mice for 2 yr. There were no differences in body weight, no clinical signs of toxicity, and no signs of dermatotoxicity. At 15 mo, relative kidney weights were slightly higher in females treated with 15 or 30 μL. There were no treatment-related changes in hematology after 15 mo. There was equivocal evidence of liver carcinogenesis. However, it was noted that this strain of mouse is prone to liver tumors. There was no evidence of tumors at the site of application (523).

103.4.1.3 Pharmacokinetics, Metabolism, and Mechanisms. DEP is readily absorbed from the skin, intestinal tract, peritoneal cavity, and lung. Following absorption, it is rapidly hydrolyzed to the monoester, monoethyl phthalate, and ethanol. Monoethyl phthalate can be absorbed through the intestinal wall (513). It may undergo further oxidation of the alcohol side chain, but further hydrolysis is minimal (515).

The pharmacokinetics of DEP after dermal application have been examined in rats. After 7 d, 34% of the dose remained at the application site. A substantial portion of the dose (25%) was unaccounted for (516). Urine was the primary source of excretion, with 24% of the dose excreted in the urine during the first day.

103.4.1.4 Reproductive and Developmental. In a study of the developmental toxicity of DEP following IP injection, Singh et al. (501) found evidence of resorptions, fetotoxicity, and skeletal anomalies after injecting >1 g/kg into the peritoneal cavity. In a standard rat developmental toxicity (teratology) study, pregnant rats were fed diets containing 0, 0.25, 2.5, or 5.0% DEP on gestational days 6–15. Maternal toxicity was observed at 2.5 and 5.0% DEP, but there was no developmental toxicity except for an increase in extra ribs (540). These data confirm earlier work in which DEP was found to produce no developmental effects in the Chernoff–Kavlock screening test in mice (496).

The reproductive toxicity of DEP was assessed in mice treated with 0, 0.25, 1.25, or 2.5% in the diet (495). There were no effects on reproduction. Studies evaluating direct effects on the testes have reported conflicting results. Jones et al. (541) treated rats with 2 g/kg DEP by gavage and found structural changes in the testes after 2 d, while Foster (498) found no adverse effects in rats after 4 d with the same dose of DEP. Fredricsson et al. (522) reported a decrease in human sperm motility following incubation with >1 mM DEP.

At a concentration of 10^{-3} M, DEP exhibited weak estrogenic activity in a recombinant yeast screen (520). Its approximate potency relative to 17β-estradiol was 0.0000005. At 10^{-5} M, it induced proliferation in estrogen-responsive breast cancer cell line MCF-7, but not ZR-75 (520). The significance of these results is unclear, given the lack of demonstrable effects *in vivo* (542).

103.4.1.5 Carcinogenesis. As mentioned, chronic dermal studies in mice and rats were conducted by the NTP (543). Doses of 0, 100, or 300 μL neat DEP were applied to the shaved skin of F344 rats for 2 y. There was no evidence of carcinogenesis, and effects on the skin at the site of application were considered to be adaptive. Doses of 0, 7.5, 15, or 30 μL neat DEP were applied to the shaved skin of $B6C3F_1$ mice for 2 y. There was equivocal evidence of liver carcinogenesis. There was no evidence of tumors at the site of application.

103.4.1.6 Genetic and Related Cellular Effects Studies. The mutagenic potential of DEP has been evaluated in a number of *in vitro* and *in vivo* assays. Studies in *S. typhimurium* strains TA98, TA100, TA1535, and TA1537 with and without S-9 metabolic activation systems showed no mutagenic activity at concentrations up to 10 mg/plate (406). In contrast, Kozumbo et al. (543) and Agarwal et al. (524) reported that DEP (at concentrations from 10–2000 μg/plate) was positive in strains TA100 and TA1535 in the absence of S-9. DEP had no effect on chromosomal aberrations in Chinese hamster ovary cells when compared to controls (526, 544).

103.4.1.7 Other: Neurological, Pulmonary, Skin Sensitization. DEP is slightly irritating to the eye (545) and skin (537). In an irritation test involving three rabbits, DEP elicited a primary irritation index score of 0.17 (out of a maximum of 8) (11).

103.4.2 Human Experience

The human skin absorption rate of DEP has been reported as 12.75 μg/cm^2/h (532) which is considered to be "slow." Using 12.75 μg/cm^2/h for the absorption rate, 720 cm^2 for the skin surface area of both hands, and 60 kg for body weight, contact with the skin of both hands for one hour can be estimated to produce an absorbed dose of approximately 0.15 mg/kg.

103.5 Standards, Regulations, or Guidelines of Exposure

The current TLV for DEP set by the ACGIH is 5 mg/m^3 (533).

103.6 Studies on Environmental Impact

DEP may enter the environment from waste sites (534). Some investigators have reported DEP in air and water (546), but recent data are not available. A method for quantification of DEP in water and sediment has been developed using HPLC (547).

104.0 *n*-Dibutyl Phthalate

104.0.1 CAS Number: *[84-74-2]*

104.0.2 Synonyms: 1, 2-Benzenedicarboxylic acid, dibutyl ester

104.0.3 Trade Names: Palatinol C and Eastman DBP Plasticizer

104.0.4 Molecular Weight: 278.34

104.0.5 Molecular Formula: $C_{16}H_{22}O_4$

104.0.6 Molecular Structure:

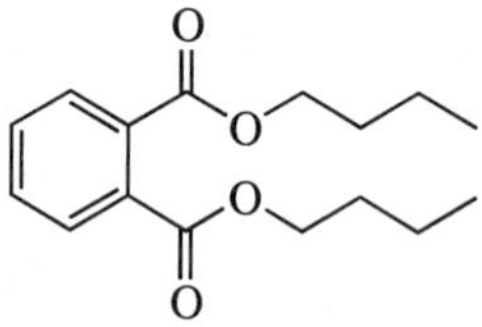

104.1 Chemical and Physical Properties

104.1.1 General

The empirical formula of *n*-dibutyl phthalate (DBP) is $C_{16}H_{22}O_4$. Its structural formula is 1,2-$(COOC_4H_9)_2C_6H_4$. DBP is a colorless to faint yellow, oily liquid. It is a liquid from −35°C (melting point) to 340°C (boiling point) with a very low vapor pressure (1.42×10^{-5} mmHg at 20°C). DBP is virtually insoluble in water (0.013 g/l) but is soluble in organic solvents. It has a specific gravity of slightly greater than water (1.0459). Physico-chemical properties are summarized in Table 80.17.

104.1.2 Odor and Warning Properties

DBP has a faint, fruity or aromatic odor.

104.2 Production and Use

DBP is manufactured by esterification of phthalic anhydride and butanol. It is used primarily as a plasticizer for polymers contained in adhesives, printing inks, lacquers, and cosmetics. It is also used as a plasticizer for cellulose plastics (480, 482).

104.3 Exposure Assessment

The levels of DBP in the workplace have generally been below 0.2 mg/m^3 (482). Since DBP is used in consumer products such as adhesives and cosmetics, some consumer exposure is expected. However, the amount of DBP used in cosmetics tends to be less that 5% of the total ingredients (480).

Methods are available to detect DBP in tissue samples (482). A survey of human adipose tissue demonstrated that DBP was detectable in fat of various age groups (551). The tissue level was not provided; however, for individuals in the southwestern and northeastern parts of the United States, the relative rank was high compared with other "semivolatile" organics such as trichlorobenzene, organic phosphates, and naphthalene.

In 1998, the Ministry of Agriculture, Fisheries and Food (MAFF) in the United Kingdom determined that DBP was detectable in 12 of 39 infant formula samples (548). The concentrations detected ($<0.05-0.09$ mg/kg) were well below the tolerable daily intake levels established by the U.K. government.

104.4 Toxic Effects

104.4.1 Experimental Studies

104.4.1.1 Acute Toxicity. The LD_{50} values for DBP show little acute toxicity. Oral LD_{50} values are >5 g/kg for rats and mice (549, 550). Inhalation exposure of rats to an aerosol of >15 mg/L (but not 12 mg/L) for 4 h produced mortality (550). The LC_{50} value for mice was 20 mg/L for a 2 h exposure (550). The dermal LD_{50} value was >20 mL/kg for rabbits (550). Additional LD_{50} data are listed in Table 80.18.

104.4.1.2 Chronic and Subchronic Toxicity. Numerous subchronic toxicity studies of DBP have been conducted in mice and rats. These studies have focused on the effect of DBP on the liver (specifically increases in liver enzyme activity associated with organelles called peroxisomes) and the testes (482). Changes in the liver or testes of rats, mice, and guinea pigs were observed after 7 d of treatment with high oral dose levels (2 g/kg). The no observable effect level (NOEL) appears to be 152–177 mg/kg for rats, with the testes being the most sensitive organ. The NOEL is higher in mice. Guinea pigs were resistant to the liver effects, and hamsters were resistant to the testicular effects of DBP. No effect on the liver or testes of rats was seen following inhalation exposure to DBP (508, 509).

Oral exposure of rats to 120 or 1200 mg/kg/d DBP for 3 mo resulted in 5% mortality and increased liver weight, but no histologic changes in the liver (575). Exposure to 0.125% DBP in the diet for 12 mo resulted in 15% mortality, but no "remarkable alterations" in the liver, kidneys, or spleen. In another 12-mo study, mortality was 50% for rats treated with 1.25%. No effect on body weight, hematology, or pathology was noted in rats treated with 0, 0.01, 0.05, 0.25, or 1.25% DBP (565). Likewise, no pathologic effects were observed in rats treated orally with 1 mL/kg twice weekly for 78 wk (576).

104.4.1.3 Pharmacokinetics, Metabolism, and Mechanisms. DBP is readily absorbed from the skin, intestinal tract, peritoneal cavity, and lung (515). Following dermal exposure, urine is the primary route of excretion in the rat. The excretion rate remains nearly constant at 10–12% of the dose excreted per day (516). Following oral administration, DBP is rapidly and completely (95%) hydrolyzed in the gastrointestinal tract to the monoester, monobutyl phthalate (MBP) and butanol. The monoester is absorbed through the intestinal wall (513, 514, 577). White et al. (578) reported that peak blood levels were obtained 2 h after oral treatment of rats. Another study employing two dose levels indicated that peak blood levels occur within 30 min for mice, and 60 min for rats and hamsters (579). High dose levels (>100 mg/kg) of DBP resulted in nonlinear kinetics due to enterohepatic circulation or decreased elimination. Excretion was primarily via the urine and was virtually complete (90%) within 48 h after oral exposure (580). Four glucuronic acid conjugates of DBP have been identified in urine. The amount of MBP-glucuronide excreted differs between rats and hamsters. This may account for the excessive testicular toxicity in rats versus hamsters (581). There appears to be little retention of DBP or MBP in tissues of rats treated with DBP for 12 wk (580).

Species differences in hepatocellular effects may be due to the fact that the mechanism for peroxisome proliferation is receptor mediated (582). Because guinea pig and human livers have fewer receptors than do rat and mouse livers guinea pigs, primates, and human livers are resistant to phthalate esters and other chemicals that act via this receptor (416, 422, 423, 583). Hamsters and primates are also resistant to the testicular effects of phthalate esters (417, 419).

104.4.1.4 Reproductive and Developmental. The reproductive toxicity of DBP has been tested under several protocols in mice and rats. Lamb et al. (511) described a continuous breeding study of DBP in which mice were treated for 60 d prior to mating, during pregnancy, and lactation. A reduction in fertility was observed in animals ingesting a diet containing 1.0%. No effects were noted in those ingesting 0.3%. Mating treated males with untreated females (crossover mating), and vice versa, indicated that females were affected rather than males. Results of crossover mating experiments performed by Wine et al. (497) indicated that ingestion of 1.0% DBP affects both males and females. In this study, exposure of dams to levels as low as 0.1% affected the developing reproductive system of male pups.

Although DBP has long been known to interfere with spermatogenesis in male rats and mice (498–500), hamsters have not shown testicular toxicity (500). Some have speculated that this difference is related to metabolism of monobutyl phthalate (581). The potential of DBP to cause similar effects in primates or humans has not been tested. However, when

DEHP (a reproductive toxicant in rats) was given to primates for 13 wk, no effects on the testes were observed (419).

The developmental toxicity of DBP has been investigated in mice and rats under the standard protocol of treatment during organogenesis (days 6–15 of gestation in rodents). Shiota et al. (502) treated mice with 0.05 to 1.0% dietary DBP throughout gestation. Craniofacial malformations (cleft palate, exencephaly) and spina bifida were observed in offspring of mice ingesting 1.0% DBP, but not 0.4% (660 mg/kg) or lower dose levels. Similar effects have been observed for rats. Cleft palate occurred in the offspring of rats treated with 630 mg/kg DBP from day 7–15 of gestation (503). The no-observable-effect level was 500 mg/kg. Subsequent studies indicated that treatment with higher dose levels early in gestation (days 7–9) resulted in craniofacial and neural tube anomalies, while treatment on days 13–15 resulted in craniofacial and skeletal anomalies (504). No anomalies were seen when animals were treated during days 10–12.

The developmental toxicity of DBP in rats during late gestation was recently investigated (497, 505, 506). Effects on the reproductive system of male pups (underdeveloped or absent epididymides, hypospadia, cryptorchidism, and aspermia resulting in loss of reproductive capacity) and "feminitation" (as judged by external landmarks at birth) were reported (497, 506). These latter effects did not appear to influence the ability of the males to reproduce or to alter the external genitalia in the mature animal. Depending on the strain of rat used and the nature of the effect, the no-observable-effect level was 50–250 mg/kg.

The effects on the developing male fetus described by Wine et al. (497) were, at first, considered to represent an estrogenic mechanism. Subsequent work by others indicated that DBP is not estrogenic *in vivo* (584–586). Mylchreest et al. (506) suggested that DBP acted as an antiandrogen to block the developmental triggers of testosterone. However, DBP does not act via the androgen receptor (587). Therefore, the mechanism for the developmental effects on the reproductive system of rats is not understood, and there is no information as to whether this mechanism is operant in humans.

In an NTP study, pregnant F344 rats were treated with 1% DBP in the diet throughout pregnancy and lactation until the pups were 6 wk of age. Offspring were then divided into groups and treated with 0, 0.25, 0.5, 1.0, 2.0, or 4.0% DBP in the diet for an additional 13 wk (588). Histopathologic effects on the liver and testes were seen in the groups receiving $\geq$1.0% dietary DBP.

Results of several assays indicate that DBP is weakly estrogenic *in vitro*. At a concentration of 10^{-4} M, DBP exhibited weak estrogenic activity in a recombinant yeast screen (520). Its approximate potency relative to 17β-estradiol was 0.0000001. At 10^{-5} M, it induced proliferation in estrogen-responsive breast cancer cell lines MCF-7 and ZR-75 (521). In an *in vitro* estrogen receptor competitive ligand-binding assay, DBP competed with 17β-estradiol for binding to the receptor with an IC_{50} of 47 μM (586). In transfected MCF-7 (but not HeLa cells), 10 μM DBP was capable of inducing estrogen-receptor mediated gene expression (586). Since oral administration of up to 2 g/kg DBP to ovariectomized rats had no effect on uterine wet weight or vaginal epithelial cell cornification (586), DBP does not appear to be estrogenic *in vivo*.

104.4.1.5 Carcinogenesis. No information found.

104.4.1.6 Genetic and Related Cellular Effects Studies. The mutagenic potential of DBP has been evaluated in a number of *in vitro* and *in vivo* assays. Studies in *S. typhimurium* strains TA98, TA100, TA1535, and TA1537 (with and without S-9 metabolic activation systems) showed that at concentrations up to 10 mg/plate, neither DBP nor diisobutyl phthalate are mutagenic (406). In contrast, Kozumbo et al. (543) and Agarwal et al. (524) reported that DBP was positive in strains TA100 and TA1535 in the absence of S-9. DBP had no effect on chromosomal aberrations (526, 577, 589) or sister chromatid exchange in Chinese hamster cells (589). There was also no effect on transformation of BALB/3T3 cells (530, 550). Positive results have been obtained in the mouse lymphoma assay with and without activation (566, 580). However, the results of these studies were confounded by poor cell survival. A recent study performed by Barber et al. indicated that DBP (0.025–0.075 μL/mL) is active in the mouse lymphoma assay in the presence, but not the absence, of metabolic activation (546). Since no evidence of genetic toxicity was seen *in vivo* in a mouse micronucleus assay (550, 588), any reactive metabolite (s) appears to be rapidly detoxified.

104.4.1.7 Other: Neurological, Pulmonary, Skin Sensitization. DBP was not irritating to the eye or skin, and is not a dermal sensitizer (550).

104.4.2 Human Experience

The human skin absorption rate of DBP has been reported as 2.40 $\mu g/cm^2/h$ (532), which is considered to be "slow." Using 2.40 $\mu g/cm^2/h$ for the absorption rate, 720 cm^2 for the skin surface area of both hands, and 60 kg for body weight, contact with the skin of both hands for 1 h can be estimated to produce an absorbed dose of approximately 0.029 mg/kg.

Cagianut (590) describes a case in which a chemical worker accidentally swallowed about 10 g of DBP. Delayed signs and symptoms included nausea, vomiting, and dizziness, followed later by headache, pain, and irritation in the eyes, lacrimation, photophobia, and conjunctivitis. Complete recovery occurred within 2 wk. There was evidence of a slight effect on the kidney, which may have been the result of systemic hydrolysis of the ester and cumulative effects of the alcohol and the acid, as well as their oxidation and decomposition products.

In 173 subjects with suspected occupational dermatoses to plastic or glue allergens, two subjects (1.2%) experienced irritation after patch testing with 5.0% *n*-dibutyl phthalate. None of the patients had allergic reactions (129).

104.5 Standards, Regulations, or Guidelines of Exposure

The current ACGIH TLV, NIOSH REL and OSHA PEL is 5 mg/m^3 (533).

104.6 Studies on Environmental Impact

Methods are available to detect DBP in environmental samples (482). DBP has been detected in runoff water (591) and sediment (482). Some have reported levels of DBP in water ranging from 0.5 to 11 μg/L (591); however, others reported values that are much lower (482).

105.0 *n*-Dihexyl Phthalate

105.0.1 CAS Number: *[84-75-3]*

105.0.2 Synonyms: 1,2-Benzenedicarboxylic acid, hexyl ester, DnC6P

105.0.3 Trade Names: NA

105.0.4 Molecular Weight: 334.46

105.1 Chemical and Physical Properties

105.1.1 General

The empirical formula of *n*-dihexyl phthalate (DnC6P) is $C_{20}H_{30}O_4$. Its structural formula is 1,2-$[COO(CH_2)_5CH_3]_2C_6H_4$. It is a clear, oily liquid, with a melting point of -58°C and a boiling point of 340–350°C. Physical data are summarized in Table 80.17.

105.1.2 Odor and Warning Properties

No information found.

105.2 Production and Use

DnC6P is manufactured by esterifying phthalic anhydride and hexanol in the presence of sulfuric acid. It is commonly used as a plasticizer for cellulose and vinyl plastics.

105.3 Exposure Assessment

No information found.

105.4 Toxic Effects

105.4.1 Experimental Studies

105.4.1.1 Acute Toxicity. DnC6P has a low order of toxicity via the oral, dermal, and inhalation exposure routes. The oral LD_{50} value for the rat is 29.6 g/kg, and the dermal LD_{50} value for the rabbit is >20 mL/kg (5). In inhalation studies in rats, exposure for 8 h at the maximum attainable vapor concentration was not acutely lethal (5). Acute toxicity data are summarized in Table 80.18.

105.4.1.2 Chronic and Subchronic Toxicity. Repeated dose studies with DnC6P have been carried out with rats for periods ranging from 21 to 90 d. The shorter studies, conducted primarily to compare hepatic effects of DnC6P to those of DEHP, demonstrated that dietary administration of 20,000 ppm produced some effects (i.e., hepatomegaly, peroxisome proliferation) in the liver (483) and thyroid (591). Rats given DnC6P at approximately 2 g/kg/d for 4 d developed testicular atrophy (499), although no testicular effects were noted after 21 d of exposure to 2% DnC6P in the diet (483). Shibko and Blumenthal list a no-effect level of 50 mg/kg/d in the rat and 125 mg/kg/d in the dog, but do not provide any data (563).

105.4.1.3 Pharmacokinetics, Metabolism, and Mechanisms. Following ingestion, phthalates are rapidly hydrolyzed to the monoester and the parent alcohol. They may undergo further oxidation of the alcohol side chain, but further hydrolysis is minimal (514). Phthalates are rapidly eliminated.

105.4.1.4 Reproductive and Developmental. In a continuous breeding study, DnC6P was administered to mice at dietary concentrations of 0, 0.3, 0.6, and 1.2% (or approximately 0, 375, 750, and 1500 mg/kg/d) for more than 100 d prior to and during mating, gestation, and lactation (495). There were dose-related decreases in the numbers of litters produced and live pups per litter in all of the treatment groups. Animals exposed to 1.2% were infertile, and animals exposed to 0.6% had severely decreased fertility.

In a screening study for developmental toxicity, pregnant mice given extremely high doses of DnC6P (approximately 10 g/kg/d) for 8 d had complete litter loss (496).

Results of *in vitro* studies investigating potential estrogenicity of DnC6P are equivocal. At a concentration up to 10^{-3} M, DnC6P was not active in a recombinant yeast screen for estrogenic activity (520). In an *in vitro* estrogen receptor competitive ligand-binding assay, DnC6P was a weak competitor of estradiol (586). In transfected MCF-7 but not HeLa cells, 10 μM DnC6P was capable of inducing estrogen-receptor mediated gene expression (586). Since oral administration of up to 2 g/kg to ovariectomized rats had no effect on uterine wet weight or vaginal epithelial cell cornification (586), DnC6P does not appear to be estrogenic *in vivo*.

105.4.1.5 Carcinogenesis. No information found.

105.4.1.6 Genetic and Related Cellular Effects Studies. At concentrations up to 10 mg/plate, DnC6P did not induce mutations in *S. typhimurium* strains TA98, TA100, TA1535, or TA1537 in the absence or presence of metabolic activation (406). It also was inactive in two other bacterial assays (571). A di(hexyl, octyl, decyl) phthalate (610P) was equivocal in a mouse lymphoma assay and inactive an *in vitro* (BALB/3T3) test for cell transformation (530).

105.4.1.7 Other: Neurological, Pulmonary, Skin Sensitization. DnC6P causes a minimal amount of irritation to rabbit skin and eyes (5).

105.4.2 Human Experience

No information found.

105.5 Standards, Regulations, or Guidelines of Exposure

No information found.

105.6 Studies on Environmental Impact

No information found.

106.0 Diisohexyl Phthalate

106.0.1 CAS Number: [68515-50-4]

106.0.2 Synonyms: 1,2-Benzenedicarboxylic acid, dihexyl ester; and DiC6P

106.0.3 Trade Names: JAYFLEX DHP

106.0.4 Molecular Weight: 334.46

106.1 Chemical and Physical Properties

106.1.1 General

The empirical formula of diisohexyl phthalate (DiC6P) is $C_{20}H_{30}O_4$. The structural formula for DiC6P varies because the iso-alcohols used in manufacture contain several isomers. DiC6P is a water white liquid with a mild odor (Exxon Chemical Company, 1998). It is a liquid from −33°C (pour point) to 340–350°C (boiling point). The vapor pressure is 0.01 mmHg (100°C). DiC6P is essentially insoluble in water, but it is soluble in organic solvents. It has a specific gravity of 1.01. These data are summarized in Table 80.17.

106.1.2 Odor and Warning Properties

No information found.

106.2 Production and Use

DiC6P is manufactured by reaction of phthalic anhydride with isohexanol in the presence of an acid catalyst. Its major use is as a plasticizer.

106.3 Exposure Assessment

Phthalates are manufactured within a closed system under vacuum. The published survey data (592–594) indicate that exposures are normally well below the recommended occupational exposure limits.

106.4 Toxic Effects

106.4.1 Experimental Studies

106.4.1.1 Acute Toxicity. No information found.

106.4.1.2 Chronic and Subchronic Toxicity. In 90-d studies, rats were given dietary DiC6P at levels of 0.1 and 0.5% (approximately 100 and 500 mg/kg/d) for 13 wk. Another group of rats was given DiC6HP in the diet at 0.05% for the first 6 wk, then 1% for weeks 7–11, and finally 3% for weeks 12–13. The effects observed, which include liver enlargement and testicular atrophy, were only in the group that was increased from 0.05% to 3% over the course of the study. There were no notable effects in the other two groups. The no-observable-effect level (NOEL) was 500 mg/kg/d (590). In another dietary study, dogs were given DiC6P at levels of 0.5 and 1% for 13 wk. A third group was started at 0.5%, and, after 57 d, was increased to 5%. The treated animals showed normal appearance, behavior, hematology, clinical chemistry, and urinalysis parameters. Increased relative liver weights were seen in the high-dose group. Both weight reduction and histologic effects were observed in the testes of the high-dose males but not in those from animals treated with the lower doses. The NOEL was approximately 125 mg/kg/d (595).

106.4.1.3 Pharmacokinetics, Metabolism, and Mechanisms. Consult Section 105.4.1.3.

106.4.1.4 Reproductive and Developmental. DiC6P was inactive in *in vitro* and *in vivo* screening tests for estrogenic activity (520, 586).

106.4.1.5 Carcinogenesis. No information found.

106.4.1.6 Genetic and Related Cellular Effects Studies. At concentrations up to 5 g/kg, DiC6P was inactive in a mouse micronucleus assay (596).

106.4.1.7 Other: Neurological, Pulmonary, Skin Sensitization. DiC6P produced slight conjunctival irritation in the eyes of rabbits that cleared within 24 h. The Draize score at 24 h was 0. There was no chemosis, and no iridial or corneal effects (597). DiC6P did not produce significant skin irritation in rabbits (598).

106.4.2 Human Experience

DiC6P did not produce skin irritation or sensitization in repeated insult patch tests with human volunteers (599). No other information on toxicity in humans was located.

106.5 Standards, Regulations, or Guidelines of Exposure

No information found.

106.6 Studies on Environmental Impact

No information found.

107.0 *n*-Diheptyl Phthalate

107.0.1 CAS Number: *[3648-21-3]*

107.0.2 Synonyms: 1,2-Benzenedicarboxylic acid, diheptyl ester; and DHP

107.0.3 Trade Names: NA

107.0.4 Molecular Weight: 362.51

107.0.5 Molecular Formula: $C_{22}H_{34}O_4$

107.0.6 Molecular Structure:

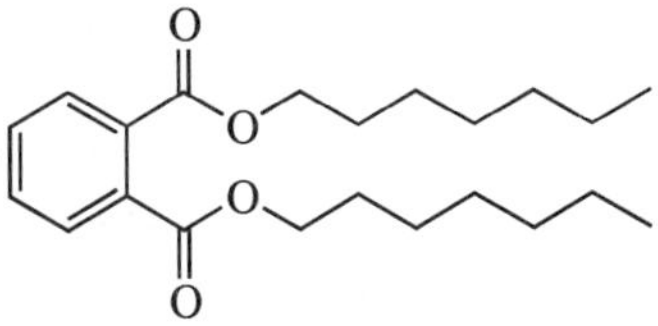

107.1 Chemical and Physical Properties

107.1.1 General

The empirical formula of *n*-diheptyl phthalate (DHP) is $C_{22}H_{34}O_4$. Its structural formula is 1,2-$(COOCH_2(CH_2)_5CH_3)_2C_6H_4$. It is a colorless liquid at ambient temperature with a boiling point of 360°C. It has a vapor pressure of 2.1×10^{-6} Mm Hg at 25°C. DHP is essentially insoluble in water (0.01%), but it is soluble in organic solvents. Table 80.17 summarizes the physical properties.

107.1.2 Odor and Warning Properties

No information found.

107.2 Production and Use

DHP is manufactured by reaction of phthalic anhydride with *n*-heptanol in the presence of an acid catalyst. Its major use is as a plasticizer for vinyl resins. DHP is often used in a mixture of heptanol, nonanol, and undecanol phthalate esters (711P, Section 118).

107.3 Exposure Assessment

No information found.

107.4 Toxic Effects

107.4.1 Experimental Studies

107.4.1.1 Acute Toxicity. No information found.

107.4.1.2 Chronic and Subchronic Toxicity. Foster and co-workers (499) treated rats with 2600 mg/kg *n*-diheptyl phthalate by oral gavage for 4 d to evaluate the effect on the testes and zinc excretion. No effects on these variables were noted.

A 28-d oral toxicity test of DHP was conducted using F344 rats treated with dose levels of 0, 200, 1000, and 5000 mg/kg/d (600). Additional groups of animals exposed to dose levels of 0 and 5000 mg/kg were used to evaluate the effect of recovery. There was no mortality, but animals in the 5000-mg/kg group had lower weight gain. Clinical chemistry revealed significant increases in albumin and A/G ratio for males treated with 200 mg/kg or more, and in albumin and total protein for females treated with 1000 mg/kg or more. In addition, urea nitrogen values, aspartate transaminase activity, alanine transaminase activity, alkaline phosphatase activity, and zinc levels were increased for the 5000-mg/kg males, and aspartate transaminase activity was increased for the 5000-mg/kg females. The increases in aspartate transaminase activity, alanine transaminase activity, and alkaline phosphatase activity also were observed in the 1000-mg/kg male group. Liver weights were increased for both sexes treated with $\geq$1000 mg/kg, and kidney weights were increased for the 5000-mg/kg male group. Testicular weight was decreased in the 5000-mg/kg group. Histopathological examination indicated liver effects in males of the 1000- and 5000-mg/kg groups, and atrophy of the seminiferous tubules for 5000-mg/kg males, with loss of spermatogenesis. Following a 2-wk recovery period, similar changes were detected in the testis, but some of the seminiferous tubules showed slight regenerative changes. In studies with dialkyl 79 phthalate (a phthalate synthesized from a mixture of C7, C8, and C9 alcohols), treatment for 21 d with 2500 mg/kg resulted in liver alterations as well as reduced testicular weight (601).

107.4.1.3 Pharmacokinetics, Metabolism, and Mechanisms. DHP may act on rodent liver via a mechanism that is similar for other phthalate esters. DHP does not increase peroxisomal enzyme activity in rat hepatocytes; however, the monoester, monoheptyl phthalate, does (602).

107.4.1.4 Reproductive and Developmental. In mice (ICR) treated on days 7–11 of gestation with dose levels of 0.94–11.3 mL/kg (~900–11000 mg/kg), embryolethality was observed at the highest dose level only (603). Fetal abnormalities were seen at the mid- and high-dose levels. A no-observed-effect level of ~900 mg/kg was determined.

107.4.1.5 Carcinogenesis. No information found.

107.4.1.6 Genetic and Related Cellular Effects Studies. No information found.

107.4.1.7 Other: Neurological, Pulmonary, Skin Sensitization. No information found.

107.4.2 Human Experience

No information found.

107.5 Standards, Regulations, or Guidelines of Exposure

No information found.

107.6 Studies on Environmental Impact

No information found.

108.0 Diisoheptyl Phthalate

108.0.1 CAS Number: *[41451-28-9], [71888-89-6]*

108.0.2 Synonyms: 1,2-Benzenedicarboxylic acid, di-isoheptyl ester; and DIHP

108.0.3 Trade Names: JAYFLEX 77

108.0.4 Molecular Weight: 362.51

108.0.5 Molecular Formula: $C_{22}H_{34}O_4$

108.0.6 Molecular Structure:

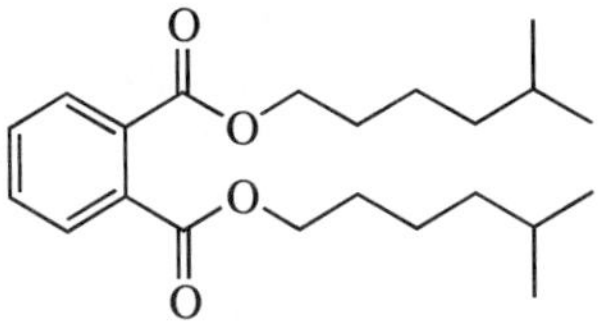

108.1 Chemical and Physical Properties

108.1.1 General

The empirical formula of diisoheptyl phthalate (DIHP) is $C_{22}H_{34}O_4$. The structural formula for DIHP varies because the iso-alcohols used in manufacture contain several isomers. It exists as a clear liquid between $-45°C$ (pour point) and $>300°C$. The vapor pressure is 0.006 mmHg at 100°C. DIHP is essentially insoluble in water, but it is soluble in organic solvents. It has a specific gravity of 0.994. Table 80.17 summarizes the physical properties.

108.1.2 Odor and Warning Properties

No information found.

108.2 Production and Use

DIHP is manufactured by reaction of phthalic anhydride with isoheptanol in the presence of an acid catalyst. Its major use is as a plasticizer.

108.3 Exposure Assessment

Similar to diisohexyl phthalate (see Section 106.3).

108.4 Toxic Effects

108.4.1 Experimental Studies

108.4.1.1 Acute Toxicity. DIHP has a low order of toxicity via the oral and dermal routes. The rat oral LD_{50} value was >10 g/kg, and the rabbit dermal LD_{50} value was >3.16 g/kg (552).

108.4.1.2 Chronic and Subchronic Toxicity. In one repeated dose toxicity test, rats and mice were given low and high dose levels of DIHP in the diet (1000 and 12,000 ppm for rats; 500 and 6,000 ppm for mice) for 4 wk. The purpose of this study was to investigate liver weight effects and the time course of other biomarkers for peroxisomal proliferation (including liver enzyme induction and elevation of DNA synthesis). These effects were similar to but less pronounced than those produced by corresponding doses of DEHP (484).

108.4.1.3 Pharmacokinetics, Metabolism, and Mechanisms. Consult Section 105.4.1.3.

108.4.1.4 Reproductive and Developmental. In an assessment of developmental toxicity, DIHP was administered by oral gavage to rats (0, 100, 300, or 750 mg/kg) on days 6–20 of gestation. Embryo/fetotoxicity and teratogenic effects were observed in the high-dose group, but there was no overt maternal toxicity. The maternal NOAEL was 750 mg/kg, and the developmental NOAEL was 300 mg/kg (552).

DIHP was inactive in both *in vitro* and *in vivo* screening assays for estrogenic activity (586).

108.4.1.5 Carcinogenesis. No information found.

108.4.1.6 Genetic and Related Cellular Effects Studies. At concentrations up to 5 mg/plate, DIHP was inactive in the Ames *Salmonella*/mammalian microsome mutagenicity assay (strains TA98, TA100, TA1535, TA1537, and TA1538). Concentrations of up to 99 μg/mL (without metabolic activation) and 4990 μg/mL (with metabolic activation) were without effect in an *in vitro* (CHO) cytogenetics test (552).

108.4.1.7 Other: Neurological, Pulmonary, Skin Sensitization. DIHP produced slight conjuctival irritation when instilled in rabbit eyes and very slight erythema when applied to rabbit skin (552). It was not sensitizing when tested in a guinea pig maximization test (552).

108.4.2 Human Experience

DIHP was neither irritating or sensitizing when applied to the skin of human volunteers (599). No other information on toxicity in humans was located.

108.5 Standards, Regulations, or Guidelines of Exposure

No information found.

108.6 Studies on Environmental Impact

No information found.

109.0 *n*-Dioctyl Phthalate

109.0.1 CAS Number: *[117-84-0]*

109.0.2 Synonyms: 1,2-Benzenedicarboxylic acid, dioctyl ester; phthaltic acid, dioctyl ester; and DnOP

109.0.3 Trade Names: Celluflex DOP; Dinopol NOP; Polycizer 162; and Vinicizer 85

109.0.4 Molecular Weight: 390.56

109.0.5 Molecular Formula: $C_{24}H_{38}O_4$

109.0.6 Molecular Structure:

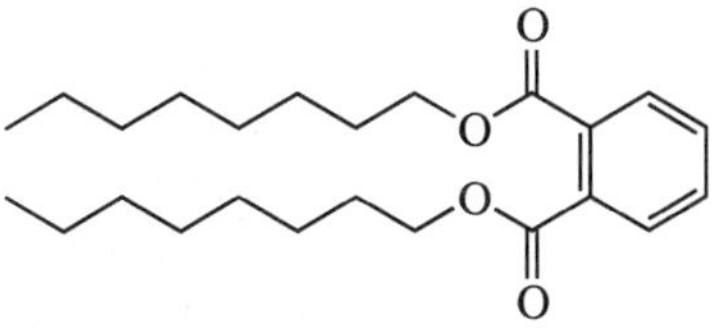

109.1 Chemical and Physical Properties

109.1.1 General

The empirical formula of *n*-dioctyl phthalate (DnOP) is $C_{24}H_{38}O_4$. Its structural formula is 1,2-$[COO(CH_2)_7CH_3]_2C_6H_4$. Its melting and boiling points are −25 and 220°C, respectively. It is essentially insoluble in water. Additional physical properties are summarized in Table 80.17.

109.1.2 Odor and Warning Properties

No information found.

109.2 Production and Use

n-Dioctyl phthalate is formed via the esterification of *n*-octanol with phthalic anhydride in the presence of a catalyst (sulfuric acid or *p*-toluenesulfonic acid) or noncatalytically at high temperature. *n*-Dioctyl phthalate is used as a plasticizer in plastics and cellulose acetate resins. Very little pure DnOP is commercially manufactured. DnOP exists primarily as a component of mixed phthalate products, such as 6,10-phthalate, which is produced by reaction of *n*-hexanol, *n*-octanol, and *n*-decanol with phthalic anhydride. DnOP comprises approximately 20% of the resultant product.

109.3 Exposure Assessment

No information found.

109.4 Toxic Effects

109.4.1 Experimental Studies

109.4.1.1 Acute Toxicity. *n*-Dioctyl phthalate has a low order of toxicity via the oral and dermal routes. The oral LD_{50} values for mice and rats are 13 and 53.7 g/kg, respectively (553, 554). The dermal LD_{50} value for guinea pigs is 75 mL/kg (12) (Table 80.18). There are no data for inhalation exposure.

109.4.1.2 Chronic and Subchronic Toxicity. The toxicology of DnOP is often confused with that of di (2-ethylhexyl) phthalate (DEHP). DEHP is often referred to as DOP in the older literature. In general, studies conducted prior to 1980 that cited "DOP" as the test substance used DEHP rather than DnOP. The two substances have different toxicological effects and different dose–response relationships. The effects of DnOP have been studied in repeated dose studies up to 13 wk of exposure.

The effect of DnOP and DEHP on liver metabolism and biochemical changes that are associated with peroxisome proliferation were studied by Lake et al. (604, 605). In this study, rats were treated with 1000 mg/kg DnOP or DEHP for 14 d. DnOP produced a 16% increase in liver weight over control, compared with a 72% increase by DEHP. However, there was no increase in peroxisomal enzyme activities with 1000 mg/kg DnOP, and a dose of 2000 mg/kg DnOP failed to produce peroxisome proliferation.

Treatment of rats with 20,000 ppm DnOP in the diet for up to 21 d resulted in significant increases in liver weight after 7 d. The increases in liver weight were associated with deposition of fat. There were no effects on kidney or testes weights (483). Oishi and Hiraga (606) also treated rats with 20,000 ppm DnOP in the diet for 1 wk and observed an increase in liver weight. They found no effect on kidney or testes weights. Testosterone and dihydrotestosterone levels in the serum were unaffected by treatment, but the zinc content was decreased. However, Foster et al. (499) found no effect on testicular zinc in rats treated with 2800 mg/kg DnOP for 4 d.

Poon et al. (607) treated rats for 13 wk with 5, 50, 500, or 5000 ppm DnOP in the diet (high-dose equivalent to 350 and 403 mg/kg for males and females, respectively). There were no effects on feed consumption, or body, liver, or kidney weights. There were slight microscopic changes in the liver, and decreased follicle size and colloid density in the thyroid (which may have been stress induced). The NOEL was 500 ppm, which is equivalent to 36.8 mg/kg.

No chronic toxicity studies for DnOP were located.

109.4.1.3 Pharmacokinetics, Metabolism, and Mechanisms. Blood levels of the monoester, MnOP, were measured in rats following administration of 2000 mg/kg DnOP by gavage (608). The biological half-life in the blood was 3.3 h, with an area under the curve (AUC) of 1066 μg h/mL. Peak blood levels were observed at 3 h following administration.

109.4.1.4 Reproductive and Developmental. The effect of DnOP on reproduction was assessed using a continuous-breeding protocol (609). Mice were fed diets containing 12500, 25000, or 50000 ppm (1800, 3600, 7500 mg/kg) DnOP prior to mating and for 98 d thereafter. There were no adverse effects on reproduction.

The effect of DnOP on development was assessed by injecting 5 or 10 mL/kg (4890 and 9780 mg/kg, respectively) IP into pregnant rats (501). There was a dose-related increase in the number of resorptions and gross abnormalities (mostly twisted legs). When administered orally to pregnant rats, a dose of 9780 mg/kg produced a slight decrease in litter size and pup weight (496), but no gross abnormalities. In this study, there was no mention of any adverse effects caused by oral administration of 4890 mg/kg DnOP.

At concentrations up to 10^{-3} M, DnOP was not active in a recombinant yeast screen for estrogenic activity (520). Results of *in vitro* estrogen receptor competitive ligand-binding and mammalian- and yeast-based gene expression assays and *in vivo* uterotrophic and vaginal cornification assays also indicated that DnOP is not estrogenic (586).

109.4.1.5 Carcinogenesis. No information found.

109.4.1.6 Genetic and Related Cellular Effects Studies. Results of studies in *S. typhimurium* strains TA98, TA100, TA1535, and TA1537 (with and without S-9 metabolic activation systems) indicated that *n*-dioctyl phthalate is not mutagenic (406).

109.4.1.7 Other: Neurological, Pulmonary, Skin Sensitization. Rats treated with 100, 300, or 600 mg/kg DnOP by IP injection for up to 90 d were evaluated for immunological responses (610). The high dose resulted in early loss of distinction between the cortex and medulla of the thyroid, reduced numbers of follicles in the lymph nodes, and loss of morphology in the adrenal glands. However, there was no effect on immune response.

Tests in rabbits indicate that DnOP is not irritating to skin or eyes (611).

109.4.2 Human Experience

In 173 subjects with suspected dermatoses to plastic or glue allergens, two subjects (1.2% of the group) experienced irritation after patch testing with 2.0% dioctyl phthalate. None of the patients had allergic reactions (129). No other information on toxicity in humans was located.

109.5 Standards, Regulations, or Guidelines of Exposure

No information found.

109.6 Studies on Environmental Impact

No information found.

110.0 Diisooctyl Phthalate

110.0.1 CAS Number: *[27554-26-3]*

110.0.2 Synonyms: 1,2-Benzenedicarboxylic acid, di-isooctyl ester; and DIOP

110.0.3 Trade Names: JAYFLEX DIOP

110.0.4 Molecular Weight: 390.56

110.0.5 Molecular Formula: $C_{24}H_{38}O_4$

110.0.6 Molecular Structure:

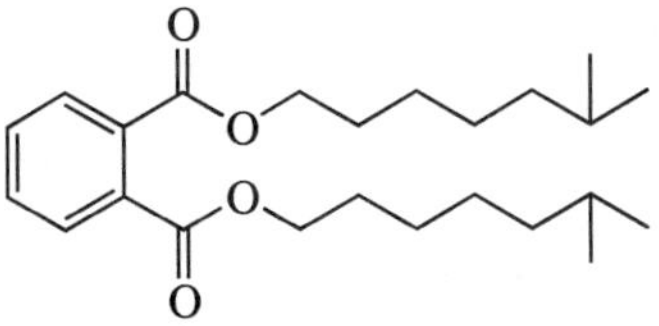

110.1 Chemical and Physical Properties

110.1.1 General

The empirical formula of diisooctyl phthalate (DIOP) is $C_{24}H_{38}O_4$. The structural formula for DIOP varies because the isoalcohols used in manufacture contain several isomers. DIOP is a colorless liquid between $-46°C$ (pour point) and $>300°C$. The vapor pressure is 1 mmHg (200°C). DIOP is insoluble in water, but it is soluble in organic solvents. It has a specific gravity of 0.99. These properties are summarized in Table 80.17.

110.1.2 Odor and Warning Properties

No information found.

110.2 Production and Use

DIOP is manufactured by reaction of phthalic anhydride with iso-octanol in the presence of an acid catalyst. Its major use is as a plasticizer.

110.3 Exposure Assessment

Similar to diisohexyl phthalate (106.3).

110.4 Toxic Effects

110.4.1 Experimental Studies

110.4.1.1 Acute Toxicity. The rat oral LD_{50} value for DIOP is >20 g/kg (555).

110.4.1.2 Chronic and Subchronic Toxicity. Summary information from the FDA indicates that 100 mg/kg body weight/d was a no-effect level for DIOP in rats and dogs in studies lasting 4-wk (563).

110.4.1.3 Pharmacokinetics, Metabolism, and Mechanisms. A study was carried out to assess the comparative pharmacokinetics of DIOP in rats, dogs, and pigs (612). The animals were given DIOP for 21–28 d before administration of ^{14}C-labeled material. DIOP was excreted most rapidly by the rats; however, after 4 d, $>95\%$ of the administered dose had been excreted by all species. There were marked differences in the excretion patterns—the rats and pigs excreted the majority of the material in the urine, and dogs excreted nearly 70% of the administered dose in the feces. Consult Section 105.4.1.3 for information on metabolism.

110.4.1.4 Reproductive and Developmental. No information found.

110.4.1.5 Carcinogenesis. No information found.

110.4.1.6 Genetic and Related Cellular Effects Studies. DIOP was inactive in the Ames *Salmonella*/mammalian microsome mutagenicity assay (613) and in an *in vitro* transformation assay in BALB/c 3T3 cells (614).

110.4.1.7 Other: Neurological, Pulmonary, Skin Sensitization. No information found.

110.4.2 Human Experience

No information on adverse effects in humans was located.

110.5 Standards, Regulations, or Guidelines of Exposure

No information found.

110.6 Studies on Environmental Impact

No information found.

111.0 Di(2-ethylhexyl) Phthalate

111.0.1 CAS Number: [117-81-7]

111.0.2 Synonyms: DEHP; 1,2-benzenedicarboxylic acid, *bis* (2-ethylhexyl) ester; *bis* (2-ethylhexyl)phthalate; and dioctyl phthalate (DOP) [*Note*: DOP is used in reference to DEHP, but it is often confused with DnOP, which properly refers to di-*n*-octyl phthalate.]

111.0.3 Trade Names: Bisoflex-81; PX-138; Palatinol AH; and Eastman DOP Plasticizer

111.0.4 Molecular Weight: 390.54

111.0.5 Molecular Formula: $C_{24}H_{38}O_4$

111.0.6 Molecular Structure:

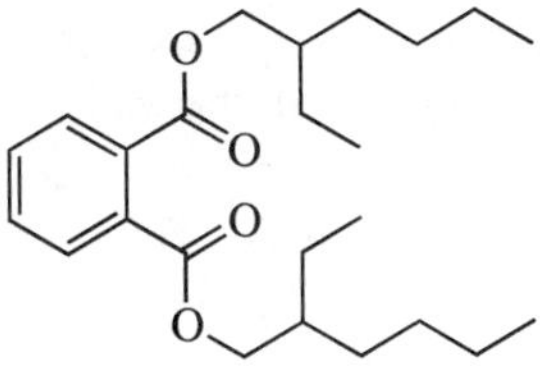

111.1 Chemical and Physical Properties

111.1.1 General

The empirical formula of di(2-ethyhexyl) phthalate (DEHP) is $C_{24}H_{38}O_4$. Its structural formula is 1,2-[$COOCH_2CH(C_2H_5)(CH_2)_3CH_3]_2C_6H_4$. It is a clear liquid with a pour point of −46°C and a boiling point of 385°C at 760 torr (615). It is essentially insoluble in water, and has a very low vapor pressure. Its specific gravity is 0.9861. These properties are summarized in Table 80.17.

111.1.2 Odor and Warning Properties

DEHP has no characteristic odor.

111.2 Production and Use

DEHP is commercially produced from esterification of phthalic acid anhydride and 2-ethylhexanol. Esterification is completed with an acid or metal catalyst or with no catalyst at high temperature. DEHP is used as a plasticizer for polyvinyl chloride resin (PVC) to make consumer products, medical devices, and cable/wire covering (616).

111.3 Exposure Assessment

DEHP has no characteristic odor or odor threshold. There are numerous analytical methods for measurement of DEHP or its metabolites in human fluids or tissues (617–620).

111.4 Toxic Effects

111.4.1 Experimental Studies

111.4.1.1 Acute Toxicity. The acute toxicity of DEHP has been evaluated following ingestion, dermal and inhalation routes of exposure. The rat, mouse, and rabbit acute oral LD_{50} values reported for DEHP range from >25.6 to >34.5 g/kg (Table 80.17), indicating that DEHP possesses very low acute toxicity following ingestion. The acute dermal toxicity of DEHP is similarly low; dermal LD_{50} values reported are >20 mL/kg in rabbits and >10 mL/kg in guinea pigs (556, 558), indicating that DEHP is virtually acutely nontoxic following dermal exposure. Inhalation studies of rats exposed to atmospheres of heated DEHP aerosol and vapor indicate that the acute inhalation LC_{50} value for DEHP is between 4 and 8.7 mg/L (556), suggesting that DEHP does not possess a high degree of acute inhalation toxicity. The techniques employed to produce this test atmosphere (heating DEHP to 170°C) resulted in animal exposure to an airborne mixture of DEHP (mixed vapor/aerosol) and probably DEHP thermal decomposition products.

111.4.1.2 Chronic and Subchronic Toxicity. There are numerous studies of the effects of DEHP treatment on rodents and nonrodents. These studies have been reviewed extensively (621–623). Target organs include the liver and testes, although there are data to suggest that the kidneys of rats are affected by high doses of DEHP. The effects are species dependent; rodents tend to be more susceptible to the systemic effects of DEHP than higher orders of mammals, with primates demonstrating no effects.

F344 rats receiving up to 2.5% DEHP in their diet (approximately 2000 mg/kg/d) for 21 d developed statistically significant increases in liver weight at daily doses of approximately 600 mg/kg and higher and decreases in testicular weight at the highest dose (624). The liver changes were accompanied by cytoplasmic eosinophilia and peroxisome proliferation. Lake (640) showed that oral gavage of 100 mg/kg DEHP for 14 d induced liver changes in Sprague–Dawley rats, whereas daily administration of 1000 mg/kg was required to produce similar changes in the Syrian hamster. Short et al. (418) found that oral administration of 100 mg/kg/d DEHP for 21 d induced hepatic peroxisome proliferation in F344 rats, whereas dose levels of up to 500 mg/kg/d for 21 d did not cause peroxisome proliferation in Cynomolgus monkeys. In marmoset monkeys, oral (2000 mg/kg/d) or intraperitoneal (1000 mg/kg/d) administration of DEHP for 14 d did not produce hepatic or testicular changes (417), and administration of up to 2500 mg/kg/d DEHP via gavage for 90 d did not produce hepatic peroxisome proliferation or testicular effects (419).

In experimental animals, oral administration of 1000–2800 mg/kg/d DEHP for up to 10 d produced testicular damage, specifically to the Sertoli cell (626, 627). The testicular effect was age and species dependent, with younger animals being more sensitive than mature animals and species such as hamsters being more resistant than rats (500). In hamsters treated with 4200 mg/kg DEHP for 9 d, minimal testicular atrophy was observed (500). Dostal et al. (628) found that 6-d old rats were more sensitive to the testicular effects of DEHP than were older animals. A recent study by Poon and co-workers (607), describes subtle testicular effects of DEHP in rats receiving 500 ppm (about 37.6 mg/kg/d) in their diet for 90 d. Testicular damage was not observed in marmoset monkeys receiving up to 2500 mg/kg/d DEHP via gavage for 90 d (419).

In dogs, intravenous administration of DEHP in plasma (0.50, 0.75, or 1.40 mg/kg/d) for a period of 21 d had no effect on behavior, clinical chemistry, hematology, body weight gain, organ weights, gross pathology, or histopathology of the liver, testes, lungs, heart, or ovaries (629).

Several chronic toxicity studies indicate that oral administration of $\geq$100 mg/kg/d DEHP causes growth retardation, noncarcinogenic liver and kidney toxicity, and testicular atrophy (generally at higher dose levels) in rodents (630–634). In general, liver tumors are observed in B6C3F1 mice and F344 rats but not Sprague–Dawley rats (see Section 111.4.1.5). In addition, F344 rats (but not Sprague–Dawley rats or B6C3F1 mice) given high dose levels of DEHP had a higher incidence of mononuclear cell leukemia.

Studies by Carpenter et al. (630) evaluated the effect of dietary administration of DEHP for 2-yr to rats, guinea pigs, and dogs. Dose levels of up to 4000 ppm caused decreased body weight gain and increased relative liver and kidney weights in rats. There was no evidence of tumorigenesis or altered hematology. There were no effects in guinea pigs fed up to 1300 ppm DEHP for 1 yr or dogs treated with up to 90 mg/kg DEHP for 1 yr. The reliability of this study is reduced because of the high mortality within all groups. Similar results were reported by Harris et al. (635) using Wistar rats and dogs fed diets up to 5000 ppm. These results also are difficult to interpret because of high mortality (> 85%).

The National Toxicology Program (631) tested DEHP in F344 rats at dietary concentrations of 6000 or 12,000 ppm (corresponding to a dose level of about 674 mg/kg/d in males and 744 mg/kg/d in females) for 2 yr. Evidence of testicular atrophy, aspermia, and a loss of germinal epithelium was observed in rats treated with the highest dose level; seminiferous tubule degeneration, chronic inflammatory changes in the kidney, and hypertrophy of the cells of the anterior pituitary occurred in high-dose male rats; and hepatic adenomas and carcinomas occurred in both sexes at all dose levels (see Section 111.4.1.5).

In Sprague–Dawley rats chronically exposed for 2-yr to DEHP (200, 2000, or 20,000 ppm), reduced body weight gain occurred after treatment with the highest dose only (632). Peroxisome proliferation in the liver occurred at all dose levels. Cessation of treatment following 1 yr of dosing returned enzyme activity to baseline after 2 wk. All dose levels produced testicular damage (inhibition of spermatogenesis and seminiferous tubule atrophy).

Concentrations of up to 6000 or 12500 ppm DEHP were administered in the diet to B6C3F1 mice or Fischer 344 rats, respectively, for 2 yr (633, 634). The studies employed four treatment levels and included a recovery group given control diet after 78 wk of exposure to the high-dose level. In mice treated with 1500 or 6000 ppm, survival (to 104 wk) was significantly decreased. At these dose levels, chronic progressive nephropathy and liver changes including hepatomegaly, peroxisome proliferation, and hepatocellular carcinoma were seen in both sexes. In addition, testis weights were significantly decreased and the incidence of hypospermia was increased. The kidney and testicular effects were reduced in severity, and all the non-neoplastic liver effects were reversible following cessation of dosing at week 78. The no-observable-adverse-effect level (NOAEL) in the mouse study was 500 ppm (approximately 98.5 mg/kg/d in males and 116.8 mg/kg/d in females). In rats, ingestion of a diet containing 2500 or 12500 ppm DEHP caused hepatic toxicity (hepatomegaly, peroxisome proliferation, increased cytoplasmic eosinophilia, increased pigment in hepatocytes and Kupffer cells, carcinomas); increased kidney weight and increased incidences of renal tubule cell pigment and chronic progressive nephropathy; a decrease in the absolute and relative testis weight accompanied by an increase in bilateral aspermatogenesis; and an increased incidence of mononuclear cell leukemia (males only). Cessation of dosing at week 78 produced a reversal of hepatic effects including neoplasms; however, treatment-related changes in tissues other than the liver did not disappear after removal of DEHP from the diet. The NOAEL for systemic and carcinogenic effects in rats was 500 ppm (28.9 and 36.1 mg/kg/d in males and females, respectively). At this dose level, there were no effects on survival, organ weights, or histopathology in any tissue (including the liver).

111.4.1.3 Pharmacokinetics, Metabolism, and Mechanisms. In most species, hydrolytic cleavage of the diester to form the monoester (MEHP) is the first step in DEHP metabolism (623). Following ingestion, MEHP is formed predominantly in the gut as a result of the action of pancreatic lipase on DEHP. Other routes of exposure to DEHP yield less MEHP because of relatively low levels of hydrolytic enzyme activity. There are differences across species in the rate of formation of MEHP following ingestion of DEHP. Intestinal hydrolytic activity in the mouse is greater than in the rat, which is greater than the guinea pig and the hamster (625, 636). The bioavailability of DEHP in rats is higher than in monkeys. Humans metabolize DEHP similarly to other primates (637).

Following hydrolysis of the diester, the ethylhexyl chain of the monoester is oxidized in the liver via several pathways (623). MEHP and its hydroxy and keto metabolites are primarily responsible for the toxicity of DEHP—these metabolites induce hepatic peroxisome proliferation in rodents (638). The differences in species and strain response to DEHP can be partially explained by differences in rodent metabolism and distribution of DEHP or its metabolites (principally MEHP) to a sensitive target site. In rats and mice, orally administered DEHP is more efficiently absorbed from the gut than in humans or nonhuman primates (417). Therefore, for any given dose of DEHP administered, the delivered dose (to target tissues) is much greater in rodents than humans. A pharmacokinetic model for DEHP in rats has been developed recently (639).

Based on results of *in vitro* testing, DEHP does not effectively penetrate intact rat or human skin (640). *In vivo* testing has been performed to evaluate percutaneous absorption of DEHP from DEHP-plasticized polyvinyl chloride (PVC) film. Male Fischer 344 rats with shaved backs were wrapped for 24 h with PVC film plasticized with [^{14}C] DEHP. Urine and feces were collected from the animals immediately after exposure and for up to

7 d postexposure. The amount of radioactivity detected in the excreta immediately after removal of the wrap and 1 wk later was 0.126% and 0.0643%, respectively (641).

111.4.1.4 Reproductive and Developmental. The results of subchronic and chronic studies (see Section 111.4.1.2) indicate that the testes are a target organ for DEHP toxicity in rodents. Therefore, it is not surprising that the results of reproductive toxicity studies have demonstrated that DEHP has adverse effects on fertility. Agarwal et al. (642) exposed male F344 rats to 320, 1250, 5000, or 20000 ppm DEHP in the diet for 60 d followed by a 5-d mating trial. A dose level of 20000 ppm reduced fertility, an effect that was reversible after 70 d of control diet. Lamb et al. (495) reported the results of a continuous breeding study in which CD-1 mice were exposed to dietary concentrations of 100, 1000, or 3000 ppm DEHP for 7 d prior to mating and during 98 d of cohabitation. Dose levels of 1000 and 3000 ppm (~ 200 and 600 mg/kg) resulted in decreased fertility. A dose level of 100 ppm (~ 20 mg/kg) had no effect. Dostal et al. (628) determined that younger rats were more susceptible than older rats to the testicular effects of DEHP, but animals recovered. In that study, 6-d-old Sprague–Dawley rats were treated with 200, 500, or 1000 mg/kg DEHP by oral gavage for 5 d, and the animals were tested for reproductive toxicity at 8 wk of age. There was no effect on fertility, although testicular effects were seen at 500 and 1000 mg/kg. No testicular effect was seen at 200 mg/kg. In addition, adverse effects were not noted in a three-generation reproduction study in albino rats of standard design that employed dietary doses of 150, 500 and 1500 ppm in the diet (643). In addition, inhalation exposure of male Wistar rats to 0.01, 0.05, or 1.0 mg/L DEHP (6 h/d for 28 d) before mating with untreated females had no adverse effect on fertility (644).

The developmental toxicity of DEHP has also been investigated using a variety of routes of administration, test species, and durations of exposure. The effects differ based on the timing of exposure, route of exposure, dose level, and species. Dietary levels of 0, 0.025, 0.05, 0.10, or 0.15% DEHP (equivalent to 0, 44, 91, 190.6, or 292.5 mg/kg) were administered to groups of CD-1 mice throughout gestation (days 0–17) (507). Compared to controls, there was a dose-related trend towards increased water and food consumption, with significantly higher food consumption in the 0.15% dose group. Maternal toxicity, indicated by reduced maternal body weight gain (due mainly to reduced uterine weight), was noted in the two highest-dose groups. Maternal relative liver weight (but not absolute liver weight) was significantly increased by 0.10% and 0.15% DEHP. There were no treatment-related effects on the number of corpora lutea, implantation sites per dam, the percentage of preimplantation loss, and sex ratio of live pups. The percentages of resorptions, late fetal deaths, and dead and malformed fetuses were significantly increased at 0.1% and 0.15% DEHP. The percentage of all fetuses with malformations and the percentage of malformed fetuses per litter were significantly increased by doses ≥0.05%. The maternal NOEL appeared to be 0.05% (91 mg/kg body weight/d), and the embryofetal NOEL was 0.025% (44 mg/kg body weight/d) DEHP. In another study, malformations were noted in offspring of ICR mice ingesting a diet containing ≥0.2 % DEHP (≥400 mg/kg) throughout gestation (645). The NOEL was 0.1% (190 mg/kg), indicating that different strains of mice have different sensitivities.

Dietary levels of 0, 0.5, 1.0, 1.5, or 2% DEHP (equivalent to 0, 357, 666, 857, or 1055 mg/kg) were given to groups of F344 rats throughout gestation (days 0–20) (507).

Food intake was significantly increased at 0.5 and 1.0% but was significantly decreased at 2% DEHP. Maternal toxicity was indicated by a dose-related decrease in gestational weight gain (weight gain minus gravid uterine weight) and an increase in maternal absolute and relative liver weights. Gravid uterine weight was significantly reduced in rats ingesting 2% DEHP. There were no treatment-related differences in the number of corpora lutea or implantation sites per dam or in the percentage of preimplantation loss. The number of resorptions, nonlive (dead plus resorbed), and affected (nonlive plus malformed) implants per litter were significantly increased by ingestion of 2% DEHP. Mean fetal body weight was significantly reduced by all dose levels. The number and percentage of malformed fetuses per litter were not significantly different from controls. In this study, the dietary NOEL for maternal and embryofetal toxicity was 0.5% DEHP (357 mg/kg body weight/d).

Pregnant female Wistar rats (20 per group) were examined for developmental toxicity after head-only inhalation exposure to aerosol concentrations of 0, 10, 50, or 300 mg/m^3 DEHP for 6 h/d from gestation day 6 to 15 (646).

Exposure 200 mg/m^3 and above was known to induce hepatic peroxisome proliferation in dams. Dams were killed on day 20 of gestation. Five additional animals per group were allowed to litter and the offspring were raised until day 21 after birth. No maternal toxicity or dose-related adverse effects in pre- or postnatal development were observed except that the body weight of the dams at day 21 postpartum was significantly decreased in the 300-mg/m^3 exposure group. The number of live fetuses/dam was significantly decreased, and the percentage of resorptions/dam increased in the 50-mg/m^3 group but not in the 300-mg/m^3 group. The percentage of abnormal fetuses per litter and the number of litters with abnormal fetuses were not different from controls. Survival rates and developmental landmarks of the pups were unaffected by treatment.

Recent investigations by Arcadi et al. (647) and Gray et al. (648) indicate that treatment of pregnant Long–Evans rats with DEHP during the last trimester of gestation affects the developing male reproductive tract and maturation process. Arcadi and co-workers exposed pregnant rats to DEHP in the drinking water (32.5 or 325 μL/L) and found that male offspring had reduced testicular weight, morphologic changes in the testes, and delayed onset of spermatogenesis. In addition, increased time to complete a beam walking test successfully was noted in rats exposed to 325 μL/L (647). More pronounced results on testes were observed in rats treated via gavage with 750 mg/kg DEHP (648).

Results of *in vitro* studies with rat ovarian granulosa cells and *in vivo* studies employing doses of up to 2 g/kg/d indicate that DEHP may alter hormone sensitivity (FSH) and estrous cyclicity in female rats (649). DEHP is capable of binding to the rainbow trout hepatic estrogen receptor (401). Using estrogen receptor competitive ligand-binding and mammalian and yeast-based gene expression *in vitro* assays, Zacharewski (586) subsequently demonstrated that neither DEHP or its metabolite MEHP interact with the estrogen receptor. Results of an *in vivo* rat uterotropic assay, in which uterine wet weight and vaginal cell cornification were assessed, confirmed the lack of estrogenicity of DEHP. This conclusion is supported by the results of Harris et al. (520) and Milligan et al. (585) using mice.

111.4.1.5 Carcinogenesis. Studies conducted by the National Toxicology Program (NTP) indicate that DEHP is a liver carcinogen in F344 rats and B6C3F1 mice (631).

An increased incidence of hepatic adenomas and carcinomas was noted in male and female F344 rats treated with dietary concentrations of 6000 and 12,000 ppm for 2 yr. In mice, dose levels of 3000 and 6000 ppm induced hepatic tumors in both sexes after 2 yr.

Results of bioassays performed by David et al. indicate that chronic ingestion of 1500 or 6000 ppm DEHP in the diet is associated with the occurrence of hepatocellular carcinomas and adenomas in male and female $B6C3F_1$ mice (481). The NOAEL for carcinogenic effects in mice was 100 ppm (approximately 19.2 mg/kg/d in males and 23.8 mg/kg/d in females). In F344 rats, chronic ingestion of 2500 or 12500 ppm DEHP induces hepatic neoplasms (carcinoma and adenoma) in both males and females and mononuclear cell leukemia in males. The NOAEL for carcinogenic effects in rats was 500 ppm (28.9 and 36.1 mg/kg/d in males and females, respectively). Some hepatic tumors were reversible upon cessation of treatment following 78 wk of exposure to 12,500 ppm.

111.4.1.6 Genetic and Related Cellular Effects Studies. A recent review of published standard *in vitro* and *in vivo* genotoxicity studies states that DEHP appears to be free of genetic toxicity (622). In a rat host-mediated assay, DEHP was not mutagenic at dose levels up to 2000 mg/kg/d for 15 d (650). In mice, IP injection of DEHP for 5 d (at doses less than 1/6 of the LD_{50} dose) produced signs of systemic toxicity but did not induce micronuclei formation in peripheral blood DNA (657). The only positive results reported for DEHP have been for nongenetic endpoints such as modification of nucleotides (652) and cellular gap junction assays (653), for which the biological significance in unclear.

Peroxisome proliferation and carcinogenesis is mediated through the peroxisome proliferator-activated receptor α (PPARα) (654), and activation of this receptor has been shown to be necessary for DEHP-induced liver toxicity (655) and hepatocarcinogenesis in rodents (421). Although human liver also contains PPARα, the receptor population is lower for humans than for rodents (423). In addition, the human response element that binds to the activated receptor to initiate the cellular responses may be inactive (656), suggesting that the lack or responsiveness in humans is related to the response element rather than to the receptor population. Two scientific panels have reviewed the mechanistic data available on peroxisome proliferator-induced rodent carcinogenesis and have concluded that these substances do not present a cancer risk to humans under likely exposure conditions (416, 657).

111.4.1.7 Other: Neurological, Pulmonary, Skin Sensitization. The ability of DEHP to alter neurobehavioral activity and function in adult rats was tested by Moser et al (658). Rats treated with single doses of up to 5000 mg/kg or repeated doses of 1500 mg/kg for 14 d showed no neurobehavioral alterations.

Dermal irritation studies in rabbits failed to produce signs of skin irritation (556). The potential of DEHP to irritate or damage the eye following direct contact with the liquid has not been evaluated.

111.4.2 Human Experience

In humans, repeated dermal application of undiluted DEHP over a period of 7 d did not produce skin irritation or sensitization (6, 556). In 144 people who were occupationally exposed to 5.0% DEHP, two people (1.4% of the group) had irritant reactions, and none had allergic reactions (129).

111.5 Standards, Regulations, or Guidelines of Exposure

The ACGIH has set a TLV, and NIOSH has set a PEL, of 5 mg/m^3, and a STEL of 10 mg/m^3. DEHP is considered a Class B2 carcinogen by the U.S. EPA and a Class 3 carcinogen by IARC and a carcinogen by NIOSH. The OSHA PEL is also 5 mg/m^3. Health Canada and the EU have not determined that DEHP should be labeled a carcinogen.

111.6 Studies on Environmental Impact

No information found.

112.0 *n*-Dinonyl Phthalate

112.0.1 CAS Number: [84-76-4]

112.0.2 Synonyms: 1,2-Benzenedicarboxylic acid dinonyl ester, di-*n*-nonyl phthalate, and phthalic acid dinonyl ester

112.0.3 Trade Names: NA

112.0.4 Molecular Weight: 418.62

112.0.5 Molecular Formula: $C_{26}H_{42}O_4$

112.0.6 Molecular Structure:

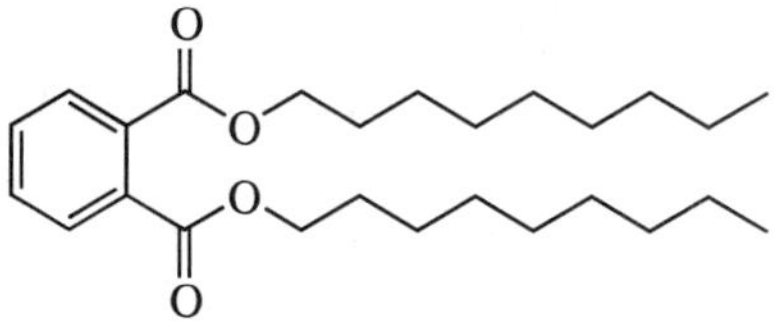

112.1 Chemical and Physical Properties

112.1.1 General

The empirical formula of *n*-dinonyl phthalate is $C_{26}H_{42}O_4$. Its structural formula is 1,2-[COO $(CH_2)_8CH_3]_2C_6H_4$. Its density and boiling point are 0.965 and 413°C, respectively. Other properties are summarized in Table 80.17.

112.1.2 Odor and Warning Properties

No information found.

112.2 Production and Use

Dinonyl phthalate is used for manufacturing vinyls that must be resistant to heat and migration and extraction by detergents.

112.3 Exposure Assessment

No information found.

112.4 Toxic Effects

112.4.1 Experimental Studies

The rat oral LD_{50} value is >2 g/kg (12). In three rats, inhalation of 5500 ppm heated (200°C) dinonyl phthalate for 1.5 h caused nasal irritation (12). Dinonyl phthalate does not cause irritation to guinea pig skin or rabbit eyes (12).

112.4.2 Human Experience

No information found.

112.5 Standards, Regulations, or Guidelines of Exposure

No information found.

112.6 Studies on Environmental Impact

No information found.

113.0 Diisononyl Phthalate

113.0.1 CAS Number: *[68515-48-0], [28553-12-0]*

113.0.2 Synonyms: 1,2-Benzenedicarboxylic acid, di-isononyl ester; and DINP

113.0.3 Trade Names: JAYFLEX DINP; Palatinol N; PX-139; and Santicizer DINP

113.0.4 Molecular Weight: 418.62

113.0.5 Molecular Formula: $C_{26}H_{42}O_4$

113.0.6 Molecular Structure:

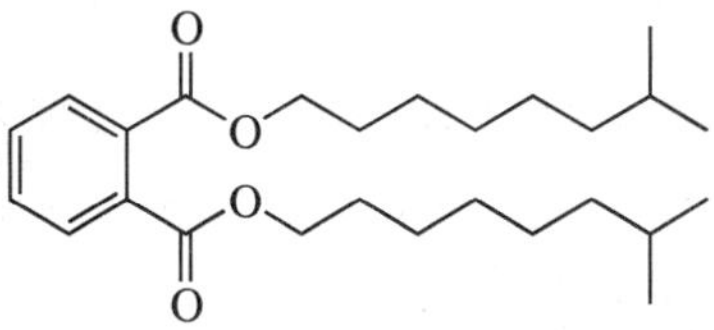

113.1 Chemical and Physical Properties

113.1.1 General

The empirical formula of diisononyl phthalate (DINP) is $C_{26}H_{42}O_4$. The structural formula for DINP varies because the isoalcohols used in manufacture contain several isomers. DINP is a clear colorless to light yellow liquid. It exists as a liquid at temperatures between −48°C (pour point) and 245–252°C (boiling point at 5 mmHg). The vapor pressure is 0.5 mmHg (at 200°C). DINP is insoluble in water, but it is soluble in organic solvents. It has a specific gravity of 0.97. Physico-chemical properties are summarized in Table 80.17.

113.1.2 Odor and Warning Properties

No information found.

113.2 Production and Use

DINP is manufactured by reaction of phthalic anhydride with iso-nonanol in the presence of an acid catalyst. Its major use is as a plasticizer.

113.3 Exposure Assessment

Similar to diisohexyl phthalate (106.3).

113.4 Toxic Effects

113.4.1 Experimental Studies

113.4.1.1 Acute Toxicity. For DINP, the rat oral LD_{50} value was >10 g/kg, and the rabbit dermal LD_{50} value was >3.16 g/kg (Table 80.18). Rats, mice, and guinea pigs exposed to the maximum attainable vapor concentration (0.067 mg/L) survived with no evidence of systemic toxicity or respiratory irritation (485).

113.4.1.2 Chronic and Subchronic Toxicity. Results of several repeated dose toxicity studies of DINP have been reviewed (500). In most of these the test material was administered in the diet. In one study rats were given 0, 50, 150, or 500 mg/kg/d DINP for 3 mo. Increased liver weight and hepatocytic hypertrophy were observed at the highest dose. In a second study, rats were given dietary DINP at levels of 0, 0.1, 0.3, 0.6, 1.0, and 2.0% (approximately 50, 150, 300, 600, and 1000 mg/kg) for 13 wk. Effects seen in those treated with 300 mg/kg or more DINP included decreased serum triglycerides and elevated liver and kidney weights. Electron microscopic examination of the liver showed a dose-related proliferation of peroxisomes and smooth endoplasmic reticulum. At necropsy, kidney lesions were seen only in the male rats. Decreased food consumption and reduced body weight gain were observed for only the high-dose animals. A mixture of commercial DINPs was given to rats at dietary levels of 0, 0.6, 1.2, and 2.5% (300, 600 and 1250 mg/kg/d) for 21 d. Changes at the 300-mg/kg/d level included increases in liver peroxisomal enzymes and total hepatic protein levels, decreased serum cholesterol levels, and elevated liver weights. Decreased body weight gain was seen in groups exposed to 600 mg/kg/d and above. Peroxisome proliferation was seen only in high-dose animals.

In lifetime studies using F344 rats, there were no biologically important effects at levels below 0.3% in the diet (or approximately 150 mg/kg/d) (486, 488). The no-effect level for liver effects (the most sensitive indicator of effects in rodents) was approximately 90 mg/kg/d. At levels of 0.3–0.6% (approximately 150–340 mg/kg/d) there was an increase in liver and kidney weights, elevated liver enzymes, and some mild effects on urinalysis/clinical chemistry parameters. At 1.2% in the diet (or approximately 600 mg/kg/d), there was a statistically significant increase in hepatocellular carcinoma in both the males and females (486). A statistically significant elevation in renal cell carcinoma was found in male F344 rats treated with 600 mg/kg/d (486). An increased incidence of mononuclear cell leukemia (MNCL) was seen in both studies at levels of 150 mg/kg/d and above.

Similar studies have been performed using the B6C3F1 mouse (487). The animals were given DINP by dietary administration at levels of 0.05, 0.15, 0.4, and 0.8% (or approximately 100, 300, 800, and 1700 mg/kg/d) for 104 wk. There were no significant effects in those treated with 100 mg/kg/d. Higher treatment levels were associated with changes in the liver and, ultimately, hepatocellular carcinoma. The dose–response relationship for hepatocellular carcinoma in the B6C3F1 mice was similar to that found in the F344 rats. However, the mice did not develop either renal cell carcinoma or mononuclear cell leukemia. The tumor response data are discussed in more detail in the carcinogenesis section.

In a 13-wk study, beagle dogs were given DINP via the diet at levels of 0, 0.125, 0.5, or 2.0% (increasing to 5.0% at week 9) or approximately 0, 37, 160, or 1000 mg/kg/d. Effects that were seen (only at the highest level) included body weight loss, decreased food consumption, elevated liver weights, and elevated liver enzyme markers. Histopathological changes included hepatocytic hypertrophy and hypertrophy of kidney tubular epithelial cells. There were no changes in gonadal weight and no histopathological abnormalities in the reproductive organs. The NOAEL was 0.5%, or approximately 160 mg/kg/d (485).

The toxicity of DINP also has been tested in two repeated dose studies in primates. In a 2-wk study in the cynomologous monkey, administration of 500 mg/kg/d DINP by nasogastric intubation had no significant effect on body weight, liver, kidney, or testes weights, urinalysis, hematology, clinical chemistry, or any other parameter that was measured (494). In a 13-wk study, DINP was given by oral administration to marmosets at levels of 100, 500, or 2500 mg/kg/d. There were no treatment-related changes in organ weights, blood chemistries, hematology, or pathology. In particular, there was no evidence of peroxisome proliferation, even at the highest dose tested. The only effects noted were changes in body weight and body weight gain in animals treated with the highest dose (659).

113.4.1.3 Pharmacokinetics, Metabolism, and Mechanisms. The dermal absorption of DINP is extremely low (660). DINP is rapidly metabolized to the monoester and parent alcohol and excreted. Repeated dosing does not result in accumulation of DINP and/or its metabolites in blood or tissue (661, 662).

113.4.1.4 Reproductive and Developmental. In a two-generation reproductive toxicity study, rats were given DINP at 0, 0.2, 0.4, or 0.8% (or approximately 170, 340, and 680 mg/kg/d) in the diet. There were no adverse effects on reproductive organs or other measures of reproductive function in either parental generation. Treatment-related microscopic changes in the liver (indicative of peroxisomal proliferation) were seen in male and female animals in all levels above 100 mg/kg/d in both parental generations. Microscopic changes in the kidney were seen in mid- and high-dose males of the second generation only. There were no differences in survival, clinical or gross observations for offspring from either the first or second generation. DINP had no effects on fertility or other classical endpoints of reproductive toxicity. A NOAEL for fertility of 1.5%

(approximately 1000 mg/kg/d) was established for both the parents and the offspring based on data from one- and two-generation reproductive toxicity studies (663). Offspring body weight was significantly reduced in both generations; however, for the most part, the decreases were not significantly different from the control group and were within the range of the historical values for the laboratory. Further, the body weight changes were largely reversible following weaning.

In a definitive developmental toxicity study in rats, DINP was administered by gavage at 0, 100, 500, and 1000 mg/kg on gestational days 6–15. There was maternal toxicity at the high level, as evidenced by decreased body weight and food consumption. In offspring of high-dose rats, there was some evidence of delayed fetal development and increased skeletal variations. DINP was not embryotoxic or teratogenic, nor was it a selective developmental toxicant. The NOAEL for both maternal and fetal effects was 500 mg/kg/d (664). Similar data were obtained by Hellwig and co-workers, in a developmental toxicity screening test of similar design (665).

A study to assess the effects of DINP administration on testicular weight of rats demonstrated that treatment for 3 wk at levels up to 2.5% (approximately 2600 mg/kg/d) did not result in significantly decreased testicular weight or produce any histological changes in F344 rats (666).

Results of *in vitro* screening tests for endocrine modulation are conflicting. At a concentration of 10^{-3} M, DINP showed weak estrogenic activity in a recombinant yeast screen (520). At 10^{-5} M, it induced proliferation of estrogen-responsive breast cancer cell lines MCF-7 and ZR-75 (520). In contrast, results of *in vitro* estrogen receptor competitive ligand-binding and mammalian- and yeast-based gene expression assays and *in vivo* uterotrophic and vaginal cornification assays indicate that DINP is not estrogenic (586).

113.4.1.5 Carcinogenesis. As stated, there have been two carcinogenesis studies in the F344 rat (486, 488) and one in the $B6C3F_1$ mouse (502). These studies establish that at dose levels of approximately 600 mg/kg/d, DINP can induce hepatocellular carcinoma in rats and mice. As there was evidence of peroxisomal proliferation at the carcinogenic doses in both species, it seems most likely that this was the mechanism for hepatocellular carcinoma induction (667). The renal cell carcinomas observed in the male rats was associated with the induction of alpha 2u-globulin (668) indicating that it was a sex- and species-specific effect. Kidney tumors that are the consequence of alpha 2u-induction are not considered to be clinically relevant to humans (669). Mononuclear cell leukemia (MNCL) is a tumor type that occurs spontaneously at a high and variable frequency in the F344 rat. Because there is no human equivalent, MNCL is not considered to be relevant to humans (488).

113.4.1.6 Genetic and Related Cellular Effects Studies. At concentrations up to 10 mg/plate, DINP was not mutagenic in *S. typhimurium* strains TA98, TA100, TA1535, TA1537, or TA1538 (with and without metabolic activation) (406, 485). It was inactive in an *in vitro* (CHO) test for chromosome aberrations at concentrations up to 160 μg/mL (485), a mouse lymphoma test at concentrations up to 8 μL/mL (without S-9) and 0.6 μL/mL (with S-9) (530), a test for unscheduled DNA synthesis in rat hepatocytes at a concentration of 10 μL/mL (500), and an *in vitro* (BALB/3T3) transformation test (546) at a concentration of

3.75 μL/mL. Treatment of rats with 5 mL/kg did not increase the incidence of chromosome aberrations in rat bone marrow (485).

Sufficiently high concentrations of DINP can induce peroxisomal proliferation in cultured rodent hepatocytes (670). In contrast, DINP does not cause changes consistent with peroxisomal proliferation in cultured human hepatocytes (690).

113.4.1.7 Other: Neurological, Pulmonary, Skin Sensitization. DINP produced slight conjunctival irritation when instilled in rabbit eyes, which completely cleared within 72 h. The maximum total Draize score in any animal at any time point was 6.8 (out of 110). DINP produced extremely mild irritation when applied to rabbit skin. The primary irritation score was 0.08, and no sign of irritation persisted beyond 24 h (485). Whereas an initial sensitization study in guinea pigs produced equivocal results, a subsequent test was negative (485).

113.4.2 Human Experience

In human volunteers, a 21-d cumulative irritation/repeat insult patch test provided no evidence of clinical sensitization or dermal irritation (599).

113.5 Standards, Regulations, or Guidelines of Exposure

No information found.

113.6 Studies on Environmental Impact

No information found.

114.0 Diisodecyl Phthalate

114.0.1 CAS Number: *[68515-49-1], [26761-40-0]*

114.0.2 Synonyms: 1,2-Benzenedicarboxylic acid, di-isodecyl ester; and DIDP

114.0.3 Trade Names: JAYFLEX DIDP, Palatinol Z, PX-120, and Vestinol DZ

114.0.4 Molecular Weight: 446

114.1 Chemical and Physical Properties

114.1.1 General

The empirical formula of diisodecyl phthalate (DIDP) is $C_{28}H_{46}O_4$. The structural formula for DIDP varies because the iso-alcohols used in manufacture contain several isomers. DIDP is a clear liquid between −48°C (pour point) and 256°C (the boiling point at 5 mmHg). The vapor pressure of DIDP is 0.34 mmHg at 200°C. DIDP is insoluble in water, but it is soluble in organic solvents. It has a specific gravity of 0.97. Table 80.17 summarizes the physical properties.

114.1.2 Odor and Warning Properties

No information found.

114.2 Production and Use

DIDP is manufactured by reaction of phthalic anhydride with isodecanol in the presence of an acid catalyst. Its major use is as a plasticizer.

114.3 Exposure Assessment

Similar to diisohexyl phthalate (106.3).

114.4 Toxic Effects

114.4.1 Experimental Studies

114.4.1.1 Acute Toxicity. DIDP is practically nontoxic following oral, dermal, or inhalation exposure. The oral LD_{50} value for rats is >6 g/kg (373, 563, 555, 559), and the dermal LD_{50} value for rabbits is >3.16 g/kg (Table 80.18). In acute inhalation studies, rats, mice and guinea pigs were exposed to a nominal concentration of 0.13 mg/L for 6 h. There was no mortality, no evidence of systemic effects, and no apparent respiratory irritation (560).

114.4.1.2 Chronic and Subchronic Toxicity. In one 90-d subchronic toxicity study in rats, DIDP was given in the feed at levels of 0.05, 0.3, and 1% (approximately 30, 200, and 650 mg/kg/d). The predominant effect was an increase in liver weight in animals treated with the highest dose. There were no effects noted in the mid-dose animals (approximately 200 mg/kg/d). Other studies ranging from 3 wk to 90 d have confirmed the liver as the target organ (489–492). In one of these studies (505), significant effects on liver weight and liver enzyme induction were noted in animals treated with approximately 350 mg/kg/d. In other studies, the lowest effect levels were higher.

Further screening studies in rats and mice (484) have demonstrated that ingestion of 6000 ppm DIDP in the diet (approximately 400 mg/kg/d) causes increases in liver weight, liver enzymes, and cell replication. These effects are considered to be hallmarks of a peroxisomal proliferation process (416).

114.4.1.3 Pharmacokinetics, Metabolism, and Mechanisms. Consult Section 105.4.1.3.

114.4.1.4 Reproductive and Developmental. In a reproductive toxicity screening test in mice, daily administration of 10 mL/kg DIDP (approximately 9650 mg/kg) had no significant effect on maternal parameters, number of live born/litter, birth weight, or survival or weight gain of pups to postnatal day 3 (496). In a definitive developmental toxicity study in rats, DIDP was administered orally (0, 100, 500, and 1000 mg/kg/d) to dams on gestational days 6–15. Maternal toxicity (as indicated by decreased body weight gain and reduced food consumption) was seen at the high-dose level. Increased fetal skeletal variations (supernumerary ribs) occurred in offspring of dams treated with the high-dose level. No effects on fetal viability or weight and no increase in the incidence of frank malformations were observed. These results indicate that DIDP is not embryotoxic or teratogenic and shows no selective developmental effects (664). Equivalent results were obtained in a developmental toxicity screening test of similar design (665).

Treatment for three weeks at levels up to 2.5% DIDP in the diet (approximately 2600 mg/kg/d) did not result in significantly decreased testicular weight or histological changes in male rats (490, 666). At concentrations up to 10^{-3} M, DIDP was inactive in a recombinant yeast screen for estrogenic activity (520). Results of *in vitro* estrogen receptor competitive ligand-binding and mammalian- and yeast-based gene expression assays and *in vivo* uterotrophic and vaginal cornification assays also indicate that DINP is not estrogenic (586).

114.4.1.5 Carcinogenesis. No information found.

114.4.1.6 Genetic and Related Cellular Effects Studies. DIDP shows no evidence of mutagenic potential. It was inactive in *S. typhimurium* strains TA98, TA100, TA1535, and TA1537 at concentrations up to 10 mg/plate (with and without metabolic activation) (406) and in a bacterial assay for 8-azaguanine resistance (562). DIDP was inactive in an *in vitro* mouse lymphoma assay at concentrations up to 10 μL/mL (without metabolic activation) and 2 μL/mL (with metabolic activation) (530), and in a bone marrow micronucleus test in mice treated with 5 g/kg (560). Additionally, 20 μL/mL DIDP did not increase transformation of BALBc/3T3 cells *in vitro* (530).

114.4.1.7 Other: Neurological, Pulmonary, Skin Sensitization. When instilled into rabbit eyes, DIDP produced slight conjunctival irritation which cleared completely in 48 h. The maximum total Draize score was 4. There was no chemosis and no iridial or corneal effects (560). When applied to rabbit skin, no evidence of irritation was produced (the primary irritation index was 0.0) (560). Tests for sensitization in guinea pigs produced inconsistent results (599).

114.4.2 Human Experience

There was no evidence of clinical sensitization or dermal irritation in human volunteers participating in a 21-d repeated insult patch test (599). In 144 subjects with suspected dermatoses to plastic or glue allergens, two subjects (1.4% of the group) experienced irritation after patch testing with 5.0% DIDP. None of the patients had allergic reactions (129).

114.5 Standards, Regulations, or Guidelines of Exposure

No information found.

114.6 Studies on Environmental Impact

No information found.

115.0 *n*-Diundecyl Phthalate

115.0.1 CAS Number: [3648-20-2]

115.0.2 Synonyms: 1,2-Benzenedicarboxylic acid, di-undecyl ester; and DUP

115.0.3 Trade Names: Eastman DUP

115.0.4 Molecular Weight: 474.72

115.0.5 Molecular Formula: $C_{30}H_{50}O_4$

115.0.6 Molecular Structure:

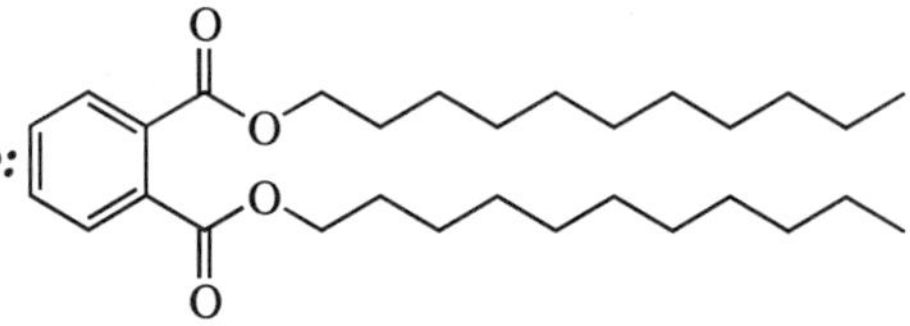

115.1 Chemical and Physical Properties

115.1.1 General

The empirical formula of diundecyl phthalate (DUP) is $C_{30}H_{50}O_4$. Its structural formula is 1,2-$[COO(CH_2)_{10}CH_3]_2C_6H_4$. It is a clear liquid between −45°C (pour point) and 270°C (the boiling point at 5 mmHg). The specific gravity of DUP is 0.96. It is fairly insoluble in water (< 0.1 g/100 mL at 21°C). The vapor pressure of DUP is < 0.15 at 392 mmHg. Table 80.17 summarizes the physical properties.

115.1.2 Odor and Warning Properties

No information found.

115.2 Production and Use

DUP is manufactured by the reaction of phthalic anhydride with undecanol in the presence of an acid catalyst. Its major use is as a plasticizer.

115.3 Exposure Assessment

No information found.

115.4 Toxic Effects

115.4.1 Experimental Studies

115.4.1.1 Acute Toxicity. DUP is practically nontoxic following a single exposure. The acute oral LD_{50} value for the rat is > 15 g/kg (561), the acute dermal LD_{50} value for the rabbit is > 7.9 g/kg, and the acute inhalation LC_{50} value (6 h) for the rat is > 1.8 mg/L (the maximum attainable vapor concentration at ambient temperature) (561). The LD_{50} value reported for intraperitoneal injection in the mouse was greater than 100 g/kg (559). These data are summarized in Table 80.18.

115.4.1.2 Chronic and Subchronic Toxicity. To determine the effect of repeated dosing, rats were given DUP in the diet at levels of 0, 0.3, 1.2, and 2.5% (approximately 280, 1100, and 2300 mg/kg/d) for 21 d. There was evidence of liver and kidney weight increases at the 1.2 and 2.5% dose levels. In addition, the activities of palmitoyl-CoA (an indicator of peroxisome proliferation) and other liver enzymes were elevated. Testis weights also were increased at these dose levels, but no pathological changes were noted during the microscopic examination. A no-observable-effect level (NOEL) of 0.3% (approximately 280 mg/kg/d) and a lowest observable-effect-level (LOEL) of 1.2% (approximately 1100 mg/kg/d) were established in this study (671).

115.4.1.3 Pharmacokinetics, Metabolism, and Mechanisms. Consult Section 105.4.1.3.

115.4.1.4 Reproductive and Developmental. No studies were found that specifically measured the effects of DUP on reproductive and/or developmental endpoints *in vivo*. However, the potential of DUP to cause estrogenic effects *in vitro* was measured. At concentrations up to 10^{-3} M, diundecyl phthalate was not active in a recombinant yeast screen for estrogenic activity (520).

115.4.1.5 Carcinogenesis. No information found.

115.4.1.6 Genetic and Related Cellular Effects Studies. DUP was negative in several types of assays measuring genotoxicity. At concentrations up to 10 mg/plate, DUP did not induce mutations in *S. typhimurium* strains TA98, TA100, TA1535, and TA1537 (with and without metabolic activation) (406). In the absence or presence of metabolic activation, concentrations of up to 10 or 8 μL/mL DUP, respectively, were inactive in a mouse lymphoma test (530). At concentrations up to 40 μL/mL, it did not induce transformation of BALB/3T3 cells (530).

115.4.1.7 Other: Neurological, Pulmonary, Skin Sensitization. DUP was not irritating to either the skin or the eyes of rabbits (559).

115.4.2 Human Experience

No information found.

115.5 Standards, Regulations, or Guidelines of Exposure

No information found.

115.6 Studies on Environmental Impact

No information found.

116.0 Diisoundecyl Phthalate

116.0.1 CAS Number: *[85507-79-5]*

116.0.2 Synonyms: 1,2-Benzenedicarboxylic acid, di-isoundecyl ester; and DIUP

116.0.3 Trade Names: JAYFLEX DIUP and Palatinol 11

116.0.4 Molecular Weight: 474

116.1 Chemical and Physical Properties

116.1.1 General

The empirical formula of diisoundecyl phthalate (DIUP) is $C_{30}H_{50}O_4$. The structural formula for DIUP varies because the iso-alcohols used in manufacture contain several isomers. It is a clear liquid between $-45°C$ (pour point) and 270°C (the boiling point at 5 mmHg). The vapor pressure of DIUP is 5.3×10^{-7} mmHg (at 298°C). DIUP is insoluble in water, but it is soluble in organic solvents. The specific gravity of DIUP is 0.96. Table 80.17 summarizes the physical properties.

116.1.2 Odor and Warning Properties

No information found.

116.2 Production and Use

DIUP is manufactured by reaction of phthalic anhydride with isoundecanol in the presence of an acid catalyst. Its major use is as a plasticizer.

116.3 Exposure Assessment

Similar to diisohexyl phthalate (106.3).

116.4 Toxic Effects

116.4.1 Experimental Studies

116.4.1.1 Acute Toxicity. No specific acute toxicity data for DIUP was located. The acute toxicity of DIUP is assumed to be similar to that of DUP.

116.4.1.2 Chronic and Subchronic Toxicity. No specific subchronic or chronic toxicity data for DIUP were located. The subchronic toxicity of DIUP is assumed to be similar to that of DUP.

116.4.1.3 Pharmacokinetics, Metabolism, and Mechanisms. Consult Section 105.4.1.3.

116.4.1.4 Reproductive and Developmental. No information found.

116.4.1.5 Carcinogenesis. No information found.

116.4.1.6 Genetic and Related Cellular Effects Studies. No information found.

116.4.1.7 Other: Neurological, Pulmonary, Skin Sensitization. DIUP was not active in skin sensitization assays in guinea pigs (599).

116.4.2 Human Experience

DIUP was not active in irritation or skin sensitization assays in human volunteers (599).

116.5 Standards, Regulations, or Guidelines of Exposure

No information found.

116.6 Studies on Environmental Impact

No information found.

117.0 Diisotridecyl Phthalate

117.0.1 CAS Number: *[27253-26-5; 68515-47-9]*

117.0.2 Synonyms: 1,2-Benzenedicarboxylic acid ditridecyl ester; and DTDP

117.0.3 Trade Names: JAYFLEX DTDP and Vestinol TD

117.0.4 Molecular Weight: 530.83

117.0.5 Molecular Formula: $C_{34}H_{58}O_4$

117.0.6 Molecular Structure:

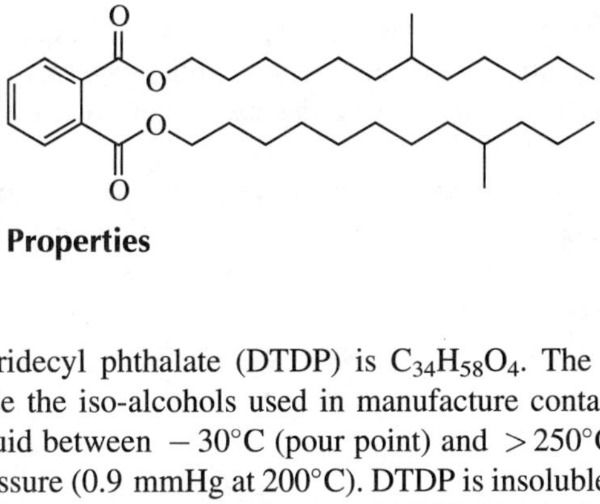

117.1 Chemical and Physical Properties

117.1.1 General

The empirical formula of diisotridecyl phthalate (DTDP) is $C_{34}H_{58}O_4$. The structural formula for DTDP varies because the iso-alcohols used in manufacture contain several isomers. DTDP is a colorless liquid between −30°C (pour point) and >250°C (boiling point). It has a very low vapor pressure (0.9 mmHg at 200°C). DTDP is insoluble in water, but it is soluble in organic solvents. It has a specific gravity of 0.957. Physical properties are summarized in Table 80.17.

117.1.2 Odor and Warning Properties

No information found.

117.2 Production and Use

DTDP is manufactured by reaction of phthalic anhydride with isotridecanol in the presence of an acid catalyst. Its major use is as a plasticizer.

117.3 Exposure Assessment

Similar to diisohexyl phthalate (106.3).

117.4 Toxic Effects

117.4.1 Experimental Studies

117.4.1.1 Acute Toxicity. DTDP is essentially nontoxic in acute toxicity studies in animals. The oral LD_{50} value for the rat is >10 g/kg (555, 672). The dermal LD_{50} value for the rabbit is >3.16 g/kg (672).

117.4.1.2 Chronic and Subchronic Toxicity. No information found.

117.4.1.3 Pharmacokinetics, Metabolism, and Mechanisms. Consult Section 105.4.1.3.

117.4.1.4 Reproductive and Developmental. When tested for estrogenic effects in a recombinant yeast screen, DTDP was initially judged to be positive. However, the effects seen were subsequently found to be due to *bis*-phenol A, which was present in the DTDP sample (520).

117.4.1.5 Carcinogenesis. No information found.

117.4.1.6 Genetic and Related Cellular Effects Studies. No information regarding the genetic toxicity of DTDP was located. However, at concentrations up to 10 mg/plate, the *n*-isomer of ditridecyl phthalate was nonmutagenic in *S. typhimurium* strains TA98, TA100, TA1535 or TA1537 (in the absence or presence of metabolic activation) (430).

117.4.1.7 Other: Neurological, Pulmonary, Skin Sensitization. DTDP produces very slight conjunctival irritation when instilled in rabbit eyes. The mean Draize scores were 1.0, 0.33, and 0 for conjunctival redness and 0.67, 0, and 0 for chemosis. There were no iridial or corneal effects (672). When applied to rabbit skin, DTDP produced a very mild erythema (672). DTDP did not produce a skin-sensitizing reaction in guinea pigs (592).

117.4.2 Human Experience

DTDP did not produce a skin-sensitizing reaction in human volunteers (599). No other information about toxicity in humans was located.

117.5 Standards, Regulations, or Guidelines of Exposure

No information found.

117.6 Studies on Environmental Impact

No information found.

118.0 Di-C7-11-Phthalate

118.0.1 CAS Number: [68515-42-4]

118.0.2 Synonyms: 1,2-Benzenedicarboxylic acid, di-C7-11-branched and linear alkyl esters; and D711P

118.0.3 Trade Names: Palatinol 711 and Santicizer 711 Plasticizer

118.0.4 Molecular Weight: 418.6 (assuming equally weighted composition)

118.1 Chemical and Physical Properties

118.1.1 General

Di-C7-11-phthalate (D711P) is a mixture of C7, C9, and C11 esters of phthalic acid where different alcohols can be esterified to the same molecule of phthalic acid. It is a colorless, oily liquid. It has a melting point of $-57°C$, a boiling point of 235–278°C, and a very low vapor pressure. D711P is only minimally soluble in water (< 0.1 mg/L). It has a neutral pH value. Physico-chemical properties are summarized in Table 80.17.

118.1.2 Odor and Warning Properties

No information found.

118.2 Production and Use

D711P is commercially produced by catalyzed esterification of phthalic anhydride with C7-9-11 alcohol blend utilizing batch or continuous processes. The individual process steps consist of (1) a reaction step, (2) excess alcohol removal, (3) drying and (4) filtration. The final product has a phthalate ester content of well above 99.5%. D711P is used as a plasticizer in vinyl formulations. Major end uses include automotive vinyl, film, sheeting, and industrial vinyl sealant.

118.3 Exposure Assessment

No information found.

118.4 Toxic Effects

118.4.1 Experimental Studies

118.4.1.1 Acute Toxicity. The acute LD_{50} values for D711P are very high, demonstrating that D711P is practically nontoxic after acute exposure. The oral LD_{50} value is > 15.8 g/kg for rats (555, 564) and > 20 g/kg for mice (555). Inhalation exposure of rats to an estimated vapor concentration of 5.89 mg/L of air for 6 h did not produce mortality (564). The dermal LD_{50} value for D711P in rabbits is > 7.94 g/kg (564).

118.4.1.2 Chronic and Subchronic Toxicity. D711P was tested as one of seven phthalate esters, representing a variety of chain lengths and degrees of branching in the alcohol moiety, for its ability to produce peroxisome proliferation in the Fischer 344 rat (624, 673). The animals were exposed to 0, 0.3, 1.2, or 2.5% D711P in the diet over 21 d. Compared to most of the other phthalate esters (including di-2-ethylhexyl phthalate), D711P was a very weak peroxisome proliferator. The low-dose group showed no signs of peroxisome proliferation.

Dose levels of 0, 250, 500, 750, 1000, or 2000 mg/kg/d D711P have been administered in feed to groups of 10 male and 10 female Fischer 344 rats for 4 wk (564). Males ingesting the highest dose level had significantly reduced mean body weights beginning at week 2, resulting in reduced terminal body weights (86% of control males). Food consumption was reduced in high-dose level males during week 1, and the two high-dose-level female groups during week 2. No deaths occurred during the course of the study. Clinical observations other than body weights were similar for treated and control animals. The only lesion considered to be treatment related was a mottled, diffuse, yellow-tan discoloration of the livers of male rats from groups receiving 750 mg/kg or more D711P. Based on this finding, the NOEL for the study was considered to be 500 mg/kg for males and greater than 2000 mg/kg for females.

In rats given 0, 1000, 3000, or 10,000 ppm D711P in the diet for 90 d, no compound-related changes in body weight or body weight gain or increases in mortality were observed (564). On days 45, 84, and 90, serum alkaline phosphatase and pyruvic transaminase activities were elevated in males receiving 10000 ppm D711P. No outstanding differences between test and control rats were noted at the time of gross pathological examination. Significant elevations in absolute and relative liver weight were observed among females fed 10,000 ppm. There were no other organ weight changes that could be attributed to ingestion of D711P. It was concluded that degenerative changes and necrosis in the livers of male but not female rats fed 10,000 ppm were compound related. Findings observed in male and female rats fed 1000 and 3000 ppm were not attributed to the ingestion of D711P.

Rats, guinea pigs, and monkeys were exposed to D711P by inhalation, 6 h/d, 5 d/wk for 6 mo (564). D711P was generated as an aerosol at concentrations of 6.5 or 22 mg/m^3. There were no treatment-related effects in monkeys or guinea pigs at either concentration. Therefore, the NOEL for these species was 22 mg/m^3. Rats in the high-exposure group exhibited significant increases in liver, kidney, pituitary, spleen, and gonad weights. The NOEL in rats was 6.5 mg/m^3.

In a chronic oral study, F344 rats received 0, 300, 1000, or 3000 ppm D711P in the diet for 2 y (674, 675). Interim sacrifices were conducted after 12 mo. No substance-related effects on organ weights, clinical pathology, gross pathology, or histopathology were observed after 12 mo. At the terminal sacrifice, slight decreases in serum total protein levels in treated males and females and higher body weights in treated males occurred, which were considered to be of equivocal toxicological relevance

118.4.1.3 Pharmacokinetics, Metabolism, and Mechanisms. Consult Section 105.4.1.3.

118.4.1.4 Reproductive and Developmental. D711P was tested for developmental toxicity in Wistar rats (665). An oily solution of the test substance was administered by stomach tube at dose levels of 40, 200, and 1000 mg/kg/d on day 6 through day 15 of gestation. Maternal toxicity (reduced body weight, impaired weight gain, reduced weight of the uterus, vaginal hemorrhage, and increased relative kidney and liver weights) was observed in the highest-dose group. Increased embryolethality, reduced fetal body weights, increased malformation rate of fetuses/litter with soft tissue and skeletal variations, skeletal retardation, and necrotic placentas also occurred at the 1000-mg/kg dose level. At 40 and 200 mg/kg/d, no substance-related adverse effects on the dams or teratogenic effects on the fetuses occurred. Therefore, the NOAEL for maternal or fetal toxicity in this study was 200 mg/kg/d.

In another developmental toxicity study, Sprague–Dawley rats were given 0, 250, 1000, or 5000 mg/kg/d D711P by gavage from gestation day 6 to 19 (564). No signs of maternal toxicity were reported. Reduced fetal body weights were observed in offspring of dams treated with the highest dose level. No other developmental or teratogenic effects were seen in any of the groups.

118.4.1.5 Carcinogenesis. In the 2-yr oral study referred to previously (644), the incidence of several types of neoplasms was increased in rats treated with 300, 1000, or 3000 ppm D711P as compared to control rats. These include pancreatic islet cell adenomas (males), mononuclear cell leukemia (males and females), and malignant mammary gland tumors (females). A review of previous 2 yr studies in F344 rats indicated that the number and distribution of these neoplasms in rats from this study were within the range of normal biological variation; therefore, the tumors were not considered to be associated with administration of D711P.

In the chronic toxicity study conducted by Hirzy (675), there was no significant increase in liver tumors following dietary treatment with levels up to 3000 ppm.

118.4.1.6 Genetic and Related Cellular Effects Studies. At concentrations up to 0.6 μL/mL (without S-9) and 0.15 μL/mL (with S-9), D711P was inactive in the mouse lymphoma assay (530, 676). In another mouse lymphoma assay (564), a positive response occurred without metabolic activation that was not reproducible. Treatment of BALBc/3T3 cells with up to 6 μL/mL D711P had no effect on cell transformation (530).

118.4.1.7 Other: Neurological, Pulmonary, Skin Sensitization. D711P was not irritating to the skin and eye as tested in rabbits and not a skin sensitizer in guinea pigs (677). The Draize scores for eye and skin irritation are 1.3 out of 110 and 0.8 out of 8.0, respectively (564).

118.4.2 Human Experience

A patch test on 57 human volunteers produced no positive reactions following the initial application of D711P, any of the 15 serial applications, or the subsequent challenge application made 10–14 d later (564). On the basis of this test, it was concluded that this product was not a primary or cumulative skin-irritant or skin-sensitizing agent.

118.5 Standards, Regulations, or Guidelines of Exposure

No information found.

118.6 Studies on Environmental Impact

D711P has an aquatic half-life of 5–8 d. Biodegradability in water and sludge is classified as intermediate (564).

119.0 Diallyl Phthalate

119.0.1 CAS Number: [131-17-9]

119.0.2 Synonyms: 1, 2-Benzenedicarboxylic acid, di-2-propenyl ester (9CI); phthalic acid, diallyl ester (6CI, 8CI); and DAP

119.0.3 Trade Names: Dapon R

119.0.4 Molecular Weight: 246.26

119.0.5 Molecular Formula: $C_{14}H_{14}O_4$

119.0.6 Molecular Structure:

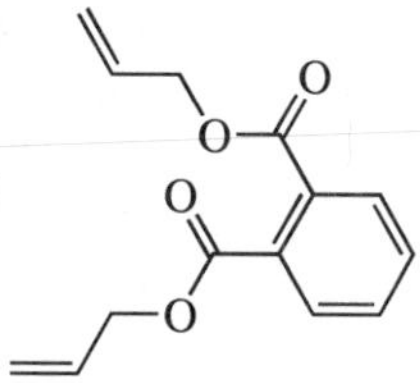

119.1 Chemical and Physical Properties

119.1.1 General

The empirical formula of diallyl phthalate is $C_{14}H_{14}O_4$. Its structural formula is 1,2-$[COOCH_2CH{:}CH_2]_2C_6H_4$. Its density is 1.121, melting point −70°C, and boiling point 290°C. Physico-chemical properties are summarized in Table 80.17.

119.1.2 Odor and Warning Properties

No information found.

119.2 Production and Use

Diallyl phthalate (DAP) is used in industry as a monomer in the processing of thermosetting plastics, polyester resins, and varnishes. It is also used for critical electronic applications, and as a dye carrier.

119.3 Exposure Assessment

No information found.

119.4 Toxic Effects

119.4.1 Experimental Studies

119.4.1.1 Acute Toxicity. DAP is more acutely toxic than many other phthalate esters. The oral LD_{50} values for rats and mice are 0.896 and 1.0–1.47 g/kg, respectively (565, 566). The dermal LD_{50} value for rabbits is approximately 3.8–3.9 g/kg (565).

119.4.1.2 Chronic and Subchronic Toxicity. In a study performed by the NTP in 1985, F344 rats were given 0, 50, 100, 200, 400, and 600 mg/kg/d DAP via gavage for 14 consecutive days (678). All rats administered 600 mg/kg died within 5 d. Body weight gains of males and females receiving 400 mg/kg and of males receiving 200 mg/kg were decreased during the second week compared to the control group. Upon necropsy, changes in the liver, lung, stomach, and spleen were observed. The livers of the animals given 200 mg/kg and above were enlarged, dark, and mottled, and had small yellowish spots on the surface. No histological examination was performed in this range-finding study.

In another subchronic study, F344 rats were given 0, 25, 50, 200, and 400 mg/kg/d via gavage for 13 wk (678). Eight out of ten males given 400 mg/kg died or were killed moribund during the study. Clinical signs of toxicity included diarrhea, rough fur or alopecia, hunched posture, and general emaciation in males and females of the two highest-dose groups. Histopathologically, periportal hepatocellular necrosis and fibrosis and bile duct hyperplasia were observed in both sexes treated with 200 mg/kg/d or more DAP. In animals treated with 50 or 100 mg/kg, mild to moderate, dose-dependent, periportal liver cell hyperplasia was found. Although no histopathology was performed in animals treated with 25 mg/kg, this level was listed as the NOAEL.

Administration of 14 daily doses of 0, 50, 100, 200, 400, or 600 mg/kg DAP via gavage had no effect on body weight and body weight gain of B6C3F$_1$ mice (686). Although deaths occurred at the two highest dose levels, no chemically related lesions were observed in these or any other treatment groups. In a longer-term study, 0, 25, 50, 100, 200, or 400 mg/kg/d DAP was administered to B6C3F1 mice by gavage 5 d/wk for 13 wk (566). Mean body weight gain was decreased by 12% in the high-dose group compared to the control group. No other substance-related clinical effects were seen, and no effects were observed during the macroscopical and histopathological examination. The NOEL was given by the authors as 200 mg/kg.

In lifetime studies, groups of 50 F344 rats/sex were given 0, 50, or 100 mg/kg/d DAP by gavage for 103 wk. No substance-related clinical signs of toxicity were observed in any of the dose groups (678). Administration of the test substance resulted in dose-related chronic liver injury, characterized by periportal necrosis and fibrosis, pigment accumulation in periportal histiocytes, and bile duct hyperplasia. In B6C3F1 mice receiving 0, 150, and 300 mg/kg/d DAP via gavage for 103 wk, no clinical signs of toxicity were observed (566). Dose-dependent inflammation and hyperplasia of the forestomach were seen in male and female mice.

119.4.1.3 Pharmacokinetics, Metabolism, and Mechanisms. Thirty minutes after rats and mice were treated orally with ^{14}C-labeled DAP, the highest levels of radioactivity were found in small intestine, liver, dermis, muscles, blood, and kidneys. After 24 h, about 6–7% of the radioactivity was present in rats and 1–3% in mice. In rats, 60% of the radioactivity was found in urine and 30% was exhaled as CO_2. In mice, 91% was present in urine, and only 8% was detected as CO_2 (678).

The following metabolic pathway is suggested for DAP. First, the diester is hydrolyzed to monoallyl phthalate (MAP) and allyl alcohol (AA). AA can be oxidized to acrolein and acrylic acid and further metabolized to CO_2. Allyl alcohol and acrolein can also react with reduced glutathione to form 3-hydroxypropylmercapturic acid (678–681). Alternatively, allyl alcohol and acrolein can be oxidized to the epoxides glycidol and glycidaldehyde.

These epoxides can be hydrolyzed to glycerin and glyceraldehyde or conjugated with reduced glutathione (678). It is not clear whether DAP is metabolized by this pathway *in vivo* (682). However, some of the products of the aforementioned reactions (e.g., monoallyl phthalate, 3-hydroxypropylmercapturic acid, allyl alcohol) as well as an unidentified polar metabolite have been detected in urine of rats and mice treated with DAP (679).

119.4.1.4 Reproductive and Developmental. No information found.

119.4.1.5 Carcinogenesis. In the 103-wk study referred to previously (678), a slight increase in mononuclear cell leukemia (MNCL) was seen in female rats treated with 50 or 100 mg/kg/d DAP. MNCL occurs in F344 control rats at a high incidence; however, the incidence of 51% in female rats at the high-dose level was above historical control data for the laboratory (29%). No significant increases in tumor incidences were seen in male rats. Based on this study, DAP was considered to have demonstrated equivocal evidence for carcinogenicity in female F344 rats according to the NTP (678).

In male and female B6C3F1 mice receiving 300 mg/kg DAP via gavage for 103 wk (5 d/wk), the incidence of forestomach papillomas was significantly greater than that of controls (566). Because of the rarity of forestomach papillomas in control B6C3F1 mice and the concomitant observation of dose-related forestomach hyperplasia, the development of these tumors was considered to be test-substance related. Compared to controls, a slight increase in the incidence of lymphomas was observed in males receiving 300 mg/kg/d DAP. Because the increase was not statistically significant compared to historical control data, this effect was not considered to be test-substance related.

119.4.1.6 Genetic and Related Cellular Effects Studies. DAP showed no mutagenic effects in a series of *S. typhimurium* reverse-mutation assays with and without metabolic activation using tester strains TA98, TA100, TA1535, TA1537, and TA1538 (406, 683, 684). A weak positive result (factor 2.1–2.27) was observed in one test with TA1535 only (685). No mutagenic effects were found in an 8-azaguanine resistance assay with *S. typhimurium* using TA100 and TA1537. DAP showed positive results in a mouse lymphoma assay with and without metabolic activation (686). The positive effects occurred only at cytototoxic concentrations. DAP induced chromosomal aberration and sister chromatid exchange in CHO cells *in vitro* with metabolic activation (687). No mutagenic effects were seen with DAP in *D. melanogaster* using the SLRL test (688).

DAP was tested for induction of bone marrow chromosomal aberrations in B6C3F1 mice after a single IP injection of 0, 75, 150, or 300 mg/kg (689). Because the percentage of cells with chromosomal aberrations was only increased in one of two trials, and no dose–response relationship was found, the results of this test were considered to be ambiguous. In an *in vivo* micronucleus assay using B6C3F1 mice 0, 43.8, 87.5, and 175 mg/kg/d DAP was administered IP for three consecutive days (689). No significant increase in micronuclei was observed in the bone marrow of treated animals compared to the control group.

119.4.1.7 Other: Neurological, Pulmonary, Skin Sensitization. DAP was nonirritating to the rabbit skin when applied under occlusive condition for 4 h (12). It was also nonirritating to the rabbit eye (690).

119.4.2 Human Experience

No information found.

119.5 Standards, Regulations, or Guidelines of Exposure

No information found.

119.6 Studies on Environmental Impact

No information found.

120.0 Butyl Benzyl Phthalate

120.0.1 CAS Number: *[85-68-7]*

120.0.2 Synonyms: 1, 2-Benzenedicarboxylic acid, butyl phenylmethylester; and BBP

120.0.3 Trade Names: Santicizer 160; Sicol 160; and Unimoll BB

120.0.4 Molecular Weight: 312.36

120.0.5 Molecular Formula: $C_{19}H_{20}O_4$

120.0.6 Molecular Structure:

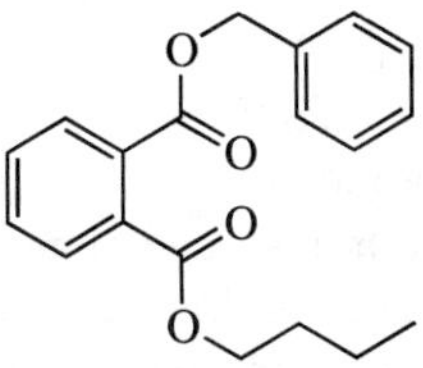

120.1 Chemical and Physical Properties

120.1.1 General

The empirical formula of butyl benzyl phthalate (BBP) is $C_{19}H_{20}O_4$. Its structural formula is 1-($COOC_4H_9$)-2-($COOCH_2C_6H_5$)C_6H_4. BBP is an oily liquid from $< -35°C$ (melting point) to 370°C (boiling point). It has a very low vapor pressure (8×10^{-5} mmHg at 25°C). BBP is slightly soluble in water (0.00269 g/L) and is soluble in organic solvents. It has a specific gravity of slightly greater than water (1.1). Physico-chemical properties are summarized in Table 80.17.

120.1.2 Odor and Warning Properties

BBP has a "characteristic aromatic odor" and a bitter taste.

120.2 Production and Use

BBP is manufactured via by the sequential addition of butanol and benzyl chloride to phthalic anhydride. It is used as a plasticizer for polyvinyl chloride plastics, particularly vinyl floor tile, vinyl leather, and cloth coating. It may be used in sealants, foams, adhesives, coating and inks, car care products, and cosmetics (691).

120.3 Exposure Assessment

As BBP is used in a variety of products used in construction, exposure can occur from contact with these materials. However, overall exposure is extremely low. Mean concentrations of BBP in indoor air samples were determined to be 35 ng/m^3 (692). BBP has been detected in certain foodstuffs with an estimate overall intake of 0.02 mg/kg/d (548).

120.4 Toxic Effects

120.4.1 Experimental Studies

120.4.1.1 Acute Toxicity. The acute oral toxicity of BBP is extremely low. Oral LD_{50} values for BBP range from 2.3 to 20 g/kg (567, 694). The oral LD_{50} value for B6C3F$_1$ mice was 6.2 g/kg for males and 4.2 g/kg for females (567). Hammond et al. (569) report a dermal LD_{50} value for rabbits of > 10 g/kg. The dermal LD_{50} value for rats and mice is estimated to be 6.7 g/kg (568).

120.4.1.2 Chronic and Subchronic Toxicity. Sprague–Dawley rats were administered BBP in the diet for 3 mo at dose levels of 0, 188, 375, 1125 or 1500 mg/kg/d (males) and 188, 375, 750, 1125, or 1500 mg/kg/d (females) (569). No compound-related lesions were observed at necropsy or on histopathological examination. Relative liver weight was increased at 750 mg/kg/d and higher in females, and at 1125 mg/kg/d and higher in males. In male rats, relative kidney weight increased at 750 mg/kg/d and higher. A NOAEL of 375 mg/kg/d was established based on liver and kidney effects at 750 mg/kg/d.

Wistar rats were fed diets administering 0, 151, 381, or 960 mg BBP/kg/d (males) and 0, 171, 422,or 1069 mg BBP/kg/d (females) for 3 mo (569). Slight anemia at the highest dose level and decreased urinary pH at the mid- and high-dose levels were observed in males. Reduced body weight gain was observed in high-dose animals. Relative liver weight was significantly increased at all dose levels in females and at the highest dose level in males. A significant increase in kidney-to-body-weight ratio occurred in a dose-related manner in both sexes at the mid- and high-dose levels. An increased incidence of red spots was observed on the liver of the mid- and high-dose males; there were also small areas of hepatocellular necrosis in high-dose males. Histopathological lesions of the pancreas were observed in males at the mid- and high-dose levels, and included islet enlargement with cell vacuolization, and peri-islet congestion. No histopathological lesions were described for females. A NOAEL of 151 mg/kg/d was established based on liver, kidney, and pancreatic lesions observed at 381 mg/kg/d.

Groups of ten F344 rats per sex were fed diets containing 0, 1600, 3100, 6300, 12500, and 25000 ppm BBP for 13 wk (567). No assessment of daily feed intake, and thus compound ingestion, was made. Reduced body weight and testicular degeneration were observed in male rats receiving 25000 ppm BBP. No other adverse effects were reported.

Male F344 rats were fed diets containing 0, 300, 900, 2800, 8300, or 25000 ppm BBP for 6 mo (693). Reduced body weight and body weight gain were observed in the high-dose group. Further, reduced testicular weight, reduced right cauda and epididymis weight, and reduced sperm concentration were also observed in the high-dose group.

A subchronic inhalation study was conducted in groups of 25 male or 25 female Sprague–Dawley rats exposed to concentrations of 0, 51, 218, or 789 mg/m^3 BBP aerosol, 6 h/d, 5 d/wk for a total of 59 exposures. Increases in relative liver and kidney weights were observed in high-concentration animals (693). A NOAEL of 218 mg/m^3 was established.

B6C3F1 mice were treated with 0, 1600, 3100, 6300, 12500, and 25000 ppm BBP in feed for 13 wk (567). At the two highest dose levels, reduced body weights were observed. No other indications of toxicity were reported.

Groups of three male and female dogs were fed diets containing 10000–50000 ppm BBP for 3 mo (567). Body weight was reduced at the highest dose levels (1852 and 1973 mg/kg/d for males and females, respectively). No changes were observed in clinical chemistry or histopathological parameters.

120.4.1.3 Pharmacokinetics, Metabolism, and Mechanisms. Oral administration of 5 g/kg to dogs resulted in approximately 10% of the dose being absorbed (694). The half-life of BBP in blood is 10 min, while the blood half-life of monoester metabolites of BBP is 5.9 h (695). Following intravenous administration of 20 mg/kg of ^{14}C-BBP, 55% of the dose was excreted into bile and 34% was excreted into the urine (695). Examination of urinary metabolites of rats following oral administration of 3.6 mmol BBP/kg/d for 3 d indicated that approximately 70% of the metabolites were unconjugated monoesters, while the remainder was conjugated. A dose-dependent change in both the route of excretion and the ratio of certain metabolites was observed, with urinary excretion predominating at lower doses and fecal excretion observed at higher doses (695). It can be concluded that BBP is rapidly metabolized to monoester components. These components are either excreted directly or are conjugated and excreted.

120.4.1.4 Reproductive and Developmental. Groups of 30 pregnant Sprague–Dawley rats ingested 0, 420, 1100 or 1640 mg/kg/d of BBP on gestational days 6–15 (695). Reduced body weight gain, increased relative liver, and increased water consumption were observed in maternal animals at the mid- and high-dose levels. Further, increased relative kidney weight was observed in high-dose dams. Fetal toxicity, which was characterized by increases in the percentage of litters resorbed and the percentage of litters with resorptions, was observed at the highest dose level. Additional indices of fetal toxicity observed in the high-dose group were decreased fetal weight and an increased percentage of litters with malformed fetuses. An increase in the percentage of variations per litter was observed in high-dose group. The most common variation was rudimentary extra lumbar ribs. The NOAEL for maternal and developmental toxicity was 420 mg/kg/d.

Developmental toxicity also was examined in Swiss Albino CD-1 mice (697). Groups of 28 or 30 pregnant mice were fed diets containing 0, 0.1%, 0.5%, or 1.25% (approximately 0, 182, 910, and 2330 mg/kg/d) BBP in diet on days 6–15 of gestation. Decreased maternal body weight gain was observed in the mid- and high-dose group, and relative liver and kidney weights were increased in the high-dose animals. Fetal toxicity was observed in the mid- and high-dose groups. Effects observed in the high-dose groups included an increase in percent resorptions per litter, percent litters with resorptions, percent litters with nonlive implants, percent fetuses with variations per litter, and decreased mean fetal body weight. The following effects were observed in both the mid- and high-dose litters: an increase in the percent nonlive implants per litter, a decrease in the number of live fetuses per litter, and an increase in the percent of malformed fetuses per litter and percent of litters with malformed fetuses. The NOAEL for maternal and developmental toxicity was 182 mg/kg/d.

Pregnant Wistar rats ingested diets containing 0%, 0.25%, 0.5%, 1.0%, or 2.0% BBP (approximately 0, 185, 375, 654, or 974 mg/kg/d) from days 0–20 of gestation (698). Maternal toxicity ranged from maternal weight loss in the high-dose group (974 mg/kg/d) to reduced body weight gain in the 375- and 654-mg/kg/d groups. At the highest level of exposure, there was complete resorption of all implanted embryos. Fetal body weights were reduced at 654 mg/kg/d compared to controls. Morphological evaluations of the fetuses revealed no evidence of teratogenesis. The developmental NOAEL was 375 mg/kg/d, while the maternal NOAEL was 185 mg/kg/d.

Pregnant Wistar rats were administered 0, 500, 750, or 1000 mg/kg BBP via gavage on gestation days 7–15 (699). Maternal body weight was reduced at the two highest dose levels, and maternal food consumption was reduced at all dose levels. The number of resorptions, dead fetuses per litter, and percent postimplantation loss per litter were increased in the 750- and 1000-mg/kg dose groups. All litters were resorbed at the highest-dose group. The mean number of live fetuses per litter and mean fetal weight were reduced in the 750-mg/kg group. Teratogenic effects were observed in the 750-mg/kg/d group, which included cleft palate, fused sternebrae, and dilation of the renal pelvis. No fetal effects were observed in the lowest-dose group; however, maternal toxicity was observed. Thus 500 mg/kg/d was the LOAEL for maternal toxicity and the NOAEL for developmental toxicity.

In a combined reproductive/developmental screening protocol (OECD Protocol 421), male and female rats were treated with 0, 250, 500, or 1000 mg BBP/kg/d via gavage (700). Males were treated 2 wk prior to and during mating, and were sacrificed after 29 d of treatment. Females were treated 2 wk prior to mating, during mating and gestation, and were sacrificed on post natal day 6. Effects on the parental animals were observed at the high-dose level. Males had reduced body weights, decreased testicular and epididymal weights, and evidence of testicular degeneration at 1000 mg/kg/d. Female rats had reduced body weight at 1000 m/kg/d. Reproductive indices were affected at 1000 mg/kg/d. Effects observed included a decreased number of pregnant females, reduced number of live pups per dam at birth, and reduced pup weight on post partum days 1 and 6. With the exception of a transient decrease in pup weight, there were no effects on the parental generation or offspring at 500 mg/kg/d.

A one-generation (two-litter) reproductive toxicity study was conducted in male and female Wistar rats administered 0, 0.2, 0.4, and 0.8% BBP in the diet (701). These levels resulted in the ingestion of approximately 0, 108, 206, and 418 mg/kg/d for males and 0, 106, 217, and 446 mg/kg/d for females. Males were treated for a 10-wk period prior to mating and females began treatment 2 wk prior to mating. Both groups continued treatment for the production of two litters. Females in the high-dose group demonstrated a reduction in mean body weight and body weight change during gestation and lactation. An increase in the relative liver weight was observed in high-dose females. A NOAEL for parental animals was established at 0.4% in the diet. In the two litters that were produced, there were no adverse effects on fertility, pregnancy, or offspring development. A NOAEL was established for reproductive effects at 0.8% in the diet.

Reproductive effects of BBP in male F344 rats have been investigated by the National Toxicology Program (702, 703). Groups of 10 males were administered 0, 0.625, 1.25, 2.5, or 5.0% (approximately, 0, 312.5, 625, 1250, or 2500 mg/kg/d) in the diet for 14 d. No deaths occurred during the study. Body weight was reduced in the two highest-dose groups. Food consumption was consistently reduced in the highest-dose group throughout the experiment. Absolute weights of testis, epididymis, prostate, and seminal vesicles were significantly reduced at the two highest-dose levels in a dose-related manner. Atrophy of the testis (both aspermatogenesis and seminiferous tubules), seminal vesicles, and prostate were observed at the two highest-dose levels. Similarly, effects on the epididymis were observed in only the two highest-dose groups. Relative weights of liver and kidneys were increased at all levels of exposure, in a dose-related manner. Although histopathological changes were described for all dose groups, atrophy was observed only in the highest-dose group. Plasma testosterone was decreased at the highest dose level. The NOAEL was 625 mg/kg/d.

Effects on the testes were further evaluated in two dietary feeding studies (705). BBP was administered in the diet to male F344/N rats for either 10 or 26 wk. In the 10-wk study, concentrations of BBP in the diet were 0, 300, 2800, or 25000 ppm, which resulted in ingestion of 0, 20, 200, or 2200 mg/kg/d. Dietary concentrations of BBP in the 26-wk study were 0, 300, 8300, or 25000 ppm, which resulted in ingestion of 0, 200, 550, or 2200 mg/kg/d. In the 10-wk study, an assessment of fertility and reproductive performance was conducted by mating treated male rats with two unexposed females. In both studies, assessments of organ weights, testicular histopathology, and epididymal sperm concentration were made at necropsy. Body weight and body weight gain of the high-dose group (2200 mg/kg/d) were significantly lower than the controls in both studies. Marked effects on the prostate, testis, cauda, and epididymis were observed in the high-dose group (2200 mg/kg/d) in both studies. Fertility indices were significantly lower at the high dose level, while no effect was observed at lower dose-levels. An analysis of sperm concentration produced conflicting results. Clear reductions in sperm concentrations were observed in the high-dose group. In the 10-wk study, a statistically significant reduction in sperm count was observed at 200 mg/kg/d; however, rats receiving 550 mg/kg/d for approximately three times as long had no reduction in sperm count. Examination of both sets of data indicate that normal variation in sperm number could be the cause of the reported reduction at 200 mg/kg/d.

Exposure to low doses of BBP during mating, pregnancy, and lactation was reported to produce effects in male offspring, reduced testicular weights, and sperm concentrations (704). However, two larger followup studies were unable to reproduce this finding (705, 706).

Though BBP has been reported to have weak estrogenic activity in several *in vitro* assays, (401, 520, 584, 586, 707), the two key metabolites, monobutyl phthalate, and monobenzyl phthalate, were not found to be estrogenic *in vitro* (520). Moreover, neither BBP nor its metabolites monobutyl phthalate and monobenzyl phthalate were uterotrophic *in vivo* in rats (585, 586, 708, 709) or mice (584).

120.4.1.5 Carcinogenesis. The NTP examined the carcinogenicity of BBP in rats and mice (567). Groups of 50 male and female rats and mice were exposed to BBP via the diet, at levels of 0, 6000, or 12,000 ppm (0, 300, and 600 mg/kg for rats and 0, 780, or 1560 mg/kg for mice). Male and female mice and female rats were exposed for 103 wk. Due to poor survival, all males were sacrificed at weeks 29–30; this part of the study was later repeated (705). No treatment-related neoplasms were observed in mice. Survival was not affected. A dose-dependent reduction in body weight in both sexes was the only treatment-related effect observed in this study. Further, non-neoplastic changes were all within the normal limits of incidence for $B6C3F_1$ mice. The NTP concluded that, under the conditions of the bioassay, BBP "was not carcinogenic for $B6C3F_1$ mice of either sex." An increased incidence of mononuclear cell leukemias was observed in the high-dose female rats. No other treatment-related findings were observed. The NTP concluded that BBP was "probably carcinogenic for female F344/N rats, causing an increased incidence of mononuclear cell leukemias" (567). The biological significance of this finding is uncertain as the background incidence of this tumor type in F344 rats is quite high (702).

A second NTP bioassay was conducted on BBP (693). This study was conducted in groups of 50 male and female F344 rats. Rats were fed diets containing 3000, 6000, and 12,000 ppm BBP for male rats (delivering approximately 0, 120, 240, or 500 mg/kg/d) and 6000, 12,000, and 24,000 ppm BBP for female rats (approximately 0, 300, 600, or 1200 mg/kg/d). In contrast to the previous study, no tumors were observed in the high-dose groups. An increased incidence of pancreatic acinar cell adenoma was observed in high-dose male rats. No other elevation in tumor incidence was observed. The authors concluded that there was "some evidence of carcinogenic activity" in male rats, based upon the increased incidences of pancreatic acinar cell adenoma and acinar cell adenoma or carcinoma (combined).

Tumor responses following chronic administration of BBP were examined under conditions of dietary restriction (710). Test groups included control and treated rats fed ad libitum, and restricted-diet treated and weight-matched control rats. An increase in the incidence of pancreatic acinar cell neoplasms was observed in BBP-treated ad libitum fed male rats compared to ad libitum fed and weight-restricted controls. Interestingly, no increase was observed in restricted-diet group receiving BBP, suggesting that diet may play an important role in the expression of this tumor following administration of BBP. In female rats, a slight increase in urinary bladder neoplasm was observed, but only at a 32-mo time point, in the restricted-feed treated group. This increase was not observed in any other groups. These data suggest that duration of study may be a predominant factor in the development of this response.

Taken as a whole, the bioassay data for BBP do not reveal any strong tumor responses. IARC has recently reevaluated BBP and reconfirmed a Group 3 classification.

120.4.1.6 Genetic and Related Cellular Effects Studies. BBP was not mutagenic in the Ames assay with and without metabolic activation (406, 543, 711–713). In the mouse lymphoma assay, BBP produced either negative (530, 714, 715), or equivocal responses (686, 716). BBP did not cause *in vitro* transformation of Balb/c-3T3 cells (530, 717). In assays for chromosomal aberrations and sister chromatid exchanges (SCE) in Chinese hamster ovary cells (409), there was slight evidence for a trend in one SCE test (without activation).

Results of *in vivo* genotoxicity assays have been negative or equivocal. BBP did not increase the incidence of sex-linked recessive lethals in *D. melanogaster* (688). Results from a mouse bone marrow tests examining induction of either sister chromatid exchanges or chromosomal aberrations indicated weak responses (693). These responses demonstrated no real dose response and were only observed at test concentrations that exceeded the LD_{50} for mice. Both of these responses, although statistically significant, were small and indicative of only weak clastogenic activity. In contrast, negative results were reported by Ashby et al. (705) in a micronucleus assay in rats. Thus the weight of the evidence indicates that BBP is not mutagenic.

120.4.1.7 Other: Neurological, Pulmonary, Skin Sensitization. Hammond et al. (703) reported that 0.5 mL of neat BBP applied on the abraded or unabraded skin of rabbits for 24 h produced essentially no irritation. Other reports indicate that BBP produces varying degrees of skin irritation; however, the test methods and route of application vary greatly (551, 718). Numerous studies in rodents and humans indicate that BBP is not a skin sensitizer (719, 720).

120.4.2 Human Experience

No significant acute or cumulative irritation was reported in human patch tests (719).

120.5 Standards, Regulations, or Guidelines of Exposure

No information found.

120.6 Studies on Environmental Impact

The prevalence of BBP in the environment has recently been reviewed (691). Low concentrations of BBP have been detected in air and water. Once released into the environment, BBP rapidly degrades.

121.0 Dimethoxyethyl Phthalate

121.0.1 CAS Number: *[117-82-8]*

121.0.2 Synonyms: 1,2-Benzenedicarboxylic acid, dimethoxyethyl ester; 1,2-benzenedicarboxylic acid, *bis*(2-methoxyethyl) ester; *bis*(2-methoxyethyl)phthalate; dimethyl glycol phthalate; di(2-methoxyethyl) phthalate; dimethyl cellosolve phthalate; DMEP; 2-methoxylethyl phthalate; methyl glycol phthalate; phthalic acid, *bis*(2-methoxyethyl) ester; phthalic acid, di (methoxyethyl)ester; di-(2-methoxyethyl) ester kyseliny flatlove [Czech]; 4-09-00-03241 [Beilstein]; and BRN 2056929

121.0.3 Trade Names: Kesscoflex MCP

121.0.4 Molecular Weight: 282.29

121.0.5 *Molecular Formula:* $C_{14}H_{18}O_6$

121.0.6 *Molecular Structure:*

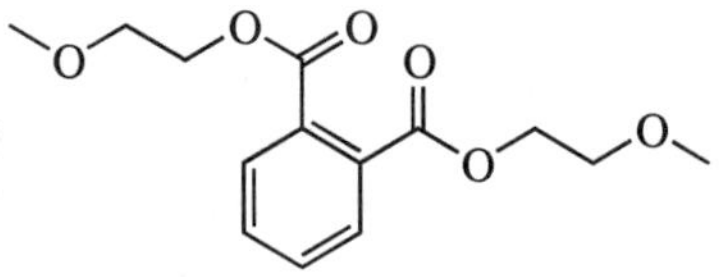

121.1 Chemical and Physical Properties

121.1.1 *General*

The empirical formula of dimethoxyethyl phthalate (DMEP) is $C_{14}H_{18}O_6$. Its structural formula is 1,2-$(COOCH_2CH_2OCH_3)_2C_6H_4$. It is a colorless, oily, liquid, with a density of 1.1708. Its melting and boiling points are −45 and 340°C, respectively. It is slightly soluble in water (8.5 mg/L). Other physico-chemical properties are listed in Table 80.17.

121.1.2 *Odor and Warning Properties*

DMEP has a mild odor.

121.2 Production and Use

DMEP is formed by the esterification of ethylene glycol monomethyl ether with phthalic anhydride in the presence of a catalyst (sulfuric acid or *p*-toluenesulfonic acid), or noncatalytically at high temperature. It is used as a plasticizer in the manufacture of polyvinylchloride devices and cellulose acetate and as an industrial solvent.

121.3 Exposure Assessment

No information found.

121.4 Toxic Effects

121.4.1 *Experimental Studies*

121.4.1.1 Acute Toxicity. The oral LD_{50} values for dimethoxyethyl phthalate in the rat, mouse, and guinea pig range from 1.6 to 6.4 g/kg (12). For the guinea pig, the dermal LD_{50} value is > 10 mL/kg (Table 80.18). DMEP produced slight dermal irritation in the guinea pig, and slight ocular irritation in the rabbit (537).

121.4.1.2 Chronic and Subchronic Toxicity. Groups of rats were treated with 0, 100, or 1000 mg/kg by oral intubation for 2 wk. Body weight and food consumption was reduced at 1000 mg/kg.

Hemoglobin and hematocrit values were reduced at 100 and 1000 mg/kg; erythrocyte and leukocyte counts were reduced at 1000 mg/kg. Absolute liver, kidney, thymus, and testes weights were decreased at 1000 mg/kg, as were relative thymus and testes weights. Thymic and testicular atrophy were observed as was a decrease in spermatogonia (537). During a 30-d study in rats, a no-effect chronic feeding level of 0.5 g/kg/d was determined. At 1.7 g/kg/d, decreased food consumption, reduced growth rate, and some histochemical changes were noted (570). In rats inhaling dimethoxyethyl phthalate 6 h/d for 62 d, a no-effect level of 145 ppm was observed (12).

In rats, ingestion of 300, 500, or 900 mg/kg DMEP for a period up to 2 yr had no effect on body weight, reproduction, or pathology (721).

121.4.1.3 Pharmacokinetics, Metabolism, and Mechanisms. Consult Section 105.4.1.3.

121.4.1.4 Reproductive and Developmental. In male rats, IP injection of 1.19–2.38 mL/kg DMEP caused a dose-dependent reduction of fertility (722). The effect of oral administration of 1, 1.5, or 2 mg/kg DMEP on fertility of male rats also was examined (723). Increased sperm head count was noted in high-dose animals, and a highly significant increase in the percentage of abnormal sperm heads was noted in mid- and high-dose groups. Although decreased testes weights were observed in all treatment groups, no histopathological abnormalities were observed.

In female rats, IP injection of 374–600 mg/kg DMEP during gestation caused abnormalities in the musculoskeletal, central nervous and/or cardiovascular systems of offspring (501, 724, 725). Similar defects were observed in offspring of rats given an oral dose of 593 mg/kg on day 10 or 13 of gestation (726). In a more traditional developmental toxicity study, pregnant Sprague–Dawley rats were treated by oral intubation with dose levels of 60, 180, or 600 mg/kg/d DMEP from gestation day 6–15. The dams were allowed to litter and the pups followed for the first few days post partum. Body weight gain was lower than controls for the 600-mg/kg group, and this dose level produced 100% embryolethality. Body weight gain and survival for the pups at 180 mg/kg were lower than controls. Individual pups from 3 of 10 litters had skeletal abnormalities. Body weight and pup survival were reduced at 60 mg/kg, but no morphologic abnormalities were seen (537). Injection of DMEP (dose unknown) into the yolk sac of developing chick embryos caused damage to the central nervous system (727). After hatching, grossly abnormal behavior, such as tremor, nonpurposeful body movement, and inability to stand or walk normally, was observed.

It has been hypothesized that the metabolites of DMEP (2-methoxyethanol or methoxyacetic acid) are responsible for the teratogenic effects of DMEP. All three of these agents produce hydronephrosis, short limbs and tails, and unusual heart defects that are not seen with other agents (728). In addition, *in vitro* treatment of post implantation rat embryos with 1–5 mM methoxyacetic acid, but not 5 mM DMEP or 2-methoxyethanol, caused developmental abnormalities (729).

121.4.1.5 Carcinogenesis. No information found.

121.4.1.6 Genetic and Related Cellular Effects Studies. When administered by IP injection at 1.19, 1.78, or 2.38 mL/kg, DMEP was mutagenic in a dominant lethal assay in mice (722).

121.4.1.7 Other: Neurological, Pulmonary, Skin Sensitization. Application to rabbit and guinea pig skin causes slight irritation (12, 730). Instillation of 100 mg into rabbit eyes causes mild irritation.

121.4.2 Human Experience

No information found.

121.5 Standards, Regulations, or Guidelines of Exposure

No information found.

121.6 Studies on Environmental Impact

DMEP is expected to exist solely as a vapor in the ambient atmosphere. Vapor-phase DMEP will be degraded in the atmosphere by reaction with photochemically produced hydroxyl radicals. The half-life for this reaction is estimated to be 20 h. If released into soil, DMEP is expected to be highly mobile and is not expected to adsorb to suspended solids and sediment in the water column. Volatilization from soil and water is not expected to be an important fate process. Although the biodegradation of DMEP had not been studied, results of studies with other phthalates suggest that it will be readily biodegraded. The potential for bioconcentration of DMEP in aquatic organisms is low (124).

122.0 Diethoxyethyl Phthalate

122.0.1 CAS Number: NA

122.0.2 Synonyms: 1,2-Benzenedicarboxylic acid, diethoxyethyl ester

122.0.3 Trade Names: NA

122.0.4 Molecular Weight: 310.35

122.1 Chemical and Physical Properties

122.1.1 General

The empirical formula of diethoxyethyl phthalate is $C_{16}H_{22}O_6$. Its structural formula is 1,2-$(COOCH_2CH_2OCH_2CH_3)_2C_6H_4$. Its melting and boiling points are -34 and 345°C, respectively. It is insoluble in water. Physico-chemical properties are summarized in Table 80.17.

122.1.2 Odor and Warning Properties

No information found.

122.2 Production and Use

No information found.

122.3 Exposure Assessment

No information found.

122.4 Toxic Effects

122.4.1 Experimental Studies

In rats, ingestion of 10% diethoxyethyl phthalate causes death in 7–15 d (570). No other information was located.

122.4.2 Human Experience

No information found.

122.5 Standards, Regulations, or Guidelines of Exposure

No information found.

122.6 Studies on Environmental Impact

No information found.

123.0 Dibutoxyethyl Phthalate

123.0.1 CAS Number: [117-83-9]

123.0.2 Synonyms: 1,2-Benzenedicarboxylic acid, dibutoxyethyl ester; 1,2-benzenedicarboxylic acid, *bis*(2-butoxyethyl) ester; *b*-butoxyethyl phthalate; *bis*(2-butoxyethyl)-phthalate; DBEP: di (2-butoxyethyl) phthalate; butyl cellosolve phthalate; dibutyl cellosolve phthalate; dibutylglycol phthalate; *n*-butyl glycol phthalate; ethanol, 2-butoxy-, phthalate; di-(2-butoxyethyl) ester kyseliny flatlove [Czech]; Dibutylcellosolve Ftalat [Czech]; AI3-00524; and 4-09-00-03242 [Beilstein], BRN 2006754

123.0.3 Trade Names: Kesscoflex BCP; Kronisol; Palatinol K; and Plasthall DBEP

123.0.4 Molecular Weight: 366.45

123.0.5 Molecular Formula: $C_{20}H_{30}O_6$

123.0.6 Molecular Structure:

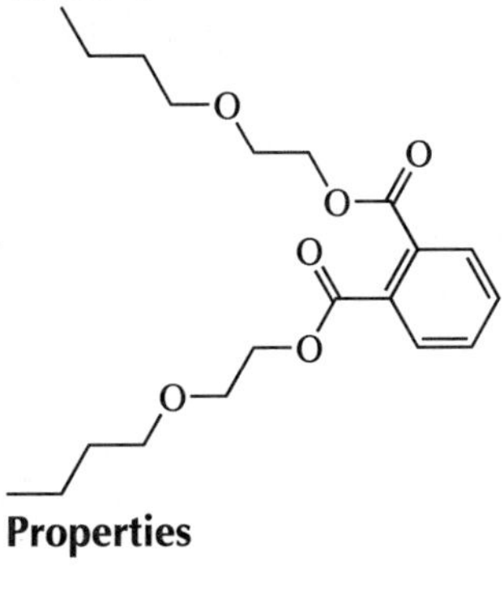

123.1 Chemical and Physical Properties

123.1.1 General

The empirical formula of dibutoxyethyl phthalate (DBEP) is $C_{20}H_{30}O_6$. Its structural formula is 1,2-$(COOCH_2CH_2O(CH_2)_3CH_3)_2C_6H_4$. It is a colorless liquid, with a density of

1.063. Its melting and boiling points are −55 and 270°C, respectively. It is soluble in organic solvents and fairly insoluble in water. Physico-chemical properties are summarized in Table 80.17.

123.1.2 Odor and Warning Properties

When heated to decomposition, DBEP emits acrid smoke and irritating fumes.

123.2 Production and Use

DBEP is formed by the esterification of glycol monobutyl ether with phthalic anhydride in the presence of a catalyst (sulfuric acid or *p*-toluenesulfonic acid) or noncatalytically at high temperature. It is used as a plasticizer for polyvinyl chloride, polyvinyl acetate, and other resins.

123.3 Exposure Assessment

No information found.

123.4 Toxic Effects

123.4.1 Experimental Studies

123.4.1.1 Acute Toxicity. The acute toxicity of dibutoxyethyl phthalate is very low. The oral LD_{50} value for the rat is 8.4 g/kg (12). Rats inhaling 290 ppm over a 6-h period showed no adverse effects (12).

123.4.1.2 Chronic and Subchronic Toxicity. No information found.

123.4.1.3 Pharmacokinetics, Metabolism, and Mechanisms. Consult Section 105.4.1.3.

123.4.1.4 Reproductive and Developmental. Injection of DBEP into the yolk sac of developing chicks causes exopthalmia, cranial bifida, blindness, and damage to the central nervous system (727).

123.4.1.5 Carcinogenesis. No information found.

123.4.1.6 Genetic and Related Cellular Effects Studies. No information found.

123.4.1.7 Other: Neurological, Pulmonary, Skin Sensitization. Application of DBEP to rabbit skin causes slight irritation (12). Instillation of 500 mg into rabbit eyes causes mild irritation (731).

123.4.2 Human Experience

No information found.

123.5 Standards, Regulations, or Guidelines of Exposure

No information found.

123.6 Studies on Environmental Impact

DBEP is expected to exist solely as a vapor in the ambient atmosphere. Vapor-phase DBEP will be degraded in the atmosphere by reaction with photochemically produced hydroxyl radicals. The half-life for this reaction is estimated to be 9.1 h. If released into soil, DBEP is expected to be slightly mobile. Volatilization from soil and water is not expected to be an important fate process. If released into soil, DBEP is not expected to adsorb to suspended solids and sediment in the water column. Although the biodegradation of DBEP has not been studied, results of studies with other phthalates suggest that it will be readily biodegraded. Theoretically, the potential for bioconcentration of DMEP in aquatic organisms is high. However, bioconcentration studies on other phthalate esters suggest that bioconcentration may be lower than indicated by theoretical equations due to the ability of aquatic organisms to metabolize phthalate esters (124).

124.0 Methyl Phthalyl Ethyl Glycolate

124.0.1 CAS Number: [85-71-2]

124.0.2 Synonyms: (Ethoxycarbonyl)methyl phthalate; ethoxycarbonylmethyl methyl phthalate; ethyl *o*-(*o*-(methoxycarbonyl)benzoyl)glycolate; ethyl *o*-(methoxycarbonyl) benzoyloxyacetate; glycolic acid ethyl ester, methyl phthalate; methyl phthalyl ether glycolate; 1,2-benzenedicarboxylic acid 2-ethoxy-2-oxoethyl methyl ester; 1,2-benzene-dicarboxylic acid methyl hydroacetic acid ethyl ester; methyl carbethoxymethyl phthalate; ethoxykarbonylmethyl-methylester kyseliny flatlove [Czech]; 4-09-00-03256 [Beilstein]; AI3-01792 [NLM]; BRN 1995396 [RTECS]; EINECS 201-625-3; and NSC 4836 [NLM]

124.0.3 Trade Names: Santicizer M-17

124.0.4 Molecular Weight: 266.25

124.0.5 Molecular Formula: $C_{13}H_{14}O_6$

124.0.6 Molecular Structure:

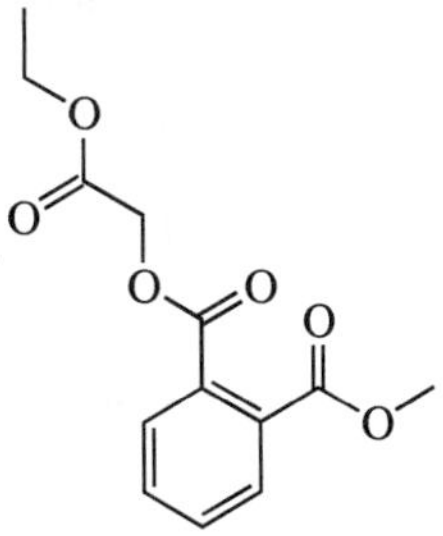

124.1 Chemical and Physical Properties

124.1.1 General

The empirical formula of methyl phthalyl ethyl glycolate is $C_{13}H_{14}O_6$. Its structural formula is 1-($COOCH_3$)-2-($COOCH_2COOC_2H_5$)C_6H_4. It is a liquid, with a density of

1.220. Its melting and boiling points are < -35 and 189°C, respectively. It is insoluble in water. Physico-chemical properties are summarized in Table 80.17.

124.1.2 Odor and Warning Properties

No information found.

124.2 Production and Use

No information found.

124.3 Exposure Assessment

No information found.

124.4 Toxic Effects

124.4.1 Experimental Studies

124.4.1.1 Acute Toxicity. The oral LD_{50} value for the rat is 9.04 g/kg (570). For the guinea pig, the dermal LD_{50} value is > 10 mL/kg (12).

124.4.1.2 Chronic and Subchronic Toxicity. When fed to rats for 30 d, the no-effect level for methyl phthalyl ethyl glycolate was 0.48 g/kg/d (733). In a 2-yr study, Hodge observed that a dietary concentration of 1.5% caused no effects on growth, life span, or hematologic, clinical, or histopathological parameters in rats (12). However, enlarged kidney tubules that were plugged with oxalate crystals were noted in rats ingesting 5.0%. Dogs tolerated 1.0 g/kg/d mixed into the diet. They showed no observable hematologic, chemical, or microscopic tissue changes. No oxalate crystal deposits were found in kidneys.

124.4.1.3 Pharmacokinetics, Metabolism, and Mechanisms. No information found.

124.4.1.4 Reproductive and Developmental. No information found.

124.4.1.5 Carcinogenesis. No information found.

124.4.1.6 Genetic and Related Cellular Effects Studies. No information found.

124.4.1.7 Other: Neurological, Pulmonary, Skin Sensitization. Application of 500 mg methyl phthalyl ethyl glycolate caused mild irritation to rabbit eyes (731).

124.4.2 Human Experience

No information found.

124.5 Standards, Regulations, or Guidelines of Exposure

No information found.

124.6 Studies on Environmental Impact

No information found.

125.0 Ethyl Phthalyl Ethyl Glycolate

125.0.1 CAS Number: [84-72-0]

125.0.2 Synonyms: Carbethoxymethyl ethyl phthalate; diethyl *o*-carboxybenzoyloxyacetate; EPEG: ethoxycarbonylmethyl ethyl phthalate; ethyl carbethoxymethyl phthalate; 1,2-benzenedicarboxylic acid, 2-ethoxy-2-oxoethyl ethyl ester; 1,2-benzenedicarboxylic acid ethyl hydroxyactetic acid ethyl ester; 4-09-00-03256; [Beilstein]; AI3-01794 [NLM]; BRN 1389422 [RTECS]; and EINECS 201-555-3

125.0.3 Trade Names: Santicizer E-15

125.0.4 Molecular Weight: 280.28

125.0.5 Molecular Formula: $C_{14}H_{16}O_6$

125.0.6 Molecular Structure:

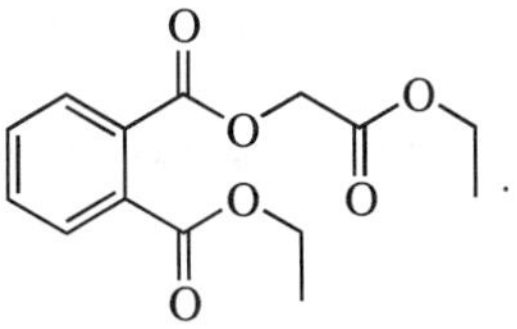

125.1 Chemical and Physical Properties

125.1.1 General

The empirical formula of ethyl phthalyl ethyl glycolate is $C_{14}H_{16}O_6$. Its structural formula is 1-($COOC_2H_5$)-2-($COOCH_2COOC_2H_5$)C_6H_4. It is a liquid, with a density of 1.180 and a boiling point is of 320°C. It is insoluble in water. Physico-chemical properties are summarized in Table 80.17.

125.1.2 Odor and Warning Properties

No information found.

125.2 Production and Use

No information found.

125.3 Exposure Assessment

No information found.

125.4 Toxic Effects

125.4.1 Experimental Studies

125.4.1.1 Acute Toxicity. The oral and intraperitoneal LD_{50} values for the mouse are 5.66 and 3.65 mL/kg, respectively (Table 80.18).

125.4.1.2 Chronic and Subchronic Toxicity. Male and female rats (25/group/sex) were administered either 0, 0.05, 0.5, or 5% ethyl phthalyl ethyl glycolate in the diet for 2 yr (732). Hematology analysis and urinalysis were performed at seven different time points, and histopathology was performed at the end of the study. No effects were seen in the two lowest-dose groups. In the 5% group, decreased growth and survival were noted. Kidneys of the high-dose animals were granular, sometimes swollen and of a pale yellow color. The renal pelvis was dilated, and there were deposits of crystalline calcium oxalate in renal tubules. In paired groups of dogs receiving 0.01, 0.05, or 0.25 g/kg/d for 1 yr, no compound-related effects were seen.

125.4.1.3 Pharmacokinetics, Metabolism, and Mechanisms. No information found.

125.4.1.4 Reproductive and Developmental. No information found.

125.4.1.5 Carcinogenesis. No information found.

125.4.1.6 Genetic and Related Cellular Effects Studies. No information found.

125.4.1.7 Other: Neurological, Pulmonary, Skin Sensitization. No information found.

125.4.2 Human Experience

No information found.

125.5 Standards, Regulations, or Guidelines of Exposure

No information found.

125.6 Studies on Environmental Impact

No information found.

126.0 Butyl Phthalyl Butyl Glycolate

126.0.1 CAS Number: *[85-70-1]*

126.0.2 Synonyms: BGBP; 1,2-benzenedicarboxylic acid 2-butoxy-2-oxoethyl butyl ester; 1,2-benzenedicarboxylic acid butyl hydroxyacetic acid butyl ester; butoxycarbonylmethyl butyl phthalate; butyl carbobutoxymethyl phthalate; butyl glycol butyl phthalate; butyl phthalate butyl glycolate; dibutyl *o*-(*o*-carboxybenzoyl) glycolate; dibutyl *o*-carboxybenzoyloxyacetate; glycolic acid butyl ester, butyl glycolate; glycolic acid, phthalate, dibutyl ester; phthalic acid, butoxycarbonylmethyl butyl ester; phthalic acid butyl ester, butyl glycolate; 4-09-00-03256; [Beilstein]; AI3-01793 [NLM]; BRN 2007363 [RTECS]; Casewell No. 131B [NLM]; CTFA 00353; EINECS 201-624-8; and HSDB 284

126.0.3 Trade Names: Reomol 4pg and Santicizer B-16

126.0.4 Molecular Weight: 336.38

126.0.5 Molecular Formula: $C_{18}H_{24}O_6$

126.0.6 Molecular Structure:

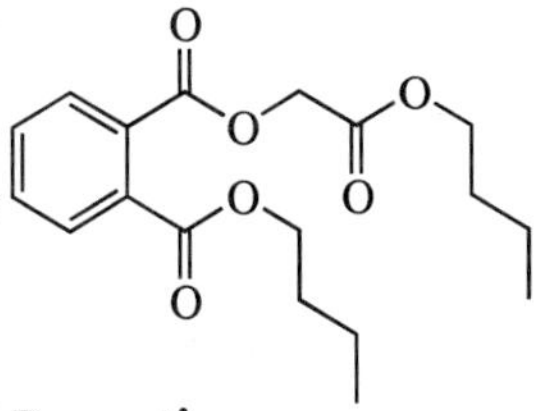

126.1 Chemical and Physical Properties

126.1.1 General

The empirical formula of butyl phthalyl butyl glycolate is $C_{18}H_{24}O_6$. Its structural formula is 1-($COOC_4H_9$)-2-($COOCH_2COOC_4H_9$)C_6H_4. It is a colorless liquid, with a specific gravity of 1.1, a melting point < -35°C, and a boiling point is of 220°C at 10 mmHg. It is relatively insoluble in water but is soluble in organic solvents. Physico-chemical properties are summarized in Table 80.17.

126.1.2 Odor and Warning Properties

Butyl phthalyl butyl glycolate is odorless in its native state. When heated to decomposition, it emits acrid smoke and irritating fumes.

126.2 Production and Use

Butyl phthalyl butyl glycolate is used as a plasticizer for polyvinyl chloride. It is used for making a variety of packaging films because it imparts very good stability to heat and light.

126.3 Exposure Assessment

No information found.

126.4 Toxic Effects

126.4.1 Experimental Studies

126.4.1.1 Acute Toxicity. Butyl phthalyl butyl glycolate is of low acute oral toxicity. The oral LD_{50} values for the rat and mouse are 7 and 12.567 g/kg, respectively (563, 572). Available LD_{50} values are summarized in Table 80.18. Doses of 2.1 and 4.7 g/kg did not cause any adverse effects in rabbits and rats (733).

126.4.1.2 Chronic and Subchronic Toxicity. Subchronic studies by Smyth and Carpenter (570) demonstrated that rats tolerated 0.45 g/kg/d (0.9%) butyl phthalyl butyl glycolate in the diet. At 1.56 g/kg/d, reduced growth and histopathologic effects were noted in a 30-d study in rats (570).

Twenty Sherman rats receiving up to 20,000 ppm (1 g/kg/d) butyl phthalyl butyl glycolate in the diet for 1 yr exhibited no significant changes in behavior, body weight, mortality, tumor incidence, hematology, and gross pathology (with special attention given to the endocrine system). Two mongrel dogs treated for 2 yr with 140 mg/d in capsule form also did not show any evidence of toxicity (734). A more recent 2-yr feeding study in which rats were fed 0.02, 0.2, or 2.0% butyl phthalyl butyl glycolate showed no toxicity or retardation of growth in any treatment group (515).

126.4.1.3 Pharmacokinetics, Metabolism, and Mechanisms. After oral administration, butyl phthalyl butyl glycolate is excreted primarily in the urine in rats (77–83%), dogs (72–75%), and pigs (68–89%) (612).

126.4.1.4 Reproductive and Developmental. When injected IP at high doses, butyl phthalyl butyl glycolate is teratogenic in the rat (516).

126.4.1.5 Carcinogenesis. In a 2-year dietary study, rats receiving up to 20,000 ppm (1 g/kg/d) butyl phthalyl butyl glycolate exhibited no significant increases in tumor incidence, when compared to control animals (734).

126.4.1.6 Genetic and Related Cellular Effects Studies. No information found.

126.4.1.7 Other: Neurological, Pulmonary, Skin Sensitization. When applied to rabbit eyes, butyl phthalyl butyl glycolate caused a slight degree of irritation (1/10 score) (735).

126.4.2 Human Experience

No information found.

126.5 Standards, Regulations, or Guidelines of Exposure

No information found.

126.6 Studies on Environmental Impact

No information found.

127.0 Di(2-ethylhexyl) Isophthalate

127.0.1 CAS Number: *[137-89-3]*

127.0.2 Synonyms: Dioctylisophthalate and DOIP

127.0.3 Trade Names: NA

127.0.4 Molecular Weight: 390.56

127.0.5 Molecular Formula: $C_{24}H_{38}O_4$

127.0.6 Molecular Structure:

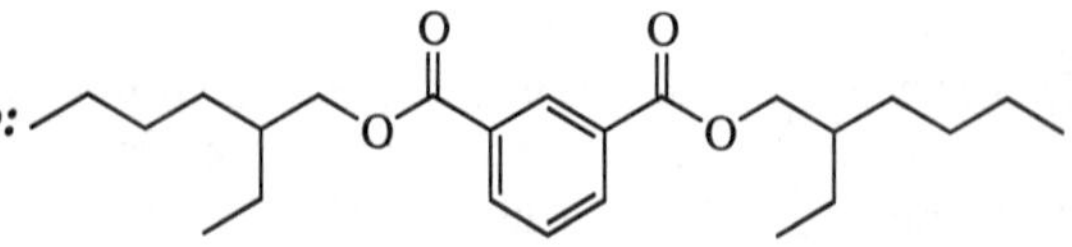

127.1 Chemical and Physical Properties

Water Solubility: < 0.1 g/100 mL at 16°C.

127.2 Production and Use

No information found.

127.3 Exposure Assessment

No information found.

127.4 Toxic Effects

127.4.1 Experimental Studies

Di(2-ethylhexyl) isophthalate is of low acute oral and dermal toxicity. The oral LD_{50} value for the rat is > 3.2 g/kg, and the dermal LD_{50} values for the rabbit and guinea pig are 7.94 and > 20 mL/kg, respectively (Table 80.18). Lethality occurs in guinea pigs inhaling saturated vapor for 8 h (373). No other experimental toxicity information was located.

127.4.2 Human Experience

No information found.

127.5 Standards, Regulations, or Guidelines of Exposure

No information found.

127.6 Studies on Environmental Impact

No information found.

128.0 Dimethyl Terephthalate

128.0.1 CAS Number: *[120-61-6]*

128.0.2 Synonyms: 1,4-Benzenedicarboxylic acid, dimethyl ester; di-Me terephthalate; dimethyl *p*-benzenedicarboxylate; dimethyl *p*-phthalate; dimethyl 4-phthalate; DMT; methyl 4-carbomethoxybenzoate; methyl *p*-(methoxycarbonyl)benzoate; terephthalic acid dimethyl ester; terephthalic acid methyl ester; dimethylester kyseliny isoftalove; dimethylester kyseliny tereftalove; AI3-02246 [NLM]; CCRIS 266; EINECS 204-411-8; HSDB 2580; NCI-C50055 [HSDB:RTECS]; and NSC 3503 [NLM]

128.0.3 Trade Names: NA

128.0.4 Molecular Weight: 194.19

128.0.5 Molecular Formula: $C_{10}H_{10}O_4$

128.0.6 Molecular Structure:

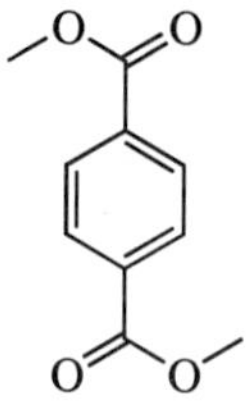

128.1 Chemical and Physical Properties

128.1.1 General

The empirical formula of dimethyl terephthalate (DMT) is $C_{10}H_{10}O_4$. Its structural formula is 1,4-$(COOCH_3)_2C_6H_4$. At room temperature, it is colorless crystals. Its melting and boiling points are 141 and 288°C, respectively. It sublimes at temperatures above 300°C. DMT is soluble in ether and chloroform, slightly soluble in ethanol, and fairly insoluble in water < 1 g/L at 13°C. Additional physical information is summarized in Table 80.17.

128.1.2 Odor and Warning Properties

No information found.

128.2 Production and Use

DMT is manufactured via oxidation of *p*-xylene to monomethyl terephthalate, followed by esterification with methanol. The compound is used for production of film, synthetic fiber, and polybutylene terephthalate and polyethylene terephthalate (PET) resins.

128.3 Exposure Assessment

No information found.

128.4 Toxic Effects

128.4.1 Experimental Studies

128.4.1.1 Acute Toxicity. DMT has a low acute oral toxicity. Oral or intraperitoneal dose levels of 0, 3000, 3900, 5020, or 6590 mg/kg/d DMT were administered to groups of 3–6 male rats as a 20% solution in corn oil. In rats given DMT orally, no mortality was observed during a 14-d post-treatment period. Clinical signs of toxicity were limited to slight to moderate weakness at all dose levels and slight tremors and ataxia at the 5020- and 6590-mg/kg dose levels. No signs of gross or histopathological changes due to systemic toxicity were noted at necropsy. For rats receiving DMT intraperitoneally, the LD_{50} value was 3.9 g/kg (736).

In six rats inhaling 6 mg/L DMT vapors (100–110°C) for 2 h, respiratory irritation was evident along with mucosal hyperemia, increased excitation upon stimulation, irregular breathing, and cyanosis. No deaths were reported (737).

Other short-term studies in rats have shown that ingestion of 500–1000 mg/kg/d DMT is associated with urinary acidosis and attendant calciuria. Bladder stones were noted in rats consuming 1500 mg/kg/d (738).

128.4.1.2 Chronic and Subchronic Toxicity. In F344 weanling rats (13–18/sex/group) fed a diet containing 0, 0.5, 1.0, 1.5, 2, or 3% DMT for 14 d, the average body weight of the animals consuming diets containing 1.5% or greater DMT was decreased on study days 6–8 and 12–14 (postnatal days 34–36 and 40–42). Decreases in body weight were accompanied by reduced feed consumption. In males ingesting 1.5, 2, or 3% DMT, the incidence of bladder calculi was 35, 72, and 100%, respectively. The incidence of bladder calculi in females from the 1.5, 2.0, and 3.0% DMT dietary groups was 0, 36, and 47%, respectively. Grossly observable irregular thickening of the bladder wall was limited to animals having bladder calculi. The composition of the bladder calculi from the DMT-treated animals was primarily calcium and terephthalic acid (TPA) with 5–7% protein. Phosphate levels were low, in contrast to bladder calculi from animals treated directly with terephthalic acid. Neither oxalate nor uric acid were found in the calculi (573).

Female F344 rats (4/group) fed diets containing 0, 1.0, or 2.0% DMT for 3 wk developed hypercalciuria and urinary acidosis, which appeared to be caused by the metabolite terephthalic acid (739).

In rats fed a diet containing 5% DMT (3.75 g/kg) for 28 d, a marked reduction in food consumption, weight loss, and high mortality were noted. No hematologic effects were found, and the cause of death was not established (12).

In male Long–Evans rats fed a diet containing 1% DMT for 96 d, a significant decrease in body weight gain was observed. Body weight gain was slightly (but not significantly) reduced in rats fed 0.25 or 0.5%. Hematologic, clinical chemistry, liver and kidney weight, and histopathologic parameters were similar in treated and control animals (736).

In another 13-wk study, rats (7–19/group) were fed higher doses of DMT (0.5, 1.6, or 3% DMT). The incidence of bladder calculi in animals fed diets containing 3% DMT for 13 wk was 12/16 and 6/16 in male and female rats, respectively. The incidence of bladder calculi in male rats fed diets containing 1.6% DMT was 1/19, and 2/19 in the 0.5% exposure group. Calculi were not noted in mid- and low-dose females. The incidence of moderate hyperplasia of the internal bladder epithelial lining in animals fed diets containing 3% DMT for 13 wk was 11/16 and 7/16 in males and females, respectively. Of the animals fed the 3% DMT diet and developing hyperplasia of the bladder lining, calculi were found in 11/11 males and 6/7 females. There were no neoplastic changes (740).

In a 13-wk study performed by the National Cancer Institute (NCI), male and female rats and mice (10 sex/group) were fed a diet containing 1750, 2500, 5000, 10,000, or 20,000 ppm DMT (741). No compound-related effects were noted in the physical appearance, behavior, or feed consumption measures of either species. No deaths occurred in the rats. Single male mice in the 2500-, 5000-, and 20,000-ppm treatment groups and two females at 20,000 ppm died during the study. Body weight gains of male and female rats fed 20,000 ppm DMT were slightly less than their respective controls. There were no differences in body weight or body weight gain in the male and female mice receiving DMT in the diet. No gross lesions were observed in rats or mice at necropsy. Microscopic examinations of the livers from both species in all dose groups revealed diffuse hepatocellular swelling. Although this change was considered compound related, it was not manifested in a dose-related manner.

The National Cancer Institute also performed a study in which mice and rats (50/sex/group) were fed a diet containing 0, 2500, and 5000 ppm DMT for 103 wk (742). A small, dose-related increase in the incidence of chronic kidney inflammation was seen in male mice and female rats. The association of this lesion with exposure to DMT is unclear since preliminary subchronic studies gave no indication of kidney toxicity.

Male Sprague–Dawley rats and Hartley guinea pigs have been exposed via inhalation to 15 mg/m^3 DMT (6 h/d, 5 d/wk) for 6 mo (743). The amount of respirable DMT was 5 mg/m^3. DMT treatment had no effect on body weight, organ weight, clinical chemistry, urinalyses, or gross or histopathology. No evidence of renal or bladder toxicity was found. Inhalation of higher concentrations (16.5 or 86.4 mg/m^3, 4 h/d, 3.5 μm mean particle size) over a 58-d period caused no toxicologically significant effects in groups of 30 male Long–Evans rats (736).

128.4.1.3 Pharmacokinetics, Metabolism, and Mechanisms. In rats fed diets containing 5% DMT for 5 d, DMT was almost completely absorbed. The majority was metabolized to terephthalic acid, which was primarily eliminated by the kidney. Only a trace amount was found to be excreted unchanged in the urine. About 15% of the unabsorbed ester appeared in the feces (744).

Eight albino rabbits were used to test the absorption and excretion of DMT following ocular administration. A single 50-mg dose of ^{14}C-labeled DMT was instilled into the conjunctival sac of one eye of each rabbit. Approximately 29% of the ^{14}C label was recovered in the urine of rabbits receiving a 5-min exposure, and 37% of the dose was recovered following a 24-h exposure. Fecal excretion was minimal. Examination of internal organs for ^{14}C label revealed only 0.1% remaining 10 d after the exposure (745).

Ten days after dermal exposure of 80 mg of ^{14}C-DMT to rats, approximately 11% of the ^{14}C label was recovered in the urine and feces. Approximately 13% of the ^{14}C label was recovered in the urine and feces of rats treated every other day over the 10-d period (745).

Twenty-four hours after intratracheal instillation of a tracer dose of ^{14}C labeled DMT to rats, 53% of the total dose was excreted in urine and feces. Similar results (62% excretion) were obtained at 48 h. ^{14}C label in the urine was 52–58% of the total dose, and the feces contained 1.8–3.2% of the total dose. In rats given ^{14}C labeled DMT every other day for 10 d, less than 1% of the total dose remained in the lungs and tracheal lymph nodes. Negligible radioactivity ($< 0.1\%$) was found in all the other organs assayed. The largest percentage of the total radioactivity was recovered from the urine (745).

In rats administered ^{14}C labeled DMT orally, 86% of the dose was excreted into the urine within 48 h. Less than 10% of the radiolabel was in the feces at that time point. After dosing five times over a 10-d period, greater than 91% of the total administered dose was recovered from feces and urine (745). Results of a study in male rats and mice indicate that urinary and fecal excretion account for 90% and 10% of an orally administered dose, respectively, in both species (746). Less than 1% remained in the carcass after 48 h. In the rat, terephthalic acid was the only compound detected in the urine. In mice, urinary metabolites consisted of monomethyl terephthalate (70%), terephthalic acid (30%), and traces of DMT. Similar metabolites were identified in the feces of both species. DMT administration did not lower the concentration of nonprotein sulfydryl groups, demonstrating DMT is not activated to form electrophilic metabolites.

128.4.1.4 Reproductive and Developmental. In male rats fed 0.25, 0.5, or 1% DMT in the diet for 115 d prior to mating and female rats fed the same dose levels for 6 d prior to mating and during gestation, parturition and lactation, no signs of toxicity were observed (751). Consumption of DMT had no effect on fertility, reproductive capacity, libido, pregnancy, gestation, litter size, or offspring viability. Compared to controls, offspring of parents fed 0.5 or 1% DMT had significantly lower body weights at weaning. No effects were noted in offspring from rats fed 0.25%. No other information on the teratogenic potential of DMT was found.

128.4.1.5 Carcinogenesis. In the NCI study referred to, there was no mention of an increased tumor incidence in rats or mice ingesting 2500 or 5000 ppm in the diet for 103 wk (742).

128.4.1.6 Genetic and Related Cellular Effects Studies. At concentrations of 0.5–5000 μg/plate, DMT was not mutagenic in *S. typhimurium* strains TA98, TA100, TA102, TA1535, TA1537, or TA1538 in the presence or absence of S-9 (406, 747, 748).

Results of DNA single-strand break assays in CO60 cells and primary rat hepatocytes, unscheduled DNA synthesis studies in HeLa cells, chromosome aberration and micronucleus assays in human peripheral blood lymphocytes, sister chromatid assays in CHO cells, and selective DNA amplification tests in CO60 and Syrian hamster embryo cells also indicate that DMT is nongenotoxic (747–749).

Results of a study by Goncharova and co-workers (1988) suggest that DMT is clastogenic *in vivo* (750). A solution of DMT in DMSO (0.20–1.00 mmol/kg) was injected intraperitoneally into male (C57Bl/6j × CBA)F_1 Mice (15/group). All dose levels of DMT tested increased the frequency of micronuclei, at the 24-h observation point. Since treatment with DMSO also was associated with an increase in micronuclei, interpretation of the study is somewhat limited. In a more recent study, B6C3F1 mice (5–7/group) were injected intraperitoneally with DMT (438–1750 mg/kg in a corn oil vehicle) over 3 consecutive days (751). Bone marrow smears (2 per mouse) were evaluated for the number of micronuclei. No differences were noted between control or treated animals (412).

128.4.1.7 Other: Neurological, Pulmonary, Skin Sensitization. The skin of rabbits (free of fur) was exposed to a suspension of DMT (concentration not reported) in 5% starch for 2 h, after which the test material was washed away with warm water. Repeated application to the skin induced a slight irritation after 3 d that was reported to have disappeared on day 4. After the tenth (final) application, a pigmentation was noted. The skin was reported to be normal by day 12 (737).

A skin-sensitization study was conducted in 20 guinea pigs. Neither primary irritation nor skin sensitization was induced by DMT (751). Instillation of DMT into rabbit eyes causes slight irritation (737, 751, 752).

128.4.2 Human Experience

An oily paste containing 80% DMT showed no irritant effects 24 h after 10 applications to human skin (753).

128.5 Standards, Regulations, or Guidelines of Exposure

AIHA WEEL TWA = 10 mg/m^3 (total) 5 mg/m^3 (respirable).

128.6 Studies on Environmental Impact

No information found.

129.0 Di-(2-ethylhexyl) Terephthalate

129.0.1. CAS Number: *[6422-86-2]*

129.0.2 Synonyms: *bis*(2-Ethylhexyl)-1,4-benzenedicarboxylate; dioctyl terephthalate; terephthalic acid, *bis*(2-ethylhexyl) ester; *bis*(2-ethylhexyl) terephthalate; and DEHT

129.0.3 Trade Names: Eastman DOTP Plasticizer

129.0.4 Molecular Weight: 390.56

129.0.5 Molecular Formula: $C_{24}H_{38}O_4$

129.0.6 Molecular Structure:

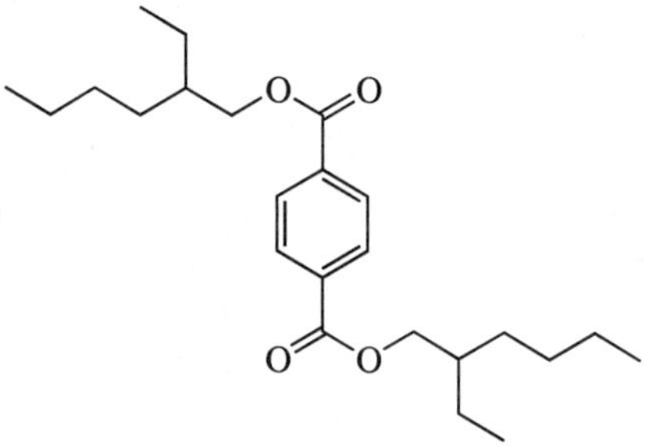

129.1 Chemical and Physical Properties

129.1.1 General

The empirical formula of di (2-ethyhexyl) terephthalate (DEHT) is $C_{24}H_{38}O_4$. Its structural formula is 1,4-$[COOCH_2CH(C_2H_5)(CH_2)_3CH_3]_2C_6H_4$. DEHT is a colorless liquid with a very low vapor pressure (1 mmHg at 217°C). It is a liquid from 30°C (melting point) to 400°C (boiling point). DEHT is virtually insoluble in water (0.00035 g/L). It has a lower specific gravity than water (0.984). Physical properties are summarized in Table 80.17.

129.1.2 Odor and Warning Properties

DEHT has a mild odor.

129.2 Production and Use

DEHT is manufactured by esterification of terephthalic acid and 2-ethylhexanol. It is used primarily as a plasticizer for polyvinyl chloride and other polymers (754).

129.3 Exposure Assessment

No information found.

129.4 Toxic Effects

129.4.1 Experimental Studies

129.4.1.1 Acute Toxicity. The LD_{50} values for DEHT show little acute toxicity. Oral LD_{50} values are >5.0 g/kg for rats and 3.2 g/kg for mice (755). The dermal LD_{50} value was >20 mL/kg for guinea pigs (755).

129.4.1.2 Chronic and Subchronic Toxicity. There have been several subchronic studies of DEHT in mice and rats. Male rats fed diets containing either 0.1 or 1.0% DEHT for 10 d gained weight and showed no evidence of toxicity based on clinical chemistry, hematology, and histopathology (755). Male and female F344 rats fed diets containing 0, 0.1, 0.5, 1.0, 1.2, or 2.5% DEHT for 21 d were evaluated for liver changes (756). Compared to controls, body weights were significantly lower and liver weights significantly higher in those ingesting 2.5%. Increases in liver weight also were seen in animals consuming 1.0 or 1.2%. Increases in liver enzyme activity (peroxisomal) were seen in animals treated with 1.2 or 2.5% DEHT.

Rats given diets containing 0, 0.1, 0.5, or 1.0% DEHT for 90 d showed no adverse effects other than a slight increase in liver weight at 1.0% (574). No effects were observed in the testes.

129.4.1.3 Pharmacokinetics, Metabolism, and Mechanisms. DEHT is hydrolyzed in the gastrointestinal tract to terephthalic acid and 2-ethylhexanol (757). Following oral administration, most of the DEHT is excreted in the feces (57%), with the remainder excreted in the urine (32%). These data suggest that DEHT is not rapidly absorbed from the GI tract. Metabolites of DEHT identified in urine were terephthalic acid and oxidized or conjugated metabolites of 2-ethylhexanol or the monoester, MEHT.

129.4.1.4 Reproductive and Developmental. There are no reproductive or developmental toxicity data for DEHT. However, based on the lack of effect on the testes of rats treated for 90 d, no effect on reproduction is anticipated.

129.4.1.5 Carcinogenesis. No information found.

129.4.1.6 Genetic and Related Cellular Effects Studies. DEHT was not mutagenic in *S. typhimurium* strains TA98, TA100, TA1535, TA1537, or TA1538 with or without S-9 metabolic activation systems, and DEHT did not cause mutations in the HGPRT assay with or without S-9 activation (758). There was no effect on chromosomal aberrations in CHO cells (758).

129.4.1.7 Other: Neurological, Pulmonary, Skin Sensitization. Male rats exposed to an airborne concentration of 46.3 mg/m^3 DEHT for 10 d showed no signs of respiratory, pulmonary, or systemic toxicity (755). DEHT was slightly irritating to the eyes and skin of rabbits, but it was not a dermal sensitizer in guinea pigs (755).

129.4.2 Human Experience

No information found.

129.5 Standards, Regulations, or Guidelines of Exposure

No information found.

129.6 Studies on Environmental Impact

No information found.

130.0 Dimethyl Tetrahydrophthalate

130.0.1 CAS Number: NA

130.0.2 Synonyms: NA

130.0.3 Trade Names: NA

130.0.4 Molecular Weight: NA

130.1 Chemical and Physical Properties

No information found.

130.2 Production and Use

No information found.

130.3 Exposure Assessment

No information found.

130.4 Toxic Effects

130.4.1 Experimental Studies

Dimethyl tetrahydrophthalate is one of the most acutely toxic phthalates. The oral LD_{50} value for the rat is 0.7 g/kg (446). The dermal LD_{50} value for the rabbit is >10 mL/kg (446). No other experimental toxicity data were located.

130.4.2 Human Experience

No information found.

130.5 Standards, Regulations, or Guidelines of Exposure

No information found.

130.6 Studies on Environmental Impact

No information found.

131.0 Di(2-ethylhexyl) Tetrahydrophthalate

131.0.1 CAS Number: *[1330-92-3]*

131.0.2 Synonyms: NA

131.0.3 Trade Names: NA

131.0.4 Molecular Weight: 394.59

131.0.5 Molecular Formula: $C_{24}H_{42}O_4$

131.0.6 Molecular Structure:

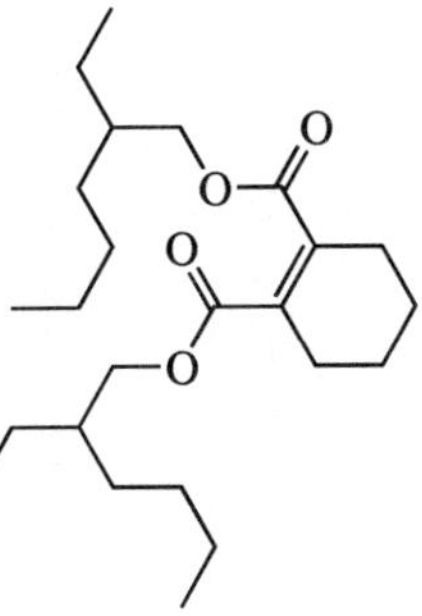

131.1 Chemical and Physical Properties

No information found.

131.2 Production and Use

No information found.

131.3 Exposure Assessment

No information found.

131.4 Toxic Effects

131.4.1 Experimental Studies

Di(2-ethylhexyl) tetrahydrophthalate is one of the least acutely toxic phthalates. The oral LD_{50} value for the rat is 114 g/kg (12). No other experimental toxicity data were located.

131.4.2 Human Experience

No information found.

131.5 Standards, Regulations, or Guidelines of Exposure

No information found.

131.6 Studies on Environmental Impact

No information found.

132.0 Didecyl Tetrahydrophthalate

132.0.1 CAS Number: NA

132.0.2 Synonyms: NA

132.0.3 Trade Names: NA

132.0.4 Molecular Weight: NA

132.1 Chemical and Physical Properties

No information found.

132.2 Production and Use

No information found.

132.3 Exposure Assessment

No information found.

132.4 Toxic Effects

132.4.1 Experimental Studies

Didecyl tetrahydrophthalate is one of the least acutely toxic phthalates. The oral LD_{50} value for the rat is 64 g/kg, and the dermal LD_{50} value for the rabbit is >20 mL/kg (759). No other toxicological information was located.

132.4.2 Human Experience

No information found.

132.5 Standards, Regulations, or Guidelines of Exposure

No information found.

132.6 Studies on Environmental Impact

No information found.

133.0 Di(1,4-hexadienyl) Tetrahydrophthalate

133.01 CAS Number: NA

133.0.2 Synonyms: NA

133.0.3 Trade Names: NA

133.0.4 Molecular Weight: NA

133.1 Chemical and Physical Properties

No information found.

133.2 Production and Use

No information found.

133.3 Exposure Assessment

No information found.

133.4 Toxic Effects

133.4.1 Experimental Studies

Di (1,4-hexadienyl) tetrahydrophthalate is one of the least acutely toxic phthalates. The oral LD_{50} value for the rat is 45.2 g/kg and the dermal LD_{50} value for the rabbit is > 16 mL/kg (759). No other toxicological information was located.

133.4.2 Human Experience

No information found.

133.5 Standards, Regulations, or Guidelines of Exposure

No information found.

133.6 Studies on Environmental Impact

No information found.

134.0 Tris(2-ethylhexyl) Trimellitate

134.0.1 CAS Number: [3319-31-1]

134.0.2 Synonyms: TOTM and TEHT

134.0.3 Trade Names: Eastman TOTM Plasticizer, Morflex 510, and Trimex T08

134.0.4 Molecular Weight: 546.8

134.0.5 Molecular Formula: $C_{33}H_{54}O_6$

134.0.6 Molecular Structure:

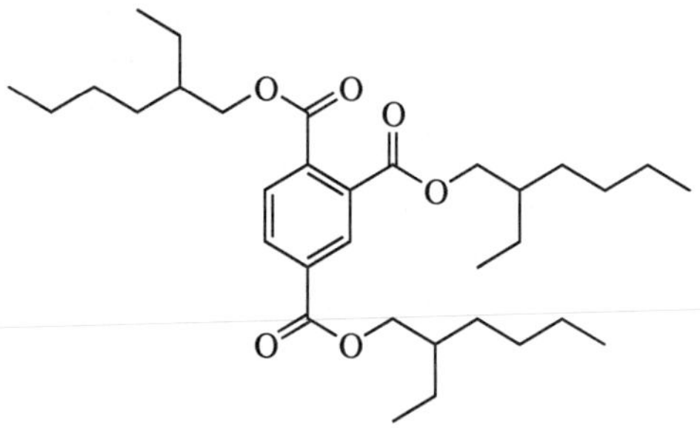

134.1 Chemical and Physical Properties

134.1.1 General

The empirical formula of trioctyl trimellitate (TOTM) is $C_{33}H_{54}O_6$. Its structural formula is 1,2,4-$[COOCH_2CH(C_2H_5)(CH_2)_3CH_3]_3C_6H_4$. It is a clear to white viscous liquid at room temperature. It has a specific gravity of 0.989, and is only slightly soluble in water (<0.1 g/100 mL at 20°C). The vapor pressure is very low ($<7.0 \times 10^{-8}$ mmHg), and the boiling point is 414°C.

134.1.2 Odor and Warning Properties

No information found.

134.2 Production and Use

TOTM is manufactured by acid-catalyzed esterification of trimellitic anhydride and 2-ethylhexanol. It is used as a plasticizer for high-temperature electrical wire coatings.

134.3 Exposure Assessment

No information found.

134.4 Toxic Effects

134.4.1 Experimental Studies

134.4.1.1 Acute Toxicity. TOTM is not acutely toxic. The oral LD_{50} values for rats and mice are >3200 mg/kg, and the dermal LD_{50} value for guinea pigs is >20 mg/kg (755). TOTM is a slight dermal and ocular irritant.

134.4.1.2 Chronic and Subchronic Toxicity. The subchronic toxicity was evaluated in male and female rats exposed to dietary concentrations of 0, 0.2, 0.67, or 2.0% TOTM for 28 d (760). There was no effect on body weight or food consumption. Liver weights for the 0.67 and 2.0% were slightly increased, but metabolic enzymes and cholesterol was moderately increased. Slight changes in hematology (lower erythrocyte count and hemoglobin concentration, higher leukocyte count) were seen at the top two dose levels, with lower erythrocyte counts seen for low-dose males, but the clinical significance of these changes in questionable. The no-observed-effect level was 0.2% (184 mg/kg) for females. This was a low-observed-effect level for males.

134.4.1.3 Pharmacokinetics, Metabolism, and Mechanisms. The absorption and metabolism of TOTM have been studied in male rats treated by oral intubation with 100 mg/kg in corn oil. Based on excretion of 75% of the dose in the feces, 85% of which was unchanged, little TOTM is absorbed. Urinary excretion accounted for 16% of the dose with 1.9% exhaled as CO_2. The ester is hydrolyzed to 2-ethylhexanol and mono-(2-ethylhexyl)trimellitate (MEHT). The 2EH is further metabolized, and those metabolites are excreted along with MEHT in the urine. Less than 1.0% of the dose was retained in the animal after 144 h (755).

134.4.1.4 Reproductive and Developmental. No information found.

134.4.1.5 Carcinogenesis. No information found.

134.4.1.6 Genetic and Related Cellular Effects Studies. TOTM was not mutagenic in *S. typhimurium* strains TA98, TA100, TA1535, TA1537, or TA1538 with or without S-9 metabolic activation systems even at concentrations of up to 1.0 mg/plate (755), and TOTM did not cause mutations in the HGPRT assay in CHO cells with or without S-9 activation (755). There was no increase in unscheduled DNA synthesis in rat hepatocytes (755).

134.4.1.7 Other: Neurological, Pulmonary, Skin Sensitization. TOTM was not a dermal sensitizer in guinea pigs (755).

134.4.2 Human Experience

No information found.

134.5 Standards, Regulations, or Guidelines of Exposure

No information found.

134.6 Studies on Environmental Impact

No information found.

ACKNOWLEDGMENT

Authors acknowledge the assistance of Laurie C. Deyo, Ph.D. in preparing this Chapter.

BIBLIOGRAPHY

1. N. I. Sax and R. J. Lewis, Sr., Eds, *Hawley's Condensed Chemical Dictionary*, 11th ed., Van Nostrand Reinhold Co., New York, 1987.
2. J. H. Draize et al., Toxicological investigations of compounds proposed for use as insect repellants, A. Local and systemic effects following topical skin application. B. Acute oral toxicity. C. Pathological Examination. *J. Pharmacol. Exper. Therapeutics* **93**, 26–39 (1948).
3. P. M. Jenner et al., Food flavorings and compounds of related structure. I. Acute oral toxicity. *Food Cosmetic Toxicol.* **2**, 327–343 (1964).
4. B. E. Graham and M. H. Kuizenga, Toxicity studies on benzyl benzoate and related benzoyl compounds. *J. Pharmacol. Experi. Therapeutics* **84**, 358 (1945).
5. H. F. Smyth, Jr. et al., Range finding toxicity data. List V. *J. Ind. Hyg. Occupational Med.* **10**, 61 (1954).
6. M. Branca et al., Macro- and microscopic alterations in 2 rabbit skin regions following topically repeated applications of benzoic acid *n*-alkyl esters. *Contact Dermatitis* **19**, 320–334 (1988).
7. H. F. Smyth, C. P. Carpenter, and C. S. Weil, Range-finding toxicity data: List IV. *Archives of Industrial Hygiene and Occupational Medicine* **4**, 119 (1951).
8. NTIS [OTS0539230]
9. D. L. J. Opdyke and C. Letizia, Monographs on fragrance raw materials. Butyl benzoate. *Food Cosmetic Toxicol.* **21**(5), 651–665 (1983).
10. D. L. J. Opdyke, Monographs on fragrance raw materials: Benzyl Benzoate. *Food and Cosmetic Toxicol.* **11**, 1015 (1973).
11. D. M. Bagley et al., Skin irritation: Reference chemicals data bank. *Toxicology in Vitro* **10**, 1–6 (1996).
12. F. A. Patty, D. W. Fassett, and D. D. Irish, Ed., *Industrial Hygiene and Toxicology*, 2nd Ed., Vol. 2, John Wiley & Sons, Inc., New York, 1963.
13. C. Alder-Hradecky and B. Kelentey, *Archives of International Pharmacodymamics* **128**, 135 (1960).
14. C. Davison, E. F. Zimmerman, and P. K. Smith, On the metabolism and toxicity of methyl salicylate. *J. Pharmacol. Exp. Therapeutics* **132**, 207 (1961).
15. D. L. J. Opdyke, Monographs on fragrance raw materials. Methyl salicylate. *Food Cosmetic Toxicol.* **16**, 821 (1978).
16. *Food Agr. Organ. UN Rep. Ser.* **44**, 63 (1967).
17. W. K. Webb and W. H. Hansen, Chronic and subacute toxicology and pathology of methyl salicylate in dogs, rats, and rabbits. *Toxicol. Appl. Pharmacol.* **5**, 576–587 (1963).
18. Monographs on fragrance raw materials. Ethyl salicylate. *Food and Cosmetic Toxicology* **16**, 751–752 (1978).
19. W. S. Spector, Ed., *Handbook of Toxicology*, Vol I., W. B. Saunders, Philadelphia, 1955, p. 138.
20. T. Sollmann and P. J. Hanzlink, *Experimental Pathology*, Saunders, Philadelphia, 1928.
21. D. L. J. Opdyke, Monographs on fragrance raw materials. Phenyl Salicylate. *Food Cosmetic Toxicol.* **14**, 837 (1976).
22. *NRC Chem. Biol. Coord. Ctr. Summ. Tables Biol. Texts* **6**, 149 (1954).

23. M. O. Amdur, J. Doull, and C. D. Klaasen, Eds., *Casarett and Doull's Toxicology*, 4th ed. Pergamon Press, New York, 1991, p. 940.
24. W. M. Grant, *Toxicology of the Eye*, 3rd ed., Charles C. Thomas, Springfield IL, 1986, p. 627.
25. A. A. Fisher, *Contact Dermatitis*, 2nd ed., Lea & Febiger, Philadelphia, 1973, p. 358.
26. J. L. Schardein, Agents Used for Pain. In *Chemically Induced Birth Defects*, Marcel Dekker, Inc., New York, 1993, Chapt. 5 pp. 126–146.
27. *WHO Food Additives Series No. 5*, World Health Organization, Geneva, 1974.
28. M. Siegel, *Antibiotics and Chemotherapy* **185**, (1969).
29. G. Angelini et al, Contact allergy to preservatives and perfumed compounds used in skin care products. *J. Appl. Cosmetology* **15**, 49–57 (1997).
30. W. I. Metzger, L. T. Wright, and J. C. DiLorenzo, *J. Amer. Med. Assoc.* **155**(4), 352 (1974).
31. P. S. Jones et al., *J. Amer. Pharm. Assoc. Sci. Ed.* **45**, 268 (1956).
32. F. N. Marzulli and H. I. Maibach, Antimicrobials: experimental contact sensitization in man. *J. Soci. Cosmetic Chemi.* **24**, 399–421 (1973).
33. E. J. Rudner, North American Group Results. *Contact Dermatitis* **3**, 208–209 (1977).
34. G. Angelini, G. A. Vena, and C. L. Meneghini, Allergic contact dermatitis to some medicaments. *Contact Dermatitis* **12**, 263–269 (1985).
35. A. G. Gilman, T. W. Rall, A. S. Nies, and P. Taylor, Eds., *Goodman and Gilman: The Pharmacological Basis of Therapeutics*, 8th ed., Pergamon Press, New York, 1990, p. 969.
36. M. Windholz and S. Budavari, Eds., *The Merck Index*, 9th ed., Merck and Company, Rahway, NJ, 1976.
37. S. Ranganithan and T. Ramasarman, Enzymatic formation of *p*-hydroxycinnamate, *Biochemical Journal* **122**, 487 (1971).
38. I. Snapper and A. Saltzman, *Archives of Biochemistry* **24**, 1 (1949).
39. D. L. J. Opdyke, Monographs on fragrance raw materials: Allyl cinnamate. *Food and Cosmetic Toxicology* **15**, 615–616 (1977).
40. W. B. Deichmann and H. W. Gerarde, *Toxicology of Drugs and Chemicals*, Academic Press, NY, 1969.
41. D. Wild, M-T. King, E. Gocke, and K. Eckhardt, Study of artificial flavouring substances for mutagenicity in the Salmonella/microsome, Basc and micronucleus tests. *Food Chem. Toxicol.* **21**, 707–719 (1983).
42. D. L. J. Opdyke, Monographs on fragrance raw materials. Methyl cinnamate. *Food Cosmetic Toxicol.* **13**, 849, 1975.
43. *Vopr. Pitan* **33** (5), 48 (1974).
44. D. L. J. Opdyke, Monographs on fragrance raw materials: Ethyl cinnamate. *Food Cosmetic Toxicol.* **12**, 721–722 (1974).
45. D. L. J. Opdyke, Monographs on fragrance raw materials: *n*-Butyl cinnamate. *Food Cosmetic Toxicol.* **18**, 655 (1980).
46. *Ann Pharm Fr.* **14**, 370 (1956).
47. D. L. J. Opdyke, Monographs on fragrance raw materials: Linalyl cinnamate. *Food Cosmetic Toxicol.* 14, 463 (1976).
48. R. H. Dreisbach, *Handbook of Poisoning*, Lange, Los Altos, CA, 1974, p. 245.
49a. L. A. Linares, T. Y. Peretz, and J. Chin, Methemoglobinemia induced by topical anesthetic (benzocaine). *Radiotherapy and Oncology* **18**(3), 267–269 (1990).

49b. N. Goluboff, Methenoglobinemia due to benzocaine. *Pediatrics* **221**, 430–431 (1955).
50. L. F. Rodriguez, L. M. Smolik, and A. J. Zbehlik, Benzocaine-induced methemoglobinemia; report of a severe reaction and review of the literature. *Annals of Pharmacotherapy* **28**, 643–649 (1994).
51. W. J. O'Donohue, L. M. Moss, and V. A. Angelillo, Acute methemoglobinemia induced by topical benzocaine and lidocaine. *Archives of Internal Medicine* **140**, 1508–1509 (1980).
52. S. E. Guerriero, Methemoglobinemia caused by topical benzocaine. *Pharmacother.* **17**(5), 1038–1040 (1997).
53. H. D. C. Peterson, Acquired methemoglobinemia in an infant due to benzocaine suppository. *New England J. Medi.* **263**, 454–455 (1960).
54. J. S. Fruton and S. Simmonds, *General Biochemistry*, 2nd ed., Wiley, New York, 1958, p. 781.
55. H. R. Mahler and E. H. Cordes, *Biological Chemistry*, Harper and Row, New York, 1966, p. 522.
56. N. I Sax, *Dangerous Properties of Industrial Materials*, 4th ed., Litton, New York, 1975.
57. F. W. Grant, W. C. Martin, and R. W. Quackenbush, Simple sensitive field-test for cocaine based on recognition of odor of methyl benzoate as a test product. *Bulletin on Narcotics* **27**, 33–35 (1975).
58. Ref. 1, p. 759.
59. *Kirk-Othmer Encyclopedia of Chemical Technology*, 3rd ed., Vol. 9, John Wiley, New York, 1980, p. 303.
60. W. H. Glaze, Reaction products of ozone: a review. *Environmental Health Perspectives* **69**, 151–157 (1986).
61. NTP Chemical Repository, Radian Corporation, Aug. 29, 1991
62. J. H. Kuney and J. N. Nullican, Eds., *Chemcyclopedia*, American Chemical Society, Washington, DC, 1988, p. 90.
63. K. Verushen, *Handbook of Environmental Data and Organic Chemistry*, Van Nostrand Reinhold, New York, 1983.
64. S. Budavari, Ed., *The Merck Index—Encyclopedia of Chemicals, Drugs, and Biologicals*, Merck and Co., Inc., Rahway, NJ, 1989, p. 950.
65. T. E. Graedel, D. T. Hawkins, and L. D. Claxton, *Atmospheric Chemical Compounds: Sources, Occurrence and Bioassay*, Academic Press, Orlando, 1986.
66. E. Zeiger et al., Salmonella mutagenicity tests. V. Results from the testing of 311 chemicals. *Environmental and Molecular Mutagenesis* **19**(Suppl. 21), 2–141 (1992).
67. D. L. J. Opdyke, Monographs on fragrance raw materials. Methyl benzoate. *Food Cosmetic Toxicol.* **12**, 937–938 (1974).
68. 21 *CFR* 172.515 (4/1/90).
69. T. E. Daubert and R. P. Danner, *Data Compilation Tables of Properties of Pure Compounds*, American Institute for Physical Properties Data, New York, 1989.
70. R. Atkinson, Structure-activity relationship for the estimation of rate constants for the gas-phase reactions of hydroxyl radicals with organic compounds. *Intern. J. Chem. Kinetics* **19**, 799–828 (1987).
71. J. A. Riddick, W. B. Bunger, and T. K. Sakano, *Techniques of Chemistry*, Vol 2 *of Organic Solvents: Physical Properties and Methods of Purification*, 3rd ed., Wiley, New York, 1986.

72. S. Arctander, Aroma Chemicals. In *Perfume and Flavor Chemicals*, Vol. 1, No. 403. S. Arctander, Montclair, NJ, 1969.
73. Ref. 1, p. 183.
74. M. Crespi-Rosell and J. Cegarra-Sanchez, Purification of the waste waters coming from dyeing with carriers. *Bol. Inst. Invest. Text. Coop. Ind.* **77**, 41–57 (1980).
75. J. M. Haas, H. W. Earhart, and A. S. Todd, Environmental guide to dye carrier selection. *Amer. Dyestuff Reporter* **64**, 34–49 (1975).
76. D. B. Harper, A. A. M. Nour, and R. H. Thompson, Volatile flavor components of dalieb (*Borassus aethipum* L.). *J. Sci. Food. Agric.* **37**, 685–688 (1986).
77. R. A. Flath and R. R. Forrey, Volatile components of papaya (Carica papaya L., Solo variety). *J. Agric. Food Chem.* **25**, 103–109 (1977).
78. M. Barbeni et al., Identification and sensory analysis of volatile constituents of babaco fruit (Carica pentagona Heliborn). *Flavour Fragrance J.* **5**, 27–32 (1990).
79. M. Aoyama, H. Fujise, and K. Yoshikawa, Residues of dye carriers in cloth and effects of residual carriers on human skin. *Igaku to Seibutsugaku* **116**, 85–88 (1988).
80. 21 *CFR* 175.105 (4/1/93).
81. E. Freese et al., Correlation between the growth inhibitory effects, partition coefficients and teratogenic effects of lipophilic acids. *Teratology* **20**, 413 (1979).
82. F. J. Lewis and R. L. Tatken, Eds., *Registry of Toxic Effects of Chemical Substances*, NIOSH, Cincinnati, OH, 1978, p. 208 (Entry No DG4925000).
83. NTP Chemical Repository, Hexyl benzoate. Radian Corporation, Aug. 29 1991.
84. W. Gerhartz, Ed., *Ullmann's Encyclopedia of Industrial Chemistry*, 5th Ed., Vol A1, VCH Publishers, Deerfield Beach, FL, 1985.
85. F. E. Furia and N. Bellanca, Eds., *Fenaroli's Handbook of Flavor Ingredients.* 2nd ed. Vol 2, The Chemical Rubber Company, Cleveland, 1975.
86. A. K. Korolkovas and J. H. Burckhalter, *Essentials of Medicinal Chemistry*, 1st Ed., Wiley, New York, 1976.
87. W. J. Hayes, Jr. and E. R. Laws Jr., Eds., Classes of Pesticides. In *Handbook of Pesticide Toxicology*, Vol. 3. Academic Press, Inc., New York, 1991, pp. 1505–1507.
88. K. E. Murray et al., The volatiles of off-flavored unblanched green peas (Pisum sativum). *J. Sci. Food Agric.* **27**, 1093–1107 (1977).
89. D. J. Humphreys, *Veterinary Toxicology*, 3rd. ed., Bailliere Tindell, London, 1988, p. 131.
90. S. Morita et al., Safety evaluation of chemicals for use in household products. 2. Teratological studies on benzyl benzoate and 2-(morpholinothio)benzothiazole in rats. *Annu. Rep. Osaka City Inst. Public Health Environ. Sci.* **43**, 90–97 (1981).
91. American Medical Association, Council on Drugs, *AMA Drug Evaluations Annual, 1994*, American Medical Association, Chicago, 1994, p. 1615.
92. C. M. Gruber, *J. Lab. Clin. Med.* **9**, 15 (1923).
93. C. M. Gruber, *J. Lab. Clin. Med.* **9**, 92 (1923).
94. D. Blanc and P. Deprez, Unusual adverse reaction to an acaricide. *Lancet* **335**, 1291–1292 (1990).
95. R. L. Bronaugh, *In vivo* percutaneous absorption of fragrance ingredients in rhesus monkeys and humans. *Food Chemical Toxicol.* **28**, 369–373 (1990).
96. 21 *CFR* 172.515. (4/1/93).

97. T. Kurihara and M. Kilcuchim, *Y. Zasshi* **95**(9), 1098 (1975).
98. M. S. Karawya, F. M. Hashim, and M. S. Hifnawy, *Bull. Frac. Pharm.* **13**(1), 183 (1974).
99. E. W. Packman et al., Chronic oral toxicity of oil sweet birch (methyl salicylate). *Pharmacologist* **3**, 62, (1961).
100. J. Lamb IV et al., Methyl salicylate. *Environmental Health Perspectives* **105**(Suppl.1), 323–324 (1997).
101. R. E. Morrissey et al., Results and evaluations of 48 continuous breeding reproduction studies conducted in mice. *Fundamental and Applied Toxicology* **13**, 747–777 (1989).
102a. G. P. Daston et al., Functional teratogens of the rat kidney. I. Colchicine, dinoseb, and methyl salicylate. *Fundamental and Applied Toxicology* **11**(3), 381–400 (1988).
102b. E. Gibson, Perinatal nephropathies. *Environmental Health Perspectives* **15**, 121–130 (1976).
103. D. O. Overman and J. A. White, Comparative teratogenic effects of methyl salicylate applied orally or topically to hamsters. *Teratology* **28**(3), 421–426 (1983).
104. R. Infurna et al., Evaluation of the dermal absorption and teratogenic potential of methyl salicylate in a petroleum based grease. *Teratology* **41**(5), 566 (1990).
105. Ref. 35, pp. 649–651.
106. *Am. J. Dis. Child.* **69**, 37 (1945).
107. *Arch. Dis. Child.* **28**, 475 (1953).
108. K. Mortelmans et al., Salmonella mutagenicity tests: II. Results from the testing of 270 chemicals. *Environmental Mutagenesis* **8**(Suppl. 7), 1–119 (1986).
109. N. Kuboyama and A. Fujii, Mutagenicity of analgesics, their derivatives, and anti-inflammatory drugs with S-9 mix of several animal species. *J. Nihon Univ. Sch. Dent.* **34**(3), 183–195 (1992).
110. *Am. J. Med. Sci.* **193**, 772 (1937).
111. *Clinical Toxicology* **6**, 189 (1973).
112. B. Bartle et al., Salicylate toxicity-from topical methylsalicylate. *On Contin. Pract.* **19**, 23–25 (1992).
113. F. Littleton, Warfarin and topical salicylates. *J. Amer. Med. Assoc.* **263**, 2888 (1990).
114. A. S. Yip et al., Adverse effect of topical methylsalicylate ointment on warfarin anti-coagulation: An unrecognized potential hazard. *Postgraduate Med. J.* **66**, 3667–3669 (1990).
115. P. Morra et al., Serum concentrations of salicylic acid following topically applied salicylate derivatives. *Annals of Pharmacotherapy* **30**, 935–940 (1996).
116. H. M. Caan and H. L. Verhulst, *J. Pediatrics* **53**, 271 (1958).
117. A. J. Collins et al., Some observations on the pharmacology of "Deep-Heat", a topical rubifacient. *Annals of Rheumatory Disease* **43**(3), 411–415 (1984).
118. Syracuse Research Corporation, unpublished report.
119. Ref. 1, p. 192.
120. N. R. Stephenson, The effect of salicylate on the thymus gland of the immature rat. *J. Pharmacy and Pharmacology* **11**, 339 (1959).
121. H. Köchling, Salizylsüureester als Masthilfsmittel. *Prakt. Tiearzt* **53**, 232 (1972).
122. G. Valetter and R. Cavier, Hydrocarbons, alcohols and esters. *Archives of International Pharmacodynamics and Therapeutics* **97**, 232 (1954).
123. M. Siddiqi and W. A. Ritschel, pH-effects on salicylate absorption through the intact rat skin. *Scientia. Pharm.* **40**, 181 (1972).

124. SQC Syracuse Research Corp., unpublished report.

125. *Osaka City Med. J.* **112**, 223 (1966).

126. E. Zeiger et al., Salmonella mutagenicity tests: III. Results from the testing of 255 chemicals. *Environmental Mutagenesis* **9**(Suppl. 9), 1–110 (1987).

127. NTP Chemical Repository, *Phenyl Salicylate.* Radian Corporation, Aug. 29, 1991.

128. M. Fimmiani, L. Casini, and S. Bocci, Contact dermatitis from phenyl salicylate in a galenic cream. *Contact Dermatitis* **22**(4), 239 (1990).

129. L. Kanerva, R. Jolanski, and T. Estlander, Allergic and irritant patch test reactions to plastic and glue allergens. *Contact Dermatitis* **37**, 301 (1997).

130. *J. Am. Pharm. Assoc. Sci. Ed.* **45**, 260 (1956).

131. *Food Agr. Organ. UN Rep. Ser.* **53**A, 81 (1974).

132. T. Sabalitschka and R. Neufeld-Crzelliter, *Arzneim. Forch.* **4**, 575 (1954).

133. K. Schuebel and J. Manger, *Arch. Exper. Pathol. Pharmakol.* **146**, 208 (1929).

134. *Clinical Toxicol.* **4**, 185 (1971).

135. H. Cremer, *Z. Lebensm. Unters.* **70**, 136 (1935).

136. C. Matthews et al., *J. Am. Pharm. Assoc.* **45**, 260 (1956).

137. A. J. Quick, *J. Biological Chem.* **97**, 403 (1932).

138. R. Derachi and J. Gourdon, Métabolisme d'un conservateur alimentaire: L'acide parahydroxylbenzoïque et ses esters. *Food Cosmetic Toxicol.* 189 (1963).

139. M. J. Prival, V. F. Simmon, and K. E. Mortelmans, Bacterial mutagenicity testing of 49 food ingredients gives very few positive results. *Mutation Research* **260**(4), 321–329 (1991).

140. B. H. Harvey, M. E. Carstens, and J. J. F. Taljaard, Central effects of the preservative, methyparaben. *In vivo* activation of cAMP-specific phosphodiesterase and reduction of cortical cAMP. *Biochemical Pharmacology* **44**(6), 1053–1057 (1992).

141. B. L. Song, L. Hai-Ying, and P. Dun-Ren, *In vitro* spermicidal activity of parabens against human spermatozoa. *Contraception* **39**(30), 331–335 (1989).

142a. A. M. Kligman, The identification of contact allergens by human assay. III. The maximization test. A procedure for screening and rating contact sensitizers. *J. Investigative Dermatology* **47**, 393 (1966).

142b. A. M. Kligman and W. Epstein, Updating the maximization test for identifying contact allergens. *Contact Dermatitis* **1**, 231 (1975).

143. S. W. Edwards et al., Inhibition of neutrophil superoxide secretion by the preservative, methylhydroxybenzoate: Effects mediated by perturbation of intracellular Ca^{2+} ?. *Free Radical Research Communications* **10**(6), 333–343 (1990).

144. L. Jian and W. Po, Kinetic evaluation of the ciliotoxicity of methyl- and propyl-p-hydroxybenzoates using factorial experiments. *J. Pharm. Pharmacol.* **45**(2), 98–101 (1993).

145. T. Menné and N. Hjorth, Routine patch testing with paraben esters. *Contact Dermatitis* **19**, 189–191 (1988).

146. 21 *CFR* 582.3490 (4/1/91).

147. J. M. Chapman et al., Subconjunctival gentamicin induction of extraocular toxic muscle myopathy. *Opthal. Res.* **24**(4), 189–196 (1992).

148. I. Moriyama, K. Hiraoka, and R. Yamaguchi Teratogenic effects of food additive ethyl-*p*-hydroxy benzoate studied in pregnant rats. *Acta Obstet. Gynaecol. Jpn.* **22**, 94–106 (1975) as cited in the Hazardous Substance Data Base.

149. *Bromatol. Chem. Toksykol.* **14**, 301 (1981).
150. H. Sokol, *Drug Stand.* **20**, 89 (1952).
151. *Nippon Eiseeigaku Zasshi* **28**, 463 (1973).
152. *Arch. Int. Pharmacodyn. Ther.* **128**, 135 (1960).
153. Ref. 24, p. 695.
154. Hinmarch K. W. et al., *J. Pharm. Sci.* **72**, 1039–1041 (1983).
155. Y. Kajimoto et al., Anaphylactoid skin reactions after intravenous regional anaesthesia using 0.5% prilocaine with or without preservative-a double-blind study. *Acta Anaesthesiol Scand.* **39**(6), 782–784 (1995).
156. T. E. Furia, Ed., *CRC Handbook of Food Additives*, 2nd ed., The Chemical Rubber Co., Cleveland, 1972, p. 126.
157. A. Osol and Hoover et al., *Remington's Pharmaceutical Sciences*, 15th Ed., Mack Publishing Co., Easton, PA, 1975, p. 1090.
158. *Zentalbl Bakteriol Parasitenkd infektionskrankh Hyg Abt II* **112**, 226 (1959).
159. M. von Bubnoff, D. Schnell, and J. Vogt-Moykoff, *Arzneim. Forsch.* **7**, 340 (1957).
160. P. H. Rossier and T. W. Wegmann, *Wien. Med. Wochenschr.* **19/20**, 358 (1953).
161. M. Levy and P. R. Ocken, Purification and properties of pig liver esterase. *Archives of Biochemistry and Biophysics* **135**, 259 (1969).
162. J. J. Child et al., Purification and properties of a phenol carboxylic acid acyl esterase from *Aspergillus flavus. Canadian Journal of Microbiology* **17**, 1455–1463 (1971).
163. G. E. Ghirardi, *Arch. Ital. Sci. Farmacol.* **9**, 282 (1959).
164. Y. Kurata et al., Structure–activity relations in promotion of rat urinary bladder carcinogenesis by phenolic antioxidants. *Jpn. J. Cancer Res.* **81**(8), 754–759 (1990).
165. M.-A. Shibata et al., Early proliferative responses of forestomach and glandular stomach of rats treated with five different phenolic antioxidants. *Carcinogenes* **11**(3), 435–439 (1990).
166. E. A. Nera et al., Short-term pathological and proliferative effects of butylated hydroxyanisole and other phenolic antioxidants in the forestomach of Fischer 344 rats. *Toxicology* **32**(3), 197–213 (1984).
167. L. E. Kier et al., The *Salmonella typhimurium*/mammalian microsomal assay. A report of the U.S. Environmental Protection Agency Gene-Tox Program. *Mutation Research* **168**, 69–240 (1986).
168. J. E. Nagel, J. T. Fuscaido, and P. Fireman, Paraben allergy. *J. Amer. Med. Assoc.* **237**, 1594 (1977).
169. *J. Soc. Cosmetic Chem.* **28**, 357 (1977).
170. P. Z. Bedoukian, *Perfumery and Flavoring Synthesies*, 2nd Ed., Elseiver Publishing Co., New York, 1967, p. 82.
171. Y. Kitamura, Effects of loral anesthetics on the peripheral nerve and the spinal cord. *Osalca City Medical Journal* **25(1)**, 7–24 (1979).
172. G. D. Clayton and F. E. Clayton, *Patty's Industrial Hygiene and Toxicology*, 3rd ed., Vol. 2, J Wiley, New York, 1982, pp. 2312–2315.
173. H. R. Peterson and A. Hall, Dermal irritating properties of perfume materials. *Drug Cosmet. Ind.* **58**, 113 (1946).
174. E. H. Silver and S. D. Murphy, Effect of carboxylesterase inhibitors on the acute hepatotoxicity of esters of allyl alcohol. *Toxicol. Appl. Pharmacol.* **45**(2), 377–389 (1978).

175. J. J. Clapp, C. M. Kaye, and L. Young, Observation on the metabolism of allyl compounds in the rat. *Biochem. J.* **114**, 6 (1969).

176. 21 *CFR* 121.1164.

177. E. C. Hagan et al., Food flavourings and compounds of related structure. II. Subacute and chronic toxicity. *Food Cosmetic. Toxicol.* **5**, 141–157 (1967).

178. R. E. Gosselin, R. P. Smith, and H. C. Hodge, *Clinical Toxicology of Commercial Products*, 5th ed., Williams and Wilkins, Baltimore, MD, 1984.

179. M. Corazza et al., Allergic contact dermatitis from benzyl alcohol. *Contact Dermatitis* **34**, 74 (1996).

180. E. V. Eremyan and K. A. Davtyan, *Prom-st. Arm.* **6**, 43–45 (1977).

181. T. J. Fitzgerald, J. Doull, and F. G. DeFeo, Radioprotective activity of *p*-aminopropiophenone. A structure-activity investigation. *J. Medicinal Chemi.* **17**, 900–902, (1974).

182. *Toksikol. Vestn.* **2**, 34 (1996).

183. *Drugs Japane* **6**, 33 (1982).

184. E. W. Schafer, The acute oral toxicity of 369 pesticidal, pharmaceutical and other chemicals to wild birds. *Toxicol. Appl. Pharmacol.* **21**, 315 (1972).

185. M. F. Brulos et al., The influence of perfumes on the sensitizing potential of cosmetic bases. I. A technique for evaluating sensitizing potential. *J. Soc. Cosmet. Chem.* **28**, 357–365 (1977).

186. *Arzneim-Forch.* **16**, 1275 (1966).

187. *Arch. Int. Pharmacodyn. Ther.* **104**, 388 (1956).

188. *Farmakol. Toksikol.* **29**, 425 (1966).

189. *Barke. Diss.*(Hannover Ger) (1936).

190. *Naunyn-Schmiedebergs Arch. Exp. Pathol. Pharmakol.* **144**, 197 (1929).

191. *Klin. Wochenscr.* **31**, 97 (1953).

192. *Res. Prog. Org. Biol. Med. Chem.* **2**, 213, (1970).

193. *Physiol. Rev.* **12**, 190 (1932).

194. *J. Am. Pharm. Assoc. Sci. Ed.* **45**, 382 (1956).

195. K. H. Beyer and A. R. Latven, Effect of 1-Cyclohexylamino-2-propylbenzoate (Cyclaine) and other local anesthetic agents administered intravenously on the cardiovascular respiratory systems of the dog. *J. of Pharmacol Exp. Therapeutics* **106**, 37 (1952).

196. *Naunyn-Schmiedebergs Arch. Exp. Pathol. Pharmakol.* **131**, 171 (1928).

197. *J. Pharmacol. Exp. Therapeutics* **54**, 137 (1935).

198. *J. Pharmacol. and Exp. Therapeutics* **45**, 291 (1932).

199. P. M. Jenner et al., Food flavourings and compounds of related structure. I. Acute oral toxicity. *Food Cosmet. Toxicol.* **2**, 327–343 (1964).

200. Anon, Aminobenzoic acid. Properties, uses as fragrance raw material, status and toxicity of methyl ester. *Food Cosmetic. Toxicol.* **12**, 935 (1974).

201. I. F. Gaunt et al., Acute and short-term toxicity of methyl-*N*-methyl anthranilate in rats. *Food and Cosmetic Toxicology* **8**, 359–368 (1970).

202. U.S. Army Armament Research & Development Command, NX #07000.

203. N. Goluboff and D. J. MacDayden, Methemoglobinemia in an infant associated with application of tar-benzocaine ointment. *J. Pediatrics* **47**, 222–226 (1955).

204. O. P. Heinonen, D. Slone, and S. Shapiro, *Birth Defects and Drugs in Pregnancy*, Publishing Sciences Group, Littleton, Mass, 1977.

205a. S. D. Prystowsky et al., Allergic contact hypersensitivity to nickel, neomycin, ethylenediamine, and benzocaine. Relationships between age, sex, history of exposure and reactivity to standard patch test and use test in a general population. *Archives of Dermatology* **115**, 959–962 (1979).

205b. K. H. Kaidbey and H. Allen, Photocontact allergy to benzocaine. *Archives of Dermatology* **117**, 77–79 (1981).

206a. S. Haworth et al., Salmonella mutagenicity test results for 250 chemicals. *Environmental Mutagenesis* **5**(suppl. 1), 3–142 (1983).

206b. H. H. M. Korsten, L. J. Hellebrekers, R. J. E. Grouls, E. W. Ackerman, A. A. J. van Zudert, H. van Herpen, and E. Gruys, Long-lasting epidural sensory blockade by *n*-Butyl *p*-Aminobenzoate in the dog: Neurotoxic of local anesthetic effect ? *Anesthesiology* **73**, 491–498 (1990).

207. M. Shulman, N. J. Joseph, and C. A. Haller, Effect of epidural and subarachnoid injections of a 10% butamben suspension. *Reg. Anest.* **15**(3), 142–146 (1990).

208. C. Bayar and N. Cumer, The effect of some local anesthetics on methemoglobin levels and erythrocyte enzymes. *Turk. J. Med. Sci.* **26**(5), 439–443 (1996).

209a. N. Suzuki et al., The effect of plain 0.5% 2-chloroprocaine on venous endothelium after intravenous regional anaesthesia in the rabbit. *Acta Anaesthesiol Scand.* **38**(7), 653–665 (1994).

209b. E. Zhivkov and L. Atanasov, [Experiments in obtaining and preventing congenital cataracts in rats]. *Opthalmologia* **2**, 105–112 (1965).

209c. V. L. Schnell et al., Effects of oocyte exposure to local anesthetics on *in vitro* fertilization and embryo development in the mouse. *Reproductive Toxicology* **6**(4), 323–327 (1992).

209d. R. H. Wainberg et al., Effects of heat and other agents on amino acid uptake in *Escherichia coli. Intern. J. Hyperthermia* **6**(3), 597–605 (1990).

209e. Z. N. Ding et al., Brainstem auditory evoked potentials during procaine toxicity in dogs. *Can. J. Anaesthesiol.* **39**(6), 600–603 (1992).

209f. O. Dehkordi et al., Cardiorespiratory effects of cocaine and procaine at the ventral brainstem. *Neurotoxicology* **17**(2), 387–395 (1996).

209g. I. Chazal et al., Prediction of drug-induced immediate hypersensitivity in guinea pigs. *Toxicology in vitro* **8**(5), 1045–1047 (1994).

210. *Poisoning Toxicology Symptom Treatment*, 2nd Ed., Vol. 2., 1970, p. 73.

211. F. Gjerris, Transitory procaine-induced Parkinsonism. *J. Neurol. Neurosurg. Psychiatry* **34**, 20–22 (1971).

212. D. A. Fulcher and C. H. Katelaris, Anaphylactoid reactions to local anesthetics despite IgE deficiency: a case report. *Asian Pacific Journal of Allergy and Immunology* **8**(2), 133–136 (1990).

213. L. Clark et al., Acute toxicity of the bird repellant, methyl anthranilate, to fry of Salmo salar, Oncorhynchus mykiss, Ictalurus punctatus and Lepmis macrochirus. *Pesticide Science* **39**(4), 313–317 (1993).

214. J. S. Fruton and S. Simmonds, *General Biochemistry*, 2nd Ed., Wiley, New York, 1958, p. 781.

215. H. R. Mahler and E. H. Cordes, *Biological Chemistry*, Harper and Row, New York, 1966, p. 522.

216. *Medicina et Pharmacologia Experimentalis* **15**, 7 (1966).

217. R. L. Clark et al., Cleft lip and palate caused by anthranilate methyl esters. *Teratology* **21**, 34–35A (1980).

218. H. Shimizu and N. Takamura, Mutagenicity of some aniline derivatives. *Occup. Health Chem. Ind. Proc. Int Congr.* **11**, 496–506 (1983).

219. Ref. 24, p. 932.

220. J. W. Bailey, M. W. Haymond, and J. M. Miles, Triacetin: a potential parenteral nutrient. *J. Parenter. Enteral Nutr.* **15**, 32–36 (1991).

221. N. Yoshimi et al., The genotoxicity of a variety of aniline derivatives in a DNA repair test with primary cultured hepatocytes. *Mutation Research* **206**(2), 183–191 (1988).

222. B. L. Oser, S. Carson, and M. Oser, Toxicological tests on flavouring materials. *Food and Cosmetic Toxicology* **3**, 563 (1965).

223. Ref. 85, p. 380.

224. Food and Drug Research Laboratories, Inc., unpublished information, 1963, as cited in I. F. Gaunt, M. Sharratt, P. Grasso and M. Wright, Acute and short-term toxicity of methyl-*N*-methyl anthranilate in rats. *Food Cosmetic Toxicol.* **8**, 359–368 (1970).

225. R. J. C. Barry, M. J. Jackson, and D. H. Smyth, Handling of glycerides of acetic acid by rat small intestine *in vitro*. *J. Physiology* **185**(3), 667 (1966).

226. A. J. Conely and J. J. Kavara, *Antimicrob. Agents Chemother.* **4**(5), 501 (1973).

227. G. D. Clayton and F. E. Clayton, Eds., *Patty's Industrial Hygiene and Toxicology*, 4th Ed., Vol. II, Part D., J Wiley, New York, 1994.

228. B. M. Hausen and W. Beyer, The sensitizing capacity of the antioxidants propyl, octyl and dodecyl gallate and some related gallic acid esters. *Contact Dermatitis* **26**(4), 253–258 (1992).

229. L. J. Casarett and J. Doull, Eds., *Toxicology: The Basic Science of Poisons*, Macmillan, New York, 1975.

230. R. C. Li, P. P. T. Sah, and H. H. Anderson, Acute toxicity of monoacetin, diacetin and triacetin. *Proc. Soc. Exp. Biol. Med.* **46**, 26 (1941).

231. R. C. Li and H. H. Anderson, Intravenous toxicity of the acetins in dogs and rabbits. *J. Pharmacol. Exp. Therapeutics* **72**, 26 (1941).

232. M. B. Chenowith et al., Factors influencing fluoroacetate poisoning. *J. Pharmacol. Exp. Therapeutics* **102**, 31 (1951).

233. F. Berte et al., *In vivo* and *in vitro* toxicity of carbitol. *Boll Chim. Farm.* **125**, 401–403 (1986).

234. A. R. Latven and H. Molitor, *J. Pharmacol. and Exp. Therapeutics* **7**, 159 (1975).

235. *Prehled. Prumsyslove Toxikol.* Org. Latky, 1986, p. 667.

236. W. F. von Oettingen, The aliphatic acids and their esters: Toxicity and potential danger. *AMA Arch. Ind. Hlth.* **21**, 28–65 (1960).

237. J. H. Gast, Some toxicity studies with triacetin. *Federation. Proceedings* **22**, 368 (1963).

238. W. H. Lawrence, M. Malik, and J. Autian, Development of a toxicity evaluation program for dental materials and products. II. Screening for systemic toxicity. *J. Biomed. Materials Res.* **8**, 11–34 (1974).

239. D. L. J, Opdyke, Monographs on fragrance raw materials. Triacetin. *Food Cosmetic. Toxicol.* **16**, 879 (1978).

240. *Acta Phsyiol. Scand.* **40**, 338 (1957).

241. *J. Pharmacol. Expe. Therapeutics* **16**, 879 (1978).
242. Eastman Kodak Company, unpublished data.
243a. A. Wretlind, *Acta Physiol. Scand.* **40**, 338 (1957).
243b. P. Mosnier et al., Effects of tributyrin on the gastric mucosa in the rat. *Nutrition Research* **13**, 23–30 (1993).
244. K. Ohta et al., Toxicity, teratogenicity and pharmacology of tricaprylin, *Oyo Yakuri.* **4**, 871–882 (1970).
245. Eastman Kodak Company, unpublished Report.
246. *J. Am. Pharm. Accoc. Sci. Ed.* **46**, 185 (1957).
247. Litton Bionetics, Inc. Mutagenic evaluation of compound FDA 71-39, propyl gallate. *National Tech. Inf. Service* [PB245-441], Dec. 31, 1974.
248. A. J. Orten, A. C. Kuyper, and A. H. Smith, *Food Technology* **2**, 308 (1948).
249. A. J. Lehman et al., *Adv. Food. Res.* **3**, 197 (1951).
250. *J. Am. Oil Chem Soc.* **54**, 239 (1977).
251. Federation of American Societies for Experimental Biology, Life Sciences Research Office, Evaluation of the health aspects of propyl gallate as a food ingredient. *National Tech. Inf. Service* [PB223-840], Jan. 1973.
252. K. J. H. von Sluis, *Food Manufactureh* **26**, 99 (1951).
253. R. Brun, [Contact eczema due to an antioxidant of margarine (gallate) and change of occupation]. *Dermatologica* **140**, 390–394 (1970).
254. G. J. van Esch and H. van Genderen, *Netherlands Inst. Publ. Health Rep. No 481*, 1954.
255. *Food Agr. Organ. U. N. Rep. Ser.* **28**, 304 (1951).
256. J. W. Bailey, R. L. Barker, and M. D. Karlstad, Total parenteral nutrition with short- and long-chain triglycerides: triacetin improves nitrogen balance in rats. *Journal of Nutrition* **122**, 1823–1829 (1992).
257. R. B. Alfin-Slater et al., *J. Am. Oil Chem. Soc.* **35**, 122 (1958).
258. W. F. Hughes, *Bull. Johns Hopkins Hosp.* **82**, 338 (1948).
259. S. L. Hem et al., Tissue irritation evaluation of potential parenteral vehicles. *Drug Dev. Commun.* **1**(5), 471–477 (1975).
260. International Labour Office, *Encylopedia of Occupational Health and Safety*, Vols I, II., International Labour Office, Geneva, Switzerland, 1983, p. 973
261. 21 *CFR*184.1901 (4/1/91).
262. R. Atkinson, Structure-activity relationship for the estimation of rate constants for the gas-phase reactions of hydroxyl radicals with organic compounds. *Int. J. Chem. Kinet.* **19**, 799–828 (1987).
263. USEPA, *PCGEMS Graphical Exposure Modeling System*, PCHYDRO, 1991.
264. H. Ogata et al., Mechanism of the intestinal absorption of drugs from oil in water emulsions. IV. Absorption from emulsions containing the higher concentration of emulsifiers. *Chem. Pharm. Bull.* **23**(4), 707–715 (1975).
265. L. R. Crum et al., Substrate specificity and inhibition characteristics of a fluoride-sensitive tributyinase. *Biochimica Biophysica Acta* **198**(2), 229–235 (1970).
266. Z. Yuan et al., Plasma pharmacokinetics (PK) of butyrate after the administration of tributyrin and Na butyrate to mice and rats. *Proc. Annu. Meet. Am. Assoc. Cancer Res.* **35**, A2556 (1994).

267. E. E. Deschner et al., Dietary butyrate (Tributyrin) does not enhance AOAM-induced colon tumorigenesis. *Cancer Lett.* **52**(1), 79–82 (1990).

268. B. A. Conley et al., Phase I study of the orally administered butyrate prodrug, tributyrin, in patients with solid tumors. *Clin. Cancer Res.* **4**(3), 629–634 (1998).

269. Z. X. Chen and T. R. Breitman, Tributryin: A prodrug of butyric acid for potential clinical application in differentiation therapy. *Cancer Res.* **54**(13), 3494–3499 (1994).

270. P. Feng, L. Ge, N. Akyhani, and G. Liau, Sodium butyrate is a potent modulator of smooth cell muscle cell proliferation and gene expression. *Cell Prolif.* **29**(5), 231–241 (1996).

271. A. Novorgrodsky et al.,, Effect of polar organic compounds on leukemic cells. Butyrate-induced partial remission of acute myelogenous leukemia in a child. *Cancer*, **51**, 9–14 (1983).

272. J. Meidani et al., Evidence for the mechanisms of formation of radiolysis products using a deutero labeled triglyceride. *J. Am. Oil Chem Soc.* **54**(11), 502–505 (1977).

273. A. A. Miller et al., Clinical pharmacology of sodium butyrate in patients with acute leukemia. *Eur. J. Cancer Clin. Oncol.* **23**, 1283–1287 (1987).

274. NTP working group, Comparative toxicology studies of corn oil, safflower oil, and tricaprylin in male F344N rats as vehicles for gavage. *National Toxicology Program Technical Report Series* **426**, (1994).

275. R. R. Fox et al., Transplacental teratogenic and carcinogenic effects in rabbits chronically treated with *N*-ethyl-*N*-nitrosourea. *J. National Cancer Institute* **65**, 607–614 (1980).

276. E. Stephens et al., Association of liver necrosis with maternal deaths in a one generation reproduction study of elemental phosphorus. *The Toxicologist* **12**, 199 (1992).

277. M. K. Smith et al., Developmental toxicity of dichloroacetonitrile: A by product of drinking water disinfection. *Fundam. Appl. Toxicol.* **12**, 765–772 (1989).

278. S. A. Christ et al., Developmental effects of tricholoroacetonitrile administered in corn oil to pregnant Long-Evans rats. *J. Toxicol. Environ. Health* **47**(3), 233–247 (1996).

279. D. A. Gordon, T. K. Wessendarp, W. Crocker, M. K. Smith, and A. C. Roth, Comparative absorption and distribution of radiolabeled trichloroacetonitrile (TCAN) in pregnant rats from corn oil (CO) and tricaprylin vehicles. *Teratology* **43**, 427 (1991).

280. L. Kanerva, R. Jolanski, and T. Estlander, Allergic and irritant patch test reactions to plastic and glue allergens. *Contact Dermatitis* **37**, 301 (1997).

281. C. A. Van Der Heijden, P. J. C. M. Janssen, and J. J. T. W. A. Strik, Toxicology of gallates: A review and evaluation. *Food Chem. Toxicol.* **24**(10–11), 1067–1070 (1986).

282. Joint FAO/WHO Expert Committee on Food Additives Toxicological Evaluation of Certain Food Additives. *WHO Fd. Add. Ser.* **10**, 45 (1976).

283a. WHO, Toxicological Evaluation of Certain Food Additives and Contaminants. *Environ. Health Criteria Number 32*, 1993, pp. 4–12.

283b. J. C. Dacre, Long-term toxicity study of *n*-propyl gallate in mice. *Food and Cosmetic Toxicology* **12**, 125 (1976).

284. K. M. Abdo et al., Carcinogenesis bioassay of propyl gallate in F344 rats and B6C3F1 mice. *Journal of the American College of Toxicology* **2**(6), 425–433 (1983).

285. J. M. Orten, A. C. Kuyper, and A. H. Smith, Studies on the toxicity of propyl gallate and of antioxidant mixtures containing propyl gallate. *Fd. Technol.* **2**, 308 (1948).

286a. A. Booth et al., The metabolic fate of gallic acid and related compounds. *Journal of Biological Chemistry* **234**, 3014 (1959).

286b. J. C. Dacre, Metabolic pathways of phenolic antioxidants. *J. N. Z. Inst. Chem.* **24**, 161 (1960).

286c. G. J. Van Esch, The toxicity of the antioxidants propyl, octyl and dodecyl gallate. *Voeding* **16**, 683 (1955).

286d. I. R. Telford, C. S. Woodruff, and R. H. Linford, Fetal resorption in the rat as influenced by certain antioxidants. *Am. J. Anat.* **110**, 29–36 (1962).

287. J. M. Desesso, Amelioration of teratogenesis. I. Modification of hydroxyurea-induced teratogenesis by the antioxidant propyl gallate. *Teratology* **24**, 19–35 (1981).

288. D. L. Madhavi and D. K. Salunkhe, Toxicological Aspects of Food Antioxidants. In D. L. Madhavi, S. S. Deshpande, and D. K. Salunkhe, ed., *Food Antioxidants: Technological, Toxicological and Health Perspectives*, Marcel Dekker, Inc., New York, 1996, Chapt. 5, pp. 267–359.

289. FDA, *NTIS PB-223816*(1973).

290. M. Hirose et al., Effects of subsequent antioxidant treatment on 7,12-dimethylbenz[a]anthracene-initiated carcinogenesis of the mammary gland ear duct and forestomach in Sprague-Dawley rats. *Carcinogenesis* **9**(1), 101–104 (1988).

291. M. M. King and P. B. McCay, Modulation of tumor incidence and possible mechanisms of inhibition of mammary carcinogenesis by dietary antioxidants. *Cancer Research* **43**, 2485S–2490S (1983).

292. H. Fujita, M. Nakano, and M. Sasaki, Mutagenicity test of food additives with Salmonella Typhimurium TA 97 and TA 102, III. *Kenkyu Nenpo-Tokyo-Toritsu Eisei Kenkynsho* **39**, 343–350 (1988).

293. J. S. Yoon et al., Chemical mutagenesis testing in Drosophila IV. Results of 45 coded compounds tested for the National Testing Program. *Environmental Mutagenesis* **7**(3), 349–367 (1985).

294. K. S. Loveday et al., Chromosome aberration and sister chromatid exchange tests in Chinese Hamster Ovary Cells *in vitro*, II. Results with 20 chemicals. *Environmental and Molecular Mutagenesis* **13**, 60 (1989).

295. M. D. Shelby and K. L. Witt, Comparison of results from mouse bone marrow chromosome aberration and micronucleus tests. *Environmental and Molecular Mutagenesis* **25**, 302–313 (1995).

296. Y. Nakagawa and S. Tayama, Cytotoxicity of propyl gallate and related compounds in rat hepatocytes. Archives of Toxicology, 69 (3), 204–208 (1995).

297. Y. Nakagawa, P. Modeus, and G. Moore, Propyl gallate-induced DNA fragmentation in isolated rat hepatocytes. Archives of Toxicology, 72 (1) 33–37 (1997).

298. T. W. Wu, K. P. Fung, L. H. Zeng, J. Wu, and H. Nakamura, Propyl gallate as a hepatoprotector *in vitro* and *in vivo. Biochemical Pharmacology* **48**(2), 419–422 (1994).

299. M. Younes, E. Kayser, and O. Strubelt, Effect of antioxidants on hypoxia/reoxygenation-induced injury in isolated perfused rat liver. *Pharmacology and Toxicology* **71**(4), 278–283 (1992).

300. B. M. Hausen and W. Beyer, The sensitizing capacity of the antioxidants propyl octyl and dodecyl gallate and some related gallic acid esters. *Contact Dermatitis* **26**(4), 253–258 (1992).

301. N. Hernandez et al., Allergic contact dermatitis from propyl gallate in desonide cream (Locapred). *Contact Dermatitis* **36**(2), 111 (1997).

302. S. M. Wilkinson and M. H. Beck, Allergic contact dermatitis from dibutyl phthalate, propyl gallate and hydrocortisone, in Timodine. *Contact Dermatitis* **27**(3), 197 (1992).

303. M. Corazza et al., Allergic contact dermatitis from propyl gallate. *Contact Dermatitis* **31**, 203 (1994).

304. G. Kahn, P. Phanuphak, and H. N. Claman, Propyl gallate- Contact sensitization and orally-induced tolerance. *Archives of Dermatology* **109**, 506–509 (1974).

305. E. Cronin, Lipstick dermatitis due to propyl gallate. *Contact Dermatitis* **6**, 213–214 (1980)

306. A. G. Wilson, I. R. White, and J. D. T. Kirby, Allergic contact dermatitis from propyl gallate in a lip balm. *Contact Dermatitis* **20**, 145–146 (1989).

307. FAO/WHO, *WHO Tech. Rep. Ser. No 21*, World Health Organization Geneva, Switzerland 1987.

308. 40 *CFR* 180.1001 (July 1, 1994).

309. Hazelton Laboratories, Inc., 13-Week dietary feedings-rats: Octyl gallate. *Project No. 458–115*(1969).

310. Hazelton Laboratories, Inc., 13-Week dietary feedings-dogs: Octyl gallate. *Project No. 458–117*(1969).

311. G. Koss and W. Koransky, Enteral absorption and biotransformation of the food additive octyl gallate in the rat. *Food Chem. Toxicol.* **20**, 591–594 (1982).

312. Hazelton Laboratories, Inc. Modified two-generation reproduction study-rats: Octyl gallate. *Project No. 458–116*(1970).

313. Industrial Bio-Test Laboratories, Inc. Three-generation reproduction study with Cold-Pro GA 8, in albino rats. *Report No. IBT P84*(1970).

314. H. Luther et al., Inhibition of rabbit erythroid 15-lipoxygenase and sheep vesicular gland prostaglandin H synthase by gallic esters. *Pharmazie* **46**, 134–136 (1991).

315. S. Christow et al., Actions of gallic esters on the arachidonic acid metabolism of human polymorphonuclear leukocytes. *Pharmazie* **46**, 282–283 (1991).

316. W. Baer-Dubowska, J. Gnojkowski, and W. Fenrych, Effect of tannic acid on benzo[a]pyrene-DNA adduct formation in mouse epidermis: comparison with synthetic gallic acid esters. *Nutr. Cancer* **29**(1), 42–47 (1997).

317. M. Pemberton et la., Allergy to octyl gallate causing stomatitis. *Br. Dent. J.* **175**(3), 106–108 (1993).

318. A. C. deGroot and F. Gerkens, Occupational airborne contact dermatitis from octyl gallate. *Contact Dermatitis* **23**(3), 184–186 (1990).

319. A. Schnuch et al., Patch testing with preservatives antimicrobials and industrial biocides. Results from a multicentre study. *Brit. J. Dermatology* **138**(3), 467–476 (1998).

320. F. D. Tollenaar, Prevention of rancidity in edible oils and fats with special reference to the use of antioxidants. *Proc. Pacific Sci. Congr.* **5**, 92 (1957).

321. H. Gali et al., Hydolyzable tannins: Potent inhibitors of hydroperoxide production and tumor production in mouse skin treated with 12-o-tetradecanoylphorbol 13-acetate *in vivo. Int. J. Cancer* **51**(3), 425–432 (1992).

322. T. Nakayama et al., The protective role of galalic acid esters in bacterial cytotoxicity and SOS responses induced by hydrogen peroxide. *Mutation Research* **303**(1), 29–34 (1993).

323. A. A. Raccagni et al., Lauryl gallate hand dermatitis in a cheese counter assistant. *Contact Dermatitis* **37**(4), 182 (1997).

324. U.S. Pat 3920883 (1975) J. Yamada, S. Kubo, and J. Hriano.

325. A. R. Singh, W. H. Lawrence, and J. Autian, Embryonic-fetal toxicity and teratogenic effects of adipic esters in rats. *J. Pharmaceut. Sci.* **62**, 1596–1600 (1973).

326. K. W. Schneider, Contact dermatitis due to diethyl sebacate. *Contact Dermatitis* **6**, 506–507 (1980)
327. K. M. Abdo et al., No evidence of carcinogenicity of D-mannitol and propyl gallate in F344 rats or B6C3F1 mice. *Food Chemi Toxicol* **24**(10–11), 1091–1094 (1986).
328. E. Browning, *Toxicity of Industrial Organic Solvents*, Chemical Publishing Company, New York, 1953.
329. E. Sasaki et al., Allergic contact dermatitis due to diethyl sebacate. *Contact Dermatitis* **36**, 172 (1997).
330. *Nat. Acad. Sci. Natl. Res. Counc. Chem-Biol Coord. Cent. Rev.* **5**, 15 (1953).
331. E. F. Dempsey et al., *Metab. Clin. Exp.* **9**, 52 (1960).
332. L. Hagler and R. H. Herman, Oxalate metabolism. IV. *American J. Clinical Nutrition* **26**, 1073–1079 (1973).
333. *Gig. Sanit.* **46**(5), 87 (1981).
334. D. L. J. Opdyke, Fragrance raw materials monographs. Dimethyl malonate. *Food Cosmetic Toxicol.* **17**, 363 (1979).
335. H. F. Smyth et al., Range-Finding Toxicity Date: List VII. *Amer. Industrial Hyg. Assoc. J.* **30**, 470–476 (1969).
336. *Neuropsychopharmacology* **1**, 286 (1959).
337. Ref. 228, p. 376.
338. *Soc. Cosmet. Chem.* **13**, 469 (1962).
339. M. T. Case, J. K. Smith, and R. A. Nelson, Acute mouse and chronic dog toxicity studies of danthron, dicotyl sodium sulfosuccinate, poloxalkol and combinations. *Drug and Chemical. Toxicology* **1**(1), 89–101 (1978).
340. *Bromatol. Chem. Toksykol.* **7**, 161 (1974).
341. *J. Am. Pharm. Assoc. Sci. Ed.* **38**, 428 (1949).
342. W. B. Deichmann and H. W. Gerarde, *Toxicology of Drugs and Chemicals*, Academic Press, New York, 1969.
343. H. Desoille, L. Truffert, and Bidegarray, Some cases of poisoning by ethyl oxalate. *Archives des Maladies Professionnelles de Medecine du Travail et de Securite Sociale* **8**, 265–268 (1947).
344. S. S. Arctander, *Perfume and Flavor Chemicals (Aroma Chemicals)*, Vol. 2. S. Arctander, Montclair NJ, 1969.
345. B. A. McLaughlin et al., Toxicity of dopamine to striatal neurons *in vitro* and potentiation of cell death by a mitochondrial inhibitor. *J. Neurochemistry* **70**(6), 2406–2415 (1998).
346. A. M. Kligman and W. Epstein, Updating the maximization test for identifying contact allergens. *Contact Dermatitis* **1**, 231 (1975).
347. R. H. Springer et al., Synthesis and enzymic activity of some novel xanthine oxidase inhibitors. 3-Substituted 5,7-dihydroxypyrazolo[1,5-a]pyrimidines. *J. Medicinal Chem.* **19**(2), 291–296 (1976).
348. E. Y. Spencer, *Guide to the Chemicals Used in Crop Protection*, 7th Ed. (Publication 1093), Research Institute Agriculture Canada, Ottawa Canada, 1982, p. 182.
349. A. E. Benaglia et al., The chronic toxicity of aerosol-DT. *J. Indust. Hyg. Toxicol.* **25**, 175 (1943).

350. C. A. Dujovne and L. W. Shoeman, Toxicity of a hepatic laxative preparation in tissue culture and excretion in bile in man. *Clin. Pharmaco. Ther.* **13**, 602–608 (1972).

351. American Medical Association, Council on Drugs, *AMA Evaluations Annual 1994*, American Medical Association, Chicago, 1994, p. 949.

352. G. K. McEvoy, Ed., *American Hospital Formulary Service- Drug Information 94*, American Society of Hospital Pharmacists Inc. Bethesda, MD, 1994, p. 1886.

353. K. MacKenzie et al., Three-generation reproduction study with dioctyl sodium sulfosuccinate in rats. *Fundamental and Applied Toxicology* **15**(1), 53–62 (1990).

354. DSS Scientific Review Panel, *Docket No. 84-N-0184*, March 1984, as cited in J. L. Schardein, *Chemically Induced Birth Defects*, Marcel Dekker, Inc, New York, Chapt. 16, 1993, pp. 445–456.

355. Ref. 24, p. 871.

356. P. Aselton et al., First-trimester drug use and congenital disorders. *Obstet. Gynecol.* **65**, 451–455 (1985).

357. O. P. Heinonen, D. Slone, and S. Shapiro, *Birth Defects and Drugs in Pregnancy*, Publishing Sciences Group, Littleton, MA, 1977.

358. J. L. Shardein, Gastrointestinal Drugs. In *Chemically Induced Birth Defects*, Marcel Dekker, Inc., New York, 1993, Chapt. 16 pp. 445–456.

359. *Goodman and Gilman's The Pharmacological Basis of Therapeutics*, 5th Ed., Macmillan Publishing Co., Inc., New York, 1975, p. 977.

360. H. Suomalainene, K. Konttinen, and E. Oura, Decarboxylation by intact yeast and pyruvate decarboxylase of some derivatives of pyruvic acid and α-ketoglutaric acid. *Arch. Mikrobiol.* **64**(3), 251–261 (1969).

361. B. A. Trela, S. R. Frame, and M. S. Bogdanffy, A microscopic and ultrastructural evaluation of dibasic esters (DBE) toxicity in rat nasal explants. *Experimental and Molecular Pathology* **56**(3), 208–218 (1992).

362. W. Mabey and T. Mill, *J. Phys. Chem. Ref. Data* **7**, 383–415 (1978).

363. Syracuse Research Corporation, unpublished report.

364. USEPA, *Graphical Exposure Modeling System. Fate of atmospheric pollutants (FAP) data base*, Office of Toxic Substances.

365. *Toksikol Vestn.* **4**, 40 (1994).

366. A. R. Singh, W. H. Lawrence, and J. Autian, Dominant lethal mutations and antifertility effects of di-2-ethylhexyl adipate and diethyl adipate in male mice. *Toxicol. Appl. Pharmacol.* **32**, 566 (1975).

367. J. M. Nikitakis, Ed., *CTFA Cosmetic Ingredient Handbook*, 1st Ed. Cosmetic Toiletry and Fragrance Association, Washington, DC, 1988, p. 178.

368. Food Chemical News, Inc., *The Food and Chemical News Guide*, Food Chemical News, Inc., Washington, DC, 1990, p. 147

369. *Gig. Sanit.* **55**(6), 86 (1990).

370. R. Lefaux, *Practical Toxicology of Plastics*, CRC Press, Inc., Cleveland, 1968, p. 358.

371. Esso Research and Engineering Company, unpublished Report, Acute oral administration in rats, 1968.

372. Esso Research and Engineering Company, unpublished Report, Acute dermal administration in rabbits, 1968.

373. H. F. Smyth, Jr. et al., Range-finding toxicity data. List VI. *Amer. Ind. Hyg. Assoc. J.* **23**, 95–107 (1962).

374. *U.S. Army Chem. Corps. Med. Labs Res. Rep.* **256**(1954).

375. *Raw Materials Data Handbook* **2**, 22 (1975).

376. *Toxic Param. Ind. Tox. Chem Under Single Exposure*, 62 (1982).

377. CIR Expert Panel, Final report on the safety assessment of dibutyl adipate. *J. Am. Coll. Toxicol.* **15**(4), 295–300 (1996).

378. B. Ekwall, C. Nordensten, and L. Albanus, Toxicity of 29 plasticizers to HeLa cells in the MIT-24 system. *Toxicol.* **24**, 199–210, (1982).

379. R. D. Ashford, *Ashford's Dictionary of Industry Chemicals*, Wavelength Publications Ltd., London, 1994.

380. D. E. Till et al., Plasticizers migration from polyvinyl chloride film to solvents and foods. DBP. *Food Chem. Toxicol.* **20**, 95–104 (1982).

381. B. Denis Page and G. M. LaCroix, The occurrence of phthalate ester and di-2-ethylhexyl adipate plasticizers in Canadian packaging and food sampled in 1985–1989: A survey. *Food Additives and Contaminants* **12** 129–151 (1995).

382. Ministry of Agriculture, Fisheries and Food, Plasticizers: Continuing surveillance. *Food Surveillance Paper No. 30*, 1990.

383. Ministry of Agriculture, Fisheries and Food, Survey of plasticiser levels in food contact materials and foods. *Food Surveillance Paper No. 21*, 1987.

384. N. Harrison, Migration of plasticizers from cling-film. *Food Additives and Contaminants* **5**, Suppl. 1, 493–499 (1988).

385. R. P. Kozyrod and J. Ziaziaris, A survey of plasticizer migration into foods. *J. Food Protection* **52**, 578–580 (1989).

386. J. H. Petersen, E. T. Naamansen, and P. A. Nielsen, PVC cling film in contact with cheese: Health aspects related to global migration and specific migration of DEHA. *Food Additives and Contaminants* **12**, 245–253 (1995).

387. N. J. Loftus et al., An assessment of the dietary uptake of di-2-(ethylhexyl)adipate (DEHA) in a limited population study. *Food Chem. Toxicol.* **32**, 1–5 (1994).

388. Kolmar Research Center. Unpublished report, *The toxicological examination of di-2-ethyl-hexyl-adipate*, (Wickenol 158), 1967.

389. Mason Research Institute, unpublished report, Single dose acute toxicity test of di(2-ethyl hexyl) adipate in Fischer 344 rats and B6C3F1 mice. *Report No. MRI-TRA 16-76-34*, 1976.

390. Mason Research Institute, unpublished report, Repeated dose acute toxicity test of di(2-ethylhexyl) adipate in Fischer 344 rats and $B6C3F_1$ mice. *Report No. MRI-TRA 31-76-54*, 1976.

391. National Toxicology Program, Carcinogenesis bioassay of di(2ethylhexyl)adipate. *Technical Report No. 212*, revised March 1982.

392. CMA unpublished report, Toxicological effects of diethylhexyl adipate. *MRI Project 7343-B*, 1982.

393. CMA unpublished report, A study of the hepatic effects of diethylhexyl adipate in the mouse and rat. *BIBRA Project 3.0709*, 1989.

394. CMA unpublished report, Studies of the hepatic effects of diethylhexyl adipate (DEHA) in the mouse and rat. *SRI Project 2759-S01-91*, 1995.

395. T. Takahashi, A. Tanaka, and T. Yamaha, Elimination, distribution and metabolism of di(2-ethylhexyl)adipate (DEHA) in rats. *Toxicology* **22**, 223–233 (1981).

396. M. C. Cornu et al., *In vivo* and *in vitro* metabolism of di-(2-ethylhexyl) adipate a peroxisome proliferator, in the rat. *Arch. Toxicol.* Suppl. **12**, 265–268 (1988).

397. CMA unpublished report, Metabolism and disposition of di(2-ethylhexyl)adipate. *MRI Project 9550-B*, 1984.

398. N. J. Loftus et al., Metabolism and pharmacokinetics of deuterium-labelled di-2-(ethylhexyl) adipate (DEHA) in humans. *Food. Chem. Toxicol.* **31**, 609–614 (1993).

399. CEFIC unpublished report, Di-(2-ethylhexyl)adipate (DEHA) fertility study in rats. *CTL Study RR0374*, 1988.

400. CEFIC unpublished report, Di(2-ethylhexyl) adipate: Teratogenicity study in the rat. *CTL Study RR0372*, 1988.

401. S. Jobling et al., A variety of environmentally persistent chemicals, including some phthalate plasticisers, are weakly oestrogenic. *Environmental Health Perspectives* **103**, 582–587 (1995).

402. Mitsubishi Chemical Safety Institute, Ltd., Report: Evaluation of adipic esters on estrogenicity by *in vivo* uterotrophy in ovariectomized rats. *Study No. 8L306* 1998.

403. National Toxicology Program, Carcinogenesis bioassay of di(2ethylhexyl)adipate. *Technical Report No. 212*, revised March 1982.

404. H. C. Hodge et al., Tests on mice for evaluating carcinogenicity. *Toxicol. Appl. Pharmacology* **9**, 583–596 (1966).

405. CMA unpublished report, Mutagenicity evaluation of di(2-ethylhexyl)adipate (DEHA) in the Ames Salmonella/microsome plate test. *LBI Project 20988*, 1982.

406. E. Zeiger et al., Mutagenicity testing of di(2-ethylhexyl)phthalate and related chemicals in Salmonella. *Environmental Mutagenesis* **7**, 213–232 (1985).

407. CMA unpublished report, Mutagenicity evaluation of DEHA in the mouse lymphoma forward mutation assay. *LBI Project 20989*, 1982.

408. D. B. McGregor et al., Responses of the L5178Y tk+/tk− Mouse lymphoma cell forward mutation assay:III. 72 Coded Chemicals. *Environmental and Molecular Mutagenesis* **12**, 85–154 (1988).

409. S. M. Galloway et al., Chromosome aberrations and sister chromatid exchanges in Chinese hamster ovary cells: Evaluations of 108 Chemicals. *Environmental and Molecular Mutagenesis* **10**, (Suppl. 10) 1–175 (1987).

410. CMA unpublished report, Evaluation of DEHA in the primary rat hepatocyte unscheduled DNA synthesis assay. *LBI Project 20991*, 1982.

411. CMA unpublished report, Mutagenicity evaluation of DEHA in the mouse micronucleus test. *LBI Project 20996*, 1982.

412. M. D. Shelby et al., Evaluation of a three-exposure mouse bone marrow micronucleus protocol: Results with 49 chemicals. *Environmental and Molecular Mutagenesis* **21**, 160–179 (1993).

413. Y. Keith et al., Peroxisome proliferation due to di(2-ethylhexyl)adipate, 2-ethylhexanol and 2-ethylhexanoic acid. *Toxicology* **66**, 321–326 (1992).

414. B. G. Lake et al., Comparison of the effects of di(2-ethylhexyl)adipate on hepatic peroxisome proliferation and cell replication in the rat and mouse. *Toxicology* **123**, 217–226 (1997).

415. M. S. Rao and J. K. Reddy, An overview of peroxisome proliferator-induced hepatocarcinogenesis. *Environmental Health Perspectives* **93**, 205–209 (1991).
416. R. Cattley et al., Do peroxisome proliferating compounds pose a hepatocarcinogenic hazard to humans?. *Reg. Toxicol. Pharmacol.* **27**, 47–60 (1998).
417. C. Rhodes et al., Comparative pharmacokinetics and subacute toxicity of di(2-ethylhexyl)-phthalate (DEHP) in rats and marmosets: Extrapolation of effects in rodents to man. *Environmental Health Perspectives* **65**, 299–308 (1986).
418. R. D. Short et al., Metabolic and peroxisome proliferation studies with di(2-ethylhexyl)-phthalate in rats and monkeys. *Toxicol. Ind. Health* **3**, 185–195 (1987).
419. Y. Kurata et al., Subchronic toxicity of di(2-ethylhexyl)phthalate in common marmosets: lack of hepatic peroxisome proliferation, testicular atrophy, or pancreatic acinar cell hyperplasia. *Toxicol Sci.* **42**, 49–56 (1998).
420. C. R. Elcombe et al., Peroxisome proliferators: Species differences in response of primary hepatocyte cultures. *Ann. N.Y. Acad. Sci.* **804**, 628–635 (1996).
421. J. M. Peters, R. C. Cattley, and F. J. Gonzalez, Role of PPARa in the mechanism of action of the nongenotoxic carcinogen and peroxisome proliferator WY-14,643. *Carcinogenesis* **18**, 2029–2033 (1997).
422. J. D. Tugwood et al., Peroxisome proliferator-activated receptors: structures and functions. *Ann. N. Y. Acad. Sci.* **804**, 252–265 (1996).
423. C. A. N. Palmer et al., Peroxisome proliferator activated receptor-a expression in human liver. *Molecularl. Pharmacology* **53**, 14–22 (1998).
424. Kolmar Research Center. unpublished report, The Toxicological Examination of Di-2-Ethyl-Hexyl-Adipate (Wickenol 158). 1967.
425. G. A. Andreeva, Toxicology of the plasticizer dioctyl adipate. *Gig. Primen. Toksikol. Pestits. Klin. Otravl.* **9**, 373–377 (1971).
426. Esso Research and Engineering Company, unpublished Report, Three-month dietary administration in rats. 1971.
427. Esso Research and Engineering Company, unpublished Report, Three-month dietary administration in dogs. 1971.
428. R. McKee, A. Lington, and K. Traul, An evaluation of the genotoxic potential of di-isononyl adipate. *Environmental Mutagenesis* **8**, 817–827 (1986).
429. Esso Research and Engineering Company, unpublished Report, Acute eye application in rabbits. 1968.
430. NIOSH, *Manual of Analytical Methods*, 3rd Ed., USDHHS/PHS/CDC/NIOSH, Cincinnati, OH, 1984.
431. K. W. Schneider, Contact dermatitis due to diethyl sebacate. *Contact Dermatitis* **6**, 506–507 (1980).
432. E. Sasaki et al., Allergic contact dermatitis due to diethyl sebacate. *Contact Dermatitis* **36**, 172 (1997).
433. C. C. Smith, *Archives of Industrial Hygiene and Occupational Medicine.* **7**, 310 (1953).
434. G. Weitzel, *Z. Physiol. Chem.* **287**, 254 (1951).
435. Y. K. Shapashnikov et al., Study of the composition of gaseous emissions from production of technical-grade acetic acid and ethyl acetate. *Prom. Sanit. Ochistka Gazov.* **1**, 22–23 (1976).
436. Ref. 370, p. 363.

437. J. F. Treon et al., *Ind. Hyg. Quart.* **16**, 183 (1955).
438. D. Oesterle and E. Deml, Promoting activity of di(2-ethylhexyl)phthalate in rat liver foci bioassay. *J. Cancer Res. Clin. Oncol.* **114**, 133–136 (1988).
439. J. D. Brain et al., Relative toxicity of di(2-ethylhexyl) sebacate and related compounds in an *in vivo* hamster bioassay. *Inhal. Toxicol.* **8**(6), 579–593 (1996).
440. B. D. Silverstein et al., Biological effects summary report on di(2-ethylhexyl) sebacate. Brookhaven National Lab Report, Issue BNL-51729, 1983.
441. J. Gootjes, A. B. H. Funcke, and W. T. Nauta, Experiments in the 5 *H*-dibenzo[a,d]cycloheptene series. IV. Synthesis and pharmacology of some 10-mono and 5,10-disubstituted 5 *H*-dibenzo[a,d]cycloheptenes. *Arzneim.-Forch.* **19**(12), 1936–1938 (1969).
442. E. Boyland and L. F. Chasseaud, The effect of some carbonyl compounds on rat liver glutathione levels. *Biochemical Pharmacology* **19**, 1526 (1970).
443. D. Gerard-Monnier, S. Fougeat, and J. Chaudiere, Glutathione and cysteine depletion in rats and mice following intoxication with diethylmaleate. *Biochemical Pharmacology* **43**(3), 451–456 (1992).
444. L. Meites, *Handbook of Analytical Chemistry*, 1st ed., Mc-Graw-Hill, New York, 1963.
445. J. Thormann, I. Hansen, and J. Misfeldt, Occupational dermatitis from dibutyl maleate. *Contact Dermatitis* **13**(5), 314–316 (1985).
446. H. F. Smyth, C. P. Carpenter, and C. S. Weil, *J. Ind. Hyg. Toxicol.* **31**, 60 (1949).
447. D. L J. Opdyke, Monographs on fragrance raw materials. Diethyl maleate. *Food Cosmeticst Toxicol.* **14**, 443–444 (1976).
448. NTIS [AD691-490].
449. L. Ullman, *Maleic acid dimethylester- acute dermal toxicity study in rats.* Research & Consulting Company, Itingen, Switzerland, Project 035910, 1985.
450. K. G. Heimann, R. Jung, and H. Kieczka Maleic acid dimethylester: Evaluation of dermal toxicity and genotoxicity. *Food. Chem. Toxicol.* **29**(8), 575–578 (1991).
451. M. E. Andersen et al., The significance of multiple detoxification pathways for reactive metabolites in the toxicity of 1,1-dichloroethylene. *Toxicol. Appl. Pharmacology* **52**, 422 (1980).
452. E. Maellaro et al., Lipid peroxidation and antioxidant systems in the liver injury produced by glutathione depleting agents. *Biochemical Pharmacol.* **39**(10),1513–1521 (1990).
453. S. K. Hanson and M. W. Anders, The effect of diethyl maleate treatment, fasting, and time of administration an allyl alcohol hepatotoxicity. *Toxicology Letter* **1**(5–6), 301–305 (1978).
454. S. C. Langley and F. J. Kelly, Depletion of pulmonary glutathione using diethylmaleic acid accelerates the development of oxygen-induced lung injury in term and preterm guinea-pig neonates. *J. Pharmacy and Pharmacol.* **46**(2), 98–102 (1994).
455. C. A. Weber et al., Depletion of tissue glutathione with diethyl maleate enhances hyperbaric oxygen toxicity. *Amer. J. Physiology* **258**(6), 308–312 (1990).
456. K. Sai et al., The protective role of glutathione, cysteine, and vitamin C against oxidative DNA damage induced in rat kidney by potassium bromate. *Jpn. J. Cancer Res.* **83**(1), 45–51 (1992).
457. F. E. Mirer, B. S. Levine, and S. D. Murphy, Effects of piperonyl butoxide and diethyl maleate on toxicity and metabolism of parathion and methyl parathion. *Toxicol. Appl. Pharmacology* **33**, 181 (1975).

458. L. G. Sultatos et al., The effect of glutathione monoethyl ester on the potentiation of the acute toxicity of methyl parathion, methyl paraoxon or fenitrothion by diethyl maleate in the mouse. *Toxicology Letters* **55**, 77–83 (1991).

459. V. K. Nonavinakere et al., Selenium lethality: Role of glutathione and metallothionein. *Toxicology Letters* **66**, 273–279 (1993).

460. M. Wong, L. M. J. Helston, and P. G. Wells, Enhancement of murine phenytion teratogenicity by the gamma-glutamylcysteine synthetase inhibitor L-buthionine-(S,R)-sulfoximine and by the glutathione depletor diethyl maleate. *Teratology* **40**, 127–141 (1989).

461. D. J. Fort, T. L. Propst, and E. L. Stover, Evaluation of the developmental toxicity of 4-bromobenzene using frog embryo teratogenesis assay-Xenopus: Possible mechanisms of action. *Teratog. Carcinog. Mutagen.* **16**(6), 307–315 (1996).

462. A. Ogata et al., Glutathione and cysteine enhance and diethyl maleate reduces thisbendazole teratogenicity in mice. *Food. Chem. Toxicol.* **27**(2), 117–123 (1989).

463. M.-A. Shibata et al., Modification of BHA forestomach carcinogenesis in rats: Inhibition by diethylmaleate or indomethacin and enhancement by a retinoid. *Carcinogenesis* **14**, 1265–1269, (1993).

464. G. Kallistratos and U. Kallistratos, Inhibition of the 3,4-benzopyrene carcinogenesis by natural and synthetic compounds. *Muench. Med. Wochenschr* **117**(10), 391–394 (1975).

465. A. H. L. Chuang, H. Mukhtar, and E. Bresnick, Effects of diethyl maleate on aryl hydrocarbon hydroxylase and on 3-methylcholanthrene-induced skin tumorigenesis in rats and mice. *Journal of the National Cancer Institute* **60**(2), 321–325 (1978).

466. T. L. McNutt and C. Harris, Glutathione and cysteine modulation in bovine oviduct epithelium cells cultured *in vitro*. *Teratology* **49**(5), 413–414 (1994).

467. C. S. Gardiner and D. J. Reed, Glutathione status and consequences of chemically-induced oxidative stress in the preimplantation mouse embryo. *Toxicologist* **30**, 2 (1996).

468. C-S. Yang et al., Effect of diethylmaleate on liver extracellular glutathione levels before and after global liver ischemia in anesthetized rats. *Biochemical Pharmacol.* **53**, 357–361 (1997).

469. L. G. Costa and S. D. Murphy, Effect of diethyl maleate and other glutathione depletors on protein synthesis. *Biochemical. Pharmacol.* **19**, 3383–3388 (1986).

470. M. W. Anders, Inhibition and enhancement of microsomal drug metabolism by diethyl maleate. *Biochemical. Pharmacol.* **27**, 1098–1101 (1978).

471. J. L. Barnhart and B. Combes, Choleresis associated with metabolism and biliary excretion of diethyl maleate in the rat and dog. *J. Pharmacol. Exp. Therapeutics* **206**, 614–623 (1978).

472. J. W. Bauman, J. M. McKim, Jr., J. Liu, and C. D. Klaassen, Induction of metallothionein by diethyl maleate. *Toxicol. Appl. Pharmacol.* **114**, 188–196 (1992).

473. C. F. Morales et al., Diethylmaleate produces diaphragmatic impairment after resistive breathing. *J. Appl. Physiology* **75**(6), 2406–2411 (1993).

474. K. E. Malten, and R. L. Sielhuis, *Industrial Toxicology and Dermatology in the Production and Processing of Plastics*, Elseiver Publishing Co., Amsterdam, 1964, p. 66.

475. V. A. Druzhinina and T. A. Kochetova, Experimental data on the toxicological characteristics of dibutyl maleate and poly(dibutyl maleate) plasticizers. *Gig. Aspekty Okhr. Okruzhayushchei Sredy* 53–58 (1976).

476. K. Chruscielska, B. Graffstein, and J. Szarapinska-Kwaszewska, Mutagenic activity of dibutyl maleate. *Oraganika* 65–69 (1991).

477. *Gig. Tr. Prof. Zabol.* **31**(1), 49 (1987).

478. Y. S. Lee and J.-J. Jang, Potent promoting activity of dimethylitaconate on gastric carcinogenesis induced by *N*-methyl nitrosourea. *Cancer Letters* **85**, 177–184 (1994).

479. P. Talay, M. DeLong, and H. J. Prochaska, Identification of a common chemical signal regulating the induction of enzymes that protect against chemical carcinogenesis. *Proceedings of the National Academy of Sciences* **85**, 8261–8265 (1988).

480. Cosmetic Ingredient Review (CIR), Final report on the safety assessment of dibutyl phthalate, dimethyl phthalate, and diethyl phthalate. *Journal of the American College of Toxicology* **4**, 267–303 (1985).

481. R. M. David et al., Chronic peroxisome proliferation and hepatomegaly associated with the hepatocellular tumorigenesis of di(2-ethylhexyl)phthalate and the effects of recovery. *Toxicol. Sci.*, 1999, in press.

482. International Programme on Chemical Safety (IPCS), Di-*n*-butyl Phthalate. *Environmental Health Criteria 189*, WHO, 1997.

483. A. Mann et al., Comparison of the short-term effects of di(2-ethylhexyl)phthalate, di(*n*-hexyl)phthalate and di(*n*-octyl)phthalate in rats. *Toxicol. Appl. Pharmacol.* **77**, 116–132 (1985).

484. J. Smith et al., The hepatic effects of di-isononyl phthalate (DINP) and related analogs in rats and mice. *The Toxicologist* **48**, 234 (1999).

485. HEDSET (Harmonized Electronic Data Set), *Di-isononyl phthalate, CAS # 68515-48-0*, prepared by Exxon Chemical Holland, BV, 1992.

486. J. Butala et al., Oncogenicity study of di(isononyl)phthalate in rats. *The Toxicologist* **30**, 202 (1996).

487. J. Butala et al., Oncogenicity study of di(isononyl)phthalate in mice. *The Toxicologist* **36**, 173 (1997).

488. A. Lington et al., Chronic toxicity and carcinogenic evaluation of diisononyl phthalate in rats. *Fundamental and Applied Toxicology* **36**, 79–89 (1997).

489. BIBRA, A 21 day feeding study of di-isodecyl phthalate to rats: effects on the liver and liver lipids. *Project No. 3.0495.5. Report No 0495/5/85 to CMA*, 1986.

490. BIBRA, An investigation of the effect of di-isodecyl phthalate (DIDP) on rat hepatic peroxisomes. *Human Experimental Toxicology* **10**, 67–68 (1990).

491. BASF, *Bericht uber den 28-tage-ratten-Futterungsversuch mit Palatinol Z*, unpublished report, 1969.

492. BASF, *Bericht uber den 90-tage-ratten-futterungsversuch mit Palatinol Z*, unpublished report, 1969.

493. W. M. Kluwe, Carcinogenic potential of phthalic acid esters and related compounds: structure activity relationships. *Environmental Health Perspectives* **65**, 271–278 (1986).

494. G. Pugh et al., Absence of liver effects in cynomolgus monkeys treated with peroxisomal proliferators. *The Toxicologist* **48**, 235 (1999).

495. J. C. Lamb, IV et al., Reproductive effects of four phthalate acid esters in the mouse. *Toxicol. Appl. Pharmacol.* **88**, 255–269 (1987).

496. B. D. Hardin et al., Evaluation of 60 chemicals in a preliminary developmental toxicity test. *Teratogenesis, Carcinogenesis, and Mutagenesis* **7**, 29–48 (1987).

497. R. N. Wine et al., Reproductive toxicity of di-*n*-butylphthalate in a continuous breeding protocol in Sprague-Dawley rats. *Environmental Health Perspectives* **105**, 102–107 (1997).

498. B. R. Cater et al., Studies on dibutyl phthalate-induced testicular atrophy in the rat: Effects on zinc metabolism. *Toxicol. Appl. Pharmacol.* **41**, 609–618 (1977).

499. P. M. D. Foster et al., Study of the testicular effects and changes in zinc excretion produced by some *n*-alkyl phthalates in the rat. *Toxicol. Appl. Pharmacol.* **54**, 392–398 (1980).

500. T. J. B. Gray et al., Species differences in the testicular toxicity of phthalate esters. *Toxicology Letters* **11**, 141–147 (1982).

501. A. R. Singh, W. H. Lawrence, and J. Autian, Teratogenicity of phthalate esters in rats. *J. Pharmaceutical Sci.* **61**, 51–55 (1972).

502. K. Shiota, M. J. Chou, and H. Nishimura, Embryotoxic effects of di-2-ethylhexyl phthalate (DEHP) and di-*n*-butyl phthalate (DBP) in mice. *Environ. Res.* **22**, 245–253 (1980).

503. M. Ema et al., Teratogenic evaluation of di-*n*-butyl phthalate in rat. *Toxicology Letters* **69**, 197–203 (1993).

504. M. Ema, H. Amano, and Y. Ogawa, Characterization of the developmental toxicity of di-*n*-butyl phthalate in rat. *Toxicol.* **86**, 163–174 (1994).

505. M. Ema, E. Miyawaki, and K. Kawashima, Further evaluation of developmental toxicity of di-*n*-butyl phthalate following administration during late pregnancy in rats. *Toxicology Letters* **98**, 87–93 (1998).

506. E. Mylchreest, R. C. Cattley, and P. M. D. Foster, Male reproductive tract malformations in rats following gestational and lactational exposure to di(*n*-butyl)phthalate: An antiandrogneic mechanism?. *Toxicol. Sci.* **43**, 47–60 (1998).

507. R. W. Tyl et al., Developmental toxicity evaluation of dietary di(2-ethylhexyl)phthalate in Fischer 344 rats and CD-1 mice. *Fundamental and Applied Toxicology* **10**, 395–412 (1988).

508. M. Kawano, Toxicological studies on phthalate esters. *Jpn. J. Hyg.* **35**, 684–692 (1980).

509. F. Walseth and O. G. Nilsen, Phthalate Esters: II. Effects of inhaled dibutylphthalate on cytochrome p-450 mediated metabolism in rat liver and lung. *Archives of Toxicology* **55**, 132–136 (1984).

510. C. R. Noller, *Chemistry of Organic Compounds*, 3rd. ed., Saunders, Philadelphia, 1966.

511. A. J. Lehman, Insect repellents. *Food and Drug Officials of the United States, Quarterly Bulletin* **19**, 87–99 (1955).

512. L. A. Timofievskaya et al. Experimental research on the effect of phthalate plasticizers on the body. *Gigiena i Sanitariya* **12**, 26–28 (1974) (English translation of abstract).

513. B. G. Lake et al., The *in vitro* hydrolysis of some phthalate diesters by hepatic and intestinal preparations from various species. *Toxicol. Appl. Pharmacol.* **39**, 239–248 (1977).

514. R. D. White et al., Absorption and metabolism of three phthalate diesters by the rat small intestine. *Food Cosmetic Toxicol.* **18**, 383–386 (1983).

515. USEPA, *Ambient Water Quality Criteria Document: Phthalate Esters*, 1980, p. C-29.

516. A. E. Elsisi, D. E. Carter, and I. G. Sipes, Dermal absorption of phthalate diesters in rats. *Fundamental and Applied Toxicology* **12**, 70–77 (1989).

517. M. Ioku et al., Studies on absorption distribution and excretion of phthalates. *Yakuji Hsido Hen* **10**, 57–62 (1976).

518. E. A. Field, Developmental toxicity evaluation of dimethyl phthalate (CAS No. 131-11-3) administered to CD rats on gestational days 6 through 15. *NTP Report NTIS/PB89-164826*, 1989, 408 pp.

519. M. R. Plasterer et al., Developmental toxicity of nine selected compounds following prenatal exposure in the mouse. *J. Toxicol. Environ. Health* **15**, 25–38 (1985).

520. C. A. Harris et al., The estrogenic activity of phthalate esters *in vitro*. *Environmental Health Perspectives* **105**, 802–811 (1997).

521. S. Oishi and H. Hiraga, Effect of phthalic acid esters on mouse testis. *Toxicology Letters* **5**, 413–416 (1980).

522. B. Fredricsson et al., Human sperm motility is affected by plasticizers and diesel particle extracts. *Pharmacol. Toxicol.* **72**, 128–133 (1993).

523. National Toxicology Program, Toxicology and carcinogenesis studies of diethylphthalate in F344/N rats and B6C3F1 mice with dermal initiation/promotion study of diethylphthalate and dimethylphthalate in male Swiss (CD-1) mice. *NTP TR 429, NIH Publication No. 95-3356*, May 1995.

524. D. K. Agarwal et al., Mutagenicity evaluation of phthalic acid esters and metabolites. *J. Toxicol. Environ. Health* **16**, 61–69 (1985).

525. V. V. Yurchenko and S. E. Gleiberman, Study of long-term effects of repellent use. Part III. Study of mutagenic properties of dimethylphthalate and phenoxyacetic acid *N,N*-dimethylamide by dominate lethal mutations. *Meditsinskaya Parazitologiya i Parazitarnye Bolezni* **49**, 58–61 (1980).

526. K. Tsuchiya and K. Hattori, Chromosomal study on human leucocyte cultures treated with phthalic acid esters. *Hokkaidoritsu Eisei Kenkyusho Ho* **26**, 114 (1976).

527. L. D. Katosova and G. I. Pavlenko, Cytogenetic examination of the workers of chemical industry. *Mutation. Research* **147**, 301–302 (1985).

528. Chemical Manufacturers Association, Evaluation of dimethyl phthalate in the *in vitro* transformation of Balb/3T3 cells assay, final report by Litton Bionetics, Inc., 1985.

529. Chemical Manufacturers Association, Mutagenicity of dimethyl phthalate in a mouse lymphoma mutation assay, final report by Hazleton Biotech Co., 1986.

530. E. Barber et al., Results in the L5178Y mouse lymphoma and the *in vitro* transformation of Balb 3T3 cell assays for eight phthalate esters. *J. Appl. Toxicol.* 1999, in press.

531. P. H. Dugard et al., Absorption of some glycol ethers through human skin *in vitro*. *Environmental Health Perspectives* **57**, 193–197 (1984).

532. R. C. Scott et al., *In vitro* absorption of some *o*-phthalate diesters through human and rat skin. *Environmental Health Perspectives* **74**, 223–227 (1987), Errata, *Environmental Health Perspectives* **79**, 323 (1989).

533. American Conference of Governmental Industrial Hygienists, *Threshold Limit Values (TLVs) for Chemical Substances and Biological Exposure Indices for 1998*, ACGIH, Cincinnati, OH, 1998.

534. ATSDR, *Toxicological Profile for Diethyl Phthalate*, U.S. Department of Health and Human Services, 1995.

535. L. J. Phillips and G. F. Birchard, Regional variations in human toxics exposure in the USA: an analysis based on the National Human Adipose Tissue Survey. *Arch. Environ. Contam. Toxicol.* **21**, 159–168 (1991).

536. N. I. Sax, *Dangerous Properties of Industrial Materials*, 6th ed., Van Nostrand, New York, 1984, p. 1026.

537. Eastman Kodak Company, unpublished data.

538. A. M. Api, *In vitro* assessment of phototoxicity. *In Vitro Toxicology* **10**, 339–350 (1997).

539. D. Brown et al., Short-term oral toxicity study of diethyl phthalate in the rat. *Food Cosmetic Toxicol.* **16**, 415–422 (1978).

540. A. E. Field et al., Developmental toxicity evaluation of diethyl and dimethyl phthalate in rats. *Teratology* **48**, 33–44 (1993).

541. H. B. Jones, D. A. Garside, R. Liu, and J. C. Roberts, The influences of phthalate esters on leydig cell structure and function *in vitro* and *in vivo. Environmental and Molecular Pathology* **58**, 179–193 (1993).

542. BASF AG, unpublished data, 1974.

543. W. J. Kozumbo and R. J. Rubin, Assessment of the mutagenicity of phthalate esters. *Environmental Health Perspectives* **45**, 103–109 (1982).

544. M. Ishidate and S. Odashima Chromosome tests with 134 compounds on Chinese Hamster Cells *in vitro*—A screening for chemical carcinogens. *Mutation. Research* **48**, 337–354 (1977).

545. Ref. 24, p. 349.

546. D. B. Peakall, Phthalate esters: Occurrence and biological effects. *Residue. Rev.* **54**, 1–41 (1975).

547. W. R. Payne and J. E. Benner, Liquid and gas chromatographic analysis of diethyl phthalate in water and sediment. *J. Assoc. Off. Anal. Chem.* **64**, 1403–1407 (1981).

548. Ministry of Agriculture, Fisheries, and Food (MAFF), Phthalates in infant formulae - follow-up survey. *UK Dept. of Health, No. 168*, 1998.

549. C. C. Smith, Toxicity of butyl stearate, dibutyl sebacate, dibutyl phthalate, and methoxyethyl oleate. *Ind. Hyg. Occup. Med.* **7**, 310–318 (1953).

550. Netherlands Organization for Applied Scientific Research (TNO) and the National Institute of Public Health and the Environment (RIVM), *IUCLID Data Set for Dibutyl Phthalate*, 1996.

551. D. Calley, J. Autian, and W. L. Guess, Toxicology of a series of phthalate esters. *J. Pharm. Sci.* **55**, 158–162 (1966).

552. HEDSET (Harmonized Electronic Data Set), *Di-isoheptyl phthalate*, CAS # [71888-89-6], prepared by Exxon Chemical Europe, Inc, 1996.

553. R. K. Dogra et al., Modification of immune response in rats by di-octyl phthalate. *Ind. Health* **25**, 97–101 (1987).

554. R. K. S. Dogra et al., Di-octyl phthalate induced altered host resistance: viral and protozoal models in mice. *Ind. Health* **27**, 83–87 (1989).

555. L. Krauskoph, Studies on the toxicity of phthalates via ingestion. *Environmental Health Perspectives* **3**, 61–72 (1973).

556. C. Shaffer et. al., Acute and subacute toxicity of DEHP with note upon its metabolism. *J. Ind. Hyg. Toxicol.* **27**, 130–135 (1945).

557. H. C. Hodge, Acute toxicity for rats and mice of 2-ethyl hexanol and 2-ethylhexyl phthalate. *Proc. Soc. Exp. Biol. Med.* **53**, 20–23.

558. Eastman Kodak Company, unpublished Report.

559. W. Lawrence et al., A toxicological investigation of some acute, short-term, and chronic effects of administering di-2-ethylhexyl phthalate (DEHP) and other phthalate esters. *Environmental Research* **9**, 1–11 (1975).

560. HEDSET (Harmonized Electronic Data Set), *Di-isodecyl phthalate.* CAS # [68515-49-1], Exxon Chemical Holland, BV, 1992.

561. European Commission, European Chemicals Bureau, International Uniform Chemical Information Database (IUCLID), Diundecyl Phthalate Data Sheet. *Section 5.1, Version 3.0.4.*(1994).

562. J. Seed, Mutagenic activity of phthalate esters in bacterial liquid suspension assays. *Environmental Health Perspectives* **45**, 111–114 (1982).

563. S. Shibko and H. Blumenthal, Toxicology of phthalic acid esters used in food-packaging materials. *Environmental Health Perspectives* **3**, 131–137 (1973).

564. Solutia Canada, Inc., Santicizer® 711 Plasticizer Material Safety Data Sheet.

565. *EPA TSCATS: OTS 0521089, Doc. ID. 86-890001475*, Food & Drug Research Labs, Inc. for FMC Corp., 6-22-78.

566. NTP, *Technical Report No. 242, NIH Publication No.* 83-1798, U.S. Department of Health and Human Services, 1983.

567. National Toxicology Program, Carcinogenesis bioassay of butyl benzyl phthalate (CAS No. 85-68-7) in F344/N rats and B6C3F$_1$ mice (Feed Study), *NTP Technical Report No. 213*, National Technical Information Service Publication Number PB83-118398, 1982.

568. N. K. Statsek, Hygienic investigations of certain esters of phthalic acid and of polyvinylchloride materials plastificated thereby. *Gig. Sanit.* **39**(6), 25–28, 1974.

569. B. G. Hammond et al., A review of the subchronic toxicity of butyl benzyl phthalate. *Toxicol. Ind. Health* **3**(2), 79–98 (1987).

570. H. Smyth and C. Carpenter, *J. Ind. Hyg. Toxicol.* **30**, 63 (1948).

571. Y. Omori, Recent progress in safety evaluation studies on plasticizers and plastics and their controlled use in Japan. *Environmental Health Perspectives* **17**, 203–209 (1976).

572. *Int. Polym. Sci. Technol.* **3**, 93 (1976).

573. T. Y. Chin et al., Chemical urolithiasis: 1. Characteristics of bladder stone induction by terephthalic acid and dimethyl terephthalate in weanling Fischer-344 Rats. *Toxicol. Appl. Pharmacol.* **58**, 307–321 (1981).

574. E. D. Barber and D. C. Topping, Subchronic 90-day oral toxicology of di(2-ethylhexyl) terephthalate in the rat. *Food Chem. Toxicol.* **33**, 971–978 (1995).

575. M. Nikinorow, H. Mazur, and H. Piekacz, Effect of orally administered plasticizers and polyvinyl chloride stabilizers in the rat. *Toxicol. Appl. Pharmacol.* **26**, 253–259 (1973).

576. J. W. Daniel, Toxicity and metabolism of phthalate esters. *Clinical Toxicol.* **13**, 257–268 (1978).

577. B. G. Lake et al., Studies on the hydrolysis *in vitro* of phthalate esters by hepatic and intestinal mucosal preparation from various species. *Biochem. Soc. Trans.* **4**, 654–655 (1976).

578. R. D. White, D. L. Earnest, and D. E. Carter, Absorption and metabolism of three phthalate diesters by the rat small intestine. *Food Cosmetics Toxicol.* **21**, 99–101 (1983).

579. National Toxicology Program, "The toxicokinetics and metabolism of di-*n*-butyl phthalate." *RTI Project 60U-4940-515*, NTP Task CHEM00324, 1995.

580. D. T. Williams and B. J. Blanchfield, The retention, distribution, and metabolism of dibutyl phthalate-7-^{14}C in the rat. *J. Agr. Food Chem.* **23**, 854–858 (1975).

581. P. M. D. Foster et al., Differences in urinary metabolic profile from di-*n*-butyl phthalate-treated rats and hamsters. A possible explanation for species differences in susceptibility to testicular atrophy. *Drug Metabolism and Disposition* **11**, 59–61 (1982).

582. I. Issemann and S. Green, Activation of a member of the steroid hormone receptor superfamily by peroxisome proliferators. *Nature* **347**, 645–650 (1990).

583. A. R. Bell et al., Molecular basis of non-responsiveness to peroxisome proliferators: the guinea pig PPARα is functional and mediates peroxisome proliferator-induced hypolipidaemia. *Biochem. J.* **332**, 689–693 (1998).

584. N. G. Coldham et al., Evaluation of a recombinant yeast cell estrogen screening assay. *Environmental Health Perspectives* **105**, 734–742 (1997).

585. S. R. Milligan, A. V. Balasubramanian, and J. C. Kalita, Relative potency of xenobiotic estrogens in an acute *in vivo* mammalian assay. *Environmental Health Perspectives* **106**, 1–9 (1998).

586. T. R. Zacharewski et al., Examination of the *in vitro* and *in vivo* estrogenic activities of eight commercial phthalate esters. *Toxicological Sciences* **42**, 282–293 (1998).

587. L. E. Gray, Jr., J. S. Ostby, E. Mylchreest, P. M. D. Foster, and W. R. Kelce, Dibutyl phthalate (DBP) induces antiandrogenic but not estrogenic *in vivo* effects in LE Hooded Rats. *Toxicological Sciences* **42**(1-S), 176 (1998).

588. National Toxicology Program, NTP technical report on toxicity studies of dibutyl phthalate administered in feed to F344/N rats and B6C3F_1 mice. *NTP TR 30, NIH Publication No. 95-3353*(1995).

589. S. Abe and M. Sasaki, Chromosome aberrations and sister chromatid exchanges in Chinese hamster cells exposed to various chemicals. *J. Natl. Cancer Inst.* **58**, 1635–1640 (1977).

590. V. P. Cagianut, *Schewiz. Med. Wochschr.* **84**, 1243 (1954).

591. ATSDR, *Toxicological Profile for Di-n-Butyl Phthalate*, US Department of Health and Human Services, TP-90-10, 1990.

591a. R. Hinton et al., Effects of phthalic acid esters on the liver and thyroid. *Environmental Health Perspectives* **70**, 195–210 (1986).

592. L. Hagmar et al., Mortality and health of workers processing polyvinylchloride. *Amer. J. Ind. Med.* **17**, 553–565 (1990).

593. J. Nielsen, B. Akesson, and S. Skerfving, Phthalate ester exposure: Air levels and health of workers processing polyvinylchloride. *Amer. Ind. Hyg. Associ. J.* **46**, 643–647 (1985).

594. S. Vainiotalo and P. Pfaffli, Air impurities in the PVC plastics processing industry. *Annals of Occupational Hygiene* **34**, 585–590 (1990).

595. Esso Research and Engineering Company, Ninety-day dietary administration in rats and dogs. unpublished report, 1962.

596. Exxon Biomedical Sciences Inc., Mutagenicity test on JAYFLEX (r) DHP in an *in vivo* mouse micronucleus assay. unpublished results, 1996.

597. Esso Research and Engineering Company, Eye irritation tests in albino rabbits. unpublished report, 1975.

598. Exxon Corporation, Evaluation of potential hazards by dermal contact. unpublished report, 1976.

599. A. Medeiros, D. Devlin, and L. Keller, Evaluation of skin sensitization response of dialkyl (C6-C13) phthalate esters. *Contact Dermatitis* 1999, in press.

600. Y. Matsushima et al., *Eisei Shikenjo Hokoku* **110**, 26–31 (1992).

601. B. Mangham, J. Foster, and B. Lake, Comparison of the hepatic and testicular effects of orally administered di(2-ethylhexyl) phthalate and dialkyl 79 phthalate in the rat. *Toxicol. Appl. Pharmacol.* **61**, 205–214 (1981).

602. T. J. B. Gray et al., Peroxisomal effects of phthalate esters in primary cultures of rat hepatocytes. *Toxicology* **28**, 167–179, 1983.

603. K. Nakashima et al., Teratogenicity of di-*n*-heptyl-phthalate in mice. *Teratology* **16**, 117 (1977).

604. B. G. Lake et al., Comparative studies of the hepatic effects of di- and mono-*n*-octyl phthalates, di-(2-ethylhexyl) phthalate and clofibrate in the rat. *Acta. Pharmacol. et Toxicol.* **54**, 167–176 (1984).

605. B. G. Lake, T. J. B. Gray, and S. D. Gangolli, Hepatic effects of phthalate esters and related compounds- *in vivo* and *in vitro* correlations. *Environmental Health Perspectives* **67**, 283–290 (1986).

606. S. Oishi and K. Hiraga, Testicular atrophy induced by phthalate acid esters: Effect on testosterone and zinc concentration. *Toxicology and Applied Pharmacology* **53**, 35–41 (1980).

607. R. Poon et al., Subchronic oral toxicity of di-*n*-octyl phthalate and di(2-ethylhexyl) phthalate in the rat. *Food Chem. Toxicol.* **35**, 225–239 (1997).

608. S. Oishi, Effects of phthalic acid ester on testicular mitochondrial function in the rat. *Archives of Toxicol.* **64**, 143–147 (1990).

609. J. J. Heindel and C. J. Powell, Reproductive toxicity of three phthalic acid esters on a continuous breeding protocol. *Fundamental Appl. Toxicol.* **12**, 508–518 (1989).

610. R. K. S. Dogra et al., Effect of dioctyl phthalate on immune system of rat. *Indian J. Experimental Biology* **23**, 315–319 (1985).

611. Ref. 220, p. 3053.

612. G. Ikeda et al., Distribution and excretion of two phthalate esters in rats, dogs, and miniature pigs. *Food Cosmetic Toxicol.* **16**, 409–413 (1978).

613. Goodyear Fiber and Polymer Products Research Division, Laboratory Report No. 81-4-7, Mutagenicity evaluation of di-isooctyl phthalate (USS Chemical). *EPA document number 878210369, Fiche no. OTS0206046*, 1981.

614. Litton Bionetics Inc., Evaluation of di-isooctyl phthalate in the *in vitro* transformation of BALB/c 3T3 cells assay. Final Report. *EPA Document No. 878101226, Fiche No. OTS0206260*, 1981.

615. Eastman Chemical Company, MSDS.

616. K. Weissermel and H. J. Arpe, *Industrial Organic Chemistry*, 3rd, ed., VCH Publishers, New York 1997, pp. 390–391.

617. L. Fishbein and P. W. Albro, Chromatographic and biological aspects of the phthalate esters. *J. Chromatogr.* **70**, 365–412 (1972).

618. E. Draviam, J. Kerkay, and K. H. Pearson, Separation and quantitation of urinary phthalates by HPLC. *Anal. Lett.* **13**, 1137–1155 (1980).

619. H. Shintani, Determination of phthalic acid mono-(2-ethylhexyl) phthalate and di-(2-ethylhexyl) phthalate in human plasma and in blood products. *J. Chromatogr. Biomed. Appl.* **337**, 279–290 (1985).

620. G. M. Liss et al., Urine phthalate determinations as an index of occupational exposure to phthalate anhydride and di(2-ethylhexyl)phthalate. *Scand. J. Work. Environ. Health.* **11**, 381–387 (1985).

621. International Programme on Chemical Safety (IPCS), *Environmental Health Criteria 131, Diethylhexyl Phthalate*, World Health Organization, Geneva, 1992.

622. U.S. Department of Health and Human Services Agency for Toxic Substances and Disease Registry, *Toxicological Profile for Di(2-ethylhexyl) phthalate*, Atlanta, 1993.

623. W. W. Huber, B. Grasl-Kraupp, and R. Schulte-Hermann, Hepatocarcinogenic potential of di(2-ethylhexyl)phthalate in rodents and its implications on human risk. *Critical Reviews in Toxicology* **26**, 365–481 (1996).

624. D. Barber et al., Peroxisome induction studies on seven phthalate esters. *Toxicology and Industrial Health* **3**(2), 7–24 (1987).

625. B. G. Lake et al., Comparative studies on di-(2-ethylhexyl)phthalate-induced hepatic peroxisome proliferation in the rat and hamster. *Toxicol. Appl. Pharmacol.* **72**, 46–60 (1984).

626. T. J. B. Gray and S. Gangolli, Aspects of the testicular toxicity of phthalate esters. *Environmental Health Perspectives* **65**, 229–235 (1986).

627. P. Sjoberg, N. G. Lindqvist, and L. Ploen, Age-dependent response of the rat testes to di(2-ethylhexyl)phthalate. *Environmental Health Perspectives* **65**, 237–242 (1986).

628. L. A. Dostal et al., Testicular toxicity and reduced Sertoli Cell numbers in neonatal rats by di(2-ethylhexyl) phthalate and the recovery of fertility as adults. *Toxicol. Appl. Pharmacol.* **95**, 104–121 (1988).

629. H. A. Rutter, *Three week intravenous administration in the dog: Di(2-ethylhexyl)phthalate*, Hazleton Laboratories, America, Project Number 542-129 for the National Heart and Lung Institute, NTIS Document Number PB 244 262, Springfield, VA, 1975.

630. C. P. Carpenter, C. S. Weil, and H. F. Smith, Jr., Chronic oral toxicity of di(2-ethylhexyl)-phthalate for rats, guinea pigs and dogs. *Amer. Med. Assoc. Arch. Ind. Hyg. Occup. Med.* **8**, 219–226 (1953).

631. National Toxicology Program., Carcinogenesis bioassay of di(2-ethylhexyl)phthalate in F-344 Rats and B6C3F1 Mice (feed study). *Tech. Rep. Series 217*, 1982.

632. A. E. Ganning, M. J. Olsson, U. Brunk, and G. Dallner, Effects of prolonged treatment with phthalate ester on rat liver. *Pharmacol. Toxicol.* **68**, 392–401 (1991).

633. R. M. David et al., Further investigations of the oncogenicity of di(2-ethylhexyl) phthalate in rats. *Fundam. Appl. Toxicol.* **30**(suppl), 204 (1996).

634. R. M. David et al., Correlation of peroxisome proliferation and oncogenicity of di(2-ethylhexyl)phthalate in mice. *Fundam. Appl. Toxicol.* **36**(Suppl), 173 (1997).

635. R. S. Harris et al., Chronic oral toxicity of di(2-ethylhexyl)phthalate in rats and dogs. *Archives of Industrial Health* **13**, 259–264 (1956).

636. P. W. Albro and S. R. Levanher, Metabolism of di-(2-ethylhexyl)phthalate. *Drug Metabolism Review* **21**, 13–34 (1989).

637. P. W. Albro et al., Pharmacokinetics, interactions with macromolecules and species differences in metabolism of DEHP. *Environmental Health Perspectives* **45**, 19–23 (1982).

638. C. R. Elcombe and A. M. Mitchell, Peroxisome proliferation due to di-(2- ethylhexyl)phthalate (DEHP): Species differences and possible mechanisms. *Environmental Health Perspectives* **70**, 211–219 (1986).

639. D. A. Keys et al., Quantitative evaluation of alternative mechanisms of blood and testes disposition of di-(2-ethylhexyl)phthalate and mono(2-ethylhexyl)phthalate in rats. *Toxicol. Sci.* **49**, 172–185 (1999).

640. P. Albro, Identification of the metabolites of DEHP by rats: Isolation and characterization of the urinary metabolites. *J. Chromatogr.* **76**, 321–330 (1981).

641. P. J. Deisinger, L. G. Perry, and D. Guest, *In vivo* percutaneous absorption of [^{14}C]DEHP-plasticized polyvinyl chloride film in male Fischer 344 rats. *Food Chem. Toxicol.* **36**, 521–527 (1998).

642. D. K. Agarwal et al., Effects of di(2-ethylhexyl) phthalate on the gonadal pathophysiology, sperm morphology, and reproductive performance of male rats. *Environmental Health Perspectives* **65**, 343–350 (1986).

643. "Final report of three generation reproduction study with di-2-ethylhexylphthalate in albino rats". *IBT No. 621-06359*, Industrial Biotest Laboratories, Northbrook, IL, 1977.

644. H.-J. Klimisch et al., Di-(2-ethylhexyl)phthalate: A short-term repeated inhalation toxicity study including fertility assessment. *Food Chem. Toxicol.* **30**, 915–1191 (1992).

645. K. Shiota, K., and H. Nishimura, Teratogenicity of di(2-ethylhexyl)phthalate (DEHP) and di-*n*-butylphthalate (DBP) in mice. *Environmental Health Perspectives* **45**, 65–70 (1982).

646. J. Merkle, H. J. Klimisch, and R. Jackh, Developmental toxicity in rats after inhalation exposure of di-2-ethylhexylphthalate (DEHP). *Toxicology Letters* **42**, 215–223 (1988).

647. F. A. Arcadi et al., Oral toxicity of *bis*(2-ethylhexyl) phthalate during pregnancy and suckling in the Long-Evans rat. *Food Chem. Toxicol.* **36**, 963–970 (1998).

648. L. E. Gray, Jr. et al., Administration of potentially antiandrogenic pesticides (Procymidone, Linuron, Iprodione, Chlozolinate, *p,p′*-DDe, and Ketoconazole) and toxic substances (dibutyl- and diethylhexyl phthalate, PCB 169, and ethane dimethane sulphonate) during sexual differentiation produces diverse profiles of reproductive malformations in the male rat. *Toxicology and Industrial Health* **15**, 94–118 (1999).

649. B. J. Davis, R. R. Maronpot, and J. J. Heindel, Di-(2-ethylhexyl)phthalate suppresses estradiol and ovulation in cycling rats. *Toxicol. Appli Pharmacology* **128**, 216–223 (1994).

650. G. D. Divincenzo, Bacterial mutagenicity testing of urine from rats dosed with 2-ethylhexyl-derived plasticizers. *Toxicology* **34**, 247–259 (1985).

651. G. R. Douglas, Genetic toxicology of phthalate esters: Mutagenic and other genotoxic effects. *Environmental Health Perspectives* **65**, 255–262 (1986).

652. H. Takagi, Significant increase of 8-hydroxydeoxyguanosine in liver DNA of rats following short-term exposure to the peroxisome proliferators DEHP and DEHA. *Jpn. J. of Cancer Res.* **81**, 213–215 (1990).

653. A. R. Malcolm, Inhibition of gap-junction intracellular communication between Chinese Hamster lung fibroblasts by DEHP and trisodium nitroloacetate monohydrate. *Cell Biol. Toxicol.* **34**, 145–153 (1989).

654. S. Green, PPAR: A mediator of peroxisome proliferation action. *Mutation Research* **333**, 101–109 (1995).

655. J. M. Ward et al., Receptor and non-receptor mediated organ-specific toxicity of di(2-ethylhexyl)phthalate (DEHP) in peroxisome proliferator-activated receptor α-null mice. *Toxicologic Pathology* **26**, 240–246 (1998).

656. N. Woodyatt et al., The peroxisome proliferator (PP) response element (PPRE) upstream of the human acyl CoA oxidase gene is inactive in a sample human population: Significance for species differences in response to PPs. *Carcinogenesis* **20**, 369–375 (1999).

657. J. Doull et al., A Cancer Risk Assessment of Di(2-ethylhexy)phthalate: Application of the New U.S. EPA Risk Assessment Guidelines. *Regulatory Toxicology and Pharmacology* **29**, 327–357 (1999).

658. V. C. Moser, B. M. Cheek, and R. C. MacPhail, A multidisciplinary approach to toxicological screening: III. Neurobehavioral toxicity. *J. Toxicol. Environ. Health* **45**, 173–210 (1995).

659. Huntingdon Life Sciences, Toxicity Study by Oral Gavage Administration to Marmosets for 13 Weeks, unpublished data, 1998.

660. Exxon Corporation, "Dermal Disposition of ^{14}C-Labelled Di-isononyl Phthalate in Rats," unpublished report, 1983.

661. M. El-Hawari et al., Disposition and metabolism of diisononyl phthalate (DINP) in Fischer 344 rats: Single dosing studies. *The Toxicologist* **5**, 237 (1985).

662. A. Lington et al., Disposition and metabolism of diisononyl phthalate (DINP) in Fischer 344 rats: Multiple dosing studies. *The Toxicologist* **5**, 238 (1985).

663. A. Nikiforov, S. Harris, and L. Keller, Two generation reproduction study in rats with di-isononyl phthalate (DINP). *Toxicology Letters*(suppl. 1/78), 61 (1995).

664. S. Waterman et al., Developmental toxicity of di-isodecyl and di-isononyl phthalates in rats. *Reproductive Toxicology* **13**, 131–136 (1999).

665. J. Hellwig, H. Freudenberger, and R. Jackh Differential prenatal toxicity of branched phthalate esters in rats. *Food Chem. Toxicol.* **35**, 501–512 (1997).

666. A. Lington et al., Short-term feeding studies assessing the testicular effects of nine plasticizers in the F344 rat. *Toxicology Letters*(suppl. 1/93), 132 (1993).

667. J. Ashby et al., Mechanistically-based human hazard assessment of peroxisome proliferator-induced hepatocarcinogenesis. *Human and Experimental Toxicology* **13**(suppl. 2), S1–S117 (1994).

668. D. Caldwell et al., Retrospective evaluation of alpha 2u globulin accumulation in male rat kidneys following high doses of diisononyl phthalate. *Toxicological Sciences* 1999, in press.

669. US EPA, *Report of the Peer Review Workshop on Alpha 2u-Globulin. Association with Renal Toxicity and Neoplasia in the Male Rat*, U.S. Environmental Protection Agency, 1991.

670. D. Benford et al., Species differences in the response of cultured hepatocytes to phthalate esters. *Food Chem. Toxicol.* **24**, 799–800 (1986).

671. BIBRA, The British Industrial Biological Research Association, *A 21-day Feeding Study of Di-undecyl phthalate to rats: effects on the Liver and Liver Lipids. EPA Document No. 40+85262007, Fiche No. OTS0509538.*

672. HEDSET (Harmonized Electronic Data Set), *Di-tridecyl phthalate. CAS # [68515-47-9]*, Exxon Chemical Europe, Inc., 1995.

673. BIBRA (The British Industrial Biological Research Association), unpublished data, 1985.

674. Monsanto Company, unpublished data, 1984.

675. J. Hirzy, Carcinogenicity of general purpose phthalates: Structure-activity relationships. *Drug Metabolism Reviews* **21**, 55–63 (1989).

676. EPA TSCATS, OTS0510526, Doc I. D. 40-8626219, 4A, 01.06, Chemical Manufacturers Association, 1986.

677. BASF AG, unpublished data, 1974.

678. NTP: *Technical Report No. 284, NIH Publication No. 85-2540*, U.S. Department of Health and Human Services, 1985.

679. D. A. Eigenberg et al., Examination of the differential hepatotoxicity of diallyl phthalate in rats and mice. *Toxicol. Appl. Pharmacology* **86**, 12–21 (1986).

680. C. Kaye and L. Young, *Biochem. J.* **127**(5), 87 (1972); Cited in *NTP: Technical Report No. 242, NIH Publication No.83-1798*, U.S. Department of Health and Human Services, 1983.

681. C. Kaye, Biosynthesis of mercapturic acids from allyl alcohol, allyl esters, and acrolein. *Biochem. J.* **134**, 1093–1101 (1973).

682. K. N. Woodward, *Phthalate Esters: Toxicity and Metabolism*, Issue 1 and 2, CRC Press, Boca Raton, FL, 1988.

683. *EPA TSCATS: OTS 0520458, Doc. ID. 86-890000411*, Ethyl Corp., 2-16-79.

684. *EPA TSCATS: OTS 0521093, Doc. ID. 86-890001479*, Microbiological Assoc. for FMC Corp., 8-10-77.

685. *EPA TSCATS: OTS 0521094, Doc. ID. 86-890001480*, Genetic Toxicol. Laboratory for FMC Corp., 5-01-86.

686. B. C. Myhr and W. J. Caspary, Chemical mutagenesis at the thymidine kinase locus in L5178Y mouse lymphoma cells: Results for 31 coded compounds in the National Toxicology Program. *Environmental and Molecular Mutagenesis* **18**, 51–83 (1991).

687. D. K. Gulati et al., Chromosome aberration and sister chromatid exchange tests in Chinese hamster ovary cells *in vitro*. III. Results with 27 chemicals. *Environmental and Molecular Mutagenesis* **13**, 133–193 (1989).

688. R. Valencia et al., Chemical mutagenesis testing in Drosophila. III. Results of 48 coded compounds tested for the National Toxicology Program. *Environmental Mutagenesis* **7**, 325–348 (1985).

689. M. D. Shelby and K. L. Witt, Comparison of results from mouse bone marrow chromosome aberration and micronucleus tests. *Environmental and Molecular Mutagenesis* **25**, 302–313 (1995).

690. W. McOmie, (1949), cited in *Information Profiles on Potential Occupational Hazards: Phthalates, 2nd draft, SRC TR 82-520 Center for Chemical Hazard Assessment*, Syracuse Research Corporation, Syracuse, New York, for NIOSH, PB89-215818, 1982.

691. International Programme on Chemical Safety, *Concise International Chemical Assessment Document Butyl Benzyl Phthalate*, 1999, in press.

692. California Environmental Protection Agency, *PTEAM: Monitoring of Phthalates and PAHS in Indoor and Outdoor Air Samples in Riverside, California*, Final Report, Volume II, prepared under contract by Research Triangle Institute, Research Triangle Park, North Carolina, Contract Number A933-144, submitted to Air Resources Board, Research Division, 1992.

693. NTP (National Toxicology Program, *Toxicology and Carcinogenesis Studies of Butyl Benzyl Phthalate (CAS No. 85-68-7) in F344/N Rats (Feed Studies)*, NTP Technical Report No. 458. NIH Publication No. 97-3374, 1997.

694. N. G. Erickson, The metabolism of diphenyl phthalate and butylbenzyl phthalate in the beagle dog. *Dissert. Abs.* **26**(5), 3014–3015 (1965).

695. D. A. Eigenberg et al., Distribution, excretion, and metabolism of butylbenzyl phthalate in the rat. *J. Toxicol. Environ. Health* **17**(4), 445–456 (1986).

696. National Toxicology Program, *Developmental Toxicity Evaluation of Butyl Benzyl Phthalate (CAS No. 85-68-7) Administered in Feed to CD Rats on Gestational Days 6 to 15.* Report No. NTP-89-246. National Technical Information Service Publication No. PB90-115346, 1989.

697. National Toxicology Program, *Final Report on the Developmental Toxicity of Butyl Benzyl Phthalate (CAS No. 85-68-7) in CD-1-Swiss Mice.* NTP Report No. 90-114. National Technical Information Service Publication No. PB91-129999, 1990.

698. M. Ema et al., Evaluation of the teratogenic potential of the plasticizer butyl benzyl phthalate in rats. *J. Appl. Toxicol.* **1**(5), 339–343 (1990).

699. M. Ema et al., Comparative developmental toxicity of *n*-butyl benzyl phthalate and di-*n*-butyl phthalate in rats. *Arch. Environ. Contam. Toxicol.* **28**(2), 223–228 (1995).

700. A. H. Piersma, A. Verhoef, and P. M. Dortant, Evaluation of the OECD 421 reproductive toxicity screening test protocol using butyl benzyl phthalate. *Toxicology* **99**(3), 191–197 (1995).

701. TNO Biotechnology and Chemistry Institute, *Dietary One-Generation Reproduction Study with Butyl Benzyl Phthalate in Rats*, contract for Monsanto Company, submitted by Monsanto to Office of Toxic Substances, U.S. Environmental Protection Agency, Document Identification Number 86-930000189, Microfiche Number OTS0538169, 1993.

702. W. Kluwe, Comprehensive evaluation of the biological effects of phthalate esters. Presented at *Int. Conf. Phthalic Acid Esters, University of Surrey, Guildford, England, 1984. In* K. Woodward, *Phthalate Esters: Toxicity and Metabolism*, Vol. 1., CRC Press, Boca Raton, 1988.

703. D. K. Agarwal et al., Adverse effects of butyl benzyl phthalate on the reproductive and hematopoietic systems of male rats. *Toxicology* **35**(3), 189–206 (1985).

704. R. M. Sharpe et al., Gestational and lactational exposure of rats to xenoestrogens results in reduced testicular size and sperm production. *Environmental Health Perspectives* **103**(12), 1136–1143 (1995).

705. J. Ashby et al., Normal sexual development of rats exposed to butyl benzyl phthalate from conception to weaning. *Regulatory Toxicology and Pharmacology* **26**, 102–118 (1997).

706. R. S. Nair et al., Lack of developmental/reproduction effects with low concentrations of butyl benzyl phthalate in drinking water in rats. *Toxicol. Sci.* **48**(1-S) 218 (1999).

707. A. M. Soto et al., The E-SCREEN assay as a tool to identify estrogens: an update on estrogenic environmental pollutants. *Environmental Health Perspectives* **103**(suppl. 7), 113–122 (1995).

708. *Report No. CTL/R/1280. Study to Evaluate the Effect of Butyl Benzyl Phthalate on Uterine Growth in Immature Female Rats after Oral Administration*, study conducted for Monsanto Europe SA by Central Toxicology Laboratory, Cheshire, UK., 1995.

709. *Report No. CTL/R/1278. Study to Evaluate the Effect of Butyl Benzyl Phthalate on Uterine Growth in Immature Female Rats after Subcutaneous Administration*, study conducted for Monsanto Europe SA by Central Toxicology Laboratory, Cheshire, UK, 1996.

710. National Toxicology Program, *Effect of Dietary Restriction on Toxicology and Carcinogenesis Studies in F344/N Rats and B6C3F$_1$ Mice*, NTP Technical Report No. 460, NIH Publication No. 97–3376, 1997.

711. Litton Bionetics Inc., *Mutagenicity Evaluation of BIO-76-17 Santicizer 160. NB 259784*, Final Report, submitted by Monsanto Company, St. Louis, Missouri, to Office of Toxic Substances, U.S. Environmental Protection Agency, Document Identification Number 87-7800282, Microfiche Number OTS200290, 1976.

712. R. J. Rubin, W. Kozumbo, and R. Kroll, Ames mutagenic assay of a series of phthalic acid esters: positive response of the dimethyl and diethyl esters in TA100 [abstract]. *Toxicol. Appl. Pharmacol.* **48**(1 part 2), A133 (1979).

713. E. Zeiger et al., Phthalate ester testing in the National Toxicology Program's Environmental Mutagenesis Test Development Program. *Environmental Health Perspectives* **45**, 99–101 (1982).

714. Litton Bionetics Inc., *Mutagenicity Evaluation of BIO-76-243 CP731 (Santicizer 160) in the Mouse Lymphoma Assay*, Final Report, submitted by Monsanto Company, St. Louis, Missouri, to Office of Toxic Substances, U.S. Environmental Protection Agency. Document Identification Number 87-7800282, Microfiche Number OTS200290, 1977.

715. Hazleton Biotechnologies Company, *Mutagenicity of 1D in a Mouse Lymphoma Mutation Assay*, Final Report, submitted by Chemical Manufacturers Association, Washington, DC, to Office of Toxic Substances, U.S. Environmental Protection Agency. Document Identification Number 40-8626225, Microfiche Number OTS0510527, 1986.

716. B. C. Myhr, L. R. Bowers, and W. J. Caspary, Results from the testing of coded chemicals in the L5178Y $TK^{+}/^{-}$ mouse lymphoma mutagenesis assay [abstract]. *Environmental Mutagenesis* **8**(suppl. 6), 58 (1986).

717. Litton Bionetics Inc., *Evaluation of 1D in the In Vitro Transformation of BALB/3T3 Cells Assay*, Final Report, Submitted by Chemical Manufacturers Association, Washington, DC, to U.S. Environmental Protection Agency. Document Identification Number 40+8526206, Microfiche Number OTS0509537, 1985.

718. L. A. Dueva and M. V. Aldyreva, Experimental assessment of sensitizing and irritating action of phthalate plasticizers [English abstract]. *Gig. Trud. Prof. Zabol.* **13**(10), 17–19 (1969).

719. Solutia study SH-79-0006, 1979.

720. K. D. Little and J. R. Little, *Investigation of Santicizer 160 (Benzyl Butyl Phthalate) as a Potential Allergen*, Washington University School of Medicine. Report to Monsanto Company, Project Number JH-81-302, 1983.

721. Ref. 372, p. 356.

722. A. R. Singh, W. H. Lawrence, and J. Autian, Mutagenic and antifertility sensitivities of mice to di-2-ethylhexylphthalate (DEHP) and dimethoxyethyl phthalate (DMEP). *Toxicol. Appl. Pharmacol.* **29**, 35–46 (1974).

723. Shell Oil Co., The effects of acute exposure of dimethoxyethyl phthalate, glycerol alpha-monochlorohydrin, epichlorohydrin, formaldehyde and methylmethanesulfonate upon testicular sperm in the rat. *EPA Doc. 878210077*(1982).

724. M. R. Parkhie, M. Webb, and M. A. Norcross, Dimoethoxyethyl phthalate: embryopathy, teratogenicity, fetal metabolism and the role of zinc in the rat. *Environmental Health Perspectives* **45**, 89–97 (1982).

725. E. J. Ritter, W. J. Scott, and J. L. Randall, Mechanistic studies of the teratogenic action of di(2-methoxyethyl)phthalate (DMEP) and 2-methoxyethanol in Wistar rats. *Teratology* **29**(2), 54A (1984).

726. E. J. Ritter et al., Computer analysis of rat teratology data following administration of phthalates and their metabolites. *Teratology* **33**, 93C (1986).

727. R. K. Bower, S. Haberman, and P. D. Minton, Teratogenic effects in the chick embryo caused by esters of phthalic acid. *J. Pharmacol. Exp. Therapeutics* **171**(2), 314–24 (1970).

728. E. J. Ritter et al., Teratogenicity of dimethoxyethylphthalate and its metabolites methoxyethanol and methoxyacetic acid in the rat. *Teratology* **32**(1), 25–31 (1985).

729. J. Yonemoto, N. A. Brown, and M. Webb, Effects of dimethoxyethylphthalate, monomethoxyethyl phthalate, 2-methoxyethanol and methoxyacetic acid on post implantation rat embryos in culture. *Toxicology Letters* **21**, 97–102 (1984).

730. Eastman Kodak Company, unpublished Report.

731. *Amer. J. Ophthalmology* **29**, 1363 (1946).

732. H. C. Hodge et al., Chronic oral toxicity of ethylphthalyl ethyl glycolate in rats and dogs. *Archives of Industrial Hygiene and Occupational Medicine* **8**, 289–295 (1953).

733. Ref. 372, p. 377.

734. B. F. Goodrich Company, *A study on the Toxicity of Butylphthalyl Butylglycolate*(Santicizer B-16), *report to Monsanto*, St. Louis, MO, 1950.

735. Ref. 24, p. 1021.

736. W. J. Krasavage, F. J. Yanno, and C. J. Terhaar, Dimethyl terephthalate (DMT): Acute toxicity, subacute feeding and inhalation studies in male rats. *Amer. Ind. Hyg. Assoc. J.* **34**(10), 455–462 (1973).

737. Y. P. Sanina and T. A. Kocketkova, *Toksikol. Novykh. Prom. Khim. Veshchestv.* **5**, 107–123 (1963).

738. H. d'A. Heck and R. W. Tyl, The induction of bladder stones by terephthalic acid, dimethyl terephthalate, and melamine (2,4,6-Triamino-*s*-triazine) and its relevance to risk assessment. *Regulatory Toxicology and Pharmacology* **5**(3), 294–313 (1985).

739. H. d'A. Heck and C. L. Kluwe, Microanalysis of urinary electrolytes and metabolites in rats ingesting dimethyl terephthalate. *J. Anal. Toxicol.* **4**(5), 222–226 (1980).

740. E. E. Vogin, Food and Drug Research Laboratories, Inc., unpublished data (1972), cited in Ref. 753.

741. National Cancer Institute Technical Report Series, *NCI-CG-TR-121, No. 121*, 1979.

742. National Cancer Institute, Bioassay of dimethyl terephthalate for possible carcinogenicity. *Technical Report Series 121, DHEW Pub No. (NIH)* 79-1376, U.S. Department of Health Education and Welfare, National Cancer Institute, Bethesda MD, 1979.

743. T. R. Lewis et al., *The Toxicologist* **2**, 7, Abstract 25 (1982).

744. Du Pont Co., Haskell Laboratory, unpublished data, MR-468-1, HL-55-58.

745. A. E. Moffitt, Jr. et al., Absorption, distribution, and excretion of terephthalic acid and dimethyl terephthalate. *J. Am. Ind. Hyg. Assoc.* **36**(8), 633–641 (1975).

746. H. d.'A., Heck, *The Toxicologist* A81 (1980).

747. S. Monarca et al., *In vitro* genotoxicity of dimethyl terephthalate. *Mutation Research* **262**(2), 85–92 (1991).

748. D. E. Lerda, Genotoxicity tests on the compounds of polyethylene glycol terephthalate (PET); dimethyl terephthalate (DMT) and terephthalic acid (TPA). *Intern. J. Environ. Health Res.* **6**(2), 125–130 (1996).

749. K. S. Loveday et al., Chromosome aberration and sister chromatid exchange tests in Chinese hamster ovary cells *in vitro*. V: Results with 46 chemicals. *Environmental and Molecular Mutagenesis* **16**, 272–303 (1990).

750. R. I. Goncharova et al., Mutagenic effects of dimethyl terephthalate on mouse somatic cells *in vivo*. *Mutation Research* **204**, 703–709 (1988).

751. Ref. 228, p. 386.

752. Z. M. Kamal' Dinova et al., *Prom. Toxikol. I Klinika Prof. Zabol. Khim. Etiol. Sb.* 159–160 (1962).

753. W. Massmann, Evaluation of the Occupational Hygiene/Toxicology of *p*-Toluic Acid Methyl Ester, Dimethyl Terephthalic and Terephthalic acid. Institute of Occupational Medicine, University of Tubingen, 1966.

754. USEPA, *Final Technical Support Document Bis(2-ethylhexyl) terephthalate* 1983.

755. Eastman Kodak Company, unpublished Report.

756. D. C. Topping et al., Peroxisome induction studies on di(2-ethylhexyl)terephthalate. *Toxicology and Industrial Health* **3**, 63–77 (1987).

757. E. D. Barber, J. A. Fox, and C. J. Giordano, Hydrolysis, absorption and metabolism of di(2-ethylhexyl)terephthalate in the rat. *Xenobiotica* **24**, 441–450 (1994).

758. E. D. Barber, Genetic toxicology testing of di(2-ethylhexyl)terephthalate. *Environmental and Molecular Mutagenesis* **23**, 228–233 (1994).

759. C. P. Carpenter, C. S. Weil, and H. F. Smyth, Jr., Range-finding toxicity data. List VIII. *Toxicol. Appl. Pharmacology* **28**(2), 313–319 (1974).

760. A 28-day toxicity study with tri(2-ethylhexyl)trimellitate in the rat. *BIBRA Project 3.0496*. CMA unpublished report, 1985.

CHAPTER EIGHTY-ONE

Esters of Carbonic and Orthocarbonic Acid, Organic Phosphorous, Monocarboxylic Halogenated Acids, Haloalcohols, and Organic Silicon

Michael S. Bisesi, Ph.D., CIH

1 INTRODUCTION

A OVERVIEW

This chapter covers (*1*) esters of carbonic and orthocarbonic acid, (*2*) esters of organic phosphorous compounds, (*3*) esters of monocarboxylic halogenated acids, alkanols, or haloalcohols, and (*4*) organic silicon esters. Other classes of esters are summarized in Chapters 79 and 80. Refer to the Introduction in Chapter 79 for a more detailed overview of general properties of esters.

Unfortunately, as shown in the two prior chapters, mainly fragmented toxicological evaluations are available for esters. Most of these esters are characterized by low toxicity.

Patty's Toxicology, Fifth Edition, Volume 6, Edited by Eula Bingham, Barbara Cohrssen, and Charles H. Powell.
ISBN 0-471-31939-2

Indeed, as expressed in Chapter 79, lethal dose (e.g., LD_{50}) values are frequently difficult or impractical to measure. Localized dermal irritation is one common effect characteristic of exposures to most organic solvents. Few esters are readily absorbed, but there are exceptions, such as tri-*o*-cresyl phosphate (TOCP). Several of the halogenated derivatives, such as ethylchloro- and ethylbromo-, are potent lacrimators. Ethyl fluoroacetate and fluoroacetic acid exhibit about the same mode of action, which may indicate that the acetate is rapidly hydrolyzed and metabolized in the mammalian system. The unsaturated carbonates are also associated with high lacrimatory activity.

TOCP is an example of an ester that can cause neuropathy in a variety of animal species. The initial weakness and paralysis are normally reversible in early stages, but repeated or massive assaults result in demyelination of the nerve fibers. The mechanism of action is not yet certain, but it appears to involve phosphorylation of proteins. Only selected phosphates exhibit neuropathic effects, including diisopropyl fluorophosphorate and *N,N′*-diisopropyl phosphorodiamidic fluoride.

B INDUSTRIAL HYGIENE EVALUTION

As was expressed in Chapter 79, industrial hygiene evaluation of esters involves collecting and analyzing air samples to determine their airborne concentrations. Published industrial hygiene air sampling and analytical methods, however, are unavailable for most esters. In relation, there are few occupational exposure and biological limits. A list of ester compounds covered in this chapter that have industrial hygiene sampling and analytical methods are presented here in Table 81.1 along with their respective occupational exposure limits, established by the American Conference of Governmental Industrial Hygienists (ACGIH), the Occupational Safety and Health Administration (OSHA), and the National Institute for Occupational Safety and Health (NIOSH) (1–4). As stated in Chapter 79, since sampling and analytical methods and occupational exposure limits are subject to periodic revision, the reader is encouraged to refer to current publications of ACGIH, OSHA, and NIOSH.

II ESTERS OF CARBONIC AND ORTHOCARBONIC ACID

A OVERVIEW

The hypothetical hydration of carbon dioxide produces two compounds, carbonic acid, $C(O)(OH)_2$, and orthocarbonic acid, $C(OH)_4$. These acids have never been isolated, although they may be formed and exist in aqueous solution. However, numerous esterified derivatives have been prepared chemically. Only very few of them have been toxicologically investigated in great detail. The majority of toxicological data reported is limited range-finding studies. Organic carbonates are chemically RO(CO)OR, the ortho esters, $RC(OR')_3$, whereby R may equal R′ or consist of alkyl, alkenyl, cyclane, or aromatic moieties. These di- and triortho esters have similar physical properties; however,

Table 81.1. Summary of Occupational Exposure Limits (OELs) and Monitoring Methods for Some Esters (1–4)

	OSHA		ACGIH			NIOSH		
Compound	PEL	STEL	TLV	STEL	Notations	REL	IDLH	NIOSH Monitoring Method
Triphenyl phosphate	3 mg/m^3	—	3 mg/m^3	—	ACGIH A4	3 mg/m^3	1000	Filter; diethyl ether; GC-FPD NIOSH #5038[a]
Methyl silicate	—	—	1 ppm	—	—	1 ppm	—	—
Ethyl silicate	100 ppm	700 ppm	10 ppm	—	—	10 ppm	700 ppm	XAD-2; CS_2; GC-FID NIOSH #5264[b]
Tri-*o*-cresyl phosphate	0.1 mg/m^3	—	0.1 mg/m^3	—	Skin; CNS cholinergic; ACGIH A4; BEI	0.1 mg/m^3	40 mg/m^3	Filter; diethyl ether; GC-FPD NIOSH #5037[a]
Trimethyl phosphite	—	—	2 ppm	—		2 ppm	—	—

[a]NIOSH MAM, 4th ed., 1994.
[b]NIOSH MAM, 2nd ed., vol. 3, 1977.

the latter resemble acetals more than carboxylates. The general toxicological impacts are also expected to be similar but nonsignificant, owing to the esters' low stability in acid solutions and their resemblance to normal mammalian metabolites. A summary of physical and chemical properties is found in Table 81.2 (5), and a summary of toxicological data is shown in Table 81.3 (6, 7).

B CARBONATES

1.0a Dimethyl Carbonate

1.0.1a CAS Number: [616-38-6]

1.0.2a Synonyms: Carbonic acid dimethyl ester

1.0.3a Trade Names: NA

1.0.4a Molecular Weight: 90.08

1.0.5a Molecular Formula: $C_3H_6O_3$

1.0.6a Molecular Structure:

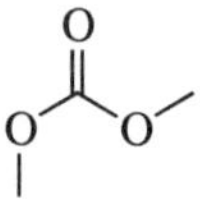

1.0b Diethyl Carbonate

1.0.1b CAS Number: [105-58-8]

1.0.2b Synonyms: Carbonic acid diethyl ester

1.0.3b Trade Names: NA

1.0.4b Molecular Weight: 118.13

1.0.5b Molecular Formula: $C_5H_{10}O_3$

1.0.6b Molecular Structure:

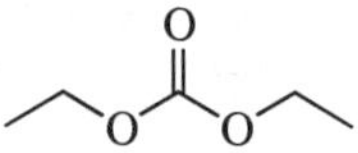

1.1 Chemical and Physical Properties

1.1.1 General

Dimethyl carbonate is miscible with most acids and alkalis, soluble in most organic solvents, but insoluble in water.

Diethyl carbonate is miscible with aromatic hydrocarbons, most organic solvents, and castor oil, but not with water.

1.1.2 Odor and Warning Properties

Dimethyl carbonate has a pleasant odor. Diethyl carbonate has a weak odor resembling that of ethyl oxybutyrate.

Table 81.2. Chemical and Physical Properties of Representative Esters of Carbonic and Orthocarbonic Acids (5)

Compound	CAS No.	Molecular Formula	MW	Boiling Point (°C) (mmHg)	Melting Point (°C)	Density	Solubility[a] in Water	Refractive Index (20°C)	Vapor density (Air = 1)	Vapor Pressure (mmHg) (°C)	Flash Pt. (°C)
Dimethyl carbonate	[*616-38-6*]	$C_3H_6O_3$	90.08	89.7	3	1.0694	i/s/s	1.3687			18
Diethyl carbonate	[*105-58-8*]	$C_5H_{10}O_3$	118.13	127	– 43	0.9752	i/s/s	1.3845	4.1	10 (23.8)	33
Ethylene carbonate	[*96-49-1*]	$C_3H_4O_3$	88.06	243(740)	36.4	1.3218	d/v/v	1.4158	3.04	0.01 (80)	145
Propylene carbonate	[*108-32-7*]	$C_4H_6O_3$	102.09	241.7	– 55	1.2057	v/v/v	1.4189		0.03 (20)	132
Triethyl orthoformate	[*122-51-0*]	$C_7H_{16}O_3$	148.20	146	– 76	0.8909	d/s/s	1.3922	5.11	10 (40.05)	30
Triethyl orthoacetate	[*78-39-7*]	$C_8H_{18}O_3$	162.23	142		0.8847	i/v/v	1.3980			36
Triethyl orthopropionate	[*115-80-0*]	$C_7H_{20}O_3$	176.26	155~160		0.886	-/v/v	1.4000			60

[a]Solubility in water: v = very soluble; s = soluble; d = slightly soluble; i = insoluble d[b]: decomposes when dissolved; – 68[c] and 84[c]: Freezing point (°C).

Table 81.3. Summary of Inhalation, Oral, Dermal, Subcutaneous, and Intraperitoneal Toxicity Data for Some Esters of Carbonic and Orthocarbonic Acids

Compound	Chemical Formula	Mode or Route of Entry	Parameter	Species	Dose or Concentration	Ref.
Dimethyl carbonate	$C_3H_6O_3$	Oral	LD_{50}	Mouse; rat	6.4–12.8 g/kg	6
		IP	LD_{50}	Mouse; rat	0.8–1.6 g/kg	6
		Dermal	LD_{50}	Guinea pig	10 mL/kg	6
		Inhale	$LD_{100/2h}$	Rat	8000 ppm	6
Ethylene carbonate	$C_3H_4O_3$	Oral	LD_{50}	Rat	10.4 g/kg	7
		Dermal	LD_{50}	Rabbit	0.20 mL/kg	7
Propylene carbonate	$C_4H_6O_3$	Oral	LD_{50}	Rabbit	20 mL/kg	6
Triethyl orthoformate	$C_7H_{16}O_3$	Oral	LD_{50}	Rat	3.2–6.4 g/kg	6
		Dermal	LD_{50}	Guinea pig	> 10 mL/kg	6
Triethyl orthoacetate	$C_8H_{18}O_3$	Oral	LD_{50}	Rat	6.4–12.8 g/kg	6
		IP	LD_{50}	Rat	12.8–25.6 g/kg	6
		Dermal	LD_{50}	Guinea pig	> 10 mL/kg	6
Triethyl orthopropionate	$C_9H_{20}O_3$	Oral	LD_{50}	Rat; rabbit	6.4–12.8 g/kg	6
		IP	LD_{50}	Rat; rabbit	6.4–12.8 g/kg	6
		Dermal	LD_{50}	Rabbit	> 10 mL/kg	6

1.2 Production and Use

The compound diethyl carbonate is manufactured via a reaction of phosgene and ethanol to produce ethyl chlorocarbonate, followed by reaction with anhydrous ethanol. It is used as a solvent for nitrocellulose, in the manufacture of radio tubes, and for fixing elements to cathodes.

1.3 Exposure Assessment

1.3.3 Workplace Methods : NA

1.4 Toxic Effects

1.4.1 Experimental Studies

1.4.1.1 Acute Toxicity. The undiluted dimethyl carbonate liquid has an oral LD_{50} in the rat and the mouse between 6.4 and 12.8 g/kg and an intraperitoneal LD_{50} in the range of 800 to 1600 mg/kg (6). Symptoms were weakness, ataxia with gasping, and unconsciousness. A dermal LD_{50} in the guinea pig was found to be greater than 10 mL/kg. Some weight loss was noted, and minimal skin absorption was suspected. However, the degree of irritation was relatively slight (6). Exposure by inhalation appeared relatively hazardous, since 8000 ppm caused rapid onset of gasping, loss of coordination, frothing from the mouth and nose, and pulmonary edema with death of all rats in a period of 2 h.

Diethyl carbonate is estimated to be moderately toxic via ingestion, dermal, and ocular contact. Intraperitoneally injected diethyl carbonate showed a slight neoplastic effect on the skin at the injection site in the mouse at about 11.4 mg, but not when administered orally at 12.5 mg (8).

1.4.1.2 Chronic and Subchronic Toxicity: NA

1.4.1.3 Pharmacokinetics, Metabolism, and Mechanisms. Metabolically, the dimethyl and diethyl carbonates may possess alkylating properties similar to those of dimethyl and ethyl sulfate.

1.4.1.4 Reproductive and Developmental. A teratogenic effect in 7.6–16.6% of 8-d pregnant hamsters was observed at a diethyl carbonate doses of 0.5–1.0 g/kg when injected intraperitoneally (9).

1.4.1.5 Carcinogenesis. A study using male and female mice treated with 0, 50, 250, or 1000 ppm (0–140 mg/kg/d) diethyl carbonate in drinking water indicated no carcinogenic effects (10).

C CYCLIC CARBONATES

2.0a Ethylene Carbonate

2.0.1a CAS Number: *[96-49-1]*

2.0.2a Synonyms: Ethylene carbonate, ethylene glycol carbonate, and 1,3-dioxolan-2-one

2.0.3a Trade Names: NA

2.0.4a Molecular Weight: 88.06

2.0.5a Molecular Formula: $C_3H_4O_3$

2.0.6a Molecular Structure: 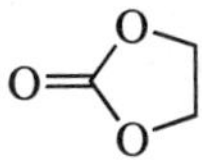

2.0b Propylene Carbonate

2.0.1b CAS Number: *[108-32-7]*

2.0.2b Synonyms: Propylene carbonate and 4-methyl-1,3-dioxolan-2-one

2.0.3b Trade Names: NA

2.0.4b Molecular Weight: 102.09

2.0.5b Molecular Formula: $C_4H_6O_3$

2.0.6b Molecular Structure: 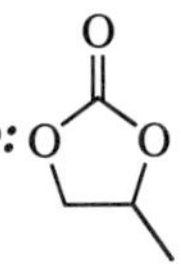

2.2 Production and Use

Used as intermediates in plastic industry.

2.3 Exposure Assessment

2.3.3 Workplace Methods : NA

2.4 Toxic Effects

2.4.1 Experimental Studies

2.4.1.1 Acute Toxicity. A range-finding study using ethylene carbonate by Smyth et al. (7) has recorded an oral LD_{50} value of 10.4 g/kg in the rat and a dermal LD_{50} of .20 mL/kg for the rabbit. Inhalation of the concentrated vapor for 8 h caused no deaths in rats. It proved to be a very low irritant to the skin, but moderately irritating to the eye of the rabbit. From a range-finding study using propylene carbonate, an oral LD_{50} in the rabbit above 20 mL/kg was determined. Inhalation of the concentrated vapor for 8 h was not lethal to rats. The undiluted material was a slight irritant to the skin and a moderate irritant to the rabbit eye.

2.4.1.2 Chronic and Subchronic Toxicity: NA

2.4.1.3 Pharmacokinetics, Metabolism, and Mechanisms. The mechanistic toxicity of ethylene carbonate was determined to be similar to the toxicity of ethylene glycol (11). Ethylene carbonate is enzymatically metabolized to ethylene glycol, and an enzyme has recently been isolated (12).

2.4.2 Human Experience

2.4.2.2.3 Pharmacokinetics, Metabolism, and Mechanisms. A dermal absorption study using living skin from human donors indicated that propylene carbonate had permeability constants of 0.7⟩0.4 g/m^2/h and 0.6⟩0.3 cm^3/m^2/h (13).

D ORTHO ACID ESTERS

3.0a Triethyl Orthoformate

3.0.1a CAS Number: *[122-51-0]*

3.0.2a Synonyms: Formic acid triethyl ester, triethoxymethane, ethyl formate (ortho), ethane, 1,1′,1″-[methylidynetris(oxy)]tris-, and ethyl orthoformate

3.0.3a Trade Names: NA

3.0.4a Molecular Weight: 148.20

3.0.5a Molecular Formula: $C_7H_{16}O_3$

3.0.6a Molecular Structure:

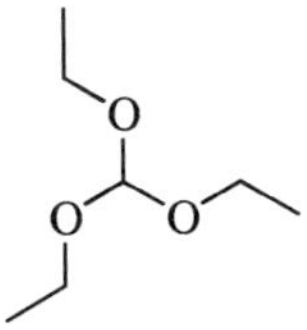

3.0b Triethyl Orthoacetate

3.0.1b CAS Number: *[78-39-7]*

3.0.2b Synonyms: Acetic acid triethyl ester and 1,1,1-triethoxyethane

3.0.3b Trade Names: NA

3.0.4b Molecular Weight: 162.23

3.0.5b Molecular Formula: $C_8H_{18}O_3$

3.0.6b Molecular Structure:

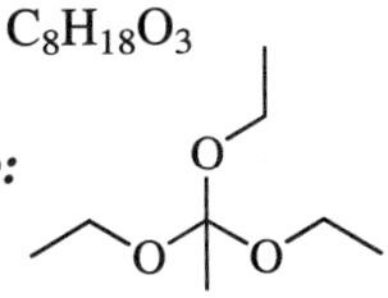

3.0c Triethyl Orthopropionate

3.0.1c CAS Number: *[115-80-0]*

3.0.2c Synonyms: Orthopropionic acid triethyl ester

3.0.3c Trade Names: NA

3.0.4c Molecular Weight: 176.26

3.0.5c Molecular Formula: $C_9H_{20}O_3$

3.0.6c Molecular Structure:

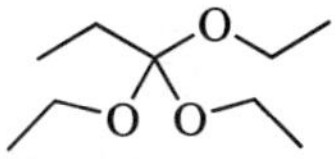

3.1 Chemical and Physical Properties

3.1.1 General

Triethyl orthoformate is a stable liquid, even in sunlight.

Triethyl orthoacetate is liquid at ambient temperature.

Triethyl orthopropionate is a highly soluble liquid.

3.3 Exposure Assessment

3.3.3 Workplace Methods : NA

3.4 Toxic Effects

3.4.1 Experimental Studies

3.4.1.1 Acute Toxicity. Oral LD_{50} values in the rat between 3.2 and 6.4 mg/kg of triethyl orthoformate have been recorded (6). Symptoms of toxicity included dyspnea and weakness. Conversely, a dermal LD_{50} in the guinea pig was found to be above 10 mL/kg, indicating that the material was practically not absorbed. No skin irritation was noted.

The oral LD_{50} for triethyl orthoacetate has been documented (6) between 6.4 and 12.8 g/kg for the rat, the intraperitoneal LD_{50} even higher, as 12.8 to 25.6 g/kg. It is not absorbed through guinea pig skin, and a dermal LD_{50} was found to be above 10 mL/kg. A very low dermal irritation was observed.

Similar to triethyl orthoacetate, the LD_{50} for triethyl orthopropionate ranges in the rat and rabbit are 6.4 to 12.8 g/kg by oral and intraperitoneal administration, and > 10 mL/kg in the rabbit when tested dermally (6).

III ESTERS OF ORGANIC PHOSPHORUS COMPOUNDS

A OVERVIEW

This class of esters includes phosphines, phosphinates, phosphonates, phosphites, and phosphates. The basic compound of the series is phosphine, PH_3, which then is successively alkylated, oxygenated, and esterified. The progressive series then contains phosphines, R_2PH or R_3P, phosphine oxide, $R_2P(O)H$ or $R_3P(O)$, phosphinic acid esters, phosphinates, $R_2P(O)OR$, phosphites, $(RO)_3P_7$, phosphonic acid esters, $RP(O)(OR')_2$, and phosphites $(RO)_3P(O)$, where R and R′ can comprise alkyl or aryl groups for any of these compounds. Some physical and chemical properties and characteristics are listed in Table 81.4 and toxicological information in Tables 81.5 (14–45) and 81.6.

Of the organic phosphorus derivatives, the aromatic phosphate esters are of greatest individual importance. They are widely used as plasticizers, flame retardants, solvents, and the lower-molecular-weight members as antifoaming agents. Phosphates and phosphites serve as stabilizers or antioxidants in the lubricant, rubber, and plastics industries. In the petroleum industry some phosphorus compounds are utilized as dispersants and detergents, such as the phospho-sulfurized hydrocarbons or nitrogen and phosphorus derivatives. The latter are primarily phosphoric acid esters. The majority of these materials are high-boiling liquids or low-melting solids. Their vapor pressures are relatively low, thus rendering the inhalation hazard almost minimal.

The stability of hydrolysis increases with increasing degree of oxygenation and esterification. Thus the acidic mono- and diesters hydrolyze more readily.

The toxicological properties of the organic phosphorous esters vary extensively, from very low to high toxicity and from neurologically inert to highly paralytic. Indeed, a review of the general physiological properties of organic phosphites and phosphates revealed groups of toxic effects. These effects range from most severe to none, as follows: (*1*) organic damage to the central nervous system with secondary flaccid or spastic paralysis, (*2*) nervous system stimulation or convulsive effects with anesthetic like action in some cases, (*3*) weak true or mainly pseudocholinesterase-inhibiting effects, (*4*) irritation to dermal or respiratory system surfaces, and (*5*) no major effects.

The initial interest in a series of these compounds stems from a widespread epidemic poisoning in the early 1930s due to tri-*o*-cresyl phosphate used in the manufacture of "Jamaica ginger," as described by Smith et al. (46). At that time, some 10,000 to 15,000

Table 81.4. Chemical and Physical Poperties of Selected Organic Phosphorous Esters

Compound	CAS No.	Molecular Formula	MW	Boiling Point (°C) (mmHg)	Melting Point (°C)	Density	Solubility[a] in Water	Refractive Index (20°C)	Vapor density (Air=1)	Vapor Pressure (mmHg) (°C)	Flash Pt. (°C)
Tributyl phosphine	*[998-40-3]*	$C_{12}H_{27}P$	202.32	240~242		0.812					40
Triisooctyl phosphine	*[10138-88-2]*	$C_{24}H_{51}P$	370.64								
Triphenyl phosphine	*[603-35-0]*	$C_{18}H_{15}P$	262.29	377	80	1.132	i	1.6358	9.0		181
Octyl *O*-butyl phosphinate		$C_{12}H_{27}O_2P$	234.32								
Phenyl *O*-ethyl phosphinate	*[2511-09-3]*	$C_8H_{11}O_2P$	170.15								
Butyl phenyl *O*-allyl phosphinate		$C_{13}H_{19}O_2P$	238.29								
O-Dimethyl phosphonate	*[868-85-9]*	$C_2H_7O_3P$	110.05	170.5		1.1029	s	1.4128			
Methane *O*-dimethyl phosphonate		$C_3H_9O_3P$	124.08			1.1507		1.4099			
O-Ethyl propyl phosphonate	*[21921-96-0]*	$C_5H_{13}O_3P$	152.13	181							
Diisopropyl phosphonate	[1809-20-7]	$C_6H_{15}O_3P$	166.16	72~75 (10)		0.997					
Octyl *O*-dibutyl phosphonate	*[5929-67-9]*	$C_{16}H_{35}O_3P$	306.42								
Dibutyl phenyl phosphonate		$C_{14}H_{23}O_3P$	222.31								
Phenyl propyl 2-propynyl phosphonate	*[18705-22-1]*	$C_{12}H_{15}O_3P$	238.22								
Vinyl, bis (2-chloroethyl) phosphonate		$C_6H_{11}Cl_2O_3P$	233.03								

Table 81.4. (*Continued*)

Compound	CAS No.	Molecular Formula	MW	Boiling Point (°C) (mmHg)	Melting Point (°C)	Density	Solubility[a] in Water	Refractive Index (20°C)	Vapor density (Air=1)	Vapor Pressure (mmHg) (°C)	Flash Pt. (°C)
Methyl ethyl chlorophenyl phosphonate	[*033232-85-8*]	$C_9H_{12}ClO_3P$	234.62								
Diisopropyl fluoro phosphonate	[*33232-85-8*]	$C_6H_{14}FO_3P$	214.20								
Ethyl *O*-methyl *O*-*p*-nitrophenyl phosphonate	[*25536-01-3*]	$C_9H_{12}NO_5P$	245.19								
Ethyl *O*-ethyl *O*-*p*-nitrophenyl phosphonate	[*546-71-4*]	$C_{10}H_{14}NO_5P$	259.22								
Methyl *O*-isopropyl *O*-*p*-nitrophenyl phosphonate	[*15536-03-5*]	$C_{10}H_{14}NO_5P$	259.22								
Ethyl *O*-isopropyl *O*-*p*-nitrophenyl phosphonate		$C_{11}H_{16}NO_5P$	273.25								
Isopropyl *O*-ethyl *O*-*p*-nitrophenyl phosphonate		$C_{11}H_{16}NO_5P$	273.25								
Isopropyl *O*-isopropyl *O*-*p*-nitrophenyl phosphonate	[*7284-60-8*]	$C_{12}H_{18}NO_5P$	287.28								
Pentyl *O*-ethyl *O*-*p*-nitrophenyl phosphonate	[*3015-75-6*]	$C_{13}H_{20}NO_5P$	301.31								
Heptyl *O*-ethyl *O*-*p*-nitrophenyl phosphonate		$C_{15}H_{24}NO_5P$	329.37								

Octyl *O*-ethyl *O*-*p*-nitrophenyl phosphonate		$C_{16}H_{26}NO_5P$	343.40								
Ethyl *O*-phenyl *O*-*p*-nitrophenyl phosphonate		$C_{14}H_{14}NO_5P$	335.32								
Phenylpropyl *O*-ethyl *O*-*p*-nitrophenyl phosphonate		$C_{17}H_{20}NO_5P$	349.32								
Phenylbutyl *O*-ethyl *O*-*p*-nitrophenyl phosphonate		$C_{18}H_{22}NO_5P$	363.35								
Phenylvinyl *O*-ethyl *O*-*p*-nitrophenyl phosphonate		$C_{16}H_{16}NO_5P$	333.30								
O-ethyl propylphenyl *O*-*p*-nitrophenyl phosphonate		$C_{17}H_{20}NO_5P$	349.32								
Ethyl butylphenyl *O*-*p*-nitrophenyl Phosphonate		$C_{18}H_{22}NO_5P$	363.35								
Diethyl phosphite	*[762-04-9]*	$C_4H_{11}O_3P$	138.10	50~51(2)		1.0720	d	1.4101	4.76		90
Cyclic ethylene phosphite		$C_2H_4O_3P$	107.72								
Dibutyl phosphite	*[1809-19-4]*	$C_8H_{19}O_3P$	194.21	118~119 (11)		0.995			6.7	1 (20)	121
Trimethyl phosphite	*[121-45-9]*	$C_3H_9O_3P$	124.08	111–112	– 78	1.046	d.i.	1.4095	4.3	24 (25)	37.8
Triethyl phosphite	*[122-52-1]*	$C_6H_{15}O_3P$	166.16	156		0.97	d^b	1.4127			52
Tripropyl phosphite	*[923-99-9]*	$C_9H_{21}O_3P$	208.24	206–207		0.9417		1.4282			
Triisopropyl phosphite	*[116-17-6]*	$C_{12}H_{21}O_3P$	208.24	63~64 (11)		0.844		1.4085			73
Tributyl phosphite	*[102-85-2]*	$C_{12}H_{27}O_3P$	250.32	118~125 (7)	– 80	0.9259	d	1.4321			121
Triisosooctyl phosphite	*[25103-12-1]*	$C_{24}H_{51}O_3P$	418.64								
Tridecyl phosphite		$C_{30}H_{63}O_3P$	502.80		< 0	0.892					235

Table 81.4. *(Continued)*

Compound	CAS No.	Molecular Formula	MW	Boiling Point (°C) (mmHg)	Melting Point (°C)	Density	Solubility[a] in Water	Refractive Index (20°C)	Vapor density (Air=1)	Vapor Pressure (mmHg) (°C)	Flash Pt. (°C)
Triphenyl phosphite	*[101-02-0]*	$C_{18}H_{15}O_3P$	310.29	360	22~24	1.1844	i	1.5900			218.3
Tri-*o*-cresyl phosphite		$C_{21}H_{21}O_3P$	352.7	238 (11)		1.1423	d	1.5740			
Tri-*p*-cresyl phosphite		$C_{21}H_{21}O_3P$	352.7	250–255 (10)		1.1313		1.5703			
2-Ethylhexyl octylphenyl phosphite	*[7346-61-4]*	$C_{22}H_{39}O_3P$									
Nonyl diphenyl phosphite		$C_{21}H_{29}O_3P$	360.43								
Tris(2-chloroethyl) phosphite	*[140-08-9]*	$C_6H_{12}Cl_{13}O_3$-P	269.50	125~135 (7)		1.3348	d[b]	1.5174	9.32	< 1 (20)	137.7
Trimethyl phosphate	*[512-56-1]*	$C_3H_9O_4P$	140.08	197.2	− 46	1.197	v	1.3967			
Triethyl phosphate	*[78-40-0]*	$C_6H_{15}O_4P$	182.16	215	− 56.4	1.072	s	1.4053	6.28	1 (39.6)	130
Tributyl phosphate	*[126-73-8]*	$C_{12}H_{27}O_4P$	266.32	289	− 80	0.976	s	1.4224	9.2	0.067 (125)	146
Trioctyl phosphate	*[1806-54-8]*	$C_{24}H_{51}O_4P$	434.64	220~230	− 74		i				
Di(2-ethylhexyl) phosphate	*[298-07-7]*	$C_{16}H_{35}O_4P$	322.42	48 (12)	− 60						137
Tri(2-ethylhexyl) phosphate	*[78-42-2]*	$C_{24}H_{51}O_4P$	434.64	190~233	− 70	0.9262	d			0.23 (150)	
3,5,5-Trimethylhexyl phosphate		$C_9H_{21}O_4P$	518.80								
Diundecyl phosphate		$C_{22}H_{47}O_4P$	406.60								
Triallyl phosphate	*[1623-19-4]*	$C_9H_{15}O_4P$	218.19								
Triphenyl phosphate	*[115-86-6]*	$C_{18}H_{15}O_4P$	326.28	245 (11)	49	1.2055	i		1.19	1 (193.5)	220
Tri-*o*-cresyl phosphate	*[78-30-8]*	$C_{21}H_{21}O_4P$	368.37	420	− 33	1.1955	i	1.5575	12.7	10 (265)	410
Tri-*m*-cresyl phosphate	*[563-04-2]*	$C_{21}H_{21}O_4P$	368.37	275(17)	25.6	1.150	i	1.5575			210

Tri-cresyl phosphate	[*1330-78-5*]	$C_{21}H_{21}O_4P$	368.37	420	– 33	1.247		12.7		410
Tri-*p*-cresyl phosphate	[*78-32-0*]	$C_{21}H_{21}O_4P$	368.37	244 (3.5)	77–78	1.16	i			410
Dimethyl benzyl phosphate		$C_9H_{13}O_4P$	216.17							
2-Ethylhexyl diphenyl phosphate	[*1241-94-7*]	$C_{20}H_{27}O_4P$	362.44	330	– 80	1.0884	i		0.5 (100)	
Cresyl diphenyl phosphate	[*26444-49-5*]	$C_{19}H_{17}O_4P$	340.31	235~255	– 38	1.2				232
Di(2-chloroethyl) phosphate		$C_4H_9Cl_2O_4P$								
Tri(2-chloroethyl) phosphate	[*115-96-8*]	$C_6H_{12}Cl_3O_4P$	285.49	330	– 55	1.39	i		0.5 (145)	232
2,2-Dichlorovinyl dimethyl phosphate	[*62-73-7*]	$C_4H_7Cl_2O_4P$	220.98	140 (20)	– 60	1.415	d		0.012 (20)	
2-Chlorovinyl diethyl phosphate	[*311-47-7*]	$C_6H_{12}ClO_4P$	214.85							

[a]Solubility in water: v = very soluble; s = soluble; d = slightly soluble; i = insoluble.
[b] Decomposes when dissolved.
[c] Freezing point (°C).

Table 81.5. Summary of Inhalation, Oral, Dermal, Subcutaneous, and Intraperitoneal Toxicity Data for Some Organic Phosphorus Esters

Compound	Chemical Formula	Route of Entry	Parameter	Species	Results (g/kg)	Ref.
Triisooctyl phosphine	$P[CH_2CH(CH_3)(CH_2)_4CH_3)]_3$	Oral	LD_{50}	Rat	21.4	7
		Dermal	LD_{50}	Rabbit	3.97	7
Triphenyl phospine	$P(C_6H_5)_3$	Oral	LD_{50}	Rat	0.80–1.6	6
		Oral	LD_{50}	Rat	4.29	14
		Oral	LD_{50}	Mouse	0.8–1.6	15
		Dermal	LD_{50}	Rabbit	> 5.0	14
		IP	LD_{50}	Rat	1.6–3.2	6
		IP	LD_{50}	Mouse	0.8–1.6	6
		Satd. inhal.	DT	Rat	8 h	14
Phosphinate						
Octyl butyl	$C_8H_{17}PH(O)OC_4H_9$	Oral	LD_{50}	Mouse	2.120	16
Phenyl ethyl	$C_6H_5PH(O)OC_2H_5$	IP	LD_{50}	Mouse	0.625	17
Allyl phenyl butyl	$(CH_2{:}CHCH_2)(C_6H_5)P(O)OC_4H_9$	Oral	LD_{50}	Mouse	0.760	16
Butyl phenyl allyl	$(C_4H_9)(C_6H_5)P(O)OCH_2CH{:}CH_2$	Oral	LD_{50}	Mouse	1.060	16
Phosphonate						
Monomethyl	$HP(O)(OH)OCH_3$	Oral	LD_{50}	Rat	1.740	18
Dimethyl	$HP(O)(OCH_3)_2$	Oral	LD_{50}	Rat	4.250	18
Propyl monoethyl	$C_3H_7P(O)(OH)OC_2H_5$	Oral	LD_{LO}	Rat	2.300	19
Octyl dibutyl	$C_8H_{17}P(O)(OC_4H_9)_2$	IP	LD_{LO}	Mouse	0.250	17
Ethyl methyl tolyl	$C_2H_5P(O)(OCH_3)(OC_6H_5CH_3)$	Oral	LD_{50}	Rat	0.475	17
Phenyl dibutyl	$C_6H_5P(O)(OC_4H_9)_2$	Unknown	LD_{50}	Mouse	1.980	16
Phenyl propyl 2-propynyl	$C_6H_5P(O)(OC_3H_7)(OCH_2C{:}CH_2)$	Oral	LD_{50}	Rat	0.362	20
		Dermal	LD_{50}	Rabbit	2.650	20
Phosphonate *p*-nitrophenyl						
Ethyl ethyl	$(C_2H_5)P(O)(OC_2H_5)(O\text{-}p\text{-}C_4H_4NO_2)$	Unknown	LD_{50}	Mouse	0.54	21
Methyl isopropyl	$(CH_3)P(O)[OCH(CH_3)_2](O\text{-}p\text{-}C_6H_4NO_2)$	Oral	LD_{50}	Mouse	5.0	19
Ethyl isopropyl	$(C_2H_5)P(O)[OCH(CH_3)_2](O\text{-}p\text{-}C_6H_4NO_2)$	Oral	LD_{LO}	Mouse	5.0	19

Isopropyl ethyl	$(C_3H_7)P(O)[OCH(C_2H_5)](O\text{-}p\text{-}C_6H_4NO_2)$	Oral	LD_{LO}	Mouse	21.0	19
Isopropyl isopropyl	$(C_3H_7)P(O)[OCH(CH_3)_2](O\text{-}p\text{-}C_6H_4NO_2)$	Oral	LD_{LO}	Mouse	300.0	19
Pentyl ethyl	$(C_5H_{11})P(O)(OC_2H_5)(O\text{-}p\text{-}C_6H_4NO_2)$	IV	LD_{50}	Rabbit	0.118	19
		IV	LD_{50}	Guinea pig	0.299	22
Heptyl ethyl	$(C_7H_{15})P(O)(OC_2H_5)(O\text{-}p\text{-}C_6H_4NO_2)$	IV	LD_{50}	Rabbit	0.550	22
		IV	LD_{50}	Guinea pig	1.601	22
Octyl ethyl	$(C_8H_{17})P(O)(OC_2H_5)(O\text{-}p\text{-}C_6H_4NO_2)$	IV	LD_{50}	Rabbit	1.202	22
		IV	LD_{50}	Guinea pig	2.603	22
Ethyl phenethyl	$(C_2H_5)P(O)(OC_6H_5C_2H_4)(O\text{-}p\text{-}C_6H_4NO_2)$	IV	LD_{50}	Rabbit	0.349	22
		IV	LD_{50}	Guinea pig	0.919	22
Phenylpropyl ethyl	$(C_6H_5C_3H_6)P(O)(OC_2H_5)(O\text{-}p\text{-}C_6H_4NO_2)$	IV	LD_{50}	Rabbit	0.350	22
		IV	LD_{50}	Guinea pig	0.880	22
Phenylbutyl ethyl	$(C_6H_5C_4H_9)P(O)(OC_2H_5)(O\text{-}p\text{-}C_6H_4NO_2)$	IV	LD_{50}	Rabbit	0.987	22
		IV	LD_{50}	Guinea pig	0.149	22
Phenylvinyl ethyl	$(C_6H_5CH{:}CH)P(O)(OC_2H_5)CO\text{-}p\text{-}C_6H_4NO_2)$	IV	LD_{50}	Rabbit	1.230	22
		IV	LD_{50}	Guinea pig	1.499	22
Phosphite						
Trimethyl	$P(OCH_3)_3$	Oral	LD_{50}	Rat	2.890	23
		Dermal	LD_{LO}	Rabbit	2.200	23
Diethyl	$(HO)P(OC_2H_5)_2$	Oral	LD_{50}	Rat	5.190	24
		Dermal	LD_{50}	Rabbit	2.020	24
Dibutyl	$(HO)P(OC_4H_9)_2$	Oral	LD_{50}	Rat	3.200	15
		Dermal	LD_{50}	Rat	2.000	15
Triethyl	$P(OCH_2CH_3)_3$	IP	LD_{50}	Mouse	>0.5	6
Tripropyl	$P[O(CH_2)_2CH_3]_3$	IP	LD_{50}	Mouse	0.250–0.500	6
Triisopropyl	$P[OCH(CH_3)_2]_3$	IP	LD_{50}	Mouse	0.500	6
Tributyl	$P(OC_4H_9)_3$	Oral	LD_{50}	Rat	3.000	6
		Unknown	LD_{50}	Rat	1.640	16
Tri-*sec*-butyl	$P[OCH(CH_3)(CH_2CH_3)]_2$	IP	LD_{50}	Mouse		6
Trihexyl	$P[O(CH_2)_5CH_3]_3$	IP	LD_{50}	Mouse		25
Triisooctyl	$P[OCH_2CH(CH_3)(CH_2)_4CH_3]_3$	Dermal	LD_{50}	Rabbit	3.970	7
Tridecyl	$P[O(CH_2)_9CH_3]_3$	Dermal	LD_{50}	Rat		23

Table 81.5. *(Continued)*

Compound	Chemical Formula	Route of Entry	Parameter	Species	Results (g/kg)	Ref.
Triphenyl	$P(OCH_6H_5)_3$	Oral	LD_{50}	Rat	1.600–3.200	6
					0.800–1.600	6
		SC	LD_{LO}	Rat	2.000	26
			LD_{LO}	Cat	0.300	26
		Oral	LD_{LO}	Chicken	1.000	26
Tri-*o*-cresyl	$P(O\text{-}2\text{-}CH_3C_6H_4)_3$	Oral	LD_{LO}	Chicken	1.000	26
		SC	LD_{LO}	Rat	0.010	26
			LD_{50}	Cat	0.100	6
Tri-*m*-cresyl	$P(O\text{-}3\text{-}CH_3C_6H_4)_3$	Oral	LD_{LO}	Chicken	2.000	26
		SC	LD_{LO}	Rat	3.000	26
			LD_{LO}	Cat	0.300	26
Tri-*p*-cresyl	$P(O\text{-}4\text{-}CH_3C_6H_4)_3$	SC	LD_{LO}	Rat	3.000	26
			LD_{LO}	Cat	0.200	26
2-Ethylhexyl octyl phenyl	$P[OCH_2CH(CH_2CH_3)(CH_2)_3CH_3](OC_8H_{17})(OC_6H_5)$	Oral	LC_{50}	Rat	7.0–10.0	6
Diphenyl nonyl	$P(OC_6H_5)_2(OC_9H_{19})$	Oral	LD_{50}	Mouse	1.0	27
bis(2-Chloroethyl)	$P(OH)(OCH_2CH_2Cl)_2$	IP	LD_{50}	Mouse	.25	17
Phosphate						
Trimethyl	$PO(OCH_3)_3$	Oral	LD_{50}	Rat		28
			LD_{50}	Rabbit		28
			LD_{50}	Guinea pig		28
		Dermal	LD_{50}	Rabbit	2.83	24
		IP	TD_{50}	Mouse	0.700	29
Triethyl	$PO(OCH_2CH_3)_3$	Oral	LD_{50}	Rat/Rabbit	~1.60	23
			LD_{50}	Mouse	> 1.5	6
		Dermal	LD_{50}	Rabbit	> 20.0	23
		IP	LD_{50}	Rat/Rabbit	0.8	23
		IP[b]	TD_{LO}	Mouse	0.660	29

Tributyl	$PO(OC_4H_9)_3$	Oral	LD_{50}	Rat	3.0	30
		Oral	LD_{50}	Hen	1.86	31
		IP	LD_{50}	Rat	0.8–1.6	6
Di(2-ethylhexyl)-hydroxypropyl	$PO[OCH_2CH(C_2H_5)(CH_2)_3CH_3]_2[O(CH_2)_3OH]$	Oral	LD_{50}	Rat		7
Tri(2-ethylhexyl)	$PO[OCH_2CH(C_2H_5)(CH_2)_3CH_3]_3$	Oral	LD_{50}	Rat	37.08	23
		Dermal	LD_{50}	Rabbit		23
		Eye	Irrit.	Rabbit	0.01–0.05[a]	23
					0.1–0.5[b]	23
Triallyl	$PO(OCH_2CH{:}CH_2)_3$	IP	LD_{50}	Mouse	0.25–0.5	6
		IV	LD_{50}	Mouse	0.071	6
Triphenyl	$PO(OC_6H_5)_3$	Oral	LD_{50}	Rat	> 6.4	6
			LD_{50}	Mouse	> 3.0	32
			LD_{50}	Mouse	1.3	33
			LD_{50}	Guinea pig	>4.0	32
			LD_{50}	Chicken	> 2.0	6
		SC	LD_{50}	Cat	0.1–0.2	6
Tri-cresyl (mixed isomers)	$PO(OC_6H_{45}CH_3)_3$	Oral	LD_{LO}	Rat	4.68	18
			LD_{LO}	Rabbit	0.10	27
			LD_{LO}	Dog	0.50	27
Tri-*o*-cresyl	$PO(OC_6H_4CH_3)_3$	Oral	LD_{50}	Rat	3–10	6
			LD_{50}	Chicken	0.1–0.2	6
		SC	LD_{50}	Cat	0.1–0.2	34
		IP	LD_{LO}	Mouse	0.050	35
Tri-*p*-cresyl		Oral	LD_{50}	Rat	> 12.8	6
			LD_{50}	Chicken	> 2.0	6
		SC	LD_{50}	Cat	> 1.0	6
Di-*o*-cresyl	$PO(OH)(OC_6H_5CH_3)_2$	IV	LD_{LO}	Rat	0.8	36
			LD_{LO}	Rabbit	0.5	36
			LD_{LO}	Cat	0.3	36
Cresyl diphenyl	$PO(OC_6H_5)_2(OC_6H_5CH_3)$	Oral	LD_{50}	Rat	6.4–12.8	6
				Mouse	6.4–12.8	6
				Guinea pig	1.6–3.2	6

Table 81.5. (*Continued*)

Compound	Chemical Formula	Route of Entry	Parameter	Species	Results (g/kg)	Ref.
2-Ethylhexyl diphenyl	$PO(OC_6H_5)_2[OCH_2CH(C_2H_5)(CH_2)_3CH_3]$	Oral	LD_{50}	Rat	> 24.0	37
				Rabbit	> 24.0	37
		IV	LD_{50}	Rabbit	0.24	37
		Dermal	LD_{50}	Rabbit	13.7	37
		Oral	LD_{00}	Rat	0.0625%	37
Tri-(2-chloroethyl)	$PO(OCH_2CH_2Cl)_3$	Oral	LD_{50}	Rat	1.410	38
2-2-Dichlorovinyl dimethyl	$PO(OCH_3)_2(OCH:CCl_2)$	Oral	LD_{50}	Dog	1090	39
			LD_{50}	Chicken	15	40
		Dermal	LD_{50}	Rat	75	41
				Rabbit	107	42
		SC	LD_{50}	Rat	12	42
		IP	LD_{LO}	Mouse	17	43
			TD_{Lo}	Mouse	15	44
2-Chlorovinyl diethyl	$PO(OCH_2CH_3)_2(OCH:CHCl)$	Oral	LD_{50}	Rat	10	45
			LD_{50}	Mouse	33	45
			LD_{50}	Rabbit	3.37	45
		Dermal	LD_{50}	Rabbit	18	45
		IP	LD_{50}	Rat	9	45
Chlorophosphoric acid diethyl ester	$P(Cl)(O)(OCH_2CH_3)_2$	Oral	LD_{50}	Rat		7
		Dermal	LD_{50}	Rabbit		38

[a] 2–yr chronic no-effect level.
[b] 8–wk feeding study.

Table 81.6. Irritant or Neurotoxic Effects of Phosphines, Phosphites, and Phosphates (6)

Compound	Species	CNS Effects	Cholinesterase Changes	Skin and Eye Effects
Triphenyl phosphine	All tested	No paralysis; only symptoms are weakness and ataxia	Not reported	Slight irritant to guinea pig skin
Phosphite				
Tributyl	All tested	No paralysis reported; no deaths after inhalation of saturated vapor for 8 h	Not reported	Slight eye and skin irritant
Triphenyl	All tested	Tremors, diarrhea, vasodilation by all routes; no paralysis in survivors	Inhibition in whole blood of mice	Skin and eye irritant; absorbed through guinea pig skin
Tri-*o*-cresyl	Rat	Tremors, but no paralysis	Inhibition in fowl plasma *in vivo*	
	Cat	Flaccid paralysis; extensor rigidity		
	Monkey	Flaccid paralysis; extensor regidity		
	Chicken	Lower effect		
Phosphate				
Trimethyl	Rabbit Cat	Oral 0.3 mL/kg/d × 6 d and dermal 2.0 mL/kg/d × 20 doses on skin caused flaccid and spastic paralysis; SC 0.2 mL/kg/d × 79 d produced loss of weight and weakness	Not reported	None on rabbit skin
Triethyl	Rat	IP 400 mg/kg × 37 d caused peritoneal irritation, ascites, anesthesia, no paralysis. Inhalation of 28,000 ppm × 6 h–3/3 deaths, weakness, gasping respirations	Weak *in vitro* rat brain inhibition	Slight irritant to guinea pig skin
Tributyl	Rat	Large oral or IP doses caused weakness, dyspnea, pulmonary edema, twitching; no paralysis in survivors; inhalation of 123 ppm × 6 h — 0/3 deaths	Weak *in vitro* inhibition human RBC and plasma CE	Strong skin and respiratory irritant
	Guinea pig	IP 1 cm^3/kg/d × 4 d — 2/3 deaths		
Triisobutyl	Rat	Same as tri-*n*-butyl phosphate, no paralysis in survivors; inhalation of 122 ppm × 6 h caused 0/3 deaths	Not reported	
Tri-2-ethyl hexyl	Mouse Rat	Delayed deaths; no paralysis	Not reported	Not a skin irritant in guinea pig; no irritation in rabbit eye

Table 81.6. *(Continued)*

Compound	Species	CNS Effects	Cholinesterase Changes	Skin and Eye Effects
Tri-2-ethyl isohexyl	All tested	No paralysis	Not reported	Moderate skin irritation in guinea pig
Triallyl	Mouse	IP 0.5 g/kg caused 5/5 deaths, ataxia, dyspnea; no paralysis; no delayed deaths	Not reported	Not reported but probably an irritant
Tri-2-chloroethyl	Rat	Oral or IP LD_{100} caused convulsions; IP 80–125 mg/kg × 37 d; no paralysis; hemorrhagic effect at high dose level	Weak *in vitro* rat inhibition	None on guinea pig
Triphenyl	Rat	No effect	Weak inhibition of RBC acetylcholine esterase, but not plasma esterases in humans	Not a skin irritant; not absorbed through skin appreciably
	Chicken	No effect		
	Cat	Flaccid paralysis, delayed onset, no cholinergic symptoms		
2-Ethylhexyl diphenyl	Rat	Chronic 2-yr feeding caused rapid loss of weight and death at 5%; retarded growth at 1%; no effects at 0.125%; no paralysis or CNS symptoms	Not reported	Not a skin irritant to human
Cresyl diphenyl	All tested	No paralysis; some dyspnea at high doses	Not reported	Not a skin irritant; not absorbed through guinea pig skin
Tri-*o*-cresyl	Human, rabbit, cat, dog, chicken, primate, calf	Flaccid or spastic paralysis; minimum oral dose in humans about 10–30 mg/kg	Probably inhibits pseudocholine esterases	Not a skin irritant; 0.1–0.4% of dose applied to human skin is absorbed
Tri-*p*-cresyl	All tested	No paralysis or neurological symptoms with pure ester	No effects on choline esterases	Not a skin irritant

people developed a neuromuscular disturbance characterized by flaccid paralysis, from which recovery was slow or which resulted in permanent disability (34). Since then, several epidemics have occurred, such as one in Switzerland due to an accidental mixing of food and hydraulic oil (47), an incident in 1959 in Morocco due to food oil adulteration (48) with hydraulic jet oil containing 3% cresyl phosphates (49), or the accidental contamination of drinking water stored in paint drums in Durban, South Africa (50).

The absorption of the phosphorous esters discussed in this section varies widely, as do the vascular translocation, tissue affinity, hydrolysis, biologic degradation, and excretion. It is apparent that some of the compounds are readily metabolized and excreted. Some of the aryl phosphites and phosphates that are more resistant to hydrolysis appear to bear the neurotoxic properties. This has been shown by the low excretion of phenols (46).

According to Fest and Schmidt (51), the phosphites and phosphates can be metabolized by several different pathways. Some possibilities might be alkylation or arylation potentials, or phosphorylation of tissue or cell components. However, it does not appear that the phosphorus esters have an affinity to nuclear material or affect cell propagation mechanisms. On the contrary, Lauwerys and Buchet (52) have shown a phosphoric diamide to reduce 7,12-dimethylbenzanthracene-induced mammary carcinogenicity. Other indications are molecular interference with nerve conduction by cholinesterase inhibition and related mechanisms. There is also the possibility of metabolic interferences with various catabolic pathway systems; an example would constitute the diisopropyl phosphofluoridate inhibition of pyruvate oxidation (53), decreased tributyrinase activity on the spinal cords of hens (54),and also retarded lipid resynthesis (55). An additional explanation for the delayed neurotoxicity may be decreased enzyme activities that affect rate-limiting enzyme conversions. Certain detoxification enzymes may occur in very small quantities or may have to be stimulated for activity. This, therefore, may present the secondary nonconverted substrate that may form the toxic metabolites and limiting conversion rates, thus causing the delayed neurotoxicity. The effects involved appear to be absorption, transport to neurological centers, transformation into active metabolites, and inhibition of various enzymes, such as cholinesterase, pseudocholinesterase, and various other esterases. There is a structure–action relationship, but it currently still appears too complex to explain the subtle structural changes affecting negative or positive neurotoxicity (56). Additional data have revealed that exposure to nontoxic doses of organophosphate esters can inhibit noncritical tissue esterases and alter metabolism and toxicity of select ester and amide compounds (57).

B ALKYL AND ARYL PHOSPHINES

4.0a Tributyl Phosphine

4.0.1a CAS Number: *[998-40-3]*

4.0.2a Synonyms: NA

4.0.3a Trade Names: NA

4.0.4a Molecular Weight: 202.32

4.0.5a Molecular Formula: $C_{12}H_{27}P$

4.0.6a Molecular Structure:

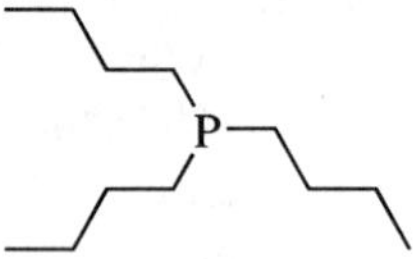

4.0b Triphenyl Phosphine

4.0.1b CAS Number: [603-35-0]

4.0.2b Synonyms: NA

4.0.3b Trade Names: NA

4.0.4b Molecular Weight: 262.29

4.0.5b Molecular Formula: $C_{18}H_{15}P$

4.0.6b Molecular Structure:

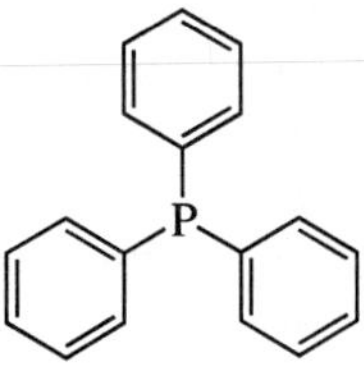

4.0c Triisooctyl Phosphine

4.0.1c CAS Number: [10138-88-2]

4.0.2c Synonyms: NA

4.0.3c Trade Names: NA

4.0.4c Molecular Weight: 370.64

4.1 Chemical and Physical Properties

4.1.1 General

4.1.2 Odor and Warning Properties

4.2 Production and Use

Tributylphosphine is used as a gasoline additive.

4.4 Toxic Effects

4.4.1 Experimental Studies

4.4.1.1 Acute Toxicity. The overall toxicity is expected to be relatively high (58); however, a moderate degree of toxic effects for triphenylphosphine and low toxicity for triisooctylphosphine has been demonstrated (6). This is also confirmed by an oral LD_{50} of

4.29 mL/kg (14). Conversely, according to Carpenter et al. (14), an inhalation exposure of rats to saturated vapor required 8 h for lethal effects. A dermal LD_{50} in the rabbit was greater than 5 mL/kg. Triisooctylphosphine appears to be less toxic by the oral route at 21.4 mL/kg, but is more readily absorbed through the skin, showing an LD_{50} of 3.97 mL/kg (7). It is also more of an irritant to the skin, but less so to the eye (7).

C PHOSPHINATES

Alkyl aryl phosphinic acid esters bear the chemical structure $R_2P(O)OR'$. A paucity of toxicological data were found. Some data are summarized in Table 81.5 for the following phosphinates: octylphosphinic acid *O*-butyl ester, $C_8H_{17}PH(O)OC_4H_9$, phenylphosphinic acid *O*-ethylester, $C_6H_5PH(O)OC_2H_5$, allylphenylphosphinic acid *O*-butyl ester, $(CH_2{:}CHCH_2)(C_6H_5)P(OH)OC_4H_9$, and butylphenylphosphinic acid *O*-allyl ester, $(C_4H_9)(C_6H_5)P(O)OCH_2CH{:}CH_2$.

D PHOSPHONATES

Phosphonates are esters of phosphoric acid of the type $RP(O)(OR')_2$, whereby R and R′ can constitute alkyl, aryl, cyclic, or heterocyclic groups. They are less acidic and hydrolyze with less ease than the corresponding phosphates (51). The phosphonates often occur as mixed products, unless they are prepared through their ester anhydrides (51). Some derivatives, especially the halogenated types, are used as pesticides.

Various aliphatic, aromatic, and substituted phosphonates appear to be moderately to highly toxic. The toxicity appears to be lowest for the alkyl derivatives, but rises with increasing aromatization and even further with halogen or nitro group substitution. Diisopropyl fluorophosphonate, $FP(O)[OCH(CH_3)_2]_2$, is neurotoxic against the housefly (59).

E ALKYL AND ARYL PHOSPHITES

Organic phosphites are esters of phosphorous acid and alkyl or aryl alcohols, of the type $(RO)_3P$. Chemically, they can undergo transesterification with alcohols (51). Such esters in the presence of allyl halides tend to convert easily from the trivalent to the pentavalent form, the thermodynamically more stable phosphonates.

The biologic properties of the organic phosphites resemble closely those of the phosphonates, possibly owing to their chemical similarity or transformation possibilities. Most animal experiments have been carried out with rats or rabbits, which are not particularly sensitive to neurotoxic stimuli. Thus no definite toxicologic conclusions can be drawn from these experiments. The differential irritancy characteristics may be due to the hydrolyzability of the individual esters.

5.0a Trimethyl Phosphite

5.0.1a CAS Number: [121-45-9]

5.0.2a Synonyms: Phosphorous acid trimethyl ester

5.0.3a Trade Names: NA

5.0.4a Molecular Weight: 124.08

5.0.5a Molecular Formula: $C_3H_9O_3P$

5.0.6a Molecular Structure:

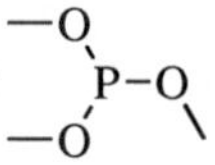

5.0b Tridecyl Phosphite

5.0.1b CAS Number: *[2929-86-4]*

5.0.2b Synonyms: Phosphorous acid tridecyl ester

5.0.3b Trade Names: NA

5.0.4b Molecular Weight: 502.80

5.0.5b Molecular Formula: $C_{30}H_{63}O_3P$

5.0.6b Molecular Structure:

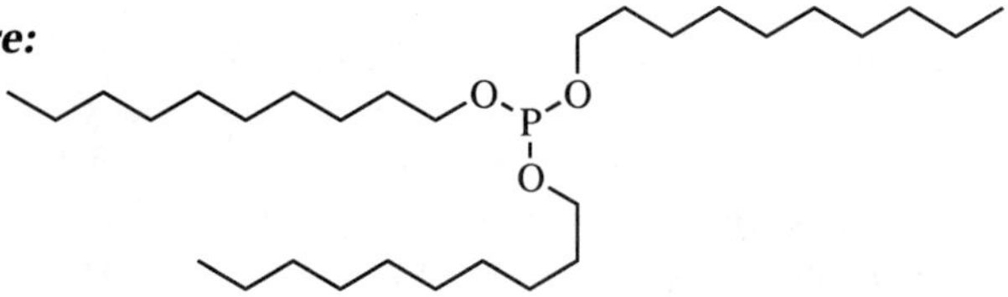

5.0c Triphenyl Phosphite

5.0.1c CAS Number: *[101-02-0]*

5.0.2c Synonyms: Phosphorous acid triphenyl ester

5.0.3c Trade Names: NA

5.0.4c Molecular Weight: 310.29

5.0.5c Molecular Formula: $C_{18}H_{15}O_3P$

5.0.6c Molecular Structure:

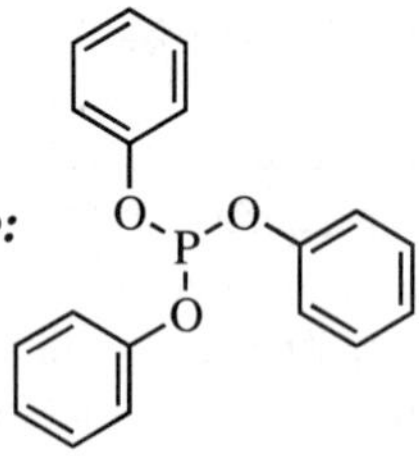

5.1 Chemical and Physical Properties

5.1.1 General

Organic phosphites are esters of phosphorous acid and alkyl or aryl alcohols, of the type $(RO)_3P$. Chemically, they can undergo transesterification with alcohols (51). The biologic properties of the organic phosphites resemble closely those of the phosphonates, possibly owing to their chemical similarity or transformation possibilities.

5.3 Exposure Assessment

5.3.3 Workplace Methods : NA

5.4 Toxic Effects

5.4.1 Experimental Studies

5.4.1.1 Acute Toxicity. The oral acute toxicity of trimethyl phosphite was found to be comparatively low and the material was not appreciably absorbed through the rabbit skin (23); however, it caused moderately severe irritation lasting several days. Introduction into the rabbit eye caused severe irritation and edema that persisted for several days. In vapor inhalation experiments, rats survived a 6 h exposure to saturated air. However, the material produced considerable discomfort, restlessness, and severe irritation of the eyes and respiratory tract (23).

According to Deichmann and Gerarde (23), the dermal LD_{50} of tridecyl phosphite in the rat is greater than 10 mL/kg. This indicates that the material is not absorbed through the skin, but still may cause irritation. It may also indicate that it has similar physical properties to those of tri(2-ethylhexyl) phosphite.

Triphenyl phosphite appears to be more toxic than the aliphatic derivatives. In addition, its neurotoxic properties, first described by Smith et al. (26), it was used for a convulsive model in experimental epilepsy (60). In addition, the authors studied the phosphite blood distribution and the motor cortex, diencephalon, and brain activity with radioactive triphenyl phosphite (60).

Aird et al. (60) using 0.3 mL/kg ^{32}P-labeled triphenyl phosphite intraperitoneally in the cat, observed considerable hydrolysis of the phosphite and only a very small quantity of original material in the central nervous system. The authors concluded that even this small quantity could have degenerative effects (60). The cresyl derivatives, similar to the phenyl, are charactersitically more toxic than the aliphatic equivalents. Nonetheless, triphenyl phosphite causes a delayed neuropathy with related ataxia and degeneration of the spinal cord in animals (61). The chemical inhibits neurotoxic esterase activity, but the structurally related compounds triphenyl phosphate, triphenyl phosphine, and trimethyl phosphite did not at the experimental doses (61). Animal studies have demonstrated that triphenyl phosphite causes potent neurotoxic effects due to cellular and axonal degeneration in the spinal cord, medulla, cerebellum, thalmus, and cerebral cortex (62).

Mixed alkyl–aryl phosphites tend to have lower toxicities than the aryl phosphites, as might be predicted. An oral LD_{50} in the rat for 2-ethylhexyl octylphenyl phosphite was7–10 g/kg (23), and an oral LD_{50} in the mouse for diphenylnonyl phosphite was 0.1 g/kg (27).

Chlorination of alkyl phosphites increases the toxicity, as demonstrated with bis(2-chloroethyl)phosphite with an intraperitoneal LD_{50} in the mouse of 250 mg/kg (17).

5.5 Standards, Regulations, or Guidelines of Exposure

Both the ACGIH TLV and the NIOSH REL for trimethyl phosphite is 2 ppm.

F ALKYL PHOSPHATES

Organic phosphates constitute a highly hazardous class of compounds. The materials most extensively investigated concern the organophosphate insecticides that are discussed in a separate chapter. The summary here is limited to the alkyl, aryl, and their lower-molecular-weight derivatives. The phosphates are chemically the most stable, and thus the most persistent phosphorus esters in biologic systems.

6.0a Trimethyl Phosphate

6.0.1a CAS Number: *[512-56-1]*

6.0.2a Synonyms: Phosphoric acid trimethyl ester, methyl phosphate

6.0.3a Trade Names: NA

6.0.3a Molecular Weight: 140.08

6.0.5a Molecular Formula: $C_3H_9O_4P$

6.0.6a Molecular Structure: 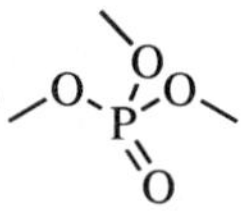

6.0b Triethyl Phosphate

6.0.1b CAS Number: *[78-40-0]*

6.0.2b Synonyms: Phosphoric acid triethyl ester

6.0.3b Trade Names: NA

6.0.4b Molecular Weight: 182.16

6.0.5b Molecular Formula: $C_6H_{15}O_4P$

6.0.6b Molecular Structure: 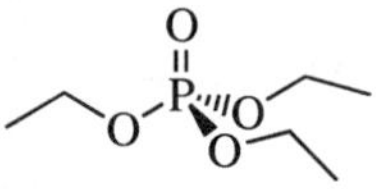

6.0c Tributyl Phosphate

6.0.1c CAS Number: *[126-73-8]*

6.0.2c Synonyms: Phosphoric acid tributyl ester, butyl phosphate

6.0.3c Trade Names: Celluphos 4

6.0.4c Molecular Weight: 266.32

6.0.5c Molecular Formula: $C_{12}H_{27}O_4P$

6.0.6c Molecular Structure: 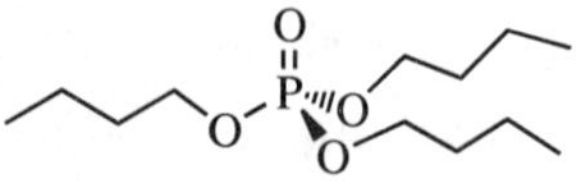

6.0d Di(2-ethylhexyl) Phosphate

6.0.1d CAS Number: [298-07-7]

6.0.2d Synonyms: Phosphoric acid bis(2-ethyhexyl)ester

6.0.3d Trade Names: NA

6.0.4d Molecular Weight: 322.42

6.0.5d Molecular Formula: $C_{16}H_{35}O_4P$

6.0.6d Molecular Structure:

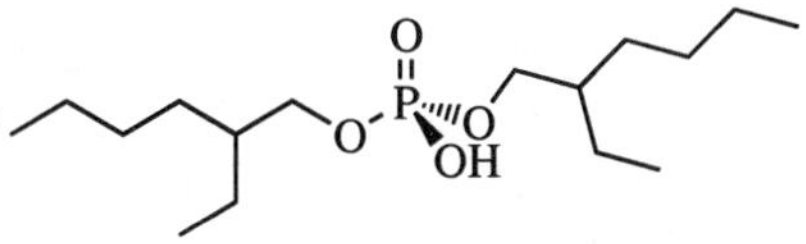

6.1 Chemical and Physical Properties

6.1.1 General

6.1.2 Odor and Warning Properties

6.2 Production and Use

The aryl phosphates are used widely in industry as plasticizers, as motor lubricants and gasoline additives (63), and as part of fire-resistant hydraulic fluid (63). Certain types of organic phosphates have found application as military defense agents (64). For the treatment of aryl phosphates poisoning, see Mieth and Beier (65). Zech et al. (64) reported that the oximes, which had raised great hopes in the field of medicine, however, do not always act as antidotes, but as synergists in certain cases. Trimethyl phosphate is used as an ethylating agent and as raw material to prepare insecticides, such as tetraethyl pyrophosphate (23). Triethyl phosphate is used as a plasticizer for cellulose esters, lacquers, plastics, especially vinyl resins, and as an antifoaming agent. The di(2-ethylhexyl) phosphate has wide industrial use.

6.3 Exposure Assessment

6.3.3 Workplace Methods

NIOSH Method 5034 is recommended for determining workplace exposures to tributyl phosphate (3a).

6.4 Toxic Effects

6.4.1 Experimental Studies

Male rats fed 0.5% of trimethyl phosphate, triethyl phosphate, tri-*n*-butyl phosphate, trioctyl phosphate, or tricresyl phosphate for 9 wk exhibited decreased body weight relative to controls (66). To date, it has been the only alkyl phosphate with documented flaccid paralytic effects (28). Orally, it is in the toxicity range of the phenyl phosphates. According to Jones and Jackson (67), trimethyl phosphate is not a teratogen. However, subtoxic oral or parenteral administration caused mutagenic effects in mice (68). Incubation experiments of TMP and DNA solutions have shown decreased sedimentation coefficients of the biopolymers. This is presumably the result of DNA alkylation, followed by depurination and scission of the phosphodiester's backbone. Gas chromatographic analysis yielded a compound, 7-methyldeoxyguanosine.

6.4.1.1 Acute Toxicity. Toxicological data in Table 81.5 show a somewhat higher degree of toxicity. Fassett and Roudabush conducted a subacute study by injecting rats intraperitoneally with 400 mg/kg TEP five times a wk for 4–6 wk (69). The animals still gained weight, and no side effects were noted. Experimental work with triethyl phosphate has so far failed to show neurotoxic effects. This is partially confirmed by the *in vitro* studies with the neurotoxin antidote pyridine-2-aldoxime methylchloride (PAM), whereby TEP was not hyrolyzed *in vitro* (70).

The acute oral toxicity of tributyl phosphate is relatively low, although it is a weak inhibitor in rats and *in vitro* human red blood cells (25). It also is a CNS stimulant. Affected rats show muscle twitching, general weakness, dyspnea, coma, and pulmonary edema. No latent paralysis has been observed. The material is a respiratory, eye, and primary skin irritant, and may be dermally absorbed (30, 55). Intraperitoneally, the dose with the lowest effect in the mouse was 63 mg/kg (30). The same toxicological properties are expected to hold for triisobutyl phosphate. Male rats exposed to the compound exhibited decreased body weight, alteration of brain cholinesterase, and increased blood coagulation rates (71).

An oral LD_{50} in the rat has been recorded as 4940 mg/kg (7), a dermal LD_{50} in the rabbit as 1250 mg/kg (7), and a lowest-effect level intraperitoneally in the mouse of 63 mg/kg (17).

6.5 Standards, Regulations, or Guidelines of Exposure

The ACGIH TLV for tributyl phosphate is 0.2 ppm.

G ARYL PHOSPHATES

Aryl phosphates contain an aromatic (benzene) ring and have the wide application in industry.

7.0a Triphenyl Phosphate

7.0.1a CAS Number: *[115-86-6]*

7.0.2a Synonyms: Phosphoric acid triphenyl ester; phenyl phosphate; TPP; triphenyl phosphoric acid ester; celluflex tpp

7.0.3a Trade Names: NA

7.0.4a Molecular Weight: 326.28

7.0.5a Molecular Formula: $C_{18}H_{15}O_4P$

7. 0.6a Molecular Structure:

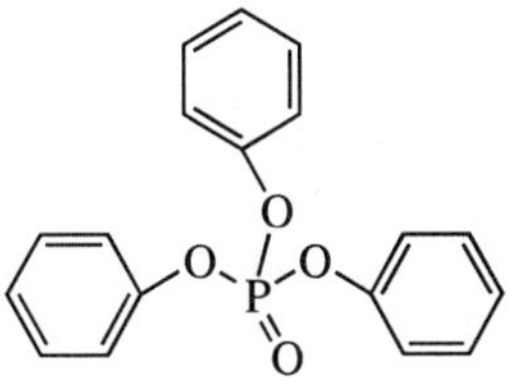

7.0b Tri-*o*-cresyl Phosphate

7.0.1b CAS Number: *[78-30-8]*

7.0.2b Synonyms: Tricresylphosphate; *o*-trioyl phosphate; TCP; TOCP; phosphoric acid tris(2-methylphenyl) ester; *o*-cresyl phosphate; phosflex 179-c; phosphoric acid, tri-*o*-cresyl ether; TOFK; *o*-tolyl phosphate; TOTP; tris(*o*-methylphenyl) phosphate; tris(*o*-tolyl) phosphate; phosphoric acid, tri-*o*-tolyl ester; triothocresyl phosphate

7.0.3b Trade Names: NA

7.0.4b Molecular Weight: 368.37

7.0.5b Molecular Formula: $C_{21}H_{21}O_4P$

7. 0.6b Molecular Structure:

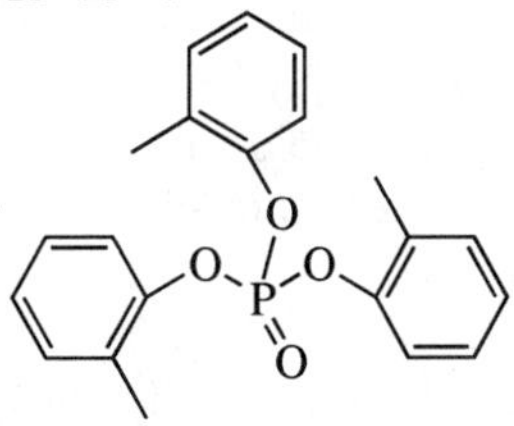

7.1 Chemical and Physical Properties

7.1.1 General

Triphenyl phosphate (TPP) is crystalline solid with a faint, aromatic odor.

Tri-*ortho*-cresyl phosphate (TOCP) is a liquid, also available as triaryl phosphate oil containing a mixture of the three isomers, the *ortho*, *meta*, and *para* forms. However, whenever possible, the portion of the *ortho* isomer is reduced or eliminated because of its paralytic properties.

7.1.2 Odor and Warning Properties

7.2 Production and Use

Triphenyl phosphate serves as a noncombustible substitute for camphor in celluloid, as a fire retardant in acetylcellulose, nitrocellulose, and airplane glue, as a plasticizer for lacquers and varnishes, and to impregnate roofing paper.

The tri-*o*-, *m*-, or *p*-cresyl phosphates, also named *o*-, *m*-, or *p*-tritolyl phosphates, phosphoric acid tri-2-, 3-, or 4-tolyl or methylphenyl phosphate esters, are widely used industrial additives. Tri-*o*-cresyl phosphate is used widely as a gasoline additive, plasticizer, fire retardant, solvent, extreme pressure additive, and as a lead scavenger in gasoline.

7.3 Exposure Assessment

7.3.3 Workplace Methods

Table 81.1 describes the method recommended for determining workplace exposures to TPP and TOCP.

7.4 Toxic Effects

7.4.1 Experimental Studies

7.4.1.1 Acute Toxicity. The basic animal experiments were carried out by Smith et al. (72) in 1930, who observed low acute toxicity in the commonly used laboratory species, and no effects in rats and mice when administered 1/10 to 1/20 of the oral LD_{50} for 3 mo. It caused generalized delayed illness and paralysis in cats and primates. The observed demyelination of the spinal cord resembled that obtained with tri-*o*-cresyl phosphate (72). Neuromuscular signs were observed above 0.2 g/kg in the cat. *In vitro* studies showed a 50 mg/kg no-effect level on whole blood cholinesterase, but a 20% decrease at the 100 mg/kg level (32). The material is poorly absorbed through the intact skin (23) and had no irritant effects in the rat skin (33). Dermal sensitization has been reported (73).

The compound studied most extensively has been tri-*o*-cresyl phosphate, since it has proved to be the causative agent in most epidemics. Some experiments by Hine et al. (58), using White Leghorn cockerels as test animals, were carried out in order to test the neurotoxic properties of a series of substituted aromatic phosphorus and related compounds for anticholinesterase activity. The authors demonstrated that in the fowl, only the *o*-tolyl phosphate produced paralytic properties either as a tri-*ortho* derivative or mono- and disubstituted compounds mixed with phenyl or *p*-tolyl phosphates. The *o*-tolyl and other substituted aryl and arylalkyl phosphates caused cholinesterase inhibition *in vivo* but not *in vitro* (58).

Of the aryl phosphates, only the tri-*o*-cresyl, the mono- and di-*o*-cresyl, and di- or monophenyl derivatives cause paralysis in the cockerel, and also have anticholinesterase activity. The fact that a series of aryl phosphates possess only anticholinesterase activity indicates that these pathological signs are due to different, nonrelated mechanisms (58). TOCP was also found to inhibit true erythrocyte cholinesterase in rats (77). This has been studied by Bleiberg and Johnson (78), who confirmed that TOCP, when administered orally, was toxic to mice and chickens and produced cholinesterase inhibition in red blood cells of the dog and the chicken. TOCP is a low dermal irritant and has been documented as the causative agent in a dermatitis case (73). *In vitro* studies with whole blood, using about 10 mg/kg TTP, depressed the acetyl cholinesterase level approximately 30% (32). Mechanistically, it appears that TOCP causes biochemical changes in the sciatic nerve of hens, marked by evidence of phosphorylation of proteins (79, 80).

Pathological effects observed in the human have been reproduced in various experimental animals, such as the cat, chicken, rabbit, dog, primate, and calf. The rat, the mouse, and the guinea pig are more resistant. Smith et al. (46, 81) report some minor fatty changes in animal liver and some degenerative deviations in the kidney. Barnes and Denz (82) studied the histopathology of the nervous fibers produced by TOCP and other phosphates. They determined that the demyelination did not appear to be directly related to anticholinesterase activity. Rayner et al. (83) have noted hyporeflexia in some workers chronically exposed to more complex organophosphate insecticides.

The chicken appears somewhat more resistant to tri-*o*-cresyl phosphate (26). The *o*-cresyl phosphate reduces the cholinesterase activity of whole blood in mice, accompanied by symptoms of tremor and diarrhea, resembling other cholinesterase inhibitory agents in the rat. A study revealed that tributyl, tributoxyethyl, and dibutylphenyl phosphate inhibited brain neurotoxic esterase activity as high as 70% (84). Rats were exposed to a fire-resistant hydraulic fluid containing dibutyl phenyl phosphate and tributyl phosphate at airborne levels ranging from 5 to 300 mg/m^3 6 h/d, 5 d/wk for 6–13 wk (85). The data revealed observations of a reddish nasal discharge and simultaneous oral salivation, and, at high exposures reduced body weights, increased liver weights, decreased number of erythrocytes, decreased hemoglobin levels, and decreased hematocrit values (85).

Injection into the yolk sac of fertile eggs prior to incubation has shown TOCP to cause reductions in survival rates with 100% deaths at 0.02 mL of undiluted material. At a concentration of 0.01 mL, 40% of the chicks hatched, but showed growth retardation and developed paralysis in 2–6 wk after hatching (86). The typical paralysis in the majority of the chicks is associated with demyelination in the spinal cord, cerebellum, and peripheral nerves (81), as observed in humans. Investigators of another study involving hens exposed to oral doses of solutions containing triphenyl phosphate, tricresyl phosphate, or butylated triphenyl phosphate additives in synthetic polyol-based lubricating oils concluded that there was low neurotoxicity and hazard under realistic conditions of exposure (87). A study by Barrett et al. in 1994 showed decreased acetylcholine esterase activity in red blood cells and plasma and decreased neurotoxic esterase activity in leukocytes due to single oral exposures of swine tri-*o*-cresyl phosphate (88).

An *in vitro* study involving use of immunocompetent cells (peritoneal cells and splenocytes) isolated from mice and subsequently exposed to triphenyl phosphate and triphenylphosphine oxide (TPPO) revealed signs of altered immune functions (89). Another study conducted *in vivo* and involving rats exposed to oral doses of TPP indicated no significant immunotoxicity (90). Another study showed that TOCP caused inhibition of several cytoplasmic enzymes (proteases) that are important in processes of intracellular protein turnover, processing hormonal peptides, and processing immune system antigenic proteins (91). Accordingly, the investigators suggest that the data may represent a previosuly unrecognized toxicity hazard.

7.4.1.3 Pharmacokinetics, Metabolism, and Mechanisms. Some studies have shown that TOCP is readily absorbed, transported by the vascular and central nervous system, and distributed into the brain, cord, and cerebellar tissue within 24 h of oral administration (58). Of 17 compounds tested, tri-*o*-cresyl phosphate is one of the most active inducers of drug-metabolizing enzymes. The compounds had in common low systemic toxicity, but high lipid solubility (92). Cohen and Murphy have shown that TOCP is capable of potentiating malathion activity (93). A study published by Wu and Leng (94) using white Leghorn hens suggested that TOCP causes phosphorylation of brain membrane proteins leading to organophosphate-induced delayed neuropathy, but treatment with the drug

Verapamil, a calcium channel blocker, eliminated enhancement and caused some inhibition of the adverse effects. Saligenin cyclic *o*-tolyl phosphate (SCOTP) is thought to be an active metabolite of TOCP and is more neurotoxic (95). Hodge and Sterner demonstrated with radioactive materials that TOCP was absorbed through the intact human palm and the abdominal skin of the dog (35). About 0.10% of the quantities administered were excreted in the urine within a few hours. Experimentally, a dermal LD_{50} of 2.8 cm^2/kg was determined for the mouse, 1/5 to 1/25 as toxic for the cat, dog, or rabbit (35). Paralysis of hind legs, however, was observed in mice (35). It also was reported that TOCP was found in the blood of the dog at 8 mg/100 cm^3 within 24 h of application and was found distributed throughout the visceral organs, muscle, brain, and bone in the dog (35).

7.4.1.4 Reproductive and Developmental. TOCP is toxic to the male reproductive system due to damage caused to the testes (96). Rats exposed to 135 d of oral doses of TCP and butylated triphenyl phosphate (BTP)-based hydraulic fluids were studied to evaluate reproductive toxicity (97). The results showed decreased fertility and number of litters born for BTP-exposed groups and decreased fertility and litter sizes in TCP-exposed groups. Rat studies also have shown that both BTP and TCP appear to be associated with endocrine toxicity (98). A study involving rats fed a diet containing various levels of triphenyl phosphate for a period from 4 wk postweaning for 91 d, through mating and gestation, indicated that there was no significant impact to the mothers or their offspring (99).

7.4.1.5 Carcinogenesis. TOCP was shown in a study by Mentzshel et al. (100) to be associated with detection of DNA adducts in tissues harvested from rats treated with the chemical. The authors concluded that a metabolite formed from biotransformation of TOCP may be associated with initiation of carcinogenesis.

7.4.2 Human Experience

7.4.2.1 General Information. Of the three isomers, the toxicities and neurotoxic character increase *para* < *meta* < *ortho* (54). The lethal oral dose for adults is 1–10 g for the *ortho* isomer; *meta* and *para* isomers are practically nontoxic and do not cause demyelination (27). Most human intoxications have involved accidental ingestion of adulterated alcoholic beverages, cooking oils (48, 50), or, as an unusual case, industrial inhalation of heated vapors (76).

Reviews of human incidents have been presented by Hunter et al. (76), Susser and Stein (50), Smith et al. (72), and Smith and Lillie (81). The minimum paralytic dose in humans is unknown, but from studies by Smith et al. (26, 72), it can be estimated as about 10–30 mg/kg of tri-*o*-cresyl phosphate for an adult. The clinical picture after ingestion has been similar in all reported cases. Immediately post ingestion usually no symptoms occur or occasionally there may be gastrointestinal disturbances lasting from a few hours to 2 d; sharp cramplike pains may occur in the calves with some numbness in the hands and feet. A few hours later weakness in legs and feet may progress to bilateral foot drop, but the cramping pain may disappear. Within another few days, weakness of fingers and wrists may develop, but not extending above elbows and thighs. There is a dispute over the extent of sensory loss; however, it is agreed that it may run parallel to the extent of dysfunction of the motor system. In mild cases, sensory changes regress rapidly (50). The progress of affected cases varies considerably, but in general, the muscular weakness may increase over a period of several weeks or even months, and then become more or less stationary. The extent to which the paralysis may be permanent depends mainly on the quantities involved, intake rates, and cumulative effects. Fatalities are quite rare and occur mainly in response to the ingestion of very large phosphate quantities in a short time.

Due to the lack of human autopsy material of the rarely fatal cases, little is known of the progressive microbiological changes. The pathology described by Smith and Lillie (81) in six human cases was characterized by some involvement of the anterior horn cells, fatty degeneration of the white substance of the cord, tigrolysis of nerve cells, displacement of the nucleus to the periphery, and particularly demyelination with marked fragmentation and fatty degeneration of the myelin sheath. Exposure of personnel for various time periods to TCP, a preparation that contains several aryl derivatives, including tri-*o*-cresyl phosphate, handled as a component of hydraulic oil on an aircraft carrier, showed no physiological effects. Exposed and control personnel were checked intensively for neurological signs, hematologic effects, and blood cholinesterase levels (74).

7.4.2.2 Clinical Studies. Tabershaw and Kleinfeld (101) have described a monitoring study of a chemical TOCP manufacturing plant. Results revealed TOCP air concentrations ranging from 0.27 to 3.40 mg/m^3, depending on the type of operation. The effect on the worker population (101) indicated some plasmic cholinesterase depression. However, no evidence of neuromuscular difficulties attributable to the industrial exposure was found.Tabershaw et al. (102) have described a triphenyl and tricresyl phosphate manufacturing plant with air concentration measurements ranging from 0.27 to 3.4 mg/m^3. The authors found no correlation between results from cholinesterase determinations and minor gastrointestinal or neuromuscular symptoms and the degree and duration of exposure (102). Occupational exposures have occurred on various occasions. Sutton et al. (32) have published a study covering the chemical, toxicological, and industrial aspects of triphenyl phosphate exposure. Fourteen employees were exposed to triphenyl phosphate vapor, mist, or dust over a period of 8–10 yr. The particle sizes of the dusts measured < 1 μm in diameter for 90%, and the average air concentration was determined as 3.5 mg/m^3 with occasional excursions to 40 mg/m^3. Results showed no signs of illnesses, but a slight statistically significant reduction in red blood cell cholinesterase activity. The human data were confirmed by Smith et al. (46, 81).

A similar report of men handling lubricating oils containing some cresyl phosphates on an aircraft carrier showed no signs of illnesses or chemical changes (74). One report by Hunter et al. (76) describes three cases of polyneuritis from the manufacture of various aryl phosphates, with the suspicion that exposure to TOCP was the causative factor. One case of permanent paralysis in a worker has been reported by Bidstrup and Bonnell (103). He had been engaged in the manufacturing of *m*- and *p*-tricresyl phosphate isomers, where the final product contained only 1% of the *ortho* form (TOCP), but 6–10% appeared during the processing. Large quantities of TPP inhibit human cholinesterase *in vitro* and *in vivo*; however, it is not considered a potent anticholinesterase agent. No adverse clinical effects were found in men exposed to TPP vapor mist and dust for 10 yr at a concentration of 3.5 mg/m^3 (23). Another study reported polyneuropathy in a mechanic exposed to hydraulic fluids containing isopropylated tricresyl phosphate (ITP) (104). The investigators concluded that this observation and results from a related cross-sectional study of eight other exposed men and eight controls suggest a possible relationship between heavy exposure to ITP and polyneuropathy. A report indicated a case of allergic contact dermatitis associated with exposure to TPP that was contained in plastic eyeglass frames (105). In relation to the observed dermatitis associated with wearing the plastic eyeglass frames, the investigators reported positive dermal patch test results for TPP and tri-*m*-cresyl phosphate, but not for tri-*p*-cresyl phosphate.

Triaryl or tricresyl phosphate (TCP) is a mixture of various aryl phosphates, including tri-*o*-cresyl phosphate. One study indicated that personnel exhibited no toxicological effects when exposed for various durations to TCP, a preparation handled as a component of hydraulic oil on an aircraft carrier and that contained several aryl derivatives including (74). Exposed and control personnel were checked intensively for neurological signs, hematologic effects, and blood cholinesterase levels (74). An epidemiological study in Italy of workers handling rubber-base materials and cements containing triaryl phosphate revealed 47 cases of motor polyneuritis (75). However, in addition, the workers had also handled a variety of other neurotoxic chemicals when previously employed in the shoe, tire, rubber, or adhesive tape industry.

7.4.2.3 Epidemiology Studies. An epidemiological study in Italy of workers handling rubber-base materials and cements revealed 47 cases of motor polyneuritis (75). One of the causative agents was suspected to be triaryl phosphate (75). However, in addition, the workers had also handled a variety of other neurotoxic chemicals when previously employed in the shoe, tire, rubber, or adhesive tape industry.

7.5 Standards, Regulations, or Guidelines of Exposure.

The ACGIH TLV, NIOSH REL and OSHA PEL for triphenyl phosphate are all 3 mg/m^3.

The ACGIH TLV, NIOSH REL and OSHA PEL for triorthocresyl phosphate are all 0.1 mg/m^3 for tri-*o*-ethylphosphate.

I HALOGENATED PHOSPHATE ESTERS

In contrast to the previously discussed phosphate esters, the chlorinated derivatives possess anesthetic-like and muscle-relaxant properties, even at relatively high doses, without pathological side effects.

8.0 Trichloroethyl Phosphate

8.0.2 Synonyms: Trichloroethyl phosphate, also known as trichloroethyl phosphate (TCEP; trichloroethyl phosphoric acid mono ester; Triclofos), $P(O)(OH)_2(OCH_2CCl_3)$ and tris(β-chloroethyl) phosphate, (tri(2-chloroethyl) phosphoric acid ethyl ester; TRCP), $P(O)(OCH_2CH_2Cl)_3$

8.4 Toxic Effects

Triclofos is a sedative or hypnotic. Tri-2-chloroethyl phosphate produces prolonged epileptiform convulsions in the rat intraperitoneally at levels of 0.28 g/kg (38). At higher doses, convulsions occur, but only weak cholinesterase inhibition (6). The material does not appear to be absorbed through the skin, nor is it a dermal irritant. A subacute study comparing triethyl phosphate and trichloroethyl phosphate by Fassett and Roudabush (69) showed a definite hemorrhagic tendency by the trichloroethyl phosphate. A variety of chlorovinyl phosphates are available, which serve as pesticides and insecticides. For example, an acute human poisoning by 2-chloro-1-(2,4-dichlorophenyl) vinyl diethyl phosphate, chlorfenvinphos, has been reported. The clinical picture was dominated by the nicotinic, mascarinic, and central nervous syndromes with severe respiratory difficulties, accompanied by acidosis and by initial hyperglycemia, which converted into hypoglycemia (106). Siggins et al. have tested a series of 2-chloroethanol esters of α-keto acids and similarly found some to possess hypoglycemic properties (107).

IV ESTERS OF MONOCARBOXYLIC HALOGENATED ACIDS, ALKANOLS, OR HALOALCOHOLS

A OVERVIEW

The halogenated acid esters are discussed in the order of prevalence of the halogen, that is, the chlorine, the fluorine, bromine, and iodine derivatives. The esters are listed with increasing molecular weight or complexity. Halogenation of hydrocarbons increases their toxicity considerably. The halogen formates are highly irritating and are used as lacrimators. The halogenated acetates serve chiefly as intermediates in various organic syntheses. They are generally colorless liquids with relatively high densities, soluble in water, and more stable toward hydrolysis than the chloroformates, although ethyl iodoacetate may release iodine on extensive heating. Physiologically, the halogenated acetates behave similarly, such as producing pulmonary edema. Especially, their lacrimatory and irritant properties can be classified in the order of potency $I > Br > Cl > F$. According to Dixon (108), the lacrimatory mechanism may involve specific reactions of the halogen with certain enzyme-sulfhydryl groups. The ester groups appear to enhance this activity. The reaction appears to occur rapidly, is reversible at low and irreversible at very high concentrations in some tissues (109, 110). Some physical and chemical properties and characteristics are listed in Table 81.7 and toxicological information in Tables 81.8, 81.9 (111–117), and 81.10 (118, 119).

B CHLOROESTERS

9.0 Chloroformates

A number of chloroformates are of industrial importance. For example, ethyl chloroformate is prepared by reacting phosgene with ethanol. The chloroformates are

Table 81.7. Chemical and Physical Properties of Representative Esters of Monocarboxylic Halogenated Acids and Haloalcohols (5)

Compound	CAS No.	Molecular Formula	MW	Boiling Point (°C) (mm Hg)	Melting Point (°C)	Density	Solubility[a] in Water	Refractive Index (20°C)	Vapor Density (Air = 1)	Vapor Pressure (mm Hg) (°C)	Flash Pt. (°C)
Methyl chloroformate	[*79-22-1*]	$C_2H_3ClO_2$	94.50	70.4–70.9 (752)	− 61	1.223	d[b]	1.3868	3.26		15
Ethyl chloroformate	[*541-41-3*]	$C_3H_5ClO_2$	108.52	93	− 81	1.135	d[b]	1.3947	3.74		10
Propyl chloroformate	[*109-61-5*]	$C_4H_7ClO_2$	122.55	105		1.0901	d[b]	1.4035	4.2		− 50
Isopropyl chloroformate	[*108-23-6*]	$C_4H_7ClO_2$	122.55	104.6–104.9		1.08	i	1.4013	4.2		15.6
Allyl chloroformate	[*2937-50-0*]	$C_4H_5ClO_2$	120.54	110		1.14	i		4.2		31.1
Benzyl chloroformate	[*501-53-1*]	$C_8H_7ClO_2$	170.60	103 (20)		1.20	d[b]	1.5160			7
Trichloromethyl chloroformate	[*503-38-8*]	$C_2Cl_4O_2$	197.83	128	− 57	1.64	i	1.4566			
Chloroethyl chloroformate	[*627-11-2*]	$C_3H_4Cl_2O_2$	142.97	155.7–156.0		1.3847	i	1.4465			
Methyl chloroacetate	[*96-34-4*]	$C_3H_5ClO_2$	108.52	131	− 33	1.238	d[b]	1.4218	1.24		47
Ethyl chloroacetate	[*105-39-5*]	$C_4H_7ClO_2$	122.55	144 (740)	− 27	1.145	i	1.4125	4.3	10 (37.5)	54
Butyl chloroacetate	[*590-02-3*]	$C_6H_{11}ClO_2$	150.60	183		1.0704	i	1.4297			
2,4,5-Trichlorophenyl chloroacetate		$C_8H_4Cl_4O_2$	273.93								
2-Chloroallylidene 3,3-diacetate chloroacetate	[*5459-90-5*]	$C_7H_9ClO_4$	192.60								
Methyl chloropropionate	[*17639-93-9*]	$C_4H_7ClO_2$	122.55								

Ethyl chloropropionate	*[535-13-7]*	$C_5H_9ClO_2$	136.58	146–149		1.072	i				38
2-Chloro-3-(4-chlorophenyl) propionic acid methyl ester		$C_{10}H_{10}Cl_2O_2$	233.09								
2-Chloroethyl acrylate	*[2206-89-5]*	$C_5H_7ClO_2$	134.56	51–53			i				
Methyl fluoroformate		$C_2H_3FO_2$	78.04	40		1.06			2.69		
Ethyl fluoroformate		$C_3H_5FO_2$	92.07	53–54	−80	0.917	s	1.3597	3.18		−20
Methyl fluoroacetate	*[453-18-9]*	$C_3H_5FO_2$	92.07								
Ethyl fluoroacetate	*[459-72-3]*	$C_4H_7FO_2$	106.10	119.2 (753)		1.098			3.7		30
Propyl fluoroacetate		$C_5H_9FO_2$	120.12								
Isopropyl fluoroacetate		$C_5H_9FO_2$	120.12								
Allyl fluoroacetate		$C_5H_7FO_2$	118.12								
2-Chloroethyl fluoroacetate		$C_4H_6ClFO_2$	140.54								
2-Fluoroethyl acetate		$C_4H_7FO_2$	106.10								
Ethyl bromoacetate	*[105-36-2]*	$C_4H_7BrO_2$	167.00	159	−38	1.5059	i	1.4489	5.8		47
Ethyl 2-bromo-propionate	*[535-11-5]*	$C_5H_9BrO_2$	181.03	156∼160		1.394	i	1.4490			51
Ethyl 2-bromo-butyrate	*[533-68-6]*	$C_6H_{11}BrO_2$	195.06	177.5 (765)		1.321	i	1.4475			58
Ethyl iodoacetate	*[623-48-3]*	$C_4H_7IO_2$	214.00	178–180		1.808		1.5079	7.4	0.54(20)	76

[a]Solubility in water: v = very soluble; s = soluble; d = slightly soluble; i = insoluble.
[b]Decomposes when dissolved.
[c]Freezing point (°C).

Table 81.8. Summary of Oral, Dermal, Ocular, and Intraperitoneal Toxicity Data for Some Alkyl and Chloroalkyl Esters

	Oral LD_{50}			IP LD_{LO}			Dermal and Eye		
Compound	Species	g/kg	Ref.	Species	mg/kg	Ref.	Species	Result	Ref.
Chloroformate									
Methyl	Rat	< 0.05	17				Guinea pig	Damage to skin	6
Ethyl	Rat	< 0.05	17	Mouse	15	17	Guinea pig	Damage to skin	6
Isopropyl	Rat	1.07	17						
	Rabbit	11.3	17						
							Rabbit	Permanent corneal opacity of eye	6
Chloroacetate									
Ethyl	Rat	0.050	17						
2-Chloroallylidene 3,3-diacetate	Rat	0.32	17				Rabbit	0.98 mL/kg (LD_{50})	6
Propionate									
(2-Chloro) methyl				Mouse	250	17			
(3-Chloro) methyl				Mouse	500	17			
(6-Chloro) *o*-tolyl				Mouse	500	17			
(2-Chloro)	Rat	1.072	17				Rabbit	0.756	6
3-(4-chloro)phenyl	Rabbit	0.5	17						
-, methyl	Guinea pig	0.5	17						
Chloroethyl acrylate	Rat	0.18	17				Rabbit	Skin and eye effects severe	6

Table 81.9. Summary of Oral, Dermal, Intraperitoneal and Subcutaneous Toxicity Data for Some Fluoroesters

	Oral LD_{50}			IP LD_{LO}			SC LD_{50}			IV and Other Routes LD_{50}			
Compound	Species	mg/kg	Ref.	Species	mg/kg	Ref.	Species	mg/kg	Ref.	Route	Species	mg/kg	Ref.
Fluoroacetate													
Methyl	Primate[a]	10	17	Mouse	7.5	111	Rat	5	112	IV	Primate	4.0	17
	Mouse[a]	5	17	Guinea pig	0.35	112	Mouse	7–10	113	IV	Rabbit	0.25	114
	Rabbit	0.5	17	Hamster	0.25		Rabbit	0.3	17	IV	Dog	0.06	112
	Guinea pig	0.5	17				Guinea pig	0.2	17	IV	Cat	0.5	112
	Dog	0.1	17				Cat	0.3	17	IV	Chicken	15.0	112
	Cat	0.3	17							Dermal	Rabbit	10.0	115
										IM	Rat	2.5	113
Ethyl	Mouse	6–10	6	Mouse	19.0	116							
Ethylene	Rat	2.2	115										
Propyl				Mouse	78.0	17							
2-Ethylhexyl										Dermal	Rabbit[a]	10.0	17
										IV	Rabbit[a]	10.0	116
Allyl							Mouse	6.0	117				
Phenyl							Mouse	6.0	115				
2-Fluoroethyl							Mouse	8.5	113				
Acetate													
Fluoroethyl	Mouse	18.0	6	Mouse	19.0	17							
Fluorobutyl				Mouse	0.96	17							
Fluoropentyl				Mouse	100.0	116							
Methyl 4-fluorobutyrate	Rat	25.0	17				Mouse	1	116				

[a] LD_{LO}.

Table 81.10. Inhalation Effects of Alkyl and Chloroalkyl Esters

Compound	Species	Parameter	Conc. (mg/m^3)	Time (min)	Effect or Mortality	Ref.
Chloroformate						
Methyl	Human		10	10	Lacrimation	109
	Human	LC_{100}	190	10	Lethal	109
Ethyl	Mouse	LC_{LO}	2260	10		17
Isopropyl	Rat	LC_{LO}	< 100	5 h		118
	Rat	LC_{100}	6450	50		6
	Rat	LC_{LO}	122,000	1 h		6
Allyl	Mouse	LC_{LO}	2000	1 h		17
Trichloromethyl	Mouse	LC_{LO}	344			17
Chloroethyl	Mouse	LC_{LO}	200			17
Chloroacetate 2-chloroallylidene 3,3-diacetate	Rat	LC_{50}	8	4 h		17
Chloroethyl acrylate	Rat	LC_{100}	125	4 h	0/6	17
	Rat	LC_{100}	250	4 h	6/6	17
Fluoroacetate						
Methyl	Primate	LC_{50}	800	10		17
	Mouse	LC_{LO}	3200	10		113
	Rabbit	LC_{LO}	100	10		119
	Guinea pig	LC_{LO}	100	10		115
	Dog	LC_{LO}	25	10		17
Ethyl	Guinea pig	LC_{LO}	100			17
Isopropyl	Guinea pig	LC_{LO}	100	10		17
Allyl	Mouse	LC_{LO}	500			17
2-Chloroethyl	Rabbit	LC_{LO}	100			117
	Guinea pig	LC_{LO}	100	10		17
2-Fluoroethyl	Rat	LC_{LO}	92			119
	Mouse	LC_{50}	0.45			113
	Rabbit	LD_{50}	50.0			117
	Guinea pig	LC_{LO}	92	10		119
4-Fluorobutyrate						
Methyl	Rat	LC_{50}	350	10		17
	Mouse	LC_{50}	120	10		17
	Rabbit	LC_{50}	35	10		17
	Guinea pig	LC_{50}	70	10		17
	Cat	LC_{50}	35	10		17
	Dog	LC_{50}	50	10		17
	Primate	LC_{50}	500	10		17
Ethyl	Rat	LC_{LO}	500	10		17
	Mouse	LC_{50}	500	10		17
	Guinea pig	LC_{50}	500	10		17
Isopropyl	Mouse	LC_{LO}	10	10		17
	Guinea pig	LC_{50}	200	10		17

Table 81.10. *(Continued)*

Compound	Species	Parameter	Conc. (mg/m^3)	Time (min)	Effect or Mortality	Ref.
2-Fluoroethyl	Primate	LC_{50}	500	10		17
	Rat	LC_{50}	200	10		17
	Mouse	LC_{50}	73	10		17
	Guinea pig	LC_{50}	35	10		17
	Cat	LC_{50}	25	10		17
	Dog	LC_{50}	25	10		17

generally corrosive, flammable, colorless liquids that have physical properties of the acid chlorides. They tend to decompose spontaneously to form hydrochloric acid and other products. They hydrolyze readily in the presence of water or moist air. Chlorination of alkyl or alkenyl carboxylates increases their toxicity drastically. The lower members are intense irritants and lacrimators, and thus have been used as warfare agents.

9.0a Methyl Chloroformate

9.0.1a CAS Number: [79-22-1]

9.0.2a–3a Synonyms and Tradenames: Methylchloroformate; carbonochloric acid, methyl ester; Carbonochloridic acid methyl ester; METHYL CHLOROFORMATE (MCF); Methyl chlorocarbonate

9.0.4a Molecular Weight: 94.50

9.0.5a Molecular Formula: $C_2H_3ClO_2$

9.0.6a Molecular Structure: 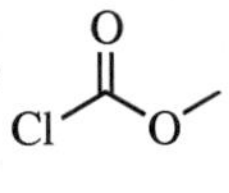

Methyl chloroformate (carbonochloridic acid methyl ester), $ClCOOCH_3$, is a clear liquid. Known as "K-stof," it was used as a warfare agent during World War I (110). Physiologically, it is highly irritating and corrosive by ingestion, inhalation, or eye or skin contact. Concentrations of 10 ppm (5.28 mg/m^3) have caused lacrimation, and a concentration of 190 ppm has been lethal in 10 min (108). Comparatively, the irritant potency may be about five times that of chlorine and one-half that of phosgene. Humans exposed to methyl chloroformate have experienced respiratory tract and eye irritation, even persisting after cessation of the exposure. Skin sensitization may occur, although employees in a chemical manufacturing plant showed no health effects nor positive laboratory findings from blood and liver function tests (6).

9.0b Propyl Chloroformate

9.0.1b CAS Number: [109-61-5]

9.0.2b–3b Synonyms and Tradenames: propyl chloroformate; chloroformic acid propyl ester; *n*-propyl chloroformate; Carbonochloridic acid propyl ester; Propyl chloroformate; NORMAL PROPYL CHLOROFORMATE (NPCF)

9.0.4b Molecular Weight: 122.5511

9.0.5b Molecular Formula: $C_4H_7ClO_2$

9.0.6b Molecular Structure:

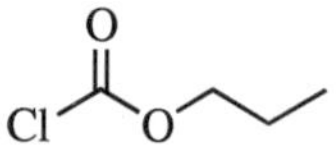

Propyl chloroformate (propyl chlorocarbonate), $ClCOOC_3H_7$, is a colorless liquid and is gradually decomposed by water. The vapors are strongly irritating to the eyes and mucous membranes.

9.0c Isopropyl Chloroformate

9.0.1c CAS Number: *[108-23-6]*

9.0.2c–3c Synonyms and Tradenames: Carbonochloridic acid 1-methylethyl ester

9.0.4c Molecular Weight: 122.5511

9.0.5c Molecular Formula: $C_4H_7ClO_2$

9.0.6c Molecular Structure:

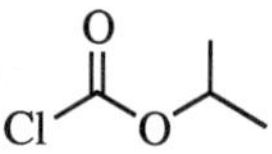

Isopropyl chloroformate (isopropyl chlorocarbonate), $ClCOOC_3H_7$, is a colorless liquid with similar toxicological properties to the propyl isomer. Inhalation exposures in elevated concentrations cause death by immediate pulmonary edema; lower concentrations cause respiration distress, prostration, and convulsions. Direct dropwise application to the guinea pig skin causes deep necrosis and eschar formation, and direct contact with the rabbit eye causes permanent corneal opacity (6).

Chloromethyl Chloroformate. Chloromethyl chloroformate, $ClCOOCH_2Cl$, is a colorless liquid with a penetrating, irritating odor. It appears to be even more corrosive than the methyl ester. A more highly chlorinated derivative, trichloromethyl chloroformate, $ClCOOCl_3$, has been called diphosgene. An inhalation LC_{LO}, a lowest lethal dose, of 344 mg/m^3 has been determined in the mouse (17).

10.0 Chloroacetates

Ethyl Chloroacetate. Ethyl chloroacetate (chloroacetic acid ethyl ester), $ClCH_2COOC_2H_5$, is a highly flammable liquid with a pungent odor. The ethyl chloroacetate is a weaker lacrimator compared with the bromo or iodo derivative. It is a moderately strong irritant to the guinea pig skin and may be absorbed percutaneously (6).

Cyclic Chloroacetate. 2-Cyclohexylethyl chloroacetate, $ClCH_2COOCH_2C_6H_{11}$, benzyl chloroacetate, $ClCH_2COOCH_2C_6H_5$, the phenethyl chloroacetate, $ClCH_2COOCH_2CH_2C_6H_5$, have shown an intraperitoneal LD_{LO} value, lowest lethal dose, of 500 mg/kg in the mouse (17).

The more highly chlorinated 2,4,5-trichlorophenyl chloroacetate indicates an oral LD_{50} in the guinea pig of 1 mg/kg (17).

Chlorodiacetate. A related compound, 2-chloroallylidene-3,3-diacetate, $CH_3CCl{:}C(OCOCH_3)_2$, has shown an oral LD_{50} of 0.32 g/kg in the rat and dermal LD_{50} of 0.98 mL/kg in the rabbit (6). This material was evidently highly toxic by inhalation, causing a lethal effect in 3 of 6 rats following an exposure of 8 ppm for 4 h (6).

11.0 Fluoroesters

Since the elucidation of the biologic action of fluoroacetic acid (120), a large number of experimental fluoro compounds have been synthesized and biologically tested. This information has served to substantiate current theories on ester hydrolysis and fatty and carboxylic acid metabolism. The fluoro esters, especially the acetates and other even-numbered flurocarboxylates, are highly toxic since they hydrolyze directly or eventually to fluoroacetate and the corresponding alcohol.

Fluoroacetates. The fluoroacetates are mainly liquids with corrosive but pleasantly aromatic properties. Methyl, ethyl, propyl, isopropyl, ethylhexyl, vinyl, allyl, phenyl, 2-chloroethyl, and 2-fluoroethyl fluoroacetate are structurally $ClCH_2COOR$, whereby the R groups can vary. The fluoroacetates are highly toxic, with a narrow range between species and routes of entry. Fluoroacetate is a potent inhibitor of the Krebs cycle, and a study using exposed mice suggested related defects in neurotubular development in mouse embryos (121). Research has shown that the compound blocks the tricarboxylic acid cycle in brain glial cells of mice (122). Dermally, the toxicity is somewhat lower, owing to slow absorption. The major toxic metabolite is fluoroacetic acid, which is also found in the South African poisonous plant *Dichapetalum cymosum* (120).

A study suggested that a series of 1-(di)halo-2-fluoroethanes, including 1,2-difluoroethane, 1-chloro-2-fluoroethane, 1-chloro-1,2,-difluoroethane, and 1-bromo-2-fluoroethane, were all highly toxic to rats exposed via inhalation for 4 h to $\leq$100 ppm (123). The researchers concluded that the 1-(di)halo-2-fluorethanes were metabolized to the active toxicant fluoroacetate.

Fluoroalkyl Acetates. Fluoroethyl, fluorobutyl, and fluoropentyl acetate, $CH_3COOCH_2CH_2F$, $CH_3COO(CH_2)_4F$, and $CH_3COO(CH_2)_5F$. Their intraperitoneal toxicities appear in a similar range, indicative of rapid metabolic hydrolysis to acetic acid and the corresponding fluoroalcohol. The latter apparently oxidizes readily to the corresponding carboxylic acid. The even-numbered acids have been shown to be more highly toxic than the odd-numbered, which ultimately catabolize to fluoropropionic acid, not exhibiting the citric acid cycle inhibiting properties of fluoroacetic acid as one of its highly toxic actions. Refer to Tables 81.9 and 81.10.

Fluoropropionates. The fluoropropionates are highly toxic, but not to the extent of the fluoroacetates, since the C-3 derivative does not appear to interfere with the mammalian enzyme systems, such as citric acid cycle *cis*-aconitase. A mammalian LD_{50} by an unknown route was 200 mg/kg (17) for ethyl 3-fluoropropionate, $FCH_2CH_2COOC_2H_5$.

Fluorobutyrates. Methyl fluorobutyrate, $F(CH_2)_3COOCH_3$ still appears highly toxic but not to the degree of the acetate. Some information is provided in Tables 81.9 and 81.10. Results indicate that fluorobutyrates may hydrolize at a lower rate or degrade only very slowly from the C_4 to the C_2 carboxylate. Information is also provided for ethylfluorobutyrate.

12.0 Bromoesters

Bromine-containing carboxylic esters are more highly odorous, lacrimatory, irritating, and toxic than the fluorine derivatives previously discussed.

Ethyl Bromoacetate. Ethyl bromoacetate (bromoacetic acid ethyl ester), $BrCH_2COOC_2H_5$, is a potent human lacrimator at a concentration as low as 3 mg/m^3 and effects at 40 mg/m^3 on the eye are intolerable. It is thus utilized in combination with other similar chemicals, such as common tear gas, but may cause blindness on direct contact (23). It can produce sensitization dermatitis in humans (6). When injected subcutaneously into the mouse at 332 mg/kg for 83 wk, some nonsignificant systemic and slight neoplastic signs were observed when an initiating agent accompanied the treatment (124). An aquatic toxicity rating Tm 96 was 1000–100 ppm. For ethyl tribromoacetate, tribromoacetic acid ethyl ester, $CBr_3COOC_2H_5$, an oral LD_{50} of 100 mg/kg has been recorded in the rat (125).

13.0 Iodoesters

Iodine-containing carboxylic esters are highly irritant, lacrimatory, and toxic.

Ethyl Iodoacetate. Ethyl iodoacetate, (odoacetic acid ethyl ester), $ICH_2COOC_2H_5$, is a dense, colorless liquid. The material is highly toxic and lacrimatory to humans at a concentration of 15 mg/m^3, about 1.5 ppm, with eye damage possible at higher concentrations. According to Vedder (109), two cases of fatal worker exposures to sudden high concentrations of ethyl iodoacetate have occurred. The autopsy findings showed pulmonary edema. Intraperitoneal injections of ethyl iodoacetate revealed an LD_{50} of 45 mg/kg in the mouse (126).

V ORGANIC SILICON ESTERS

A OVERVIEW

The alkyl silicon esters can be classified into several categories, namely, the silanes, which are alkyl-substituted silicon tetrahydrides (R_4S) and the next more highly oxygenated derivatives only, silanols (R_3SiOR' $R_2Si(OR')_3$). Some silanols can be polymerized to form silicones of the type $[R_xSiO_{(4-x/2)}]_n$. The most highly oxygenated compound is silicic acid and its esters, the silicates, which are chemically $Si(OR)_4$.

Some of these compounds are important industrial raw materials used to produce a variety of widely used end products, such as the silicones that serve as lubricating fluids, oil baths, resins, and plastic copolymers. The toxicological effects vary from inert to caustic, depending on the oxidation potential and the molecular size. Some chemical and physical properties and characteristics are listed in Table 81.11 and toxicological information in Table 81.12 (127–130).

Table 81.11. Chemical and Physical Properties of Representative Esters of Organic Silicon (5)

Compound	CAS No.	Molecular Formula	MW	Boiling Point (°C) (mmHg)	Melting Point (°C)	Density	Solubility[a] in Water	Refractive Index (20°C)	Vapor density (Air=1)	Vapor Pressure (mmHg) (°C)	Flash Pt. (°C)
Tetrachloro silane	[*10026-04-7*]	Cl_4Si	169.90	57.6	− 70	1.483	d[b]				
Methyltrichloro silane	[*75-79-6*]	CH_3Cl_3Si	149.48	66	− 77	1.27	d[b]		5.17		− 13
Dimethyldichloro silane	[*75-78-5*]	$C_2H_6Cl_2Si$	129.06	70.0	− 16	1.07	d[b]		4.45		− 9
Ethyltrichloro silane	[*115-21-9*]	$C_2H_5Cl_3Si$	163.51	168	− 105.6	1.2381	d	1.4257	5.6		22.2
Diethyldichloro silane	[*1719-52-5*]	$C_4H_{10}Cl_2Si$	157.12	128–130	− 96.5	1.0504	d[b]		5.14		21.1
Dimethyldiethoxy silane	[*78-62-6*]	$C_6H_{16}O_2Si$	148.28	114		0.865					11
Methyltriethoxy silane	[*2031-67-6*]	$C_7H_{18}O_3Si$	178.30	141–143		0.8925		1.3835	6.14		23
Ethyltriethoxy silane	[*78-07-9*]	$C_8H_{20}O_3Si$	192.33	158.9		0.8594	i	1.3955			
Amyltriethoxy silane	[*2761-24-2*]	$C_{11}H_{26}O_3Si$	234.41	198		0.889					
Vinyltriethoxy silane	[*78-08-0*]	$C_8H_{18}O_3Si$	190.31	160–161		0.903					34
Tri(2-chloroethoxy) silane		$C_6H_{13}Cl_3O_3Si$	267.61								
Methyl silicate	[*681-84-5*]	$C_4H_{12}O_4Si$	152.22	121–122	− 2°[c]	1.0232				10(25)	45
Ethyl silicate	[*78-10-4*]	$C_8H_{20}O_4Si$	208.30	168	− 85	0.933	i, d[b]		7.2	2(20)	37.2
Hexamethyl disiloxane	[*107-46-0*]	$C_6H_{18}OSi_2$	162.38	101	− 59	0.764	i	1.3818	7.22	1.5 (25)	− 1
Dodecamethyl pentasiloxane	[*141-63-9*]	$C_{12}H_{36}O_4Si_5$	384.84	229 (710)	− 84[c]	0.8755	i	1.3925			− 30

[a]Solubility in water: v = very soluble; s = soluble; d = slightly soluble; i = insoluble.
[b]Decomposes when dissolved.
[c]Freezing point (°C).

Table 81.12. Summary of Oral, Inhalation, and Contact Toxicity Data for Some Organic Silicon Esters

	Oral LC_{50}		LD_{50}		LD_{50}		Inhalation (Rat)						
Compound	Species	g/kg	Species	g/kg	Species	g/kg	Con (ppm)	Exposure (h)	Mortality or Effects	Vap.	Inhal. Mortality Satd. Dermal	Eye Irritation (Rabbit)	Ref.
Silane													
Methyltrichloro	Rat	0.8			Rat	~0.06							64
Dimethyldichloro	Rat	0.8			Rat	~0.06							64
Ethyltrichloro	Rat	0.8			Rat	~1.0							64
Diethyldichloro	Rat	2.0			Rat	~0.6							64
Trimethylethoxy	Rat	9.33											64
Methyltrimethoxy	Rat	12.5	Rabbit	>10							Low to mod.	Very low–low	127
Trimethylethoxy	Rat	12.5	Rabbit	>10			2000	8	0/5 LC_{00}				128
							4000	8	4/5 LC_{80}				128
Dimethyldiethoxy	Rat	2.5					500	10 × 7	Sl. inc. heart wt.				128
							4000	8	0/5 LC_{00}				128
							2000	7	0/5 LC_{00}				128
Methyltriethoxy	Rat	5.0					4000	8	4/5 LC_{80}				128
							125	5–30 × 7	Sl. effects				128
Tetraethoxy							250	4–10	Wt. loss, renal and lung damage				128
							500	3–5 × 5 d	Decr. wt., renal effects, lung irrit.				128
							1000	3 × 7	Decr. wt., other signs				128
Amyltrimethoxy	Rat	4.92	Rabbit	10.0							Mod.	Very low	127
Amyltriethoxy	Rat	19.6	Rabbit	7.13						8 h–none	Low to mod.	Low	7
Vinyltrimethoxy	Rat	11.3	Rabbit	3.54			4000	4	1/6				127
Vinyltriethoxy	Rat	22.5	Rabbit	10.0							Low	Low	7
Tri(2-chloroethoxy)	Rat	0.19	Rabbit	0.089 mL/kg									6

Silicate													
Tetramethoxy	Rat	0.7[a]					1000	8	4/5				128
Tetraethoxy	Rat	0.2–0.4[b]			Rat	0.9–3.5	2000	4	3/5				128
							4000	4	5/5				128
							23–88[c]		None				129
							500[d]	Several	None				130
							2000[d]	1	None				130
										8 h–all	Low to mod.	Very low	127
2-Ethylbutyl	Rat	19.7	Rabbit	> 10									
Silicones													
Siloxane													
Hexamethyldi-	Guinea pig	> 50[e]											128
Dodecamethylpenta-	Guinea pig	40[f]											128

[a]Minimal lethal dose with renal damage.
[b]No-effect level.
[c]Guinea pig and rat.
[d]Guinea pig.
[e]No symptoms
[f]10–30 mL/kg have laxative effect 8 h post administration.

B SILANES

14.0a Tetrachloro Silane

14.0.1a CAS Number: *[10026-04-7]*

14.0.2a Synonyms: SIC-L(TM); silicon chloride; Tetrachlorosilane; Tetrachlorosilicon; Silicon(IV) chloride

14.0.3a Trade Names: NA

14.0.4a Molecular Weight: 169.90

14.0.5a Molecular Formula: Cl_4Si

14.0.6a Molecular Structure:

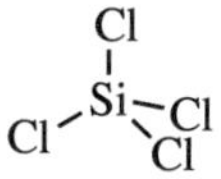

14.0b Methyltrichloro Silane

14.0.1b CAS Number: *[75-79-6]*

14.0.2b Synonyms: Trichloromethylsilane

14.0.3b Trade Names: NA

14.0.4b Molecular Weight: 149.48

14.0.5b Molecular Formula: CH_3Cl_3Si

14.0.6b Molecular Structure:

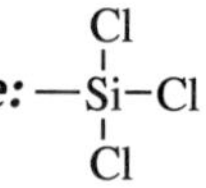

14.0c Dimethyldichloro Silane

14.0.1c CAS Number: *[75-78-5]*

14.0.2c Synonyms: Dimethyl dichlorosilane

14.0.3c Trade Names: NA

14.0.4c Molecular Weight: 129.06

14.0.5c Molecular Formula: $C_2H_6Cl_2Si$

14.0.6c Molecular Structure:

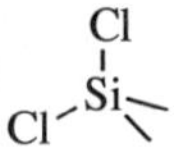

14.0d Ethyltrichloro Silane

14.0.1d CAS Number: *[115-21-9]*

14.0.2d Synonyms: Trichloroethylsilane, ethyl silicon trichloride, and trichloroethylsilicane

14.0.3d Trade Names: NA

14.0.4d Molecular Weight: 163.51

14.0.5d Molecular Formula: $C_2H_5Cl_3Si$

14.0.6d Molecular Structure: 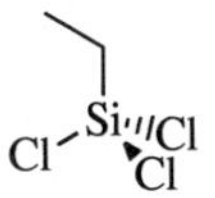

14.0e Diethyldichloro Silane

14.0.1e CAS Number: *[1719-53-5]*

14.0.2e Synonyms: Dichlorodiethylsilane

14.0.3e Trade Names: NA

14.0.4e Molecular Weight: 157.12

14.0.5e Molecular Formula: $C_4H_{10}Cl_2Si$

14.0.6e Molecular Structure: 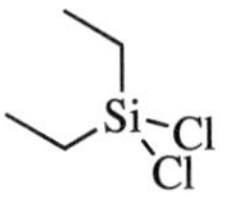

For comparative purposes, some physical properties and range-finding toxicological data are presented in Tables 81.11 and 81.12. Toxicological data show high oral and intraperitoneal toxicities for the methyl and ethyl chlorosilanes. Tetrachlorosilane, $SiCl_4$, an irritant gas, has been used as a warfare agent and to prepare smoke screens (128). The chlorinated derivatives appear to be most toxic of this series, as shown with tri(2-chloroethoxy)silane, $HSI(OCH_2CH_2Cl)_3$.

C SILANOL ESTERS

15.0a Trimethylethoxy Silane

15.0.1a CAS Number: *[1825-62-3]*

15.0.2a Synonyms: Ethoxytrimethyl silane

15.0.3a Trade Names: NA

15.0.4a Molecular Weight: 118

15.0.5a Molecular Formula: $C_5H_{14}OSi$

15.0.6a Molecular Structure: 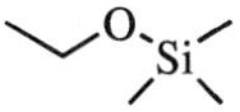

15.0b Dimethyldiethoxy Silane

15.0.1b CAS Number: *[78-62-6]*

15.0.2b Synonyms: Diethoxydimethyl silane

15.0.3b Trade Names: NA

15.0.4b Molecular Weight: 148.28

15.0.5b Molecular Formula: $C_6H_{16}O_2Si$

15.0.6b Molecular Structure: 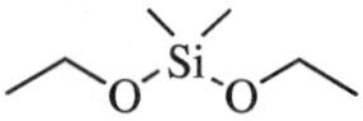

15.0c Methyltriethoxy Silane

15.0.1c CAS Number: *[2031- 67-6]*

15.0.2c Synonyms: Methyltriethoxy silanate

15.0.3c Trade Names: NA

15.0.4c Molecular Weight: 178.30

15.0.5c Molecular Formula: $C_7H_{18}O_3Si$

15.0.6c Molecular Structure: 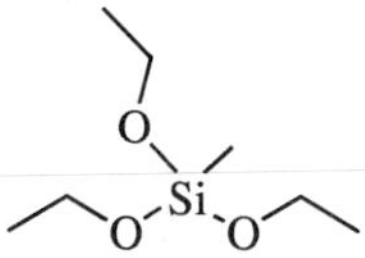

15.0d Amyltriethoxy Silane

15.0.1d CAS Number: *[2761-24-2]*

15.0.2d Synonyms: Amyltriethoxy silanate

15.0.3d Trade Names: NA

15.0.4d Molecular Weight: 234.41

15.0.5d Molecular Formula: $C_{11}H_{26}O_3Si$

15.0.6d Molecular Structure: 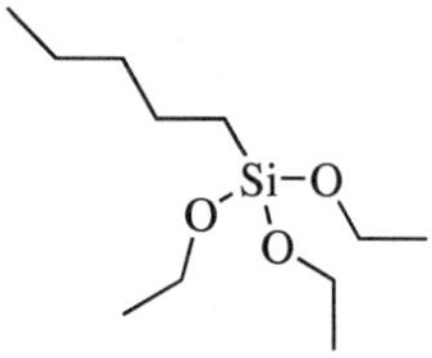

15.0e Vinyltriethoxy Silane

15.0.1e CAS Number: *[78-08-0]*

15.0.2e Synonyms: NA

15.0.3e Trade Names: NA

15.0.4e Molecular Weight: 190.31

15.0.5e Molecular Formula: $C_8H_{18}O_3Si$

15.0.6e Molecular Structure: 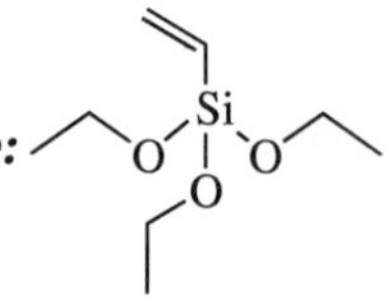

The silanol esters are commonly called silanes, but should more correctly be called silanates. They represent the basic raw materials for the commercial production of silicones. Range-finding toxicological data are shown in Table 81.12. Trimethylethoxysilane (ethoxytrimethylsilane), $(CH_3)_3SiOC_2H_5$, is a liquid at ambient temperatures. Intraperitoneal injections of 1000 mg/kg tetramethoxysilane (TMOS), tetraethoxysilane (TEOS), tetrapropoxysilane (TPOS), and tetrabutoxysilane (TBOS) caused renal toxicologic effects in mice. TMOS and TEOS caused acute tubular necrosis; TEOS, TPOS, and TBOS caused elevated creatinine and blood urea nitrogen; and TMOS exposed mice died exhibiting cytolysis suggestive of spleen damage (131). Mice exposed via inhalation to 100 ppm TEOS 6 h/d, 5 d/wk, for 2–4 wk developed tubulo-interstitial nephritis (132). The mice exposed to 50 ppm TEOS for the same period did not develop nephritis; however, histopathological changes developed in their nasal mucosa. Dimethyldiethoxysilane (diethoxydimethylsilane), $(CH_3)_2Si(OC_2H_5)_2$, is a liquid at ambient temperatures. A mixture of dimethyldiethoxysilane and glycerol has successfully been applied in the rabbit as a model for the prevention of arterial wall and metabolic disorders (133). Amyltriethoxysilane (amyltriethoxysilanate), $C_5H_{11}Si(OC_2H_5)_3$, is of relatively low toxicity.The vinyl triethoxy derivative appears in the same toxicity range.

D SILICATES

16.0a Methyl Silicate

16.0.1a CAS Number: *[681-84-5]*

16.0.2a Synonyms: Tetramethyl orthosilicate, silicic acid tetramethyl ester; methyl orthosilicate, Tetramethoxysilane, tetramethyl silicate, and TMOS

16.0.3a Trade Names: Dynasil M

16.0.4a Molecular Weight: 155.22

16.0.5a Molecular Formula: $C_4H_{12}O_4Si$

16.0.6a Molecular Structure:

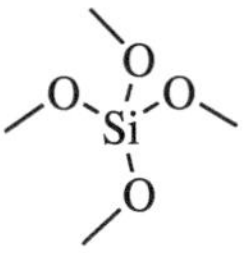

16.0b Ethyl Silicate

16.0.1b CAS Number: *[78-10-4]*

16.0.2b Synonyms: Silicic acid tetraethyl ester

16.0.3b Trade Names: NA

16.0.4b Molecular Weight: 208.30

16.0.5b Molecular Formula: $C_8H_{20}O_4Si$

16.0.6b Molecular Structure:

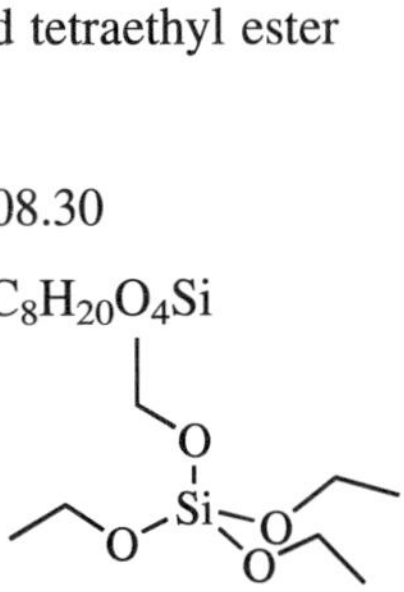

16.3 Exposure Assessment

16.3.3 Workplace Methods

NIOSH Method S264 is recommended for determining workplace exposures to ethyl silicate.

Alkyl silicates, tetraalkyloxysilanes, are organic derivatives of hydrocarbons esterified with silicic acid of the type $Si(OR)_4$. The methyl and ethyl silicates have industrial uses in protective coatings and as preservatives or waterproofing agents for stone and concrete. Range-finding toxicological data are shown in Table 81.12.

Methyl silicate (tetramethyl orthosilicate; silicic acid tetramethyl ester), $Si(OCH_3)_4$, is a liquid under ambient conditions. Methyl silicate has been used in the ceramic industry for closing pores, including those in concrete and cement, for coating metal surfaces and as a bonding agent in paints and lacquers. It is of moderate toxicity, and under certain humid conditions effects progressive necrosis of the cornea.

Ethyl silicate (silicic acid tetraethyl ester), $Si(OC_2H_5)_2$, is a high-boiling liquid. Ethyl silicate is used as a preservative for stone, brick, concrete, and plaster. It is used in water, weather- and acid-proofing processes, heat- and chemical-resistant paints, and protective coatings. Orally, the ethyl silicate is moderately toxic, but may be narcotic in high concentrations. Injected into the skin of the rabbit, it produced transient erythema, edema, and slight necrosis at the injection site (128). In the rabbit eye, it produced transient irritation (128). Inhalation of 400 ppm by rats for 7 h/d for 30 d caused mortalities and lung, liver, and kidney pathological effects. Under similar conditions, 88 ppm caused no effects. Inhalation exposure of the guinea pig to ethyl silicate revealed that humid air was related to more severe effects than dry air (130).

6.5 Standards, Regulations, or Guidelines of Exposure

Both the ACGIH TLV and the NIOSH REL are 1 ppm for methyl silicate. The ACGIH TLV and the NIOSH REL for ethyl silicate is 10 ppm while the OSHA PEL is 100 ppm for ethyl silicate.

E SILICONES (SILOXANES)

17.0a Hexamethyl Disiloxane (HMS)

17.0.1a CAS Number: [107-46-0]

17.0.5a Molecular Formula: $C_6H_{18}OSi_2$

17.0.6a Molecular Structure:

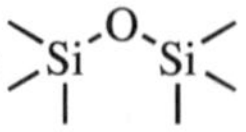

17.0b Dodecamethylpentasiloxane

17.0.1b CAS Number: [141-63-9]

17.0.5b Molecular Formula: $C_{12}H_{36}O_4Si_5$

17.0.6b Molecular Structure:

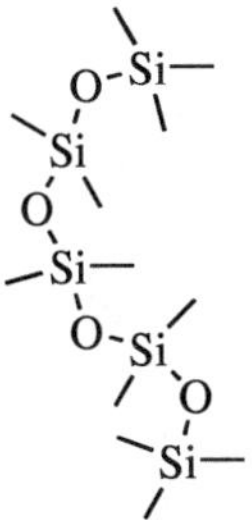

The silicones, organopolysiloxanes, may be divided into the commercial materials, the fluids, and the resins, structurally $R_2[R_2SiO_{n-1}]_n$ for the linear compounds. The final products are water repellent, insoluble in most solvents, and resistant toward oxidation and chemical attack (128). Silicones are used medically and for cosmetic prosthetic devices.

In chronic feeding experiments, rats on HMS showed widespread systemic irritation. Rabbits injected intradermally with HMS showed irritation with edema and necrosis at the injection site (128). Siloxanes injected into the rabbit eye resulted in transient irritation with complete clearing after 48 h (128). When inhaled at 4400 ppm for 19–26 d, HMS caused slight depression in the rat and the guinea pig, with a very slight increase in rat liver and kidney weights (128).

Silicone resins had no influence on health when fed for 94 d to rats, and did not result in irritation to the rabbit skin or eye or when injected into rats intraperitoneally (128). It has been postulated and reported in the literature, however, that implanted silicone prostheses may cause granulomas, lymphadenopathy, cancer, and various autoimmune diseases in humans (134–136). Other reports indicate, however, that a definite correlation has not yet been confirmed by scientific data (137, 138).

BIBLIOGRAPHY

1. American Conference of Governmental and Industrial Hygienists (ACGIH), *1999 TLVs and BEIs*, ACGIH, Cincinnati, OH, 1999.
2. Occupational Safety and Health Administration (OSHA), *Code of Federal Regulations*, Title 29 1910.1000, 1999.
3. National Institute for Occupational Safety and Health (NIOSH), *Manual of Analytical Methods*, 4th ed., U.S. Department of Health and Human Services, Public Health Service, Centers for Disease Control, Cincinnati, OH, 1994 and 2nd ed., Vol. 3, 1977.
4. National Institute for Occupational Safety and Health (NIOSH), *Pocket Guide to Chemical Hazards*, U.S. Department of Health and Human Services, Public Health Service, Centers for Disease Control, Cincinnati, OH, 1997.
5. D. R. Lide, ed., *Handbook of Chemistry and Physics*, CRC Press, Boca Rotan, FL, 1998.
6. F. A. Patty, *Industrial Hygiene and Toxicology*, 2nd rev. ed., Vol. 3, Wiley, New York, 1963.
7. H. F. Smyth, Jr. et al., *Am. Ind. Hyg. Assoc. J.*, **23**, 95 (1962).

8. R. O. B. Wijeskera, A. L. Jayewardene, and L. S. Rajapakse, *J. Sci. Food Agric.* **24**(10), 1211 (1974).
9. J. A. DiPaolo and J. Elis, *Cancer. Res.* **27** 1696 (1967).
10. D. Brown et al., *Toxicology*, **10**(3), 291–295 (1978).
11. T. R. Hanley et al., *Toxicol. Appl. Pharmacol.* **100**, 24–31 (1989).
12. Y. I. Yang, S. G. Ramaswamy, and W. B. Jakoby. *J. Biol. Chem.* **273**(14), 7814–7817 (1998).
13. C. Ursin et al., *Am. Ind. Hyg. Assoc. J.* **56**, 651–660.
14. C. P. Carpenter, C. S. Weil, and H. F. Smyth, Jr., *Toxicol. Appl. Pharmacol.* **28**, 313 (1974).
15. H. F. Smyth, C. P. Carpenter, and C. S. Weil, *J. Ind. Hyg. Toxicol.* **31**, 60 (1949).
16. J. K. Kodama et al., *AMA Arch. Ind. Health* **11**, 487 (1955).
17. National Institute for Occupational Safety and Health (NIOSH), *Registry of Toxic Effects of Chemical Substances*, Vol. II, NIOSH, Cincinnati, OH, 1977.
18. J. V. Marhold, Institut Pro Vychovu Vedousisn Pracovniku Chemickeho Prumyclu Praha, 1972.
19. D. E. Seizinger and B. Dimitriades, *J. Air Pollut. Conrol Assoc.* **22**, 47–51 (1972).
20. E. H. Frear, *Pestic. Index* **4**, 332 (1969).
21. B. O. Holmstedt, *Pharmacol. Rev.* **11**, 567 (1959).
22. E. L. Becker, C. L. Punte, and J. F. Barbaro, *Biochem. Pharmacol.* **13**, 1229 (1964).
23. W. B. Deichmann and H. W. Gerarde, *Toxicology of Drugs and Chemicals*, Academic Press, New York, 1969.
24. H. F. Smyth et al., *Toxicol. Appl. Pharmacol.* **24**, 340 (1969).
25. J. C. Sabine and F. N. Hayes, *Arch. Ind. Hyg. Occup. Med.* **6**, 174 (1952).
26. M. I. Smith et al., *J. Pharmacol. Exp. Ther.* **49**, 79 (1933).
27. R. LeFaux, *Practical Toxicology of Plastics*, 1968.
28. W. B. Deichmann and S. Witherup, *J. Pharmacol. Exp. Ther.* **88**, 338 (1946).
29. S. S. Epstein et al., *Toxicol. Appl. Pharmacol.* **23**, 288 (1972).
30. H. F. Smyth, Jr. and C. P. Carpenter, *J. Ind. Health* **26**, 269 (1944).
31. C. D. Carrington et al., *Int. J. Occup. Med. Immunol. Toxicol.* **5**(1) 61–68 (1996).
32. W. L. Sutton et al., *AMA Arch. Environ. Health* **1**, 33 (1960).
33. O. K. Antonyuk, *Gig. Sanit.* **8**, 98 (1974).
34. J. G. Kidd and O. R. Langworthy, *Bull. Johns Hopkins Hosp.* **52**, 39 (1933).
35. H. C. Hodge and J. H. Sterner, *J. Pharmacol. Exp. Ther.* **79**, 225 (1943).
36. M. I. Smith and E. F. Stohlman, *J. Pharmacol. Exp. Ther.* **51**, 217 (1934).
37. J. F. Treon, F. R. Dutra, and F. P. Cleveland, *Arch. Ind. Hyg. Occup. Med.* **8**, 170 (1957).
38. H. F. Smyth, C. P. Carpenter, and C. S. Weil, *Arch. Ind. Hyg. Occup. Med.* **4**, 119 (1951).
39. E. I. Goldenthal, *Toxicol. Appl. Pharmacol.* **18**, 185 (1971).
40. M. Sherman and E. Ross, *Toxicol. Appl. Pharmacol.* **3**, 521 (1961).
41. T. B. Gaines, *Toxicol. Appl. Pharmacol.* **2**, 88 (1960).
42. *Bull. Entomol. Soc. Am.* **12**, 117 (1966), through Ref. 18.
43. I. L. Natoff and B. Reiff, *Br. J. Pharmacol.* **40**, 124 (1970).
44. R. D. Kimbrough and T. B. Gaines, *Arch. Environ. Health* **16**, 805 (1968).
45. J. K. Kodama et al., *AMA Arch. Ind. Hyg. Occup. Med.* **9**, 45 (1954).

46. M. I. Smith, E. W. Engel, and E. F. Stohlman, *Natl. Inst. Health Bull.* **160** (1932).
47. R. Staehelin, *Schweiz. Med. Wochenschr.* **1** (1941).
48. H. V. Smith and J. M. K. Spalding, *Lancet*, 1019 (1959).
49. R. E. Gosselin et al., *Clinical Toxicology of Commercial Products*, 4th ed., Williams & Wilkins, Baltimore, MD, 1976.
50. M. Susser and Z. Stein, *Br. J. Ind. Med.* **14**, 111 (1957).
51. C. Fest and K. J. Schmidt, *The Chemistry of Organo-Phosphorous Pesticides*, Springer, New York and Berlin, 1973.
52. R. Lauwerys and J. P. Bushet, *J. Eur. Toxicol.* **5**, 163 (1972).
53. K. P. Strickland, R. H. S. Thompson, and G. R. Webster, *Biochemistry* **62**, 512 (1956).
54. C. J. Earl, R. H. S. Thompson, and G. R. Webster, *Br. J. Pharmacol.* **8**, 110 (1953).
55. L. Austin, *Br. J. Pharmacol.* **12**, 356 (1957).
56. W. N. Aldridge, J. M. Barnes, and M. K. Johnson, *Ann. N.Y. Acad. Sci.* **160**, 314 (1969).
57. S. D. Cohen, *Fundam. Appl. Toxicol.* **4**(3,1), 315–324 (1984).
58. C. H. Hine et al., *J. Pharmacol. Exp. Ther.* **116**, 227 (1956).
59. C. Yu and G. M. Booth, *Life Sci.* **10**(6), 337 (1971).
60. R. B. Aird, W. E. Cohn, and S. Weiss, *Proc. Soc. Exp. Biol. Med.* **45**, 306 (1940).
61. S. S. Padilla, T. B. Grizzle, and D. Lyerly, *Toxicol. Appl. Pharmacol.* **87**(2), 249–256 (1987).
62. D. Tanaka, Jr. et al., *Brain Research* **531**(1–2), 294–298 (1990).
63. Y. B. Guthrie, ed., *Petroleum Products Handbook*, McGraw-Hill, New York, 1960.
64. R. Zech, W. D. Erdmann, and H. Engelhard, *Arzneim. Forsch.* **17**(2), 1196 (1967).
65. K. Mieth and D. Beier, *Monatsh. Veterinaer. Med.* **28**(2), 45 (1973).
66. H. Oishi, S. Oishi, and K. Hiraga, *Toxicol. Lett.* **13** (1–2), 29–34 (1982).
67. P. Jones and H. Jackson, *J. Reprod. Fertil.* **38**, 347 (1974).
68. S. S. Epstein et al., *Science* **168**, 584 (1970).
69. D. W. Fassett and R. L. Roudabush, *Arch. Ind. Hyg. Occup. Med.* **6**, 525 (1952).
70. F. Bergmann and H. Govrin, *Biochimie* **55**, 515 (1973).
71. H. Oishi, S. Oishi, and K. Hiraga, *Toxicol. Lett.* **6**(2), 81–85 (1980).
72. M. I. Smith, E. Elvove, and W. H. Frazier, *Public. Health Rep.* **45**, 2509 (1930).
73. J. S. Pegum, *Br. J. Dermatol.* **78**, 626 (1966).
74. H. D. Baldridge et al., *AMA Arch. Ind. Health* **20**, 258 (1959).
75. E. C. Vigliani, *Minerva Med.* **59**(78), 4075 (1968).
76. D. Hunter, K. M. A. Perry, and R. B. Evans, *Br. J. Ind. Med.* **1**, 227 (1944).
77. M. M. Coursey, M. K. Dunlap, and C. H. Hine, *Proc. Soc. Exp. Biol. Med.* **96**, 673 (1957).
78. M. J. Bleiberg and H. Johnson, *Toxicol. Appl. Pharmacol.* **7**, 227 (1965).
79. E. S. Lapadula, D. M. Lapadula, and M. B. Abou-Donia, *Neurochem. Int.* **20**(2), 247–255 (1992).
80. W. E. Luttrell, E. J. Olajos, and P. A. Pleban, *Environ Res.* **60**, 290–294 (1993).
81. M. I. Smith and R. D. Lillie, *Arch. Neurol. Psychiatr* **26**, 976 (1931).
82. J. M. Barnes and F. A. Denz, *J. Pathol. Bacteriol.* **65**, 587 (1953).
83. M. C. Rayner et al., *Res. Commun. Chem. Pathol. Pharmacol.* **4**, 595 (1972).
84. C. D. Carrington et al., *Toxicol. Ind. Health* **6**(3–4), 415–423 (1990).

85. C. E. Healy et al., *J.Am. Ind. Health Assoc.* **53**(3), 175–180 (1992).
86. H. N. MacFarland and C. L. Punte, *Arch. Environ. Health* **13**, 13 (1966).
87. W. Daughtrey et al., *Fundam. Appl. Toxicol.* **32**(20), 244–249 (1996).
88. D. S. Barrett et al., *Vet. Hum. Toxicol.* **36**(2), 103–109 (1994).
89. R. Fautz and H. G. Miltenburger, *Toxicol. In Vitro* **8**(5), 1027–1031 (1994).
90. D. M. Hinton et al., *Int. J. Occup. Med. Immunol. Toxicol.* **5**(1), 43–60 (1996).
91. M. A. Saleem et al., *J. Environ. Pathol. Toxicol. & Oncol.* **17**(1), 69–73 (1998).
92. I. Gruebner, W. Klinger, and H. Ankermann, *Arch. Int. Pharmacodyn. Ther.* **196**(2), 288 (1972).
93. S. D. Cohen and S. D. Murphy, *Biochem. Pharmacol.* **20**(3), 575 (1971).
94. Y.-J. Wu and X.-F. Leng, *Pestic. Biochem. Physiol.* **58**, 7–12 (1997).
95. L. T. Burka and R. E. Chapin, *Reprod. Toxicol.* **7**, 81–86 (1993).
96. S. G. Somkuti et al., *Toxicol. Appl. Pharmacol.* **89**, 64–72 (1987).
97. J. R. Latendresse et al., *Fundam. Appl. Toxicol.* **22**(3), 392–399 (1994).
98. J. R. Latendresse, C. L. Brooks, and C. C. Capen, *Vet. Pathol.* **32**(4), 394–402 (1995).
99. J. J. Welsh et al., *Toxicol. Ind. Health* **3**(30), 357–369 (1987).
100. A. Mentzschel et al., *Carcinogenesis (London)* **14**(10), 2039–2043 (1993).
101. I. R. Tabershaw and M. Kleinfeld, *AMA Arch. Ind. Health* **15**, 541 (1957).
102. I. R. Tabershaw, M. Kleinfeld, and B. Feiner, *AMA Arch. Ind. Health* **15**, 537 (1957).
103. P. A. Bidstrup and J. A. Bonnell, *Chem. Ind. (London)*, p. 674 (1954).
104. B. Jarvholm et al., *Am. J. Ind. Med.* **9**(6), 561–566 (1986).
105. L. Carlsen, K. E. Andersen, and H. Egsgaard, *Contact Dermatitis* **15**(5), 274–277 (1986).
106. F. Jaros and R. Kratinova, *Prac. Lek.* **25**(4), 152 (1973).
107. J. E. Siggins, H. R. Harding, and G. O. Potts, *J. Med. Chem.* **12**(25), 941 (1969).
108. M. Dixon, *Biochem. Soc. Symp.* **2** (1948).
109. E. B. Vedder, *Medical Aspects of Chemical Warfare*, Williams & Wilkins, Baltimore, MD, 1925.
110. A. M. Prentiss, *Chemicals in War*, 1st ed., McGraw-Hill, New York, 1937.
111. F. L. M. Pattison, S. B. D. Hunt, and J. B. Strotheus, *J. Pharmacol. Exp. Toxicol.* **21**, 883 (1956).
112. M. G. Chenoweth and A. Gilman. *J. Pharmacol. Exp. Toxicol.* **87**, 90 (1946).
113. B. C. Saunders, *Nature (London)* **160**, 179 (1947).
114. B. C. Saunders, *J. Chem. Soc.*, p. 1279 (1949).
115. B. C. Saunders and G. J. Stacey, *J. Chem. Soc.*, p. 773. (1948).
116. J. H. Simons ed., *Fluorine Chemistry*, Vol. 7, Academic Press, New York, 1963.
117. B. C. Saunders and G. J. Stacey, *J. Chem. Soc.* **9**, 916 (1949).
118. J. C. Gage, *Br. J. Ind. Med.* **27**(1), 1 (1970).
119. H. McCombie and B. C. Saunders, *Nature (London)* **158**, 382 (1946).
120. F. L. M. Pattison, *Toxic Aliphatic Fluorine Compounds*, Elsevier, Princeton, NJ, 1959.
121. E. S. Hunter, *Teratology* **49**(5), 394. (1994).
122. B. Hassel et al., *J. Cereb Blood Flow Metab.* **17**(11), 1230–1238 (1997).
123. D. A. Keller, D. C. Roe, and P. H. Lieder, *Fundam. Appl. Toxicol.* **30**, 213–219 (1996).

124. B. L. Van Duuren et al., *J. Natl. Cancer Inst.* **53**, 695 (1974).
125. R. R. Burtner and G. Lehmann, *J. Pharmacol. Exp. Ther.* **63**, 183 (1938).
126. S. P. Kramer et al., *J. Natl. Cancer Inst.* **31**(2), 297 (1963).
127. H. F. Smyth et al., *Am. Ind. Hyg Assoc. J.* **30**, 470 (1969).
128. V. K. Rowe, H. C. Spencer, and S. L. Bass, *J. Ind. Hyg. Toxicol.* **30**, 332 (1948).
129. U. C. Pozzani and C. P. Carpenter, *Arch. Ind. Hyg. Occup. Med.* **14**, 465 (1951).
130. H. F. Smyth and J. Seaton, *J. Ind. Hyg. Toxicol.* **22**, 288 (1940).
131. H. Nakashima et al., *Arch. Toxicol.* **68**(5), 277–283 (1994).
132. K. Omae et al., *Sangyo Eiseigaku Zasshi* **37**(1), 1–4 (1995).
133. J. B. Fourtillan and J. P. Dupin, *Chim. Ther.* **8**(2), 207 (1973).
134. W. Kaiser and J. Zazgornik, *Urologe , Ausgabe A* **30**(2), 302–305 (1991).
135. H. P. Frey, G. Emperie, and K. Exner, *Handchir. Mikrochir. Plast. Chir.* **24**(4), 171–178 (1992).
136. T. Randall, *J. Am. Med. Assoc.* **268**(1), 12–13 (1992).
137. H. Berkel, D. C. Birdsell, and H. Jenkins, *N. Engl. J. Med.* **326**(25), 1649–1653 (1992).
138. F. Samdal et al., *Tidssk. Nor. Laegeforen.* **112**(15), 1971–1973 (1992).

CHAPTER EIGHTY-TWO

Epoxy Compounds—Olefin Oxides, Aliphatic Glycidyl Ethers and Aromatic Monoglycidyl Ethers

John M. Waechter, Jr., Ph.D., DABT, Lynn H. Pottenger, Ph.D., and Gauke E. Veenstra, Ph.D.

INTRODUCTION

An epoxy compound is defined as any compound containing one or more oxirane rings. An oxirane ring (epoxide) consists of an oxygen atom linked to two adjacent (vicinal) carbon atoms as follows for the example compound, ethylene oxide:

$$H_2C\overset{O}{\diagup\diagdown}CH_2$$

The term alpha-epoxide is sometimes used for this structure to distinguish it from rings containing more carbon atoms. The alpha does not indicate where in a carbon chain the oxirane ring occurs.

The oxirane ring is highly strained and is thus the most reactive ring of the oxacyclic carbon compounds. The strain is sufficient to force the four carbon atoms nearest the oxygen atom in 1,2-epoxycyclohexane into a common plane, whereas in cyclohexane the carbon atoms are in a zigzag arrangement or boat structure (1). As a result of this strain, epoxy compounds are attacked by almost all nucleophilic substances to open the ring and

Patty's Toxicology, Fifth Edition, Volume 6, Edited by Eula Bingham, Barbara Cohrssen, and Charles H. Powell.
ISBN 0-471-31939-2

form addition compounds. For example,

$$\mathrm{R{-}NH_2 + H_2C\overset{O}{\overbrace{{-}}}CH{-}R_1 \longrightarrow R{-}NH{-}CH_2{-}\overset{OH}{\overset{|}{C}}H{-}R_1}$$

Among agents reacting with epoxy compounds are halogen acids, thiosulfate, carboxylic acids, hydrogen cyanide, water, amines, aldehydes, and alcohols.

A major portion of this chapter presents information on the two olefin oxides, ethylene oxide and propylene oxide, which are produced in high volume and are largely used as intermediates in the production of the glycol ethers. In addition, these compounds are used in the production of several other important products (e.g., polyethylene glycols, ethanolamines, and hydroxypropylcellulose) and have minor uses as fumigants for furs and spices and as medical sterilants. The other olefin oxides discussed are used as chemical intermediates (e.g., vinylcyclohexene mono- and dioxide), as gasoline additives, acid scavengers, and stabilizing agents in chlorinated solvents (butylene oxide) or in limited quantities as reactive diluents for epoxy resins. The discussion of the toxicology of certain olefinic oxides may be pertinent to their respective olefin precursors. However, it must be pointed out that the olefinic precursors of these different oxides demonstrate widely varying degrees of toxicity in mammalian models, mostly attributable to pharmacokinetic/metabolism differences in metabolic conversion of olefins to their respective oxide metabolites. For example, chronic bioassay results range from repeated negatives (ethylene, propylene) to clear positives (butadiene). A major use of the glycidyl ethers discussed in this chapter are as reactive diluents in epoxy resin mixtures. However, some of these materials are also used as intermediates in chemical synthesis as well as in other industrial applications.

The concept that epoxides, through their binding to nucleophilic biopolymers such as DNA, RNA, and protein, can produce toxic effects is well established. However, the magnitude and nature of physiological disruption depend on the reactivity of the particular epoxide (2–4), its molecular weight, and its solubility (5), all of which may control its access to critical molecular targets. In addition, the number of epoxide groups present, the dose and dose rate, the route of administration, and the affinity for the enzymes that can detoxify or activate the compound may affect the degree and nature of the physiologic response. A key enzyme for epoxide detoxification is microsomal epoxide hydrolase (EH), which is widely distributed throughout the body, but it is organ, species, and even strain variant (6). It should be noted that mouse tissues have a much lower level of EH activity than human tissue; in fact, at least two strains of mice, C57BL/6N and DBA/2N, have no EH activity in their skins (7). Hence, the relevance of risk assessment based on the toxicity findings from studies of epoxides in mice is moot. Epoxy compounds may also be metabolized by the cytoplasmic enzyme glutathione-*S*-transferase (GST), which conjugates epoxides with its co-substrate, glutathione (GSH), leading to formation of 2-alkylmercapturic acids. This enzyme, because it is in the aqueous phase, may play a minor role in the detoxification of larger more lipophilic epoxides but is active against low-molecular-weight epoxides (8, 9). GST also exhibits organ, species, and strain differences in expression and activity as well as genetic polymorphisms.

Acute toxic effects most commonly observed in animals have been dermatitis (either primary irritation or secondary to induction of sensitization), eye irritation, pulmonary irritation, and gastric irritation, which are found in these tissues after direct contact with the epoxy compound. Skin irritation is usually manifested by more or less sharply localized lesions that develop rapidly on contact, more frequently on the arms and hands. Signs and symptoms usually include redness, swelling, and intense itching. In severe cases, secondary infections may occur. Workers show marked differences in sensitivity.

Most of the glycidyl ethers in this chapter have shown evidence of delayed contact skin sensitization, in either animals or humans. The animal and human data available on skin sensitization of epoxy compounds do not assist in determining the structural requirements necessary to produce sensitization, but do provide some practical guidance for industrial hygiene purposes. Specifically, of the alkyl glycidyl ethers, only the C8–C10 alkyl glycidyl ether appears to be a human sensitizer. Despite equivocal results in tests for delayed contact sensitization in guinea pigs, *n*-butyl glycidyl ether and cresyl glycidyl ether do produce dermal sensitization in some humans. Skin sensitization reactions can be elicited from much less agent than is required for an irritative response. Because this condition is difficult to treat, sensitized individuals may require transfer to other working areas.

Animals exposed to vapors of gaseous or volatile epoxy compounds, primarily ethylene oxide and propylene oxide, have shown pulmonary irritation. Sequelae of this effect may include pulmonary edema, cardiovascular collapse, and pneumonia. However, this route of exposure is unlikely for some of the other epoxy compounds owing to their lower volatility.

For those epoxy compounds for which repeat exposures have been conducted, respiratory epithelium or nasal mucosa (when inhalation was the route of exposure) and stomach (when given by gavage) appear to be the major target organs. In some cases liver and kidney have been target organs. These effects on the liver and kidney have been relatively nonspecific or adaptive, as indicated by an increased organ weight without accompanying histopathology. Exceptions include ethylene oxide, which caused renal tubular degeneration and necrosis in mice; vinylcyclohexene dioxide, which produced kidney tubule cell necrosis; *n*-butyl glycidyl ether, which produced liver necrosis; and phenyl glycidyl ether, which produced atrophic liver and kidney effects in rats. Ovarian toxicity and depression of hematopoesis have also been observed in laboratory animals for butadiene dioxide and vinylcyclohexene dioxide. Respiratory epithelium and nasal mucosa effects have been responses typical of irritation, such as flattening or destruction of epithelial cells.

Although all of the compounds described in this chapter were mutagenic to bacteria (excluding epoxidized glycerides) as well as positive in other *in vitro* genotoxicity assays, not all have produced genotoxicity in *in vivo* studies. Ethylene oxide was positive in the mouse micronucleus assay and mouse dominant lethal assay. In contrast, propylene oxide, although positive in all the *in vitro* assays in which it was tested, was negative in all of the *in vivo* mammalian assays where propylene oxide was administered via the relevant inhalation route. These negative mammalian studies include a mouse micronucleus assay (although positive by IP injection of high doses), mouse sperm cell analysis, and a rat dominant lethal assay. In addition, propylene oxide failed to cause chromosomal changes (SCE and chromosomal aberrations) in monkey lymphocytes following chronic exposures to 300 ppm. Other compounds showing positive or equivocal effects *in vitro* but negative effects *in vivo* are styrene oxide, and many of the glycidyloxy compounds used in epoxy resin formulations.

There has been no evidence of teratogenicity for glycidyl ethers or olefin oxides, except ethylene oxide (EO), when tested by oral or inhalation exposure in conventional developmental toxicity studies. Fetal toxicity has been observed at maternally toxic doses for ethylene oxide, propylene oxide by inhalation in rats, and 1,2-epoxybutane by inhalation in rabbits. No evidence for fetal toxicity, in some instances even at maternally toxic doses, has been observed for phenyl glycidyl ether by inhalation in rats, 1,2-epoxybutane by inhalation in rats, or propylene oxide or styrene oxide by inhalation in rabbits. Additionally, repeated intravenous infusion of ethylene oxide was teratogenic in mice. Inhalation of extremely high levels of EO (600–1200 ppm compared to the ACGIH-recommended 8-h TWA–TLV of 1 ppm) in mice at the time of fertilization or early zygote development has led to fetal deaths or malformations in some survivors. However, no teratological effects have been demonstrated by inhalation exposures up to 150 ppm in rats or rabbits.

A number of these epoxide compounds have been found to be carcinogenic in rodents, although there has been no clear epidemiologic evidence for cancer in the workplace. In rats and/or mice, many epoxy compounds produce a carcinogenic response in the tissues of first contact. These compounds include ethylene oxide, butylene oxide, propylene oxide, styrene oxide allyl glycidyl ether, phenyl glycidyl ether, and neopentyl glycol diglycidyl ether. A few of them, such as ethylene oxide, butadiene dioxide, and vinylcyclohexene dioxide, have produced tumors at sites other than the "portal of entry."

A OLEFIN OXIDES

1.0 Ethylene Oxide

1.0.1 CAS Number: [75-21-8]

1.0.2 Synonyms: 1,2-epoxyethane, oxirane, dimethylene oxide, oxacyclopropane, dihydrooxirene, epoxyethane, oxane, qazi-ketcham, ethene oxide, alpha, beta-oxidoethane, oxidoethane, amprolene, anprolene, anproline, ethox, fema no. 2433, merpol, oxyfume, oxyfume 12, and T-GAS. Abbreviations in common use are EO, EtO, and ETO

1.0.3 Trade Names: NA

1.0.4 Molecular Weight: 44.05

1.0.5 Molecular Formula: C_2H_4O

1.0.6 Molecular Structure: 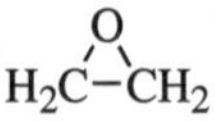 and 

1.1 Chemical and Physical Properties

1.1.1 General

The physical and chemical properties of ethylene oxide are summarized as follows.

Specific gravity:
- 0/4°C: 0.8966
- 20/20°C: 0.8711

Freezing point: − 111.3°C

Boiling point (760 mmHg): 10.7°C

Vapor density (40°C): 1.49

Refractive index (4°C): 1.3614

Solubility: Miscible with water, acetone, methanol, ether, benzene, carbon tetrachloride

Flash point: − 18°C (open cup)

Flammability limits: 3 to 100%

Color: Colorless

Odor: Ethereal—characteristic sweet olefinic odor

Odor threshold: 260 ppm for perception and 500–700 ppm for recognition

1 ppm = 1.80 mg/m^3 at 25°C, 760 mmHg; 1 mg/L = 555 ppm at 25°C, 760 mmHg

It is a highly flammable gas at room temperature and normal pressure, condensing to a liquid at 10°C. Polymerization may occur violently if initiated by acids, bases, or heat. Ethylene oxide is highly reactive toward molecules with active hydrogen atoms and toward nucleophilic agents such as amines and alcohols. For instance, it readily forms ethylene glycol with water, ethylene chlorohydrin with HCl, and hydroxyethylamines with primary or secondary amines. It also reacts with sulfhydryl compounds such as cysteine and glutathione. The major residue that occurs in products such as sterilized foodstuffs and medical equipment is 2-chloroethanol, as a result of reaction with chloride ions. In spite of its generally reactive nature, ethylene oxide is more slowly removed from the atmosphere or water at a neutral pH.

The high chemical reactivity and exothermic nature of the reactions of ethylene oxide present several problems in storage, handling, and use. Although liquid ethylene oxide is relatively stable, the vapor in concentrations ranging from 3 to 100% is highly flammable and subject to explosive decomposition. Ignition may result from many common sources of heat, and the resulting pressure rise may cause the violent rupture of containing equipment. Ethylene oxide vapor is heavier than air and will travel rapidly as a layer to the lowest points. It reacts with materials containing a labile hydrogen. Polymerization is catalyzed by acids, alkalis, some carbonates, oxides of iron and aluminum, and boron. No acetylide-forming metals such as copper or copper alloys should be allowed to come into contact with ethylene oxide. When boron trifluoride is used as a catalyst for ethylene oxide, very toxic organofluorine compounds may be produced (10).

1.1.2 Odor and Warning Properties

Ethylene oxide is reported to have a characteristic sweet, ethereal odor. The reported perception threshold for ethylene oxide is 200 ppm; the odor recognition threshold is 500–700 ppm (11, 12). Because of the high odor threshold for ethylene oxide, sensory recognition does not provide an adequate indication of hazardous exposures.

1.2 Production and Use

Ethylene oxide is currently produced by the direct oxidation of ethylene with oxygen or air over a catalyst. Ethylene is approximately 60% converted to the oxide at temperatures in the range of 100–150°C. In the past, an indirect but more general and more specific synthesis path consisted of adding hypochlorous acid to olefins to form the chlorohydrins. Subsequent treatment with strong bases results in dehydrochlorination and the formation of the epoxide. Although this methodology is no longer used, many epidemiology studies published on ethylene oxide–exposed cohorts included workers who used this production methodology. Industrial production for 1990 in the United States has been estimated at 6.2 billion lb, with worldwide production estimated at over 16.5 billion lb.

About 60% of the ethylene oxide produced in the United States is converted into monoethylene glycol for use in antifreeze, polyester resins, industrial chemicals, and solvents. About 15% is used to produce various nonionic surfactants. Most of the remainder is used to produce polyethylene glycols, glycol ethers, ethanolamines, and mixed polyglycols. A small amount of the total production is used as a sterilant for medical and hospital supplies, and as a fumigant for furs and certain foods, such as spices.

1.3 Exposure Assessment

1.3.2 Background Levels

Ethylene oxide is formed in the body from exogenous and endogenous sources of ethylene. The latter source may be formed as a result of several different metabolic processes, including lipid peroxidation (14); oxidation of free methionine (15, 16); and intestinal microbial metabolism (14, 17). Ethylene is a natural plant hormone; thus exogenous exposure to ethylene also occurs. Ethylene oxide is present in cigarette smoke and automobile exhaust, which provide exogenous sources of background exposure to ethylene oxide itself (14, 18). Several studies have compared the level of EO-derived hemoglobin adducts found in smokers and nonsmokers, and generally reported elevated levels in smokers (18, 19). It has been estimated that about 2% of inhaled ethylene is converted to ethylene oxide in humans (120).

1.3.3 Workplace Methods

Because of the high odor threshold for ethylene oxide, sensory recognition does not provide an adequate indication of hazardous exposures. The current NIOSH recommended method (13) for monitoring potential exposures in the workplace involves adsorption on activated charcoal that has been treated with hydrobromic acid. The ethylene oxide reacts to form bromoethanol, which is desorbed with dimethylformamide. The sample is derivatized to a heptafluorobutyrate ester by reaction with heptafluorobutylimidazole and analyzed by gas chromatography, using an electron capture detector.

An alternative method involves adsorption on activated charcoal, desorption with cold carbon disulfide, and analysis by gas chromatography. Care must be taken to keep the sample tube and the desorbed sample cold to avoid loss of ethylene oxide.

Ambient air monitoring may be done directly, using gas chromatography with flame ionization or photoionization detection. Grab samples may be taken for this type of analysis, but inert sampling bags (such as Tedlar®) must be used to avoid loss of ethylene oxide.

Colorimetric detector tubes may also be used to give an approximation of ethylene oxide levels but should not be relied on for quantitative data.

1.3.5 Biomonitoring/Biomarkers

Several areas of research related to ethylene oxide are particularly active; among these are the use of hemoglobin adducts and lymphocytic chromosomal changes as biomarkers of exposure in human populations and in experimental animals. DNA adducts have also been quantified in experimental animals as potential biomarkers of exposure. Epidemiological investigations of worker populations, either in combination with biomarkers or not, also have been actively pursued in recent years.

As mentioned earlier, studies employing such biomonitoring techniques in animals and humans have demonstrated that there is a low level of endogenous production of ethylene oxide from ethylene formed in the body. It has also been established that environmental exposure to both ethylene and ethylene oxide occurs. Thus there is a quantifiable background level of hydroxyethyl adducts (both hemoglobin and DNA) present in animals and humans that can vary depending on lifestyle and environmental and individual factors (21).

1.3.5.1 Blood. Quantitation of various biomarkers of exposure for ethylene oxide has been reported over the past decade. Hemoglobin adducts and certain cytogenetic biomarkers identified in peripheral blood lymphocytes, such as sister chromatid exhanges (SCE), micronuclei (MN), and chromosomal aberrations (CA), have been the ones most often measured, both in experimental animals and in humans. The cytogenetic biomarkers are discussed at length in Sections 1.4.1.6 and 1.4.2.6.

It has been found that the nucleophilic amino acids cysteine, histidine, and valine become ethoxylated and can be used as biomarkers of exposure. Erythrocytes have a measured lifetime in humans and animals and, because they lack suitable repair enzymes, the altered amino acids in hemoglobin act as a cumulative indicator of exposure with a predictable lifetime. These properties have been explored in animal models, and also studied in an effort to monitor occupational exposure. Many reports have attempted to establish a quantitative relationship between measured atmospheric exposures and alkylated hemoglobin bases ("adducts") in workers or other humans. Lars Ehrenberg of Sweden has been a leading pioneer of this approach (22), and now there are dozens of monitoring studies reported by many researchers available in the published literature. Several have been tied to exogenous exposure levels through extensive monitoring of ambient air levels of ethylene oxide before, during, and/or after collection of blood samples (23–26).

There is need for caution in interpreting the results of these endpoints as to their accuracy for biomarkers of exposure. It should be noted that erythrocytes with high levels of adducts may have a shorter half-life than unexposed cells. Erythrocytes from different species have very different survival curves. The data summarized by Rhomberg et al. (27) indicated that the relationship between hemoglobin adduct formation and exposure in parts per million–hours is rather similar for mice, rats, rabbits, and humans when these parameters are plotted on a log–log scale. However, these data may need to be reexamined in light of the recently demonstrated mouse specific nonlinear kinetics (see Metabolism discussion below).

Walker et al. (28, 29) quantified both hemoglobin and DNA adducts in mice and rats exposed over a 4-wk period to various levels of ethylene oxide and found an accumulation of hemoglobin adducts over the 4-wk period. However, there was not a linear relationship between the levels of hemoglobin adducts and the amount of DNA adducts (represented by levels of 7-hydroxyethylguanine, 7-HEG) in either target (i.e., brain or spleen) or nontarget (i.e., liver) tissues. The ratio between the 7-HEG and the hemoglobin adducts changed over time during multiple exposures to ethylene oxide, and was dependent on many factors. Thus the use of hemoglobin adducts as a systemic dosimeter does not appear to be an accurate estimate of the target tissue dose, even for a direct-acting compound like ethylene oxide (30) EHP. This work was extended by Wu et al. (21), who used a more sensitive and more specific gas chromatography–high-resolution mass spectroscopy (GC-HRMS) to quantify structurally the level of endogenous hydroxyethyl adducts (hemoglobin and DNA) in unexposed animals and ones exposed at very low levels (3 ppm for 4 wk). This work also demonstrated tissue- and species-specific dose–response relationships, including linear and nonlinear effects.

Hemoglobin adducts were determined in Cynomologus monkeys following IV administration of ethylene oxide in combination with propylene oxide at equimolar concentrations (31). "Blood doses" were determined based on hemoglobin adducts, and EO blood levels and half-life were linear across the fourfold increase in administered dose, while propylene oxide (PO) blood levels demonstrated evidence of saturation of detoxication. The small number of animals treated ($n = 2$) and the toxicity of the high dose complicated analysis of the data.

A recent report examined the relationship between hemoglobin adducts in workers exposed to EO and to PO, in conjunction with extensive air and personnel monitoring data during a three-part study (26). In a first study, where single blood samples were taken in conjunction with air level monitoring, there was no correlation between adduct levels and measured air levels. In a second and third study, they examined the difference in hemoglobin levels at the start and finish of a work shift. By taking blood samples before and immediately after work shifts, along with continuous personal monitoring of air levels, they were able to derive an equation that related additional hemoglobin adduct levels to air exposure times time ($mg/m^3/hr$). This gave them the basis for proposing a tentative biological exposure limit (BEL) for ethylene oxide that would correspond with a specific air exposure level. Monitoring exposure to ethylene oxide by measuring the difference in hemoglobin adducts over a work shift seems to offer a useful gauge for determining what level of exposure to ethylene oxide has occurred. However, hemoglobin adducts do not seem to offer a reliable surrogate for internal dose to target tissues.

1.3.5.2 Urine. One published report included measurement of urinary 2-hydroxyethyl-mercapturic acid (HEMA), a predicted urinary metabolite of ethylene oxide, in EO-exposed and in control worker population (32). Although the mean level of urinary HEMA was elevated for exposed workers, there was a wide variation in urinary concentration of HEMA, and no correlation with ethylene oxide exposure levels. This biomarker was not considered suitable for the biomonitoring of exposure.

1.3.5.3 Other. Several groups have utilized DNA adduct formation as a biomarker of exposure to study the tissue dose of ethylene oxide at specific target organ sites in experimental animals (21, 28, 29, 33). The major reaction product of ethylene oxide with DNA is 7-(2-hydroxethyl)guanine, or 7-HEG. This represents about 90% of the alkylated sites in DNA and is detectable in all tissues investigated from treated rats and mice (28). As was discussed, adduct levels were quantified in different tissues from mice and rats during and following up to 4 wk of inhalation exposure to 100 or 300 ppm ethylene oxide. Rat tissues had higher levels than mouse, by two- to threefold, with lung and brain having the highest levels for both species, and testis having the lowest. However, there was no great difference between tissues with increased incidence of tumors and those without. The half-life of these adducts varied between tissues, with mouse liver the shortest (1 d) and mouse kidney the longest (7 d). These values suggest that 7-HEG is removed either by chemical depurination alone (7 d) or by a combination of chemical depurination and normal DNA repair processes (30, 33, 34).

As discussed, the relationship between systemic dose indicated by hemoglobin adducts and tissue dose indicated by DNA adducts is not always straightforward, even for a direct-acting chemical like ethylene oxide. The kinetics of the interaction of ethylene oxide with hemoglobin differ from that with DNA, because the latter is more protected and also has repair enzymes. However, there is some parallelism of results, with the alkylated levels of DNA being lower than those of hemoglobin. The rate constant for protein binding is three to four times greater than that for DNA. Determination of the ratio of hemoglobin adducts to DNA adducts following *in vivo* exposures to ethylene oxide was summarized in Pauwels and Veulemann (35), and ranged from 50 to 1254 for different species, tissues, and exposure routes.

The potential impact of genetic polymorphisms in metabolic enzymes on adduct formation has been investigated, and there are no clear conclusions as of yet. Based on *in vitro* testing, Fost et al. (36) have suggested that a genetic polymorphism may exist in human red blood cell GSTs, which affects their capability to metabolize ethylene oxide, resulting in different levels of hemoglobin adduct for different genotypes. However, they did not see any significant interindividual differences in levels of DNA adducts. Other studies have identified genotypes in conjunction with quantifying hemoglobin adduct levels and have not found any major impact of GST genotype on hydroxyethylvaline hemoglobin adduct levels. Comparison between hydroxyethylvaline hemoglobin adducts in $GSTT_1$-positive or $GSTT_1$-null smokers only showed 30–50% increase in adduct levels for the null genotype, while $GSTM_1$ genotype had little effect on adduct levels (37).

It must be re-emphasized that these biomonitoring techniques should be regarded as indicators of exposure rather than as a direct prediction of a toxicologic effect (biomarkers of effect) at this stage of the science. This is confirmed by the fact that tissue concentration does not appear to be related to the incidence of adverse responses in specific tissues.

1.4 Toxic Effects

Ethylene oxide must be handled with the full appropriate precautions. As a reactive substance, ethylene oxide is highly irritating to exposed tissues. Dilute solutions may cause irritation or necrosis of the eyes and irritation, blistering, and necrosis of the skin.

Excessive exposure to the vapor may also cause irritation of the eyes, skin, and respiratory tract, including the lungs. Systemic effects may include depression of the central nervous system, nausea, vomiting, convulsive seizures, and profound limb weakness. Tissue damage may result in secondary infections such as pneumonia. Other toxicologic end points are discussed in more detail in sections to follow. The aim of this section is to provide an overview of this and also provide sources for additional information (38, 39).

1.4.1 Experimental Studies

Currently active areas of animal model research on ethylene oxide include the study of mutagenicity in mice and the pharmacokinetics and metabolic disposition of ethylene oxide.

1.4.1.1 Acute Toxicity. The oral LD_{50} of ethylene oxide administered in water or corn oil to rats and mice was in the range of 250–350 mg/kg. The dose–response curve for mortality was quite steep (39).

The most likely route of exposure is by inhalation. A recent study in Sprague–Dawley rats provided a 1-h LC_{50} of 5029 ppm (40). The 4-h LC_{50} has been estimated to be approximately 1460 ppm for rats, 835 ppm for mice, and 960 ppm for dogs (41). More recently, the 4-h LC_{50} was confirmed in Sprague–Dawley rats, as 1741 ppm (42). Again, the dose–response curve for mortality appears to be steep.

The first signs of inhalation toxicity in animals relate to respiratory and eye irritation. These include scratching of the nose, nasal discharge, lacrimation, and salivation. Respiration may become labored or gasping as the lungs become congested and edematous. Secondary lung infections and pneumonia can ensue, leading to delayed deaths. There may also be signs of central nervous system effects, including ataxia, prostration, convulsions, and vomiting. Corneal opacities have been seen in some species, including guinea pigs and monkeys.

Skin Effects. Liquid ethylene oxide does not adversely affect skin unless in sufficient quantities to cause "frostbite" by evaporative cooling; however, liquid ethylene oxide can cause severe chemical burns and blistering if left in contact with human skin, such as inside a leather shoe or underneath clothing. Even dilute aqueous solutions can produce irritation, edema, or skin burns in rabbits. Similar effects may be seen in humans, if contact is prolonged, for example, by the continued wearing of contaminated clothing. The vapor may also be irritating, particularly if the skin is wet or oily. The data are mixed concerning skin sensitization in humans. Guinea pig tests have been negative. Certainly occupational dermatitis can occur in people handling materials containing high residues of ethylene oxide. Ethylene oxide will "off-gas" from materials that have been fumigated. The reduction in the levels of residues is a function of time and ventilation, with the residual ethylene oxide dissipating over a period of a few hours up to 2 or 3 d, depending on the initial level and type of material. Thus the incidence of occupational dermatitis would reflect this decay period in specific contact situations.

Eye Irritation. Liquid ethylene oxide can cause severe irritation and corneal injury. Immediate flushing with water or saline may reduce the severity of these effects. High concentrations of the vapor are irritating to the eyes of both animals and humans. There is also an indication that cataracts may occur in exposed monkeys (43).

1.4.1.2 Chronic and Subchronic Toxicity. Subchronic studies have been conducted in rats, mice, rabbits, guinea pigs, and Rhesus monkeys. The concentrations studied range from 50 ppm to as high as 830 ppm (39). These species generally tolerate 50 ppm for 7 h/d, 5 d/w. 830 ppm is lethal in these species after repeated exposures. There are some specie differences in susceptibility, but at doses in the order of 110–330 ppm there is increasing evidence of lung damage, including hemorrhages, edema, collapse, and pneumonia. At lower levels of exposure there appears to be minimal irritation or evidence of frank tissue damage or increased cell turnover. In addition to pulmonary effects there is also evidence of nervous system involvement, which is discussed more fully in Section 1.4.1.7.

1.4.1.3 Pharmacokinetics, Metabolism, and Mechanisms. Ethylene oxide is very soluble in water and also in blood. Uptake of the vapor via the lungs into the body is rapid, although not complete (44); estimates of absorbed vapor are between 40 and 60% (46). Absorbed ethylene oxide is widely distributed among all the tissues. Tissue levels have been quantified in rats and mice, and have shown that testes get from one-half (mouse) to one-fifth (rat) the levels of systemic exposure, based on blood levels (44, 45). Estimated half-life of absorbed ethylene oxide in blood is short (3–12 min in mouse; 9 min in rat; 42 min in humans; (20, 44, 46).

There are two metabolic detoxication pathways for ethylene oxide, namely, glutathione-*S*-transferase–mediated conjugation with glutathione, which leads to the formation of hydroxyethyl cysteine derivatives, and hydrolysis via epoxide hydrolase (EH), which results in formation of ethylene glycol (1,2-ethanediol) (48–50). Metabolites from both of these pathways are mainly excreted in urine. Some ethylene glycol may be further metabolized to hydroxyacetic acid, oxoacetic acid, ethanioic acid, and formic acid. Further metabolism to CO_2 occurs with the metabolic products from both pathways. Another potential metabolite, namely, 2-chloroethanol, which is formed by the reaction with chloride ions, has not been demonstrated in animals, although it is found in materials of plant origin such as foodstuffs fumigated with ethylene oxide. Goldberg (38) suggested that the alcohol is an intermediate in the formation of other metabolites, but there is not general agreement on this (33). More recently, glycolaldehyde has been proposed as a possible metabolite of ethylene oxide via further metabolism of the EO metabolite, ethylene glycol (52).

Although all species investigated seem to have qualitatively similar metabolic pathways for ethylene oxide, detailed investigation into mouse and rat metabolism has shown clear quantitative specie differences, and *in vitro* work with human tissues indicates that human metabolism also differs quantitatively from the mouse and rat (37, 44, 45). In particular, the mouse relies mainly on GST-mediated conjugation of ethylene oxide with GSH as the main metabolic route, with about 75% of absorbed ethylene oxide metabolized through GSH-dependent pathways. In contrast, the rat metabolism is equally split between GST conjugation and EH hydrolysis of ethylene oxide. Based on *in vitro* metabolism studies conducted with human liver samples, humans appear to rely mainly on hydrolysis via EH to ethylene glycol for ethylene oxide metabolism.

These quantitative specie differences in metabolism result in nonlinearity in the elimination of ethylene oxide from the mouse at exposures above 200 ppm, while no evidence of non-linearity was found for overall elimination from rats (37, 44, 53). High or repeated exposures can deplete tissue glutathione levels in both species (54, 55), but the mouse is particularly sensitive. The mechanistic explanation of mouse nonlinearity lies in their increased tendency for depletion of tissue GSH. Once the GSH is depleted to about 20% of control values, there is no longer sufficient GSH present to maintain GST-mediated conjugation at a high rate. With the major detoxication pathway no longer operating at maximal rate, blood and tissue levels of ethylene oxide increase in the mouse. This does not occur with rats, where GST-mediated conjugation accounts for only 50% of ethylene oxide detoxication, nor is it predicted to occur in humans, where GST-mediated metabolism is predicted to account for only a small fraction of total detoxication.

The pharmacokinetic data collected over the past 10 yr have been used to develop physiologically based pharmacokinetic (PB-PK) models to describe the metabolism and fate of ethylene oxide. Such models permit cross-species extrapolation, predicting internal doses of ethylene oxide in rats, mice, and humans for different exposure scenarios. The models can use different dosimeters to describe internal dose, such as blood and/or tissue levels of ethylene oxide, or hemoglobin and/or DNA adduct formation and elimination (37, 53, 56, 57). These models are expected to play an important role in understanding the ever-increasing body of biological monitoring (biomarker) data that are being collected. In addition, validated PB-PK models are expected to offer important perspective on potential risks from occupational exposures.

Ethylene oxide that is not metabolized is either exhaled unchanged or reacts with cellular macromolecules, such as proteins, RNA, and DNA. Formation of hemoglobin and DNA adducts were discussed previously (see Section 1.3.5).

1.4.1.4 Reproductive and Developmental. There is ample evidence that ethylene oxide reaches the testes of exposed animals, and it can be assumed that this is also true for the ovaries. Testicular tubular degeneration and atrophy have been reported in rats and guinea pigs exposed to ethylene oxide. Maximum alkylation of sperm occurs at the middle to late spermatid stage. In rats, abnormal sperm heads are also seen. Some of these testicular effects are prevented in rats by the administration of methylcobalamin. Mori et al. and Sega et al. (58–60) reported that in mice, the alkylation of sperm DNA was greater when a given dose was administered in a short time frame than when the same total dose (concentration times time) was given over a longer period. This is consistent with the corresponding mutagenicity data. The dose–rate effect of ethylene oxide represents a very active area of research, which is discussed briefly in the following.

In a one-generation F344 rat reproductive toxicity study, a reduction in the fertility index and litter sizes was seen at 100 ppm, whereas 33 ppm was considered a no-observed-effect exposure (NOEL) (61). A more recent two-generation reproductive toxicity study in Sprague–Dawley rats investigated effects of EO exposure at the same levels as the previously described study (431). The NOEL for reproductive effects was 10 ppm, and, based on an increase in postimplantation loss for the F_0 generation and reductions in fetal weight gain, the LOEL was 33 ppm ethylene oxide.

No teratological effects were demonstrated by inhalation exposures up to 100–150 ppm in rats and rabbits, although effects have been seen after the repeated intravenous infusion of up to 150 mg/kg in mice (63). A developmental toxicity study was conducted using Sprague–Dawley rats exposed to 50, 125, and 225 ppm ethylene oxide (62). This study did not establish a NOEL, based on a 4% reduction in fetal body weights at 50 ppm. However, this effect was considered to be of little or no biological significance; hence 50 ppm can be considered a no observed adverse effect level (NOAEL). There was no increased incidence in malformations at any exposure.

There are several studies indicating that exposure of female mice to 600–1200 ppm ethylene oxide at the time of fertilization or early zygote development leads to midgestational or late fetal deaths and malformations in some of the survivors Katoh et al. and Rutledge and Generoso (64, 65) showed that these effects were most obvious when exposure was 1 or 6 h after mating and were marginal by 9 h. Generoso et al. (66) have suggested that altered gene expression is responsible rather than a mutation *per se*. The no-effect concentration for these effects is 600 ppm.

Developmental toxicity has been used to investigate dose–rate effects of exposure to ethylene oxide. Two recent studies clearly demonstrate that exposure to high levels of ethylene oxide over short time periods can result in more noticeable effects than an equivalent total dose administered at a lower exposure level over a longer time period. Repeated brief exposures (3 times/day) compared with single daily exposure showed that the result of the product of concentration and time ($c \times t$) is not necessarily a constant for ethylene oxide–induced developmental effects in rats (67). Whereas fetal toxicity was produced in rats exposed for 30 min to 1200 ppm, no fetal toxicity was seen in rats exposed for three 30 min exposures to 400 ppm. Generally, higher exposures resulted in more severe effects than repeated exposures to lower concentrations, although the total $c \times t$ was kept equivalent. A similar result was reported for mice by Weller et al. (68), who showed that, for several different $c \times t$ equivalent totals, short, high exposures to ethylene oxide on day 7 resulted in an increased incidence of developmental effects compared with longer, lower exposures. These apparent dose–rate effects have been repeatedly demonstrated using different endpoints, and may be related to nonlinear pharmacokinetics, as discussed in the section on pharmacokinetics and metabolism (1.4.1.3).

1.4.1.5 Carcinogenesis. Ethylene oxide has been tested for carcinogenic potential by several routes of exposure. Dermal carcinogenesis has not been demonstrated, although subcutaneous injection has produced tumors in mice. Intragastric administration was associated with stomach tumors in rats.

There have been two chronic inhalation bioassays conducted with rats (38, 69, 70) and one with mice (71). The rat studies showed an increased incidence of mononuclear cell leukemias, peritoneal mesotheliomas, and brain tumors. The study in mice showed a statistical excess of tumors of the lung and Harderian gland. An increased incidence of tumors of the uterus and mammary gland and malignant lymphoma also occurred. These bioassays have been subjected to several intensive analyses (38, 72). Given the similarity in tissue concentrations between organs and the range of tumors found, there is some ambiguity as to what are the "target organs." In the rat studies the incidence of tumors increased toward the end of the study.

Preneoplastic foci have been demonstrated in the livers of rats. These foci have not been demonstrated to lead to hepatocarcinogenesis in rats exposed to ethylene oxide (73).

1.4.1.6 Genetic and Related Cellular Effects Studies. Ethylene oxide has been shown to be mutagenic in a wide range of test systems, ranging from bacteria, fungi, and yeasts to mammalian cells including human lymphocytes (74). Although the *in vivo* mutagenic potency of ethylene oxide is generally weak, requiring high inhalation exposure levels to obtain clear positive responses, it is often used as a positive control substance in such tests. Ethylene oxide is regarded as a direct mutagen; that is, metabolic activation is not required. Molecular characterization of mutations induced *in vitro* (75) and *in vivo* (76) indicate that large deletions and modifications at guanine and adenine comprise important aspects in the mechanism of ethylene oxide mutagenicity.

New techniques available for studying mutagenicity have been applied to ethylene oxide, including the use of transgenic rodents to identify mutations in target and nontarget tissue and studies of mutational spectra. One study, conducted using Big Blue® mice, compared mutation frequency in the transgene, *lacI* (77), to mutation frequency in an endogenous gene, *hprt* (78) in splenic and/or thymus-derived T-cells and lymphocytes, following 4 wk of exposure to ethylene oxide (50–200 ppm). Other tissues were investigated for *lacI* mutation frequency also, including lung, bone marrow, and germ cells (77). The transgenic system demonstrated an increased mutation frequency only for lung, only at 200 ppm, while HPRT mutations in thymic and splenic T-lymphocytes increased with dose following 4 wk exposure to 100 or 200 ppm ethylene oxide. The difference in sensitivity between these two systems was explained based on the inability of the *lacI* system to identify large deletions (77).

There are two particular aspects of the mutagenicity of ethylene oxide that require detailed evaluation, namely, ability to induce mutations in germ cells and the demonstration of cytogenetic and other mutational endpoints in human lymphocytes.

Inhalation exposure to high levels of ethylene oxide has been shown to produce both dominant lethal mutations and chromosome aberrations, such as reciprocal translocations that may be seen in the sperm of treated mice Generoso et al. and Wiencke et al. (79, 80) have shown that there is a gradient of susceptibility during the spermatogenic germ-cell stages, with late spermatids being particularly susceptible. Protamine, which is protein unique to sperm cells, is also alkylated. Potential transmissable effects of exposure to ethylene oxide were investigated in *Drosophila* (51); mating of exposed males to excision repair-deficient females resulted in a significant increase in sex-linked recessive lethality.

Rhomberg et al. (27) have examined the available mouse data and have concluded that the dose–response relationship for ethylene oxide–induced genotoxic events is nonlinear. The shape of the ethylene oxide dose–response curve was also addressed in a review on genetic risk assessment of ethylene oxide (82). Those authors agreed that ethylene oxide presented a nonlinear dose–response curve. They proposed that the dose–response curve must be a function of the square of the dose to account for a two-hit mechanism required for the formation of reciprocal translocations. In fact, the hockey-stick curve that a dose-squared relationship would predict does fit the reciprocal translocation and dominant lethal data reported by Generoso et al. (79) very well. This would indicate that there is a nonlinear dose–response for the murine dominant lethal effects of ethylene oxide. This dose–response relationships could be based on GSH depletion, as discussed, especially since the positive dominant lethal effects were identified at exposure levels approximately at or above the 200 ppm breakpoint for GSH depletion in mouse.

Based on these animal data, Rhomberg et al. (27) estimated that for human males, the risk of fathering a translocation carrier when exposed to an average of 0.5 ppm for 60 h was less than 1 in 10,000 (ppm concentration was multiplied by 120 h exposures); exposure to 100 ppm for 12,000 ppm h might result in a more than 1 in 10 risk. Because sperm have a limited life span, the time sequence of exposure and fertilization is a key factor; that is, there is a relatively narrow window for an effect to occur. A more recent genetic risk review (82) estimated that any genetic risk to males from low exposures to ethylene oxide would be negligible, given the proposed shape of the low dose–response curve. They further proposed that the frequency of translocation in female germ cells would also be negligible at low exposures, given the very high probability of DNA repair prior to the next S-phase in a resting oocyte. The biological model for risk assessment described in this report resulted in a predicted genetic risk from low exposures to ethylene oxide of at least an order of magnitude less than the previous effort.

There are many reports of ethylene oxide producing chromosomal-type mutagenic effects in the blood cells of animals and humans. For instance, sister chromatid exchanges (SCEs) can be quantified in the chromosomes of peripheral lymphocytes. Micronuclei may also be seen in bone marrow or red blood cells. Mutations have also been demonstrated in the *hprt* gene locus, which controls a specific enzyme. A review by Natarajan et al. (83) summarizes both animal and human genotoxicity data on ethylene oxide, and estimates a risk for dominantly inherited disease in offspring based on a parallelogram approach. These authors concluded that humans are around 3 times less sensitive than mice to induction of somatic mutations, and predicted a 4×10^{-4} increase over background risk for 1 working year exposed to the 1 ppm TWA. This is in contrast with the estimate discussed previously, where negligible risk was predicted for exposure at 1 ppm (82).

In general, these chromosomal-type effects have been seen in populations with regular exposures in the range of 10 ppm or more. The background rates are such that studies must be conducted on exposed and control populations to measure any effect and that little weight can usually be placed on the results of single samples. Several studies are discussed under Section 1.4.2.2.6. Mechanistic considerations of these cytogenetic changes, discussed in Preston et al. (84), indicate that these alterations are nontransmissable *in vivo*, and as such, should be considered as biomarkers of exposure. In fact, with no known health effects associated with these alterations that are based on *in vitro* DNA replication errors, both SCE and chromosome aberrations are considered as markers of exposure rather than effect. In addition, Preston concludes that measuring transmissable effects (i.e., reciprocal translocations, inversions), using techniques such as fluorescence *in situ* hybridization (FISH) are necessary to be able to provide a mechanistic link between cytogenetic effects and prediction of carcinogenicity or other *in vivo* genetic effects.

1.4.1.7 Other: Neurological, Pulmonary, Skin Sensitization. Many of the clinical signs of exposure to high concentrations of ethylene oxide suggest observable pharmacological effects on the nervous system. In several species (rat, mouse, rabbit, monkey, dog) repeated high exposures have a characteristic peripheral effect on the lumbosacral nerves. This results in some paralysis, muscle atrophy, and decrease in pain perception and reflexes in the hind limbs. Exposures sufficient to cause these effects lie above 200 ppm, with 100 ppm being regarded as a no-effect level. It is possible that more subtle neurobehavioral effects will be detected in animals at somewhat lower exposures.

Onishi et al. (85) reported that subchronic, high-level exposures of rats to ethylene oxide (500 ppm for 13 wk) resulted in clinical signs of neuropathy (awkward gait at weeks 5–8, followed by slight to moderate hindlimb ataxia by week 9 or 10), which were confirmed by degenerative changes in peripheral nerves. Chronic exposure of monkeys, reported by Setzer et al. (86) demonstrated that, following 2 yr of 7 h/d, 5 d/wk inhalation exposures to 50 and 100 ppm EO, there were no neurological effects identified in monkeys that were considered compound related, including no neurohistopathology. A few monkeys were sacrificed 7 yr after termination of exposure, and no treatment-related effects were detected. A neurotoxicity study was conducted recently in Sprague–Dawley rats, and resulted in no evidence of treatment-related subchronic neurotoxicity after 13 wk of whole-body inhalation exposure up to 200 ppm ethylene oxide (87).

1.4.2 Human Experience

1.4.2.1 General Information. No case histories of deaths in humans exposed to high concentrations of ethylene oxide have been located. There is no evidence of cardiovascular effects such as electrocardiogram changes following exposure to ethylene oxide (38). Renal effects have not been reported in humans, although high exposures in mice have led to tubular degeneration and necrosis.

1.4.2.2 Clinical Cases

1.4.2.2.1 Acute Toxicity. Acute exposure to high levels of ethylene oxide can cause irritation to eyes and skin, including corneal injury. In addition, CNS-related symptoms such as nausea, vomiting, dizziness, and even unconsciousness have been reported.

1.4.2.2.2 Chronic and Subchronic Toxicity. Clinical observations suggest that continued overexposure (possibly several hundred parts per million) in humans may lead to nystagmus, ataxia, incoordination, weakness, and slurred speech. Sensorimotor neuropathy may occur in both the upper and lower limbs. Full or partial recovery may occur over a several-month period.

Cataracts have been reported in men after working on a leaking sterilizer (38), and more recently, in a hospital sterilizing unit workers exposed to levels up to 90 ppm, well above the 1 ppm TLV, without any use of personal protective equipment (88). In that study, 14 of 16 workers had lens opacities, of which 3 were possibly related to occupational exposure to ethylene oxide and the others were deemed age related. Deschamps et al. (89, 90) examined a population of 55 persons exposed to ethylene oxide and found that the incidence of lens opacities was similar to controls. A significant percentage of people over 45 in both exposed and nonexposed groups demonstrated ocular opacities. However, in terms of cataracts, there was an increased incidence in the exposed population. Eye protection (chemical goggles) should be worn when handling ethylene oxide or its solutions.

1.4.2.2.3 Pharmacokinetics, Metabolism, and Mechanisms. Few data were located on pharmacokinetics and metabolism of ethylene oxide in humans. Brugnone et al. (91) showed that, at exposures of 0.11–12.3 ppm ethylene oxide, workers absorbed 75–80%

of inhaled ethylene oxide at a steady state. Alveolar air concentrations ranged from 0.03 to 3.8 ppm, and the absorption following an 8-h exposure to 1 ppm was estimated to be 7.2–7.7 mg ethylene oxide. As mentioned previously, Filser et al. (20) used these data to calculate a mean half-life of 42 min for ethylene oxide in humans. The *in vitro* data and PB-PK model were discussed in Section 1.4.1.3.

1.4.2.2.4 Reproductive and Developmental. There is an Eastern European report of a range of gynecological and pregnancy disorders resulting from the exposure of women to ethylene oxide. This study is difficult to evaluate without further data (38). Other studies have not confirmed these effects but warn that potential reproductive effects from ethylene oxide should not be disregarded (92). A more recent report investigated potential effects on pregnancies of dental hygienists using ethylene oxide gas to sterilize instruments while working during pregnancy (93). The authors reported an increase in spontaneous abortions in the potentially exposed group in both preterm and postterm births, compared with the controls. The number of subjects was small, and the controls demonstrated a much lower than expected rate of spontaneous abortions. In addition, the data were based on self-reported recollection of outcomes several years after the fact, without validation of reported outcomes (94). Other potential confounders included incomplete information on potential of co-exposures for the majority of exposed subjects reporting spontaneous abortions, and no information on birth weight to validate reported outcome for preterm and postterm births. Although ethylene oxide does not appear to be a highly active reproductive toxin as judged by classical reproductive toxicologic techniques, the potential for mutagenic events to affect offspring adversely should be considered; see also the following section on genetic toxicity.

1.4.2.2.6 Genetic and Related Cellular Effects Studies. Many studies have investigated potential impact of ethylene oxide exposure on cytogenetic and other biomarkers in worker and other populations. Wiencke et al. (80) suggested that there may be a bimodal distribution of SCEs in the unexposed population. Tates et al. (96) concluded that in terms of sensitivity for detection, hemoglobin adducts were more readily detected than SCEs > chromosomal aberrations > micronuclei > HPRT mutations. This work has been extended by Mayer et al. (97), who showed in a relatively small population of hospital workers exposed to ethylene oxide at or below the current OSHA exposure limits that, although control smokers had more hemoglobin adducts and SCEs than exposed nonsmokers, there was an exposure-related increase when nonsmokers and smokers were analyzed separately (see also Reference 98). Other endpoints such as chromosomal aberrations, micronuclei, DNA repair index, and single-strand breakage did not show these trends. This information also confirms that at low exposure levels, extraneous sources must also be considered.

De Jong et al. (99) explored the use of cytogenetic monitoring at low levels of exposure to several compounds, including ethylene oxide, and concluded that the methods were not sufficiently sensitive for the typical current occupational exposures.

Several more recent studies monitoring cytogenetic changes as potential biomarkers of effect in workers exposed to ethylene oxide have been published, including both sterilization and manufacturing industry populations (98, 100–105).

These have been critically reviewed (84). Based on the criteria applied in this review, many of the studies of the human population exposed to EO contained shortcomings that limited their usefulness in establishing cytogenetics as a definitive biomarker of effect. In particular, the review discusses the implications of the assessment of unstable, nontransmissable alterations such as SCEs, micronuclei, and chromosomal aberrations in peripheral lymphocytes. These cytogenetic alterations represent effects that occur in the first *in vitro* S phase, that is, once the cells are removed from the body, and are not alterations produced *in vivo*. Hence, the frequency of cytogenetic alteration measured by these methods will be proportional to the damage remaining at the time of *in vitro* DNA replication. Because DNA damage is rapidly repaired in cells in G_0 phase, elevated frequencies in these parameters measured months and even years after ethylene oxide exposure are most probably due to other factors not adequately controlled for in the study. Additional difficulties were identified in some of these studies due to insufficient subjects (exposed and appropriate controls) and confounding exposures (smoking, radon, etc).

The overall conclusion from this critical review was that, although acute, high-level exposures to ethylene oxide, with sampling shortly following exposures can result in observable increases in unstable chromosome aberrations and SCE, the human cytogenetic monitoring studies currently available do not provide convincing evidence that chronic exposure to 25 ppm or less ethylene oxide results in increased chromosome aberrations or SCE (84).

1.4.2.2.7 Other: Neurological, Pulmonary, Skin Sensitization, etc. The available evidence does not suggest that any neurotoxic hazard will exist at the current exposure limits, although there is some evidence of effects within the older occupational exposure limits (106, 107).

The potential for skin sensitization was discussed previously (see Section 1.4.1.1 on skin effects).

There have been reports of allergic and/or asthmatic reactions in sterilization function workers (108, 109), and in dialysis patients, (110), with some immunological evidence for formation of anti-ethylene oxide antibodies in the published literature. Contact dermatitis has been the subject of several reports over the years, but is not a common occurrence (111). There is not good agreement on the causality of ethylene oxide in these effects.

1.4.2.3 Epidemiology Studies. There is a rich epidemiological database available on ethylene oxide, comprising about 33,000 workers in both manufacturing and sterilizing populations, with 10 different cohorts and some large studies with as much as 27 yr average follow up. Recent publications include a meta-analysis and an update of that meta-analysis of all the EO epidemiology studies (112, 113). Even with this unusually extensive database, the human evidence relating to carcinogenicity remains uncertain. There is one epidemiologic report from 1979 that suggested ethylene oxide exposure may be related to leukemias. Three cases versus 0.2 expected (acute myelogenous and "associated with macroglobulinemia") were found in a population exposed to methyl formate and ethylene oxide (114). Over the ensuing years, several other studies have failed to confirm this conclusion (115–117), neither in manufacturing nor in sterilizer populations. Generally, additional studies have not confirmed the leukemia conclusion. Olsen et al. (118) reported on a cohort of 1361 workers in a chlorohydrin manufacturing setting, with a followup period of 24 yr on average, that demonstrated no increases in leukemia. Other recent studies have found either similar or inconclusive results, with no clear significant relationship between ethylene oxide exposure and leukemia (119–121).

Cohorts have been investigated for other cancers, too. Contradictory results exist for breast cancer, for example, where the largest study, with 9,000 female subjects (116) showed a deficit of breast cancer mortality. This is in contrast with a smaller study, which reported an increased incidence of breast cancer in workers potentially exposed to EO, but which diminished over time (122).

Based on the meta-analyses conducted on the overall data, there is some evidence of an association between exposure to EO and non-Hodgkins lymphoma and possibly leukemia that requires additional study (112, 113). The current assessment is that the epidemiologic study of high-exposure populations has not revealed a consistent cancer response in humans and, given the reductions in exposure that have occurred, it seems unlikely that studies on current or future workers will show any response (113).

A more recent epidemiology study and update included data on hemoglobin adducts to validate exposure to ethylene oxide and to attempt to quantify internal dose (123, 124). The original Hagmar et al. study (123) indicated that, in the sterilizer plants examined, exposures about 30 years ago would be 75 ppm but for the past 10 years were usually well below 1 ppm (<0.2 ppm). The more recent update on this cohort demonstrated only a nonsignificantly increased risk estimate for leukemia, and only for cumulative exposures above the median.

1.5 Standards, Regulations, or Guidelines of Exposure

Generally, international workplace standards range between 0.5 and 1 ppm as an 8h TWA. The Occupational Safety and Health Administration (OSHA) has established a permissible exposure limit (PEL) for ethylene oxide of 1 ppm in air, as an 8h time-weighted average. An excursion limit of 5 ppm, as a 15-min STEL, has also been established. An action level of 0.5 ppm, as an 8-h TWA, has been set that, if met or exceeded for 30 or more days per year, triggers certain requirements such as ongoing workplace monitoring, medical surveillance, and annual employee training. For further information on the OSHA requirements, refer to the Ethylene Oxide Standard, 29 CFR 1910.1047.

The American Conference of Governmental Industrial Hygienists (ACGIH) has established a threshold limit value TLV for ethylene oxide of 1 ppm, as an 8-h TWA, and has assigned an A2 designation (suspected human carcinogen).

The National Institute for Occupational Safety and Health (NIOSH) has classified ethylene oxide as a potential human carcinogen and recommends maintaining workplace exposure below 0.1 ppm, as an 8-h TWA.

The International Agency for Research on Cancer (IARC) reviewed the database for ethylene oxide in 1994, and classified EO as a Group 1, known human carcinogen. IARC justified the classification based on mechanistic considerations, as the epidemiology data were judged to be limited (125).

2.0 Propylene Oxide

2.0.1 CAS Number: [75-56-9]

2.0.2 Synonyms: 1,2-Epoxypropane, propene oxide, methyl ethylene oxide, methyl oxirane, 1,2-propylene oxide; epoxypropane; propylene epoxide; 2,3-epoxypropane; ad 6 (suspending agent); PO and AD 6

2.0.3 Trade Names: NA

2.0.4 Molecular Weight: 58.080

2.0.5 Molecular Formula: C_3H_6O

2.0.6 Molecular Structure: 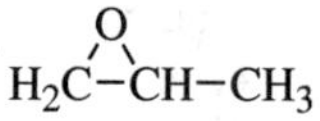 or 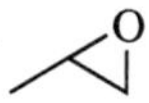

2.1 Chemical and Physical Properties

2.1.1 General

Propylene oxide is a colorless liquid that is highly reactive chemically. It is also extremely flammable. The liquid is relatively stable but may react violently with materials having a labile hydrogen, particularly in the presence of catalysts such as acids, alkalis, and certain salts. It polymerizes exothermically. No acetylide-forming metals such as copper or copper alloys should be in contact with propylene oxide. The same general handling procedures as described for ethylene oxide (126) should be employed.

Physical and chemical properties of propylene oxide:

Molecular formula: C_3H_6O
Specific gravity:
  20/20°C: 0.8304
  25/25°C: 0.826
Freezing point: – 112°C
Boiling point (760 mmHg): 34.2°C
Vapor density (air = 1): 2.0
Refractive index (25°C): 1.363
Solubility: 40.5% by wt. in water at 20°C, 59% by wt. in water at 25°C; miscible with acetone, benzene, carbon tetrachloride, ether, and methanol
Flash point: – 30°C (open)
  – 37°C (closed cup)
Flammability limits: 2.1 to 38.5% by vol in air

1 ppm = 2.376 mg/m^3 at 25°C, 760 mmHg; 1 mg/L = 421 ppm at 25°C, 760 mmHg

2.1.2 Odor and Warning Properties

Propylene oxide is reported to have a sweet, alcoholic odor. Reported ranges for the odor threshold for propylene oxide vary widely. The median detectable odor concentration was calculated to be 200 ppm (127). Amoore and Hautala (128) reported an odor threshold of 44 ppm, and Hellman and Small (129) reported a threshold value of 10 ppm for detection of an odor, and 35 ppm for when the chemical odor was recognized.

2.2 Production and Use

Propylene oxide is synthesized commercially from propylene through the intermediate propylene chlorohydrin. It also can be made by direct oxidation of propylene with either air or oxygen, but this method produces co-products as well. The largest use of propylene oxide is in the production of polyurethane foams and resins. The second largest use is for the production of propylene glycol and its derivatives. Substantial quantities are used also in the preparation of hydroxypropylcelluloses and sugars, surface-active agents, and isopropanolamine. Minor applications include use as a fumigant, herbicide, preservative, and solvent (130).

2.3 Exposure Assessment

2.3.2 Background Levels

Background/endogenous exposure to propylene oxide is judged to occur based on the background levels of hydroxypropylvaline (HPV) adducts in hemoglobin from unexposed control animals and humans (131, 132). The source of these background levels in hemoglobin adducts is not determined. Environmental exposure to propylene and propylene oxide also occurs, for example, from tobacco smoke (14, 18). No detectable background/endogenous levels of DNA adducts (hydroxypropylguanine, HPG) have been reported in the literature.

2.3.3 Workplace Methods

A NIOSH method 1612 for the determination of propylene oxide in air is available (13). This method describes the collection of propylene oxide on coconut shell charcoal solid sorbent tubes, desorption with CS_2, and analysis by gas chromatography with flame ionization detection.

2.3.5 Biomonitoring/Biomarkers

Several areas of research related to propylene oxide are particularly active; among these are investigations of hemoglobin and DNA adducts as measures of internal dose in experimental animals. In addition, there are a few reports of using hydroxypropyl–hemoglobin adducts as a biomarker of exposure in propylene oxide worker populations.

As mentioned earlier, studies employing such biomonitoring techniques in animals and humans have demonstrated that there is a low level of background/endogenous exposure to propylene oxide in the body. It is known that environmental exposure to propylene oxide occurs, for example, from tobacco smoke, because the level of hydroxypropyl adducts is higher in smokers than nonsmokers. Thus there is a quantifiable background level of hydroxypropyl–hemoglobin adducts present in animals and humans that can vary depending on lifestyle and other environmental factors (18, 131, 132).

2.3.5.1 Blood. The first reports of hydroxpropyl–hemoglobin adducts were from exposing mice to ^{14}C-labeled propylene oxide by IP injection (133). They demonstrated a linear relationship between administered dose and blood dose (determined as HPV adducts). Recent work in rats has reported steady state levels of about 78 nmol HPV/g globin following 4 wk of inhalation exposure (6 h/d; 5 d/wk) to 500 ppm propylene oxide (131).

A few reports of attempts to correlate air exposure levels with hemoglobin adducts in worker populations have been published. Some earlier reports did not always include appropriate controls, and therefore are difficult to interpret (134, 135). More recent work included extensive personnel air monitoring and sampling blood at the beginning and end of maintenance projects (26). By taking blood samples before and at the conclusion of the project, along with continuous personal monitoring of air levels, the authors reported they were able to derive an equation that related additional hemoglobin adduct levels to air exposure times time (mg/m^3 h). This gave them the basis for proposing a tentative biological exposure limit for propylene oxide that would correspond with a specific air exposure level. Monitoring exposure to propylene oxide by measuring the difference in hemoglobin adducts over a work shift seems to offer a useful gauge for determining what level of exposure to propylene oxide has occurred. Nonetheless, hemoglobin adducts do not seem to offer a reliable surrogate for internal dose to target tissues, as is discussed in the following.

2.3.5.2 Urine. No data identified. It seems unlikely that a urinary metabolite of propylene oxide would offer a reliable biomarker of exposure since present-day exposures are in the low ppm levels, and propylene oxide is likely extensively metabolized to propylene glycol, an endogenous product of intermediary metabolism that is further metabolized to carbon dioxide in humans.

2.3.5.3 Other. Investigations of hydroxypropyl DNA adducts have demonstrated that the major adduct formed is the N^7-hydroxypropylguanine (HPG), both *in vitro* (137–139) and *in vivo* (132, 138, 139). HPG comprises about 90+% of total DNA adducts identified from rats exposed to propylene oxide by inhalation (132, 138). The minor adducts so far identified include the 1-/N^{6}HP-adenine and 3-HP-cytosine/uracil, which comprise about 2 and 0.02% of the HPG in the target tissue in rat nasal epithelium (139). The 3-HP-cytosine/uracil was only quantifiable in the nose.

Although propylene oxide is expected to distribute evenly throughout the body, quantitation of HPG adducts in different tissues demonstrated a clear titration effect on adduct levels following 4 wk of inhalation exposure to 500 ppm (132, 141). In fact, tissues distant from the site of contact (respiratory tree) demonstrated adduct levels that were up to 25-fold and 42-fold less than the nasal respiratory epithelium (NRE). Thus there was not a linear relationship between the level of hemoglobin adducts (the systemic dosimeter) and DNA adducts in various tissues (internal dose dosimeter).

Segerback et al. (142) evaluated specie comparisons in hydroxypropyl hemoglobin and liver DNA adducts as biomarkers of internal dose among mice, rats, and beagle dogs exposed to 2-^{14}C-propylene oxide vapors. The purpose of the experiment was to evaluate whether or not surface-area-based dose extrapolations (vs. body-weight-based extrapolations) were warranted in risk assessment modeling. Although sensitivity concerns due to the low specific activity of the radiolabel limit the usefulness of this study, the authors reported that hemoglobin and DNA alkylation data correlated with calculated administered dose. It must be re-emphasized that these biomonitoring techniques should be regarded as indicators of exposure rather than as a direct prediction of a toxicologic effect (biomarkers of effect) at this stage of the science. This is confirmed by the fact that tissue concentration does not demonstrate a consistent relationship with an adverse response in specific tissues.

2.4 Toxic Effects

2.4.1 Experimental Studies

2.4.1.1 Acute Toxicity. Oral LD_{50} values by gavage of a 5% aqueous solution were 1.14 g/kg in rats and 0.69 g/kg in guinea pigs (143). When gavaged as a 10% olive oil solution, rats survived 0.3 g/kg and died at 1.0 g/kg (144). The dermal LD_{50} in rabbits was 1.3 g/kg (145). Four-hour inhalation LC_{50} of propylene oxide were 4126 mg/m^3 (1740 ppm) in mice and 9486 mg/m^3 (4000 ppm) in rats (143). At 7200 ppm in rats, deaths were 0/10, 2/10, and 5/10 after exposures of 0.25, 0.5, and 1.0 h, respectively. There was no evidence of systemic toxicity at necropsy in rats exposed by inhalation to 4000 ppm for 0.5 h, 2000 ppm for 2 h, or 1000 ppm for 7 h. Sensitivity to the acute lethal effects of propylene oxide appears to be in the following order, from most to least sensitive: dog > mouse > rat > guinea pig (143, 144).

Skin Effects. Undiluted propylene oxide has low irritation potential if the skin is uncovered to allow for evaporation. However, when the skin is covered, as when contaminated clothing or shoes are worn, even dilute (10%) water solutions can cause severe irritation.

Eye Irritation. Propylene oxide vapor is irritating to the eyes of animals, and liquid caused severe eye irritation in rabbits (145).

2.4.1.2 Chronic and Subchronic Toxicity. When inhaled repeatedly, propylene oxide vapor appears to be about one-third as toxic as ethylene oxide (143, 144). The primary toxic effect is lung irritation. When vapor exposures were conducted for 5 d/wk for approximately 6–7 mo, at a vapor concentration of 102 ppm, 9/40 rats died, but there were 0/16 deaths in guinea pigs, and there were no pathological findings at necropsy in either species. At 195 ppm, 7/40 rats died, there were no deaths in 16 guinea pigs, four rabbits, or two monkeys, and only female guinea pig; lungs increased in weight. At 457 ppm, no adverse effects were seen in rabbits and monkeys; growth depression, lung damage, and eye irritation were seen in rats and guinea pigs; and deaths occurred in rats (144). In a more recent National Toxicology Program (NTP) inhalation study, however, no treatment-related gross or microscopic pathology was observed in mice or rats exposed to up to 500 ppm propylene oxide vapor for 6 h/d, 5 d/wk, for 13 wk (146).

Eldridge et al. (147), using more sensitive methods of cell damage detection (BrdU uptake), reported increased cell turnover in nasal epithelium but not in liver from rats exposed to 500 ppm propylene oxide, 6 h/d for a total of 20 exposures. Nasal respiratory epithelium (NRE) was more severely affected than was olfactory epithelium. More recently, a dose–response study reported that exposure to $\geq$300 ppm propylene oxide for either 3 or 20 d (6 h/d; 5 d/wk) resulted in a statistically significant increase in cell proliferation (> four fold difference) in the anterior portion of the nasal passages (NRE) (148). A significant increase in cell proliferation in the nasopharyngeal duct was only evident at 500 ppm, which suggests a lesser cytotoxic effect induced by propylene oxide in this region. No exposure-related changes in cell proliferation were observed in the liver. These studies demonstrate that the target organ for carcinogenesis (nose) is a sensitive site for toxicity-induced compensatory cell proliferation after repeated (3 or 20 d) inhalation exposure to levels of propylene oxide that induced tumors. They also demonstrate that tissues that do not receive sufficient delivered dose to induce toxicity also lack any compensatory response in increased cell proliferation. This finding is consistent with the limitation of propylene oxide–induced effects to the site of contact.

Repeated gavage doses of 0.2 g/kg as a 10% solution in olive oil given five times a week until 18 doses had been given failed to produce significant toxicity in rats; a dose of 0.3 g/kg caused loss of body weight, gastric irritation, and slight liver injury (144).

2.4.1.3 Pharmacokinetics, Metabolism, and Mechanics. Research on the pharmacokinetics and metabolism of propylene oxide has been actively pursued in recent years. Propylene oxide is detoxified either by hydrolysis to propylene diol (propylene glycol) via epoxide hydratase (EH) or by conjugation with glutathione via glutathione-*S*-transferase activity. Propylene glycol can be further metabolized to pyruvic and lactic acid, eventually leading to formation of CO_2 or other normal intermediary metabolites.

An early study demonstrated depletion of liver GSH levels and nonlinear increases in PO blood levels in rats following inhalation exposure to propylene oxide (155). These data have been confirmed and extended in work that has been presented mostly as abstracts up to the present. Rats exposed to propylene oxide by inhalation demonstrated nonlinear effects for PO blood levels and tissue GSH depletion (156). The gas-uptake data for these *in vivo* exposures were described using a two-compartment pharmacokinetic model.

Tissue and blood partition coefficients were determined for several different tissues and showed that propylene oxide was equally soluble in lipid and water, indicating that absorbed PO would be expected to distribute evenly throughout the body (157). These values were collected to be used in the development of a physiologically based pharmacokinetic model to describe propylene oxide uptake, metabolism, and elimination in rats. *In vitro* metabolism studies were conducted to determine the *in vitro* enzyme constants for GST and for EH-mediated metabolism and detoxication of propylene oxide (158). Values were obtained for liver and lung using subcellular fractions from mouse, rat, and humans. In addition, values were obtained for the target tissue, rat nasal respiratory epithelium. The reported V_{max}/K_m indicated that GST-mediated metabolism predominates in mouse liver, while GST- and EH-mediated metabolism are about equal in rat liver. Samples from two different human livers indicated that EH-mediated metabolism predominates in humans. Results from lung tissue were different, with GST-mediated metabolism contributing a larger fraction of detoxication than EH-mediated hydrolysis in mouse and rat, but both contributing similar amounts to human lung metabolism based on lung samples from two humans. Again, these data will serve in the development of a PB-PK model capable of cross-species extrapolation for predicting internal dose of propylene oxide.

Additional *in vivo* kinetic data have been reported in the past, including an estimated blood half-life of 3 min in rats, based on gas-uptake studies (159). Blood levels in male F344/N rats were measured following inhalation exposure to 14 ppm propylene for 60 min (160). These data showed increasing blood propylene oxide concentrations during the first 10 min of exposure, with concentrations leveling off at 3 ng/g blood for the remainder of the exposure. The authors reported a similar plateau-type curve for propylene oxide blood levels following exposure to propylene. Steady-state levels of propylene oxide in rats were determined following 20 d (6 h/d; 5 d/wk) inhalation exposures to 500 ppm. The average blood level of 37 μmol propylene oxide/l blood agreed with the predicted value of 37.1 μmol/l blood, based on a two-compartment pharmacokinetic model (131, 157).

2.4.1.4 Reproductive and Developmental. Sprague–Dawley rats and New Zealand White rabbits were exposed by inhalation to 500 ppm propylene oxide in a developmental toxicity study (161). There was no evidence of fetal toxicity or developmental defects in rabbits. In rats, however, fetal toxicity was evident by decreased fetal body weight and crown-rump length. An increase in wavy ribs and reduced ossification in fetal vertebrae and ribs were also noted in rats. Reduction in number of corpora lutea, implantation sites, and live fetuses occurred in rats exposed during the pregestational period (total of 3 wk of exposures), suggesting an effect on fertility and survival of offspring. Because maternal toxicity was also evident in the dams, clear interpretation of the rat developmental effects in this study is difficult. The data have been extensively reviewed (162, 163).

In another subsequent inhalation developmental toxicity study in F344 rats, exposures were for 6 h/d on gestational days 6 to 15 at concentrations of 0, 100, 300, and 500 ppm. Despite maternal toxicity at the high exposure level of 500 ppm, there was no evidence of fetal toxicity. Only at the maternally toxic level of 500 ppm was there evidence for any developmental toxicity, expressed as an increased frequency of seventh cervical ribs in fetuses. The no observed adverse effect level was considered to be 300 ppm (164).

Inhalation exposure to propylene oxide at levels up to 300 ppm over two generations did not produce any adverse effects on reproductive function, including a lack of effect on fertility and survival of offspring (165).

2.4.1.5 Carcinogenesis. In an NTP inhalation bioassay in F344 rats and $B6C3F_1$ mice, propylene oxide vapor was administered for 6 h/d, 5 d/wk, for 103 wk at concentrations of 0, 200, or 400 ppm. The NTP concluded there was some evidence of carcinogenicity in rats, as indicated by increased incidences of papillary adenomas of the nasal turbinates in high-dose males and females, and clear evidence of carcinogenicity in mice, as indicated by increased incidences of hemangiomas or hemangiosarcomas of the nasal turbinates at 400 ppm. Suppurative inflammation, hyperplasia, and squamous metaplasia in rats and inflammation in mice were non-neoplastic histopathology also observed in the respiratory epithelium of the nasal turbinates (146). These studies have been extensively reviewed (149, 150).

A second bioassay, conducted by NIOSH in male F344 rats, examined the effect of chronic exposure to 100 and 300 ppm propylene oxide (151). This study demonstrated again that nasal respiratory epithelium was the major target of chronic effects of inhaled propylene oxide. Similar pathology of complex epithelial hyperplasia was found, as had been described for the NTP study. Development of proliferative lesions in nasal mucosa was influenced by an intercurrent Mycoplasma infection. A minimal, site-of-contact tumorigenesis response occurred, with two adenomas in nasal passages. There were no treatment-related systemic tumors identified from this study.

In a 28-mo inhalation bioassay in Wistar rats (152), rats were exposed at concentrations of 0, 30, 100, or 300 ppm. Nasal and combined-site respiratory tract tumors were observed at the high dose. Hyperplasia of the respiratory epithelium and degeneration of the olfactory epithelium were also observed in the high-dose group. Although there was a statistically significant increase identified in mammary tumors in females for the high-dose group, the incidence (67%) was barely outside of the highly variable historical control incidence for the testing laboratory (19–61%). Both the longer duration of exposures (28 mo) and the different strain of rats were proposed as possible contributors to this increased incidence. Increased incidence of mammary tumors was not identified in either of the other two, well-conducted bioassays where F344 rats were exposed to higher levels of PO (400 ppm) for up to 2 yr (146, 151). F344 rats, like Wistar rats, are highly prone to the development of mammary tumors, and therefore would be expected to respond with an increased incidence in mammary tumors if that tissue were truly a target for PO-induced carcinogenesis.

Using a different study design, Sellakumar et al. (153) investigated carcinogenicity of propylene oxide in Sprague–Dawley rats following 30 d (6 h/d; 5 d/wk) inhalation exposure to 435, 870, and 1740 ppm. The 1740-ppm exposures resulted in significant lethality after 8 d and were discontinued. The remaining groups were exposed over the 30 d, and then held until they died spontaneously (about 645 d for propylene oxide–exposed rats). Nasal histopathology similar to other reports, including rhinitis and squamous metaplasia, was described for the propylene oxide–exposed rats. No nasal tumors were found in any animals exposed to propylene oxide, although other treatment groups did demonstrate increases in nasal tumors (e.g., methylmethansulfonate).

In an oral carcinogenicity study in Sprague–Dawley rats, propylene oxide was administered twice weekly by gavage in an edible oil for 109.5 wk at doses of 0, 15, or 60 mg/kg. Treatment resulted in a dose-dependent increase in the incidence of squamous cell carcinomas of the forestomach. Tumor incidence at other sites was not increased compared to controls. Propylene oxide–treated rats also developed papillomas, hyperplasia, or hyperkeratosis of the forestomach (154).

A dose-related increase in local sarcomas was observed in female NMRI mice given subcutaneous injections of 0.1, 0.3, 1.0, or 2.5 mg/mouse once weekly for 95 wk (154).

2.4.1.6 Genetic and Related Cellular Effects Studies. Propylene oxide is a direct-acting chemical, and its mutagenicity data have been reviewed extensively (166–168). Propylene oxide was mutagenic *in vitro* to *Salmonella* strains TA100 and TA1535 in both the presence and absence of a metabolizing system in the Ames assay (169, 170), and was positive in the *E. coli* WP2 strain (171), indicating it is a base substitution mutagen. It also induced forward mutations in the yeast *Schizosaccharomyces pombe* (172).

Propylene oxide was positive in all the *in vitro* mammalian cell assays investigated. It induced chromosomal aberrations in an epithelial-like cell line derived from rat liver (173) and in human lymphocytes in cultures established from peripheral blood (170, 174). It also was positive in the CHO/HGPRT mutation assay (175), induced chromosomal aberrations and sister chromatid exchanges in Chinese hamster ovary cells (176), and was mutagenic in the mouse lymphoma forward mutation assay (177).

Sex-linked recessive lethal mutation was observed in *Drosophila* exposed to 645 ppm propylene oxide vapor for 24 h, with spermatocytes and mature sperm being sensitive stages (178).

Evidence for propylene oxide mutagenicity following *in vivo* exposure is not conclusive. All the *in vivo* mammalian cytogenetic and mutagenic data are negative, except when propylene oxide is administered by IP injection at high doses. There was no evidence of either chromosome aberrations or SCEs in lymphocytes from monkeys exposed for 2 yr to propylene oxide vapor at concentrations as high as 300 ppm (179). Orally administered propylene oxide at a dose as high as 500 mg/kg was negative in a mouse micronucleus assay, and gave no evidence of mutagenic action on sperm at a dose of 250 mg/kg/d for 14 d in a mouse dominant lethal test. However, when it was injected intraperitoneally at a dose of 300 mg/kg, propylene oxide was positive in the mouse micronucleus assay; lower doses were without effect (170). In a similar, more recent study, IP injection of 350 mg/kg propylene oxide resulted in increased incidences of SCE, micronuclei and chromosomal aberrations (180). In male rats or mice exposed to 300 ppm propylene oxide vapors for 7 h/d on five consecutive days, there were no sperm head abnormalities in mice, and no dominant lethal mutation was observed in rats following mating with virgin females weekly for 6 wk (178).

2.4.1.7 Other: Neurological, Pulmonary, Skin Sensitization. Propylene oxide has been reported to cause central-peripheral distal axonopathy in rats exposed five times a week for 7 wk to a vapor concentration of 1500 ppm, a nearly lethal level (182). However, in a separate study at lower, nonlethal concentrations, there was no evidence of neurotoxicity associated with inhalation exposure in rats at concentrations as high as 300 ppm for 6 h/d for 24 wk (183). Monkeys exposed to 100 or 300 ppm propylene oxide for 2 yr did not demonstrate any physiological or functional alternations compared with control monkeys (86). Exposed monkeys sacrificed 7 yr following the end of exposures did not demonstrate any neuropathological effects.

2.4.2 Human Experience

2.4.2.1 General Information. Propylene oxide is reported to cause eye irritation at exposure levels considerably higher than current workplace permissible exposure levels. No reports of human fatalities following acute exposure to propylene oxide were located in the literature.

2.4.2.2 Clinical Cases

2.4.2.2.1 Acute Toxicity. Human experiential data from the late 1960s were described in the NAC/EPA Acute Exposure Guideline Level (AEGL) for propylene oxide (185).

Those data indicate that exposures for 30–170 min of up to 1500 ppm propylene oxide resulted in "strong odor, however the irritation was not intolerable."

2.4.2.2.2 Chronic and Subchronic Toxicity. Additional human experimental data, from the late 1940s and also discussed in the NAC/AEGL document (185), describe "eye irritation after about two weeks of steady operation" under conditions of exposures measured at between about 350 and 900 ppm over two 30-min sampling periods. No other symptoms were reported.

2.4.2.2.3 Pharmacokinetics, Metabolism, and Mechanisms. The *in vitro* data from human tissues were described in Section 2.4.1.4. No additional data were identified in the literature.

2.4.2.2.4 Reproductive and Developmental. No data on humans were identified in the literature.

2.4.2.2.6 Genetic and Related Cellular Effects Studies. A few studies have attempted to correlate genotoxic effects with propylene oxide exposure. A series of publications, using hemoglobin adducts as an internal dosimeter, found that lymphocytes from individuals exposed to propylene oxide and ethylene oxide demonstrated a reduced capacity for unscheduled DNA synthesis (186–188). The authors interpreted these results as indicating a reduced ability to repair DNA damage, but this study has several drawbacks, including wide inter- and intraindividual variability in hemoglobin adduct values, limited numbers of controls, no adjustment for smoking (smokers: 100% of PO-exposed subjects; 38% of controls), and overlapping values for control and exposed subjects. Thus it is difficult to draw conclusions from these data. Exposures were for 1–20 years to time-weighted-average air concentrations of 0.6 to 12 ppm with short-term exposures to air levels as high as 1000 ppm (186, 188). As part of an epidemiology study described, an increase in chromosome aberrations in peripheral blood lymphocytes was reported in a subset of 43 male workers exposed for more than 20 years to both propylene oxide and ethylene oxide (189). The mixed exposures make it difficult to interpret these results. Hogstedt et al. (134) reported values for cytogenetic alterations (chromosomal aberrations and micronuclei) in workers exposed to propylene oxide; however, there were no unexposed control values for comparison; so no conclusions can be drawn from these data.

2.4.2.2.7 Other: Neurological, Pulmonary Skin Sensitization, etc. Contact dermatitis (190, 191) and corneal burns (192) have been reported in humans exposed to propylene oxide.

2.4.2.3 Epidemiology Studies. No significant excess of mortality could be found due to any cause in a cohort of 602 workers exposed occupationally to both propylene oxide and ethylene oxide over the period 1928–1980 (193). Other studies have also been reported, but are confounded by exposures to several chemicals in addition to propylene oxide (194).

2.5 Standards, Regulations, or Guidelines of Exposure

The ACGIH proposed 8 h TWA TLV is 5 ppm (2000). The majority of countries with OELs have similar values (UK = 5 ppm; NL = 2.9 ppm; DE = 2.5 ppm). Propylene oxide is considered a carcinogen by NIOSH. The OSHA PEL is 100 ppm. IARC reviewed the available data on propylene oxide in 1994 and downgraded its classification to a Group 2B, a possible human carcinogen, based on sufficient animal and insufficient human data (195).

3.0a 1,2-Butylene Oxide

3.0b 2,3-Butylene Oxide

3.0.1a CAS Number: *[106-88-7]*

3.0.1b CAS Number: *[3266-23-7]*

3.0.2a Synonyms: 1-Butene oxide, 1,2-butylene oxide, 1,2-butene oxide, 1,2-butylene epoxide, 1-butylene oxide, epoxybutane, ethyl ethylene oxide, 2-ethyloxirane, ethyloxirane and 1,2-epoxybutane

3.0.2b Synonyms: 2,3-Epoxybutane, beta-butylene oxide, 2,3-dimethylethylene oxide, 2,3-butylene oxide, and 2,3-epoxybutane (*cis,trans* mixture)

3.0.3a Trade Names: NA

3.0.3b Trade Names: NA

3.0.4a Molecular Weight: 72.107

3.0.4b Molecular Weight: 72.107

3.0.5a Molecular Formula: C_4H_8O

3.0.5b Molecular Formula: C_4H_8O

3.0.6a Molecular Structure: $H_2C{-}CH{-}CH_2{-}CH_3$ (O bridging the first two carbons)

3.0.6b Molecular Structure: $H_3C{-}HC{-}CH{-}CH_3$ (O bridging the two central carbons)

3.1 Chemical and Physical Properties

1,2-Butylene oxide (1,2-epoxybutane) and butylene oxide(s), a mixture of 1,2- (80–90%) and 2,3-epoxybutane (10–20%), are water-white liquids with the following properties:

	1,2-Butylene Oxide	Butylene Oxides
Specific gravity (25°C)	0.826	0.815
Freezing point (°C)	Below −60	Below −50
Boiling point (°C)(760 mmHg)	63.3	54
Refractive index (25°C)	1.381	1.378
Density of satd. air (air = 1)	~0.977	~1.36
Vapor density	~177	~183
Solubility at 25°C g/100 g H_2O	~8.24	>=10
Other common solvents	Miscible with common aliphatic and aromatic solvents	Miscible with aliphatic and aromatic solvents
Flash point (°F/closed cup)	−15	−16
Flammability limits		1.5–18.3% by volume in air

1 ppm = 2.94 mg/m^3 at 25°C, 760 mmHg; 1 mg/L = 340 ppm at 25°C, 760 mmHg

3.1.2 *Odor and Warning Properties*

The odor of the mixed straight-chain isomers of butylene oxide may be described as sweetish, somewhat like butyric acid, and disagreeable. It is doubtful that the odor can be relied on to prevent excessive exposure.

3.2 Production and Use

Butylene oxide is available commercially as the mixed isomers or as the single 1,2-isomer. The oxides are prepared commercially from butylene through the intermediate butylene chlorohydrin. The butylene oxides are used for the production of the corresponding butylene glycols and their derivatives, such as polybutylene glycols, mixed polyglycols, and glycol ethers and esters. They are also used to make butanolamines, surface-active agents, and gasoline additives, and as acid scavengers and stabilizers for chlorinated solvents.

The butylene oxides are highly flammable and highly reactive chemically, but are less reactive than ethylene or propylene oxide. The liquids are relatively stable, but they may react violently with materials having a labile hydrogen, particularly in the presence of catalysts such as acids, alkalis, and certain salts. They are capable of polymerizing exothermically. The same general precautions should be taken when handling the butylene oxides as when handling ethylene oxide (196).

3.3 Exposure Assessment

There was no standard method found for the analysis of butylene oxides in air. However, the NIOSH standard method available for the determination of propylene oxide (13) appears suitable and are subject to limitations.

3.4 Toxic Effects

Toxicologic data are available mostly for the 1,2 isomer, which is the primary component of the commercial butylene oxides. They tend to be generally less reactive than ethylene oxide and propylene oxide. Butylene oxides are moderately acutely toxic and are substantial irritants, which react with "portal of entry" tissues, such as nasal epithelium and lung when inhaled. The isomers are direct-acting alkylating agents and have been shown to be genotoxic in several *in vitro* bacterial and mammalian cell assays and *in vivo* in *Drosophila.* There is limited evidence for carcinogenicity in rodents, and no data for humans. IARC recently reviewed butylene oxide and classified it as a Group 2B, possibly carcinogenic to humans (197). Because fetal toxicity was observed only in rabbits at maternally toxic doses of 1,2-epoxybutane vapor, NIOSH concluded that the fetus would be protected at maternally safe exposure levels. 1,2-Epoxybutane is extensively metabolized and rapidly eliminated.

3.4.1 Experimental Studies

3.4.1.1 Acute Toxicity. The LD_{50} of 1,2-epoxybutane in rats was 1.17 g/kg orally and 1.76 g/kg dermally in rabbits (195). The 4-h inhalation LC_{50} was reported to be >2000 ppm in rat but between 400 and 1400 ppm in mouse; clinical signs of eye and respiratory tract irritation were observed (198).

Skin Effects. 1,2-Epoxybutane was not a skin sensitizer in guinea pigs (199), but elicited marked skin irritation when applied occluded in rabbits (145).

Eye Irritation. Marked irritation with corneal injury resulted in rabbits when liquid chemical was instilled in the eye (145).

3.4.1.2 Chronic and Subchronic Toxicity. In rats inhaling 400–600 ppm 1,2-epoxybutane for 6 h/d, 5 d/wk for 2 wk, there was inflammation of the nasal mucosa and lungs, bone marrow hyperplasia, and elevated white blood cell count in the high-dose animals. Surviving mice exposed to the same levels had similar effects on the respiratory tract (200). In an NTP study (198) at the same doses, compound-related lesions included pulmonary hemorrhage and rhinitis in rats at 1600 ppm and nephrosis in mice at 800 and 1600 ppm.

In a 13-wk subchronic inhalation study (200) at doses of 75 to 600 ppm, inflammatory and degenerative changes in the nasal mucosa were observed in mice and rats, and myeloid hyperplasia in bone marrow was observed in male rats. In a comparable NTP study (198) at doses of 50–800 ppm, inflammation of nasal turbinates was seen in rats, and renal tubular necrosis was seen in mice at 800 ppm. Rhinitis was observed in mice at levels as low as 100 ppm.

3.4.1.3 Pharmacokinetics, Metabolism, and Mechanisms. 1,2-Epoxybutane is extensively metabolized and rapidly eliminated following either inhalation exposure or gavage in male rats. Acute exposures of males to vapor concentrations of 2000, 1000, or 400 ppm caused a dose-related depletion of nonprotein sulfhydryl groups in liver and kidney tissue (201). Steady-state uptake rates of 1,2-epoxybutane were determined to be 0.0433 mg/kg/min at 50 ppm and 0.720 mg/kg/min at 1000 ppm. These rates correspond to an estimated uptake of 15.6 and 252 mg/kg during a 6-h exposure. It appears that the physical and biologic processes involved in absorption, metabolism, and elimination of 1,2-epoxybutane are essentially linear throughout the exposure range of 50–1000 ppm (201).

3.4.1.4 Reproductive and Developmental. Inhalation teratology studies were conducted in rats and rabbits at vapor concentrations of 0, 250, or 1000 ppm 1,2-epoxybutane (202). In rats, although maternal mortality and depressed body weight gain were observed at the high dose, fetal growth and viability were not affected, and there was no evidence of teratogenicity. In rabbits, litter size was reduced and fetal mortality was increased in the high-dose group; however, maternal mortality in this group was high (14/24).

3.4.1.5 Carcinogenesis. In a dermal carcinogenicity study, 10% 1,2-epoxybutane in acetone was applied to the backs of mice three times a week for 77 wk and the animals were killed in week 85. No visible skin reactions or tumors were observed in treated or 100% acetone control mice (203).

Chronic 2-yr inhalation exposures to 1,2-epoxybutane vapor were conducted by NTP in mice and rats (198). Exposure concentrations were 0, 200, or 400 ppm in rats and 0, 50, or 100 ppm in mice. Inflammatory, degenerative, and proliferative lesions occurred in the nasal cavity of both rats and mice. No exposure-related neoplastic lesions were seen in mice. However, in high-dose rats, nasal papillary adenomas occurred in males (7/50) and females (2/50) compared with none in controls, and alveolar/bronchiolar adenomas or carcinomas (combined) occurred with increased incidence in exposed males (5/49) relative to controls (0/50) (150).

3.4.1.6 Genetic and Related Cellular Effects Studies. The genetic toxicity of 1,2-epoxybutane has been reviewed (197, 204). It was positive for gene mutation in the Ames assay in *Salmonella* strains TA100 and TA1535 with or without metabolic activation, but was negative in strains TA98 and TA1537, indicative of a direct-acting mutagen that induces base-pair substitutions. Forward mutations were also induced *in vitro* in the mouse lymphoma assay with and without activation. Both chromosomal aberrations and SCEs were induced in cultured Chinese hamster ovary cells. When fed to male *Drosophila*, 1,2-epoxybutane caused significant increases in the number of sex-linked recessive lethal mutations and reciprocal translocations in the germ cells (198). Negative results have been reported for rat dominant lethal and mouse sperm morphology studies (205). It did not induce unscheduled DNA synthesis in the hepatocyte rat primary culture/DNA repair test (206).

3.4.1.7 Other: Neurological, Pulmonary, Skin Sensitization. Twenty weeks of repeated exposures to 2000 ppm of butylene oxide resulted in slight peripheral neurotoxicity in rats (207). Data reported for 13-wk inhalation exposure studies found no evidence of peripheral neurotoxicity at exposure concentrations of 800 and 600 ppm for rat and mouse, respectively (198).

3.4.2 Human Experience

No data were available in humans; however, excessive exposure to vapors would likely result in disagreeable respiratory irritation.

3.5 Standards, Regulations, or Guidelines of Exposure

The American Industrial Hygiene Association (AIHA) has established a TWA workplace environmental exposure limit (WEEL) of 2 ppm. The International Agency for the Research on Cancer has placed 1,2-epoxybutane in Group 2B (possibly carcinogenic to humans) (197).

4.0 Butadiene Diepoxide

4.0.1 CAS Number: *[1464-53-5]*

4.0.2 Synonyms: Diepoxybutane, butadiene dioxide, BDE DEB; 2,2′-bioxirane, 1,2:3,4-diepoxybutane, 3,4-diepoxybutane, erythritol anhydride, bioxirane, and 1,3-butadiene diepoxide

4.0.3 Trade Names: NA

4.0.4 Molecular Weight: 86.090

4.0.5 Molecular Formula: $C_4H_6O_2$

4.0.6 Molecular Structure:  and 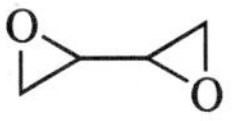

4.1 Chemical and Physical Properties

Diepoxybutane (BDE) is a water-white, low-viscosity liquid with the following properties (208):

Molecular weight: 86.09
Specific gravity (25/4°C): 0.962
Boiling point (760 mmHg): 138°C
Vapor pressure (20°C): 3.9 mmHg
Refractive index (20°C): 1.435
Solubility: Miscible with water

1 ppm = 3.52 mg/m^3 at 25°C, 760 mmHg; 1 mg/L = 284 ppm at 25°C, 760 mmHg

4.1.2 Odor and Warning Properties

No information on odor, sensory threshold limits, or warning properties was found.

4.2 Production and Use

Diepoxybutane is prepared by chlorination of butadiene followed by epoxidation with peracetic acid, with subsequent hydrolysis of the epoxide group and final reepoxidation with caustic. It is suggested as a chemical intermediate, cross-linking agent, and in the preparation of erythritol and pharmaceuticals.

4.3 Exposure Assessment

No standard method for the determination of diepoxybutane was found. The pyridinium chloride-chloroform method for epoxy groups (209) and GLC is suggested.

4.4 Toxic Effects

Diepoxybutane is a severe pulmonary irritant. It is classed as highly toxic on inhalation, and moderately toxic following ingestion or skin absorption. It causes severe eye and skin irritation. The *meso* form appears to have half the acute fatal toxicity to mice of the DL form. Repeated inhalation of BDE is irritating to the upper respiratory tract. In rodents BDE causes ovarian toxicity. The compound produced local skin tumors in skin painting studies, local sarcomas upon repeated subcutaneous injections, and lung tumors following repeated intraperitoneal injections. It is a potent genotoxin in *in vitro* and *in vivo* systems.

Butadiene diepoxide (BDE) is a metabolite of 1,3-butadiene. Butadiene (BD) is a multisite rodent carcinogen; the carcinogenic potency in mice is significantly higher than in rats. Extensive research has been conducted into the dosimetry, metabolism, kinetics, and genotoxicity of the BD metabolites in an attempt better to understand their roles in the carcinogenicity of BD. BDE has been detected in blood and tissues of rodents exposed to BD; in mice levels of BDE are considerably higher than in rats. The mutagenic potential of BDE is approximately 100 times that of other butadiene epoxides. These data and other information had led to the suspicion that BDE is the proximate carcinogen in mice. The toxicology, metabolism, and genotoxicity of 1,3-butadiene has been reviewed (210). The information presented here is a summary of the toxicology of BDE.

4.4.1 Experimental Studies

4.4.1.1 Acute Toxicity. The LD_{50} value for oral in the rat was 78 mg/kg and for dermal in the rabbit was 89 mg/kg. Inhalation of concentrated vapors killed all of six rats in 15 min; the 4-h LC_{50} in rats was 90 ppm (211).

Skin Effects. The undiluted material is a severe primary skin irritant resulting in necrosis in rabbits (211).

Eye Irritation. Diepoxybutane is a severe eye irritant, causing corneal necrosis in rabbit eyes (211).

4.4.1.2 Chronic and Subchronic Toxicity. The toxicity of BDE was studied in rats and mice following a single 6-h exposure to 18 ppm or repeated exposures (6 h/d) up to 20 ppm for 7 d or 5 ppm for 5 d (mice only). The toxicity was assessed immediately after the last exposure and up to 10 d post-exposure. A single exposure of 18 ppm to BDE was without effect in the lungs of rats. In mice there were noninflammatory responses in the lungs consisting mainly of alveolar hyperplasia. Both rats and mice showed signs of irritation of the respiratory and olfactory epithelium of nasal mucosa; the lesions in mice were more severe than in rats. There was no evidence of adverse changes in bronchial lavage fluid or, microscopically, in several organs and tissues, including the ovaries (212).

Repeated inhalation of 20 ppm was severely toxic to rats; mice did not tolerate repeated exposures to 20 ppm beyond 12 d. The primary toxic effect in rats was degeneration, necrosis, and inflammation of the respiratory, transitional, and olfactory epithelium of the nasal passages; no laryngeal effects were noted. No significant toxicity was found in other tissues. In mice the effects in the nasal epithelium were qualitatively similar to those found in rats; however, the lesions in mice were more severe. In addition, some mice showed myocardial degeneration, bone marrow atrophy, or focal hepatic necrosis. There were no toxic effects in the respiratory tract or in any other tissue or organ in mice inhaling 5 ppm for 5 d (212).

In a study designed to establish the ovarian toxicity of BDE, rats and mice received daily intraperitoneal doses of BDE in the range 0.2–25 mg/kg/d for a period of 30 d. Significant toxicity and mortality was observed in rats receiving 25 doses of 25 mg/kg BDE. In mice ovarian and uterine weights were reduced and follicle counts were severely depressed. Rats showed also evidence of ovotoxicity; however, the severity was significantly less than in mice (213).

4.4.1.3 Pharmacokinetics, Metabolism, and Mechanisms. *In vitro* experiments with liver and lung preparations from mouse, rat, and human (liver only) showed that BDE was rapidly metabolized. Two deactivation mechanisms have been recognized: enzyme-mediated conjugation with glutathione and enzymatic hydrolysis. The order of the GSH conjugation rate was mice > rat > humans. Hydrolysis by epoxide hydrolase was highly efficient, with human samples generally being more efficient than rat or mouse. The fraction of BDE hydrolyzed through nonenzymatic mechanisms was negligible. Overall, the epoxide hydrolase activity was shown to be the major pathway for elimination of BDE in human tissues; in mice the glutathione conjugation is expected to be more prominent. The hydrolysis products identified in these experiments were erythritol and 1,4-anhydroerythritol (214, 215). The monoglutathione conjugate of BDE was also formed by incubation of *Salmonella* TA1535 transfected with rat GST (216).

Partition coefficients measured in blood and several tissues showed that BDE is hydrophilic in nature, suggesting that BDE has a potential for wide distribution within the body. These data, together with the kinetic parameters of the enzymatic conversion of BDE, were used to develop several different physiologically based pharmacokinetic (PB–PK) models for inhaled butadiene; the most recent models include a pathway for the formation and removal of BDE (217, 218).

A single dose of 26 mg/kg BDE was administered intraperitoneally (IP) or intratracheally (IT) to rats and mice in order to determine if parent BDE could be detected in the general circulation and in tissues after absorption. BDE was rapidly absorbed and could be detected in blood within 5 min after the IT dose; by the IP route BDE was detected after 30 min of dosing (first sampling time point), with blood levels in rats and mice being quantitatively similar. BDE was cleared rapidly from the blood but was still detectable 2 h after administration. In a subsequent experiment, rats and mice inhaled an atmosphere of 12 ppm BDE for 6 h. Immediately after termination, concentrations of BDE in mouse blood were about 1000 pmol/g; BDE blood levels in rats were about 50% of the concentrations in mice; lung tissue levels of BDE were about 2–3 times higher than in blood (212). Similarly, intravenous administration of BDE to rats resulted in very rapid clearance from the blood (217).

4.4.1.5 Carcinogenesis. Both the DL and *meso* isomers of BDE produced significant numbers of skin tumors in mouse dermal bioassays. Skin applications (10 mg/animal in 0.1 mL acetone), repeated three times weekly for 1 yr, also resulted in sebaceous gland suppression, hyperkeratosis, and hyperplasia (219). Repeated once-weekly subcutaneous injections of the DL isomer in mice (0.1 or 1.1 mg/animal) and rats (1 mg/animal) resulted in local fibrosarcomas in 9/50 rats and 5/50 and 5/30 mice, respectively, compared to none in controls (220, 221). In mice given 12 thrice-weekly intraperitoneal injections of the L isomer dissolved in water or in tricaprylin, the incidence of lung tumors was increased significantly at doses in water of 27, 108, and 192 mg/kg (222). Intragastric administration in rats once weekly for a year was nontoxic, presumably because of rapid acid hydrolysis of the epoxide in the stomach (221).

4.4.1.6 Genetic and Related Cellular Effects Studies. The genotoxic effects of BDE were reviewed in 1988 (223) and, more recently, as part of the IARC evaluation of butadiene (224) and a review on butadiene (210).

Diepoxybutane is genotoxic in a wide variety of test systems and organisms, including bacteria, yeast, cultured mammalian cells, *Drosophila*, mice and rats. BDE is a bifunctional compound and has been shown to act as an alkylating agent and as an interstrand cross linker. It can induce point mutations and small and large deletions in *in vitro* and *in vivo* systems; the concentrations necessary to induce the effects *in vitro* are similar to those occurring *in vivo* following exposure to butadiene. The summary to follow is not a comprehensive listing of all studies, but presents representative examples of the significant research effort on the genotoxicity of diepoxybutane.

In vitro. Diepoxybutane was mutagenic in *Salmonella* strains TA98, TA100, TA1535 (225–227), and TA1530 (228) with or without metabolic activation, and *E. coli* (229, 230). In mammalian cell assays, *in vitro*, diepoxybutane was mutagenic in the mouse lymphoma cell forward mutation assay (231), induced chromosome aberrations in Chinese hamster ovary cells (232, 233), induced SCEs in Chinese hamster V79 cells, and in human lymphocytes at concentrations as low as 0.1 or 0.5 μM, respectively (234–239). It caused chromosome damage in cultured rat liver cells (240).

In human TK6 lymphoblasts BDE caused large deletions and rearrangements of the *hprt* locus or loss of heterozygosity at the *tk* locus (241–243). BDE alkylates DNA and forms specific DNA cross links, similar to other bifunctional alkylkating agents (244). However, the compound was negative in the rat hepatocyte DNA-repair test (245) and did not induce unscheduled DNA synthesis (UDS) in rat or mouse hepatocytes *in vitro* (246).

The susceptibility of human lymphocytes to SCE formation upon incubation with BDE had a bimodal distribution, with one group being twice as sensitive as other individuals. Glutathione transferase polymorphisms may have an impact on the individual susceptibility to genotoxic effects such as SCEs; however, the data show that the increased sensitivity to SCEs due to BDE is not linked with GST genotypes (238, 239, 248–250). GSH conjugation modulates the mutagenicity of BDE as was demonstrated by the increased response in *Salmonella* tester strains expressing GST (216).

In vivo. Diepoxybutane was found to induce several clastogenic effects such as strand breakages, deletions, chromosomal aberrations, sister chromatid exchanges, and micronuclei. Structural chromosomal abnormalities and SCEs in bone marrow cells of mice and Chinese hamsters were reported, when BDE was inhaled from an aerosol during a 2-h head-only exposure or administered as a single intraperitoneal injection; inhalation doses, calculated from blood concentrations, were 102 mg/kg in mice and 335 mg/kg in hamsters; intraperitoneal doses were 32 and 147 mg/kg (247). BDE also induced SCEs in bone marrow, alveolar macrophages, and regenerating liver cells of mice following intraperitoneal injection (251). BDE induced micronuclei in bone marrow, splenocytes, and germ cells of rats and mice (252). BDE induced a dose–related response in bone marrow cells of mice and rats in a Comet and a UDS assay; however, no effects were observed in haploid testicular cells (253).

Mutation frequencies (Mf) were significantly increased at the *htrp* locus in splenic T cells of preweaning mice treated IP with doses up to 21 mg BDE/kg; in adult mice and rats no increase in Mf was observed. Also no increases in Mf in *lacI* were observed in bone marrow and spleen of transgenic mice and rats after inhalation of 5 ppm BDE for 2 wk. Inhalation of 4 ppm BDE for 4 wk (6 h/d; 5 d/wk) caused increases in Mf at the *htrp* locus in splenic T cells of rats and mice, with higher Mf observed in rats compared to mice (241, 254, 255). In a number of studies, BDE was mutagenic in *Drosophila*, which at the *rosy* locus were shown to be related to point mutations, deletions of a small number of base pairs, or large-scale deletions (241, 421–426).

4.4.2 Human Experience

No documented reports of human exposure to BDE were available. However, BDE would be expected to be severely irritating acutely. The compound has been used to detect Fanconi anemia by virtue of its cytogenetic action (427, 428).

4.5 Standards, Regulations, or Guidelines of Exposure

No standards of permissible occupational exposure to BDE were found.

5.0a Epoxidized Linseed Oil

5.0b Epoxidized Soya Bean Oil

5.0c Epoxidized Tall Oil

5.0.1a CAS Number: *[8016-11-3]*

5.0.1b CAS Number: *[8013-07-8]*

5.0.1c CAS Number: *[8002-26-4]*

5.0.1b Synonyms: Soybean oil epoxide, ESBO

5.0.1c Synonyms: Tall oil rosin and fatty acids

5.1 Chemical and Physical Properties

The properties of epoxidized linseed oil and epoxidized soya bean oil (ESBO) are provided below. No data on epoxidized tall oil could be found. These products are similar in composition and are based on naturally occurring mixtures of triglycerides, containing variable amounts of unsaturated fatty acids (e.g., linoleic, oleic, and linolenic acid).

	Epoxidized Linseed Oil	Epoxidized Soya Bean Oil
Oxirane oxygen (%)	9.3	7.0–7.2
Specific gravity (25/25°C)	1.03	0.992–0.996
Color (Gardner)	1	1
Viscosity, Stokes (25°C)	3.3	3.2–4.2
Fire point (°F)	650	600
Flash point (°F)	590	590
Odor	Very low	Mild
Vapor pressure (mmHg, 25°C)	0.1	0.1
Freezing point (°C)	0	0

5.1.2 Odor and Warning Properties

Epoxidized glycerides have a nondistinctive odor.

5.2 Production and Use

These epoxidized oils are made by epoxidizing the unsaturated bonds of unsaturated carboxylic acid–glycerin esters (triglycerides) with peracids (peracetic acid or its equivalent, hydrogen peroxide in acetic acid). Because the molecular weights of these esters approach 900 before epoxidation and because the unsaturated esters are diluted with inert palmitates and stearates, there are sufficient epoxy groups to bind them into polymers, but too few to constitute much of a handling hazard even though there may be a di- or triepoxide content. The epoxidized glycerides are primarily reactive diluents and find use in coatings of food cans. They are recommended for PVC homo- and copolymer stabilization and plasticization for rigid, flexible, extruded, calendered, and molded compounds. Applications are found in intravenous tubing, blood bags, food wrap film, cap liners and seals, meat trays, upholstery, pipe, and construction materials.

5.3 Exposure Assessment

No OSHA nor NIOSH standard method for the determination of these materials was found. These products are nonvolatile until heated to about 500°F, when acrolein is released from glyceride decomposition. The pyridinium chloride–chloroform method for epoxy groups (209) is suggested.

5.4 Toxic Effects

Epoxidized glycerides are of a low order of acute oral and dermal toxicity. They are slightly irritant to skin and eyes but have no skin sensitization potential. Repeated oral administration at high-dose levels caused mild effects in liver, kidney, testis, and uterus.

Studies on reproduction and development did not show any adverse effects, and there was no evidence of a carcinogenic response in long-term studies in rats. Epoxidized glycerides do not have a mutagenic potential. There are sporadic reports of asthmatic symptoms following exposure to vapors of epoxidized glycerides. This toxicological assessment has been based on data available for ESBO.

The toxicological properties of ESBO have been reviewed in 1997, and the original sources of the information presented here can be found in (256).

5.4.1 Experimental Studies

5.4.1.1 Acute Toxicity. Acute toxicity studies with ESBO have shown acute oral LD_{50} values in rats in the range of $>5-40$ g/kg. In rabbits the acute dermal LD_{50} was $>$ 20 mL/kg. The intraperitoneal LD_{50} in mice was >25 g/kg. Signs of toxicity included difficult breathing, lethargy and diarrhea. There were no gross abnormalities at necropsy (256).

Skin Irritation. A 24-h occluded exposure to intact rabbit skin caused mild redness and minimal swelling; uncovered skin contact caused slight irritation. No skin sensitization potential was identified in guinea pigs in studies in using repeated intracutaneous injections during the induction phase (256).

Eye Irritaton. ESBO was a nonirritant when instilled (0.5 mL) into rabbit eyes. In another study slight conjuctival reddening was observed upon instillation of 0.1 mL in a rabbit eye (256).

5.4.1.2 Chronic and Subchronic Toxicity. Several subchronic and chronic toxicity studies with ESBO have been conducted; these were reviewed in Ref. 256. In a subchronic study conducted to current testing guidelines, rats received diets containing up to 5% ESBO. Reduced growth, liver and kidney enlargement, and fatty changes in the liver were observed. Similar findings were observed in the 2.5% exposure group, except for the effects on body weight. Exposure of rats to dietary concentrations up to 2.5% for 2 yr caused minimal toxicity. No changes were recorded in survival, blood parameters, and macroscopic and microscopic pathology of a comprehensive range of tissues and organs, with the exception of slight changes in the uterus at 2.5%. In males the no-observed adverse effect level (NOAEL) was 2.5% (ca. 1000 mg/kg bw/d); in females a NOAEL of 0.25% (ca. 140 mg/kg) was observed. Several older studies in rats with exposures up to 10% covering periods up to 2 yr were reported. The results of these older studies were broadly consistent with the more recent studies, with the exception of uterine effects in the latest study (reported 1986) and testicular degeneration in male rats in a single study reported in 1963 (256). Concentrations of 5% ESBO in the diet of dogs was well tolerated for 1 yr. Food intake and body weight were adversely affected. Microscopically, fatty infiltration of the liver was observed (256).

5.4.1.4 Reproductive and Developmental. ESBO did not affect parental reproductive parameters and development of the offspring in a one-generation reproduction study in rats exposed to daily oral doses of 1 g/kg bw/d. Animals were exposed during all phases of the reproductive cycle up to weaning, including a 10-wk premating period. In a developmental toxicity study, pregnant rats received oral daily doses of ESBO at levels up to 1 g/kg bw/d during days 6–15 of gestation. There were no treatment-related effects on a wide range of parameters, including examination of external, skeletal, and soft tissue abnormalities (256).

5.4.1.5 Carcinogenesis. In a study conducted to current testing guidelines, no evidence of carcinogenic response was observed in rats exposed to dietary concentrations up to 2.5% ESBO. Older studies also showed a lack of a carcinogenic effect in orally exposed rats. Other studies using dermal, intraperitoneal, or unspecified exposure routes were inadequate or inconclusive regarding the carcinogenic potential of ESBO (256).

5.1.4.6 Genetic and Related Cellular Effects Studies. ESBO was negative in *Salmonella* strains TA98 and TA100 (257) and the CHO/HGPRT gene mutation assay (258). In a mouse lymphoma assay and a chromosomal aberration study in cultured human lymphocytes ESBO tested negative both in absence and the presence of a rat liver S9 fraction (256). Additionally, stomach and intestine contents of rats dosed orally with ESBO were tested in the Ames assay (TA98 and TA100) at 6 h after dosing, and were found negative (259). ESBO and epoxidized linseed oils were also tested for cytotoxicity in the HeLa cell MIT-24 *in vitro* test system; both materials demonstrated low toxicity (260).

5.4.2 Human Experience

Exposure to vapors of heated polyvinyl chloride film containing ESBO caused asthma in a worker. Challenge with ESBO vapor rapidly produced asthmatic symptoms. A similar report of occupational asthma in response to exposure to epoxidized tall oil was described in the rubber industry (261).

5.5 Standards, Regulations, or Guidelines of Exposure

The EU Scientific Committee on Food allocated a temporary Tolerable Daily Intake (TDI) of 1 mg/kg for ESBO. No occupational standards have been established for these materials.

6.0 Vinylcyclohexene Monoxide

6.0.1 CAS Number: [106-86-5]

6.0.2 Synonyms: 7-Oxabicyclo(4.1.0)heptane, 3-vinyl-, 1,2-epoxy-4-vinylcyclohexane, 1-vinyl-3,4-epoxycyclohexane, 3-vinyl-7-oxabicyclo(4.1.0)heptane, 4-vinylcyclohexane, 1,2-epoxide 4-vinylcyclohexene monoxide, and 4-vinylcyclohexene-1,2-epoxide.

6.0.3 Trade Names: EP-101; Epoxide 101

6.0.4 Molecular Weight: 124.18

6.0.5 Molecular Formula: $C_8H_{12}O$

6.0.6 Molecular Structure:

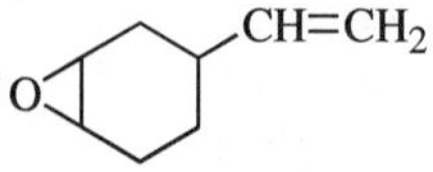

6.1 Chemical and Physical Properties

Vinylcyclohexene monoxide is a colorless, mobile liquid combining readily with water, alcohols, phenols, and other agents containing active hydrogens. It has the following properties:

Specific gravity (20/20°C): 0.9598
Freezing point: Sets to a glass below – 100°C
Boiling point (760 mmHg): 169°C
Vapor density: 3.75
Vapor pressure (20°C): 2.0 mmHg
Refractive index (20°C): 1.4700
Solubility: 0.5% in water
Concentration in "saturated" air: 0.263%

1 ppm @ 5.07 mg/m^3 at 25°C, 760 mmHg; 1 mg/L @ 197 ppm at 25°C, 760 mmHg

6.1.2 Odor and Warning Properties

No information on odor or warning properties were found. It is irritating to the respiratory tract.

6.2 Production and Use

Vinylcyclohexene monoxide is a chemical intermediate; it can be copolymerized with other epoxides to yield polyglycols having unsaturation available for further reaction.

6.3 Exposure Assessement

No OSHA nor NIOSH standard methods for the determination of vinylcyclohexene monoxide were found. The pyridinium chloride–chloroform method for epoxy groups (209) and GLC are suggested, but the HCl–dioxane method for epoxides (209) might also be used.

6.4 Toxic Effects

The LD_{50} in rats given the material by mouth was 2 mL/kg. The percutaneous LD_{50} for rabbits was 2.83 g/kg. Rats exposed to saturated vapors for 2 h survived, but three of six died after a 4-h exposure. The compound produced moderate skin irritation in rabbits, although eye irritation was not marked. Inhalation exposure caused pulmonary irritation in rats (262). The ovarian toxicity of vinylcyclohexene monoxide as compared to vinylcyclohexene diepoxide has been studied (263).

6.5 Standards, Regulations, or Guidelines of Exposure

No standards for permissible occupational exposure were found.

7.0 Vinylcyclohexene Dioxide

7.0.1 CAS Number: [106-87-6]

7.0.2 Synonyms: 1,2-Epoxy-4-(epoxyethyl)cyclohexane, 1-epoxyethyl-3,4-epoxycyclohexane, 1-ethyleneoxy-3,4-epoxycyclohexane, 1-vinyl-3-cyclohexene dioxide, 3-(1,2-epoxyethyl)-7-oxabicyclo(4.1.0)heptane, 3-oxiranyl-7-oxabicyclo(4.1.0)heptane, 4-(1,2-epoxyethyl)-7-oxabicyclo(4.1.0)heptane, 4-vinlycyclohexene dioxide, 4-vinyl-1,2-cyclohexene diepoxide, 4-vinyl-1-cyclohexene diepoxide, 4-vinyl-1-cyclohexene dioxide, 4-vinylcyclohexene diepoxide, vinyl cyclohexene diepoxide, vinyl cyclohexene dioxide, VCHD

7.0.3 Trade Names: NA

7.0.4 Molecular Weight: 140.18

7.0.5 Molecular Formula: $C_8H_{12}O_2$

7.0.6 Molecular Structure:

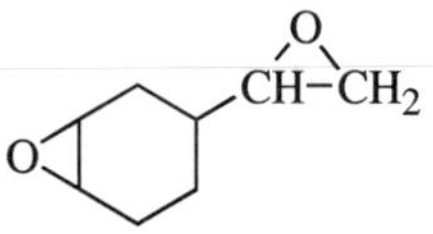

7.1 Chemical and Physical Properties

VCHD is a clear, water-soluble liquid with a faintly olefinic odor and the following properties (264):

Specific gravity (20/20°C): 1.0986
Freezing point: Sets to glass at -55°C
Boiling point (760 mmHg): 227°C
Vapor density: 4.07
Vapor pressure (20°C): 0.1 mmHg
Refractive index (20°C): 1.4787
Solubility: 18.3% in water at 20°C

1ppm = 5.73 mg/m^3 at 25°C, 760 mmHg; 1 mg/L = 1.74 ppm at 25°C, 760 mmHg

7.1.2 Odor and Warning Properties

Although no odor threshold data were found, the irritating properties of VCHD may serve to warn of overexposure.

7.2 Production and Use

VCHD is manufactured by epoxidation of 4-vinylcyclohexene with peroxyacetic acid (265). VCHD is used as a chemical intermediate and as a monomer for preparation of polyglycols containing unreacted epoxy groups.

7.3 Exposure Assessment

No OSHA nor NIOSH standard methods for the determination of vinylcyclohexene dioxide were found. The pyridinium chloride–chloroform method for epoxy groups (209) or GLC is suggested, but the HCl–dioxane method for epoxides (209) might also be used.

7.4 Toxic Effects

VCHD is moderately acutely toxic and is severely irritating. With repeated dermal exposure, it produces marked skin lesions, which progress to skin cancer in rodents; additionally, ovarian and lung cancer are seen in mice. Systemically, VCHD causes reproductive organ toxicity in mice and forestomach and kidney lesions in rodents when given orally. It is mutagenic in the Ames assay, and there is evidence that it is immunosuppressive in mice. VCHD is a contact allergen in humans and has the potential to induce respiratory sensitivity.

7.4.1 Experimental Studies

7.4.1.1 Acute Toxicity. A summary of lethal dose data from animals studies is as follows: LD_{50} rat, oral—2.8 g/kg (266); LD_{50} rabbit, dermal—0.06 mL/kg (266); LD_{50} rat, oral—2130 mg/kg (267); LC_{50} rat, inhalation—800 ppm/4 h (267); LD_{50} rabbit, dermal—620 mg/kg (267).

Skin Effects. Undiluted VCHD causes severe skin irritation in the rabbit.

Eye Irritation. The compound would be expected to be a severe eye irritant.

7.4.1.2 Chronic and Subchronic Toxicity. In 14-d dermal toxicity studies, rats receiving 139 mg/rat (males) or 112 mg/rat (females) had congestion and/or hypoplasia of the bone marrow; most had acute nephrosis; skin lesions included necrosis and ulceration, epidermal hyperplasia, and hyperkeratosis. Similar skin effects were seen in mice receiving 5 mg/mouse (268).

In subsequent 13-wk NTP studies (269), VCHD toxicity was evaluated by both the oral and dermal routes of exposure. In the dermal studies, groups of 10 rats of each sex received 0.3 mL of VCHD in acetone daily 5 d/wk at concentrations of 3.75–60 mg/mL, and mice received 0.1 mL at concentrations of 0.625–10 mg/mL. At the two highest dose levels tested, skin lesions consisted of hyperkeratosis of the epidermis and sebaceous gland hyperplasia, and follicular atrophy of the ovary was seen in mice. In oral studies, mice and rats (10 of each sex) were given 62.5–1000 mg/kg VCHD in corn oil by gavage daily 5 d/wk. In rats and mice, systemic effects included hyperplasia and hyperkeratosis of the forestomach. In rats, there was also kidney tubule cell necrosis, and in mice there were also reproductive organ effects, which included follicular atrophy in ovaries and degeneration of germinal epithelium in testes.

7.4.1.3 Pharmacokinetics, Metabolism, and Mechanisms. VCHD is absorbed by rodents exposed dermally, orally, or by inhalation (145). In dermal absorption studies conducted by NTP in rats and mice, 30% of a dose of radiolabeled VCHD was absorbed over a 24-h period in both species. By 24 h, 70–80% of the absorbed dose was eliminated from the body, mostly in the urine. Of the radioactivity remaining in the body, no tissue contained more than 1% of the applied dose, and tissue/blood ratios ranged from 0.3 to 1.5 in rats and 0.8 to 2.8 in mice (268). *In vitro* studies with rabbit liver microsomal preparations showed that 4-vinyl-1-cyclohexene diepoxide can be metabolized to mono-epoxymonoglycols, 1,2-hydroxy-4-vinylcyclohexane oxide, and 4-(1′,2′-dihydroxyethyl)-1-cyclohexane oxide. Formation of these products is catalyzed by epoxide hydrolase. Conjugation with glutathione is another pathway for metabolism of 4-vinyl-1-cyclohexene. Depletion of reduced glutathione was reported in the liver of mice given IP injections of 500 mg/kg 4-vinyl-1-cyclohexene diepoxide (268). The metabolism of 4-vinylcyclohexene to VCHD has also been studied due the potential for occupational exposure to 4-vinylcyclohexene in the rubber industry. The rate of epoxidation of 4-vinylcyclohexene (1 mM) to 4-vinylcyclohexene-1,2-epoxide was 6.5-fold greater in mouse liver microsomes than that in rat liver microsomes. The species difference in the rate of epoxide formation by the liver may be an important factor in the species difference in susceptibility to 4-vinylcyclohexene induced ovarian tumors (263).

7.4.1.5 Carcinogenesis. NTP conducted dermal cancer bioassays in F344 rats and $B6C3F_1$ mice by administering VCHD in acetone 5 d/wk for 105 wk to rats at 0, 15, or 30 mg/animal or for up to 103 wk to mice at 0, 2.5, 5, or 10 mg/animal. There was an increased incidence of squamous cell papillomas and carcinomas of the skin in both species, and mid- and high-dose female mice had benign and malignant ovarian tumors. The incidence of lung tumors in female mice was also marginally increased (270).

7.4.1.6 Genetic and Related Cellular Effects Studies. There are numerous studies that demonstrate the mutagenicity of VCHD in the Ames assay in *Salmonella* strains TA100 and TA1535 in the presence or absence of S9 metabolic activation (3, 222, 271–274).

7.4.1.7 Other: Neurological, Pulmonary, Skin Sensitization. The NTP has carried out immunotoxicity studies in $B6C3F_1$ mice. Immune function tests indicated that VCHD was an immunosuppressive at 10 mg/mouse and, to a lesser extent, at 5 mg/mouse when applied dermally daily for 5 d and at 2.5 and 5 mg/mouse when applied dermally for 14 d (268).

7.4.2 Human Experience

Allergic contact dermatitis has been reported in an individual exposed occupationally to VCHD (275). A study of occupational exposure to sensitizing chemicals including VCHD was initiated because of numerous complaints of eye, nasal, skin, and respiratory irritation among workers at an industrial facility (276). Serum samples were obtained from 31 workers and assayed for immunoglobulin G and immunoglobulin E using ELISA. Immunoglobulin G and immunoglobulin E specific to four chemicals were found: an aliphatic diisocyanate, VCHD, trimellitic anhydride, and an unknown substance that was a component of *n*-octyl-*n*-decyl-trimellitate. The presence or absence of antibodies could not be correlated with the presence or absence of symptoms; however, the highest incidence of symptoms occurred in workers who were classified as having the highest exposures. It was concluded that there was no occupational allergic disease in this worker population. The lack of correlation between the antibody responses and symptoms suggests that an immunologic factor was not the cause of the reported symptoms.

7.5 Standards, Regulations, or Guidelines of Exposure

The 2000 ACGIH TLV for VCHD is 0.1 ppm (skin) and was classified as A3 (confirmed animal carcinogen with unknown relevance to humans). NIOSH REL TWA 10 ppm (60 mg/m^3). The International Agency for Research on Cancer (277) has classified VCHD as 2B (possibly carcinogenic to humans).

8.0 Styrene Oxide

8.0.1 CAS Number: *[96-09-3]*

8.0.2 Synonyms: Styrene-7,8,-oxide, styrene epoxide, 1,2-epoxyethylbenzene, 2-phenyloxirane, and 1,2-epoxy-1-phenylethane, Styrene Oxide; 1-Phenyl-1,2-Epoxy-ethane; Phenyloxirane; epoxyethylbenzene; epoxystyrene; alpha, beta-epoxystyrene; phenethylene oxide; phenylethylene oxide; 2-phenyloxirane; styryl oxide; Phenyloxirane, d8; Styrene oxide-d8

8.0.3 Trade Names: NA

8.0.4 Molecular Weight: 120.15

8.0.5 Molecular Formula: C_8H_8O

8.0.6 Molecular Structure:

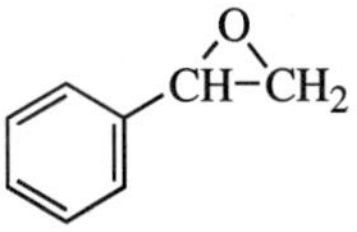

8.1 Chemical and Physical Properties

Styrene oxide is a colorless liquid, existing in two optical isomers, and the commercial product is a racemic mixture with the following properties:

Specific gravity (25/25°C): 1.054
Freezing point: – 36.8°C
Vapor density: 4.30
Boiling point (760 mmHg): 194.1°C
Vapor pressure (20°C): 0.3 mmHg
Refractive index (25°C): 1.533
Solubility: < 0.12 /100 mL in water at 25°C; miscible with methanol, ether, carbon tetrachloride, benzene, and acetone
Flash point (TOC): 74°C
Flammability limits: Precautions necessary at elevated temperature

1 ppm @ 4.91 mg/m^3 at 25°C, 760 mmHg; 1 mg/L @ 203.6 ppm at 25°C, 760 mmHg

8.1.2 Odor and Warning Properties

No information on odor or warning properties was found. Irritation to the respiratory tract might serve as a warning property for overexposure.

8.2 Production and Use

Styrene oxide is made commercially from styrene through the intermediate chlorohydrin. It exists largely as an intermediate in the production of styrene glycol and its derivatives, for example, 2-phenylethanol (oil of roses), which is used as a perfume base. Styrene oxide can also be polymerized to polystyrene glycols, and limited amounts may be used as a liquid diluent in the epoxy resin industry.

Styrene oxide has a flash point of about 80°C (175°F) as determined by the open-cup procedure. On this basis, it presents a hazard of flammability similar to that encountered with such well-known chemical products as *o*-cresol, *o*-dichlorobenzene, naphthalene, phenol, and dimethylaniline. A definite hazard exists whenever styrene oxide is heated to temperatures at and above the flash point.

Styrene oxide will polymerize exothermally or react vigorously with compounds having a labile hydrogen, including water, in the presence of catalysts such as acids, bases, and certain salts. Experience with the chemical reactivity of styrene oxide has shown that styrene oxide is not as hazardous as ethylene oxide; however, precautions should be taken to prevent excessive pressure under storage or reaction conditions and to relieve such pressure should it occur.

8.3 Exposure Assessment

No OSHA nor NIOSH standard methods for the determination of styrene oxide were found. A method for measuring styrene oxide in the blood using gas chromatographic/mass spectrometry has been described (278); a similar approach may be applicable for the determination of styrene oxide in air. The pyridinium chloride–chloroform method for epoxy groups (209) and GLC have been suggested, but the HCl–dioxane method for epoxides (209) might also be used.

8.4 Toxic Effects

The acute oral and dermal toxicity of styrene oxide is low. Acute inhalation toxicity studies have shown lethality in laboratory animals at concentrations close to saturation. Styrene oxide may produce severe eye irritation, moderate skin irritation, and skin sensitization. Styrene oxide has been shown to be genotoxic in several *in vitro* assays, although only weakly mutagenic or negative *in vivo* assays. Several investigators have shown that styrene oxide produced malignant tumors in the forestomach of rodents when administered by gavage without producing tumors at other sites. Dermal application of styrene oxide has not resulted in an increased incidence of tumors in mice, although it has been reported to induce malignant lymphoma in mice after administration by an unspecified route.

8.4.1 Experimental Studies

8.4.1.1 Acute Toxicity (279). The LD_{50} for guinea pigs and rats is about 2.0 g/kg. Single 24-h skin exposures in rabbits gave an LD_{50} value of 2.83 g/kg. Rats exposed to air saturated with vapors (theoretically calculated to be 395 ppm) survived a 2-h exposure, but three of six died following a 4-h exposure. Forty-four of 106 rats exposed to 300 ppm styrene oxide for a single 7-h inhalation exposure died within a 3-d period postexposure (280).

Eye Irritation. Undiluted styrene oxide may cause relatively severe irritation and pain to the eyes, but it is not apt to cause serious burns with permanent loss of vision. Solutions as dilute as 1% may have some irritating action.

Skin Irritation. Tests with laboratory animals and human subjects indicate that styrene oxide is capable of causing moderate skin irritation and skin sensitization. These effects may result from single or repeated contact with undiluted material and with solutions as dilute as 1%. Experience indicates that persons who have become hypersensitive may react rather severely to contact with the vapor as well as with the liquid material. There is some evidence that styrene oxide is absorbed slowly through the skin. This absorption could be significant only from exposures that produced extensive and serious injury to the skin.

8.4.1.3 Pharmacokinetics, Metabolism, and Mechanisms. The metabolism of styrene to styrene 7,8-oxide and the subsequent metabolism of styrene oxide has been widely described and summarized (281, 282). Styrene oxide is metabolized by two different pathways to mandelic and hippuric acid; these can be conjugated with glucuronic acid or mandelic acid can be dehydrogenated to phenylglyoxylic acid. Styrene oxide is also directly conjugated with glutathione and subsequently metabolized to mercapturic acids. The metabolites of styrene oxide are excreted primarily in the urine. Mendrala et al., conducted *in vitro* studies with mouse, rat, and human liver homogenates to compare the metabolism of styrene and styrene oxide and reported that the affinity (inverse K_m value) of styrene oxide for epoxide hydrolase was greatest in humans suggesting humans would detoxify styrene oxide more readily than these other species. Subsequently, a physiologically based pharmacokinetic model describing the disposition and metabolism of styrene oxide in mice, rat and humans after multiple routes of exposure has been developed (283). The model indicated that small changes in the elimination of styrene oxide could cause large changes in the body burden of styrene oxide in mice, but not in rats or humans, which may explain the greater sensitivity of mice to the toxicity of styrene oxide.

8.4.1.4 Reproductive and Developmental. No fetal malformations were observed in the offspring of female rats exposed to 100 ppm styrene oxide for 7 h/d, 5 d/wk for 3 wk prior to mating and then exposed to 100 ppm styrene oxide for 7 h/d daily during days 0–18 of gestation even though this exposure regimen resulted in severe maternal toxicity (285). Exposure of pregnant rabbits to 15 or 50 ppm styrene oxide for 7 h/d on days 1–24 of gestation resulted in maternal toxicity but no fetotoxicity or teratogenicity (285, 286). Injection of styrene oxide into the air space of fertilized chicken eggs resulted in embryo death (LD_{50} = 1.5 mmol/egg) and a 7% incidence of malformations as compared to none in controls (284).

A statistically significant decrease in the fraction of mated female rats that were found pregnant following exposure to styrene oxide (100 ppm) has been reported (287). Because corpora lutea but no implantation sites were found in the nonpregnant animals, it was suggested by the authors that styrene oxide may cause preimplantation loss of fertilized ova.

8.4.1.5 Carcinogenesis. Several chronic toxicity/carcinogenicity studies have been conducted on styrene oxide; five studies involved administration of the material at high doses by gavage; in two others styrene oxide was applied to the skin. Maltoni et al. (288) and Conti et al. (289) administered styrene oxide to Sprague–Dawley rats by gavage at dose levels of 50 or 250 mg/kg/d in olive oil, 4–5 d/wk, for 52 wk. The surviving animals were then held until spontaneous death for up to 52 additional weeks. A dose-related increase in the incidence of forestomach neoplasias, papillomas, and precursor lesions (acanthosis and dysplasia) was observed without any other indications of an oncogenic response in any other organ or tissue.

In another study (290), styrene oxide was given in olive oil to pregnant rats at a dose of 200 mg/kg on the seventeenth day of pregnancy and the offspring dosed with styrene oxide by gavage for 96 weekly doses at a dose level of 100–150 mg/kg. A statistically significant increase in forestomach tumors together with papillomas and other early changes indicative of chronic irritation (hyperkeratosis, hyperplasia, and dysplasia) were observed in the treated animals. The incidence of tumors in the styrene oxide groups was not increased relative to controls for any other tissues.

Lijinski (291) reported in a study where styrene oxide was administered in corn oil by gavage three times a week for up to 104 wk to rats and mice at dose levels of 275 or 550 mg/kg/d and 375 or 750 mg/kg/d, respectively. Consistent with the results of previous gavage studies, a high incidence of squamous cell carcinomas, papillomas, and other non-neoplastic lesions indicative of irritation were found in the forestomach of styrene oxide-treated animals. As in other studies, the incidence of tumors in the styrene oxide groups was not increased relative to controls for any other tissues.

Long-term dermal studies in which styrene oxide was applied three times per week to the skin as a 5 or 10% solution in acetone or benzene showed no indication of an oncogenic response in mice (145, 219), but it has been reported to induce malignant lymphoma in mice after administration by an unspecified route (292).

8.4.1.6 Genetic and Related Cellular Effects Studies. Styrene oxide and styrene have been tested in several *in vitro* and *in vivo* assays for genotoxicity; the results have been reviewed (293–296).

In *in vitro* assays, styrene oxide shows evidence of producing both gene mutations and chromosomal effects that appear to be reduced or eliminated by the presence of glutathione or exogenous metabolic systems (S9). Consistent with the species differences in metabolism described above, Herrero et al. (297) have reported that recombinant expression of human microsomal epoxide hydrolase in V79 Chinese hamster ovary cells significantly protected them from DNA damage by styrene oxide; this effect was blocked by valpromide, an inhibitor of epoxide hydrolase. In cultured lymphocytes, styrene oxide has been reported to induce DNA strand breaks (298) and sister chromatid exchange (296, 299). Genetic polymorphism in the expression glutathione *S*-transferases in the human population has been shown to affect the degree of sister chromatid exchange that occurred in cultured human lymphocytes exposed to styrene oxide (300). Styrene oxide can also form adducts with DNA (296), the molecular analysis of these adducts has been the subject of several more recent investigations (301–306).

In vivo, the responses to styrene oxide are more variable. Styrene oxide was weakly genotoxic or showed no genotoxicity depending on the species and test system, possibly due to metabolic inactivation prior to its distribution from the site of administration to the target cells. For example, sex-linked recessive lethal mutations and also somatic mutations have been reported in *D. melanogaster* (307, 308), but a dominant lethal test in mice was negative following an IP dose of 250 mg/kg (309). More recently, IP doses of styrene oxide given to mice at doses up to 200 mg/kg were reported to cause approximately a three- to 15-fold increase in DNA damage as measured by the Comet assay, depending on the tissue examined (310). However, caution has been advised in the conduct and interpretation of the Comet assay, as methodological factors such as time allowed for DNA unwinding has been shown dramatically to affect the results of the assay (311). Cantoreggi and Lutz (312) have reported that despite a relatively long half-life of styrene oxide in animals, the chemical reactivity of styrene oxide appears to be too low to result in a detectable production of DNA adducts *in vivo*. A comparison of these results to the DNA binding of other carcinogens suggests it is unlikely that the tumorigenic effect of styrene oxide results from a purely genotoxic mechanism. In summary, styrene oxide is a direct genotoxin *in vitro*, but appears to have a lower genotoxic potential *in vivo*, which is most likely due to the its rapid inactivation by metabolism.

8.4.1.7 Other: Neurological, Pulmonary, Skin Sensitization. Styrene oxide has been shown *in vitro* to form adducts with mammalian hemoglobin (35, 313), but a high background levels of these adducts in exposed individuals may preclude their use as biomarkers (314).

8.4.2 Human Experience: NA

8.5 Standards, Regulations, or Guidelines of Exposure

No current recommendations for exposure limits to styrene oxide could be found; however, current recommended threshold limits established by various nations for styrene range from 20 to 50 ppm (295). The International Agency for Research on Cancer (IARC) has indicated that there is sufficient evidence to establish the carcinogenicity of styrene oxide in animals and have also classified styrene oxide as a probable human carcinogen (Group 2A) (296). It is recommended that every precaution be taken when handling or using styrene oxide to prevent contact with the person.

B GLYCIDYL ETHERS

9.0 Allyl Glycidyl Ether

9.0.1 CAS Number: [106-92-3]

9.0.2 Synonyms: [(2-propenyloxy)methyl]oxirane, 1-(allyloxy)-2,3-epoxypropane, glycidyl allyl ether, AGE, 1,2-epoxy-3-allyloxypropane, allyl 2,3-epoxypropyl ether, and allyl glycidyl ether–ethylene glycol prepolymer (18/1)

9.0.3 Trade Names: NA

9.0.4 Molecular Weight: 114.14

9.0.5 Molecular Formula: $C_6H_{10}O_2$

9.0.6 Molecular Structure:

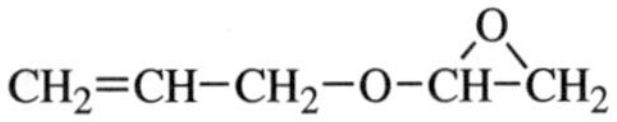

9.1 Chemical and Physical Properties

AGE is a colorless liquid with the following properties:

Specific gravity (20/4°C): 0.9698
Freezing point: Forms glass at – 100°C
Boiling point (760 mmHg): 153.9°C
Vapor density (25°C): 3.32
Vapor pressure (25°C): 4.7 mmHg
Refractive index (20°C): 1.4348
Solubility: 14.1% in water; miscible with acetone, toluene, octane
Flash point (Tag open cup): 57.22°F
Concentration in "saturated" air (25°C): 0.62%

1 ppm = 4.66 mg/m^3 at 25°C, 760 mmHg; 1 mg/L = 214 ppm at 25°C, 760 mmHg

9.1.2 Odor and Warning Properties

AGE has a characteristic, but not unpleasant, odor. The odor threshold is 10 ppm.

9.2 Production and Use

AGE is manufactured through the condensation of allyl alcohol and epichlorohydrin with subsequent dehydrochlorination with caustic to form the epoxy ring. It is a commercial chemical of primary interest as a resin intermediate, and is also used as a stabilizer of chlorinated compounds, vinyl resins, and rubber.

9.3 Exposure Assessment

9.3.3 Workplace Methods

A NIOSH standard method (No. 2545) for the determination of AGE in air has been developed (13).

9.4 Toxic Effects

AGE is moderate to low in acute oral and dermal toxicity, with the liver and kidney as main target organs. AGE is a CNS depressant and irritant to the respiratory tract upon single inhalation. Severe lung damage results from repeated inhalation at lethal doses in rats. AGE is appreciably irritating to the skin, severely irritating to the eyes and the upper respiratory tract. It causes occupational dermatitis and skin sensitization in humans. It appears to be a weak carcinogen in mice and is mutagenic *in vitro* and *in vivo*.

9.4.1 Experimental Studies

9.4.1.1 Acute Toxicity. Oral administration of AGE to rats and mice resulted in labored breathing and central nervous system depression, preceded by incoordination, ataxia, and reduced motor activity. Piloerection, diarrhea, and coma immediately preceded death. Oral LD_{50} values were 390 mg/kg in mice and 830–1600 mg/kg in rats. Necropsy of rats surviving a dose of 500 mg/kg AGE revealed irritation of the nonglandular stomach, including hyperkeratosis, erosion, and ulceration. Occasionally the liver showed focal areas of necrosis, and adhesions of the stomach wall to adjacent tissues were noted. The dermal LD_{50} in rabbits was 2550 mg/kg (315, 316).

The LC_{50} following 7-h inhalation exposures in rats to atmospheric concentrations of 100–2600 ppm AGE was calculated to be 308 ppm. Rats exposed to 300 ppm and higher had dyspnea, irritation of the nasal turbinates and lungs, nasal discharge, and discoloration, and gross pathological effects were noted in the liver and kidneys. No visible lesions were found at necropsy in rats exposed to 100 or 250 ppm (316). LC_{50}s of 670 ppm (8 h) in rats and 270 ppm (4 h) in mice have also been reported. In these latter studies, the most common finding was lung irritation, and pneumonitis was confirmed by microscopic examination. Discoloration of liver and kidneys was also noted frequently at necropsy, but tissue damage was not always confirmed microscopically. Occasionally, hepatic focal inflammatory cells and moderate congestion of the central zones were observed (315). Although intramuscular injection has been reported to affect the hematopoietic system at lethal doses (317), subchronic and chronic inhalation studies have not demonstrated any effect on hematopoiesis even at concentrations that result in increased mortality (315, 318).

Skin Irritation. Undiluted AGE was slightly irritating to intact rabbit skin, but moderate to severe irritation was observed on abraded skin (319).

Eye Irritation. Undiluted compound was severely irritating to the rabbit eye in the Draize test (319). Corneal opacity was seen in some rats after a single 8-h whole-body inhalation exposure (concentration not specified) (315) and following a 7-h exposure to air concentrations of 300–500 ppm; no corneal effects were seen at 250 ppm or less (316).

9.4.1.2 Chronic and Subchronic Toxicity. Rats were exposed, in groups of 10, to 260, 400, 600, or 900 ppm AGE vapor for 7 h/d, 5 d/wk for 10 wk. At 600 and 900 ppm, seven and eight animals in each group, respectively, died between the seventh and twenty-first exposure; rats in these groups had bronchopneumonic consolidation, severe emphysema, bronchiectasis, and inflammation of the lungs. Necrotic spleens were found in two of the rats exposed at 900 ppm. At 400 ppm, one rat died after the eighteenth exposure; the kidney-to-body weight ratio was significantly increased in animals exposed at this concentration, and necropsy revealed a decrease in peritoneal fat, severe emphysema, mottled liver, and enlarged and congested adrenal glands. At 260 ppm, slight eye irritation and respiratory distress persisting throughout the exposure period were observed. Decreased body weight gain was seen at all concentrations (315).

Repeated exposure (6 h/d, 5 d/wk) of mice to AGE to concentrations of 7.1 ppm (RD_{50} value) caused very severe lesions in the respiratory and olfactory epithelium of the nasal turbinates after four exposures. After 9 or 14 exposures, the severity of the lesion had decreased. Microscopic examination of the trachea and lungs showed no changes (320).

9.4.1.4 Reproductive and Developmental. Testicular degeneration was reported in rats after intramuscular injection of AGE; however, the results of the study were not statistically significant (317). Male rats were dosed with 400 mg/kg on days 1, 2, 8, and 9, and sacrificed on day 12. Focal necrosis of the testis was observed in one of the three surviving rats. Inhalation exposure of rats to 300 ppm AGE vapor 7 h/d, 5 d/wk for a total of 50 exposures also resulted in testicular atrophy in 5 of 10 rats, and 1 of 10 had small testes (315). In an 8-wk inhalation study of reproductive effects in rats and mice of both sexes, the NTP exposed rats to 0 to 200 ppm and mice to 0 to 30 ppm AGE (318). Although the mating performance of exposed male rats was markedly reduced, there was no effect on sperm morphology, motility, or number. Exposed female rats and male and female mice showed no effect on reproductive performance.

9.4.1.5 Carcinogenesis. In a 2-yr inhalation carcinogenicity study in Osborne Mendel rats and $B6C3F_1$ mice (50 of each sex at each exposure level), animals were exposed to concentrations of 0, 5, or 10 ppm AGE, 6 h/d, 5 d/wk. Although occasional respiratory epithelial tumors were observed, the NTP concluded the data provided only equivocal evidence of carcinogenicity in male rats and female mice. No evidence was obtained to support a carcinogenic effect in female rats. Some evidence was provided for a carcinogenic response in male mice, which included three adenomas of the respiratory epithelium, dysplasia in four mice, and focal basal cell hyperplasia of the respiratory epithelium in the nasal passages of seven mice (318, 321).

9.4.1.6 Genetic and Related Cellular Effects Studies. AGE has been shown to be positive in the Ames test in strains TA100 and TA1535 with and without activation and negative in strains TA1537 and TA98 (318, 322, 323). AGE induced SCEs and chromosomal aberrations in Chinese hamster ovary cells in both the presence and absence of metabolic activation (318). In the same study, it was reported that AGE induced a significant increase in sex-linked recessive lethal mutations in *Drosophila*, but did not induce reciprocal translocations.

In vitro incubation of AGE with nucleosides and DNA showed nucleoside alkylation, and DNA adduct formation with 7-AGE-guanine and 1-AGE-adenine was expected to be suitable for monitoring AGE exposures *in vivo*. (324). ^{32}P-postlabeling analysis confirmed the presence of the 7-guanine adduct in skin and liver of mice treated topically or intraperitoneally with AGE (325).

9.4.1.7 Other: Neurological, Pulmonary, Skin Sensitization. Oronasal exposure of mice to 1.9–8.6 ppm AGE for 15 min produced a concentration-dependent expiratory bradypnea, indicative of irritation of the nasal mucosa (326). The RD_{50} (airborne concentration producing a 50% decrease in respiratory rate) was 5.7 ppm. In another study an RD_{50} of 7.1 ppm was determined in mice (320). When mice were exposed via tracheal cannulation to 105–185 ppm AGE for 120 min, there was a concentration-dependent decrease in the respiratory rate due to pulmonary toxicity. The RD_{50} via tracheal cannulation was 134 ppm (326).

Incubation of AGE with mouse and human erythrocytes or IP injection of AGE to mice resulted in formation of N-terminal valine of hemoglobin adducts. In mice these hemoglobin adducts appeared to be stable over the lifetime of the erythrocyte. In unexposed individuals these adducts could not be detected (327).

9.4.2 Human Experience

Twenty-three cases of occupational dermatitis were reported in one study in which four of the workers developed sensitivity reactions to AGE; in one instance there was eye irritation from exposure to AGE vapor (326). Of 20 patients tested for allergenic properties of AGE, two had allergic reactions (328). More recently incidental cases of occupational contact allergy to AGE were described (329, 330).

9.5 Standards, Regulations, and Guidelines of Exposure

The ACGIH TLV is 1 ppm. The NIOSH recommended exposure limit (REL) is 5 ppm, with a short-term exposure limit (STEL) of 10 ppm. The concentration immediately dangerous to life and health (IDLH) is 50 ppm. The OSHA ceiling concentration is 10 ppm.

10.0 Isopropyl Glycidyl Ether

10.0.1 CAS Number: *[4016-14-2]*

10.0.2 Synonyms: [(1-Methylethoxy)methyl]-oxirane, 2,3-epoxypropyl isopropyl ether, and IGE, 1,2-epoxy-3-isopropoxypropane

10.0.3 Trade Names: NA

10.0.4 Molecular Weight: 116.18

10.0.5 Molecular Formula: $C_6H_{12}O_2$

10.0.6 Molecular Structure: $H_3C-CH(CH_3)-O-CH_2-CH(-O-)CH_2$

10.1 Chemical and Physical Properties

Isopropyl glycidyl ether is a mobile, colorless liquid with the following properties:

Specific gravity (20/4°C): 0.9186
Boiling point (760 mmHg): 137°C

Relative vapor density: 4.0
Vapor pressure (25°C): 9.4 mmHg
Solubility: 18.8% in water, soluble in ketones and alcohols
Concentration in "saturated" air: 1.237%

1 ppm = 3.74 mg/m^3 at 25°C, 760 mmHg; 1 mg/L = 211 ppm at 25°C, 760 mmHg

10.1.2 Odor and Warning Properties

No information on odor or warning properties was found.

10.2 Production and Use

Isopropyl glycidyl ether is manufactured through the condensation of isopropyl alcohol and epichlorohydrin with subsequent dehydrochlorination with caustic to form the epoxy ring. It is used as a reactive diluent for epoxy resins, stabilizer for organic compounds, and intermediate for synthesis of ethers and esters.

10.3 Exposure Assessment

A NIOSH standard method (1620) for the analysis of isopropyl glycidyl ether by absorption on charcoal, displacement with solvent, and determination by GLC has been described (13).

10.4 Toxic Effects

Isopropyl glycidyl ether was slightly toxic following acute oral or inhalation exposure, and practically nontoxic via dermal exposure. It is moderately irritating to the eyes and skin and capable of causing skin sensitization. Slight systemic toxicity occurred following repeated inhalation exposure in rats.

10.4.1 Experimental Studies

10.4.1.1 Acute Toxicity. Oral LD_{50} values were 1.3 and 4.2 g/kg, respectively, in mice and rats. The dermal LD_{50} in rabbits was 9.65 g/kg. Inhalation LC_{50} values were 1500 ppm in the mouse after 4-h exposure and 1100 ppm in the rat after 8-h exposure. Signs of toxicity following oral and inhalation exposures were mainly depressed motor activity, incoordination, and respiratory depression; irritation of the lungs was observed at necropsy following inhalation exposure (316).

Skin Irritation. Isopropyl glycidyl ether was moderately irritating to rabbit skin, with a Draize score of 4.3/8.0 (316).

Eye Irritation. The compound was moderately irritating to rabbit eyes, with a Draize score of 40/110 (316).

10.4.1.2 Chronic and Subchronic Toxicity. Rats exposed to 400 ppm for 10 wk (7-h/d; 5 d/wk) had a slight retardation of weight gain, slight ocular irritation, and elevated hemoglobin but no other evidence of cumulative toxicity. Some animals had patchy bronchopneumonia at autopsy (315). Repeated dermal exposure of a single rabbit to daily doses of about 90 mg/kg for 7 d caused local erythema and growth retardation (315).

10.4.1.6 Genetic and Related Cellular Effects Studies. Isopropyl glycidyl ether was positive in the Ames assay in *Salmonella* strains TA100 and TA1535 both with and without S9 activation (323). It also induced SOS repair in *E. coli* PQ37 (228). In an *in vitro* mammalian assay, it induced sister chromatid exchanges in Chinese hamster V79 cells (235). It has been shown to be a direct alkylating agent (331). *In vivo*, isopropyl glycidyl ether was positive in the sex-linked recessive lethal assay and the reciprocal translocation assay in *Drosophila melanogaster* (332).

10.4.2 Human Experience

No untoward effects, other than slight skin irritation on repeated contact, have been reported.

10.5 Standards, Regulations, or Guidelines of Exposure

The ACGIH TLV is 50 ppm; the STEL is 75 ppm; OSHA PEL TWA 50 ppm. The NIOSH 15-min ceiling value is 50 ppm; the concentration "immediately dangerous to life and health" (IDLH) is 400 ppm.

11.0 *n*-Butyl Glycidyl Ether

11.0.1 CAS Number: *[2426-08-6]*

11.0.2 Synonyms: 1-Butoxy-2,3-epoxypropane, BGE, 1,2-epoxy-3-butoxypropane, butyl glycidyl ether, *n*-butyl 2,3-epoxypropyl ether, (butoxymethyl)oxirane, 2,3-epoxypropyl butyl ether, butyl 2,3-epoxypropyl ether, glycidyl butyl ether, ageflex bge, glycidyl *n*-butyl ether, and Araldite RD-1

11.0.3 Trade Names: NA

11.0.4 Molecular Weight: 130.21

11.0.5 Molecular Formula: $C_7H_{14}O_2$

11.0.6 Molecular Structure: $C_4H_9{-}O{-}CH_2{-}\overset{O}{CH{-}CH_2}$

11.1 Chemical and Physical Properties

BGE is a colorless liquid with the following properties:

Specific gravity (25/4°C): 0.9087
Boiling point (760 mmHg): 164°C
Vapor pressure (25°C): 3.2 mmHg

Relative vapor density (25°C): 3.78 (air = 1)
Concentration in "saturated" air (25°C): 0.42%
Flash point: 54.5°C
Solubility: 2% in water at 20°C

1 ppm = 5.32 mg/m^3 at 25°C, 760 mmHg; 1 mg/L = 188 ppm at 25°C, 760 mmHg

11.1.2 Odor and Warning Properties

Although an odor threshold has not been measured, the odor is not unpleasant but is a slight irritant.

11.2 Production and Use

n-Butyl glycidyl ether (BGE) is made by the condensation of *n*-butyl alcohol and epichlorohydrin with subsequent dehydrochlorination with caustic to form the epoxy ring. Uses include the roles of viscosity-reducing agent for easier handling of conventional epoxy resins, acid acceptor for stabilizing chlorinated solvents, and chemical intermediate. Some curing agents may produce hazardous polymerizations in large quantities.

11.3 Exposure Assessment

A NIOSH standard method (Method 1616) for the analysis of BGE by absorption on charcoal, displacement with solvent, and determination by GLC has been described (13).

11.4 Toxic Effects

BGE is a CNS depressant and causes irritation of the respiratory tract when inhaled. It is of a low order of acute toxicity. It is appreciably irritating to the skin and eyes and causes occupational dermatitis and skin sensitization in humans. Inhalation of vapors causes irritation of the upper respiratory tract. It appears to be mutagenic *in vitro*, but mixed results have been obtained *in vivo*. A brief summary of the limited data on toxicological properties of *t*-BGE is also provided.

11.4.1 Experimental Studies

11.4.1.1 Acute Toxicity. The oral LD_{50} of BGE was reported to be 1.53 g/kg for mice and 2.26 g/kg for rats. Clinical signs prior to death were incoordination, ataxia, agitation, and excitement (315). The acute oral toxicity of *t*-BGE was comparable (316). Subsequent studies were fairly consistent with these studies with oral LD_{50} values reported in the range of 1800–3400 mg/kg (317, 319). The dermal LD_{50} of BGE in rabbits ranged from 0.788 g/kg (333) to 4.93 g/kg (316). In 4-h inhalation studies, air saturated with BGE vapors (ca. 2300 ppm) did not kill any mice (326), but one of six rats died at an air concentration of 4000 ppm (145). The LC_{50} for an 8-h exposure in rats was 1030 ppm (316). Intraperitoneal injection caused essentially the same pattern of signs as oral gavage, with LD_{50} values of 1.14 and 0.70 g/kg for rats and mice, respectively (315).

Skin Irritation. BGE was found to be a mild skin irritant in rabbits, with a Draize score of 2.8 out of 8 possible (316). BGE was a skin sensitizer in guinea pigs and in mice using the local lymph node assay (145, 334) Repeated dermal application in humans has resulted in appreciable skin irritation and sensitization (333).

Eye Irritation. BGE appeared to be a slight to moderate eye irritant in rabbits, with a Draize score of 23/110; there were corneal effects in 3/3 rabbits (335). In another study, BGE was reported to cause slight pain, slight conjunctival irritation, and slight corneal injury, all healing within 48 h (319).

11.4.1.2 Chronic and Subchronic Toxicity. Male rats inhaled BGE vapors for 50 7-h exposures to air concentrations of 38, 75, 150, or 300 ppm (336). There were no signs of treatment-related toxicity at the two lowest doses. At 150 ppm, 1 of 10 rats died, and survivors were significantly retarded in growth. At 300 ppm there was 50% mortality with additional signs of toxicity in the survivors, such as emaciation, unkempt fur, liver necrosis, and a significant increase in kidney/body and lung/body weight ratios. Testicular atrophy was noted in four out of five of the surviving rats at the lethal level of 300 ppm and in one rat exposed to 75 ppm; however, the rats used in this study were juveniles (57–97 g) in which the testes were probably immature at the start of the study, and there was a high incidence of bronchopneumonia, which may have also contributed to their general poor health. Thus it is difficult to identify this effect clearly as treatment related.

In a 28-d inhalation study, rats were exposed to BGE for 6 h/d, 5 d/wk at levels of 0.1, 0.5, or 1.0 mg/L air (equivalent to 18, 94, and 188 ppm) (337). No testicular toxicity was observed even though systemic toxicity was clearly evident. The exposures resulted in decreased body weights in the high-dose group, changes in fasting glucose in a high-dose reversibility group, elevated aspartate transferase levels in serum of high-dose males, and slightly increased hemoglobin in high-dose males, which was reversible. Histopathological examination revealed a degeneration of the olfactory mucosa and metaplasia of the ciliated respiratory epithelium. These latter changes were evident at the middle-dose level but not in the low-dose group.

In a repeated intramuscular injection study (6 d over a 10-d period) in rats, BGE did not suppress bone marrow or significantly alter leukocyte counts (318).

In a 90-d vapor inhalation study with the structural analogue *t*-BGE, male and female rats, mice, and rabbits were exposed at dose levels of 0, 25, 75, or 225 ppm for 6 h/d, 5 d/wk for 13 wk. There were no deaths, and all animals appeared normal and healthy during the course of the study. Microscopic examination revealed that the only tissue affected was the nasal mucosa. Hyperplasia and/or flattening of the nasal respiratory epithelium and inflammation of the nasal mucosa were seen in the 225-ppm group. Decreased body weight gain and concomitant decreases in organ weights were observed in all three species at this highest dose. Minimal effects, primarily in the nasal respiratory epithelium, were observed in most rats and mice exposed to 75 ppm, and no adverse effects were found in animals exposed to 25 ppm (338).

11.4.1.3 Pharmacokinetics, Metabolism, and Mechanisms. When ^{14}C-BGE was administered orally to male rats and rabbits (20 mg/kg), it was rapidly absorbed and metabolized. Most of the compound, 87% in the rat and 78% in the rabbit, was eliminated in the

0- to 24-h urine. In both species, a major route of biotransformation was via the hydrolytic opening of the epoxide ring followed by oxidation of the resulting diol to 3-butoxy-2-hydroxypropionic acid and subsequent oxidative decarboxylation to yield butoxyacetic acid (339). However, 23% of the dose administered to the rats was excreted in the urine as 3-butoxy-2-acetylaminopropionic acid, a metabolite that was not found in rabbits.

Incubation of BGE with mouse and human erythrocytes or after IP injection to mice has been shown to form adducts with the *N*-terminal valine of hemoglobin. In unexposed individuals these adducts could not be detected (327).

11.4.1.6 Genetic and Related Cellular Effects Studies. BGE has been found positive in a number of *in vitro* genetic toxicity assays, including the Ames assay (323, 340–343) and unscheduled DNA synthesis assay in cultured human blood lymphocytes (341, 342, 344) and W138 cells (345), with and without metabolic activation.

t-BGE was mutagenic in *S. typhimurium* TA100 and TA1535 in the absence and presence of an activation (S9) system. No response was observed in strains TA98, TA1537, TA1538 (343). An increase in unscheduled DNA synthesis (UDS) was observed in cultured human lymphocytes (344).

In vivo, mixed results have been obtained. BGE was negative in the mouse micronucleus assay when it was dosed by gavage (342) and negative in the host-mediated assay in mice when injected intraperitoneally (341). However, positive results were seen in a micronucleus assay in mice exposed intraperitoneally to >225 mg/kg/d for 2 d and $>$ 675 mg/kg for 1 d (346).

A dominant lethal test was conducted in mice in which males were treated topically three times a week for 8 wk with doses of 375, 750, or 1500 mg/kg BGE and then mated with untreated females (347). Although the results suggested a dominant lethal effect, the results were tentative, because the control fetal death rate was as high as the BGE fetal death rate at the top dose in the second experiment of the study. There was no gross pathology or histopathology in liver, lungs, or testes. A separate dominant lethal test at similar doses was negative (342).

t-BGE was negative in a chromosome aberration test and a micronucleus assay in the bone marrow of mice receiving 5 daily oral doses up to 400 mg/kg (346). Negative results were observed in a dominant lethal test in mice topically exposed to doses up to 1500 mg/kg *t*-BGE for 8 wk (applied 3 times weekly) (345).

11.4.2 Human Experience

Appreciable skin irritation and sensitization have been reported in humans (328, 348–350).

11.5 Standards, Regulations, or Guidelines of Exposure

The ACGIH TLV for 2000 is 25 ppm. The OSHA PEL is 50 ppm. The NIOSH ceiling limit is 5.6 ppm. The concentration "immediately dangerous to life and health" (IDLH) is 250 ppm.

12.0 Ethylhexyl Glycidyl Ether

12.0.1 CAS Number: [2461-15-6]

12.0.2 Synonyms: [(2-Ethylhexyl)oxy]methyloxirane, [(2-ethylhexyl)oxy]1,2-epoxy-3-propane, 1-glycidyloxy-2-ethylhexane, 1,2-epoxy-3-((2-ethylhexyl)oxy)propane, glycidyl 2-ethylhexyl ether, and EHGE

12.0.3 Trade Names: NA

12.0.4 Molecular Weight: 186.29

12.0.5 Molecular Formula: $C_{11}H_{22}O_2$

12.0.6 Molecular Structure: $CH_3{-}(CH_2)_3{-}CH(CH_2{-}CH_3){-}CH_2{-}O{-}CH_2{-}CH{-}CH_2$ (epoxide ring, O bridging the last two carbons)

12.1 Chemical and Physical Properties

2-Ethylhexyl glycidyl ether (EHGE) is a colorless liquid. Its properties are as follows:

Color, Gardner–Holdt: 1–2
Specific gravity, 25°C: 0.891
Viscosity, 25°C: 3 cm/sec
Epoxide equivalent weight: 230
Boiling point (°C, 11 mBar): 118–120
Flash Point (°C): 95–97

12.1.2 Odor and Warning Properties

It has a weak rancid odor; the odor threshold is unknown.

12.2 Production and Use

2-Ethylhexyl glycidyl ether is made by condensation of 2-ethylhexanol with epichlorohydrin followed by dehydrochlorination with caustic to form the epoxy ring. It constitutes a reactive diluent and may be used as a relatively nonvolatile chloride-scavenging agent and stabilizer for vinyl resins and rubber.

12.3 Exposure Assessment

No NIOSH or OSHA standard method for the determination of 2-ethylhexyl glycidyl ether has been developed was found. The pyridinium chloride–chloroform method for epoxy groups and GLC are suggested, but the HCl–dioxane method for epoxides (122, 209) might also be used.

12.4 Toxic Effects

EHGE has a low acute toxicity and is slightly irritating to the skin and eyes. It is a skin sensitizer. No cumulative toxicity was observed upon repeated inhalation of vapors of EHGE. It was found to be mutagenic in *in vitro* systems.

12.4.1 Experimental Studies

12.4.1.1 Acute Toxicity. The acute oral toxicity in rats of EHGE is >7500 mg/kg; the dermal LD_{50} in rats was >3600 mg/kg. Inhalation of vapors at concentrations ranging between 1.75 and 11.2 ppm for 19 d induced no clinical signs of toxicity or adverse effects growth and behavior (351).

Skin Effects. EHGE was moderately irritating when applied to rabbit skin for 24 h under occluded conditions. Signs of irritation persisted up to 7 d after applications. It was a moderate skin sensitizer in a maximization test in guinea pigs (351).

Eye Irritation. A eye irritation test in rabbits indicated that EHGE was a mild irritant, with irritation persisting for 24 h (351).

12.4.1.6 Genetic and Related Cellular Effects Studies. EHGE was tested in the Ames assay in the presence or absence of rat liver S9. EHGE increased the mutation frequency in the bacterial strains *S. typhimurium* TA1538, TA98, TA92, only in the presence of S9. In strain TA100, EHGE caused an increase in the mutation frequency with and without S9, but the increase was less in the nonactivated tests (352). In other studies, EHGE was mutagenic in TA100 and TA1535 only in the presence of S9 (rat liver); no response was seen in TA97 and TA98 (323, 353). The mitotic gene conversion was increased in *Saccharomyces cerevisiae* cultures exposed to EHGE in the presence of S9 (352). No significant chromosomal damage was observed in cultured rat liver cells exposed to EHGE/L (352).

12.5 Standards, Regulations, or Guidelines of Exposure

No standards of permissible occupational exposure were found.

13.0 Alkyl Glycidyl Ethers

13.0.1 CAS Numbers: (C_{10}–C_{16}) *[6881-84-5]*, (C_8–C_{10}) *[68609-96-1]*, (C_{12}–C_{14}) *[68609-97-2]*, (C_{12}–C_{13}) *[120547-52-6]*

13.0.2 Synonyms: Alkyl (C_{10}–C_{16}) glycidyl ether, alkyl (C_8–C_{10}) glycidyl ether, alkyl (C_{12}–C_{14}) glycidyl ether, and alkyl (C_{12}–C_{13}) glycidyl ether

13.0.3 Trade Names: Heloxy 7, (C_8–C_{10}) glycidyl ether, Heloxy 8, Epoxide 8, Araldite DY 025, Neodol glycidyl ether (C_{12}–C_{14}).

13.0.4 Molecular Weight: C_8–C_{10} 229, C_{12}–C_{14} 286

13.0.6 Molecular Structure:

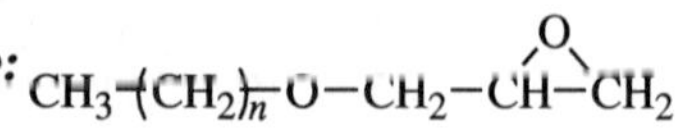

13.1 Chemical and Physical Properties

The properties of two of the epoxides are summarized as follows:

	C_8–C_{10} (Epoxide 7)	C_{12}–C_{14} (Epoxide 8)
Epoxide equivalent weight	230	291
Specific gravity (4/25%)	0.9	0.89
Melting point (°F)	10	35
Boiling point (°F)(100 mmHg)	283	420
Vapor pressure at 70°F (mmHg)	0.08	0.06
Flash point, ASTM D-1393-59 (°F)	245	310
Viscosity at 77°F (cm/sec)	10	10

These materials, because of their high epoxide equivalent weight, are claimed not to produce explosively hazardous polymerizations no matter what the quantity of curing agent added. This is in contrast to more reactive diluents.

13.1.2 Odor and Warning Properties

The alkyl glycidyl ethers have a mild fatty alcohol citrus-like odor. There is no room temperature vapor hazard.

13.2 Production and Use

Three fractions of straight-chain alcohols derived from reduction of fats are converted to their respective glycidyl ethers by reaction with epichlorohydrin followed by dehydrohalogenation. The glycidyl derivatives are the C_8–C_{10} fraction, the C_{12}–C_{14} fraction, and the C_{10}–C_{16} fraction, and the C_{12}–C_{13} fraction. The products are used mainly as reactive diluents for epoxy resins. The manufacturers claim low volatility hazard, improved substrate wettability (better adhesion), reduction of surface tension, and a lower tendency to promote resin crystallization. Although vaporization hazard is low, hot mixing should still be done with adequate ventilation and adequate precautions should be taken to prevent skin contact (354).

13.3 Exposure Assessment

Alkyl glycidyl ethers are essentially nonvolatile at room temperature. No OSHA nor NIOSH standard methods for their determination were found.

13.4 Toxic Effects

The alkyl glycidyl ethers are very low in acute toxicity, are moderate skin and mild eye irritants, and are skin sensitizers in guinea pigs. Only the C_8–C_{10} fraction appears to be a human skin sensitizer. No evidence of significant systemic toxicity or developmental toxicity has been found in repeated dermal exposure studies. Although mutagenic in some strains of *Salmonella* in the Ames assay, the alkyl glycidyl ethers are negative in other *in vitro* and *in vivo* genetic toxicity assays.

13.4.1 Experimental Studies

13.4.1.1 Acute Toxicity. Acute toxicity studies of alkyl glycidyl ethers by oral, dermal, and inhalation routes have demonstrated very low acute toxicity, with oral LD_{50}s for C_8–C_{10}, C_{12}–C_{14}, and C_{16}–C_{18} alkyl glycidyl ethers of 10.4, 19.2, and >31.6 mL/kg, respectively (355). Thus toxicity becomes less with increasing carbon chain length. In an acute dermal toxicity study, undiluted C_{12}–C_{14} alkyl glycidyl ether was applied to the skin of 10 rabbits at 1 mL/kg. Exposure produced moderate erythema without edema at all test sites; no other effects were observed (356). In a similar dermal study at doses as high as 4.5 mL/kg, no treatment-related toxic effects were noted in major organs at necropsy (357).

Skin Effects. Alkyl glycidyl ethers are considered to be moderate skin irritants with Draize scores in rabbit irritation studies of 3 or 4 out of a possible 8 (358, 359). Heloxy 9 was considered noncorrosive when applied undiluted to the intact skin of albino rabbits (360). Alkyl glycidyl ethers have been reported to be significant skin sensitizers in guinea pig assays.

Eye Irritation. Alkyl glycidyl ethers produced only mild eye irritation in rabbit Draize tests (361–363). There was slight conjunctivitis, which cleared within 1 d, and no corneal involvement.

13.4.1.2 Chronic and Subchronic Toxicity. In a 20-d dermal toxicity test in rabbits, C_8–C_{10} alkyl glycidyl ether was dosed at 2 mL/kg (5% solution in dimethyl phthalate). At necropsy, body weight and hematologic values were within the normal range, and there was no histological evidence of toxicity in any major organ (364). In a 90-d dermal toxicity study in rabbits with a mixture of C_{16}–C_{18} and C_8–C_{10} alkyl glycidyl ethers dosed at 2 mL/kg (5% solution in mineral oil), there was also no evidence of systemic toxicity (365).

A 2-wk study in which two rats/sex/dose level were dosed dermally at 0, 10, 100, or 1000 mg/kg/d, 5 d/wk with a blend of C_{12}–C_{13} alkyl glycidyl ether (~49% *n*-dodecyl glycidyl ether and ~39% *n*-tridecyl glycidyl ether, remainder unidentified, in acetone) found that after four doses at 1000 mg/kg/d the skin was scaly, thickened, and ulcerated (366). Microscopically, the response consisted of epidermal hyperplasia with hyperkeratosis and parakeratosis, along with hyperplasia of the sebaceous glands with ulceration, inflammation, and edema. Epidermal response at 100 mg/kg was similar to the 1000 mg/kg, probably due to the longer period of treatment; however, the epidermis remained intact. At 10 mg/kg/d only slight scaling was found at the site of application, and there were no histopathological effects of treatment to the skin. Based on the results of this 2-wk study, nonoccluded dermal doses of 0, 1, 10, or 100 mg/kg/d C_{12}–C_{13} alkyl glycidyl ether in acetone were administered to ten rats/sex/dose level 5 d/wk for 13 wk. No treatment effects were found on body weight, food consumption, ophthalmology, hematology, clinical chemistry, or urinalysis. Necropsy and complete histopathology revealed that treatment-related gross and histopathological lesions were limited to the skin of animals dosed at 100 mg/kg/d. The effects observed were epidermal and sebaceous gland hyperplasia, hyperkeratosis, and a mild inflammatory response. Treatment-related effects at 10 mg/kg/d were limited to slight scaling of the skin at the application site, although histopathologically the skin appeared normal. The no-observed effect level was 1 mg/kg/d.

13.4.1.3 Pharmacokinetics, Metabolism, and Mechanisms. *In vitro*, full thickness mouse skin was more permeable to ^{14}C-C_{12} AGE (1-dodecyl glycidyl ether) than was dermatomized rat or human skin with human skin showing only an average of 0.87% of applied dose absorbed within 24 h with an average lag time of about 3.8 h required for penetration (367). During skin penetration in all the species studied, ^{14}C-C_{12} AGE was extensively metabolized to its *bis*-diol.

In vitro metabolism studies by Boogaard, (367) have shown that ^{14}C-C_{12} AGE was very rapidly hydrolyzed by either cytosolic or microsomal fractions of liver and lung derived from rats, mice, and human tissues. The cytosol and microsomes derived from human tissues had a higher hydrolytic efficiency towards ^{14}C-C_{12} AGE than did those derived from rat or mouse tissues, with the exception being mouse liver cytosol and mouse lung microsomes, which both showed about the same activity as found in the corresponding derivatives from humans. Microsomal activity was greater than cytosolic activity for both tissues for all species except the mouse where microsomal and cytosolic activities were about equivalent. No detectable chemical reaction or enzyme-catalyzed conjugation of ^{14}C-C_{12} AGE with glutathione was observed.

13.4.1.4 Reproductive and Developmental. No adverse effects on fetal development were found in a study where eight pregnant rats/sex/dose level were dosed with 0, 1, 10, 50, 100, or 200 mg/kg/d, on days 6 through 15 of gestation with a blend of C_{12}–C_{13} alkyl glycidyl ether (~49% *n*-dodecyl glycidyl ether and ~39% *n*-tridecyl glycidyl ether, remainder unidentified, in acetone) (368). Intrauterine growth and survival were also unaffected at all dose levels as measured by the number of fetuses, early and late resorptions, total implantations, and corpora lutea. Dermal irritation at the application site was observed in the 50, 100, and 200 mg/kg/d groups with more severe signs of dermal irritation in the 100 and 200 mg/kg/d groups.

13.4.1.6 Genetic and Related Cellular Effects Studies. The C_8–C_{10}, C_{12}, and C_{14} alkyl glycidyl ethers were found to be weakly mutagenic in *S. typhimurium*, strains TA1535 and TA100 only in the presence of metabolic activation (369). More recently, a blend of C_{12}–C_{13} alkyl glycidyl ether (~49% *n*-dodecyl glycidyl ether and ~39% *n*-tridecyl glycidyl ether, remainder unidentified) was tested for mutagenicity both with and without metabolic activation in *S. typhimurium* strains TA97, TA98, TA100, TA1535, TA1537, and TA1538 and *E. coli* WP2uvrA. This material was found to mutagenic in TA1535 in three of five tests and in TA100 in one of five tests but not mutagenic in any other the other strains of bacteria (370). This same C_{12}–C_{13} alkyl glycidyl ether did not cause gene mutations in mammalian cells (CHO/HPRT mutation assay) in either the absence or presence of Aroclor-induced rat liver S9 (371). Negative results have been obtained for C_{12}–C_{14} alkyl glycidyl ether in the mouse lymphoma assay and unscheduled DNA synthesis assay using a human cell line (369). Negative results were also obtained for this compound in a host-mediated assay in which mice were dosed orally once a day for 4 d and urine was collected and tested with TA1535 for mutagenicity and in a dominant lethal assay in which compound was applied dermally at a dose of 2 g/kg, three times a week, for 8 wk (341). C_{12}–C_{13} alkyl glycidyl ether (~49% *n*-dodecyl glycidyl ether and ~39% *n*-tridecyl glycidyl ether, remainder unidentified) did not produce an increase in micronucleated polychromatic erythrocytes when administered to male and female mice intraperitoneally at doses up to 4000 mg/kg (372). In summary, alkyl glycidyl ethers appear to be a weak bacterial mutagen but have not been shown to cause mutations or chromosomal effects in mammalian test systems.

13.4.1.7 Other: Neurological, Pulmonary, Skin Sensitization. Mattsson et al. (430) have studied the potential of C_{12}–C_{13} alkyl glycidyl ether to produce neurotoxicity following dermal exposure using the same test material described for the studies of McGuirk and Johnson (366) Nonoccluded dermal doses of 0, 1, 10, or 100 mg/kg/d C_{12}–C_{13} alkyl glycidyl ether in acetone were administered to ten rats/sex/dose level 5 d/wk for 14 wk. Treatment-related effects on the skin, which were the same as described were observed, but there were no treatment-related effects noted on the functional observational battery, motor activity, clinical examination or the comprehensive histopathological examination of perfused tissues. Also evaluated were the evoked nerve potentials, which include tests of the visual pathway (flash-evoked potential; FEP), auditory pathway somatosensory pathway, and caudal nerves. Statistically significant differences occurred in the FEPs, but not on any of the other evoked potentials. FEPs from all dose levels were very "normal" in appearance, and treatment-related differences were complex. In general, FEPs from male mid- and high-dose rats were smaller than controls, and FEPs from female high-dose rats were larger than controls; male rat FEPs collected 5 wk postexposure showed a pattern of effects similar to the earlier FEPs. The alteration in the FEPs suggested that the input from the eyes or optic tracts might be altered and electroretinograms (ERGs) collected on control and high-dose males found the ERGs of high-dose males were smaller than controls. However, the histopathological evaluation the found no treatment-related differences. The no observed effect level for the study was 1 mg/kg/application, based on mild skin effects and mild FEP alterations at the 10-mg/kg dose.

13.4.2 Human Experience

Although alkyl glycidyl ethers appear to be skin sensitizers in guinea pig assays, only the C_8–C_{10} fraction appears to be a human skin sensitizer under the conditions of the repeated insult patch test (373).

13.5 Standards, Regulations, or Guidelines of Exposure

None has been established.

14.0 Diglycidyl Ether

14.0.1 CAS Number: [2238-07-5]

14.0.2 Synonyms: Di(2,3-epoxy)propyl ether, diallyl ether dioxide, glycidyl ether, oxirane, 2,2′-(oxybis(methylene))bis-, bis(2,3-epoxypropyl) ether, ether, bis(2,3-epoxypropyl), diallyl ether dioxide, DGE, di(epoxypropyl)ether, NSV 54739, and 2,2′-[oxybis(methylene)]bisoxirane

14.0.3 Trade Names: NA

14.0.4 Molecular Weight: 130.14

14.0.5 Molecular Formula: $C_6H_{10}O_3$

14.0.6 Molecular Structure:

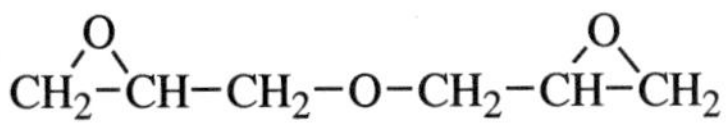

14.1 Chemical and Physical Properties

The properties are as follows:

Specific gravity (20/4°C): 1.1195
Boiling point (760 mmHg): 260°C
Vapor pressure (25°C): 0.09 mmHg
Vapor concentration in saturated air (25°C): 0.0121%
Vapor density (air = 1): 3.78 at 25°C

14.1.2 Odor and Warning Properties (315)

The odor threshold is not established, but the odor is recognizable at approximately 5 ppm. The odor is strong and irritating (374). It is irritating to eyes at 10 ppm (315).

14.2 Production and Use

Diglycidyl ether is significant only because it is a possible trace component of epoxy compounds derived from epichlorohydrin.

14.3 Exposure Assessment

No OSHA nor NIOSH standard methods for the determination of diglycidyl ether were found. The pyridinium chloride–chloroform method for epoxy groups and GLC are suggested, but the HCl–dioxane method for epoxides might also be used (209).

14.4 Toxic Effects

Diglycidyl ether is a severe irritant to skin, eyes, and respiratory tract. It is a strong skin sensitizer. It exhibits radiomimetic effects following acute and chronic exposure as evidenced by depression of bone marrow and other rapidly growing cells. It is genotoxic in lower organisms, and has been shown to be carcinogenic in mice following repeated skin application. At high levels of exposure it may cause adverse kidney and liver effects.

14.4.1 Experimental Studies

14.4.1.1 Acute Toxicity. (315). The acute oral LD_{50} ranged from 170 to 192 mg/kg in mice and from 450 to 510 mg/kg in rats. At gross necropsy, pulmonary hemorrhage, hyperemia and irritation of the enteric tract, liver and renal changes, and "inflammation" of the adrenal gland were found (315).

The intravenous LD_{50} in rabbits was 141 mg/kg. Autopsy showed severe congestion of the lungs, congestion of patchy ischemia of the liver, slight ischemia of the kidneys, and ascites. Single injections of 50, 100, and 200 mg/kg had an effect on the peripheral blood cell counts and cell morphology. These consisted of leukopenia due to a decrease of polynuclear cells, the duration of which increased with the dose. In animals given 100 mg/kg, an increase was observed in nucleated red blood cells which appeared 7 d after injection. In surviving animals there was eventually a full recovery.

The percutaneous LD_{50} ranged from 1000 to 1500 mg/kg in either rats or rabbits. The application induced severe skin irritation in both species and also caused systemic effects similar to those following intravenous injection, that is, weight loss and leukopenia on the third day after application, which after a few days changed into a leukocytosis. In rabbits, a decrease of the hemoglobin concentration also occurred.

The LC_{50} values by inhalation are listed as follows:

Species	Time of Exposure (h)	LC_{50} (ppm)
Rabbit	24	13.3
Mouse	4	86
Rat	4	200
Mouse	8	30
Rat	8	68

The immediate effects of vapor were few and due to irritation of the mucosal membranes; however, after removal from the chamber within 24 h the rabbits showed cloudy corneas as well as irritation of the mucous membranes and the skin. High vapor exposure (113 ppm) in rats also caused dyspnea. Necropsy showed lung congestion, granular, discolored livers, enlarged kidneys, and prominent adrenals. At single 24-h vapor exposures of rabbits (to 3, 6, 12, and 24 ppm) the morphology of the blood cells was not affected at the 3 and 6 ppm level; at 12 ppm a possible thrombocytosis was noted; and at 24 ppm there was a marked leukocytosis preterminally. At 3 ppm there was still clear evidence of mucous membrane irritation, which became very severe at the highest level.

Skin Effects. Single and repeated application to the intact or scarified skin of rabbits induced severe irritation and chemical burns. Repeated application led to necrosis of subcutaneous tissues. Diglycidyl ether was a strong skin sensitizer in guinea pigs (315).

Eye Irritation. In the eyes, instillation of liquid diglycidyl ether induced severe irritation and corneal necrosis. A 15% solution in propylene glycol was rated as severely irritating (315).

14.4.1.2 Chronic and Subchronic Toxicity (315). Rats received daily percutaneous applications (5 d/wk, 4 wk) of 15, 30, 60, 125, 250, and 500 mg/kg. At the two highest dose levels there was a high mortality and at 125 mg/kg and higher there were pronounced symptoms of systemic toxicity, necrosis of the skin, and corneal opacity. The symptoms of systemic toxicity were weight loss and leukopenia due to a decrease in polynuclear cells. Autopsy of these animals showed general bone marrow aplasia, atrophy of the thymus, focal lymphoid necrosis, necrosis of the proximal convoluted tubules of the kidneys, focal necrosis of the testes, and hemorrhage of the adrenal medulla. At 30 and 60 mg/kg there were only weight loss and a decrease in polynuclear cells in the blood but no leukopenia. The bone marrow showed no changes. The thymus only was decreased in weight in the 60-mg/kg animals; the testes were all normal. At the 15-mg/kg level no effect was observed.

Dogs receiving a few repeated intravenous injections at weekly intervals of 25 mg/kg developed leukopenia. Severe inflammation in the muscles occurred when the injections were given intramuscularly. Several dogs died with secondary pulmonary infections. Also a weekly intravenous dose of 12.5 mg/kg induced leukopenia. In surviving dogs, killed a month after the last injection, the bone marrow appeared normal.

Rats were exposed to DGE by inhalation of vapors with concentrations of 0.0, 0.3, or 3.0 ppm 4 h/d, 5 d/wk. The rats exposed to 3.0 ppm received only 19 exposures; the exposures to 0.3 ppm amounted to a maximum of 60. The animals at 3.0 ppm showed mortality due to pulmonary infection and showed evidence of systemic intoxication: diminished weight gain, blood changes appearing after the seventh exposure, and bone marrow changes at autopsy. In 10 animals surviving for 1 yr after the final exposure to 3.0 ppm the signs of systemic intoxication had disappeared and the bone marrow was normal, but several rats showed peribronchiolitis and one a fatty dystrophy of the liver. Thirty animals were exposed to 0.3 ppm (estimated to be approximately 0.2 ppm actual value), of which 10 were sacrificed after 20 exposures. These animals appeared normal. Ten other animals were sacrificed after 60 exposures and 10 others kept for 1 yr after the final exposure and then sacrificed. There was no effect on the peripheral blood or the bone marrow in these two groups. The group sacrificed immediately after the final exposure showed "poorly defined focal degeneration of the germinal epithelium." Thus it appears that any changes occurring at low levels of exposure are reversible. However, a clear no-effect level has not been demonstrated.

14.4.1.5 Carcinogenesis. In mice, diglycidyl ether has been shown to produce epithelioma following repeated skin application. A total dose of 100 mg produced these tumors in 4/20 animals; a dose of 33 mg produced only 1/20 (315).

14.4.1.6 Genetic and Related Cellular Effects Studies. Diglycidyl ether has been shown to have a mutagenic effect in bacteriophage T2 and *Neurospora* and to induce chromosome aberrations in *Vicia faba* and several other plant cells. It has a mutagenic action in bacterial systems such as *Salmonella typhimurium* (315).

14.4.2 Humans Experience

No published reports could be found on adverse effects from industrial handling of DGE.

14.5 Standards, Regulations, or Guidelines of Exposure

The 2000 ACGIH TLV is 0.1 ppm (0.53 mg/m^3) with an A4 classification (not classifiable as a human carcinogen). NIOSH recommends TWA 0.1 ppm (0.5 mg/m^3), IDLH value = 10 ppm, and diglycidyl ether be regulated as a potential human carcinogen. The OSHA ceiling concentration is 0.5 ppm.

15.0 1,4-Butanediol Diglycidyl Ether

15.0.1 CAS Number: [2425-79-8]

15.0.2 Synonyms: Butanediol diglycidyl ether, 1,4-bis(2,3-epoxypropoxy)butane, 2,2′-[1,4-butanediylbis(oxymethylene)]bisoxirane, 1,4-bis(oxiranylmethyloxy)butane, 1,4-bis(glycidyloxy)butane, 1,4-butane diglycidyl ether, butane-1:4-diol diglycidyl ether, and Araldite RD-2

15.0.3 Trade Names: NA

15.0.4 Molecular Weight: 202.25

15.0.5 Molecular Formula: $C_{10}H_{18}O_4$

15.0.6 Molecular Structure:

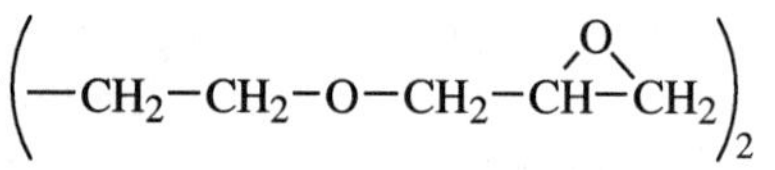

15.1 Chemical and Physical Properties

1,4-Butanediol diglycidyl ether is a clear to pale yellow liquid with the following properties:

Boiling point: 266°C
Specific gravity: 1.049

15.1.2 Odor and Warning Properties

No information on odor or warning properties was found.

15.2 Production and Use

The compound results from condensation of butanediol with epichlorohydrin followed by dehydrochlorination with caustic. Its primary use is as a reactive diluent in epoxy resin systems to reduce viscosity and to allow higher filler loading.

15.3 Exposure Assessment

No OSHA or NIOSH standard methods for the determination of 1,4-BDGE were found. The pyridinium chloride–chloroform method for epoxy groups (209) and GLC are suggested, but the HCl–dioxane method for epoxides (209) might also be used (209).

15.4 Toxic Effects

The acute toxicity of 1,4-BDGE by the oral and dermal routes is low; however, it is a severe skin and eye irritant and skin sensitizer. 1,4-BDGE was mutagenic in *in vitro* bacterial and mammalian cell mutagenicity assays and in an *in vivo* mutagenicity assay; however, no carcinogenic potential was identified negative in a 2-yr dermal cancer bioassay in mice.

15.4.1 Experimental Studies

15.4.1.1 Acute Toxicity. The oral LD_{50} of 1,4-BDGE was 1.41–1.88 g/kg in rats (375, 376) and 3.61 g/kg in hamsters (376). The dermal LD_{50} was >2.15 g/kg in rats; no evidence of systemic toxicity was apparent (377).

Skin Effects. A single dermal application to rabbits resulted in marked skin irritation with a Draize score of 4.3 out of a possible 8.0 (375); five daily consecutive applications produced extreme irritation with a maximum score of 8.0 (376). 1,4-BDGE is a severe sensitizer in guinea pigs (378, 379).

Eye Irritation. Draize scores in rabbits have been 44–80 out of a possible 110, indicating severe irritation (375).

15.4.1.5 Carcinogenesis. A 2-yr dermal cancer bioassay was conducted in CF_1 mice at doses of 0, 0.05, and 0.2% in acetone (380). Treatment did not adversely affect survival, did not increase the incidence of skin tumors, and did not result in significant skin irritation. There were no statistically significant increases in incidence of any systemic tumors, except for lymphatic tumors in females. Because there was a high background incidence of this tumor type in CF_1 mice used in the testing laboratory (381), there was no clear evidence of 1,4-BDGE induced carcinogenicity.

15.4.1.6 Genetic and Related Cellular Effects Studies. 1,4-BDGE was mutagenic in several *Salmonella* strains in the Ames assay (323) and was positive in a mouse lymphoma assay (376), in both instances with and without activation. When 1,4-BDGE was gavaged in Chinese hamsters at daily doses of 0.6 to 3.0 g/kg for 2 d, there was a significant increase in the percentage of bone marrow cells with nuclear anomalies at 24 h after the second dose (382).

15.4.2 Human Experience

1,4-BDGE is a skin sensitizer in exposed workers (383).

15.5 Standards, Regulations, or Guidelines of Exposure

No standards of permissible occupational exposure were found.

16.0 Neopentyl Glycol Diglycidyl Ether

16.0.1 CAS Number: *[17557-23-2]*

16.0.2 Synonyms: 2,2-((Dimethyl-1,3-propanediyl)bis(oxymethylene))bisoxirane, 1,3-Bis(2,3-epoxypropoxy)-2,2-dimethylpropane, Diglycidyl ether of neopentyl glycol,

Heloxy wc68, 2,2′-((2,2-Dimethyl-1,3-propanediyl)bis(oxymethylene))bis-oxirane, NPGDE

16.0.3 Trade Names: NA

16.0.4 Molecular Weight: 216.28

16.0.5 Molecular Formula: $C_{11}H_{20}O_4$

16.0.6 Molecular Structure:

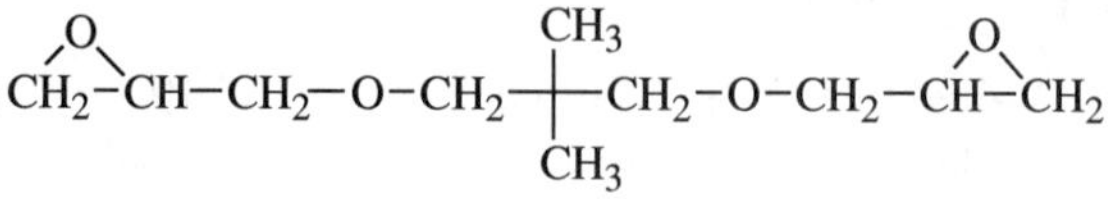

16.1 Chemical and Physical Properties

The properties are as follows:

Specific gravity (25/4°C): 1.07
Flash point (open cup): 190°F min
Viscosity (25°C): 10–16 cm/sec
Color, Gardner-Holdt: 3 max
Epoxide equivalent weight: 135–155

Hazardous polymerization may occur with aliphatic amines in masses greater than 1 lb.

16.1.2 Odor and Warning Properties

No information was found on odor or warning properties.

16.2 Production and Use

The ether is made by condensing neopentyl glycol with epichlorohydrin followed by dehydrochlorination with caustic. This may be used as a bifunctional reactive diluent for viscosity reduction of resins with minimum change in cured resin properties and to allow increased filler incorporation.

16.3 Exposure Assessment

No OSHA or NIOSH standard methods for the determination of NPGDGE were found. The pyridinium chloride–chloroform method for epoxy groups (209) and GLC are suggested, but the HCl–dioxane method for epoxides (209) might also be used.

16.4 Toxic Effects

Acute toxicity of NPGDGE is low by both the oral and dermal routes of exposure. The compound is slightly irritating to the rabbit eye but is a severe skin irritant and skin sensitizer. Systemic toxicity was not apparent following repeated dermal exposure. NPGDGE appears to be a weak genotoxin and weak carcinogen in mice.

16.4.1 Experimental Studies

16.4.1.1 Acute Toxicity. In rats, the oral LD_{50} was 4.5 g/kg, and the dermal LD_{50} was >2.15 g/kg (379).

Skin Effects. NPGDGE was a moderate skin irritant in rabbits, with a Draize irritation score of 2.3 out of a possible 8.0 (384). In a 5-d repeated dermal exposure study, skin irritation became severe with necrosis (385). NPGDGE was a potent skin sensitizer in guinea pigs (379, 386).

Eye Irritation. NPGDGE was only slightly irritating to the washed rabbit eye (379).

16.4.1.2 Chronic and Subchronic Toxicity. In two separate 5-d repeated dermal application studies in rabbits, in which 0.5 mL of NPGDGE was applied occluded for 24 h, daily, there was no evidence of systemic toxicity at necropsy carried out at 5 d after the last exposure (386).

16.4.1.5 Carcinogenesis. A 2-yr dermal carcinogenicity study of NPGDGE in C3H mice was conducted at dose levels of 0.94, 1.87, and 3.75 mg/mouse (387). Skin tumor incidences were 10/50, 6/50, and 0/50 at 3.75, 1.87, and 0.94 mg/mouse; no tumors occurred in the 300 acetone-treated controls. Tumor potency was calculated to be 1/700th that of benzo[a]pyrene, which was tested concurrently.

16.4.1.6 Genetic and Related Cellular Effects Studies. NPGDGE was positive in the Ames test with strain TA1535 without activation, but was negative with activation; it was negative in strain TA98 with or without activation. In a host-mediated assay in mice, minimal positive results were obtained in strain TA1535 with the addition of β-glucuronidase, but not without activation. Positive results were obtained with induction of DNA repair in cultured human leukocytes. Negative results were seen in a micronucleus test and in a dominant lethal test in mice with dermal application of 1.5 g/kg three times a week for 8 wk to the males (341).

16.5 Standards, Regulations, or Guidelines of Exposure

No standards of permissible occupational exposure were found.

17.0a Phenyl Glycidyl Ether

17.0.1a CAS Number: [122-60-1]

17.0.2a Synonyms: PGE, 1,2-epoxy-3-phenoxypropane, 2-(phenoxymethyl) oxirane, glycidyl phenyl ether, phenyl 2,3-epoxypropyl ether, (phenoxymethyl)oxirane, benzene, (2,3-epoxypropoxy)-, 2,3-epoxypropyl phenyl ether, 3-phenoxy-1,2-epoxypropane, phenoxypropene oxide, (+/−)-1,2-epoxy-3-phenoxypropane, and 1,2-epoxy-3-phenoxypropane

17.0.3a Trade Names: NA

17.0.4a Molecular Weight: 150.17

17.0.5a Molecular Formula: $C_9H_{10}O_2$

17.0.6a Molecular Structure:

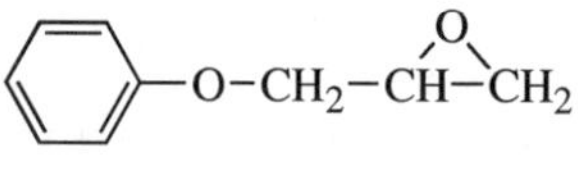

17.0b *p*-(*t*-Butyl)phenyl Glycidyl Ether

17.0.1b CAS Number: *[3101-60-8]*

17.0.2b Synonyms: TBPGE, [4-(1,1-dimethylethyl)phenoxy]methyl oxirane, 1-(*p-tert*-butylphenoxy)-2,3-epoxy propane, [[4-(1,1-dimethylethyl)phenoxy]methyl]-oxirane, *p-tert*-butylphenyl, 2,3-epoxypropyl ether, and Araldite 6005

17.0.3b Trade Names: NA

17.0.4b Molecular Weight: 206.28

17.0.5b Molecular Formula: $C_{13}H_{18}O_2$

17.0.6b Molecular Structure:

17.1 Chemical and Physical Properties

PGE is a relatively high boiling, colorless liquid with the following properties

Specific gravity (20/4°C): 1.1092
Melting point: 3.5°C
Boiling point (760 mmHg): 245°C
Relative vapor density (25°C): 4.37
Vapor pressure (20°C, mmHg): 0.01
Viscosity (25°C): 6 cP
Refractive index: 1.5314
Solubility: 0.24% in water; 12.9% in octane; completely soluble in acetone and toluene
Conversion factor 1 ppm = 6.23 mg/m^3

17.1.2 Odor and Warning Properties

It has an unpleasant sweet odor. No warning properties were found.

17.2 Production and Use

PGE is synthesized by condensation of phenol with epichlorohydrin, with subsequent dehydrochlorination with caustic to form the epoxy ring. As an acid acceptor it is very effective as a stabilizer of halogenated compounds. Its high solvency for halogenated materials offers many possibilities as an intermediate.

17.3 Exposure Assessment

NIOSH standard method 1619 for the analysis of PGE by absorption on charcoal, displacement with solvent, and determination by GLC has been described (13).

17.4 Toxic Effects

The acute oral and dermal toxicity of PGE is low, with the liver and kidney as target organs for systemic effects. PGE is an appreciable skin and eye irritant and a skin sensitizer. These acute effects were comparable for TBPGE. PGE is irritant to the respiratory tract upon single or repeated inhalation exposure. PGE induced nasal tumors in a 2-yr rat bioassay and is classified by IARC as possibly carcinogenic to humans (2B). PGE did not affect development and reproduction in rats. The compound was a direct-acting mutagen in *in vitro* assays but no activity was observed *in vivo*. TBPGE was not mutagenic *in vitro*.

17.4.1 Experimental Studies

17.4.1.1 Acute Toxicity. Oral LD_{50}s reported for PGE ranged from 2.5 to 6.4 g/kg in the rat (145, 211, 316, 388) and 1.4 g/kg in the mouse (316). Incoordination, ataxia, and decreased motor activity were followed by coma and death in rats dosed orally. Extensive congestion of the liver and kidney were also observed at necropsy. Dermal LD_{50}s in the rabbit were 2.28–2.99 g/kg (145, 211, 316). Although irritation of the lungs occurred, no deaths were produced in mice exposed for 4 h or rats exposed for 8 h to saturated vapors at room temperature. The acute toxic properties of TBPGE and CGE were comparable (384, 389).

Skin Effects. PGE was moderately to severely irritating to rabbit skin, particularly upon prolonged or repeated treatment (145, 211, 316). In a 4 h test under semioccluded conditions PGE was slightly irritant (390). PGE was a strong skin sensitizer in guinea pigs after topical and intradermal administration (145, 390–392). Skin effects of TBPGE were similar (389).

Eye Irritation. PGE has produced irritation ranging from mild to severe in the rabbit eye (145, 211, 316, 335, 388, 390).

17.4.1.2 Chronic and Subchronic Toxicity. When rats inhaled aerosols of PGE at a concentration of 29 ppm for 4 h/d, 5 d/wk for 2 wk, the animals exhibited weight loss, atrophic changes in the liver, kidneys, spleen, thymus, and testes, depletion of hepatic glycogen, and chronic catarrhal tracheitis (393). In a subsequent study (394) rats and dogs were exposed to concentrations of 1, 5, and 12 ppm for 6 h/d, 6 d/wk for 3 mo. In rats at the high-dose there was alopecia and hyperkeratosis and inflammatory atrophy of hair follicles; this change was not seen in dogs. There were no compound-related changes in the histopathology of major organs or in blood, urine, or biochemical indexes at any exposure concentration in either rats or dogs. Six intramuscular injections of 400 mg/kg PGE in rats failed to produce any effects on hematopoiesis (317).

17.4.1.3 Pharmacokinetics, Metabolism, and Mechanisms. Studies with PGE in rats and rabbits have shown evidence for two key mechanisms in the metabolism of PGE: hydrolysis of the epoxide group and glutathione conjugation. In rats, the capacity for glutathione (GSH) conjugation is limited and the ratio of GSH conjugation and hydrolysis decreases with increasing exposure to PGE. Urinary metabolites identified for rat following subcutaneous administration of PGE were 3-phenyloxylacetic acid (up to 94% of the administered dose) and small quantities of the mercapturic acid of phenyl glycidyl ether [*N*-acetyl-*S*-(2-hydroxy-3-phenoxypropyl)- L-cysteine]. Recently, a novel metabolite (*N*-acetyl-*O*-phenylserine) derived from transamination from the epoxide hydrolysis pathway was described (395, 396). When administered dermally, the absorption rates were 4.2 $mg/cm^2/h$ for rats and 13.6 $mg/cm^2/h$ for rabbits (388).

Incubation of PGE with mouse and human erythrocytes or IP injection of PGE to mice resulted in formation of *N*-terminal valine hemoglobin adducts. In unexposed individuals these adducts could not be detected (327).

17.4.1.4 Reproductive and Developmental (393). Male rats were exposed to PGE at concentrations of 0, 2, 6, or 11 ppm for 6 h/d for 19 consecutive days and were mated with untreated females for 6 consecutive weeks. Offspring from those matings were mated with the treatment group. There were no significant effects on reproduction. Although one of eight males from each of the three treatment groups showed testicular atrophy upon histological examination, there did not appear to be any compromise in their functional capacity to reproduce. In the same study, pregnant rats were exposed to PGE by inhalation on the fourth and fifteenth days of gestation at concentrations of 1, 5, and 12 ppm for 6 h/d. No clinical signs of toxicity were observed, and there was no evidence of developmental toxicity (393).

17.4.1.5 Carcinogenesis. An inhalation carcinogenicity study in rats at concentrations of 0, 1, and 12 ppm for 2 yr resulted in nasal carcinomas, 11% in males and 4.4% in females at 12 ppm, and dose-related squamous cell metaplasia, which corresponded to the increased incidence of nasal tumors (397).

17.4.1.6 Genetic and Related Cellular Effects Studies. PGE was mutagenic in *Salmonella* strains TA100 and TA1535 with and without metabolic activation, suggesting it is a direct-acting mutagen causing base substitutions. Generally, no response was observed in strains TA98, TA1537, TA1538 (4, 323, 398–400). Mutagenic activity was also observed in *Klebsiella* (3) and *E. coli*, including repair deficient strains, suggesting that PGE may cause DNA damage that can be repaired by recombination. (331, 401, 402). In mammalian cells, PGE did not induce chromosomal aberrations or gene mutations in cultured Chinese hamster ovary cells (398, 400), but did induce transformation of hamster embryo cells in culture (398). No unscheduled DNA synthesis was induced by PGE in cultured rat hepatocytes (230). TBPGE was not mutagenic *in vitro* in bacteria or yeast (4, 389).

Inconsistent, nonspecific results were seen in mice given PGE orally or intraperitoneally in a host-mediated assay (398). No effects on DNA synthesis in mouse testes was observed after oral administration of 0.5 g/kg PGE (402). PGE was also negative in the mouse micronucleus assay following an oral dose of 1 g/kg (400). There was also no increase in the incidence of chromosomal aberrations in bone marrow cells from male rats exposed to PGE by inhalation at concentrations of 0, 1, 5 or 12 ppm for 6 h/d for 19 consecutive days (402). A dominant lethal assay in rats inhaling PGE at concentrations of 0, 2, 6 and 11 ppm for 19 d (6 h/d) was also negative (402).

PGE has been reported to alkylate nucleic acid bases *in vitro*; however PGE did not bind to DNA in *E. coli* with or without metabolic activation (403). Incubation of PGE with deoxynucleosides or calf thymus DNA resulted in formation of phosphate alkylated and base alkylated adducts (404, 405).

17.4.2 Human Experience

Contact allergy to PGE has been observed in workers handling epoxy resin systems, glues, and plastics. There is some evidence suggesting that cross-sensitization with other glycidyloxy compounds may also occur. No adverse systemic effects have been described in any of the case reports (316, 328, 329, 406).

17.5 Standards, Regulations, or Guidelines of Exposure

The ACGIH TLV is 0.1 ppm with a skin designation; OSHA PEL TWA 10 ppm (60 mg/m^3. The NIOSH considers PGE a potential carcinogen with a ceiling limit of 1 ppm; the immediately dangerous to health concentration is 100 ppm. The International Agency for Research on Cancer (IARC) classified PGE in Group 2B (possibly carcinogenic to humans).

18.0 Cresyl Glycidyl Ethers

18.0.1 CAS Number: *[26447-14-3]* (mixed isomers; in this section referred to CGE). The isomers of CGE include *o*-CGE CAS Number *[2210-79-9]* *m*-CGE CAS Number *[2186-25-6]*, and *p*-CGE CAS Number *[2186-24-5]*

18.0.2 Synonyms: For the ortho isomer: (2-methylphenoxy) methyl oxirane; 1,2-epoxy-3-(*o*-tolyloxy) propane; cresyl 2-(2,3-epoxypropyl) ether, *o*-CGE

18.0.3 Trade Names: Heloxy Modifier 62, Grilonit RV1805

18.0.4 Molecular Weight: 164.28

18.0.5 Molecular Formula: $C_{10}H_{12}O_2$

18.0.6 Molecular Structure:

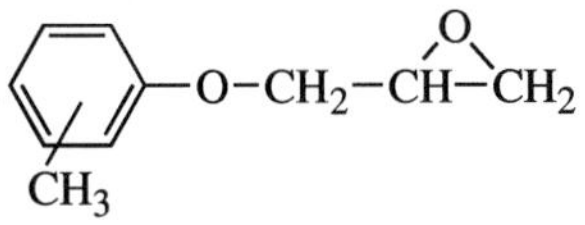

18.1 Chemical and Physical Properties

The properties of CGE are summarized as follows:

Appearance: Clear, Colorless liquid
Viscosity (25°C): 5–25 cm/sec
Color, Gardner-Holdt: 4 max
Relative molecular mass: 164.2
Epoxide equivalent weight: 190–195
Specific gravity (25°C): 1.14
Boiling Range (100 mbar): 170–195°C
Flash point (open cup): 250°F min
Conversion Factor 1 mg/m^3 = 0.15 ppm:

The substance is incompatible with strong oxidizing agents, strong acids, or bases.

18.1.2 Odor and Warning Properties

No information on odor or warning properties was found.

18.2 Production and Use

CGE may be made from cresols by reaction with allyl chloride followed by epoxidation or with epichlorohydrin followed by dehydrochlorination. Suggested uses are as a reactive diluent for viscosity reduction of liquid epoxy resin systems, to increase the level of filler loading of such systems, and to reduce the tendency of the resins to crystallize.

18.3 Exposure Assessment

No OSHA standard methods for the determination of CGEs were found. The pyridinium chloride–chloroform method for epoxy groups and GLC are suggested, but the HCl–dioxane method for epoxides might also be used (209).

18.4 Toxic Effects

CGE and *o*-CGE are of a low order of acute oral and dermal toxicity. They are moderate to severe skin irritants and potent skin sensitizers. CGE and *o*-GCE are slightly irritating to the eye. *o*-CGE causes no adverse effects upon inhalation of the highest attainable vapors, however, inhalation of higher concentrations causes significant irritation in the upper respiratory tract. *In vitro* skin penetration studies have identified a potential for skin penetration of *o*-CGE. *In vitro*, *o*-CGE is a direct-acting, weak genotoxin; however, *in vivo* the weight of evidence indicates there was no genotoxic activity in tissues directly or indirectly exposed to *o*-CGE. There was no evidence of chromosomal aberrations in exposed workers.

18.4.1 Experimental Studies

18.4.1.1 Acute Toxicity. The oral LD_{50} of CGE in the rat was 5.8 g/kg (376). Signs of toxicity included dyspnea and lacrimation, and congested liver was observed at necropsy. The acute dermal LD_{50} in rats was > 2.15 g/kg, and no toxic effects were observed (376). In a 4-h inhalation study in rats, the LC_{50} was 1220 ppm (389).

For *o*-CGE the acute oral LD_{50} was > 5000 mg/kg; the dermal $LD_{50} > 2000$ mg/kg and the inhalational 4-h $LC_{50} > 6.1$ ppm in the rat (412). Signs of intoxication were ataxia and general depression. Dermal exposure resulted in severe dermal irritation.

Skin Effects. CGE is a moderate skin irritant in rabbits, with a Draize score of 5.2 out of a possible 8 (384). Undiluted *o*-CGE produced severe skin irritation at 24 h, which progressed to necrosis by 14 d (413). Following a 4-h semioccluded exposure *o*-CGE caused slight irritation to the skin with a complete recovery of the irritant effects within 7 d (390).

CGE is a potent skin sensitizer in guinea pigs (414). In a maximization test in guinea pigs according to Magnusson and Kligman *o*-CGE caused a response in 80% of the treated animals (415).

Eye Irritation CGE and *o*-CGE are slightly irritating to the rabbit eye; recovery is rapid (335, 376, 390).

18.4.1.2 Chronic and Subchronic Toxicity. Rats inhaled aerosols of *o*-CGE at concentrations of 0, 53, 152, or 305 mg/m^3 for 3 wk (5 d/wk; 6 h/d) (389). There was significant mortality in the highest exposure group. Survivors and animals of the 53-mg/m^3 group had significant inflammation and ulceration of the nasal passages; within 3 wk these effects had largely recovered. At all exposure levels reduced food consumption and growth, dyspnea, and exophthalmos were observed.

In a 4-wk (6 h/d, 5 d/wk) study rats were exposed to vapors of *o*-CGE at the highest practically attainable concentration (26.9 mg/m^3, 4 ppm). There were no adverse effects observed in any of the in-life parameters, hematology, urinalysis, clinical chemistry, and macroscopic and microscopic pathology (412).

18.4.1.3 Pharmacokinetics, Metabolism, and Mechanisms. *In vitro*. *o*-CGE was converted rapidly to the corresponding diol compounds when incubated with guinea pig liver homogenate *in vitro* (416). Studies in preparations of microsomal and cytosolic fractions of liver and lung derived from human, rat, and mouse showed that *o*-CGE is a good substrate for glutathione tranferase, with mice being the most efficient in enzymatic glutathione conjugation. Overall, enzymatic hydrolysis of the epoxide group is the most important route of *in vitro* biotransformation of *o*-CGE, with the highest activity located in the microsomes. Human samples generally had a higher efficiency for hydrolysis than mice or rats (367).

In vitro skin penetration studies with dermatomized human and rat skin and whole mouse skin indicated that *o*-CGE has a potential to penetrate the skin. Permeability was in the order: mouse > rat > human. *o*-CGE is rapidly hydrolyzed after penetration into the skin; however, a minimum of 10% of the applied dose may be present as unchanged material after absorption through the skin (367).

The urinary metabolite profile of *o*-CGE in rats, following IP administration of single doses up to 164 mg/kg, indicated that glutathione (GSH) conjugation and epoxide hydrolysis were key biotransformation pathways of *o*-CGE. At low doses levels GSH conjugation and epoxide hydrolysis accounted each for ca. 25% of urinary metabolites; however at higher doses epoxide hydrolysis became the most prominent route of detoxification (40%) (396).

18.4.1.6 Genetic and Related Cellular Effects Studies. *In vitro*. *o*-CGE was a direct-acting mutagen in *Salmonella typhimurium* strains TA1535 and TA100 in the Ames assay. *p*-CGE was active in these strains both in the presence and absence of a rat and hamster liver S9 fraction. No activity was observed in strain TA98 (4, 323). Similarly, CGE caused reverse mutations in strains TA100 and TA1535 in the absence of rat liver S9; however, a weak activity was also recorded inTA1535 in the presence of S9. CGE did not respond in strains TA98, TA1537, and TA1538 and in *Saccharomyces cerevisiae* D4 (389). The data indicate that CGE exerts its mutagenic potential by causing base-pair mutations. In an unscheduled DNA synthesis assay in human lymphocytes, *o*-CGE produced significant increases in unscheduled DNA synthesis at 10 and 100 ppm; it was cytotoxic at 1000 ppm (341). In a host-mediated assay mice were exposed orally to 4 daily doses of 125 mg/kg *o*-CGE; after last dose the animals were inoculated with *S. typhimurium* TA1535. Urine of mice (ICR or B6D2F1 strain) treated similarly with *o*-CGE was also assayed for mutagenic activity in TA1535, either in presence or absence of β-glucuronidase. There was no evidence of mutagenicity, with the exception of a weak response of the urine of ICR mice treated with β-glucuronidase. (341).

In vivo. In a dominant lethal assay, *o*-CGE was administered topically to male B6D2F1 mice at a dose of 1.5 g/kg, three times a week, for a minimum of 8 wk. A reduction in the pregnancy rate and the number of implantations was observed, although the number of resorptions was not increased, a typical indication of dominant lethality. Although the authors interpreted these data as evidence of a dominant lethal effect, the observed pregnancy rate in animals prior to treatment was lower than the post-treatment rates, indicating that exposure to CGE had not adversely affected this parameter. Since the number of implants varied widely in control animals, the interpretation of treatment-related decrease in implantations in the CGE animals is tenous (341). In a micronucleus assay in female B6D2F1 mice 5 daily oral doses of 125 mg/kg *o*-CGE did not induce an increase in micronuclei in bone marrow extracted 4 h following the last dose (341). Another micronucleus assay in mice conducted to current regulatory guidelines also gave negative results (412). *o*-GCE was tested for its ability to induce mutations in male transgenic mice (Muta^{TM} mice). Animals were exposed dermally to 5 daily doses of 500 mg/kg *o*-CGE. After 7 and 28 d, DNA was extracted from samples of skin—at the site of treatment—liver and bone marrow and assayed for mutations in *LacZ* transgenes. No increases in mutation frequency were recorded in the treated skin or in any of the distant tissues (417).

18.4.1.7 Other: Neurological, Pulmonary, Skin Sensitization. *o*-CGE has been shown to form adducts with the *N*-terminal valine of hemoglobin following incubation with mouse and human erythrocytes or after IP injection to mice. In unexposed individuals these adducts could not be detected (327).

18.4.2 Human Experience

CGE and *o*-CGE are potent skin sensitizers in humans (257, 328, 329, 350, 407, 409, 418). No biologically significant increase in the frequency of chromosomal aberrations was found in peripheral blood lymphocytes of exposed workers (419).

18.5 Standards, Regulations, or Guidelines of Exposure

A TWA value of 10 ppm was established by Denmark.

BIBLIOGRAPHY

1. B. Ottar, *Acta Chem. Scand.* **1**, 283 (1947).
2. D. R. Wade, S. C. Airy, and J. E. Sinsheimer, Mutagenicity of aliphatic epoxides. *Mutat. Res.* **58**, 217–223 (1978).
3. E. Voogd, J. J. van der Stel, and J. J. J. A. A. Jacobs, The mutagenic action of aliphatic epoxides. *Mutat. Res.* **89**, 269–282 (1981).
4. S. H. Neau et al., Substituent effects on the mutagenicity of phenyl glycidyl ethers in *Salmonella typhimurium*. *Mutat. Res.* **93**, 297–304 (1982).
5. H. Glatt, R. Jung, and F. Oesch, Bacterial mutagenicity investigation of epoxides: drugs, drug metabolites, steroids and pesticides. *Mutat. Res.* **11**, 99–118 (1983).
6. F. Oesch et al., *Biochem. Pharmacol.* **26**, 603 (1977).
7. D. W. Nebert et al., *Mol. Pharmacol.* **8**, 374 (1972).
8. F. Gesch et al., *Arch. Toxicol. (Berlin)* **39**, 97 (1977).
9. T. Hayakawa et al., *Arch. Biochem. Biophys.* **170**, 438 (1975).
10. C. T. Bedford, D. Blair, and D. E. Stevenson, *Nature* **267**, 335 (1977).
11. T. M. Hellman and F. H. Small. Characterization of the odor properties of 101 petrochemicals using sensory methods. *J. Air Pollut. Contr. Assoc.* **24**, 979–982 (1974).
12. K. H. Jacobson, E. B. Hackley, and L. Feinsilver, The toxicity of inhaled ethylene oxide and propylene oxide vapors. *AMA Arch. Ind. Health* **13**, 237–244 (1956).
13. NIOSH. *Manual of Analytical Methods*., 4th ed., P. M. Eller and M. E. Cassinelli eds., Method No. 1614, Washington, DC, 1994.
14. M. Tornqvist, et al., *Carcinogenesis* **10**, 39–41 (1989).
15. M. Lieberman and L. W. Mapson, Genesis and biogenesis of ethylene. *Nature*, **204**, 343 (1964).
16. W. Kessler and H. Remmer, Generation of volatile hydrocarbons from amino acids and proteins by an iron/ascorbate/GSH system. *Biochem. Pharmacol.* **39**, 1347 (1990).
17. R. A. Gelmont, J. F. Stein, and Mead, The bacterial origin of rat breath pentane. *Biochem. Biophys. Res. Commun.*, **102**(3), 932–936 (1981).
18. M. Törnqvist et al., Monitoring of environmental cancer initiators through hemoglobin adducts by a modified Edman degradation method. *Anal. Biochem.* **154**, 255–266 (1986).
19. E. Bailey et al., Hydroxyethylvaline adduct formation in hemoglobin as a biological monitor of cigarette smoking intake. *Arch. Toxicol.* **62**, 247–253 (1988).
20. J. G. Filser et al., *Arch. Toxicol.* **66**, 157–163 (1992).

21. K. Y. Wu et al., Molecular dosimetry of endogenous and ethylene oxide–induced N7-(2-hydroxyethyl) guanine formation in tissues of rodents. *Carcinogenesis* **20**(9) 1787–1792 (1999).
22. H. Bartsch, K. Hemminki, and I. K. O'Neill, Methods for Detecting DNA Damaging Agents in Humans. *IARC Scientific Publication No. 89*, Lyon, 1988.
23. J. Mayer et al., *Mutat. Res.* **248**, 163–176 (1991).
24. N. van Sittert et al., Monitoring occupational exposure to ethylene oxide by the determination of hemoglobin adducts. *Environmental Health Perspectives* **99**, 217–220 (1993).
25. A. Tates et al., Biological effect monitoring in industrial workers following incidental exposure to high concentrations of ethylene oxide. *Mutat. Res.* **329**, 63–77 (1995).
26. P. J. Boogaard, P. S. J. Rocchi, and N. J. van Sittert, Biomonitoring of exposure to ethylene oxide and propylene oxide by determination of hemoglobin adducts: correlations between airborne exposure and adduct levels. *Int. Arch. Occup. Environ. Health* **72**, 142–150 (1999).
27. L. Rhomberg et al., *Environ. Mol. Mutagen.* **16**, 104–125 (1990).
28. V. E. Walker et al., Molecular dosimetry of ethylene oxide: formation and persistence of 7-(2-hydroxyethyl)guanine in DNA following repeated exposures of rats and mice. *Cancer Research* **52**, 4328–4334 (1992).
29. V. E. Walker et al., Molecular dosimetry of ethylene oxide: formation and persistence of *N*-(2-hydroxyethyl)valine in hemoglobin following repeated exposures of rats and mice. *Cancer Research*, **52**, 4320–4327 (1992).
30. V. E. Walker et al., Molecular dosimetry of DNA and hemoglobin adducts in mice and rats exposed to ethylene oxide. *Environmental Health Perspectives* **99**, 11–17 (1993).
31. R. Couch et al., *in vivo* dosimetry of ethylene oxide and propylene oxide in the cynomolgus monkey. *Mutat. Res.* **357**(1–2), 17–23 (1996).
32. W. Popp et al., DNA-protein cross-links and sister chromatid exchange frequencies in lymphocytes and hydroxyethyl mercapturic acid in urine of ethylene oxide–exposed hospital workers. *Int. Arch. Occup. Environ. Health* **66**, 325–332 (1994).
33. V. E. Walker et al., *Mutat. Res.* **233**, 151–164 (1990).
34. D. Segerbaeck, *Carcinogenesis* **11**, 307–312 (1990).
35. W. Pauwels and H. Veulemans, Comparison of ethylene, propylene and styrene 7,8-oxide *in vitro* adduct formation on *N*-terminal valine in human haemoglobin and on *N*-7-guanine in human DNA. *Mutat. Res.* **418**, 21–33 (1998).
36. U. Fost et al., *Human Exp. Toxicol.* **10**, 25–31 (1991).
37. T. R. Fennell, Inter-species extrapolation of ethylene oxide dosimetry. *Proceedings of The Toxicology Forum 22nd Annual Winter Meeting*, pp. 541–562 (1997).
38. L. Golberg, *Hazard Assessment of Ethylene Oxide*, CRC Press, Boca Raton, FL, 1986.
39. World Health Organization (WHO), *International Programme on Chemical Safety: Environment Health Criteria 55, Ethylene Oxide*, Geneva, 1985.
40. D. J. Nachreiner, *Ethylene Oxide: Acute Vapor Inhalation Toxicity Testing According to D.O.T. Regulations (One-Hour Test)*. Bushy Run Research Center, Export, PA, Project ID 54–593, 1992.
41. K. H. Jacobsen, E. B. Hackley, and L. Feinsilver, *AMA Arch. Ind. Health* **13**, 237 (1956).
42. D. J. Nachreiner, *Ethylene Oxide: Acute Vapor Inhalation Toxicity Test in Rats (Four-Hour Test)*, Bushy Run Research Center, Export, PA, Project ID 54–76, 1991.

43. D. W. Lynch et al., Chronic inhalation toxicity of ethylene oxide in monkeys—lens opacities at termination of exposure and 10-year follow-up., *The Toxicologist* **12**(1), Abstract No. 1384 (1992).

44. C. D. Brown et al., Ethylene oxide dosimetry in the mouse. *Toxicol. Appl. Pharmacol.* **148**, 215–221 (1998).

45. C. D. Brown, B. A. Wong, and T. R. Fennell, *in vivo* and *in vitro* kinetics of ethylene oxide metabolism in rats and mice. *Toxicol. Appl. Pharmacol.* **136**, 8–19 (1996).

46. J. G. Filser and H. M. Bolt, Inhalation pharmacokinetics based on gas uptake studies. VI. Comparative evaluation of ethylene oxide and butadiene monoepoxide as exhaled reactive metabolites of ethylene and 1,3-butadiene in rats. *Arch. Toxicol.* **55**, 219–223 (1984).

47. D. Segerback, Alkylation of DNA and hemoglobin in the mouse following exposure to ethene and ethene oxide. *Chem-Biol. Interact.* **45**, 139–151 (1983).

48. A. R. Jones and G. Wells. The comparative metabolism of 2-bromoethanol and ethylene oxide in the rat. *Xenobiotica* **11**, 763–770 (1981).

49. L. Martis et al., Disposition kinetics of ethylene oxide, ethylene glycol and 2-chloroethanol in the dog. *J. Toxicol. Environ. Health* **10**, 847–856 (1982).

50. R. Tardif et al., Species differences in the urinary disposition of some metabolites of ethylene oxide. *Fundam. Appl. Toxicol.* **9**, 448–453 (1987).

51. E. W. Vogel and M. J. Nivard, Genotoxic effects of inhaled ethylene oxide, propylene oxide and butylene oxide on germ cells: sensitivity of genetic endpoints in relation to dose and repair status. *Mutat. Res.* **405**, 259–271 (1998).

52. J. G. Hengstler et al., Glycolaldehyde causes DNA—protein crosslinks: a new aspect of ethylene oxide genotoxicity. *Mutat. Res.* **304**, 229–234 (1994).

53. K. Krishnan, M. L. Gargas, T. R. Fennell, M. E. Andersen, A physiologically based description of ethylene oxide dosimetry in the rat. *Toxicology and Industrial Health* **8**(3) (1992).

54. T. Katoh, K. Higashi, N. Inoue, et al., *Toxicology* **58**, 1–9 (1989).

55. J. A. McKelvey and M. A. Zemaitis. The effects of ethylene oxide (EO) exposure on tissue glutathione levels in rats and mice. *Drug Chem. Toxicol.* **9**, 51–66 (1986).

56. T. R. Fennell, S. C. J. Sumner, and V. E. Walker, A model for the formation and removal of hemoglobin adducts. *Cancer Epidemiology, Biomarkers and Prevention* **1**, 213–219 (1992).

57. Gy. A. Csanády et al., A physiological toxicokinetic model for exogenous and endogenous ethylene and ethylene oxide in rat, mouse and human. Modeling of tissue concentrations, hemoglobin and DNA adducts. *Toxicol. Appl. Pharmacol.*, in press, 2000.

58. K. Mori, M. Kaido, K. Fujishiro, et al., *Arch. Toxicol.* **65**, 369–401 (1991).

59. K. Mori, M. Kaido, K. Fujishiro, et al., *Brit. J. Ind. Med.* **48**, 270–274 (1991).

60. G. A. Sega, P. A. Brimer, and E. E. Generoso, *Mutat. Res.* **249**, 339–349 (1991).

61. W. M. Snellings, J. P. Zelenak, and C. S. Weil. Effects on reproduction in Fischer 334 rats exposed to ethylene oxide by inhalation for one generation. *Toxicol. Appl. Pharmacol.* **63**, 382–388 (1982).

62. T. L. Neeper-Bradley and M. F. Kubena, Ethylene oxide: Developmental toxicity study of maternally inhaled vapor in Cd^{e} rats. *Balchem Corp. and Praxair Inc.*, 1993, unpublished report.

63. J. B. LaBorde and C. A. Kimmel, The teratogenicity of ethylene oxide administered intravenously to mice. *Toxicol. Appl. Pharmacol.* **56**, 16–22 (1980).

64. M. Katoh et al., *Mutat. Res.* **210**, 337–344 (1989).

65. J. C. Rutledge and W. M. Generoso, *Teratology* **39**, 563–572 (1989).

66. W. M. Generoso et al., *Mutat. Res.* **250** 439–446 (1991).

67. A. M. Saillenfait et al., Developmental toxicity of inhaled ethylene oxide in rats following short-duration exposure, *Fundamental and Applied Toxicology* **34**, 223–227 (1996).

68. E. Weller et al., Dose-rate effects of ethylene oxide exposure on developmental toxicity. *Toxicological Sciences* **50**, 259–270 (1999).

69. D. W. Lynch, T. R. Lewis, and W. J. Moorman, et al., *Toxicol. Appl. Pharmacol.* **76**, 85 (1984).

70. W. M. Snellings, C. S. Weil, and R. R. Maronpot, A two-year inhalation study of carcinogenic potential of ethylene oxide in Fischer 344 rats. *Toxicol. Appl. Pharmacol.* **75**, 105–117 (1984).

71. National Toxicology Program, *Toxicology and Carcinogenesis Studies of Ethylene Oxide in B6C3F$_1$* Mice, NTP TR 326, NIH Publication No. 88–2582, U.S. Department of Health and Human Services, National Institutes of Health, 1987.

72. S. G. Austin and R. L. Sielken, Jr., *J. Occup. Med.* **30**, 236–245 (1988).

73. B. Denk, J. G. Filser, D. Oesterle, et al., *J. Cancer Res. Clin. Oncol.* **114**, 35–38 (1988).

74. V. L. Dellarco et al., *Environ. Mol. Mutagen.* **16**, 85–103 (1990).

75. T. Bastlova et al., Molecular analysis of ethylene oxide–induced mutations at the HPRT locus in human diploid fibroblasts, *Mutat. Res.* **287**(2), 283–292 (1993).

76. V. Walker and R. Skopek, A mouse model for the study of *in vivo* mutational spectra: Sequence specificity of ethylene oxide at the *hprt* locus. *Mutat. Res.* **288**, 151–162 (1993).

77. S. Sisk et al., Assessment of the *in vivo* mutagenicity of ethylene oxide in the tissues of B6C3F$_1$ *lacI* transgenic mice following inhalation exposure. *Mutat. Res.* **391**, 153–164 (1997).

78. V. Walker et al., *In vivo* mutagenicity of ethylene oxide at the *hprt* locus in lymphocytes of B6C3F$_1$ *lacI* transgenic mice following inhalation exposure. *Mutat. Res.* **392**, 211–222 (1997).

79. W. M. Generoso et al., Concentration-response curves for ethylene-oxide–induced heritable translocations and dominant lethal mutations. *Environ. Mol. Mutagen.* **16**, 126–131 (1990).

80. J. K. Wiencke et al., *Cancer Res.* **51**, 5266–5269 (1991).

81. L. Rhomberg et al., *Environ. Mol. Mutagen.* **16**, 104–125 (1990).

82. R. J. Preston et al., Reconsideration of the genetic risk assessment for ethylene oxide exposures, *Environmental and Molecular Mutagenesis* **26**, 189–202 (1995).

83. A. T. Natarajan et al., Ethylene oxide: evaluation of genotoxicity data and an exploratory assessment of genetic risk, *Mutat. Res.* **330**, 55–70 (1995).

84. R. J. Preston, Cytogenetic effects of ethylene oxide, with an emphasis on population monitoring. *Critical Reviews in Toxicology* **29**(3), 263–282 (1999).

85. A. Ohnishi et al., Ethylene oxide induces central-peripheral distal axonal degeneration of the lumbar primary neurones in rats. *Br. J. Ind. Med.* **42**, 373–379 (1985).

86. J. Setzer et al., Neurophysiological and neuropathological evaluation of primates exposed to ethylene oxide and propylene oxide. *Toxicology and Industrial Health* **12**(5), 667 (1996).

87. R. C. Mandella, A 13-week inhalation neurotoxicity study of ethylene oxide (498–95A) in the rat via whole-body exposures with recovery. Huntingdon Life Sciences, East Millstone, New Jersey, Aug. 12, 1997, pp. 1–1089.

88. A. Sobaszek et al., Working conditions and health effects of ethylene oxide exposure at hospital sterilization sites. *J. Occupational Environ. Med.* **41**(6), 492–499 (June 1999).

89. D. Deschamps, M. Leport, and S. Cordier, *J. Fr. Ophthalmol.* **13**, 189–197 (1990).

90. D. Deschamps et al., *Brit. J. Indust. Med.* **47**, 308–313 (1990).

91. F. Brugnone et al., *Int. Arch. Occup. Environ. Health* **58**, 105–112 (1986).
92. E. I. M. Florack and G. A. Zielhaus, *Int. Arch. Occup. Environ. Health.* **62**, 273–277 (1990).
93. A. Rowland et al., Ethylene oxide exposure may increase the risk of spontaneous abortion, preterm birth, and postterm birth. *Epidemiology* **7**(4), 363–368 (1996).
94. G. W. Olsen, L. Lucas, and J. Teta, Letter to the Editor: Ethylene oxide exposure and risk of spontaneous abortion, preterm birth and postterm birth. *Epidemiology* **8**(4) (1997).
95. J. K. Wiencke et al., *Cancer Res.* **51**, 5266–5269 (1991).
96. A. D. Tates et al., *Mutat. Res.* **250**, 483–497 (1991).
97. J. Mayer et al., *Mutat. Res.* **248**, 163–176 (1991).
98. P. Schulte et al., Biologic markers in hospital workers exposed to low levels of ethylene oxide. *Mutat. Res.* **278**, 237–251 (1992).
99. G. de Jong, N. J. Van Sittert, and A. T. Natarajan, *Mutat. Res.* **204**, 451–464 (1988).
100. W. Popp et al., *Int. Arch. Occup. Environ. Health* **66**, 325–332 (1994).
101. D. Lerda and R. Rizzi, Cytogenetic study of persons occupationally exposed to ethylene oxide. *Mutat. Res.* **281**, 31–37 (1992).
102. D. J. Tomkins et al., A study of sister chromatid exchange and somatic cell mutation in hospital workers exposed to ethylene oxide. *Environ. Health Perspectives* **101** (Suppl. 3), 159–164 (1993).
103. J. Major, M. Jakab, and A. Tompa, Genotoxicological investigation of hospital nurses occupationally exposed to ethylene-oxide: I. Chromosome aberrations, sister-chromatid exchanges, cell cycle kinetics, and UV-Induced DNA synthesis in peripheral blood lymphocytes. *Environmental and Molecular Mutagenesis* **27**, 84–92 (1996).
104. L. Ribeiro et al., Biological monitoring of workers occupationally exposed. *Mutat. Res.* **313**, 81–87 (1994).
105. A. Tates et al., Biological effect monitoring in industrial workers following incidental exposure to high concentrations of ethylene oxide. *Mutat. Res.* **329**, 63–77 (1995).
106. H. A. Crystall, H. H. Schaumberg, E. Cooper, et al., *Neurology* **38**, 567–569 (1988).
107. W. J. Estrin, R. M. Bowler, A. Lash et al., *J. Toxicol. Clin. Toxicol.* **28**, 1–20 (1990).
108. J. Bousquet and F.-B. Michel, Allergy to formaldehyde and ethylene oxide. *Clin. Rev. Allergy* **9**, 357–370 (1991).
109. S. Verraes and O. Michel, Occupational asthma induced by ethylene oxide (Letter). *Lancet* **346**, 1434–1435 (1995).
110. J. Bommer and E. Ritz, Ethylene oxide (ETO) as a major cause of anaphylactoid reactions in dialysis (a review). *Artif. Organs* **11**, 111–117 (1987).
111. C. Romaguera and J. Vilaplana, Airborne occupational contact dermatitis from ethylene oxide. *Contact Dermatitis* **39**, 85 (1997).
112. R. E. Shore, M. J. Gardner, and B. Pannett, Ethylene oxide: An assessment of the epidemiological evidence on carcinogenicity. *Brit. J. Ind. Med.* **50**, 971–997 (1993).
113. M. J. Teta, R. L. Sielken Jr., and C. Valdez-Flores, Ethylene oxide cancer risk assessment based on epidemiological data: Application of revised regulatory guidelines. *Risk Analysis* **19**, 1135–1155 (1999).
114. C. Hogstedt et al., *Brit. J. Ind. Med.* **36**, 766–780 (1979).
115. M. J. Gardner et al., *Brit. J. Ind. Med.* **46**, 860–865 (1989).
116. K. Steenland et al., *N. Engl. J. Med.* **324**, 1402–1407 (1991).

117. H. L. Greenberg, M. G. Ott, and R. E. Shore, *Brit. J. Ind. Med.* **47**, 221–230 (1990).

118. G. W. Olsen et al., Mortality from pancreatic and lymphopoietic cancer among workers in ethylene and propylene chlorohydrin production. *Occupational and Environmental Medicine* **54**, 592–598 (1997).

119. M. J. Teta, L. O. Benson, and J. N. Vitale, Mortality study of ethylene oxide workers in chemical manufacturing: a ten-year update. *Br J. Ind. Med.* **50**(8), 704–709 (1993).

120. L. Hagmar, Z. Mikoczy, and H. Welinder, Cancer incidence in Swedish sterilant workers exposed to ethylene oxide. *Occupational and Environmental Medicine* **52**, 154–156 (1995).

121. L. Bisanti et al., Cancer mortality in ethylene oxide workers. *Brit. J. Ind. Med.* **50**, 317–324 (1993).

122. S. Norman et al., Cancer incidence in a group of workers potentially exposed to ethylene oxide. *International Journal of Epidemiology* **24**(2) (1995).

123. L. Hagmar et al., An epidemiological study of cancer risk among workers exposed to ethylene oxide using hemoglobin adducts to validate environmental exposure assessments. *Int. Arch. of Occupational and Environmental Health* **63**, 271–278 (1991).

124. L. Hagmar, Z. Mikoczy, and H. Welinder, Cancer incidence in Swedish sterilant workers exposed to ethylene oxide. *Occupational and Environmental Medicine* **52**, 154–156 (1995).

125. IARC (International Agency for Research on Cancer), *Ethylene Oxide*, IARC Monographs, Vol. 60, 1994, pp. 73–159.

126. Manufacturing Chemists' Association, *Chemical Safety Data Sheet SD-38*, 1951, (revised 1971).

127. K. H. Jacobson, E. B. Hackley, and L. Feinsilver. The toxicity of inhaled ethylene oxide and propylene oxide vapors. *AMA Arch. Ind. Health* **13**, 237–244 (1956).

128. J. E. Amoore and E. Hautala. Odor as an aid to chemical safety: odor thresholds compared with threshold limit values and volatilities for 214 industrial chemicals in air and water dilution. *J. Appl. Toxicol.* **3**, 272–290 (1983).

129. T. M. Hellman, and F. H. Small. Characterization of the odor properties of 101 petrochemicals using sensory methods. *J. Air Pollut. Contr. Assoc.* **24**, 979–982 (1974).

130. G. O. Curme, Jr., and F. Johnston, Eds., *Glycols, ACS Monograph Series No. 114*, Reinhold, New York, 1952.

131. S. Osterman-Golkar, Methods for biological monitoring of propylene oxide exposure in Fischer 344 rats. *Toxicology* **134**, 1–8 (1999).

132. M. N. Rìos-Blanco et al., Propylene Oxide: mutagenesis, carcinogenesis and molecular dose. *Mutat. Res.* **380**, 179–197 (1997).

133. K. Svensson, K. Olofsson, and S. Osterman-Golkar. Alkylation of DNA and hemoglobin in the mouse following exposure to propene and propylene oxide. *Chem. Biol. Interact* **78**, 55–66 (1991).

134. B. Hogstedt et al., Chromosomal aberrations and micronuclei in lymphocytes in relation to alkylation of hemoglobin in workers exposed to ethylene oxide and propylene oxide. *Hereditas* **111**, 133–138 (1990).

135. R. W. Pero, S. Osterman-Golkar, and B. Hogstedt. Unscheduled DNA synthesis correlated to alkylation of hemoglobin in individuals occupationally exposed to propylene oxide. *Cell Biol. Toxicol.* **1**, 309–314 (1985).

136. P. J. Boogaard, P. S. J. Rocchi, and N. J. van Sittert, Biomonitoring of exposure to ethylene oxide and propylene oxide by determination of hemoglobin adducts: Correlations between

airborne exposure and adduct levels. *Int. Arch. Occup. Environ. Health* **72**, 142–150 (1999).

137. J. J. Solomon, F. Mukai, J. Fedyk, et al., *Chem. Biol. Interact.* **67**, 275 (1988).
138. D. Segerbäck et al., Tissue distribution of DNA adducts in male Fischer rats exposed to 500 ppm of propylene oxide: quantitative analysis of 7-(2-hydroxypropyl)guanine by ^{32}P-postlabelling. *Chem.-Biol. Interact* **115**, 229–246 (1998).
139. K. Plna et al., ^{32}P-:Postlabelling of propylene oxide 1- and N^{6}-substituted adenine and 3-substituted cytosine/uracil: formation and persistence *in vitro* and *in vivo*. *Carcinogenesis* **20**, 2025–2032 (1999).
140. K. Svensson, K. Olofsson, and S. Osterman-Golkar. Alkylation of DNA and hemoglobin in the mouse following exposure to propene and propylene oxide. *Chem. Biol. Interact* **78**, 55–66 (1991).
141. M. N. Rìos-Blanco et al., DNA and hemoglobin adducts of propylene oxide in F344 rats and the quantitation of DNA apurinic/apyrimidinic sites in tissues after inhalation exposure. submitted.
142. D. Segerback et al., *In vivo* tissue dosimetry as a basis for cross-species extrapolation in cancer risk assessment of propylene oxide. *Regul. Toxicol. Pharmacol.* **20**, 1–14 (1994).
143. K. H. Jacobson, E. B. Hackley, and L. Feinsilver. The toxicity of inhaled ethylene oxide and propylene oxide vapors. *AMA Arch. Ind. Health* **13**, 237–244 (1956).
144. V. K. Rowe, R. L. Hollingsworth, et al., *AMA Arch. Ind. Health* **13**, 228 (1956).
145. C. S. Weil, N. Condra, C. Haun, and J. A. Striegel, *Am. Ind. Hyg. Assoc. J.* **24**, 305 (1963).
146. National Toxicology Program, *TR 267*, 1985.
147. S. R. Eldridge et al., Effects of propylene oxide on nasal epithelial cell proliferation in F344 rats. *Fundam. Appl. Toxicol.* **27**, 25–32 (1995).
148. M. N. Rìos-Blanco et al., Changes in cell proliferation in rat nasal respiratory epithelium after inhalation exposure to propylene oxide. *Toxicology Letters*, vol. 109 (Suppl. 1), Abstract # 37, (1999).
149. R. A. Renne et al., *J. Natl. Cancer Inst.* **77**, 573 (1986).
150. J. K. Dunnick et al., *Toxicology* **50**, 69 (1988).
151. D. W. Lynch et al., Carcinogenic and toxicologic effects of inhaled ethylene oxide and propylene oxide in F344 rats. *Toxicol. Appl. Pharmacol.* **76**, 69–84 (1984).
152. C. F. Kuper, P. G. J. Reuzel, and V. J. Feron, *Food Chem. Toxicol.* **26**, 159 (1988).
153. A. R. Sellakumar, C. A. Snyder, and R. E. Albert. Inhalation carcinogenesis of various alkylating agents, *J. National Cancer Institute* **79**, 285–289 (1987).
154. H. Dunkelberg, *Brit. J. Cancer* **46**, 924 (1982).
155. R. J. Nolan et al., Effects of Single 6-h exposures to various concentrations of propylene oxide on liver non-protein sulfhydryls in male Wistar/Lewis rats. Unpublished report of The Dow Chemical Company, Midland, MI, 1980.
156. M. S. Lee et al., Propylene oxide in blood and glutathione depletion in nose, lung and liver of rats exposed to propene oxide. *Arch. Pharmacol.* **357**(Suppl. 4), R 172, (1998).
157. R. Schmidbauer et al., Determination of the partition coefficients of propylene oxide in water, olive oil and fat. *Arch. Pharmacol.* **353**(Suppl. 4), R 111, (1996).
158. T. H. Faller et al., Kinetics of propylene oxide in cytosol and microsomes of liver and lung of mouse, rat and man, and of respiratory and olfactory nasal mucosa of rat. *Toxicol. Letters*, (Suppl. 1/95), 217–218 (1998).

159. K. Golka et al., Pharmacokinetics of propylene and its reactive metabolite propylene oxide in Sprague–Dawley rats. *Arch. Toxicol.* **13**(Suppl), 240–242 (1989).
160. K. R. Maples and A. R. Dahl. Levels of epoxides in blood during inhalation of alkenes and alkene oxides. *Inhal. Toxicol.* **5**, 43–54 (1993).
161. P. L. Hackett, M. G. Brown, and R. L. Buschbom, et al., NIOSH, U.S. Department of Health and Human Services, 1982.
162. B. D. Hardin, R. W. Niemeier, M. R. Sikov, et al., *Scand. J. Work Environ. Health* **9**, 94 (1983).
163. C. A. Kimmel, J. B. LaBorde, and B. D. Hardin, *Toxicology Newborn* (1984).
164. S. B. Harris, J. L. Schardein, C. E. Ulrich, et al., *Fundam. Appl. Toxicol.* **13**, 323 (1989).
165. W. C. Hayes, H. D. Kirk, T. S. Gushow, et al., *Fundam. Appl. Toxicol.* **10**, 82 (1988).
166. IARC (International Agency for Research on Cancer), *Propylene Oxide*, IARC Monographs, Vol. 60, 1994, pp. 181–213.
167. W. Meylan et al., *Toxicol. Ind. Health* **2**, 219 (1986).
168. A. K. Giri. Genetic toxicology of propylene oxide and trichloropropylene oxide- a review. *Mutat. Res.* **277**, 1–9 (1992).
169. E. H. Pfeiffer and H. Dunkelberg, *Food Cosmet. Toxicol.* **18**, 115 (1980).
170. J. D. Bootman, C. Lodge, and H. E. Whalley, *Mutat. Res.* **67**, 101 (1979).
171. R. E. McMahon, J. C. Cline, and C. Z. Thompson, *Cancer Res.* **39**, 682 (1979).
172. L. Migliore, A. M. Rossi, and N. Loprieno, *Mutat. Res.* **102**, 425 (1982).
173. B. J. Dean and G. Hodson-Walker, *Mutat. Res.* **64**, 329 (1979).
174. J. D. Tucker, J. Xu, J. Stewart, et al., *Teratog. Carcinog. Mutagen.* **6**, 15 (1986).
175. P. O. Zamora, J. M. Benson, A. P. Li, et al., *Environ. Mutagen.* **5**, 795 (1983).
176. D. K. Gulati, K. Witt, B. Anderson, et al., *Environ. Mol. Mutagen.* **13**, 133 (1989).
177. D. McGregor, A. G. Brown, P. Cattanach, et al., *Environ. Mol. Mutagen.* **17**, 122 (1991).
178. B. D. Hardin, R. L. Schuler, P. M. McGinnis, et al., *Mutat. Res.* **177**, 337 (1983).
179. D. W. Lynch, T. R. Lewis, W. J. Moorman, et al., *Toxicol. Appl. Pharmacol.* **76**, 85 (1984).
180. Z. Farooqui et al., Genotoxic effects of ethylene oxide and propylene oxide in mouse bone marrow cells. *Mutat. Res.* **288**, 223–228 (1993).
181. B. D. Hardin, R. L. Schuler, P. M. McGinnis, et al., *Mutat. Res.* **177**, 337 (1983).
182. A. Ohnishi, T. Yamamoto, Y. Murai, et al., *Arch. Environ. Health* **43**, 353 (1988).
183. J. T. Young et al., unpublished report, The Dow Chemical Company, 1985.
184. J. Setzer et al., Neurophysiological and Neuropathological evaluation of primates exposed to ethylene oxide and propylene oxide. *Toxicology and Industrial Health* **12**(5), 667, (1996).
185. C. Troxel, Acute Exposure Guideline Levels for Propylene Oxide NAC//Pro Draft 3, Nov. 1998.
186. R. W. Pero, T. Bryngelsson, B. Widegren, et al., *Mutat. Res.* **104**, 193 (1982).
187. R. W. Pero, S. Osterman-Golkar, and B. Hogstedt. Unscheduled DNA synthesis correlated to alkylation of hemoglobin in individuals occupationally exposed to propylene oxide. *Cell Biol. Toxicol.* **1**, 309–314 (1985).
188. S. Osterman-Golkar, E. Bailey, P. B. Farmer, et al., *Scand. J. Work Environ. Health* **10**, 99 (1984).
189. A. M. Theiss et al., *J. Occup. Med.* **23**, 343 (1981).
190. R. Jolanki, Riitta, Estlander, et al., *ASTM Spec. Tech. Publ.* (1988).

191. O. Jensen, *Contact Dermatitis* **7**, 148 (1981).
192. R. L. McLaughlin, *Am. J. Ophthalmol.* **29**, 1355 (1946).
193. A. M. Theiss, et al., *Occupational Safety & Health Series*, 1981.
194. M. G. Ott, M. J. Teta, and H. L. Greenberg. Lymphatic and hematopoietic tissue cancer in a chemical manufacturing environment. *Am. J. Ind. Med.* **16**, 631–643.
195. IARC (International Agency for Research on Cancer), *Propylene Oxide*, IARC Monographs Vol. 60, 1994, pp. 181–213.
196. Manufacturing Chemists' Association, *Chemical Safety Data Sheet SD-38, 1951* (revised 1971).
197. IARC Monographs on the Evaluation of Carcinogenic Risks to Humans, International Agency for Research on Cancer (IARC) *Re-evaluation of Some Organic Chemicals, Hydrazine and Hydrogen Peroxide. Monograph No. 71*, 1999, pp. 629–640.
198. *National Toxicology Program (NTP) Technical Report Series No. 329*, Toxicology and Carcinogenicity of 1,2-epoxybutane in F344/N rats and $B6C3F_1$ mice (inhalation studies) 1988.
199. The Dow Chemical Company, unpublished report, 1984.
200. R. R. Miller et al., *Fundam. Appl. Toxicol.* **1**, 319 (1981).
201. R. H. Reitz, T. R. Fox, and E. A. Hermann, unpublished report, The Dow Chemical Company, 1983.
202. M. R. Sikov et al., *NIOSH Technical Report 81–124*, 1981.
203. B. L. Van Duuren et al., *J. Natl. Cancer Inst.* **39**, 1217 (1967).
204. L. Ehrenberg and S. Hussain, *Mutat. Res.* **86**, 1–113 (1981).
205. *National Toxicology Program (NTP) Annual Plan*, U.S. National Institute of Environmental Health Sciences, 1983.
206. G. M. Williams, H. Mori, and C. A. McQueen, *Mutat. Res.* **221**, 263–286 (1989).
207. Ohnishi and Y. Muria, Polyneuropathy due to ethylene oxide, propylene oxide and butylene oxide. *Environ. Res.* **60**, 242–247 (1993).
208. B. Phillips, Union Carbide Chemicals Co., New York, 1992.
209. J. L. Jungnickel et al., *Organic Analysis*, Vol. 1, Wiley-Interscience, New York, 1963, p. 127.
210. M. W. Himmelstein et al., *Critical Reviews in Toxicology* **27**, 1 (1997).
211. H. F. Smyth et al., *AMA Arch. Indust. Hyg. Occup. Med.* **10**, 61 (1954).
212. R. F. Henderson et al., *Toxicol. Sciences* **51**, 146 (1999).
213. J. K. Doerr, E. A. Hollis, and I. G. Sipes, *Toxicology* **113**, 128 (1996).
214. P. J. Boogaard and J. A. Bond, *Toxicol. Appl. Pharmacol.* **141**, 617 (1996).
215. P. J. Boogaard, S. C. J. Sumner, and J. A. Bond, *Toxicol. Appl. Pharmacol.* **136**, 307 (1996).
216. R. Thier et al., *Chem. Res. Toxicol.* **8**, 465 (1995).
217. L. M. Sweeney et al., *Carcinogenesis* **18**, 611 (1997).
218. Gy. A. Csanady et al., *Toxicology* **113**, 300 (1996).
219. B. L. Van Duuren et al., *J. Natl. Cancer Inst.* **31**, 41 (1963).
220. B. L. Van Duuren, L. Orris, and N. Nelson, *J. Natl. Cancer Inst.* **35**, 707 (1965).
221. B. L. Van Duuren et al., *J. Natl. Cancer Inst.* **37**, 825 (1966).
222. M. B. Shimkin et al., *J. Natl. Cancer Inst.* **36**, 915 (1966).
223. C. De Meester, *Mutat. Res.* **195**, 273 (1988).

224. IARC, Monograph on the Evaluation of Carcinogenic Risk to Humans, *Re-Evaluation Of Some Organic Chemicals, Hydrazine And Hydrogen Peroxide.*, Vol. 71, Part 1, 109, 1999.
225. M. J. Wade, J. W. Moyer, and C. H. Hine, *Mutat. Res.* **66**, 367 (1979).
226. P. G. Gervasi et al., *Mutat. Res.* **156**, 77 (1985).
227. V. C. Dunkel et al., *Environ. Mutagen.* **6**, 1 (1984).
228. C. DeMeester, M. Mercier, and F. Poncelet, *Mutat. Res.* **97**, 204 (1982).
229. D. Ichinotsubo et al., *Mutat. Res.* **46**, 53 (1977).
230. W. von der Hude, A. Seelbach, and A. Basler, *Mutat. Res.* **231**, 205 (1990).
231. D. B. McGregor et al., *Environ. Molc. Mutag.* **11**, 91 (1988).
232. F. Abka'i et al., *Cell Biol. Toxicol.* **3**, 285 (1987).
233. F. Darroudi, A. T. Natarajan, and P. H. M. Lohman, *Mutat. Res.* **212**, 103 (1989).
234. Y. Nishi et al., *Cancer Res.* **44**, 3270 (1984).
235. W. von der Hude, S. Carstensen, and G. Obe, *Mutat. Res.* **249**, 55 (1991).
236. M. Sasiadek, H. Jarventaus, and M. Sorsa, *Mutat. Res.* **263**, 47 (1991).
237. M. Sasiadek, H. Norppa, and M. Sorsa, *Mutat. Res.* **261**, 117 (1991).
238. J. K. Wiencke, D. C. Christiani, and K. T. Kelsey, *Mutat. Res.* **248**, 17 (1991).
239. K. T. Kelsey, D. C. Christiani, and J. K. Wiencke, *Mutat. Res.* **248**, 27 (1991).
240. B. J. Dean and G. Hodson-Walker, *Mutat. Res.* **64**, 329 (1979).
241. J. E. Cochrane and T. R. Skopek, *Carcinogenesis* **15**, 713 (1994).
242. C. Y. Li, D Yandell, and J. B. Little, *Somatic Cell and Mol. Genet.* **18**, 77 (1992).
243. A. M. Steen, K. G. Meyer, and L. Reccio, *Mutagenesis* **12**, 61 (1997).
244. J. T. Millard and M. M. White, *Biochemistry* **32**, 2120 (1993).
245. G. M. Williams, H. Mori, and C. A. McQueen, *Mutat. Res.* **221**, 263 (1989).
246. G. T. Arce et al., *Environ. Health Perspect.* **86**, 75 (1990).
247. R. A. Walk et al., *Mutat. Res.* **182**, 333 (1987).
248. J. K. Wiencke et al., *Cancer Res.* **51**, 5266 (1991).
249. J. K. Wiencke et al., *Cancer Epidemiol. Biomark. Prevent.* **4**, 253 (1995).
250. H. Norppa et al., *Carcinogenesis* **16**, 1261 (1995).
251. M. K. Conner, J. E. Luo, and O. Gutierrez de Gotera, *Mutat. Res.* **108**, 251 (1983).
252. Y. Xiao and A. D. Tates, *Environ. Mol. Mutagen.* **26**, 97 (1995).
253. D. Anderson et al., *Mutat. Res.* **391**, 233 (1997).
254. A. D. Tates et al., *Mutat. Res.* **397**, 21 (1998).
255. Q. Meng et al., *Mutat. Res.* **429**, 127 (1999).
256. *BIBRA*, Toxicity Profile, 1997.
257. D. P. Ward, unpublished report of Monsanto Proprietary, 1986.
258. A. P. Li, unpublished report of Monsanto Proprietary, 1987.
259. D. P. Ward, unpublished report of Monsanto Proprietary, 1987.
260. B. Ekwall, C. Nordensten, and L. Albanus, *Toxicology* **24**, 199 (1982).
261. S. M. Tarlo, *Clin. Exp. Allergy* **22**, 99 (1992).
262. C. H. Hine et al., Chapter 32, Epoxy compounds. In G. D. Clayton and F. E. Clayton, Eds., *Patty's Industrial Hygiene and Toxicology*, 3rd ed., Vol. IIa, J Wiley, New York, 1981.

263. B. J. Smith, D. R. Mattison, and I. G. Sipes, *Tox. Appl. Pharmacol.* **105**, 372–381 (1990).
264. Industrial Medical and Toxicology Department, Union Carbide Corp., New York, 1958.
265. J. G. Wallace, Epoxidation. In R. E. Kirk and D. F. Othmer, Eds., *Encyclopedia of Chemical Technology*, 2nd ed., Vol. 8, Wiley, New York, 1964, pp. 249–265.
266. ACGIH (American Conference of Governmental Industrial Hygienists), *Documentation of the Threshold Limit Values and Biological Exposure Indices*, 5th ed. American Conference of Governmental Industrial Hygienists, Cincinnati, 1986, p. 627.
267. N. I. Sax, *Dangerous Properties of Industrial Materials*, 6th ed. Van Nostrand Reinhold, New York, 1984, p.2729.
268. *National Toxicology Program (NTP) Technical Report No. 362*, Toxicology and carcinogenicity studies of 4-vinyl-1-cyclohexene diepoxide (CAS 106–87–6) in F344/N rats and $B6C3F_1$ mice (Dermal Administration) 1989.
269. R. S. Chhabra, M. R. Elwell, and A. Peters, *Fundam. Appl. Toxicol.* **14**, 745 (1990).
270. R. S. Chhabra, M. R. Elwell, and A. Peters, *Fundam. Appl. Toxicol.* **14**, 745 (1990).
271. M. P. Murray and J. E. Cummins, *Environ. Mutagen.* **1**, 307 (1979).
272. M. A. El-Tantawy and B. D. Hammock, *Mutat. Res.* **79** (1980).
273. G. Gervasi, L. Citti, and G. Turchi, *Mutat. Res.* **74**, 202 (1980).
274. S. W. Frantz and J. E. Sinsheimer, *Mutat. Res.* **90** (1981).
275. C. J. Dannaker, *J. Occup. Med.* **30**, 641 (1988).
276. R. Patterson, K. E. Harris, W. Stopford, et al., *Int. Arch. Allergy Appl. Immunol.* **85**, 467 (1988).
277. IARC Monographs On The Evaluation Of Carcinogenic Risks To Humans, Volume 60, *Some Industrial Chemicals*, World Health Organization, International Agency for Research on Cancer, 1994, pp. 361–372.
278. P. W. Langvardt and R. J. Nolan, *J. Chromatogr.* **567**, 93 (1991).
279. Industrial Medicine and Toxicology Department, Union Carbide Corp., New York, 1958.
280. M. R. Sikov et al., *J. Appl. Toxicol.* **6**, 155 (1986).
281. *IPCS Environmental Health Criteria 26, Styrene*, United Nations Environment Programme, International Labour Organization and World Health Organization, Geneva, 1987.
282. J. A. Bond, *CRC Critic. Rev. Toxicol.* **19**, 227 (1989).
283. Gy. A. Csanady et al., *Arch. Toxicol.* **68**, 143–157, (1994).
284. H. Vaino, K. Hemminki, and E. Elovaara, *Toxicology* **8**, 319 (1977).
285. M. R. Sikov et al., *J. Appl. Toxicol.* **6**, 155 (1986).
286. B. D. Hardin et al., *Scand. J. Work Environ. Health* **7**, (Suppl. 4), 66 (1981).
287. B. D. Hardin, R. W. Niemeier, M. R. Sikov, et al., *Scand. J. Work Environ. Health* **9**, 94 (1983).
288. C. Maltoni, A. Ciliberti, and D. Carrietti, *Ann. NY Acad. Sci.* **381**, 216 (1982).
289. B. Conti, C. Maltoni, G. Perino, and A. Ciliberti, *Ann. NY Acad. Sci.* **534**, 203 (1988).
290. V. Ponomarkov and L. Tomatis, *Cancer Lett.* **24**, 95 (1984).
291. W. Lijinski, J. Natl. *Cancer Inst.* **77**, 471 (1986).
292. P. Kotin et al., *Radiat. Res.* (Suppl 3), 193 (1963).
293. R. J. Preston, *S.I.R.C. Review* **1**(1), 23 (1990); Styrene Information and Research Center, Washington, DC.
294. R. J. Preston, *S.I.R.C. Review* **1**(2), 24 (1990); Styrene Information and Research Center, Washington, DC.

295. R. Barale, *Mutat. Res.* **257**, 107 (1991).
296. IARC Monographs On The Evaluation Of Carcinogenic Risks To Humans, Volume 60, *Some Industrial Chemicals*, World Health Organization, International Agency for Research on Cancer, 1994, pp. 321–346.
297. M. E. Herrero et al., *Env. Mol. Mutag.* **30**, 429–439 (1997).
298. B. Maarczynski, M. Peel, and X. Baur, *Toxicology* **120**, 111–117 (1997).
299. S. Charkrabarti, X. Zhang, and C. Richer, *Mut. Res.* **395**, 37–45 (1997).
300. T. Ollikainen, A. Hirvonen, and H. Norppa, *Env. Mol. Mutag.* **31**, 311–315 (1998).
301. P. Vodicka et al., *Carcinogenesis* **17**, 801–808 (1996).
302. T. Bastlova and A. Padlutsky, *Mutagenesis* **11**, 581–591 (1996).
303. W. Schrader and M. Linscheid, *Arch. Toxicol.* **71**, 588–595 (1997).
304. R. Kumar et al., *Carcinogenesis* **18**, 407–414 (1997).
305. F. R. Setayesh et al., *Chem. Res. Toxicol.* **11**, 766–777 (1998).
306. M. Otteneder, E. Eder, and W. K. Lutz, *Chem. Res. Toxicol.* **12**, 93–99 (1999).
307. M. Donner, M. Sorsa, and H. Vainio *Mutat. Res.* **67**, 373–376 (1979).
308. R. Rodriguez-Arnaiz, *Env. Mol. Mutag.* **31**, 390–401 (1998).
309. L. Fabry, A. Leonard, and M. Roberfroid, *Mutat. Res.* **51**, 377–381 (1978).
310. H. Vaghef and B. Hellman, *Pharmacol. Toxicol.* **83**, 69–74, (1998).
311. Yendle et al., *Mutat. Res.* **375**, 125–136 (1997).
312. S. Cantoreggi and W. K. Lutz, *Carcinogenesis* **13**, 193 (1992).
313. K. Yeowell-O'Connell, W. Pauwels, M. Severi, Z. Jin, M. R. Walker, S. M. Rappaport, and H. Veulemans, *Chem. Biol. Inter.* **106**, 67–85 (1997).
314. S. Fustinoni, C. Colosio, A. Colombi, L. Lastrucci, K. Yeowell-O'Connell, and S. M. Rappaport, *Int. Arch. Occup. Environ. Health* **71**, 35–41 (1998).
315. C. J. Hine et al., *AMA Arch. Ind. Health* **14**, 250 (1956).
316. J. W. Henck, D. D. Lockwood, and H. O. Yakel, unpublished report of The Dow Chemical Company, 1978.
317. J. K. Kodama et al., *J. Arch. Environ. Health* **2**, 50 (1961).
318. National Toxicology Program, *NTP TR 376*, 1990.
319. K. J. Olson, unpublished report of The Dow Chemical Company, 1957.
320. D. Zissu, *J. Appl. Toxicol.* **15**, 207 (1995).
321. R. A. Renne, H. R. Brown, and M. P. Jokinen, *Toxicol. Pathol.* **20**, 416 (1992).
322. M. J. Wade, J. W. Moyer, and C. J. Hine, *Mutat. Res.* **66**, 367 (1979).
323. D. A. Canter et al., *Mutat. Res.* **172**, 105 (1986).
324. K. Plna, D. Segerbaeck, and E. K. H. Schweda, *Carcinogenesis* **17**, 1465 (1996).
325. K. Plna and D. Segerbaeck, *Carcinogenesis* **18**, 1457 (1997).
326. F. Gagnaire et al., *Toxicol. Lett.* **39**, 139 (1987).
327. H. Licea Perez, K. Plna, and S. Osterman-Golkar, *Chem.-Biol. Interactions* **103**, 1 (1997).
328. S. Fregert and H. Rorsman, *Acta Allerg.* **19**, 296 (1964).
329. G. Angelini, L. Rigano, C. Fote et al., *Contact Dermatitis* **35**, 11 (1996).
330. M. A. Dooms-Goossens et al., *Contact Dermatitis* **33**, 17 (1995).
331. K. K. Hemminki, K. Falck, and H. Vainio, *Arch. Toxicol.* **46**, 277 (1980).

332. P. Foureman, J. M. Mason, R. Valencia, and S. Zimmering, *Environ. Mol. Mutagen.* **23**, 51 (1994).
333. D. D. Lockwood and H. W. Taylor, unpublished report of The Dow Chemical Company, 1982.
334. D. A. Basketer, E. W. Scholes, and I. Kimber, *Food Chem. Toxicol.* **32**, 543 (1994).
335. Procter and Gamble Company, unpublished report, 1973.
336. M. Andersen, P. Kiel, and H. Larsen, unpublished data, 1957.
337. Ciba-Geigy Limited, unpublished data, 1985.
338. J. F. Quast and R. R. Miller, *Toxicologist* **5**, 1 (1985).
339. L. V. Eadsforth et al., *Xenobiotica* **15**, 579 (1985).
340. T. H. Connor et al., *Environ. Mutat.* **2**, 284 (1980).
341. T. G. Pullin, unpublished report to The Dow Chemical Company, 1977.
342. Reichhold Chemicals, Inc., studies submitted under Section 8(d) of *TSCA (40-7840108)*, 1978.
343. E. D. Thompson et al., *Mutat. Res.* **90**, 213 (1981).
344. A. F. Frost and M. S. Legator, *Mutat. Res.* **102**, 193 (1982).
345. B. J. Dabney, unpublished data, The Dow Chemical Company, 1979.
346. T. H. Connor et al., *Environ. Mutat.* **2**, 521 (1980).
347. E. B. Whorton et al., *Mutat. Res.* **124**, 225 (1983).
348. M. A. Wolf and V. K. Rowe, unpublished data, Dow Chemical Company, 1958.
349. R. Jolanki et al., *Contact Dermatitis* **23**, 172 (1990).
350. A. Tosti et al., *Toxiol. Indust. Health* **9**, 493 (1993).
351. A. D. Coombs, D. G. Clark, confidential report of Shell Chemicals Ltd., 1977.
352. B. J. Dean and T. M. Brooks, confidential report of Shell Chemicals Ltd., 1978.
353. E. Zeiger et al., *Environ. Mol. Mutagen.* **11**, 1 (1988).
354. Procter & Gamble, *Tech. Inf. Sheets DUP 2660-128A, DUP 2660-129.*
355. *EPA/OTS Document 878212388 (V-1146-153, BTS 171)*, 1982.
356. Procter and Gamble Miami Valley Laboratories, unpublished data 1979.
357. Springborn Institute for Bioresearch, Inc., unpublished data 1980.
358. Industrial Bio-Test Laboratories, Inc., unpublished data 1973.
359. International Bio-Research, Inc., unpublished data 1975.
360. J. O. Kuhn and S. Cagen, unpublished report of Shell Chemical Company, 1994.
361. *EPA/OTS Document 878212424 (Y152-116, BTS 1393)*, 1982.
362. *EPA/OTS Document 878212429 (Y154-103, BTS 1463)*, 1982.
363. *EPA/OTS Document 878212408 (V200043)*, 1982.
364. *EPA/OTS Document 878212396 (V1439-143)*, 1982.
365. *EPA/OTS Document 878212410 (V1927-43, BTS 345S and 368)*, 1982.
366. R. J. McGuirk and K. A. Johnson, Alkyl glycidyl ether: 2-week range finding and 13-week repeated dose dermal toxicity study in Fischer 344 rats. Unpublished report of The Dow Chemical Company, 1997.
367. P. J. Boogaard et al., *Xenobiotica* **30**, 469, 485 (2000).
368. D. G. Stump, A dermal developmental toxicity screening study of alkyl glycidyl ethers in rats. Report of WIL Research Laboratories to The Society of the Plastics Industry, Inc., 1997.
369. E. D. Thompson et al., *Mutat. Res.* **90**, 213 (1981).

370. V. O. Wagner, III, Bacterial reverse mutation assay with an independent repeat assay. Report of Microbiological Associates to The Society of the Plastics, Industry, Inc., 1997.
371. R. H. C. San and J. J. Clarke, *In vitro* mammalian cell gene mutation test with an independent repeat assay. Report of Microbiological Associates to The Society of the Plastics, Industry, Inc., 1997.
372. R. Gudi, Alkyl glycidyl ether: Micronucleus cytogenetic assay in mice. Report of Microbiological Associates to The Society of the Plastics, Industry, Inc., 1997.
373. Hill Top Research Institute, Inc., unpublished data 1964.
374. N. I. Sax and R. J. Lewis, *Hawley's Condensed Chemical Dictionary* 11th ed. Van Noostrand Reinhold Company, New York, 1987 p. 398.
375. Ciba-Geigy, Limited, unpublished report, 1981.
376. Ciba-Geigy, Limited, unpublished report, 1983.
377. Ciba-Geigy, Limited, unpublished report, 1972.
378. A. Thorgeirsson, *Acta Dermatol.* **58**, 219 (1978).
379. S. Clemmensen, *Drug Chem. Toxicol.* **7**, 527 (1977).
380. E. Thorpe et al., unpublished report of Shell Research Limited, 1980.
381. G. C. Peristianis, M. A. Doak, P. N. Cole, and R. W. Hend, *Food Chem. Toxicol.* **26**, 611 (1988).
382. *EPA/OTS Document 878210058*, 1987.
383. R. Jolanki et al., *Contact Dermatitis* **16**, 87 (1987).
384. Ciba-Geigy, Limited, unpublished report, 1975.
385. Ciba-Geigy, Limited, unpublished report, 1982.
386. Ciba-Geigy, Limited, unpublished report, 1976.
387. J. M. Holland et al., *Oak Ridge National Laboratory Report No. ORNL 5762*, 1981.
388. T. Czajkowska and J. Stetkiewica, *Med. Pharm.* **23**, 363 (1972).
389. Ciba-Geigy Limited, 1978, unpublished report.
390. L. Ullman and Porricello, 1988, report to CIBA Geigy Ltd.
391. E. Zschunke and P. Behrbohm, *Dermatol. Wochenschr.* **151**, 480 (1965).
392. J. E. Betso, R. E. Carreon, P. W. Langvardt, and E. Martin, 1986, unpublished report of the Dow Chemical Toxicology Research Laboratory.
393. J. B. Terrill and K. P. Lee, *Toxicol. Appl. Pharmacol.* **42**, 263 (1977).
394. K. P. Lee, J. B. Terrill, and N. W. Henry, *J. Toxicol. Environ. Health* **3**, 859 (1977).
395. S. P. James, A. E. Pheasant, and E. Solheim, *Xenobiotica* **8**, 219 (1976).
396. B. M. de Rooij et al., *Chem. Res. Toxicol.* **11**, 111 (1998).
397. K. P. Lee, P. W. Schneider, and H. J. Trochimowica, *Am. J. Pathol.* **3**, 140 (1983).
398. E. J. Greene et al., *Mutat. Res.* **67**, 9 (1979).
399. G. W. Mvie, *Mutat. Res.* **79**, 73 (1980).
400. J. P. Seiler, *Mutat. Res.* **135**, 159 (1984).
401. H. Ohtani and H. Nishioka, *Sci. Eng. Rev.* **21**, 247 (1981).
402. J. B. Terrill et al., *Toxicol. Appl. Pharmacol.* **64**, 204 (1982).
403. K. Hemminki and H. Vainio, *Dev. Toxicol. Environ. Sci.* **8**, 241 (1980).
404. F. Lemiere et al., *J. Chromatogr.* **647**, 211 (1993).

405. D. L. D. Deforee et al., *Carcinogenesis* **19**, 1077 (1998).
406. A. Tosti et al., *Tox. Industrial Health* **9**, 493 (1993).
407. K. Tarvainen, *Contact Dermatitis* **32**, 346 (1995).
408. R. Jolanki et al., *Contact Dermatitis* **34**, 390 (1996).
409. L. Kanerva, R. Jolanski, and T. Estlander, *Contact Dermatitis* **37**, 301 (1997).
410. L. Kanerva and T. Estlander, *Contact Dermatitis* **38**, 274 (1998).
411. IARC Monographs On The Evaluation Of Carcinogenic Risks To Humans, Volume 71, *Re-evaluation of Some Industrial Chemicals, Hydrazine and Hydrogen Peroxide*, World Health Organization, International Agency for Research on Cancer, Lyon, France, 1999.
412. APME, 1991, unpublished report to the Association of the Plastics Manufacturers in Europe.
413. J. R. Albert, H. C. Wimberly, and N. L. Wilburn, 1983, unpublished report of Shell Oil Company.
414. M. N. Pinkerton and R. L. Schwebel, 1977, unpublished report of the Dow Chemical Company.
415. L. Ullman, 1991, unpublished report of RCC Research and Consulting Company.
416. K. Sollner and Irrgana, *K. Arzneim. Forsch.* **15**, 1355 (1965).
417. M. Ballantyne, unpublished report to the Association of the Plastics Manufacturers in Europe, 2000.
418. C. Chierregato et al., *Contact Dermatitis* **30**, 120 (1994).
419. G. De Jong, N. J. Van Sittert, and A. T. Natarajan, *Mutat. Res.* **204**, 451 (1988).
420. A. L. Mendrala et al., *Arch. Toxicol.* **67**, 18–27 (1993).
421. O. G. Fahmy and M. J. Fahmy, *Cancer Res.* **30**, 195 (1970).
422. R. E. Denell, M. C. Lim, and C. Auerbach, *Mutat. Res.* **49**, 219 (1978).
423. M. L. Goldberg, R. A. Colvin, and A. F. Mellin, *Genetics* **123**, 145 (1989).
424. J. T. Reardon et al., *Genetics* **115**, 323 (1987).
425. O. A. Olson and M. M. Green, *Mutat. Res.* **92**, 107 (1982).
426. P. Gay and D. Contamine, *Mol. Gen. Genet.* **239**, 361 (1993).
427. A. D. Auerbach, B. Adler, and R. S. K. Chagani, *Pediatrics* **67**, 128 (1981).
428. A. D. Auerbach, *Exp. Hematol.* **21**, 731 (1993).
429. R. F. Henderson et al., *Toxicol. Sci.* **52**, 33–44 (2000).
430. J. L. Mattsson et al., Unpublished report of The Dow Chemical Company, 1997.
431. T. L. Neeper, Reproductive study of inhaled ethylene oxide vapor in CD® Rats. Praxair & Balchem, 1993.

CHAPTER EIGHTY-THREE

Epoxy Compounds — Aromatic Diglycidyl Ethers, Polyglycidyl Ethers, Glycidyl Esters, and Miscellaneous Epoxy Compounds

John M. Waechter, Jr., Ph.D., DABT, and
Gauke E. Veenstra, Ph.D.

INTRODUCTION

The principal focus of this chapter is on the epoxy compounds frequently encountered in industrial use as uncured epoxy resins. These resins are marketed in a variety of physical forms from low-viscosity liquids to tack-free solids and require admixture with curing agents to form hard and nonreactive cross-linked polymers. They are in demand because of their toughness, high adhesive properties (polarity), low shrinkage in molds, and chemical inertness.

It is the uncured resins that are of main interest to toxicologists, for a well-cured resin should have few or no unreacted epoxide groups remaining in it. The toxicology of the curing agents is not treated in this chapter. They are most frequently bi- or trifunctional amines, di- or tricarboxylic acids and their anhydrides, polyols, and compounds containing mixed functional groups, such as aminols and amino acids, as well as other resins containing such groups (1).

Some of the other epoxide compounds described in this chapter are used as reactive diluents in epoxy resin mixtures; others are of commercial importance for their multiple

Patty's Toxicology, Fifth Edition, Volume 6, Edited by Eula Bingham, Barbara Cohrssen, and Charles H. Powell.
ISBN 0-471-31939-2

uses in the synthesis of other compounds (specifically, epichlorohydrin). As reactive diluents, monomeric epoxides are added to epoxy resins to reduce viscosity and modify the handling characteristics of the uncured materials. The epoxide functionalities of these diluents react with the resin curing agents in the same manner as the resin to become part of the finished polymer. Epoxy resins have found application as protective coatings, adhesives for most substrates (metals included), caulking compounds, flooring and special road paving, potting and encapsulation resins, low-pressure molding mixtures, and binding agents for fiber glass products. Uncured, they are used as plasticizers and stabilizers for vinyl resins.

Epoxy resin coating formulations can generally be limited to one of three forms: solution coatings, high-solids formulations, and epoxy powder coatings. Solid epoxies are used in coating applications, as solid solutions or heat-converted coating. Solution coatings are often room-temperature applications, and typically there is little potential for vapor exposure. The potential clearly exists for skin contact during application of coatings of this type. Heat-converted coatings are usually applied and cured by mechanical means and exposure to vapors or contact with skin is minimal.

Solid resins are used for other applications such as electrical molding powders and decorative or industrial powder coatings. For applications of this kind, exposure to vapors and dust can occur and is greatest during formulating and grinding.

Considering that epoxides can react with nucleophiles, particularly basic nitrogens, one might expect the epoxides to react with cellular biomolecules such as glutathione, proteins, and nucleic acids, and indeed this has been demonstrated for some of these epoxide compounds. The nature and magnitude of these interactions with these biomolecules is most likely related to the toxicity and observed for any given molecule in this class of compounds. However, the potential for any epoxide to react with cellular nucleophilic biomolecules is dependent on several factors, including the reactivity of the particular epoxide, the dose and dose rate, as well as the molecular weight, and solubility, these latter two influencing access to molecular targets within the cell. In addition, the efficiency of metabolism via epoxide hydrolase or other metabolic routes of detoxication may significantly influence the toxicologic potential and potency of these materials.

Epoxide hydrolase activity is widely distributed throughout the body, but it is organ, species, and even strain variant (2). The liver, testes, lung, and kidney have considerable epoxide hydrolase activity; the activities in the skin and gut, however, are considerably lower. In this regard it should be noted that mouse tissues have a much lower level of EH activity than does human tissue; in fact, at least two strains of mice, C57BL/6N and DBA/2N, have no EH activity in their skins (3). Therefore, it may be questioned if toxicity or treatment-related effects observed in dermal mouse studies are relevant for hazard evaluation. Epoxy compounds may also be metabolized by the cytoplasmic enzyme glutathione-*S*-transferase, which converts epoxides to 2-alkylmercapturic acids. This enzyme because it is in the aqueous phase, may play a minor role in the detoxification of large lipophilic epoxides, but is active against low-molecular-weight epoxides (4, 5). Due to differences in physio-chemical properties and the effectiveness and nature of the detoxification of these materials through metabolism, the toxicity of these compounds ranges from the highly active, electrophilic, low-molecular-weight mono- and diepoxides to the nontoxic and inert cured materials, which possess only a few epoxy groups per molecule.

In general, the acute toxicity of epoxy resin compounds as observed in laboratory animals can be considered low; oral and dermal LD_{50} values generally range from about 2000 to greater than 10,000 mg/kg in rodents; there are not marked differences in acute toxicity among the structurally diverse categories of epoxy resin compounds. The acute toxicity of low-molecular-weight epoxides such as epichlorohydrin and glycidol is significantly greater (oral LD_{50} values range from 90 to about 500 mg/kg). Lung irritation following inhalation or gastrointestinal irritation following gavage has also been observed in animals. It is generally difficult to achieve acutely toxic levels of epoxy compounds by dermal exposure. Usually the irritating properties of epoxy liquids or vapors limit significant exposure to produce systemic toxicity.

Effects most commonly observed in animals have been dermatitis (either primary irritation or secondary to induction of sensitization), eye irritation, pulmonary irritation, and gastric irritation, which are typically found in the tissues that are the first to come into contact with the epoxy compound. In general, it appears that epoxy compounds of higher molecular weight (e.g., epoxy novolac resins and diglycidyl ether of bisphenol A) produce less dermal irritation than those of lower molecular weight. In some instances, liquid epoxy compounds splashed directly into the eye may cause pain and, in severe cases, corneal damage. Skin irritation is usually manifested by more or less sharply localized lesions that develop rapidly on contact, more frequently on the arms and hands. Signs and symptoms usually include redness, swelling, and intense itching. In severe cases, secondary infections may occur.

Workers show marked differences in sensitivity. Devices made from epoxy resins have produced severe dermatitis when not properly cured and when in prolonged contact with the skin (6). Skin irritation also has been reported from exposure to epoxy vapors (7). Most of the epoxy compounds have the ability to produce delayed contact skin sensitization, although there are notable exceptions, such as the advanced bisphenol A/epichlorohydrin resins. The higher molecular weight of these resins may be responsible for the absence of dermal sensitization (8, 9). Skin sensitization reactions can be elicited from much less agent than is required for a primary irritation response. Because this condition is difficult to treat, sensitized individuals may require transfer to other working areas. Particular attention should be paid to vapors and fine airborne dusts.

Animals exposed to vapors of epichlorohydrin have shown pulmonary irritation. Sequelae of this effect may include pulmonary edema, cardiovascular collapse, and pneumonia. However, this route of exposure is unlikely for many of these epoxy compounds owing to their low volatility. For the glycidyloxy compounds for which LC_{50}s have been determined, it appears that none of these compounds can be considered highly acutely toxic by the inhalation route.

For those epoxy compounds for which repeat dosing studies have been conducted, generally the liver, kidneys, respiratory epithelium or nasal mucosa (when inhalation was the route of exposure), and stomach (when given by gavage) appear to be the major target organs. Respiratory epithelium and nasal mucosa effects have been responses typical of irritation, such as flattening or destruction of epithelial cells. Disruption of hematopoesis, primarily leukopenia, has also been demonstrated in laboratory animals with a polyglycidyl ether of substituted glycerin and resorcinol diglycidyl ether, but similar changes have not been observed in workers as a result of occupational exposures. The testes have been found to a target organ for glycidol and epichlorohydrin. Glycidol was also embryotoxic in laboratory animals when administered by a route not relevant to occupational exposure (intra-amniotic injection). However, no developmental toxicity has been observed for any other compounds in this chapter where there are data available. In addition, a two-generation reproduction study in rats on the diglycidyl ether of bisphenol A also indicated that this material did not produce adverse effects on either male or female reproduction.

There is evidence in rats from a National Toxicology Program study that glycidol produces neurotoxicity (10), and this finding suggests that glycidyl esters, if metabolized to glycidol, could have this effect. However, glycidyl ethers have shown no evidence of neurotoxic effects in numerous acute or repeated dosing subchronic studies in rodents. A recently completed neurotoxicity study of diglycidyl ether of bisphenol found no evidence of neurotoxic effects in rats dosed dermally.

Generally, *in vitro* genetic toxicity testing of the epoxide compounds has resulted in positive (genotoxic) responses; the majority of the studies of genotoxic potential have been carried out using bacteria. These results are not surprising because many of these compounds have been tested in strains TA1535 and TA100 of *S. typhimurium* or in other gene mutation assays that are specifically sensitive to base-pair substitution, Metabolic activation was not required for most of the epoxides, which showed mutagenic in these tests. Many other *in vitro* assays examining both gene mutation and chromosomal effects have been employed to test the epoxy compounds, including assays in *E. coli*, yeast, Chinese hamster ovary cells (CHO/HPGRT), mouse lymphoma cells, and cultured human lymphocytes; the results have usually been mixed or positive. Fewer epoxy compounds have been tested using *in vivo* assays for genotoxic effects, although some have been extensively studied. Glycidol was positive in the *Drosophila* sex-linked recessive lethal assay and mouse micronucleus assay and produced chromosomal aberrations in the bone marrow of mice dosed orally or intraperitoneally. Glycidaldehyde was positive in the *Drosophila* sex-linked recessive lethal assay. In contrast, epichlorohydrin, also a low-molecular-weight epoxy compound, was negative in both the mouse micronucleus test following intraperitoneal administration and the mouse dominant lethal assay following oral or intraperitoneal administration, although it was positive in many of the *in vitro* assays. Another compound showing positive or equivocal effects *in vitro* but negative effects *in vivo* is the diglycidyl ether of bisphenol A.

Of the compounds in this chapter, four (resorcinol diglycidyl ether, epichlorohydrin, glycidaldehyde, and glycidol) were the subject of studies in which there was clear evidence of tumorigenic effects in rodents. Larger-molecular-weight glycidyloxy compounds such as castor oil glycidyl ether, the diglycidyl ether of bisphenol A, and advanced bisphenol A/epichlorohydrin epoxy resins have been negative in dermal bioassays. Epidemiology studies have not provided any evidence for an association between workplace exposure and cancer to any of the materials in this chapter.

AROMATIC DIGLYCIDYL ETHERS

1.0 Bisphenol A Diglycidyl Ether

1.0.1 CAS Numbers: [1675-54-3, 25068-38-6, 25085-99-8, 25036-25-3]

1.0.2 Synonyms: Diglycidyl ether of bisphenol A; 2,2′-[(1-methylethylindene)*bis*(4,1-phenyleneoxymethylene)]*bis*-oxirane; diglycidyl ether of 4,4′-isopropylidenediphenol; 2,2-*bis*[parp-2,3-epoxypropoxy)phenyl]propane; and BADGE

1.0.3 Trade Names: Various trademarks of epoxy resins based on bisphenol A diglycidyl ether are EPON® resin series, D.E.R® series, Epotuff® series, Araldite® series, EPI-Rez® series and the ERL Bakelite® epoxy series

1.0.4 Molecular Weight: 340.42

1.0.5 Molecular Formula: $C_{21}H_{24}O_4$

1.0.6 Molecular Structure:

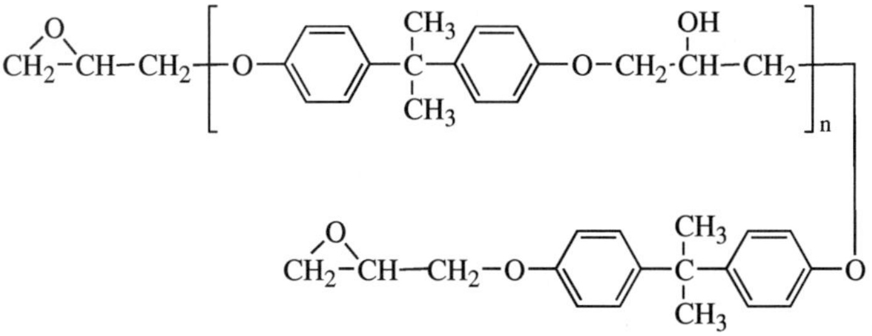

1.1 Chemical and Physical Properties

These compounds include the diglycidyl ether of bisphenol A and the condensation products of their further reaction or advancement with bisphenol A. The resins are usually mixtures and may contain homologues of higher weight, isomers, branched-chain homologues, and occasionally, monoglycidyl ethers.

The general formula may be written as indicated for the molecular formula above, where *n* is the number of repeating units in the resin chain.

	$n = 0$	$n = 2$	$n = 9$
Molecular weight (approx.)	340.42	900	2900
Specific gravity	1.16	1.204	1.146
Melting point (°C)	8–12	64–76	127–133
Epoxy equivalent	190–210	450–525	1650–2050

The lower-molecular-weight resins are liquids, and as the molecular weight increases, they become increasingly viscous and finally solids. Hazardous polymerizations may occur with aliphatic amines in masses greater than one pound.

1.1.2 Odor and Warning Properties

There are no odor and warning properties; the resins are sticky when handled.

1.2 Production and Use

The synthesis of the basic epoxy resin molecule involves the reaction of epichlorohydrin with bisphenol A, the latter requiring two basic intermediates for synthesis, acetone and phenol. Theoretically, the production of the bisphenol A diglycidyl ether (BADGE) requires 2 mol of epichlorohydrin for each mole of the phenol.

Epoxy resins of higher molecular weight are obtained by reducing the epichlorohydrin/bisphenol A ratio. This reaction involves consumption of the initial epoxy groups in the epichlorohydrin and of some of the groups formed by dehydrohalogenation.

The properties of epoxy resins make them ideally suited for sealing, encapsulating, making castings and pottings, and formulating lightweight foams. Castings may be used for patterns, molds, and finished products. These resins are used as binders in the preparation of laminates of paper, polyester cloth, fiberglass cloth, and wood sheets. The epoxy resins have outstanding adhesive properties.

The amines were the first materials to gain general acceptance as curing agents for epoxy resins. Polyfunctional primary aliphatic amines give fast cures and provide overall properties satisfactory for a wide variety of applications. These materials are usually considerably more active physiologically than the epoxy resins and more volatile, and skin and eye irritation may occur. Other curing agents, including the acid anhydrides and organic acids, have given fewer problems in handling. A number of diluents are also physiologically more active than the resins themselves. Some of these are the epoxy esters, ethers, and aliphatic compounds of low molecular weight.

Resin modifiers include phenolic substances, aniline formaldehyde resins, furfural, isocyanates, and silicone resins, all of which may contribute to handling problems.

1.3 Exposure Assessment

Inhalation exposure to vapors from these resins is unlikely because these materials are essentially nonvolatile. Dust from the solid resins may be trapped according to standard filter paper and liquid entrapment methods and determined by the HCl–dioxane method (11).

1.4 Toxic Effects

The acute systemic toxicity of pure BADGE and the low-molecular-weight BADGE-based resins is low by either dermal or oral administration. A single dermal dose of these materials produces only slight irritation to the skin of rabbits, but repeated dermal application may produce greater irritation. Dermal exposure with liquid BADGE-based resins or pure BADGE has produced skin sensitization in guinea pigs. Thus contact with the skin presents the greatest potential problem for occupational exposures, and it is well known that the lower-molecular-weight epoxy resins can cause dermatitis and skin sensitization in certain individuals upon prolonged or repeated exposure. Manufacturers of epoxy resins recommend that precautionary action be taken to avoid all skin contact with uncured resins as well as other components of epoxy resins sytstems. Owing to their sticky nature, resins on the skin are difficult to remove, and there often is a tendency to use solvents to remove them. This is not recommended, because the solvents may facilitate penetration of the resin through the skin. These materials produce minimal irritation to the eye. Inhalation exposure to these resins is unlikely owing to the very low vapor pressure of these materials. Oral subchronic studies on these materials have shown that the systemic toxicity of these materials is very low. In addition, four separate studies have indicated that BADGE or BADGE-based resins are not teratogenic. No adverse effects on reproductive parameters or reproductive organs were noted in either male or female rats gavaged with a low-molecular-weight BADGE-based resin or with pure BADGE. BADGE and various BADGE-based epoxy resins have been tested in several *in vitro* and *in vivo* assays for genotoxicity. Although some of the *in vitro* mutagenicity assays for these materials have given positive results, the results of these and other *in vitro* tests have been negative, and all the *in vivo* tests for mutagenicity have been negative. *In vitro* and *in vivo* studies have shown that BADGE is only slowly absorbed through the skin, and is detoxified by metabolism to the *bis*-diol of BADGE by the enzyme epoxide hydrolase. *In vitro* studies found human skin to be the least permeable to BADGE as compared to the rat and mouse. There is also some evidence to suggest that very little parent compound would be systemically available following the oral administration of BADGE. A number of carcinogenicity studies involving the topical application of pure BADGE, as well as EPON® Resin 828, and other commercial BADGE-based resins, have been carried out in experimental animals. Viewing the studies as a whole, the weight of evidence does not show that BADGE or BADGE-based epoxy resins are carcinogenic.

The acute systemic toxicity of solid, that is, the advanced bisphenol A/epichlorohydrin epoxy resins (advanced BADGE-based) is low by either dermal or oral routes. Inhalation of these materials is unlikely because of their low volatility. Instillation into the eyes of rabbits produced only slight irritation. Dermal contact may result in slight irritation, especially with repeated or prolonged exposure. However, unlike BADGE-based resins of lower molecular weight, these materials do not appear to cause delayed contact hypersensitivity on the skin. Subchronic oral administration of an advanced BADGE-based epoxy resin did not result in demonstrable toxicity. Mutagenicity testing has produced mixed results. Chronic dermal administration did not induce tumor formation in mice.

1.4.1 Experimental Studies

1.4.1.1 Acute Toxicity. Single-dose oral toxicity of BADGE is very low; early limit tests resulted in estimations of the LD_{50} at values greater than 2000 or 4000 mg/kg (12) reported "exact" oral LD_{50} values of 11,400 mg/kg in rats, 15,600 mg/kg in mice, and 19,800 mg/kg in rabbits for a commercial BADGE-based epoxy resin. Weil et al. (13) reported an oral LD_{50} of 19.6 mL/kg ($\sim$19,600 mg/kg) for rats with a commercial BADGE-based epoxy resin. More recent studies with pure BADGE or commercial BADGE-based resins have produced results consistent with those previously reported; the single-dose oral LD_{50} value was reported as >1000 mg/kg in the rat and >500 mg/kg in the mouse (14, 15). Lockwood and Taylor (16) found the single-dose oral LD_{50} value for rats to be >2000 mg/kg. The pharmacological effects observed even in lethal doses were not remarkable. Moderate antemortem depression occurs, and loss of body weight and diarrhea are often observed in surviving animals. Intraperitoneal toxicity is greater than oral toxicity by five to tenfold (2400 and 4000 mg/kg in rats and mice, respectively) (12).

The potential for absorption through the skin in acutely toxic amounts is low; the single-dose dermal LD_{50} value in rabbits has been reported to be 20 mL/kg for a BADGE-based commercial resin (13). Lockwood and Taylor (16) reported 100% survival with no adverse effects for rabbits treated with a single dermal dose of BADGE-based commercial resin at a dose level of 2000 mg/kg. In other species, studies show dermal LD_{50} values of pure BADGE to be >800 and >1600 mg/kg for mice and rats, respectively (14). The acute dermal toxicity of a commercial BADGE-based resin was similar, with single-dose dermal LD_{50} values of >800 and >1200 mg/kg for mice and rats, respectively (14, 15).

The inhalation toxicity of BADGE or BADGE-based resins has not been studied because vapor exposures are unlikely owing to the low vapor pressure of the material. Nolan et al. (17) reported difficulty in generating an atmosphere at a respirable temperature (~22°C) that contained sufficient BADGE to conduct a rodent inhalation study even when using large surface areas and high temperatures initially to generate an atmosphere prior to cooling to respirable temperatures.

Skin Effects. Single prolonged (24-h) application to the skin of rabbits showed BADGE-based resin to be only slightly irritating at most, even when occluded or if skin was abraded (12, 16, 18–22). Repeated applications were reported to be more irritating (12, 23). Application of the same liquid resin for 4 h/d for 20 d resulted in Draize scores of 0 to 8, indicating that prolonged and repeated skin contact with liquid resins may cause severe irritation.

When representative solid resins (mol. wt. = 900 and 2900) were applied for 4 h/d for 20 d to rabbits, Draize scores of 0 were seen, which indicates that the solid resins are less likely to cause primary irritation even with repeated to prolonged skin contact. However, it is still strongly recommended that skin contact with either solid or liquid resins be avoided (12).

Skin Sensitization. Several studies have reported lower-molecular-weight BADGE-based resins to produce skin sensitization in guinea pigs (14, 15, 24–28). Pure BADGE was also found to be a skin sensitizer in guinea pigs when tested with the Magnusson–Kligmann maximization test (29). Higher-molecular-weight, solid BADGE-based resins have not produced this effect (30, 31). For additional information, see the section on effects in humans.

Eye Irritation. BADGE-based resins have been reported to cause only minimal eye irritation (12, 15, 18, 19, 21, 22). Eye irritation in an industrial setting usually results from inadvertent transfer of resin from the hands when rubbing the eyes, and may occur even if the hands are protected with gloves. The use of goggles is an effective way to prevent accidental eye exposure.

Solid resins of higher molecular weight (~900 and above) are capable of causing moderate eye irritation by the Draize test, and scores as high as 41/110 have been reported (32). The greater degree of eye irritation potential with the solid resins is due to their ability to form dust. The small particles of resin can thus cause mechanical abrasion of the eye and surrounding tissue. Again, use of goggles is effective in preventing exposure to resin dusts.

1.4.1.2 Chronic and Subchronic Toxicity. *Oral Administration.* Incorporation of 1% undiluted uncured liquid resin of molecular weight 450 (EPON® resin 834) or as 75 and 50% in dioctyl phthalate into the diet of rats for 26 d resulted in body weight loss and decreased food consumption, but no gross or histopathological lesions (33). Incorporation of a semisolid resin of molecular weight about 600 (EPON® resin 836) as 1% undiluted, uncured resin, or as 50% in dioctyl phthalate in the diet of rats for 26 d gave the same results (33). A lower-molecular-weight resin (mol. wt. = 350, EPON® resin 828) was cured separately with three different curing agents (boron fluoride, metaphenylenediamine, and EPON® curing agent E); 1, 5, and 10% of each cured resin was added to the diet of rats for 6 wk (33). There was no evidence of behavioral change during treatment, and no diarrhea or other abnormalities were seen. Rats in the 10% groups ate more than controls, but actual "food" intake was about equivalent to controls. There were no statistically significant differences from controls for growth rate or for liver/body and kidney/body weight ratios. The average amount of cured resin consumed by each rat was 2.4 g. In one subchronic dietary study, rats were fed BADGE in their diets for 3 mo at concentrations up to 3% (33a). Rats at the highest level rejected the diets and failed to gain weight; these rats showed effects upon gross and histopathological examination that were consistent with under-nutrition. There was no evidence of systemic toxicity at any levels. In another subchronic study, BADGE was fed to rats at dietary concentrations of 0.2, 1, or 5% for 26 wk (12). All rats at the highest dose died by the end of 20 wk, but gross and histopathological examination did not reveal evidence of systemic toxicity at any dose (12). A 28-d study with a low-molecular-weight BADGE-based resin (Araldite GY250) in which rats were gavaged daily with the resin in an aqueous solution of 0.5% carboxymethylcellulose/0.1% Tween 80 at doses of 0, 50, 200, or 1000 mg/kg/d did not alter any of the following parameters relative to controls: body weight, food consumption, water consumption, food conversion, mortality, clinical observations, eyes or hearing, hematology, blood chemistries, organ weights, and gross pathology or histopathology of the spleen, heart, liver, kidney, or adrenal gland (34).

The effects of incorporating solid resins into the diet of rats have also been studied. A resin of molecular weight 950 (EPON® resin 1001) was cured separately with two different curing agents (diethylenetriamine and Versamide 115), and 1, 5, and 10% of each cured resin was added to the diet of rats for 6 wk (33). There were no adverse effects noted in any group with respect to behavior, toxic effects, growth rate, or liver/body and kidney/body weight ratio. There was increased food consumption in the 10% groups. No gross or microscopic lesions were found. A resin of molecular weight 3000 (EPON® resion 1007) was added at 2% to the diet of rats for 12 wk as the uncured resin, cured with diethylenetriamine, and cured with urea–formaldehyde resin (33). An unspecified number of deaths occurred in each group, including controls, which were described as being unrelated to treatment. The growth rate of treated groups was not statistically different from control, and no other toxic effects were described.

Dermal Administration. A BADGE-based resin (Epidian® 5) was applied at 6.8 mg/d for 17 d to pregnant and nonpregnant rats (35, 36). Local effects (erythema, edema, and erosions) were seen on the skin of the treated sites with all groups. Nonpregnant rats had reduced body weight, hyperemia of livers and kidneys, and elevated kidney and liver enzymes. There was a decrease in the brain acetylcholinesterase activity of nonpregnant (but not pregnant) rats treated with Epidian® 5 reported in one study (35), but in the other study by these same investigators neither pregnant nor nonpregnant rats had reduced acetylcholinesterase activity (36). In pregnant rats, there was a slight body weight gain, atrophy of adipose tissue, reduced liver and kidney enzymes, hyperemia of liver and kidney, elevation of other enzymes, and slight changes in amniotic fluid compared to controls. In other related studies by these investigators, pregnant guinea pigs treated dermally with Epidian® 5 (36, 37) appeared more susceptible to the actions of the resin than the nonpregnant ones.

BADGE was also evaluated for its potential to produce systemic toxicity and neurotoxicity in Fischer 344 rats following approximately 13 wk of repeated dermal application in two studies conducted concurrently (38–40). BADGE was applied to the skin of male and female rats five times per week at doses of 0, 10, 100 or 1000 mg/kg/d. BADGE caused no apparent systemic toxicity with the exception of decreased body weight in males and females at 1000 mg/kg/application, accompanied by a decrease in food consumption. Serum cholesterol values were increased in a dose-related manner in the mid- and high-dose groups of both sexes but were not correlated with any changes in histopathology. The increases in serum cholesterol were likely a result of the decreases in body weight occurring at these doses as this effect has also been observed in animals with body weight loss due to feed restriction (41). Dermal application of BADGE did not alter any of the following parameters relative to controls: mortality, clinical observations and behavior, gross pathology or histopathology, with exception of a chronic dermatitis at the site of BADGE application. This dermatitis was characterized by very slight to moderate epidermal hyperplasia with very slight to slight chronic inflammation, and, in some animals, very slight to slight follicular or epidermal spongiosis. Males tended to be affected slightly more severely than females, perhaps due to the higher concentrations of BADGE applied in order to achieve the same mg/kg dose in a fixed volume. BADGE showed no evidence of neurotoxicity in the treated animals, which included evaluation in a functional observational battery (hand-held and open field observations, grip strength, and landing foot splay), motor activity, evoked potential testing (visual, auditory, somatosensory systems and conduction velocity of caudal nerves), and a comprehensive neuropathological examination.

BADGE in acetone applied to the skin male $B6C3F_1$ mice three times per week for 13 wk at dosages of 1, 10 or 100 mg/kg/application caused no apparent systemic toxicity (42). Data were collected on clinical appearance and behavior, gross dermal irritation at the dose site, body weights, food consumption, clinical pathology, gross pathology, and histopathology. Mild to moderate chronic active dermatitis with a weak dose–response relationship was observed histopathologically. Spongiosis (intraepithelial edema in the epidermis or at the neck of the hair follicle) and epidermal microabscess formation were also observed at the 100 mg/kg dose.

Intramuscular Administration. An undiluted liquid resin of molecular weight 350 (EPON® 828) was injected intramuscularly in rats at 800 mg/kg once a day for 12 d (43). The only significant effect was a reduction in the rate of body weight gain.

1.4.1.3 Pharmacokinetics, Metabolism, and Mechanisms. *In vivo*, BADGE was very slowly absorbed through the skin of mice (44). *In vitro*, full-thickness mouse skin was more permeable to BADGE than was dermatomized rat or human skin, with human skin showing only about 0.1 to 0.2% of applied dose absorbed within 24 h with an average lag time of about 6 h required for penetration (45). During skin penetration in all the species studied, BADGE was extensively metabolized to its *bis*-diol (45).

Following a single oral administration of ^{14}C-BADGE to mice, the dose was relatively rapidly excreted as metabolites in the urine and feces and the profile of fecal and urinary metabolites was independent of the route of exposure (44). The major metabolite was the bis-diol of BADGE formed by hydrolysis of epoxides by epoxide hydrolase; bisphenol A was not found as a metabolite of BADGE (16). The *bis*-diol was excreted in both free and conjugated forms and was also further metabolized to various carboxylic acids (46). BADGE did not appear to be metabolized to phenyl glycidyl ether by mice. Metabolic pathways for BADGE in the rabbit appear similar to those described for the mouse (47).

Route-dependent differences in the plasma ^{14}C concentration–time profiles, tissue/plasma ^{14}C ratios, and urinary excretion following the intravenous or oral administration of ^{14}C-BADGE to rats have been observed (48). These data suggest that very little BADGE is absorbed unchanged following oral administration (48). The primary route of excretion in the rat was the feces after either intravenous or oral administration, although the plasma data suggest that only 13% of the orally administered radioactivity was absorbed so that some of the fecal radioactivity following oral administration may represent unabsorbed material (48). Consistent with the observations of Climie et al. (46), no unchanged BADGE was excreted in the urine or bile following oral or intravenous administration.

Bentley et al. (49) have investigated the hydrolysis of the epoxide functionalities of BADGE by the microsomal and cytosolic fractions of mouse liver and skin. These investigators reported that BADGE was rapidly hydrolyzed by the eposide hydrolase of both tissues, with skin microsomal activity toward BADGE about 10 times greater than that found in the cytosol of skin.

In vitro metabolism studies by Boogaard (45) have shown that BADGE was very rapidly hydrolyzed by either cytosolic or microsomal fractions of liver and lung derived from rats, mice, and human. The cytosol and microsomes derived from human tissues had a higher hydrolytic efficiency towards BADGE than did those derived from rat or mouse tissues, with the exception being mouse liver cytosol, which showed about twice the activity found in humans. Microsomal activity was greater than cytosolic activity for both tissues for all species except the mouse, where microsomal and cytosolic activity was about equivalent. Only a small degree of chemical reaction and some inconsistent enzyme-catalyzed conjugation of BADGE with glutathione were observed.

Bentley et al. (49) also reported that the dermal administration of ^{14}C-BADGE (radiolabeled in the glycidyl side chain) resulted in radioactivity being covalently associated with the protein, DNA, and RNA purified from the skin at the site of application. Most of this radioactivity appeared to be a result of the metabolism of the glycidyl side chain to glyceraldehyde, a normal endogenous product of intermediary metabolism. Glyceraldehyde was subsequently metabolized to single carbon units that entered the one-carbon pool and were then incorporated into tissue macromolecules via normal catabolic pathways. These findings are consistent with those of Climie et al. (46) who could find no evidence for the *in vivo* metabolism of BADGE to glycidaldehyde in mice. However, Bentley et al. (49) did report the formation of small amounts of a DNA adduct that was tentatively identified as the reaction product of glycidyladehyde and deoxyguanosine when doses of 0.8 or 2 mg/mouse of ^{14}C-BADGE were applied to the skin of mice (49). However, Bentley et al. based their identification on the liquid chromatography retention volume of the unknown peak as compared to a known standard and did not conclusively identify this "adduct" using other analytical techniques such as mass spectrometry. No adducts were found at the dose level of 0.4 mg BADGE/mouse, the lowest dose level used. Subsequently, the work of Steiner et al. (50) conclusively identified hydroxymethylethenodeoxyadenosine-3′-monophosphate as an adduct formed in the DNA isolated from the epidermis of mice dosed topically with BADGE dissolved in acetone at 2 mg/kg. This adduct appears to be formed at levels of 0.14–0.8 adducts per million nucleotides, about 1000-fold lower than the production of this adduct from an equivalent dose of glycidaldehyde.

1.4.1.4 Reproductive and Developmental. *Developmental Toxicity.* Low-molecular-weight BADGE-based epoxy resin was not teratogenic in rats or a chick embryo assay but was embryo toxic at doses of 10% of the oral LD_{50} (51). The reference cited for this statement is an abstract and does not provide any information regarding specific chemicals tested. In a dermal teratology probe study, rabbits were administered doses of 0, 100, 300, or 500 mg/kg/d on days 6–18 (52). No embryo toxicity was observed at any dose. The full teratology study, conducted at dermal doses of 0, 30, 100, or 300 mg/kg in the rabbit, showed no evidence of embryo/fetal toxicity or teratogenicity (23). Gavage teratology studies using both rats and rabbits with a low-molecular-weight BADGE-based epoxy resin (Araldite GY250 or TK10490) have also been conducted (53, 54). Dose levels of 0, 60, 180, and 540 mg/kg/d were used for rats, and dose levels of 0, 20, 60, and 180 mg/kg/d were used for rabbits. The test material in both studies was suspended in an aqueous solution of 0.5% carboxymethylcellulose and 0.1% Tween 80 (polysorbate 80). Treatment at the top dose levels resulted in signs of material toxicity in both studies, but there were no adverse effects on mean litter size, pre- and postimplantation losses, or any evidence of a teratogenic or embryotoxic effect at any dose level.

Reproductive Toxicity. A one-generation reproduction study in rats has been conducted in which a BADGE-based epoxy resin (Araldite GY250 or TK 10490) was administered by gavage at dose levels of 0, 20, 60, 180, and 540 mg/kg/d (55). The vehicle used was an aqueous solution of 0.5% carboxymethylcellulose, 0.1% Tween 80. Oral administration of this resin to males for 10 wk and females for 2 wk prior to mating produced a lower mean body weight in males at 540 mg/kg/d, but did not affect mating performance, gestation period, or the ability of females to rear their offspring successfully to weaning (55). No treatment-related macroscopic changes, differences in mean organ weights, or histological changes to the reproductive and alimentary tracts (top dose only) in either sex of the F_0 generation were observed.

Pure BADGE (>99%) was studied to examine its potential to adversely affect reproduction when given orally to laboratory rats over two generations (56, 57). Both males and females were dosed with solutions of BADGE by gavage, once a day, 7 d/wk at doses of 0, 50, 540 or 750 mg/kg/d. The first generation was dosed for a total of 34 wk (a long treatment period for a study of this type), and the highest dose used was about 15,000 times greater than the potential human exposures from consumer applications. At the two higher doses, body weight was decreased approximately 8–11% in the adult males of both generations; adult females of both generations showed a similar decrease in body weight but only at the highest dose. Despite the extremely high doses used in this study, there were no treatment-related effects on reproductive parameters in any dose group. The ability successfully to mate, conceive, deliver, and rear their offspring (two sets of litters produced by the P_0 adults, one by the F_1 generation) as well as the growth and survival of the offspring were unaffected by BADGE treatment. In addition to looking at reproductive endpoints, the study included the microscopic examination of all the tissues from ten females and ten males of the first generation adults by a veterinary pathologist. No effects were found in any of the tissues, with the exception of some irritation and inflammation in nasal tissue, which could be attributed to gastric reflux of the dosing solution as a result of repeated gavage administration.

In summary, studies of the potential of BADGE to produce reproductive toxicity have not shown adverse effects on reproduction.

1.4.1.5 Carcinogenesis. Older studies in which cured and uncured, solid and liquid BADGE-based epoxy resins were orally administered at concentrations up to 10% in the diet produced no adverse effects in rats with exposures up to 26 wk (12). Neither did the oral administration in the diet or intraperitoneal or subcutaneous injection of solid and liquid epoxy resins into mice susceptible to lung tumors cause any statistically significant increase in incidence of lung tumors compared to negative controls (58). The subcutaneous implantation of 1 × 1 cm disks of cured resin in mice for 575 d did not result in tumors (59). Other, older studies where EPON® resin 828 was repeatedly injected subcutaneously into rats caused 4 of 30 animals to develop fibrosarcoma at the injection site (12), but this result is not felt to be surprising because other materials not considered to be carcinogenic also produce this same type of tumor under the same conditions (e.g., nylon, table salt, glucose).

Information more relevant to the potential occupational exposure to epoxy resins is provided by the numerous studies involving long-term skin application of BADGE-based resins. In an early study, these materials were applied to the skin of mice as acetone solutions (0.3 to 5%), three times weekly for 2 yr. A solvent control as well as a positive control group were also included. There was no increase in the incidence of grossly detectable skin tumors in any of the resin-treated groups (12).

In another skin painting study in mice, "one brushful" of undiluted resin was applied to the skin of C3H mice three times for up to 23 mo. A skin papilloma was detected in a single mouse after 16 mo of treatment, at which time 32 of the 40 mice started on the study were still alive (13). When the study was repeated, twice for 24 mo and once for 27 mo, no skin tumors were found (160).

Holland et al. (62) investigated the carcinogenic potential of a modified commercial EPON® 828–type resin. The test material was applied as a 50% solution in acetone to the skin of C3H and C57BL/6 mice of both sexes, three times weekly for 2 yr at a dose of 15 or 75 mg/kg/wk. No skin tumors were found in the C3H mice, but a weak carcinogenic response was noted in the C57BL/6 strain. However, it was subsequently found that the resin sample used in this test contained a typically high levels of several active contaminants, including epichlorohydrin (1500 ppm), phenyl glycidyl ether (830 ppm), and diglycidyl ether (3400 ppm), as well as about 10% of a presumed diluent identified only as an epoxidized polyglycol. In view of the presence of the contaminants, and particularly the high level of the epoxidized polyglycol, the weak carcinogenic response noted in the one mouse strain cannot be clearly ascribed to the resin.

In a subsequent study by the group at Oak Ridge National Laboratory (62), a low-molecular-weight BADGE-based epoxy resin and two comparable commercial resins from other manufacturers were evaluated in the C3H mouse following the protocol used in the earlier study. None of the three resins elicited skin or systemic tumors in the test animals.

Of potentially greater interest is the fact that a 1:1 mixture of BADGE-based resins and *bis*-(2,3-epoxycyclopentyl) ether used as an epoxy diluent produced a clear increase in the incidence of skin tumors despite the fact that neither compound gave positive results when applied individually (62). A recent study has provided evidence that mechanisms of this effect may be related to the ability of *bis*-(2,3-epoxycyclopentyl) ether to inhibit microsomal epoxide hydrolase activity, which results in DNA binding by a metabolite of BADGE (glycidaldehyde) formed via a different metabolic pathway (49). These data add further emphasis to recommendations for the proper handling of epoxy resins (particularly with respect to diluted resins), and they should be considered in establishing industrial hygienic practices, recommending protective equipment, and advising remedial action in case of overexposure.

The carcinogenic potential and chronic dermal toxicity of three commercially available BADGE-based resins were also investigated by Agee et al. (63). The test materials were dissolved in acetone and 50 mL applied topically, twice a week for 94 wk, to the backs of C3H/HeJ male mice, 50 per treatment group. The three BADGE-based resins tested were 42% BADGE (Code C-618), 76% BADGE (Code C-621), and 27% BADGE (Code C-660) (64) and were tested in acetone at concentrations of 50%, 25%, and undiluted, respectively. Thus the actual concentrations of BADGE applied were 21% (C-618), 19% (C-621), and 27% (C-660). Two groups of 50 mice each were treated twice weekly with 50 mL of acetone or 0.025% benzo[*a*]pyrene in acetone to serve as negative and positive control groups, respectively. An additional group of 50 mice received no treatment as a negative control group. The skin from all animals was examined by light microscopy for non-neoplastic and neoplastic lesions, and histopathological examination of internal organs was conducted on half of the mice from each group. Forty-eight of the mice in the positive control group developed skin tumors with an average latent period of 32.4 wk, whereas no skin neoplasms were observed in either of the negative control groups or in the groups treated with resins in acetone at a final concentration of 19 or 27% BADGE (i.e., C-621 and C-660). Three mice of the 50 treated with C-618 in acetone (21% BADGE) had microscopically detected skin papillomas, but no malignant neoplasms of the skin were present in any of the animals in this treatment group. The incidence of hepatocellular carcinoma observed in the treated and control groups was within the range of those detected in historical control animals from the same laboratory and below values reported by the animal supplier for this strain of mouse. Thus under the conditions of this study, dermal application of these BADGE-based resins did not produce a carcinogenic response in mice.

Zakova et al. (65) evaluated the carcinogenic potential of Araldite GY250 a BADGE-based epoxy resin in CF_1 mice. Groups of 50 male and 50 female mice were treated for 2 yr by repeated epidermal application of a 1 or 10% (v/v) solution in acetone. Controls were treated with acetone alone. The treatment had no effect on survival, and no excess incidence of skin or systemic neoplasis occurred.

Comprehensive studies on the carcinogenic potential of the low-molecular-weight BADGE-based resins applied dermally have been reported. Groups of CF_1 mice of each sex were exposed to pure BADGE, EPON® resin 828, or another comparable commercial resin. The test materials were applied as a 1 or 10% solution in acetone, 0.2 mL twice weekly, for 2 yr. Solvent (acetone) and positive (*b*-propiolactone) groups were also included (166).

The animals treated with *b*-propiolactone showed a high incidence of skin tumors in comparison to the solvent control groups, demonstrating the susceptibility of the CF_1 mouse to a chemical known to produce skin cancer. In the EPON® resin 828–treated mice, the incidence of cutaneous tumors of the treated site or of the skin at all sites was not statistically significantly different from controls. In the two other treatment groups, some skin tumors were observed, but statistical analysis of this tumor data revealed the incidence was not significantly different from controls. This skin tumor data were compared to the incidence of skin tumors in control CF_1 mice from two other chronic studies conducted in the Shell laboratory. Based on the low incidence of skin tumors in these "historical" control mice, the authors suggested that BADGE and one of the commercial resins, but not EPON® resin 828, exhibited a low order of carcinogenic potential to the skin of CF_1 mice. However, the historical control data that were used for comparison by the authors were very limited, with only 100 males and 200 females tested for 2 yr. The study of Zakova et al. (65) with a similar BADGE-based epoxy resin (Araldite GY 250), conducted by another laboratory at the same time using the same protocol and with CF_1 mice supplied by the Shell laboratory, did not result in an excess of either skin or systemic tumors (see above). When the results of these two studies were combined (including additional historical control data from Zakova et al.), Peristianis and co-workers concluded that there was no evidence of carcinogenic activity of these resins in mouse skin (66).

With regard to systemic neoplasia in the mice treated with acetone solutions of EPON® resin 828, there was a statistically significant linear trend in the dose response for renal tumors in male mice when the data were analyzed by the method of Peto et al. (67). The renal tumor incidences were 6, 0, and 12% in the control, low-dose, and high-dose groups, respectively. This finding is not considered to be related to treatment because the authors state that renal neoplasms in male CF_1 mice are common in the testing laboratory, and the absence of this tumor in the 1% EPON® resin 828 was rather unusual. Furthermore, if BADGE was truly a causative agent for renal neoplasia, one would have expected kidney tumor incidence to have shown a statistically significant increase in the test statistic for trend in dose response in all the groups treated with BADGE-containing resins. Yet there was not a statistically significant trend in dose response for renal neoplasia in male or female mice treated with EPIKOTE® resin 828, pure BADGE (66), or Araldite GY 250 (65), indicating the evidence for linking BADGE to renal neoplasia in mice is tenuous.

Peristianis and co-workers (66) also reported a statistically significant increase in the dose–response trend for lymphoreticular/hematopoietic tumors in female mice treated with pure BADGE or EPIKOTE® resin 828 when the data were analyzed by the method of Peto et al. (67). However, it is likely that this finding is not treatment related, because the CF_1 mice raised in this testing laboratory have a relatively high background incidence of these lesions. It was considered likely that the mice were susceptible to the development of lymphoreticular/hematopoietic tumors as a result of the presence of virus and/or a genetic tendency to viral infection. There was no statistically significant increase in the dose–

response trend for lymphatic tumors in female or male CF_1 mice treated with other BADGE-based resins, neither EPON® resin 828 (66) nor Araldite GY 250 (65), suggesting that BADGE was not the causative agent for these lesions.

Pure BADGE (>99%) was further studied as to its potential to chronic toxicity and oncogenicity when applied to the skin of female F344 rats and male $B6C3F_1$ mice for 2 yr (68, 69). Mice were dosed 3 d/wk with BADGE in acetone at concentrations of 0, 0.005%, 0.5% or 5% w/v; rats were dosed 5 d/wk with BADGE in acetone at concentrations of 0, 0.6%, 6.0% or 60% (w/v). The cumulative dose in rats was higher than any dose previously studied (70 animals per dose level) approximately 0, 1, 100, or 1000 mg/kg/application. Mice (70 animals per dose level) received approximately 0, 0.1, 10, or 100 mg/kg/application. Histopathologically, BADGE was found to cause a mild chronic dermatitis in rats at the dermal test site in the 100 and 1000 mg/kg/application groups. In mice, BADGE produced slight to severe epidermal hyperplasia, slight to moderate chronic or chronic active dermal inflammation, and epidermal crusts were observed histopathologically at the 10- or 100-mg/kg/application doses. In a few mice at the two highest doses, epidermal ulcers were observed, indicating the maximum tolerated dose had been exceeded (70, 71). The irritation and inflammation of the skin persisted throughout the course of the study in both rats and mice and was dependent upon the dose of BADGE applied to the skin. Specifically, no irritation was seen at a dose of 1 mg/kg/d in rats and at a dose 0.1 mg/kg/d in mice. In rats, there was a very slight increase in the incidence of centrilobular hepatocyte hypertrophy at 1000 mg/kg/d and an increase in several types of foci of cellular alteration in the liver at 100 and 1000 mg/kg/d. Serum alanine aminotransferase, alkaline phosphatase, and aspartate aminotransferase tended to be slightly increased at the 1000-mg/kg/d dose throughout the study and taken together with the histological observations in liver suggest a chronic low-level hepatotoxicity at the 1000-mg/kg/d dose. One mouse given 10 mg/kg/application had a squamous cell carcinoma of the skin at the test material application site, which was not treatment related, as no tumors were found at the site of treatment in the highest-dose group and the incidence for this study was within the historical control incidence for $B6C3F_1$ mice. In summary, BADGE did not produce neoplasia in any tissue or organ in either rats or mice at any of the doses applied in either of these studies.

In summary, a number of carcinogenicity studies involving the topical application of pure BADGE, as well as EPON® resin 828 and other commercial BADGE-based resins have been carried out in experimental animals. Viewing the studies as a whole, the weight of evidence does not show that BADGE or BADGE-based epoxy resins are carcinogenic.

1.4.1.6 Genetic and Related Cellular Effects Studies. Both liquid and solid BADGE-based resins have been tested in several *in vitro* and *in vivo* assays for genotoxicity; the results of these tests have been reviewed (72). *In vitro* mutagenicity assays for these materials have given mixed results, whereas all the *in vivo* tests for mutagenicity have been negative.

1.4.2 Human Experience

Occupational dermatitis from epoxy resins has been described by a number of workers, including Hine et al. (12), Pluss (73), and Grandjean (74). The usual lesion is typical of contact dermatitis, the early manifestations being redness and edema, with weeping followed by crusting and scaling. Following initial contact, there is usually an erythematous, discrete area confined to the point of contact. Because this frequently involves the face, it is likely that it is caused by the vapors of the hardener or active diluent, though contact with contaminated gloves or droplets may also play a part, as do occasionally the vapors of the liquid-type epoxy resin. The initial lesion usually persists for 48 h to 10 d, after which the erythema fades and gives way to a macular rash followed by scaling.

Sensitization may follow the initial contact, resulting in the development of a papular, vesicular eczema. This is accompanied by considerable itching, and extension into areas beyond the point of original contact. Only occasionally, areas other than the backs of the hands, the forearms, the face, and the neck are involved. Recommended treatment consists of bland ointments and soaps. The worker is withdrawn from further contact, and the lesion usually subsides in 10–14 d. However, it may recur on further contact. If the worker is not withdrawn from contact, the dermatitis usually persists for longer periods, but usually does not become more intense. The lesions may assume a brownish color, and scaling is frequently noted.

Fregert and Thorgeirsson (75) reported that of 34 individuals previously experiencing skin sensitization as a result of occupational exposure to epoxy resins, all demonstrated a positive skin sensitization response to a BADGE-based epoxy resin of average molecular weight (MW) 340 following patch testing. Twenty-three of these individuals patch tested with resins of average MW 624 and MW 908 did not experience sensitization (75), and seven of these individuals patch tested with a resin of average MW 1192 did not react. However, eight patients tested with commercial mixtures of epoxy resins with average MW 1280 and MW 1850 reacted to these mixtures, which contained the MW 340 oligomer as determined by gel permeation chromatography (75). The authors concluded that the MW 340 oligomer is the component responsible for contact allergy to epoxy resins in humans. These results in humans are consistent with the absence of skin sensitization in guinea pigs for higher-molecular-weight resins (76, 77).

Because of the persistence of dermatitis and the possibility of sensitization, preventive measures are especially important when handling these compounds. The most important measure for combating dermatitis is good personal hygiene. This requires strong administrative controls, supervisory instruction of personnel, good work habits, and provision of adequate facilities for removing the material periodically, together with a designated cleanup period prior to "break" and "quitting" times. In addition, protective devices offer considerable aid in minimizing personal contact. Gloves should be worn, and contamination of the skin should be scrupulously avoided. Protective clothing and personal protective creams are of help. When possible, only persons with no history of allergic conditions or eczematous eruptions should be selected for epoxy resins applications. Procedures should be carefully supervised to ensure that mixing of volatile agents is carried out in properly ventilated area. Bench and floor areas should be protected with disposable paper.

Remedial measures following skin contact should include thorough cleansing with soap and water, followed by a waterless hand cleanser when absolutely necessary. The use of solvents may promote epidermal penetration of materials that would otherwise not penetrate the skin. Accidental eye contamination is unlikely, but treatment consists of the usual measures. The source of contact should be identified when dermatitis develops, and the improper work condition corrected.

Reference has been made to occupational asthma in workers exposed to "fumes" of epoxy resins (78). However, a closer examination of the literature revealed that the occupational asthma attributed to an "epoxy resin system" was due to fumes of phthalic anhydride, trimellitic anhydride, or triethylenetetramine, and that sensitized workers did not respond to the epoxy resin when exposed to that material alone (79). It should be further noted that it is unsure whether the epoxy resin contained unreacted BADGE. Kanerva et al. (80) have published a case report describing two individuals with immediate and delayed allergy to epoxy resins based on BADGE, one of these patients showed a positive response in the prick and patch tests for allergy to the phthalic anhydrides used for curing the resins.

1.5 Standards, Regulations, or Guidelines of Exposure

No standards of permissible occupational exposure to BADGE have been established.

2.0 Diglycidyl Ethers of Brominated Bisphenol A

2.0.1 CAS Number: *[26265-087-7, 40039-93-8]*

2.0.2 Synonyms: Tetrabromobisphenol A based epoxy resins (TBBAER). The lowest-molecular-weight component of this class of resins is tetrabromobisphenol A diglycidyl ether (TBBADGE).

2.0.3 Trade Names: Some of the trademarks for the polymeric brominated resins are D. E. R.® epoxy resin 500 series, EPON® resin 1000 series, Araldite LT 8000 series, and Epikote 1183.

2.0.4 Molecular Weight: 800–>4000

2.0.5 Molecular Formula: NA

2.0.6 Molecular Structure:

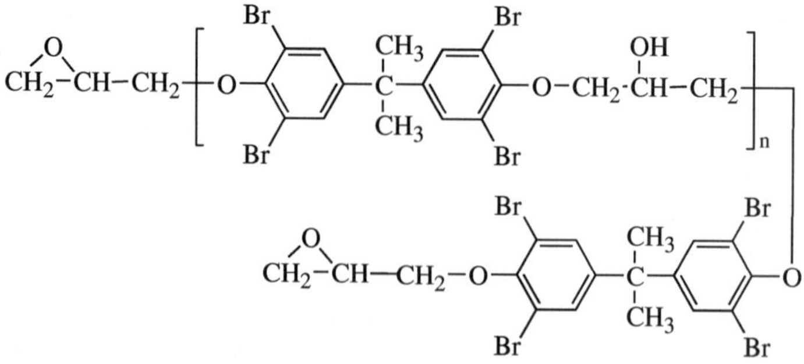

2.1 Chemical and Physical Properties

Epoxy resin content: 2500– < 1000
Appearance: yellowish powder
Specific gravity: 1.8
Bromine content: 50–52%
Melting/softening point: 60– >200°C

2.1.2 Odor and Warning Properties

TBBAER are solids and they have no odor nor warning properties.

2.2 Production and Use

These resins are the condensation product of 2,2-*bis*(3,5-dibromo-4-hydroxyphenyl) propane and epichlorohydrin. Bromine has been incorporated into BADGE-type resins to increase the ignition resistance of such resins by the use of tetrabrominated bisphenol A for the synthesis of the glycidyl ethers. Because bromine substitution reduces the thermal stability of the resins, the products are usually used with or condensed with bisphenol A. They therefore contain less than the theoretical amount of bromine indicated by the chemical formula (18–48% Br). Brominated epoxy resins are used as reactive flame retardants and are applied in electrical and electronics parts and equipment.

2.3 Exposure Assessment

Like the BADGE-based resins, inhalation exposure to vapors from these resins is unlikely because these materials are essentially nonvolatile. In the event of combustion, bromine and hydrogen bromide vapors may be released from these resins.

2.4 Toxic Effects

The acute systemic toxicity of resins based on the diglycidyl ether of tetrabromobisphenol A (TBBAER) is very low. Only slight skin and eye irritation was observed in rabbits treated with TBBAER. Like several other epoxy resin materials, the lower-molecular-weight TBBAER caused delayed contact skin sensitization in guinea pigs; higher-molecular-weight resins (epoxy resin content ca. 1400 mmol/kg) have no skin sensitization potential. In *in vitro* genotoxicity tests low-molecular-weight TBBAER have shown a genotoxic potential; for higher-molecular-weight resins epoxy resin content ca. 1400 mmol/kg) no mutagenic response in bacteria was observed. TBBAER did not induce chromosomal aberrations in the bone marrow of rats given repeated dermal doses of 1000 mg/kg.

2.4.1 Experimental Studies

2.4.1.1 Acute Toxicity. The acute toxicity of TBBAER has been observed to be very low. The acute oral and dermal LD_{50}s have been reported to be >12,000 and 6000 mg/kg, respectively (81). In other studies, TBBAER had an oral and dermal LD_{50} of >2000 mg/kg in the rat or the rabbit; no signs of systemic intoxication were observed (82–85). A 1-h exposure of four female rats to an atmosphere of dust generated at room temperature (calculated to contain 0.33 mg TBBAER/L air) did not produce any apparent adverse effects with no visible lesions upon pathological examination 14 d postexposure (82).

Skin Effects. TBBAER was found to be essentially nonirritating to the skin of rabbits (82, 84–86).

In studies to determine the skin sensitization potential of TBBAER, eight female Hartley strain guinea pigs received a total of 10 intradermal injections of TBBAER. Fourteen days after the final injection, the animals were challenged with an intradermal dose of TBBAER. None of the animals exhibited any reaction to the challenge dose at 24 h postchallenge (81). In addition, TBBAER was not found to be a skin sensitizer in guinea pigs when tested topically according to a modified Buehler test (87). In a guinea pig maximization test according to Magnusson and Kligman, 50% of the animals responded after an epicutaneous challenge with low-molecular-weight TBBAER (epoxy resin content in excess of ca. 1800 mmol/kg) (84, 88). Using the same test protocol, guinea pigs induced with TBBAER (epoxy resin content below 1400 mmol/kg) did not exhibit any responses following an epicutaneous challenge (85).

Eye Irritation. TBBAER was found to be essentially nonirritating to the eyes of rabbits (82, 84–86).

2.4.1.6 Genetic and Related Cellular Effects Studies. TBBAER was mutagenic in *S. typhimurium* strain TA100, but not in strains TA98, TA1535, TA1537, or TA1538. In the presence of metabolic activation, the mutagenic response in TA100 was eliminated at all concentrations tested (89). In another bacterial assay, TBBAER (epoxy resin content ca. 2500 mmol/kg) tested positive in *S. typhimurium* TA100 and TA1535 in the presence of a rat liver activation system (S9); however, in the absence of S9 no response was observed. There was no evidence of mutagenic activity in *S. typhimurium* TA98, TA1537, TA1538 and *E. Coli* WP2 uvrA pKM101 either in the absence or the presence of S9 (90). A negative response was obtained when TBBAER with an epoxy resin content ca. 1400 mmol/kg was tested in *S. typhimurium* TA98, TA1535, TA1537, TA1538 in the absence or the presence of S9 (91). Another study also reported a lack of activity in *S. typhimurium* when tested with and without an activation system (92). TBBAER was tested in the presence and absence of metabolic activation for its potential to induce chromosomal aberrations in cultured Chinese hamster ovary cells. A dose-related increase in the percent of cells with chromosomal aberrations was observed, with a greater response found in the presence of metabolic activation (93). TBBAER was negative when tested in BALB/C-3T3 cells to determine its potential to induce morphological transformation (94). Five daily dermal doses of 1000 mg/kg of TBBAER did not induce chromosomal aberrations in the bone marrow of rats (95).

2.4.2 Human Experience

Nine individuals known to be sensitized to bisphenol A diglycidyl ether (BADGE) were patch tested with TBBAER known to be free of residual low-molecular-weight BADGE; all subjects had negative reactions (96). Kanerva et al. (97) have reported one individual had a positive "patch test" for dermal sensitization to a brominated epoxy resin (Rutapox 0451) but showed a negative prick test and negative RAST (specific IgE determination).

2.5 Standards, Regulations, or Guidelines of Exposure

No standards of permissible exposure to TBBAER were found.

3.0 Resorcinol Diglycidyl Ether

3.0.1 CAS Number: [101-90-6]

3.0.2 Synonyms: 2,2′-[1,3-Phenylenebis(oxymethylene)]*bis*-oxirane; 1,3-*bis*(2,3-epoxy propoxy) benzene; *m*-bis(glycidyloxy)benzene; and RDGE

3.0.3 Trade Names: NA

3.0.4 Molecular Weight: 222.24

3.0.5 Molecular Formula: $C_{12}H_{14}O_4$

3.0.6 Molecular Structure:

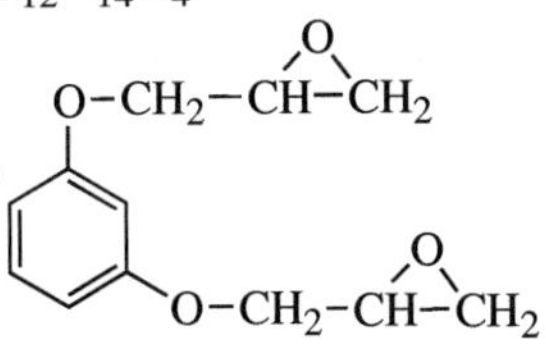

3.1 Chemical and Physical Properties

Resorcinol diglycidyl ether (RDGE) is a straw yellow liquid with a slight phenolic odor and following properties:

Molecular Structure: $C_{12}H_{14}O_4$
Specific gravity: 1.21
Melting point: 36.5–46°C
Boiling point:
  0.05 mmHg: 172°C–160°C
  12 mmHg: 208–210°C
Vapor density: 7.95
Refractive index (20°C): 1.5409
Epoxy number: 110

3.1.2 Odor and Warning Properties

The phenolic odor is easily perceptible.

3.2 Production and Use

Resorcinol diglycidyl ether is made by the reaction of epichlorohydrin and resorcinol in the presence of caustic. It is suggested for use as an epoxy resin, as a stabilizer of organic chemicals, as a curing agent for "Thiokol" rubber, and for the solubilizing of protein adhesives.

3.3 Exposure Assessment

Inhalation exposure to vapors from these resins is unlikely because these materials are essentially nonvolatile. Although dust conditions have not been encountered, sampling could be carried out by entrapment of airborne particles by standard filter paper or impinger techniques. Subsequent analysis may be carried out through the HCl–dioxane method (11).

3.4 Toxic Effects

RDGE is low in acute oral, dermal, and inhalation toxicity but moderately to severely irritating on skin contact. RDGE is severely irritating and damaging to the eye. RDGE is a skin sensitizer. Repeated exposure to RDGE causes significant irritation at the site of contact. Chronic oral exposure resulted in severe irritation of stomach and formation of local tumors in rodents. RDGE was mutagenic in *in vitro* systems, but mixed responses were observed *in vivo*.

3.4.1 Experimental Studies

3.4.1.1 Acute Toxicity. The single-dose oral LD_{50} values for rats, mice, and rabbits, respectively, were 2.57, 0.98, and 1.24 g/kg (12). Intraperitoneally the LD_{50} values were 0.178 and 0.243 for rats and mice, respectively (12).

The percutaneous LD_{50} in the rabbit was 2.0 mL/kg when RDGE (60% in xylene) was applied to the skin but not occluded (98). When RDGE remained in continuous contact with the skin, the percutaneous LD_{50} was 0.64 mL/kg (98).

Air "saturated" with RDGE vapors did not produce death to mice or rats after 8 h of exposure (12). Rats exposed to a concentrated aerosol of 44.8 mg RDGE (60% in xylene) per liter of air for 4 h died within 5 d postexposure (98).

Skin Effects. Single applications of RDGE to the skin of rabbits resulted in moderate irritation with a Draize score of 5 out of a possible 8 (12). Repeated applications were severely irritating with a leathery appearance of the skin and a Draize score of 7 out of 8 (12). Topical application of 0.01 mL of a 10% solution of RDGE in acetone to the skin of five rabbits produced a definite erythema and edema; 0.5 mL of RDGE (60% in xylene) applied topically to rabbits for 24 h produced severe irritation which progressed to necrosis (98).

Eye Irritation. The score by the Draize method was 45/110; the resin was severely irritating (12). In another study, installation of 0.5 mL RDGE (60% in xylene) in the rabbit eye resulted in severe inflammation and corneal necrosis (98).

3.4.1.2 Chronic and Subchronic Toxicity. In an early investigation, there was no evidence of toxicity as measured by body weight and organ weights in rats exposed for 50 7-h exposures to air saturated with RDGE vapors (12). However, seven repeated skin applications of 1 mL total dose caused mortality in rabbits (12).

Monkeys receiving 100–200 mg/kg intravenously once monthly showed a progressively increasing depression of the white blood cell count (99). Fourteen-day and 13-wk repeated dose gavage toxicity studies in rats and mice have been conducted by NTP. The test material was 81% pure and was gavaged in a corn oil vehicle. Dose levels were sufficient to result in increased mortality for both rats and mice in the 14-d study. Mean body weight was depressed in nearly all groups, and gross pathological examination of both species revealed lesions of the stomachs and renal medulla (reddening) with papillary growths in the stomachs of many of the dosed rats. In the 13-wk study, partial mortality (rats 1/20; mice, 16/20) occurred at the top dose levels of 200 and 400 mg/kg/d for rats and mice, respectively. Mean body weight was depressed in male rats dosed with 100 mg/kg and above and females dosed at 200 mg/kg; mice of the 400-mg/kg/d group depressed body weights. Compound-related observations in the nonglandular stomach of both rats and mice included inflammation, ulceration, squamous cell papilloma, hyperkeratosis, and basal cell hyperplasia at dose levels of 12.5 mg/kg and above in rats 25 mg/kg/d and above in mice. Some histopathological changes in the liver occurred in both rats and mice, including necrosis and fatty metamorphosis at the top dose levels only (100).

3.4.1.3 Pharmacokinetics, Metabolism, and Mechanisms. Intraperitoneal administration of RDGE results in an approximately 10-fold lower LD_{50} than observed following oral administration (12) suggesting RDGE has a low oral bioavailability. Following oral administration of RDGE in an aqueous 10% DMSO solution, it was shown to be metabolized in part to the *bis*-diol derivative (104). RDGE has also been shown to conjugate with glutathione via gluthathione-*S*-transferase *in vitra* (105).

3.4.1.5 Carcinogenesis. Older studies in which RDGE was applied repeatedly to the skin of mice or subcutaneously to rats for periods of 1 yr or longer were inconclusive due the small group sizes and lack of control groups (101, 102). In a more recent lifetime dermal study, 1% RDGE in benzene was applied to the skin of 30 female Swiss–Mellerton mice at a dose level of 100 mg dosing solution/mouse three times per week. No benign or malignant skin tumors were observed even though moderate to severe crusting and/or scarring and hair loss occurred at the site of application. The median survival time was 70 wk for the treated group, 71 wk for the negative control group (benzene-treated), and 63 wk for the untreated control group (103).

A 2-yr oral carcinogenicity study on RDGE has been conducted using rats and mice by NTP (100). The test material was an 81%-pure commercial product, dosed via gavage in corn oil five times a week for 2-yr. The dose levels for mice were 0, 50, or 100 mg/kg/d; the dose levels for rats were 0, 12, 25, or 50 mg/kg/d. At the 50-mg/kg/d dose level there was a significant decrease in body weights and a significant increase in mortality for rats. At the 25-mg/kg dose level there was a transient decrease in body weight in rats (at weeks 80–100), and the survival was significantly decreased. There were no effects on body weight at the 12 mg/kg/d dose level, but male rats had a significantly lower survival rate (46%) than controls (78%) at week 104. Histologically, RDGE produced hyperkeratosis, hyperplasia, and neoplasia of the squamous epithelium of the nonglandular stomach in all treated groups. Squamous cell carcinomas were observed in the nonglandular stomach; the respective incidence in the 0-, 12-, and 25-mg/kg/d group males was 0/100, 39/50, and 38/50; and the incidence in females was 0/99, 27/50, and 34/50.

3.1.4.6 Genetic and Related Cellular Effects Studies. RDGE was mutagenic in *S. typhimurium* strains TA100 and TA1535 with or without metabolic activation (104, 107). Seiler (104) also showed RDGE to produce chromosomal aberrations in Chinese hamster ovary cells *in vitro* at a concentration of 8 to 25 mg/mL, but reported that RDGE was negative in the mouse micronucleus test following oral gavage doses of 300 or 600 mg/kg. RDGE did not increase the number of micronuclei in the bone marrow following administration of intraperitoneal doses up to 91.2 mg/kg for 3 d, however IP administration of a single dose of RDGE at levels up to 270 mg/kg showed an increase in micronuclei (109).

3.4.2 Human Experience

Severe burns and skin sensitization on local contact have been observed (33). Leukopenia and the appearance of atypical monocytes in the peripheral blood have been reported in humans exposed in RDGE (110).

3.5 Standards, Regulations, or Guidelines of Exposure

No current recommendations for occupational exposure standards were found.

POLYGLYCIDYL ETHERS

4.0 Polyglycidyl Ethers of Phenolic Novolacs and Polyglycidyl Ethers of Cresolic Novolacs

4.0.1 CAS Numbers: Polyglycidyl ethers of phenolic novolacs: *[9003-36-5, 28064-14-4, 40216-08-8, 92183-42-1]*; Polyglycidyl ethers of cresolic novolacs: *[29690-82-2, 37382-79-9, 64425-89-4, 68609-31-4]*

4.0.2 Synonyms: The lowest-molecular-weight glycidated phenolic novolac is also referred to as bisphenol F diglycidyl ether (BFDGE).

4.0.3 Trade Names: These ethers and the condensation products of their reaction are marketed as Ciba-Geigy ECN® 1200 series, Dow D.E.N.® series 400, and Epikote® 155 or 180

4.0.6 Molecular Structure:

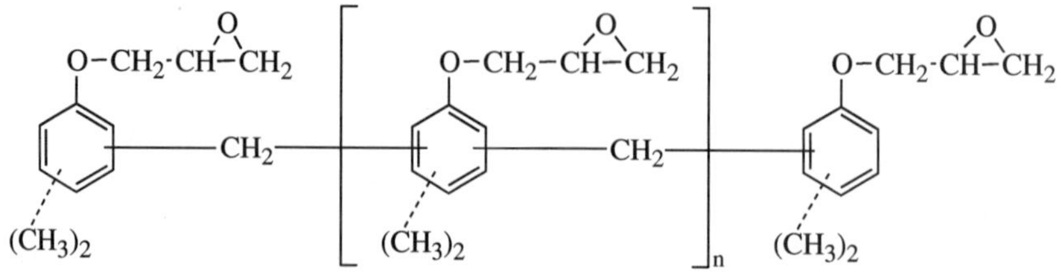

4.1 Chemical and Physical Properties

The properties of this type of resin are as follows:

Specific gravity (25°C): 122
Water solubility: Insoluble
Flash point (ASTM-D-1310-67): 490°F
Color, Gardner, Max.: 2
Viscosity (52°C): 20,000–50,000 cm/sec
Epoxide equivalent weight: 178*

Hazardous polymerization can occur with aliphatic amines in quantities greater than 1 lb.

4.1.2 Odor and Warning Properties

No information on odor or warning properties was found.

4.2 Production and Use

Ther term *novolac* refers to the reaction products of phenolic compounds and formaldehyde. Glycidation of the phenolic hydroxyl groups by epichlorohydrin under similar conditions to those used for forming the diglycidyl ether of bisphenol A produces the corresponding epoxy novolac resins. The average *n* in the above formula varies from 0.2 to 1.8, depending on the ultimate use for the resin. The epoxy novolac resins are used and cured in ways similar to the resins based on the diglycidyl ether of bisphenol A. These resins differ from bisphenol A epoxy resins in that whereas a bisphenol A polymeric chain can contain a maximum of two epoxide group regardless of chain length, cresolic and phenolic novolacs can contain an epoxide group for each phenol or cresol molecule incorporated in the chain.

Because of their high functionality, epoxy novolac resins, when cured, produce tightly cross-linked systems with improved high-temperature performance, chemical resistance, and adhesion over the resins based on the diglycidyl ether of bisphenol A. The thermal stability of epoxy novolac resins has made them useful for structural and electrical laminates and as coatings and castings for elevated temperature service. Owing to the chemical resistance of these resins, they are used for lining storage tanks, pumps, and other process equipment as well as for corrosion-resistant coatings.

4.3 Exposure Assessment

No standards of permissible occupational exposure were found. Like the diglycidyl ether of bisphenol A, these resins are essentially nonvolatile. Dusts and vapors from high-temperature reactions may be trapped by filter or liquid entrapment methods and determined by the HCl–dioxane method (11).

*Fairly constant for all resins unless modified by partial esterification.

4.4 Toxic Effects

The phenolic and cresolic novolac resins have a low order of acute oral and dermal toxicity potential. Both are minor irritants to the skin upon single contact. The phenolic resins have been studied for irritation potential following repeated application, and the results indicate appreciable irritation can develop following repeated or prolonged skin contact. Both resins can cause minor transient eye irritation. The phenolic novolac resins were shown to be weak skin sensitizers in humans. There are no sensitization data available for the cresolic novolac resins. The lowest-MW member of the class of phenolic novolac resins (BFDGE) showed only slight toxicity in rats when administered orally at levels up to 250 mg/kg/d for 13 wk. BFDGE was mutagenic in *in vitro* systems; however, BFDGE was negative in a mouse micronucleus assay. Cresolic novolac resin has been reported to cause bacterial gene mutaions.

4.4.1 Experimental Studies

4.4.1.1 Acute Toxicity. The phenolic novolac resins (EPNRs) have been shown to have low acute toxicity potential by oral route. Acute oral toxicity testing in rats shows that the oral LD_{50} was >4000 mg/kg for a liquid resin with molecular weight of 427 (111), whereas a solid EPNR advanced with pisphenol A of unknown molecular weight had an oral LD_{50} >2000 mg/kg (112).

Ciba-Geigy (Unpublished report of Ciba-Geigy, Limited, 1974) reported an an LD_{50} of >10,000 mg/kg in rats. The acute dermal toxicity was low, with LD_{50} values of >4 mL/kg and 3000 mg/kg in New Zealand white rabbits (108).

In an acute inhalation study, rats were exposed for 4 h to a 1.7-mg/L dust concentration, which was the highest level attainable in the test system. All 10 animals survived the 14-d observation period. There were no behavioral reactions, adverse body weight effects, or gross pathological findings (115).

Skin Effects. EPNRs were practically nonirritating to rabbit skin. A solution of partially hydrolyzed EPNR (85% in methyl ethyl ketone) did not produce any effect in rabbits when applied dermally at a dose of 2000 mg/kg (16). The liquid resin was slightly to moderately irritating to the skin of rabbits (111), but essentially no skin irritation resulted from contact with solid resin (112) or with partially hydrolyzed EPNR (16). A study by Ciba-Geigy (108) showed pale red erythema and slight edema after repeated application of EPNR. Continued observations showed circumscribed elevations of the skin through day 7 and finally a more pronounced lesion containing yellowish exudate (pus) on day 14. These studies showed that the phenolic novolac resins were minimal primary irritants, but that repeated application could cause appreciable irritation. The studies also showed that the resins were not absorbed in sufficient quantities to cause any deaths or other signs of toxicity.

It was initially reported that EPNR did not produce skin sensitization in guinea pigs using a modified Buehler assay (113), but subsequent testing showed delayed contact skin sensitization in 3 of 20 guinea pigs tested (114). When patch tested in humans, the resins were irritating to the skin but were not found to have sensitizing potential (see below).

Eye Irritation. Eye irritation studies in rabbits indicated only minor transient irritation with no corneal injury produced by either liquid or solid EPNRs (108, 115) or a partially hydrolyzed EPNR (16).

4.4.1.2 Chronic and Subchronic Toxicity. Rats receiving daily oral doses of 0, 100, 300, and 1000 mg/kg/d bisphenol F diglycidyl ether (BFDGE) in coconut oil for 14 d showed decreased body weight gain and food consumption, increased liver weight, and thickening of the mucosal surfaces of the stomach in animals receiving 1000 mg/kg. Slight changes in liver weight were also observed in animals receiving 300 mg/kg/d.

In a subsequent 13-wk toxicity study wherein rats received daily doses of BFDGE in coconut oil at levels of 0 (control), 10, 50, and 250 mg/kg/d, BFDGE produced only slight toxicity. Clinically, an abnormal gait was observed immediately after dosing in the high-dose group. Food consumption and body weight gain was decreased in males receiving 250 mg/kg/d. Adjusted liver weights were increased in both sexes receiving 250 mg/kg/d. No biochemical, hematological, or microscopic findings were recorded. The no observed adverse effect level (NOEL) was 50 mg/kg/d (116).

4.4.1.6 Genetic and Related Cellular Effects Studies. BFDGE was assayed for mutations in cultures of *S. typhimurium* TA98, TA1535, and TA102 both in the presence and the absence of a rat liver metabolic fraction (S9). Results showed that BFDGE induced mutations in strains TA100 and TA1535 both in the presence and absence of S9. In strain TA102 a response was observed only in the presence of S9 (117).

In a chromosome aberration study in cultured human lymphocytes BFDGE induced increased frequencies of cells with sturctural aberrations both in the absence and presence of S9 (Burman, 1998).

No induction of micronuclei in the polychromatic erythrocytes of the bone marrow was observed 24 or 48 h after administration of BFDGE in coconut oil to mice at levels of 0, 500, 1000, and 2000 mg/kg for 2 d (118).

4.4.2 Human Experience

A study using a solid epoxy novolac resin in 50 male and female valunteers showed that the material was moderately to severely irritating when applied as a 10% solution in sesame oil. A 1% solution produced slight to moderate irritation. The incidence of irritation in the volunteers was 7–9%. None of the volunteers responded to a 5% challenge application made 2 wk following the ninth application, suggesting the absence of skin sensitization potential (119).

The cresolic novolac resins were found to be practically nontoxic by the oral route. The acute oral LD_{50} in rats was greater than 10,000 mg/kg (120). The acute dermal LD_{50} in rabbits was greater than 3.98 g/kg, indicating very low toxicity potential by the dermal exposure route. The cresolic novolac resins were minimally to mildly irritating to rabbit skin and eyes (121). Eye irritation response involved only the conjectivas, and no corneal injury was seen. There are no data available on sensitization potential. A cresolic novolac resin was positive in strains TA1535 and TA100 when tested for mutagenic potential in the Ames assay using *S. typhimurium* strains TA98, 100, 1535, and 1537 at concentrations of 0.9, 3.4, 10.1, 30.4, and 91.2 mg/mL (122).

5.0 Polyglycidyl Ether of Substituted Glycerin

5.0.1 CAS Number: [25038-04-4]

5.0.2 Synonyms: NA

5.0.3 Trade Names: NA

5.0.4 Molecular Weight: ~300

5.0.5 Molecular Formula: NA

5.0.6 Molecular Structure:

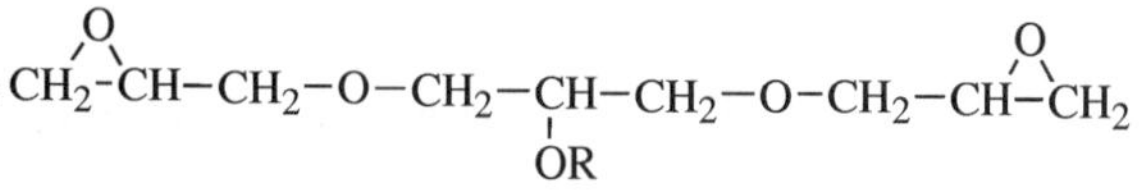

5.1 Chemical and Physical Properties

The polyglycidyl ether of substituted glycerin is the reaction product of epichlorohydrin and glycerin. Although the exact molecular arrangement in not known, it is of uniform composition and has the following general properties:

Specific gravity (20/4°C): 1.023
Refractive index (25°C): 1.478
Solubility: Miscible with water, soluble in ketones
Epoxy equivalent*: 140 to 165

5.1.2 Odor and Warning Properties

There is no characteristic odor and no irritation from the vapors.

5.2 Production and Use

The polyglycidyl ether of substituted glycerine is obtained as a reaction mixture of epichlorohydrin and glycerin. It is used in conjunction with the epoxy resins in the manufacture of adhesives. Because the compound is generally used in conjunction with a curing agent, frequently active amines, the hazardous properties of these substances must also be considered.

5.3 Exposure Assessment

No OSHA standard method was found for this meterial. The HCI–dioxane method (11) is suggested.

5.4 Toxic Effects

Systemic toxicity is low. The compound being practically nontoxic by ingestion and percutaneous absorption. No vapor hazard exists based on results of animal studies and extremely low vapor pressure. Repeated skin contact causes irritation and occasionally sensitization. Slight effects on hematopoesis have been reported.

*Grams of resin containing 1 g-equivalent of epoxide.

5.4.1 Experimental Studies

5.4.1.1 Acute Toxicity. The oral LD_{50} was 5 g/kg in rats and 1.87 g/kg in mice. The dermal LD_{50} in rabbits was 14.4 g/kg, and the intraperitoneal LD_{50} was 0.38 g/kg in rats and 0.30 g/kg in mice. No effects other than slight irriation were observed in mice and rats exposed for 8 h air saturated with the polyglycidyl ether of substituted glycerine.

Skin Irritation. A score of 0 was obtained by the Draize test for a single application. The compound is classified as nonirritating by this method. Applications of larger amounts or repeated applications give rise to severe skin injury on rabbits.

Eye Irritation. A score of 82/110 was obtained by the Draize test. The compound is classified as severely irritating.

5.4.1.2 Chronic and Subchronic Toxicity. Groups of 10 male rats exposed for 7 h/d, 5 d/wk, for a total of 50 exposures to air saturated with vapors of the polyglycidyl ether of subsituted glycerine. Aside from a slight excrustation of the eyelids in some animals, none of the rats showed any signs of toxicity or irritation attributable to the exposure.

Rabbits received 20 dermal applications of 0.2 g total dose of the resin, which was allowed to remain for 1 or 7 h. Scoring according to the Draize method gave high scores of 8 in both cases, with a final mean of 7.6–7.8, indicating the compounds to be highly irritating on repeated applications. In the second experimental, rabbits receiving 1 g/kg cutaneously on five successive days developed severe subcutaneous hemorrhage; two of four died. The uncured resin was fed in the diet of rats for 26 wk at a level of 0.04, 0.2, and 1%. No mortalities occurred at any of the levels, but at the highest level there there was retardation of weight gain. No significant pathology occurred.

5.4.1.5 Carcinogenesis. Carcinomas were produced on the skin of mice painted with the material from once to thrice weekly over a period of a year. No tumors were produced on rabbits ears in a similar test. Sarcomas were produced in rats by subcutaneous injection. When fed to strain A mice at a concentration of 0.2% in the diet, there was no increase in the incidence of spontaneous pulmonary adenomas as compared to controls (12).

5.4.1.6 Genetic and Related Cellular Effects Studies. In the Ames test, polyglycidyl ether of substituted glycerin showed negative response with *S. typhimurium* TA1535 and TA1533, both with and without activation, and positive response in strains TA98 and TA100, with and without activation (33).

5.4.1.5 Other: The slight effects of repeated exposure on the hematopoietic system are summarized in Table 83.1.

5.4.2 Human Experience

Primary irritation and occasionally sensitization occur in humans. In three of eight persons receiving the material intravenously, hematological changes occurred. These consisted of decreased in the white cell count, the total number of lymphocytes and monocytes, and the platelet count. No impairment in liver or kidney function was observed. The effects were reversible (33).

Table 83.1. Effects on the Hematopoietic System of Exposure to the Polyglycidyl Ether of Substituted Glycerin (161)

Route	Species	Dose (g/kg)	No. of Doses	Response[a]
Respiratory	Rat	Saturated vapors	50 (8 h each)	No effect noted
Intramuscular	Rat	0.10	6	No effect
		0.20	6	Depression of WBC count and bone marrow nucleated cell count
Intramuscular	Dog	0.2	2 (1/week)	Marked depression of WBC count; neutropenia; leukocytosis; ulceration and abscess of injection site.
Intravenous	Dog	0.2	1	Progressive decline in WBC count Death from overwhelming infection
Intravenous	Rabbit	0.1	2	Decrease in total WBC
Percutaneous	Rat	1.0	20	No effect
		2.0	20	No effect
		4.0	20	Depression of bone marrow nucleated cell count (only)

[a]WBC, white blood cells.

5.5 Standards, Regulations, or Guidelines of Exposure

No standard of permissible occupational exposure were found.

6.0 Castor Oil Glycidyl Ether

6.0.1 CAS Number: *[74398-71-3]*

6.0.2 Synonyms: 1,2,3-Propanetriyl ether of 12-(oxiranylmethoxy)-9-octadecanoic acid (COGE)

6.0.3 Trade Names: NA

6.0.5 Molecular Formula: $C_{66}H_{116}O_{12}$

6.1.2 Odor and Warning Properties

No information on odor or warning properties of COGE was found.

6.3 Exposure Assessment

No OSHA standard method was found for this material. The HCl–dioxane method (11) is suggested.

6.4 Toxic Effects

The acute toxicity of COGE is extremely low by both the oral and dermal routes of exposure, and the compound is not significantly irritating to the skin and eyes. No systemic toxicity or carcinogenicity were apparent following repeated dermal exposure for up to nearly 2 y in mice.

6.4.1 Experimental Studies

6.4.1.1 Acute Toxicity (124). The oral LD_{50} of COGE in rats was >5 g/kg, and the dermal LD_{50} in rabbits was >2 g/kg. There was no evidence of systemic toxicity by either route of exposure.

Skin Irritation. There was only very slight edema, with an irritation score of 0.7 out of a possible 8.0, in the Draize test in rabbits (124).

Eye Irritation. COGE was not irritating to the rabbit eye (Bionetics Research Laboratories, 1970).

6.4.1.2 Chronic and Subchronic Toxicity. COGE was tested in a 90-d skin-painting study in C3H mice in which solutions of 12.5 or 25% COGE in acetone or 100% COGE were applied twice weekly (125). No treatment-related effects were observed with regard to clinical signs, body weight, blood or urine parameters, or histopathology of the skin or major organs.

6.4.1.5 Carcinogenesis. There was no evidence of skin or systemic carcinogenicity in a cancer bioassay in mice treated topically with 50 mL of 50% COGE in acetone twice weekly for 64 wk (125).

6.5 Standards, Regulations or Guidelines of Exposure

No standards of permissible occupational exposure were found.

GLYCIDYL ESTERS

7.0 Phthalic Acid Diglycidyl Ester

7.0.1 CAS Number: *[7195-45-1]*

7.0.2 Synonyms: 1,2-Benzenedicarboxylic acid, *bis*(oxiranylmethyl) ester; 1-propanol,2,3-epoxy-, phthlate; diglycidyl *o*-phthalate; and PADGE

7.0.3 Trade Names: NA

7.0.4 Molecular Weight: 278

7.0.5 Molecular Formula: NA

7.0.6 Molecular Structure:

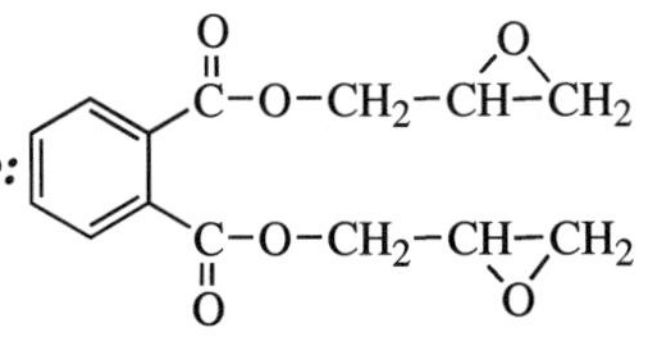

7.1 Chemical and Physical Properties

Selected physical–chemical properties of PADGE are:

Boiling point (760 mmHg): 167
Refractive index (20°C): 1.523

1 ppm 11.37 mg/m^3 at 25°C, 760 mmHg; 1 mg/L 87.9 ppm at 25°C, 760 mmHg

7.1.2 Odor and Warning Properties

No information was found on odor or warning properties of PADGE.

7.2 Production and Use

It is used in epoxy-based industrial coatings and paints.

7.3 Exposure Assessment

No OSHA standard methods for PADGE were found.

7.4 Toxic Effects

The acute oral and dermal toxicity of PADGE was low. It was moderately irritating to the skin of rabbits following a single topical application. Severe injury was produced in rabbits following the instillation of PADGE into the eye. PADGE has been shown to be mutagenic in bacteria and has been reported to cause skin sensitization in humans exposed occupationally.

7.4.1 Experimental Studies

7.4.1.1 Acute Toxicity. An acute oral toxicity test conducted with PADGE in male rats showed the single-dose oral LD_{50} to be between 500 and 5000 mg/kg (Bionetics Research Laboratories, 1970). PADGE administered topically to male and female rabbits showed the dermal LD_{50} to be greater than 2000 mg/kg (126).

Skin Irritation Undiluted PADGE (0.5 mL) applied topically to rabbit skin resulted in a well-defined to moderate erythema and edema; the primary irritation score was 2.3/8 (126).

Eye Irritation. Instillation of 0.1 mL of PADGE into the eyes of rabbits produced conjunctivitis, redness, chemosis, and lesions to the cornea and iris (126).

7.4.1.6 Genetic and Related Cellular Effect Studies. The mutagenic potential of PADGE was examined in a series of *in vitro* microbial assays using *S. typhimurium* and *Saccharomyces cerevisiae*. PADGE was tested with and without metabolic activation at dose levels from 0.001 to 5.0 mL per plate. Results demonstrated that PADGE was mutagenic to *S. typhimurium* strains TA1535 and TA100 in the presence and absence of metabolic activation (127).

7.4.2 *Human Experience*

Contact allergy tests were conducted in a aircraft factory using resin composite materials. Five of six workers patch tested with 1% solution of PADGE showed a positive response (128).

7.5 Standards, Regulations, or Guidelines of Exposure

No standards of permissible exposure of PADGE have been established.

8.0 Hexahydrophthalic Acid Diglycidyl Ester

8.0.1 CAS Number: *[5493-45-8]*

8.0.2 Synonyms: 1,2-Cyclohexanedicarboxylic acid, *bis*(oxiranylmethyl)ester; 1-propanol, 2,3-epoxy-, 1,2-cyclohexanedicarboxylate; and diglycidyl 1,2-cyclohexanedicarboxylate

8.0.3 Trade Names: NA

8.0.4 Molecular Weight: 284

8.0.5 Molecular Formula: $C_{14}H_{20}O_6$

8.0.6 Molecular Structure:

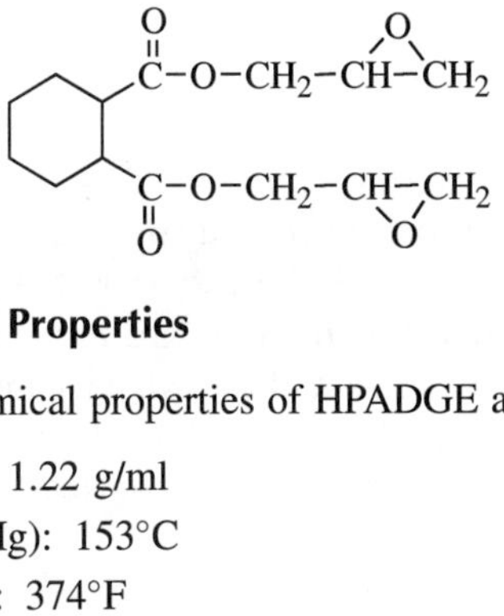

8.1 Chemical and Physical Properties

Some of the physical and chemical properties of HPADGE are as follows:

Specific gravity (25°C): 1.22 g/ml
Boiling point (at 2 mmHg): 153°C
Flash point (closed cup): 374°F
Appearance: Clear pale yellow liquid
Vapor pressure (25°C): >0.2 mmHg

1 ppm 11.62 mg/m^3 at 25°C, 760 mmHg; 1 mg/L 86.1 ppm at 25°C, 760 mmHg. Hexahydrophthalic acid diglycidyl ester (HPADGE) is a stable, clear, amber liquid with a pH of approximately 10 and is insoluble in water.

8.1.2 *Odor and Warning Properties*

No information was found on odor or warning properties of HPADGE.

8.2 Production and Use

It is used in epoxy-based industrial coatings and paints.

8.3 Exposure Assessment

No OSHA standard methods for HPADGE were found.

8.4 Toxic Effects

Acute toxicity studies of HPADGE by oral, dermal, and inhalation routes have demonstrated very low acute toxicity. HPADGE produced no irritation to the skin of rabbits after prolonged (24 h) exposure but more severe irritation upon repeated exposure. It has also been reported that HPADGE produced delayed contact skin sensitization in guinea pigs. HPADGE produced slight to moderate discomfort, moderate conjunctival redness and swelling, and slight reddening of the iris. In the eye of one rabbit it produced corneal injury that persisted 21-d postexposure. HPADGE has been shown to be mutagenic when tested *in vitro* in bacteria and mouse lymphoma cells. *In vivo*, HPADGE has been reported to produce an increase in SCEs in the bone marrow cells from Chinese hamsters treated orally and also an increased number of micronuclei in the mouse bone marrow micronucleus test.

8.4.1 Experimental Studies

8.4.1.1 Acute Toxicity. The acute oral LD_{50} in rats was reported to be 1030 mg/kg in the rat (72). Other studies have reported the acute oral LD_{50} in rats to be between 500 and 2000 mg/kg (129). The single-dose dermal LD_{50} has been reported to be greater than 2000 mg/kg (129) and greater than 4600 mg/kg (72). Nose-only exposures of rats to a liquid aerosol of HPADGE (1.52 mg/L, with a mass median aerodynamic diameter of 3.39 mm for 4 h) revealed no treatment-related effects other than transient signs of stress; all animals survived until sacrifice at 14 postexposure (130).

Skin Irritation. Topical application of HPADGE to the skin of a rabbit resulted in no irritation within 24 h, but moderate to marked erythema, moderate edema, and very slight exfoliation and necrosis on the abdominal skin of rabbits was observed after three applications (129). HPADGE produced delayed contact skin sensitization in guinea pigs (129, 131).

Eye Irritation. HPADGE produced slight to moderate discomfort, moderate conjunctival redness and swelling, slight reddening of the iris, and in the eye of one rabbit, corneal injury that persisted 21 d postexposure (129).

8.4.1.5 Carcinogenesis. HPADGE was tested *in vitro* for gene mutation effects in *S. typhimurium* at concentrations up to 231 mg/mL and was positive in strains TA100 and TA1535 (131). HPADGE has also been reported as positive in the mouse lymphoma assay (132). Negative results were found in the BALB/3T3 fibroblast transformation assay at concentrations up to 30 mg/mL (388, Ciba-Geigy, Limited, 1985). *In vivo*, HPADGE administered orally in a single dose of 1250, 2500, or 5000 mg/kg to Chinese hamsters produced a statistically significant increase in SCEs in bone marrow cells at the two higher dose levels (133). Mice given a single oral dose of 500 or 1000 mg/kg had a statistically significant increase in the frequency of micronucleated polychromatic erythrocytes as compared to negative controls (134).

8.4.2 Human Experience

Allergic contact dermatitis was observed in five workers operating machinery using a cutting fluid containing HPADGE. Positive patch test reactions were obtained in all five workers to HPADGE and were negative in 25 controls (135). Other cases of contact dermatitis have been described (136).

8.5 Standards, Regulations, or Guidelines of Exposure

No standards of permissible occupational exposure to HPADGE were found.

9.0 Glycidyl Ester of Neodecanoic Acid

9.0.1 CAS Number: *[26761-45-5]*

9.0.2 Synonyms: *tert*-Decanoic acid oxiranylmethyl ester; neodecanoic acid, 2,3-epoxypropyl ester; 2,3-epoxypropyl ester of mixed trialkyl acetic; and NA6E

9.0.3 Trade Names: CARDURA E10

9.0.4 Molecular Weight: 228

9.0.5 Molecular Formula: $C_{13}H_{24}O_3$

9.0.6 Molecular Structure:

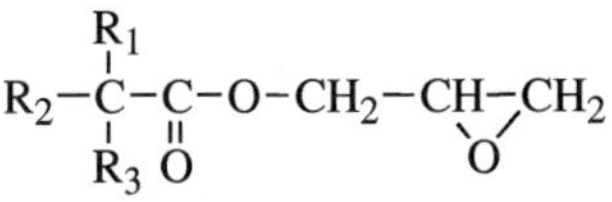

and

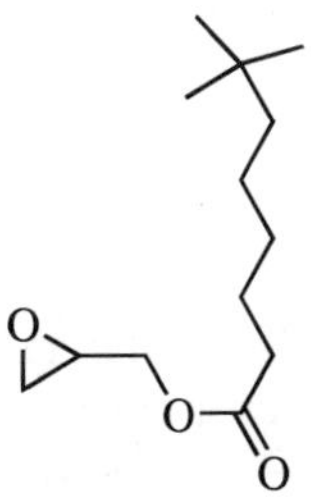

9.1 Chemical and Physical Properties

The material consists of mixture of branched aliphatic C_{10} carboxylic acids, with R_1, R_2, and R_3 in the structure above representing a distribution of aliphatic carbon chains, one of which will always be a methyl group; the total number of carbons in the other two R groups will be seven. The tertiary configuration of the acid radical confers extreme chemical stability to the ester linkage.

Typical properties of this glycidyl ester are as follows:

Epoxide equivalent Weight: 240–250

Density: 960 mg/kg (20°C)

Color: Water white

Melting point: < -60°C

Boiling point: 260°C

Flash point (Plensky Martens closed cup): 129°C

Miscibility with water (water in the ester): 0.7%(wt.)

It reacts readily with active hydrogen compounds and polymerizes readily in the presence of strong Lewis acids.

9.1.2 Odor and Warning Properties

The product has a low musty odor.

9.2 Production and Use

The glycidyl ester of neodecanoic acid (NAGE) is produced by the reaction of neodecanoic acid with epichlorohydrin in the presence of base:

$$\underset{\substack{| \\ R_3}}{\overset{\substack{R_1 \\ |}}{R_2-C-C-OH}} \;+\; CH_2-CH-CH_2-Cl \;\xrightarrow[OH^-]{}\; \underset{\substack{| \\ R_3 \;\; O}}{\overset{\substack{R_1 \\ |}}{R_2-C-C-O-CH_2-CH-CH_2}}$$
$$+ \; Cl^-$$

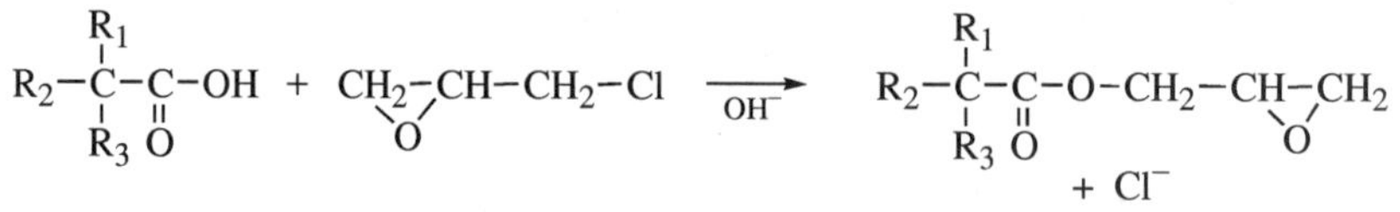

This material may be used an epoxy resin reactive diluent, as a component of epoxy coatings systems, and as an ingredient in alkyl resins to modify film properties. Improved chemical resistance, weatherability, film hardness, and acid number control are the properties enhanced in epoxy resin applications by the use of this material.

9.3 Exposure Assessment

No OSHA standard methods for NAGE were found. Because of the high boiling point and low volatility, it is not likely that detectable amounts could develop in the atmosphere.

9.4 Toxic Effects

The acute oral and dermal toxicity of NAGE is low. NAGE was only mildly irritating to the skin of rabbits but produced delayed contact skin sensitization in guinea pigs. NAGE was practically nonirritating to the eyes of rabbits. Rats fed high doses of NAGE for 5 wk showed decreased body weights, decreased hematocrit, decreased plasma alkaline phosphatase, increased plasms urea, protein, and sodium, increased urinary ketones, and increased relative liver and kidney weights with degenerative histopathological changes in the kidneys. Incubation of NAGE in liver, lung, and skin preparations showed a rapid and efficient hydrolysis of the epoxide group and the ester function in rat and human tissue samples; in mouse tissues, gluthione conjugation of the epoxide group was shown to be the major mechanism for deactivation of NAGE. In vitro, NAGE produced a weak increase in mutation frequency in bacteria but not yeast. In other in vitro studies, NAGE did not produce an increase in transformation of cultured baby hamster kidney cells but did induce a small increase in the frequency of chromosomal aberrations in cultured rat hepatocytes. NAGE did not cause DNA single-strand damage in the liver of rats given single oral doses of 5 mL/kg. Skin penetration studies with mouse, rat, and human skin showed that NAGE was extensively metabolized during skin penetration.

9.4.1 Experimental Studies

9.4.1.1 Acute Toxicity. The single-dose oral and dermal LD_{50} in rats was greater than 2000 mg/kg (137, 138). In rabbits, the acute (24 h) dermal LD_{50} was >10 mL/kg. The 4-h acute inhalation LC_{50} in rats for NAGE was greater than 240 mg/m^3 (139).

Skin Irritation. The compound is moderately irritating to rat and rabbit at 24 h postapplication (137). In a 4-h test in rabbits a slight irritation response was observed

(138). When tested by the Magnusson and Kligman maximization test in the guinea pig, NAGE was a skin sensitizer (137, 138).

Eye Irritation. NAGE was found to be nonirritant in the nonwashed eyes of rabbits (137, 138).

9.4.1.2 Chronic and Subchronic Toxicity. Rats were fed NAGE at dietary concentrations of 0, 100, 500, 1000, and 10,000 ppm for 5 wk. There were no intermediate deaths and no effect on general health or behavior in any dose group. At termination, rats receiving 10,000 ppm had decreased body weight and feed intake, decreased erythrocyte count and hematocrit, increased plasma urea, protein, and sodium, decreased plasma alkaline phosphatase, increased urinary ketones, and increased relative liver and kidney weights. Degenerative, occlusive, and regenerative changes were seen in the proximal renal tubules of male and to a much lesser extent female rats. Similar changes were seen in the 5000-ppm dose group, except that the erythrocyte count and hematocrit were normal. No treatment-related effects were observed at a dose level of 1000 ppm or below (140).

9.4.1.3 Pharmacokinetics, Metabolism, and Mechanisms. The *in vitro* kinetics of the glutathione-*S*-transferase-catalyzed conjugation of glutathione (GSH) and hydrolysis by epoxide hydrolase (EH) or carboxyesterase (CE) was studied by incubation of NAGE in microsomal and cytosolic fractions of liver and lung of mouse, rat, and human (141). Although very stable and resistant to chemical hydrolysis, NAGE was very rapidly and efficiently metabolized in all tissues of all species. In mouse tissues, significant GSH conjugation and CE activity was observed, whereas in rat and human tissues, EH and CE activity represented the major detoxification pathways. The overall rate of the *in vitro* metabolism of this glycidyl ester was at least one order of magnitude higher than for aromatic or aliphatic glycidyl ethers (45).

The transdermal penetration and metabolism of NAGE was determined *in vitro* in freshly dermatomized human and rat skin and full-thickness mouse skin (141). Penetration of NAGE was very slow, with intact mouse and dermatomized rat skin being more permeable than dermatomized human skin. NAGE was extensively metabolized in skin of all species; $< 0.2\%$ of the applied dose was recovered unchanged.

9.4.1.5 Carcinogenesis. The mutagenic activity of glycidyl ester of neodecanoic acid was investigated in cultures of *S. typhimutium* TA98, TA100, TA102, TA1535, TA1537, and TA1538, *E. coli* WP2 and WP2 uvr A, and in cultures of *S. cerevisiae* JD1, both with and without the incorporation of rat liver microsomal enzymes (107, 142, 143). The results indicated that glycidyl ester of neodecanoic acid induced an increase in mutation frequency, only after metabolic activation in *Salmonella* strains TA100, TA1535, and TA1538. The mutation frequency was within the spontaneous frequency range in the absence of S9 fractions for these strains; however, in some experiments weak responses were observed in the absence of S9 (107, 143). In TA98, TA102, TA1537, and TA1538, there was no increase in mutation frequency either with or without S9. There was no increase in gene mutation and frequency in either of the *E. coli* strains with or without activation. Similarly, in *Saccharomyces* no changes in gene conversion frequency were reported. In general, no significant effects were detected below 100 mg of material per plate, suggesting that the mutagenic activity was relatively weak. The bacterial results indicate that mutagenicity was expressed by both base-pair substitution and frameshift mechanisms.

In other studies, monolayer slides of cultures of rat liver (RL1) cells were exposed to culture medium containing glycidyl ester of neodecanoic acid, and after 24 h incubation the slides were processed for metaphase chromosome analysis (142). In the RL1 cells, NAGE induced a low frequency of chromatid aberrations at concentrations just below the cytotoxic dose (50 mg/mL).

When suspension cultures of baby hamster kidney cells were exposed to NAGE at concentrations of approximately 44, 88, and 350 mg/mL, no increased frequency of transformed cells was observed (144).

In an *in vivo* study, a single oral dose of 5 mL/kg of NAGE did not cause significant DNA single-strand damage in rat liver cells when measured at 6 h postdosing (145).

9.4.2 Human Experience

Contact dermatitis has been observed in workers exposed to NAGE (146, 147).

9.5 Standards, Regulations, or Guidelines of Exposure

No standards of permissible occupational exposure to NAGE were found.

MISCELLANEOUS EPOXY COMPOUNDS

10.0 Epichlorohydrin

10.0.1 CAS Number: *[106-89-8]*

10.0.2 Synonyms: 1-Chloro-2,3-epoxypropane; γ-chloropropylene oxide; α-epichlorohydrin (ECH); chloromethyloxirane; 1,2-epoxy-3-chloropropane; and glycidyl chloride; DL-a-epichlorohydrin; (Chloromethyl) Ethylene Oxide; 3-Chloropropylene Oxide; Ech; Glycol Epichlorohydrin; (chloromethyl)oxirane; 2-(Chloromethyl)oxirane; alpha-epichlorohydrin; 2-3-epoxypropyl chloride; skekhg; Allyl chloride oxide; 3-Chloro-1,2-epoxypropane; 3-Chloro-1,2-propylene oxide; 3-Chloropropene-1,2-oxide; 3-Chloropropyl epoxide; EPI; Epoxy-3-chloropropane; Epoxypropyl chloride; (RS)-3-Chloro-1,2-epoxypropane; Cardolite NC-513; Epichlorohydrin

10.0.3 Trade Names: NA

10.0.4 Molecular Weight: 92.53

10.0.5 Molecular Formula: C_3H_5OCl

10.0.6 Molecular Structure: 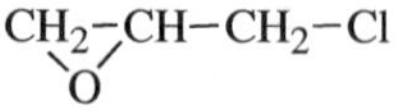 and 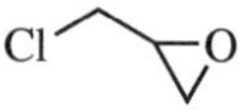

10.1 Chemical and Physical Properties

ECH is a colorless, mobile liquid that is flammable and reactive. ECH has a sweet, pungent, or chloroformlike odor and the following properties:

Specific gravity (20/4°C): 1.1812
Freezing point: −57.2°C

Boiling point (760 mmHg): 117.9°C
Vapor density: 3.21
Vapor pressure (20°C): 13 mmHg
Refractive index (20°C): 1.43805
Solubility: 6.59% in water; miscible with ethers, alcohols, CCl_4, and benzene
Flash point (Tag open cup): 40°C
Flash point (Tag closed cup): 33.89°C
Concentration in "Saturated air" (20°C): 1.7%
Flammability limits: 3.8% vol. to 21% vol. in air

1 ppm = 3.78 mg/m^3 at 25°C, 760 mmHg; 1 mg/L = 265 ppm at 25°C, 760 mmHg

10.1.2 Odor and Warning Properties

The odor is generally perceived as a slightly irritating chloroformlike odor. Sensory perception studies have indicated that the mean threshold for odor recognition is approximately 10 ppm, and that at 25 ppm it is recognized by the majority of persons. Marked nose and eye irritation occurs only at levels exceeding 100 ppm. It is concluded that local irritation of the eyes is not severe enough to force workers to evacuate potentially harmful areas. Eye irritation may be accepted as indicating an undersirable atmospheric contamination.

10.2 Production and Use

ECH is available in large-scale commercial quantities through the discovery and development of processes for its production from propylene. It is employed as a raw material for the manufacture of a number of glycerol and glycidol derivatives, in the manufacture of epoxy resins, as a stabilizer in chlorine-containing materials, and as an intermediate in preparation of condensates with polyfunctional substances.

10.3 Exposure Assessment

10.3.3 Workplace Methods

Standard methods for air sampling and analysis of the ECH by GLC with flame ionization detection have been described NIOSH Method 1010 (66, 148, 149).

10.3.4 Biomonitoring/Biomarkers

The measurement of 3-chloro-2-hydroxypropylmercapturic acid and α-chlorohydrin as biomarkers of occupational exposure to epichlorohydrin has been described (150).

10.4 Toxic Effects

ECH has been found to be moderately toxic to laboratory animals following a single oral, percutaneous, intravenous, or subcutaneous dose. By inhalation the 1-h LC_{50} has been determined to be 3617 and 2165 ppm in male and female rats, respectively (151). Because the theoretical saturated vapor concentration is 17,105 ppm, excessive vapor concentrations are readily attainable and may cause unconsciousness and death. Skin contact may cause severe irritation and sensitization. ECH vapors have been reported to be irritating to the mucous membranes of the eye and respiratory tract. The subchronic inhalation toxicity observed in laboratory animals was destruction of the nasal turbinate epithelium with slight nonprogressive kidney effects and slight nondegenerative liver effects. Repeated exposure to 25 or 50 ppm of ECH vapor or repeated oral administration of 15 mg ECH mg/kg/d also produced reversible sterility in male rats. The available data provide good evidence of gene and chromosomal mutagenicity of ECH in several experimental systems; however, cytogenetic studies of workers exposed to ECH have yielded equivocal evidence for a clastogenic effect on lymphocytes. Chronic inhalation exposure to 100 ppm ECH for 30 d produced tumors in the nasal epithelium of rats. Some indication of the potential of ECH to produce nasal tumors was also observed in rats inhaling 30 ppm for their entire lifetime. The oral administration of ECH by gavage or in drinking water resulted in hyperplasia, papillomas, and carcinomas in the nonglandular stomach of rats. ECH alone did not induce tumors when applied to the skin of mice but appeared to act as an initiator in groups treated subsequently with phorbol myristate acetate as a promotor. Tumor formation appears to be confined to tissues that are the site of initial ECH contact. Epidemiology studies have not provided definitive evidence for an association between occupational ECH exposure and cancer or any other adverse health effects.

10.4.1 Experimental Studies

10.4.1.1 Acute Toxicity. The doses or exposures of ECH producing lethality have been defined following administration by the oral, dermal, intravenous, subcutaneous, or inhalation routes of exposure. ECH has been found to be moderately toxic to laboratory animals following administration of a single oral, percutaneous, intravenous, or subcutaneous dose. Acute inhalation studies in laboratory animals have shown that lethal vapor concentrations are readily attainable at room temperature. A summary of some of the key acute toxicity data appears in Table 83.2.

Skin Effects. Undiluted epichlorohydrin (0.5 mL) was intensely irritating and necrotic to the depilated skin of laboratory rabbits when allowed to remain in contact with the skin for 24 h (152). Smaller volumes of undiluted ECH (0.1–0.2 mL) applied to rabbit skin for 2 h produced less severe irritation over a smaller area (152). A much smaller volume (0.01 mL) of undiluted ECH applied for 2 h to rabbit skin produced only a trace of redness and capillary injection (13). Weaker solutions (0.2 mL of a 0.3% solution of ECH in cottonseed oil) applied to rabbit skin for 24 h under an occlusive dressing produced no irritation, whereas a 5% solution resulted in a marked irritation of the skin (153). Repeated applications may lead to widespread necrosis. Skin sensitization tests in guines pigs have produced mixed or equivocal results, although from the information available it appears that ECH clearly has the potential to produce delayed contact skin sensitization (13, 25, 153, 154).

Eye Irritation. Undiluted material or concentrated solutions (80% ECH in cottonseed oil) were markedly irritating to the eye and produced corneal damage on local contact (13, 154). Less concentrated solutions (20% ECH in cottonseed oil) produced conjunctival

Table 83.2. Summary of Acute Toxicity Data on Epichlorohydrin

Route	Species	Dose	Parameter of Toxicity
Oral	Rats	0.09 g/kg	LD_{50}
Oral	Guinea pigs	0.178 g/kg	LD_{50}
Oral	Mice	0.238 g/kg	LD_{50}
Intravenous	Rats	0.154 g/kg	LD_{50}
Intravenous	Mice	0.178 g/kg	LD_{50}
Percutaneous	Rabbits	0.88 ml/kg	LD_{50}
Percutaneous	Rats (3 applications)	0.5 mL/kg	LD_{50}
Inhalation	Mice	2370 ppm	0/30[a]
Inhalation	Mice	8300 ppm	20/20[a]
Inhalation	Rats	250 ppm (8 h)	LC_{50}
Inhalation	Rats	500 ppm (4 h)	LC_{50}
Inhalation	Guinea pigs	561 ppm (4 h)	LC_{50}
Inhalation	Rabbits	445 ppm (4 h)	LC_{50}
Inhalation	Rats (males)	3617 ppm (1 h)	LC_{50}
Inhalation	Rats (females)	2165 ppm (1 h)	LC_{50}

[a]Number of deaths over the number of animals exposed.

irritation and edema, whereas 5% in cottonseed oil produced no eye irritation (153). The eye irritation potential of ECH appeared to be increased in rabbits if the ECH was in a solvent that is more readily soluble in water than is ECH itself (155). ECH vapors may also give rise to eye irritation.

10.4.1.2 Chronic and Subchronic Toxicity. In well-documented studies, repeated exposure of rats, mice, or rabbits to concentrations of ECH vapor up to 120 ppm (studies ranged from 4 to 13 weeks) resulted in substantial degenerative changes in the nasal turbinate mucosa (156–158). Eye irritation, slight nonprogressive kidney effects (rats), and slight nondegenerative liver effects (decreased glycogen in rats and mice) and minimal changes in the content of the epididymides (rats) were also noted with transient infertility observed in male rats (but not rabbits) (157, 158). The no observed effect level for all subchronic toxicologic effects was 5 ppm.

Other previously reported and poorly documented studies have confirmed nasal irritation and some liver and kidney effects at higher concentrations and also reported some effects not repeated in later studies. A single 4-h inhalation exposure of rats to 100 ppm ECH or exposure for 4 consecutive days for 4 h/d to 100 ppm did not result in marked changes in several endpoints used to assess hepatotoxicity and kidney toxicity (159). Inhalation exposure of rats for 1 h at 150 ppm produced decreases in renal glutathione and histologic changes in the kidney when examined by light and electron microscopy (160).

Gavage administration of ECH to rats for ten consecutive days at doses of 0, 3, 7, 19, or 46 mg/kg/d or for 90 d at doses of 0, 1, 5, and 25 mg/kg/d resulted in decreased body weight gain and reduced food and water consumption (161). Dose-related increases in kidney and liver weights were observed in both sexes at 25 mg/kg/d in the 90-d study.

Histopathology examinations revealed the forestomach as the primary target organ with dose-related increases in mucosal hyperplasia (acanthosis and hyperkeratosis). The NOAEL for this effect was 1 mg/kg/d.

10.4.1.3 Pharmacokinetics, Metabolism, and Mechanisms. Studies in rats and mice have demonstrated that ECH is readily metabolized and rapidly excreted either as urinary metabolites (approximately 50% of the absorbed dose) or as CO_2 in the expired air (162–166). The majority of the absorbed dose was eliminated within the first 12–24 h postdosing. Hydrolysis of the epoxide ring and glutathione conjugation are two metabolic routes whereby ECH is metabolized and eliminated from the body, although all the exact details of the metabolic fate of ECH have not been fully elucidated. In a more recent study (167) it has been reported that ECH was measurable in the kidney and blood of rats over 1 h after intraperitoneal administration. However, ECH rapidly disappeared from the liver and was no longer detectable at 30 min postdosing. In contrast, the metabolite α-chlorohydrin was measurable up to 5 h postdosing in the liver, kidney, and blood after the administration of ECH. These results indicate that the metabolism of ECH is rapid and that α-chlorohydrin is relatively more persistent in the tissues.

ECH did not inhibit or induce the activity of either microsomal or glutathione-*S*-transferases in mice dosed intraperitoneally for three consecutive days with doses of 40 mg/kg/d, suggesting that the metabolism of ECH is neither impaired or induced by repeated ECH exposures (168).

Since testicular toxicity following ECH treatment was similar as that induced by α-monochlorohydrin, it has been proposed that these materials are metabolized to a common gonadal toxicant (169). Indeed, two oral studies in rats and mice have detected α-monochlorohydrin as metabolite of epichlorohydrin (165, 169). Additional studies with α-monochlorohydrin have shown that α-monochlorohydrin can induce infertility by a direct interaction with spermatogenesis. The mechanism of action is through an inhibition of the glycolysis, decreasing the available ATP required for the motility of spermatozoa.

10.4.1.4 Reproductive and Developmental Toxicity. *Developmental Toxicity.* No embryo toxic, fetotoxic, or teratogenic effects were noted in the offspring of pregnant rats and rabbits inhaling 0, 2.5, or 25 ppm ECH during gestation (170). A teratology study in which rats and mice were administered daily oral doses of 0, 80, or 120 mg/kg ECH by gavage did not result in an increase in the frequency of fetal malformations in either species (171).Some evidence of fetotoxicity was noted in mice at the higher dose levels of 80 and 120 mg/kg; however, this may have been a reflection of maternal toxicity because the highest dose level was lethal to 3/32 dams (171).

Reproductive Toxicity. ECH has been found to have an effect on fertility in male rats. Rats receiving an oral dose of 15 mg/kg/d for 7 d in a 12-d period became infertile within 1 wk. This effect was reversible within 1 wk after treatment was discontinued. No histological changes were found in the testes, epididymis, prostate, and seminal vesicles. Libido and ejaculation capacity were not adversely affected (172). ECH caused a decreased sperm count when administered orally to rats at 50 mg/kg (173) and decreased sperm count, increased abnormal sperm, and testicular and epididymal effects histopathologically following a single subcutaneous dose to rats (169). More recently, a single lower subcutaneous dose of 31.5 mg/kg was found to have no effect on sperm count, sperm morphology, testicular or epididymal weights, or testicular histopathology (174).

Male and female Long–Evans rats dosed by gavage with ECH (males: 12.5, 25, and 50 mg/kg/d; females: 25, 50, and 100 mg/kg/d) for 21 and 14 d, respectively, found fertility in the high-dose males to be totally impaired (lower-dose males were not examined). No measured parameters of female fertility nor the mean number and survival of the offspring were changed. Sperm morphology and percentage motile sperm were not statistically different from control values in both ejaculated and cauda epididymal sperm count was slightly decreased in males at the 50-mg ECH/kg dose level. Mean curvilinear velocity, straight-line velocity, and amplitude of lateral head displacement of cauda epididymal sperm were significantly reduced by ECH at 12.5 mg/kg/d and above. Sperm track linearity was also reduced, but only at 50 mg/kg/d. Beat/cross frequency of sperm was significantly increased at 12.5 mg/kg/d and above. All the above sperm motion parameters showed dose-dependent trends. The effects were consistent with the spermatozoal metabolic lesions reported for α-chlorohydrin, a metabolite of ECH (175) rats exposed once for 4 h by inhalation to 100 ppm ECH, showed no effects on sperm motility over 14 d postexposure, but transient changes in sperm velocity were found (176).

Repeated inhalation exposures (5 d/wk, 6 h/d for 10 wk) to 25 or 50 ppm of ECH vapor resulted in reduced fertility in male rats (170). Males exposed to 25 ppm were able to impregnate females, but marked preimplantation losses were observed. The effects were reversible as early as 2 wk postexposure, and 5 ppm did not affect fertility. No adverse effects on fertility in female rats were observed following ECH inhalation.

Klinefelter et al. (177) reported that decreases in male fertility induced with ECH were highly correlated with in quantitative decreases in specific (SP22), although the mechanism of this decrease has not been determined. Recently, ECH was shown to have no activity in *in vitro* receptor-mediated assays to examine estrogenic and antiestrogenic activity (178). Hence the mechanism of action of ECH in inducing male antifertility effects appears to be through its metabolism to α-chlorohydrin, as discussed.

10.4.1.5 Carcinogenesis. Tumorigenic effects in laboratory animals treated with ECH have been reported, although only tissues that were the first tissues in direct contact with ECH following either inhalation or oral administration, and only at doses that caused significant chronic irritation. Specifically, Laskin et al. (179) reported that tumors in the nasal cavity (primarily squamous cell carcinomas) developed in 15 of 140 rats exposed for 30 d to 100 ppm ECH and observed for their lifetime, as compared to no nasal tumors in 150 concurrent controls. Severe inflammation of the nasal turbinates that preceded tumor development was also produced by ECH in this study. In addition, lifetime exposure of 100 rats to 30 ppm resulted in one squamous cell carcinoma of the nasal cavity and one nasal papilloma; no tumors were produced by lifetime exposure at 10 ppm. Rats administered ECH by gavage for 2 y at dose levels of 2 or 10 mg/kg/d, or by drinking water for 81 wk (29, 52, or 89 mg/kg/d), developed hyperplasia, papillomas, and carcinomas of the nonglandular stomach at all dose levels (180, 181). The dose levels in the gavage study were high enough to produce 7.7, 22.4, and 44.9% reductions in body weight for the lowest to highest dose levels, respectively.

The tumorigenic effects produced at the "portals of entry," that is, the nasal turbinates following inhalation or the nonglandular stomach following oral administration, are consistent with the chemical reactivity of ECH, a bifunctional alkylating agent that may covalently bind to many cell constituents to produce toxicity. This alkylating ability of ECH is likely responsible for the genotoxic effects observed in various *in vitro* genotoxicity test systems; these have been reviewed (182).

Chronic percutaneous application of ECH alone did not result in skin tumors in either mice or rats (13, 183). However, 9 of 30 mice developed either skin papillomas or, in one case, carcinoma when treated with a single application of 2 mg of ECH followed 2 wk later by three applications per week with the active component of croton oil, phorbol myristate acetate (PMA) (183). This result compared to 3/30 mice developing papillomas in the PMA-only treated group suggests ECH may act as a tumor initiator in mice.

10.4.1.6 Genetic and Related Cellular Effects Studies. The genetoxic potential of ECH has been extensively investigated using both *in vitro* and *in vivo* assays; the results have been summarized and reviewed (182, 184–186). As expected for a bifunctional alkylating agent, ECH had genotoxic activity in several *in vitro* assays, including those conducted in bacteria, fungi, and mammalian cells in culture, and appears to act as a direct-active mutagen as the addition of metabolic capacity to these assays typically resulted in a reduction or elimination of genotoxic effects.

In vivo studies examining the genotoxicity of ECH have produced more mixed results. Mice receiving a single dose of 100 mg/kg, or five doses at 20 mg/kg on five consecutive days by gavage or intraperitoneal injection, showed an increase in the frequency of chromosomal aberrations in bone marrow (187). Subsequent studies found that ECH did not cause an increase in chromosomal aberrations in the bone marrow of rats inhaling up to 50 ppm of ECH, 6 h/d, 5 d/wk (188) or in mice given single oral doses up to 200 mg/kg (165). Five separate studies examining the potential of ECH to produce micronuclei in the bone marrow of mice found negative results (189–193) with one positive (194) ECH produced sex-linked recessive mutations in *D. melanogaster* (195), but negative results were obtained in dominant lethal tests in mice by both the oral intraperitoneal routes of administration (187, 196). In an 8-wk inhalation study, an increase in the number of chromosomal aberrations in spermatogonia, as well as morphologically abnormal sperm, were observed in mice exposed to an atmosphere of 125 mg ECH/m^3 in air (184). Prodi et al. (197) reported that ECH binds to nucleic acids and proteins in the stomach, liver, kidney, and lungs of rats and mice following intraperitoneal administration of a single dose at approximately 0.5 mg/kg. The covalent binding index in these studies that was calculated according to Lutz (198) indicated that ECH should be considered a weak initiator. Deuterium labeling of ECH did not influence the rate of single-strand breaks in DNA from rat hepatocytes treated *in vitro*, which is evidence that metabolism is not required for ECH to alkylate DNA (199). *In vitro*, the epoxide group of ECH has been found to react with the N^{-7} position of deoxyguanosine and with the N^{-6} position of deoxyadenosine, and this work has been reviewed (200). The formation of adducts with hemaglobin and DNA *in vivo* following intraperitoneal administration to rats has also been reported (201).

10.4.2 Human Experience

10.4.2.2 Clinical Cases. Occasional cases of skin sensitization have been reported (204–207).

10.4.2.2.1 Acute Toxicity. A few days after direct skin contact with epichlorohydrin, erythema, edema, and papules associated with a burning, itching sensation were reported (202). In more severe cases blisters developed. Recovery was complete. It should be mentioned that ECH can penetrate leather shoes or rubber gloves and so may cause chemical burns. A single case was reported of chronic asthmatic bronchitis, which developed after a single, though severe, overexposure. At 2 yr postexposure the exposed individual was reported to have a pronounced fatty liver, attributed by the authors to the prior ECH exposure, although a causal relationship is questionable (203).

10.4.2.3 Epidemiology Studies. A number of studies have investigated the potential relationships between repeated exposure to ECH and health effects in workers, including specific organ toxicity, reproductive effects, genotoxicity, and carcinogenicity; these have been reviewed (185, 208). In general, none of these studies has provided definitive evidence for an association between occupational ECH exposure and an increased incidence of organ injury or disease, decrease in fertility, or other biologic effects, such as alterations in cytogenetics. Nevertheless, the evidence for the ability of ECH to produce toxicity in laboratory animals dictates that appropriate handling precautions and hygienic standards be followed during the manufacture and use of ECH.

10.5 Standards, Regulations, or Guidelines of Exposure

The 2000 ACGIH TLV is 0.5 ppm or 1.9 mg/m^3 (8 h) with skin notation, indicating that there is potential for cutaneous absorption and classification as A3, animal carcinogen. The OSHA PEL is 5 ppm. NIOSH REL recommends the lowest feasible concentration with an IDLH value of 75 ppm. The United Kingdom has maximum exposure limit (MEL) of 0.5 ppm (8-h TWA) and short-term exposure limit (STEL) of 1.5 ppm (10 min). The International Agency for the Research of Cancer (IARC) has classified ECH as Group 2A, that is probably carcinogenic to humans (209).

Skin, eye, and respiratory contact should be avoided. Protective equipment, which should be routinely used even where exposure is not expected, includes chemical goggles, safety showers, and eye wash stations in the immediate working area, and respiratory protection equipment available for use during escape. Additional personal protective equipment should be used when the potential for exposure is high, but should not be considered a substitute for proper handling and engineering controls. Protective clothing (suits, boots, and gloves) made of neoprene or butyl rubber is preferred. Protective clothing made of PVC and nitrile is penetrated more readily. For short-term exposure or low concentrations, a NIOSH-approved, fitted, full-facepiece cartridge-type respirator with organic vapor cartridge or organic vapor/acid gas combination cartridge should be used. Where there is potential for longer-term exposure, or exposure to higher concentrations can be expected, supplied air or a positive pressure breathing apparatus should be used. Leather articles should not be worn, as they cannot be decontaminated and must be cut up and burned.

11.0 Glycidol

11.0.1 CAS Number: [556-52-5]

11.0.2 Synonyms: 3-Hydroxypropylene oxide; 2,3-Epoxy-1-propanol; 2-hydroxymethyloxirane; epoxypropyl alcohol: Oxiranemethanol; 1,2-Epoxy-3-Hydroxy Propane; 2,3-epoxypropanol; epihydrin alcohol; glycide; glycidyl alcohol; 3-hydroxy-1,2-epoxypropane; oxiranylmethanol; allyl alcohol oxide; (*RS*)-3-Hydroxy-1,2-epoxypropane

11.0.3 Trade Names: NA

11.0.4 Molecular Weight: 74.08

11.0.5 Molecular Formula: $C_3H_6O_2$

11.0.6 Molecular Structure:  and 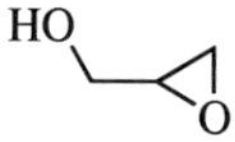

11.1 Chemical and Physical Properties

Glycidol is a colorless, slightly viscous liquid with the following properties:

Specific gravity (20/4°C): 1.115
Boiling point (760 mmHg): 166.11°C
Vapor density: 2.15
Vapor pressure (25°C): 0.9 mmHg
Solubility: Completely water soluble
Concentration in "saturated" air: 0.118%

1 ppm = 3.03 mg/m^3 at 25°C, 760 mmHg; 1 mg/L = 330 ppm at 25°C, 760 mmHg

11.1.2 Odor and Warning Properties

Eye and respiratory irritation should help prevent excessive overexposure.

11.2 Production and Use

Glycidol is made through dehydrochlorination of glycerol monochlorohydrin with caustic. It is commercially available. It is suggested for use in preparation of glycerol and glycidyl ethers, esters, and amines, in the pharmaceutical industry, and in sanitation chemicals.

11.3 Exposure Assessment

A NIOSH standard method 1608 for the analysis of glycidol by absorption on charcoal, displacement with tetrahydrofuran, and determination by GLC-FID has been described (210).

11.4 Toxic Effects

Glycidol is severely irritating to the eyes and skin, and exposure to vapors may also cause corneal damage. The brain, kidney, and testes have been identified as target organs in subchronic and chronic toxicity studies; effects include testicular atrophy and antifertility effects in males and tumorigenesis in multiple tissues.

11.4.1 Experimental Studies

11.4.1.1 Acute Toxicity. The oral LD_{50} of a 10% solution of glycidol in propylene glycol is 850 mg/kg for the rat and 450 mg/kg for the mouse (211). Symptoms of intoxication were initial depression of the CNS, followed by hypersensitivity to sound with muscular tremors and facial muscular fibrillation; some rats showed terminal convulsions. All animals had dyspnea and showed lacrimation. Neat glycidol had an oral LD_{50} in rats of 640 mg/kg in females and 760 mg/kg in males; the LD_{50} following intraperitoneal injection in rats was 210 mg/kg in females and 350 mg/kg in males (212). The dermal LD_{50} in rabbits after a 7-h exposure was 1980 mg/kg. Minimal signs of systemic toxicity were observed, but skin irritation was present. By inhalation, the 4-h LC_{50} was 580 ppm in the rat and 450 ppm in the mouse. Signs of severe irritation included dyspnea, lacrimation, salivation, and nasal discharge. Signs of CNS stimulation were also apparent. Death usually resulted from pulmonary edema; emphysema was also detected. There was some discoloration of the kidneys and liver. An 8-h vapor exposure caused corneal opacity (211).

Skin Effects. Glycidol is rated a moderate skin irritant, with a Draize score of 4.5/8.0, for a single application on occluded rabbit skin. It was severely irritating, with necrosis, when applied repeatedly over 4 d. Although data are not available, glycidol would be expected to be a skin sensitizer (211).

Eye Irritation. Glycidol is severely irritating to the rabbit eye, with a Draize score of 68/110. However, there were no permanent corneal defects after a single instillation (211).

11.4.1.2 Chronic and Subchronic Toxicity. Rats exposed to 400 ppm glycidol, 7 h/d for 50 d, showed no signs of systemic toxicity. Only slight irritation of mucous membranes was noted. Repeated intramuscular injections failed to affect hematopoiesis in the rat (211). Following gavage of 300 mg/kg glycidol in rats for 16 d, there was edema and degeneration of the epididymal stroma, atrophy of the testes, and granulomatous inflammation of the epididymis in males. Focal demyelination in the medulla and thalamus of the brain occurred in all female mice that received 300 mg/kg glycidol for 16 d. In a subsequent 13-wk gavage study, count and motility were reduced in male rats at doses of 100 or 200 mg/kg. Necrosis of the cerebellum, demyelination in the medulla of the brain, tubular degeneration and necrosis of the kidney, and testicular atrophy and degeneration occurred in rats given a 400-mg/kg dose. In a 13-wk mouse gavage study, there was demyelination of the brain at doses of 150 or 300 mg/kg, testicular atrophy at 19 mg/kg, and renal tubular cell degeneration in males that received 300 mg/kg (10).

11.4.1.4 Reproductive and Developmental. The teratogenic potential of glycidol was assessed in rats by intra-amniotic injections on day 13 of gestation. There was a 50%

resorption rate with 10, 100, and 1000 mg/fetus. Of the surviving fetuses treated with 1000 mg glycidol, 44% were malformed, with limb defects being the most frequent malformation (213). By contrast, in a mouse developmental toxicity study, glycidol gavaged at 200 mg/kg showed no evidence of teratogenicity; there was a significant increase in the number of stunted fetuses at this dose, but all these were present in a single litter, and this dose killed 5 of 30 dams (171).

11.4.1.5 Carcinogenesis. In a dermal carcinogenicity study, 20 mice were topically administered 100 mg of a 5% solution of glycidol in acetone, three times weekly, for 520 d. No tumors of any type resulted (214). Another lifetime skin carcinogenicity study in mice was negative (158). However, an oral gavage cancer study in mice and rats conducted by NTP was positive with multiple tumors (10); these data have also been published (215). NTP concluded there was clear evidence for carcinogenic activity in both species. Non-neoplastic lesions observed in the cancer study included hyperkeratosis and epithelial dysplasia of the forestomach, fibrosis of the spleen in rats, and cysts of the preputial gland and kidney in male mice.

11.4.1.6 Genetic and Related Cellular Effects Studies (10). Glycidol was mutagenic in several *in vitro* tests. Mutagenic activity was observed with and without metabolic activation in *Salmonella* strains TA97, TA98, TA100, TA1535, and TA1537 in the Ames assay. It was also positive without activation in the mouse lymphoma assay; it was not tested with activation. In the Chinese hamster ovary cell assay, glycidol induced both SCEs and chromosomal aberrations in the presence and absence of metabolic activation. Glycidol was also genotoxic in *in vivo* assays. It induced sex-linked recessive lethal mutations and reciprocal translocations in the germ cells of male *Drosophila* exposed by feeding. It was also positive in the micronucleus test following intraperitoneal injection in male B6C3F_1 mice, and induced chromosome aberrations in bone marrow cells from rats dosed orally or intraperitoneally (216). DNA adduct formation, specifically 3-dihydroxypropyl-deoxyuridine, has also been reported (217).

11.4.2 Human Experience

No effects have been reported. Effective hygiene practices should be taken to prevent the irritative and potential systemic effects of glycidol.

11.5 Standards, Regulations, or Guidelines of Exposure

The ACGIH TLV–TWA 8-h value for glycidol is 2 ppm with classification as A3, animal carcinogen. OSHA PEL TWA is 50 ppm (150 mg/m^3), NIOSH REL TWA is 25 ppm (75 mg/m^3), with IDLH value of 150 ppm.

12.0 Glycidaldehyde

12.0.1 CAS Number: *[765-34-4]*

12.0.2 Synonyms: Oxiranecarboxaldehyde, 2,3-epoxypropanol, epihydrinaldehyde, epoxypropanol, fomyloxirane, and glycidal

12.0.3 Trade Names: NA

12.0.4 Molecular Weight: 72.1

12.0.5 Molecular Formula: $C_3H_4O_2$

12.0.6 Molecular Structure:

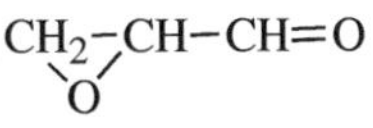 and 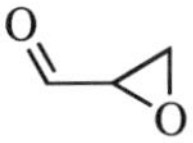

12.1 Chemical and Physical Properties

Glycidaldehyde is a mobile, colorless liquid with a pungent odor, having the following properties:

Specific gravity: 1.1403
Freezing point: – 61.8°C
Boiling point:
  760 mmHg: 112–113°C
  100 mmHg: 57–58°C
Vapor density: 2.58
Refractive index (20°C): 1.4200
Solubility: Completely soluble in all common solvents; insoluble in petroleum ether
Flash point (Tag open cup): 88°F

1 ppm = 2.94 mg/m^3 at 25°C, 760 mmHg; 1 mg/L = 339 ppm at 25°C, 760 mmHg

12.1.2 Odor and Warning Properties

There is a pronounced aldehydelike odor at low levels. Voluntary exposure to serious lung-irritating levels is unlikely.

12.2 Production and Use

Glycidaldehyde is prepared from the hydrogen peroxide epoxidation of acrolein. It is suggested as a bifunctional chemical intermediate and as a cross-linking agent for textile treatment, leather tanning, and protein insolubilization.

12.3 Exposure Assessment

No standard methods for the determination of glycidaldehyde in air were found. NIOSH method 1608 described for glycidol (adsorption on charcoal, displacement with tetrahydrofuran, and determination by gas–liquid chromatography) would likely be suitable (210).

12.4 Toxic Effects

Glycidaldehyde is moderately toxic following oral and dermal exposure, and is extremely irritating to the skin. It is a moderate eye and respiratory tract irritant. It is mutagenic *in vitro* in nonmammalian assays, and is a skin carcinogen in rodents.

12.4.1 Experimental Studies

12.4.1.1 Acute Toxicity. Following ingestion in rats, transitory excitement occurs, succeeded by depression and labored breathing. The LD_{50} in rats is 0.23 g/kg. Respiratory exposure causes marked pulmonary tract irritation and tearing. The LC_{50} in rats exposed 4 h is 251 ppm. Percutaneous absorption gives an approximately lethal dose of 0.2 g/kg in rabbits. At a dermal dose of 0.04 g/kg, severe local injury to the skin occurred, but none of the three rabbits died.

Eye Irritation. Glycidaldehyde is classified as moderately irritating to the rabbit eye, with a Draize score of 3.6/8.0.

12.4.1.5 Carcinogenesis. Glycialdehyde was a skin carcinogen in mice following topical application (219) and subcutaneous injection (220) and in rats by subcutaneous injection (214, 220). It also actively alkylated proteins and DNA (221).

Glycidaldehyde was mutagenic in bacteria, yeast *Klebsiella*, and in the recessive lethal test in *D. melanogaster* (222). It was negative in the mouse lymphoma HGPRT test (222). Glycidaldehyde was also mutagenic in a host-mediated assay, induced unscheduled DNA synthesis in human fibroblasts, and gave positive responses in two different cell-transformation systems (223). The mechanisms of adduct formation from reactions of glycidaldehyde with 2′-deoxyguanosine and/or guanosine have been studied (224).

12.4.2 Human Experience

Glycidaldehyde will produce marked skin irritation and sensitizaion; effective hygiene practices should prevent exposure.

12.5 Standards, Regulations, or Guidelines of Exposure

No standards for permissible occupational exposure were found. The International Agency for the Research of Cancer has classified glycidaldehyde as 2B, that is, a possible human carcinogen (209). The U.S. EPA has established a reference dose (RfD) for glycidaldehyde of 0.4 μg/kg/d (225).

13.0 Bis(2,3-epoxycyclopentyl) Ether

13.0.1 CAS Number: *[2386-90-5]*

13.0.2 Synonyms: 6-Oxabicyclo[3.1.0]hexane, 2,2′-oxybis-

13.0.3 Trade Names: NA

13.0.4 Molecular Weight: 182.22

13.0.5 Molecular Formula: $C_{10}H_{14}O_3$

13.0.6 Molecular Structure:

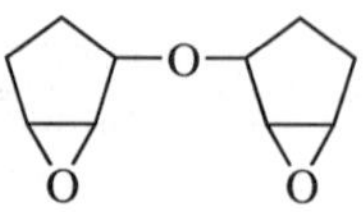

13.1 Chemical and Physical Properties

Boiling Point: 203°C
Viscosity (25°C): 35 cm/sec
Epoxy equivalent weight (g/equiv.): 96
Density: 1.17 g/cm^3

13.1.1 Odor and Warning Properties

No information on odor or warning properties was found.

13.2 Production and Use

This product was developed as a high-performance epoxy resin and epoxy resin diluent.

13.3 Exposure Assessment

No OSHA standard methods for the analysis of this material were found. A method of analysis by GLC has been described (226).

13.4 Toxic Effects

No acute toxicity data were available. The compound was positive in *in vitro* genetic toxicity assays, but mixed results have been reported in cancer bioassays in mice.

13.4.1.5 Carcinogenesis. Dermal application in C3H/JAX mice for 18–20 mo at 30% in acetone resulted in no mice with tumors. All mice had died by 20 mo. When applied in C3H/HeJ mice for up to 24 mo at 50% in acetone for the first 15 mo and 25% in acetone for the rest of the study, there were again no skin tumors (227). In another test in C3H mice at 50 and 10% in acetone, there were 5/80 with skin tumors at the 50% level and 0/40 with skin tumors at 10% (162). When a 1:1 mixture of this compound and ERL 2774 was applied at 50 and 10% in acetone to the same two strains of mice, there were 51/80 and 19/40, respectively, C57BL/6 mice with skin tumors. The sample of ERL 2774 tested was shown to have 10% of an epoxidized polyglycol (mol. wt. 7500) and up to 5000 ppm phenyl glycidyl ether present. Nothing is known of the toxicity potential of the epoxidized polyglycol, but phenyl glycidyl ether is known to cause nasal carcinoma in rats following chronic inhalation (228).

The compound has been shown to be positive in the Ames test for bacterial mutagenesis, positive in a cell transformation assay in Chinese hamster ovary cells, and negative for induction of DNA repair in cultured human leulocytes (33).

13.5 Standards, Regulations, or Guidelines of Exposure

No standards of permissible occupational exposure were found.

14.0 3,4-Epoxycyclohexylmethyl-3,4-Epoxycyclohexanecarboxylate

14.0.1 CAS Number: [2386-87-0]

14.0.6 Molecular Structure:

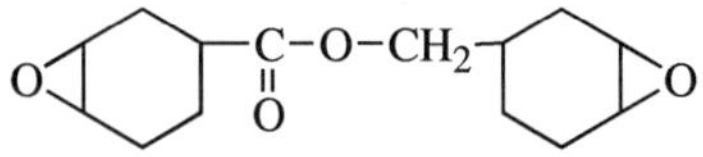

14.1 Chemical and Physical Properties

The properties are as follows:

Viscosity (25°C): 350–450 cm/sec
Specific gravity (25/25°C): 1.175
Color, Gardner, 1933 max.: 1
Epoxy equivalent weight, g/g-mol oxirane equiv.: 131–143
Boiling point (760 mmHg): 354°C
Vapor pressure (20°C): 0.1 mmHg
Freezing point: – 20°C
Solubility in water, 25°C: 0.03% wt.
Solubility of water in, 25°C: 2.8% wt.

Hazardous polymerizations with amines may occur with masses of 1 lb.

14.1.2 Odor and Warning Properties

No information on odor on warning properties was found.

14.2 Production and Use

This compound can be made by treatment of the corresponding *bis*-unsaturated ester with peracetic acid. It is used as a reactive diluent, for filament winding, as an acid scavenger, or as a plasticizer. Vapor hazard is slight at ambient temperature because of its moderately high boiling point. Skin contact is to be avoided.

14.3 Exposure Assessment

No OSHA standard methods for the determination of this material in air were found. The HCl–dioxane method may be used on trapped condensed phases (11) or GLC.

14.4 Toxic Effects

The oral LD_{50} in rats is 4.49 mL/kg. Saturated vapors at ambient temperature for 8 h did not kill rats. The compound is slightly irritating to rabbit skin and has an LD_{50} dermally of 20 g/kg. It produces moderate eye irritation in rabbits. Lifetime mouse skin painting studies were negative for carcinogenesis (133).

14.5 Standards, Regulations, or Guidelines of Exposure

No standards for permissible occupational exposure were found.

BIBLIOGRAPHY

1. C. A. May and Y. Tanaka, *Epoxy Resins*, Dekker, New York, 1973, p. 16.
2. F. Oesch et al., *Biochem. Pharmacol.* **26**, 603 (1977).
3. D. W. Nebert et al., *Mol. Pharmacol.* **8**, 374 (1972).
4. F. Gesch et al., *Arch. Toxicol.* (Berlin) **39**, 97 (1977).
5. T. Hayakawa et al., *Arch. Biochem. Biophys.* **170**, 438 (1975).
6. L. B. Bourne et al., *Brit. J. Ind. Med.* **16**, 81 (1959).
7. A. Welker, *Dermatol. Wochenschr. dd*, 871 (1955).
8. A. Thorgeirsson, S. Fregert, and O. Ramnas, *Acta Dermatol.* **57**, 253 (1977).
9. D. C. Mensik and D. D. Lockwood, unpublished report of The Dow Chemical Company, 1987.
10. R. Irwin, *National Toxicology Program Technical Report 374*, NIH Publication No. 90-2829, 1990.
11. J. L. Jungnickel et al., *Organic Analysis*, Vol. 1, Wiley-Interscience, New York, 1963.
12. C. H. Hine et al., *AMA Arch. Ind. Health* **17**, 129 (1958).
13. C. S. Weil et al., *Am. Ind. Hyg. Assoc. J.* **24**, 305 (1963).
14. R. W. Hend, D. G. Clark, and A. D. Coombs, unpublished report of Shell Chemical Company, 1977.
15. D. G. Clark and S. L. Cassidy, unpublished report of Shell Chemical Company, 1978.
16. D. D. Lockwood and H. W. Taylor, unpublished report of The Dow Chemical Company, 1982.
17. R. J. Nolan, S. Unger, and L. S. Chatterton, unpublished report of the Dow Chemical Company, 1981.
18. V. K. Rowe, unpublished report of The Dow Chemical Company, 1948.
19. M. A. Wolf, unpublished report of The Dow Chemical Company, 1956.
20. M. A. Wolf, unpublished repoft of The Dow Chemical Company, 1957.
21. K. Olson, unpublished report of The Dow Chemical Company, 1958.
22. V. K. Rowe, unpublished report of The Dow Chemical Company, 1958.
23. W. J. Breslin, H. D. Kirk, and K. A. Johnson, *Fundam. Appl. Tox.* **10**, 736 (1988).
24. S. Fregert and B. Lundin, *Sensitization Capacity of Epoxy Resin Compounds*, report to The Swedish Plastics Federation and Swedish Environmental Fund, 1977.
25. A. Thorgeirsson and S. Fregert, *Acta Dermatovener (Stockholm).* **53**, 253 (1977).
26. H. P. Til, Report of Central Institute for Nutrition and Food Research (Civo, Switzerland) to The Dow Chemical Company, Europe, 1977.
27. D. D. Lockwood, unpublished report of The Dow Chemical Company, 1978.
28. M. N. Pinkerton, unpublished report of The Dow Chemical Company, 1979.
29. F. Edgar and E. Donald, unpublished report of Inveresk Research to the Association of Plastic Manufacturers in Europe, 1999.
30. A. Thorgeirsson and S. Fregert, *Acta Dermatovener (Stockholm)* **58**, 17 (1978).

31. D. C. Mensik and D. D. Lockwood, unpublished report of The Dow Chemical Company, 1987.
32. C. S. Weil, personal communication, 1978.
33. C. H. Hine et al., Epoxy compounds. In G. D. Clayton and F. E. Clayton, Eds., Patty's Industrial Hygiene and Toxicology, 3rd ed., Vol. IIa, Wiley, New York, 1981.
33a. M. A. Wolf, unpublished report of the Dow Chemical Company, 1958.
34. W. Basler, W. Gfeller, F. Zak, and V. Skorpil, unpublished report of Ciba-Geigy Ltd., Basel, Switzerland, 1984.
35. W. Dobryszycka et al., *Arch. Toxicol.* **33**, 73 (1974).
36. J. Woyton et al., *Toxicol. Appl. Pharmacol.* **32**, 5 (1975).
37. W. Dobryszyka et al., *Arch. Immunol. Ther. Exp.* **22**, 135 (1974).
38. J. M. Redmond and J. W. Crissman, unpublished report of The Dow Chemical Company, 1996.
39. P. J. Spencer, R. R. Albee, and J. W. Crissman, unpublished report of The Dow Chemical Company, 1996.
40. P. J Spencer et al., *The Toxicologist* **36**, 65 (1997).
41. J. E. Phillips et al., Routine Toxicologic examination of feed restricted 344 male rats unpublished report of The Dow Chemical Company, 1988.
42. J. M. Redmond and J. W. Crissman, unpublished report of The Dow Chemical Company, 1996.
43. J. K. Kodama et al., *Arch. Environ. Health* **2**, 50 (1961).
44. I. J. G. Climie, D. H. Huston, and G. Stoydin, *Xenobiotica* **11**, 391 (1981).
45. P. J., Boogaard, unpublished report of Shell International Chemicals, 1999.
46. I. J. G. Climie, D. H. Huston, and G. Stoydin, *Xenobiotica* **11**, 401 (1981).
47. P. C. Coveney, unpublished report of Shell Chemical Company (SBGR.83.073), 1983.
48. R. J. Nolan, S. Unger, and L. S. Chatterton, unpublished report of the Dow Chemical Company, 1981.
49. P. Bentley, *Carcinogenesis* **10**, 321 (1989).
50. S. Steiner, G. Honger, and P. Sagelsdorff, *Carcinogenesis* **13**, 169–972 (1992).
51. S. M. Roche and C. H. Hine, *Toxicol. Appl. Pharmacol* **12**, 327 (1968).
52. W. J. Breslin et al., unpublished report of The Dow Chemical Company, 1986.
53. J. A. Smith, R. E. Masters, and I. S. Dawe, Report to Ciba-Geigy Limited, Huntingdon Research Centre Ltd., 1988.
54. J. A. Smith, D. A. John, and I. S. Dawe, report to Ciba-Geigy Limited, Huntingdon Research Centre Ltd., 1988.
55. J. A. Smith et al., report to Ciba-Geigy Limited, Huntingdon Research Centre Ltd., 1989.
56. T. R. Hanley et al., unpublished report of The Dow Chemical Company, 1996.
57. T. R. Hanley et al., *The Toxicologist* **36**, 358 (1997).
58. C. H. Hine et al., *Cancer Res.* **18**, 20 (1958).
59. W. L. Kydd and L. M. Sreenby, *J. Natl. Cancer Inst.* **25**, 749 (1960).
60. C. S. Weil, personal communication, through Ref. 33.
61. J. M. Holland, D. G. Gosslee, and N. J. Williams, *Cancer Res.* **39**, 1718 (1979).
62. J. M. Holland et al., Oak Ridge National Laboratory Report No. ORNL 5762, 1981.

63. J. Agee, et al., Report of Kettering Laboratory, Department of Environmental Health, University of Cincinnati Medical Center, Cincinnati, OH, to Celanese Corporation, 1987.
64. W. T. Solomon, Zeon Chemicals, Louisville KY, personal communication, 1990.
65. N. Zakova et al., *Food Chem. Toxicol.* **23**, 1081 (1985).
66. G. C. Peristianis et al., *Food Chem. Toxicol.* **26**, 611 (1988).
67. R. Peto et al., *A Critical Appraisal*, IARC Monographs on the Evaluation of the Carcinogenic Risk of Chemicals to Humans, Suppl. 2, 1980, p. 311.
68. J. W. Crissman and T. K. Jeffries, unpublished report of The Dow Chemical Company, 1998.
69. K. E. Stebbins and P. C. Baker, unpublished report of The Dow Chemical Company, 1998.
70. EPA, United States Environmental Protection Agency, *Summary of EPA Workshop on Carcinogenesis Bioassay via the Dermal Route PB90-146309*, NTIS Springfield, VA 1987.
71. EPA, United States Environmental Protection Agency, *Summary of the Second EPA Workshop on Carcinogenesis Bioassay via the Dermal Route PB90-146358*, NTIS Springfield, VA 1988.
72. T. H. Gardiner et al., *Reg. Pharmacol. Toxicol.* **15**, S1 (1992).
73. J. Pluss, *Z. Unfallmed.* **47**, 83 (1954).
74. E. Grandjean, *Brit J. Ind. Med.* **14**, 1 (1957).
75. S. Fregert and A. Thorgeirsson, *Contact Dermatitis* **3**, 301 (1977).
76. S. Clemmensen, *Drug Chem. Toxicol.* **7**, 527 (1977).
77. D. C. Mensick and D. D. Lockwood, unpublished report of The Dow Chemical Company, 1987.
78. R. S. Cotran, V. Kumar, and S. L. Robbins, *Robbins Pathologic Basis of Disease*, 4th ed., W. B. Saunders, Philadelphia, 1989, p. 774.
79. I. W. Fawcett, A. J. Newman Taylor, and J. Pepys, *Clin. Allergy* **7**, 1 (1977).
80. L. Kanerva, et al., *Scand. J. Work Environ. Health* **17**, 208–215, (1991).
81. Unpublished report of Ciba-Geigy, Limited, 1969.
82. D. D. Lockwood and V. Borrego, unpublished report of The Dow Chemical Company, 1981.
83. W. L. Wilborn, C. M. Parker, and D. R. Patterson, unpublished report of Shell Chemical Company, 1982.
84. J. R. Gardner, unpublished report of Shell Chemical Company, 1992.
85. J. A. Wilson, unpublished report to Shell Chemical Company, 1995.
86. V. A. Jud and C. M. Parker, unpublished report of Chemical Company, 1982.
87. A. S. Lam and C. M. Parker, unpublished report of Shell Chemical Company, 1982.
88. J. A. Heath, unpublished report to Shell Chemical Company, 1996.
89. D. M. Glueck, unpublished report of Shell Chemical Company, 1982.
90. T. M. Brooks. unpublished report of Shell Chemical Company, 1991.
91. S. E. Willington, unpublished report to Shell Chemical Company, 1995.
92. JETOC, Ministry of Labour Japan, 1996.
93. V. L. Sawin and W. M. Smith, unpublished report of Shell Chemical Company, 1983.
94. J. O. Rundell, B. S. Tang, and V. L. Sawin, unpublished report of Shell Chemical Company, 1984.
95. W. M. Smith, B. S. Tang, and V. L. Sawin, unpublished report of Shell Chemical Company, 1984.

96. S. Fregert and L. Trulsson, *Contact Dermatitis* **10**, 112–113 (1984).
97. L. Kanerva, R. Jolanki, and T. Estlander, *Contact Dermatitis* **24**, 293–300 (1991).
98. M. L. Westrick and P. Gross, Industrial Hygiene Foundation of America, Inc., 1960.
99. Wilmington Chemical Corporation, Tech Data Sheet No. 100-774.
100. National Toxicology Program, National Institutes of Health, *Technical Report Series No. 257, NIH Publication 87-2513*, 1987.
101. C. J. McCammon et al., *Proc. Am. Assoc. Cancer Res.* **2**, 229 (1957).
102. Kettering Laboratory, University of Cincinnati, 1958 through Ref. 32.
103. B. L. Van Duuren, L. Orris, and N. Nelson, *J. Nat. Cancer Inst.* **35**, 707 (1965).
104. J. P. Seiler, *Chem. Biol. Interact.* **51**, 347 (1984).
105. E. Boyland and K. Williams, *Biochemistry* **94**, 190 (1965).
106. J. P. Seiler, *Mutat. Res.* **135**, 159 (1984).
107. D. A. Canter et al., *Mutat. Res.* **172**, 105 (1986).
108. D. D. Lockwood and H. W. Taylor, unpublished report of The Dow Chemical Company (1982).
109. M.. D. Shelby et al., *Environ. Mo. Mutag.* **21**, 160 (1993).
110. Imperial Chemical Industries, Limited, unpublished report, 1959.
111. M. A. Wolf, unpublished report of The Dow Chemical Company, 1959.
112. D. D. Lockwood and V. Borrego, unpublished report of The Dow Chemical Company (1979).
113. J. R. Jones, unpublished report of The Dow Chemical Company, 1990.
114. J. R. Jones, unpublished report of The Dow Chemical Company, 1991.
115. Ciba-Geigy, Limited, unpublished report, 1969.
116. P. Chowdhury, unpublished report to the Association of the Plastic Manufacturers in Europe, 1999.
117. N. Dawkes, unpublished report to the Association of the Plastic Manufacturers in Europe, 1998.
118. M. Burman, unpublished report to the Association of the Plastic Manufacturers in Europe, 1998.
119. M. A. Wolf, unpublished report of The Dow Chemical Company, 1958.
120. Ciba-Geigy, Limited, unpublished report, 1971.
121. Ciba-Geigy, Limited, unpublished report, 1972.
122. Ciba-Geigy, Limited, unpublished report, 1979.
123. M. P. Murray and J. E. Cummins, *Environ. Mutagen.* **1**, 307 (1979).
124. Bionetics Research Laboratories, 1970.
125. Kettering Laboratory, University of Cincinnati, 1987.
126. Bionetics Research Laboratories, unpublished report to Ciba-Geigy, 1970.
127. Litton Bionetics, Inc., unpublished report to Ciba-Geigy, 1977.
128. D. Burrows, et al., *Contact Dermatitis* **11**, 80 (1984).
129. J. W. Lacher and J. W. Crissman, unpublished report of The Dow Chemical Company, 1990.
130. M. J. Beekman and J. W. Crissman, unpublished report of The Dow Chemical Company, 1992.
131. Ciba-Geigy, Limited, unpublished report, 1986.
132. Ciba-Geigy, Limited, unpublished report, 1985.

133. Ciba-Geigy, Limited, unpublished report, 1984.
134. B. B. Gollapudi and Y. E. Samson, unpublished report of The Dow Chemical Company, 1992.
135. J. S. C. English, I. et al., *Contact Dermatitis* **15**, 66 (1986).
136. A. D. Morris, et al., *Contact Dermatitis* **38**, 57 (1998).
137. D. G. Clark and A. D. Coombs, unpublished report of Shell Chemical Company, 1977.
138. S. Denton, confindential report of Shell Chemicals Europe Ltd., 1998.
139. D. Blair, unpublished report of Shell Chemical Company, 1983.
140. R. G. Pickering, unpublished report of Shell Chemical Company, 1981.
141. P. Boogaard, confidential report of Shell International Chemicals B. V., 1998.
142. B. J. Dean, et al., unpublished report of Shell Chemical Company, 1979.
143. N. Dawkes, confidential report of Shell Chemicals Europe Ltd., 1998.
144. A. L. Meyer, unpublished report of Shell Chemical Company, 1980.
145. M. F. Wooder and C. L. Creedy, unpublished report of Shell Chemical Company, 1980.
146. I. Dahlquist and S. Fregert, *Contact Dermatitis* **5**, 121 (1979).
147. C. R. Lovell, R. J. G. Rycroft, and J. Matood, *Contact Dermatitis* **11**, 190 (1984).
148. G. R. Schultz, Ed., *OSHA Analytical Methods Manual* (Organic) Part I, Method 07, 1989.
149. NIOSH. *Manual of Analytical Methods*, 4th ed., P. M. Eller and M. E. Cassinelli, Eds., United States Government Printing Office, Supt of Docs, 1994.
150. B. M. de Rooij, et al., *Environ. Toxicol. Pharmacol* **3**, 175–185 (1997).
151. F. K. Dietz, M. Grandjean, J. T. Young, unpublished report of The Dow Chemical Company, 1985.
152. S. Pallade et al., *Arch. Mal. Prof. Med. Trav. Secur. Soc.* **28**, 505 (1967).
153. W. H. Lawrence et al., *J. Pharm. Sci.* **61**, 1712 (1972).
154. K. S. Rao, J. E. Besto, and K. J. Olson, *Drug Chem. Toxicol.* **4**, 331 (1981).
155. A. Kandel, unpublished report of The Dow Chemical Company, 1953.
156. J. C. Gage, *Brit. J. Ind. Med* **16**, 11 (1959).
157. J. F. Quast et al., unpublished report of The Dow Chemical Company, 1979.
158. J. A. John et al., *Toxicol. Appl. Pharmacol.* **68**, 415 (1983).
159. B. L. Robinson, et al., The *Toxicologist* **9**, (Abstr. #909) 227 (1989).
160. A. Ito et al., *Toho Igakkai Zasshi.* **42**, 321–330 (1995).
161. F. B. Daniel et al., *Drug Chem. Toxicol.* **19**, 41–58 (1996).
162. W. E. Weigel, H. B. Plótnick, and W. L. Conner, *Res. Commun. Chem. Pathol. Pharmacol.* **20**, 275 (1978).
163. F. A. Smith, P. W. Langvardt, and J. D. Young, *Toxicol. Appl. Pharmacol.* **48**, A116 (1979).
164. G. Fakhouri and A. R. Jones, *Aust. J. Pharm. Sci.* **8**, 11 (1975).
165. A. M. Rossi et al., *Mutat. Res.* **118**, 213–226 (1983).
166. R. Gingell et al., *Drug Metab. Disp.* **13**, (1985).
167. A. Ito and Y. Feng, *Jpn. J. Toxicol.* **12**, 51–58 (1999).
168. D. E. Moody et al., *Biochem. Pharmacol.* **41**, 1625–1637 (1991).
169. W. M. Kluwe, B. N. Gupta, and J. C. Lamb, *Tox. Appl. Pharmacol.* **70**, 67–87 (1983).
170. J. A. John et al., *Fundam. Appl. Toxicol.* **3**, 437 (1983).

171. T. A. Marks, F. S. Gerling, and R. E. Staples, *J. Toxicol. Environ. Health* **9**, 87 (1982).
172. J. D. Hahn, *Nature* 226, 87 (1970).
173. S. L. Cassidy, K. M. Dix, T. Jenkins, unpublished Report of Shell Chemical, 1982.
174. M. Omura et al., *Bull. Environ. Contam. Toxicol.* **55**, 1–7 (1995).
175. G. P. Toth, H. Zenick, M. K. Smith, *Fund. Appl. Toxicol.* **13**, 16–25 (1989).
176. V. L. Slott and B. F. Hales, *Teratology* **32**, 65 (1985).
177. G. R. Klinefelter et al., *J. Androl.* **18**, 139–150 (1997).
178. K. Saito et al., unpublished report of Sumitomo Chemical Company, Ltd., 1999.
179. S. Laskin et al., *J. Natl. Cancer Inst.* **65**, 751 (1980).
180. Y. Konishi et al., *Gann* 71, 922 (1980).
181. P. W. Wester, et al., *Toxicology* **36**, 325 (1985).
182. Dynamac Corporation, Health Assessment Document for Epichlorohydrin, Draft Report #1, submitted to the US-EPA, Contract No. 68-03-3111, 1982.
183. B. L. Van Duuren et al., *J. Natl. Cancer Inst.* **53**, 695 (1974).
184. R. J. Sram et al., *Mutat. Res.* **87**, 299 (1981).
185. *Health and Safety Executive UK, Toxicity Review 24*: Ammonia, 1-Chloro-2,3-epoxypropane (Epichlorohydrin), Carcinogenicity of Cadmium and Its Compounds, 1991.
186. A. K. Giri, *Mutat. Res.* **386**, 25–38 (1997).
187. R. J. Sram, M. Cerna, and M. Kucerova, *Biol. Zentralbl.* **95**, 451 (1976).
188. B. J. Dabney et al., unpublished Report of The Dow Chemical Company, 1979.
189. T. Tsuchimoto and B. E. Matter, *Prog. Mutation Res.* **1**, 705–711 (1981).
190. B. Kirkhat, *Prog. Mutation Res.* **1**, 698–704 (1981).
191. M. F. Salamone, J. A. Heddle, and M. Katz, *Prog. Mutation Res.* **1**, 686–697 (1981).
192. M. Terada et al., *Mutati. Res.* **272**, 288 (1992).
193. A. O. Asita et al., *Mutati. Res.* **271**, 29–37 (1992).
194. Y. F. Wang and C. H. Hine, *Chin. Med. J. Engl.* **99**, 461–464 (1986).
195. P. G. N. Kramers et al., *Mutati. Res.* **252**, 17–33 (1991).
196. S. S. Epstein et al., *Toxicol. Appl. Pharmacol.* **23**, 288 (1972).
197. G. Prodi et al., *Tox. Pathol.* **14**, 438–444 (1986).
198. W. K. Lutz, *Mutati. Res.* **65**, 289–356 (1979).
199. E. V. Sargent et al., *Mutati. Res.* **263**, 9–12 (1991).
200. M. Uzeil et al., *Mutati. Res.* **277**, 35–90 (1992).
201. H. Hindo-Landin et al., *Chem. Biol. Inter.* **117**, 49–64 (1999).
202. H. Ippen and V. Mathias, *Berufsdermatosen* **18**, 144 (1970).
203. C. Schultz, *Dtsch. Med. Wochenschr.* **89**, 1342 (1964).
204. E. Epstein, *Contact Dermatitis Newsl.* **16**, 475 (1974).
205. D. Lambert et al., *Ann. Dermatol. Venerol.* **105**, 521 (1978).
206. M. H. Beck and C. M. King, *Contact Dermatitis* **9**, 315 (1983).
207. T. Van Joost, *Contact Dermatitis* **19**, 278 (1988).
208. G. W. Olsen et al., *Am. J. Industrial Med.* **25**, 205–218 (1994).
209. *IARC (International Agency for Research on Cancer) Monograph 71*, 1998.

210. NIOSH. *Manual of Analytical Methods*, 4th ed., P. M. Eller and M. E. Cassinelli, Eds., United States Government Printing Office, Supt of Docs, Washington, DC, 1994.
211. J. W. Henck, D. D. Lockwood, and H. O. Yakel, unpublished report of The Dow Chemical Company, 1978.
212. E. D. Thompson and D. P. Gibson, *Food Chem. Toxicol.* **22**, 665 (1984).
213. V. L. Slott and B. F. Hales, *Teratology* 32, 65 (1985).
214. B. L. Van Duuren et al., *J. Natl. Cancer Inst.*, **39**, 1217 (1967).
215. R. D. Irwin et al., *J. Appl. Toxicol.* **16**, 201–209 (1996).
216. E. D. Thompson and D. P. Gibson, *Food Chem. Toxicol.* **22**, 665 (1984).
217. O. S. Bhanot, U. S. Singh, and J. J. Solomon, *Proc. Am. Assoc. Cancer Res.* **38**, 40 (1997).
218. ACGIH, American Conference of Governmental Industrial Hygienists TLVs® and BELs® Cincinnati, OH, 2000.
219. B. L. VanDuuren, L. Orris, and N. Nelson, *J. Natl. Cancer Inst.* **35**, 707–717 (1965).
220. B. L. Van Duuren et al., *J. Natl. Cancer Inst.* **37**, 825 (1966).
221. S. Steiner and W. P. Watson, *Carcinogenesis* **13**, 119 (1992).
222. A. G. A. C. Knaap, C. E. Voogd, and P. G. N. Kramers, *Mutat. Res.* **101**, 199 (1982).
223. Dutch MAC Commission, 1995.
224. B. T. Golding, P. K. Slaich, and W. P. Watson, *J. Chem. Soc. Chem. Commun.* **7**, 515–517 (1996).
225. US EPA Integrated Risk Information System (IRIS), 1991.
226. J. M. Holland, et al., Oak Ridge Natl. Lab. Rep. ORNL-5375, 1978.
227. C. S. Weil, unpublished report of the Carnegie-Mellon Institute of Research, 1978.
228. K. P. Lee, P. W. Schneider, and H. J. Trochimowicz, *Am. J. Pathol.* **3**, 140 (1983).

CHAPTER EIGHTY-FOUR

Organic Peroxides

Jon B. Reid, Ph.D., DABT

INTRODUCTION

A review of the literature for these compounds has resulted in very little new information. Consequently, the content of this chapter is drawn heavily from the previous authors, including the tables. The only compounds where there has been publishing activity are: dibenzoyl peroxide, methyl ethyl ketone peroxide, *t*-butyl hydroperoxide, isopropylbenzene hydroperoxide, and peroxyacetic acid. A recent review compared the skin tumor promoting activity of different organic peroxides in SENCAR mice (especially *t*-butyl peroxide, dicumyl peroxide, and also dibenzoyl peroxide) (1). Much of this literature is, however, irrelevant to health. The introduction presented in the previous edition is entirely correct and relevant.

Peroxides are highly reactive molecules due to the presence of an oxygen–oxygen linkage. Under activating conditions, the oxygen–oxygen bond may be cleaved to form highly reactive free radicals. These highly reactive radicals can be used to initiate polymerization or curing. Consequently, organic peroxides are used as initiators for free-radical polymerization, curing agents for thermoset resins, and cross-linking agents for elastomers and polyethylene. In some cases they can be used as antiseptic agents.

These materials must be handled and stored with caution. If free radicals are formed during storage in concentrated form, an accelerated decomposition could result, leading to the release of considerable heat and energy. It has been determined that decompositions of commercially available peroxides are generally low-order deflagrations rather than detonations (2).

There have been several investigations into the types of physical hazards represented by organic peroxides (2–5). These compounds may possess the combination of thermal

Patty's Toxicology, Fifth Edition, Volume 6, Edited by Eula Bingham, Barbara Cohrssen, and Charles H. Powell.
ISBN 0-471-31939-2

instability, sensitivity to shock, and/or friction, as well as flammability. Organic peroxides tend to be unstable, with the instability increasing with greater concentrations. Because of their instability, many peroxides are stored/handled in inert vehicles (6). It has been shown by Tamura (3) that the ignition sensitivity and the violence of deflagration for each organic peroxide may have a tendency to increase with increasing active oxygen content among the same type of organic peroxide, with a few exceptions. The ignition sensitivity and the violence of deflagration for each type of organic peroxide may decrease in the following order, given the same active oxygen content: diacyl peroxides > peroxyesters > dialkyl peroxides > hydroperoxides (3).

Basically only acute health testing has been performed on organic peroxides. Exposures should be well controlled, primarily owing to the decomposition or deflagration hazard of the organic peroxide. The health data presented in Tables 84.1–84.9 were collected and furnished by the Organic Peroxide Producers Safety Division of the Society of the Plastics Industry to the previous authors of this chapter and are presented again in this edition. This represents an effort by industry to evaluate their products and provide that information to the public. Most of the information in the table has been previously published (7) in an industry bulletin.

The analytic method should also be specific for each organic peroxide. NIOSH has fully validated a high-performance liquid chromatography/ultraviolet light method for benzoyl peroxide (8). Very few of the other organic peroxides have fully validated analytic methods. One of the first conventional methods to determine concentrations of organic peroxides was the titration of iodine from sodium iodide. However, it was not specific for organic peroxides. The polarographic method came into use with the visible-recording polarograph because hydroperoxides could be distinguished from other peroxides. This method could identify the functional groups and also quantify mixtures. Di-*t*-butyl peroxide is an exception because it is not reduced polarographically (9).

Many analytical methods that have not been subjected to review by consensus standard organizations or regulatory agencies are in use. Some examples include gas chromatography (dialkyl peroxides such as di-*t*-butyl peroxide), high-pressure liquid chromatography (peroxyketals), and iodometric titration (peroxyesters, diacyl peroxides, hydroperoxides, and peroxydicarbonates). Some degree of selectivity in iodometric titrations may be obtained by variation of the reducing agent employed and the reaction conditions.

A PEROXYDICARBONATES

1.0 Diisopropyl Peroxydicarbonate

1.0.1 CAS Number: [105-64-6]

1.0.2 Synonyms: Diisopropyl perdicarbonate and isopropyl perdicarbonate

1.0.3 Trade Names: IPP

1.0.4 Molecular Weight: 206.22

Table 84.1. Toxic Properties of Peroxydicarbonates (7)*

Peroxydicarbonate	CAS No.	Oral LD_{50} (mg/kg)	Dermal LD_{50} (mg/kg)	Primary Skin Irritation[a]	Eye Irritation[a]	LC_{50}, ppm (mg/L)[b]	*Salmonella typhimurium* Assay
Diisopropyl peroxydicarbonate	[*105-64-6*]						
100%		2140		Mod. irritant	Ext. irritating		
30% in toluene		3720	2025	Ext. irritating	Mod. irritating		
45% in Soltrol 130		8500	1.7[c]	V. severe irritant	Irritant	Irritant gas	
45% in cyclohexane–benzene		6500	4.0[c]	V. severe irritant	Irritant		
Di-*n*-propyl peroxydicarbonate	[*16066-38-9*]						
100%		3400	> 6800	Ext. irritating	Ext. irritating	> 22.7 (> 0.19) 1 h	
85% in methylcyclohexane		4600	3500	Ext. irritating	Ext. irritating	> 1433 (> 12) 1 h	
Di-*sec*-butyl peroxydicarbonate	[*19910-65-7*]						
100%		7600				172 ppm — no adverse effects (1 h)	
75% in odorless mineral spirits		> 4640	Not toxic at 2000[d]		Irritant		
75% in Soltrol 130		9300	1200				
Di-(2-ethylhexyl) peroxydicarbonate	[*16111-62-9*]						
97% min			Not toxic at 2000[d]		Irritant		
75% in Soltrol 130		1020					
40% in Soltrol 130		20,800					
40% in dimethyl phthalate		3690					
Di-(4-*t*-butylcyclohexyl)peroxydicarbonate	[*15520-11-3*]	> 5000		Not an irritant	Not an irritant		Negative
Di-*n*-butyl peroxydicarbonate	[*16215-49-9*]						
50% in aromatic-free mineral spirits		10[c]		V. servere irritant	Slight irritant		

Table 84.1. (*Continued*)

Peroxydicarbonate	CAS No.	Oral LD_{50} (mg/kg)	Dermal LD_{50} (mg/kg)	Primary Skin Irritation[a]	Eye Irritation[a]	LC_{50}, ppm (mg/L)[b]	*Salmonella typhimurium* Assay
Di-(3-methylbutyl) peroxydi-carbonate	[*4113-14-8*]						
20% in white spirits						1.7 ppm — no toxic signs; slight nose irritation	
Diacetyl peroxydicarbonate 75% wet	[*26322-14-5*]	5000		Not an irritant	Not an irritant		Negative
Di-(2-phenoxyethyl) peroxy-dicarbonate	[*41935-39-1*]	> 20,000	> 20,000		Mild irritant		Negative
Di-(2-chloroethyl) peroxydi-carbonate	[*6410-72-6*]	4000 400					
Di-(3-chloropropyl) peroxydi-carbonate	[*34037-78-0*]	1500					
Di-(4-chlorobutyl) peroxidi-carbonate	[*14245-74-0*]	5200					
Di-(2-butoxyethyl) peroxidi-carbonate	[*6410-72-6*]	4000					

[a] V = Very; ext = extremly; mod = moderately.
[b] All LC_{50} tests lasted 4 h unless noted otherwise.
[c] LD_{50} reported in ml/kg.
[d] According to the Federal Hazardous Substances Act.
* All studies used rats.

Table 84.2. Toxic Properties of Diacyl Peroxides (7)*

Diacyl peroxide	CAS No.	Oral LD_{50} (mg/kg)	Dermal LD_{50} (mg/kg)	Primarry Skin Irritation	Eye Irritation	LC_{50}(mg/L)[a]	*Salmonella typhimurium* Assay (Unless Specified)
Dibenzoyl proxide	*[94-36-0]*	Not toxic at 5000[b]		Not an irritant	Not irritanting (5 min wash)	> 24.3	Negative
78% wet					Strongly irritating, but not corrosive (24 h wash)		Negative[c]
Di-(2,4-dichlorobenzoyl) peroxide	*[133-14-2]*						
50% in silicone fluid		> 12.918	> 8000		Not an irritant		
Di-*p*-chlorobenzoyl peroxide	*[94-17-7]*	500(IP)					Negative[c]
Di-(2-methylbenzoyl) peroxide	*[3034-79-5]*						
78% wet		> 5000		Severe irritant	Irritant (unwashed)		Negative
Didecanoyl peroxide	*[762-12-9]*	> 5000		Mod. irritating	Slight irritation		Negative
Dilauroyl peroxide	*[105-74-8]*	> 5000[b]		Not an irritant	Not an irritant	Toxic at 200	Negative
					RTECS-moderate[f]		Negative[d]
Diacetyl peroxide	*[110-22-5]*				Severe		
Dipproprionyl peroxide 22.7% in white spirits	*[3248-28-0]*					Saturated — 1.5 h — all animals died; 100 ppm — nose and eye irritation, respiratory difficulty, 1 death	Negative

Table 84.2. (*Continued*)

Diacyl peroxide	CAS No.	Oral LD_{50} (mg/kg)	Dermal LD_{50} (mg/kg)	Primarry Skin Irritation	Eye Irritation	LC_{50}(mg/L)[a]	*Salmonella typhimurium* Assay (Unless Specified)
Di-*n*-octanoyl peroxide	[*762-16-3*]			Severe irritation	Slight irritation		Negative
50% in Shellsol T		> 5000					
Di-(3,5,5-trimethyl hexanoyl) peroxide	[*3851-87-4*]						
75% in isododecane		12.7[e]		Very sereve irritation	Irritant		Negative

[a]All LC_{50} tests lasted 4 h unless noted otherwise.
[b]According to the Federal Hazardous Substances Act.
[c]Tumor cell growth assay.
[d]Mouse lymphoma forward mutation assay, with/without metabolic activation.
[e]LD_{50} reported in mL/kg.
[f]Registry of Toxic Effects of Chemical Substances Classified it as moderate irritant.
*All studies used rats.

Table 84.3. Toxic Properties of Peroxyesters (7)[*]

Peroxyester	CAS No.	Oral LD_{50} (mg/kg)	Dermal LD_{50}	Primary Skin Irritation	Eye Irritation	LD_{50} (mg/L)[a]	*Salmonella typhimurium* Assay (Unless Specified)[b]
t-Butyl peroxyacetate	[*107-71-1*]						
75% in OMS[c]		2562	4757	Slight irritant	Irritant	6.1	
70% in benzene (mice)		1900					
50% in Shellsol T					450 (8 hr)		
t-Butyl peroxypivalate	[*927-07-1*]						
75% in OMS[c]		4169–4640	2500	Mod to severe	Not an irritant	7.79	
t-Amyl perioxypivalate	[*29240-17-3*]						
75% in OMS[c]		4270	> 2000	Not an irritant	Not an irritant	> 9.5	
t-Butyl peroxybenzoate	[*614-45-9*]	3639–4838	3817	Not an irritant or sensitizer	Slight irritation	> .26	Slightly positive/ negative + CHO ± ML
Mice		914–2500					
t-Butyl peroxy-2-ethyl-hexonate	[*3006-82-4*]	> 10,000	16,818	Not an irritant	Not an irritant	42.2	
t-Amyl peroxy-2-ethyl-hexonoate	[*686-31-7*]	> 5000	> 2000	Slightly irritating	Not an irritant		
t-Butyl peroxy-3,5,5-tri-methylhexanoate	[*13122-18-4*]	17.4[d]		Mod. to severe	Not an irritant	> 0.8	Negative
t-Butyl peroxyneode-canoate	[*26748-41-4*]						
50% Shellosol T		> 12,918	> 8000	Mod. to severe	Not an irritant	50.0	
t-Butyl peroxy-2-ethyl-hexylcarbonate	[*34443-12-4*]	> 5,000	> 2000	Mildly irritating	Not an irritant		
t-Butyl peroxycronate	[*23474-91-1*]	4100		Moderate irritant			
t-Amyl peroxybenzoste	[*4511-39-1*]						Negative
Cumyl peroxyneode-canoate	[*26748-47-0*]						
90%		5126	> 7940 < 19,800	Not an irritant		20.2	

Table 84.3. Toxic Properties of Peroxyesters (7)*

Peroxyester	CAS No.	Oral LD_{50} (mg/kg)	Dermal LD_{50}	Primary Skin Irritation	Eye Irritation	LD_{50} (mg/L)[a]	*Salmonella typhimurium* Assay (Unless Specified)[b]
75% in OMS[c]						> 20.4 slight dyspnea, eye squint and wt. loss	
2.5-dimethyl-2,5-di-(2-ethylhexonoyl-peroxy)hexane	[*13052-09-0*]	> 12,918	> 8000	Not an irritant	Not an irritant	> 800	
Di-*t*-butyl diperoxy-azelate	[*16580-06-6*]						
75% in OMS[c]		> 5000	> 2000	Severe irritant	Not an irritant		
1,1,3,3-Tetramethylbutyl peroxyphenoxyacetate	[*59382-51-3*]						
30% in Shellosol T (OMS[c])		> 12.0[d]		Severe irritant	Not an irritant	> 24 ppm	
t-Butylperoxyisopropyl carbonate	[*2372-21-6*]	5.0[d]	> 10,000	Not an irritant	Conjunctivitis		
t-Butyl monoperoxy-maleate	[*1931-62-0*]	16(Ip)					
Di-*t*-butyl diproxyphylate	[*15042-77-0*]	128 (Ip)					
Cumyl peroxyneohep-tanoate	[*130097-36-8*]	~5000		Mildly irritating			
	[*104852-44-0*]						
75% in OMS[c]							

[a]All LC_{50} test lasted 4 h unless noted otherwise.
[b]± CHO indicates a positive or negative finding in Chinese hamster ovary cells. The following +/− indicating with/without metabolic activation. ± ML indicates a positive or negative finding in the mouse lymphoma forward mutation assay.
[c]Odorless mineral spirits.
[d]LD_{50} reported in mL/kg.
*All studies used rats unless otherwise specified.

Table 84.4. Toxic Properties of Ketone Peroxides (7)*

Ketone Peroxide	CAS No.	Oral LD_{50} (mg/kg)	Dermal LD_{50} (mg/kg)	Primary Skin Irritation	Eye Irritation	LC_{50} (mg/L)[a]	*Salmonella typhimurium* Assay
Methyl ethyl ketone peroxide	[*1338-23-4*]						Positive
OPPSD composite		> 500 < 5000		Moderate irritant	Corrosive	> 200	Negative
Lupersol DDM		681					
Lucidol & Cadet		484		Irritant	Irritant	200 ppm	
Noury(40% in DMP[b])		1017	4000		Corrosive	17	
Noury & Lucidol						33	
Methyl isobutyl Ketone peroxides	[*37206-20-5*]	1.77[d]		Very severe irritant	Sereve irritant	1.5	Negative
Acetyl acetone peroxide	[*37187-22-7*]	2870		Not an irritant	Severe irritant	Not a hazard at 13.1 mg/L for 1 h	
Diacetone alcohol peroxide	[*54693-46-8*]	2.68[d]		Very severe irritant	Severe irritant	0.54	Negative
1,1-Dihydroperoxycyclo-hexane	[*2699-11-9*]						
21% in DMP[b]		1.08[d]		Very severe irritant	Very severe irritant		Slightly positive
Cyclohexanone peroxide	[*12262-58-7*]				Severe irritant		
Di-(1-hydroxycyclohexyl)-peroxide	[*2699-12-9*]						
100% (mice)		900		Irritating	Irritating		
60% in DBP[c] (mice)		850					
1-Hydroperoxy-1-hydroxy-diclohexyl peroxide	[*78-18-2*]						
100% (mice)		880		Irritating	Irritating		
60% in DBP[c](mice)		740					

[a] All LC_{50} tests lasted 4 h unless noted otherwise.
[b] DMP dimethyl phthalate.
[c] DBP dibutyl phthalate.
[d] LD_{50} reported in ml/kg.
* All studies used rats unless otherwise specified.

Table 84.5. Toxic Properties of Dialkyl Peroxides (7)[*]

Dialkyl Peroxide	CAS. No	Oral LD_{50} (mg/kg)	Dermal LD_{50} (mg/kg)	Primary Skin Irritation	Eye Irritation	LC_{50}[a]	*Salmonella typhimurium* Assay
Di-*t*-butyl peroxide	[*110-05-4*]	> 25,000		Not an irritant	Not an irritant	> 4103 ppm	Negative
Mice		> 20.0[b]	> 10,000				
Mice		> 50.0[b]					
2,5-Dimethyl-2,5-di-(*t*-butylperoxy)hexane	[*78-63-7*]	> 32,000	4100				
2,5-Dimethyl-2,5-di-(*t*-butylperoxy)hexyne-3	[*1068-27-5*]						
In dodecane		> 7680		Moderate irritatant		Nontoxic[c]	
90%				Not an irritant			
Dicumyl peroxide	[*80-43-3*]						
96% min		4100		Mild irritation, no sensitizer			Negative
Dust from 40% on Filter						2.24 mg/L no effect (6 h)	
20% in corn oil		~4000					
50% in corn oil					Mild (unwahed) conjuctivities		

α,α′-*Bis*(*t*-butylperoxy)di-isopropylbenzenes 96% min	[*25155-25-3*] [*2781-00-2*]						
		> 23,100		Slight irritation, not a sensitizer	Minimal irritation	> 6000 ppm — vapor > 180 mg/m^3 — dust	Negative
						> 180 mg/m^3 — dust	
Mice		> 4500					
t-Butyl cumyl peroxide 92%	[*3457-61-2*]	5.18[b]		Severe irritant	Not an irritant	> 140 ppm (1.2 mg/L)	Negative
4-(*t*-Butylperoxy)-4-methyl-2-pentanone	[*26394-04-7*]	3949	> 20,000		Slight irritation (unwashed)	> 2.3 mg/L (1 h)	

[a]All LC_{50} tests lasted 4 hr unless noted otherwise.
[b]LD_{50} reported in ml/kg.
[c]According to the Fedral Hazardous Substances Act.
*All studies used rats unless rats unless otherwise specified.

Table 84.6. Toxic Properties of Peroxyketals (7)*

Peroxyketal	CAS No.	Oral LD_{50} (mg/kg)	Dermal LD_{50} (mg/kg)	Primary Skin Irritation	Eye Irritation	LC_{50} (mg/L)[a]	*Salmonella typhimurium* Assay
1,1-Di-(*t*-butylperoxy)-3,3,5-Trimethylcyclohexane	*[6731-36-8]*						
75% in DBP[b]		> 12,918	> 8000		Not an irritant	~800	
1,1-Di(*t*-butylperoxy)-cyclohexane	*[3006-86-8]*						
65% in DBP[b]		16,653		Not an irritant	Not an irrtation	> 207.2	
2,2-Di(*t*-butylperoxy)butane	*[2167-23-9]*						
50% in DBP[b]		23.2[c]		Mod. irritating	Slight irritation	> 2.42	Negative
50% in mineral oil		> 30.0[c]			Slight irritation		
n-Butyl 4,4-di-(*t*-butylperoxy)valerate	*[995-33-5]*						
40% in chalk		> 5000		Not an irritant	Slight irrtation		
2,2-Di-(cumylperoxy)-propane	*[4202-02-2]*						
50% in odorless mineral spirits		11.5[c]		Severe irritant	Not an irritant		Negative

[a]All LC_{50} tests lasted 4 h unless noted otherwise.
[b]Dibutyl phthalate.
[c]LD_{50} reporterd in mL/kg.
*All studies used rats.

Table 84.7. Toxic Properties of Hydroperoxides (7)*

Hydroperoxide	CAS No.	Oral LD_{50} (mg/kg)	Dermal LD_{50} (mg/kg)	Primary Skin Irritation	Eye Irritation	LC_{50}, ppm (mg/L)[a]	*Salmonella typhimurium* Assay
t-Butyl Hydroperoxide	*[75-91-2]*						
70%		560	0.5[b]	Extreme irritant	Extreme irritant	502 (1.85)	Positive (see text)
80% 20% di-*t*-butyl peroxide		406	> 10,000	Irritating	Irritant	500	
a-Cumyl hydroperoxide	*[80-15-9]*						
80–83% in corn oil		800–1600	> 200	Severe irritation corrosive (DOT)	Irritant	700 (4.3) (6 h)	
73%		382		Irritating	Irritating	220	Inconclusive
1-Phenylethyl hydroperoxide	*[3071-32-7]*						
30% in ethylbenzene		800	1700	Severe irritant	Severe irritant	20–33 mg/L	
1,1,3,3-Tetramethylbutyl hydroperoxide	*[5809-08-5]*	0.92[b]		Very severe irritant	Very severe irritant	> 480 (2.85)	Negative
1,2,3,4-Tetrahydro-1-naphthyl hydroperoxide	*[771-29-9]*	250 (unk.route in mice)					
1-Vinyl-3-cyclohexen-1-yl hydroperoxide	*[3736-26-3]*		1440				
Diisopropylbenzene hydroperoxide	*[26762-93-6]*						
53%		6200		Severe irritation (immidiate) corrosive (DOT)	Severe irritant	4.5 mg/L (6 h)	
p-Menthyl Hydroperoxide	*[26762-92-5]*						
55%		3700		Severe irritation (immediate) corrosive (DOT)	Severe irritation	9.2 mg/L (6 h)	Positive

[a] All LC_{50} tests lasted 4 h unless noted otherwise.
[b] LD_{50} reported in mL/kg.
* All studies used rats.

Table 84.8. Toxic Properties of Peroxyacids (7)*

Peroxyacids	CAS No.	Oral LD_{50} (mg/kg)	Dermal LD_{50}	Primary Skin Irritation	Eye Irritation	LC_{50} (ppm)	*Salmonella typhimurium* Assay (Unless Specified)
Peroxyacetic acid	*[79-21-0]*						
100%		1540	1410 mg/kg				Negative
40% in acetic acid		1230	0.71 mL/kg	Severe irritant	Severe irritant	> 500- < 1000 (4 h)	
p-Nitroperoxybenzoic acid	*[943-39-5]*						Subcutaneous sarcomas

*All studies used rats.

Table 84.9. Toxic Properties of Silyl Peroxides (7)*

Silyl Peroxide	CAS No.	Oral LD_{50} (mg/kg)	Dermal LD_{50} (mg/kg)	Primary Skin irritation	Eye Irritation	LC_{50} (ppm)[a]	*Salmonella typhimurium* Assay
Vinyltri-(*t*-butylperoxy)silane	*[15188-09-7]*						
100%		$> 100 < 250$	$> 10{,}000$	No toxic signs	Inflammation	0.68 (9 mg/L)	Negative
40% in *n*-hexane		~2.5	$> 10{,}000$	No toxic signs	Slight irritation		
5% in odorless mineral spirits						0.006–0.009 mg/L	
Cumylperoxytrimethyl silane	*[18057-16-4]*					> 22.3	

[a]All LC_{50} tests lasted 4 h unless noted otherwise.
*All studies used rats.

1.0.5 Molecular Formula: $C_8H_{14}O_6$

1.0.6 Molecular Structure:

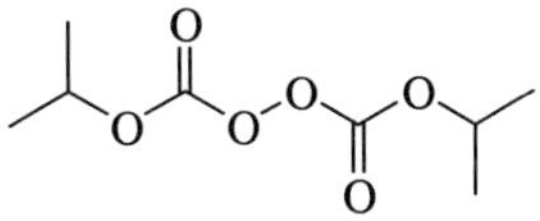

1.1 Chemical and Physical Properties

1.1.1 General

No new information was found on this compound. Diisopropyl peroxydicarbonate is a colorless, coarse granular crystalline solid (12). Diisopropyl peroxydicarbonate is one of the few commercially available organic peroxides to be classified as a high-order deflagration hazard. It has a low decomposition temperature, and its volatile decomposition products can be ignited in air. Comparably, however, IPP is much less easily ignited than black powder or dry benzoyl peroxide. IPP in the frozen state is not sensitive to friction and is less sensitive to impact.

1.2 Production and Use

This peroxide is produced by careful reaction of isopropyl chloroformate with aqueous sodium peroxide at low temperature. IPP is commercially available in greater than 99% purity with an active oxygen content of 7.8%. It has attained commercial importance as an efficient polymerization catalyst at low reaction temperature. It is particularly useful for the polymerization of unsaturated esters such as diethylene glycol bis(allyl carbonate) and has found application as an initiator for the polymerization of ethylene and vinyl chloride (13).

1.3 Exposure Assessment

1.3.3 Workplace Methods

At present there is no method for monitoring the workplace air for the concentration of IPP; however, it liberates iodine from acidified solutions of potassium iodide, providing the basis for an analytic method.

1.4 Toxic Effects

Animal toxicity test data are found in Table 84.1. IPP, however, has a low vapor pressure, which should reduce the risk in the workplace unless an operation or process causes this compound to be vaporized or aerosolized.

1.4.1 Experimental Studies

1.4.1.1 Acute Toxicity. For information on oral LD_{50}, dermal LD_{50}, primary skin irritation, eye irritation, and LC_{50}, see Table 84.1. It should be emphasized that rats inhaling vapor at 9–13 ppm on regular basis over a 3 wk period produced manifestations of an irritant gas.

1.4.2 Human Experience

1.4.2.1 General Information. No ill effects have been observed among laboratory or plant personnel, except for isolated individual cases of dermatitis and sensitivity to the characteristic odor of IPP (13).

1.5 Standards, Regulations, or Guidelines of Exposure

The National Fire Protection Association (NFPA) (12) recommends storing this material in a freezer with a temperature alarm in a detached, noncombustible building.

2.0 Di-*n*-Propyl Peroxydicarbonate

2.0.1 CAS Number: [16066-38-9]

2.0.2 Synonyms: NA

2.0.3 Trade Names: NA

2.0.4 Molecular Weight: 206.22

2.0.5 Molecular Formula: $C_8H_{14}O_6$

2.0.6 Molecular Structure:

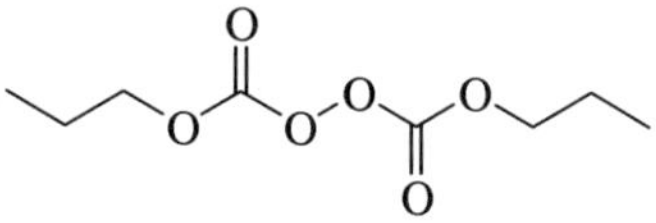

2.1 Chemical and Physical Properties

2.1.1 General

No information was located for this compound.

2.2 Production and Use

No information was located for this compound.

2.3 Exposure Assessment

2.3.1 Air

No air collection method or analytic method was located for this compound.

2.4 Toxic Effects

The data elements provided in Table 84.1 were the only toxicity information located on this chemical.

2.4.1 *Experimental Studies*

2.4.1.1 Acute Toxicity. For information on oral LD_{50}, dermal LD_{50} primary skin irritation, eye irritation, and LC_{50}, see Table 84.1.

3.0 Di(*sec*-butyl) Peroxydicarbonate

3.0.1 *CAS Number:* *[19910-65-7]*

3.0.2 *Synonyms:* Di-*sec*-butyl ester peroxydicarbonic acid, *sec*-Butyl Peroxydicarbonate

3.0.3 *Trade Names:* NA

3.0.4 *Molecular Weight:* 234.28

3.0.5 *Molecular Formula:* $C_{10}H_{18}O_6$

3.0.6 *Molecular Structure:*

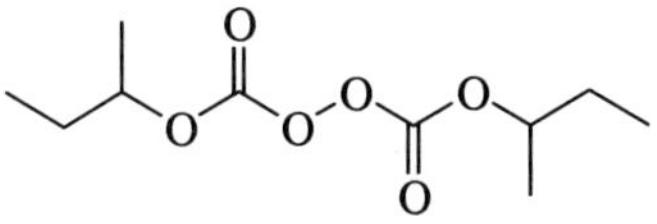

3.1 Chemical and Physical Properties

3.1.1 *General*

No additional information was located for this compound.

3.1.2 *Odor and Warning Properties*

3.2 Production and Use

No information was located for this compound.

3.3 Exposure Assessment

3.3.1 *Air*

No air collection method or analytic method was located for this compound.

3.4 Toxic Effects

The data elements provided in Table 84.1 were the only toxicity information on this chemical.

3.4.1 *Experimental Studies*

3.4.1.1 Acute Toxicity. For information on oral LD_{50}, dermal LD_{50}, eye irritation, and LC_{50}, see Table 84.1.

4.0 Di(2-ethylhexyl) Peroxydicarbonate

4.0.1 CAS Number: [16111-62-9]

4.0.2 Synonyms: Peroxydicarbonic acid and di(2-ethylhexyl) ester

4.0.3 Trade Names: NA

4.0.4 Molecular Weight: 346.46

4.0.5 Molecular Formula: $C_{18}H_{34}O_6$

4.0.6 Molecular Structure:

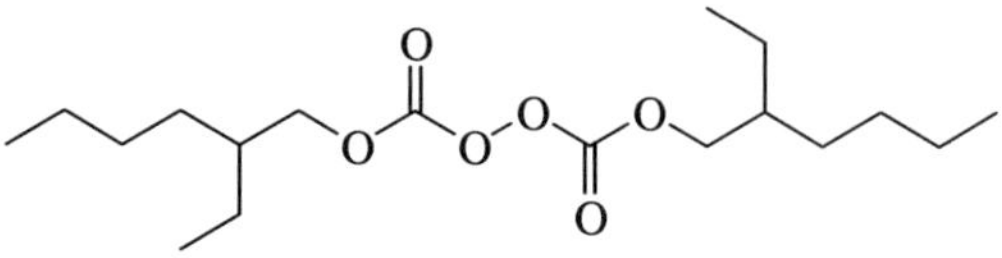

4.1 Chemical and Physical Properties

Flash point: 62°C
Insoluble in water:

4.1.1 General

No (additional) information was located for this compound.

4.1.2 Odor and Warning Properties

4.2 Production and Use

No information was located for this compound.

4.3 Exposure Assessment

4.3.1 Air

No air collection method or analytic method was located for this compound.

4.4 Toxic Effects

The data elements provided in Table 84.1 were the only toxicity information located on this chemical.

4.4.1 Experimental Studies

4.4.1.1 Acute Toxicity. For information on oral LD_{50}, dermal LD_{50}, and eye irritation, see Table 84.1.

5.0 Di-(4-*t*-butylcyclohexyl) Peroxydicarbonate

5.0.1 CAS Number: [15520-11-3]

5.0.2 Synonyms: NA

5.0.3 Trade Names: NA

5.0.4 Molecular Weight: 398.52

5.0.5 Molecular Formula: $C_{22}H_{38}O_6$

5.0.6 Molecular Structure:

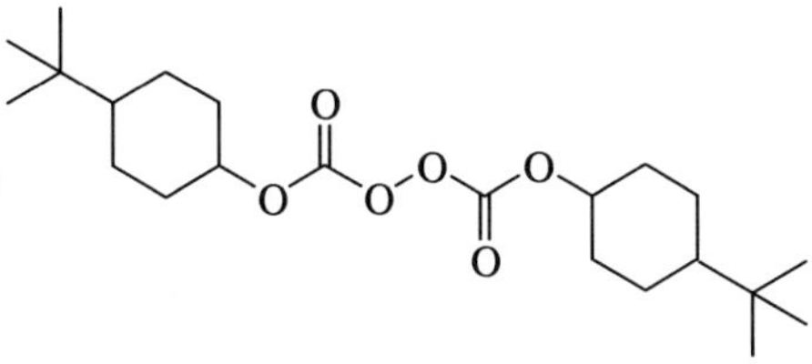

5.1 Chemical and Physical Properties

No information was located for this compound.

5.1.1 General

5.1.2 Odor and Warning Properties

5.2 Production and Use

No information was located for this compound.

5.3 Exposure Assessment

5.3.1 Air

No air collection method or analytic method was located for this compound.

5.4 Toxic Effects

The data elements provided in Table 84.1 were the only available toxicity information on this chemical.

5.4.1 Experimental Studies

5.4.1.1 Acute Toxicity. For information on oral LD_{50}, primary skin irritation, and eye irritation, see Table 84.1.

5.4.1.6 Genetic and Related Cellular Effects Studies. For information see Table 84.1.

6.0 Di-*n*-butyl Peroxydicarbonate

6.0.1 CAS Number: *[16215-49-9]*

6.0.2 Synonyms: Dibutyl ester peroxydicarbonic acid, butyl peroxydicarbonate, *n*-butyl peroxydicarbonate

6.0.3 Trade Names: NA

6.0.4 Molecular Weight: 234.28

6.0.5 Molecular Formula: $C_6H_{10}O_6$

6.0.6 Molecular Structure:

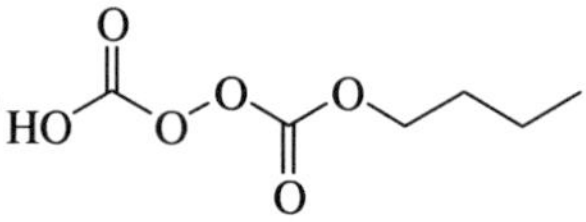

6.1 Chemical and Physical Properties

6.1.1 General

No Information was found on this compound.

6.3 Exposure Assessment

6.3.1 Air

No air collection method or analytic method was located for this compound.

6.4 Toxic Effects

The data elements provided in Table 84.1 were the only toxicity information located on this chemical.

6.4.1 Experimental Studies

6.4.1.1 Acute Toxicity. For information on oral LD_{50}, primary skin irritation, and eye irritation, see Table 84.1.

6.5 Standards, Regulations, or Guidelines of Exposure

It is forbidden by DOT to ship this peroxide at greater than 52% in solution (14).

7.0 Di-(3-methylbutyl) peroxydicarbonate

7.0.1 CAS Number: *[4113-14-8]*

7.0.2 Synonyms: Diisoamyl peroxydicarbonate

7.0.3 Trade Names: NA

7.0.4 Molecular Weight: 238.28

7.1 Chemical and Physical Properties

7.1.1 General

No (additional) information was located for this compound.

7.2 Production and Use

No information was located for this compound.

7.3 Exposure Assessment

7.3.1 Air

No air collection method or analytic method was located for this compound.

7.4 Toxic Effects

The data elements provided in Table 84.1 were the only acute toxicity information located on this chemical.

7.4.1 Experimental Studies

7.4.1.1 Acute Toxicity. A subacute inhalation study by Gage (15) used the peroxide, 20% by weight in white spirits, at 1–7 ppm to expose unidentified animals for two 6-h exposures. There were no toxic signs observed other than slight nose irritation attributed to the white spirits. Necropsy revealed all organs were normal. In the same report, Gage (15) exposed four male rats to a mist of this peroxide at a concentration of 44 mg/m^3 for eight 5-h exposures. This exposure produced nose and eye irritation, respiratory difficulty, and weight loss in the animals. The necropsy revealed thickened alveolar walls in the lungs with peribronchiolar leukocytic infiltration. When the exposure level was increased to 140 mg/m^3 pneumonia was produced after three 4-h exposures.

For information on LC_{50}, see Table 84.1.

8.0 Diacetyl Peroxydicarbonate

8.0.1 CAS Number: *[26322-14-5]*

8.0.2 Synonyms: Dihexadecyl peroxydicarbonate

8.0.3 Trade Names: NA

8.0.4 Molecular Weight: 571.00

8.0.5 Molecular Formula: $C_{34}H_{66}O_6$

8.0.6 Molecular Structure:

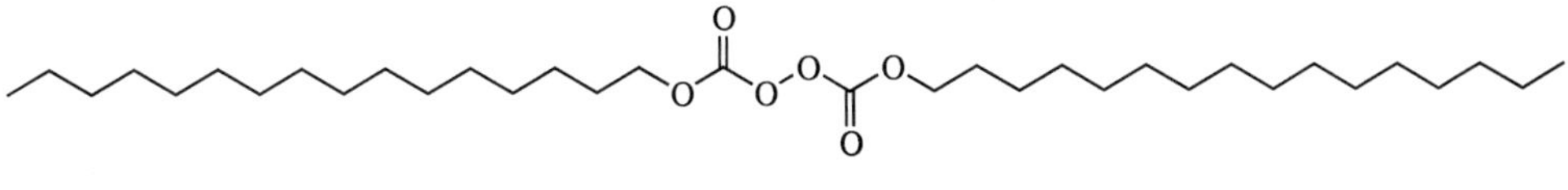

8.1 Chemical and Physical Properties

8.1.1 General

No (additional) information was located for this compound.

8.2 Production and Use

No information was located for this compound.

8.3 Exposure Assessment

8.3.1 Air

No air collection method or analytic method was located for this compound.

8.4 Toxic Effects

The data elements provided in Table 84.1 were the only toxicity information located on this chemical.

8.4.1 Experimental Studies

8.4.1.1 Acute Toxicity. For information on oral LD_{50}, primary skin irritation, and eye irritation, see Table 84.1.

9.0 Di-(2-phenoxyethyl) peroxydicarbonate

9.0.1 CAS Number: *[41935-39-1]*

9.0.2 Synonyms: NA

9.0.3 Trade Names: NA

9.0.4 Molecular Weight: 362.32

9.1 Chemical and Physical Properties

9.1.1 General

No information was located for this compound.

9.2 Production and Use

No information was located for this compound.

9.3 Exposure Assessment

9.3.1 Air

No air collection method or analytic method was located for this compound.

9.4 Toxic Effects

The data elements provided in Table 84.1 were the only toxicity information located on this chemical.

9.4.1 Experimental Studies

9.4.1.1 Acute Toxicity. For information on oral LD_{50}, dermal LD_{50}, and eye irritation, see Table 84.1.

9.4.1.6 Genetic and Related Cellular Effects Studies. For information see Table 84.1.

10.0 Di-(2-chloroethyl) Peroxydicarbonate

10.0.1 CAS Number: *[6410-72-6]* and *[34037-78-0]*

10.0.2 Synonyms: NA

10.0.3 Trade Names: NA

10.0.4 Molecular Weight: 247.03

10.1 Chemical and Physical Properties

10.1.1 General

This peroxide is nonexplosive and has a half-life of 5.7 h at 55°C.

10.1.2 Odor and Warning Properties

It also has a very weak odor associated with its use (16).

10.2 Production and Use

Di-(2-chloroethyl) peroxydicarbonate is an excellent catalyst for the polymerization of vinyl compounds and particularly for the suspension copolymerization of vinylidene chloride and vinyl chloride. It imparts superior properties with respect to odor and thermal stability as compared with those produced from conventional catalysts.

10.3 Exposure Assessment

10.3.1 Air

No air collection method or analytic method was located for this compound.

10.4 Toxic Effects

The data elements provided in Table 84.1 were the only toxicity information located on this chemical.

10.4.1 Experimental Studies

10.4.1.1 Acute Toxicity. For information on oral LD_{50}, see Table 84.1.

11.0 Di-(3-chloropropyl) Peroxydicarbonate

11.0.1 CAS Number: Undetermined

11.0.2 Synonyms: NA

11.0.3 Trade Names: NA

11.0.4 Molecular Weight: 275.09

11.1 Chemical and Physical Properties

11.1.1 General

This peroxide is nonexplosive and has a half-life of 4.0 h at 55°C.

11.1.2 Odor and Warning Properties

It also has a very weak odor associated with its use (16).

11.2 Production and Use

This peroxide is an excellent catalyst for the polymerization of vinyl compounds and particularly for the suspension copolymerization of vinylidene chloride and vinyl chloride. It imparts superior properties with respect to odor and thermal stability as compared with those produced from conventional catalysts.

11.3 Exposure Assessment

11.3.1 Air

No air collection method or analytic method was located for this compound.

11.4 Toxic Effects

The data elements provided in Table 84.1 were the only toxicity information located on this chemical.

11.4.1 Experimental Studies

11.4.1.1 Acute Toxicity. For information on oral LD_{50}, see Table 84.1.

12.0 Di-(4-chlorobutyl) Peroxydicarbonate

12.0.1 CAS Number: *[14245-74-0]*

12.0.2 Synonyms:

12.0.3 Trade Names:

12.0.4 Molecular Weight: 303.14

12.1 Chemical and Physical Properties

12.1.1 General

This peroxide is nonexplosive and has a half-life of 4.3 h at 55°C.

12.1.2 Odor and Warning Properties

It also has a very weak odor associated with its use (16).

12.2 Production and Use

This peroxide is an excellent catalyst for the polymerization of vinyl compounds and particularly for the suspension copolymerization of vinylidene chloride and vinyl chloride. It imparts superior properties with respect to odor and thermal stability as compared with those produced from conventional catalysts.

12.3 Exposure Assessment

12.3.1 Air

No air collection method or analytic method was located for this compound.

12.4 Toxic Effects

The data elements provided in Table 84.1 were the only toxicity information located on this chemical.

12.4.1 Experimental Studies

12.4.1.1 Acute Toxicity. For information on oral LD_{50}, see Table 84.1.

13.0 Di-(2-butoxyethyl) Peroxydicarbonate

13.0.1 CAS Number: *[6410-72-6]*

13.0.2 Synonyms: Butoxyethyl peroxydicarbonate

13.0.3 Trade Names: NA

13.0.4 Molecular Weight: 322.35

13.1 Chemical and Physical Properties

13.1.1 General

It has a short half-life, even at low temperatures. This catalyst also has the problem of safe handling because there is the danger of decomposition accompanied by explosion (16).

13.1.2 Odor and Warning Properties

Di-(2-butoxyethyl) peroxydicarbonate has a strong odor.

13.3 Exposure Assessment

13.3.1 Air

No air collection method or analytic method was located for this compound.

13.4 Toxic Effects

The data elements provided in Table 84.1 were the only toxicity information located on this chemical.

13.4.1 Experimental Studies

13.4.1.1 Acute Toxicity. For information on oral LD_{50} see Table 84.1.

B DIACYL PEROXIDES

14.0 Dibenzoyl Peroxide

14.0.1 CAS Number: *[94-36-0]*

14.0.2 Synonyms: Benzoyl peroxide, benzoic acid, peroxide, benzoyl superoxide, benzoperoxide, Novadelox, Acetoxyl, Acnegel, Benzac, Benzaken, Debroxide, Desanden, Benzagel 10, Benoxyl, Lucidol, Nericur, Oxy-5, Oxy 10; PanOxyl, Peroxydex, Persadox, Persa-gel, sanoxit, Theraderm, Xerac BP 5, Xerac BP 10, BPO, TCBA, Tribac, 2,3,6-TBA, diphenylglyoxal peroxide, aztec bpo, benzaknew, BZF-60, Cadet, cadox bs, dry and clear, epi-clear, fostex, Garox, incidol, loroxide,luperco, luperox fl, nayper b and bo, norox bzp-250, norox bzp-c-35, OXY-5, Oxy-10, oxylite, oxy wash, quinolor compound, superox, Topex, vanoxide, Xerac and Benzoyl peroxide, remainder water

14.0.3 Trade Names: BPO

14.0.4 Molecular Weight: 242.22

14.0.5 Molecular Formula: $C_{14}H_{10}O_4$

14.0.6 Molecular Structure:

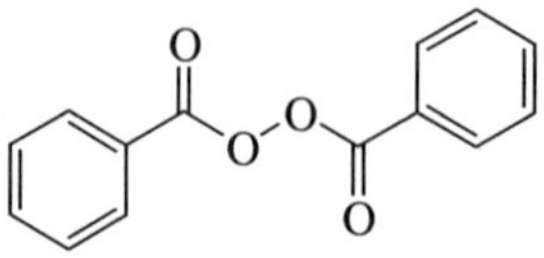

14.1 Chemical and Physical Properties

Specific gravity: 1.334
Vapor pressure: <0.1 torr at 20°C
Melting point: 103~106°C
Boiling point: (Explodes)
Flash point: 80°C
Water solubility: slightly soluble <0.1 g/100 ml at 26°C.

14.1.1 General

This material is a tasteless, white, granular, crystalline solid. Consequently this material is generally sold in a wetted or diluted (phlegmatized) form.

14.1.2 Odor and Warning Properties

Has a slight almond like odor. Dry benzoyl peroxide is a friction- and shock-sensitive material that must be handled with care. Handling procedures must avoid spark generation, heat, flame, or contamination. This peroxide should not be added to hot reaction mixtures, for this could result in explosive decomposition.

14.2 Production and Use

The compound is formed by the reaction of sodium peroxide with benzoyl chloride in water. Benzoyl peroxide is used in industry for bleaching flour and edible oils and as an additive in the self-curing of plastics, such as acrylic dentures (17). Benzoyl peroxide is widely used as an initiator for the polymerization of vinyl monomers and styrene-unsaturated polyester resin compositions. It has been used in the vulcanization of various natural and synthetic rubbers including saturated and unsaturated hydrocarbon, polyester, and silicone rubber stocks. It has also been used as a source of free radicals in organic synthesis involving addition and deletion reactions, for example, the chlorination and the addition of hydrogen halides to unsaturated compounds. It has been described as a therapeutic aid in the treatment of burns, ulcerations, and various infected cutaneous and mucous membrane lesions. Recently topical preparations containing this organic peroxide have been most effectively used to treat acne (18, 19). This successful use of benzoyl peroxide has been attributed to its antibacterial action.

14.3 Exposure Assessment

There is a recent report for allergic dermatitis due to phenol–formaldehyde resin and benzoyl peroxide in swimming goggles (20).

14.3.1 Air

The air sampling method uses a mixed cellulose ester membrane (0.8 mm) and is analyzed using high-pressure liquid chromatography (NIOSH Method 5009) (8).

14.4 Toxic Effects

14.4.1 Experimental Studies

14.4.1.1 Acute Toxicity. Animal experiments have shown that BPO is well tolerated by rats in oral dosages of 950 mg/kg (21). Other acute toxicity data are found in Table 84.2. Benzoyl peroxide has a large volume of experimental animal data surrounding it. However, most of these data concern BPO's ability to promote cancer but not initiate it in multiple strains and species, mutagenicity, and short-term studies (22–26). The relevance of these types of studies in relation to human risk has been questioned.

For more information on oral LD_{50}, primary skin irritation, eye irritation and LC_{50}, see Table 84.2.

14.4.1.4 Reproductive and Developmental. Benzoyl peroxide is one of the few organic peroxides that has been tested for embryo toxicity. The test method was to add solutions of the organic peroxide diluted in acetone directly onto the inner-shell membrane in the air chamber of the egg, focusing it exactly on the embryo visible under the membrane (36). This organic peroxide showed 13–33% malformed embryos over a range of concentrations from 0.05 to 1.7 mmol/egg, which showed no dose response. However, it is unlikely these data can be accurately extrapolated to humans owing to the direct delivery of the dose to the embryo.

14.4.1.5 Carcinogenesis. When repeatedly applied to the skin of mice, BPO was not carcinogenic (23). However, benzoyl peroxide is a tumor promoter in mice and hamsters but has shown no complete carcinogenic or tumor-initiating activity (27). There has been one controversial Japanese report (28) that was interpreted as BPO being a complete carcinogen. However, when the data were critically evaluated it was found consistent with BPO acting as a skin tumor promoter and not as a carcinogen. Additionally, there are other animal and *in vitro* studies that continue to support the lack of carcinogenic or mutagenic properties for BPO (26, 30–34). The International Agency for Research on Cancer (IARC) has evaluated the carcinogenicity of benzoyl peroxide. They classified it as Group 3. This means there is limited or inadequate evidence of carcinogenicity for animals and inadequate or absent information for humans (29).

Gimmez-Conti et al. compared the tumor-promoting potential of different organic peroxides in SENCAR mice (1). The compounds tested were benzoyl peroxide (known to be a well-characterized tumor promoter), di-*t*-butyl peroxide, dicumyl peroxide), and di-*m*-chlorobenzoyl peroxide. The authors state that tumor-promoting efficacy generally showed an inverse association with thermal stability suggesting that the rate of formation of free radicals is a k_e factor contributing to tumor production by organic peroxides. Several recent papers have investigated the mechanism of action of this compound in tumor promotion (30–34).

14.4.1.6 Genetic and Related Cellular Effects Studies. For information see Table 84.2.

14.4.1.7 Other: Neurological, Pulmonary, Skin Sensitization. Applied experimentally to the eyes of animals, it produces superficial opacities in the cornea and inflammation of the conjunctiva, but according to another report, no injury results from a single application (35).

14.4.2 Human Experience

14.4.2.1 General Information. Systemic toxicity in humans has not been reported.

14.4.2.2.7 Other: Neurological, Pulmonary, Skin Sensitization, etc. Human exposure up to 12.2 mg/m^3 has been reported to cause pronounced irritation of the nose and throat (36). Skin sensitization has been reported by acne cream users (37) and in volunteers patch tested on the upper lateral arm (38).

15.0 Di-(2,4-dichlorobenzoyl) Peroxide

15.0.1 CAS Number: [133-14-2]

15.0.2 Synonyms: NA

15.0.3 Trade Names: NA

15.0.4 Molecular Weight: 380.0

15.0.5 Molecular Formula: $C_{14}H_6C_{14}O_4$

15.0.6 Molecular Structure:

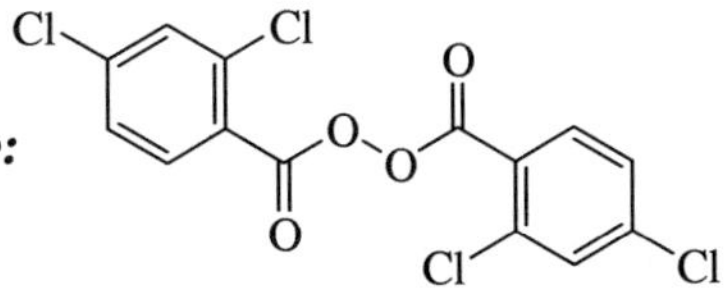

15.1 Chemical and Physical Properties

15.1.1 General

No additional information was found.

15.3 Exposure Assessment

15.3.1 Air

No air collection method or analytic method was located for this compound.

15.4 Toxic Effects

15.4.1 Experimental Studies

15.4.1.1 Acute Toxicity. The data in Table 84.2 detail the acute toxicity of this compound. This compound appears to be of low-order toxicity based on these limited data. Mice received a single intraperitoneal injection with 500 mg/kg of di-(2,4-dichlorobenzoyl) peroxide in butyl succinate, which was reported to be the maximum tolerated dose

(39). The maximum tolerated dose is defined as that dose that slightly suppresses body weight gain. However, a later report found the LD_{50} dose to be 225 mg/kg when given intraperitoneally (40).

For more information on oral LD_{50}, dermal LD_{50}, and eye irritation, see Table 84.2.

15.5 Standards, Regulations, or Guidelines of Exposure

The Department of Transportation (DOT) forbids it to be shipped in more than a 77% wetted solid in water because of its hazardous physical properties (14).

16.0 Di(*p*-chlorobenzoyl) Peroxide

16.0.1 CAS Number: [94-17-7]

16.0.2 Synonyms: *p*-Chlorobenzoyl Peroxide, bis(*p*-chlorobenzoyl)peroxide

16.0.3 Trade Names: NA

16.0.4 Molecular Weight: 311.12

16.0.5 Molecular Formula: $C_{14}H_8C_{12}O_4$

16.0.6 Molecular Structure:

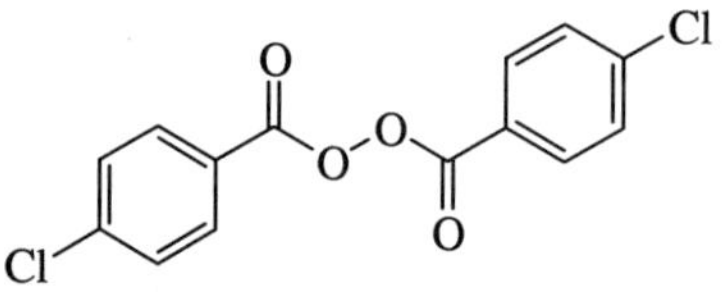

16.1 Chemical and Physical Properties

16.1.1 General

No additional information was found.

16.3.1 Air

No air collection method or analytic method was located for this compound.

16.4 Toxic Effects

16.4.1 Experimental Studies

16.4.1.1 Acute Toxicity. This compound is structurally very similar to the preceding one, which has two chlorines each in the *para* position on the benzene rings. When comparing toxicity following an intraperitoneal injection, this compound was considerably less toxic, with an LD_{50} of 500 mg/kg in butyl succinate.

For more information on oral LD_{50}, see Table 84.2.

16.4.1.5 Carcinogenesis. The only other data available for this organic peroxide show that it had no effects on tumor cells when injected intraperitoneally at 125 mg/kg in the chest wall of mice (41). The data are presented in Table 84.2.

16.4.1.6 Genetic and Related Cellular Effects Studies. For information see Table 84.2.

16.5 Standards, Regulations, or Guidelines of Exposure

The Department of Transportation (DOT) forbids this material to be shipped in more than a 77% wetted solid in water because of its hazardous physical properties (14).

17.0 Di-2-methylbenzoyl Peroxide

17.0.1 CAS Number: *[3034-79-5]*

17.0.2 Synonyms: *o*-Toluoyl peroxide, bis-(2-methylbenzoyl) peroxide, di-(*o*-methylbenzoyl) peroxide

17.0.3 Trade Names: NA

17.0.4 Molecular Weight: 270.30

17.0.5 Molecular Formula: $C_{16}H_{14}O_4$

17.0.6 Molecular Structure:

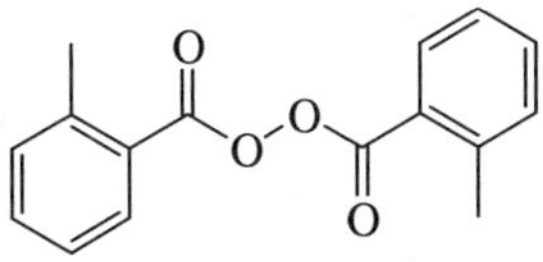

17.1 Chemical and Physical Properties

17.1.1 General

No information was located for this compound.

17.2 Production and Use

No information was located for this compound.

17.3 Exposure Assessment

17.3.1 Air

No air collection method or analytic method was located for this compound.

17.4 Toxic Effects

The data elements provided in Table 84.2 were the only toxicity information located on this chemical.

17.4.1 Experimental Studies

17.4.1.1 Acute Toxicity. For information on oral LD_{50}, primary skin irritation, and eye irritation, see Table 84.2.

17.4.1.6 Genetic and Related Cellular Effects Studies. For information see Table 84.2.

18.0 Didecanoyl Peroxide

18.0.1 CAS Number: *[762-12-9]*

18.0.2 Synonyms: Decanoyl peroxide, bis(1-oxodexyl) peroxide

18.0.3 Trade Names: NA

18.0.4 Molecular Weight: 342.58

18.1 Chemical and Physical Properties

18.1.1 General

The deflagration hazard was measured for this organic peroxide (3). The didecanoyl peroxide in powder form showed no pressure rise using 5 g of igniter at 98% purity when using the revised time–pressure test. This measurement determined that didecanoyl peroxide is not a deflagration hazard.

18.3 Exposure Assessment

18.3.1 Air

No air collection method or analytic method was located for this compound.

19.0 Dilauroyl peroxide

19.0.1 CAS Number: *[105-74-8]*

19.0.2 Synonyms: Dodecanoyl peroxide, lauroyl peroxide, laurydol, bis(1-oxododecyl) peroxide, didodecanoyl peroxide

19.0.3 Trade Names: NA

19.0.4 Molecular Weight: 398.70

19.0.5 Molecular Formula: $C_{24}H_{46}O$

19.0.6 Molecular Structure:

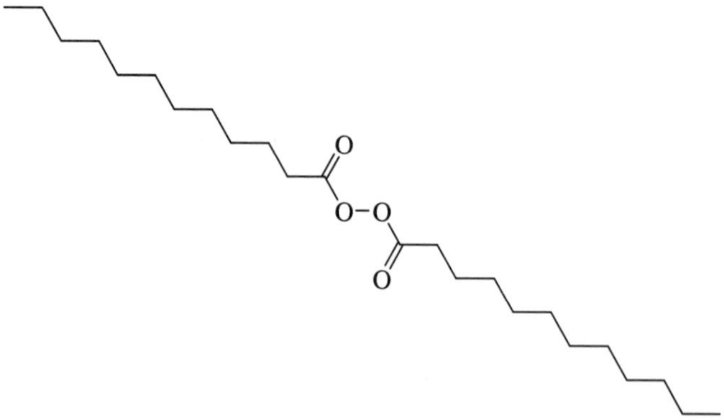

19.1 Chemical and Physical Properties

19.1.1 General

Dilauroyl peroxide is a tasteless, coarse, white powder (40). Dilauroyl peroxide is not a deflagration hazard. However, it has been judged an intermediate fire hazard by Noller (2). When all the physical tests available for this peroxide are evaluated collectively, it actually ranks as a low physical hazard (1).

19.1.2 Odor and Warning Properties

Faint pungent, soapy odor.

19.2 Production and Use

It is produced by reaction of lauroyl chloride with sodium peroxide. Its major use is as an initiator for vinyl chloride. It is used as a polymerization agent in the plastics industry and as a curing agent for rubber. It has also been used as a burn-out agent for acetate yarns. The pharmaceutical industry uses it in topical creams in combination with antibiotics for acne treatment (6).

19.3 Exposure Assessment

19.3.1 Air

A simple colorimetric method has been used to determine microgram amounts of dilauroyl peroxide in solution (42).

19.4 Toxic Effects

Studies in male rats indicate that the following parameters should be used to detect poisoning from lauroyl peroxide: hemoglobin, methemoglobin, erythrocytes, reticulocytes, and blood peroxidase activity (43).

19.4.1 Experimental Studies

19.4.1.1 Acute Toxicity. For information on oral LD_{50}, primary skin irritation, eye irritation, and LC_{50}, see Table 84.2.

19.4.1.4 Reproductive and Developmental. Dilauroyl peroxide is one of the few organic peroxides that has been tested for embryo toxicity. The test method was to add solutions of the organic peroxide diluted in acetone directly onto the inner-shell membrane in the air chamber of the egg, focusing it exactly on the embryo visible under the membrane (44). This organic peroxide showed a maximum of 25% malformed embryos over a range of concentrations from 0.25 to 0.50 mmol/egg, which showed no dose response. However, it is unlikely these data can be accurately extrapolated to humans owing to the direct delivery of the dose to the embryo.

19.4.1.6 Genetic and Related Cellular Effects Studies. For information see Table 84.2.

19.4.2.2.5 Carcinogenesis. The International Agency for Research on Cancer (IARC) has evaluated the carcinogenicity of lauroyl peroxide. They classified it as a Group 3 material, which means there is limited or inadequate evidence for animals and inadequate or absent information for humans (45). The carcinogenicity of this peroxide has primarily been studied using skin applications. After a single topical application of 10, 20, or 40 mg of lauroyl peroxide, the epidermal thickness increased markedly. This hyperplasia was characterized by a sustained production of dark basal keratinocytes (46). This peroxide is inactive as a tumor initiator or as a complete carcinogen. However, it is as effective as benzoyl peroxide as a skin tumor promoter.

20.0 Diacetyl Peroxide

20.0.1 CAS Number: [110-22-5]

20.0.2 Synonyms: Acetyl peroxide

20.0.3 Trade Names: NA

20.0.4 Molecular Weight: 118.10

20.0.5 Molecular Formula: $C_4H_6O_4$

20.0.6 Molecular Structure:

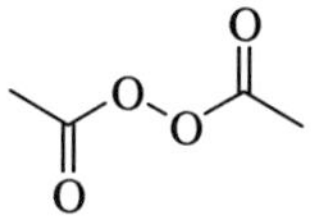

20.1 Chemical and Physical Properties

Melting point: 30°C
Boiling point: 63°C
Water solubility: slightly soluble.

20.1.1 General

Diacetyl peroxide is a colorless crystal. It must be kept in solution because of its shock sensitivity and high explosion risk.

20.1.2 Odor and Warning Properties

Has a strong pungent odor.

20.3 Exposure Assessment

20.3.1 Air

No air collection method or analytic method was located for this compound.

20.4.1 *Experimental Studies*

20.4.1.1 Acute Toxicity. For information on eye irritation, see Table 84.2.

20.4.2.2.7 Other: Neurological, Pulmonary, Skin Sensitization, etc. Human exposure has produced eye, skin, and mucous membrane irritation after inhalation or ingestion (airborne concentrations unknown) (47). Two drops of this material (30% in dimethyl phthalate) applied to the eyes of rabbits caused severe corneal damage (35).

21.0 Succinyl Peroxide

21.0.1 CAS Number: [108-30-5]

21.0.2 Synonyms: Di-(3-carboxypropionyl) peroxide, disuccinic acid peroxide, disuccinoyl peroxide, peroxydisuccinic acid, succinic acid peroxide, and succinoyl peroxide

21.0.3 Trade Names: NA

21.0.4 Molecular Weight: 100.07

21.0.5 Molecular Formula: $C_4H_4O_3$

21.0.6 Molecular Structure:

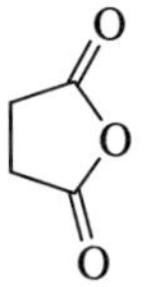

21.1 Chemical and Physical Properties

Melting point: 119.6°C
Boiling point: 261°C
Vapor density: 3.7
Water solubility: <0.1 g/100 ml at 21°C.

21.1.1 General

The odorless, fine white crystalline powder has a tart taste (48).

21.1.2 Odor and Warning Properties

Odorless.

21.2 Production and Use

The compound is produced from succinic anhydride and hydrogen peroxide (49). This material must be stabilized by using a dehydrating agent such as disodium sulfate or magnesium sulfate. It is used for disinfecting, sterilizing, as a polymerization catalyst in deodorants, and for use in antiseptic agents (50). This compound mixed with a dehydrating agent does not have the friction sensitivity of a dry powder (51). It has also been reported to be stable at ambient temperature for 14 mo.

21.3 Exposure Assessment

21.3.1 Air

No air collection method or analytic method was located for this compound.

21.4 Toxic Effects

21.4.1 Experimental Studies

21.4.1.1 Acute Toxicity. The only available toxicity information is for intraperitoneal (IP) injection. One reference (52) states that the maximum dose given IP once daily to three animals for 5–7 successive days causing no adverse toxic effect was 15.6 mg/kg (in butyl succinate). Another reference cites the IP LD_{50} to be 10–12 mm/mouse in CBA mice (52). Morphological evaluations following the IP injections produced necrotic changes in the spleen and mesenteric ganglia (53). There are no acute toxicity data presented in Table 84.2.

21.4.1.6 Genetic and Related Cellular Effects Studies. This material has been found lacking mutagenic effectiveness in the *Drosophila melanogaster* (54). Even though not mutagenic, other reports have shown disuccinyl peroxide to induce inactivating DNA alterations (55).

21.4.2.2.7 Other: Neurological, Pulmonary, Skin Sensitization, etc. The DOT (56) reports that contact may cause burns to skin and eyes, but other reports call it only a skin irritant (50).

22.0 Diproprionyl Peroxide

22.0.1 CAS Number: *[3248-28-0]*

22.0.2 Synonyms: NA

22.0.3 Trade Names: NA

22.0.4 Molecular Weight: 146.14

22.0.5 Molecular Formula: $C_6H_{10}O_4$

22.0.6 Molecular Structure: 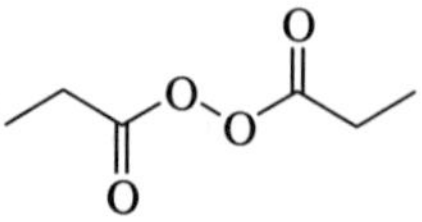

22.1 Chemical and Physical Properties

22.1.1 General

No information was located for this compound.

22.2 Production and Use

No information was located for this compound.

22.3 Exposure Assessment

22.3.1 Air

No air collection method or analytic method was located for this compound.

22.4 Toxic Effects

The only available toxicity information on this compound was a series of inhalation studies that used very limited exposures and numbers of animals. Table 84.2 outlines the most severe reactions. Exposures of two male and two female rats to 30 ppm four times for 5 h each exposure caused nose irritation only. As the number of exposures increased to 19 5-h exposures and the concentration was dropped to 10 ppm, more severe effects were observed, that is, lethargy and retarded weight gain. Necropsy found all organs normal. When the concentration was decreased to 7 ppm, no toxic signs were observed or found at necropsy when four male and four female rats were exposed to 14 5-h exposures (7).

22.4.1 Experimental Studies

22.4.1.1 Acute Toxicity. For information on LC_{50}, see Table 84.2.

22.4.1.6 Genetic and Related Cellular Effects Studies. For information see Table 84.2.

23.0 Di-*n*-Octanoyl Peroxide

23.0.1 CAS Number: *[762-16-3]*

23.0.2 Synonyms: NA

23.0.3 Trade Names: NA

23.0.4 Molecular Weight: 286.40

23.0.5 Moleculer Formula: $C_{16}H_{30}O_4$

23.0.6 Molecular Structure:

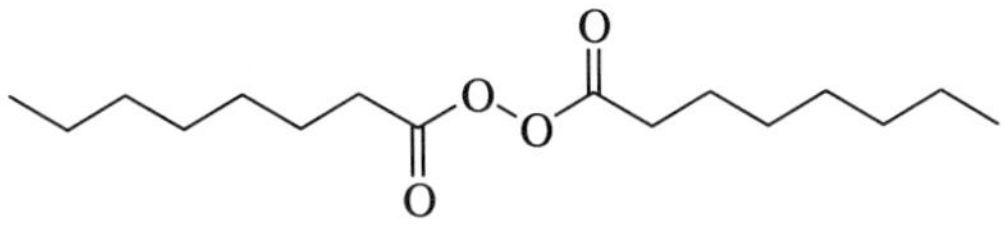

23.1 Chemical and Physical Properties

23.1.1 General

No information was located for this compound.

23.2 Production and Use

No information was located for this compound.

23.3 Exposure Assessment

23.3.1 Air

No air collection method or analytic method was located for this compound.

23.4 Toxic Effects

The data elements provided in Table 84.2 were the only toxicity information located for this chemical.

23.4.1 Experimental Studies

23.4.1.1 Acute Toxicity. For information on oral LD_{50}, primary skin irritation, and eye irritation, see Table 84.2.

24.0 Di-(3,5,5-trimethylhexanoyl) Peroxide

24.0.1 CAS Number: *[3851-87-4]*

24.0.2 Synonyms: Diisononanoyl Peroxide

24.0.4 Molecular Weight: 314.45

24.0.5 Molecular Formula: $C_{18}H_{34}O_4$

24.1 Chemical and Physical Properties

24.1.1 General

This material is not a deflagration hazard; the revised time–pressure test showed no pressure rise with 5 g of igniter. The peroxide tested was 75% pure (3).

24.3 Exposure Assessment

24.3.1 Air

No air collection method or analytic method was located for this compound.

24.4 Toxic Effects

The data elements provided in Table 84.2 were the only toxicity information located on this chemical.

24.4.1 Experimental Studies

24.4.1.1 Acute Toxicity. For information on oral LD_{50}, primary skin irritation, and eye irritation, see Table 84.2.

24.4.1.6 Genetic and Related Cellular Effects Studies. For information see Table 84.2.

25.0 *t*-Butyl Peroxyacetate

25.0.1 CAS Number: [107-71-1]

25.0.2 Synonyms: *t*-Butyl peracetate, ethaneperoxic acid, 1,1-dimethyl ester, tertiary-butyl peracetate, and 75% in mineral spirits

25.0.4 Molecular Weight: 132.18

25.0.5 Molecular Formula: $C_6H_{12}O_3$

25.0.6 Molecular Structure: 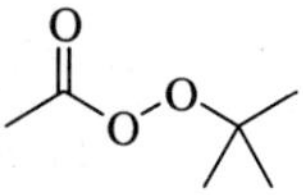

25.1 Chemical and Physical Properties

25.1.1 General

t-Butyl peroxyacetate is a clear, waterlike liquid. At 74–76% purity it has an active oxygen content of 8.9–9.2%. It has a half-life in benzene of 10.0 h at 217°F (103°C). This material decomposes to *t*-butoxy and acetoxy radicals. The *t*-butoxy radicals may be used to initiate the polymerization of vinyl monomers such as ethylene and styrene and in the cross-linking of styrene-unsaturated polyester compositions, silicone rubber, and other unsaturated resins (57).

25.2 Production and Use

This material decomposes to *t*-butoxy and acetoxy radicals. The *t*-butoxy radicals may be used to initiate the polymerization of vinyl monomers such as ethylene and styrene and in the cross-linking of styrene-unsaturated polyester compositions, silicone rubber, and other unsaturated resins (57).

25.3 Exposure Assessment

25.3.1 Air

A nonvalidated method used for air sampling employed two tubes in series of porous aromatic polymer (SKC, Catalog 226-30-02, 1989) at flow rates of 200 cm^3/min.

Immediately following sampling these tubes were desorbed with carbon disulfide and analyzed using gas chromatography (58).

25.4 Toxic Effects

25.4.1 Experimental Studies

25.4.1.1 Acute Toxicity. An 8-h inhalation study at concentrations of 3, 15, 29, 305, 400, 460, 507, and 622 ppm (50% in dimethyl phthalate) was performed. No toxic effects were observed up to 29 ppm. At 305 ppm respiration difficulties appeared during the second half of the exposure period. At the four highest exposure levels animals appeared restless briefly, then laid down, showing lacrimation, nasal discharge, and slight narcosis. Labored respiration accompanied by mouth breathing occurred after 1–2 h. At all levels, deaths, if any, occurred within 24 h after the start of the exposure (59). The inhalation of this peroxide at 50% in dimethyl phthalate produced lethal concentrations for 50% of the animals at 450 ppm after an 8-h exposure. The OSHA criteria for labeling a chemical toxic by inhalation is an $LC_{50} > 200$ at 2000 ppm. However, the time of the exposure specified by OSHA is 1 h (60). The regulation does not clarify how the difference in time of exposure is to be rectified. This product would probably not be labeled as toxic by inhalation; however, a later study performed for 4 h determined the LC_{50} to be 6.1 mg/l or approximately 1128 ppm (61). Care should be taken when evaluating these data, because the second study used 75% peroxide in odorless mineral spirits, whereas the first used 50% in dimethyl phthalate. Both inhalation studies revealed that male rats were more sensitive to the peroxide than were female rats. It is unknown if this is true in the human population. Special care should be taken to protect employees from any process that might aerosolize this material. A no observed effect level was documented at 29 ppm (~156.7 mg/m^3). See Table 84.3 for additional toxicity information.

25.4.1.7 Other: Neurological, Pulmonary, Skin Sensitization. *t*-Butyl peroxyacetate is the one member of the peroxyesters that is a primary eye irritant.

26.0 *t*-Butyl Peroxypivalate

26.0.1 CAS Number: *[927-07-1]*

26.0.2 Synonyms: *t*-Butyl perpivalate, *t*-butyl trimethylperoxyacetate, propaneperoxoic acid, 2,2-dimethyl-, 1,1-dimethylethyl ester; and *tert*-butyl peroxypivalate, 75%n mineral spirits

26.0.3 Trade Names: NA

26.0.4 Molecular Weight: 174.24

26.0.5 Molecular Formula: $C_9H_{18}O_3$

26.0.6 Molecular Structure: 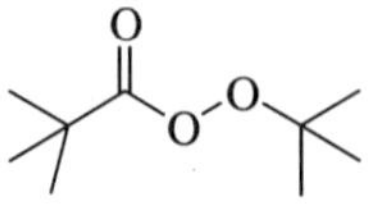

26.1 Chemical and Physical Properties

Melting point: − 17°C
Water solubility: insoluble.

26.1.1 General

t-Butyl peroxypivalate has 6.43–6.97 percent active oxygen and is commercially available at 70–76% purity (3). This organic peroxide is a highly efficient low-temperature initiator for the polymerization of numerous commercially important monomers including ethylene, vinyl chloride, vinyl acetate, styrene, acrylonitrile, and methyl methacrylate. In benzene it has a half-life of 10.0 h at 49°C (120°F) (62). Noller (2) classified *t*-butyl peroxypivalate, 75 and 50% in mineral spirits, as a deflagration hazard after testing rate and/or violence of decomposition and ease of ignition and/or decomposition. Tamura (3), on the other hand, found no pressure rise with 5 g of igniter in the revised time–pressure test to evaluate deflagration.

26.1.2 Odor and Warning Properties

No information found.

26.3.1 Air

No air collection method or analytic method was located for this compound.

26.4 Toxic Effects

26.4.1 Experimental Studies

26.4.1.1 Acute Toxicity. All the toxicity data located are presented in Table 84.3. However, additional subchronic data noted that no adverse effects were found after rats inhaled 50 ppm for 20 6-h exposures. After a single 5-h exposure of 200 ppm rats gave the appearance of nasal irritation, respiratory difficulty, and lethargy. Weight loss was observed, but at necropsy all organs appeared normal (15).

27.0 *t*-Amyl Peroxypivalate

27.0.1 CAS Number: [29240-17-3]

27.0.2 Synonyms: Propaneperoxoic acid, 2,2-dimethyl-, 1,1-dimethylethyl ester; and *tert*-butyl peroxypivalate, 75%n mineral sprits

27.0.3 Trade Names: NA

27.0.4 Molecular Weight: 188.3

27.0.5 Molecular Formula: $C_{10}H_{20}O_3$

27.0.6 Molecular Structure:

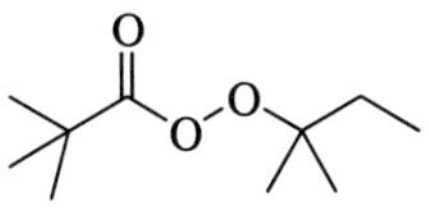

27.1 Chemical and Physical Properties

27.1.1 General

t-Amyl peroxypivalate is a colorless to slightly yellow liquid. It has an active oxygen content of 6.29–6.46% at 74–76% purity. In benzene at 0.2–mol/L it has a half-life of 10.0 h at 58°C (137°F). It is soluble in most organic solvents and is used as a low-temperature initiator for the polymerization of ethylene, vinyl chloride, vinyl acetate, styrene, acrylonitrile, and methyl methacrylate (63).

27.1.2 Odor and Warning Properties

No information found.

27.3 Exposure Assessment

27.3.1 Air

No air collection method or analytic method was located for this compound.

28.0 Perbenzoic Acid

28.0.1 CAS Number: *[614-45-9]*

28.0.2 Synonyms: *t*-Butyl perbenzoate, *t*-butyl ester, and *t*-butyl peroxybenzoate

28.0.3 Trade Names: NA

28.0.4 Molecular Weight: 194.25

28.0.5 Molecular Formula: $C_{11}H_{14}O_3$

28.0.6 Molecular Structure:

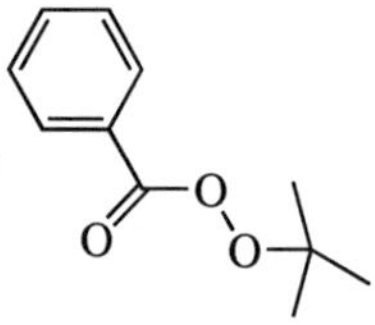

28.1 Chemical and Physical Properties

28.1.1 General

t-Butyl peroxybenzoate is produced by reacting *t*-butyl hydroperoxide with benzoyl chloride in the presence of a base. It is a colorless to slight yellow liquid. It has a half-life in benzene of 10 h at 224°F (107°C). It is a liquid peroxyester catalyst of low volatility and high purity, effective as a medium temperature initiator for the polymerization and/or cross-linking of methyl methacrylate, acrylonitrile, isoprene, styrene, and butadiene. Low-density polyethylene of excellent mechanical properties is obtained when using this peroxide as an initiator. This peroxide is also used in various organic syntheses involving coupling reactions of olefins and paraffinic compounds as well as phenolic derivatives (64). The physical properties of *t*-butyl perbenzoate are that of an intermediate fire hazard (1); however, other sources refer to it as a fire hazard (2) with 8.16% active oxygen (3). The liquid at 99% purity showed a slow pressure rise with only 1 g of igniter; therefore, it is not a deflagration hazard (3). It also has low shock or impact sensitivity (65).

28.1.2 Odor and Warning Properties

It has a mild aromatic odor (48).

28.3 Exposure Assessment

28.3.1 Air

No air collection method or analytic method was located for this compound.

28.4 Toxic Effects

The National Toxicology Program released a report in March 1991 that *t*-butyl perbenzoate induced negligible systemic toxicity after dosing rats orally with 500 mg/kg.

28.4.1 Experimental Studies

28.4.1.1 Acute Toxicity. Table 84.3 reports the acute toxicity data available on this compound. Acute inhalation of this material is reported to produce no adverse effects at the maximum attainable vapor concentration of 0.26 mg/L (65). Repeated daily inhalations of 0.006 mg/L for 4 h (number of days not specified) brought about a decrease in body weights and urinary output. An increase in adrenal ascorbic acid and a decrease in ^{131}I uptake denote adrenal and thyroid hypofunction. No morphological alterations were found in the liver or blood.

28.4.1.4 Reproductive and Developmental. *t*-Butyl peroxybenzoate is one of the few organic peroxides that has been tested for embryo toxicity. The test method was to add solutions of the organic peroxide dissolved in acetone directly onto the inner-shell membrane in the air chamber of the egg, focusing it exactly on the embryo visible under the membrane (44). This organic peroxide showed a maximum of 13% malformed embryos over a range of concentrations from 0.13 to 1.0 mmol/egg, which showed no dose response. However, it is unlikely these data can be accurately extrapolated to humans owing to the direct delivery of the dose to the embryo.

28.4.1.6 Genetic and Related Cellular Effects Studies. Mutagenicity tests performed by Hazleton Laboratories showed this organic peroxide to be nonmutagenic in the *Salmonella typhimurium* assay. In addition, these tests showed a slightly positive response in the Chinese hamster ovary cell and the mouse lymphoma forward mutation assays. The extrapolation of these tests to the human experience has become controversial in recent years. Spermatogenesis, the content of nucleic acids, and testicular depolymerase activity were not affected (66).

28.4.2.2.5 Carcinogenesis. Other studies have determined the possibility that *t*-butyl peroxybenzoate can be metabolized to free radicals by human carcinoma skin keratinocytes. Free radicals are suggested to be involved in the cascade of events occurring during tumor promotion (67).

29.0 *t*-Butyl Peroxy-2-ethylhexanoate

29.0.1 CAS Number: [3006-82-4]

29.0.2 Synonyms: *t*-Butyl peroctoate

29.0.3 Trade Names: NA

29.0.4 Molecular Weight: 216.3

29.0.5 Molecular Formula: $C_{12}H_{24}O_3$

29.0.6 Molecular Structure:

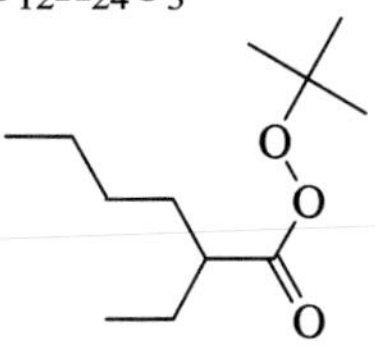

29.1 Chemical and Physical Properties

29.1.1 General

t-Butyl peroxy-2-ethylhexanoate is a colorless liquid. It has a half-life in benzene of 10.0 h at 162°F (72°C). It is used as a medium temperature initiator for the polymerization of vinyl monomers and the curing of styrene-unsaturated polyester resins. It is regarded as an intermediate fire hazard; however, it has low impact sensitivity (68).

29.1.2 Odor and Warning Properties

No information found.

29.3.1 Air

No air collection method or analytic method was located for this compound.

29.4 Toxic Effects

29.4.1 Experimental Studies

29.4.1.1 Acute Toxicity. All located acute toxicity data are presented in Table 84.3. The acute inhalation study in rats used four dose levels, 103.4, 46.5, 20.8, and 9.2 mg/L. During the 4-h exposures, nasal discharge and slight dyspnea were observed in all groups of rats. Redness of ears and paws developed shortly after exposure in all rats exposed to 103.4 and 46.5 mg/L. Deaths occurred 1 or 2 d after exposure. Necropsy revealed red patches in the lungs (69).

30.0 Peroxyhexanoic Acid

30.0.1 CAS Number: [686-31-7]

30.0.2 Synonyms: *t*-Amylperoctoate, *t*-amyl peroxy-2-ethylhexanoate, 2-ethyl-*t*-pentyl ester, 2-ethylperoxyhexanoic acid, *t*-pentyl ester

30.0.3 Trade Names: NA

30.0.4 Molecular Weight: 230.39

30.0.5 Molecular Formula: $C_{13}H_{26}O_3$

30.0.6 Molecular Structure:

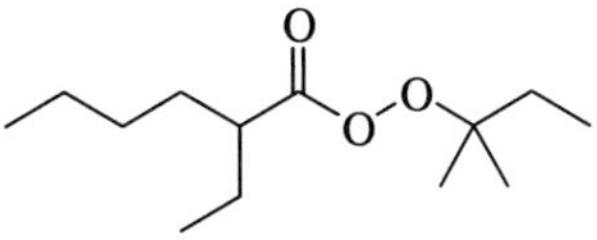

30.1 Chemical and Physical Properties

30.1.1 General

No information was located for this compound.

30.1.2 Odor and Warning Properties

30.2 Production and Use

No information was located for this compound.

30.3 Exposure Assessment

30.3.1 Air

No air collection method or analytic method was located for this compound.

30.4 Toxic Effects

There is very little information available for *t*-amyl peroxy-2-ethylhexanoate. It has been tested *in vivo* (whole animal) and in cell culture for its antimalarial potency, but it was inactive in the whole animal (70).

30.4.1 Experimental Studies

30.4.1.1 Acute Toxicity. All located acute toxicity data are presented in Table 84.3.

31.0 *t*-Butyl Peroxy-3,5,5-trimethylhexanoate

31.0.1 CAS Number: [13122-18-4]

31.0.2 Synonyms: *t*-Butyl peroxyisononanoate, peroxyhexanoic acid, and 3,5,5-trimethyl-*t*-butyl ester

31.0.3 Trade Names: NA

31.0.4 Molecular Weight: 230.35

31.0.5 Molecular Formula: $C_{13}H_{26}O_3$

31.0.6 Molecular Structure:

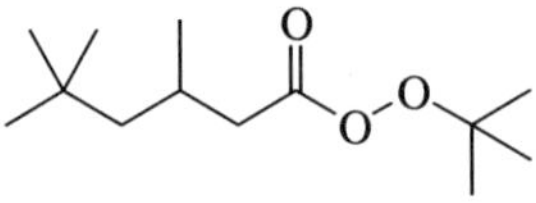

31.1 Chemical and Physical Properties

31.1.1 General

No information was located for this compound.

31.2 Production and Use

No information was located for this compound.

31.3 Exposure Assessment

31.3.1 Air

No air collection method or analytic method was located for this compound.

31.4 Toxic Effects

The data elements provided in Table 84.3 were the only toxicity information located on this chemical.

31.4.1 Experimental Studies

31.4.1.1 Acute Toxicity. The acute inhalation study used a finely dispersed aerosol. The particle size was measured using a Cascade impactor; the diameter of the droplets ranged from 1.2 to 4.7 mm, fully in the respirable range. The rats were somewhat restless during the first half-hour of exposure to 0.8 mg/L; however, gradually all the animals fell asleep for the remainder of the 4-h exposure. No mortality occurred; therefore, the LC_{50} is considered greater than 0.8 mg/L (71).

32.0 *t*-Butyl Peroxyneodecanoate

32.0.1 CAS Number: *[26748-41-4]*

32.0.2 Synonyms: *t*-Butyl perneodecanoate, peroxyneodecanoic acid, and *t*-butyl ester

32.0.3 Trade Names: NA

32.0.4 Molecular Weight: 244.42

32.1 Chemical and Physical Properties

32.1.1 General

t-Butyl peroxyneodecanoate is a colorless liquid with 4.85–4.98% active oxygen at 75% purity. This peroxide is normally used dissolved 75% in an organic solvent. It has a half-life in benzene of 10 h at 122°F (50°C). It is considered a combustible oxidizing liquid and should be handled with care (72).

32.2 Production and Use

It is useful as a low-temperature initiator for radical-catalyzed polymerization of vinyl monomers.

32.3 Exposure Assessment

32.3.1 Air

No air collection method or analytic method was located for this compound.

32.4 Toxic Effects

All toxicity data located are presented in Table 84.3.

33.0 *t*-Butyl Peroxy-2-Ethylhexylcarbonate

33.0.1 CAS Number: [34443-12-4]

33.0.2 Synonyms: NA

33.0.3 Trade Names: NA

33.0.4 Molecular Weight: 246.34

33.1 Chemical and Physical Properties

33.1.1 General

No information was located for this compound.

33.2 Production and Use

No information was located for this compound.

33.3 Exposure Assessment

33.3.1 Air

No air collection method or analytic method was located for this compound.

33.4 Toxic Effects

The data elements provided in Table 84.3 were the only toxicity information located on this chemical.

34.0 *t*-Butyl Peroxycrotonate

34.0.1 CAS Number: *[23474-91-1]*

34.0.2 Synonyms: *t*-Butyl percrotonate

34.0.3 Trade Names: NA

34.0.4 Molecular Weight: 158.2

34.0.5 Molecular Formula: $C_8H_{14}O_3$

34.0.6 Molecular Structure: 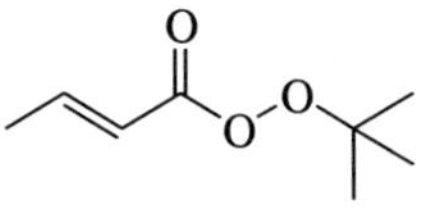

34.1 Chemical and Physical Properties

34.1.1 General

No information was located for this compound.

34.2 Production and Use

No information was located for this compound.

34.3 Exposure Assessment

34.3.1 Air

No air collection method or analytic method was located for this compound.

34.4 Toxic Effects

The data elements provided in Table 84.3 were the only toxicity information located on this chemical.

35.0 *t*-Amyl Peroxybenzoate

35.0.1 CAS Number: [4511-39-1]

35.0.2 Synonyms: *t*-Amyl Perbenzoate

35.0.3 Trade Names: NA

35.0.4 Molecular Weight: 208.25

35.0.5 Molecular Formula: $C_{12}H_{16}O_3$

35.0.6 Molecular Structure:

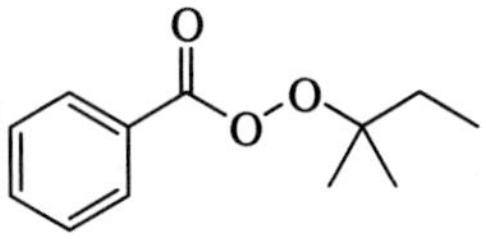

35.1 Chemical and Physical Properties

35.1.1 General

No information was located for this compound.

35.2 Production and Use

No information was located for this compound.

35.3 Exposure Assessment

35.3.1 Air

No air collection method or analytic method was located for this compound.

35.4 Toxic Effects

The data elements provided in Table 84.3 were the only toxicity information located on this chemical.

36.0 Cumyl Peroxyneodecanoate

36.0.1 CAS Number: [26748-47-0]

36.0.2 Synonyms: NA

36.0.3 Trade Names: NA

36.0.4 Molecular Weight: 306.43

36.1 Chemical and Physical Properties

36.1.1 General

No information was located for this compound.

36.2 Production and Use

No information was located for this compound.

36.3 Exposure Assessment

36.3.1 Air

No air collection method or analytic method was located for this compound.

36.4 Toxic Effects

The data elements provided in Table 84.3 were the only toxicity information located on this chemical.

36.4.1 Experimental Studies

36.4.1.1 Acute Toxicity. The acute inhalation study in rats was only conducted for 1 h. The "metered" concentration of 20.4 mg/L resulted in slight dyspnea, eye squint, and slight body weight loss measured 1 d after exposure (73).

37.0 2,5-Dimethyl-2,5-di-(2-ethylhexanoylperoxy)hexane

37.0.1 CAS Number: *[13052-09-0]*

37.0.2 Synonyms: 2-ethylperoxyhexanoic acid, 1,1,4,4-tetramethyltetramethylene ester, 2-ethylhexaneperoxoic acid, 1,1,4,4-tetramethyl-1,4-butanediyl ester

37.0.3 Trade Names: NA

37.0.4 Molecular Weight: 430.70

37.0.5 Molecular Formula: $C_{24}H_{46}O_6$

37.0.6 Molecular Structure:

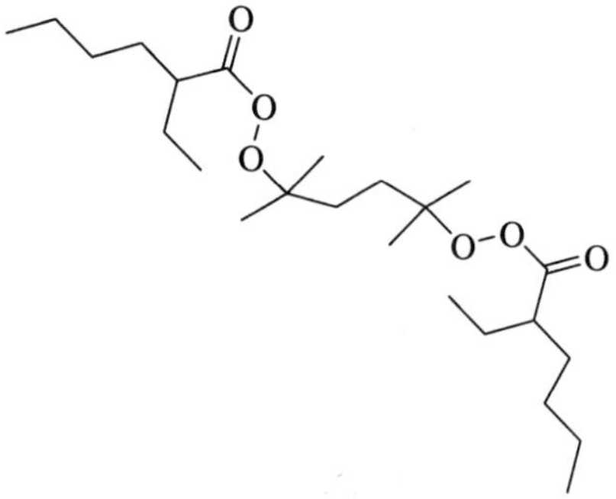

37.1 Chemical and Physical Properties

37.1.1 General

No information was located for this compound.

37.2 Production and Use

No information was located for this compound.

37.3 Exposure Assessment

37.3.1 Air

No air collection method or analytic method was located for this compound.

37.4 Toxic Effects

The data elements provided in Table 84.3 were the only toxicity information located on this chemical.

38.0 Di-*t*-butyl Diperoxyazelate

38.0.1 CAS Number: *[16580-06-6]*

38.0.2 Synonyms: Di-*t*-butyl diperazelate

38.0.3 Trade Names: NA

38.0.4 Molecular Weight: 332.4

38.1 Chemical and Physical Properties

38.1.1 General

No information was located for this compound.

38.2 Production and Use

No information was located for this compound.

38.3 Exposure Assessment

38.3.1 Air

No air collection method or analytic method was located for this compound.

38.4 Toxic Effects

The data elements provided in Table 84.3 were the only toxicity information located on this chemical.

39.0 1,1,3,3-Tetramethylbutyl Peroxyphenoxyacetate

39.0.1 CAS Number: *[59382-51-3]*

39.0.2 Synonyms: 2,4,4-Trimethylpentyl 2-peroxyphenoxyacetate

39.0.3 Trade Names: NA

39.0.4 Molecular Weight: 280.35

39.1 Chemical and Physical Properties

39.1.1 General

No information was located for this compound.

39.2 Production and Use

No information was located for this compound.

39.3 Exposure Assessment

39.3.1 Air

No air collection method or analytic method was located for this compound.

39.4 Toxic Effects

The data elements provided in Table 84.3 were the only toxicity information located on this chemical.

39.4.1 Experimental Studies

39.4.1.1 Acute Toxicity. The acute inhalation study exposed rats for 4 h to an aerosol of 30% solution of the substance in Shellsol T at a concentration of 24 ppm. During the first half-hour of the exposure period the rats were slightly restless. This symptom gradually disappeared and after 1 h all rats were asleep. No mortalities occurred; therefore, the LC_{50} is greater than 24 ppm (74).

40.0 *t*-Butyl Peroxyisopropylcarbonate

40.0.1 CAS Number: *[2372-21-6]*

40.0.2 Synonyms: peroxycarbonic acid, *OO-t*-butyl *O*-isopropyl ester, and *OO-t*-butyl *O*-isopropyl monoperoxycarbonate

40.0.3 Trade Names: NA

40.0.4 Molecular Weight: 176.21

40.0.5 Molecular Formula: $C_8H_{16}O_4$

40.0.6 Molecular Structure:

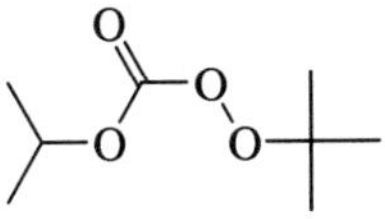

40.1 Chemical and Physical Properties

40.1.1 General

This liquid is not a deflagration hazard with 8.9% active oxygen at 98% purity. It gives a slow pressure rise using 4 g of igniter in the revised time–pressure test (3).

40.3 Exposure Assessment

40.3.1 Air

No air collection method or analytic method was located for this compound.

40.4 Toxic Effects

The data elements provided in Table 84.3 were the only toxicity information located on this chemical.

41.0 *t*-Butyl Monoperoxymaleate

41.0.1 CAS Number: [1931-62-0]

41.0.2 Synonyms: Peroxymaleic acid, *O-t*-butyl ester, and *t*-butyl peroxymaleic acid

41.0.3 Trade Names: NA

41.0.4 Molecular Weight: 188.18

41.0.5 Molecular Formula: $C_8H_{12}O_5$

41.0.6 Molecular Structure:

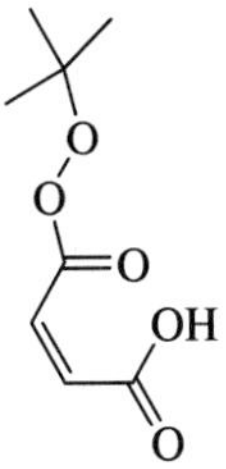

41.1 Chemical and Physical Properties

41.1.1 General

This powder with 8.4% active oxygen and 98.8% purity showed small pressure rises in the time–pressure assay when using only 1 g of igniter (3). This indicates that this material poses a deflagration hazard.

41.3 Exposure Assessment

41.3.1 Air

No air collection method or analytic method was located for this compound.

41.4 Toxic Effects

The data elements provided in Table 84.3 were the only toxicity information located on this chemical.

42.0 Di-*t*-Butyl Diperoxyphthalate

42.0.1 CAS Number: [15042-77-0]

42.0.2 Synonyms: NA

42.0.3 Trade Names: NA

42.0.4 Molecular Weight: 310.00

42.1 Chemical and Physical Properties

42.1.1 General

This peroxide powder with 8.45% active oxygen and 82% purity was mixed 20% in water and showed small pressure rises in the time–pressure assay when using 1 g of igniter (3). This indicates that this material poses a deflagration hazard.

42.3 Exposure Assessment

42.3.1 Air

No air collection method or analytic method was located for this compound.

42.4 Toxic Effects

The data elements provided in Table 84.3 were the only toxicity information located on this chemical.

43.0 Cumyl Peroxyneoheptanoate

43.0.1 CAS Number: [130097-36-8, 104852-44-0]

43.0.2 Synonyms: Cumyl perneoheptanoate

43.0.3 Trade Names: NA

43.0.4 Molecular Weight: 264.35

43.1 Chemical and Physical Properties

43.1.1 General

No information was located for this compound.

43.2 Production and Use

No information was located for this compound.

43.3 Exposure Assessment

43.3.1 Air

No air collection method or analytic method was located for this compound.

43.4 Toxic Effects

The data elements provided in Table 84.3 were the only toxicity information located on this chemical.

C KETONE PEROXIDES

44.0 Methyl Ethyl Ketone Peroxide

44.0.1 CAS Number: *[1338-23-4]*

44.0.2 Synonyms: Butanone peroxide, MEK peroxide

44.0.3 Trade Names: NA

44.0.4 Molecular Weight: 176.24

44.0.5 Molecular Formula: $C_8H_{16}O_4$

44.1 Chemical and Physical Properties

Specific Gravity: 1.12 at 15°C
Melting point: 60°C
Boiling point: 118°C (decomposes)
Water solubility: insoluble. 0.1∼0.5 g/100 ml at 22°C.

44.1.1 General

This compound is a colorless liquid with a fragrant, mintlike, moderately sharp odor. Commercially available forms are 9.0% maximum active oxygen and are not considered a deflagration hazard.

44.1.2 *Odor and Warning Properties*

It has fragrant, mintlike, moderately sharp odor.

44.2 Production and Use

MEK peroxide is used as a curing agent with polyester resins for adhesives, plastics, lacquers, and fiber glass resin kits for boat and automobile body repair. The peroxide is usually one component used in a two-component mixture referred to as "epoxies." However, this is a misnomer because they do not contain epoxy groups (75).

44.3 Exposure Assessment

A recent report (76) has indicated contact allergy from methyl ethyl ketone peroxide and cobalt in the manufacture of fiberglass reinforced plastics, esophageal stenosis in a child caused by ingestion of methyl ethyl ketone peroxide (77), contact sensitivity to methyl ethyl ketone peroxide in a paint sprayer (78), occular injury (79), peripheral zonal heaptic necrosis caused by accidental ingestion of methyl ethyl ketone peroxide (80).

44.3.1 *Air*

The air sampling method uses a standard size XAD-4 tube and is analyzed using high-pressure liquid chromatography (81).

44.4 Toxic Effects

44.4.1 *Experimental Studies*

44.4.1.1 Acute Toxicity. Acute intoxication with this organic peroxide is accompanied by a drop in blood pressure. The cause of this effect is unknown, but may involve lipid peroxidation of cell membranes (82).

For more information on oral LD_{50}, dermal LD_{50}, primary skin irritation, eye irritation, and LC_{50}, see Table 84.4.

44.4.1.2 Chronic and Subchronic Toxicity. Subchronic oral exposures three times a week for 7 wk at one-fifth the LD_{50} caused all five rats to die. Necropsy revealed mild liver damage with glycogen depletion (83). These data suggest accumulation of the material in the body.

44.4.1.4 Reproductive and Developmental. A MEK peroxide technical product (50% MEKP in phosphoric acid ester and phthalic acid ester) was tested for embryo toxicity in 3-d-old chicken embryos. A dose of 0.29 mmol/egg caused early deaths in 50% of treated eggs and malformations in 40% of the survivors (44). The authors classified MEK peroxide as a moderately potent embryo toxin. However, the relevance of these data in relation to humans is unknown. The test method applied solutions directly to the inner membrane, focusing it on the embryo. It is unlikely these data can be accurately extrapolated to humans owing to the direct delivery of the dose.

44.4.1.5 Carcinogenesis. This material is currently being assessed for carcinogenicity by the National Toxicology Program (NTP) in skin painting studies.

44.4.1.6 Genetic and Related Cellular Effects Studies. Short term *in vitro* mutagenic studies have given mixed results (84).

For more information see Table 84.4.

44.4.1.7 Other: Neurological, Pulmonary, Skin Sensitization. It is a respiratory tract irritant and a severe eye and skin irritant. The maximum nonirritating strength for skin was 1.5%, for eyes 0.6% (9). Skin irritation was delayed, with erythema and edema appearing within 2–3 d. Eyes washed within 4 sec after exposure resulted in no adverse effects. However, it has been noted that single MEK peroxide exposures to the eye can exacerbate preexisting corneal and limbal disease (85).

44.4.2 Human Experience

44.4.2.2.1 Acute Toxicity. Human exposure attests to the corrosive nature of this compound, causing chemical burns of the gastrointestinal tract with residual scarring and stricture of the esophagus (64). Death 2–3 d later was due to hepatic failure after ingestion of this material. An oral dose of 50–100 mL is toxic and potentially lethal to adults (86).

44.4.2.2.7 Other: Neurological, Pulmonary, Skin Sensitization, etc. Human exposure attests to the corrosive nature of this compound, causing chemical burns of the gastrointestinal tract with residual scarring and stricture of the esophagus (83).

45.0. Methyl Isobutyl Ketone Peroxide

45.0.1 CAS Number: *[37206-20-5]*

45.0.2 Synonyms: 4-Methyl-2-pentanone peroxide, and isobutyl methyl ketone peroxide

45.0.3 Trade Names: NA

45.0.4 Molecular Weight: Various

45.1 Chemical and Physical Properties

45.1.1 General

45.1.2 Odor and Warning Properties

45.2 Production and Use

Methyl isobutyl ketone peroxide (MIKP) is obtained from the reaction of methyl isobutyl ketone with hydrogen peroxide. During the production of this peroxide, a mixture of MIKP is obtained (87).

45.3 Exposure Assessment

45.3.1 Air

No air collection method or analytic method was located for this compound.

45.4 Toxic Effects

The data elements provided in Table 84.4 were the only toxicity information located on this chemical.

45.4.1 Experimental Studies

45.4.1.1 Acute Toxicity. The acute inhalation study used 60% methyl isobutyl ketone peroxides and 40% diisobutyl phthlate at 2.56, 1.77, 1.55, 1.38, and 1.30 mg/L (g/m^3). The particle size was determined using a Cascade impactor. Eighty percent of the mist consisted of droplets with a diameter of 1.7–3.3 mm. The maximum droplet size was 6.7 mm. During the first half-hour of exposure the rats were restless. During the whole exposure period the animals kept their eyes closed. After the exposure period most of the rats of the higher-dose groups showed mouth breathing. The exposed animals lost weight during the first week postexposure, but regained the weight the following week. Based on these data the LC_{50} of this material was determined to be 1.5 mg/L (g/m^3), with 1.4 and 1.61 mg/L (g/m^3) representing the 95% confidence limits (88).

For more information on Oral LD_{50}, primary skin irritation, eye irritation, and LC_{50}, see Table 84.4.

45.4.1.6 Genetic and Related Cellular Effects Studies. For information see Table 84.4.

46.0. 3,5-Dimethyl-3,5-dihydroxy-1,2-dioxolone

46.0.1 CAS Number: [37187-22-7]

46.0.2 Synonyms: 2,4-Pentanedione peroxide and acetyl acetone peroxide

46.0.3 Trade Names: NA

46.0.4 Molecular Weight: 134.13

46.1 Chemical and Physical Properties

46.1.1 General

No information was located for this compound.

46.2 Production and Use

No information was located for this compound.

46.3 Exposure Assessment

46.3.1 *Air*

No air collection method or analytic method was located for this compound.

46.4 Toxic Effects

All located toxicity data are presented in Table 84.4.

46.4.1 *Experimental Studies*

46.4.1.1 Acute Toxicity. For information on oral LD_{50}, primary skin irritation, eye irritation, and LC_{50}, see Table 84.4.

46.4.1.4 Reproductive and Developmental. The embryo toxicity of this peroxide was studied using 3-d chicken embryos. The peroxide was delivered directly onto the inner membrane overlaying the embryo. The median effective dose (ED_{50})/egg was 0.34 mmol, producing 23% malformed embryos (44). Because of the obvious difference in delivery of the dose in this study versus humans, the relevance of this information to humans is unknown.

47.0. Diacetone Alcohol Peroxide

47.0.1 *CAS Number: [54693-46-8]*

47.0.2 Synonyms: 2,4-Dihydroxy-2-methyl-4-hydroperoxypentane and 4-hydroxy-4-methyl-2-pentanone peroxide

47.0.3 Trade Names: NA

47.0.4 Molecular Weight: 150.17

47.1 Chemical and Physical Properties

47.1.1 *General*

No information was located for this compound.

47.2 Production and Use

No information was located for this compound.

47.3 Exposure Assessment

47.3.1 *Air*

No air collection method or analytic method was located for this compound.

47.4 Toxic Effects

The data elements provided in Table 84.4 were the only toxicity information located on this chemical.

47.4.1 Experimental Studies

47.4.1.1 Acute Toxicity. The acute inhalation study used 51–52% diacetone alcohol peroxide, with the balance being diacetone alcohol, water, and hydrogen peroxide at 0.75, 0.51, 0.37, and 0.27 mg/L (g/m^3). The particle size was determined using a Cascade impactor. Ninety percent of the mist consisted of droplets with a diameter of 1.7–2.4 mm. The maximum droplet size was 3.3 mm. During the whole exposure period the animals kept their eyes closed. Mouth breathing and labored respiration occurred during the second half of the exposure period. The exposed animals lost weight during the first 3 d postexposure, but regained the weight the following week. Based on these data the LC_{50} of this material was determined to be 0.54 mg/L (g/m^3), with 0.46 and 0.64 mg/L (g/m^3) representing the 95% confidence limits (88).

For more information on oral LD_{50}, primary skin irritation, eye irritation, and LC_{50}, see Table 84.4.

47.4.1.6 Genetic and Related Cellular Effects Studies. For information see Table 84.4.

48.0. 1,1-Dihydroperoxycyclohexane

48.0.1 CAS Number: *[2699-11-9]*

48.0.2 Synonyms: NA

48.0.3 Trade Names: NA

48.0.4 Molecular Weight: 148.16

48.1 Chemical and Physical Properties

48.1.1 General

No information was located for this compound.

48.2 Production and Use

No information was located for this compound.

48.3 Exposure Assessment

48.3.1 Air

No air collection method or analytic method was located for this compound.

48.4 Toxic Effects

The data elements provided in Table 84.4 were the only toxicity information located on this chemical.

48.4.1 Experimental Studies

48.4.1.1 Acute Toxicity. For information on oral LD_{50}, primary skin irritation, and eye irritation, see Table 84.4.

48.4.1.6 Genetic and Related Cellular Effects Studies. For information see Table 84.4.

49.0. Cyclohexanone Peroxide

49.0.1 CAS Number: *[12262-58-7]*

49.0.2 Synonyms: NA

49.0.3 Trade Names: NA

49.0.4 Molecular Weight: 228.28

49.1 Chemical and Physical Properties

49.1.1 General

No information was located for this compound.

49.2 Production and Use

No information was located for this compound.

49.3 Exposure Assessment

49.3.1 Air

No air collection method or analytic method was located for this compound.

49.4 Toxic Effects

All other toxicity data located are presented in Table 84.4.

49.4.1 Experimental Studies

49.4.1.1 Acute Toxicity. For information on eye irritation, see Table 84.4.

49.4.1.4 Reproductive and Developmental. The embryo toxicity of this peroxide was studied using 3-d chicken embryos. The peroxide was delivered directly onto the inner membrane overlaying the embryo. The median effective dose (ED_{50})/egg was 0.13 mmol, producing 27% malformed embryos (44). Because of the obvious difference in delivery of the dose in this study versus humans, the relevance of this information to humans is unknown.

50.0 Di(1-hydroxycyclohexyl) Peroxide

50.0.1 CAS Number: [2407-94-5]

50.0.2 Synonyms: Dihydroxydicyclohexyl peroxide, bis(1-hydroxycyclohexyl) peroxide, 1,1′-dioxybiscyclohexanol, 1,1′-peroxydicyclohexanol

50.0.3 Trade Names: NA

50.0.4 Molecular Weight: 230.34

50.0.5 Molecular Formula: $C_{12}H_{22}O_4$

50.1 Chemical and Physical Properties

50.1.1 General

No information was located for this compound.

50.2 Production and Use

No information was located for this compound.

50.3 Exposure Assessment

50.3.1 Air

No air collection method or analytic method was located for this compound.

50.4 Toxic Effects

A Russian article (English abstract) reports in mice a single oral dose of 65 mg/kg as the threshold dose for nerve–muscle irritation, perturbation in hemoglobin/methemoglobin content, and catalase and peroxidase activity in blood (89). All other available toxicity data are presented in Table 84.4.

50.4.1 Experimental Studies

50.4.1.1 Acute Toxicity. For information on oral LD_{50}, primary skin irritation, and eye irritation, see Table 84.4.

51.0. Di-(1-hydroperoxycyclohexyl) Peroxide

51.0.1 CAS Number: [2699-12-9]

51.0.2 Synonyms: Cyclohexyl peroxide dihydroperoxide and dioxydicyclohexylidene bis-hydroperoxide

51.0.3 Trade Names: NA

51.0.4 Molecular Weight: 262.34

51.1 Chemical and Physical Properties

51.1.1 General

No information was located for this compound.

51.2 Production and Use

No information was located for this compound.

51.3 Exposure Assessment

51.3.1 Air

The photocolorimetric determination has been used to analyze for this peroxide in water (90).

51.4 Toxic Effects

A Russian article (English abstract) reports in mice a single oral dose of 18 mg/kg as the threshold dose for nerve–muscle irritation, perturbation in hemoglobin/methemoglobin content, and catalase and peroxidase activity in blood (91). All other available toxicity data are presented in Table 84.4.

51.4.1 Experimental Studies

51.4.1.1 Acute Toxicity. For information on oral LD_{50}, primary skin irritation, and eye irritation, see Table 84.4.

52.0 1-Hydroperoxy-1′-hydroxydicyclohexyl Peroxide

52.0.1 CAS Number: *[78-18-2]*

52.0.2 Synonyms: 1-(1-Hydroperoxycyclohexyl)dioxycyclohexanol, cyclohexanone peroxide, 1-hydroperoxycyclohexyl-1-hydroxycyclohexyl peroxide, cyclohexyl peroxide dihydroperoxide

52.0.3 Trade Names: NA

52.0.4 Molecular Weight: 246.34

52.0.5 Molecular Formula: $C_{12}H_{22}O_5$

52.0.6 *Molecular Structure:*

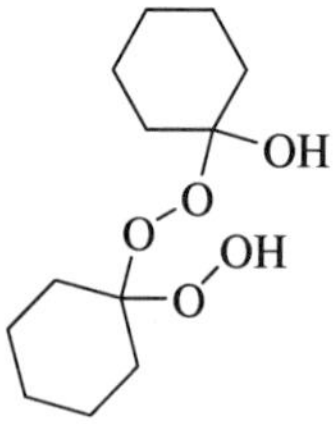

52.1 Chemical and Physical Properties

52.1.1 *General*

This peroxide is an off-white thick paste (48).

52.2 Production and Use

It is used as a catalyst for hardening certain fiberglass resins.

52.3 Exposure Assessment

52.3.1 *Air*

The photocolorimetric determination has been used to analyze for this peroxide in water (90).

52.4 Toxic Effects

52.4.1 *Experimental Studies*

52.4.1.1 Acute Toxicity. For information on oral LD_{50}, primary skin irritation, and eye irritation, see Table 84.4.

52.4.1.4 Reproductive and Developmental. When this peroxide was administered to rats orally at 72 mg/kg (10% of the LD_{50}) on a daily basis for 30 d it decreased the mobility of spermatozoids. Peroxide treatment of the males increased embryonic death and decreased the average weight of the first generation offspring. It did not affect the percentage of females impregnated nor the number of embryos per female (91).

52.4.1.7 Other: Neurological, Pulmonary, Skin Sensitization. A Russian article (English abstract) reports in mice a single oral dose of 1.028 g/kg as the threshold dose for nerve–muscle irritation, perturbation in hemoglobin/methemoglobin content, and catalase and peroxidase activity in blood (91). In the workplace the peroxide has caused eye irritation and skin sensitization (92). Prolonged inhalation of vapors result in headache and throat irritation. Prolonged skin contact with clothing contaminated with peroxides may cause irritation and blistering (6).

D DIALKYL PEROXIDES

53.0 Di-*t*-butyl Peroxide

53.0.1 CAS Number: [110-05-4]

53.0.2 Synonyms: *t*-butyl peroxide, *bis*(*t*-butyl) peroxide, di-*tert*-butyl peroxide, TBP, DTBP, bis(1,1-dimethyl)peroxide, di-tertiary-butyl peroxide, *tert*-butyl peroxide, 98% - Carc.

53.0.3 Trade Names: DTBT

53.0.4 Molecular Weight: 146.26

53.0.5 Molecular Formula: $C_8H_{18}O_2$

53.0.6 Molecular Structure:

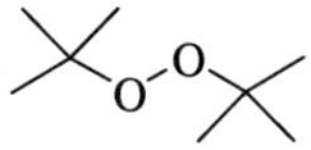

53.1 Chemical and Physical Properties

Melting point: $-40°C$;
Boiling point: 109~110°C (explodes when heated)
Flash point: 18.3
Water solubility: <0.1 g/100 mL at 21°C.

53.1.1 General

This peroxide is a clear to yellow liquid. It is produced by reacting sulfated *t*-butyl alcohol or isobutylene with hydrogen peroxide. Di-*t*-butyl peroxide is classified a deflagration hazard by Noller (2); however, it has low impact sensitivity (65). In the revised time–pressure test this peroxide gave a relatively slow rise in pressure when 3 g of igniter was used. Therefore, Tamura (3) did not classify it as a deflagration hazard. Di-*t*-butyl peroxide has been reported to increase the degree of purification in the combustion gas of power plants by removing toxic impurities (93).

53.2 Production and Use

It is an important initiator for high-temperature, high-pressure polymerizations of ethylene and halogenated ethylenes. It is used in curing resins or allyl acetate and allyl phthalate types and is also used in the synthesis of polyketones from carbon monoxide and ethylene (6).

53.3 Exposure Assessment

53.3.1 Air

One method reported for analyzing dialkyl peroxides is high-pressure liquid chromatography (94).

53.3.3 *Workplace Methods*

No air collection method exists for sampling this peroxide in the breathing zone of the worker.

53.4 Toxic Effects

A 4-h exposure to 24.5 g/m^3 of di-*t*-butyl peroxide caused excitability and breathing difficulties in 10 mice, but there were no deaths. Similar exposure of six rats caused tremors of the head and neck, weakness of the limbs, and prostration. One animal that became hyperactive died after 7 d. Autopsies of rats and mice that died during this treatment indicated lung damage was the cause of death (9).

53.4.1 *Experimental Studies*

53.4.1.1 Acute Toxicity. For information on oral LD_{50}, Dermal LD_{50}, primary skin irritation, eye irritation, and LC_{50}, see Table 84.5.

53.4.1.5 Carcinogenesis. A single exposure (route unspecified, but probably subcutaneous) of 14.6 mg ($\sim$365 mg/kg) produced unconvincing evidence for carcinogenicity owing to the lack of controls in 50 mice observed for more than 80 wk. Of 35 survivors, seven (20%) had malignant blood tumors (lymphomas) and one had a benign lung tumor (pulmonary adenoma) (94). Owing to its poor design, this study should be judged inadequate to determine carcinogenicity.

53.4.1.6 Genetic and Related Cellular Effects Studies. Di-*t*-butyl peroxide was not mutagenic in the bacterial *Salmonella typhimurium* test with or without a liver metabolic activation fraction (95, 96). No DNA damage was seen in tests with the bacterium *Pneumococcus* (97). but an equivocal result was obtained using *Escherichia coli* bacteria (SOS chromotest) (98). Mutagenicity tests performed in the fungus *Neurospora* have given both negative (99) and reportedly positive results (107), although no results were reported to support the positive findings.

For more information see Table 84.5.

53.4.1.7 Other: Neurological, Pulmonary, Skin Sensitization. Peroxide caused no eye effects when instilled at 0.1 g, 0.5 g, or an unstated amount into rabbit's eyes (9, 101), and only mild irritation was reported in two additional studies (102).

54.0 2,5-Dimethyl-2,5-di-(*t*-butylperoxy)hexane

54.0.1 *CAS Number:* *[78-63-7]*

54.0.2 *Synonyms:* 2,5-Dimethyl-2,5-di-(tertiary-butylperoxy)-hexane, and 2,5-dimethyl-2,5-*bis*(*tert*-butylperoxy)hexane, 90%

54.0.3 *Trade Names:* NA

54.0.4 *Molecular Weight:* 290.44

54.0.5 Molecular Formula: $C_{16}H_{34}O_4$

54.0.6 Molecular Structure:

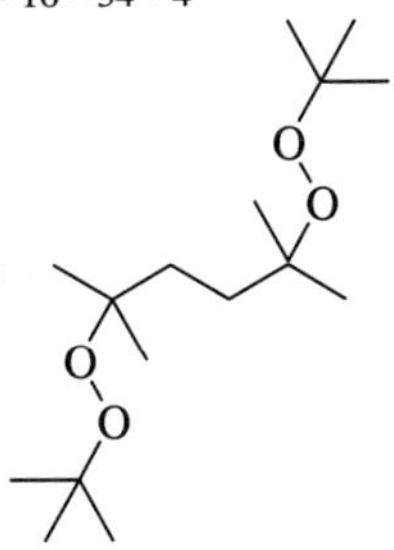

54.1 Chemical and Physical Properties

Melting point: 8°C
Boiling point: 50~52°C
Flash point: 80°C
Water solubility: < 0.1 g/100 mL at 23°C

54.1.1 General

This peroxide is a colorless to slightly yellow liquid used as a melt-flow modifier for polyolefins. Its half-life in benzene is 10 h at 248°F (120°C) (63). The deflagration properties of this material was tested by Tamura (3). There was no rise in pressure using 5 g of igniter; therefore, it is not a deflagration hazard. The thermal decomposition products of this peroxide heated at 125°C in several different organic solvents were basically acetone, *t*-butyl alcohol, *t*-amyl alcohol, methane, ethane, ethylene, *t*-amyl-*t*-butyl peroxide, and some residual parent compound (103).

54.3 Exposure Assessment

54.3.1 Air

One method reported for analyzing dialkyl peroxides is high-pressure liquid chromatography (94).

54.3.3 Workplace Methods

No air collection method exists for sampling this peroxide in the breathing zone of the worker.

54.4 Toxic Effects

The data elements provided in Table 84.5 were the only toxicity information located on this chemical.

54.4.1 Experimental Studies

54.4.1.1 Acute Toxicity. For information on oral LD_{50} and dermal LD_{50}, see Table 84.5.

55.0 2,5-Dimethyl-2,5-di-(*t*-butylperoxy)hexyne-3

55.0.1 CAS Number: *[1068-27-5]*

55.0.2 Synonyms: NA

55.0.3 Trade Names: NA

55.0.4 Molecular Weight: 286.46

55.0.5 Molecular Formula: $C_{16}H_{30}O_4$

55.0.6 Molecular Structure:

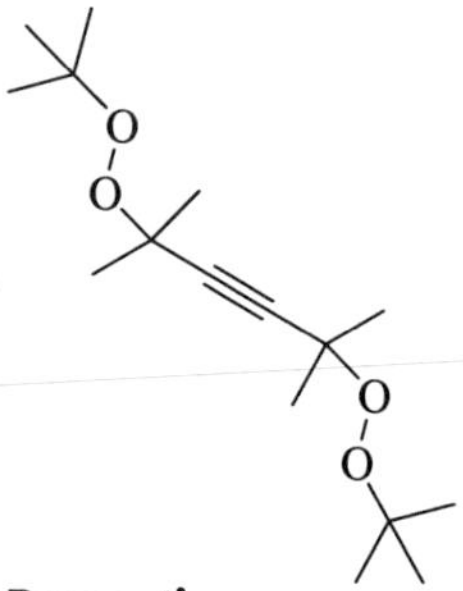

55.1 Chemical and Physical Properties

55.1.1 General

This peroxide is a yellow liquid used in the cross-linking of various polyolefins. Its half-life in benzene is 10 h at 267°F (130°C) (104).

55.3 Exposure Assessment

55.3.1 Air

One method reported for analyzing dialkyl peroxides is high-pressure liquid chromatography (94).

55.3.3 Workplace Methods

No air collection method exists for sampling this peroxide in the breathing zone of the worker.

55.4 Toxic Effects

The data elements provided in Table 84.5 were the only toxicity information located on this chemical.

55.4.1 Experimental Studies

55.4.1.1 Acute Toxicity. The information reviewed on the acute inhalation study stated that it was a nontoxic compound. The concentration tested was not specified (105).

For more information on oral LD_{50}, primary skin irritation, and LC_{50}, see Table 84.5.

56.0. Dicumyl Peroxide

56.0.1 CAS Number: [80-43-3]

56.0.2 Synonyms: Cumyl peroxide, bis(α,α-dimethylbenzyl) peroxide, diisopropylbenzene peroxide, isopropylbenzene peroxide; bis(1-methyl-1phenylethyl) peroxide, bis(α, α-dimethylbenzyl) peroxide, di-cumyl peroxide, di-cup, and cumene peroxide

56.0.3 Trade Names: NA

56.0.4 Molecular Weight: 270.37

56.0.5 Molecular Formula: $C_{18}H_{22}O_2$

56.0.6 Molecular Structure:

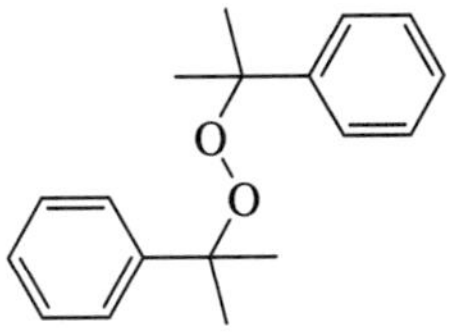

56.1 Chemical and Physical Properties

Boiling point: 130°C
Water solubility: < 0.1 g/100 mL at 23°C

56.1.1 General

Dicumyl peroxide is a crystalline solid that melts at 42°C. It is insoluble in water and soluble in vegetable oil and organic solvents (106). It is used as a high-temperature catalyst in production of polystyrene plastics. The deflagration hazard potential of this peroxide was tested using 5 g of igniter in the revised time–pressure test, but no pressure rise was produced (3). Noller (2) found it to be an intermediate fire hazard.

56.3 Exposure Assessment

56.3.1 Air

An analytic method exists that consists of using a liquid chromatograph equipped with a 254-nm ultraviolet detector (107). However, the most recent method uses high-pressure liquid chromatography (108).

56.3.3 Workplace Methods

No air collection method exists for sampling this peroxide in the breathing zone of the worker.

56.4 Toxic Effects

All other available toxicity data are presented in Table 84.5.

56.4.1 *Experimental Studies*

56.4.1.1 Acute Toxicity. For information on oral LD_{50}, primary skin irritation, eye irritation, and LC_{50}, see Table 84.5.

56.4.1.4 Reproductive and Developmental. The embryo toxicity of this peroxide was studied using 3-d chicken embryos. The peroxide was delivered directly onto the inner membrane overlaying the embryo (44). The median effective dose (ED_{50})/egg was 1.24 mmol producing 57% malformed embryos. Because of the obvious difference in delivery of the dose in this study versus that in humans, the relevance of this information to humans is unknown.

56.4.1.6 Genetic and Related Cellular Effects Studies. No evidence of mutagenicity was seen in the *S. typhimurium* assay with or without metabolic activation (86, 109–111). However, this material was reported to be mutagenic in *E. coli* in the absence of a liver metabolic activation fraction (112).

56.4.1.7 Other: Neurological, Pulmonary, Skin Sensitization. Mild conjunctivitis was produced in the eyes of rabbits after instillation of 0.1 mL of 50% dicumyl peroxide in corn oil. Guinea pigs receiving intradermal injections of 0.1% in saline produced no sensitization (107). Rabbits' nostrils instilled with 50 mL of 0.001 or 0.0025% in saline suffered slight inflammation of the nasal mucosa within 1 h. Repeated applications of 0.001%, three times a day, 5 d/wk for up to 4 wk caused increased vascularization in the nose (87). No overt signs of toxicity were seen in groups of 10 rats and four rabbits exposed for 6 h to atmospheres containing 21–224 mg dust/m^3 (8–90 mg dicumyl peroxide/m^3) (113).

56.4.2 *Human Experience*

56.4.2.2.7 Other: Neurological, Pulmonary, Skin Sensitization, etc. Dicumyl peroxide has been reported to cause slight skin irritation, with no skin sensitization in 200 human volunteers patch tested with a technical grade of 90% of this material (107). However, there has been one report of one female worker allegedly contracting occupational asthma after inhaling the vapors from a heated polyethylene repair tape (114). It was reported that the vapors most likely consisted of dicumyl peroxide and its breakdown products (113). A group of 18 workers exposed to dicumyl peroxide in the atmosphere (concentration unspecified) exhibited nasal "crusting" and visible blood vessels (115).

57.0 α,α′-*Bis*(*t*-butylperoxy)diisopropylbenzene

57.0.1 *CAS Number:* *[25155-25-3, 2781-00-2]*

57.0.2 *Synonyms:* NA

57.0.3 *Trade Names:* NA

57.0.4 *Molecular Weight:* 338.47

57.0.5 Molecular Formula: $C_{20}H_{34}O_4$

57.0.6 Molecular Structure:

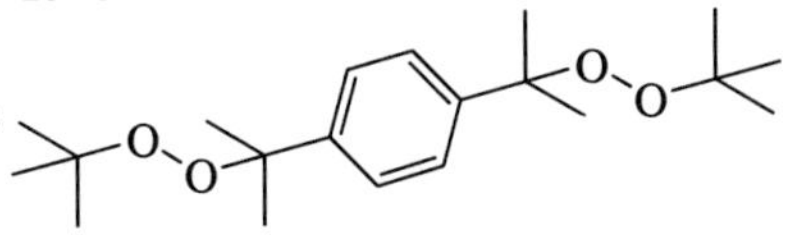

57.1 Chemical and Physical Properties

57.1.1 General

No information was located for this compound.

57.2 Production and Use

No information was located for this compound.

57.3 Exposure Assessment

57.3.1 Air

One method reported for analyzing dialkyl peroxides is high-pressure liquid chromatography (94).

57.3.3 Workplace Methods

No air collection method exists for sampling this peroxide in the breathing zone of the worker.

57.4 Toxic Effects

The data elements provided in Table 84.5 were the only toxicity information located on this chemical.

57.4.1 Experimental Studies

57.4.1.1 Acute Toxicity. An acute dust inhalation study was conducted using rats, guinea pigs, and mice. The animals were exposed to 180 mg/m^3 for 6 h. No deaths or untoward behavioral reactions or effects on body weight were noted during the exposure or during the 5-d postexposure observation period (116). An acute vapor inhalation study was carried out on two rats, two guinea pigs, and two mice. The animals were exposed to an average nominal vapor concentration of 0.1 mg/L (100 mg/m^3) for 4 h. No adverse effects were reported. The LC_{50} was considered to be greater than 0.1 mg/L (116).

For more information on oral LD_{50}, primary skin irritation, eye irritation, and LC_{50}, see Table 84.5.

57.4.1.6 Genetic and Related Cellular Effects Studies. For information see Table 84.5.

58.0 Cumyl *t*-Butyl Peroxide

58.0.1 CAS Number: [3457-61-2]

58.0.2 Synonyms: *t*-Butyl cumyl peroxide, *t*-butyl αα-dimethylbenzyl peroxide, *t*-butyl cumyl peroxide

58.0.3 Trade Names: NA

58.0.4 Molecular Weight: 208.33

58.0.5 Molecular Formula: $C_{13}H_{20}O_2$

58.0.6 Molecular Structure: 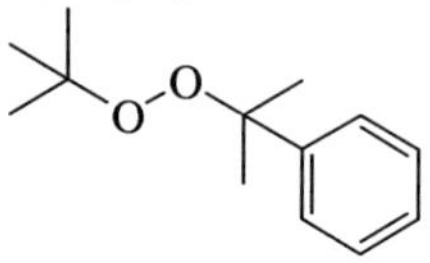

58.1 Chemical and Physical Properties

58.1.1 General

t-Butyl cumyl peroxide is a pale yellow liquid.

58.3 Exposure Assessment

58.3.1 Air

Gas chromatography has been used to determine the concentration of this peroxide (160). An "Intersmat" gas chromatograph with a flame ionization detector with a stainless steel column (06 × mm; 0.30 m) packed with QF-1 (14.8% diglycerol, 0.13% on chromosorb GAW-DMCS). Column, injector, and detector temperatures were 80, 95, and 100°C, respectively. The gas chromatograph was calibrated by injecting samples of a known concentration of the peroxide in ethanol.

58.4 Toxic Effects

The data elements provided in Table 84.5 were the only toxicity information located on this chemical.

58.4.1 Experimental Studies

58.4.1.1 Acute Toxicity. An acute vapor inhalation study in rats showed the LC_{50} to be greater than 140 ppm (1.2 mg/L) (117). A later acute vapor inhalation study was conducted using 13 rats. The 4-h exposure at a nominal concentration of 152 mg/L resulted in no deaths. After the first 15 min of exposure the animals showed reduced activity and partially closed eyes. During the exposure lacrimation and salivation were observed. In addition, coarse intermittent body tremors were observed the last 2 h of exposure. Slight lacrimation continued through day 9. The LC_{50} was judged greater than 152 mg/L (118).

For more information on oral LD_{50}, primary skin irritation, eye irritation, and LC_{50}, see Table 84.5.

58.4.1.6 Genetic and Related Cellular Effects Studies. For information see Table 84.5.

59.0 4-(*t*-Butylperoxy)-4-methyl-2-pentanone

59.0.1 CAS Number: [26394-04-7]

59.0.2 Synonyms: NA

59.0.3 Trade Names: NA

59.0.4 Molecular Weight: 188.26

59.1 Chemical and Physical Properties

59.1.1 General

No information was located for this compound.

59.2 Production and Use

No information was located for this compound.

59.3 Exposure Assessment

59.3.1 Air

One method reported for analyzing dialkyl peroxides is high-pressure liquid chromatography (71).

59.3.3 Workplace Methods

No air collection method exists for sampling this peroxide in the breathing zone of the worker.

59.4 Toxic Effects

The data elements provided in Table 84.5 were the only toxicity information located on this chemical.

59.4.1 Experimental Studies

59.4.1.1 Acute Toxicity. For information on oral LD_{50}, dermal LD_{50}, eye irritation, and LC_{50}, see Table 84.5.

E PEROXYKETALS

60.0. 1,1-Di-(*t*-butylperoxy)-3,3,5-trimethylcyclohexane

60.0.1 CAS Number: [6731-36-8]

60.0.2 Synonyms: (3,3,5-Trimethylcyclohexylidene)*bis*(*t*-butyl) peroxide and (3,3,5-trimethylcyclohexylidene)*bis*(1,1-dimethylethyl) peroxide

60.0.3 Trade Names: NA

60.0.4 Molecular Weight: 302.5

60.0.5 Molecular Formula: $C_{17}H_{34}O_4$

60.0.6 Molecular Structure:

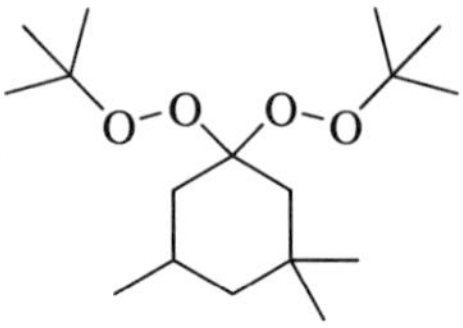

60.1 Chemical and Physical Properties

60.1.1 General

This chemical is a clear liquid with a half-life in benzene of 10 h at 202°F (94°C).

60.2 Production and Use

It is used as an effective source of free radicals in the cross-linking of elastomers. It resists the decomposition effects of most fillers and pigments. This commercially available product is a deflagration hazard.

60.3 Exposure Assessment

The extraction and photometric analysis of this peroxide in water was reported by Sevast'yanova and Smirnova (90).

60.4 Toxic Effects

A foreign study reports that this peroxide is of low toxicity. However, when used at concentrations exceeding maximum permissible levels (unspecified), it caused local skin and eye irritation, as well as irritation of the respiratory tract. It did not show any cumulative effects (time unspecified) (119).

60.4.1 Experimental Studies

60.4.1.1 Acute Toxicity. For information on oral LD_{50}, Dermal LD_{50}, eye irritation, and LC_{50}, see Table 84.6.

61.0. 1,1-Di-(*t*-butylperoxy)cyclohexane

61.0.1 CAS Number: *[3006-86-8]*

61.0.2 Synonyms: NA

61.0.3 Trade Names: NA

61.0.4 Molecular Weight: 260.36

61.0.5 Molecular Formula: $C_{14}H_{28}O_4$

61.0.6 Molecular Structure:

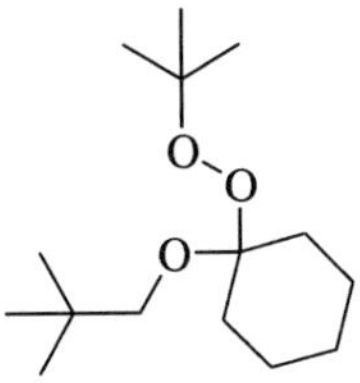

61.1 Chemical and Physical Properties

61.1.1 General

No information was located for this compound.

61.2 Production and Use

No information was located for this compound.

61.3 Exposure Assessment

61.3.1 Air

No air collection method or analytic method was located for this compound.

61.4 Toxic Effects

The data elements provided in Table 84.6 were the only toxicity information located on this chemical.

61.4.1 Experimental Studies

61.4.1.1 Acute Toxicity. An acute inhalation study using rats exposed them to 207.2 or 20.8 mg/L for 4 h. Salivation, nasal discharge, and dyspnea were noted during the study. Slight dyspnea persisted at the lower dose for 1–2 d and at the higher dose for up to 7 d. No deaths occurred and no morphological changes were noted at necropsy (120). The LC_{50} is considered to be greater than 207.2 mg/L.

62.0 2,2-Di-(*t*-butylperoxy)butane

62.0.1 CAS Number: *[2167-23-9]*

62.0.2 Synonyms: NA

62.0.3 Trade Names: NA

62.0.4 Molecular Weight: 234.1

62.0.5 *Molecular Formula:* $C_{12}H_{26}O_4$

62.0.6 *Molecular Structure:*

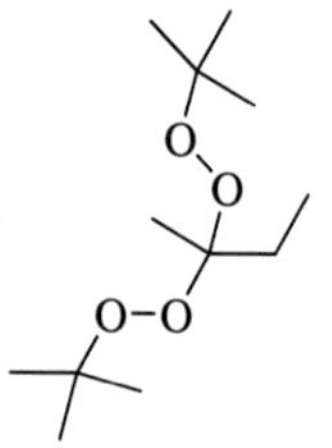

62.1 Chemical and Physical Properties

62.1.1 *General*

This peroxide dissolved in dibutyl phthalate is a clear, colorless liquid.

62.3 Exposure Assessment

62.3.1 *Air*

No air collection method or analytic method was located for this compound.

62.4 Toxic Effects

The data elements provided in Table 84.6 were the only toxicity information located on this chemical.

62.4.1 *Experimental Studies*

62.4.1.1 Acute Toxicity. An acute inhalation study used the maximum attainable concentration of 2.42 g/m^3 of air. The particle size was determined using a Cascade impactor. Ninety percent of the mist consisted of droplets with a diameter of 1.2–3.3 mm. The maximum droplet size was 6.7 mm. No information on the behavior of the rats during the exposure is available because the animals were completely invisible as a consequence of the dense mist prevailing in the inhalation chamber. No mortality occurred. The LC_{50} is considered to be greater than 2.42 g/m^3 (88).For more information on oral LD_{50}, primary skin irritation, eye irritation, and LC_{50}, see Table 84.6.

62.4.1.6 Genetic and Related Cellular Effects Studies. For information see Table 84.6.

63.0 *n*-Butyl 4,4-Di(*t*-butylperoxy)valerate

63.0.1 *CAS Number:* *[995-33-5]*

63.0.2 *Synonyms:* 4,4-Di-(*t*-butylperoxy) *n*-butyl valerate

63.0.3 *Trade Names:* NA

63.0.4 *Molecular Weight:* 334.45

63.0.5 ***Molecular Formula:*** $C_{17}H_{34}O_6$

63.0.6 ***Molecular Structure:***

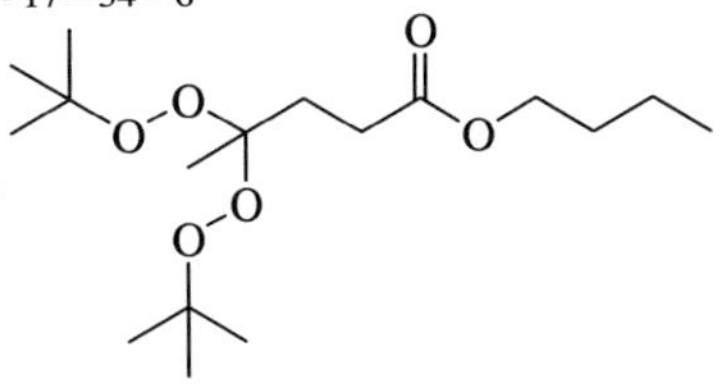

63.1 Chemical and Physical Properties

63.1.1 General

No information was located for this compound.

63.2 Production and Use

No information was located for this compound.

63.3 Exposure Assessment

63.3.1 Air

No air collection method or analytic method was located for this compound.

63.4 Toxic Effects

The data elements provided in Table 84.6 were the only toxicity information located on this chemical.

63.4.1 Experimental Studies

63.4.1.1 Acute Toxicity. For information on oral LD_{50}, primary skin irritation, and eye irritation, see Table 84.6.

64.0 2,2-Di-(Cumylperoxy)propane

64.0.1 ***CAS Number:*** *[4202-02-2]*

64.0.2 ***Synonyms:*** Biscumylperoxypropane

64.0.3 ***Trade Names:*** NA

64.0.4 ***Molecular Weight:*** 344.43

64.1 Chemical and Physical Properties

64.1.1 General

No information was located for this compound.

64.2 Production and Use

No information was located for this compound.

64.3 Exposure Assessment

64.3.1 Air

No air collection method or analytic method was located for this compound.

64.4 Toxic Effects

The data elements provided in Table 84.6 were the only toxicity information located on this chemical.

64.4.1 Experimental Studies

64.4.1.1 Acute Toxicity. For information on oral LD_{50}, primary skin irritation, and eye irritation, see Table 84.6.

64.4.1.2 Chronic and Subchronic Toxicity. For information see Table 84.6.

F HYDROPEROXIDES

65.0 *t*-Butyl Hydroperoxide

65.0.1 CAS Number: [75-91-2]

65.0.2 Synonyms: 1,1-Dimethyl ethyl hydroperoxide

65.0.3 Trade Names: TBHP

65.0.4 Molecular Weight: 90.12

65.0.5 Molecular Formula: $C_4H_{10}O_2$

65.0.6 Molecular Structure:

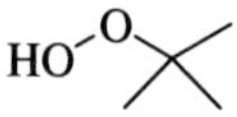

65.1 Chemical and Physical Properties

Specific gravity: 0.94
Melting point: 50°C
Boiling point: 156~160°C
Flash point: 62°C
Water solubility: moderately sol. >= 10 g/100 mL at 22°C

65.1.1 General

TBHP vapor can burn in the absence of air and may be flammable at either elevated temperature or reduced pressure. Fine mist/spray may be combustible at temperatures below the normal flash point. When evaporated the residual liquid will concentrate the TBHP content and may reach an explosive concentration (>90%). Closed containers may generate internal pressure through the degradation of TBHP to oxygen (121).

65.1.2 Odor and Warning Properties

The odor threshold for this peroxide is reported to be 0.17 mg/m^3. The warning threshold concentration is 1 ppm (122).

65.2 Production and Use

TBHP is produced by the liquid-phase reaction of isobutane and molecular oxygen or by mixing equimolar amounts of *t*-butyl alcohol and 30–50% hydrogen peroxide. This chemical has many industrial uses such as an intermediate in the production of propylene oxide and *t*-butyl alcohol from isobutane and propylene (123) TBHP, however, is primarily used as an initiator and finishing catalyst in the solution and emulsion polymerization methods for polystyrene and polyacrylates. It is used in the polymerization of vinyl chloride and vinyl acetate. It is used as an oxidation and sulfonation catalyst in bleaching and deodorizing operations. Two other possible industrial uses are as an antislime agent in cooling systems and as a settling agent in aqueous slurries of various mineral tailings (124, 125).

65.3 Exposure Assessment

65.3.1 Air

Several analytic methods have been used over the years such as high-pressure liquid chromatography ((126, 127), chromatography using a flame ionization detector (128, 129), and colorimetric detection (130).

65.3.3 Workplace Methods

No air collection method was located for sampling this peroxide in the breathing zone of the worker.

65.4 Toxic Effects

65.4.1 Experimental Studies

65.4.1.1 Acute Toxicity. Oral studies, as shown in Table 84.7 have shown TBHP is moderately toxic if ingested with an LD_{50} of 560 mg/kg. It should be noted that even though only 1 of 10 rats died when given 0.6 mL/kg, the animals in this dose group exhibited signs of depression and lacrimation. As the dose levels increased, 0.8, 1.0, 1.2,

1.4, and 1.6 mL/kg, the animals showed signs of loss of righting, hypothermia, and hematuria.

For more information on oral LD_{50}, dermal LD_{50}, primary skin irritation, eye irritation, and LC_{50}, see Table 84.7.

65.4.1.2 Chronic and Subchronic Toxicity. Zavodnik et al. (131). investigated the structural and functional transitions of the drug-metabolizing systems under oxidate injury. They noted that ionizing radiation damage to cells is mainly due to the action of very reactive hydroxyl radicals, excited states, and free radicals of macromolecules. They studied the radiation-induced damage of rat liver microsomal membranes comparing the susceptibility of microsomal membranes to chemically induced oxidative stress (with TBH). The postmortal liver microsomal membrane treatment by TBH drastically changed the membrane structure and enzymatic activities. The microsomal membrane rigidity increased after TBH treatment up to 0.5 mM and slowly decreased at higher oxidant concentrations. The microsomal NADPH oxidase and Fe(3+)-NADPH oxidoreductase activities decreased after TBH treatment of the microsomes of nonirradiated animals and either increased or remained unchanged for irradiated rats. They conclude that low-dose-rate irradiation as well as TBH significantly changed the membrane functional properties. The preliminary irradiation increased the membrane susceptibility to chemically induced oxidative stress.

Kapahi et al. (132) investigated the relationship between mammalian life span and cellular resistance to stress. They suggested that identifying the mechanisms determining species-specific life spans is a central challenge in understanding the biology of aging. Cellular stresses produce damage that may accumulate and cause aging. Evolution theory predicts that long-lived species secure their longevity through investment in a more durable soma, including enhanced cellular resistance to stress. They compared cellular resistance of primary skin fibroblasts from 8 mammalian species with a range of life spans. Cell survival was measured by the thymidine incorporation assay following stresses induced by several agents including TBH. Significant positive correlations between cell LD_{90} and maximum life span were found for all these stresses. Similar results were obtained when cell survival was measured by the MTT assay, and when lymphocytes from different species were compared. Cellular resistance to a variety of oxidative and nonoxidative stresses as positively correlated with mammalian longevity. They conclude that their research supports the concept that the gene network regulating the cellular response to stress is functionally important in aging and longevity.

Cai et al. (133) studied the oxidant-induced apoptosis in cultured human retinal pigment epithelial cells using the chemical oxidant, TBH. Their results indicate that TBH can induce apoptosis in retinal pigment epithelium, probably by triggering the mitochondrial permeability transition, which results in swelling and release of mitochondrial intermembrane proteins.

Nowak et al. (134) investigated recovery of cellular function following oxidant injury by examining the recovery of renal proximal tubule cellular (RPTC) functions following oxidant-induced sublethal injury with TBH. Their results indicate that oxidant-induced sublethal injury to RPTC may contribute to renal dysfunction and that RPTC can repair and regain cellular functions following oxidant injury.

Lavoie and Chessex (135) studied the gender-related response to TBH-induced oxidation in human neonatal tissue. They concluded that the prostaglandin concentration was not proportional to the oxidative stimulus, suggesting that a critical level of TBH exists at which the oxidative state differs in tissues derived from boys and girls.

Adams et al. (136) investigated brain oxidative stress induced by TBH. They suggest that many diseases and aging may be associated with oxidative stress in the brain. They performed intracerebroventricular injection of TBH. Oxidizing glutathione levels in the brain increased by as much as 90-fold during TBH induced oxidative stress. At the same time, brain GSH levels decreased. The brain appears to retain GSSG and not reduce it or export it efficiently. Vitamin E levels in the striatum increased during TBH-induced oxidative stress. Aging alters the ability of the brain to detoxify an oxidative stress, in that 8-mo-old mice retain GSSG in their brains much more than 2-mo-old mice. Eight-mo-old mice were much more susceptible to TBH-induced toxicity than 2-mo-old mice. This may indicate that aging makes the brain more susceptible to oxidative damage.

65.4.1.4 Reproductive and Developmental. The teratogenicity of TBHP has been investigated. Female rats inhaled 226 mg/m^3 for 4 h during day 19 of gestation. This dose caused nonspecific impairment in the development of their fetuses in addition to general maternal toxicity (137).

Kumar and Muralidhara (138) investigated the propensity for prooxidant treatment to induce dominant lethal (DL) type mutations in a randomly bred closed colony of CFT-Swiss mice. They examined TBH and cumene hydroperoxide (CHP) giving IP doses to adult males. The LD_{50} was determined as 1500 and 3000 μmol/kg for CHP and TBH, respectively. Major variables observed were implantations, live embryos, and dead implants (DI). TBHP induced a marginal increase only during the first week. They conclude that peroxidants induce DL-type effects only in specific postmeiotic stages of spermatogenesis and stress the need to further investigate the implications of chronic oxidative stress on the male reproductive system.

65.4.1.5 Carcinogenesis. A study performed to evaluate the carcinogenicity of TBHP found it was not carcinogenic when applied to the skin of mice at 16.6% of the peroxide six times a week for 45 wk. However, if its application was preceded by 0.05 mg of 4-nitroquinoline-1-oxide as a 0.25% solution in benzene applied 20 times over 7 wk followed by TBHP (16.6% in benzene), then malignant skin tumors appeared between days 390 and 405 of the experiment (139). This supports the theory that peroxides are not complete carcinogens, but may act as promoters (140). The effects of TBHP on promotable and nonpromotable mouse epidermal cell culture lines were reported by Muehlematter (141).

65.4.1.6 Genetic and Related Cellular Effects Studies. Several mutagenicity studies have been carried out with *t*-butyl hydroperoxide (96, 142, 143). The first two articles cited a positive result with TBHP in test strains of bacteria. The latter study was an *in vivo* experiment that showed no mutagenesis in the bone marrow cells after rats inhaled 100 ppm for 6 h/d for 5 d. ARCO Chemical Company (121) concerned about sorting out this information, conducted other tests:

	Mutagenicity of TBHP		
Test	Result	Comment	Dose
Salmonella, typhimurium, – 1537, necessary assay	Positive	TA-98, – 100, Liver S-9 fraction	
Mouse lymphoma	Positive	See below	0.018–0.0013 mL/mL
Cell transformation	Negative	C3H/10T-1/2	0.0049–0.0003 mL/mL
Rat bone marrow	Negative	Whole animal	5-d exposure to 100 ppm for 6/d

Of special interest in the mouse lymphoma test was the indication that TBHP is several times less active when the enzyme preparation is added than without the microsomal preparation. This may indicate that TBHP can be readily metabolized in a whole animal to a much less genotoxic material. To support this theory, the whole animal experiments showed no indication of genotoxic potential. Also, an *in vitro* assay looking at cell transformation was negative. This assay has been interpreted to be an *in vitro* estimation of carcinogenic potential. In conclusion, TBHP is mutagenic in bacterial and mammalian cells in culture, but is nonmutagenic in whole animals.

In cells, TBHP is reduced by glutathione peroxidase to *t*-butyl alcohol and water. TBHP is not destroyed by catalase, an enzyme that destroys other peroxides (94). TBHP has been reported to cause lipid peroxidation and to produce other evidence of oxidative stress, resulting in adverse effects in cells in culture and other *in vitro* systems (144–147).

For more information see Table 84.7.

65.4.1.7 Other: Neurological, Pulmonary, Skin Sensitization. In a routine primary skin irritation assay, three out of the six rabbits died. The peroxide is a severe dermal irritant, by DOT standards, a corrosive. It can also be absorbed through the skin in toxic amounts to cause cyanosis, depression, loss of righting, blanching of the treated skin, convulsions, and death.

65.4.2 Human Experience

65.4.2.2.7 Other: Neurological, Pulmonary, Skin Sensitization, etc. A single inhalation study reported a respiratory irritation threshold for humans of 45 mg/m^3. TBHP poisoning in humans causes severe depression, incoordination, cyanosis, and death due to respiratory arrest. However, levels producing such effects would not normally be expected to occur (48).

65.6 Studies on Environmental Impact

TBHP is not be subject to biomagnification in an aquatic environment owing to its low partition coefficient (octanol/water, – 1.30) (148) and high reactivity (149).

66.0. Isopropylbenzene Hydroperoxide

66.0.1 CAS Number: [80-15-9]

66.0.2 Synonyms: Cumene hydroperoxide, α,α-dimethylbenzyl hydroperoxide, α-cumyl hydroperoxide

66.0.3 Trade Names: CHP

66.0.4 Molecular Weight: 152.18

66.0.5 Molecular Formula: $C_9H_{12}O_2$

66.0.6 Molecular Structure:

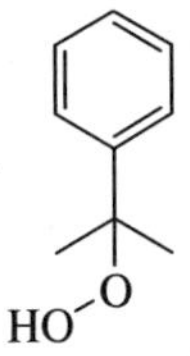

66.1 Chemical and Physical Properties

Specific gravity: 1.024
Water solubility: slightly soluble <0.01 g/100 mL at 18°C
Melting point: < -40°C
Boiling point: 153°C
Flash point: 79°C

66.1.1 General

This 90% pure, colorless liquid has a flash point of 175°F and is relatively insensitive to shock as indicated by impact and detonation tests (107, 113).

66.3 Exposure Assessment

66.3.1 Air

Several analytic methods have been used over the years such as high-pressure liquid chromatography (126, 127), chromatography using a flame ionization detector (128, 129), and colorimetric detection (130).

66.3.3 Workplace Methods

No air collection method was located for sampling this peroxide in the breathing zone of the worker.

66.4 Toxic Effects

Based on data presented in Table 84.7, this peroxide is moderately toxic when ingested or inhaled. Prolonged inhalation of vapors of CHP causes headache and throat irritation.

Necropsy of rats used in the inhalation LC_{50} test showed severe inflammation of the trachea and lungs.

66.4.1 Experimental Studies

66.4.1.1 Acute Toxicity. For information on oral LD_{50}, dermal LD_{50}, primary skin irritation, eye irritation, and LC_{50}, see Table 84.7.

66.4.1.2 Chronic and Subchronic Toxicity. Repeated sublethal doses (one-fifth the LD_{50}) of CHP given rats three times a week for 7 wk either orally or intraperitoneally resulted in cumulative effects. Specific effects were not reported (9).

The review by Kumar and Muralidhara (138) regarding dominant lethal mutations is reviewed in the previous section (TBH).

66.4.1.6 Genetic and Related Cellular Effects Studies. For information see Table 84.7.

66.4.1.7 Other: Neurological, Pulmonary, Skin Sensitization. This peroxide causes burning and throbbing when in contact with the skin. Edema, erythema, and vesiculation may take 2–3 d to appear. The maximal concentration that produced no irritation to rabbit skin was 7%. High concentrations of CHP applied directly to the eyes of rabbits affected the cornea, iris, and conjunctiva extensively. Washing the eyes with water for 4 sec after application prevented any adverse reactions. The maximal nonirritating concentration to rabbits' eyes was 1% (107, 113).

Inhalation experiments conducted in six female rats exposed to 16 ppm 12 times, each exposure lasting 4 to 5 h, produced salivation and nose irritation. All organs were normal at necropsy. When the concentration was increased to 31.5 ppm for six rats for seven exposures, 5 h each, respiratory difficulty was noted along with salivation, tremor, weight loss, and hyperemia of eyes and tail. The lungs were the target organ showing emphysema and thickening of the alveolar walls. Two rats were exposed to 50 ppm; one died of congested lungs and kidneys (150).

67.0 Ethylbenzene Hydroperoxide

67.0.1 CAS Number: [3071-32-7]

67.0.2 Synonyms: 1-Phenylethyl hydroperoxide

67.0.3 Trade Names: NA

67.0.4 Molecular Weight: 138.16

67.0.5 Molecular Formula: $C_8H_{10}O_2$

67.0.6 Molecular Structure:

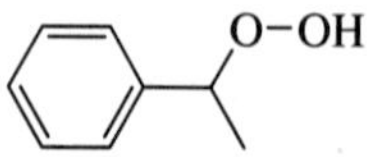

67.1 Chemical and Physical Properties

67.1.1 General

No information was located for this compound.

67.2 Production and Use

No information was located for this compound.

67.3 Exposure Assessment

67.3.1 Air

Several analytic methods have been used over the years such as high-pressure liquid chromatography (126, 127), chromatography using a flame ionization detector (128, 129), and colorimetric detection (130).

67.3.3 Workplace Methods

No air collection method was located for sampling this peroxide in the breathing zone of the worker.

67.4 Toxic Effects

The data elements provided in Table 84.7 were the only toxicity information located on this chemical.

67.4.1 Experimental Studies

67.4.1.1 Acute Toxicity. Two acute inhalation studies have been performed using this material; however, neither study critically evaluated the actual composition of the chamber atmosphere. Therefore, the LC_{50} ranges from 20 to 33 mg/L.

For more information on oral LD_{50}, Dermal LD_{50}, primary skin irritation, eye irritation, and LC_{50}, see Table 84.7.

68.0 1,1,3,3-Tetramethylbutyl Hydroperoxide

68.0.1 CAS Number: *[5809-08-5]*

68.0.2 Synonyms: 2,4,4-Trimethylpentyl-2-hydroperoxide

68.0.3 Trade Names: NA

68.0.4 Molecular Weight: 146.22

68.0.5 Molecular Formula: $C_8H_{18}O_2$

68.0.6 Molecular Structure:

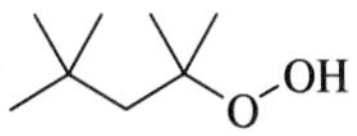

68.1 Chemical and Physical Properties

68.1.1 General

No information was located for this compound.

68.2 Production and Use

No information was located for this compound.

68.3 Exposure Assessment

68.3.1 Air

Several analytic methods have been used over the years such as high-pressure liquid chromatography (126, 127), chromatography using a flame ionization detector (128, 129), and colorimetric detection (130).

68.3.3 Workplace Methods

No air collection method was located for sampling this peroxide in the breathing zone of the worker.

68.4 Toxic Effects

The data elements provided in Table 84.7 were the only toxicity information located on this chemical.

68.4.1 Experimental Studies

68.4.1.1 Acute Toxicity. An acute inhalation study using rats indicated that the LC_{50} was greater than 480 ppm. However, the severity of effects expressed during exposure indicates the LC_{50} will not be exceedingly greater than this value. The effects were mouth breathing and incoordination; severity increased as the exposure progressed (54).

For more information on oral LD_{50}, primary skin irritation, eye irritation, and LC_{50}, see Table 84.7.

68.4.1.6 Genetic and Related Cellular Effects Studies. For information see Table 84.7.

69.0 Tetraline Hydroperoxide

69.0.1 CAS Number: *[771-29-9]*

69.0.2 Synonyms: 1,2,3,4-Tetrahydro-1-naphthyl hydroperoxide

69.0.3 Trade Names: NA

69.0.4 Molecular Weight: 164.22

69.0.5 Molecular Formula: $C_{10}H_{12}O_2$

69.0.6 Molecular Structure:

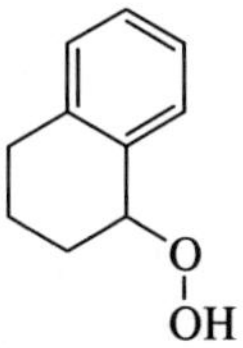

69.1 Chemical and Physical Properties

69.1.1 General

No information was located for this compound.

69.2 Production and Use

No information was located for this compound.

69.3 Exposure Assessment

69.3.1 Air

Several analytic methods have been used over the years such as high-pressure liquid chromatography (126, 127), chromatography using a flame ionization detector (128, 129), and colorimetric detection (130).

69.3.3 Workplace Methods

No air collection method was located for sampling this peroxide in the breathing zone of the worker.

69.4 Toxic Effects

The data elements provided in Table 84.7 were the only toxicity information located on this chemical.

69.4.1 Experimental Studies

69.4.1.1 Acute Toxicity. For information on Oral LD_{50}, see Table 84.7.

70.0 1-Vinyl-3-cyclohexen-1-yl Hydroperoxide

70.0.1 CAS Number: *[3736-26-3]*

70.0.2 Synonyms: 1-Hydroperoxy-1-vinylcyclohex-3-ene

70.0.3 Trade Names: NA

70.0.4 Molecular Weight: 140.20

70.1 Chemical and Physical Properties

70.1.1 General

No information was located for this compound.

70.2 Production and Use

No information was located for this compound.

70.3 Exposure Assessment

70.3.1 Air

Several analytic methods have been used over the years such as high-pressure liquid chromatography (126, 127), chromatography using a flame ionization detector 128, 129), and colorimetric detection (130).

70.3.3 Workplace Methods

No air collection method was located for sampling this peroxide in the breathing zone of the worker.

70.4 Toxic Effects

When mice were exposed to this material (route unknown) at 1440 mg/kg, there was an equivocal increase in incidences of tumors (151).

The data elements provided in Table 84.7 were the only acute toxicity information located on this chemical.

70.4.1 Experimental Studies

70.4.1.1 Acute Toxicity. For information on dermal LD_{50}, see Table 84.7.

71.0. Isopropyl Cumyl Hydroperoxide

71.0.1 CAS Number: *[26762-93-6]*

71.0.2 Synonyms: Diisopropylbenzene hydroperoxide and DIBHP

71.0.3 Trade Names: NA

71.0.4 Molecular Weight: 194.26

71.0.5 Molecular Formula: $C_{12}H_{18}O_2$

71.0.6 Molecular Structure:

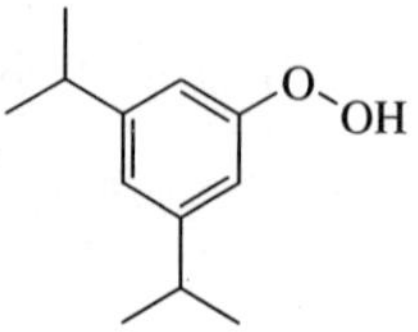

71.1 Chemical and Physical Properties

71.1.1 General

Diisopropylbenzene hydroperoxide is a clear, yellow liquid.

71.3 Exposure Assessment

71.3.1 Air

Several analytic methods have been used over the years such as high-pressure liquid chromatography (126, 127), chromatography using a flame ionization detector (128, 129), and colorimetric detector (130).

71.3.3 Workplace Methods

No air collection method was located for sampling this peroxide in the breathing zone of the worker.

71.4 Toxic Effects

The data elements provided in Table 84.7 were the only toxicity information located on this chemical.

71.4.1 Experimental Studies

71.4.1.1 Acute Toxicity. An acute inhalation study using rats exposed to the nominal concentrations of 2.18, 4.26, and 5.59 mg/L caused red nasal discharge at even the lowest dose. The animals also showed signs of shortness of breath, reduced activity, and vasodilitation. Fifty percent of the animals exposed to the 4.26 mg/L level died showing similar symptoms (152).

For more information on oral LD_{50}, primary skin irritation, eye irritation, and LC_{50}, see Table 84.7.

71.4.1.7 Other: Neurological, Pulmonary, Skin Sensitization. A 4-d subacute skin irritation test in rabbits was performed with this hydroperoxide in propylene glycol. After four applications of one a day, 1.75% wt/vol of this chemical caused mild to moderate dermatitis; 3.5, 7.0, and 14.0% caused severe dermatitis at the end of 4 d. In fact, for the two highest concentrations the fourth application was discontinued due to necrosis (143).

72.0 *p*-Menthane Hydroperoxide

72.0.1 CAS Number: [26762-92-5]

72.0.2 Synonyms: *p*-Menthyl hydroperoxide

72.0.3 Trade Names: NA

72.0.4 Molecular Weight: 172.30

72.1 Chemical and Physical Properties

72.1.1 General

No information was located for this compound.

72.2 Production and Use

No information was located for this compound.

72.3 Exposure Assessment

72.3.1 Air

Several analytic methods have been used over the years such as high-pressure liquid chromatography (126, 127), chromatography using a flame ionization detector (128, 129), and colorimetric detection (130).

72.3.3 Workplace Methods

No air collection method was located for sampling this peroxide in the breathing zone of the worker.

72.4 Toxic Effects

This peroxide was tested in five strains of bacteria in the *S. typhimurium* assay. It was negative in four, but weakly positive in TA97 (a frameshift detector) with and without metabolic activation (111).

72.4.1 Experimental Studies

72.4.1.1 Acute Toxicity. The data elements provided in Table 84.7 were the only acute toxicity information located on this chemical. It should be noted that corneal opacity was still present 14 d after instilling this material in the eyes of rabbits.

A 4-d subacute skin irritation test in rabbits was performed with this hydroperoxide in propylene glycol. All concentrations, 1.75, 3.5, 7.0, and 14.0% wt/vol, caused severe dermatitis after two applications. Therefore, the last two applications were not made for the three highest concentrations. The 1.75% dose was applied for 4 d, resulting in second-degree burns and focal third-degree burns (143).

An acute inhalation study using 18 rats exposed to the nominal concentrations of 5.37, 8.99, and 10.64 mg/L found red nasal discharge at even the lowest dose. The animals also showed signs of shortness of breath, reduced activity, and vasodilation. Fifty percent of the animals exposed to the 8.99 mg/L level died showing similar symptoms (152).

For more information on oral LD_{50}, primary skin irritation, eye irritation, and LC_{50}, see Table 84.7.

72.4.1.6 Genetic and Related Cellular Effects Studies. For information see Table 84.7.

73.0 Peroxyacetic Acid

73.0.1 CAS Number: [79-21-0]

73.0.2 Synonyms: Acetyl hydroperoxide, peracetic acid, peracetic acid solution

73.0.3 Trade Names: NA

73.0.4 Molecular Weight: 76.052

73.0.5 Molecular Formula: $C_2H_4O_3$

73.0.6 Molecular Structure:

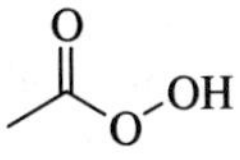

73.1 Chemical and Physical Properties

Melting point: 0.1°C
Boiling point: 105°C
Flash point: 40.5°C
Water solubility: ≥10 g/100 ml at 19°C

73.2 Production and Use

Peracetic acid has several uses. It is often used as a cold-temperature sterilant because it is an efficient bactericide, fungicide, and viricide (153). Dilute solutions have been recommended for preoperative sterilization of the hands before surgery.

73.3 Exposure Assessment

73.3.1 Air

Several analytic methods have been used over the years, such as a gas chromatograph equipped with a flame ionization detector (129).

73.3.3 Workplace Methods

No air collection method was located for sampling this peroxide in the breathing zone of the worker.

73.4 Toxic Effects

There were several reports of the oral toxicity of this peroxide being quite low; for example, in guinea pigs, the oral LD_{50} was 10 mg/kg (7) and in an unspecified species it was 17.5 mg/kg, 40% peracetic acid in acetic acid (154). The differences from those reported in Table 84.8 could be due to species differences.

73.4.1 Experimental Studies

73.4.1.1 Acute Toxicity. For information on oral LD_{50}, Dermal LD_{50}, primary skin irritation, eye irritation, and LC_{50}, see Table 84.8.

73.4.1.6 Genetic and Related Cellular Effects Studies. Peracetic acid is an oxidized by-product of ethanol. Experiments have shown it may play a role in ingested ethanol toxicity by interfering with arachidonic acid incorporation into the membranes of erythrocytes. It was suggested it directly inhibits the transferase enzyme (155).

For more information see Table 84.8.

73.4.1.7 Other: Neurological, Pulmonary, Skin Sensitization. Bock (117) has shown that this peroxide in acetone is highly toxic owing to dermal absorption at 2% when painted onto the backs of mice. The study was intended to study skin tumors and cocarcinogenicity. Subsequently, the peroxide was applied in water at 3 and 1% and was tolerated for the duration of the experiment, which was five times a week for 66 wks. The peroxide was a potent promoter when its application was preceded with one skin application of dimethylbenz[a]anthracene (DMBA) of 125 mg in 0.125 mL of acetone.

73.4.2 Human Experience

73.4.2.2.7 Other: Neurological, Pulmonary, Skin Sensitization, etc. When this peroxide escapes into the air, it is intensely irritating to human nasal passages (156).

73.6 Studies on Environmental: Impact

Muela et al. (157) investigated discharge of disinfected wastewater in recipient aquatic systems and the fate of autochthonous bacterial and autochthonous protozoa populations. They noted that chlorine provokes a decrease in the number of sprotozoa and delay in the bacterivorous ability. The discharge of ozonated and peracetic acid–treated wastewater provokes only an initial slight decrease in bacterivorous ability. After the disinfection processes, recipient systems (fresh and marine water) have different effects on the survival of *E. coli* populations discharged into them. The effect of the freshwater recipient system is less negative than seawater.

74.0 4-Nitroperoxybenzoic Acid

74.0.1 CAS Number: *[943-39-5]*

74.0.2 Synonyms: *p*-Nitroperoxybenzoic acid

74.0.3 Trade Names: NA

74.0.4 Molecular Weight: 183.12

74.0.5 Molecular Formula: $C_7H_5NO_5$

74.0.6 Molecular Structure:

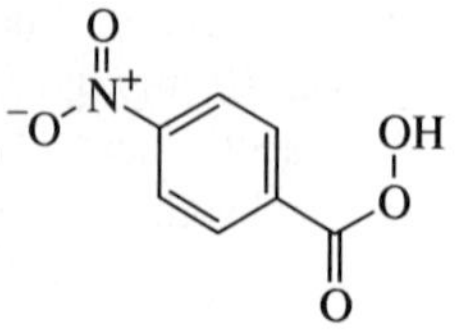

74.2 Production and Use

This peroxide can be prepared from *p*-nitrobenzoic acid and 90% hydrogen peroxide in methanesulfonic acid medium (158).

74.3 Exposure Assessment

74.3.1 Air

Several analytic methods have been used over the years, such as a gas chromatograph equipped with a flame ionization detector (129).

74.3.3 Workplace Methods

No air collection method was located for sampling this peroxide in the breathing zone of the worker.

74.4 Toxic Effects

Subcutaneous injections given mice three times a week for 4 wk caused subcutaneous sarcomas after a total dose of 114 mg of peroxide. However, this is not a relevant route of exposure and stearic acid, used as a control, caused subcutaneous sarcomas with this dosing regimen. No other toxicity data were located for this chemical.

74.4.1 Experimental Studies

74.4.1.6 Genetic and Related Cellular Effects Studies. For information see Table 84.8.

G SULFONYL PEROXIDES

75.0 Acetyl Cyclohexanesulfonyl Peroxide

75.0.1 CAS Number: [3179-56-4]

75.0.2 Synonyms: NA

75.0.3 Trade Names: NA

75.0.4 Molecular Weight: 222.2

75.0.5 Molecular Formula: $C_8H_{14}O_5S$

75.0.6 Molecular Structure:

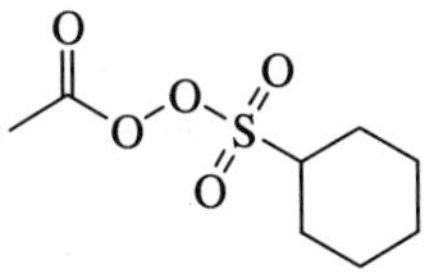

75.1 Chemical and Physical Properties

75.1.1 General

No information was located for this compound.

75.2 Production and Use

No information was located for this compound.

75.3 Exposure Assessment

75.3.1 Air

No air collection method or analytic method was located for this compound.

75.4 Toxic Effects

The only toxicity data found for acetyl cyclohexanesulfonyl peroxide were as follows (7): It had an oral LD_{50} of >4640 mg/kg, a dermal LD_{50} of >2000 mg/kg, and was classified as an eye irritant. At a concentration of 29% in dimethyl phthalate, its oral LD_{50} was 1710 mg/kg.

An acute inhalation study was performed exposing 40 rats to 25, 50, 100, and 200 mg/L of this chemical. At the lowest concentration, 25 mg/L, bloody nasal discharge and congested lungs were noted. One death was noted. The LC_{50} was judged to be 58.3 mg/L with 95% confidence limits of 46–74 mg/L (159).

H SILYL PEROXIDES

76.0 1-Vinyltri-(*t*-butylperoxy)silane

76.0.1 CAS Number: *[15188-09-7]*

76.0.2 Synonyms: VTBS

76.0.3 Trade Names: NA

76.0.4 Molecular Weight: 322.47

76.0.5 Molecular Formula: $C_{14}H_{30}O_6Si$

76.0.6 Molecular Structure:

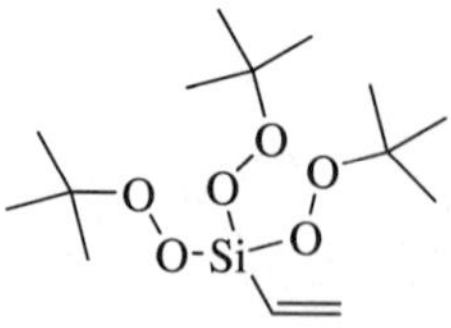

76.1 Chemical and Physical Properties

76.1.1 General

Colorless liquid.

76.2 Production and Use

It is used to promote adhesion in polymer-to-metal bonding applications (159).

76.3 Exposure Assessment

76.3.1 Air

No air collection method or analytic method was located for this compound.

76.4 Toxic Effects

Data indicate that this peroxide is highly toxic by inhalation and toxic by ingestion. The acute inhalation study used rats and was performed using 4, 6, 9, and 18 mg/m^3 of this material diluted 50% in Shellsol T. The exposure concentrations were calculated for pure undiluted VTBS. A control experiment exposing the test animals to saturated Shellsol T vapor produced no effects. However, the peroxide exposures caused salivation and nasal discharge. Mouth breathing was observed in the latter part of the experiment. Deaths occurred within 48 h after the exposure. The LC_{50} was calculated to be 9 mg/m^3 (0.68 ppm) with 11 (0.85 ppm) and 7 mg/m^3 (0.55 ppm) as the 95% confidence limits (160).

76.4.1 Experimental Studies

76.4.1.1 Acute Toxicity. For information on Oral LD_{50}, Dermal LD_{50}, primary skin irritation, eye irritation, and LC_{50}.

77.0. Cumylperoxytrimethylsilane

77.0.1 CAS Number: [18057-16-4]

77.0.2 Synonyms: NA

77.0.3 Trade Names: NA

77.0.4 Molecular Weight: 224.37

77.0.5 Molecular Formula: NA

77.0.6 Molecular Structure: NA

77.1 Chemical and Physical Properties

77.1.1 General

No information was located for this compound.

77.2 Production and Use

No information was located for this compound.

77.3 Exposure Assessment: NA

77.4 Toxic Effects: NA

BIBLIOGRAPHY

1. I. B. Gimenez-Conti et al., *Toxicol Applied Pharmacol* **149**(1), 73–79 (March 1998).
2. D. Noller et al., *Ind. Eng. Chem.* **56**(12) 18–27 (1964).
3. M. Tamura et al., *J. Hazardous Mater.* **17** 89–98 (1987).
4. J. Cywinski, *Reinforced Plastics* (May, 1962).
5. J. Martin, *Ind. Eng. Chem.* **52**(4), 65 (1960).
6. *Encyclopedia of Occupational Health and Safety*, Vols. I and II, International Labour Office, McGraw-Hill, New York, 1983, p. 1612.
7. SPI Bulletin, *Commercial Organic Peroxide Toxicological Data*, Organic Peroxide Producers Safety Division of the Society of Plastics Industry, Publication #19-B, 9/82.
8. NIOSH Method 5009, *Manual of Analytical Methods*, 4th ed., 1994.
9. E. Floyd and H. Stokinger, *Am. Ind. Hyg. Assoc. J.* **19**, 205–212 (1958).
10. A. G. McFarland, Elf ATOCHEM North America, Inc. May 6, 1992, personal communication.
11. M. O. Funk and W. J. Baker, *J. Liq. Chromatogr.* **8**(4), 663–675 (1985).
12. NFPA, *Fire Protection Guide Hazardous Materials*, 49–135 (1978).
13. W. Strong, *Ind. Eng. Chem.* **33**, 38 (1964).
14. *Code of Federal Regulations* 49, 173.225, Department of Transportation, 1992.
15. J. C. Gage, *Brit. J. Ind. Med.* **27**, 1–18 (1970).
16. Patent Specification L040826 No. 20504/65 Application made in Japan (No. 27097), May 14, 1964.
17. W. Eaglstein, *Arch. Dermatol.* **97**, 527 (1968).
18. J. W Fluhr et. al., *Dermatology*, **198**(3), 273–237 (1999)
19. H. Gollnick, M. Schramm, *J. Euro. Academy of Dermatol. Venereology* **11** (Suppl 1), S8–12; discussion S28–9 (Sept. 1998).
20. R. H. Azurdia and C.M King, *Contact Dermatitis* **38**(4), 234–235 (April 1998).
21. R. L. Roudabush, *Toxicity and Health Hazard Summary*, Laboratory of Industrial Medicine, Eastman Kodak Co., Rochester, NY, 1964.
22. M. Sharrat et al., *Food Cosmet. Toxicol.* **2**, 527 (1964).
23. B. VanDuuren et al., *J. Natl. Cancer Inst.* **31**, 41 (July 1963).
24. W. C. Hueper, *NCI Monograph No. 10*, 1963, pp. 349–359.
25. J. H. Epstein, *J. Photochem. Photobiol.* **37**, S-38 (1983).
26. Litton Bionetics, FDA OTC Vol. 070294, June, 1975.
27. T. Slaga et al., *Science* 213–228 (Aug. 1981).
28. Y. Kurokawa et al., *Cancer Lett.* **24**, 299 (1984).
29. IARC Monographs on the Evaluation of the Carcinogen Risk of Chemicals to Humans No. 36, 1985, p. 267.

30. N. Rudra and N. Singh, *Indian Journal of Physiology and Pharmacology* **41**(2) 109–15 (April 1997).
31. N. Singh and S. Aggarwal *Indian Journal of Experimental Biology* **34**(7) 647–51 (July 1996).
32. C. Hazelwood and M. J Daview *Archives of Biochemistry and Biophysics* **332**(1) 79–91 (Aug. 1996).
33. R. L Binder, M. J Aardema, and E. D. Thompson *Progress in Clinical and Biological Research*, **391**, 245–94 (1995)
34. A. L. Kraus et al., *Regulatory Toxicology and Pharmacology.* **21**(1) 87–107 (Feb. 1995).
35. W. M. Grant, *Toxicology of the Eye*, 3rd ed., Charles C Thomas, Springfield, IL, 1986.
36. American Conference of Governmental Industrial Hygienists, TLV Documentation, 1991.
37. J. R. Tkach, *Cutis* **29**(2), 187 (1982).
38. F. N. Marzulli and H. L. Maibach, *Food Cosmet. Toxicol.* **12**(2), 219 (1974).
39. National Research Council, Summary Table Biological Tests **2**, 241, 1950.
40. *Int. Polymer Sci. Technol.* 3, 93 (1976).
41. Ref 39. vol 4, 1952, p 110.
42. P. R. Dugan, *Anal. Chem.* **33**(6), 696 (1961).
43. F. Orlova, *Sb. Nauch. Tr. Kuibyshev Nauch-Issled Inst. Gig.* **6**, 101 (1971).
44. A. Korhonen et al., *Environ. Res.*, **33**, 54 (1984).
45. Ref. 29, p 315.
46. A. Klein-Szanto and T. Slaga, *J. Invest. Dermatol.* **79**(1), 30 (1982).
47. National Research Council, *Prudent Practices for Handling Hazardous Chemicals in Laboratories*, National Academy Press, Washington, DC, 1981, p. 106.
48. N. I. Sax, *Dangerous Properties of Industrial Materials*, 5th ed., Van Nostrand Rheinhold, New York, 1979.
49. *(The) Merck Index,* 10th ed., Merck Co., Rahway , NJ, 1983.
50. G. G. Hawley, *The Condensed Chemical Dictionary,* 10th ed., Van Nostrand Reinhold, New York, 1981.
51. U. S. Pat. 4402853, (Sept. 6, 1983) (to Sterling Drug, Inc.); *Chem abstr.*/099/181508J.
52. V. J. Horgan et al., *Biochem. J.* **67**, 551 (1957).
53. V. J. Horgan et al., *Organic Peroxides in Radiobiology*, London, 1958, p. 50.
54. E. Altenburg and L. S. Altenburg, *Genetics* **42**, 357–358 (1957).
55. E. B. Freese et al., *Mutat. Res.* **4**(5), 517–531 (1967).
56. Department of Transportation, *Emergency Response Guidebook,* DOT P 5800.4, U.S. Government Printing Office G-49, Washington, DC, 1987.
57. Catalyst Resources Incorporated, Technical Bulletin No. 1.1, July, 1983.
58. M. King, Catalyst Resources, Inc., Elyria, OH, 1992, personal communication.
59. Central Instituut voor Voedingsonderzoek TNO (CIVO), Zeist, Holland-Reports for Akzo Chemie No. R4707, 1975; No. R5519, 1977.
60. Code of Federal Regulations 29 *CFR* 1910.1200 Appendix A 6(c), Occupational Safety and Health Administration, July 1, 1988.
61. International Research and Development Corp., unpublished data sponsored by Aztec Chemicals, 1979.
62. Ref. 57, No. 5.1, June 1985.

63. Ref. 57, No. 6.1, June 1985.
64. Ref. 57, No. 4.1, July 1983.
65. D. Walrod et al., *Plastics Compounding* **52** (Jan./Feb. 1979).
66. I. V. Sanotskii et al., *Toksikol. Nov. Prom. Khim. Veshchestv.* **10**, 55–63 (1968).
67. M. Athar et al., *Carcinogenesis* **10**(8), 1499–1503 (1989).
68. Ref. 57, No. 7.1, July 1983.
69. Ref. 61, No 376-8, 1978.
70. J. Vennerstrom et al., *Drug Design Delivery* **4**, 45 (1989).
71. Ref. 59, No R 5624, 1978.
72. Ref. 57, No. 12.1, April 1990.
73. Ref. 61, No. 164-76, 1978.
74. Ref. 59, No R 5619, 1978.
75. R. Hall, Adhesives, National Clearinghouse for Poison Control Centers Bulletin, Jan–Feb. 1969.
76. M. Bhushan, N. M. Craven, and M. H. Bech, *Contact Dermatitis* **39**(4), 203 (Oct. 1998).
77. A. Prez-Martnez et al., *European Journal of Pediatrics* **156**(12), 976–977 (Dec 1997)
78. L. Stewart and M.H. Beck, Contact Dermatitis, **26**(1), 52–53 (Jan. 1992).
79. F. T. Fraunfelder et al., *American Journal of Ophthalmology* **110**(6), 635–40 (Dec. 15 1990).
80. P. J. Karhunen et al., *Human and Experimental Toxicology* **9**(3), 197–200 (May 1990).
81. NIOSH Method 3508, *Manual of Analytical Methods*, 4th ed., 1994.
82. L. A. Tiunov, Toxicology of organic peroxides. *Gig. Sanit.* **29**(10), 82–87 (1964).
83. M. Sittig, *Handbook of Toxic and Hazardous Chemicals and Carcinogens*, Noyes Publications, Park Ridge, NJ, 1985.
84. National Institute of Environmental Health Sciences, private communication to Lucidol Div., Pennwalt Corp., Dec. 1981. Results also published in *NTP Tech. Bulletin of NTP Program,* (6), Table 3 (Jan 1982)
85. F. Fraunfelder et al., *Am. J. Ophthalmol.* **110**(6), 635–640 (1990).
86. R. Mittleman et al., *J. Forensic Sci.* **31**(1), 312–320 (1986).
87. G. Luft et al., *Angew. Makromol. Chem.* **141** 207 (1986).
88. Ref. 59, No R 6143, 1979.
89. A. N. Klimina, *Sb. Nauch. Tr. Keibyshev Nauch.-Issled. Inst. Epidemiol. Gig.* **5**, 98–100, (1968).
90. E. M. Sevast'yanova and Z. S. Smirnova, Gig. Sanit. **0**(2), 48–49 (1987); 90–91, (aug. 8 1990).
91. Ref. 89, vol. 6, 1971, pp. 98–100.
92. R. E. Gosselin et al., *Clinical Toxicology of Commercial Products,* 4th ed., Williams & Wilkins, Baltimore, 1976.
93. USSR Pat. 841659, (June 30, 1981) K. I. Ivanov et al., (to All-Union, Scientific-Research Thermotechnical Institute).
94. L. A. Cornish et al., *J. Chromatogr.* **19**, 85–87 (1981).
95. Ref. 59, No R 6141, 1979.
96. T. Yamaguchi and Y. Yamashita, *Agric. Biol. Chem.* **44**(7), 1675 (1980).

97. R. Latarjet et al., In M. Haissinsky, Ed., *Organic Peroxides in Radiobiology*, Pergamon Press, London, 1958.
98. E. Eder et al., *Toxicol. Lett.* **48**, 225 (1989).
99. K. A. Jensen et al., *Cold Spring Harbor Symp. Quant. Biol.* **16**, 245 (1951).
100. L. Fishbein, Pesticidal, industrial, food additive and drug mutagens. In H. E. Sutton and M. I. Harris, Eds., *Mutagenic Effects of Environmental Contaminants*, Academic Press, New York, 1972.
101. Nippon Kayaka K. K., unpublished study sponsored by Kayaka Nory Corp., 1971.
102. H. J. Kuchle, Zentralbl. *ArbMed. ArbSchutz.* **8**, 25, (1958).
103. F. Tang and E. S. Huyser, *J. Org. Chem.* **42**(12), 2160–2163 (1976).
104. Ref. 57, No. 52.1, March 1987.
105. Ref. 61, No 164-16, 1970.
106. BIBRA Toxicology International, Toxicity Profile for Dicumyl Peroxide, 1990.
107. Hercules Rubber Chemicals, Hercules Powder Company, Wilmington, DE, 1964 Bulletin T-104, Revision I.
108. *ASTM Standard Test Method E 755–90*, American Society for Testing and Materials, Philadelphia, 1990.
109. Ref. 59, No R4934, Feb. 1976.
110. A. Rannug et al., *Industrial Hazards of Plastics and Synthetic Elastomers*, Alan R. Liss Inc., New York, 1984, p. 407.
111. E. Zeiger et al., *Environ. Mol. Mutagen.* **11**, (Suppl. 12), 1–158 (1988).
112. M. R. Chevallier and D. Luzzati, *Compt. Rend.* **250**, 1572 (1960).
113. Ref 107, Bulletin ORC-207C.
114. S. C. Stenton et al., *J. Soc. Occup. Med.* **39**, 33 (1989).
115. B. Petruson and B. Jauarvholm, *Acta Oto-Lar.* **95**, 333 (1983).
116. Industrial Bio-Test Laboratories, Inc., unpublished data sponsored by Hercules, Inc. No. N8328, 1970.
117. Ref. 59, No V81. 337/2/1424, 1981.
118. Biodynamics, Inc., unpublished data sponsored by Hercules, Inc. No. 81-7514, 1981; unpublished data supported by ARCO Chemical Co. No. 81-7532, 1981.
119. Anonymous, Flammability and toxic properties of the diperoxides DIGIF-40 and BPIB-40, *Kauch. Rezina* **12**, 28–29 (1985).
120. Ref. 61, No 378-11, 1978.
121. ARCO Chemical Company, Summaries of Acute Toxicity and Mutagenicity Tests, June, 1982.
122. G. Leonardos, A. D. Little, Inc., memorandum report to Oxirane Corp., dated Dec. 10, 1979.
123. S.R.I. International, Stanford Research Institute International, Inc., Chemical Economics Handbook, Menlo Park, CA, 1978, 1979.
124. R. Brink (to Pentz Laboratories, Inc.), U.S. Patent No. 514278, 1968.
125. U.S. Pat. 4017392, M. Hamer and O. Petzen (to International Minerals and Chemicals Corporation).
126. W. J. M. VanTilborg, *J. Chromatogr.* **115**, 616–620 (1975).
127. B. B. Jones et al., *J. Chromatogr.* **202**, 127–130 (1980).
128. L. Cerveny et al., *J. Chromatogr.* **74**, 118–120 (1972).

129. G. T. Cairns et al., *J. Chromatogr.* **103**, 381–384 (1975).
130. R. S. Deelder and M. G. F. Kroll, *J. Chromatogr.* **125**, 307–314 (1976).
131. L. Zavodnik et al., Experimental and Toxicologic Pathology **51**(4–5) 446–450 (July 1999).
132. P. Kapahi, M. E. Boulton, and T. B. Kirkwood *Free Radical Biology and Medicine* **26**(5–6) 495–500 (March 1999).
133. J. Cai et al., *Investigatigve Opthalmology and Visual Science* **40**(5) 959–66 (April 1999).
134. G. Nowak et al., *Americal Journal of Physiology*, **274**(3 Pt. 2) F509–515 (March 1998).
135. J. C. Lavoie and P. Chessex, *Free Radical Biology and Medicine*, **16**(3), 307–313 (Mar 1994).
136. J. D. Adams Jr., et al., *Free Radical Biology and Medicine*, **15**(2) 195–202 (Aug. 1993).
137. G. A. Shevelava, *Gig. Tr. Prof. Zabol.* **12**, 40–48 (1976).
138. T. R. Kumar and Muralidhara, *Mutation Res.* **444**(1), 145–9 (Jul 21, 1999).
139. H. Hoshino et al., *Gann* **61**(2), 121–124 (1970).
140. P. Cerrutti, paper presented at the 14th Annual Cancer Research Workshop, University of New Orleans, Feb. 12, 1981.
141. D. Muehlematter et al., *Chem.-Biol. Interact.* **71**, 339–362 (1989).
142. National Toxicology Program, unpublished reports on Ames testing of BPO (1985) and multistrain comparison of DMBA/MNNG/BPO promotion (1986); *t*-butyl hydroperoxide.
143. Ref. 116, No 76-13, 1976.
144. R. J. Trotta et al., *Biochim. Biophys. Acta* **679**, 230–237 (1981).
145. S. A. Jewell et al., *Science* **217**, 1257–1259 (1982).
146. G. Rush and D. Alberts, *Toxicol. Appl. Pharmacol.* **85**, 324–331 (1986).
147. D. P. Jones et al., *Adv. Biosci.* **76** 13–19 (1989).
148. EPA, *Review of Environmental Fate of Selected Chemicals*, EPA 560/5-77-3, Office of Toxic Substances, May, 1977.
149. H. Seis et al., *Fed. Fur. Biochem. Soc.* **27**, 171–175 (1972).
150. *Brit. J. Ind. Med.* **27** 11–12 (1970).
151. Mycotoxins in foodstuffs. *Proceedings of the Symposium held at Massachusetts Institute of Technology*, March, 1964.
152. Ref. 116, No 8562-9495, 1977.
153. L. B. Kline and R. N. Hull, *Am. J. Clin.* Pathol. **33** 30–33 (1960).
154. FMC Corporation, data supplied by letter, dated April 1972 from H. M. Castrantas, FMC, to Organic Peroxide Producers Safety Division of the Society of Plastics Industry.
155. D. W. Allen et al., *Biochem. Biophys. Acta. 1081*(3), 267–273 (1991).
156. F. G. Bock et al., *J. Natl. Cancer Inst.* **55**(6), 1359–1361 (1975).
157. A. Muelal et al., *J. Appl. Microbiology* **85**(2), 263–270 (1998).
158. L. Silbert et al., *J. Org. Chem.* **27**, 133–142 (1962).
159. Ref. 61, No 378-5, 1976.
160. Ref. 59, No R 5170, 1976.

Subject Index

Chemical Index